Prefixes and Suffixes in Anatomical and Medical Terminology

Element	Definition and Example
mito-	thread: *mitochondrion*
mono-	alone, one, single: *monocyte*
morph-	form, shape: *morphology*
multi-	many, much: *multinuclear*
myo-	muscle: *myology*
narc-	numbness, stupor: *narcotic*
necro-	corpse, dead: *necrosis*
neo-	new, young: *neonatal*
nephr-	kidney: *nephritis*
neuro-	nerve: *neurolemma*
noto-	back: *notochord*
ob-	against, toward, in front of: *obturator*
oc-	against: *occlusion*
-oid	resembling, likeness: *sigmoid*
oligo-	few, small: *oligodendrocyte*
-oma	tumor: *lymphoma*
oo-	egg: *oocyte*
or-	mouth: *oral*
orchi-	testis: *orchiectomy*
ortho-	straight, normal: *orthopnea*
-ory	pertaining to: *sensory*
-ose	full of: *adipose*
osteo-	bone: *osteoblast*
oto-	ear: *otolith*
ovo-	egg: *ovum*
par-	give birth to, bear: *parturition*
para-	near, beyond, beside: *paranasal*
path-	disease, that which undergoes sickness: *pathology*
-pathy	abnormality, disease: *neuropathy*
ped-	children: *pediatrician*
pen-	need, lack: *penicillin*
-penia	deficiency: *thrombocytopenia*
per-	through: *percutaneous*
peri-	near, around: *pericardium*
phag-	to eat: *phagocyte*
-phil	have an affinity for: *neutrophil*
phleb-	vein: *phlebitis*
-phobia	abnormal fear, dread: *hydrophobia*
-plasty	reconstruction of: *rhinoplasty*
platy-	flat, side: *platysma*
-plegia	stroke, paralysis: *paraplegia*
-pnea	to breathe: *apnea*
pneumo(n)-	lung: *pneumonia*
pod-	foot: *podiatry*
-poiesis	formation of: *hemopoiesis*
poly-	many, much: *polyploid*
post-	after, behind: *postnatal*
pre-	before in time or place: *prenatal*
pro-	before in time or place: *prophase*

Element	Definition and Example
proct-	anus: *proctology*
pseudo-	false: *pseudostratified*
psycho-	mental: *psychology*
pyo-	pus: *pyorrhea*
quad-	fourfold: *quadriceps femoris*
re-	back, again: *repolarization*
rect-	straight: *rectus abdominis*
ren-	kidney: *renal*
rete-	network: *rete testis*
retro-	backward, behind: *retroperitoneal*
rhin-	nose: *rhinitis*
-rrhagia	excessive flow: *menorrhagia*
-rrhea	flow or discharge: *diarrhea*
sanguin-	blood: *sanguineous*
sarc-	flesh: *sarcoma*
-scope	instrument for examining a part: *stethoscope*
-sect	cut: *dissect*
semi-	half: *semilunar*
-sis	process or action: *dialysis*
steno-	narrow: *stenosis*
-stomy	surgical opening: *tracheostomy*
sub-	under, beneath, below: *subcutaneous*
super-	above, beyond, upper: *superficial*
supra-	above, over: *suprarenal*
syn- (sym-)	together, joined, with: *synapse*
tachy-	swift, rapid: *tachycardia*
tele-	far: *telencephalon*
tens-	stretch: *tensor tympani*
tetra-	four: *tetrad*
therm-	heat: *thermogram*
thorac-	chest: *thoracic cavity*
thrombo-	lump, clot: *thrombocyte*
-tomy	cut: *appendectomy*
tox-	poison: *toxemia*
tract-	draw, drag: *traction*
trans-	across, over: *transfuse*
tri-	three: *trigone*
trich-	hair: *trichology*
-trophy	a state relating to nutrition: *hypertrophy*
-tropic	turning toward, changing: *gonadotropic*
ultra-	beyond, excess: *ultrasonic*
uni-	one: *unicellular*
-uria	urine: *polyuria*
uro-	urine, urinary organs or tract: *uroscope*
vas-	vessel: *vasoconstriction*
viscer-	organ: *visceral*
vit-	life: *vitamin*
zoo-	animal: *zoology*
zygo-	union, join: *zygote*

Kent M. Van De Graaff
Weber State University

Stuart Ira Fox
Pierce College

Concepts of HUMAN ANATOMY & PHYSIOLOGY

FIFTH EDITION

Boston Burr Ridge, IL Dubuque, IA Madison, WI New York San Francisco St. Louis
Bangkok Bogotá Caracas Lisbon London Madrid
Mexico City Milan New Delhi Seoul Singapore Sydney Taipei Toronto

WCB/McGraw-Hill

A Division of The **McGraw-Hill** *Companies*

CONCEPTS OF HUMAN ANATOMY AND PHYSIOLOGY, FIFTH EDITION

This book is printed on acid-free paper.

1 2 3 4 5 6 7 8 9 0 VNH/VNH 9 3 2 1 0 9 8

ISBN 0–697–28425–5

Vice president and editorial director: *Kevin T. Kane*
Publisher: *Colin H. Wheatley*
Sponsoring editor: *Kristine Tibbetts*
Developmental editor: *Patrick F. Anglin*
Marketing manager: *Heather K. Wagner*
Senior project manager: *Peggy J. Selle*
Senior production supervisor: *Mary E. Haas*
Designer: *K. Wayne Harms*
Photo research coordinator: *John C. Leland*
Art editor: *Brenda A. Ernzen*
Compositor: *Shepherd, Inc.*
Typeface: *10.5/12 Goudy*
Printer: *Von Hoffman Press, Inc.*

Cover/interior design: *Jamie O'Neal*
Cover illustration: *Rictor Lew*

Library of Congress Cataloging-in-Publication Data

Van De Graaff, Kent M. (Kent Marshall), 1942–
Concepts of human anatomy and physiology / Kent M. Van De Graaff, Stuart Ira Fox. — 5th ed.
p. cm.
Includes bibliographical references and index.
ISBN 0–697–28425–5 — ISBN 0–07–115865–0 (ISE)
1. Human physiology. 2. Human anatomy. I. Fox, Stuart Ira. II. Title.
QP36.V36 1999
612—dc21 98–6292
CIP

www.mhhe.com

Brief Contents

Contents

Chapter Five

Membrane Transport and the Membrane Potential 110

Chapter Six

Histology 130

UNIT II

Support and Movement of the Human Body 159

Chapter Seven

Integumentary System 159

Chapter Eight

Skeletal System: Bone Tissue and Bone Development 182

Chapter Nine

Skeletal System: Axial Skeleton 200

Chapter Ten

Skeletal System: Appendicular Skeleton 225

Chapter Eleven

Articulations 247

Chapter Twelve

Muscle Tissue and Muscle Physiology 280

Chapter Thirteen

Muscular System 306

UNIT III

Integration and Control Systems of the Human Body 371

Chapter Fourteen

Functional Organization of the Nervous System 371

Chapter Fifteen

Central Nervous System 407

UNIT IV

Regulation and Maintenance of the Human Body 591

Chapter Twenty

Circulatory System: Blood 591

Chapter Twenty-One

Circulatory System 610

Chapter Twenty-Two

Circulatory System: Cardiac Output and Blood Flow 655

CHAPTER TWENTY-THREE

Lymphatic System and Immunity 692

CHAPTER TWENTY-FOUR

Respiratory System 728

CHAPTER TWENTY-FIVE

Urinary System: Fluid, Electrolyte, and Acid-Base Balance 778

CHAPTER TWENTY-SIX

CHAPTER TWENTY-SEVEN

UNIT V

Continuance of the Human Species 891

CHAPTER TWENTY-EIGHT

CHAPTER TWENTY-NINE

CHAPTER THIRTY

APPENDIXES 985

Preface

While the fifth edition of *Concepts of Human Anatomy and Physiology* has taken on a fresh, new look and has changed in other significant ways, we have made every effort to retain those features that have contributed to the great popularity of this text over the years. Of major importance, the fifth edition is consistent with previous editions in its focus on unifying concepts as a means of integrating factual information. Just as importantly, a clear and interesting narrative, carefully rendered and attractive illustrations, and numerous pedagogical devices continue to be central in enabling students to assimilate a large body of information and to place what they have learned in a meaningful context.

As in previous editions, the material is organized so that instructors may tailor required text readings to their individual course needs. Because the text is designed for students who do not have extensive science backgrounds but who plan to enter health or other careers that require considerable knowledge of anatomy and physiology, the chapters in the opening unit present basic chemical, cellular, biological, and anatomical concepts. The chapters in the remaining four units then take a detailed approach to the anatomy and physiology of organs and systems. Throughout the text, we continue to promote the view of anatomy and physiology as dynamic sciences that serve as foundations for the health professions.

Having said this, we have no doubt that the fifth edition is the strongest by far. We are confident that it can be of immense value in helping students achieve learning objectives, in fostering in them a love of and respect for the science of human anatomy and physiology, and in persuading them to continue in the field.

What's New, Revised, or Improved

Followers of previous editions will quickly note the major physical improvements in the fifth edition. We have also added new material in light of recent scientific findings, taking care to connect new developments to basic principles. Listed below are some of the major fifth-edition changes.

New Content

A great deal of new information has been incorporated into this edition. These include the following changes:

Some of the New Topics Added

- Skin cancer
- ICE proteins
- Telomeres and telomerase
- Peroxisomes
- Aquaporins
- Serotonin and neuropeptide Y as neurotransmitters
- Nicotinic ACh receptors in brain
- Retrograde dendritic potentials
- Thermal receptors and nociceptors
- Mesolimbic and nigrostriatal dopamine pathways
- Nuclear receptor proteins
- Tyrosine kinase second messenger system
- Adhesion molecules
- Dendritic cells as antigen presenters
- FAS and FAS ligand
- Function of the enterochromaffin-like (ECL) cells of the gastric mucosa
- Guanylin and uroguanylin
- Retinoic acid nuclear receptors
- Adipose tissue physiology
- Leptin and its actions
- Obesity
- Brown fat and β_3-adrenergic receptors
- Effect of weightlessness on calcium balance and bones
- Male contraception
- The human sexual response

Some of the Revised Sections

- Chemiosmotic presentation and number of ATP generated
- Symport and antiport concepts added to coupled transport discussion
- Voltage-gated ion channels
- Neurotransmitter release from axons
- Alzheimer's disease
- REM sleep, reticular formation, and medial temporal lobe
- Paracrine and autocrine regulators
- Cross-bridge cycle
- Muscle metabolism during exercise
- Cardiovascular adaptations to exercise
- Mechanisms of smooth muscle relaxation
- Functions of neutrophils
- Stem cell differentiation, with added discussion of cytokines
- Physiology of lymphatic vessels
- Frank–Starling law
- Capillary dynamics (Starling equilibrium)
- Hypertension and congestive heart failure
- Nonspecific immune recognition and function
- AIDS and AIDS treatments

- Central regulation of breathing
- Respiration during exercise
- Structure and function of the vasa recta
- ADH action and diabetes insipidus
- Atrial natriuretic peptide
- Regulation of gastric acid secretion
- Catecholamine regulation of metabolism
- Thyroxine regulation of metabolism
- Regulation of human parturition

Some Expanded Topics

- p53, with additional information about p21
- Dopaminergic receptors
- Physiology of taste, including role of gustducin
- Physiology of olfaction
- Effects of urea in concentrating the urine
- Benefits of breastfeeding the neonate
- Enteric nervous system
- Body weight homeostasis
- Nutrition and fatty acids
- Vitamins
- Regulation of eating
- Cholecystokinin physiology
- Mast cell function
- Regulation of insulin secretion and mode of insulin action
- Treatment for diabetes mellitus
- Regulation of peristaltic contractions

New Clinical Investigations

Case Studies appear at the beginning of most of the chapters. These hypothetical situations are indicative of the type of clinical material that will be presented in the chapters. The solution to the case study is presented at the end of the chapter following the last major section.

New Critical Thinking Questions

Following each chapter summary, sets of objective, essay, and critical thinking questions give students the opportunity to obtain feedback as to the depth of their understanding and learning. They challenge students to use the chapter information in novel ways toward the solution of practical problems. The correct responses to the objective questions are provided in Appendix A (page 985).

New and Revised Illustrations

An already outstanding illustration program has been greatly improved in this edition, which features nearly 40% new or revised illustrations. The figure legends have been modified to enhance the identification of a figure and improve the readability of the legend.

Figure 30.16

The formation of the umbilical cord and other extraembryonic structures. These structures are depicted in sagittal sections of a gravid uterus from week 4 to week 22. (*a*) A connecting stalk forms as the developing amnion expands around the embryo, finally meeting ventrally. (*b*) The umbilical cord begins to take form as the amnion ensheathes the yolk sac. (*c*) A cross section of the umbilical cord showing the embryonic vessels, mucoid connective tissue, and the tubular connection to the yolk sac. (*d*) By week 22, the amnion and chorion have fused, and the umbilical cord and placenta have become well-developed structures.

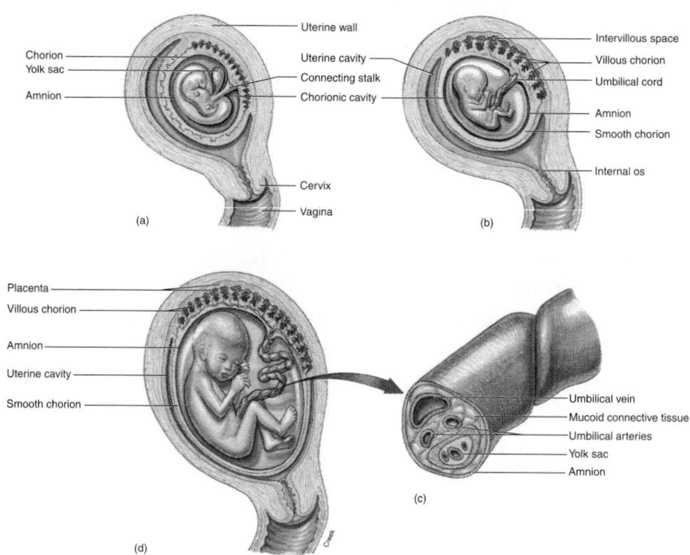

NEXUS

Toward the end of each chapter, or group of chapters, on a particular body system, newly designed and revised interrelationship charts, called NEXUS, tie the functional aspects of one body system to each of the other systems, underscoring the concept of homeostasis. Each listed interaction has a page reference in blue for students to read for additional information. This is analogous to the hyperlinks of an Internet web page, and can be used in a similar manner to pursue related concepts of interest.

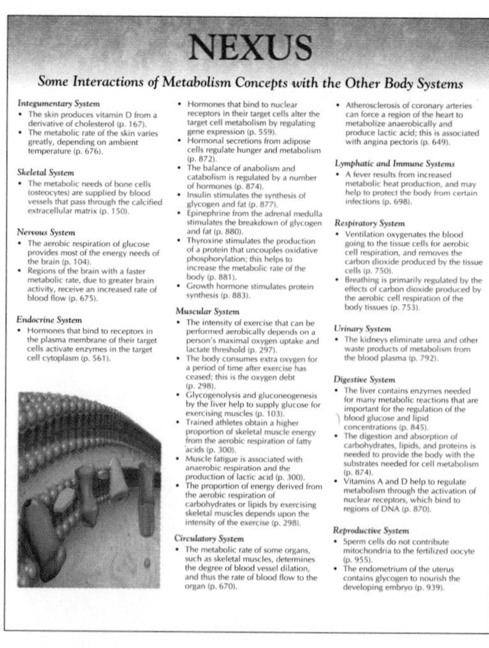

NEXUS

Some Interactions of Metabolism Concepts with the Other Body Systems

Integumentary System
- The skin produces vitamin D from a derivative of cholesterol (p. 167).
- The metabolic rate of the skin varies greatly, depending on ambient temperature (p. 676).

Skeletal System
- The metabolic needs of bone cells (osteocytes) are supplied by blood vessels that pass through the calcified extracellular matrix (p. 150).

Nervous System
- The aerobic respiration of glucose provides most of the energy needs of the brain (p. 104).
- Regions of the brain with a faster metabolic rate, due to greater brain activity, receive an increased rate of blood flow (p. 675).

Endocrine System
- Hormones that bind to receptors in the plasma membrane of their target cells activate enzymes in the target cell cytoplasm (p. 561).
- Hormones that bind to nuclear receptors in their target cells alter the target cell metabolism by regulating gene expression (p. 559).
- Hormonal secretions from adipose cells regulate hunger and metabolism (p. 872).
- The balance of anabolism and catabolism is regulated by a number of hormones (p. 874).
- Insulin stimulates the synthesis of glycogen and fat (p. 877).
- Epinephrine from the adrenal medulla stimulates the breakdown of glycogen and fat (p. 880).
- Thyroxine stimulates the production of a protein that uncouples oxidative phosphorylation; this helps to increase the metabolic rate of the body (p. 881).
- Growth hormone stimulates protein synthesis (p. 883).

Muscular System
- The intensity of exercise that can be performed aerobically depends on a person's maximal oxygen uptake and lactate threshold (p. 297).
- The body consumes extra oxygen for a period of time after exercise has ceased; this is the oxygen debt (p. 298).
- Glycogenolysis and gluconeogenesis by the liver help to supply glucose for exercising muscles (p. 103).
- Trained athletes obtain a higher proportion of skeletal muscle energy from the aerobic respiration of fatty acids (p. 300).
- Muscle fatigue is associated with anaerobic respiration and the production of lactic acid (p. 300).
- The proportion of energy derived from the aerobic respiration of carbohydrates or lipids by exercising skeletal muscles depends upon the intensity of the exercise (p. 298).

Circulatory System
- The metabolic rate of some organs, such as skeletal muscles, determines the degree of blood vessel dilation, and thus the rate of blood flow to the organ (p. 670).
- Atherosclerosis of coronary arteries can force a region of the heart to metabolize anaerobically and produce lactic acid; this is associated with angina pectoris (p. 649).

Lymphatic and Immune Systems
- A fever results from increased metabolic heat production, and may help to protect the body from certain infections (p. 698).

Respiratory System
- Ventilation oxygenates the blood going to the tissue cells for aerobic cell respiration, and removes the carbon dioxide produced by the tissue cells (p. 750).
- Breathing is primarily regulated by the effects of carbon dioxide produced by the aerobic cell respiration of the body tissues (p. 753).

Urinary System
- The kidneys eliminate urea and other waste products of metabolism from the blood plasma (p. 792).

Digestive System
- The liver contains enzymes needed for many metabolic reactions that are important for the regulation of the blood glucose and lipid concentrations (p. 845).
- The digestion and absorption of carbohydrates, lipids, and proteins is needed to provide the body with the substrates needed for cell metabolism (p. 874).
- Vitamins A and D help to regulate metabolism through the activation of nuclear receptors, which bind to regions of DNA (p. 870).

Reproductive System
- Sperm cells do not contribute mitochondria to the fertilized oocyte (p. 955).
- The endometrium of the uterus contains glycogen to nourish the developing embryo (p. 939).

New Design

A conscious effort has been made to make this book more readable with a fresh, clean design with fewer interruptions in the narrative.

Under Development

Near the end of most system chapters is a discussion that includes exhibits and explanations of the morphogenic events involved in the development of a body system. Placement near the end of a chapter ensures that the terminology needed to understand the embryonic structures has been introduced. In a few chapters, an Under Development feature follows the relevant discussion of a specific body part or region; this occurs, for example, in sections on the skull, brain and spinal cord, ear and eye, and pituitary gland.

A More Personal Approach

It has been our experience that beginning students in anatomy and physiology are often intimidated by a very formal, academic writing style. In this edition, the language has been relaxed to engage the reader and make learning more enjoyable. Simple analogies are frequently used to promote understanding of concepts. The level of difficulty has been carefully controlled, recognizing the wide variation in motivation and background that typifies a broad spectrum of students.

Learning Aids: A Guide to the Student

The pedagogical devices in this text are designed to help you learn anatomy and physiology. Don't just read this text as you would a novel. Interact with it, using the pedagogical devices as tools. The more you use these tools, the more effective and enjoyable your study will become.

Chapter Introductions

The opening page of each chapter contains an overview of the contents of the chapter in outline form. Page numbers are indicated to guide you to the major sections. Learning objectives are also included, and should be checked both before and after studying each section of a chapter.

Concept Statements

One of the unique attributes of this text is the way in which major sections are introduced. Each of these sections is prefaced by a concept statement—a succinct expression of the main idea presented in the section. These concept statements will help you gain an overview before encountering the details.

Terminology Aids

The first time each technical term appears in the narrative, it is set off by boldface or italic type and is often followed by a phonetic pronunciation in parentheses. In this fifth edition, many new phonetic pronunciations have been added.

Word Derivations

The derivations of many terms are provided in footnotes at the bottom of the page on which the terms are introduced. Don't skip over these footnotes; they are often interesting in themselves. Furthermore, if you know how a word was derived, it becomes more meaningful and is easier to remember. You can identify the roots of each term by referring to the glossary of prefixes and suffixes on the inside front cover of the text.

Clinical Applications

Set off from the text narrative are short paragraphs highlighted by accompanying topic icons. This interesting information is relevant to the discussion that precedes it, but more importantly, it demonstrates how basic scientific knowledge is applied. New clinical applications—some of topical interest—have been added to the fifth edition, and others have been updated. The two icons used are as follows:

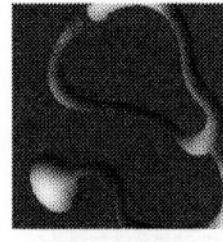

Clinical information is indicated by a stethoscope. These sections explore selected medical applications of the preceeding anatomical and physiological concepts.

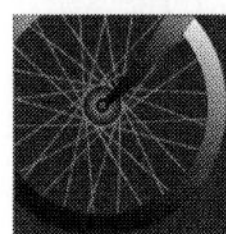

Exercise physiology is indicated by a bicycle wheel. These sections explore how the preceeding anatomical and physiological concepts can be applied to understanding the physiology of exercise.

Illustrations

Because anatomy is a descriptive science, great care has been taken to provide an outstanding illustration program that maximizes students' learning. These illustrations represent a collaborative effort between author and illustrator, often involving dissection of cadavers to ensure accuracy. In addition to being aesthetically pleasing, each illustration has been checked and rechecked for conceptual clarity and precision of the linework, labels, and caption. All of the figures are integrated with the text narrative, and most are original full-color art. In addition to the anatomical renderings, color graphics are used to clarify complex physiological processes. Light and scanning electron photomicrographs are also used where appropriate, and carefully selected photographs appear throughout the text. Color-coding is used in certain art sequences as a technique to aid learning. For example, the bones of the skull in chapter 9 are color-coded so that each bone can be readily identified in the many renderings included in the chapter. In chapters 9, 10, and 11 on the skeletal system and articulations, new orientation diagrams have been added to highlight the location of specific bones and joints relative to the body as a whole or to a particular region of the body. Following chapter 13 is a set of reference plates, including photographs of human cadaver dissections and several full-page illustrations of the male and female trunk. The photographs of dissected human cadavers illustrate the complexity of structural relationships that can be fully appreciated only when seen in a human specimen. Elsewhere in the text, photographs of specific organs from cadavers are used to augment the illustrations.

This fifth edition features many new and revised physiology figures to accompany the new topics and updated discussions. We believe that, for such figures to be understandable and useful to the beginning student, they should be relatively simple. A complex, summary figure that incorporates a great deal of information may be brilliantly conceived and attractive to an instructor, but it is deadly to a beginning student who can be overwhelmed by the complexity. Therefore, in this text, each figure is designed to illustrate a single concept or to summarize only a limited number of concepts. In total, they present all of the information covered in the text, but they do so in digestible bites.

Chapter Summaries

At the end of each chapter the material is summarized for you in outline form, following the sequence of the text narrative. Review each summary after studying the chapter to be sure that you have not missed any points. In addition, use the chapter summaries in preparing for examinations.

Review Activities

A series of objective, essay questions, and new critical thinking questions provide you with feedback as to the depth of your learning and understanding. The answers to the objective questions are provided in Appendix A.

Glossary

The glossary of terms at the end of the text has been updated and expanded, and continues to be particularly noteworthy for its comprehensiveness. Phonetic pronunciations are included for most of the terms, and an easy-to-use pronunciation guide appears at the beginning of the glossary. Synonyms, including eponymous terms, are indicated, and for some terms antonyms are given as well. The majority of the terms in the glossary are accompanied by a page number indicating where the term is discussed in the text narrative. (Adjectival terms and general terms are not page referenced.) Look to the glossary as you review to check your understanding of the technical terminology.

Multimedia Correlations

This fifth edition introduces the Dynamic Human, Version 2.0, 3-D Visual Guide to Anatomy and Physiology CD-ROM, which interactively illustrates the complex relationships between anatomical structures and their functions in the human body. This program covers each body system, demonstrating clinical concepts, histology, and physiology. The Dynamic Human (dancing man) icon appears in appropriate figure legends to alert the reader to the corresponding information. A list of correlating figures to specific sections of The Dynamic Human, Version 2.0, follows this preface.

A set of five videotapes contains nearly 53 animations of physiological processes integral to the study of human anatomy and physiology. Entitled "WCB's Life Science Animation (LSA) Videotape Series," these videotapes cover such topics as cell division, genetics, and reproduction. A new LSA 3-D Videotape with 42 key biological processes is included in these correlations. Videotape icons appear in appropriate figure legends to alert the reader to these animations. A list of the figures that relate to the animations follows this preface.

World Wide Web

Concepts of Human Anatomy and Physiology has a home page on the World Wide Web. The address is www.mhhe.com/biosci/abio, and is listed at the end of each chapter in a section called "Related Web Sites." At this home page, instructors and students can find up-to-date addresses and hot links to related sites for each chapter.

Supplementary Materials

The supplementary materials that accompany the text are designed to help students in their learning activities and to guide instructors in planning course work and presentations. Following are brief descriptions of these supplements.

1. *Laboratory Manual to accompany Concepts of Human Anatomy and Physiology*, fifth edition (0–697–28428–X) by Kent M. Van De Graaff, Stuart Ira Fox, and Laurence G. Thouin, Jr., has been thoroughly revised. It can be brought to the laboratory and used as a stand-alone manual. It can also be used as a source of quiz and test questions, without recourse to the textbook. However, this manual is closely tied to the textbook, so that students can maximize their laboratory experience by studying the referenced portions of the textbook in conjunction with the laboratory exercises. The exercises have been carefully refined and updated to keep pace with continuous changes in laboratory technology, vendor supply sources, and address updates in computer-assisted instruction and biohazard health concerns.

2. *Instructor's Manual for the Laboratory Manual* (0–697–28429–8) by Laurence G. Thouin, Jr., provides the answers to the questions that appear in the laboratory reports in the Laboratory Manual.
3. *Student Study Guide* (0–697–28430–1) by Kent M. Van De Graaff features the concept statements from the text, focus questions, mastery quizzes, study activities, and answer keys with explanations.
4. *Instructor's Manual and Test Item File* (0–697–28427–1) by Jeffrey and Karianne Prince provides instructional support in the use of the textbook. It also contains a test item file with approximately 70 items for each chapter to aid instructors in creating examinations.
5. *MicroTest III* (0–697–28440–9 Macintosh or 0–697–28439–5 Windows) is a computerized test generator, available free to qualified adopters, which enables instructors to generate tests from questions in the Instructor's Manual.
6. *Visual Resource Library* (0–697–42202–X) is a CD-ROM containing all of the color line art in this textbook that can be incorporated into computer-assisted lecture presentations.
7. *Transparencies* (0–697–28431–X) include 200 color illustrations from this book reprinted as overhead lecture transparencies, packaged in a 3-ring binder.
8. *The Dynamic Human CD-ROM, Version 2.0* (0–697–38935–9) consists of 3-D and other visualizations of relationships between human structure and function.
9. *The Dynamic Human Videodisc* (0–697–38937–5) contains all of the CD-ROM animations, with a bar code directory.
10. *Virtual Physiology Lab CD-ROM* (0–697–37994–9) has 10 simulations of animal-based experiments common in the physiology component of a laboratory course; allows students to repeat experiments for improved mastery.
11. *WCB Anatomy and Physiology Videodisc* (0–697–27716–X) has more than 30 physiological animations, line art, and photomicrographs, with a bar code directory.
12. *WCB's Life Science Animations (LSA)* contains 53 animations on VHS videocassettes; Chemistry, The Cell, and Energetics (0–697–25068–7); Cell Division, Heredity, Genetics, Reproduction, and Development (0–697–25069–5); Animal Biology No. 1 (0–697–25070–9); Animal Biology No. 2 (0–697–25071–7); and Plant Biology, Evolution, and Ecology (0–697–26600–1). Another available videotape is Physiological Concepts of Life Science (0–697–21512–1). A new 3-D videotape (0–07–290652–9) is also available with 42 key biological processes all narrated and animated in vibrant color with dynamic three-dimensional graphics.
13. *WCB Anatomy and Physiology Videotape Series* consists of four videotapes, free to qualified adopters, including Blood Cell Counting, Identification and Grouping (0–697–11629–8); Introduction to the Human Cadaver and Prosection (0–697–11177–6); Introduction to Cat Dissection: Cat Musculature (0–697–11630–1); and Internal Organs and Circulatory System of the Cat (0–697–13922–0).
14. *Human Anatomy and Physiology Study Cards*, third edition (0–697–26447–5) by Kent Van De Graaff, Ward Rhees, and Christopher Creek is a boxed set of 300 illustrated cards (3 × 5 in.), each of which concisely summarizes a concept of structure or function, defines a term, and provides a concise table of related information.
15. *Coloring Guide to Anatomy and Physiology* (0–697–17109–4) by Robert and Judith Stone consists of outline drawings and text that emphasize learning through color association. Students retain information through a meditative exercise in color-coding structures and correlated labels. This can be an especially effective aid for students who more easily remember visual concepts than verbal ones.
16. *An Atlas to Human Anatomy* (0–697–38793–3) by Dennis Strete and Christopher Creek is a new full-color atlas that contains over 200 full-color photographs and over 150 black-and-white illustrations that accompany and portray the necessary detail of human anatomy.
17. *Atlas of the Skeletal Muscles*, second edition (0–697–13790–2) by Robert and Judith Stone illustrates each skeletal muscle in a diagram that the student can color, and provides a concise table of the origin, insertion, action, and innervation of each muscle.
18. *Laboratory Atlas of Anatomy and Physiology*, second edition (0–697–39480–8) by Douglas Eder et al. is a full-color atlas containing histology, human skeletal anatomy, human muscular anatomy, dissections, and reference tables.
19. *Case Histories in Human Physiology*, third edition (0–697–34234–4) by Donna Van Wynsberghe and Gregory Cooley is a web-based workbook that stimulates analytical thinking through case studies and problem solving; includes an instructor's answer key.
20. *Explorations in Human Biology CD-ROM* (0–697–37907–8 Macintosh and 0–697–37906–X Windows) by George Johnson consists of 16 interactive animations of human biology.
21. *Explorations in Cell Biology and Genetics CD-ROM* (0–697–37908–6) by George Johnson contains 17 animations that afford an engrossing way for students to delve into these often-challenging topics.
22. *Life Science Living Lexicon CD-ROM* (0–697–37993–0) by William Marchuk provides interactive vocabulary-

building exercises. It includes the meanings of word roots, prefixes, and suffixes with illustrations and audio pronunciations.

23. *Survey of Infectious and Parasitic Diseases* (0–697–27535–3) by Kent Van De Graaff is a booklet of essential information on 100 of the most significant infectious diseases.

Acknowledgments

This book could not have been written without the enduring patience and support of our wives, Karen Van De Graaff and Ellen Fox, to whom this book is gratefully dedicated.

Many of the improvements in the fifth edition of *Concepts of Human Anatomy and Physiology* came about through comments that we received from the many users of previous editions. Although it would be impossible in this space to acknowledge them individually, we are deeply grateful to each one. As in the past, our colleagues at our respective institutions were very supportive and helpful. In particular, we would like to thank Michael J. Shively, Laurence G. Thouin, Jr., R. Ward Rhees, James Rikel, Samuel I. Zeveloff, and J. Ronald Galli.

We also wish to thank physicians who assisted in specific ways. Drs. Kyle M. Van De Graaff, Eric J. Van De Graaff, and Ryan L. Van De Graaff provided professional advice. Dr. Brent C. Chandler provided many of the radiographs used in the text. Drs. James N. Jones, Harrihar A. Pershadsingh, and Paul Urie assisted in updating the clinical information.

Quality illustrations for this text were provided by a number of talented artists. We are especially grateful for their tremendous contributions. Many of the renderings new to this edition were contributed by Christopher H. Creek and Rictor Lew.

The editorial and production staffs at WCB/McGraw-Hill inspired, guided, and shaped this enormous project, and they were superb to work with. We owe a large debt of gratitude to Sponsoring Editor Kris Tibbetts, Developmental Editor Pat Anglin, Senior Editorial Assistant Darlene Schueller, Senior Project Manager Peggy Selle, Art Editor Brenda Ernzen, and Photo Editor John Leland, and many other talented individuals at WCB/McGraw-Hill. We are also especially appreciative of Ann Mirels and Jane Matthews who laboriously copyedited the manuscript and provided numerous helpful suggestions.

Reviewers

The forthright criticisms and helpful suggestions of a knowledgeable and hard-working panel of reviewers added immeasurably to the quality of the final draft. The review panel for the fifth edition included

L. Amini-Sereshki-Kormi
Worchester State College

Mary A. Anderson
Gustavus Adolphus College

Len M. Archer
Florida Hospital College of Health Sciences

Timothy A. Ballard
University of North Carolina–Wilmington

Steven Bassett
Southeast Community College

Clinton Benjamin
Lower Columbia College

Joanna D. Borucinska
University of Hartford

Julie Harrill Bowers
East Tennessee State University

James M. Britton
Arkansas State University–Beebe

Jennifer Carr Burtwistle
Northeast Community College

Larry I. Crawshaw
Portland State University

John R. Crooks
Iowa Wesleyan College

Weldon “Tex” Davis

Teresa DeGolier
Bethel College

Larry Delay
Waubonsee Community College

Danielle Desroches
William Paterson University of New Jersey

Mike Eoff
Marian College

Kathy McCann Evans
Reading Area Community College

Brian D. Feige
Mott Community College

Gregory R. Garman
Centralia College

William A. Gibson
University of New Orleans

Clare Hays
Metropolitan State College of Denver

Robert W. Hays
University of Pittsburgh School of Medicine

Karen R. Hickman
University of Mary Hardin-Baylor

Robert E. Hillis
Okaloosa-Walton Community College

Melanie W. Jenkins
Formerly of Fayetteville Technical Community College

Ronald K. Jyring
Bismarck State College

Robert J. Keating
Houston Community College–Northwest

Robert Knudsen
San Joaquin Delta College

Ibrahim Y. Mahmoud
University of Wisconsin–Oshkosh

J. Ray Marak, Jr.
San Jacinto College–North

Wilma Jo McNamara
Milwaukee Area Technical College

Melissa Meador
Arkansas State University–Beebe

Anne D. Merkel
Eastern Maine Technical College

Alfredo Munoz
University of Texas
Texas Southmost College

Richard A. Nyhof
Calvin College

Valerie Dean O'Loughlin
Indiana University–Bloomington

John G. Osborne
East Tennessee State University

Glenn Perrigo
Texas A&M University

Jon C. Pigage
University of Colorado at Colorado Springs

Elizabeth Rayhel
Missouri Baptist College

Louis Reed
Lamar University Port Arthur

John M. Ripper
Butler County Community College

Laura H. Ritt
Burlington County College

Steven C. Roschke
School of Nursing
Good Samaritan Hospital

Tim Roye
San Jacinto College–South

Traci M. Santos
Stonehill College

Timothy P. Scott
Texas A&M University

Eileen Kennedy Shull
Scott Community College

Deborah K. Smith
Meredith College

Xavier Stewart
Wilmington College
Wesley College
University of Maryland

Mark F. Taylor
Baylor University

Kent R. Thomas
Wichita State University

M. Thomas
Houston Community College System–Southeast College

Ed W. Thompson
Winona State University

Patricia Turner
Howard Community College

Itzick Vatnick
Widener University

John J. Wielgus
Washington and Lee University

Joseph B. Williams
Ohio State University

Gary R. Wilson
McMurry University

Jerry Woolpy
Earlham College

Harry E. Womack
Salisbury State University

Life Science 3D Animations Correlation Guide

Chapter 2

2.2	Module 1	Atomic Structure and Covalent and Ionic Bonding
2.5	Module 1	Atomic Structure and Covalent and Ionic Bonding
2.8	Module 1	Atomic Structure and Covalent and Ionic Bonding

Chapter 3

3.16	Module 13	Structure of DNA
3.19	Module 18	Transcription
3.22	Module 18	Transcription
	Module 19	Translation
3.28	Module 14	DNA Replication
3.31	Module 10	Mitosis
3.34	Module 11	Meiosis
3.35	Module 12	Crossing Over

Chapter 4

4.2	Module 7	Enzyme Action
4.11	Module 8	Photosynthesis
4.26	Module 9	Electron Transport Chain

Chapter 5

5.1	Module 4	Diffusion
5.4	Module 5	Osmosis
5.16	Module 6	Sodium/Potassium Pump

Chapter 12

12.10	Module 40	Muscle Contraction

Chapter 14

14.13	Module 39	Action Potential

Chapter 19

19.5	Module 41	Hormone Action
19.5	Module 41	Hormone Action

Chapter 23

23.14	Module 33	Complement System
23.20	Module 35	Clonal Selection
23.24	Module 34	How T Lymphocytes Work
23.25	Module 34	How T Lymphocytes Work

Chapter 28

28.16	Module 11	Meiosis

Chapter 29

29.12	Module 11	Meiosis

Life Science Animations Correlation Guide

Chapter 2

2.5	Tape 1	Concept 1	Formation of an Ionic Bond

Chapter 3

3.1	Tape 1	Concept 2	Journey into a Cell
3.4	Tape 1	Concept 3	Endocytosis
3.19	Tape 2	Concept 16	Transcription of a Gene
3.22	Tape 2	Concept 16	Transcription of a Gene
	Tape 2	Concept 17	Protein Synthesis
3.24	Tape 2	Concept 17	Protein Synthesis
3.28	Tape 2	Concept 15	DNA Replication
3.31	Tape 2	Concept 12	Mitosis
3.34	Tape 2	Concept 13	Meiosis
3.34	Tape 2	Concept 14	Crossing Over

Chapter 4

4.2	Tape 6	Concept 1	Lock and Key Model of Enzyme Action
4.11	Tape 1	Concept 8	Photosynthetic Electron Transport Chain and the Production of ATP
	Tape 6	Concept 5	Electron Transport Chain and Oxidative Phosphorylation
4.15	Tape 1	Concept 11	ATP as an Energy Carrier
4.18	Tape 1	Concept 5	Glycolysis
4.19	Tape 1	Concept 5	Glycolysis
4.25	Tape 1	Concept 6	Oxidative Phosphorylation
4.26	Tape 1	Concept 6	Oxidative Phosphorylation
	Tape 1	Concept 7	Electron Transport Chain and the Production of ATP

Chapter 5

5.4	Tape 6	Concept 2	Osmosis
5.15	Tape 6	Concept 3	Active Transport

Chapter 12

12.2	Tape 3	Concept 29	Levels of Muscle Structure
12.10	Tape 3	Concept 30	Sliding Filament Model of Muscle Contraction
12.14	Tape 3	Concept 31	Regulation of Muscle Contraction

Chapter 14

14.5	Tape 3	Concept 22	Formation of Myelin Sheath
14.16	Tape 6	Concept 6	Conduction of Nerve Impulses
14.17	Tape 6	Concept 23	Saltatory Nerve Conduction
14.21	Tape 6	Concept 8	Synaptic Transmission
14.29	Tape 6	Concept 7	Temporal and Spatial Summation

Chapter 16

16.28	Tape 3	Concept 25	Reflex Arcs

Chapter 18

18.16	Tape 3	Concept 26	Organ of Static Equilibrium
18.27	Tape 3	Concept 27	Organ of Corti
18.41	Tape 6	Concept 9	Visual Accommodation

Chapter 19

19.5	Tape 6	Concept 10	Action of Steroid Hormone on Target Cells
19.6	Tape 6	Concept 11	Action of T_3 in Target Cells
19.8	Tape 6	Concept 12	Cyclic AMP Action
	Tape 3	Concept 28	Peptide Hormone Action (cAMP)

Chapter 20

20.5	Tape 4	Concept 40	A, B, O Blood Types

Chapter 21

21.4	Tape 4	Concept 37	Blood Circulation
21.6	Tape 4	Concept 32	Cardiac Cycle and Production of Heart Sounds
21.12	Tape 4	Concept 38	Production of Electrocardiogram
21.13	Tape 4	Concept 32	Cardiac Cycle and Production of Heart Sounds
21.39	Tape 4	Concept 39	Common Congenital Defects of the Heart

Chapter 23

23.8	Tape 6	Concept 13	Life Cycle of HIV
23.10	Tape 4	Concept 41	B-Cell Immune Responses
23.12	Tape 4	Concept 42	Structure and Function of Antibodies
23.22	Tape 4	Concept 43	Types of T-Cells
23.24	Tape 4	Concept 44	Relationship of Helper T Cells and Killer T Cells
23.25	Tape 4	Concept 44	Relationship of Helper T Cells and Killer T Cells

Chapter 26

26.14	Tape 4	Concept 33	Peristalsis
26.39	Tape 4	Concept 35	Digestion of Proteins
26.40	Tape 4	Concept 33	Peristalsis
26.42	Tape 4	Concept 34	Digestion of Carbohydrates
26.44	Tape 4	Concept 36	Digestion of Lipids
26.45	Tape 4	Concept 36	Digestion of Lipids

Chapter 27

27.10	Tape 6	Concept 12	Cyclic AMP Action

Chapter 28

28.16	Tape 2	Concept 13	Meiosis
	Tape 2	Concept 19	Spermatogenesis

Chapter 29

29.12	Tape 2	Concept 13	Meiosis
	Tape 2	Concept 20	Oogenesis

Chapter 30

30.18	Tape 2	Concept 21	Human Embryonic Development
30.19	Tape 2	Concept 21	Human Embryonic Development
30.20	Tape 2	Concept 21	Human Embryonic Development

Dynamic Human 2.0 Correlation Guide

Chapter 1

1.7	Skeletal/Gross Anatomy/Axial Skeleton/Vertebral Column/Intervertebral Disc
1.9	Skeletal/Clinical Concepts/Fractured Femur Digestive/Clinical Concepts/Gallstones
1.10	Human Body/Clinical Concepts/Clinical Imaging
1.12	Human Body/Explorations/Anatomical Orientation/Planes
1.13	Human Body/Explorations/Anatomical Orientation/Planes
TA Table 1.2	Human Body/Explorations/Anatomical Orientation/Directional Terminology

Chapter 3

3.8	Human Body/Anatomy/Cell Components
3.9	Human Body/Anatomy/Cell Components
3.10	Human Body/Anatomy/Cell Components
3.11	Human Body/Anatomy/Cell Components
3.12	Human Body/Anatomy/Cell Components
3.27	Human Body/Anatomy/Cell Components
3.31	Human Body/Explorations/Mitosis

Chapter 6

6.2	Human Body/Histology/Simple Squamous Epithelium
6.4	Human Body/Histology/Simple Columnar Epithelium
6.6	Human Body/Histology/Pseudostratified Ciliated Columnar Epithelium
6.7	Human Body/Histology/Stratified Squamous Epithelium
6.9	Human Body/Histology/Transitional Epithelium
6.17	Human Body/Histology/Dense Irregular Connective Tissue
6.21	Human Body/Histology/Hyaline Cartilage
6.22	Human Body/Histology/Fibrocartilage
6.23	Human Body/Histology/Elastic Cartilage
6.24	Skeletal/Histology/Compact Bone
6.26	Muscular/Histology/Skeletal Muscle (cross section) Muscular/Histology/Skeletal Muscle (longitudinal) Muscular/Histology/Smooth Muscle Muscular/Histology/Cardiac Muscle
6.27	Nervous/Histology/Dorsal Root Ganglion Neurons

Chapter 8

8.1	Skeletal/Anatomy/Gross Anatomy
8.2	Skeletal/Explorations/Cross Section of a Long Bone
8.4	Skeletal/Explorations/Cross Section of a Long Bone
8.6	Skeletal/Histology/Compact Bone

Chapter 9

9.1	Skeletal/Anatomy/Gross Anatomy/Axial Skeleton/Skull Skeletal/Anatomy/3D Viewer: Cranial Anatomy
9.2	Skeletal/Anatomy/Gross Anatomy/Axial Skeleton/Skull Skeletal/Anatomy/3D Viewer: Cranial Anatomy
9.3	Skeletal/Anatomy/Gross Anatomy/Axial Skeleton/Skull Skeletal/Anatomy/3D Viewer: Cranial Anatomy
9.4	Skeletal/Anatomy/Gross Anatomy/Axial Skeleton/Skull Skeletal/Anatomy/3D Viewer: Cranial Anatomy
9.5	Skeletal/Anatomy/Gross Anatomy/Axial Skeleton/Skull Skeletal/Anatomy/3D Viewer: Cranial Anatomy
9.19	Skeletal/Anatomy/Gross Anatomy/Axial Skeleton/Vertebral Column
9.21	Skeletal/Anatomy/Gross Anatomy/Axial Skeleton/Vertebral Column
9.22	Skeletal/Anatomy/Gross Anatomy/Axial Skeleton/Vertebral Column
9.23	Skeletal/Anatomy/Gross Anatomy/Axial Skeleton/Vertebral Column
9.24	Skeletal/Anatomy/Gross Anatomy/Axial Skeleton/Vertebral Column
9.25	Skeletal/Anatomy/Gross Anatomy/Axial Skeleton/Thoracic Cage
9.26	Skeletal/Anatomy/Gross Anatomy/Axial Skeleton/Thoracic Cage

Chapter 10

10.1	Skeletal/Anatomy/Gross Anatomy/Appendicular Skeleton/Pectoral Girdle
10.2	Skeletal/Anatomy/Gross Anatomy/Appendicular Skeleton/Pectoral Girdle

continued

Dynamic Human 2.0 Correlation Guide (*continued*)

10.4	Skeletal/Anatomy/Gross Anatomy/Appendicular Skeleton/Upper Limb
10.5	Skeletal/Anatomy/Gross Anatomy/Appendicular Skeleton/Upper Limb
10.7	Skeletal/Anatomy/Gross Anatomy/Appendicular Skeleton/Upper Limb
10.10	Skeletal/Anatomy/Gross Anatomy/Appendicular Skeleton/Pelvic Girdle
10.12	Skeletal/Anatomy/Gross Anatomy/Appendicular Skeleton/Pelvic Girdle
10.15	Skeletal/Anatomy/Gross Anatomy/Appendicular Skeleton/Lower Limb
10.17	Skeletal/Anatomy/Gross Anatomy/Appendicular Skeleton/Lower Limb
10.18	Skeletal/Anatomy/Gross Anatomy/Appendicular Skeleton/Lower Limb
10.22	Skeletal/Clinical Concepts/Fractured Femur

Chapter 11

11.1	Skeletal/Explorations/Fibrous Joint
11.2	Skeletal/Explorations/Fibrous Joint
11.4	Skeletal/Explorations/Cartilaginous Joint
11.6	Skeletal/Explorations/Synovial Joints
11.8	Skeletal/Explorations/Synovial Joints
11.9	Skeletal/Explorations/Synovial Joints
11.10	Skeletal/Explorations/Synovial Joints
11.11	Skeletal/Explorations/Synovial Joints
11.12	Skeletal/Explorations/Synovial Joints
11.13	Skeletal/Explorations/Synovial Joints
11.17	Skeletal/Explorations/Synovial Joints
11.28	Skeletal/Clinical Concepts/MRI of Knee
11.32	Skeletal/Clinical Concepts/Joint Disorders (Arthritis)

Chapter 12

12.1	Muscular/Anatomy/Skeletal Muscle
12.2	Muscular/Histology/Skeletal Muscle
12.4	Muscular/Explorations/Isometric vs. Isotonic Contraction
12.5	Muscular/Explorations/Neuromuscular Junction
12.10	Muscular/Explorations/Sliding Filament Theory

Chapter 13

13.1	Muscular/Anatomy/Body Regions
13.2	Muscular/Explorations/Muscle Action around Joints

Chapter 15

15.2	Human Body/Clinical Concepts/Clinical Imaging
15.4	Nervous/Anatomy/Gross Anatomy
15.5	Nervous/Anatomy/Gross Anatomy
15.7	Nervous/Explorations/Motor and Sensory Pathways
15.9	Nervous/Anatomy/Gross Anatomy
15.23	Nervous/Anatomy/Spinal Cord Anatomy Nervous/Histology/Spinal Cord
15.24	Nervous/Explorations/Motor and Sensory Pathways
15.25	Nervous/Explorations/Motor and Sensory Pathways

Chapter 16

16.28	Nervous/Explorations/Reflex Arc

Chapter 18

18.11	Nervous/Explorations/Motor and Sensory Pathways
18.12	Nervous/Explorations/Taste
18.13	Nervous/Explorations/Zones of Taste
18.14	Nervous/Explorations/Olfaction
18.16	Nervous/Explorations/Static Equilibrium Nervous/Explorations/Dynamic Equilibrium
18.18	Nervous/Explorations/Static Equilibrium Nervous/Explorations/Dynamic Equilibrium
18.26	Nervous/Explorations/Hearing
18.27	Nervous/Histology/Organ of Corti
18.33	Nervous/Histology/Eye
18.38	Nervous/Explorations/Vision
18.42	Nervous/Histology/Retina
18.55	Nervous/Explorations/Nearsightedness vs. Farsightedness

Chapter 19

19.1	Endocrine/Anatomy/Gross Anatomy
19.5	Endocrine/Explorations/Endocrine Function
19.6	Endocrine/Explorations/Endocrine Function
19.12	Endocrine/Anatomy/Gross Anatomy/Hypothalamus and Pituitary Gland
19.16	Endocrine/Explorations/Hypothalamo-Pituitary-Thyroid Axis
19.18	Endocrine/Anatomy/Gross Anatomy/Adrenal Glands Endocrine/Histology/Adrenal Cortex Endocrine/Histology/Adrenal Medulla
19.21	Endocrine/Anatomy/Gross Anatomy/Thyroid
19.22	Endocrine/Histology/Thyroid
19.27	Endocrine/Anatomy/Gross Anatomy/Pancreas

Chapter 21

21.1	Cardiovascular/Anatomy/3D Viewer: Thoracic Anatomy
21.2	Cardiovascular/Anatomy/Gross Anatomy of the Heart
21.6	Cardiovascular/Explorations/Heart Dynamics/Heart Sounds
21.10	Cardiovascular/Explorations/Heart Dynamics/Conduction System Cardiovascular/Explorations/Heart Dynamics/Electrocardiogram
21.12	Cardiovascular/Explorations/Heart Dynamics/Conduction System Cardiovascular/Explorations/Heart Dynamics/Electrocardiogram
21.13	Cardiovascular/Explorations/Heart Dynamics/Cardiac Cycle
21.14	Cardiovascular/Explorations/Generic Vasculature
21.35	Cardiovascular/Explorations/Generic Portal System

Chapter 22

22.9	Immune & Lymphatic/Anatomy/Gross Anatomy

Chapter 23

23.1	Immune & Lymphatic/Anatomy/Gross Anatomy
23.2	Immune & Lymphatic/Explorations/Lymph Formation and Movement
23.3	Immune & Lymphatic/Anatomy/Microscopic Components
23.5	Immune & Lymphatic/Histology/Spleen
23.7	Immune & Lymphatic/Explorations/Non-Specific Immunity/Phagocytosis
23.8	Immune & Lymphatic/Clinical Concepts/HIV-AIDS
23.10	Immune & Lymphatic/Explorations/Specific Immunity/Antibody-Mediated Immunity
23.16	Immune & Lymphatic/Explorations/Non-Specific Immunity/Inflammation
23.22	Immune & Lymphatic/Explorations/Specific Immunity
23.24	Immune & Lymphatic/Explorations/Specific Immunity/T-Helper Cell Function
23.25	Immune & Lymphatic/Explorations/Specific Immunity/Cytotoxic T-Cell Function
23.26	Immune & Lymphatic/Explorations/Specific Immunity

Chapter 24

24.1	Respiratory/Anatomy/Gross Anatomy
24.2	Respiratory/Anatomy/Gross Anatomy
24.3	Respiratory/Histology

24.7	Respiratory/Anatomy/Gross Anatomy
24.8	Respiratory/Histology/Trachea
24.11	Human Body/Explorations/Visible Human/Male/Thorax Human Body/Explorations/Visible Human/Female/Thorax
24.19	Respiratory/Explorations/Mechanics of Breathing
24.21	Respiratory/Clinical Concepts/Spirometry
24.22	Respiratory/Clinical Concepts/Spirometry
24.25	Respiratory/Explorations/Gas Exchange
24.37	Respiratory/Explorations/Oxygen Transport

Chapter 25

25.1	Urinary/Anatomy/Gross Anatomy
25.3	Urinary/Anatomy/Kidney Anatomy
25.10	Urinary/Explorations/Urine Formation
25.12	Urinary/Explorations/Urine Formation
25.14	Urinary/Explorations/Urine Formation
25.17	Urinary/Explorations/Urine Formation
25.20	Urinary/Explorations/Urine Formation
25.30	Urinary/Anatomy/Gross Anatomy

Chapter 26

26.2	Digestive/Anatomy/Gross Anatomy Digestive/Anatomy/3D Viewer: Digestive Anatomy
26.6	Digestive/Explorations/Oral Cavity
26.7	Digestive/Explorations/Oral Cavity
26.16	Digestive/Histology/Fundic Stomach
26.19	Digestive/Histology/Duodenum Digestive/Histology/Duodenal Villi
26.21	Digestive/Histology/Duodenum Digestive/Histology/Duodenal Villi
26.28	Digestive/Clinical Concepts/Barium Radiograph
26.29	Human Body/Explorations/Visible Human/Male/Abdomen Human Body/Explorations/Visible Human/Female/Abdomen
26.30	Digestive/Histology/Liver
26.39	Digestive/Explorations/Digestion
26.42	Digestive/Explorations/Digestion
26.44	Digestive/Explorations/Digestion
26.45	Digestive/Explorations/Digestion

Chapter 28

28.10	Reproductive/Anatomy/3D Viewer: Male Reproductive Anatomy Reproductive/Anatomy/Male Reproductive Anatomy
28.12	Reproductive/Histology/Testis
28.16	Reproductive/Explorations/Male: Spermatogenesis
28.17	Reproductive/Explorations/Male: Spermatogenesis
28.29	Reproductive/Clinical Concepts/Male: Vasectomy

Chapter 29

29.1	Reproductive/Anatomy/3D Viewer: Female Reproductive Anatomy Reproductive/Anatomy/Female Reproductive Anatomy
29.2	Reproductive/Histology/Uterine Tube
29.4	Reproductive/Anatomy/3D Viewer: Female Reproductive Anatomy Reproductive/Anatomy/Female Reproductive Anatomy
29.8	Reproductive/Histology/Ovarian Follicle
29.10	Reproductive/Histology/Ovary
29.12	Reproductive/Explorations/Female: Oogenesis
29.15	Reproductive/Explorations/Female: Ovarian Cycle Reproductive/Explorations/Female: Menstrual Cycle
29.23	Reproductive/Clinical Concepts/Female: Tubal Ligation
29.25	Reproductive/Clinical Concepts/Female: Breast Cancer

Chapter 30

30.5	Reproductive/Explorations/Female: Ovarian Cycle
30.13	Reproductive/Explorations/Female: Amniocentesis

ONE

Introduction to Anatomy and Physiology

OBJECTIVES

- List some of the historic events that helped to define the sciences of anatomy and physiology.
- Explain what is meant by the scientific method and discuss its significance in furthering our understanding of anatomy and physiology.
- Describe the taxonomic classification of humans.
- List the characteristics that identify humans as chordates and mammals.
- Describe the different levels of organization in the human body.
- Describe the general function of each body system.
- Identify the planes of reference used to depict the structural arrangement of the human body.
- Describe the anatomical position and use the descriptive and directional terms that refer to body structures, surfaces, and regions.
- List the regions of the body and the localized areas within each region.
- Identify the body cavities and list the organs found within each.
- Define *homeostasis* and explain how negative feedback helps to maintain homeostasis.

Clinical Investigation

A student, enjoying a ski vacation during winter break, was brought into the emergency room of a hospital with a high fever. She stated that severe pain in her lower right abdomen caused her to fall during a ski run. While down, she was hit by an out-of-control skier. The tip of his ski rammed into her right popliteal fossa, and his ski pole punctured her left, lateral thoracic cavity, just below her axillary fossa. This caused collapse of the left lung, but she was still able to breathe. Laboratory analysis revealed a white blood cell count that was higher than normal. In layperson's terms, what happened to this student?

Clues: Study the section on directional terms, body regions, and body cavities to locate the sites of the injuries. Check the discussion of the thoracic cavity to learn why our patient could still breathe even though one of her lungs had collapsed, and examine figure 1.15 to determine which organ located in the lower right abdomen might be inflamed to cause a fever and high white blood cell count.

The Sciences of Anatomy and Physiology

Anatomy and physiology are integrated, dynamic sciences with an exciting heritage. These sciences provide the foundation for personal health and clinical applications.

Human anatomy and physiology are sciences concerned with the structure and function of the body. The term **anatomy** (*ă-nat′ŏ-me*) is derived from a Greek word meaning "to cut up"; indeed, in ancient times, the word *anatomize* was more common than the word *dissect*. Dissection of human *cadavers* (*kă-dav′erz*) is the basis for understanding the structure of the human body. The science of **physiology** (*fiz″e-ol′ ŏ-je*) is concerned with explaining how the body functions through physical and chemical processes. The term *physiology* is derived from another Greek word—this one meaning the "study of nature." The "nature" of an organism is its function. Much of the knowledge of physiology is gained through experimentation.

Anatomy and physiology are both subdivisions of the science of *biology*, the study of living organisms. Frequently, anatomy and physiology are studied as separate disciplines, in which case the anatomy of the body is learned before its physiology. However, since anatomical structures are adapted to perform specific functions, a proper understanding of structure and function is best achieved through an integrated study.

anatomy: Gk. *ana*, up; *tome*, a cutting
cadaver: L. *cadere*, to fall
physiology: Gk. *physis*, nature; *logos*, study

Figure 1.1

A fourteenth-century painting of Hippocrates.
The famous Greek physician Hippocrates is referred to as the father of medicine; his creed is immortalized as the Hippocratic oath.

Historical Development

The study of human anatomy and physiology has had a rich, long, and frequently troubled heritage. Its history parallels that of medicine because interest in the structure and function of the body frequently grew out of a desire to understand and treat body dysfunctions.

Grecian Period

Anatomy and physiology first found wide acceptance as sciences in ancient Greece. *Hippocrates* (460–337 B.C.) was a famous Greek physician who is regarded as the father of medicine (fig. 1.1). Perhaps Hippocrates' greatest contribution was his attribution of disease to natural causes rather than to the displeasure of the gods. His application of logic and reason to medical study was the beginning of observational medicine.

The field of medicine at the time of Hippocrates held to the notion that an individual's health depended upon the balance of four body fluids, or **humors.** This so-called *humoral theory* suggested that, if blood, bile, black bile, and phlegm were balanced, the person would be healthy and have an even disposition. If blood was the predominant humor, one would have a *sanguine* personality—courageous and passionate. If there was too much bile, one would be *choleric*—angry and mean. A *melancholic* personality—moody and depressed—resulted from an overproduction of black bile. Too much phlegm resulted in a *phlegmatic* personality—sluggish and apathetic. Although medicine has long since abandoned this explanation of health and personality, it is interesting that these terms are still used in our language. The term *sanguine*, however, has evolved to refer simply to cheerfulness and optimism, and no longer refers directly to the humoral theory.

Aristotle (384–332 B.C.) made careful investigations of all kinds of animals, including humans, and pursued a limited type of scientific method in obtaining data (fig. 1.2). He wrote the first known account of embryology, in which he described the development of the heart in a chick embryo. His best-known zoological works are *History of Animals*, *Parts of Animals*, and *Generation of Animals*.

Despite his tremendous accomplishments, Aristotle perpetuated some erroneous theories regarding human anatomy. For example, he disagreed with Plato, who had described the brain as the seat of feeling and thought, and proclaimed the heart to be the seat of intelligence. Aristotle thought that the function of the brain, which was bathed in fluid, was to cool the blood that was pumped from the heart and thus maintain body temperature.

The Greek scientist *Erasistratus* (about 300 B.C.) was more interested in body functions than structure, and is therefore frequently referred to as the father of physiology. Erasistratus authored a book on the causes of diseases, in which he included observations on the heart, vessels, brain, and cranial nerves. He noted the toxic effects of snake venom on various internal organs and described changes in the liver resulting from certain metabolic diseases. Although some of the writings of Erasistratus were scientifically accurate, others were based on primitive and mystical concepts. For instance, he thought that the cranial nerves carried animal spirits and that muscles contracted because of distention by spirits.

Both Erasistratus and another Greek philosopher, *Herophilus* (about 325 B.C.), were greatly criticized later in history for their use of *vivisection* (*viv″ ĭ-sek′shun*), the dissection of living animals. Herophilus was described as a "butcher of men" who had dissected as many as 600 living human beings, some of them in public demonstrations.

humor: L. *humor*, fluid
sanguine: L. *sanguis*, bloody
choler: Gk. *chole*, bile
melancholic: Gk. *melan*, black; *chole*, bile
phlegm: Gk. *phlegm*, inflammation
vivisection: L. *vivus*, living; *sectus*, to cut

Figure 1.2

A sculpture of Aristotle, the famous Greek philosopher.

Roman Era

In many respects, the Roman Empire stifled scientific advancements and set the stage for the Dark Ages. The interest and emphasis of science shifted from the theoretical to the practical under Roman rule. Few dissections of cadavers were performed other than at autopsies in attempts to determine the cause of death in criminal cases. Medicine was not preventive but was limited, almost without exception, to the treatment of soldiers injured in battle.

Claudius Galen (A.D. 130–201) was a Greek living under Roman domination. Galen was the most famous physician of his time and the most influential writer to date on medical subjects. For nearly 1,500 years, the writings of Galen represented the ultimate authority on anatomy and medical treatment. Galen probably dissected no more than two or three human cadavers during his career, of necessity limiting his anatomical descriptions to animal dissections. He compiled nearly 500 medical papers (of which 83 have been preserved)

autopsy: Gk. *autopsia*, seeing with one's own eyes

from earlier works of others, as well as from his personal studies. Galen believed in the humors of the body and perpetuated this concept. He also gave authoritative explanations for nearly all body functions.

Galen's works contain many errors, primarily because of his desire to draw definitive conclusions regarding human body functions on the basis of data obtained largely from nonhuman animals. He did, however, provide some astute and accurate anatomical details that are still regarded as classics. He proved to be an experimentalist, demonstrating that the heart of a pig would continue to beat when the spinal nerve was transected so that nerve impulses could not reach the heart. He showed that the squealing of a pig stopped when the particular nerve that innervated its vocal cords was cut. He also proved that arteries contained blood rather than air.

Middle Ages

The Middle Ages (Dark Ages) came with the fall of the Roman Empire in A.D. 476 and lasted nearly 1,000 years. Dissections of cadavers were totally prohibited during this period, and molesting a corpse was a criminal act that was frequently punished by burning at the stake. If mysterious deaths occurred only examinations by inspection and palpation were allowed. During the plague epidemic in the sixth century, however, a few *necropsies* (*nek'rop-sēz*) and dissections were performed in hopes of determining the cause of this dreaded disease.

Renaissance

The period known as the Renaissance was characterized by a rebirth of science. Lasting roughly from the fourteenth through the sixteenth century, the Renaissance was a period of transition between the Middle Ages and the modern age of science. The development of movable type in about 1450 revolutionized the production of printed books and helped to usher in the Renaissance.

The major advancements in anatomy that occurred during the Renaissance were in large part due to the artistic and scientific ability of *Andreas Vesalius* (1514–64). By the time he was 28 years old, Vesalius had completed the masterpiece of his life, *De Humani Corporis Fabrica*, in which he beautifully illustrated and described the various body systems and individual organs (fig. 1.3). Because of the eventual impact of this book, Vesalius is often called the father of anatomy. His book was especially important because it boldly challenged Galen's erroneous teachings. Vesalius wrote of his surprise upon finding numerous anatomical errors that were being taught as fact, and he refused to accept Galen's explanations on faith. Because he was so outspokenly opposed to Galen's writings, he incurred the wrath of many of the traditional anatomists, including his former teacher Sylvius (Jacques Dubois) of Paris. Sylvius even went so far as to give him the nickname Vesanus (for "madman"). Vesalius became so unnerved by these relentless attacks that he destroyed much of his unpublished work and ceased his dissections.

Figure 1.3

Plates from *De Humani Corporis Fabrica* published in 1543. Completed by Vesalius at the age of 28, this book revolutionized anatomy and physiology.

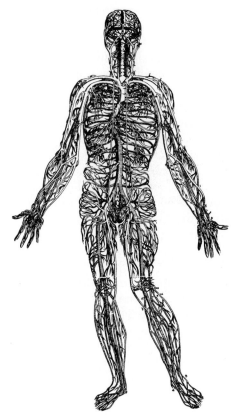

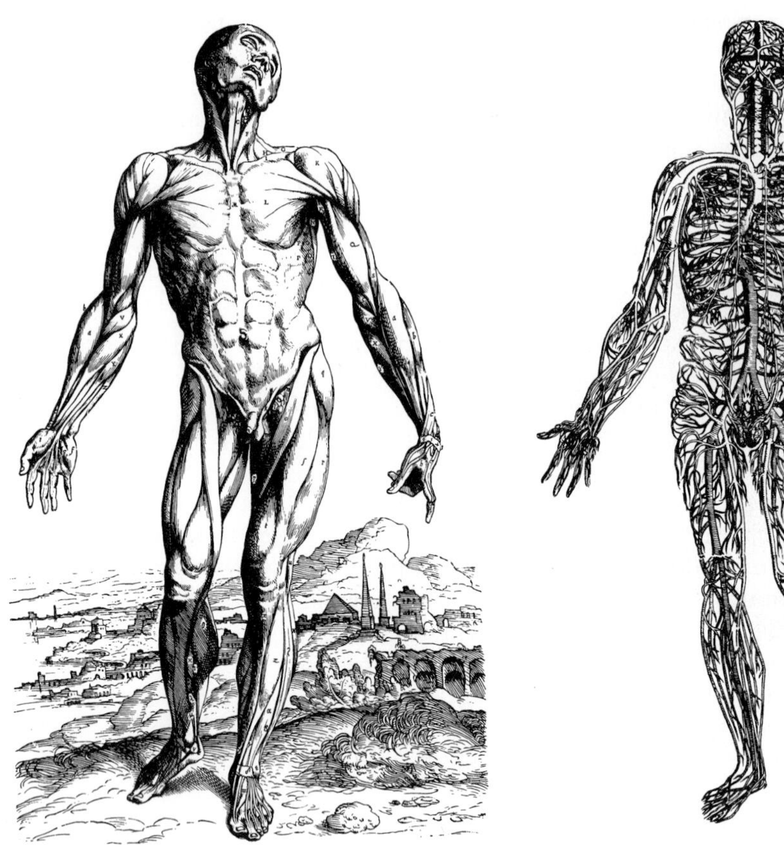

Seventeenth and Eighteenth Centuries

Two of the most significant contributions to anatomy and physiology of the seventeenth and eighteenth centuries were the explanation of blood flow and the development and use of the microscope.

In 1628, the English physician *William Harvey* (1578–1657) published his pioneering work *On the Movement of the Heart and Blood in Animals*. Not only did this brilliant research establish proof of the continuous circulation of blood within vessels, it also provided a classic example of the scientific method of investigation (fig. 1.4). Harvey is widely

necropsy: Gk. *nekros*, corpse; *opsy*, view

Figure 1.4

A painting of the English physician William Harvey. In the early seventeenth century, the English physician William Harvey demonstrated that blood circulates and does not flow back and forth through the same vessels.

regarded as the father of modern physiology, although like Vesalius, he was severely criticized in his time for his departure from Galenic philosophy. The controversy over the circulation of the blood raged for 20 years until other anatomists finally repeated Harvey's experiments and confirmed his observations.

Antoni van Leeuwenhoek (1632–1723) was a Dutch lens grinder who so improved the microscope that he achieved a magnification of 270 times. His many contributions included developing techniques for examining tissues and describing blood cells, spermatozoa, and the striated appearance of skeletal muscle.

The development of the microscope added an entirely new dimension to anatomy and physiology and eventually led to explanations of basic body functions. In addition, the improved microscope was invaluable for understanding the etiologies (causes) of diseases, and thus for discovering cures for many of them.

Nineteenth and Twentieth Centuries

The major contribution in the nineteenth century was the formulation of the biological principle known as the **cell theory** and the implications it had for a clearer understanding of the structure and functioning of the body. According to the cell theory, all living organisms are composed of cells and the products of cells.

Johannes Müller (1801–58), a German physiologist and comparative anatomist, is noted for applying the sciences of physics, chemistry, and psychology to the study of the human body. *Claude Bernard* (1813–78), a French physiologist, extended the experimental study of physiology and is often regarded as the father of experimental medicine. The study of anatomy and physiology during the twentieth century has become highly specialized, and the research more detailed and complex. In response to the increased technology and depths of understanding, new disciplines and specialties have emerged (fig. 1.5). The explosive growth of knowledge in these specialties in recent times has made it impossible for any one individual to remain expert in them all.

Human anatomy and physiology are dynamic, applied sciences that are constantly changing as new discoveries are made. Keeping up with the changes requires a solid and broad understanding of these sciences; hence, an objective of this text is to provide such a foundation for the student. An excellent way to remain informed about anatomy and physiology is to read scientific magazines like *Science*, *Scientific American*, *Discover*, and *Science News*. To be an educated contributor to society, becoming and staying informed is essential.

Scientific Method

Researchers in anatomy and physiology employ the processes of scientific inquiry in conducting their investigations and making discoveries. Although the boundaries of scientific investigations leave room for considerable creativity, they progress in well-defined, orderly ways. Such a disciplined approach to gaining information about the world is referred to as the **scientific method.** Simply stated, the scientific method depends on a systematic search for information and a continual checking and rechecking to see whether previous ideas still hold up in the light of new information. Actually, the scientific method may be used in seeking answers to questions encountered in everyday life, as well as in methodical research.

All of the information presented in this text has been gained through the application of the scientific method. Although many different techniques are involved, all share three attributes: (1) confidence that natural phenomena are ultimately explainable in terms we can understand; (2) descriptions and explanations of the natural world that are honestly based on observations and that are subject to modification or refutation as a result of other observations; and (3) humility, or the willingness to accept the fact that we could be wrong. If further study should yield conclusions that refuted all or part of an idea, the idea would have to be modified accordingly. In short, the scientific method is based on a confidence in our rational ability, honesty, and humility. Practicing scientists may not always display these attributes, but the validity of the large body of scientific knowledge that has been accumulated—as evidenced by technological applications and the predictive value of scientific hypotheses—are ample testimony to the fact that the scientific method works.

Figure 1.5

Some of the subdivisions, or specialties, of human anatomy and physiology.

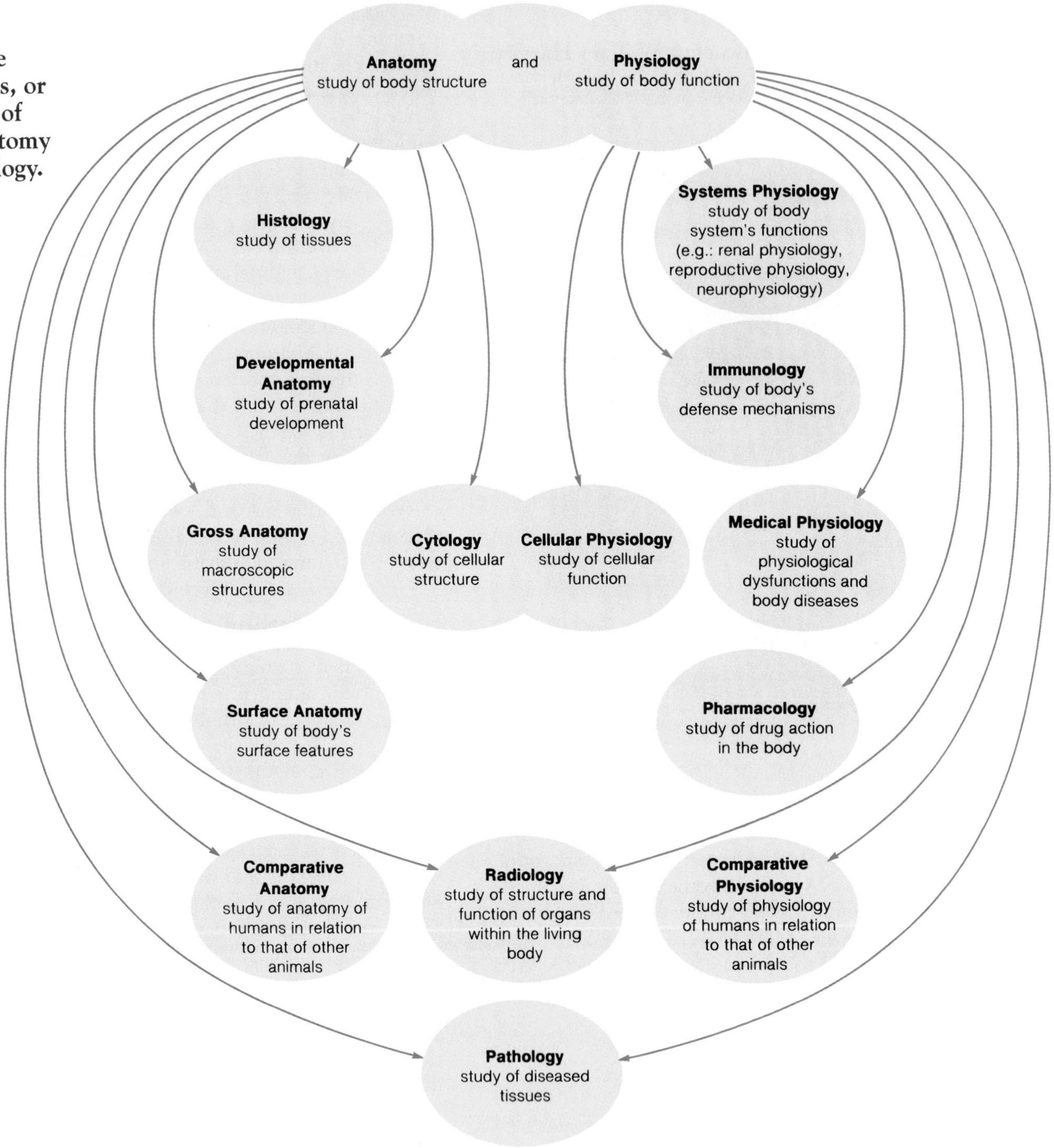

Anatomy study of body structure and **Physiology** study of body function

Histology study of tissues

Systems Physiology study of body system's functions (e.g.: renal physiology, reproductive physiology, neurophysiology)

Developmental Anatomy study of prenatal development

Immunology study of body's defense mechanisms

Gross Anatomy study of macroscopic structures

Cytology study of cellular structure

Cellular Physiology study of cellular function

Medical Physiology study of physiological dysfunctions and body diseases

Surface Anatomy study of body's surface features

Pharmacology study of drug action in the body

Comparative Anatomy study of anatomy of humans in relation to that of other animals

Radiology study of structure and function of organs within the living body

Comparative Physiology study of physiology of humans in relation to that of other animals

Pathology study of diseased tissues

The scientific method involves specific steps. In the first step, a **hypothesis** is formulated. In order for this hypothesis to be scientific, it must be *testable;* that is, it must be open to possible refutation by experiments or other observations of the natural world. For example, one might hypothesize that people who exercise regularly have a lower resting pulse rate than people who don't. Experiments are conducted, or other observations are made, and the results are analyzed. Conclusions are then drawn as to the validity of the hypothesis. If the hypothesis survives such testing, it might be incorporated into a more general **theory.** Scientific theories are statements about the natural world that incorporate a number of proven hypotheses. They serve as a logical framework by which these hypotheses can be interrelated and provide the basis for predictions that may as yet be untested.

The hypothesis in the preceding example is scientific because it can be tested. The pulse rates of 100 athletes and 100 sedentary people can be measured to see whether there are statistically significant differences. If the pulse rates do vary, the statement that athletes, on the average, have lower resting pulse rates than sedentary people is justified *based on these data.* But one must still keep in mind that this conclusion could be wrong. Before the discovery could become generally accepted as fact, other scientists would have to consistently replicate the results. Scientific theories are based on *reproducible data*.

It is quite possible that in attempting to replicate the experiment, other scientists will obtain slightly different results. They may, for example, construct scientific hypotheses that the differences in resting pulse rate also depends on the nature of the exercise performed, or on other variables. When scientists attempt to test these hypotheses, they will likely encounter new problems, requiring new hypotheses, which then must be tested by additional experiments.

In this way, a large body of highly specialized information is gradually accumulated and a more generalized explanation (a scientific theory) can be formulated. This explanation will almost always be different from preconceived notions. People who follow the scientific method will then appropriately modify their concepts, realizing that their new ideas will probably have to be changed again as additional experiments are performed.

Classification and Characteristics of Humans

Humans are biological organisms belonging to the phylum Chordata within the kingdom Animalia and to the family Hominidae within the class Mammalia and the order Primates.

Taxonomic Scheme

In the classification, or taxonomic, system established by biologists to organize the structural and evolutionary relationships of living organisms, each category of classification is referred to as a *taxon*. The highest taxon is the kingdom, and the most specific taxon is the species. Humans are species belonging to the **animal kingdom.** *Phylogeny* (*fi-loj'ĕ-ne*) is the science that studies relatedness on the basis of taxonomy. A phylogenetic tree of animal taxa can be constructed, much like a family tree, to show genealogy and relationships of different animal groups.

Human beings belong to the **phylum Chordata** (*fi'lum kor-dă-tă*) and **subphylum vertebrata,** along with fish, amphibians, reptiles, birds, and other mammals. All chordates have three structures in common: a **notochord** (*no'tŏ-kord"*), a **dorsal hollow nerve cord,** and **pharyngeal** (*fă-rin'je-al*) **pouches** (fig. 1.6). These chordate characteristics are well expressed during the embryonic period of development, and to a certain extent, are present in an adult.

The notochord is a flexible rod of tissue that extends the length of the back of an embryo. A portion of the notochord persists in the adult as the *nucleus pulposus*, which is the gelatinous center located within each intervertebral disc (fig. 1.7). The dorsal hollow nerve cord is positioned above the notochord and develops into the *brain* and *spinal cord*. Pharyngeal pouches form gill openings in fish and some amphibians. In other chordates, such as humans, embryonic pharyngeal pouches develop, but only one of the pouches persists,

Figure 1.6

A schematic diagram of a chordate embryo.
The three diagnostic chordate characteristics are indicated in boldface type.

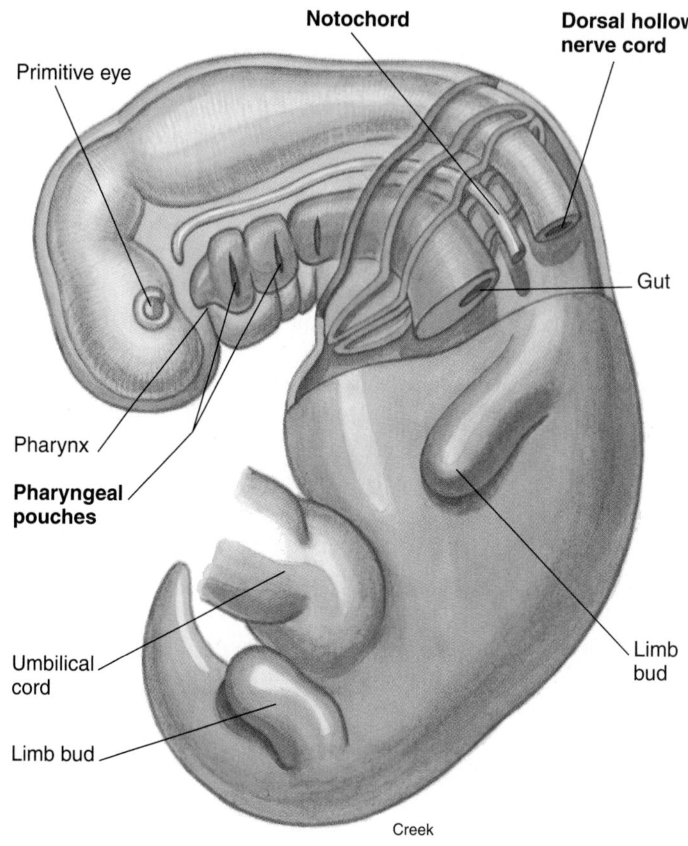

Figure 1.7

Vertebrae and intervertebral discs.
A lateral view of three vertebrae from the vertebral column showing the intervertebral discs and, to the right, a superior view of an intervertebral disc showing the nucleus pulposus.

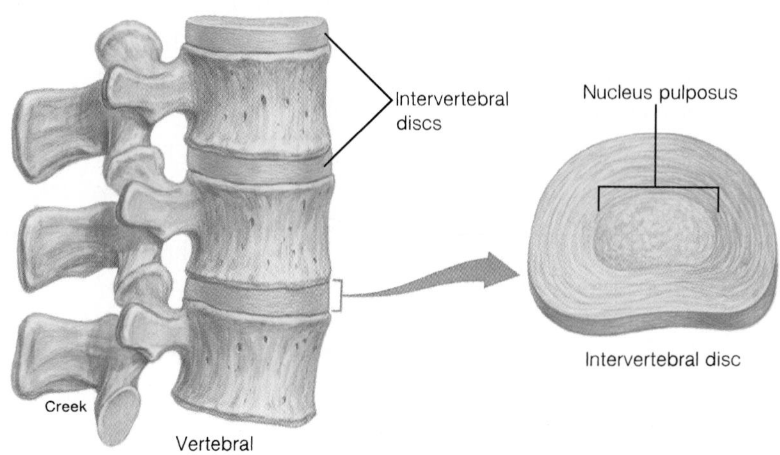

becoming the middle-ear cavity. The *auditory* (eustachian) (*yoo-sta'shun*) *tube* is a persisting connection between the middle-ear cavity and the pharynx (*far'ingks*) (throat area).

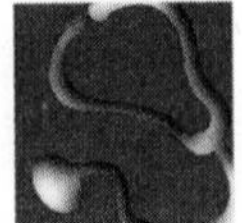

The function of an intervertebral disc and its nucleus pulposus is to allow flexibility between vertebrae for movement of the entire spinal column while preventing compression. Spinal nerves exit between vertebrae, and the discs maintain the spacing to avoid nerve damage. A "slipped disc," resulting from straining the back, is a misnomer. What actually occurs is a herniation, or rupture, because of a weakened wall of the nucleus pulposus. This may cause severe pain as a nerve is compressed.

Humans are in the **class mammalia,** which are vertebrates that possess hair and mammary glands. Hair is a thermoregulatory protective covering for most mammals, and mammary glands serve for suckling the young. Other characteristics of mammals include three small auditory ossicles (ear bones), a fleshy outer ear (auricle), heterodont dentition (teeth, such as incisors and molars, that are shaped differently), a temporomandibular joint (a joint between the lower jaw and skull), usually seven cervical vertebrae, an attached placenta, well-developed facial muscles, a muscular diaphragm, and a four-chambered heart with a left aortic arch.

Humans are in the **order primates,** along with monkeys and apes. Members of this order have prehensile hands (with the digits modified for grasping) and relatively large, well-developed brains.

Humans are the sole living members of the **family Hominidae.** *Homo sapiens* is included within this family, to which all the varieties or ethnic groups of humans belong. Our classification pedigree is presented in table 1.1.

Human Characteristics

As human beings, we possess certain anatomical characteristics that are so specialized that they distinguish us from other animals, and even from other closely related mammals. We also have characteristics that are equally well developed in other animals, but when these function with the human brain, they provide remarkable and unique capabilities. Our anatomical characteristics include the following:

1. **A large, well-developed brain.** The adult human brain weighs between 1,350 and 1,400 grams (3 pounds). This gives us a large brain-to-body-weight ratio. But more important is the development of portions of the brain. Certain extremely specialized regions and structures within the brain account for emotion, thought, reasoning, memory, and even precise, coordinated movement.

heterodont: Gk. *heteros*, other; *odontos*, tooth
placenta: L. *placenta*, flat cake
primates: L. *primas*, first
prehensile: L. *prehensus*, to grasp

Table 1.1
Classification of Human Beings

Taxon	Designated Grouping	Characteristics
Kingdom	Animalia	Eucaryotic cells without cell walls, plastids, or photosynthetic pigments
Phylum	Chordata	Dorsal hollow nerve cord; notochord; pharyngeal pouches
Subphylum	Vertebrata	Vertebral column
Class	Mammalia	Mammary glands; hair
Order	Primates	Well-developed brain; prehensile hands
Family	Hominidae	Large cerebrum; bipedal locomotion
Genus	*Homo*	Flattened face; prominent chin and nose with inferiorly positioned nostrils
Species	*sapiens*	Largest cerebrum

2. **Bipedal locomotion.** Because humans stand and walk on two appendages, our style of locomotion is said to be *bipedal. Upright posture* imposes other diagnostic structural features, such as the *sigmoid* (S-shaped) *curvature* of the spine, the anatomy of the hip and thighs, and arched feet. Some of these may cause clinical problems in older individuals.
3. **An opposable thumb.** The human thumb is structurally adapted for tremendous versatility in grasping objects. The saddle joint at the base of the thumb allows a wide range of movement (see fig. 11.12). All primates have opposable thumbs.
4. **Well-developed vocal structures.** Humans, like no other animals, have developed articulated speech. The anatomical structure of our vocal organs (larynx, tongue, and lips) and our well-developed brain have made this possible.
5. **Stereoscopic vision.** Although this characteristic is well developed in several other animals, it is also keen in humans. Our eyes are directed forward so that when we focus on an object, we view it from two angles. Stereoscopic vision gives us depth perception, or a three-dimensional image.

We also differ from other animals in the number and arrangement of vertebrae (vertebral formula), the kinds and number of teeth (tooth formula), the degree of development of our facial muscles (allowing for a wide range of facial expression), and the structural organization of various body organs.

bipedal: L. *bi*, two; *pedis*, foot

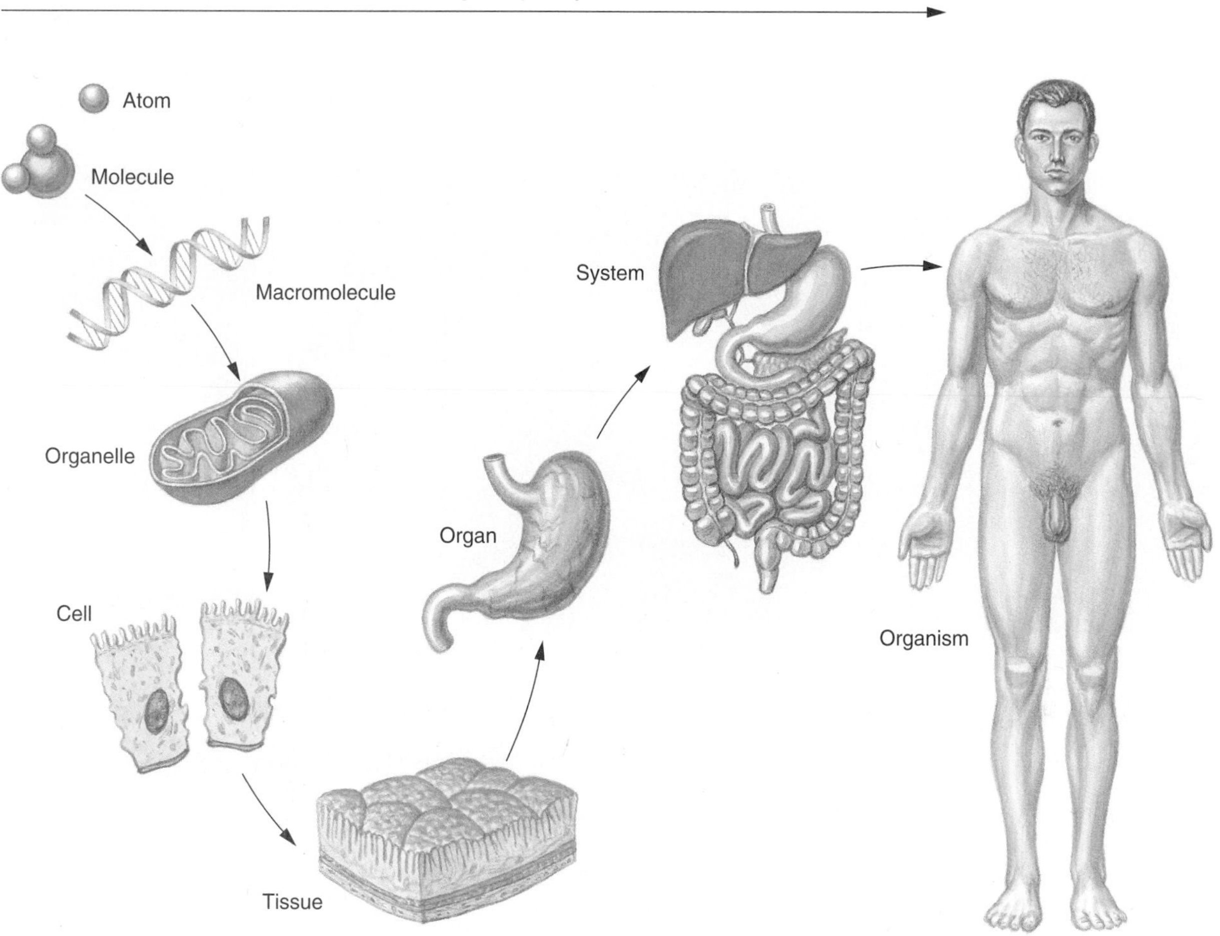

Figure 1.8
The levels of structural organization and complexity within the human body.

Body Organization

Structural and functional levels of organization characterize the human body, and each of its parts contributes to the total organism.

Cellular Level

The **cell** is the basic structural and functional component of life. Humans are multicellular organisms composed of 60 to 100 trillion cells. At this microscopic cellular level (fig. 1.8), vital functions of life such as metabolism, growth, irritability (responsiveness to stimuli), repair, and replication are carried on. All cells contain a semifluid substance called **protoplasm** (*pro'tŏ-plaz'em*). Certain protoplasmic structures and molecules are arranged into small functional units called **organelles** (*or"gă-nelz*). Each organelle carries out a specific function within the cell.

The human body contains many distinct kinds of cells, each specialized to perform specific functions. Examples of specialized cells are bone cells, muscle cells, fat cells, blood cells, and nerve cells. The unique structure of each of these cell types is directly related to its function.

Tissue and Organ Levels

Tissues are layers or aggregations of similar cells that perform a common function. The entire body is composed of only four primary tissue types: *epithelial*, *connective*, *muscular*, and *nervous* tissue. The outer layer of skin is a tissue (epithelium) because it is composed of similar cells that together serve as a protective shield for the body. *Histology* is the science concerned with the microscopic study of tissues, which will be discussed fully in chapter 6.

An **organ** is an aggregate of two or more tissue types that performs a specific function. Organs occur throughout the body and vary greatly in size and function. Examples of organs are the heart, spleen, pancreas, ovary, skin, and even any of the bones within the body. Each organ usually has one or more primary tissues and several secondary tissues. In the stomach, for example, the inside epithelial lining is considered the primary tissue because the basic functions of secretion and absorption occur within this layer. Secondary tissues of the stomach are the supporting connective tissue and vascular, nervous, and muscle tissues.

cell: L. *cella*, small room
protoplasm: Gk. *protos*, first; *plassein*, to mold

tissue: Fr. *tissu*, woven; from L. *texo*, to weave
organ: Gk. *organon*, instrument

Figure 1.9

Examples of radiographic images.
The versatility of radiology makes this technique one of the most important tools in diagnostic medicine and provides a unique way of observing specific anatomical structures within the body. (*a*) A radiograph of a healing fracture, (*b*) a radiograph of gallstones within a gallbladder, and (*c*) a radiograph of a stomach filled with a radiopaque contrast medium.

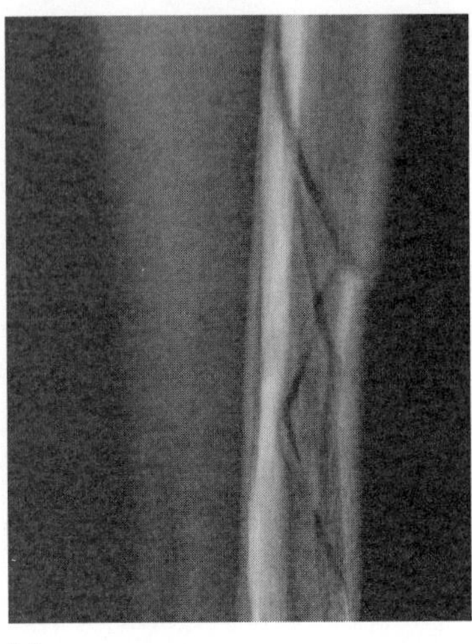

(a)

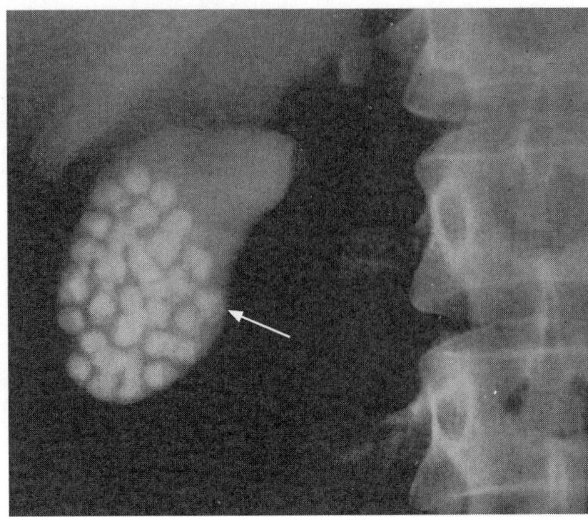

(b)

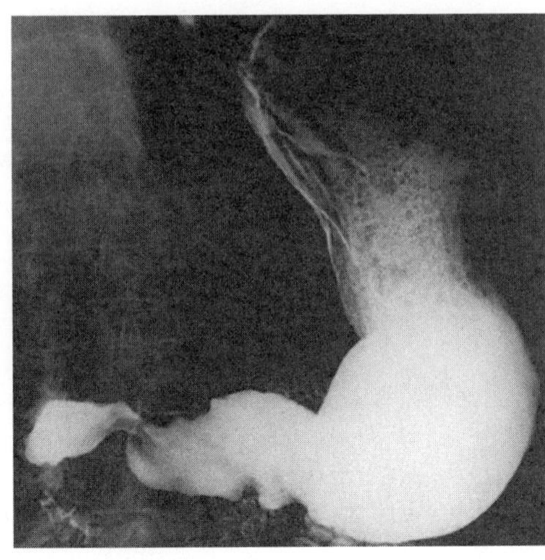

(c)

System Level

The **systems** of the body constitute the next level of structural organization. A body system consists of various organs that have similar or related functions. Examples of systems are the digestive system, nervous system, circulatory system, and endocrine system. Some organs serve two systems. For example, the pancreas functions with both the digestive and endocrine systems, and the pharynx serves both the digestive and respiratory systems. All of the systems of the body are interrelated and function together, making up the **organism.**

A *systemic approach* to studying anatomy and physiology emphasizes the functional relationships of various organs within a system. For example, the functional role of the digestive system can be better understood when all of the organs within that system are studied together. Another approach to anatomy, the *regional approach,* has merit in professional schools because the structural relationships of portions of several systems can be observed simultaneously. This is important for surgeons, who must be familiar with all the systems within a particular region. Dissections of cadavers are usually conducted on a regional basis. By means of radiographs (images produced by X rays) (fig. 1.9) and newer radiographic techniques, including computerized tomography (CT), magnetic resonance imaging (MRI) scans, and PET (positron emission tomography) scans (fig. 1.10), the organs within different body regions can be safely visualized in a patient.

This text uses a systematic approach to anatomy and physiology. In the chapters that follow, you will become acquainted system by system with the structural and functional aspects of the entire body. An overview of each of the body systems is presented in figure 1.11.

system: Gk. *systema,* being together

Planes of Reference and Descriptive Terminology

All of the descriptive planes of reference and terms of direction used in the science of anatomy and physiology are standardized because of their reference to the body in anatomical position.

Planes of Reference

In order to visualize and study the structural arrangements of various organs, the body may be sectioned (cut) and diagrammed according to three fundamental planes of reference: a sagittal (*saj'-ĭ-tal*) plane, a coronal plane, and transverse plane (see figs. 1.12 and 1.13).

A **sagittal plane** extends vertically through the body dividing it into right and left portions. A *midsagittal plane* is a sagittal plane that passes lengthwise through the midplane of the body, dividing it equally into right and left halves. **Coronal,** or **frontal, planes** also extend vertically and divide the body into anterior (front) and posterior (back) portions. **Transverse planes,** also called **horizontal,** or **cross-sectional, planes,** divide the body into superior (upper) and inferior (lower) portions.

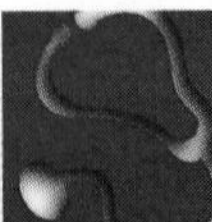

The value of the computerized tomographic X-ray (CT) scan is that it displays an image along a transverse plane similar to that which could otherwise be obtained only by actually sectioning the body. Prior to the development of this technique, the vertical plane of conventional radiographs (X-ray images) made it difficult, if not impossible, to assess the extent of body irregularities.

Figure 1.10

Different techniques in radiography.
Unique perspectives of the anatomy of the human head are seen in (*a*) a CT (computerized tomography) scan, (*b*) an MRI (magnetic resonance image) scan, and (*c*) a PET (positron emission tomography) scan.

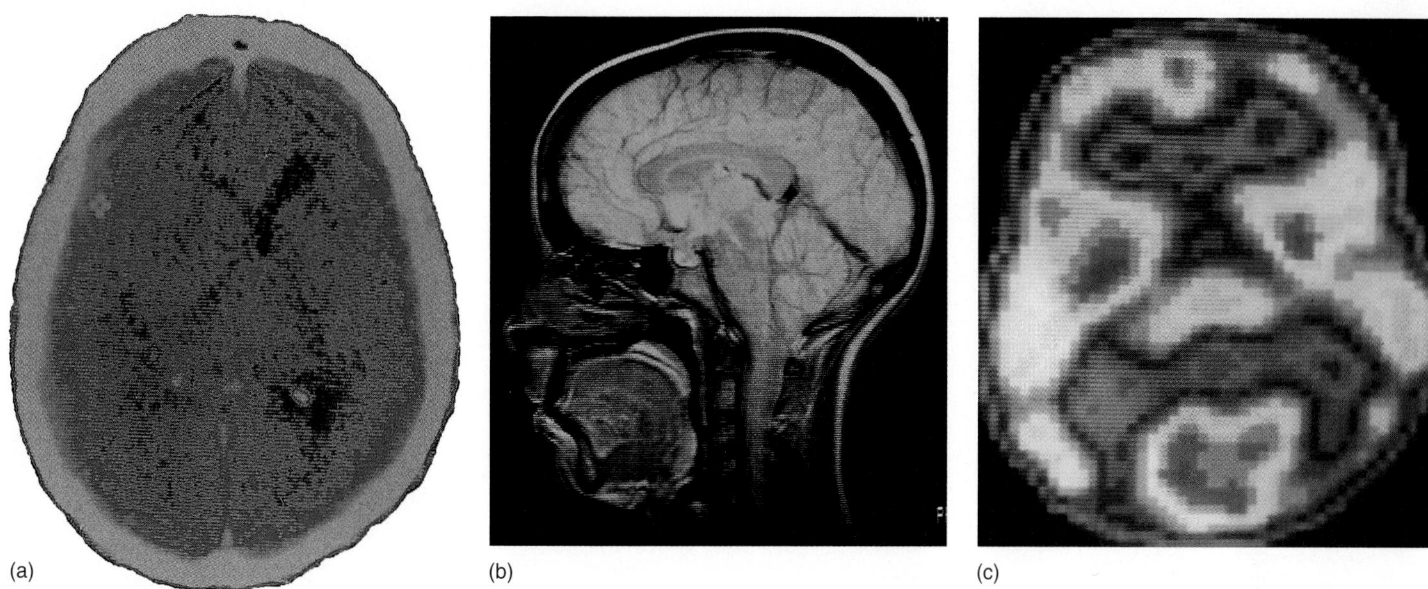

Anatomical Position and Directional Terms

All terms of direction that describe the relationship of one body part to another are made in reference to the **anatomical position.** In the anatomical position, the body is erect, the feet are parallel to one another and flat on the floor, the eyes are directed forward, and the arms are at the sides of the body with the palms of the hands turned forward and the fingers pointed straight down (see the photograph on the left in table 1.2).

Directional terms are used to locate the position of structures, surfaces, and regions of the body. These terms are always relative to the anatomical position. For example, if a person is in the anatomical position, the thumb is always lateral to the little finger and the nose is anterior to the ears. A summary of directional terms is presented in table 1.2.

Word Derivations

Analyzing anatomical and physiological terminology can be a rewarding experience. Not only is an understanding of the roots of words of academic interest, but a familiarity with technical terms reinforces the learning process. The majority of scientific terms are of Greek or Latin derivation, but some are German, French, and Arabic. Some anatomical and medical terms have been coined in honor of various anatomists or physicians; unfortunately, such terms have no descriptive basis and must simply be memorized.

Because many Greek and Latin terms were coined more than 2,000 years ago, we can gain a glimpse into our medical heritage by deciphering the meanings of these terms. For example, many terms refer to common plants or animals. Thus, the term *vermis* means "worm"; *cochlea*, "snail shell"; *cancer*, "crab"; and *uvula*, "grape." Other terms provide a clue to the warlike environment of the Greek and Latin era. *Thyroid*, for example, means "shield"; *xiphos* (*zi'fos*), "sword"; and *thorax*, "breastplate." *Sella* means "saddle," and *stapes* (*sta'pēz*) means "stirrup." Various tools or instruments were also referred to. The malleus (hammer) and incus (anvil) of the ear, for example, visually resemble miniatures of a blacksmith's implements.

Certain clinical procedures are important in determining body structure and function. *Palpation* is feeling with firm pressure for surface landmarks, lumps, tender spots, or pulsations. *Percussion* is tapping sharply at points on the thorax or abdomen to determine fluid concentrations and organ densities. *Auscultation* is listening to the sounds that various organs make as they perform their functions.

You will encounter many new terms throughout your study of anatomy and physiology, and learning these terms will be easier if you understand their prefixes and suffixes. A glossary of prefixes and suffixes (on the inside front cover) and a guide to the relationship between the singular and plural forms of words (table 1.3) have been provided to aid you. Pronouncing these terms as you learn them will also help you recall them later.

Body Regions and Body Cavities

The human body is divided into regions and specific local areas that can be identified on the surface. The head and trunk contain internal organs housed in distinct body cavities.

Learning the terminology used in reference to the body regions now will help you learn the names of underlying structures

Figure 1.11
The body systems.

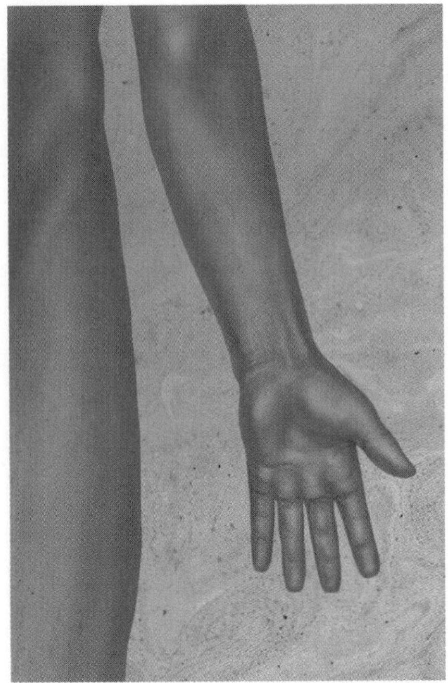

Integumentary system
Function: external support and protection of body

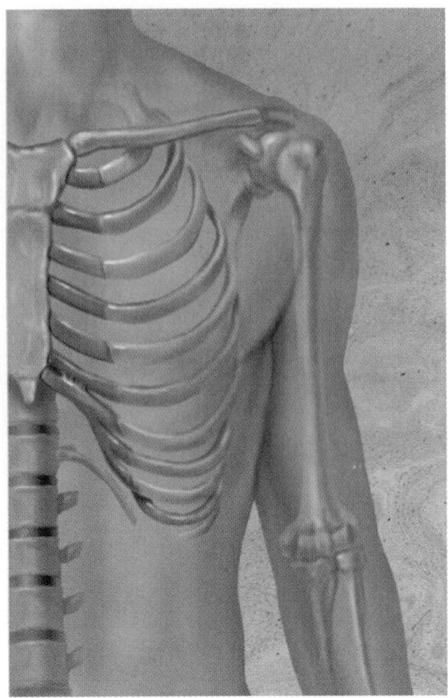

Skeletal system
Function: internal support and flexible framework for body movement; production of blood cells

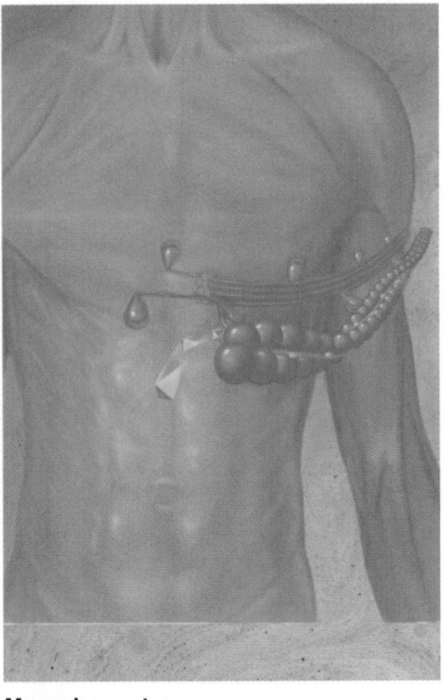

Muscular system
Function: body movement; production of body heat

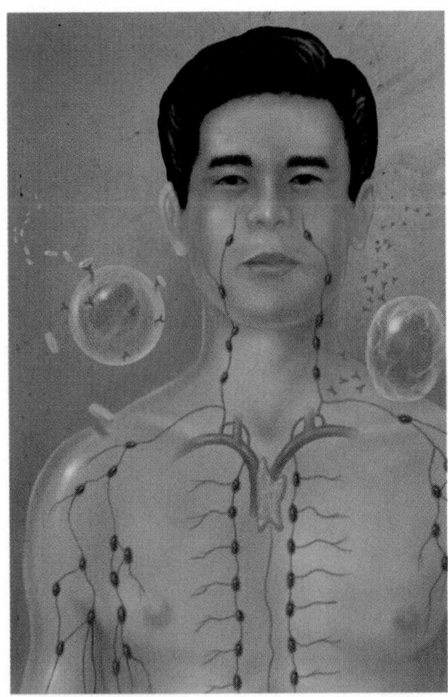

Lymphatic system
Function: body immunity; absorption of fats; drainage of tissue fluid

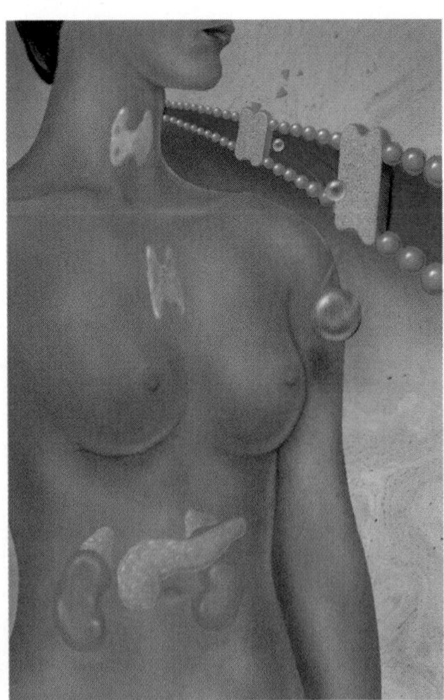

Endocrine system
Function: secretion of hormones for chemical regulation

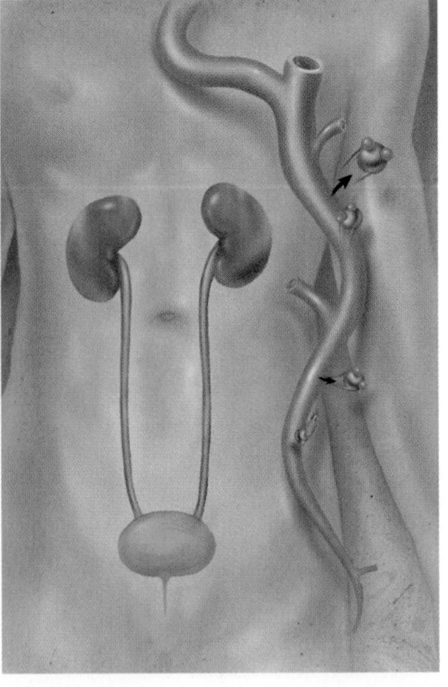

Urinary system
Function: filtration of blood; maintenance of volume and chemical composition of blood; removal of metabolic wastes from body

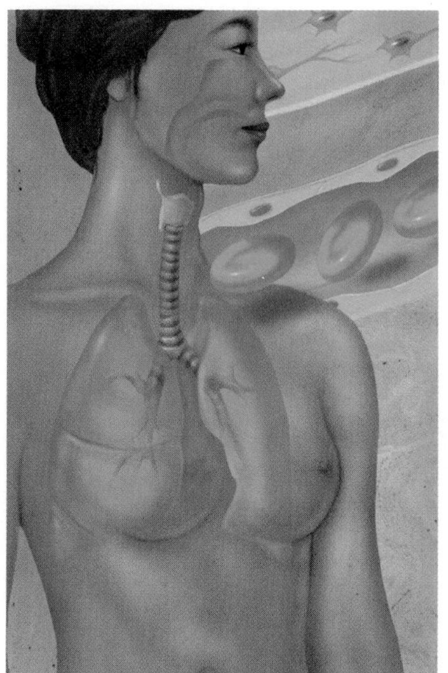

Respiratory system
Function: gaseous exchange between external environment and blood

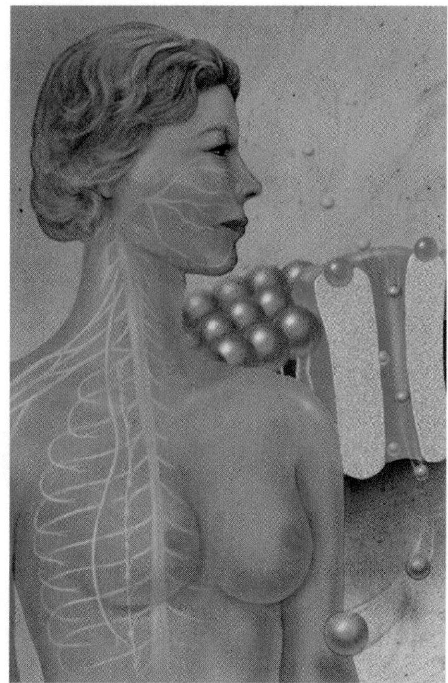

Nervous system
Function: control and regulation of all other systems of the body

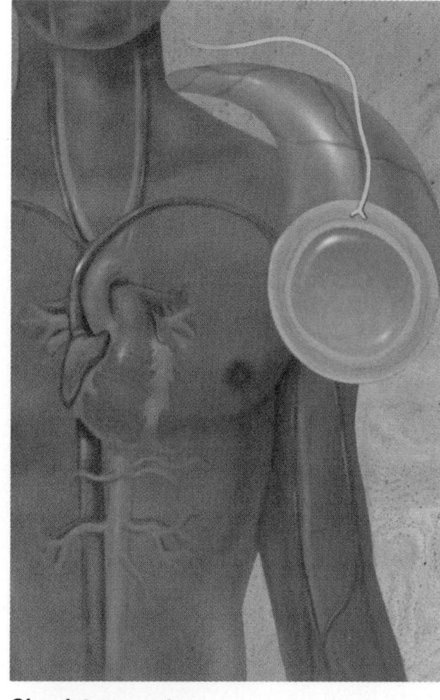

Circulatory system
Function: transport of life-sustaining materials to body cells; removal of metabolic wastes from cells

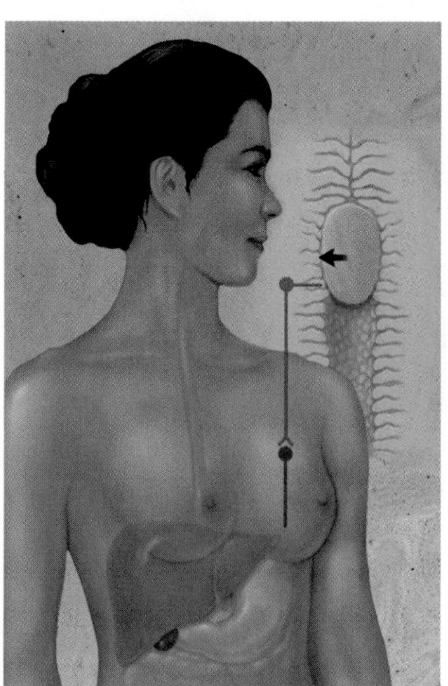

Digestive system
Function: breakdown and absorption of food materials

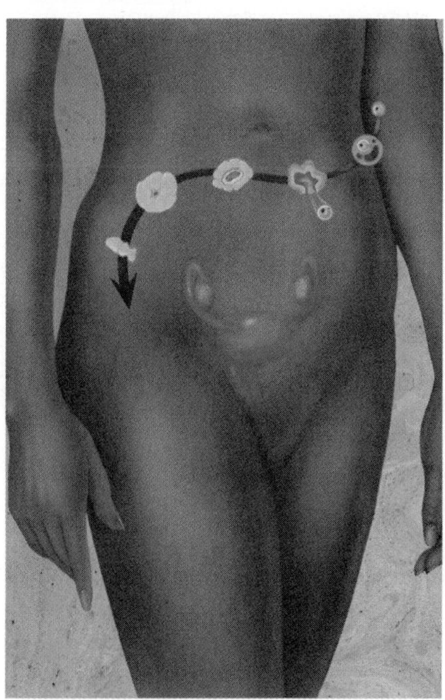

Female reproductive system
Function: production of female sex cells (ova); receptacle for sperm from male; site for fertilization of ovum, implantation, and development of embryo and fetus; delivery of fetus

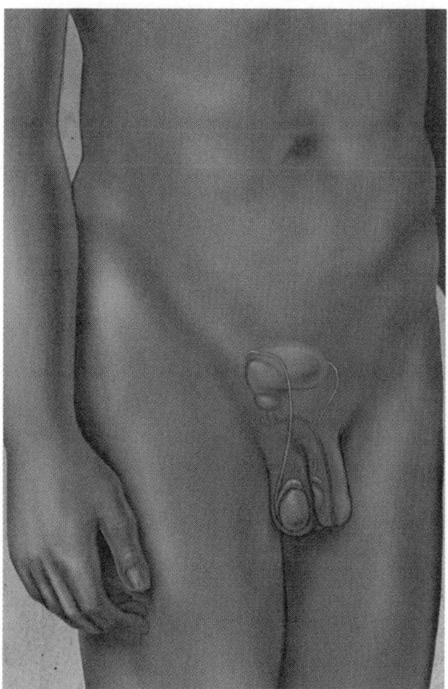

Male reproductive system
Function: production of male sex cells (sperm); transfer of sperm to reproductive system of female

Figure 1.12

Planes of reference through the body.

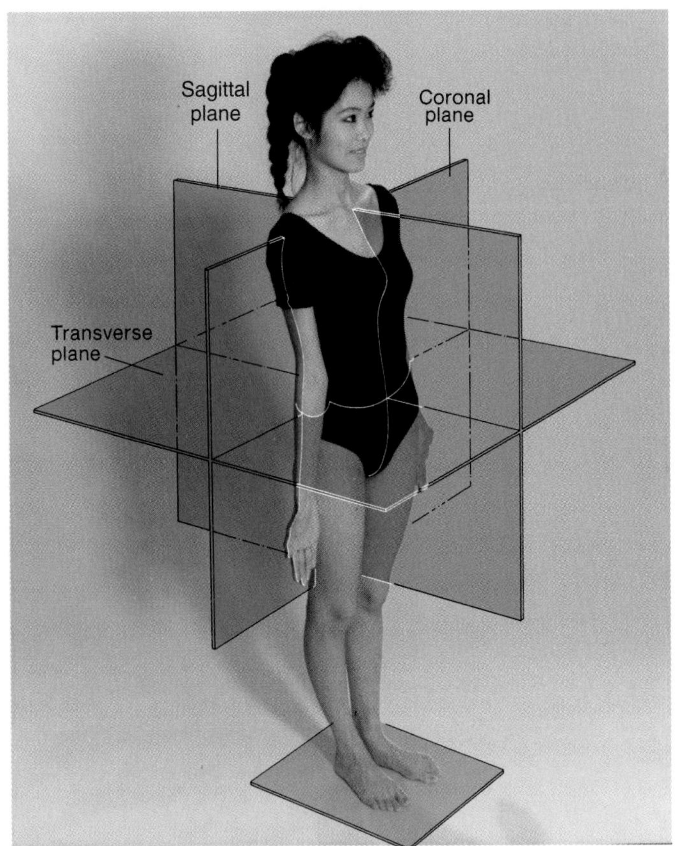

Figure 1.13

Planes of reference through the human brain. The brain sectioned along (*a*) a transverse plane, (*b*) a coronal plane, and (*c*) a sagittal plane.

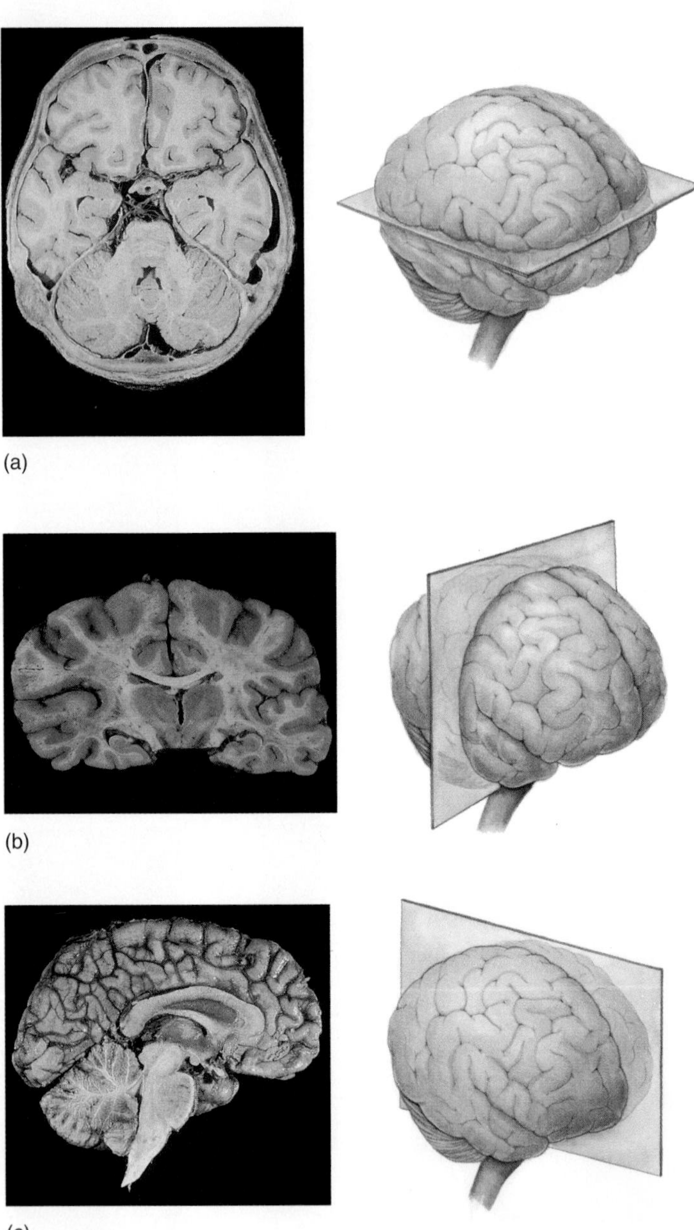

later. The major body regions are the **head, neck, trunk, upper extremity,** and **lower extremity.** The trunk is frequently divided into the thorax and abdomen. Figure 1.14 presents an outline of the major body regions.

Head and Neck

The **head** is divided into a **facial region,** which includes the eyes, nose, and mouth, and a **cranial region,** or *cranium* (*kra′ne-um*), which covers and supports the brain. The identifying names for specific surface regions of the face are based on associated organs—for example, the orbital (eye), nasal (nose), oral (mouth), and auricular (ear) regions—or underlying bones—such as the frontal, zygomatic, temporal, parietal, and occipital regions. The neck, referred to as the **cervical region,** or *cervix* (*ser′viks*), supports the head and permits it to move.

Thorax

The **thorax,** or **thoracic** (*thŏ-ras′ik*) **region,** is commonly referred to as the chest. The **mammary region** of the thorax surrounds the nipple, and in sexually mature females, is enlarged as the breast. Between the mammary regions is the **sternal region.** The armpit is called the **axillary fossa,** or simply the **axilla,** and the surrounding area, the **axillary region.** Paired **scapular regions** (shoulder blades) can be identified from the back of the thorax. The **vertebral region** extends the length of the back, following the vertebral column. On either lateral side of the thorax are the **pectoral regions.**

thorax: L. *thorax*, chest
mammary: L. *mamma*, breast

axillary: L. *axilla*, armpit

Table 1.2
Directional Terms for the Human Body

Term	Definition	Example
Superior (cranial, cephalic)	Toward the head; toward the top	The thorax is superior to the abdomen.
Inferior (caudal)	Away from the head; toward the bottom	The legs are inferior to the trunk.
Anterior (ventral)	Toward the front	The navel is on the anterior side of the body.
Posterior (dorsal)	Toward the back	The kidneys are posterior to the intestine.
Medial	Toward the midline of the body	The heart is medial to the lungs.
Lateral	Away from the midline of the body	The ears are lateral to the nose.
Internal (deep)	Away from the surface of the body	The brain is internal to the cranium.
External (superficial)	Toward the surface of the body	The skin is external to the muscles.
Proximal	Toward the trunk of the body	The knee is proximal to the foot.
Distal	Away from the trunk of the body	The hand is distal to the elbow.

The heart and lungs are contained within the thoracic cavity. Easily identified surface landmarks are helpful in assessing the condition of these organs. A physician must know, for example, where the valves of the heart can best be detected and where to listen for respiratory sounds. The axilla becomes important in examining for infected lymph nodes. When fitting a patient for crutches, a physician will instruct the patient to avoid supporting the weight of the body on the axillary region because of the possibility of damaging the underlying nerves and vessels.

Abdomen

The **abdomen** (*ab'do-men*) is located below the thorax. Centered on the front of the abdomen, the **umbilicus (navel)** is an obvious landmark. The abdomen has been divided into nine regions to describe the location of internal organs. The subdivisions of the abdomen are diagrammed in figure 1.15.

The **pelvic region** forms the lower portion of the trunk. Within the pelvic region is the **pubic area,** which is covered with pubic hair in sexually mature individuals. The **perineum**

Table 1.3

Examples of Singular and Plural Word Endings

Singular Ending	Plural Ending	Examples
-a	-ae	Axilla, axillae
-ax	-aces	Thorax, thoraces
-en	-ina	Lumen, lumina
-ex	-ices	Cortex, cortices
-is	-es	Diagnosis, diagnoses
-is	-ides	Epididymis, epididymides
-ix	-ices	Appendix, appendices
-ma	-mata	Carcinoma, carcinomata
-on	-a	Mitochondrion, mitochondria
-um	-a	Cilium, cilia
-us	-i	Tarsus, tarsi
-us	-ora	Corpus, corpora
-us	-era	Viscus, viscera
-x	-ges	Pharynx, pharynges
-y	-ies	Ovary, ovaries

Courtesy of Dr. Kenneth S. Saladin, Georgia College and State University.

(*per″ĭ-ne′um*) is the region where the external sex organs and the anal opening are located. The center of the back side of the abdomen, commonly called the small of the back, is the **lumbar region.** The **sacral region** is located farther down, at the point where the vertebral column terminates. The large hip muscles form the **buttock,** or **gluteal region.** This region is a common injection site for hypodermic needles.

Upper and Lower Extremities

The upper extremity is anatomically divided into the **shoulder, brachium** (*bra′ke-um*) (upper arm), **antebrachium** (forearm), and **manus** (hand). The shoulder is the region between the pectoral girdle and the brachium in which the shoulder joint is located. The shoulder is referred to as the **deltoid region** (omos). The **cubital region** is the area between the arm and forearm that contains the elbow joint. The **cubital fossa** is the depressed anterior portion of the cubital region. It is a common site for intravenous injections or the withdrawal of blood. The wrist is the flexible junction between the forearm and the hand. The front of the hand is referred to as the **palmar region** (palm), and the back of the hand is called the **dorsum of the hand.**

The lower extremity consists of the **thigh, knee, leg,** and **pes** (foot). The thigh is commonly called the **upper leg,** or **femoral region.** The knee has two surfaces: the front surface is the **patellar region** (kneecap); the back of the knee is called the **popliteal** (*pop″lĭ-te′al*) **fossa.** The shin is a prominent bony ridge extending longitudinally along the **anterior crural region,** and the calf is the thickened muscular mass of the **posterior crural region.** The ankle is the junction between the leg and the foot. The **heel** is the back of the foot, and the **sole** of the foot is referred to as the **plantar surface.** The **dorsum of the foot** is the top surface.

Body Cavities and Associated Membranes

Body Cavities

Body cavities are confined spaces within the body that contain organs that are protected, compartmentalized, and supported by associated membranes. There are two principal body cavities: the **posterior** (dorsal) **body cavity** and the larger **anterior** (ventral) **body cavity.** The posterior body cavity contains the brain and the spinal cord. During development, the anterior body cavity forms from a cavity within the trunk called the **coelom** (*se′lom*). The coelom is lined with a membrane that secretes a lubricating fluid. As development progresses, the coelom is partitioned by the muscular *diaphragm* into an upper **thoracic cavity,** or chest cavity, and a lower **abdominopelvic cavity** (figs. 1.16 and 1.17). Organs within the coelom are collectively called **viscera,** or **visceral** (*vis′er-al*) **organs.** Within the thoracic cavity are two **pleural** (*ploor′al*) **cavities** surrounding the right and left lungs and a **pericardial** (*per″ĭ kar′-de-al*) **cavity** surrounding the heart. The area between the two lungs is known as the **mediastinum** (*me″de-ă-sti′num*).

The abdominopelvic cavity consists of an upper **abdominal cavity** and a lower **pelvic cavity.** The abdominal cavity contains the stomach, small intestine, large intestine, liver, gallbladder, pancreas, spleen, and kidneys. The pelvic cavity is occupied by the terminal portion of the large intestine, the urinary bladder, and certain reproductive organs (uterus, uterine tubes, and ovaries in the female; seminal vesicles and prostate in the male).

Body Membranes

Body membranes are composed of thin layers of connective and epithelial tissue that cover, separate, and support viscera and line body cavities. There are two basic types of body membranes: mucous (*myoo′kus*) membranes and serous (*ser′us*) membranes.

Mucous membranes secrete a thick liquid substance called *mucus*. Generally, mucus lubricates or protects the associated organs. Mucous membranes line cavities and tubes that enter or exit from the body, such as the oral (mouth) and nasal cavities and tubes of the digestive and respiratory systems.

Serous membranes line the thoracic and abdominopelvic cavities and cover visceral organs, secreting a watery lubricant called *serous fluid*. **Pleurae** (singular, *pleura* [*ploor′a*]) are serous

---

cubital: L. *cubitis*, elbow

popliteal: L. *poples*, ham (hamstring muscles) of the knee
coelom: Gk. *koiloma*, cavity

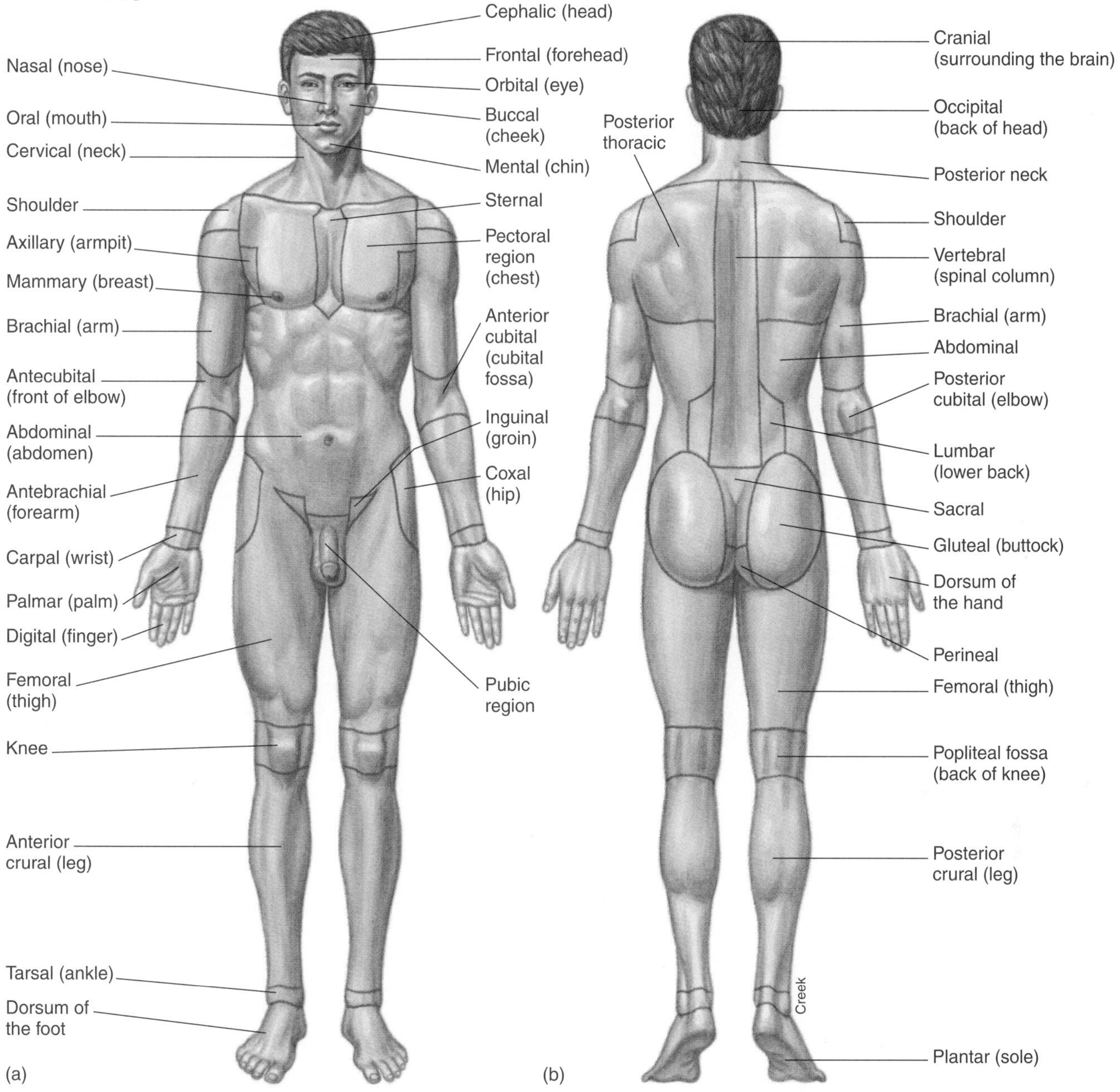

Figure 1.14

Body regions.
(*a*) An anterior and (*b*) posterior view.

membranes associated with the lungs (fig. 1.18*a*). Each pleura (pleura of right lung and pleura of left lung) has two parts. The **visceral pleura** is attached to the outer surface of the lung, whereas the **parietal pleura** lines the thoracic walls and the thoracic side of the diaphragm. The moistened space between the two pleurae is the **pleural cavity.** Each lung is surrounded by its own pleural cavity.

Pericardial membranes are the serous membranes of the heart (fig. 1.18*b*). The thin **visceral pericardium** (*per″i-kar′de-um*) is the outer layer of tissue attached to the heart, and a thicker **parietal pericardium** is the durable covering that surrounds the heart. The space between these two membranes is the **pericardial cavity.** Like the pleural cavity, the pericardial cavity contains a small amount of fluid.

Serous membranes of the abdominal cavity are called peritoneal membranes (fig. 1.19). The parietal peritoneum attaches to the abdominal wall, and the visceral peritoneum (*per″i-to-ne′um*) attaches to the visceral organs. The peritoneal cavity is the fluid-filled space within the abdominopelvic cavity between the parietal and visceral peritoneal membranes. A **mesentery** (*mes′en-ter″e*) is a fused double layer of parietal peritoneum that connects the parietal peritoneum to the visceral peritoneum. Mesenteries help to hold many of the visceral organs in place, while at the same time permitting involuntary digestive movement.

---

peritoneum: Gk. *peritonaion*, stretched over

Figure 1.15

The nine regions of the abdomen.
The vertical planes are positioned lateral to the rectus abdominis muscles. The upper horizontal plane is positioned at the level of the rib cage, and the lower horizontal plane is even with the upper border of the hipbones.

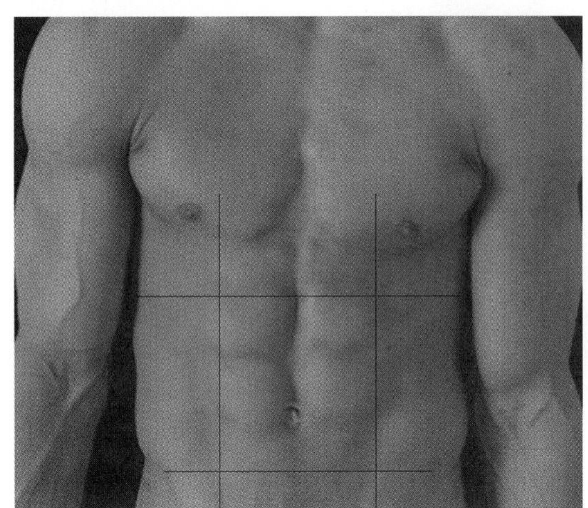

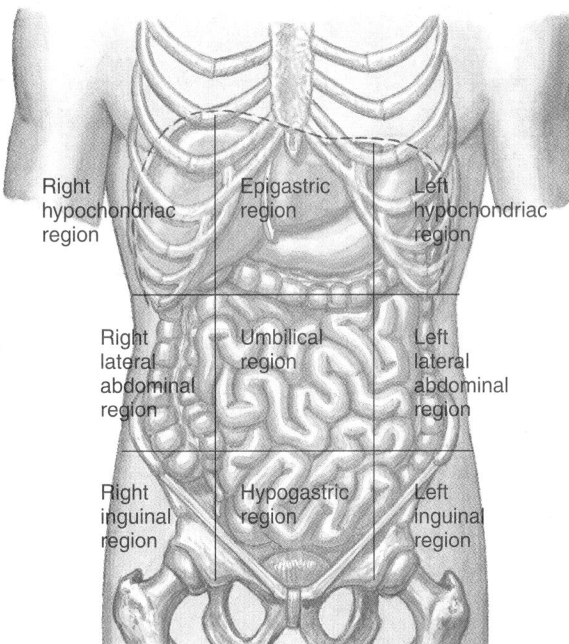

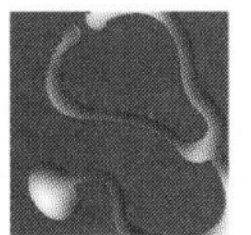

Body cavities serve to confine organs and systems that have related functions. The major portion of the nervous system occupies the posterior cavity; the principal organs of the respiratory and circulatory systems are in the thoracic cavity; the primary organs of digestion are in the abdominal cavity; and the reproductive organs are in the pelvic cavity. Not only do these cavities house and support various body organs, the associated membranes also effectively compartmentalize them so that infections and diseases cannot spread from one compartment to another. For example, pleurisy of one lung membrane does not usually spread to the other, and an injury to the thoracic cavity will usually result in the collapse of one lung rather than both.

Figure 1.16

A midsagittal (median) section showing the body cavities.

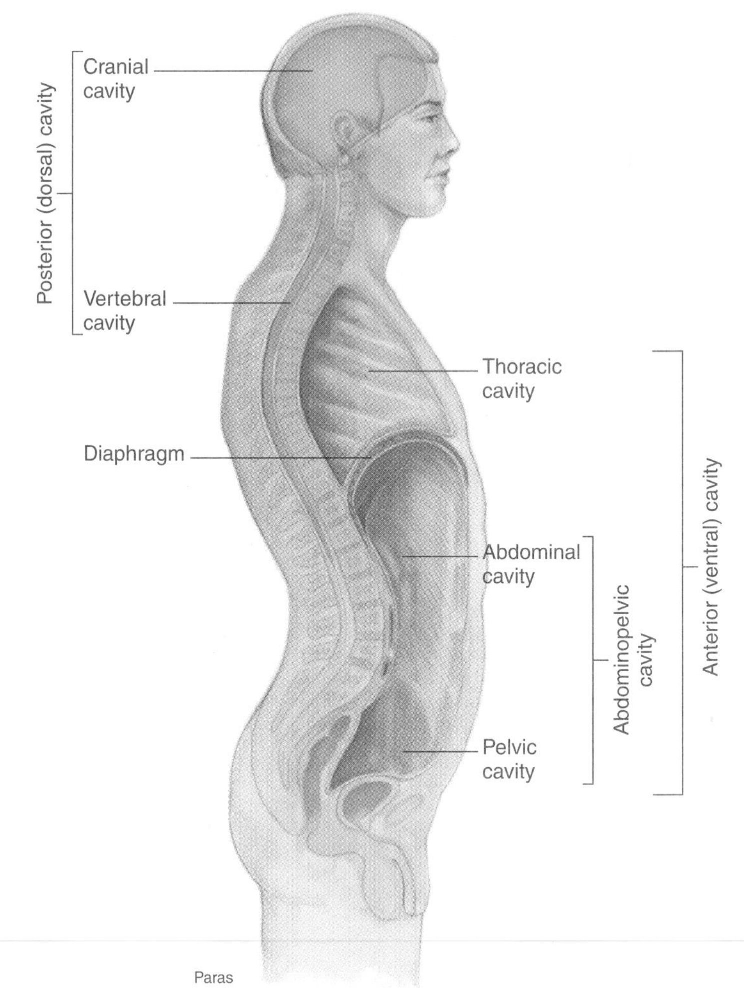

Homeostasis and Feedback Control

The regulatory mechanisms of the body can be understood in terms of a single, shared function: that of maintaining a dynamic constancy of the internal environment, or homeostasis. Homeostasis is maintained by effectors, which are regulated by sensory information from the internal environment.

Over a century ago, the French physiologist Claude Bernard observed that the *milieu interieur* (internal environment) remains remarkably constant despite changing conditions in the external environment. In a book entitled *The Wisdom of the Body*, published in 1932, the American physiologist Walter Cannon coined the term **homeostasis** to describe this internal constancy. Cannon suggested that mechanisms of physiological regulation exist for one purpose—the maintenance of internal constancy.

Figure 1.17

An anterior view showing the cavities of the trunk.

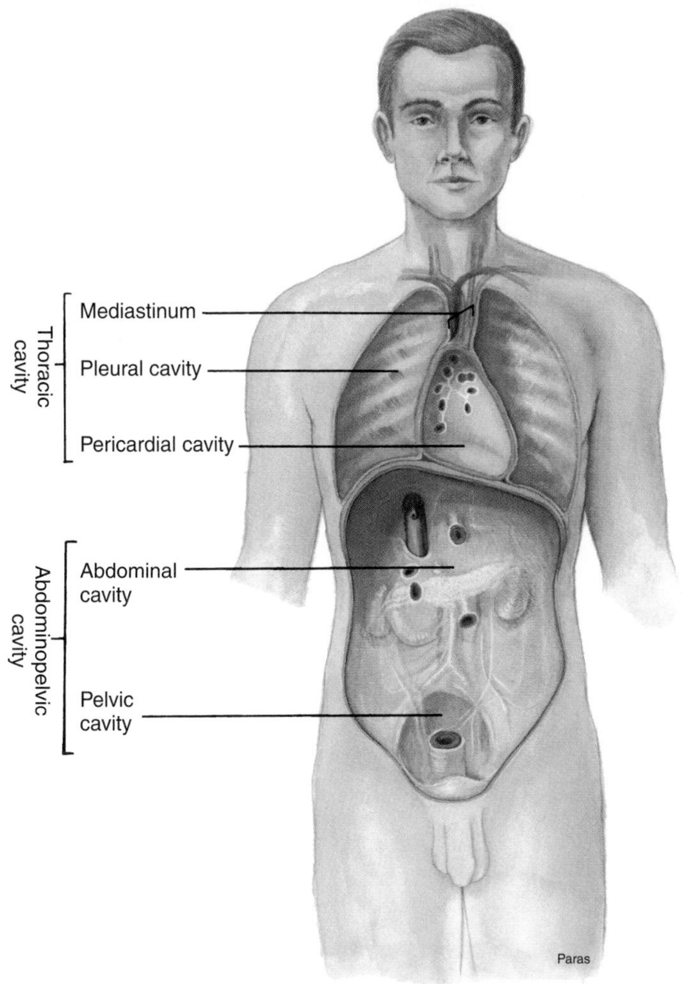

Figure 1.18

Serous membranes of the thorax.

(*a*) Serous membranes surrounding the lungs and (*b*) serous membranes surrounding the heart.

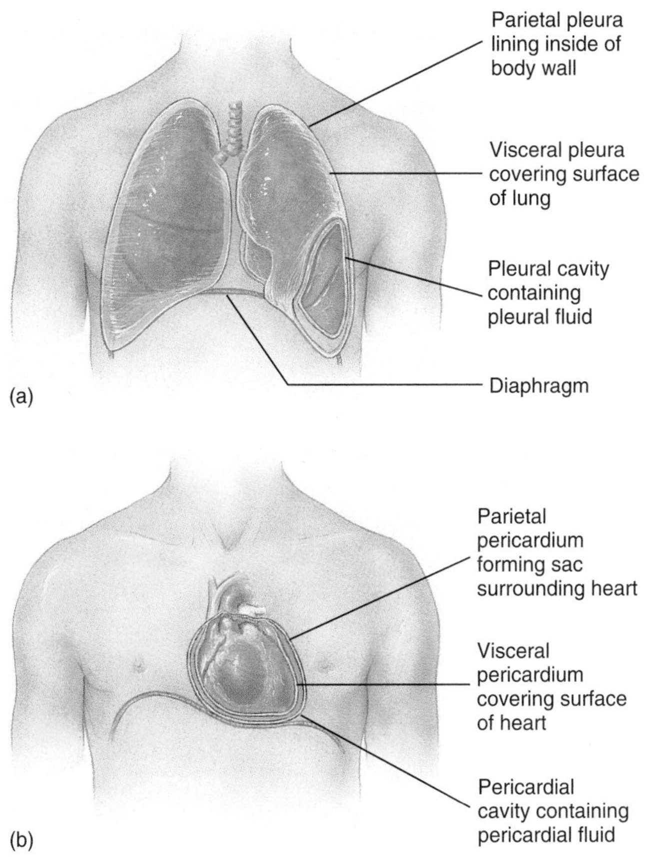

The concept of homeostasis has been immensely valuable in the study of anatomy and physiology because it allows diverse regulatory mechanisms to be understood in terms of their "why" as well as their "how." The concept of homeostasis also provides a foundation for medical diagnostic procedures. When a particular measurement of the internal environment deviates significantly from the normal range, we can conclude that homeostasis of a specific body function is not being maintained. A number of such measurements, together with clinical observations, may allow a particular defective mechanism, or disease, to be identified.

The systems of the body are regulated by negative feedback mechanisms that operate to maintain homeostasis. Positive feedback mechanisms are rare in the human body, and act to amplify changes.

Negative Feedback Mechanisms

In order for internal constancy to be maintained, the body must have *sensors* that are able to detect deviations from a *set point,* which is analogous to the temperature set on a house thermostat (fig. 1.20). In a similar manner, there is a set point for body temperature, blood glucose concentration, tension on a tendon, and so on. When a sensor detects a deviation from a set point, it must relay this information to an *integrating center* (fig. 1.21), which usually receives information from many different sensors. The integrating center is often a particular region of the brain or spinal cord, but in some cases it can also be cells of endocrine glands. The relative strengths of different sensory inputs are weighed in the integrating center, and, in response, the integrating center either increases or decreases the activity of particular *effectors,* which are generally muscles or glands.

If, for example, your body temperature exceeds the set point of 37° C (98.6° F), sensors in a part of the brain detect this deviation and, acting via an integrating center (also in the brain), stimulate activities of effectors (including sweat glands) that lower the temperature. As another example, if your blood glucose concentration falls below normal, the effectors (endocrine glands) increase the blood glucose. One can think of the effectors as "defending" the set points against deviations. Since the activity of the effectors is influenced by the effects they produce, and since this regulation is in a negative, or reverse, direction, this type of control system is known as a **negative feedback mechanism.** (Notice that in

Figure 1.19

Visceral organs of the abdominal cavity and supporting serous membranes.

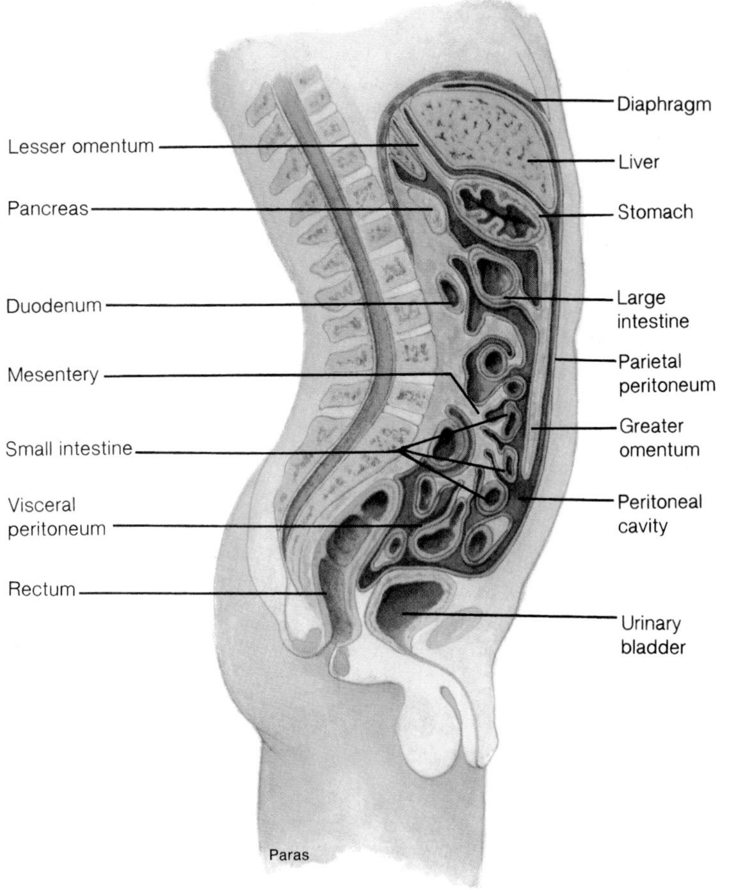

Figure 1.20

A negative feedback loop in response to a rise in the internal environment.
A rise in some factor of the internal environment (↑X) is detected by a sensor. Acting through an integrating center, this caused an effector to produce a change in the opposite direction (↓X). The initial deviation is thus reversed, completing a negative feedback loop (shown by the dashed arrow and negative sign). The numbers indicate the sequence of changes.

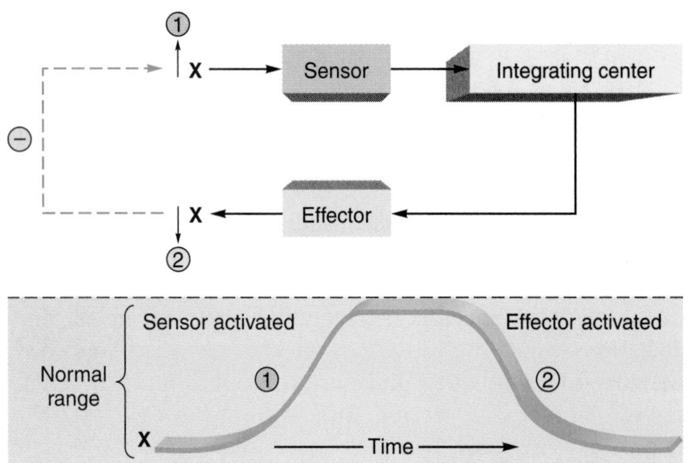

Figure 1.21

A negative feedback loop in response to a fall in the internal environment.
(Compare this figure with figure 1.20.)

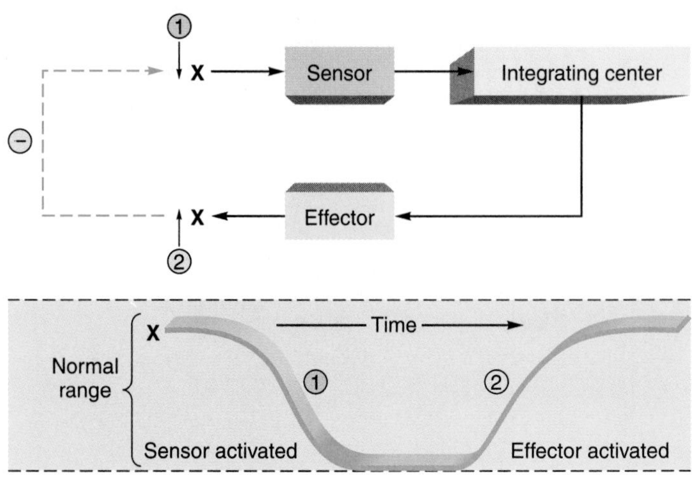

Figure 1.22

Negative feedback loops.
Indicated by negative signs, negative feedback loops maintain a state of dynamic constancy within the internal environment.

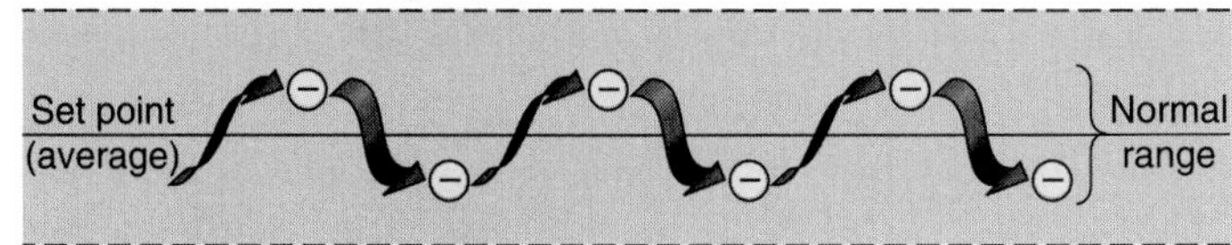

figs. 1.20 and 1.21 negative feedback is illustrated by a dashed line and a negative sign.)

Homeostasis is best conceived as a state of *dynamic constancy* rather than as a state of absolute constancy. The values of particular measurements of the internal environment fluctuate above and below the set point, which can be taken as the average value within the normal range of measurements (fig. 1.22). This state of dynamic constancy results from a greater or a lesser degree of activation of effectors in response to sensory feedback and from the competing actions of antagonistic effectors.

Antagonistic Effectors

Most factors in the internal environment are controlled by several effectors, which often display antagonistic activity. Control by antagonistic effectors is sometimes described as push-pull, where the increasing activity of one effector is accompanied by decreasing activity of an antagonistic effector. This affords a finer degree of control than could be achieved by simply switching one effector on and off. Normal body temperature, for example, is maintained at a set point of about 37° C by the antagonistic effects of sweating, shivering, and other mechanisms (fig. 1.23).

Figure 1.23

Regulation of body temperature.
A simplified scheme by which body temperature is maintained within the normal range (with a set point of 37° C) by two antagonistic mechanisms—shivering and sweating. Shivering is induced when the body temperature falls too low and gradually subsides as the temperature rises. Sweating occurs when the body temperature is too high and diminishes as the temperature falls. Most aspects of the internal environment are regulated by the antagonistic actions of different effector mechanisms.

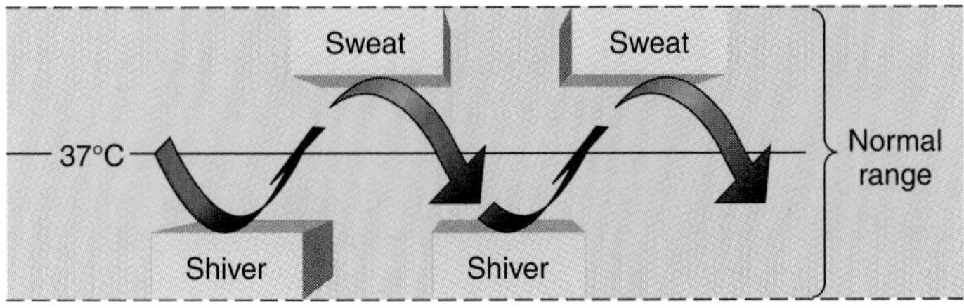

The blood concentrations of glucose, calcium, and other substances are regulated by negative feedback loops that involve chemical regulators called hormones that produce opposite effects. Although insulin, for example, lowers blood glucose, other hormones raise the blood glucose concentration. The heart rate, similarly, is controlled by impulses through nerve fibers that produce opposite effects. Stimulation of one group of nerve fibers increases the heart rate, whereas stimulation of another group slows the heart rate.

Positive Feedback Mechanisms

Constancy of the internal environment is maintained by effectors that act to compensate for changes that served as stimuli for their activation; in short, by negative feedback mechanisms. A thermostat, for example, maintains a constant temperature by increasing heat production when a room is cold and decreasing heat production when a room is warm. The opposite occurs during a **positive feedback mechanism**—in this case, the action of effectors *amplifies* those changes that stimulated the effectors. A thermostat that worked by positive feedback, for example, would increase heat production in response to a rise in temperature.

It is clear that homeostasis must ultimately be maintained by negative rather than by positive feedback mechanisms. The effectiveness of some negative feedback mechanisms, however, is increased by positive feedback mechanisms that amplify the actions of a negative feedback response. Blood clotting, for example, occurs as a result of a sequential activation of clotting factors. The activation of one clotting factor results in activation of many in a positive feedback, avalanchelike, manner. In this way, a single change is amplified to produce a blood clot. Formation of the clot, however, can prevent further loss of blood and thus represents the completion of a negative feedback mechanism.

Neural and Endocrine Regulation

The effectors of most negative feedback mechanisms include the actions of nerves and hormones. In both neural and endocrine regulation, particular chemical regulators released by nerve fibers or endocrine glands stimulate target cells by interacting with specific receptor proteins in these cells. The mechanisms by which this regulation is achieved will be described in later chapters.

Homeostasis is maintained by two general categories of regulatory mechanisms: (1) those that are *intrinsic*, or "built-in," to the organs that produce them and (2) those that are *extrinsic*, as in regulation of an organ by the nervous and endocrine systems.

The endocrine system functions closely with the nervous system in regulating and integrating body processes and maintaining homeostasis. The nervous system controls the secretion of many endocrine glands, and some hormones in turn affect the function of the nervous system. Together, the nervous and endocrine system regulate the activities of most of the other systems of the body.

Regulation by the endocrine system is achieved by the secretion of chemical regulators called **hormones** into the blood. Since hormones are secreted into the blood, they are carried by the blood to all organs in the body. Only specific organs can respond to a particular hormone, however; these are known as the *target organs* of that hormone.

Nerve fibers are said to *innervate* the organs that they regulate. When stimulated, these fibers produce electrochemical nerve impulses that are conducted from the origin of the fiber to the target organ innervated by that fiber. These target organs can be muscles or glands that may function as effectors in the maintenance of homeostasis.

Feedback Control of Hormone Secretion

We will discuss the details of the nature of the endocrine glands, the interaction of the nervous and endocrine systems, and the actions of hormones in later chapters. For now it is sufficient to describe the regulation of hormone secretion very broadly, since it so superbly illustrates the principles of homeostasis and negative feedback regulation.

Hormones are secreted in response to specific chemical stimuli. A rise in the plasma glucose concentration, for example, stimulates insulin secretion from the pancreatic islets (islets of Langerhans) in the pancreas. Hormones are also secreted in response to nerve stimulation and to stimulation by other hormones.

The secretion of a hormone can be inhibited by its own effects, in a negative feedback manner. Insulin, for example, lowers blood glucose. Since a rise in blood glucose stimulates insulin secretion, a lowering of blood glucose caused by insulin's action inhibits further insulin secretion. This closed-loop control system is called **negative feedback inhibition** (fig. 1.24).

Figure 1.24

The negative feedback control of insulin secretion and blood glucose concentration.

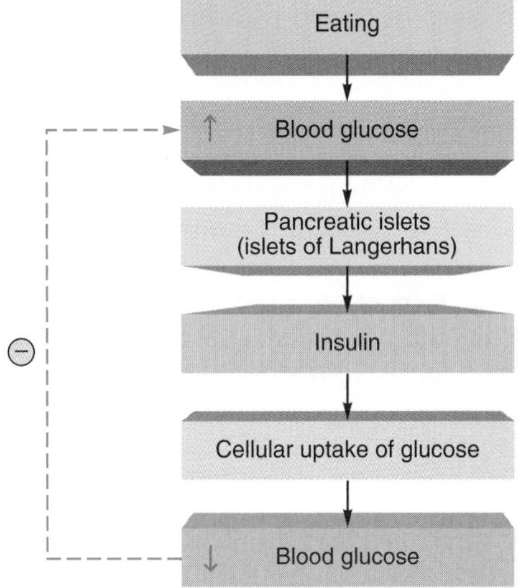

Clinical Investigation Answer

The pain in the student's lower right abdomen may have been caused by appendicitis. This possibility is supported by the observations that she had a high fever and that her white blood cell count was high. The out-of-control skier rammed the tip of his ski into her right popliteal fossa, which is on the posterior surface of the lower extremity in the knee region. His ski pole punctured the lateral side of her thoracic cavity just inferior to her axillary fossa, causing the collapse of her left lung. She was still able to breathe because her right lung is in a separate anatomical compartment surrounded by its own pleural cavity.

Chapter Summary

The Sciences of Anatomy and Physiology (pp. 2–7)

1. The history of anatomy and physiology parallels that of medicine; Hippocrates, the famous Greek physician, is regarded as the father of medicine.
2. Aristotle established a type of scientific method for obtaining data.
3. Erasistratus is frequently referred to as the father of physiology.
4. The writings of the Roman physician Galen were the ultimate authority on anatomy and medical treatment for nearly 1,500 years.
5. Often regarded as the father of anatomy, Vesalius challenged Galen's teachings and accelerated research in anatomy and physiology during the Renaissance.
6. Two major contributions of the seventeenth and eighteenth centuries were the explanation of blood flow by Harvey and the development and utilization of the microscope by Leeuwenhoek.
7. The scientific method is a disciplined approach considered necessary for scientific investigation. It includes observation, data collection, and formulation of a hypothesis that is testable.

Classification and Characteristics of Humans (pp. 7–9)

1. Humans are biological organisms belonging to the phylum Chordata within the kingdom Animalia and to the family Hominidae within the class Mammalia and the order Primates.
2. Humans belong to the phylum Chordata because of the presence of a notochord, a dorsal hollow nerve cord, and pharyngeal pouches during the embryonic period of development.
3. Some of the characteristics of humans include a large, well-developed brain, bipedal locomotion, an opposable thumb, well-developed vocal structures, and stereoscopic vision.

Body Organization (pp. 9–10)

1. The cell is the structural and functional component of life.
2. Tissues are aggregations of similar cells that perform specific functions.
3. An organ is an aggregate of two or more tissues that performs specific functions.
4. Body systems consist of various organs that have similar or interrelated functions.

Planes of Reference and Descriptive Terminology (pp. 10–11)

1. In the anatomical position, the subject stands erect and faces forward with arms at the sides and palms turned forward.
2. Directional terms are used to describe the location of one body part with respect to another part.
3. The majority of anatomical and physiological terms are of Greek or Latin derivation.

Body Regions and Body Cavities (pp. 11–18)

1. The human body is divided into regions, which can be identified on the surface. In each region are internal organs, the locations of which are anatomically, physiologically, and clinically important.
2. For functional and protective purposes, the viscera are compartmentalized and supported in specific body cavities by connective and epithelial membranes.

Homeostasis and Feedback Control (pp. 18–21)

1. Homeostasis is the dynamic constancy of the internal environment. This concept is the central theme of anatomy and physiology, and even of medicine.
2. Deviations from a set point are detected by sensors, and changes are instituted by effectors that act to compensate for the deviations in a negative feedback fashion.
3. Neural and endocrine effectors regulate most of the organs of the body. The secretion of hormones is controlled by negative feedback mechanisms.

Review Activities

Objective Questions

1. Which of the following men would be most likely to disagree with the concept of body humors?
   (a) Galen (c) Vesalius
   (b) Hippocrates (d) Aristotle
2. The most important contribution of William Harvey was his research on
   (a) the continuous circulation of blood.
   (b) the microscopic structure of spermatozoa.
   (c) the detailed structure of the kidney.
   (d) the striated appearance of skeletal muscle.
3. The taxonomic scheme from specific to general is
   (a) species, class, order, phylum.
   (b) genus, family, kingdom, phylum.
   (c) species, family, class, kingdom.
   (d) genus, phylum, class, kingdom.
4. Which of the following is (are) *not* a principal chordate characteristic?
   (a) dorsal hollow nerve cord
   (b) distinct head, thorax, and abdomen
   (c) notochord
   (d) pharyngeal pouches
5. The cubital fossa is located in
   (a) the thorax.
   (b) the upper extremity.
   (c) the abdomen.
   (d) the lower extremity.

6. Which of the following is *not* a fundamental plane?
   (a) coronal plane
   (b) transverse plane
   (c) vertical plane
   (d) sagittal plane
7. In the anatomical position,
   (a) the arms are extended away from the body.
   (b) the palms of the hands face posteriorly.
   (c) the body is erect and the palms face anteriorly.
   (d) the body is in a fetal position.
8. Which of the following statements about homeostasis is true?
   (a) The internal environment is maintained absolutely constant.
   (b) Negative feedback mechanisms act to correct deviations from a normal range within the internal environment.
   (c) Homeostasis is maintained by switching effector actions on and off.
   (d) All of the above are true.
9. In a negative feedback mechanism, the effector organ produces changes that are
   (a) similar in direction to that of the initial stimulus.
   (b) opposite in direction to that of the initial stimulus.
   (c) unrelated to the initial stimulus.
10. A hormone called parathyroid hormone acts to help raise the blood calcium concentration. According to the principles of negative feedback, an effective stimulus for parathyroid hormone secretion would be
   (a) a fall in blood calcium.
   (b) a rise in blood calcium.

Essay Questions

1. What is meant by the humoral theory of body organization? Which great anatomists were influenced by this theory? When was the humoral theory discarded?
2. Discuss the impact Galen had on the advancement of anatomy and physiology and medicine. What ideological circumstances permitted the philosophies of Galen to survive for so long?
3. What role did the development of the microscope play in the advancement of the sciences of anatomy and physiology and medicine? What specialties of anatomical and physiological study have emerged since the introduction of the microscope?
4. Describe the scientific method and comment on its immense importance in anatomical and physiological research.
5. Explain the role of antagonistic negative feedback processes in the maintenance of homeostasis.
6. Explain, using examples, how the secretion of a hormone is controlled by the effects of that hormone's actions.

Related Web Sites

In a listing of the most current web sites related to this chapter, please visit the *Concepts of Human Anatomy and Physiology* home page at http://www.mhhe.com/biosci/abio/.

Two

Chemical Composition of the Body

Objectives

- Describe the structure of an atom and define the terms *atomic mass* and *atomic number*.
- Describe the different types of chemical bonds, noting their relative strengths.
- Define the terms *acid*, *base*, *pH*, and *ion*.
- Discuss the properties of water and explain why compounds may be either hydrophilic or hydrophobic.
- Describe the structures of some organic molecules and identify different functional groups.
- List the subcategories of carbohydrates and give examples for each subcategory.
- Describe dehydration synthesis and hydrolysis reactions and indicate where they occur in the body.
- Identify the subclasses of lipids and explain why they are all classified as lipids.
- Distinguish between saturated and unsaturated fats and describe the chemical reactions in which triglycerides are formed and broken down.
- Describe the structures of phospholipids and prostaglandins and explain the functions of these molecules in the body.
- Describe the structure of amino acids and explain how one type of amino acid differs from another.
- Explain what is meant by the primary, secondary, tertiary, and quaternary structure of proteins.
- List some of the functions of different proteins and explain why protein structure is so diverse.

Clinical Investigation

A student, believing it immoral to eat plants or animals, decided to eat only artificial food. After raiding a chemistry laboratory, he placed himself on a diet consisting only of the D-amino acids and L-sugars he obtained in his raid. After a couple of weeks he began to feel weak and sought medical attention. Laboratory analysis of his urine revealed very high concentrations of ketone bodies (ketonuria). What might be the cause of his weakness and ketonuria?

Clues: See the description of stereoisomer in the section on organic molecules and the description of ketone bodies in the section on lipids.

Atoms, Ions, and Chemical Bonds

An understanding of the structure and function of the human body requires some familiarity with the basic concepts and terminology of chemistry. A knowledge of atomic and molecular structure, the nature of chemical bonds, and the nature of pH and associated concepts provides the foundation for much of human physiology.

The structures and physiological processes of the body are based, to a large degree, on the properties and interactions of atoms, ions, and molecules. Water is the major solvent in the body and accounts for 65% to 75% of the total weight of an average adult. Of this amount, two-thirds is contained within the body cells, or in the *intracellular compartment*; the remainder is contained in the *extracellular compartment*, a term that refers to the blood and tissue fluids. Dissolved in this water are many organic molecules (carbon-containing molecules such as carbohydrates, lipids, proteins, and nucleic acids), as well as inorganic molecules and ions (atoms with a net charge). Before describing the structure and function of organic molecules within the body, it is useful to consider some basic chemical concepts, terminology, and symbols.

Atoms

Atoms are the smallest units of matter that can undergo chemical change. They are much too small to be seen individually, even with the most powerful electron microscope. Through the efforts of generations of scientists, however, atomic structure is now understood. At the center of an atom is its **nucleus.** The nucleus contains two types of subatomic particles—**protons,** which bear a positive charge, and **neutrons,** which carry no charge. The mass of a proton is equal to the mass of a neutron, and the sum of the protons and neutrons in an atom is equal to the **atomic mass** of the atom. For example, an atom of carbon, which contains six protons and six neutrons, has an atomic mass of 12 (table 2.1).

The number of protons in an atom is given as its **atomic number.** Carbon has six protons and thus has an atomic number of 6. Outside the positively charged nucleus are negatively charged subatomic particles called **electrons.** Since the number of electrons in an atom is equal to the number of protons, atoms have a net charge of zero.

Although it is often convenient to think of electrons as orbiting the nucleus like planets orbiting the sun, this simplified model of atomic structure is no longer believed to be correct. A given electron can occupy any position in a certain volume of space called the *orbital* of the electron. The orbital is like a "shell," or energy level, beyond which the electron usually does not pass.

There are potentially several such orbitals surrounding a nucleus, with each successive orbital being farther from the nucleus. The first orbital, closest to the nucleus, can contain only two electrons. If an atom has more than two electrons (as do all atoms except hydrogen and helium), the additional electrons must occupy orbitals that are more distant from the nucleus. The second orbital can contain a maximum of eight electrons, and higher orbitals can contain still more electrons that possess more energy the farther they are from the nucleus. Most elements of biological

Table 2.1
Atoms Commonly Present in Organic Molecules

Atom	Symbol	Atomic Number	Atomic Mass	Orbital 1	Orbital 2	Orbital 3	Number of Chemical Bonds
Hydrogen	H	1	1	1	0	0	1
Carbon	C	6	12	2	4	0	4
Nitrogen	N	7	14	2	5	0	3
Oxygen	O	8	16	2	6	0	2
Sulfur	S	16	32	2	8	6	2

significance (other than hydrogen), however, require eight electrons to complete the outermost orbital. (The orbitals are filled from the innermost outward.) Carbon, with six electrons, has two electrons in its first orbital and four electrons in its second orbital (fig. 2.1).

Only electrons in the outermost orbital, if this orbital is incomplete, participate in chemical reactions and form chemical bonds. These outermost electrons are known as the **valence electrons** of the atom.

Isotopes

A particular atom with a given number of protons in its nucleus may exist in several forms that differ from one another in their number of neutrons. The atomic number of these forms is thus the same, but their atomic mass is different. These different forms are called **isotopes** (*ī′sŏ-tōps*). All of the isotopic forms of a given atom are included in the term **chemical element.** The element hydrogen, for example, has three isotopes. The most common of these has a nucleus consisting of only one proton. Another isotope of hydrogen (called *deuterium*) has one proton and one neutron in the nucleus, whereas the third isotope (*tritium*) has one proton and two neutrons. Tritium is a radioactive isotope that is commonly used in physiological research and in many clinical laboratory procedures.

Chemical Bonds, Molecules, and Ionic Compounds

Molecules are formed through interaction of the valence electrons between two or more atoms. These interactions, such as the sharing of electrons, produce **chemical bonds** (fig. 2.2). The number of bonds that each atom can have is determined by the number of electrons needed to complete the outermost orbital. Hydrogen, for example, must obtain only one more electron—and can thus form only one chemical bond—to complete the first orbital of two electrons. Carbon, by contrast, must obtain four more electrons—and can thus form four chemical bonds—to complete the second orbital of eight electrons (fig. 2.3, *left*).

Covalent Bonds

Covalent bonds result when atoms share their valence electrons. Covalent bonds that are formed between identical atoms, as in oxygen gas (O_2) and hydrogen gas (H_2), are the strongest because their electrons are equally shared. Since the electrons are equally distributed between the two atoms, these molecules are said to be **nonpolar,** and the bonds between them are called nonpolar covalent bonds. Such bonds are also important in living systems. The unique nature of carbon atoms and the organic molecules formed through covalent bonds between carbon atoms provide the chemical foundation of life.

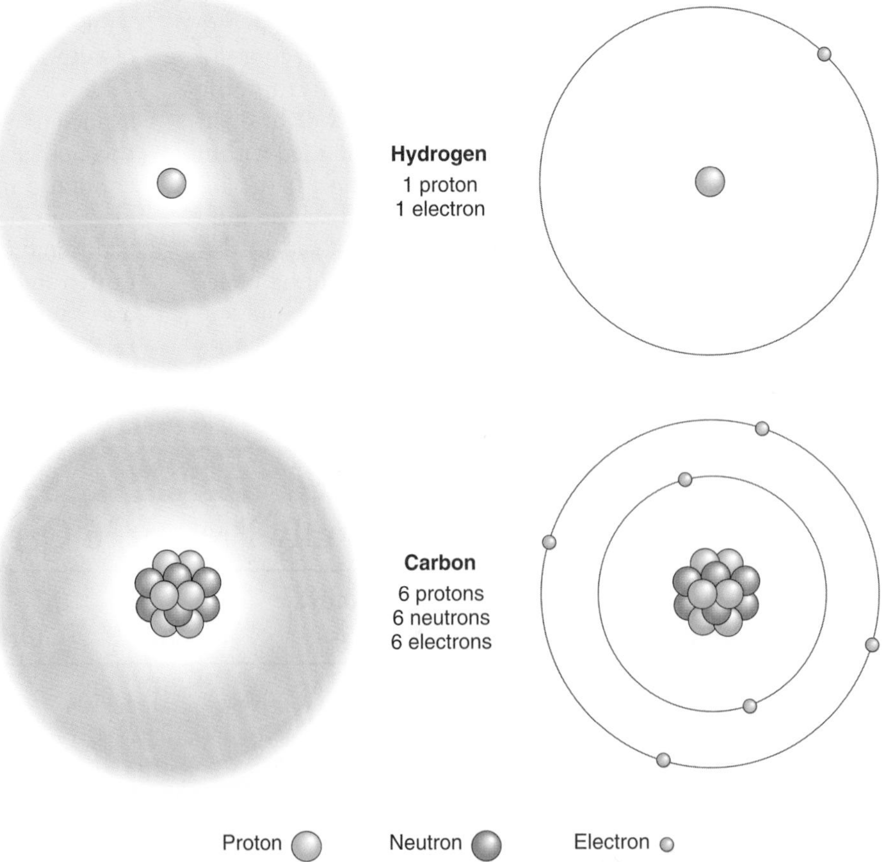

Figure 2.1

Diagrams of the hydrogen and carbon atoms.
The electron orbitals on the left are represented by shaded spheres indicating probable positions of the electrons. The orbitals on the right are represented by concentric circles.

When covalent bonds are formed between two different atoms, the electrons may be pulled more toward one atom than the other. The end of the molecule toward which the electrons are pulled is electrically negative compared to the other end. Such a molecule is said to be **polar** (has a positive and negative pole). Atoms of oxygen, nitrogen, and phosphorus have a particularly strong tendency to pull electrons toward themselves when they bond with other atoms; thus, they tend to form polar molecules.

Figure 2.2

A hydrogen molecule showing the covalent bonds between hydrogen atoms.
These bonds are formed by the equal sharing of electrons.

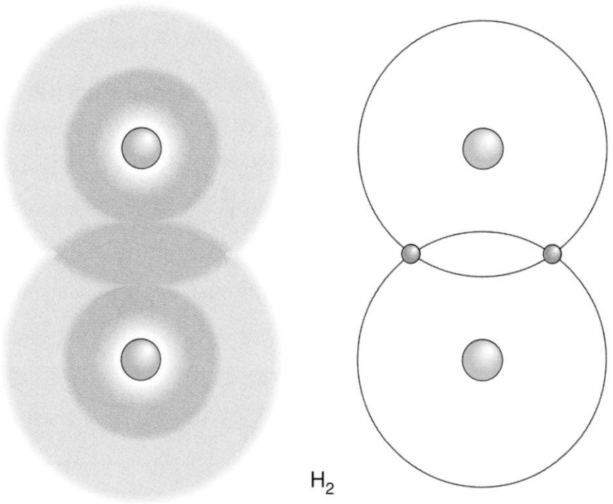

Water is the most abundant molecule in the body and serves as the solvent for body fluids. Water is a good solvent because it is polar; the oxygen atom pulls electrons from the two hydrogens toward its side of the water molecule, so that the oxygen side is more negatively charged than the hydrogen side of the molecule (fig. 2.4). The significance of the polar nature of water in its function as a solvent is discussed in the next section.

Ionic Bonds

Ionic bonds result when one or more valence electrons from one atom are completely transferred to a second atom. Thus, the electrons are not shared at all. The first atom loses electrons, so that its number of electrons becomes smaller than its number of protons; it becomes positively charged. Atoms or molecules that have positive or negative charges are called **ions** (*i'onz*). Positively charged ions are called *cations* (*kat'i-onz*) because they move toward the negative pole, or cathode, in an electric field. The second atom now has more electrons than it has protons and becomes a negatively charged ion, or *anion* (*an'i-on*) (so called because it moves toward the positive pole, or anode, in an electric field). The cation and anion then attract each other to form an **ionic compound.**

Common table salt, sodium chloride (NaCl), is an example of an ionic compound. Sodium, with a total of eleven electrons, has two in its first orbital, eight in its second orbital, and only one in its third orbital. Chlorine, conversely, is one electron short of completing its outer orbital of eight electrons. The lone electron in sodium's outer orbital is attracted to chlorine's outer orbital. This creates a chloride ion (represented as Cl^-) and a sodium ion (Na^+). Although table salt is shown as NaCl, it is actually composed of Na^+Cl^- (fig. 2.5).

Figure 2.3

The molecules methane and ammonia represented in three different ways.
Notice that a bond between two atoms consists of a pair of shared electrons (the electrons from the outer orbital of each atom).

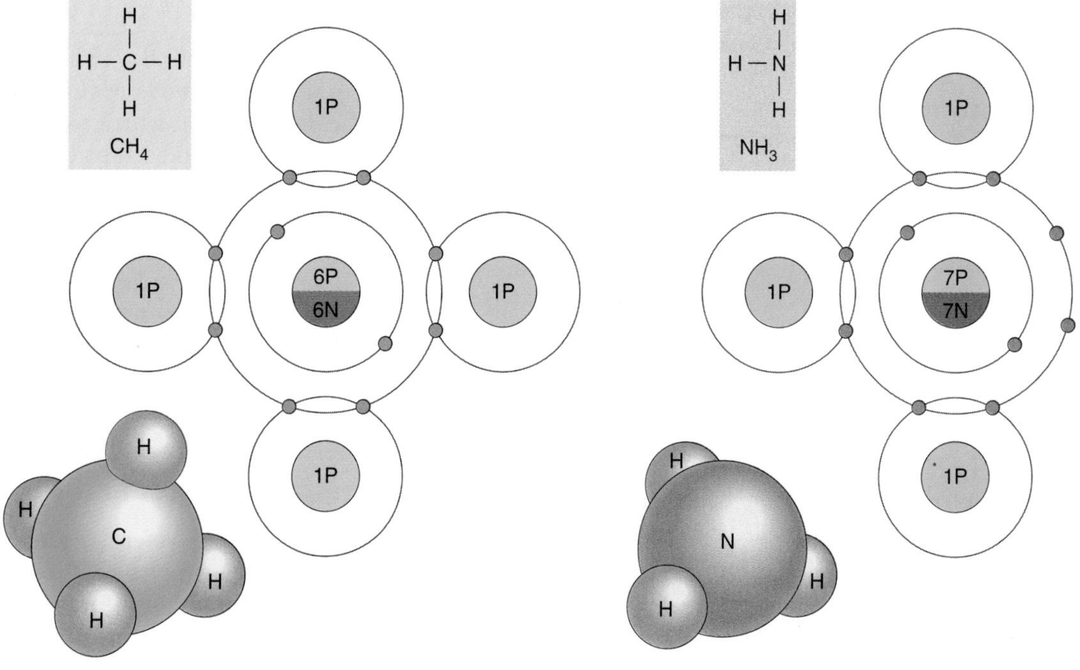

Figure 2.4

A model of a water molecule showing its polar nature. Notice that the oxygen side of the molecule is negative, whereas the hydrogen side is positive. Polar covalent bonds are weaker than nonpolar covalent bonds. As a result, some water molecules ionize to form a hydroxyl ion (OH^-) and a hydrogen ion (H^+). The H^+ combines with water molecules to form hydronium (H_3O^+) ions (not shown).

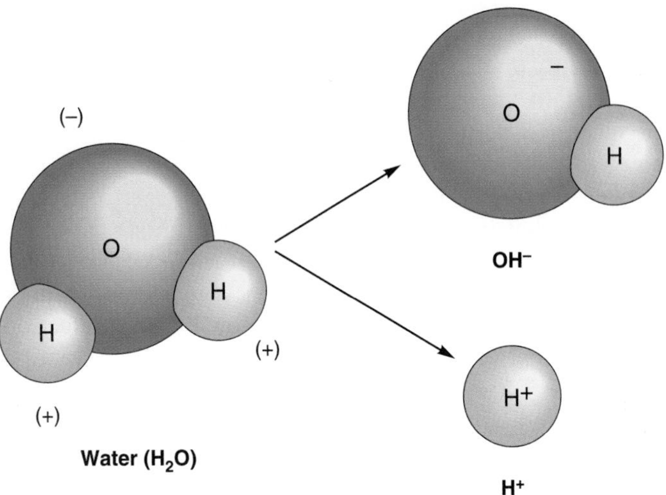

Figure 2.5

The dissociation of sodium and chlorine to produce sodium and chloride ions. The positive sodium and negative chloride ions attract each other, producing the ionic compound sodium chloride (NaCl). The transferred electron is indicated in red.

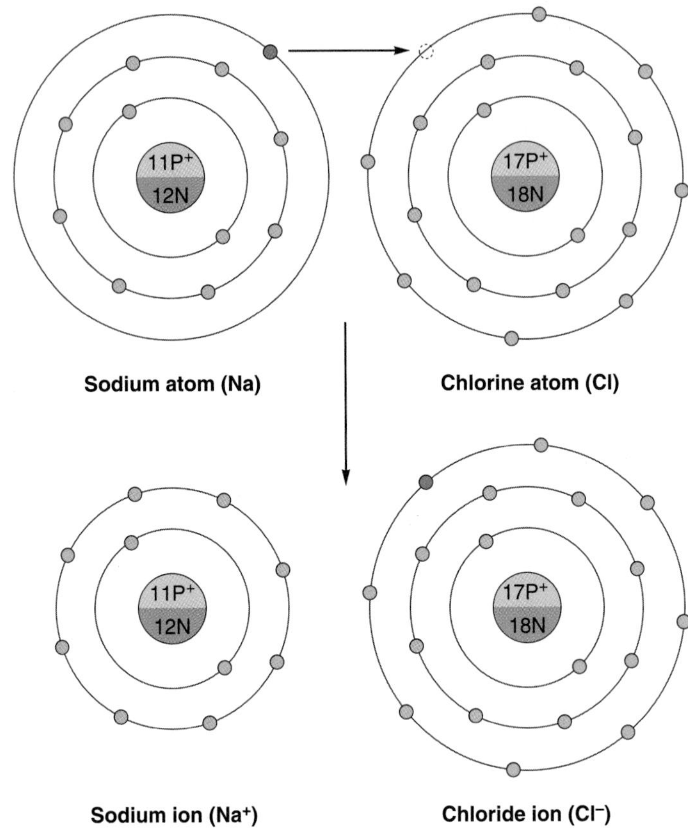

Ionic bonds are weaker than polar covalent bonds, and therefore ionic compounds easily separate (dissociate) when dissolved in water. Dissociation of NaCl, for example, yields Na^+ and Cl^-. Each of these ions attracts polar water molecules; the negative ends of water molecules are attracted to the Na^+, and the positive ends of water molecules are attracted to the Cl^- (fig. 2.6). The water molecules that surround these ions in turn attract other molecules of water to form *hydration spheres* around each ion.

The formation of hydration spheres makes an ion or a molecule soluble in water. Glucose, amino acids, and many other organic molecules are water-soluble because hydration spheres can form around atoms of oxygen, nitrogen, and phosphorus, which are joined by polar covalent bonds to other atoms in the molecule. Such molecules are said to be **hydrophilic**(*hi″drŏ-fil′ik*). By contrast, molecules composed primarily of nonpolar covalent bonds, such as the hydrocarbon chains of fat molecules, have few charges and thus cannot form hydration spheres. They are insoluble in water, and in fact are repelled by water molecules. For this reason, nonpolar molecules are said to be **hydrophobic** (*hi″drŏ-fo′bik*).

Hydrogen Bonds

When a hydrogen atom forms a polar covalent bond with an atom of oxygen or nitrogen, the hydrogen gains a slight positive charge as the electron is pulled toward the other atom. This other atom is thus described as being *electronegative*. Since the hydrogen atom has a slight positive charge, it will have a weak attraction for a second electronegative atom (oxygen or nitrogen) that may be located near it. This weak attraction is called a **hydrogen bond.** Hydrogen bonds are usually shown with dashed or dotted lines (fig. 2.7) to distinguish them from strong covalent bonds, which are shown with solid lines.

Although each hydrogen bond is relatively weak, the sum of their attractive forces is largely responsible for the folding of a protein and for the holding together of the two strands of a DNA molecule (see chapter 3). Hydrogen bonds can also be formed between adjacent water molecules (fig. 2.7). The hydrogen bonding between water molecules is responsible for many of the biologically important properties of water, including its *surface tension* (see chapter 24) and its ability to be pulled as a column through narrow channels in a process called *capillary action*.

Acids, Bases, and the pH Scale

The bonds in water molecules that join hydrogen and oxygen atoms are, as previously discussed, polar covalent bonds. Although these bonds are strong, a small proportion of them break as the electron from the hydrogen atom is completely transferred to oxygen. When this occurs, the water molecule ionizes to form a *hydroxyl* (*hi-drok′sil*) *ion* (OH^-) and a hydrogen ion (H^+), which is simply a free proton (see fig. 2.4). A proton released in this way does not remain free for long,

hydrophilic: Gk. *hydro*, water; *philos*, fond
hydrophobic: Gk. *hydro*, water; *phoos*, fear

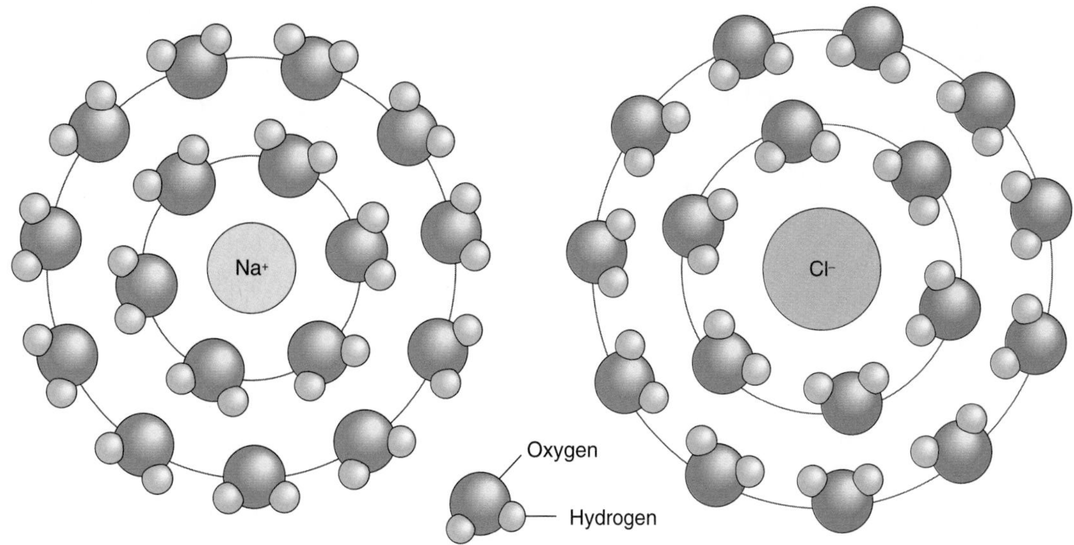

Figure 2.6

How NaCl dissolves in water. The negatively charged oxygen-ends of water molecules are attracted to the positively charged Na^+, whereas the positively charged hydrogen-ends of water molecules are attracted to the negatively charged Cl^-. Other water molecules are attracted to this first concentric layer of water, forming hydration spheres around the sodium and chloride ions.

Figure 2.7

Hydrogen bonds between water molecules. The oxygen atoms of water molecules are weakly joined together by the attraction of the electronegative oxygen for the positively charged hydrogen. These weak bonds are called hydrogen bonds.

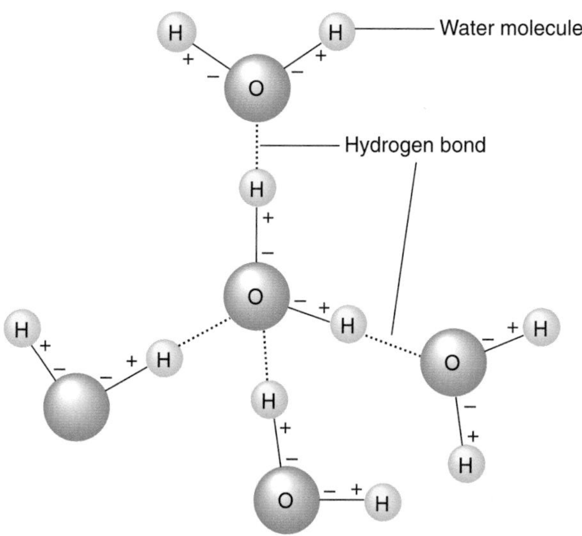

however, because it is attracted to the electrons of oxygen atoms in water molecules. This forms a *hydronium ion*, shown by the formula H_3O^+. For the sake of clarity in the following discussion, however, H^+ will be used to represent the ion resulting from the ionization of water.

Ionization of water molecules produces equal amounts of OH^- and H^+. Since only a small proportion of water molecules ionize, the concentrations of H^+ and OH^- are each equal to only 10^{-7} molar (the term *molar* is a unit of concentration, described in chapter 5; for hydrogen, one molar equals one gram per liter). A solution with 10^{-7} molar hydrogen ion, which is produced by the ionization of water molecules in which the H^+ and OH^- concentrations are equal, is said to be **neutral.**

Table 2.2

Common Acids and Bases

Acid	Symbol	Base	Symbol
Hydrochloric acid	HCl	Sodium hydroxide	$NaOH$
Phosphoric acid	H_3PO_4	Potassium hydroxide	KOH
Nitric acid	HNO_3	Calcium hydroxide	$Ca(OH)_2$
Sulfuric acid	H_2SO_4	Ammonium hydroxide	NH_4OH
Carbonic acid	H_2CO_3		

A solution that has a higher H^+ concentration than that of water is called *acidic;* one with a lower H^+ concentration is called *basic*, or *alkaline* (*al'kă-lin*). An **acid** is defined as a molecule that can release protons (H^+) into a solution; it is a "proton donor." A **base** is a negatively charged ion (anion), or a molecule that ionizes to produce the anion, that can combine with H^+ and thus remove the H^+ from solution; it is a "proton acceptor." Most strong bases release OH^- into a solution; the OH^- combines with H^+ to form water, thus lowering the H^+ concentration. Examples of common acids and bases are shown in table 2.2.

pH

The H^+ concentration of a solution is usually indicated in pH units on a pH scale that runs from 0 to 14. The pH number is equal to the logarithm of 1 over the H^+ concentration:

$$\mathbf{pH = \log \frac{1}{[H^+]}}$$

where $[H^+]$ = molar H^+ concentration. This can also be expressed as $pH = -\log [H^+]$.

Table 2.3
The pH Scale

	H^+ Concentration (Molar)	pH	OH^- Concentration (Molar)
	1.0	0	10^{-14}
	0.1	1	10^{-13}
	0.01	2	10^{-12}
Acids	0.001	3	10^{-11}
	0.0001	4	10^{-10}
	10^{-5}	5	10^{-9}
	10^{-6}	6	10^{-8}
Neutral	10^{-7}	7	10^{-7}
	10^{-8}	8	10^{-6}
	10^{-9}	9	10^{-5}
	10^{-10}	10	0.0001
Bases	10^{-11}	11	0.001
	10^{-12}	12	0.01
	10^{-13}	13	0.1
	10^{-14}	14	1.0

Pure water has a H^+ concentration of 10^{-7} molar at 25° C, and thus has a pH of 7 (neutral). Because of the logarithmic relationship, a solution with 10 times the hydrogen ion concentration (10^{-6} M) has a pH of 6, whereas a solution with one-tenth the H^+ concentration (10^{-8} M) has a pH of 8. The pH number is easier to write than the molar concentration, but it is admittedly confusing because it is *inversely related* to the H^+ concentration: a solution with a higher H^+ concentration has a lower pH number; one with a lower H^+ concentration has a higher pH number. A strong acid with a high H^+ concentration of 10^{-2} molar, for example, has a pH of 2, whereas a solution with only 10^{-10} molar has a pH of 10. **Acidic solutions,** therefore, have a pH of less than 7 (that of pure water), whereas **basic (alkaline) solutions** have a pH between 7 and 14 (table 2.3).

Buffers

A **buffer** is a system of molecules and ions that prevents changes in H^+ concentration and thus stabilizes the pH of a solution. In blood plasma, for example, the pH is stabilized by the following reversible reaction involving the bicarbonate ion (HCO_3^-) and carbonic acid (H_2CO_3):

$$HCO_3^- + H^+ \rightleftharpoons H_2CO_3$$

The double arrows indicate that the reaction could go either to the right or to the left; the net direction depends on the concentration of molecules and ions on each side. If an acid (such as lactic acid) should release H^+ into the solution, for example, the increased concentration of H^+ would drive the equilibrium to the right, and the following reaction would be promoted:

$$HCO_3^- + H^+ \rightarrow H_2CO_3$$

Blood pH

Lactic acid and other organic acids are produced by the cells of the body and secreted into the blood. Despite the release of H^+ by these acids, the arterial blood pH normally does not decrease but remains remarkably constant at pH 7.40 ± 0.05. This constancy is achieved, in part, by the buffering action of bicarbonate shown in the preceding equation. Bicarbonate serves as the major buffer of the blood.

Certain conditions could cause an opposite change in pH. For example, excessive vomiting that results in loss of gastric acid could cause the concentration of free H^+ in the blood to fall and the blood pH to rise. In this case, the reaction previously described could be reversed:

$$H_2CO_3 \rightarrow H^+ + HCO_3^-$$

The dissociation of carbonic acid yields free H^+, which helps to prevent an increase in pH. Bicarbonate ions and carbonic acid thus act as a *buffer pair* to prevent either decreases or increases in pH, respectively. This buffering action normally maintains the blood pH within the narrow range of 7.35 to 7.45.

If the arterial blood pH falls below 7.35, the condition is called *acidosis*. A blood pH of 7.20, for example, represents significant acidosis. Notice that acidotic blood need not be acidic. An increase in blood pH above 7.45, conversely, is known as *alkalosis*. Acidosis and alkalosis are normally prevented by the action of the bicarbonate/carbonic acid buffer pair and by the functions of the lungs and kidneys. Regulation of blood pH is discussed in more detail in chapters 24 and 25.

Organic Molecules

Organic molecules are those molecules that contain carbon and hydrogen atoms. Since the carbon atom has four electrons in its outer orbital, it must share four additional electrons by covalently bonding with other atoms to fill its outer orbital with eight electrons. The unique bonding requirements of carbon enable it to join with other carbon atoms to form chains and rings, while still allowing the carbon atoms to bond with hydrogen and other atoms.

Most organic molecules in the body contain hydrocarbon chains and rings, as well as other atoms bonded to carbon. Two adjacent carbon atoms in a chain or ring may share one or two pairs of electrons. If the two carbon atoms share one pair of electrons, they have a *single covalent bond;* this leaves each carbon atom free to bond with as many as three other atoms. If the two carbon atoms share two pairs of electrons, they have a *double covalent bond,* and each carbon atom can bond with a maximum of only two additional atoms (fig. 2.8).

Figure 2.8

Single and double covalent bonds.
Two carbon atoms may be joined by a single covalent bond (*left*) or a double covalent bond (*right*). In both cases, each carbon atom shares four pairs of electrons (has four bonds) to complete the eight electrons required to fill its outer orbital.

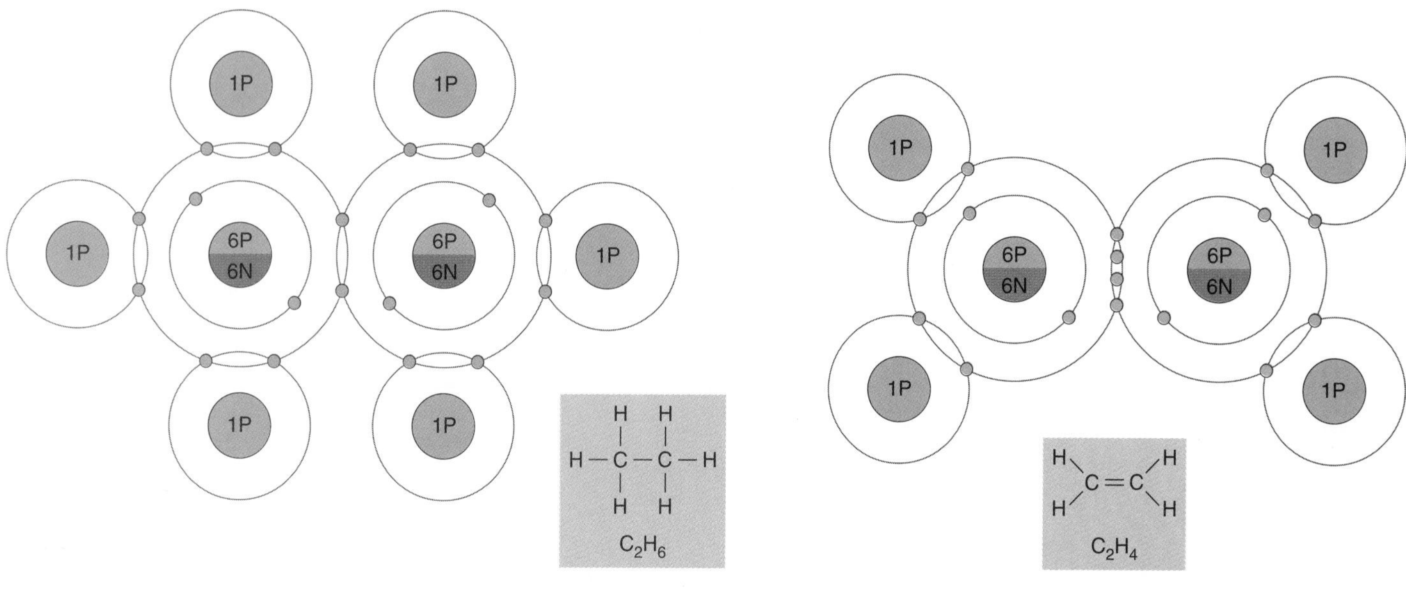

The ends of some hydrocarbons join to form rings. In the shorthand structural formulas of these molecules, the carbon atoms are not shown but are understood to be located at the corners of the ring. Some of these cyclic molecules have a double bond between two adjacent carbon atoms. Benzene and related molecules are shown as a six-sided ring with alternating double bonds. Such compounds are called **aromatic.** Since all of the carbons in an aromatic ring are equivalent, double bonds can be shown between any two adjacent carbons in the ring (fig. 2.9), or even as a circle within the hexagonal structure of carbons.

The hydrocarbon chain or ring of many organic molecules provides a relatively inactive molecular "backbone" to which more reactive groups of atoms are attached. Known as *functional groups* of the molecule, these reactive groups usually contain atoms of oxygen, nitrogen, phosphorus, or sulfur. They are largely responsible for the unique chemical properties of the molecule (fig. 2.10).

Classes of organic molecules can be named according to their functional groups. **Ketones** (*ke'tonz*), for example, have a carbonyl group within the carbon chain. An organic molecule is an **alcohol** if it has a hydroxyl group bound to a hydrocarbon chain. All **organic acids** (acetic acid, citric acids, lactic acid, and others) have a *carboxyl* (*kar-bok'sil*) *group* (fig. 2.11).

A carboxyl group can be abbreviated COOH. This group is an acid because it can donate its proton (H^+) to the solution. Ionization of COOH forms COO^- and H^+ (fig. 2.12). The ionized organic acid is designated with the suffix *-ate*. For example, when the carboxyl group of lactic acid ionizes, the molecule is called *lactate*. Since both ionized and un-ionized forms of the molecule exist together in

Figure 2.9

Different shapes of hydrocarbon molecules.
Hydrocarbon molecules can be (*a*) linear, (*b*) cyclic, and (*c*) have aromatic rings.

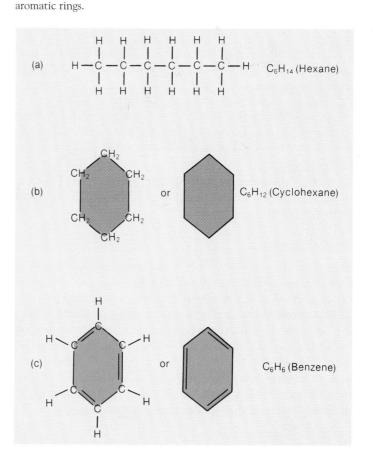

(a) C_6H_{14} (Hexane)

(b) CH_2 or C_6H_{12} (Cyclohexane)

(c) or C_6H_6 (Benzene)

Figure 2.10

Various functional groups of organic molecules.

The general symbol for a functional group is R.

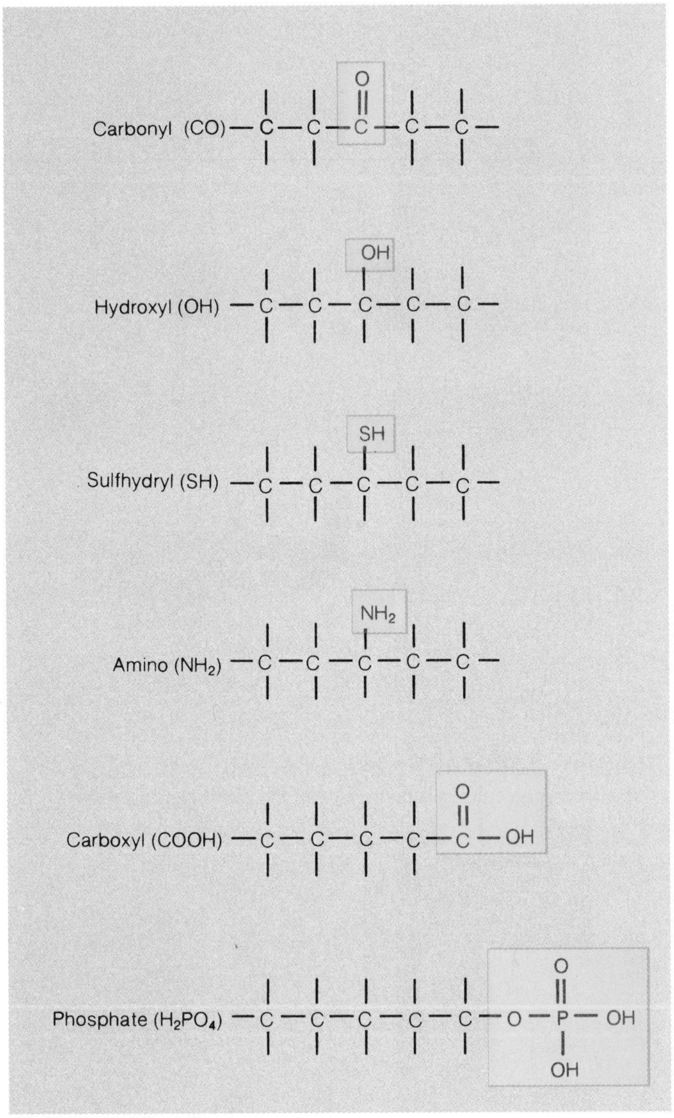

Carbonyl (CO)

Hydroxyl (OH)

Sulfhydryl (SH)

Amino (NH_2)

Carboxyl (COOH)

Phosphate (H_2PO_4)

Figure 2.11

Categories of organic molecules based on functional groups.

Acids, alcohols, and other types of organic molecules are characterized by specific functional groups.

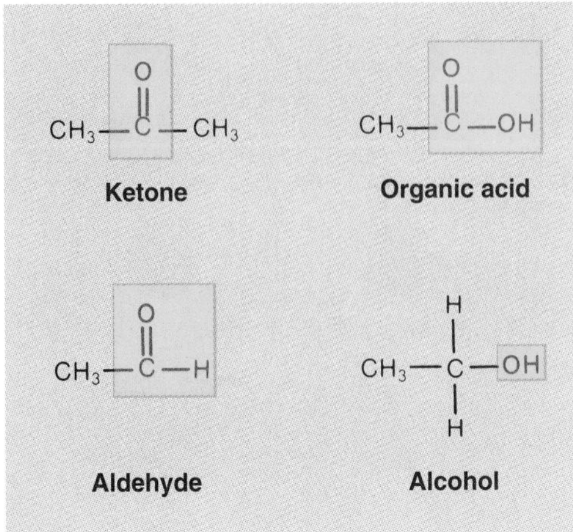

Figure 2.12

The carboxyl group of an organic acid.

This group can ionize to yield a free proton, or hydrogen ion (H^+). This process is shown for lactic acid, and the double arrows indicate that the reaction is reversible.

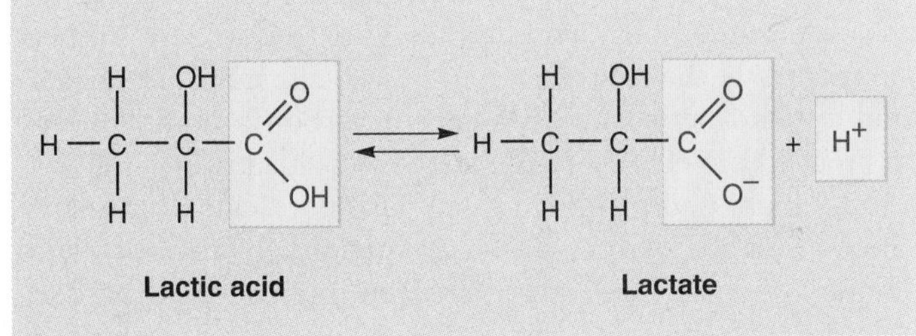

a solution (the proportion of each depends on the pH of the solution), one can correctly refer to the molecule as either lactic acid or lactate.

Stereoisomers

Two molecules may have exactly the same atoms arranged in exactly the same sequence yet differ with respect to the spatial orientation of a key functional group. Such molecules are called **stereoisomers** (*ster″e-o-i′sŏ-merz*) of each other. Depending upon the direction in which the key functional group is oriented with respect to the molecules, stereoisomers are called either *D-isomers* (for *dextro*, or right-handed) or *L-isomers* (for *levo*, or left-handed). Their relationship is similar to that between a right and left glove—if the palms are both pointing in the same direction, the two cannot be superimposed.

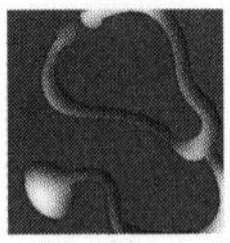

Severe birth defects often resulted when pregnant women were prescribed the sedative *thalidomide* in the early 1960s to alleviate morning sickness. The drug available at the time contained a mixture of both right-handed (D) and left-handed (L) forms. This tragic circumstance emphasizes the clinical importance of stereoisomers. It has since been learned that the L-stereoisomer is a potent tranquilizer, but the right-handed version causes disruption of fetal development and the resulting birth defects. Interestingly, thalidomide is now being used in the treatment of AIDS patients and others with cachexia (prolonged ill health and malnutrition).

These subtle differences in structure are extremely important biologically. They ensure that enzymes—which interact with such molecules in a stereo-specific way in chemical reactions—cannot combine with the "wrong" stereoisomer. For example, the enzymes of all cells (human and others) can combine only with L-amino acids and D-sugars. The opposite stereoisomers (D-amino acids and L-sugars) cannot be absorbed into the body from the small intestine or be used by any enzyme in metabolism.

Carbohydrates and Lipids

Carbohydrates are a class of organic molecules that includes monosaccharides, disaccharides, and polysaccharides. All of these molecules are based on a characteristic ratio of carbon, hydrogen, and oxygen atoms. Lipids are a category of diverse organic molecules that share the physical property of being nonpolar and thus insoluble in water.

Carbohydrates and lipids are similar in many ways. Both groups of molecules consist primarily of the atoms carbon, hydrogen, and oxygen, and both serve as major sources of energy in the body (accounting for most of the calories consumed in food). Carbohydrates and lipids differ, however, in some important aspects of their chemical structures and physical properties. Such differences significantly affect the functions of these molecules in the body.

Carbohydrates

Carbohydrates are organic molecules that contain carbon, hydrogen, and oxygen in the ratio described by their name—*carbo* (carbon) and *hydrate* (water, H_2O). The general formula for a carbohydrate molecule is thus CH_2O; the molecule contains twice as many hydrogen atoms as carbon or oxygen atoms.

Monosaccharides, Disaccharides, and Polysaccharides

Carbohydrates include simple sugars, or **monosaccharides** (*mon″ŏ-sak′ă-rīdz*), and longer molecules that contain a nu of monosaccharides joined together. The suffix *-ose* denotes a sugar molecule; the term *hexose*, for example, refers to a six-carbon monosaccharide with the formula $C_6H_{12}O_6$. This formula is adequate for some purposes, but it does not distinguish between related hexose sugars, which are *structural isomers* of each other. The structural isomers glucose, galactose, and fructose, for example, are monosaccharides that have the same ratio of atoms arranged in slightly different ways (fig. 2.13).

Two monosaccharides can be joined covalently to form a **disaccharide** (*di-sak′ă-rīd*), or double sugar. Common disaccharides include table sugar, or *sucrose* (composed of glucose and fructose); milk sugar, or *lactose* (composed of glucose and galactose); and malt sugar, or *maltose* (composed of two glucose molecules). When numerous monosaccharides join, the resulting molecule is called a **polysaccharide.** *Starch*, for example, a polysaccharide found in many plants, is formed by the bonding of thousands of glucose subunits. **Glycogen** (*gli′kŏ-jen*), found in the liver and muscles, likewise consists of repeating glucose molecules, but it is more highly branched than plant starch (fig. 2.14).

monosaccharide: Gk. *monos*, single; *sakcharon*, sugar

Figure 2.13

The structural formulas of three hexose sugars. These are (*a*) glucose, (*b*) galactose, and (*c*) fructose. All three have the same ratio of atoms—$C_6H_{12}O_6$.

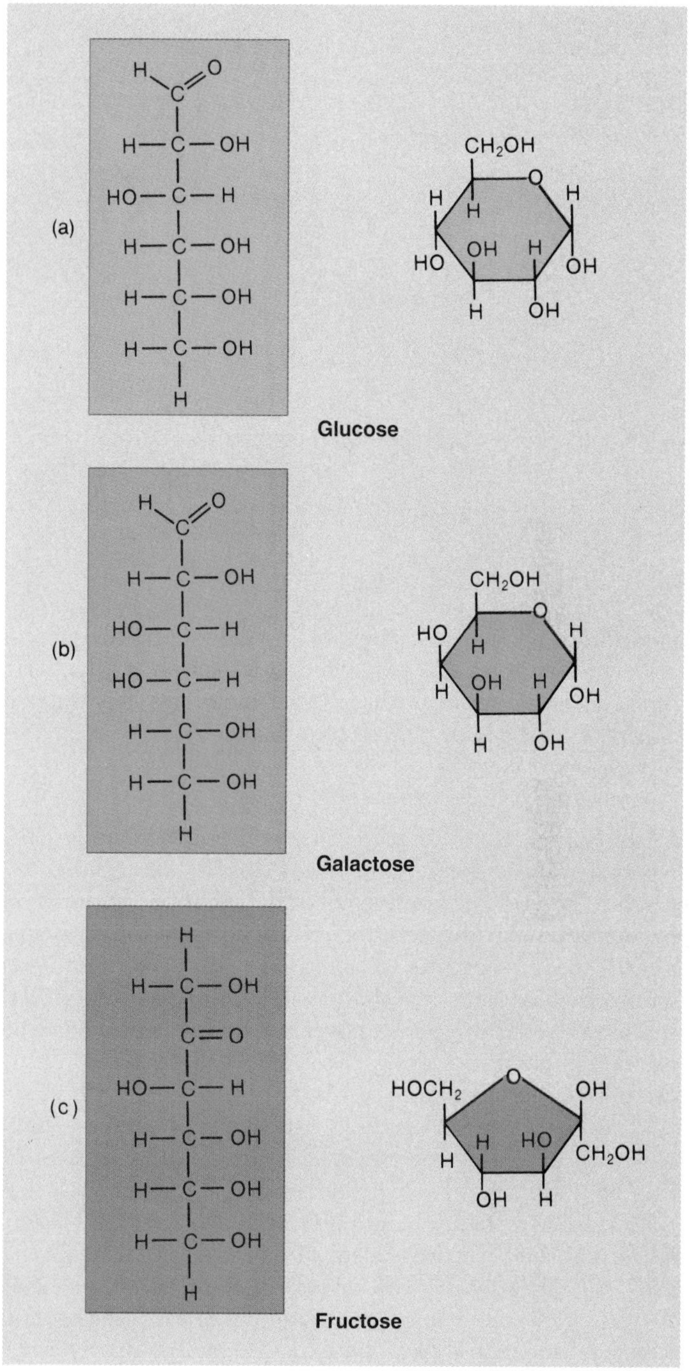

Many cells store carbohydrates for use as an energy source, as described in chapter 4. If a cell were to store many thousands of separate monosaccharide molecules, however, their high concentration would draw an excessive amount of water into the cell, damaging or even killing it. The net movement of water through membranes is called *osmosis*, and is discussed in chapter 5. Cells that store carbohydrates for energy

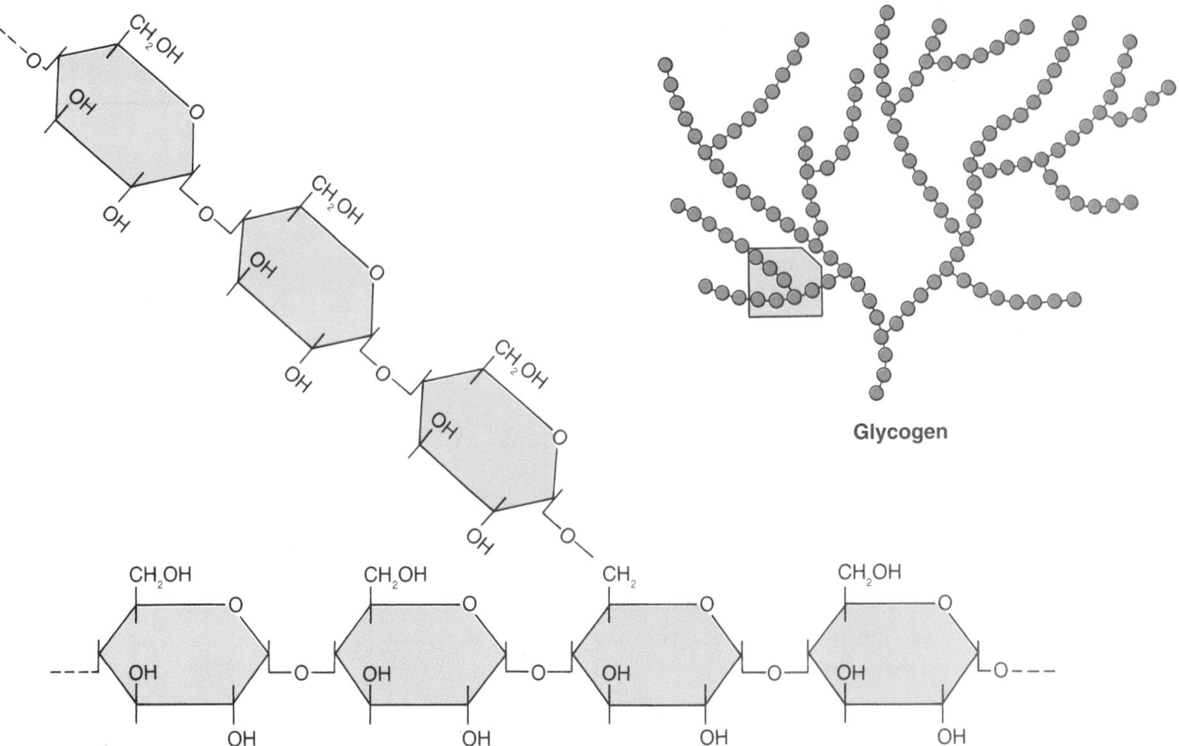

Figure 2.14

The structure of glycogen. Glycogen is a polysaccharide composed of glucose subunits joined together to form a large, highly branched molecule.

minimize this osmotic damage by instead joining the glucose molecules to form the polysaccharides starch or glycogen. Since there are fewer of these larger molecules, less water is drawn into the cell by osmosis (see chapter 5).

Dehydration Synthesis and Hydrolysis

In the formation of disaccharides and polysaccharides, the separate subunits (monosaccharides) are bonded together covalently by a type of reaction called **dehydration synthesis,** or **condensation.** In this reaction, which requires the participation of specific enzymes (chapter 4), a hydrogen atom is removed from one monosaccharide and a hydroxyl group (OH) is removed from another. A covalent bond is formed between the two monosaccharides as water (H_2O) is produced. Dehydration synthesis reactions are illustrated in figure 2.15.

When a person eats disaccharides or polysaccharides, or when the stored glycogen in the liver and muscles is to be used by tissue cells, the covalent bonds that join monosaccharides into disaccharides and polysaccharides must be broken. These *digestion reactions* occur by means of **hydrolysis** (*hi-drol'ĭ-sis*). Hydrolysis is the reverse of dehydration synthesis. A water molecule is split, and the resulting hydrogen atom is added to one of the free glucose molecules as the hydroxyl group is added to the other (fig. 2.16).

When a person eats a potato, the starch within the potato is hydrolyzed into separate glucose molecules within the small intestine. This glucose is absorbed into the blood and carried to the tissues. Some tissue cells may use this glucose for energy. Liver and muscles, however, can store excess glucose in the form of glycogen by dehydration synthesis reactions in these cells. During fasting or prolonged exercise, the liver can add glucose to the blood through hydrolysis of its stored glycogen.

Dehydration synthesis and hydrolysis reactions do not occur spontaneously; they require the action of specific enzymes. Similar reactions, in the presence of other enzymes, build and break down lipids, proteins, and nucleic acids. In general, therefore, hydrolysis reactions digest molecules into their subunits, and dehydration synthesis reactions build larger molecules by the bonding of their subunits.

Lipids

The category of molecules known as **lipids** includes several types of molecules that differ greatly in chemical structure. These diverse molecules are all in the lipid category by virtue of a common physical property—they are all *insoluble in polar solvents* such as water. This is because lipids consist primarily of hydrocarbon chains and rings, which are nonpolar and therefore hydrophobic. Although lipids are insoluble in water, they can be dissolved in nonpolar solvents such as ether, benzene, and related compounds.

Triglycerides

Triglycerides are the subcategory of lipids that includes fat and oil. These molecules are formed by the condensation of one molecule of *glycerol* (*glis'ĕ-rol*) (a three-carbon alcohol) with three molecules of *fatty acids*. Each fatty acid molecule consists of a nonpolar hydrocarbon chain with a carboxylic acid group (abbreviated COOH) on one end. If the carbon atoms within the hydrocarbon chain are joined by single covalent bonds so

hydrolysis: Gk. *hydro*, water; *lysis*, break

Figure 2.15

Dehydration synthesis of disaccharides. The two disaccharides formed here are (*a*) maltose and (*b*) sucrose. Notice that as the disaccharides are formed, a molecule of water is produced.

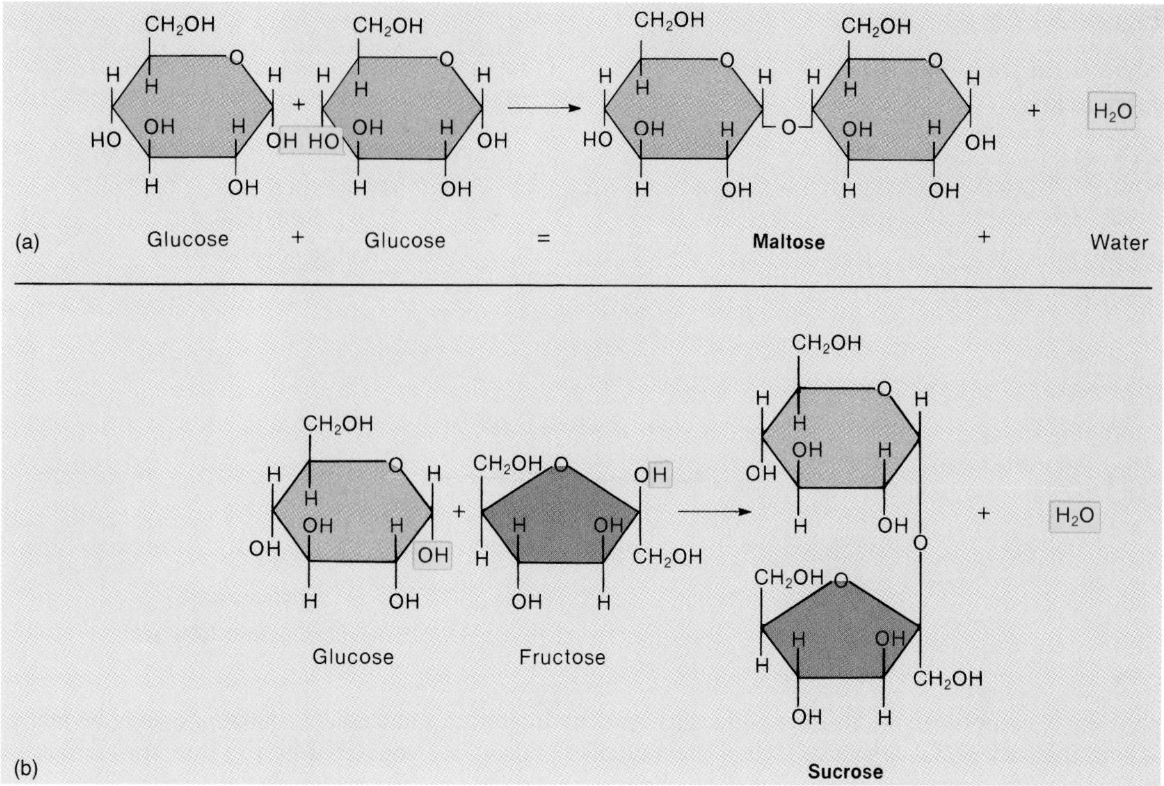

Figure 2.16

The hydrolysis of starch. The polysaccharide is first hydrolyzed (*a*) into disaccharides (maltose) and then (*b*) into monosaccharides (glucose). Notice that as the covalent bond between the subunits breaks, a molecule of water is split. In this way, the hydrogen atom and hydroxyl group from the water are added to the ends of the released subunits.

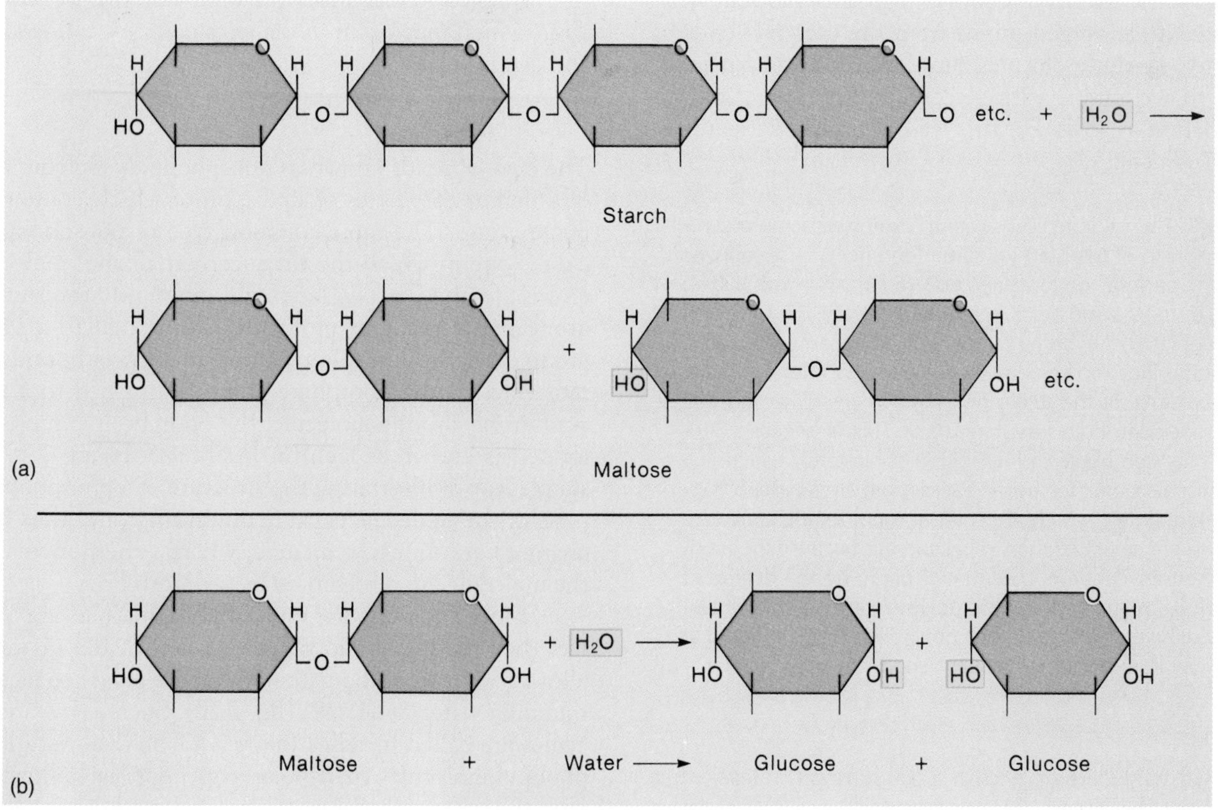

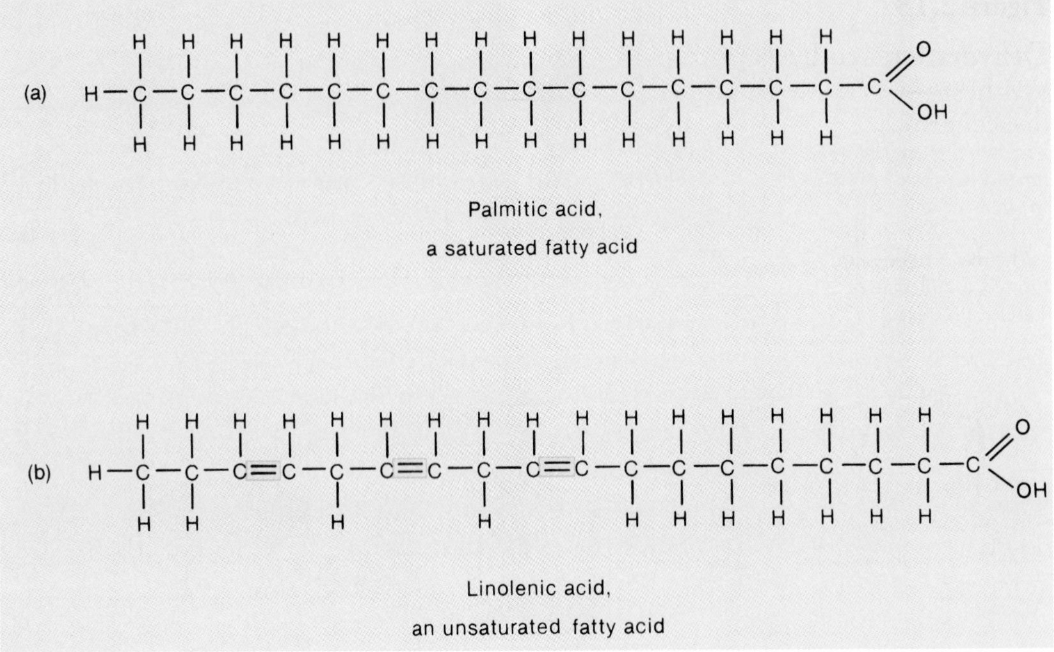

(a)

Palmitic acid,
a saturated fatty acid

(b)

Linolenic acid,
an unsaturated fatty acid

Figure 2.17

Structural formulas for fatty acids.
(*a*) Formula for saturated fatty acids and (*b*) unsaturated fatty acids.

that each carbon atom can also bond with two hydrogen atoms, the fatty acid is *saturated.* If there are a number of double covalent bonds within the hydrocarbon chain so that each carbon atom can bond with only one hydrogen atom, the fatty acid is *unsaturated.* Triglycerides that contain saturated fatty acids are **saturated fats;** those that contain unsaturated fatty acids are **unsaturated fats** (fig. 2.17).

Within the adipose cells of the body, triglycerides are formed as the carboxylic acid ends of fatty acid molecules condense with the hydroxyl groups of a glycerol molecule (fig. 2.18). Since the hydrogen atoms from the carboxyl ends of fatty acids form water molecules during dehydration synthesis, fatty acids that combine with glycerol can no longer release H^+ and function as acids. For this reason, triglycerides are described as *neutral fats*.

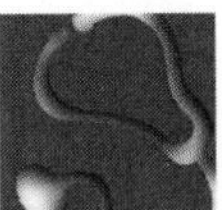

The saturated fat content (expressed as a percentage of total fat) for some food items is as follows: canola, or rapeseed, oil (6%); olive oil (14%); margarine (17%); chicken fat (31%); palm oil (51%); beef fat (52%); butter fat (66%); and coconut oil (77%). Health authorities recommend that a person's total fat intake not exceed 30% of the total energy intake per day, and that saturated fat contribute less than 10%. This is because saturated fat may contribute to high blood cholesterol, which is a significant risk factor for heart disease and stroke (see chapter 21). Animal fats, which are solid at room temperature, are generally more saturated than vegetable oils because the hardness of the triglyceride is determined partly by the degree of saturation. Palm and coconut oil, however, are notable exceptions. Though very saturated, they nonetheless remain liquid at room temperature because they have short fatty acid chains.

Ketone Bodies

Hydrolysis of triglycerides within adipose tissue releases *free fatty acids* into the blood. Free fatty acids can be used as an immediate source of energy by many organs; they can also be converted by the liver into derivatives called **ketone bodies.** These include four-carbon-long acidic molecules (acetoacetic acid and β-hydroxybutyric acid) and acetone (the solvent in nail-polish remover). A rapid breakdown of fat, as occurs during dieting and in uncontrolled diabetes mellitus, results in elevated levels of ketone bodies in the blood. This is a condition called **ketosis** (*ke-to'sis*). If there are sufficient amounts of ketone bodies in the blood to lower the blood pH, the condition is called **ketoacidosis** (*ke"to-as"ĭ-do'sis*). Severe ketoacidosis, which may occur in diabetes mellitus, can lead to coma and death.

Phospholipids

The class of lipids known as **phospholipids** includes a number of different categories of lipids, all of which contain a phosphate group. The most common type of phospholipid molecule is one in which the three-carbon alcohol molecule glycerol is attached to two fatty acid molecules; the third carbon atom of the glycerol molecule is attached to a phosphate group, and the phosphate group in turn is bound to other molecules. If the phosphate group is attached to a nitrogen-containing choline molecule, the phospholipid molecule thus formed is known as **lecithin** (*les'ĭ-thin*). Figure 2.19 shows a simple way of illustrating the structure of a phospholipid—the parts of the molecule capable of ionizing (and thus becoming charged) are shown as a circle, whereas the nonpolar parts of the molecule are represented by sawtooth lines.

Since the nonpolar ends of phospholipids are hydrophobic, they tend to group together when mixed in water. This allows the hydrophilic parts (which are polar) to face the surrounding water molecules (fig. 2.20). Such aggregates of molecules are called **micelles** (*mi-selz'*). The dual nature of phospholipid molecules (part polar, part nonpolar) allows them to alter the interaction of water molecules and thus decrease the

Figure 2.18

Dehydration synthesis forms triglyceride from glycerol and fatty acids.
A molecule of water is produced as an ester bond forms between each fatty acid and the glycerol. Sawtooth lines represent hydrocarbon chains, which are symbolized by an R.

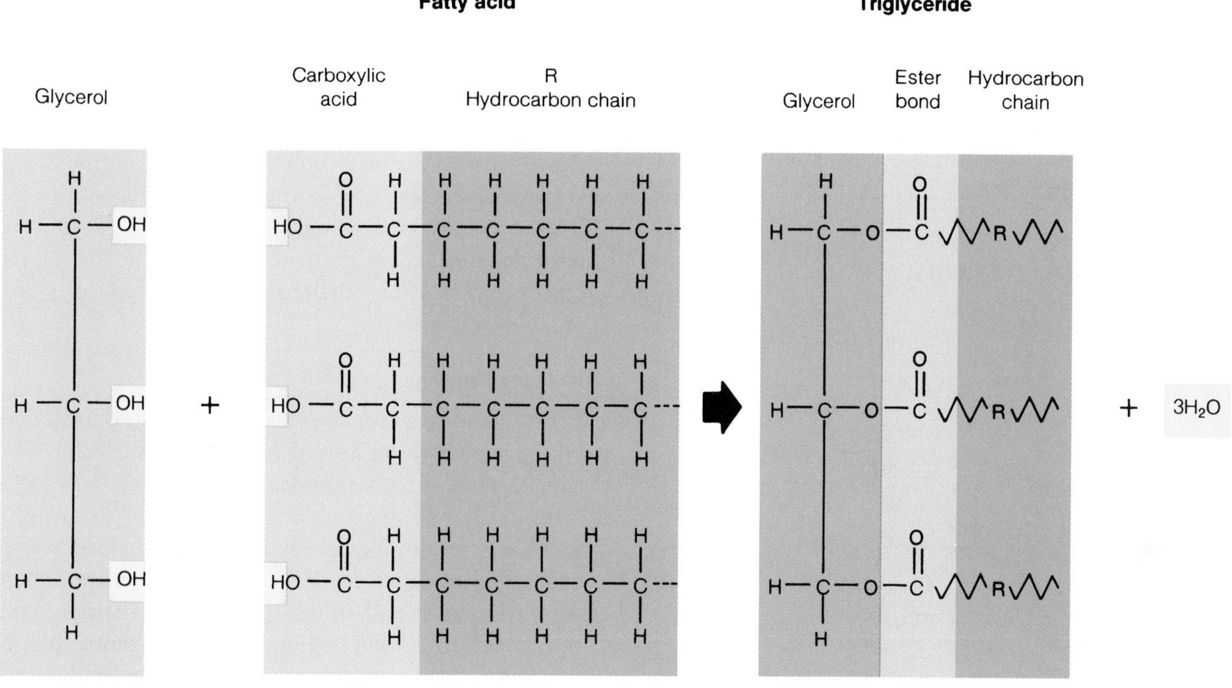

Figure 2.19

The structure of lecithin.
Lecithin (*top*) is a typical phospholipid, which is often represented in a more simplified manner (*bottom*).

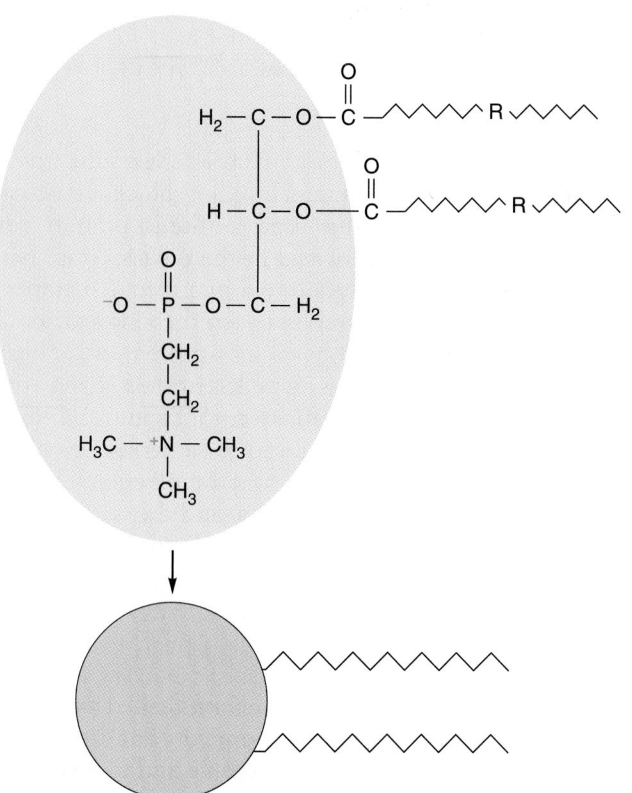

Figure 2.20

The formation of a micelle structure by phospholipids such as lecithin.
The outer layer of the micelle is hydrophilic, and faces the aqueous environment.

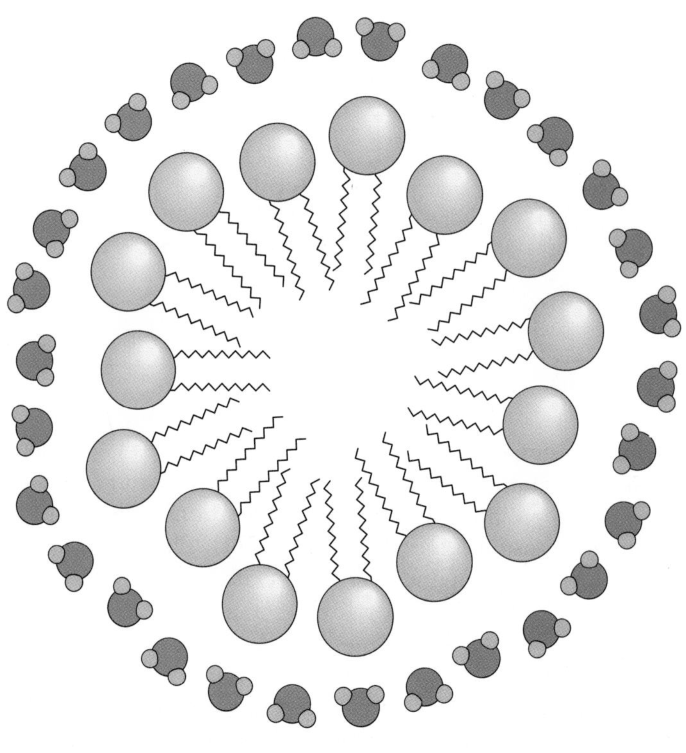

Figure 2.21

Cholesterol and some of the steroid hormones derived from cholesterol.
The steroid hormones are secreted by the gonads and the adrenal cortex.

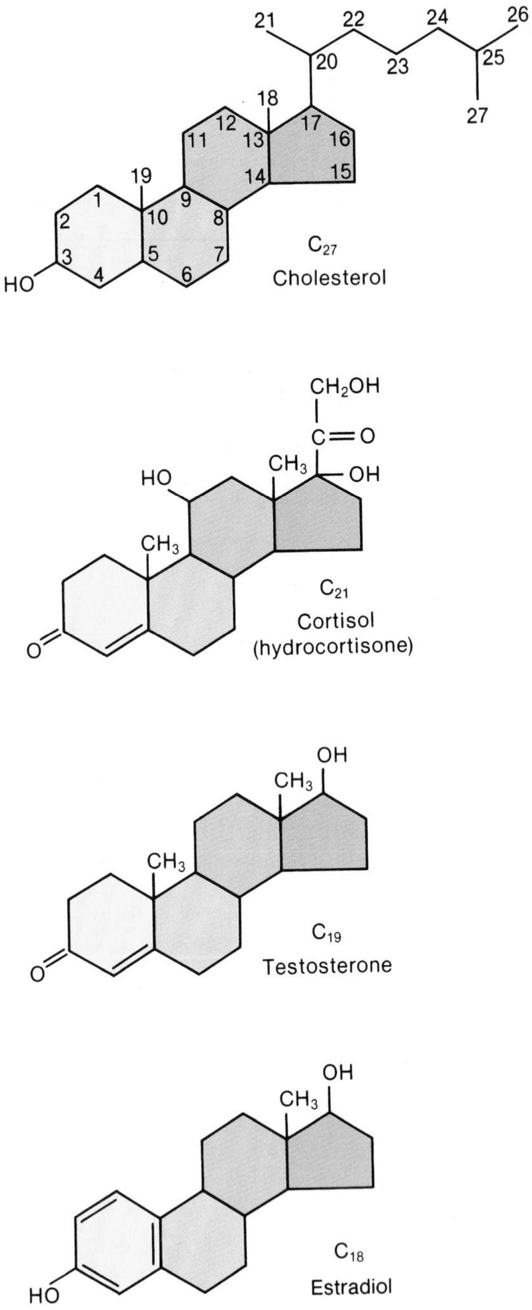

surface tension of water. This function of phospholipids makes them **surfactants** (surface-active agents). The surfactant effect of phospholipids prevents the lungs from collapsing due to surface tension forces (see chapter 24). Phospholipids are also the major component of cell membranes, as will be described in chapter 3.

Steroids

In terms of structure, **steroids** (*ster'oidz*) differ considerably from triglycerides or phospholipids, and yet steroids are still included in the lipid category of molecules because they are nonpolar and insoluble in water. All steroid molecules have the same basic structure: three six-carbon rings joined to one five-carbon ring (fig. 2.21). However, different kinds of steroids have different functional groups attached to this basic structure, and they vary in the number and position of the double covalent bonds between the carbon atoms in the rings.

Cholesterol (*kŏ-les'ter-ol*) is an important molecule in the body because it serves as the precursor (parent molecule) for the steroid hormones produced by the gonads and adrenal cortex. The testes and ovaries (collectively called the *gonads*) secrete **sex steroids,** which include estradiol and progesterone from the ovaries and testosterone from the testes. The adrenal cortex secretes the **corticosteroids,** including hydrocortisone and aldosterone.

Prostaglandins

Prostaglandins are a type of fatty acid with a cyclic hydrocarbon group. Although their name is derived from the fact that they were originally noted in the semen as a secretion of the prostate, it has since been shown that they are produced by and are active in almost all organs, where they serve a variety of regulatory functions. Prostaglandins are implicated in the regulation of blood vessel diameter, ovulation, uterine contraction during labor, inflammation reactions, blood clotting, and many other functions. Structural formulas for different types of prostaglandins are shown in figure 2.22.

Proteins

Proteins are large molecules composed of amino acid subunits. Since there are 20 different types of amino acids that can be used in constructing a given protein, the variety of protein structures is immense. This variety allows each type of protein to perform very specific functions.

The enormous diversity of protein structure results from the fact that there are 20 different building blocks—the *amino* (*ă-me'no*) *acids*—that can be used to form a protein. These amino acids, as will be described in the next section, join to form a chain that can twist and fold in a specific manner because of chemical interactions between the amino acids. The specific sequence of amino acids in a protein, and thus the specific structure of the protein, is determined by genetic information. This genetic information for protein synthesis is contained in another category of organic molecules, the *nucleic acids*, which includes the macromolecules DNA and RNA. The structure of nucleic acids, and the mechanisms by which the genetic information they encode directs protein synthesis, is described in chapter 3.

Structure of Proteins

Proteins consist of long chains of subunits called **amino acids.** As the name implies, each amino acid contains an *amino group* (NH_2) on one end of the molecule and a *carboxylic acid*

Figure 2.22

Structural formulas of various prostaglandins.
Prostaglandins are a family of regulatory compounds derived from a membrane lipid known as arachidonic acid.

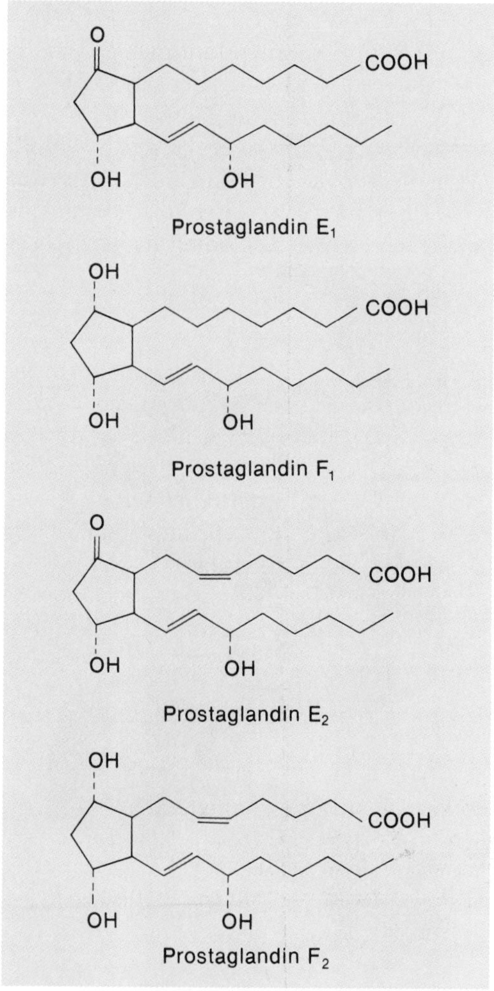

Prostaglandin E_1

Prostaglandin F_1

Prostaglandin E_2

Prostaglandin F_2

Figure 2.23

Representative amino acids.
The figure depicts different types of functional (R) groups. There are over twenty different amino acids that have different functional groups.

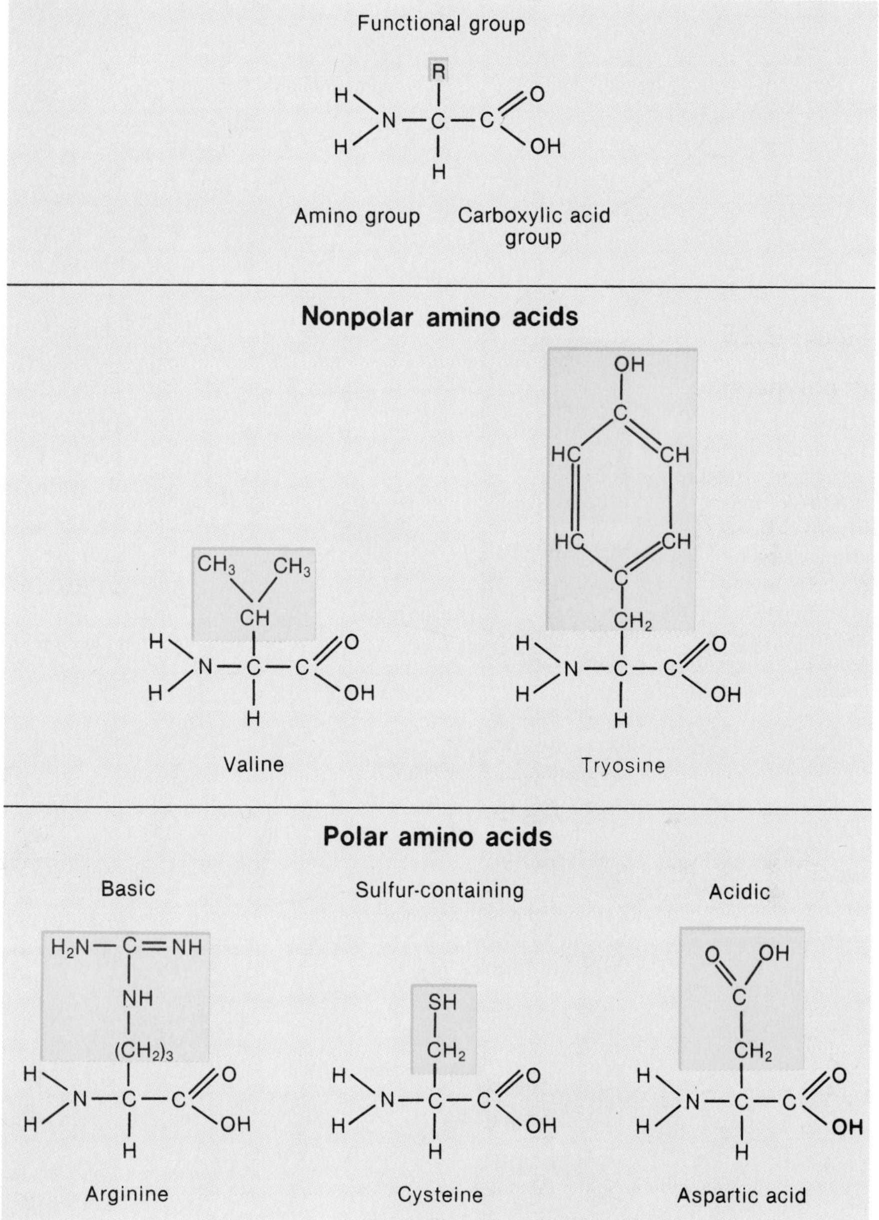

Functional group

Amino group Carboxylic acid group

Nonpolar amino acids

Valine

Tryosine

Polar amino acids

Basic

Sulfur-containing

Acidic

Arginine

Cysteine

Aspartic acid

group (COOH) on another end. There are 20 different amino acids, each with a distinct structure and chemical properties, that are used to build proteins. The differences between the amino acids are due to differences in their *functional groups*. "R" is the abbreviation for *functional group* in the general formula for an amino acid (fig. 2.23). The R symbol actually stands for the word *residue*, but it can be thought of as indicating the "*rest* of the molecule."

When amino acids join by dehydration synthesis, the hydrogen from the amino end of one amino acid combines with the hydroxyl group of the carboxylic acid end of another amino acid. As a covalent bond forms between the two amino acids, water is produced (fig. 2.24). The bond between adjacent amino acids is called a **peptide bond,** and the compound formed is called a *peptide*. Two amino acids bound together is called a *dipeptide;* three, a *tripeptide*. When numerous amino acids are joined in this way, a chain of amino acids, or a **polypeptide,** is produced.

The lengths of polypeptide chains vary widely. A hormone called *thyrotropin-releasing hormone*, for example, is only three amino acids long, whereas *myosin*, a muscle protein, contains about 4,500 amino acids. When the length of a polypeptide chain becomes very long (containing more than about 100 amino acids), the molecule is called a **protein.**

Figure 2.24

The formation of peptide bonds by dehydration synthesis reactions.
Water molecules are split off as the peptide bonds are produced between the amino acids.

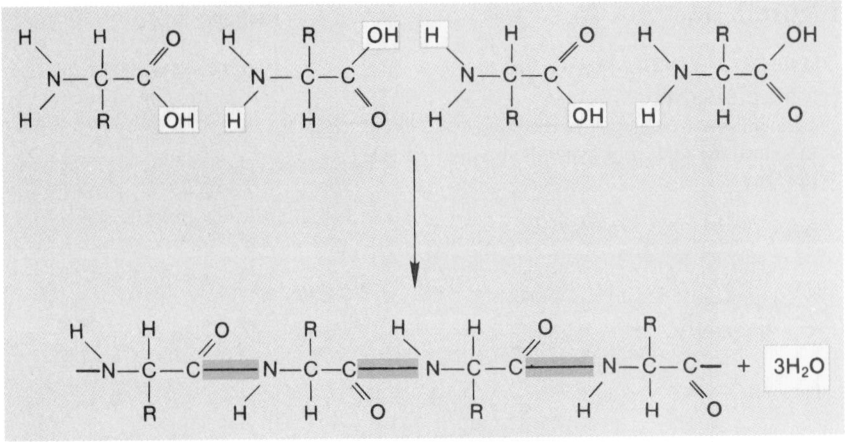

Figure 2.25

A polypeptide chain.
The figures depict its (*a*) primary structure and (*b*) secondary structure. The primary structure refers to its amino acid sequence, and its secondary structure refers to its helical coiling.

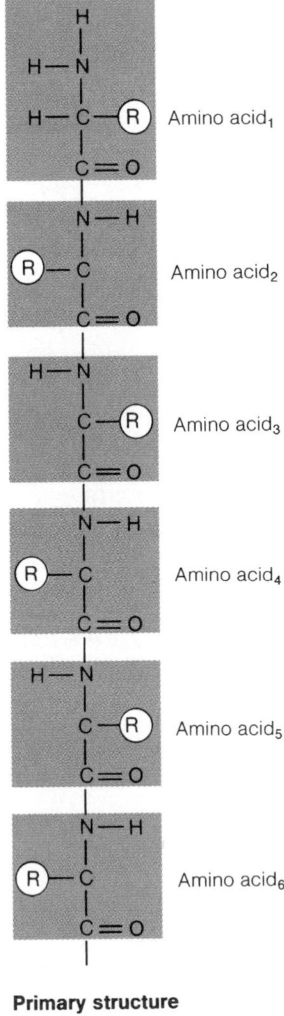

Primary structure

(a)

Hydrogen bond

(b) **Secondary structure** (α helix)

The structure of a protein can be described at four different levels. At the first level, the sequence of amino acids in the protein is described; this is called the **primary structure** of the protein. Each type of protein has a different primary structure. All of the billions of *copies* of a given type of protein in a person have the same structure, however, because the structure of a given protein is coded by the person's genes. The primary structure of a protein is illustrated in figure 2.25*a*.

Weak hydrogen bonds may form between the hydrogen atom of an amino group and an oxygen atom from a different amino acid nearby. These weak bonds cause the polypeptide chain to twist into a *helix*. The extent and location of the

Figure 2.26

The tertiary structure of a protein.
This is a model of the tertiary (three-dimensional) structure of myoglobin, a protein in muscle that serves to store oxygen.

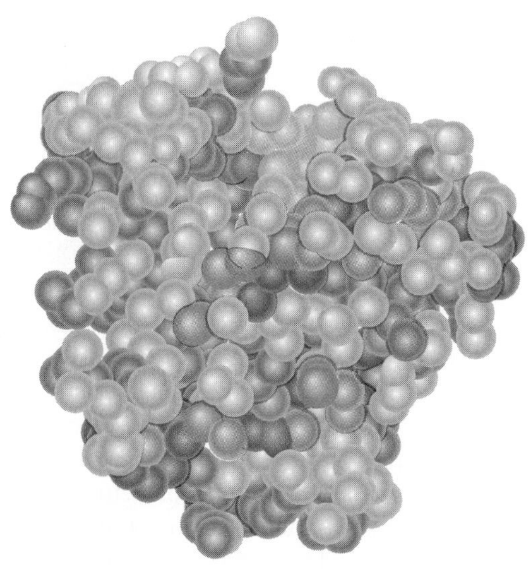

helical structure is different for each protein because of differences in amino acid composition. A description of the helical structure of a protein is termed its **secondary structure** (fig. 2.25*b*).

Most polypeptide chains bend and fold upon themselves to produce complex three-dimensional shapes called the **tertiary** (*ter'she-ar-e*) **structure** of the protein (fig. 2.26). Each type of protein has its own characteristic tertiary structure. This is because the folding and bending of the polypeptide chain is produced by chemical interactions between particular amino acids located in different regions of the chain.

Most of the tertiary structure of proteins is formed and stabilized by weak chemical bonds (such as hydrogen bonds) between the functional groups of widely spaced amino acids. Since most of the tertiary structure is stabilized by weak bonds, this structure can easily be disrupted by high temperature or by changes in pH. Irreversible changes in the tertiary structure of proteins that occur by these means are referred to as **denaturation** (*de-na"chur-a'shun*) of the proteins. The tertiary structure of some proteins, however, is made more stable by strong covalent bonds between sulfur atoms (called *disulfide bonds* and abbreviated S—S) in the functional group of an amino acid known as cysteine (fig. 2.27).

Figure 2.27

The bonds responsible for the tertiary structure of a protein.
The tertiary structure of a protein is held in place by a variety of different bonds. These include relatively weak bonds, such as hydrogen bonds, ionic bonds, and hydrophobic bonds, as well as the strong covalent disulfide bonds.

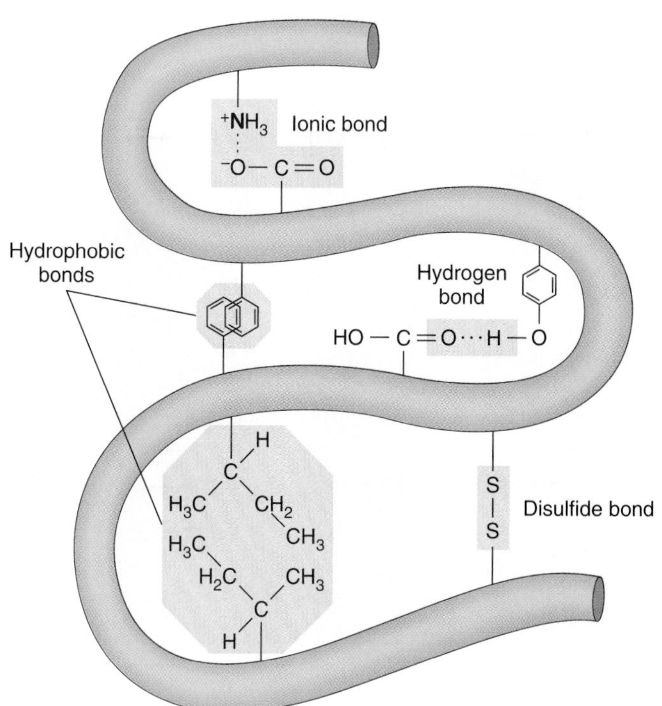

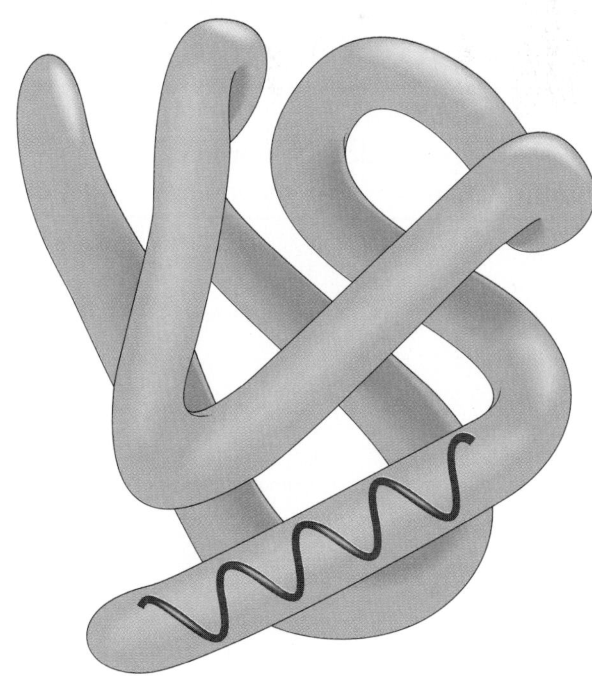

Denatured proteins retain their primary structure (the peptide bonds are not broken) but have altered chemical properties. Cooking a pot roast, for example, alters the texture of the meat proteins—it doesn't result in an amino acid soup. Denaturation is most dramatically demonstrated by frying an egg. Egg albumin proteins are soluble in their native state, in which they form the clear, viscous fluid of a raw egg. When denatured by cooking, these proteins change shape, cross-bond with each other, and by this means form an insoluble white precipitate—the egg white.

Some proteins (such as hemoglobin and insulin) are composed of a number of polypeptide chains covalently bonded together. This is the **quaternary** (*kwot'er-nar"e*) **structure** of these proteins. Insulin, for example, is composed of two polypeptide chains—one that is 21 amino acids long, the other that is 30 amino acids long. Hemoglobin (the protein in red blood cells that carries oxygen) is composed of four separate polypeptide chains (see chapter 24, fig. 24.36). The composition of various body proteins is shown in table 2.4.

Many proteins in the body are normally found combined, or *conjugated,* with other types of molecules. **Glycoproteins** are proteins conjugated with carbohydrates. Examples of such molecules include certain hormones and some proteins found in the cell membrane. **Lipoproteins** are proteins conjugated with lipids. These are found in cell membranes and in the plasma (the fluid portion of the blood). Proteins may also be conjugated with pigment molecules. These include hemoglobin, which transports oxygen in red blood cells, and the cytochromes, which are needed for oxygen utilization and energy production within cells.

Functions of Proteins

Because of their tremendous structural diversity, proteins serve a wider variety of functions than any other type of molecule in the body. Many proteins, for example, contribute significantly to the structure of different tissues and in this way play a passive role in the functions of these tissues. Examples of such *structural proteins* include collagen (fig. 2.28) and keratin. Collagen is a fibrous protein that provides tensile strength to connective tissues, such as tendons and ligaments. Keratin is found in the outer layer of dead cells in the epidermis, where it prevents water loss through the skin.

Many proteins play a more active role in the body, where specificity of structure and function is required. *Enzymes* and *antibodies*, for example, are proteins—no other type of molecule could provide the vast array of different structures needed for their tremendously varied functions. As another example, proteins in cell membranes may serve as *receptors* for specific regulator molecules (such as hormones) and as *carriers* for transport of specific molecules across the membrane. Proteins provide the diversity of shape and chemical properties required by these functions.

Table 2.4

Composition of Selected Body Proteins

Protein	Number of Polypeptide Chains	Nonprotein Component	Function
Hemoglobin	4	Heme pigment	Carries oxygen in the blood
Myoglobin	1	Heme pigment	Stores oxygen in muscle
Insulin	2	None	Hormone that regulates metabolism
Luteinizing hormone	1	Carbohydrate	Hormone that stimulates gonads
Fibrinogen	1	Carbohydrate	Aids in blood clotting
Mucin	1	Carbohydrate	Forms mucus
Blood group proteins	1	Carbohydrate	Produces blood types
Lipoproteins	1	Lipids	Transports lipids in blood

Figure 2.28

A photomicrograph of collagen fibers within connective tissue.
Collagen proteins give strength to the connective tissues.

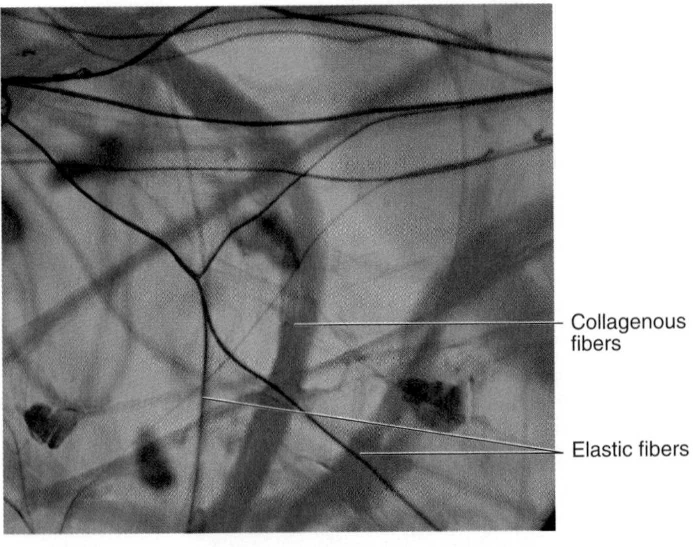

Clinical Investigation Answer

Since our enzymes can recognize only L-amino acids and D-sugars, the opposite stereoisomer that the student was eating could not be used by his body. He was weak because he was literally starving. The ketonuria also may have contributed to his malaise. Since he was starving, his stored fat was rapidly being hydrolyzed into glycerol and fatty acids for use as energy sources. The excessive release of fatty acids from his adipose tissue resulted in the excessive production of ketone bodies by his liver; hence, his ketonuria.

Chapter Summary

Atoms, Ions, and Chemical Bonds (pp. 25–33)

1. Covalent bonds are formed by atoms that share electrons. They are the strongest type of chemical bond.
   (a) Electrons are equally shared in nonpolar covalent bonds and unequally shared in polar covalent bonds.
   (b) Atoms of oxygen, nitrogen, and phosphorus strongly attract electrons and become electrically negative compared to the other atoms sharing electrons with them.
2. Ionic bonds are formed by atoms that transfer electrons. These weak bonds join atoms together in an ionic compound.
   (a) The atom in an ionic compound that takes the electron from another atom gains a net negative charge; the other atom becomes positively charged.
   (b) Ionic bonds break easily when the ionic compound is dissolved in water. Dissociation of the ionic compound yields charged atoms called ions.
3. When hydrogen is bonded to an electronegative atom, it gains a slight positive charge and is weakly attracted to another electronegative atom. This weak attraction is a hydrogen bond.
4. Acids donate hydrogen ions to a solution, whereas bases lower the hydrogen ion concentration of a solution.
   (a) The pH scale is a negative function of the logarithm of the hydrogen ion concentration.
   (b) In a neutral solution, the concentration of H^+ is equal to the concentration of OH^-, and the pH is 7.
   (c) Acids raise the H^+ concentration, and thus lower the pH below 7. Bases lower the H^+ concentration, and thus raise the pH above 7.
5. Organic molecules contain atoms of carbon joined by covalent bonds. Atoms of nitrogen, oxygen, phosphorus, or sulfur may be present as specific functional groups in the organic molecule.

Carbohydrates and Lipids (pp. 33–38)

1. Carbohydrates contain carbon, hydrogen, and oxygen, usually in a ratio of 1:2:1.
   (a) Carbohydrates consist of simple sugars (monosaccharides), disaccharides, and polysaccharides (such as glycogen).
   (b) Covalent bonds between monosaccharides are formed by dehydration synthesis, or condensation. Bonds are broken by hydrolysis reactions.
2. Lipids are organic molecules that are insoluble in water.
   (a) Triglycerides (fat and oil) consist of three fatty acid molecules joined to a molecule of glycerol.
   (b) Ketone bodies are smaller derivatives of fatty acids.
   (c) Phospholipids (such as lecithin) are phosphate-containing lipids that have a hydrophilic polar group. The rest of the molecule is hydrophobic.
   (d) Steroids (including the hormones of the gonads and adrenal cortex) are lipids with a characteristic five-ring structure.
   (e) Prostaglandins are a family of cyclic fatty acids that serve a variety of regulatory functions.

Proteins (pp. 38–42)

1. Proteins are composed of long chains of amino acids bonded together by covalent peptide bonds.
   (a) Each amino acid contains an amino group, a carboxyl group, and a functional group that is different for each of the more than 20 different amino acids.
   (b) The polypeptide chain may be twisted into a helix (secondary structure) and bent and folded to form the tertiary structure of the protein.
   (c) Proteins that are composed of two or more polypeptide chains are said to have a quaternary structure.
   (d) Proteins may be combined with carbohydrates, lipids, or other molecules.
   (e) Because they are so diverse structurally, proteins serve a wider variety of specific functions than any other type of molecule.

Review Activities

Objective Questions

1. Which of the following statements about atoms is *true?*
   (a) They have more protons than electrons.
   (b) They have more electrons than protons.
   (c) They are electrically neutral.
   (d) They have as many neutrons as they have electrons.
2. The bond between oxygen and hydrogen in a water molecule is
   (a) a hydrogen bond.
   (b) a polar covalent bond.
   (c) a nonpolar covalent bond.
   (d) an ionic bond.
3. Which of the following is a nonpolar covalent bond?
   (a) the bond between two carbons
   (b) the bond between sodium and chloride
   (c) the bond between two water molecules
   (d) the bond between nitrogen and hydrogen
4. Solution A has a pH of 2, and solution B has a pH of 10. Which of the following statements about these solutions is *true?*
   (a) Solution A has a higher H^+ concentration than solution B.
   (b) Solution B is basic.
   (c) Solution A is acidic.
   (d) All of the above are true.
5. Glucose is
   (a) a disaccharide.
   (b) a polysaccharide.
   (c) a monosaccharide.
   (d) a phospholipid.
6. Digestion reactions occur by means of
   (a) dehydration synthesis.
   (b) hydrolysis.
7. Carbohydrates are stored in the liver and muscles in the form of
   (a) glucose.
   (b) triglycerides.
   (c) glycogen.
   (d) cholesterol.
8. Lecithin is
   (a) a carbohydrate.
   (b) a protein.
   (c) a steroid.
   (d) a phospholipid.

9. Which of the following lipids have regulatory roles in the body?
   (a) steroids
   (b) prostaglandins
   (c) triglycerides
   (d) both *a* and *b*
   (e) both *b* and *c*
10. The tertiary structure of a protein is *directly* determined by
    (a) the genes.
    (b) the primary structure of the protein.
    (c) enzymes that "mold" the shape of the protein.
    (d) the position of peptide bonds.
11. The type of bond formed between two molecules of water is
    (a) a hydrolytic bond.
    (b) a polar covalent bond.
    (c) a nonpolar covalent bond.
    (d) a hydrogen bond.
12. The carbon-to-nitrogen bond that joins amino acids is called
    (a) a glycosidic bond.
    (b) a peptide bond.
    (c) a hydrogen bond.
    (d) a double bond.

Essay Questions

1. Compare and contrast nonpolar covalent bonds, polar covalent bonds, and ionic bonds.
2. Define *acid* and *base* and explain how acids and bases influence the pH of a solution.
3. Using dehydration synthesis and hydrolysis reactions, explain the relationships between starch in an ingested potato, liver glycogen, and blood glucose.
4. "All fats are lipids, but not all lipids are fats." Explain why this is an accurate statement.
5. What are the similarities and differences between a fat and an oil? Comment on the physiological and clinical significance of the degree of saturation of fatty acid chains.

Critical Thinking Questions

1. Explain the relationship between the primary structure of a protein and its secondary and tertiary structures. What do you think would happen to the tertiary structure if some amino acids were substituted for others in the primary structure? What physiological significance might this have?
2. Suppose you try to discover a hormone by homogenizing an organ in a fluid, filtering the fluid to eliminate the solid material, and then injecting the extract into an animal to see the effect. If an aqueous (water) extract does not work but one using benzene as the solvent does have an effect, what might you conclude about the chemical nature of the hormone? Explain.
3. From the ingredients listed on a food wrapper, it would appear that the food contains high amounts of fat. Yet on the front of the package is the large slogan, "Cholesterol Free!" In what sense is this slogan chemically correct? In what way is it misleading?

Related Web Sites

For a listing of the most current web sites related to this chapter, please visit the *Concepts of Human Anatomy and Physiology* home page at http://www.mhhe.com/biosci/abio/.

THREE

Cell Structure and Genetic Regulation

Objectives

- Describe the structure of the cell membrane.
- Discuss the processes of amoeboid motion and phagocytosis.
- Describe the structure and functions of cilia and flagella.
- Describe the structure and functions of the cytoskeleton and of lysosomes.
- Describe the structure of mitochondria and the endoplasmic reticulum and discuss their significance.
- Describe the structure of DNA and RNA nucleotides and explain complementary base pairing.
- Discuss the process of RNA synthesis and distinguish between the different types of RNA.
- Explain how RNA directs the synthesis of proteins.
- Describe the functions of mRNA, tRNA, and rRNA.
- Explain how codons and anticodons function in protein synthesis.
- Describe the functions of the rough endoplasmic reticulum and the Golgi complex in the packaging and secretion of proteins.
- Explain how DNA replicates itself and state the events that occur during each phase of the cell cycle.
- Explain the function of mitosis and describe the events that occur during the different phases of this process.
- Explain how meiosis differs from mitosis and state the function of meiotic cell division.

Clinical Investigation

A liver biopsy was taken from a teenage boy with apparent liver disease, and different microscopic techniques for viewing the sample were employed. The biopsy revealed an unusually extensive smooth endoplasmic reticulum, and the patient admitted to a history of substance abuse. In addition, an abnormally large amount of glycogen granules were found, and many intact glycogen granules were seen within secondary lysosomes. Laboratory analysis revealed a lack of the enzyme that hydrolyzes glycogen. What is the relationship between these observations?

Clues: See the sections on lysosomes and the endoplasmic reticulum, paying particular attention to the boxed information in these sections.

Cell Membrane and Associated Structures

The cell is the basic unit of structure and function in the body. Many of the functions of cells are performed by particular subcellular structures known as organelles. The cell membrane allows selective communication between the intracellular and extracellular compartments and aids cellular movement.

Cells look so small and simple when viewed with the ordinary (light) microscope that it is difficult to think of each one as a living entity unto itself. Equally amazing is the fact that the physiology of our organs and systems derives from the complex functions of the cells of which they are composed. Complexity of function demands complexity of structure, even at the subcellular level.

As the basic functional unit of the body, each cell is a highly organized molecular factory. Cells come in a wide variety of shapes and sizes. This great diversity, which is also apparent in the subcellular structures within different cells, reflects the diversity of function of different cells in the body. All cells, however, share certain characteristics; for example, they are all surrounded by a cell membrane, and most of them possess the structures listed in table 3.1. Thus, although no single cell can be considered "typical," the general structure of cells can be indicated by a single illustration (fig. 3.1).

Figure 3.1

A generalized human cell showing the principal organelles.
Most cells of the body are highly specialized, and so have structures that differ from this general model.

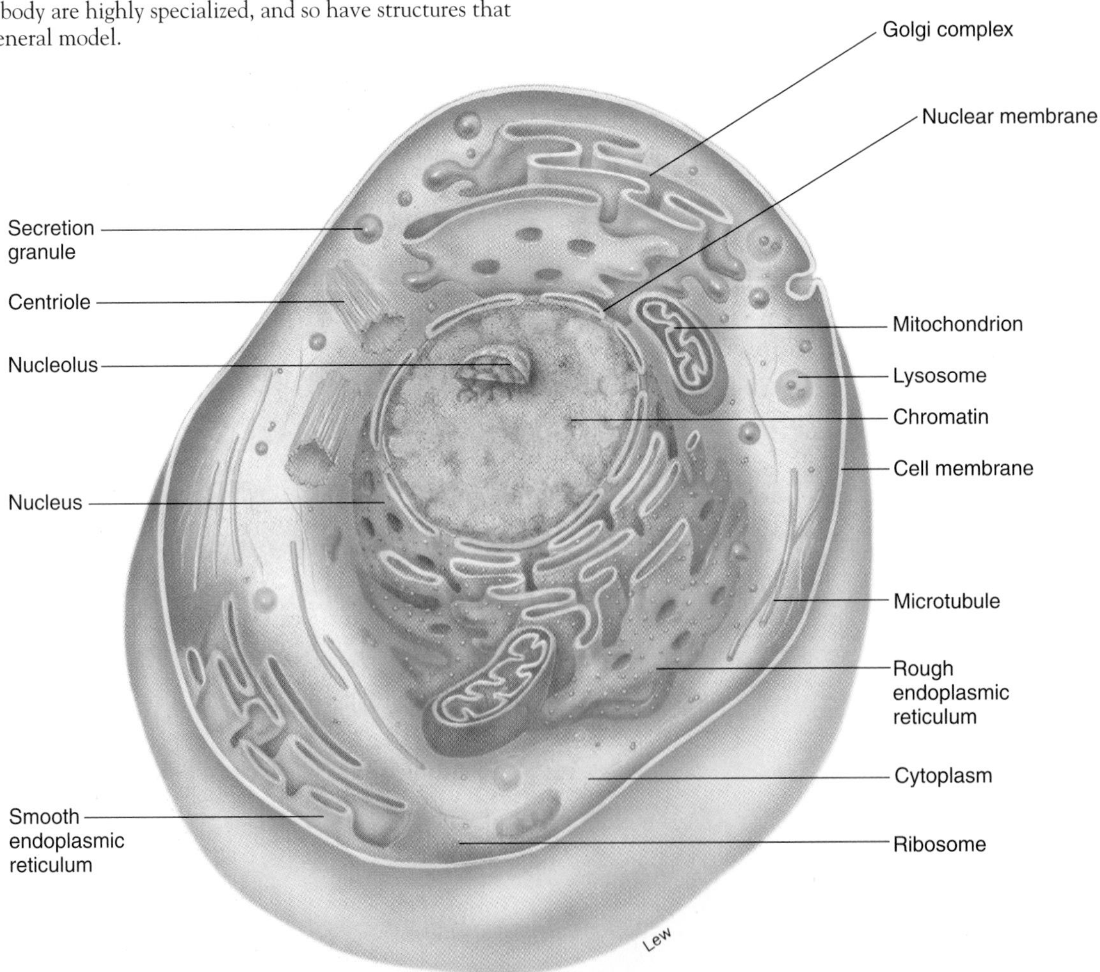

Table 3.1
Cellular Components: Structure and Function

Component	Structure	Function
Cell (plasma) membrane	Membrane composed of double layer of phospholipids in which proteins are embedded	Gives form to cell and controls passage of materials in and out of cell
Cytoplasm	Fluid, jellylike substance between the cell membrane and the nucleolus in which organelles are suspended	Serves as matrix substance in which chemical reactions occur
Endoplasmic reticulum	System of interconnected membrane-forming canals and tubules	Smooth endoplasmic reticulum metabolizes nonpolar compounds and stores Ca^{2+} in striated muscle cells; rough endoplasmic reticulum assists in protein synthesis
Ribosomes	Granular particles composed of protein and RNA	Synthesize proteins
Golgi complex	Cluster of flattened membranous sacs	Synthesizes carbohydrates and packages molecules for secretion; secretes lipids and glycoproteins
Mitochondria	Double-walled membranous sacs with folded inner partitions	Release energy from food molecules and transform energy into usable ATP
Lysosomes	Single-walled membranous sacs	Digest foreign molecules and worn and damaged cells
Peroxisomes	Spherical membranous vesicles	Contain enzymes that detoxify harmful molecules and break down hydrogen peroxide
Centrosome	Nonmembranous mass of two rodlike centrioles	Helps organize spindle fibers and distribute chromosomes during mitosis
Vacuoles	Membranous sacs	Store and release various substances within the cytoplasm
Fibrils and microtubules	Thin, hollow tubes	Support cytoplasm and transport materials within the cytoplasm
Cilia and flagella	Minute cytoplasmic projections that extend from the cell surface	Move particles along cell surface or move the cell
Nuclear membrane	Double-layered membrane composed of protein and lipid molecules that surrounds the nucleus	Supports nucleus and controls passage of materials between nucleus and cytoplasm
Nucleolus	Dense nonmembranous mass composed of protein and RNA molecules	Forms ribosomes
Chromatin	Fibrous strands composed of protein and DNA molecules	Contains genetic code that determines which proteins (especially enzymes) will be manufactured by the cell

For descriptive purposes, a cell can be divided into three principal parts:

1. **Cell (plasma) membrane.** The selectively permeable cell membrane surrounds the cell, gives it form, and separates the cell's internal structures from the extracellular environment. The cell membrane also participates in intercellular communication.
2. **Cytoplasm and organelles** (*or″ga-nelz′*). The cytoplasm is the aqueous content between the nucleus of a cell and the cell membrane. Organelles (excluding the nucleus) are subcellular structures within the cytoplasm of a cell that perform specific functions. The term **cytosol** is frequently used to describe the soluble portion of the cytoplasm; that is, the part that cannot be removed by centrifugation.
3. **Nucleus** (*noo′kle-us*). The nucleus is a large, generally spheroid body within a cell. The largest of the organelles, it contains the DNA, or genetic material, of the cell and thus directs the cell's activities. The nucleus also contains one or more *nucleoli* (*noo-kle′ŏ-li*). Nuclei are centers for the production of ribosomes, which are the sites of protein synthesis.

aqueous: L. *aqua*, water

Structure of the Cell Membrane

Because both the intracellular and extracellular environments (or "compartments") are aqueous, a barrier must be present to prevent the loss of enzymes, nucleotides, and other cellular molecules that are water-soluble. Since this barrier surrounding the cell cannot itself be composed of water-soluble molecules, it is instead composed of lipids.

The **cell membrane** (also called the **plasma membrane,** or **plasmalemma**), and indeed all of the membranes surrounding

Figure 3.2

The fluid-mosaic model of the cell membrane.

The membrane consists of a double layer of phospholipids, with the polar phosphates (shown by spheres) oriented outward and the nonpolar hydrocarbons (wavy lines) oriented toward the center. Proteins may completely or partially span the membrane. Carbohydrates are attached to the outer surface.

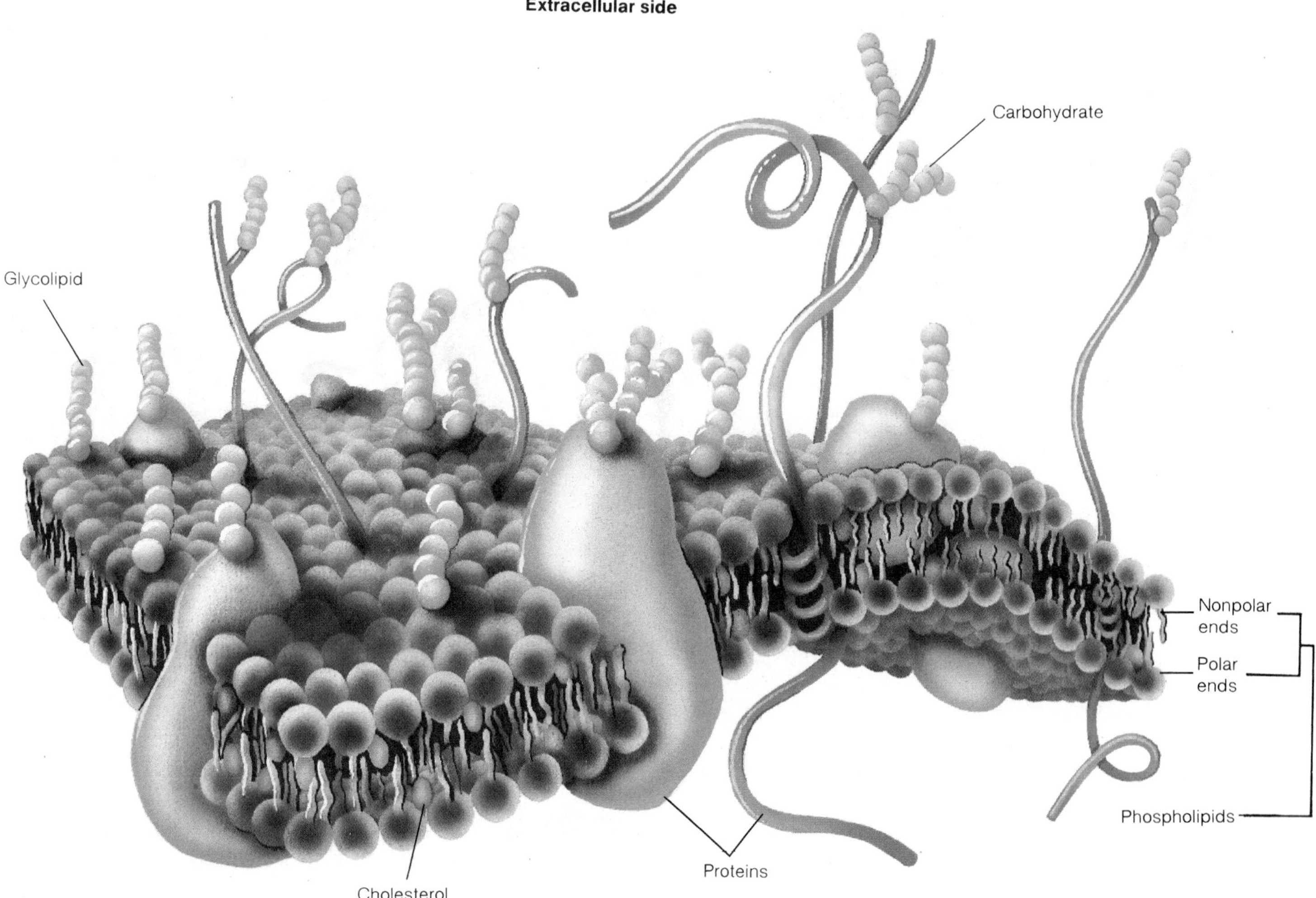

organelles within the cell, are composed primarily of phospholipids and proteins. Phospholipids, described in chapter 2, are polar on the end that contains the phosphate group and nonpolar (and hydrophobic) throughout the rest of the molecule. Since the environment on each side of the membrane is aqueous, the hydrophobic parts of the molecules "huddle together" in the center of the membrane, leaving the polar ends exposed to water on both surfaces. This results in the formation of a double layer of phospholipids in the cell membrane.

The hydrophobic middle of the membrane restricts the passage of water and water-soluble molecules and ions. Certain polar compounds, however, do pass through the membrane. The specialized functions and selective transport properties of the membrane are believed to be due to its protein content. Some proteins are found partially submerged on each side of the membrane; other proteins span the membrane completely from one side to the other. Since the membrane is not solid—phospholipids and proteins are free to move laterally—the proteins within the phospholipid "sea" are not uniformly distributed. Rather, they present a constantly changing mosaic pattern, an arrangement known as the **fluid-mosaic model** of membrane structure (fig. 3.2).

The proteins found in the cell membrane serve a variety of functions, including structural support, transport of molecules across the membrane, and enzymatic control of chemical reactions at the cell surface. Some proteins function as receptors for hormones and other regulatory molecules that arrive at the outer surface of the membrane. Receptor proteins are usually specific for one particular messenger, much like an enzyme that is specific for a single substrate. Other cellular proteins serve as "markers" (antigens) that identify the blood and tissue type of an individual.

In addition to lipids and proteins, the cell membrane also contains carbohydrates, which are primarily attached to the outer surface of the membrane as glycoproteins and glycolipids. These surface carbohydrates have numerous negative

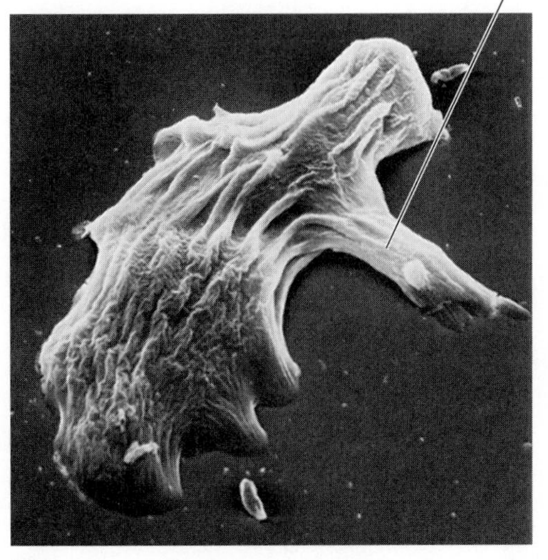

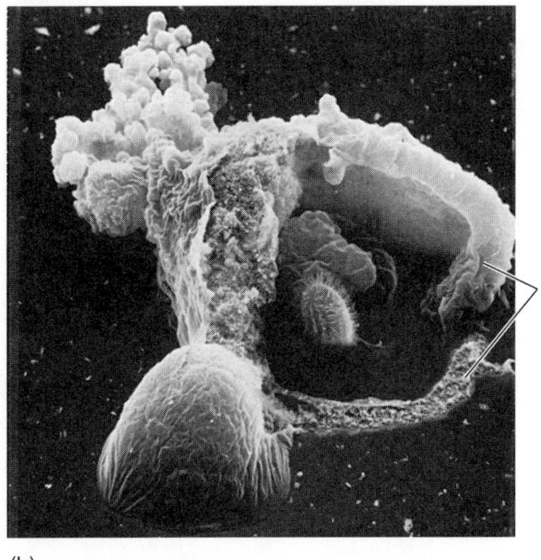

Figure 3.3

Scanning electron micrographs of phagocytosis. The photographs show (*a*) the formation of pseudopods and (*b*) the entrapment of the prey within a food vacuole.

charges and, as a result, affect the interaction of regulatory molecules with the membrane. The negative charges at the surface also affect interactions between cells—they help keep red blood cells apart, for example. Stripping the carbohydrates from the outer red blood cell surface results in their more rapid destruction by the liver, spleen, and bone marrow.

Phagocytosis

Most of the movement of molecules and ions between the intracellular and extracellular compartments involves passage through the cell membrane (see chapter 5). However, the cell membrane also participates in the **bulk transport** of larger portions of the extracellular environment. Bulk transport includes the processes of *phagocytosis* and *endocytosis*.

Some body cells—including certain white blood cells and macrophages in connective tissues—are able to transport themselves in the manner of an amoeba (a single-celled animal). They perform this **amoeboid movement** by extending parts of their cytoplasm to form *pseudopods* (*soo'dŏ-podz*), which attach to a substrate and pull the cell along. This process depends on the bonding of membrane-spanning proteins called *integrins* with proteins outside the membrane in the extracellular *matrix* (generally, an extracellular gel of proteins and carbohydrates).

Cells that move by amoeboid motion—as well as certain liver cells, which are not mobile—use pseudopods to surround and engulf particles of organic matter (such as bacteria). This process is a type of cellular "eating" called **phagocytosis** (*fag"ŏ-si-to'sis*). It serves to protect the body from invading microorganisms and to remove extracellular debris.

Phagocytic cells surround their victim with pseudopods, which join and fuse (fig. 3.3). After the inner membrane of the pseudopods has become a continuous membrane surrounding the ingested particle, it pinches off from the cell membrane. The ingested particle is now contained in an organelle called a *food vacuole* (*vak'yoo-ōl*) within the cell. The food vacuole will subsequently fuse with an organelle called a lysosome (described later), and the particle will be digested by lysosomal enzymes.

Endocytosis

Endocytosis is a process in which the cell membrane furrows inwards, instead of extending outward with pseudopods. One form of endocytosis, **pinocytosis** (*pin"ŏ-si-to'sis*), is a nonspecific process performed by many cells. The cell membrane invaginates to produce a deep, narrow furrow. The membrane near the surface of this furrow then fuses, and a small vesicle containing the extracellular fluid pinches off and enters the cell. Pinocytosis allows a cell to engulf large molecules such as proteins, as well as any other molecules that may be present in the extracellular fluid.

Another type of endocytosis involves a smaller area of cell membrane, and it occurs only in response to specific molecules in the extracellular environment. Since the extracellular molecules must bind to very specific *receptor proteins* in the cell membrane, this process is known as **receptor-mediated endocytosis.**

In receptor-mediated endocytosis, the interaction of specific molecules in the extracellular fluid with specific membrane receptor proteins causes the membrane to invaginate, fuse, and pinch off to form a vesicle (fig. 3.4). Vesicles formed in this way contain extracellular fluid and molecules that could not have passed by other means into the cell. Cholesterol attached to specific proteins, for example, is taken up into artery cells by receptor-mediated endocytosis. This is in part responsible for atherosclerosis, as described in chapter 21. Hepatitis, polio, and AIDS viruses also exploit the process of receptor-mediated endocytosis to invade cells.

pseudopod: Gk. *pseudes*, false; *pod*, foot
phagocytosis: Gk. *phagein*, to eat; *kytos*, hollow body
vacuole: L. *vacuus*, empty
pinocytosis: Gk. *pinein*, to drink; *kytos*, hollow body

Figure 3.4

Receptor-mediated endocytosis.
Stages 1–4 are shown, during which specific bonding of extracellular particles with membrane receptor proteins results in the formation of endocytotic vesicles.

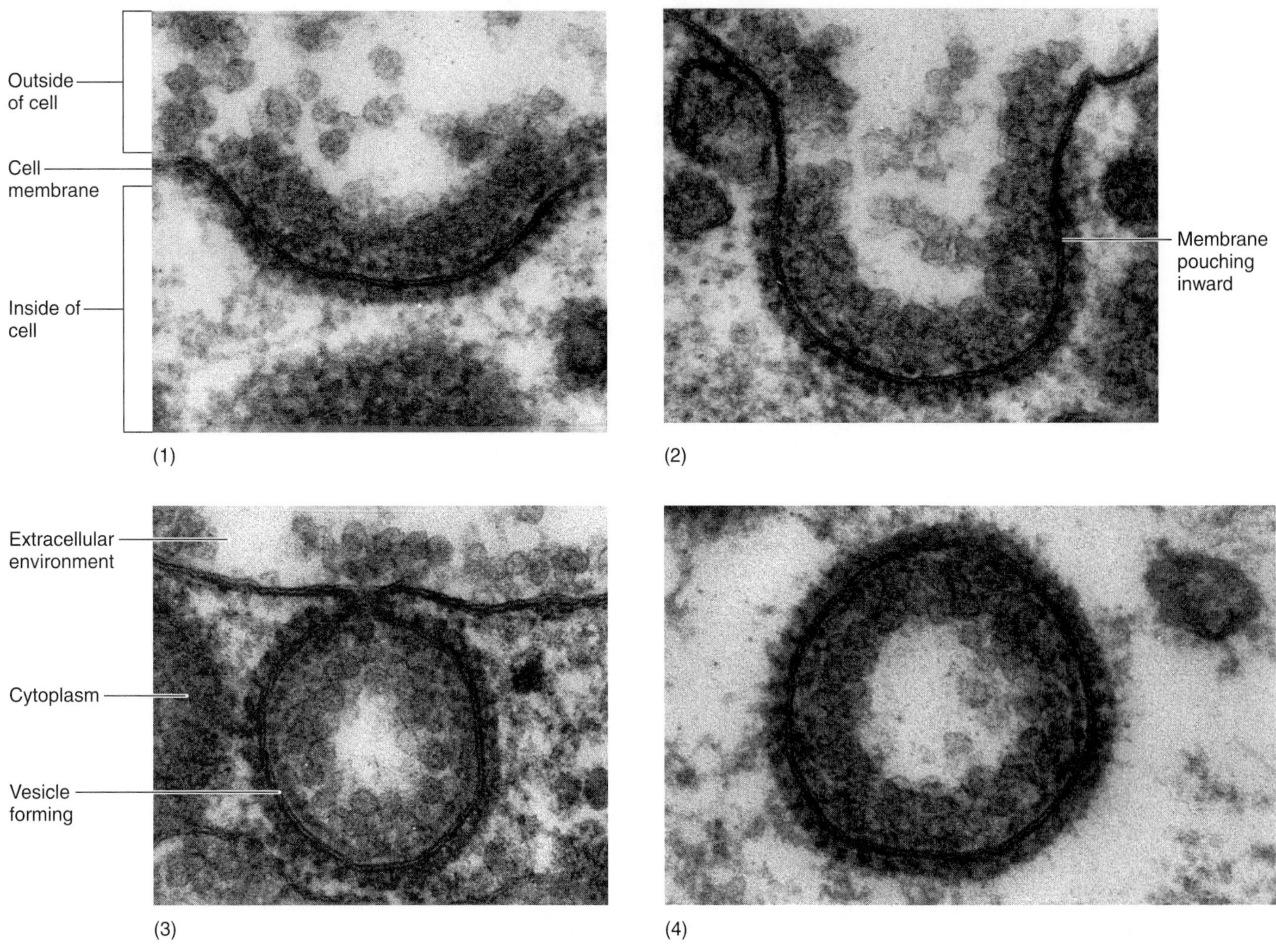

Exocytosis

Exocytosis is a process by which cellular products are secreted into the extracellular environment. Proteins and other molecules produced within the cell that are destined for export (secretion) are packaged within vesicles by an organelle known as the Golgi complex. In the process of exocytosis, these secretory vesicles fuse with the cell membrane and release their contents into the extracellular environment (see fig. 3.27). Nerve endings, for example, release their chemical neurotransmitters in this manner (see chapter 14).

When the vesicle containing the secretory products of the cell fuses with the cell membrane during exocytosis, the total amount of cell membrane is increased. This process replaces material that was lost from the cell membrane during endocytosis.

Cilia and Flagella

Cilia (*sil′e-ă*) are tiny hairlike structures that project from the surface of a cell and, like the coordinated action of rowers in a boat, stroke in unison. Cilia in the human body are found on the apical surface (the surface facing the lumen, or cavity) of stationary epithelial cells in the respiratory and female reproductive tracts. In the respiratory system, the cilia transport strands of mucus to the pharynx (throat), where the mucus can either be swallowed or expectorated. In the female reproductive tract, ciliary movements in the epithelial lining of the uterine tube draw the ovum (egg) into the tube and move it toward the uterus.

---

cilia: L. *cili*, small hair

Figure 3.5

Electron micrographs of cilia.
The cilia can be seen in a scanning electron micrograph of entire cilia and a transmission electron micrograph of cross sections of cilia. Note the characteristics "9 + 2" arrangement of microtubules in the cross sections.

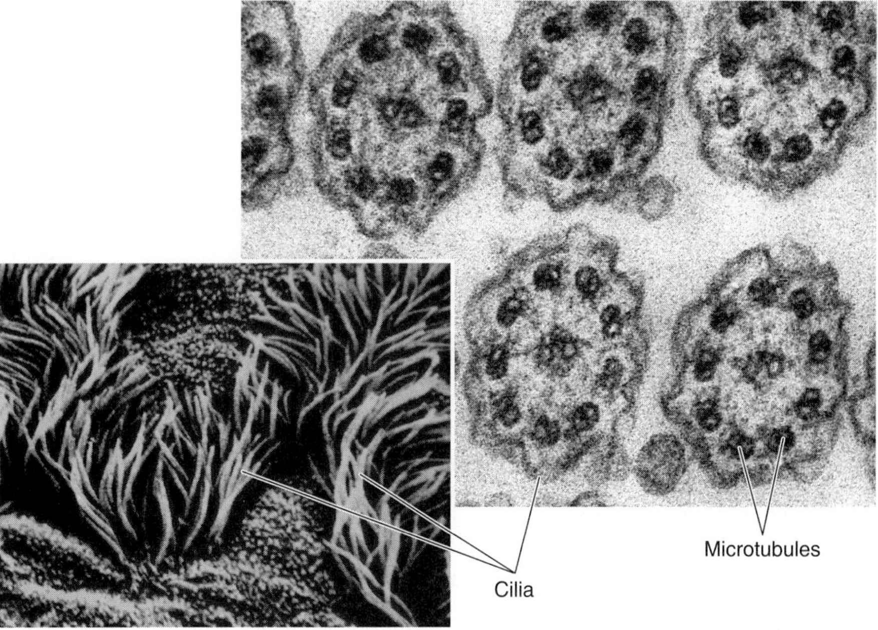

Figure 3.6

Microvilli in the small intestine.
The microvilli (Mv), are seen here with (*a*) the transmission and (*b*) the scanning electron microscope. (TW is the terminal web, a protein mesh to which the microvilli are anchored.)

Reproduced from R. G. Kessel and R. H. Kardon, Tissues and Organs: A Text Atlas of Scanning Electron Microscopy, *W. H. Freeman and Co., 1979.*

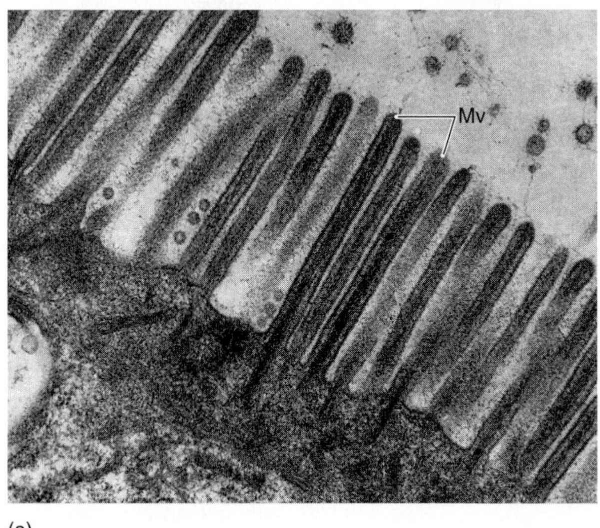

(a)

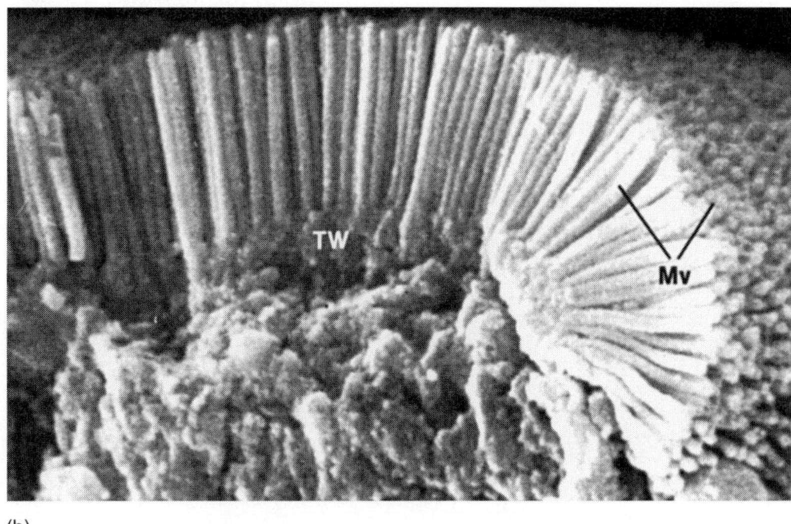

(b)

Sperm cells are the only cells in the human body that have **flagella** (*flă-jel'ă*). The flagellum is a single whiplike structure that propels the sperm cell through its environment. Both cilia and flagella are composed of *microtubules* (thin cylinders formed from proteins) arranged in a characteristic way. One pair of microtubules in the center of a cilium or flagellum is surrounded by nine other pairs of microtubules, to produce what is often described as a "9 + 2" arrangement (fig. 3.5).

Microvilli

In areas of the body that are specialized for rapid diffusion, the surface area of the cell membranes may be increased by numerous folds called **microvilli** (*mi"kro-vil'i*). The rapid passage of the products of digestion across the epithelial membranes in the small intestine, for example, is aided by these structural adaptations. The surface area of the apical membranes (the part facing the lumen) in the small intestine is increased by the numerous tiny fingerlike projections (fig. 3.6). Similar

flagellum: L. *flagrum*, whip

microvillus: Gk. *mikros*, small; L. *villus*, shaggy hair

microvilli are found in the epithelium of the kidney tubule, which must reabsorb various molecules that are filtered out of the blood.

Cytoplasm and Its Organelles

Many of the functions of a cell that are performed in the cytoplasmic compartment result from the activity of specific structures called organelles. Among these are the lysosomes, which contain digestive enzymes, and the mitochondria, where most of the cellular energy is produced. Other organelles participate in the synthesis and secretion of cellular products.

Cytoplasm and Cytoskeleton

The jellylike matrix within a cell (exclusive of that within the nucleus) is known as **cytoplasm** (*si'tŏ-plaz"em*). When viewed in a microscope without special techniques, the cytoplasm appears to be uniform and unstructured. According to recent evidence, however, the cytoplasm is not a homogenous solution; it is, rather, a highly organized structure in which protein fibers—in the form of *microtubules* and *microfilaments*—are arranged in a complex latticework surrounding the membrane-bound organelles. Using fluorescence microscopy, these structures can be visualized with the aid of antibodies against their protein components (fig. 3.7). The interconnected microfilaments and microtubules are believed to provide structural organization for cytoplasmic enzymes and support for various organelles.

The latticework of microfilaments and microtubules is said to function as a **cytoskeleton** (fig. 3.8). The structure of this "skeleton" is not rigid; it is capable of quite rapid movement and reorganization. Contractile proteins—including actin and myosin, which are responsible for muscle contraction—are microfilaments found in most cells. Such microfilaments aid in amoeboid movement, for example, so that the cytoskeleton is

Figure 3.7

Microtubules forming the cytoskeleton of a cell.
The microtubules in this photograph are visualized with the aid of fluorescent antibodies against tubulin, the major protein component of the microtubules.

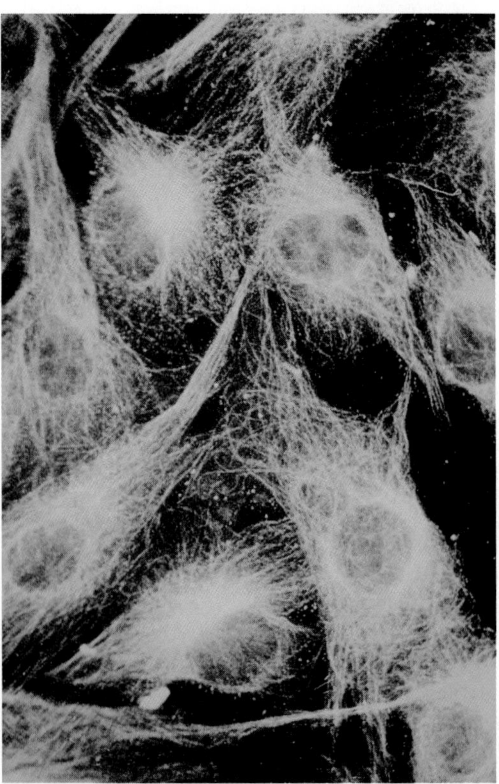

Figure 3.8

Microtubules form the cytoskeleton.
Microtubules are also important in the motility (movement) of the cell and the movement of materials within the cell.

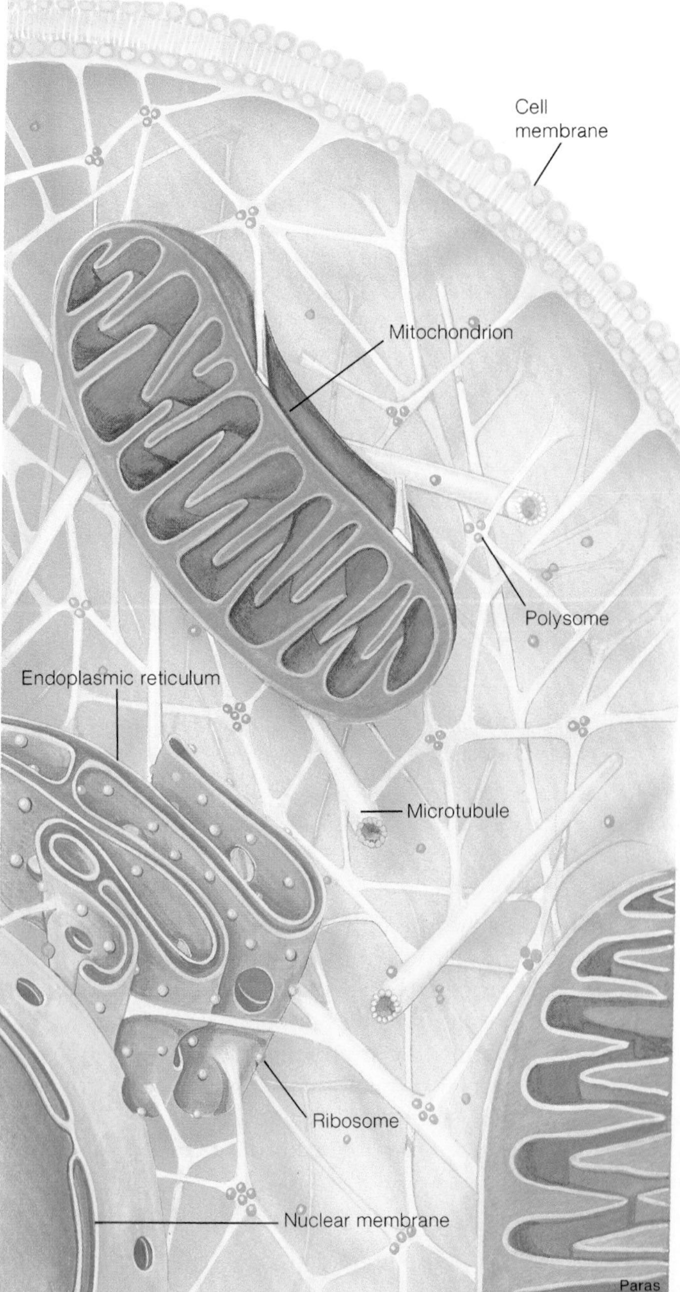

also the cell's "musculature." Microtubules, as another example, form the *spindle apparatus* that pulls chromosomes away from each other in cell division. Microtubules also form the central parts of cilia and flagella and contribute to the structure and movements of these projections from the cells.

The cytoplasm of some cells contains stored chemicals in aggregates called **inclusions.** Examples are *glycogen granules* in the liver, skeletal muscles, and in some other tissues; *melanin granules* in the melanocytes of the skin; and *triglycerides* within adipose cells.

Lysosomes

After a phagocytic cell has engulfed the proteins, polysaccharides, and lipids present in a particle of "food" (such as a bacterium), these molecules are still kept isolated from the cytoplasm by the membranes surrounding the food vacuole. The large molecules of proteins, polysaccharides, and lipids must first be digested into their smaller subunits (including amino acids, monosaccharides, and fatty acids) before they can cross the vacuole membrane and enter the cytoplasm.

The digestive enzymes of a cell are isolated from the cytoplasm and concentrated within membrane-bound organelles called **lysosomes** (*li'sŏ-sōmz*) (fig. 3.9). A *primary lysosome* is one that contains only digestive enzymes (about 40 different types) within an environment that is considerably more acidic than the surrounding cytoplasm. A primary lysosome may fuse with a food vacuole (or with another cellular organelle) to form a *secondary lysosome* in which worn-out organelles and the products of phagocytosis can be digested. Thus, a secondary lysosome contains partially digested remnants of other organelles and ingested organic material. A lysosome that contains undigested wastes is called a *residual body*. Residual bodies may eliminate their wastes by exocytosis, or the wastes may accumulate within the cell as the cell ages.

Partly digested membranes of various organelles and other cellular debris are often observed within secondary lysosomes. This is a result of **autophagy** (*aw-tof'ă-je*), a process that destroys worn-out organelles so that they can be continuously replaced. Lysosomes are thus aptly characterized as the "digestive system" of the cell.

Lysosomes have also been called "suicide bags" because a break in their membranes would release their digestive enzymes and thus destroy the cell. This happens normally in *programmed cell death* (or *apoptosis*), described in a later section. An example is the destruction of tissues that occurs during the remodeling processes in embryological development.

Peroxisomes

Peroxisomes (*pĕ-roks'ĭ-sōmz*) are membrane-enclosed organelles containing several specific enzymes that promote oxidative reactions. Although peroxisomes are present in most cells, they are particularly large and active in the liver.

All peroxisomes contain one or more enzymes that promote reactions in which hydrogen is removed from particular organic molecules and transferred to molecular oxygen (O_2), thereby oxidizing the molecule and forming hydrogen peroxide (H_2O_2) in the process. The oxidation of toxic molecules by peroxisomes in this way is an important function of liver and kidney cells. For example, much of the alcohol ingested in alcoholic drinks is oxidized into acetaldehyde by liver peroxisomes.

---

autophagy: Gk. *autos*, self; *phagein*, to eat

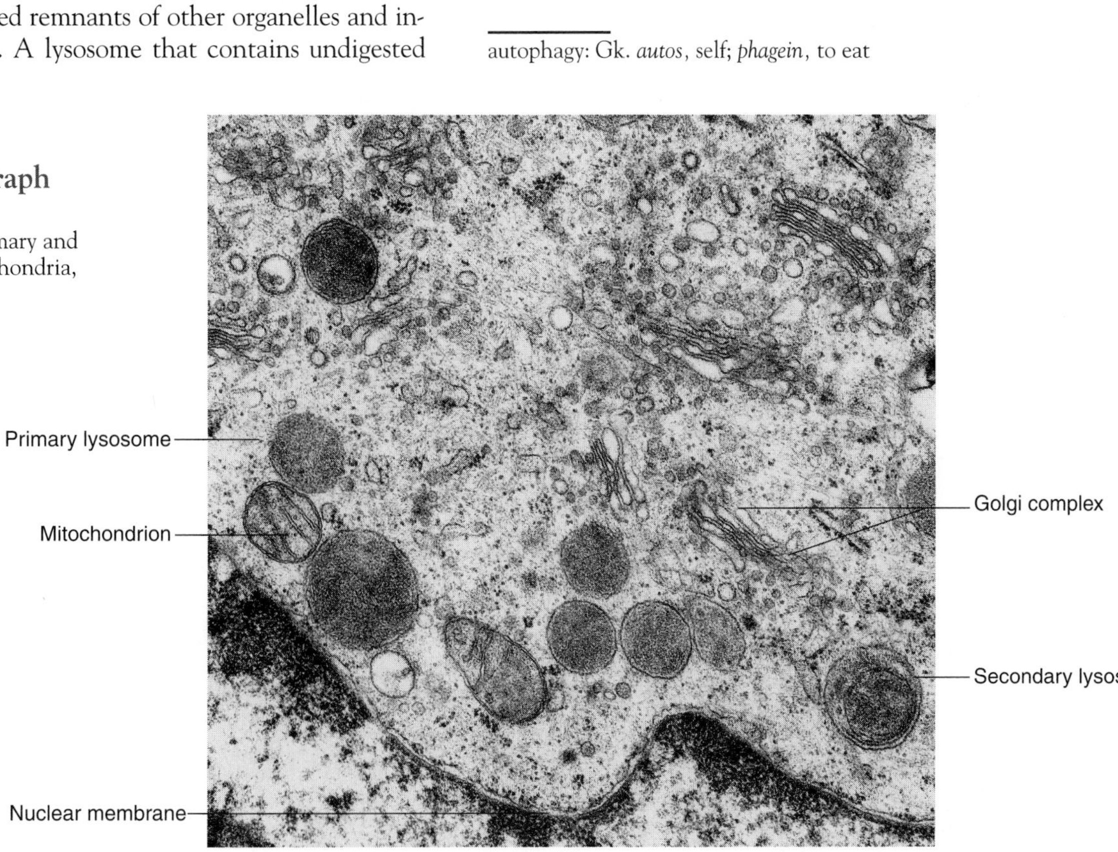

Figure 3.9

An electron micrograph of lysosomes.
This photograph shows primary and secondary lysosomes, mitochondria, and the Golgi complex.

The enzyme *catalase* within the peroxisomes prevents the excessive accumulation of hydrogen peroxide by catalyzing the reaction $2\ H_2O_2 \rightarrow 2\ H_2O + O_2$. Catalase is one of the fastest acting enzymes known (see chapter 4), and it is this reaction that produces the characteristic fizzing when hydrogen peroxide is poured on a wound.

Mitochondria

All cells in the body, with the exception of mature red blood cells, have from a hundred to a few thousand organelles called **mitochondria** (*mi″tŏ-kon′dre-ă*). Mitochondria serve as sites for the production of most of the energy of cells (see chapter 4).

Mitochondria vary in size and shape, but all have the same basic structure (fig. 3.10). Each mitochondrion is surrounded by an inner and outer membrane, separated by a narrow intermembranous space. The outer mitochondrial membrane is smooth, but the inner membrane is characterized by many folds, called *cristae* (*kris′te*), which project like shelves into the central area (or *matrix*) of the mitochondrion. The cristae and the matrix compartmentalize the space within the mitochondrion and have different roles in the generation of cellular energy. The structure and functions of mitochondria will be described in more detail in the context of cellular metabolism in chapter 4.

Mitochondria can migrate through the cytoplasm of a cell and are able to reproduce themselves. Indeed, mitochondria contain their own DNA. This is a more primitive form of DNA (consisting of a circular, relatively small, double-stranded molecule) than that found within the cell nucleus. For this and other reasons, many scientists believe that mitochondria evolved from separate organisms, related to bacteria, that invaded the ancestors of animal cells and remained in a state of symbiosis.

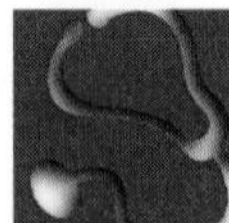

An ovum (egg cell) contains mitochondria; the head of a sperm cell contains none. Therefore, all of the mitochondria in a fertilized egg are derived from the mother. The mitochondrial DNA replicates itself and the mitochondria divide, so that all of the mitochondria in the fertilized ovum and the cells derived from it during embryonic and fetal development are genetically identical to those in the original ovum. This provides a unique form of inheritance that is passed only from mother to child. A rare cause of blindness known as *Leber's hereditary optic neuropathy,* as well as several other disorders, are inherited only along the maternal lineage and are known to be caused by defective mitochondrial DNA.

Endoplasmic Reticulum

Most cells contain a system of membranes known as the **endoplasmic reticulum** (*en-do-plaz′mik rĕ-tik′yŭ-lum*), or **ER.** The ER may be either of two types: (1) a **rough,** or **granular, endoplasmic reticulum** and (2) a **smooth endoplasmic reticulum** (fig. 3.11). A rough endoplasmic reticulum bears ribosomes (discussed in a later section) on its surface, whereas a smooth endoplasmic reticulum does not. The smooth endoplasmic reticulum serves a variety of purposes in different cells; it provides a site for enzyme reactions in steroid hormone production and inactivation, for example, and a site for the storage of Ca^{2+} in skeletal and cardiac muscle cells. The rough endoplasmic reticulum is found in cells that are active in protein synthesis and secretion, such as those of many exocrine and endocrine glands.

mitochondrion: Gk. *mitos*, a thread; *chondros*, grain

symbiosis: Gk. *syn*, together with; *bios*, life
Leber's hereditary optic neuropathy: from Theodor Leber, German ophthalmologist, 1840–1917.

Figure 3.10

The structure of a mitochondrion.
(*a*) An electron micrograph of a mitochondrion. The outer mitochondrial membrane and the infoldings of the inner membrane—the cristae—are clearly seen. The fluid in the center is the matrix. (*b*) A diagram of the structure of a mitochondrion.

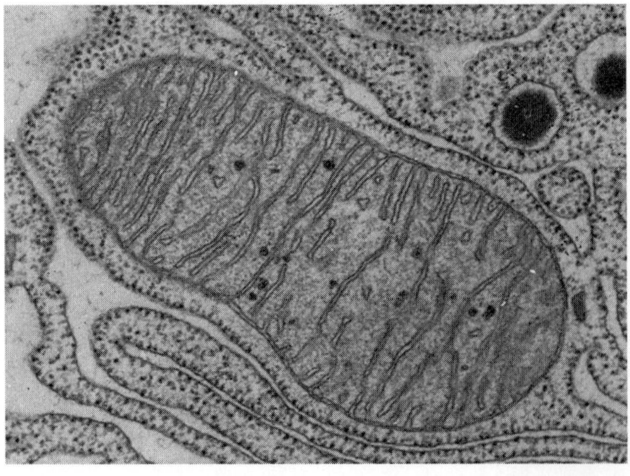

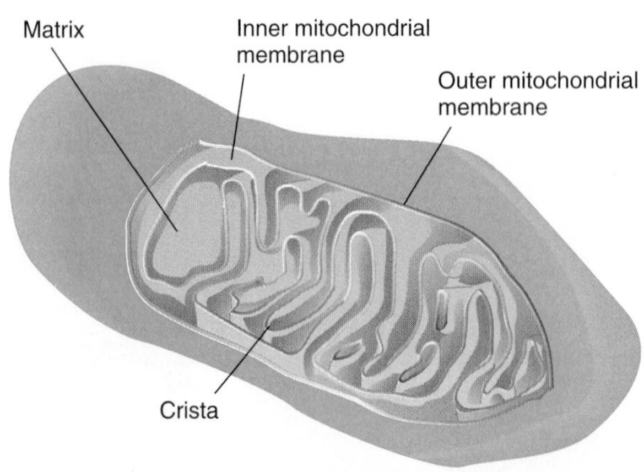

The function of the rough endoplasmic reticulum and that of another organelle, the Golgi complex, will be described in a later section on protein synthesis. The structure of centrioles and the spindle apparatus, which are involved in DNA replication and cell division, will also be described in a separate section.

Figure 3.11

The endoplasmic reticulum.
(*a*) An electron micrograph of the endoplasmic reticulum (about 100,000×). The rough endoplasmic reticulum (*b*) has ribosomes attached to its surface, whereas the smooth endoplasmic reticulum (*c*) lacks ribosomes.

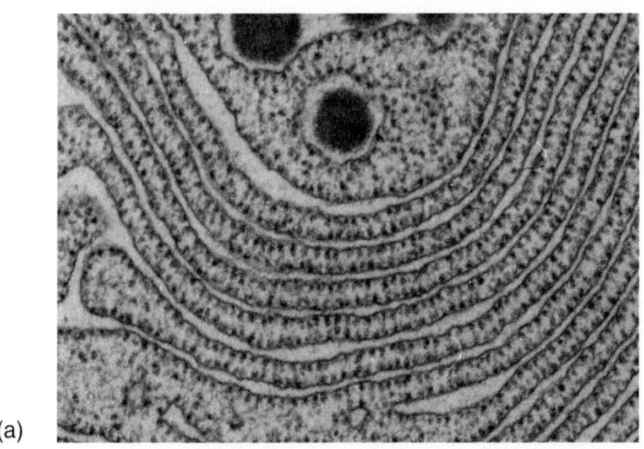

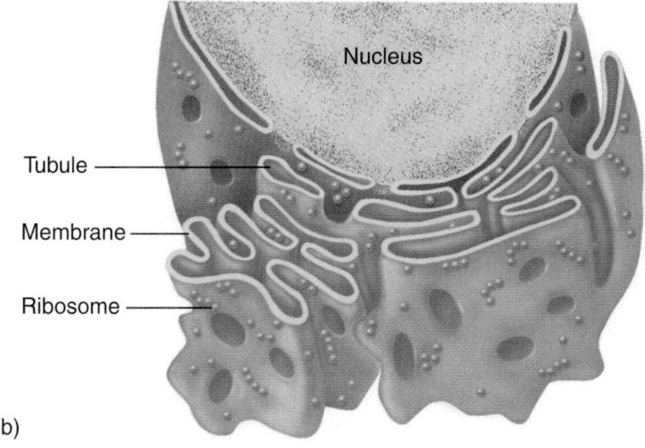

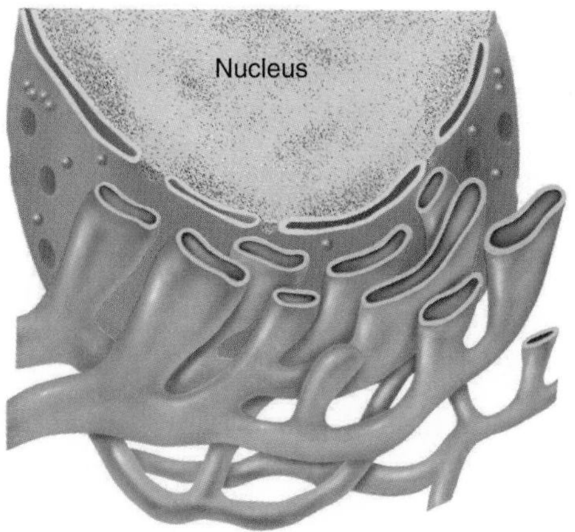

Cell Nucleus and Nucleic Acids

The genetic code is based on the structure of DNA and expressed through the structure and function of RNA. DNA and RNA are composed of subunits called nucleotides, and together these molecules are known as nucleic acids. The sequences of DNA nucleotides direct the synthesis of RNA molecules, and it is through the RNA-directed synthesis of proteins that the genetic code is expressed.

Most cells in the body have a single **nucleus.** Exceptions include skeletal muscle cells, which have two or more nuclei, and mature red blood cells, which have none. The nucleus is enclosed by two membranes—an inner membrane and an outer membrane—that together are called the **nuclear membrane,** or **nuclear envelope** (fig. 3.12). The outer membrane is continuous with the endoplasmic reticulum in the cytoplasm. At various points, the inner and outer membranes are fused together by structures called *nuclear pore complexes*. These structures function as rivets, holding the two membranes together. Each nuclear pore complex has a central opening, the *nuclear pore* (fig. 3.13), surrounded by interconnected rings and columns of proteins. Small molecules may pass through the complexes by diffusion, but movement of protein and RNA through the nuclear pores is a selective, energy-requiring process.

Transport of specific proteins from the cytoplasm, through the nuclear pores, and into the nucleus may serve a variety of functions, including regulation of gene expression by hormones (see chapter 19). Transport of RNA out of the nucleus, where it is formed, is required for gene expression. As described in this section, *genes* are regions of the DNA within

Figure 3.12

The structure of a nucleus.
The nucleus of a liver cell with its nuclear membrane and nucleolus is shown in this electron micrograph.

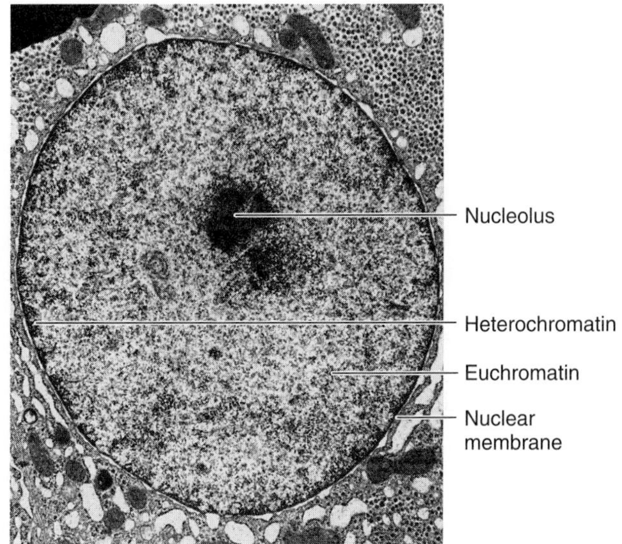

Figure 3.13

The nuclear pores.
(*a*) An electron micrograph of a freeze-fractured nuclear envelope showing the nuclear pores. (*b*) A diagram showing the nuclear pore complexes.

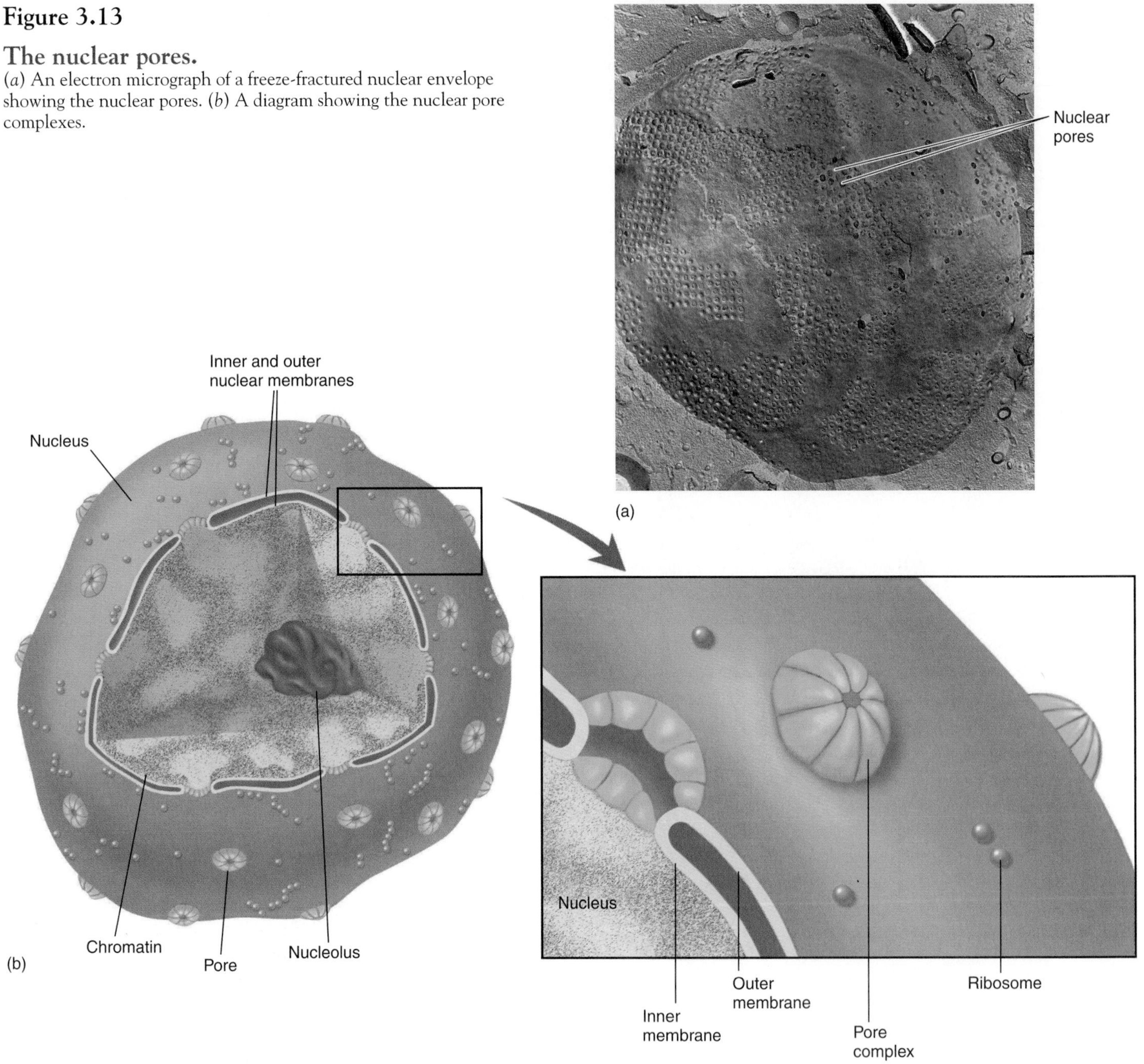

the nucleus. Each gene contains the code for the production of a particular type of RNA called messenger RNA (mRNA). As an mRNA is transported through the nuclear pore, it becomes associated with ribosomes that are either free in the cytoplasm or associated with the rough endoplasmic reticulum. The mRNA then provides the code for the production of a specific type of protein.

The primary structure of the protein (its amino acid sequence) is determined by the sequence of bases in mRNA. The base sequence of mRNA has been previously determined by the sequence of bases in the region of the DNA (the gene) that codes for the mRNA. Genetic expression therefore occurs in two stages: first **transcription** (synthesis of RNA) and then **translation** (synthesis of protein).

Each nucleus contains one or more dark areas (see fig. 3.12). These regions, which are not surrounded by membranes, are called **nucleoli** (singular, *nucleolus*). The DNA within the nucleoli contains the genes that code for the production of ribosomal RNA (rRNA), an essential component of ribosomes.

Nucleic Acids

Nucleic acids include the macromolecules of **DNA** and **RNA,** which are critically important in genetic regulation, and the subunits from which these molecules are formed. These subunits are known as **nucleotides** (*noo'kle-ŏ-tīdz*).

Figure 3.14

The general structure of a nucleotide.
A polymer of nucleotides, or polynucleotide (shown below) is formed by sugar-phosphate bonds between nucleotides.

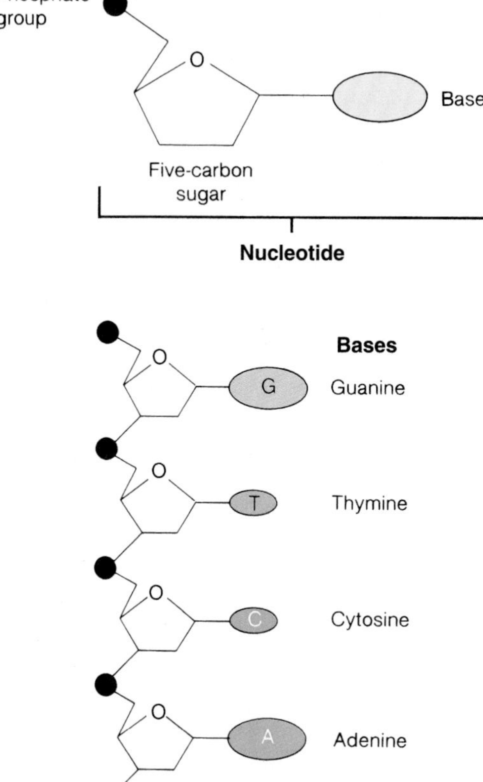

Figure 3.15

The four nitrogenous bases in deoxyribonucleic acid (DNA).
Notice that hydrogen bonds can form between guanine and cytosine and between thymine and adenine.

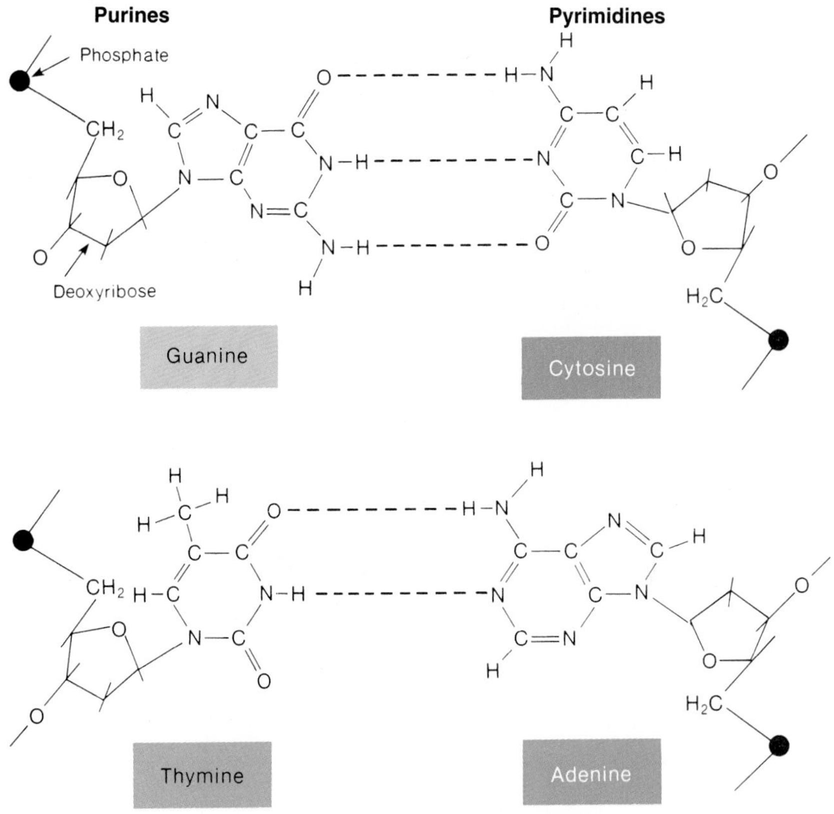

Nucleotides are used as subunits in the formation of long polynucleotide chains. Each nucleotide, however, is composed of three smaller subunits: a five-carbon sugar, a phosphate group attached to one end of the sugar, and a *nitrogenous base* attached to the other end of the sugar (fig. 3.14). The nitrogenous bases are cyclic nitrogen-containing molecules of two kinds: pyrimidines and purines. The *pyrimidines* (*pi-rim'ĭ-dēnz*) contain a single ring of carbon and nitrogen, whereas the *purines* have two such rings (fig. 3.15).

Deoxyribonucleic Acid

The structure of **DNA (deoxyribonucleic acid)** serves as the basis for the genetic code. For this reason, it might seem logical that DNA should have an extremely complex structure. DNA is indeed larger than any other molecule in the cell, but its structure is actually simpler than that of most proteins. This simplicity of structure deceived some of the early scientists into believing that the protein content of chromosomes, rather than their DNA content, provided the basis for the genetic code.

Sugar molecules in the nucleotides of DNA are a type of pentose (five-carbon) sugar called **deoxyribose** (*de-ok"se-ri'bōs*). Each deoxyribose can be covalently bonded to one of four possible bases. These bases include the two purines **adenine** (*ad'n-ēn*) and **guanine** (*gwă'nēn*) and the two pyrimidines **cytosine** (*si'tŏ-sēn*) and **thymine** (*thi'mēn*). There are thus four different types of nucleotides that can be used to produce the long DNA chains. If you remember that there are 20 different amino acids used to produce proteins, you can now understand why many scientists were deceived into thinking that genes were composed of proteins rather than nucleic acids.

When nucleotides combine to form a chain, the phosphate group of one condenses with the deoxyribose sugar of another nucleotide. This forms a sugar-phosphate chain as water is removed in dehydration synthesis. Since the nitrogenous bases are attached to the sugar molecules, the sugar-phosphate chain looks like a "backbone" from which the bases project. Each of these bases can form hydrogen bonds with other bases, which are in turn joined to a different chain of nucleotides. Such hydrogen bonding between bases thus produces a *double-stranded* DNA molecule; the two strands are like a staircase, with the paired bases as steps (fig. 3.15).

Actually, the two chains of DNA twist about each other to form a **double helix,** so that the molecule resembles a spiral staircase (fig. 3.16). It has been shown that the number of purine bases in DNA is equal to the number of pyrimidine bases. The reason for this is explained by the **law of**

Figure 3.16

The double-helix structure of DNA.
The two strands are held together by hydrogen bonds between complementary bases in each strand.

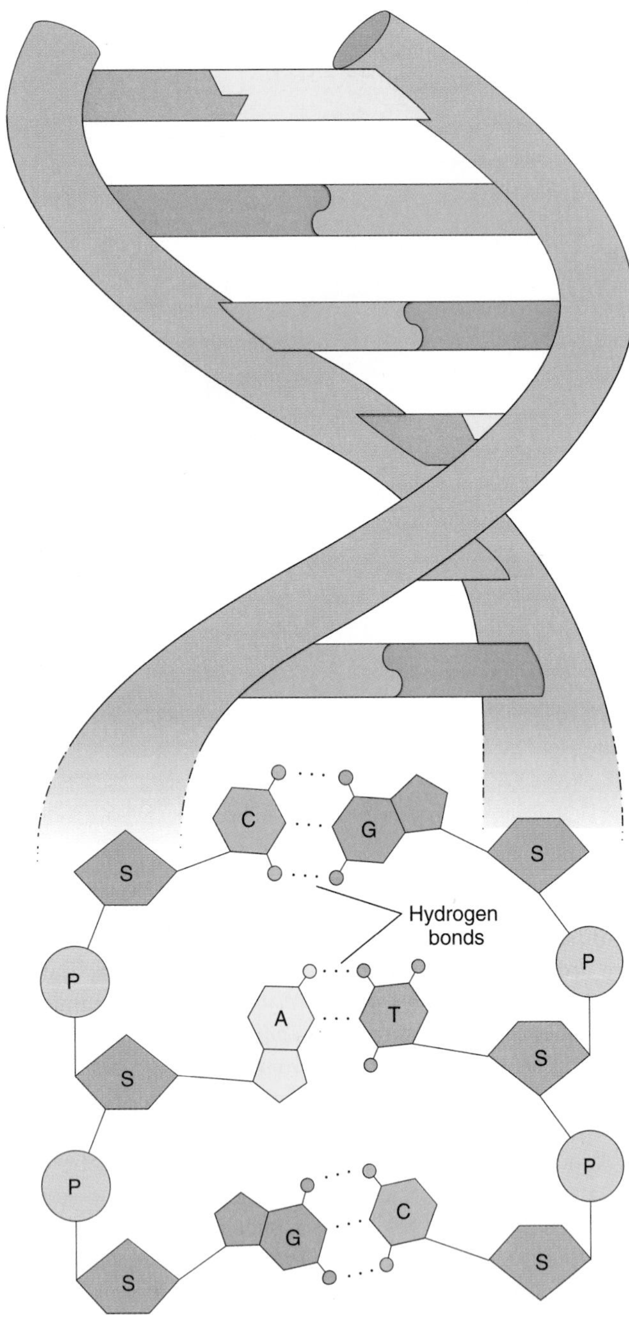

Figure 3.17

The structure of chromatin.
Part of the DNA is wound around complexes of histone proteins, forming particles known as nucleosomes.

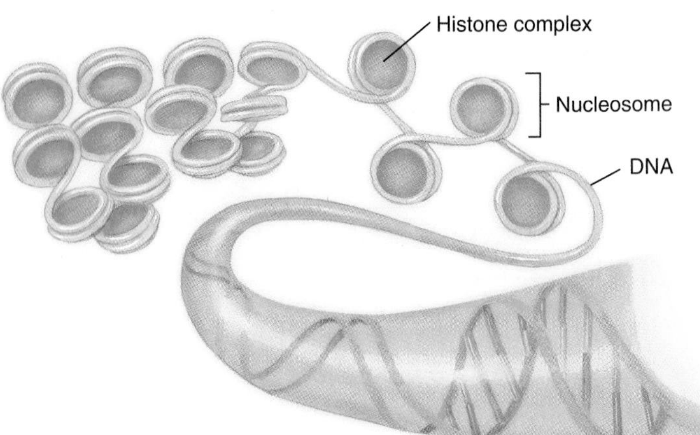

complementary base pairing: *adenine can pair only with thymine* (through two hydrogen bonds), whereas *guanine can pair only with cytosine* (through three hydrogen bonds). With knowledge of this rule, we could predict the base sequence of one DNA strand if we knew the sequence of bases in the complementary strand.

Although we can predict which base is opposite a given base in DNA, we cannot predict which bases will be above or below that particular pair within a polynucleotide chain. Although there are only four bases, the number of possible base sequences along a stretch of several thousand nucleotides (the length of most genes) is almost infinite. To gain perspective, it is useful to realize that the total human **genome** (*je'nōm*)—all of the genes in a cell—consists of 3.3 billion base pairs that would extend over a meter if the DNA molecules were unraveled and stretched out.

Yet, even with this amazing variety of possible base sequences, almost all of the billions of copies of a particular gene in a person are identical. The mechanisms by which identical DNA copies are made and distributed to the daughter cells when a cell divides will be described in a later section.

Chromatin

The DNA within the cell nucleus is combined with protein to form **chromatin** (*kro'mă-tin*), the material that makes up the chromosomes. Much of the protein content of the chromatin is of a type known as *histones*. Histone proteins (of which there are several different forms) are positively charged and organized to form spools about which the negatively charged strands of DNA are wound. Each spool consists of two turns of DNA, comprising 146 base pairs, wound around a core of histone proteins. This spooling creates particles known as **nucleosomes** (fig. 3.17).

Spooled DNA cannot be transcribed, so that the spooling, and other levels of "packaging" of DNA by histone proteins, serves to repress genes. Some genes are permanently repressed in particular cells; others are repressed much of the time but are activated under specific conditions. According to one model, a different kind of protein that functions as a gene activator may work by causing dissociation of histone proteins from the "promotor" region of a gene's DNA. Removal of the repressive effects of the histones, then, would activate the gene. Active regions of chromatin, called

chromatin: Gk. *chroma*, color

Figure 3.18

The differences between the nucleotides and sugars in DNA and RNA.
DNA has deoxyribose and thymine; RNA has ribose and uracil. The other three bases are the same in DNA and RNA.

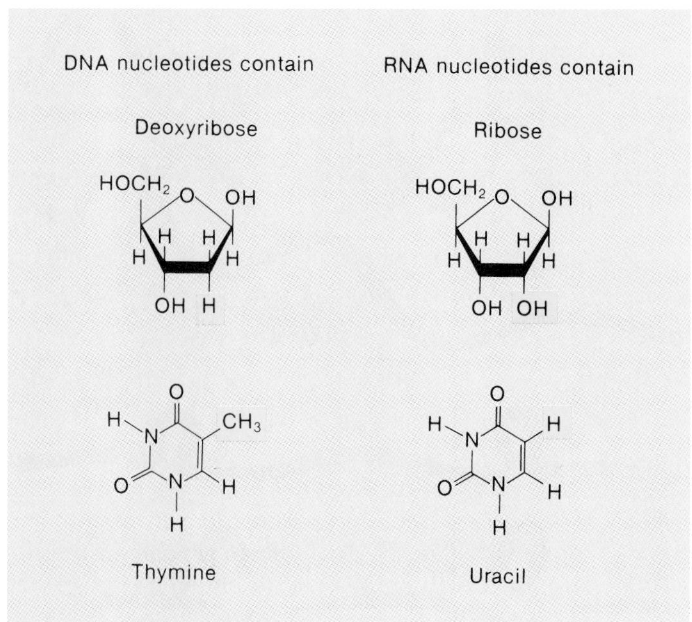

euchromatin (*yoo-kro'mă-tin*), have a threadlike appearance in the electron microscope, whereas inactive chromatin appears as blotches called *heterochromatin*.

Ribonucleic Acid

DNA can direct the activities of the cell only by means of another type of nucleic acid—**RNA (ribonucleic acid).** Like DNA, RNA consists of long chains of nucleotides joined together by sugar-phosphate bonds. Nucleotides in RNA, however, differ from those in DNA (fig. 3.18) in three ways: (1) a **ribonucleotide** (*ri"bo-noo'kle-ŏ-tīd*) contains the sugar **ribose** (instead of deoxyribose), (2) the base **uracil** (*yoor'ă-sil*) is present in place of thymine, and (3) RNA is composed of a single polynucleotide strand (it is not double-stranded like DNA).

There are three types of RNA molecules that function in the cytoplasm of cells: *messenger RNA* (*mRNA*), *transfer RNA* (*tRNA*), and *ribosomal RNA* (*rRNA*). All three types are made within the cell nucleus by using information contained in DNA as a guide.

RNA Synthesis

One gene codes for one polypeptide chain. Each gene is a stretch of DNA that is several thousand nucleotide pairs long. The DNA in a human cell contains over 3 billion base pairs—enough to code for at least 3 million proteins. Since the average human cell contains less than this amount (30,000 to 150,000 different proteins), it follows that only a fraction of the DNA in each cell is used to code for proteins. The remainder of the DNA may be inactive or redundant. Also, some segments of DNA serve to regulate those regions that do code for proteins.

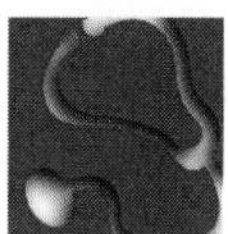

The *Human Genome Project* was launched by Congress in 1988, with the goal of completely mapping the human genome by September 30, 2005. That allowed just 17 years to determine the exact sequences of bases with which the 3 billion base pairs are arranged to form the 50,000 to 100,000 genes in the haploid human genome. (The haploid genome is the genome of a sperm cell or oocyte.) Such a detailed map will greatly aid the diagnosis and treatment of the 4,000 different genetic diseases that are directly caused by particular abnormal genes. Other diseases—the majority—have a genetic component but are caused by a variety of factors. The diagnosis and treatment of diseases such as these also may be greatly aided by research utilizing a complete map of the human genome.

In order for the genetic code to be translated into the synthesis of specific proteins, the DNA code first must be copied onto a strand of RNA. This is accomplished by DNA-directed RNA synthesis—the process of **transcription.**

In RNA synthesis, the enzyme **RNA polymerase** (*pol-im'er-ās*) breaks the weak hydrogen bonds between paired DNA bases. This does not occur throughout the length of DNA, but only in the regions that are to be transcribed. There are base sequences that code for "start" and "stop," and there are regions of DNA that function as *promotors*. Specific regulatory molecules, such as hormones, act as **transcription factors** by binding to the promotor region of a particular gene and thereby activating the gene. The double-stranded DNA separates in the region to be transcribed, so that the freed bases can pair with the complementary RNA nucleotide bases in the nucleoplasm.

This pairing of bases, like that which occurs in DNA replication (described in a later section), follows the law of complementary base pairing: *guanine bonds with cytosine* (and vice versa), and *adenine bonds with uracil* (because uracil in RNA is equivalent to thymine in DNA). Unlike DNA replication, however, only *one* of the two freed strands of DNA serves as a guide for RNA synthesis (fig. 3.19). Once an RNA molecule has been produced, it detaches from the DNA strand on which it was formed. This process can continue indefinitely, producing many thousands of RNA copies of the DNA strand that is being transcribed. When the gene is no longer to be transcribed, the separated DNA strands can then go back together again.

Types of RNA

There are four types of RNA produced within the nucleus by genetic transcription: (1) **precursor messenger RNA (pre-mRNA),** which is altered within the nucleus to form mRNA; (2) **messenger RNA (mRNA),** which contains the code for the synthesis of specific proteins; (3) **transfer RNA (tRNA),** which is needed for decoding the genetic message contained in mRNA; and (4) **ribosomal RNA (rRNA),**

Figure 3.19

RNA synthesis (genetic transcription).
Notice that only one of the two DNA strands is used to form a single-stranded molecule of RNA.

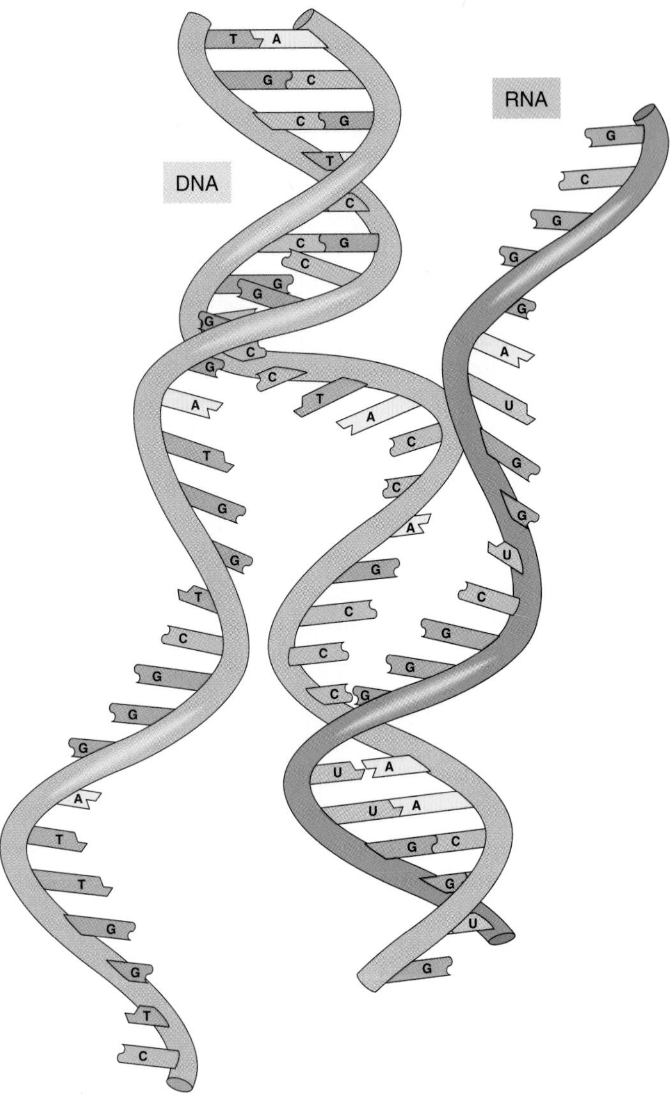

Figure 3.20

The processing of pre-mRNA into mRNA.
Noncoding regions of the genes, called introns, produce excess bases within the pre-mRNA. These excess bases are removed, and the coding regions of mRNA are spliced together.

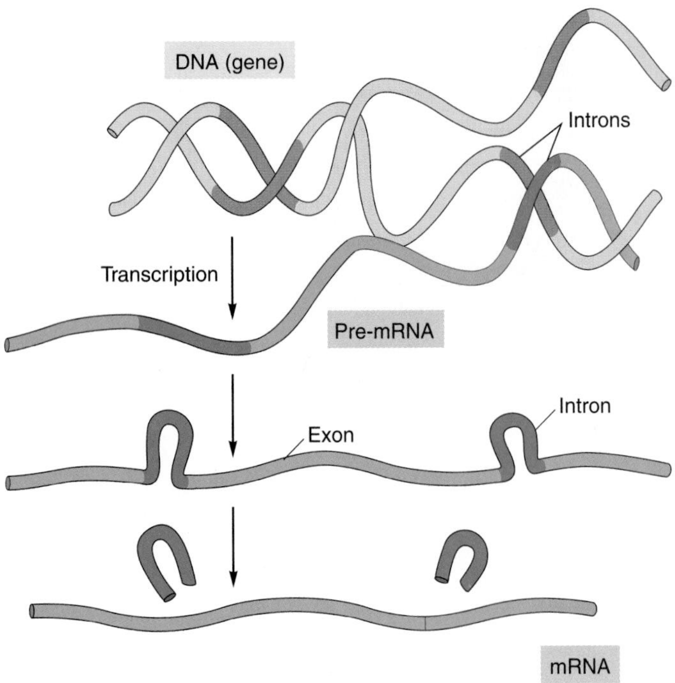

which forms part of the structure of ribosomes. The DNA that codes for rRNA synthesis is located in the part of the nucleus called the nucleolus. The DNA that codes for pre-mRNA and tRNA synthesis is located elsewhere in the nucleus.

In bacteria, where the molecular biology of the gene is best understood, a gene that codes for one type of protein produces an mRNA molecule that begins to direct protein synthesis as soon as it is transcribed. This is not the case for higher organisms, including humans. In more complex cells, a pre-mRNA is produced that must be modified within the nucleus before it can enter the cytoplasm as mRNA and direct protein synthesis.

Precursor mRNA is much larger than the mRNA it forms. Surprisingly, this large size of pre-mRNA is not due to excess bases at the ends of the molecule that must be trimmed; rather, the excess bases are located *within* the pre-mRNA. The genetic code for a particular protein, in other words, is split up by stretches of base pairs that do not contribute to the code. These regions of noncoding DNA within a gene are called *introns;* the coding regions are known as *exons*. Consequently, pre-mRNA must be cut and spliced to make mRNA (fig. 3.20). This cutting and splicing can be quite extensive—a single gene may contain up to 50 introns, all of which must be removed from the pre-mRNA in order to convert it to mRNA. Introns are cut out of the pre-mRNA, and the ends of the exons spliced, by macromolecules called *SnRNPs* (pronounced "snurps"), producing the functional mRNA that leaves the nucleus and enters the cytoplasm. SnRNPs stands for *small nuclear ribonucleoproteins*. These are small, ribosomelike aggregates of RNA and protein that form a body called a *spliceosome* that splices the exons together.

Protein Synthesis and Secretion

In order for a gene to be expressed, it first must be used as a guide, or template, in the production of a complementary strand of messenger RNA. This mRNA is then itself used as a guide to produce a particular type of protein whose sequence of amino acids is determined by the sequence of base triplets (codons) in the mRNA.

When mRNA enters the cytoplasm, it attaches to **ribosomes** (*ri'bo-sōmz*), which appear in the electron microscope as numerous small particles. A ribosome is composed

Figure 3.21

An electron micrograph of polyribosomes.
An RNA strand (arrow) joins the ribosomes together.

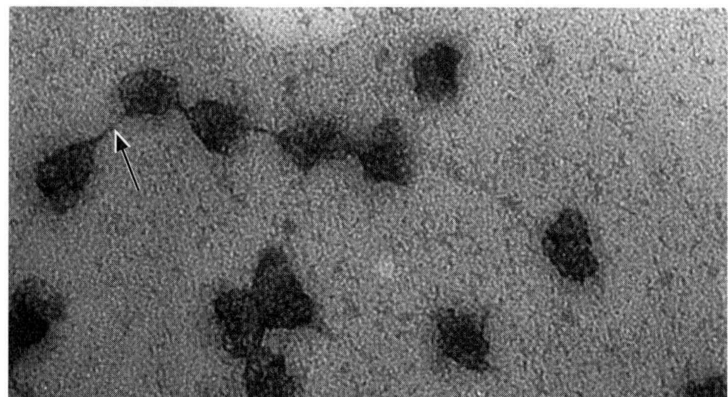

of four molecules of ribosomal RNA and 82 proteins, arranged to form two subunits of unequal size. The mRNA passes through a number of ribosomes to form a "string-of-pearls" structure called a *polyribosome* (or *polysome*, for short), as shown in figure 3.21. The association of mRNA with ribosomes is needed for the process of **translation**—the production of specific proteins according to the code contained in the mRNA base sequence.

Each mRNA molecule contains several hundred or more nucleotides, arranged in the sequence determined by complementary base pairing with DNA during transcription (RNA synthesis). Every three bases, or *base triplet*, is a code word—called a **codon** (*ko'don*)—for a specific amino acid. Sample codons and their amino acid "translations" are listed in table 3.2 and illustrated in figure 3.22. As mRNA moves through the ribosome, the sequence of codons is translated into a sequence of specific amino acids within a growing polypeptide chain.

Transfer RNA

Translation of the codons is accomplished by tRNA and particular enzymes. Each tRNA molecule, like mRNA and rRNA, is single-stranded. Although tRNA is single-stranded, it bends in on itself to form a cloverleaf structure (fig. 3.23*a*), which is believed to be further twisted into an upside down "L" shape (fig. 3.23*b*). One end of the "L" contains the **anticodon**—three nucleotides that are complementary to a specific codon in mRNA.

Enzymes in the cell cytoplasm called *aminoacyl-tRNA synthetase* enzymes join specific amino acids to the ends of tRNA, so that a tRNA with a given anticodon can bind to only one specific amino acid. There are 20 different varieties of synthetase enzymes, one for each type of amino acid. Each synthetase must not only recognize its specific amino acid, it must be able to attach this amino acid to the particular tRNA that has the correct anticodon for that amino acid. The cytoplasm of a cell thus contains tRNA molecules that are each bonded to a specific amino acid, and each of these tRNA molecules is capable of bonding with a specific codon in mRNA via its anticodon base triplet.

Table 3.2

Selected DNA Base Triplets and mRNA Codons

DNA Triplet	RNA Codon	Amino Acid
TAC	AUG	"Start" (Methionine)
ATC	UAG	"Stop"
AAA	UUU	Phenylalanine
AGG	UCC	Serine
ACA	UGU	Cysteine
GGG	CCC	Proline
GAA	CUU	Leucine
GCT	CGA	Arginine
TTT	AAA	Lysine
TGC	ACT	Tyrosine
CCG	GGC	Glycine
CTC	GAG	Aspartic acid

Formation of a Polypeptide

The anticodons of tRNA bind to the codons of mRNA as the mRNA moves through the ribosome. Since each tRNA molecule carries a specific amino acid, the joining of these amino acids by peptide bonds creates a polypeptide whose amino acid sequence has been determined by the sequence of codons in mRNA.

The first and second tRNA bring the first and second amino acids close together. The first amino acid then detaches from its tRNA and is enzymatically transferred to the amino acid on the second tRNA, forming a dipeptide. When the third tRNA binds to the third codon, the amino acid it brings forms a peptide bond with the second amino acid (which detaches from its tRNA). A tripeptide is now attached by the third amino acid to the third tRNA. The polypeptide chain thus grows as new amino acids are added to its growing tip (fig. 3.24). This growing polypeptide chain is always attached by means of only one tRNA to the strand of mRNA, and this tRNA molecule is always the one that has added the latest amino acid to the growing polypeptide.

As the polypeptide chain grows in length, interactions between its amino acids cause the chain to twist into a helix (secondary structure) and to fold and bend upon itself (tertiary structure). At the end of this process, the new protein detaches from the tRNA as the last amino acid is added.

Figure 3.22

Genetic transcription and translation.

The genetic code is first transcribed into base triplets (codons) in mRNA and then translated into a specific sequence of amino acids in a polypeptide.

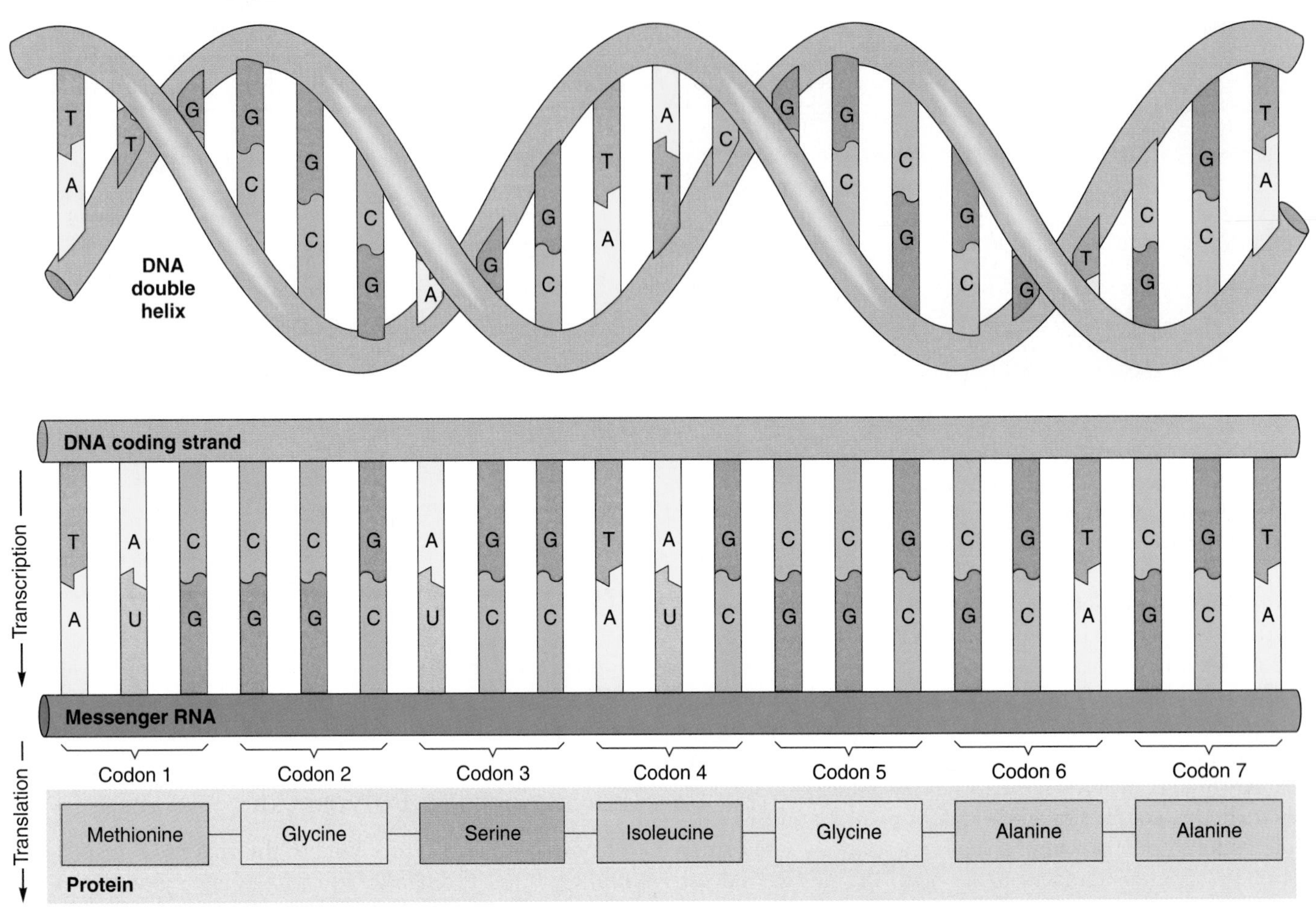

Figure 3.23

The structure of transfer RNA (tRNA).

(*a*) A simplified cloverleaf representation and (*b*) the three-dimensional structure of tRNA.

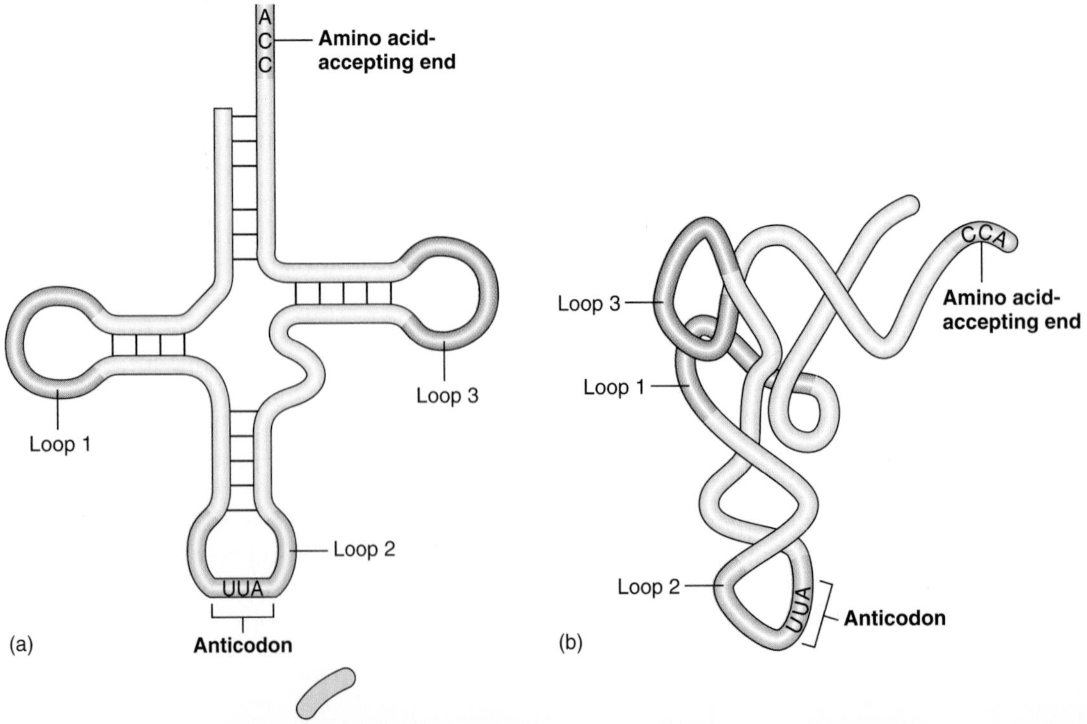

Figure 3.24

The translation of messenger RNA (mRNA).
As the anticodon of each new aminoacyl-tRNA bonds with a codon on the mRNA, new amino acids are joined to the growing tip of the polypeptide chain.

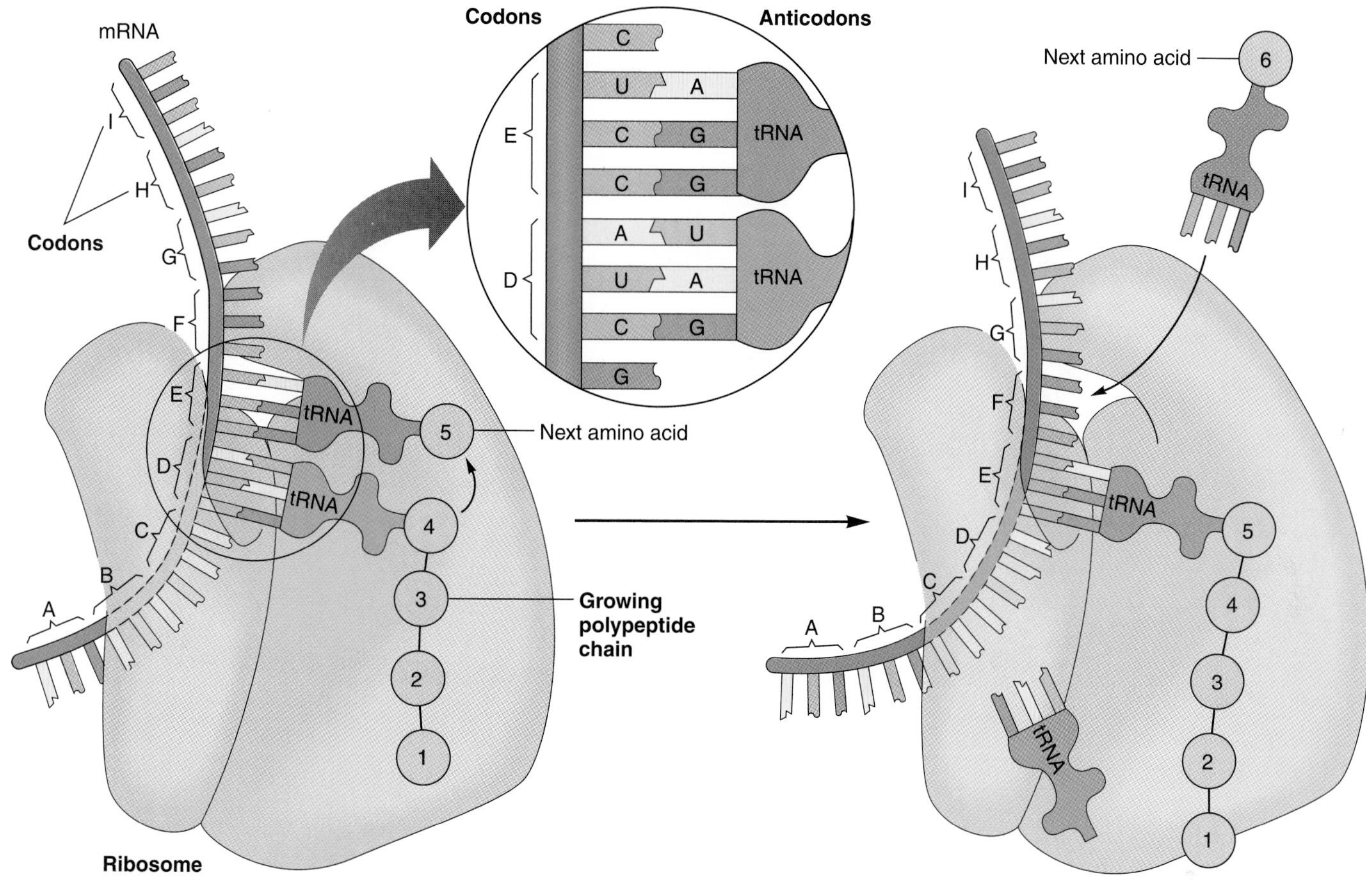

Many proteins are further modified after they are formed; these modifications occur in the rough endoplasmic reticulum and Golgi complex.

Function of the Rough Endoplasmic Reticulum

Proteins that are to be used within the cell are likely to be produced by free polyribosomes in the cytoplasm. If a protein is to be secreted by the cell, however, it is more likely to be made by mRNA-ribosome complexes located on the rough endoplasmic reticulum. The membranes of this system enclose fluid-filled spaces called *cisternae* (*sis-ter'ne*), which the newly formed proteins may enter. Once in the cisternae, the structure of these proteins is modified in specific ways.

When proteins destined for secretion are produced, the first 30 or so amino acids are primarily hydrophobic. This *leader sequence* is attracted to the lipid component of the membranes of the endoplasmic reticulum. As the polypeptide chain elongates, it is "injected" into the cisterna within the endoplasmic reticulum. The leader sequence is, in a sense, an "address" that directs secretory proteins into the endoplasmic reticulum. Once the proteins are in the cisterna, the leader sequence is enzymatically removed so that the protein cannot reenter the cytoplasm (fig. 3.25).

The processing of the hormone insulin can serve as an example of the changes that occur within the endoplasmic reticulum. The original molecule enters the cisterna as a single polypeptide composed of 109 amino acids. This molecule is called *preproinsulin*. The first 23 amino acids serve as a leader sequence that allows the molecule to be injected into the cisterna within the endoplasmic reticulum. The leader sequence is then quickly removed, producing a molecule called *proinsulin*. The remaining chain folds within the cisterna so that the first and last amino acids in the polypeptide are brought close together. The central region is then enzymatically removed, producing two chains—one of them, 21 amino acids long; the other, 30 amino acids long—which are subsequently joined together by disulfide bonds (fig. 3.26). It is this form of insulin that is normally secreted from the cell.

Function of the Golgi Complex

Secretory proteins do not remain trapped within the rough endoplasmic reticulum; they are transported to another organelle within the cell—the **Golgi** (*gol'je*) **complex,** also called the **Golgi apparatus.** This organelle serves three interrelated

Golgi complex: from Camilio Golgi, Italian histologist, 1843–1926

Figure 3.25

Secretory proteins enter the endoplasmic reticulum.
A protein destined for secretion begins with a leader sequence that enables it to be inserted into the cisterna (cavity) of the endoplasmic reticulum. Once it has been inserted, the leader sequence is removed, and carbohydrate is added to the protein.

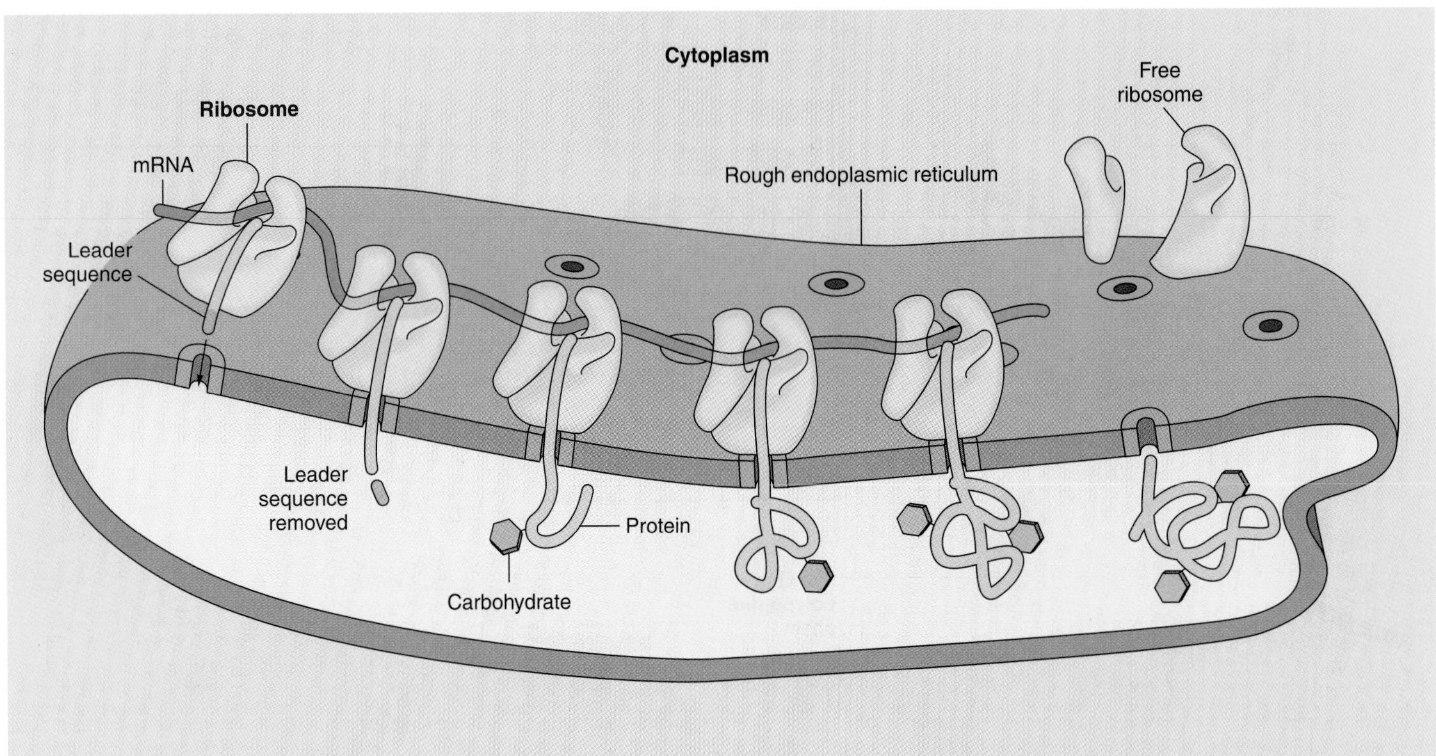

functions: (1) further modifications of proteins (such as the addition of carbohydrates to form *glycoproteins*) occur in the Golgi complex; (2) different types of proteins are separated according to their function and destination; and (3) the final products are packaged and shipped to their destinations. In the Golgi complex, for example, proteins that are to be secreted are separated from those that will be incorporated into the cell membrane and from those that will be introduced into lysosomes, and these different proteins are packaged into separate membrane-enclosed vesicles.

The Golgi complex consists of a stack of flattened sacs with bulges at their edges. One side of the sac serves as the site of entry for cellular products that arrive in vesicles from the endoplasmic reticulum. After specialized modifications of the proteins are made within one sac, the modified proteins are passed by means of vesicles to the next sac until the finished products leave the Golgi complex from the opposite side of the stack (fig. 3.27). Depending on the nature of the specific product, the vesicles that leave the Golgi complex may become lysosomes, storage granules of secretory products, or additions to the cell membrane.

DNA Synthesis and Cell Division

When a cell is going to divide, each strand of the DNA within its nucleus acts as a template for the formation of a new complementary strand. Organs grow and repair themselves through a type of cell division known as mitosis. The two daughter cells produced by mitosis contain the same genetic information as the parent cell. Gametes contain only half the number of chromosomes as their parent cell and are formed by a type of cell division called meiosis.

Genetic information is required for the life of the cell and for the ability of the cell to perform its functions in the body. Each cell obtains this genetic information from its parent cell through the process of DNA replication and cell division. DNA is the only type of molecule in the body capable of replicating itself, and mechanisms exist within the dividing cell to ensure that the duplicate copies of DNA will be properly distributed to the daughter cells.

DNA Replication

When a cell is going to divide, each DNA molecule replicates itself, and each of the identical DNA copies thus produced is distributed to the two daughter cells. Replication of DNA requires the action of a complex composed of many enzymes and proteins. As this complex moves along the DNA molecule, certain enzymes (*DNA helicases*) break the weak hydrogen bonds between complementary bases to produce two free strands at a fork in the double-stranded molecule. As a result, the bases of each of the two freed DNA strands can bond with new complementary bases (which are part of nucleotides) that are available in the surrounding environment.

Figure 3.26

The conversion of proinsulin into insulin.
The long polypeptide chain called proinsulin is converted into the active hormone insulin by enzymatic removal of a length of amino acids. The insulin molecule produced in this way consists of two polypeptide chains (colored circles) joined by disulfide bonds.

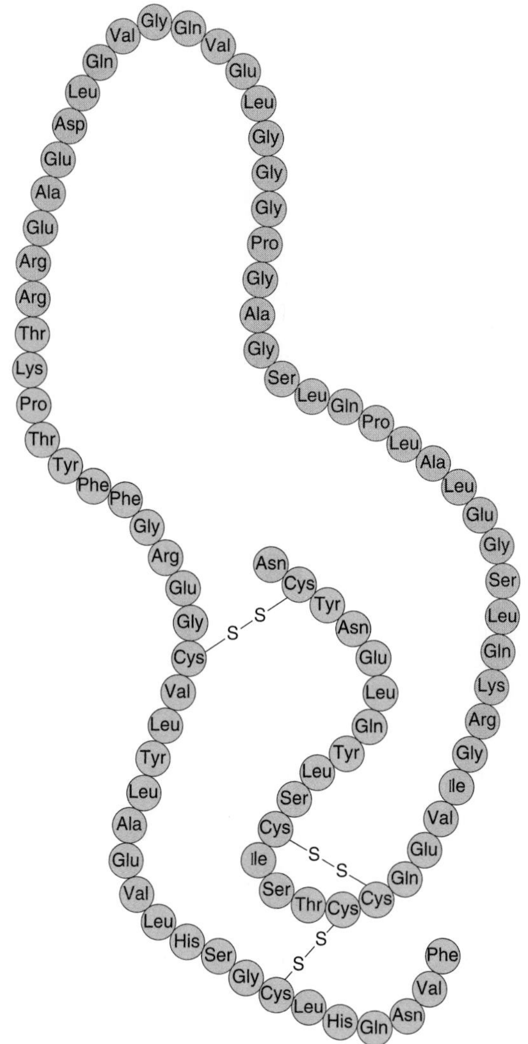

According to the rules of complementary base pairing, the bases of each original strand bond with the appropriate free nucleotides; adenine bases pair with thymine-containing nucleotides; guanine bases pair with cytosine-containing nucleotides; and so on. Enzymes called **DNA polymerases** join the nucleotides together to form a second polynucleotide chain in each DNA that is complementary to the first DNA strands. In this way, two new molecules of DNA, each containing two complementary strands, are formed. Thus, two new double-helix DNA molecules are produced that contain the same base sequence as the parent molecule (fig. 3.28).

When DNA replicates, therefore, each copy is composed of one new strand and one strand from the original DNA molecule. Replication is said to be **semiconservative** (half of the original DNA is "conserved" in each of the new DNA molecules). Through this mechanism, the sequence of bases in DNA—the basis of the genetic code—is preserved from one cell generation to the next.

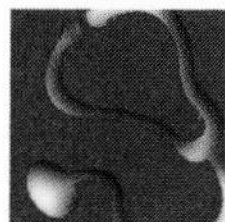

Advances in the identification of human genes, methods of cloning (replicating) isolated genes, and other technologies have made gene therapy a realistic possibility. The first federally approved human genetic engineering experiment began in 1989 when patients with advanced cancer were treated with special white blood cells from their own body. These "tumor infiltrating lymphocytes" had been given an exogenous (foreign) genetic marker. The exogenous gene didn't aid the treatment but served only to test whether or not the procedure was workable and safe. The first approved gene therapy of a disease began testing in 1990. This involved the attempted correction of a genetic defect for an enzyme called *adenosine deaminase (ADA)* that causes failure of the immune system. Clinical human trials of gene therapy for diseases of the blood-forming cells, liver, lungs, clotting system, and other diseases are currently in progress.

The Cell Cycle

Unlike the life of an organism, which can be pictured as a linear progression from birth to death, the life of a cell follows a cyclical pattern. Each cell is produced as a part of its "parent" cell; when the daughter cell divides, it in turn becomes two new cells. In a sense, then, each cell is potentially immortal as long as its progeny can continue to divide. Some cells in the body divide frequently; the epidermis of the skin, for example, is renewed approximately every 2 weeks, and the stomach lining is renewed about every 2 or 3 days. Other cells, however, such as nerve and striated muscle cells in the adult, do not divide at all. All cells in the body, of course, live only as long as the person lives (some cells live longer than others, but eventually all cells die when vital functions cease).

The nondividing cell is in a part of its life cycle known as **interphase** (fig. 3.29), which is subdivided into G_1, S, and G_2 phases, as will be described shortly. The chromosomes are in their extended form, and their genes actively direct the synthesis of RNA. Through their direction of RNA synthesis, genes control the metabolism of the cell. During this time the cell may be growing, and this part of interphase is known as the *G_1 phase*. Although sometimes described as "resting," cells in the G_1 phase perform the physiological functions characteristic of the tissue in which they are found. The DNA of resting cells in the G_1 phase thus produces mRNA and proteins as previously described.

If a cell is going to divide, it replicates its DNA in a part of interphase known as the *S phase* (S stands for *synthesis*). Once DNA has replicated in the S phase, the chromatin condenses in the *G_2 phase*, forming short, thick, rodlike structures by the end of G_2. This is the more familiar form of chromosomes because they are easily seen in the ordinary (light) microscope.

Figure 3.27

The Golgi complex.
(*a*) An electron micrograph of a Golgi complex. Notice the formation of vesicles at the ends of some of the flattened sacs. (*b*) An illustration of the processing of proteins by the rough endoplasmic reticulum and Golgi complex.

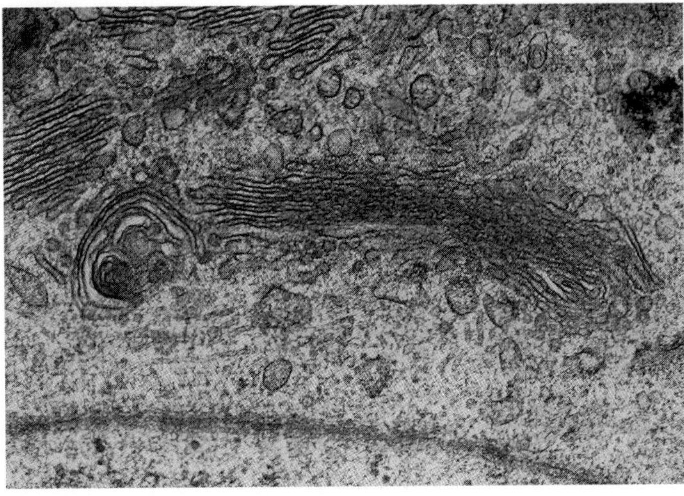
(a)

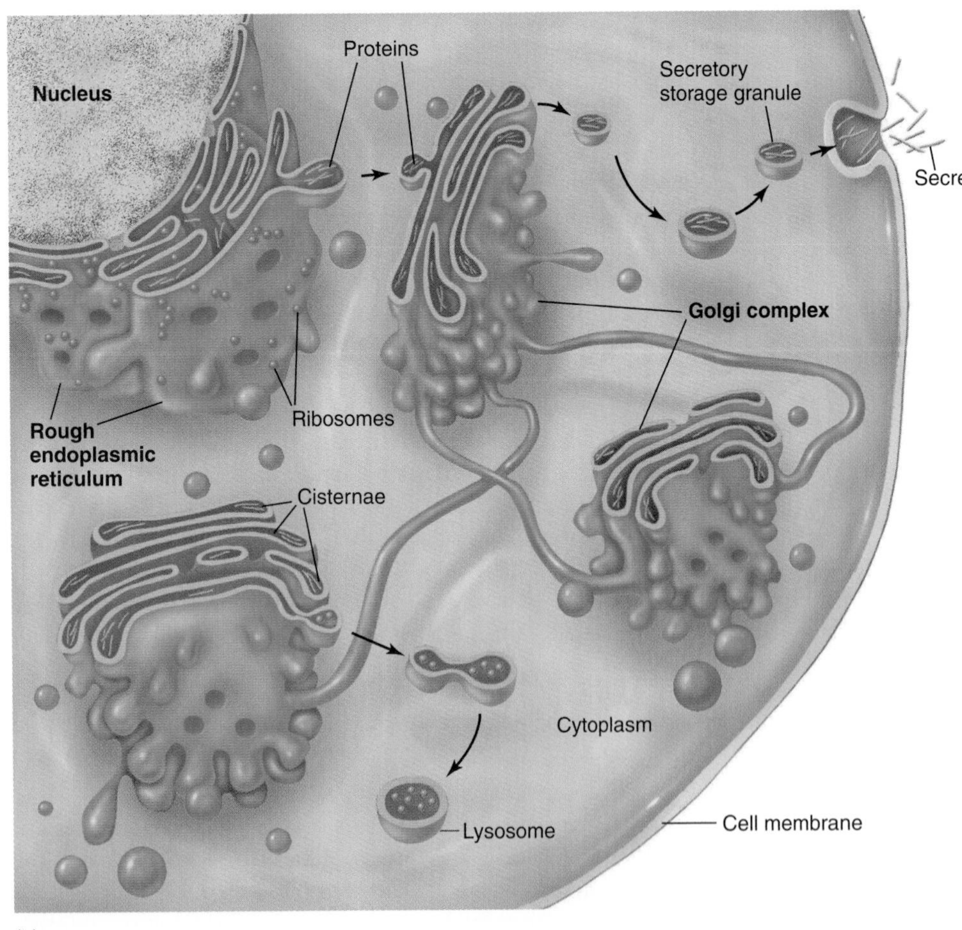

(b)

Cell Death

Cell death occurs both pathologically and naturally. Pathologically, cells deprived of a blood supply may swell, rupture their membranes, and burst. Such cellular death, leading to tissue death, is known as **necrosis** (*nĕ-kro'sis*). In certain cases, however, a different pattern is observed. Instead of swelling, the cells shrink. The membranes remain intact but become bubbled, and the nuclei condense. This pattern has been named **apoptosis** (*ap"o-to'sis* or *ă-pop-to'sis*), from a Greek term describing the shedding of leaves from a tree.

Apoptosis occurs normally as part of programmed cell death—a process described previously in the section on lysosomes. Programmed cell death refers to the physiological process responsible for the remodeling of tissues during embryonic development and for tissue turnover in the adult body. The epithelial cells lining the digestive tract, for example, are programmed to die 2–3 days after they are produced, and epidermal cells of the skin live only for about 2 weeks until they die and become completely cornified. Apoptosis is also important in the functioning of the immune system. A neutrophil (a type of white blood cell), for example, is programmed to die by apoptosis 24 hours after its creation in the bone marrow. A killer T lymphocyte (another type of white blood cell) destroys targeted cells by triggering their apoptosis.

Mitosis

At the end of the G_2 phase of the cell cycle, which is generally shorter than G_1, each chromosome consists of two strands called **chromatids** (*kro'mă-tidz*) that are joined together by a *centromere* (fig 3.30). The two chromatids within a chromosome contain identical DNA base sequences because each is produced by the semiconservative replication of DNA. Each chromatid, therefore, contains a complete double-helix DNA molecule that is a copy of the single DNA molecule existing prior to replication. Each chromatid will become a separate chromosome once mitotic cell division has been completed.

The G_2 phase completes interphase. The cell next proceeds through the various stages of cell division, or **mitosis** (*mi-to'sis*). This is the M *phase* of the cell cycle. Mitosis is subdivided into four stages:

mitosis: Gk. *mitos*, thread

Figure 3.28

The replication of DNA.
Each new double helix is composed of one old and one new strand. The base sequence of each of the new molecules is identical to that of the parent DNA because of complementary base pairing.

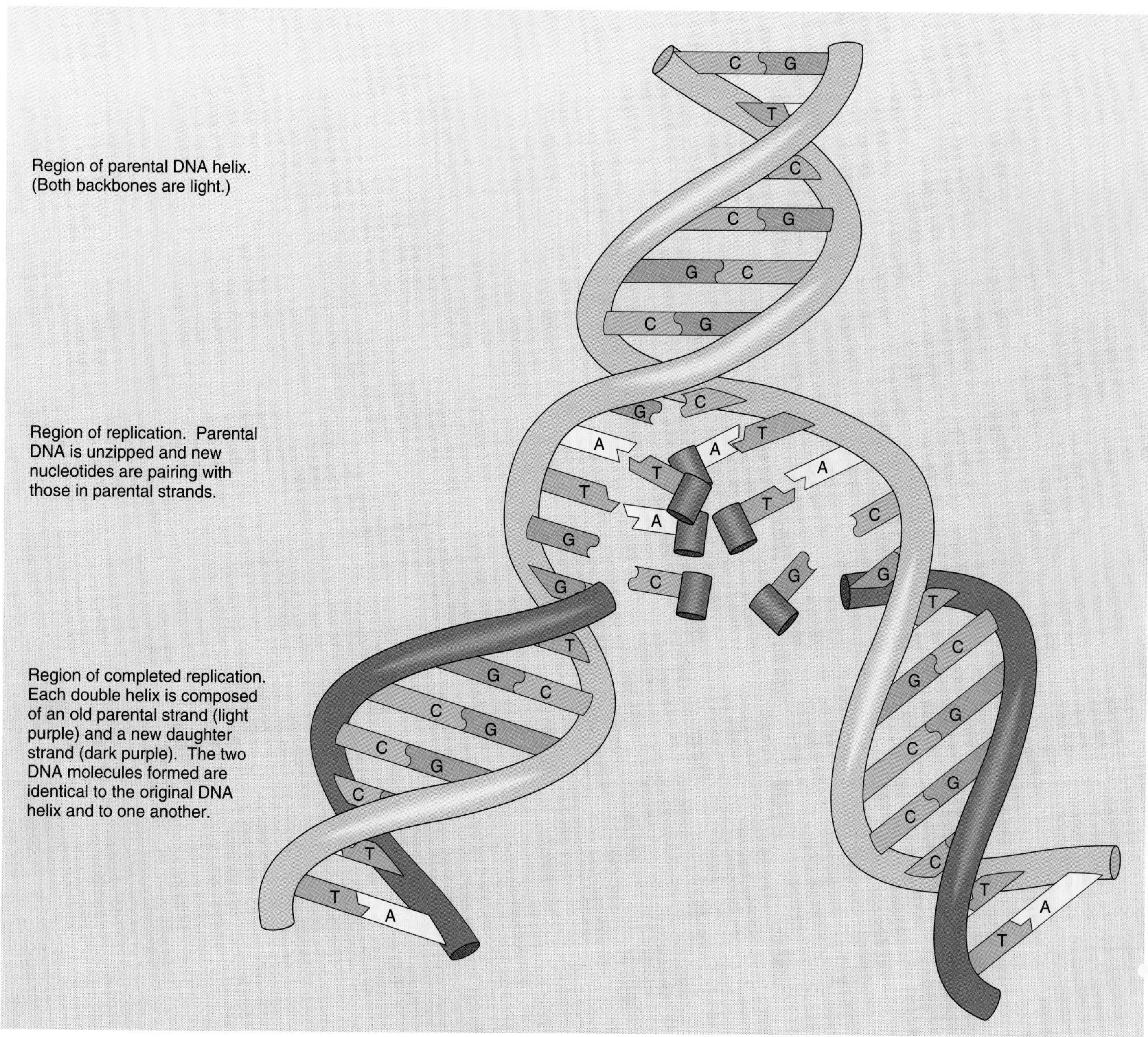
Region of parental DNA helix. (Both backbones are light.)
Region of replication. Parental DNA is unzipped and new nucleotides are pairing with those in parental strands.
Region of completed replication. Each double helix is composed of an old parental strand (light purple) and a new daughter strand (dark purple). The two DNA molecules formed are identical to the original DNA helix and to one another.

prophase, *metaphase*, *anaphase*, and *telophase* (fig. 3.31). In prophase, chromosomes become visible as distinctive structures. In metaphase of mitosis, the chromosomes line up single file along the equator of the cell. This aligning of chromosomes at the equator is believed to result from the action of **spindle fibers,** which are attached to a protein structure called the *kinetochore* (*kĭ-net'ŏ-kor*) at the centromere of each chromosome (fig. 3.31).

Anaphase begins when the centromeres split apart and the spindle fibers shorten, pulling the two chromatids in each chromosome to opposite poles. Each pole therefore gets one copy of each of the 46 chromosomes. During early telophase, division of the cytoplasm, called *cytokinesis* (*si"to-kĭ-ne'sis*), results in the production of two daughter cells that are genetically identical to each other and to the original parent cell.

Figure 3.29

The life cycle of a cell.
The different stages of mitotic division are shown, but it should be noted that not all cells undergo mitosis.

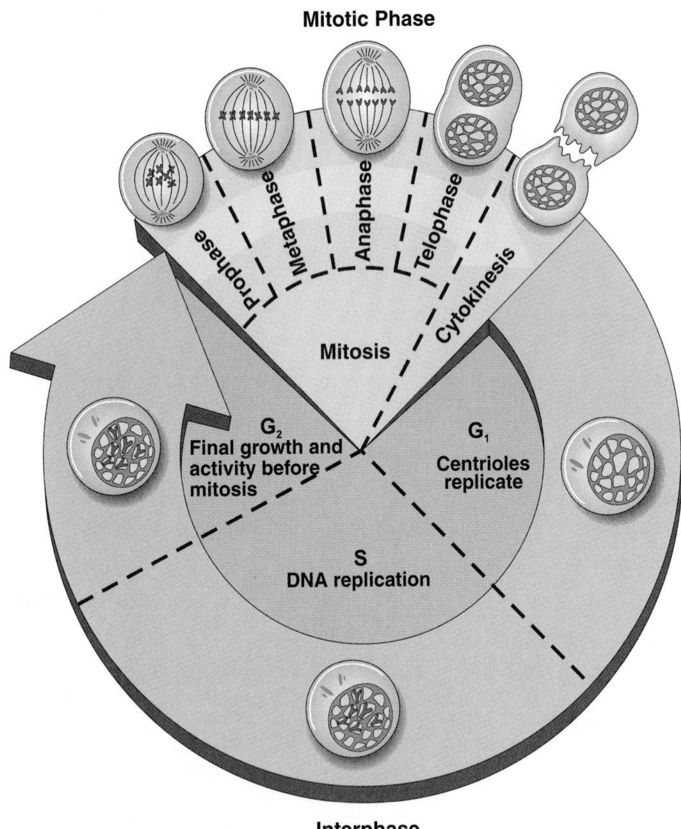

Figure 3.30

The structure of a chromosome after DNA replication.
At this stage, a chromosome consists of two identical strands, or chromatids.

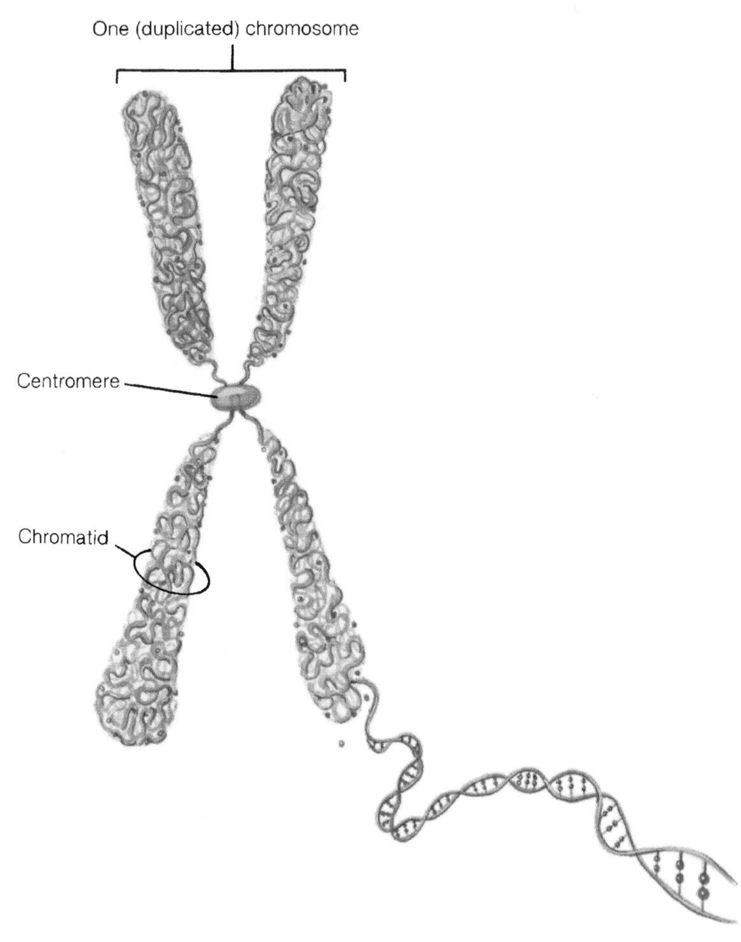

Role of the Centrosome

All animal cells have a **centrosome,** located near the nucleus in a nondividing cell. At the center of the centrosome are two **centrioles** (*sen'trĭ-ōlz*), which are positioned at right angles to each other. Each centriole is composed of nine evenly spaced bundles of microtubules, with three microtubules per bundle (fig. 3.32). Surrounding the two centrioles is an amorphous mass of material called the *pericentriolar material.* Microtubules grow out of the pericentriolar material, which is believed to function as the center for the organization of microtubules in the cytoskeleton.

Through a mechanism that is still not understood, the centrosome replicates itself during interphase if a cell is going to divide. The two identical centrosomes then move away from each other during prophase of mitosis and take up positions at opposite poles of the cell by metaphase. At this time, the centrosomes produce new microtubules. These new microtubules are very dynamic, rapidly growing and shrinking as if they were "feeling out" randomly for chromosomes. A microtubule becomes stabilized when it finally binds to the proper region of a chromosome. In this way, the microtubules from both centrosomes form the spindle fibers that are attached to each of the replicated chromosomes at metaphase.

The spindle fibers pull the chromosomes to opposite poles of the cell during anaphase, so that at telophase, when the cell pinches inward, two identical daughter cells will be produced. This also requires the centrosomes, which somehow organize a ring of contractile filaments halfway between the two poles. These filaments are attached to the cell membrane, and when they contract, the cell is pinched in two. The filaments consist of actin and myosin proteins, the same contractile proteins present in muscle.

Hypertrophy and Hyperplasia

The growth of an individual from a fertilized egg into an adult involves an increase in the number of cells and an increase in the size of cells. Growth that is due to an increase in cell number results from an increased rate of mitotic cell division and is termed **hyperplasia** (*hi"per-pla'ze-ă*). Growth of a tissue or organ that is due to an increase in cell size is termed **hypertrophy** (*hi"per'trŏ-fe*).

Most growth is due to hyperplasia. A callus on the palm of the hand, for example, involves thickening of the skin by

Figure 3.31 The stages of mitosis.

(a) **Interphase**

- The chromosomes are in an extended form and seen as chromatin in the electron microscope.
- The nucleus is visible.

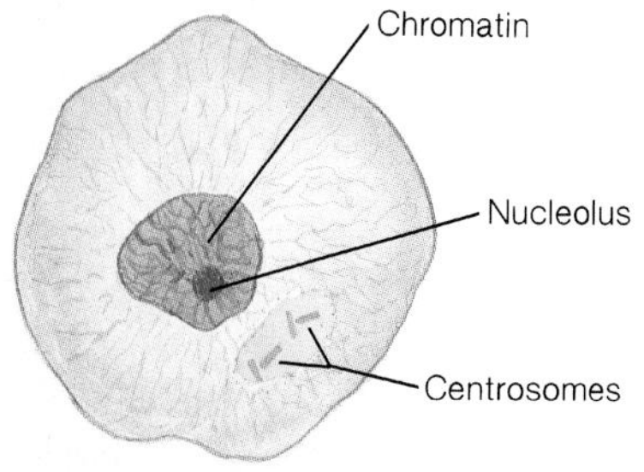

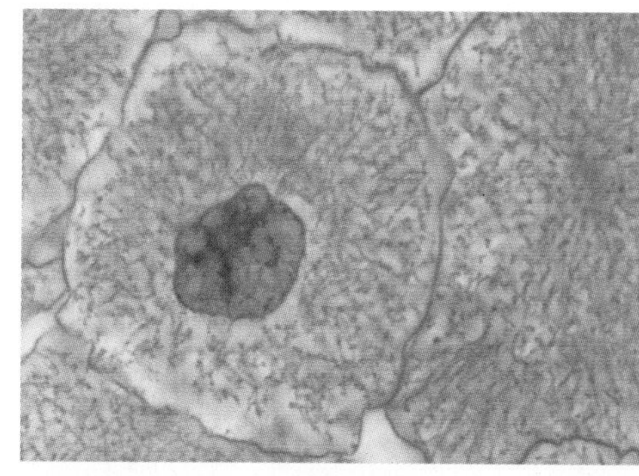

(b) **Prophase**

- The chromosomes are seen to consist of two chromatids joined by a centromere.
- The centrioles move apart toward opposite poles of the cell.
- Spindle fibers are produced and extended from each centrosome.
- The nuclear membrane starts to disappear.
- The nucleolus is no longer visible.

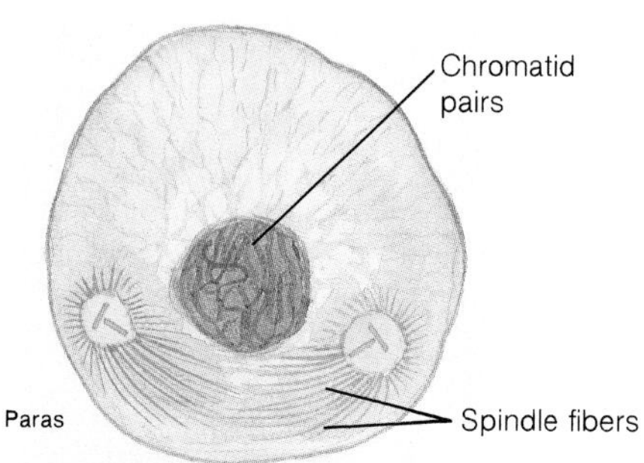

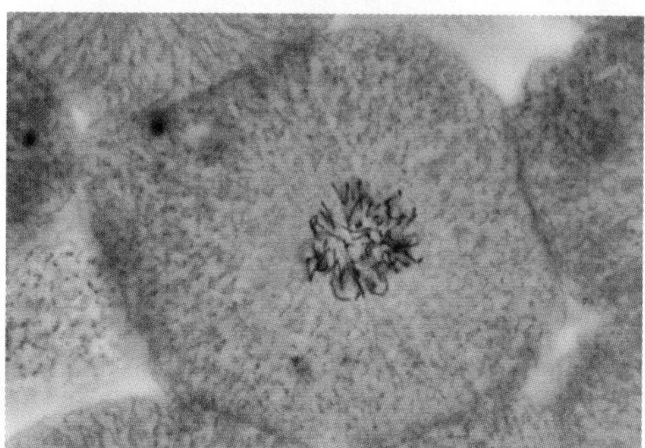

(c) **Metaphase**

- The chromosomes are lined up at the equator of the cell.
- The spindle fibers from each centriole are attached to the centromeres of the chromosomes.
- The nuclear membrane has disappeared.

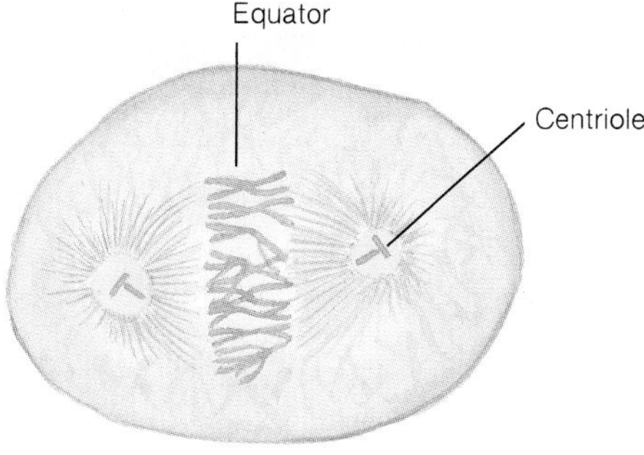

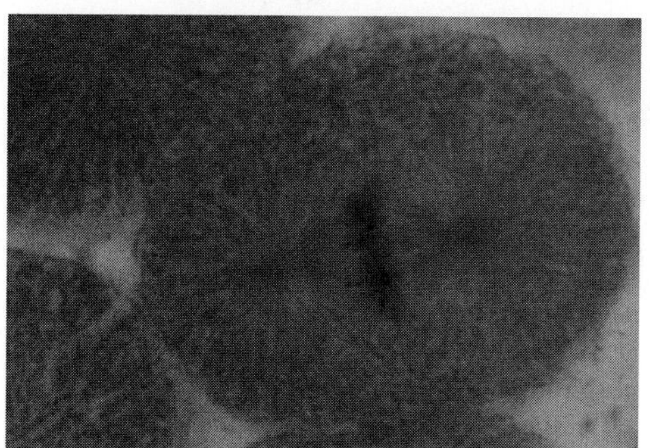

(d) **Anaphase**

- The centromeres split, and the sister chromatids separate as each is pulled to an opposite pole.

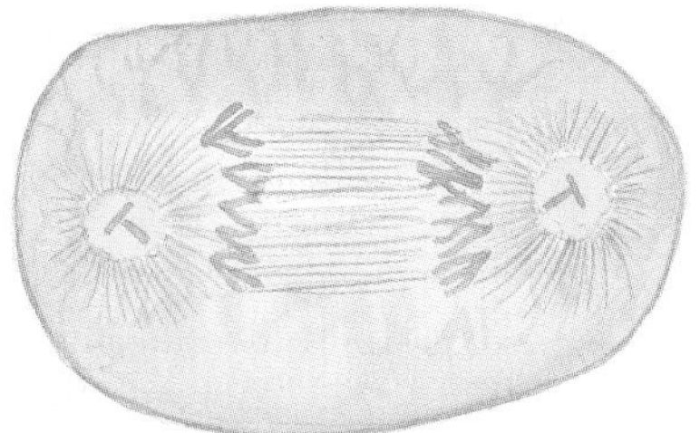

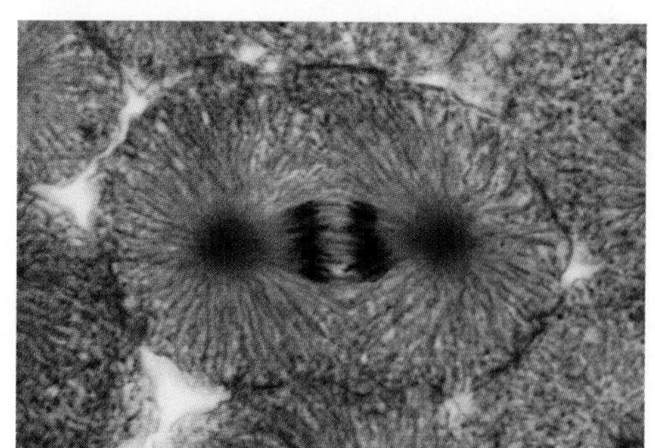

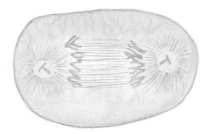

(e) **Telophase**

- The chromosomes become longer, thinner, and less distant.
- New nuclear membranes form.
- The nucleolus reappears.
- Cell division is nearly complete.

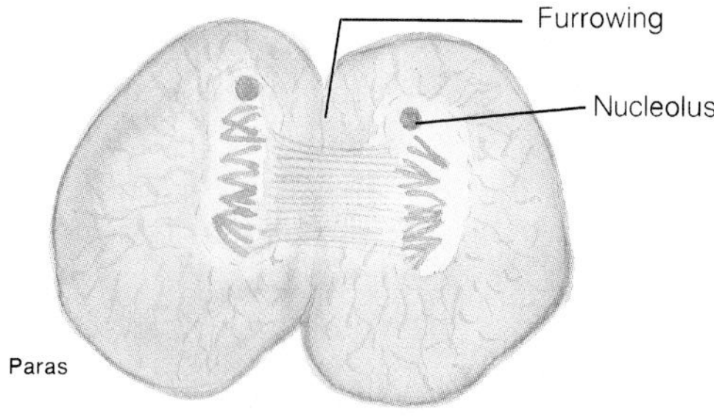

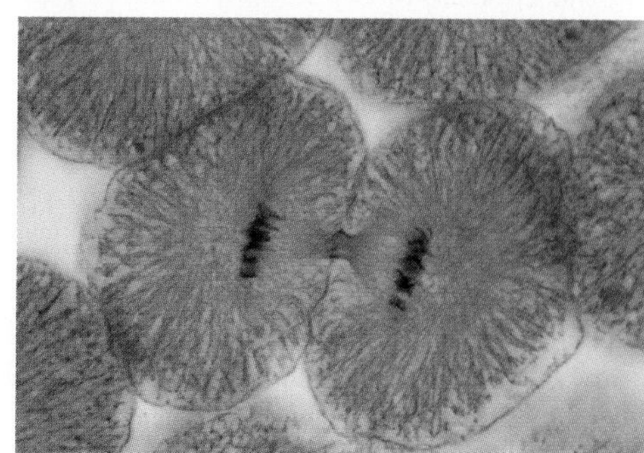

Figure 3.32

The centrioles.
(*a*) A micrograph of the two centrioles in a centrosome (14,200×).
(*b*) A diagram showing that the centrioles are positioned at right angles to each other.

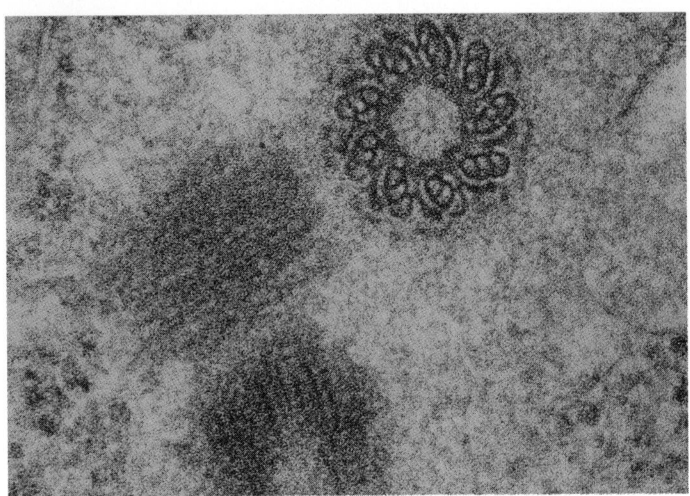

(a)

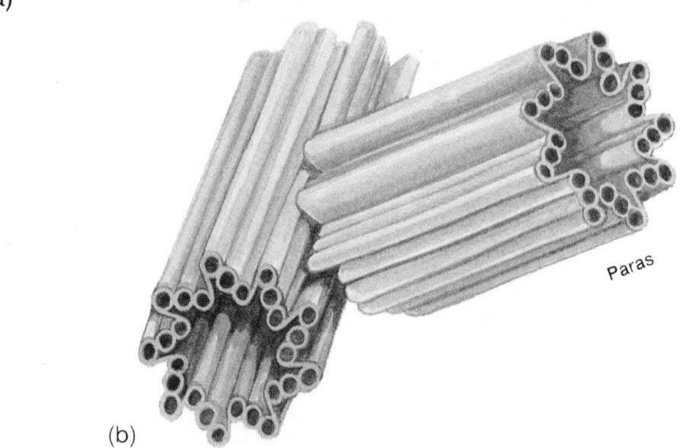

(b)

Figure 3.33

Chromosomes arranged in homologous pairs.
A false-color light micrograph of chromosomes from a male arranged in numbered homologous pairs.

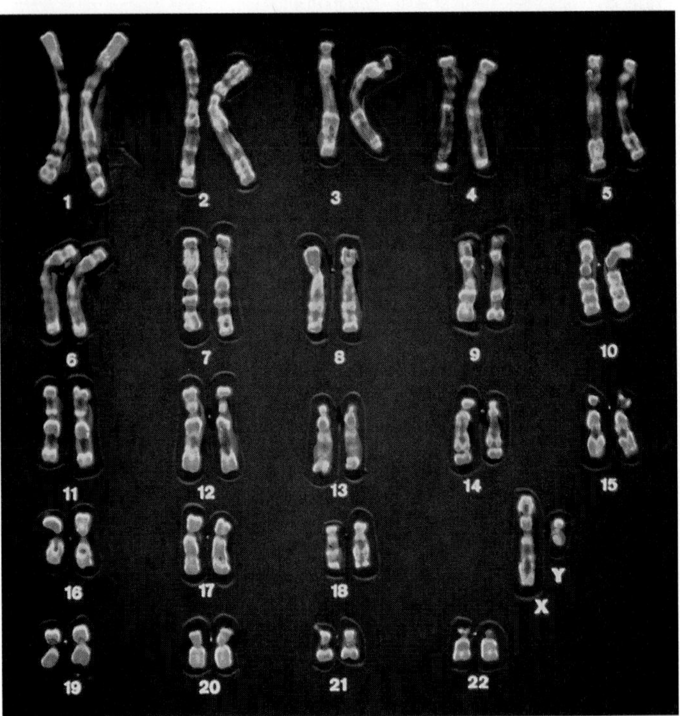

hyperplasia as a result of frequent abrasion. An increase in skeletal muscle size as a result of exercise, by contrast, is produced by hypertrophy.

Skeletal muscle and cardiac (heart) muscle can grow only by hypertrophy. When growth occurs in skeletal muscles in response to an increased workload—during weight training, for example—it is called *compensatory hypertrophy.* The heart muscle may also demonstrate compensatory hypertrophy when its workload increases because of hypertension (high blood pressure). The opposite of hypertrophy is *atrophy* (*at′rŏ-fe*), the wasting or decrease in size of a cell, tissue, or organ. This may result from the disuse of skeletal muscles, as occurs in prolonged bed rest, various diseases, or advanced age.

Meiosis

When a cell is going to divide, either by mitosis or meiosis, the DNA is replicated (forming chromatids) and the chromosomes become shorter and thicker, as previously described. At this point the cell has 46 chromosomes, each of which consists of two duplicate chromatids.

The short, thick chromosomes seen at the end of the G_2 phase can be matched as pairs, the members of each pair appearing to be structurally identical. These matched chromosomes are called **homologous** (*hŏ-mol′ŏ-gus*) **chromosomes.** One member of each homologous pair is derived from a chromosome inherited from the father, and the other member is a copy of one of the chromosomes inherited from the mother. Homologous chromosomes do not have identical DNA base sequences; one member of the pair may code for blue eyes, for example, and the other for brown eyes. There are 22 homologous pairs of *autosomal* (*aw″to-so′mal*) *chromosomes* and one pair of *sex chromosomes*, described as X and Y. Females have two X chromosomes, whereas males have one X and one Y chromosome (fig. 3.33).

Meiosis (*mi-o′sis*), which has two divisional sequences, is a special type of cell division that occurs only in the gonads (testes and ovaries), where it is used only in the production of gametes—sperm cells and ova. (Gamete production is described in detail in chapters 28 and 29.) In the first division of meiosis, the homologous chromosomes line up side by side, rather than single file, along the equator of the cell. The spindle fibers then pull one member of a homologous pair to one pole of the cell, and the other member of the pair to the other pole. Each of the two daughter cells thus acquires only one chromosome from each of the 23 homologous pairs contained in the parent. The daughter cells,

meiosis: Gk. *meioun*, lessen

Table 3.3
Stages of Meiosis

Stage	Events
First Meiotic Division	
Prophase I	Chromosomes appear double-stranded. Each strand, called a chromatid, contains duplicate DNA joined together by a structure known as a centromere. Homologous chromosomes pair up side by side.
Metaphase I	Homologous chromosome pairs line up at the equator. Spindle apparatus is complete.
Anaphase I	Homologous chromosomes separate; the members of a homologous pair move to opposite poles.
Telophase I	Cytoplasm divides to produce two haploid cells.
Second Meiotic Division	
Prophase II	Chromosomes appear, each containing two chromatids.
Metaphase II	Chromosomes line up single file along the equator as spindle formation is completed.
Anaphase II	Centromeres split and chromatids move to opposite poles.
Telophase II	Cytoplasm divides to produce two haploid cells from each of the haploid cells formed at telophase I.

in other words, contain 23 rather than 46 chromosomes. For this reason meiosis is also known as **reduction division.**

At the end of this cell division, each daughter cell contains 23 chromosomes—but *each of these consists of two chromatids*. (Since the two chromatids per chromosome are identical, this does not make 46 chromosomes; there are still only 23 *different* chromosomes per cell at this point.) The chromatids are separated by a second meiotic division. Each of the daughter cells from the first cell division itself divides, with the duplicate chromatids going to each of two new daughter cells. A grand total of four daughter cells can thus be produced from the meiotic cell division of one parent cell. This occurs in the testes, where one parent cell produces four sperm cells. In the ovaries, one parent cell also produces four daughter cells, but three of these die and only one progresses to become a mature egg cell (as will be described in chapter 29).

The stages of meiosis are subdivided according to whether they occur in the first or the second meiotic cell division. These stages are designated as prophase I, metaphase I, anaphase I, telophase I; and then prophase II, metaphase II, anaphase II, and telophase II (table 3.3 and fig. 3.34).

The reduction of the chromosome number from 46 to 23 is obviously necessary for sexual reproduction, where the sex cells join and add their content of chromosomes together to produce a new individual. The significance of meiosis, however, goes beyond the reduction of chromosome number. At metaphase I, the pairs of homologous chromosomes can line up with either member facing a given pole of the cell. (Recall that each member of a homologous pair came from a different parent.) Maternal and paternal members of homologous pairs are thus randomly shuffled. Hence, when the first meiotic division occurs, each daughter cell will obtain a complement of 23 chromosomes that are randomly derived from the maternal or paternal contribution to the homologous pairs of chromosomes of the parent cell.

In addition to this "shuffling of the deck" of chromosomes, exchanges of parts of homologous chromosomes can occur at prophase I. That is, pieces of one chromosome of a homologous pair can be exchanged with the other homologous chromosome in a process called *crossing-over* (fig. 3.35). These events together result in **genetic recombination** and ensure that the gametes produced by meiosis are genetically unique. This provides additional genetic diversity for organisms that reproduce sexually, and genetic diversity is needed to promote survival of species over evolutionary time.

Clinical Considerations

Some Functions of Lysosomes and the Smooth Endoplasmic Reticulum

Lysosomes

Most, if not all, molecules in the cell have a limited life span. They are continuously destroyed and must be continuously replaced. Glycogen and some complex lipids in the brain, for example, are digested at a particular rate by lysosomes. If a person does not have the proper amount of lysosomal enzymes because of some genetic defect, the resulting abnormal accumulation of glycogen and lipids could destroy the tissues. Examples of such diseases include *glycogen storage disease*, *Tay–Sach's disease*, and *Gaucher's disease*.

Smooth Endoplasmic Reticulum

The smooth endoplasmic reticulum in liver cells and other cells contains enzymes used for the inactivation of steroid hormones and many toxic compounds. This inactivation is generally achieved by reactions that convert these compounds to forms that are more water-soluble and less active, and thus more easily excreted by the kidneys. When people take certain drugs (such as alcohol and phenobarbital) for a long period of time, increasingly larger doses are required to produce the effect achieved initially. This phenomenon, called *drug tolerance*, is accompanied by an increase in the smooth endoplasmic reticulum, and thus an increase in the enzymes charged with inactivation of these drugs.

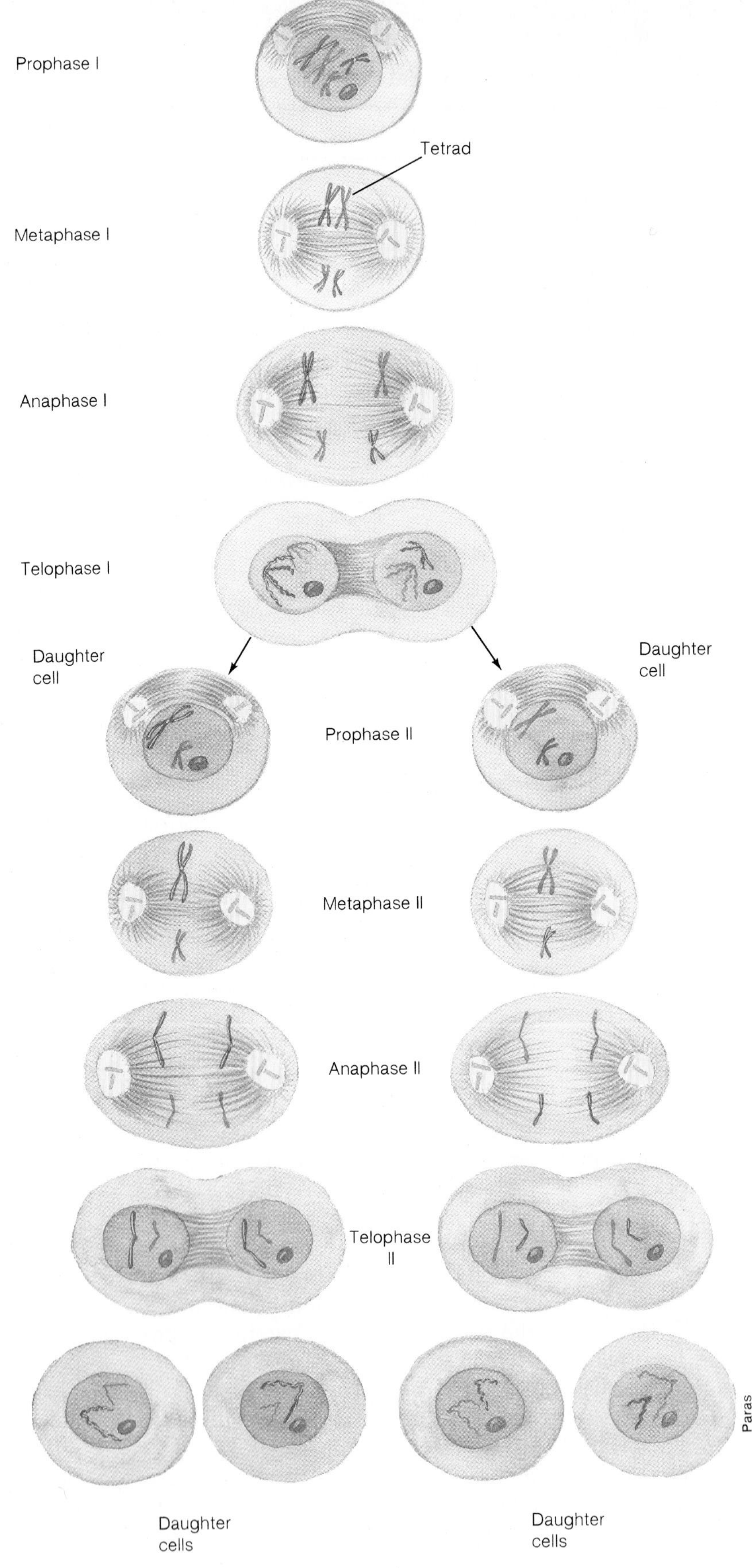

Figure 3.34

Meiosis, or reduction division.
In the first meiotic division, the homologous chromosomes of a diploid parent cell are separated into two haploid daughter cells. Each of these chromosomes contains duplicate strands, or chromatids. In the second meiotic division, these chromosomes are distributed to two new haploid daughter cells.

The Cell Cycle and Cancer

Cyclins and p53

A group of proteins known as the **cyclins** (*si'klinz*) promote different phases of the cell cycle. The concentration of **cyclin D** proteins within the cell, for example, rises during the G_1 phase of the cycle and acts to move the cell quickly through this phase. Cyclin D proteins do this by activating a group of otherwise inactive enzymes known as *cyclin-dependant kinases*. Therefore, overactivity of a gene that codes for a cyclin D might be predicted to cause uncontrolled cell division, as occurs in a cancer. Indeed, overexpression of the gene for cyclin D has been shown to occur in some cancers, including those of the breast and esophagus. Genes that contribute to cancer are called **oncogenes** (*ong'kŏ-jēnz*).

Although oncogenes promote cancer, other genes—called **tumor suppressor genes**—inhibit its development. One very important tumor suppressor gene is known as **p53.** This name refers to the protein coded by the gene, which has a molecular weight of 53,000. The normal gene protects against cancer by indirectly blocking the ability of cyclins to stimulate cell division. In part, p53 accomplishes this by inducing the expression of another gene, called *p21*, which produces a protein that binds to and inactivates the cyclin-dependant kinases. The p21 protein thus inhibits cell division as it promotes cell differentiation (specialization). For these reasons, cancer is likely to develop if the p53 gene becomes mutated and thus ineffective as a tumor suppressor gene. Indeed, mutated p53 genes are found in over 50% of all cancers. Mice whose p53 genes were "knocked out" through an exciting new technology in genetic engineering all developed tumors. (*Knockout mice* are strains of mice in which a specific targeted gene has been inactivated by developing the mice from embryos injected with specifically mutated cells.) These important discoveries have obvious relevance to cancer diagnosis and treatment.

Skin Cancer

Using mice with their gene for p53 knocked out, scientists have learned that p53 is also needed for the apoptosis that occurs when a cell's DNA is damaged. The damaged DNA, if it is not repaired, activates

Figure 3.35

Crossing-over.
(*a*) Genetic variation results from the crossing-over of tetrads, which occurs during the first meiotic prophase. (*b*) A diagram depicting the recombination of chromosomes that occurs as a result of crossing-over.

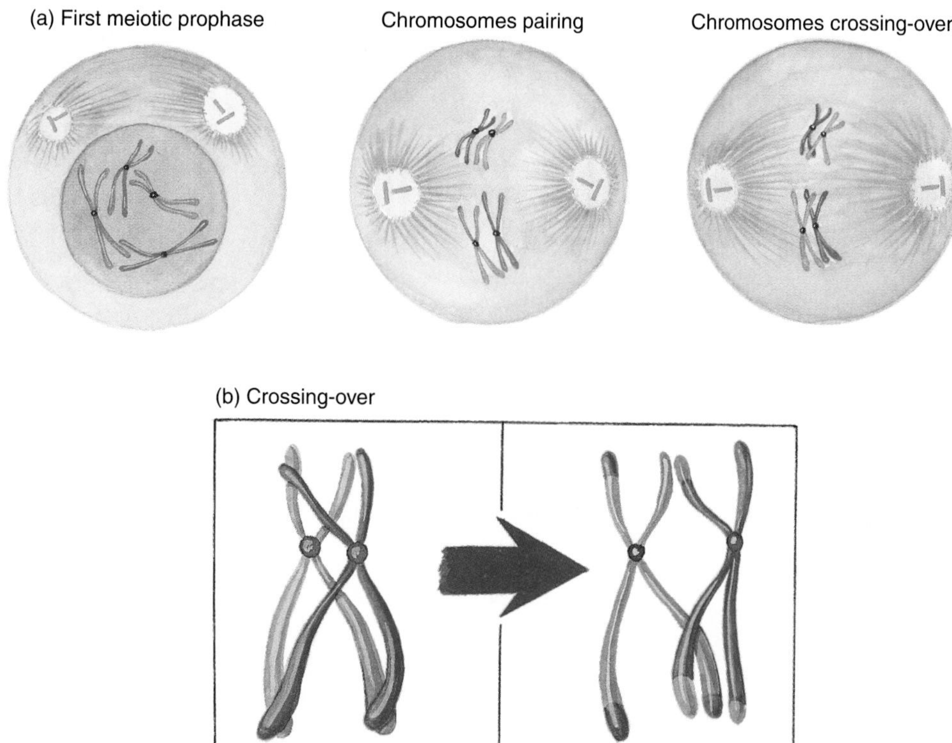

p53, which in turn causes the cell to be destroyed. If the p53 gene has mutated to an ineffective form, however, the cell will not be destroyed by apoptosis as it should; rather, it will divide to produce daughter cells with damaged DNA. This may be one mechanism responsible for the development of a cancer.

As described in chapter 7, there are three forms of **skin cancer**—squamous cell carcinoma, basal cell carcinoma, and melanoma—all of which are promoted by the damaging effects of the ultraviolet portion of sunlight. Ultraviolet light promotes a characteristic type of DNA mutation in which either of two pyrimidines (cytosine or thymine) is affected. In squamous cell and basal cell carcinoma (but not melanoma), the cancer is believed to involve mutations that affect the p53 gene, among others. Whereas cells with normal p53 genes may die by apoptosis when their DNA is damaged, and are thus prevented from replicating themselves and perpetuating the damaged DNA, those cells with a mutated p53 gene survive and divide, producing the cancer.

Telomeres, Aging, and Cancer

Certain types of cells can be removed from the body and grown in nutrient solutions (outside the body, or *in vitro*). Under these artificial conditions, the potential longevity of different cell lines can be studied. For unknown reasons, normal connective tissue cells (called fibroblasts) stop dividing in vitro after a certain number of population doublings. Cells from a newborn will divide 80–90 times, whereas those from a 70-year-old will stop after 20–30 divisions. The decreased ability to divide is thus an indicator of *senescence* (aging). Cells that become transformed into cancer, however, apparently do not age and continue dividing indefinitely in culture.

This senescent decrease in the ability of cells to replicate may be related to a loss of DNA sequences at the ends of chromosomes, in regions called **telomeres** (*tel'ŏ-mērz*). Scientists discovered that DNA polymerase copies only part of the DNA sequence in the end-regions. Each time a chromosome replicates, it loses 50–100 base pairs in its telomeres. Cell division may ultimately stop when there is too much loss of DNA in the telomeres, and the cell dies because of damage it sustains in the course of aging.

Germinal cells that give rise to gametes (sperm cells and ova) can continue to divide indefinitely. This may be because they produce an enzyme called **telomerase** (*tĕ-lom'er-ās*), which duplicates the telomere DNA. Telomerase is also found in hematopoietic stem cells (those in bone marrow that produce blood cells) and other stem cells that must divide continuously. Similarly, telomerase is produced by cancer cells and may be responsible for their ability to divide indefinitely.

---

in vitro: L. *in vitro*, in a glass

telomere: Gk. *telos*, end

Clinical Investigation Answer

The substance abuse could have resulted in the development of an extensive smooth endoplasmic reticulum, which contains many of the enzymes required to metabolize drugs. Liver disease could have been caused by the substance abuse, but there is an alternative explanation. The deficiency of the enzyme that breaks down glycogen signals the presence of glycogen storage disease, a genetic condition in which a key lysosomal enzyme is lacking. This enzymatic evidence is supported by the observations of large amounts of glycogen granules and the lack of partially digested glycogen granules within secondary lysosomes. (In reality, such a genetic condition would more likely be diagnosed in early childhood.)

Chapter Summary

Cell Membrane and Associated Structures (pp. 46–52)

1. The structure of the cell, or plasma, membrane is described by a fluid-mosaic model.
   (a) The membrane is predominately composed of a double layer of phospholipids.
   (b) The membrane also contains proteins, some of which span its entire width.
2. Some cells move by extending pseudopods. Cilia and flagella protrude from the cell membrane of some specialized cells.
3. In the process of endocytosis, invaginations of the cell membrane allow the cells to take up molecules from the external environment.
   (a) In phagocytosis, the cell extends pseudopods that eventually fuse together to create a food vacuole; pinocytosis involves the formation of a narrow furrow in the membrane that eventually fuses.
   (b) Receptor-mediated endocytosis requires the interaction of a specific molecule in the extracellular environment with a specific receptor protein in the cell membrane.
   (c) Exocytosis, the reverse of endocytosis, is a process that allows the cell to secrete its products.

Cytoplasm and Its Organelles (pp. 52–55)

1. Microfilaments and microtubules produce a cytoskeleton, which aids movements of organelles within a cell.
2. Lysosomes contain digestive enzymes and are responsible for the elimination of structures and molecules within the cell and for digestion of the contents of phagocytic food vacuoles.
3. Mitochondria serve as the major sites for energy production in the cell. They have an outer membrane with a smooth contour and an inner membrane with numerous infoldings called cristae.
4. The endoplasmic reticulum is a system of membranous tubules in the cell.
   (a) The rough endoplasmic reticulum is covered with ribosomes and is involved in protein synthesis.
   (b) The smooth endoplasmic reticulum provides a site for many enzymatic reactions and, in skeletal muscles, serves to store calcium ions.

Cell Nucleus and Nucleic Acids (pp. 55–60)

1. The cell nucleus is surrounded by a double-layered nuclear membrane. At some points, the two layers are fused by nuclear pore complexes that allow for passage of molecules.
2. Nucleic acids include DNA, RNA, and their nucleotide subunits.
   (a) The DNA nucleotides contain the sugar deoxyribose, whereas the RNA nucleotides contain the sugar ribose.
   (b) There are four different types of DNA nucleotides that contain one of four possible bases: adenine, guanine, cytosine, and thymine. In RNA, the base uracil substitutes for the base thymine.
   (c) DNA consists of two long polynucleotide strands twisted into a double helix. The two strands are held together by hydrogen bonds between specific bases—adenine pairs with thymine, and guanine pairs with cytosine.
   (d) RNA is single-stranded. Four types are produced within the nucleus: ribosomal RNA, transfer RNA, precursor messenger RNA, and messenger RNA.
3. Active euchromatin directs the synthesis of RNA in a process called transcription.
   (a) The enzyme RNA polymerase causes separation of the two strands of DNA along the region of the DNA that constitutes a gene.
   (b) One of the two separated strands of DNA serves as a template for the production of RNA. This occurs by complementary base pairing between the DNA bases and ribonucleotide bases.

Protein Synthesis and Secretion (pp. 60–64)

1. Messenger RNA leaves the nucleus and attaches to the ribosomes.
2. Each transfer RNA, with a specific base triplet in its anticodon, bonds to a specific amino acid.
   (a) As the mRNA moves through the ribosomes, complementary base pairing between tRNA anticodons and mRNA codons occurs.
   (b) As each successive tRNA molecule bonds with its complementary codon, the amino acid it carries is added to the end of a growing polypeptide chain.
3. Proteins destined for secretion are produced in ribosomes located in the rough endoplasmic reticulum and enter the cisternae of this organelle.
4. Secretory proteins move from the rough endoplasmic reticulum to the Golgi complex, which consists of a stack of membranous sacs.
   (a) The Golgi complex modifies the proteins it contains, separates different proteins, and packages them in vesicles.
   (b) Secretory vesicles from the Golgi complex fuse with the cell membrane and release their products by exocytosis.

DNA Synthesis and Cell Division (pp. 64–71)

1. Replication of DNA is semiconservative. Each DNA strand serves as a template for the production of a new strand.
   (a) The strands of the original DNA molecule gradually separate along their entire length and, through complementary base pairing, form a new complementary strand.
   (b) In this way, each DNA molecule consists of one old and one new strand.
2. During the G_1 phase of the cell cycle, the DNA directs the synthesis of RNA, and hence that of proteins.

3. During the S phase of the cycle, DNA directs the synthesis of new DNA and replicates itself.
4. After a brief rest (G_2), the cell begins mitosis (the M stage of the cycle).
   (a) Mitosis consists of the following phases: prophase, metaphase, anaphase, and telophase.
   (b) In mitosis, the homologous chromosomes line up single file and are pulled by spindle fibers to opposite poles.
   (c) This results in the production of two daughter cells that each contain 46 chromosomes, just like the parent cell.
5. Meiosis is a special type of cell division that results in the production of gametes in the gonads.
   (a) The homologous chromosomes line up side by side, so that only one of each pair is pulled to each pole.
   (b) This results in the production of two daughter cells, each containing only 23 chromosomes.
   (c) The duplicate chromatids in each of the 23 chromosomes go to each of two new daughter cells in the second meiotic cell division.

Review Activities

Objective Questions

1. According to the fluid-mosaic model of the cell membrane,
   (a) protein and phospholipids form a regular, repeating structure.
   (b) the membrane is a rigid structure.
   (c) phospholipids form a double layer, with the polar parts facing each other.
   (d) proteins are free to move within a double layer of phospholipids.
2. After the DNA molecule has replicated itself, the duplicate strands are called
   (a) homologous chromosomes.
   (b) chromatids.
   (c) centromeres.
   (d) spindle fibers.
3. Nerve and skeletal muscle cells in the adult, which do not divide, remain in the
   (a) G_1 phase.
   (b) S phase.
   (c) G_2 phase.
   (d) M phase.
4. The phase of mitosis in which the chromosomes line up at the equator of the cell is called
   (a) interphase.
   (b) prophase.
   (c) metaphase.
   (d) anaphase.
   (e) telophase.
5. The phase of mitosis in which the chromatids separate is called
   (a) interphase.
   (b) prophase.
   (c) metaphase.
   (d) anaphase.
   (e) telophase.
6. The RNA nucleotide base that pairs with adenine in DNA is
   (a) thymine.
   (b) uracil.
   (c) guanine.
   (d) cytosine.
7. Which of the following statements about RNA is *true?*
   (a) It is made in the nucleus.
   (b) It is double-stranded.
   (c) It contains the sugar deoxyribose.
   (d) It is a complementary copy of the entire DNA molecule.
8. Which of the following statements about mRNA is *false?*
   (a) It is produced as a larger pre-mRNA.
   (b) It forms associations with ribosomes.
   (c) Its base triplets are called anticodons.
   (d) It codes for the synthesis of specific proteins.
9. The organelle that combines proteins with carbohydrates and packages them within vesicles for secretion is
   (a) the Golgi complex.
   (b) the rough endoplasmic reticulum.
   (c) the smooth endoplasmic reticulum.
   (d) the ribosome.
10. The organelle that contains digestive enzymes is
    (a) the mitochondrion.
    (b) the lysosome.
    (c) the endoplasmic reticulum.
    (d) the Golgi complex.
11. If four bases in one DNA strand are A (adenine), G (guanine), C (cytosine), and T (thymine), the complementary bases in the RNA strand made from this region are
    (a) T,C,G,A.
    (b) C,G,A,U.
    (c) A,G,C,U.
    (d) U,C,G,A.
12. Which of the following statements about tRNA is *true?*
    (a) It is made in the nucleus.
    (b) It is looped back on itself.
    (c) It contains the anticodon.
    (d) There are over 20 different types.
    (e) All of the above are true.
13. The step in protein synthesis during which tRNA, rRNA, and mRNA are all active is known as
    (a) transcription.
    (b) translation.
    (c) replication.
    (d) RNA polymerization.
14. The anticodons are located in
    (a) tRNA.
    (b) rRNA.
    (c) mRNA.
    (d) ribosomes.
    (e) endoplasmic reticulum.

Essay Questions

1. Give some specific examples that illustrate the dynamic nature of the cell membrane.
2. Explain how one DNA molecule serves as a template for the formation of another DNA and why DNA synthesis is said to be semiconservative.
3. What is the genetic code, and how does it affect the structure and function of the body?
4. Why may tRNA be considered the "interpreter" of the genetic code?
5. Compare the processing of cellular proteins with that of proteins secreted by a cell.
6. Explain the interrelationship between the endoplasmic reticulum and the Golgi complex. What becomes of vesicles released from the Golgi complex?
7. Explain the functions of centrioles in nondividing and dividing cells.
8. Describe the phases of the cell cycle and explain how this cycle may be regulated.
9. Distinguish between oncogenes and tumor suppressor genes and give examples of how such genes may function.
10. Define *apoptosis* and explain the physiological significance of this process.

Critical Thinking Questions

1. Discuss the role of chromatin proteins in regulating gene expression. How does the three-dimensional structure of the chromatin affect genetic regulation? How do hormones influence genetic regulation?

2. Explain how p53 functions as a tumor suppressor gene. Discuss how mutations in p53 could lead to cancer, and how gene therapy or other drug interventions might be used to inhibit the growth of a tumor.
3. Release of lysosomal enzymes from white blood cells during a local immune attack can contribute to the symptoms of inflammation. Suppose, to alleviate inflammation, you develop a drug that destroys all lysosomes. Would this drug have negative side effects? Explain.
4. Antibiotics can have different mechanisms of action. An antibiotic called puromycin blocks genetic translation. One called actinomycin D blocks genetic transcription. These drugs can be used to determine how regulatory molecules, such as hormones, work. For example, if a hormone's effects on a tissue were blocked immediately by puromycin but not by actinomycin D, what would that tell you about the mechanism of action of the hormone?

Related Web Sites

For a listing of the most current web sites related to this chapter, please visit the *Concepts of Human Anatomy and Physiology* home page at http://www.mhhe.com/biosci/abio/.

NEXUS

Some Interactions of Basic Cell Concepts with the Other Body Systems

Integumentary System

- Mitotic cell division in the epidermis replaces cells lost from the surface (p. 162).

Skeletal System

- Mitotic cell division in the epiphyseal discs of growing bones is stimulated by growth hormone (p. 884).

Nervous System

- Different forms (alleles) of a gene produce different forms of receptors for particular neurotransmitter chemicals (p. 396).
- Microglia, located in the brain and spinal cord, are cells that move by amoeboid motion (p. 375).
- The insulating material around nerve fibers, called a myelin sheath, is derived from the cell membrane of particular types of cells in the nervous system (p. 376).

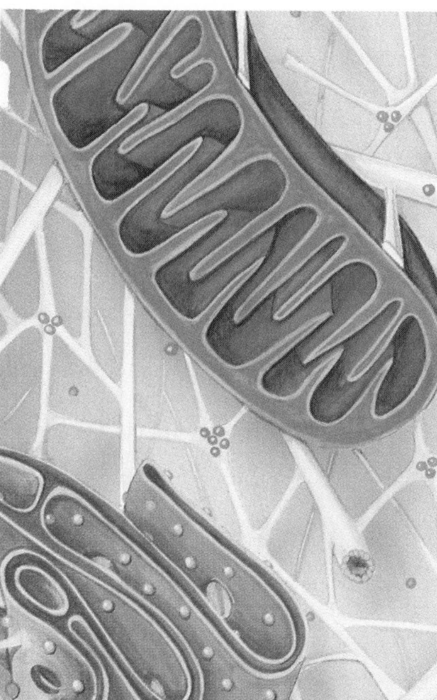

Endocrine System

- Many hormones act on their target cells by regulating gene expression (p. 558).
- The endoplasmic reticulum of some cells stores Ca^{2+}, which is released in response to hormone action (p. 563).
- Liver and adipose cells store glycogen and triglycerides, respectively, which can be mobilized for energy needs by the action of particular hormones (p. 874).
- The sex of an individual is determined by the presence of a particular region of DNA in the Y chromosome (p. 893).

Muscular System

- Muscle cells have cytoplasmic proteins called actin and myosin, which are needed for contraction (p. 290).
- Muscular dystrophy is caused by a defective gene that produces a protein called dystrophin, associated with the muscle cell membrane (p. 284).
- The endoplasmic reticulum of skeletal muscle fibers stores Ca^{2+}, which is needed for muscle contraction (p. 294).

Circulatory System

- Blood cells are formed in the bone marrow (p. 596).
- Mature red blood cells lack nuclei and mitochondria (p. 595).
- The different white blood cells are distinguished by the shape of their nuclei and the presence of cytoplasmic granules (p. 595).

Lymphatic and Immune System

- Some white blood cells and tissue macrophages destroy bacteria by phagocytosis (p. 697).
- When a B lymphocyte is stimulated by a foreign molecule (antigen), its endoplasmic reticulum becomes more developed to produce more antibody proteins (p. 702).
- Apoptosis is responsible for the destruction of T lymphocytes after an infection has been cleared (p. 716).

Respiratory System

- The air sacs of the lungs (pulmonary alveoli) are composed of cells that are very thin, minimizing the separation between air and blood (p. 736).
- The epithelial cells lining the respiratory airways have cilia (p. 735).
- The alveolar cells produce a protein-phospholipid secretion that helps to prevent the collapse of the lungs by lowering their surface tension (p. 741).

Urinary System

- Parts of the renal tubules have microvilli to increase the rate of reabsorption (p. 780).
- Some regions of the renal tubules have water channels; these are produced by the Golgi complex and inserted by means of vesicles into the cell membrane (p. 791).

Digestive System

- The mucosa of the GI tract has unicellular glands called goblet cells, which secrete mucus (p. 821).
- The cells of the small intestine have microvilli, which increases the rate of absorption (p. 835).
- The liver contains phagocytic cells (p. 840).

Reproductive System

- Males have an X and a Y chromosome, whereas females have two X chromosomes per diploid cell (p. 893).
- Gametes are produced by meiotic cell division (p. 904).
- Follicles degenerate (undergo atresia) in the ovaries by means of apoptosis (p. 932).
- Sperm are motile through the action of flagella (p. 908).
- The uterine tubes are lined with cilia, to help move the ovulated egg towards the uterus (p. 926).

Four

Enzymes, Energy, and Metabolism

Objectives

- Explain how catalysts function in chemical reactions and how enzymes function as catalysts.
- Describe the effects of pH and temperature on enzyme activity.
- Describe the effects of cofactors and coenzymes on enzyme activity and the effects of substrate and enzyme concentrations.
- Explain how end-product inhibition affects the direction of a branched metabolic pathway.
- Use the first and second laws of thermodynamics to explain why some molecules have more chemical-bond energy than others.
- Describe the coupling of energy-releasing and energy-requiring reactions and discuss the significance of ATP.
- Describe the nature of oxidation-reduction reactions.
- Describe glycolysis in terms of its initial substrate and its products.
- Describe the pathway of anaerobic respiration and discuss the significance of lactic acid formation.
- Define *gluconeogenesis* and discuss its significance.
- Describe the fate of pyruvic acid in aerobic respiration and discuss the nature of the Krebs cycle, naming the products that result from it.
- Explain the function of the electron-transport system and the role of oxygen in aerobic respiration.
- Define *oxidative phosphorylation*, state where it occurs, and discuss its significance.
- Explain how glucose and glycogen can be interconverted and how the liver can secrete free glucose derived from its stored glycogen.
- Define *lipolysis* and *β-oxidation* and explain how these processes function in cellular energy production.
- Explain how ketone bodies are formed and discuss their significance.
- Describe the processes of oxidative deamination and transamination of amino acids and explain how they contribute to energy production in the cell.
- Explain, in terms of the metabolic pathways involved, how carbohydrates or protein can be converted into fat.

Clinical Investigation

A 77-year-old man was brought to the hospital after experiencing severe chest pains. Analysis of blood samples revealed an abnormally high plasma concentration of the MB isoform of creatine phosphokinase. During his hospital stay, the patient complained of difficulty in urination, and an additional blood test revealed a high concentration of acid phosphatase. What do these blood tests suggest?

Clues: Read the section "Naming of Enzymes" and the boxed information in this section. Examine table 4.5.

Enzymes as Catalysts

Enzymes are biological catalysts that increase the rate of chemical reactions. Most enzymes are proteins, and their catalytic action results from their complex structure. The great diversity of protein structure allows different enzymes to be specialized in their action.

The ability of yeast cells to make alcohol from glucose (a process called *fermentation*) had been known since antiquity, yet even as late as the mid-nineteenth century no scientist had been able to duplicate the transformation in the absence of living yeast. Also, a vast array of chemical reactions occurred in yeast and other living cells at body temperature that could not be duplicated in the chemistry laboratory without adding substantial amounts of heat energy. These observations led many mid-nineteenth-century scientists to believe that chemical reactions in living cells were aided by a "vital force" that operated beyond the laws of the physical world. This *vitalist concept* was squashed along with the yeast cells when a pioneering biochemist, Eduard Buchner, demonstrated that juice obtained from yeast could ferment glucose to alcohol. The yeast juice was not alive—evidently some chemicals in the cells were responsible for fermentation. Buchner didn't know what these chemicals were, so he simply named them with the Greek term for "in yeast": **enzymes.**

As a general rule, enzymes are proteins. In the few special cases in which RNA demonstrates enzymatic activity, the term *ribozymes* is often used. Recent experiments, for example, suggest that the RNA component of ribosomes (chapter 3) acts as an enzyme that helps form peptide bonds within the growing polypeptide. Regardless of its chemical nature, an enzyme acts as a *biological catalyst.* A catalyst is a chemical that (1) increases the rate of a reaction, (2) is not itself changed at the end of the reaction, and (3) does not change the nature of the reaction or its final result. The same reaction would have occurred to the same degree in the absence of the catalyst, but it would have progressed at a much slower rate.

In order for a given reaction to occur, the reactants must have sufficient energy. The amount of energy required for a reaction to proceed is called the **activation energy.** By analogy, a match will not burn and release heat energy unless it is first "activated" by striking the match or by placing it in a flame.

In a large population of molecules, only a small fraction will possess sufficient energy for a reaction. Adding heat will raise the energy level of all the reactant molecules, thus increasing the percentage of the population that has the activation energy. Heat makes reactions go faster, but it also produces undesirable side effects in cells. Catalysts make reactions go faster at lower temperatures by lowering the activation energy required, thus ensuring that a larger percentage of the reactant molecules will have sufficient energy to participate in a reaction (fig. 4.1).

Since a small fraction of reactants will have the activation energy required for a reaction even in the absence of a catalyst, the reaction could theoretically occur spontaneously at a slow rate. This rate, however, would be much too slow for the needs of a cell. So, from a biological standpoint, the presence or absence of a specific enzyme catalyst acts as a switch—the reaction will occur if the enzyme is present and will not occur if the enzyme is absent.

Mechanism of Enzyme Action

The ability of enzymes to lower the activation energy of a reaction is a result of their structure. Enzymes are large proteins with complex, three-dimensional shapes produced by physical and chemical interactions between their amino acid subunits. Each type of enzyme has a characteristic three-dimensional shape, or *conformation*, with ridges, grooves, and pockets lined with specific amino acids. The particular pockets that are active in catalyzing a reaction are called the *active sites* of the enzyme.

The reactant molecules, which are the *substrates* of the enzyme, have specific shapes that allow them to fit into the active sites. The enzyme can thus be thought of as a lock into which only a specifically shaped key—the substrate—can fit. This **lock-and-key model** of enzyme activity is illustrated in figure 4.2.

In some cases, the fit between an enzyme and its substrate may not be perfect at first. A perfect fit may be induced, however, as the substrate gradually slips into the active site. This induced fit, together with temporary bonds that form between the substrate and the amino acids lining the active sites of the enzyme, weaken the existing bonds within the substrate molecules and allows them to be more easily broken. New bonds are more easily formed as substrates are brought close together in the proper orientation. This model of enzyme activity, in which the enzyme undergoes a slight structural change to better fit the substrate, is called the **induced-fit model.** The *enzyme-substrate complex*, formed temporarily in the course of the reaction, then dissociates to yield *products* and the free unaltered enzyme.

Since enzymes are very specific as to their substrates and activity, the concentration of a specific enzyme in a sample of

---

Buchner, Eduard: German biochemist, 1860–1917

Figure 4.1

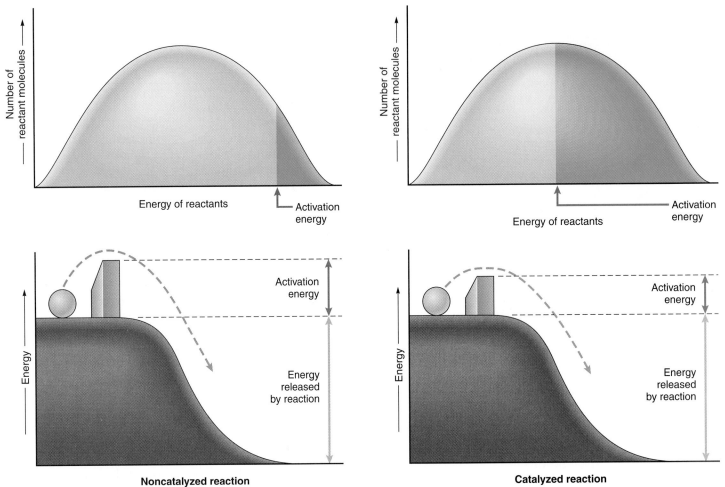

Comparison between a noncatalyzed and a catalyzed reaction.

The upper figures compare the proportion of reactant molecules that have sufficient activation energy to participate in the reaction (blue = insufficient energy; green = sufficient energy). This proportion is increased in the enzyme-catalyzed reaction because enzymes lower the activation energy required for the reaction (shown as a barrier on top of an energy "hill" in the lower figures). Reactants that can overcome this barrier are able to participate in the reaction, as shown by arrows pointing to the bottom of the energy hill.

Figure 4.2

The lock-and-key model of enzyme action.

(*a*) Substrates A and B fit into active sites in the enzyme, forming an enzyme-substrate complex (*b*). This then dissociates (*c*), releasing the products of the reaction and the free enzyme.

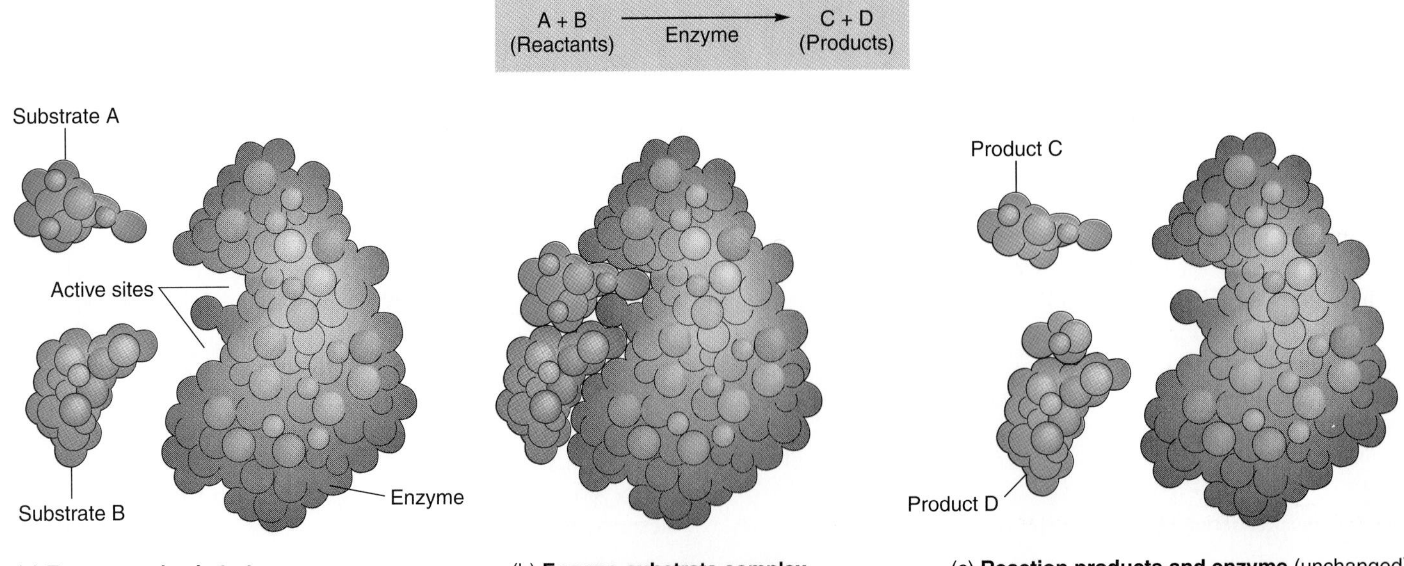

fluid can be measured relatively easily. This is usually done by measuring the rate of conversion of the enzyme's substrates into products under specified conditions. The presence of an enzyme in a sample can thus be detected by the job it does, and its concentration can be measured by how rapidly it performs its job.

Naming of Enzymes

In the past, enzymes were given names that were somewhat arbitrary. The modern system of naming enzymes, established by an international committee, is more orderly and informative. With the exception of some older enzyme names (such as pepsin, trypsin, and renin), all enzyme names end with the suffix *-ase* (table 4.1) and classes of enzymes are named according to their activity, or "job category." *Hydrolases* (*hi'drŏ-lās-ez*), for example, promote hydrolysis reactions. Other enzyme categories include *phosphatases* (*fos'fĕ-tās-ez*), which catalyze the removal of phosphate groups; *synthases* and *synthetases* (*sin'thĭ-tās-ez*), which catalyze dehydration synthesis reactions; *dehydrogenases* (*de"hi-droj'ĕ-nās-ez*), which remove hydrogen atoms from their substrates; and *kinases* (*ki'nsā-ez*), which add a phosphate group to (phosphorylate) particular molecules. Enzymes called *isomerases* (*i-som'ĕ-rās-ez*) rearrange atoms within their substrate molecules to form structural isomers, such as glucose and fructose.

The names of many enzymes specify both the substrate and the job category of the enzyme. Lactic acid dehydrogenase, for example, removes hydrogens from lactic acid. Enzymes that do exactly the same job (that catalyze the same reaction) in different organs have the same name, since the name describes the activity of the enzyme. Different organs, however, may make slightly different "models" of the enzyme that differ in one or a few amino acids. These different models of the same enzyme are called **isoenzymes** (*i-so-en'zīmz*). The differences in structure do not affect the active sites (otherwise the enzymes would not catalyze the same reaction), but they do alter the structure of the enzymes at other locations, so that the different isoenzymatic forms can be separated by standard biochemical procedures. These techniques are useful in the diagnosis of diseases (see "Clinical Considerations" at the end of this chapter).

Table 4.1

Selected Enzymes and the Reactions They Catalyze

Enzyme	Reaction Catalyzed
Catalase	$2\ H_2O_2 \rightarrow 2\ H_2O + O_2$
Carbonic anhydrase	$H_2CO_3 \rightarrow H_2O + CO_2$
Amylase	starch + $H_2O \rightarrow$ maltose
Lactate dehydrogenase	lactic acid $\rightarrow$ pyruvic acid + H_2
Ribonuclease	RNA + $H_2O \rightarrow$ ribonucleotides

Control of Enzyme Activity

The rate of an enzyme-catalyzed reaction depends on numerous factors, including the concentration of the enzyme and the pH and temperature of the solution. Genetic control of enzyme concentration, for example, affects the rate of progress along particular metabolic pathways and thus regulates cellular metabolism.

The activity of an enzyme, as measured by the rate at which its substrates are converted to products, is influenced by such factors as (1) the temperature and pH of the solution; (2) the concentration of cofactors and coenzymes, which are needed by many enzymes as "helpers" for their catalytic activity; (3) the concentration of enzyme and substrate molecules in the solution; and (4) the stimulatory and inhibitory effects of some products of enzyme action on the activity of the enzymes that helped to form these molecules.

Effects of Temperature and pH

An increase in temperature will increase the rate of non-enzyme-catalyzed reactions. A similar relationship between temperature and reaction rate occurs in enzyme-catalyzed reactions. At a temperature of 0° C the reaction rate is immeasurably slow. As the temperature is raised above 0° C the reaction rate increases, but only up to a point. At a few degrees above body temperature (which is 37° C) the reaction rate reaches a plateau; further increases in temperature actually *decrease* the rate of the reaction (fig. 4.3). This decrease is due to the fact that the tertiary structure of enzymes becomes altered at higher temperatures.

A similar relationship is observed when the rate of an enzymatic reaction is measured at different pH values. Each enzyme characteristically exhibits peak activity in a very narrow pH range, which is the **pH optimum** for the enzyme. If

Figure 4.3

The effect of temperature on enzyme activity. This is measured by the rate of the enzyme-catalyzed reaction under standardized conditions as the temperature of the reaction is varied.

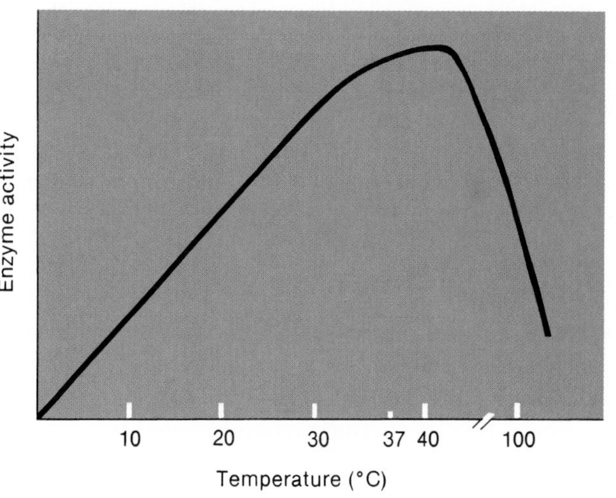

Figure 4.4

The effect of pH on the activity of three digestive enzymes.

Pepsin is found in acidic gastric juice; salivary amylase is found in saliva, which has a pH close to neutral; and trypsin is found in alkaline pancreatic juice.

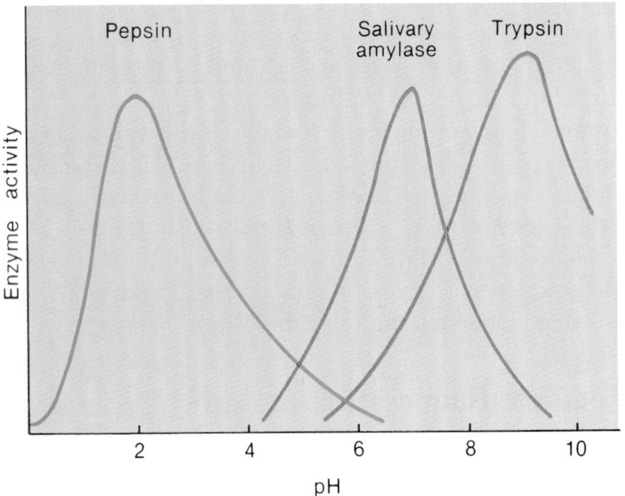

the pH is changed so that it is no longer within the enzyme's optimum range, the reaction rate will decrease (fig. 4.4). This decreased enzyme activity is due to changes in the conformation of the enzyme and in the charges of the R groups of the amino acids lining the active sites.

The pH optimum of an enzyme usually reflects the pH of the body fluid in which the enzyme is found. The acidic pH optimum of the protein-digesting enzyme *pepsin*, for example, allows it to be active in the strong hydrochloric acid of gastric juice. Similarly, the neutral pH optimum of *salivary amylase* (*am'ĭ-lās*) and the alkaline pH optimum of *trypsin* (*trip'sin*) in pancreatic juice allow these enzymes to digest starch and protein, respectively, in other parts of the digestive tract.

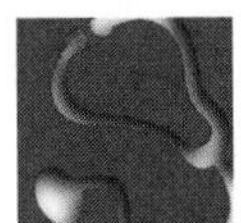

Although the pH of other body fluids shows less variation than that of the fluids of the digestive tract, the pH optima of different enzymes found throughout the body do show significant differences (table 4.2). Some of these differences can be exploited for diagnostic purposes. Disease of the prostate, for example, may be associated with elevated blood levels of a prostatic phosphatase with an acidic pH optimum (descriptively called acid phosphatase). Bone disease, on the other hand, may be associated with elevated blood levels of alkaline phosphatase, which has a higher pH optimum than the similar enzyme released from the diseased prostate.

Cofactors and Coenzymes

Many enzymes are completely inactive when they are isolated in a pure state. Evidently some of the ions and smaller organic molecules that are removed in the purification procedure play an essential role in enzyme activity. These ions and smaller organic molecules needed for the activity of specific enzymes are called *cofactors* and *coenzymes*.

Table 4.2

pH Optima of Selected Enzymes

Enzyme	Reaction Catalyzed	pH Optimum
Pepsin (stomach)	Digestion of protein	2.0
Acid phosphatase (prostate)	Removal of phosphate group	5.5
Salivary amylase (saliva)	Digestion of starch	6.8
Lipase (pancreatic juice)	Digestion of fat	7.0
Alkaline phosphatase (bone)	Removal of phosphate group	9.0
Trypsin (pancreatic juice)	Digestion of protein	9.5
Monoamine oxidase (nerve endings)	Removal of amine group from norepinephrine	9.8

Figure 4.5

The roles of cofactors in enzyme function.

In (*a*) the cofactor changes the conformation of the active site, allowing for a better fit between the enzyme and its substrates. In (*b*) the cofactor participates in the temporary bonding between the active site and the substrates.

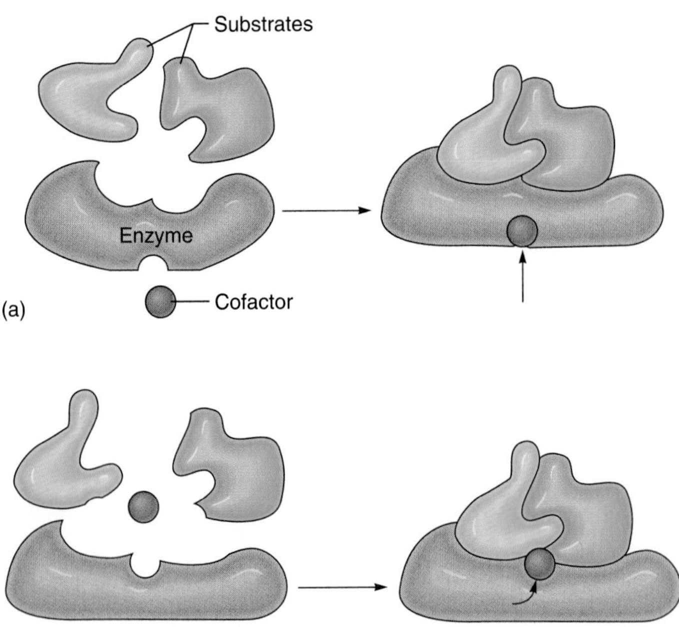

Cofactors include metal ions such as Ca^{2+}, Mg^{2+}, Mn^{2+}, Cu^{2+}, Zn^{2+}, and selenium. Some enzymes with a cofactor requirement do not have a properly shaped active site in the absence of the cofactor. In these enzymes, the attachment of cofactors causes a conformational change in the protein that allows it to combine with its substrate. The cofactors of other enzymes participate in the temporary bonds between the enzyme and its substrate when the enzyme-substrate complex is formed (fig. 4.5).

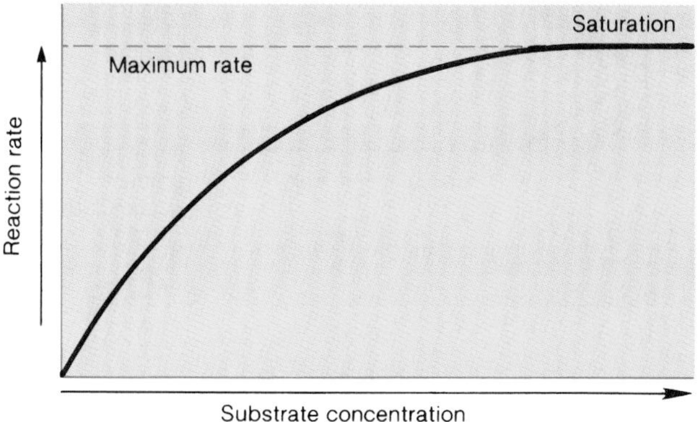

Figure 4.6

The effect of substrate concentration on the rate when the enzyme concentration is constant.
When the reaction rate is maximal, the enzyme is said to be saturated.

Other cofactors, called **coenzymes** (*ko-en'zīmz*), are organic molecules derived from niacin, riboflavin, and other water-soluble vitamins. Coenzymes participate in enzyme-catalyzed reactions by transporting hydrogen atoms and small molecules from one enzyme to another. Examples of the actions of cofactors and coenzymes in specific reactions will be given in the context of their roles in cellular metabolism later in this chapter.

Substrate Concentration and Reversible Reactions

At a given level of enzyme concentration, the rate of product formation will increase as the substrate concentration increases. Eventually, however, a point will be reached where additional increases in substrate concentration do not result in comparable increases in reaction rate. When the relationship between substrate concentration and reaction rate reaches a plateau of maximum velocity, the enzyme is *saturated*. If we think of enzymes as workers and substrates as jobs, there is 100% employment when the enzyme is saturated; further availability of jobs (substrate) cannot further increase employment (conversion of substrate to product). This concept is illustrated in figure 4.6.

Some enzymatic reactions within a cell are reversible, with both the forward and backward reactions catalyzed by the same enzyme. The enzyme *carbonic anhydrase*, for example, is named because it can catalyze the following reaction:

$$H_2CO_3 \rightarrow H_2O + CO_2$$

The same enzyme, however, can also catalyze the reverse reaction:

$$H_2O + CO_2 \rightarrow H_2CO_3$$

The two reactions can be more conveniently illustrated by a single equation:

$$H_2O + CO_2 \rightleftharpoons H_2CO_3$$

The direction of the reversible reaction depends, in part, on the relative concentrations of the molecules to the left and right of the arrows. If the concentration of CO_2 is very high (as it is in the tissues), the reaction will be driven to the right. If the concentration of CO_2 is low and that of H_2CO_3 is high (as it is in the lungs), the reaction will be driven to the left. The principle that reversible reactions will be driven from the side of the equation where the concentration is higher to the side where the concentration is lower is known as the **law of mass action.**

Although some enzymatic reactions are not directly reversible, the net effects of the reactions can be reversed by the action of different enzymes. Some of the enzymes that convert glucose to pyruvic acid, for example, are different from those that reverse the pathway and produce glucose from pyruvic acid. Likewise, the formation and breakdown of glycogen (a polymer of glucose) are catalyzed by different enzymes.

Figure 4.7

The general pattern of a metabolic pathway.
In metabolic pathways, the product of one enzyme becomes the substrate of the next.

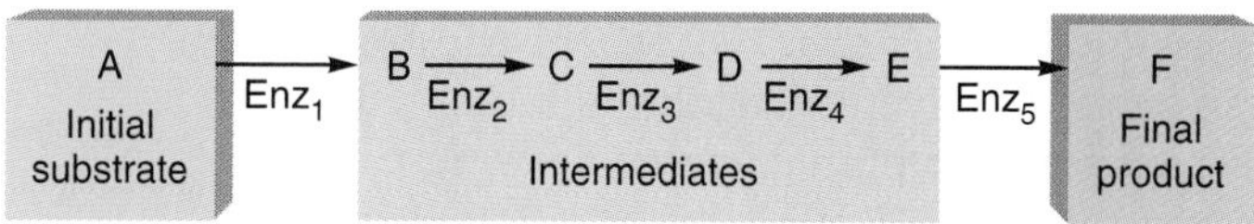

Figure 4.8

A branched metabolic pathway.
Two or more different enzymes can work on the same substrate at the branch point of the pathway, catalyzing two or more different reactions.

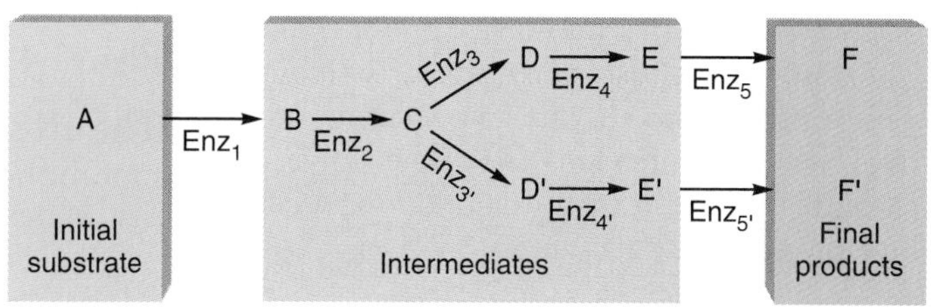

Metabolic Pathways

The many thousands of different types of enzymatic reactions within a cell do not occur independently of each other. Rather, they are all linked by intricate webs of interrelationships, the total pattern of which constitutes cellular metabolism. A sequence of enzymatic reactions that begins with an *initial substrate*, progresses through a number of *intermediates*, and ends with a *final product* is known as a **metabolic pathway.**

The enzymes in a metabolic pathway cooperate in a manner analogous to workers on an assembly line, where each contributes a part to the final product. In this process, the product of one enzyme in the line becomes the substrate of the next enzyme, and so on (fig. 4.7).

Few metabolic pathways are completely linear. Most are branched so that one intermediate at the branch point can serve as a substrate for two different enzymes. Two different products can thus be formed that serve as intermediates of two pathways (fig. 4.8).

End-Product Inhibition

The activities of enzymes at the branch points of metabolic pathways are often regulated by a process called **end-product inhibition,** which is a form of negative feedback inhibition. In this process, one of the final products of a divergent pathway inhibits the activity of the branch-point enzyme that began the path toward the production of this inhibitor. This inhibition prevents that final product from accumulating excessively and results in a shift toward the final product of the alternate pathway (fig. 4.9).

The mechanism by which a final product inhibits an earlier enzymatic step in its pathway is known as **allosteric** (*al″ŏ-ster′ik*) **inhibition.** The allosteric inhibitor combines with a part of the enzyme at a location other than the active site. This causes the active site to change shape so that it can no longer combine properly with its substrate.

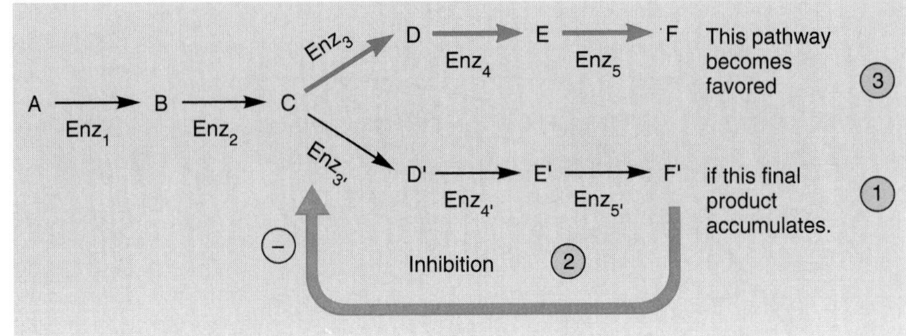

Figure 4.9

End-product inhibition in a branched metabolic pathway. Inhibition is shown by the arrow in step 2.

Inborn Errors of Metabolism

Since each different polypeptide in the body is coded by a different gene (chapter 3), each enzyme protein that participates in a metabolic pathway is coded by a different gene. An inherited defect in one of these genes may result in a disease known as an **inborn error of metabolism.** In this type of disease, the quantity of intermediates formed *prior* to the defective enzymatic step *increases*, and the quantity of intermediates and final products formed *after* the defective step *decreases*. Diseases may result from deficiencies of the normal end product or from excessive accumulation of intermediates formed prior to the defective step. If the defective enzyme is active at a step that follows a branch point in a pathway, the intermediates and final products of the alternate pathway will increase (fig. 4.10). An abnormal increase in the production of these products can be the cause of some metabolic diseases.

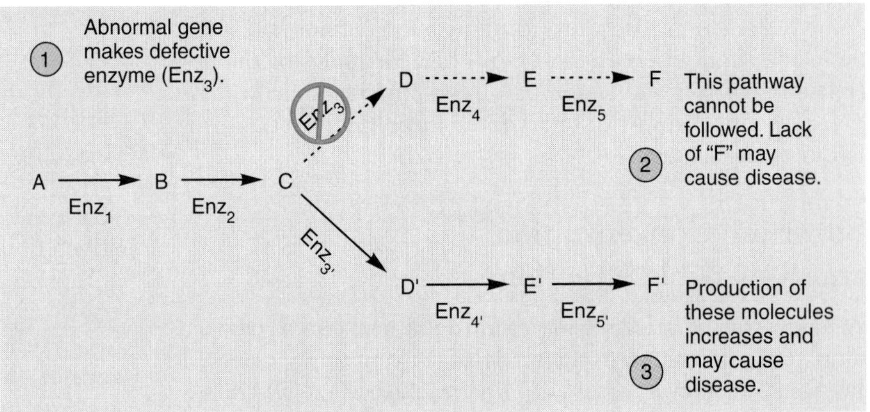

Figure 4.10

The effects of an inborn error of metabolism on a branched metabolic pathway. The defective gene produces a defective enzyme, indicated here by a line through its symbol.

Bioenergetics

Living organisms require the constant expenditure of energy to maintain their complex structures and processes. Central to life processes are chemical reactions that are coupled, so that the energy released by one reaction is incorporated into the products of another reaction. The transformation of energy in living systems is largely based on reactions involving ATP and on oxidation-reduction reactions.

Bioenergetics refers to the flow of energy in living systems. Organisms maintain their highly ordered structure and life-sustaining activities through the constant expenditure of energy obtained ultimately from the environment. The energy flow in living systems obeys the first and second laws of a branch of physics known as *thermodynamics*.

According to the **first law of thermodynamics,** energy can be transformed but it can neither be created nor destroyed. This is sometimes called the *law of conservation of energy*. As a result of energy transformations, according to the **second law of thermodynamics,** the universe and its parts (including living systems) become increasingly disorganized. The term *entropy* (*en′trŏ-pe*) is used to describe the degree of disorganization of a system. Energy transformations thus increase the amount of entropy of a system. Only energy that is in an organized state—called *free energy*—can be used to do work. Thus, since entropy increases in every energy transformation, the

allosteric: Gk. *allos*, other; *stereos*, position

bioenergetics: Gk. *bios*, life; *energeia*, work
thermodynamics: Gk. *therme*, heat; *dynamis*, force

Figure 4.11

A simplified diagram of photosynthesis.
Some of the sun's radiant energy is captured by plants and used to produce glucose from carbon dioxide and water. As the product of this endergonic reaction, glucose has more free energy than the initial reactants.

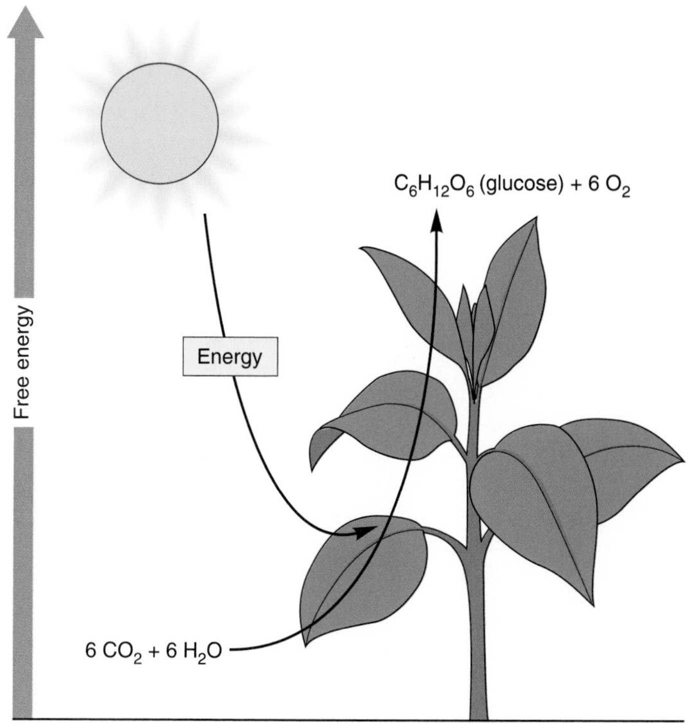

amount of free energy available to do work decreases. As a result of the increased entropy described by the second law, systems tend to go from states of higher free energy to states of lower free energy.

The chemical bonding of atoms into molecules obeys the laws of thermodynamics. A complex organic molecule such as glucose, for example, has more free energy (less entropy) than six separate molecules each of carbon dioxide and water. Therefore, in order to convert carbon dioxide and water to glucose, energy must be added. Plants perform this feat using energy from the sun in the process of *photosynthesis* (fig. 4.11).

Endergonic and Exergonic Reactions

Chemical reactions that require an input of energy are known as **endergonic** (*en″der-gon′ik*) **reactions.** Since energy is added to make these reactions "go," the products of endergonic reactions must contain more free energy than the reactants. A portion of the energy added, in other words, is contained within the product molecules. This follows from the fact that energy cannot be created or destroyed (first law of thermodynamics) and from the fact that a more-organized state of matter contains more free energy, or less entropy, than a less-organized state (second law of thermodynamics).

The fact that glucose contains more free energy than carbon dioxide and water can easily be proven by combusting glucose to CO_2 and H_2O. This reaction releases energy in the form of heat. Reactions that convert molecules with more free energy to molecules with less—and, therefore, that release energy as they proceed—are called **exergonic reactions.**

As illustrated in figure 4.12, the amount of energy released by an exergonic reaction is the same whether the energy is released in a single combustion reaction or in the many small, enzymatically controlled steps that occur in tissue cells. The energy that the body obtains from consuming particular foods can therefore be measured as the amount of heat energy released when these foods are combusted.

Heat is measured in units called *calories*. One calorie is defined as the amount of heat required to raise the temperature of one cubic centimeter of water one degree on the Celsius scale. The caloric value of food is usually indicated in *kilocalories* (1 kilocalorie = 1,000 calories), which are often called large calories and spelled with a capital C.

Coupled Reactions: ATP

In order to remain alive, a cell must maintain its highly organized, low-entropy state at the expense of free energy in its environment. Accordingly, the cell contains many enzymes that catalyze exergonic reactions using substrates that come ultimately from the environment. The energy released by these exergonic reactions drives the energy-requiring processes (endergonic reactions) in the cell. Since cells cannot use heat energy to drive energy-requiring processes, the chemical-bond energy that is released in exergonic reactions must be directly transferred to chemical-bond energy in the products of endergonic reactions. Energy-liberating reactions are thus *coupled* to energy-requiring reactions. This relationship is like that of two meshed gears; the turning of one (the energy-releasing exergonic gear) causes turning of the other (the energy-requiring endergonic gear). This relationship is illustrated in figure 4.13.

The energy released by most exergonic reactions in the cell is used, either directly or indirectly, to drive one particular endergonic reaction (fig. 4.14): the formation of **adenosine** (*ă-den′o-sēn*) **triphosphate (ATP)** from adenosine diphosphate (ADP) and inorganic phosphate (abbreviated P_i).

The formation of ATP requires the input of a fairly large amount of energy. Since this energy must be conserved (first law of thermodynamics), the bond produced by joining P_i to ADP must contain a part of this energy. Thus, when enzymes reverse this reaction and convert ATP to ADP and P_i, a large amount of energy is released. Energy released from the breakdown of ATP powers the energy-requiring processes in all cells. As the **universal energy carrier,** ATP serves to more

photosynthesis: Gk. *phos*, light; *synthesis*, a putting together
endergonic: Gk. *endon*, within; *ergon*, work
exergonic: Gk. *exo*, outside; *ergon*, work
calorie: L. *calor*, heat

Figure 4.12

A comparison of combustion with cell respiration.
Since glucose contains more energy than six separate molecules each of carbon dioxide and water, the combustion of glucose is an exergonic reaction. The same amount of energy is released when glucose is broken down stepwise within the cell.

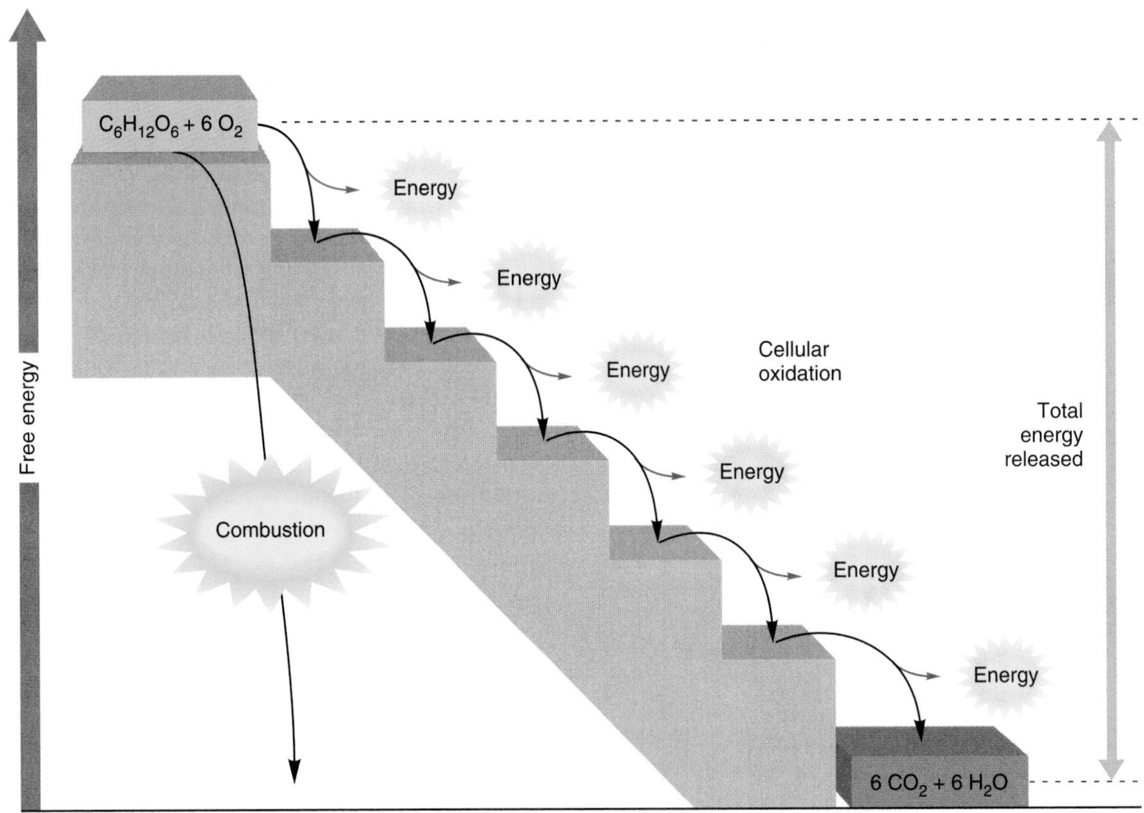

Figure 4.13

A model of the coupling of exergonic and endergonic reactions.
The reactants of the exergonic reaction (represented by the larger gear) have more free energy than the products of the endergonic reaction because the coupling is not 100% efficient—some energy is lost as heat.

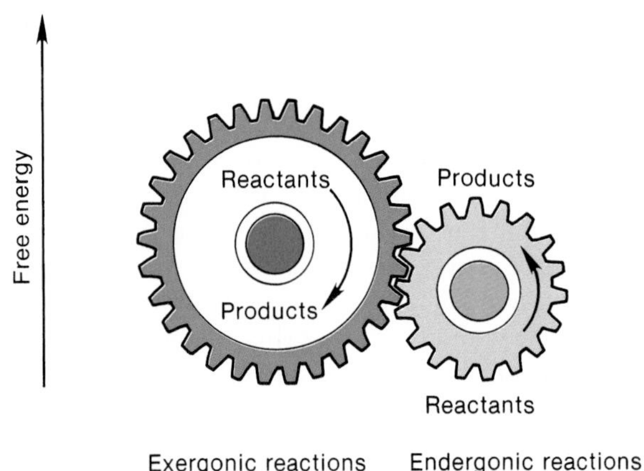

efficiently couple the energy released by the breakdown of food molecules to the energy required by the diverse endergonic processes in the cell (fig. 4.15).

Coupled Reactions: Oxidation-Reduction

When an atom or a molecule gains electrons, it is said to become **reduced;** when it loses electrons, it is said to become **oxidized.** Reduction and oxidation are always coupled reactions: an atom or a molecule cannot become oxidized unless it donates electrons to another, which therefore becomes reduced. The atom or molecule that donates electrons *to* another is a **reducing agent,** and the one that accepts electrons *from* another is an **oxidizing agent.** It is important to understand that a particular atom (or molecule) can play both roles; it may function as an oxidizing agent in one reaction and as a reducing agent in another reaction. When atoms or molecules play both roles, they gain electrons in one reaction and pass them on in another reaction to produce a series of coupled oxidation-reduction reactions—like a bucket brigade, with electrons in the buckets.

Notice that the term *oxidation* does not imply that oxygen participates in the reaction. This term is derived from the fact that oxygen has a great tendency to accept electrons; that is, to act as a strong oxidizing agent. This property of oxygen is exploited by cells; oxygen acts as the final electron acceptor in a chain of oxidation-reduction reactions that provides energy for ATP production.

Oxidation-reduction reactions in cells often involve the transfer of hydrogen atoms rather than of free electrons. Since a hydrogen atom contains one electron and one proton in the nucleus, a molecule that loses hydrogen becomes oxidized and one that gains hydrogen becomes reduced. In many oxidation-reduction reactions, pairs of electrons—either as free electrons or as a pair of hydrogen atoms—are transferred from the reducing agent to the oxidizing agent.

Two molecules that serve important roles in the transfer of hydrogens are **nicotinamide adenine dinucleotide** (*nik″ŏ-tin′ă-mīd ad′n-ēn di-noo′kle-ŏ-tīd*) **(NAD),** derived from the vitamin niacin (vitamin B_3), and **flavin** (*fla′vin*) **adenine dinucleotide (FAD),** derived from the vitamin riboflavin (vitamin B_2). These molecules (fig. 4.16) are coenzymes that function as *hydrogen carriers* because they accept hydrogens (becoming reduced) in one enzyme reaction and donate hydrogens (becoming oxidized) in a different enzyme reaction (fig. 4.17). The oxidized forms of these molecules may be written simply as NAD (or NAD^+) and FAD.

Each FAD can accept two electrons and can bind two protons. Therefore, the reduced form of FAD is combined with the equivalent of two hydrogen atoms and may be written as $FADH_2$. Each NAD can also accept two electrons but

Figure 4.14

The formation and structure of adenosine triphosphate (ATP).
ATP is the universal energy carrier of the cell.

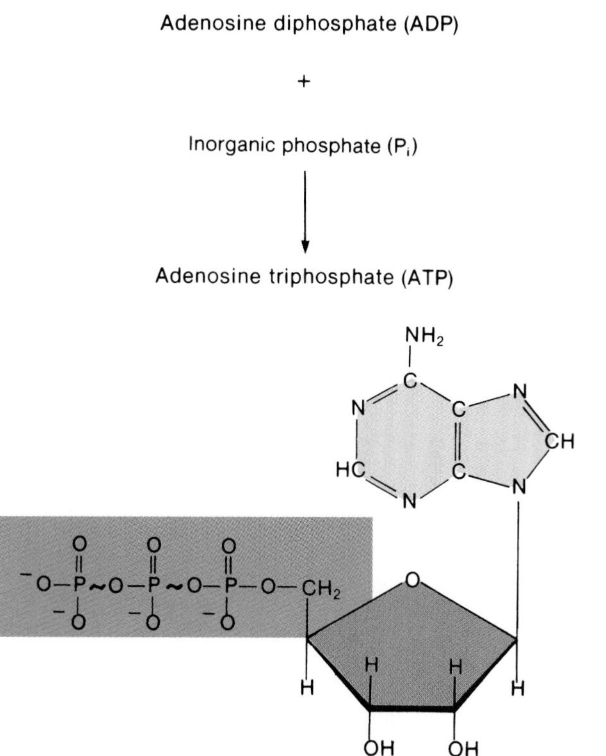

Figure 4.15

A model of ATP as the universal energy carrier of the cell.
Exergonic reactions are shown as gears with arrows going down (these reactions produce a decrease in free energy); endergonic reactions are shown as gears with arrows going up (these reactions produce an increase in free energy).

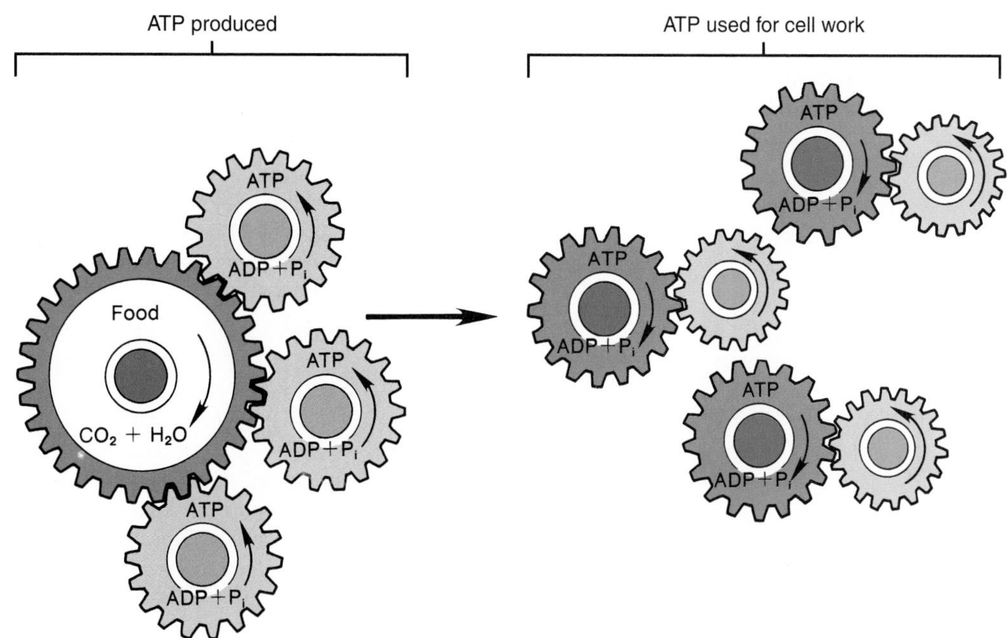

Figure 4.16

The structures NAD and $FADH_2$.
(*a*) The oxidized form of NAD (nicotinamide adenine dinucleotide) and (*b*) the reduced form of FAD (flavin adenine dinucleotide). Notice the two additional hydrogen atoms (shown in color) that reduce FAD to $FADH_2$.

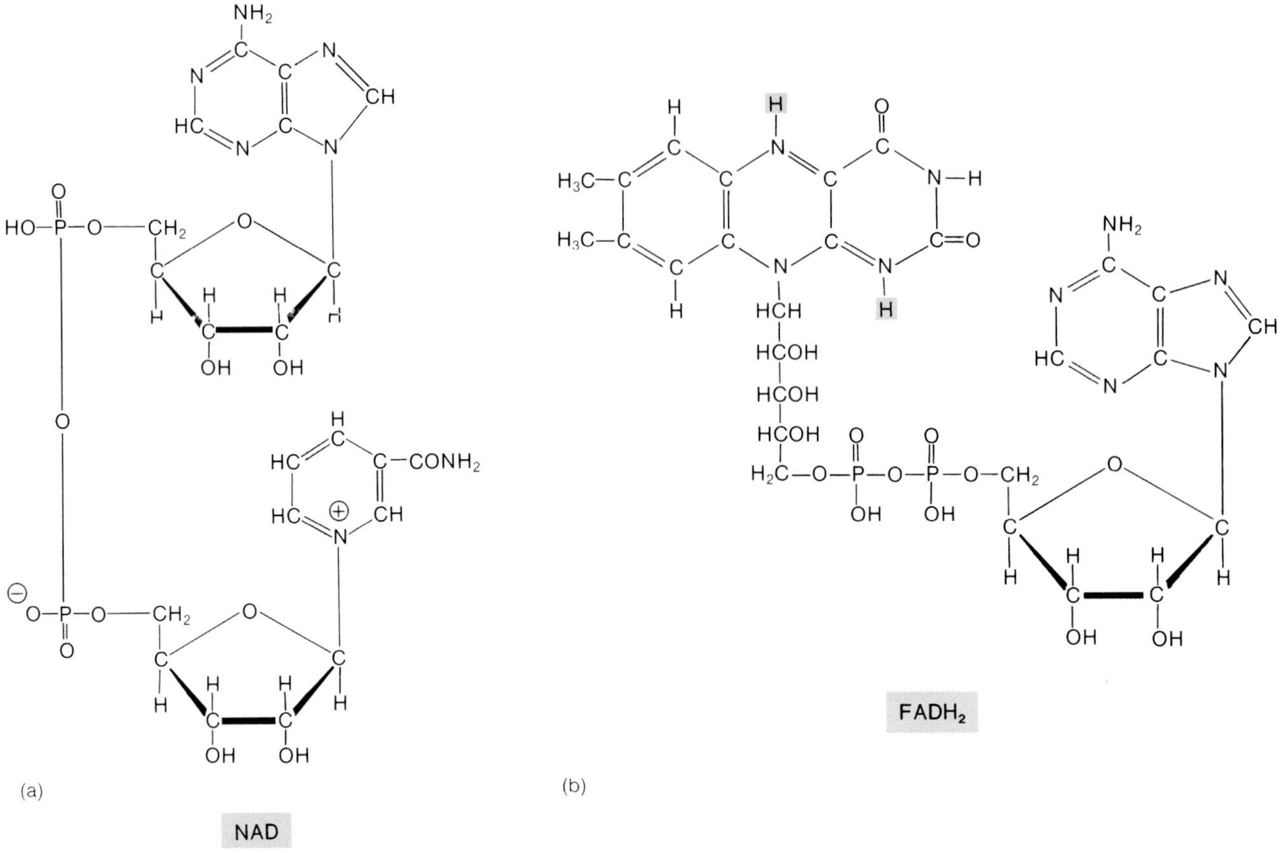

(a) NAD

(b) $FADH_2$

Figure 4.17

The action of NAD.
NAD is a coenzyme that transfers pairs of hydrogen atoms from one molecule to another. In the first reaction, NAD is reduced (acts as an oxidizing agent); in the second reaction, NADH is oxidized (acts as a reducing agent).

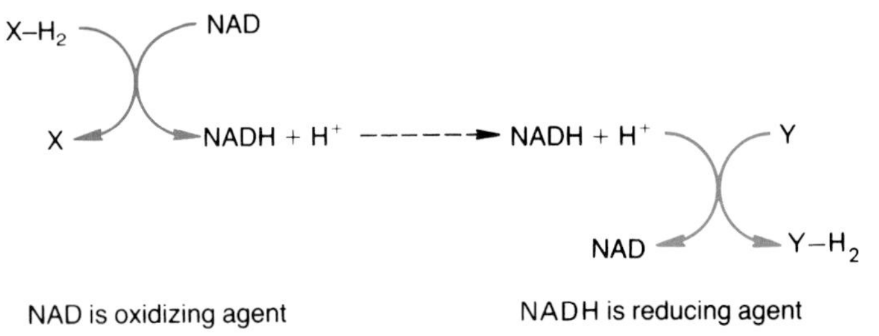

$X{-}H_2$ + NAD → X + NADH + H^+ - - - → NADH + H^+ + Y → NAD + $Y{-}H_2$

NAD is oxidizing agent — NADH is reducing agent

can bind only one proton. The reduced form of NAD is therefore indicated by NADH + H^+ (the H^+ represents a free proton). When the reduced forms of these two coenzymes participate in an oxidation-reduction reaction, they transfer two hydrogen atoms to the oxidizing agent (fig. 4.17).

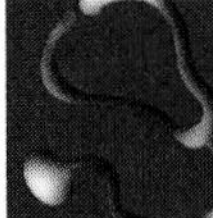

Production of the coenzymes NAD and FAD is the major reason that we need the vitamins niacin and riboflavin in our diet. As described later in this chapter, NAD and FAD are required to transfer hydrogen atoms in the chemical reactions that provide energy for the body. Niacin and riboflavin do not themselves provide the energy, although this is often claimed in misleading advertisements. Nor can eating extra amounts of niacin and riboflavin provide extra energy. Once the cells have obtained sufficient NAD and FAD, the excess amounts of these vitamins are simply eliminated in the urine.

Glycolysis and the Lactic Acid Pathway

In cellular respiration, energy is released by the stepwise breakdown of glucose and other molecules, and some of this energy is used to produce ATP. The complete combustion of glucose requires the presence of oxygen and yields 30 ATP per glucose. However, some energy can be obtained in the absence of oxygen by the pathway that leads to the production of lactic acid. This process results in a net gain of 2 ATP per glucose.

All of the reactions in the body that involve energy transformation are collectively termed **metabolism** (*mĕ-tab'ŏ-liz"em*). Metabolism may be divided into two categories: **anabolism** (*ă-nab'ŏ-liz"em*) and **catabolism** (*ka-tab'o-liz"em*). Catabolic reactions release energy, usually by the breakdown of larger organic molecules into smaller molecules. Anabolic reactions require the input of energy and include the synthesis of large energy-storage molecules, including glycogen, fat, and protein.

The catabolic reactions that break down glucose, fatty acids, and amino acids serve as the primary sources of energy for the synthesis of ATP. For example, this means that some of the chemical-bond energy in glucose is transferred to the chemical-bond energy in ATP. Since energy transfers can never be 100% efficient (according to the second law of thermodynamics), some of the chemical-bond energy from glucose is lost as heat.

This energy transfer involves oxidation-reduction reactions. Oxidation of a molecule occurs when the molecule loses electrons, and this must be coupled to the reduction of another atom or molecule, which accepts the electrons. In the breakdown of glucose and other molecules for energy, some of the electrons initially present in these molecules are transferred to intermediate carriers and then to a *final electron acceptor*. When a molecule is completely broken down to carbon dioxide and water within an animal cell, the final electron acceptor is always an atom of oxygen. Because of the involvement of oxygen, the metabolic pathway that converts molecules such as glucose or fatty acid to carbon dioxide and water (transferring some of the energy to ATP) is called **aerobic** (*ă-ro'bik*) **cell respiration.** The oxygen for this process is obtained from the blood. The blood, in turn, obtains oxygen from air in the lungs through the process of breathing, or ventilation, as described in chapter 24. Ventilation also serves the important function of eliminating the carbon dioxide produced by aerobic cell respiration.

Unlike the process of burning, or combustion, which quickly releases the energy content of molecules as heat, the conversion of glucose to carbon dioxide and water within the cells occurs in small, enzymatically catalyzed steps. Oxygen is used only at the last step. Since a small amount of the chemical-bond energy of glucose is released at early steps in the metabolic pathway, some tissue cells can obtain energy for ATP production in the temporary absence of oxygen. This process is described in the next two sections.

Glycolysis

The breakdown of glucose for energy involves a metabolic pathway in the cytoplasm known as **glycolysis** (*gli"kol'ĭ-sis*). Glycolysis is the metabolic pathway by which glucose—a six-carbon (hexose) sugar—is converted into two molecules of *pyruvic* (*pi-roo'vik*) *acid*, or *pyruvate*. Even though each pyruvic acid molecule is roughly half the size of a glucose, glycolysis is not simply the breaking in half of glucose. Glycolysis is a metabolic pathway involving many enzymatically controlled steps.

Each pyruvic acid molecule contains three carbons, three oxygens, and four hydrogens. The number of carbon and oxygen atoms in one molecule of glucose—$C_6H_{12}O_6$—can thus be accounted for in the two pyruvic acid molecules. Since the two pyruvic acids together account for only eight hydrogens, however, it is clear that four hydrogen atoms are removed from the intermediates in glycolysis. Each pair of these hydrogen atoms is used to reduce a molecule of NAD. In this process, each pair of hydrogen atoms donates two electrons to NAD, thus reducing it. The reduced NAD binds one proton from the hydrogen atoms, leaving one proton unbound as H^+. Starting from one glucose molecule, therefore, glycolysis results in the production of two molecules of NADH and two H^+. The H^+ will follow the NADH in subsequent reactions, so for simplicity we can simply refer to reduced NAD as NADH.

Glycolysis is exergonic, and a portion of the energy that is released drives the endergonic reaction ADP + $P_i \rightarrow$ ATP. At the end of the glycolytic pathway, there is a net gain of two ATP molecules per glucose molecule, as indicated in the overall equation for glycolysis:

$$\textbf{Glucose + 2 NAD + 2 ADP + 2 } \mathbf{P_i} \rightarrow \textbf{2 pyruvic acid + 2 NADH + 2 ATP}$$

Although the overall equation for glycolysis is exergonic, glucose must be "activated" at the beginning of the pathway before energy can be obtained. This activation requires the addition of two phosphate groups derived from two molecules of ATP. Energy from the reaction ATP $\rightarrow$ ADP + P_i is therefore consumed at the beginning of glycolysis. This is shown as an "up-staircase" in figure 4.18. Notice that the P_i is not shown in these reactions in figure 4.18; this is because the phosphate is not released from, but instead is added to, the intermediate molecules of glycolysis. The addition of a phosphate group is known as *phosphorylation* (*fos"for-ĭ-la'shun*). Besides being essential for glycolysis, the phosphorylation of glucose (to glucose-6-phosphate) has an important side benefit: it traps the glucose within the cell. This is because *phosphorylated organic molecules cannot cross cell membranes*.

At later steps in glycolysis, four molecules of ATP are produced (and two molecules of NAD are reduced) as energy is liberated (the "down-staircase" in fig. 4.18). The two molecules of ATP used in the beginning, therefore, represent an energy investment; the net gain of two ATP and two NADH by the end of the pathway represents an energy profit. The overall equation for glycolysis obscures the fact that this is a metabolic pathway consisting of nine separate steps. The individual steps in this pathway are shown in figure 4.19.

metabolism: Gk. *metabole*, change
anabolism: Gk. *anabole*, a raising up
catabolism: Gk. *katabole*, a casting down
glycolysis: Gk. *glyco*, sugar; *lysis*, breaking

Figure 4.18

The energy expenditure and gain in glycolysis.
Notice that there is a "net profit" of two ATP and two NADH molecules for every molecule of glucose that enters the glycolytic pathway. Molecules listed by number are (*1*) fructose-1,6-diphosphate, (*2*) 1,3-diphosphoglyceric acid, and (*3*) 3-phosphoglyceric acid (see fig. 4.19).

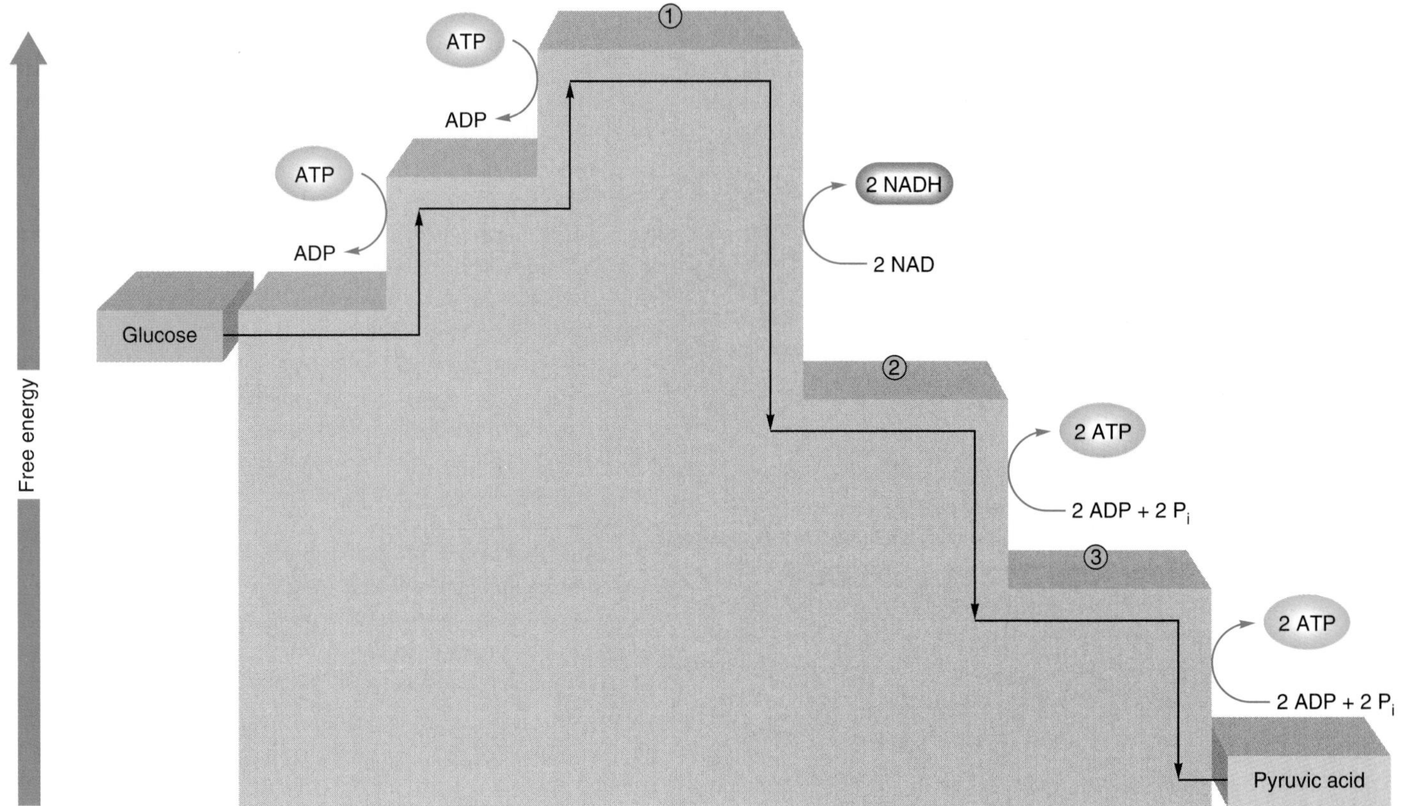

The Lactic Acid Pathway

In order for glycolysis to continue, there must be adequate amounts of NAD available to accept hydrogen atoms. Therefore, the NADH produced in glycolysis must become oxidized by donating its electrons to another molecule. (In aerobic respiration this other molecule is located in the mitochondria and ultimately passes its electrons to oxygen.)

When oxygen is *not* available in sufficient amounts, the NADH (+ H^+) produced in glycolysis is oxidized in the cytoplasm by donating its electrons to pyruvic acid. This results in the reformation of NAD and the addition of two hydrogen atoms to pyruvic acid, which is thus reduced. This addition of two hydrogen atoms to pyruvic acid produces **lactic acid** (fig. 4.20).

The metabolic pathway by which glucose is converted to lactic acid is frequently referred to by physiologists as **anaerobic** (*an-ă-ro'bik*) **respiration.** "Anaerobic" describes the fact that oxygen is not used in the process. However, many biologists prefer that "anaerobic respiration" be reserved for the pathways in certain bacteria that employ sulfur or iron as a final electron acceptor in place of oxygen. In that sense, the term "respiration" refers to the use of an inorganic atom as the final electron acceptor. If anaerobic respiration is used to describe this bacterial metabolism, then another term is needed for the production of lactic acid. In this case the term **lactic acid fermentation** may be used, since the metabolic pathway is analogous to that used by yeast cells to produce ethyl alcohol. In both the production of alcohol by yeast and of lactic acid by human cells, the electron acceptor is an organic molecule. In this text, anaerobic respiration and lactic acid fermentation will be used synonymously.

The lactic acid pathway yields a net gain of two ATP molecules (produced by glycolysis) per glucose molecule. A cell can survive without oxygen as long as it can produce sufficient energy for its needs in this way, and as long as lactic acid concentrations do not become excessive. Some tissues are better adapted to anaerobic conditions than others—skeletal muscles survive longer than cardiac muscle, which in turn survives under anaerobic conditions longer than the brain.

Red blood cells, which lack mitochondria, can use only the lactic acid pathway; therefore (for reasons described in the next section), they cannot use oxygen. This spares the oxygen they carry for delivery to other cells. Except for red blood cells, anaerobic respiration occurs only for a limited period of time in tissues that have energy requirements in excess of their aerobic ability. Anaerobic respiration occurs in the

anaerobic: Gk. *an*, without; *aer*, air; *bios*, life

Figure 4.19

Glycolysis.

In glycolysis, one glucose is converted into two pyruvic acids in nine separate steps. In addition to two pyruvic acids, the products of glycolysis include two molecules of NADH and four molecules of ATP. Since two ATP molecules were used at the beginning, however, the net gain is two ATP molecules per glucose. Dashed arrows indicate reverse reactions that may occur under other conditions.

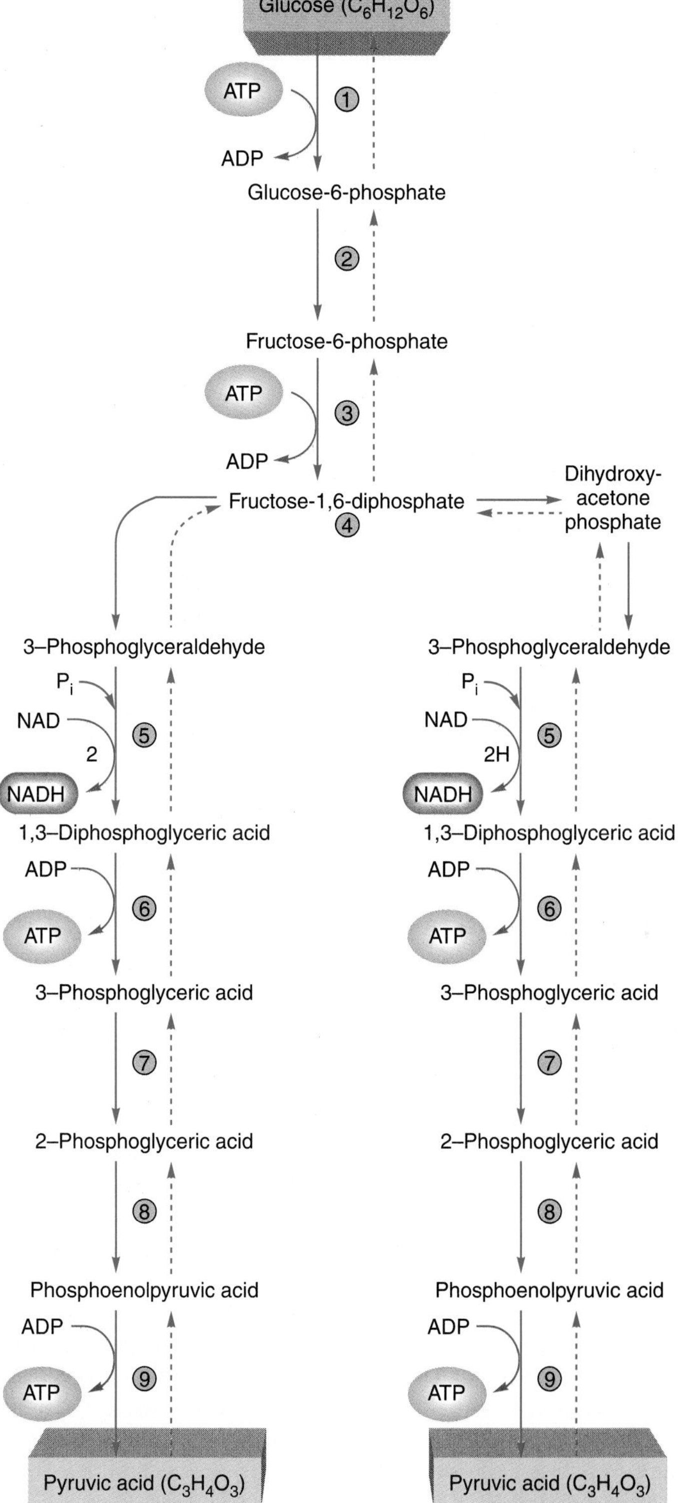

Figure 4.20

The formation of lactic acid.

The addition of two hydrogen atoms (colored boxes) from reduced NAD to pyruvic acid produces lactic acid and oxidized NAD. This reaction is catalyzed by lactic acid dehydrogenase (LDH), and is reversible under the proper conditions.

NADH + H^+ → NAD

Pyruvic acid —LDH→ Lactic acid

skeletal muscles and heart when the *ratio of oxygen supply to oxygen need* (related to the concentration of NADH) falls below a critical level. Anaerobic respiration is, in a sense, an emergency procedure that provides some ATP until the emergency (oxygen deficiency) has passed.

It should be noted, though, that there is no real "emergency" in the case of skeletal muscles, where anaerobic respiration is a normal, daily occurrence that does not harm muscle tissue or the individual. Excessive lactic acid production by muscles, however, is associated with pain and muscle fatigue. (The metabolism of skeletal muscles is discussed in chapter 12.) In contrast to skeletal muscles, the heart normally respires only aerobically. When anaerobic conditions do occur in the heart, a potentially dangerous situation may be present.

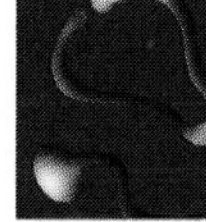

Ischemia (*ĭ-ske′me-ă*) refers to inadequate blood flow to an organ, such that the rate of oxygen delivery is insufficient to maintain aerobic respiration. Inadequate blood flow to the heart, or *myocardial ischemia,* may occur if the coronary blood flow is occluded by atherosclerosis, a blood clot, or an artery spasm. People with myocardial ischemia often experience *angina pectoris* (*an-ji′nă pek′tor-is*)—severe pain in the chest and left (or sometimes, right) arm area. This pain is associated with increased blood levels of lactic acid that are produced by the ischemic heart muscle. If the ischemia is maintained, the cells may die and produce an area called an *infarct.* The degree of ischemia and angina can be decreased by vasodilator drugs such as nitroglycerin and amyl nitrite, which improve blood flow to the heart and also decrease the work of the heart by dilating peripheral blood vessels.

Glycogenesis and Glycogenolysis

Cells cannot accumulate many separate glucose molecules, because abundance of these would exert an osmotic pressure (see chapter 5) that would draw a dangerous amount of water into the cells. Instead, many organs—particularly the liver, skeletal muscles, and heart—store carbohydrates in the form of glycogen.

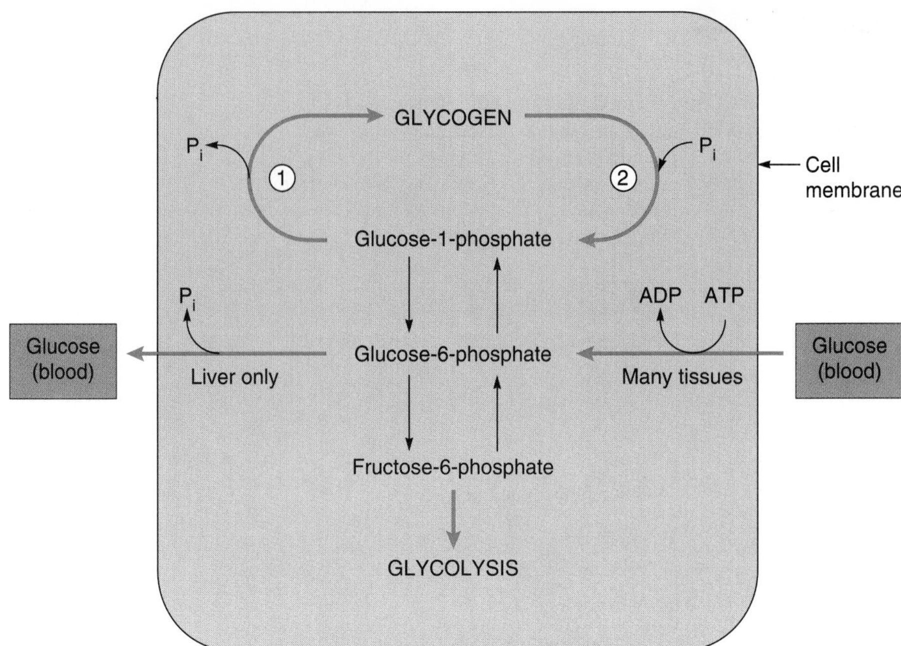

Figure 4.21

Blood glucose that enters tissue cells is converted to glucose-6-phosphate. This intermediate can be metabolized for energy in glycolysis, or it can be converted to glycogen (*1*) in a process called glycogenesis. Glycogen represents a storage form of carbohydrates, which can be used as a source for new glucose-6-phosphate (*2*) in a process called glycogenolysis. The liver contains an enzyme that can remove the phosphate from glucose-6-phosphate; liver glycogen thus serves as a source for new blood glucose.

The formation of glycogen from glucose is called **glycogenesis** (*gli″kŏ-jen′ĭ-sis*). In this process, glucose is converted to glucose-6-phosphate by utilizing the terminal phosphate group of ATP. Glucose-6-phosphate is then converted into its isomer, glucose-1-phosphate. Finally, the enzyme *glycogen synthase* removes the phosphate groups as it polymerizes glucose to form glycogen.

The reverse reactions are similar. The enzyme *glycogen phosphorylase* catalyzes the breakdown of glycogen to glucose-1-phosphate. (The phosphates are derived from inorganic phosphate, not from ATP, so glycogen breakdown does not require metabolic energy.) Glucose-1-phosphate is then converted to glucose-6-phosphate, a process called **glycogenolysis** (*gli″ko-jĕ-nol′ĭ-sis*). In most tissues, glucose-6-phosphate can then be respired for energy (through glycolysis) or used to resynthesize glycogen. Only in the liver, for reasons that will now be explained, can the glucose-6-phosphate also be used to produce free glucose for secretion into the blood.

As mentioned earlier, organic molecules with phosphate groups cannot cross cell membranes. Since the glucose derived from glycogen is in the form of glucose-1-phosphate and then glucose-6-phosphate, it cannot leak out of the cell. Similarly, glucose that enters the cell from the blood is "trapped" within the cell by conversion to glucose-6-phosphate. Skeletal muscles, which have large amounts of glycogen, can generate glucose-6-phosphate for their own glycolytic needs, but they cannot secrete glucose into the blood because they lack the ability to remove the phosphate group.

Unlike skeletal muscles, the liver contains an enzyme—known as *glucose-6-phosphatase*—that can remove the phosphate groups and produce free glucose (fig. 4.21). This free glucose can then be transported through the cell membrane. The liver, then, can secrete glucose into the blood, whereas skeletal muscles cannot. Liver glycogen can thus supply blood glucose for use by other organs, including exercising skeletal muscles that may have depleted much of their own stored glycogen during exercise.

The Cori Cycle

In humans and other mammals, much of the lactic acid produced in anaerobic respiration is later eliminated by aerobic respiration of the lactic acid to carbon dioxide and water. However, some of the lactic acid produced by exercising skeletal muscles is delivered by the blood to the liver. Within the liver cells under these conditions, the enzyme *lactic acid dehydrogenase* (*LDH*) converts lactic acid to pyruvic acid. This is the reverse of the step of anaerobic respiration shown in figure 4.20, and in the process NAD is reduced to NADH + H^+. Unlike most other organs, the liver contains the enzymes needed to take pyruvic acid molecules and convert them to glucose-6-phosphate, a process that is essentially the reverse of glycolysis.

Glucose-6-phosphate in liver cells can then be used as an intermediate for glycogen synthesis or converted to free glucose that is secreted into the blood. The conversion of noncarbohydrate molecules (not just lactic acid, but also amino acids and glycerol) through pyruvic acid to glucose is an extremely important process called **gluconeogenesis** (*gloo″ko-ne″o-jen′ĭ-sis*). The significance of this process in exercise and starvation will be discussed later in this chapter.

During exercise, some of the lactic acid produced by skeletal muscles may be transformed through gluconeogenesis in the liver to blood glucose. This new glucose can serve as an energy source during exercise and can be used after exercise to help replenish the depleted muscle glycogen. This two-way traffic between skeletal muscles and the liver is called the

glycogenesis: Gk. *glyco*, sugar; *genesis*, production

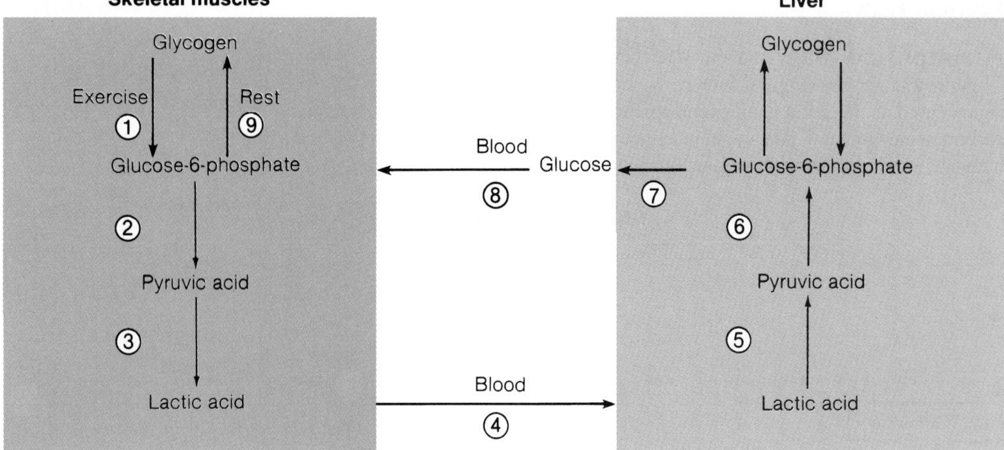

Figure 4.22

The Cori cycle.
The sequence of steps is indicated by numbers 1 through 9.

Cori cycle (fig. 4.22). Through the Cori cycle, gluconeogenesis in the liver allows depleted skeletal muscle glycogen to be restored within 48 hours.

Aerobic Respiration

In the aerobic respiration of glucose, pyruvic acid is formed by glycolysis and then converted into acetyl coenzyme A. This begins a cyclic metabolic pathway called the Krebs cycle. As a result of these pathways, a large amount of reduced NAD and FAD (NADH and $FADH_2$) is generated. These reduced coenzymes provide electrons for an energy-generating process that drives the formation of ATP.

Aerobic respiration is equivalent to combustion in terms of its final products (CO_2 and H_2O) and in terms of the total amount of energy liberated. In aerobic respiration, however, the energy is released in small, enzymatically controlled oxidation reactions, and a portion (38% to 40%) of the energy released in this process is captured in the high-energy bonds of ATP.

The aerobic respiration of glucose begins with glycolysis. Glycolysis in both anaerobic and aerobic respiration results in the production of two molecules of pyruvic acid, two molecules of ATP, and two molecules of NADH + H^+ per glucose. In aerobic respiration, however, the electrons in NADH are *not* donated to pyruvic acid and lactic acid is not formed, as happens in anaerobic respiration. Instead, the pyruvic acids will move to a different cellular location and undergo a different reaction; the NADH produced by glycolysis will eventually be oxidized, but that occurs later in the story.

In aerobic respiration, pyruvic acid leaves the cell cytoplasm and enters the interior (the matrix) of mitochondria. Once pyruvic acid is inside a mitochondrion, carbon dioxide is enzymatically removed from each three-carbon-long pyruvic acid to form a two-carbon-long organic acid—acetic acid.

Cori cycle: from Carl F. Cori, American biochemist, 1896–1984

Figure 4.23

The formation of acetyl coenzyme A in aerobic respiration.
Notice that NAD is reduced to NADH in this process.

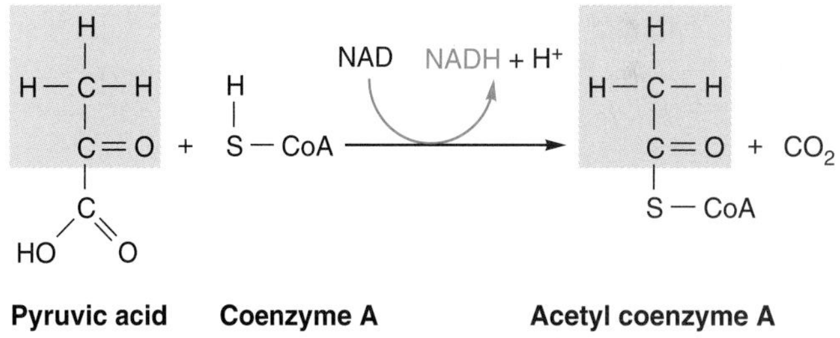

Pyruvic acid + Coenzyme A (H–S–CoA) —(NAD → NADH + H^+)→ Acetyl coenzyme A + CO_2

Pyruvic acid **Coenzyme A** **Acetyl coenzyme A**

The enzyme that catalyzes this reaction combines the acetic acid with a coenzyme (derived from the vitamin pantothenic acid) called *coenzyme* A. The combination thus produced is called **acetyl** (*as'ĭ-tl* or *ă-sēt'l*) **coenzyme A,** abbreviated **acetyl CoA** (fig. 4.23).

Glycolysis converts one glucose molecule into two molecules of pyruvic acid. Since each pyruvic acid molecule is converted into one molecule of acetyl CoA and one CO_2, two molecules of acetyl CoA and two molecules of CO_2 are derived from each glucose. These acetyl CoA molecules serve as substrates for mitochondrial enzymes in the aerobic pathway, whereas the carbon dioxide is a waste product that is carried by the blood to the lungs for elimination. It is important to note that the oxygen in CO_2 is derived from pyruvic acid, not from oxygen gas.

The Krebs Cycle

Once acetyl CoA has been formed, the acetic acid subunit (two carbons long) combines with oxaloacetic acid (four carbons long) to form a molecule of citric acid (six carbons long). Coenzyme A acts only as a transporter of acetic acid from one enzyme to another, similar to the transport of hydrogen by NAD. The formation of citric acid begins a cyclic metabolic pathway known as the **citric acid cycle,** or **TCA cycle** (for tricarboxylic acid; citric acid has three carboxylic

Figure 4.24

A simplified diagram of the Krebs cycle.
This shows how the original four-carbon-long oxaloacetic acid is regenerated at the end of the cyclic pathway. Only the numbers of carbon atoms in the Krebs cycle intermediates are shown; the numbers of hydrogens and oxygens are not accounted for in this simplified scheme.

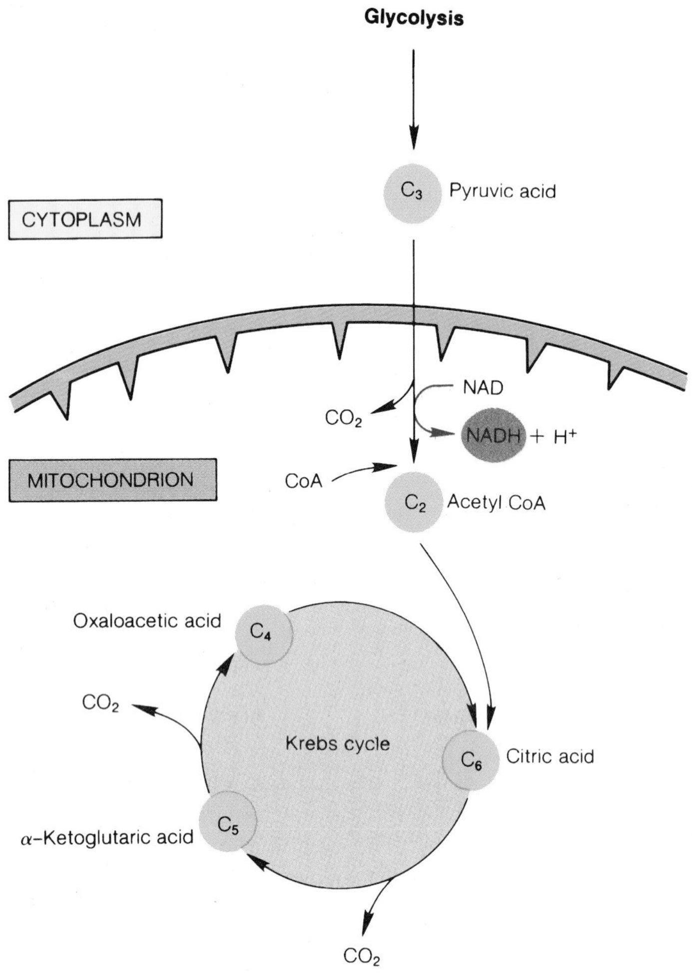

acid groups). More commonly, however, this cyclic pathway is called the **Krebs cycle** after its principal discoverer, Sir Hans Krebs. A simplified illustration of this pathway is shown in figure 4.24.

Through a series of reactions involving the elimination of two carbons and four oxygens (as two CO_2 molecules) and the removal of hydrogens, citric acid is eventually converted to oxaloacetic acid, which completes the cyclic metabolic pathway. In this process the following events occur: (1) one guanosine triphosphate (GTP) is produced (step 5 of fig. 4.25), which donates a phosphate group to ADP to produce one ATP; (2) three molecules of NAD are reduced to NADH (steps 4, 5, and 8 of fig. 4.25); and (3) one molecule of FAD is reduced to $FADH_2$ (step 6).

The production of NADH and $FADH_2$ by each "turn" of the Krebs cycle is far more significant, in terms of energy production, than the single GTP (converted to ATP) produced directly by the cycle. This is because NADH and $FADH_2$ eventually donate their electrons to an energy-transferring process that results in the formation of a large number of ATP.

Krebs cycle: from Hans A. Krebs, German biochemist, 1900–1981

Electron Transport and Oxidative Phosphorylation

Built into the foldings, or cristae, of the inner mitochondrial membrane are a series of molecules that serve as an **electron-transport system** during aerobic respiration. This electron-transport chain of molecules consists of a protein containing *flavine mononucleotide* (abbreviated FMN and derived from the vitamin riboflavin), *coenzyme Q*, and a group of iron-containing pigments called *cytochromes* (*si'tŏ-krōmz*). The last of these cytochromes is cytochrome a_3, which donates electrons to oxygen in the final oxidation-reduction reaction (as will be described). These molecules of the electron-transport system are fixed in position within the inner mitochondrial membrane in such a way that they can pick up electrons from NADH and $FADH_2$ and transport them in a definite sequence and direction.

In aerobic respiration, NADH and $FADH_2$ become oxidized by transferring their pairs of electrons to the electron-transport system of the cristae. It should be noted that the protons (H^+) are not transported together with the electrons; their fate will be described a little later. The oxidized forms of NAD and FAD are thus regenerated and can continue to "shuttle" electrons from the Krebs cycle to the electron-transport chain. The first molecule of the electron-transport chain in turn becomes reduced when it accepts the electron pair from NADH. When the cytochromes receive a pair of electrons, two ferric ions (Fe^{3+}) become reduced to two ferrous ions (Fe^{2+}).

The electron-transport chain thus acts as an oxidizing agent for NAD and FAD. Each element in the chain, however, also functions as a reducing agent; one reduced cytochrome transfers its electron pair to the next cytochrome in the chain (fig. 4.26). In this way, the iron ions in each cytochrome alternately become reduced (from Fe^{3+} to Fe^{2+}) and oxidized (from Fe^{2+} to Fe^{3+}). This is an exergonic process, and the energy derived is used to phosphorylate ADP to ATP. The production of ATP in this manner is thus appropriately termed **oxidative phosphorylation** (*ok"sĭ-da'tiv fos"for-ĭ-la'shun*).

Coupling of Electron Transport to ATP Production

Under very high magnification with the electron microscope, lollipop-like structures can be seen protruding from the inner mitochondrial membrane into the matrix. Indeed, the cristae are covered with these structures (fig. 4.27). The "lollipops" are called *respiratory assemblies* and are critically important in the generation of ATP.

According to the **chemiosmotic** (*kem-e"os-mot'ik*) **theory,** the electron-transport system, powered by the transport of

Figure 4.25

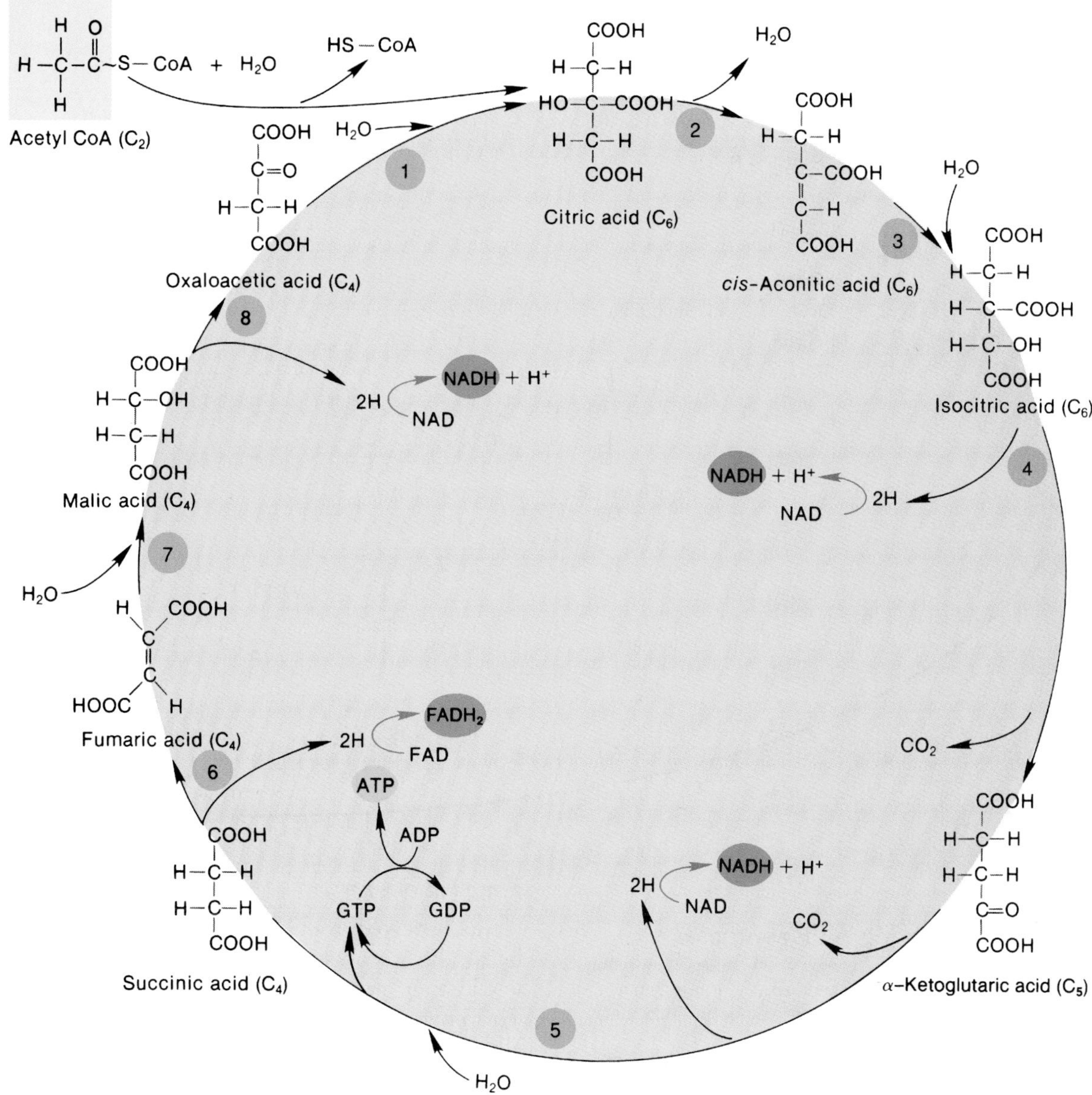

The complete Krebs cycle.

Notice that, per each cycle, 1 ATP, 3 NADH, and 1 $FADH_2$ are produced.

electrons, pumps protons (H^+) from the mitochondrial matrix into the space between the inner and outer mitochondrial membranes. The electron-transport system is grouped into three complexes that serve as **proton pumps** (figure 4.28). The first pump (the NADH-Coenzyme Q reductase complex) transports four H^+ from the matrix to the intermembrane space for every pair of electrons moved along the electron-transport system. The second pump (the cytochrome C reductase complex) also transports four protons into the intermembrane space, and the third pump (the cytochrome C oxidase complex) transports two protons into the intermembrane space. As a result, there is a higher concentration of H^+ in the intermembrane space than in the matrix, favoring the diffusion of H^+ back out into the matrix. The inner mitochondrial membrane, however, does not permit diffusion of H^+ except through the respiratory assemblies.

The respiratory assemblies consist of a group of proteins that form a "stem" and a globular subunit. The stem contains a channel through the inner mitochondrial membrane that permits the passage of protons (H^+). The globular subunit, which protrudes into the matrix, contains an **ATP synthase** enzyme that is capable of catalyzing the reaction ADP + $P_i \rightarrow$ATP when it is activated by the diffusion of protons through the respiratory assemblies into the matrix (fig. 4.28). In this way, phosphorylation (the addition of phosphate to ADP) is coupled to oxidation (the transport of electrons) in oxidative phosphorylation.

Figure 4.26

Electron transport and oxidative phosphorylation.
Each element in the electron-transport chain alternately becomes reduced and then oxidized as it transports electrons to the next member of the chain. This process provides energy for the formation of ATP. At the end of the electron-transport chain the electrons are donated to oxygen, which becomes reduced (by the addition of two hydrogen atoms) to water.

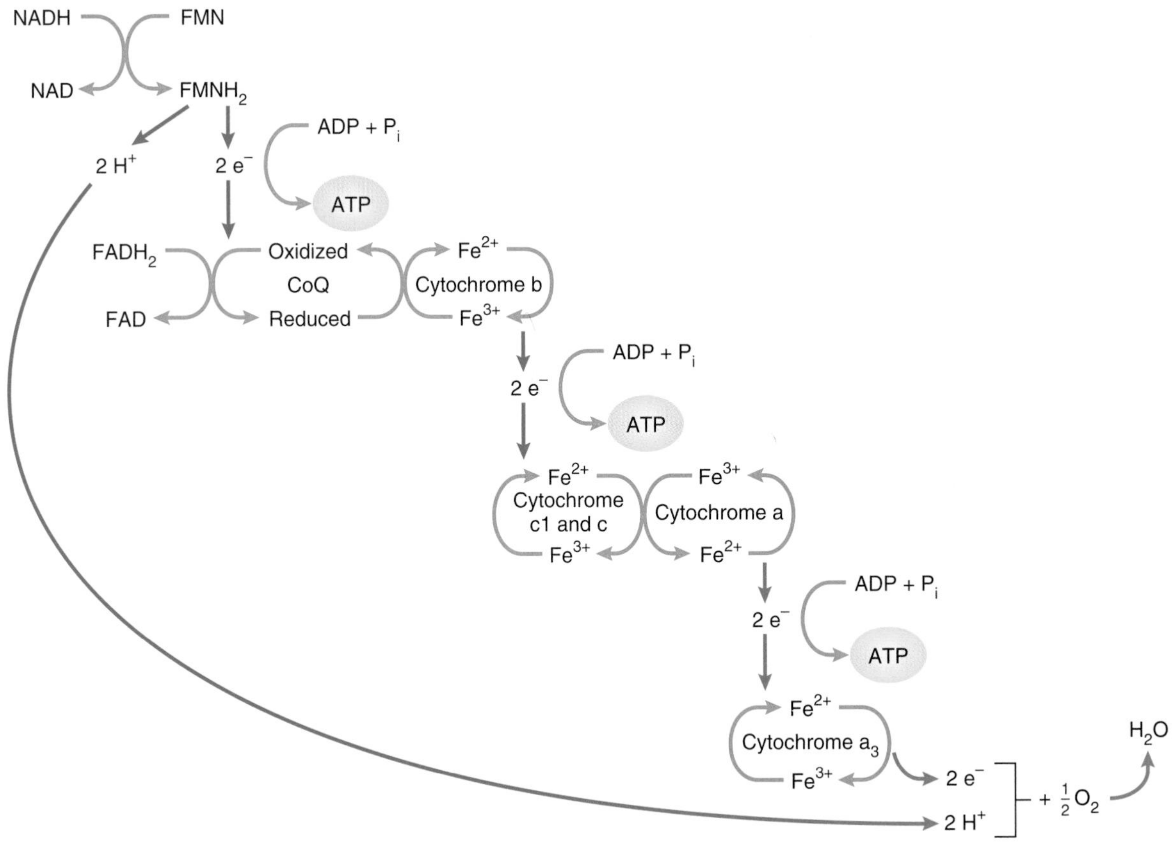

Figure 4.27

Electron micrograph showing ATP synthase.
The ATP synthase (arrow) at the end of the electron-transport system is attached to the cristae by a stalk. This unit produces ATP using energy from electron transport, in a process called oxidative phosphorylation.

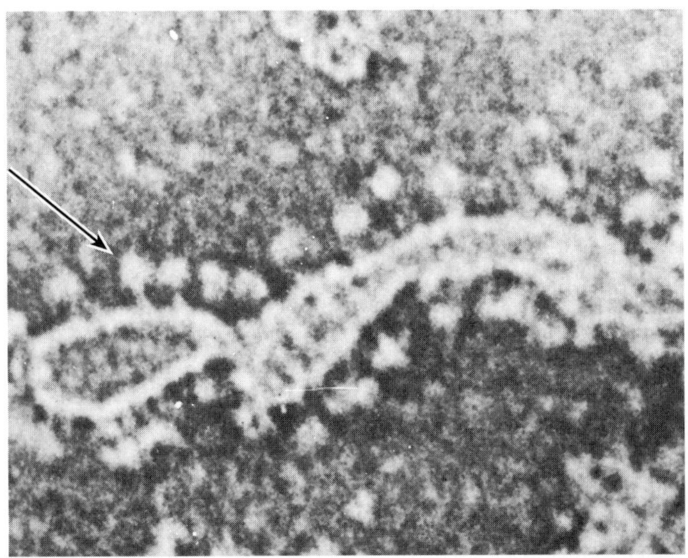

Function of Oxygen

If the last cytochrome remained in a reduced state, it would be unable to accept more electrons. Electron transport would then progress only to the next-to-last cytochrome. This process would continue until all of the elements of the electron-transport chain remained in the reduced state. At this point, the electron-transport system would stop functioning and no ATP could be produced in the mitochondria. With the electron-transport system incapacitated, NADH and $FADH_2$ could not become oxidized by donating their electrons to the chain and, through inhibition of Krebs cycle enzymes, no more NADH and $FADH_2$ could be produced in the mitochondria. The Krebs cycle would stop and respiration would become anaerobic.

Oxygen, from the air we breathe, allows electron transport to continue by functioning as the **final electron acceptor** of the electron-transport chain. This oxidizes cytochrome a_3, allowing electron transport and oxidative phosphorylation to continue. At the very last step of aerobic respiration, therefore, oxygen becomes reduced by the two electrons that were passed to the chain from NADH and $FADH_2$. This reduced oxygen binds two protons to form a molecule of water.

Figure 4.28

A schematic representation of the chemiosmotic theory.
(*a*) A mitochondrion. (*b*) The matrix and the compartment between the inner and outer mitochondrial membrane showing how the electron-transport system functions as H^+ pumps. This results in a steep H^+ gradient between the intermembrane space and the cytoplasm of the cell. The diffusion of H^+ through ATP synthase results in the production of ATP.

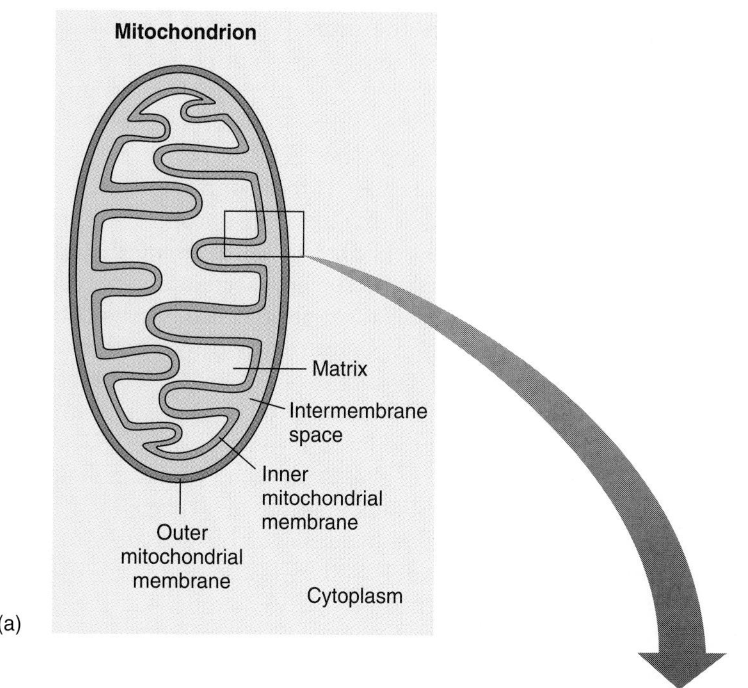

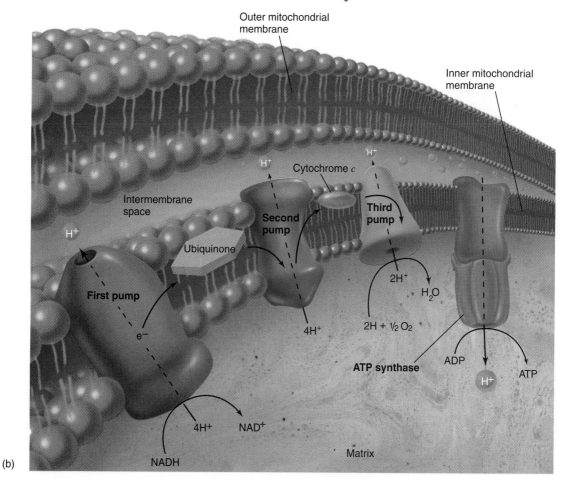

Since the oxygen atom is part of a molecule of oxygen gas (O_2), this last reaction can be shown as follows:

$$O_2 + 4\ e^- + 4\ H^+ \rightarrow 2\ H_2O$$

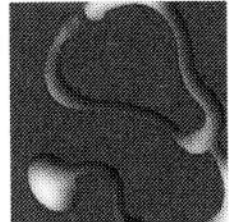

Cyanide is a fast-acting lethal poison that produces such symptoms as rapid heart rate, tiredness, seizures, and headache. Cyanide poisoning can result in coma and ultimately death in the absence of quick treatment. The reason that cyanide is so deadly is that it has one very specific action: it blocks the transfer of electrons from cytochrome a_3 to oxygen. The effects are thus the same as would occur if oxygen were completely removed—aerobic cell respiration and the production of ATP by oxidative phosphorylation comes to a halt.

ATP Balance Sheet

There are two different methods of ATP formation in cell respiration. One method is the **direct** (also called **substrate-level) phosphorylation** that occurs in glycolysis (producing a net gain of 2 ATP) and the Krebs cycle (producing 1 ATP per cycle). These numbers are certain and constant. The second method of ATP formation, **oxidative phosphorylation,** produces numbers of ATP that are less certain and constant because they vary under different conditions and for different kinds of cells. For many years, it was believed that 1 NADH yielded 3 ATP and that 1 $FADH_2$ yielded 2 ATP by oxidative phosphorylation. This gave a grand total of 36 to 38 ATP per glucose through cell respiration (see the footnote in table 4.3). Newer biochemical information, however, suggests that these numbers may be overestimates, because although 36 to 38 ATP are produced per glucose in the mitochondrion, only 30 to 32 ATP actually enter the cytoplasm of the cell.

Three protons must pass through the respiratory assemblies and activate ATP synthase to produce 1 ATP. However, the newly formed ATP is located in the mitochondrial matrix and must be moved into the cytoplasm; this transport also uses the proton gradient and costs one more proton. The ATP and H^+ are transported into the cytoplasm in exchange for ADP and P_i, which are transported into the mitochondrion. Thus, it effectively takes four protons to produce 1 cytoplasmic ATP.

Each NADH formed in the mitochondrion donates two electrons to the electron-transport system at the first proton pump (fig. 4.28). The electrons are then passed to the second and third proton pumps, activating each of them in turn until the two electrons are ultimately passed to oxygen. The first and second pumps transport four protons each, and the third pump transports two protons, for a total of ten. Dividing ten protons by the four it takes to produce an ATP gives 2.5 ATP that are produced for every pair of electrons donated by an NADH. (There is no such thing as half an ATP; the decimal fraction simply indicates an average.)

Three molecules of NADH are formed with each Krebs cycle, and 1 NADH is also produced when pyruvate is converted into acetyl CoA (see fig. 4.23). Starting from one glucose, two Krebs cycles (producing 6 NADH) and two pyruvates converted to acetyl CoA (producing 2 NADH), yield 8 NADH. Multiplying by 2.5 ATP per NADH gives 20 ATP.

Electrons from $FADH_2$ are donated later in the electron-transport system than those donated by NADH; consequently, these electrons activate only the second and third proton

Table 4.3
ATP Yield per Glucose in Aerobic Respiration

Phases of Respiration	ATP Made Directly	Reduced Coenzymes	ATP Made by Oxidative Phosphorylation*
Glucose to pyruvate (in cytoplasm)	**2 ATP** (net gain)	2 NADH, but usually goes into mitochondria as 2 $FADH_2$	1.5 ATP per $FADH_2$ × 2 = **3 ATP**
Pyruvate to acetyl coA (× 2 because one glucose yields 2 pyruvates)	None	1 NADH (× 2) = 2 NADH	2.5 ATP per NADH × 2 = **5 ATP**
Krebs cycle (× 2 because one glucose yields 2 Krebs cycles)	1 ATP (× 2) = **2 ATP**	3 NADH (× 2) 1 $FADH_2$ (× 2)	2.5 ATP per NADH × 3 = 7.5 ATP × 2 = **15 ATP**
			1.5 ATP per $FADH_2$ × 2 = **3 ATP**
Subtotals	4 ATP		26 ATP
Total		**30 ATP**	

*Older estimates of ATP production from oxidative phosphorylation are 2 ATP per $FADH_2$ and 3 ATP per NADH. If these numbers are used, a total of 32 ATP will be calculated as arising from oxidative phosphorylation. This is increased to 34 ATP if the cytoplasmic NADH remains as NADH when it is shuttled into the mitochondrion. Adding these numbers to the four ATP made directly gives a total of 38 molecules of ATP produced per glucose molecule.

pumps. Since the first proton pump is bypassed, the electrons passed from $FADH_2$ result in the pumping of only six protons (four by the second pump and two by the third pump). Since 1 ATP is produced for every four protons pumped, electrons derived from $FADH_2$ result in the formation of $6 \div 4 = 1.5$ ATP. Each Krebs cycle produces 1 $FADH_2$, and we get two Krebs cycles from one glucose, so there are 2 $FADH_2$ that give 2×1.5 ATP = 3 ATP.

The 23 ATP subtotal from oxidative phosphorylation we have at this point includes only the NADH and $FADH_2$ produced in the mitochondrion. Remember that glycolysis, which occurs in the cytoplasm, also produces 2 NADH. These cytoplasmic NADH molecules cannot directly enter the mitochondrion, but their electrons can be "shuttled" in. The net effect of the most common shuttle is that a molecule of NADH in the cytoplasm is translated into a molecule of $FADH_2$ in the mitochondrion. The 2 NADH produced in glycolysis, therefore, usually become 2 $FADH_2$ and yield 2×1.5 ATP = 3 ATP by oxidative phosphorylation. (Less commonly, the cytoplasmic NADH may become mitochondrial NADH and produce 2×2.5 ATP = 5 ATP.)

We now have a total 26 ATP (or, less commonly, 28 ATP) produced by oxidative phosphorylation from glucose. Adding the 2 ATP made by direct (substrate-level) phosphorylation in glycolysis and the 2 ATP made directly by the two Krebs cycles yields a grand total of 30 ATP (or, less commonly, 32 ATP) produced by the aerobic respiration of glucose (table 4.3).

Metabolism of Lipids and Proteins

Triglycerides can be hydrolyzed into glycerol and fatty acids. The latter are of particular importance because they can be converted into numerous molecules of acetyl CoA that can enter Krebs cycles and generate a large amount of ATP. Amino acids derived from proteins also may be used for energy. This involves deamination (removal of the amine group) and the conversion of the remaining molecule into either pyruvic acid or one of the Krebs cycle molecules.

Energy can be derived from the cellular respiration of lipids and proteins using the same aerobic pathway previously described for the metabolism of pyruvic acid. Indeed, some organs preferentially use molecules other than glucose as an energy source. Pyruvic acid and the Krebs cycle acids also serve as common intermediates in the interconversion of glucose, lipids, and amino acids.

When food energy is taken into the body faster than it is consumed, the concentration of ATP within the body cells rises. Cells, however, do not store extra energy in the form of extra ATP. When cellular ATP concentrations rise because more energy (from food) is available than can be immediately used, high ATP concentrations inhibit glycolysis. Under conditions of high ATP concentrations, when glycolysis is inhibited, glucose is instead converted into glycogen and fat (fig. 4.29).

Lipid Metabolism

When glucose is going to be converted into fat, glycolysis occurs and pyruvic acid is converted into acetyl CoA. Some of the glycolytic intermediates—phosphoglyceraldehyde and dihydroxyacetone phosphate—do not complete their conversion to pyruvic acid, however, and acetyl CoA does not enter a Krebs cycle. The acetic acid subunits of these acetyl CoA molecules can instead be used to produce a variety of lipids, including cholesterol (used in the synthesis of bile salts and steroid hormones), ketone bodies, and fatty acids (fig. 4.30). Acetyl CoA may thus be considered a branch point from which a number of different possible metabolic pathways may progress.

In the formation of fatty acids, a number of acetic acid (two-carbon) subunits join to form the fatty acid chain. Six acetyl CoA molecules, for example, will produce a fatty acid that is twelve carbons long. When three of these fatty acids condense with one glycerol (derived from phosphoglyceraldehyde), a triglyceride molecule is produced. The formation of fat, or **lipogenesis** (*lip″ŏ-jen′ĕ-sis*), occurs primarily in adipose tissue and in the liver when the concentration of blood glucose is elevated following a meal.

Fat represents the major form of energy storage in the body. One gram of fat contains 9 kilocalories of energy, compared to 4 kilocalories for a gram of carbohydrates or protein. In a nonobese 70-kilogram man, 80% to 85% of the body's energy is stored as fat, which amounts to about 140,000 kilocalories. Stored glycogen, by contrast, accounts for less than 2,000 kilocalories, most of which (about 350 g) is stored in skeletal muscles and is available for use only by the muscles. The liver contains from 80 to 90 grams of glycogen, which can be converted to glucose and used by other organs. Protein accounts for 15% to 20% of the stored calories in the body, but protein is usually not used extensively as an energy source because that would involve the loss of muscle mass.

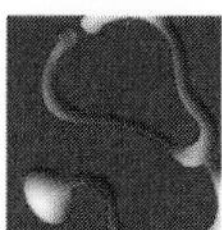

It is common experience that the ingestion of excessive calories in the form of carbohydrates (cakes, ice cream, candy, and so on) increases fat production. The rise in blood glucose that follows carbohydrate-rich meals stimulates insulin secretion, and this hormone, in turn, promotes the entry of blood glucose into adipose cells. Increased availability of glucose within adipose cells, under conditions of high insulin secretion, promotes the conversion of glucose to fat (see fig. 4.29). The hormonal control of carbohydrate and fat metabolism is discussed in chapter 27.

Breakdown of Fat (Lipolysis)

When fat stored in adipose tissue is going to be used as an energy source, *lipase* enzymes hydrolyze triglycerides into glycerol and free fatty acids in a process called **lipolysis** (*lĭ-pol′ĭ-sis*). These molecules (primarily the free fatty acids) serve as *bloodborne energy carriers* that can be used by the liver, skeletal muscles, and other organs for aerobic respiration.

Figure 4.29

The conversion of glucose into glycogen and fat.

This conversion occurs when there is inhibition of respiratory enzymes when the cell has adequate amounts of ATP. Favored pathways are indicated by blue arrows.

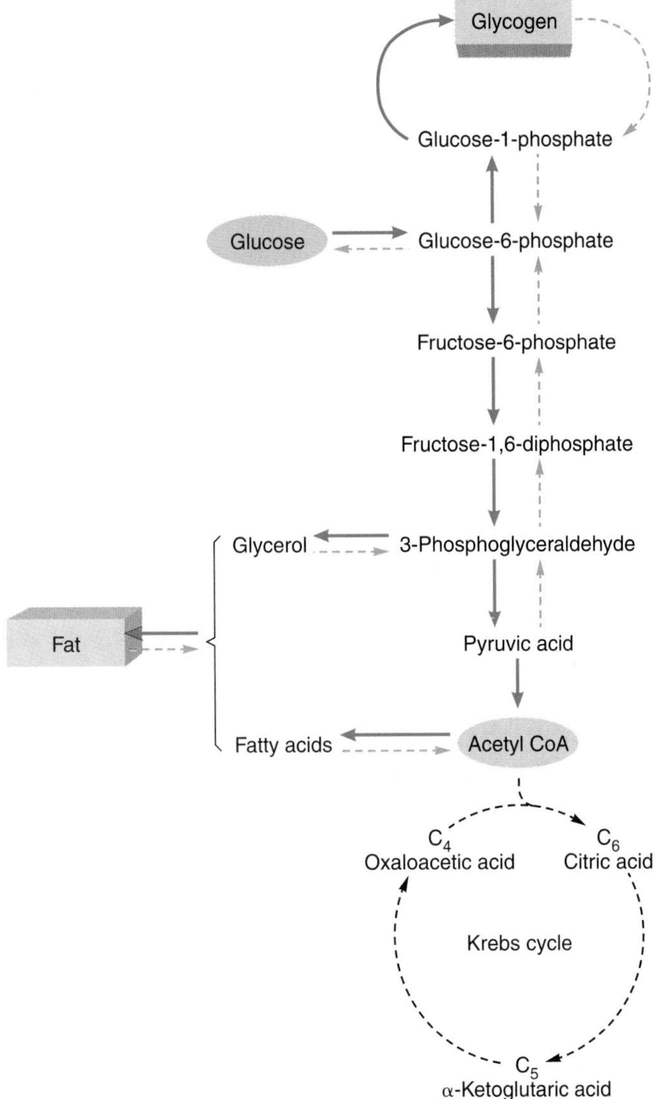

A few organs can utilize glycerol for energy by virtue of an enzyme that converts glycerol to phosphoglyceraldehyde. Free fatty acids, however, serve as the major energy source derived from triglycerides. Most fatty acids consist of a long hydrocarbon chain with a carboxylic acid group (COOH) at one end. In a process known as **β-oxidation** (β is the Greek letter *beta*), enzymes remove two-carbon acetic acid molecules from the acid end of a fatty acid chain. This results in the formation of acetyl CoA, as the third carbon from the end oxidizes to produce a new carboxylic acid group. The fatty acid chain is thus decreased in length by two carbons. The process of β-oxidation continues until the entire fatty acid molecule is converted to acetyl CoA (fig. 4.31).

A 16-carbon-long fatty acid, for example, yields eight acetyl CoA molecules. Each of these can enter a Krebs cycle and produce ten ATP per turn of the cycle, so that 8×10, or 80, ATP are produced. In addition, each time an acetyl CoA is formed and the end carbon of the fatty acid chain is oxidized, one NADH and one $FADH_2$ are produced. Oxidative phosphorylation produces 2.5 ATP per NADH and 1.5 ATP per $FADH_2$. For a 16-carbon-long fatty acid, these four ATP molecules would be formed seven times (producing 4×7, or 28, ATP). Not counting the single ATP used to start β-oxidation (fig. 4.31), this fatty acid could yield a grand total of 28 + 80, or 108, ATP molecules!

Figure 4.30

Divergent metabolic pathways for acetyl coenzyme A.

Acetyl CoA is a common substrate that can be used to produce a number of chemically related products.

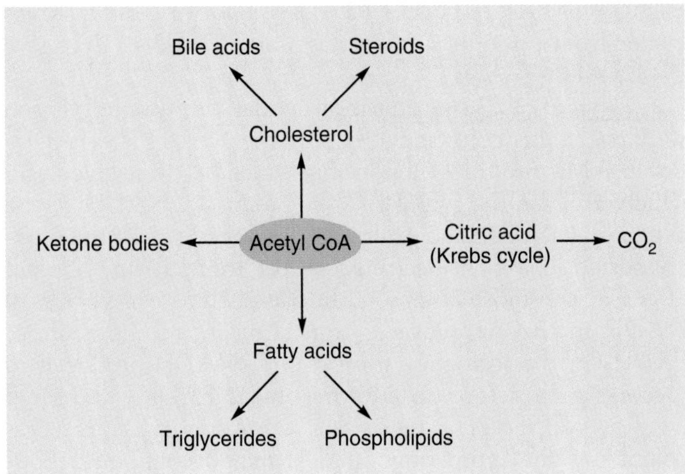

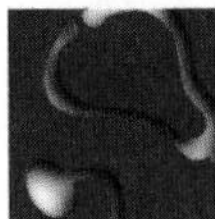

The amount of *brown fat* in the body is greatest at the time of birth. Brown fat is the major site for thermogenesis (heat production) in the newborn, and is especially prominent around the kidneys and adrenal glands. Smaller amounts are found around the blood vessels of the chest and neck. In response to regulation by thyroid hormone (see chapter 19) and norepinephrine from sympathetic nerves (see chapter 17), brown fat produces a unique uncoupling protein. This protein causes H^+ to leak out of the inner mitochondrial membrane, so that less H^+ is available to pass through the respiratory assemblies and drive ATP synthase activity. Therefore, less ATP is made by the electron-transport system than would otherwise be the case. Lower ATP concentrations cause the electron-transport system to be more active and generate more heat from the respiration of fatty acids. Experimental evidence in other mammals suggests that this extra heat is needed to prevent hypothermia (low body temperature) in newborns.

Ketone Bodies

Even when a person is not losing weight, the triglycerides in adipose tissue are continuously being broken down and resynthesized. New triglycerides are produced, while others are hydrolyzed into glycerol and fatty acids. This turnover ensures that the blood will normally contain a sufficient level

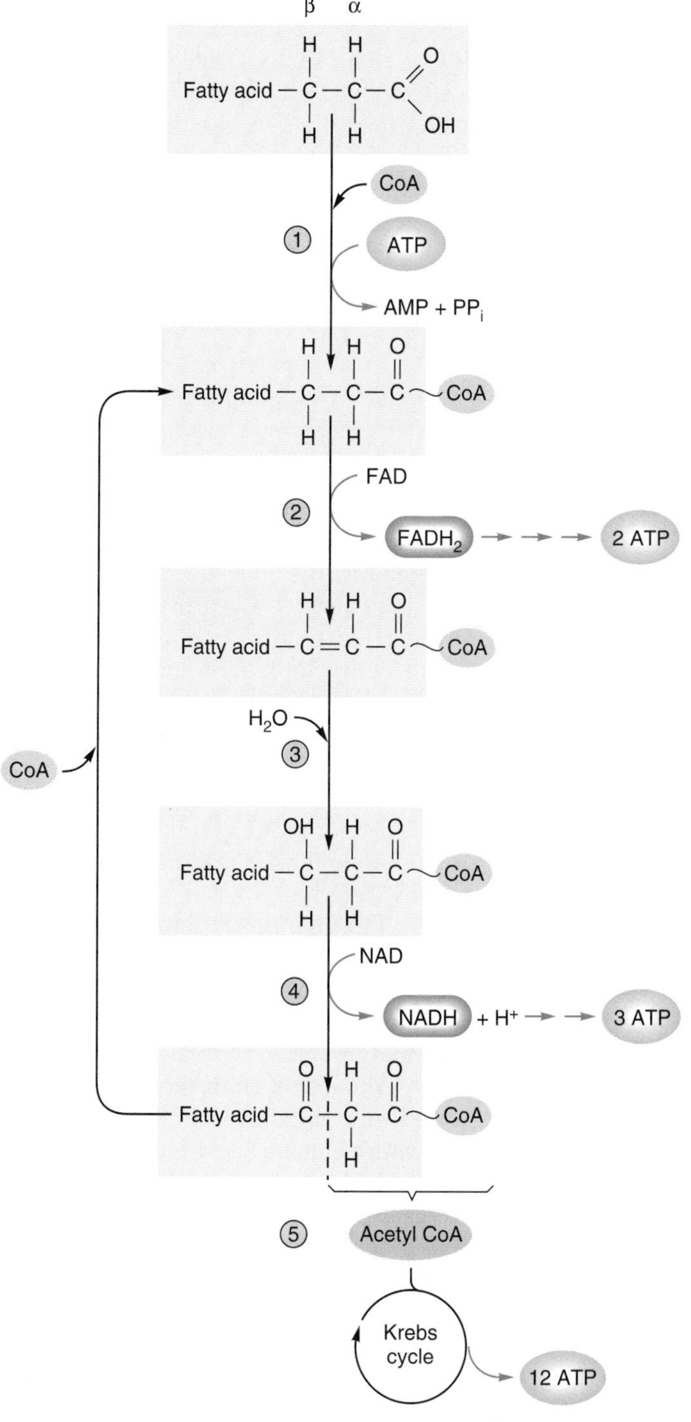

Figure 4.31

Beta-oxidation of a fatty acid.
After the attachment of coenzyme A to the carboxylic acid group (*step 1*), a pair of hydrogens is removed from the fatty acid and used to reduce one molecule of FAD (*step 2*). When this electron pair is donated to the cytochrome chain, 2 ATP are produced. The addition of a hydroxyl group from water (*step 3*), followed by the oxidation of the β carbon (*step 4*), results in the production of 3 ATP from the electron pair donated by NADH. The bond between the α and β carbons in the fatty acid is broken (*step 5*), releasing acetyl coenzyme A and a fatty acid chain that is two carbons shorter than the original. With the addition of a new coenzyme A to the shorter fatty acid, the process begins again (*step 2*), as acetyl CoA enters the Krebs cycle and generates 12 ATP.

of fatty acids for aerobic respiration by skeletal muscles, the liver, and other organs. When the rate of lipolysis exceeds the rate of fatty acid utilization—as it may in starvation, dieting, and in diabetes mellitus—the blood concentration of fatty acids increases.

If the liver cells contain sufficient amounts of ATP so that further production of ATP is not needed, some of the acetyl CoA derived from fatty acids is channeled into an alternate pathway. This pathway involves the conversion of two molecules of acetyl CoA into four-carbon-long acidic derivatives, *acetoacetic acid* and *β-hydroxybutyric* acid. Together with *acetone*, which is a three-carbon-long derivative of acetoacetic acid, these products are known as **ketone bodies.**

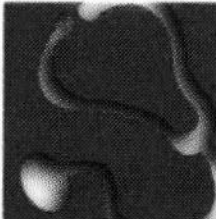

Ketone bodies can be used for energy by many organs and are found in the blood under normal conditions. Under conditions of fasting or of diabetes mellitus, however, the increased liberation of free fatty acids from adipose tissue results in the increased production of ketone bodies by the liver. The secretion of abnormally high amounts of ketone bodies into the blood produces *ketosis,* which is one of the signs of fasting or an uncontrolled diabetic state. A person in this condition may also have a sweet-smelling breath due to the presence of acetone, which is volatile and leaves the blood in the exhaled air.

Amino Acid Metabolism

Nitrogen is primarily ingested as proteins, enters the body as amino acids, and is mainly excreted as urea in the urine. In childhood, the amount of nitrogen excreted may be less than the amount ingested because amino acids are incorporated into proteins during growth. Growing children are thus said to be in a state of *positive nitrogen balance.* People who are starving or suffering from prolonged wasting diseases, by contrast, are in a state of *negative nitrogen balance;* they excrete more nitrogen than they ingest because they are breaking down their tissue proteins.

Healthy adults maintain a state of nitrogen balance, in which the amount of nitrogen excreted is equal to the amount ingested. This does not imply that the amino acids ingested are unnecessary; on the contrary, they are needed to replace the protein that is "turned over" each day. When more amino acids are ingested than are needed to replace proteins, the excess amino acids are not stored as additional protein (one cannot build muscles simply by eating large amounts of protein). Rather, the amine groups can be removed, and the "carbon skeletons" of the organic acids that are left can be used for energy or converted to carbohydrate and fat.

Transamination

An adequate amount of all 20 amino acids is required to build proteins for growth and to replace the proteins that are turned over. However, only eight of these (in adults) or nine (in children) cannot be produced by the body and must be obtained in the diet. These are the **essential amino acids** (table 4.4). The remaining amino acids are "nonessential" only in the

Table 4.4

The Essential and Nonessential Amino Acids

Essential Amino Acids	Nonessential Amino Acids
Lysine	Aspartic acid
Tryptophan	Glutamic acid
Phenylalanine	Proline
Threonine	Glycine
Valine	Serine
Methionine	Alanine
Leucine	Cysteine
Isoleucine	Arginine
Histidine (children)	Asparagine
	Glutamine
	Tyrosine

sense that the body can produce them if provided with a sufficient amount of carbohydrates and the essential amino acids.

Pyruvic acid and the Krebs cycle acids are collectively termed *keto acids* because they have a ketone group; these should not be confused with the ketone bodies (derived from acetyl CoA) discussed in the previous section. Keto acids can be converted to amino acids by the addition of an amine (NH_2) group. This amine group is usually obtained by "cannibalizing" another amino acid; in this process, a new amino acid forms as the one that was cannibalized converts to a new keto acid. This type of reaction, in which the amine group transfers from one amino acid to form another, is called **transamination** (*trans"am-ĭ-na'shun*) (fig. 4.32).

Each transamination reaction is catalyzed by a specific enzyme (a transaminase) that requires vitamin B_6 (pyridoxine) as a coenzyme. The amine group from glutamic acid, for example, may be transferred to either pyruvic acid or oxaloacetic acid. The former reaction is catalyzed by the enzyme alanine transaminase (ALT); the latter reaction is catalyzed by aspartate transaminase (AST). These enzyme names reflect the fact that the addition of an amine group to pyruvic acid produces the amino acid alanine; the addition of an amine group to oxaloacetic acid produces the amino acid known as aspartic acid (fig. 4.32).

Figure 4.32

Two important transamination reactions.
The areas shaded in blue indicate the parts of the molecules that are changed. (AST = aspartate transaminase; ALT = alanine transaminase. The amino acids are identified in boldface type.)

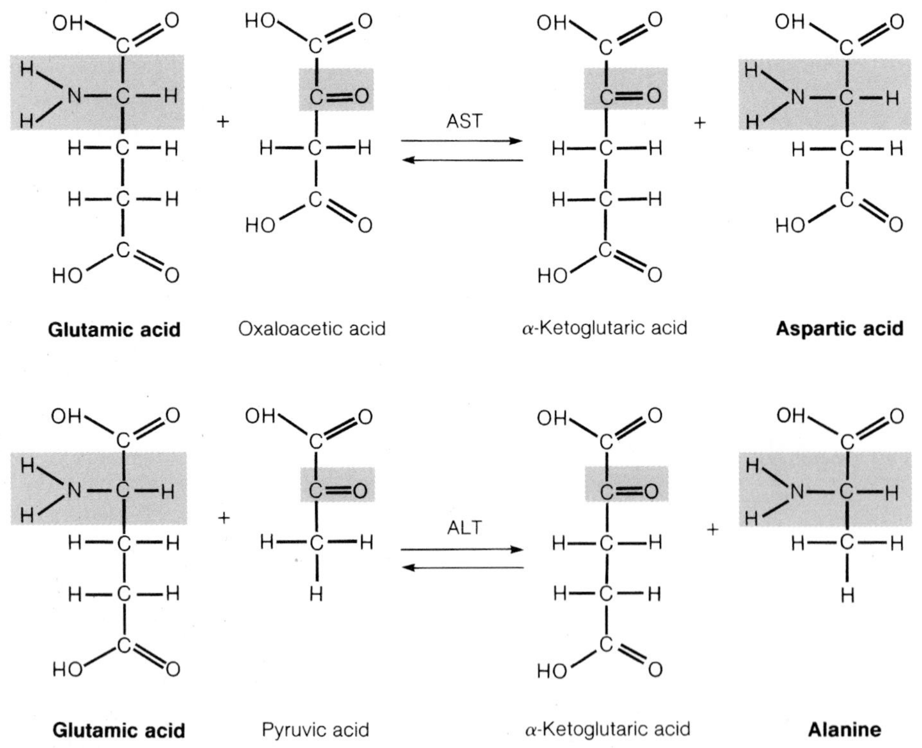

Oxidative Deamination

As shown in figure 4.33, glutamic acid can be formed through transamination by the combination of an amine group with α-ketoglutaric acid. Glutamic acid is also produced in the liver from the ammonia that is generated by intestinal bacteria and carried to the liver in the hepatic portal vein. Since free ammonia is very toxic, its removal from the blood and incorporation into glutamic acid is an important function of the healthy liver.

If there are more amino acids than are needed for protein synthesis, the amine group from glutamic acid may be removed and excreted as *urea* in the urine (fig. 4.33). The metabolic pathway that removes amine groups from amino acids—leaving a keto acid and ammonia (which is converted to urea)—is known as **oxidative deamination** (*ok"sĭ-da'tiv de-am"ĭ-na'shun*).

A number of amino acids can be converted into glutamic acid by transamination. Since glutamic acid can donate amine groups to urea (through deamination), it serves as a channel through which other amino acids can be used to produce keto acids (pyruvic acid and Krebs cycle acids). These keto acids may then be used in the Krebs cycle as a source of energy (fig. 4.34).

Figure 4.33

Oxidative deamination.

Glutamic acid is converted to α-ketoglutaric acid as it donates its amine group to the metabolic pathway that results in the formation of urea.

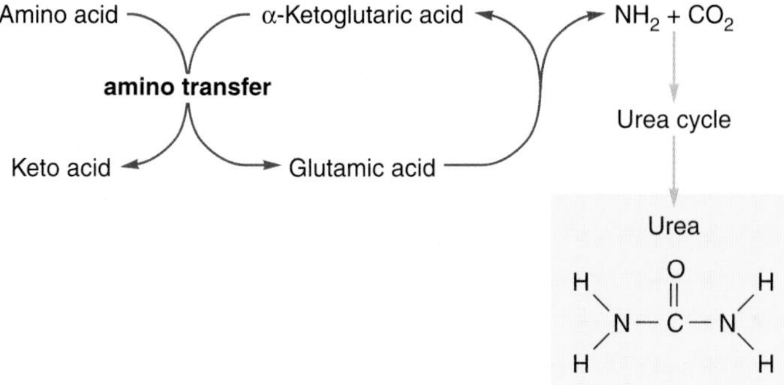

Depending upon which amino acid is deaminated, the keto acid left over may be either pyruvic acid or one of the Krebs cycle acids. These can be respired for energy, converted to fat, or converted to glucose. In the last case, the amino acids eventually change to pyruvic acid, which is used to form glucose. This process—the formation of glucose from amino acids or other noncarbohydrate molecules—is called *gluconeogenesis*, as mentioned previously in connection with the Cori cycle.

The main substrates for gluconeogenesis are the three-carbon-long molecules of alanine (an amino acid), lactic acid, and glycerol. This illustrates the interrelationship between amino acids, carbohydrates, and fat, as shown in figure 4.35. Recent experiments in humans have suggested that, even in the early stages of fasting, most of the glucose secreted by the liver is derived through gluconeogenesis. Findings indicate that hydrolysis of liver glycogen (glycogenolysis) contributes only 36% of the glucose secreted during the early stages of a fast. By 42 hours of fasting, all of the glucose secreted by the liver is produced by gluconeogenesis.

Figure 4.34

Pathways by which amino acids can be catabolized for energy.

These pathways are indirect for some amino acids, which first must be transaminated into other amino acids before being converted into keto acids by deamination.

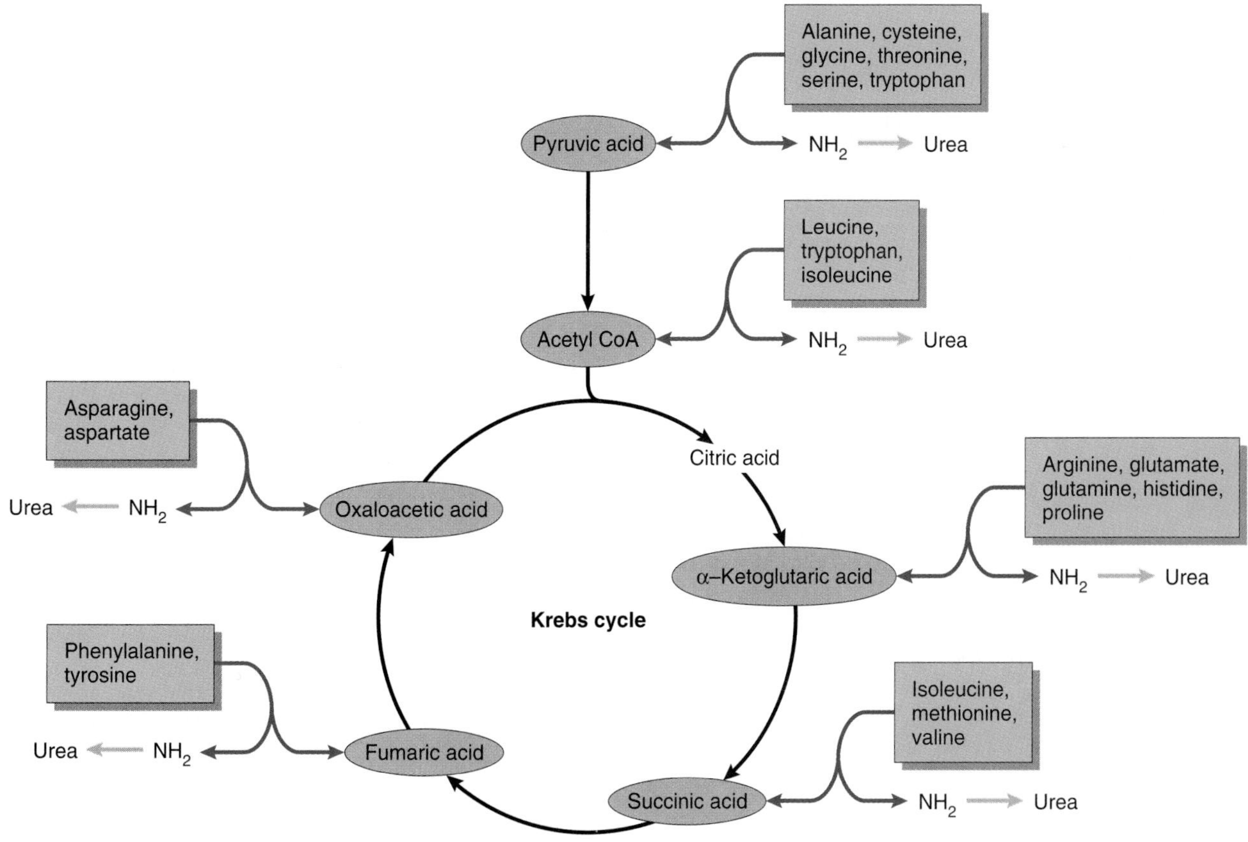

Figure 4.35

The interconversion of glycogen, fat, and protein.
These simplified metabolic pathways show how glycogen, fat, and protein can be interconverted. Note that most reactions are reversible, but the reaction from pyruvic acid to acetyl CoA is not. This is because a CO_2 is removed in the process. (Only plants, in a phase of photosynthesis called the dark reaction, can use CO_2 to produce glucose.)

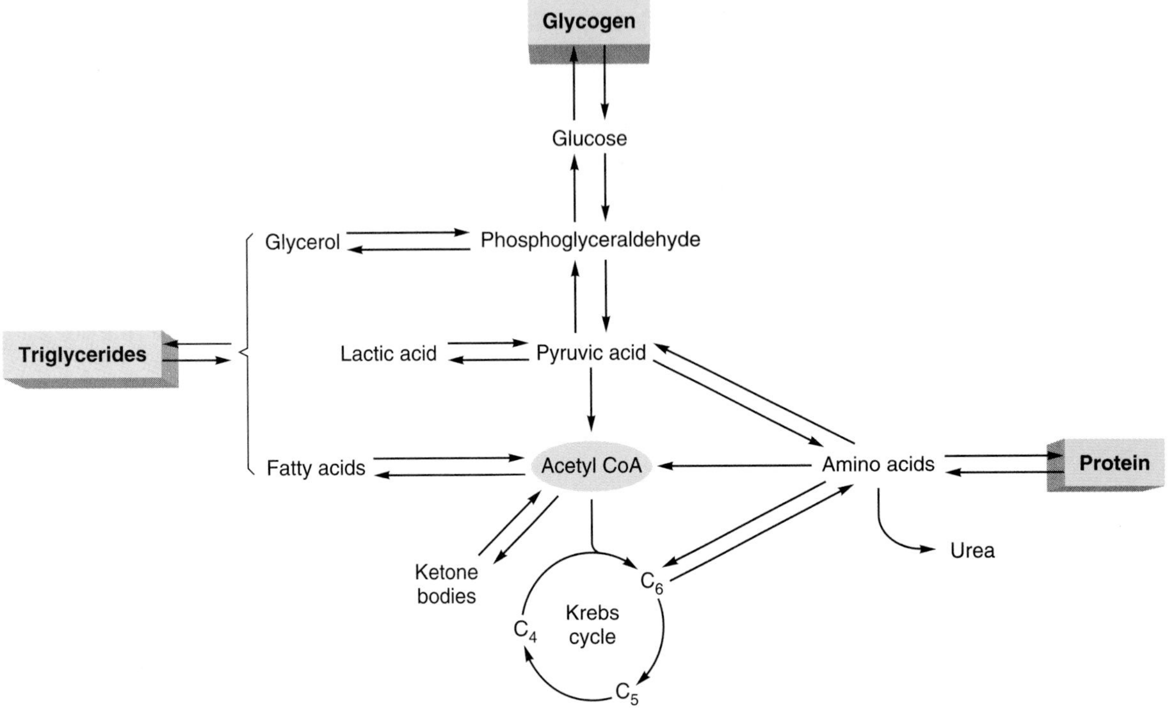

Uses of Different Energy Sources

The blood serves as a common trough from which all the cells in the body are fed. If all cells used the same energy source, such as glucose, this source would quickly be depleted and cellular starvation would occur. However, the blood normally contains a variety of energy sources from which to draw: glucose and ketone bodies from the liver, fatty acids from adipose tissue, and lactic acid and amino acids from muscles. Some organs preferentially use one energy source more than the others, so that energy sources may be "spared" for organs with strict energy needs.

The brain uses blood glucose as its major energy source. Under fasting conditions, blood glucose is primarily supplied by the liver through glycogenolysis and gluconeogenesis. In addition, the blood glucose concentration is maintained because many organs spare glucose by using fatty acids, ketone bodies, and lactic acid as energy sources. During severe starvation, the brain also gains some ability to metabolize ketone bodies for energy.

As mentioned earlier, lactic acid produced anaerobically during exercise can be used for energy following exercise. The lactic acid is reconverted to pyruvic acid, which then enters the aerobic respiratory pathway. The extra oxygen required to metabolize lactic acid contributes to the *oxygen debt* following exercise (see chapter 12).

Clinical Considerations

Clinical Enzyme Measurements

Assays of Enzymes in Plasma

When tissues become damaged as a result of diseases, some of the dead cells disintegrate and release their enzymes into the blood. Most of these enzymes are not normally active in the blood because their specific substrates are not available, but their enzymatic activity can be measured in a test tube by the addition of the appropriate substrates to samples of plasma. Such measurements are clinically useful because abnormally high plasma concentrations of particular enzymes are characteristic of certain diseases (table 4.5).

Identification of Isoenzymes

Different organs, when they are diseased, may liberate different isoenzymatic forms of an enzyme that can be measured in a clinical laboratory. For example, the enzyme **creatine phosphokinase** (*kre′ă-tin fos-fo-ki′nās*), abbreviated either **CPK** or **CK,** exists in three isoenzymatic forms. These forms are identified by two letters that indicate two components of this enzyme. One form is identified as MM and is liberated from diseased skeletal muscle; the second is BB, released by a damaged

Table 4.5

Diseases Associated with Abnormal Plasma Concentrations of Selected Enzymes

Enzyme	Associated Disease(s)
Alkaline phosphatase	Obstructive jaundice, Paget's disease (osteitis deformans), carcinoma of bone
Acid phosphatase	Benign hypertrophy of prostate, cancer of prostate
Amylase	Pancreatitis, perforated peptic ulcer
Aldolase	Muscular dystrophy
Creatine kinase (or creatine phosphokinase—CPK)	Muscular dystrophy, myocardial infarction
Lactate dehydrogenase (LDH)	Myocardial infarction, liver disease, renal disease, pernicious anemia
Transaminases (GOT and GPT)	Myocardial infarction, hepatitis, muscular dystrophy

brain; and the third is MB, released from a diseased heart. Clinical tests utilizing antibodies that can bind to the M and B components are now available to specifically measure the level of the MB form in the blood when heart disease is suspected.

Metabolic Disturbances

Phenylketonuria (PKU)

The branched metabolic pathway that begins with phenylalanine as the initial substrate is subject to a number of inborn errors of metabolism (fig. 4.36). When the enzyme that converts this amino acid to tyrosine is defective, the final products of a divergent pathway accumulate and can be detected in the blood and urine. This disease, **phenylketonuria** (*fen"il-kēt"n-oor'e-ă*) (**PKU**), can result in severe mental retardation and a shortened life span. Although inborn errors of metabolism are relatively rare, the incidence of PKU is high enough and the defect is so easy to detect that all newborn babies are routinely tested for it. If this disease is detected early, brain damage can be prevented by placing the child on an artificial diet low in the amino acid phenylalanine.

Figure 4.36

Metabolic pathways for the degradation of the amino acid phenylalanine.
Defective $enzyme_1$ produces phenylketonuria (PKU); defective $enzyme_2$ produces alcaptonuria (not a clinically significant condition); and defective $enzyme_3$ produces albinism.

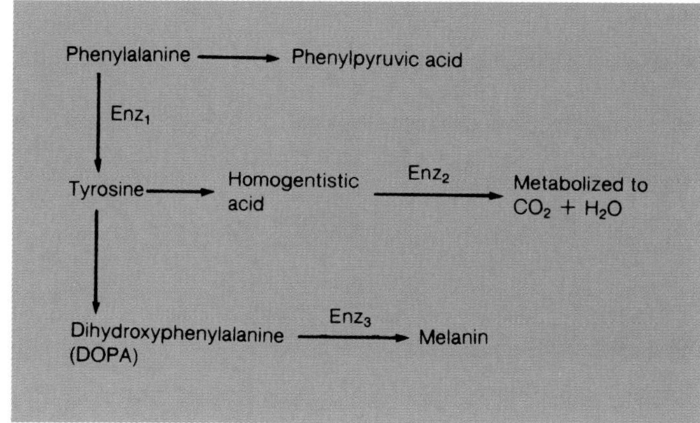

Albinism and Other Defects

One of the conversion products of phenylalanine is a molecule called *DOPA*, an acronym for dihydroxyphenylalanine (fig. 4.36). DOPA is a precursor of the pigment molecule melanin, which gives skin, eyes, and hair their normal coloration. An inherited defect in the enzyme that catalyzes the formation of melanin from DOPA results in the lack of normal pigmentation characteristic of **albinism** (*al"bĭ-niz"em*). Besides PKU and albinism, there are many other inherited defects of amino acid metabolism, as well as inborn errors in the metabolism of carbohydrates and lipids.

Endocrine Disorders and Metabolism

Since metabolism is largely regulated by hormones (chemicals secreted by endocrine glands into the blood), endocrine diseases can produce metabolic disorders. For example, **diabetes mellitus**—characterized by high blood glucose and the presence of glucose in the urine—results from the inadequate secretion or action of the hormone insulin. This disease may also be associated with the excessive production of ketone bodies, which can alter blood pH and produce **ketoacidosis.** Abnormally low blood glucose, **hypoglycemia,** may be produced by excessive insulin secretion. Other metabolic disorders can result from diseases of the pituitary, thyroid, and adrenal glands, as described in chapter 27.

Clinical Investigation Answer

The high blood concentrations of the MB isoenzyme form of creatine phosphokinase (CPK) following severe chest pains suggest that this man experienced a myocardial infarction (heart attack—see chapter 21). His difficulty in urination, together with his high blood levels of acid phosphatase, suggest prostate disease. (The relationship between the prostate and the urinary system is described in chapter 28.) Further tests—including one for *prostate-specific antigen* (*PSA*)—can be performed to confirm this diagnosis.

Chapter Summary

Enzymes as Catalysts (pp. 79–81)

1. Enzymes are biological catalysts, acting to increase the rate of chemical reactions.
2. Most enzymes are proteins, and the tertiary structures of proteins grant specificity to the actions of the enzymes.
3. Substrates are the reactant molecules that fit into the active sites of an enzyme. The process by which the enzyme-substrate products are formed leaves the enzyme unaltered, so that it is able to act again.

Control of Enzyme Activity (pp. 81–84)

1. The activity of an enzyme is affected by a variety of factors, including temperature and pH.
   (a) An enzyme functions best at a particular pH, called its pH optimum.
   (b) At too high a temperature, the enzyme proteins denature and no longer function effectively.
2. Metabolic pathways involve a number of enzyme-catalyzed reactions, in which enzymes cooperate in a stepwise fashion.
   (a) The product of one enzyme becomes the substrate of the next enzyme in the pathway.
   (b) If an enzyme is defective, its product is not made. The products formed prior to that step and at branch points in the pathway may accumulate.

Bioenergetics (pp. 84–88)

1. Reactions that liberate energy may be coupled to those that require energy.
2. ATP is the universal energy carrier of the cells.
   (a) Exergonic reactions provide the energy for the formation of ATP, in which some of the liberated energy is trapped in the bond formed between ADP and the last phosphate.
   (b) The hydrolysis of ATP provides the energy that powers all of the energy needs of the cells.
3. Oxidation-reduction reactions are coupled and usually involve the transfer of hydrogen atoms.
   (a) Atoms or molecules that gain hydrogens (or electrons) are reduced; those that lose hydrogens (or electrons) are oxidized.
   (b) In many oxidation-reduction reactions, pairs of electrons are transferred from the reducing agent to the oxidizing agent.

Glycolysis and the Lactic Acid Pathway (pp. 88–93)

1. Glycolysis refers to the metabolic pathway that converts glucose to two molecules of pyruvic acid.
   (a) Two molecules of ATP are hydrolyzed in the process, but four molecules of ATP are produced, for a net gain of two ATP.
   (b) Two molecules of the oxidized form of NAD are reduced to NADH + H^+ during glycolysis.
2. In anaerobic respiration, pyruvic acid is converted into lactic acid.
   (a) The two NADH + H^+ formed during glycolysis are oxidized to NAD when pyruvic acid is reduced to lactic acid.
   (b) Anaerobic respiration often occurs during skeletal muscle contraction. The lactic acid thus produced can cause muscle pain and fatigue.
3. Hydrolysis of glycogen yields glucose-6-phosphate, which can be used for energy in glycolysis or secreted (by the liver) after removal of the phosphate group.
4. Conversion of noncarbohydrate molecules (lactic acid, amino acids, and glycerol) through pyruvic acid into glucose is called gluconeogenesis.

Aerobic Respiration (pp. 93–99)

1. Pyruvic acid enters into a mitochondrion and, through the loss of carbon dioxide, is converted into a two-carbon molecule that binds to coenzyme A to form acetyl CoA.
2. The Krebs cycle begins when coenzyme A donates acetic acid to an enzyme reaction that combines it with oxaloacetic acid to form citric acid.
   (a) As the reactions of the Krebs cycle proceed, carbon dioxide is lost, three NAD are converted to NADH + H^+, and one FAD is reduced to $FADH_2$.
   (b) Through an intermediate reaction, the Krebs cycle produces one ATP in addition to the NADH and $FADH_2$.
3. NADH and $FADH_2$ donate their electrons to an electron-transport chain of molecules, located in the cristae.
   (a) Iron ions in the oxidized state (Fe^{3+}) gain an electron and are reduced to Fe^{2+}. They then give up that electron as they pass it to the next molecule in the electron-transport chain.
   (b) The last cytochrome donates its electron to oxygen, which functions as the final electron acceptor. One atom of oxygen gains two electrons and two protons to form H_2O.
4. Electron transport provides energy for the formation of ATP.
   (a) The coupling of electron transport to the production of ATP is called oxidative phosphorylation.
   (b) The aerobic respiration of one glucose molecule to carbon dioxide and water yields 30 to 32 ATP.

Metabolism of Lipids and Proteins (pp. 99–104)

1. Triglycerides can be hydrolyzed to glycerol and fatty acids for energy.
   (a) Glycerol can be converted into phosphoglyceraldehyde, and fatty acids into acetyl CoA molecules, through the process of β-oxidation.
   (b) Fatty acids can also be converted, through acetyl CoA, into ketone bodies.
2. Proteins can be hydrolyzed into amino acids, which may serve as an energy source.
   (a) One amino acid can be converted into another through transfer of its amine group to a keto acid. This process is called transamination.
   (b) In oxidative deamination, amino acids are converted into keto acids for energy, and the amino groups are incorporated into molecules of urea.
3. Different organs have different preferred energy sources. During fasting conditions,

some organs spare glucose for the brain by using fatty acids, ketone bodies, and lactic acid as energy sources.

Review Activities

Objective Questions

1. Which of the following statements about enzymes is *true?*
   (a) Most proteins are enzymes.
   (b) Most enzymes are proteins.
   (c) Enzymes are changed by the reactions they catalyze.
   (d) The active sites of enzymes have little specificity for substrates.
2. Which of the following statements about enzyme-catalyzed reactions is *true?*
   (a) The rate of reaction is independent of temperature.
   (b) The rate of all enzyme-catalyzed reactions is decreased when the pH is lowered from 7 to 2.
   (c) The rate of reaction is independent of substrate concentration.
   (d) Under given conditions of substrate concentration, pH, and temperature, the rate of product formation varies directly with enzyme concentration up to a maximum, at which point the rate cannot be increased further.
3. Which of the following represents an endergonic reaction?
   (a) ADP + $P_i \rightarrow$ ATP
   (b) ATP $\rightarrow$ ADP + P_i
   (c) glucose + $O_2 \rightarrow CO_2 + H_2O$
   (d) $CO_2 + H_2O \rightarrow$ glucose
   (e) both *a* and *d*
   (f) both *b* and *c*
4. Which of the following statements about ATP is *true?*
   (a) The bond joining ADP and the third phosphate is a high-energy bond.
   (b) The formation of ATP is coupled to energy-liberating reactions.
   (c) The conversion of ATP to ADP and P_i provides energy for biosynthesis, cell movement, and other cellular processes that require energy.
   (d) ATP is the "universal energy carrier" of cells.
   (e) All of the above are true.
5. When oxygen is combined with two hydrogens to make water,
   (a) oxygen is reduced.
   (b) the molecule that donated the hydrogens becomes oxidized.
   (c) oxygen acts as a reducing agent.
   (d) both *a* and *b* apply.
   (e) both *a* and *c* apply.
6. Enzymes increase the rate of chemical reactions by
   (a) increasing the body temperature.
   (b) decreasing the blood pH.
   (c) increasing the affinity of reactant molecules for each other.
   (d) decreasing the activation energy of the reactants.
7. In anaerobic respiration in humans, the oxidizing agent for NADH (that is, the molecule that removes electrons from NADH) is
   (a) pyruvic acid.
   (b) lactic acid.
   (c) citric acid.
   (d) oxygen.
8. When skeletal muscles lack sufficient oxygen, there is an increased blood concentration of
   (a) pyruvic acid.
   (b) glucose.
   (c) lactic acid.
   (d) ATP.
9. Which of the following statements about the oxygen in the air we breathe is *true?*
   (a) It functions as the final electron acceptor of the electron-transport chain.
   (b) It combines with hydrogen to form water.
   (c) It combines with carbon to form CO_2.
   (d) Both *a* and *b* are true.
   (e) Both *a* and *c* are true.
10. In terms of the number of ATP molecules directly produced, the major energy-yielding process in the cell is
    (a) glycolysis.
    (b) the Krebs cycle.
    (c) oxidative phosphorylation.
    (d) gluconeogenesis.
11. Ketone bodies are derived from
    (a) fatty acids.
    (b) glycerol.
    (c) glucose.
    (d) amino acids.
12. The formation of glucose from pyruvic acid derived from lactic acid, amino acids, or glycerol is called
    (a) glycogenesis.
    (b) glycogenolysis.
    (c) glycolysis.
    (d) gluconeogenesis.
13. Which of the following organs has an almost absolute requirement for blood glucose as its energy source?
    (a) liver
    (b) brain
    (c) skeletal muscles
    (d) heart
14. When amino acids are used as an energy source,
    (a) oxidative deamination occurs.
    (b) pyruvic acid or one of the Krebs cycle acids (keto acids) is formed.
    (c) urea is produced.
    (d) all of the above occur.

Essay Questions

1. Explain the relationship between the chemical structure and the function of an enzyme and describe how both structure and function may be altered in various ways.
2. The coenzymes NAD and FAD can "shuttle" hydrogens from one reaction to another. How does this process serve to couple oxidation and reduction reactions?
3. Using albinism and phenylketonuria as examples, explain what is meant by inborn errors of metabolism.
4. State the advantages and disadvantages of anaerobic respiration.
5. What purpose is served by the formation of lactic acid during anaerobic respiration? How is this accomplished during aerobic respiration?
6. Describe the effect of cyanide on oxidative phosphorylation and on the Krebs cycle. Why is cyanide deadly?
7. Describe the metabolic pathway by which glucose can be converted into fat. How can end-product inhibition by ATP favor this pathway?
8. Describe the metabolic pathway by which fat can be used as a source of energy and explain why the metabolism of fatty acids can yield more ATP than the metabolism of glucose.
9. Explain how energy is obtained from the metabolism of amino acids. Why does a starving person have a high concentration of urea in the blood?
10. Explain why the liver is the only organ able to secrete glucose into the blood. What are the possible sources of hepatic glucose?
11. Explain the two possible meanings of the term *anaerobic respiration*. Why is the production of lactic acid sometimes termed a "fermentation" pathway?
12. Explain the function of brown fat. What does its mechanism imply about the effect of ATP concentrations on the rate of cell respiration?
13. What three molecules serve as the major substrates for gluconeogenesis? Describe the situations in which each one would be involved in this process. Why can't fatty acids be used as a substrate for gluconeogenesis? (*Hint:* Count the carbons in acetyl CoA and pyruvic acid.)

Critical Thinking Questions

1. A friend, wanting to lose weight, eliminates all fat from her diet. How would this help her to lose weight? Could she possibly gain weight on this diet? How? Discuss the health consequences of such a diet.
2. Suppose a drug is developed that promotes the channeling of H^+ out of the intermembrane space into the matrix of the mitochondria of adipose cells. How would this drug affect ATP production, body temperature, and body weight?
3. For many years, the total number of ATP produced per glucose in aerobic respiration was given as 38. Later, it was estimated to be closer to 36, and now it is believed to be closer to 30. What factors must be considered in estimating the yield of ATP molecules? Why are the recent numbers considered to be approximate values?

Related Web Sites

For a listing of the most current web sites related to this chapter, please visit the *Concepts of Human Anatomy and Physiology* home page at http://www.mhhe.com/biosci/abio/.

NEXUS

Some Interactions of Metabolism Concepts with the Other Body Systems

Integumentary System

- The skin produces vitamin D from a derivative of cholesterol (p. 167).
- The metabolic rate of the skin varies greatly, depending on ambient temperature (p. 676).

Skeletal System

- The metabolic needs of bone cells (osteocytes) are supplied by blood vessels that pass through the calcified extracellular matrix (p. 150).

Nervous System

- The aerobic respiration of glucose provides most of the energy needs of the brain (p. 104).
- Regions of the brain with a faster metabolic rate, due to greater brain activity, receive an increased rate of blood flow (p. 675).

Endocrine System

- Hormones that bind to receptors in the plasma membrane of their target cells activate enzymes in the target cell cytoplasm (p. 561).
- Hormones that bind to nuclear receptors in their target cells alter the target cell metabolism by regulating gene expression (p. 559).
- Hormonal secretions from adipose cells regulate hunger and metabolism (p. 872).
- The balance of anabolism and catabolism is regulated by a number of hormones (p. 874).
- Insulin stimulates the synthesis of glycogen and fat (p. 877).
- Epinephrine from the adrenal medulla stimulates the breakdown of glycogen and fat (p. 880).
- Thyroxine stimulates the production of a protein that uncouples oxidative phosphorylation; this helps to increase the metabolic rate of the body (p. 881).
- Growth hormone stimulates protein synthesis (p. 883).

Muscular System

- The intensity of exercise that can be performed aerobically depends on a person's maximal oxygen uptake and lactate threshold (p. 297).
- The body consumes extra oxygen for a period of time after exercise has ceased; this is the oxygen debt (p. 298).
- Glycogenolysis and gluconeogenesis by the liver help to supply glucose for exercising muscles (p. 103).
- Trained athletes obtain a higher proportion of skeletal muscle energy from the aerobic respiration of fatty acids (p. 300).
- Muscle fatigue is associated with anaerobic respiration and the production of lactic acid (p. 300).
- The proportion of energy derived from the aerobic respiration of carbohydrates or lipids by exercising skeletal muscles depends upon the intensity of the exercise (p. 298).

Circulatory System

- The metabolic rate of some organs, such as skeletal muscles, determines the degree of blood vessel dilation, and thus the rate of blood flow to the organ (p. 670).
- Atherosclerosis of coronary arteries can force a region of the heart to metabolize anaerobically and produce lactic acid; this is associated with angina pectoris (p. 649).

Lymphatic and Immune Systems

- A fever results from increased metabolic heat production, and may help to protect the body from certain infections (p. 698).

Respiratory System

- Ventilation oxygenates the blood going to the tissue cells for aerobic cell respiration, and removes the carbon dioxide produced by the tissue cells (p. 750).
- Breathing is primarily regulated by the effects of carbon dioxide produced by the aerobic cell respiration of the body tissues (p. 753).

Urinary System

- The kidneys eliminate urea and other waste products of metabolism from the blood plasma (p. 792).

Digestive System

- The liver contains enzymes needed for many metabolic reactions that are important for the regulation of the blood glucose and lipid concentrations (p. 845).
- The digestion and absorption of carbohydrates, lipids, and proteins is needed to provide the body with the substrates needed for cell metabolism (p. 874).
- Vitamins A and D help to regulate metabolism through the activation of nuclear receptors, which bind to regions of DNA (p. 870).

Reproductive System

- Sperm cells do not contribute mitochondria to the fertilized oocyte (p. 955).
- The endometrium of the uterus contains glycogen to nourish the developing embryo (p. 939).

FIVE

Membrane Transport and the Membrane Potential

Objectives

- Describe how nonpolar molecules and small inorganic ions penetrate the cell membrane and explain how net diffusion occurs.
- Define *osmosis* and describe how the osmotic pressure of solutions affects the direction of osmosis.
- Discuss the significance of osmolality measurements and define the terms *isotonic*, *hypertonic*, and *hypotonic solutions*.
- Explain the mechanisms that help to maintain a constant plasma osmolality.
- Describe the characteristics of carrier-mediated transport.
- Distinguish between simple diffusion and facilitated diffusion and explain why both are passive transport processes.
- Distinguish between facilitated diffusion and active transport and describe the characteristics of active transport.
- Distinguish between primary and secondary active transport.
- Explain how the membrane potential is determined by the permeability characteristics of the cell membrane.
- Explain why the true membrane potential is close to, but less than, the theoretical potassium equilibrium potential.
- Describe the role of the Na^+/K^+ pumps in the establishment of the membrane potential.

Clinical Investigation

A student complained that he was constantly thirsty. During a laboratory exercise at school involving urinalysis, he discovered that his urine contained a significant amount of glucose. After a medical examination, he later learned that he had hyperglycemia, hyperkalemia, and a high plasma osmolality. He was also told that his electrocardiogram showed some abnormalities. How might this student's symptoms and medical findings be interrelated?

Clues: Refer to the boxed information on hyperglycemia in the section "Carrier-Mediated Transport" and review the characteristics of carrier-mediated transport. Think of glucose in urine as exerting an osmotic pressure and, using deductive reasoning, determine what effect this would have on the amount of water excreted in the urine and on the resulting blood volume and concentration. Also read the information on the effect of hyperkalemia on the heart in "Clinical Considerations."

Diffusion and Osmosis

Net diffusion of a molecule or ion through a cell membrane always occurs in the direction of its lower concentration. Nonpolar molecules can penetrate the phospholipid barrier, and small inorganic ions can pass through channels in the membrane. The net diffusion of water through a membrane is known as osmosis.

The cell (plasma) membrane separates the intracellular environment from the extracellular environment. Proteins, nucleotides, and other molecules needed for the structure and function of the cell cannot penetrate, or "permeate," this membrane. The cell membrane is, however, selectively permeable to certain molecules and many ions; this allows two-way traffic in nutrients and wastes needed to sustain metabolism and provides electrical currents created by the movements of ions through the membrane.

The mechanisms involved in the transport of molecules and ions through the cell membrane may be divided into two categories: (1) transport that requires the action of specific *carrier proteins* in the membrane **(carrier-mediated transport)** and (2) transport through the membrane that is not carrier-mediated. Carrier-mediated transport may further be subdivided into *facilitated diffusion* and *active transport*, both of which will be described later. Membrane transport that does not use carrier proteins involves the *simple diffusion* of ions, lipid-soluble molecules, and water through the membrane. *Osmosis* is the net diffusion of solvent (water) through a membrane.

Membrane transport processes may also be categorized on the basis of their energy requirements. **Passive transport** is the net movement of molecules and ions across a membrane from higher to lower concentration (down a concentration gradient); it does not require metabolic energy. Passive transport includes simple diffusion, osmosis, and facilitated diffusion. **Active transport** is net movement across a membrane that occurs against a concentration gradient (to the region of higher concentration). Active transport requires the expenditure of metabolic energy (ATP) and involves specific carrier proteins.

Figure 5.1

Diffusion of a solute.
(*a*) Net diffusion occurs when there is a concentration difference (or concentration gradient) between two regions of a solution, provided that the membrane separating these regions is permeable to the diffusing substance. (*b*) Diffusion tends to equalize the concentrations of these regions, and thus to eliminate the concentration differences.

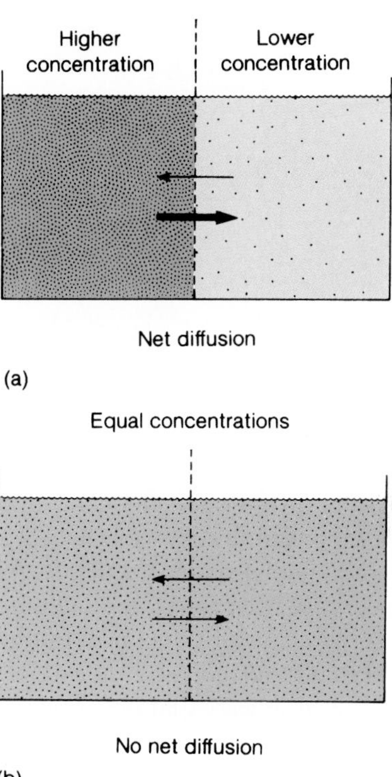

Diffusion

Molecules in a gas and molecules and ions dissolved in a solution are in a constant state of random motion as a result of their thermal (heat) energy. This random motion, called **diffusion,** tends to scatter the molecules evenly, or diffusely, within a given volume. Whenever a *concentration difference*, or *concentration gradient*, exists between two regions of a solution, therefore, random molecular motion tends to eliminate the gradient and distribute the molecules uniformly (fig. 5.1). In terms of the second law of thermodynamics, the concentration difference represents an unstable state of high organization (low entropy) that changes to produce a uniformly distributed solution with maximum disorganization (high entropy).

diffusion: L. *dis-*, apart; *fundere*, to pour

Figure 5.2

Gas exchange occurs by diffusion.
The colored dots, which represent oxygen and carbon dioxide molecules, indicate relative concentrations inside the cell and in the extracellular environment. Gas exchange between the intracellular and extracellular compartments thus occurs by diffusion.

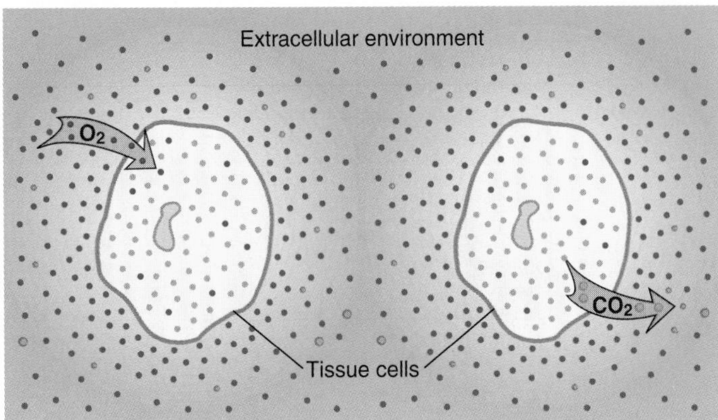

As a result of random molecular motion, molecules in the part of the solution with a higher concentration will enter the area of lower concentration. Molecules will also move in the opposite direction, but not as frequently. As a result, there will be a *net movement* from the region of higher to the region of lower concentration until the concentration difference no longer exists. This net movement is called **net diffusion.** Net diffusion is a physical process that occurs whenever there is a concentration difference across a membrane that is permeable to the diffusing substance.

Diffusion through the Cell Membrane

Since the cell membrane consists primarily of a double layer of phospholipids, molecules that are nonpolar and thus lipid-soluble can easily pass from one side of the membrane to the other. The cell membrane, in other words, does not present a barrier to the diffusion of nonpolar molecules such as oxygen gas (O_2) or steroid hormones. Small molecules that have polar covalent bonds that are uncharged, such as CO_2 (as well as ethanol and urea), are also able to penetrate the phospholipid bilayer. Net diffusion of these molecules can thus easily occur between the intracellular and extracellular compartments when concentration gradients exist.

For example, the oxygen concentration is relatively high in the extracellular fluid because oxygen is carried from the lungs to the body tissues by the blood. Since oxygen is converted to water in aerobic cell respiration, the oxygen concentration within the cells is lower than in the extracellular fluid. The concentration gradient for carbon dioxide is in the opposite direction because cells produce CO_2. *Gas exchange* thus occurs by diffusion between the tissue cells and their extracellular environments (fig. 5.2).

Although water is not lipid-soluble, water molecules can diffuse through the cell membrane because of their small size and lack of net charge. In certain membranes, however, the passage of water is aided by specific channels that can open or

Figure 5.3

Ion pores through the cell membrane.
Inorganic ions (such as Na^+ and K^+) are able to penetrate the membrane through pores within integral proteins that span the thickness of the double layer of phospholipid molecules.

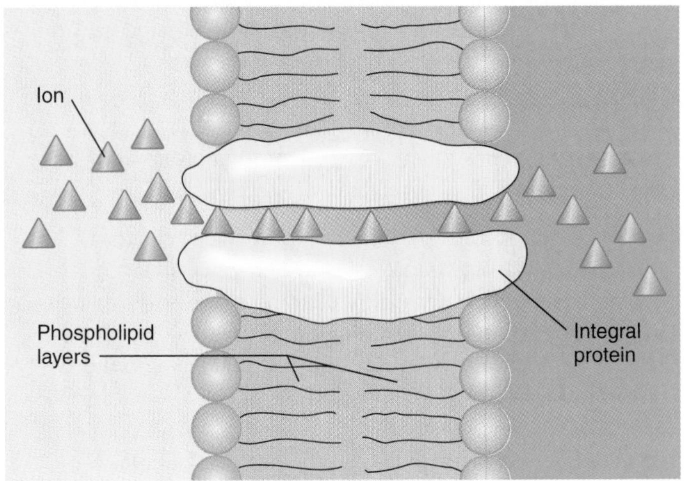

close in response to physiological regulation. The net diffusion of water molecules (the solvent) across the membrane is known as *osmosis*. Since osmosis is the simple diffusion of solvent instead of solute, a unique terminology (discussed in a later section) is used to describe it.

Larger polar molecules, such as glucose, cannot pass through the double layer of phospholipid molecules and thus require special *carrier proteins* in the membrane for transport (described later). The phospholipid portion of the membrane is similarly impermeable to charged inorganic ions, such as Na^+ and K^+. Tiny **ion channels** through the membrane, which are too small to be seen even with an electron microscope, may permit passage of these ions. The ion channels are provided by some of the proteins that span the thickness of the membrane (fig. 5.3).

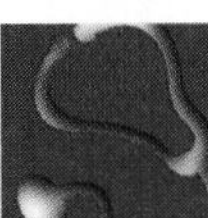

Cystic fibrosis occurs about once in every 2,500 births in the Caucasian population. As a result of a genetic defect, abnormal NaCl and water movement occurs across wet epithelial membranes. Where such membranes line the pancreatic ductules and small respiratory airways they produce a dense, viscous mucus that cannot be properly cleared, leading to pancreatic and pulmonary disorders. The genetic defect involves a particular glycoprotein that forms chloride (Cl^-) channels in the apical membrane of the epithelial cells. This protein, known as *CFTR* (cystic fibrosis transmembrane conductance regulator), is formed as normal in the endoplasmic reticulum. It does not move into the Golgi complex for processing, however, and therefore it doesn't get correctly processed and inserted into vesicles that would introduce it into the cell membrane (chapter 3). The gene for CFTR has been identified and cloned. Recently, in a preliminary test of gene therapy for cystic fibrosis, the gene for CFTR was inserted into an adenovirus (cold virus) and introduced into the nasal passages of human volunteers. More research is required, however, before gene therapy for this disease becomes a real possibility.

Rate of Diffusion

The rate of diffusion, measured by the number of diffusing molecules passing through the membrane per unit time, depends on (1) the magnitude of the concentration difference across the membrane (the "steepness" of the concentration gradient), (2) the permeability of the membrane to the diffusing substances, (3) the temperature of the solution, and (4) the surface area of the membrane through which the substances are diffusing.

The magnitude of the concentration difference across a membrane serves as the driving force for diffusion. Regardless of this concentration difference, however, the diffusion of a substance across a membrane will not occur if the membrane is not permeable to that substance. With a given concentration difference, the rate at which a substance diffuses through a membrane will depend on how permeable the membrane is to it. In a resting neuron, for example, the cell (plasma) membrane is about 20 times more permeable to potassium (K^+) than to sodium (Na^+); consequently, K^+ diffuses much more rapidly than Na^+. Changes in the protein structure of the membrane channels, however, can change the permeability of the membrane. This occurs during the production of a nerve impulse (see chapter 14), when specific stimulation opens Na^+ channels temporarily and allows a faster diffusion rate for Na^+ than for K^+.

In areas of the body that are specialized for rapid diffusion, the surface area of the cell membranes may be increased by numerous folds. The rapid passage of the products of digestion across the epithelial membranes in the small intestine, for example, is aided by tiny fingerlike projections called *microvilli* (discussed in chapter 3). Similar microvilli are found in the kidney tubule epithelium, which must reabsorb various molecules that are filtered out of the blood.

Osmosis

Osmosis (*oz-mo'sis*) is the net diffusion of water (the solvent) across a membrane. For osmosis to occur, the membrane must be *semipermeable;* that is, it must be more permeable to water molecules than to at least one species of solute. Therefore, there are two requirements for osmosis: (1) there must be a difference in the concentration of a solute on the two sides of a semipermeable membrane; and (2) the membrane must be relatively impermeable to the solute. Solutes that cannot freely pass through the membrane are said to be **osmotically active.**

Like the diffusion of solute molecules, the diffusion of water occurs when the water is more concentrated on one side of the membrane than on the other side (that is, when one solution is more dilute than the other—fig. 5.4). The more dilute solution has a higher concentration of water molecules and a lower concentration of solute. Although the terminology associated with osmosis can be awkward (because we are describing water instead of solute), the principles of osmosis are the same as those governing the diffusion of solute molecules through a membrane. Remember that, during osmosis, there is a net movement of water molecules from the side of higher water concentration to the side of lower water concentration.

Imagine a cylinder divided into two equal compartments by an artificial membrane partition that can freely move. One compartment initially contains 180 g/L (grams per liter) of glucose and the other compartment contains 360 g/L of glucose. If the membrane is permeable to glucose, glucose will diffuse from the 360 g/L compartment to the 180 g/L compartment until both compartments contain 270 g/L of glucose. If the membrane is not permeable to glucose but is permeable to water, the same result (270 g/L solutions on both sides of the membrane) will be achieved by the diffusion of water. As water diffuses from the 180 g/L compartment to the 360 g/L compartment, the former solution becomes more concentrated while the latter becomes more dilute. This is accompanied by volume changes, as illustrated in figure 5.5. Osmosis ceases when the concentrations become equal on both sides of the membrane.

Figure 5.4

A model of osmosis.
The diagram illustrates the net movement of water from the solution of lesser solute concentration (greater water concentration) to the solution of greater solute concentration (lower water concentration).

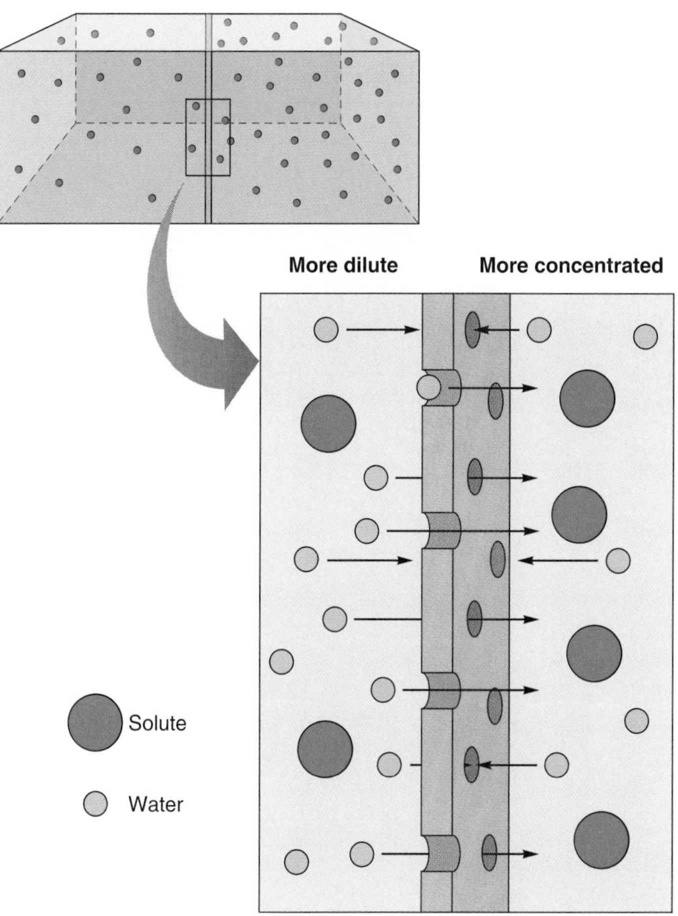

osmosis: Gk. *osmos*, a thrust

Figure 5.5

The effects of osmosis.
(*a*) A movable semipermeable membrane (permeable to water but not to glucose) separates two solutions of different glucose concentration. As a result, water moves by osmosis into the solution of greater concentration until (*b*) the volume changes equalize the concentrations on both sides of the membrane.

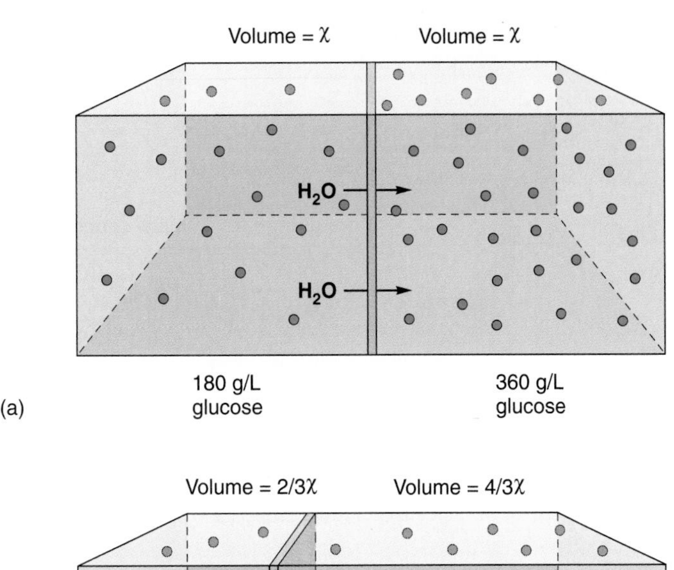

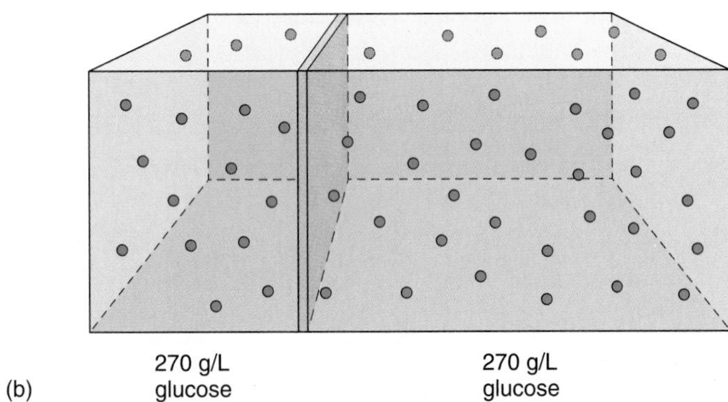

Figure 5.6

A model illustrating osmotic pressure.
If a semipermeable membrane separates pure water from a 180 g/L glucose solution, water will tend to move by osmosis into the glucose solution, thus creating a hydrostatic pressure that will push the membrane to the left and expand the volume of the glucose solution. The amount of pressure that must be applied to just counteract this volume change is equal to the osmotic pressure of the glucose solution.

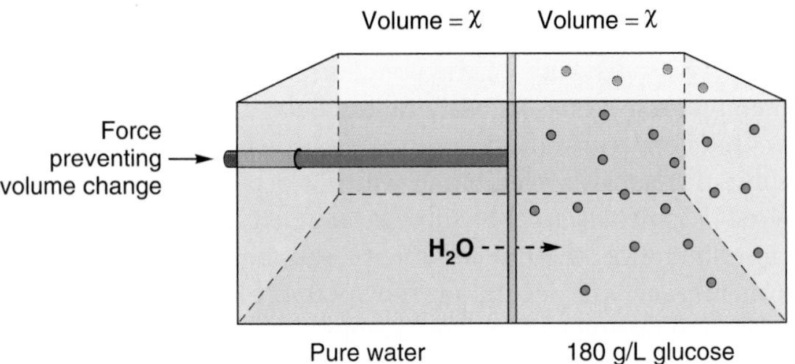

Cell membranes behave in a similar manner because water is able to move to some degree through the lipid component of most cell membranes. The membranes of some cells, however, have special water channels that allow water to move through more rapidly. These channels are known as **aquaporins** (*ak″wă-por′inz*). Some of the aquaporin channels are always open, and some are open only when stimulated by a regulatory molecule. Such regulation is particularly important in the functioning of the kidneys, as will be described in chapter 25.

Osmotic Pressure

Osmosis and the movement of the membrane partition could be prevented by an opposing force. If one compartment contained 180 g/L of glucose and the other compartment contained pure water, the osmosis of water into the glucose solution could be prevented by pushing against the membrane with a certain force (in this case, equal to 22.4 atmospheres pressure). This concept is illustrated in figure 5.6.

The force that would have to be exerted to prevent osmosis in the situation just described is the **osmotic pressure** of the solution. This backward measurement indicates how strongly the solution "draws" water into it by osmosis. The greater the solute concentration of a solution, the greater its osmotic pressure. Thus, pure water has an osmotic pressure of zero, and a 360 g/L glucose solution has twice the osmotic pressure of a 180 g/L glucose solution.

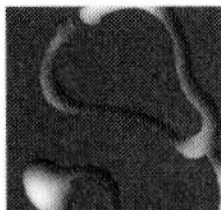

Water returns from tissue fluid to blood capillaries because the protein concentration of blood plasma is higher than the protein concentration of tissue fluid. Plasma proteins, in contrast to other plasma solutes, cannot pass from the capillaries into the tissue fluid. Therefore, plasma proteins are osmotically active. If a person has an abnormally low concentration of plasma proteins, excessive accumulation of fluid in the tissues—called *edema*—will result. This may occur when a damaged liver (as in cirrhosis) is unable to produce sufficient amounts of albumin, the major protein in the blood plasma.

Molarity and Molality

Glucose is a monosaccharide with a molecular weight of 180 (the sum of its atomic weights). Sucrose is a disaccharide of glucose and fructose, which have molecular weights of 180 each. When glucose and fructose join by dehydration synthesis to form sucrose, a molecule of water (molecular weight = 18) splits off. Therefore, sucrose has a molecular weight of 342 (the sum of 180 + 180 minus 18). Since the molecular weights of sucrose and glucose are in a ratio of 342/180, it follows that 342 grams of sucrose must contain the same number of molecules as 180 grams of glucose.

Notice that an amount of any compound equal to its molecular weight in grams must contain the same number of molecules as an amount of any other compound equal to its molecular weight in grams. This unit of weight, a *mole*, always contains 6.02×10^{23} molecules **(Avogadro's** (*av″ŏ-gad′rōs*) **number).** One mole of solute dissolved in water to make one

Avogadro's number: from Amadeo Avogadro, Italian chemist and physicist, 1766–1856

Figure 5.7

Molar and molal solutions.
The diagrams illustrate the difference between (*a*) a one-molar (1.0 M) and (*b*) a one-molal (1.0 *m*) glucose solution.

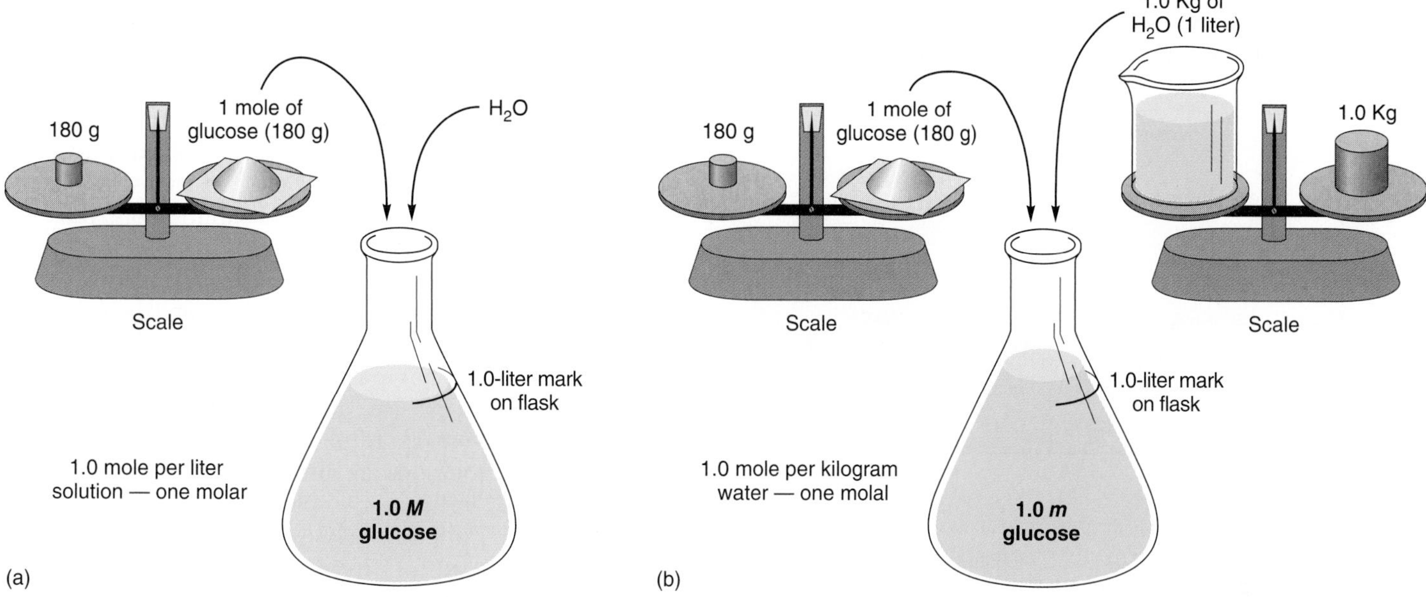

liter of solution is described as a **one-molar solution** (abbreviated 1.0 M). Although this unit of measurement is commonly used in chemistry, it is not completely desirable in discussions of osmosis because the exact ratio of solute to water is not specified. For example, more water is needed to make a 1.0 M NaCl solution (where a mole of NaCl weighs 58.5 grams) than is needed to make a 1.0 M glucose solution, since 180 grams of glucose take up more volume than 58.5 grams of salt.

Since the ratio of solute to water molecules is critically important in osmosis, a more desirable measurement of concentration is **molality.** In a one-molal solution (abbreviated 1.0 *m*), 1 mole of solute (180 grams of glucose, for example) is dissolved in 1 kilogram of water (equal to 1 liter at 4° C). Therefore, a 1.0 *m* NaCl solution and a 1.0 *m* glucose solution both contain a mole of solute dissolved in exactly the same amount of water (fig. 5.7).

Osmolality

If 180 grams of glucose and 180 grams of fructose were dissolved in the same kilogram of water, the osmotic pressure of the solution would be the same as that of a 360 g/L glucose solution. Osmotic pressure depends on the ratio of solute to solvent, *not* on the chemical nature of the solute molecules. The expression for the total molality of a solution is **osmolality** (*oz"mo-lal'ĭ-te*) **(Osm).** Thus, the solution of 1.0 *m* glucose plus 1.0 *m* fructose has a total molality, or *osmolality*, of 2.0 osmol/L (abbreviated 2.0 Osm). This osmolality is the same as that of the 360 g/L glucose solution, which has a concentration of 2.0 *m* and 2.0 Osm (fig. 5.8).

Unlike glucose, fructose, and sucrose, electrolytes such as NaCl ionize when they dissolve in water. One molecule of NaCl dissolved in water yields two ions (Na^+ and Cl^-); 1 mole of

Figure 5.8

The osmolality of a solution.
The osmolality (Osm) is equal to the sum of the molalities of each solute in the solution. If a semipermeable membrane separates two solutions with equal osmolalities, no osmosis will occur.

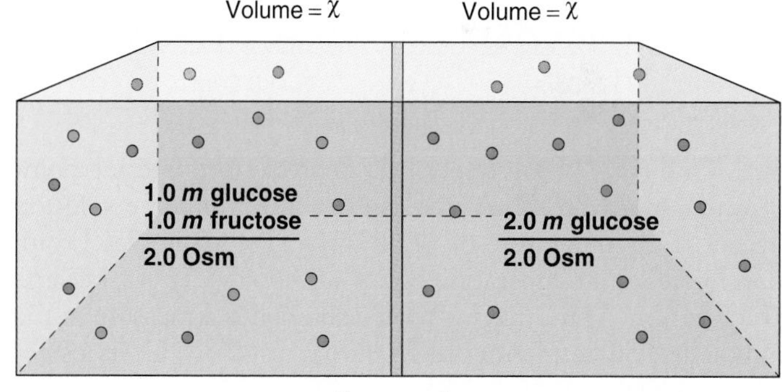

NaCl ionizes to form 1 mole of Na^+ and 1 mole of Cl^-. Thus, a 1.0 *m* NaCl solution has a total concentration of 2.0 Osm. The effect of this ionization on osmosis is illustrated in figure 5.9.

Measurement of Osmolality

Plasma and other biological fluids contain many organic molecules and electrolytes. The osmolality of such complex solutions can only be estimated by calculations. Fortunately, however, there is a relatively simple method for measuring osmolality. This method is based on the fact that the freezing point of a solution, like its osmotic pressure, is affected by the total concentration of the solution and not by the chemical nature of the solute.

Figure 5.9

The effect of ionization on the osmotic pressure. (*a*) If a semipermeable membrane (permeable to water but not to glucose, Na^+, or Cl^-) separates a 1.0 *m* glucose solution from a 1.0 *m* NaCl solution, water will move by osmosis into the NaCl solution. Osmosis occurs because NaCl can ionize to yield one-molal Na^+ plus one-molal Cl^-. (*b*) After osmosis, the total concentration, or osmolality, of the two solutions is equal.

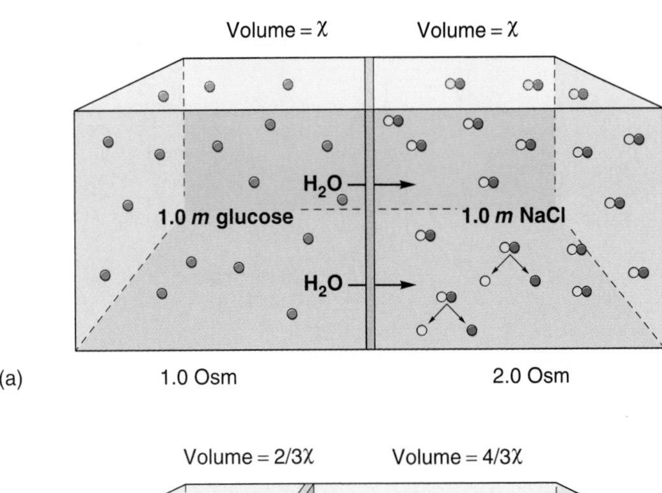

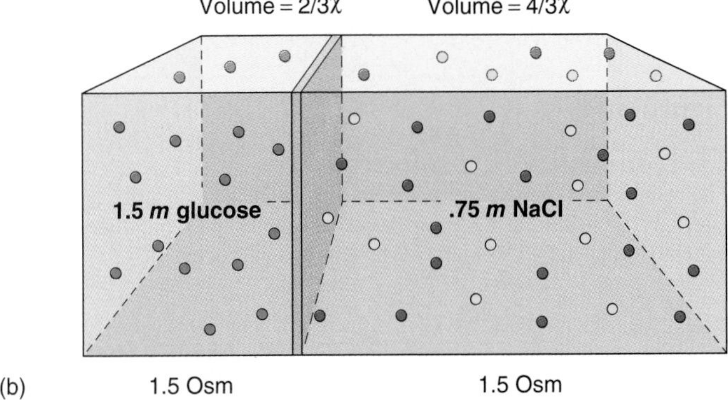

One mole of solute per liter depresses the freezing point of water by –1.86° C. Accordingly, a 1.0 *m* glucose solution freezes at a temperature of –1.86° C, and a 1.0 *m* NaCl solution freezes at a temperature of 2 × –1.86 = –3.72° C, because of ionization. Thus, *freezing-point depression* is a measure of the osmolality. Since plasma freezes at about –0.56° C, its osmolality is equal to 0.56 ÷ 1.86 = 0.3 Osm, which is more commonly indicated as 300 milliosmolal (or 300 mOsm).

Tonicity

A 0.3 *m* glucose solution, which is 0.3 Osm or 300 milliosmolal (300 mOsm), has the same osmolality and osmotic pressure as plasma. The same is true of a 0.15 *m* NaCl solution, which ionizes to produce a total concentration of 300 mOsm. Both of these solutions are used clinically as intravenous infusions, labeled 5% *dextrose* (5 g of glucose per 100 ml, which is 0.3 *m*) and *normal saline* (0.9 g of NaCl per 100 ml, which is 0.15 *m*). Since 5% dextrose and normal saline have the same osmolality as plasma, they are said to be **isosmotic** (*i″sos-mot′ik*) to plasma.

The term *tonicity* is used to describe the effect of a solution on the osmotic movement of water. For example, if an isosmotic glucose or saline solution is separated from plasma by a membrane that is permeable to water but not to glucose or NaCl, osmosis will not occur. In this case, the solution is said to be **isotonic** (*i″so-ton′ik*) to plasma.

Red blood cells placed in an isotonic solution will neither gain nor lose water. It should be noted that a solution may be isosmotic but not isotonic; such is the case whenever the solute in the isosmotic solution can freely penetrate the membrane. A 0.3 *m* urea solution, for example, is isosmotic but not isotonic because the cell membrane is permeable to urea. When red blood cells are placed in a 0.3 *m* urea solution, the urea diffuses into the cells until its concentration on both sides of the cell membrane becomes equal. Meanwhile, the solutes within the cells that cannot exit—and which are, therefore, osmotically active—cause osmosis of water into the cells. Red blood cells placed in 0.3 *m* urea will thus eventually burst.

Solutions that have a lower total concentration of solutes than that of plasma, and therefore a lower osmotic pressure, are *hypo-osmotic* to plasma. If the solute is osmotically active, such solutions are also **hypotonic** to plasma. Red blood cells placed in hypotonic solutions gain water and may burst—a process called *hemolysis* (*he-mol′ĭ-sis*). When red blood cells are placed in a **hypertonic** solution (such as sea water), which contains osmotically active solutes at a higher osmolality and osmotic pressure than plasma, they shrink as a result of the osmosis of water out of the cells. This process is called *crenation* (*krĭ-na′shun*) because the cell surface takes on a scalloped appearance (fig. 5.10).

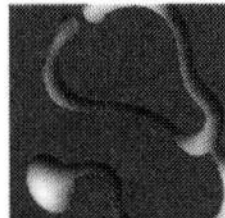

Fluids delivered intravenously must be isotonic to blood in order to maintain the correct osmotic pressure and prevent cells from either expanding or shrinking from the gain or loss of water. Common fluids used for this purpose are normal saline and 5% dextrose, which, as previously described, have about the same osmolality as normal plasma (approximately 300 mOsm). Another isotonic solution frequently used in hospitals is *Ringer's lactate.* This solution contains glucose and lactic acid in addition to a number of different salts. Isotonic solutions are also used in heart-lung machines, which take the place of the heart and lungs during open-heart surgery.

Regulation of Blood Osmolality

The osmolality of the blood plasma is normally maintained within very narrow limits by a variety of regulatory mechanisms. When a person becomes dehydrated, for example, the blood becomes more concentrated as the total blood volume is reduced. The increased blood osmolality and osmotic pressure stimulate *osmoreceptors* (*oz″mŏ-re-cep′torz*), which are neurons located in a part of the brain called the hypothalamus.

isotonic: Gk. *isos*, equal; *tonus*, tension
hypotonic: Gk. *hypo*, under; *tonus*, tension
hypertonic: Gk. *hyper*, over; *tonus*, tension
crenation: L. *crena*, a notch
Ringer's lactate: from Sidney Ringer, English physiologist, 1835–1910

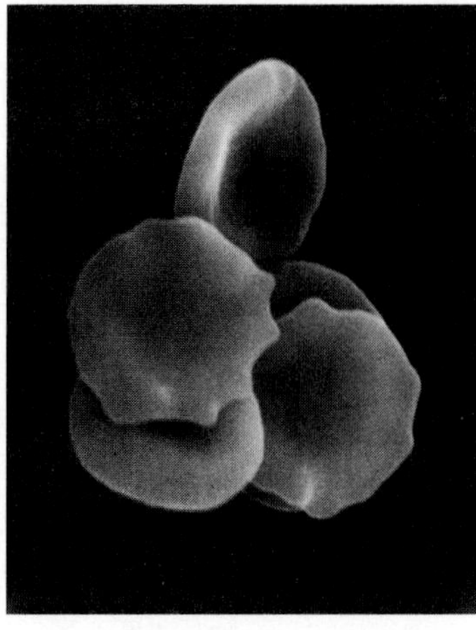

Figure 5.10

A scanning electron micrograph of normal and crenated red blood cells.
Notice that the cell membrane of the crenated cells is notched, or crenulated, as a result of shrinkage due to the loss of water by osmosis.

Figure 5.11

Homeostasis of plasma concentration.
An increase in plasma osmolality (increased concentration and osmotic pressure) due to dehydration stimulates thirst and increased ADH secretion. These effects cause the person to drink more and urinate less. The blood volume, as a result, is increased while the plasma osmolality is decreased. These effects help to bring the blood volume back to the normal range and complete the negative feedback loop (indicated by a negative sign).

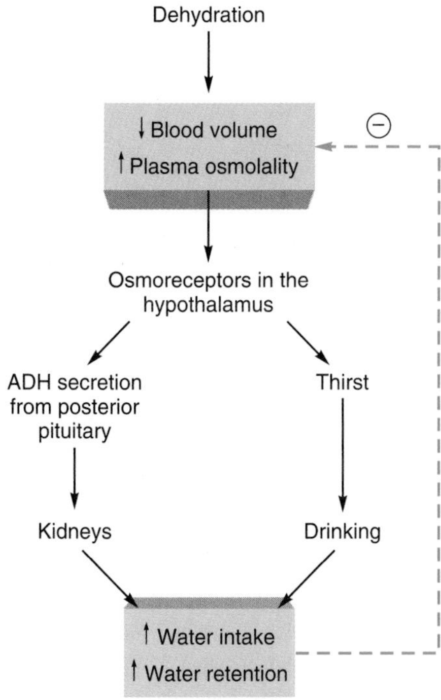

As a result of increased osmoreceptor stimulation, the person becomes thirsty and, if water is available, drinks. Along with increased water intake, a person who is dehydrated excretes a lower volume of urine. This occurs as a result of the following sequence of events: (1) increased plasma osmolality stimulates osmoreceptors in the hypothalamus of the brain; (2) the osmoreceptors, by means of a tract of axons, stimulate the posterior pituitary to release **antidiuretic** (*an″te-di″yŭ-ret′ik*) **hormone (ADH);** and (3) ADH acts on the kidneys to promote water retention, so that a lower volume of more concentrated urine is excreted.

Therefore a person who is dehydrated drinks more and urinates less. This represents a negative feedback loop (fig. 5.11) that maintains homeostasis of the plasma concentration (osmolality) and, in the process, helps to maintain a proper blood volume.

A person with a normal blood volume who eats salty food will also get thirsty and have an increased secretion of ADH. By drinking more and excreting less water in the urine, the salt from the food will become diluted to restore the normal blood concentration, but at a higher blood volume. The opposite occurs in salt deprivation. With a lower plasma osmolality, the osmoreceptors are not stimulated as much, and the posterior pituitary releases less ADH. Consequently, more water is excreted in the urine to again restore the proper range of plasma concentration, but at a lower blood volume. Low blood volume and pressure as a result of prolonged salt deprivation can be fatal (refer to the discussion of blood volume and pressure in chapter 22).

Carrier-Mediated Transport

Molecules such as glucose are transported across cell membranes by special protein carriers. Carrier-mediated transport in which the net movement is down a concentration gradient, and which is therefore passive, is called facilitated diffusion. Carrier-mediated transport that occurs against a concentration gradient, and which therefore requires metabolic energy, is called active transport.

In order to sustain metabolism, cells must take up glucose, amino acids, and other organic molecules from the extracellular environment. Molecules such as these, however, are too large and polar to pass through the lipid barrier of the cell membrane by a process of simple diffusion. The transport of such molecules is mediated by **protein carriers** within the membrane. Although such carriers cannot be directly observed, their presence has been inferred by the observation that this transport has characteristics in common with enzyme activity. These characteristics include (1) *specificity*, (2) *competition*, and (3) *saturation*.

Like enzyme proteins, carrier proteins interact only with specific molecules. Glucose carriers, for example, can interact only with glucose and not with closely related monosaccharides. As a further example of specificity, particular carriers for amino acids transport some types of amino acids but not others. Two amino acids that are transported by the same carrier compete with each other, so that the rate of transport for each is lower when they are present together than it would be if each were present alone (fig. 5.12).

As the concentration of a transported molecule is increased, its rate of transport will also be increased—but only up to a maximum. Beyond this rate, called the **transport maximum (T_m),** subsequent increases in concentration do

Figure 5.12

Characteristics of carrier-mediated transport.
Carrier-mediated transport displays the characteristics of saturation (illustrated by the transport maximum) and competition. Molecules X and Y compete for the same carrier, so that when they are present together the rate of transport of each is lower than when either is present separately.

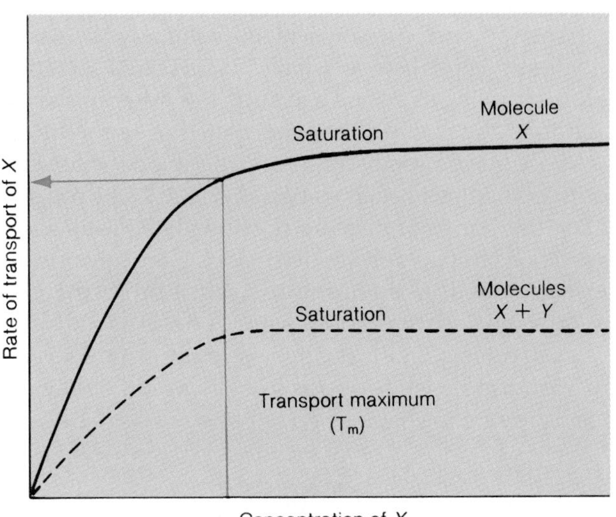

Figure 5.13

A model of the facilitated diffusion of glucose.
A carrier—with characteristics of specificity and saturation—is required for this transport, which occurs from the blood into the cells of such tissues as muscles, liver, and adipose tissue. This is passive transport because the net movement is to the region of lower concentrations, and ATP is not required.

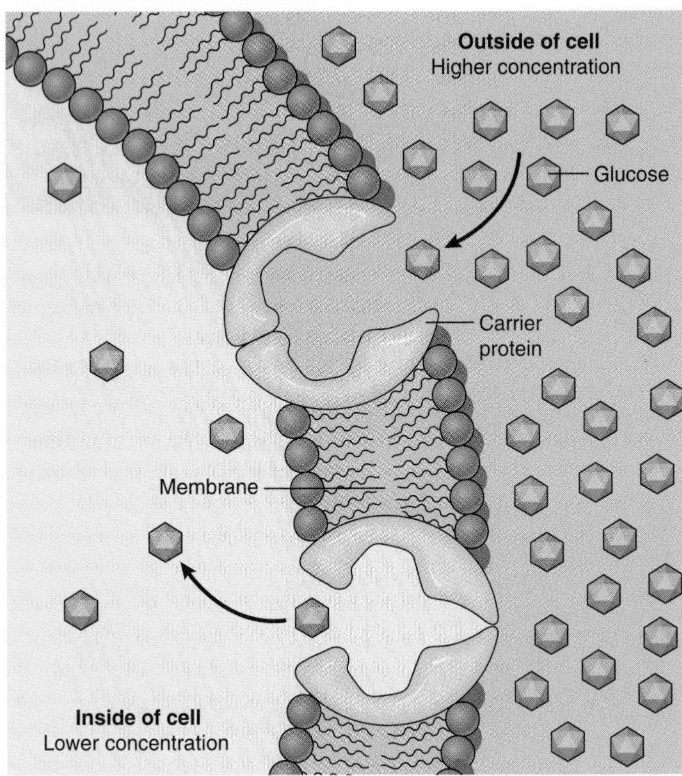

not further increase the transport rate. This indicates that the carriers have become saturated (fig. 5.12).

As an example of saturation, imagine a bus stop that is serviced once an hour by a bus that can hold a maximum of 40 people (its "transport maximum"). If there are 10 people waiting at the bus stop, 10 will be transported each hour. If 20 people are waiting, 20 will be transported each hour. This linear relationship will hold up to a maximum of 40 people; if there are 80 people at the bus stop, the transport rate will still be 40 per hour.

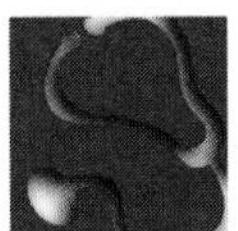

The kidneys transport a number of molecules from the blood filtrate (which eventually becomes urine) back into the blood. Glucose, for example, is normally completely reabsorbed so that urine is normally free of glucose. However, if the glucose concentration of the blood and filtrate is too high (a condition called *hyperglycemia*), the transport maximum will be exceeded. In this case, glucose will be found in the urine (a condition called *glycosuria*). This may result from the consumption of too much sugar or from inadequate action of the hormone *insulin* in the disease *diabetes mellitus.*

Facilitated Diffusion

The transport of glucose from the blood across the cell membranes of tissue cells occurs by **facilitated diffusion.** Facilitated diffusion, like simple diffusion, is powered by the thermal energy of the diffusing molecules and involves the net transport of substances through a cell membrane from the side of higher to the side of lower concentration. ATP is not required for either facilitated or simple diffusion.

Unlike simple diffusion of nonpolar molecules, water, and inorganic ions through a membrane, the diffusion of glucose through the cell membrane displays the properties of carrier-mediated transport: specificity, competition, and saturation. The diffusion of glucose through a cell membrane must therefore be mediated by carrier proteins. In the conceptual model shown in figure 5.13, each transport carrier is composed of two protein subunits that interact with glucose in such a way as to create a channel through the membrane, thus enabling the movement of glucose down its concentration gradient.

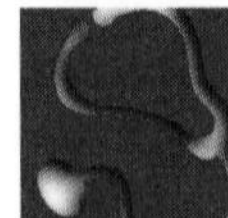

The rate of the facilitated diffusion of glucose into tissue cells depends directly on the plasma glucose concentration. When the plasma glucose concentration is abnormally low—a condition called *hypoglycemia*—the rate of transport of glucose into brain cells may be inadequate for the metabolic needs of the brain. Severe hypoglycemia, as may be produced in a diabetic person by an overdose of insulin, can thus result in loss of consciousness or even death.

Like the isoenzymes described in chapter 4, carrier proteins that do the same job may exist in various tissues in

Figure 5.14

Insertion of carrier proteins into the cell membrane. In the unstimulated state, carrier proteins (e.g., for glucose) may be located in the membrane of intracellular vesicles. In response to stimulation, the vesicle fuses with the cell membrane, and the carriers are inserted into the cell membrane.

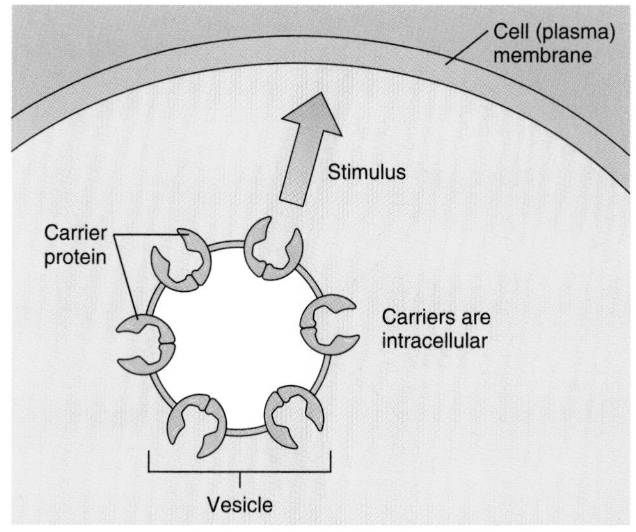

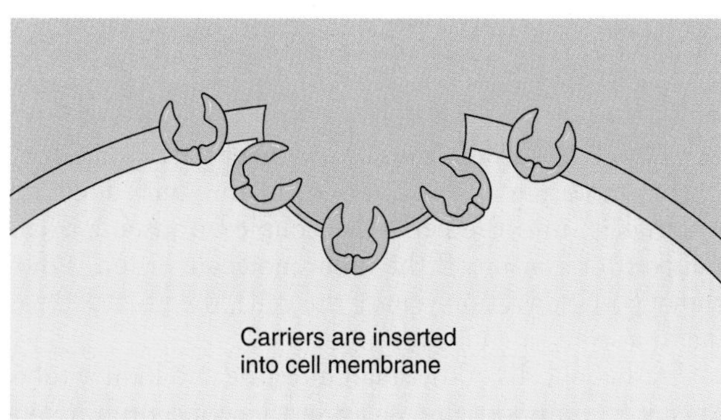

slightly different forms. The transport carriers for the facilitative diffusion of glucose are designated with the letters **GLUT,** followed by a number for the isoform. The carrier for glucose in skeletal muscles, for example, is designated *GLUT4*. In unstimulated muscles, GLUT4 proteins are located within the membrane of cytoplasmic vesicles. In response to exercise of a muscle, these vesicles fuse with the cell membrane. This is similar to exocytosis (chapter 3; also see fig. 5.19), except that no cellular product is secreted. Instead, the transport carriers are inserted into the cell membrane (fig. 5.14). During exercise, as a result, more glucose is able to enter the skeletal muscle cells from the blood plasma.

Active Transport

Some aspects of cell transport cannot be explained by simple or facilitated diffusion. The epithelial linings of the small intestine and kidney tubules, for example, move glucose from the side of lower to the side of higher concentration—from the space within the tube (*lumen*) to the blood. Similarly, all cells extrude Ca^{2+} into the extracellular environment and, by this means, maintain an intracellular Ca^{2+} concentration that is about 10,000 times lower than the extracellular Ca^{2+} concentration. This steep concentration gradient sets the stage for Ca^{2+} to be used as a regulatory signal. The opening of plasma membrane Ca^{2+} channels, and the rapid diffusion of Ca^{2+} that results, provides a signal for neurotransmitter release, muscle contraction, and many other cellular activities.

Active transport is the movement of molecules and ions against their concentration gradients, from lower to higher concentrations. This transport requires the expenditure of cellular energy obtained from ATP; if a cell is poisoned with cyanide (which inhibits oxidative phosphorylation), active transport will stop. Passive transport, by contrast, can continue even if metabolic poisons kill the cell by preventing the formation of ATP.

Primary Active Transport

Primary active transport occurs when the hydrolysis of ATP is directly required for the function of the carriers. These carriers are composed of proteins that span the thickness of the membrane. The following sequence of events is believed to occur: (1) the molecule or ion to be transported binds to a specific "recognition site" on one side of the carrier protein; (2) this bonding stimulates the breakdown of ATP, which in turn results in phosphorylation of the carrier protein; (3) as a result of phosphorylation, the carrier protein undergoes a conformational (shape) change; and (4) a hingelike motion of the carrier protein releases the transported molecule or ion on the opposite side of the membrane. This model of active transport is illustrated in figure 5.15.

The Sodium-Potassium Pump

Primary active transport carriers are often referred to as *pumps*. Although some of these carriers transport only one molecule or ion at a time, other carriers exchange one molecule or ion for another. The most important of the latter type of carriers is the **Na^+/K^+ pump.** This carrier protein, which is also an ATPase enzyme that converts ATP to ADP and P_i, actively extrudes three sodium ions (Na^+) from the cell as it transports two potassium ions (K^+) into the cell. This transport is energy-dependent because Na^+ is more highly concentrated outside the cell and K^+ is more concentrated within the cell. Both ions, in other words, are moved against their concentration gradients (fig. 5.16).

All cells have numerous Na^+/K^+ pumps that are constantly active. For example, there are about 200 Na^+/K^+ pumps per red blood cell, about 35,000 per white blood cell, and several million per cell in a part of the tubules within the kidney. This represents an enormous expenditure of energy used to maintain a steep gradient of Na^+ and K^+ across the cell membrane. This steep gradient serves four functions: (1) the steep Na^+ gradient provides energy for the "coupled transport" of other molecules; (2) the activity of

Figure 5.15

A model of active transport.
This model (a mental construct, consistent with the scientific evidence) features a hingelike motion of the integral protein subunits.

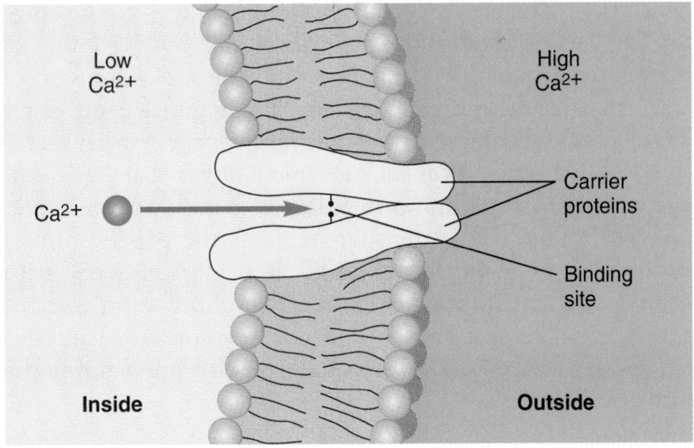

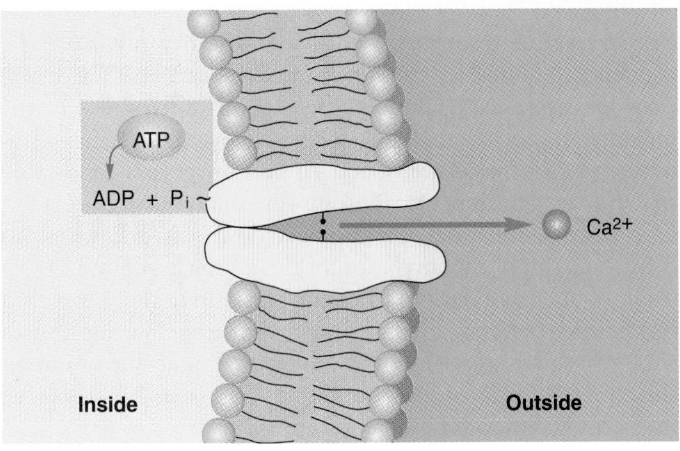

Figure 5.16

The Na^+/K^+ pump exchanges intracellular Na^+ for K^+.
The active transport carrier itself is an ATPase that breaks down ATP for energy. Dashed arrows indicate the direction of passive transport (diffusion); solid arrows indicate the direction of active transport.

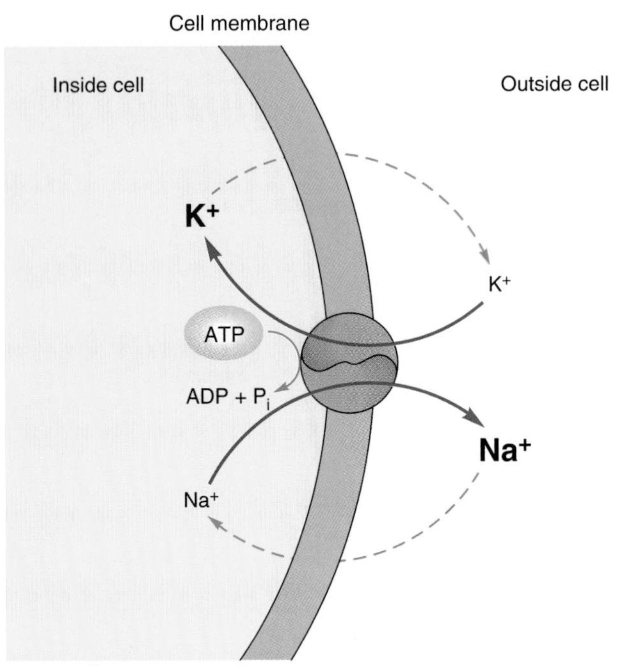

the Na^+/K^+ pumps can be adjusted (primarily by thyroid hormones) to regulate the resting calorie expenditure and basal metabolic rate of the body; (3) the Na^+ and K^+ gradients across the cell membranes of nerve and muscle cells produce electrochemical impulses needed for the functions of nerves and muscles, including the heart; and (4) the active extrusion of Na^+ is important for osmotic reasons. If the pumps stopped functioning, the increased Na^+ concentration within cells would promote the osmotic inflow of water and damage the cells.

Secondary Active Transport (Coupled Transport)

In **secondary active transport,** or **coupled transport,** the energy needed for the "uphill" movement of a molecule or ion is obtained from the "downhill" transport of Na^+ into the cell. Hydrolysis of ATP by the action of the Na^+/K^+ pumps is required indirectly, in order to maintain low intracellular Na^+ concentrations. The diffusion of Na^+ down its concentration gradient into the cell can then power the movement of a different ion or molecule against its concentration gradient. If the other molecule or ion is moved in the same direction as Na^+—that is, into the cell—the coupled transport is called *cotransport,* or *symport.* If the other molecule or ion is moved in the opposite direction (out of the cell), the process is called *countertransport,* or *antiport.*

Epithelial cells of the small intestine and kidney tubules, for example, transport glucose against its concentration gradient by a carrier that requires the simultaneous bonding of Na^+ (fig. 5.17). Glucose and Na^+ cotransport into the cell as a result of the Na^+ gradient created by the Na^+/K^+ pumps. Because of the distribution of the Na^+/K^+ pumps and glucose carriers in the epithelial cell membrane, the Na^+ and glucose move from the lumina of the intestine and kidney tubules into the blood (fig. 5.18).

An example of countertransport is the uphill extrusion of Ca^{2+} from a cell by a type of pump that is coupled to the passive diffusion of Na^+ into the cell. Cellular energy, obtained from ATP, is not used to move Ca^{2+} directly out of the cell in this case, but energy is constantly required to maintain the steep Na^+ gradient. Another example of countertransport is the exchange of chloride (Cl^-) for bicarbonate (HCO_3^-) across the red blood cell membrane. Diffusion of bicarbonate out of the cell powers the entry of chloride (this is discussed together with red blood cell function in chapter 24).

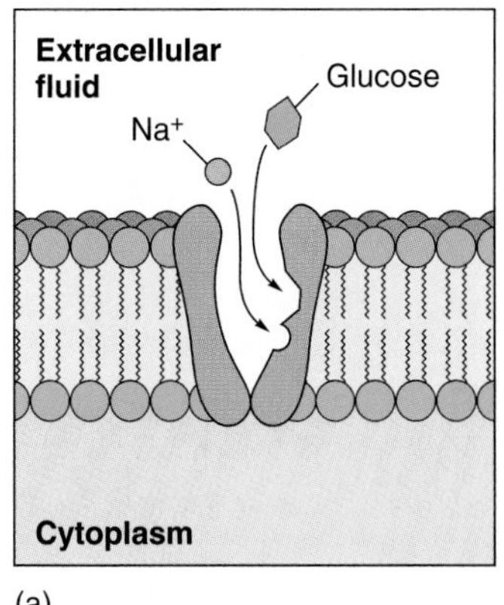

(a)

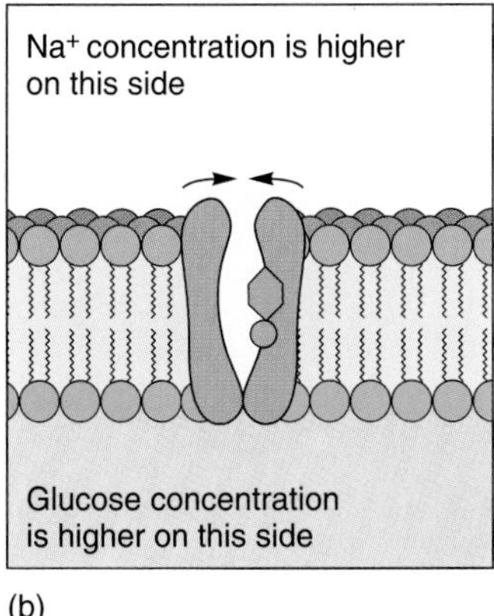

(b)

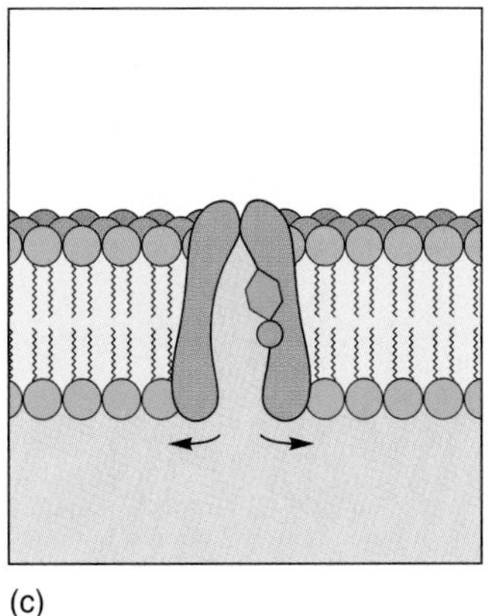
(c)

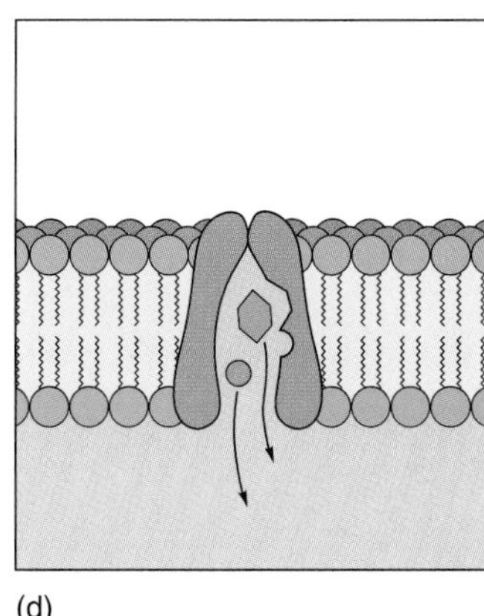
(d)

Figure 5.17

A model for the cotransport of Na^+ and glucose into a cell.
The sequence of events is illustrated in (*a*) through (*d*). This is secondary active transport because it is dependent upon the diffusion gradient for Na^+ created by the Na^+/K^+ pumps.

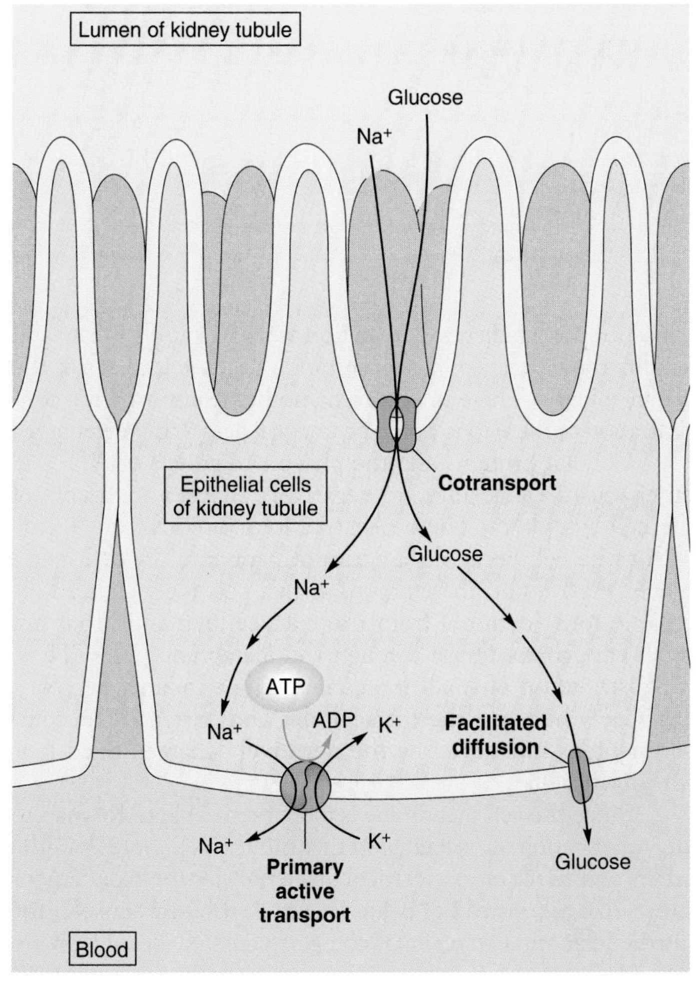

Figure 5.18

The membrane transport of glucose.
This diagram illustrates the transport of glucose from the fluid in the kidney tubules, through the epithelial cells of the tubule wall, and into the blood. All three types of carrier-mediated transport are used in this process. A similar transport process occurs in the absorption of glucose from the intestine.

Severe diarrhea is responsible for almost half of all deaths worldwide of children under the age of 4 (amounting to about 4 million deaths per year). Because rehydration through intravenous therapy is often not practical, the World Health Organization (WHO) developed a simpler, more economical treatment called *oral rehydration therapy.* The therapy is effective because the absorption of water by osmosis across the intestine is proportional to the absorption of Na^+, and the intestinal epithelium cotransports Na^+ and glucose. The WHO provides those in need with a mixture (which can be diluted with tap water in the home) containing both glucose and Na^+, as well as other ions. The glucose in the mixture promotes the cotransport of Na^+, and the Na^+ transport promotes the osmotic movement of water from the intestine into the blood. It has been estimated that oral rehydration therapy saves the lives of more than a million small children each year.

Bulk Transport

Polypeptides and proteins, as well as many other molecules, are too large to be transported through a membrane by the carriers described in previous sections. Yet many cells do secrete these molecules—for example, as hormones or neurotransmitters—by the process of **exocytosis.** As described in chapter 3, this involves the fusion of a membrane-bound vesicle with the cell membrane, so that the membranes become continuous (fig. 5.19).

The process of **endocytosis** (see fig. 3.4) resembles exocytosis in reverse. In receptor-mediated endocytosis, specific molecules, such as protein-bound cholesterol, can be taken into the cell because of the interaction between the cholesterol transport protein and a protein receptor on the cell membrane. Cholesterol is removed from the blood by the liver and by the walls of blood vessels through this mechanism (see chapters 21 and 26 for a more complete discussion).

Together, exocytosis and endocytosis provide **bulk transport** out of and into the cell, respectively. (The term "bulk" is used because many molecules are moved at the same

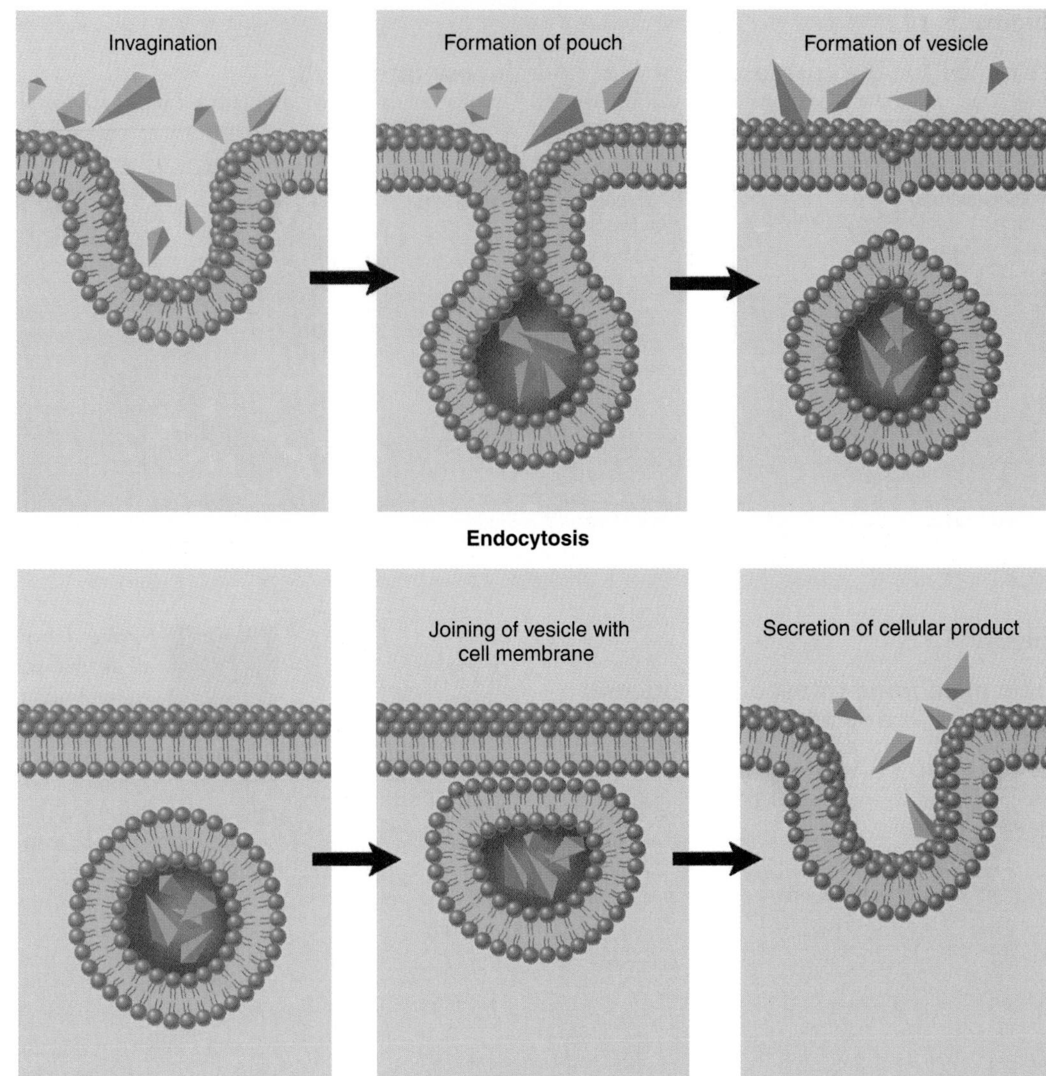

Figure 5.19

Endocytosis and exocytosis.
Endocytosis and exocytosis are responsible for the bulk transport of molecules into and out of a cell, respectively.

time.) It should be noted that molecules taken into a cell by endocytosis are still separated from the cytoplasm by the cell membrane of the endocytotic vesicle. Some of these molecules, such as membrane receptors, will be moved back to the cell membrane; the rest will end up in lysosomes.

The Membrane Potential

As a result of the permeability properties of the cell membrane, the presence of nondiffusible negatively charged molecules inside the cell, and the action of the Na^+/K^+ pumps, there is an unequal distribution of charges across the membrane. Consequently, the inside of the cell is negatively charged compared to the outside. This difference in charge, or potential difference, is known as the membrane potential.

In the preceding section the action of the Na^+/K^+ pumps was discussed in conjunction with the topic of active transport, and it was noted that these pumps move Na^+ and K^+ against their concentration gradients. This action would, by itself, create and amplify the difference in concentration of these ions across the cell membrane. There is, however, another reason why the concentration of these ions would be unequal across the membrane.

Cellular proteins and the phosphate groups of ATP and other organic molecules are negatively charged at the pH of the cell cytoplasm. These negative ions (anions) are "fixed" within the cell because they cannot penetrate the cell membrane. As a result, these anions attract positively charged inorganic ions (cations) from the extracellular fluid that are small enough to diffuse through the membrane pores. Thus, the distribution of small inorganic cations (mainly K^+, Na^+, and Ca^{2+}) between the intracellular and extracellular compartments is influenced by the negatively charged fixed ions within the cell.

Since the cell membrane is more permeable to K^+ than to any other cation, K^+ accumulates within the cell more than the others as a result of its electrical attraction for the fixed anions (fig. 5.20). So, instead of being evenly distributed between the intracellular and extracellular compartments, K^+ becomes more

Figure 5.20

Effect of fixed anions on the distribution of cations.
Proteins, organic phosphates, and other organic anions that cannot leave the cell create a fixed negative charge on the inside of the membrane. This negative charge attracts positively charged inorganic ions (cations), which therefore accumulate within the cell at a higher concentration than is found in the extracellular fluid. The amount of cations that accumulate within the cell is limited by the fact that a concentration gradient builds up, which favors the diffusion of the cations out of the cell.

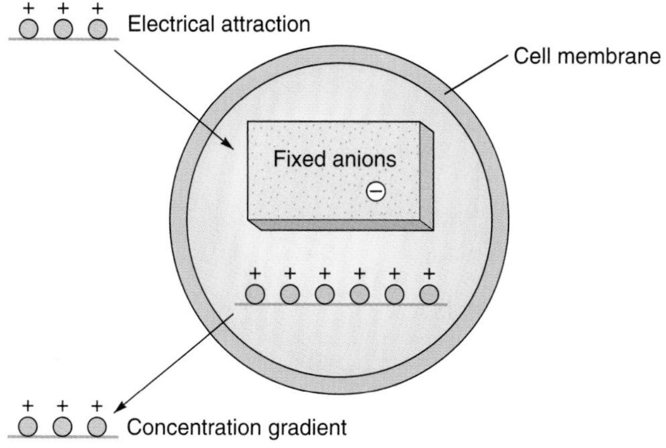

Figure 5.21

The equilibrium potential for K^+.
If K^+ were the only ion able to diffuse through the cell membrane, it would distribute itself between the intracellular and extracellular compartments until an equilibrium was established. At equilibrium, the K^+ concentration within the cell would be higher than outside the cell due to the attraction of K^+ for the fixed anions. Not enough K^+ would accumulate within the cell to neutralize these anions, however, so the inside of the cell would be –90 millivolts compared to the outside of the cell. This membrane voltage is the equilibrium potential (E_K) for potassium.

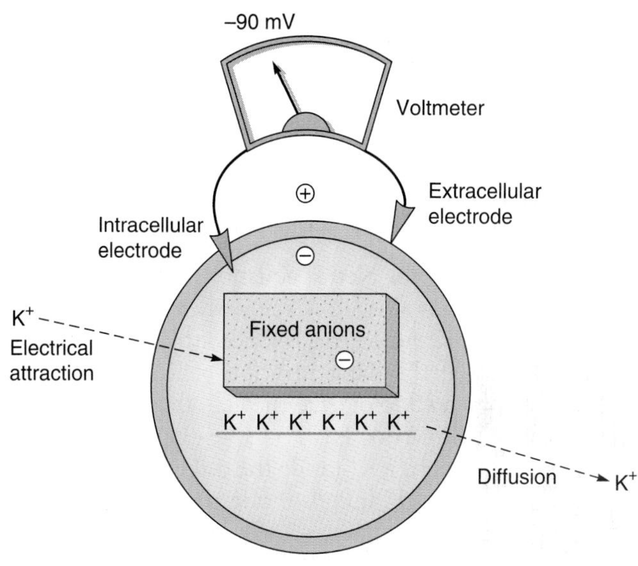

highly concentrated within the cell. The intracellular K^+ concentration is 150 mEq/L in the human body compared to an extracellular concentration of 5 mEq/L (mEq = milliequivalents, which is the millimolar concentration multiplied by the valence of the ion—in this case, by 1).

As a result of the unequal distribution of charges between the inside and outside of cells, each cell is a tiny battery with the positive pole outside the cell membrane and the negative pole inside. The magnitude of this charge difference is measured in *voltage*. Although the voltage of this battery is very small (less than a tenth of a volt), it is of critical importance in such physiological processes as muscle contraction, the regulation of the heartbeat, and the generation of nerve impulses. In order to understand these processes, then, we must first examine the electrical properties of cells.

Equilibrium Potentials

An **equilibrium potential** is a theoretical voltage that would be produced across a cell membrane if only one ion were able to diffuse through the membrane. Since the membrane is most permeable to K^+, we can construct a theoretical approximation of what would happen if K^+ were the *only* ion able to cross the membrane. If this were the case, K^+ would diffuse until its concentration inside and outside of a cell became stable, thus establishing an *equilibrium*. In this condition, if a certain amount of K^+ were to move inside the cell (by electrical attraction for the fixed anions), an identical amount of K^+ would diffuse out of the cell (down its concentration gradient). At equilibrium, the forces of electrical attraction and of the diffusion gradient are equal and opposite.

At this equilibrium, the concentration of K^+ would be higher inside the cell than outside the cell; a concentration difference would exist across the cell membrane that was stabilized by the attraction of K^+ to the fixed anions. At this point we could ask, Are the fixed anions neutralized? Are the charges balanced? The answer depends on how much K^+ gets into the cell, which in turn depends on the K^+ concentration in the extracellular fluid. At the K^+ concentrations that are, in fact, found in the body, the answer to our question is no. Not enough K^+ is present in the cell to neutralize the fixed anions (fig. 5.21).

At equilibrium, therefore, the inside of the cell membrane would have a higher concentration of negative charges than the outside. There is a difference in charge, as well as a difference in concentration, across the membrane. Under these conditions, the magnitude of the difference in charge, or **potential difference,** on the two sides of the membrane is 90 millivolts (mV). A sign (+ or –) placed in front of this number indicates the polarity within the cell. This is shown with a negative sign (as –90 mV) to indicate that the inside of the cell is the negative pole. The potential difference of –90 mV, which would be developed if K^+ were the only diffusible ion, is called the **K^+ equilibrium potential** (abbreviated $\mathbf{E_K}$).

Nernst Equation

There is another way to look at the equilibrium potential: it is the membrane potential that would *exactly balance* the diffusion gradient and prevent the net movement of a particular ion. Since the diffusion gradient depends on the difference in concentration of the ion, the value of the equilibrium potential must depend on the ratio of the concentrations of the ion on the two sides of the membrane. The **Nernst equation** allows this theoretical equilibrium potential to be calculated for a particular ion when its concentrations are known. The following simplified form of the equation is valid at a temperature of 37° C:

$$E_x = \frac{61}{z} \log \frac{[X_o]}{[X_i]}$$

where

E_x = equilibrium potential in millivolts (mV) for ion x
X_o = concentration of the ion outside the cell
X_i = concentration of the ion inside the cell
z = valence of the ion (+1 for Na^+ or K^+)

Note that, using the Nernst equation, the equilibrium potential for a cation has a negative value when X_i is greater than X_o. If we substitute K^+ for X, this is indeed the case. As a hypothetical example, if the concentration of K^+ were 10 times higher inside compared to outside the cell, the equilibrium potential would be 61 mV (log 1/10) = 61 × (–1) = –61 mV. In reality, the concentration of K^+ inside the cell is actually 30 times greater than outside (150 mEq/L inside compared to 5 mEq/L outside). Since the log of 1/30 is –1.477, the equilibrium potential for K^+ given realistic concentration values is –90 mV.

If we wish to calculate the equilibrium potential for Na^+, different values must be used. The concentration of Na^+ in the extracellular fluid is 145 mEq/L, whereas its concentration inside cells is only 12 mEq/L. The diffusion gradient thus promotes the movement of Na^+ into the cell, and, in order to oppose this diffusion, the membrane potential would have to have a positive polarity on the inside of the cell. This is indeed what the Nernst equation would provide, since $[X_o]/[X_i]$ is greater than 1. Using this equation, the equilibrium potential for Na^+ can be calculated to be +60 mV.

Resting Membrane Potential

A membrane potential of +60 mV would prevent the diffusion of Na^+ into the cell, while a membrane potential of –90 mV would prevent the diffusion of K^+ out of the cell. It is clear that the membrane potential cannot be both values at the same time; indeed, it is seldom either value but instead is somewhere between these two extremes. We will call this the **resting membrane potential** to distinguish it from the theoretical equilibrium potentials. The actual value of the resting membrane potential depends on the permeability of the membrane to each ion and on the equilibrium potential of each diffusible ion.

Nernst equation: from Walther Hermann Nernst, German physicist and chemist, 1864–1941

Figure 5.22

The resting membrane potential.
Because some Na^+ leaks into the cell by diffusion, the actual resting membrane potential is lower than the K^+ equilibrium potential. As a result, some K^+ diffuses out of the cell, as indicated by the dashed lines.

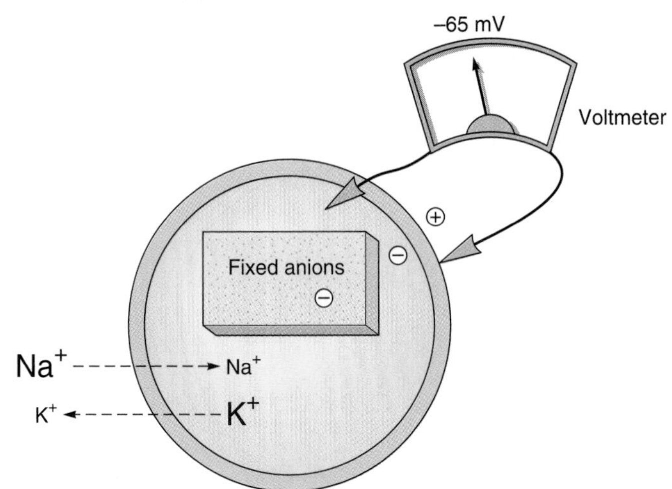

The resting membrane potential of most cells ranges from –65 mV to –85 mV (in neurons it averages –70 mV). This value is very close to the one that we had predicted if K^+ were the only ion able to move through the cell membrane and establish an equilibrium. The resting membrane potential, however, is quite different from the Na^+ equilibrium potential. The resting membrane potential is closer to the K^+ than to the Na^+ equilibrium potential because, as previously discussed, the membrane is much more permeable to K^+ than to Na^+.

The actual value of the membrane potential thus depends on the differential permeability of the membrane. When neurons produce nerve impulses, the permeability to Na^+ increases dramatically, sending the membrane potential toward the Na^+ equilibrium potential (nerve impulses are discussed in chapter 14). This is the reason that the term *resting* is used to describe the membrane potential when it is not producing impulses.

Role of the Na^+/K^+ Pumps

Since the membrane potential is less negative than E_K as a result of some Na^+ entry, some K^+ leaks out of the cell (fig. 5.22). The cell is *not* at equilibrium with respect to K^+ and Na^+ concentrations. Nonetheless, the concentrations of K^+ and Na^+ are maintained constant because of the constant expenditure of energy in active transport by the Na^+/K^+ pumps. The Na^+/K^+ pumps act to counter the leaks and thus maintain the membrane potential.

Actually, the Na^+/K^+ pump does more than simply work against the ion leaks; since it transports *three* Na^+ out of the cell for every *two* K^+ that it moves in, it has the net effect of contributing to the negative intracellular charge (fig. 5.23). This *electrogenic effect* of the pumps adds a few more millivolts

Figure 5.23

The contribution of the Na^+/K^+ pumps to the membrane potential.
The concentrations of Na^+ and K^+ both inside and outside the cell do not change as a result of diffusion (dashed arrows) because of active transport (solid arrows) by the Na^+/K^+ pump. Since the pump transports three Na^+ for every two K^+, the pump itself helps to create a charge separation (a potential difference, or voltage) across the membrane.

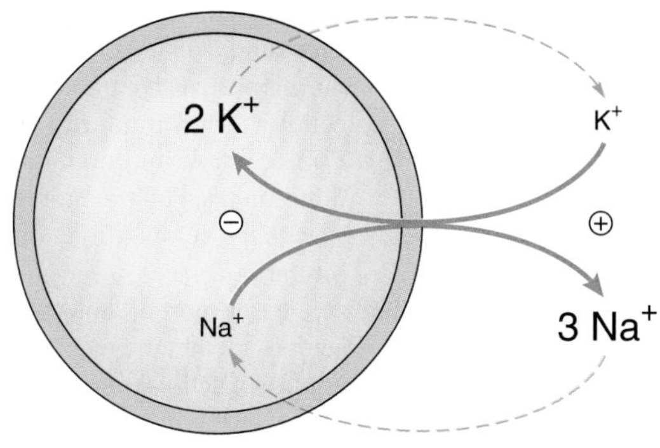

to the membrane potential. As a result of all of these activities, a real cell has (1) a relatively constant intracellular concentration of Na^+ and K^+ and (2) a constant membrane potential (in the absence of stimulation) in nerves and muscles of –65 mV to –85 mV.

Clinical Considerations

Dialysis

Normally functioning kidneys remove waste products from the blood. After the blood is filtered through pores in capillary walls that are large enough to permit the passage of wastes and other molecules, the molecules needed by the body are reabsorbed back into the blood. The wastes generally remain in the filtrate and are excreted in the urine. If the kidneys are not functioning properly, waste molecules can be removed from the blood artificially by a process called **dialysis** (*di-al'ĭ-sis*). Dialysis involves the removal of particular molecules from a solution by having them pass, by means of diffusion, through an artificial porous membrane (see fig. 25.31). Since the pores in this dialysis membrane are large enough to permit the passage of some molecules but too small to permit the passage of others (the blood plasma proteins), small waste molecules can be removed from the blood by this technique.

Inherited Defects in Membrane Carriers

Since membrane transport carriers are proteins that are coded by specific genes, inherited defects in these carriers can result if there is an alteration in the genetic code. Defective protein carriers in the cell membranes of epithelial cells that line the small intestine may produce diseases that result from an inadequate absorption of ingested molecules. *Pernicious anemia,* for example, is due to the inability to absorb vitamin B_{12}. Defects in transport carriers within the epithelial cells of kidney tubules may result in the abnormal excretion of particular molecules in the urine. For example, the inability of kidney tubules to transport glucose is responsible for a rare form of *diabetes mellitus.*

Hyperkalemia and the Membrane Potential

Although changes in the extracellular concentration of many ions can affect the membrane potential, this potential is particularly sensitive to changes in blood K^+. Since the maintenance of a particular membrane potential is critical for the generation of electrical events in nerves and muscles (including the heart), the body has a variety of mechanisms that serve to maintain blood K^+ concentrations within very narrow limits. As described in chapter 25, these mechanisms act primarily through the kidneys, which can excrete K^+ in the urine or reabsorb it into the blood. The excretion of K^+ is stimulated by hormones of the adrenal cortex—particularly by the steroid hormone *aldosterone* (*al-dos-ter'ōn*). Indeed, if the adrenal glands of an experimental animal are removed, the animal may die as a result of an accumulation of K^+ in the blood. An abnormal increase in the blood concentration of K^+ is called **hyperkalemia** (*hi"per-kă-le'me-a*).

When hyperkalemia occurs, more K^+ can enter the cell and neutralize more of the fixed negative charges. In terms of the Nernst equation, the ratio $[K^+_o]/[K^+_i]$ is decreased. This reduces the membrane potential (brings it closer to zero) and thus interferes with the proper function of the heart. For these reasons, the blood electrolyte concentrations are monitored very carefully in patients with heart or kidney disease.

hyperkalemia: Gk. *hyper*, over; L. *kalium*, potash; Gk, *haima*, blood

Clinical Investigation Answer

The student's hyperglycemia caused his renal carrier proteins to become saturated, resulting in glycosuria (glucose in the urine). The elimination of glucose in the urine and its subsequent osmotic effects caused the urinary excretion of an excessive amount of water, resulting in dehydration. This raised the plasma osmolality, stimulating the thirst center in the hypothalamus. (Hyperglycemia and excessive thirst and urination are cardinal signs of diabetes mellitus.) Further, the loss of plasma water (increased plasma osmolality) caused the concentration of plasma solutes, including K^+, to be raised. The resulting hyperkalemia affected the membrane potential of myocardial cells in the heart, producing electrical abnormalities that were revealed in the student's electrocardiogram.

Chapter Summary

Diffusion and Osmosis (pp. 111–117)

1. Diffusion is the net movement of molecules or ions from regions of high concentration to regions of low concentration.
   (a) This is a type of passive transport—energy is provided by the thermal energy of the molecules, not by cellular metabolism.
   (b) Net diffusion stops when the concentration is equal on both sides of the membrane.
2. The rate of diffusion is dependent on the magnitude of the concentration difference between the two sides of the membrane, the degree of permeability of the cell membrane to the diffusing substance, the temperature of the solution, and the extent of the surface area of the membrane.
3. Simple diffusion is the type of passive transport in which small molecules and inorganic ions, such as Na^+ and K^+, move through the cell membrane.
   (a) Inorganic ions pass through specific channels in the membrane.
   (b) Lipids (such as steroid hormones) and dipolar water molecules can pass directly through the phospholipid layers of the membrane by simple diffusion.
4. Osmosis is the simple diffusion of solvent (water) through a membrane that is more permeable to the solvent than it is to the solute.
5. Water moves from the solution that is more dilute to the solution that has a higher solute concentration.
6. Osmosis depends on a difference in total solute concentration, not on the chemical nature of the solute.
   (a) The concentration of total solute (in moles) per kilogram (liter) of water is measured in osmolality units.
   (b) The solution with the higher osmolality has the higher osmotic pressure.
   (c) Water moves by osmosis from the solution of lower osmolality and osmotic pressure to the solution of higher osmolality and osmotic pressure.
7. Solutions containing osmotically active solutes that have the same osmotic pressure as plasma (such as 0.9% NaCl and 5% glucose) are said to be isotonic.
   (a) Solutions with a lower osmotic pressure are hypotonic; those with a higher osmotic pressure are hypertonic.
   (b) Cells in a hypotonic solution gain water and swell; those in a hypertonic solution lose water and shrink (crenate).
8. The osmolality and osmotic pressure of the plasma is detected by osmoreceptors in the hypothalamus of the brain and maintained within a normal range by the action of antidiuretic hormone (ADH) secreted by the posterior pituitary.
   (a) Increased osmolality inhibits the electrical activity of the osmoreceptors.
   (b) Stimulation of the osmoreceptors causes thirst and triggers the secretion of antidiuretic hormone (ADH) from the pituitary.
9. ADH stimulates water retention by the kidneys, which serves to maintain a normal blood volume and osmolality.

Carrier-Mediated Transport (pp. 117–122)

1. The passage of glucose, amino acids, and other polar molecules through the cell membrane is mediated by carrier proteins in the cell membrane.
   (a) Carrier-mediated transport exhibits the properties of specificity, competition, and saturation.
   (b) The transport rate of molecules such as glucose reaches a maximum when the carriers are saturated. This maximum rate is called the transport maximum, or T_m.
2. The transport of molecules such as glucose from the side of higher to the side of lower concentration by means of membrane carriers is called facilitated diffusion.
   (a) Like simple diffusion, facilitated diffusion is passive transport—cellular energy is not required.
   (b) Unlike simple diffusion, facilitated diffusion displays the properties of specificity, competition, and saturation.
3. The active transport of molecules and ions across a membrane requires the expenditure of cellular energy (ATP).
   (a) In active transport, carriers move molecules or ions from the side of lower to the side of higher concentration.
   (b) One example of active transport is the action of the Na^+/K^+ pump.
4. Sodium is more concentrated on the outside of the cell, whereas potassium is more concentrated on the inside of the cell.
   (a) The Na^+/K^+ pump helps to maintain these concentration differences by transporting Na^+ out of the cell and K^+ into the cell.
   (b) Three sodium ions are transported out of the cell for every two potassium ions that are transported into the cell.

The Membrane Potential (pp. 122–125)

1. The cytoplasm of the cell contains negatively charged organic ions (anions) that cannot leave the cell—they are "fixed anions."
   (a) These fixed anions attract K^+, which is the inorganic ion that can most easily pass through the cell membrane.
   (b) As a result of this electrical attraction, the concentration of K^+ within the cell is greater than the concentration of K^+ in the extracellular fluid.
2. If K^+ were the only diffusible ion, the concentrations of K^+ on the inside and outside of the cell would reach an equilibrium.
   (a) At this point, the rate of K^+ entry (due to electrical attraction) would equal the rate of K^+ exit (due to diffusion).

(b) At this equilibrium, there would still be a higher concentration of negative charges within the cell (because of the fixed anions) than outside the cell.
(c) At this equilibrium, the inside of the cell would be 90 millivolts negative (–90 mV) compared to the outside of the cell. This potential difference is called the K^+ equilibrium potential (E_K).

3. The resting membrane potential is less than E_K (usually –65 mV to –85 mV) because some Na^+ can also enter the cell.
   (a) Na^+ is more highly concentrated outside than inside the cell, and the inside of the cell is negative. These forces attract Na^+ into the cell.
   (b) The rate of Na^+ entry is generally slow because the membrane is usually not very permeable to Na^+.
4. The slow rate of Na^+ entry is accompanied by a slow rate of K^+ leakage out of the cell.
   (a) The Na^+/K^+ pump counters this leakage, thus maintaining constant concentrations and a constant resting membrane potential.
   (b) All cells in the body contain numerous Na^+/K^+ pumps that require a constant expenditure of energy.
   (c) The Na^+/K^+ pump itself contributes to the membrane potential because it pumps more Na^+ out than it pumps K^+ in (by a ratio of three to two).

Review Activities

Objective Questions

1. The movement of water across a cell membrane occurs by
   (a) active transport.
   (b) facilitated diffusion.
   (c) simple diffusion (osmosis).
   (d) all of the above.
2. Which of the following statements about the facilitated diffusion of glucose is *true?*
   (a) There is a net movement from the region of lower to the region of higher concentration.
   (b) Carrier proteins in the cell membrane are required for this transport.
   (c) This transport requires energy obtained from ATP.
   (d) It is an example of cotransport.
3. If a poison such as cyanide stopped the production of ATP, which of the following transport processes would cease?
   (a) the movement of Na^+ out of a cell
   (b) osmosis
   (c) the movement of K^+ out of a cell
   (d) all of the above
4. Red blood cells crenate in
   (a) hypotonic solution.
   (b) an isotonic solution.
   (c) a hypertonic solution.
5. Plasma has an osmolality of about 300 mOsm. The osmolality of isotonic saline is equal to
   (a) 150 mOsm.
   (b) 300 mOsm.
   (c) 600 mOsm.
   (d) none of the above.
6. Which of the following statements comparing a 0.5 *m* NaCl solution and a 1.0 *m* glucose solution is *true?*
   (a) They have the same osmolality.
   (b) They have the same osmotic pressure.
   (c) They are isotonic to each other.
   (d) All of the above are true.
7. The most important diffusible ion in the establishment of the membrane potential is
   (a) K^+.
   (b) Na^+.
   (c) Ca^{2+}.
   (d) Cl^-.
8. Which of the following statements regarding an increase in blood osmolality is *true?*
   (a) It can occur as a result of dehydration.
   (b) It causes a decrease in blood osmotic pressure.
   (c) It is accompanied by a decrease in ADH secretion.
   (d) All of the above are true.
9. In hyperkalemia, the resting membrane potential
   (a) moves farther from 0 millivolts.
   (b) moves closer to 0 millivolts.
   (c) remains unaffected.
10. Which of the following statements about the Na^+/K^+ pump is *true?*
    (a) Na^+ is actively transported into the cell.
    (b) K^+ is actively transported out of the cell.
    (c) An equal number of Na^+ and K^+ ions are transported with each cycle of the pump.
    (d) The pumps are constantly active in all cells.
11. Which of the following statements about carrier-mediated facilitated diffusion is *true?*
    (a) It uses cellular ATP.
    (b) It is used for cellular uptake of blood glucose.
    (c) It is a form of active transport.
    (d) None of the above are true.
12. Which of the following is *not* an example of cotransport?
    (a) movement of glucose and Na^+ through the apical epithelial membrane in the intestinal epithelium
    (b) movement of Na^+ and K^+ through the action of the Na^+/K^+ pumps
    (c) movement of Na^+ and glucose across the kidney tubules
    (d) movement of Na^+ into a cell while Ca^{2+} moves out

Essay Questions

1. Describe the conditions required to produce osmosis and explain why osmosis occurs under these conditions.
2. Explain how simple diffusion can be distinguished from facilitated diffusion and how active transport can be distinguished from passive transport.
3. Compare the theoretical membrane potential that occurs at K^+ equilibrium with the true resting membrane potential. Explain why these values differ.
4. Explain how the Na^+/K^+ pump contributes to the resting membrane potential.
5. Describe the cause-and-effect sequence whereby a genetic defect results in improper cellular transport and the symptoms of cystic fibrosis.
6. Using the principles of osmosis, explain why movement of Na^+ through a cell membrane is followed by movement of water. Use this concept to explain the rationale on which oral rehydration therapy is based.
7. Distinguish between primary active transport and secondary active transport, and between cotransport and countertransport. Give examples of each.

Critical Thinking Questions

1. Mannitol is a sugar that does not pass through the walls of blood capillaries in the brain (does not cross the "blood-brain barrier," as described in chapter 14). It also does not cross the walls of kidney tubules, the structures that transport blood filtrate to become urine (chapter 25).

Explain why mannitol can be described as osmotically active. How might its clinical administration help to prevent swelling of the brain in head trauma? Also, explain the effect it might have on the water content of urine.

2. Discuss carrier-mediated transport. How could you experimentally distinguish between the different types of carrier-mediated transport?
3. Remembering the effect of cyanide (described in chapter 4), explain how you might determine the extent to which the Na^+/K^+ pumps contribute to the resting membrane potential. Using a measurement of the resting membrane potential as your guide, how could you experimentally determine the relative permeability of the cell membrane to Na^+ and K^+?

Related Web Sites

For a listing of the most current web sites related to this chapter, please visit the *Concepts of Human Anatomy and Physiology* home page at http://www.mhhe.com/biosci/abio/.

NEXUS

Some Interactions of Membrane Transport Concepts with the Other Body Systems

Skeletal System

- Osteoblasts secrete Ca^{2+} and PO_4^{3-} into the extracellular matrix to form calcium phosphate crystals for bone hardness (p. 185)

Nervous System

- Glucose enters neurons by facilitative diffusion (p. 118).
- Voltage-gated ion channels produce action potentials, or nerve impulses (p. 380).
- Ion channels in particular regions of a neuron open in response to binding to a chemical ligand known as a neurotransmitter (p. 386).
- Neurotransmitters are released by axons through a process of exocytosis (p. 386).
- Some cell membranes are electrically coupled by means of connections called gap junctions (p. 386).

Endocrine System

- Lipophilic hormones pass through the cell membrane of their target cells, where they then bind to receptors in the cytoplasm or nucleus (p. 558).
- Active transport Ca^{2+} pumps and the passive diffusion of Ca^{2+} are important in the mechanism of action of some hormones (p. 563).
- Insulin stimulates the facilitative diffusion of glucose into skeletal muscle cells (p. 877).

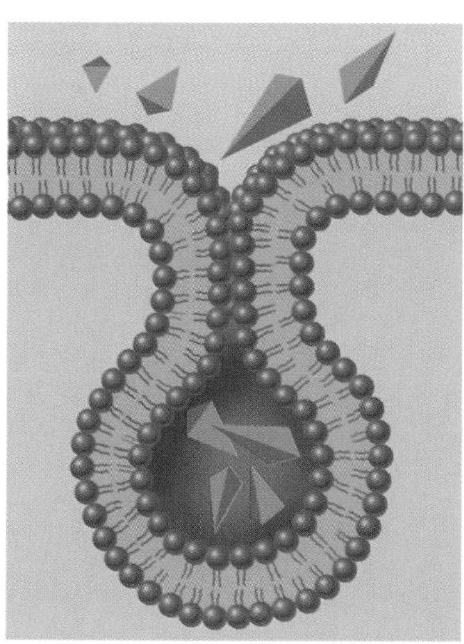

Muscular System

- Exercise increases the number of carriers for the facilitative diffusion of glucose in the muscle cell membrane (p. 887).
- Ca^{2+} transport processes in the endoplasmic reticulum of skeletal muscle fibers are important in the regulation of muscle contraction (p. 294).
- Voltage-gated Ca^{2+} channels in the cell membrane of smooth muscle open in response to depolarization, producing contraction of the muscle (p. 302).

Circulatory System

- Ion diffusion across the cell membrane of myocardial cells is responsible for the electrical activity of the heart and the heartbeat (p. 656).
- The LDL carriers for blood cholesterol are taken into arterial smooth muscle cells by receptor-mediated endocytosis (p. 651).
- Sympathetic nerves increase the rate of spontaneous depolarization in the pacemaker cells of the heart (p. 656).

Lymphatic and Immune Systems

- B lymphocytes secrete antibody proteins that function in humoral immunity (p. 701).
- T lymphocytes secrete polypeptides called cytokines, which promote the cell-mediated immune response (p. 713).
- Antigen-presenting cells engulf foreign proteins by pinocytosis, modify these proteins, and present them to T lymphocytes (p. 714).

Respiratory System

- Oxygen and carbon dioxide pass through the cells of the pulmonary alveoli (air sacs) by simple diffusion (p. 750).
- CO_2, diffusing into red blood cells, converts to bicarbonate, which passively diffuses into the plasma (p. 762).

Urinary System

- Urine is produced as a filtrate of blood plasma, but most of the filtered water is reabsorbed into the blood by osmosis (p. 785).
- Osmosis across the wall of the renal tubules is promoted by membrane pores known as aquaporins (p. 791).
- Urea transport occurs passively across particular regions of the renal tubules (p. 791).
- Antidiuretic hormone stimulates the permeability of the renal tubule to water (p. 791).
- Aldosterone stimulates Na^+ transport in a region of the renal tubule (p. 798).
- Glucose and amino acids are reabsorbed by secondary active transport (p. 798).
- The active transport of many foreign molecules into the renal tubules allow them to be rapidly cleared from the blood (p. 797).

Digestive System

- Cells in the stomach have a membrane H^+/K^+ ATPase active transport pump that creates an extremely acidic gastric juice (p. 830).
- Water is absorbed in the intestine by osmosis following the absorption of sodium chloride (p. 839).
- An intestinal membrane carrier protein transports dipeptides and tripeptides from the intestinal lumen into the epithelial cells (p. 854).

Six

Histology

Objectives

- Define *tissue* and discuss the importance of histology.
- Describe the functional relationship between cells and tissues.
- Classify the tissues of the body according to their four principal types and list the distinguishing characteristics of each type.
- Compare and contrast the various types of epithelia.
- Define *exocrine gland* and compare and contrast the various types of exocrine glands in the body.
- Describe the general characteristics, locations, and functions of connective tissues.
- List the various ground substance fiber types and cells that constitute connective tissue and explain their functions.
- Describe the structure, location, and function of the three types of muscle tissue.
- Describe the basic characteristics and functions of nervous tissue.
- Distinguish between neurons and neuroglia.

Clinical Investigation

In a nutrition class, the professor assigned the students to prepare a short report on why scurvy was such a dreaded disease in the past, especially among sailors. As a student studying anatomy and physiology, reply to the same question. What is the cause of scurvy and why is it a concern to the health of connective tissue? Is scurvy a disease that we should be concerned about today?

Definition and Classification of Tissues

Histology is the specialty of anatomy that involves study of the microscopic structure of tissues. Tissues are assigned to four basic categories on the basis of their cellular composition and histological appearance.

Although cells are the structural and functional units of the body, the cells of a multicellular organism are so specialized that they do not function independently. *Tissues* are aggregations of similar cells and cell products that perform specific functions. The various types of tissues are established during early embryonic development. As the embryo grows, organs form from specific arrangements of tissues. Many adult organs, including the heart and muscles, contain the original cells and tissues that were formed prenatally, although some functional changes occur as they are acted upon by hormones or as their effectiveness diminishes with age.

The study of tissues is referred to as **histology** (*his-tol'ŏ-je*). It provides a foundation for understanding the structure and functions of the organs discussed in the chapters that follow. Many diseases profoundly alter the tissues within an affected organ; therefore, by knowing the normal tissue structure, a physician can recognize the abnormal. In medical schools a course in histology is usually followed by a course in *pathology*, the study of abnormal tissues in diseased organs.

Although histologists employ many different techniques for preparing, staining, and sectioning tissues, only two basic kinds of microscopes are used to view the prepared tissues. The *light microscope* is used to observe overall tissue structure (fig. 6.1*a*), and the *electron microscope* is used to observe the fine details of tissue and cellular structure (fig. 6.1*b*). Most of the histological photomicrographs in this text are at the light microscopic level. However, where fine structural detail is needed to understand a particular function, electron micrographs are used.

Tissue cells are surrounded and bound together by a nonliving intercellular *matrix* (*ma'triks*) that the cells secrete.

histology: Gk. *histos*, web (tissue); *logos*, study
pathology: Gk. *pathos*, suffering, disease; *logos*, study
matrix: L. *matris*, mother

Figure 6.1

The histological appearance of skin.
(*a*) The skin magnified 25 times, as seen through a compound light microscope, and (*b*) the skin magnified 280 times, as seen through a scanning electron microscope (SEM).

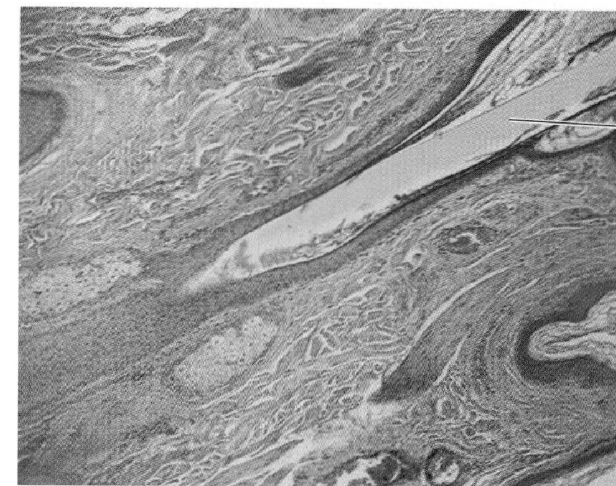

(a)

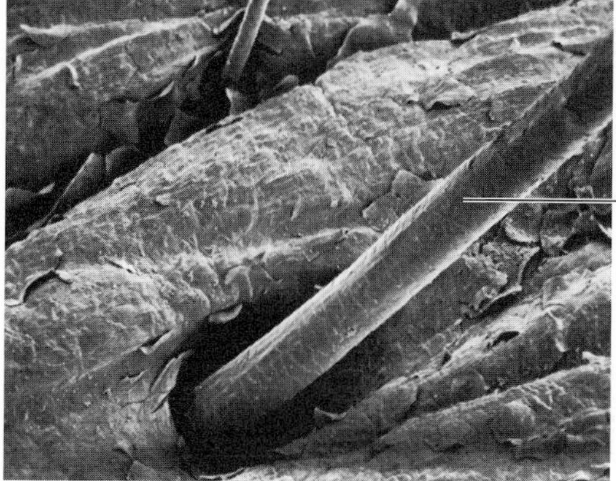

(b)

Matrix varies in composition from one tissue to another and may take the form of a liquid, semisolid, or solid. Blood, for example, has a liquid matrix, permitting this tissue to flow through vessels. By contrast, bone cells are separated by a solid matrix, permitting this tissue to support the body.

The tissues of the body are classified into four basic kinds on the basis of structure and function.

1. **Epithelial** (*ep"ĭ-the'le-al*) **tissue.** Epithelial tissue (*epithelium*; plural, *epithelia*) consists of one or more layers of cells that cover body surfaces and line body cavities and ducts. Tightly clustered epithelial tissue in certain locations forms glands.
2. **Connective tissue.** Connective tissue has abundant matrix that enables it to bind, support, and protect body parts.

epithelium: Gk. *epi*, upon; *thelium*, to cover

UNDER DEVELOPMENT

Development of Tissues

Human prenatal development is initiated by the fertilization of an ovulated ovum (egg) from a female by a sperm cell from a male. The chromosomes within the nucleus of a zygote (*zi'gōt*) (fertilized egg) contain all the genetic information necessary for the differentiation and development of all body structures.

Within 30 hours after fertilization, the zygote undergoes a mitotic division as it moves through the uterine tube toward the uterus (see fig. 30.5). After several more cellular divisions, the embryonic mass consists of 16 or more cells and is called a morula (*mor'yŭ-lă*) (fig. 1). Three or 4 days after conception, the morula enters the uterine cavity where it remains unattached for about 3 days. During this time, the center of the morula fills with fluid passing in from the uterine cavity. As the fluid-filled space develops inside the morula, two distinct groups of cells form. The single layer of cells forming the outer wall is known as the trophoblast and the inner aggregation of cells is known as the embryoblast. After further development, the trophoblast becomes a portion of the placenta and the embryoblast becomes the embryo. With the establishment of these two groups of cells, the morula becomes known as a blastocyst (*blas'tŏ-sist*). Implantation of the blastocyst begins between the fifth and seventh day (see fig. 30.8).

zygote: Gk. *zygotos*, yolked
morula: Gk. *morus*, mulberry
trophoblast: Gk. *trophe*, nourishment; *blastos*, germ
embryoblast: Gk. *embryon*, to be full, swell; *blastos*, germ

Figure 1 The early stages of embryonic development. (*a*) Fertilization and the formation of the zygote, (*b*) the morula at about the third day, (*c*) the early blastocyst at the time of implantation between the fifth and seventh day, (*d*) a blastocyst at 2 weeks, and (*e*) a blastocyst at 3 weeks showing the three primary germ layers that constitute the embryonic disc.

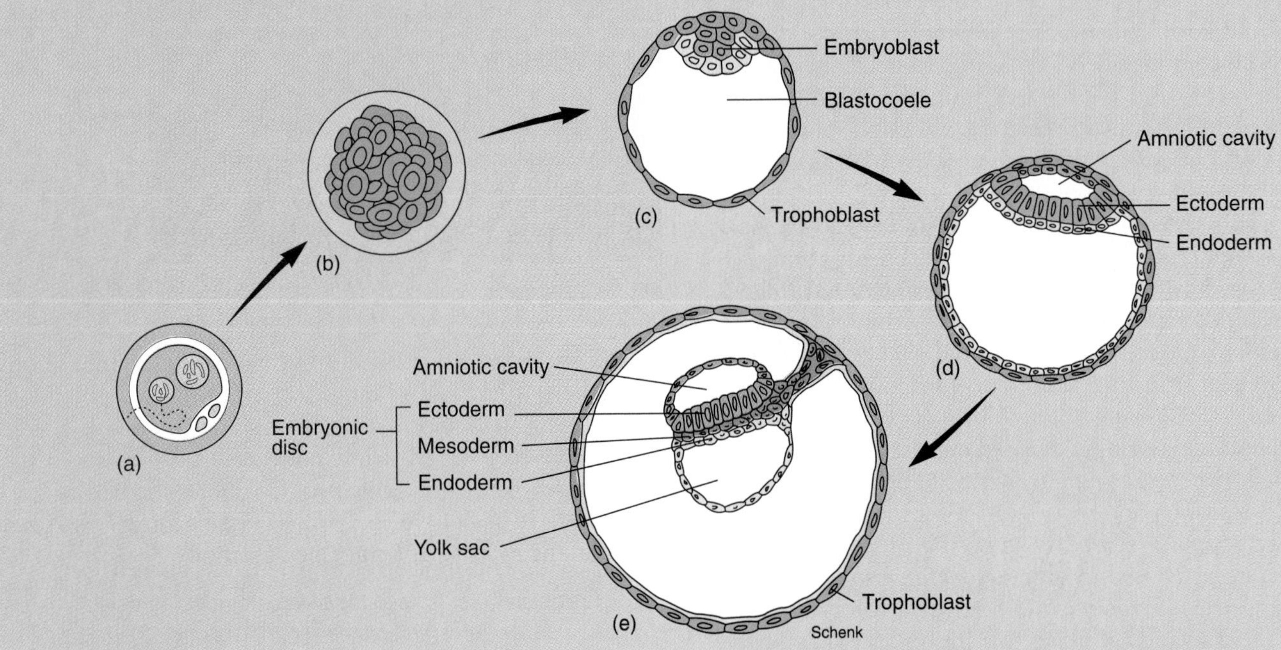

3. **Muscle tissue.** Muscle tissue is composed of elongated cells that contract when stimulated producing movement.
4. **Nervous tissue.** Nervous tissue consists of highly branched, elongated cells called neurons (*noo'ronz*) and other supportive cells called neuroglia (*noo-rog'le-ah*). Neurons initiate and transmit nerve impulses from one body part to another.

Epithelial Tissue

There are two major categories of epithelia: membranous and glandular. Membranous epithelia are located throughout the body and form such structures as the outer layer of the skin, the inner lining of body cavities and lumina, and the covering of visceral organs. Glandular epithelia are specialized tissues that form the secretory portion of glands.

As the blastocyst completes implantation during the second week of development, the embryoblast undergoes marked differentiation. A slitlike space called the amniotic (*am'ne-ot-ic*) cavity forms within the embryoblast, adjacent to the trophoblast. The embryoblast now consists of two layers: an upper ectoderm, which is closer to the amniotic cavity, and a lower endoderm, which borders the blastocyst cavity. A short time later, a third layer called the mesoderm forms between the endoderm and ectoderm. These three layers constitute the primary germ layers.

ectoderm: Gk. *ecto*, outside; *derm*, skin
endoderm: Gk. *endo*, within; *derm*, skin
mesoderm: Gk. *meso*, middle; *derm*, skin

The primary germ layers are important because all the cells and tissues of the body are derived from them (see fig. 30.9). Ectodermal cells form the nervous system; the outer layer of skin (epidermis), including hair, nails, and skin glands; and portions of the sensory organs. Mesodermal cells form the skeleton, muscles, blood, reproductive organs, dermis of the skin, and connective tissue. Endodermal cells give rise to the lining of the gastrointestinal (GI) tract, the digestive organs, the respiratory tract and lungs, and the urinary bladder and urethra. Table 1 lists the derivatives of the primary germ layers.

Table 1
Derivatives of the Germ Layers

Ectoderm	Mesoderm	Endoderm
Epidermis of skin and epidermal derivatives: hair, nails, glands of the skin; linings of oral, nasal, anal, and vaginal cavities	Muscle: smooth, cardiac, skeletal	Epithelium of pharynx, auditory canal, tonsils, thyroid, parathyroid, thymus, larynx, trachea, lungs, GI tract, urinary bladder and urethra, and vagina
Nervous tissue; sense organs	Connective tissue: embryonic, connective tissue proper, cartilage, bone, blood	Liver and pancreas
Lens of eye; enamel of teeth	Dermis of skin; dentin of teeth	
Pituitary gland	Epithelium of blood vessels, lymphatic vessels, body cavities, joint cavities	
Adrenal medulla	Internal reproductive organs	
	Kidneys and ureters	
	Adrenal cortex	

Characteristics of Membranous Epithelia

Membranous epithelia always have one free surface exposed to a body cavity, a lumen (hollow portion of a body tube), or the skin surface. Some membranous epithelia are derived from ectoderm, such as the outer layer of the skin; some from mesoderm, such as the inside lining of blood vessels; and others from endoderm, such as the inside lining of the gastrointestinal (GI) tract (digestive tract).

Membranous epithelia may be one layer or several layers thick. The upper surface of epithelia may be exposed to gases, as in the case of epithelium in the integumentary and respiratory systems; to liquids, as in the circulatory and urinary systems; or to semisolids, as in the GI tract. The deep surface of most membranous epithelia is bound to underlying supportive tissue by a **basement membrane,** consisting of glycoprotein from the epithelial cells and a meshwork of collagenous and reticular fibers from the underlying connective tissues. With few exceptions, membranous epithelia are avascular (without blood vessels) and must be nourished by diffusion from underlying connective tissues. Cells that make up membranous epithelia are tightly packed together, with little intercellular matrix between them.

Some of the functions of membranous epithelia are quite specific, but certain generalities can be made. Epithelia that cover or line surfaces provide *protection* from pathogens, physical injury, toxins, and desiccation (drying). Epithelia lining the lumen of the GI tract function in *absorption*. The epithelium of the kidneys provides *filtration*, whereas that within the pulmonary alveoli (air sacs) of the lungs allows for *diffusion*. Highly specialized neuroepithelium in the taste buds and in the nasal region have a chemoreceptor function.

Many membranous epithelia are exposed to friction or harmful substances from the outside environment. For this reason, epithelial tissues have remarkable regenerative abilities. The mitotic replacement of the outer layer of skin and the lining of the GI tract, for example, is a continuous process.

Membranous epithelia are histologically classified by the number of layers of cells and the shape of the cells along the exposed surface. Epithelial tissues that are composed of a single layer of cells are called *simple;* those that are layered are said to be *stratified*. Squamous cells are flattened; *cuboidal* cells are cube-shaped; and *columnar* cells are taller than they are wide.

Figure 6.2

Simple squamous epithelium.
(*a*) This epithelium lines the lumina of vessels, where it permits diffusion. A photomicrograph of the tissue is shown in (*b*) and a labeled diagram in (*c*).

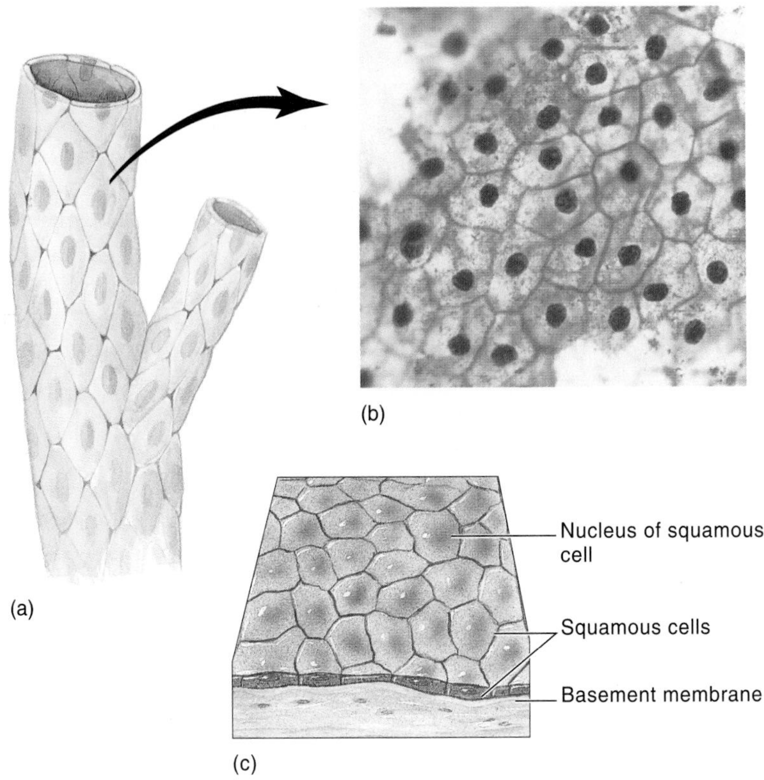

Simple Epithelia

Simple epithelial tissue is a single layer thick and is located where diffusion, absorption, filtration, and secretion are principal functions. The cells of simple epithelial tissue range from thin, flattened cells to tall, columnar cells. Some of these cells have cilia that create currents for the movement of materials across cell surfaces. Others have microvilli that increase the surface area for absorption.

Simple Squamous Epithelium

Simple squamous (*skwa'mus*) epithelium is composed of flattened, irregularly shaped cells that are tightly bound together in a mosaiclike pattern (fig. 6.2). Each cell contains an oval centrally located nucleus. This epithelium is adapted for diffusion and filtration. It occurs in the pulmonary alveoli within the lungs (where gaseous exchange occurs), in portions of the kidney (where blood is filtered), on the inside lining of blood vessels, in the lining of body cavities, and in the covering of the viscera (internal body organs). The simple squamous epithelium forming the inner lining of blood and lymphatic vessels is termed **endothelium.** That which covers visceral organs and lines body cavities is called **mesothelium** (*mes"o-the' le-um*).

Simple Cuboidal Epithelium

Simple cuboidal epithelium is composed of a single layer of tightly fitted cube-shaped cells (fig. 6.3). This type of epithelium is found lining lumina of small ducts and tubules that have excretory, secretory, or absorptive functions. It occurs on the surface of the ovaries, forms a portion of the tubules within the kidneys, and lines the ducts of the salivary glands and the pancreas.

Simple Columnar Epithelium

Simple columnar epithelium is composed of tall, columnar cells (fig. 6.4). The height of the cells varies, depending on the site and function of the tissue. Each cell contains a single nucleus, which is usually located near the basement membrane. Specialized unicellular glands called **goblet cells** are scattered through this tissue at most locations. Goblet cells secrete a lubricative

squamous: L. *squamosus*, scaly

endothelium: Gk. *endon*, within; *thelium*, to cover
mesothelium: Gk. *meso*, middle; *thelium*, to cover

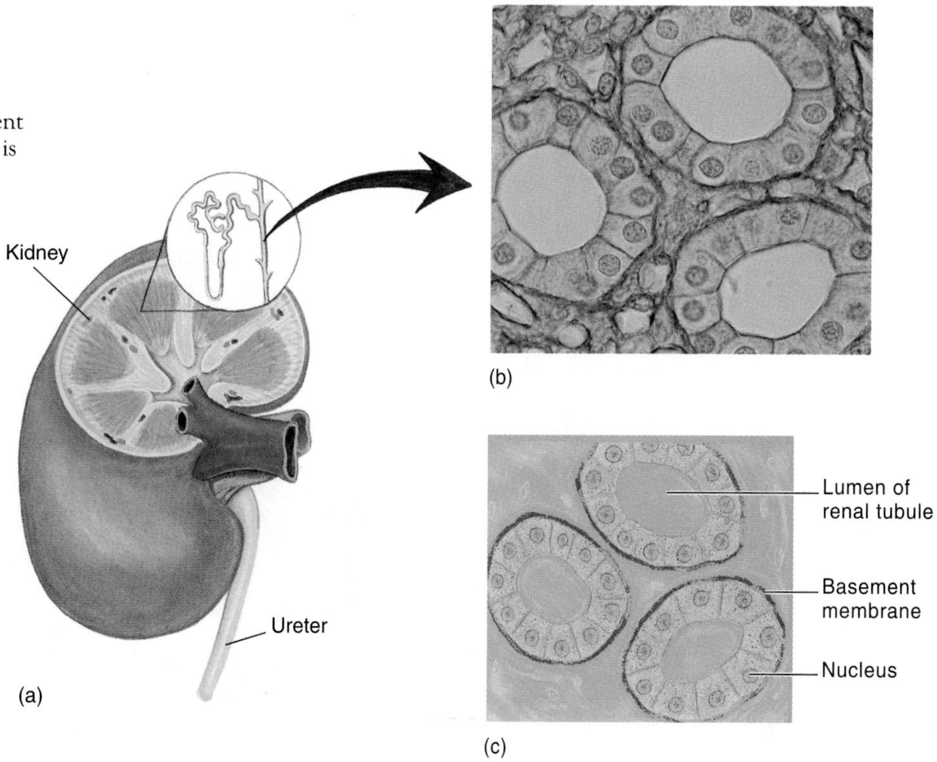

Figure 6.3

Simple cuboidal epithelium.
(*a*) This epithelium lines the lumina of ducts (for example, in the kidneys), where it permits movement of fluids and ions. A photomicrograph of the tissue is shown in (*b*) and a labeled diagram in (*c*).

Kidney
Ureter
(a)
(b)
Lumen of renal tubule
Basement membrane
Nucleus
(c)

Figure 6.4

Simple columnar epithelium.
(*a*) This epithelium lines the lumen of the GI tract, where it permits secretion and absorption. A photomicrograph of the tissue is shown in (*b*) and a labeled diagram in (*c*).

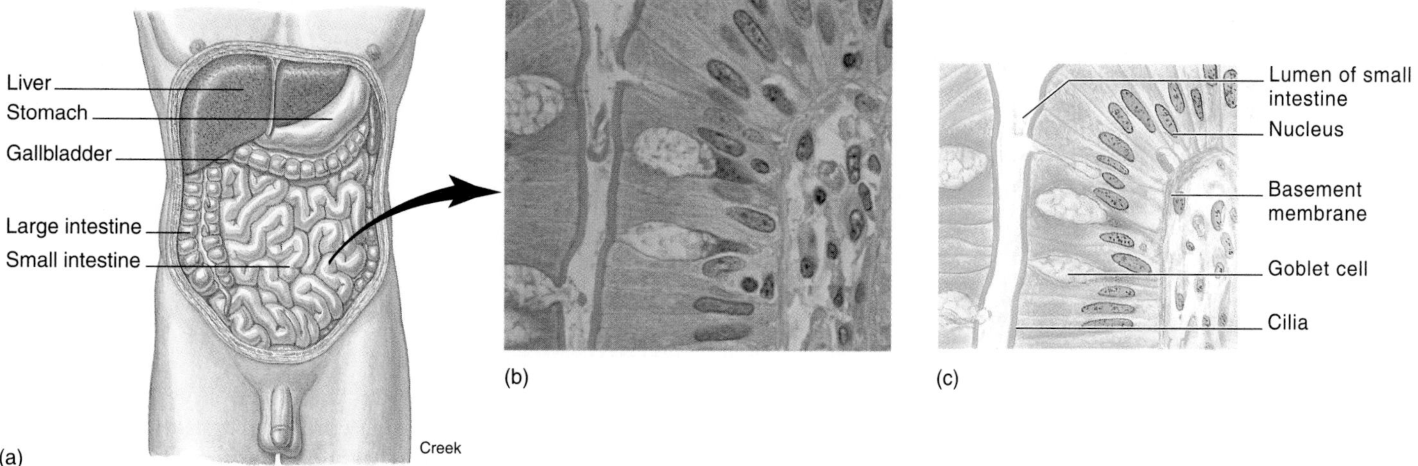

and protective mucus along the free surfaces of the cells. Simple columnar epithelium is found lining the lumen of the stomach and intestine. In the digestive system, it forms a highly absorptive surface and also secretes certain digestive chemicals. Within the stomach, simple columnar epithelium has a tremendous mitotic rate. It replaces itself every 2 or 3 days.

Simple Ciliated Columnar Epithelium

Simple ciliated columnar epithelium is characterized by the presence of cilia along its free surface (fig. 6.5). By contrast, the simple columnar type is unciliated. Cilia produce wavelike movements that transport materials through tubes or passageways. This type of epithelium occurs in the uterine tubes

Figure 6.5

Simple ciliated columnar epithelium.
(*a*) This epithelium lines the lumen of the uterine tube, where currents generated by the cilia propel the egg cell toward the uterus. A photomicrograph of the tissue is shown in (*b*) and a labeled diagram in (*c*).

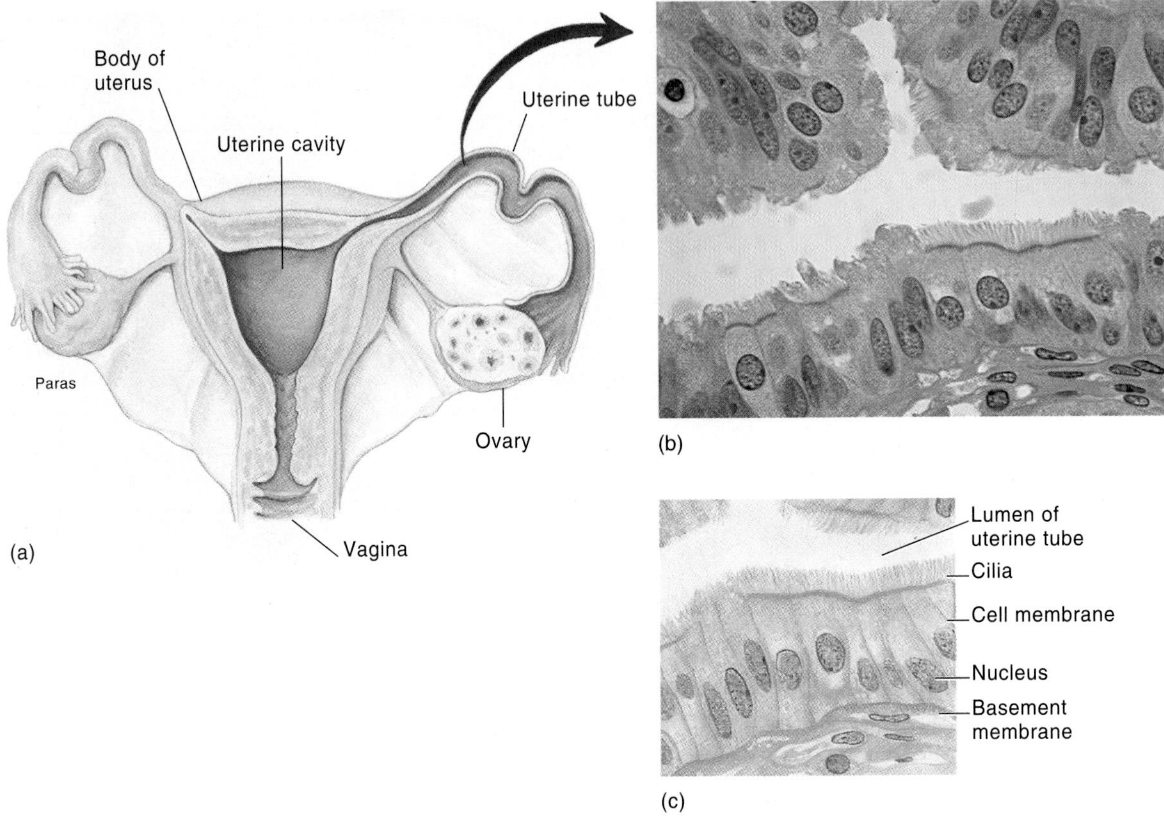

of the female, where the currents generated by the cilia propel the ovum (egg cell) toward the uterus. Furthermore, recent evidence indicates that sperm introduced during sexual intercourse may be moved along the return currents, or eddies, generated by ciliary movement. This greatly enhances the likelihood of fertilization.

Pseudostratified Ciliated Columnar Epithelium

As the name implies, this type of epithelium has a layered appearance (*strata* = layers). Actually, it is not multilayered (*pseudo* = false), since each cell is in contact with the basement membrane. Not all cells are exposed to the surface, however (fig. 6.6). The epithelium has a stratified appearance because the nuclei of these cells are located at different levels. Numerous goblet cells and a ciliated, exposed surface are characteristic of this epithelium. The lumina of the trachea and the bronchial tubes are lined with this tissue; hence, it is frequently called *respiratory epithelium*. Its function is to remove foreign dust and bacteria entrapped in mucus from the lower respiratory system.

Coughing, sneezing, or simply "clearing the throat" are protective reflex mechanisms for clearing the respiratory passages of obstructions or inhaled particles that have been trapped in the mucus along the ciliated lining. The coughed-up material consists of the mucus-entrapped particles.

Stratified Epithelia

Stratified epithelia are tissues consisting of two or more layers of cells. In contrast to simple epithelia, stratified epithelia are poorly suited for absorption and secretion because of their thickness. Stratified epithelia have a primarily protective function that is enhanced by rapid cell divisions. Stratified epithelia are classified according to the shape of the surface layer of cells, since the layer in contact with the basement membrane is cuboidal or columnar in shape.

Stratified Squamous Epithelium

Stratified squamous epithelium is composed of a variable number of cell layers that are flattest at the surface (fig. 6.7).

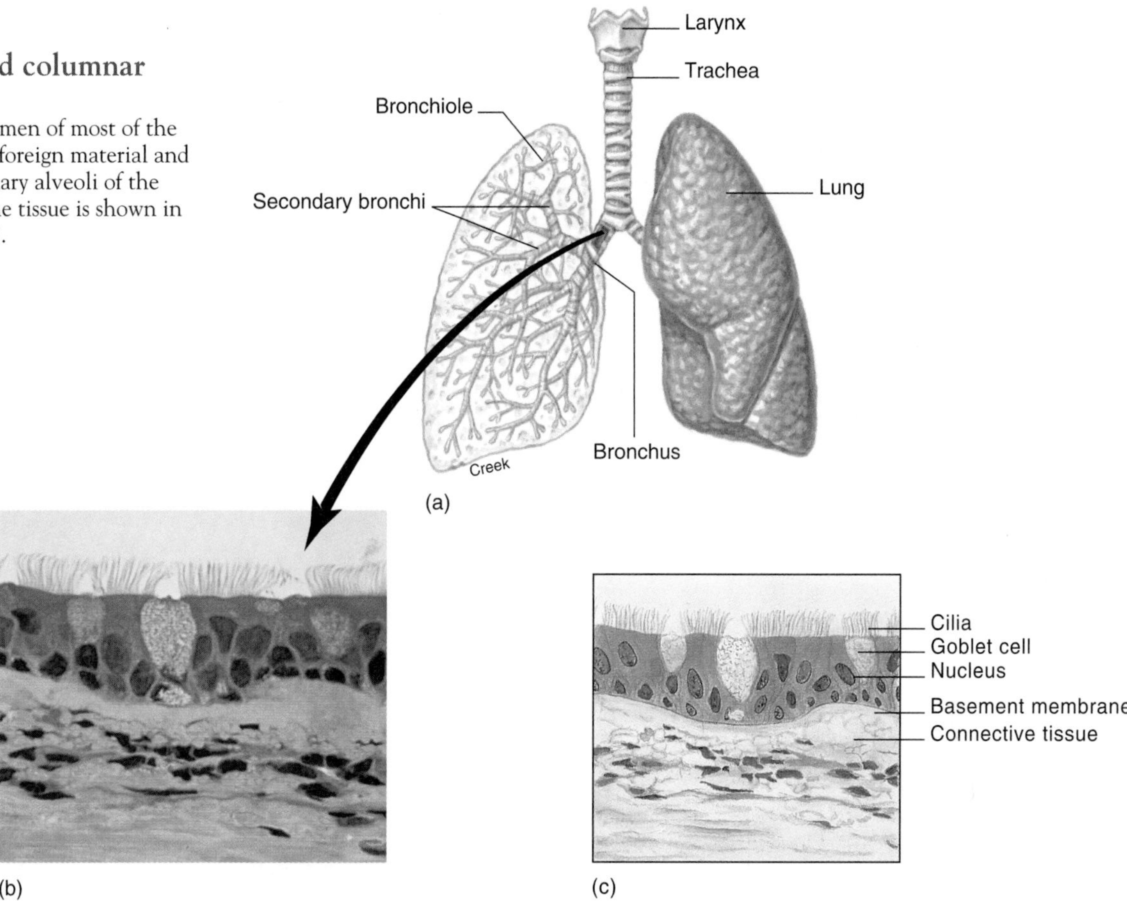

Figure 6.6

Pseudostratified ciliated columnar epithelium.

(*a*) This epithelium lines the lumen of most of the respiratory tract, where it traps foreign material and moves it away from the pulmonary alveoli of the lungs. A photomicrograph of the tissue is shown in (*b*) and a labeled diagram in (*c*).

Mitosis occurs only at the deepest layers (see table 1). The mitotic rate approximates the rate at which cells are sloughed off at the surface. As the newly produced cells grow, they are pushed toward the surface where they replace the cells that are sloughed off. Movement of the epithelial cells away from the supportive basement membrane is accompanied by the production of the protein keratin (described below), progressive dehydration, and flattening.

There are two types of stratified squamous epithelia: keratinized and nonkeratinized.

1. **Keratinized stratified squamous epithelium** contains *keratin* (*ker'ă-tin*), a protein that strengthens the tissue. Keratin makes the epidermis (outer layer) of the skin somewhat waterproof and protects it from bacterial invasion. The outer layers of the skin are dead, but glandular secretions keep them soft (see chapter 7).
2. **Nonkeratinized stratified squamous epithelium** lines the oral cavity and pharynx, nasal cavity, vagina, and anal canal. This type of epithelium, called *mucosa*, is well adapted to withstand moderate abrasion but not fluid loss. The cells on the free surface are alive and are always moistened.

Stratified squamous epithelium is the first line of defense against the entry of living organisms into the body. Stratification, along with rapid mitotic activity and keratinization within the epidermis of the skin, are important protective features. An acidic pH along the surfaces of this tissue also helps to prevent disease. The pH of the skin ranges from 4 to 6.8. The pH in the oral cavity ranges from 5.8 to 7.1, which tends to retard the growth of microorganisms. The pH of the anal region is about 6, and the pH along the surface of the vagina is 4 or lower.

Stratified Cuboidal Epithelium

Stratified cuboidal epithelium usually consists of only two or three layers of cuboidal cells (fig. 6.8). This type of epithelium is confined to the linings of the large ducts of sweat glands, salivary glands, and the pancreas, where its stratification probably provides a more robust lining than would simple epithelium.

keratin: Gk. *keras*, horn

Figure 6.7

Stratified squamous epithelium. This epithelium forms the outer layer of skin and the lining of body openings. In the moistened areas, such as in the vagina (*a*), it is nonkeratinized, whereas in the epidermis of the skin it is keratinized. A photomicrograph of the tissue is shown in (*b*) and a labeled diagram in (*c*).

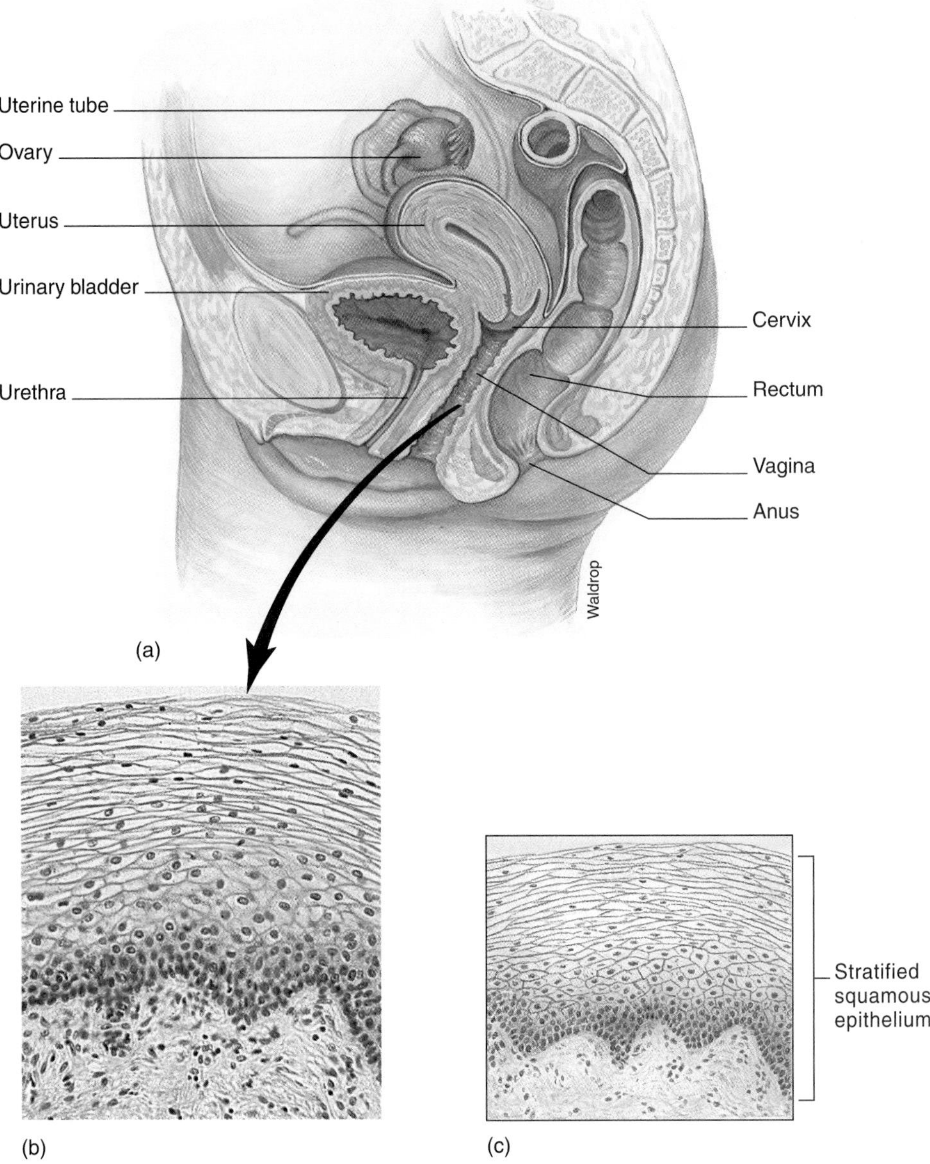

Transitional Epithelium

Transitional epithelium is similar to nonkeratinized stratified squamous epithelium except that the surface cells of the former are large and round rather than flat, and some may have two nuclei (fig 6.9). Transitional epithelium is located only in the urinary system, particularly lining the cavity of the urinary bladder and lining the lumina of the ureters. This tissue is specialized to permit distension (stretching) of the urinary bladder as it fills with urine.

A summary of membranous epithelial tissue is presented in table 6.1.

Body Membranes

Body membranes are composed of thin layers of epithelial tissue and, in certain locations, epithelial tissue coupled with supporting connective tissue. Body membranes cover, separate, and support visceral organs and line body cavities. There are two basic types of body membranes: mucous (*myoo'kus*) membranes and serous (*se'rus*) membranes.

1. **Mucous membranes** secrete a thick, viscid substance called *mucus*. Generally, mucus lubricates or protects the associated organs where it is secreted. Mucous

Figure 6.8

Stratified cuboidal epithelium.

(*a*) This epithelium lines the lumina of large ducts like the parotid duct, which drains saliva from the parotid gland. A photomicrograph of the tissue is shown in (*b*) and a labeled diagram in (*c*).

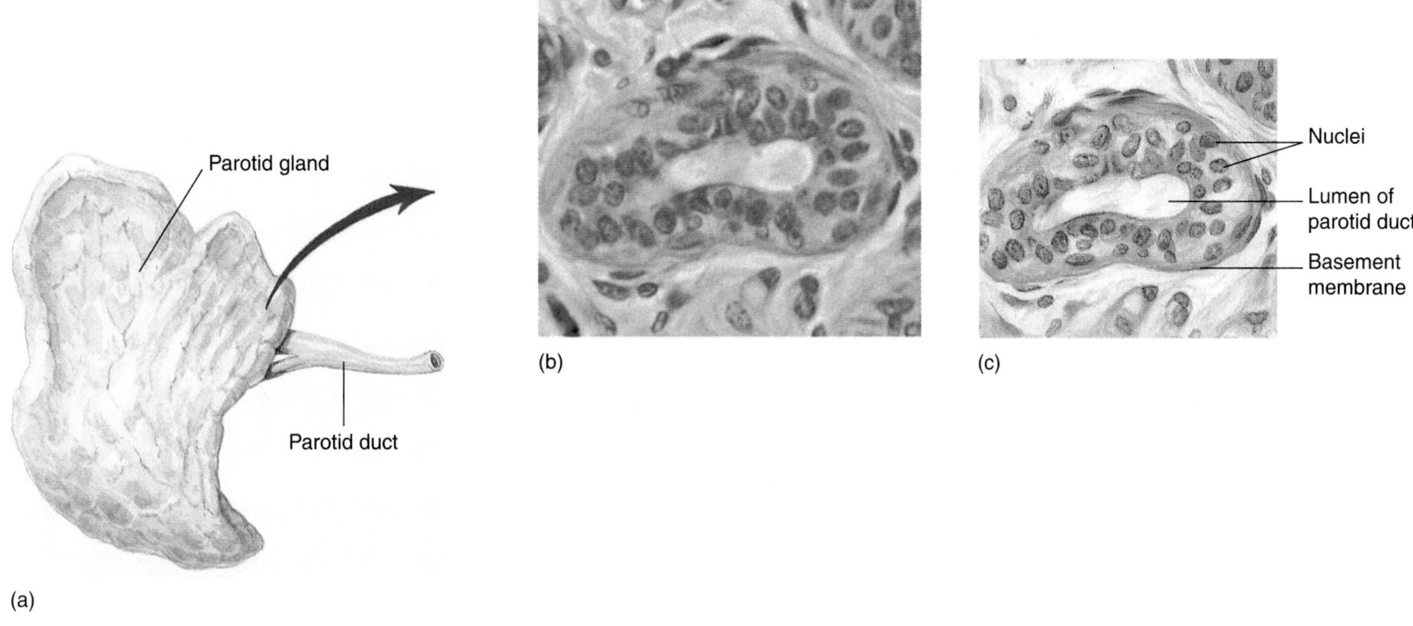

Figure 6.9

Transitional epithelium.

(*a*) This epithelium lines the lumina of the ureters and the cavity of the urinary bladder, where it permits distension. A photomicrograph of the tissue is shown in (*b*) and a labeled diagram in (*c*).

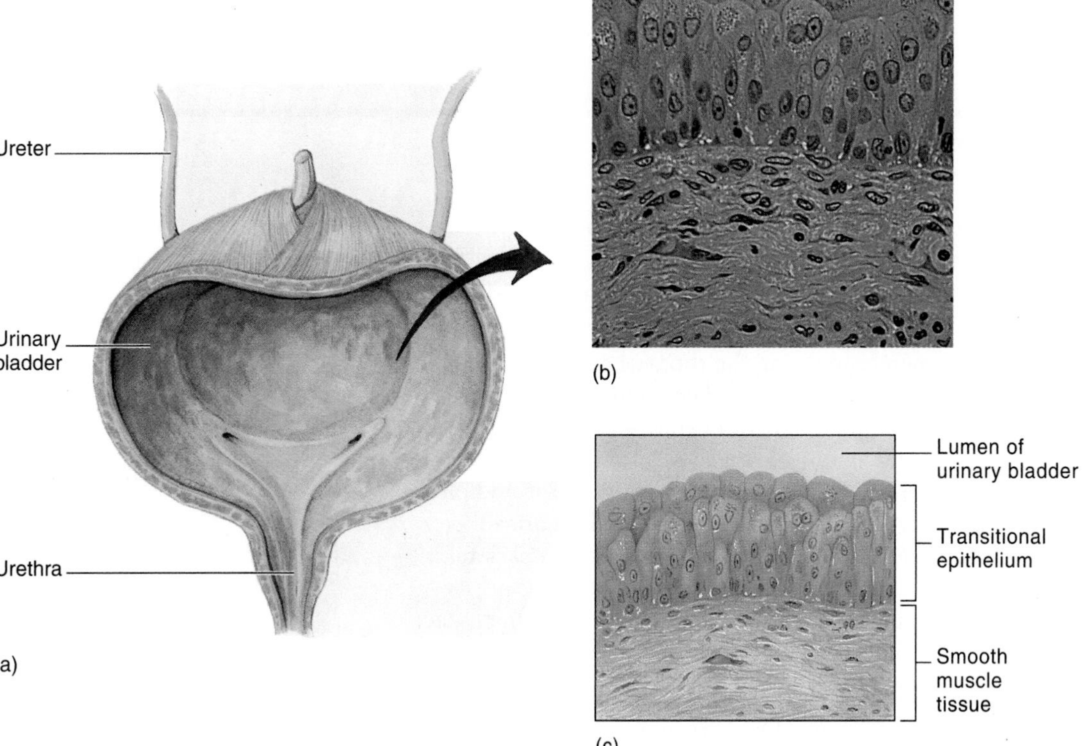

Table 6.1
Summary of Membranous Epithelial Tissue

Type	Structure and Function	Location
Simple Epithelia	Single layer of cells; function varies with type	Covering visceral organs; linings of body cavities, tubes, and ducts
Simple squamous epithelium	Single layer of flattened, tightly bound cells; diffusion and filtration	Capillary walls; pulmonary alveoli of lungs, covering visceral organs; linings of body cavities
Simple cuboidal epithelium	Single layer of cube-shaped cells; excretion, secretion, or absorption	Surface of ovaries; linings of renal tubules, salivary ducts, and pancreatic ducts
Simple columnar epithelium	Single layer of nonciliated, tall, column-shaped cells; protection, secretion, and absorption	Lining of most of GI tract
Simple ciliated columnar epithelium	Single layer of ciliated, column-shaped cells; protection, secretion, and absorption	Lining of uterine tubes
Pseudostratified ciliated columnar epithelium	Single layer of ciliated, irregularly shaped cells; many goblet cells; protection, secretion, ciliary movement	Lining of respiratory passageways
Stratified Epithelia	Two or more layers of cells; function varies with type	Epidermal layer of skin; linings of body openings, ducts, and urinary bladder
Stratified squamous epithelium (keratinized)	Numerous layers containing keratin, with outer layers flattened and dead; protection	Epidermis of skin
Stratified squamous epithelium (nonkeratinized)	Numerous layers lacking keratin, with outer layers moistened and alive; protection and pliability	Linings of oral and nasal cavities, vagina, and anal canal
Stratified cuboidal epithelium	Usually two layers of cube-shaped cells	Large ducts of sweat glands, salivary glands, and pancreas
Transitional epithelium	Numerous layers of rounded, nonkeratinized cells; distension	Walls of ureters, part of urethra; and urinary bladder

membranes line various cavities and tubes that enter or exit from the body, such as the oral and nasal cavities and the tubes of the respiratory, reproductive, urinary, and digestive systems.

2. **Serous membranes** line the thoracic and abdominopelvic cavities and cover visceral organs, secreting a watery lubricant called *serous fluid.* **Pleurae** are serous membranes associated with the lungs (see chapter 24). Each pleura (pleura of right lung and pleura of left lung) has two parts. The **visceral pleura** adheres to the outer surface of the lung, whereas the **parietal** (*pă-ri′ĕ-tal*) **pleura** lines the thoracic wall and the thoracic surface of the diaphragm. The moistened space between the two pleurae is know as the **pleural cavity.**

Pericardial membranes are the serous membranes of the heart (see chapter 21). A thin **visceral pericardium** covers the surface of the heart and a thicker **parietal pericardium** is the durable covering that surrounds the heart. The space between these two membranes is called the **pericardial cavity.**

Serous membranes of the abdominal cavity are called **peritoneal** (*per″ĭ-tŏ-ne′al*) **membranes** (see chapter 26). The **parietal peritoneum** lines the abdominal wall, and the **visceral peritoneum** covers the visceral organs (fig. 6.10). The **peritoneal cavity** is the potential space within the abdominopelvic cavity between the parietal and visceral peritoneal membranes. Certain organs, such as the kidneys, adrenal glands, and a portion of the pancreas, which are within the abdominal cavity, are positioned behind the parietal peritoneum and therefore are said to be **retroperitoneal.** **Mesenteries** (*mes′en-ter″ēz*) are double folds of peritoneum that connect the parietal to the visceral peritoneum.

Figure 6.10

Abdominal visceral organs and their associated serous membranes.
The serous membranes are labeled in boldface type.

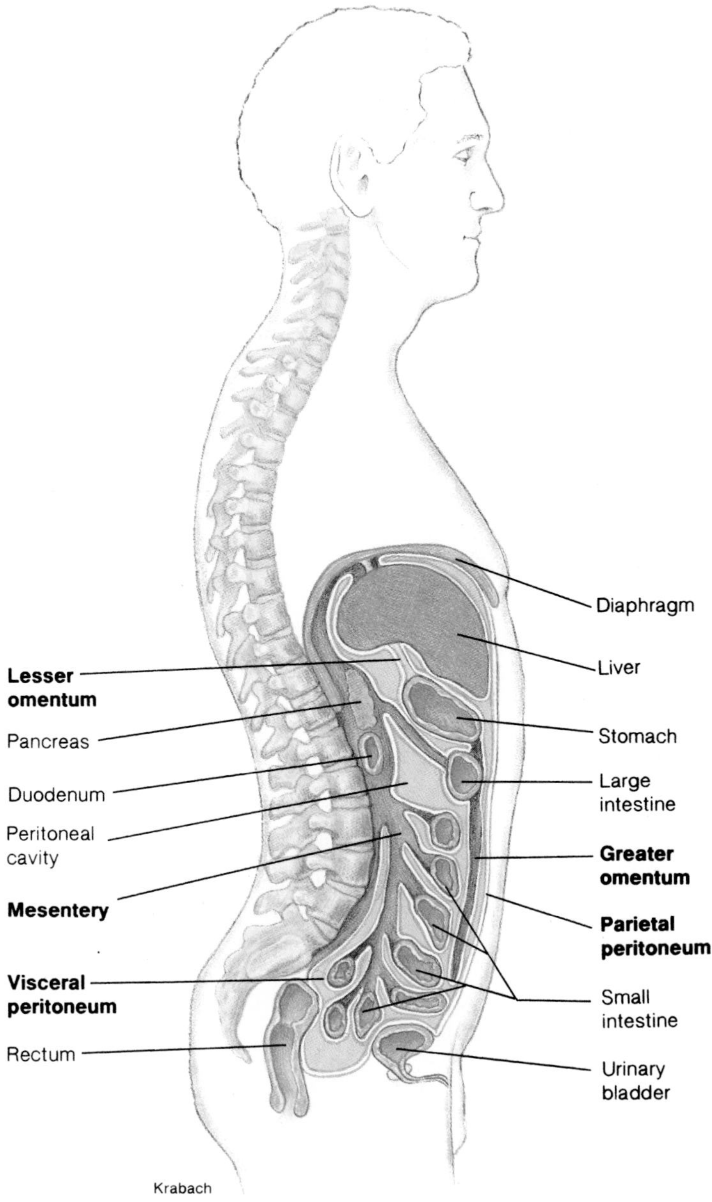

Glandular Epithelia

As tissues develop in the embryo, tiny invaginations (infoldings) of membranous epithelia give rise to specialized secretory structures called **exocrine** (*ek'sŏ-krin*) **glands.** These glands remain connected to the epithelium by ducts, and their secretions pass through the ducts onto body surfaces or into body cavities. Exocrine glands should not be confused with endocrine glands, which are ductless, and which secrete their products (hormones) into the blood or surrounding intercellular fluid. Exocrine glands within the skin include oil (sebaceous) glands, sweat glands, and mammary glands. Exocrine glands within the digestive system include the salivary and pancreatic glands.

Exocrine glands are classified according to their structure and how they discharge their products. Classified according to structure, there are two types of exocrine glands: unicellular and multicellular glands.

1. **Unicellular glands.** Unicellular glands are single-celled exocrine glands, such as *goblet cells* (fig. 6.11). They are modified columnar cells that occur within most epithelial tissues. Goblet cells are found in the epithelial linings of the respiratory and digestive systems. The mucus secretion of these cells lubricates and protects the surface linings.
2. **Multicellular glands.** Multicellular glands, as their name implies, are composed of both secretory cells and cells that form the walls of the ducts. Multicellular glands are classified as *simple* or *compound glands*. The ducts of the simple glands do not branch, whereas those of the compound type do (fig. 6.12). Multicellular glands are also classified according to the shape of the secretory portion. They are identified as *tubular glands* if the secretory portion resembles a tube and as *acinar glands* if the secretory portion resembles a flask. Multicellular glands with a secretory portion that resembles both a tube and a flask are termed *tubuloacinar glands*.

Multicellular exocrine glands are also organized according to the means by which they release their product (fig. 6.13). Glands that secrete their products by exocytosis through the cell membrane of the secretory cells are called **merocrine** (*mer'ŏ-krin*) **glands.** Salivary glands, pancreatic glands, and certain sweat glands are of this type. In **apocrine** (*ap'ŏ-krin*) **glands** the secretion accumulates near the surface of the secretory cell, and then a portion of the cell, along with the secretion, pinches off to be discharged. Mammary glands and certain sweat glands are apocrine glands. In a **holocrine** (*hol'ŏ-krin*) **gland,** the entire secretory cell is discharged along with the secretory product. An example of a holocrine gland is a sebaceous, or oil-secreting, gland of the skin.

Glandular epithelia are summarized in table 6.2.

---

exocrine: Gk. *exo*, outside; *krinein*, to separate

merocrine: Gk. *meros*, part; *krinein*, to separate
apocrine: Gk. *apo*, off; *krinein*, to separate
holocrine: Gk. *holos*, whole; *krinein*, to separate

Figure 6.11

A goblet cell.

Goblet cells are unicellular glands that secrete mucus, which lubricates and protects surface linings. (*a*) Goblet cells are abundant in the columnar epithelium lining the lumen of the small intestine. A photomicrograph of the tissue is shown in (*b*) and a labeled diagram in (*c*).

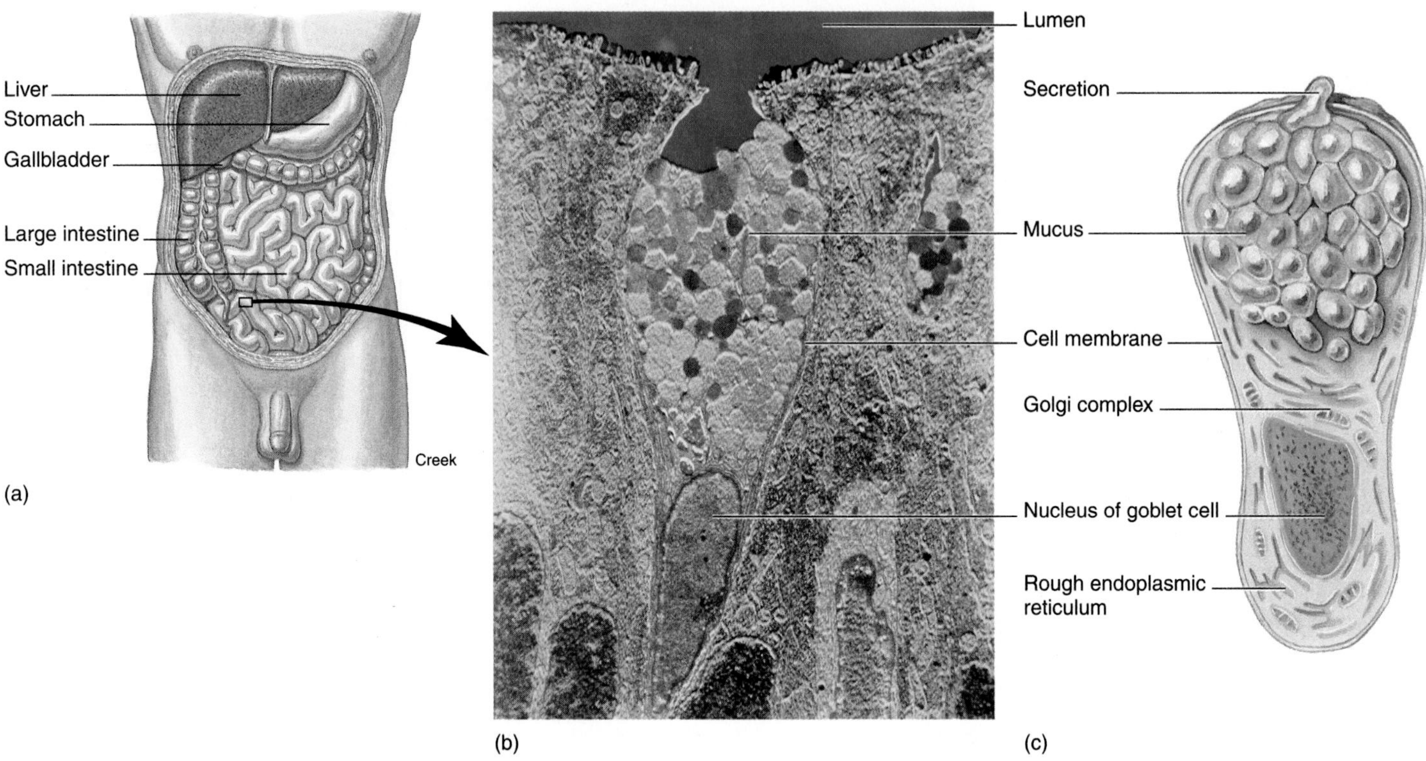

Figure 6.12

Structural classification of multicellular exocrine glands.

The ducts of the simple glands do not branch or have few branches, whereas those of the compound glands have multiple branches.

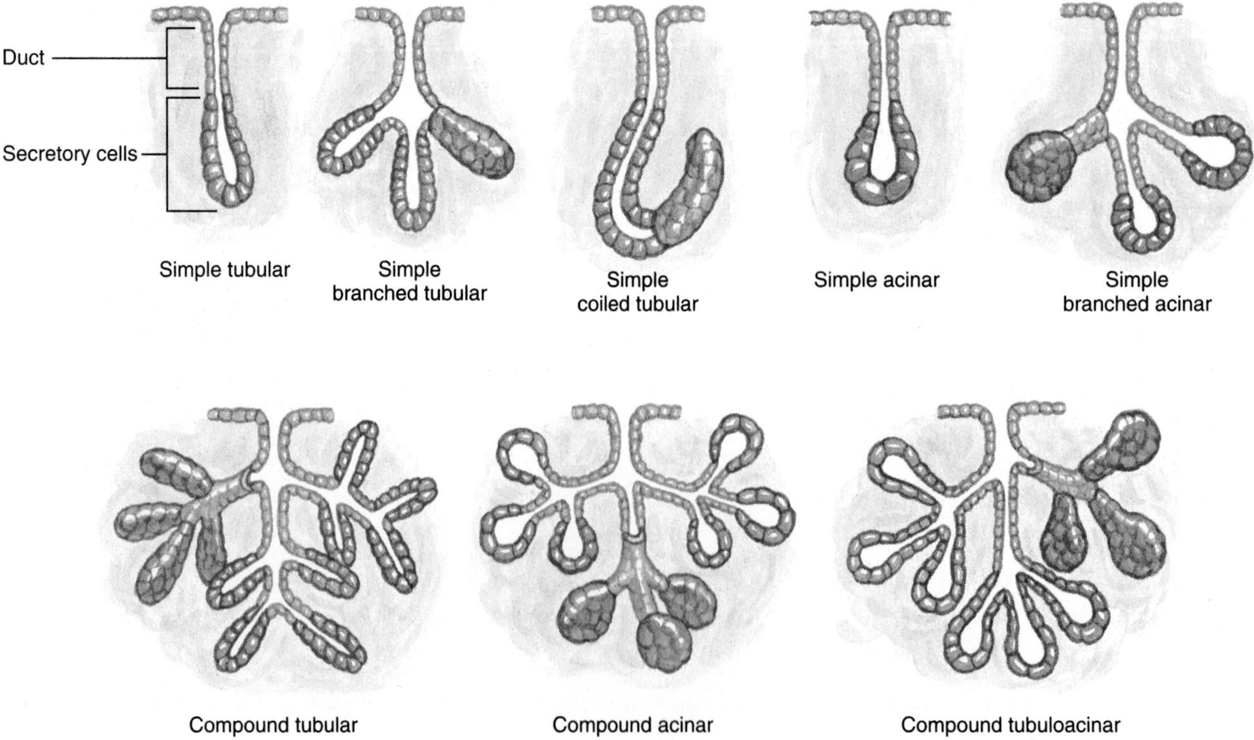

Figure 6.13
Examples of multicellular exocrine glands.

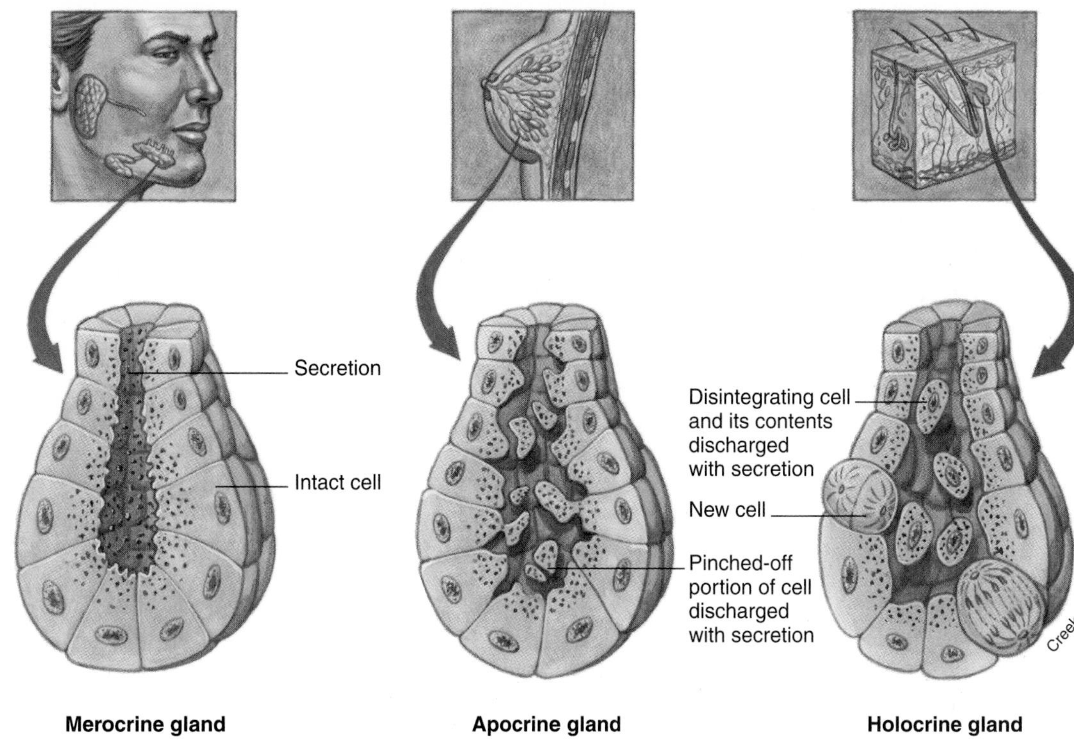

Table 6.2
Summary of Glandular Epithelial Tissue

Classification of Exocrine Glands by Structure

Type	Function	Example
I. Unicellular	Lubricate and protect	Goblet cells of digestive, respiratory, urinary, and reproductive systems
II. Multicellular	Protect, cool body, lubricate, aid in digestion, maintain body homeostasis	Sweat glands, digestive glands, liver, mammary glands, sebaceous glands
A. Simple		
1. Tubular	Aid in digestion	Intestinal glands
2. Branched tubular	Protect, aid in digestion	Uterine glands, gastric glands
3. Coiled tubular	Regulate temperature	Certain sweat glands
4. Acinar	Provide additive for spermatozoa	Seminal vesicles of male reproductive system
5. Branched acinar	Condition skin	Sebaceous glands of skin
B. Compound		
1. Tubular	Lubricate urethra of male, assist body digestion	Bulbourethral glands of male reproductive system, liver
2. Acinar	Provide nourishment of infant, aid in digestion	Mammary glands, salivary glands (sublingual and submandibular)
3. Tubuloacinar	Aid in digestion	Salivary gland (parotid), pancreas

Classification of Exocrine Glands by Mode of Secretion

Type	Description of Secretion	Example
Merocrine glands	Watery secretion for regulating temperature or enzymes that promote digestion	Salivary and pancreatic glands, certain sweat glands
Apocrine glands	Portion of secretory cell and secretion are discharged; provide nourishment for infant, assist in regulating temperature	Mammary glands, certain sweat glands
Holocrine glands	Entire secretory cell with enclosed section is discharged; condition skin	Sebaceous glands of the skin

Figure 6.14

Mesenchyme.
This embryonic connective tissue migrates and gives rise to other kinds of connective tissue. (*a*) It is found within an early developing embryo and (*b*) consists of irregularly shaped cells lying in a jellylike homogeneous matrix.

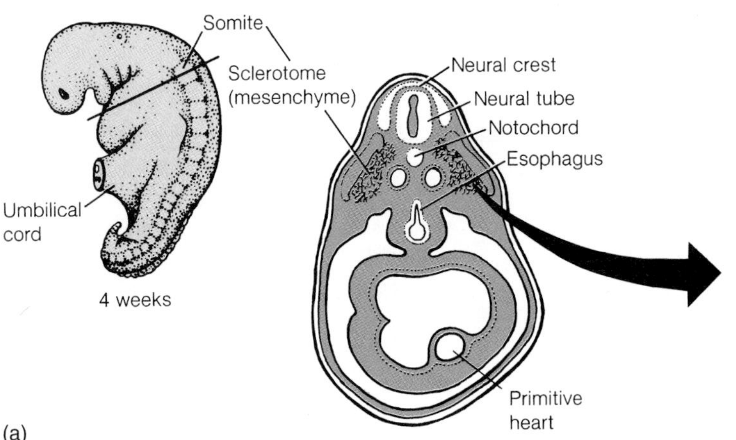

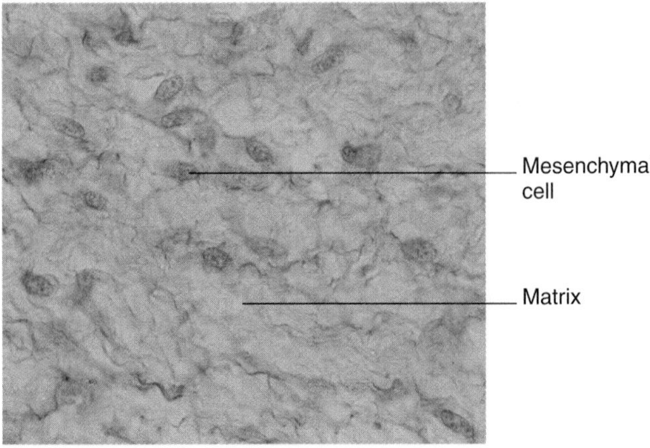

Connective Tissue

Connective tissue is divided into subtypes according to the matrix that binds the cells. Connective tissue provides structural and metabolic support for other tissues and organs of the body.

Characteristics and Classification of Connective Tissues

Connective tissue is abundant throughout the body. It supports other tissues or binds them together and provides for the metabolic needs of all body organs. Certain types of connective tissue store nutritional substances; other types manufacture protective and regulatory materials.

Although connective tissue varies widely in structure and function, all types of connective tissues have similarities. With the exception of mature cartilage, connective tissue is highly vascular and well nourished. It is able to replicate and, by so doing, is responsible for the repair of body organs. Unlike epithelial tissue, which is composed of tightly fitted cells, connective tissue contains considerably more matrix (intercellular material) than cells. Connective tissue does not occur on free surfaces of body cavities or on the surface of the body, as does epithelial tissue. Furthermore, connective tissue is embryonically derived from mesoderm, whereas epithelial tissue derives from ectoderm, mesoderm, and endoderm.

The classification of connective tissues is not exact; in fact, several schemes have been devised. In general, however, the various types are named according to the kind and arrangement of the matrix. The following are the basic kinds of connective tissues.

A. Embryonic connective tissue
B. Connective tissue proper
   1. Loose connective (areolar) tissue
   2. Dense regular connective tissue
   3. Dense irregular connective tissue
   4. Elastic connective tissue
   5. Reticular connective tissue
   6. Adipose connective tissue
C. Cartilage tissue
   1. Hyaline cartilage
   2. Fibrocartilage
   3. Elastic cartilage
D. Bone tissue
E. Blood (vascular tissue)

Embryonic Connective Tissue

The embryonic period of development, which lasts 6 weeks (from the beginning of the third to the end of the eighth week), is characterized by extensive tissue differentiation and organ formation. At the beginning of the embryonic period, all connective tissue appears the same and is referred to as **mesenchyme** (*mez'en-kīm*). Mesenchyme is undifferentiated embryonic connective tissue that is derived from mesoderm. It consists of irregularly shaped cells surrounded by large amounts of a homogeneous, jellylike matrix (fig. 6.14). In certain periods of development, mesenchyme migrates to predisposed sites where it interacts with other tissues to form organs. Once mesenchyme has completed its embryonic migration to predetermined sites, it differentiates into all other kinds of connective tissue.

Some mesenchymal-like tissue persists past the embryonic period in certain locations within the body. Examples are the undifferentiated cells that surround blood vessels and form fibroblasts if the vessels are traumatized. Fibroblasts assist in healing wounds (see chapter 7).

Another kind of prenatal connective tissue exists only in the fetus (the fetal period is from 9 weeks to birth) and is called *mucous connective tissue*, or *Wharton's jelly*. It gives a turgid consistency to the umbilical cord.

Wharton's jelly: from Thomas Wharton, English anatomist, 1614–73

Figure 6.15

Loose connective tissue.
This packing and binding tissue surrounds muscles (*a*), nerves, and vessels and binds the skin to the underlying muscles. A photomicrograph of the tissue is shown in (*b*) and a labeled diagram in (*c*).

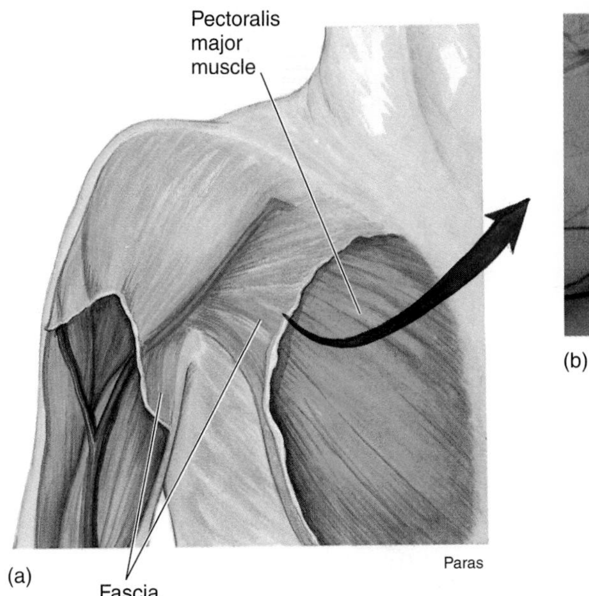

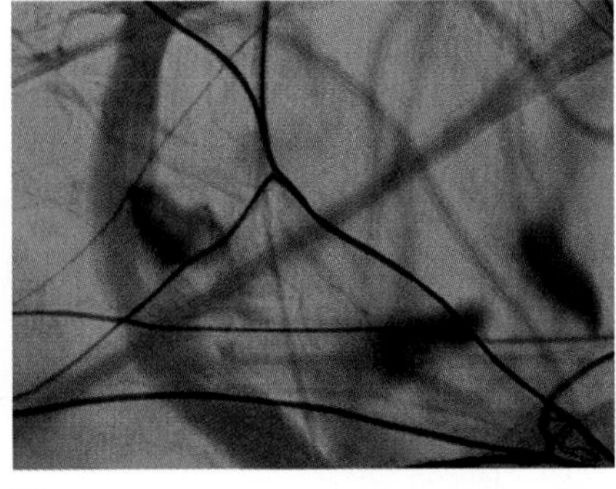

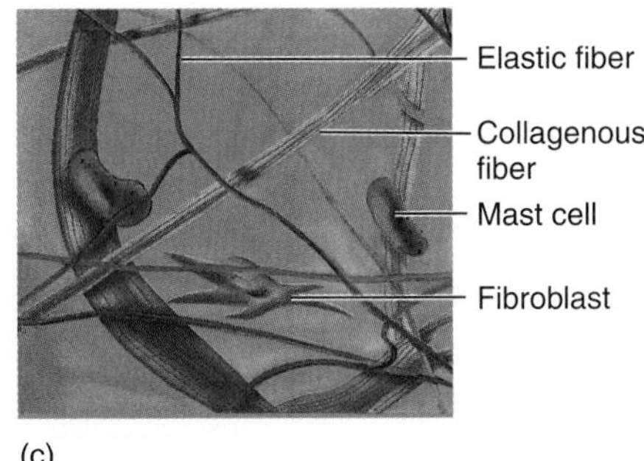

Connective Tissue Proper

Connective tissue proper has a loose, flexible matrix, frequently called *ground substance*. The most common cell within connective tissue proper is called a **fibroblast** (*fi'bro-blast*). Fibroblasts are large, star-shaped cells that produce collagenous, elastic, and reticular (*rĕ-tik'yŭ-lar*) fibers. **Collagenous** (*kŏ-laj'ĕ-nus*) **fibers** are composed of a protein called *collagen* (*col'ă-jen*); they are flexible, yet they have tremendous strength. **Elastic fibers** are composed of a protein called *elastin*, which provides certain tissues with elasticity. Collagenous and elastic fibers may be either sparse and irregularly arranged, as in loose connective tissue, or tightly packed, as in dense connective tissue. Tissues with loosely arranged fibers generally form packing material that cushions and protects various organs, whereas those that are tightly arranged form the binding and supportive connective tissues of the body.

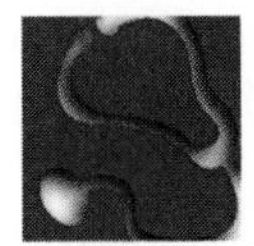

Resilience in tissues that contain elastic fibers is extremely important for a number of physical body functions. Consider, for example, that elastic fibers are found in the walls of arteries and in the walls of the lower respiratory passageways. As these walls are expanded by blood moving through vessels or by inspired air, the elastic fibers must first stretch and then recoil. This maintains the pressures of the fluid or air moving through the lumina, thus ensuring adequate flow rates and rates of diffusion through capillary and lung surfaces.

Reticular fibers reinforce by branching and joining to form a delicate lattice or reticulum. Reticular fibers are common in lymphatic glands, where they form a meshlike center called the *stroma*.

Six basic types of connective tissue proper are generally recognized. These tissues are distinguished by the consistency of the ground substance and the type and arrangement of the reinforcement fibers.

Loose Connective (Areolar) Tissue

Loose connective tissue is distributed throughout the body as a binding and packing material. It binds the skin to the underlying muscles and is highly vascular, providing nutrients to the skin. Loose connective tissue surrounding muscle fibers and muscle groups is known as **fascia** (*fash'e-ă*). It also surrounds blood vessels and nerves, where it provides both protection and nourishment. Specialized cells called **mast cells** are dispersed throughout the loose connective tissue surrounding blood vessels. Mast cells produce *heparin* (*hep'ă-rin*), an anticoagulant that prevents blood from clotting within the vessels. They also produce *histamine*, which is released during inflammation and acts as a powerful vasodilator.

The cells of loose connective tissue are predominantly fibroblasts, with collagenous and elastic fibers dispersed throughout the ground substance (fig. 6.15). The irregular arrangement of this tissue provides flexibility, yet strength, in any direction. It is this tissue layer, for example, that permits the skin to move when a part of the body is rubbed.

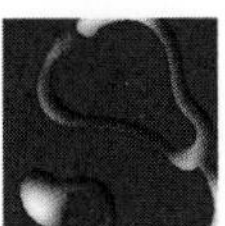

Much of the fluid of the body is found within loose connective tissue and is called interstitial fluid (tissue fluid). Sometimes excessive interstitial fluid accumulates, causing a swelled condition called *edema* (*ĕ-de'mă*). Edema is symptomatic of a variety of disease processes.

---

reticular: L. *rete*, net or netlike
collagen: Gk. *kolla*, glue
elastin: Gk. *elasticus*, to drive

stroma: Gk. *stroma*, a couch or bed
fascia: L. *fascia*, band or girdle
heparin: Gk. *hepatos*, the liver

Figure 6.16

Dense regular connective tissue.
This tissue forms the strong and highly flexible tendons (*a*) and ligaments. A photomicrograph of the tissue is shown in (*b*) and a labeled diagram in (*c*).

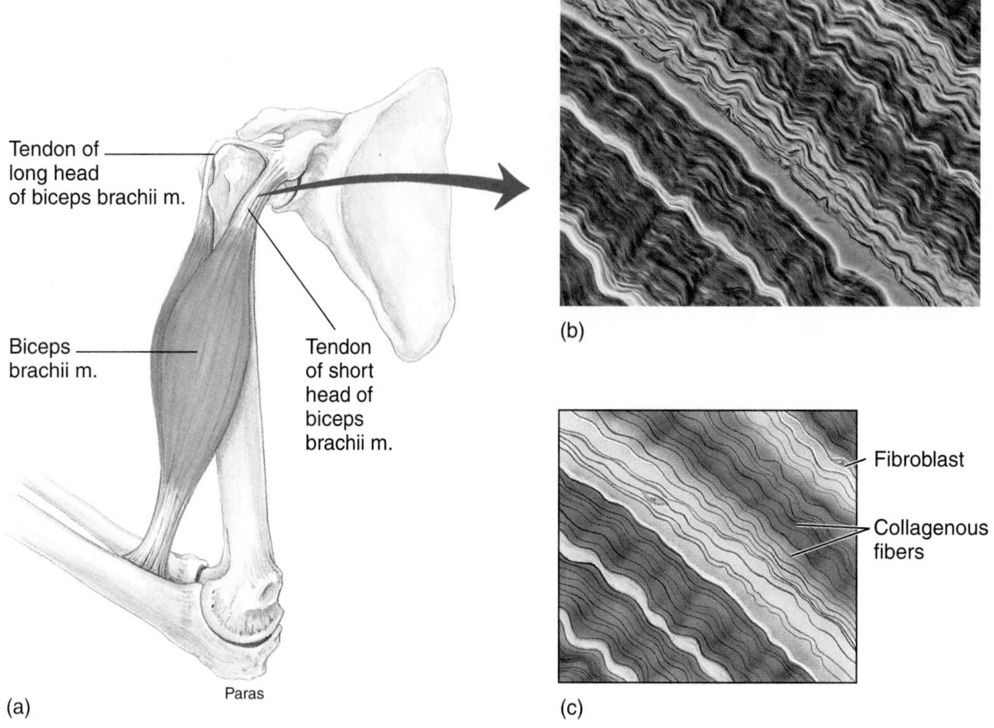

Dense Regular Connective Tissue

Dense regular connective tissue is characterized by large amounts of densely packed collagenous fibers that run parallel to the direction of force placed on this tissue during body movement. Because this tissue is silvery white in appearance, it is sometimes called white fibrous connective tissue.

Dense regular connective tissue occurs where strong, flexible support is needed (fig. 6.16). **Tendons,** which attach muscles to bones and transfer the forces of muscle contractions, and **ligaments,** which connect bone to bone across articulations, are composed of this type of tissue.

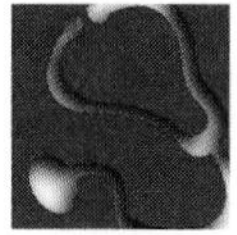

Trauma to ligaments, tendons, and muscles are common sports-related injuries. A strain is an excessive stretch of the tissue composing the tendon or muscle, with no serious damage. A sprain is a tearing of the tissue of a ligament and may be slight, moderate, or complete. A complete tear of a major ligament is especially painful and disabling. Ligamentous tissue does not heal well because it has a poor blood supply. Surgical reconstruction is generally needed for the treatment of a severed ligament.

tendon: L. *tendere*, to stretch
ligament: L. *ligare*, bind

Dense Irregular Connective Tissue

Dense irregular connective tissue is characterized by large amounts of densely packed collagenous fibers that are interwoven to provide tensile strength in any direction. This tissue is found in the dermis of the skin and the submucosa of the GI tract. It also forms the fibrous capsules of organs and joints (fig. 6.17).

Elastic Connective Tissue

Elastic connective tissue is composed primarily of elastic fibers that are irregularly arranged and yellowish in color (fig. 6.18). They can be stretched to one-and-a-half times their original lengths and will snap back to their former size. Elastic connective tissue is found in the walls of large arteries, in portions of the larynx, and in the trachea and bronchial tubes of the lungs. It is also present between the arches of the vertebrae that make up the vertebral column.

Reticular Connective Tissue

Reticular connective tissue is characterized by a network of reticular fibers woven through a jellylike matrix (fig. 6.19). Certain specialized cells within reticular tissue are *phagocytic* (*fag″ŏ-sit′ik*), and therefore can ingest foreign materials. The liver, spleen, lymph nodes, and bone marrow contain reticular connective tissue.

Figure 6.17

Dense irregular connective tissue.

This tissue forms joint capsules (*a*) that contain synovial fluid for lubricating movable joints. A photomicrograph of the tissue is shown in (*b*) and a labeled diagram in (*c*).

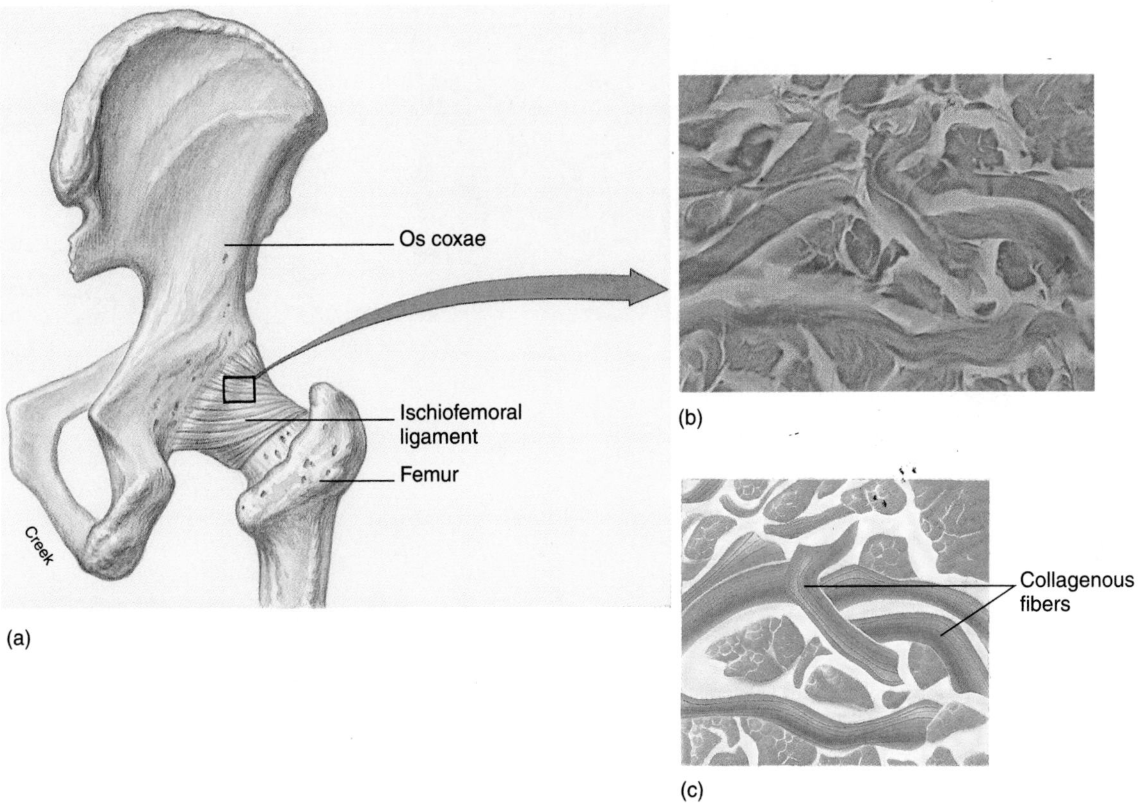

Figure 6.18

Elastic connective tissue.

This tissue permits stretching of a large artery (*a*) as blood flows through. A photomicrograph of the tissue is shown in (*b*) and a labeled diagram in (*c*).

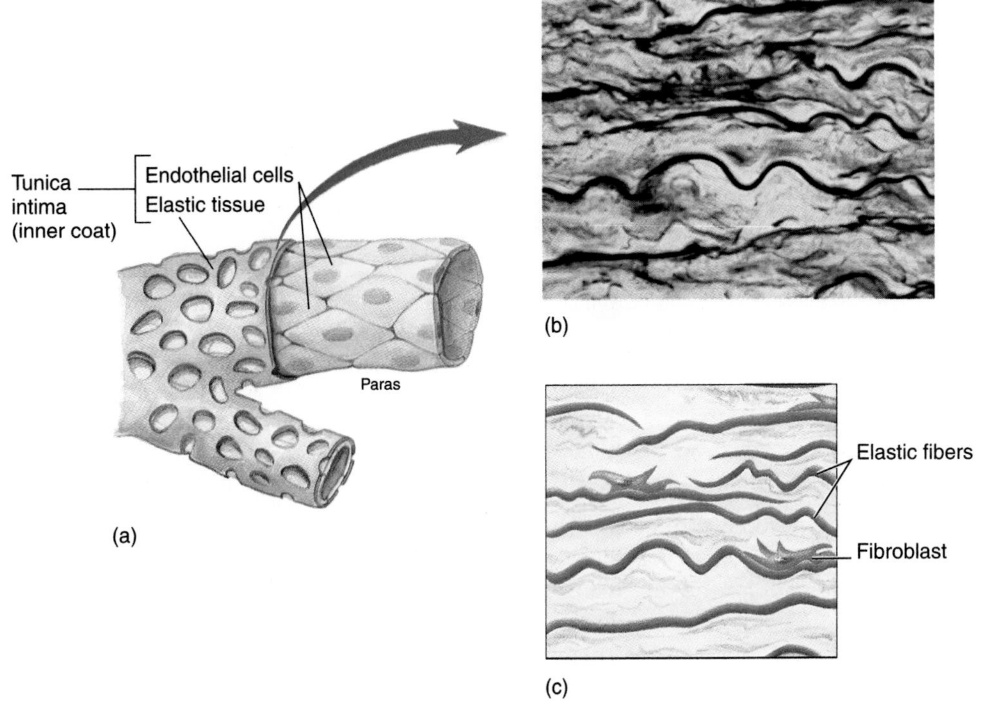

Figure 6.19

Reticular connective tissue.
This tissue forms the stroma, or framework, of such organs as the spleen (*a*), liver, thymus, and lymph nodes. A photomicrograph of the tissue is shown in (*b*) and a labeled diagram in (*c*).

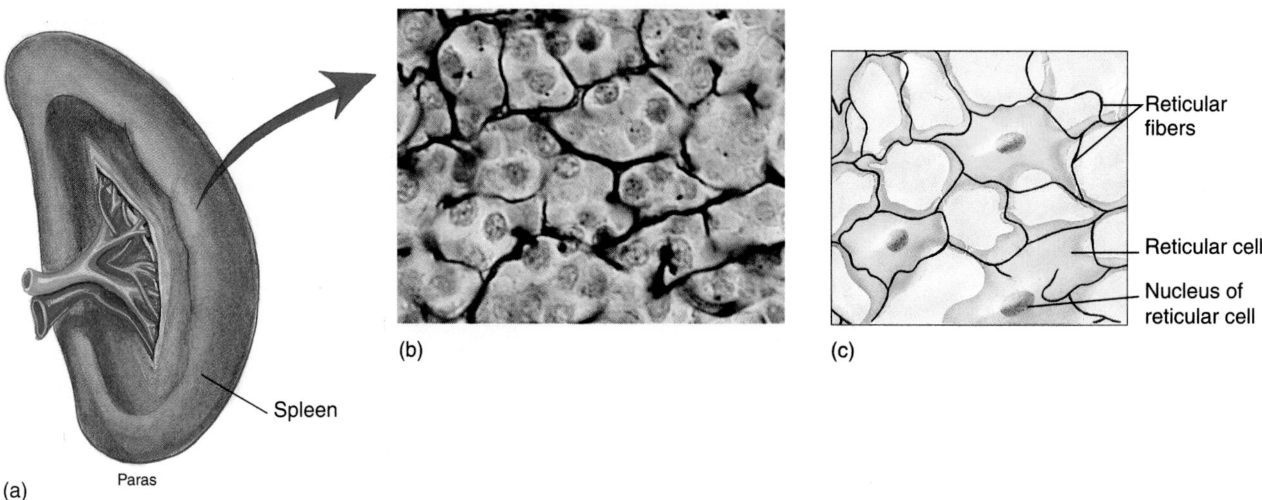

Figure 6.20

Adipose tissue.
This tissue is abundant in the hypodermis of the skin (*a*) and around various internal organs. A photomicrograph of the tissue is shown in (*b*) and a labeled diagram in (*c*).

Dermis
Hypodermis
(a)
Paras
(b)
Nucleus of adipocyte
Fat globule
Cytoplasm
Cell membrane
(c)

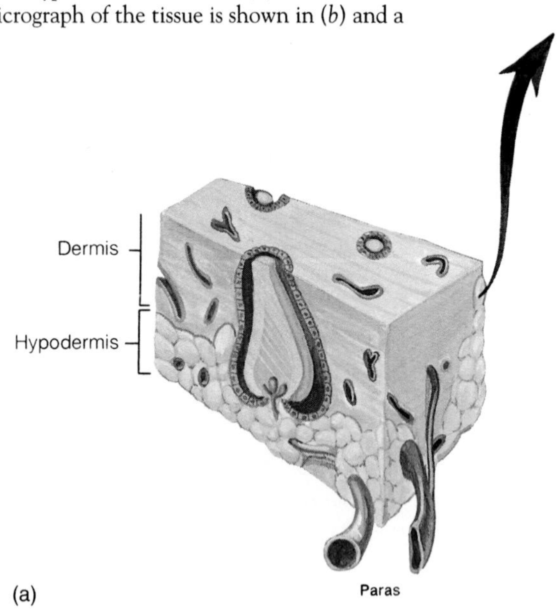

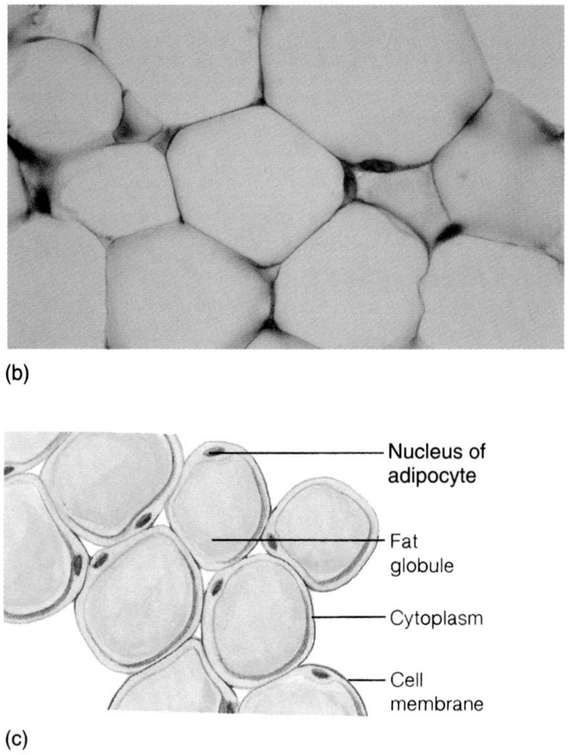

Adipose Connective Tissue

Adipose tissue is a specialized type of loose connective tissue that contains large quantities of adipose cells, or **adipocytes.** Adipocytes form from mesenchyme and, for the most part, are formed prenatally and during the first year of life. Adipocytes store droplets of fat within their cytoplasm, causing them to swell and forcing their nuclei to one side (fig. 6.20).

Adipose tissue is found throughout the body but is concentrated around the kidneys, in the hypodermis of the skin, on the surface of the heart, surrounding joints, and in the breasts of sexually mature females. Fat functions not only as a food reserve but also as support and protection for various organs. Fat is a good insulator against cold because it is a poor conductor of heat.

Table 6.3
Summary of Connective Tissue Proper

Type	Structure and Function	Location
Loose connective (areolar) tissue	Predominantly fibroblast cells with lesser amounts of collagen and elastic proteins; binds organs, holds interstitial fluid	Surrounding nerves and vessels, between muscles, beneath the skin
Dense regular connective tissue	Densely packed collagenous fibers that run parallel to the direction of force; provides strong, flexible support	Tendons, ligaments
Dense irregular connective tissue	Densely packed collagenous fibers arranged in a tight interwoven pattern; provides tensile strength in any direction	Dermis of skin, fibrous capsules of organs and joints, periosteum of bone
Elastic connective tissue	Predominantly irregularly arranged elastic fibers; supports, provides framework	Large arteries, lower respiratory tract, between the arches of vertebrae
Reticular connective tissue	Reticular fibers that form a supportive network; stores, performs phagocytic function	Lymph nodes, liver, spleen, thymus, bone marrow
Adipose tissue	Adipocytes; protects, stores fat, insulates	Hypodermis of skin, surface of heart, omentum, around kidneys, back of eyeball, surrounding joints

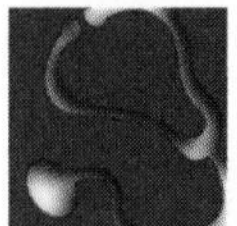

The excessive fat of obesity is a significant risk factor in cardiovascular disease, in diabetes mellitus, and in endometrial and breast cancer. For these reasons, good exercise programs and sensible diets are extremely important. Adipose tissue can also retain environmental pollutants that are ingested or absorbed through the skin. Dieting eliminates the fat stored within the tissue but not the tissue itself.

The surgical procedure of suction lipectomy may be used to remove small amounts of adipose tissue from certain localized body areas, such as the breasts, abdomen, buttocks, or thighs. Suction lipectomy is used for cosmetic purposes rather than as a treatment for obesity, and the risks for potentially detrimental side-effects need to be seriously considered. Potential candidates should be between 30 and 40 years old, only about 15 to 20 pounds overweight, and also have good skin elasticity.

The characteristics, functions, and locations of connective tissue proper are summarized in table 6.3.

Cartilage Tissue

Cartilage tissue consists of cartilage cells, or **chondrocytes** (*kon'dro-sĭtz*), and a semisolid matrix that imparts marked elastic properties to the tissue. It is a supportive and protective connective tissue that is frequently associated with bone. Cartilage forms a precursor to one type of bone and persists at the articular surfaces on the bones of all movable joints.

The chondrocytes within cartilage may occur singly but are frequently clustered. Chondrocytes occupy cavities, called **lacunae** (*lă-kyoo'ne*), within the matrix. Most cartilage tissue is surrounded by a dense irregular connective tissue called **perichondrium** (*per"ĭ-kon'dre-um*). Cartilage at the articular surfaces of bones (articular cartilage) lacks a perichondrium. Because mature cartilage is avascular, it must receive nutrients through diffusion from the perichondrium and the surrounding tissue. For this reason, cartilaginous tissue has a slow rate of mitotic activity; if damaged, it heals with difficulty.

There are three kinds of cartilage: *hyaline cartilage*, *fibrocartilage*, and *elastic cartilage*. Each is distinguished by the type and amount of fibers embedded within the matrix.

Hyaline Cartilage

Hyaline (*hi' ă-lĭn*) cartilage, commonly called "*gristle*," has a homogeneous, bluish-staining matrix in which the collagenous fibers are so fine that they can be observed only with an electron microscope. When viewed through a microscope, hyaline cartilage has a clear, glassy appearance (fig. 6.21).

Hyaline cartilage is the most abundant cartilage within the body. It covers the articular surfaces of bones, supports the tubular trachea and bronchi of the respiratory system, reinforces the nose, and forms the flexible bridge, called the **costal cartilage,** between the anterior end of each of the first 10 ribs and the sternum. Most of the bones of the body form first as hyaline cartilage and later become bone in a process called *endochondral ossification*.

lacuna: L. *lacuna*, hole or pit
hyaline: Gk. *hyalos*, glass

Figure 6.21

Hyaline cartilage.
This tissue is the most abundant cartilage within the body. It occurs in places such as the larynx (*a*), trachea, rib cage, and embryonic skeleton. A photomicrograph of the tissue is shown in (*b*) and a labeled diagram in (*c*).

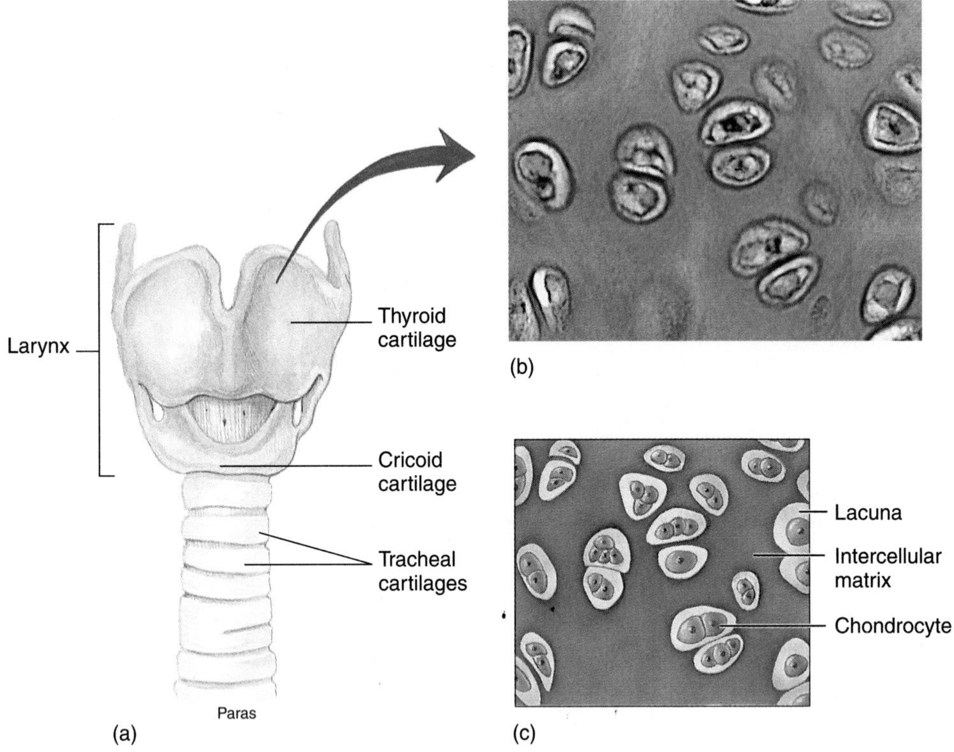

Fibrocartilage

Fibrocartilage has a matrix that is reinforced with numerous collagenous fibers (fig. 6.22), making it durable and able to withstand tension and compression. It is found at the symphysis pubis, where the two pelvic bones articulate, and between the vertebrae as intervertebral discs. It also forms the cartilaginous wedges within the knee joint, called *menisci* (see chapter 11).

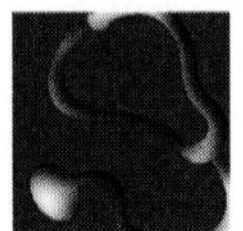

By the end of the day, the intervertebral discs of the vertebral column are somewhat compacted; therefore, a person is actually slightly shorter in the evening than in the morning following a recuperative rest. Aging, however, brings with it a gradual compression of the intervertebral discs that is irreversible.

Elastic Cartilage

Elastic cartilage is similar to hyaline except for the presence of abundant elastic fibers that make elastic cartilage very flexible without compromising its strength (fig. 6.23). The numerous elastic fibers also give it a yellowish appearance. This tissue is found in the outer ear, portions of the larynx, and in the auditory canal.

The three types of cartilage are summarized in table 6.4.

Bone Tissue

Bone tissue is the most rigid of the connective tissues. Unlike cartilage, bone tissue has a rich vascular supply and is the site of considerable metabolic activity. The hardness of bone is largely due to the inorganic calcium phosphate (calcium hydroxyapatite) deposited within the intercellular matrix. Numerous collagenous fibers, also embedded within the matrix, give bone some flexibility.

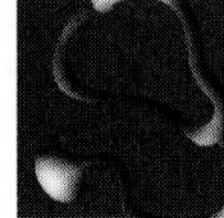

When a bone is placed in a weak acid, the calcium salts dissolve away and it becomes pliable. The bone retains its basic shape but can be easily bent and twisted. In calcium deficiency diseases, such as rickets, the bone tissue becomes pliable and bends under the weight of the body (see fig. 7.8).

Based on porosity, bone tissue is classified as either compact or spongy, and most bones have both types (fig. 6.24). **Compact** (*dense*) **bone tissue** comprises the hard, outer portion of a bone, and **spongy** (*cancellous*) **bone tissue** comprises the porous, highly vascular, inner portion. The outer surface of a bone is covered by a connective tissue layer called the *periosteum* that serves as a site of attachment for ligaments and tendons, provides protection, and gives durable strength to the bone. Spongy bone tissue makes the bone lighter and provides a space for red bone marrow, where blood cells are produced.

In compact bone tissue, mature bone cells, called **osteocytes,** are arranged in concentric layers around a **central** (haversian) **canal,** which contains a vascular and nerve supply. Each osteocyte occupies a cavity called a **lacuna.** Radiating from each lacuna are numerous minute canals, called **canaliculi,** which traverse the dense matrix of the bone tissue to adjacent lacunae. Nutrients diffuse through the canaliculi to reach each osteocyte. The inorganic matrix is deposited in concentric layers called **lamellae.** Bone tissue is described further in chapter 8.

Blood (Vascular Tissue)

Blood, or **vascular tissue,** is a highly specialized, fluid connective tissue that plays a vital role in maintaining homeostasis. The cells, or **formed elements,** of blood are suspended in

haversian canal: from Clopton Havers, English anatomist, 1650–1702

Figure 6.22

Fibrocartilage.

This tissue is located at the symphysis pubis, within the knee joints, and between the vertebrae as the intervertebral discs (*a*). A photomicrograph of the tissue is shown in (*b*) and a labeled diagram in (*c*).

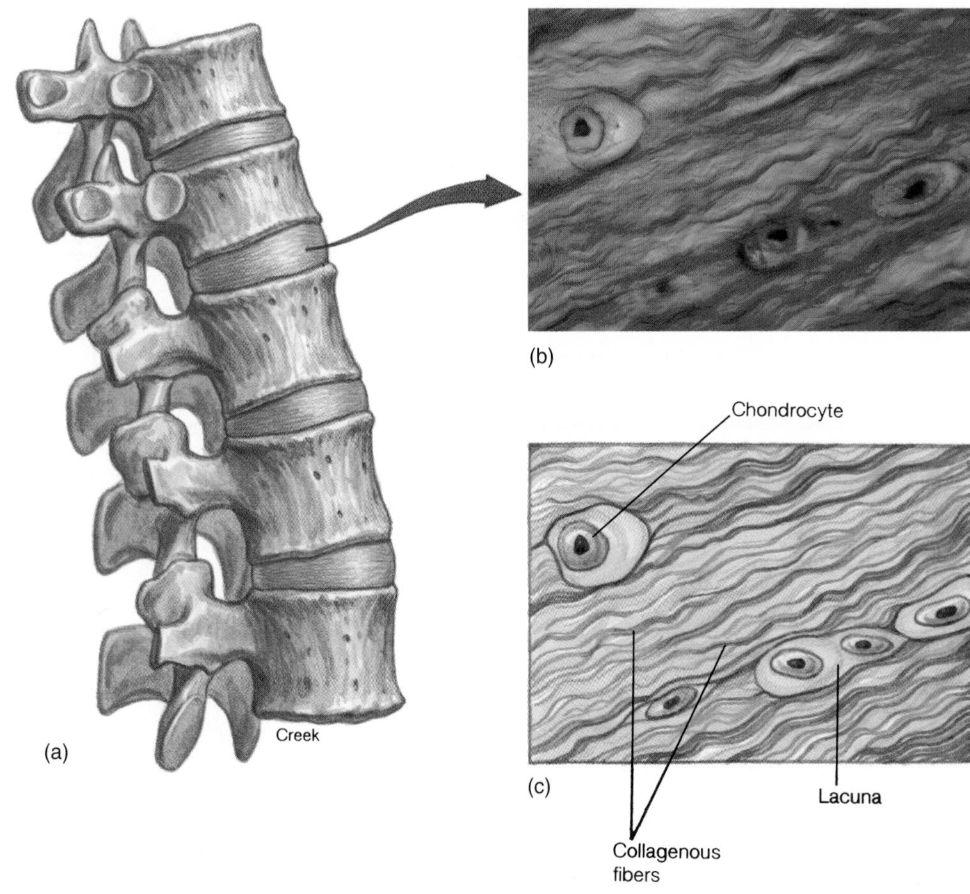

Figure 6.23

Elastic cartilage.

This tissue gives support to the outer ear (*a*), auditory canal, and parts of the larynx. A photomicrograph of the tissue is shown in (*b*) and a labeled diagram in (*c*).

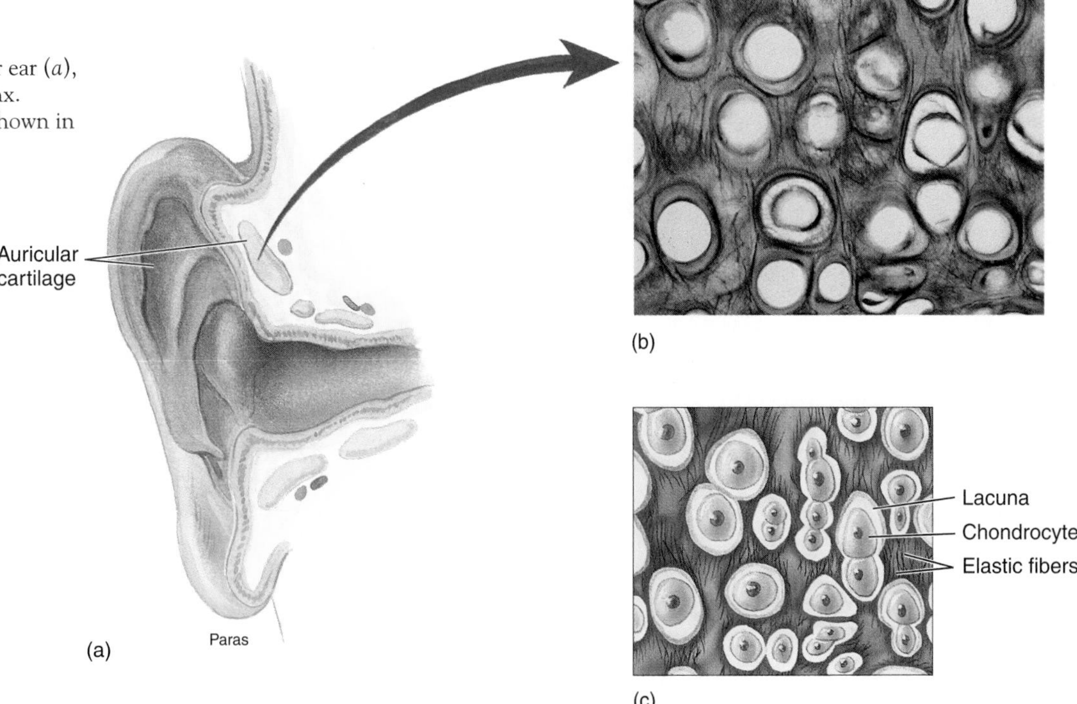

Table 6.4
Summary of Cartilage

Type	Structure and Function	Location
Hyaline cartilage	Homogeneous matrix with extremely fine collagenous fibers; provides flexible support, protects, is precursor to bone	Articular surfaces of bones, nose, walls of respiratory passages, fetal skeleton
Fibrocartilage	Abundant collagenous fibers within matrix; supports, withstands compression	Symphysis pubis, intervertebral discs, knee joint
Elastic cartilage	Abundant elastic fibers within matrix; supports, provides flexibility	Framework of outer ear, auditory canal, portions of larynx

Figure 6.24
The histology of bone.
(*a*) Bone consists of compact and spongy bone tissues. A photomicrograph of the tissue is shown in (*b*) and a labeled diagram in (*c*).

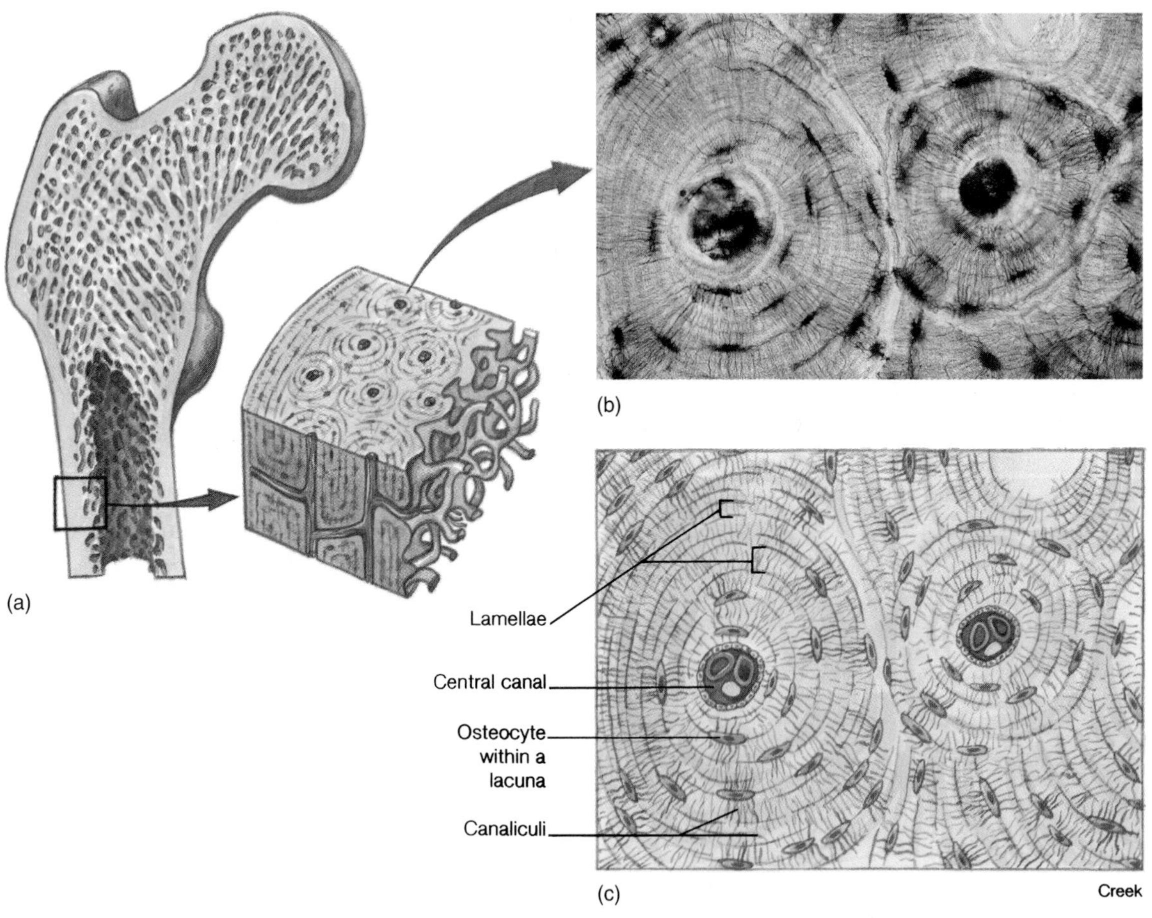

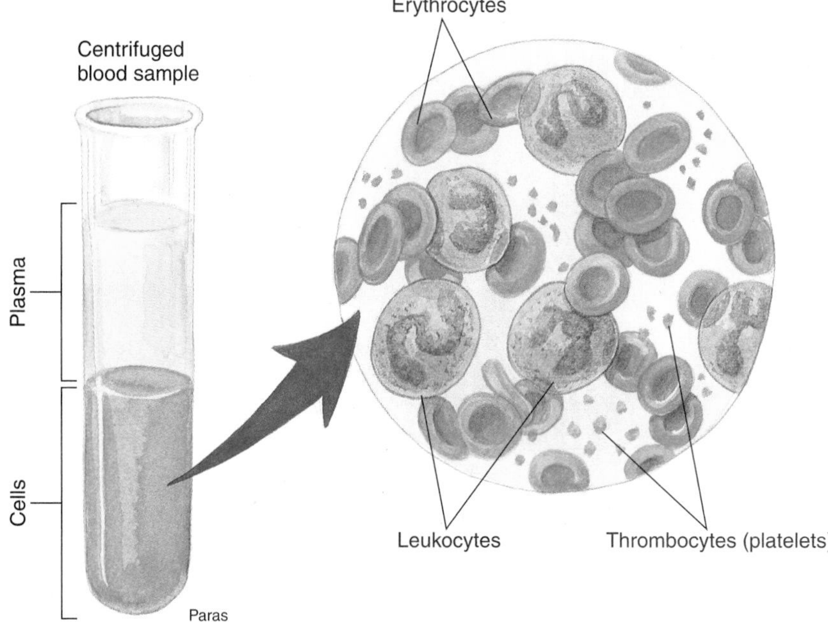

Figure 6.25

The histology of blood.
Blood consists of formed elements—erythrocytes (red blood cells), leukocytes (white blood cells), and thrombocytes (platelets)—suspended in a liquid plasma matrix.

the liquid matrix called **blood plasma** (fig. 6.25). The three types of formed elements are **erythrocytes** (red blood cells), **leukocytes** (white blood cells), and **thrombocytes** (platelets). Blood is discussed fully in chapter 20.

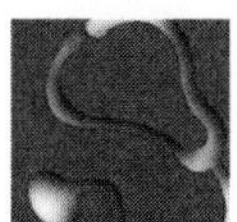

An injury to a portion of the body may stimulate tissue repair activity, usually involving connective tissue. A minor scrape or cut results in thrombocyte and plasma activity of the exposed blood and in the formation of a scab. The epidermis of the skin regenerates beneath the scab. A severe open wound heals through connective tissue granulation. In this process, collagenous fibers form from surrounding fibroblasts to strengthen the traumatized area. The healed area is known as a *scar.*

Muscle Tissue

Muscle tissue is responsible for the movement of materials through the body, the movement of one part of the body with respect to another, and locomotion. Fibers in the three kinds of muscle tissue are adapted to contract in response to stimuli.

Muscle tissue is unique in its ability to contract, and thus make movement possible. The muscle cells, or *fibers*, are elongated in the direction of contraction, and movement is accomplished through the shortening of the fibers in response to a stimulus. Muscle tissue is derived from mesoderm. There are three types of muscle tissue in the body: *smooth*, *cardiac*, and *skeletal muscle tissue* (fig. 6.26).

Smooth Muscle

Smooth muscle tissue is common throughout the body, occurring in many of the systems. For example, in the wall of the GI tract it provides the contractile force for the peristaltic movements involved in the mechanical digestion of food. Smooth muscle is also found in the walls of arteries, in the walls of respiratory passages, and in the urinary and reproductive ducts. The contraction of smooth muscle is under autonomic (involuntary) nervous control, and is discussed in more detail in chapter 12.

Smooth muscle fibers are long, spindle-shaped cells. They contain a single nucleus and lack striations. These cells are usually grouped together in flattened sheets, forming the muscular portion of a wall around a lumen.

Cardiac Muscle

Cardiac muscle tissue makes up most of the wall of the heart. This tissue is characterized by bifurcating (branching) fibers, each with a centrally positioned nucleus, and by transversely positioned **intercalated** (*in-ter'kă-lāt-ed*) **discs.** Intercalated discs help to hold adjacent cells together and transmit electrical impulses from cell to cell. Like skeletal muscle, cardiac muscle is striated, but unlike skeletal muscle it experiences rhythmical involuntary contractions. Cardiac muscle is further discussed in chapter 12.

Skeletal Muscle

Skeletal muscle tissue attaches to the skeleton and is responsible for voluntary body movements. Each elongated, multinucleated fiber has distinct transverse striations. Fibers of this muscle tissue are grouped into parallel fasciculi (bundles) that can be seen without a microscope in fresh muscle. Both cardiac and skeletal muscle fibers cannot replicate once tissue formation has been completed shortly after birth. Skeletal muscle tissue is discussed further in chapter 12. The three types of muscle tissue are summarized in table 6.5.

---

erythrocyte: Gk. *erythros*, red; *kytos*, hollow (cell)
leukocyte: Gk. *leukos*, white; *kytos*, hollow (cell)
thrombocyte: Gk. *thrombos*, a clot; *kytos*, hollow (cell)

Figure 6.26

The histology of muscle.

There are three types of muscle tissue: (*a*) smooth, (*b*) cardiac, and (c) skeletal.

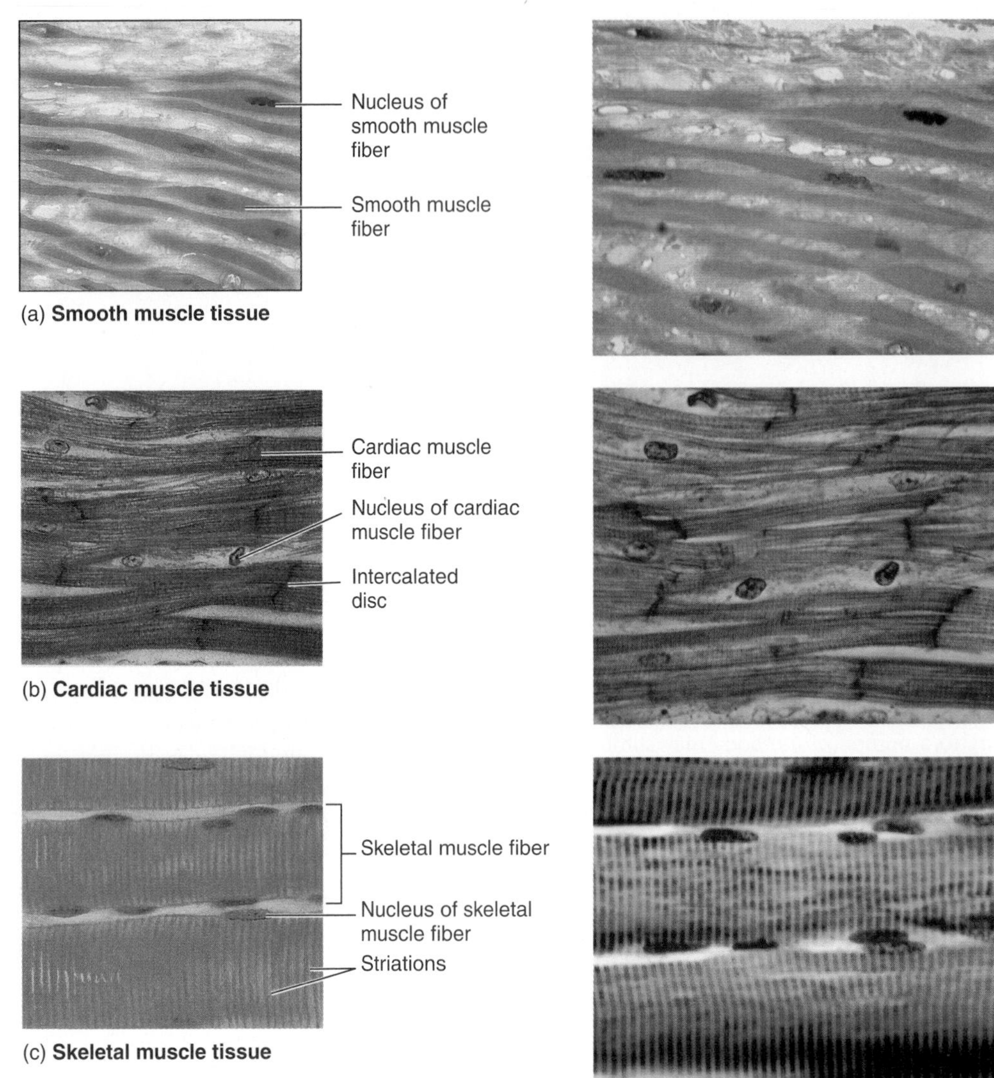

Table 6.5

Summary of Muscle Tissue

Type	Structure and Function	Location
Smooth	Elongated, spindle-shaped fiber with single nucleus; involuntary movements of internal organs	Walls of hollow internal organs
Cardiac	Branched, striated fiber with single nucleus and intercalated discs, involuntary rhythmic contraction	Heart wall
Skeletal	Multinucleated, striated, cylindrical fiber that occurs in fasciculi; voluntary movement of skeletal parts	Associated with skeleton; spans joints of skeleton via tendons

Nervous Tissue

Nervous tissue is composed of neurons, which respond to stimuli and conduct impulses to and from all body organs, and neuroglia, which functionally support and physically bind neurons.

Neurons

Although there are several kinds of neurons (*noo'ronz*) in nervous tissue, they all have three principal components: (1) a cell body, or perikaryon; (2) dendrites; and (3) an axon (fig. 6.27*b*). **Dendrites** receive stimuli and conduct impulses to the cell body. The **cell body,** or **perikaryon** (*per"ĭ-kar'e-on*), contains the nucleus and specialized organelles and microtubules. The **axon** is a cytoplasmic extension that conducts impulses away from the cell body. The term **nerve fiber** refers to any process extending from the cell body of a neuron and the myelin sheath that surrounds it (see fig. 14.2).

Neurons derive from ectoderm and are the basic structural and functional units of the nervous system. They are specialized to respond to physical and chemical stimuli, convert stimuli into nerve impulses, and conduct these impulses to other neurons, muscle fibers, or glands. Of all the body's cells, neurons are probably the most specialized. As with muscle cells, the number of neurons is established shortly after birth; thereafter, they lack the ability to undergo mitosis, although under certain circumstances a severed portion can regenerate.

neuron: Gk. *neuron,* sinew or nerve
perikaryon: Gk. *peri,* around; *karyon,* nut or kernel

Neuroglia

In addition to neurons, nervous tissue is composed of neuroglia (*noo-rog'le-ah*)(fig. 6.27*c*). Neuroglial cells, sometimes called glial cells, are about five times as abundant as neurons and have limited mitotic abilities. They do not transmit impulses but support and bind neurons together. Certain neuroglial cells are phagocytic; others assist in providing sustenance to the neurons.

Neurons and neuroglia are discussed in detail in chapter 14.

Clinical Considerations

As stated at the beginning of this chapter, the study of tissues is extremely important in understanding the structure and function of organs and systems. Histology has immense clinical importance as well because many diseases are diagnosed

neuroglia: Gk. *neuron,* nerve; *glia,* glue

Figure 6.27

Nervous tissue.
This tissue is found in the brain (*a*), spinal cord, nerves, and ganglia. It consists of two principal kinds of cells: (*b*) neurons and (*c*) neuroglia.

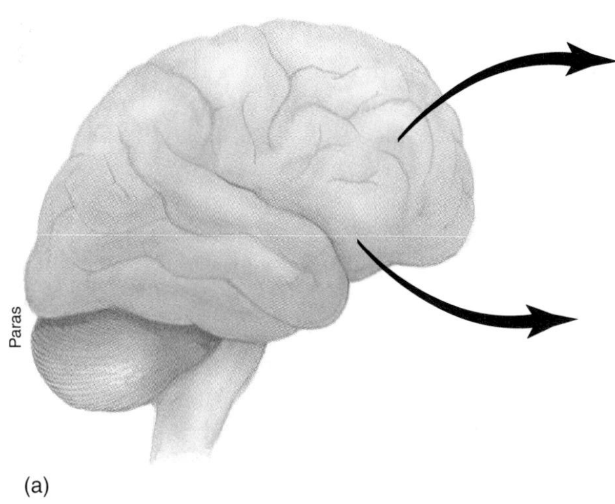

(a)

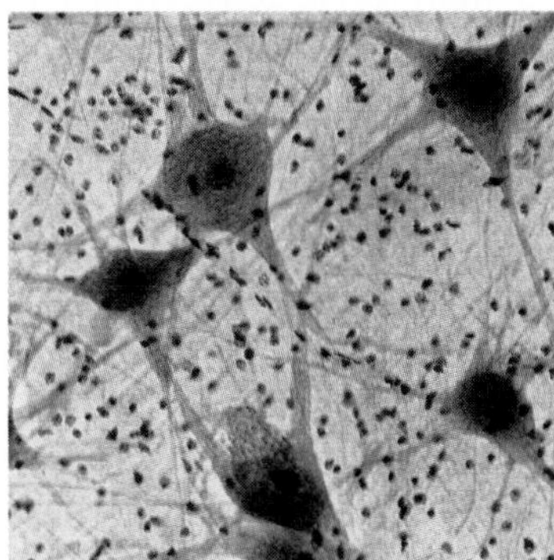
(b) Neurons

(c) Neuroglia

through microscopic examination of tissue sections. Even in performing an autopsy, an examination of various tissues is vital in establishing the cause of death.

Several sciences are concerned with specific aspects of tissues. *Histopathology* is the study of diseased tissues. *Histochemistry* is concerned with the physiology of tissues as they maintain homeostasis. *Histotechnology* explores the ways in which tissues can be better stained and observed. In all of these disciplines, a thorough understanding of normal, or healthy, tissues is imperative for recognizing altered, or abnormal, tissues.

Changes in Tissue Composition

Most diseases alter tissue structure *locally* where the disease is prevalent. Some diseases, however, called *general conditions*, cause changes that are far removed from the locus of the disease. **Atrophy** (wasting of body tissue), for example, may be limited to a particular organ where the disease interferes with the metabolism of that organ; or it may involve an entire limb if nourishment or nerve impulses are impaired. *Muscle atrophy*, for example, can be caused by a disease of the nervous system such as polio, or it can be the result of a diminished blood supply to a muscle. *Senescence* (*sĕ-nes′ens*) *atrophy*, or simply *senescence*, is the natural aging of tissues and organs within the body. *Disuse atrophy* is a local atrophy that results from the inactivity of a tissue or organ. Muscular dystrophy causes a disuse atrophy that decreases muscle size and strength due to the loss of sarcoplasm within the muscle.

Necrosis (*nĕ-kro′sis*) is cellular or tissue death within the living body. Recognizable by physical changes in the dead tissues, necrosis can be caused by a number of factors, such as severe injury, physical agents (trauma, heat, radiant energy, chemical poisons), or interference with the nutrition of tissues. When examined histologically, the necrotic tissue usually appears opaque, and a whitish or yellow cast. **Gangrene** is a massive necrosis of tissue accompanied by an invasion of microorganisms that live on decaying tissues.

Somatic death is the death of the body as a whole. Following somatic death, tissues undergo irreversible changes such as **rigor mortis** (muscular rigidity), clotting of the blood, and cooling of the body. Postmortem (after death) changes occur under varying conditions at predictable rates, which is useful in estimating the approximate time of death.

Tissue Analysis

In diagnosing a disease, it is frequently important to examine tissues from a living person histologically. When this is necessary, a **biopsy** (*bi′op-se*) (removal of a section of living tissue) is performed. There are several techniques for obtaining a biopsy. *Surgical removal* is usually done on large masses or tumors. *Curettage* involves cutting and scraping tissue, as may be done in examining for uterine cancer. In a *percutaneous needle biopsy*, a biopsy needle is inserted through a small skin incision, and tissue samples are aspirated. Both normal and diseased tissues are removed for purposes of comparison.

Preparing tissues for examination involves a number of steps. **Fixation** is fundamental for all histological preparation. It is the rapid killing, hardening, and preservation of tissue to maintain its existing structure. **Embedding** the tissue in a supporting medium such as paraffin wax usually follows fixation. The next step is **sectioning** the tissue into extremely thin slices, followed by **mounting** the specimen on a slide. Some tissues are fixed by rapid freezing and then sectioned while frozen, making embedding unnecessary. Frozen sections enable the pathologist to make a quick diagnosis during a surgical operation. These are done frequently, for example, in cases of suspected breast cancer. **Staining** is the next step. Hematoxylin and eosin (H & E) stains, which give a differential blue and red color to the basic and acidic structures within the tissue, are routinely used on all tissue specimens. Other dyes may be needed to stain for specific structures.

Examination is first done with the unaided eye and then with a microscope. Practically all histological conditions can be diagnosed with low magnification (40×). Higher magnification is used to clarify specific details. Further examination may be performed with an electron microscope, which makes visible the cellular structures that are the morphological bases of metabolic processes. Histological observation is the foundation of subsequent diagnosis, prognosis, treatment, and reevaluation.

Tissue Transplantation

In the last two decades, medical science has made tremendous advancements in tissue transplants. Tissue transplants are necessary for replacing nonfunctional, damaged, or lost body parts. The most successful transplant is one where tissue is taken from one place on a person's body and moved to another place, such as a skin graft from the thigh to replace burned tissue of the hand. This type of transplant is termed an **autograft. Isografts** are transplants between individuals who are closely related genetically. Identical twins have the best acceptance success in this type of transplant. **Homotransplants,** or **allografts,** are grafts between individuals of the same species but of different genotype, and **heterografts,** or **xenografts,** are grafts between individuals of different species. Both allografts and xenografts present the problem of possible *tissue-rejection reaction*. When this occurs, the recipient's immune mechanisms are triggered, and the donor's tissue is identified as foreign and is destroyed. The reaction can be minimized by "matching" recipient and donor tissue. Immunosuppressive drugs also may lessen the rejection rate. These drugs act by interfering with the recipient's immune mechanisms. Unfortunately, immunosuppressive drugs may also lower the recipient's resistance to infections. New techniques involving blood transfusions from donor to recipient before transplant are proving successful. In any event, tissue transplants are an important aspect of medical research, and significant breakthroughs are on the horizon.

---

atrophy: Gk. *a*, without; *trophe*, nourishment
necrosis: Gk. *nekros*, corpse
gangrene: Gk. *gangraina*, gnaw or eat

Clinical Investigation Answer

Vitamin deficiency diseases that were once quite common seldom occur now because many foods are fortified with vitamins, nutritious foods are available year-round, and dietary supplements are widely used.

Scurvy is characterized by a loss of collagen, the main structural protein in many connective tissues. Scurvy is caused by a dietary deficiency of vitamin C, which is a necessary factor in the formation of collagenous fibers. Without vitamin C, these fibers break up and cannot form to support the tissue. The resulting symptoms include skin sores, spongy gums, loose teeth, weak blood vessels, and poor healing of wounds.

Chapter Summary

Definition and Classification of Tissues (pp. 131–132)

1. Tissues are aggregations of similar cells that perform specific functions. The study of tissues is called histology.
2. Cells are separated and bound together by an intercellular matrix, the composition of which varies from solid to liquid.
3. The four principal types of tissues are epithelial tissue, connective tissue, muscle tissue, and nervous tissue.

Epithelial Tissue (pp. 132–144)

1. Epithelia are derived from all three germ layers and may be one or several layers thick. The lower surface of most membranous epithelia is supported by a basement membrane.
   (a) Simple epithelia vary in shape and surface characteristics. They are located where diffusion, filtration, and secretion occur.
   (b) Stratified epithelia consist of two or more layers of cells and are adapted for protection.
   (c) Transitional epithelium lines the urinary bladder and is adapted for distension.
2. The body has two principal types of membranes: mucous membranes, which secrete protective mucus, and serous membranes, which line the thoracic and abdominopelvic cavities and cover visceral organs. Serous membranes secrete a lubricating serous fluid.
3. Glandular epithelia derive from developing epithelial tissue and function as secretory exocrine glands.

Connective Tissue (pp. 144–153)

1. Connective tissues are derived from mesoderm and, with the exception of cartilage, are highly vascular.
2. Connective tissue proper contains fibroblasts, collagenous fibers, and elastic fibers within a flexible ground substance.
3. Cartilage tissue provides a flexible framework for many organs. It consists of a semisolid matrix of chondrocytes and various fibers.
4. Bone tissue consists of osteocytes, collagenous fibers, and a durable matrix of mineral salts.
5. Blood (vascular tissue) consists of formed cellular elements (erythrocytes, leukocytes, and thrombocytes) suspended in a fluid plasma matrix.

Muscle Tissue (pp. 153–154)

1. Muscle tissues (smooth, cardiac, and skeletal) are responsible for the movement of materials through the body, the movement of one part of the body with respect to another, and for locomotion.
2. Fibers in muscle tissues are adapted to contract in response to stimuli.

Nervous Tissue (p. 155)

1. Neurons are the functional units of the nervous system. They respond to stimuli and conduct impulses to and from all body organs.
2. Neuroglia support and bind neurons. Some are phagocytic; others provide sustenance to neurons.

Review Activities

Objective Questions

1. Which of the following is *not* a principal type of body tissue?
   (a) nervous tissue
   (b) integumentary tissue
   (c) connective tissue
   (d) muscular tissue
   (e) epithelial tissue
2. Which of the following statements regarding tissues is *false?*
   (a) Tissues are aggregations of similar kinds of cells that perform specific functions.
   (b) All tissues are microscopic and are studied within the science of histology.
   (c) All tissues are stationary within the body at the location of their developmental origin.
   (d) A body organ is composed of two or more kinds of tissues.
3. Connective tissues, muscle tissues, and the dermis of the skin derive from embryonic
   (a) mesoderm.
   (b) endoderm.
   (c) ectoderm.
4. Which statement is *false* regarding epithelia?
   (a) They are derived from mesoderm, ectoderm, and endoderm.
   (b) They are strengthened by elastic and collagenous fibers.
   (c) One side is exposed to the lumen, cavity, or external environment.
   (d) They have very few intercellular matrix-binding cells.
5. A gastric ulcer of the stomach would involve
   (a) simple cuboidal epithelium.
   (b) transitional epithelium.
   (c) simple ciliated columnar epithelium.
   (d) simple columnar epithelium.
6. Which structural and secretory designation describes mammary glands?
   (a) acinar, apocrine
   (b) tubular, holocrine
   (c) tubular, merocrine
   (d) acinar, holocrine
7. Dense regular connective tissue is found in
   (a) blood vessels.
   (b) the spleen.
   (c) tendons.
   (d) the wall of the uterus.
8. The phagocytic connective tissue found in the lymph nodes, liver, spleen, and bone marrow is
   (a) reticular connective tissue.
   (b) loose connective tissue.
   (c) mesenchyme.
   (d) elastic connective tissue.
9. Cartilage is slow in healing following an injury because
   (a) it is located in body areas that are under constant physical strain.
   (b) it is avascular.
   (c) its chondrocytes cannot reproduce.
   (d) it has a semisolid matrix.
10. Cardiac muscle tissue has
    (a) striations.
    (b) intercalated discs.
    (c) rhythmical involuntary contractions.
    (d) all of the above.

Essay Questions

1. Define *tissue*. What are the differences between cells, tissues, glands, and organs?
2. What physiological functions are epithelial tissues adapted to perform?
3. Identify the epithelial tissue
   (a) in the pulmonary alveoli
   (b) lining the lumen of the GI tract;
   (c) in the outer layer of skin;
   (d) in the urinary bladder;
   (e) in the uterine tube; and
   (f) lining the lumina of the lower respiratory tract.
   Describe the function of the tissue in each case.
4. Why are both keratinized and nonkeratinized epithelia found within the body?
5. Describe how epithelial glands are classified according to structural complexity and secretory function.
6. Identify the connective tissue
   (a) on the surface of the heart and surrounding the kidneys;
   (b) lining the lumen of the aorta;
   (c) forming the symphysis pubis;
   (d) supporting the outer ear;
   (e) forming the lymph nodes; and
   (f) forming the tendo calcaneus.
   Describe the function of the tissue in each case.
7. Compare and contrast the following: reticular fibers, collagenous fibers, elastin, fibroblasts, and mast cells.
8. What is the relationship between adipose cells and fat? Discuss the function of fat and explain the potential danger of excessive fat.
9. Discuss the mitotic abilities of each of the four principal types of tissues.
10. Define the following terms: *atrophy*, *necrosis*, *gangrene*, and *somatic death*.

Critical Thinking Questions

1. The function of a tissue is actually a function of its cells. And the function of a cell is a function of its organelles. Knowing this, what type of organelles would be particularly abundant in cardiac muscle tissue that requires a lot of energy; in reticular tissue within the liver, where cellular debris and toxins are ingested; and in dense regular connective tissue that consists of tough protein strands?
2. Your aunt was recently diagnosed as having brain cancer. In talking with your aunt's physician, she indicated that the cancer was actually a neuroglioma, and went on to say that cancer of neurons and muscle cells is a rare occurrence. Explain why neuroglial cells are much more susceptible to cancer than are neurons or muscle cells.
3. Compare the vascular supply of bones and ligaments and discuss how this may be relevant to the clinical course of an ankle sprain and an ankle fracture.
4. The connective tissue diseases are a group of disorders most likely caused by an abnormal immune response to a person's own connective tissue. The best known of these is rheumatoid arthritis, in which small joints of the body become inflamed and the articulating surfaces erode away. Knowing where connective tissue is found in the body, can you predict which organs might be involved in other connective tissue diseases?

Related Web Sites

For a listing of the most current web sites related to this chapter, please visit the *Concepts of Human Anatomy and Physiology* home page at http://www.mhhe.com/biosci/abio/.

SEVEN

Integumentary System

Objectives

- Explain why the skin is considered an organ and a component of the integumentary system.
- Describe some common clinical conditions of the skin that result from nutritional deficiencies or body dysfunctions.
- Describe the histological characteristics of each layer of the skin.
- Summarize the transitional events that occur within each of the epidermal layers.
- Discuss the role of the skin in the protection of the body from disease and external injury, and the regulation of body fluids and temperature, absorption, synthesis, sensory reception, and communication.
- Describe the structure of hair and list the three principal types.
- Describe the structure and function of nails.
- Compare and contrast the structure and function of the three principal kinds of integumentary glands.

Clinical Investigation

A 27-year-old male was involved in a gasoline explosion and sustained burns to his face, neck, chest, and arms. Upon arrival at the emergency room, he complained of intense pain in his face and neck, both of which exhibited extensive blistering and erythema (redness). His burned chest and arms, however, appeared pale and waxy, and examination revealed this skin area to be leathery and lacking sensation. The emergency room physician commented to an observing medical student that the patient had third-degree burns on his chest and arms and that excision of the burn eschar (traumatized tissue) with subsequent skin grafting would be required.

Why would the areas that sustained second-degree burns be red, blistered, and painful, while the third-degree burns were pale and insensate (without sensation, including pain)? Why would the chest and arms require skin grafting, but probably not the face and neck?

Clues: Think in terms of functions of the skin and survival of the germinal cells in functioning skin. Carefully examine figures 7.1 and 7.14.

The Skin as an Organ

The skin is the largest organ of the body. Together with its accessory structures (hair, glands, and nails), it constitutes the integumentary system. In certain areas of the body, skin has adaptive modifications that accommodate protective or metabolic functions. In its role as a dynamic interface between the continually changing external environment and the body's internal environment, the skin helps to maintain homeostasis.

We are more aware of and concerned with our integumentary system than perhaps any other system of our body. One of the first things we do in the morning is look in a mirror and see what we have to do to make our skin and hair presentable. Periodically, we examine our skin for wrinkles and our scalp for gray hairs as signs of aging. We recognize other people to a large extent by features of their skin.

The appearance of our skin frequently determines our self-image and the initial impression we make on others. Unfortunately, it may also determine whether or not we succeed in gaining social acceptance. For example, social rejection during teenage years, imagined or real, can be directly associated with skin problems such as acne. Even clothing styles are somewhat determined by how much skin we, or the designers, want to expose. But our skin is much more than a showpiece. It helps to regulate certain body functions and protect certain body structures.

The skin, or *integument* (*in-teg'yoo-ment*), and its accessory structures (hair, glands, and nails) constitute the integumentary system. Included in this system are the millions of sensory receptors of the skin, and its extensive vascular network. The skin is a dynamic interface between the body and the external environment. It protects the body from the environment even as it allows for communication with the environment.

The skin is considered an organ, since it consists of several kinds of tissues that are structurally arranged to function together. It is the largest organ of the body, covering over 7,600 sq cm (3,000 sq in.) in the average adult, and accounts for approximately 7% of a person's body weight. The skin is of variable thickness, averaging 1.5 mm. It is thickest on the parts of the body exposed to wear and abrasion, such as the soles of the feet and palms of the hand, where it is about 6 mm thick. It is thinnest on the eyelids, external genitalia, and tympanic membrane (eardrum), where it is approximately 0.5 mm thick. Even its appearance and texture varies from the rough, callous skin covering the elbows and knuckles to the soft, sensitive areas of the eyelids, nipples, and genitalia.

The general appearance and condition of the skin are clinically important because they provide clues to certain body conditions or dysfunctions. Pale skin may indicate shock, whereas red, flushed, overwarm skin may indicate fever and infection. A rash may indicate allergies or local infections. Abnormal textures of the skin may be the result of glandular or nutritional problems (table 7.1). Even chewed fingernails may be a clue to emotional problems.

integument: L. *integumentum*, a covering

Layers of the Skin

The skin consists of two principal layers. The outer epidermis is stratified into four or five structural layers, and the thick and deeper dermis consists of two layers. The hypodermis (subcutaneous tissue) connects the skin to underlying organs.

Epidermis

The **epidermis** (*ep"i-der'mis*) is the superficial protective layer of the skin. It is composed of stratified squamous epithelium that varies in thickness from 0.007 to 0.12 mm. All but the deepest layers of the epidermis are composed of dead cells. Either four or five layers may be present, depending on where the epidermis is located (figs. 7.1 and 7.2). The epidermis of the palms and soles has five layers because these areas are exposed to the most friction. In all other areas of the body, the epidermis has only four layers. The names and characteristics of the epidermal layers are as follows:

1. **Stratum basale (basal layer).** The stratum basale is composed of a single layer of cells in contact with the dermis. Four types of cells compose the stratum basale: keratinocytes (*ker'ă-tin'o-sītz*), melanocytes (*mel'ă-no-sītz*),

epidermis: Gk. *epi*, on; *derma*, skin
stratum: L. *stratum*, something spread out
basale: Gk. *basis*, base

Table 7.1

Conditions of the Skin and Associated Structures Indicating Nutritional Deficiencies or Body Dysfunctions

Condition	Deficiency	Comments
General dermatitis	Zinc	Redness and itching
Scrotal or vulval dermatitis	Riboflavin	Inflammation in genital region
Hyperpigmentation	Vitamin B_{12}, folic acid, or starvation	Dark pigmentation on backs of hands and feet
Dry, stiff, brittle hair	Protein, calories, and other nutrients	Usually occurs in young children or infants
Follicular hyperkeratosis	Vitamin A, unsaturated fatty acids	Rough skin due to keratotic plugs from hair follicles
Pellagrous dermatitis	Niacin and tryptophan	Lesions on areas exposed to sun
Thickened skin at pressure points	Niacin	Noted at belt area at the hips
Spoon nails	Iron	Thin nails that are concave or spoon-shaped
Dry skin	Water or thyroid hormone	Dehydration, hypothyroidism, rough skin
Oily skin (acne)		Hyperactivity of sebaceous glands

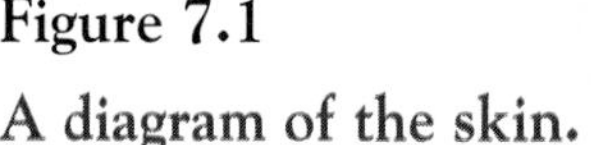

Figure 7.1
A diagram of the skin.

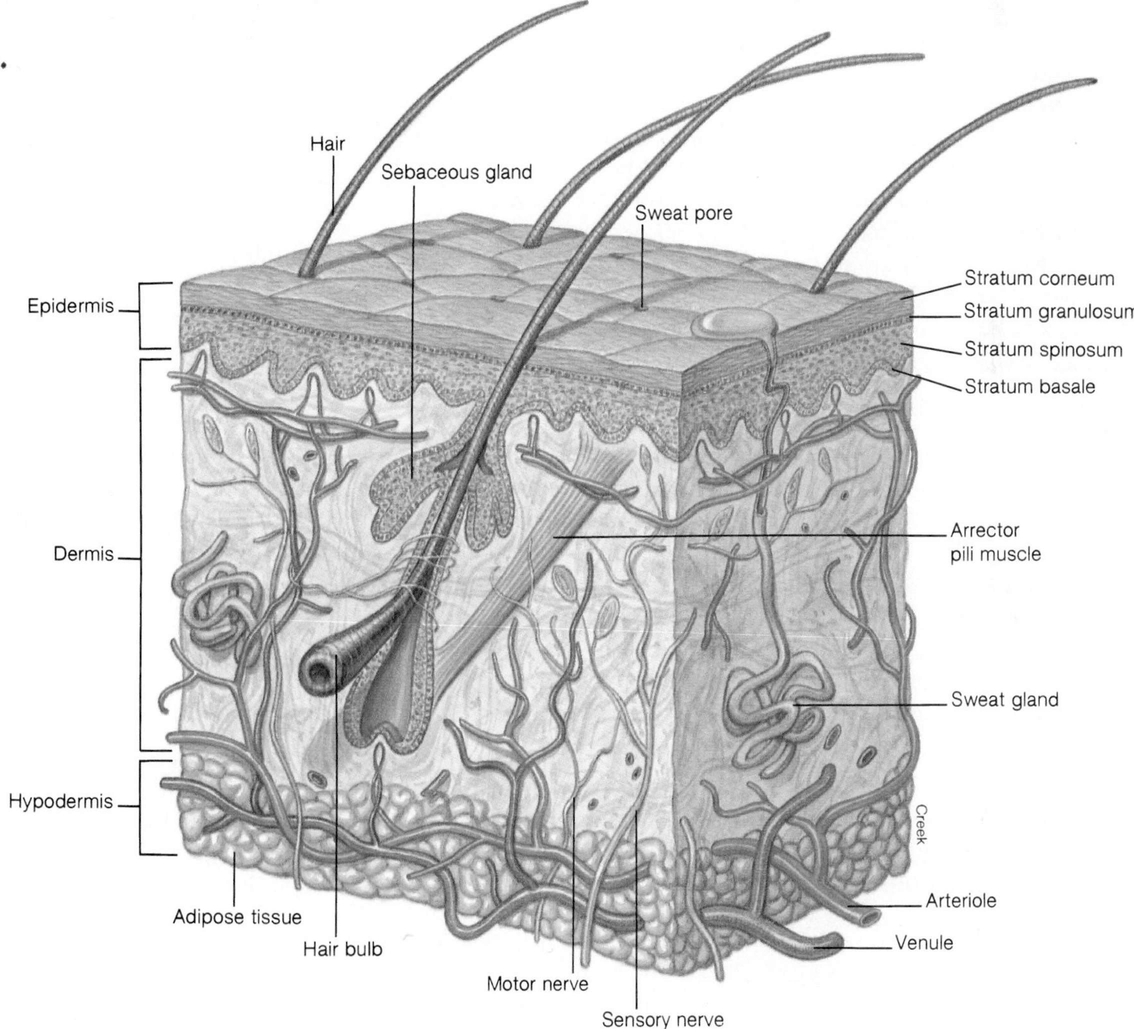

Figure 7.2

A photomicrograph of the epidermis (250×).

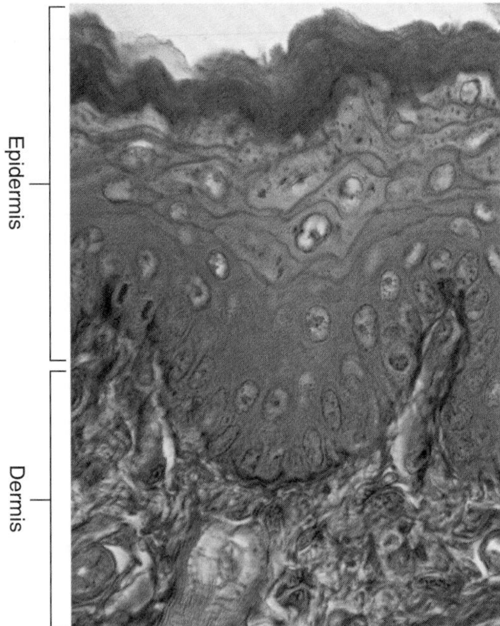

Figure 7.3

A scanning electron micrograph of the surface of the skin.
Note the opening of a sweat gland. The fragmented-appearing particles are bacteria, which are present throughout the body on the surface of the skin.

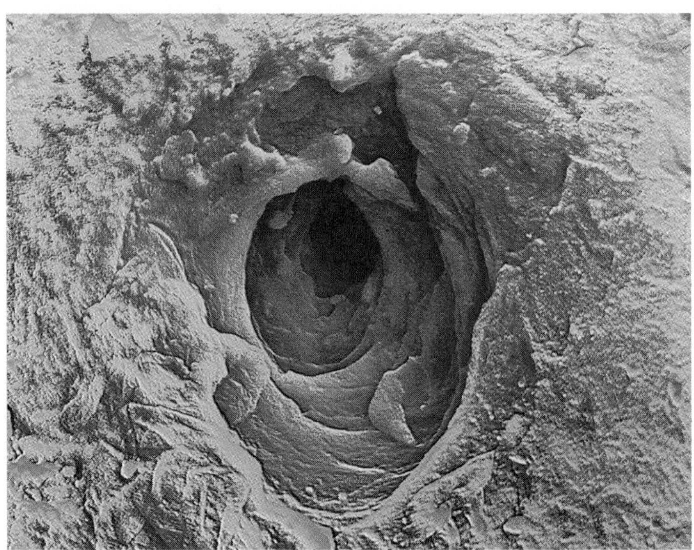

tactile cells (Merkel cells), and nonpigmented granular dendrocytes (dendritic cells, or Langerhans cells). With the exception of tactile cells, these cells are constantly dividing mitotically and moving outward to renew the epidermis. It usually takes between 6 and 8 weeks for the cells to move from the stratum basale to the surface of the skin.

Keratinocytes are specialized cells that produce **keratin** (*ker'ă-tin*), which toughens and waterproofs the skin. As keratinocytes are pushed away from the vascular nutrient and oxygen supply of the dermis, their nuclei degenerate, their cellular content is dominated by keratin, and the process of *keratinization* is completed. By the time keratinocytes have reached the surface of the skin, they resemble flat dead scales. **Melanocytes** are specialized epithelial cells that synthesize the pigment **melanin** (*mel'ă-nin*), providing a protective barrier to the ultraviolet radiation in sunlight. **Tactile cells** are sparse compared to keratinocytes and melanocytes. These cells aid in tactile (touch) reception. **Nonpigmented granular dendrocytes** are scattered throughout the stratum basale. They are protective *macrophagic cells* that ingest bacteria and other foreign debris and thereby present foreign molecules (antigens) to the immune system (see chapter 23).

2. **Stratum spinosum (spiny layer).** The stratum spinosum contains several stratified layers of cells. The spiny appearance of this layer is due to the changed shape of the keratinocytes when the tissue is fixed for microscopic examination. Since there is limited mitosis in the stratum spinosum, this layer and the stratum basale are collectively referred to as the **stratum germinativum** (*jer-mĭ"nă-tĭ'vum*).
3. **Stratum granulosum (granular layer).** The stratum granulosum (*gran"yoo-lo'sum*) consists of only three or four flattened rows of cells. These cells contain granules that are filled with *keratohyalin*, a chemical precursor to keratin.
4. **Stratum lucidum (clear layer).** The nuclei, organelles, and cell membranes are no longer visible in the cells of the stratum lucidum, and so histologically this layer appears clear. It exists only in the lips and in the thickened skin of the soles and palms.
5. **Stratum corneum (hornlike layer).** The stratum corneum (*kor'ne-um*) is composed of 25 to 30 layers of flattened, scalelike cells. Thousands of these dead cells shed from the skin surface each day, only to be replaced by new ones from deeper layers. This surface layer is cornified; it is the layer that actually protects the skin (fig. 7.3). *Cornification*, brought on by keratinization, is the drying and flattening of the cells comprising the stratum corneum and is an important protective adaptation of the skin. Friction at the surface of the skin

keratinocyte: Gk. *keras*, hornlike; *kytos*, cell
melanocyte: Gk. *melas*, black; *kytos*, cell
macrophagic: Gk. *makros*, large; *phagein*, to eat
spinosum: L. *spina*, thorn

germinativum: L. *germinare*, sprout or growth
granulosum: L. *granum*, grain
lucidum: L. *lucidus*, light
corneum: L. *corneus*, hornlike

Table 7.2
Layers of the Epidermis

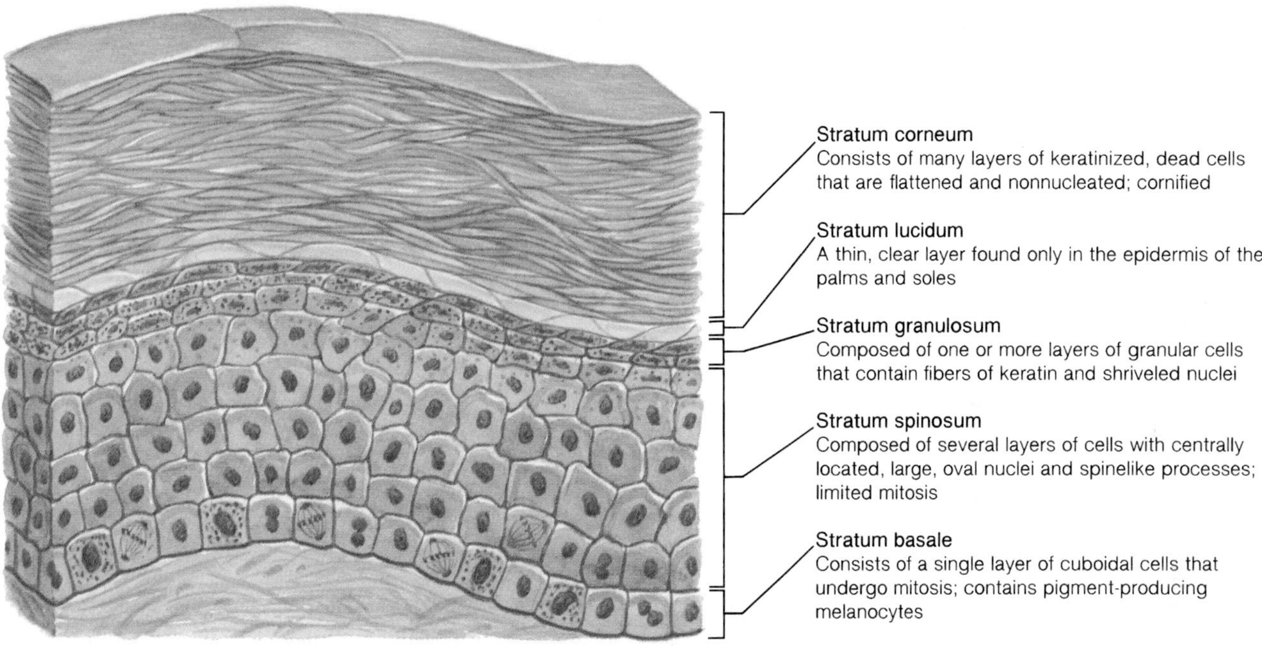
Stratum corneum
Consists of many layers of keratinized, dead cells that are flattened and nonnucleated; cornified
Stratum lucidum
A thin, clear layer found only in the epidermis of the palms and soles
Stratum granulosum
Composed of one or more layers of granular cells that contain fibers of keratin and shriveled nuclei
Stratum spinosum
Composed of several layers of cells with centrally located, large, oval nuclei and spinelike processes; limited mitosis
Stratum basale
Consists of a single layer of cuboidal cells that undergo mitosis; contains pigment-producing melanocytes

stimulates additional mitotic activity of the stratum basale, which may result in the formation of a *callus* for additional protection.

The specific characteristics of each epidermal layer are described in table 7.2.

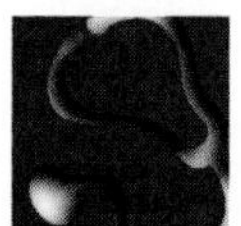

Tattooing colors the skin permanently because pigmented dyes are injected below the mitotic basal layer into the dermis. In nonsterile conditions, infectious organisms may be introduced along with the dye. Small tattoos can be removed by skin grafting; for large tattoos, mechanical abrasion of the skin is preferred.

Coloration of the Skin

Normal skin color is caused by the expression of a combination of three pigments: melanin, carotene, and hemoglobin. **Melanin** is a brown-black pigment produced in the melanocytes of the stratum basale (fig. 7.4). All individuals of similar size have approximately the same number of melanocytes, but the amount of melanin produced and the distribution of melanin determine racial variations in skin color, such as black, brown, yellow, and white. Melanin protects the basal layer against the damaging effect of the ultraviolet (UV) rays of the sun. A gradual exposure to the sunlight promotes the increased production of melanin within the melanocytes, thereby tanning the skin. The skin of a person with albinism (*al'bi-niz-em*) has the normal number of melanocytes in the epidermis but lacks the enzyme *tyrosinase*, which converts the amino acid tyrosine to melanin. Albinism is a hereditary condition.

Figure 7.4

Melanocytes.
These pigment-producing cells are found throughout the stratum basale (see arrow), where they produce melanin.

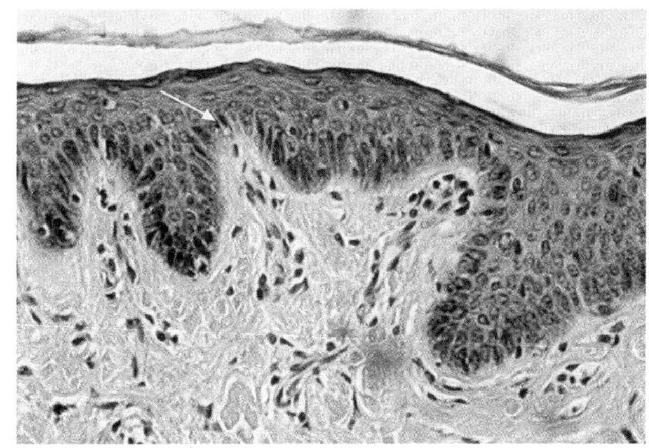

Other genetic expressions of melanocytes are more common than albinism. *Freckles*, for example, are caused by aggregated patches of melanin. A lack of melanocytes in localized areas of the skin causes distinct white spots in the condition called *vitiligo* (*vit-ĭ-li'gō*). After the age of 50, brown plaquelike growths, called *seborrheic* (*seb"o-re'ik*) *hyperkeratoses*, may

vitiligo: L. *vitiatio*, blemish

appear on the skin, particularly on exposed portions. Commonly called "liver spots," these pigmented patches are benign growths of pigment-producing melanocytes. Usually no treatment is required, unless for cosmetic purposes.

Carotene (*kar'o-t-en*) is not produced naturally in the body. Rather, it is a yellowish pigment produced by a number of plants, including carrots, that tends to accumulate in the epidermal cells and fatty parts of the dermis. Carotene was once thought to account for the yellow-tan skin of people of Asian descent, but this coloration is now known to be caused by variations in melanin.

Hemoglobin (*he'mŏ-glo"bin*) is not a pigment of the skin; rather, it is the oxygen-binding pigment found in red blood cells. Oxygenated blood flowing through the dermis gives the skin its pinkish tones.

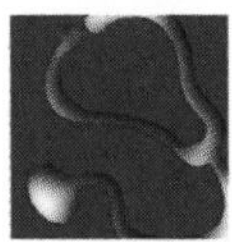

Certain physical conditions or diseases cause symptomatic discoloration of the skin. *Cyanosis* is a bluish discoloration of the skin that appears in people with certain cardiovascular or respiratory diseases. People also become cyanotic during an interruption of breathing. In *jaundice,* the skin appears yellowish because of an excess of bile pigment in the bloodstream. Jaundice is usually symptomatic of liver dysfunction and sometimes of liver immaturity, as in a jaundiced newborn. *Erythema* is a redness of the skin generally due to vascular trauma, such as a sunburn.

Surface Patterns

The exposed surface of the skin has recognizable patterns that are either present at birth or develop later. **Fingerprints** (*friction ridges*) are congenital patterns on the finger and toe pads, as well as on the palms and soles. The designs formed by these lines have basic similarities but are not identical in any two individuals. They are formed by the pull of elastic fibers within the dermis and are well-established prenatally. The ridges of fingerprints prevent slippage when grasping objects. Because they are precise and easy to reproduce, fingerprints are customarily used for identifying individuals in the science known as *dermatoglyphics*. All primates have fingerprints, and even dogs have a characteristic "nose print" that is used for identification in the military canine corps and in certain dog kennels.

Acquired lines include the deep **flexion creases** on the palms and the shallow **flexion lines** on the knuckles and on the surface of other joints. Furrows on the forehead and face are acquired from continual contraction of facial muscles, such as from smiling or squinting in bright light or against the wind. Facial lines become more strongly delineated as a person ages.

carotene: L. *carota*, carrot (referring to orange coloration)
hemoglobin: Gk. *haima*, blood; *globus*, globe
cyanosis: Gk. *kyanosis*, dark-blue color
jaundice: L. *galbus*, yellow
erythema: Gk. *erythros*, red; *haima*, blood

Figure 7.5

Stretch marks on the abdomen of a pregnant woman.
Stretch marks (lineae albicantes) generally fade with time but frequently leave permanent integumentary markings.

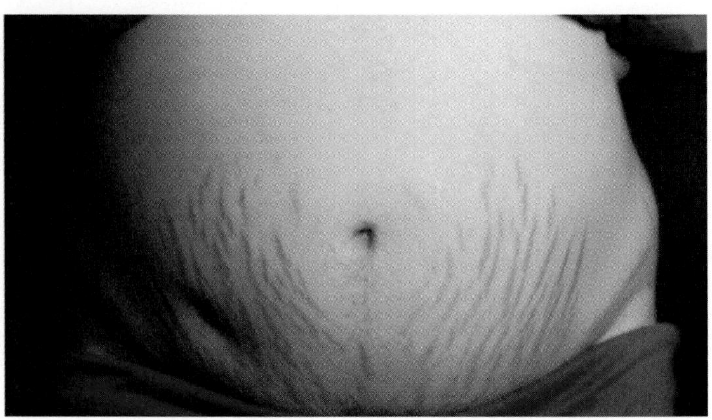

Dermis

The **dermis** is deeper and thicker than the epidermis (see fig. 7.1). Elastic and collagenous fibers within the dermis are arranged in definite patterns, producing lines of tension on the surface of the skin and providing skin tone. There are many more elastic fibers in the dermis of a young person than in an elderly one, and a decrease in number is apparently associated with aging. The extensive network of blood vessels in the dermis provides nourishment to the living portion of the epidermis. The dermis also contains many sweat glands, oil-secreting glands, nerve endings, and hair follicles.

Layers of the Dermis

The dermis is composed of two layers. The upper layer, called the **stratum papillarosum** (papillary layer), is in contact with the epidermis and accounts for about one-fifth of the entire dermis. Numerous projections, called *papillae,* extend from the upper portion of the dermis into the epidermis. Papillae form the base for the friction ridges on the fingers and toes.

The deeper and thicker layer of the dermis is called the **stratum reticularosum** (reticular layer). In fact, it is this layer that corresponds to the hide of an animal used to make leather and suede. Fibers within this layer are more dense and regularly arranged to form a tough, flexible meshwork. It is quite distensible, as is evident in pregnant women or obese individuals, but it can be stretched too far, causing "tearing" of the dermis. The repair of a strained dermal area leaves a white streak called a stretch mark, or *linea albicans*. Lineae albicantes frequently develop on the buttocks, thighs, abdomen, and breasts (fig. 7.5).

papilla: L. *papula*, swelling or pimple

Innervation of the Skin

The dermis of the skin has extensive innervation. Specialized integumentary *effectors* consist of muscles or glands within the dermis that respond to motor impulses transmitted from the central nervous system to the skin by autonomic nerve fibers.

Several types of **sensory receptors** respond to various stimuli including tactile (touch), pressure, temperature, tickle, or pain. Some are free nerve endings, some form a network around hair follicles, and some extend into the papillae of the dermis. Certain areas of the body, such as the palms, soles, lips, and external genitalia, have a greater concentration of sensory receptors and are therefore more sensitive to touch. Chapter 18 includes a detailed discussion of the structure and function of the various sensory receptors.

Vascular Supply of the Skin

Blood vessels within the dermis supply nutrients to the mitotically active stratum basale of the epidermis and to the cellular structures of the dermis, such as the glands and hair follicles. Dermal blood vessels play an important role in regulating body temperature and blood pressure. Autonomic vasoconstriction or vasodilation responses can either shunt the blood away from the superficial dermal arterioles or permit it to flow freely throughout dermal vessels. Fever or shock can be detected by the color and temperature of the skin. Blushing is the result of involuntary vasodilation of dermal blood vessels.

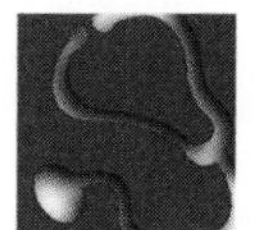

People who are bedridden need to maintain good blood circulation to prevent bedsores, or *decubitus ulcers.* When a person lies in one position for an extended period, the dermal blood flow is restricted where the body presses against the bedding. As a consequence, cells die and open wounds may develop (fig. 7.6). Frequently changing the position of the patient and periodically massaging the skin to stimulate blood flow can help prevent decubitus ulcers.

Hypodermis

The **hypodermis,** or *subcutaneous tissue*, is not actually a part of the skin, but it binds the dermis to underlying organs. The hypodermis is composed primarily of loose connective tissue and adipose cells interlaced with blood vessels (see fig. 7.1). Collagenous and elastic fibers reinforce the hypodermis—particularly on the palms and soles, where the skin is firmly attached to underlying structures. The amount of adipose tissue in the hypodermis varies with the region of the body and the sex, age, and nutritional state of the individual. Females generally have hypodermis that is about 8% thicker than males. This layer stores lipids, insulates and cushions the body, and regulates temperature.

decubitus: L. *decumbere*, lie down
ulcer: L. *ulcus*, sore
hypodermis: Gk. *hypo*, under; *derma*, skin

Figure 7.6

A decubitus ulcer on the medial surface of the ankle. Sites for decubitus ulcers occur most frequently in the skin overlying a bony projection, such as at the hip, ankle, heel, shoulder, or elbow.

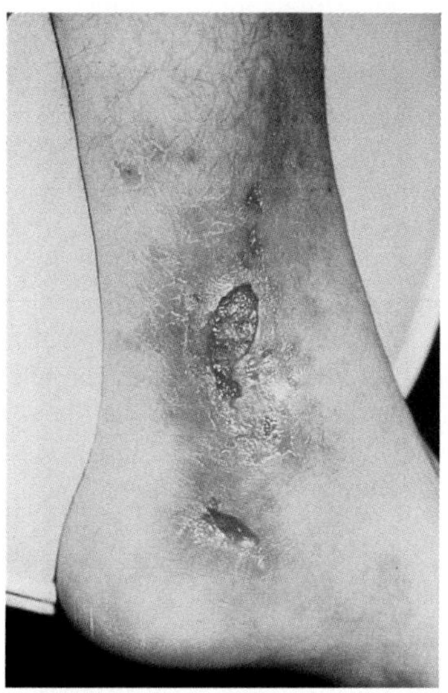

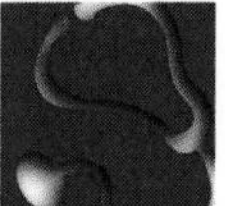

The hypodermis is the site for subcutaneous injections. Through a hypodermic needle, medicine can be administered to patients who are unconscious or uncooperative, and when oral medications are not practical. Subcutaneous devices to administer slow-release, low-dosage medications are now available. For example, insulin may be administered in this way to treat some forms of diabetes. Even a subcutaneous birth-control device (Norplant) is currently being marketed (see fig. 29.26)

Physiology of the Skin

The skin not only protects the body from pathogens and external injury, it is a highly dynamic organ that plays a key role in maintaining body homeostasis.

Physical Protection

The skin is a physical barrier to most microorganisms, water, and excessive sunlight (UV light). Oily secretions onto the surface of the skin form an acidic (pH 4.0–6.8) protective film that waterproofs the body and retards the growth of most pathogens. The protein keratin in the epidermis also waterproofs the skin, and the cornified outer layer (stratum corneum) resists scraping and keeps out microorganisms. As mentioned previously, exposure to UV light stimulates the melanocytes in the stratum basale to synthesize melanin, which absorbs and disperses sunlight. In addition, surface friction causes the epidermis to thicken by increasing the rate of mitosis in the cells of the stratum basale and stratum spinosum, resulting in the formation of a protective callus.

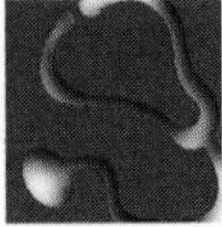

Regardless of skin pigmentation, everyone is susceptible to skin cancer if his or her exposure to sunlight is sufficiently intense. In fact, an estimated 800,000 people are diagnosed with skin cancer yearly in the United States, and approximately 9,300 of these people have the potentially life-threatening melanoma (cancer of melanocytes). Melanomas are usually termed malignant because they may spread rapidly. Sunscreens are advised for people who must be in direct sunlight for long periods of time.

The skin absorbs two wavelengths of ultraviolet rays from the sun: UV-A and UV-B. The DNA within the basal skin cells may be damaged as the sun's more dangerous UV-B rays penetrate the skin. Although it was once believed that UV-A rays were harmless, recent findings indicate that excessive exposure to UV-A rays may inhibit the DNA repair process that follows exposure to UV-B. Therefore, individuals who are exposed solely to UV-A rays in tanning salons still risk melanomas, since they will later be exposed to UV-B rays of sunlight when they are out-of-doors.

Hydroregulation

The thickened, keratinized, and cornified epidermis of the skin is adapted for continuous exposure to the air. In addition, the outer layers are dead and scalelike, and a protein-polysaccharide basement membrane adheres the stratum basale to the dermis. Human skin is virtually waterproof, protecting the body from desiccation (dehydration) on dry land and even from water absorption when immersed in water.

Thermoregulation

The skin plays a crucial role in the regulation of body temperature. Body heat comes from cellular metabolism, particularly in muscle cells as they maintain tone or a degree of tension. A normal body temperature of 37° C is maintained by the antagonistic effects of sweating and shivering, which involve feedback mechanisms. Excess heat is actually lost from the body in three ways, all involving the skin: (1) through radiation from dilated blood vessels, (2) through secretion and the evaporation of perspiration, and (3) through convection and the conduction of heat directly through the skin (fig. 7.7). The volume increases approximately 100–150 ml/day for each 1° C in body temperature. For each hour of hard physical work out-of-doors in the summertime, a person may produce 1 to 10 L of perspiration.

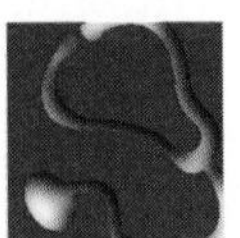

A serious danger of continued exposure to heat and excessive water and salt loss is heat exhaustion, characterized by nausea, weakness, dizziness, headache, and a decreased blood pressure. Heat stroke is similar to heat exhaustion, except that in heat stroke, sweating is inhibited (for reasons that are not clear), and body temperature rises. Convulsions, brain damage, and death may follow.

Excessive heat loss triggers a shivering response in muscles, which increases cellular metabolism and consequent heat production. Not only do skeletal muscles contract, but tiny smooth muscles called **arrector pili** (*ă-rek'tor pi'li*), which are attached to hair follicles (see fig. 7.9), contract involuntarily, causing goose bumps.

Figure 7.7

A thermogram of the hand showing differential heat radiation.

Hair and body fat are good insulators. Red and yellow indicate the warmest parts of the body. Blue, green, and white indicate the coolest.

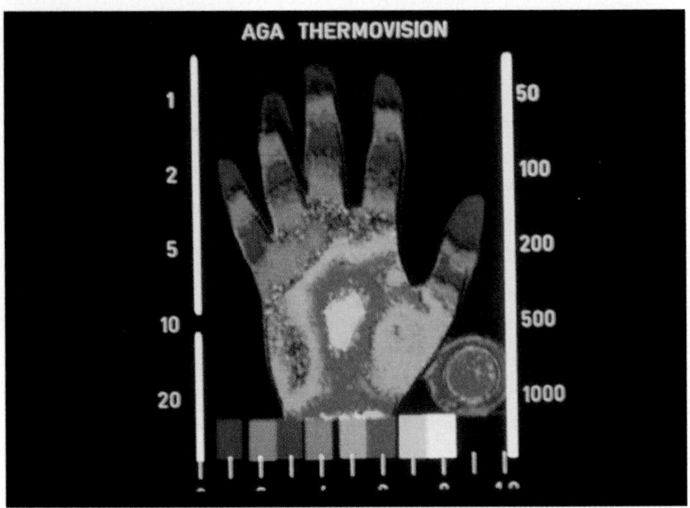

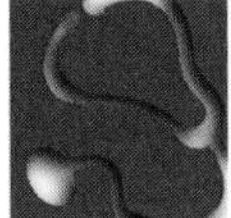

When the body's heat-producing mechanisms cannot keep pace with heat loss, *hypothermia* results. A lengthy exposure to temperatures below 20° C and to dampness may lead to this condition. This is why it is so important that a hiker, for example, dress appropriately for the weather conditions, especially on cool, rainy spring or fall days. The initial symptoms of hypothermia are numbness, paleness, delirium, and uncontrolled shivering. If the core temperature falls below 32° C (90° F), the heart loses its ability to pump blood and will go into fibrillation (erratic contractions). If the victim is not warmed, extreme drowsiness, coma, and death follow.

Cutaneous Absorption

Because of the effective protective barriers of the skin already described, cutaneous (through the skin) absorption is limited. Some gases, such as oxygen and carbon dioxide, may pass through the skin and enter the bloodstream. Small amounts of UV light, necessary for synthesis of vitamin D, are absorbed readily. The skin is no barrier to steroid hormones, such as cortisol, and to fat-soluble vitamins (A, D, E, and K). Of clinical consideration is the fact that certain chemicals such as lipid-soluble toxins and pesticides can easily enter the body through the skin.

---

pili: L. *pilus*, hair

(a)

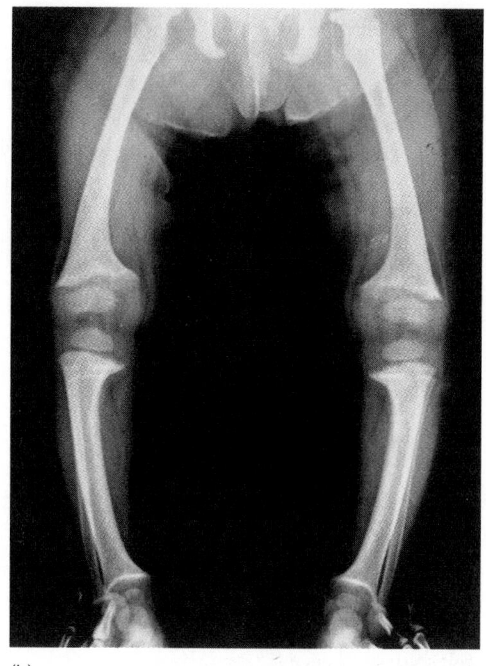

(b)

Figure 7.8

Rickets.
(*a*) A case of rickets in a child who lives in a village in Nepal, where the people reside in windowless huts. During the 5-to-6-month rainy season, the children are kept indoors. (*b*) A radiograph (X ray) of rickets in a 10-month-old child. Rickets develop from improper diets and also from lack of UV light, which is needed to synthesize vitamin D.

Synthesis

The integumentary system synthesizes melanin and keratin, which remain in the skin, and vitamin D, which is used elsewhere in the body. The integumentary cells contain a compound called *dehydrocholesterol* (*de-hi″dro-kŏ-les′tă-rol*), from which they synthesize vitamin D in the presence of UV light. Only small amounts of UV light are necessary for vitamin D synthesis, but these amounts are very important to a growing child (fig. 7.8). Synthesized vitamin D enters the blood and helps regulate the metabolism of calcium and phosphorus, which are important for development of strong and healthy bones. *Rickets* is a disease caused by vitamin D deficiency.

Sensory Reception

Highly specialized sensory receptors (see chapter 18) that respond to thermal (heat and cold), mechanical (pressure, touch, and vibration), and noxious (pain) stimuli are located throughout the dermis and hypodermis of the integument. These receptors, referred to as **cutaneous receptors,** are abundant in the skin in parts of the face, the palms and fingers of the hands, the soles of the feet, and the genitalia. They are less abundant along the back and on the back of the neck and are sparse in the skin over joints, especially the elbow. Generally speaking, the thinner the skin, the greater the sensitivity.

Communication

Humans are highly social animals, and the skin plays an important role in communication. Various emotions, such as anger or embarrassment, may be reflected in changes of skin color. The contraction of specific facial muscles produces facial expressions that convey an array of emotions, including surprise, happiness, sadness, and despair. Secretions from certain integumentary glands have odors that frequently elicit subconscious responses from others who detect them.

Accessory Structures of the Skin

Hair, nails, and integumentary glands form from the epidermal layer and are therefore of ectodermal derivation. Hair and nails are structural features of the integument and have a limited functional role. By contrast, integumentary glands are extremely important in body defense and the maintenance of homeostasis.

Hair

The presence of **hair** is characteristic of all mammals, but its distribution, function, density, and texture varies across mammalian species. Humans are relatively hairless, with only the scalp, face, pubis, and axillae being densely haired. Men and women have about the same density of hair on their bodies, but it is generally more obvious on men due to male hormones (see chapters 19 and 28). Certain structures and regions of the body are hairless, such as the palms, soles, lips, nipples, penis, and parts of the female genitalia.

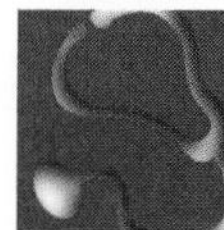

Hirsutism (*her′soo-tiz″em*) is a condition of excessive body and facial hair, especially in women. It may be a genetic expression, as in certain ethnic groups, or result from a metabolic disorder, usually endocrine in nature. Hirsutism occurs in some women as they experience hormonal changes during menopause (see chapter 29). Various treatments for hirsutism include hormonal injections and electrolysis to permanently destroy selected hair follicles.

hirsutism: L. *hirsutus*, shaggy

The primary function of hair is protection, even though its effectiveness is limited. Hair on the scalp and eyebrows protects against sunlight. The eyelashes and the hair in the nostrils protect against airborne particles. Hair on the scalp may also protect against mechanical injury. Some secondary functions of hair are to distinguish individuals and to serve as a sexual attractant.

Each hair consists of a diagonally positioned **shaft, root,** and **bulb** (fig. 7.9). The shaft is the visible, but dead, portion of the hair projecting above the surface of the skin. The bulb is the enlarged base of the root within the **hair follicle.** Each hair develops from stratum basale cells within the bulb of the hair, where nutrients are received from dermal blood vessels. As the cells divide, they are pushed away from the nutrient supply toward the surface, and cellular death and keratinization occur. In a healthy person, hair grows at the rate of approximately 1 mm every 3 days. As the hair becomes longer, however, it goes through a resting period, during which there is minimal growth.

The life span of a hair varies from 3 to 4 months for an eyelash to 3 to 4 years for a scalp hair. Each hair is replaced by a new hair that grows from the base of the follicle and pushes the old hair out. Between 10 and 100 hairs are lost daily. Baldness results when hair is lost and not replaced. This condition may be disease-related, but it is generally inherited and most frequently occurs in males because of genetic influences combined with the action of the male sex hormone *testosterone* (*tes-tos'tĕ-rōn*). No treatment is effective in reversing genetic baldness; however, flaps or plugs of skin containing healthy follicles from hairy parts of the body can be grafted onto hairless regions.

Three layers can be observed in hair that is cut in cross section. The inner **medulla** (*mĕ-dul'ă*) is composed of loosely arranged cells separated by many air cells. The thick **cortex** surrounding the medulla consists of hardened, tightly packed cells. The **cuticle** covers the cortex and forms the toughened outer layer of the hair. Cells of the cuticle have serrated edges that give a hair a scaly appearance when observed under a dissecting scope.

---

medulla: L. *medulla*, marrow
cortex: L. *cortex*, bark
cuticle: L. *cuticula*, small skin

Figure 7.9

The structure of hair and the hair follicle.

(*a*) A photomicrograph (63×) of the bulb and root of a hair within a hair follicle. (*b*) A scanning electron micrograph (280×) of a hair as it extends from a follicle. (*c*) A diagram of hair, a hair follicle, a sebaceous gland, and an arrector pili muscle.

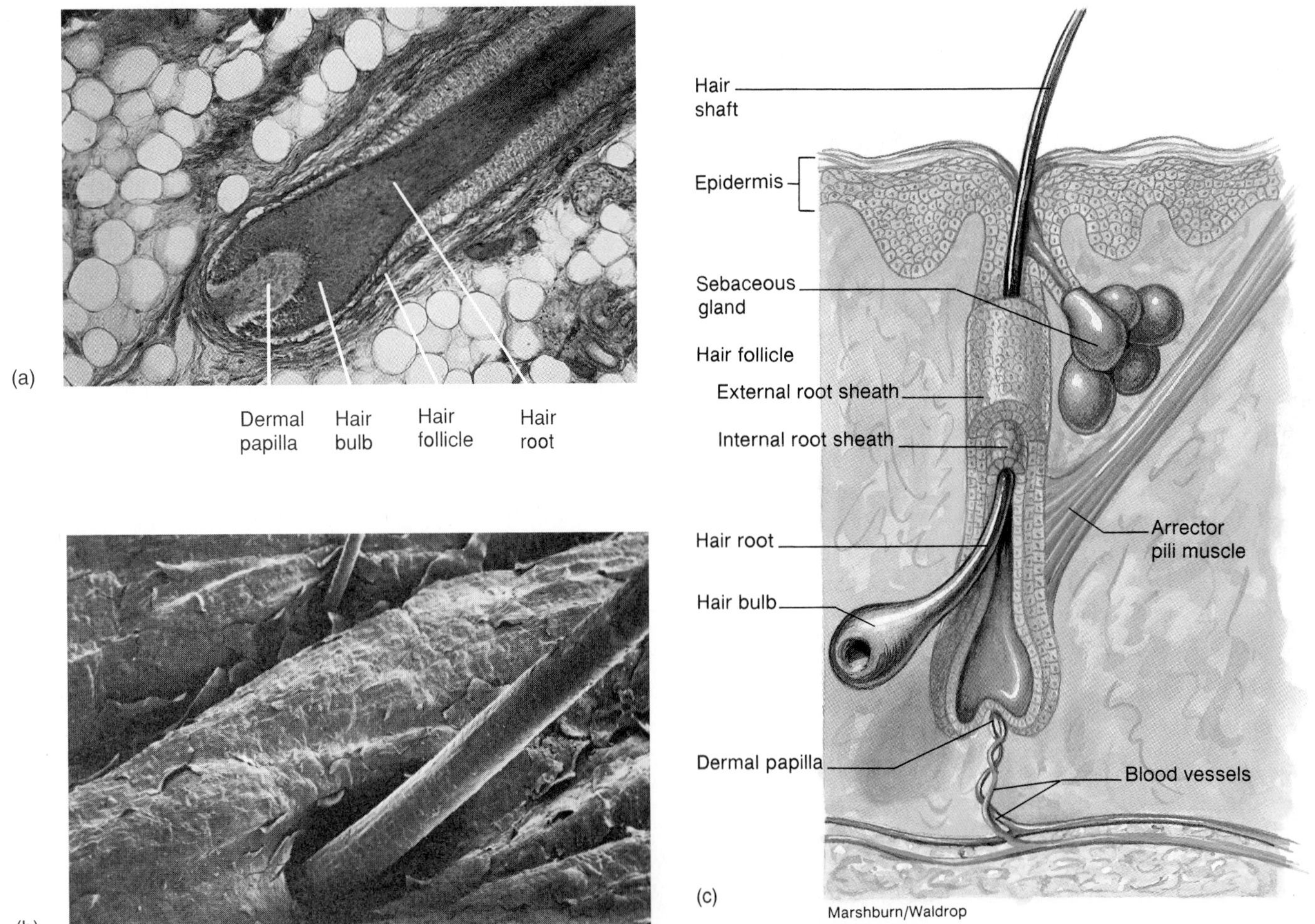

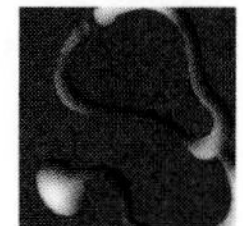

People exposed to heavy metals, such as lead, mercury, arsenic, or cadmium, will have concentrations of these metals in their hair that are 10 times as great as those found in their blood or urine. Because of this, hair samples can be extremely important in certain diagnostic tests.

Even evidence of certain metabolic diseases or nutritional deficiencies may be detected in hair samples. For example, the hair of children with cystic fibrosis will be deficient in calcium and display excessive sodium. There is a deficiency of zinc in the hair of malnourished individuals.

Hair color is determined by the type and amount of pigment produced in the stratum basale at the base of the hair follicle. Varying amounts of melanin produce hair ranging in color from blond to brunette to black. The more abundant the melanin, the darker the hair. A pigment with an iron base (*trichosiderin*) produces red hair. Gray or white hair is the result of a lack of pigment production and air spaces within the layers of the shaft of the hair. The texture of hair is determined by its cross-sectional shape: straight hair is round, wavy hair is oval, and kinky hair is flat.

Sebaceous glands and arrector pili muscles are attached to the hair follicles (see fig. 7.9c). The arrector pili muscles are involuntary, responding to thermal or psychological stimuli. When they contract, the hair is pulled into a more vertical position, causing goose bumps.

Humans have three distinct kinds of hair:

1. **Lanugo.** Lanugo (*la-noo'go*) is a fine, silky fetal hair that appears during the last trimester of development. It is usually seen only on premature infants.
2. **Angora.** Angora hair grows continuously. It is found on the scalp and on the male face.
3. **Definitive.** Definitive hair grows to a certain length and then stops. It is the most common type of hair. Eyelashes, eyebrows, and pubic and axillary hair are examples.

lanugo: L. *lana*, wool

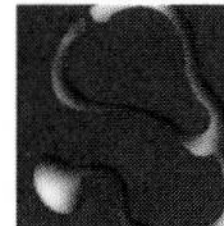

Anthropologists have referred to humans as the naked apes because of our relative hairlessness. The clothing that we wear over the exposed surface areas of our bodies insulates and protects us, just as hair or fur does for other mammals. However, the nakedness of our skin does lead to some problems. Skin cancer occurs frequently in humans, particularly in regions of the skin exposed to the sun. *Acne,* another problem unique to humans, is partly related to the fact that hair is not present to dissipate the oily secretion from the sebaceous glands.

Nails

The **nails** on the ends of the fingers and toes are formed from the compressed outer layer (stratum corneum) of the epidermis. The hardness of the nail is due to the dense keratin fibrils running parallel between the cells. Both fingernails and toenails protect the digits, and fingernails also aid in grasping and picking up small objects.

Each nail consists of a **body, free border,** and **hidden border** (fig. 7.10). The platelike body of the nail rests on a **nail bed,** which is actually the stratum spinosum of the epidermis. The body and nail bed appear pinkish because of the underlying vascular tissue. The sides of the nail body are protected by a **nail fold,** and the furrow between the sides and body is the **nail groove.** The free border of the nail extends over a thickened region of the stratum corneum called the

Figure 7.10

The fingertip and the associate structures of the nail.
(*a*) A diagram of a dissected nail, and (*b*) a photomicrograph of a nail from a fetus (3.5×).

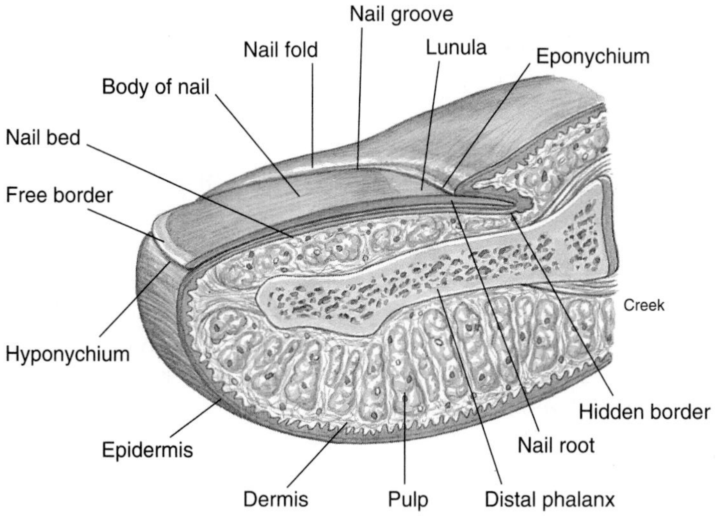

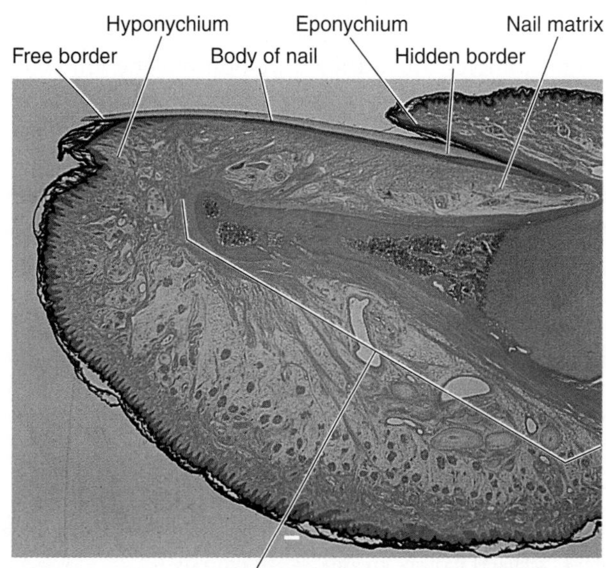

hyponychium (*hi″pŏ-ni′kē-um*). The hidden border of the nail is attached at the base.

An **eponychium** (cuticle) covers the hidden border of the nail. The eponychium frequently splits, causing a hangnail. The growth area of the nail is the **nail matrix.** A small part of the nail matrix, the **lunula** (*loo′nyoo-la*), can be seen as a white, half-moon–shaped area near the eponychium of the nail.

The nail grows by the transformation of the superficial cells of the nail matrix into nail cells. These harder, transparent cells are then pushed forward over the strata basale and spinosum of the nail bed. Fingernails grow at the rate of approximately 1 mm per week. The growth rate of toenails is somewhat slower.

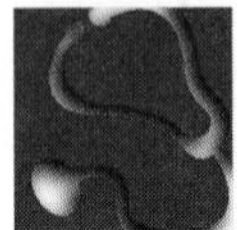

The condition of nails can be an indication of a person's general health and well-being. Nails should appear pinkish, showing the rich vascular capillaries beneath the translucent nail. A yellowish hue may indicate certain glandular dysfunctions or nutritional deficiencies. Split nails may also be caused by nutritional deficiencies. A prominent bluish tint may indicate improper oxygenation of the blood. Spoon nails (concave body) may be the result of iron-deficiency anemia, and "clubbing" at the base of the nail may be caused by lung cancer. Dirty or ragged nails may indicate poor personal hygiene, and chewed nails may suggest emotional problems.

Glands

Although they originate in the epidermal layer, all of the glands of the skin are located in the dermis, where they are physically supported and receive nutrients. Glands of the skin are referred to as *exocrine* because they are externally excreting glands that either release their secretions directly or through ducts. The glands of the skin are of three basic types: *sebaceous*, (*sĕ-ba′shus*), *sudoriferous* (*soo′dor-if′er-us*), and *ceruminous* (*sĕ-roo′mĭ-nus*).

Sebaceous Glands

Commonly called oil glands, sebaceous glands are associated with hair follicles, since they develop from the follicular epithelium of the hair. They are holocrine glands (chapter 6) that secrete **sebum** (*se′bum*) onto the shaft of the hair (fig. 7.11). Sebum, which consists mainly of lipids, is dispersed along the shaft of the hair to the surface of the skin, where it lubricates and waterproofs the stratum corneum and also prevents the hair from becoming brittle. If the ducts of sebaceous glands become blocked for some reason, the glands may become infected, resulting in acne. Sex hormones regulate the production and secretion of sebum, and hyperactivity of sebaceous glands can result in serious acne problems, particularly during teenage years.

hyponychium: Gk. *hypo*, under; *onyx*, nail
lunula: L. *lunula*, small moon
sebum: L. *sebum*, tallow or grease

Figure 7.11
Types of skin glands.

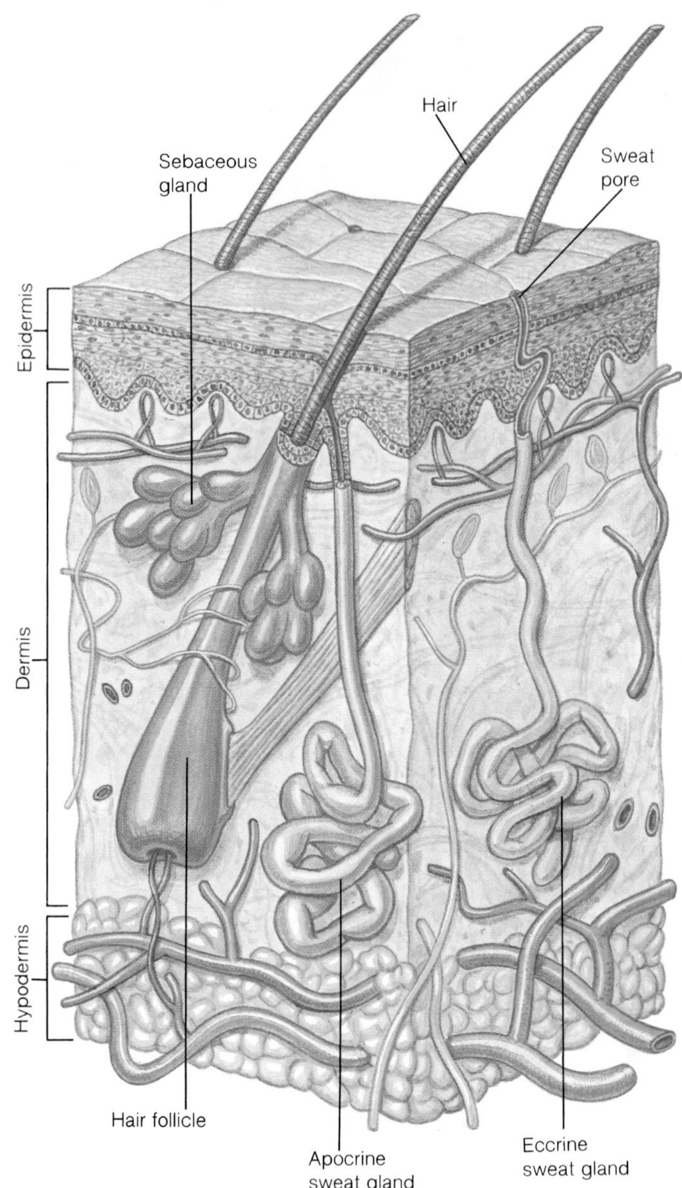

Sudoriferous Glands

Commonly called sweat glands, **sudoriferous glands** secrete perspiration, or sweat, onto the surface of the skin. Perspiration is composed of water, salts, urea, and uric acid. It serves not only for evaporative cooling, but also for the excretion of certain wastes. Sweat glands are most numerous on the palms, soles, axillary and pubic regions, and the forehead. They are coiled and tubular (fig. 7.11) and are of two types: eccrine (*ek′rin*) and apocrine (*ap′ŏ-krin*) sweat glands.

sudoriferous: L. *sudorifer*, sweat; *ferre*, to bear

1. **Eccrine sweat glands** are widely distributed over the body, especially on the forehead, back, palms, and soles. These glands are formed before birth and function in evaporative cooling (fig. 7.11).
2. **Apocrine sweat glands** are much larger than eccrine glands. They are found in the axillary and pubic regions, where they secrete into hair follicles. Apocrine glands are not functional until puberty, and their odoriferous secretion is thought to act as a sexual attractant.

Mammary glands, found within the breasts, are specialized sudoriferous glands that secrete milk during lactation (see chapter 29). The breasts of the human female reach their greatest development during the childbearing years under the stimulus of pituitary and ovarian hormones.

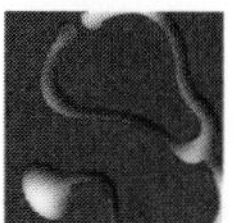

Good routine hygiene is very important for health and social reasons. Washing away the dried residue of perspiration and sebum eliminates dirt. Excessive bathing, however, can wash off the natural sebum and dry the skin, causing it to itch or crack. The commercial lotions used for dry skin are, for the most part, refined and perfumed lanolin, which is sebum from sheep.

Ceruminous Glands

These highly specialized glands are found only in the external acoustic canal (ear canal). They secrete **cerumen** (*sĕ-roo'men*), or earwax. Cerumen is a water and insect repellent, and also keeps the tympanic membrane (eardrum) pliable. Excessive amounts of cerumen may interfere with hearing.

Clinical Considerations

The skin is a buffer against the external environment and is therefore subject to a variety of disease-causing microorganisms and physical assaults. A few of the many diseases and disorders of the integumentary system are briefly discussed here.

Inflammatory Conditions (Dermatitis)

Inflammatory skin disorders are caused by immunologic hypersensitivity, infectious agents, poor circulation, or exposure to environmental assaults such as wind, sunlight, or chemicals. Some people are allergic to certain foreign proteins and, because of this inherited predisposition, experience such hypersensitive reactions as asthma, hay fever, hives, drug and food allergies, and eczema. **Lesions,** as applied to inflammatory conditions, are defined as more or less circumscribed pathologic changes in the tissue. Some of the more common inflammatory skin disorders and their usual sites are illustrated in figure 7.12.

There are also a number of *infectious diseases* of the skin, which is not surprising considering the highly social and communal animals we are. Most of these diseases can now be prevented, but too frequently people fail to take appropriate precautionary measures. Infectious diseases involving the skin include childhood viral infections (measles and chicken pox); bacteria, such as staphylococcus (impetigo); sexually transmitted diseases; leprosy; fungi (ringworm, athlete's foot, candida); and mites (scabies).

Neoplasms

Both benign and malignant neoplastic conditions or diseases are common in the skin. *Pigmented moles* (nevi), for example, are a type of benign neoplastic growth of melanocytes. *Dermal cysts* and *benign viral infections* are also common. *Warts* are virally caused abnormal growths of tissue that frequently occur on the hands and feet. These warts are usually treated effectively with liquid nitrogen or acid. A different type of wart, called a *venereal wart,* occurs in the anogenital region of affected sexual partners. Risk factors for cervical cancer may be linked to venereal warts, so these warts are treated aggressively with chemicals, cryosurgery, cautery, or laser therapy.

Skin cancer is the most common malignancy in the United States. As shown in figure 7.13, there are three frequently encountered types. **Basal cell carcinoma,** the most common skin cancer, accounts for about 70% of total cases. It usually occurs on the body where exposure to sunlight is the greatest—on the face and arms. This type of cancer arises from cells in the stratum basale. It appears first on the surface of the skin as a small, shiny bump. As the bump enlarges, it often develops a central crater that erodes, crusts, and bleeds. Fortunately, there is little danger that it will spread (metastasize) to other body areas. These carcinomas are usually treated by excision (surgical removal).

Squamous cell carcinoma arises from cells immediately superficial to the stratum basale. Normally, these cells undergo very little division, but in squamous cell carcinoma they continue to divide as they produce keratin. The result is usually a firm, red keratinized tumor, confined to the epidermis. If untreated, however, it may invade the dermis and metastasize. Treatment usually consists of excision and radiation therapy.

Malignant melanoma, the most life-threatening form of skin cancer, arises from the melanocytes located in the stratum basale. Often, it begins as a small molelike growth, which enlarges, changes color, becomes ulcerated, and bleeds easily. Metastasis occurs quickly, and unless treated early—usually by extended excision and radiation therapy—this cancer is often fatal.

Burns

A burn is an epithelial injury caused by contact with thermal, radioactive, chemical, or electrical agents. Burns generally

cerumen: L. *cera,* wax

neoplasm: Gk. *neo,* new; *plasma,* something formed
benign: L. *benignus,* good-natured
malignant: L. *malignus,* acting from malice

Figure 7.12

Common inflammatory skin disorders and their usual site of occurrence.

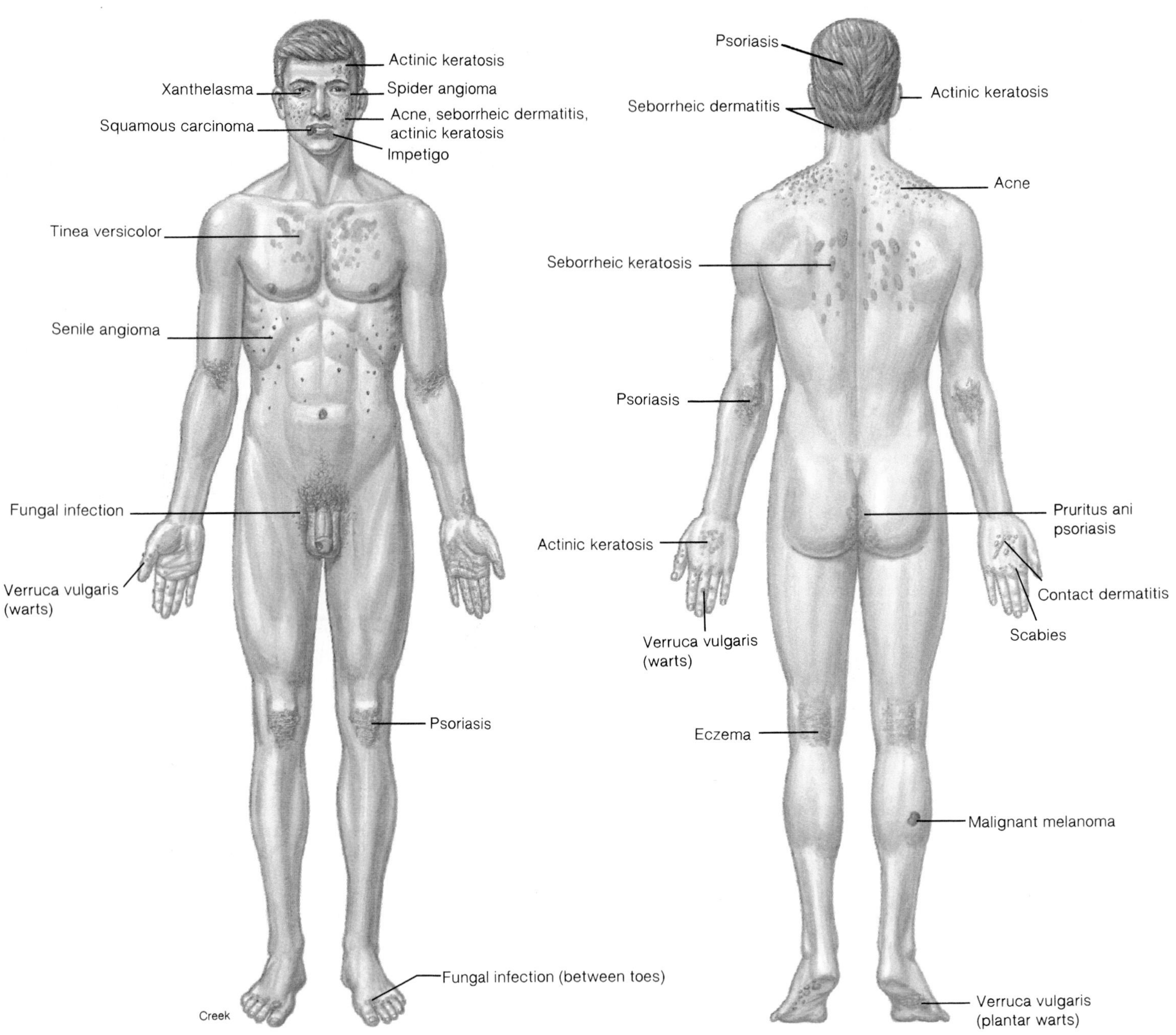

occur on the skin, but they can involve the linings of the respiratory and GI tracts. The extent and location of a burn is frequently less important than the degree to which it disrupts body homeostasis. Burns that have a **local effect** (local tissue destruction) are not as serious as those that have a **systemic effect.** Systemic effects directly or indirectly involve the entire body and are a threat to life. Possible systemic effects include body dehydration, shock, reduced circulation and urine production, and bacterial infections.

Burns are classified as first degree, second degree, and third degree based on their severity (fig. 7.14). In **first-degree burns,** the epidermal layers of the skin are damaged and symptoms are restricted to local effects such as redness, pain, and edema (swelling). A shedding of the surface layers (desquamation) generally follows in a few days. A sunburn is an example. **Second-degree burns** involve both the epidermis and dermis. Blisters appear and recovery is usually complete, although slow. **Third-degree burns** destroy the entire thickness of the skin and frequently some of the underlying connective tissue and muscle. The skin appears waxy or charred and is insensitive to touch. As a result, ulcerating wounds develop, and the body attempts to heal itself by forming scar tissue. Skin grafts are frequently used to assist recovery.

Figure 7.13

Types of skin cancer.

(*a*) Squamous cell carcinoma, (*b*) malignant melanoma, and (*c*) basal cell carcinoma.

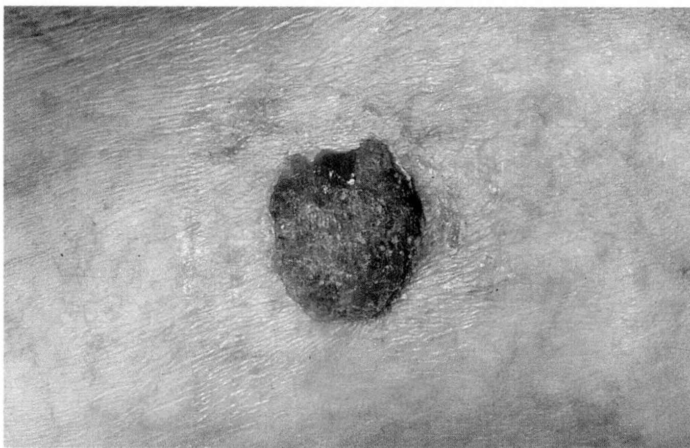

(a)

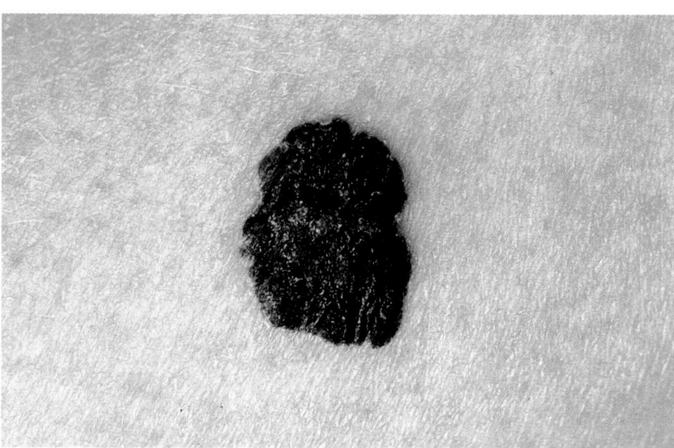

(b)

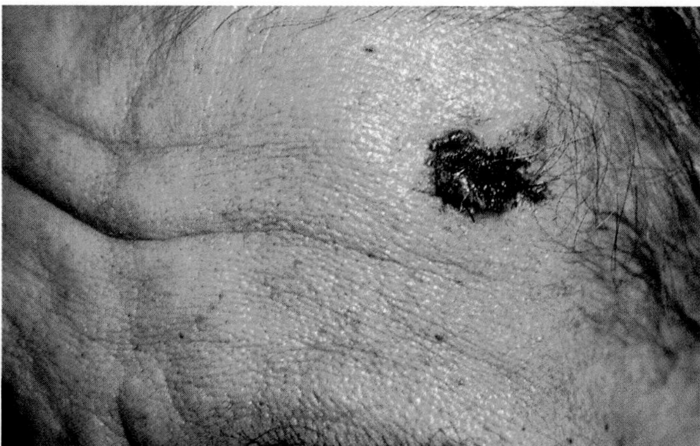

(c)

Figure 7.14

The classification of burns.

(*a*) First-degree burns involve the epidermis and are characterized by redness, pain, and edema—such as with a sunburn. (*b*) Second-degree burns involve the epidermis and dermis and are characterized by intense pain, redness, and blistering. (*c*) Third-degree burns destroy the entire skin and frequently expose the underlying organs. The skin is charred and numb and does not protect against fluid loss.

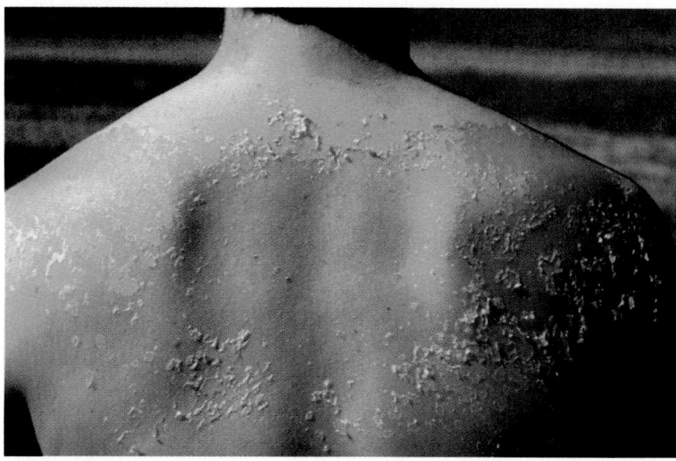

(a)

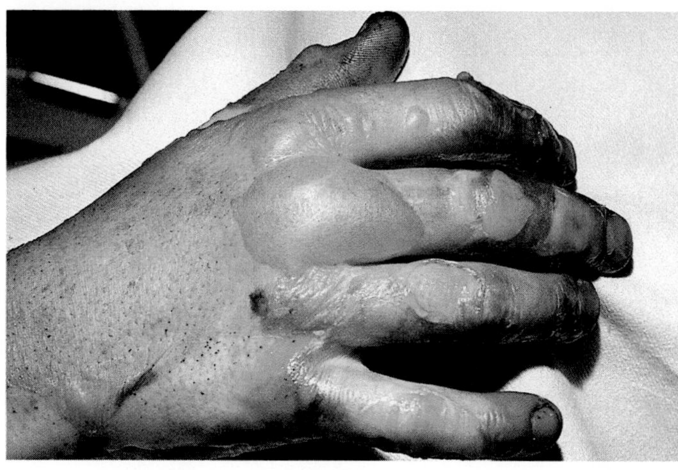

(b)

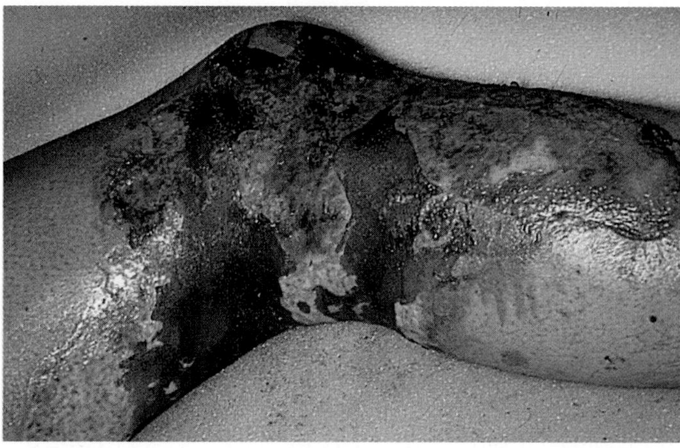

(c)

Under Development

Development of the Integumentary System

Both the ectodermal and mesodermal germ layers (see chapter 6) function in the formation of the structures of the integumentary system. The epidermis and the hair, glands, and nails of the skin develop from the ectodermal germ layer (figs. 1 and 2). The dermis develops from a thickened layer of undifferentiated mesoderm called **mesenchyme** (*mez'en-kim'*).

By 6 weeks, the ectodermal layer has differentiated into an outer flattened periderm and an inner, cuboidal **germinal** (*basal*) **layer** in contact with the mesenchyme. The periderm eventually sloughs off, forming the **vernix caseosa** (*ka"se-o'să*), a cheeselike protective coat that covers the skin of the fetus.

By 11 weeks, the mesenchymal cells below the germinal cells have differentiated into the distinct collagenous and elastic connective tissue fibers of the dermis. The tensile properties of these fibers cause a buckling of the epidermis and the formation of dermal papillae. During the early fetal period (about 10 weeks), specialized neural crest cells called **melanoblasts** migrate into the developing dermis and differentiate into **melanocytes.** The melanocytes soon migrate to the germinal layer of the epidermis, where they produce the pigment **melanin,** which colors the epidermis.

Before hair can form, a **hair follicle** must be present. Each hair follicle begins to develop at about 12 weeks (fig. 2) as a mass of germinal cells, called a **hair bud,** proliferates into the underlying mesenchyme. As the hair bud becomes club-shaped, it is referred to as a **hair bulb.** The hair follicle, which physically supports and provides nourishment to the hair, is derived from specialized mesenchyme called the **hair papilla,** which is localized around the hair bulb, and from the epithelial cells of the hair bulb called the **hair matrix.** Continuous mitotic activity in the epithelial cells of the hair bulb results in the growth of the hair.

Sebaceous glands and sweat glands are the two principal types of integumentary glands. Both develop from the germinal layer of the epidermis (fig. 2). Sebaceous glands develop as proliferations from the sides of the developing hair follicle. Sweat glands become coiled as the secretory portion of the developing gland proliferates into the dermal mesenchyme. Mammary glands are modified sweat glands that develop in the skin of the anterior thoracic region.

mesenchyme: Gk. *mesos*, middle; *enchyma*, infusion
periderm: Gk. *peri*, around; *derm*, skin
vernix caseosa: L. *vernix*, varnish; *caseus*, cheese

As a way of estimating the extent of damaged skin suffered in burned patients, the *rule of nines* (fig. 7.15) is often applied. The surface area of the body is divided into regions, each of which accounts for about 9% (or a multiple of 9%) of the total skin body surface. An estimation of the percentage of surface area damaged is important in treating with intravenous fluid, which replaces the fluids lost from tissue damage.

Frostbite

Frostbite is a local destruction of the skin resulting from freezing. Like burns, frostbite is classified by its degree of severity: first degree, second degree, and third degree. In **first-degree frostbite** the skin will appear cyanotic (bluish) and swollen. Vesicle formation and hyperemia (engorgement with blood) are symptoms of **second-degree frostbite.** As the effected area is warmed, there will be further swelling, and the skin will redden and blister. In **third-degree frostbite,** there will be severe edema, some bleeding, and numbness, followed by intense throbbing pain and necrosis of the affected tissue. Gangrene will result from untreated third-degree frostbite.

Skin Grafts

If extensive areas of the stratum basale of the epidermis are destroyed in second-degree or third-degree burns or frostbite, new skin cannot grow back. In order for this type of wound to heal, a skin graft must be performed. A **skin graft** is a segment of skin that has been excised from a *donor site* and transplanted to the *recipient site*, or *graft bed*. As stated in chapter 6, an *autograft* is the most successful type of tissue transplant. It involves taking a thin sheet of healthy epidermis from a donor site of the burn or frostbite patient and moving it to the recipient site. A *heterotransplant* (*xenograft*—between two

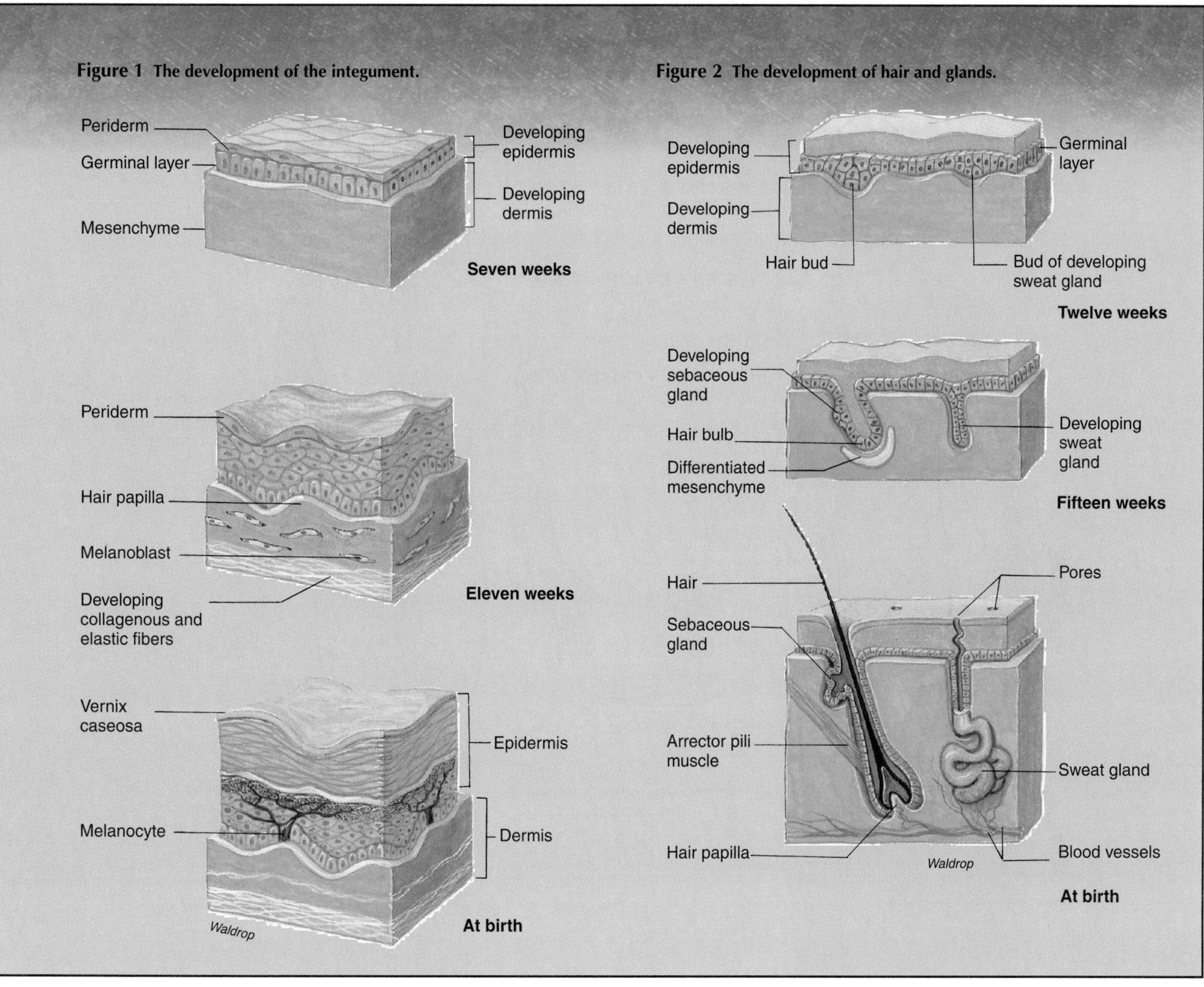

different species) can serve as a temporary treatment to prevent infection and fluid loss.

Synthetic skin fabricated from animal tissue bonded to a silicone film may be used on a patient who is extensively burned. The process includes seeding the synthetic skin with basal skin cells obtained from healthy locations on the patient. This treatment eliminates some of the problems of skin grafting—for example, additional trauma, widespread scarring, and rejection, as in the case of skin obtained from a cadaver.

Wound Healing

The skin effectively protects against many abrasions, but if a wound does occur, a sequential chain of events promotes rapid healing. The process of wound healing depends on the extent and severity of the injury. Trauma to the epidermal layers stimulates an increased mitotic activity in the stratum basale, whereas injuries that extend to the dermis or subcutaneous tissue elicit activity throughout the body as well as within the wound itself. General body responses include a temporary elevation of temperature and pulse rate.

In an open wound (fig. 7.16), blood vessels are broken and bleeding occurs. Through the action of blood **platelets** (*plāt′letz*) and protein molecules, called **fibrinogen** (*fi-brin′ŏ-jen*), a clot forms and soon blocks the flow of blood and entry of pathogens. A scab forms and covers and protects the damaged area. Mechanisms are activated to destroy bacteria, dispose of dead or injured cells, and isolate the injured area. These responses are collectively referred to as *inflammation* and are characterized by redness, heat, edema, and pain. Inflammation is a response that confines the injury and promotes healing.

The next step in healing is the differentiation of binding **fibroblasts** from connective tissue, forming **fibrin** at the

Figure 7.15

The extent of burns as estimated by the rule of nines.
(*a*) Anterior and (*b*) posterior.

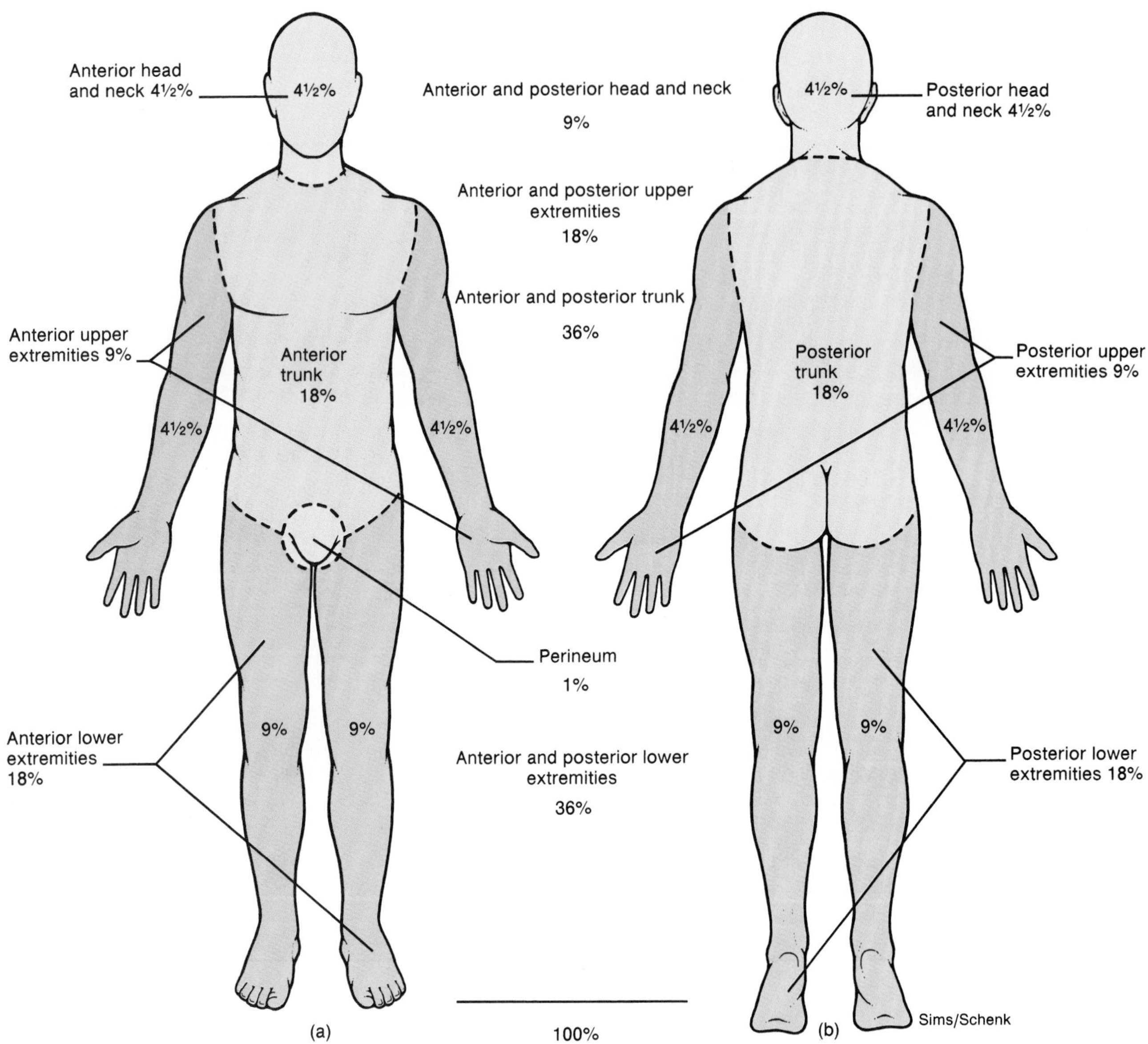

wound margins. Together with new branches from surrounding blood vessels, **granulation tissue** is formed. Phagocytic cells migrate into the wound and ingest dead cells and foreign debris. Eventually the damaged area is repaired and the protective scab is sloughed off.

If the wound is severe enough, the granulation tissue may develop into **scar tissue.** The collagenous fibers of scar tissue are more dense than those of normal tissue, and scar tissue has no stratified squamous or epidermal layer. Scar tissue also has fewer blood vessels than normal skin, and may lack hair, glands, and sensory receptors. The closer the edges of a wound, the less granulation tissue develops and the less obvious a scar. This is one reason for suturing a large break in the skin.

Aging

As the skin ages, it becomes thin, dry, and begins to lose its elasticity. Collagenous fibers in the dermis become thicker and stiffer, and the amount of adipose tissue in the hypodermis diminishes, making it thinner. Skinfold measurements indicate that the diminution of the hypodermis begins at about

Figure 7.16

The process of wound healing.

(*a*) A penetrating wound into the dermis ruptures blood vessels. (*b*) Blood cells, fibrinogen, and fibrin flow out of the wound. (*c*) Vessels constrict, and a clot blocks the flow of blood. (*d*) A protective scab is formed from the clot, and granulation tissue forms within the site of the wound. (*e*) The scab sloughs off as the epidermal layers regenerate.

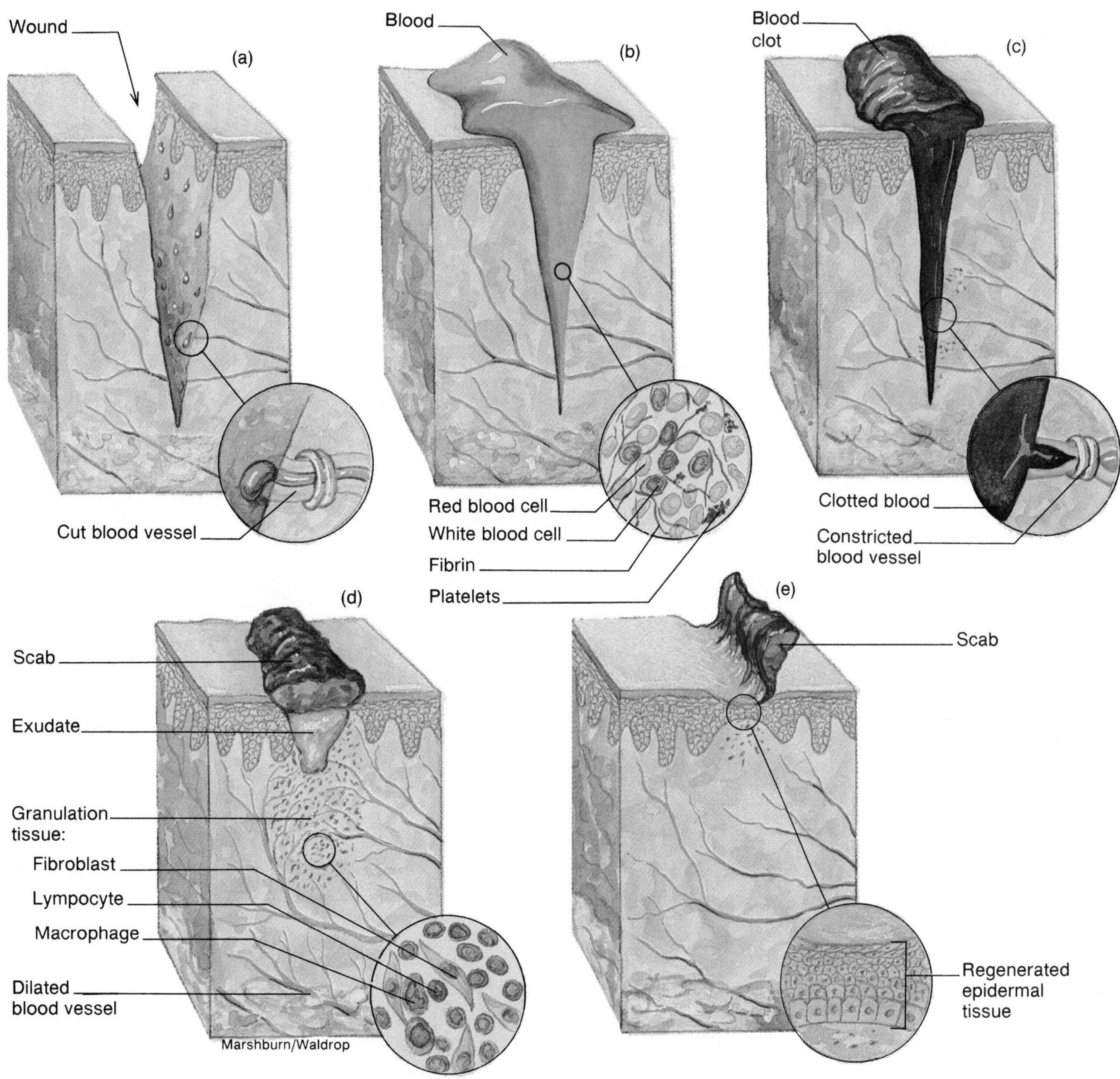

the age of 45. With a loss of elasticity and a reduction in the thickness of the hypodermis, wrinkling, or the permanent infolding of the skin, becomes apparent (fig. 7.17).

During the aging of the skin, the number and activity of hair follicles, sweat glands, and sebaceous glands also diminish. Consequently, there is a marked thinning of scalp hair and hair on the extremities, reduced sweating, and decreased sebum production. Since elderly people cannot perspire as freely as they once did, they are more likely to complain about heat and are at greater risk for heat exhaustion. They also become more sensitive to cold because of the loss of insulating adipose tissue and diminished circulation. A decrease in the production of sebum causes the skin to dry and crack frequently.

The integument of an elderly person is not well protected from the sun because of thinning, and melanocytes that produce melanin gradually atrophy. The loss of melanocytes accounts for graying of the hair and pallor of the skin.

Figure 7.17

Aging of the skin.
As the skin ages, elasticity diminishes and wrinkles appear.

Figure 7.18

Albinism.
The albino individual in this photograph has melanocytes within his skin, but he lacks the ability to synthesize melanin as a result of a mutant gene.

Clinical Investigation Answer

The blistering and erythema characteristic of second-degree burns is a manifestation of intact and functioning blood vessels, which exist in abundance within the spared dermis. In third-degree burns, the entire dermis and its vasculature is destroyed, thus explaining the absence of these findings. In addition, nerve endings and other nerve end organs that reside in the dermis are destroyed in third-degree burns, resulting in a desensitized area. By contrast, significant numbers of these structures are spared and functional in second-degree burns, thus preserving sensation—including pain. The third-degree burn areas will all require skin grafting in order to prevent infection, one of the skin's most vital functions. In second-degree burns, the spared dermis serves somewhat of a barrier to bacteria. Consequently, skin grafting is usually unnecessary, especially if sufficient numbers of skin adnexa (hair follicles, sweat glands, and so forth), which generally lie deep within the dermis, are spared. These structures serve as starting points for regeneration of surface epithelium and skin organs.

Important Clinical Terminology

acne An inflammatory condition of sebaceous glands. Acne is affected by gonadal hormones and is therefore most common during puberty and adolescence. Pimples and blackheads on the face, chest, and back are expressions of this condition.

albinism (*al'bi-niz"em*) A congenital condition in which the pigment of the skin, hair, and eyes is deficient as a result of a metabolic block in the synthesis of melanin (fig. 7.18).

alopecia (*al"o-pe'she-a*) Loss of hair; baldness. Male pattern baldness is genetically determined and irreversible. Other types of hair loss may respond to treatment.

athlete's foot (tinea pedis) A fungus disease of the skin of the foot.

blister A collection of fluid between the epidermis and dermis caused by excessive friction or a burn.

boil (furuncle) A localized bacterial infection originating in a hair follicle or skin gland.

carbuncle A bacterial infection similar to a boil, except that a carbuncle infects the subcutaneous tissues.

cold sore (fever blister) A lesion on the lip or oral mucous membrane caused by type I herpes simplex virus (HSV) and transmitted by oral or respiratory exposure.

comedo (*kom'e-do*) A plug of sebum and epithelial debris in the hair follicle and excretory duct of the sebaceous gland; also called a *blackhead* or *whitehead*.

corn A type of callus that is localized on the foot, usually over toe joints.

dandruff Common dandruff is the continual shedding of epidermal cells of the scalp; it can be removed by normal washing and brushing of the hair. Abnormal dandruff may be caused by certain skin diseases, such as seborrhea or psoriasis.

decubitus (*de-kyoo'bi-tus*) ulcer A bedsore—an exposed ulcer caused by a continual pressure that restricts dermal blood flow to a localized portion of the skin (see fig. 7.6).

dermabrasion A procedure for removing tattoos or acne scars by high-speed sanding or scrubbing.

dermatitis An inflammation of the skin.

dermatology A specialty of medicine concerned with the study of the skin—its anatomy, physiology, histopathology, and the relationship of cutaneous lesions to systemic disease.

eczema (*eg'zĕ-mă*) A noncontagious inflammatory condition of the skin producing red, itching, vesicular lesions that may be crusty or scaly.

erythema (*er"ĭ-the'mă*) Redness of the skin, generally as a result of vascular trauma.

furuncle A boil—a localized abscess resulting from an infected hair follicle.

gangrene Necrosis of tissue resulting from the obstruction of blood flow. It may be localized or extensive and may be infected secondarily with anaerobic microorganisms.

hives (urticaria) (*ur"ti-ka're'a*) A skin eruption of reddish wheals usually accompanied by extreme itching. It may be caused by drugs, food, insect bites, inhalants, emotional stress, or exposure to heat or cold.

impetigo A contagious skin infection that results in lesions followed by scaly patches. It generally occurs on the face and is caused by staphylococci or streptococci.

keratosis Any abnormal growth and hardening of the stratum corneum of the skin.

melanoma (*mel-a-no'ma*) A cancerous tumor originating from proliferating melanocytes within the epidermis of the skin.

nevus (*ne'vus*) A mole or birthmark—a congenital pigmentation of a limited area of the skin.

papilloma (*pap-i-loma*) A benign epithelial neoplasm, such as a wart or a corn.

papule A small inflamed elevation of the skin, such as a pimple.

pruritus (*proo-ri'tus*) Itching. It may be symptomatic of systemic disorders but is generally due to dry skin.

psoriasis (*sŏ-ri"ă-sis*) An inherited inflammatory skin disease, usually expressed as circular scaly patches of skin.

pustule A small, localized pus-filled elevation of the skin.

seborrhea (*seb"ŏ-re'ă*) A disease characterized by an excessive activity of the sebaceous glands and accompanied by oily skin and dandruff. It is known as "cradle cap" in infants.

wart A roughened projection of epidermal cells caused by a virus.

Chapter Summary

The Skin as an Organ (p. 160)

1. The skin is considered an organ because it consists of several kinds of tissues that are structurally arranged to function together.
2. The appearance of the skin is clinically important because it provides clues to certain body conditions or dysfunctions.

Layers of the Skin (pp. 160–165)

1. The stratified squamous epithelium of the epidermis is divisible into five structural and functional layers: the stratum basale, stratum spinosum, stratum granulosum, stratum lucidum, and stratum corneum.
   (a) Normal skin color is the result of a combination of melanin and carotene in the epidermis and hemoglobin in the blood of the dermis and hypodermis.
   (b) Fingerprints on the surface of the epidermis are individually unique; flexion creases and flexion lines are acquired.
2. The thick dermis of the skin is composed of fibrous connective tissue interlaced with elastic fibers. The two layers of the dermis are the upper papillary layer and the deeper reticular layer.
3. The hypodermis, composed of adipose and fibrous connective tissue, binds the dermis to underlying organs.

Physiology of the Skin (pp. 165–167)

1. Structural features of the skin protect the body from disease and external injury.
   (a) Keratin and an acidic oily secretion on the surface protect the skin from water and microorganisms.
   (b) Cornification of the skin protects against abrasion.
   (c) Melanin is a barrier to UV light.
2. The skin regulates body fluids and temperatures.
   (a) Fluid loss is minimal due to keratinization and cornification.
   (b) Temperature regulation is maintained by radiation, convection, and the antagonistic effects of sweating and shivering.
3. The skin permits the absorption of UV light, respiratory gases, steroids, fat-soluble vitamins, and certain toxins and pesticides.
4. The integument synthesizes melanin and keratin, which remain in the skin, and vitamin D, which is used elsewhere in the body.
5. Sensory reception in the skin is provided through cutaneous receptors throughout the dermis and hypodermis. Cutaneous receptors respond to precise sensory stimuli and are more sensitive in thin skin.
6. Certain emotions are reflected in changes in the skin.

Accessory Structures of the Skin (pp. 167–171)

1. Hair is characteristic of all mammals, but its distribution, function, density, and texture varies across mammalian species.
   (a) Each hair consists of a shaft, root, and bulb. The bulb is the enlarged base of the root within the hair follicle.
   (b) The three layers of the hair shaft are the medulla, cortex, and cuticle.
   (c) Lanugo, angora, and definitive are the three kinds of human hair.
2. Hardened, keratinized nails are protective on the ends of each digit; fingernails aid in grasping and picking up small objects.
   (a) Each nail consists of a nail body, free border, and hidden border.
   (b) The hyponychium, eponychium, and nail fold support the nail on the nail bed.
3. Integumentary glands are exocrine, since they either secrete or excrete substances through ducts.
   (a) Sebaceous glands secrete sebum onto the shaft of the hair.
   (b) The two types of sudoriferous (sweat) glands are eccrine and apocrine.
   (c) Mammary glands are specialized sudoriferous glands that secrete milk during lactation.
   (d) Ceruminous glands secrete cerumen (earwax).

Review Activities

Objective Questions

1. Hair, nails, integumentary glands, and the epidermis of the skin are derived from embryonic
   (a) ectoderm. (c) mesoderm.
   (b) endoderm. (d) mesenchyme.
2. Spoon-shaped nails may result when a person has a dietary deficiency of
   (a) zinc. (c) iron.
   (b) niacin. (d) vitamin B_{12}.
3. The epidermal layer *not* present in the thin skin of the face is the stratum
   (a) granulosum. (c) spinosum.
   (b) lucidum. (d) corneum.
4. Which of the following does *not* contribute to skin color?
   (a) hair papillae (c) melanin
   (b) carotene (d) hemoglobin

5. Which of the following is *not* true of the epidermis?
   (a) It is composed of stratified squamous epithelium.
   (b) As the epidermal cells die, they undergo keratinization and cornification.
   (c) Rapid mitotic activity (cell division) within the stratum corneum accounts for the thickness of this epidermal layer.
   (d) In most areas of the body, the epidermis lacks blood vessels and nerves.
6. Integumentary glands that empty their secretions into hair follicles are
   (a) sebaceous glands.
   (b) endocrine glands.
   (c) eccrine glands.
   (d) ceruminous glands.
7. Fetal hair that is present during the last trimester of development is referred to as
   (a) angora. (c) lanugo.
   (b) definitive. (d) replacement.
8. Which of these conditions is potentially life threatening?
   (a) acne (c) eczema
   (b) melanoma (d) seborrhea
9. The skin of a burn victim has been severely damaged through the epidermis and into the dermis. Integumentary regeneration will be slow, with some scarring, but it will be complete. Which kind of burn is this?
   (a) first degree
   (b) second degree
   (c) third degree
10. The technical name for a blackhead or whitehead is
   (a) carbuncle. (c) melanoma.
   (b) nevus. (d) comedo.

Essay Questions

1. List the functions of the skin. Which of these functions occurs passively due to the structure of the skin, and which occurs dynamically due to physiological processes?
2. Why is the skin considered an organ? What types of tissues are found in each of the three layers of the skin?
3. Discuss the growth process and regeneration of the epidermis.
4. What are some physical and chemical features of the skin that make it an effective protective organ?
5. Define the following: *lines of tension*, *friction ridges*, and *flexion lines*. What causes each of these to develop?
6. Distinguish between a hair follicle and a hair. What other accessory structures are associated with hair follicles and hair?
7. Compare and contrast the structure and function of sebaceous, sudoriferous, mammary, and ceruminous glands.
8. Discuss the development of the skin and its associated hair and glands. What role do the ectoderm and mesoderm play in integumentary development?
9. Discuss what is meant by an inflammatory lesion. What are some frequent causes of skin lesions?
10. Describe each of the three degrees of burns and discuss the physiological danger of burns.

Critical Thinking Questions

1. Why is it important that the epidermis serve as a barrier against UV rays, yet not block them out completely?
2. Review the structure and function of the skin by explaining (*a*) the mechanisms involved in thermoregulation; (*b*) variations in skin color; (*c*) abnormal coloration of the skin (for example, cyanosis, jaundice, and pallor); and (*d*) the occurrence of acne.
3. Do you think that humans derive any important benefit from contraction of the arrector pili muscles? Justify your answer.
4. The relative hairlessness of humans is unusual among mammals. Why should it be that we have any hair at all?
5. Compounds such as lead, zinc, and arsenic may accumulate in the hair and nails. Chemical toxins from pesticides and pollutants may accumulate in the adipose tissue (subcutaneous fat) of the hypodermis. Discuss some of the possible clinical situations where this knowledge would be of importance.
6. Pale or gray skin that is cool and clammy to the touch is one of the early symptoms of shock. Explain the physiological changes in the body that account for these symptoms.

Related Web Sites

For a listing of the most current web sites related to this chapter, please visit the *Concepts of Human Anatomy and Physiology* home page at http://www.mhhe.com/biosci/abio/.

NEXUS

Some Interactions of the Integumentary System with the Other Body Systems

Skeletal System

- The skin provides protection for all body systems, including the skeletal system (p. 165)
- The skin produces vitamin D, needed for maintenance of plasma calcium concentrations to ensure proper calcification of the skeletal system (p. 194)
- Bones store calcium phosphate, needed for the normal functioning of all organs, including the skin (p. 193)

Nervous System

- The skin provides protection for all body systems, including the nervous system (p. 165).
- The skin houses cutaneous (tactile) receptors that convey sensory sensations to the brain (p. 501).
- The nervous system provides autonomic regulation of cutaneous vessels and glands (p. 481).

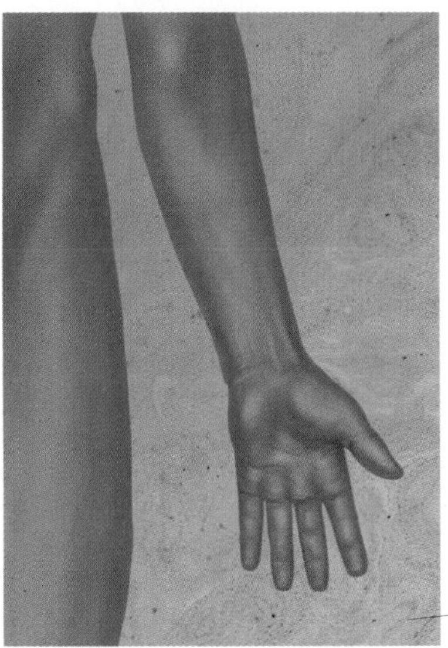

Endocrine System

- The skin provides protection for all body systems, including the endocrine system (p. 165).
- Sex hormones cause integumentary features to change during puberty (p. 897).
- Parathyroid hormone and calcitonin act together with the vitamin D secreted by the skin to regulate calcium balance (p. 194).

Muscular System

- The skin provides protection for all body systems, including the muscular system (p. 165).
- The skin produces vitamin D, needed for maintenance of plasma calcium concentrations for muscle contraction (p. 167).
- The skin provides radiant and evaporative heat loss (through sweating) of heat produced by muscle metabolism during exercise (p. 166).
- Skeletal muscle contractions pull on skin, producing facial expressions (p. 281).

Circulatory System

- The skin provides protection for all body systems, including the circulatory system (p. 165).
- Blood vessels in the skin help to maintain constant body temperature (p. 676).
- The keratinized epidermis helps to prevent fluid loss (p. 166).
- The circulatory system transports O_2 and CO_2, nutrients, and wastes to and from the skin (p. 616).

Lymphatic and Immune Systems

- The skin serves as the first line of defense against invasion by pathogens (p. 696).
- The dermis serves as a site for local inflammation reactions (p. 706).
- Immune cells protect the skin from infection (p. 707).

Respiratory System

- The skin provides protection for all body systems, including the respiratory system (p. 165).
- The respiratory system provides the skin with O_2 and eliminates CO_2 (p. 729).

Urinary System

- The skin provides protection for all body systems, including the urinary system (p. 165).
- The kidneys eliminate wastes and regulate homeostasis of the intercellular fluid volume and composition, required by all body systems (p. 792).
- Enzymes in the kidneys are required to convert vitamin D to an active derivative (p. 195).

Digestive System

- Vitamin D, produced by the skin, stimulates the intestinal absorption of calcium (p. 195).
- The digestive system provides nutrients for all body systems, including the integumentary system (p. 817).

Reproductive System

- Cutaneous receptors respond to erotic stimuli (p. 912).
- Mammary glands, which are modified integumentary glands, produce milk (p. 940).
- Gonads produce sex hormones, which affect the skin and sebaceous glands (p. 897).

EIGHT

Skeletal System: Bone Tissue and Bone Development

OBJECTIVES

- Describe the structural organization of the skeletal system and list the bones of the axial and appendicular portions.
- Discuss the principal functions of the skeletal system and identify the body systems served by these functions.
- Classify bones according to their shapes and give an example of each type.
- Describe the various markings on the surfaces of bones.
- Describe the gross features of a typical long bone and list the functions of each feature.
- Identify the five types of bone cells and list the functions of each.
- Distinguish between spongy and compact bone tissues.
- Describe the process of endochondral ossification as it relates to bone growth.
- Describe the relationships between bone and plasma calcium levels and the importance of maintaining normal blood calcium concentrations.
- Describe how parathyroid hormone and calcitonin secretions are regulated and how these hormones regulate the plasma calcium and phosphate concentrations.
- Explain how vitamin D functions as a prehormone and describe its effects on calcium and phosphate balance.

Clinical Investigation

A 72-year-old woman made an appointment to see her physician because her daughter was concerned about her mother's "changing posture." The daughter expressed that "mom used to be so poised and stately and had excellent carriage, but now she seemed to slouch frequently and had trouble walking." The elderly woman had sustained a hip fracture two years prior that healed with considerable difficulty. Upon receiving a thorough physical exam, which included a series of radiographs (X rays) of the vertebral column, the woman was told by her physician that she had moderate osteoporosis. This information was very frightening to the patient and she had many questions about the disease, such as, "What is osteoporosis? What actually happens within the bones that caused this condition? Can anything be done to stop the progress of the disease? Will my daughter be afflicted with osteoporosis when she is my age?"

How should the physician answer these questions?

Clues: The patient sought medical attention early in the progression of the disease and she was physically active prior to her hip fracture. Carefully study the section on bone tissue and understand the physiology of bone. In addition, the clinical section on nutritional and hormonal disorders is very informative.

Organization of the Skeletal System

The axial and appendicular components of the skeletal system of an adult human consist of 206 individual bones arranged to form a strong, flexible body framework.

Although the adult skeletal system consists of approximately 206 bones, the exact number of bones differs from person to person depending on age and genetic factors. For example, at birth, the skeleton consists of approximately 270 bones. As further bone development (ossification) occurs during infancy, the number increases. During adolescence, however, the number of bones decreases, as separate bones gradually fuse. Each bone is actually an organ that plays a part in the total functioning of the skeletal system. The science concerned with the study of bones is called *osteology*.

Some adults have extra bones within the sutures (joints) of the skull called **sutural** (*wormian*) **bones.** Additional bones may develop in tendons in response to stress as the tendons repeatedly move across a joint. Bones formed this way are called **sesamoid** (*ses'ă-moid*) **bones.** Sesamoid bones, like the sutural bones, vary in number. The patellae ("kneecaps") are two sesamoid bones all people have.

ossification: Gk. *os*, bone; L. *facio*, to make
sesamoid: Gk. *sesamon*, like a sesame seed

For the convenience of study, the skeleton is divided into *axial* and *appendicular portions*, as shown in figure 8.1 and summarized in table 8.1. The **axial skeleton** consists of the bones that form the axis of the body and support and protect the organs of the head, neck, and trunk. The components of the axial skeleton are as follows:

1. **Skull.** The skull consists of two sets of bones: the *cranial bones* that form the cranium, or braincase, and the *facial bones* that support the eyes and nose and form the bony framework of the oral cavity.
2. **Auditory ossicles.** Three auditory ossicles ("ear bones") are present in the middle-ear chamber of each ear and serve to transmit sound impulses.
3. **Hyoid bone.** The hyoid (*hi'oid*) bone is located above the larynx ("voice box") and below the mandible ("jawbone"). It supports the tongue and assists in swallowing.
4. **Vertebral column.** The vertebral column ("backbone") consists of 26 individual bones (*vertebrae*) separated by cartilaginous *intervertebral discs*. In the pelvic region, several vertebrae are fused to form the *sacrum*, which is the attachment portion of the pelvic girdle. The terminal vertebrae are fused to form the *coccyx* ("tailbone").
5. **Rib cage.** The rib cage, or thoracic cage, forms the bony and cartilaginous framework of the thorax. It articulates posteriorly with the thoracic vertebrae and includes the 12 pairs of *ribs*, the flattened *sternum*, and the *costal cartilages* that connect the ribs to the sternum.

The **appendicular skeleton** is composed of the bones of the upper and lower extremities and the bony girdles that anchor the appendages to the axial skeleton. The components of the appendicular skeleton are as follows:

1. **Pectoral girdle.** The paired *scapulae* ("shoulder blades") and *clavicles* ("collarbones") are the appendicular components of the pectoral girdle, and the sternum ("breastbone") is the axial component. The primary function of the pectoral girdle is to provide attachment for the muscles that move the brachium (arm) and antebrachium (forearm).
2. **Upper extremities.** Each upper extremity contains a proximal *humerus* within the brachium, an *ulna* and *radius* within the forearm, the *carpal bones* ("wrist bones"), the *metacarpal bones*, and *phalanges* ("finger bones") of the hand.
3. **Pelvic girdle.** The two *ossa coxae* ("hipbones") are the appendicular components of the pelvic girdle, and the sacrum is the axial component. The ossae coxae are united anteriorly by the *symphysis* (*sim'fĭ-sis*) *pubis* and posteriorly by the sacrum. The pelvic girdle supports the weight of the body through the vertebral column and protects the viscera within the pelvic cavity.

ossicle: L. *ossiculum*, little bone

Figure 8.1

The human skeleton.

(*a*) An anterior view and (*b*) a posterior view. The axial portion is colored light blue.

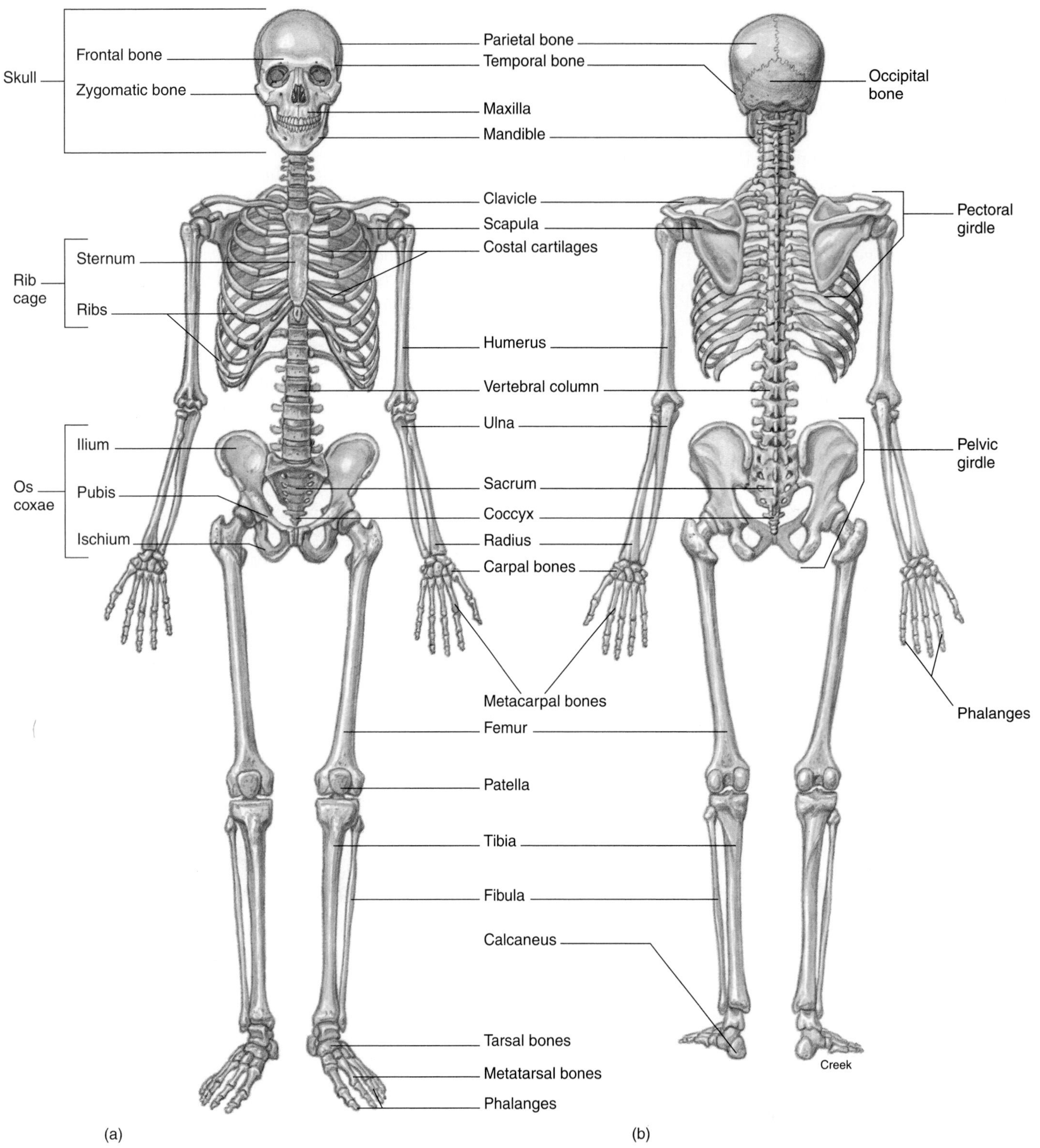

Table 8.1
Bones of the Adult Skeleton

Axial Skeleton			Appendicular Skeleton	
Skull—22 bones		**Auditory Ossicles**—6 bones	**Pectoral Girdle**—5 bones	
14 Facial Bones	*8 Cranial Bones*	malleus (2)	sternum* (1)	
maxilla (2)	frontal (1)	incus (2)	scapula (2)	
palatine (2)	parietal (2)	stapes (2)	clavicle (2)	
zygomatic (2)	occipital (1)	**Hyoid**—1 bone	**Upper Extremities**—60 bones	
lacrimal (2)	temporal (2)	**Vertebral Column**—26 bones	humerus (2)	carpal bones (16)
nasal (2)	sphenoid (1)	cervical vertebra (7)	radius (2)	metacarpal bones (10)
vomer (1)	ethmoid (1)	thoracic vertebra (12)	ulna (2)	phalanges (28)
inferior nasal concha (2)		lumbar vertebra (5)	**Pelvic Girdle**—3 bones	
mandible (1)		sacrum (1) (4 or 5 fused bones)	sacrum* (1)	
		coccyx (1) (3–5 fused bones)	os coxae (2) (each contains 3 fused bones)	
		Rib Cage—25 bones	**Lower Extremities**—60 bones	
		rib (24)	femur (2)	tarsal bones (14)
		sternum (1)	tibia (2)	metatarsal bones (10)
			fibula (2)	phalanges (28)
			patella (2)	

*Although the sternum and sacrum are bones of the axial skeleton, technically speaking they are also considered bones of the pectoral and pelvic girdles, respectively.

4. **Lower extremities.** Each lower extremity contains a proximal *femur* ("thighbone") within the thigh, a *tibia* ("shinbone") and *fibula* within the leg, the *tarsal bones* ("ankle bones"), the *metatarsal bones* and *phalanges* ("toe bones") of the foot. In addition, the *patella* (pă-tel′ă: "kneecap") is located on the anterior surface of the knee joint, between the thigh and leg.

Functions of the Skeletal System

The bones of the skeleton perform the mechanical functions of support, protection, and leverage for body movement and the metabolic functions of hemopoiesis and mineral storage.

The strength of bone comes from its inorganic components, of such durability that they resist decomposition even after death. Much of what we know of prehistoric animals, including humans, has been determined from preserved skeletal remains. When we think of bone, we frequently think of a hard, dry structure. In fact, the term *skeleton* comes from a Greek word meaning "dried up." Living bone, however, is not inert material; it is dynamic and adaptable. It performs many body functions, including support, protection, leverage for body movement, hemopoiesis, and mineral storage.

1. **Support.** The skeleton forms a rigid framework to which the softer tissues and organs of the body are attached. Interestingly, the mass of muscles and organs that the skeleton's 206 bones support may weigh five times as much as the bones themselves.
2. **Protection.** The skull and vertebral column enclose the brain and spinal cord; the rib cage protects the heart, lungs, great vessels, liver, and spleen; and the pelvic cavity supports and protects the pelvic viscera. Even the site where blood cells are produced is protected within the central hollow portion of certain bones.
3. **Body movement.** Bones serve as anchoring attachments for most skeletal muscles. In this capacity, the bones act as levers (the joints functioning as pivots) when muscles contract and cause body movement.
4. **Hemopoiesis.** The process of blood cell formation is called hemopoiesis (*he″mŏ-poi-e′sis*). It takes place in tissue called red bone marrow located internally in some bones. In an infant, the spleen and liver produce red blood cells, but as the bones mature, the bone marrow takes over this formidable task. An estimated 2.5 million red blood cells are produced every second by the bone marrow to replace those that are worn out and destroyed by the liver.
5. **Mineral storage.** The inorganic matrix of bone is composed primarily of the minerals calcium and phosphorus. These minerals, which account for approximately two-thirds of the weight of bone, give

UNDER DEVELOPMENT

Development of the Skeletal System

Bone formation, or *ossification,* begins about the fourth week of embryonic development, but ossification centers cannot be readily observed until about the tenth week (fig. 1). Bone tissue derives from specialized migratory cells of mesoderm (chapter 6) known as mesenchyme. Some of the embryonic mesenchymal cells will transform into *chondroblasts* (*kon'dro-blasts*) and develop a cartilage matrix that is later replaced by bone in a process known as **endochondral** (*en"dŏ-kon'dral*) **ossification.** Most of the skeleton is formed in this fashion—first it goes through a hyaline cartilage stage and then it is ossified as bone.

A smaller number of mesenchymal cells develop directly into bone without first going through a cartilage stage. This type of bone-formation process is referred to as **intramembranous** (*in"tră-mem'bră-nus*) **ossification.** Facial bones and certain bones of the cranium are formed this way. *Sesamoid bones* are specialized intramembranous bones that develop in tendons. The patella is an example of a sesamoid bone.

chondroblast: Gk. *chondros*, cartilage; *blastos*, offspring or germ

Figure 1 Ossification centers of the skeleton of a 10-week-old fetus. (*a*) The diagram depicts endochondrial ossification in red and intramembranous ossification in a stippled pattern. The cartilaginous portions of the skeleton are shown in grey. (*b*) The photograph shows the ossification centers stained with a red indicator dye.

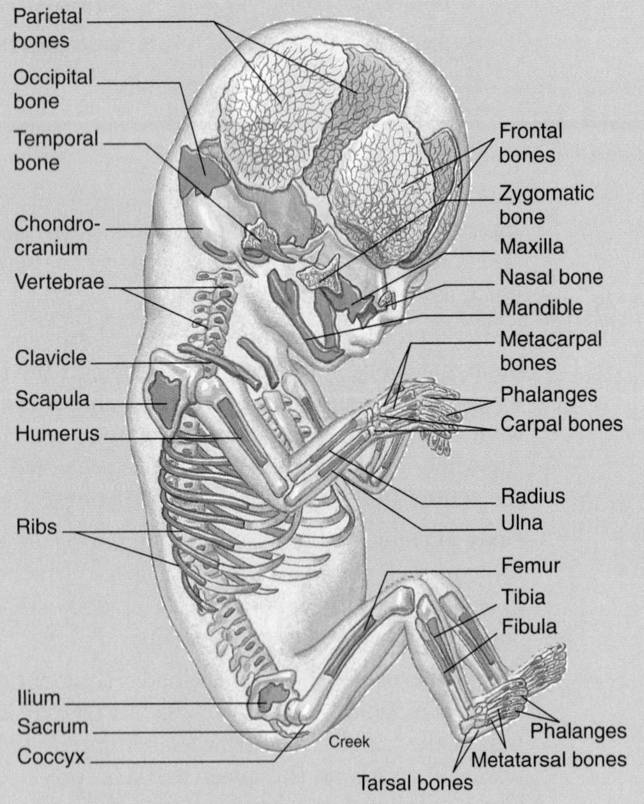

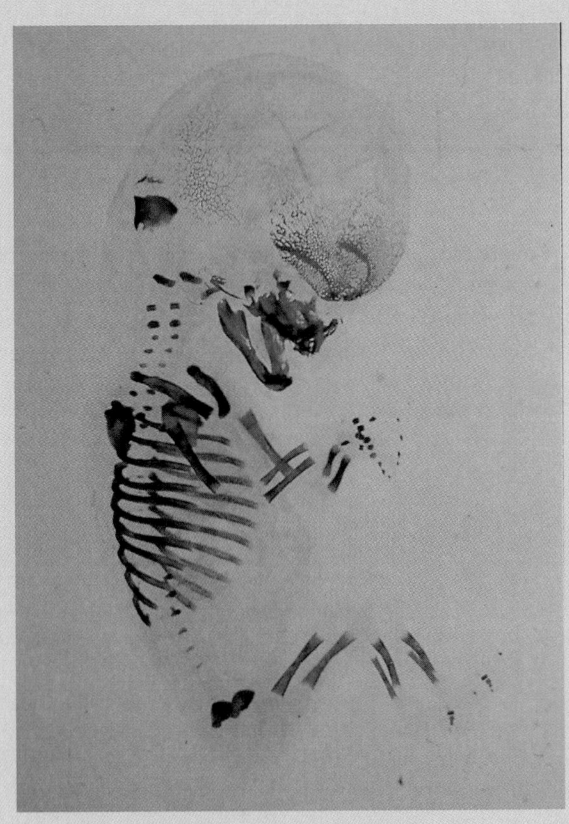

bone its firmness and strength. About 95% of the calcium and 90% of the phosphorus within the body are deposited in the bones and teeth. Although the concentration of these inorganic salts within the blood is kept within narrow limits, both of these mineral salts are essential for other body functions. Calcium is necessary for muscle contraction, blood clotting, and the movement of ions and nutrients across cell membranes. Phosphorus is required for the activities of the nucleic acids DNA and RNA, as well as for ATP utilization. If mineral salts are not present in the diet in sufficient amounts, they may be withdrawn from the bones until they are replenished through proper nutrition. In addition to calcium and phosphorus, lesser amounts of magnesium, sodium, fluorine, and strontium are stored in bone tissue.

Table 8.2
Surface Features of Bone

Structure	Description and Example
Articulating Surfaces	
Condyle (*kon'dil*)	A large, rounded articulating knob (the occipital condyle of the occipital bone)
Facet	A flattened or shallow articulating surface (costal face of a thoracic vertebra)
Head	A prominent, rounded articulating end of a bone (the head of the femur)
Depressions and Openings	
Alveolus (*al-ve'o-lus*)	A deep pit or socket (the dental alveolus [tooth socket] in the maxilla and mandible)
Fissure	A narrow, slitlike opening (the superior orbital fissure of the sphenoid bone)
Foramen (*fo-ra'men*; plural, *foramina*)	A rounded opening through a bone (foramen magnum of the occipital bone)
Fossa (*fos'a*)	A flattened or shallow surface (the mandibular fossa of the temporal bone)
Sinus	A cavity or hollow space in a bone (the frontal sinus of the frontal bone)
Sulcus	A groove that accommodates a vessel, nerve, or tendon (the intertubercular sulcus of the humerus)
Nonarticulating Prominences	
Crest	A narrow, ridgelike projection (the iliac crest of the os coxae)
Epicondyle	A projection adjacent to a condyle (the medial epicondyle of the femur)
Process	Any marked bony prominence (the mastoid process of the temporal bone)
Spine	A sharp, slender process (the spine of the scapula)
Trochanter	A massive process found only on the femur (the greater trochanter of the femur)
Tubercle (*too'ber-k'l*)	A small, rounded process (the greater tubercle of the humerus)
Tuberosity	A large, roughened process (the radial tuberosity of the radius)

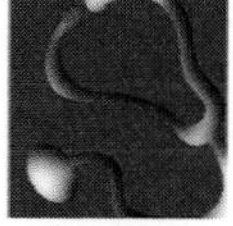

Vitamin D assists in the absorption of calcium and phosphorus from the small intestine into the blood. As bones develop in a child, it is extremely important that the child's diet contain an adequate amount of these two minerals and vitamin D. If the diet is deficient in these essentials, the blood level falls below that necessary for calcification, and a condition known as *rickets* develops (see fig. 7.8). Rickets is characterized by soft bones that may result in bowlegs and malformation of the head, chest, and pelvic girdle.

In summary, the skeletal system is not an isolated body system. It functions with the muscle system in storing calcium needed for muscular contraction and providing attachments for muscles as they span the movable joints. The skeletal system serves the circulatory system by producing blood cells in protected sites. Directly or indirectly, the skeletal system supports and protects all of the systems of the body (see Nexus, p. 244).

Bone Structure

Each bone has a characteristic shape and diagnostic surface features that indicate its functional relationship to other bones, muscles, and the body structure as a whole.

The shape and surface features of each bone indicate its functional role in the skeleton (table 8.2). Bones that are long, for example, provide body support and function as levers during body movement. Bones that support the body are massive and have large articular surfaces and processes for muscle attachment. Roughened areas on these bones may serve for the attachment of ligaments, tendons, or muscles. A flattened surface provides an attachment site for a large muscle or may provide protection. Grooves around an articular end of a bone are where tendons or nerves pass, and openings through a bone permit the passage of nerves or blood vessels.

Shapes of Bones

The bones of the skeleton are classified into four principal types on the basis of shape rather than size. The four classes are long bones, short bones, flat bones, and irregular bones (fig. 8.2).

facet: Fr. *facette*, little face
trochanter: Gk. *trochanter*, runner
tuberosity: L. *tuberosus*, lump

Figure 8.2

Examples of bone types, as classified by shape.

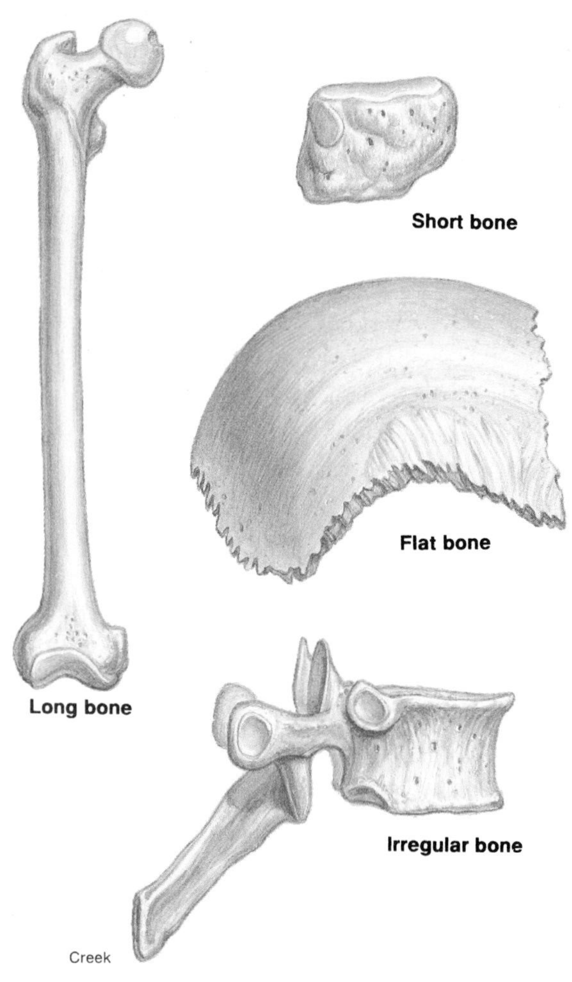

Figure 8.3

A section through the skull showing diploe.
Diploe is a layer of spongy bone sandwiched between two surface layers of compact bone. It is extremely strong yet light in weight.

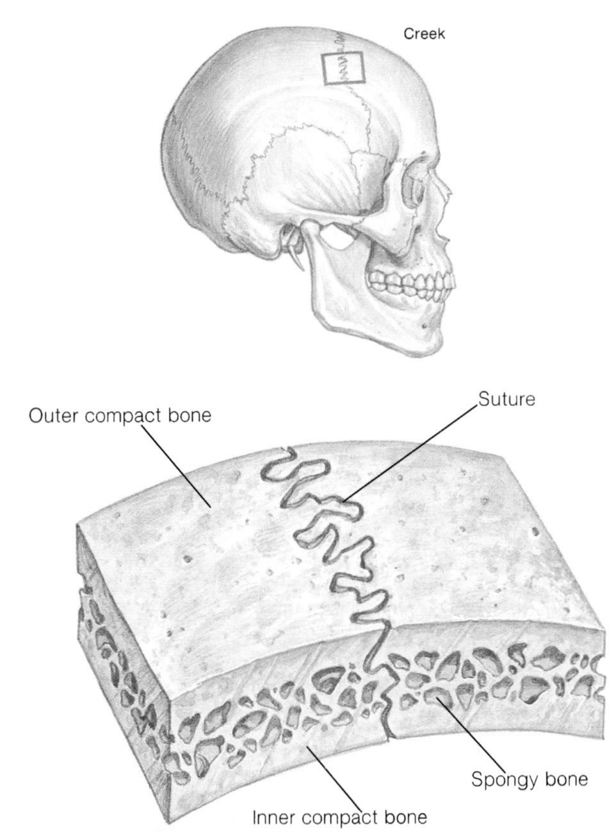

1. **Long bones.** Long bones are longer than they are wide and function as levers. Most of the bones of the upper and lower extremities are of this type (e.g., the humerus, radius, ulna, metacarpal bones, femur, tibia, fibula, metatarsal bones, and phalanges).
2. **Short bones.** Short bones (e.g., the wrist and ankle bones) are somewhat cube-shaped and are found in confined spaces, where they transfer forces.
3. **Flat bones.** Flat bones (e.g., the cranium, ribs, and bones of the shoulder girdle) have a broad, dense surface for muscle attachment or protection of underlying organs.
4. **Irregular bones.** Irregular bones (e.g., the vertebrae and certain bones of the skull) have varied shapes and many surface markings for muscle attachment or articulation.

Structure of a Typical Long Bone

Bone tissue is organized as *compact (dense) bone* or *spongy (cancellous) bone*, and most bones have both types. Compact bone is hard and dense, and is the protective exterior portion of all bones. The spongy bone, when it occurs, is deep to the compact bone and is quite porous. The microscopic structure of compact and spongy bone will be considered shortly.

In a flat bone of the skull, the spongy bone is sandwiched between the compact bone and is called a *diploe* (*dip'lo-e*) (fig. 8.3). Because of this protective layering of bone tissue, a blow to the head may fracture the outer compact bone layer without harming the inner compact bone layer and the brain.

The long bones of the skeleton have a descriptive terminology all their own. In a long bone from an appendage, the body of the bone (shaft), or **diaphysis** (*di-af'ĭ-sis*), consists of a cylinder of compact bone surrounding a central cavity called the **medullary** (*med"yoo-lar-e*) **cavity** (fig. 8.4). The medullary cavity is lined with a thin layer of connective tissue called the **endosteum** (*en-dos'te-um*). In an adult, the cavity contains **yellow bone marrow,** so-named because it contains a large amount of fat. On each end of the diaphysis is an **epiphysis** (*ĕ-pif'ĭ-sis*), consisting of spongy bone surrounded by a layer of

diploe: Gk. *diplous*, double
diaphysis: Gk. *dia*, throughout; *physis*, growth
epiphysis: Gk. *epi*, upon; *physis*, growth

Figure 8.4

A diagram of a long bone (the humerus) shown in a partial longitudinal section.

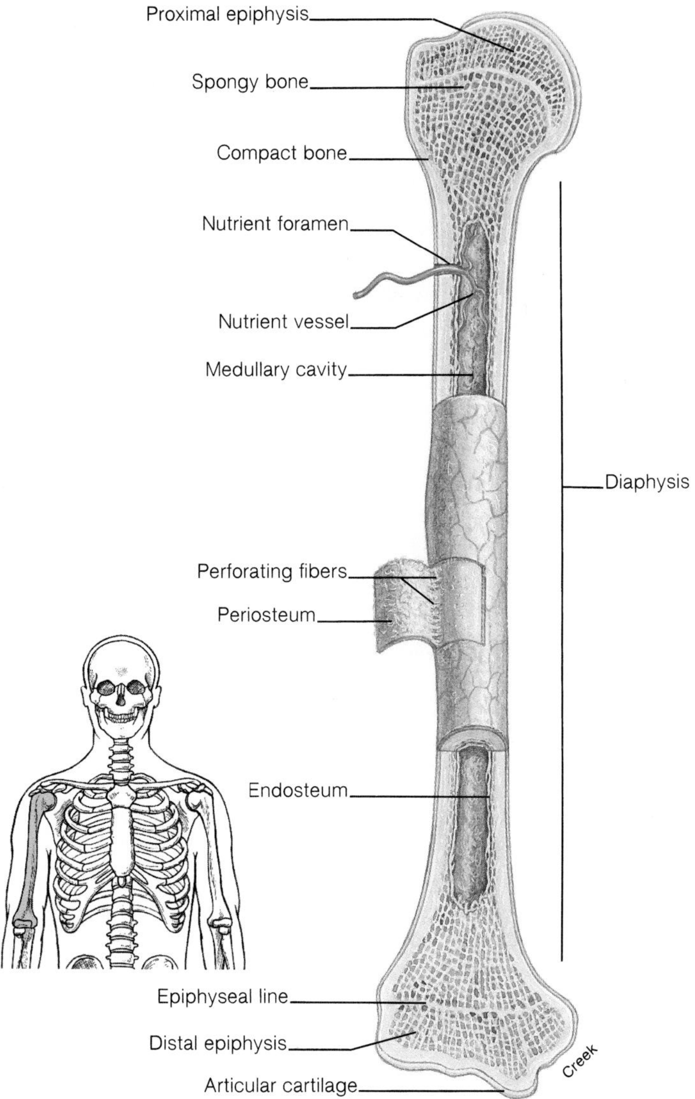

compact bone. **Red bone marrow** is found within the porous chambers of spongy bone. In an adult, erythropoiesis (the production of red blood cells; see chapter 20) occurs in the red bone marrow, especially that of the sternum, vertebrae, portions of the ossa coxae, and the proximal epiphyses of the femora and humeri. The red bone marrow is also responsible for the formation of white blood cells and platelets and for the phagocytosis of worn-out red blood cells. **Articular cartilage,** which is composed of thin hyaline cartilage, caps each epiphysis and facilitates joint movement. Along the diaphysis are **nutrient foramina**—small openings into the bone that allow for passage of nutrient vessels for nourishment of the living tissue.

Between the diaphysis and epiphysis is an **epiphyseal** (*ep"ĭ-fiz'e-al*) **plate** of cartilage—a region of mitotic activity that is responsible for *linear bone growth.* As bone growth is completed, an **epiphyseal line** replaces the plate, and ossification occurs between the epiphysis and the diaphysis (see fig. 8.8). A **periosteum** (*per"e-os'te-um*) of dense regular connective tissue covers the surface of the bone, except at the articulating surfaces. This highly vascular layer serves as a place for a tendon-muscle attachment and is responsible for *appositional bone growth* (increase in width). The periosteum is secured to the bone by **perforating** (Sharpey's) **fibers** (fig. 8.4), composed of bundles of collagenous fibers.

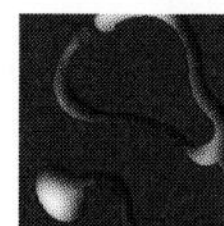

Fracture of a long bone in a young person may be especially serious if it damages an epiphyseal plate. If such an injury goes untreated, or is not treated properly, linear bone growth may be arrested or slowed, resulting in permanent shortening of the affected limb.

Bone Tissue

Bone tissue is composed of several types of bone cells embedded in a matrix of ground substance, inorganic salts (calcium and phosphorus), and collagenous fibers. Bone cells and ground substance give bone flexibility and strength; the inorganic salts give it hardness.

Bone Cells

There are five principal types of bone cells contained within bone tissue. **Osteogenic** (*os"te-ŏ-jen'ik*) **cells** are found in the bone tissues in contact with the endosteum and the periosteum. These cells respond to trauma, such as a fracture, by giving rise to bone-forming cells (osteoblasts) and bone-destroying cells (osteoclasts). **Osteoblasts** (*os'te-ŏ-blasts*) are bone-forming cells (fig. 8.5*a*) that synthesize and secrete nonmineralized ground substance. They are abundant in areas of high metabolism within bone, such as under the periosteum and bordering the medullary cavity. **Osteocytes** (*os'te-ŏ-sītz*) are mature bone cells (figs. 8.5 and 8.6) derived from osteoblasts that have secreted bone tissue around themselves. Osteocytes maintain healthy bone tissue by secreting enzymes and influencing bone mineral content. They also regulate the calcium release from bone tissue to blood. **Osteoclasts** (*os'te-ŏ-klasts*) are large, multinuclear cells (fig. 8.5*b*) that enzymatically break down bone tissue. These cells are important in bone growth, in remodeling, and in healing. **Bone-lining cells** are derived from osteoblasts along the surface of most bones in the adult skeleton. These cells are thought to regulate the movement of calcium and phosphate into and out of bone matrix.

periosteum: Gk. *peri*, around; *osteon*, bone
Sharpey's fibers: from William Sharpey, Scottish physiologist and histologist, 1802–80
osteoblast: Gk. *osteon*, bone; *blastos*, offspring or germ
osteoclast: Gk. *osteon*, bone; *klastos*, broken

Figure 8.5

Types of bone cells.
(*a*) Osteoblasts are important in secreting nonmineralized ground substance. Osteocytes derive from osteoblasts and play a regulatory role in maintaining bone tissue. (*b*) Osteoclasts are bone-destroying cells that help to maintain the dynamic state of bone tissue.

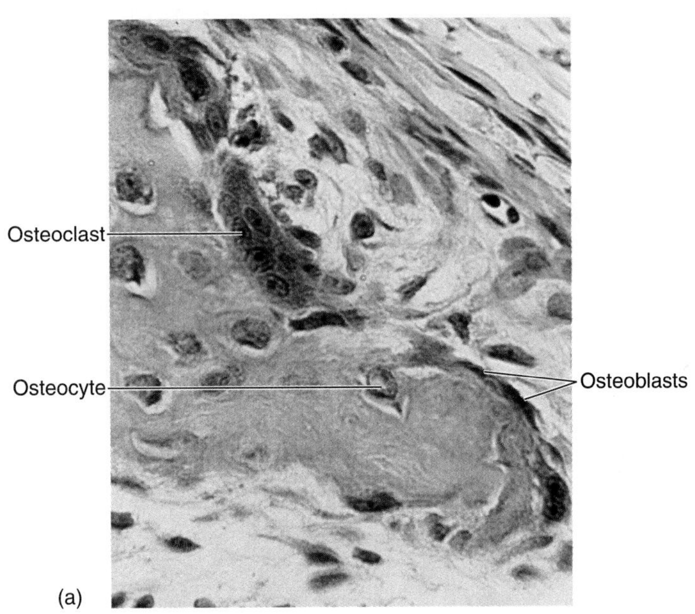

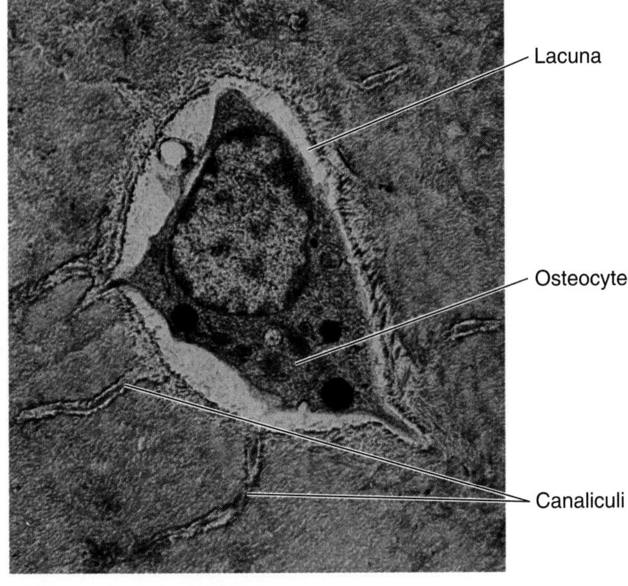

Spongy and Compact Bone Tissues

As mentioned earlier, most bones contain both spongy and compact bone tissues (fig. 8.6). **Spongy bone tissue** is located deep to the compact bone tissue, and is quite porous. Minute spikes of bone tissue called **trabeculae** (*tra-bek'yu-le*) give spongy bone a latticelike appearance. Spongy bone is highly vascular and provides great strength with minimal weight.

Compact bone tissue forms the external portion of a bone and is very hard and dense. It consists of precise arrangements of microscopic cylindrical structures oriented parallel to the long axis of the bone (fig. 8.6). These columnlike structures are the **osteons,** or *haversian systems*, of the bone tissue. The matrix of an osteon is laid down in concentric rings, called **lamellae,** that surround a **central** (haversian) **canal** (fig. 8.7). The central canal contains minute nutrient vessels and a nerve. Osteocytes within spaces called **lacunae** are regularly arranged between the lamellae. The lacunae are connected by tiny channels called **canaliculi** (*kan"a-lik'yu-li*), through which nutrients diffuse. Metabolic activity within bone tissue occurs at the osteon level. Between osteons are incomplete remnants of osteons called **interstitial systems. Perforating** (Volkmann's) **canals** penetrate compact bone, connecting osteons with blood vessels and nerves.

haversian system: from Clopton Havers, English anatomist, 1650–1702
Volkmann's canal: from Alfred Volkmann, German physiologist, 1800–1877

Bone Growth

The development of bone from embryo to adult depends on the orderly processes of cell division, growth, and ongoing remodeling. Bone growth is influenced by genetics, hormones, and nutrition.

In most bone development, a cartilaginous model is gradually replaced by bone tissue during *endochondral bone formation*. As the cartilage model grows, the *chondrocytes* (cartilage cells) in the center of the diaphysis hypertrophy, and minerals are deposited within the matrix in a process called *calcification* (fig. 8.8). Calcification restricts the passage of nutrients to the chondrocytes, causing them to die. At the same time, some cells of the perichondrium (dense regular connective tissue surrounding cartilage) differentiate into osteoblasts. These cells secrete **osteoid** (*os'te-oid*), the hardened inorganic component of bone. As the perichondrium calcifies, it gives rise to a thin plate of compact bone called the **periosteal bone collar.** The periosteal bone collar is surrounded by the periosteum.

A **periosteal bud,** consisting of osteoblasts and blood vessels, invades the disintegrating center of the cartilage model from the periosteum. Once in the center, the osteoblasts secrete osteoid, and a **primary ossification center** is established. Ossification then expands into the deteriorating cartilage. This process is repeated in both the proximal and distal epiphyses, forming **secondary ossification centers** where spongy bone develops.

Figure 8.6

Compact bone tissue.

(*a*) A diagram of the femur showing a cut through the compact bone into the medullary cavity. (*b*) The arrangement of the osteons within the diaphysis of the bone. (*c*) An enlarged view of an osteon showing the osteocytes within lacunae and the concentric lamellae. (*d*) An osteocyte within a lacuna.

Figure 8.7

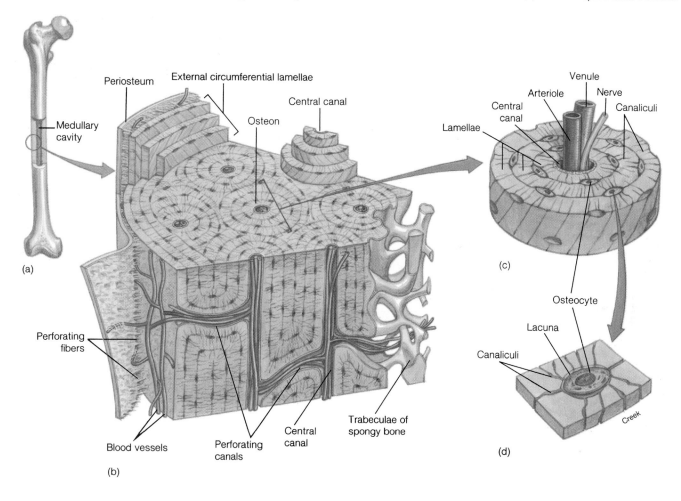

Bone tissue as seen in (*a*) a scanning electron micrograph and (*b*) a photomicrograph.

The lacunae (LA) provide spaces for the osteocytes, which are connected to one another by canaliculi (CA). Note the divisions between the lamellae (arrows).

From: Tissues and Organs: A Text Atlas of Scanning Electron Microscopy by R. G. Kessel and R. Kardon. © 1979, W. H. Freeman and Company.

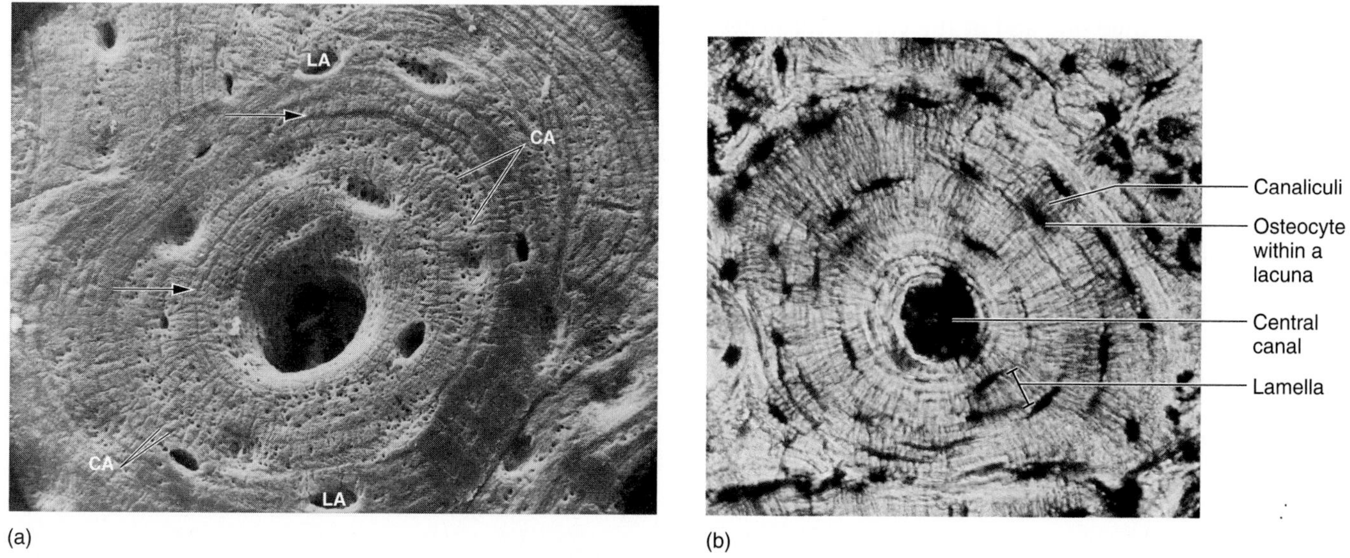

Figure 8.8

The process of endochondral ossification, beginning with (*a*) the cartilaginous model as it occurs in an embryo at 6 weeks.
The bone develops (*b–e*) through intermediate stages to (*f*) adult bone.

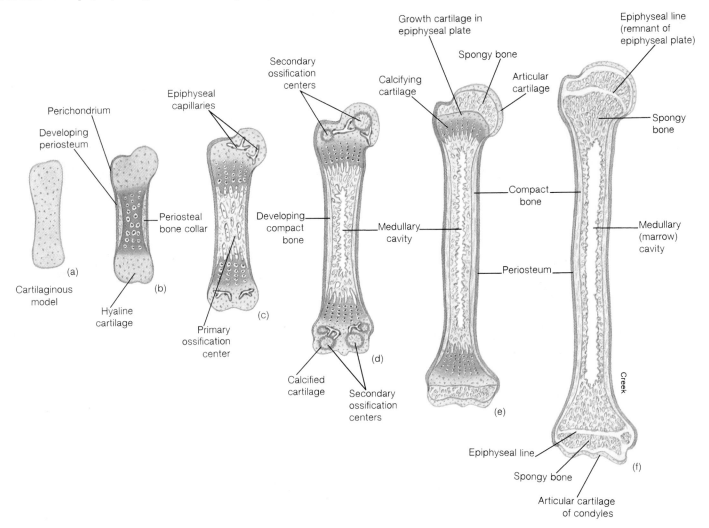

Once the secondary ossification centers have been formed, bone tissue totally replaces cartilage tissue, except at the articular ends of the bone and at the epiphyseal plates. An epiphyseal plate contains five histological zones (fig. 8.9). The *reserve zone* borders the epiphysis and consists of small chondrocytes irregularly dispersed throughout the intercellular matrix. The chondrocytes in this zone anchor the epiphyseal plate to the bony epiphysis. The *proliferating zone* consists of larger, regularly arranged chondrocytes that are constantly dividing. The *hypertrophic zone* consists of very large chondrocytes that are arranged in columns. The linear growth of long bones is due to the cellular proliferation at the proliferating zone and the growth and maturation of these new cells within the hypertrophic zone. The *resorption zone* is the area where a change in mineral content occurs. The *ossification zone* is a region of transformation from cartilage tissue to bone tissue. The chondrocytes within this zone die because the intercellular matrix surrounding them becomes calcified. Osteoclasts then break down the calcified matrix, and the area is invaded by osteoblasts and capillaries from the bone tissue of the diaphysis. As the osteoblasts mature, osteoid is secreted and bone tissue is formed. The result of this process is a gradual increase in the length of bone at the epiphyseal plates.

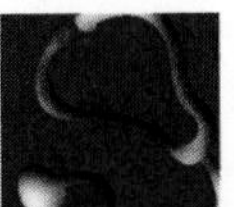

The time at which epiphyseal plates ossify varies widely from bone to bone but generally it occurs between the ages of 18 and 20 within the long bones (table 8.3). Because ossification of the epiphyseal cartilages within each bone occurs at predictable times, radiologists can determine the ages of people who are still growing by examining radiographs of their bones (fig. 8.10). Large discrepancies between bone age and chronological age may indicate a genetic or endocrine abnormality.

Bone is continually being remodeled over the course of a person's life. Bony prominences develop as stress is applied

Figure 8.9

A photomicrograph from an epiphyseal plate (63×).

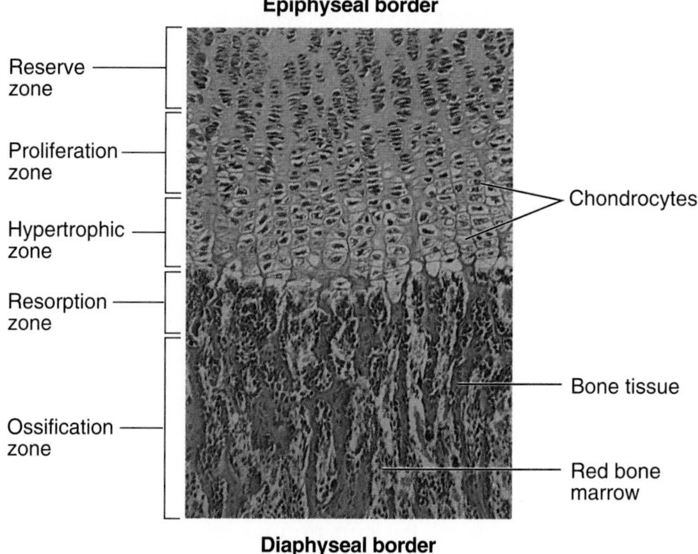

Table 8.3

Average Age of Completion of Bone Ossification

Bone	Chronological Age of Fusion
Bones of upper extremity	17–20
Scapula	18–20
Bones of lower extremity	18–22
Os coxae	18–23
Sternum (body)	23
Sacrum	23–25
Clavicle	23–31
Vertebra	25
Sternum (manubrium, xiphoid)	30+

to the periosteum, causing the osteoblasts to secrete osteoid and form new bone tissue. The greater trochanter of the femur, for example, develops in response to forces of stress applied to the periosteum of the bone where the tendons of muscles attach. Even though a person has stopped growing in height, bony processes may continue to enlarge somewhat if he or she remains physically active.

As new bone layers are deposited on the outside surface of the bone, osteoclasts dissolve bone tissue adjacent to the medullary cavity. In this way, the size of the cavity keeps pace with the increased growth of the bone.

Figure 8.10

Epiphyseal plates.
The presence of epiphyseal plates, as seen in a radiograph of a child's hand, indicates that bones are still growing in length.

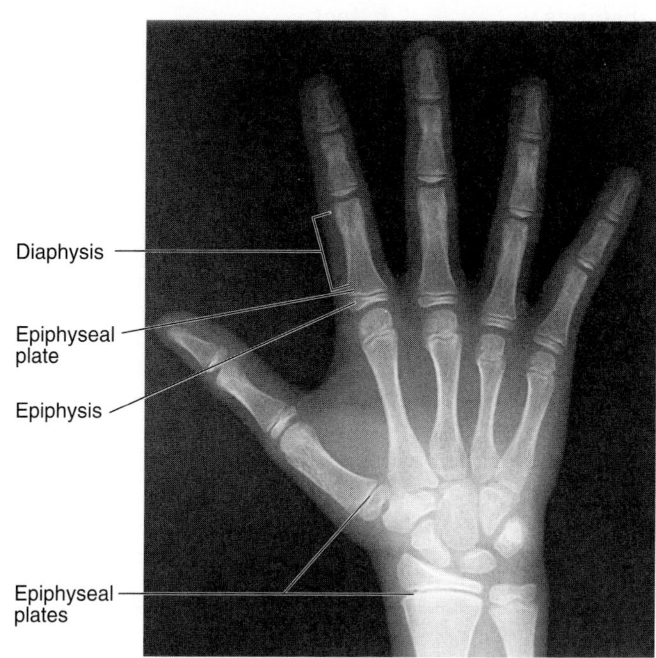

Even the absence of stress causes a remodeling of bones. This effect can best be seen in the bones of bedridden or paralyzed individuals. Radiographs of their bones reveal a marked loss of bone tissue. The absence of gravity that accompanies space flight may result in mineral loss from bones if an exercise regimen is not followed.

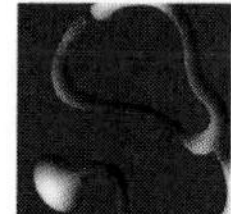

The movement of teeth in orthodontics involves bone remodeling. The dental alveoli (tooth sockets) are reshaped through the activity of osteoclast and osteoblast cells as stress is applied with braces. The use of traction in treating certain skeletal disorders has a similar effect.

Physiology of Bone Tissue

Bone deposition and bone resorption maintain homeostasis of calcium and phosphate concentrations within the plasma of the blood. These processes are regulated by parathyroid hormone, 1,25-dihydroxyvitamin D_3, and calcitonin.

One of the functions of the skeletal system is mineral storage, and bone tissue is highly dynamic in maintaining homeostasis of calcium and phosphate concentrations. Calcium and phosphate are stored in bone as *hydroxyapatite* (*hi-drok″se-ap′ĭ-tīt*) *crystals*, which have the formula $Ca_{10}(PO_4)_6(OH)_2$. The calcium phosphate in these hydroxyapatite crystals is derived from the blood by the action of bone-forming cells, or osteoblasts (see fig. 8.5). The osteoblasts secrete an organic

Table 8.4

Factors Affecting Bone Physiology

Substance	Effect
Growth hormone	Stimulates osteoblast activity and collagen synthesis
Thyroid hormones	Stimulate osteoblast activity, collagen synthesis, and formation of ossification centers
Parathyroid hormone (PTH)	Stimulates resorption
1,25-dihydroxyvitamin D_3	Stimulates intestinal absorption of calcium and phosphate, bone resorption
Calcitonin	Stimulates deposition
Sex hormones (especially androgens)	Stimulate osteoblast activity and bone growth
Adrenocorticoid hormones	Stimulate osteoclast activity
Vitamin A	Promotes chondrocyte function; synthesis of lysosomal enzymes for osteoclast activity
Vitamin C	Promotes collagen synthesis

matrix, composed largely of collagenous fibers, which becomes hardened by deposits of hydroxyapatite. This process is called **bone deposition** or **ossification. Bone resorption** (dissolution), produced by the action of osteoclasts, results in the return of bone calcium and phosphate to the blood.

The formation and resorption of bone occur constantly at rates determined by a number of physiological factors (table 8.4). Body growth during the first two decades of life occurs because bone formation proceeds at a faster rate than bone resorption. By age 50 or 60, the rate of bone resorption often exceeds the rate of bone deposition. The constant activity of osteoblasts and osteoclasts allows bone to be remodeled throughout life.

Despite the changing rates of bone formation and resorption, the blood plasma concentrations of calcium and phosphate are maintained by hormonal control of the intestinal absorption and urinary excretion of these ions. These hormonal control mechanisms are very effective at maintaining the plasma calcium and phosphate concentrations within narrow limits.

The maintenance of normal blood plasma calcium concentrations is important because of the wide variety of effects that calcium has in the body. In addition to its role in bone formation, calcium is essential for muscle contraction (see chapter 12), hormonal action (see chapter 19), and maintenance of proper membrane permeability. An abnormally low blood plasma calcium concentration increases the permeability of the cell membranes to Na^+ and other ions. Hypocalcemia (low blood plasma calcium concentration) enhances the excitability of nerves and muscles and can result in muscle spasm (tetany).

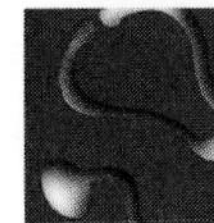

The rate of bone deposition equals the rate of bone resorption in healthy people on earth. In the *microgravity* (essentially, weightlessness) of space, however, astronauts have suffered from a slow, progressive loss of calcium from the weight-bearing bones of the legs and spine. For reasons that are not presently understood, about 100 mg of calcium is lost per day, which has reduced bone mineral density up to 20% in some astronauts who have been in space for several months. This loss cannot be countered by simply giving astronauts calcium, because hypercalcemia may cause kidney stones and other problems. The exercise machines that have been used in space have helped to prevent loss of muscle mass in astronauts, but have not been successfully able to counter the problem of bone resorption.

Parathyroid Hormone

Whenever the blood plasma concentration of calcium ions begins to fall, the parathyroid glands are stimulated to secrete increased amounts of *parathyroid hormone* (*PTH*), which acts to raise the blood Ca^{2+} back to normal levels. As might be predicted from this action of PTH, people who have their parathyroid glands removed will experience hypocalcemia. This can cause severe muscle tetany (spasm) and serves as a dramatic reminder of the importance of PTH.

Parathyroid hormone helps to raise the blood Ca^{2+} concentration primarily by stimulating the activity of osteoclasts to resorb bone. In addition, PTH stimulates the kidneys to retain blood Ca^{2+} while promoting PO_4^{-3} (phosphate ion) excretion. This raises blood Ca^{2+} levels without promoting the deposition of calcium phosphate crystals in bone. Finally, PTH promotes the formation of 1,25-dihydroxyvitamin D_3 (as described in the next section), and so it also helps to raise the blood calcium levels indirectly through the effects of this other hormone.

1,25-Dihydroxyvitamin D_3

The production of **1,25-dihydroxyvitamin D_3** begins in the skin, where vitamin D_3 is produced from its precursor molecule (7-dehydrocholesterol) under the influence of sunlight. When the skin does not make sufficient vitamin D_3 because of insufficient exposure to sunlight, this compound must be ingested in the diet—that is why it is called a vitamin. Whether this compound is secreted into the blood from the skin or enters the blood after being absorbed from the small intestine, vitamin D_3 functions as a *prehormone*; thus, it must be chemically changed in order to be biologically active (see chapter 19).

An enzyme in the liver adds a hydroxyl group (—OH) to carbon 25, which converts vitamin D_3 into 25-hydroxyvitamin D_3. In order to be active, however, another hydroxyl group must be added to the first carbon. Hydroxylation of the first carbon is accomplished by an enzyme in the kidneys, which converts the molecule to 1,25-dihydroxyvitamin D_3 (fig. 8.11). The activity of this enzyme in the kidneys is stimulated by parathyroid hormone (fig. 8.12). Increased secretion of PTH, stimulated by low blood Ca^{2+}, is thus accompanied by the increased production of 1,25-dihydroxyvitamin D_3.

Figure 8.11

The pathway for the production of the hormone 1,25-dihydroxyvitamin D_3.
This hormone is produced in the kidneys from the inactive precursor, 25-hydroxyvitamin D_3 (formed in the liver). This latter molecule is produced from vitamin D_3 secreted by the skin.

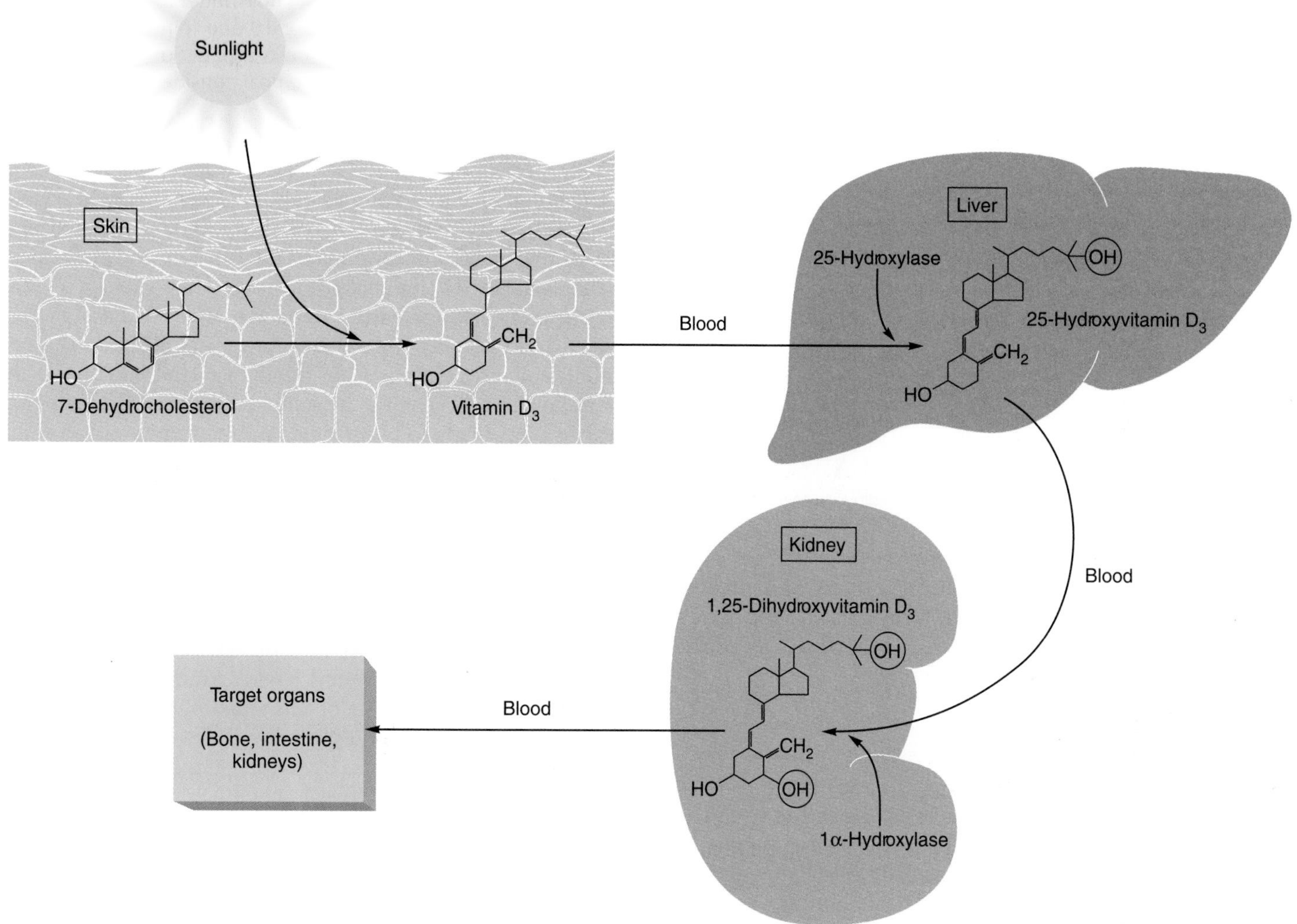

The hormone 1,25-dihydroxyvitamin D_3 helps to raise the blood plasma concentrations of calcium and phosphate by stimulating (1) the intestinal absorption of calcium and phosphate, (2) the resorption of bones, and (3) the renal reabsorption of calcium and phosphate so that less is excreted in the urine. Notice that 1,25-dihydroxyvitamin D_3, but not parathyroid hormone, directly stimulates intestinal absorption of calcium and phosphate. The effect of simultaneously raising the blood concentrations of Ca^{2+} and PO_4^{-3} results in the increased tendency of these two ions to precipitate as hydroxyapatite crystals in bone.

Since 1,25-dihydroxyvitamin D_3 directly stimulates bone resorption, it seems paradoxical that this hormone is needed for proper bone deposition and, in fact, that inadequate amounts of 1,25-dihydroxyvitamin D_3 result in the bone demineralization of osteomalacia and rickets. This apparent paradox may be explained logically by the fact that the primary function of 1,25-dihydroxyvitamin D_3 is stimulation of intestinal Ca^{2+} and PO_4^{-3} absorption. When calcium intake is adequate, the major result of 1,25-dihydroxyvitamin D_3 action is the availability of Ca^{2+} and PO_4^{-3} in sufficient amounts to promote bone deposition. Only when calcium intake is inadequate does the direct effect of 1,25-dihydroxyvitamin D_3 on bone resorption become significant, acting to ensure proper blood Ca^{2+} levels.

Negative Feedback Control of Calcium and Phosphate Balance

The secretion of parathyroid hormone is controlled by the blood plasma calcium concentrations. Its secretion is stimulated by low-calcium concentrations and inhibited by high-calcium concentrations. Since parathyroid hormone stimulates the final hydroxylation step in the formation of 1,25-dihydroxyvitamin D_3,

Figure 8.12

The effect of plasma Ca^{2+}.
A decrease in plasma Ca^{2+} directly stimulates the secretion of parathyroid hormone (PTH). The production of 1,25-dihydroxyvitamin D_3 also rises when Ca^{2+} is low because PTH stimulates the final hydroxylation step in the formation of this compound in the kidneys.

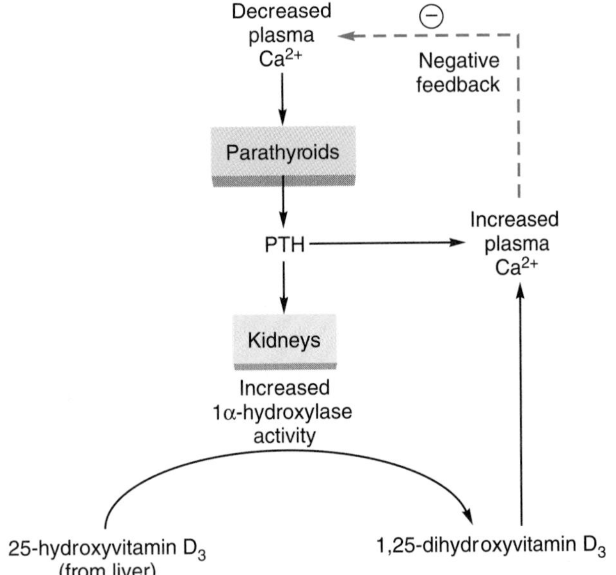

a rise in parathyroid hormone results in an increase in production of 1,25-dihydroxyvitamin D_3. Low blood calcium can thus be corrected by the effects of increased parathyroid hormone and 1,25-dihydroxyvitamin D_3.

Calcitonin

Experiments in the 1960s revealed that high blood calcium in dogs could be lowered by a hormone secreted from the thyroid gland. This hormone thus has an effect opposite to that of parathyroid hormone and 1,25-dihydroxyvitamin D_3. The calcium-lowering hormone, called **calcitonin,** was found to be a 32-amino-acid polypeptide secreted by the thyroid gland (see chapter 19).

The secretion of calcitonin is stimulated by high blood plasma calcium levels and acts to lower calcium levels by (1) inhibiting the activity of osteoclasts, thus reducing bone resorption, and (2) stimulating the urinary excretion of calcium and phosphate by inhibiting their reabsorption in the kidneys.

Although it is attractive to think that calcium balance is regulated by the effects of antagonistic hormones, the significance of calcitonin in human physiology remains unclear. Patients who have had their thyroid gland surgically removed (as for thyroid cancer) are *not* hypercalcemic, as one would expect if calcitonin were needed to lower blood calcium levels. The ability of very large, pharmacological doses of calcitonin to inhibit osteoclast activity and bone resorption, however, is clinically useful in the treatment of Paget's disease, in which osteoclast activity causes softening of bone.

Paget's disease: from Sir James Paget, English surgeon, 1814–99

Clinical Considerations

Each bone is a dynamic living organ that is influenced by hormones, diet, aging, and disease. Since the development of bone is genetically controlled, congenital abnormalities may occur. The hardness of bones gives them strength, yet they lack the resiliency to avoid fracture when they undergo severe trauma. (Fractures are discussed in chapter 10 and joint injuries are discussed in chapter 11.) All of these aspects of bone make for some important and interesting clinical considerations.

Developmental Disorders

Congenital malformations account for several types of skeletal deformities. Certain bones may fail to form during osteogenesis, or they may form abnormally. **Cleft palate** and **cleft lip** are malformations of the palate and face. They vary in severity and seem to involve both genetic and environmental factors. **Spina bifida** (*spi'nă bif'ĭ-dă*) is a congenital defect of the vertebral column resulting from a failure of the laminae of the vertebrae to fuse, leaving the spinal cord exposed. The lumbar area is most likely to be affected, and frequently only a single vertebra is involved.

Nutritional and Hormonal Disorders

Several bone disorders result from nutritional deficiencies or from excessive or deficient amounts of the hormones that regulate bone development and growth. Vitamin D has a tremendous influence on proper bone structure and function. When there is a deficiency of this vitamin, the body is unable to metabolize calcium and phosphorus. Vitamin D deficiency in children is linked to rickets, which causes the bones to remain soft and structurally weak, bending under the weight of the body (see fig. 7.8).

A vitamin D deficiency in adults causes the bones to demineralize, or to give up stored calcium and phosphorus. This demineralization results in a condition called **osteomalacia** (*os"te-o-mă-la'shă*). Osteomalacia occurs most often in malnourished women who have repeated pregnancies and who experience relatively little exposure to sunlight.

The consequences of endocrine disorders are described in chapter 19. Since hormones exert a strong influence on bone development, however, a few endocrine disorders will be briefly mentioned here. Hypersecretion of the growth hormone from the pituitary gland leads to **gigantism** in young people, if it begins before ossification of their epiphyseal plates. In adults, it leads to **acromegaly** (*ak"ro-meg'ă-le*), which is characterized by hypertrophy of the bones of the face, hands, and feet. In a child, growth hormone deficiency results in slowed bone growth—a condition called **dwarfism.**

Paget's disease, a bone disorder that mainly affects older adults, occurs more frequently in males than in females. It is characterized by disorganized metabolic processes within the bone tissue. The activity of osteoblasts and osteoclasts becomes irregular, resulting in thick bony deposits in some areas of the skeleton and fragile, thin bones in other areas.

Figure 8.13

A bone scan of the legs of a patient suffering from arthritis in the left knee joint.
In a bone scan, arthritis is depicted as a brighter image than a normal joint.

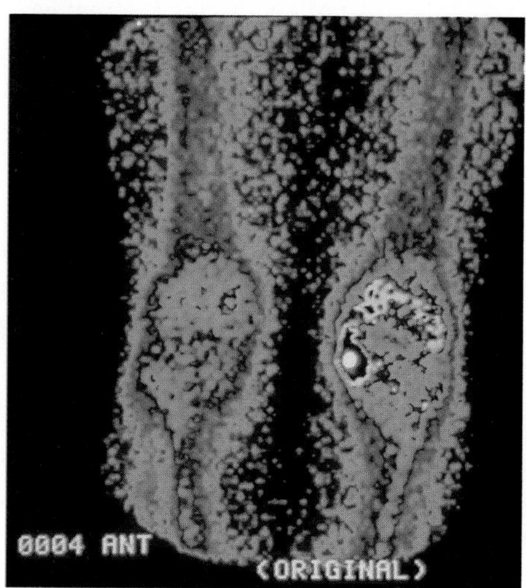

The vertebral column, pelvis, femora, and skull are most often involved, and become increasingly painful and deformed. Bowed leg bones, abnormal curvature of the spine, and enlargement of the skull may develop. The cause of Paget's disease is currently not known.

Neoplasms of Bone

Malignant bone tumors are three times more common than benign tumors. Pain is the usual symptom of either type of osseous neoplasm, although benign tumors may not have accompanying pain.

Two types of benign bone tumors are **osteomas,** which are the more frequent and which often involve the skull, and **osteoid osteomas,** which are painful neoplasms of the long bones, usually in children.

Osteogenic sarcoma (*sar-ko'ma*) is the most virulent type of bone cancer. It frequently metastasizes through the blood to the lungs. This disease usually originates in the long bones and is accompanied by aching and persistent pain.

A **bone scan** (fig. 8.13) is a diagnostic procedure frequently done on a person who has had a malignancy elsewhere in the body that may have metastasized to the bone. The patient receiving a bone scan may be injected with a radioactive substance that accumulates more rapidly in malignant tissue than normal tissue, causing the malignant bone areas to appear as intensely dark dots.

Aging of the Skeletal System

Senescence affects the skeletal system by decreasing skeletal mass and density and increasing porosity and erosion (fig. 8.14). Bones become more brittle and susceptible to fracture. Articulating surfaces also deteriorate, contributing to arthritic conditions. Arthritic diseases are second to heart disease as the most common debilitation in the elderly.

Osteoporosis (*os"te-o-pŏ-ro'sis*) is a weakening of the bones, primarily as a result of calcium loss. The causes of osteoporosis include aging, inactivity, poor diet, and an imbalance in hormones or other chemicals in the blood. It is most common in older women because low levels of estrogens after menopause lead to increased bone resorption, and the formation of new bone is not sufficient to keep pace. People with osteoporosis are prone to bone fracture, particularly at the pelvic girdle and vertebrae, as the bones become too brittle to support the weight of the body. Complications of hip fractures often lead to permanent disability, and vertebral compression fractures may produce a permanent curved deformity of the spine.

Figure 8.14

Scanning electron micrographs of spongy bone biopsy specimens from the ilium.
(*a*) A normal specimen and (*b*) a specimen from a person with osteoporosis.

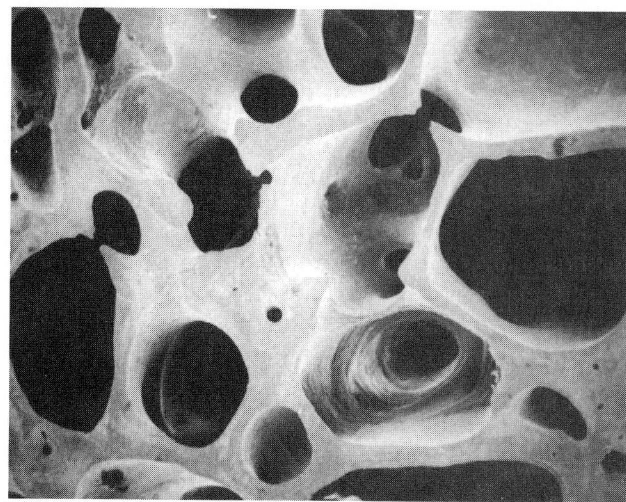

(a)

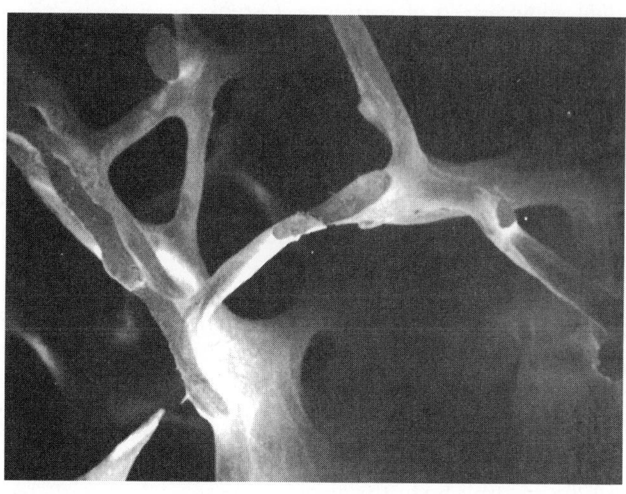

(b)

Although there is no known cure for osteoporosis, good eating habits and a regular program of exercise, established at an early age and continued throughout adulthood, can minimize its effects. Treatment in women through dietary calcium, exercise, and estrogens has had limited positive results. In addition, a drug called *alendronate* (Fosamax), approved by the FDA in 1995, has been shown to be effective in managing osteoporosis. This drug works without hormones to block osteoclast activity, making it useful for women who choose not to be treated with estrogen replacement therapy.

Clinical Investigation Answer

Osteoporosis is the deterioration or wasting of bone tissue. In healthy bone tissue, there is a balance between the breakdown of old bone through the action of osteoclasts and the production of new bone through the action of osteoblasts. In osteoporosis, the breakdown occurs at a faster rate than the replacement, causing the bones to become weak and brittle. Because it carries the weight of the body and maintains posture, the vertebral column is particularly susceptible to the effects of osteoporosis. Abnormal curvature of the vertebral column may develop due to the gradual compression of vertebrae. Osteoporosis in elderly persons frequently follows prolonged immobilization such as from a fracture. Although there is no specific treatment for osteoporosis, exercise and a diet that includes a daily calcium intake of about 1,500 mg generally slows its progress. In addition, some women respond favorably to estrogen supplements. A person with osteoporosis must take precautions to avoid falls and should use a cane if unstable while walking. Evidence indicates that, to a certain extent, osteoporosis has a genetic basis. There is also abundant evidence, however, that maintaining a healthy body throughout one's life greatly reduces the chance of being afflicted with this disease.

Chapter Summary

Organization of the Skeletal System (pp. 183–185)

1. The axial skeleton consists of the skull, auditory ossicles, hyoid bone, vertebral column, and rib cage.
2. The appendicular skeleton consists of the bones within the pectoral girdle, upper extremities, pelvic girdle, and lower extremities.

Functions of the Skeletal System (pp. 185–187)

1. The mechanical functions of bones include the support and protection of softer body tissues and organs; in addition, certain bones function as levers during body movement.
2. The metabolic functions of bones include hemopoiesis and mineral storage.

Bone Structure (pp. 187–189)

1. Bone structure includes the shape and surface features of each bone, along with gross internal components.
2. Structurally speaking, bones may be classified as long, short, flat, or irregular.
3. The surface features of bones can be broadly classified into articulating surfaces, nonarticulating prominences, and depressions and openings.
4. A typical long bone has a diaphysis filled with marrow in the medullary cavity, epiphyses, epiphyseal plates for linear bone growth, and a covering of periosteum for appositional bone growth and the attachments of ligaments and tendons.

Bone Tissue (pp. 189–190)

1. The five types of bone cells are osteogenic cells, in contact with the endosteum and periosteum; osteoblasts (bone-forming cells); osteocytes (mature bone cells); osteoclasts (bone-destroying cells); and bone-lining cells, along the surface of most bones.
2. Compact bone consists of precise arrangements of osteons. The osteons contain osteocytes and lamellae.

Bone Growth (pp. 190–193)

1. Bone growth from embryonic to adult size is an orderly process determined by genetics, hormonal secretions, and nutritional supply.
2. Most bones develop through endochondral ossification.
3. Bone remodeling is a continual process that involves osteoclasts in bone resorption and osteoblasts in the formation of new bone tissue.

Physiology of Bone Tissue (pp. 193–196)

1. Bone tissue contains a reserve supply of calcium and phosphate for the blood in the form of hydroxyapatite crystals.
2. Parathyroid hormone stimulates bone resorption and calcium reabsorption in the kidneys, and thus raises the blood calcium concentration.
3. 1,25-dihydroxyvitamin D_3 is derived from vitamin D by hydroxylation reactions in the liver and kidneys.
4. A rise in parathyroid hormone, accompanied by the increased production of 1,25-dihydroxyvitamin D_3, helps to maintain proper blood levels of calcium and phosphate in response to a fall in calcium levels.
5. Calcitonin is secreted by the thyroid gland and lowers blood calcium by inhibiting bone resorption and stimulating the urinary excretion of calcium and phosphate.

Review Activities

Objective Questions

1. A bone is considered to be
   (a) a tissue. (c) a cell.
   (b) an organ. (d) a system.
2. Which of the following statements is *false?*
   (a) Bones are important in the synthesis of vitamin D.
   (b) Bones and teeth contain about 95% of the body's calcium.
   (c) Red bone marrow is the primary site for hemopoiesis.
   (d) Most bones develop through endochondral ossification.
3. Bone tissue derives from specialized migratory mesodermal cells called
   (a) dermatomes. (c) mesenchyme.
   (b) myotomes. (d) somites.
4. As a structural feature of certain bones, a fovea is
   (a) a rounded opening through a bone.
   (b) a small rounded process.
   (c) a deep pit or socket.
   (d) a small pit or depression.

5. The periosteum is secured to bone by
   (a) the epiphyseal plate.
   (b) perforating fibers.
   (c) interosseous ligaments.
   (d) diploe.
6. Columnlike structures within compact bone tissue are called
   (a) osteons. (c) lamellae.
   (b) lacunae. (d) diaphyses.
7. Specialized bone cells that enzymatically reabsorb bone tissue are called
   (a) osteoblasts. (c) osteocytes.
   (b) osteons. (d) osteoclasts.
8. The increased intestinal absorption of calcium is stimulated directly by
   (a) parathyroid hormone.
   (b) 1,25-dihydroxyvitamin D_3.
   (c) calcitonin.
   (d) all of the above.
9. A rise in blood calcium levels directly stimulates
   (a) parathyroid hormone secretion.
   (b) calcitonin secretion.
   (c) 1,25-dihydroxyvitamin D_3 formation.
   (d) all of the above.
10. The bone disorder common in elderly people, particularly if they are subject to prolonged inactivity, malnutrition, or an unbalanced secretion of hormones, is
    (a) osteitis. (c) osteoporosis.
    (b) osteonecrosis. (d) osteomalacia.

Essay Questions

1. Distinguish between the axial skeleton and the appendicular skeleton. List the bones that compose the pectoral and pelvic girdles.
2. Sketch a typical long bone. Label the diaphysis, epiphyses, articular cartilages, periosteum, and medullary cavity.
3. Define *osteon*. Sketch an osteon and label the osteocytes, lacunae, lamellae, central canal, and canaliculi.
4. Describe how bones grow in length and in circumference. How are these processes similar, and how do they differ? Explain how radiographs can be used to determine normal bone growth.
5. Describe the process of endochondral ossification of a long bone. Why is it important that a balance be maintained between osteoblast activity and osteoclast activity?
6. Explain why a proper balance of vitamins, hormones, and minerals is essential in maintaining healthy bone tissue. Give examples of diseases or skeletal conditions that may occur if there is an imbalance of any of these three essential substances.
7. Why is vitamin D considered to be both a vitamin and a prehormone? Explain why people with osteoporosis might be helped by taking controlled amounts of vitamin D.

Critical Thinking Questions

1. Your kid brother is convinced that the bones in our bodies are dead—understandable considering that many people associate bones with graveyards, Halloween, and leftover turkey from a Thanksgiving dinner. What information could you use to try to get him to change his mind?
2. Over the past century, the average adult heights of both men and women in developing countries have significantly increased from one generation to the next. What accounts for this change? Will the trend continue?
3. Why are there more bones in a child than in an adult? Which bones undergo fusion during aging? What is the possible clinical importance of this knowledge?
4. Explain why a proper balance of vitamins, hormones, and minerals is essential in maintaining healthy bone tissue. Give examples of diseases or skeletal conditions that may occur in the event of an imbalance of any of these three essential substances.

Related Web Sites

For a listing of the most current web sites related to this chapter, please visit the *Concepts of Human Anatomy and Physiology* home page at http://www.mhhe.com/biosci/abio/.

Nine

Skeletal System: Axial Skeleton

Objectives

- Identify the cranial and facial bones of the skull and describe their structural characteristics.
- Describe the location of each of the bones of the skull and identify the articulations that affix them to each other.
- Identify the bones of the five regions of the vertebral column and describe the characteristic curves of each region.
- Describe the structure of a typical vertebra.
- Identify the parts of the rib cage and compare and contrast the various types of ribs.

Clinical Investigation

A 68-year-old man visited his family doctor for his first physical examination in 30 years. Upon sensing a disgruntled patient, the doctor gently tried to determine the reason. In response to the doctor's inquiry, the patient blurted out, "The nurse who measured my height is incompetent! I know for a fact I used to be six feet even when I was in the navy, but she tells me I'm 5′10″!" The doctor then performed the measurement himself, noting that although the patient's posture was excellent, he was indeed 5′10″, just as the nurse had said. He explained to the patient that the spine contains some nonbony tissue, which shrivels up a bit over the years. The patient interrupted, stating indignantly that he knew anatomic terms and principles and would like a detailed explanation.

How would you explain the anatomy of the vertebral column and the changes it undergoes during the aging process?

Clues: The patient's normal posture and the fact that he had no complaints of pain indicated good health for his age. Examine figure 9.19 and carefully read the accompanying caption. Also see the Clinical Considerations section at the end of chapter 8.

Skull

The human skull, consisting of 8 cranial and 14 facial bones, contains several cavities that house the brain and sensory organs. Each bone of the skull articulates with the adjacent bones and has diagnostic and functional processes, surface features, and foramina.

The skull consists of *cranial bones* and *facial bones*. The 8 bones of the cranium join firmly with one another to enclose and protect the brain and sense organs. The 14 facial bones form the framework for the facial region and support the teeth. Variation in size, shape, and density of the facial bones is a major contributor to the individuality of each human face. The facial bones, with the exception of the mandible ("jawbone"), are also firmly interlocked with one another and the cranial bones.

The skull has several cavities. The **cranial cavity** is the largest, with a capacity of about 1,300–1,350 cc. The **nasal cavity** is formed by both cranial and facial bones and is partitioned into two chambers, or **nasal fossae,** by a **nasal septum** of bone and cartilage. Four sets of **paranasal sinuses,** located within the bones surrounding the nasal area, communicate via ducts into the nasal cavity. **Middle-** and **inner-ear cavities** are positioned inferior to the cranial cavity and house the organs of hearing and balance. The two **orbits** for the eyeballs are formed by facial and cranial bones. The **oral,** or **buccal** (*buk′al*) **cavity** (mouth), which is only partially formed by bone, is completely within the facial region.

The bones of the skull contain numerous foramina to accommodate nerves, vessels, and other structures. A summary of the foramina of the skull is presented in table 9.1. Figures 9.1 through 9.8 show various views of the skull. Radiographs of the skull are shown in figure 9.9.

Although the hyoid bone and the three paired auditory ossicles are not considered part of the skull, they are located within the axial skeleton and are described immediately following the discussion of the skull.

cranium: Gk. *kranion*, skull

Cranial Bones

The cranial bones enclose and protect the brain and associated sensory organs. They consist of one *frontal*, two *parietals*, two *temporals*, one *occipital*, one *sphenoid*, and one *ethmoid*.

Frontal Bone

The frontal bone forms the anterior roof of the cranium, the forehead, the roof of the nasal cavity, and the superior arch of the orbits, which contain the eyeballs. The bones of the orbit are summarized in table 9.2. The frontal bone develops in two halves that grow together. Generally, they are completely fused by age 5 or 6. A suture sometimes persists between these two portions beyond age 6 and is referred to as a *metopic* (*mĕ-top′ik*) *suture*. The **supraorbital margin** is a prominent bony ridge over the orbit. Slightly medial to its midpoint is an opening called the **supraorbital foramina,** which provides passage for a nerve, artery, and vein.

The frontal bone also contains **frontal sinuses,** which are connected to the nasal cavity (fig. 9.9). These sinuses, along with the other paranasal sinuses, lessen the weight of the skull and act as resonance chambers for voice production.

Parietal Bone

The two parietal bones form the upper sides and roof of the cranium (figs. 9.2 and 9.4). The **coronal** (*kă-ro′nal*) **suture** separates the frontal bone from the parietal bones, and the **sagittal** (*saj′ĭ-tal*) **suture** along the superior midline separates the right and left parietals from each other. The inner concave surfaces of each parietal bone, as well as the inner concave surfaces of other cranial bones, is marked by shallow impressions from convolutions of the brain and vessels serving the brain.

metopic suture: Gk. *metopon*, forehead; L. *sutura*, sew

Table 9.1

Major Foramina of the Skull

Foramen	Location	Structure Transmitted
Carotid canal	Petrous part of temporal bone	Internal carotid artery and sympathetic nerves
Greater palatine foramen	Palatine bone of hard palate	Greater palatine nerve and descending palatine vessels
Hypoglossal canal	Anterolateral edge of occipital condyle	Hypoglossal nerve and branch of ascending pharyngeal artery
Incisive foramen	Anterior region of hard palate, posterior to incisors	Branches of descending palatine vessels and ansopalatine nerve
Jugular foramen	Between part of temporal and occipital bones, posterior to carotid canal	Internal jugular vein; vagus, glossopharyngeal, and accessory nerves
Foramen lacerum	Between petrous part of temporal and sphenoid bones	Branch of ascending pharyngeal artery and internal carotid artery
Lesser palatine foramen	Posterior to greater palatine foramen in hard palate	Lesser palatine nerves
Foramen magnum	Occipital bone	Union of medulla oblongata and spinal cord, meningeal membranes, and accessory nerves, vertebral and spinal arteries
Mandibular foramen	Medial surface of ramus of mandible	Inferior alveolar nerve and vessels
Mental foramen	Below second premolar on lateral side of mandible	Mental nerve and vessels
Nasolacrimal canal	Lacrimal bone	Nasolacrimal (tear) duct
Cribriform foramina	Cribriform plate of ethmoid bone	Olfactory nerves
Optic foramen	Back of orbit in lesser wing of sphenoid bone	Optic nerve and ophthalmic artery
Foramen ovale	Greater wing of sphenoid bone	Mandibular nerve of trigeminal nerve
Foramen rotundum	Within body of sphenoid bone	Maxillary nerve of trigeminal nerve
Foramen spinosum	Posterior angle of sphenoid bone	Middle meningeal vessels
Stylomastoid foramen	Between styloid and mastoid processes of temporal bone	Facial nerve and stylomastoid artery
Superior orbital fissure	Between greater and lesser wings of sphenoid bone	Four cranial nerves (oculomotor, trochlear, ophthalmic nerve of trigeminal, and abducens)
Supraorbital foramen	Supraorbital ridge of orbit	Supraorbital nerve and artery
Zygomaticofacial foramen	Anterolateral surface of zygomatic bone	Zygomaticofacial nerve and vessels

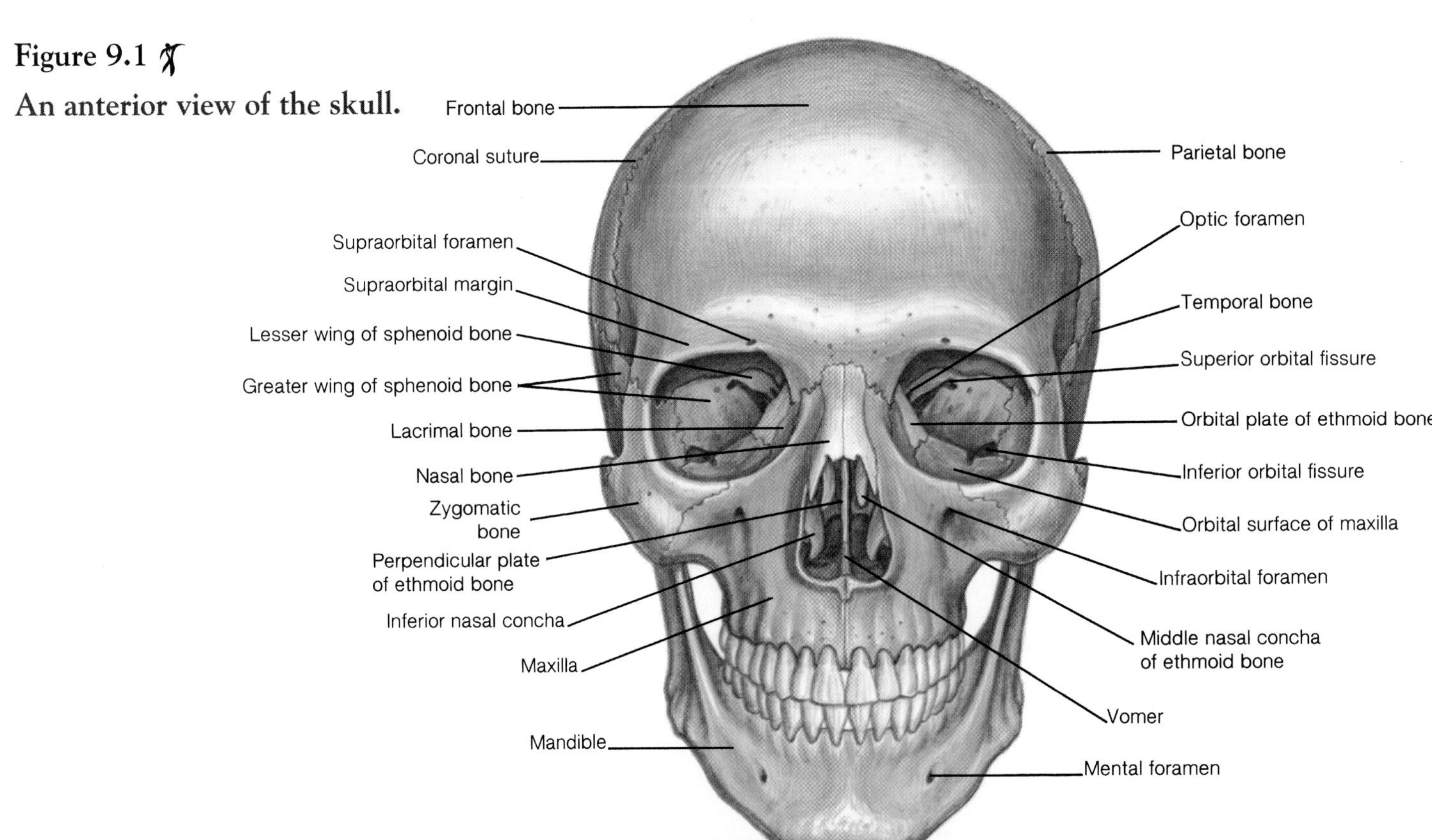

Figure 9.1 An anterior view of the skull.

Figure 9.2
A lateral view of the skull.

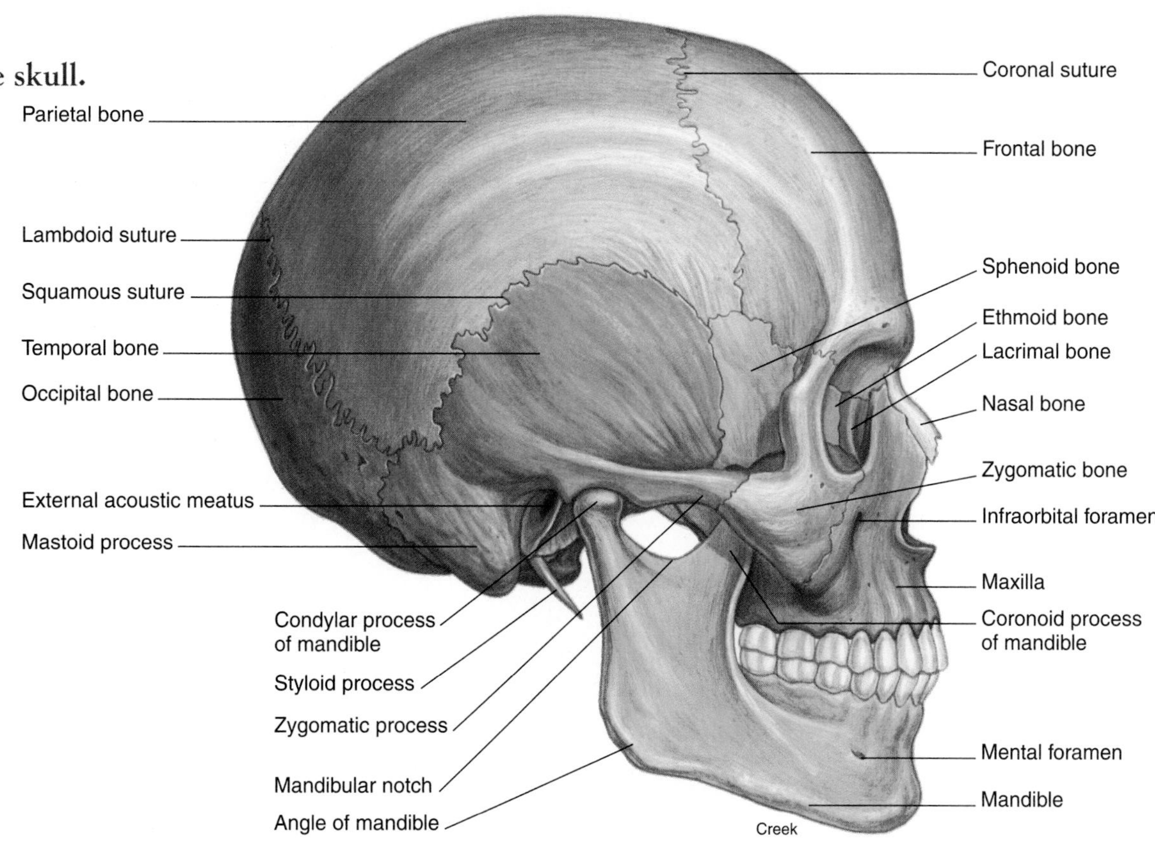
Parietal bone
Lambdoid suture
Squamous suture
Temporal bone
Occipital bone
External acoustic meatus
Mastoid process
Condylar process of mandible
Styloid process
Zygomatic process
Mandibular notch
Angle of mandible
Coronal suture
Frontal bone
Sphenoid bone
Ethmoid bone
Lacrimal bone
Nasal bone
Zygomatic bone
Infraorbital foramen
Maxilla
Coronoid process of mandible
Mental foramen
Mandible
Creek

Figure 9.3
An inferior view of the skull.

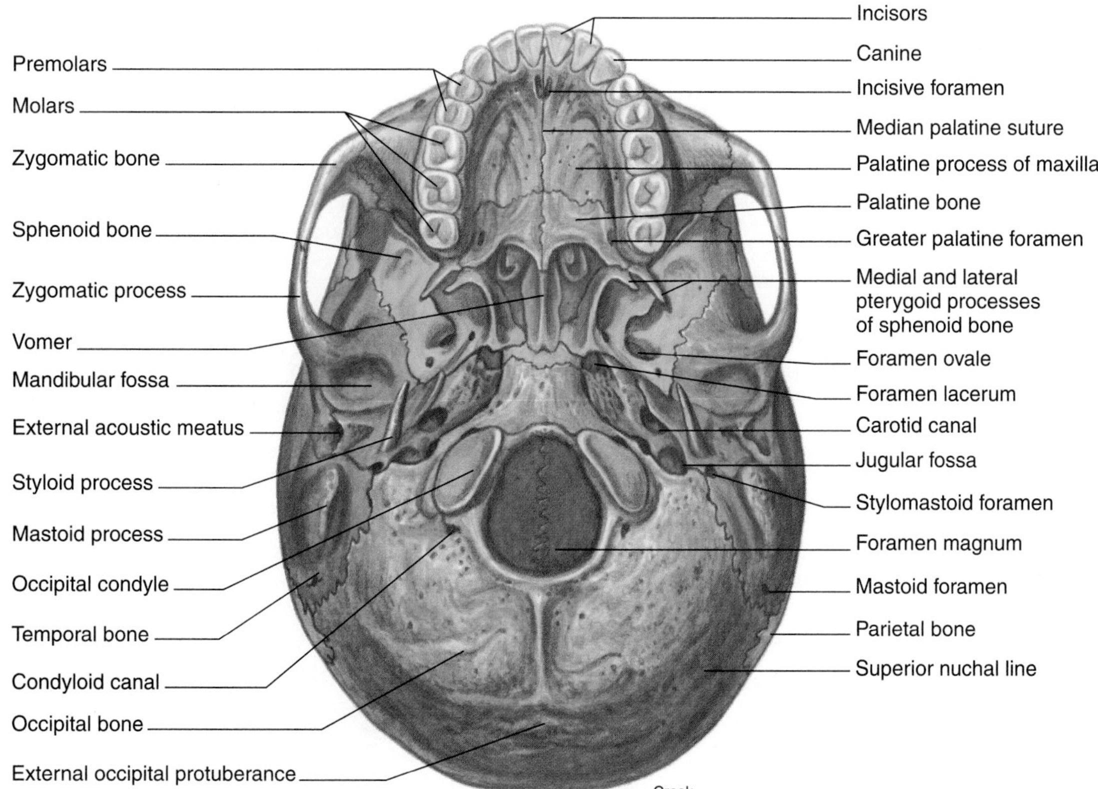
Premolars
Molars
Zygomatic bone
Sphenoid bone
Zygomatic process
Vomer
Mandibular fossa
External acoustic meatus
Styloid process
Mastoid process
Occipital condyle
Temporal bone
Condyloid canal
Occipital bone
External occipital protuberance
Incisors
Canine
Incisive foramen
Median palatine suture
Palatine process of maxilla
Palatine bone
Greater palatine foramen
Medial and lateral pterygoid processes of sphenoid bone
Foramen ovale
Foramen lacerum
Carotid canal
Jugular fossa
Stylomastoid foramen
Foramen magnum
Mastoid foramen
Parietal bone
Superior nuchal line
Creek

Figure 9.4

A sagittal view of the skull.

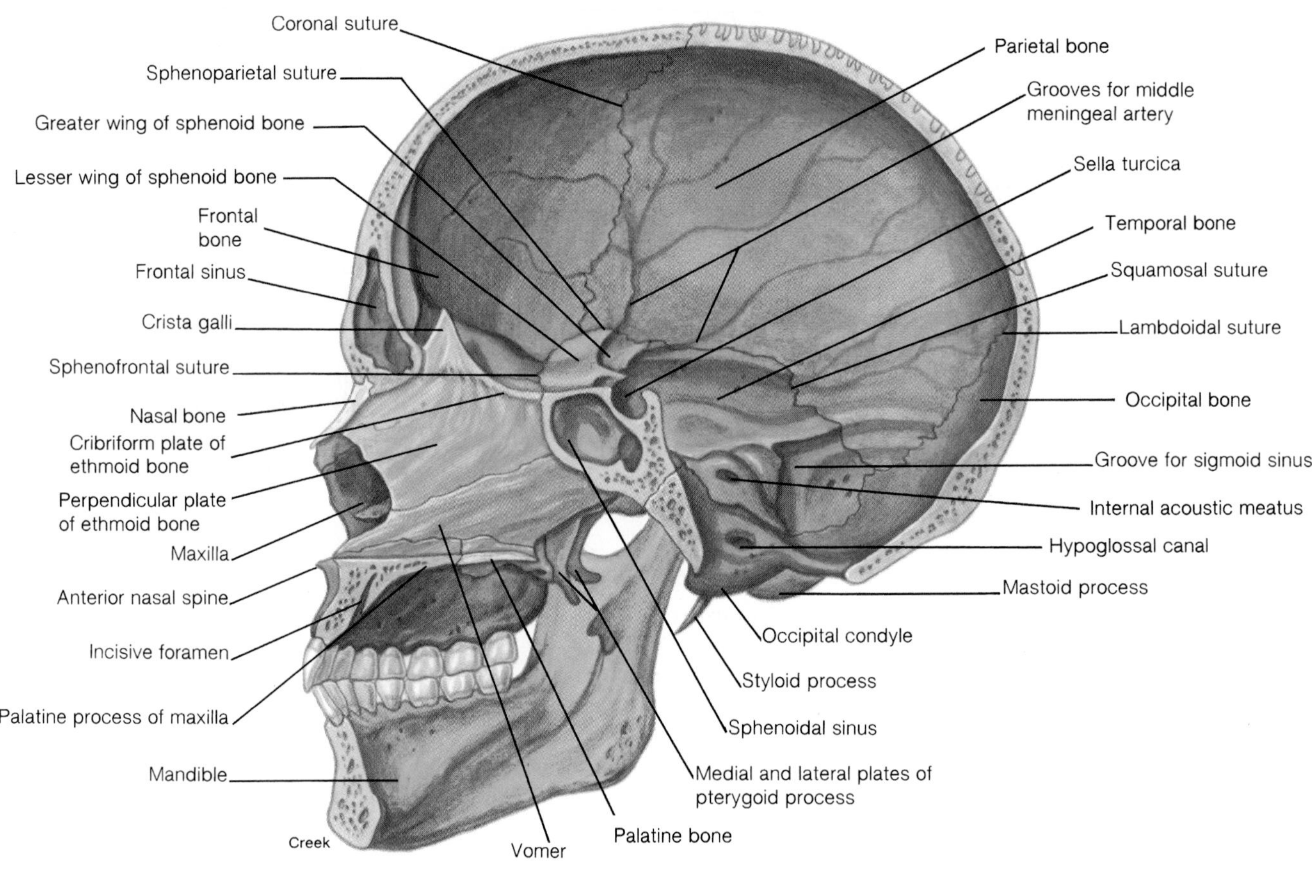
Coronal suture
Sphenoparietal suture
Greater wing of sphenoid bone
Lesser wing of sphenoid bone
Frontal bone
Frontal sinus
Crista galli
Sphenofrontal suture
Nasal bone
Cribriform plate of ethmoid bone
Perpendicular plate of ethmoid bone
Maxilla
Anterior nasal spine
Incisive foramen
Palatine process of maxilla
Mandible
Creek
Vomer
Palatine bone
Parietal bone
Grooves for middle meningeal artery
Sella turcica
Temporal bone
Squamosal suture
Lambdoidal suture
Occipital bone
Groove for sigmoid sinus
Internal acoustic meatus
Hypoglossal canal
Mastoid process
Occipital condyle
Styloid process
Sphenoidal sinus
Medial and lateral plates of pterygoid process

Figure 9.5

An inferolateral view of the skull.

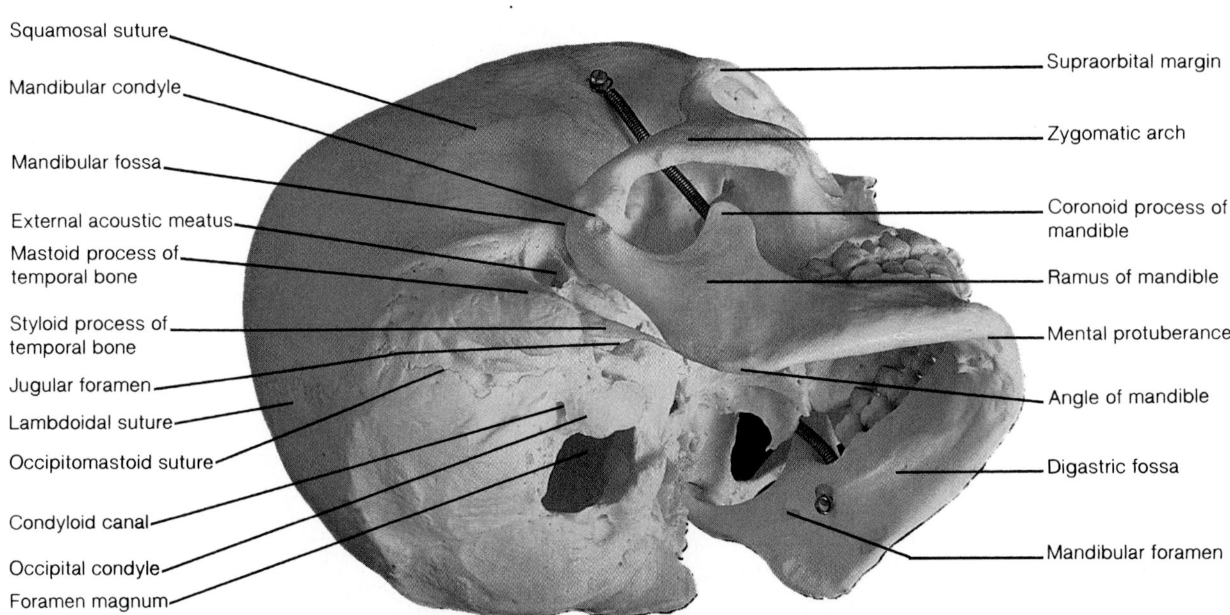
Squamosal suture
Mandibular condyle
Mandibular fossa
External acoustic meatus
Mastoid process of temporal bone
Styloid process of temporal bone
Jugular foramen
Lambdoidal suture
Occipitomastoid suture
Condyloid canal
Occipital condyle
Foramen magnum
Supraorbital margin
Zygomatic arch
Coronoid process of mandible
Ramus of mandible
Mental protuberance
Angle of mandible
Digastric fossa
Mandibular foramen

Figure 9.6
The floor of the cranial cavity.

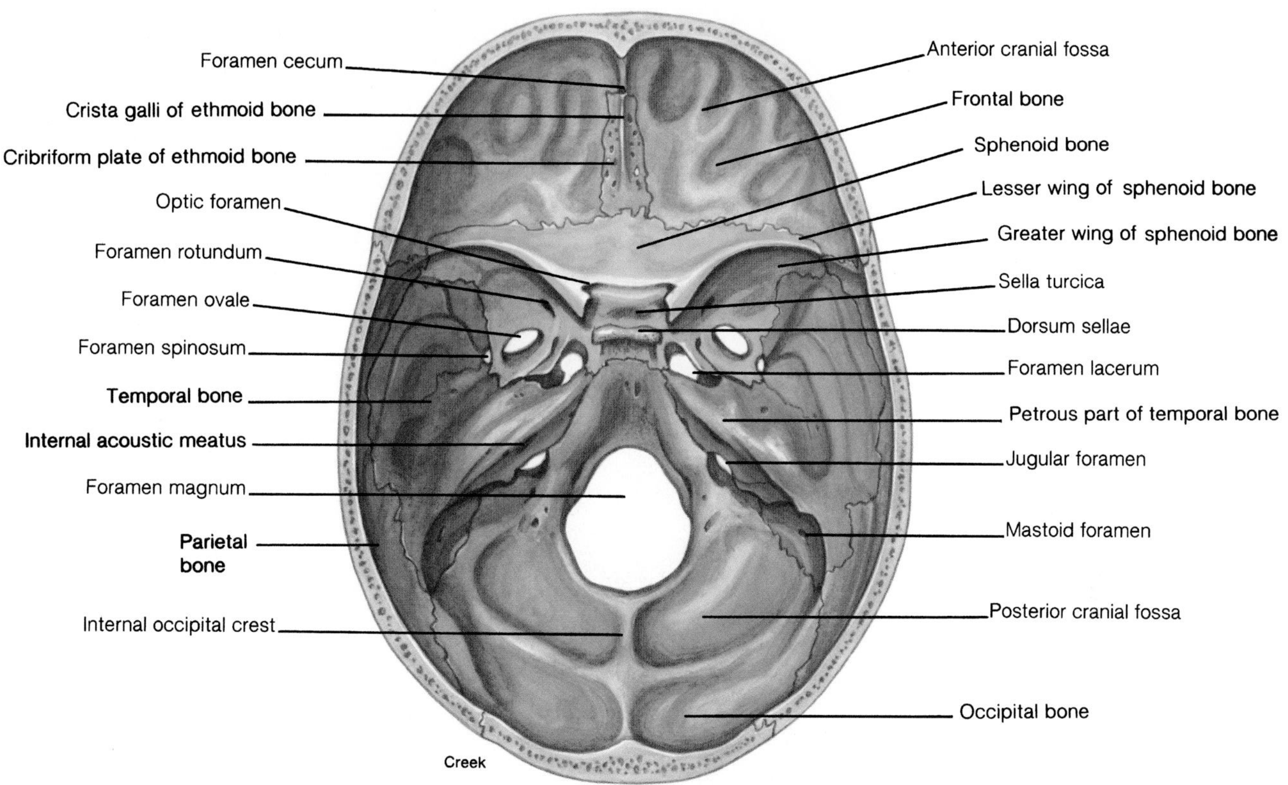

Figure 9.7
A posterior view of a frontal (coronal) section of the skull.

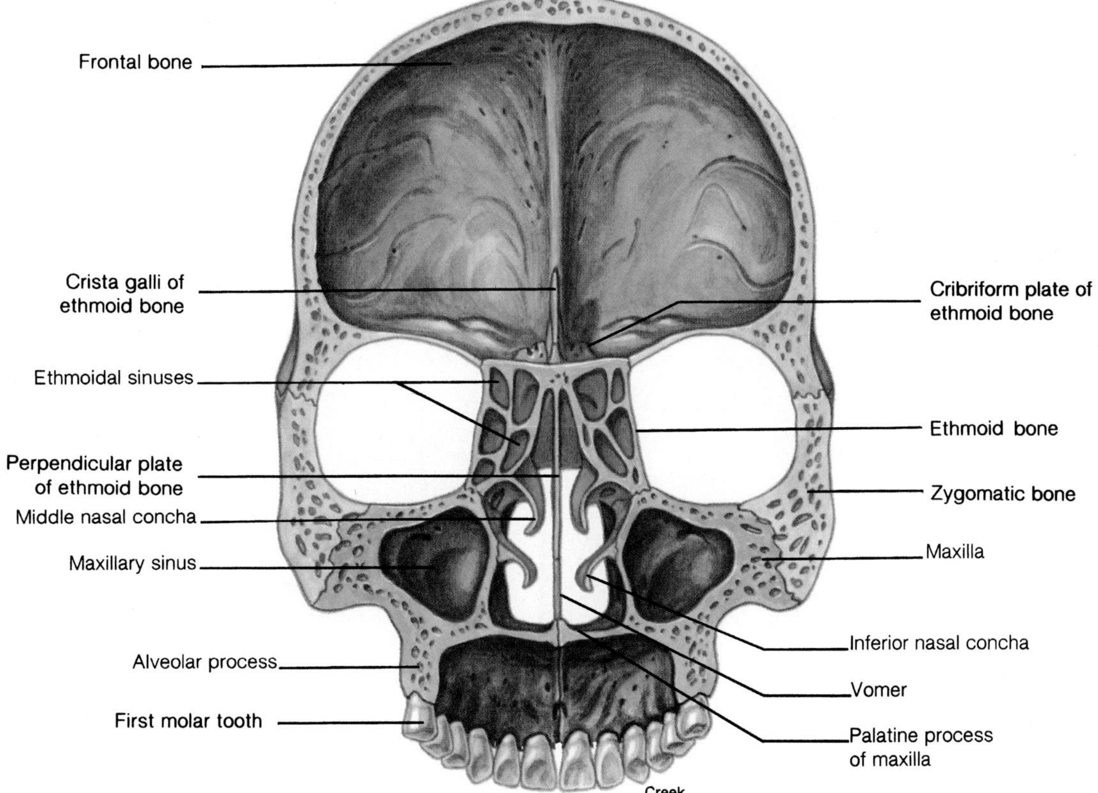

Figure 9.8

Bones of the orbit.

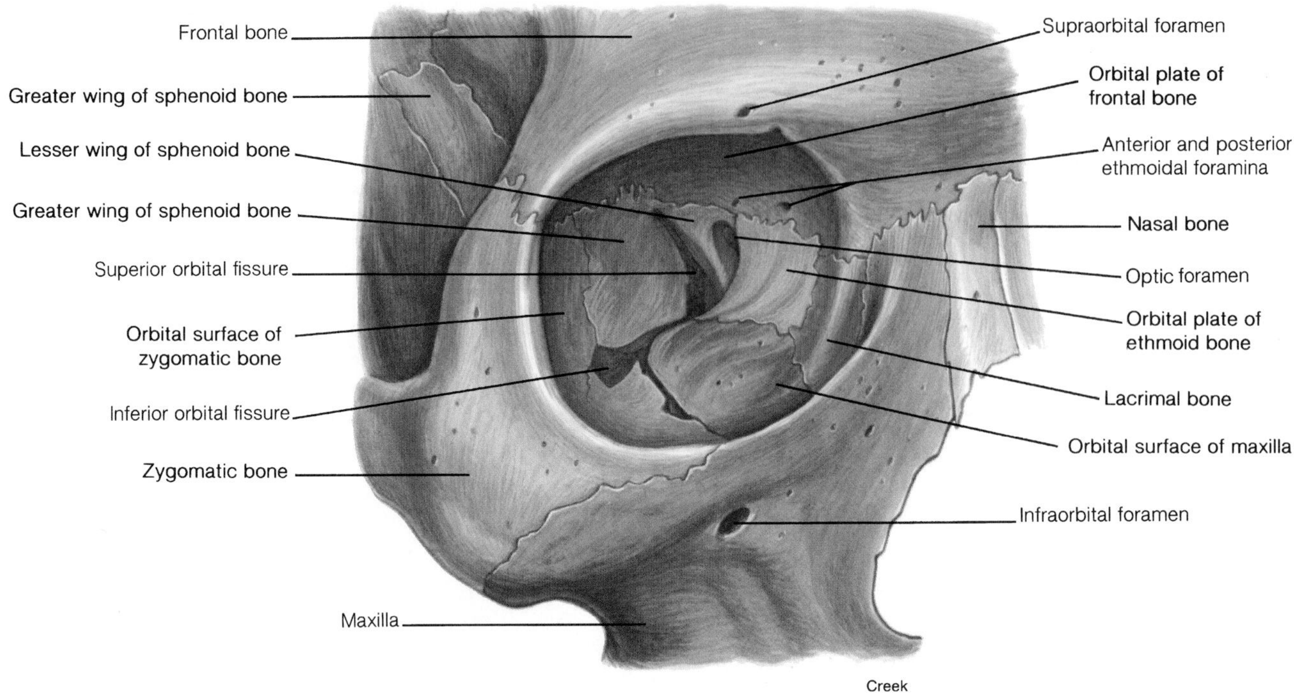

Figure 9.9

Radiographs (X rays) of the skull showing paranasal sinuses.
(*a*) An anteroposterior view and (*b*) a right lateral view.

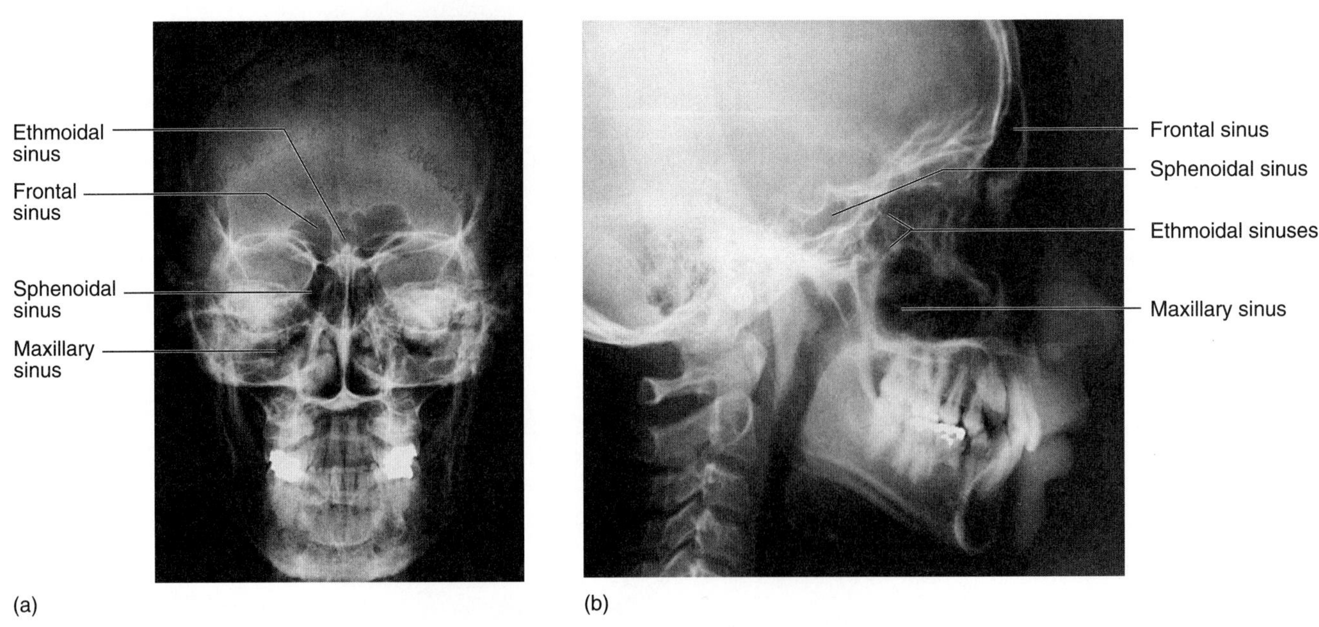

Temporal Bone

The two temporal bones form the lower sides of the cranium (figs. 9.2, 9.3, 9.4, and 9.10). Each temporal bone is joined to its adjacent parietal bone by the **squamosal** (*skwă-mo'sal*) **suture.** Structurally, each temporal bone has four parts.

1. **Squamous part.** The squamous part is the flattened plate of bone at the sides of the skull. Projecting forward is a **zygomatic** (*zi"go-mat'ik*) **process** that forms the posterior portion of the **zygomatic arch.** On the inferior surface of the squamous part is the cuplike **mandibular fossa,** which forms a joint with the condyle of the mandible. The articulation is the *temporomandibular* (*tem"pŏ-ro-man-dib'yŭ-lar*) *joint*.
2. **Tympanic part.** The tympanic part of the temporal bone contains the **external acoustic meatus** (*me-a'tus*), or ear canal, which is posterior to the mandibular fossa. A thin, pointed **styloid process** (figs. 9.3 and 9.4) projects inferiorly from the tympanic part.
3. **Mastoid part.** The **mastoid process,** a rounded projection posterior to the external acoustic meatus, accounts for the mass of the mastoid part. The **mastoid foramen** (fig. 9.3) is directly posterior to the mastoid process. The **stylomastoid foramen,** located between the mastoid and styloid processes (fig. 9.3), provides the passage for part of the facial nerve.
4. **Petrous part.** The petrous (*pet'rus*) part can be seen in the floor of the cranium (figs. 9.6 and 9.10*b*). The structures of the middle ear and inner ear are housed in

Table 9.2
Bones Forming the Orbit

Region of the Orbit	Contributing Bones
Roof (superior)	Frontal; lesser wing of sphenoid bone
Floor (inferior)	Maxilla; zygomatic bone; palatine bone
Lateral wall	Zygomatic bone
Posterior wall	Greater wing of sphenoid bone
Medial wall	Maxilla; lacrimal bone; ethmoid bone
Superior rim	Frontal bone
Lateral rim	Zygomatic bone
Medial rim	Maxilla

zygomatic: Gk. *zygoma,* yolk
styloid: Gk. *stylos,* pillar
mastoid: Gk. *mastos,* breast
petrous: Gk. *petra,* rock

Figure 9.10

The temporal bone.
(*a*) A lateral view and (*b*) a medial view.

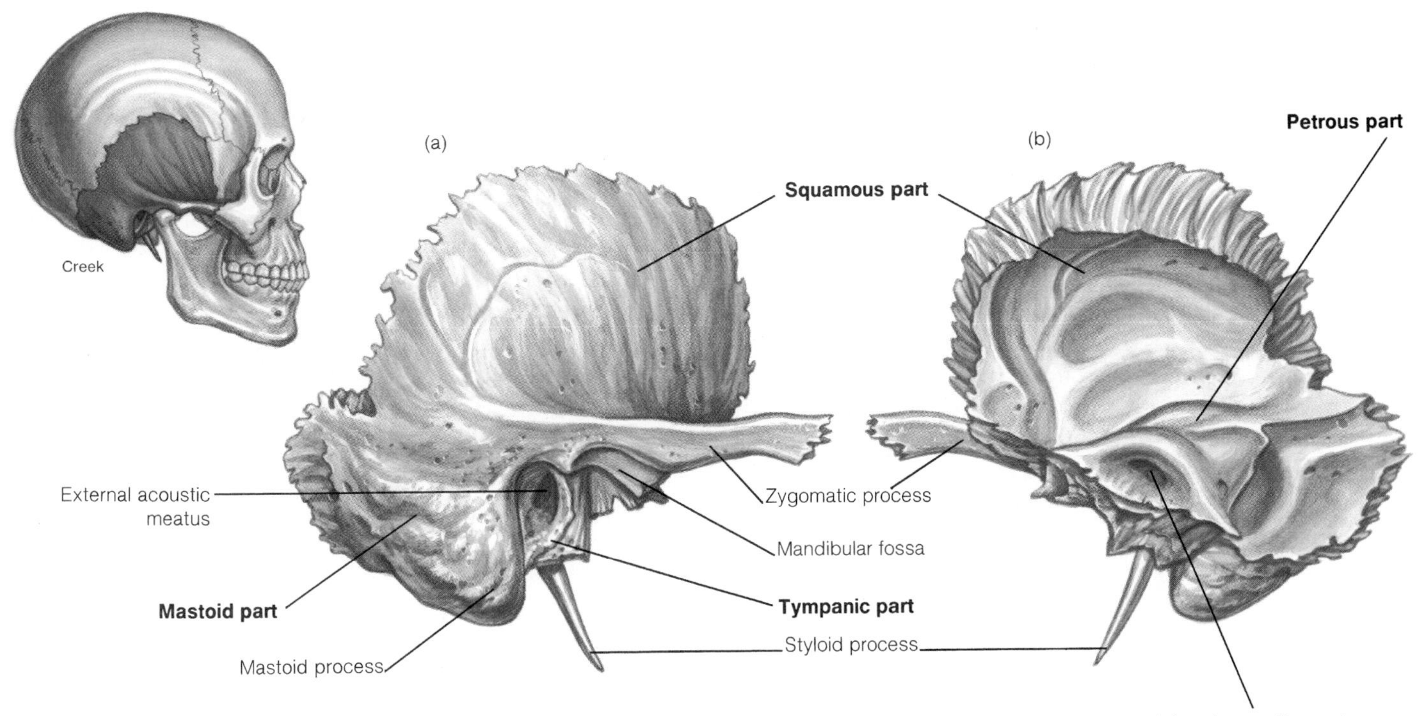

Figure 9.11

The sphenoid bone.
(*a*) A superior view and (*b*) a posterior view.

this dense, part of the temporal bone. The **carotid** (*că-rot'id*) **canal** and the **jugular foramen** border on the medial side of the petrous part at the junction of the temporal and occipital bones. The carotid canal allows blood into the brain via the internal carotid artery, and the jugular foramen lets blood drain from the brain via the internal jugular vein. Three cranial nerves also pass through the jugular foramen (see table 9.1).

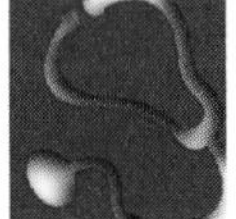

The mastoid process of the temporal bone can be easily palpated as a bony knob immediately behind the earlobe. This process contains a number of small air-filled spaces called *mastoid cells* that can become infected in *mastoiditis,* as a result, for example, of a prolonged middle-ear infection.

Occipital Bone

The occipital bone forms the posterior and most of the base of the skull. It articulates with the parietal bones at the **lambdoidal suture** (fig. 9.5). The **foramen magnum** is the large hole in the occipital bone through which the spinal cord attaches to the brain stem. On each side of the foramen magnum are the **occipital condyles** (fig. 9.3), which articulate with the first vertebra (the atlas) of the vertebral column. At the anterolateral edge of the occipital condyle is the **hypoglossal canal** (fig. 9.4), through which the hypoglossal nerve passes. A **condyloid** (*kon'dĭ-loid*) **canal** lies posterior to the occipital condyle (fig. 9.3). The **external occipital protuberance** is a prominent posterior projection on the occipital bone that can be felt as a definite bump just under the skin. The **superior nuchal** (*noo'kal*) **line** is a ridge of bone extending laterally from the occipital protuberance to the mastoid part of the temporal bone. **Sutural bones** are small clusters of irregularly shaped bones that frequently occur along the lambdoidal suture.

Sphenoid Bone

The sphenoid (*sfe'noid*) bone forms part of the anterior base of the cranium and can be viewed laterally and inferiorly (figs. 9.2 and 9.3). This bone has a somewhat mothlike shape (fig. 9.11). It consists of a **body** with laterally projecting **greater** and **lesser wings** that form part of the orbit. The wedgelike body contains the **sphenoidal** (*sfe-noi'dal*) **sinuses** and a prominent saddlelike depression, the **sella turcica** (*sel'ă tur'sĭ-kă*) (figs. 9.4 and 9.6). Commonly called "Turk's saddle," the sella turcica houses the pituitary gland. A pair of **pterygoid** (*ter'ĭ-goid*) **plates** (processes) project inferiorly from the sphenoid bone and help form the lateral walls of the nasal cavity.

Several foramina (figs. 9.3, 9.6, and 9.11) are associated with the sphenoid bone.

lambdoidal: Gk. *lambda*, letter L in Greek alphabet
magnum: L. *magnum*, great

nuchal: Fr. *nuque*, nape of neck
sphenoid: Gk. *sphenoeides*, wedgelike

1. The **optic canal** is a large opening through the lesser wing into the back of the orbit that provides passage for the optic nerve and the ophthalmic artery.
2. The **superior orbital fissure** is a triangular opening between the wings of the sphenoid bone that provides passage for the ophthalmic nerve, a branch of trigeminal nerve, and the oculomotor, trochlear, and abducens nerves.
3. The **foramen ovale** is an opening at the base of the lateral pterygoid plate, through which the mandibular nerve passes.
4. The **foramen spinosum** is a small opening at the posterior angle of the sphenoid bone that provides passage for the middle meningeal vessels.
5. The **foramen lacerum** (*las'er-um*) is an opening between the sphenoid bone and the petrous part of the temporal bone, through which the internal carotid artery and the meningeal branch of the ascending pharyngeal artery pass.
6. The **foramen rotundum** is an opening located just posterior to the superior orbital fissure, at the junction of the anterior and medial portions of the sphenoid bone. The maxillary nerve passes through this foramen.

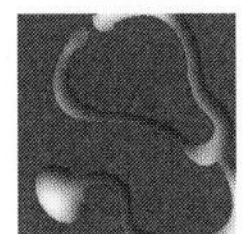

Located on the inferior side of the cranium, the sphenoid bone would seem to be well protected from trauma. Actually just the opposite is true—and in fact, the sphenoid bone is the most frequently fractured bone of the cranium. It has several broad, thin, platelike extensions that are perforated by numerous foramina. A blow to almost any portion of the skull causes the buoyed, fluid-filled brain to rebound against the vulnerable sphenoid bone, often causing it to fracture.

Ethmoid Bone

The ethmoid bone is located in the anterior portion of the floor of the cranium between the orbits, where it forms the roof of the nasal cavity (figs. 9.4, 9.7, and 9.12). An inferior projection of the ethmoid bone, called the **perpendicular plate,** forms the superior part of the nasal septum that separates the nasal cavity into two chambers. Each chamber of the nasal cavity is referred to as a **nasal fossa.** Flanking the perpendicular plate on each side is a large but delicate mass of bone riddled with ethmoidal air cells, collectively constituting the **ethmoid sinus.** A spine of the perpendicular plate, the **crista galli** (*kris'ta gal'e*), projects superiorly into the cranial cavity and serves as an attachment for the meninges covering the brain. On both lateral walls of the nasal cavity are two scroll-shaped plates of the ethmoid bone, the **superior** and

ethmoid: Gk. *ethmos*, sieve
crista galli: L. *crista*, crest; *galli*, cock's comb

Figure 9.12

An anterior view of the ethmoid bone.

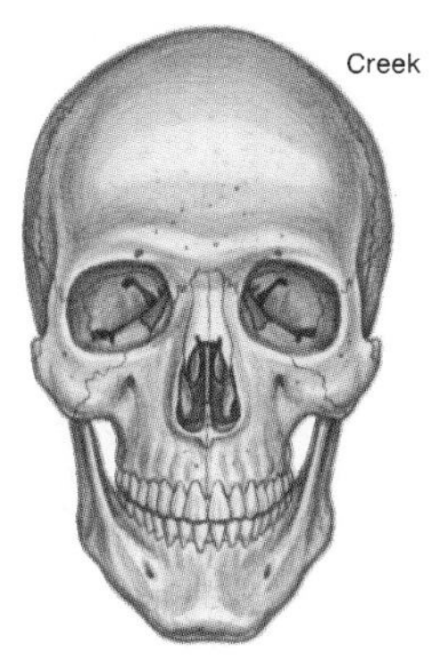

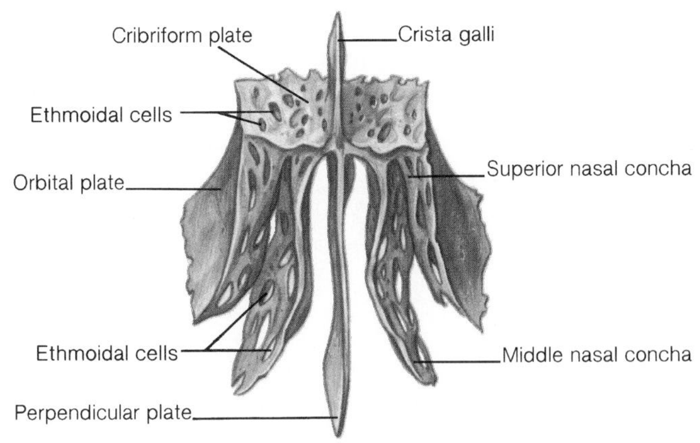

middle nasal conchae (*kong'ke*; singular, *concha*) (fig. 9.13), also known as *turbinates*. At right angles to the perpendicular plate, within the floor of the cranium, is the **cribriform** (*krib'ri-form*) **plate,** which has numerous **cribriform foramina** for the passage of olfactory nerves from the nasal cavity. The bones of the nasal cavity are summarized in table 9.3.

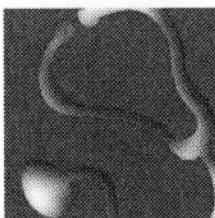

The moist, warm vascular lining within the nasal cavity is susceptible to infections, particularly if a person is not in good health. Unfortunately, these infections can spread to several surrounding areas, especially the paranasal sinuses, since they connect to the nasal cavity. During a nasal infection, the eyes may become reddened and swollen because of the connection of the nasolacrimal duct, through which tears drain from the orbit to the nasal cavity. Organisms may spread via the auditory tube from the nasopharynx to the middle ear. When untreated, infectious organisms may even ascend to the meninges covering the brain via the sheaths of the olfactory nerves and pass through the cribriform plate to cause *meningitis*.

conchae: L. *conchae*, shells
cribriform: L. *cribrum*, sieve; *forma*, like

Under Development

Development of the Skull

The formation of the skull is a complex process that begins during the fourth week of embryonic development and continues well beyond the birth of the baby. Three factions are involved in the formation of the skull: the **chondrocranium,** the **neurocranium,** and the **viscerocranium** (fig. 1). The chondrocranium is the portion of the skull that undergoes endochondral ossification to form the bones supporting the brain. The neurocranium is the portion of the skull that develops through membranous ossification to form the bones covering the brain and facial region. The viscerocranium (splanchnocranium) is the portion that develops from the embryonic visceral arches and forms the auditory ossicles, the hyoid bone, and specific processes of the skull.

During fetal development and infancy, the bones of the neurocranium covering the brain are separated by fibrous sutures. There are also six large membranous "soft spots" of the skull that provide spaces between the developing bones (fig. 2). Because the baby's pulse can be felt surging in these areas, they are called **fontanels** (*fon″tă-nelz*), meaning "little fountains." They permit the skull to undergo changes of shape, called molding, during parturition (childbirth), and they also allow for rapid growth of the brain during infancy. Ossification of the fontanels is normally complete by 20 to 24 months of age. A description of the six fontanels follows.

1. **Anterior (frontal) fontanel.** The anterior fontanel is diamond-shaped and is the most prominent of the six. It is located on the anteromedian portion of the skull.
2. **Posterior (occipital) fontanel.** The posterior fontanel is positioned at the back of the skull on the median line.
3. **Anterolateral (sphenoidal) fontanels.** The paired anterolateral fontanels are found on both sides of the skull, lateral to the anterior fontanel.
4. **Posterolateral (mastoid) fontanels.** The paired posterolateral fontanels are located on the posterolateral sides of the skull.

chondrocranium: Gk. *chondros*, cartilage; *kranion*, skull
viscerocranium: L. *viscera*, soft parts; Gk. *kranion*, skull

fontanel: Fr. *fontaine*, little fountain

Figure 1 The embryonic skull. At 12 weeks, the embryonic skull is composed of bony elements from three developmental sources: the chondrocranium (colored blue-gray), the neurocranium (colored light yellow), and the viscerocranium (colored salmon).

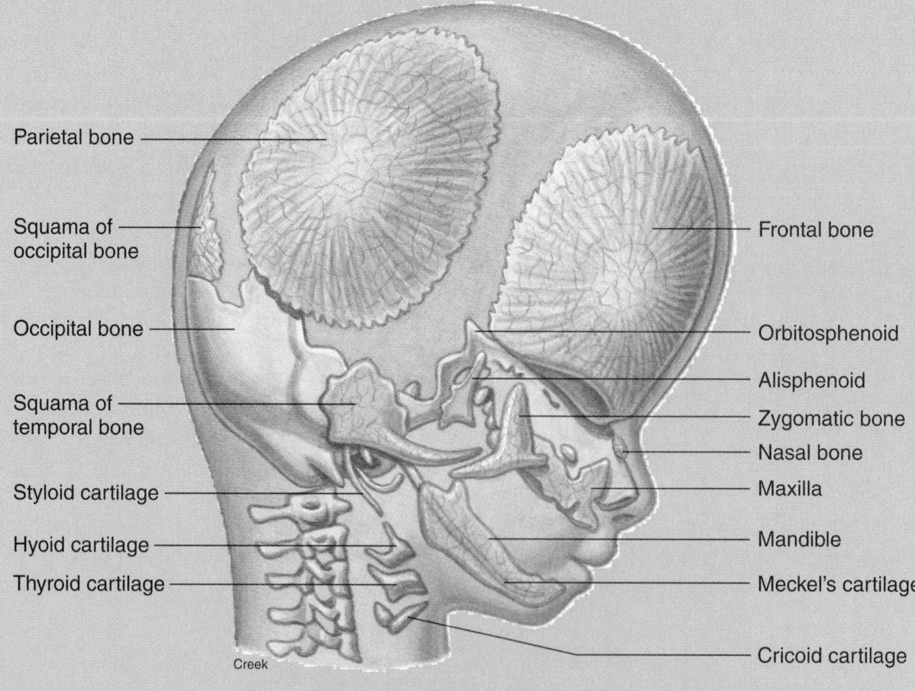

A prominent **sagittal suture** extends the anteroposterior median length of the skull between the anterior and posterior fontanels. A **coronal suture** extends from the anterior fontanel to the anterolateral fontanel. A **lambdoidal suture** extends from the posterior fontanel to the posterolateral fontanel. A **squamosal suture** connects the posterolateral fontanel to the anterolateral fontanel.

During normal parturition, the molding of the fetal skull is such that the occipital bone is usually pressed under the two parietal bones. In addition, one parietal bone overlaps the other, with the depressed one against the promontory of the mother's sacrum. If a baby is born breech (buttocks first), molding does not occur and delivery is more difficult.

Figure 2 The fetal skull showing the six fontanels and the sutures. (*a*) A right lateral view and (*b*) a superior view.

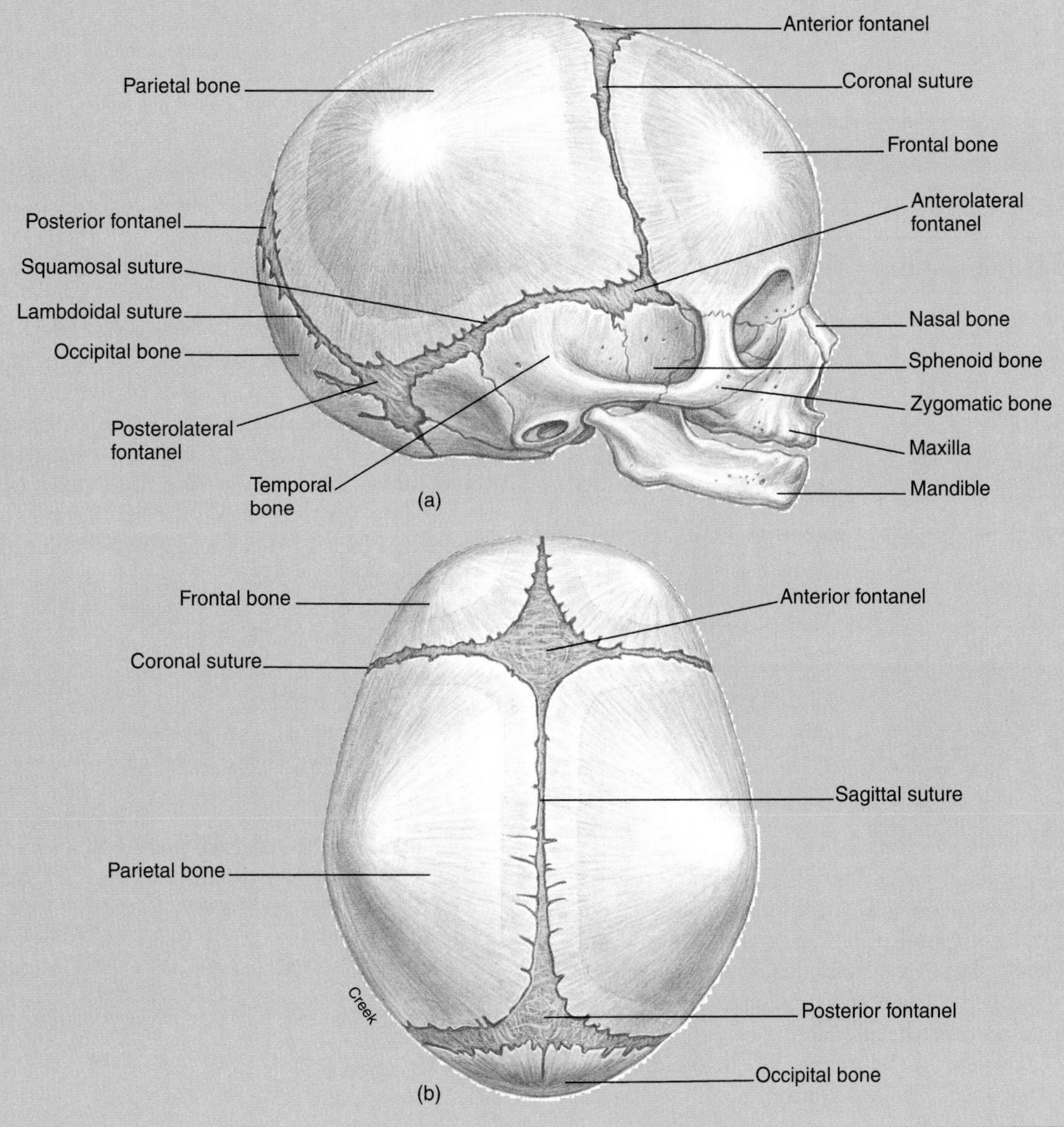

Figure 9.13

A lateral wall of the nasal cavity.

Table 9.3

Bones That Enclose the Nasal Cavity

Region of the Nasal Cavity	Contributing Bones
Roof (superior)	Ethmoid bone (cribriform plate); frontal bone
Floor (inferior)	Maxilla; palatine bone
Lateral wall	Maxilla; palatine bone
Nasal septum (medial)	Ethmoid bone (perpendicular plate); vomer
Bridge	Nasal bones
Conchae	Ethmoid bone (superior and middle conchae); inferior nasal concha

Facial Bones

The 14 bones of the skull not in contact with the brain are called **facial bones.** These bones, together with certain cranial bones (frontal bone and portions of the ethmoid and temporal bones), give shape and individuality to the face. Facial bones also support the teeth and provide attachments for various muscles that move the jaw and cause facial expressions. With the exception of the vomer and mandible, all of the facial bones are paired. The articulated facial bones are illustrated in figures 9.1 through 9.8.

Maxilla

The two maxillae (*mak-sil'e*) unite at the midline to form the upper jaw, which supports the upper teeth. **Incisors** (*in-si'zorz*), **canines** (cuspids), **premolars,** and **molars** are anchored in **dental alveoli** (tooth sockets) within the **alveolar** (*al-ve'ŏ-lar*) **process** of the maxilla (fig. 9.14). The palatine (*pal'a-tin*) process, a horizontal plate of the maxilla, forms the greater portion of the **hard palate** (*pal'it*), or roof of the mouth. The **incisive foramen** (fig. 9.3) is located in the anterior region of the hard palate, behind the incisors. An **infraorbital foramen** is located under each orbit and serves as a passageway for the infraorbital nerve and artery to the nose (figs. 9.1, 9.2, 9.8, and 9.14). A final opening within the maxilla is the **inferior orbital fissure.** It is located between the maxilla and the greater wing of the sphenoid bone (fig. 9.1) and is the external opening for the maxillary nerve of the trigeminal nerve and infraorbital vessels. The large **maxillary sinus** located within the maxilla is one of the four paranasal sinuses (figs. 9.7, 9.9, and 9.14*b*).

If the two palatine processes fail to join during early prenatal development (about 12 weeks), a *cleft palate* results. A cleft palate may be accompanied by a *cleft lip* lateral to the midline. These conditions can be surgically treated with excellent cosmetic results. An immediate problem, however, is that a baby with a cleft palate may have a difficult time nursing because it is unable to create the necessary suction within the oral cavity to swallow effectively.

incisor: L. *incidere*, to cut
canine: L. *canis*, dog
molar: L. *mola*, millstone
alveolus: L. *alveus*, little cavity

Figure 9.14

The maxilla.
(*a*) A lateral view and (*b*) a medial view.

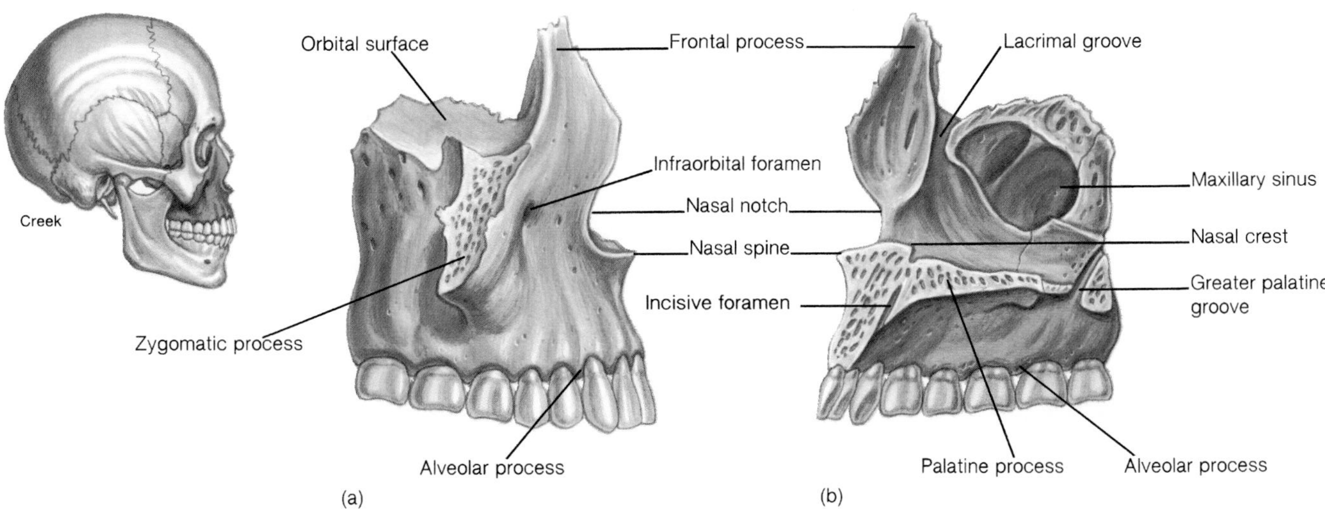

Palatine Bone

The L-shaped palatine bones form the posterior third of the hard palate, a portion of the orbits, and a part of the supporting wall of the nasal cavity. The **horizontal plates** of the palatines contribute to the formation of the hard palate (fig. 9.15). At the posterior angle of the hard palate is the large **greater palatine foramen** that provides passage for the greater palatine nerve and descending palatine vessels (fig. 9.3). Two or more smaller **lesser palatine foramina** are positioned posterior to the greater palatine foramen. Branches of the lesser palatine nerve pass through these openings.

Zygomatic Bone

The two zygomatic bones ("cheekbones") are the lateral margins of the orbits. A posteriorly extending **zygomatic process** of the zygomatic bone unites with the zygomatic process of the temporal bone to form the **zygomatic arch** (figs. 9.3 and 9.5). A small **zygomaticofacial** (*zi"gŏ-mat"ĭ-kŏ-fa'shal*) **foramen,** located on the anterolateral surface of this bone, allows passage of the zygomatic nerves and vessels.

Lacrimal Bone

The thin lacrimal (*lak'rĭ-mal*) bones form the anterior part of the medial wall of each orbit (fig. 9.8). These are the smallest of the facial bones. Each one has a **lacrimal sulcus**—a groove that helps to form the **nasolacrimal canal.** This opening permits the tears of the eye to drain into the nasal cavity.

Nasal Bone

The small, rectangular nasal bones (fig. 9.1) join in the midline to form the bridge of the nose. The nasal bones support the flexible cartilaginous plates, which are a part of the framework of the nose. Fractures of the nasal bones or fragmentation of the associated cartilages are common facial injuries.

Inferior Nasal Concha

The two inferior nasal conchae are fragile, scroll-like bones that project horizontally and medially from the lateral walls of the nasal cavity (figs. 9.1 and 9.7). They extend into the nasal cavity just below the superior and middle nasal conchae, which are part of the ethmoid bone (fig. 9.12). The inferior nasal conchae are the largest of the three paired conchae, and, like the other two, are covered with a mucous membrane to warm, moisten, and cleanse inhaled air.

Vomer

The vomer (*vo'mer*) is a thin, flattened bone that forms the lower part of the nasal septum (figs. 9.3, 9.4, and 9.7). Along with the perpendicular plate of the ethmoid bone, it supports the septal cartilage that forms most of the anterior part of the nasal septum.

Mandible

The mandible ("jawbone") is the largest, strongest bone in the face. It is attached to the skull by the temporomandibular articulations, and is the only movable bone of the skull. The horseshoe-shaped front and horizontal lateral sides of the mandible are referred to as the **body** (fig. 9.16). Extending vertically from the posterior part of the body are two **rami** (*ra'mi;* singular, *ramus*). At the superior margin of each

vomer: L. *vomer*, plowshare
mandible: L. *mandere*, to chew
ramus: L. *ramus*, branch

Figure 9.15

The palatine bone.
(*a*) A medial view. (*b*) The two palatine bones viewed posteriorly. The two palatine bones form the posterior portion of the hard palate.

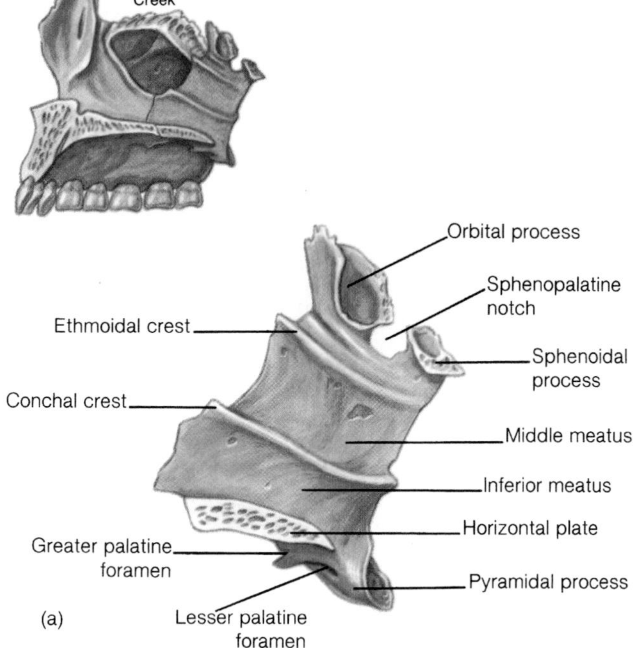

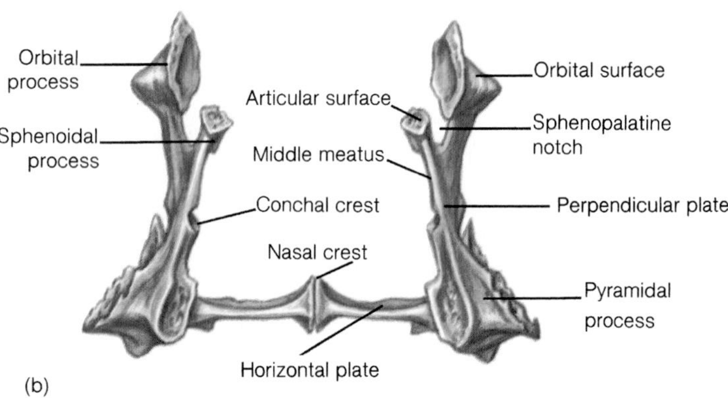

ramus is a knoblike **condylar process,** which articulates with the mandibular fossa of the temporal bone, and a pointed **coronoid process** for the attachment of the temporalis muscle. The depressed area between these two processes is called the **mandibular notch.** The angle of the mandible is where the horizontal body and vertical ramus meet at the corner of the jaw.

Two sets of foramina are associated with the mandible: the **mental foramen** on the anterolateral aspect of the body of the mandible below the first molar and the **mandibular foramen,** on the medial surface of the ramus. The mental nerve and vessels pass through the mental foramen, and the inferior alveolar nerve and vessels are transmitted through the mandibular foramen. Several muscles that close the jaw extend from the skull to the mandible (see chapter 13). The mandible of an adult supports 16 teeth within dental alveoli, which occlude with the teeth of the maxilla.

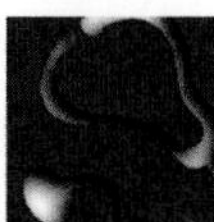

Dentists use bony landmarks of the facial region to locate the nerves that traverse the foramina in order to inject anesthetics. For example, the trigeminal cranial nerve is composed of three large nerves, the lower two of which convey sensations from the teeth, gums, and jaws. The mandibular teeth can be desensitized by an injection near the mandibular foramen called a *third-division,* or *lower, nerve block.* An injection near the foramen rotundum of the skull, called a *second-division nerve block,* desensitizes all the upper teeth on one side of the maxilla.

Hyoid Bone

The single **hyoid** (*hi'oid*) **bone** is a unique part of the skeleton in that it does not attach directly to any other bone. It is located in the neck region, below the mandible, where it is suspended from the styloid process of the temporal bone by the stylohyoid muscles and ligaments. The hyoid bone has a **body,** two **lesser cornua** (*kor'nyoo-a*; singular, *cornu*) extending anteriorly, and two **greater cornua** (fig. 9.17), which project posteriorly to the stylohyoid ligaments.

The hyoid bone supports the tongue and provides attachment for some of its muscles (see fig. 13.7). It may be palpated by placing a thumb and a finger on either side of the upper neck under the lateral portions of the mandible and firmly squeezing medially. This bone is carefully examined in an autopsy when strangulation is suspected, since during strangulation it frequently fractures.

Auditory Ossicles

Three small paired bones, called **auditory ossicles,** are located within the middle-ear chambers in the petrous part of the temporal bones (fig. 9.18). From outer to inner, these bones are the **malleus** ("hammer"), **incus** ("anvil"), and **stapes** ("stirrup"). As described in chapter 18, their movements transmit and amplify sound impulses through the middle-ear chamber.

condyloid: L. *condylus*, knucklelike
coronoid: Gk. *korone*, like a crow's beak

malleus: L. *malleus*, hammer
incus: L. *incus*, anvil
stapes: L. *stapes*, stirrup

Figure 9.16

The mandible.
(*a*) A lateral view and (*b*) a posterior view.

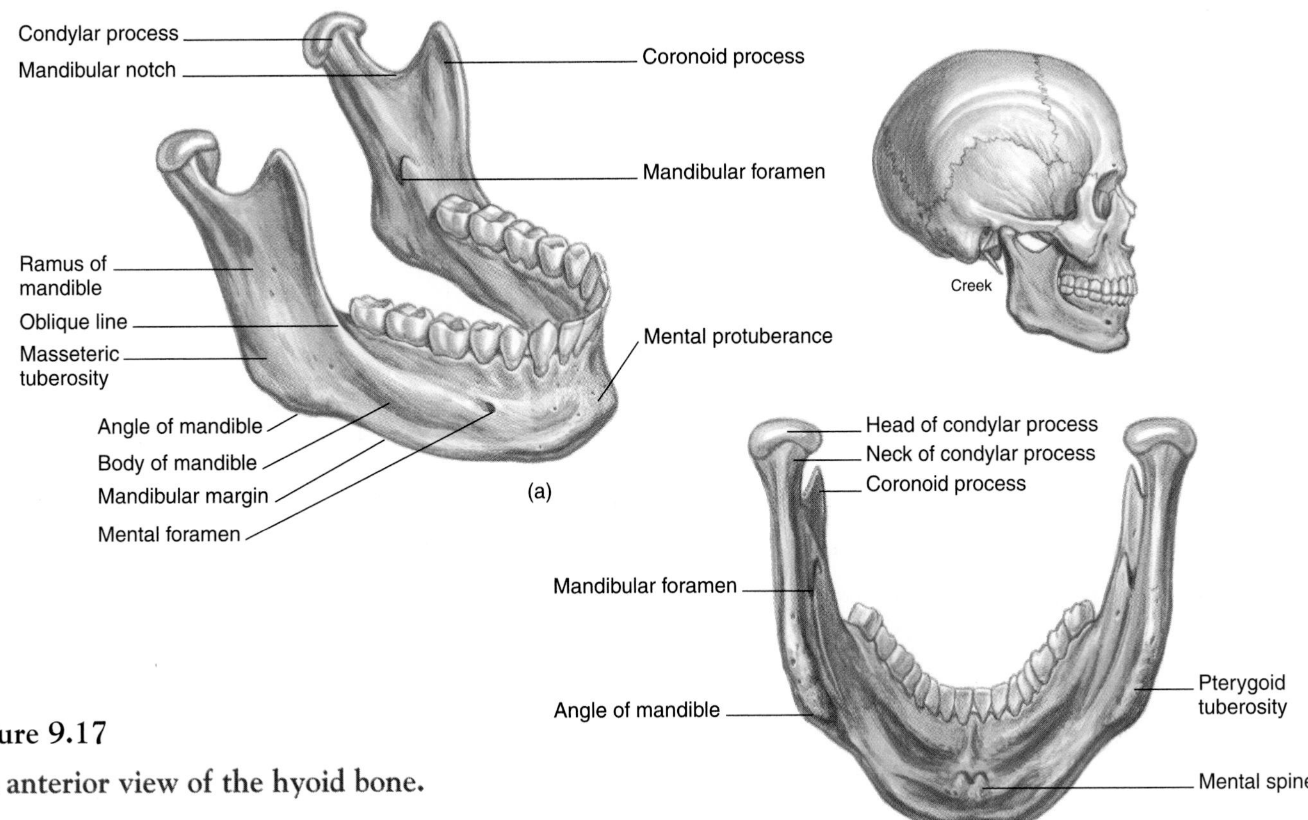

Figure 9.17

An anterior view of the hyoid bone.

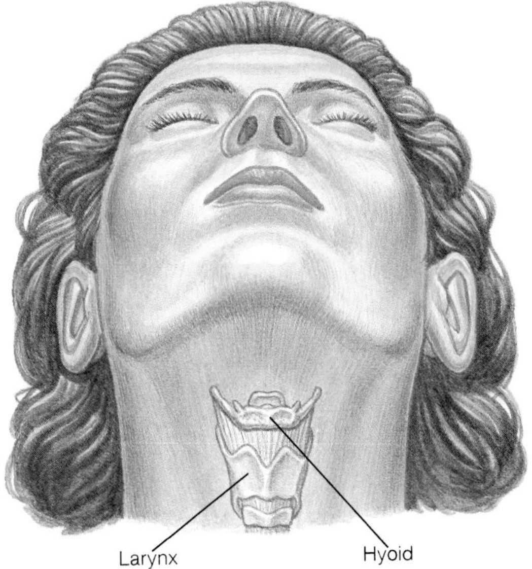

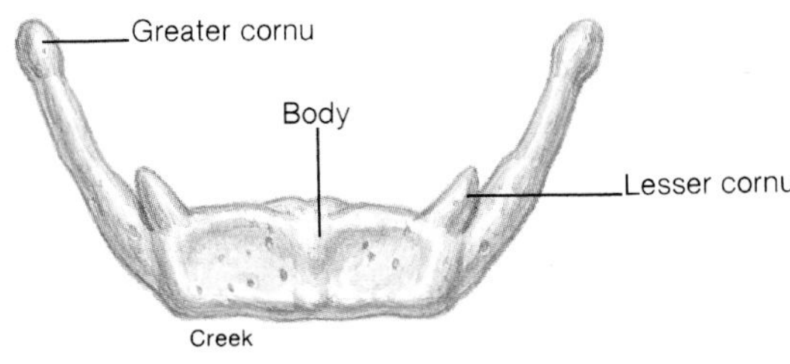

Vertebral Column

The vertebral column consists of a series of irregular bones called vertebrae, separated from each other by fibrocartilaginous intervertebral discs. Vertebrae enclose and protect the spinal cord, support the skull and allow for its movement, articulate with the rib cage, and provide for the attachment of trunk muscles. The intervertebral discs lend flexibility to the vertebral column and absorb vertical shock.

The **vertebral column** ("backbone") and the spinal cord of the nervous system constitute the *spinal column*. The vertebral column has three functions:

1. to support the head and upper extremities while permitting freedom of movement;
2. to provide attachment for various muscles, ribs, and visceral organs; and
3. to protect the spinal cord and permit passage of the spinal nerves.

The vertebral column is typically composed of 33 individual vertebrae, some of which are fused. There are 7 **cervical,** 12 **thoracic** (*thŏ-ras'ik*), 5 **lumbar,** 5 fused **sacral,** and 4 or 5 fused

Figure 9.18

The three auditory ossicles (in boldface type) within the middle-ear cavity.

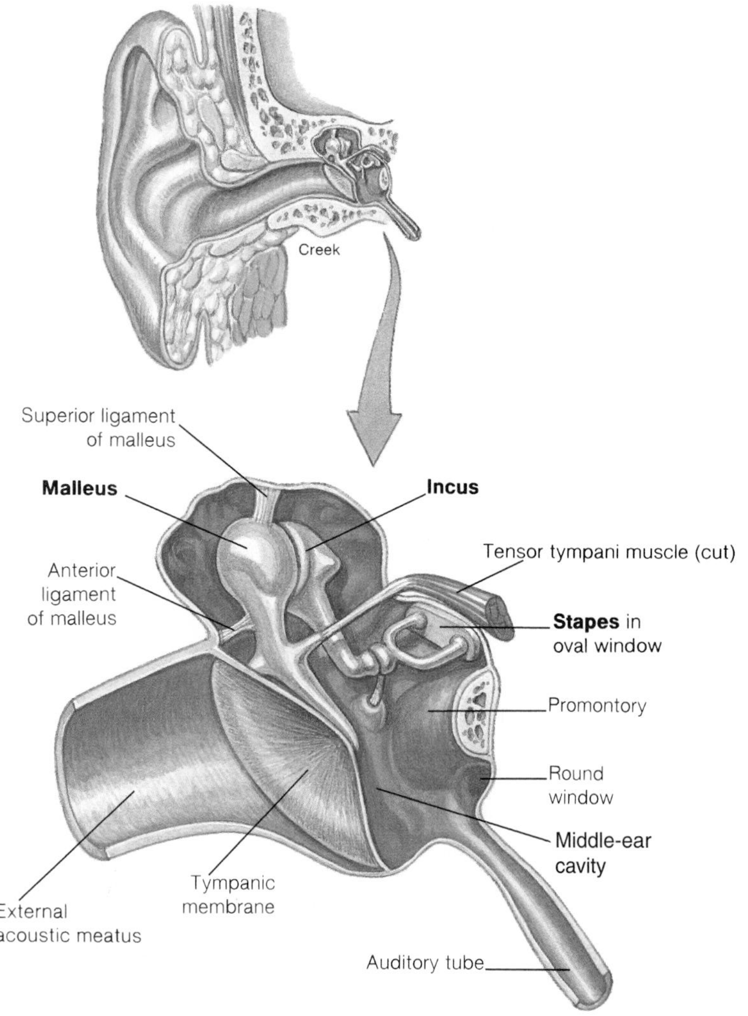

coccygeal (*kok-sij'e-al*) **vertebrae;** thus, the vertebral column is composed of a total of 26 movable parts. Vertebrae are separated by fibrocartilaginous intervertebral discs and are secured to each other by interlocking processes and binding ligaments. This structural arrangement provides limited movements between adjacent vertebrae but extensive movements for the vertebral column as a whole. Between the vertebrae are openings called **intervertebral foramina** that allow passage of spinal nerves.

When viewed from the side, four curvatures of the vertebral column of an adult can be identified (fig. 9.19). The **cervical, thoracic,** and **lumbar curves** are identified by the type of vertebrae they include. The **pelvic curve** (sacral curve) is formed by the shape of the sacrum and coccyx (*kok'siks*). The curves of the vertebral column play an important role in increasing the strength and maintaining the balance of the upper part of the body; they also make possible a bipedal stance.

Figure 9.19

The curves of the vertebral column.
The vertebral column of an adult has four curves named according to the region in which they occur. The bodies of the vertebrae are separated by intervertebral discs, which allow flexibility.

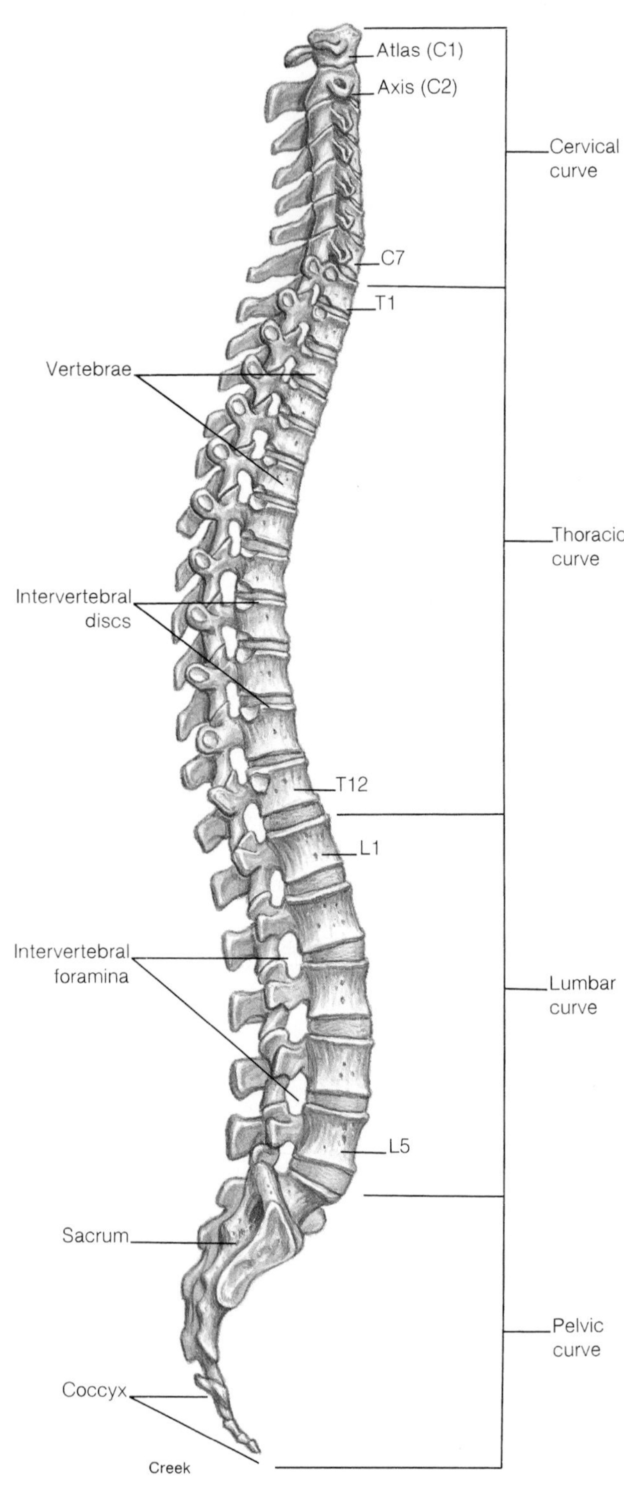

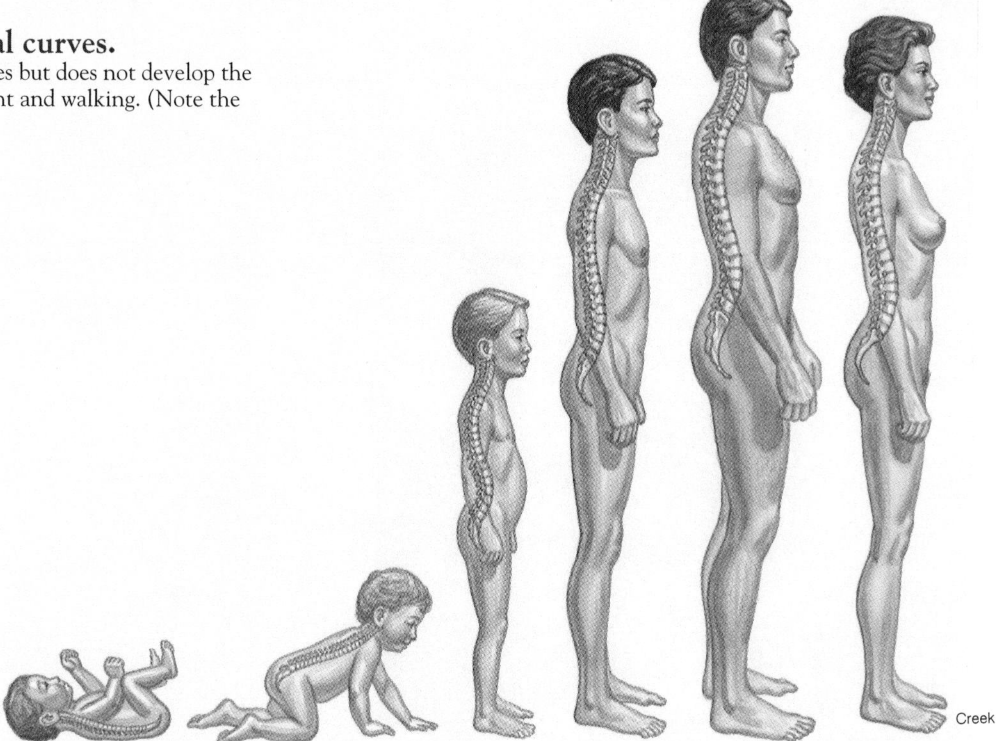

Figure 9.20

The development of the vertebral curves.
An infant is born with the two primary curves but does not develop the secondary curves until it begins sitting upright and walking. (Note the differences in the curves between the sexes.)

The four vertebral curves are not present in an infant. The cervical curve begins to develop at about 3 months as the baby begins holding up its head, and it becomes more pronounced as the baby learns to sit up (fig. 9.20). The lumbar curve develops as a child begins to walk. The thoracic and pelvic curves are called *primary curves* because they retain the shape of the fetus. The cervical and lumbar curves are called *secondary curves* because they are modifications of the fetal shape.

General Structure of Vertebrae

Vertebrae are similar in their general structure from one region to another. A typical vertebra consists of an anterior drum-shaped **body,** which is in contact with an intervertebral disc above and below (fig. 9.21). The **vertebral arch** is attached to the posterior surface of the body and is composed of two supporting **pedicles** (*ped'i-kulz*) and two arched **laminae** (*lam'i-ne*). The space formed by the vertebral arch and body is the **vertebral foramen,** through which the spinal cord passes. Between the pedicles of adjacent vertebrae are the **intervertebral foramina,** through which spinal nerves emerge as they branch off the spinal cord.

Seven processes arise from the vertebral arch of a typical vertebra: the **spinous** (*spi'nus*) **process,** two **transverse processes,** two **superior articular processes,** and two **inferior articular processes** (fig. 9.22). The spinous process and transverse processes serve for muscle attachment, and the superior and inferior articular processes limit twisting of the vertebral column. The spinous process protrudes posteriorly and inferiorly from the vertebral arch. The transverse process extends laterally from each side of a vertebra at the point where the lamina and pedicle join. The superior articular processes of a vertebra interlock with the inferior articular processes of the bone above.

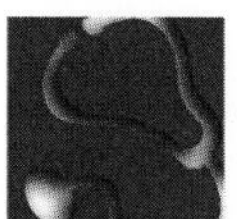

A *laminectomy* is the surgical removal of the spinous processes and their supporting vertebral laminae in a particular region of the vertebral column. A laminectomy may be performed to relieve pressure on the spinal cord or nerve root caused by a blood clot, a tumor, or a herniated disc. It may also be performed on a cadaver to expose the spinal cord and its surrounding meninges.

Regional Characteristics of Vertebrae

Cervical Vertebrae

The seven cervical vertebrae form the flexible framework of the neck and support the head. The bone tissue of cervical vertebrae is more dense than that found in the other vertebral regions, and, except for those in the coccygeal region, the cervical vertebrae are smallest. Cervical vertebrae are distinguished by the presence of a **transverse foramen** in each transverse process (fig. 9.21). The vertebral artery and vein pass through this opening as they contribute to the blood flow

pedicle: L. *pediculus,* small foot
lamina: L. *lamina,* thin layer

Figure 9.21

Cervical vertebrae.
(*a*) A radiograph (X ray) of the cervical region, (*b*) a superior view of a typical cervical vertebra, and (*c*) the articulated atlas and axis.

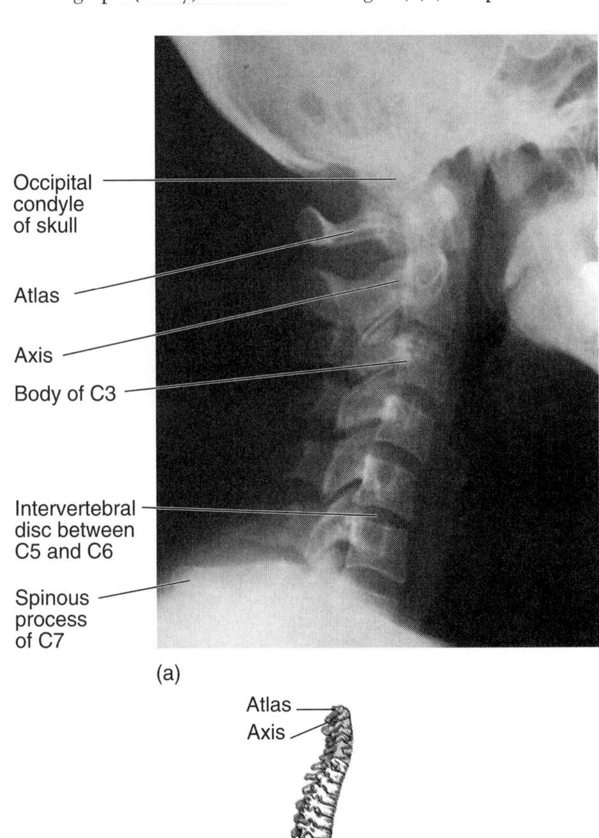

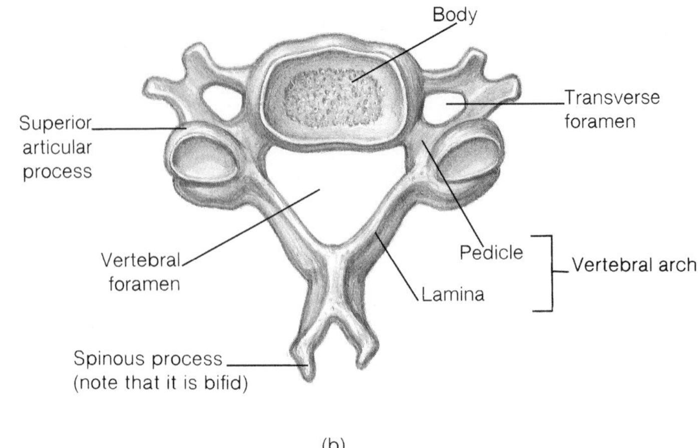

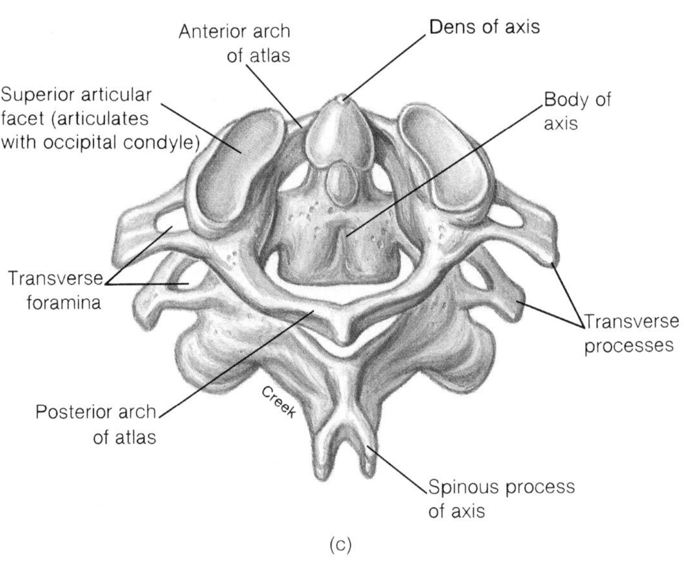

associated with the brain. Cervical vertebrae C2–C6 generally have a *bifid* (*bi'fid*), or notched, spinous process. The bifid spinous processes increase the surface area for attachment of the strong *nuchal ligament* that attaches to the back of the skull. The first cervical vertebra has no spinous process, and the process of C7 is not bifid and is larger than those of the other cervical vertebrae.

The **atlas** is the first cervical vertebra (sometimes called cervical 1, or C1). The atlas lacks a body, but it does have a short rounded spinous process called the **posterior tubercle.** It also has cupped **superior articular surfaces** that articulate with the oval occipital condyles of the skull. This *atlanto-occipital joint* supports the skull and permits the nodding of the head in a "yes" movement.

The **axis** is the second cervical vertebra (C2). It has a peglike **dens** (*odontoid process*) for rotation with the atlas in turning the head from side to side, as in a "no" movement.

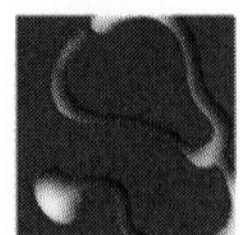

Whiplash is a common term for any injury to the neck. Muscle, bone, or ligament injury in this portion of the spinal column is relatively common in individuals involved in automobile accidents and sports injuries. Joint dislocation occurs commonly between the fourth and fifth or fifth and sixth cervical vertebrae, where neck movement is greatest. Bilateral dislocations are particularly dangerous because of the probability of spinal cord injury. Compression fractures of the first three cervical vertebrae are common and follow abrupt forced flexion of the neck. Fractures of this type may be extremely painful because of pinched spinal nerves.

atlas: from Gk. mythology, *Atlas* (the Titan who supported the heavens)

axis: L. *axis*, axle
odontoid: Gk. *odontos*, tooth

Figure 9.22

Thoracic vertebrae.
Representative vertebrae in (*a*) a lateral view and (*b*) a superior view.

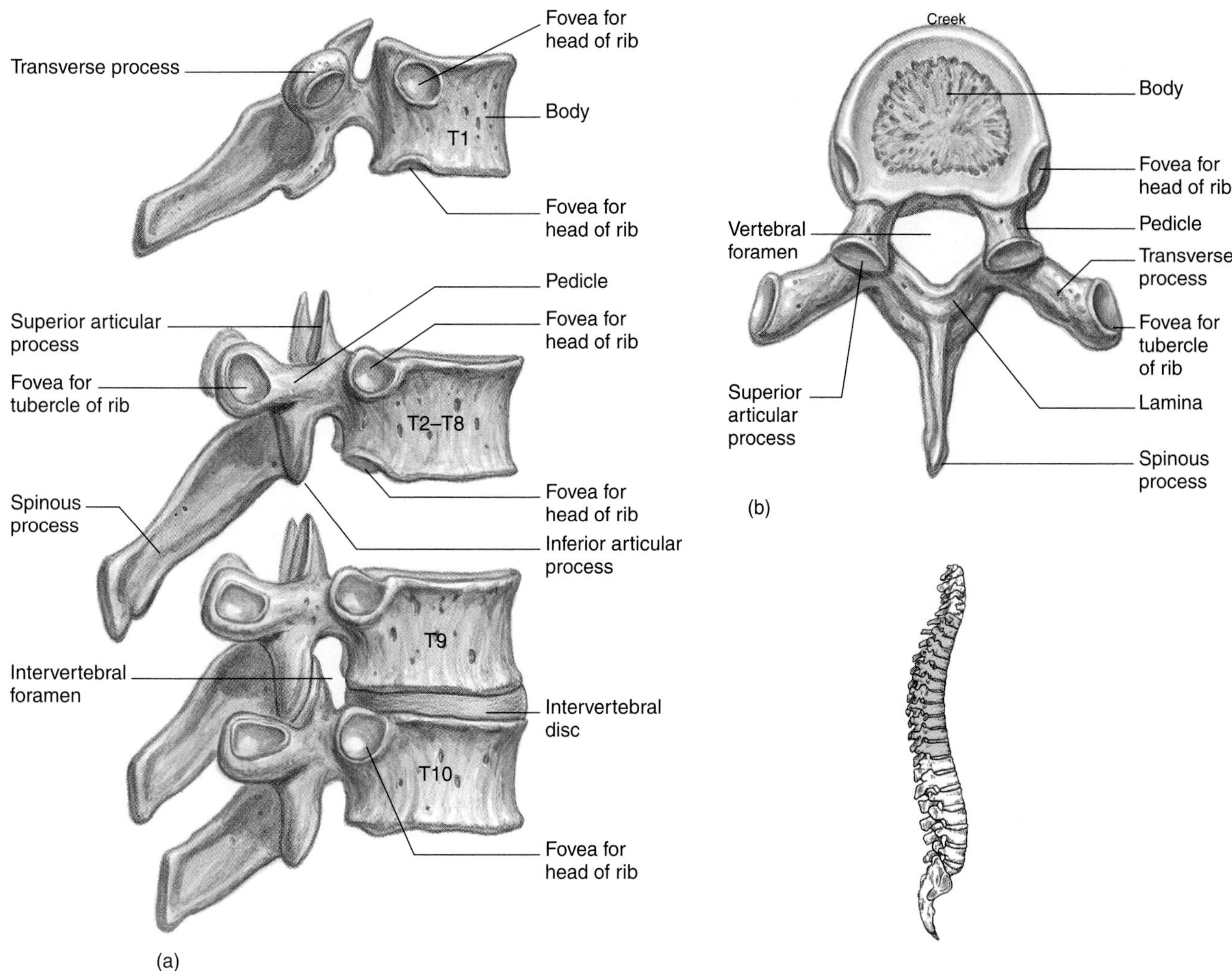

Thoracic Vertebrae

Twelve thoracic vertebrae articulate with the ribs to form the posterior anchor of the rib cage. Thoracic vertebrae are larger than cervical vertebrae and increase in size from superior (T1) to inferior (T12). Each thoracic vertebra has a long spinous process, which slopes obliquely downward, and **foveae** (*facets*) for articulation with the ribs (fig. 9.22).

Lumbar Vertebrae

The five lumbar vertebrae are easily identified by their heavy bodies and thick, blunt spinous processes (fig. 9.23) for attachment of powerful back muscles. They are the largest vertebrae of the vertebral column. Their articular processes are also distinctive in that the foveae of the superior pair are directed medially instead of posteriorly, and the foveae of the inferior pair are directed laterally instead of anteriorly.

lumbar: L. *lumbus*, loin

Sacrum

The wedge-shaped sacrum provides a strong foundation for the pelvic girdle. It consists of four or five sacral vertebrae (fig. 9.24) that become fused after age 26. The sacrum has an extensive **auricular surface** on each lateral side for the formation of a slightly movable **sacroiliac** (*sak″ro-il′e-ak*) **joint** with the ilium of the hip. A **median sacral crest** is formed along the posterior surface by the fusion of the spinous processes. **Posterior sacral foramina** on either lateral side of the crest allow for the passage of nerves from the spinal cord. The **sacral canal** is the tubular cavity within the sacrum that is continuous with the vertebral canal. Paired **superior articular processes,** which articulate with the fifth lumbar vertebra, arise from the roughened **sacral tuberosity** along the posterior surface.

The smooth anterior surface of the sacrum forms the posterior surface of the pelvic cavity. It has four **transverse**

sacrum: L. *sacris*, sacred

Figure 9.23

Lumbar vertebrae.
(*a*) A radiograph (X ray), (*b*) a superior view, and (*c*) a lateral view.

L3
Transverse process
Spinous process
L4
Superior articular process
L5
Intervertebral disc
Sacrum
Coccyx
(a)

Body
Vertebral foramen
Pedicle
Transverse process
Superior articular process
Lamina
Spinous process
(b)

Body
Superior articular process
Transverse process
Spinous process
Inferior articular process
Intervertebral foramen
Intervertebral disc
Creek
(c)

Figure 9.24

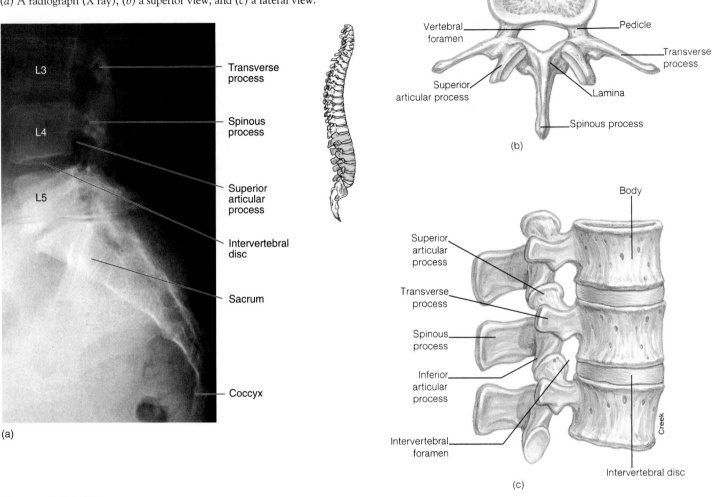

The sacrum and coccyx.
(*a*) An anterior view and (*b*) a posterior view.

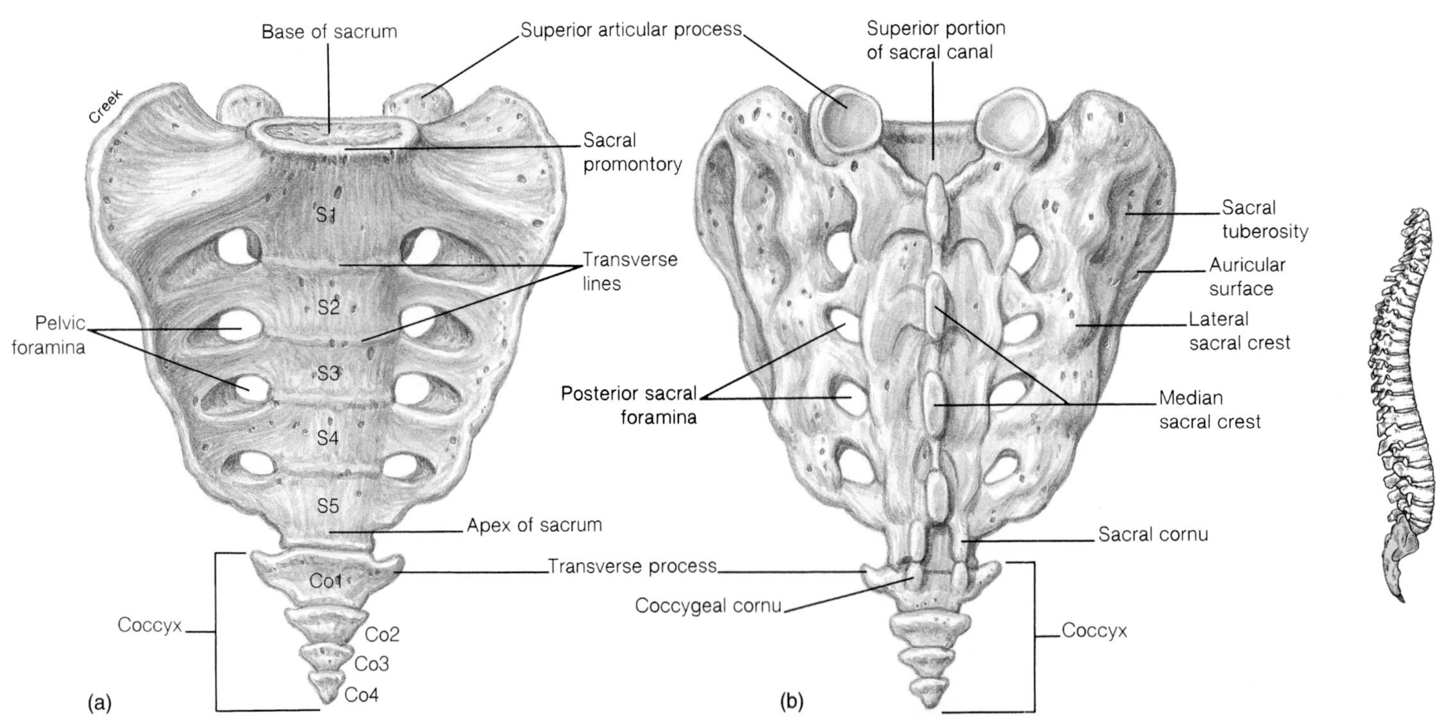

lines denoting the fusion of the vertebral bodies. At the ends of these lines are the paired **pelvic foramina (anterior sacral foramina).** The superior border of the anterior surface of the sacrum, called the **sacral promontory** (*prom'on-tor"e*), is an important obstetric landmark for pelvic measurements.

Coccyx

The triangular coccyx ("tailbone") is composed of three to five fused coccygeal vertebrae. The first vertebra of the fused coccyx has two long **coccygeal cornua,** which are attached by ligaments to the sacrum (fig. 9.24). Lateral to the cornua are the transverse processes.

The regions of the vertebral column are summarized in table 9.4.

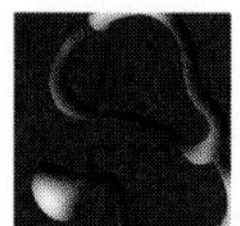

When a person sits, the coccyx flexes anteriorly, acting as a shock absorber. An abrupt fall on the coccyx, however, may cause a painful subperiosteal bruising, fracture, or fracture-dislocation of the sacrococcygeal joint. An especially difficult childbirth can even injure the coccyx of the mother. Coccygeal trauma is painful and may require months to heal.

Table 9.4

Regions of the Vertebral Column

Region	Number of Bones	Diagnostic Features
Cervical	7	Transverse foramina, superior facets of atlas articulate with occipital condyle; dens of axis; spinous processes of third through sixth vertebrae are generally bifid
Thoracic	12	Long spinous processes that slope obliquely downward; fovea for articulation with ribs
Lumbar	5	Large bodies, prominent transverse processes; short, thick spinous processes
Sacrum	4 or 5 fused vertebrae	Extensive auricular surface; median sacral crest; posterior sacral foramina; sacral promontory; sacral canal
Coccyx	3 to 5 fused vertebrae	Small and triangular; coccygeal cornua

coccyx: Gk. *kokkyx*, like a cuckoo's beak

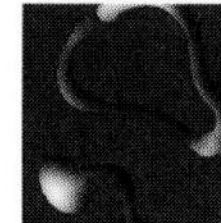

Distinct losses in height occur during middle and old age. Between the ages of 50 and 55, the body shortens by 0.5 to 2.0 cm (0.25 to 0.75 in.) because of compression and shrinkage of the intervertebral discs. Elderly individuals may suffer a further loss of height because of osteoporosis (see Clinical Considerations at the end of chapter 8).

Rib Cage

The cone-shaped, flexible rib cage consists of the thoracic vertebrae, 12 paired ribs, costal cartilages, and the sternum. It encloses and protects the thoracic viscera and is directly involved in the mechanics of breathing.

The sternum, ribs, costal cartilages, and the previously described thoracic vertebrae form the **rib cage** (*thoracic cage*) (fig. 9.25). The rib cage is anteroposteriorly compressed and more narrow superiorly than inferiorly. It supports the pectoral girdle and upper extremities, protects and supports the thoracic and upper abdominal viscera, and plays a major role in breathing (see fig. 13.10). Certain bones of the rib cage contain active sites in the bone marrow for the production of blood cells.

Sternum

The **sternum** ("breastbone") is an elongated, flattened bony plate consisting of three separate bones: the upper **manubrium** (*ma-noo'bri-um*), the central **body,** and the lower **xiphoid** (*zif'oid; zi'foid*) **process.** On the lateral sides of the sternum are **costal notches** where the costal cartilages attach. A **jugular notch** is formed at the superior end of the manubrium, and a **clavicular** (*klă-vik'yŭ-lar*) **notch** for articulation with the clavicle is present on both lateral sides of the jugular notch. The manubrium articulates with the costal cartilages of the first and second ribs. The body of the sternum attaches to the costal cartilages of the second through the tenth ribs. The xiphoid process does not attach to ribs but is an attachment for abdominal muscles. The costal cartilages of the eighth, ninth, and tenth ribs fuse to form the **costal margin** of the rib cage (see fig. 4, p. 353). A **costal angle** is formed where the two costal margins come together at the xiphoid process. The **sternal angle** (angle of Louis) may be palpated as an elevation between the manubrium and body of the sternum at the level of the second rib (fig. 9.25). The costal angle, costal margins, and sternal angle are important surface landmarks of the thorax and abdomen.

sternum: Gk. *sternon*, chest
manubrium: L. *manubrium*, a handle
xiphoid: Gk. *xiphos*, sword
costal: L. *costa*, rib

Figure 9.25
The rib cage.

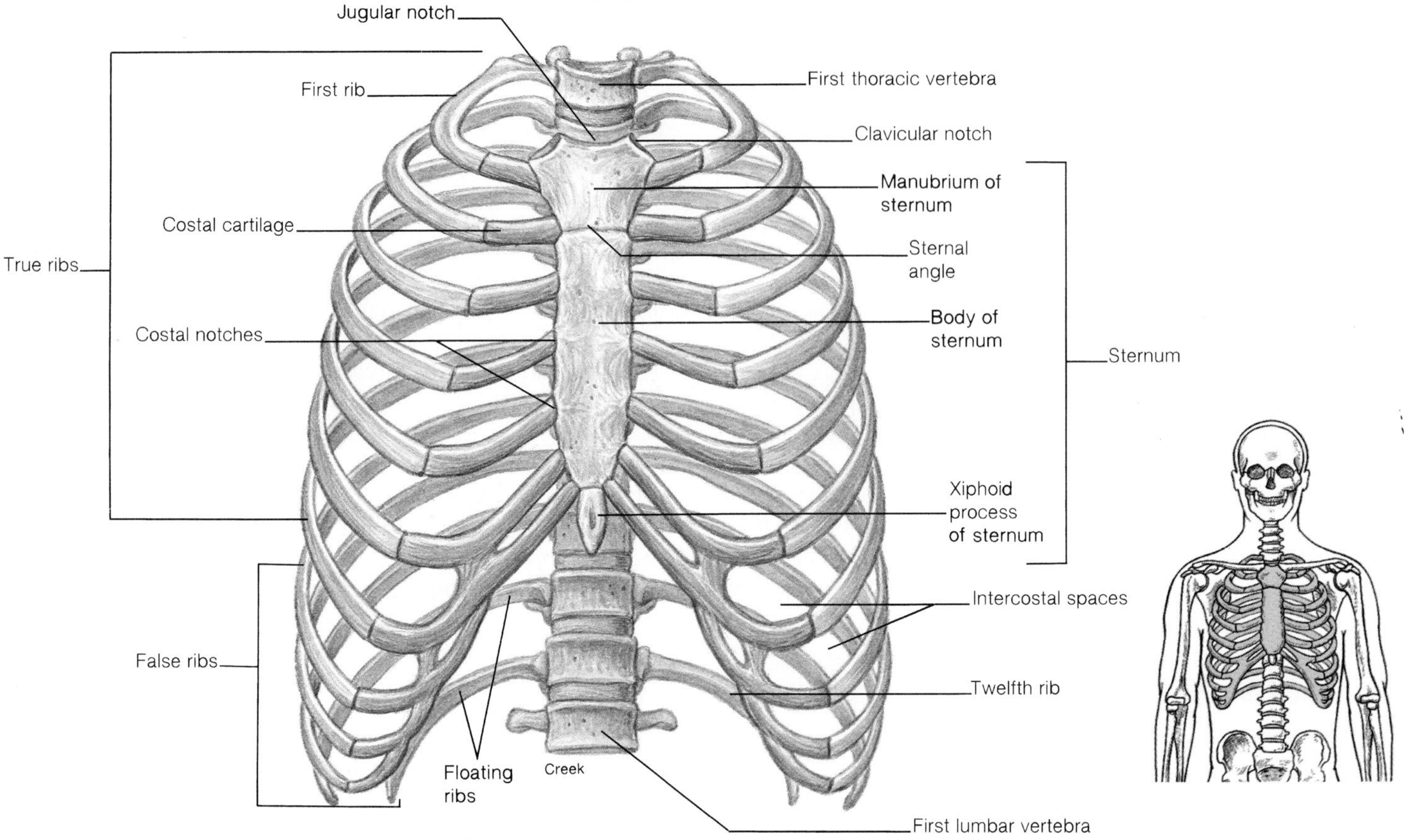

Figure 9.26
The structure of a rib.

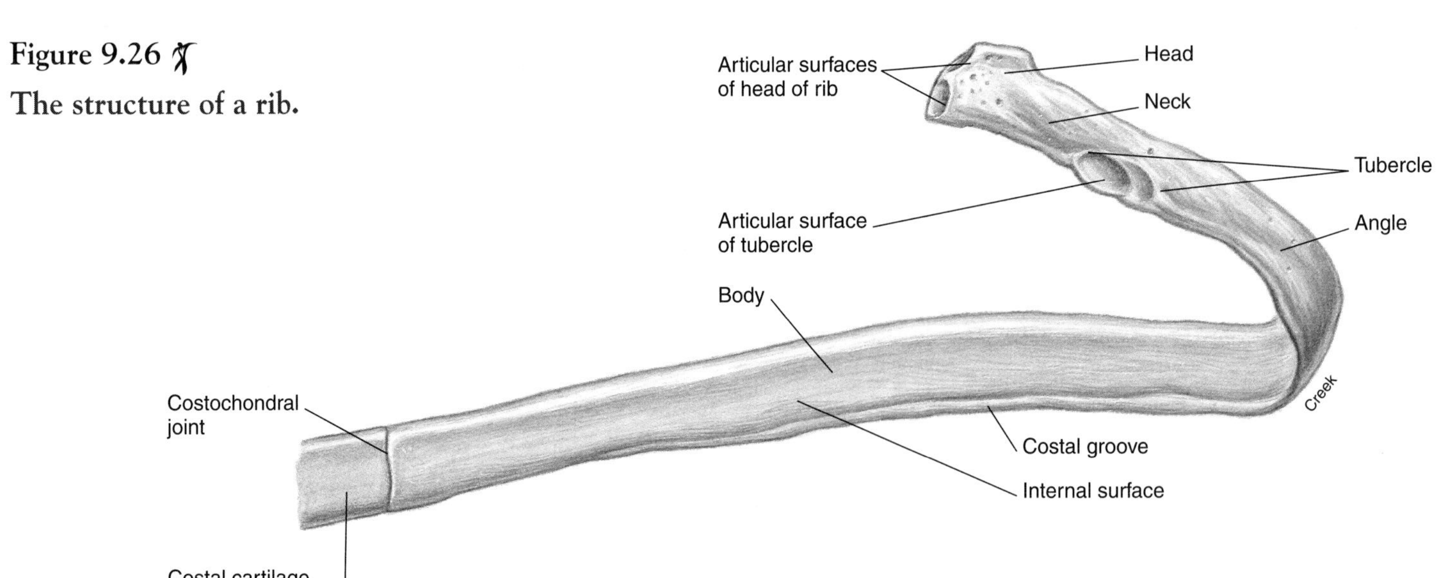

Ribs

Embedded in the muscles of the body wall are twelve pairs of **ribs,** each pair attached posteriorly to a thoracic vertebra. Anteriorly, the first seven pairs are anchored to the sternum by individual **costal cartilages;** these ribs are called **true ribs.** The remaining five pairs (ribs 8, 9, 10, 11, and 12) are termed **false ribs.** Because the last two pairs of false ribs do not attach to the sternum at all, they are referred to as **floating ribs.**

Although the structure of ribs vary structurally, each of the first ten pairs has a **head** and a **tubercle** for articulation with a vertebra. The last two have a head but no tubercle. In addition, each of the twelve pairs has a **neck, angle,** and **body** (fig. 9.26). The head projects posteriorly and articulates with the body of a thoracic vertebra (fig. 9.27). The tubercle is a knoblike process, just lateral to the head, that articulates with the fovea on the transverse process of a thoracic vertebra. The neck is the constricted area between the head and the tubercle. The body is the curved main part of the rib. Along the inner surface of the body is a depressed canal called the **costal groove** that protects the costal vessels and nerve. Spaces between the ribs are called **intercostal spaces** and are occupied by the intercostal muscles.

Figure 9.27

Articulation of a rib with a thoracic vertebra as seen in a superior view.

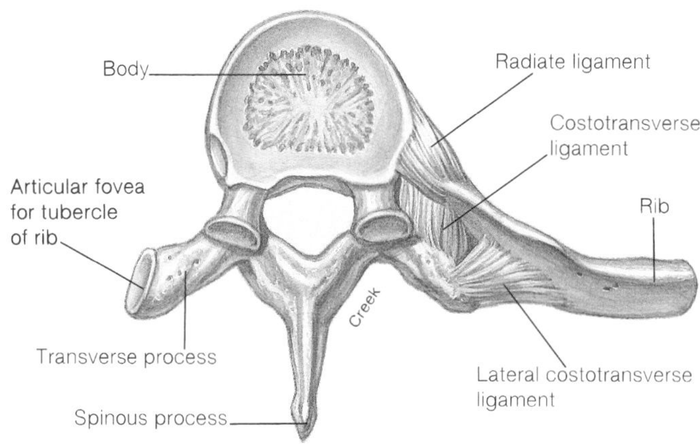

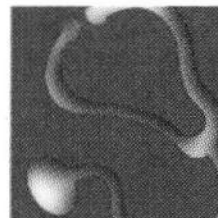

Fractures of the ribs are relatively common, and most frequently occur between ribs 3 and 10. The first two pairs of ribs are protected by the clavicles; the last two pairs move freely and will give with an impact. Little can be done to assist the healing of broken ribs other than binding them tightly to limit movement.

Clinical Investigation Answer

The height (overall length) of the vertebral column is equal to the sum of the thicknesses of the vertebrae plus the sum of the thicknesses of the intervertebral discs. The body of a vertebra consists of outer compact bone and inner spongy bone. An intervertebral disc consists of a fibrocartilage sheath called the *anulus fibrosus* and a mucoid center portion called the *nucleus pulposus*. The intervertebral discs generally change their anatomical configuration as one ages. In early adulthood, the nucleus pulposus is spongy and moist. With advanced age, however, it desiccates, resulting in a flattening of the intervertebral disc. Collectively, the intervertebral discs account for 25% of the height of the vertebral column. As they flatten with age, there is a gradual decrease in a person's overall height. Height loss may also result from undetectable compression fractures of the vertebral bodies, which are common in elderly people. This phenomenon, however, is considered pathological and is not an aspect of the normal aging process. In a person with osteoporosis, there is often a marked decrease in height and perhaps more serious clinical problems as well, such as compression of spinal nerves.

Chapter Summary

Skull (pp. 201–215)

1. The eight cranial bones include the frontal (1), parietals (2), temporals (2), occipital (1), sphenoid (1), and ethmoid (1).
   (a) The cranium encloses and protects the brain and provides for the attachment of muscles.
   (b) Sutures are fibrous joints between cranial bones.
2. The 14 facial bones include the nasals (2), maxillae (2), zygomatics (2), mandible (1), lacrimals (2), palatines (2), inferior nasal conchae (2), and vomer (1).
   (a) The facial bones form the basic shape of the face, support the teeth, and provide for the attachment of the facial muscles.
   (b) Sutures are the joints between facial bones with the exception of the freely movable joints between the mandible and the temporal bones.
3. The hyoid bone is located in the neck, between the mandible and the larynx.
4. The three paired auditory ossicles (malleus, incus, and stapes) are located within the middle-ear chambers of the petrous part of the temporal bones.

Vertebral Column (pp. 215–221)

1. The vertebral column consists of 7 cervical, 12 thoracic, 5 lumbar, 4 or 5 fused sacral, and 3 to 5 fused coccygeal vertebrae.
2. Cervical vertebrae have transverse foramina; thoracic vertebrae have foveae for articulation with ribs; lumbar vertebrae have large bodies; sacral vertebrae are triangularly fused and articulate with the pelvic girdle; and the coccygeal vertebrae form a small triangular bone.

Rib Cage (pp. 221–223)

1. The sternum consists of a manubrium, body, and xiphoid process.
2. There are seven pairs of true ribs and five pairs of false ribs. The inferior two pairs of false ribs (pairs 11 and 12) are called floating ribs.

Review Activities

Objective Questions

Match the following foramina to the correct bone in which it occurs.

1. rotundum (a) ethmoid bone
2. mental (b) occipital bone
3. carotid canal (c) sphenoid bone
4. olfactory (d) mandible
5. magnum (e) temporal bone
6. With respect to the hard palate, which of the following statements is *false?*
   (a) The hard palate is composed of two maxillae and two palatine bones.
   (b) The hard palate separates the oral cavity (mouth) from the nasal cavity.
   (c) The mandible articulates with the posterolateral angles of the hard palate.
   (d) The median palatine suture, incisive fossae, and greater palatine foramina are structural features of the hard palate.
7. The location of the sella turcica is immediately
   (a) superior to the sphenoidal sinus.
   (b) inferior to the frontal sinus.
   (c) medial to the petrous part of the temporal bones.
   (d) superior to the perpendicular plate of the ethmoid bone.
8. Which is the most prominent of the six fontanels?
   (a) anterior (c) posterior
   (b) anterolateral (d) posterolateral
9. The parietal bone articulates with the occipital bone at
   (a) the coronal suture.
   (b) the squamosal suture.
   (c) the posterolateral suture.
   (d) the lambdoidal suture.
10. Which of the following is *not* a cranial bone?
   (a) sphenoid bone (c) vomer
   (b) ethmoid bone (d) frontal bone
11. Which of the following is *not* one of the four parts of the temporal bone?
   (a) squamous part (d) petrous part
   (b) auricular part (e) mastoid part
   (c) tympanic part
12. The mandibular fossa is located in which structural part of the temporal bone?
   (a) squamous part (d) petrous part
   (b) auricular part (e) mastoid part
   (c) tympanic part
13. The facial nerve passes through the foramen
   (a) magnum. (c) stylomastoid.
   (b) ovale. (d) spinosum.
14. The crista galli is a structural feature of which bone?
   (a) sphenoid bone (c) palatine bone
   (b) ethmoid bone (d) temporal bone
15. Thoracic vertebrae are distinguished by the presence of
   (a) transverse foramina.
   (b) bifid spinous processes.
   (c) fovea.
   (d) auricular surfaces.
16. The usual number of false ribs is
   (a) two pairs. (c) three pairs.
   (b) five pairs. (d) seven pairs.

Essay Questions

1. Describe the development of the skull. What are fontanels, where are they located, and what are their functions?
2. List the bones of the skull that are paired. Which are unpaired? Identify the bones of the skull that can be palpated.
3. Which facial bones contain foramina? What structures traverse these openings?
4. List the bones that form the cranial cavity, the orbit, and the nasal cavity. Describe the location of the paranasal sinuses, the mastoid sinus, and the inner-ear cavity.
5. Describe the curvature of the vertebral column. What is meant by primary curves as compared to secondary curves?
6. List two or more characteristics by which vertebrae from each of the five regions of the vertebral column can be identified.
7. Identify the bones that form the rib cage. What functional role do the bones and the costal cartilages have in respiration?

Critical Thinking Questions

1. The sensory organs involved with sight, smell, and hearing are protected by bone. Describe the locations of each of these sensory organs and list the associated bones that provide protection.
2. The most common surgical approach to a pituitary gland tumor is through the nasal cavity. With the knowledge that the pituitary gland is supported by the sella turcica of the brain case, list the bones that would be involved in the removal of the tumor.
3. The contour of a child's head is distinctly different from that of an adult. Which skull bones exhibit the greatest amount of change as a child grows to adulthood?
4. Describe the structural features of the vertebral column that permit movement between adjacent vertebrae. Which structures restrict movement? Why is it undesirable for the vertebral column to not have even more movement?

Related Web Sites

For a listing of the most current web sites related to this chapter, please visit the *Concepts of Human Anatomy and Physiology* home page at http://www.mhhe.com/biosci/abio/.

TEN

Skeletal System: Appendicular Skeleton

Objectives

- Describe the bones of the pectoral girdle and the positions of articulations.
- Identify the bones of the upper extremity and list the diagnostic features of each bone.
- Describe the structure of the pelvic girdle and list its functions.
- Describe how the male and female pelves differ structurally.
- Identify the bones of the lower extremity and list the diagnostic features of each bone.
- Describe the structural features and functions of the arches of the foot.

Clinical Investigation

A 12-year-old boy was hit by a car while crossing a street. He was brought to the emergency room in stable condition, complaining of severe pain in his right leg. Radiographs revealed a 4-inch fracture extending inferiorly from the surface of the tibial plateau into the anterior body of the tibia. The fragment of bone created by the fracture was moderately displaced. With the radiographs in hand, the orthopedic surgeon went into the waiting room and conferred with the boy's parents. He told them that this kind of injury was more serious in children and growing adolescents than in adults. He went on to say that future growth of the bone might be jeopardized and that surgery, although recommended, could not guarantee normal growth. The parents asked, "What is it about this particular fracture that threatens future growth?"

If you were the surgeon, how would you respond?

Clues: Review the section on bone growth in chapter 8. Carefully examine figures 8.4 and 8.8 in chapter 8 and figures 10.16 and 10.22 in this chapter.

Pectoral Girdle and Upper Extremity

The structure of the pectoral girdle and upper extremities is adaptive for freedom of movement and extensive muscle attachment.

Pectoral Girdle

The two *clavicles* and two *scapulae* make up the **pectoral** (*shoulder*) **girdle** (see fig. 8.1). It is not a complete girdle, having only an anterior attachment to the axial skeleton at the sternum. As an axial bone, the sternum was described in chapter 9 (see fig. 9.25). Lacking a posterior attachment to the axial skeleton, the pectoral girdle has a wide range of movement. Because it is not weight-bearing, it is structurally more delicate than the pelvic girdle. The primary function of the pectoral girdle is to provide attachment areas for the numerous muscles that move the shoulder and elbow joints.

Clavicle

The slender S-shaped clavicle (*klav'i-kul;* "collarbone") connects the upper extremity to the axial skeleton and holds the shoulder joint away from the trunk for freedom of movement. The articulation of the medial **sternal extremity** (fig. 10.1) of the clavicle to the manubrium is referred to as the *sternoclavicular joint*. The lateral **acromial** (*ă-kro'me-al*) **extremity** of the clavicle articulates with the acromion of the scapula. This articulation is referred to as the *acromioclavicular joint*. A **conoid tubercle** is present on the acromial extremity of the clavicle, and a **costal tuberosity** is present on the inferior surface of the sternal extremity. Both processes serve as attachments for ligaments.

Figure 10.1

The right clavicle.
(*a*) A superior view and (*b*) an inferior view.

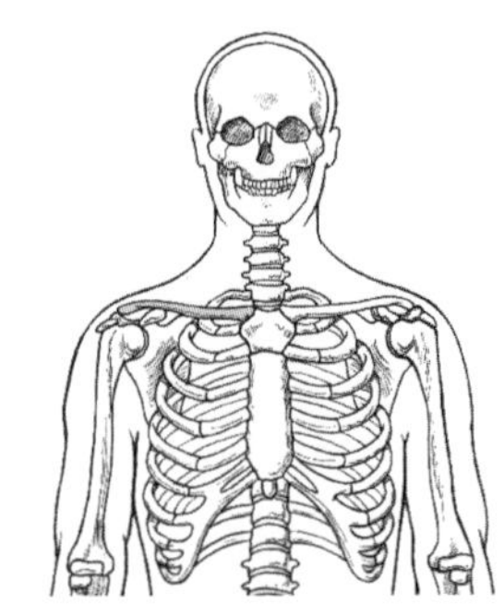

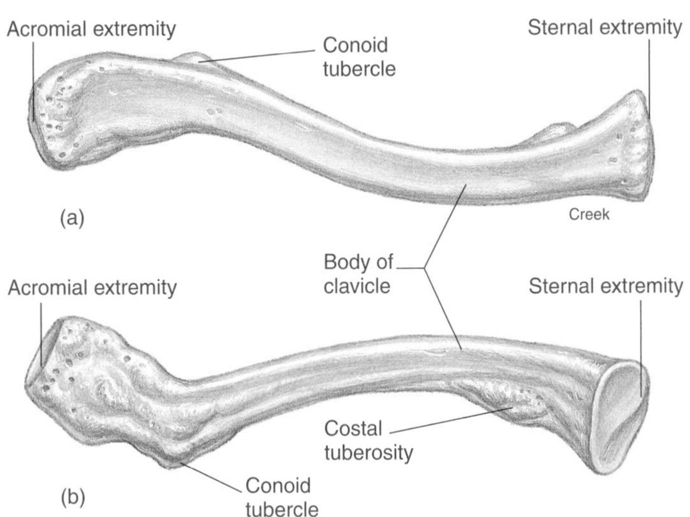

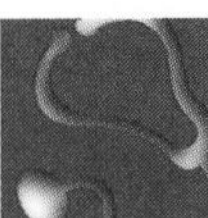

The long, delicate clavicle is the most commonly broken bone in the body. When a person receives a blow to the shoulder, or attempts to break a fall with an outstretched hand, the force is transmitted to the clavicle, possibly causing it to fracture. The most vulnerable area for a fracture of this bone is through its center, immediately proximal to the conoid tubercle. Because the clavicle is directly beneath the skin and not covered with muscle, a fracture can easily be palpated, and frequently seen.

Scapula

The **scapula** (*skap'yoo-la;* "shoulder blade") is a large, triangular flat bone on the posterior side of the rib cage, overlying ribs 2 through 7. The **spine** of the scapula is a prominent diagonal bony ridge seen on the posterior surface (figs. 10.2 and 10.3). Above the spine is the **supraspinous fossa,** and below

clavicle: L. *clavicula*, a small key

conoid tubercle: Gk. *konos*, cone; L. *tuberculum*, a small swelling
scapula: L. *scapula*, shoulder

Figure 10.2

The right scapula.
(*a*) An anterior view and (*b*) a posterior view.

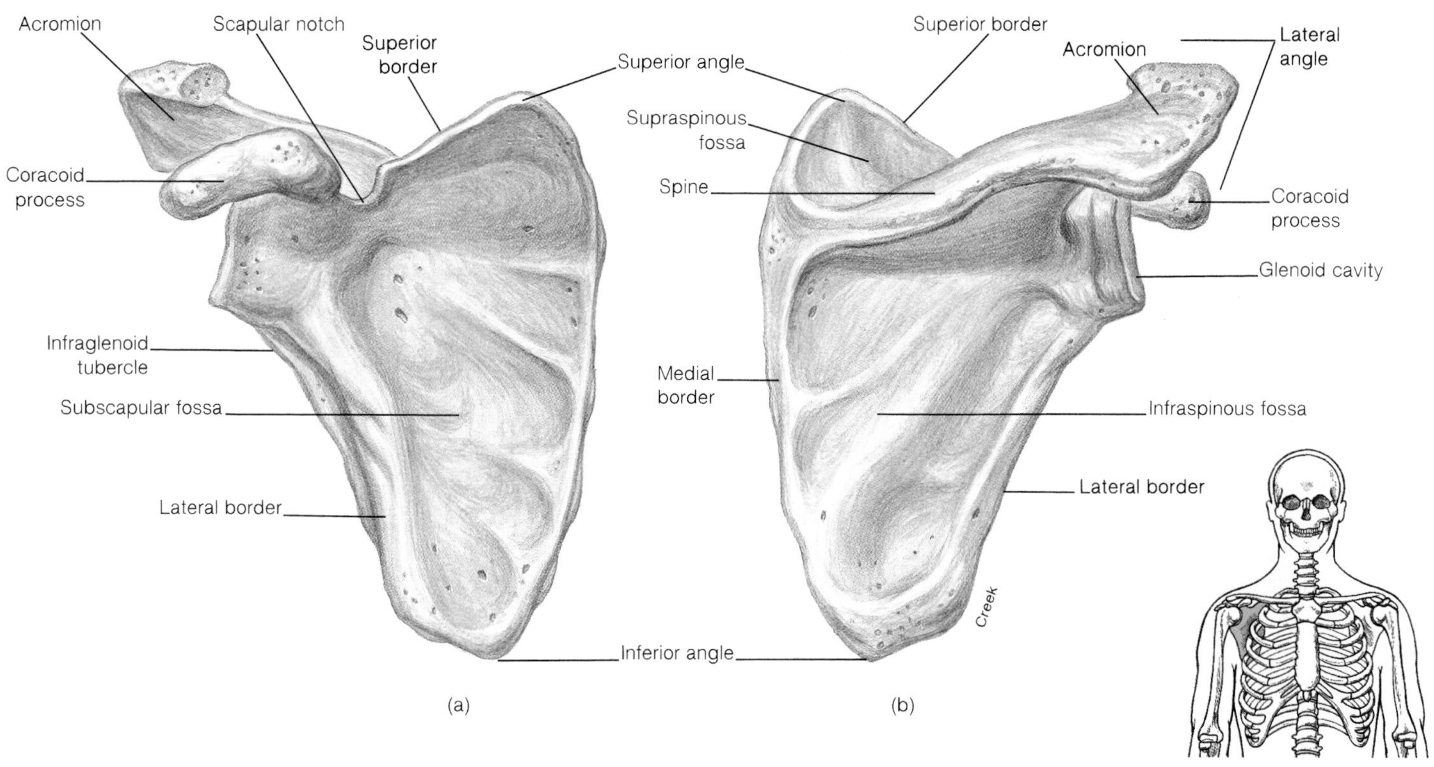

Figure 10.3

A radiograph of the right shoulder.
This X ray shows the positions of the clavicle, scapula, and humerus.

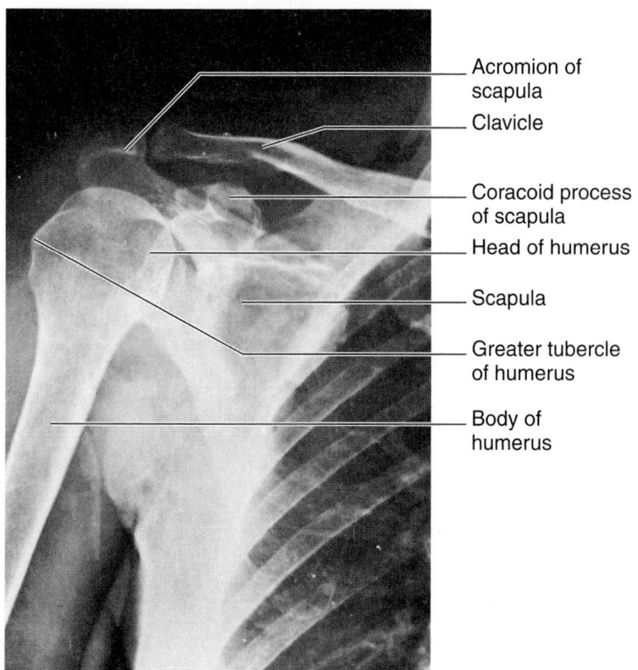

the spine is the **infraspinous fossa.** The spine broadens toward the shoulder as the **acromion** (*ă-kro'me-on*). This process serves for the attachment of several muscles, as well as for articulation with the clavicle. Inferior to the acromion is a shallow depression, the **glenoid** (*gle'noid*) **cavity,** into which the head of the humerus fits. The **coracoid** (*kor'a-koid*) **process** is a thick upward projection lying superior and anterior to the glenoid cavity. On the anterior surface of the scapula is a slightly concave area known as the **subscapular fossa.**

The scapula has three borders delimited by three angles. The superior edge is called the **superior border.** The **medial border** is nearest to the vertebral column, and the **lateral border** is directed toward the arm. The **superior angle** is located between the superior and medial borders; the **inferior angle,** at the junction of the medial and lateral borders; and the **lateral angle,** at the junction of the superior and lateral borders. The scapula articulates with the head of the humerus at the lateral angle. Along the superior border, a distinct depression called the **scapular notch** is a passageway for the suprascapular nerve.

---

acromion: Gk. *akros,* peak; *amos,* shoulder
glenoid: Gk. *glenoeides,* shallow form
coracoid process: Gk. *korakodes,* like a crow's beak

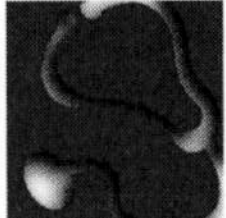

The scapula has numerous surface features for the attachment of 15 muscles. Clinically, the pectoral girdle is significant because the clavicle and acromion of the scapula are frequently fractured in trying to break a fall. The acromion is palpated as a landmark for identifying the site for an injection in the arm. This site is chosen because the musculature of the shoulder is quite thick and contains few nerves.

Brachium (Arm)

The **brachium** (*bra'ke-um*) extends from the shoulder to the elbow. In a strict anatomical usage, *arm* refers only to this portion of the upper limb. The brachium contains a single bone—the humerus.

Humerus

The humerus (fig. 10.4) is the longest bone of the upper extremity. It consists of a proximal **head,** which articulates with the glenoid cavity of the scapula; a **body** (shaft); and a distal end, which is modified to articulate with the two bones of the forearm. Surrounding the margin of the head is a slightly indented groove denoting the **anatomical neck.** The **surgical neck,** the constriction just below the head, is a frequent fracture site. The **greater tubercle** is a large knob on the lateral proximal portion of the humerus. The **lesser tubercle** is slightly anterior to the greater tubercle and is separated from the greater by an **intertubercular** (bicipital) **groove.** The tendon of the biceps brachii muscle passes through this groove. Along the lateral midregion of the body of the humerus is a roughened area, the **deltoid tuberosity,** for the attachment of the deltoid muscle. Small openings in the bone along the body are called **nutrient foramina.**

The *humeral condyle* on the distal end of the humerus has two articular surfaces. The **capitulum** (*kă-pich'ŭ-lum*) is the lateral rounded part that articulates with the radius. The **trochlea** (*trok'le-ă*) is the pulleylike medial part that articu-

deltoid tuberosity: Gk. *deltoeides*, shaped like the letter D
capitulum: L. *caput*, little head
trochlea: Gk. *trochilia*, a pulley

Figure 10.4

The right humerus.
(*a*) An anterior view and (*b*) a posterior view.

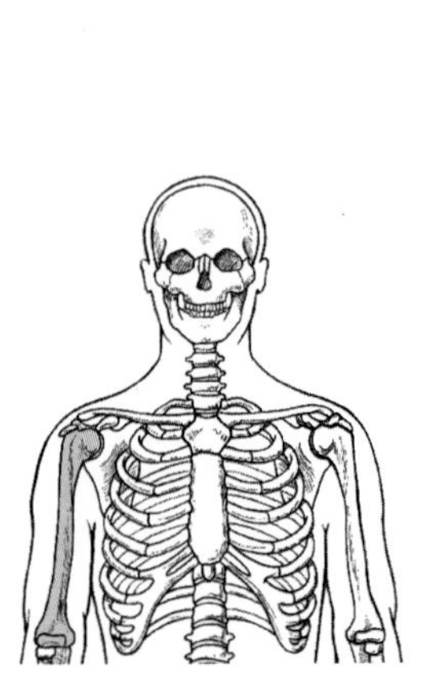

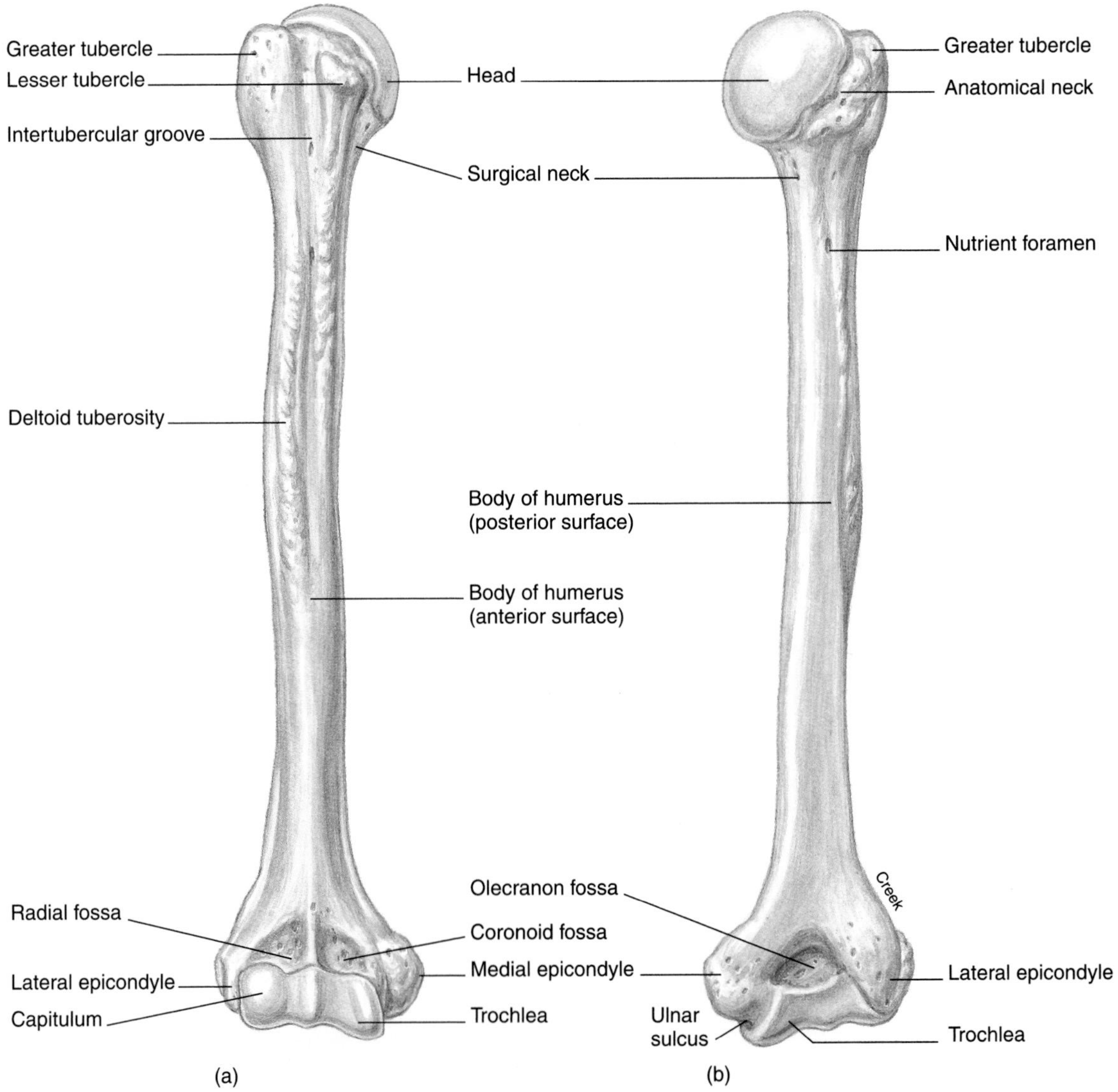

lates with the ulna. On either side above the condyles are the **lateral** and **medial epicondyles.** The large medial epicondyle protects the ulnar nerve that passes posteriorly through a depression on the back of the elbow called the **ulnar sulcus.** This region is popularly known as the "funny bone" because striking the elbow on the edge of a table, for example, stimulates the ulnar nerve causing a tingling sensation. The **coronoid fossa** is a depression above the trochlea on the anterior surface. The **olecranon** (*o-lek'ră-non*) **fossa** is a depression on the distal posterior surface. Both fossae are adapted to work with the ulna during movement of the forearm.

olecranon: Gk. *olene*, ulna; *kranion*, head

The medical term for tennis elbow is *lateral epicondylitis,* which means an inflammation of the tissues surrounding the lateral epicondyle of the humerus. Six muscles that control backward (extension) movement of the wrist and finger joints originate on the lateral epicondyle. Repeated strenuous contractions of these muscles, as in stroking with a tennis racket, may strain the periosteum and tendinous muscle attachments, resulting in swelling, tenderness, and pain around the epicondyle. Binding usually eases the pain, but only rest can eliminate the causative factor, and recovery generally follows.

Antebrachium (Forearm)

The skeletal structures of the antebrachium are the *ulna* on the medial side and the *radius* on the lateral (thumb) side (figs. 10.5 and 10.6). The ulna is more firmly connected to the humerus than the radius, and it is longer than the radius.

Figure 10.5

An anterior view of the right radius and ulna.

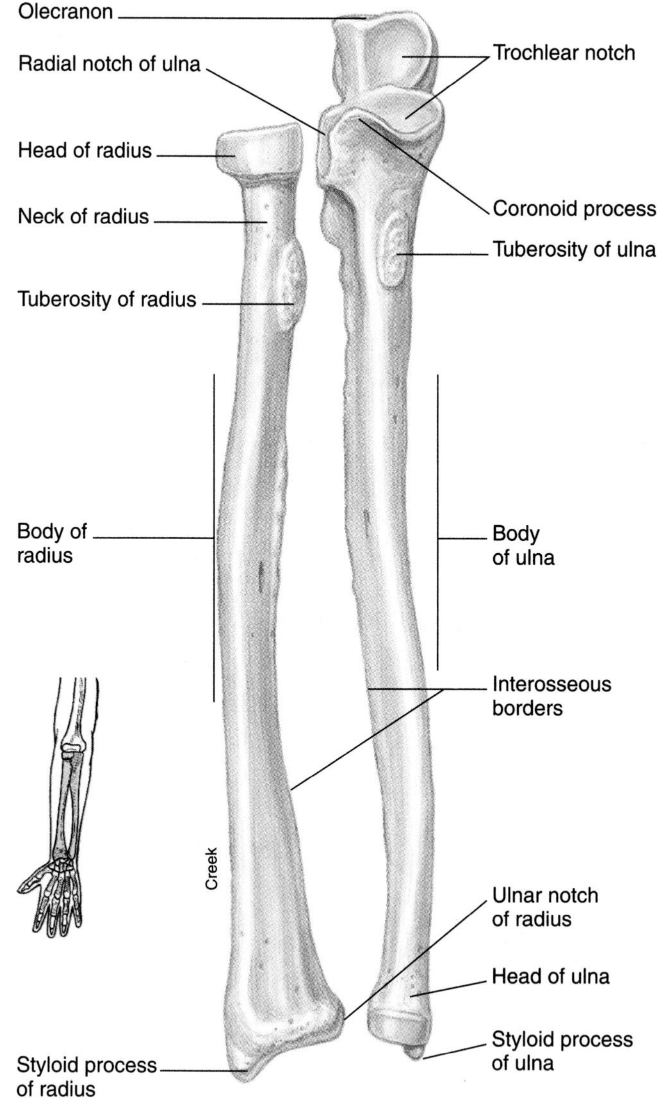

Figure 10.6

A posterior view of the right radius and ulna.

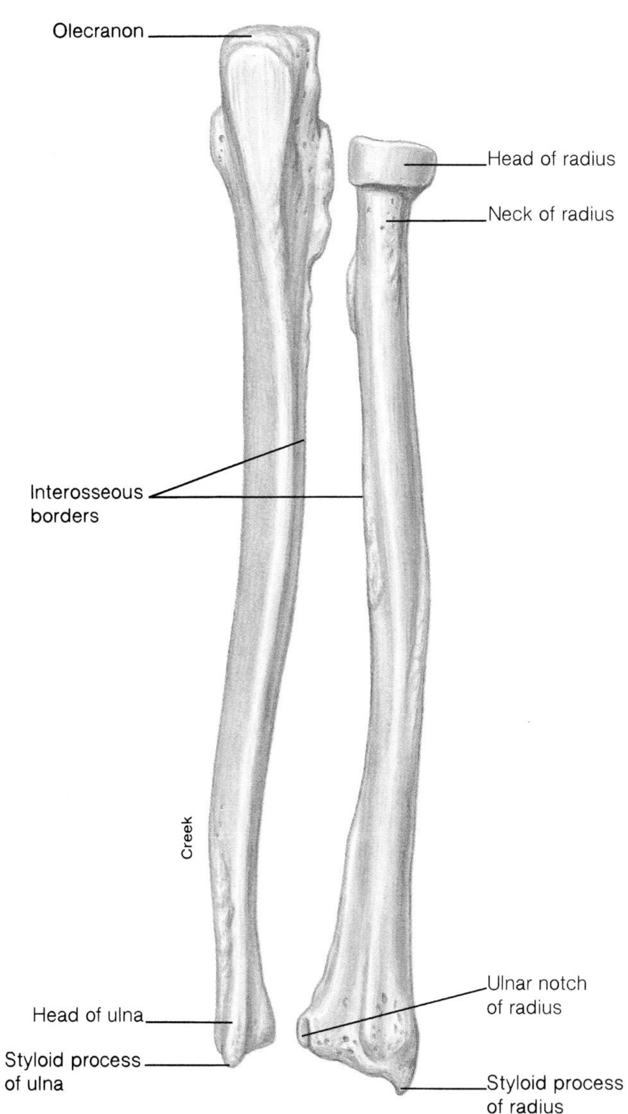

The radius, however, contributes more significantly to the articulation at the wrist joint than the ulna.

Ulna

The proximal end of the ulna articulates with the humerus and radius. A distinct depression, the **trochlear notch,** articulates with the trochlea of the humerus. The **coronoid process** forms the anterior lip of the trochlear notch, and the **olecranon** (commonly called the "elbow") forms the posterior portion (fig. 10.5). Lateral and inferior to the coronoid process is the **radial notch,** which accommodates the head of the radius.

On the tapered distal end of the ulna is a knobbed portion, the **head,** and a knoblike projection, the **styloid process.** The ulna articulates at both ends with the radius.

Radius

The radius consists of a **body** (shaft) with a small proximal end and a large distal end. A proximal disc-shaped **head** articulates with the capitulum of the humerus and the radial notch of the ulna. The prominent **tuberosity of radius** (radial tuberosity), for attachment of the biceps brachii muscle, is located on the medial side of the body just below the head. On the distal end of the radius is a double-faceted surface for articulation with the proximal carpal bones. The distal end of the radius also has a **styloid process** on the lateral tip and the **ulnar notch** on the medial side that receives the distal end of the ulna. The styloid processes on the ulna and radius provide lateral and medial stability for articulation at the wrist.

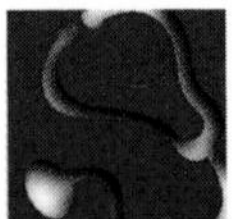

When a person falls, the natural tendency is to extend the hand to break the fall. This reflexive movement frequently causes fractured bones. Common fractures of the radius include a fracture of the head as it is driven forcefully against the capitulum, a fracture of the neck, or a fracture of the distal end (*Colles' fracture*) caused by landing on an outstretched hand.

When falling, it is actually less traumatic to the body to withdraw the appendages, bend the knees, and let the entire body hit the surface. Athletes learn that this is the safest way to fall.

Manus (Hand)

The hand contains 27 bones, grouped into the carpus, metacarpus, and phalanges (figs. 10.7–10.9).

Figure 10.7

A posterior view of the skeleton of the right hand.
(*a*) A drawing and (*b*) a photograph. Each digit (finger) is indicated by a Roman numeral, the first digit, or thumb, being Roman numeral I.

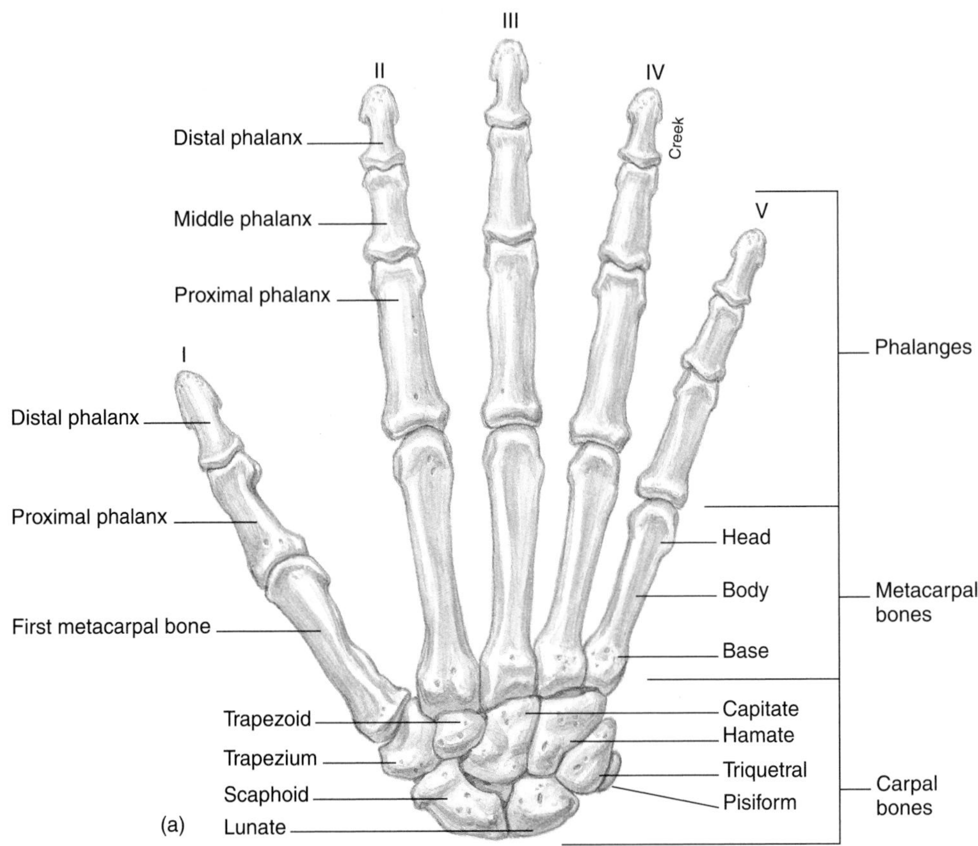

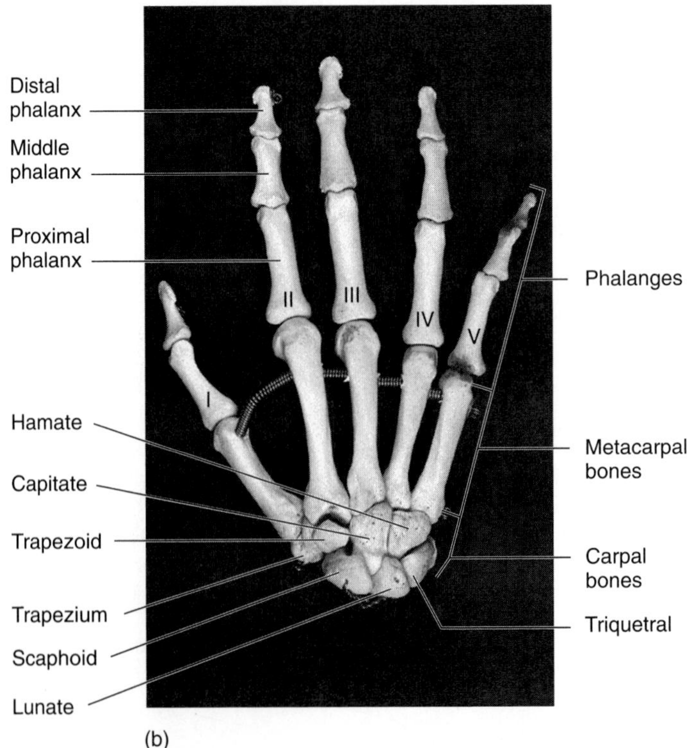

styloid process: Gk. *stylos*, pillar; *eidos*, resemblance

Figure 10.8

An anterior view of the bones of the right hand.

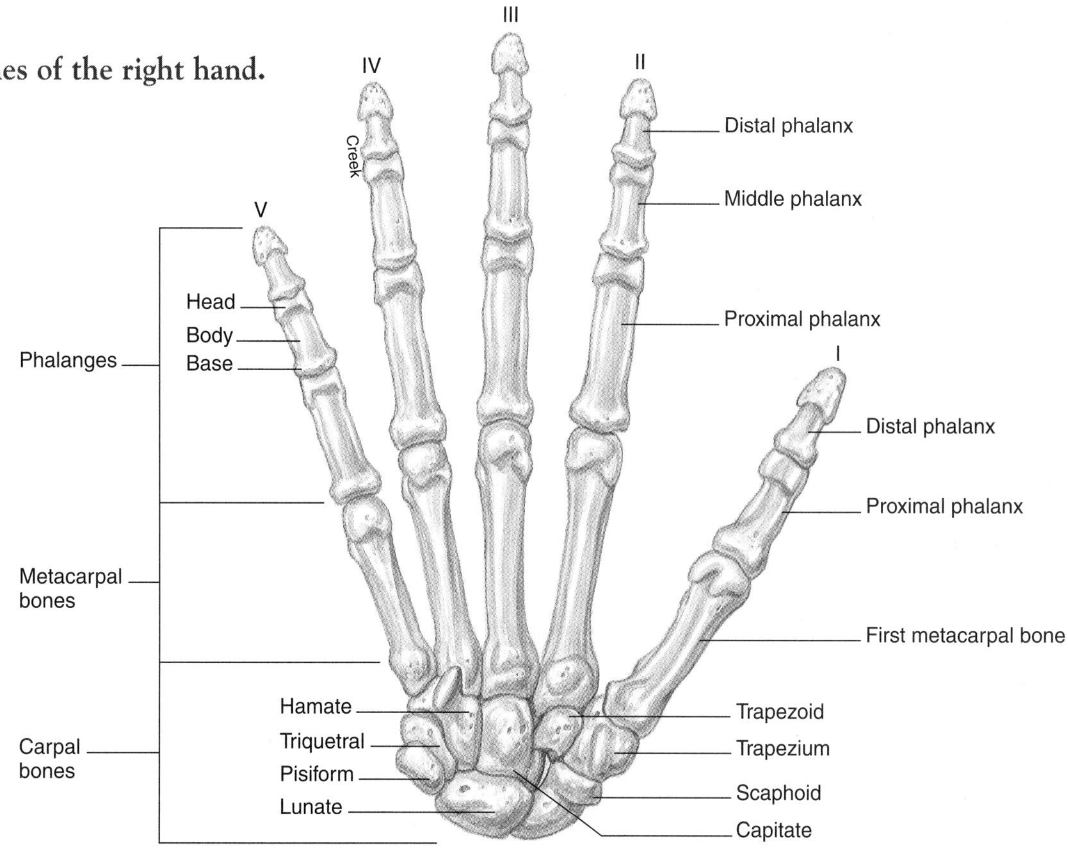

Figure 10.9

A radiograph (X ray) of the right hand shown in an anteroposterior view.
Note the presence of a sesamoid bone at the thumb joint.

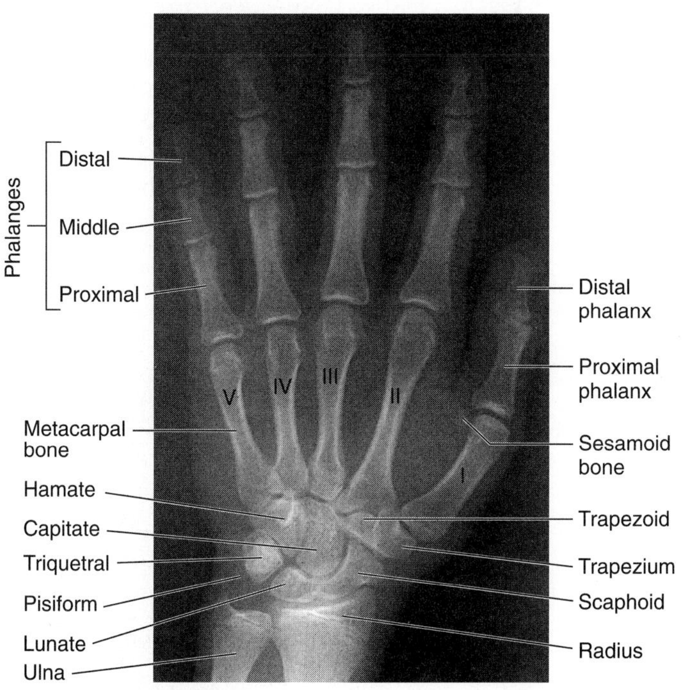

Carpus

The carpus, or wrist, contains eight carpal bones arranged in two transverse rows of four bones each. The proximal row, naming from the lateral (thumb) to medial side, consists of the **scaphoid** (navicular), **lunate, triquetral** (*tri-kwe'tral*), and **pisiform** (*pi'sĭ-form*) **bone.** The pisiform bone forms in a tendon as a sesamoid bone. The distal row, from lateral to medial, consists of the **trapezium** (greater multangular), **trapezoid** (lesser multangular), **capitate,** and **hamate** (*ham'at*). The scaphoid and lunate of the proximal row articulate with the distal end of the radius.

Metacarpus

The metacarpus, or palm of the hand, contains five metacarpal bones. Each metacarpal bone consists of a proximal **base,** a **body** (shaft), and a distal **head** that is rounded for articulation with the base of each proximal phalanx. The heads of the metacarpal bones are distally located and form the knuckles of a clenched fist.

---

carpus: Gk. *karpos*, wrist
scaphoid: Gk. *skaphe*, boat; *eidos*, resemblance
lunate: L. *lunare*, crescent- or moon-shaped
triquetrum: L. *triquetrus*, three-cornered
pisiform: Gk. *pisos*, pea
trapezium: Gk. *trapesion*, small table
capitate: L. *capitatus*, head
hamate: L. *hamatus*, hook

Table 10.1

Bones of the Pectoral Girdle and Upper Extremities

	Location	Major Distinguishing Features
Clavicle (2)	Anterior base of neck, between sternum and scapula	S-shaped, sternal and acromial extremities, conoid tubercle; costal tuberosity
Scapula (2)	Upper back forming part of the shoulder	Triangular; spine; subscapular, supraspinous, and infraspinous fossae; glenoid cavity; coracoid process; acromion
Humerus (2)	Brachium, between scapula and elbow	Longest bone of upper extremity; greater and lesser tubercles; intertubercular groove; surgical neck, deltoid tuberosity; capitulum; trochlea; lateral and medial epicondyles; coronoid and olecranon fossae
Ulna (2)	Medial side of forearm	Trochlear notch; olecranon; coronoid and styloid processes; radial notch
Radius (2)	Lateral side of forearm	Head; radial tuberosity; styloid process; ulnar notch
Carpal bone (16)	Wrist	Short bones arranged in two rows of four bones each
Metacarpal bone (10)	Palm of hand	Long bones, each aligned with a digit
Phalanx (28)	Digits	Three in each digit, except two in thumb

Phalanges

The 14 phalanges are the bones of the digits. A single finger bone is called a **phalanx** (*fa'langks*). The phalanges of the fingers are arranged in a proximal row, a middle row, and a distal row. The thumb, or *pollex* (adjective, *pollicis*), lacks a middle phalanx. The digits are sequentially numbered from I to V starting with the thumb—the lateral side, in reference to anatomical position.

A summary of the bones of the upper extremities is presented in table 10.1.

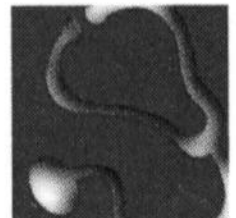

The hand is a marvel of structural complexity that can withstand considerable abuse. Other than sprained ligaments of the fingers and joint dislocations, the most common bone injury is a fracture to the scaphoid—a wrist bone that accounts for about 70% of carpal fractures. When immobilizing a fractured carpal bone, the wrist is positioned in the plane of relaxed function. This is the position in which the hand is about to grasp an object between the thumb and index finger.

Pelvic Girdle and Lower Extremity

The structure of the pelvic girdle and lower extremities is adaptive for support and locomotion. Extensive processes and surface features on certain bones of the pelvic girdle and lower extremities accommodate massive muscle use in body movements and in maintaining posture.

Pelvic Girdle

The **pelvic girdle** is formed by two *ossa coxae* (*os'a kuk'se*; "hipbones"), united anteriorly at the *symphysis pubis* (fig. 10.10). It is attached posteriorly to the sacrum of the vertebral column. The sacrum, a bone of the axial skeleton, was described in chapter 9 (see fig. 9.24). The deep, basinlike structure formed by the ossa coxae, together with the sacrum and coccyx, is called the **pelvis** (plural, *pelves* or *pelvises*). The pelvic girdle and its associated ligaments support the weight of the body from the vertebral column. The pelvic girdle also supports and protects the lower viscera, including the urinary bladder, the reproductive organs, and in a pregnant woman, the developing fetus.

The pelvis is divided into a **greater** (false) **pelvis** and a **lesser** (true) **pelvis** (see fig. 10.14). These two components are divided by the **pelvic brim,** a curved bony rim passing inferiorly from the sacral promontory to the upper margin of the symphysis pubis. The greater pelvis is the expanded portion of the pelvis, superior to the pelvic brim. The pelvic brim not only divides the two portions but surrounds the **pelvic inlet** of the lesser pelvis. The lower circumference of the lesser pelvis bounds the pelvic outlet.

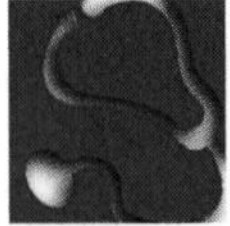

During parturition, a child must pass through its mother's lesser pelvis for a natural delivery. *Pelvimetry* measures the dimension of the lesser pelvis to determine whether a cesarean delivery might be necessary. Diameters may be determined by vaginal palpation or by radiographic measurements (fig. 10.11).

Each os coxae ("hipbone") actually consists of three separate bones: the *ilium*, the *ischium* (*is'ke-um*), and the *pubis* (figs. 10.12 and 10.13). These bones are fused together in the adult. On the lateral surface of the os coxae, where the three

phalanx: Gk. *phalanx*, finger bone or toe bone
coxae: L. *coxae*, hips

ilium: L. *ilia*, loin
ischium: Gk. *ischion*, hip joint
pubis: L. *pubis*, genital area

Figure 10.10

An anterior view of the pelvic girdle.

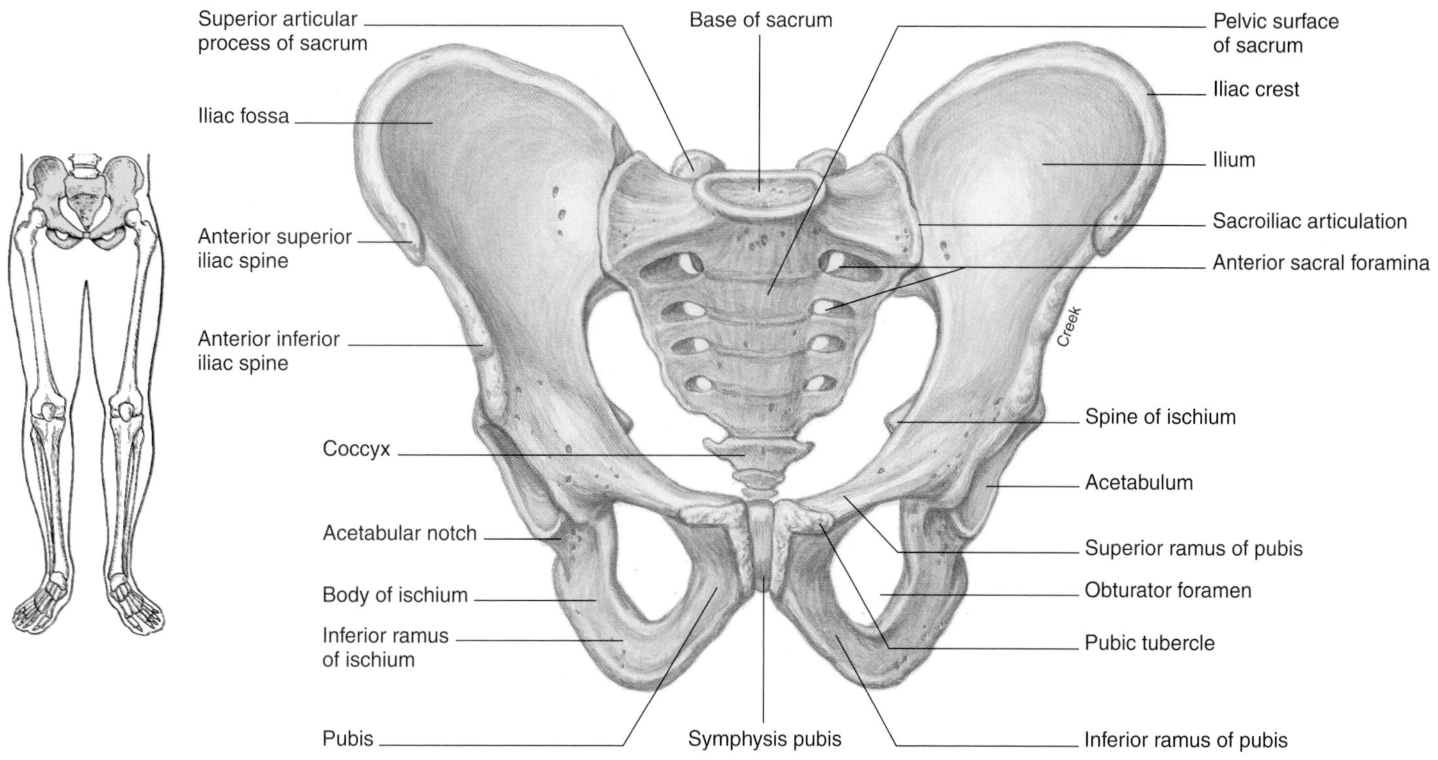

Figure 10.11

A radiograph of the pelvic girdle and the articulating femurs.

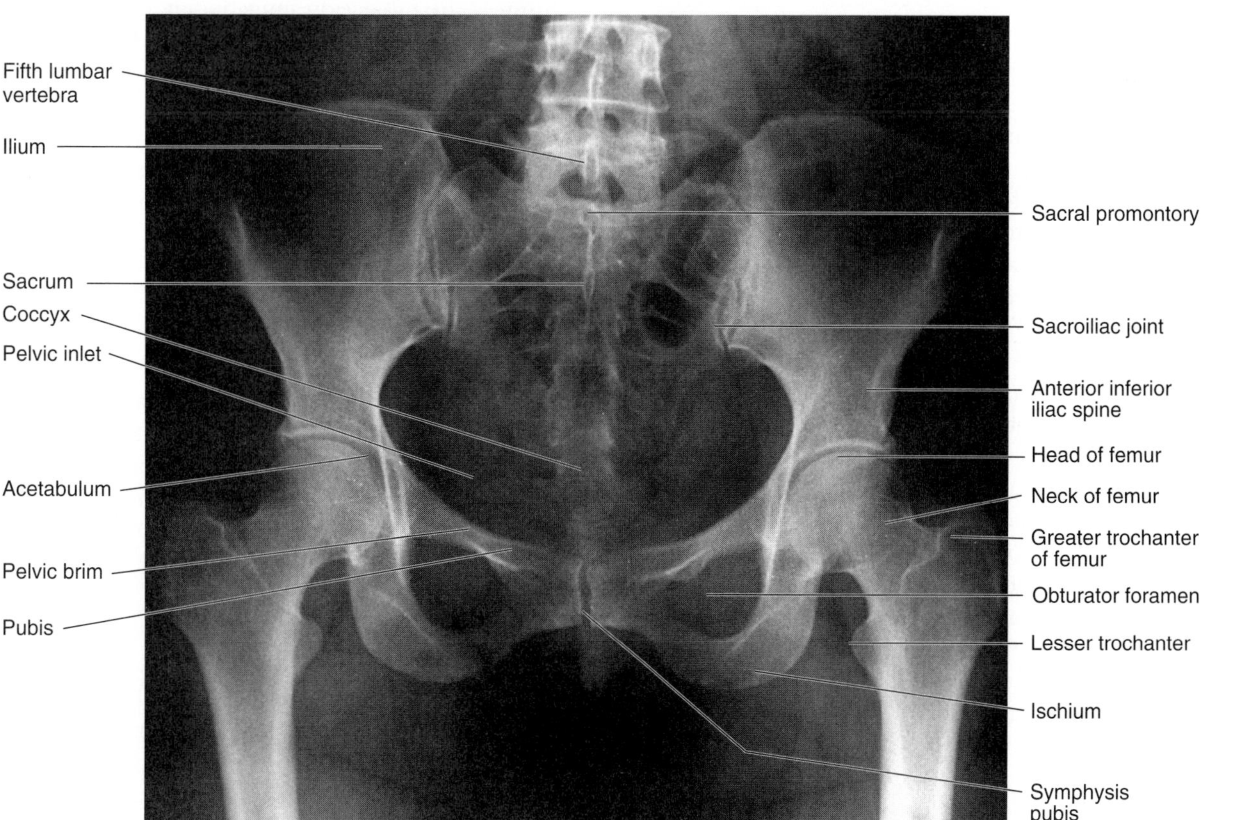

Figure 10.12

The lateral aspect of the right os coxae.
The three bones comprising the os coxae are labeled in boldface type.

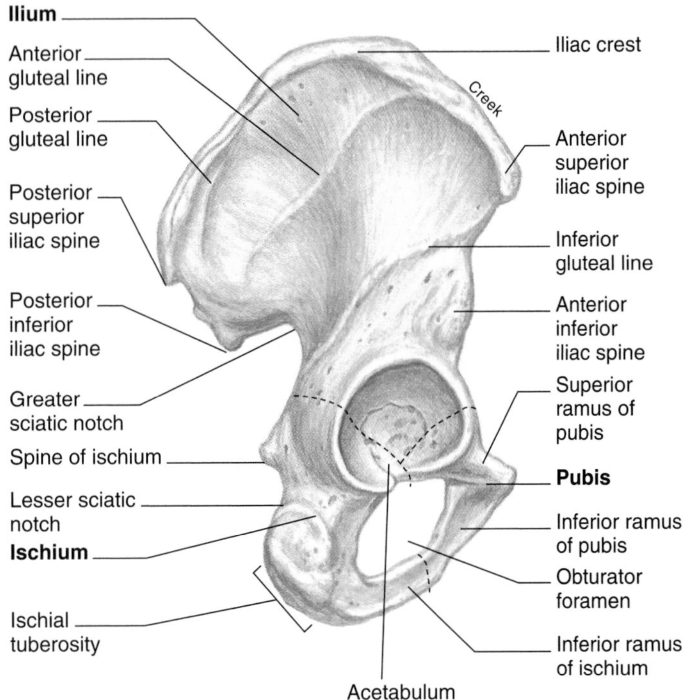

Figure 10.13

The medial aspect of the right os coxae.
The three bones comprising the os coxae are labeled in boldface type.

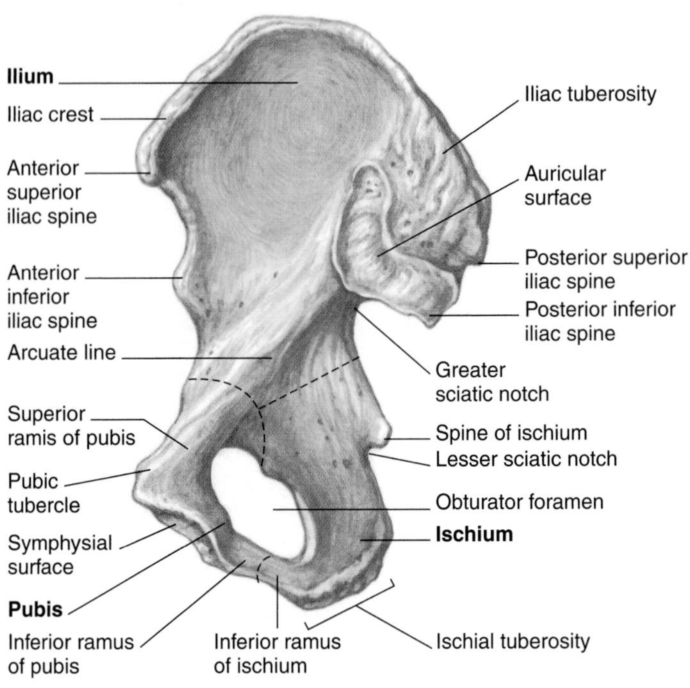

bones ossify, is a large circular depression, the **acetabulum** (*as″ĕ-tab′yŭ-lum*), which receives the head of the femur. Although both ossa coxae are single bones in the adult, the three components of each one are considered separately for descriptive purposes.

Ilium

The ilium is the largest and uppermost of the three pelvic bones. It has a crest and four angles, or spines—important surface landmarks that serve for muscle or ligament attachment. The **iliac crest** forms the prominence of the hip. This crest terminates anteriorly as the **anterior superior iliac spine.** Just below this spine is the **anterior inferior iliac spine.** The posterior termination of the iliac crest is the **posterior superior iliac spine,** and just below this is the **posterior inferior iliac spine.**

Below the posterior inferior iliac spine is the **greater sciatic** (*si-at′ik*) **notch,** through which the sciatic nerve passes. On the medial surface of the ilium is the roughened **auricular surface,** which articulates with the sacrum. The **iliac fossa** is the smooth concave surface on the anterior portion of the ilium. The iliacus muscle originates from this fossa. The **iliac tuberosity,** for the attachment of the sacroiliac ligament, is positioned posterior to the iliac fossa. Three roughened ridges are present on the **gluteal surface** of the posterior aspect of the ilium. These ridges, which serve to attach the gluteal muscles, are the **inferior, anterior,** and **posterior gluteal lines.**

---

acetabulum: L. *acetabulum*, vinegar cup

Ischium

The ischium is the posteroinferior bone of the os coxae. This bone has several significant features. The **spine of the ischium** is the projection immediately posterior and inferior to the greater sciatic notch of the ilium. Inferior to this spine is the **lesser sciatic notch** of the ischium. The **ischial tuberosity** is the bony projection that supports the weight of the body in the sitting position. A deep **acetabular** (*as″e-tab′yu-lar*) **notch** is present on the inferior portion of the acetabulum. The large **obturator** (*ob′tu-ra″tor*) **foramen** is formed by the **ramus of the ischium** together with the pubis. The obturator foramen is covered by the obturator membrane, to which several muscles attach.

Pubis

The pubis is the anterior bone of the os coxae. This bone consists of a **superior ramus** and an **inferior ramus** that support the **body** of the pubis. The body contributes to the formation of the symphysis pubis—the joint between the two ossa coxae. At the lateral end of the anterior border of the body is the **pubic tubercle,** one of the attachments for the inguinal ligament.

Sex-Related Differences in the Pelvis

Structural differences between the pelvis of an adult male and that of an adult female (fig. 10.14 and table 10.2) reflect the female's role in pregnancy and parturition. In addition to the osseous differences listed in table 10.2, the symphysis pubis and sacroiliac joints stretch during pregnancy and parturition.

Figure 10.14

A comparison of (*a*) the male and (*b*) the female pelvic girdle.

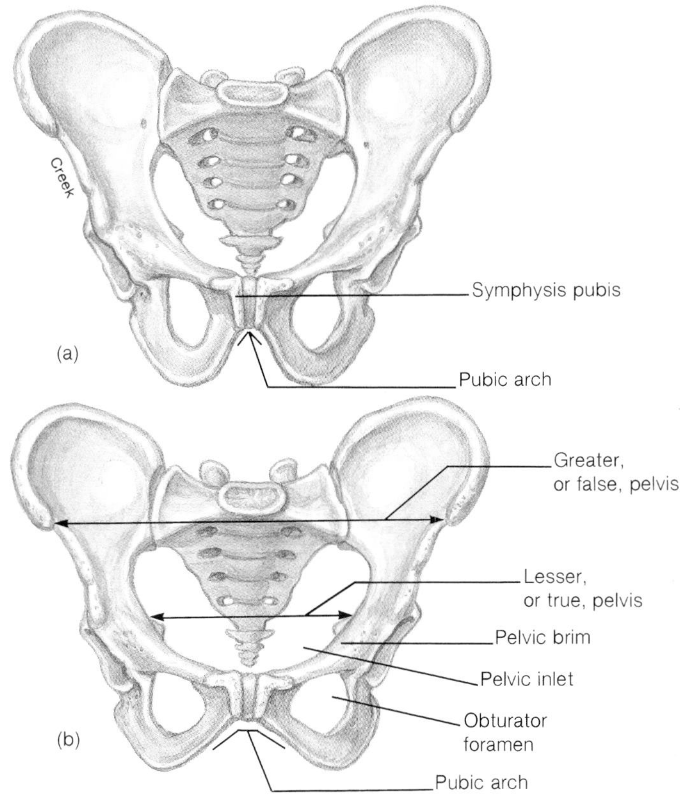

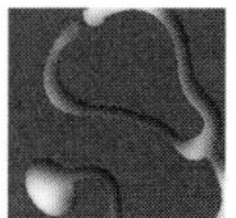

The structure of the human pelvis, in its attachment to the vertebral column, permits an upright posture and locomotion on two appendages (bipedal locomotion). An upright posture may cause problems, however. The sacroiliac joint may weaken with age, causing lower back pains. The weight of the viscera may weaken the walls of the lower abdominal area and cause hernias. Some of the problems of childbirth are related to the structure of the mother's pelvis. Finally, the hip joint tends to deteriorate with age, so that many elderly people suffer from fractured hips.

Thigh

The *femur* (*fe'mur;* "thighbone") is the only bone of the thigh. In the following discussion, however, the *patella* (*pa-tel'a;* "kneecap") will also be discussed.

Femur

The femur is the longest, heaviest, and strongest bone in the body (fig. 10.15). The proximal rounded **head** of the femur articulates with the acetabulum of the os coxae. A shallow pit, called the **fovea capitis femoris** is present in the lower center of the head of the femur. The fovea capitis femoris provides the point of attachment for the ligamentum teres, which helps to support the head of the femur against the acetabulum. The constricted region supporting the head is called the **neck** and is a common site for fractures in the elderly.

The **body** (shaft) of the femur has a slight medial curve to bring the knee joint in line with the body's plane of gravity. The degree of curvature is even greater in the female because of the wider pelvis. The body of the femur has several distinguishing features for muscle attachment. On the proximolateral side of the body of the femur is the **greater trochanter,** and on the medial side is the **lesser trochanter.** On the anterior side, between the trochanters, is the **intertrochanteric** (*in"ter-tro"kan-ter'ik*) **line.** On the posterior side, between the trochanters, is the **intertrochanteric crest.** The **linea aspera** (*lin'e-a as'per-a*) is a roughened vertical ridge on the posterior surface of the body of the femur.

The distal end of the femur is expanded for articulation with the tibia. The **medial** and **lateral condyles** are the articular processes for this joint. The depression between the condyles on the posterior aspect is called the **intercondylar fossa.** The **patellar surface** is located between the condyles on the anterior side. Above the condyles on the lateral and medial sides are the **epicondyles,** which serve for ligament and tendon attachment.

Patella

The patella is a large, triangular sesamoid bone positioned on the anterior side of the distal femur (figs. 10.16 and 10.17). It develops in response to strain in the tendon of the quadriceps femoris muscle. The patella has a broad **base** and an inferiorly pointed **apex.** Articular facets on the **articular surface** of the patella articulate with the medial and lateral condyles of the femur.

femur: L. *femur*, thigh

linea aspera: L. *linea*, line; *asperare*, rough

Table 10.2

Comparison of the Male and Female Pelves

Characteristics	Male Pelvis	Female Pelvis
General structure	More massive; prominent processes	More delicate; process not so prominent
Pelvic inlet	Heart-shaped	Round or oval
Pelvic outlet	Narrower	Wider
Anterior superior iliac spine	Not as far apart	Farther apart
Obturator foramen	Oval	Triangular
Acetabulum	Faces laterally	Faces more anteriorly
Symphysis pubis	Deeper, longer	Shallower, shorter
Pubic arch	Angle less than 90°	Angle greater than 90°

Figure 10.15

The right femur.
(*a*) An anterior view and (*b*) a posterior view.

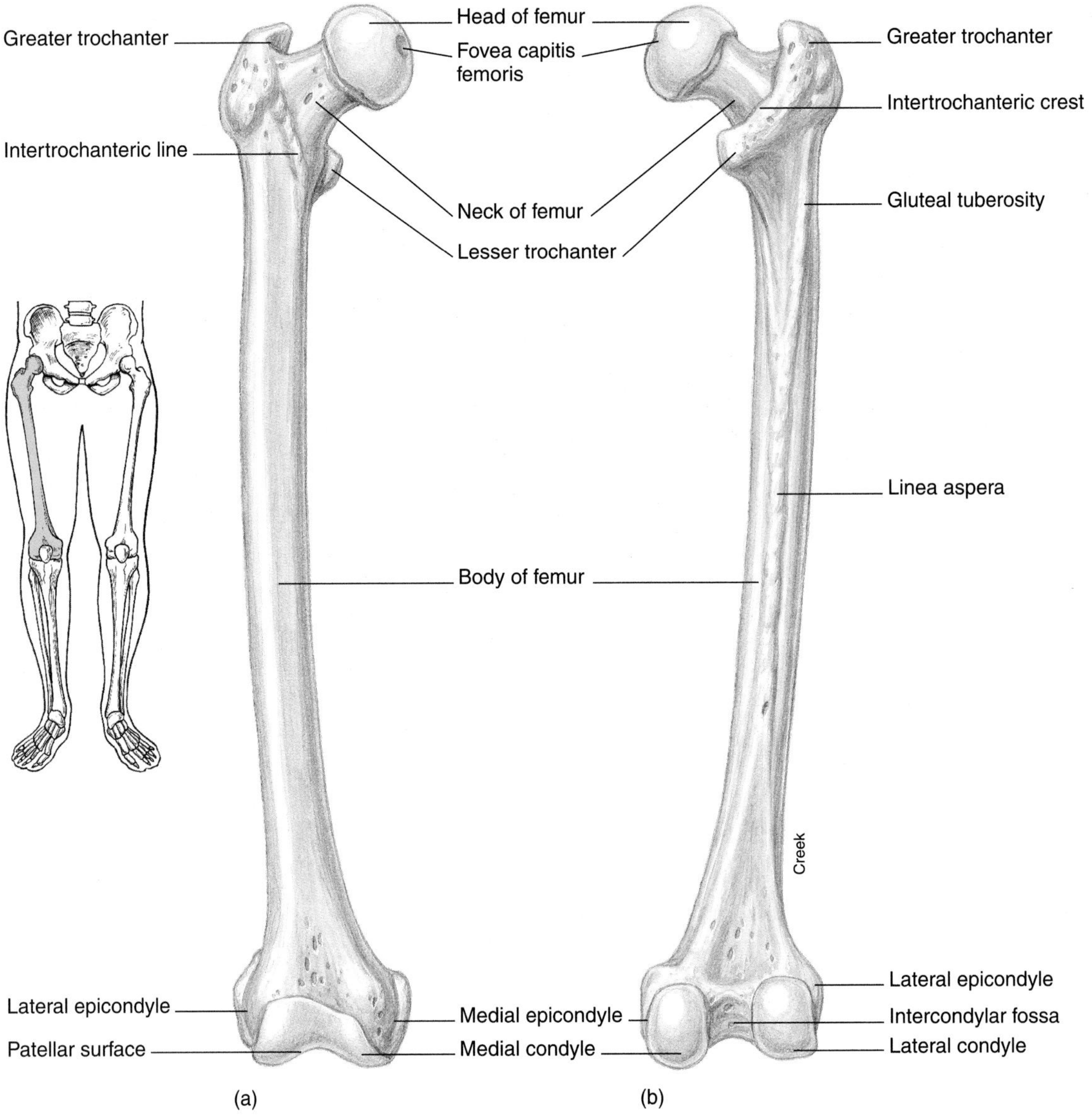

Figure 10.16

A radiograph (X ray) of the right knee.

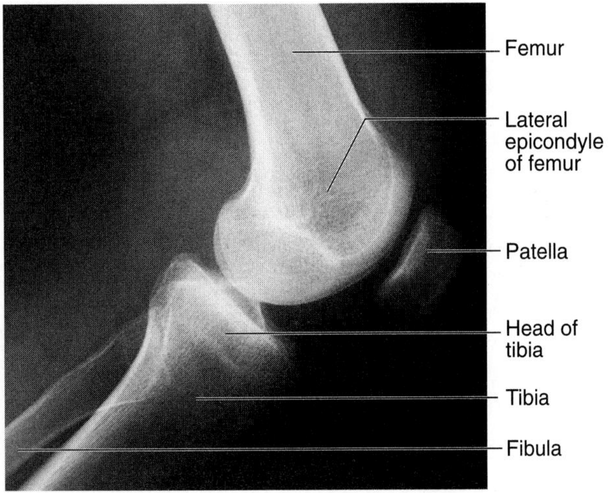

The functions of the patella are to protect the knee joint and to strengthen the tendon of the quadriceps femoris muscle. It also increases the leverage of this muscle as it extends (straightens) the knee joint.

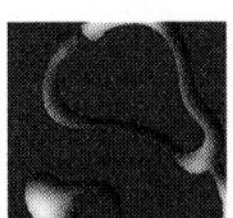

The patella can be fractured by a direct blow. It usually does not fragment, however, because it is confined within the tendon. Dislocations of the patella may result from injury or from underdevelopment of the lateral condyle of the femur.

Leg

Technically speaking, "leg" refers only to that portion of the lower limb between the knee and foot. The *tibia* (*tib'e-a;* "shinbone") and *fibula* (*fib'yu-la*) are the bones of the leg. The tibia, which is the weight-bearing bone of the leg, is the larger and more medial of the two bones.

---

tibia: L. *tibia*, shinbone, pipe, or flute

Figure 10.17 **The right tibia, fibula, and patella.** (*a*) An anterior view and (*b*) a posterior view.

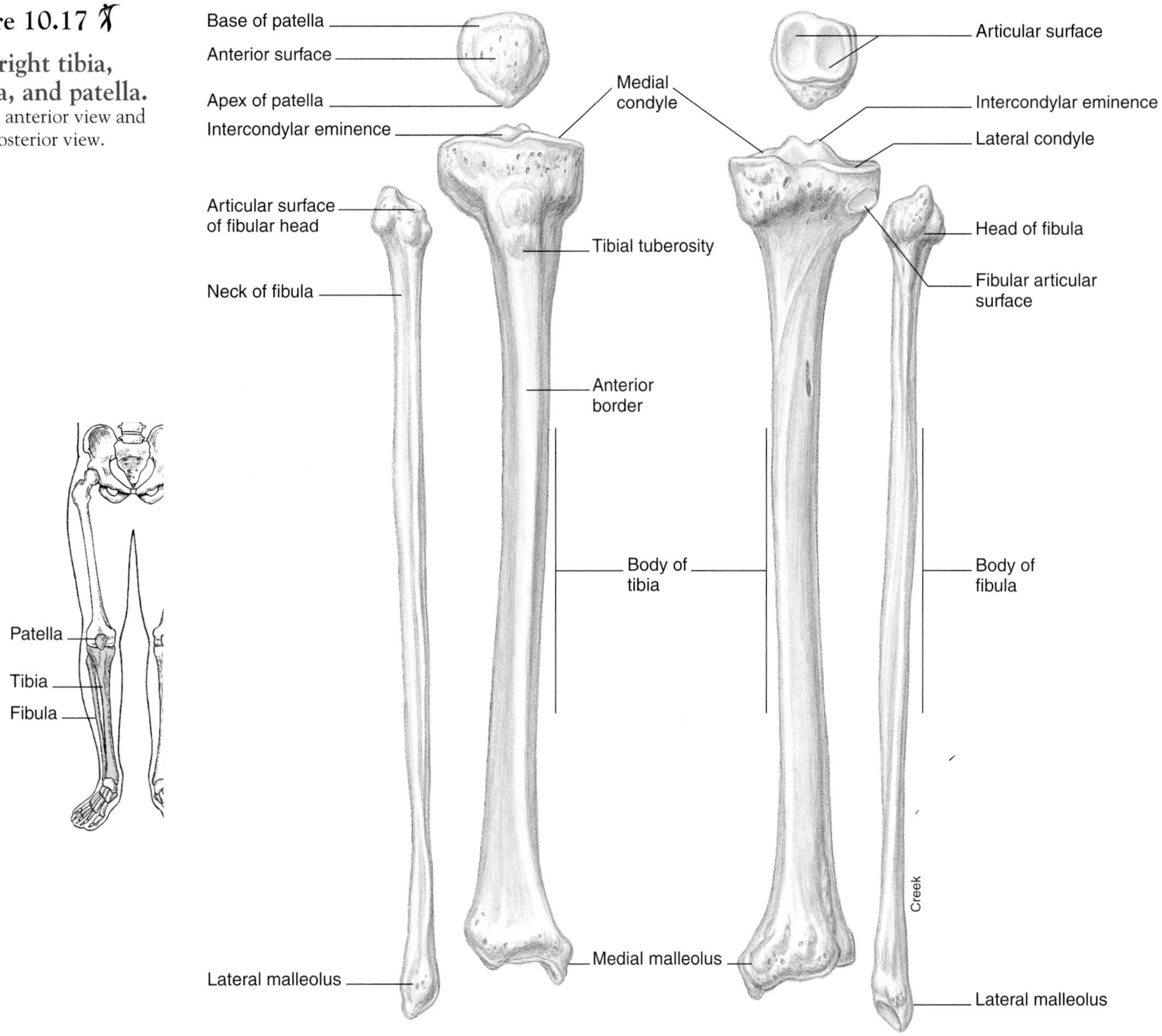

Tibia

The tibia articulates proximally with the femur at the knee and distally with the talus of the ankle. It also articulates both proximally and distally with the fibula. Two slightly concave surfaces on the proximal end of the tibia, the **medial** and **lateral condyles** (fig. 10.17), articulate with the condyles of the femur. The condyles are separated by a slight upward projection called the **intercondylar eminence.** The **tibial tuberosity,** for attachment of the patellar ligament, is located on the proximoanterior portion of the body of the tibia. The **anterior crest** is a sharp ridge along the anterior surface of the body.

The **medial malleolus** (*ma-le'o-lus*) is a prominent medial knob of bone located on the distomedial end of the tibia. A **fibular notch,** for articulation with the fibula, is located on the distolateral end.

Fibula

The fibula is a long, narrow bone that is more important for muscle attachment than for support. The **head** of the fibula articulates with the proximolateral end of the tibia. The distal end has a prominent knob called the **lateral malleolus.**

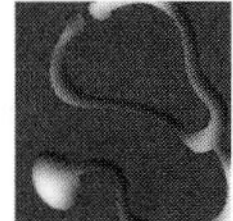

The lateral and medial malleoli are positioned on either side of the talus and help stabilize the ankle joint. Both processes can be seen as prominent surface features and are easily palpated. Fractures of the fibula above the lateral malleolus are common among skiers. Clinically referred to as a *Pott's fracture,* it is caused by a shearing force occurring at a vulnerable spot on the leg.

---

malleolus: L. *malleolus*, small hammer

fibula: L. *fibula*, clasp or brooch

Pes (Foot)

The **foot** contains 26 bones grouped into the *tarsus*, *metatarsus*, and *phalanges* (figs. 10.18 and 10.19). Although similar to the bones of the hand, the bones of the foot have distinct structural differences in order to support the weight of the body and to provide leverage and mobility during walking.

Tarsus

There are seven tarsal bones. The most superior in position is the **talus,** which articulates with the tibia and fibula to form the ankle joint. The **calcaneus** (*kal-ka'ne-us*) is the largest of the tarsal bones and provides skeletal support for the heel of the foot. It has a large posterior extension, called the **tuberosity of the calcaneus,** for the attachment of the calf muscles. Anterior to the talus is the block-shaped **navicular** (*nă-vik'yŭ-lar*) **bone.** The remaining four tarsal bones form a distal series that articulate with the metatarsal bones. They are, from the medial to lateral side, the **medial, intermediate,** and **lateral cuneiform** (*kyoo-ne'ĭ-form*) **bones,** and the **cuboid bone.**

Metatarsus

The metatarsal bones and phalanges are similar in name and number to the metacarpals and phalanges of the hand. They differ in shape, however, because of their load-bearing role. The metatarsal bones and phalanges are numbered I to V, starting with the medial (great toe) side of the foot. The first metatarsal bone is larger than the others because of its major role in supporting body weight.

The metatarsal bones each have a **base, body** (shaft), and **head.** The proximal bases of the first, second, and third metatarsal bones articulate proximally with the cuneiform bones. The heads of the metatarsal bones articulate distally with the proximal phalanges. The proximal joints are called *tarsometatarsal joints*, and the distal joints are called *metatarsophalangeal joints*. The ball of the foot is formed by the heads of the first two metatarsal bones.

Phalanges

The 14 phalanges are the skeletal elements of the toes. As with the fingers of the hand, the phalanges of the toes are arranged in a proximal row, a middle row, and a distal row. The great toe, or *hallux* (adjective, *hallucis*) has only a proximal and a distal phalanx.

Arches of the Foot

The foot has two arches that support the weight of the body and provide leverage when walking. These arches are formed by the structure and arrangement of the bones held in place by ligaments and tendons (fig. 10.19). The arches are not rigid; they "give" when weight is placed on the foot, and they spring back as the weight is lifted. A weakening of the ligaments and tendons of the foot may cause the arches to "fall"—a condition known as *pes planus*, or, more commonly, "flatfoot."

The **longitudinal arch** is divided into medial and lateral parts. The medial part is more elevated of the two. The talus is the keystone of the medial part, which originates at the calcaneus, rises at the talus, and descends to the first three metatarsal bones. The shallower lateral part consists of the calcaneus, cuboid, and fourth and fifth metatarsal bones. The cuboid is the keystone bone of this part.

The **transverse arch** extends across the width of the foot and is formed by the calcaneus, navicular, and cuboid bones posteriorly and the bases of all five metatarsal bones anteriorly.

The bones of the lower extremities are summarized in table 10.3.

tarsus: Gk. *tarsos*, flat of the foot
talus: L. *talus*, ankle
calcaneus: L. *calcis*, heel

Clinical Considerations

Developmental Disorders

Minor defects of the extremities are relatively common malformations. Extra digits, a condition called **polydactyly** (*pol″e-dak'tĭ-le*) (fig. 10.20), is the most common limb deformity. Usually an extra digit is incompletely formed and does not function. **Syndactyly** (*sin-dak'tĭ-le*), or webbed digits, is likewise a relatively common limb malformation. Polydactyly is inherited as a dominant trait, whereas syndactyly is a recessive trait.

Talipes (*tal'ĭ-pēz*), or clubfoot (fig. 10.21), is a congenital malformation in which the sole of the foot is twisted medially. It is not certain if abnormal positioning or restricted movement in utero causes this condition, but both genetics and environmental factors are involved in most cases.

Trauma and Injury

The most common type of bone injury is a **fracture**—the cracking or breaking of a bone. Radiographs are often used to diagnose the position and extent of a fracture. Fractures may be classified in several ways, and the type and severity of the fracture are often related to the age and the general health of the individual. **Pathologic fractures,** for example, result from diseases that weaken the bones. Most fractures, however, are called **traumatic fractures** because they are caused by injuries. The following are descriptions of several kinds of traumatic fractures (fig. 10.22).

polydactyly: Gk. *polys*, many; *daktylos*, finger
syndactyly: Gk. *syn*, together; *daktylos*, finger
talipes: L. *talus*, heel; *pes*, foot

Figure 10.18

The bones of the right foot.

(*a*) A photograph of a superior view, (*b*) a radiograph (X ray) of a medial view, (*c*) a superior view, and (*d*) an inferior view. Each digit (toe) is indicated by a Roman numeral, the first digit, or great toe, being Roman numeral I.

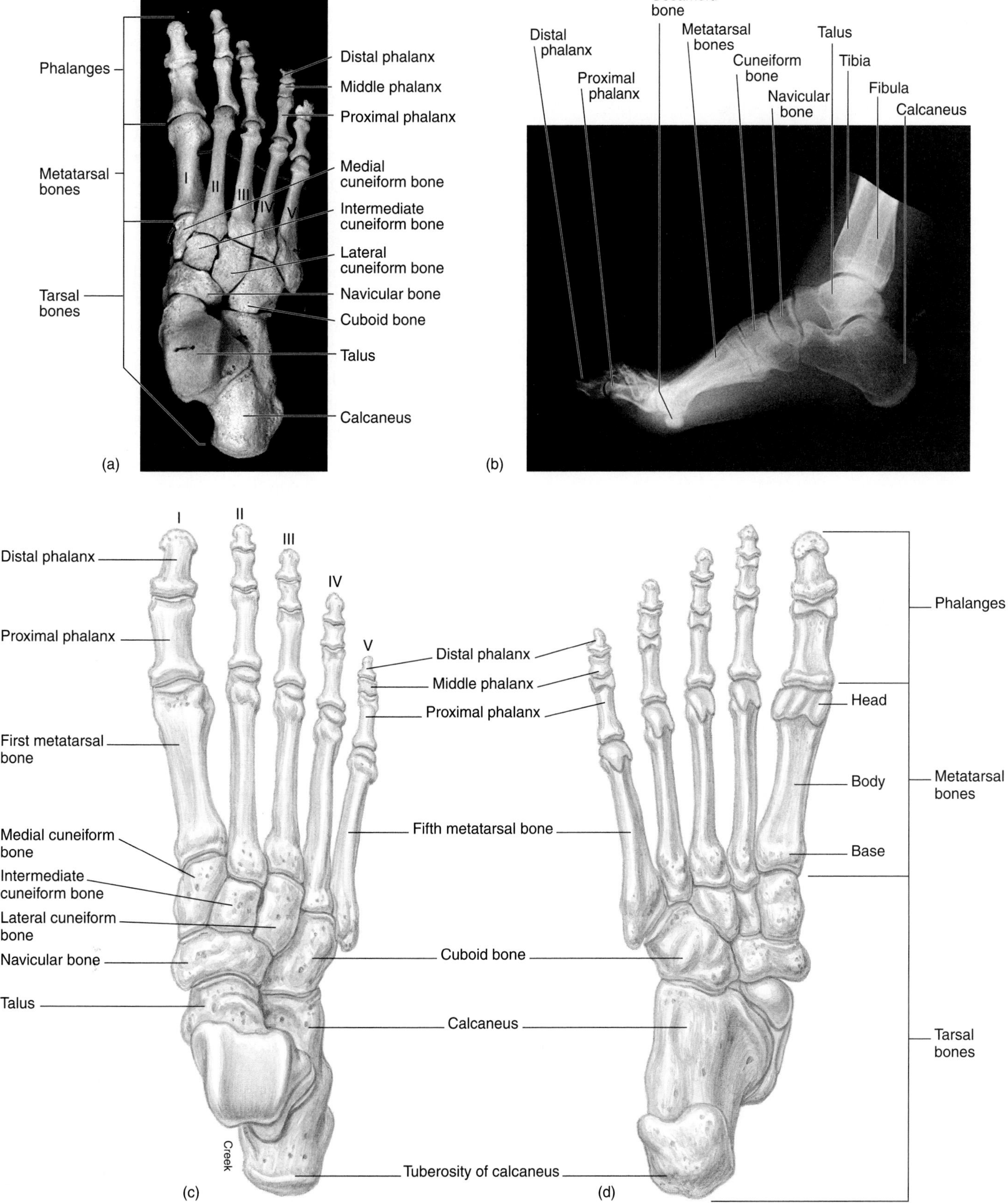

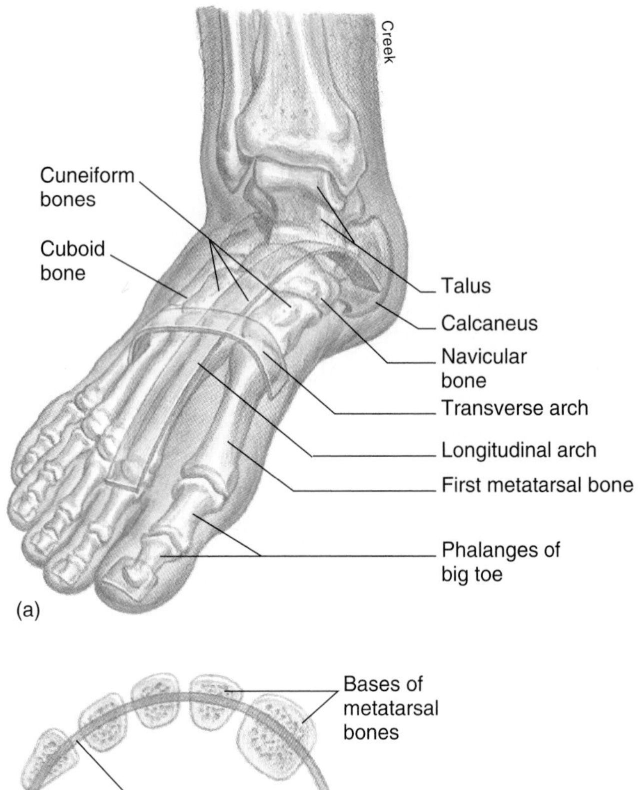

Figure 10.19

The arches of the foot.
(*a*) A medial view of the right foot showing both arches and (*b*) a transverse view through the bases of the metatarsal bones showing a portion of the transverse arch.

1. **Simple,** or **closed.** The fractured bone does not break through the skin.
2. **Compound,** or **open.** The fractured bone protrudes to the outside through an opening in the skin.
3. **Partial (fissured).** The bone incompletely breaks.
4. **Complete.** The fracture separates the bone into two parts.
5. **Comminuted** (*kom′ĭ-noot″ed*). The bone splinters into several fragments.
6. **Spiral.** The fracture line twists as it is broken.
7. **Greenstick.** An incomplete break, in which one side of the bone breaks, and the other side bows.
8. **Impacted.** One broken end of a bone impales the other.
9. **Transverse.** The fracture occurs across the bone at a right angle to the long axis.
10. **Oblique.** The fracture occurs across the bone at an oblique angle to the long axis.
11. **Colles'.** A fracture of the distal portion of the radius.
12. **Pott's.** A fracture of the distal end of the fibula at the level of the lateral malleolus, with eversion (outer displacement) of the foot.
13. **Avulsion.** A portion of a bone tears off.
14. **Depressed.** The broken portion of the bone impales inward, as in certain skull fractures.
15. **Displaced.** A fracture in which the bone fragments are not in anatomical alignment.
16. **Nondisplaced.** A fracture in which the bone fragments remain in anatomical alignment.

Table 10.3

Bones of the Pelvic Girdle and Lower Extremities

Name and Number	Location	Major Distinguishing Features
Os coxae (2)	Hip, part of the pelvic girdle, composed of the fused ilium, ischium, and pubis	Iliac crest; acetabulum; anterior superior iliac spine; greater sciatic notch of the ilium; ischial tuberosity; lesser sciatic notch of the ischium; obturator foramen, pelvic tubercle
Femur (2)	Bone of the thigh, between hip and knee	Head; fovea capitis femoris; neck; greater and lesser trochanters; linea aspera; lateral and medial condyles; lateral and medial epicondyles
Patella (2)	Anterior surface of distal femur	Triangular sesamoid bone
Tibia (2)	Medial side of leg, between knee and ankle	Medial and lateral condyles; intercondylar eminence, tibial tuberosity; anterior crest; medial malleolus; fibular notch
Fibula (2)	Lateral side of leg, between knee and ankle	Head; lateral malleolus
Tarsal bone (14)	Ankle	Large talus and calcaneus to receive weight of leg; five other wedge-shaped bones to help form arches of foot
Metatarsal bone (10)	Sole of foot	Long bones, each in line with a digit
Phalanx (28)	Digits	Three in each digit, except two in great toe

Figure 10.20

Congenital anomalies of the digits.

(a) Polydactyly is the condition in which there are extra digits. It is the most common congenital deformity of the foot, although it also occurs in the hand. (b) Syndactyly is the condition in which two or more digits are webbed together. It is a common congenital deformity of the hand, although it also occurs in the foot. Both conditions can be surgically corrected.

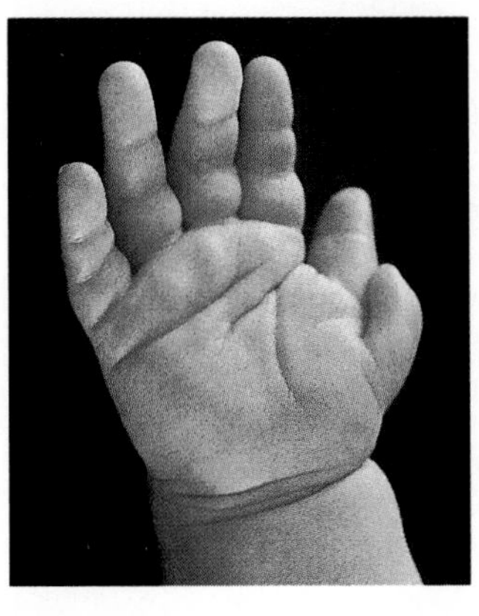

(a)

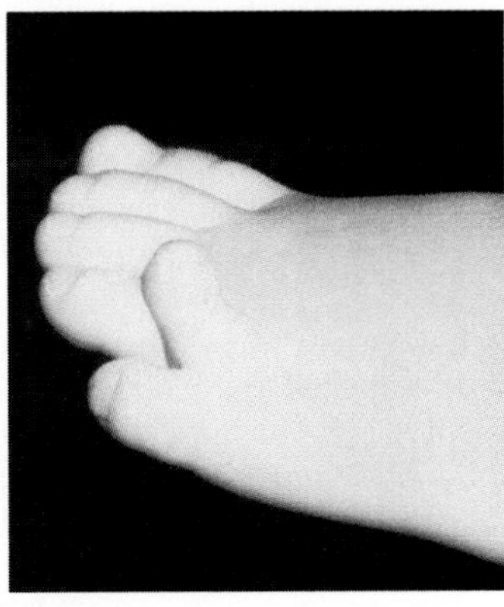

(b)

Figure 10.21

Talipes, or clubfoot, is a congenital malformation of a foot or both feet.

The condition can be effectively treated surgically if the procedure is done at an early age.

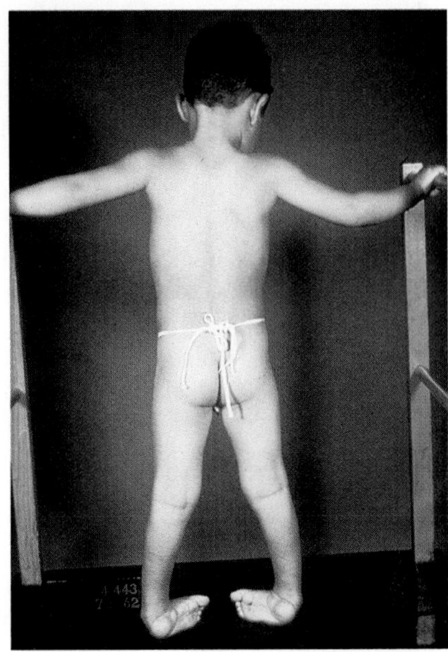

Figure 10.22

Examples of types of fractures.

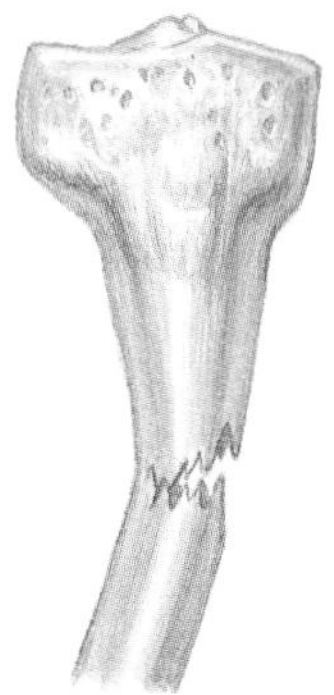

A *greenstick* fracture is incomplete, and the break occurs on the convex surface of the bend in the bone.

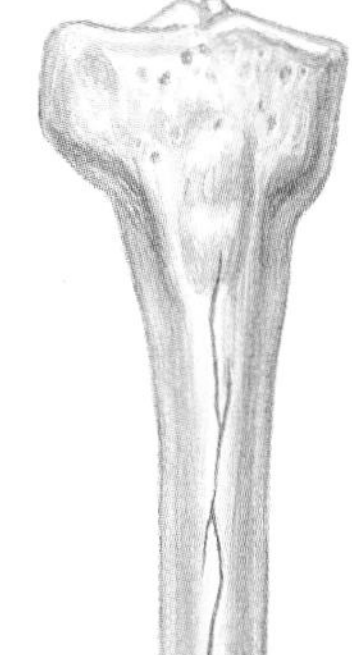

A *partial* (*fissured*) fracture involves an incomplete break.

A *comminuted* fracture is complete and results in several bony fragments.

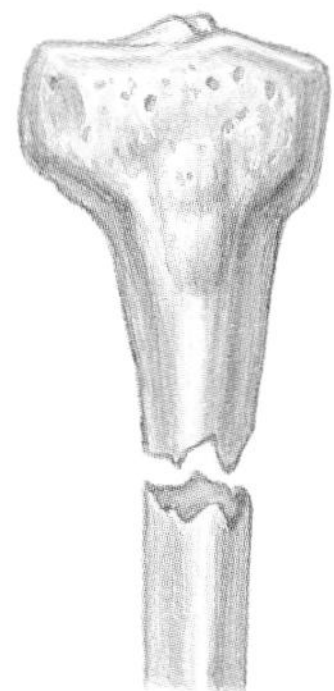

A *transverse* fracture is complete, and the fracture line is horizontal.

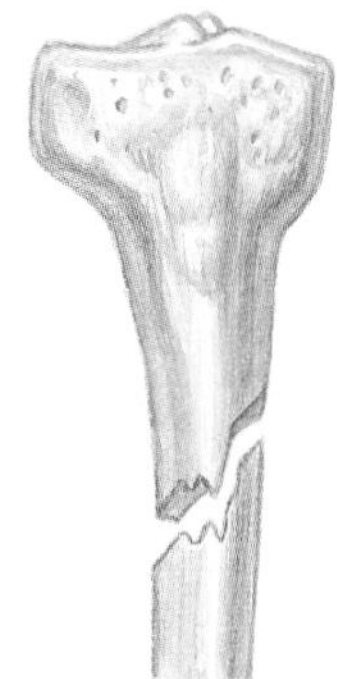

An *oblique* fracture is complete, and the fracture line is at an angle to the long axis of the bone.

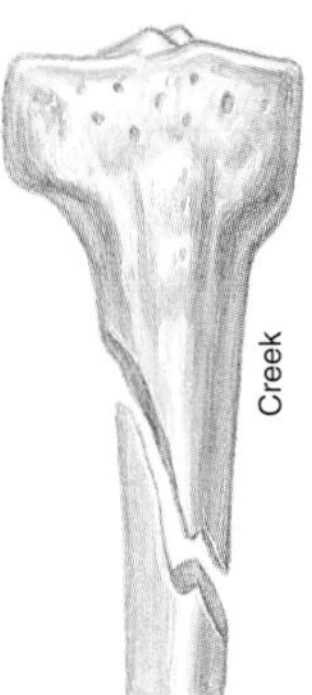

A *spiral* fracture is caused by twisting a bone excessively.

UNDER DEVELOPMENT

Development of the Extremities

The development of the upper and lower extremities is initiated toward the end of the fourth week with the appearance of four small elevations called **limb buds** (fig. 1). The superior pair are the arm buds, which precede the development of the inferior pair of leg buds by a few days. Each limb bud consists of a mass of undifferentiated mesoderm partially covered with a layer of ectoderm. This **apical** (*a'pĭ-kal*) **ectodermal ridge** promotes bone and muscle development.

As the limb buds elongate, migrating mesenchymal tissues differentiate into specific cartilaginous bones. Primary ossification centers soon form in each bone, and the hyaline cartilage tissue is gradually replaced by a bony tissue in the process of endochondral ossification (see chapter 8).

Initially, the developing limbs are directed caudally, but later there is a lateral rotation in the upper extremity and a medial rotation in the lower extremity. As a result, the elbows are directed backward and the knees directed forward.

Digital rays that will form the hands and feet are apparent by the fifth week, and the individual digits separate by the end of the sixth week.

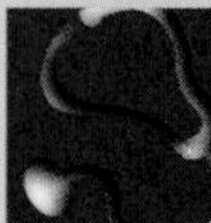

A large number of limb deformities occurred in children born between 1957 and 1962 as a result of mothers ingesting thalidomide during early pregnancy to relieve morning sickness. It is estimated that 7,000 infants were malformed by thalidomide. The malformations ranged from *micromelia* (short limbs) to *amelia* (absence of limbs).

micromelia: Gk. *mikros*, small; *melos*, limb
amelia: Gk. *a*, without; *melos*, limb

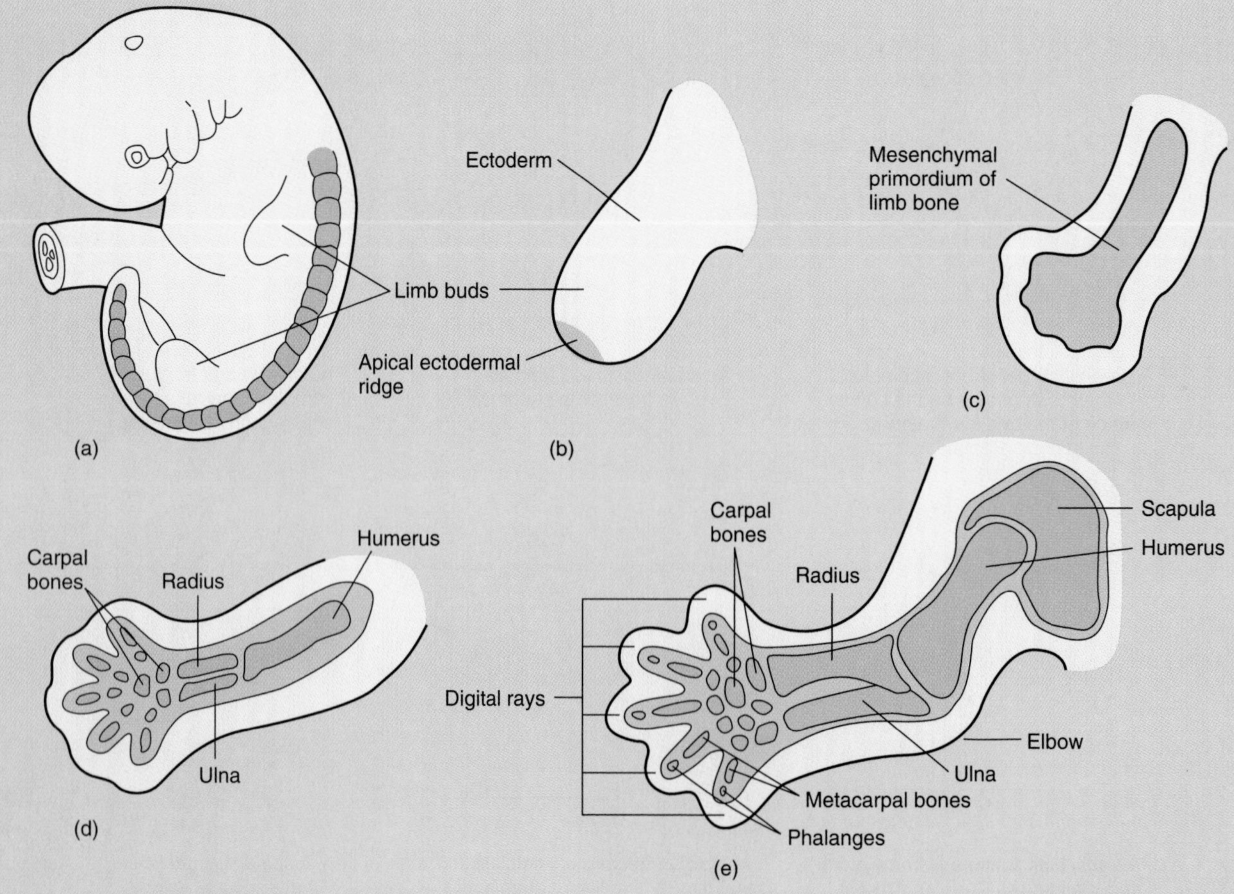

Figure 1 The development of the extremities. (*a*) Limb buds are apparent in an embryo by 28 days, and (*b*) an ectodermal ridge is the precursor of the skeletal and muscular structures. (*c*) Mesenchymal primordial cells are present at 33 days. (*d*) Hyaline cartilaginous models of individual bones develop early in the sixth week. (*e*) Later in the sixth week, the cartilaginous skeleton of the upper extremity is well formed.

Figure 10.23

The repair of a fracture.

Stages *a–d* of the repair of a fracture. (*e*) A radiograph X ray of a healing fracture.

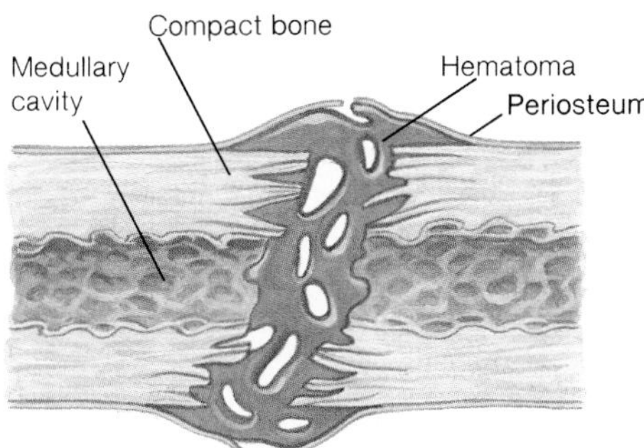

(a) Blood escapes from ruptured blood vessels and forms a hematoma.

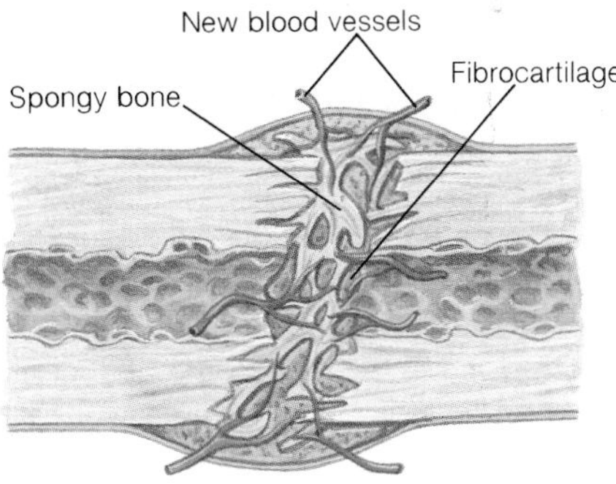

(b) Spongy bone forms in regions close to developing blood vessels; fibrocartilage forms in more distant regions.

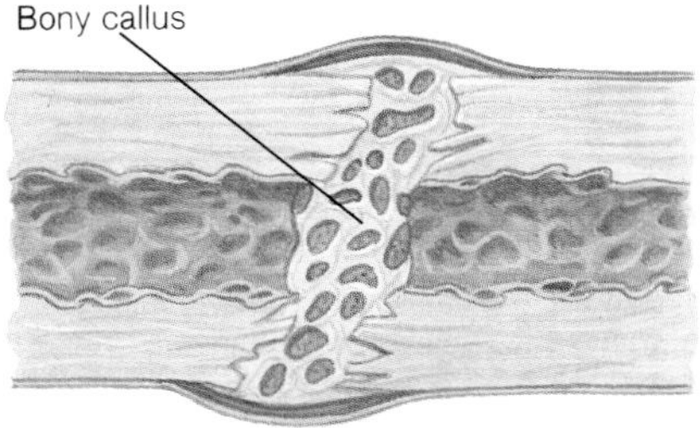

(c) Fibrocartilage is replaced by a bony callus.

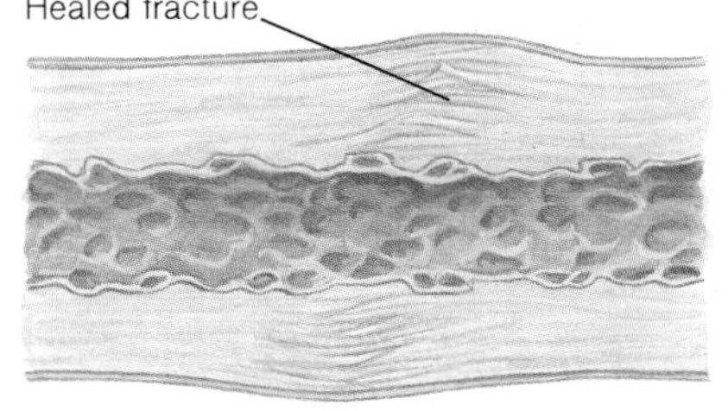

(d) Osteoclasts remove excess bony tissue, making new bone structure much like the original.

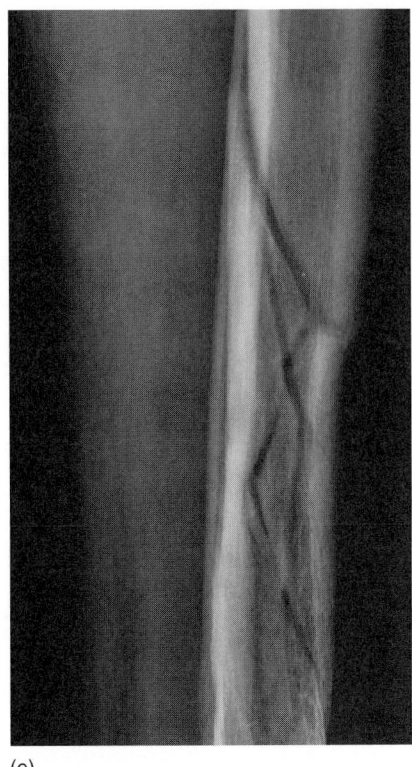

(e)

When a bone fractures, medical treatment involves realigning the broken ends and then immobilizing them until new bone tissue forms and the fracture heals. The site and severity of the fracture and the age of the patient will determine the type of immobilization. The methods of immobilization include tape, splints, casts, straps, wires, screws, plates, and steel pins. Certain fractures seem to resist healing, however, even with this array of treatment options. New techniques for treating fractures include applying weak electrical currents to fractured bones. This method has shown promise in promoting healing and significantly reducing the time of immobilization.

Physicians can realign and immobilize a fracture, but the ultimate repair of the bone occurs naturally within the bone itself. Several steps are involved in this process (fig. 10.23).

1. When a bone fractures, the surrounding periosteum usually tears, and blood vessels in both tissues rupture. A blood clot called a **fracture hematoma** (*hēm″ă-to′mă*) soon forms throughout the damaged area. A disrupted blood supply to osteocytes and periosteal cells at the fracture site causes localized cellular death. This is followed by swelling and inflammation.
2. The traumatized area is "cleaned up" by the activity of phagocytic cells within the blood and osteoclasts that resorb bone fragments. As the debris is removed, fibrocartilage fills the gap within the fragmented bone, and a cartilaginous mass, called a **bony callus** forms. The bony callus becomes the precursor of bone formation in much the same way that hyaline cartilage is the precursor of developing bone.
3. The remodeling of the bony callus is the final step in the healing process. The cartilaginous callus breaks down, a new vascular supply is established, and compact bone develops around the periphery of the fracture. A healed fracture line is frequently undetectable in a radiograph, except that the bone in this area may be slightly thicker.

fracture hematoma: Gk. *hema*, blood; *oma*, tumor

callus: L. *callosus*, hard

Clinical Investigation Answer

The injury involves the cartilaginous epiphyseal growth plate, which is the site of linear growth in long bones. At cessation of growth, this plate disappears as the epiphysis and diaphysis fuse. Until this occurrence, however, disruption of the growth plate can adversely affect growth of the bone.

Chapter Summary

Pectoral Girdle and Upper Extremity (pp. 226-232)

1. The pectoral girdle is composed of two scapulae and two clavicles. The clavicles attach the pectoral girdle to the axial skeleton at the sternum.
   (a) Diagnostic features of the clavicle include the conoid tubercle and acromial and sternal extremities.
   (b) Diagnostic features of the scapula include the spine, acromion, and coracoid process; the supraspinous, infraspinous, and subscapular fossae; the glenoid cavity; superior, medial, and lateral borders; and superior, inferior, and lateral angles.
2. The brachium contains the humerus, which extends from the scapula to the elbow.
   (a) Proximally, diagnostic features of the humerus include a rounded head, greater tubercle, anatomical neck, and an intertubercular groove. Distally, they include medial and lateral epicondyles, coronoid and olecranon fossae, a capitulum, and a trochlea.
   (b) The head of the humerus articulates proximally with the glenoid cavity of the scapula; distally, the trochlea and capitulum articulate with the ulna and radius, respectively.
3. The antebrachium contains the medial ulna and the lateral radius.
   (a) Proximally, diagnostic features of the ulna include the olecranon, coronoid process, and trochlear notch. Distally, they include the styloid process and head of the ulna.
   (b) Proximally, diagnostic features of the radius include the head and neck of the radius and the tuberosity of the radius. Distally, they include the styloid process and ulnar notch.
4. The hand contains 27 bones arranged as the carpal bones, metacarpal bones, and phalanges.

Pelvic Girdle and Lower Extremity (pp. 232–238)

1. The pelvic girdle is formed by two ossa coxae united anteriorly by the symphysis pubis.
2. The pelvis is divided into a greater pelvis, which helps to support the pelvic viscera, and a lesser pelvis, which forms the walls of the birth canal.
3. Each os coxae consists of an ilium, ischium, and pubis. Diagnostic features of the os coxae include an obturator foramen and an acetabulum, the latter of which is the socket for articulation with the head of the femur.
   (a) Diagnostic features of the ilium include an iliac crest, iliac fossa, anterior superior iliac spine, and anterior inferior iliac spine.
   (b) Diagnostic features of the ischium include the body, ramus, and ischial tuberosity.
   (c) Diagnostic features of the pubis include the ramus and pubic tubercle. The two pubic bones articulate at the symphysis pubis.
4. The thigh contains the femur, which extends from the hip to the knee where it articulates with the tibia and the patella.
   (a) Proximally, diagnostic features of the femur include the head, neck, and greater and lesser trochanters. Distally, they include the lateral and medial epicondyles, the lateral and medial condyles, and the patellar surface. The linea aspera is a roughened ridge positioned vertically along the posterior aspect of the body (shaft) of the femur.
   (b) The head of the femur articulates proximally with the acetabulum of the os coxae and distally with the condyles of the tibia and the articular surfaces of the patella.
5. The leg contains the medial tibia and the lateral fibula.
   (a) Diagnostic features of the tibia include the intercondylar eminence and tibial tuberosity proximally and medial malleolus distally. The anterior crest is a sharp ridge extending the anterior length of the tibia.
   (b) Diagnostic features of the fibula include the head proximally and the lateral malleolus distally.
6. The foot contains 26 bones arranged as the tarsal bones, metatarsal bones, and phalanges.

Review Activities

Objective Questions

1. When in anatomical position, the subscapular fossa of the scapula faces
   (a) anteriorly. (c) posteriorly.
   (b) medially. (d) laterally.
2. The clavicle articulates with
   (a) the scapula and humerus.
   (b) the humerus and manubrium.
   (c) the manubrium and scapula.
   (d) the manubrium, scapula, and humerus.
3. Which of the following bones has a conoid tubercle?
   (a) scapula (d) clavicle
   (b) humerus (e) ulna
   (c) radius
4. The "elbow" of the ulna is formed by
   (a) the lateral epicondyle.
   (b) the olecranon.
   (c) the coronoid process.
   (d) the styloid process.
   (e) the medial epicondyle.
5. Which of the following statements concerning the carpus is *false?*
   (a) There are eight carpal bones arranged in two transverse rows of four bones each.
   (b) All of the carpal bones are considered sesamoid bones.
   (c) The scaphoid and the lunate bones articulate with the radius.
   (d) The trapezium, trapezoid, capitate, and hamate bones articulate with the metacarpal bones.
6. In pelvimetry, which of the following is measured?
   (a) os coxae (c) pelvic brim
   (b) symphysis pubis (d) lesser pelvis
7. Which of the following is *not* a structural feature of the os coxae?
   (a) obturator foramen
   (b) acetabulum
   (c) auricular surface
   (d) greater sciatic notch
   (e) linea aspera
8. A fracture across the intertrochanteric line would involve which bone?
   (a) ilium (d) fibula
   (b) femur (e) patella
   (c) tibia
9. As compared to the male pelvis, the female pelvis
   (a) is more massive.
   (b) is narrower at the pelvic outlet.
   (c) is tilted backward.
   (d) has a shallower symphysis pubis.
10. Clubfoot is a congenital malformation that is medically referred to as
    (a) talipes. (c) pes planus.
    (b) syndactyly. (d) polydactyly.

Essay Questions

1. Explain the significance of the limb buds, apical ectodermal ridges, and digital rays in limb development. When does limb development begin, and when is it completed?
2. Compare the pectoral and pelvic girdles in structure, articulation to the axial skeleton, and function.
3. Explain why the clavicle is more frequently fractured than the scapula.
4. List the processes of the bones of the upper and lower extremities that can be palpated. Why is it important to be able to recognize these bony landmarks?
5. There are basic similarities and specific differences between the bones of the hands and those of the feet. Compare and contrast these appendages, taking into account the functional role of each.
6. Define *bipedal locomotion* and discuss the adaptations of the pelvic girdle and lower extremities that permit this type of movement.
7. What are the structural differences between the male and female pelves?
8. What is meant by a congenital skeletal malformation? Give two examples of such abnormalities that occur within the appendicular skeleton.
9. How do spontaneous and traumatic fractures differ? Give some examples of traumatic fractures.
10. How does a fractured bone repair itself? Why is it important that the fracture be immobilized?

Critical Thinking Questions

1. James Smithson, benefactor of the Smithsonian Institution, died in 1829 at the age of 64. Although his body was buried in Italy, it was reinterred in 1904 near the front entry of the Smithsonian in Washington, D.C. Before the reburial, scientists at the Smithsonian carefully examined Smithson's skeleton to learn more about him. They concluded that Smithson was rather slightly built but athletic—he had a large chest and powerful arms and hands. His teeth were worn on the left side from chewing a pipe. The scientists also reported that "certain peculiarities of the right little finger suggest that he may have played the harpsichord, piano, or a stringed instrument such as a violin."

   Preserved bones can serve as a storehouse of information. Considering current technology, what other types of information might be gleaned from examination of a preserved skeleton?
2. Which would you say has been more important in human evolution—adaptation of the hand or adaptation of the foot? Explain your reasoning.
3. Speculate as to why a single bone is present in both the brachium and the thigh, whereas the antebrachium and leg each have two bones.
4. Compare the tibia and fibula with respect to structure and function. Which would be more debilitating: a compound fracture of the tibia or a compound fracture of the fibula?
5. It is not uncommon for a person to develop a sesamoid bone at one or more joints in the hand (see fig. 10.9). Explain what may cause this to occur and its possible consequence.

Related Web Sites

For a listing of the most current web sites related to this chapter, please visit the *Concepts of Human Anatomy and Physiology* home page at http://www.mhhe.com/biosci/abio/.

NEXUS

Some Interactions of the Skeletal System with the Other Body Systems

Integumentary System

- The skeletal system provides structural support for the body, including the skin (p. 185).
- The skin produces vitamin D, which stimulates the intestinal absorption of the Ca^{2+} required for proper ossification (p. 167).

Muscular System

- The skeleton provides attachment sites for muscles (p. 281).
- The skeleton serves as a reservoir of Ca^{2+}, which is required for muscle contraction (p. 294).
- Skeletal muscle contraction causes bones to move at joints (p. 256).

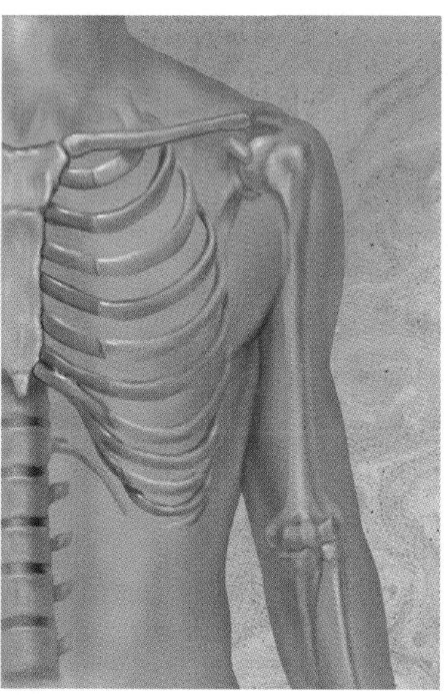

Nervous System

- The bones of the cranium and vertebrae protect the central nervous system (p. 201).
- The skeleton serves as a source of Ca^{2+}, required for neural function (p. 193).
- Proprioceptors provide sensory information regarding the position of bones at their joints (p. 504).

Endocrine System

- The skeleton serves as a reservoir of blood Ca^{2+}, which is required for the action of some hormones (p. 563).
- A variety of hormones regulate bone growth and maintenance (p. 884).

Circulatory System

- The blood transports O_2 and CO_2, nutrients, and hormones to and from the skeletal system (p. 592).
- The bone marrow produces blood cells (p. 596).
- Bones serves as a reservoir of Ca^{2+}, which is required for cardiac muscle contraction (p. 294).

Lymphatic System

- Bone marrow produces and stores lymphocytes and other cells of the immune system (p. 596).
- Lymphatic vessels maintain a balanced amount of interstitial fluid within bone tissue (p. 662).
- Lymphocytes protect bone tissue following trauma (p. 701).

Respiratory System

- The skeletal system forms a respiratory passageway through the nasal cavity (p. 729).
- The rib cage protects the lungs and is required for the respiratory movements of ventilation (p. 742).
- The respiratory system provides O_2 for all body systems, including the skeletal system (p. 729).

Digestive System

- The skeletal system provides the organs of the GI tract with physical support and protection (p. 185).
- The digestive system provides nutrients for growth, maintenance, and repair of bone tissue (p. 817).

Urinary System

- The skeletal system provides the organs of the urinary system with physical support and protection (p. 185).
- The urinary system helps to regulate homeostasis of blood Ca^{2+}, which is required by the skeletal system for proper ossification (p. 194).
- The kidneys have enzymes that convert vitamin D into an active derivative that is required for regulation of blood Ca^{2+} concentration (p. 195).

Reproductive System

- The skeletal system provides the organs of the reproductive system with physical support and protection (p. 185).
- Gonads produce sex steroid hormones that promote growth and development of the skeletal system (p. 897).

Eleven

Articulations

Objectives

- Define *arthrology* and *kinesiology*.
- Compare and contrast the three principal kinds of joints.
- Describe the structure of a suture and indicate where sutures are located.
- Describe the structure of a syndesmosis and indicate where syndesmoses are located.
- Describe the structure and note the location of gomphoses. Also, discuss the importance of these joints to the profession of dentistry.
- Describe the structure of a symphysis and indicate where symphyses occur.
- Describe the structure of a synchondrosis and indicate where synchondroses occur.
- Describe the structure of a synovial joint.
- Discuss the various kinds of synovial joints, noting where they occur and the movements they permit.
- List and discuss the various kinds of movements that are possible at synovial joints.
- Describe the components of a lever and explain the role of synovial joints in lever systems.
- Compare the structure of first-, second-, and third-class levers.
- Describe the structure, functions, and possible clinical importance of the following joints: temporomandibular, glenohumeral, elbow, metacarpophalangeal, interphalangeal, coxal, tibiofemoral, and talocrural.

Clinical Investigation

A 20-year-old college football player sustained injury to his right knee during the opening game of the season. Because of rapid swelling and intense pain, he was taken to the emergency room of the local hospital. When the attending physician asked him to describe how the injury occurred, the athlete responded, "I was carrying the ball on an end run left on third down and two. As I planted my right foot just before I was going to make my cut, I was hit in the knee from the side. I felt my knee give way, and then I felt a stabbing pain on the inside of my knee."

Close examination by the physician revealed marked swelling on the medial part of the knee. The doctor determined that *valgus stress* (an inward bowing stress on the knee) caused the medial aspect of the joint to "open."

Which stabilizing structure is most likely injured? Which cartilaginous structure is frequently injured in association with the previously mentioned structure? Is there an anatomical explanation? What are some other stabilizing structures within the knee that are frequently injured in sports?

Clues: An impact to one side of the knee generally results in greater trauma to the other side. Carefully read the sections in this chapter on the tibiofemoral (knee) joint and trauma to joints. In addition, examine figures 11.28 and 11.29.

Classification of Joints

On the basis of anatomical structure, the articulations between the bones of the skeleton are classified as fibrous joints, cartilaginous joints, or synovial joints. Fibrous joints firmly join bones with fibrous connective tissue. Cartilaginous joints firmly join bones with cartilage. Synovial joints are freely movable joints; they are enclosed by joint capsules that contain synovial fluid.

One of the functions of the skeletal system is to permit body movement. However, it is not the bones themselves but the unions between them that allow movement. These unions are called **articulations,** or **joints.** Although the joints of the body are actually part of the skeletal system, this chapter is devoted entirely to them.

The structure of a joint determines its direction and range of movement. Not all joints are flexible, however, and as one part of the body moves, other joints remain rigid to stabilize the body and maintain balance. The coordinated activity of all of the joints permits the sinuous, elegant movements of a gymnast or ballet dancer, just as it permits all of the commonplace actions associated with walking, eating, writing, and speaking.

Arthrology is the science concerned with the study of joints, including their structure, classification, function, and any dysfunctions that may develop. *Kinesiology* (*kĭ-ne′se-ol′ŏ-je*), a more applied and dynamic science, is concerned with the mechanics of human motion—the functional relationship of the bones, muscles, and joints as they work together to produce coordinated movement. Kinesiology is a subdiscipline of biomechanics, which deals with a broad range of mechanical processes, including the forces that govern blood circulation and respiration.

In studying the joints, a kinetic approach allows for the greatest understanding. The student should be able to demonstrate the various movements permitted at each of the movable joints. Additionally, he or she should be able to explain the adaptive advantage, as well as the limitation, of each type of movement.

The articulations of the body are grouped by their structure into three principal categories.

1. **Fibrous joints.** In fibrous joints, the articulating bones are held together by fibrous connective tissue. These joints lack joint cavities.
2. **Cartilaginous joints.** In cartilaginous joints, the articulating bones are held together by cartilage. These joints also lack joint cavities.
3. **Synovial joints.** In synovial (*sĭ-no′ve-al*) joints, the articulating bones are capped with cartilage, and ligaments frequently help to support them. These joints are distinguished by fluid-filled joint cavities.

Fibrous Joints

As the name suggests, the articulating bones in fibrous joints are tightly bound by fibrous connective tissue. Fibrous joints range from rigid and relatively immovable joints to those that are slightly movable. The three kinds of fibrous joints are sutures, syndesmoses, and gomphoses.

Sutures

Sutures are found only within the skull. They are characterized by a thin layer of dense irregular connective tissue that binds the articulating bones (fig. 11.1). Sutures form at about 18 months of age and replace the pliable fontanels of an infant's skull (see fig. 2, p. 211).

Different types of sutures can be distinguished on the appearance of the articulating edge of bone. A **serrate suture** is characterized by interlocking sawlike articulations. This is the most common type of suture, an example of which is the sagittal suture between the two parietal bones. In a **squamous** (*lap*) **suture,** the edge of one bone overlaps that of the articulating bone. The squamous suture formed between the temporal and parietal bones is an example (see fig. 9.2). In a **plane** (*butt*) **suture,** the edges of the articulating bones are fairly

arthrology: Gk. *arthron*, joint; *logos*, study

kinesiology: Gk. *kinesis*, movement; *logos*, study
suture: L. *sutura*, sew

Figure 11.1

A section of the skull showing a suture.

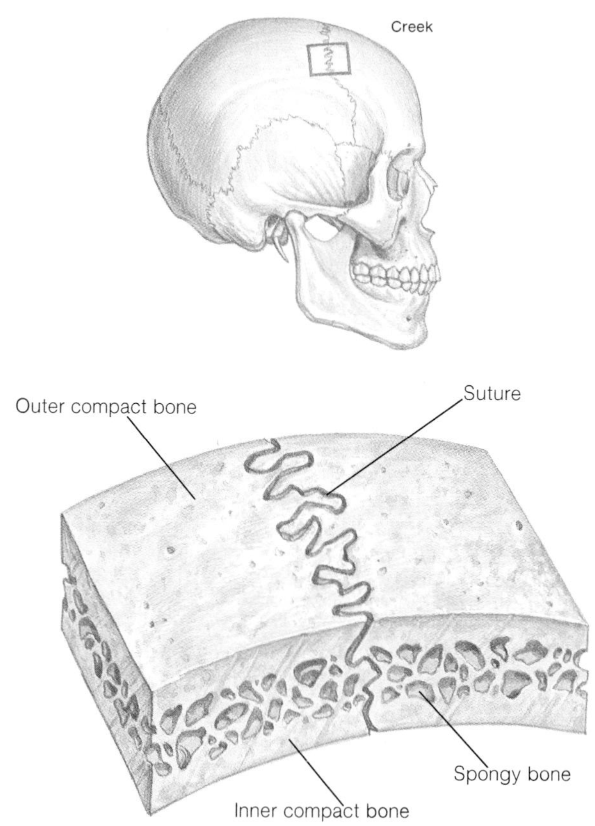

Figure 11.2

A syndesmosis.
The side-to-side articulation of the ulna and radius forms a syndesmotic joint. An interosseous ligament tightly binds these bones and permits only slight movement between them.

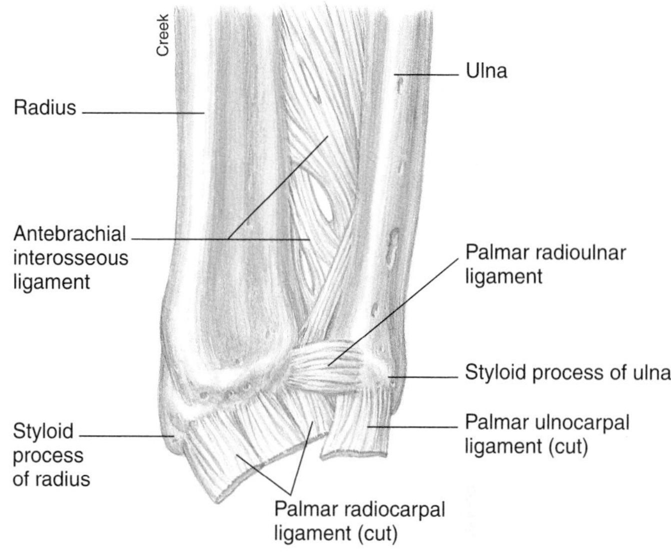

smooth and do not overlap. An example is the median palatine suture, where the maxillae and palatine bones articulate to form the hard palate (see fig. 9.3).

A **synostosis** (*sin"os-to'sis*) is a unique sutural joint. It is present during growth of the skull, but in the adult the suture becomes totally ossified. For example, the frontal bone forms as two separate components but the separation becomes obscured in most individuals as the skull completes its growth.

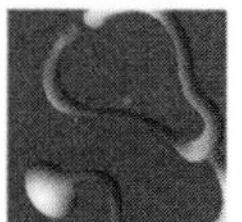

Fractures of the skull are fairly common in an adult but much less so in a child. The skull of a child is resilient to blows because of the nature of the bone and the layer of fibrous connective tissue within the sutures. The skull of an adult is much like an eggshell in its lack of resilience. It will frequently splinter on impact.

The nomenclature in human anatomy is extensive and precise. There are over 30 named sutures in the skull even though just a few of them are mentioned by name in figures 9.2, 9.3, and 9.4. Review these illustrations and make note of the bones that articulate to form the sutures identified.

Syndesmoses

Syndesmoses (*sin"des-mo'sez*) are fibrous joints held together by collagenous fibers or sheets of fibrous tissue called *interosseous ligaments*. The tympanostapedial joint in the middle-ear cavity is a syndesmosis. This type of joint also occurs in the antebrachium (forearm) between the distal parts of the radius and ulna (fig. 11.2) and in the leg between the distal parts of the tibia and fibula. Slight movement is permitted at these joints as the antebrachium or leg is rotated.

Gomphoses

Gomphoses (*gom-fo'sez*) are fibrous joints that occur between the teeth and the supporting bones of the jaws. More specifically, a gomphosis, or a dentoalveolar joint, is where the root of a tooth is anchored to the periodontal ligament of the dental alveolus (tooth socket) of the bone (fig. 11.3).

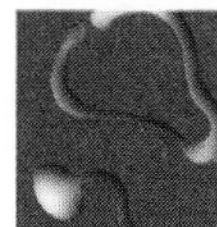

Periodontal disease occurs at gomphoses. It refers to the inflammation and degeneration of the gum, periodontal ligaments, and alveolar bone tissue. With this condition, the teeth become loose and plaque accumulates on the roots. Periodontal disease may be caused by poor oral hygiene, compacted teeth (poor alignment), or local irritants, such as impacted food, chewing tobacco, or cigarette smoke.

---

syndesmosis: Gk. *syndesmos*, binding together
gomphosis: Gk. *gompho*, nail or bolt

Figure 11.3

A gomphosis.
This is a fibrous joint in which a tooth is held in its socket (dental alveolus).

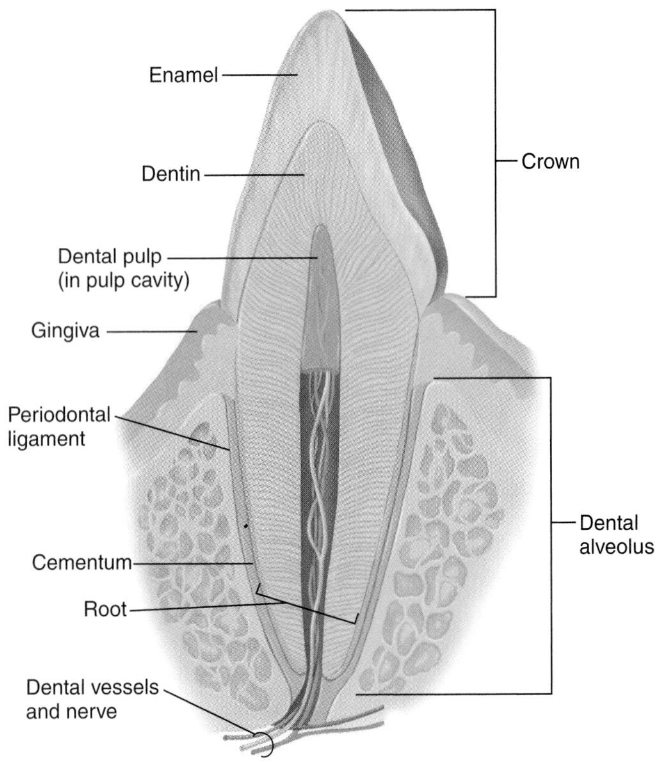

Cartilaginous Joints

Cartilaginous joints allow limited movement in response to twisting or compression. The two types of cartilaginous joints are symphyses and synchondroses.

Symphyses

The adjoining bones of a **symphysis** (*sim'fĭ-sis*) joint are covered with hyaline cartilage, which becomes infiltrated with collagenous fibers to form an intervening pad of fibrocartilage. This pad cushions the joint and allows limited movement. The symphysis pubis and the intervertebral joints formed by the intervertebral discs (fig. 11.4) are examples of symphyses. Although only limited motion is possible at each intervertebral joint, the combined movement of all the joints of the vertebral column results in extensive spinal action.

Synchondroses

Synchondroses (*sin"kon-dro'sēz*) are cartilaginous joints that have hyaline cartilage between the articulating bones. Some of these joints are temporary, forming the epiphyseal plates (growth lines) between the diaphyses and epiphyses in the

symphysis: Gk. *symphysis*, growing together
synchondrosis: Gk. *syn*, together; *chondros*, cartilage

Figure 11.4

Examples of symphyses.
(*a*) The symphysis pubis and (*b*) the intervertebral joints between vertebral bodies.

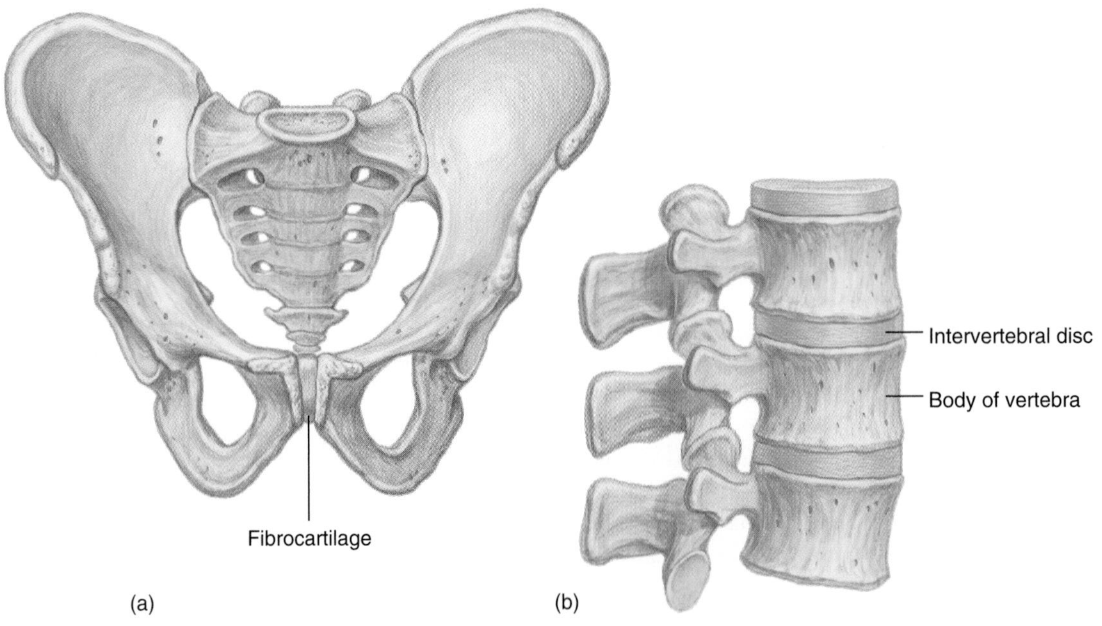

Figure 11.5

A synchondrosis.
Synchondrotic joints can be seen in this radiograph (X ray) of the left humerus of a 10-year-old child. In a long bone, this type of joint occurs at both the proximal and distal epiphyseal plates. The mitotic activity at synchondrotic joints is responsible for bone growth in length.

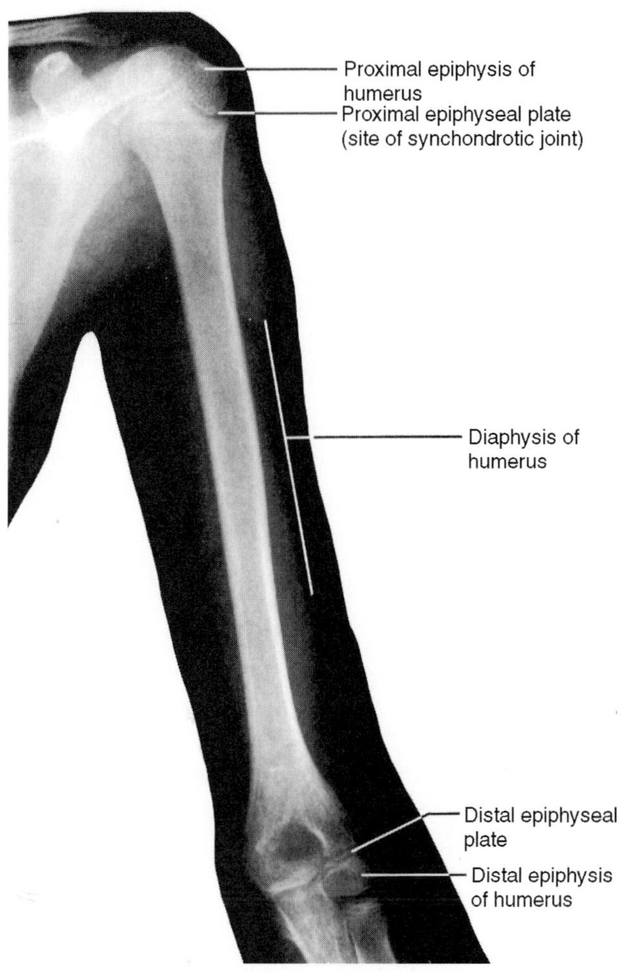

long bones of children (fig. 11.5). When growth is complete, these synchondrotic joints ossify. A totally ossified synchondrosis may also be referred to as a *synostosis*.

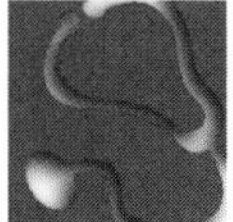

A fracture of a long bone in a child may be extremely serious if it involves the mitotically active epiphyseal plate of a synchondrotic joint. If such an injury is left untreated, bone growth is usually retarded or arrested, so that the appendage will be shorter than normal.

Synchondroses that do not ossify as a person ages are those between the occipital, sphenoid, temporal, and ethmoid bones of the skull. In addition, the costochondral articulations between the ends of the ribs and the costal cartilages that attach to the sternum are examples of synchondroses. Elderly people often exhibit some ossification of the costal cartilages of the rib cage. This may restrict movement of the rib cage and obscure an image of the lungs in a thoracic radiograph.

Synovial Joints

The freely movable synovial joints are enclosed by joint capsules containing synovial fluid. Based on the shape of the articular surfaces and the kinds of motion they permit, synovial joints are categorized as gliding, hinge, pivot, condyloid, saddle, or ball-and-socket.

The most obvious type of articulation in the body is the freely movable synovial joint. The function of synovial joints is to provide a wide range of precise, smooth movements, at the same time maintaining stability, strength, and, in certain aspects, rigidity in the body.

Synovial joints are the most complex and varied of the three major types of joints. A synovial joint's range of motion is determined by three factors:

1. the structure of the bones involved in the articulation (for example, the olecranon of the ulna prevents hyperextension of the elbow joint);
2. the strength of the joint capsule and the strength and tautness of the associated ligaments and tendons; and
3. the size, arrangement, and action of the muscles that span the joint.

Range of motion at synovial joints is characterized by tremendous individual variation, most of which is related to body conditioning. Although some people can perform remarkable contortions and are said to be "double-jointed," they have no extra joints that help them do this. Rather, through conditioning, they are able to stretch the ligaments that normally inhibit movement.

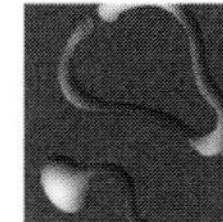

Arthroplasty is the surgical repair or replacement of joints. Advancements in this field continue as devices are sought to restore lost joint function and to permit movement that is free of pain. A recent advancement in the repair of soft tissues involves the use of *artificial ligaments.* A material consisting of carbon fibers coated with a plastic called polylactic acid is sewn in and around torn ligaments and tendons. This reinforces the traumatized structures while providing a scaffolding on which the body's collagenous fibers can grow. As the healing progresses, the polylactic acid is absorbed and the carbon fibers break down.

Structure of a Synovial Joint

Synovial joints are enclosed by a fibroelastic **joint capsule** (articular capsule), which is filled with lubricating **synovial fluid**

synostosis: Gk. *syn*, together; *osteon*, bone

arthroplasty: Gk. *arthron*, joint; *plasso*, to form

UNDER DEVELOPMENT

Development of Synovial Joints

The sites of developing synovial joints (freely movable joints) are discernible at 6 weeks as mesenchyme becomes concentrated in the areas where precartilage cells differentiate (fig. 1). At this stage, the future joints appear as intervals of less-concentrated mesenchymal cells. As cartilage cells develop within a forming bone, a thin flattened sheet of cells forms around the cartilaginous model to become the **perichondrium.** These same cells are continuous across the gap between the adjacent developing bone. Surrounding the gap, the flattened mesenchymal cells differentiate to become the **joint capsule.**

During the early part of the third month of development, the mesenchymal cells still remaining within the joint capsule begin migrating toward the epiphyses of the adjacent developing bones. The cleft eventually enlarges to become the **joint cavity.** Thin pads of hyaline cartilage develop on the surfaces of the epiphyses that contact with the joint cavity. These pads become the **articular cartilages** of the functional joint. As the joint continues to develop, a highly vascular **synovial membrane** forms on the inside of the joint capsule and begins secreting a watery *synovial fluid* into the joint cavity.

In certain developing synovial joints, the mesenchymal cells do not migrate away from the center of the joint cavity. Rather, they give rise to cartilaginous wedges called menisci, as in the knee joint, or to complete cartilaginous pads, called articular discs, as in the sternoclavicular joint.

Most synovial joints have formed completely by the end of the third month. Shortly thereafter, fetal muscle contractions, known as *quickening,* cause movement at these joints. Joint movement enhances the nutrition of the articular cartilage and prevents the fusion of connective tissues within the joint.

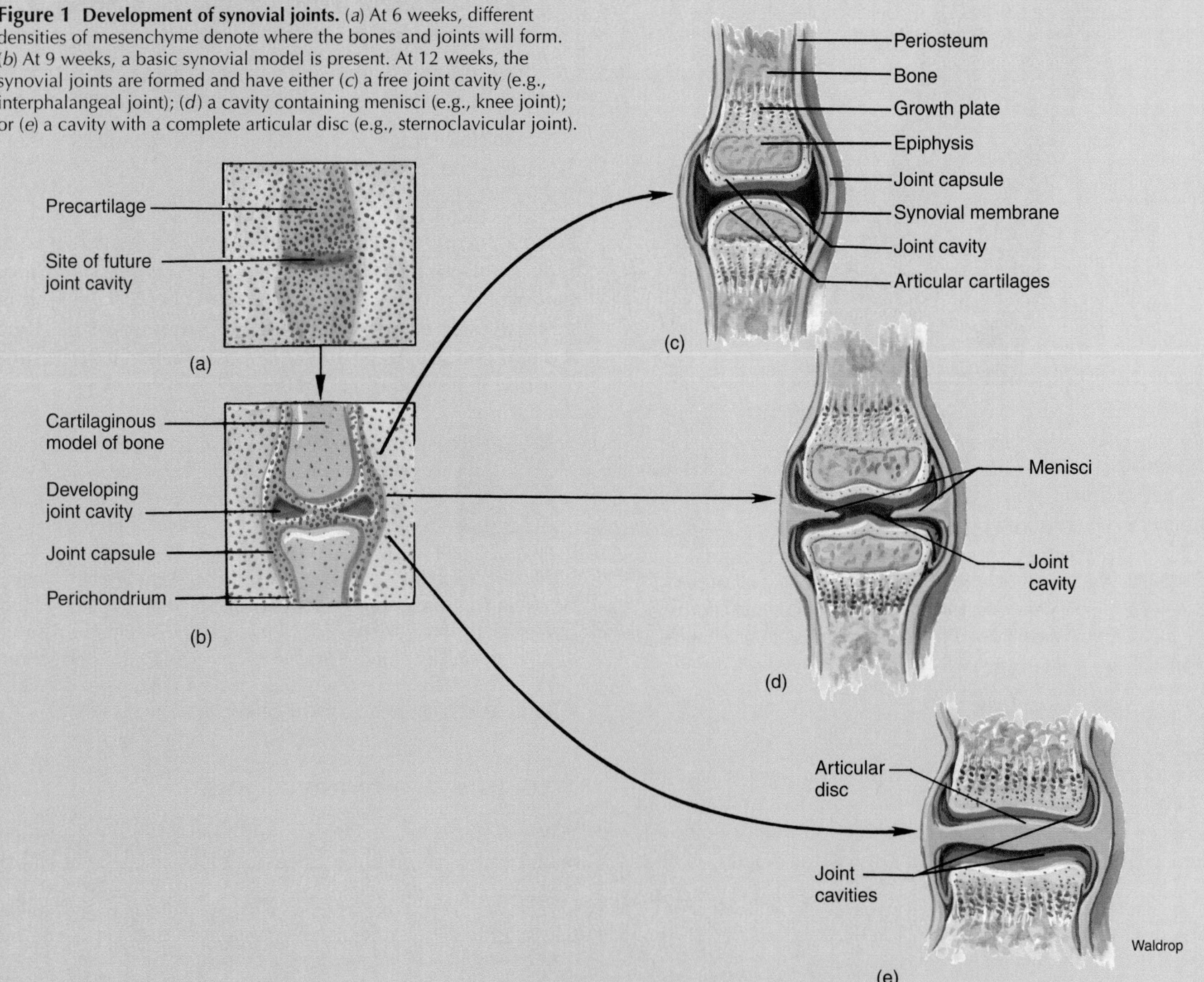

Figure 1 Development of synovial joints. (*a*) At 6 weeks, different densities of mesenchyme denote where the bones and joints will form. (*b*) At 9 weeks, a basic synovial model is present. At 12 weeks, the synovial joints are formed and have either (*c*) a free joint cavity (e.g., interphalangeal joint); (*d*) a cavity containing menisci (e.g., knee joint); or (*e*) a cavity with a complete articular disc (e.g., sternoclavicular joint).

Figure 11.6

A synovial joint.
This freely movable joint is represented by the knee joint, shown here in a sagittal view.

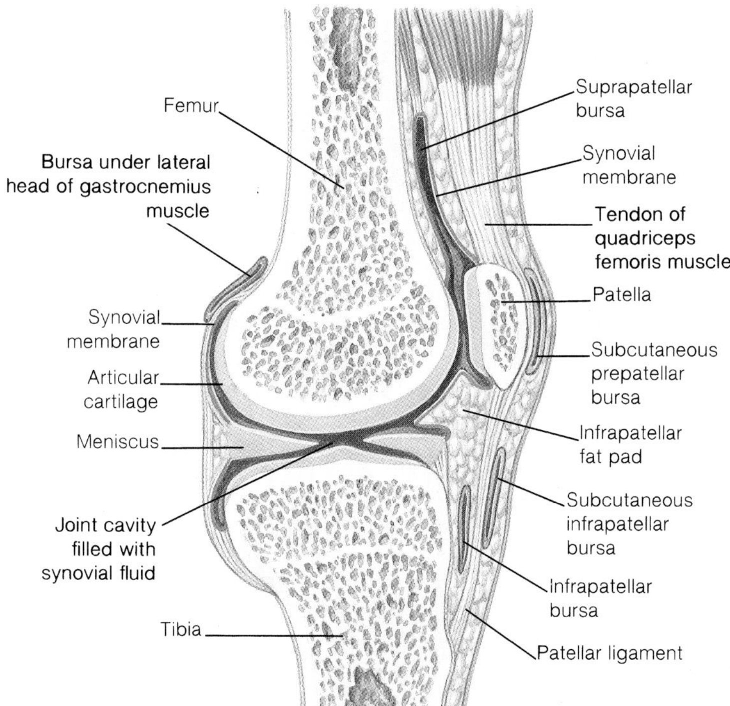

(fig. 11.6). The term *synovial* is derived from a Greek word meaning "*egg white*," which this fluid resembles. It is secreted by a thin **synovial membrane** that lines the inside of the joint capsule. Synovial fluid is similar to interstitial fluid (fluid between the cells) and has a high concentration of hyaluronic acid and albumin, and also contains phagocytic cells that eliminates tissue debris resulting from wear on the joint cartilages. The bones that articulate in a synovial joint are capped with a smooth layer of hyaline cartilage called the **articular cartilage.** Articular cartilage is only about 2 mm thick. Because articular cartilage lacks blood vessels, it has to be nourished by the movement of synovial fluid during joint activity.

Ligaments are flexible connective tissue bands or cords that connect from bone to bone as they help bind synovial joints. Ligaments may be located within the joint cavity or on the outside of the capsule. Tough, fibrous cartilaginous pads called **menisci** (*mĕ-nis′ki;* singular, *meniscus*) are unique to the knee joint, where they cushion and guide the articulating bones. A few other synovial joints, such as the temporomandibular joint (see fig. 11.21), have a fibrocartilaginous pad called an **articular disc** that provides functions similar to menisci.

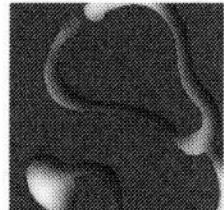

Many people are concerned about the cracking sounds they hear as joints move, or the popping sounds that result from "popping" or "cracking" the knuckles by forcefully pulling on the fingers. These sounds are actually quite normal. When a synovial joint is pulled upon, its volume is suddenly expanded and the pressure of the joint fluid is decreased, causing a partial vacuum within the joint. As the joint fluid is displaced and hits against the articular cartilage, a popping or cracking sound is heard. Similarly, displaced water in a sealed vacuum tube makes this sound as it hits against the glass wall. Popping your knuckles does not cause arthritis, but it can lower one's social standing.

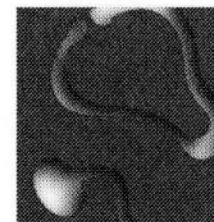

The articular cartilage that caps the articular surface of each bone and synovial fluid that circulates through the joint during movement are protective features of synovial joints. They serve to minimize friction and to cushion the articulating bones. Should trauma or disease render either of them nonfunctional, the two articulating bones will come in contact. Bony deposits will then form, and a type of arthritis will develop within the joint.

Closely associated with some synovial joints are flattened, pouchlike sacs called **bursae** (*bur′se*; singular, *bursa*) that are filled with synovial fluid. (fig. 11.7*a*). They function to cushion certain muscles and assist the movement of tendons or muscles over bony or ligamentous surfaces. A **tendon sheath** (fig. 11.7*b*) is a modified bursa that surrounds and lubricates the tendons of certain muscles, particularly those that cross the wrist and ankle joints.

Wearing improperly fitted shoes or inappropriate shoes can cause joint-related problems. For example, people who perpetually wear high-heeled shoes often have backaches and leg aches because their posture has to counteract the forward tilt of their bodies when standing or walking. Their knees are excessively flexed and their spine thrusts forward at the lumbar curvature in order to maintain balance. Tightly fitted shoes, especially ones with pointed toes, may result in the development of *hallux valgus*—a lateral deviation of the hallux (great toe) in the direction of the other toes. Hallux valgus is generally accompanied by the formation of a *bunion* at the medial base of the proximal phalanx of the hallux. A bunion is an inflammation and an accompanying callus that develops in response to pressure and the rubbing of a shoe.

Kinds of Synovial Joints

Synovial joints are classified into six main categories based on their structure and the motion they permit. The six categories are *gliding*, *hinge*, *pivot*, *condyloid*, *saddle*, and *ball-and-socket*.

meniscus: Gk. *meniskos*, small moon

bursa: Gk. *byrsa*, bag or purse

Figure 11.7

Bursae and tendon sheaths.

These friction-reducing structures are found in conjunction with synovial joints. (*a*) A bursa is a closed sac filled with synovial fluid. Bursae are commonly located between muscles or between tendons and joint capsules. (*b*) A tendon sheath is a double-layered sac of synovial fluid that completely envelops a tendon.

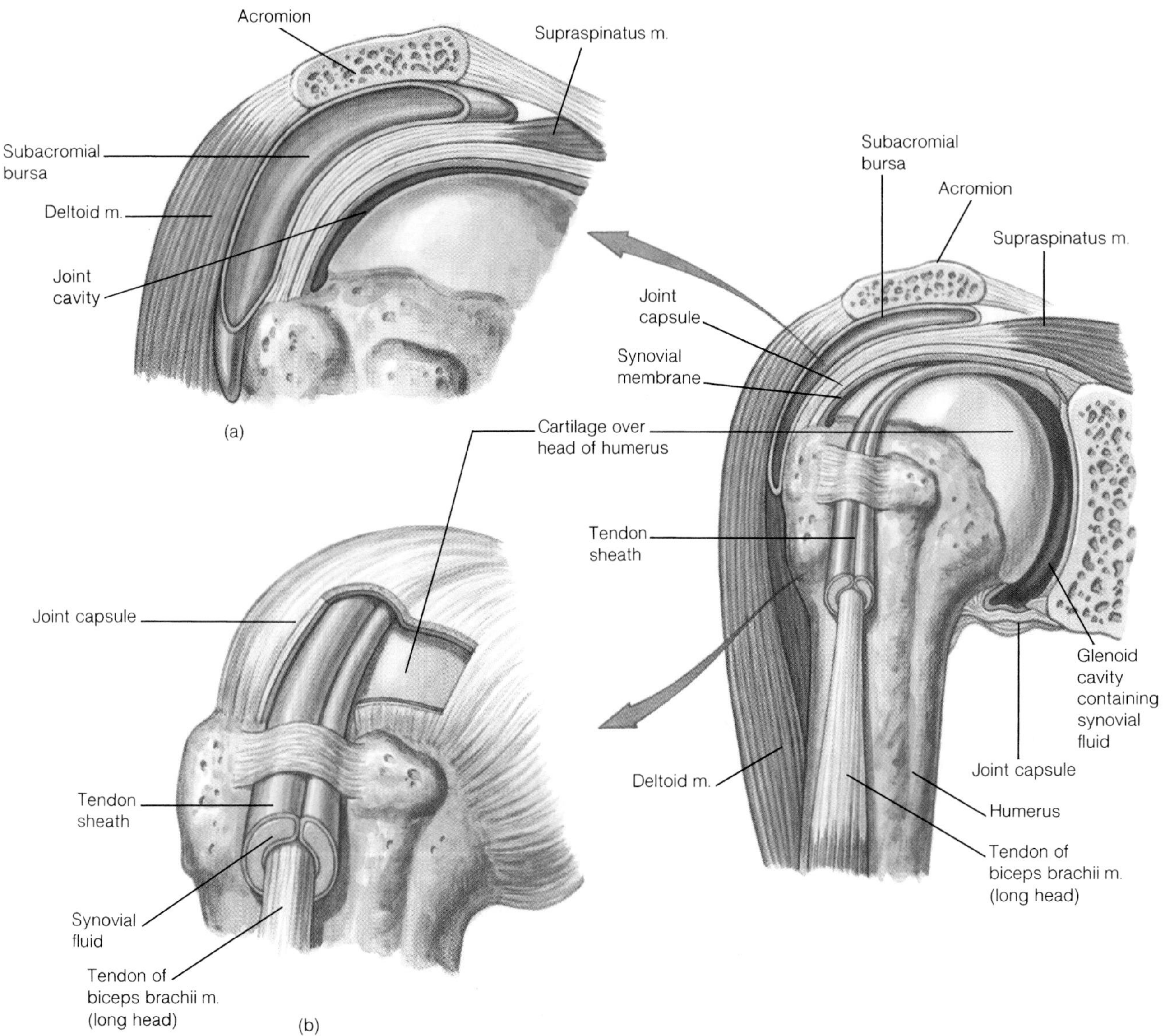

Gliding

Gliding joints allow only side-to-side and back-and-forth movements with some slight rotation. This is the simplest type of joint movement. The articulating surfaces are nearly flat, or one may be slightly concave and the other slightly convex (fig. 11.8). The intercarpal and intertarsal joints, the sternoclavicular joint, and the joint between the articular process of adjacent vertebrae are examples.

Hinge

Hinge joints are *monaxial*—like the hinge of a door, they permit movement in only one plane. In this type of articulation, the surface of one bone is always concave, and the other convex (fig. 11.9). Hinge joints are the most common type of synovial joints. Examples include the knee, the humeroulnar articulation within the elbow, and the joints between the phalanges.

Pivot

The movement in a pivot joint is limited to rotation about a central axis. In this type of articulation, the articular surface on one bone is conical or rounded and fits into a depression on another bone (fig. 11.10). Examples are the proximal articulation of the radius and ulna for rotation of the forearm, as

Figure 11.8

A gliding joint.

The intercarpal articulations in the wrist are examples of gliding joints in which the articulating surfaces of the adjacent bones are flattened or slightly curved. Note the diagrammatic representation showing the direction of possible movement.

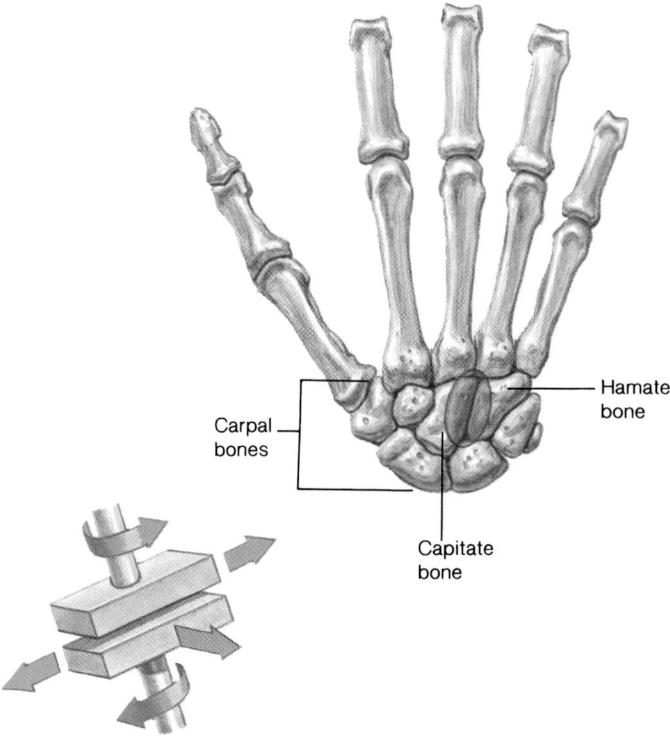

Figure 11.9

A hinge joint.

This type of synovial joint permits only a bending movement (flexion and extension). The hinge joint of the elbow involves the articulation of the distal end of the humerus with the proximal end of the ulna. Note the diagrammatic representation showing the direction of possible movement.

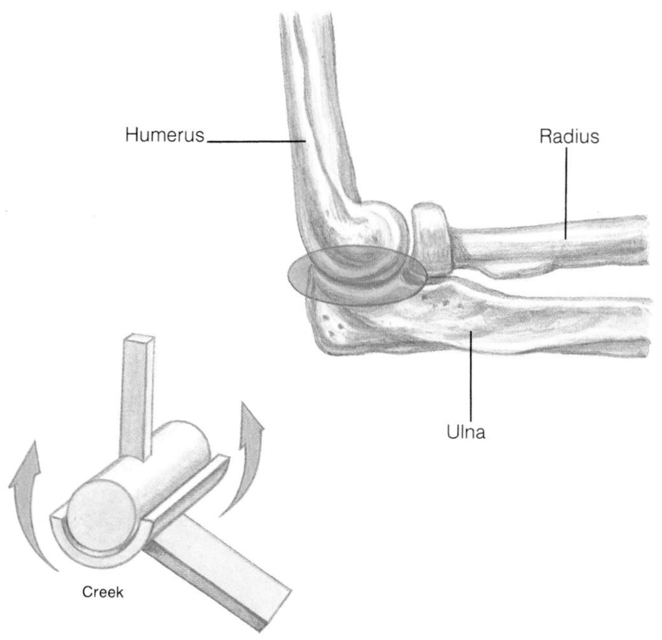

Figure 11.10

A pivot joint.

The articulation of the atlas with the axis forms a pivot joint that permits rotation. Note the diagrammatic representation showing the direction of possible movement. Refer to figure 11.9 and determine which articulating bones of the elbow region form a pivot joint.

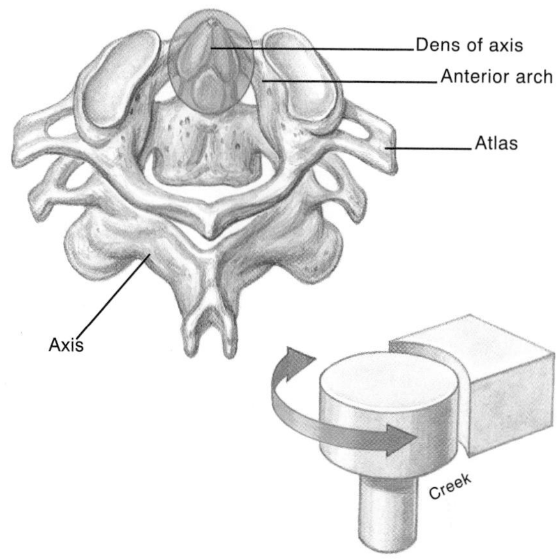

in turning a doorknob, and the articulation between the atlas and axis that allows rotational movement of the head.

Condyloid

A condyloid articulation is structured so that an oval, convex articular surface of one bone fits into a concave depression on another bone (fig. 11.11). This permits angular movement in two directions, as in up-and-down and side-to-side motions. Condyloid joints are therefore said to be *biaxial* joints. The radiocarpal joint of the wrist and the metacarpophalangeal joints are examples.

Saddle

Each articular process of a saddle-shaped joint has a concave surface in one direction and a convex surface in another. This articulation is a modified condyloid joint that allows a wide range of movement. There are two places in the body where a saddle joint occurs. One is at the articulation of the trapezium of the carpus with the first metacarpal bone (fig. 11.12). This carpometacarpal joint is the one responsible for the opposable thumb—a hallmark of primate anatomy. The other is at the articulation between the malleus and incus, two of the auditory ossicles of the middle ear (see fig. 18.24).

Ball-and-Socket

Ball-and-socket joints are formed by the articulation of a rounded convex surface with a cuplike cavity (fig. 11.13). This *multiaxial* type of articulation provides the greatest range of movement of all the synovial joints. Examples are the hip and shoulder joints.

A summary of the various types of joints is presented in table 11.1.

Figure 11.11

A condyloid joint.

The metacarpophalangeal articulations of the hand are examples of condyloid joints in which the oval condyle of one bone articulates with the cavity of another. Note the diagrammatic representation showing the direction of possible movement.

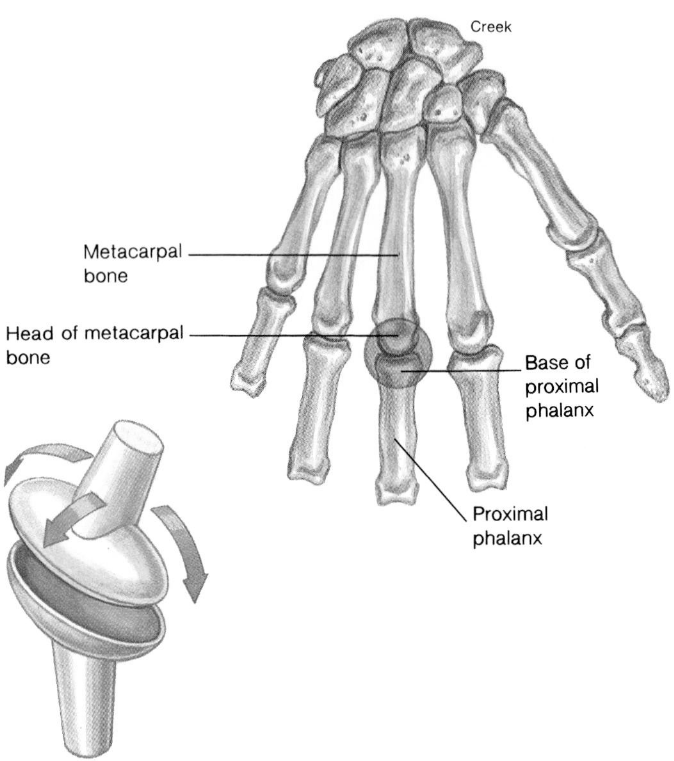

Figure 11.12

A saddle joint.

This type of synovial joint is formed as the trapezium articulates with the base of the first metacarpal bone. Note the diagrammatic representation showing the direction of possible movement.

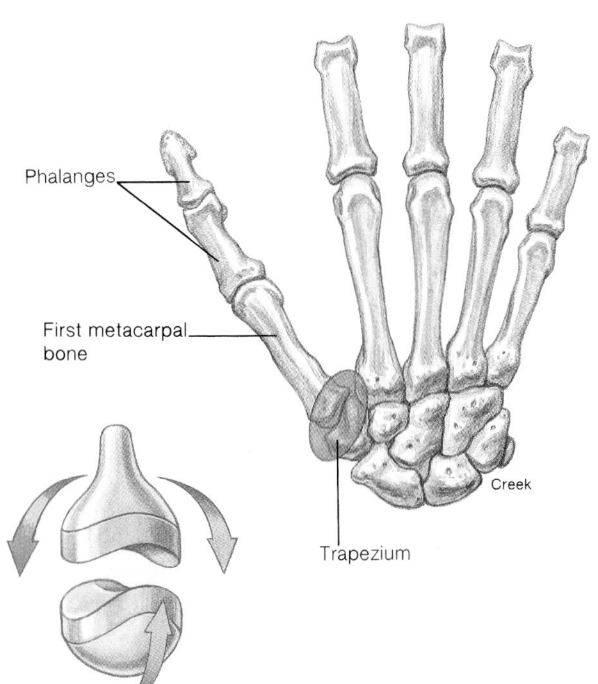

Figure 11.13

A ball-and-socket joint.

This highly movable synovial joint is formed at the hip where the head of the femur articulates with the acetabulum of the os coxae. Note the diagrammatic representation showing the direction of possible movement.

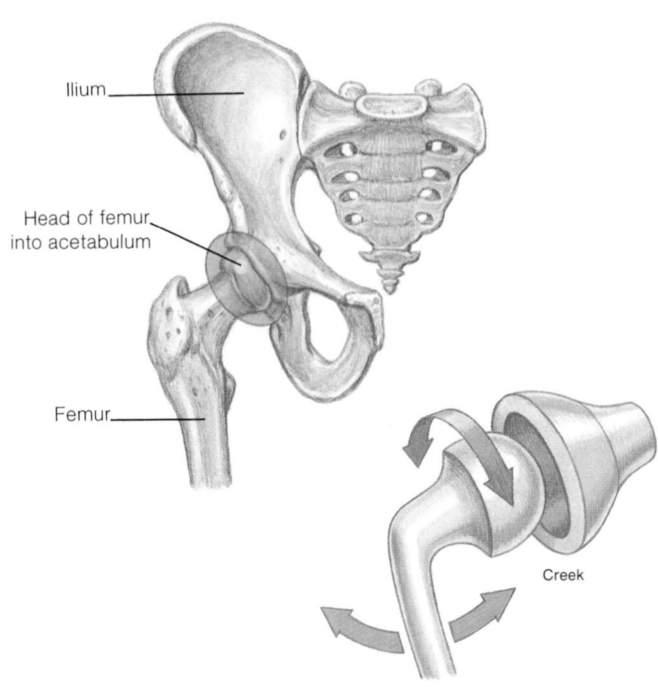

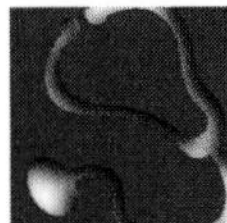

When a synovial joint is traumatized or injured, excessive synovial fluid is produced to cushion and immobilize the joint. This leads to joint swelling and discomfort. The most frequent type of joint injury is a *sprain,* in which the supporting ligaments or the joint capsule are damaged to varying degrees.

Movements at Synovial Joints

Movements at synovial joints are produced by the contraction of skeletal muscles that span the joints and attach to or near the bones forming the articulations. In these actions, the bones act as levers, the muscles provide the force, and the joints are the fulcra, or pivots.

As previously mentioned, the range of movement at a synovial joint is determined by its bony structure and the arrangement of the associated muscles, tendons, and ligaments. The movement at a hinge joint, for example, occurs in only one plane, whereas the structure of a ball-and-socket joint permits movement around many axes. Joint movements are broadly classified as *angular* and *circular.* Each of these categories includes specific types of movements, and certain special movements may involve several of the specific types. The description of joint movements are in reference to anatomical position (see table 11.2).

Table 11.1
Types of Articulations

Type	Structure	Movements	Example
Fibrous joints	Skeletal elements joined by fibrous connective tissue		
1. Suture	Edges of articulating bones frequently jagged; separated by thin layer of fibrous tissue	None	Sutures between bones of the skull
2. Syndesmoses	Articulating bones bound by interosseous ligament	Slightly movable	Joints between tibia-fibula and radius-ulna
3. Gomphoses	Teeth bound into dental alveoli of bone by periodontal ligament	Slightly movable	Dentoalveolar joints (teeth secured in dental alveoli)
Cartilaginous joints	Skeletal elements jointed by fibrocartilage of hyaline cartilage		
1. Symphyses	Articulating bones separated by pad of fibrocartilage	Slightly movable	Intervertebral joints; symphysis pubis
2. Synchondroses	Mitotically active hyaline cartilage located between skeletal elements	None	Epiphyseal plates within long bones; costal cartilages of rib cage
Synovial joints	Joint capsule containing synovial membrane and synovial fluid		
1. Gliding	Flattened or slightly curved articulating surfaces	Sliding	Intercarpal and intertarsal joints
2. Hinge	Concave surface of one bone articulates with convex surface of another	Bending motion in one plane (monaxial)	Knee; elbow; joints of phalanges
3. Pivot	Conical surface of one bone articulates with depression of another	Rotation about a central axis	Atlantoaxial joint; proximal radioulnar joint
4. Condyloid	Oval condyle of one bone articulates with elliptical cavity of another	Movement in two planes (biaxial)	Radiocarpal joint; metacarpophalangeal joint
5. Saddle	Concave and convex surface of each articulating bone	Wide range of movements	Carpometacarpal joint of thumb
6. Ball-and-socket	Round convex surface of one bone articulates with cuplike socket of another	Movement in all planes and rotation (multiaxial)	Shoulder and hip joints

Angular Movements

Angular movements increase or decrease the joint angle produced by the articulating bones. The four types of angular movements are *flexion*, *extension*, *abduction*, and *adduction*.

Flexion

Flexion is movement that decreases the joint angle on an anteroposterior plane (fig. 11.14*a*). Examples of flexion are the bending of the elbow or knee. Flexion of the elbow joint is a forward movement, whereas flexion of the knee is a backward movement. Flexion of the ankle and shoulder joints is a bit more complicated. In the ankle joint, flexion occurs as the top surface (dorsum) of the foot is elevated. This movement is frequently call **dorsiflexion** (fig. 11.14*b*). Pressing the foot downward (as in rising on the toes) is called **plantar flexion.** Flexion of the shoulder joint consists of raising the arm anterior from anatomical position, as if to point forward.

Extension

In extension, which is the reverse of flexion, the joint angle is increased (fig. 11.14*a*). Extension returns a body part to the anatomical position. In an extended joint, the angle between the articulating bones is 180°. An exception is the ankle joint, in which there is a 90° angle between the foot and the leg in the anatomical position. Examples of extension are straightening of the elbow or knee joints from flexion positions. **Hyperextension** occurs when a part of the body is extended beyond the anatomical position so that the joint angle is greater than 180°. An example of hyperextension is bending the neck to tilt the head backward, as in looking at the sky.

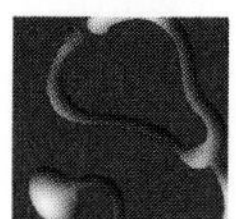

A common injury in runners is patellofemoral stress syndrome, commonly called "runner's knee." This condition is characterized by tenderness and aching pain around or under the patella. During normal knee movement, the patella glides up and down the patellar groove between the femoral condyles. In patellofemoral stress syndrome, the patella rubs laterally, causing irritation to the membranes and articular cartilage within the knee joint. Joggers frequently experience this condition from prolonged running on the slope of a road near the curb.

flexion: L. *flectere*, to bend

extension: L. *ex*, out, away from; *tendere*, stretch

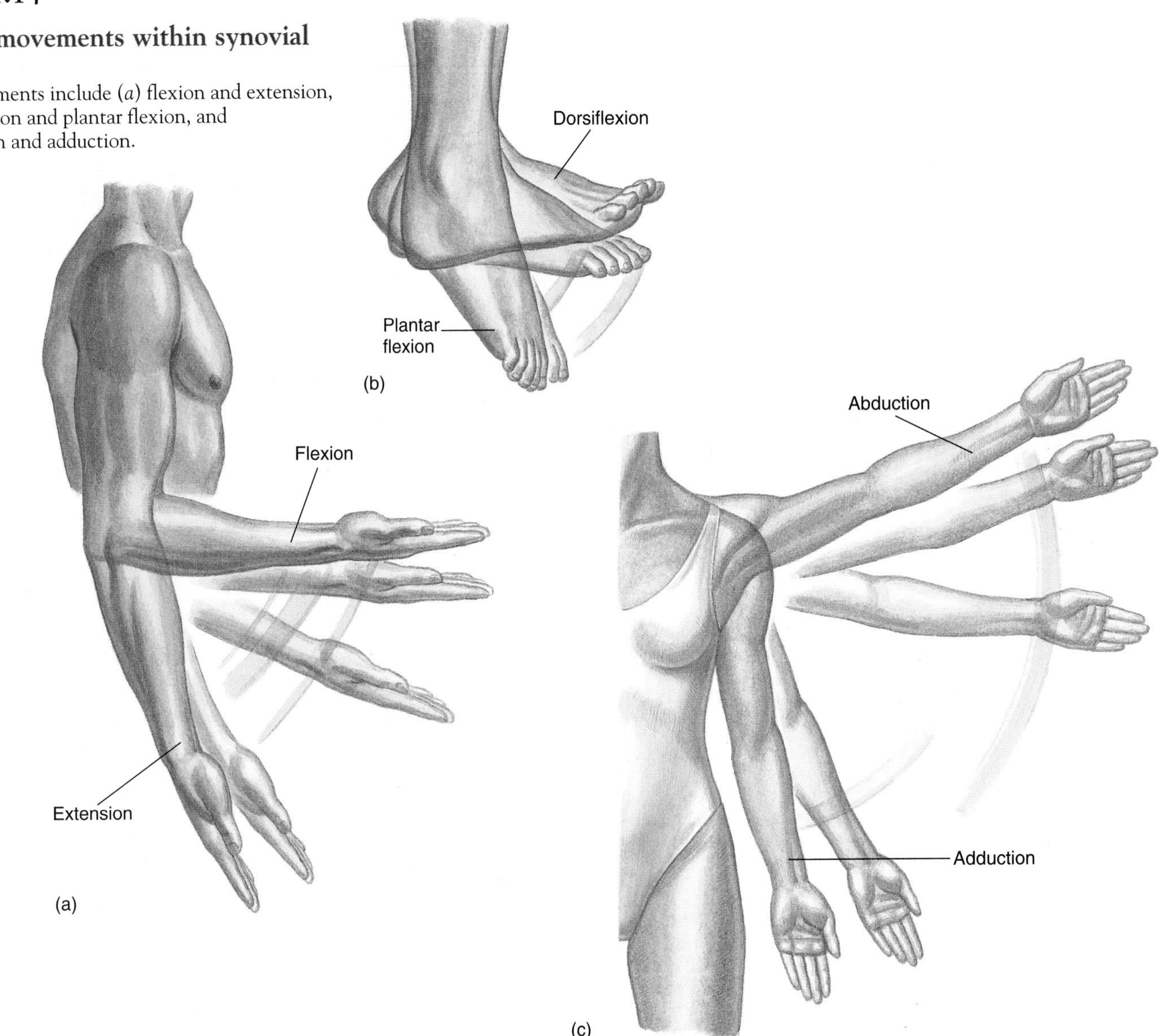

Figure 11.14

Angular movements within synovial joints.
These movements include (*a*) flexion and extension, (*b*) dorsiflexion and plantar flexion, and (*c*) abduction and adduction.

Abduction

Abduction is the movement of a body part away from the main axis of the body, or away from the midsagittal plane, in a lateral direction (fig. 11.14*c*). This term usually applies to the arm or leg but can also apply to the fingers or toes, in which case the line of reference is the longitudinal axis of the limb. An example of abduction is moving the arms sideward, away from the body. Spreading the fingers apart is another example.

Adduction

Adduction, the opposite of abduction, is the movement of a body part toward the main axis of the body (fig. 11.14*c*). In anatomical position, the arms and legs have been adducted toward the midplane of the body.

abduction: L. *abducere*, lead away
adduction: L. *adductus*, bring to

Circular Movements

In joints that permit **circular movement,** a bone with a rounded or oval surface articulates with a corresponding depression on another bone. The two basic types of circular movements are *rotation* and *circumduction*.

Rotation

Rotation is the movement of a body part around its own axis (fig. 11.15*a*). There is no lateral displacement during this movement. Examples are turning the head from side to side as if gesturing "no," and twisting at the waist.

Supination (*soo"pi-na'shun*) is a specialized rotation of the forearm so that the palm of the hand faces forward (anteriorly) or upward (superiorly). In anatomical position, the

rotation: L. *rotare*, a wheel

Figure 11.15

Circular movements within synovial joints.
These movements include (*a*) rotation and (*b*) circumduction.

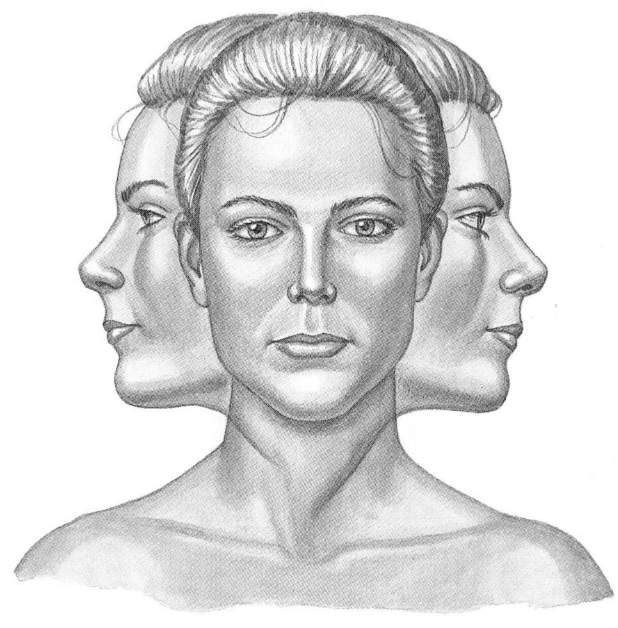

(a) Rotation

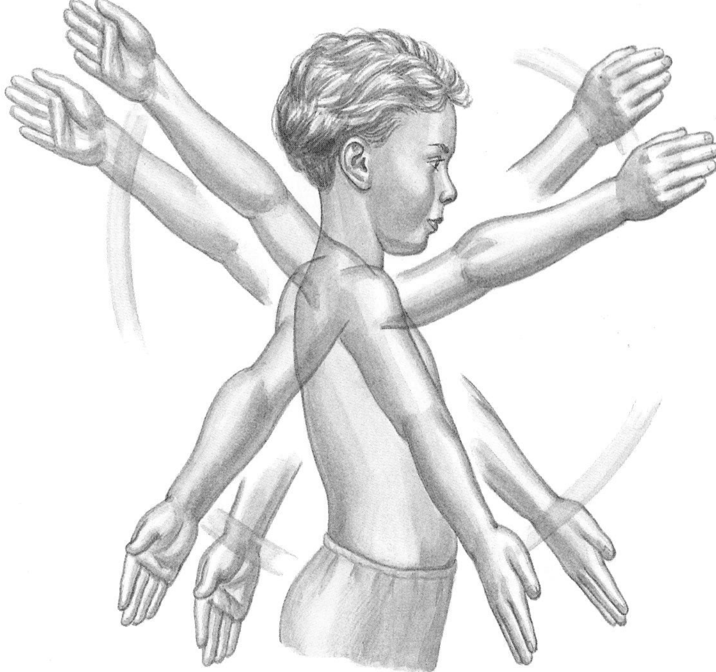

(b) Circumduction

forearm is already supine. **Pronation** (*pro-na'shun*) is the opposite of supination. It is a rotational movement of the forearm so that the palm is directed backward (posteriorly) or downward (inferiorly).

Applied to the foot, the term *pronation* describes a combination of eversion and abduction movements that result in a lowering of the longitudinal arch.

Circumduction

Circumduction is the circular movement of a body part so that a cone-shaped airspace is traced. The distal extremity performs the circular movement, and the proximal attachment serves as the pivot (fig. 11.15*b*). This type of motion is possible at the trunk, shoulder, wrist, metacarpophalangeal, hip, ankle, and metatarsophalangeal joints.

Special Movements

Because the terms used to describe generalized movements around axes do not apply to movement at certain joints or areas of the body, other terms must be used. **Inversion** is movement of the sole of the foot inward or medially (fig. 11.16*a*). **Eversion,** the opposite of inversion, is movement of the sole of the foot outward or laterally. The pivot axes for these movements are at the ankle and intertarsal joints. Both inversion and eversion are clinical terms that are usually used to describe developmental abnormalities.

The condition of the heels of your shoes can tell you whether you invert or evert your foot as you walk. If the heel is worn down on the outer side, you tend to invert your foot as you walk. If the heel is worn down on the inside, you tend to evert your foot.

Protraction is the movement of part of the body forward, on a plane parallel to the ground. The thrusting out of the lower jaw (fig. 11.16*b*) and the movement of the shoulder and upper extremity forward are examples. **Retraction,** the opposite of protraction, is the pulling back of a protracted part of the body on a plane parallel to the ground. Retraction of the mandible brings the lower jaw back in alignment with the upper jaw, so that the teeth occlude.

Elevation is a movement that raises a body part. Examples include elevating the mandible to close the mouth and lifting the shoulders to shrug (fig. 11.16*c*). **Depression** is the opposite of elevation. Both the mandible and shoulders are depressed when moved downward.

Many of the movements permitted at synovial joints are visually summarized in figures 11.17 through 11.19.

Biomechanics of Body Movement

A lever is any rigid structure that turns about a fulcrum when force is applied. Levers are generally associated with machines but can also apply to other mechanical structures, such as the human body. There are four basic elements in the function of

Figure 11.16

Special movements within synovial joints.
These movements include (*a*) inversion and eversion, (*b*) protraction and retraction, and (*c*) elevation and depression.

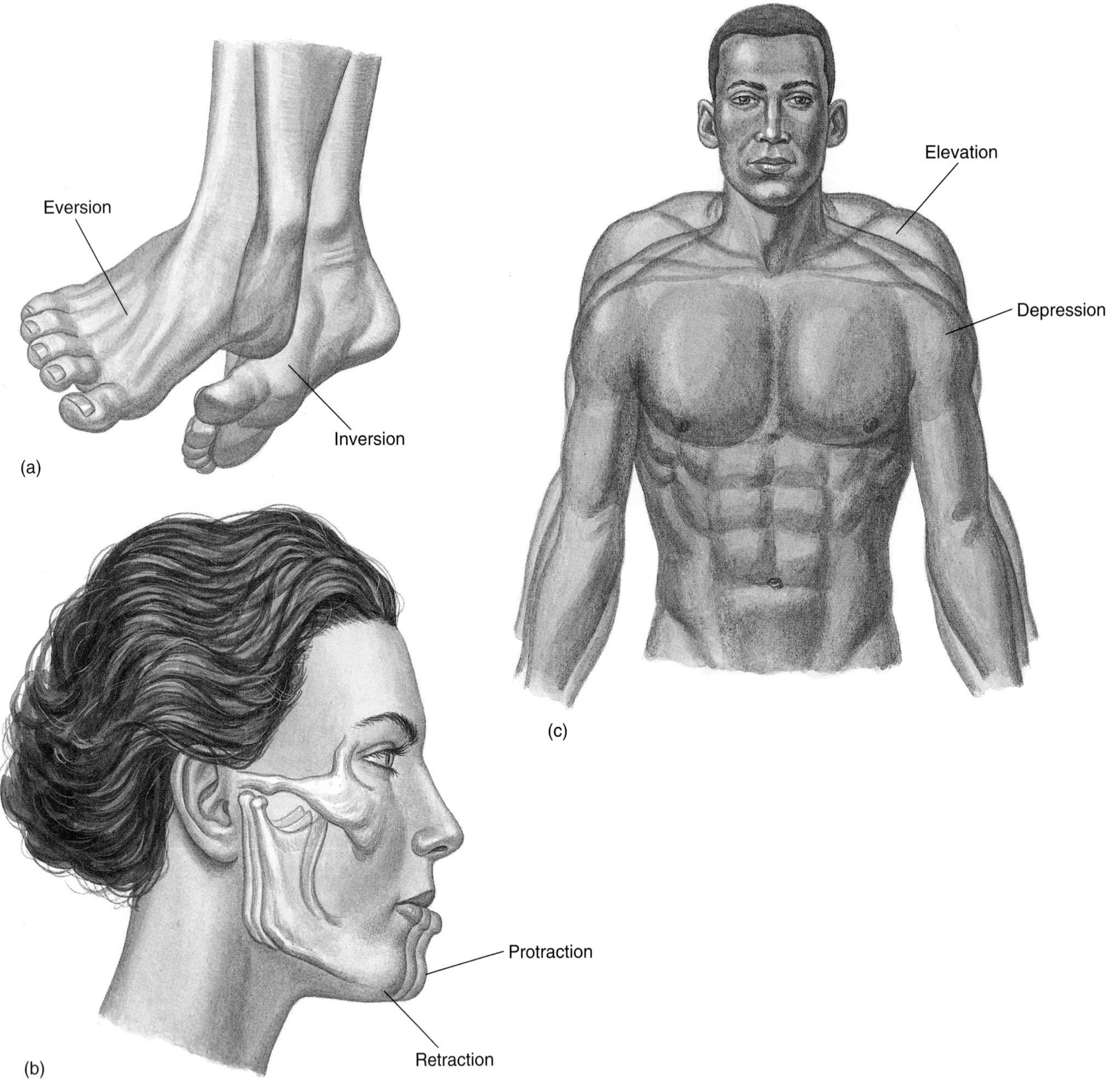

a lever: (1) the lever itself—a rigid bar or other such structure; (2) a pivot or fulcrum; (3) an object or resistance to be moved; and (4) a force that is applied to one portion of the rigid structure. In the body, synovial joints usually serve as the fulcra (*F*), the muscles provide the force, or effort (*E*), and the bones act as the rigid lever arms that move the resisting object (*R*).

There are three kinds of levers, determined by the arrangement of their parts (fig. 11.20).

1. In a **first-class lever,** the fulcrum is positioned between the effort and the resistance. The sequence of elements in a first-class lever is much like that of a seesaw—a sequence of resistance–pivot–effort. Scissors and hemostats are mechanical examples of first-class levers. In the body, the head at the atlantooccipital (*at-lan″to-ok-sip′ĭ-tal*) joint is a first-class lever. The weight of the skull and facial portion of the head is the resistance, and the posterior neck muscles that contract to oppose the tendency of the head to tip forward provide the effort.
2. In a **second-class lever,** the resistance is positioned between the fulcrum and the effort. The sequence of elements is pivot–resistance–effort, as in a wheelbarrow

Figure 11.17

A photographic summary of joint movements.
(*a*) Adduction of shoulder, hip, and carpophalangeal joints; (*b*) abduction of shoulder, hip, and carpophalangeal joints; (*c*) rotation of vertebral column; (*d*) lateral flexion of vertebral column; (*e*) flexion of vertebral column; (*f*) hyperextension of vertebral column; (*g*) flexion of shoulder, hip, and knee joints of right side of body, and extension of elbow and wrist joints; and (*h*) hyperextension of shoulder and hip joints on right side of body, and plantar flexion of right ankle joint.

or the action of a crowbar when one end is placed under a rock and the other end lifted. Contraction of the calf muscles (*E*) to elevate the body (*R*) on the toes, with the ball of the foot acting as the fulcrum, is another example.

3. In a **third-class lever,** the effort lies between the fulcrum and the resistance. The sequence of elements is pivot–effort–resistance, as in the action of a pair of forceps in grasping an object. The third-class lever is the most common type in the body. The flexion at the elbow is an example. The effort occurs as the biceps brachii muscle is contracted to move the resistance of the forearm, with the elbow joint forming the fulcrum.

Each bone-muscular interaction at a synovial joint forms some kind of lever system, and each lever system confers an advantage. Certain joints are adapted for power at the expense of speed, whereas most are clearly adapted for speed. The specific attachment of muscles that span a joint plays an

Figure 11.18

A photographic summary of some angular movements at synovial joints.

(*a*) Flexion, extension, and hyperextension in the cervical region; (*b*) flexion and extension at the knee joint and dorsiflexion and plantar flexion at the ankle joint; (*c*) flexion and extension at the elbow joint, and flexion, extension, and hyperextension at the wrist joint; (*d*) flexion, extension, and hyperextension at the hip joint, and flexion and extension at the knee joint; (*e*) adduction and abduction of the arm and fingers; and (*f*) abduction and adduction of the wrist joint (posterior view). Note that the range of abduction at the wrist joint is less extensive than the range of adduction as a result of the length of the styloid process of the radius.

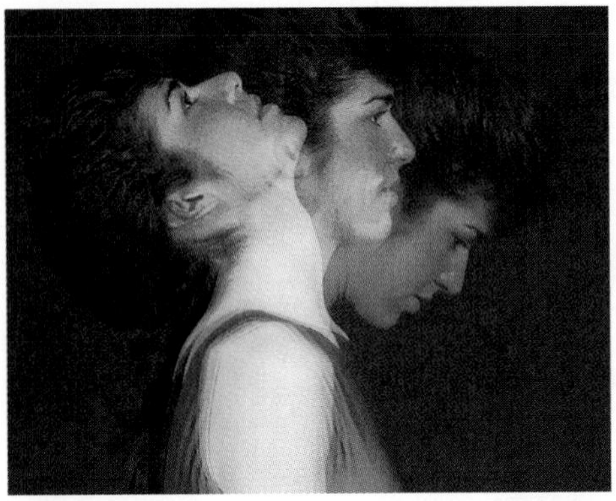

(a) 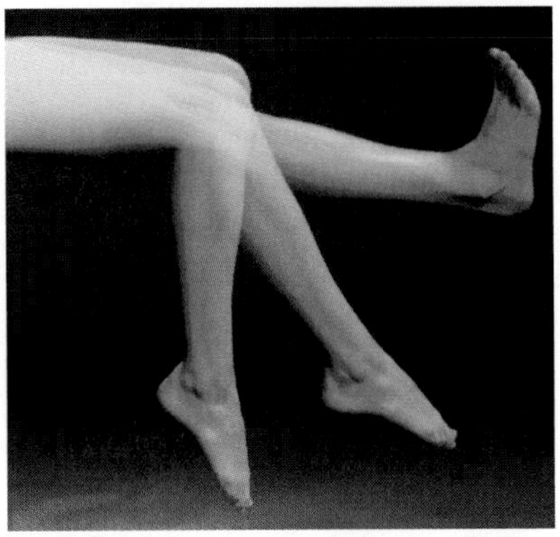 (b)  (c)

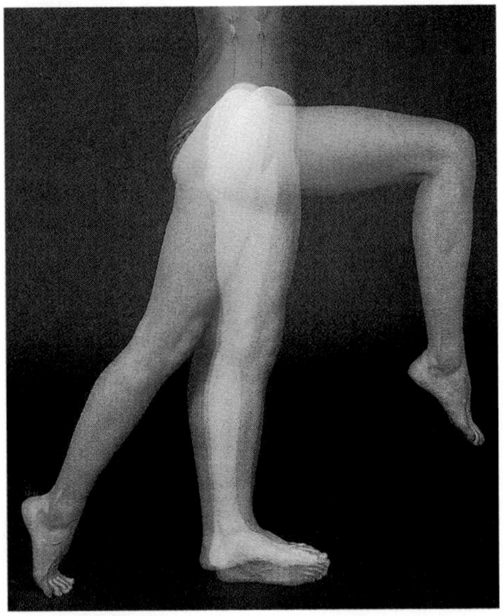

(d)  (e) 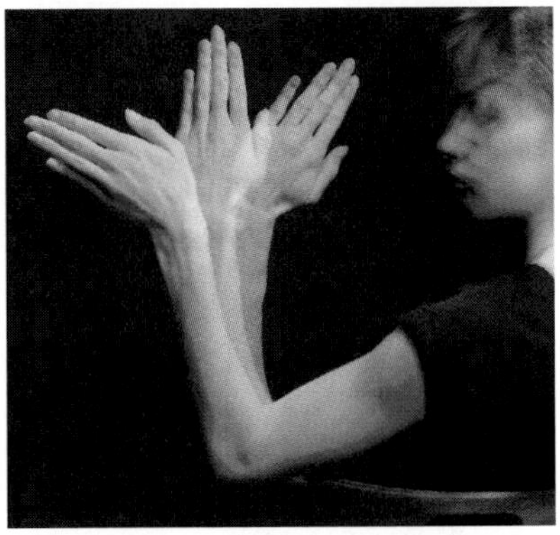 (f)

Figure 11.19

A photographic summary of some rotational movements at synovial joints.

(*a*) Rotation of the head at the cervical vertebrae—especially at the atlantoaxial joint, and (*b*) rotation of the (forearm) antebrachium at the proximal radioulnar joint.

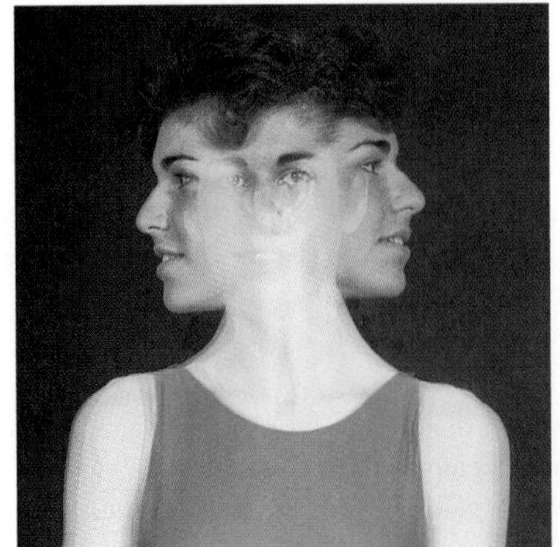

(a)

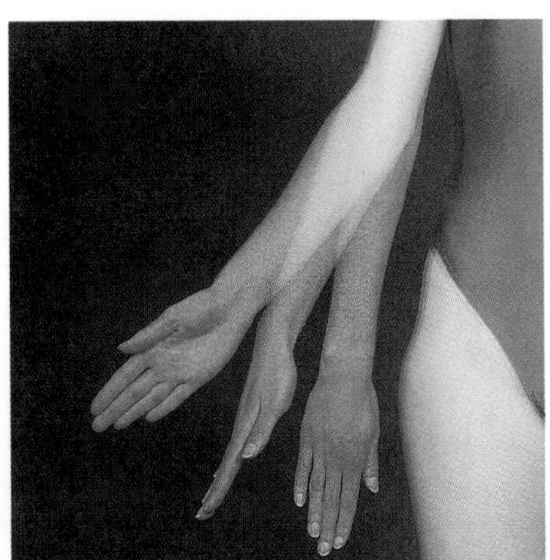

(b)

Figure 11.20

The three classes of levers.
(*a*) In a first-class lever, the fulcrum (*F*) is positioned between the resistance (*R*) and the effort (*E*). (*b*) In a second-class lever, the resistance is between the fulcrum and the effort. (*c*) In a third-class lever, the effort is between the fulcrum and the resistance.

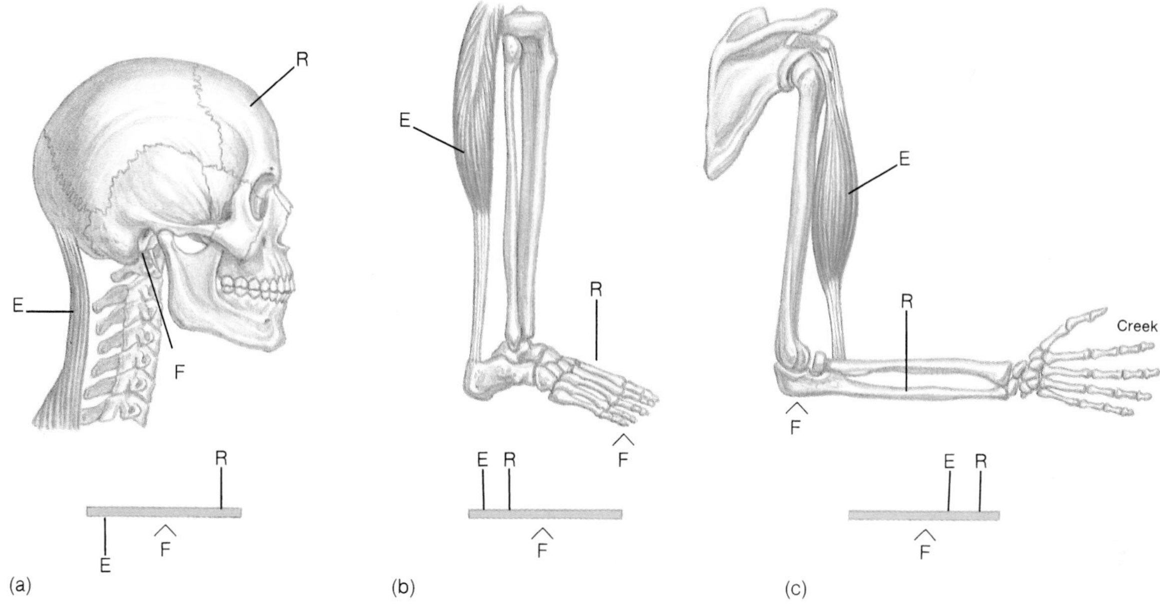

extremely important role in determining the mechanical advantage. The position of the insertion of a muscle relative to the joint is an important factor in the biomechanics of the contraction. An insertion close to the joint (fulcrum), for example, will produce a faster movement and greater range of movement than an insertion that is farther away from the joint. An attachment far from the joint capitalizes on the length of the lever arm (bone), and increases power at the sacrifice of speed and range of movement.

Specific Joints of the Body

Of the numerous joints in the body, some have special structural features that enable them to perform particular functions. These joints are also somewhat vulnerable to trauma and are therefore clinically important.

Temporomandibular Joint

The **temporomandibular joint** represents a unique combination of a hinge joint and a gliding joint (fig. 11.21). It is formed by the condylar process of the mandible and the mandibular fossa and the articular tubercle of the temporal bone. An *articular disc* separates the joint cavity into superior and inferior compartments.

Three major ligaments support and reinforce the temporomandibular joint. The **lateral ligament of the temporomandibular joint** is positioned on the lateral side of the joint capsule and is covered by the parotid gland. This ligament prevents the head of the mandible from being displaced posteriorly and fracturing the tympanic plate when the chin suffers a severe blow. The **stylomandibular ligament** is not directly associated with the joint but extends inferiorly and anteriorly from the styloid process to the posterior border of the ramus of the mandible. On the medial side by the joint, the **sphenomandibular** (*sfe″no-man-dib′yŭ-lar*) **ligament** extends from the spine of the sphenoid bone to the ramus of the mandible.

The movements of the temporomandibular joint include depression and elevation of the mandible as a hinge joint, protraction and retraction of the mandible as a gliding joint, and lateral rotatory movements. The lateral motion is made possible by the articular disc.

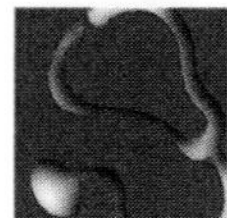

The temporomandibular joint can be easily palpated by applying firm pressure to the area in front of your ear and opening and closing your mouth. This joint is most vulnerable to dislocation when the mandible is completely depressed, as in yawning. Relocating the jaw is usually a simple task, however, and is accomplished by pressing down on the molars while pushing the jaw backward.

Temporomandibular joint (TMJ) syndrome is an ailment that may afflict an estimated 75 million Americans. The apparent cause of TMJ syndrome is a malalignment of one or both temporomandibular joints. The symptoms of the condition range from moderate and intermittent facial pain to intense and continuous pain in the head, neck, shoulders, or back. Clicking sounds in the jaw and limitation of jaw movement are common symptoms. Some vertigo (dizziness) and tinnitus (ringing in the ears) may also occur.

Figure 11.21

The temporomandibular joint.
(*a*) A lateral view, (*b*) a medial view, and (*c*) a sagittal view.

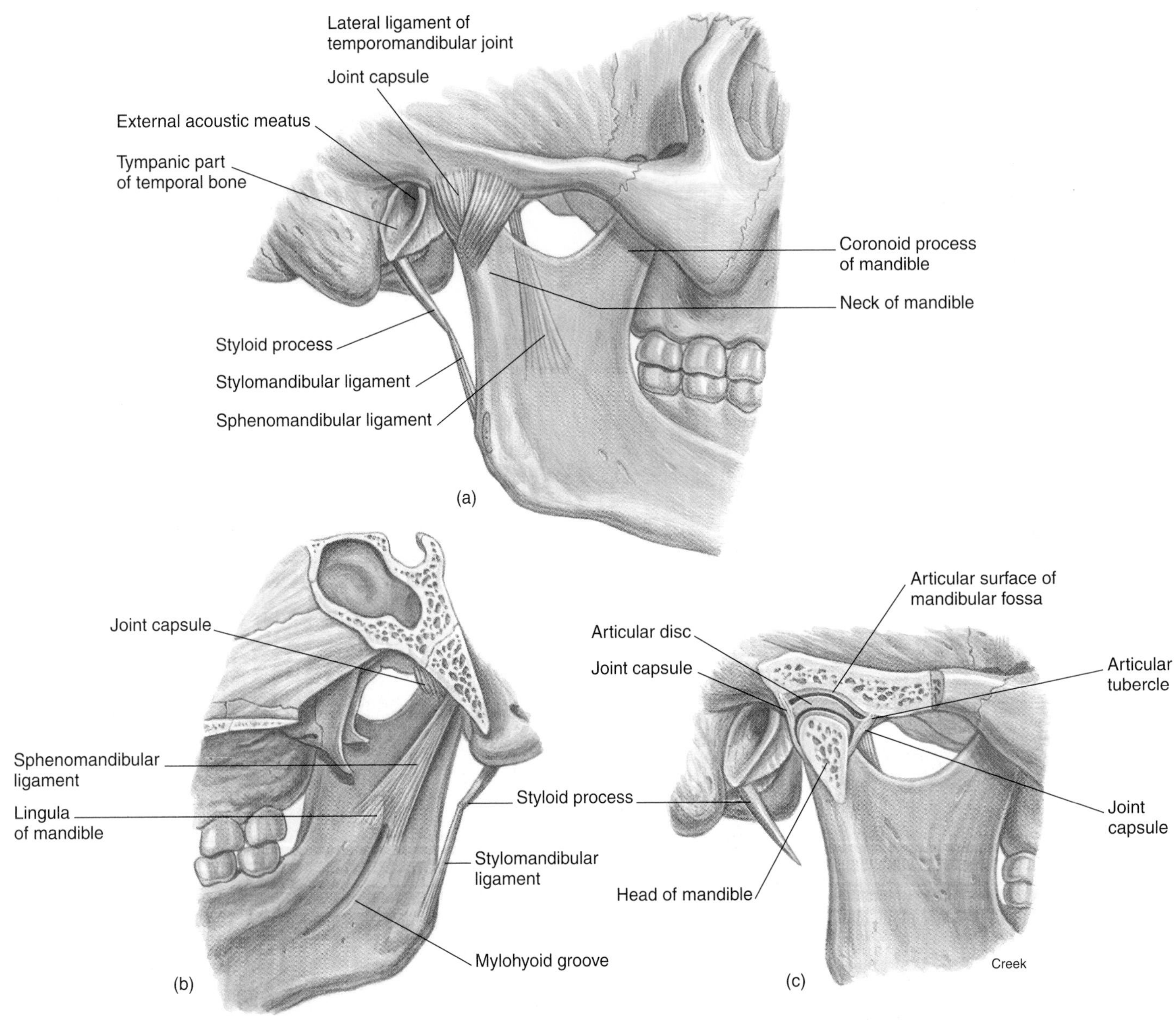

Glenohumeral (Shoulder) Joint

The **shoulder joint** is formed by the head of the humerus and the glenoid fossa of the scapula (figs. 11.22 and 11.23). It is a ball-and-socket joint and the most freely movable joint in the body. A circular band of fibrocartilage called the **glenoid labrum** passes around the rim of the shoulder joint and deepens the concavity of the glenoid fossa. The shoulder joint is protected from above by an arch formed by the acromion and coracoid process of the scapula and by the clavicle.

Although two ligaments and one retinaculum surround and support the shoulder joint, most of the stability of this joint depends on the powerful muscles and tendons that cross over it. Thus, it is an extremely mobile joint in which stability has been sacrificed for mobility. The **coracohumeral** (*kor″ă-ko-hyoo′-mer-al*) **ligament** extends from the coracoid process of the scapula to the greater tubercle of the humerus. The joint capsule is reinforced with three ligamentous bands called the **glenohumeral ligaments.** The final support of the shoulder joint is the **transverse humeral retinaculum,** a thin band that extends from the greater tubercle to the lesser tubercle of the humerus.

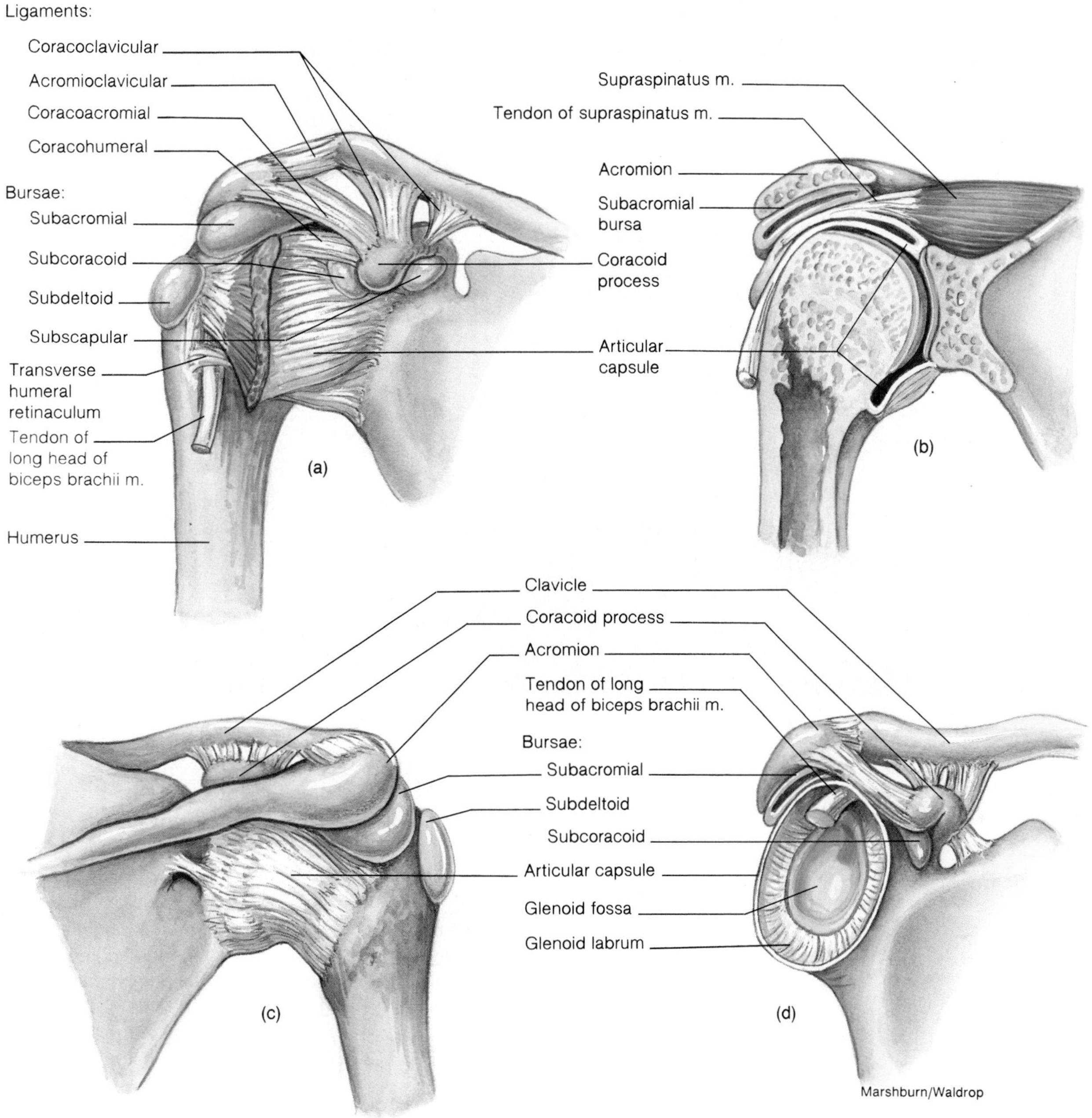

Figure 11.22

The glenohumeral (shoulder) joint.
(*a*) An anterior view, (*b*) a coronally sectioned anterior view, (*c*) a posterior view, and (*d*) a lateral view with the humerus removed.

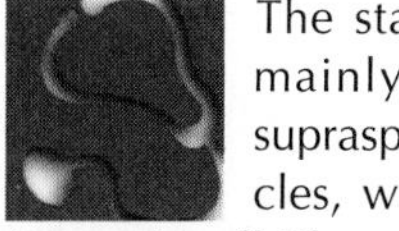

The stability of the shoulder joint is provided mainly by the tendons of the subscapularis, supraspinatus, infraspinatus, and teres minor muscles, which together form the *musculotendinous (rotator) cuff.* The cuff is fused to the underlying capsule, except in its inferior aspect. Because of the lack of inferior stability, most dislocations (subluxations) occur in this direction. The shoulder is most vulnerable to trauma when the joint is fully abducted and a sudden force from a superior direction is applied to the appendage—as for example, when the outstretched arm is struck by a heavy object falling from a shelf. Degenerative changes in the musculotendinous cuff produce an inflamed, painful condition known as *pericapsulitis.*

Two major and two minor bursae are associated with the shoulder joint. The larger bursae are the **subdeltoid bursa,** located between the deltoid muscle and the joint capsule, and the **subacromial bursa,** located between the acromion and joint capsule. The **subcoracoid bursa,** which lies between the coracoid process and the joint capsule, is frequently considered an extension of the subacromial bursa. A small **subscapular bursa** is located between the tendon of the subscapularis muscle and the joint capsule.

Figure 11.23

A posterior view of a dissected glenohumeral joint.
An incision has been made into the joint capsule and the humerus has been retracted laterally and rotated posteriorly.

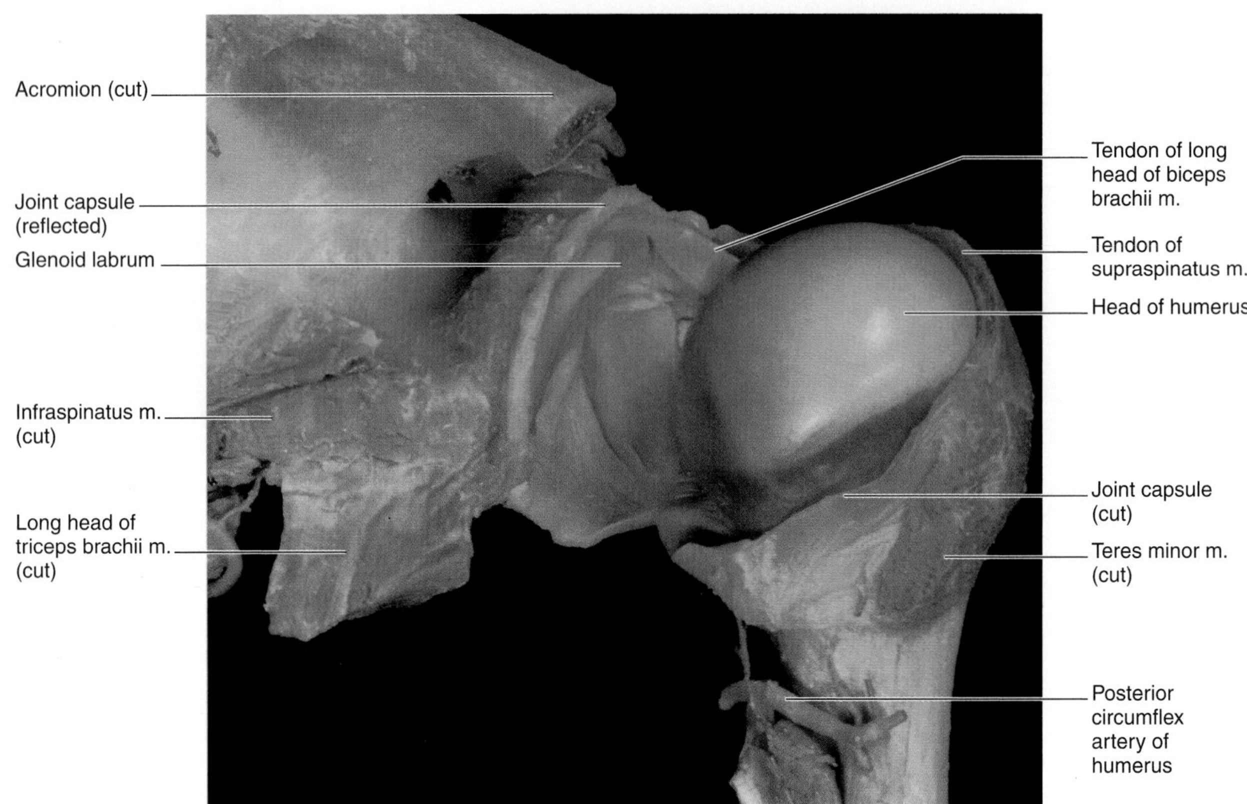

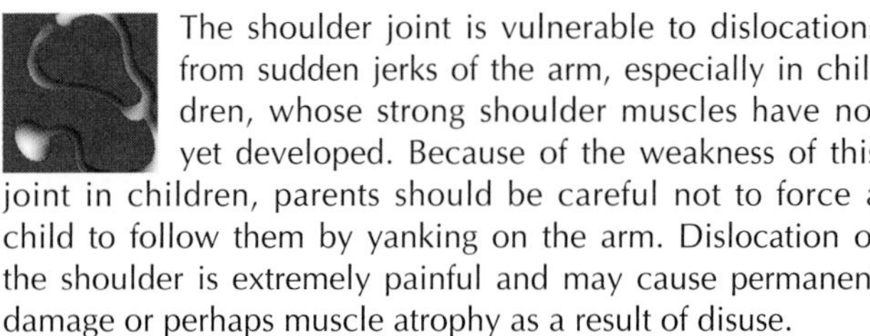

The shoulder joint is vulnerable to dislocations from sudden jerks of the arm, especially in children, whose strong shoulder muscles have not yet developed. Because of the weakness of this joint in children, parents should be careful not to force a child to follow them by yanking on the arm. Dislocation of the shoulder is extremely painful and may cause permanent damage or perhaps muscle atrophy as a result of disuse.

Elbow Joint

The elbow joint is a hinge joint composed of two articulations—the **humeroulnar joint,** formed by the trochlea of the humerus and the trochlear notch of the ulna, and the **humeroradial joint,** formed by the capitulum of the humerus and the head of the radius (figs. 11.24 and 11.25). Both of these articulations are enclosed in a singe joint capsule. On the posterior side of the elbow, there is a large **olecranon bursa** to lubricate the area. A **radial (lateral) collateral ligament** reinforces the elbow joint on the lateral side, and an **ulnar (medial) collateral ligament** strengthens the medial side.

A third joint occurs in the elbow region—the **proximal radioulnar joint**—but it is not part of the hinge. At this joint, the head of the radius fits into the radial notch of the ulna and is held in place by the **annular ligament.**

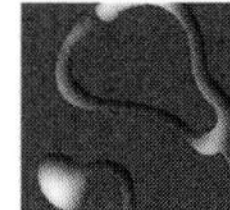

Because so many muscles originate or insert near the elbow, it is a common site of localized tenderness, inflammation, and pain. *Tennis elbow* is a general term for musculotendinous soreness in this area. The structures most generally strained are the tendons attached to the lateral epicondyle of the humerus. The strain is caused by repeated extension of the wrist against some force, as occurs during the backhand stroke in tennis.

Metacarpophalangeal Joints and Interphalangeal Joints

The **metacarpophalangeal joints** are condyloid joints, and the **interphalangeal joints** are hinge joints. The articulating bones of the former are the metacarpal bones and the proximal phalanges; those of the latter are adjacent phalanges (fig. 11.26). Each joint in both joint types has three ligaments. A **palmar ligament** spans each joint on the palmar, or anterior, side of the joint capsule. Each joint also has two **collateral ligaments,** one on the lateral side and one on the medial side, to

Figure 11.24

The right elbow region.
(*a*) An anterior view, (*b*) a posterior view, (*c*) a sagittal view, (*d*) a lateral view, and (*e*) a medial view.

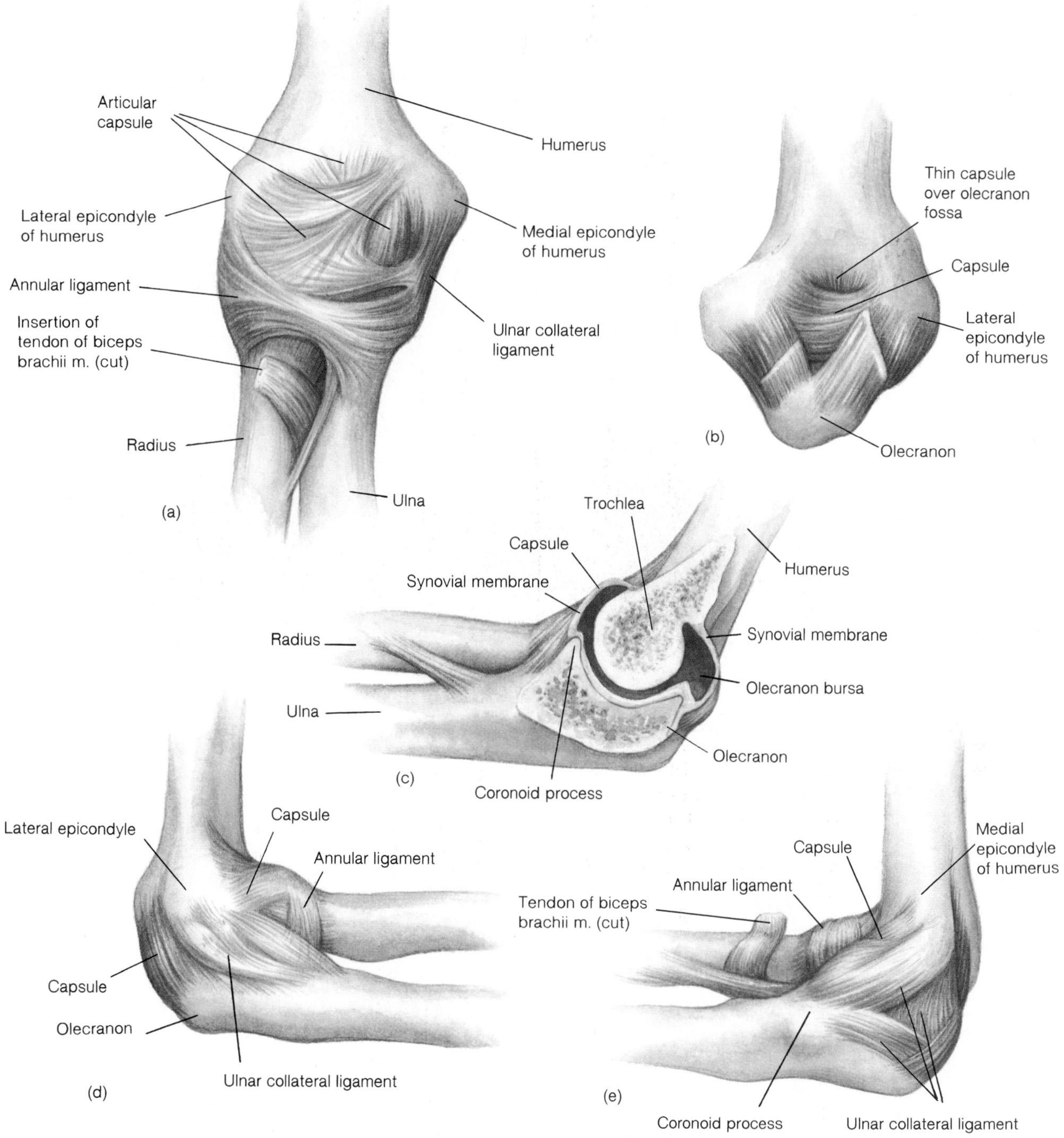

Figure 11.25

A posterior view of a dissected elbow joint.
A portion of the joint capsule has been removed to show the articular surface of the humerus.

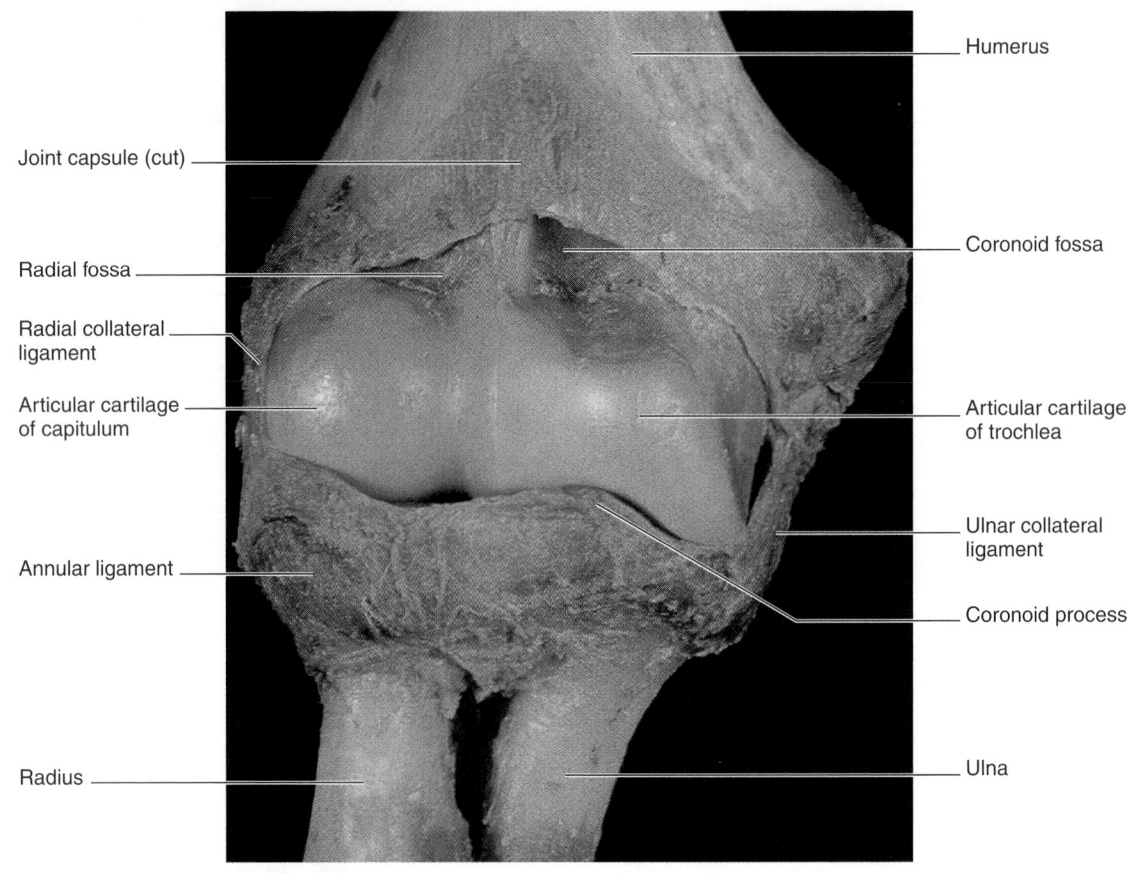

Figure 11.26

Metacarpophalangeal and interphalangeal joints.
(*a*) A lateral view, (*b*) an anterior (palmar) view, and (*c*) a posterior view.

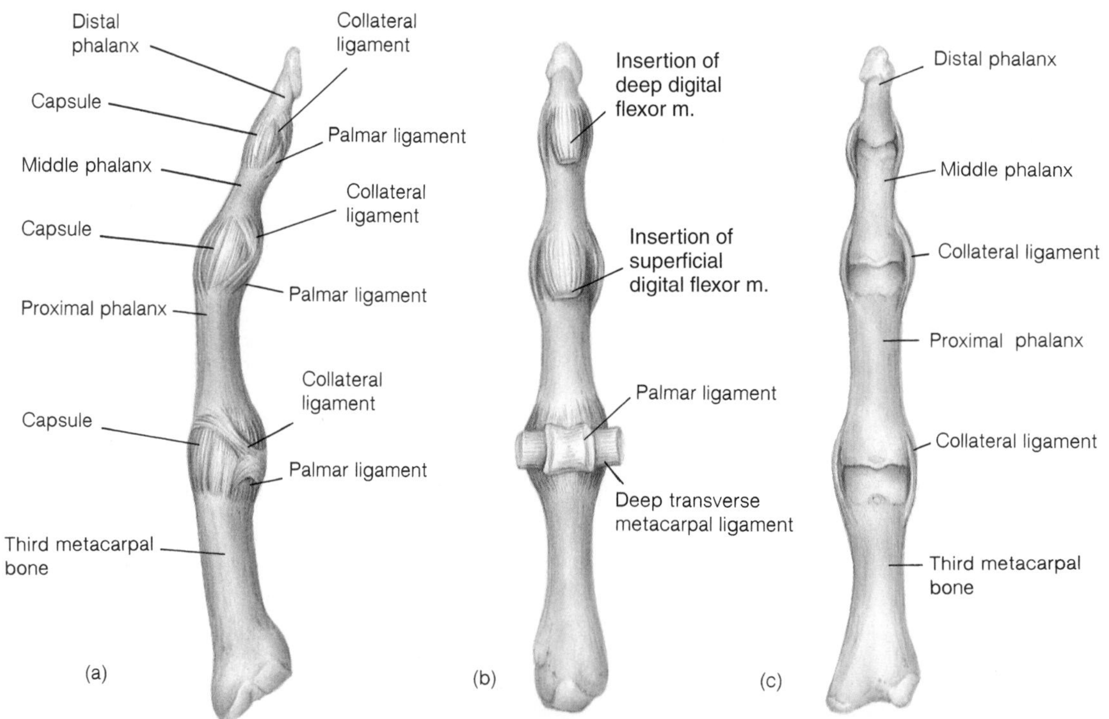

further reinforce the joint capsule. There are no supporting ligaments on the posterior side.

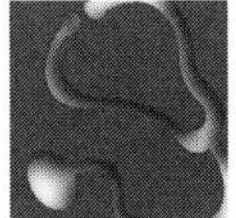

Athletes frequently jam their fingers. It occurs when a ball forcefully strikes a distal phalanx as the fingers are extended, causing a sharp flexion at the joint between the middle and distal phalanges. No ligaments support the joint on the posterior side, but there is a tendon from the digital extensor muscles of the forearm. It is this tendon that is damaged when the finger is jammed. Treatment involves splinting the finger for a period of time. If splinting is not effective, surgery is generally performed to avoid a permanent crook in the finger.

Coxal (Hip) Joint

The ball-and-socket **hip joint** is formed by the head of the femur articulating with the acetabulum of the os coxae (fig. 11.27). It bears the weight of the body and is therefore much stronger and more stable than the shoulder joint. The hip joint is secured by a strong fibrous joint capsule, several ligaments, and a number of powerful muscles.

The primary ligaments of the hip joint are the anterior **iliofemoral** (*il"ĕ-o-fem'or-al*) and **pubofemoral ligaments** and the posterior **ischiofemoral** (*is"ke-o-fem'or-al*) **ligament.** The **ligamentum capitis femoris** is located within the articular capsule and attaches the head of the femur to the acetabulum.

Figure 11.27

The right coxal (hip) joint.
(*a*) An anterior view, (*b*) a posterior view, and (*c*) a coronal view.

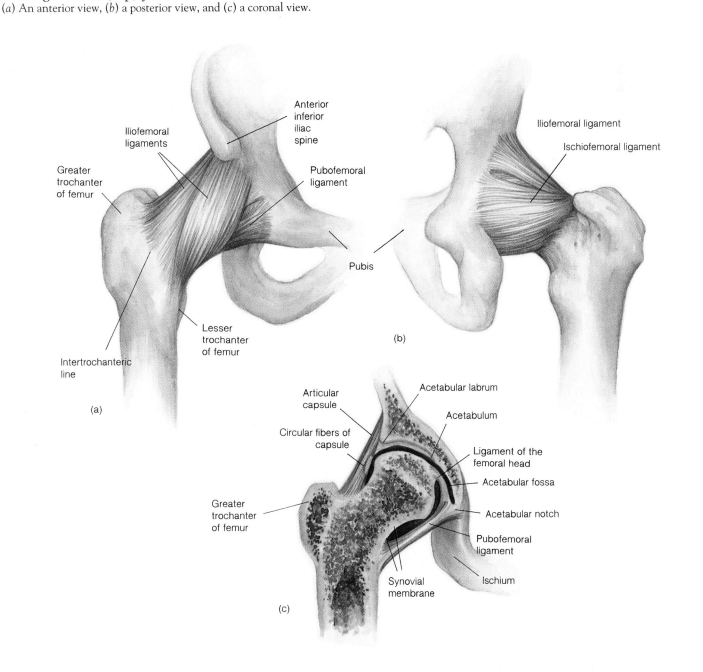

This is a relatively slack ligament, and does not play a significant role in holding the femur in place. However, it does contain a small artery that supplies blood to the head of the femur. The **transverse acetabular** (*as″ĕ-tab′yŭ-lar*) **ligament** crosses the acetabular notch and connects to the joint capsule and the ligamentum capitis femoris. The **acetabular labrum,** a fibrocartilaginous rim that rings the head of the femur as it articulates with the acetabulum, is attached to the margin of the acetabulum.

Tibiofemoral (Knee) Joint

The **knee joint,** located between the femur and tibia, is the largest, most complex, and probably the most vulnerable joint in the body. It is a complex hinge joint that permits limited rolling and gliding movements, in addition to flexion and extension. On the anterior side, the knee joint is stabilized and protected by the patella and the **patellar ligament,** forming a gliding **patellofemoral joint.**

Because of the complexity of the knee joint, only the relative positions of the ligaments, menisci, and bursae will be covered here. Although the attachments will not be discussed in detail, the locations of these structures can be seen in figures 11.28 and 11.29.

In addition to the patella and the patellar ligament on the anterior surface, the tendinous insertion of the quadriceps femoris muscle forms two supportive bands called the **lateral** and **medial patellar retinacula** (*ret″ĭ-nak′yŭ-lă*). Four bursae are associated with the anterior aspect of the knee: the **subcutaneous prepatellar bursa,** the **suprapatellar bursa,** the **cutaneous prepatellar bursa,** and the **deep infrapatellar bursa.**

The posterior aspect of the knee is referred to as the **popliteal** (*pop″li-te′al*) **fossa.** The broad **oblique popliteal ligament** and the **arcuate** (*ar′kyoo-at*) **popliteal ligament** are superficial in position, whereas the **anterior** and **posterior cruciate** (*kroo′she-at*) **ligaments** are deep within the joint. The **popliteal bursa** and the **semimembranosus bursa** are the two bursae associated with the back of the knee.

Strong **collateral ligaments** support both the medial and lateral sides of the knee joint. Two fibrocartilaginous discs, the **lateral** and **medial menisci,** are located within the knee joint interposed between the distal femoral and proximal tibial condyles. The two menisci are connected by a **transverse ligament.** In addition to the four bursae on the anterior side and the two on the posterior side, there are seven bursae on the lateral and medial sides, for a total of thirteen.

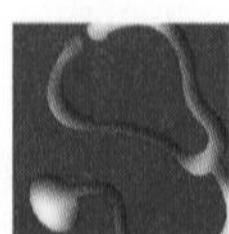

During normal walking, running, and supporting the body, the knee joint functions superbly. It can tolerate considerable stress without tissue damage. However, the knee lacks bony support to withstand sudden forceful stresses, which frequently occur in athletic competition. Knee injuries often require surgery, and they heal with difficulty because of the avascularity of the cartilaginous tissue. Knowledge of the anatomy of the knee provides insight as to its limitations. The three C's—the anterior cruciate ligament, the collateral ligaments, and the cartilage—are the most likely sites of crippling injury.

Talocrural (Ankle) Joint

There are actually two principal articulations within the **ankle joint,** both of which are hinge joints. One is formed as the distal end of the tibia, and its medial malleolus articulates with the talus; the other is formed as the lateral malleolus of the fibula and articulates with the talus (fig. 11.30).

One joint capsule surrounds the articulations of the three bones, and four ligaments support the ankle joint on the outside of the capsule. The strong **deltoid ligament** is associated with the tibia, whereas the **lateral collateral ligaments, anterior talofibular** (*ta″lo-fib′yŭ-lar*) **ligament, posterior talofibular ligament,** and **calcaneofibular** (*kal-ka″ne-o-fib′yŭ-lar*) **ligament** are associated with the fibula.

The malleoli form a cap over the upper surface of the talus that prohibits side-to-side movement at the ankle joint. Unlike the condyloid joint at the wrist, the movements of the ankle are limited to flexion and extension. Dorsiflexion of the ankle is checked primarily by the tendo calcaneus, whereas plantar flexion, or ankle extension, is checked by the tension of the extensor tendons on the front of the joint and the anterior portion of the joint capsule.

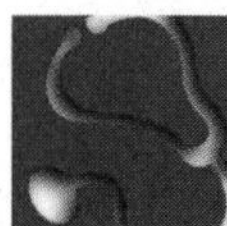

Ankle sprains are a common type of locomotor injury. They vary widely in seriousness but tend to occur in certain locations. The most common cause of ankle sprain is excessive inversion of the foot, resulting in partial tearing of the anterior talofibular ligament and the calcaneofibular ligament. Less commonly, the deltoid ligament is injured by excessive eversion of the foot. Torn ligaments are extremely painful and are accompanied by immediate local swelling. Reducing the swelling and immobilizing the joint are about the only treatments for moderate sprains. Extreme sprains may require surgery and casting of the joint to facilitate healing.

A summary of the principal joints of the body and their movement is presented in table 11.2.

Clinical Considerations

A synovial joint is a remarkable biologic system. Its self-lubricating action provides a shock-absorbing cushion between articulating bones and enables almost frictionless movement under tremendous loads and impacts. Under normal circumstances and in most people, the many joints of the body perform without problems throughout life. Joints are not indestructible, however, and are subject to various forms of trauma and disease. Although not all of the diseases of joints are fully understood, medical science has made remarkable progress in the treatment of arthrological problems.

Trauma to Joints

Joints are well adapted to withstand compression and tension forces. Torsion or sudden impact to the side of a joint, however, can be devastating. These types of injuries frequently occur in athletes.

Figure 11.28

The right tibiofemoral (knee) joint.
(*a*) An anterior view, (*b*) a superficial posterior view, (*c*) a lateral view showing the bursae, (*d*) an anterior view with the knee slightly flexed and the patella removed, and (*e*) a deep posterior view.

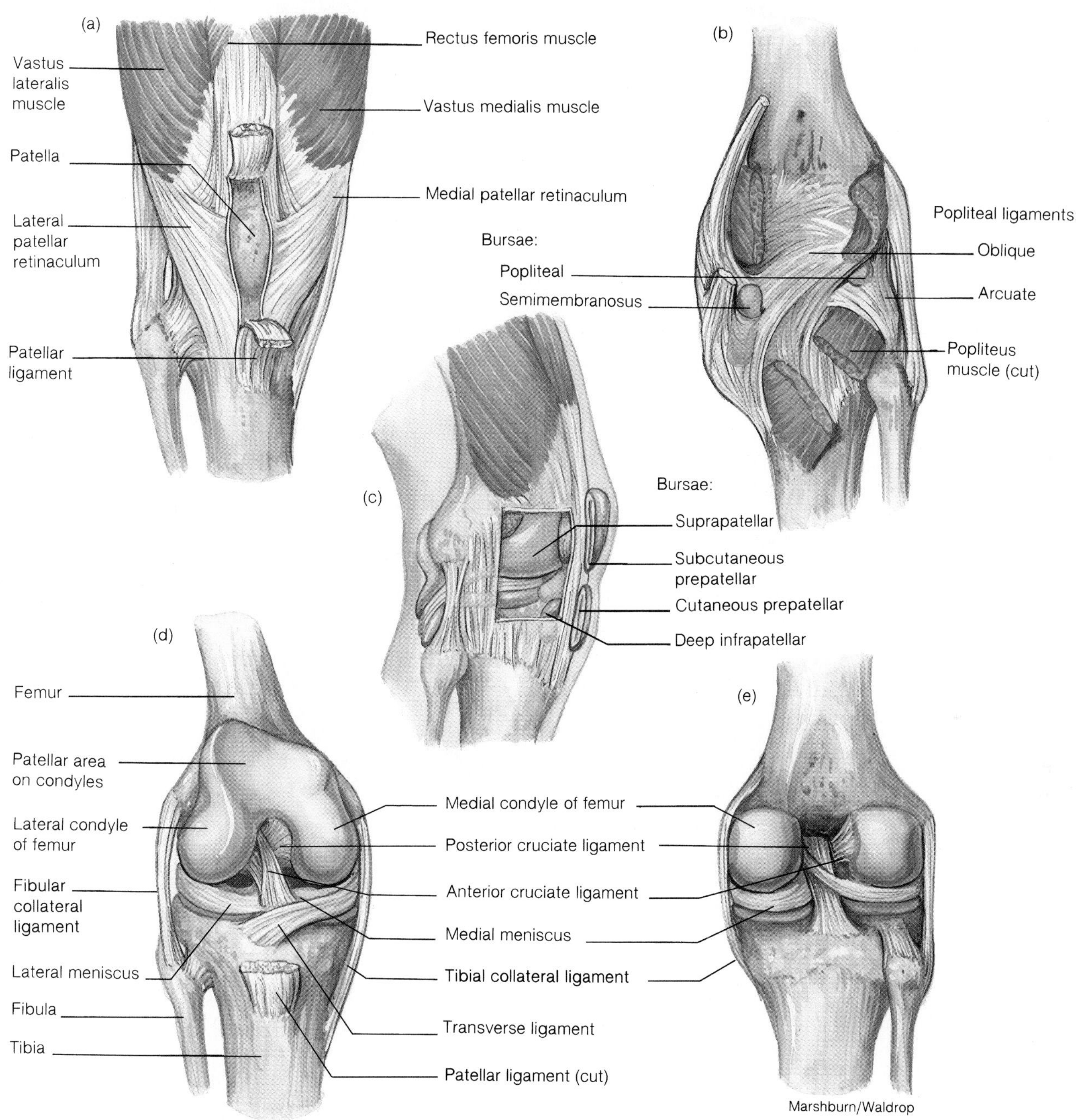

In a **strained joint,** unusual or excessive exertion stretches the tendons or muscles surrounding a joint. The damage is not serious. Strains are frequently caused by not warming up the muscles and not stretching the joints prior to exercise. A **sprain** is a tearing of the ligaments or tendons surrounding a joint. There are various grades of sprains, and the severity will determine the treatment. Severe sprains damage articular cartilages and may require surgery. Sprains are usually accompanied by **synovitis** (*sin″ŏ-vi′tis*), an inflammation of the joint capsule.

Luxation, or **joint dislocation,** is derangement of the articulating bones that compose the joint. Joint dislocation is more serious than a sprain and is usually accompanied by sprains. The shoulder and knee joints are the most vulnerable to dislocation. Self-healing of a dislocated joint may be incomplete, leaving the person with a "trick knee," for example, that may unexpectedly give way.

Subluxation is partial dislocation of a joint. Subluxation of the hip joint is a common type of birth defect that can be

Figure 11.29

A posterior view of a dissected tibiofemoral joint.
The joint capsule has been removed to expose the cruciate ligaments and the menisci.

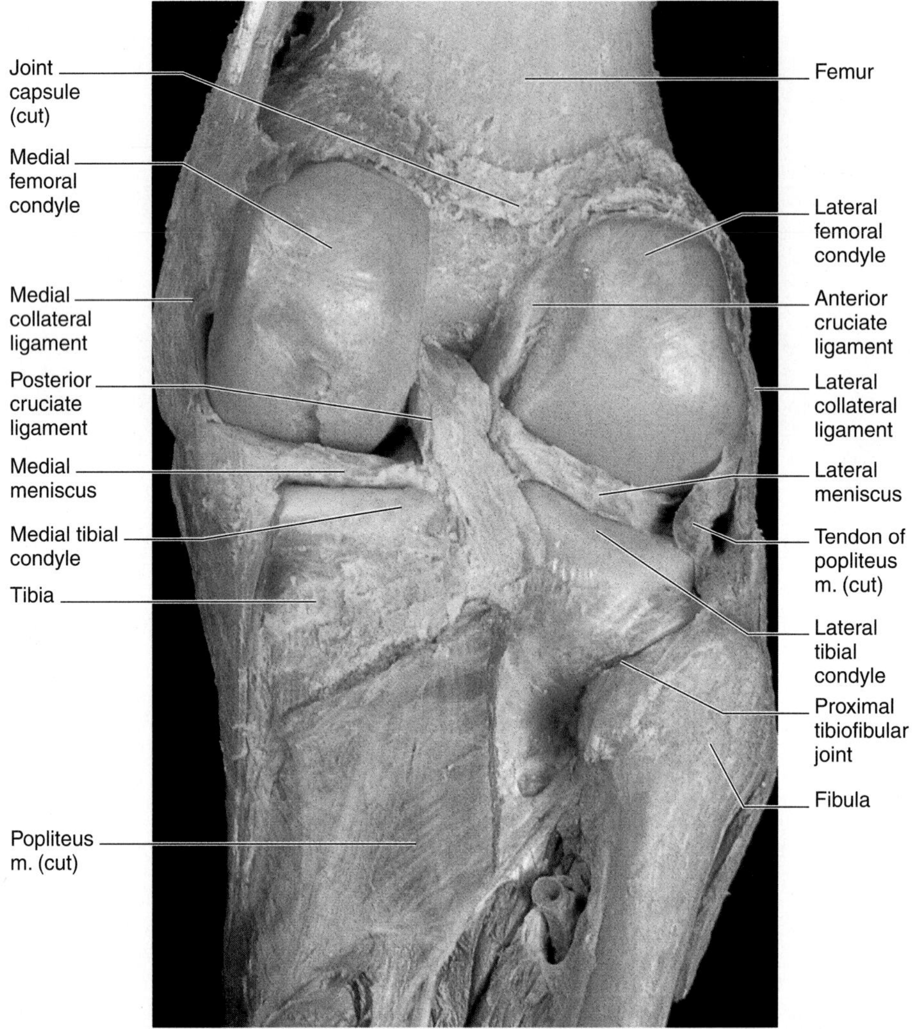

treated by bracing or casting the hip joints to promote suitable bone development.

Bursitis (*bur-si'tis*) is an inflammation of a bursa associated with a joint. Because bursae are close to joints, bursitis may affect a joint capsule as well. Bursitis may be caused by excessive stress on the bursa from overexertion, or it may be a local or systemic inflammatory process. As a bursa swells, the surrounding muscles become sore and stiff. **Tendonitis** involves inflammation of a tendon; it usually comes about in the same way as bursitis.

The flexible vertebral column is a marvel of mechanical engineering. Not only do the individual vertebrae articulate one with another, but together they form the portion of the axial skeleton with which the head, ribs, and ossa coxae articulate. The vertebral column also encloses the spinal cord and provides exits for 31 pairs of spinal nerves. Considering all the articulations in the vertebral column and the physical abuse it takes, it is no wonder that back ailments are second only to headaches as our most common physical complaint. Our way of life causes many of the problems associated with the vertebral column. Improper shoes, athletic exertion, sudden stops in vehicles, or improper lifting can all cause the back to go awry. Body weight, age, and general physical condition influence a person's susceptibility to back problems.

The most common cause of back pain is *strained muscles*, generally the result of overexertion. The second most frequent back ailment is a *herniated disc*. The dislodged nucleus pulposus of a disc may push against a spinal nerve and cause excruciating pain. The third most frequent back problem is a *dislocated articular facet* between two vertebrae, caused by sudden twisting of the vertebral column. The treatment of back ailments varies from bed rest to spinal manipulation to extensive surgery.

Figure 11.30

The right talocrural (ankle) joint.
(*a*) A lateral view, (*b*) a medial view, and (*c*) a posterior view.

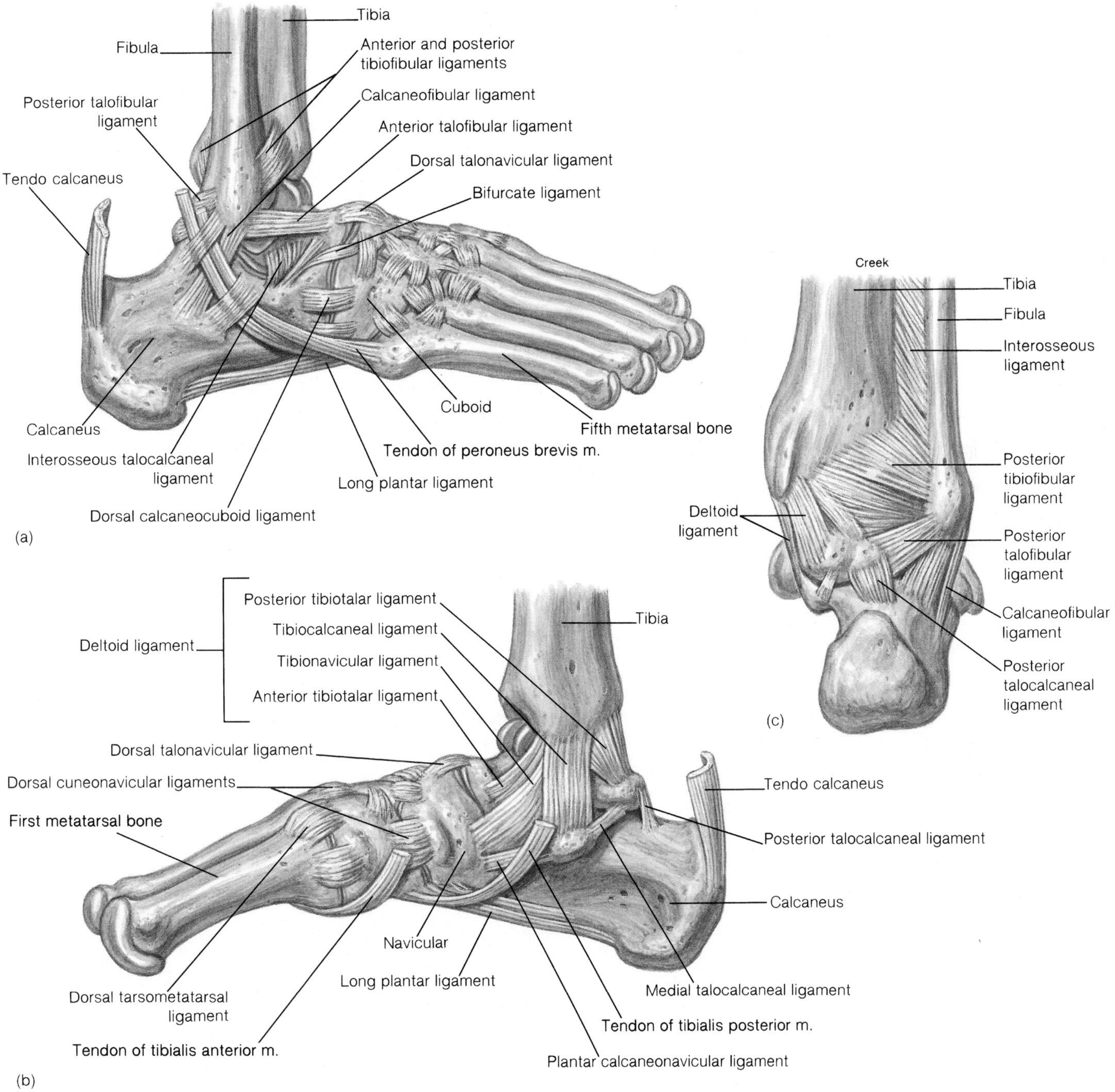

Curvature disorders are another problem of the vertebral column. **Kyphosis** (*ki-fo'sis*) (hunchback) is an exaggeration of the thoracic curve. **Lordosis (swayback) is an abnormal anterior convexity of the lumbar curve.** Scoliosis (*sko-le-o'sis*) (crookedness) is an abnormal lateral curvature of the vertebral column (fig. 11.31). It may be caused by abnormal vertebral structure, unequal length of the legs, or uneven muscular development on the two sides of the vertebral column.

kyphosis: Gk. *kyphos*, hunched
lordosis: Gk. *lordos*, curving forward
scoliosis: Gk. *skoliosis*, crookedness

Diseases of Joints

Arthritis is a generalized term for over 50 different joint diseases (fig. 11.32), all of which have the symptoms of edema, inflammation, and pain. The causes are unknown, but certain

Table 11.2

Principal Articulations

Joints	Type	Movement
Most skull joints	Fibrous (suture)	Immovable
Temporomandibular	Synovial (hinge; gliding)	Elevation, depression; protraction, retraction
Atlantooccipital	Synovial (condyloid)	Flexion, extension, circumduction
Atlantoaxial	Synovial (pivot)	Rotation
Intervertebral		
Bodies of vertebrae	Cartilaginous (symphysis)	Slight movement
Articular processes	Synovial (gliding)	Flexion, extension, slight rotation
Sacroiliac	Cartilaginous (gliding)	Slight gliding movement; may fuse in adult
Costovertebral	Synovial (gliding)	Slight movement during breathing
Sternocostal	Synovial (gliding)	Slight movement during breathing
Sternal	Cartilaginous (symphysis)	Slight movement during breathing
Acromioclavicular	Synovial (gliding)	Protraction, retraction; elevation, depression
Glenohumeral (shoulder)	Synovial (ball-and-socket)	Flexion, extension; adduction, abduction; rotation; circumduction
Elbow	Synovial (hinge)	Flexion, extension
Proximal radioulnar	Synovial (pivot)	Rotation
Distal radioulnar	Fibrous (syndesmosis)	Slight side-to-side movement
Radiocarpal (wrist)	Synovial (condyloid)	Flexion, extension; adduction, abduction; circumduction
Intercarpal	Synovial (gliding)	Slight movement
Carpometacarpal		
Fingers	Synovial (condyloid)	Flexion, extension; adduction, abduction
Thumb	Synovial (saddle)	Flexion, extension; adduction, abduction
Metacarpophalangeal	Synovial (condyloid)	Flexion, extension; adduction, abduction
Interphalangeal	Synovial (hinge)	Flexion, extension
Symphysis pubis	Fibrous (symphysis)	Slight movement
Coxal (hip)	Synovial (ball-and-socket)	Flexion, extension; adduction, abduction; rotation; circumduction
Tibiofemoral (knee)	Synovial (hinge)	Flexion, extension; slight rotation when flexed
Proximal tibiofibular	Synovial (gliding)	Slight movement
Distal tibiofibular	Fibrous (syndesmosis)	Slight movement
Talocrural (ankle)	Synovial (hinge)	Dorsiflexion, plantar flexion; slight circumduction; inversion, eversion
Intertarsal	Synovial (gliding)	Inversion, eversion
Tarsometatarsal	Synovial (gliding)	Flexion, extension; adduction, abduction

types follow joint trauma or bacterial infection. Some types are genetic, and others result from hormonal or metabolic disorders. The most common forms are *rheumatoid* (*roo'mă-toid*) *arthritis*, *osteoarthritis*, and *gouty arthritis*.

Rheumatoid arthritis results from an autoimmune attack against the joint tissues. The synovial membrane thickens and becomes tender, and synovial fluid accumulates. This is generally followed by deterioration of the articular cartilage, which eventually exposes bone tissue. When bone tissue is unprotected, joint ossification produces the crippling effect of this disease (fig. 11.32). Females are afflicted more often than males, and the disease usually begins between the ages of 30 and 50. Rheumatoid arthritis tends to occur bilaterally; if the right wrist or hip develops the disease, so does the left.

rheumatoid: Gk. *rheuma*, a flowing

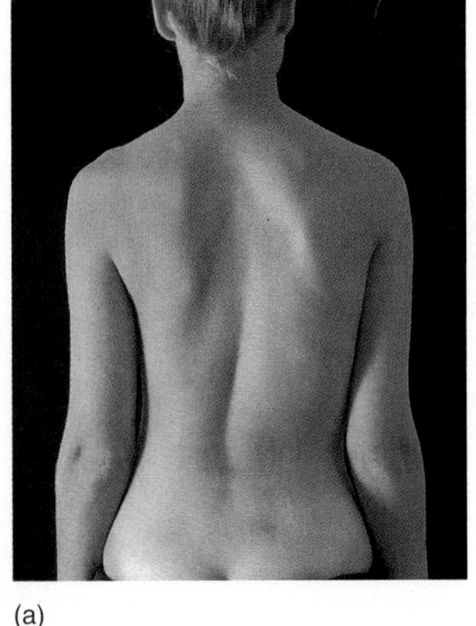
(a)

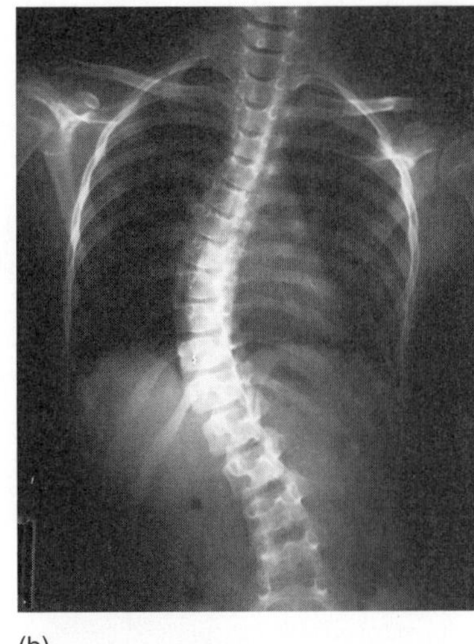
(b)

Figure 11.31

Scoliosis.
This condition is a lateral curvature of the spine, usually in the thoracic region. It may be congenital, acquired, or disease related. (*a*) A posterior view of a 19-year-old woman and (*b*) a radiograph.

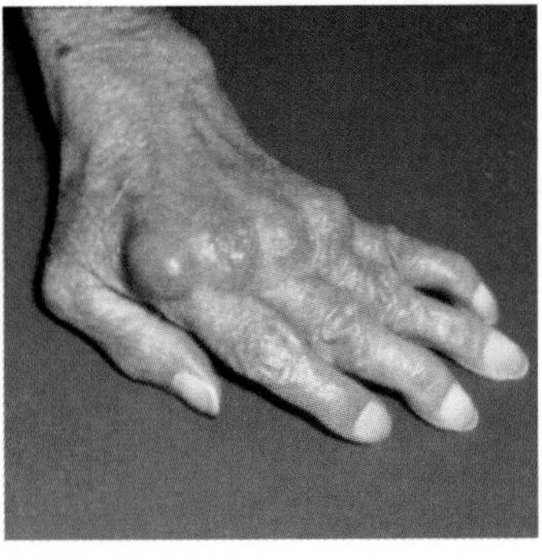
(a)

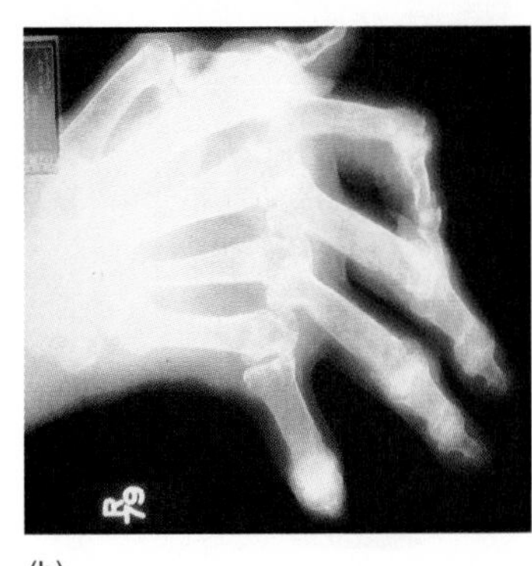
(b)

Figure 11.32

Rheumatoid arthritis.
This condition may eventually cause joint ossification and debilitation as seen in (*a*) a photograph of a patient's hand and (*b*) a radiograph.

Osteoarthritis is a degenerative joint disease that results from aging and irritation of the joints. Although osteoarthritis is far more common than rheumatoid arthritis, it is usually less damaging. Osteoarthritis is a progressive disease in which the articular cartilages gradually soften and disintegrate. The affected joints seldom swell, and the synovial membrane is rarely damaged. As the articular cartilage deteriorates, ossified spurs are deposited on the exposed bone, causing pain and restricting joint movement. Osteoarthritis most frequently affects the knee, hip, and intervertebral joints.

Gouty arthritis results from a metabolic disorder in which an abnormal amount of uric acid is retained in the blood, and sodium urate crystals are deposited in the joints. The salt crystals irritate the articular cartilage and synovial membrane, causing swelling, tissue deterioration, and pain. If gout is not treated, the affected joint fuses. Males have a greater incidence of gout than females, and apparently the disease is genetically determined. About 85% of gout cases affect the joints of the legs and feet. The most common joint affected is the metatarsophalangeal joint of the hallux (great toe).

Treatment of Joint Disorders

Arthroscopy (*ar-thros'kŏ-pe*) is widely used in diagnosing and, to a limited extent, treating joint disorders. Arthroscopic inspection involves making a small incision into the joint capsule and inserting a tubelike instrument called an arthroscope. In arthroscopy of the knee, the articular cartilage, synovial membrane, menisci, and cruciate ligaments can be observed. Samples can be extracted, and pictures can be taken for further evaluation.

Remarkable advancements have been made in the last 15 years in **joint prostheses** (*pros-the'sēz*) (fig. 11.33). These artificial articulations do not take the place of normal, healthy joints, but they are a valuable option for chronically disabled arthritis patients. They are now available for finger, shoulder, and elbow joints, as well as for hip and knee joints.

gout: L. *gutta*, a drop (thought to be caused by "drops of viscous humors")

prosthesis: Gk. *pros*, in addition to; *thesis*, a setting down

Figure 11.33

Examples of joint prostheses.
(*a*,*b*) The coxal (hip) joint and (*c*,*d*) the tibiofemoral (knee) joint.

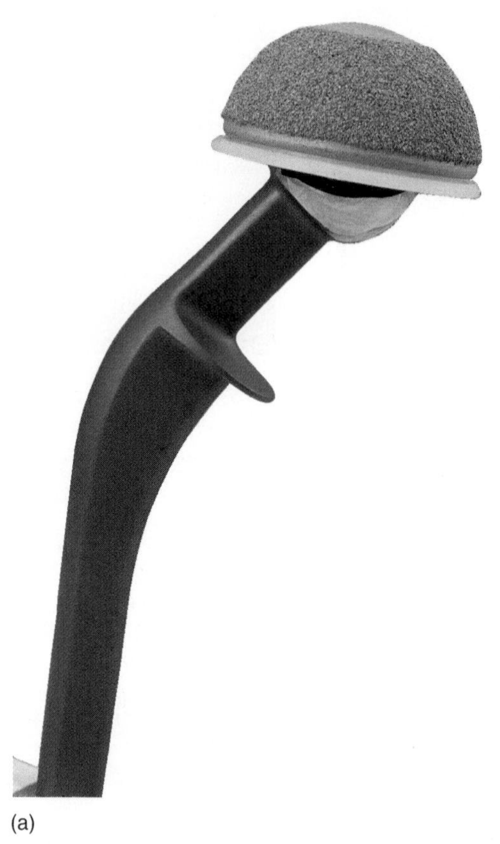
(a)

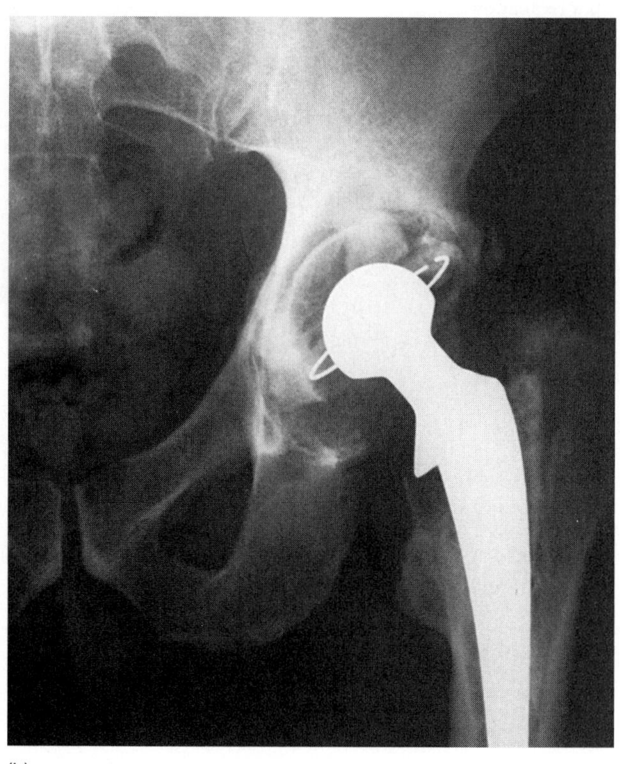
(b)

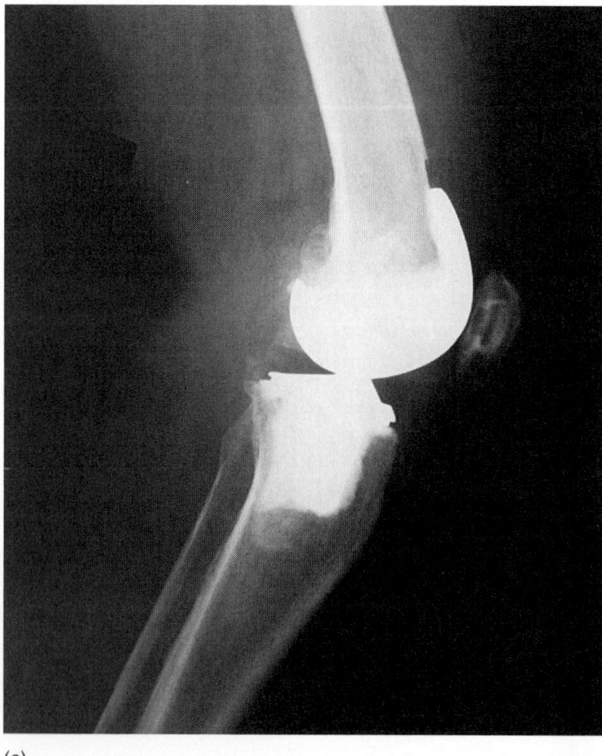
(c)

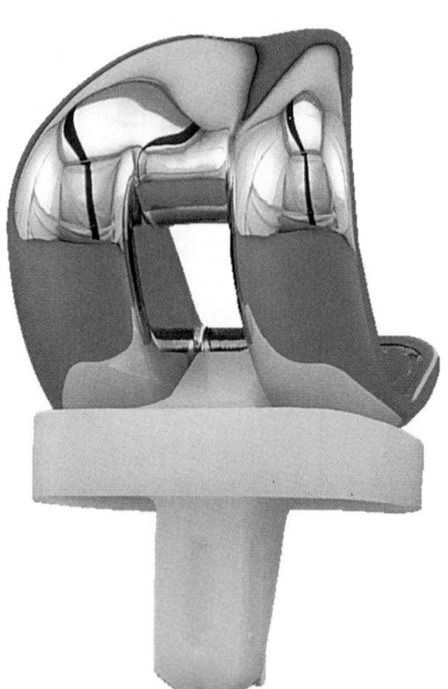
(d)

Important Clinical Terminology

ankylosis (*ang"ki-lo'sis*) Stiffening of a joint, resulting in severe or complete loss of movement.

arthralgia (*ar-thral'je-a*) Severe pain within a joint; also called *arthrodynia*.

arthrolith (*ar'thro-lith*) A gouty deposit in a joint.

arthrometry (*ar-throm'e-tre*) Measurement of the range of movement in a joint.

arthroncus (*ar-thron'kus*) Swelling of a joint as a result of trauma or disease.

arthropathy (*ar-throp'a-the*) Any disease affecting a joint.

arthroplasty (*ar'thro-plas"te*) Surgical repair of a joint.

arthrosis (*ar-thro'sis*) A joint or an articulation; also, a degenerative condition of a joint.

arthrosteitis (*ar"thros-te-itis*) Inflammation of the bony structure of a joint.

chondritis (*kon-dri'tis*) Inflammation of the articular cartilage of a joint.

coxarthrosis (*koks"ar-thro'sis*) A degenerative condition of the hip joint.

hemarthrosis (*hem-ar-thro'sis*) An accumulation of blood in a joint cavity.

rheumatology (*roo"ma-tol'o-je*) The medical speciality concerned with the diagnosis and treatment of rheumatic diseases.

spondylitis (*spon-dil-i'tis*) Inflammation of one or more vertebrae.

synovitis (*sin"o-vi'tis*) Inflammation of the synovial membrane lining the inside of a joint capsule.

Clinical Investigation Answer

The way in which the knee was injured and the location of the pain, taken together with the exam findings, indicate a complete or near-complete tear of the medial collateral ligament. Because the medial meniscus is attached to this ligament, it is frequently torn as well in an injury of this sort. Other ligaments susceptible to athletic injury are the anterior cruciate ligament (most common) and the lateral collateral and posterior cruciate ligaments. Complete tears of these ligaments usually require surgical repair for acceptable results. Incomplete tears can often be managed by nonsurgical means.

Chapter Summary

Classification of Joints (p. 248)

1. Joints are formed as adjacent bones articulate. Arthrology is the science concerned with the study of joints; kinesiology is the study of movements involving certain joints.
2. Joints are classified as fibrous, cartilaginous, or synovial.

Fibrous Joints (pp. 248–249)

1. Articulating bones in fibrous joints are tightly bound by fibrous connective tissue. Fibrous joints are of three types: sutures, syndesmoses, and gomphoses.
2. Sutures are found only in the skull; they are classified as serrate, lap, or plane.
3. Syndesmoses are found in the vertebral column, middle ear, antebrachium, and leg. The articulating bones of syndesmoses are held together by interosseous ligaments, which permit slight movement.
4. Gomphoses are found only in the skull, where the teeth are bound into their sockets by the periodontal ligaments.

Cartilaginous Joints (pp. 250–251)

1. The fibrocartilage or hyaline cartilage of cartilaginous joints allows limited motion in response to twisting or compression. The two types of cartilaginous joints are symphyses and synchondroses.
2. The symphysis pubis and the joints involving the intervertebral discs are examples of symphyses.
3. Some synchondroses are temporary joints formed in the growth plates (epiphyseal plates) between the diaphyses and epiphyses in the long bones of children. Other synchondroses are permanent (for example, the joints between the ribs and the costal cartilages of the rib cage).

Synovial Joints (pp. 251–256)

1. The freely movable synovial joints are enclosed by joint capsules that contain synovial fluid. Synovial joints include gliding, hinge, pivot, condyloid, saddle, and ball-and-socket types.
2. Synovial joints contain a joint cavity, articular cartilages, and synovial membranes that produce the synovial fluid. Some also contain articular discs, accessory ligaments, and associated bursae.
3. The movement of a synovial joint is determined by the structure of the articulating bones, the strength and tautness of associated ligaments and tendons, and the arrangement and tension of the muscles that act on the joint.

Movements at Synovial Joints (pp. 256–263)

1. Movements at synovial joints are produced by the contraction of the skeletal muscles that span the joints and attach to or near the bones forming the articulations. In these actions, the bones act as levers, the muscles provide the force, and the joints are the fulcra, or pivots.
2. Angular movements increase or decrease the joint angle produced by the articulating bones. Flexion decreases the joint angle on an anterior-posterior plane; extension increases the same joint angle. Abduction is the movement of a body part away from the main axis of the body; adduction is the movement of a body part toward the main axis of the body.
3. Circular movements can occur only where the rounded surface of one bone articulates with a corresponding depression on another bone. Rotation is the movement of a bone around its own axis. Circumduction is a conelike movement of a body part.
4. Special joint movements include inversion and eversion, protraction and retraction, and elevation and depression.
5. Synovial joints and their associated bones and muscles can be classified as first-, second-, or third-class levers. In a first-class lever, the fulcrum is positioned between the effort and the resistance. In a second-class lever, the resistance is positioned between the fulcrum and the effort. In a third-class lever, the effort lies between the fulcrum and the resistance.

Specific Joints of the Body (pp. 263–270)

1. The temporomandibular joint, a combined hinge and gliding joint, is of clinical importance because of temporomandibular joint (TMJ) syndrome.
2. The glenohumeral (shoulder) joint, a ball-and-socket joint, is vulnerable to dislocations from sudden jerks of the arm, especially in children before strong shoulder muscles have developed.

3. There are two sets of articulations at the elbow joint as the distal end of the humerus articulates with the proximal ends of the ulna and radius. It is a hinge joint that is subject to strain during certain sports.
4. The metacarpophalangeal joints (knuckles) are condyloid joints, and the interphalangeal joints (between adjacent phalanges) are hinge joints.
5. The ball-and-socket coxal (hip) joint is adapted for weight bearing. Its capsule is extremely strong and is reinforced by several ligaments.
6. The hinged tibiofemoral (knee) joint is the largest, most vulnerable joint in the body.
7. There are two hinged articulations within the talocrural (ankle) joint. Sprains are frequently associated with this joint.

Review Activities

Objective Questions

1. Which of the following statements regarding joints is *false?*
   (a) They are places where two or more bones articulate.
   (b) All joints are movable.
   (c) Arthrology is the study of joints; kinesiology is the study of the biomechanics of joint movement.
2. Synchondroses are a type of
   (a) fibrous joint.
   (b) synovial joint.
   (c) cartilaginous joint.
3. An interosseous ligament is characteristic of
   (a) a suture. (c) a symphysis.
   (b) a synchondrosis. (d) a syndesmosis.
4. Which of the following joint type–function word pairs is *incorrect?*
   (a) synchondrosis/growth at the epiphyseal plate
   (b) symphysis/movement at the intervertebral joint
   (c) suture/strength and stability in the skull
   (d) syndesmosis/movement of the jaw
5. Which of the following statements is *false?*
   (a) Synchondroses occur only in children and young adults.
   (b) Sutures occur only in the skull.
   (c) Saddle joints occur in the thumb and in the neck, where rotational movement is possible.
   (d) Syndesmoses occur in the antebrachium and leg.
6. Which of the following is *not* characteristic of all synovial joints?
   (a) articular cartilage
   (b) synovial fluid
   (c) a joint capsule
   (d) a meniscus
7. The atlantoaxial and the proximal radioulnar synovial joints are specifically classified as
   (a) hinge. (c) pivotal.
   (b) gliding. (d) condyloid.
8. Which of the following joints can be readily and comfortably hyperextended?
   (a) an interphalangeal joint
   (b) a coxal joint
   (c) a tibofemoral joint
   (d) a sternocostal joint
9. Which of the following joints is most vulnerable to luxation?
   (a) the elbow joint
   (b) the glenohumeral joint
   (c) the coxal joint
   (d) the tibiofemoral joint
10. A thickening and tenderness of the synovial membrane and the accumulation of synovial fluid are signs of the development of
   (a) arthroscopitis.
   (b) gouty arthritis.
   (c) osteoarthritis.
   (d) rheumatoid arthritis.

Essay Questions

1. What is meant by a structural classification of joints, as compared to a functional classification?
2. Why is the anatomical position so important in explaining the movements that are possible at joints?
3. What are the structural components of a synovial joint that determine the range of movement at that joint?
4. What are the advantages of a hinge joint over a ball-and-socket type? If ball-and-socket joints allow a greater range of movement, why are not all the synovial joints of this type?
5. What is synovial fluid? Where is it produced, and what are its functions?
6. Describe a bursa and discuss its function. What is bursitis?
7. Identify four types of synovial joints found in the wrist and hand region and state the types of movement permitted by each.

Figure 11.34
Joint flexion.
Which joints of the body are flexed as a person assumes a fetal position?

8. Discuss the articulations of the pectoral and pelvic regions to the axial skeleton with regard to range of movement, ligamentous attachments, and potential clinical problems.
9. What is meant by a sprained ankle? How does a sprain differ from a strain or a luxation?
10. What occurs within the joint capsule in rheumatoid arthritis? How does rheumatoid arthritis differ from osteoarthritis?

Critical Thinking Questions

1. Refer to figure 11.34 and identify the joints being flexed in the upper and lower extremities. In your own body, which are larger and stronger, the flexor muscles or the extensor muscles? Why?
2. Considering the type of synovial joint at the hip and the location of the gluteal muscles of the buttock, explain why this type of lever system is adapted for rapid, wide-ranging movements.
3. The star runningback of a local high school football team was taken to the emergency room of the local hospital following a knee injury during the championship game. The injury resulted from a hard blow to the back of his right knee ("clipping") as it was supporting the weight of his body. Suspecting a rupture of the anterior cruciate ligament, the ER physician informed the football player that this diagnosis could be confirmed by pulling the tibia forward as the knee was flexed. He explained that if the tibia slipped forward at the knee ("bureau drawer sign"), it could be assumed that the anterior cruciate ligament was ruptured.

   In terms of the anatomy of the knee joint, explain the occurrence of bureau drawer sign. What structure most likely would be traumatized if the tibia could be displaced backward?
4. In what ways do the anatomical differences between the jaw, shoulder, elbow, hip, knee, and ankle joints relate to their differences in function?

Related Web Sites

For a listing of the most current web sites related to this chapter, please visit the *Concepts of Human Anatomy and Physiology* home page at http://www.mhhe.com/biosci/abio/.

TWELVE

Muscle Tissue and Muscle Physiology

OBJECTIVES

- Explain how muscle fibers and connective tissue are arranged within skeletal muscles and describe the banding pattern of skeletal muscle fibers.
- Describe the nature of a muscle twitch and explain how summation and tetanus are produced.
- Distinguish between isometric and isotonic contractions and discuss the significance of the series-elastic component of muscles.
- Discuss the relationship between somatic motor neurons and skeletal muscles, noting the significance of motor units.
- Describe the sliding filament mechanism of contraction, noting how the bands in a muscle fiber change during contraction.
- Describe the cross-bridge cycle and the role of ATP in muscle contraction and muscle relaxation.
- Describe the function of actin, myosin, troponin, and tropomyosin in muscle contraction.
- Discuss the role of Ca^{2+} in muscle contraction and explain how electrical stimulation influences the availability of Ca^{2+}.
- Define *maximal oxygen uptake* and *oxygen debt*.
- Describe the role of phosphocreatine in muscle contraction.
- Distinguish between fast-twitch and slow-twitch fibers and explain how muscles adapt to exercise training.
- Compare and contrast smooth muscle and skeletal muscle with respect to structure and contractile mechanisms.
- Explain how the mechanism of contraction is regulated in smooth muscle.

Clinical Investigation

A woman who had been athletically active for most of her life complained that she was experiencing fatigue and muscle pain. Upon exercise testing, it was found that she had a high maximal oxygen uptake. Her muscles were not large, but appeared to be well toned—perhaps excessively so. Laboratory tests revealed that this woman had a normal blood concentration of creatine phosphokinase but an elevated blood Ca^{2+} concentration. She had a history of hypertension, and was currently taking a calcium-channel-blocking drug for this condition. What might be responsible for her fatigue and muscle pain?

Clues: Review the material on maximal oxygen uptake and on oxygen debt, and study the section "Adaptation of Muscles to Exercise Training." Look at the boxed information on calcium channel blockers in the section on smooth muscles. Also read the boxed calcium channel information under the heading "Phosphocreatine" and study the information on the role of Ca^{2+} in muscle contraction.

Structure and Actions of Skeletal Muscles

Skeletal muscles are composed of individual muscle fibers that contract when stimulated by motor neurons. Each motor neuron branches to innervate a number of muscle fibers. Activation of varying numbers of motor neurons results in gradations in the strength of muscle contraction.

As described in chapter 6 and summarized in table 12.1, there are three types of muscle tissue: skeletal, cardiac, and smooth. Skeletal and cardiac muscle cells are striated, in contrast to the nonstriated cells of smooth muscle. Our focus in this chapter is on muscle structure and function; a detailed description of the anatomy of the muscular system is presented in chapter 13.

Attachment of Muscles

Skeletal muscles are usually attached to bone on each end by tough connective tissue *tendons*. When a muscle contracts it shortens, and this places tension on its tendons and attached bones. The muscle tension causes movement of the bones at a joint, where one of the articulating bones generally moves more than the other. The more movable bony attachment of the muscle, known as its **insertion,** is pulled toward its **origin,** which is its less-movable attachment.

Specialized tendons are identified by specific names. Flattened, sheetlike tendons, for example, are called **aponeuroses** (*ap″ŏ-noo-ro′sēz*). An example is the galea aponeurotica over the skull (see fig. 13.4). In certain places, tendons are enclosed by protective sheaths that lubricate the tendons with synovial fluid (see fig. 11.6). In the ankle (see fig. 13.33) and in the wrist, the entire group of tendons is contained in place by a thin but strong band of connective tissue called a **retinaculum** (*ret″ĭ-nak′yoo-lum*).

Associated Connective Tissue

Contracting muscle fibers would be ineffective if they worked as isolated units. Each fiber is bound to adjacent fibers to form bundles, and the muscle bundles in turn are bound to other muscle bundles. In this arrangement, the contraction of muscle fibers in one area of a muscle works in conjunction with contracting fibers elsewhere in the muscle. The binding structures within muscles are the *associated connective tissues*.

A fibrous connective tissue called **fascia** (*fash′e-ă*) is found under the skin and binds adjacent muscles together. Fascia may be categorized as superficial or deep. *Superficial fascia* is the tissue that secures the hypodermis of the skin to the underlying muscles, and it varies in thickness throughout the body. For example, superficial fascia over the buttock and anterior abdominal wall is thick and laced with adipose tissue. By contrast, the superficial fascia under the skin of the dorsum of the hand and facial region is thin. *Deep fascia* is an inward extension of the superficial fascia. It occurs between individual muscles and also surrounds adjacent muscles to bind them into functional groups. Deep fascia generally lacks adipose tissue.

Surrounding each muscle is a connective tissue sheath known as the **epimysium** (*ep″ĭ-mis′e-um*) (fig. 12.1). The fibers of this sheath are continuous with those of the tendons. Additionally, the connective tissue fibers from the epimysium extend into the body of the muscle, subdividing it into bundles. These subdivisions within the muscle are known as **fasciculi** (*fă-sik′yŭ-li*), and are the "strings" in stringy meat. Each fasciculus is surrounded by its own connective tissue sheath, known as the **perimysium.**

Dissection of a muscle fasciculus under a microscope reveals that it, in turn, is composed of many **muscle fibers** (or *myofibers*) surrounded by wisps of connective tissue called **endomysium** (fig. 12.1). Since the connective tissue of the tendons, epimysium, perimysium, and endomysium is continuous, muscle fibers do not normally pull out of the tendons when they contract.

Skeletal Muscle Fibers

The muscle fibers are actually the cells of the muscle. Despite their unusual elongated shape, muscle fibers have the same organelles that are present in other cells: mitochondria,

---

aponeurosis: Gk. *aponeurosis*, change into a tendon

retinaculum: L. *retinere*, to hold back (retain)
fascia: L. *fascia*, band or girdle
epimysium: Gk. *epi*, upon; *myos*, muscle
fasciculus: L. *fascis*, bundle
perimysium: Gk. *peri*, around; *myos*, muscle

Table 12.1

Summary of Muscle Tissue

Type	Structure and Function	Location
Smooth	Elongated, spindle-shaped fiber with single nucleus; involuntary movements of internal organs	Walls of hollow internal organs 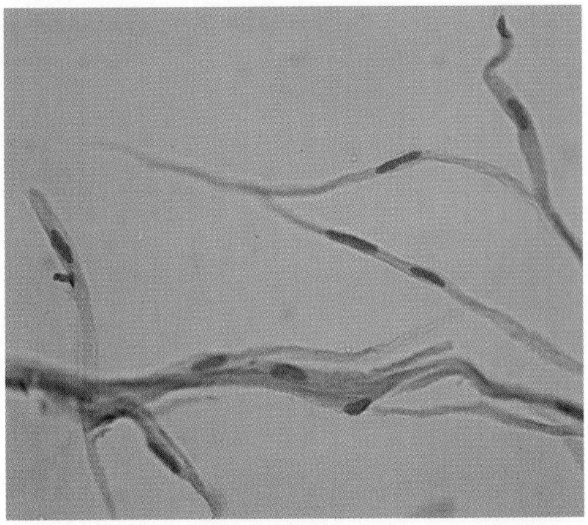
Cardiac	Branched, striated fiber with single nucleus and intercalated discs; involuntary rhythmic contraction	Heart muscle 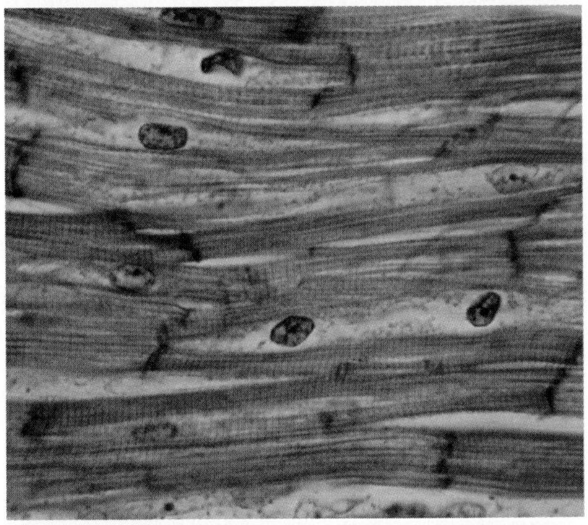
Skeletal	Multinucleated, striated, cylindrical fiber; voluntary movement of skeletal parts	Spanning joints and attached to bones of the skeleton 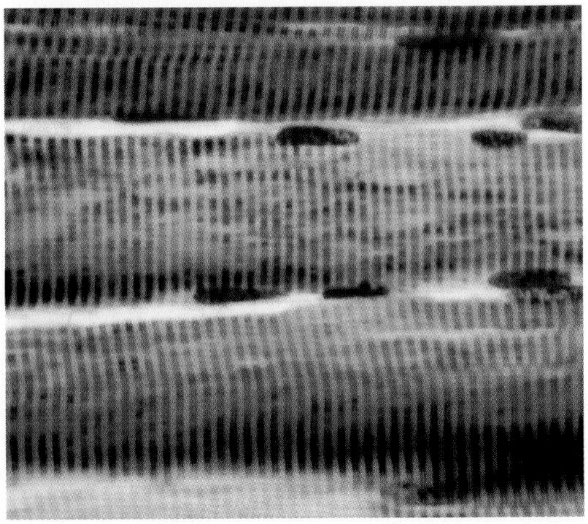

Figure 12.1

The structure of a skeletal muscle.
The relationship between muscle fibers and the connective tissues of the tendon, epimysium, perimysium, and endomysium is depicted in the upper figure (*a*, *b*, *c*). Below (*d*) is a close-up of a single muscle fiber.

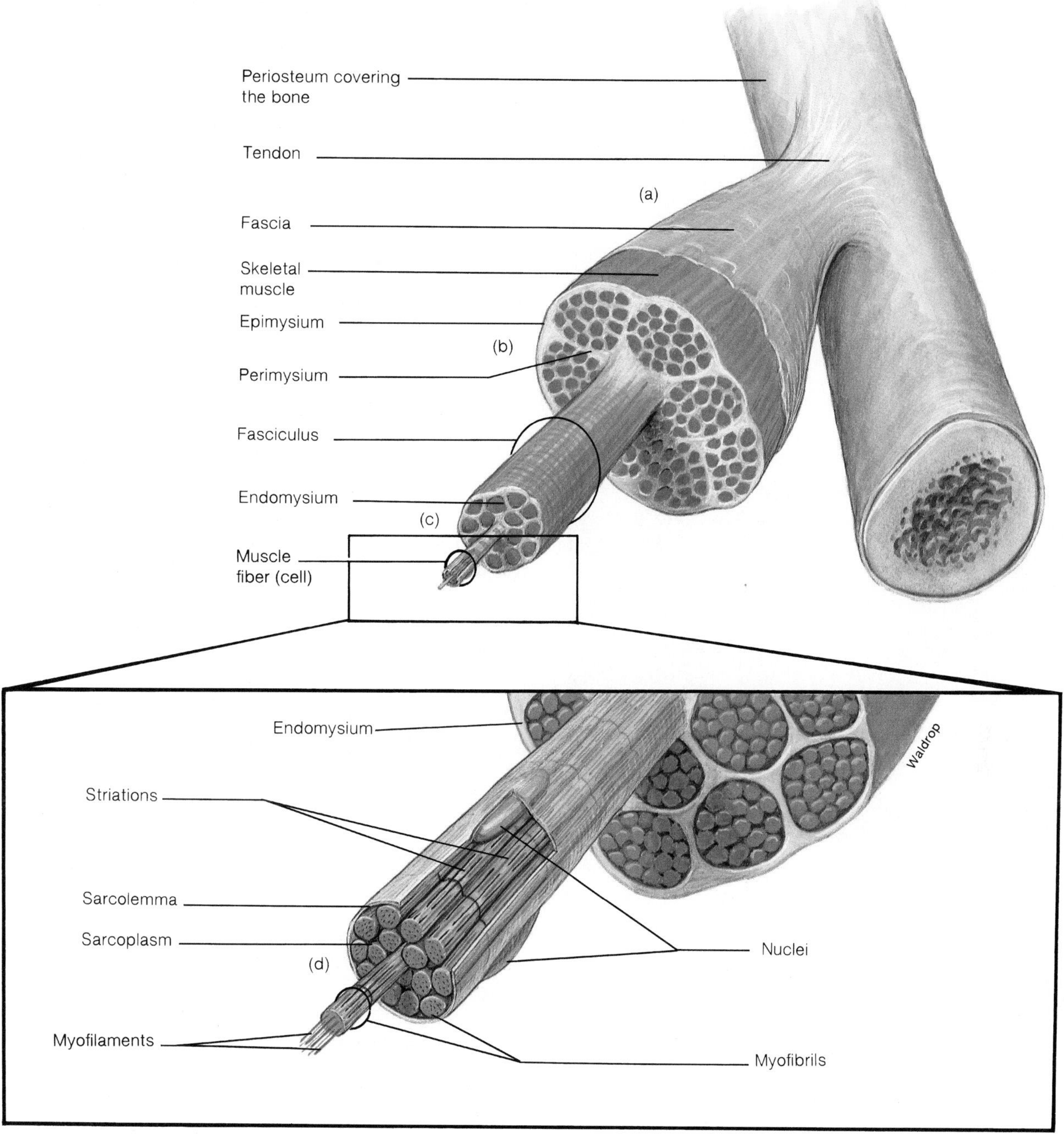

intracellular membranes, glycogen granules, and others. The most distinctive feature of skeletal muscle fibers, however, is their *striated appearance* when viewed microscopically (fig. 12.2). The striations (stripes) are produced by alternating dark and light bands that appear to span the width of the fiber.

The dark bands are called **A bands** and the light bands are called **I bands.** At high magnification in an electron microscope, thin dark lines can be seen in the middle of the I bands. These are called **Z lines.** The labels A, I, and Z are useful for describing the functional architecture of muscle fibers, and were derived during the history of muscle research. The letters A and *I* stand for *anisotropic* and *isotropic*, respectively, which indicate the behavior of polarized light as it passes through these regions; the letter Z comes from the German word *Zwischenscheibe*, which translates to "between disc." These derivations are of historical interest only.

Figure 12.2

The structure of a skeletal muscle fiber.
(*a*) A skeletal muscle fiber is composed of numerous threadlike strands of myofibrils that contain myofilaments of actin and myosin. Skeletal muscle fibers are striated and multinucleated. (*b*) A light micrograph of skeletal muscle fibers showing the striations and the peripheral location of the nuclei.

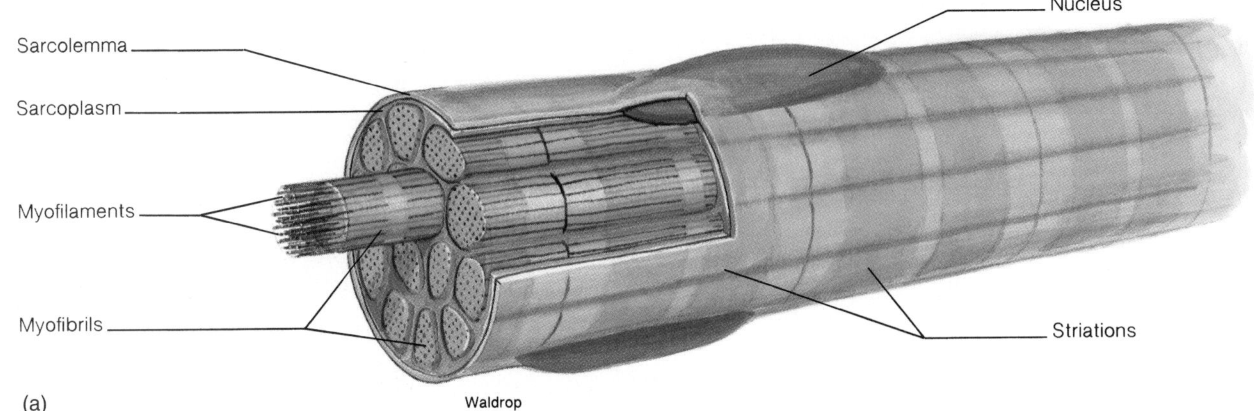

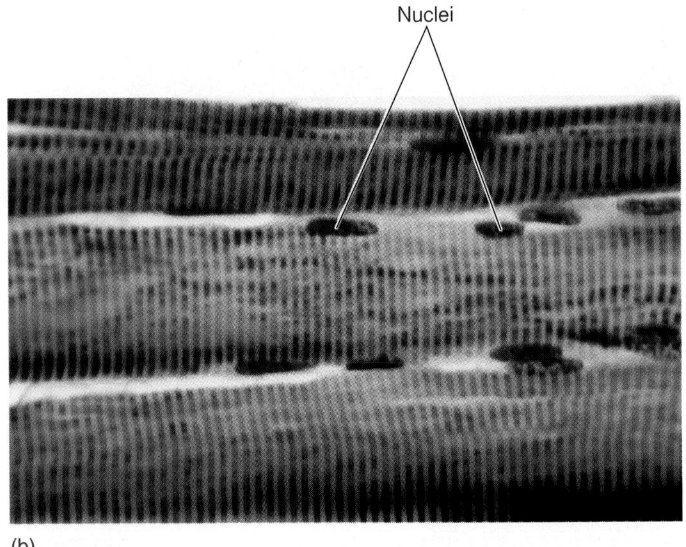

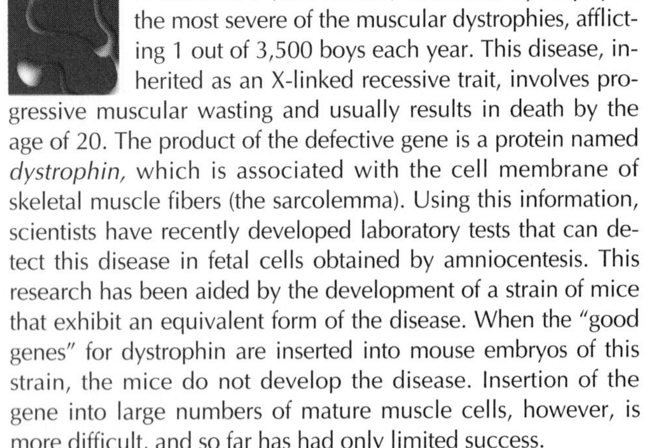

Duchenne's (*doo-shenz'*) *muscular dystrophy* is the most severe of the muscular dystrophies, afflicting 1 out of 3,500 boys each year. This disease, inherited as an X-linked recessive trait, involves progressive muscular wasting and usually results in death by the age of 20. The product of the defective gene is a protein named *dystrophin,* which is associated with the cell membrane of skeletal muscle fibers (the sarcolemma). Using this information, scientists have recently developed laboratory tests that can detect this disease in fetal cells obtained by amniocentesis. This research has been aided by the development of a strain of mice that exhibit an equivalent form of the disease. When the "good genes" for dystrophin are inserted into mouse embryos of this strain, the mice do not develop the disease. Insertion of the gene into large numbers of mature muscle cells, however, is more difficult, and so far has had only limited success.

Duchenne's muscular dystrophy: from Guillaume B. A. Duchenne, French neurologist, 1806–75.

Types of Muscle Contractions

The contractile behavior of skeletal muscles is more easily studied *in vitro* (outside the body) than *in vivo* (within the body). When a muscle—for example, the gastrocnemius (calf muscle) of a frog—is studied in vitro, it is usually mounted so that one end is fixed and the other is movable. The mechanical force of the muscle contraction is transduced (changed) into an electric current, which can be amplified and displayed as pen deflections in a multichannel recorder (fig. 12.3). In this way, the contractile behavior of the whole muscle in response to experimentally administered electric shocks can be studied.

Twitch, Summation, and Tetanus

When the muscle is stimulated with a single electric shock of sufficient voltage, it quickly contracts and relaxes. This response is called a **twitch.** Increasing the stimulus voltage increases the strength of the twitch up to a maximum. The strength of a muscle contraction can thus be *graded*, or varied—an obvious requirement for the proper control of skeletal movements. If a second electric shock is delivered immediately after the first, it will produce a second twitch that may partially "ride piggyback" on the first. This response is called **summation.**

Stimulation of fibers within a muscle in vitro with an electric stimulator, or in vivo by motor axons, usually results in the full contraction of the individual fibers. Stronger muscle contractions are produced by the stimulation of greater numbers of muscle fibers. Skeletal muscles can thus produce **graded contractions,** the strength of which depends on the number of fibers stimulated rather than on the strength of the contractions of individual muscle fibers.

If the stimulator is set to deliver an increasing frequency of electric shocks automatically, the relaxation time between successive twitches will get shorter and shorter as the strength of contraction increases in amplitude. This effect is known as **incomplete tetanus** (*tet'n-us*). Finally, at a particular "fusion

UNDER DEVELOPMENT

Development of Skeletal Muscles

The formation of skeletal muscle tissue begins during the fourth week of embryonic development as specialized mesodermal cells, called **myoblasts,** begin rapid mitotic division (fig. 1). The proliferation of new cells continues while the myoblast cells migrate and fuse together into **syncytial** (*sin-sish'al*) **myotubes.** A *syncytium* is a multinucleated protoplasmic mass formed by the union of originally separate cells. At 9 weeks, primitive myofilaments course through the myotubes, and the nuclei of the contributing myoblasts are centrally located. Growth in length continues through the addition of myoblasts.

It is not certain when skeletal muscle is sufficiently developed to sustain contractions, but by week 17 fetal movements known as *quickening* are strong enough to be recognized by the mother. The individual muscle fibers have now thickened, the nuclei have moved peripherally, and the myofilaments can be recognized as alternating dark and light bands. Growth in length still continues through the addition of myoblasts. Shortly before a baby is born, the formation of myoblast cells ceases. At this time, all of the muscle cells have been determined.

---

myoblast: Gk. *myos*, muscle; *blastos*, germ
syncytial: Gk. *syn*, with; *cyto*, cell

Figure 1 The development of skeletal muscle fibers. (*a*) At 5 weeks, the myotube is formed as individual cell membranes are broken down. Myotubes grow in length by incorporating additional myoblasts; each adds an additional nucleus. (*b*) Muscle fibers are distinct at 9 weeks, but the nuclei are still centrally located, and growth in length continues through the addition of myoblasts. (*c*) At 5 months, thin (actin) and thick (myosin) filaments are present, and moderate growth in length is still occurring. (*d*) By birth, the striated myofilaments have aggregated into bundles, the fiber has thickened, and the nuclei have shifted to the periphery. Myoblast activity ceases, and all the muscle fibers a person will have are formed.

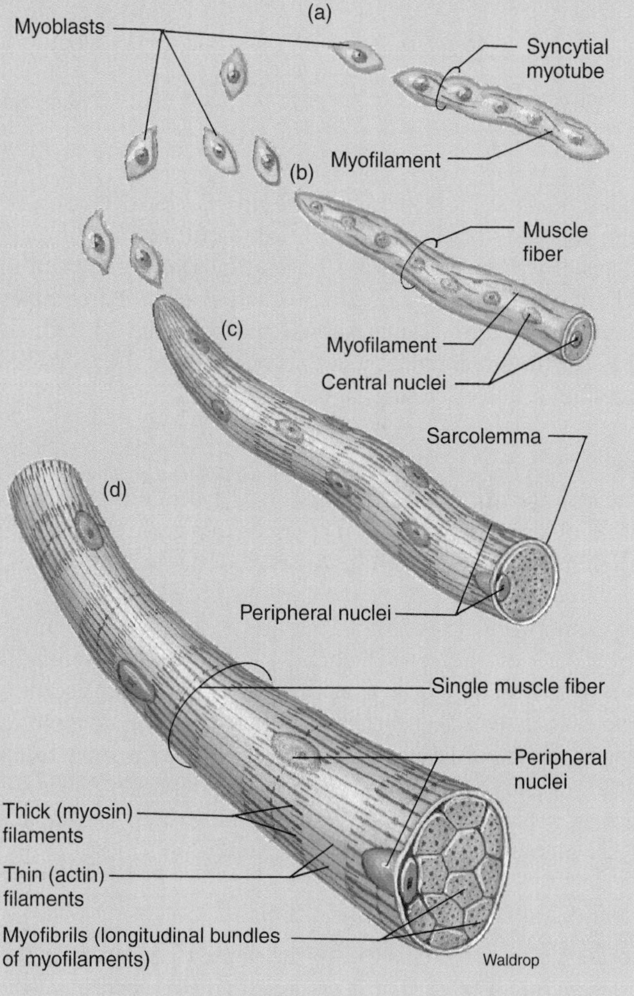

Figure 12.3

Recording muscle contractions.
(*a*) A physiograph recorder. (*b*) A photograph and (*c*) an illustration of the behavior of an isolated gastrocnemius muscle of a frog in response to electrical shocks.

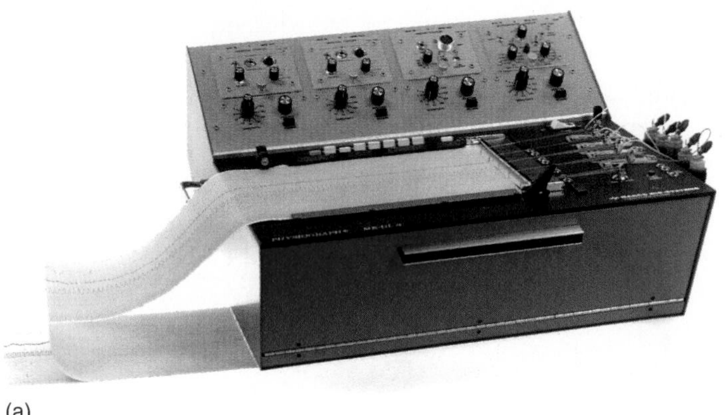

(a)

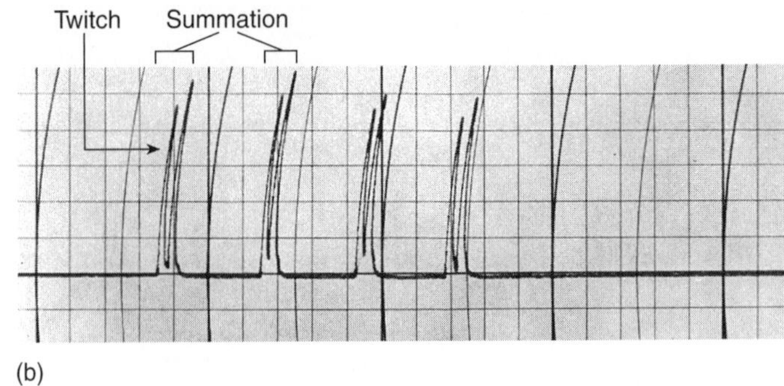

(b)

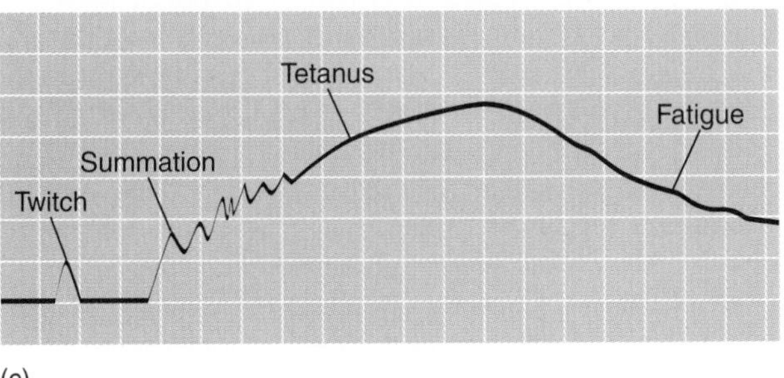

(c)

frequency" of stimulation, there is no visible relaxation between successive twitches (fig. 12.3). Contraction is smooth and sustained, as it is during normal muscle contraction in vivo. This smooth, sustained contraction is called **complete tetanus.** (The term *tetanus* should not be confused with the disease of the same name, which is accompanied by a painful state of muscle contracture, or *tetany*.)

Treppe

If the voltage of the electrical shocks delivered to an isolated muscle in vitro is gradually increased from zero, the strength of the muscle twitches will increase accordingly, up to a maximal value at which all of the muscle fibers are stimulated. This demonstrates the graded nature of the muscle contraction. If a series of electrical shocks at this maximal voltage is given to a fresh muscle so that each shock produces a separate twitch, each of the twitches evoked will be successively stronger, up to a higher maximum. This demonstrates **treppe** (*trep'e*), or the *staircase effect*. Treppe may represent a warmup effect and is believed to be due to an increase in intracellular Ca^{2+}, which is needed for muscle contraction.

Isotonic and Isometric Contractions

In order for muscle fibers to shorten when they contract, they must generate a force that is greater than the opposing forces that act to prevent movement of the muscle's insertion. When a weight is lifted by flexing the elbow joint, for example, the force produced by contraction of the biceps brachii muscle is greater than the force of gravity on the object being lifted (fig. 12.4). The tension produced separately by the contraction of each muscle fiber is insufficient to overcome the opposing force, but the combined contractions of numerous muscle fibers may be sufficient to overcome the opposing

Figure 12.4

Photograph of isometric and isotonic contractions.
(*a*) An isometric contraction, in which the muscle stays the same length and (*b*) an isotonic contraction, in which the muscle shortens.

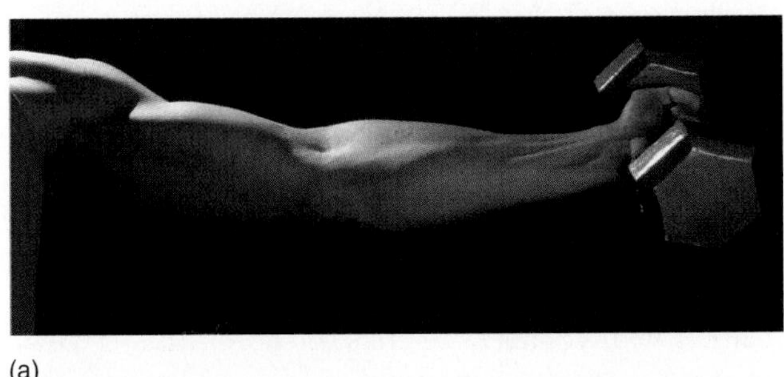

(a)

(b)

force and flex the forearm. In this case, the muscle and all of its fibers shorten in length.

Contraction that results in muscle shortening is called **isotonic contraction,** so-called because the force of contraction remains relatively constant throughout the shortening process. If the opposing forces are too great, however, or if the number of muscle fibers activated is too few to shorten the muscle, the contraction is called an **isometric** (literally, "same length") **contraction.**

Isometric contraction can be voluntarily produced, for example, by lifting a weight and maintaining the forearm in a partially flexed position. We can then increase the amount of muscle tension produced by recruiting more muscle fibers until the muscle begins to shorten; at this point, isometric contraction is converted to isotonic contraction.

Series-Elastic Component

In order for a muscle to shorten when it contracts, and thus to move its insertion toward its origin, the noncontractile parts of the muscle and the connective tissue of its tendons must first be pulled tight. These structures, particularly the tendons, have elasticity—they resist distension, and when the distending force is released, they tend to spring back to their resting lengths. Tendons provide what is called a **series-elastic component** because they are somewhat elastic and in line (in series) with the force of muscle contraction. The series-elastic component absorbs some of the tension as a muscle contracts, and it must be pulled tight before muscle contraction can result in muscle shortening.

When the gastrocnemius muscle was stimulated with a single electric shock as described earlier, the amplitude of the twitch was reduced because some of the force of contraction was used to stretch the series-elastic component. Quick delivery of a second shock thus produced a greater degree of muscle shortening than the first shock, culminating at the fusion frequency of stimulation with complete tetanus, in which the strength of contraction was greater than that of individual twitches.

Some of the energy used to stretch the series-elastic component during muscle contraction is released by elastic recoil when the muscle relaxes. This elastic recoil, which helps the muscles return to their resting length, is of particular importance for the muscles involved in breathing. As we will see in chapter 24, inspiration is produced by muscle contraction, and expiration is produced by the elastic recoil of the thoracic structures that were stretched during inspiration.

Motor Units

In vivo, each muscle fiber receives a single axon terminal from a somatic motor neuron. The motor neuron stimulates the muscle fiber to contract by liberating a chemical, *acetylcholine* (*ă-sēt″l-ko′lēn*), at the neuromuscular junction (described in chapter 14). The specialized region of cell membrane of the muscle fiber at the neuromuscular junction is known as a **motor end plate** (fig. 12.5). The cell body of a somatic motor neuron is located in the ventral horn of the gray matter of the spinal cord and gives rise to a single axon that emerges in the ventral root of a spinal nerve (see chapter 16). Each axon, however, can produce a number of collateral branches to innervate an equal number of muscle fibers. Each somatic motor neuron, together with all of the muscle fibers that it innervates, is known as a **motor unit** (fig. 12.6).

Whenever a somatic motor neuron is activated, all of the muscle fibers that it innervates are stimulated to contract with all-or-none twitches. In vivo, graded contractions of whole muscles are produced by variations in the number of motor units that are activated. In order for these graded contractions to be smooth and sustained, as in complete tetanus, different motor units must be activated by rapid, asynchronous stimulation.

Fine neural control over the strength of muscle contraction is optimal when there are many small motor units involved. In the extraocular muscles that position the eyes, for example, the *innervation ratio* (motor neuron:muscle fibers) of an average motor unit is 1 neuron per 23 muscle fibers. This affords a fine degree of control. The innervation ratio of the gastrocnemius, by contrast, averages 1 neuron per 1,000 muscle fibers. Stimulation of these motor units results in more powerful contractions at the expense of finer gradations in contraction strength.

All of the motor units controlling the gastrocnemius, however, are not the same size. Innervation ratios vary from 1:100 to 1:2,000. A neuron that innervates fewer muscle fibers has a smaller cell body and is stimulated by lower levels of excitatory input than a larger neuron that innervates a greater number of muscle fibers. The smaller motor units, as a result, are the ones that are used most often. When contractions of greater strength are required, larger and larger motor units are activated in a process known as **recruitment** of motor units.

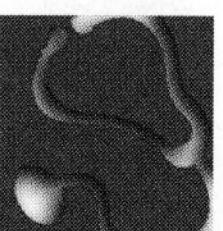

The disease known as *amyotrophic lateral sclerosis (ALS)* involves degeneration of the somatic motor neurons, leading to muscle paralysis. This disease is sometimes called *Lou Gehrig's disease,* named after the baseball player who suffered from it, and also includes the famous physicist Steven Hawking among its victims. Scientists have recently learned that the inherited form of this disease is caused by a defect in the gene for a specific enzyme—*superoxide dismutase.* This enzyme is responsible for eliminating superoxide free radicals, which are highly toxic products that can damage the motor neurons. The mutant gene produces an enzyme that has a different, and in fact destructive, action. Although most cases of ALS are not inherited, the hereditary and nonhereditary forms of the disease are clinically the same and may therefore be treatable through the same means.

---

isotonic: Gk. *isos*, equal; *tonos*, tension
isometric: Gk. *isos*, equal; *metron*, measure

Figure 12.5

Motor end plates at the neuromuscular junction.

A neuromuscular junction is the site where the nerve fiber and a muscle fiber meet. The motor end plate is the specialized portion of the sarcolemma of a muscle fiber surrounding the terminal end of the axon. (*a*) An illustration of the neuromuscular junction. Notice the slight gap between the membrane of the axon and that of the muscle fiber. (*b*) A photomicrograph of muscle fibers and neuromuscular junctions. A motor neuron and the muscle fibers it innervates constitute a motor unit.

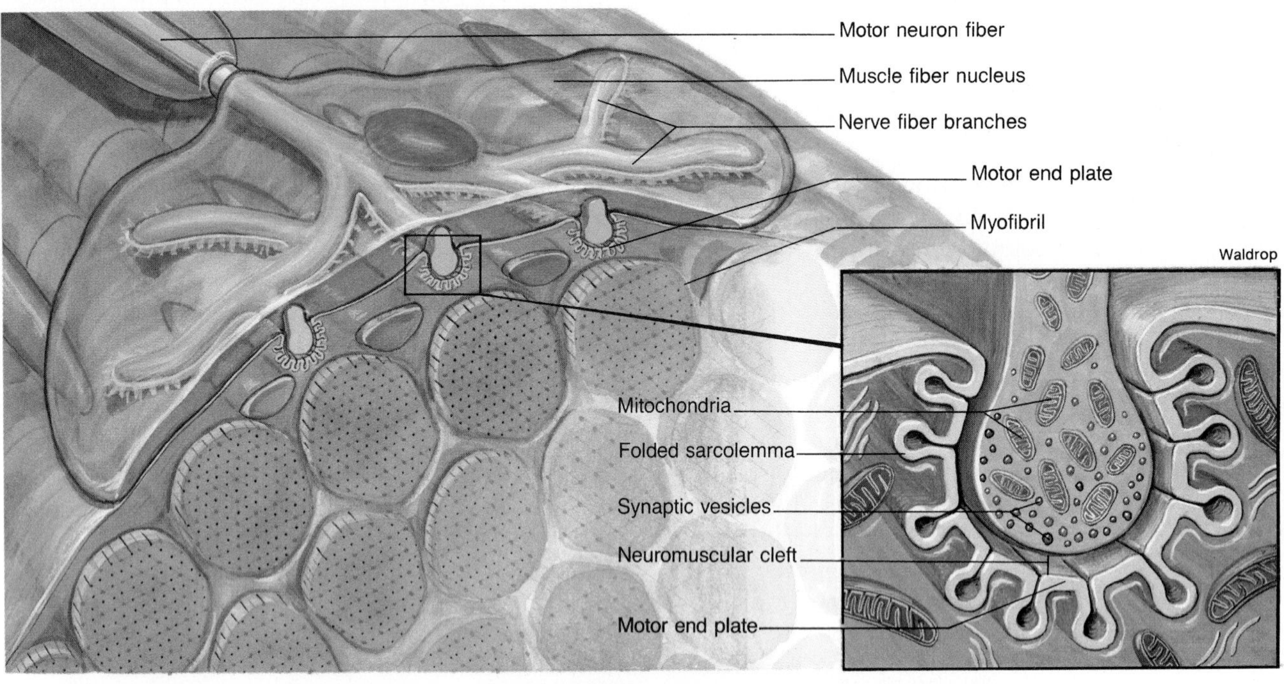

(a)

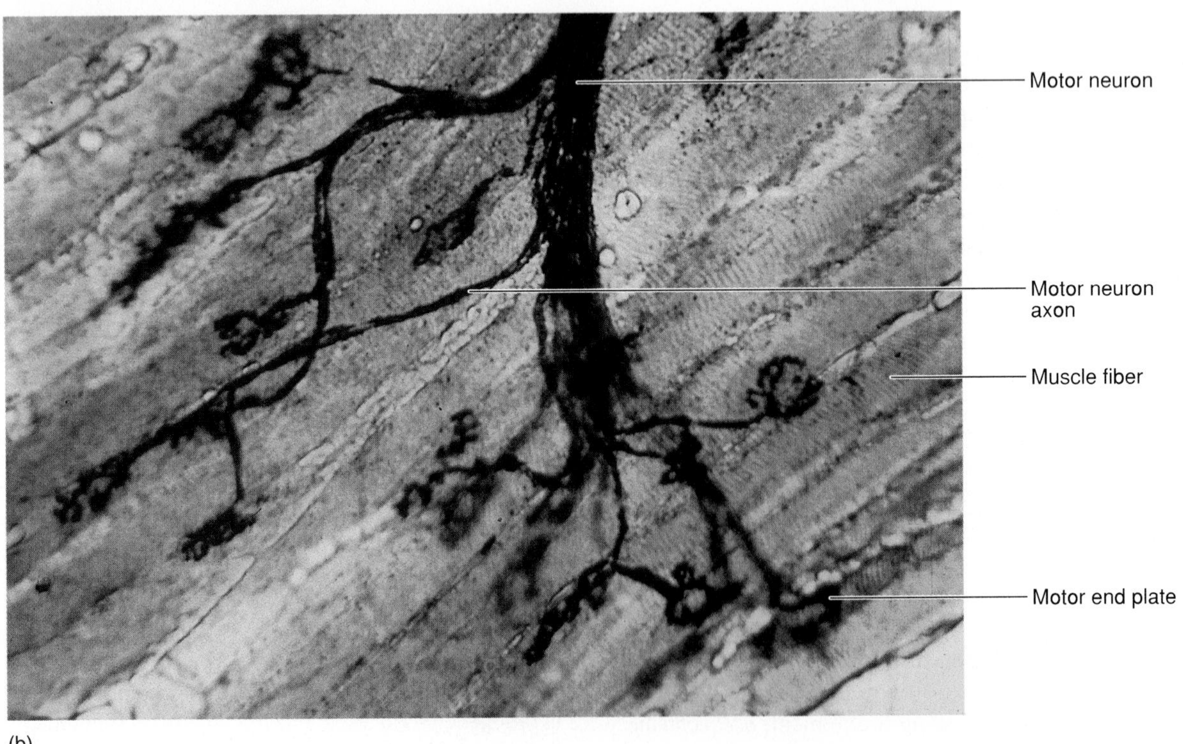

(b)

Figure 12.6

Motor units.
A motor unit is composed of a motor neuron and the muscle fibers it innervates. This diagram illustrates the innervation of muscle fibers by different motor units. (Actually, many more muscle fibers would be included in a single motor unit than are shown in this drawing.)

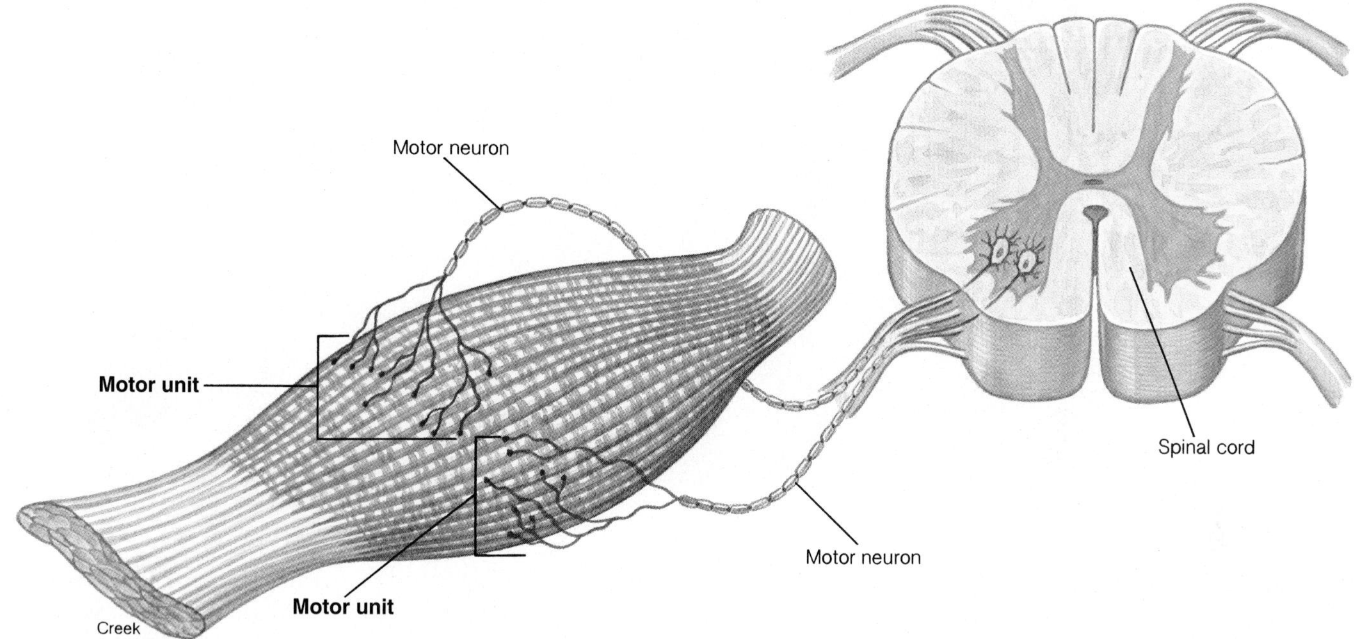

Mechanisms of Contraction

The A bands within each muscle fiber are composed of thick filaments, and the I bands contain thin filaments. Movement of cross bridges that extend from the thick to the thin filaments causes sliding of the filaments, and thus muscle tension and shortening. The activity of the cross bridges is regulated by the availability of Ca^{2+}, which is increased by electrical stimulation of the muscle fiber. Electrical stimulation produces contraction of the muscle through the binding of Ca^{2+} to regulatory proteins within the thin filaments.

When muscle cells are viewed in the electron microscope, which can produce images at several thousand times the magnification possible in an ordinary light microscope, each cell is seen to be composed of many subunits known as **myofibrils** (*mi″ŏ-fi′brilz*) (fig. 12.7). Myofibrils are approximately 1 micrometer (1 μm) in diameter and extend in parallel rows from one end of the muscle fiber to the other. The myofibrils are so densely packed that other organelles, such as mitochondria and intracellular membranes, are restricted to the narrow cytoplasmic spaces that remain between adjacent myofibrils.

The muscle fiber does not have striations that extend from one side of the fiber to the other when viewed in an electron microscope. Rather, it is the myofibrils that are striated with dark A bands and light I bands (fig. 12.8). The striated appearance of the entire muscle fiber when seen with a light microscope is an illusion created by the alignment of the dark and light bands of the myofibrils from one side of the fiber to the other. Since the separate myofibrils are not clearly seen at low magnification, the dark and light bands appear to be continuous across the width of the fiber.

Figure 12.7

The components of a skeletal muscle fiber.
A muscle fiber (muscle cell) is like a cable, containing many myofibrils. The sarcolemma is the cell membrane of the muscle fiber, and the sarcoplasm is its cytoplasm. A skeletal muscle fiber contains many nuclei (is multinucleate).

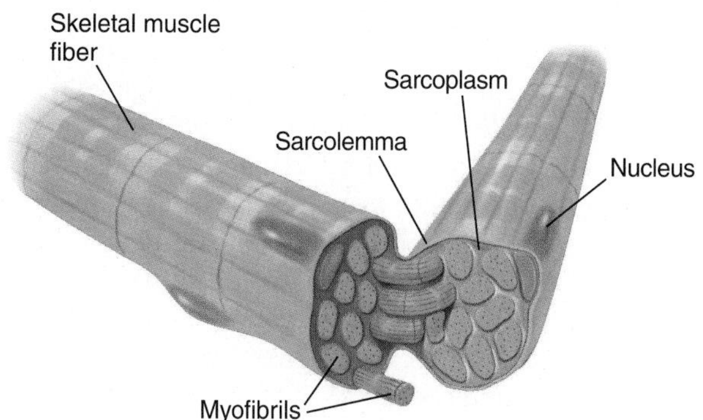

Each myofibril contains even smaller structures called **filaments,** or *myofilaments*. When a myofibril is observed at high magnification in longitudinal section (side view), the A bands are seen to contain **thick filaments.** These are about 110 angstroms thick (110 Å, where 1 Å = 10^{-10} m) and are stacked in register. It is these thick filaments that give the A band its dark appearance. The lighter I band, by contrast,

Figure 12.8

Electron micrograph of a longitudinal section of myofibrils.
The A, H, and I bands are clearly seen. Notice how the dark and light bands of each myofibril are stacked in register.

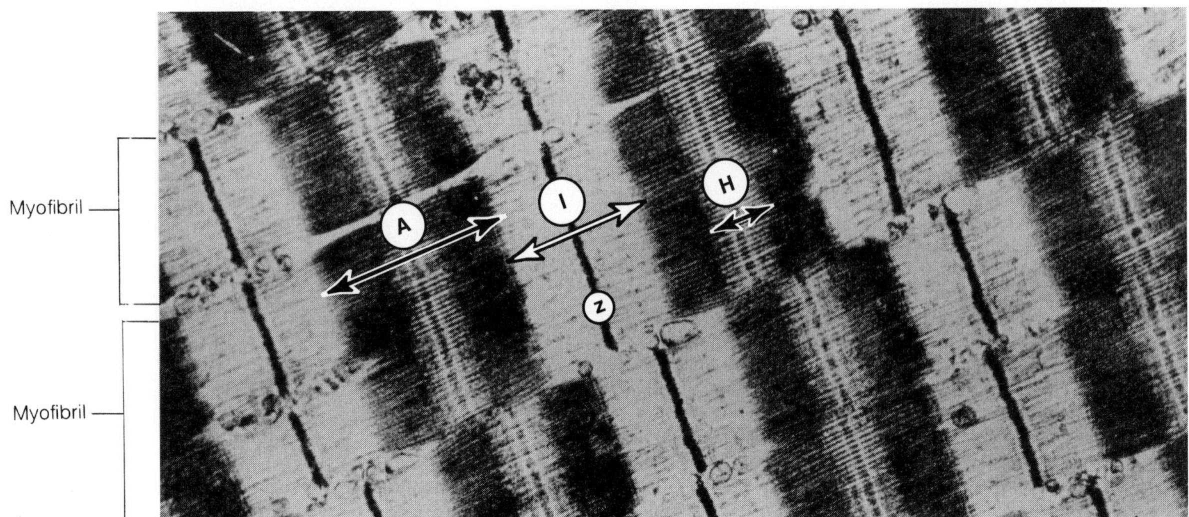

contains **thin filaments** (from 50–60 Å thick). The thick filaments are composed of the protein **myosin** (*mi′ŏ-sin*), and the thin filaments are composed primarily of the protein **actin** (*ak′tin*).

The I bands within a myofibril are the lighter areas that extend from the edge of one stack of thick myosin filaments to the edge of the next stack of thick filaments. They are light in appearance because they contain only thin filaments. The thin filaments, however, do not end at the edges of the I bands. Instead, each thin filament extends partway into the A bands on each side (between the stack of thick filaments on each side of an I band). Since thick and thin filaments overlap at the edges of each A band, the edges of the A band are darker in appearance than the central region. These central lighter regions of the A bands are called the *H bands* (for *helle,* a German word meaning "bright"). The central H bands thus contain only thick filaments that are not overlapped by thin filaments.

In the center of each I band is a thin dark Z line. The arrangement of thick and thin filaments between a pair of Z lines forms a repeating pattern that serves as the basic subunit of striated muscle contraction. These subunits, from Z to Z, are known as **sarcomeres** (*sar′kŏ-mērz*) (fig. 12.9*b*). A longitudinal section of a myofibril thus presents a side view of successive sarcomeres.

This side view is, in a sense, misleading; there are numerous sarcomeres within each myofibril that are out of the plane of the section (and out of the picture). A better appreciation of the three-dimensional structure of a myofibril can be obtained by viewing the myofibril in cross section. In this view, it can be seen that the Z lines are actually **Z discs,** and that the thin filaments that penetrate these Z discs surround the thick filaments in a hexagonal arrangement (fig. 12.9*c*). If we concentrate on a single row of dark thick filaments in this cross section, the alternating pattern of thick and thin filaments seen in longitudinal section becomes apparent.

myosin: L. *myosin,* within muscle
actin: L. *actus,* motion, doing

Sliding Filament Theory of Contraction

When a muscle contracts isotonically, it decreases in length as a result of the shortening of its individual fibers. Shortening of the muscle fibers, in turn, is produced by shortening of their myofibrils, which occurs as a result of the shortening of the distance from Z line to Z line. As the sarcomeres shorten in length, however, the A bands do *not* shorten but instead move closer together. The I bands—which represent the distance between A bands of successive sarcomeres—decrease in length (table 12.2).

The thin actin filaments composing the I band, however, do not shorten. Close examination reveals that the thick and thin filaments remain the same length during muscle contraction. Shortening of the sarcomeres is produced not by shortening of the filaments, but rather by the *sliding* of thin filaments over and between the thick filaments. In the process of contraction, the thin filaments on either side of each A band slide deeper and deeper toward the center, producing increasing amounts of overlap with the thick filaments. The I bands (containing only thin filaments) and H bands (containing only thick filaments) thus get shorter during contraction (fig. 12.10).

Cross Bridges

Sliding of the filaments is produced by the action of numerous **cross bridges** that extend from the myosin toward the actin. These cross bridges are part of the myosin proteins that

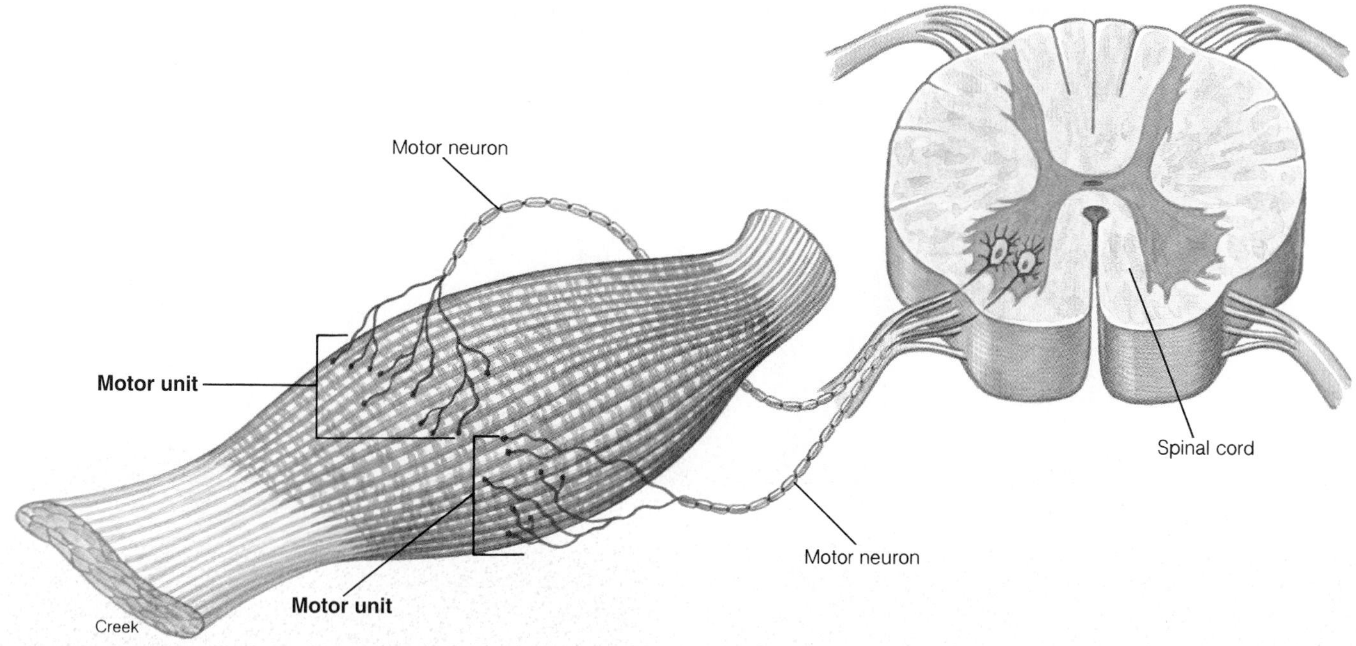

Figure 12.6

Motor units.
A motor unit is composed of a motor neuron and the muscle fibers it innervates. This diagram illustrates the innervation of muscle fibers by different motor units. (Actually, many more muscle fibers would be included in a single motor unit than are shown in this drawing.)

Mechanisms of Contraction

The A bands within each muscle fiber are composed of thick filaments, and the I bands contain thin filaments. Movement of cross bridges that extend from the thick to the thin filaments causes sliding of the filaments, and thus muscle tension and shortening. The activity of the cross bridges is regulated by the availability of Ca^{2+}, which is increased by electrical stimulation of the muscle fiber. Electrical stimulation produces contraction of the muscle through the binding of Ca^{2+} to regulatory proteins within the thin filaments.

When muscle cells are viewed in the electron microscope, which can produce images at several thousand times the magnification possible in an ordinary light microscope, each cell is seen to be composed of many subunits known as **myofibrils** (*mi″ŏ-fi′brilz*) (fig. 12.7). Myofibrils are approximately 1 micrometer (1 µm) in diameter and extend in parallel rows from one end of the muscle fiber to the other. The myofibrils are so densely packed that other organelles, such as mitochondria and intracellular membranes, are restricted to the narrow cytoplasmic spaces that remain between adjacent myofibrils.

The muscle fiber does not have striations that extend from one side of the fiber to the other when viewed in an electron microscope. Rather, it is the myofibrils that are striated with dark A bands and light I bands (fig. 12.8). The striated appearance of the entire muscle fiber when seen with a light microscope is an illusion created by the alignment of the dark and light bands of the myofibrils from one side of the fiber to the other. Since the separate myofibrils are not clearly seen at low magnification, the dark and light bands appear to be continuous across the width of the fiber.

Figure 12.7

The components of a skeletal muscle fiber.
A muscle fiber (muscle cell) is like a cable, containing many myofibrils. The sarcolemma is the cell membrane of the muscle fiber, and the sarcoplasm is its cytoplasm. A skeletal muscle fiber contains many nuclei (is multinucleate).

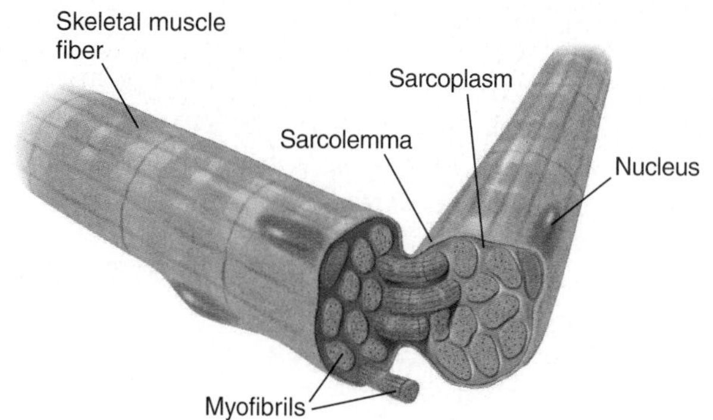

Each myofibril contains even smaller structures called **filaments,** or *myofilaments*. When a myofibril is observed at high magnification in longitudinal section (side view), the A bands are seen to contain **thick filaments.** These are about 110 angstroms thick (110 Å, where 1 Å = 10^{-10} m) and are stacked in register. It is these thick filaments that give the A band its dark appearance. The lighter I band, by contrast,

Figure 12.8

Electron micrograph of a longitudinal section of myofibrils.
The A, H, and I bands are clearly seen. Notice how the dark and light bands of each myofibril are stacked in register.

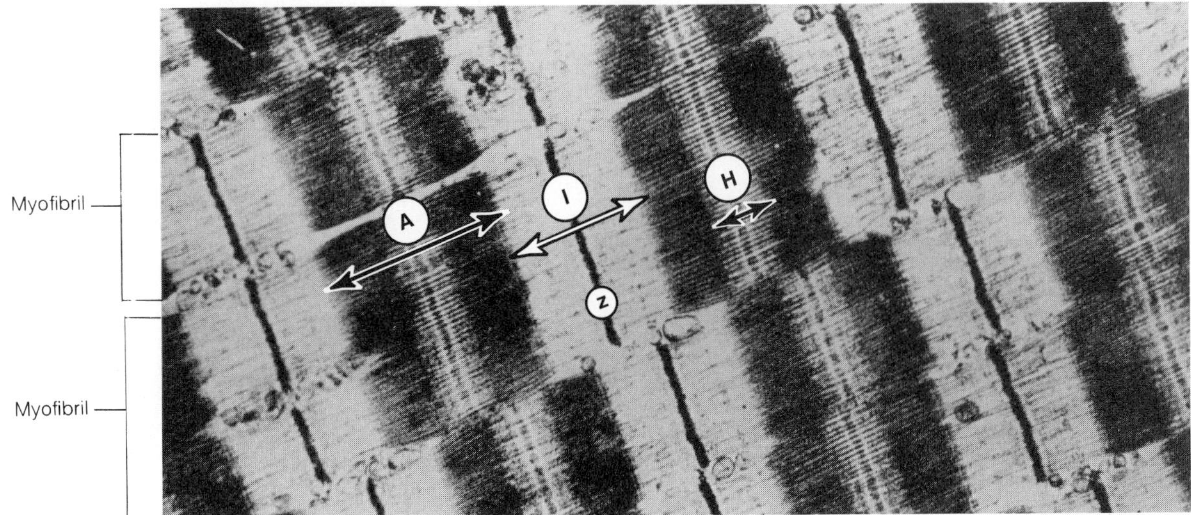

contains **thin filaments** (from 50–60 Å thick). The thick filaments are composed of the protein **myosin** (*mi'ŏ-sin*), and the thin filaments are composed primarily of the protein **actin** (*ak'tin*).

The I bands within a myofibril are the lighter areas that extend from the edge of one stack of thick myosin filaments to the edge of the next stack of thick filaments. They are light in appearance because they contain only thin filaments. The thin filaments, however, do not end at the edges of the I bands. Instead, each thin filament extends partway into the A bands on each side (between the stack of thick filaments on each side of an I band). Since thick and thin filaments overlap at the edges of each A band, the edges of the A band are darker in appearance than the central region. These central lighter regions of the A bands are called the *H bands* (for *helle*, a German word meaning "bright"). The central H bands thus contain only thick filaments that are not overlapped by thin filaments.

In the center of each I band is a thin dark Z line. The arrangement of thick and thin filaments between a pair of Z lines forms a repeating pattern that serves as the basic subunit of striated muscle contraction. These subunits, from Z to Z, are known as **sarcomeres** (*sar'kŏ-mērz*) (fig. 12.9*b*). A longitudinal section of a myofibril thus presents a side view of successive sarcomeres.

This side view is, in a sense, misleading; there are numerous sarcomeres within each myofibril that are out of the plane of the section (and out of the picture). A better appreciation of the three-dimensional structure of a myofibril can be obtained by viewing the myofibril in cross section. In this view, it can be seen that the Z lines are actually **Z discs,** and that the thin filaments that penetrate these Z discs surround the thick filaments in a hexagonal arrangement (fig. 12.9*c*). If we concentrate on a single row of dark thick filaments in this cross section, the alternating pattern of thick and thin filaments seen in longitudinal section becomes apparent.

Sliding Filament Theory of Contraction

When a muscle contracts isotonically, it decreases in length as a result of the shortening of its individual fibers. Shortening of the muscle fibers, in turn, is produced by shortening of their myofibrils, which occurs as a result of the shortening of the distance from Z line to Z line. As the sarcomeres shorten in length, however, the A bands do *not* shorten but instead move closer together. The I bands—which represent the distance between A bands of successive sarcomeres—decrease in length (table 12.2).

The thin actin filaments composing the I band, however, do not shorten. Close examination reveals that the thick and thin filaments remain the same length during muscle contraction. Shortening of the sarcomeres is produced not by shortening of the filaments, but rather by the *sliding* of thin filaments over and between the thick filaments. In the process of contraction, the thin filaments on either side of each A band slide deeper and deeper toward the center, producing increasing amounts of overlap with the thick filaments. The I bands (containing only thin filaments) and H bands (containing only thick filaments) thus get shorter during contraction (fig. 12.10).

Cross Bridges

Sliding of the filaments is produced by the action of numerous **cross bridges** that extend from the myosin toward the actin. These cross bridges are part of the myosin proteins that

myosin: L. *myosin*, within muscle
actin: L. *actus*, motion, doing

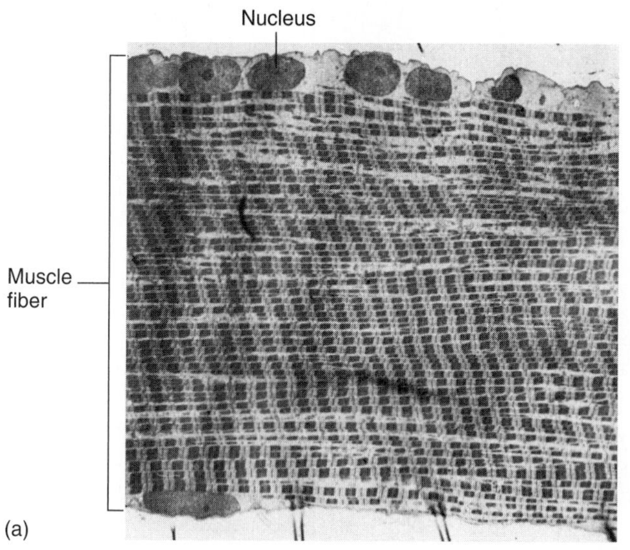

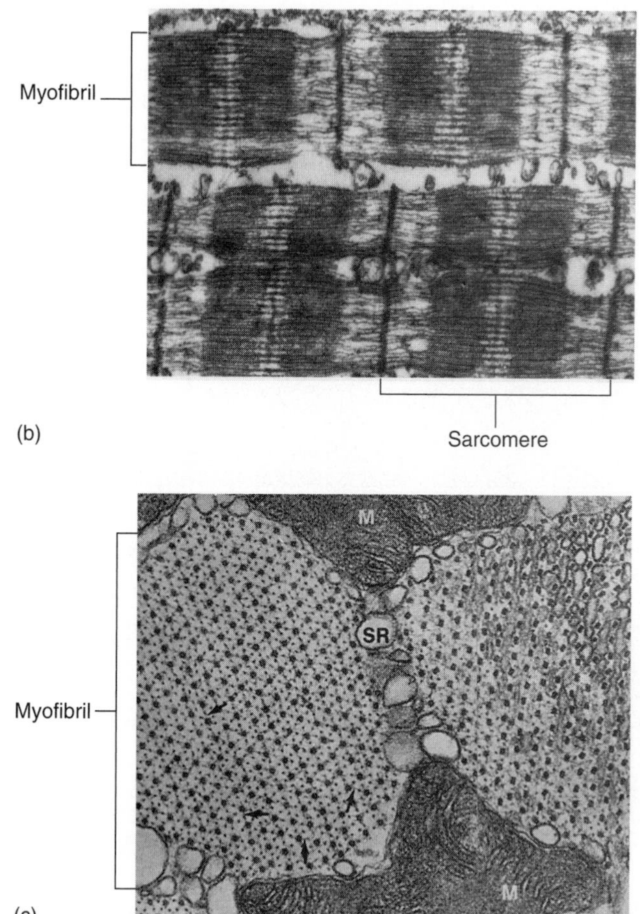

Figure 12.9

Electron micrographs of myofibrils in a muscle fiber. (*a*) At low power (1,600×), a single muscle fiber is seen to contain numerous myofibrils. (*b*) The structure of these myofibrils is more clearly seen in *b* at high power (53,000×). Notice the sarcomeres and overlapping thick and thin filaments in this longitudinal section. (*c*) The myofibrils are cut in cross section, where the hexagonal arrangement of thick and thin filaments is clearly seen (arrows point to cross bridges; SR = sarcoplasmic reticulum; M = mitochondria).

Part [c] reproduced from: R. G. Kessel and R. H. Kardon, Tissues and Organs: A Text-Atlas of Scanning Electron Microscopy, *W. H. Freeman and Co., 1979.*

Table 12.2

The Sliding Filament Theory of Contraction

1. A myofiber, together with all its myofibrils, shortens by movement of the insertion toward the origin of the muscle.
2. Shortening of the myofibrils is caused by shortening of the sarcomeres—the distance between Z lines (or discs) is reduced.
3. Shortening of the sarcomeres is accomplished by sliding of the myofilaments—each myofilament remains the same length during contraction.
4. Sliding of the myofilaments is produced by asynchronous power strokes of myosin cross bridges, which pull the thin myofilaments (actin) over the thick myofilaments (myosin).
5. The A bands remain the same length during contraction, but are pulled toward the origin of the muscle.
6. Adjacent A bands are pulled closer together as the I bands between them shorten.
7. The H bands shorten during contraction as the thin filaments from each end of the sarcomeres are pulled toward the middle.

extend from the axis of the thick filaments to form "arms" that terminate in globular "heads" (fig. 12.11). A myosin protein has two globular heads that serve as cross bridges. The orientation of the myosin heads on one side of a sarcomere is opposite to that on the other side, so that, when the myosin heads form cross bridges by attaching to actin on each side of the sarcomere, they can pull the actin from each side toward the center.

Isolated muscles in vitro are easily stretched (although this is opposed in vivo by the stretch reflex, described in chapter 18), demonstrating that the myosin heads are not attached to actin when the muscle is at rest. Each globular myosin head of a cross bridge contains an ATP-binding site closely associated with an actin-binding site (fig. 12.11). The globular heads function as **myosin ATPase** (*ā″te-pe′as*) enzymes, splitting ATP into ADP and P_i. This reaction occurs before the myosin heads combine with actin, and indeed is required for activating the myosin heads so that they can attach to actin. The ADP and P_i remain bonded to the myosin heads until the cross bridges attach to the actin.

Figure 12.10

The sliding filament model of muscle contraction.
(*a*) An electron micrograph and (*b*) a diagram of the sliding filament model of contraction. As the filaments slide, the Z lines are brought closer together and the sarcomeres get shorter. (*1*) Relaxed muscle, (*2*) partially contracted muscle, and (*3*) fully contracted muscle.

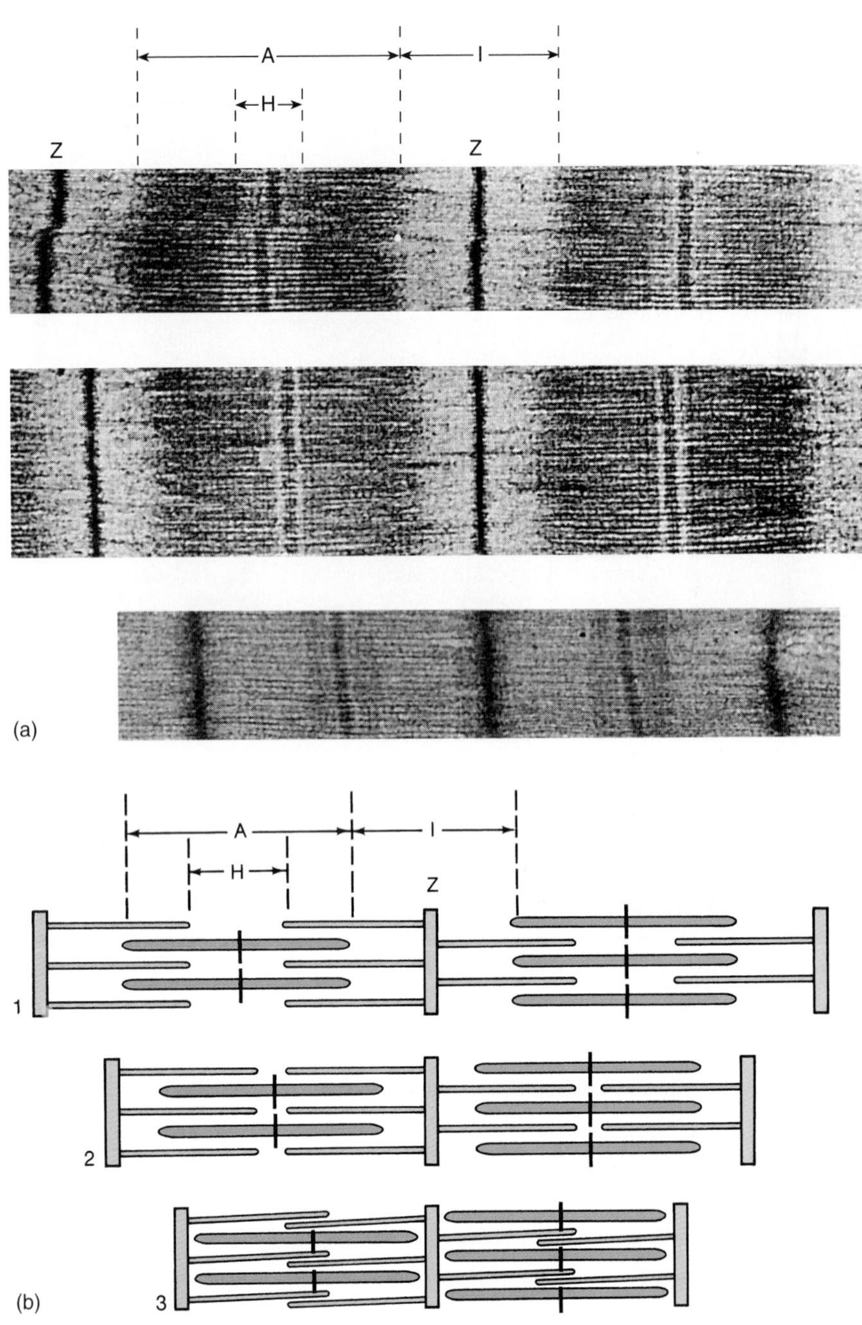

The myosin heads are able to bind to specific attachment sites in the actin subunits. When the cross bridges bind to actin, they release the P_i. This causes a conformation change in the myosin protein, resulting in a *power stroke* that pulls the thin filaments toward the center of the A bands. The ADP is released when the cross bridges bind to a fresh ATP at the end of the power stroke. This release of ADP upon binding to a new ATP is required for the cross bridges to break their bond with actin at the end of the power stroke. The myosin ATPase will then split ATP and become activated as in the previous cycle. Note that the splitting of ATP is required *before* a cross bridge can attach to actin and undergo a power stroke, and that the attachment of a *new ATP* is needed for the cross bridge to release from actin at the end of a power stroke (fig. 12.12).

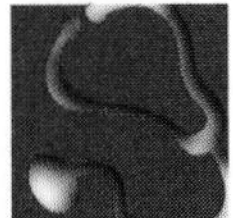

The detachment of a cross bridge from actin at the end of a power stroke requires that a new ATP molecule bind to the myosin ATPase. The importance of this process is illustrated by the muscular contracture called *rigor mortis* that occurs due to lack of ATP when the muscle dies. Without ATP, the ADP remains bound to the cross bridges, and the cross bridges remain tightly bound to actin. This results in the formation of "rigor complexes" between myosin and actin that cannot detach. In rigor mortis, the muscles remain stiff until the myosin and actin begin to decompose.

Because the cross bridges are quite short, a single contraction cycle and power stroke of all the cross bridges in a muscle would shorten the muscle by only about 1% of its resting length. Since muscles can shorten up to 60% of their resting lengths, it is obvious that the contraction cycles must be repeated many times. In order for this to occur the cross bridges must detach from the actin at the end of a power stroke, reassume their resting orientation, and then reattach to the actin and repeat the cycle.

During normal contraction, however, only a fraction of the cross bridges are attached at any given time. The power strokes are thus not in synchrony, as the strokes of a competitive rowing team would be. Rather, they are like the actions of a team engaged in tug-of-war, where the pulling action of the members is asynchronous. Some cross bridges are engaged in power strokes at all times during the contraction.

Regulation of Contraction

When the cross bridges attach to actin they undergo power strokes and cause muscle contraction. In order for a muscle to relax, therefore, the attachment of myosin cross bridges to actin must be prevented. The regulation of cross-bridge attachment to actin is a function of two proteins that are associated with actin in the thin filaments.

Figure 12.15

The sarcoplasmic reticulum.
This figure depicts the relationship between myofibrils, the transverse tubules, and the sarcoplasmic reticulum. The sarcoplasmic reticulum (green) stores Ca^{2+}, and is stimulated to release it by action potentials arriving in the transverse tubules (yellow).

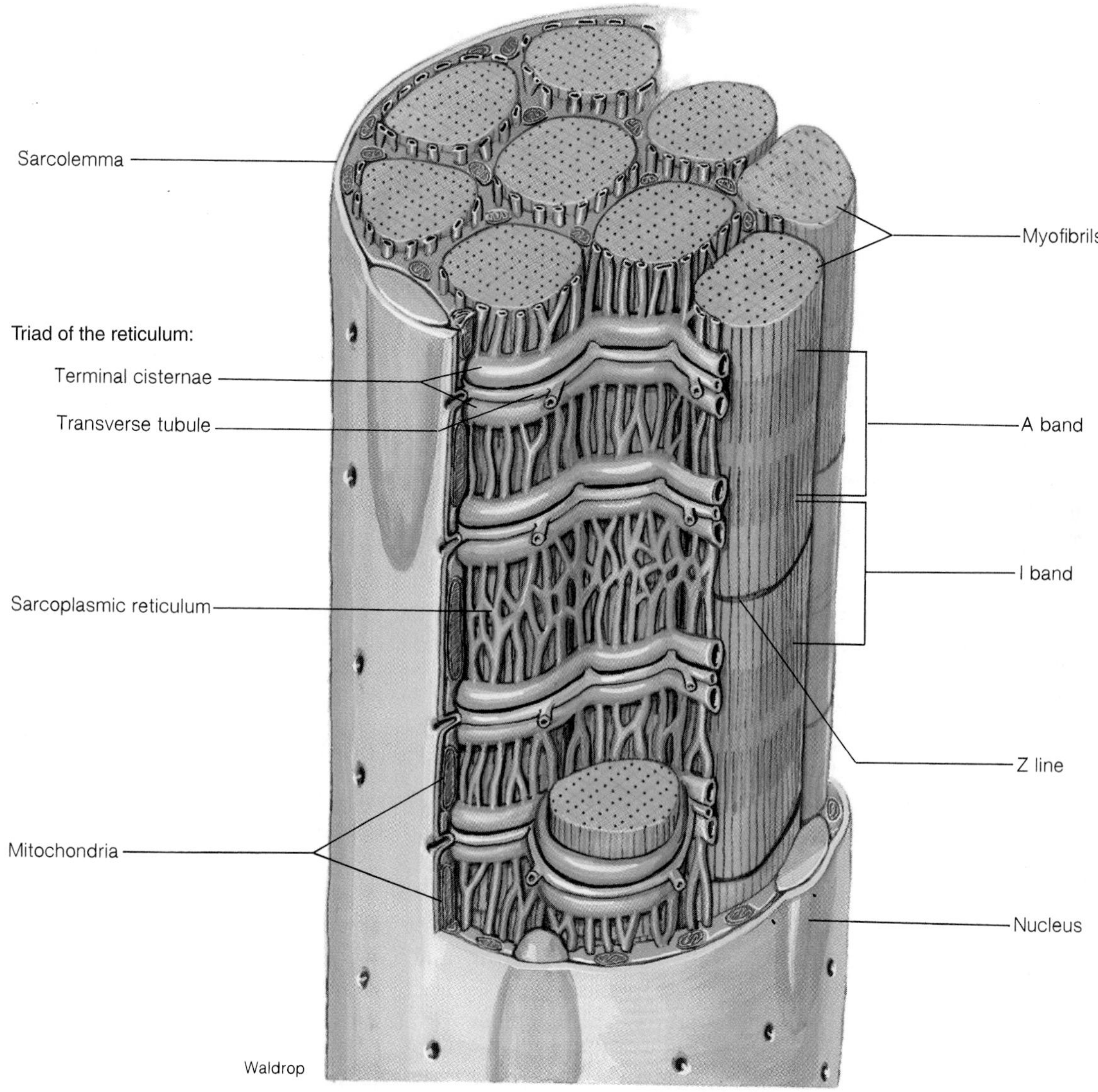

sarcoplasm into the **sarcoplasmic reticulum (SR)** (fig. 12.15). The sarcoplasmic reticulum is a modified endoplasmic reticulum, consisting of interconnected sacs and tubes that surround each myofibril within the muscle cell.

Most of the Ca^{2+} in a relaxed muscle fiber is stored within expanded portions of the sarcoplasmic reticulum known as *terminal cisternae*. When a muscle fiber is stimulated to contract by either a motor neuron in vivo or electric shocks in vitro, the stored Ca^{2+} is released from the sarcoplasmic reticulum so that it can attach to troponin. When a muscle fiber is no longer stimulated, the Ca^{2+} from the sarcoplasm is actively transported back into the sarcoplasmic reticulum. Now, in order to understand how the release and uptake of Ca^{2+} is regulated, one more organelle within the muscle fiber must be described.

The terminal cisternae of the sarcoplasmic reticulum are separated only by a very narrow gap from **transverse tubules** (or **T tubules**), which are narrow membranous "tunnels" formed from and continuous with the sarcolemma (muscle cell membrane). The transverse tubules thus open to the extracellular environment through pores in the cell surface and are capable of conducting action potentials. The stage is now set to explain exactly how a motor neuron stimulates a muscle fiber to contract.

The release of the chemical *acetylcholine* from the axon terminals at the neuromuscular junctions causes electrical activation of the skeletal muscle fibers. This generates impulses called *action potentials* (see chapter 14), which are similar to nerve impulses. Action potentials in muscle cells, like those in nerve cells, are events that are regenerated along the cell

Table 12.3
Summary of Events in Excitation-Contraction Coupling

1. Impulses in a somatic motor neuron cause the release of acetylcholine neurotransmitter at the myoneural junction (one myoneural junction per myofiber).
2. Acetylcholine, through its interaction with receptors in the muscle cell membrane (sarcolemma), produces impulses that are regenerated across the sarcolemma.
3. The membranes of the transverse tubules (T tubules) are continuous with the sarcolemma and conduct impulses deep into the muscle fiber.
4. Impulses in the T tubules, by a mechanism that is poorly understood, stimulate the release of Ca^{2+} from the terminal cisternae of the sarcoplasmic reticulum.
5. Ca^{2+} released into the sarcoplasm binds to troponin, causing a change in its structure.
6. The shape change in troponin causes its attached tropomyosin to shift position in the actin filament, thus exposing binding sites for the myosin cross bridges.
7. Myosin cross bridges, previously activated by the hydrolysis of ATP, attach to actin.
8. Once the previously activated cross bridges attach to actin, they undergo a power stroke and pull the thin filaments over the thick filaments.
9. Attachment of fresh ATP allows the cross bridges to detach from actin and repeat the contraction cycle as long as Ca^{2+} remains attached to troponin.
10. When impulses stop being produced, the sarcoplasmic reticulum actively accumulates Ca^{2+}, and tropomyosin returns to its inhibitory position.

membrane. As discussed in chapter 14, action potentials involve the flow of ions between the extracellular and intracellular environments across a cell membrane that separates these two compartments. In muscle cells, therefore, action potentials can be conducted into the interior of the fiber across the membrane of the transverse tubules.

Action potentials in the transverse tubules cause the release of Ca^{2+} from the sarcoplasmic reticulum. This process is known as **excitation-contraction coupling** (table 12.3). Since the transverse tubules are not physically continuous with the sarcoplasmic reticulum, however, there must be some mechanism to permit communication between these two organelles. One possibility is that there may be a direct coupling, on a molecular level, between voltage-regulated Ca^{2+} channels in the transverse tubules and the Ca^{2+} release channels in the sarcoplasmic reticulum. The Ca^{2+} release channel proteins of the sarcoplasmic reticulum have a part that extends into the cytoplasm. This part, which has a footlike appearance in the electron microscope, may be able to interact directly with the Ca^{2+} channel proteins of the transverse tubules. Another possibility is that Ca^{2+} diffusion through channels in the transverse tubules may stimulate the opening of Ca^{2+} channels in the sarcoplasmic reticulum.

Through whichever mechanism it is accomplished, action potentials in the transverse tubules cause the release of Ca^{2+} from the sarcoplasmic reticulum. The released Ca^{2+} then diffuses into the sarcomeres and binds to troponin, causing the displacement of tropomyosin and allowing the actin to bind to the myosin cross bridges. Muscle contraction is thus stimulated.

As long as action potentials continue to be produced—which is as long as the neural stimulation of the muscle is maintained—Ca^{2+} will remain attached to troponin, and cross bridges will be able to undergo contraction cycles. When neural activity and action potentials in the muscle fiber cease, the sarcoplasmic reticulum actively accumulates Ca^{2+} and muscle relaxation occurs. Note that the return of Ca^{2+} to the sarcoplasmic reticulum involves active transport, and thus requires the hydrolysis of ATP. ATP is therefore needed for muscle relaxation as well as for muscle contraction.

Length–Tension Relationship

The strength of a muscle's contraction is influenced by a variety of factors. These include the number of fibers within the muscle that are stimulated to contract, the frequency of stimulation, the thickness of each muscle fiber (thicker fibers have more myofibrils and thus can exert more power), and the initial length of the muscle fibers when they are at rest.

There is an "ideal" resting length for striated muscle fibers. This is the length at which they can generate maximum force. When the resting length exceeds this ideal, the overlap between actin and myosin is so small that few cross bridges can attach. When the muscle is stretched to the point that there is no overlap of actin with myosin, no cross bridges can attach to the thin filaments and the muscle cannot contract. When the muscle is shortened to about 60% of its resting length, the Z lines abut the thick filaments so that further contraction cannot occur.

The strength of a muscle's contraction can be measured by the force required to prevent it from shortening. Under these isometric conditions, the strength of contraction, or *tension*, can be measured when the muscle length at rest is varied. Maximum tension of skeletal muscle is produced when the muscle is at its normal resting length in vivo (fig. 12.16). If the muscle were any shorter or longer than its normal length, in other words, its strength of contraction would be reduced. This resting length is maintained by reflex contraction in response to passive stretching, as described in a later section of this chapter.

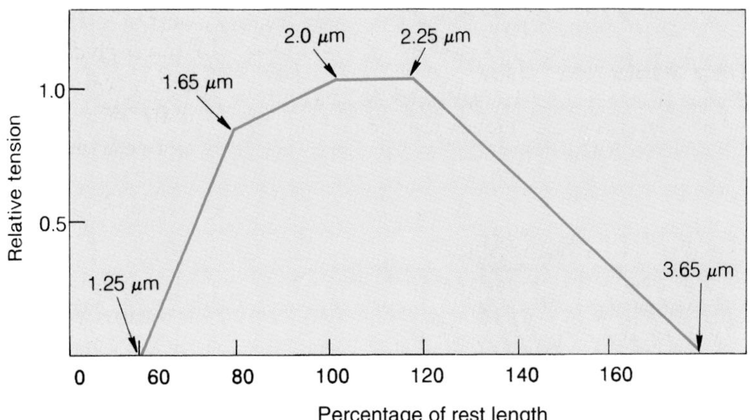

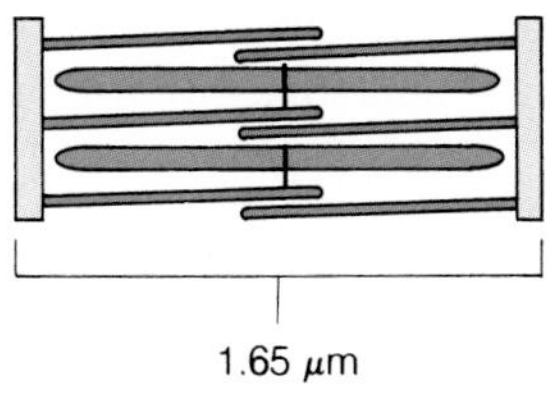

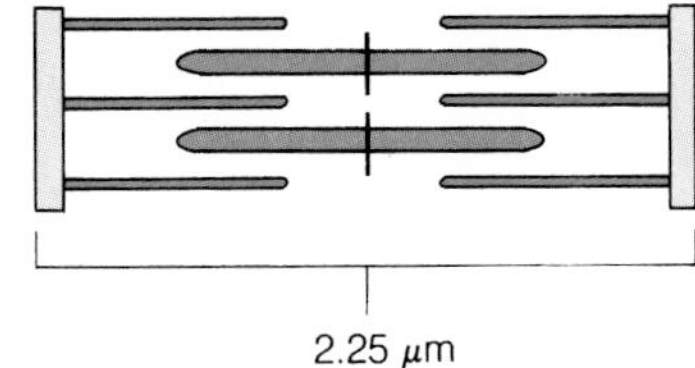

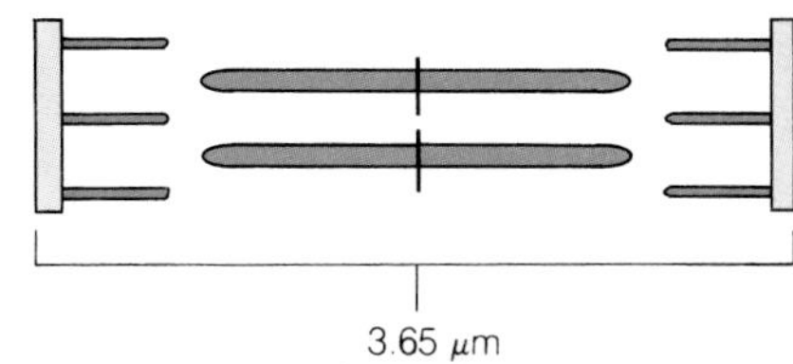

Figure 12.16

The length–tension relationship in skeletal muscles.
Maximum relative tension (1.0 on the Y axis) is achieved when the muscle is 100% to 120% of its resting length (sarcomere lengths from 2.0 to 2.25 μm). Increases or decreases in muscle (and sarcomere) lengths result in rapid decreases in tension.

Energy Requirements of Skeletal Muscles

Skeletal muscles generate ATP through aerobic and anaerobic respiration and through the use of phosphate groups donated by creatine phosphate. The aerobic and anaerobic abilities of skeletal muscle fibers differ according to muscle fiber type. Slow-twitch (type I) fibers are adapted for aerobic respiration; fast-twitch (type II) fibers are adapted for anaerobic respiration.

Skeletal muscles at rest obtain most of their energy from the aerobic respiration of fatty acids. During exercise, muscle glycogen and blood glucose are also used as energy sources (fig. 12.17). Energy obtained by cell respiration is used to make ATP, which serves as the immediate source of energy for (1) the movement of the cross bridges for muscle contraction and (2) the pumping of Ca^{2+} into the sarcoplasmic reticulum for muscle relaxation.

Metabolism of Skeletal Muscles

Skeletal muscles respire anaerobically for the first 45 to 90 seconds of moderate-to-heavy exercise, because the cardiopulmonary system requires this amount of time to sufficiently increase the oxygen supply to the exercising muscles. If exercise is moderate, aerobic respiration contributes the major portion of the skeletal muscle energy requirements following the first 2 minutes of exercise.

Maximal Oxygen Uptake

Whether exercise is light, moderate, or heavy for a given person depends on that person's maximal capacity for aerobic exercise. The maximum rate of oxygen consumption (by aerobic respiration) in the body is called the **maximal oxygen uptake,** or the **aerobic capacity,** and is often given in abbreviated form as the **$\dot{V}_{O_2}$max.** The maximal oxygen uptake is determined primarily by a person's age, size, and sex. It is 15% to 20% higher for males than for females and highest at age 20 for both sexes. The $\dot{V}_{O_2}$max ranges from about 12 ml of O_2 per minute per kilogram body weight for older, sedentary people to about 84 ml per minute per kilogram for young, elite male athletes. Some world-class athletes have maximal oxygen uptakes that are twice the average for their age and sex—this appears to be due largely to genetic factors, but training can increase the maximum oxygen uptake by about 20%.

The intensity of exercise can also be defined by the **lactate** (or **anaerobic**) **threshold.** This is the percentage of the maximal oxygen uptake at which a significant rise in blood lactate levels occurs. For average healthy people, for example, a significant amount of blood lactate appears when exercise is performed at about 50% to 70% of the $\dot{V}_{O_2}$max.

During light exercise (at about 25% of the $\dot{V}_{O_2}$max), most of the exercising muscle's energy is derived from the aerobic respiration of fatty acids. These are obtained mainly from stored fat in adipose tissue, and to a lesser extent from triglycerides stored in the muscle (fig. 12.17). When a person exercises just below the lactate threshold, where the exercise can be described as moderately intense (at 50% to 70% of the $\dot{V}_{O_2}$max), the energy is derived almost equally from fatty acids and glucose (obtained from stored muscle glycogen and the blood plasma). By contrast, glucose from these sources supplies two-thirds of the energy for muscles during heavy exercise above the lactate threshold.

During exercise, the carrier protein for the facilitated diffusion of glucose (GLUT4—chapter 5) is moved into the

Figure 12.17

Muscle fuel consumption during exercise.
The relative contributions of plasma glucose, plasma free fatty acids, muscle glycogen, and muscle triglycerides to the energy consumption of exercising muscles during mild exercise (25% of $\dot{V}_{O_2}$max), moderate exercise (65% of $\dot{V}_{O_2}$max), and heavy exercise (85% of $\dot{V}_{O_2}$max).

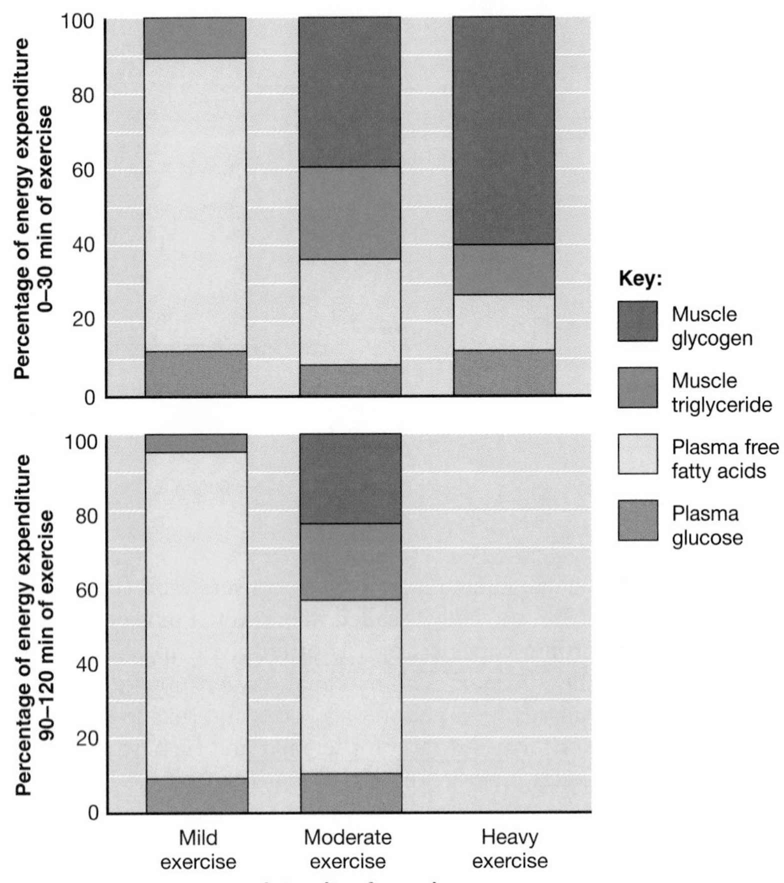

Figure 12.18

The production and utilization of phosphocreatine in muscles.
Phosphocreatine serves as a muscle reserve of high-energy phosphate, used for the rapid formation of ATP.

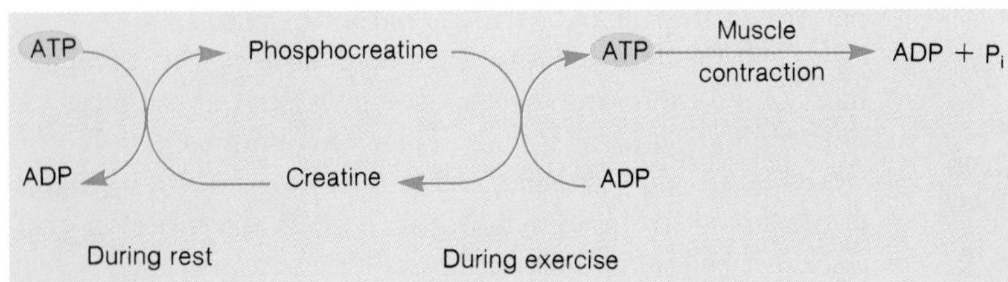

muscle cell membrane, so that the cell can take up an increasing amount of blood glucose. The uptake of plasma glucose contributes 15% to 30% of the muscle's energy needs during moderate exercise and up to 40% of the energy needs during heavy exercise. This would produce hypoglycemia if the liver failed to increase its output of glucose. The liver increases its output of glucose primarily through hydrolysis of its stored glycogen, but gluconeogenesis (the production of glucose from amino acids, lactate, and glycerol) contributes increasingly to the liver's glucose production as exercise is prolonged.

Oxygen Debt

When a person stops exercising, the rate of oxygen uptake does not immediately return to pre-exercise levels; it returns slowly (the person continues to breathe heavily for some time afterward). This extra oxygen is used to repay the **oxygen debt** incurred during exercise. The oxygen debt includes oxygen that was withdrawn from savings deposits—hemoglobin in blood and myoglobin in muscle (see chapter 24); the extra oxygen required for metabolism by tissues warmed during exercise; and the oxygen needed for the metabolism of the lactic acid produced during anaerobic respiration.

Phosphocreatine

During sustained muscle activity, ATP may be used faster than it can be produced through cell respiration. At these times the rapid renewal of ATP is extremely important. This is accomplished by combining ADP with phosphate derived from another high-energy phosphate compound called **phosphocreatine** (*fox″fo-kre′ă-tin*), or **creatine phosphate.**

Within muscle cells, the phosphocreatine concentration is more than three times the concentration of ATP and represents a ready reserve of high-energy phosphate that can be donated directly to ADP (fig. 12.18). During times of rest, the depleted reserve of phosphocreatine can be restored by the reverse reaction—phosphorylation of creatine with phosphate derived from ATP.

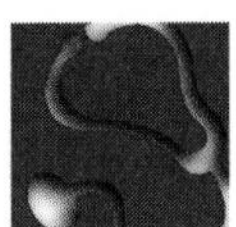

The enzyme that transfers phosphate between creatine and ATP is called *creatine kinase* (*ki′nās*), or *creatine phosphokinase*. Skeletal muscle and heart muscle have two different forms of this enzyme (they have different isoenzymes). The skeletal muscle isoenzyme is found to be elevated in the blood of people with muscular dystrophy (degenerative disease of skeletal muscles). The plasma concentration of the isoenzyme characteristic of heart muscle is elevated as a result of myocardial infarction (damage to heart muscle). Measurements of this enzyme are thus used for diagnostic purposes (see "Clinical Considerations" in chapter 4).

Slow- and Fast-Twitch Fibers

Skeletal muscle fibers can be divided on the basis of their contraction speed (time required to reach maximum tension) into **slow-twitch,** or **type I, fibers,** and **fast-twitch,** or **type II, fibers.** These differences are associated with different

Figure 12.19

Skeletal muscle (of a cat) stained to indicate the activity of myosin ATPase.
ATPase activity is greater in the type II fibers than in the type I fibers.

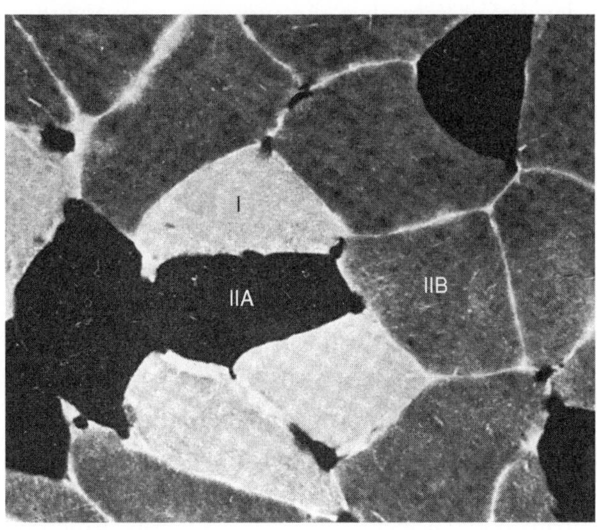

Table 12.4

Characteristics of Red, Intermediate, and White Muscle Fibers

	Red (Type I)	Intermediate (Type IIA)	White (Type IIB)
Diameter	Small	Intermediate	Large
Z-line thickness	Wide	Intermediate	Narrow
Glycogen content	Low	Intermediate	High
Resistance to fatigue	High	Intermediate	Low
Capillaries	Many	Many	Few
Myoglobin content	High	High	Low
Respiration	Aerobic	Aerobic	Anaerobic
Twitch rate	Slow	Fast	Fast
Myosin ATPase content	Low	High	High

Figure 12.20

A comparison of the rates with which maximum tension is developed in three muscles.
These are (*a*) the relatively fast-twitch extraocular, (*b*) the gastrocnemius muscles, and (*c*) the slow-twitch soleus muscle.

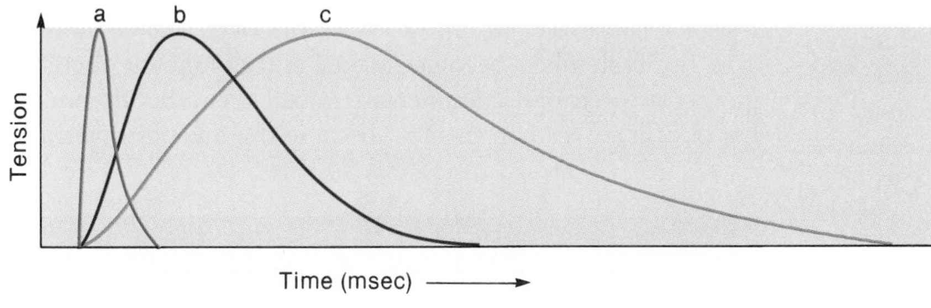

myosin ATPase isoenzymes, which can also be designated as "slow" and "fast." The two fiber types can be distinguished by their ATPase isoenzyme when they are appropriately stained (fig. 12.19). The extraocular muscles that position the eyes, for example, have a high proportion of fast-twitch fibers and reach maximum tension in about 7.3 msec (milliseconds—thousandths of a second); the soleus muscle in the leg, by contrast, has a high proportion of slow-twitch fibers and requires about 100 msec to reach maximum tension (fig. 12.20).

Muscles like the soleus are *postural muscles;* they are able to sustain a contraction for a long period of time without fatigue. The resistance to fatigue demonstrated by these muscles is aided by other characteristics of slow-twitch (type I) fibers that endow them with a high oxidative capacity for aerobic respiration. Slow-twitch fibers have a rich capillary supply, numerous mitochondria and aerobic respiratory enzymes, and a high concentration of *myoglobin* (*mi″ŏ-glo′bin*). Myoglobin is a red pigment, similar to the hemoglobin in red blood cells, that improves the delivery of oxygen to the slow-twitch fibers. Because of their high myoglobin content, slow-twitch fibers are also called *red fibers*.

The thicker, fast-twitch (type II) fibers have fewer capillaries and mitochondria than slow-twitch fibers and not as much myoglobin; hence, these fibers are also called *white fibers*. Fast-twitch fibers are adapted to respire anaerobically by a large store of glycogen and a high concentration of glycolytic enzymes. In addition to the type I (slow-twitch) and type II (fast-twitch) fibers, human muscles also have an intermediate form of fibers. These intermediate fibers are fast-twitch but also have a high oxidative capacity, and therefore they are relatively resistant to fatigue. They are called **type IIA fibers,** to distinguish them from the anaerobically adapted fast-twitch **type IIB fibers,** which have a low oxidative capacity and fatigue rapidly. The three fiber types are compared in table 12.4.

Interestingly, the conduction rate of motor neurons that innervate fast-twitch fibers is faster (80–90 meters per second) than the conduction rate to slow-twitch fibers (60–70 meters per second). The fiber type indeed seems to be determined by the motor neuron. When the motor neurons to different fiber types are switched in experimental animals, the previously fast-twitch fibers become slow, and the slow-twitch fibers become fast. As expected from these observations, all of the muscle fibers innervated by the same motor neuron (that are part of the same motor unit) are of the same type.

A muscle such as the gastrocnemius contains both fast- and slow-twitch fibers, although fast-twitch fibers predominate. A given somatic motor axon, however, innervates muscle fibers of only one type. The size of these motor units differ; the motor units composed of slow-twitch fibers tend to be smaller (have fewer fibers) than the motor units of fast-twitch fibers. As mentioned earlier, motor units are recruited from smaller to larger when increasing effort is required; thus, the smaller motor units with slow-twitch fibers would be used most often in routine activities. Larger motor units with fast-twitch fibers, which can exert a great deal of force but which respire anaerobically and thus fatigue quickly, would be used relatively infrequently and for only short periods of time.

Muscle Fatigue

Muscle fatigue may be defined as the inability to maintain a particular muscle tension when the contraction is sustained or to reproduce a particular tension during rhythmic contractions. Fatigue during a sustained maximal contraction, when all the motor units are used and the rate of neural firing is maximal—as when lifting an extremely heavy weight—appears to be due to an accumulation of extracellular K^+. (As will be explained in chapter 14, K^+ leaves axons and muscle fibers during the course of action potentials.) This reduces the membrane potential of muscle fibers and interferes with their ability to produce action potentials. Fatigue under these circumstances lasts only a short time, and maximal tension can again be produced after less than a minute's rest.

Fatigue during moderate exercise occurs as the slow-twitch fibers deplete their reserve glycogen and fast-twitch fibers are increasingly recruited. Fast-twitch fibers obtain their energy through anaerobic respiration, converting glucose to lactic acid, and this results in a rise in intracellular H^+ and a fall in pH. The decrease in muscle pH, in turn, promotes muscle fatigue, but the exact physiological mechanisms by which this occurs are not well understood. One possibility is that there may be a reduced ability of the sarcoplasmic reticulum to accumulate Ca^{2+} by active transport, or there may be a reduced ability of the sarcoplasmic reticulum to release Ca^{2+} in response to stimulation. By either mechanism, the decrease cellular pH would produce muscle fatigue by interfering with excitation-contraction coupling.

Adaptation of Muscles to Exercise Training

The maximal oxygen uptake obtained during very strenuous exercise averages 50 ml of O_2 per minute per kilogram body weight in males between the ages of 20 and 25 (females average 25% lower). For trained endurance athletes (such as swimmers and long-distance runners), maximal oxygen uptakes can be as high as 86 ml of O_2 per minute per kilogram. These considerable differences affect the lactate threshold, and thus the amount of exercise that can be performed before lactic acid production contributes to muscle fatigue. In addition to having a higher aerobic capacity, well-trained athletes also have a lactate threshold that is a higher percentage of their $\dot{V}_{O_2}$max. The lactate threshold of an untrained person, for example, might be 60% of the $\dot{V}_{O_2}$max, whereas the lactate threshold of a trained athlete can be up to 80% of the $\dot{V}_{O_2}$max. These athletes thus produce less lactic acid at a given level of exercise than the average person, and therefore they are less subject to fatigue than the average person.

Since the depletion of muscle glycogen places a limit on exercise, any adaptation that spares muscle glycogen will improve physical endurance. This is achieved in trained athletes by an increased proportion of energy that is derived from the aerobic respiration of fatty acids, resulting in a slower depletion of their muscle glycogen. The greater the level of physical training, the higher the proportion of energy derived from the oxidation of fatty acids during exercise below the $\dot{V}_{O_2}$max.

All fiber types adapt to endurance training by an increase in mitochondria, and thus in aerobic respiratory enzymes. In fact, the maximal oxygen uptake can be increased by as much as 20% through endurance training. There is a decrease in type IIB fibers, which have a low oxidative capacity, accompanied by an increase in type IIA fibers, which have a high oxidative capacity. Although the type IIA fibers are still classified as fast-twitch, they show an increase in the slow myosin ATPase isoenzyme form, indicating that they are in a transitional state between type II and type I fibers. A summary of the changes that occur as a result of endurance training is presented in table 12.5.

Endurance training does not increase the size of muscles. Muscle enlargement is produced only by frequent periods of high-intensity exercise in which muscles work against a high resistance, as in weight lifting. As a result of resistance training, type II muscle fibers become thicker, and the muscle therefore grows by hypertrophy (an increase in cell size rather than number of cells). This happens first because the myofibrils within a

Table 12.5

The Effects of Endurance Training* on Skeletal Muscles

1. Improved ability to obtain ATP from oxidative phosphorylation
2. Increased size and number of mitochondria
3. Less lactic acid produced per given amount of exercise
4. Increased myoglobin content
5. Increased intramuscular triglyceride content
6. Increased lipoprotein lipase (enzyme needed to utilize lipids from blood)
7. Increased proportion of energy derived from fat; less from carbohydrates
8. Lower rate of glycogen depletion during exercise
9. Improved efficiency in extracting oxygen from blood

*Long-distance running, swimming, bicycling, etc.

muscle fiber thicken as a result of the synthesis of actin and myosin proteins and the addition of new sarcomeres. Then, after a myofibril has attained a certain thickness it may split into two myofibrils, each of which may become thicker as a result of the addition of sarcomeres. Muscle hypertrophy, in short, is associated with an increase in the size of the muscle fibers and then in the number of myofibrils within the muscle fibers.

The decline in physical strength of older people is associated with a reduced muscle mass, which is due to a decrease in the size of fast-twitch (type IIA and IIB) muscle fibers. Aging is also associated with a reduced density of blood capillaries surrounding the muscle fibers, leading to a decrease in oxidative capacity. These changes are partially caused by a more sedentary lifestyle and can be largely reversed through physical training. Resistance training has been shown to increase muscle mass in older people; endurance training increases the density of blood capillaries in the muscles. The muscle glycogen of older people also can be increased by endurance training, but it cannot be raised to the levels present in youth.

Cardiac and Smooth Muscle

Cardiac muscle, like skeletal muscle, is striated and contains sarcomeres that shorten by sliding of thin and thick filaments. But while skeletal muscle requires nervous stimulation to contract, cardiac muscle can produce impulses and contract spontaneously. Smooth muscles lack sarcomeres, but they do contain actin and myosin that produce contractions in response to a unique regulatory mechanism.

Unlike skeletal muscles, which are voluntary effectors regulated by somatic motor neurons, cardiac and smooth muscles are involuntary effectors regulated by autonomic motor neurons. Although there are important differences between skeletal muscle and cardiac and smooth muscle, there are also significant similarities. All types of muscle are believed to contract by means of sliding of thin filaments over thick filaments. The sliding of the filaments is produced by the action of myosin cross bridges in all types of muscles, and excitation-contraction coupling in all types of muscles involves Ca^{2+}.

Cardiac Muscle

Like skeletal muscle cells, cardiac (heart) muscle cells, or **myocardial cells,** are striated; they contain actin and myosin filaments arranged in the form of sarcomeres, and they contract by means of the sliding filament mechanism. The long, fibrous skeletal muscle cells, however, are structurally and functionally separated from each other, whereas the myocardial cells are short, branched, and interconnected. Adjacent myocardial cells are joined by electrical synapses, or **gap junctions** (described in chapter 14). Gap junctions in cardiac muscle have an affinity for stain that makes them appear as dark lines between adjacent cells when viewed in the light microscope. These dark-staining lines are known as **intercalated** (*in-ter'kă-lāt-ed*) **discs** (see table 12.1).

Electrical impulses that originate at any point in a mass of myocardial cells, called a **myocardium,** can spread to all cells in the mass that are joined by gap junctions. Because all cells in a myocardium are electrically joined, a myocardium behaves as a single functional unit. Thus, unlike skeletal muscles that produce contractions that are graded depending on the number of cells stimulated, a myocardium contracts to its full extent each time because all of its cells contribute to the contraction. The ability of the myocardial cells to contract, however, can be increased by the hormone epinephrine (adrenalin) and by stretching of the heart chambers. The heart contains two distinct myocardia (atria and ventricles), as will be described in chapter 21.

Unlike skeletal muscles, which require external stimulation by somatic motor nerves before they can produce action potentials and contract, cardiac muscle is able to produce action potentials automatically. Cardiac action potentials normally originate in a specialized group of cells called the *pacemaker*. However, the rate of this spontaneous electrical activity, and thus the rate of the heartbeat, are regulated by autonomic innervation. Regulation of the cardiac rate is described more fully in chapter 22.

Smooth Muscle

Smooth (visceral) muscles are arranged in circular layers around the walls of blood vessels and bronchioles (small air passages in the lungs). Both circular and longitudinal smooth muscle layers occur in the tubular digestive tract, the ureters (which transport urine), the ductus deferentia (which transport sperm), and the uterine tubes (which transport ova). The alternate contraction of circular and longitudinal smooth muscle layers in the intestine produces **peristaltic** (*per"ĭ-stal'tik*) **waves,** which propel the contents of these tubes in one direction.

Although smooth muscle cells do not contain sarcomeres (which produce striations in skeletal and cardiac muscle), they do contain a great deal of actin and some myosin. The ratio of thin-to-thick filaments in smooth muscle is about 16:1 (in striated muscles the ratio is 2:1). Unlike striated muscles, in which the thick filaments are short and stacked between Z discs in sarcomeres, the myosin filaments in smooth muscle cells are quite long (fig. 12.21). These filaments are attached at each end of the cell to **dense bodies,** which are analogous to the Z discs of striated muscles.

The long length of myosin filaments and the fact that they are not organized into sarcomeres may be of advantage in smooth muscle function. Smooth muscles have the ability to contract even when greatly stretched—in the urinary bladder, for example, the smooth muscle cells may be stretched up to two-and-a-half times their resting length. The smooth muscle cells of the uterus may be stretched up to eight times their original length by the end of pregnancy. Striated muscles, because of their structure, lose their ability to contract when the sarcomeres are stretched to the point where actin and myosin no longer overlap.

Figure 12.21

Electron micrograph of smooth muscle.

The thick and thin filaments can be seen in this smooth muscle cell. A longitudinal section of a complete long myosin filament is shown between the arrows (32,000×).

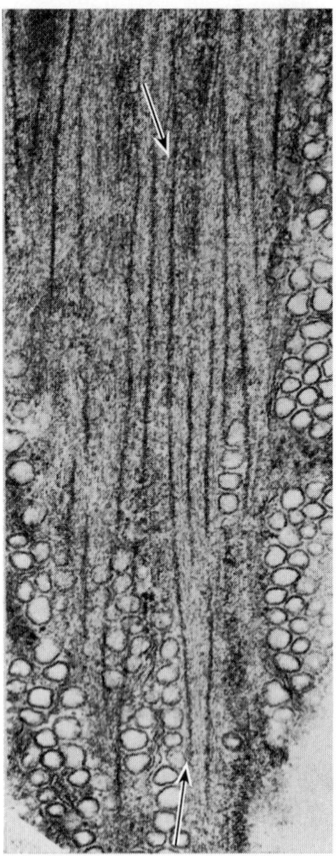

As in striated muscles, the contraction of smooth muscles is triggered by a sharp rise in the Ca^{2+} concentration within the cytoplasm of the muscle cells. However, the sarcoplasmic reticulum of smooth muscles is less developed than that of skeletal muscles, and Ca^{2+} released from this organelle may account for only the initial phase of smooth muscle contraction. Extracellular Ca^{2+} diffusing into the smooth muscle cell through its cell membrane is responsible for sustained contractions. This Ca^{2+} enters primarily through voltage-regulated Ca^{2+} channels in the cell membrane. The opening of these channels is graded by the amount of depolarization; the greater the depolarization, the more Ca^{2+} will enter the cell and the stronger the smooth muscle contraction.

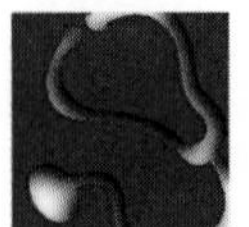

Drugs such as *nifedipine* and related newer compounds are *calcium channel blockers.* These drugs block Ca^{2+} channels in the membrane of smooth muscle cells within the walls of blood vessels, causing the muscles to relax and the vessels to dilate. This effect, called vasodilation, may be helpful in treating some cases of hypertension (high blood pressure). Calcium-channel-blocking drugs are also used when spasm of the coronary arteries (vasospasm) produces angina pectoris, which is pain caused by insufficient blood flow to the heart.

The events that follow the entry of Ca^{2+} into the cytoplasm are somewhat different in smooth muscles than in striated muscles. In striated muscles, Ca^{2+} combines with troponin. Troponin, however, is not present in smooth muscle cells. In smooth muscles, Ca^{2+} combines with a protein in the cytoplasm called *calmodulin* (*kal-mod'yŭ-lin*), which is structurally similar to troponin. The calmodulin-Ca^{2+} complex thus formed combines with and activates **myosin light-chain kinase,** an enzyme that catalyzes the phosphorylation of (addition of phosphate groups to) the myosin cross bridges. In smooth muscle cells, the cross bridges must be phosphorylated before they can bind to actin, which is not the case in striated muscles.

Relaxation of the smooth muscle follows the closing of the Ca^{2+} channels and lowering of the cytoplasmic Ca^{2+} concentrations. Under these conditions, calmodulin dissociates from the myosin light-chain kinase and thereby inactivates this enzyme. The phosphate groups that were added to the myosin are then removed by a different enzyme—a phosphatase. Dephosphorylation inhibits the cross bridge from binding to actin and undergoing another power stroke.

It is the concentration of Ca^{2+} in the cytoplasm of a smooth muscle cell that determines the number of cross bridges that will combine with actin, and thus the strength of smooth muscle contraction. The concentration of Ca^{2+} is in turn regulated by the degree of depolarization. Unlike the situation in striated muscle cells, smooth muscle cells can produce graded depolarizations and contractions without producing action potentials. (This is discussed in more detail in conjunction with intestinal contractions in chapter 26.) Indeed, only these graded depolarizations are conducted from cell to cell in many smooth muscles.

In addition to being graded, the contractions of smooth muscle cells are slow and sustained. The slowness of contraction is related to the fact that myosin ATPase in smooth muscle is slower in its action (splitting ATP for the cross-bridge cycle) than it is in striated muscle. The sustained nature of smooth muscle contraction is explained by the theory that cross bridges in smooth muscles can enter a *latch state*.

The latch state allows smooth muscle to maintain its contraction in a very energy-efficient manner, hydrolyzing less ATP than would otherwise be required. This ability is obviously important for smooth muscles, given that they encircle the walls of hollow organs and must sustain contractions for long periods of time. The mechanisms by which the latch state is produced, however, are complex and poorly understood.

The three muscle types—skeletal, cardiac, and smooth—are compared in table 12.6.

Single-Unit and Multiunit Smooth Muscles

Smooth muscles are often grouped into two functional categories: **single-unit** and **multiunit.** Single-unit smooth muscles have numerous gap junctions (electrical synapses) between adjacent cells that weld them together electrically; thus, they behave as a single unit, much like cardiac muscle. Most smooth muscles—including those in the digestive tract and uterus—are single-unit.

Table 12.6
Comparison of Skeletal, Cardiac, and Smooth Muscle

Skeletal Muscle	Cardiac Muscle	Smooth Muscle
Striated; actin and myosin arranged in sarcomeres	Striated; actin and myosin arranged in sarcomeres	Not striated; more actin than myosin; actin inserts into dense bodies and cell membrane
Well-developed sarcoplasmic reticulum and transverse tubules	Moderately developed sarcoplasmic reticulum and transverse tubules	Poorly developed sarcoplasmic reticulum; no transverse tubules
Contains troponin in the thin filaments	Contains troponin in the thin filaments	Contains calmodulin, a protein that, when bound to Ca^{2+}, activates the enzyme myosin light-chain kinase
Ca^{2+} released into cytoplasm from sarcoplasmic reticulum	Ca^{2+} enters cytoplasm from sarcoplasmic reticulum and extracellular fluid	Ca^{2+} enters cytoplasm from extracellular fluid, sarcoplasmic reticulum, and perhaps mitochondria
Cannot contract without nerve stimulation; denervation results in muscle atrophy	Can contract without nerve stimulation; action potentials originate in pacemaker cells of heart	Maintains tone in absence of nerve stimulation; visceral smooth muscle produces pacemaker potentials; denervation results in hypersensitivity to stimulation
Muscle fibers stimulated independently; no gap junctions	Gap junctions present as intercalated discs	Gap junctions in most smooth muscles

Only some cells of single-unit smooth muscles receive autonomic innervation, but the acetylcholine released by the axon can diffuse to a number of smooth muscle cells. Such stimulation, however, only modifies the automatic behavior of single-unit smooth muscles. Single-unit smooth muscles display pacemaker activity, in which certain cells stimulate others in the mass. This is similar to the situation in cardiac muscle. Single-unit smooth muscles also display intrinsic, or *myogenic*, electrical activity and contraction in response to stretch. For example, the stretch induced by an increase in the volume of a ureter or a section of the digestive tract can stimulate myogenic contraction. Such contraction does not require stimulation by autonomic nerves.

Contraction of multiunit smooth muscles, by contrast, requires nerve stimulation. Multiunit smooth muscles have few, if any, gap junctions. The cells must thus be stimulated individually by nerve fibers. Examples of multiunit smooth muscles are the arrector pili muscles in the skin and the ciliary muscles attached to the lens of the eye.

Clinical Investigation Answer

Since this woman has a high maximal oxygen uptake, she should have good endurance with little fatigue and pain during exercise. The fact that her muscles are not large but have good tone supports her statement that she frequently engages in endurance-type exercise. The normal concentration of creatine phosphokinase suggests that her skeletal muscles and heart may not be damaged. Further tests should be done to confirm this, however, particularly since she has a history of hypertension. Excessive workouts could account for her fatigue and muscle pain, but the high blood Ca^{2+} concentration suggests another possibility. The high blood Ca^{2+} could be responsible for her excessively high muscle tone; this inability of her muscles to relax might, in fact, be responsible for the pain and fatigue. This person, therefore, should undergo an endocrinological workup (for parathyroid hormone, for example) to determine the cause of her high blood Ca^{2+} levels.

Chapter Summary

Structure and Actions of Skeletal Muscles (pp. 281–288)

1. Skeletal muscles are attached to bones by means of tendons.
   (a) Muscles are separated by fascia, and each muscle is covered by an epimysium.
   (b) Muscle fibers are grouped into fasciculi that are surrounded by a connective tissue perimysium. Each fasciculus is composed of muscle fibers that are surrounded by an endomysium.
2. Skeletal muscle fibers originate from a number of myoblasts that eventually join together to form the multinucleated skeletal muscle fiber.
3. Skeletal and cardiac muscle fibers are striated, whereas smooth muscle fibers are not.
4. Muscles in vitro can exhibit twitch, summation, tetanus, and tonus.
   (a) The rapid contraction and relaxation of muscle fibers is called a twitch.
   (b) The stronger the electric shock, the stronger the muscle twitch. Whole muscles are capable of graded contractions because the number of fibers participating in the contraction varies.

(c) The summation of fiber twitches can occur so rapidly that the muscle produces a smooth, sustained contraction known as tetanus.
(d) When a muscle exerts tension without shortening, the contraction is termed isometric; when shortening does occur, the contraction is isotonic.

5. The contraction of muscle fibers in vivo is stimulated by somatic motor neurons.
   (a) Each somatic motor axon branches to innervate a number of muscle fibers.
   (b) A motor unit consists of a single motor neuron and the muscle fibers it innervates.

Mechanisms of Contraction (pp. 289–296)

1. The banding pattern of skeletal muscle fibers is produced by an orderly arrangement of myofilaments.
   (a) The A bands contain thick filaments, composed of the protein myosin; the edges of each A band also contain thin filaments overlapped by the thick filaments.
   (b) The central regions of the A bands contain only thick filaments; these regions are called the H bands.
   (c) The I bands contain only thin filaments, composed primarily of the protein actin.
   (d) The filaments slide, not shorten, during muscle contraction.
   (e) The lengths of the H and I bands decrease during contraction, whereas the A bands stay the same length.
2. Myosin cross bridges extend from the thick filaments to the thin filaments.
3. The activity of the cross bridges causes the thin filaments to slide toward the centers of the sarcomeres.
   (a) The cross-bridge heads function as myosin ATPase enzymes.
   (b) ATP is split into ADP and P_i, activating the cross bridge, prior to attachment of the cross bridge to actin.
4. When the activated cross bridges attach to actin, they undergo a power stroke.
5. At the end of a power stroke, the cross bridge binds to a new ATP. This allows the cross bridge to detach from actin and repeat the cycle.
6. A protein known as tropomyosin is located at intervals within the thin myofilaments. Another protein, troponin, is attached to the tropomyosin.
7. When a muscle is at rest, the Ca^{2+} concentration of the sarcoplasm is very low. Cross bridges are prevented from attaching to actin by the position of tropomyosin in the thin filaments.
   (a) Ca^{2+} is actively transported into the sarcoplasmic reticulum when a muscle is at rest.
   (b) Electrical impulses, conducted by transverse tubules into the muscle fiber, stimulate the release of Ca^{2+} from the sarcoplasmic reticulum.
   (c) Ca^{2+} binds to troponin, and this causes tropomyosin to shift position so that the myosin cross bridges can bind to actin.
8. When electrical impulses cease, Ca^{2+} is removed from the sarcoplasm by active transport and returned to the sarcoplasmic reticulum.

Energy Requirements of Skeletal Muscles (pp. 297–301)

1. Aerobic cell respiration is ultimately required for the production of ATP needed for cross-bridge activity.
2. New ATP can be quickly produced, however, from the combination of ADP with phosphate derived from phosphocreatine.
3. Resting muscles and muscles performing light exercise derive most of their energy from the aerobic respiration of free fatty acids.
   (a) During moderate exercise, muscles obtain energy both from fatty acids and from glucose derived from muscle glycogen and blood plasma.
   (b) During heavy exercise, muscles use carbohydrates to a greater extent, although the muscles of endurance trained athletes can use more fatty acids than the muscles of other people. Because their muscle glycogen is spared, these athletes have a higher aerobic capacity than most people.
4. Muscle fibers are of three types.
   (a) Slow-twitch red fibers are adapted for aerobic respiration and are resistant to fatigue.
   (b) Fast-twitch white fibers are adapted for anaerobic respiration.
   (c) Intermediate fibers are fast-twitch but adapted for aerobic respiration.
5. Endurance training increases the aerobic capacity of all muscle fiber types, thus decreasing their reliance on anaerobic respiration and their susceptibility to fatigue.

Cardiac and Smooth Muscle (pp. 301–303)

1. Cardiac muscle is striated and contains sarcomeres. Smooth muscle lacks striations but does contain actin and myosin.
   (a) Electrical impulses in cardiac muscle originate in myocardial fibers. These impulses can cross from one myocardial fiber to another through gap junctions.
   (b) Electrical impulses are produced spontaneously in the heart.
2. Smooth muscle fibers contain myosin and actin, but they are not arranged in sarcomeres.
   (a) Myosin myofilaments are very long; consequently, smooth muscle fibers can contract even when they are greatly stretched.
   (b) When electrically stimulated, Ca^{2+} enters smooth muscle fibers and combines with calmodulin. The calmodulin-Ca^{2+} complex activates an enzyme that phosphorylates myosin cross bridges.

Review Activities

Objective Questions

1. A graded whole muscle contraction is produced in vivo primarily by variations in
   (a) the strength of the fiber's contraction.
   (b) the number of fibers that are contracting.
   (c) both of the above.
   (d) neither of the above.
2. The series-elastic component of muscle contraction is responsible for
   (a) increased muscle shortening to successive twitches.
   (b) a time delay between contraction and shortening.
   (c) the lengthening of muscle after contraction has ceased.
   (d) all of the above.
3. Which of the following muscles have motor units with the highest innervation ratio?
   (a) leg muscles
   (b) arm muscles
   (c) muscles that move the fingers
   (d) muscles of the trunk
4. When a skeletal muscle shortens during contraction, which of the following statements is *false?*
   (a) The A bands shorten.
   (b) The H bands shorten.

(c) The I bands shorten.
(d) The sarcomeres shorten.

5. Electrical excitation of a muscle fiber *most directly* causes
   (a) movement of tropomyosin.
   (b) attachment of the cross bridges to actin.
   (c) release of Ca^{2+} from the sarcoplasmic reticulum.
   (d) splitting of ATP.
6. The energy for muscle contraction is *most directly* obtained from
   (a) phosphocreatine.
   (b) ATP.
   (c) anaerobic respiration.
   (d) aerobic respiration.
7. Which of the following statements about cross bridges is *false?*
   (a) They are composed of myosin.
   (b) They bind to ATP after they detach from actin.
   (c) They contain an ATPase.
   (d) They split ATP before they attach to actin.
8. When a muscle is stimulated to contract, Ca^{2+} binds to
   (a) myosin.
   (b) tropomyosin.
   (c) actin.
   (d) troponin.
9. Which of the following statements about muscle fatigue is *false?*
   (a) It may result when ATP is no longer available for the cross-bridge cycle.
   (b) It may be caused by a loss of muscle cell Ca^{2+}.
   (c) It may be caused by the accumulation of extracellular K^{+}.
   (d) It may be a result of lactic acid production.
10. Which of the following types of muscle cells are *not* capable of spontaneous depolarization?
    (a) single-unit smooth muscle
    (b) multiunit smooth muscle
    (c) cardiac muscle
    (d) skeletal muscle
    (e) both *b* and *d*
    (f) both *a* and *c*
11. Which of the following muscle types is striated and contains gap junctions?
    (a) single-unit smooth muscle
    (b) multiunit smooth muscle
    (c) cardiac muscle
    (d) skeletal muscle
12. In an isotonic muscle contraction,
    (a) the length of the muscle remains constant.
    (b) the muscle tension remains constant.
    (c) both muscle length and tension are changed.
    (d) movement of bones does not occur.

Essay Questions

1. Using the concept of motor units, explain how skeletal muscles in vivo produce graded and sustained contractions.
2. Describe how an isometric contraction can be converted into an isotonic contraction using the concepts of motor unit recruitment and the series-elastic component of muscles.
3. Trace the sequence of events in which the cross bridges attach to the thin filaments when a muscle is stimulated by a nerve. Why don't the cross bridges attach to the thin filaments when a muscle is relaxed?
4. Using the sliding filament theory of contraction, explain why the contraction strength of a muscle is maximal at a particular muscle length.
5. Explain the role of ATP in muscle contraction and muscle relaxation.
6. Why are all the muscle fibers of a given motor unit of the same type? Why are smaller motor units and slow-twitch muscle fibers used more frequently than larger motor units and fast-twitch fibers?
7. What changes occur in muscle metabolism as the intensity of exercise increases? Describe the changes that occur as a result of endurance training and explain how these changes raise the level of exercise that can be performed before the onset of muscle fatigue.
8. Compare the mechanism of excitation-coupling in striated muscle with that in smooth muscle.
9. Compare cardiac muscle, single-unit smooth muscle, and multiunit smooth muscle in terms of the regulation of their contraction.

Critical Thinking Questions

1. In the sixteenth century, Andreas Vesalius demonstrated that cutting a muscle along its length has very little effect on its function; on the other hand, a transverse cut puts a muscle out of action. How would you explain Vesalius's findings?
2. Compare muscular dystrophy and amyotrophic lateral sclerosis (ALS) in terms of their causes and their effects on muscles.
3. Why is it important to have a large amount of stored high-energy phosphates in the form of creatine phosphate for the function of muscles during exercise? What might happen to a muscle in your body if it ever ran out of ATP?
4. How is electrical excitation of a skeletal muscle fiber coupled to muscle contraction? Speculate why the exact mechanism of this coupling has been difficult to determine.
5. How would a rise in the extracellular Ca^{2+} concentration affect the beating of a heart? Explain the mechanisms involved. Lowering the blood Ca^{2+} concentration can cause muscle spasms. What might be responsible for this effect?

Related Web Sites

For a listing of the most current web sites related to this chapter, please visit the *Concepts of Human Anatomy and Physiology* home page at http://www.mhhe.com/biosci/abio/.

THIRTEEN

Muscular System

Objectives

- Explain how muscles are described according to their location and cooperative function.
- Explain what is meant by synergistic and antagonistic muscle groups.
- Describe the various ways in which muscle fibers are arranged and discuss the advantage of each of these arrangements.
- Use examples to describe the various ways in which muscles are named.
- Locate the major muscles of the axial skeleton. Describe the action of synergistic and antagonistic muscles, using specific muscles as examples.
- Locate the major muscles of the appendicular skeleton. Identify synergistic and antagonistic muscles and describe their action.

Clinical Investigation

A 66-year-old man went to a doctor for a routine physical exam. The man's medical history revealed that he had been treated surgically for cancer of the oropharynx 6 years earlier. The patient stated that the cancer had spread to the lymph nodes in the left side of his neck. He pointed to the involved area, explaining that lymph nodes, a vein, and a muscle, among other things, had been removed. On the right side, only lymph nodes had been removed, and they were not cancerous. The patient then stated that he had difficulty turning his head to the right. Obviously perplexed, he commented, "It seems to me Doc, that if they took the muscle out of the left side of my neck, I would be able to turn my head only to the right."

Does the patient have a valid point? If not, how would you explain the reason for his disability in terms of neck musculature?

Clues: The action of a muscle can always be explained on the basis of its points of attachment and the joint or joints it spans. Carefully examine the muscles shown in figure 13.9 and described in table 13.6.

Organization of the Muscular System

Skeletal muscles are arranged in functional groups that are adaptive in causing particular movements. Within each muscle, the fibers are arranged in a specific pattern that provides specific functional capabilities.

More than 600 skeletal muscles make up the muscular system, and technically each one is an organ—it is composed of skeletal muscle tissue, connective tissue, and nervous tissue. Each muscle also has a particular function, such as moving a finger or blinking an eyelid. Collectively, the skeletal muscles account for approximately 40% of the body weight.

In describing the various muscles, they are usually grouped according to anatomical location and cooperative function. The *muscles of the axial skeleton* have their attachments to the bones of the axial skeleton and include facial muscles, neck muscles, and the anterior and posterior trunk muscles. The *muscles of the appendicular skeleton* include those that act on the pectoral and pelvic girdles and those that cause movement at the joints of the upper and lower extremities.

The principal superficial muscles are shown in figure 13.1.

Muscle Groups

Just as individual muscle fibers seldom contract independently, muscles generally do not contract separately but work as functional groups. Muscles that contract together in accomplishing a particular movement are said to be *synergistic* (*sin″er-jis′tik*) (fig. 13.2). *Antagonistic* muscles perform opposite functions and are generally located on the opposite sides of the joint. For example, the two heads of the biceps brachii muscle, together with the brachialis muscle, contract to *flex* the elbow joint. The triceps brachii muscle, the antagonist to the biceps brachii and brachialis muscles, *extends* the elbow as it is contracted.

Seldom does the action of a single muscle cause a movement at a joint. Utilization of several synergistic muscles rather than one massive muscle allows for a division of labor. One muscle may be an important postural muscle, for example, whereas another may be adapted for rapid, powerful contraction.

Muscle Architecture

Skeletal muscles may be classified on the basis of fiber arrangement as parallel, convergent, sphincteral (circular), or pennate (table 13.1). Each type of fiber arrangement provides the muscle with distinct capabilities.

Muscle fiber architecture can be observed on a cadaver or other dissection specimen. If you have the opportunity to learn the muscles of the body from a cadaver, observe the fiber architecture of specific muscles and try to determine the advantages afforded to each muscle by its location and action.

Naming of Muscles

Skeletal muscles are named on the basis of shape, location, attachment, orientation of fibers, relative position, or function.

One of the tasks of a student of anatomy and physiology is to learn the names of the principal muscles of the body. Although this may seem overwhelming, keep in mind that most of the muscles are paired; that is, the right side is the mirror image of the left. To help you further, most muscles have names that are descriptive.

As you study the muscles of the body, consider how each was named. Identify the muscle on the figure referenced in the text narrative and locate it on your own body. Use your body to act out its movement that occurs at the joint. Learning the muscles in this way will simplify the task and make it more meaningful.

The following are some criteria by which the names of muscles have been logically derived:

1. **Shape:** rhomboideus (like a rhomboid); trapezius (like a trapezoid); or denoting the number of heads of origin: triceps (three heads), biceps (two heads)
2. **Location:** pectoralis (in the chest, or pectus); intercostal (between ribs); brachia (arm)

---

synergistic: Gk. *synergein*, cooperate
antagonistic: Gk. *antagonistes*, struggle against

Figure 13.1

The principal superficial skeletal muscles.
(*a*) An anterior view and (*b*) a posterior view.

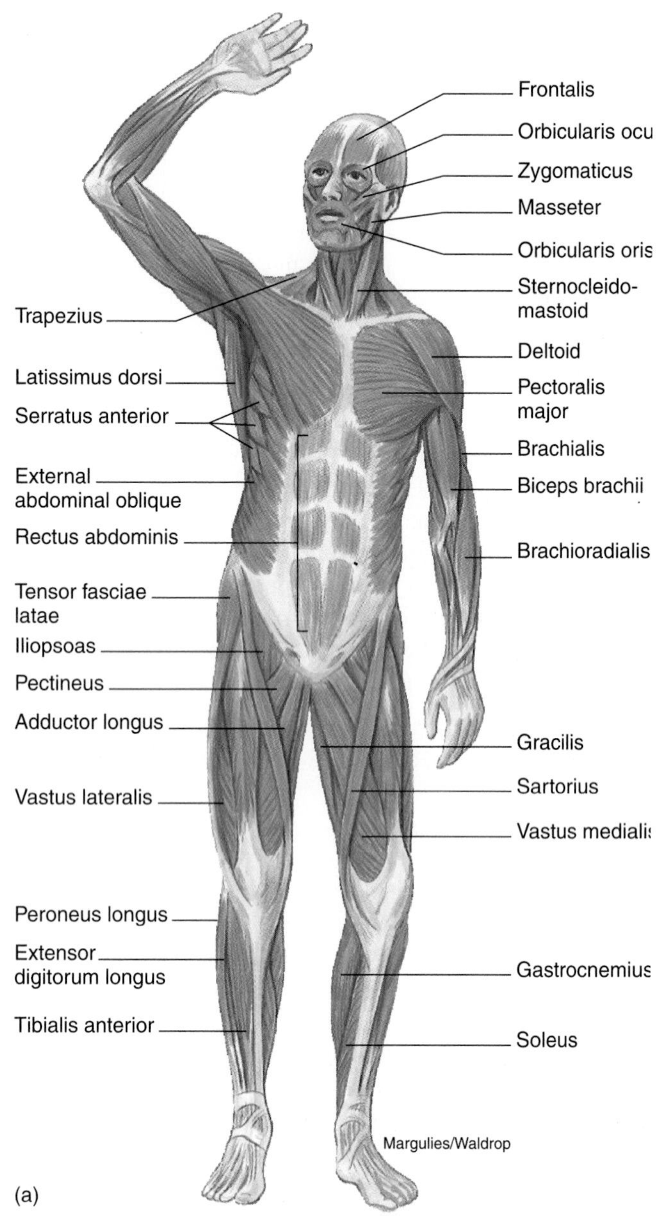

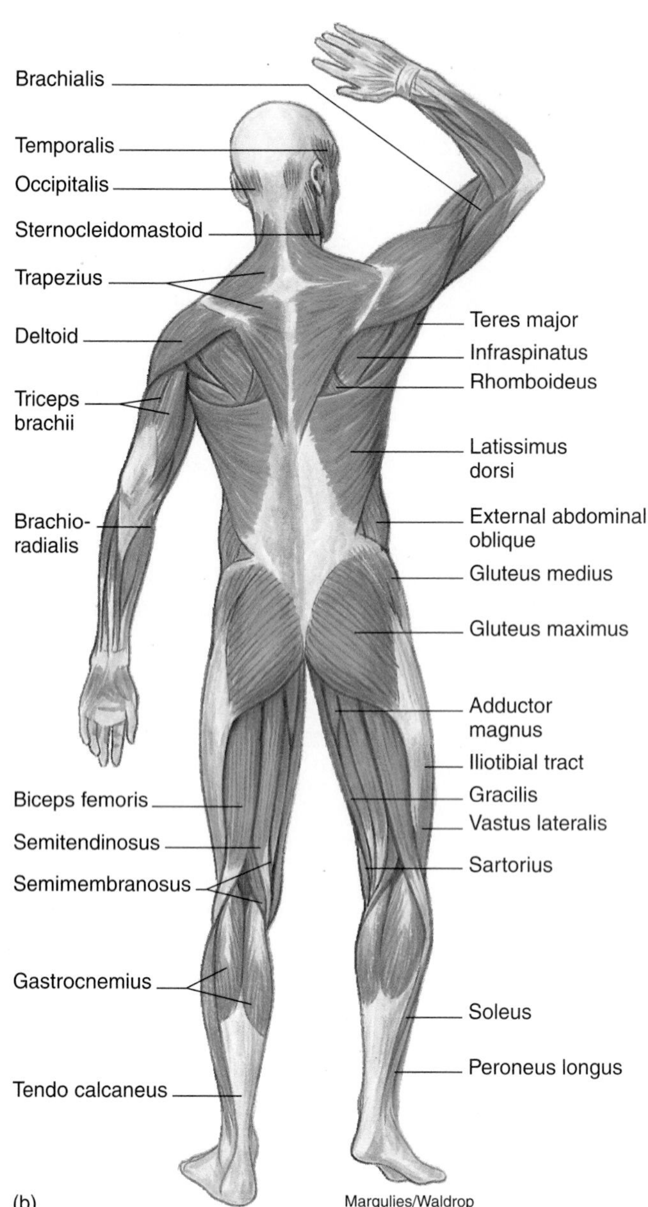

3. **Attachment:** many facial muscles (zygomaticus, temporalis, nasalis); sternocleidomastoid (sternum, clavicle, and mastoid process of the temporal bone)
4. **Size:** maximus, (larger, largest); minimus (smaller, smallest); longus (long); brevis (short)
5. **Orientation of fibers:** rectus (straight); transverse (across); oblique (in a slanting or sloping direction)
6. **Relative position:** lateral, medial, internal, and external
7. **Function:** adductor, flexor, extensor, pronator, and levator (lifter)

Muscles of the Axial Skeleton

Muscles of the axial skeleton include those responsible for facial expression, mastication, eye movement, tongue movement, neck movement, and respiration, and those of the abdominal wall, the pelvic outlet, and the vertebral column.

Muscles of Facial Expression

Humans have a well-developed facial musculature (figs. 13.3 and 13.4) that allows for complex facial expression as a means

Figure 13.2

Examples of synergistic and antagonistic muscles. The two heads of the biceps brachii muscle and the brachialis muscle are synergistic to each other, as are the three heads of the triceps brachii muscle. The biceps brachii and the brachialis are antagonistic to the triceps brachii, and the triceps brachii is antagonistic to the biceps brachii and the brachialis muscles. When one antagonistic group contracts, the other one must relax; otherwise, movement does not occur.

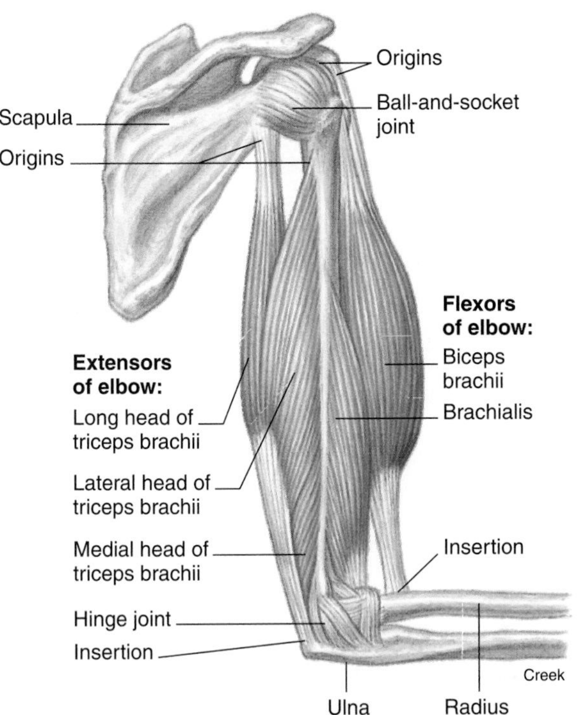

Table 13.1

Muscle Architecture

Type and Description	Appearance
Parallel—straplike; long excursion (contract over a great distance); good endurance; not especially strong; examples: sartorius and rectus abdominis muscles	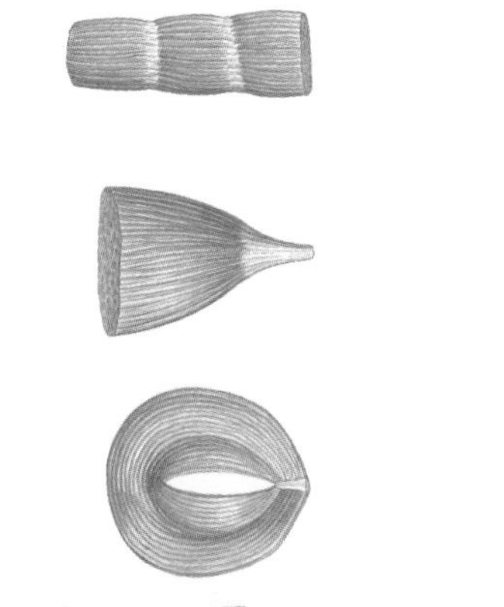
Convergent—fan-shaped; force of contraction focused onto a single point of attachment; stronger than parallel type; examples: deltoid and pectoralis major	
Sphincteral—fibers concentrically arranged around a body opening (*orifice*); act as a sphincter when contracted; examples: orbicularis oculi and orbicularis oris	
Pennate—many fibers per unit area; strong muscles; short excursions; highly dexterous; tire quickly; three types: (a) unipennate, (b) bipennate, and (c) multipennate	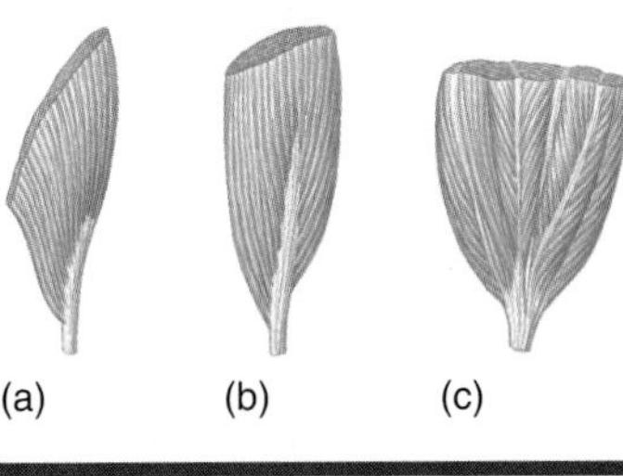

orifice: L. *orificium*, mouth; *facere*, to make
pennate: L. *pennatus*, feather

Figure 13.3

An anterior view of the superficial facial muscles involved in facial expression.

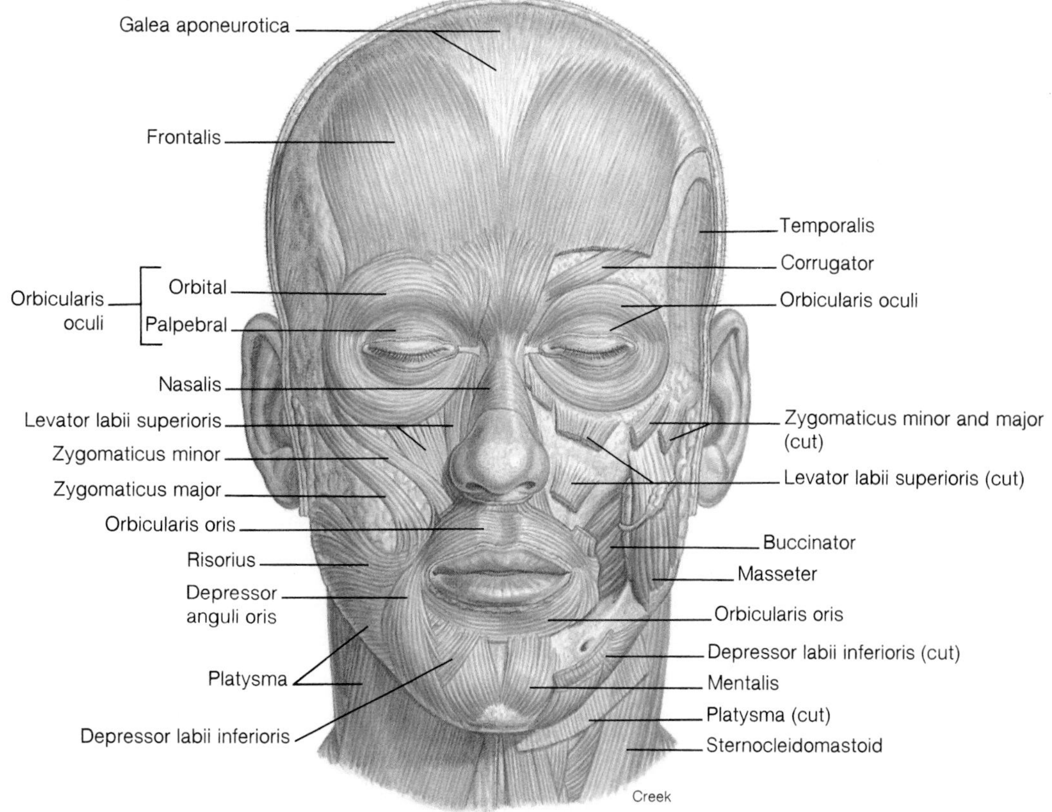

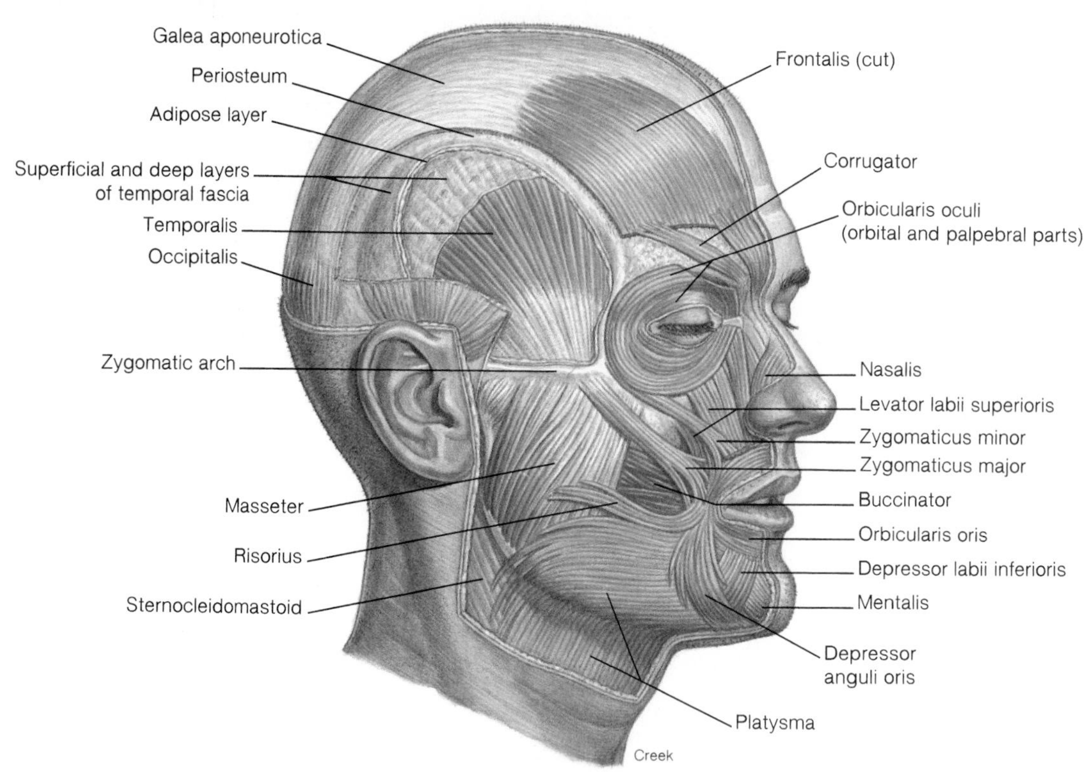

Figure 13.4

A lateral view of the superficial facial muscles involved in facial expression.

Table 13.2

Muscles of Facial Expression

Muscle	Origin	Insertion	Action
Epicranius	Galea aponeurotica and occipital bone	Skin of eyebrow and galea aponeurotica	Wrinkles forehead and moves scalp
Frontalis	Galea aponeurotica	Skin of eyebrow	Wrinkles forehead and elevates eyebrow
Occipitalis	Occipital bone and mastoid process	Galea aponeurotica	Moves scalp backward
Corrugator	Fascia above eyebrow	Root of nose	Draws eyebrow toward midline
Orbicularis oculi	Bones of medial orbit	Tissue of eyelid	Closes eye
Nasalis	Maxilla and nasal cartilage	Aponeurosis of nose	One part widens nostrils; another part depresses nasal cartilages and compresses nostrils
Orbicularis oris	Fascia surrounding lips	Mucosa of lips	Closes and purses lips
Levator labii superioris	Upper maxilla and zygomatic bone	Orbicularis oris and skin above lips	Elevates upper lip
Zygomaticus	Zygomatic bone	Superior corner of orbicularis oris	Depresses corner of mouth
Depressor labii inferioris	Mandible	Oribularis oris at corner of mouth	Draws angle of mouth laterally
Depressor anguli oris	Mandible	Inferior corner of orbicularis oris	Depresses corner of mouth
Depressor labii inferioris	Mandible	Orbicularis oris and skin of lower lip	Depresses lower lip
Mentalis	Mandible (chin)	Orbicularis oris	Elevates and protrudes lower lip
Platysma	Fascia of neck and chest	Inferior border of mandible	Depresses mandible and lower lip
Buccinator	Maxilla and mandible	Orbicularis oris	Compresses cheek

corrugator: L. *corrugo*, a wrinkle
risorius: L. *risor*, a laughter
mentalis: L. *mentum*, chin
platysma: Gk. *platys*, broad
buccinator: L. *bucca*, cheek

of social communication. Very often we let our feelings be known without speaking a word.

The muscles of facial expression are located in a superficial position on the scalp, face, and neck. Although highly variable in size and strength, these muscles all originate on the bones of the skull or in the fascia and insert into the skin (table 13.2). They are all innervated by the facial nerves. The locations and points of attachments of most of the facial muscles are such that, when contracted, they cause movements around the eyes, nostrils, or mouth.

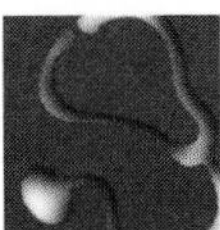

The muscles of facial expression are of clinical concern for several reasons, all of which involve the facial nerve. Located right under the skin, the many branches of the facial nerve are vulnerable to trauma. Facial lacerations and fractures of the skull frequently damage branches of this nerve. The extensive pattern of motor innervation (see fig. 16.8) becomes apparent in stroke victims and persons suffering from *Bell's palsy,* whose facial muscles on one side of the face are affected, causing that side of the face to sag.

Muscles of Mastication

The large **temporalis** and **masseter** (*mă-se′ter*) muscles (fig. 13.5) are powerful elevators of the mandible in conjunction with the **medial pterygoid** (*ter′ĭ-goyd*) muscle. The primary function of the medial and lateral pterygoid muscles is to provide grinding movements of the teeth. The **lateral pterygoid** muscle also protracts the mandible (table 13.3).

Ocular Muscles

The movements of the eyeball are controlled by six extrinsic ocular (eye) muscles (fig. 13.6 and table 13.4). Five of these muscles arise from the margin of the optic foramen at the back of the orbital cavity and insert on the outer layer (sclera) of the eyeball. Four **rectus muscles** maneuver the eyeball in the direction indicated by their names (**superior, inferior, lateral,** and **medial**), and two **oblique muscles** (**superior** and **inferior**) rotate the eyeball on its axis. The medial rectus on one side contracts with the medial rectus of the opposite eye when focusing on close objects. When looking to the side, the lateral rectus of one eyeball works with the medial rectus of the opposite eyeball to keep both eyes functioning together. The superior oblique muscle passes through a pulleylike cartilaginous loop, the *trochlea,* before attaching to the eyeball.

Another muscle, the **levator palpebrae** (*le-va′tor pal′pĕ-bre*) **superioris** muscle (fig. 13.6*b*), is located in the ocular region but is not attached to the eyeball. It extends into the upper eyelid and raises the eyelid when contracted.

Muscles That Move the Tongue

The tongue is a highly specialized muscular organ that functions in speaking, manipulating food, cleansing the teeth, and swallowing. The *intrinsic tongue muscles* are located within the tongue and are responsible for its mobility and changes of shape. The *extrinsic tongue muscles* are those that originate on structures away from the tongue and insert onto it to cause gross tongue movement (fig. 13.7 and table 13.5). The four paired extrinsic muscles are the **genioglossus** (*je-ne″o-glos′us*), **styloglossus, hyoglossus,** and **palatoglossus.** When the anterior portion of the genioglossus muscle contracts, the tongue is depressed and thrust forward. If both genioglossus muscles contract together along their entire lengths, the superior surface of the tongue becomes transversely concave. This muscle is extremely important in nursing infants; the tongue positions itself around the nipple with a concave groove channeled toward the pharynx.

Muscles of the Neck

Muscles of the neck either support and move the head or are attached to structures within the neck region, such as the hyoid bone and larynx. Only the more obvious neck muscles will be considered in this chapter. You can observe on your own body the location of many of the muscles in this section and those that follow. These muscles are illustrated in figures 13.8 and 13.9 and are summarized in table 13.6.

Posterior Muscles

The posterior muscles include the sternocleidomastoid (originates anteriorly), trapezius, splenius capitis, semispinalis capitis, and longissimus capitis.

As the name implies, the **sternocleidomastoid** (*ster″no-kli″dŏ-mas′toid*) muscle originates on the sternum and clavicle and inserts on the mastoid process of the temporal bone (fig. 13.9). When contracted on one side, it turns the head sideways in a direction opposite the side on which the muscle is located. If both sternocleidomastoid muscles are contracted, the head is pulled forward and down. The sternocleidomastoid is covered by the platysma muscle (see fig. 13.4 and table 13.6).

Although a portion of the **trapezius** muscle extends over the posterior neck region, it is primarily a superficial muscle of the back and will be described later.

The **splenius capitis** (*sple′ne-us kap′i-tis*) is a broad muscle, positioned deep to the trapezius (fig. 13.8). It originates on the ligamentum nuchae and the spinous processes of the seventh cervical and first three thoracic vertebrae. It inserts on the back of the skull below the superior nuchal line and on the mastoid process of the temporal bone. When the splenius capitis contracts on one side, the head rotates and extends to one side. Contracted together, these muscles extend the head at the neck. Further contraction causes hyperextension of the neck and head.

The broad, sheetlike **semispinalis capitis** muscle extends upward from the seventh cervical and first six thoracic vertebrae to insert on the occipital bone (fig. 13.8). When the two semispinalis capitis muscles contract together, they extend the head at the neck, along with the splenius capitis muscle. If one of the muscles acts alone, the head is rotated to the side.

Figure 13.5

Muscles of mastication.
(*a*) A superficial view, (*b*) a deep view, and (*c*) the deepest view, showing the pterygoid muscles. (The muscles of mastication are labeled in boldface type.)

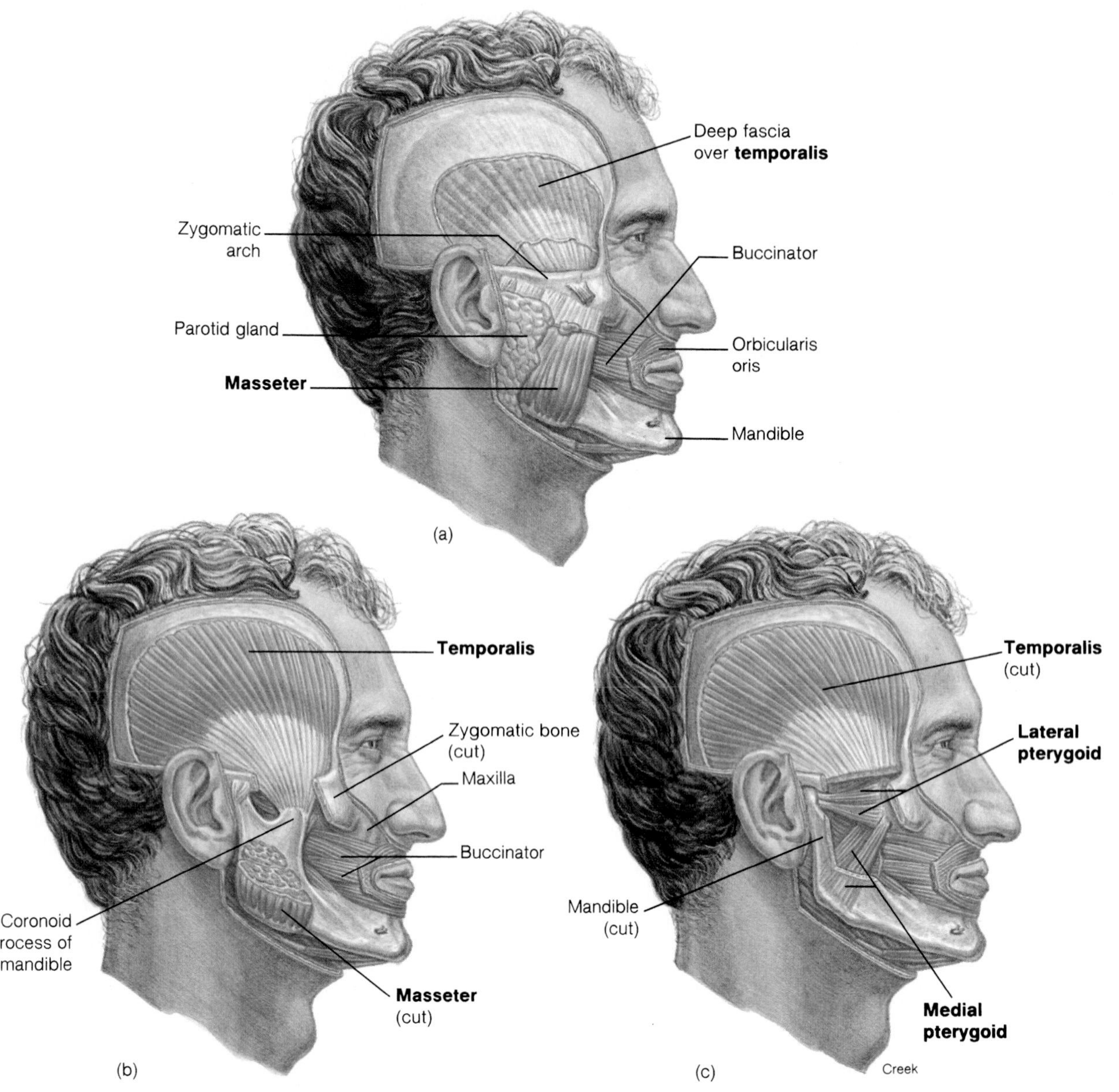

Table 13.3
Muscles of Mastication

Muscle	Origin	Insertion	Action
Temporalis	Temporal fossa	Coronoid process of mandible	Elevates mandible
Masseter	Zygomatic arch	Lateral part of ramus of mandible	Elevates mandible
Medial pterygoid	Sphenoid bone	Medial aspect of mandible	Elevates mandible and moves mandible laterally
Lateral pterygoid	Sphenoid bone	Anterior side of mandibular condyle	Protracts mandible

masseter: Gk. *maseter*, chew
pterygoid: Gk. *pteron*, wing

Figure 13.6

Extrinsic ocular muscles of the left eyeball.
(*a*) An anterior view and (*b*) a lateral view. (The extrinsic occular muscles are labeled in boldface type.)
(*a*) *From* The Color Atlas of Human Anatomy, *translated by Richard T. Jolly. Copyright © 1980 Fabbri Representative Offices, Inc., New York, NY. Reprinted by permission.*

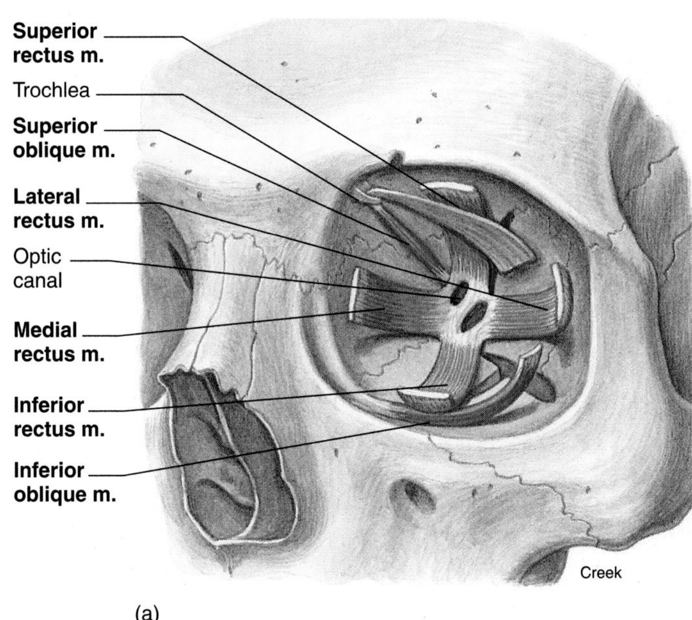

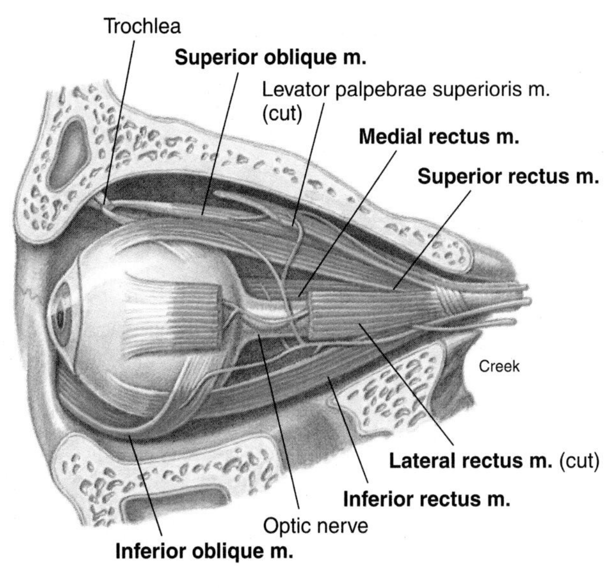

Table 13.4
Ocular Muscles

Muscle	Cranial Nerve Innervation	Movement of Eyeball
Lateral rectus	Abducens n.	Lateral
Medial rectus	Oculomotor n.	Medial
Superior rectus	Oculomotor n.	Superior and medial
Inferior rectus	Oculomotor n.	Inferior and medial
Inferior oblique	Oculomotor n.	Superior and lateral
Superior oblique	Trochlear n.	Inferior and lateral

n. = "nerve."

The narrow, straplike **longissimus** (*lon-jis′ĭ-mus*) **capitis** muscle ascends from processes of the lower four cervical and upper five thoracic vertebrae and inserts on the mastoid process of the temporal bone (fig. 13.8). This muscle extends the head at the neck, bends it to the one side, or rotates it slightly.

Suprahyoid Muscles

The group of suprahyoid muscles located above the hyoid bone includes the digastric, mylohyoid, and stylohyoid muscles (fig. 13.9).

The **digastric** is a two-bellied muscle of double origin that inserts on the hyoid bone. The anterior origin is on the mandible at the point of the chin, and the posterior origin is near the mastoid process of the temporal bone. The digastric muscle can open the mouth or elevate the hyoid.

The **mylohyoid** forms the floor of the mouth. It originates on the inferior border of the mandible and inserts on the median raphe and hyoid bone. As this muscle contracts, the floor of the mouth elevates. It aids swallowing by forcing the food toward the back of the mouth.

The slender **stylohyoid** muscle extends from the styloid process of the temporal bone to the hyoid bone, which it elevates as it contracts. The secondary effect of this muscle on tongue movement has already been described.

Infrahyoid Muscles

The thin, straplike infrahyoid muscles are located below the hyoid bone. They are individually named on the basis of their origin and insertion and include the sternohyoid, sternothyroid, thyrohyoid, and omohyoid muscles (fig. 13.9).

The **sternohyoid** muscle originates on the manubrium of the sternum and inserts on the hyoid bone. It depresses the hyoid bone as it contracts.

The **sternothyroid** muscle also originates on the manubrium but inserts on the thyroid cartilage of the larynx. When this muscle contracts, the larynx is pulled downward.

The short **thyrohyoid** muscle extends from the thyroid cartilage to the hyoid bone. It elevates the larynx and lowers the hyoid bone.

The long, thin **omohyoid** muscle originates on the superior border of the scapula and inserts on the clavicle bone and on the hyoid bone. It acts to depress the hyoid bone.

Figure 13.7

Extrinsic muscles of the tongue and deep structures of the neck.
(The extrinsic muscles of the tongue are labeled in boldface type.)

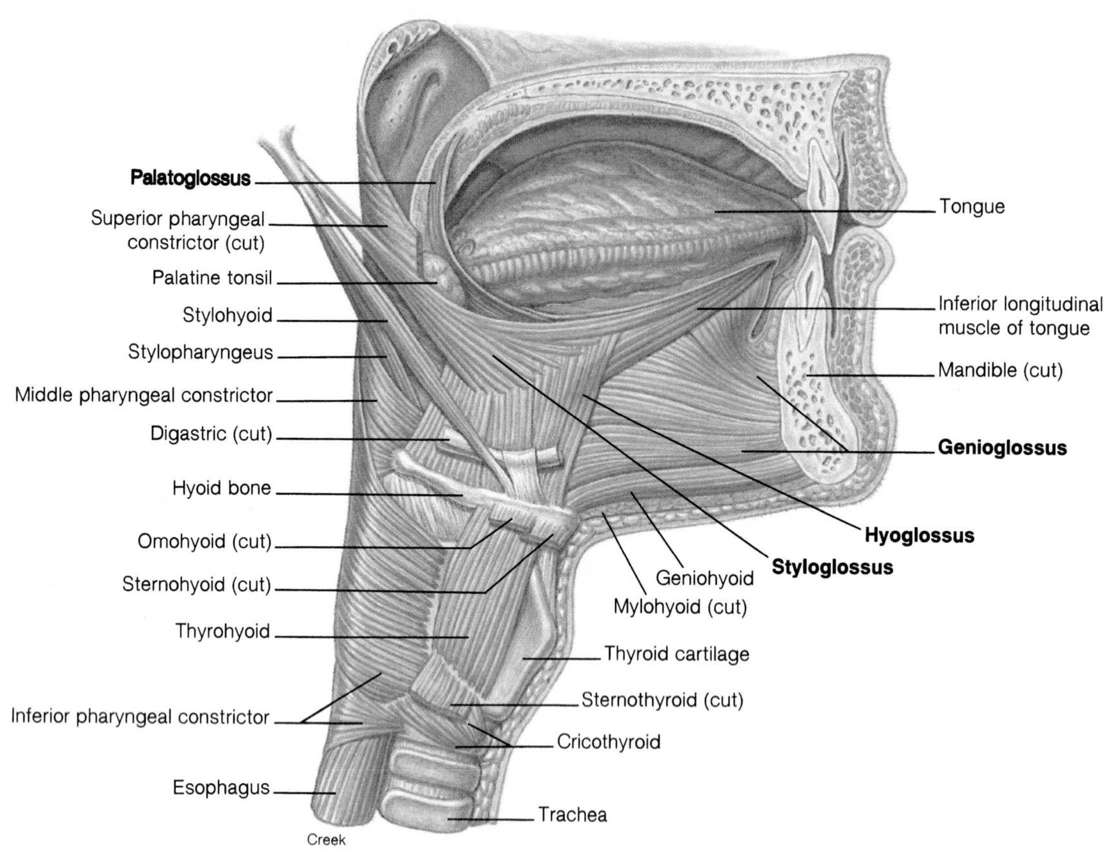

Table 13.5

Extrinsic Tongue Muscles

Muscle	Origin	Insertion	Action
Genioglossus	Mental spine of mandible	Undersurface of tongue	Depresses and protracts tongue
Styloglossus	Styloid process of temporal bone	Lateral side and undersurface of tongue	Elevates and retracts tongue
Hyoglossus	Body of hyoid bone	Side of tongue	Depressed sides of tongue
Palatoglossus	Soft palate	Side of tongue	Elevates posterior tongue; constricts fauces (opening from oral cavity to pharynx)

genioglossus: L. *geneion*, chin; *glossus*, tongue

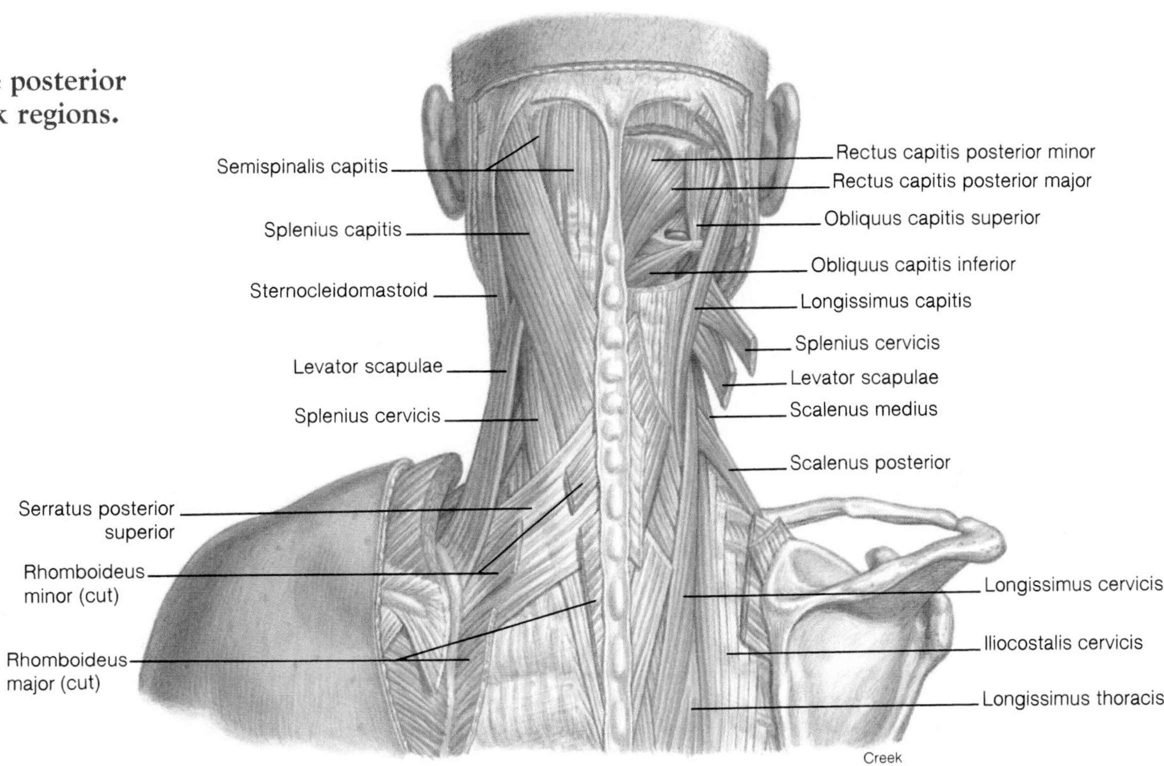

Figure 13.8

Deep muscles of the posterior neck and upper back regions.

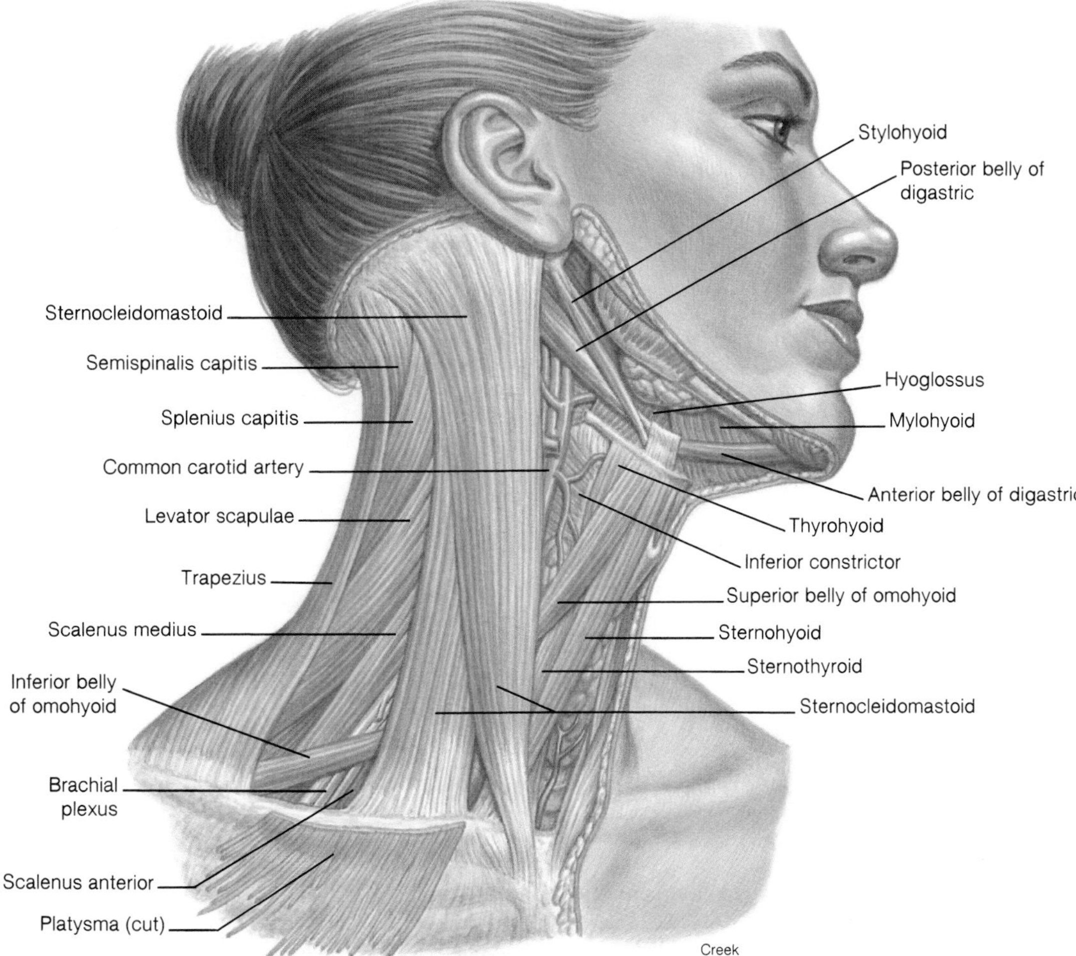

Figure 13.9

Muscles of the anterior and lateral neck regions.

Table 13.6
Muscles of the Neck

Muscle	Origin	Insertion	Action	Innervation
Sternocleidomastoid	Sternum and clavicle	Mastoid process of temporal bone	Turns head to side; flexes neck	Accessory n.
Digastric	Inferior border of mandible and mastoid process of temporal bone	Hyoid bone	Opens mouth; elevates hyoid bone	Trigeminal n. (ant. belly); facial n. (post. belly)
Mylohyoid	Inferior border of mandible	Body of hyoid bone and median raphe	Elevates hyoid bone and floor of mouth	Trigeminal n.
Geniohyoid	Medial surface of mandible at chin	Body of hyoid bone	Elevates hyoid bone	Spinal n. (C1)
Stylohyoid	Styloid process of temporal bone	Body of hyoid bone	Elevates and retracts tongue	Facial n.
Sternohyoid	Manubrium	Body of hyoid bone	Depresses hyoid bone	Spinal nn. (C1–C3)
Sternothyroid	Manubrium	Thyroid cartilage	Depresses thyroid cartilage	Spinal nn. (C1–C3)
Thyrohyoid	Thyroid cartilage	Great cornu of hyoid bone	Depresses hyoid bone; elevates thyroid	Spinal nn. (C1–C3)
Omohyoid	Superior border of scapula	Body of hyoid bone	Depresses hyoid bone	Spinal nn. (C1–C3)

n. = "nerve"; *nn.* = "nerves."
digastric: L. *di*, two; Gk. *gaster*, belly
mylohyoid: Gk. *mylos*, akin to; *hyoeides*, pertaining to hyoid bone
omohyoid: Gk. *omos*, shoulder

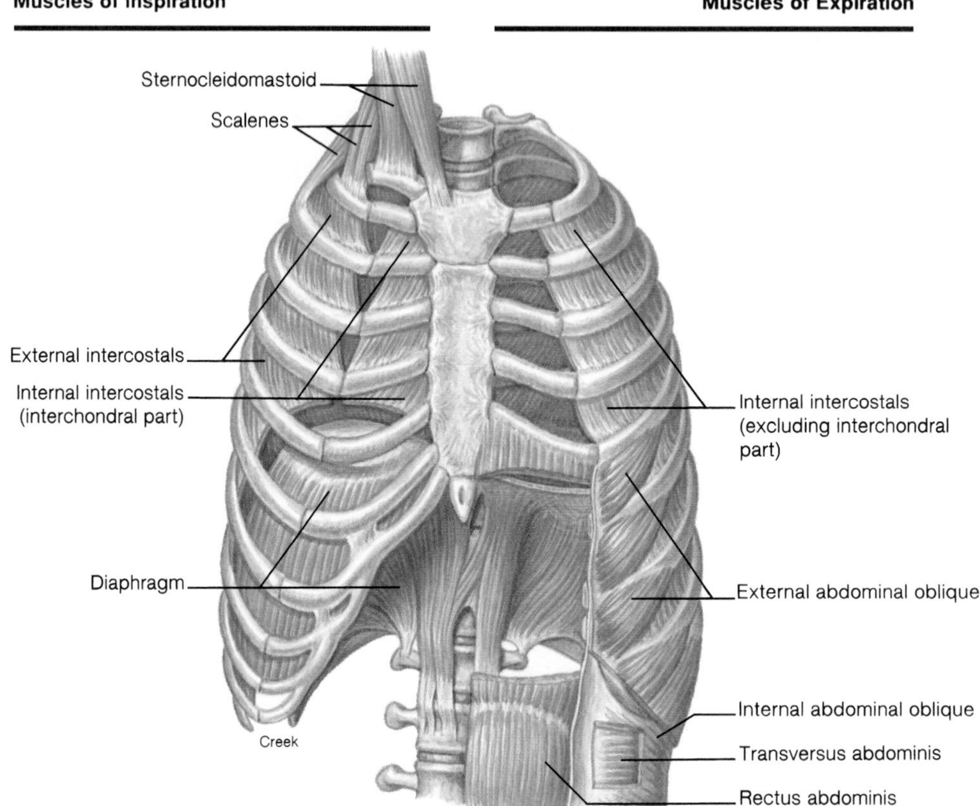

Figure 13.10
Muscles of respiration.

Muscles of Respiration

The muscles of respiration are skeletal muscles that continually contract rhythmically, usually involuntarily. Breathing, or *pulmonary ventilation*, is divided into two phases: *inspiration* (*inhalation*) and *expiration* (*exhalation*).

During normal, relaxed inspiration, the contracting muscles are the **diaphragm** (*di'ă-fram*), the **external intercostal** muscles, and the interchondral portion of the **internal intercostal** muscles (fig. 13.10). A downward contraction of the dome-shaped diaphragm causes a vertical increase in tho-

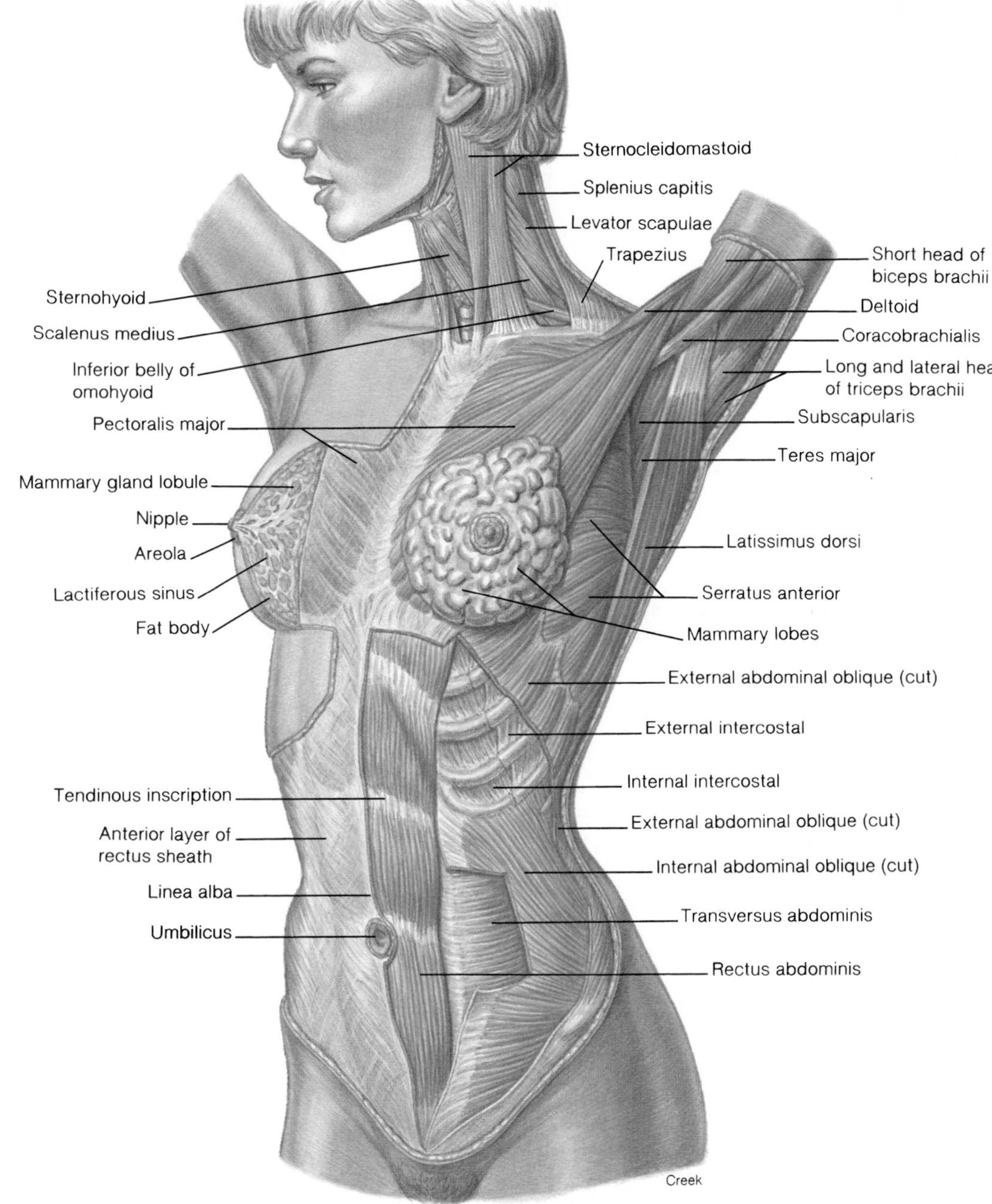

Figure 13.11

Muscles of the anterolateral neck, shoulder, and trunk regions.
The mammary gland is an integumentary structure positioned over the pectoralis major muscle.

racic dimension. A simultaneous contraction of the external intercostal muscles and the interchondral portion of the internal intercostal muscles produces an increase in the lateral dimension of the thorax. In addition, the **sternocleidomastoid** and **scalene** (*ska'len*) muscles may assist in inspiration through elevation of the first and second ribs, respectively. The intercostal muscles are innervated by the intercostal nerves, and the diaphragm receives its stimuli through the phrenic nerves.

Expiration is primarily a passive process, occurring as the muscles of inspiration relax and the rib cage recoils to its original position. During forced expiration, the interosseous portion of the **internal intercostal** muscles contracts, causing the rib cage to depress. This portion of the internal intercostal muscles lies under the external intercostal muscles, and its fibers are directed downward and backward. The *abdominal muscles* may also contract during forced expiration, which increases pressure within the abdominal cavity and forces the diaphragm superiorly, squeezing additional air out of the lungs.

Muscles of the Abdominal Wall

The anterolateral abdominal wall is composed of four pairs of flat, sheetlike muscles: the external abdominal oblique, internal abdominal oblique, transversus abdominis, and rectus abdominis muscles (fig. 13.11). These muscles support and protect the organs of the abdominal cavity and aid in breathing. When they contract, the pressure in the abdominal cavity increases, which can aid in defecation and in stabilizing the spine during heavy lifting.

Table 13.7

Muscles of the Abdominal Wall

Muscle	Origin	Insertion	Action
External abdominal oblique	Lower eight ribs	Iliac crest and linea alba	Compresses abdomen; lateral rotation; draws thorax downward
Internal abdominal oblique	Iliac crest, inguinal ligament, and lumbodorsal fascia	Linea alba and costal cartilage of last three or four ribs	Compresses abdomen; lateral rotation; draws thorax downward
Transversus abdominis	Iliac crest, inguinal ligament, lumbar fascia, and costal cartilage of last six ribs	Xiphoid process, linea alba, and pubis	Compresses abdomen
Rectus abdominis	Pubic crest and symphysis pubis	Costal cartilage of fifth to seventh ribs and xiphoid process of sternum	Flexes vertebral column

rectus abdominis: L. *rectus*, straplike; *abdomino*, belly

Table 13.8

Muscles of the Pelvic Outlet

Muscle	Origin	Insertion	Action
Levator ani	Spine of ischium and pubic bone	Coccyx	Supports pelvic viscera; aids in defecation
Coccygeus	Ischial spine	Sacrum and coccyx	Supports pelvic viscera; aids in defecation
Transversus perinei	Ischial tuberosity	Central tendon	Supports pelvic viscera
Bulbospongiosus	Central tendon	Males: base of penis; females: root of clitoris	Constricts urethra; constricts vagina
Ischiocavernosus	Ischial tuberosity	Males: pubic arch and crus of the penis; females: pubic arch and crus of the clitoris	Aids erection of penis and clitoris

The **external abdominal oblique** is the strongest and most superficial of the three layered muscles of the lateral abdominal wall. Its fibers are directed inferiorly and medially. The **internal abdominal oblique** lies deep to the external abdominal oblique, and its fibers are directed at right angles to those of the external abdominal oblique. The **transversus abdominis** is the deepest of the abdominal muscles; its fibers are directed horizontally across the abdomen. The long, straplike **rectus abdominis** muscle is entirely enclosed in a fibrous sheath formed from the aponeuroses of the other three abdominal muscles. The *linea alba* is a band of connective tissue on the midline of the abdomen that separates the two rectus abdominis muscles. *Tendinous inscriptions* transect the rectus abdominis muscles at several points, causing the abdominal region of a well-muscled person with low body fat to appear segmented.

Refer to table 13.7 for a summary of the muscles of the abdominal wall.

Muscles of the Pelvic Outlet

Any sheet that separates cavities may be termed a diaphragm. The **pelvic outlet**—the entire muscular wall at the bottom of the pelvic cavity—contains two: the pelvic diaphragm and the urogenital diaphragm. The urogenital diaphragm lies immediately deep to the external genitalia; the pelvic diaphragm is situated closer to the internal viscera. Together, these sheets of muscle provide support for pelvic viscera and help to regulate the passage of urine and feces.

The pelvic diaphragm consists of the levator ani and the coccygeus muscles (table 13.8). The **levator ani** (*le-va'tor a'ni*) (fig. 13.12) is a thin sheet of muscle that helps support the pelvic viscera and constrict the lower part of the rectum, pulling it forward and aiding defecation. The deeper, fan-shaped **coccygeus** (*kok-sij'e-us*) aids the levator ani in its functions. Either or both of these muscles stretch and occasionally tear during parturition.

Figure 13.12

Muscles of the pelvic outlet.
(*a*) Male and (*b*) female. (*c*) A superior view of the internal muscles of the female pelvic outlet.

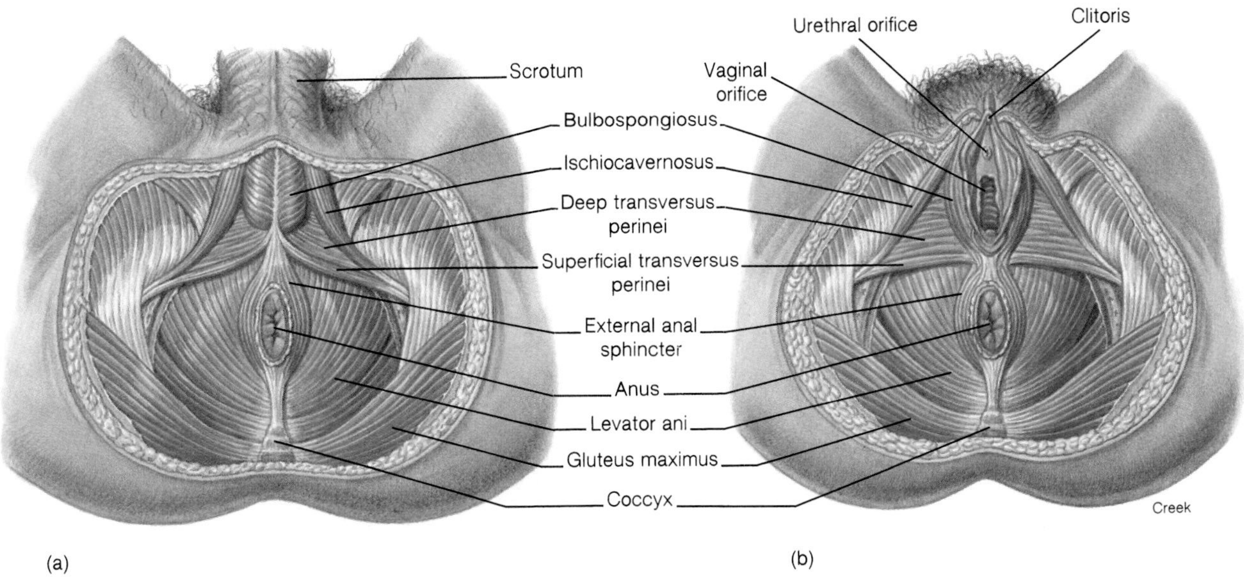

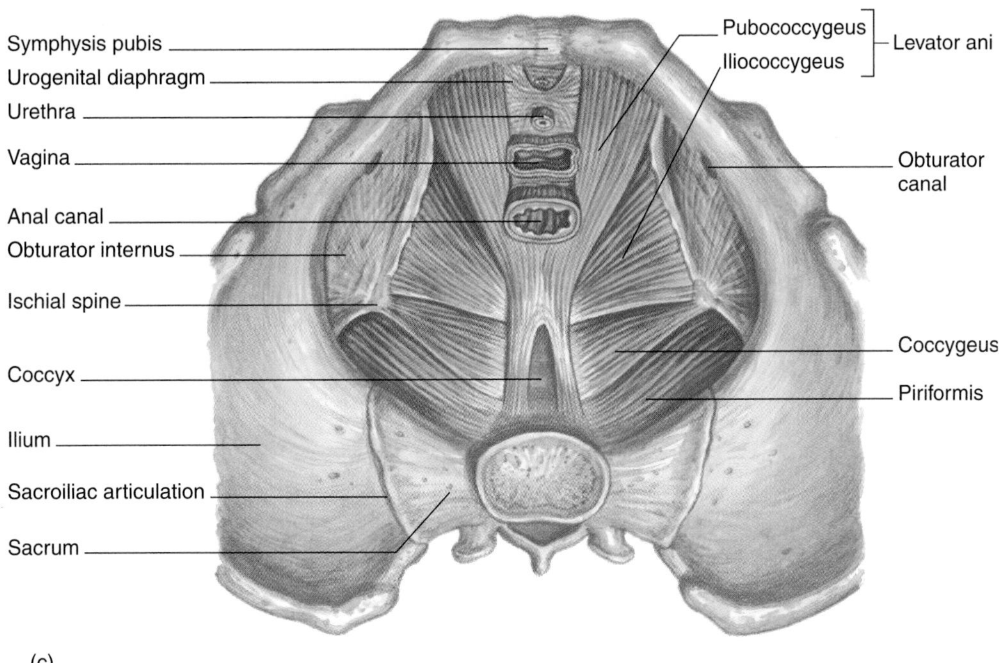

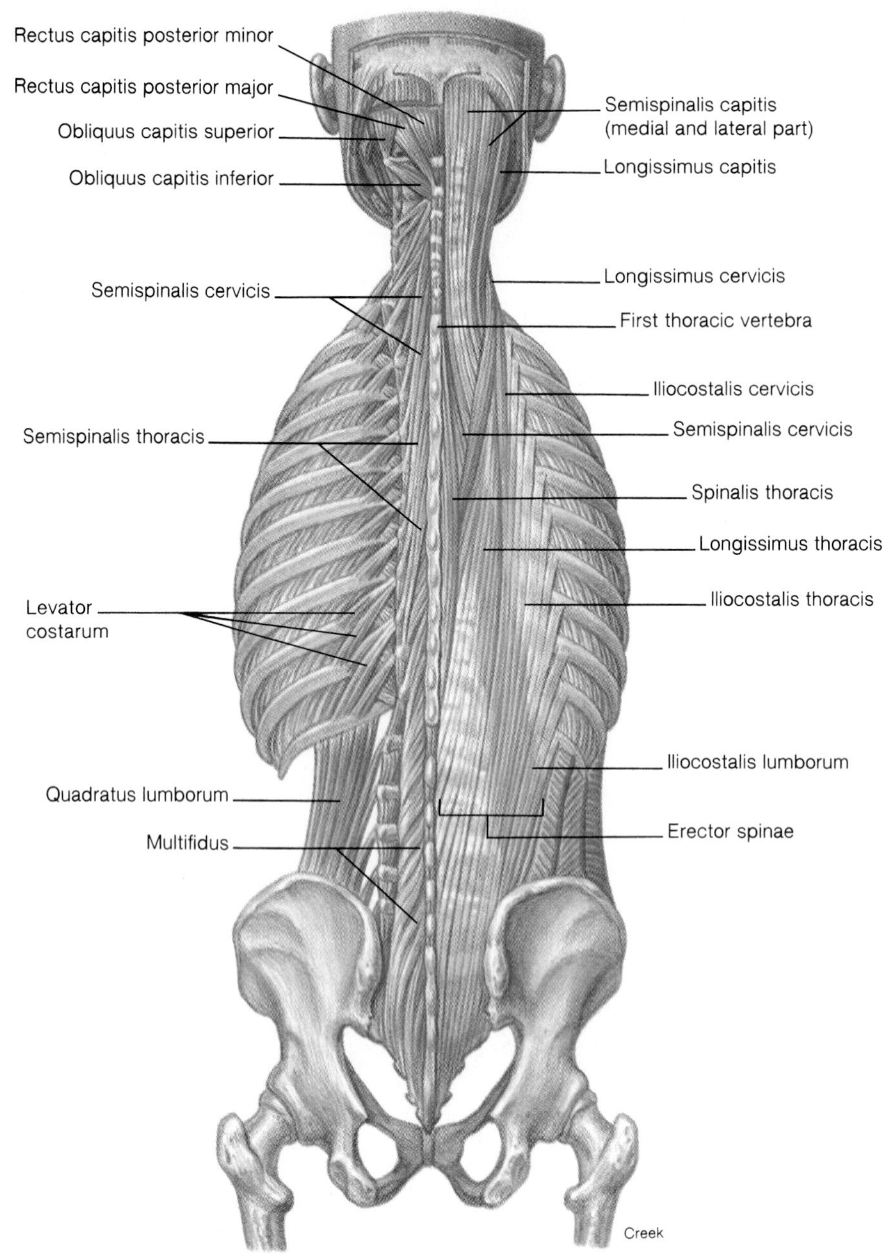

Figure 13.13

Muscles of the vertebral column.
The superficial neck muscles and erector spinae group of muscles are illustrated on the right, and the deep neck and back muscles are illustrated on the left.

The urogenital diaphragm consists of the deep, sheetlike **transversus perinei** (*per-i-ne'i*) muscle and the associated **external anal sphincter** muscle. The external anal sphincter is a funnel-shaped constrictor muscle that surrounds the anal canal.

Inferior to the pelvic diaphragm are the *perineal muscles*, which provide the skeletal muscular support to the genitalia. They include the *bulbocavernosus*, *ischiocavernosus*, and the *transversus perinei superficialis* muscles (fig. 13.12). The muscles of the pelvic diaphragm and the urogenital diaphragm are similar in the male and female, but the perineal muscles exhibit marked sex-based differences. In males, the **bulbospongiosus** (*bul"bo-spon"je-o'sus*) muscle of one side unites with that of the opposite side to form a muscular constriction surrounding the base of the penis. When contracted, the two muscles constrict and assist in emptying the urethra. In females, these muscles are separated by the vaginal orifice, which they constrict as they contract. The **ischiocavernosus** muscle inserts onto the pubic arch and crus of the penis in the male and the pubic arch and crus of the clitoris of the female. This muscle aids the erection of the penis in the male and that of the clitoris in the female.

The levator ani muscle and the transversus perinei muscle are innervated by the pudendal plexus. The coccygeus, bulbospongiosus, and ischiocavernosus muscles are innervated by the perineal branch of the pudendal nerve.

The muscles of the pelvic outlet are illustrated in figure 13.12 and summarized in table 13.8.

Table 13.9
Muscles of the Vertebral Column

Muscle	Origin	Insertion	Action	Innervation
Quadratus lumborum	Iliac crest and lower three lumbar vertebrae	Twelfth rib and upper four lumbar vertebrae	Extends lumbar region; laterally flexes vertebral column	Intercostal nerve T12 and lumbar nerves L2–L4
Erector spinae	Consists of three groups of muscles: iliocostalis, longissimus, and spinalis. The iliocostalis and longissimus are further subdivided into three groups each on the basis of location along the vertebral column.			
Iliocostalis lumborum	Crest of ilium	Lower six ribs	Extends lumbar region	Posterior rami of lumbar nerves
Iliocostalis thoracis	Lower six ribs	Upper six ribs	Extends thoracic region	Posterior rami of thoracic nerves
Iliocostalis cervicis	Angles of third to sixth rib	Transverse processes of fourth to sixth cervical vertebrae	Extends cervical region	Posterior rami of cervical nerves
Longissimus thoracis	Transverse processes of lumbar vertebrae	Transverse processes of all the thoracic vertebrae and lower nine ribs	Extends thoracic region	Posterior rami of spinal nerves
Longissimus cervicis	Transverse processes of upper four or five thoracic vertebrae	Transverse processes of second to sixth cervical vertebrae	Extends and laterally flexes cervical region	Posterior rami of spinal nerves
Longissimus capitis	Transverse processes of upper five thoracic vertebrae and articular processes of lower three cervical vertebrae	Posterior margin of cranium and mastoid process of temporal bone	Extends head; acting separately, turns face toward that side	Posterior rami of middle and lower cervical nerves
Spinalis thoracis	Spinous processes of upper lumbar and lower thoracic vertebrae	Spinous processes of upper thoracic vertebrae	Extends vertebral column	Posterior rami of spinal nerves
Semispinal thoracis	Transverse processes of T6–T10	Spinous processes of C6–T4	Extends vertebral column	Posterior rami of spinal nerves
Semispinalis cervicis	Transverse processes of T1–T6	Spinous processes of C2–C5	Extends vertebral column	Posterior rami of spinal nerves
Semispinalis capitis	Transverse processes of C7–T7	Nuchal line of occipital bone	Extends head	Posterior rami of spinal nerves

Muscles of the Vertebral Column

The muscles that move the vertebral column are strong and complex because they have to provide support and movement in resistance to the effect of gravity.

The vertebral column can be flexed, extended, abducted, adducted, and rotated. The muscle that flexes the vertebral column, the rectus abdominis, has already been described as a paired, straplike muscle of the anterior abdominal wall. The extensor muscles, located on the posterior side of the vertebral column, have to be stronger than the flexors because extension (such as lifting an object) is in opposition to gravity. The extensor muscles consist of a superficial group and a deep group. Only some of the muscles of the vertebral column will be described here.

The **erector spinae** (*spi'ne*) muscles constitute a massive superficial muscle group that extends from the sacrum to the skull. It actually consists of three groups of muscles: the **iliocostalis, longissimus,** and **spinalis** muscles (fig. 13.13 and table 13.9). Each of these groups, in turn, consists of overlapping slips of muscle. The iliocostalis is the most lateral group, the longissimus is intermediate in position, and the spinalis, in medial position, is in contact with the spinous processes of the vertebrae.

The erector spinae muscles are frequently strained through improper lifting of objects. A heavy object should not be lifted with the vertebral column flexed; instead, the hip and knee joints should be flexed so that the pelvic and leg muscles can aid in the task.

Pregnancy may also put a strain on the erector spinae muscles. Pregnant women will try to counterbalance the effect of a protruding abdomen by hyperextending the vertebral column. This results in an exaggerated lumbar curvature, strained muscles, and a peculiar gait.

The deep **quadratus lumborum** (*kwad-ratus lum-bor'um*) muscle originates on the iliac crest and the lower three lumbar vertebrae. It inserts on the transverse processes of the first four lumbar vertebrae and the inferior margin of the twelfth rib. When the right and left quadratus lumborum muscle contract together, the vertebral column in the lumbar region extends. Separate contraction causes lateral flexion of the spine.

Figure 13.14

Muscles of the anterior trunk and shoulder regions.
The superficial muscles are illustrated on the right, and the deep muscles are illustrated on the left.

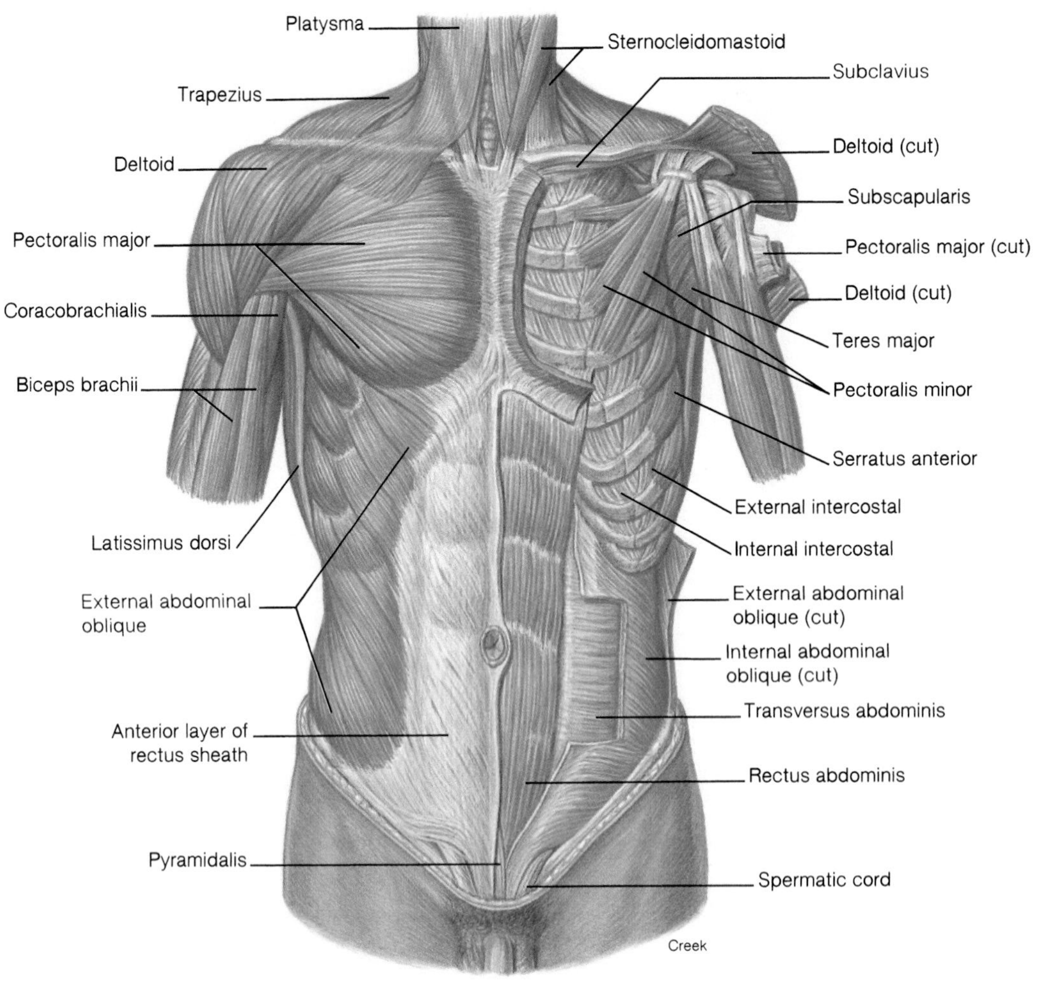

Muscles of the Appendicular Skeleton

The muscles of the appendicular skeleton include those of the pectoral girdle, arm, forearm, wrist, hand, and fingers, and those of the pelvic girdle, thigh, leg, ankle, foot, and toes.

Muscles That Act on the Pectoral Girdle

The shoulder is attached to the axial skeleton only at the sternoclavicular joint; therefore, strong, straplike muscles are necessary in this region. Furthermore, muscles that move the brachium originate on the scapula, and during brachial movement the scapula has to be held stationary. The muscles that act on the pectoral girdle originate on the axial skeleton and can be divided into anterior and posterior groups.

The anterior group of muscles that act on the pectoral girdle includes the **serratus** (*ser-a'tus*) **anterior, pectoralis** (*pek"to-ra'lis*) **minor,** and **subclavius** (*sub-kla've-us*) muscles (fig. 13.14 and table 13.10). The posterior group includes the **trapezius, levator scapulae** (*skap-yu'le*), and **rhomboideus** (*rom-boid'e-us*) muscles (fig. 13.15). These muscles are positioned so that one of them does not cause an action on its own. Rather, several muscles contract synergistically to result in any movement of the girdle.

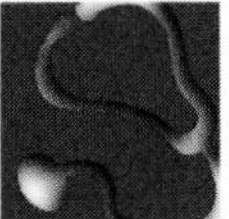

Treatment of advanced stages of *breast cancer* requires the surgical removal of both pectoralis major and pectoralis minor muscles in a procedure called a *radical mastectomy.* Postoperative physical therapy is primarily geared toward strengthening the synergistic muscles of this area. As you learn the muscles that act on the brachium, determine which are synergists with the pectoralis major muscle.

Table 13.10

Muscles That Act on the Pectoral Girdle and That Move the Shoulder Joint

Muscle	Origin	Insertion	Action	Innervation
Serratus anterior	Upper eight or nine ribs	Anterior vertebral border of scapula	Rotates scapula superiorly and laterally	Long thoracic n.
Pectoralis minor	Sternal ends of third, fourth, and fifth ribs	Coracoid process of scapula	Pulls scapula anteriorly and inferiorly	Medial and lateral pectoral nn.
Subclavius	First rib	Subclavian groove of clavicle	Draws clavicle inferiorly	Spinal nerves C5, C6
Trapezius	Occipital bone and spines of seventh cervical and all thoracic vertebrae	Clavicle, spine of scapula, and acromion	Elevates, depresses, and adducts scapula; hyperextends neck; braces shoulder	Accessory nerve
Levator scapulae	First to fourth cervical vertebrae	Medial border of scapula	Elevates scapula	Dorsal scapular n.
Rhomboideus major	Spines of second to fifth thoracic vertebrae	Medial border of scapula	Elevates and adducts scapula	Dorsal scapular n.
Rhomboideus minor	Seventh cervical and first thoracic vertebrae	Medial border of scapula	Elevates and adducts scapula	Dorsal scapular n.
Pectoralis major	Clavicle, sternum, and costal cartilages of second to sixth rib; rectus sheath	Crest of greater tubercle of humerus	Flexes, adducts, and rotates shoulder joint medially	Medial and lateral pectoral nn.
Latissimus dorsi	Spines of sacral, lumbar, and lower thoracic vertebrae; iliac crest and lower four ribs	Intertubercular groove of humerus	Extends, adducts, and rotates shoulder joint medially; adducts shoulder joint	Thoracodorsal n.
Deltoid	Clavicle, acromion and spine of scapula	Deltoid tuberosity of humerus	Abducts, extends, or flexes shoulder joint	Axillary n.
Supraspinatus	Fossa—superior to spine of scapula	Greater tubercle of humerus	Abducts and laterally rotates shoulder joint	Suprascapular n.
Infraspinatus	Fossa—inferior to spine of scapula	Greater tubercle of humerus	Rotates shoulder joint laterally	Suprascapular n.
Teres major	Inferior angle and lateral border of scapula	Crest of lesser tubercle of humerus	Extends shoulder joint, or adducts and rotates shoulder joint medially	Lower subscapular n.
Teres minor	Axillary border of scapula	Greater tubercle and groove of humerus	Rotates shoulder joint laterally	Axillary n.
Subscapularis	Subscapular foss	Lesser tubercle of humerus	Rotates shoulder joint medially	Subscapular nn.
Coracobrachialis	Coracoid process of scapula	Body of humerus	Flexes and adducts shoulder joint	Musculocutaneous n.

serratus: L. *serratus*, saw-shaped
pectoralis: L. *pectus*, chest
trapezius: Gk. *trapezeoides*, trapezoid-shaped
rhomboideus: Gk. *rhomboides*, rhomboid-shaped
latissimus: L. *latissimus*, widest
deltoid: Gk. *delta*, triangular
teres: L. *teres*, rounded

Muscles That Move the Humerus at the Shoulder Joint

Of the nine muscles that span the shoulder joint to insert on the humerus, only two—the pectoralis major and latissimus dorsi—do not originate on the scapula (table 13.10). These two are designated as axial muscles, whereas the remaining seven are scapular muscles. The muscles of this region are shown in figures 13.14 and 13.15, and the attachments of all the muscles that either originate or insert on the scapula are shown in figure 13.16.

In terms of their development, the pectoralis major and the latissimus dorsi muscles are not axial muscles at all. They develop in the forelimb and extend to the trunk secondarily. They are considered axial muscles only because their origins are on the axial skeleton.

Axial Muscles

The **pectoralis major** is a large, fan-shaped chest muscle (see fig. 13.14) that binds the humerus to the pectoral girdle. It is the principal flexor muscle of the shoulder joint. The large,

Figure 13.15

Muscles of the posterior neck, shoulder, trunk, and gluteal regions.
The superficial muscles are illustrated on the left, and the deep muscles are illustrated on the right.

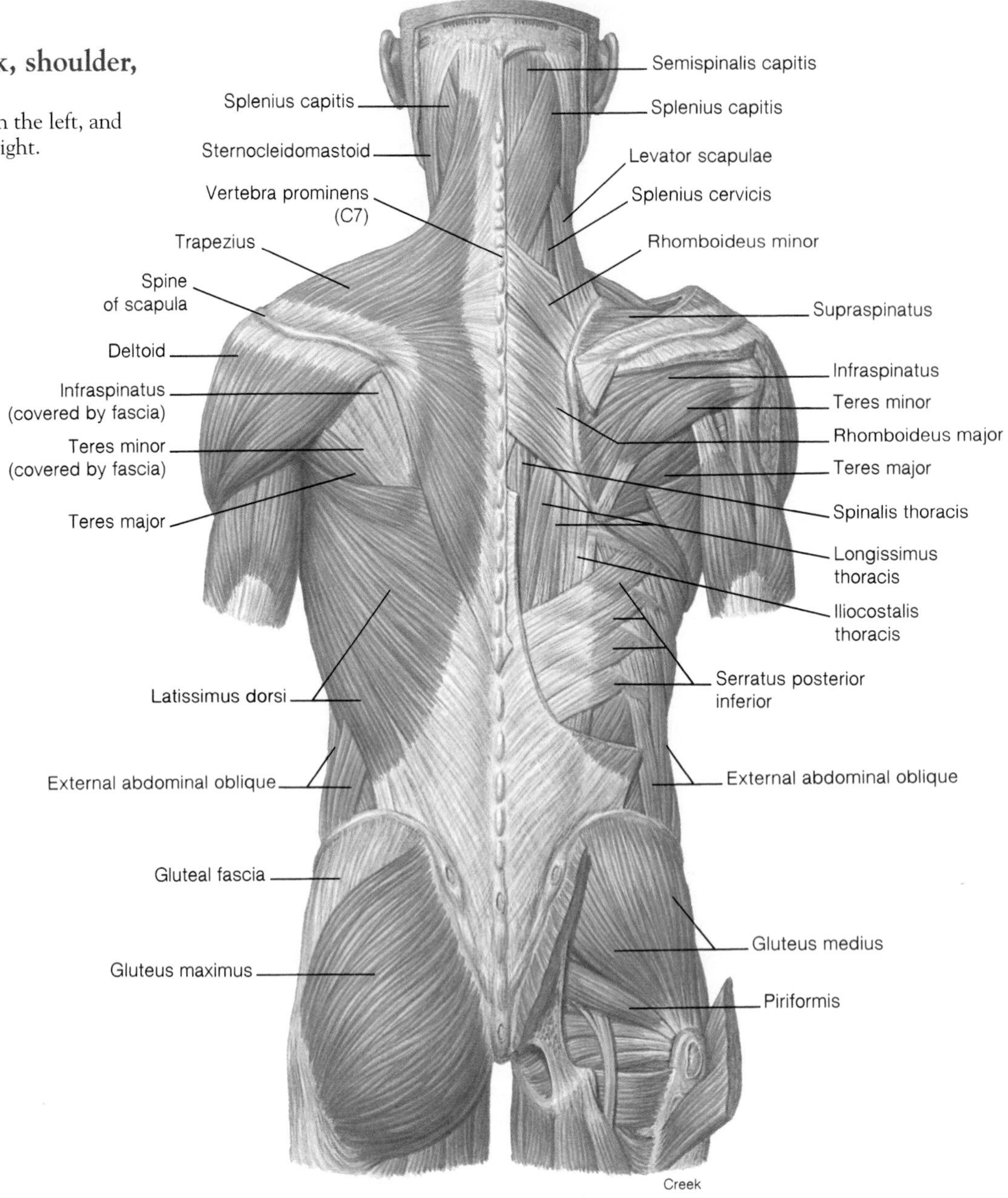

flat, triangular **latissimus dorsi** (*la-tis'i-mus dor'si*) muscle covers the inferior half of the thoracic region of the back (see fig. 13.15) and is the antagonist to the pectoralis major muscle. The latissimus dorsi is frequently called the "swimmer's muscle" because it powerfully extends the shoulder joint, drawing the arm downward and backward while it rotates medially. Extension of the shoulder joint is in reference to anatomical position and is therefore a backward, retracting (increasing the shoulder joint angle) movement of the arm.

Scapular Muscles

The nonaxial scapular muscles include the deltoid, supraspinatus, infraspinatus, teres major, teres minor, subscapularis, and coracobrachialis muscles.

The **deltoid** is a thick, powerful muscle that caps the shoulder joint (figs. 13.17 and 13.18). Although it has several functions (table 13.10), its principal action is abduction of the shoulder joint. Functioning together, both the pectoralis major and the latissimus dorsi muscles are antagonists to the deltoid muscle in that they cause adduction of the shoulder joint. The deltoid muscle is a common site for intramuscular injections.

The remaining six scapular muscles also help stabilize the shoulder and have specific actions at the shoulder joint (table 13.10). The **supraspinatus** (*soo"pra-spi-na'tus*) muscle laterally rotates the arm and is synergistic with the deltoid muscle in abducting the arm at the shoulder joint. The **infraspinatus** muscle rotates the arm laterally. The action of the **teres** (*te'rez*) **major** muscle is similar to that of the latissimus dorsi, adducting and medially rotating the shoulder joint. The **teres minor** muscle works with the infraspinatus in laterally rotating the arm at the shoulder joint. The **subscapularis**

Figure 13.16

A posterior view of the scapula and humerus showing the areas of attachment of the associated muscles.
(Points of origin are color coded red, and points of insertion are color coded blue.)

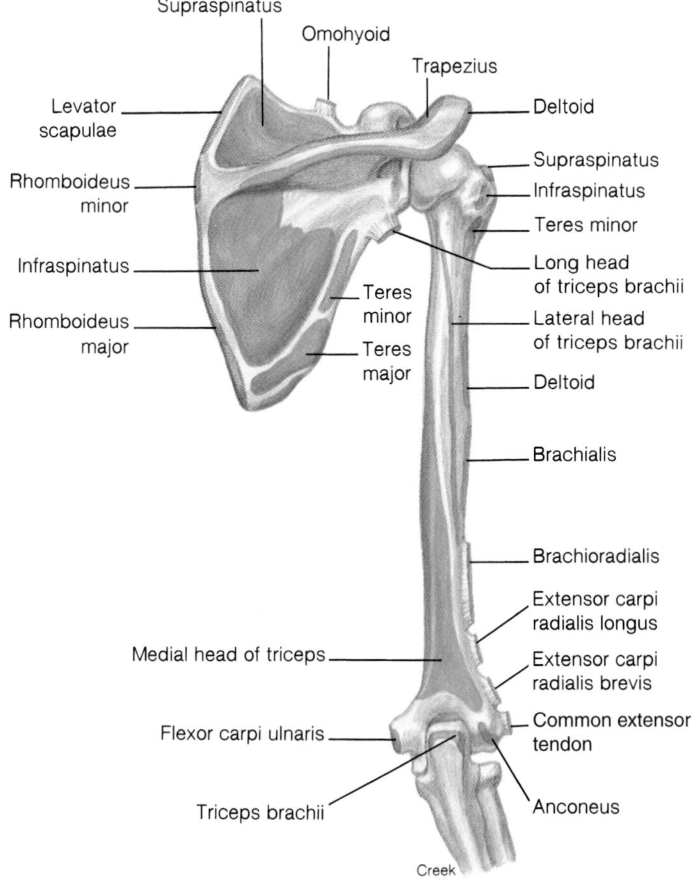

muscle is a strong stabilizer of the shoulder and also aids in medially rotating the arm at the shoulder joint. The **coraco-brachialis** (*kor″a-ko-bra″ke-al′is*) muscle is a synergist to the pectoralis major muscle in flexing and adducting the arm at the shoulder joint.

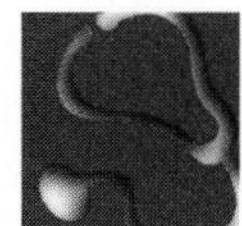

Four of the nine muscles that cross the shoulder joint, the supraspinatus, infraspinatus, teres minor, and subscapularis, are commonly called the *musculotendinous cuff,* or *rotator cuff.* Their distal tendons blend with and reinforce the fibrous capsule of the shoulder joint en route to their points of insertion on the humerus. This structural arrangement plays a major role in stabilizing the shoulder joint. Musculotendinous cuff injuries are common among baseball players. When throwing a baseball, an abduction of the shoulder is followed by a rapid and forceful rotation and flexion of the shoulder joint, which may strain the musculotendinous cuff.

Muscles That Move the Forearm at the Elbow Joint

The powerful muscles of the brachium are responsible for flexion and extension of the elbow joint. These muscles are the biceps brachii, brachialis, brachioradialis, and triceps brachii (figs. 13.17 and 13.18). In addition, a short triangular muscle, the *anconeus,* is positioned over the distal end of the triceps brachii muscle, near the elbow.

The powerful **biceps brachii** (*bi′ceps bra′ke-i*) muscle, positioned on the anterior surface of the humerus, is the most familiar muscle of the arm, yet it has no attachments on the humerus. This muscle has a dual origin: a medial tendinous head, the **short head,** arises from the coracoid process of the scapula, and the **long head** originates on the superior tuberosity of the glenoid cavity, passes through the shoulder joint, and descends in the intertubercular groove on the humerus (see fig. 10.4). Both heads of the biceps brachii muscle insert on the radial tuberosity. The **brachialis** (*bra″ke-al′is*) muscle is located on the distal anterior half of the humerus, deep to the biceps brachii muscle. It is synergistic to the biceps brachii muscle in flexing the elbow joint.

The prominent **brachioradialis** (*bra″ke-o-ra″de-a′lis*) is the prominent muscle positioned along the lateral (radial) surface of the forearm. It, too, flexes the elbow joint.

The **triceps brachii** muscle, located on the posterior surface of the brachium, extends the forearm at the elbow joint, in opposition to the action of the biceps brachii muscle. Thus, these two muscles are antagonists. The triceps brachii muscle has three heads, or origins. Two of the three, the **lateral head** and **medial head,** arise from the humerus, whereas the **long head** arises from the infraglenoid tuberosity of the scapula. A common tendinous insertion attaches the triceps brachii muscle to the olecranon of the ulna. The small **anconeus** (*an-ko′ne-us*) muscle is a synergist of the triceps brachii muscle in elbow extension.

Refer to table 13.11 for a summary of the muscles that act on the forearm at the elbow joint.

Muscles of the Forearm That Move the Joints of the Wrist, Hand, and Fingers

The muscles that cause wrist, hand, and most finger movements are positioned along the forearm (figs. 13.19 and 13.20). Several of these muscles act on two joints—the elbow and wrist. Others act on the joints of the wrist, hand, and digits. Still others produce rotational movement at the radioulnar joint. The precise actions of these muscles are complex, so only the basic movements will be described here. Most of these muscles perform four primary actions on the hand and digits: supination, pronation, flexion, and extension. Other actions of the hand include adduction and abduction.

Figure 13.17
Muscles of the right anterior shoulder and brachium.

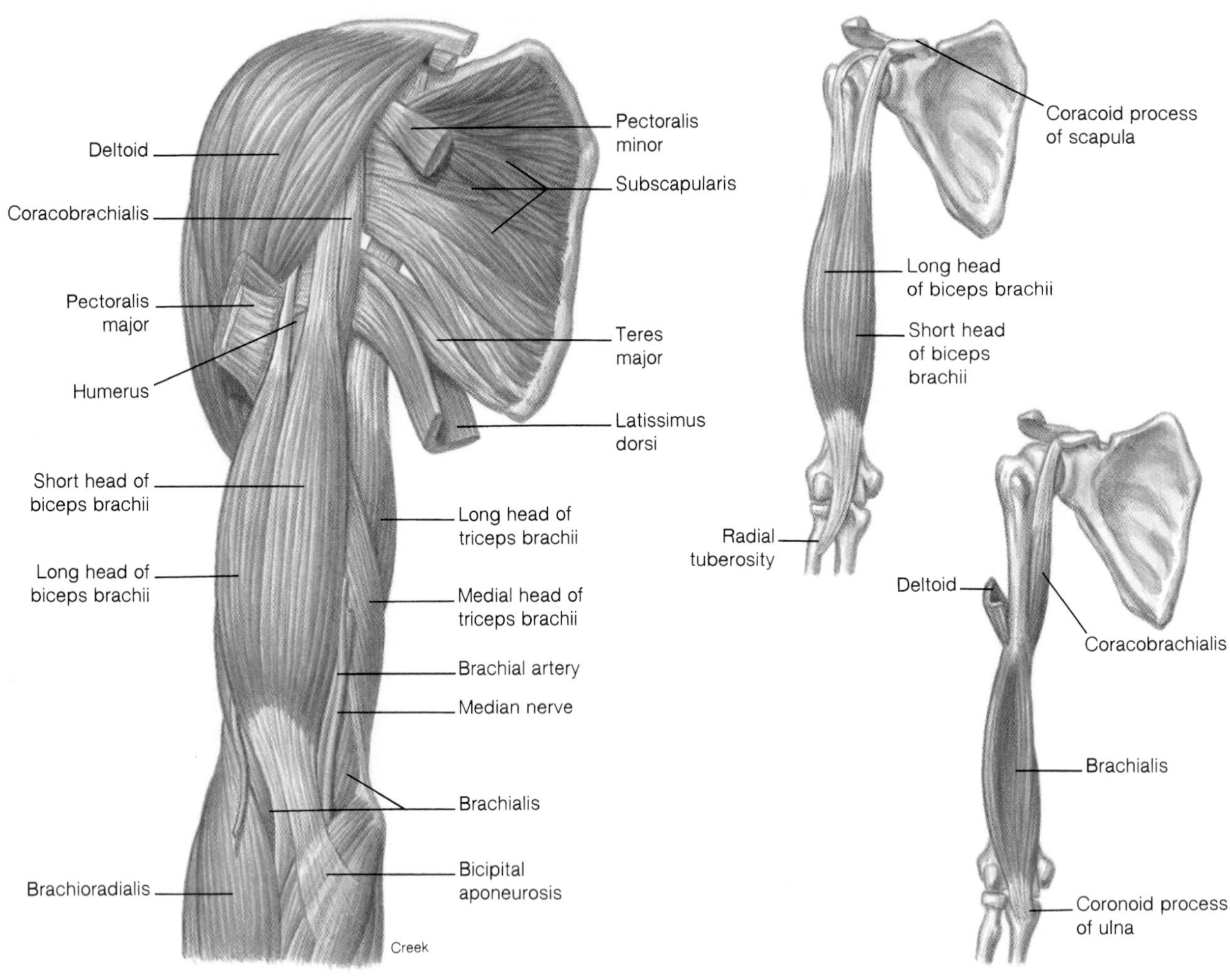

Supination and Pronation of the Hand

The **supinator** (*soo″pi-na′tor*) muscle wraps around the upper posterior portion of the radius (fig. 13.20), where it works synergistically with the biceps brachii muscle to supinate the hand. Two muscles are responsible for pronating the hand—the pronator teres and pronator quadratus. The **pronator teres** muscle is located on the upper medial side of the forearm, whereas the deep, anteriorly positioned **pronator quadratus** muscle extends between the ulna and radius on the distal fourth of the forearm. These two muscles work synergistically to rotate the palm of the hand posteriorly and position the thumb medially.

Flexion of the Wrist, Hand, and Fingers

Six of the muscles that flex the joints of the wrist, hand, and fingers will be described from lateral to medial and from superficial to deep (figs. 13.19 and 13.20). Although four of the six arise from the medial epicondyle of the humerus (table 13.11), their actions on the elbow joint are minimal. The brachioradialis, already described, is an obvious reference muscle for locating the muscles of the forearm that flex the joints of the hand.

The **flexor carpi radialis** muscle extends diagonally across the anterior surface of the forearm, and its distal cordlike tendon crosses the wrist under the *flexor retinaculum*. This muscle is an important landmark for locating the radial artery, where the pulse is usually taken.

The narrow **palmaris longus** muscle is superficial in position on the anterior surface of the forearm. It has a long, slender tendon that attaches to the *palmar aponeurosis*, where it assists in flexing the wrist joints. The palmaris longus is the most variable muscle in the body. It is totally absent in approximately 8% of all people, and in 4%, it is absent in one or the other forearm. Furthermore, it is absent more often in females than males, and on the left side in both sexes. Because of the superficial position of the palmaris longus muscle, you can readily determine whether or not it is present in your own forearm by flexing the wrist while touching the thumb and little finger, and examining for its tendon just proximal to the wrist (see fig. 6 on page 354).

Figure 13.18
Muscles of the right posterior shoulder and brachium.

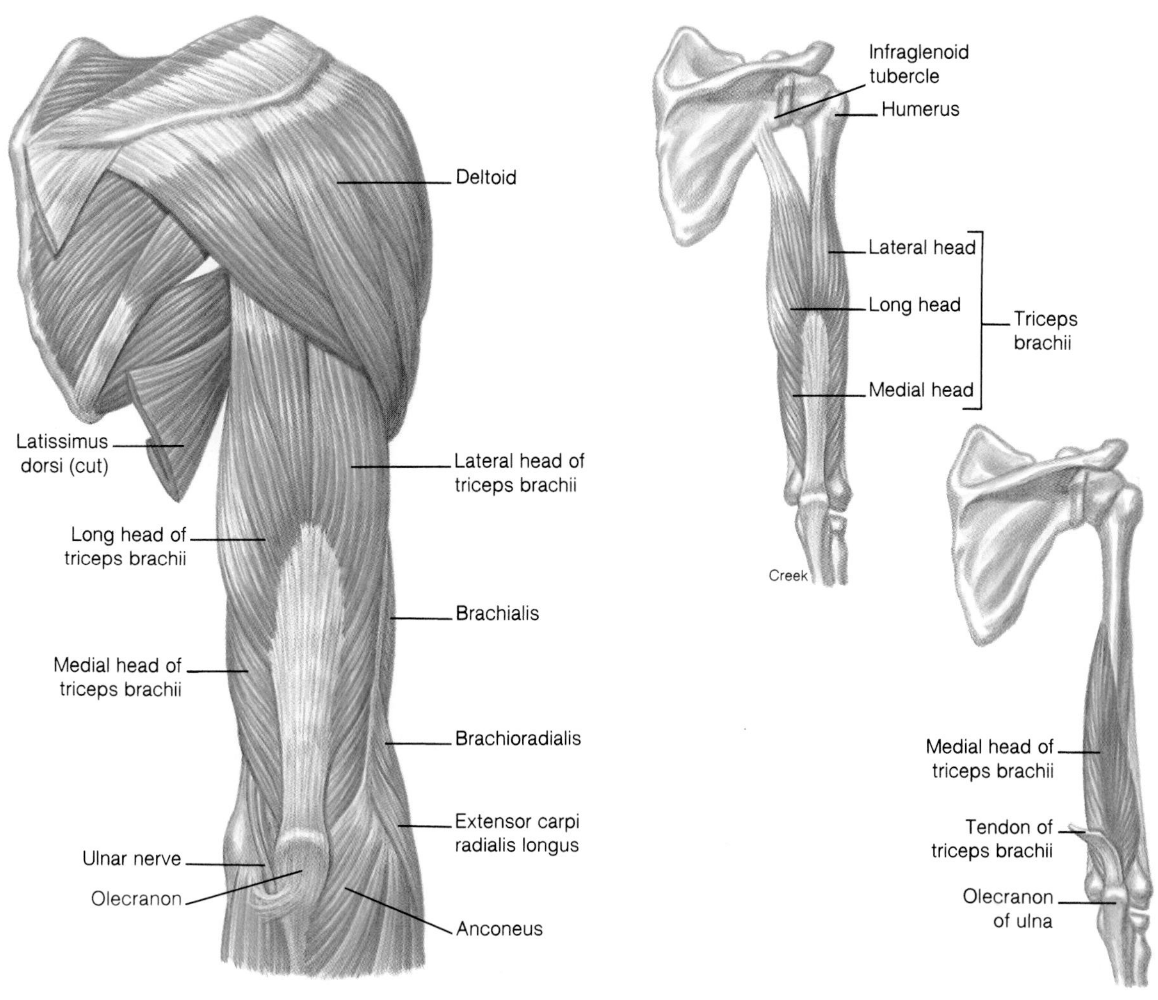

Table 13.11
Muscles That Act on the Forearm at the Elbow Joint

Muscle	Origin	Insertion	Action	Innervation
Biceps brachii	Coracoid process and tuberosity above glenoid cavity of scapula	Radial tuberosity	Flexes elbow joint; supinates forearm and hand at radioulnar joint	Musculocutaneous n.
Brachialis	Anterior body of humerus	Coronoid process of ulna	Flexes elbow joint	Musculocutaneous n.
Brachioradialis	Lateral supracondylar ridge of humerus	Proximal to styloid process of radius	Flexes elbow joint	Radial n.
Triceps brachii	Tuberosity below glenoid cavity; lateral and medial surfaces of humerus	Olecranon of ulna	Extends elbow joint	Radial n.
Anconeus	Lateral epicondyle of humerus	Olecranon of ulna	Extends elbow joint	Radial n.

biceps: L. *biceps*, two heads
triceps: L. *triceps*, three heads
anconeus: Gk. *ancon*, elbow

Figure 13.19

Superficial muscles of the right forearm.
(*a*) An anterior view and (*b*) a posterior view.

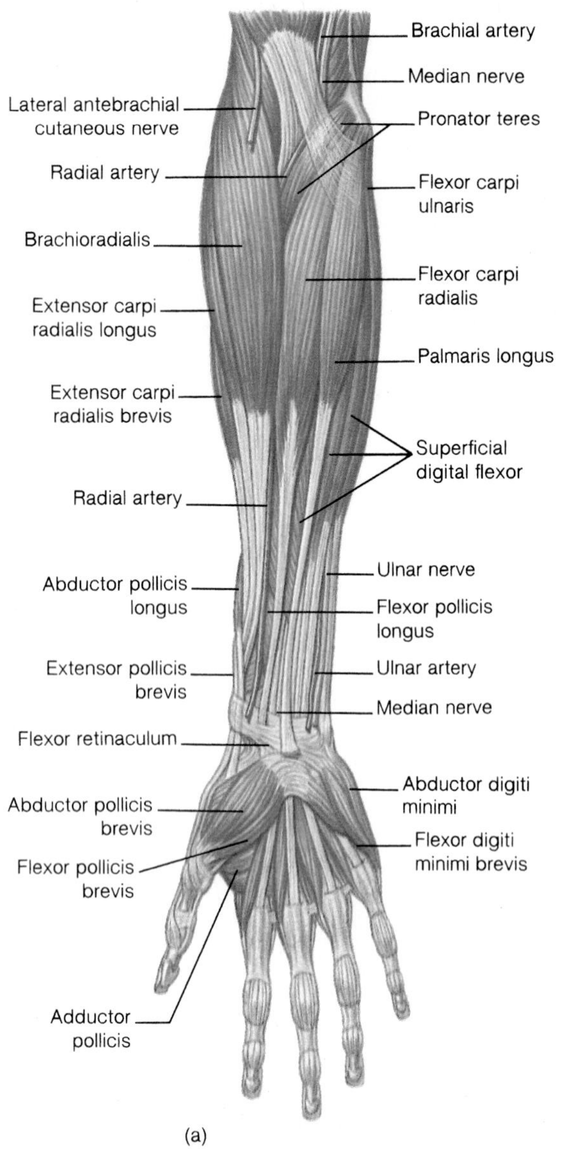

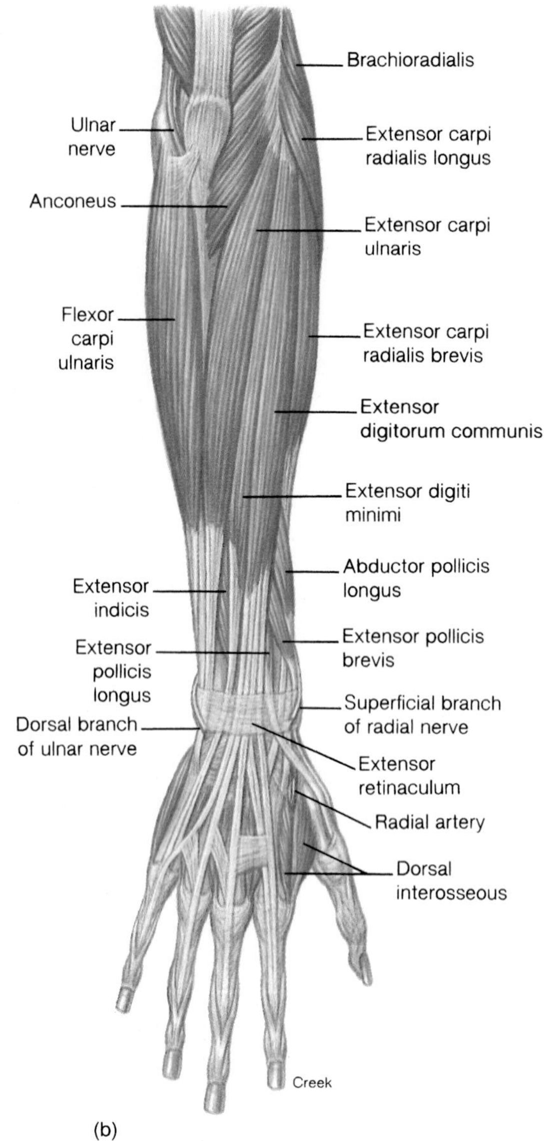

The **flexor carpi ulnaris** muscle is positioned on the medial anterior side of the forearm, where it assists in flexing the wrist joints and adducting the hand.

The broad **superficial digital flexor** muscle lies directly beneath the three flexor muscles just described (figs. 13.19 and 13.20). It has an extensive origin, involving the humerus, ulna, and radius (table 13.12). The tendon at the distal end of this muscle is united across the wrist joint but then splits to attach to the middle phalanx of digits two (II) through five (V).

The **deep digital flexor** muscle lies deep to the superficial digital flexor muscle. It inserts on the distal phalanges two (II) through five (V). These two muscles flex the joints of the wrist, hand, and the second, third, fourth, and fifth digits.

The **flexor pollicis longus** muscle is a deep, lateral muscle of the forearm. It flexes the joints of the thumb, assisting the grasping mechanism of the hand.

The tendons of the muscles that flex the joints of the hand can be seen on the wrist as a fist is made. These tendons are securely positioned by the *flexor retinaculum* (fig. 13.19*a*), which crosses the wrist area transversely.

Extension of the Hand

The muscles that extend the joints of the hand are located on the posterior side of the forearm. Most of the primary extensor muscles can be seen superficially in figure 13.19*b* and will be discussed from lateral to medial.

The long, tapered **extensor carpi radialis longus** muscle is medial to the brachioradialis muscle. It extends the joints of the wrist and abducts the hand at the wrist. Immediately medial to the extensor carpi radialis longus muscle is the **extensor carpi radialis brevis** muscle, which performs approxi-

Figure 13.20

Deep muscles of the right forearm.
(*a*) Rotators, (*b*) flexors, and (*c*) extensors.

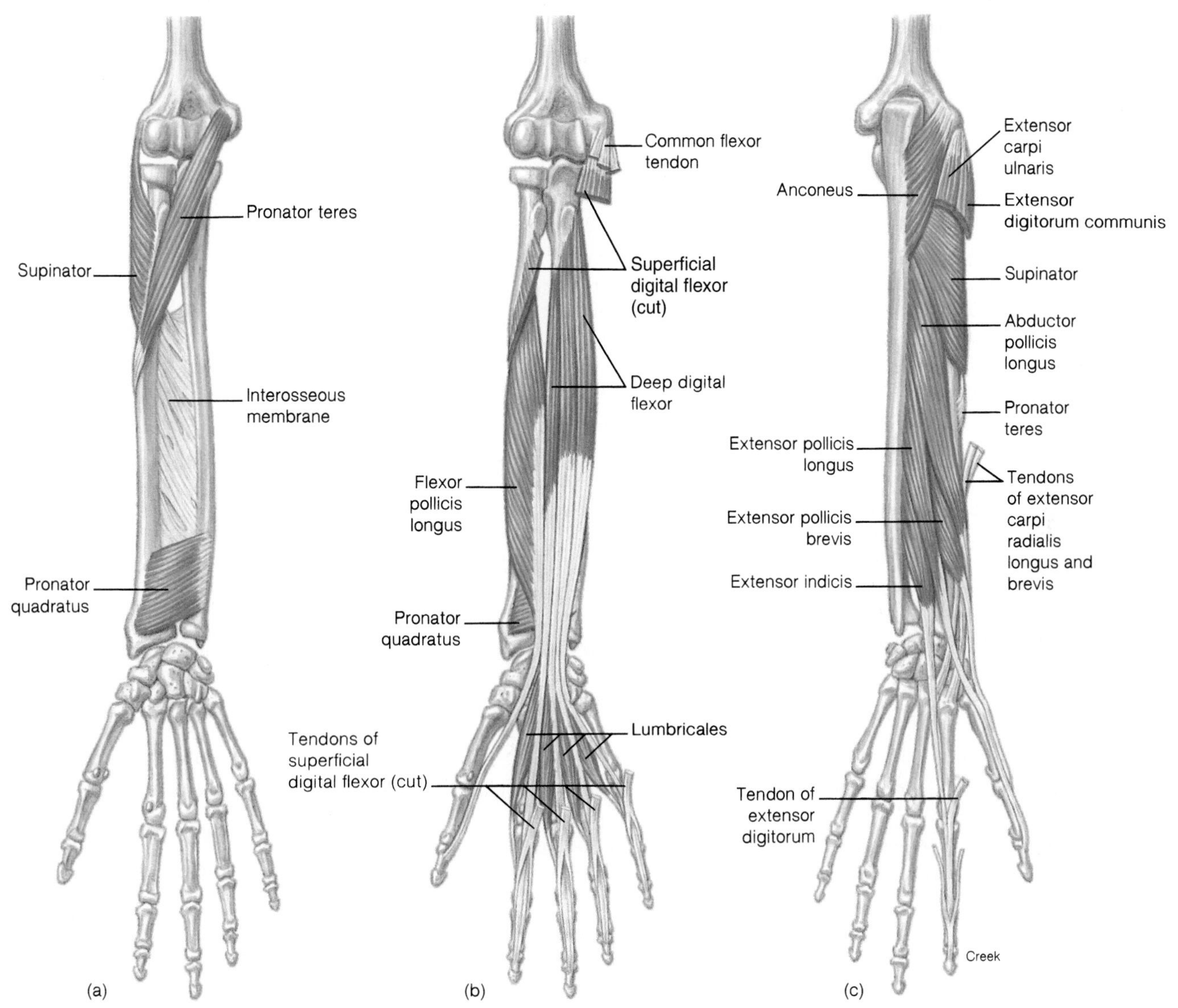

mately the same functions. The origin and insertion of the latter muscle are different, however (table 13.12).

The **extensor digitorum communis** muscle is positioned in the center of the forearm, along the posterior surface. It originates on the lateral epicondyle of the humerus. Its tendon of insertion divides at the wrist, beneath the extensor retinaculum, into four tendons that attach to the distal tip of the medial phalanges of digits two (II) through five (V).

The **extensor digiti minimi** is a long, narrow muscle located on the ulnar side of the extensor digitorum communis muscle. Its tendinous insertion fuses with the tendon of the extensor digitorum communis muscle going to the fifth digit.

The **extensor carpi ulnaris** is the most medial muscle on the posterior surface of the forearm. It inserts on the base of the fifth metacarpal bone, where it functions to extend and adduct the joints of the hand.

The **extensor pollicis longus** muscle arises from the mid-ulnar region, crosses the lower two-thirds of the forearm, and inserts onto the base of the distal phalanx of the thumb (fig. 13.20). It extends the joints of the thumb and abducts the hand. The **extensor pollicis brevis** muscle arises from the lower midportion of the radius and inserts on the base of the proximal phalanx of the thumb (fig. 13.20). The action of this muscle is similar to that of the extensor pollicis longus.

As its name implies, the **abductor pollicis longus** muscle abducts the joints of the thumb and hand. It originates on the interosseous ligament, between the ulna and radius, and inserts on the base of the first metacarpal bone.

The muscles that act on the joints of the wrist, hand, and digits are summarized in table 13.12.

Table 13.12

Muscles of the Forearm That Move the Joints of the Wrist, Hand, and Digits

Muscle	Origin	Insertion	Action	Innervation
Supinator	Lateral epicondyle of humerus and crest of ulna	Lateral surface of radius	Supinates forearm and hand	Radial n.
Pronator teres	Medial epicondyle of humerus	Lateral surface of radius	Pronates forearm and hand	Median n.
Pronator quadratus	Distal fourth of ulna	Distal fourth of radius	Pronates forearm and hand	Median n.
Flexor carpi radialis	Medial epicondyle of humerus	Base of second and third metacarpal bones	Flexes and abducts hand at wrist	Median n.
Palmaris longus	Medial epicondyle of humerus	Palmar aponeurosis	Flexes wrist	Median n.
Flexor carpi ulnaris	Medial epicondyle and olecranon	Carpal and metacarpal bones	Flexes and adducts wrist	Ulnar n.
Superficial digital flexor	Medial epicondyle, coronoid process, and anterior border of radius	Middle phalanges of digits II–V	Flexes wrist and digits at metacarpophalangeal and interphalangeal joints	Median n.
Deep digital flexor	Proximal two-thirds of ulna and interosseous ligament	Distal phalanges of digits II–V	Flexes wrist and digits at metacarpophalangeal and interphalangeal joints	Median and ulnar nn.
Flexor pollicis longus	Body of radius, interosseous membrane, and coronoid process of ulna	Distal phalanx of thumb	Flexes joints of thumb	Median n.
Extensor carpi radialis longus	Lateral supracondylar ridge of humerus	Second metacarpal bone	Extends and abducts wrist	Radial n.
Extensor carpi radialis brevis	Lateral epicondyle of humerus	Third metacarpal bone	Extends and abducts wrist	Radial n.
Extensor digitorum communis	Lateral epicondyle of humerus	Posterior surfaces of digits II–V	Extends wrist and phalanges at joints of carpophalangeal and interphalangeal joints	Radial n.
Extensor digiti minimi	Lateral epicondyle of humerus	Extensor aponeurosis of fifth digit	Extends joints of fifth digit and wrist	Radial n.
Extensor carpi ulnaris	Lateral epicondyle of humerus and olecranon	Base of fifth metacarpal bone	Extends and adducts wrist	Radial n.
Extensor pollicis longus	Middle of body of ulna, lateral side	Base of distal phalanx of thumb	Extends joints of thumb; abducts joints of hand	Radial n.
Extensor pollicis brevis	Distal body of radius and interosseous ligament	Base of first phalanx of thumb	Extends joints of thumb; abducts joints of hand	Radial n.
Abductor pollicis longus	Distal radius and ulna and interosseous ligament	Base of first metacarpal bone	Abducts joints of thumb and joints of hand	Radial n.

supinator: L. *supin*, bend back
pronator: L. *pron*, bend forward
palmaris: L. *palma*, flat of hand
pollicis: L. *pollex*, thumb

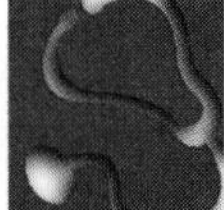

Notice that the joints of your hand are partially flexed even when the hand is relaxed. The muscles that extend the hand are not as strong as the muscles that flex it. This is why people who receive strong electrical shocks through the arms from a cord or wire will tightly flex their hands and cling to it. All the muscles of the arm are stimulated to contract, but the flexors, being stronger, cause the hands to close tightly.

Muscles of the Hand

The hand is a marvelously complex structure, adapted to permit an array of intricate movements. Flexion and extension movements of the hand and phalanges are accomplished by the muscles of the forearm just described. Precise finger movements that require coordinating abduction and adduction with flexion and extension are the function of the small intrinsic muscles of the hand. These muscles and associated structures of the hand are depicted in figure 13.21. The position and actions of the muscles of the hand are listed in table 13.13.

Figure 13.21

Muscles of the hand.

(*a*) An anterior view and (*b*) a lateral view of the second digit (index finger).

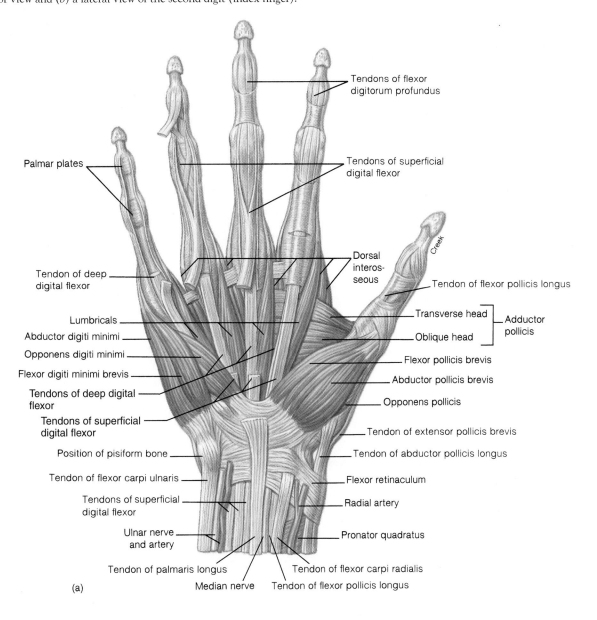

Tendons of flexor digitorum profundus
Palmar plates
Tendons of superficial digital flexor
Dorsal interosseous
Tendon of deep digital flexor
Tendon of flexor pollicis longus
Lumbricals
Transverse head
Oblique head
Adductor pollicis
Abductor digiti minimi
Opponens digiti minimi
Flexor pollicis brevis
Flexor digiti minimi brevis
Abductor pollicis brevis
Tendons of deep digital flexor
Opponens pollicis
Tendons of superficial digital flexor
Tendon of extensor pollicis brevis
Position of pisiform bone
Tendon of abductor pollicis longus
Tendon of flexor carpi ulnaris
Flexor retinaculum
Tendons of superficial digital flexor
Radial artery
Ulnar nerve and artery
Pronator quadratus
Tendon of palmaris longus
Tendon of flexor carpi radialis
Median nerve
Tendon of flexor pollicis longus
Creek
(a)

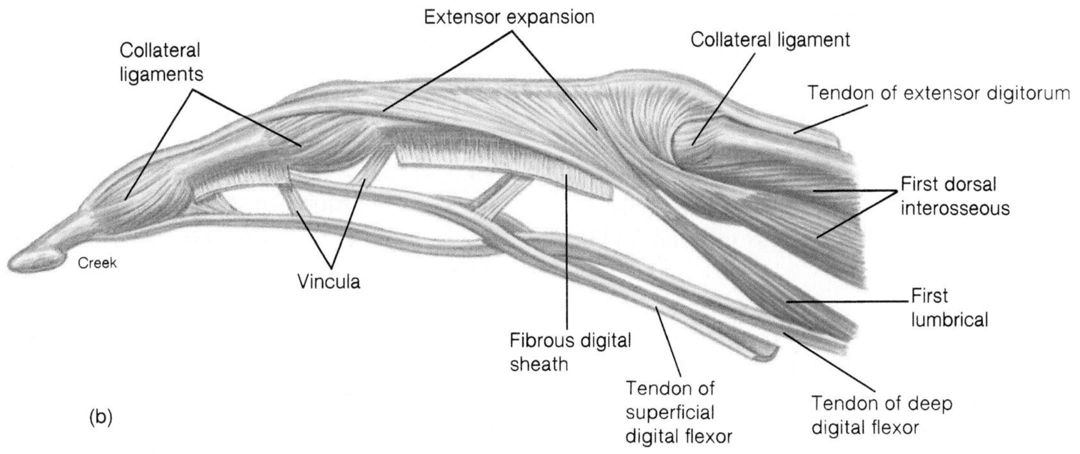

Table 13.13
Intrinsic Muscles of the Hand

Muscle	Origin	Insertion	Action	Innervation
Thenar Muscles				
Abductor pollicis brevis	Flexor retinaculum, scaphoid, and trapezium	Proximal phalanx of thumb	Abducts joints of thumb	Median n.
Flexor pollicis brevis	Flexor retinaculum and trapezium	Proximal phalanx of thumb	Flexes joints of thumb	Median n.
Opponens pollicis	Trapezium and flexor retinaculum	First metacarpal bone	Opposes joints of thumb	Median n.
Intermediate Muscles				
Adductor pollicis (oblique and transverse heads)	Oblique head, capitate; transverse head, second and third metacarpal bones	Proximal phalanx of thumb	Adducts joints of thumb	Ulnar n.
Lumbricales (4)	Tendons of flexor digitorum profundus	Extensor expansions of digits II–V	Flexes digits at metacarpophalangeal joints; extends digits at interphalangeal joints	Median and ulnar nn.
Palmar interossei (3)	Medial side of second metacarpal bone; lateral sides of fourth and fifth metacarpal bones	Proximal phalanges of index, ring, and little fingers and extensor digitorum communis	Adducts fingers toward middle finger at metacarpophalangeal joints	Ulnar n.
Dorsal interossei (4)	Adjacent sides of metacarpal bones	Proximal phalanges of index and middle fingers (lateral sides) plus proximal phalanges of middle and ring fingers (medial sides) and extensor digitorum communis	Abducts fingers away from middle finger at metacarpaophalgeal joints	Ulnar n.
Hypothenar Muscles				
Abductor digiti minimi	Pisiform and tendon of flexor carpi ulnaris	Proximal phalanx of digit V	Abducts joints of digit	Vulnar n.
Flexor digiti minimi	Flexor retinaculum and book of hamate	Proximal phalanx of digit V	Flexes joints of digit V	Ulnar n.
Opponens digiti minimi	Flexor retinaculum and hook of hamate	Fifth metacarpal bone	Opposes joints of digit V	Ulnar n.

opponens: L. *opponens*, against

The muscles of the hand are divided into **thenar** (*the′nar*), **hypothenar** (*hi-poth′ē-nar*), and **intermediate** groups. The *thenar eminence* is the fleshy base of the thumb and is formed by three muscles: the **abductor pollicis brevis,** the **flexor pollicis brevis,** and the **opponens pollicis.** The most important of the thenar muscles is the opponens pollicis, which opposes the thumb to the palm of the hand.

The *hypothenar eminence* is the elongated, fleshy bulge at the base of the little finger. It also is formed by three muscles: the **abductor digiti minimi,** the **flexor digiti minimi,** and the **opponens digiti minimi.**

Muscles of the intermediate group are positioned between the metacarpal bones in the region of the palm. This group includes the **adductor pollicis muscle,** the **lumbricales** (*lum′brĭ-kalz*), and the **palmar** and **dorsal interossei** muscles.

Muscles That Move the Thigh at the Hip Joint

The muscles that move the thigh at the hip joint originate from the pelvic girdle and the vertebral column, and insert on various places on the femur. These muscles stabilize a highly movable hip joint and provide support for the body during bipedal stance and locomotion. The most massive muscles of the body are found in this region, along with some extremely small muscles. The muscles that move the thigh at the hip joint are divided into anterior, posterior, and medial groups.

Anterior Muscles

The anterior muscles that move the thigh at the hip joint are the iliacus and psoas major (figs. 13.22 and 13.23). The

Figure 13.22

Muscles of the right anterior pelvic and thigh regions.

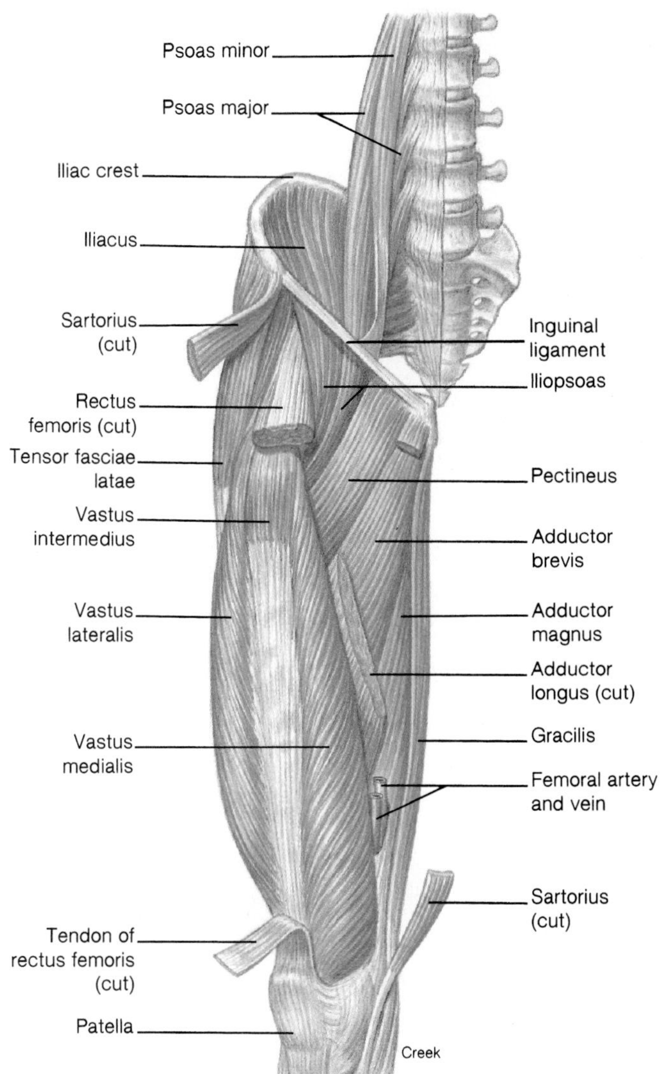

Figure 13.23

Anterior pelvic muscles that move the hip joint.

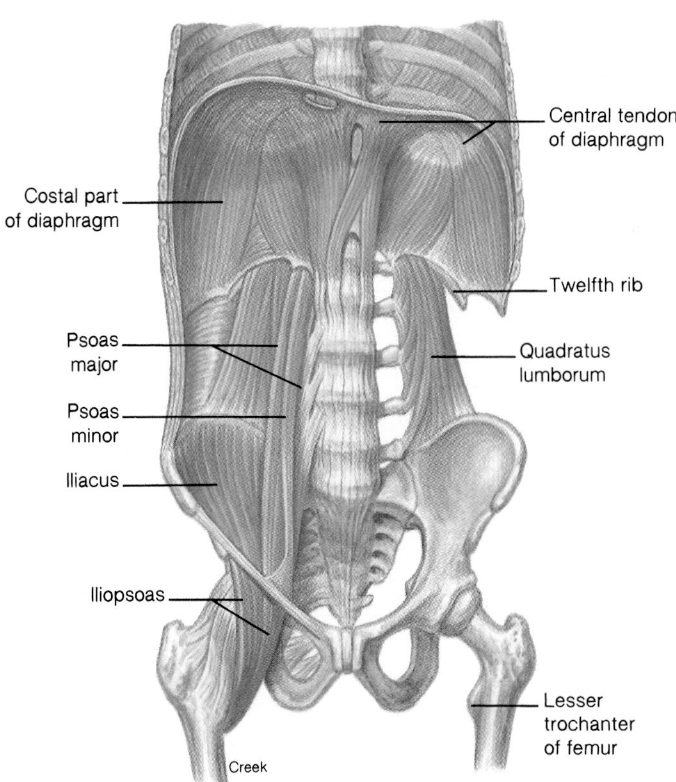

triangular **iliacus** (*il″e-ak′us;* i-l′a-kus) muscle arises from the iliac fossa and inserts on the lesser trochanter of the femur. The long, thick **psoas** (*so′as*) **major** muscle originates on the bodies and transverse processes of the lumbar vertebrae; it inserts, along with the iliacus muscle, on the lesser trochanter (fig. 13.23). The psoas major and the iliacus work synergistically in flexing and rotating the hip joint and flexing the vertebral column. These two muscles are collectively called the **iliopsoas** (*il″e-ŏ-so′as*) muscle.

Posterior and Lateral (Buttock) Muscles

The posterior muscles that move the thigh include the gluteus maximus, gluteus medius, gluteus minimus, and tensor fasciae latae.

The large **gluteus** (*gloo′te-us*) **maximus** muscle forms much of the prominence of the buttock (fig. 13.24). It is a powerful extensor muscle of the hip joint and is very important for bipedal stance and locomotion. The gluteus maximus originates on the ilium, sacrum, coccyx, and aponeurosis of the lumbar region. It inserts on the gluteal tuberosity of the femur and the *iliotibial tract,* a thickened tendinous region of the fascia lata extending down the thigh (see fig. 13.26).

The **gluteus medius** muscle is located immediately deep to the gluteus maximus (fig. 13.24). It originates on the lateral surface of the ilium and inserts on the greater trochanter of the femur. The gluteus medius abducts and medially rotates the hip joint. The mass of this muscle is of clinical significance as a site for intramuscular injections.

The **gluteus minimus** muscle is the smallest and deepest of the gluteal muscles (fig. 13.24). It also arises from the lateral surface of the ilium, and it inserts on the lateral surface of the greater trochanter, where it acts synergistically with the gluteus medius to abduct and medially rotate the hip joint.

The quadrangular **tensor fasciae latae** (*fash′e-e la′te*) muscle is positioned superficially on the lateral surface of the hip (see fig. 13.26). It originates on the iliac crest and inserts on a broad lateral fascia of the thigh called the iliotibial tract. The tensor fasciae latae muscle and the gluteus medius muscle are synergistic abductor muscles of the hip joint.

A deep group of six lateral rotators of the hip joint is positioned directly over the posterior aspect of the hip. These muscles are not discussed here but are identified in figure 13.24 from superior to inferior as the **piriformis** (*pi-ri-for′mis*), **superior gemellus** (*je-mel′us*), **obturator internus, inferior gemellus, obturator externus,** and **quadratus femoris** muscles.

The anterior and posterior group of muscles that move the thigh at the hip joint are summarized in table 13.14.

Figure 13.24
Deep gluteal muscles.

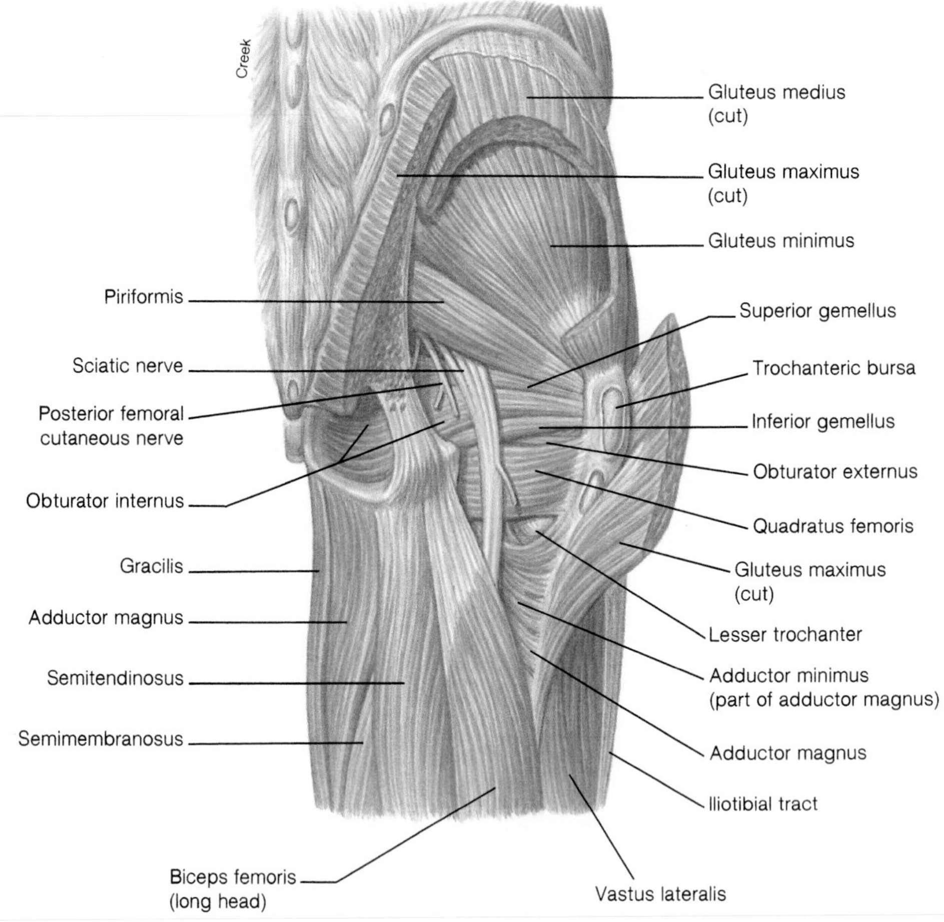

Table 13.14

Anterior and Posterior Muscles That Move the Thigh at the Hip Joint

Muscle	Origin	Insertion	Action	Innervation
Iliacus	Iliac fossa	Lesser trochanter of femur, along with psoas major	Flexes and rotates thigh laterally at the hip joint; flexes joints of vertebral column	Femoral n.
Psoas major	Transverse processes of all lumbar vertebrae	Lesser trochanter, along with iliacus	Flexes and rotates thigh laterally at the hip joint; flexes joints of vertebral column	Spinal nerves L2, L3
Gluteus maximus	Iliac crest, sacrum, coccyx, and aponeurosis of the lumbar region	Gluteal tuberosity and iliotibial tract	Extends and rotates thigh laterally at the hip joint	Inferior gluteal n.
Gluteus medius	Lateral surface of ilium	Greater trochanter	Abducts and rotates thigh medially at the hip joint	Superior gluteal n.
Gluteus minimus	Lateral surface of lower half of ilium	Greater trochanter	Abducts and rotates thigh medially at the hip joint	Superior gluteal n.
Tensor fasciae latae	Anterior border of ilium and iliac crest	Iliotibial tract	Abducts thigh at the hip joint	Superior gluteal n.

psoas: Gk. *psoa*, loin
gluteus: Gk. *gloutos*, rump

Figure 13.25

Adductor muscles of the right thigh.

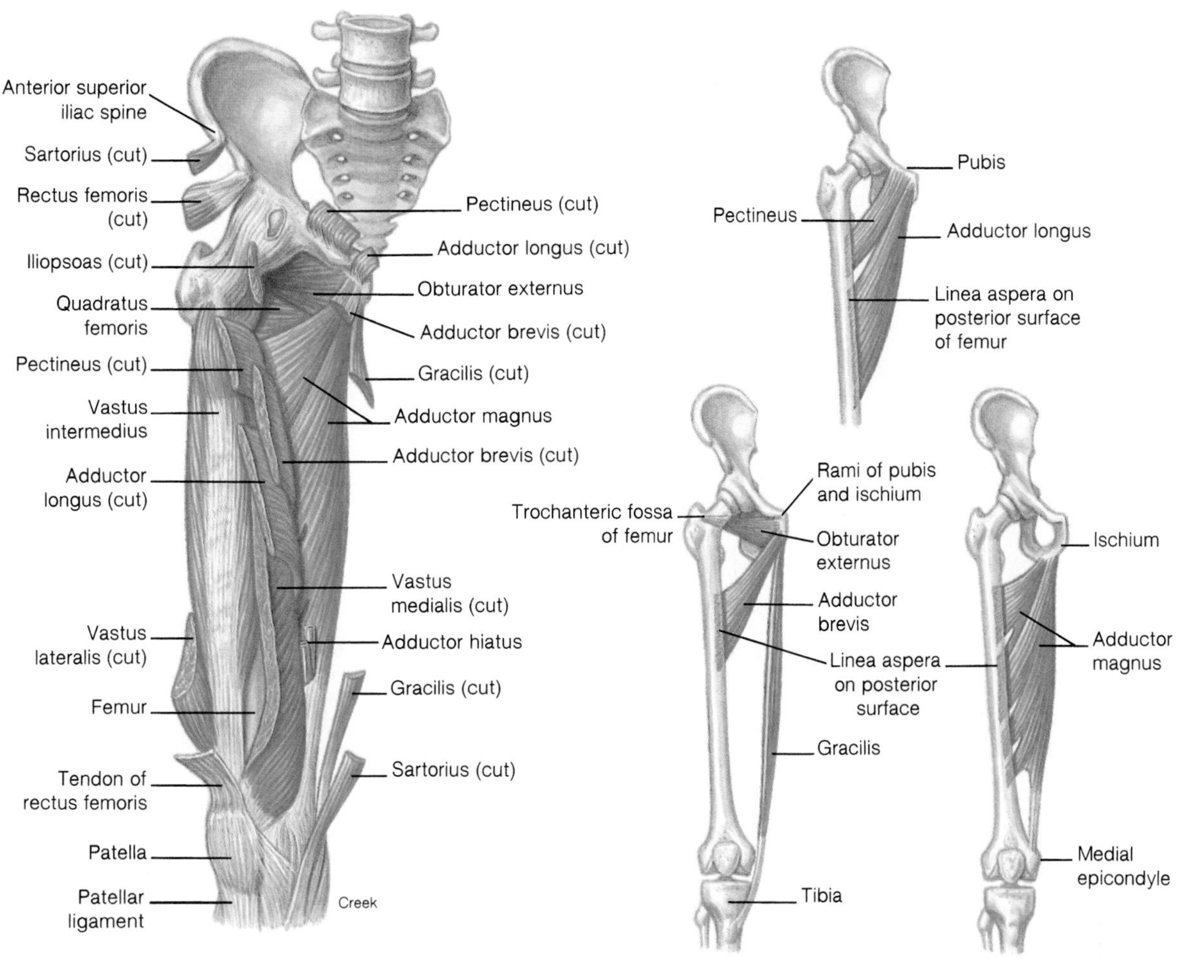

Medial, or Adductor, Muscles

The medial muscles that move the hip joint include the gracilis, pectineus, adductor longus, adductor brevis, and adductor magnus muscles (figs. 13.25, 13.26, 13.27, and 13.28).

The long, thin **gracilis** (*gras'il-is*) muscle is the most superficial of the medial thigh muscles. It is a two-joint muscle and can adduct the hip joint or flex the knee.

The **pectineus** (*pek-tin'e-us*) muscle is the uppermost of the medial muscles that move the hip joint. It is a flat, quadrangular muscle that flexes and adducts the hip.

The **adductor longus** muscle is located immediately lateral to the gracilis on the upper third of the thigh; it is the most anterior of the adductor muscles. The **adductor brevis** is a triangular muscle located deep to the adductor longus and pectineus muscles, which largely conceal it. The **adductor magnus** muscle is a large, thick muscle, somewhat triangular in shape. It is located deep to the other two adductor muscles. The adductor longus, adductor brevis, and the adductor magnus are synergistic in adducting, flexing, and laterally rotating the hip joint.

The muscles that adduct the hip joint are summarized in table 13.15.

Muscles of the Thigh That Move the Knee Joint

The muscles that move the knee originate on the pelvic girdle or thigh. They are surrounded and compartmentalized by tough fascial sheets, which are a continuation of the iliotibial tract. These muscles are divided according to function and position into two groups: *anterior extensors* and *posterior flexors*.

Anterior, or Extensor, Muscles

The anterior muscles that move the knee joint are the sartorius and quadriceps femoris muscles (figs. 13.25 and 13.26).

The long, straplike **sartorius** (*sar'to're-us*) muscle obliquely crosses the anterior aspect of the thigh. It can act on both the hip and knee joints to flex and rotate the hip laterally, and also to assist in flexing the knee joint and rotating it medially. The sartorius is the longest muscle of the

Figure 13.26
Muscles of the right anterior thigh.

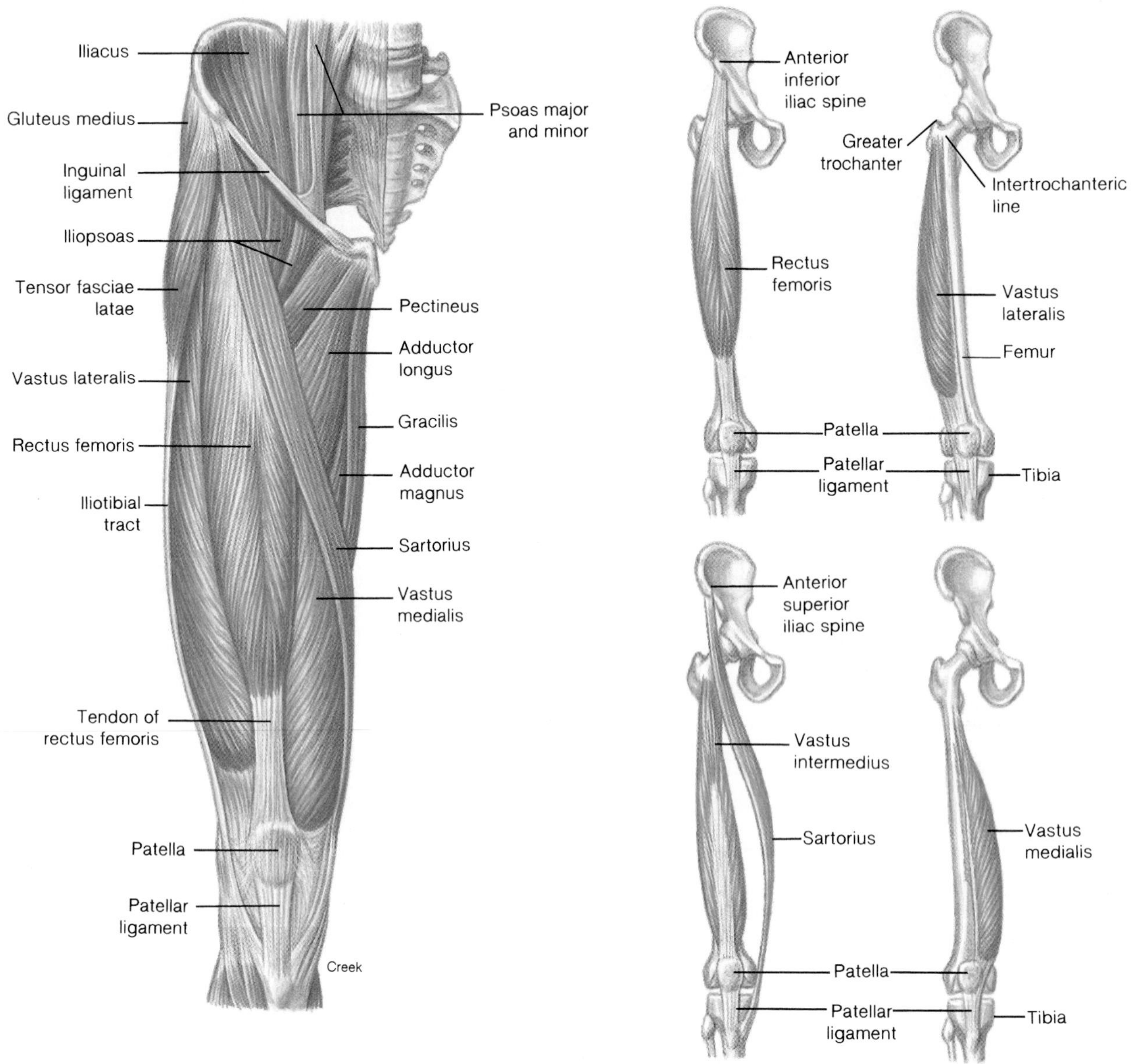

body. It is frequently called the "tailor's muscle" because it helps to effect the cross-legged sitting position in which tailors are often depicted.

The **quadriceps femoris** muscle is actually a composite of four distinct muscles that have separate origins but a common insertion on the patella via the tendon of the rectus femoris. The tendon of the rectus femoris is continuous over the patella and becomes the *patellar ligament* as it attaches to the tibial tuberosity (fig. 13.26). These muscles function synergistically to extend the knee joint, as in kicking a football. The four muscles of the quadriceps femoris muscle are the rectus femoris, vastus lateralis, vastus medialis, and vastus intermedius.

The **rectus femoris** muscle occupies a superficial position and is the only one of the four quadriceps that functions in both the hip and knee joints. The laterally positioned **vastus lateralis** is the largest muscle of the quadriceps femoris. It is a common intramuscular injection site in infants who have small, underdeveloped buttock and shoulder muscles. The **vastus medialis** muscle occupies a medial position along the thigh. The **vastus intermedius** muscle lies deep to the rectus femoris.

The anterior thigh muscles that move the leg at the knee joint are summarized in table 13.16.

Figure 13.27

Muscles of the right posterior thigh.

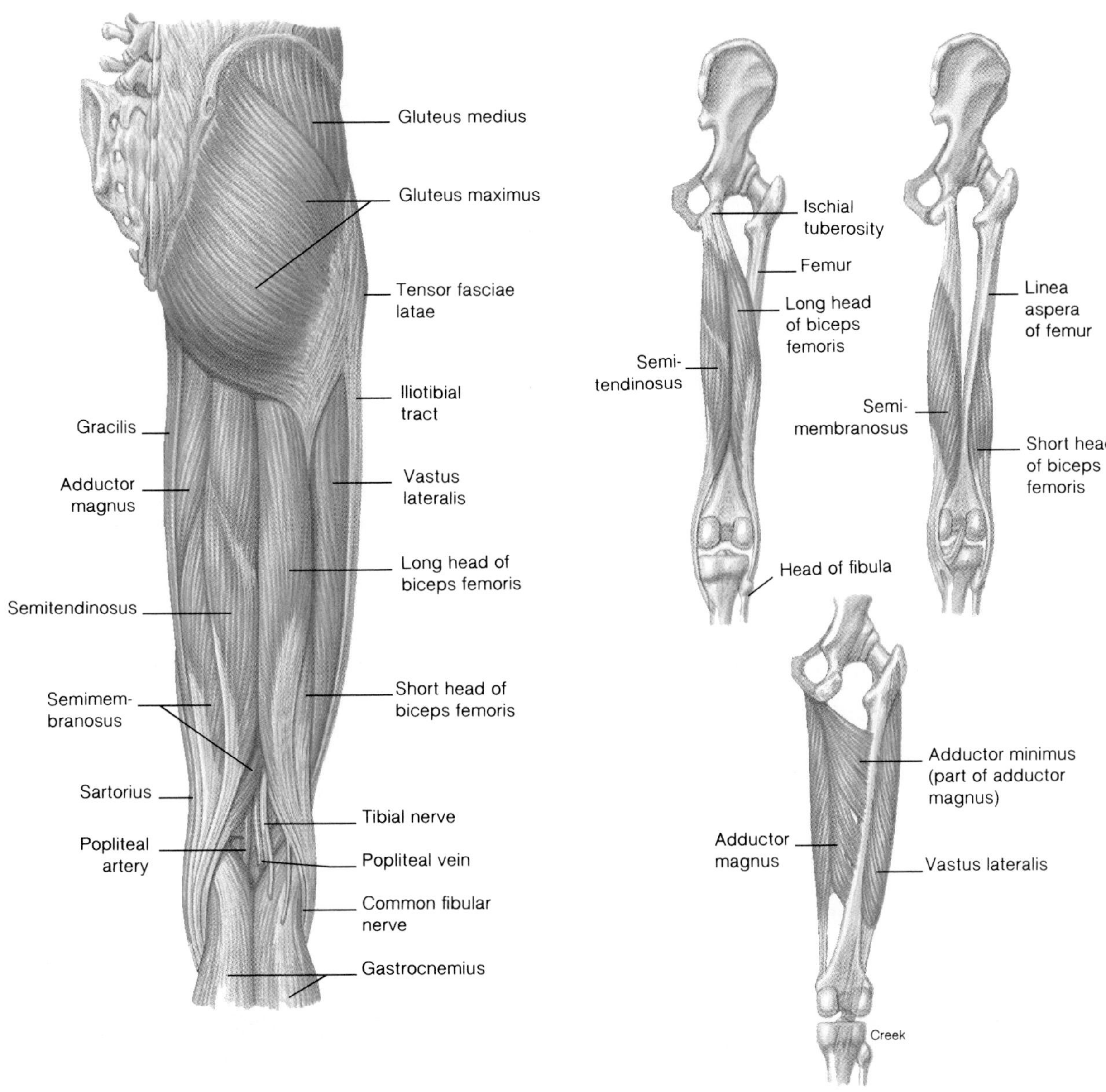

Posterior, or Flexor, Muscles

There are three posterior thigh muscles, which are antagonistic to the quadriceps femoris muscles in flexing the knee joint. These muscles are known as the **hamstrings** (fig. 13.27). The name derives from the butchers' practice of curing a ham by hanging the hog by the tendons of these knee muscles.

The **biceps femoris** muscle occupies the posterior lateral aspect of the thigh. It has a superficial long head and a deep short head and causes movement at both the hip and knee joints. The superficial **semitendinosus** muscle is fusiform and is located on the posterior medial aspect of the thigh. It also works over two joints. The flat **semimembranosus** muscle lies deep to the semitendinosus muscle on the posterior medial aspect of the thigh.

The posterior thigh muscles that move the leg at the knee joint are summarized in table 13.17. The relative positions of the muscles of the thigh are illustrated in figure 13.29.

Figure 13.28

Muscles of the right medial thigh.

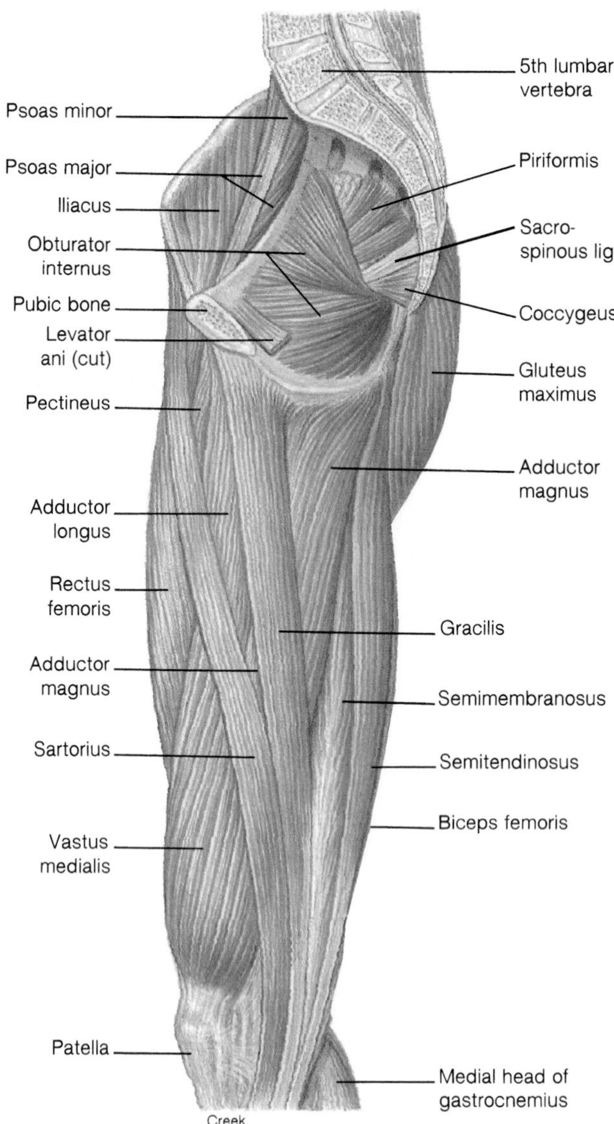

Table 13.15

Medial Muscles That Move the Thigh at the Hip Joint

Muscle	Origin	Insertion	Action	Innervation
Gracilis	Inferior edge of symphysis pubis	Proximal medial surface of tibia	Adducts thigh at hip joint; flexes and rotates leg at knee joint	Obturator n.
Pectineus	Pectineal line of pubis	Distal to lesser trochanter of femur	Adducts and flexes thigh at hip joint	Femoral n.
Adductor longus	Pubis—below pubic crest	Linea aspera of femur	Adducts, flexes, and laterally rotates thigh at hip joint	Obturator n.
Adductor brevis	Inferior ramus of pubis	Linea aspera of femur	Adducts, flexes, and laterally rotates thigh at hip joint	Obturator n.
Adductor magnus	Inferior ramus of ischium and pubis	Linea aspera and medial epicondyle of femur	Adducts, flexes, and laterally rotates thigh at hip joint	Obturator and tibial nn.

gracilis: Gk. *gracilis*, slender

Table 13.16
Anterior Thigh Muscles That Move the Knee Joint

Muscle	Origin	Insertion	Action	Innervation
Sartorius	Anterior superior iliac spine	Medial surface of tibia	Flexes knee and hip joints, abducts hip joint, rotates thigh laterally at hip joint, and rotates thigh medially at hip joint	Femoral n.
Quadriceps femoris		Patella by common tendon, which continues as patellar ligament to tibial tuberosity	Extends leg at knee joint; rectus femoris alone also flexes hip joint	Femoral n.
Rectus femoris	Anterior superior iliac spine and lip of acetabulum			
Vastus lateralis	Greater trochanter and linea aspera of femur			
Vastus medialis	Medial surface and linea aspera of femur			
Vastus intermedius	Anterior and lateral surfaces of femur			

sartorius: L. *sartor*, a tailor (muscle used to cross legs in a tailor's position)

Table 13.17
Posterior Thigh Muscles That Move the Knee Joint

Muscle	Origin	Insertion	Action
Biceps femoris	Long head—ischial tuberosity; short head—linea aspera of femur	Head of fibular and lateral epicondyle of tibia	Flexes knee joint; extends and laterally rotates thigh at hip joint
Semitendinosus	Ischial tuberosity	Proximal portion of medial surface of body of tibia	Flexes knee joint; extends and medially rotates thigh at hip joint
Semimembranosus	Ischial tuberosity	Medial epicondyle of tibia	Flexes knee joint; extends and medially rotates thigh at hip joint

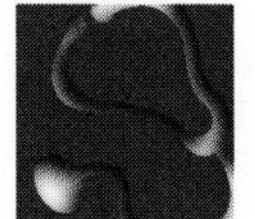

Hamstring injuries are a common occurrence in some sports. The injury usually occurs when sudden lateral or medial stress to the knee joint tears the muscles or tendons. Because of its structure and the stress applied to it in competition, the knee joint is also highly susceptible to injury. Altering the rules in contact sports could reduce the incidence of knee injury. At the least, additional support and protection should be provided for this vulnerable joint.

Muscles of the Leg That Move the Joints of the Ankle, Foot, and Toes

The muscles of the leg, the **crural** muscles, are responsible for the movements of the foot. There are three groups of crural muscles: anterior, lateral, and posterior. The anteromedial aspect of the leg along the body of the tibia lacks muscle attachment.

Anterior Crural Muscles

The anterior crural muscles include the tibialis anterior, extensor digitorum longus, extensor hallucis longus, and peroneus tertius muscles (figs. 13.30, 13.31, and 13.32).

The large, superficial **tibialis anterior** muscle can be easily palpated on the anterior lateral portion of the tibia (fig. 13.30). It parallels the prominent anterior crest of the tibia. The **extensor digitorum longus** muscle is positioned lateral to the tibialis anterior on the anterolateral surface of the leg. The **extensor hallucis** (*ha-loo'sis*) **longus** muscle is positioned deep between the tibialis anterior muscle and the extensor digitorum longus muscle. The small **peroneus tertius** muscle is continuous with the distal portion of the extensor digitorum longus muscle.

Lateral Crural Muscles

The lateral crural muscles are the peroneus longus and peroneus brevis (figs. 13.30 and 13.31). The long, flat **peroneus**

Figure 13.29

A transverse section of the right thigh as seen from above.
(Note the position of the vessels and nerves.)

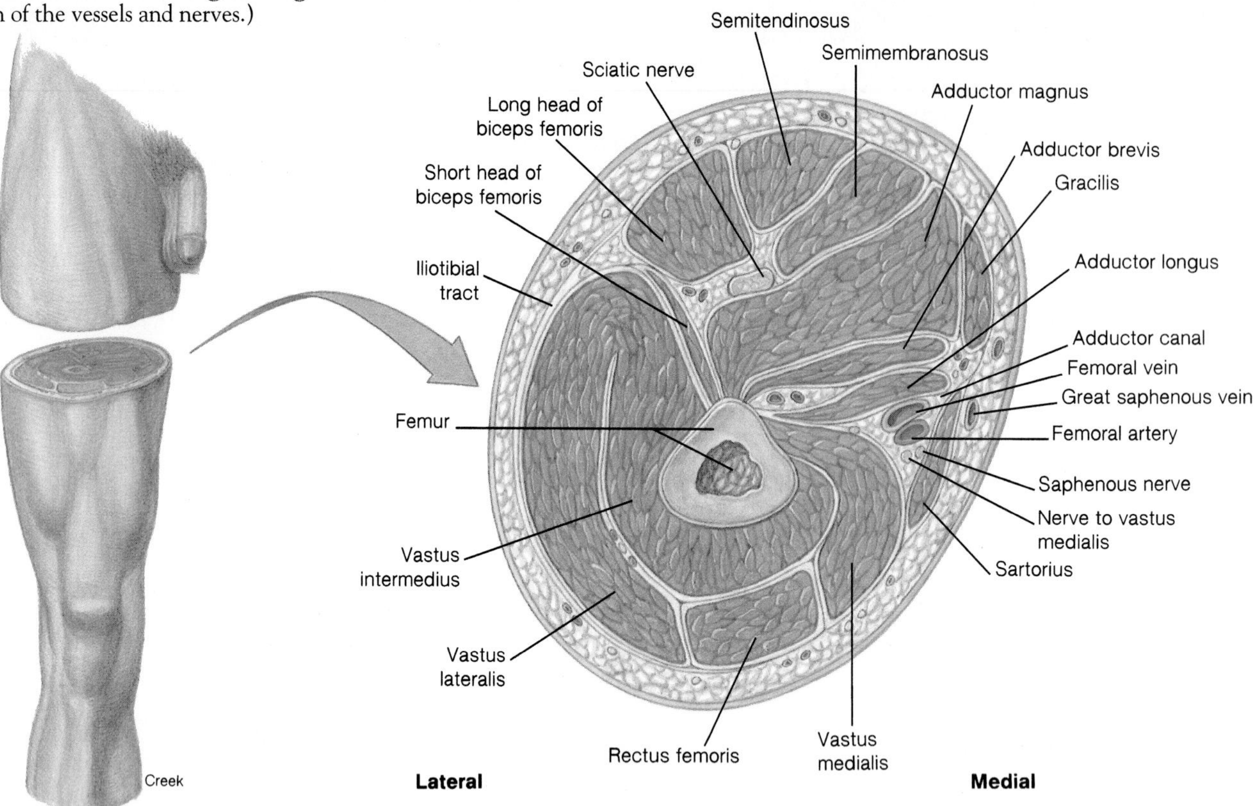

Figure 13.30

Anterior crural muscles.

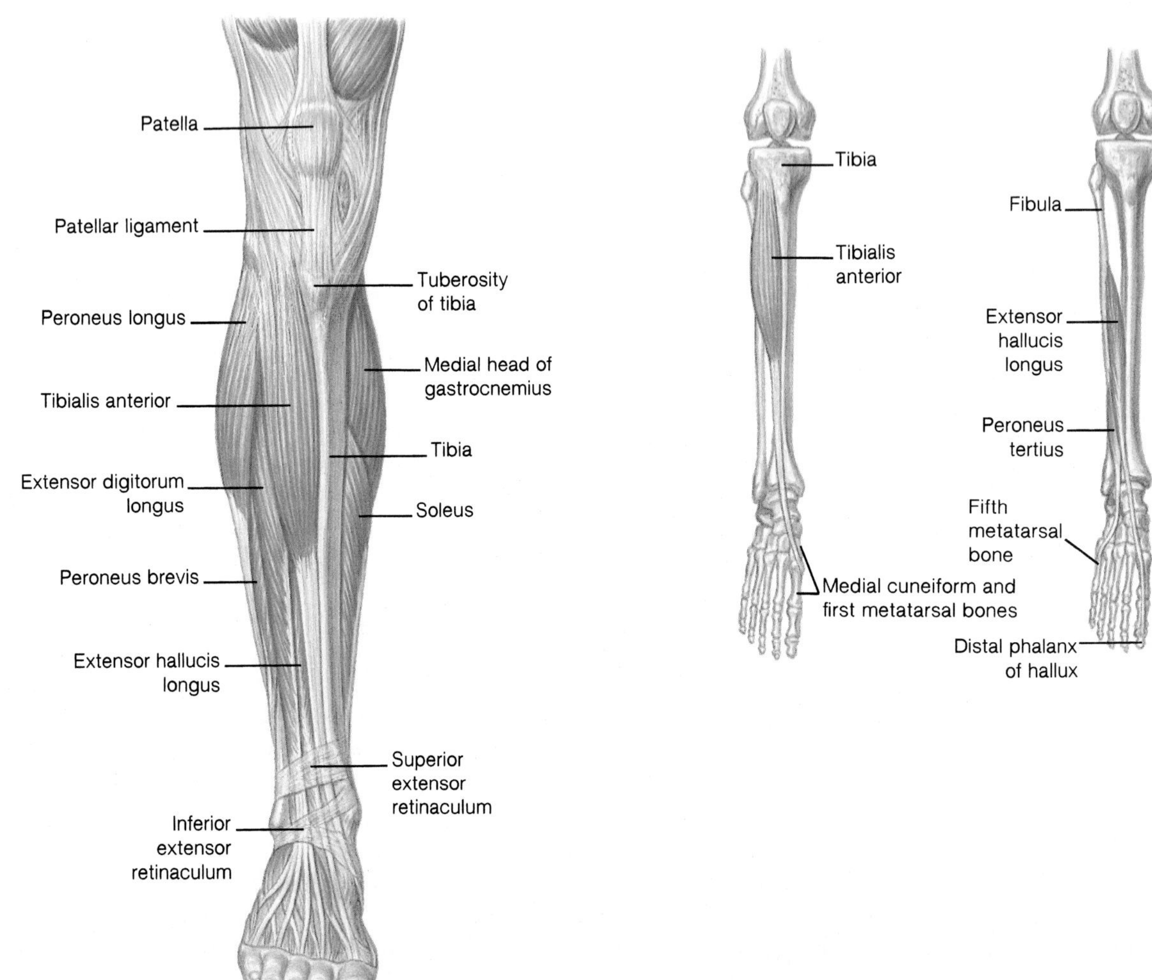

Figure 13.31
Lateral crural muscles.

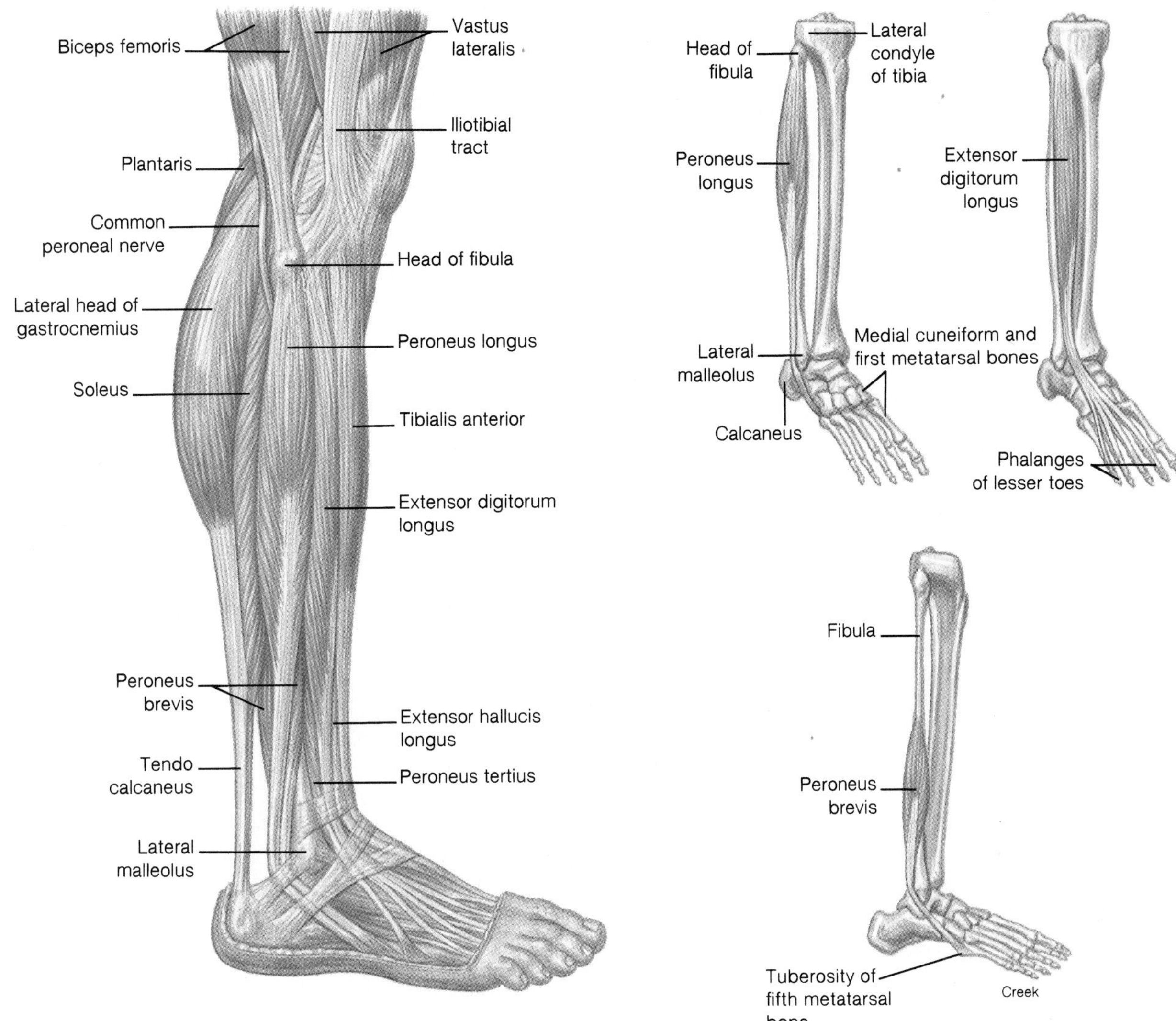

longus muscle is a superficial lateral muscle that overlies the fibula. The **peroneus brevis** muscle lies deep to the peroneus longus muscle and is positioned closer to the foot. These two muscles are synergistic in flexing the ankle joint and everting the foot (table 13.18).

Posterior Crural Muscles

The seven posterior crural muscles can be grouped into a superficial and a deep group. The superficial group is composed of the gastrocnemius, soleus, and plantaris muscles (fig. 13.33). The four deep posterior crural muscles are the popliteus, flexor hallucis longus, flexor digitorum longus, and tibialis posterior muscles (fig. 13.34).

The **gastrocnemius** (*gas″trok-ne′me-us*) muscle is a large, superficial muscle that forms the major portion of the calf of the leg. It consists of two distinct heads that arise from the posterior surfaces of the medial and lateral epicondyles of the femur. This muscle and the deeper soleus muscle insert onto the calcaneus via the common *tendo calcaneus* (*tendon of Achilles*). This is the strongest tendon of the body, but it is frequently ruptured from sudden stress during athletic competition. The gastrocnemius acts over two joints to cause flexion of the knee joint and plantar flexion of the foot at the ankle joint.

The **soleus** muscle lies deep to the gastrocnemius. These two muscles are frequently referred to as a single muscle, the **triceps surae** (*sur′e*). The soleus and gastrocnemius muscles have a common insertion, but the soleus acts only on the ankle joint, in plantar flexing the foot.

The small **plantaris** muscle arises just above the origin of the lateral head of the gastrocnemius muscle on the lateral supracondylar ridge of the femur. It has a very long, slender tendon of insertion onto the calcaneus. The tendon of this muscle is frequently mistaken for a nerve by those dissecting it for the first time. The plantaris is a weak muscle, with limited ability to flex the knee joint and plantar flex the ankle joint.

Figure 13.32
A medial view of crural muscles.

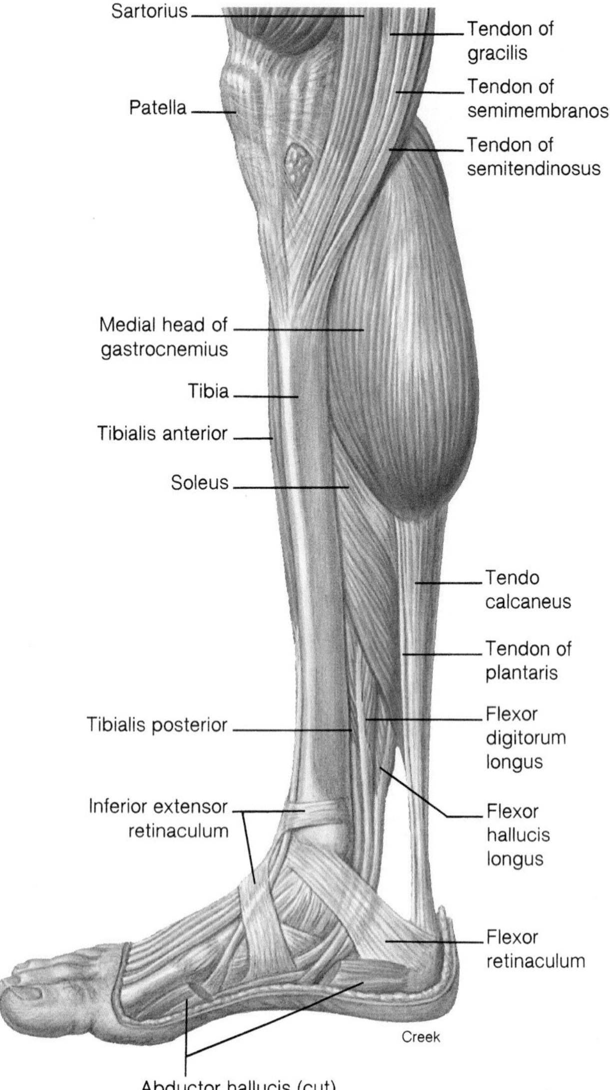

The thin, triangular **popliteus** (*pop-lit'e-us*) muscle is situated deep to the heads of the gastrocnemius muscle, where it forms part of the floor of the *popliteal fossa*—the depression on the back side of the knee joint (fig. 13.35). The popliteus muscle is a medial rotator of the tibia at the knee joint. The bipennate **flexor hallucis longus** muscle lies deep to the soleus muscle on the posterolateral side of the leg. It flexes the joints of the great toe (hallux) and assists in plantar flexing the ankle joint and inverting the foot.

The **flexor digitorum longus** muscle also lies deep to the soleus muscle, and it parallels the flexor hallucis longus muscle on the medial side of the leg. Its distal tendon passes posterior to the medial malleolus and continues along the plantar surface of the foot, where it branches into four tendinous slips that attach to the bases of the distal phalanges of the second, third, fourth, and fifth digits (fig. 13.36). The flexor digitorum longus works over several joints, flexing the joints in four of the digits and assisting in plantar flexing the ankle joint and inverting the foot.

The **tibialis posterior** muscle is located deep to the soleus muscle, between the posterior flexors. Its distal tendon passes posterior to the medial malleolus and inserts on the plantar surfaces of the navicular, cuneiform and cuboid bones, and the second, third, and fourth metatarsal bones (fig. 13.36). The tibialis posterior plantar flexes the ankle joint, inverts the foot, and supports the arches of the foot.

The crural muscles are summarized in table 13.18.

Muscles of the Foot

With the exception of one additional intrinsic muscle, the **extensor digitorum brevis,** the muscles of the foot are similar in name and number to those of the hand. The functions of the muscles of the foot are different, however, because the foot is adapted to provide support while bearing body weight rather than to grasp objects.

The muscles of the foot can be grouped into four layers (fig. 13.36), but these are difficult to dissociate, even in dissection. The muscles function either to move the toes or to support the arches of the foot through their contraction. Because of their complexity, the muscles of the foot will be presented only in illustrations (see figs. 13.36 and 13.37).

Clinical Considerations

Compared to the other systems of the body, the muscular system is extremely durable. If properly conditioned, the muscles of the body can adequately serve a person for a lifetime. Muscles are capable of doing incredible amounts of work; through exercise, they can become even stronger.

Evaluation of Muscle Condition

The clinical symptoms of muscle diseases are usually weakness, loss of muscle mass (atrophy), and pain. The most obvious diagnostic procedure is a clinical examination of the patient. Following this, it may be necessary to test muscle function using **electromyography (EMG)** to measure conduction rates and motor unit activity within a muscle. Laboratory tests may include serum enzyme assays or muscle biopsies. A muscle biopsy is perhaps the most definitive diagnostic tool. Progressive atrophy, polymyositis, and metabolic diseases of muscles can be determined through a biopsy.

Functional Conditions in Muscles

Muscles depend on systematic, periodic contraction to maintain optimal health. Obviously, overuse or disease will cause a change in muscle tissue. The immediate effect of overexertion on muscle tissue is the accumulation of lactic acid, which results in fatigue and soreness. Excessive contraction of a muscle can also damage the fibers or associated connective tissue, resulting in a **strained muscle.**

Table 13.18

Muscles of the Leg That Move the Joints of the Ankle, Foot, and Toes

Muscle	Origin	Insertion	Action	Innervation
Tibialis anterior	Lateral epicondyle and body of tibia	First metatarsal bone and first cuneiform	Dorsiflexes ankle and inverts foot at ankle	Deep fibular n.
Extensor digitorum longus	Lateral epicondyle of tibia and anterior surface of fibular	Extensor expansions of digits II–V	Extends joints of digits II–V and dorsiflexes foot at ankle	Deep fibular n.
Extensor hallucis longus	Anterior surface of fibular and interosseous membrane	Distal phalanx of digit I	Extends joints of great toe and assists dorsiflexion of foot at ankle	Deep fibular n.
Peroneus tertius	Anterior surface of fibular and interosseous membrane	Dorsal surface of fifth metatarsal bone	Dorsiflexes and everts foot at ankle	Deep fibular n.
Peroneus longus	Lateral epicondyle of tibia and head and body of fibular	First cuneiform and metatarsal bone I	Plantar flexes and everts foot at ankle	Superficial fibular n.
Peroneus brevis	Lower aspect of fibular	Metatarsal bone V	Plantar flexes and everts foot at ankle	Superficial fibular n.
Gastrocnemius	Lateral and medial epicondyle of femur	Posterior surface of calcaneus	Plantar flexes foot at ankle; flexes knee joint	Tibial n.
Soleus	Posterior aspect of fibular and tibia	Calcaneus	Plantar flexes foot at ankle	Tibial n.
Plantaris	Lateral supracondylar ridge of femur	Calcaneus	Plantar flexes foot at ankle	Tibial n.
Popliteus	Lateral condyle of femur	Upper posterior aspect of tibia	Flexes and medially rotates leg at knee joint	Tibial n.
Flexor hallucis longus	Posterior aspect of fibular	Distal phalanx of great toe	Flexes joint of distal phalanx of great toe	Tibial n.
Flexor joints of digitorum longus	Posterior surface of tibia	Distal phalanges of digits II–V	Flexes joints of distal phalanges of digits II–V	Tibial n.
Tibialis posterior	Tibia and fibula and interosseous membrane	Navicular, cuneiform, cuboid, and metatarsal bones II–IV	Plantar flexes and inverts foot at ankle; supports arches	Tibial n.

hallucis: L. *hallus*, great toe
peroneus tertius: Gk. *perone*, fibula; *tertius*, third
gastrocnemius: Gk. *gaster*, belly; *kneme*, leg
soleus: L. *soleus*, sole of foot
popliteus: L. *poples*, ham of the knee

A **cramp** within a muscle is an involuntary, painful, prolonged contraction. Cramps can occur while muscles are in use or at rest. The precise cause of cramps is unknown, but evidence indicates that they may be related to conditions within the muscle. They may result from general dehydration, deficiencies of calcium or oxygen, or excessive stimulation of the motor neurons.

Torticollis (*tor″ti-kol′is*), or **wryneck,** is an abnormal condition in which the head is inclined to one side as a result of a contracted state of muscles on that side of the neck. This disorder may be either inborn or acquired.

A condition called **rigor mortis** (rigidity of death) affects skeletal muscle tissue several hours after death, as depletion of ATP within the fibers causes a state of muscle contracture and stiffness of the joints. After a few days, however, as the muscle proteins decompose, the rigidity of the corpse disappears.

When skeletal muscles are not contracted, either because the motor nerve supply is blocked or because the limb is immobilized (as when a broken bone is in a cast), the muscle fibers **atrophy** (*at′ro-fe*), or diminish in size. Atrophy is reversible if exercise is resumed, as after a healed fracture, but tissue death is inevitable if the nerves cannot be stimulated.

The fibers in healthy muscle tissue increase in size, or **hypertrophy,** when a muscle is systematically exercised. This increase in muscle size and strength is due not to an increase in the number of muscle cells but rather to the increased production of myofibrils, accompanied by a strengthening of the associated connective tissue.

Figure 13.33

Posterior crural muscles and the popliteal region.

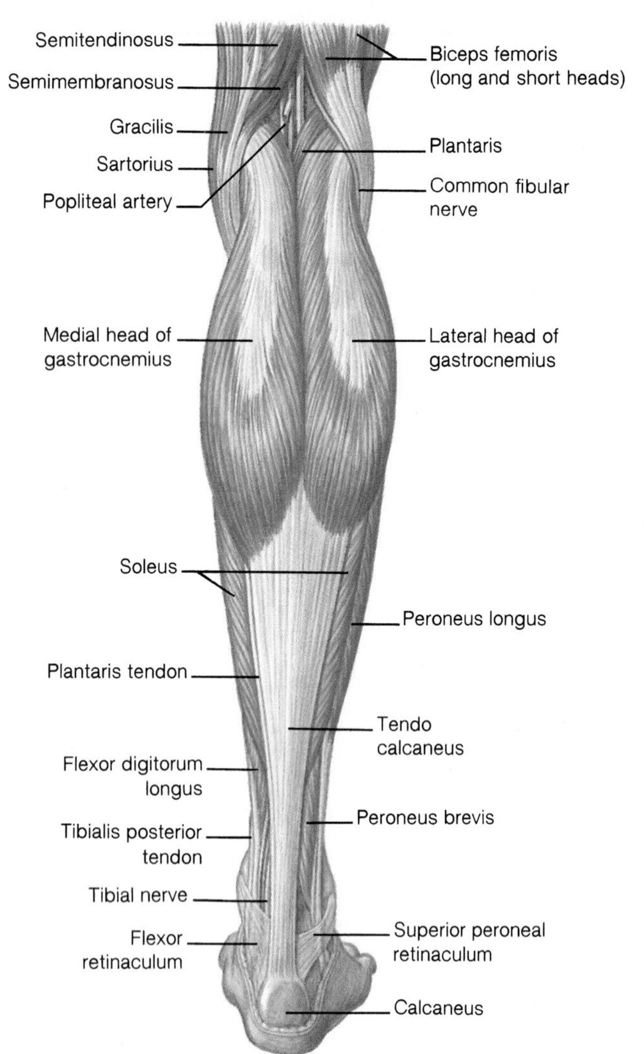

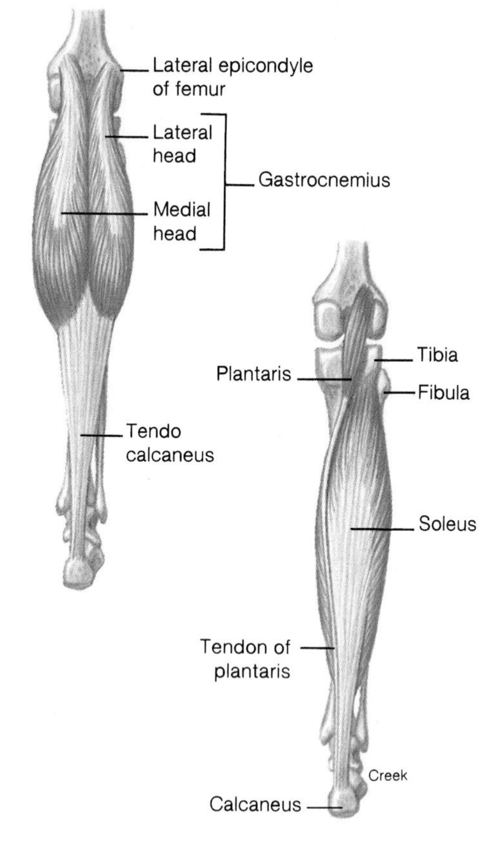

Diseases of Muscles

Fibromyositis (*fi″bro-mi″o-si′tis*) is an inflammation of both skeletal muscular tissue and the associated connective tissue. Its causes are not fully understood. Pain and tenderness frequently occur in the extensor muscles of the lumbar region of the spinal column where there are extensive aponeuroses. Fibromyositis of this region is called **lumbago** (*lum-ba′go*), or **rheumatism.**

Muscular dystrophy is a genetic disease characterized by a gradual atrophy and weakening of muscle tissue. There are several kinds of muscular dystrophy, none of whose etiology is completely understood. The most frequent type affects children and is sex-linked to the male child. As muscular dystrophy progresses, the muscle fibers atrophy and are replaced by adipose tissue.

The disease **myasthenia gravis** (*mi″as-the′ne-a grav′is*) is characterized by extreme muscle weakness and low endurance. It results from a defect in the transmission of impulses at the neuromuscular junction. Myasthenia gravis is believed to be an autoimmune disease, and it typically affects women between the ages of 20 and 40.

Poliomyelitis (polio) is actually a viral disease of the nervous system that causes muscle paralysis. The viruses are usually localized in the anterior (ventral) horn of the spinal cord, where they affect the motor nerve impulses to skeletal muscles.

Neoplasms (abnormal growths of new tissue) are rare in muscle, but when they do occur, they are usually malignant. **Rhabdomyosarcoma** (*rab″do-mi″ŏ-sar-ko′mă*) is a malignant tumor of skeletal muscle. It can arise in any skeletal muscle, and most often afflicts young children and elderly people.

lumbago: L. *lumbus*, loin

myasthenia: Gk. *myos*, muscle; *astheneia*, weakness
poliomyelitis: Gk. *polios*, gray; *myolos*, marrow
rhabdomyosarcoma: Gk. *rhabdos*, rod; *myos*, muscle; *oma*, a growth

Figure 13.34

Deep posterior crural muscles.

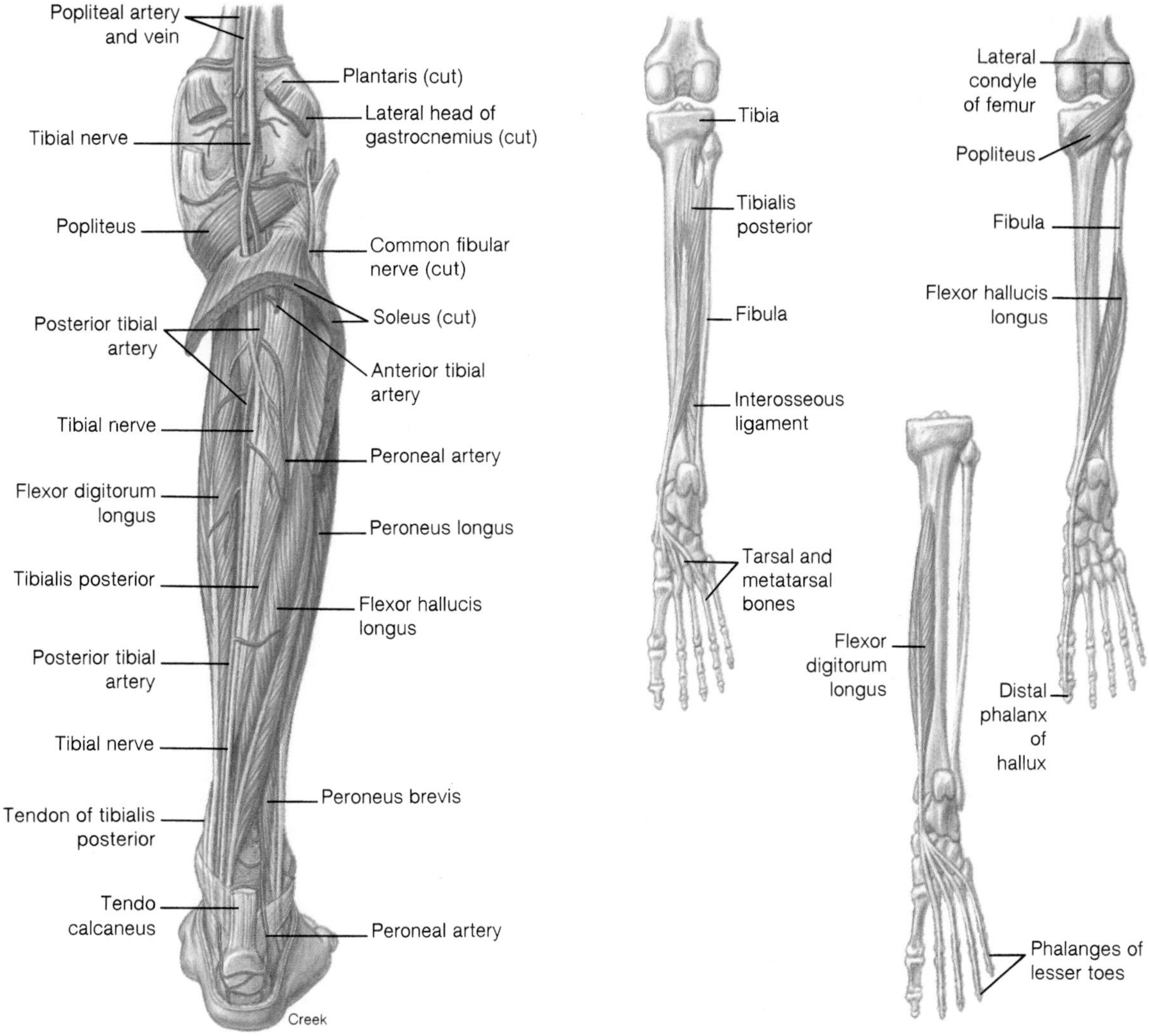

Popliteal artery and vein
Plantaris (cut)
Lateral head of gastrocnemius (cut)
Tibial nerve
Popliteus
Common fibular nerve (cut)
Soleus (cut)
Posterior tibial artery
Anterior tibial artery
Tibial nerve
Peroneal artery
Flexor digitorum longus
Peroneus longus
Tibialis posterior
Flexor hallucis longus
Posterior tibial artery
Tibial nerve
Peroneus brevis
Tendon of tibialis posterior
Tendo calcaneus
Peroneal artery
Creek
Tibia
Tibialis posterior
Fibula
Interosseous ligament
Tarsal and metatarsal bones
Lateral condyle of femur
Popliteus
Fibula
Flexor hallucis longus
Distal phalanx of hallux
Flexor digitorum longus
Phalanges of lesser toes

Figure 13.35

Muscles that surround the popliteal fossa.

Semitendinosus
Semimembranosus
Gracilis
Great saphenous vein
Medial superior genicular artery
Popliteal artery and vein
Sartorius
Saphenous nerve
Medial head of gastrocnemius
Small saphenous vein
Long head of biceps femoris
Short head of biceps femoris
Vastus lateralis
Tibial nerve
Lateral superior genicular artery
Common fibular nerve
Lateral sural cutaneous nerve
Plantaris
Medial sural cutaneous nerve
Lateral head of gastrocnemius
Communicating branch of fibular nerve
Creek

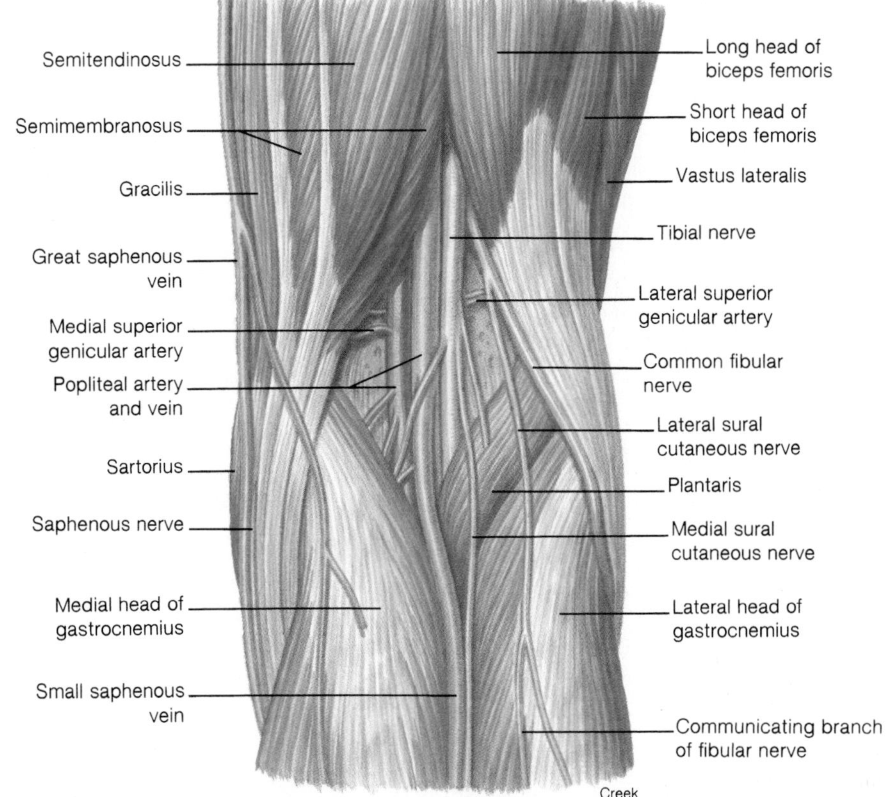

Figure 13.36

The four musculotendinous layers of the plantar aspect of the foot.
(*a*) Superficial layer, (*b*) second layer, (*c*) third layer, and (*d*) deep layer.

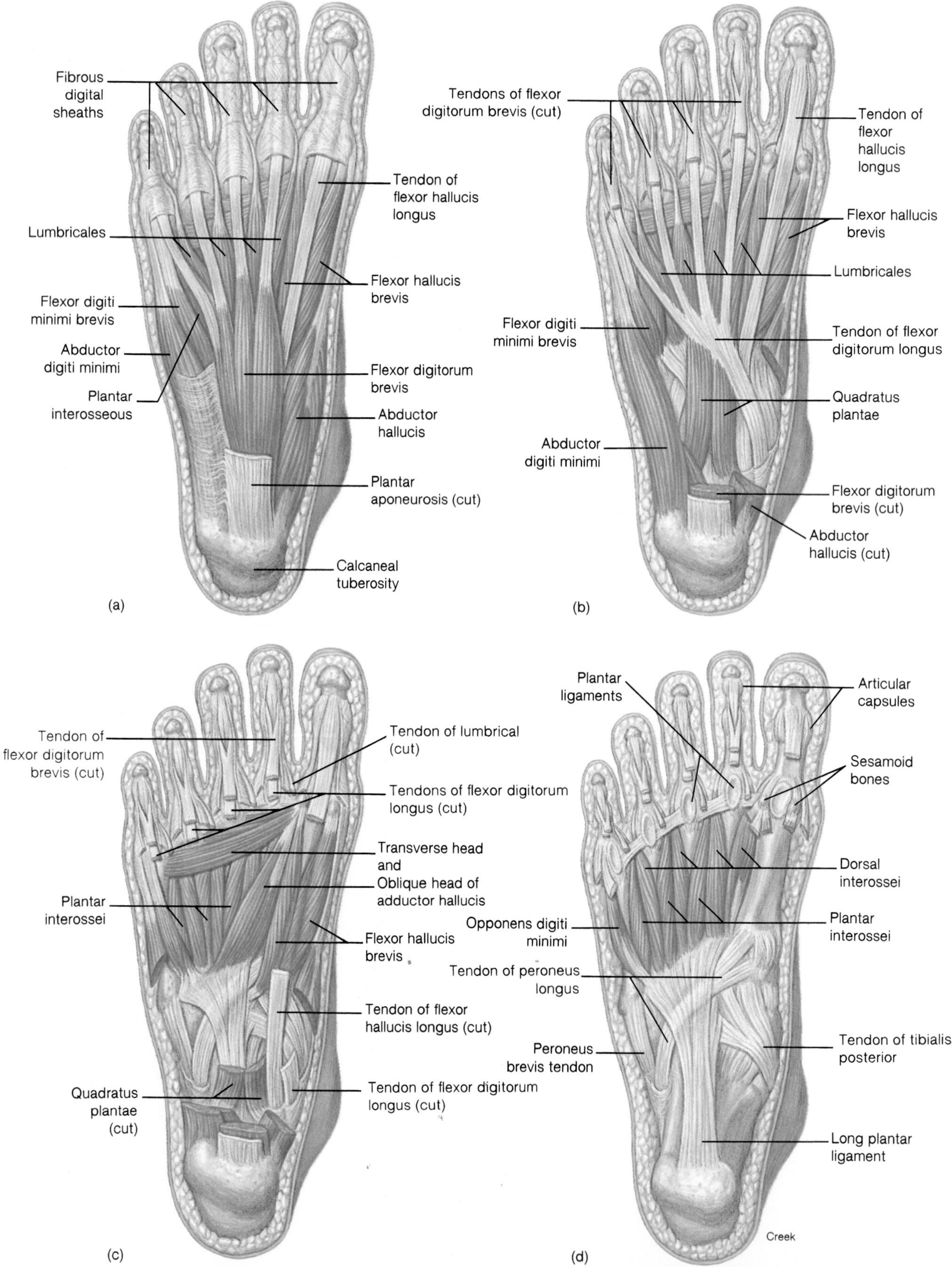

Figure 13.37

An anterior view of the dorsum of the foot.

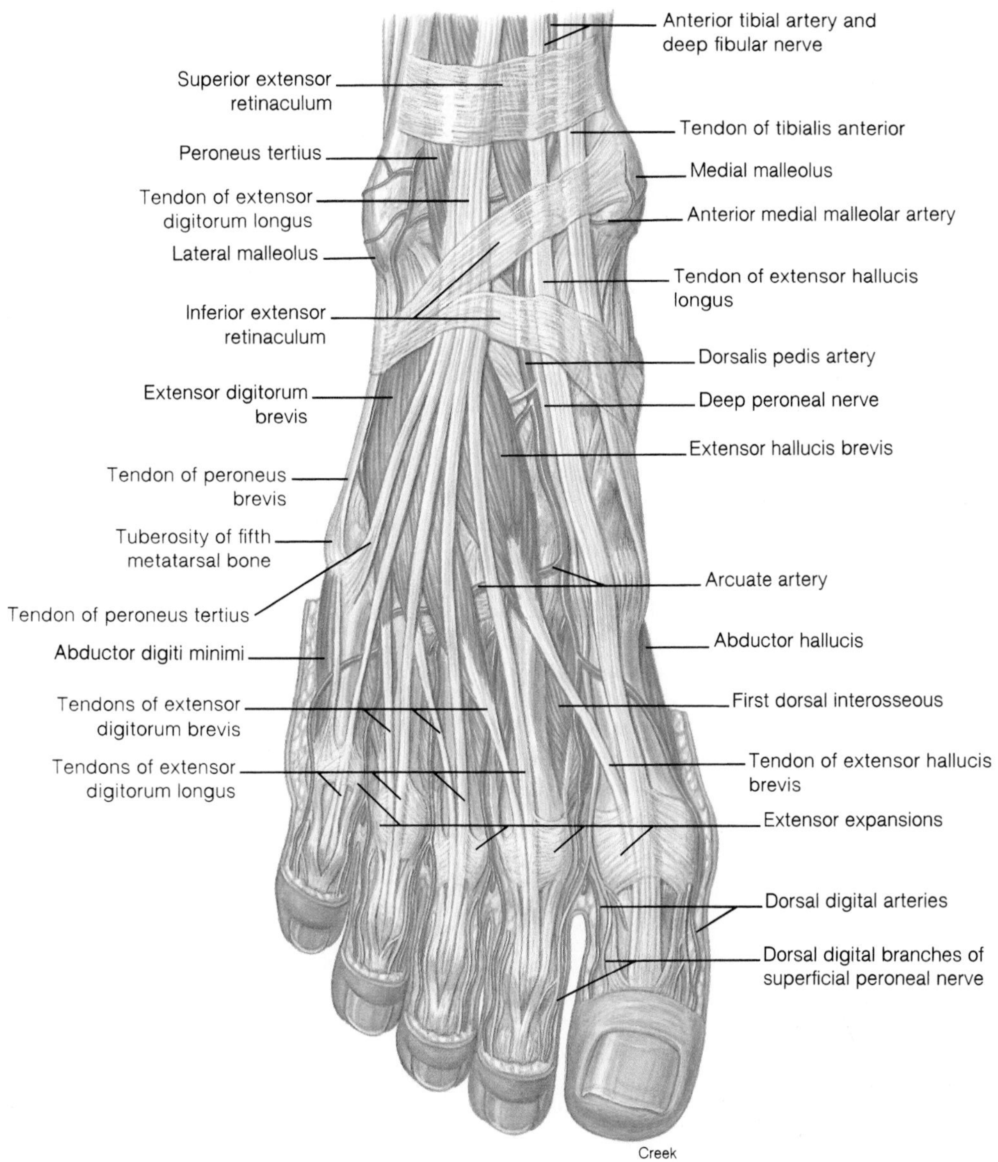

Important Clinical Terminology

convulsion An involuntary, spasmodic contraction of skeletal muscle.

fibrillation (fib-rĭ-la′shun) A series of rapid, uncoordinated, and spontaneous contractions involving individual motor units of a muscle.

hernia The rupture or protrusion of a portion of the underlying viscera through muscle tissue. Most hernias occur in the normally weak places of the abdominal wall. There are four common abdominal hernia types:

1. **femoral**—viscera descending through the femoral ring
2. **hiatal**—the superior portion of the stomach protruding through the esophageal opening of the diaphragm
3. **inguinal**—viscera descending through the inguinal canal
4. **umbilical**—a hernia occurring at the navel

intramuscular injection A hypodermic injection into a heavily muscled area to avoid damaging nerves. The most common site is the buttock.

myalgia (*mi-al′je-a*) Pain within a muscle resulting from any muscular disorder or disease.

myokymia (*mi-o-ki′me-a*) Twitching of isolated segments of muscle; also called *kymatism*.

myoma (*mi-o′ma*) A tumor of muscle tissue.

myopathy (*mi-op′athe*) Any muscular disease.

myotomy (*mi-ot′ŏ-me*) Surgical cutting or anatomical dissection of muscle tissue.

myotonia (*mi-ot′ŏ-ne-a*) A prolonged muscular spasm.

paralysis The loss of nervous control of a muscle.

shinsplints Tenderness and pain on the anterior surface of the leg generally caused by straining the tibialis anterior or extensor digitorium longus muscle.

Clinical Investigation Answer

When cancer of the head or neck involves lymph nodes in the neck, a number of structures on the affected side are removed surgically. This procedure usually includes the sternocleidomastoid muscle. This muscle, which originates on the sternum and clavicle, inserts on the mastoid process of the temporal bone. When it contracts, the mastoid process is pulled forward, causing the chin to rotate away from the contracting muscle. This explains why the patient who had his left sternocleidomastoid muscle removed would have difficulty turning his head to the right.

Chapter Summary

Organization of the Muscular System (p. 307)

1. Synergistic muscles contract together. Antagonistic muscles perform in opposition to a particular group of muscles.
2. Muscles may be classified according to fiber arrangement as parallel, convergent, pennate, or sphincteral.
3. Axial muscles include facial muscles, neck muscles, and trunk muscles; appendicular muscles include those that act on the girdles and those that move the segments of the appendages.

Naming of Muscles (pp. 307–308)

1. Skeletal muscles are named on the basis of shape, location, attachment, size, orientation of fibers, relative position, and function.
2. Most of the muscles are paired; that is, one side of the body is an image of the other.

Muscles of the Axial Skeleton (pp. 308–321)

1. The muscles of the axial skeleton include those responsible for facial expression, mastication, eye movement, tongue movement, neck movement, and respiration and those of the abdominal wall, the pelvic outlet, and the vertebral column.

Muscles of the Appendicular Skeleton (pp. 322–342)

1. The muscles of the appendicular skeleton include those of the pectoral girdle, humerus, forearm, wrist, hand, and fingers and those of the pelvic girdle, thigh, leg, ankle, foot, and toes.

Review Activities

Objective Questions

1. Which of the following muscles is a synergist to the biceps brachii muscle in flexing the elbow joint?
   (a) deltoid muscle
   (b) triceps brachii muscle
   (c) brachialis muscle
   (d) anconeus muscle
2. Muscles that are strong, fatigue quickly, and have short excursions are classified as
   (a) parallel. (c) pennate.
   (b) convergent. (d) sphincteral.
3. Which of the following is *not* used as a means of naming muscles?
   (a) location
   (b) action
   (c) shape
   (d) attachment
   (e) strength of contraction
4. Which of the following muscles originates on the zygomatic arch, inserts on the lateral portion of the mandible, and, when contracted, elevates the mandible?
   (a) masseter muscle
   (b) lateral pterygoid muscle
   (c) temporalis muscle
   (d) zygomaticus muscle
5. Which of the following muscles is a flexor of the shoulder joint?
   (a) pectoralis major muscle
   (b) supraspinatus muscle
   (c) teres major muscle
   (d) trapezium muscle
   (e) latissimus dorsi muscle
6. Which of the following muscles is an antagonist to the longissimus thoracis muscle?
   (a) spinalis thoracis muscle
   (b) quadratus lumborum muscle
   (c) longissimus cervicis muscle
   (d) rectus abdominis muscle
7. Which of the following muscles does *not* have either an origin or insertion upon the humerus?
   (a) teres minor muscle
   (b) biceps brachii muscle
   (c) supraspinatus muscle
   (d) brachialis muscle
   (e) pectoralis major muscle
8. Which of the following muscles in a female constricts the urethra canal and the vagina?
   (a) transversus perinei muscle
   (b) bulbospongiosus muscle
   (c) coccygeus muscle
   (d) ischiocavernosus muscle
   (e) levator ani muscle
9. Which of the following muscles does *not* contract over both the hip and knee joints (is not a two-joint muscle)?
   (a) gracilis muscle
   (b) sartorius muscle
   (c) rectus femoris muscle
   (d) semitendinosus muscle
   (e) vastus medialis muscle
10. The genetic disease of muscle that is most common in male children and is characterized by gradual atrophy and weakening of muscle tissue is
    (a) poliomyelitis.
    (b) myasthenia gravis.
    (c) muscular dystrophy.
    (d) rhabdomyosarcoma.

Essay Questions

1. What are the advantages and disadvantages of pennate-fibered muscles?
2. List the muscles of the neck that either originate or insert on the hyoid bone.
3. Diagram a posterior view of the scapula and indicate the locations of the muscles that originate and insert upon it.
4. Discuss the position of flexor and extensor muscles relative to the shoulder, elbow, and wrist joint.
5. Based on function, describe exercises that would strengthen the following muscles: (a) the pectoralis major muscle; (b) the deltoid muscle; (c) the triceps brachii muscle; (d) the pronator teres muscle; (e) the rhomboideus major muscle; (f) the trapezius muscle; (g) the serratus anterior muscle; and (h) the latissimus dorsi muscle.

6. Give three examples of synergistic muscle groups within the lower extremity and identify the antagonistic muscle group of each.
7. Describe the actions of the muscles of inspiration. Which muscles participate in forced expiration?
8. Which muscles of the pelvic outlet support the floor of the pelvic cavity and which are associated with the genitalia?

Critical Thinking Questions

1. As a result of a severe head trauma sustained in an automobile accident, a 17-year-old male lost function of his right oculomotor nerve. Explain what will happen to the function of the affected eye.
2. Discuss the position of flexor and extensor muscles relative to the shoulder, elbow, and wrist joints.
3. Based on function, describe exercises that would strengthen the following muscles: (a) the pectoralis major, (b) the deltoid, (c) the triceps brachii, (d) the pronator teres, (e) the rhomboideus major, (f) the trapezius, (g) the serratus anterior, and (h) the latissimus dorsi.
4. Why is it necessary to have dual (sensory and motor) innervation to a muscle? Give an example of a disease that results in loss of motor innervation to specific skeletal muscles, and describe the effects of this denervation.
5. Compare muscular dystrophy and myasthenia gravis as to causes, symptoms, and the effect they have on muscle tissue.

Related Web Sites

For a listing of the most current web sites related to this chapter, please visit the *Concepts of Human Anatomy and Physiology* home page at http://www.mhhe.biosci/abio/.

NEXUS

Some Interactions of the Muscular System with the Other Body Systems

Integumentary System

- The skin helps to protect all organs of the body from invasion by pathogens (p. 697).
- The smooth muscles of cutaneous blood vessels are needed for the regulation of cutaneous blood flow (p. 676).
- The arrector pili muscles in the skin produce goose bumps (p. 168).

Skeletal System

- Bones store calcium, which is needed for the control of muscle contraction (p. 193).
- The skeleton provides attachment sites for muscles (p. 185).
- Joints of the skeleton provide levers for movement (p. 256).
- Muscle contractions maintain the health and strength of bone (p. 193).

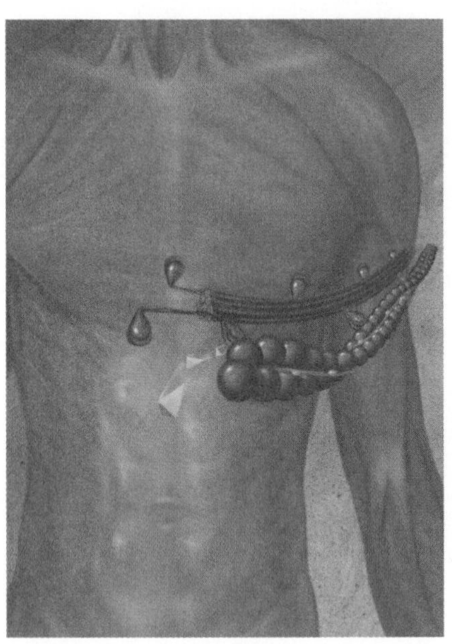

Nervous System

- Somatic motor neurons stimulate contraction of skeletal muscles (p. 287).
- Autonomic neurons stimulate smooth muscle contraction or relaxation (p. 478).
- Autonomic nerves increase cardiac output during exercise (p. 672).
- Sensory neurons from muscles monitor muscle length and tension (p. 504).

Endocrine System

- Sex hormones promote muscle development and maintenance (p. 897).
- Parathyroid hormone and other hormones regulate blood calcium and phosphate concentrations (p. 194).
- Epinephrine and norepinephrine influence contractions of cardiac and smooth muscles (p. 488).
- Insulin promotes glucose entry into skeletal muscles (p. 877).
- Adipose tissue secretes hormones that regulate the sensitivity of muscles to insulin (p. 873).

Circulatory System

- Blood transports O_2 and nutrients to muscles and removes CO_2 and lactic acid (p. 592).
- Contractions of skeletal muscles serve as a pump to assist blood movement within veins (p. 627).
- Cardiac muscle enables the heart to function as a pump (p. 618).
- Smooth muscle enables blood vessels to constrict and dilate (p. 667).

Lymphatic and Immune Systems

- Lymphatic vessels help to maintain a balanced amount of interstitial fluid within muscle tissue (p. 662).
- Lymphocytes provide defense against infection (p. 701).
- Muscle contractions assist lymph movement (p. 693).

Respiratory System

- The lungs provide O_2 for muscle metabolism and eliminate CO_2 (p. 736).
- Respiratory muscles enable ventilation of the lungs (p. 743).

Urinary System

- The kidneys eliminate creatinine and other metabolic wastes from muscle (p. 795).
- The kidneys help to regulate the blood calcium and phosphate concentrations (p. 194).
- Muscles of the urinary tract are needed for the control of urination (p. 806).

Digestive System

- The GI tract provides nutrients for all body organs, including muscles (p. 817).
- Smooth muscle contractions push digestion products along the GI tract (p. 851).
- Muscular sphincters of the GI tract help to regulate the passage of food (p. 829).

Reproductive System

- Testicular androgen promotes growth of skeletal muscle (p. 898).
- Muscle contractions contribute to orgasm in both sexes (p. 912).
- Uterine muscle contractions are required for vaginal delivery of the fetus (p. 973).

Figure 1

The surface anatomy of the facial region.

(Also refer to figure 13.3 for the underlying muscles.)

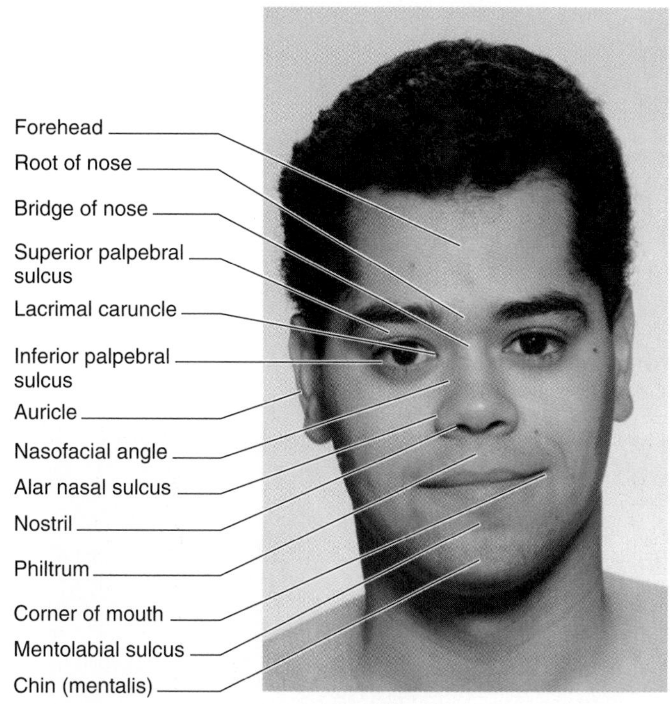

Figure 2

An anterolateral view of the neck.

(Also refer to figure 13.9 for the underlying muscles.)

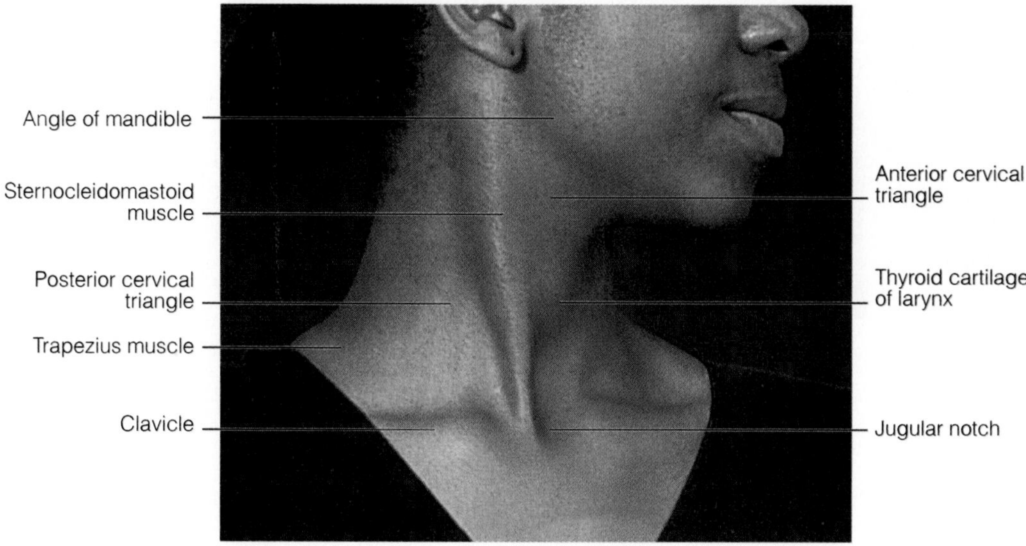

Figure 3

The surface anatomy of the back.
(Also refer to figure 13.16 for the underlying muscles.)

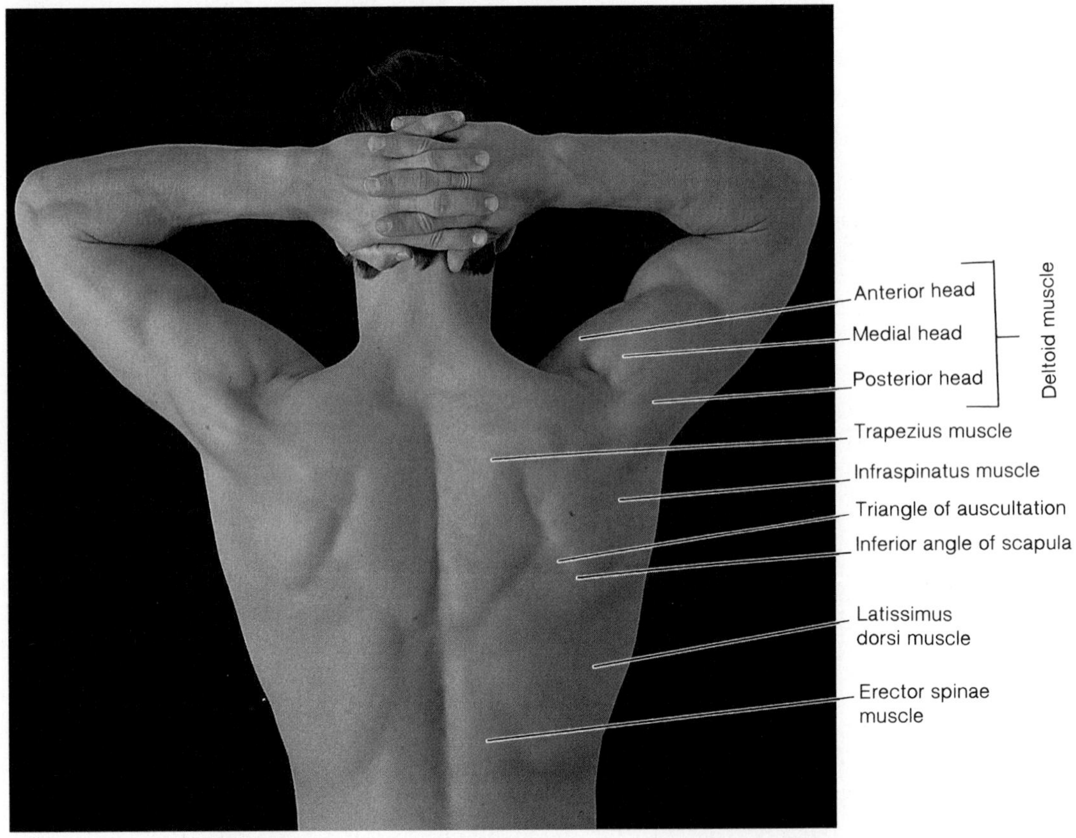

Figure 4

An anterolateral view of the trunk and axilla.
(Also refer to figures 13.11, 13.17, and 13.18 for the underlying muscles.)

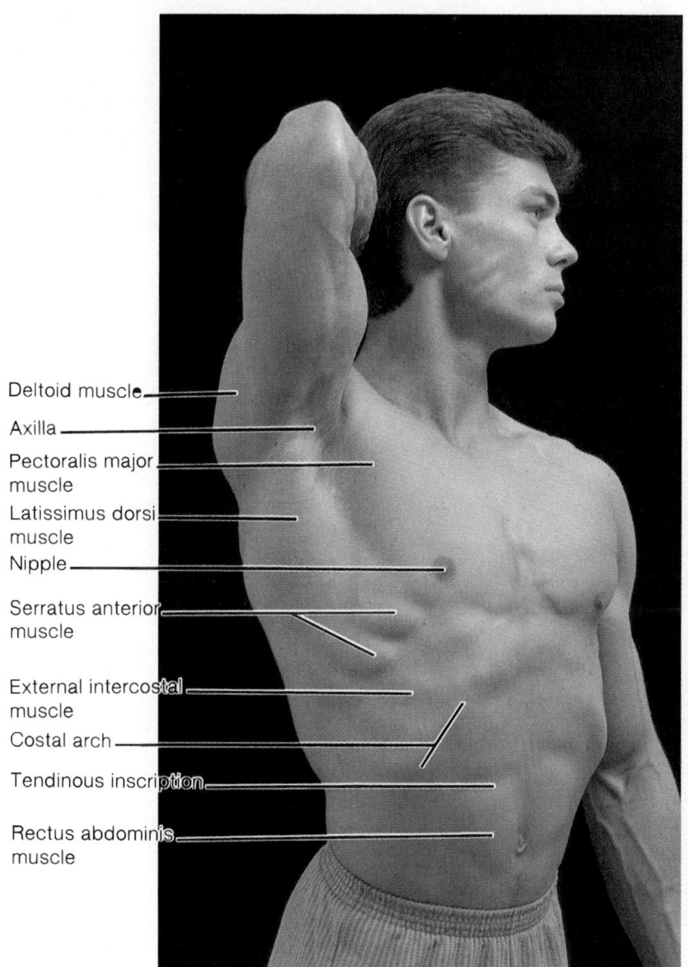

Figure 5

The anatomical snuffbox.
(Also refer to figure 13.19*b* for the underlying muscles and tendons.)

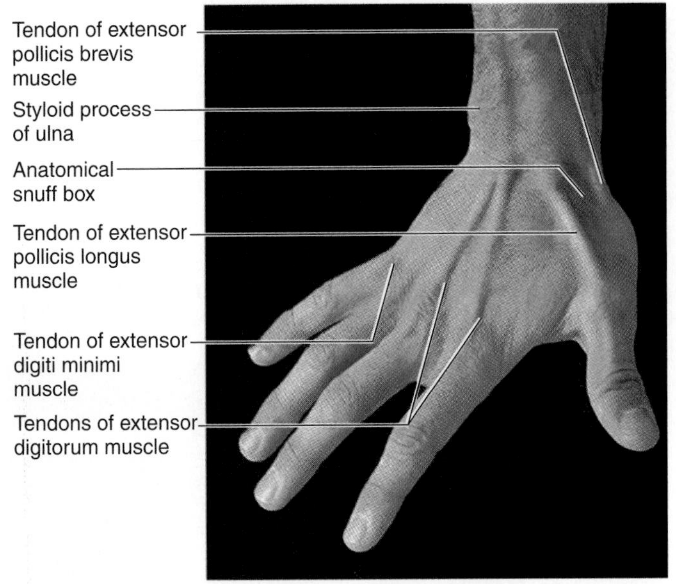

Figure 6

An anterior view of the forearm and hand.
(Also refer to figure 13.19*a* for the underlying muscles and tendons.)

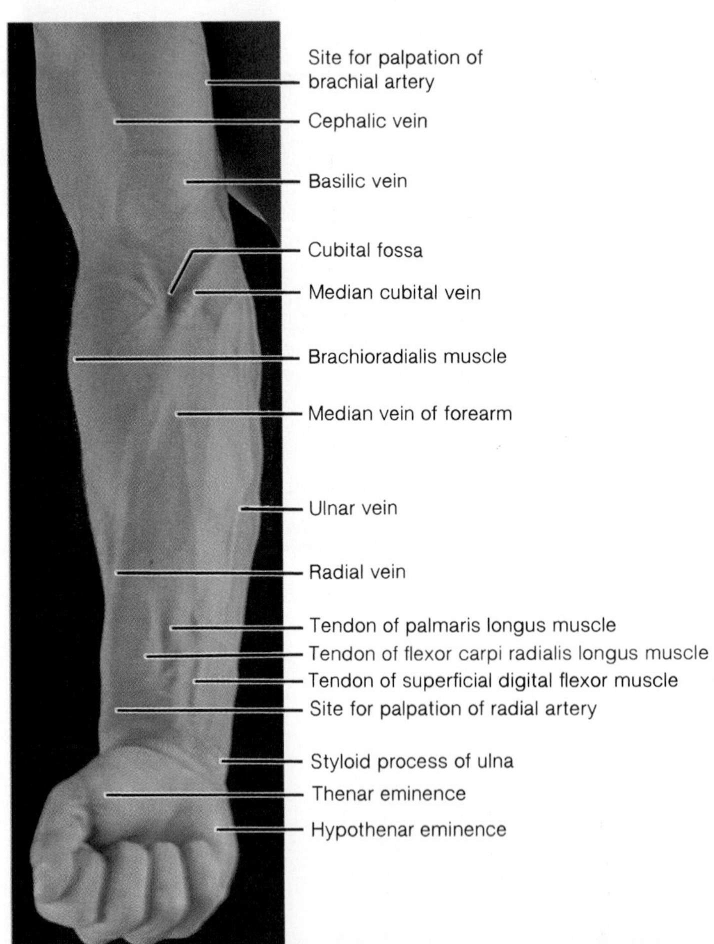

Figure 7

The (*a*) lateral, (*b*) posterior, and (*c*) medial surfaces of the leg.
(Also refer to figures 13.31, 13.32, and 13.33 for the underlying muscles.)

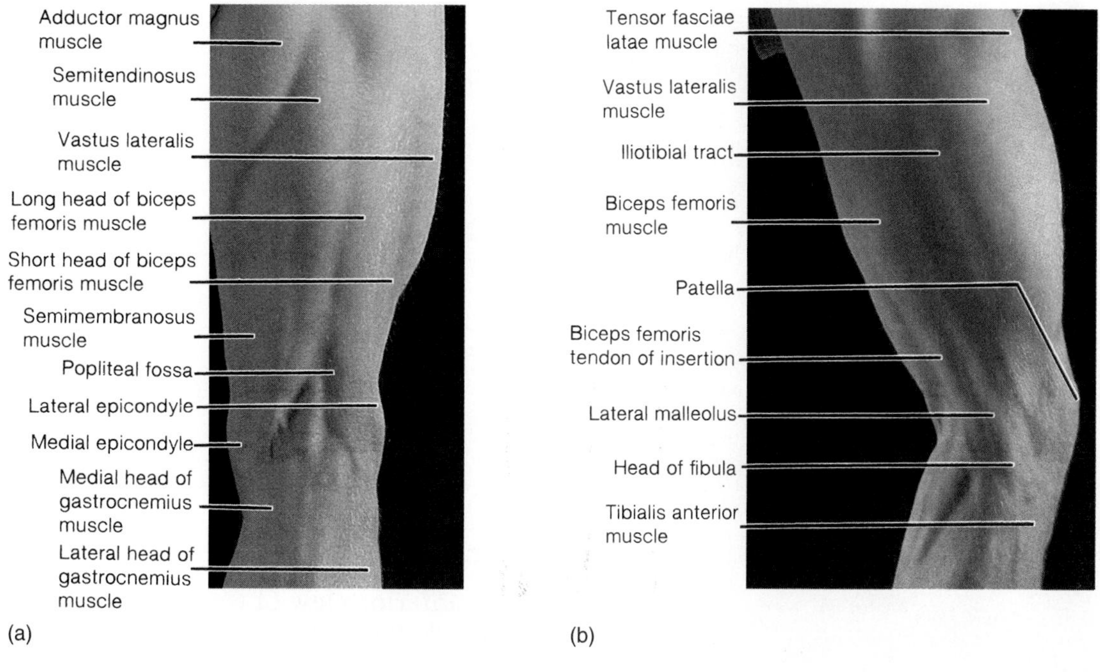

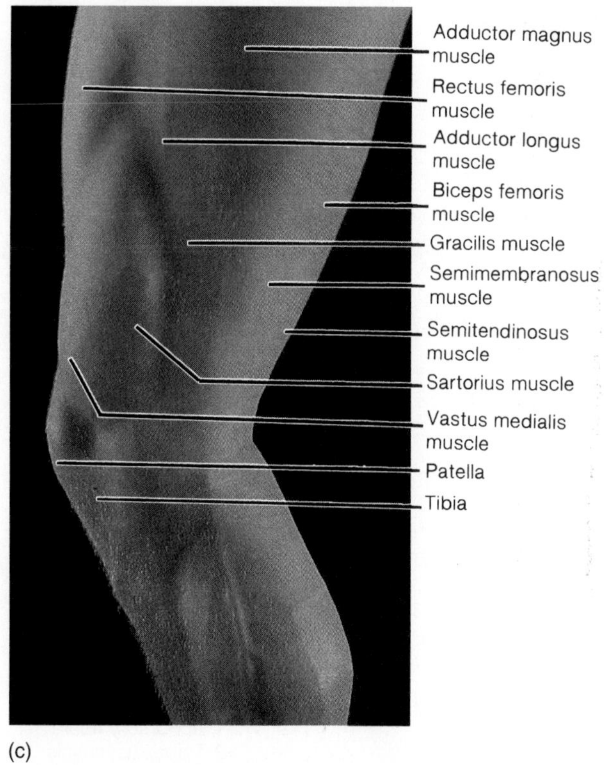

Figure 8

An anterior view of the muscles of the head.

(Also refer to figures 13.3and 13.4.)

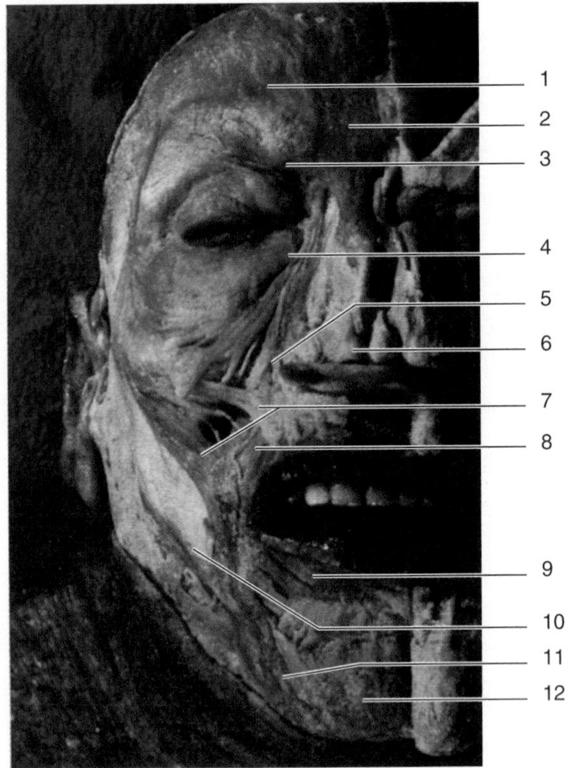

1 Frontalis muscle
2 Supratrochlear artery
3 Corrugator muscle
4 Orbicularis oculi muscle
5 Levator labii superioris muscle
6 Alar cartilage
7 Zygomaticus muscles
8 Facial artery
9 Orbicularis oris muscle
10 Risorius muscle
11 Depressor angularis oris muscle
12 Mentalis muscle

Figure 9

An anterior view of the right cervical region.

(Also refer to figure 13.9.)

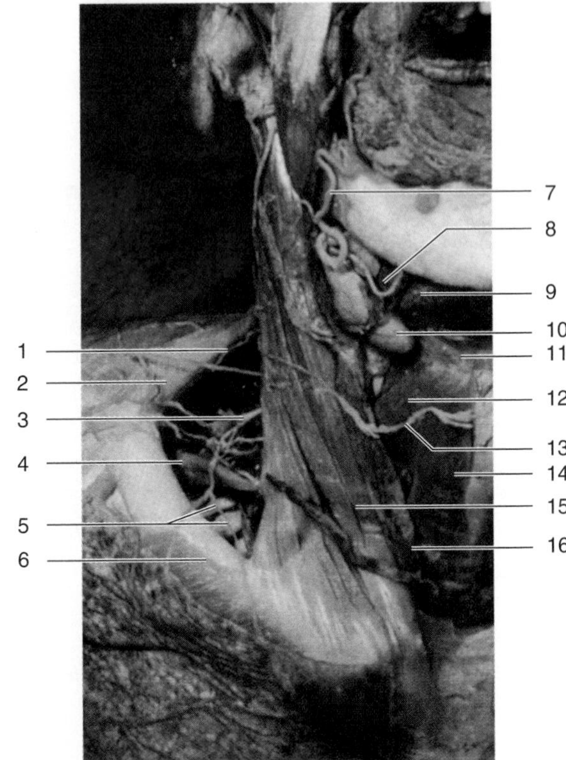

1 Accessory nerve
2 Trapezius muscle
3 Supraclavicular nerve
4 Omohyoid muscle
5 Brachial plexus
6 Clavicle
7 Facial artery
8 Mylohyoid muscle
9 Digastric muscle
10 Submandibular gland
11 Hyoid bone
12 Omohyoid muscle
13 Transverse cervical nerve
14 Sternohyoid muscle
15 Sternocleidomastoid muscle
16 External jugular vein

Figure 10

A sagittal section of the head and neck.

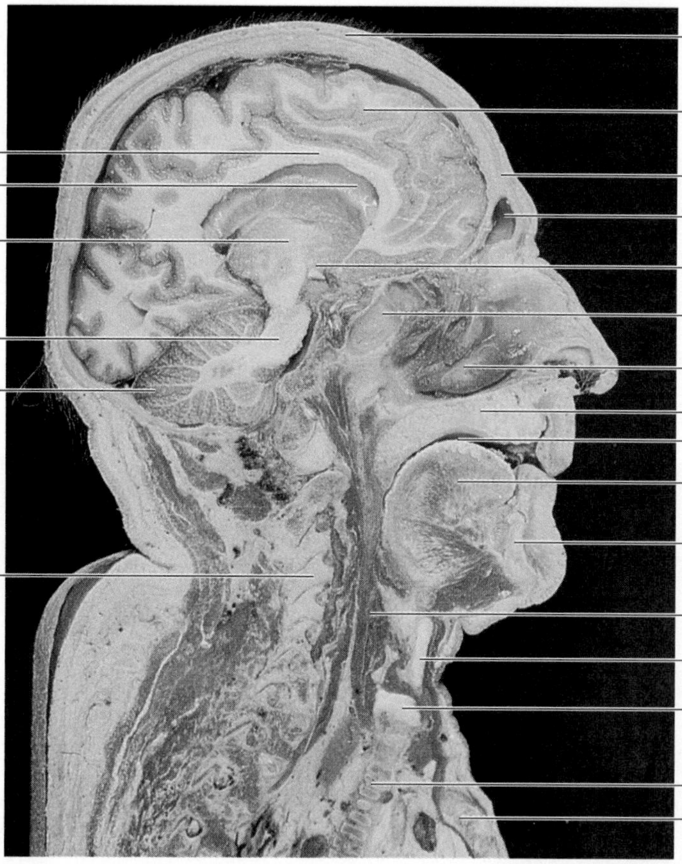

Corpus callosum
Lateral ventricle
Thalamus
Brain stem
Cerebellum
Cervical vertebra
Scalp
Cerebrum
Frontal bone
Frontal sinus
Hypothalamus
Sphenoidal sinus
Inferior nasal concha
Maxilla
Oral cavity
Tongue
Mandible
Esophagus
Thyroid cartilage
Cricoid cartilage
Trachea
Manubrium

Figure 11

An anterior view of the right thorax, shoulder, and brachium.
(Also refer to figures 13.14 and 13.17.)

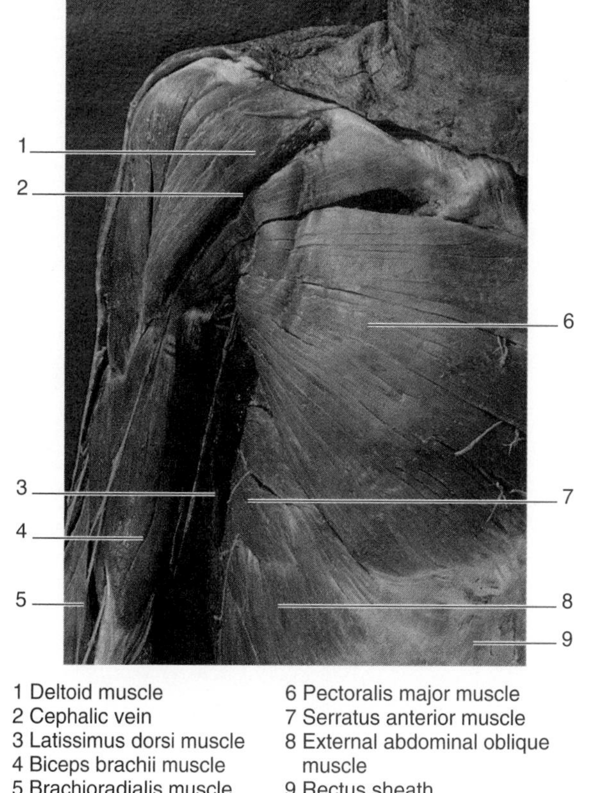

1 Deltoid muscle
2 Cephalic vein
3 Latissimus dorsi muscle
4 Biceps brachii muscle
5 Brachioradialis muscle
6 Pectoralis major muscle
7 Serratus anterior muscle
8 External abdominal oblique muscle
9 Rectus sheath

Figure 12

The thoracic cavity with the heart and lungs removed.

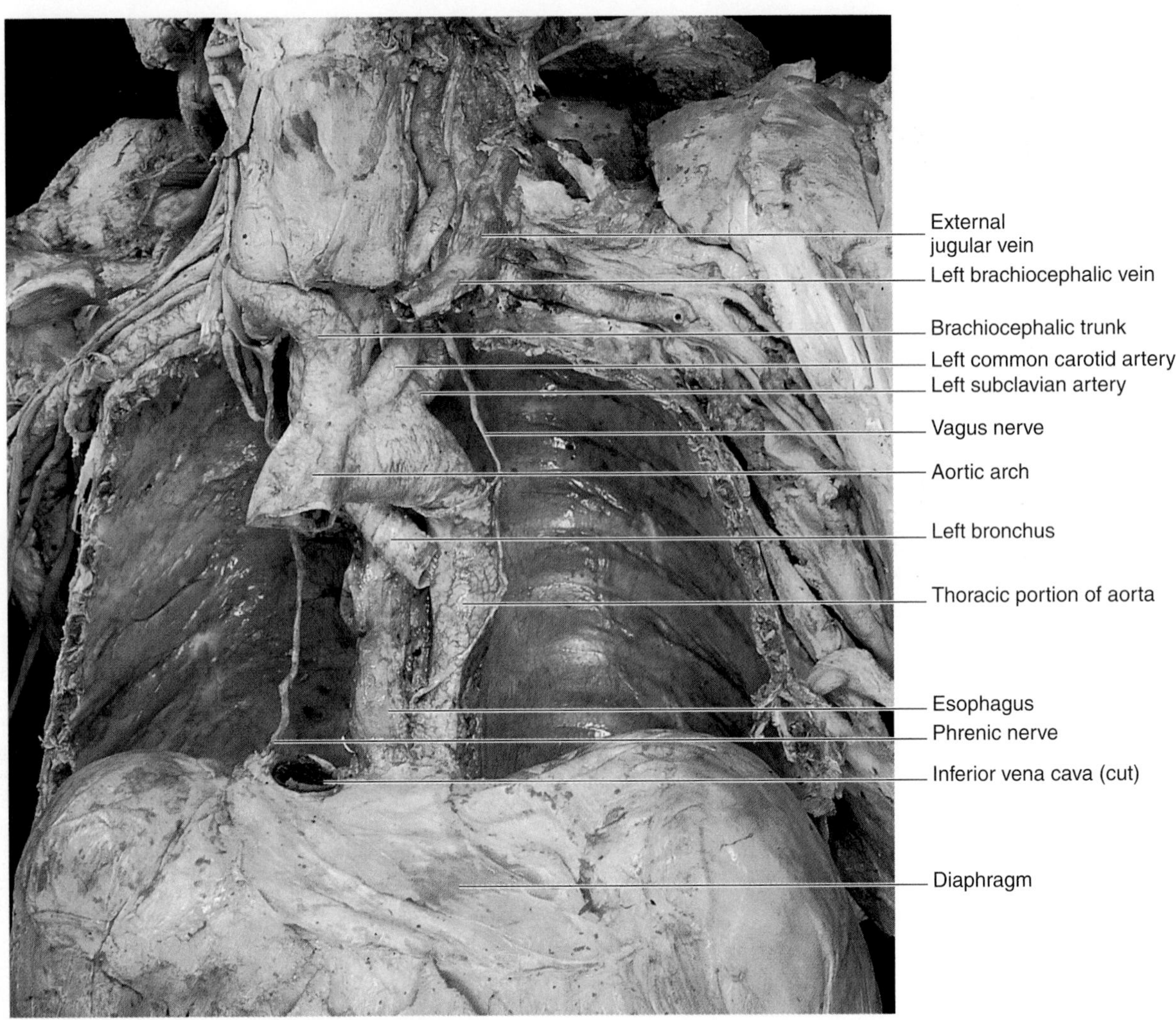

Figure 13

A posterior view of the right thorax and neck.

(Also refer to figure 13.15.)

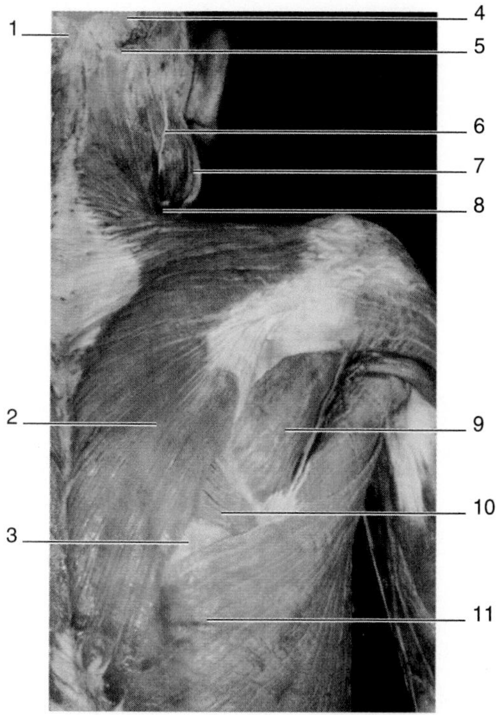

1 External occipital protuberance
2 Trapezius muscle
3 Triangle of auscultation
4 Occipital artery
5 Greater occipital nerve
6 Lesser occipital nerve
7 Sternocleidomastoid muscle
8 Great auricular nerve
9 Infraspinatus muscle
10 Rhomboideus major muscle
11 Latissimus dorsi muscle

Figure 14

An anterior view of the muscles of the abdominal wall.

(Also refer to figures 13.11 and 13.14.)

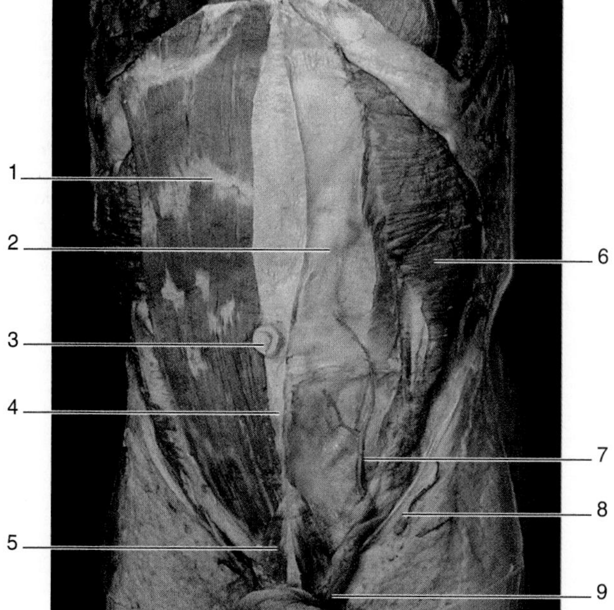

1 Rectus abdominis muscle
2 Rectus sheath
3 Umbilicus
4 Linea alba
5 Pyramidalis muscle
6 Transverse abdominis muscle
7 Inferior epigastric artery
8 Inguinal ligament
9 Spermatic cord

Figure 15

Viscera of the abdomen.

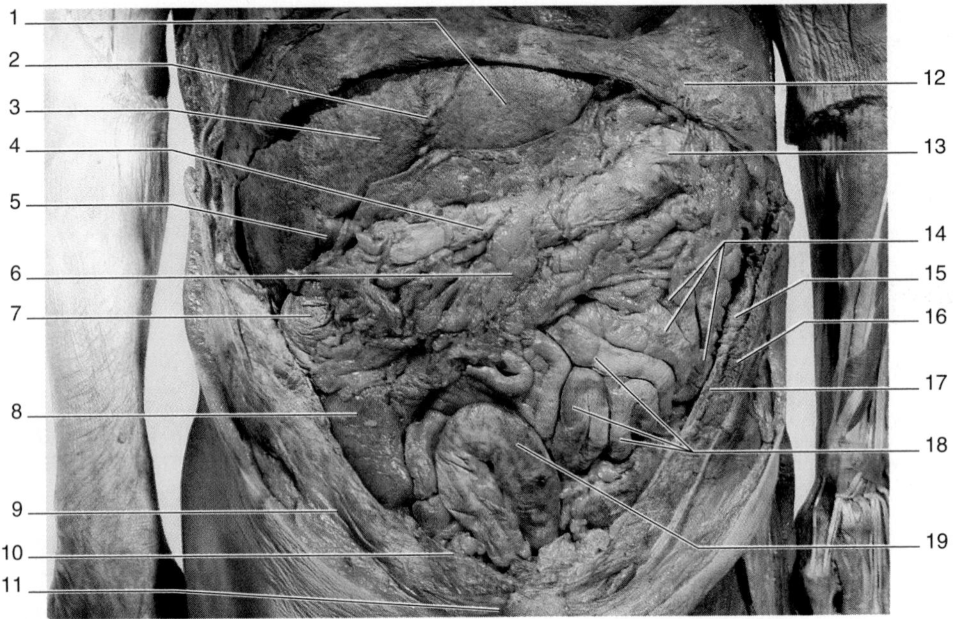

1 Left lobe of liver
2 Falciform ligament
3 Right lobe of liver
4 Transverse colon
5 Gallbladder
6 Greater omentum
7 Hepatic (right colic) flexure
8 Fat deposit within greater omentum
9 Aponeurosis of internal abdominal oblique muscle
10 Rectus abdominis muscle (cut)
11 Rectus sheath (cut)
12 Diaphragm
13 Splenic (left colic) flexure
14 Jejunum
15 Transversus abdominis muscle (cut)
16 Internal and external abdominal oblique muscles (cut)
17 Parietal peritoneum (cut)
18 Ileum
19 Sigmoid colon

Figure 16

A posterior view of the left forearm and hand.
(Also refer to figures 13.19 and 13.20 for views of the right forearm.)

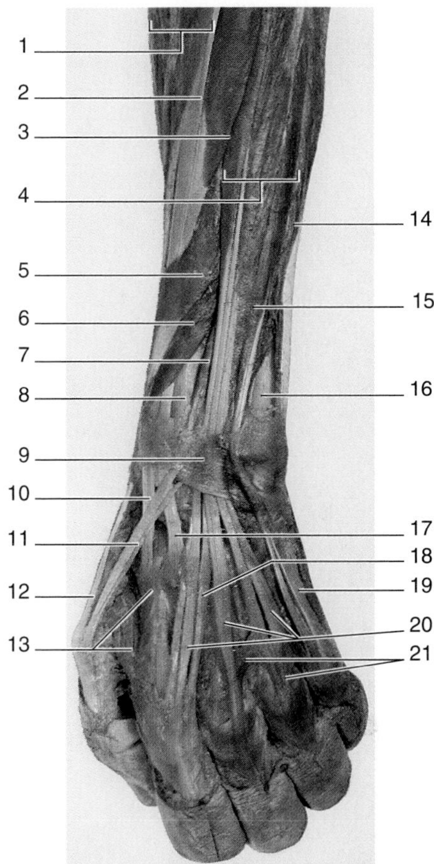

1 Brachioradialis muscle
2 Extensor carpi radialis longus tendon
3 Extensor carpi radialis brevis muscle
4 Extensor digitorum communis muscle
5 Abductor pollicis longus muscle
6 Extensor pollicis brevis muscle
7 Extensor pollicis longus muscle
8 Radius
9 Extensor retinaculum
10 Tendon of extensor carpi radialis longus muscle
11 Tendon of extensor pollicis longus muscle
12 Tendon of extensor pollicis brevis muscle
13 First dorsal interosseous muscle
14 Extensor carpi ulnaris muscle
15 Extensor digiti minimi muscle
16 Ulna
17 Tendon of extensor carpi radialis brevis muscle
18 Tendon of extensor indicis muscle
19 Tendon of extensor digiti minimi muscle
20 Tendons of extensor digitorum muscle
21 Intertendinous connections

Figure 17

A posterior view of the deep muscles of the right abdominal and gluteal regions.
(Also refer to figures 13.13, 13.14, and 13.15.)

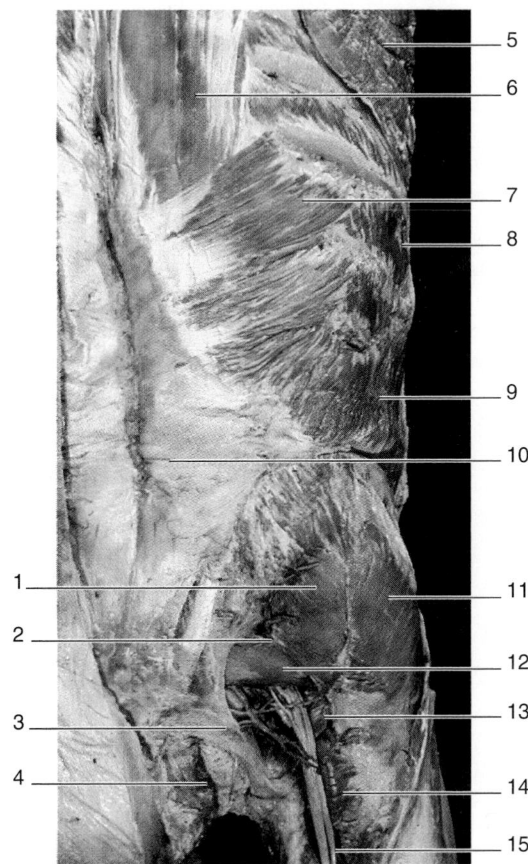

1 Superior gluteal vessels
2 Inferior gluteal vessels
3 Sacrotuberous ligament
4 Levator ani muscle
5 Serratus anterior muscle
6 Erector spinae muscle
7 Serratus posterior muscle
8 External intercostal muscle
9 Internal abdominal oblique muscle
10 Lumbar aponeurosis
11 Gluteus medius muscle
12 Piriformis muscle
13 Obturator internus muscle
14 Quadratus femoris muscle
15 Sciatic nerve

Figure 18

An anterior view of the right thigh.
(Also refer to figures 13.22 and 13.26.)

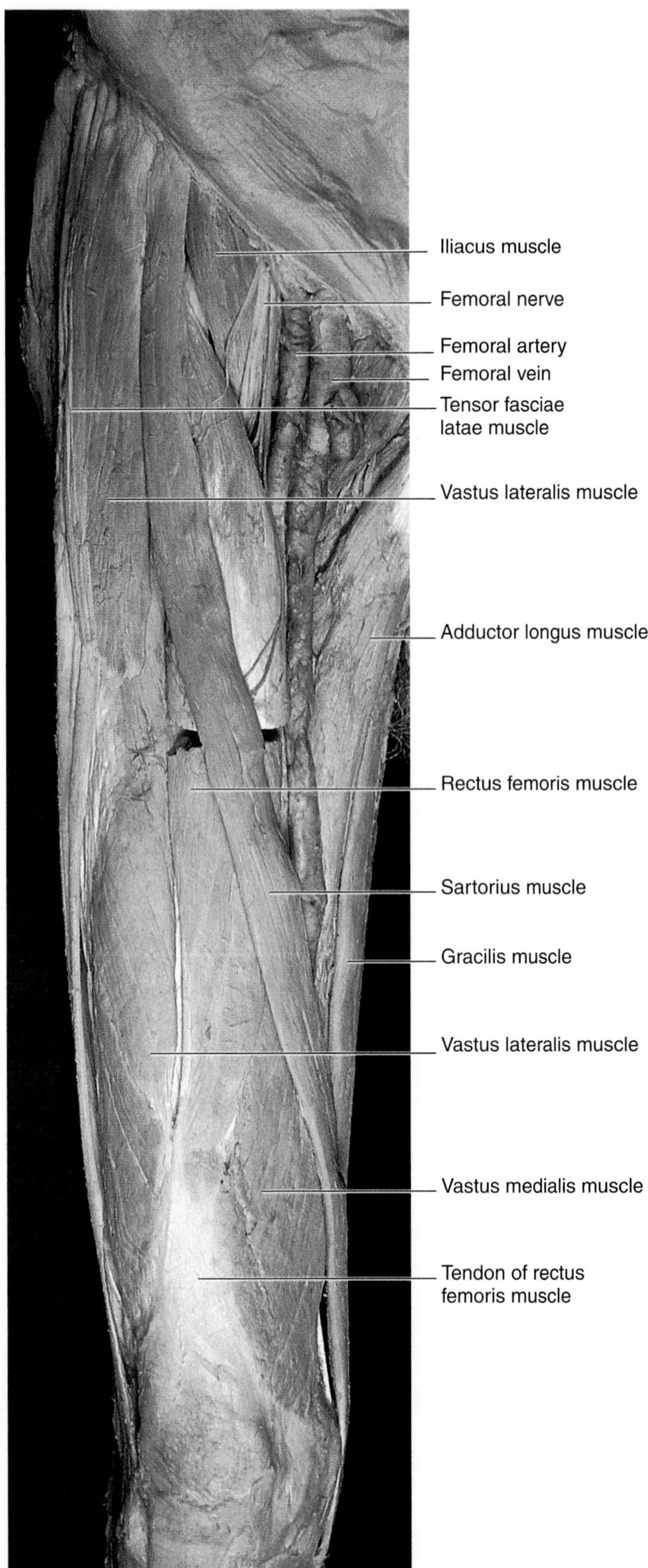

Figure 19

A posterior view of the right hip and thigh.
(Also refer to figures 13.24 and 13.27.)

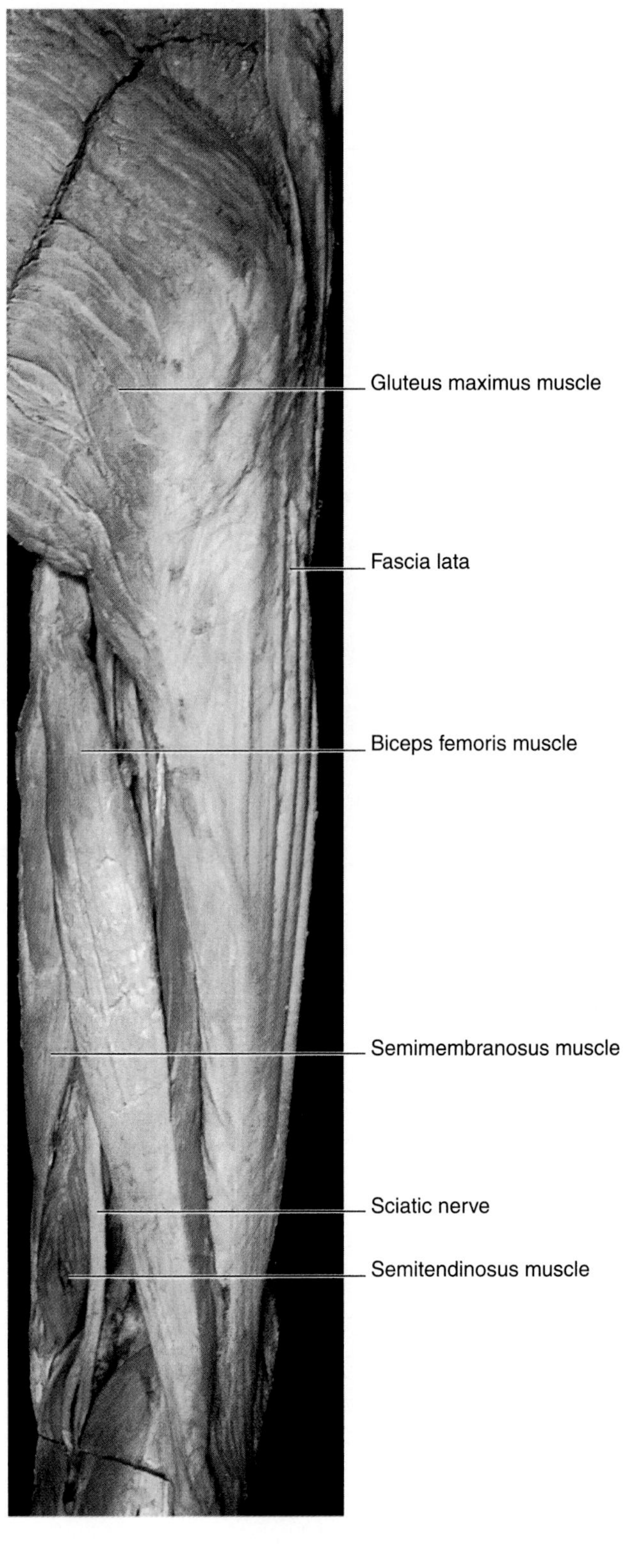

Figure 20

An anterior view of the right leg.
(Also refer to figure 13.30.)

Patella
Patellar ligament
Tibialis anterior muscle
Soleus muscle
Tibia
Tendon of extensor digitorum longus muscle
Tendon of peroneus longus muscle
Superior extensor retinaculum
Inferior extensor retinaculum
Tendons of extensor digitorum longus muscle
Tendon of extensor hallucis longus muscle

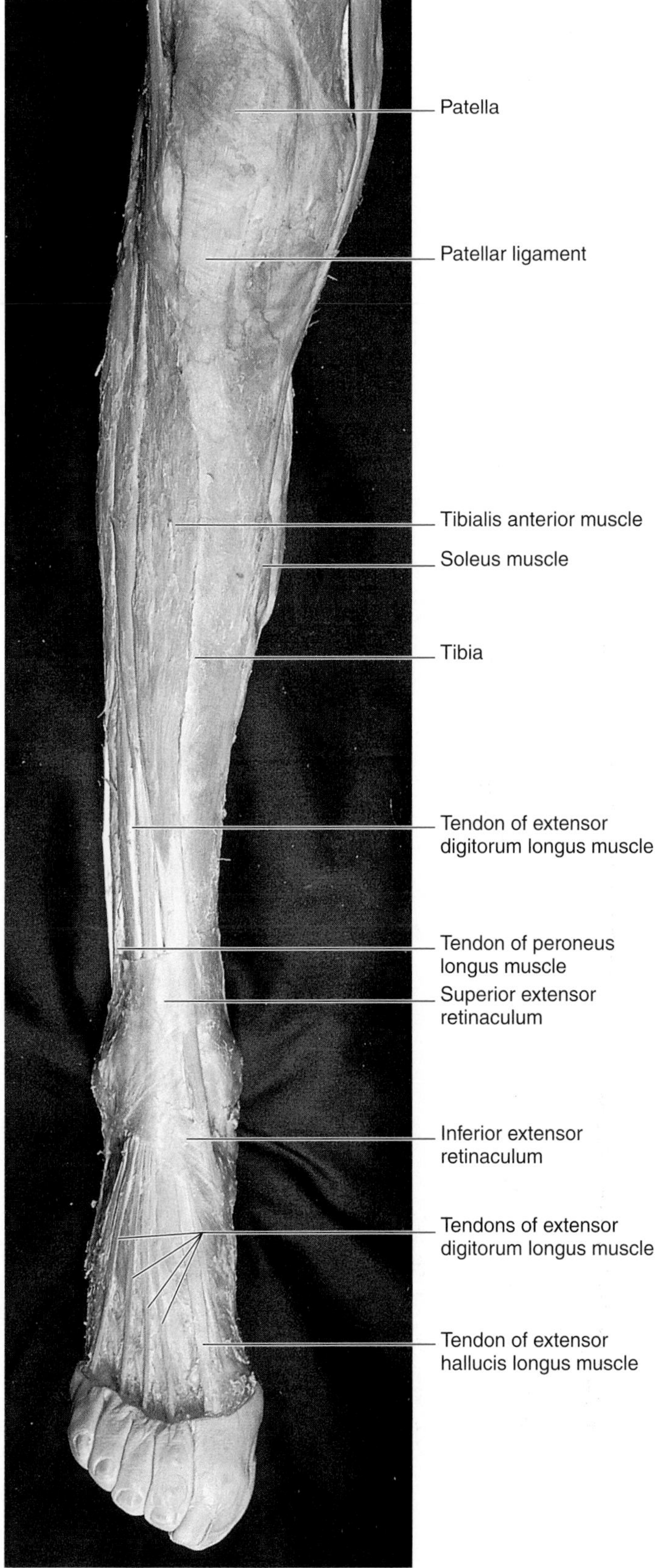

Figure 21

A posterior view of the right leg.
(Also refer to figure 13.33.)

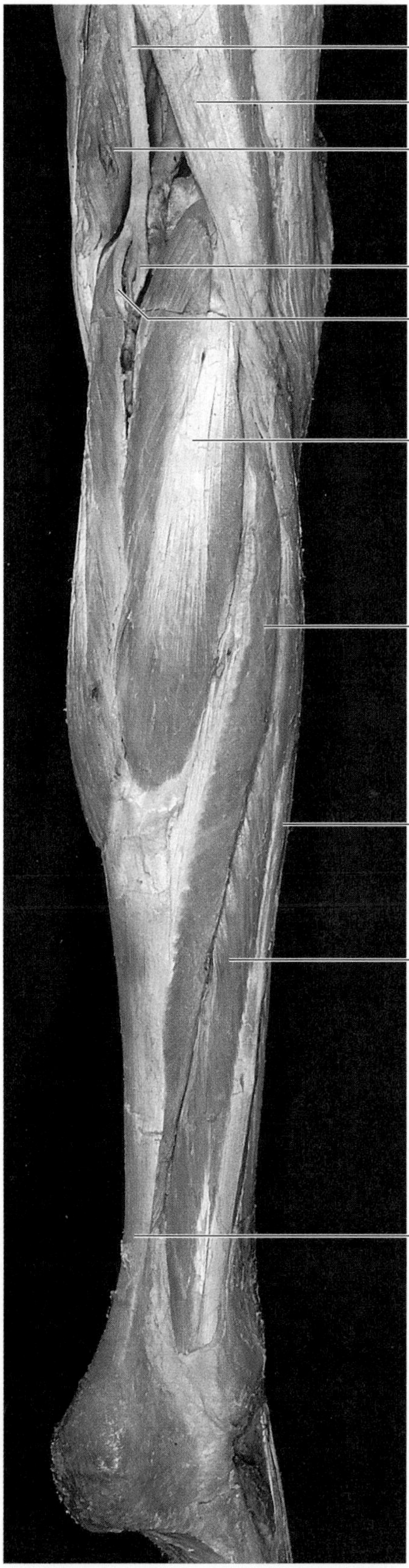

Sciatic nerve
Biceps femoris muscle
Semitendinosus muscle
Common peroneal nerve
Tibial nerve
Gastrocnemius muscle
Soleus muscle
Peroneus longus muscle
Peroneus brevis muscle
Tendo calcaneus

Figure 22

An anterior view of the female trunk with the superficial muscles exposed on the left side.
(*m.* stands for *muscle; v.* stands for *vein.*)

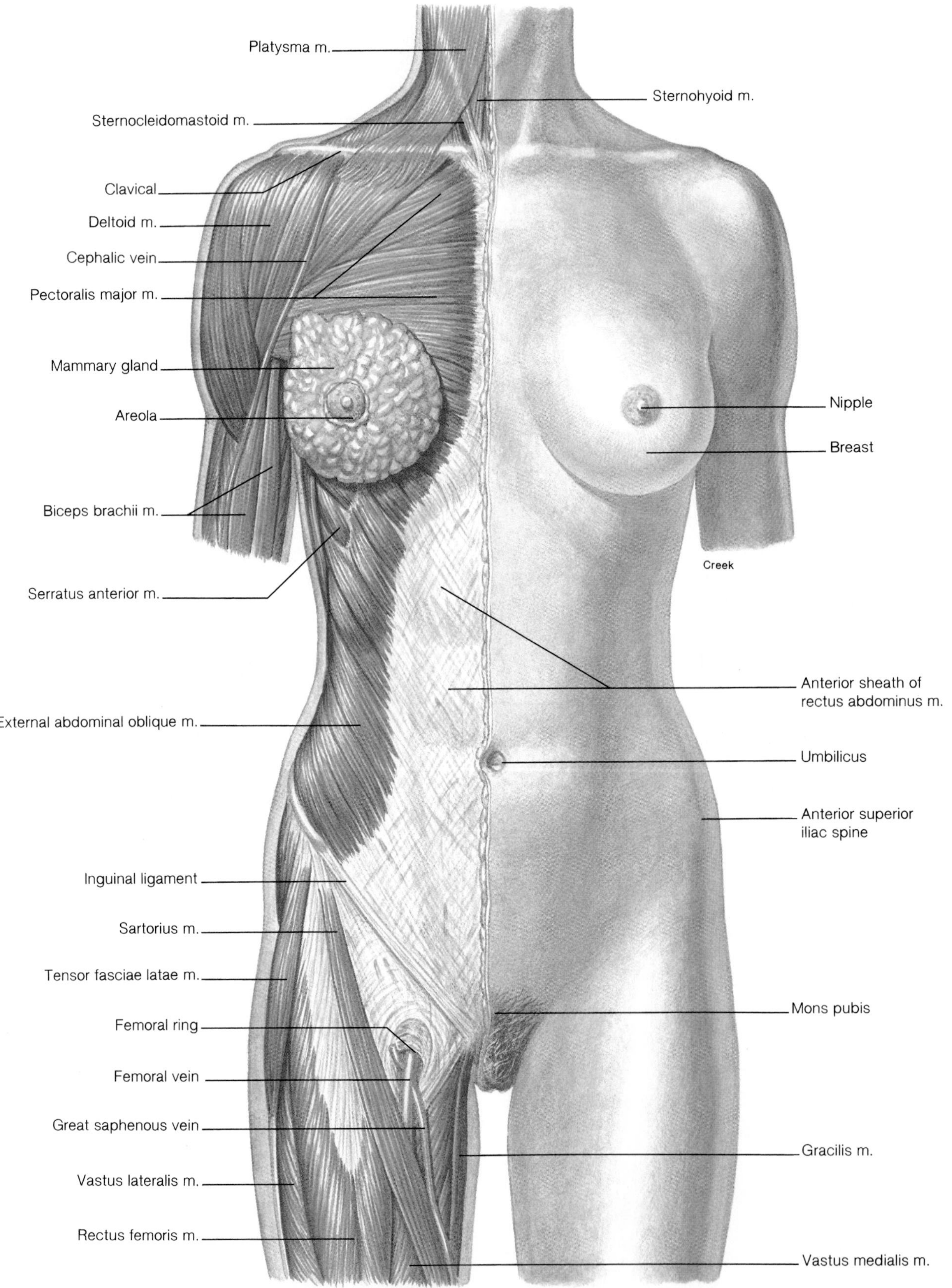

Figure 23

An anterior view of the male trunk with the deeper muscle layers exposed.
(*n.* stands for *nerve; a.* stands for *artery*.)

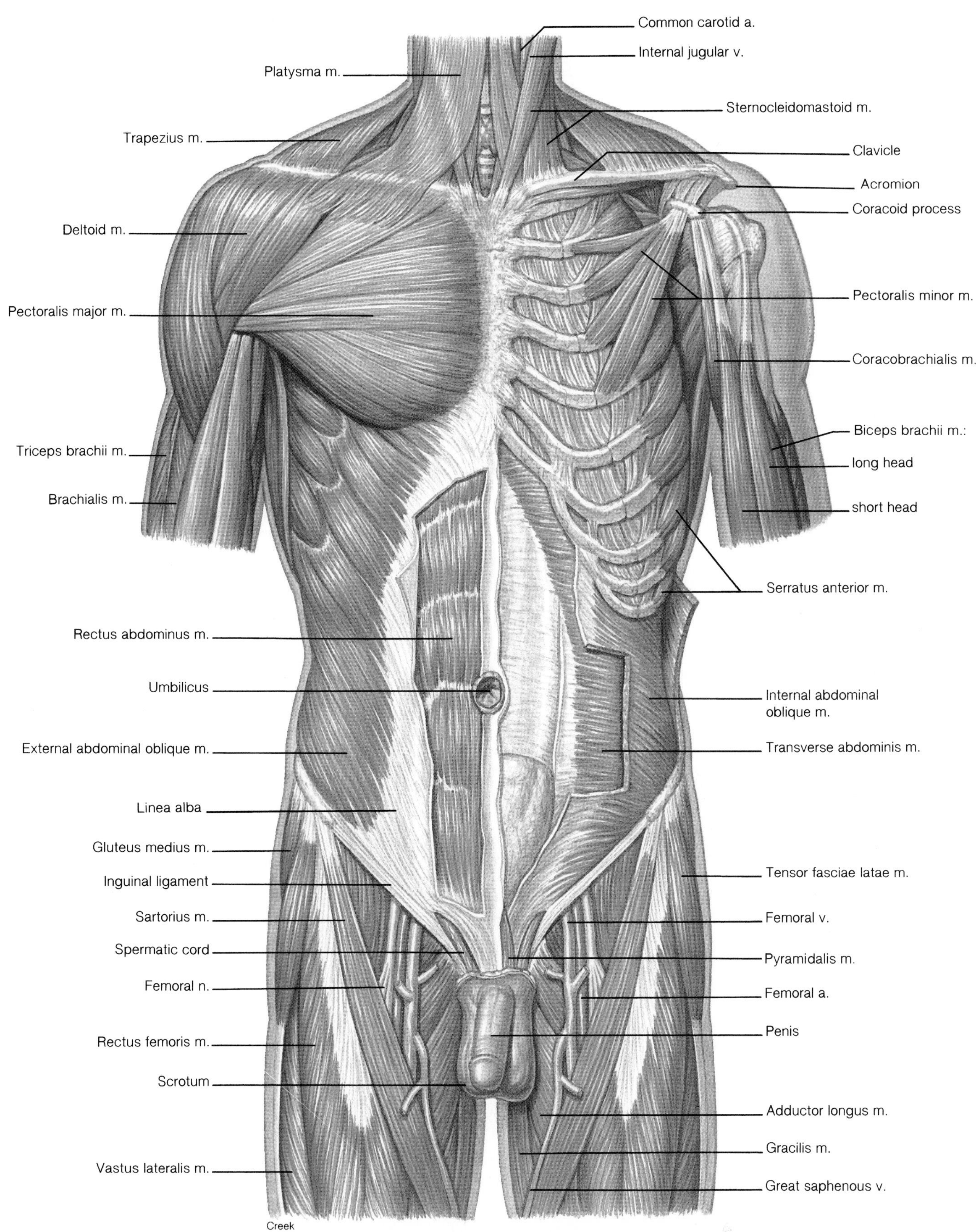

Figure 24

An anterior view of the male trunk with the deep muscles removed and the abdominal viscera exposed.
(*a.* stands for *artery.*)

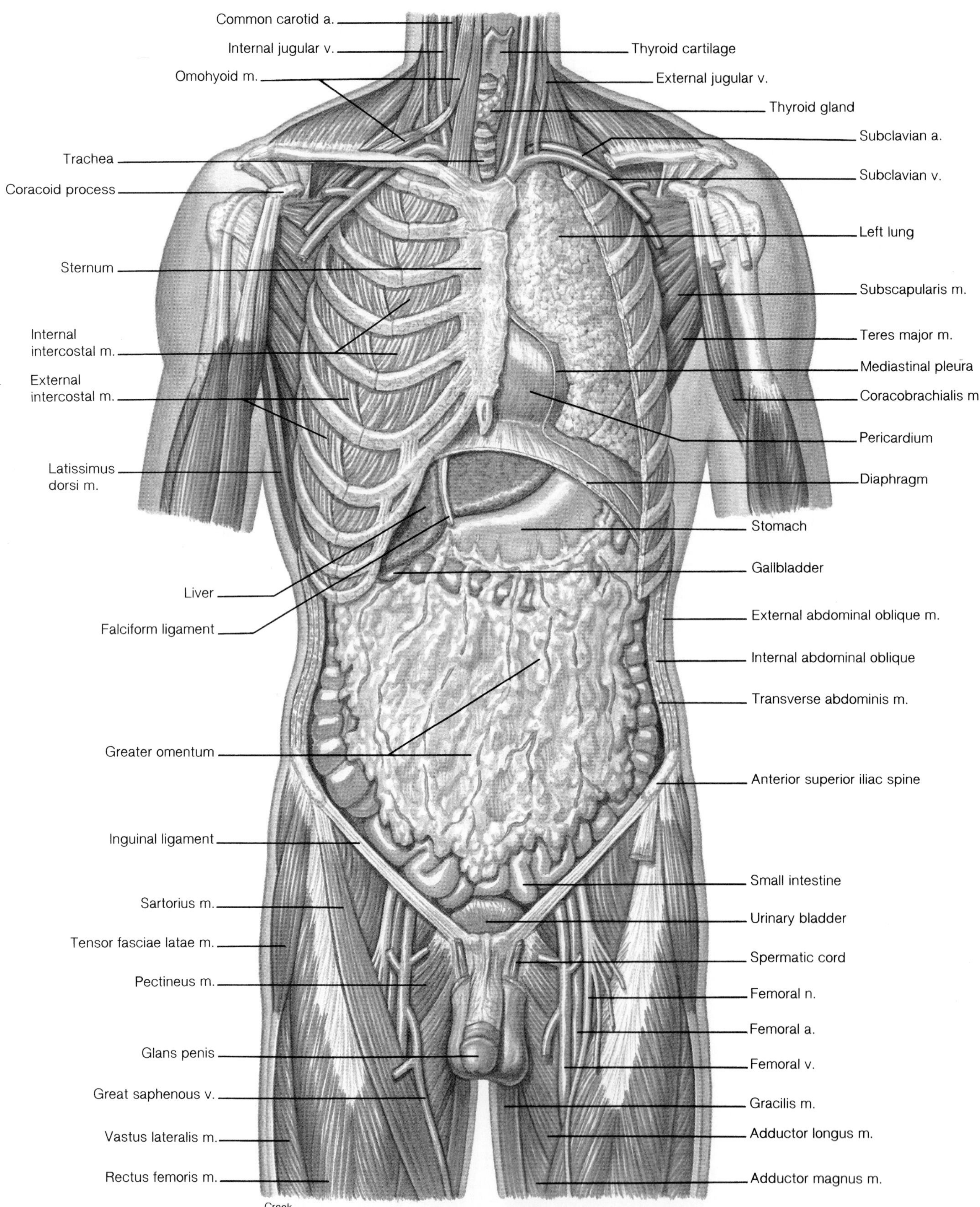

Figure 25

An anterior view of the male trunk with the thoracic and abdominal viscera exposed.

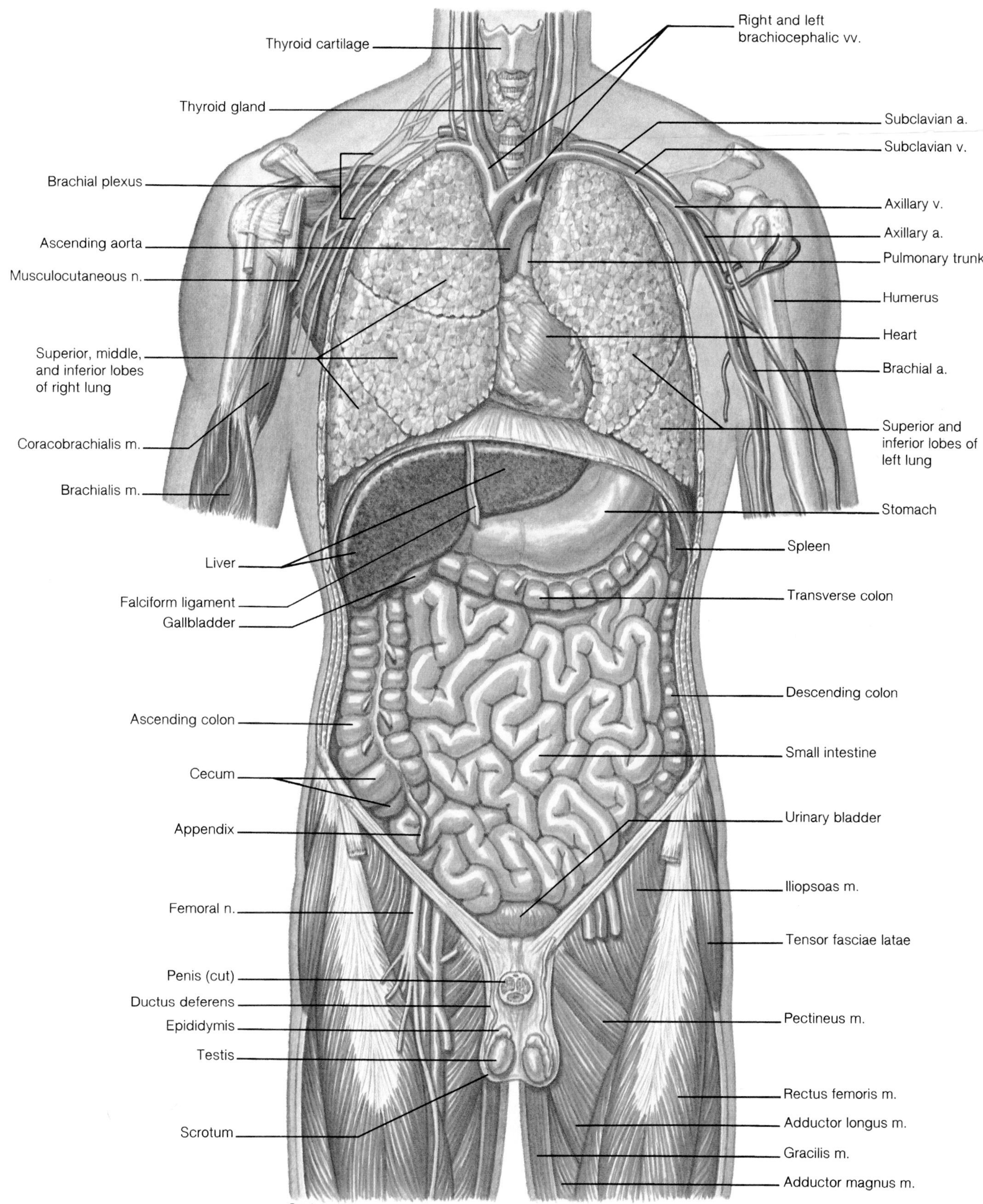

Figure 26

An anterior view of the female trunk with the lungs, heart, and small intestine sectioned.

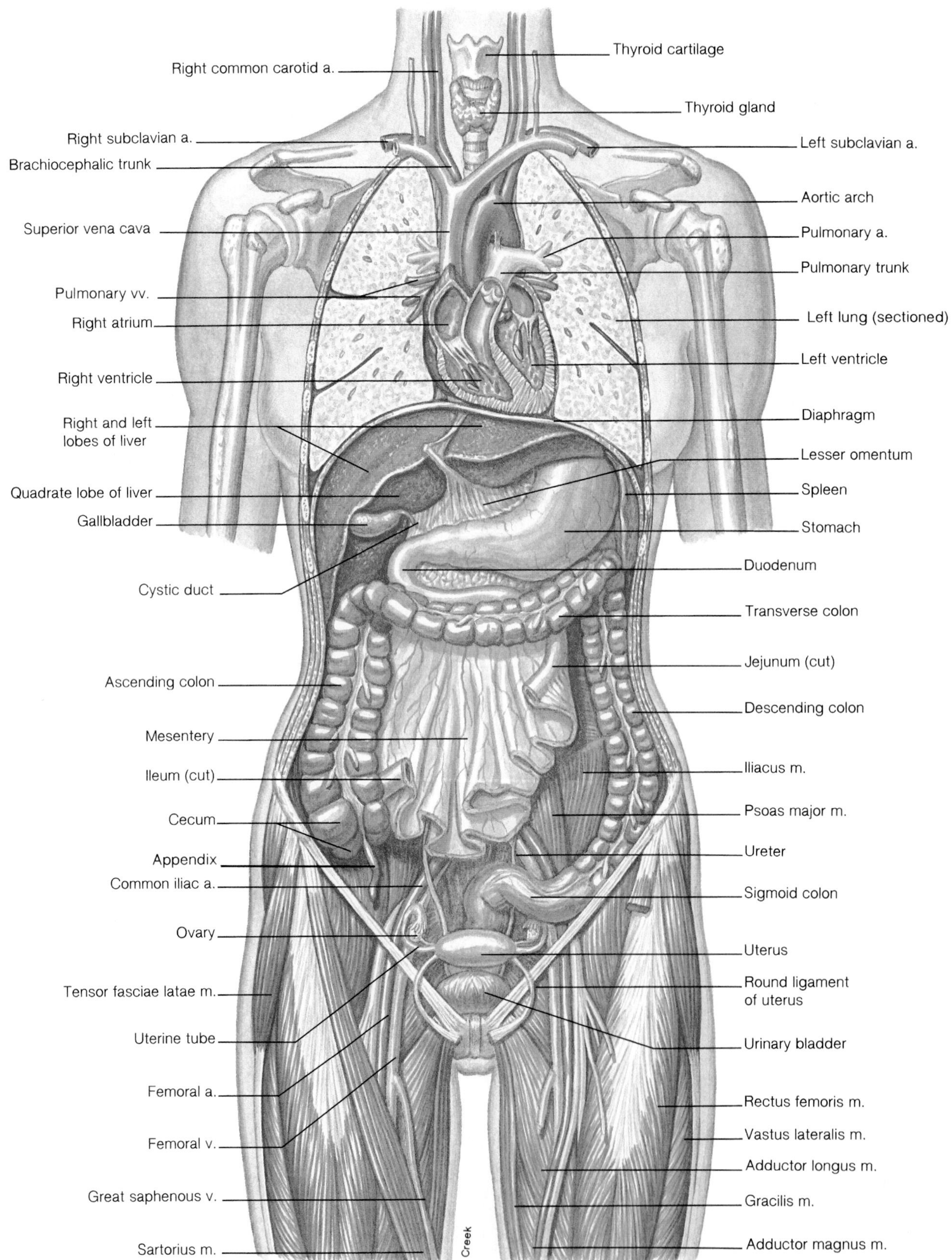

Figure 27

An anterior view of the female trunk.

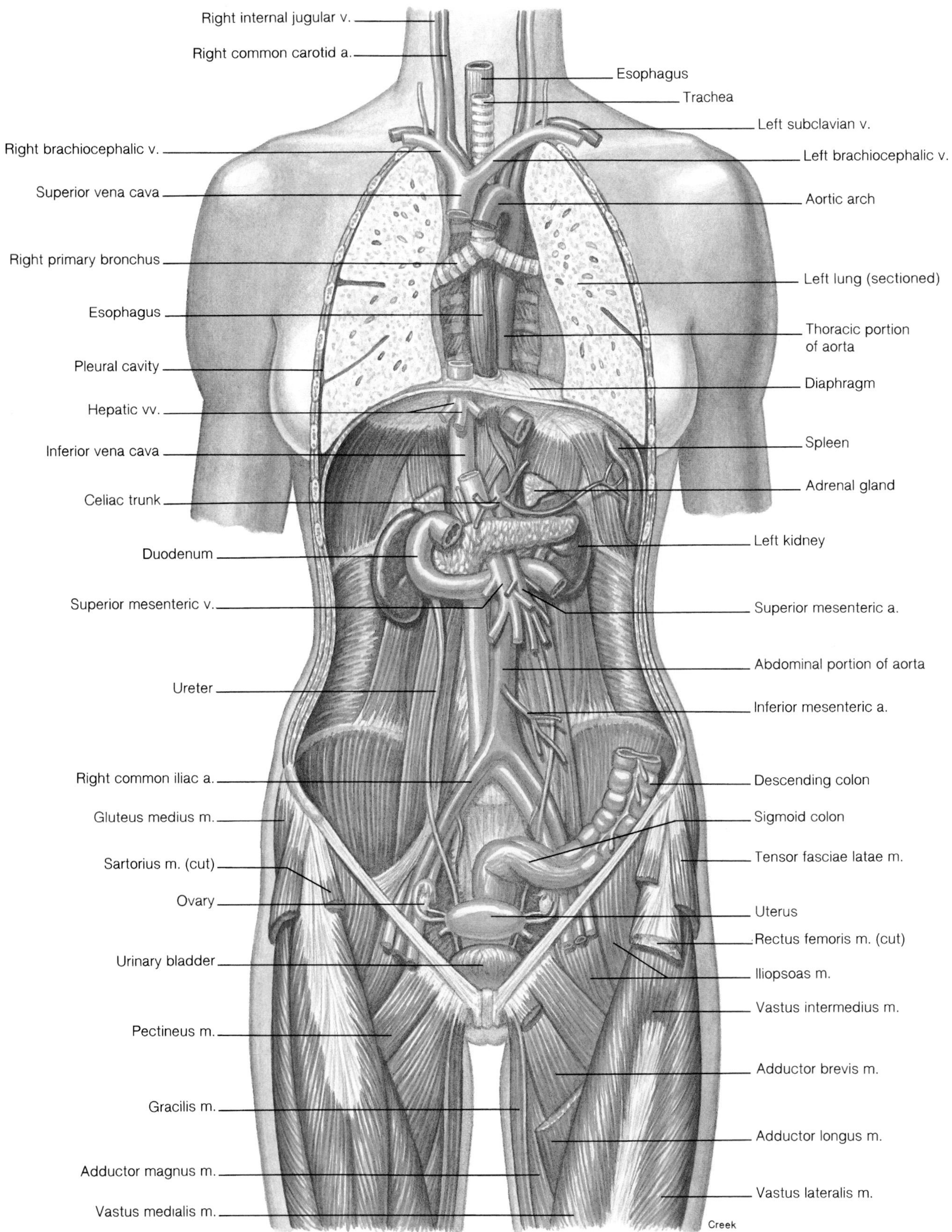
Right internal jugular v.
Right common carotid a.
Esophagus
Trachea
Left subclavian v.
Right brachiocephalic v.
Left brachiocephalic v.
Superior vena cava
Aortic arch
Right primary bronchus
Left lung (sectioned)
Esophagus
Thoracic portion of aorta
Pleural cavity
Diaphragm
Hepatic vv.
Spleen
Inferior vena cava
Adrenal gland
Celiac trunk
Left kidney
Duodenum
Superior mesenteric v.
Superior mesenteric a.
Abdominal portion of aorta
Ureter
Inferior mesenteric a.
Right common iliac a.
Descending colon
Gluteus medius m.
Sigmoid colon
Tensor fasciae latae m.
Sartorius m. (cut)
Ovary
Uterus
Rectus femoris m. (cut)
Urinary bladder
Iliopsoas m.
Vastus intermedius m.
Pectineus m.
Adductor brevis m.
Gracilis m.
Adductor longus m.
Adductor magnus m.
Vastus lateralis m.
Vastus medialis m.
Creek

Figure 28

An anterior view of the female trunk with the thoracic, abdominal, and pelvic visceral organs removed.

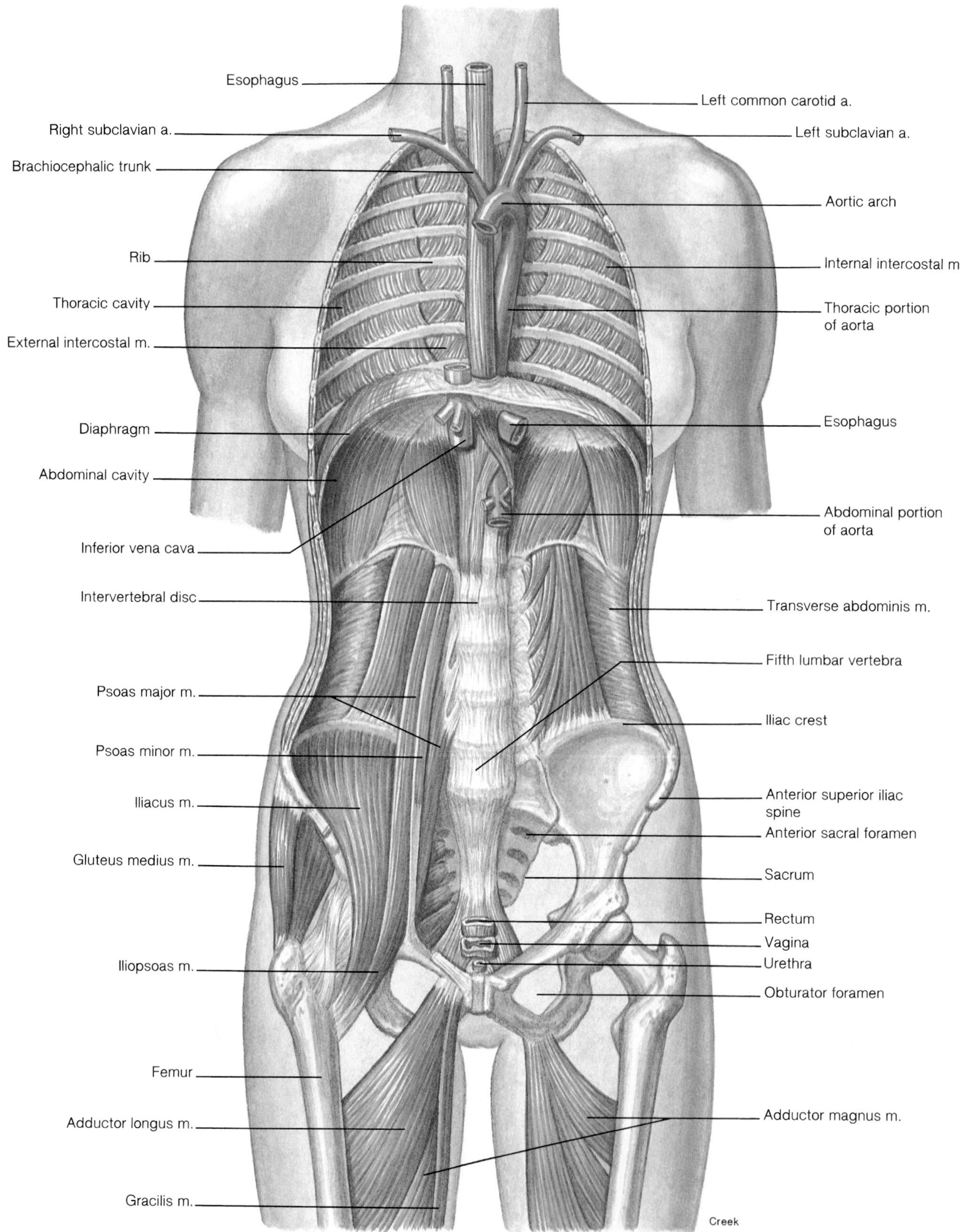

FOURTEEN

Functional Organization of the Nervous System

Objectives

- Define the basic terms used to describe the nervous system and describe the different categories of neurons.
- Explain how myelin sheaths are formed in the PNS and CNS.
- Describe the process of axon regeneration in the PNS.
- Describe the nature of the blood-brain barrier and discuss its significance.
- Explain why an action potential is an all-or-none phenomenon and describe the nature of the refractory period.
- Explain how action potentials code for the strength of a stimulus.
- Compare the conduction of action potentials in unmyelinated and myelinated axons.
- Describe the nature of gap junctions.
- Describe the structure and function of chemical synapses.
- Explain how acetylcholine stimulates the production of EPSPs.
- Compare the characteristics of EPSPs and action potentials.
- Explain how EPSPs cause action potentials to be produced.
- Explain the actions of acetylcholine in the PNS and CNS.
- Identify the monoamines and describe their actions.
- Explain how GABA, the endogenous opioids, and nitric oxide function as neurotransmitters.
- Describe spatial and temporal summation and explain how EPSPs and IPSPs can interact in the process of postsynaptic inhibition.

Clinical Investigation

A student whose clinical depression was causing her grades to fall decided to treat herself to a dinner at a seafood restaurant. After eating a meal of mussels and clams, which had been gathered from the local shore, she fell to the floor. Examination revealed that she had flaccid paralysis of her muscles and was experiencing difficulty breathing. Fortunately, quick emergency medical care saved her life. While this emergency care was being administered a prescription bottle containing a monoamine oxidase inhibitor was found in her purse. Laboratory tests later revealed that her blood contained this drug at a concentration consistent with its prescribed therapeutic use. What might have happened to cause this student's medical emergency?

Clues: Read the section "Monoamines as Neurotransmitters," including the box on MAO inhibitors, and examine table 14.5.

Neurons and Supporting Cells

The nervous system is composed of neurons, which produce and conduct electrochemical impulses, and supporting cells, which assist the functions of neurons. Neurons are classified according to structure or function; the various types of supporting cells perform specialized functions.

The nervous system is divided into the **central nervous system (CNS),** which includes the brain and spinal cord, and the **peripheral nervous system (PNS),** which includes the *cranial nerves* arising from the brain and the *spinal nerves* arising from the spinal cord.

The nervous system is composed of only two principal types of cells—neurons and supporting cells. **Neurons** (*noor'onz*) are the basic structural and functional units of the nervous system. They are specialized to respond to physical and chemical stimuli, conduct electrochemical impulses, and release chemical regulators. Through these activities, neurons enable the perception of sensory stimuli, learning, memory, and the control of muscles and glands. Neurons cannot divide by mitosis, although some neurons can regenerate a severed portion or sprout small new branches under some conditions.

Supporting cells aid the functions of neurons and are about five times more abundant than neurons. Supporting cells are collectively called **neuroglia** (*noo-rog'le-ă*), or simply **glial** (*gle'al*) **cells.** Unlike neurons, glial cells retain limited mitotic abilities (brain tumors that occur in adults are usually composed of glial cells rather than neurons).

Neurons

Although neurons vary considerably in size and shape, they generally have three principal regions: (1) a cell body, (2) dendrites, and (3) an axon (figs. 14.1 and 14.2). Dendrites and axons can be referred to generically as *processes*, or extensions from the cell body.

neuroglia: Gk. *neuron*, nerve; *glia*, glue

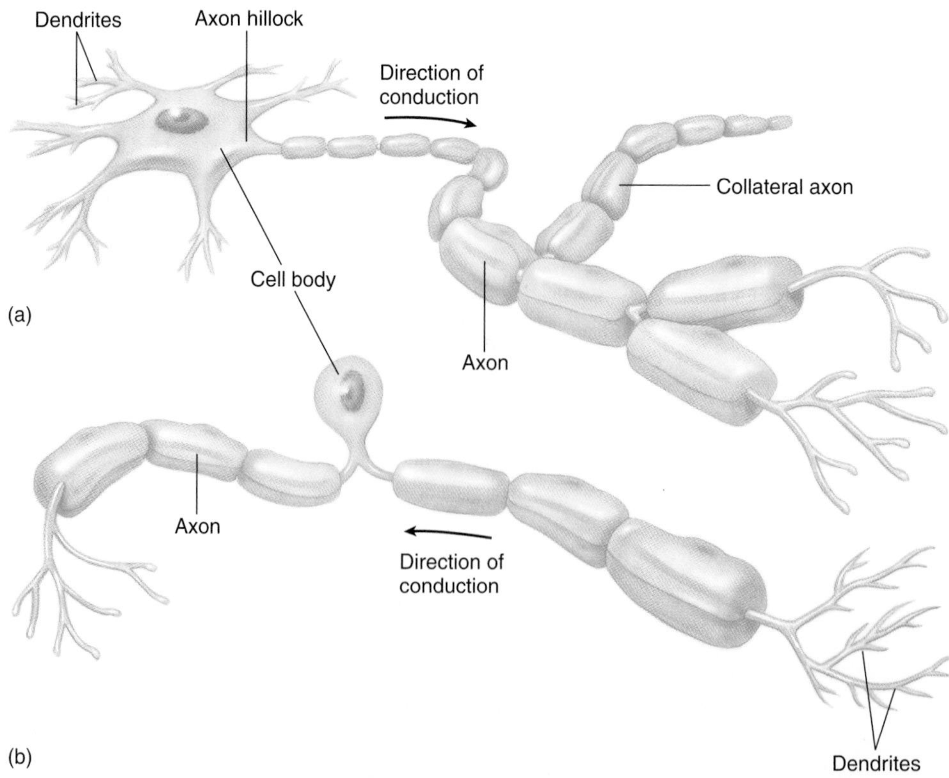

Figure 14.1

The structure of two kinds of neurons.
These are (*a*) a motor neuron and (*b*) a sensory neuron.

Figure 14.2

The parts of a neuron.
The axon of this neuron is wrapped by Schwann cells, which form a myelin sheath.

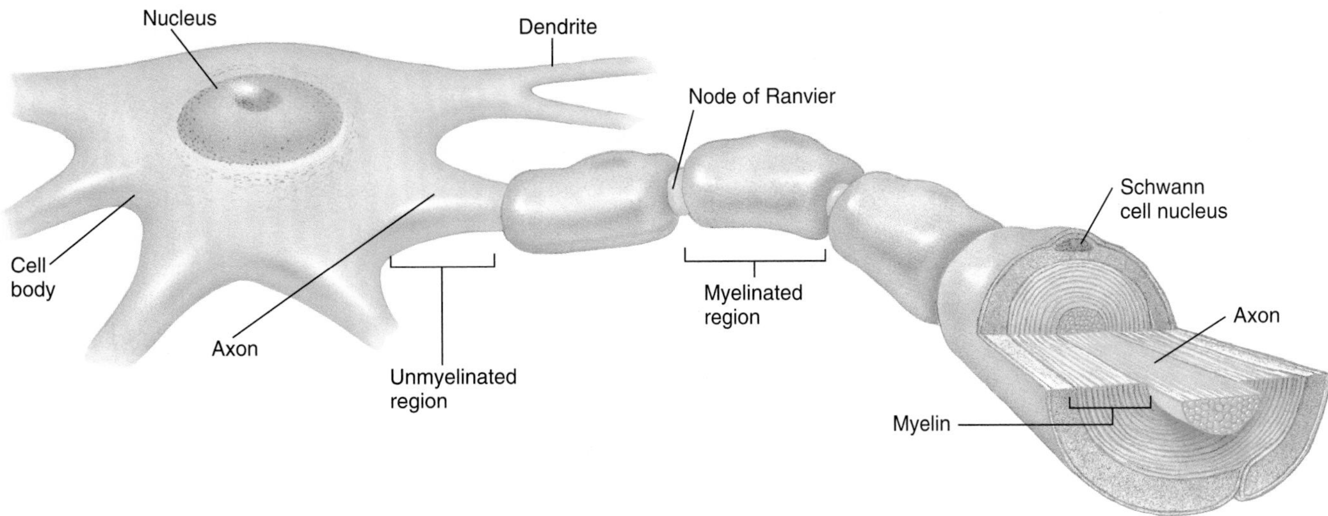

Table 14.1

Terminology Pertaining to the Nervous System

Term	Definition
Central nervous system (CNS)	Brain and spinal cord
Peripheral nervous system (PNS)	Nerves, ganglia, and nerve plexuses (outside of the CNS)
Association neuron (interneuron)	Multipolar neuron located entirely within the CNS
Sensory neuron (afferent neuron)	Neuron that transmits impulses from a sensory receptor into the CNS
Motor neuron (efferent neuron)	Neuron that transmits impulses from the CNS to an effector organ; for example, a muscle
Nerve	Cablelike collection of many axons; may be "mixed" (contain both sensory and motor fibers)
Somatic motor nerve	Nerve that stimulates contraction of skeletal muscles
Autonomic motor nerve	Nerve that stimulates contraction (or inhibits contraction) of smooth muscle and cardiac muscle and that stimulates glandular secretion
Ganglion	Grouping of neuron cell bodies located outside the CNS
Nucleus	Grouping of neuron cell bodies within the CNS
Tract	Grouping of nerve fibers that interconnect regions of the CNS

The **cell body,** or **perikaryon** (*per″ĭ-kar′e-on*), is the enlarged portion of the neuron that contains the nucleus. It is the "nutritional center" of the neuron, where macromolecules are produced. The cell body also contains densely staining areas of rough endoplasmic reticulum known as *chromatophilic substance* (*Nissl* [*nis′l*] *bodies*) that are not found in the dendrites or axon. The cell bodies within the CNS are frequently clustered into groups called *nuclei* (not to be confused with the nucleus of a cell). Cell bodies in the PNS usually occur in clusters called *ganglia* (*gang′gle-ă*) (table 14.1).

Dendrites (*den′drīts*) are thin, branched processes that extend from the cytoplasm of the cell body. Dendrites provide a receptive area that transmits electrical impulses to the cell body. The **axon** (*ak′son*) is a longer process that conducts impulses away from the cell body. Axons vary in length from only a millimeter long up to a meter or more (for those that extend from the CNS to the foot). The origin of the axon near the cell body is an expanded region called the *axon hillock*; it is here that nerve impulses originate. Side branches called *axon collaterals* may extend from the axon.

Proteins and other molecules are transported through the axon at faster rates than could be achieved by simple diffusion. This rapid movement is produced by two different mechanisms: axoplasmic flow and axonal transport. **Axoplasmic flow,** the

perikaryon: Gk. *peri*, around; *karyon*, nucleus
chromatophilic: Gk. *khroma*, color; *philus*, loving
Nissl body: from Franz Nissl, German neuroanatomist, 1860–1919
ganglion: Gk. *ganglion*, swelling
dendrite: Gk. *dendron*, tree branch
axon: Gk. *axon*, axis

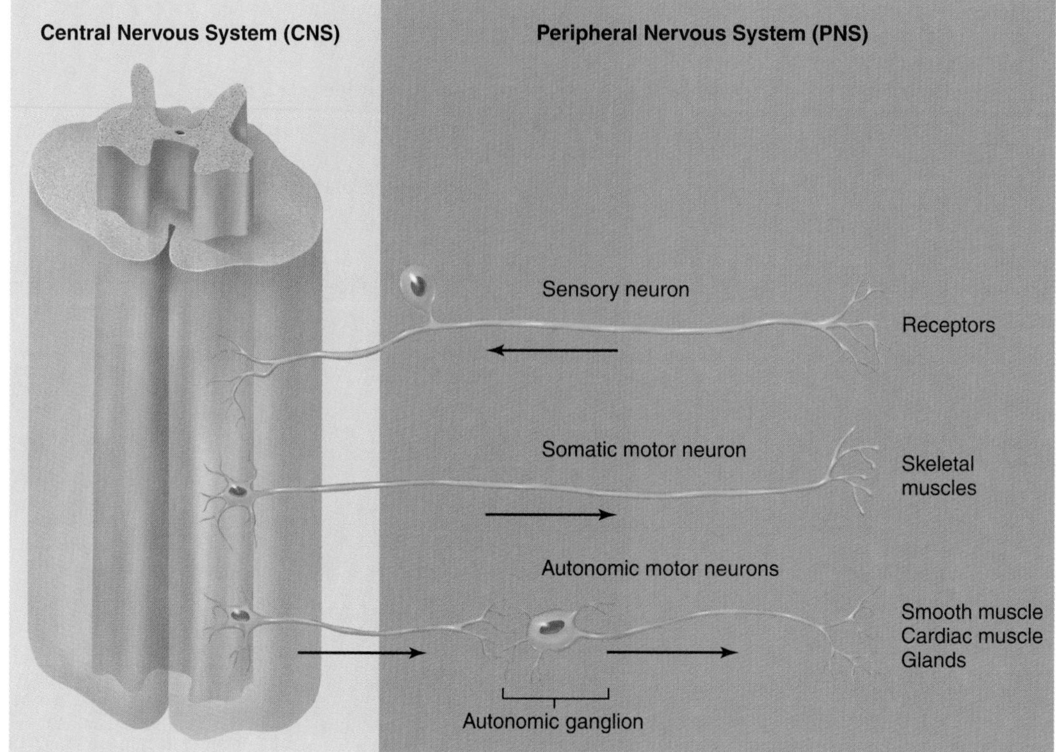

Figure 14.3

The relationship between the CNS and PNS. The PNS consists of sensory and motor neurons, which carry information into and out of, respectively, the CNS (brain and spinal cord).

slower of the two, results from rhythmic waves of contraction that push the cytoplasm from the axon hillock to the nerve endings. **Axonal** (*ak'sŏ-nal* or *ak-son'al*) **transport,** which employs microtubules and is more rapid and more selective, may occur in a reverse (retrograde) direction as well as in a forward (orthograde) direction. Indeed, retrograde transport may be responsible for the movement of herpes virus, rabies virus, and tetanus toxin from the nerve terminals into cell bodies.

Classification of Neurons and Nerves

Neurons may be classified according to their function or structure. The functional classification is based on the direction in which they conduct impulses, as indicated in figure 14.3. **Sensory,** or **afferent, neurons** conduct impulses from sensory receptors *into* the CNS. **Motor,** or **efferent, neurons** conduct impulses *out of* the CNS to effector organs (muscles and glands). **Association neurons,** or **interneurons,** are located entirely within the CNS and serve the associative, or integrative, functions of the nervous system.

There are two types of motor neurons: somatic and autonomic. **Somatic motor neurons** are responsible for both reflex and voluntary control of skeletal muscles. **Autonomic motor neurons** innervate the involuntary effectors—smooth muscle, cardiac muscle, and glands. The cell bodies of the autonomic neurons that innervate these organs are located outside the CNS in autonomic ganglia (fig. 14.3). There are two subdivisions of autonomic neurons: *sympathetic* and *parasympathetic*. Autonomic motor neurons, together with their central control centers, constitute the *autonomic nervous system*, the focus of chapter 17.

The structural classification of neurons is based on the number of processes that extend from the cell body of the neuron (fig. 14.4). **Bipolar neurons** have two processes, one at either end; this type is found in the retina of the eye. **Multipolar neurons,** the most common type, have several dendrites and one axon extending from the cell body; motor neurons are good examples of this type. **Pseudounipolar neurons** have a single short process that branches like a T to form a pair of longer processes. They are called pseudounipolar (*pseudo* = false) because they originate as bipolar neurons, but during early embryonic development their two processes converge and partially fuse. Sensory neurons are pseudounipolar—one of the branched processes receives sensory stimuli and produces nerve impulses; the other delivers these impulses to synapses within the brain or spinal cord. Anatomically, the part of the process that conducts impulses toward the cell body can be considered a dendrite, and the part that conducts impulses away from the cell body can be considered an axon. Functionally, however, the two branched processes behave as a single long axon and may be covered with a myelin sheath (discussed later), as are axons. Only the small projections at the receptive end of the process function as typical dendrites.

A **nerve** is a bundle of axons located outside the CNS. Most nerves are composed of both motor and sensory fibers and are thus called *mixed nerves*. Some of the cranial nerves, however, contain only sensory fibers. These are the nerves that serve the special senses of sight, hearing, taste, and smell.

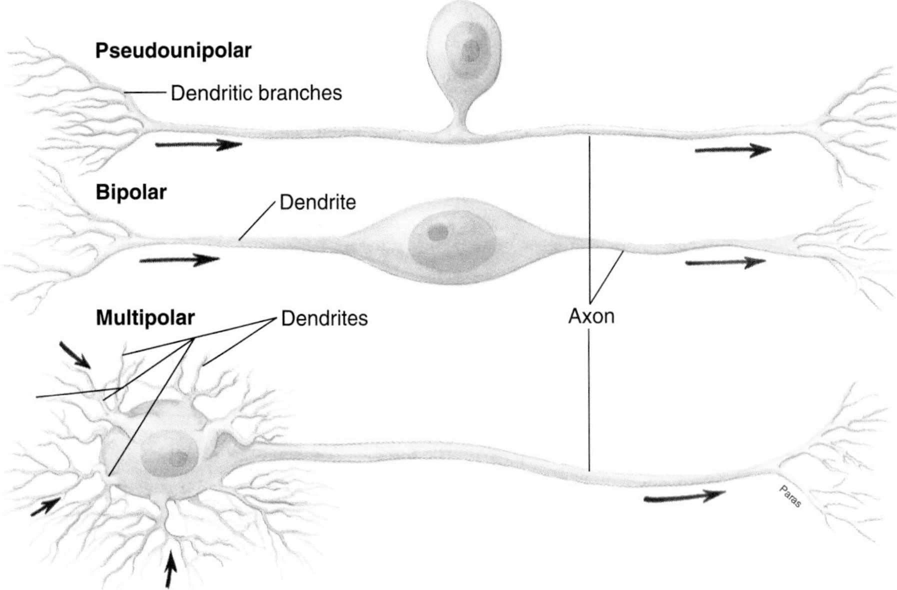

Figure 14.4

Three different types of neurons. Pseudounipolar neurons have one process, which splits; bipolar neurons have two processes; and multipolar neurons have many processes (one axon and many dendrites).

Supporting Cells

Unlike other organs that are "packaged" in connective tissue derived from mesoderm (the middle layer of embryonic tissue), the supporting cells of the nervous system are derived from the same embryonic tissue layer (ectoderm) that produces neurons. There are six categories of supporting cells: (1) **Schwann cells,** or **neurolemmocytes,** which form myelin sheaths around peripheral axons; (2) **satellite cells,** or **ganglionic gliocytes,** which support neuron cell bodies within the ganglia of the PNS; (3) **oligodendrocytes** (*ol″ĭ-go-den′drŏ-sītz*), which form myelin sheaths around axons of the CNS; (4) **microglia** (*mi-krog′le-ă*), which migrate through the CNS and phagocytize foreign and degenerated material; (5) **astrocytes,** which help to regulate the external environment of neurons in the CNS; and (6) **ependymal** (*ĕ-pen′dĭ-mal*) **cells,** which line the ventricles of the brain and the central canal of the spinal cord. A summary of the supporting cells is presented in table 14.2.

Sheath of Schwann and Myelin Sheath

Some axons in the CNS and PNS are surrounded by a myelin sheath and are known as *myelinated axons*. Other axons do not have a myelin sheath and are *unmyelinated*. The myelin sheaths in the PNS are formed by Schwann cells, whereas the myelin sheaths in the CNS are formed by oligodendrocytes.

All axons in the PNS are surrounded by a living sheath of Schwann cells, known as the **sheath of Schwann.** The outer surface of this layer of Schwann cells is encased in a glycoprotein basement membrane called the *neurilemma* (*nu″rĭ-lem′mah*), which is analogous to the basement membrane that underlies epithelial membranes. The axons of the CNS, by contrast, lack a sheath of Schwann (Schwann cells are found only in the PNS) and also lack a continuous basement membrane. This is significant in terms of nerve regeneration, as will be described in a later section.

Axons that are smaller than 2 micrometers (2 μm) in diameter are usually unmyelinated. Larger axons are generally surrounded by a **myelin sheath,** which is composed of successive wrappings of the cell membrane of Schwann cells (in the PNS) or oligodendrocytes (in the CNS).

Schwann cell: from Theodor Schwann, German histologist, 1810–82
oligodendrocytes: Gk. *oligos*, few; L. *dens*, tooth; Gk. *kytos*, hollow (cell)
microglia: Gk. *micros*, small; *glia*, glue
astrocyte: Gk. *aster*, star; *kytos*, hollow (cell)

Table 14.2

Supporting Cells and Their Functions*

Cell Type	Functions
Schwann cells	Surround axons of all peripheral nerve fibers, forming a neurolemmal sheath, or sheath of Schwann; wrap around many peripheral fibers to form myelin sheaths; also called *neurolemmocytes*
Satellite cells	Support ganglia within the PNS; also called *ganglionic gliocytes*
Oligodendrocytes	Form myelin sheaths around axons, producing white matter of the CNS
Astrocytes	Vascular processes cover capillaries within the brain and contribute to the blood-brain barrier
Microglia	Phagocytize pathogens and cellular debris within the CNS
Ependymal cells	Form the epithelial lining of brain cavities (ventricles) and the central canal of the spinal cord; cover tufts of capillaries to form choroid plexuses—structures that produce cerebrospinal fluid

*Supporting cells in the CNS are also known as neuroglia.

Figure 14.5

The formation of a myelin sheath around a peripheral axon.

The myelin sheath is formed by successive wrappings of the Schwann cell membranes, leaving most of the Schwann cell cytoplasm outside the myelin. The sheath of Schwann is thus external to the myelin sheath.

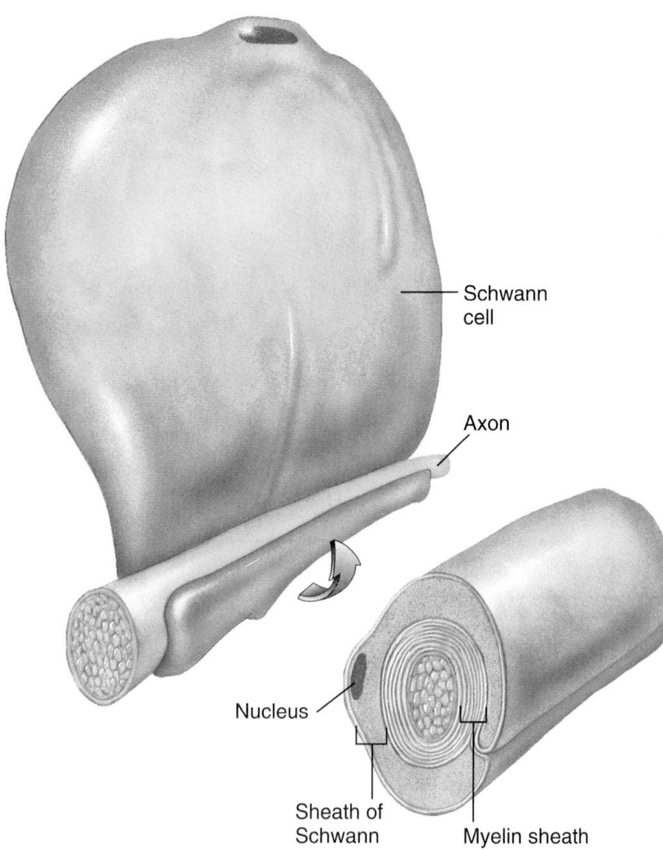

In the process of myelin formation in the PNS, Schwann cells roll around the axon, much like a roll of electrician's tape is wrapped around a wire. Unlike electrician's tape, however, the Schwann cell wrappings are made in the same spot, so that each wrapping overlaps the previous layers. The cytoplasm, meanwhile, is forced into the outer region of the Schwann cell, much as toothpaste is squeezed into the top of the tube as the bottom is rolled up (fig. 14.5). Each Schwann cell wraps only about a millimeter of axon, leaving gaps of exposed axon between the adjacent Schwann cells. These gaps in the myelin sheath are known as the **nodes of Ranvier** (*ran've-a*), also called *neurofibril nodes*. The successive wrappings of Schwann cell membrane provide insulation around the axon, leaving only the nodes of Ranvier exposed to produce nerve impulses.

nodes of Ranvier: from Louis A. Ranvier, French pathologist, 1835–1922

Figure 14.6

Electron micrograph of unmyelinated and myelinated axons.

Notice that myelinated axons have Schwann cell cytoplasm to the outside of their myelin sheath, and that Schwann cell cytoplasm also surrounds unmyelinated axons.

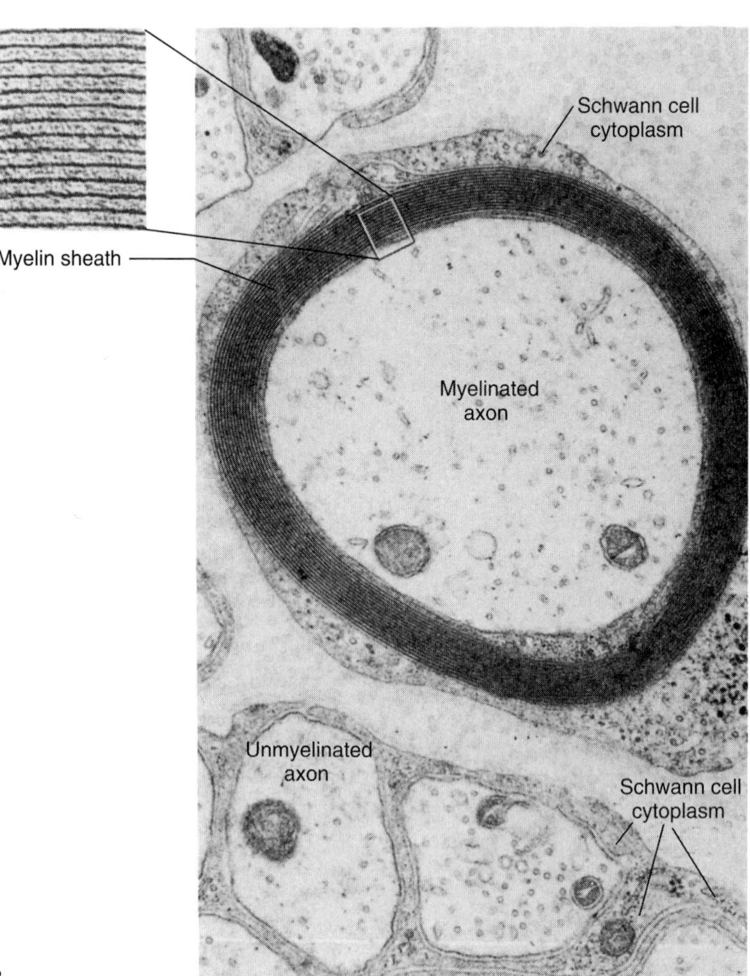

The Schwann cells remain alive as their cytoplasm is forced to the outside of the myelin sheath. As a result, myelinated axons of the PNS, like their unmyelinated counterparts, are surrounded by a living sheath of Schwann (fig. 14.6).

The myelin sheaths of the CNS are formed by oligodendrocytes. Unlike a Schwann cell, which forms a myelin sheath around only one axon, each oligodendrocyte has extensions like the tentacles of an octopus that form myelin sheaths around several axons (fig. 14.7). The myelin sheaths around axons of the CNS give this tissue a white color; areas of the CNS that contain a high concentration of axons thus form the **white matter.** The **gray matter** of the CNS is composed of high concentrations of cell bodies and dendrites, which lack myelin sheaths.

Figure 14.7

The formation of myelin sheaths in the CNS by an oligodendrocyte.
One oligodendrocyte forms myelin sheaths around several axons.

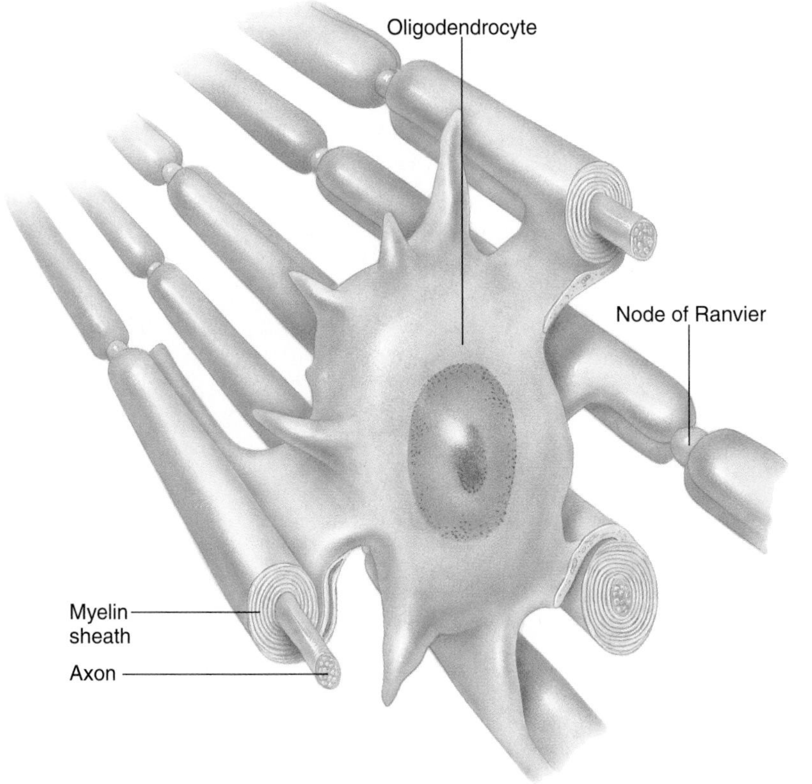

Regeneration of a Cut Axon

When an axon in a peripheral nerve is cut, the distal portion of the axon that was severed from the cell body degenerates and is phagocytosed by Schwann cells. The Schwann cells, surrounded by the neurilemma, then form a *regeneration tube* (fig. 14.8) as the part of the axon that is connected to the cell body begins to grow and exhibit amoeboid movement. The Schwann cells of the regeneration tube are believed to secrete chemicals that attract the growing axon tip, and the regeneration tube helps to guide the regenerating axon to its proper destination. Even a severed major nerve may be surgically reconnected—and the function of the nerve largely reestablished—if the surgery is performed before tissue death.

Injury in the CNS stimulates growth of axon collaterals, but central axons have a much more limited ability to regenerate than peripheral axons. This may be due in part to the absence of a continuous neurilemma (as is present in the PNS), which precludes the formation of a regeneration tube, and to molecules on the membrane of oligodendrocytes that act to inhibit neural growth. In addition to the limited ability of CNS neurons to regenerate, injury to the spinal cord has recently been shown to actually evoke apoptosis (cell suicide—chapter 3) in neurons that were not directly damaged by the injury. The causes of this are not presently understood.

Neurotrophins

In a developing fetal brain, chemicals called **neurotrophins** promote neuron growth. *Nerve growth factor* (NGF) was the first neurotrophin to be identified; others include *brain-derived neurotrophic factor* (BDNF), *glial-derived neurotrophic factor* (GDNF), *neurotrophin-3*, and *neurotrophin-4/5* (the number depends on the species). NGF and neurotrophin-3 are known to be particularly important in the embryonic development of sensory neurons and sympathetic ganglia.

Neurotrophins also have important functions in the adult nervous system. NGF is required for the maintenance of sympathetic ganglia, and there is evidence that neurotrophins are required in order for mature sensory neurons to regenerate after injury. In addition, GDNF may be needed in the adult to maintain spinal motor neurons and to sustain neurons in the brain that use the chemical dopamine as a neurotransmitter (discussed later in this chapter).

Experiments suggest that neurons of the CNS can regenerate if they are provided with the appropriate environment. Although neurotrophins promote neuron growth, some chemicals, including *myelin-associated inhibitory proteins*, have been shown to inhibit axon regeneration. In one experiment, axon regeneration in a rat spinal cord was improved by simultaneously blocking the inhibitory proteins with antibodies while providing a neurotrophin. In another experiment, limited regeneration and restored function was produced in the cut spinal cord of rats using a neurotrophin and grafted peripheral nerves that bridged the gap between the two cut ends of the spinal cord.

Astrocytes and the Blood-Brain Barrier

Astrocytes (fig. 14.9) are large stellate cells with numerous cytoplasmic processes that radiate outward. They are the most abundant of the glial cells in the CNS, constituting up to 90% of the nervous tissue in some areas of the brain.

Astrocytes are known to interact with neurons in two different ways. First, they have been shown to take up potassium ions from the extracellular fluid. Since K^+ is released from active neurons during the production of nerve impulses, this action of astrocytes may be very important in maintaining a proper ionic environment for the neurons. Second, astrocytes have been shown to take up specific neurotransmitter chemicals (which are released from the axon endings, as described in a later section). These neurotransmitters, glutamic acid and gamma-aminobutyric acid (GABA), are broken down within the astrocytes. The molecule produced from this breakdown—glutamine—is released from the astrocytes and made available to the neurons, which may then resynthesize these particular neurotransmitters.

Astrocytes have also been shown to interact with blood capillaries within the brain. Indeed, the brain capillaries are

Figure 14.8

The process of peripheral neuron regeneration.
(*a*) If a neuron is severed through a myelinated axon, the proximal portion may survive, but (*b*) this distal portion will degenerate through phagocytosis. The myelin sheath provides a pathway (*c* and *d*) for the regeneration of an axon, and (*e*) innervation is restored.

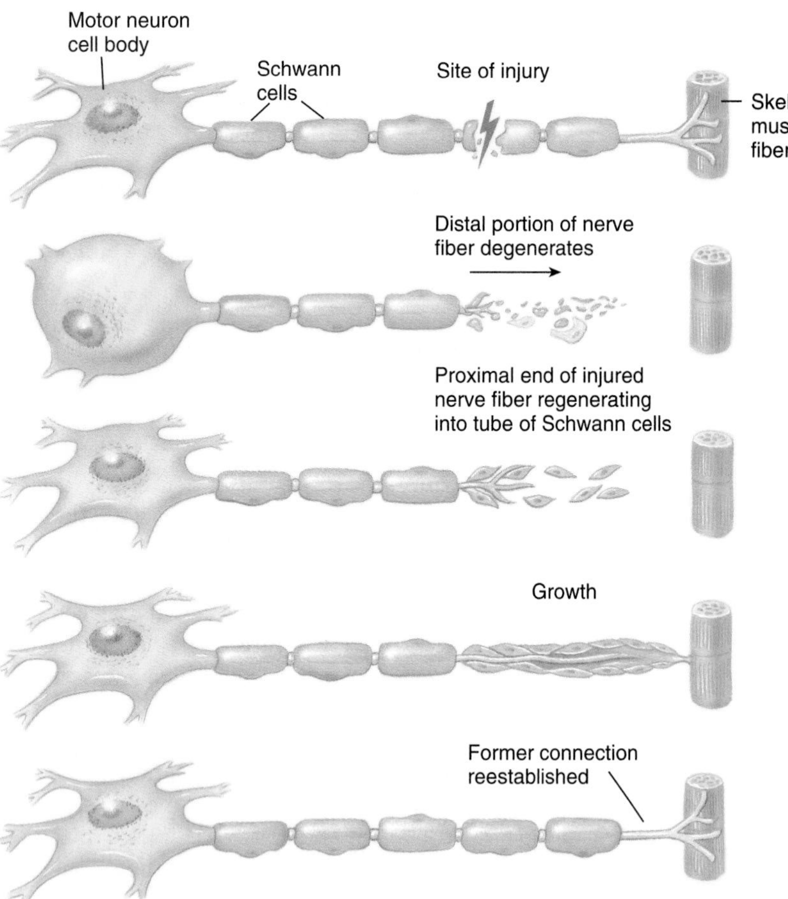

Figure 14.9

Astrocytes.
Astrocyte processes cover most of the surface area of brain capillaries.

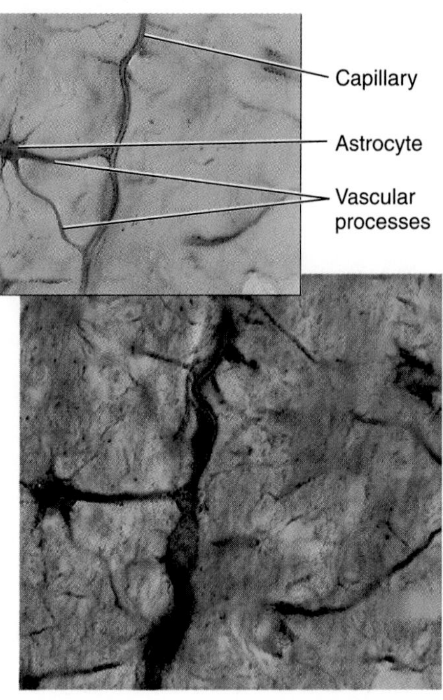

almost entirely surrounded by cytoplasmic extensions of the astrocytes called *vascular processes*. This association between astrocytes and brain capillaries has important physiological consequences.

Capillaries in the brain, unlike those of most other organs, do not have pores between adjacent endothelial cells (the cells that compose the walls of capillaries). Instead, the endothelial cells of brain capillaries are joined together by tight junctions. Unlike other organs, therefore, the brain cannot obtain molecules from the blood plasma by a nonspecific filtering process. Instead, molecules within brain capillaries must be moved through the endothelial cells by diffusion and active transport, as well as by endocytosis and exocytosis. This feature of brain capillaries imposes a very selective **blood-brain barrier.** There is evidence to suggest that the development of tight junctions between adjacent endothelial cells in brain capillaries, and thus the development of the blood-brain barrier, results from the effects of astrocytes on the brain capillaries.

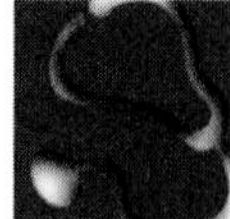

The blood-brain barrier presents difficulties in the chemotherapy of brain diseases because drugs that could enter other organs may not be able to enter the brain. In the treatment of *Parkinson's disease,* for example, patients who need a chemical called dopamine in the brain must be given a precursor molecule called levodopa (L-dopa) because L-dopa can cross the blood-brain barrier but dopamine cannot. Also, many antibiotics cannot cross the blood-brain barrier; therefore, in treating infections such as meningitis, only those antibiotics that can cross the blood-brain barrier are used.

Electrical Activity in Axons

The permeability of the axon membrane to Na^+ and K^+ is regulated by gates, which open in response to stimulation. Net diffusion of these ions occurs in two stages: first Na^+ moves into the axon, then K^+ moves out. This flow of ions and the changes in the membrane potential that result constitute an event called an action potential, or a nerve impulse.

All cells in the body maintain a potential difference (voltage) across the membrane, or **resting membrane potential,** in which the inside of the cell is negatively charged in comparison to the outside of the cell (for example –70 mV). As explained in chapter 5, this potential difference is largely the result of the permeability properties of the cell membrane. The membrane traps large, negatively charged organic molecules within the cell and permits only limited diffusion of positively charged inorganic ions. These properties result in an unequal distribution of these ions across the membrane. The action of the Na^+/K^+ pumps also helps to maintain a potential difference because they pump out three sodium ions for every two potassium ions that they transport into the cell. Partly as a result of these pumps, Na^+ is more highly concentrated in the extracellular fluid than in the cell, whereas K^+ is more highly concentrated within the cell.

Although all cells have a membrane potential, only a few types of cells have been shown to alter their membrane potential in response to stimulation. Such alterations in membrane potential are achieved by varying the membrane permeability to specific ions in response to stimulation. A central aspect of the physiology of neurons and muscle cells is their ability to produce and conduct these changes in membrane potential. Such an ability is termed *excitability* or *irritability*.

An increase in membrane permeability to a specific ion results in the diffusion of that ion down its concentration gradient, either into or out of the cell. These *ion currents* occur only across limited patches of membrane (located fractions of a millimeter apart), where specific ion channels are located. Changes in the potential difference across the membrane at these points can be measured by the voltage developed between two electrodes—one placed inside the cell and the other placed outside the cell membrane at the region being recorded. The voltage between these two recording electrodes can be visualized by connecting them to an oscilloscope (fig. 14.10).

In an oscilloscope, electrons from a cathode-ray "gun" are sprayed across a fluorescent screen, producing a line of light. This line deflects upward or downward in response to a potential difference between the two electrodes. The oscilloscope can be calibrated in such a way that an upward deflection of the line indicates that the inside of the membrane has become less negative (or more positive) compared to the outside of the membrane. A downward deflection of the line, conversely, indicates that the inside of the cell has become more negative. The oscilloscope can thus function as a fast-responding voltmeter with an ability to display voltage changes as a function of time.

Figure 14.10

Observing depolarization and hyperpolarization. The difference in potential (in millivolts) between an intracellular and extracellular recording electrode is displayed on an oscilloscope screen. The resting membrane potential (rmp) of the axon may be reduced (depolarization) or increased (hyperpolarization). Depolarization is seen as a line going up from the rmp; hyperpolarization is represented by a line going down from the rmp.

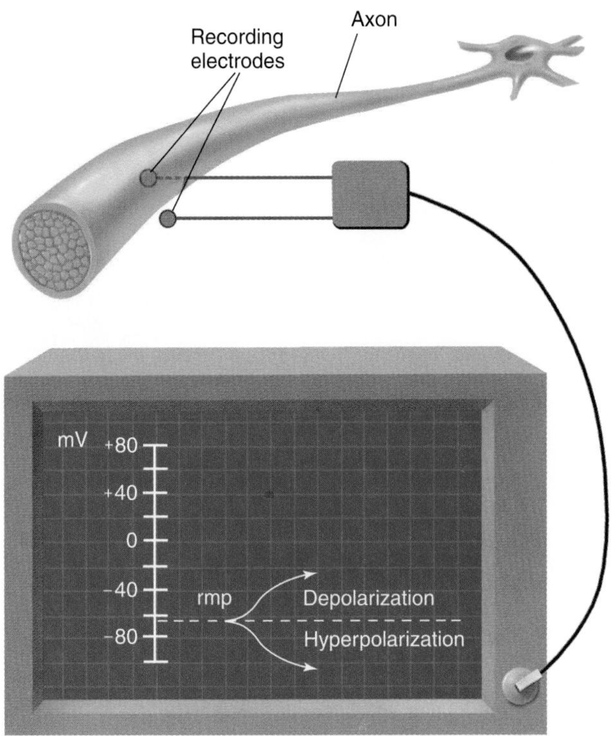

If both recording electrodes are placed outside of the cell, the potential difference between the two will be zero (because there is no charge separation). When one of the two electrodes penetrates the cell membrane, the oscilloscope will indicate that the intracellular electrode is electrically negative with respect to the extracellular electrode; a membrane potential is recorded. We will call this the **resting membrane potential (rmp)** to distinguish it from events described in later sections. All cells have a resting membrane potential, but its magnitude can be different in different types of cells. Neurons maintain an average rmp of –70 mV, for example, whereas heart muscle cells may have an rmp of –85 mV.

If appropriate stimulation causes positive charges to flow into the cell, the line will deflect upward. This change is called **depolarization,** since the potential difference between the two recording electrodes is reduced. A return to the resting membrane potential is known as **repolarization.** If stimulation causes negative charges to flow into the cell so that the inside of the cell becomes more negative than the resting membrane potential, the line on the oscilloscope will deflect downward. This change is called **hyperpolarization** (fig. 14.10).

Figure 14.11

Model of a voltage-gated ion channel.
The channel is closed at the resting membrane potential but opens in response to a threshold level of depolarization. This permits the diffusion of ions required for action potentials. After a brief period of time, the channel is inactivated by a "ball-and-chain" portion of a polypeptide chain.

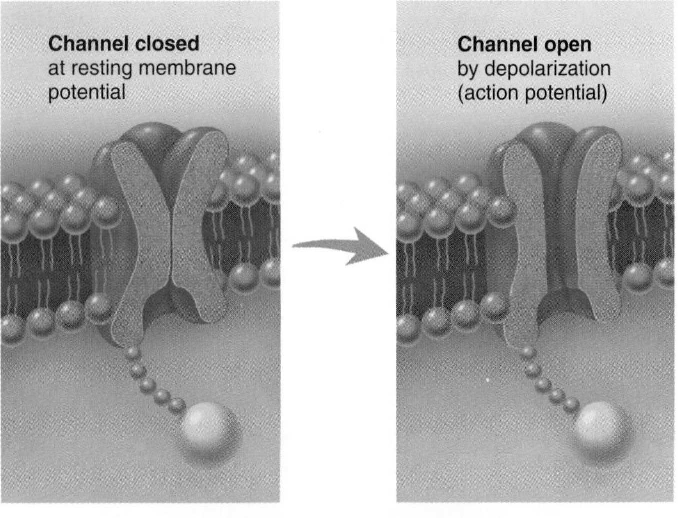

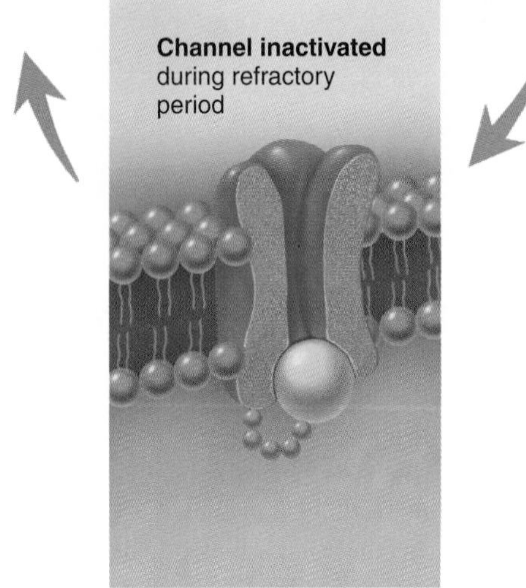

Ion Gating in Axons

The changes in membrane potential just described—depolarization, repolarization, and hyperpolarization—are caused by changes in the net flow of ions through ion channels in the membrane. The permeability of the membrane to Na^+, K^+, and other ions is regulated by parts of the ion channels called **gates.** These polypeptide chains can open or close a membrane channel according to specific conditions. When the gates of specific ion channels are closed, the membrane is not very permeable to that ion, and when the channel gates are opened, the permeability to that ion is greatly increased (fig. 14.11).

The ion channels for Na^+ and K^+ are fairly specific for each of these ions. It is believed that there are two types of channels for K^+; one type lacks gates and is always open, whereas the other type has gates that are closed in the resting cell. Channels for Na^+, by contrast, always have gates, and these gates are closed in the resting cell. The resting cell is thus more permeable to K^+ than to Na^+. (As described in chapter 5, some Na^+ does leak into the cell; this leakage may occur in a nonspecific manner through open K^+ channels.) The resting membrane potential is thus close to, but slightly less than, the equilibrium potential for K^+ (chapter 5).

Depolarization of a small region of an axon can be experimentally induced by a pair of stimulating electrodes that act as if they were injecting positive charges into the axon. If two recording electrodes are placed in the same region (one electrode within the axon and one outside), an upward deflection of the oscilloscope line will be observed as a result of this depolarization. If a certain level of depolarization is achieved (from –70 mV to –55 mV, for example) by this artificial stimulation, a sudden and very rapid change in the membrane potential will be observed. This is because *depolarization to a threshold level causes the Na^+ gates to open.* Now the permeability properties of the membrane are changed, and Na^+ diffuses down its concentration gradient into the cell.

A fraction of a second after the Na^+ gates open, they close again. Just before they do, *the depolarization stimulus causes the K^+ gates to open.* This makes the membrane more permeable to K^+ than it is at rest, and K^+ diffuses down its concentration gradient out of the cell. The K^+ gates will then close and the permeability properties of the membrane will return to what they were at rest.

Since opening of the gated Na^+ and K^+ channels is stimulated by depolarization, these ion channels in the axon membrane are said to be **voltage regulated.** The channel gates are closed at the resting membrane potential of –70 mV and open in response to depolarization of the membrane to a threshold value.

Action Potentials

We will now consider the events that occur at one point in an axon, when a small region of axon membrane is stimulated artificially and responds with changes in ion permeabilities. The resulting changes in membrane potential at this point are detected by recording electrodes placed in this region of the axon. The nature of the stimulus in vivo (in the body), and the manner by which electrical events are conducted to different points along the axon will be described in later sections.

When the axon membrane has been depolarized to a threshold level—in the previous example, by stimulating electrodes—the Na^+ gates open and the membrane becomes permeable to Na^+. This permits Na^+ to enter the axon by diffusion, which further depolarizes the membrane (makes the inside less negative, or more positive). Since the gates for the Na^+ channels of the axon membrane are voltage regulated, this additional depolarization opens more Na^+ channels and makes the membrane even more permeable to Na^+. As a result, more Na^+ can enter the cell and induce a depolarization that opens even more voltage-regulated Na^+ gates. A *positive feedback loop* (fig. 14.12) is thus created, causing

Figure 14.12

Depolarization of an axon affects Na^+ and K^+ diffusion in sequence.

(*1*) NA^+ gates open and NA^+ diffuses into the cell, then the NA^+ gates close. (*2*) After a brief period, K^+ gates open and K^+ diffuses out of the cell. An inward diffusion of Na^+ causes further depolarization, which in turn causes further opening of Na^+ gates in a positive feedback (+) fashion. The opening of K^+ gates and outward diffusion of K^+ makes the inside of the cell more negative, and thus has a negative feedback effect (–) on the initial depolarization.

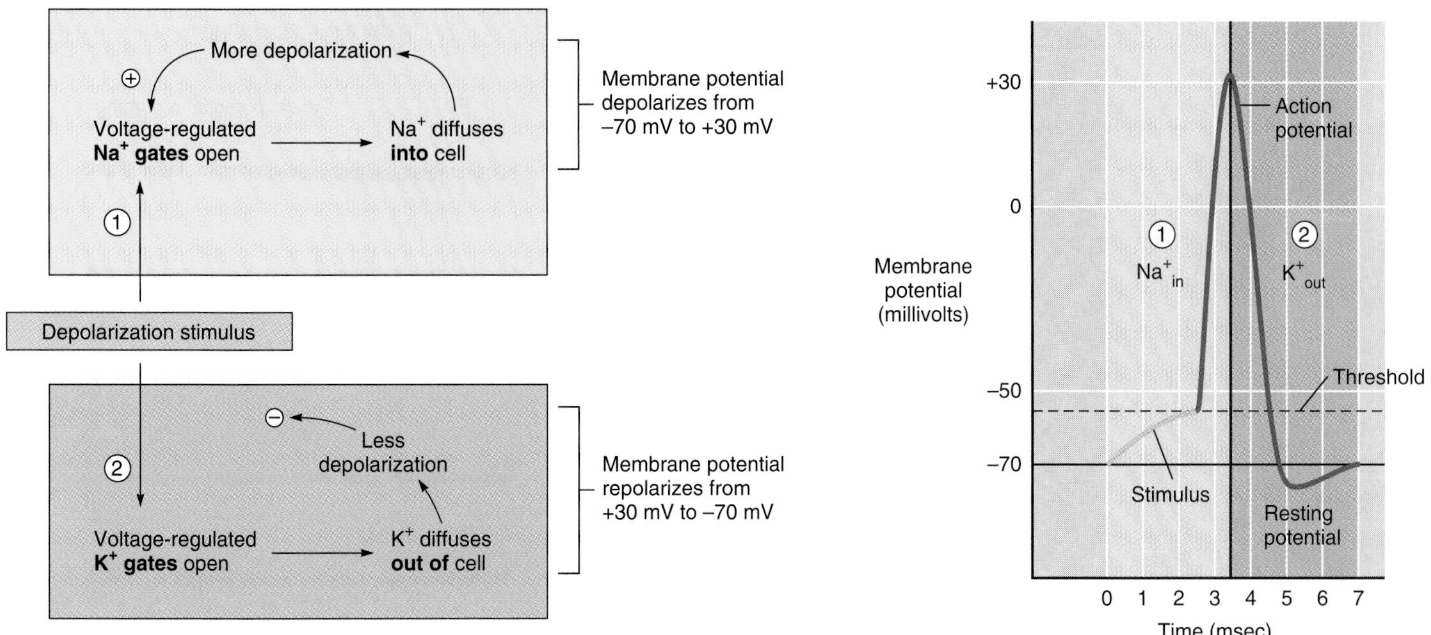

the rate of Na^+ entry and depolarization to accelerate in an explosive fashion.

After a slight delay following the opening of the Na^+ gates, depolarization of the axon membrane also causes the opening of voltage-regulated K^+ gates and the diffusion of K^+ out of the cell. Since K^+ is positively charged, the diffusion of K^+ out of the cell makes the inside of the cell less positive, or more negative, and acts to restore the original resting membrane potential. This process is called **repolarization** and represents the completion of a *negative feedback loop* (fig. 14.12). These changes in Na^+ and K^+ diffusion and the resulting changes in the membrane potential that they produce constitute an event called the **action potential,** or **nerve impulse.**

The correlation between ion movements and changes in membrane potential is shown in figure 14.13. The bottom portion of this figure illustrates the movement of Na^+ and K^+ through the axon membrane in response to a depolarization stimulus. Notice that the explosive increase in Na^+ diffusion causes rapid depolarization to 0 mV and then *overshoot* of the membrane potential so that the inside of the membrane actually becomes positively charged (almost +30 mV) compared to the outside (top portion of fig. 14.13). The greatly increased permeability to Na^+ thus drives the membrane potential toward the equilibrium potential for Na^+ (chapter 5). The Na^+ permeability then rapidly decreases and the diffusion of K^+ increases, resulting in repolarization to the resting membrane potential.

Once an action potential has been completed, the Na^+/K^+ pumps extrude the extra Na^+ that has entered the axon and recovers the K^+ that has diffused out of the axon. This active transport of ions occurs very quickly because the events described occur across only a very small area of membrane. Only a relatively small amount of Na^+ and K^+ actually diffuse through the membrane during the production of an action potential, and so the total concentrations of Na^+ and K^+ in the axon and in the extracellular fluid are not significantly changed.

Notice that active transport processes are not directly involved in the production of an action potential; both depolarization and repolarization are produced by the diffusion of ions down their concentration gradients. A neuron poisoned with cyanide, so that it cannot produce ATP, can still produce action potentials for a period of time. After awhile, however, the lack of ATP for active transport by the Na^+/K^+ pumps results in a decline in the concentration gradients, and therefore in the ability of the axon to produce action potentials. This shows that the Na^+/K^+ pumps are not directly involved; rather, they are required to maintain the concentration gradients needed for the diffusion of Na^+ and K^+ during action potentials.

All-or-None Law

Once a region of axon membrane has been depolarized to a threshold value, the positive feedback effect of depolarization on Na^+ permeability and of Na^+ permeability on depolarization causes the membrane potential to shoot toward about +30 mV. It does not normally become more positive because the Na^+ gates quickly close and the K^+ gates open. The length

Figure 14.13

Membrane potential changes and ion movements during an action potential.

An action potential (*top*) is produced by an increase in sodium diffusion that is followed, with a short time delay, by an increase in potassium diffusion (*bottom*). This drives the membrane potential first toward the sodium equilibrium potential and then toward the potassium equilibrium potential.

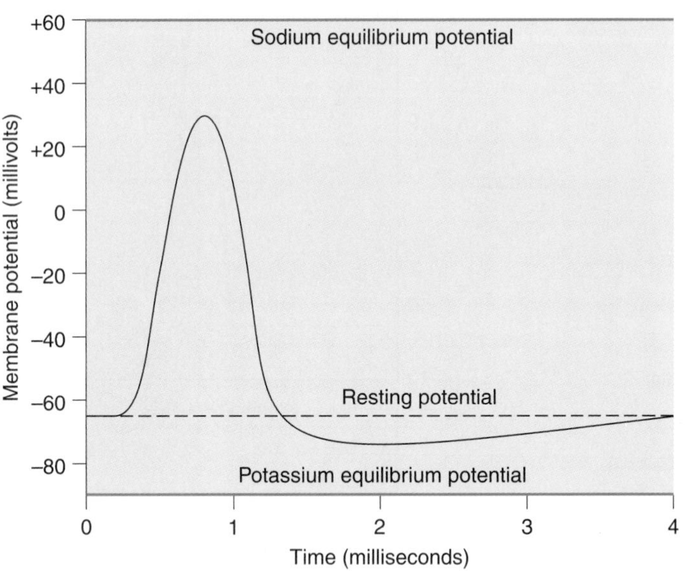

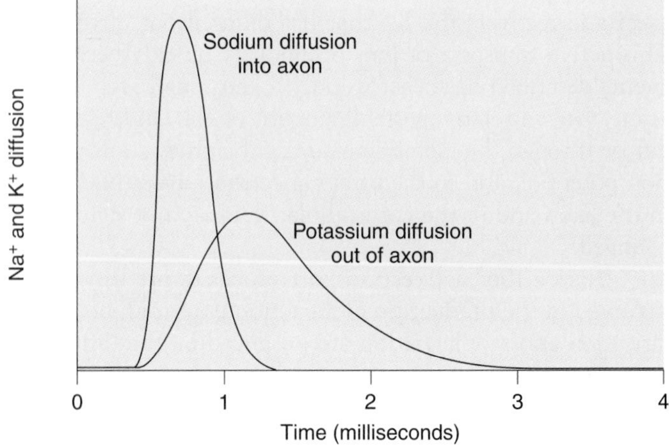

Figure 14.14

The effect of stimulus strength on action potential frequency.

These are recordings from a single sensory fiber of a sciatic nerve of a frog stimulated by varying degrees of stretch of the gastrocnemius muscle. Notice that increasing degrees of stretch (indicated by increasing weights attached to the muscle) result in a higher frequency of action potentials.

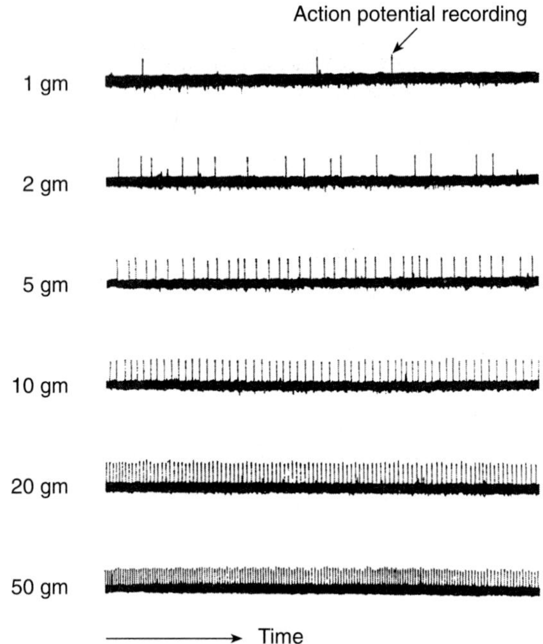

of time that the Na^+ and K^+ gates stay open is independent of the strength of the depolarization stimulus.

The amplitude of action potentials is therefore **all or none.** When depolarization is below a threshold value, the voltage-regulated gates are closed; when depolarization reaches threshold, a maximum potential change (the action potential) is produced. Since the change from −70 mV to +30 mV and back to −70 mV lasts only about 3 msec, the image of an action potential on an oscilloscope screen looks like a spike. Action potentials are therefore sometimes called *spike potentials*.

The gates are only open for a fixed period of time because they soon become inactivated. This occurs automatically, as will be described shortly. Because of this automatic inactivation, the duration of one action potential is about the same as that of another. Likewise, since the concentration gradient for Na^+ is relatively constant, the amplitudes of the action potentials are about the same in all axons at all times (from −70 mV to +30 mV, or about 100 mV in total amplitude).

Coding for Stimulus Intensity

If one depolarization stimulus is greater than another, the greater stimulus strength is not coded by a greater amplitude of action potentials (because action potentials are all-or-none events). The code for stimulus strength in the nervous system is not amplitude modulated (AM). When a greater stimulus strength is applied to a neuron, identical action potentials are produced more frequently (more are produced per second). Therefore, the code for stimulus strength in the nervous system is frequency modulated (FM). This concept is illustrated in figure 14.14.

When an entire collection of axons (in a nerve) is stimulated, different axons will be stimulated at different stimulus intensities. A weak stimulus will activate only those few fibers with low thresholds, whereas stronger stimuli can activate fibers with higher thresholds. As the intensity of stimulation increases, more and more fibers become activated. This process, called **recruitment,** represents another mechanism by which the nervous system can code for stimulus strength.

Refractory Periods

If a stimulus of a given intensity is maintained at one location of an axon and depolarizes it to threshold, action potentials will be produced at that point at a given frequency (number per second). As the stimulus strength increases, the frequency of action potentials produced at that location increases accordingly. As action potentials are produced with increasing frequency, the time between successive action potentials decreases—but only up to a minimum time interval. The interval between successive action potentials will never become so short as to allow a new action potential to be produced before the preceding one has finished.

During the time that a patch of axon membrane is producing an action potential, it is incapable of responding—or *refractory*—to further stimulation. If a second stimulus is applied during most of the time that an action potential is being produced, the second stimulus will have no effect on the axon membrane. The membrane is thus said to be in an **absolute refractory period;** it cannot respond to any subsequent stimulus.

The cause of the absolute refractory period is now understood at a molecular level. In addition to the voltage-regulated gates that open and close the channel, each ion channel has a polypeptide that may function as a "ball and chain" apparatus dangling from its cytoplasmic side (see fig. 14.11). After a voltage-regulated channel is opened by depolarization for a set time, it enters an *inactive state*. An inactivated channel is different from one that is simply closed. The inactivated channel cannot be opened by depolarization because it is blocked by the molecular ball attached to the chain! After a fixed period of time, the ball leaves the mouth of the channel and the Na^+ channel enters its resting state—a state in which it is closed by voltage-regulated gates.

If a second stimulus is applied while the K^+ gates are open (and the membrane is in the process of repolarizing), the membrane is said to be in a **relative refractory period.** During this time only a very strong depolarization can overcome the repolarization effects of the open K^+ channels and produce a second action potential (fig. 14.15).

Because the cell membrane is refractory during the time it is producing an action potential, each action potential remains a separate, all-or-none event. In this way, as a continuously applied stimulus increases in intensity, its strength can be coded strictly by the frequency of the action potentials it produces at each point of the axon membrane.

After a large number of action potentials have been produced, one might think that the relative concentrations of Na^+ and K^+ would be changed in the extracellular and intracellular compartments. This is not the case. In a typical mammalian axon that is 1 mm in diameter, for example, only one intracellular K^+ in 3,000 is exchanged for a Na^+ to produce an action potential. Since a typical neuron has about 1 million Na^+/K^+ pumps that can transport nearly 200 million ions per second, these small changes can be quickly corrected.

Cable Properties of Neurons

If a pair of stimulating electrodes produces a depolarization that is too weak to cause the opening of voltage-regulated

Figure 14.15

Absolute and relative refractory periods.
While a segment of axon is producing an action potential, the membrane is absolutely or relatively resistant (refractory) to further stimulation.

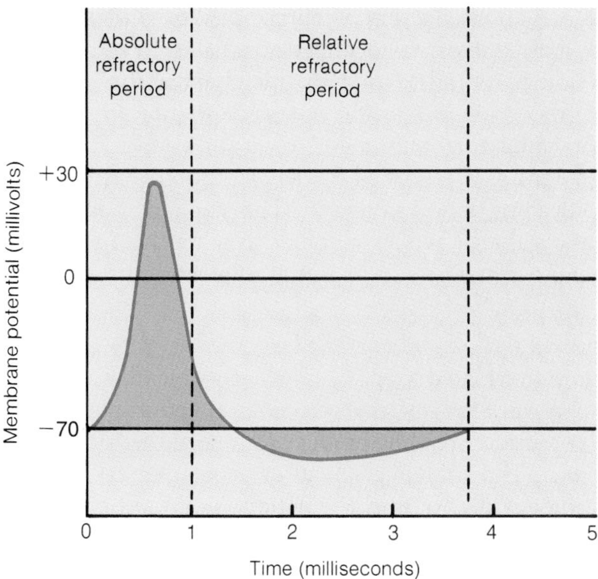

Na^+ gates—that is, if the depolarization is below threshold (about –55 mV)—the change in membrane potential will be *localized* to within 1 to 2 mm of the point of stimulation. For example, if the stimulus causes depolarization from –70 mV to –60 mV at one point, and the recording electrodes are placed only 3 mm away from the stimulus, the membrane potential recorded will remain at –70 mV (the resting potential). The axon is thus a very poor conductor compared to a metal wire.

The term *cable properties* refers to the ability of a neuron to transmit charges through its cytoplasm. These cable properties are quite poor because there is a high internal resistance to the spread of charges and because many charges leak out of the axon through its membrane. If an axon had to conduct only through its cable properties, therefore, no axon could be more than a millimeter in length. The fact that some axons are a meter or more in length suggests that the conduction of nerve impulses does not rely on the cable properties of the axon.

Conduction of Nerve Impulses

When stimulating electrodes artificially depolarize one point of an axon membrane to a threshold level, voltage-regulated gates open and an action potential is produced at that small region of axon membrane containing those gates. For about the first millisecond of the action potential, when the membrane voltage changes from –70 mV to +30 mV, a current of Na^+ enters the cell by diffusion because of the opening of the Na^+ gates. Each action potential thus "injects" positive charges (sodium ions) into the axon.

These positively charged sodium ions are conducted by the cable properties of the axon to an adjacent region that still has a membrane potential of –70 mV. Within the limits of the cable properties of the axon (1 to 2 mm), this helps to depolarize the adjacent region of axon membrane. When this adjacent region of membrane reaches a threshold level of depolarization, it too produces an action potential as its voltage-regulated gates open.

Each action potential thus acts as a stimulus for the production of another action potential at the next region of membrane that contains voltage-regulated gates. In the description of action potentials earlier in this chapter, the stimulus for their production was artificial—depolarization produced by a pair of stimulating electrodes. Now it can be seen that each action potential is produced by depolarization that results from the preceding action potential. This explains how all action potentials along an axon are produced after the first action potentials are generated at the initial segment of the axon.

Conduction in an Unmyelinated Axon

In an unmyelinated axon, every patch of membrane that contains Na^+ and K^+ gates can produce an action potential. Action potentials are thus produced along the entire length of the axon. The cablelike spread of depolarization induced by the influx of Na^+ during one action potential helps to depolarize the adjacent regions of membrane—a process that is also aided by movements of ions on the outer surface of the axon membrane (fig. 14.16). This process would depolarize the adjacent membranes on each side of the region to produce an action potential, but the area that had previously produced one cannot produce another at this time because it is still in its refractory period.

It is important to recognize that action potentials are not really conducted, although it is convenient to use that word. Each action potential is a separate, complete event that is repeated, or *regenerated,* along the axon's length. This is analogous to the "wave" performed by spectators in a stadium. One person after another gets up (depolarization) and then sits down (repolarization); it is thus the "wave" (spread of action potentials) that travels, not the people (individual action potentials).

The action potential produced at the end of the axon is thus a completely new event that was produced in response to depolarization from the previous action potential. The last action potential has the same amplitude as the first. Therefore, action potentials are said to be **conducted without decrement** (without decreasing in amplitude).

The spread of depolarization by the cable properties of an axon is fast compared to the time it takes to produce an action potential. Thus, the more action potentials along a given stretch of axon that have to be produced, the slower the conduction. Since action potentials must be produced at every fraction of a micrometer in an unmyelinated axon, the conduction rate is relatively slow. This conduction rate is somewhat faster if the unmyelinated axon is thicker, since the ability of fibers to conduct charges by cable properties improves with increasing diameter. The conduction rate is substantially faster if the axon is myelinated because fewer action potentials are produced along a given length of myelinated axon.

Figure 14.16

The conduction of action potentials in an unmyelinated axon.

Each action potential "injects" positive charges that spread to adjacent regions. The region that has just produced an action potential is refractory. The next region, not having been stimulated previously, is partially depolarized. As a result, its voltage-regulated Na^+ gates open, and the process is repeated. Successive segments of the axon thereby regenerate, or "conduct," the action potential.

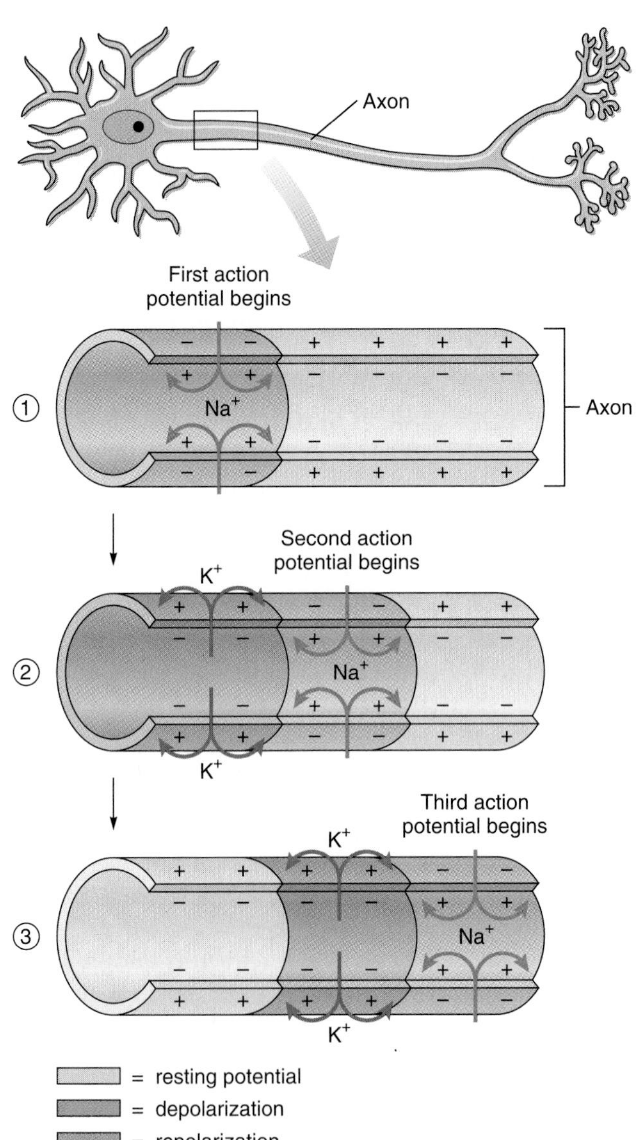

Conduction in a Myelinated Axon

The myelin sheath provides insulation for the axon, preventing movements of Na^+ and K^+ through the membrane. If the myelin sheath were continuous, therefore, action potentials could not be produced. The myelin thus has interruptions—the nodes of Ranvier, as previously described.

Because the cable properties of axons can conduct depolarizations only over a very short distance (1–2 mm), the nodes

Figure 14.17

The conduction of a nerve impulse in a myelinated axon.
Since the myelin sheath prevents inward Na^+ current, action potentials can be produced only at gaps in the myelin sheath, called the nodes of Ranvier. This "leaping" of the action potential from node to node is known as saltatory conduction.

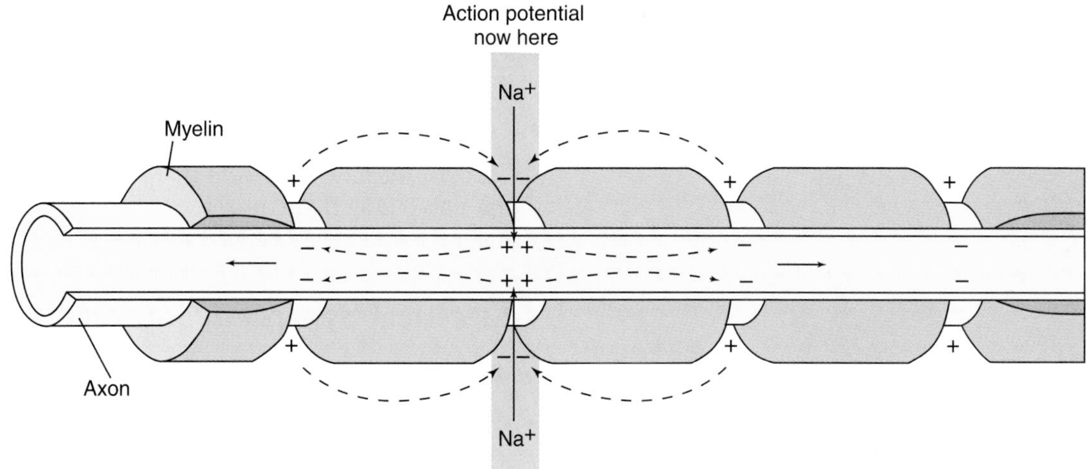

Table 14.3

Conduction Velocities and Functions of Mammalian Nerves of Different Diameters

Diameter (µm)	Conduction Velocity (m/sec)	Examples of Functions Served
12–22	70–120	Sensory: muscle position
5–13	30–90	Somatic motor fibers
3–8	15–40	Sensory: touch, pressure
1–5	12–30	Sensory: pain, temperature
1–3	3–15	Autonomic fibers to ganglia
0.3–1.3	0.7–2.2	Autonomic fibers to smooth and cardiac muscles

of Ranvier cannot be separated by more than this distance. Studies have shown that Na^+ channels are highly concentrated at the nodes (estimated at 10,000 per square micrometer) and almost absent in the regions of axon membrane between the nodes. Action potentials, therefore, occur only at the nodes of Ranvier (fig. 14.17) and seem to "leap" from node to node—a process called **saltatory conduction.** The leaping is, of course, just a metaphor; the action potential at one node depolarizes the membrane at the next node to threshold, so that a new action potential is produced at the next node of Ranvier.

saltatory: L. *saltario,* leap

Since the cablelike spread of depolarization between the nodes is very fast and fewer action potentials need to be produced per given length of axon, saltatory conduction allows a *faster rate of conduction* than is possible in an unmyelinated fiber. Conduction rates in the human nervous system vary from 1.0 m/sec—in thin, unmyelinated fibers that mediate slow, visceral responses—to faster than 100 m/sec (225 miles per hour)—in thick, myelinated fibers involved in quick stretch reflexes in skeletal muscles (table 14.3).

The Synapse

Axons end close to, or in some cases at the point of contact with, another cell. Once action potentials reach the end of an axon, they directly or indirectly stimulate (or inhibit) the other cell. In specialized cases, action potentials can directly pass from one cell to another. In most cases, however, the action potentials stop at the axon ending, where they stimulate the release of a chemical neurotransmitter that affects the next cell.

A **synapse** (*sin'aps*) is the functional connection between a neuron and a second cell. In the CNS, this other cell is also a neuron. In the PNS the other cell may be either a neuron or an *effector cell* within a muscle or gland. Although the physiology of neuron-neuron synapses and neuron-muscle synapses is similar, the latter synapses are often called **myoneural** (*mi"ŏ-noor'al*), or **neuromuscular, junctions.**

Neuron-neuron synapses usually involve a connection between the axon of one neuron and the dendrites, cell body, or axon of a second neuron. These are called, respectively, *axodendritic, axosomatic,* and *axoaxonic synapses* (fig. 14.18). In almost all synapses, transmission is in one direction only—from the axon of the first (or **presynaptic**) neuron to the second (or **postsynaptic**) neuron. Most commonly, the synapse

Figure 14.18

Different types of synapses.
Depicted here are (*a*) axodendritic, (*b*) axoaxonic, (*c*) dendrodendritic, and (*d*) axosomatic synapses.

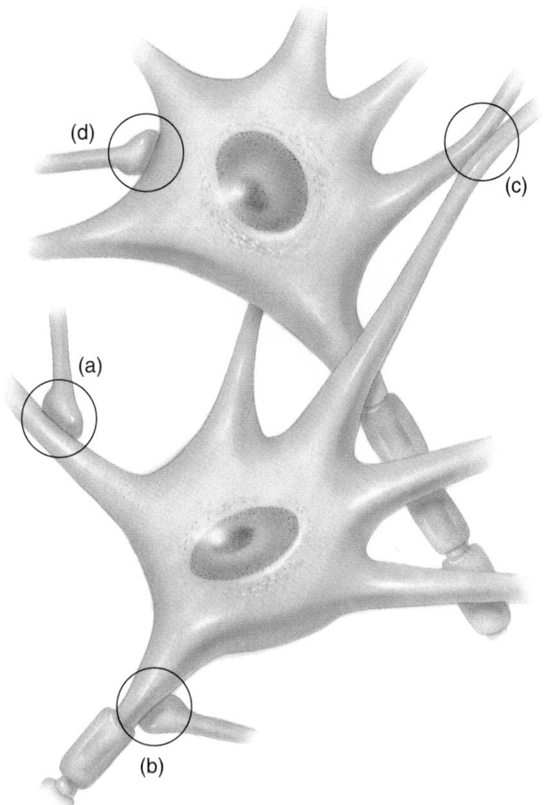

occurs between the axon of the presynaptic neuron and the dendrites or cell body of the postsynaptic neuron.

In the early part of the twentieth century most physiologists believed that synaptic transmission was *electrical*—that is, that action potentials were conducted directly from one cell to the next. This was a logical assumption given that nerve endings appeared to touch the postsynaptic cells and that the delay in synaptic conduction was extremely short (about 0.5 msec). Improved histological techniques, however, revealed tiny gaps in the synapses, and experiments demonstrated that the actions of autonomic nerves could be duplicated by certain chemicals. This led to the hypothesis that synaptic transmission might be *chemical*—that the presynaptic nerve endings might release chemicals called **neurotransmitters** that stimulated action potentials in the postsynaptic cells.

In 1921, a pharmacologist named Otto Loewi published the results of experiments suggesting that synaptic transmission was indeed chemical, at least at the junction between a branch of the vagus nerve (see chapter 16) and the heart. He had isolated the heart of a frog and, while stimulating the branch of the vagus that innervates the heart, perfused the heart with an isotonic salt solution. Stimulation of this nerve slowed the heart rate, as expected. More importantly, application of this salt solution to the heart of a second frog caused the second heart to also slow its rate of beat.

Loewi concluded that the nerve endings of the vagus must have released a chemical—which he called *vagusstoff*—that inhibited the heart rate. This chemical was subsequently identified as **acetylcholine**(*a-sēt"l-ko'lēn*), or **ACh.** In the decades following Loewi's discovery many other examples of chemical synapses were discovered, and the theory of electrical synaptic transmission fell into disrepute. More recent evidence, ironically, has shown that electrical synapses do exist in the nervous system (though they are the exception), within smooth muscles, and between cardiac cells in the heart.

Electrical Synapses: Gap Junctions

In order for two cells to be electrically coupled, they must be approximately equal in size and they must be joined by areas of contact with low electrical resistance. In this way the impulses can be regenerated from one cell to the next without interruption. Adjacent cells that are electrically coupled are joined together by **gap junctions.** In gap junctions, the membranes of the two cells are separated by only 2 nanometers (1 nanometer = 10^{-9} meter). A surface view of gap junctions in the electron microscope reveals hexagonal arrays of particles that are believed to be channels through which ions and molecules may pass from one cell to the next (fig. 14.19).

Gap junctions are present in cardiac muscle and some smooth muscles, where they allow excitation and rhythmic contraction of large masses of muscle cells. Gap junctions have also been observed in various regions of the brain. Although their functional significance in the brain is unknown, it has been speculated that they may allow a two-way transmission of impulses (in contrast to chemical synapses, which are always one-way). Gap junctions have also been observed between glial cells; these may act as channels for the passage of informational molecules between cells. It is interesting in this regard that gap junctions are present in many embryonic tissues, and that these gap junctions disappear as the tissue becomes more specialized.

Chemical Synapses

Transmission across the majority of synapses in the nervous system is one-way and occurs through the release of chemical neurotransmitters from presynaptic axon endings. These presynaptic endings, called **axon terminals,** also called *terminal boutons* (*boo-tonz'*) because of their swollen appearance, are separated from the postsynaptic cell by a **synaptic cleft** so narrow that it can be seen clearly only with an electron microscope (fig. 14.20).

Neurotransmitter molecules within the presynaptic neuron endings are contained within many small, membrane-enclosed **synaptic vesicles.** In order for the neurotransmitter within these vesicles to be released into the synaptic cleft, the vesicle membrane must fuse with the axon membrane in the process of *exocytosis* (chapter 3). The neurotransmitter is released in multiples of the amount contained in one vesicle,

Otto Loewi, Austrian pharmacologist (1873–1961)

Figure 14.19

The structure of gap junctions.

(*a*) An electron micrograph showing a gap junction. (*b*) An illustration of a gap junction, with the arrow indicating a channel through which ions and molecules may pass. Gap junctions function as electrical synapses.

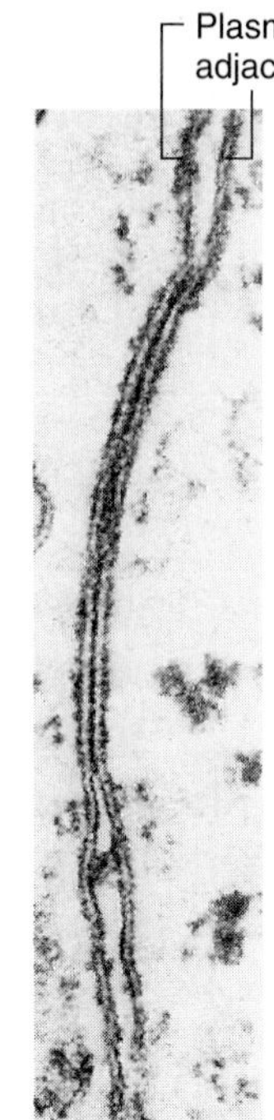

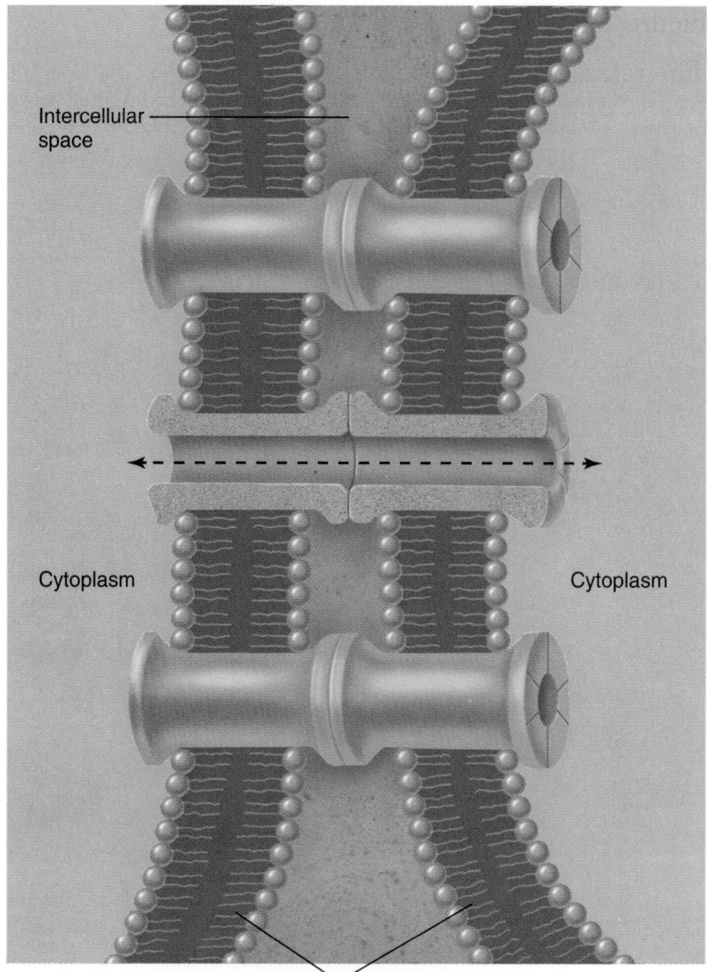

Figure 14.20

Electron micrograph of a chemically transmitting synapse.

This is the synapse between an axon of a somatic motor neuron and a skeletal muscle cell, showing synaptic vesicles at the end of the axon and the synaptic cleft. The synaptic vesicles contain the neurotransmitter chemical.

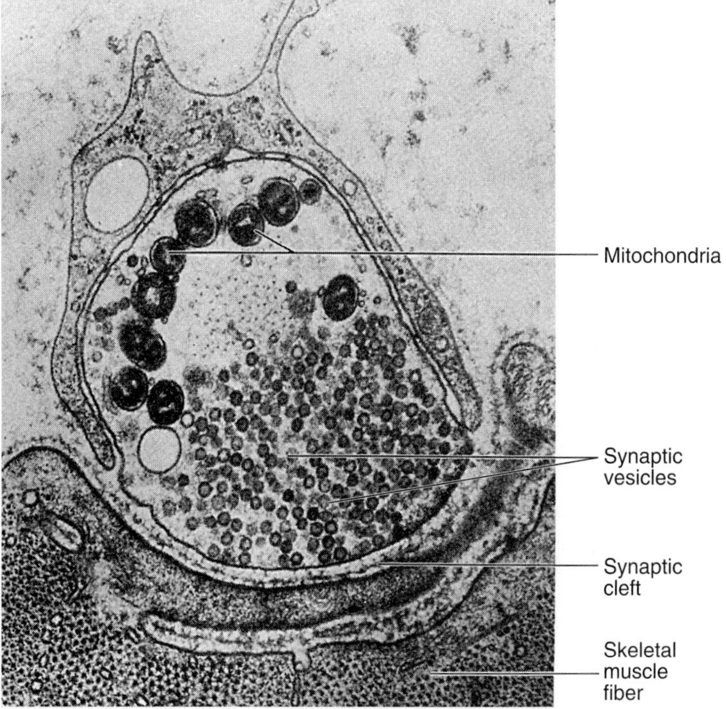

Figure 14.21

The release of neurotransmitter.
Action potentials, by opening Ca^{2+} channels, stimulate the fusion of docked synaptic vesicles with the cell membrane of the axon terminals. This leads to exocytosis and the release of neurotransmitter. The activation of protein kinase by Ca^{2+} may also contribute to this process.

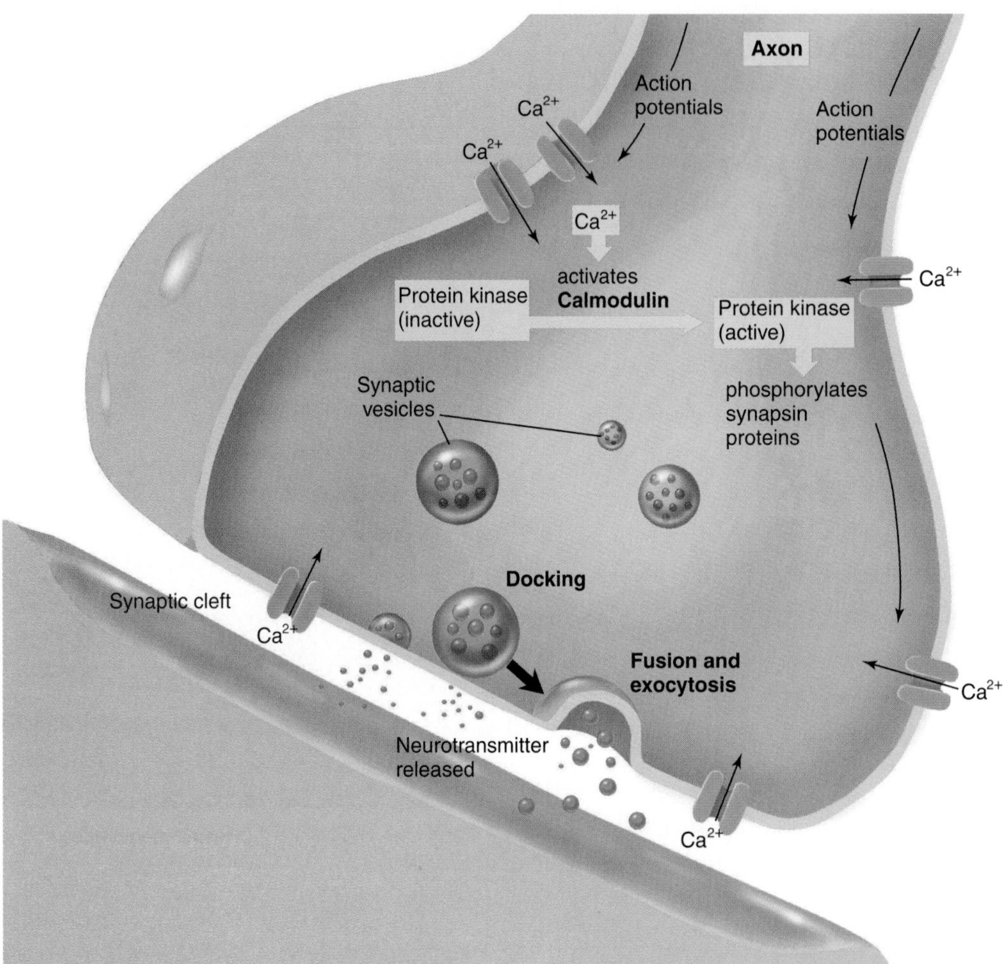

and the number of vesicles that undergo exocytosis depends on the frequency of action potentials produced at the presynaptic axon ending. Therefore, when stimulation of the presynaptic axon is increased, more of its vesicles will release their neurotransmitters to more greatly stimulate the postsynaptic cell.

Action potentials that arrive at the axon terminal trigger the release of neurotransmitter quite rapidly. The release is rapid because many synaptic vesicles are already "docked" at the correct areas of the presynaptic membrane before the arrival of the action potentials. At these docking sites, the vesicles are attached by proteins that form a *fusion complex* associated with the presynaptic membrane. The fusion complex attaches the vesicle to the docking site, but actual fusion of the vesicle membrane and the axon membrane is prevented until the arrival of action potentials.

Tetanus toxin and *botulinum toxin* are bacterial products that cause paralysis by preventing neurotransmission. These neurotoxins function as *proteases* (protein-digesting enzymes), digesting particular components of the fusion complex and thereby inhibiting the exocytosis of synaptic vesicles and preventing the release of neurotransmitter. Botulinum toxin prevents the release of ACh, causing flaccid paralysis; tetanus toxin blocks inhibitory synapses (discussed later), causing spastic paralysis.

Voltage-regulated calcium (Ca^{2+}) channels are located in the axon terminal adjacent to the docking sites. The arrival of action potentials at the axon terminal opens these voltage-regulated calcium channels, and it is the inward diffusion of Ca^{2+} that triggers the rapid fusion of the synaptic vesicle with the axon membrane and the release of neurotransmitter through exocytosis (fig. 14.21).

In addition, Ca^{2+} diffusing into the axon terminal activates a regulatory protein within the cytoplasm known as **calmodulin,** (*kal-mod'ŭ-lin*), which in turn activates an enzyme called **protein kinase.** This enzyme phosphorylates (adds a phosphate group to) specific proteins known as *synapsins* in the membrane of the synaptic vesicle. This action may influence the amount of neurotransmitter released when the neuron is repetitively stimulated. The Ca^{2+}–calmodulin–protein kinase regulatory mechanism is also important in the action of some hormones, and is therefore discussed in more detail in chapter 19.

Once the neurotransmitter molecules have been released from the presynaptic axon terminals, they diffuse rapidly across the synaptic cleft and reach the membrane of the postsynaptic cell. The neurotransmitters then bind to specific **receptor proteins** that are located within the postsynaptic membrane. Receptor proteins have high specificity for their neurotransmitter, which is the **ligand** (*li'gand* or *lig'and*) of the receptor protein. The term *ligand* in this case refers to a smaller molecule (the neurotransmitter) that binds to and forms a complex with a larger protein molecule (the receptor). Binding of the neurotransmitter ligand to its receptor protein causes ion channels to open in the postsynaptic membrane. The gates that regulate these channels, therefore, can be called **chemically regulated** (or **ligand regulated**) **gates** because they open in response to chemical changes in the postsynaptic cell membrane.

Note that two broad categories of gated ion channels have been described: *voltage-regulated* and *chemically regulated*. Voltage-regulated channels are found primarily in the axons; chemically regulated channels are found in the postsynaptic membrane. Voltage-regulated channels open in response to depolarization; chemically regulated channels open in response to the binding of postsynaptic receptor proteins to their neurotransmitter ligands.

The chemically regulated channels are opened by a number of different mechanisms, and the effects of opening these channels vary. Opening of ion channels often produces a depolarization—the inside of the postsynaptic membrane becomes less negative. This depolarization is called an **excitatory postsynaptic potential (EPSP)** because the membrane potential moves closer toward threshold. In other cases, a hyperpolarization occurs—the inside of the postsynaptic membrane becomes more negative. This hyperpolarization is called an **inhibitory postsynaptic potential (IPSP)** because the membrane potential moves farther from threshold. The mechanisms by which EPSPs and IPSPs are produced will be described in the following sections that deal with different types of neurotransmitters.

Excitatory postsynaptic potentials, as their name implies, stimulate the postsynaptic cell to produce action potentials, and inhibitory postsynaptic potentials antagonize this effect. In synapses between the axon of one neuron and the dendrites of another, the EPSPs and IPSPs are produced at the dendrites and must propagate to the initial segment of the axon to stimulate (or inhibit) action potentials (fig. 14.22). Once the first action potentials are produced, they will regenerate themselves along the axon as previously described.

Figure 14.22

The functional specialization of different regions in a multipolar neuron.
Integration of input (EPSPs and IPSPs) generally occurs in the dendrites and cell body, whereas the axon is specialized for the conduction of impulses (action potentials).

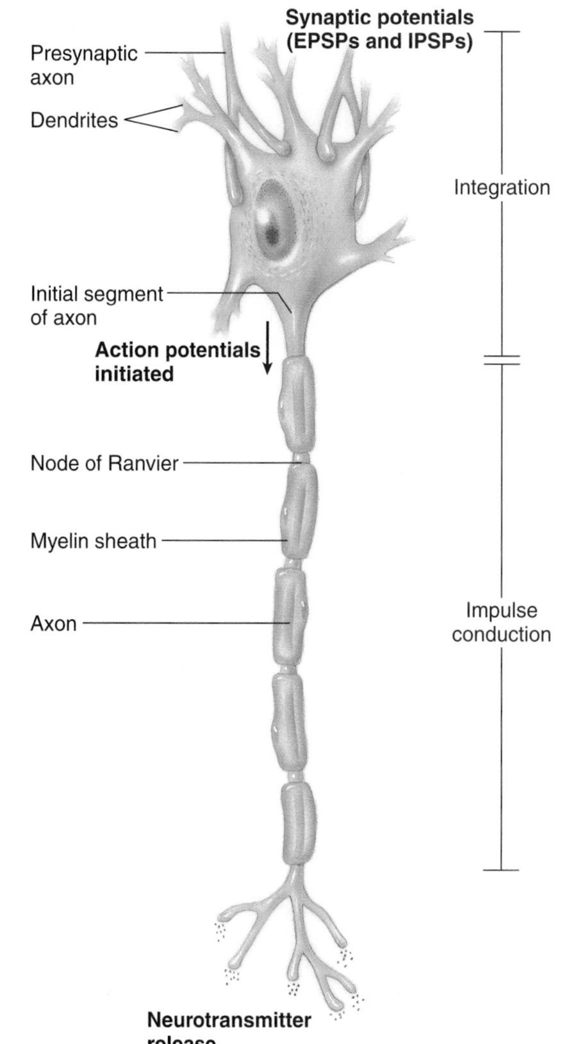

Acetylcholine as a Neurotransmitter

When acetylcholine (ACh) binds to its receptor it directly or indirectly causes the opening of chemically regulated gates. In most cases this produces a depolarization called an excitatory postsynaptic potential, or EPSP. In some cases, however, ACh causes a hyperpolarization known as an inhibitory postsynaptic potential, or IPSP.

Acetylcholine (ACh) is used as an excitatory neurotransmitter by some neurons in the CNS and by somatic motor neurons at the neuromuscular junction. At autonomic nerve endings, ACh may be either excitatory or inhibitory, depending on the organ involved.

The varying responses of postsynaptic cells to the same chemical can be explained, in part, by the fact that different postsynaptic cells have different subtypes of ACh receptors. These receptor subtypes can be specifically stimulated by particular toxins, and are named for these toxins. The stimulatory effect of ACh on skeletal muscle cells is produced by the binding of ACh to **nicotinic** (*nik″ŏ-tin′ik*) **ACh receptors,** so-named because they can also be activated by nicotine. Effects of ACh on other cells are produced when ACh binds to receptors called **muscarinic** (*mus″kă-rin′ik*) **ACh receptors,** because these effects can also be produced by muscarine (derived from poisonous mushrooms).

Chemically Regulated Gated Channels

The binding of a neurotransmitter to its receptor protein can cause the opening of ion channels through two different mechanisms. These two mechanisms can be illustrated by the actions of ACh on the nicotinic and muscarinic subtypes of the ACh receptors.

Ligand-Operated Channels

This is the most direct mechanism by which chemically regulated gates can be opened. In this case, the ion channel runs through the receptor itself. The ion channel is opened by the binding of the receptor to the neurotransmitter ligand.

This is what happens when ACh binds to its nicotinic ACh receptor. This receptor consists of five polypeptide subunits that enclose the ion channel. Two of these subunits contain ACh binding sites, and the channel opens when both sites are bound to ACh (fig. 14.23). The opening of this channel permits the inward diffusion of Na^+, which produces a depolarization in the postsynaptic membrane. This depolarization serves as an excitatory postsynaptic potential (EPSP), as will be described in a separate section.

G-Protein–Operated Channels

The muscarinic ACh receptors are formed from only a single subunit that can bind to one ACh molecule. Unlike the nicotinic receptors, these receptors do not contain ion channels. The ion channels are separate proteins located at some distance from the muscarinic receptors. Binding of ACh (the ligand) to the muscarinic receptor causes it to activate a complex of proteins in the cell membrane known as **G-proteins**—so-named because their activity is influenced by guanosine nucleotides (GDP and GTP).

There are three G-protein subunits, designated alpha, beta, and gamma. In response to the binding of ACh to its receptor, the alpha subunit dissociates from the other two subunits, which stay together to form a beta-gamma complex. Depending on the specific case, either the alpha subunit or the beta-gamma complex then diffuses through the membrane

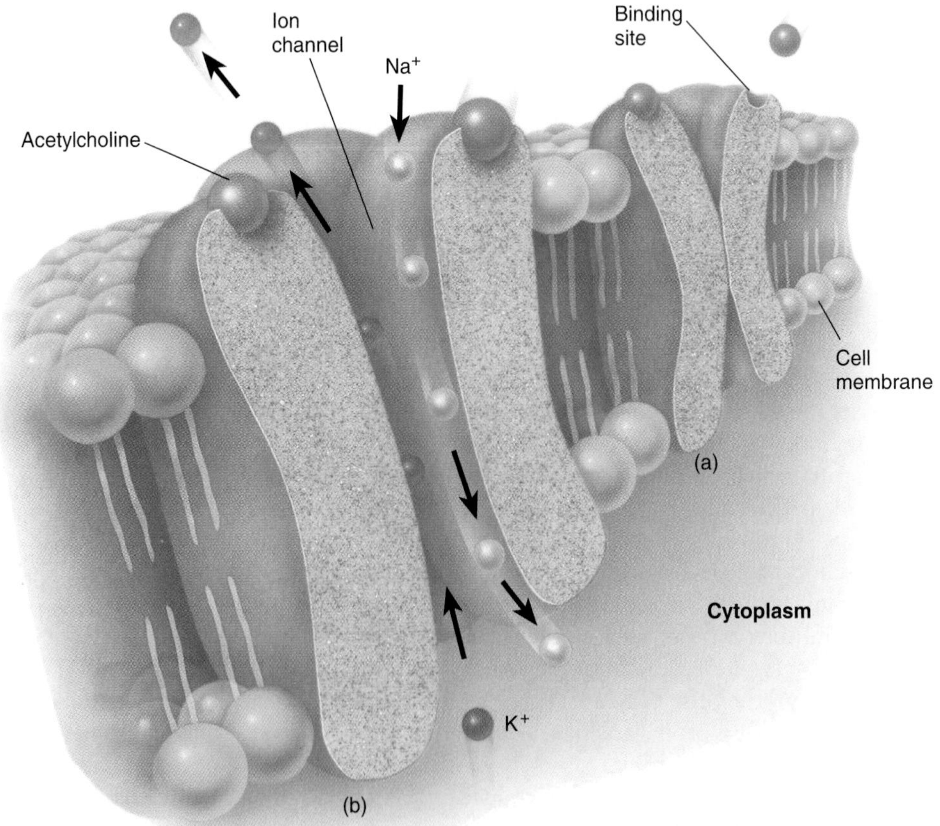

Figure 14.23

Nicotinic acetylcholine (ACh) receptors also function as ion channels.
The nicotinic acetylcholine receptor contains a channel that is closed (*a*) until the receptor binds to ACh. (*b*) Na^+ and K^+ diffuse simultaneously, and in opposite directions, through the open ion channel.

until it binds to an ion channel, causing the channel to open (fig. 14.24). A short time later, the G-protein alpha subunit (or beta-gamma complex) dissociates from the channel and moves back to its previous position. This causes the ion channel to close.

Scientists have recently learned that it is the beta-gamma complex that binds to the K^+ channels in the heart muscle cells and causes these channels to open (fig. 14.24). This leads to the diffusion of K^+ out of the postsynaptic cell (in the direction of its concentration gradient). As a result the cell becomes hyperpolarized, producing an inhibitory postsynaptic potential (IPSP). Such an effect is produced in the heart, for example, when autonomic nerve fibers (part of the vagus nerve) synapse with pacemaker cells and slow the rate of beat. It should be noted that inhibition also occurs in the CNS in response to other neurotransmitters, but those IPSPs are produced by a different mechanism that will be described in later sections.

There are cases in which the alpha subunit is the effector, and examples where its effects are substantially different from that shown in figure 14.24. In the smooth muscle cells of the stomach, the binding of ACh to its muscarinic receptors causes a different type of G-protein alpha subunit to dissociate and bind to the K^+ channels. In this case, however, the binding of the G-protein subunit to the K^+ channels causes the channels to close rather than to open. As a result, the outward diffusion of K^+, which occurs at an ongoing rate in the resting cell, is reduced to below resting levels. Since the resting membrane potential is maintained by a balance between cations flowing into the cell and cations flowing out, a reduction in the outward flow of K^+ produces a depolarization. This depolarization produced in smooth muscle cells results in contractions of the stomach (chapter 12).

Acetylcholinesterase (AChE)

The bond between ACh and its receptor protein exists for only a brief instant. The ACh-receptor complex quickly dissociates but can be quickly re-formed as long as free ACh is in the vicinity. In order for activity in the postsynaptic cell to be controlled, free ACh must be inactivated very soon after it is released. The inactivation of ACh is achieved by means of an enzyme called **acetylcholinesterase** (*ă-sēt″l-ko″lĭ-nes′tĕ-rās*), or **AChE,** which is present on the postsynaptic membrane or immediately outside the membrane with its active site facing the synaptic cleft (fig. 14.25).

Nerve gas exerts its odious effects by inhibiting AChE in skeletal muscles. Since ACh is not degraded, it can continue to combine with receptor proteins and can continue to stimulate the postsynaptic cell, leading to spastic paralysis. Clinically, cholinesterase inhibitors (such as neostigmine) are used to enhance the effects of ACh on muscle contraction when neuromuscular transmission is weak, as in the disease myasthenia gravis.

Excitatory Postsynaptic Potentials (EPSPs)

Binding of ACh to the nicotinic receptors causes opening of channels that permit the simultaneous diffusion of Na^+ and K^+ into and out of the postsynaptic cell, respectively, through the same membrane channel (see fig. 14.23). The two ion

Figure 14.24

Muscarinic ACh receptors require the mediation of G-proteins.
The figure depicts the effects of ACh on the pacemaker cells of the heart. Binding of ACh to its muscarinic receptor causes the beta-gamma subunits to dissociate from the alpha subunit. The beta-gamma complex of G-proteins then binds to a K^+ channel, causing it to open. Outward diffusion of K^+ results, slowing the heart rate.

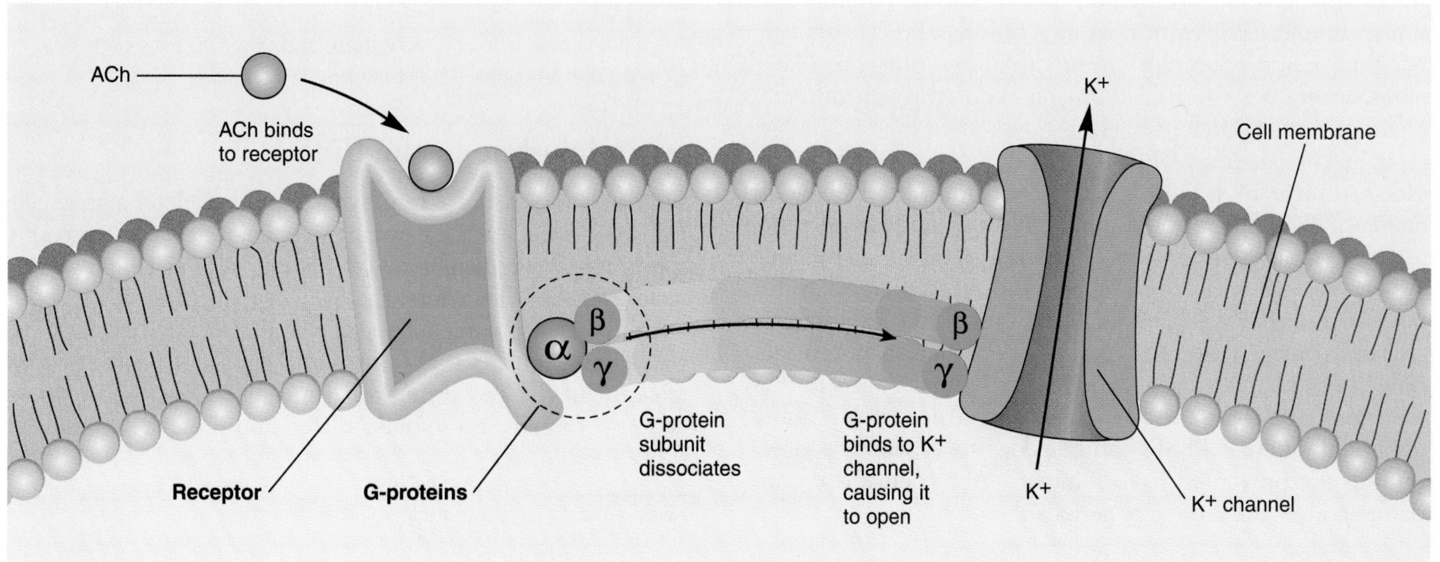

Figure 14.25

The action of acetylcholinesterase (AChE).
The AChE in the postsynaptic cell membrane inactivates the ACh released into the synaptic cleft. This prevents continued stimulation of the postsynaptic cell unless more ACh is released by the axon.

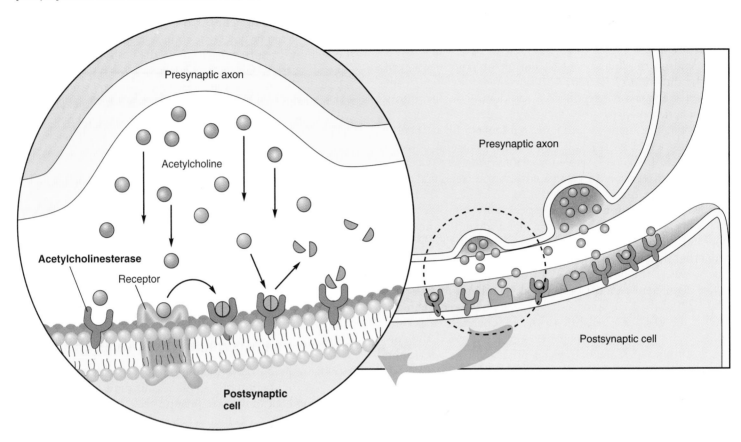

movements do not cancel each other out, however. The inward Na^+ diffusion predominates because of its greater electrochemical gradient, and a net depolarization is produced. The outflow of K^+ does, however, prevent the depolarization from overshooting 0 mV. Therefore, the membrane polarity does not reverse in EPSPs as it does in action potentials. (Remember that action potentials are produced by voltage-regulated gates, where Na^+ and K^+ diffusion occur through different channels and at different stages.)

A comparison of EPSPs and action potentials is provided in table 14.4. Unlike action potentials, EPSPs have *no threshold;* the ACh released from a single synaptic vesicle produces a tiny depolarization of the postsynaptic membrane. When more vesicles are stimulated to release their ACh, the depolarization is correspondingly greater. EPSPs are therefore *graded* in magnitude, unlike all-or-none action potentials. Since EPSPs can be graded, and have *no refractory period,* they are capable of *summation.* This means that the depolarizations of several different EPSPs can be added together. Action potentials are prevented from summating by their all-or-none nature and by the refractory periods they exhibit.

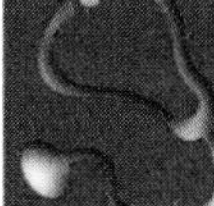

Muscle weakness in the disease *myasthenia gravis* is due to the fact that ACh receptors are blocked and destroyed by antibodies secreted by the immune system of the affected person. Paralysis in people who eat shellfish poisoned with saxitoxin, or pufferfish containing tetrodotoxin, results from the blockage of Na^+ gates. The effects of these and other poisons on neuromuscular transmission are summarized in table 14.5.

Acetylcholine in the PNS

Somatic motor neurons form synapses with skeletal muscle cells (muscle fibers). At these synapses, or neuromuscular junctions, the postsynaptic membrane of the muscle fiber is known as a *motor end plate.* Therefore, the EPSPs produced by ACh acting at this postsynaptic membrane are often called **end-plate potentials.** This depolarization opens voltage-regulated gates that are adjacent to the end plate. Voltage-regulated channels produce action potentials in the muscle fiber, and these are reproduced

myasthenia: Gk. *myos,* muscle; *asthenia,* weakness

Table 14.4

Comparison of Action Potentials with Excitatory Postsynaptic Potentials (EPSPs)

Characteristic	Action Potential	Excitatory Postsynaptic Potential
Stimulus for opening of ionic gates	Depolarization	Acetylcholine (ACh)
Initial effect of stimulus	Na^+ gates open	Na^+ and K^+ gates open
Cause of repolarization	Opening of K^+ gates	Loss of intracellular positive charges with time and distance
Conduction distance	Not conducted—regenerated over length of axon	1–2 mm; a localized potential
Positive feedback between depolarization and opening of Na^+ gates	Yes	No
Maximum depolarization	+40 mV	Close to zero
Summation	No summation—all-or-none phenomenon	Summation of EPSPs, producing graded depolarizations
Refractory period	Yes	No
Effect of drugs	Inhibited by tetrodotoxin, not by curare	Inhibited by curare, not by tetrodotoxin

Table 14.5

Drugs That Affect the Neural Control of Skeletal Muscles

Drug	Derivation	Effect
Botulinus toxin	Produced by *Clostridium botulinum* (bacteria)	Inhibits release of acetylcholine (ACh)
Curare	Resin from a South American tree	Prevents interaction of ACh with the postsynaptic receptor protein
α-Bungarotoxin	Venom of *Bungarus* snakes	Binds to ACh receptor proteins
Saxitoxin	Red tide (*Gonyaulax*) protozoa	Blocks voltage-regulated Na^+ channels
Tetrodotoxin	Pufferfish	Blocks voltage-regulated Na^+ channels
Nerve gas	Artificial	Inhibits acetylcholinesterase in postsynaptic cell
Prostigmine	Nigerian bean	Inhibits acetylcholinesterase in postsynaptic cell
Strychnine	Seeds of an Asian tree	Prevents IPSPs in spinal cord that inhibit contraction of antagonistic muscles

by other voltage-regulated channels along the muscle cell membrane. This conduction is analogous to that of action potentials by axons; it is significant because action potentials in muscle fibers stimulate muscle contraction (as described in chapter 12).

If any stage in the process of neuromuscular transmission is blocked, muscle weakness—sometimes leading to paralysis and death—may result. The drug *curare* (*koo-ră're*), for example, competes with ACh for attachment to the nicotinic ACh receptors and thus reduces the size of the depolarizations (table 14.5). This drug was first used on blow-gun darts by South American Indians because it produced flaccid paralysis in their victims. Clinically, curare is used in surgery as a muscle relaxant and in electroconvulsive shock therapy to prevent muscle damage.

Autonomic motor neurons innervate cardiac muscle, smooth muscles in blood vessels and visceral organs, and glands. As previously mentioned, there are two classifications of autonomic nerves: sympathetic and parasympathetic. Most of the parasympathetic axons that innervate the effector organs use ACh as their neurotransmitter. In some cases, these axons have an inhibitory effect on the organs they innervate through the binding of ACh to muscarinic ACh receptors. The action of the vagus nerve in slowing the heart rate is an example of this inhibitory effect. In other cases, ACh released by autonomic neurons produces stimulatory effects as previously described. The structures and functions of the autonomic system are described in chapter 17.

Figure 14.26

Excitatory postsynaptic potentials (EPSPs) are graded.
This means that stimuli of increasing strength produce increasing amounts of depolarization. At a threshold level of depolarization, action potentials are generated in the axon.

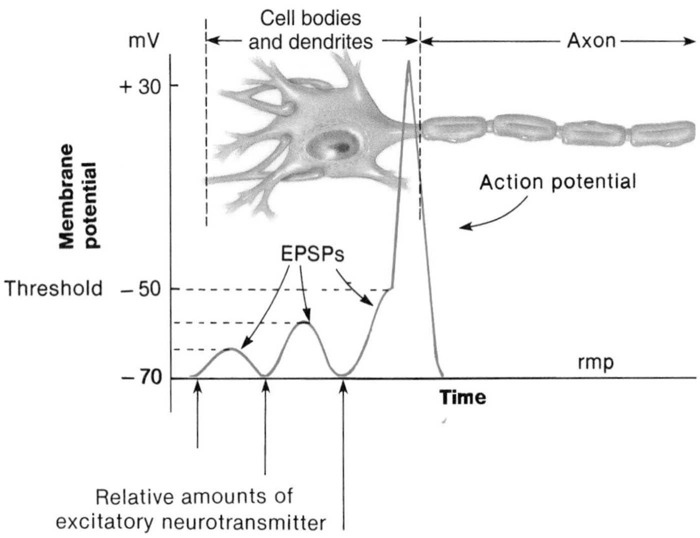

Acetylcholine in the CNS

There are many **cholinergic** (*ko″lĭ-ner′jik*) **neurons** (those that use ACh as a neurotransmitter) in the CNS, where the axon terminals of one neuron typically synapse with the dendrites or cell body of another. The dendrites and cell body thus serve as the receptive area of the neuron, and it is in these regions that receptor proteins for neurotransmitters and chemically regulated gated channels are located. Voltage-regulated gated channels are located at the *axon hillock,* a cone-shaped elevation on the cell body from which the axon arises. The *initial segment* of the axon, which is the unmyelinated region of the axon around the axon hillock, has a high concentration of voltage-regulated gated channels. It is here that action potentials are first produced (see fig. 14.22).

Depolarizations—EPSPs—in the dendrites and cell body spread by cable properties to the initial segment of the axon in order to stimulate action potentials. If the depolarization is at or above threshold by the time it reaches the initial segment of the axon, the EPSP will stimulate the production of action potentials, which can then regenerate themselves along the axon. If, however, the EPSP is below threshold at the initial segment, no action potentials will be produced in the postsynaptic cell (fig. 14.26). Gradations in the strength of the EPSP above threshold determine the frequency with which action potentials will be produced at the axon hillock, and at each point in the axon where the impulse is conducted. The action potentials that begin at the initial segment of the axon are conducted without loss of amplitude towards the axon terminals.

Earlier in this chapter, the action potential was introduced by describing the events that occurred when a depolarization stimulus was artificially produced by stimulating electrodes. Now it is apparent that EPSPs, conducted from the dendrites and cell body, serve as the normal stimuli for the production of action potentials at the axon hillock, and that the action potentials at this point serve as the depolarization stimulus for the next region, and so on. This chain of events ends at the axon terminal where neurotransmitter is released.

Monoamines as Neurotransmitters

A variety of chemicals in the CNS function as neurotransmitters. Among these are the monoamines, a chemical family that includes dopamine, norepinephrine, and serotonin. Although these molecules have similar mechanisms of action, they are used by different neurons for different functions.

The regulatory molecules epinephrine, norepinephrine, dopamine, and serotonin are in the chemical family known as **monoamines.** Serotonin is derived from the amino acid tryptophan. Epinephrine, norepinephrine, and dopamine are derived from the amino acid tyrosine and form a subfamily of monoamines called the **catecholamines** (*kat′ĕ-kol′ă-mēnz*). Epinephrine is not a neurotransmitter; it is the hormone widely known as adrenalin, secreted by the adrenal glands. Norepinephrine is likewise secreted as a hormone, but it also functions as a neurotransmitter. Dopamine functions exclusively as a neurotransmitter.

Like ACh, monoamine neurotransmitters are released by exocytosis from presynaptic vesicles, diffuse across the synaptic cleft, and interact with specific receptor proteins in the membrane of the postsynaptic cell. The stimulatory effects of these monoamines, like those of ACh, must be quickly inhibited so as to maintain proper neural control. The inhibition of monoamine action is due to (1) **reuptake** of monoamines into the presynaptic neuron endings, (2) enzymatic degradation of monoamines in the presynaptic neuron endings by *monoamine oxidase* (MAO), and (3) the enzymatic degradation of catecholamines in the postsynaptic neuron by *catechol-O-methyltransferase* (COMT). This process is illustrated in figure 14.27. Drugs that inhibit MAO thus promote the effects of monoamine action (see the boxed information on page 396).

The monoamine neurotransmitters do not directly cause opening of ion channels in the postsynaptic membrane. Instead, these neurotransmitters act by means of an intermediate regulator known as a **second messenger.** In the case of some synapses that use catecholamines for synaptic transmission, this second messenger is a compound known as **cyclic adenosine monophosphate (cAMP).** Although other synapses can use other second messengers, only the function of cAMP as a second messenger will be considered here.

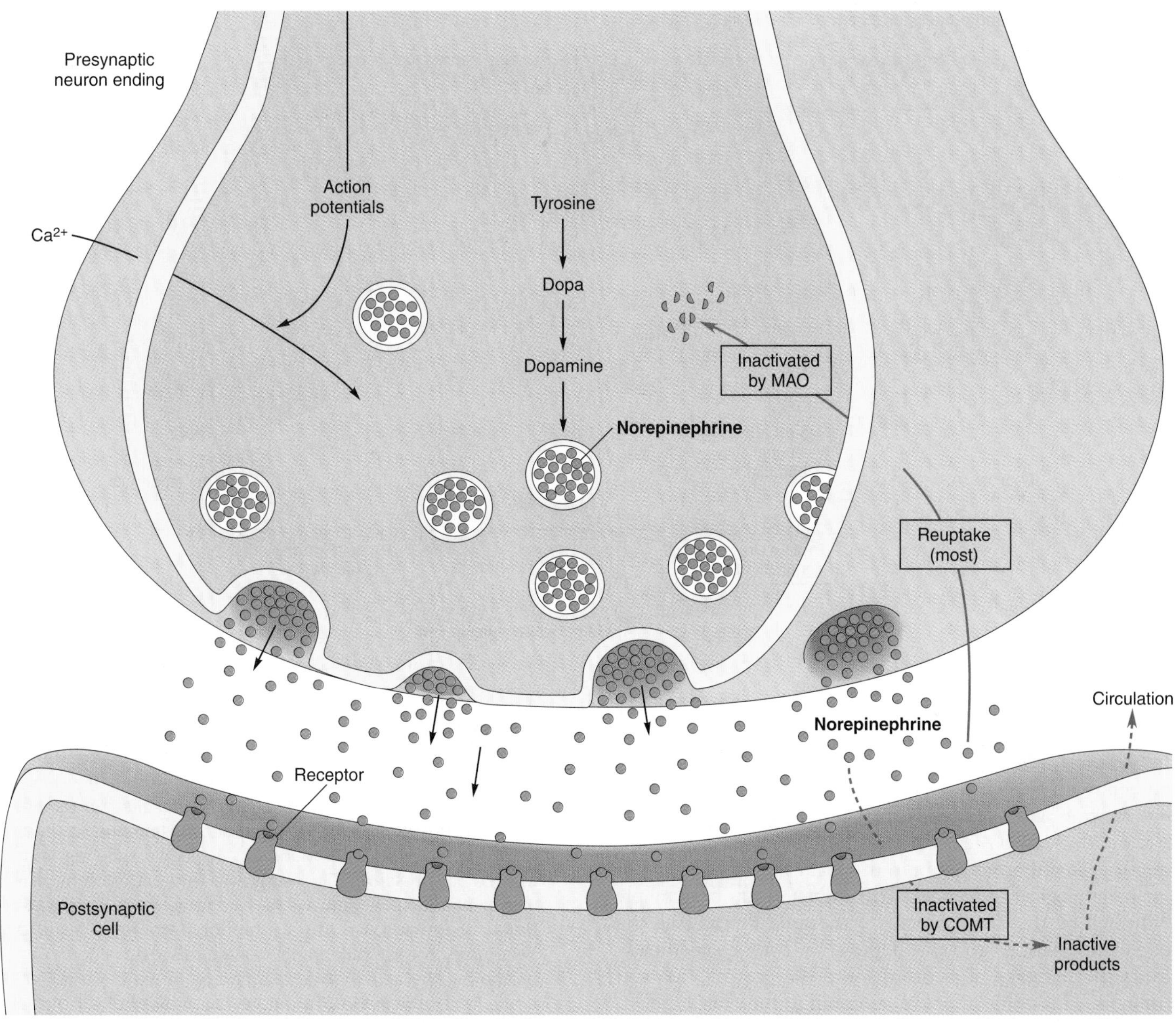

Figure 14.27

The production, release, and reuptake of catecholamine neurotransmitters. The transmitters combine with receptor proteins in the postsynaptic membrane. (COMT = catechol-O-methyltransferase; MAO = monoamine oxidase.)

Other second-messenger systems are discussed in conjunction with hormone action in chapter 19.

Binding of norepinephrine, for example, to its receptor in the postsynaptic membrane stimulates the dissociation of the G-protein alpha subunit from the others in its complex (fig. 14.28). This subunit diffuses in the membrane until it binds to an enzyme known as *adenylate cyclase* (also called *adenylyl cyclase*). This enzyme converts ATP to cyclic AMP (cAMP) and pyrophosphate (two inorganic phosphates) within the postsynaptic cell cytoplasm. Cyclic AMP in turn activates another enzyme, *protein kinase,* which phosphorylates (adds a phosphate group to) other proteins (fig. 14.28). It is through this action that ion channels are opened in the postsynaptic membrane.

Serotonin as a Neurotransmitter

Serotonin (*ser″ŏ-to′nin*), or *5-hydroxytryptamine,* is used as a neurotransmitter by neurons with cell bodies in the *raphe* (*ra′fe*) *nuclei* that are located along the midline of the brain stem (see chapter 15). Serotonin is derived from the amino acid L-tryptophan, and variations in the amount of this amino acid in the diet (tryptophan-rich foods include milk and turkey) can affect the amount of serotonin

Figure 14.28

Norepinephrine action requires G-proteins.
The binding of norepinephrine to its receptor (*1*) causes the dissociation of G-proteins (*2*). Binding of the alpha G-protein subunit to the enzyme adenylate cyclase (*3*) activates this enzyme, leading to the production of cyclic AMP (*4*). Cyclic AMP, in turn, activates protein kinase (*5*), which can open ion channels (*6*) and produce other effects.

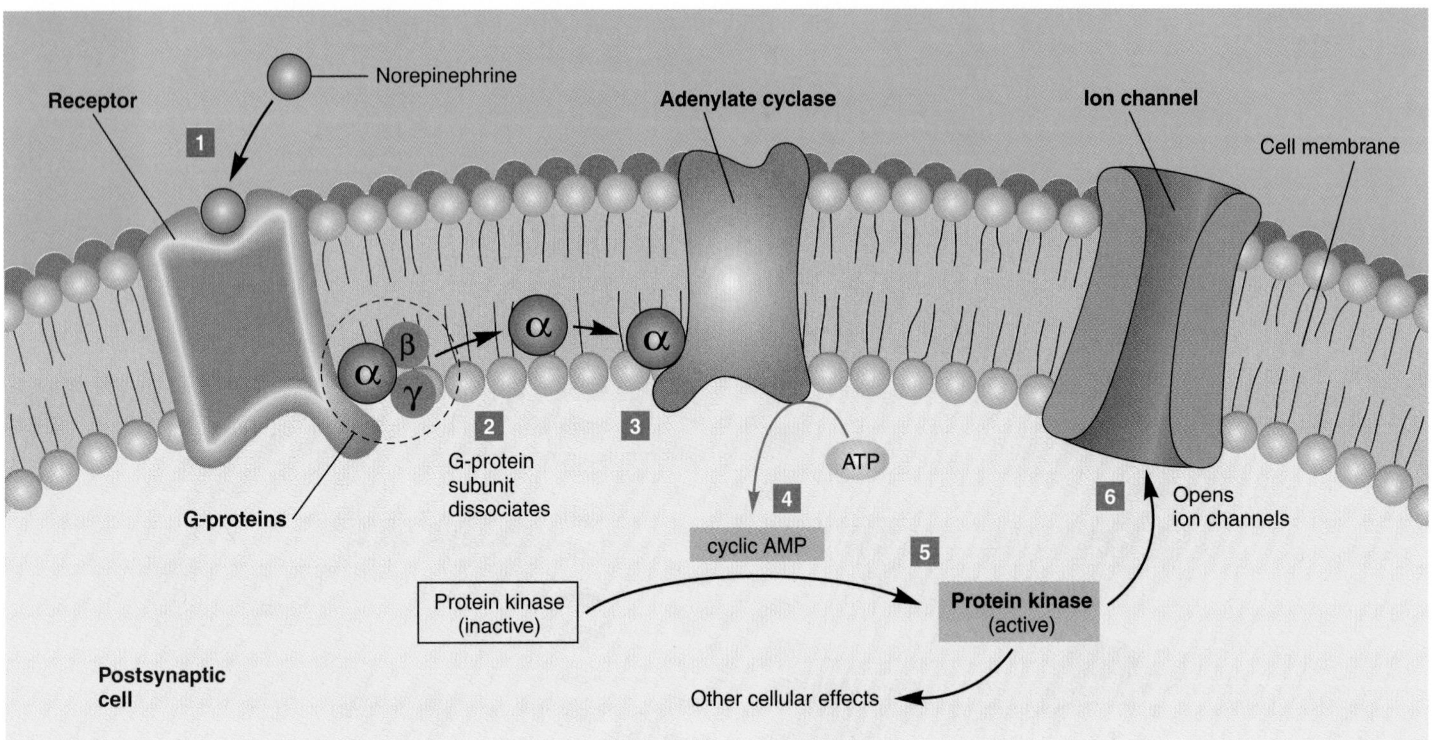

produced by these neurons. Physiological functions attributed to serotonin include a role in the regulation of mood and behavior, appetite, and cerebral circulation.

Since LSD (a powerful hallucinogen) mimics the structure, and thus likely the function, of serotonin, scientists have long suspected that serotonin in the brain should have important psychological influences. This suspicion has been amply confirmed by the development of the drug *Prozac* (see the boxed information) to treat depression. Prozac specifically blocks the reuptake of serotonin into the presynaptic axon terminals. This results in more serotonin in the synaptic cleft, which thereby stimulates those neural pathways that use serotonin as a neurotransmitter.

Serotonin's diverse functions are related, at least in part, to the finding that there are a large number of different subtypes of serotonin receptors—over a dozen are currently known. Thus, whereas Prozac may be given to relieve depression, another drug that promotes serotonin action is sometimes given to reduce the appetite of obese patients. A different drug that may activate a different serotonin receptor is used to treat anxiety, and yet another drug that promotes serotonin action is given to relieve migraine headaches. It should be noted that the other monoamine neurotransmitters, dopamine and norepinephrine, also influence mood and behavior in a way that complements the actions of serotonin.

Monoamine oxidase (MAO) is an enzyme in the endings of presynaptic axons that breaks down catecholamines and serotonin after they have been taken up from the synaptic cleft. Drugs that act as *MAO inhibitors* thus increase transmission at these synapses and have been found to aid people suffering from clinical depression. This suggests that a deficiency in monoamine neural pathways may contribute to severe emotional depression. An MAO inhibitor (*Deprenyl*) has also been used to treat Parkinson's disease (discussed shortly) by promoting the activity of dopamine as a neurotransmitter. Drugs that inhibit MAO promote the activity of all of the monoamines, and thus can produce undesired side effects. A newer drug, fluoxetine hydrochloride (Prozac), blocks the reuptake of serotonin into presynaptic axons. This drug thus promotes serotonin action, and it is because of this action that Prozac is effective in the treatment of depression. The widespread use of this drug, however, has engendered considerable controversy.

Dopamine as a Neurotransmitter

Neurons that use **dopamine** as a neurotransmitter are called **dopaminergic** (*do″pă-men-er′jik*) **neurons.** Neurons that have dopamine receptor proteins on the postsynaptic membrane, and therefore respond to dopamine, have been identified in postmortem brain tissue. More recently, the location of these receptors has been observed in the living brain using the technique of positron emission tomography (PET). These investigations have been spurred by the great clinical interest in the effects of dopaminergic neurons.

The cell bodies of dopaminergic neurons are highly concentrated in the midbrain. Their axons project to different parts of the brain and can be divided into two systems. The *nigrostriatal dopamine system* is involved in motor control and the *mesolimbic dopamine system* is involved in emotional reward.

Nigrostriatal Dopamine System

The cell bodies of the **nigrostriatal** (*ni″gro-stri-a′tal*) **dopamine system** are located in a part of the midbrain called the *substantia nigra* (*sub-stan′she-ă ni′gra*) ("dark substance") because it contains melanin pigment. Neurons in the substantia nigra send fibers to the caudate nucleus and putamen (also known as the corpus striatum because of its striped appearance—hence the term *nigrostriatal system*). These regions are centers for motor control that are part of the *basal nuclei*—large masses of cell bodies deep in the cerebrum involved in the coordination of skeletal movements (see chapter 15). There is much evidence that in **Parkinson's disease** there is degeneration of these dopaminergic neurons in the substantia nigra. Parkinson's disease is a major cause of neurological disability in people over the age of 60 and is associated with such symptoms as muscle tremors and rigidity, difficulty in initiating movements and speech, and other severe motor problems. These patients are treated with L-dopa to increase the production of dopamine in the brain, as described previously in this chapter.

The cause of the degeneration of dopaminergic neurons in Parkinson's disease is not well understood. Some scientists believe that neural destruction might be caused by free radicals (superoxide and nitric oxide), perhaps released by overactive microglia, that produce oxidative damage. Animal experiments demonstrate that the neurotrophin known as *glial-derived neurotrophic factor* (*GDNF*) acts to maintain the viability of dopaminergic neurons in the midbrain. Further, injections of GDNF into the brains of rodents and monkeys who were previously made to have a Parkinson's-like condition resulted in alleviation of the Parkinson's symptoms.

Mesolimbic Dopamine System

The **mesolimbic dopamine system** involves neurons that originate in the midbrain and send axons to structures in the forebrain that are part of the limbic system (structures in the forebrain involved in emotion—see fig. 15.12). The dopamine released by these neurons may be involved in behavior and reward. For example, several studies involving human twins separated at birth and reared in different environments, and other studies involving the use of rats, have implicated the gene that codes for one subtype of dopamine receptor (designated D_2) in alcoholism. Other addictive drugs, including cocaine, morphine, and amphetamines, are also known to activate dopaminergic pathways.

Recent studies demonstrate that alcohol, amphetamines, cocaine, marijuana, and morphine promote the activity of dopaminergic neurons that arise in the midbrain and terminate in a particular location, the *nucleus accumbens*, of the forebrain. Interestingly, nicotine has recently been shown to also promote the release of dopamine by axons that terminate in this very location. This suggests that the physiological mechanism for nicotine addiction in smokers is similar to that for other abused drugs.

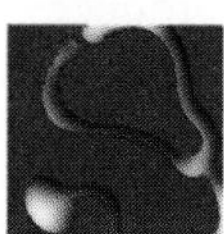

A side effect of L-dopa treatment in some patients with Parkinson's disease is the appearance of symptoms characteristic of *schizophrenia*. This effect is not surprising in view of the fact that the drugs used to treat schizophrenic patients (drugs called *neuroleptics,* and including chlorpromazine and haloperidol) act as antagonists of the D_2 subtype of dopamine receptors. As might be predicted from these observations, schizophrenic patients treated with these drugs often develop symptoms of Parkinson's disease. Based on this evidence, it seems that schizophrenia may be caused, at least in part, by overactivity of the mesolimbic dopaminergic system. This is supported by studies that show a link between schizophrenia and increased amounts of the D_2 dopamine receptors in the forebrain.

Norepinephrine as a Neurotransmitter

Norepinephrine (*nor″ep-ĭ-nef′rin*), like ACh, is used as a neurotransmitter in both the PNS and the CNS. Sympathetic neurons of the PNS use norepinephrine as a neurotransmitter at their synapse with smooth muscles, cardiac muscle, and glands. Some neurons in the CNS also use norepinephrine as a neurotransmitter; these neurons seem to be involved in general behavioral arousal. This would help to explain the mental arousal elicited by *amphetamines* (*am-fet′ă-mēnz*) and other such drugs that stimulate pathways in which norepinephrine is used as a neurotransmitter. Such drugs also stimulate the PNS pathways that use norepinephrine, however, and this duplicates the effects of sympathetic nerve activation. A rise in blood pressure, constriction of arteries, and other effects similar to the deleterious consequences of cocaine use can thereby be produced.

Other Neurotransmitters

A surprisingly large number of diverse molecules appear to function as neurotransmitters. These include some amino acids and their derivatives, many polypeptides, and even the gas nitric oxide.

Amino Acids as Neurotransmitters

The amino acids **glutamic acid** and **aspartic acid** function as excitatory neurotransmitters in the CNS. Glutamic acid (or *glutamate*) is indeed the major excitatory neurotransmitter in

---

Parkinson's disease: from James Parkinson, English physician, 1755–1824

the brain, producing excitatory postsynaptic potentials (EPSPs). Experiments using chemicals that differ slightly from glutamate but that mimic its actions have revealed different subtypes of glutamate receptors in brain neurons.

One subtype of glutamate receptor is named after the glutamate analogue *N-methyl-D-aspartate* (*NMDA*). The NMDA receptors for glutamate have been implicated in the physiology of memory and will be described in more detail in a later section. Interestingly, the NMDA receptors are a major site of action for the street drug known as angel dust.

The amino acid **glycine** is inhibitory; instead of depolarizing the postsynaptic membrane and producing an EPSP, it hyperpolarizes the postsynaptic membrane and produces an inhibitory postsynaptic potential (IPSP). The binding of glycine to its receptor proteins causes the opening of chloride (Cl^-) channels in the postsynaptic membrane. As a result, Cl^- diffuses into the postsynaptic neuron and produces the hyperpolarization. This inhibits the neuron by making the membrane potential even more negative than it is at rest, and therefore farther from the threshold depolarization required to stimulate action potentials.

The inhibitory effects of glycine are very important in the spinal cord, where they help in the control of skeletal movements. Flexion of an arm, for example, involves stimulation of the flexor muscles by motor neurons in the spinal cord. The motor neurons that innervate the antagonistic extensor muscles are inhibited by IPSPs produced by glycine released from other neurons. The importance of the inhibitory actions of glycine is revealed by the deadly effects of *strychnine* (*strik'nīn*), a poison that causes spastic paralysis by specifically blocking the glycine receptor proteins. Animals poisoned with strychnine die from asphyxiation because they are unable to relax the diaphragm.

The neurotransmitter **gamma-aminobutyric** (*ă-me"no-byoo-tir'ik*) **acid (GABA)** is a derivative of another amino acid, glutamic acid. GABA is the most prevalent neurotransmitter in the brain; in fact, as many as one-third of all the neurons in the brain use GABA as a neurotransmitter. Like glycine, GABA is inhibitory—it hyperpolarizes the postsynaptic membrane. Also, the effects of GABA, like those of glycine, are involved in motor control. For example, large *Purkinje* (*pur-kin'je*) *cells* mediate the motor functions of the cerebellum by producing IPSPs in their postsynaptic neurons. A deficiency of GABA-releasing neurons is responsible for the uncontrolled movements seen in people with *Huntington's chorea*.

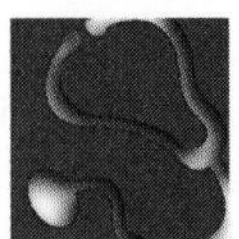

Benzodiazepines (*ben"zo-di-az'ĕ-pēnz*) are drugs that act to increase the ability of GABA to activate its receptors in the brain and spinal cord. Since GABA inhibits the activity of spinal motor neurons that innervate skeletal muscles, the intravenous infusion of benzodiazepines acts to inhibit the muscular spasms in epileptic seizures and seizures resulting from drug overdose and poisons. Probably as a result of its general inhibitory effects on the brain, GABA also functions as a neurotransmitter involved in mood and emotion. Benzodiazepines such as *Valium* are thus given orally as tranquilizers.

Polypeptides as Neurotransmitters

Many polypeptides of various sizes are found in the synapses of the brain. These are often called **neuropeptides** and are believed to function as neurotransmitters. Interestingly, some of the polypeptides that function as hormones secreted by the small intestine and other endocrine glands are also produced in the brain, where they may function as neurotransmitters (table 14.6). For example, *cholecystokinin* (*CCK*), which is secreted as a hormone from the small intestine, is also released from neurons and used as a neurotransmitter in the brain. Recent evidence suggests that CCK, acting as a neurotransmitter, may promote feelings of satiety in the brain following meals. Another polypeptide found in many organs, *substance P*, functions as a neurotransmitter in pathways in the brain that mediate sensations of pain.

Table 14.6

Examples of Chemicals That Are Either Proven or Supposed Neurotransmitters

Category	Chemicals
Amines	Acetylcholine
	Histamine
	Serotonin
Catecholamines	Dopamine
	Epinephrine
	Norepinephrine
Amino acids	Aspartic acid
	GABA (gamma-aminobutyric acid)
	Glutamic acid
	Glycine
Polypeptides	Glucagon
	Insulin
	Somatostatin
	Substance P
	ACTH (adrenocorticotrophic hormone)
	Angiotensin II
	Endogenous opioids (enkephalins and endorphins)
	LHRH (luteinizing hormone-releasing hormone)
	TRH (thyrotrophin-releasing hormone)
	Vasopressin (antidiuretic hormone)
	CCK (cholecystokinin)
Gases	Nitric oxide
	Carbon monoxide

Synaptic Plasticity

Although some of the polypeptides released from neurons may function as neurotransmitters in the traditional sense (that is, by stimulating the opening of ionic gates and causing changes in the membrane potential), others may have more subtle and poorly understood effects. **Neuromodulators** has been proposed as a name for compounds with such alternative effects. An exciting recent discovery is that some neurons in both the PNS and CNS produce both a classical neurotransmitter (ACh or a catecholamine) and a polypeptide neurotransmitter. These are contained in different synaptic vesicles that can be distinguished using the electron microscope. The neuron can thus release either the classical neurotransmitter or the polypeptide neurotransmitter under different conditions.

Discoveries such as the one just described indicate that synapses have a greater capacity for alteration at the molecular level than was previously believed. This attribute has been termed **synaptic plasticity.** Synapses are also more plastic at the cellular level. There is evidence that sprouting of new axon branches can occur over short distances to produce a turnover of synapses, even in the mature CNS. This breakdown and re-forming of synapses may occur within a time span of only a few hours. The physiological significance of these interesting discoveries is not yet fully understood.

Endogenous Opioids

The ability of opium and its analogues—that is, the *opioids*—to relieve pain (promote analgesia) has been known for centuries. Morphine, for example, has long been used for this purpose. The discovery in 1973 of opioid receptor proteins in the brain suggested that the effects of these drugs might be due to the stimulation of specific neuron pathways. This implied that opioids—along with LSD, mescaline, and other mind-altering drugs—might resemble neurotransmitters produced by the brain.

The analgesic effects of morphine are blocked in a specific manner by a drug called *naloxone* (*nal-oks'ōn*). In the same year that opioid receptor proteins were discovered, it was found that naloxone also blocked the analgesic effect of electrical brain stimulation. Subsequent evidence suggested that the analgesic effects of hypnosis and acupuncture could also be blocked by naloxone. These experiments indicated that the brain might be producing its own morphinelike analgesic compounds that served as the natural ligands of the opioid receptors in the brain.

These compounds have been identified as a family of polypeptides produced by the brain and pituitary gland. One member is called **β-endorphin** (for "endogenously produced morphinelike compounds"). Another consists of a group of 5-amino-acid peptides called **enkephalins** (*en-kef'ă-linz*), and a third is a polypeptide neurotransmitter called **dynorphin.**

The endogenous opioid system is inactive under normal conditions, but when activated by stressors it can block the transmission of pain. Current evidence for this effect includes results obtained both from neurophysiological studies—in which endogenous opioids blocked the release of substance P (the chemical transmitter believed to mediate pain pathways)—and from behavioral studies. The pain threshold of pregnant rats, for example, was found to decrease when they were treated with naloxone, which blocks the opioid receptors. Also, a burst in β-endorphin secretion was shown to occur in pregnant women during parturition (childbirth).

Exogenous opioids such as opium and morphine can produce euphoria, and so endogenous opioids may mediate reward or positive reinforcement pathways. This is consistent with the observation that overeating in genetically obese mice can be blocked by naloxone. It has also been suggested that the feeling of well-being and reduced anxiety following exercise (the "joggers' high") may be an effect of endogenous opioids. Blood levels of β-endorphin increase when exercise is performed at greater than 60% of the maximal oxygen uptake (see chapter 12) and peak 15 minutes after the exercise ends. Although obviously harder to measure, an increased level of opioids in the brain and cerebrospinal fluid has also been found to result from exercise. The opioid antagonist drug naloxone, however, does not block the exercise-induced euphoria, suggesting that the joggers' high is not primarily an opioid effect. Use of naloxone, however, does demonstrate that the endogenous opioids are involved in the effects of exercise on blood pressure, and that they are responsible for the ability of exercise to raise the pain threshold.

Neuropeptide Y

Neuropeptide Y is the most abundant neuropeptide in the brain. It has been shown to have a variety of physiological effects, including a role in the response to stress, in the regulation of circadian rhythms, and in the control of the cardiovascular system. Neuropeptide Y has been shown to inhibit the release of the excitatory neurotransmitter glutamate in a part of the brain called the hippocampus. This is significant because excessive glutamate released in this area can cause convulsions. Indeed, frequent seizures were a symptom of a recently developed strain of mice with the gene for neuropeptide Y "knocked out." (Knockout strains of mice have specific genes inactivated, as described in chapter 3.)

Neuropeptide Y is the most powerful stimulator of appetite. When injected into a rat's brain it can cause the rat to eat until it becomes obese. Conversely, inhibitors of neuropeptide Y that are injected into the brain inhibit eating. This research has become particularly important in light of the recent discovery of *leptin*, a satiety factor secreted by the adipose tissue. Leptin suppresses appetite by acting, at least in part, to inhibit neuropeptide Y release. This topic is discussed in more detail in chapter 27.

Nitric Oxide as a Neurotransmitter

Nitric oxide (NO) was the first gas to be identified as a neurotransmitter. Produced by nitric oxide synthetase in the cells of many organs from the amino acid L-arginine, nitric oxide's actions are very different from those of the more familiar nitrous oxide (N_2O), or laughing gas, sometimes used by dentists.

Nitric oxide has a number of different roles in the body. Within blood vessels, it acts as a local tissue regulator that causes the smooth muscles of those vessels to relax, so that the blood vessels dilate. This role will be described in conjunction with the circulatory system in chapter 22. Within macrophages and other cells, nitric oxide helps to kill bacteria. This activity is described in conjunction with the immune system in chapter 23. In addition, nitric oxide is a neurotransmitter of certain neurons in both the PNS and CNS. It diffuses out of the presynaptic axon and into neighboring cells by simply passing through the lipid portion of the cell membranes. Once in the target cells, NO exerts its effects by stimulating the production of cyclic guanosine monophosphate (cGMP), which acts as a second messenger.

In the PNS, nitric oxide is released by some neurons that innervate the gastrointestinal tract, penis, respiratory passages, and cerebral blood vessels. These are autonomic neurons that cause smooth muscle relaxation in their target organs. This can produce, for example, the engorgement of the spongy tissue of the penis with blood. In fact, scientists now believe that erection of the penis results from the action of nitric oxide released by specific parasympathetic nerves (see chapter 28). As a neurotransmitter in the brain, nitric oxide has been implicated in the processes of learning and memory. This will be discussed in more detail in the following section.

In addition to nitric oxide, another gas—**carbon monoxide (CO)**—may function as a neurotransmitter. Certain neurons, including those of the cerebellum and olfactory epithelium, have been shown to produce carbon monoxide (derived from the conversion of one pigment molecule, heme, to another, biliverdin). Also, carbon monoxide, like nitric oxide, has been shown to stimulate the production of cGMP within the neurons. Experiments suggest that carbon monoxide may promote odor adaptation in olfactory neurons, contributing to the regulation of olfactory sensitivity. Other physiological functions of neuronal carbon monoxide have also been suggested, including neuroendocrine regulation in the hypothalamus.

Synaptic Integration

The summation of numerous EPSPs may be needed to produce a depolarization of sufficient magnitude to stimulate the postsynaptic cell. The net effect of EPSPs on the postsynaptic neuron is reduced by hyperpolarization (IPSPs), which is produced by inhibitory neurotransmitters. The activity of neurons within the central nervous system is thus the net result of both excitatory and inhibitory effects.

Since voltage-regulated Na^+ and K^+ channels are absent in the dendrites and cell bodies, changes in membrane potential induced by neurotransmitters in these areas do not have the all-or-none characteristics of action potentials. Synaptic potentials are graded and can add together, or summate. **Spatial summation** occurs because numerous presynaptic nerve fibers (up to a thousand, in some cases) converge on a single postsynaptic neuron. In spatial summation, synaptic depolarizations

Figure 14.29

Spatial and temporal summation.
EPSPs can summate over distance (spatial summation) and time (temporal summation). When summation results in a threshold level of depolarization at the axon hillock, voltage-regulated NA^+ gates are opened, and an action potential is produced.

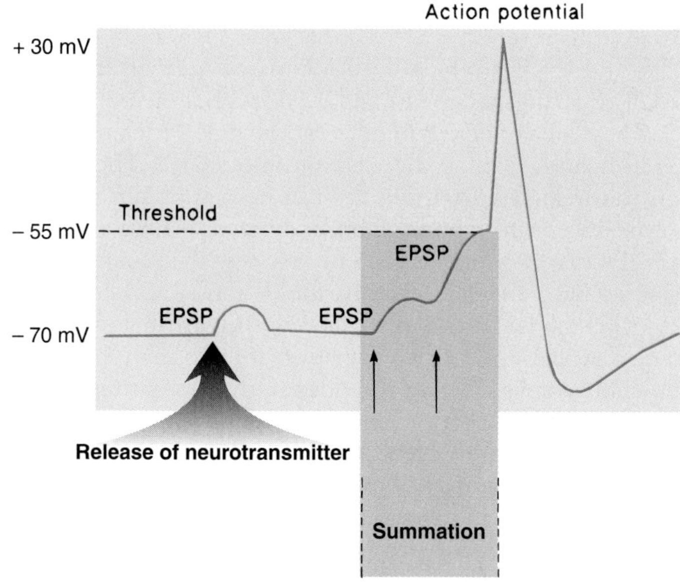

(EPSPs) produced at different synapses may summate in the postsynaptic dendrites and cell body. In **temporal summation,** the successive activity of presynaptic axon terminals, causing successive waves of transmitter release, may result in the summation of EPSPs in the postsynaptic neuron. The summation of EPSPs helps to ensure that the depolarization that reaches the axon hillock will be of sufficient magnitude to generate new action potentials in the postsynaptic neuron (fig. 14.29).

Long-Term Potentiation

When a presynaptic neuron is experimentally stimulated at a high frequency, even for just a few seconds, the excitability of the synapse is enhanced—or "potentiated"—when this neuron pathway is subsequently stimulated. The improved efficacy of synaptic transmission may last for hours or even weeks and is called **long-term potentiation (LTP).** Long-term potentiation may favor transmission along frequently used neural pathways, and thus may represent a mechanism of neural "learning." It is interesting in this regard that LTP has been observed in the hippocampus of the brain, an area implicated in memory storage (see chapter 15).

Most of the neural pathways in the hippocampus use glutamate as their neurotransmitter. There are two subclasses of glutamate receptors: NMDA receptors (as previously mentioned) and another type called *AMPA receptors*. Binding of glutamate to NMDA receptors is not required for normal synaptic transmission. It is the binding of glutamate to the AMPA receptors that produces the EPSP depolarization. However, when this depolarization occurs at a time when other glutamate molecules are binding to the NMDA receptors, channels for Ca^{2+} open in the postsynaptic membrane. The diffusion of Ca^{2+} into the postsynaptic neuron results in an increase in the number of AMPA receptors on the postsynaptic membrane, thereby increasing its ability to depolarize. There is also recent evidence that the AMPA receptors become phosphorylated. This helps to induce LTP by making the neuron more sensitive to subsequent stimulation so that it produces stronger EPSPs.

Another factor that may contribute to LTP is the recent observation that action potentials can travel backward from the axon into the dendrites! (EPSPs go from the dendrites to the cell body and axon; the action potentials are always first produced at the initial segment of the axon, as previously described, but can spread in a backward as well as forward direction.) This backward transmission of action potentials into the dendrites, when paired with an EPSP, has been shown to cause an increased entry of Ca^{2+} into the postsynaptic cell and thereby enhance LTP.

The diffusion of Ca^{2+} into the dendrites of the postsynaptic neuron also activates nitric oxide synthetase, causing the postsynaptic neuron to produce and release nitric oxide. The nitric oxide released from the postsynaptic neuron may then act as a "retrograde transmitter" and diffuse to the presynaptic neuron, where it may stimulate the release of more glutamate transmitter. In this way, synaptic transmission during LTP could be strengthened first postsynaptically and then presynaptically.

Synaptic Inhibition

Although many neurotransmitters depolarize the postsynaptic membrane (produce EPSPs), some transmitters do just the opposite. They hyperpolarize the postsynaptic membrane; that is, they make the inside of the membrane more negative than it is at rest. Since hyperpolarization (from –70 mV to, for example, –85 mV) drives the membrane potential farther from the threshold depolarization required to stimulate action potentials, this inhibits the activity of the postsynaptic neuron. Hyperpolarizations produced by neurotransmitters are therefore called *inhibitory postsynaptic potentials* (*IPSPs*), as previously described. The inhibition produced in this way is called **postsynaptic inhibition.**

Excitatory and inhibitory inputs (EPSPs and IPSPs) to a postsynaptic neuron can summate in an algebraic fashion (fig. 14.30). The effects of IPSPs in this way reduce, or may even preclude, the ability of EPSPs to generate action potentials in the postsynaptic cell. Considering that a given neuron may receive as many as 1,000 presynaptic inputs, the interactions of EPSPs and IPSPs can vary greatly.

In **presynaptic inhibition** (fig. 14.31), the amount of excitatory neurotransmitter released at the end of an axon is decreased by a second neuron, whose axon makes a synapse with the axon of the first neuron (axoaxonic synapse). The neurotransmitter released at this axoaxonic synapse partially depolarizes the axon of the first neuron, bringing it closer to

Figure 14.30

An IPSP hyperpolarizes the postsynaptic membrane. An inhibitory postsynaptic potential (IPSP) makes the inside of the postsynaptic membrane more negative than the resting potential—it hyperpolarizes the membrane. Subsequent or simultaneous excitatory postsynaptic potential (EPSPs), which are depolarizations, must thus be stronger to reach the threshold required to generate action potentials at the axon hillock.

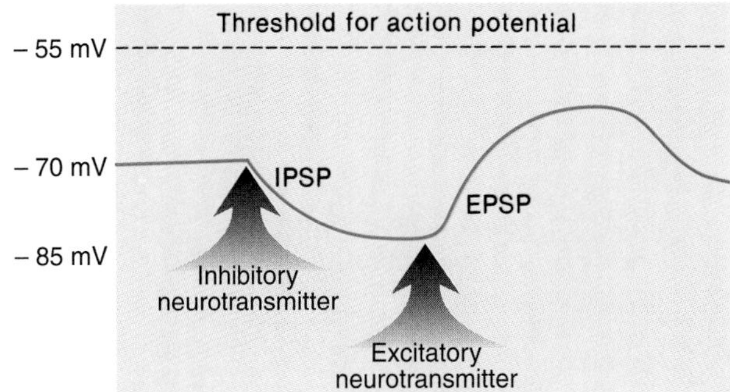

threshold and making it easier to "fire" action potentials. The amplitude of these action potentials (subtracting the new lower potential from the +30-mV "top" of the action potential), however, is less than normal. The smaller action potentials that result cause the release of lesser amounts of excitatory neurotransmitter by the first neuron, resulting in inhibition of its effect on its postsynaptic cell.

Clinical Considerations

The clinical aspects of the nervous system are extensive and usually complex. Numerous diseases and developmental problems directly involve the nervous system, and it is indirectly involved in most diseases because of the perception of pain. In this section, we will consider only a few of the diseases involving developmental problems of the nervous system and clinical aspects of nerve conduction and synaptic transmission. Other clinical aspects of neurology will be discussed in later chapters.

Developmental Problems

Congenital malformations of the CNS are common and frequently involve overlying bone, muscle, and connective tissue. The more severe abnormalities make life impossible; those that are less severe frequently result in functional disability.

Most congenital malformations of the nervous system occur during the sensitive embryonic period. Neurological malformations are generally caused by genetic abnormalities but also may result from environmental factors, such as anoxia, infectious agents, drugs, and ionizing radiation.

Spina bifida (*spi'nă bif'ĭ-dă*) is a defective fusion of the vertebral elements and may or may not involve the spinal cord. *Spina bifida occulta* is the most common and least serious type of spina bifida. This defect usually involves few vertebrae, is not externally apparent except for perhaps a pigmented spot

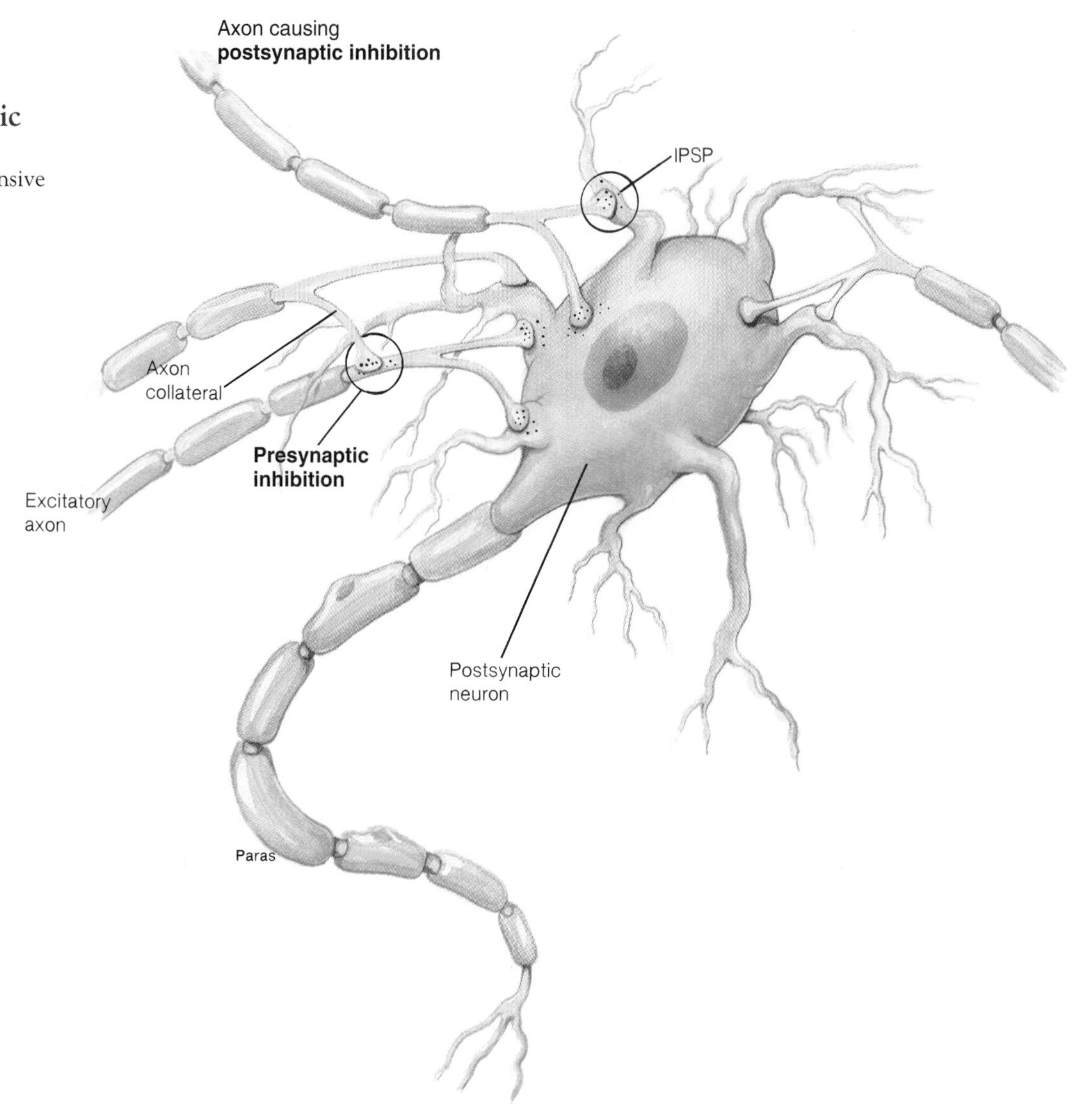

Figure 14.31

A diagram illustrating postsynaptic and presynaptic inhibition.
These and other processes allow extensive integration within the CNS.

with a tuft of hair, and usually does not cause neurological disturbances. *Spina bifida cystica,* a severe type of spina bifida, is a saclike protrusion of skin and underlying meninges that may contain portions of the spinal cord and nerve roots. Spina bifida cystica is most common in the lower thoracic, lumbar, and sacral regions. The position and extent of the defect determines the degree of neurological impairment.

Anencephaly (*an″en-sef′ă-le*) is a markedly defective development of the brain and the surrounding cranial bones. Anencephaly occurs in 1 per 1,000 births and makes sustained extrauterine life impossible. This congenital defect apparently results from the failure of the neural folds at the cranial portion of the neural plate to fuse and form the prosencephalon.

Microcephaly is an uncommon condition in which brain development is not completed. If enough neurological tissue is present, the infant will survive but will be severely mentally retarded.

Diseases of the Myelin Sheath

Multiple sclerosis (MS) is a relatively common neurological disease in people between the ages of 20 and 40. It is a chronic, degenerating, remitting, and relapsing disease that progressively destroys the myelin sheaths of neurons in multiple areas of the CNS. Initially, lesions form on the myelin sheaths and soon develop into hardened *scleroses,* or scars (hence the name). Destruction of the myelin sheaths prohibits the normal conduction of impulses, resulting in a progressive loss of functions. Because myelin degeneration is widely distributed and affects different areas of the nervous system in different people, MS has a wider variety of symptoms than any other neurological disease. This characteristic, coupled with remissions, frequently causes misdiagnosis of the disease.

---

multiple sclerosis: L. *multiplus,* many parts; Gk. *skleros,* hardened

During the early stages of MS, many patients are considered neurotic because the symptoms they report vary so widely and come and go quickly. As the disease progresses, the symptoms may include double vision (diplopia), spots in the visual field, blindness, tremor, numbness of the appendages, and locomotor difficulty. Eventually the patient becomes bedridden, and death may occur anytime from 7 to 30 years after the initial onset of symptoms.

In **Tay–Sachs disease,** the myelin sheaths are destroyed by the excessive accumulation of one of the lipid components of the myelin. This results from an enzyme defect caused by the inheritance of genes carried by the parents in a recessive state. The disease is inherited primarily by individuals of Eastern European Jewish descent and appears before the infant is a year old. It causes blindness, loss of mental and motor ability, and ultimately death by the age of 3. Potential parents can tell if they are carriers for Tay–Sachs by having a special blood test for the defective enzyme.

Problems of Neural Transmission

The muscle weakness associated with the disease **myasthenia gravis** (*mi″as-the′ne-ă grav′is*) results because ACh receptors are blocked by antibodies secreted by the immune system. Paralysis in people who eat shellfish poisoned with *saxitoxin*, produced by the unicellular organisms that cause the red tides, results from inhibition of the chemically regulated Na^+ gates at the neuromuscular junction. A similar inhibition and paralysis is produced by *tetrodotoxin*, a poison found in the pufferfish.

Alzheimer's disease is the most common cause of senile dementia, which often begins in middle age and produces progressive mental deterioration. The cause of Alzheimer's disease is not known, but there is evidence that it is associated with a loss of neurons that use acetylcholine as a neurotransmitter. These axons terminate in the hippocampus and cerebral cortex of the brain, which are areas concerned with memory storage.

Alzheimer's is associated with a deficiency in the enzyme responsible for producing acetylcholine from acetyl coenzyme A and choline. Treatment with different drugs that inhibit the activity of acetylcholinesterase has been reported to be fairly effective. By inhibiting this enzyme, the breakdown of ACh in synapses is reduced, so that the action of ACh is improved.

Autopsies of people who have died of Alzheimer's disease reveal tangled masses of fibrils within the cell bodies of many neurons. Also evident are "neuritic plaques" consisting of degenerating neurons and deposits of amyloid protein. Similar plaques are seen in the brains of people with Down syndrome, a genetic disease caused by an extra chromosome number 21. Recently, scientists have discovered that the gene that codes for the amyloid protein in people with Alzheimer's disease is located in chromosome number 21, suggesting that Alzheimer's may be caused by genetic defects located on this chromosome.

Cocaine—a stimulant related to the amphetamines in its action—is currently widely abused in the United States. Although early use of this drug produces feelings of euphoria and social adroitness, continued use leads to social withdrawal, depression, dependence upon ever-higher dosages, and serious cardiovascular and renal disease that can result in heart and kidney failure. The numerous effects of cocaine on the central nervous system appear to be mediated by one primary mechanism: cocaine binds to the dopamine reuptake transporter and blocks the reuptake of dopamine into the presynaptic axon endings. This results in overstimulation of those neural pathways that use dopamine as a neurotransmitter.

Tay–Sachs disease: from Warren Tay, English physician, 1843–1927, and Bernard Sachs, American neurologist, 1858–1944
Alzheimer's disease: from Alois Alzheimer, German neurologist, 1864–1915

Clinical Investigation Answer

The muscular paralysis and difficulty in breathing (due to paralysis of the diaphragm) could have been caused by saxitoxin poisoning from the shellfish, which may have been gathered during a red tide. A positive chemical analysis of the student's blood and of the shellfish for saxitoxin would confirm this diagnosis. The monoamine oxidase (MAO) inhibitor was probably prescribed to treat the student's depression. It turns out, however, that there are significant drug-food interactions with MAO inhibitors—in fact, shellfish is specifically contraindicated! Other drugs are now available to treat depression that have fewer side effects.

Chapter Summary

Neurons and Supporting Cells (pp. 372–378)

1. Every neuron contains a cell body, dendrites, and an axon.
2. On the basis of the number of processes extending from the cell body, neurons can be classified as pseudounipolar, bipolar, or multipolar.
3. Neurons in the PNS that conduct impulses into the CNS are sensory; those that conduct impulses out of the CNS are motor.
4. Motor neurons that innervate skeletal muscles are somatic; those that innervate the heart, smooth muscles, and glands are autonomic.
5. Supporting cells include Schwann cells (neurolemmocytes) and satellite cells in the PNS; in the CNS they include oligodendrocytes, astrocytes, microglia, and ependymal cells.
   (a) In the PNS, Schwann cells surround axons to form a sheath of Schwann that provides a continuous basement membrane around the axon.
   (b) Many axons of the PNS have a myelin sheath, formed by successive wrappings of the Schwann cell membrane.
   (c) In the CNS, myelin sheaths are formed by oligodendrocytes.
   (d) The sheath of Schwann allows damaged peripheral axons to regenerate; the absence of a sheath

of Schwann in the CNS hinders regeneration of central axons.
   (e) Astrocytes send out vascular processes that surround capillaries in the brain and help to establish the blood-brain barrier.

Electrical Activity in Axons (pp. 379–385)

1. The permeability of the axon membrane to Na^+ and K^+ is regulated by gates at the openings of ion channels.
   (a) When the membrane is depolarized to a threshold level, the Na^+ gates open first, allowing Na^+ to flow into the axon by diffusion.
   (b) This is followed quickly by the opening of K^+ gates, allowing K^+ to diffuse out of the axon.
2. The opening of voltage-regulated gates and the resulting flow of ions produce an action potential.
   (a) The inward diffusion of Na^+ causes a reversal of the membrane potential from –70 mV to +30 mV.
   (b) The opening of K^+ gates and outward diffusion of K^+ causes the reestablishment of the resting membrane potential in a process called repolarization.
3. Action potentials are all-or-none events.
   (a) A depolarization stimulus that is lower than threshold has no effect.
   (b) Any depolarization stimulus higher than a threshold level will cause an action potential of maximal amplitude.
   (c) Stronger stimuli thus do not produce stronger action potentials; rather, they cause action potentials to be produced at a greater frequency (more per second).
4. The axon membrane is in a refractory period while producing an action potential.
   (a) While in the refractory period, the membrane cannot be stimulated (absolute refractory period) or can only be stimulated by a very large stimulus (relative refractory period).
   (b) This prevents action potentials from being able to summate or interfere with each other.
5. One action potential serves as the depolarization stimulus for production of the next action potential in the axon.
   (a) In unmyelinated axons, action potentials are produced fractions of a micrometer apart.
   (b) In myelinated axons, action potentials are produced only at the nodes of Ranvier; this saltatory conduction is faster than conduction in unmyelinated axons.

The Synapse (pp. 385–389)

1. Gap junctions are electrical synapses and are found in cardiac muscle, smooth muscle, and in some regions of the brain.
2. In chemical synapses, neurotransmitters are packaged in synaptic vesicles and released into the synaptic cleft.
   (a) Action potentials arriving at the axon terminal stimulate the entry of Ca^{2+}, which causes fusing of the synaptic vesicle with the axon membrane and the release of neurotransmitter by exocytosis.
   (b) The greater the frequency of action potentials that arrive at the axon terminal, the more neurotransmitter is released.

Acetylcholine as a Neurotransmitter (pp. 389–394)

1. The binding of ACh to a nicotinic ACh receptor in the postsynaptic membrane causes chemically regulated gates to open, which produces a depolarization known as an excitatory postsynaptic potential (EPSP).
   (a) Chemically regulated gates allow Na^+ and K^+ to diffuse simultaneously, resulting in a graded depolarization with a maximum amplitude of 0 volts.
   (b) The extent of the depolarization depends on the amount of transmitter released; EPSPs are therefore graded.
   (c) EPSPs have no refractory period; they are therefore capable of summation.
   (d) Unlike action potentials, EPSPs cannot be self-regenerated; they therefore decrease in amplitude with distance as they are conducted.
2. The binding of ACh to a muscarinic ACh receptor activates a system of G-proteins that dissociate to produce subunits that diffuse through the membrane to the chemically gated ion channels.
   (a) In some cases, binding of a G-protein subunit to an ion channel causes the channel to open; in other cases, it causes the channel to close.
   (b) As a result of its action via muscarinic receptors, ACh can inhibit the heart, causing it to slow its beat, or it can stimulate the stomach to contract.
3. ACh is inactivated by acetylcholinesterase. This acts to stop the stimulation and is essential for normal synaptic function.
4. EPSPs produced at synapses in the dendrites or cell body travel to the axon hillock.
   (a) Here, they stimulate the opening of voltage-regulated gates and generate action potentials in the axon.
   (b) Summation of EPSPs, through both spatial and temporal summation, occurs at the axon hillock and influences the frequency of action potentials produced at the axon of the postsynaptic neuron.

Monoamines as Neurotransmitters (pp. 394–397)

1. Monoamines include serotonin and a family of compounds called catecholamines. The catecholamines include dopamine, norepinephrine, and epinephrine.
2. These neurotransmitters act via second messengers, such as cyclic AMP.
3. The monoamine neurotransmitters are inactivated by their reuptake into the presynaptic nerve endings and by their degradation via the action of monoamine oxidase.
4. The monoamine neurotransmitters have diverse functions in the brain and are of great clinical significance.

Other Neurotransmitters (pp. 397–400)

1. Aspartic acid and glutamic acid are amino acids that serve as excitatory neurotransmitters. Glutamate is the major excitatory neurotransmitter in the brain.
2. Glycine and GABA exert inhibitory effects via the production of hyperpolarizations, or inhibitory postsynaptic potentials (IPSPs).
3. Many polypeptides serve as neurotransmitters. These include the endogenous opioids, which reduce sensations of pain, and neuropeptide Y, which stimulates appetite.
4. Nitric oxide is a gas that serves as a neurotransmitter in both the PNS and CNS.

Synaptic Integration (pp. 400–401)

1. Spatial and temporal summation of EPSPs allows a depolarization of sufficient

magnitude to cause the stimulation of action potentials in the postsynaptic neuron.

2. In long-term potentiation (LTP), bursts of stimulation along a synapse will increase the excitability of the postsynaptic neuron as a form of neuronal learning.
3. Neurotransmitters that cause hyperpolarization of the postsynaptic membrane produce inhibitory postsynaptic potentials (IPSPs).
   (a) The production of IPSPs is called postsynaptic inhibition.
   (b) IPSPs and EPSPs from different synaptic inputs can summate.
   (c) Presynaptic inhibition occurs in axoaxonic synapses. The inhibited neuron releases less neurotransmitter, thus reducing the stimulation of the postsynaptic cell.

Review Activities

Objective Questions

1. The supporting cells that form myelin sheaths in the peripheral nervous system are
   (a) oligodendrocytes.
   (b) satellite cells.
   (c) Schwann cells.
   (d) astrocytes.
   (e) microglia.
2. A collection of neuron cell bodies located outside the CNS is called
   (a) a tract.
   (b) a nerve.
   (c) a nucleus.
   (d) a ganglion.
3. Which of the following neurons are pseudounipolar?
   (a) sensory neurons
   (b) somatic motor neurons
   (c) neurons in the retina
   (d) autonomic motor neurons
4. Depolarization of an axon is produced by
   (a) inward diffusion of Na^+.
   (b) active extrusion of K^+.
   (c) outward diffusion of K^+.
   (d) inward active transport of Na^+.
5. Repolarization of an axon during an action potential is produced by
   (a) inward diffusion of Na^+.
   (b) active extrusion of K^+.
   (c) outward diffusion of K^+.
   (d) inward active transport of Na^+.
6. As the strength of a depolarizing stimulus to an axon is increased,
   (a) the amplitude of action potentials increases.
   (b) the duration of action potentials increases.
   (c) the speed with which action potentials are conducted increases.
   (d) the frequency with which action potentials are produced increases.
7. The conduction of action potentials in a myelinated nerve fiber is
   (a) saltatory.
   (b) without decrement.
   (c) faster than in an unmyelinated fiber.
   (d) all of the above.
8. Which of the following is *not* a characteristic of synaptic potentials?
   (a) They are all or none in amplitude.
   (b) They decrease in amplitude with distance.
   (c) They are produced in dendrites and cell bodies.
   (d) They are graded in amplitude.
   (e) They are produced by chemically regulated gates.
9. Which of the following is *not* a characteristic of action potentials?
   (a) They are produced by voltage-regulated gates.
   (b) They are conducted without decrement.
   (c) Na^+ and K^+ gates open at the same time.
   (d) The membrane potential reverses polarity during depolarization.
10. A drug that inactivates acetylcholinesterase
    (a) inhibits the release of ACh from presynaptic endings.
    (b) inhibits the attachment of ACh to its receptor protein.
    (c) increases the ability of ACh to stimulate muscle contraction.
    (d) does all of the above.
11. Postsynaptic inhibition is produced by
    (a) depolarization of the postsynaptic membrane.
    (b) hyperpolarization of the postsynaptic membrane.
    (c) axoaxonic synapses.
    (d) long-term potentiation.
12. Hyperpolarization of the postsynaptic membrane in response to glycine or GABA is produced by the opening of
    (a) Na^+ gates.
    (b) K^+ gates.
    (c) Ca^{2+} gates.
    (d) Cl^- gates.
13. The absolute refractory period of a neuron
    (a) is due to the high negative polarity of the inside of the neuron.
    (b) occurs only during the repolarization phase.
    (c) occurs only during the depolarization phase.
    (d) occurs during depolarization and the first part of the repolarization phase.
14. Which of the following statements about catecholamines is *false?*
    (a) They include norepinephrine, epinephrine, and dopamine.
    (b) Their effects are increased by action of the enzyme catechol-O-methyltransferase.
    (c) They are inactivated by monoamine oxidase.
    (d) They are inactivated by reuptake into the presynaptic axon.
    (e) They may stimulate the production of cyclic AMP in the postsynaptic axon.
15. The summation of EPSPs from numerous presynaptic nerve fibers converging onto one postsynaptic neuron is called
    (a) spatial summation.
    (b) long-term potentiation.
    (c) temporal summation.
    (d) synaptic plasticity.
16. Which of the following statements about ACh receptors is *false?*
    (a) Skeletal muscles contain nicotinic ACh receptors.
    (b) The heart contains muscarinic ACh receptors.
    (c) G-proteins are needed to open ion channels for nicotinic receptors.
    (d) Stimulation of nicotinic receptors results in the production of EPSPs.
17. Hyperpolarization is caused by all of the following neurotransmitters *except:*
    (a) glutamic acid in the CNS.
    (b) ACh in the heart.
    (c) glycine in the spinal cord.
    (d) GABA in the brain.
18. Which of the following may be produced by the action of nitric oxide?
    (a) dilation of blood vessels
    (b) erection of the penis
    (c) relaxation of smooth muscles in the digestive tract
    (d) LTP among neighboring synapses in the brain
    (e) all of the above

Essay Questions

1. Compare the characteristics of action potentials with those of synaptic potentials.
2. Explain how voltage-regulated gates produce an all-or-none action potential.
3. Explain how action potentials are regenerated along an axon.
4. Explain why conduction in a myelinated axon is faster than in an unmyelinated axon.
5. Describe the structure of nicotinic ACh receptors. Explain how ACh causes the production of an EPSP and relate this

process to the neural stimulation of skeletal muscle contraction.

6. Describe the nature of muscarinic ACh receptors and the role of G-proteins in the action of these receptors. How does stimulation of these receptors cause the production of hyperpolarization or a depolarization?
7. Trace the course of events in the interval between the production of an EPSP and the generation of action potentials at the axon hillock. Describe the effect of spatial and temporal summation on this process.
8. Explain how an IPSP is produced and how IPSPs can inhibit activity of the postsynaptic neuron.
9. List the endogenous opioids in the brain and describe some of their proposed functions.
10. Explain what is meant by long-term potentiation and discuss the significance of this process. What may account for LTP, and what role might nitric oxide play?

Critical Thinking Questions

1. Grafting peripheral nerves onto the two parts of a cut spinal cord in rats was found to restore some function in the hind limbs. Apparently, when the white matter of the peripheral nerve was joined to the gray matter of the spinal cord, some regeneration of central neurons occurred across the two spinal cord sections. What component of the peripheral nerve probably contributed to the regeneration? Discuss the factors that promote and inhibit central neuron regeneration.
2. Discuss the different states of a voltage-gated ion channel and distinguish between these states. How has molecular biology/biochemistry aided our understanding of the physiology of the voltage-gated channels?
3. Suppose you are provided with an isolated nerve-muscle preparation in order to study synaptic transmission. In one of your experiments, you give this preparation a drug that blocks voltage-regulated Ca^{2+} channels; in another, you give tetanus toxin to the preparation. How will synaptic transmission be affected in each experiment?
4. What functions do G-proteins serve in synaptic transmission? Speculate on the advantages of having G-proteins mediate the effects of a neurotransmitter.
5. Studies indicate that alcoholism may be associated with a particular allele (form of a gene) for the D_2 dopamine receptor. Suggest some scientific investigations that might further explore these possible genetic and physiological relationships.

Related Web Sites

For a listing of the most current web sites related to this chapter, please visit the *Concepts of Human Anatomy and Physiology* home page at http://www.mhhe.com/biosci/abio/.

FIFTEEN

Central Nervous System

OBJECTIVES

- Describe the general characteristics of the brain and spinal cord.
- Discuss the basic metabolic demands of the brain.
- Describe the structure of the cerebrum and list the functions of the cerebral lobes.
- Define the term *electroencephalogram* and discuss its clinical importance.
- Describe the locations of the language centers and compare short-term and long-term memory.
- Discuss the functions of the limbic system and list its parts.
- List the autonomic functions of the thalamus and the hypothalamus.
- Describe the location and structure of the pituitary gland.
- List the structures of the mesencephalon and describe their function.
- Describe the location and structure of the pons and cerebellum and list their functions.
- Describe the location and structure of the medulla oblongata and list its functions.
- Define the term *reticular formation* and explain its function.
- Describe the position of the meninges as they protect the CNS.
- Describe the locations of the ventricles of the brain.
- Discuss the formation, function, and flow of cerebrospinal fluid.
- Describe the structure of the spinal cord.
- Describe the arrangement of ascending and descending tracts.

Clinical Investigation

A 56-year-old woman visited her family doctor for evaluation of a headache that had persisted for nearly a month. Upon questioning the patient, the doctor learned that her left arm, as she put it, "was a bit unwieldy, hard to control, and weak." Through examination, the doctor determined that the entire left upper extremity was generally weak. He also found weakness of the left lower extremity, although less significant. Sensation in the limbs seemed to be normal, although mild rigidity and hyperactive reflexes were present. Expressing concern, the doctor told the patient that she needed a MRI of her head, and explained that there could be a problem within the brain, possibly a tumor or other lesion. The doctor then picked up the phone and contacted a radiologist. After explaining the patient's case, the doctor remarked parenthetically that he believed he knew where the problem was located.

Why did the doctor suggest to the patient that there might be a problem within her brain when the symptoms were weakness of the extremities, and just on one side of her body? Also, how would he know the location of the suspected brain tumor? In which side of the brain and in which lobe would it be? Explain the muscle weakness in terms of neuronal pathways from the brain to the periphery.

Clues: Remember the controlling and integrating functions of the brain. Carefully study the information and accompanying figures concerning the structures and functions of the brain and the neuronal tract.

Characteristics of the Central Nervous System

The central nervous system, composed of gray and white matter, is covered with meninges and is bathed in cerebrospinal fluid. The tremendous metabolic rate of the brain requires a continuous flow of blood amounting to approximately 20% of the total resting cardiac output.

The central nervous system (CNS) consists of the brain and spinal cord. The entire delicate CNS is protected by a bony encasement—the cranium surrounding the brain (fig. 15.1) and the vertebral column surrounding the spinal cord. The meninges (*mĕ-nin'jēz*) are connective tissue encasements that form a protective membrane between the bone and the soft tissue of the CNS. The CNS is bathed in cerebrospinal (*ser"ĕ-bro-spi'nal*) fluid that circulates within the hollow ventricles of the brain, the central canal of the spinal cord, and the subarachnoid (*sub"ă-rak'noid*) space surrounding the entire CNS.

Figure 15.1

The central nervous system.
This division of the nervous system consists of the brain and the spinal cord, both of which are covered with meninges and bathed in cerebrospinal fluid. (*a*) A sagittal section showing the brain within the cranium of the skull and the spinal cord within the vertebral canal. (*b*) The spinal cord, shown in a posterior view, extends from the level of the foramen magnum to the first lumbar vertebra (L1).

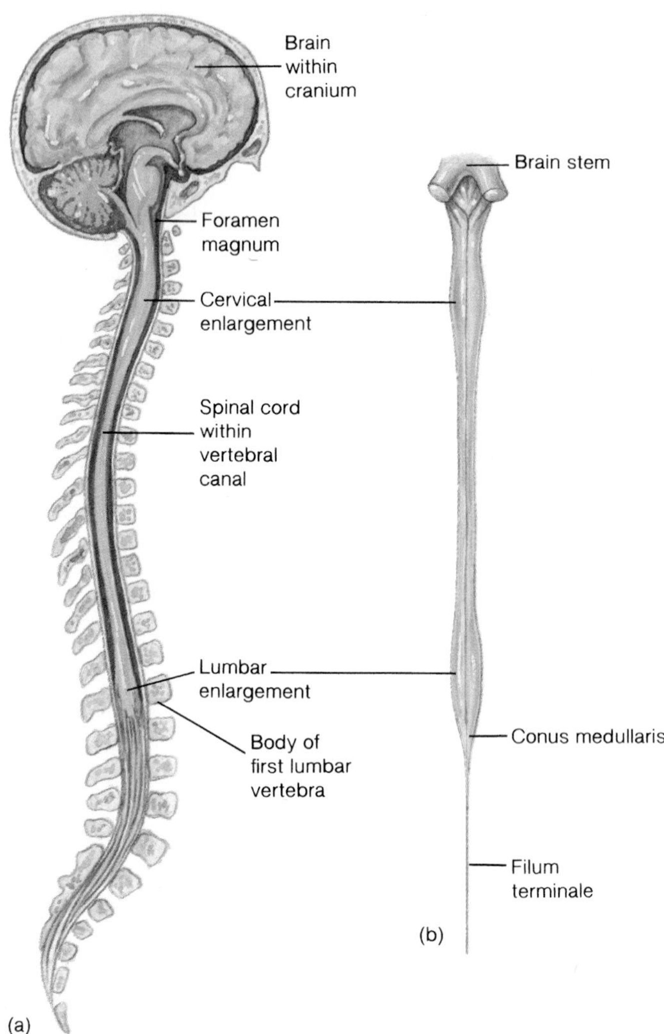

The CNS is composed of gray and white matter. **Gray matter** consists of either nerve cell bodies and dendrites or bundles of unmyelinated axons and neuroglia. The gray matter of the brain exists as the outer convoluted **cortex layer** of the cerebrum and cerebellum. In addition, specialized gray matter clusters of nerve cells called **nuclei** are found deep within the white matter. **White matter** forms the tracts within the CNS and consists of aggregations of dendrites and myelinated axons, along with associated neuroglia.

The brain of an adult weighs nearly 1.5 kg (3–3.5 lb) and is composed of an estimated 100 billion (10^{11}) neurons. As described in the previous chapter, neurons communicate with one another by means of innumerable synapses between the axons and dendrites within the brain. Neurotransmission within the brain is regulated by specialized neurotransmitter chemicals that are found in specific brain regions and tracts.

The brain has a tremendous metabolic rate and needs a continuous supply of oxygen and nutrients. Although it accounts for only 2% of a person's body weight, the brain receives approximately 20% of the total resting cardiac output. This amounts to a flow of about 750 ml of blood per minute. The volume remains relatively constant even with changes in physical or mental activity. This continuous flow is so crucial that a failure of cerebral circulation for as short an interval as 10 seconds causes unconsciousness.

The brain is composed of perhaps the most sensitive tissue of the body. Because of its high metabolic rate, it not only requires continuous oxygen, but also a continuous nutrient supply and the rapid removal of wastes. It is also very sensitive to certain toxins and drugs. The cerebrospinal fluid aids the metabolic needs of the brain by serving as a medium of exchange of nutrients and waste products between the blood and nervous tissue. Cerebrospinal fluid also maintains a protective homeostatic environment within the brain. The blood-brain barrier (chapter 14) and the secretory activities of neural tissue also help to maintain homeostasis. The brain has an extensive vascular supply through the paired internal carotid and vertebral arteries that unite at the cerebral arterial circle (circle of Willis) (see chapter 21 and fig. 21.21).

The brain of a newborn is especially sensitive to oxygen deprivation or to excessive oxygen. If complications arise during childbirth and the oxygen supply from the mother's blood to the baby is interrupted while still in the birth canal, the infant may be stillborn or suffer brain damage that can result in cerebral palsy, epilepsy, paralysis, or mental retardation. Excessive oxygen administered to a newborn may cause blindness.

Measurable increases in regional blood flow and in glucose and oxygen metabolism within the brain accompany mental functions, including perception and emotion. These metabolic changes can be assessed through the use of *positron emission tomography* (*PET*). The technique of a PET scan (fig. 15.2) is based on injecting radioactive tracer molecules labeled with carbon-11, fluorine-18, and oxygen-15 into the bloodstream and photographing the gamma rays that are subsequently emitted from the patient's brain through the skull. PET scans are of value in studying neurotransmitters and neuroreceptors, as well as the substrate metabolism of the brain.

The development of the five basic regions of the brain—telencephalon, diencephalon, mesencephalon, metencephalon,

Figure 15.2

A positron emission tomographic (PET) scan of the brain.
This PET scan shows a transverse section of the brain from an unmedicated patient with schizophrenia. Red areas indicate high glucose use (uptake of 18-F-deoxyglucose). The scan shows highest glucose uptake in the posterior region, where the brain's visual center is located.

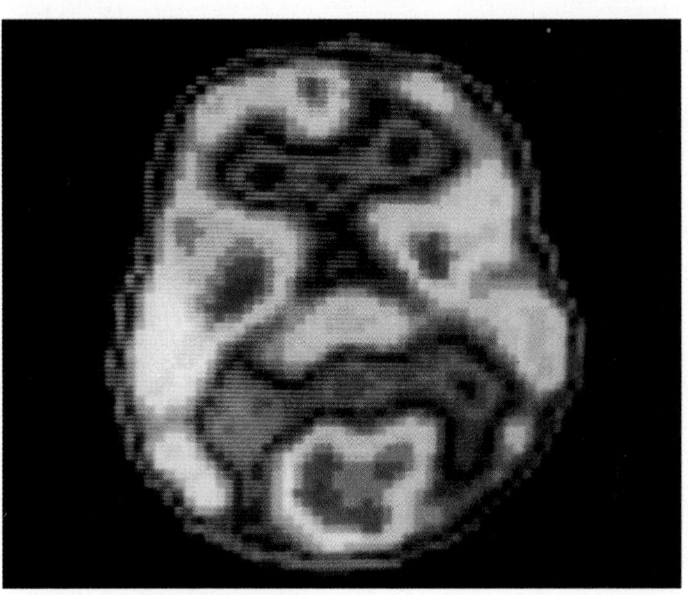

and myelencephalon—is discussed in the boxed developmental material on pages 410 and 411. From each of these regions, distinct functional structures are formed. These structures are summarized in table 15.1 and will be discussed in greater detail in the following sections.

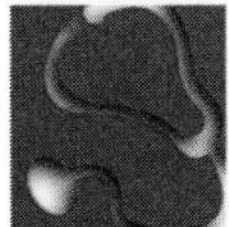

Mitotic activity within nervous tissue is completed during prenatal development. Thus, a person is born with all the neurons he or she is capable of producing. However, nervous tissue continues to grow and to specialize after a person is born, particularly in the first several years of postnatal life. The extent to which the nervous tissue is altered during the aging process is not known. It has been estimated that as many as 100,000 neurons die each day of our adult life. Recent studies, however, show that such estimates are unfounded, indicating that relatively few neural cells are lost during the normal aging process. Neurons are, however, extremely sensitive and susceptible to various drugs or interruptions of blood flow, such as those caused by strokes or other cardiovascular diseases.

There is evidence that aging alters neurotransmitters. Age-related conditions such as depression or specific diseases such as Alzheimer's disease may be caused by an imbalance of neurotransmitter chemicals. Changes in sleeping patterns in elderly people also probably result from neurotransmitter problems.

UNDER DEVELOPMENT

Development of the Brain

The first indication of nervous tissue development occurs about 17 days following conception, when a thickening appears along the entire dorsal length of the embryo. This thickening, called the **neural plate** (fig. 1), differentiates and eventually gives rise to all of the **neurons** and to most of **neuroglia** that support the neurons. As development progresses, the midline of the neural plate invaginates to become the neural groove. At the same time, cells proliferate along the lateral margins of the neural plate, forming the thickened **neural folds.** The neural groove continues to deepen as the neural folds elevate. By day 20, the neural folds have met and fused at the midline, and the neural groove has become a **neural tube.** For a short time, the neural tube is open both cranially and caudally. These openings, called neuropores, close during the fourth week. Once formed, the neural tube separates from the surface ectoderm and eventually develops into the central nervous system (brain and spinal cord). The **neural crest** forms from the neural folds as they fuse longitudinally along the dorsal midline. Most of the peripheral nervous system (cranial

Figure 1 The early development of the nervous system from embryonic ectoderm. (*a*) A dorsal view of an 18-day-old embryo showing the formation of the neural plate and the position of a transverse cut indicated in (a_1). (*b*) A dorsal view of a 22-day-old embryo showing cranial and caudal neuropores and the positions of three transverse cuts indicated in (b_1–b_3). (Note the amount of fusion of the neural tube at the various levels of the 22-day-old embryo. Note also the relationship of the notochord to the neural tube.)

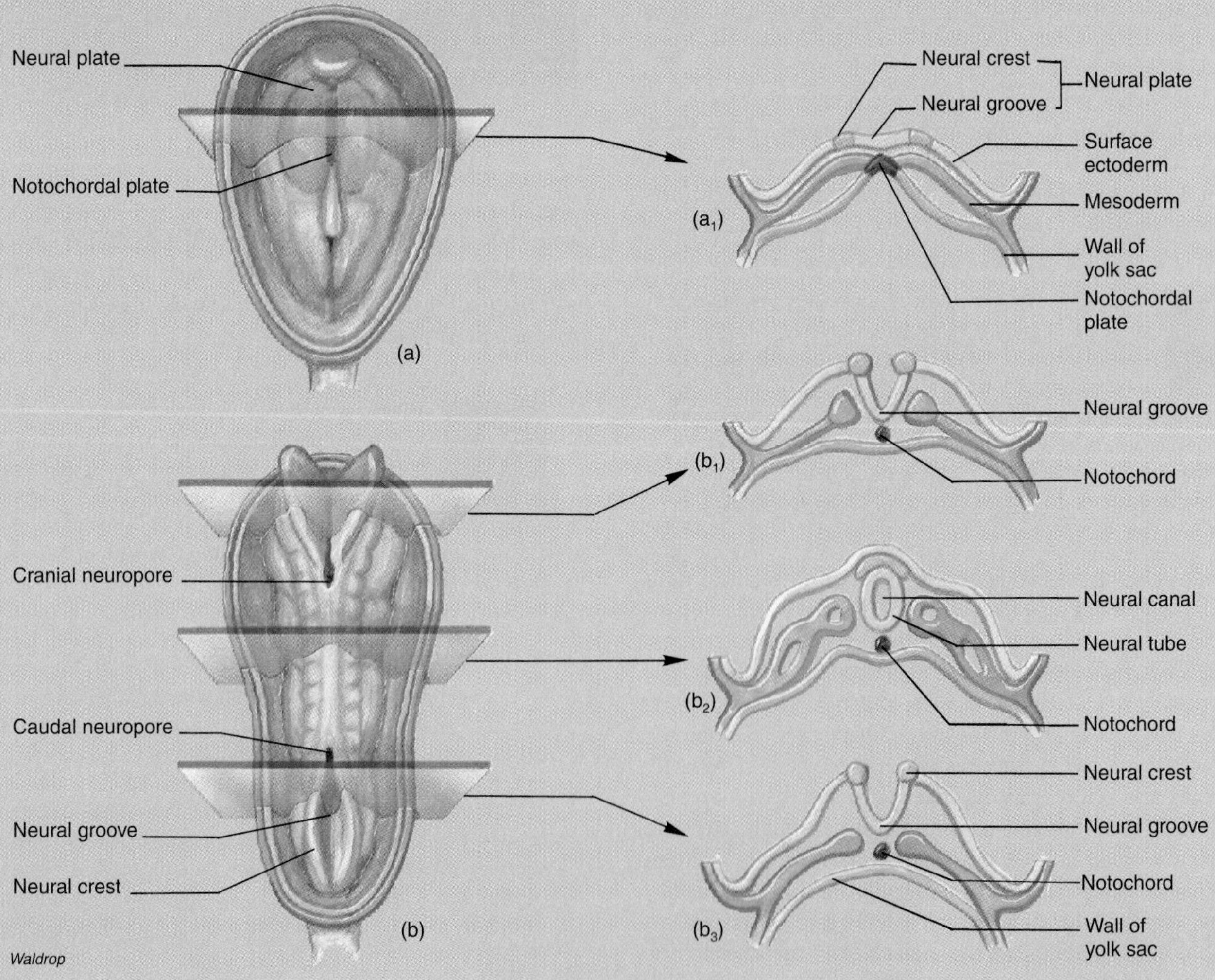

and spinal nerves) forms from the neural crest. Some neural crest cells break away from the main tissue mass and migrate to other locations, where they differentiate into motor nerve cells of the sympathetic division of the autonomic nervous system or into *Schwann cells* (neurolemmocytes), which are a type of glial cell important in the peripheral nervous system.

The brain begins its embryonic development as the cephalic end of the neural tube starts to grow rapidly and to differentiate (fig. 2). By the middle of the fourth week, three distinct swellings are evident: the **prosencephalon** (*pros″en-sef′ă-lon*) (forebrain), the **mesencephalon** (midbrain), and the **rhombencephalon** (hindbrain). Further development during the fifth week results in the formation of five specific regions: The telencephalon and the **diencephalon** (*di″en-sef′ă-lon*) derive from the forebrain, the mesencephalon remains unchanged, and the **metencephalon** and **myelencephalon** form from the hindbrain. The caudal portion of the myelencephalon is continuous with and resembles the spinal cord.

Figure 2 The developmental sequence of the brain. During the fourth week, the three principal regions of the brain are formed. During the fifth week, a five-regioned brain develops and specific structures begin to form.

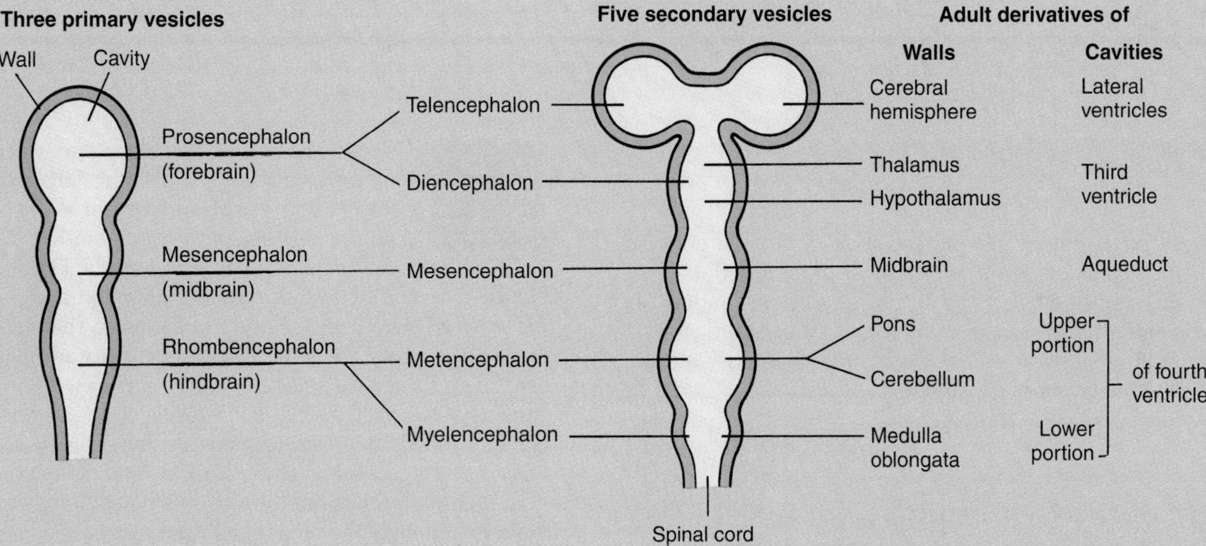

Table 15.1

Derivation and Functions of the Major Brain Structures

	Region	Structure	Function
Prosencephalon (forebrain)	Telencephalon	Cerebrum	Control of most sensory and motor activities; reasoning, memory, intelligence, etc.; instinctual and limbic functions
	Diencephalon	Thalamus	Relay center; all impulses (except olfactory) going into the cerebrum synapse here; some sensory interpretation; initial autonomic response to pain
		Hypothalamus	Regulation of food and water intake, body temperature, heartbeat, etc.; control of secretory activity in anterior pituitary; instinctual and limbic functions
		Pituitary gland	Regulation of other endocrine glands
Mesencephalon (midbrain)	Mesencephalon	Superior colliculi	Visual reflexes (eye-hand coordination)
		Inferior colliculi	Auditory reflexes
		Cerebral peduncles	Reflex coordination; contain many motor fibers
Rhombencephalon (hindbrain)	Metencephalon	Cerebellum	Balance and motor coordination
		Pons	Relay center; contains nuclei (pontine nuclei)
	Myelencephalon	Medulla oblongata	Relay center; contains many nuclei; visceral autonomic center (e.g., respiration, heart rate, vasoconstriction)

Cerebrum

The cerebrum, consisting of five paired lobes within two convoluted hemispheres, is concerned with higher brain functions, including the perception of sensory impulses, the instigation of voluntary movement, the storage of memory, thought processes, and reasoning ability. The cerebrum is also concerned with instinctual and limbic (emotional) functions.

Structure of the Cerebrum

The **cerebrum** (*ser'ĕ-brum*), located in the region of the telencephalon, is the largest and most obvious portion of the brain (fig. 15.3). It accounts for about 80% of the mass of the brain and is responsible for the higher mental functions, including memory and reason. The cerebrum consists of the **right** and **left hemispheres,** which are incompletely separated by a **longitudinal cerebral fissure** (fig. 15.4*b*). Portions of the two hemispheres are connected internally by the **corpus callosum** (*că-lo'sum*), a large tract of white matter (see fig. 15.3*c*). A portion of the meninges, called the **falx** (*falks*) **cerebri** extends into the longitudinal fissure. Each cerebral hemisphere contains a central cavity called the **lateral ventricle** (fig. 15.5), which is lined with ependymal cells and filled with cerebrospinal fluid.

cerebrum: L. *cerebrum*, brain

The two cerebral hemispheres carry out different functions. In most people, the left hemisphere controls analytical and verbal skills, such as reading, writing, and mathematics. The right hemisphere is the source of spatial and artistic kinds of intelligence. The corpus callosum unifies attention and awareness between the two hemispheres and permits a sharing of learning and memory. Severing the corpus callosum is a radical treatment to control severe epileptic seizures. Although this surgery has proven successful, it results in the cerebral hemispheres functioning as separate structures, each with its own information, competing for control. A more recent and effective technique of controlling epileptic seizures is a precise laser treatment of the corpus callosum.

The cerebrum consists of two layers. The surface layer, referred to as the **cerebral cortex,** is composed of gray matter that is 2–4 mm (0.08–0.16 in.) thick (fig. 15.5). Beneath the cerebral cortex is the thick white matter of the cerebrum, which constitutes the second layer. The cerebral cortex is characterized by numerous folds and grooves called **convolutions.** Convolutions form during early fetal development, when brain size increases rapidly and the cortex enlarges out of proportion to the underlying white matter. The elevated folds of the convolutions are the **cerebral gyri** (*ji'ri;* singular, *gyrus*), and the depressed grooves are the **cerebral sulci** (*sul'si;* singular, *sulcus*). The convolutions effectively triple the area of the gray matter, which is composed of cell bodies of neurons.

gyrus: Gk. *gyros*, circle
sulcus: L. *sulcus*, a furrow or ditch

Figure 15.3

The brain.

(*a*) a lateral view, (*b*) an inferior view, and (*c*) a sagittal view. (*d*) A left/right magnetic resonance image (MRI) of the skull, brain, and cervical portion of the spinal cord.

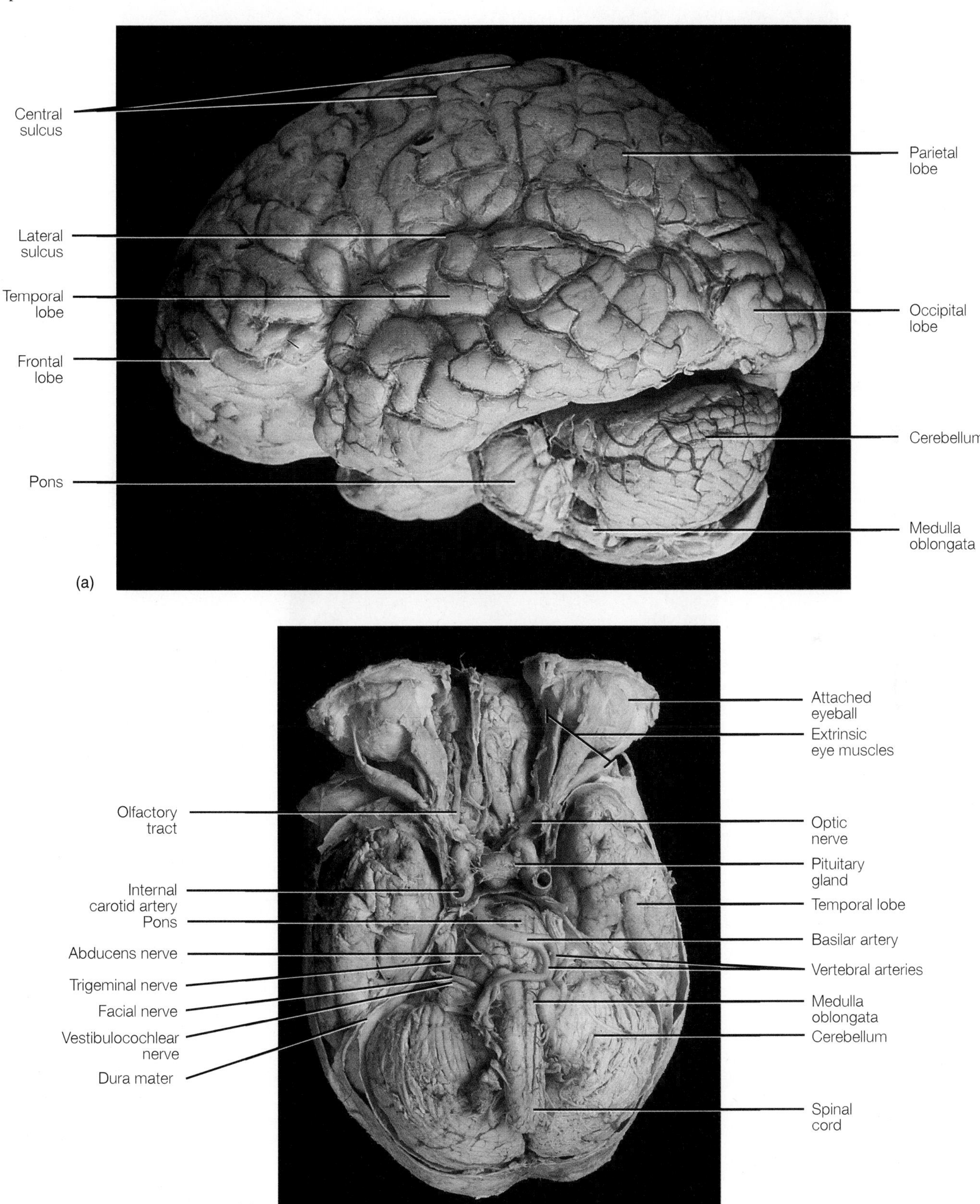

Figure 15.3—*Continued*

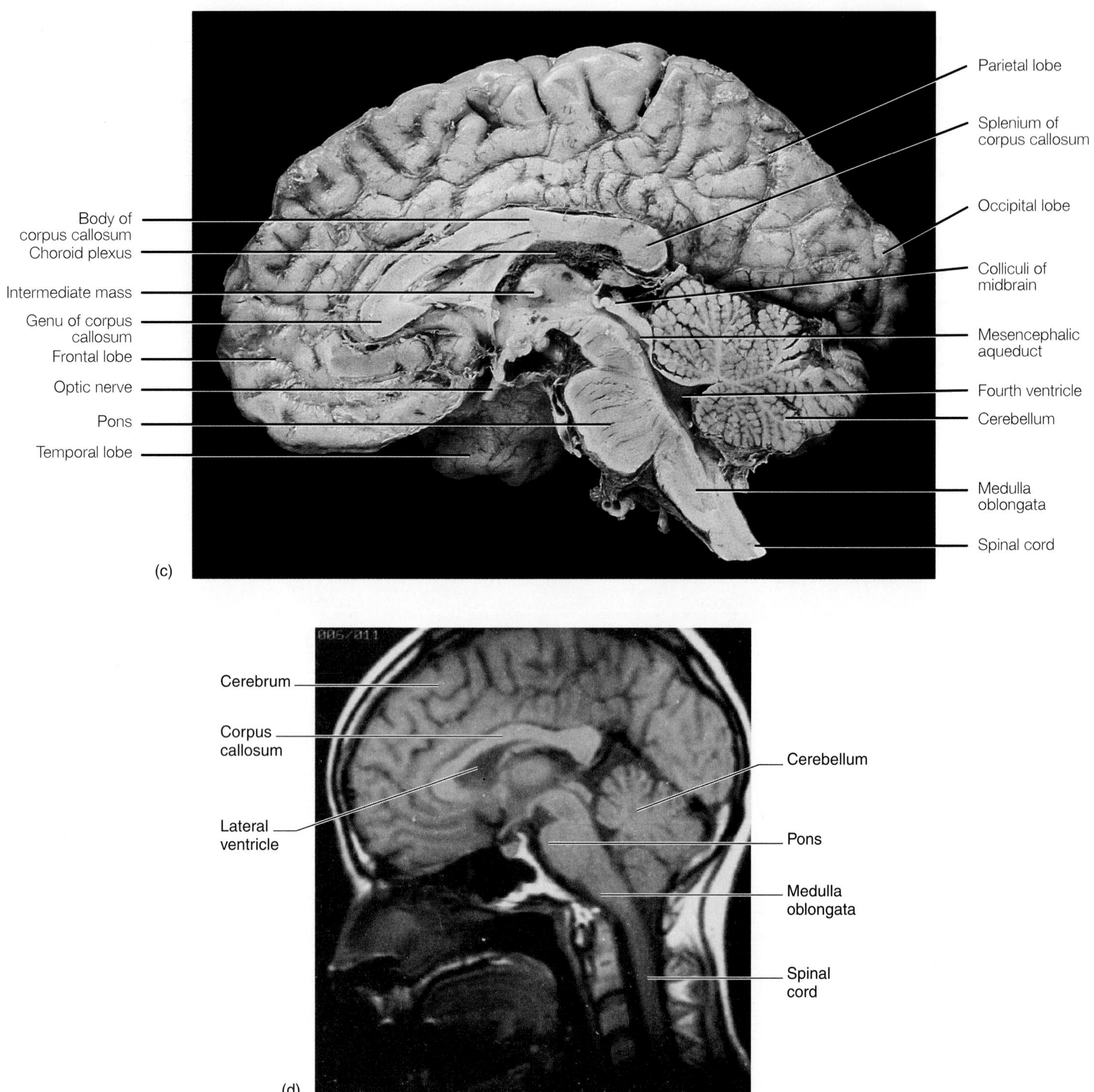
Parietal lobe
Splenium of corpus callosum
Occipital lobe
Colliculi of midbrain
Mesencephalic aqueduct
Fourth ventricle
Cerebellum
Medulla oblongata
Spinal cord
Body of corpus callosum
Choroid plexus
Intermediate mass
Genu of corpus callosum
Frontal lobe
Optic nerve
Pons
Temporal lobe
(c)
Cerebrum
Corpus callosum
Lateral ventricle
Cerebellum
Pons
Medulla oblongata
Spinal cord
(d)

Figure 15.4

The cerebrum.
(*a*) A lateral view and (*b*) a superior view.

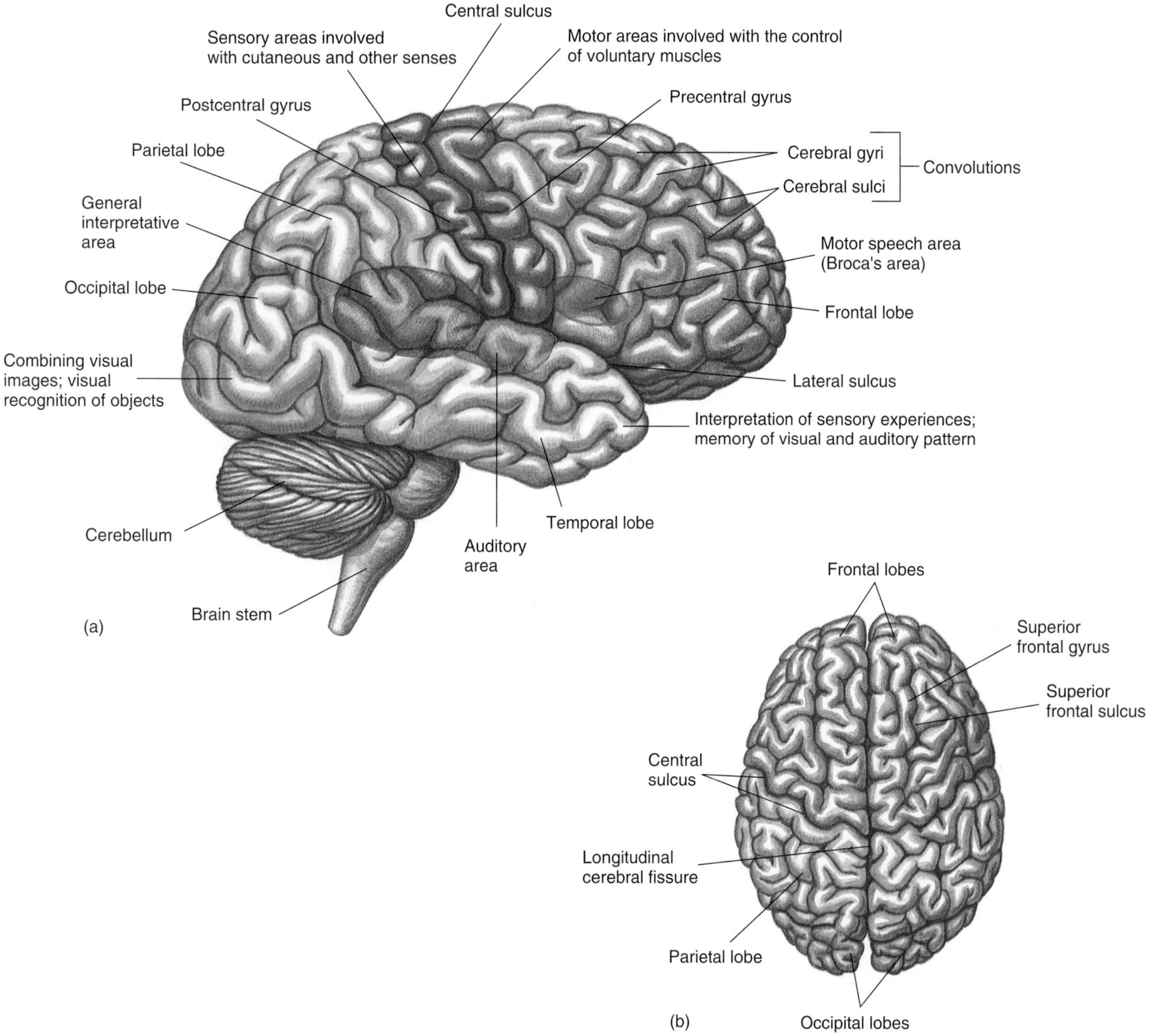

Recent studies indicate that increased learning is accompanied by an increase in the number of synapses between neurons within the cerebrum. Although the number of neurons is established during prenatal development, the number of synapses is variable depending upon the learning process. The number of cytoplasmic extensions from the cell body of a neuron determines the extent of nerve impulse conduction and the associations that can be made to cerebral areas already containing stored information.

Lobes of the Cerebrum

Each cerebral hemisphere is subdivided into five lobes by especially deep sulci. Four of these lobes appear on the surface of the cerebrum and are named according to the overlying cranial bones (fig. 15.6). The reasons for the separate cerebral lobes, as well as two cerebral hemispheres, have to do with specificity of function (table 15.2).

Figure 15.5

Sections through the cerebrum and diencephalon.
(*a*) A coronal section and (*b*) a cross section.

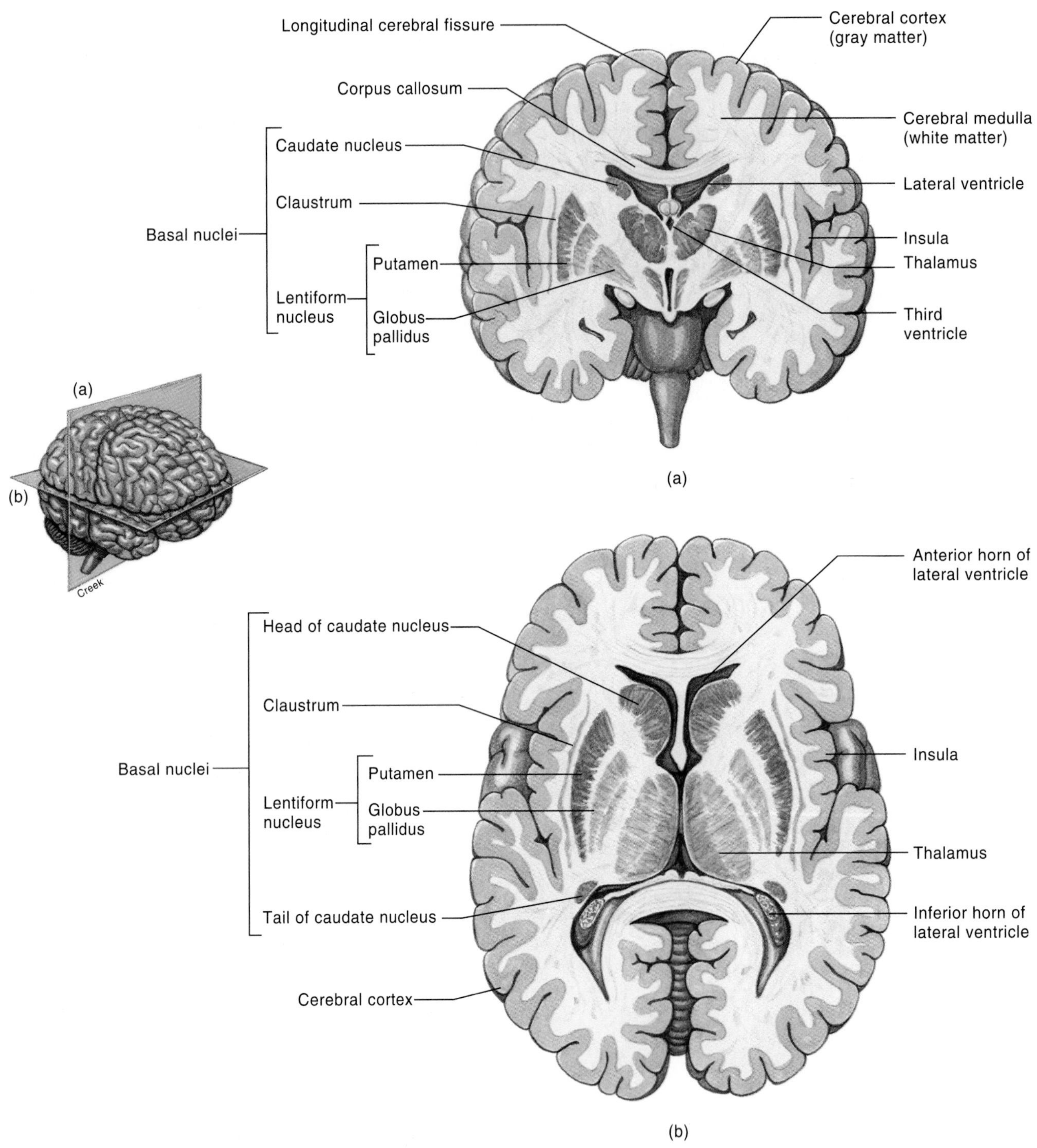

Frontal Lobe

The frontal lobe forms the anterior portion of each cerebral hemisphere (fig. 15.6). A prominent, deep furrow called the **central sulcus** (fissure of Rolando) separates the frontal lobe from the parietal lobe. The central sulcus extends at right angles from the longitudinal fissure to the lateral sulcus. The **lateral sulcus** (fissure of Sylvius) extends laterally from the inferior surface of the cerebrum to separate the frontal and temporal lobes. The **precentral gyrus** (see fig. 15.4), an important motor area, is positioned immediately in front of the central

fissure of Rolando: from Luigi Rolando, Italian anatomist, 1773–1831

fissure of Sylvius: from Franciscus Sylvius de la Boe, Dutch anatomist, 1614–72

Figure 15.6

The lobes of the left cerebral hemisphere showing the principal motor and sensory areas of the cerebral cortex.

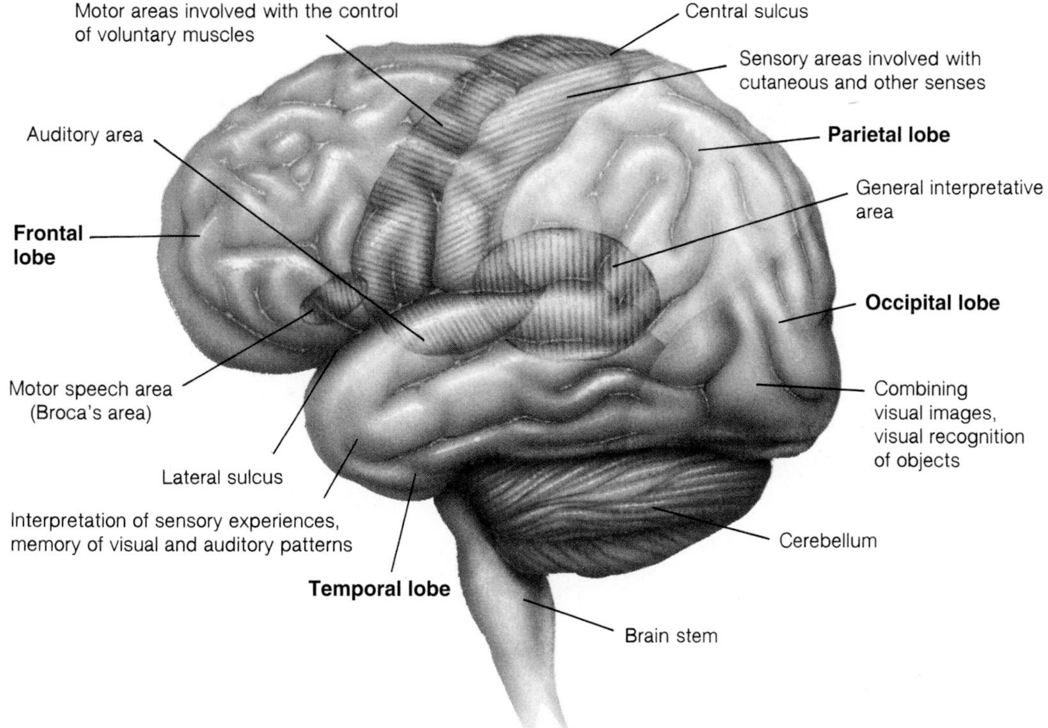

Table 15.2

Functions of the Cerebral Lobes

Lobe	Functions
Frontal	Voluntary motor control of skeletal muscles; personality; higher intellectual processes (e.g., concentration, planning, and decision making); verbal communication
Parietal	Somatesthetic interpretation (e.g., cutaneous and muscular sensations); understanding speech and formulating words to express thoughts and emotions; interpretation of textures and shapes
Temporal	Interpretation of auditory sensations; storage (memory) of auditory and visual experiences
Occipital	Integration of movements in focusing the eye; correlation of visual images with previous visual experiences and other sensory stimuli; conscious perception of vision
Insula	Memory; integration of other cerebral activities

sulcus. Functions of the frontal lobes include initiating voluntary motor impulses for the movement of skeletal muscles, analyzing sensory experiences, and providing responses relating to personality. The frontal lobes also mediate responses related to memory, emotions, reasoning, judgment, planning, and verbal communication.

Parietal Lobe

The parietal lobe is posterior to the central sulcus of the frontal lobe. An important sensory area called the **postcentral gyrus** (fig. 15.4) is positioned immediately behind the central sulcus. The postcentral gyrus is designated as a *somatesthetic area* because it responds to stimuli from cutaneous and muscular receptors throughout the body.

Portions of the precentral gyrus responsible for motor movement and portions of the postcentral gyrus that respond to sensory stimuli do not correspond in size to the part of the body being served but rather to the number of motor units activated or the density of receptors (fig. 15.7). For example, because the hand has many motor units and sensory receptors, larger portions of the precentral and postcentral gyri serve it than serve the thorax, even though the thorax is much larger.

In addition to responding to somatesthetic stimuli, the parietal lobe functions in understanding speech and in articulating thoughts and emotions. The parietal lobe also interprets the textures and shapes of objects as they are handled.

Temporal Lobe

The temporal lobe is located below the parietal lobe and the posterior portion of the frontal lobe. It is separated from both by the lateral sulcus (see fig. 15.6). The temporal lobe contains auditory centers that receive sensory neurons from the cochlea of the ear. This lobe also interprets some sensory experiences and stores memories of both auditory and visual experiences.

Figure 15.7

Motor and sensory areas of the cerebral cortex.
Motor areas control skeletal muscles, and sensory areas receive somatesthetic sensations.

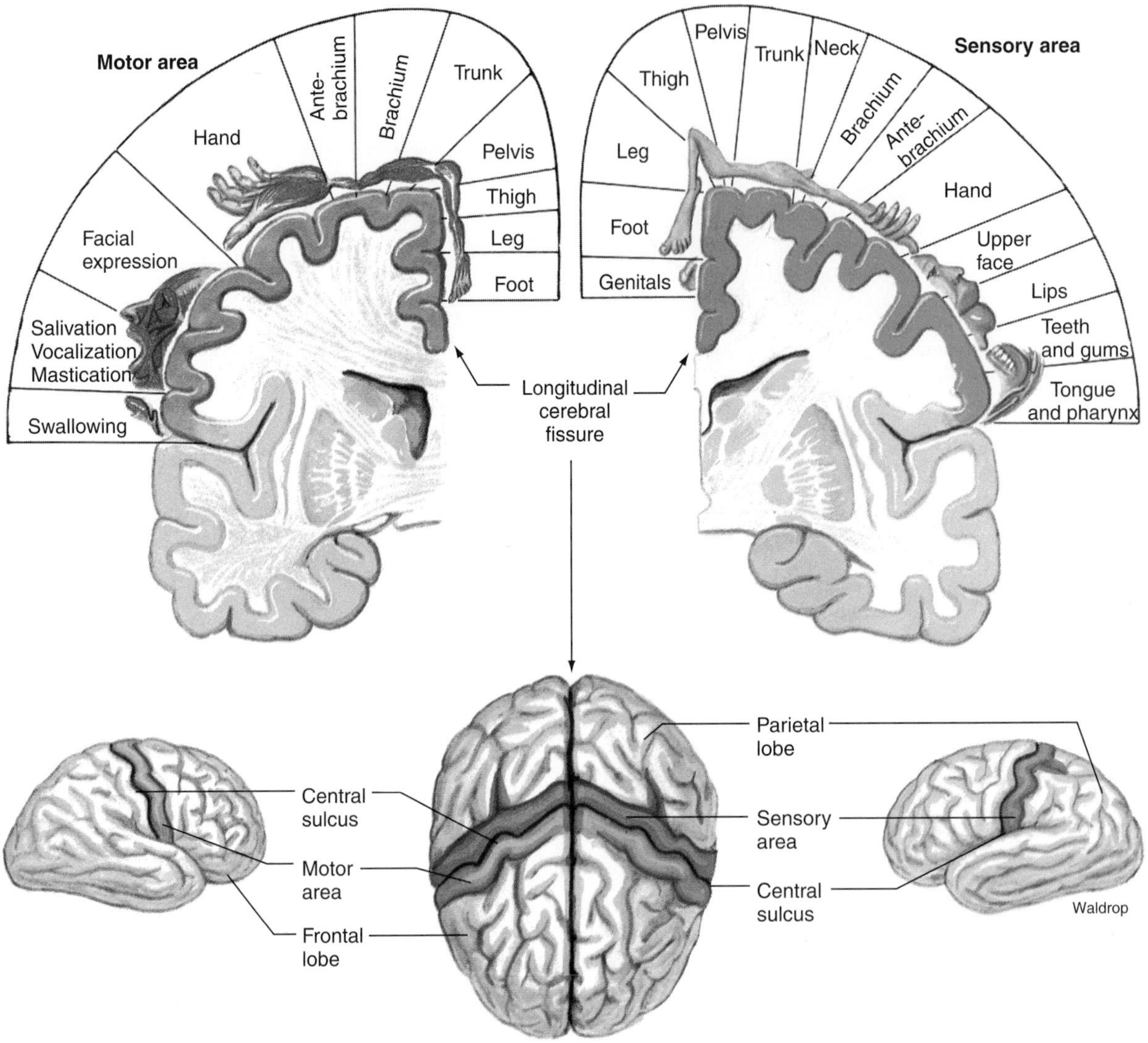

Occipital Lobe

The occipital lobe forms the posterior portion of the cerebrum and is not distinctly separated from the temporal and parietal lobes (see fig. 15.6). It is superior to the cerebellum and is separated from it by an infolding of the meningeal layer called the **tentorium cerebelli** (*ten-to're-um ser"e-bel'i*). The principal functions of the occipital lobe concern vision. It integrates eye movements by directing and focusing the eye. It is also responsible for visual association—correlating visual images with previous visual experiences and other sensory stimuli.

Insula

The insula is a deep lobe of the cerebrum that cannot be viewed on the surface (see fig. 15.5). It lies deep to the lateral sulcus and is covered by portions of the frontal, parietal, and temporal lobes. Little is known of the function of the insula except that it integrates other cerebral activities. It is also thought to have some function in memory.

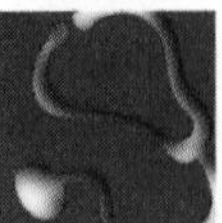

Because of its size and position, portions of the cerebrum frequently suffer brain trauma. A *concussion* to the brain may cause a temporary or permanent impairment of cerebral functions. Much of what is known about cerebral function comes from observing body dysfunctions when specific regions of the cerebrum are traumatized.

Brain Waves

Neurons within the cerebral cortex continuously generate electrical activity. This activity can be recorded by electrodes

insula: L. *insula*, island

Table 15.3
EEG Patterns

Alpha waves—recorded on the scalp over the parietal and occipital regions while a person is awake and relaxed, but with the eyes closed; rhythmic oscillation at about 10 to 12 cycles/sec; in a child under the age of 8, at 4 to 7 cycles/sec.	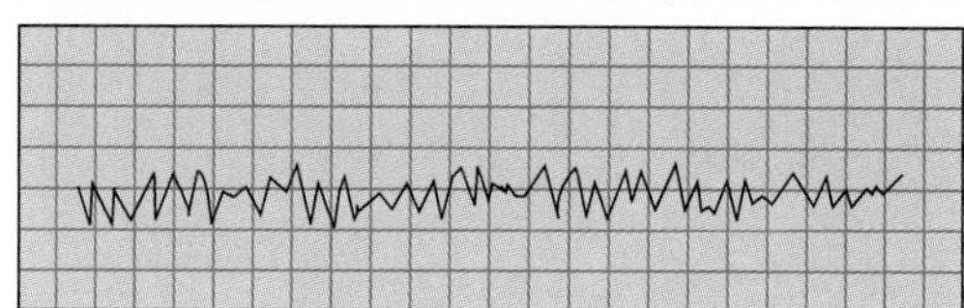
Beta waves—recorded on the scalp over the precentral gyrus of the frontal region while a person is experiencing visual and mental activity; rhythmic oscillation at about 13 to 25 cycles/sec.	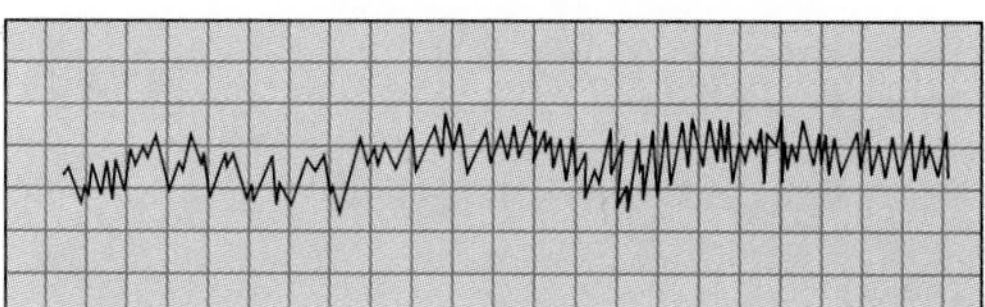
Theta waves—recorded on the scalp over the temporal and occipital lobes while a person is awake and relaxed; rhythmic oscillation at about 5 to 8 cycles/sec; typical in newborns, but its presence in an adult generally indicates severe emotional stress and can be a forewarning of a nervous breakdown.	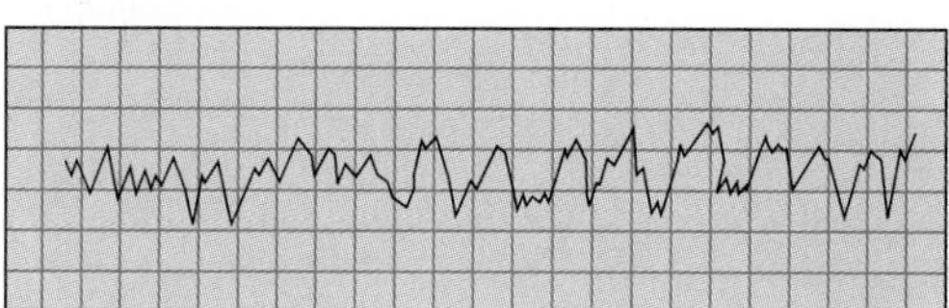
Delta waves—recorded on the scalp over all the cerebral lobes while a person is asleep; rhythmic oscillation at about 1 to 5 cycles/sec; typical in an awake infant, but its presence in an awake adult indicates brain damage.	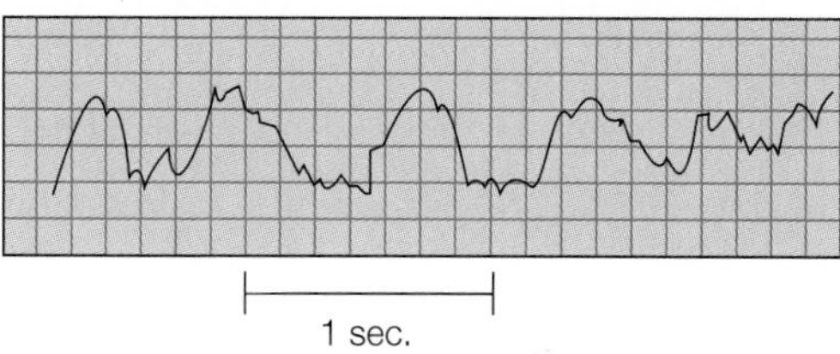

attached to precise locations on the scalp, producing an **electroencephalogram** (*e-lek"tro-en-sef'a-lo-gram*) (EEG). An EEG pattern, commonly called *brain waves,* is the collective expression of millions of action potentials from cerebral neurons.

Brain waves are first emitted from a developing brain during early fetal development and continue throughout a person's life. The cessation of brain-wave patterns (a "flat EEG") may be a decisive factor in the legal determination of death.

Certain distinct EEG patterns signify healthy mental functions. Deviations from these patterns are of clinical significance in diagnosing trauma, mental depression, hematomas, tumors, infections, and epileptic lesions. Normally, there are four kinds of EEG patterns (table 15.3).

White Matter of the Cerebrum

The thick white matter of the cerebrum is deep to the cerebral cortex (see fig. 15.5) and consists of dendrites, myelinated axons, and associated neuroglia. These fibers form the billions of connections within the brain by which information is transmitted to the appropriate places in the form of electrical impulses. The three types of fiber tracts within the white matter are named according to location and the direction in which they conduct impulses (fig. 15.8).

1. **Association fibers** are confined to a given cerebral hemisphere and conduct impulses between neurons within that hemisphere.
2. **Commissural** (*kă-mĭ-shur'al*) **fibers** connect the neurons and gyri of one hemisphere with those of the other. The **corpus callosum** and **anterior commissure** (fig. 15.9) are composed of commissural fibers.
3. **Projection fibers** form the ascending and descending tracts that transmit impulses from the cerebrum to other parts of the brain and spinal cord and from the spinal cord and other parts of the brain to the cerebrum.

Figure 15.8

Types of fiber tracts within the white matter associated with the cerebrum.
(*a*) Association fibers of a given hemisphere. (*b*) Commissural fibers connecting the hemispheres and projection fibers connecting the hemispheres with other structures of the CNS. (Note the decussation [crossing-over] of projection fibers within the medulla oblongata.)

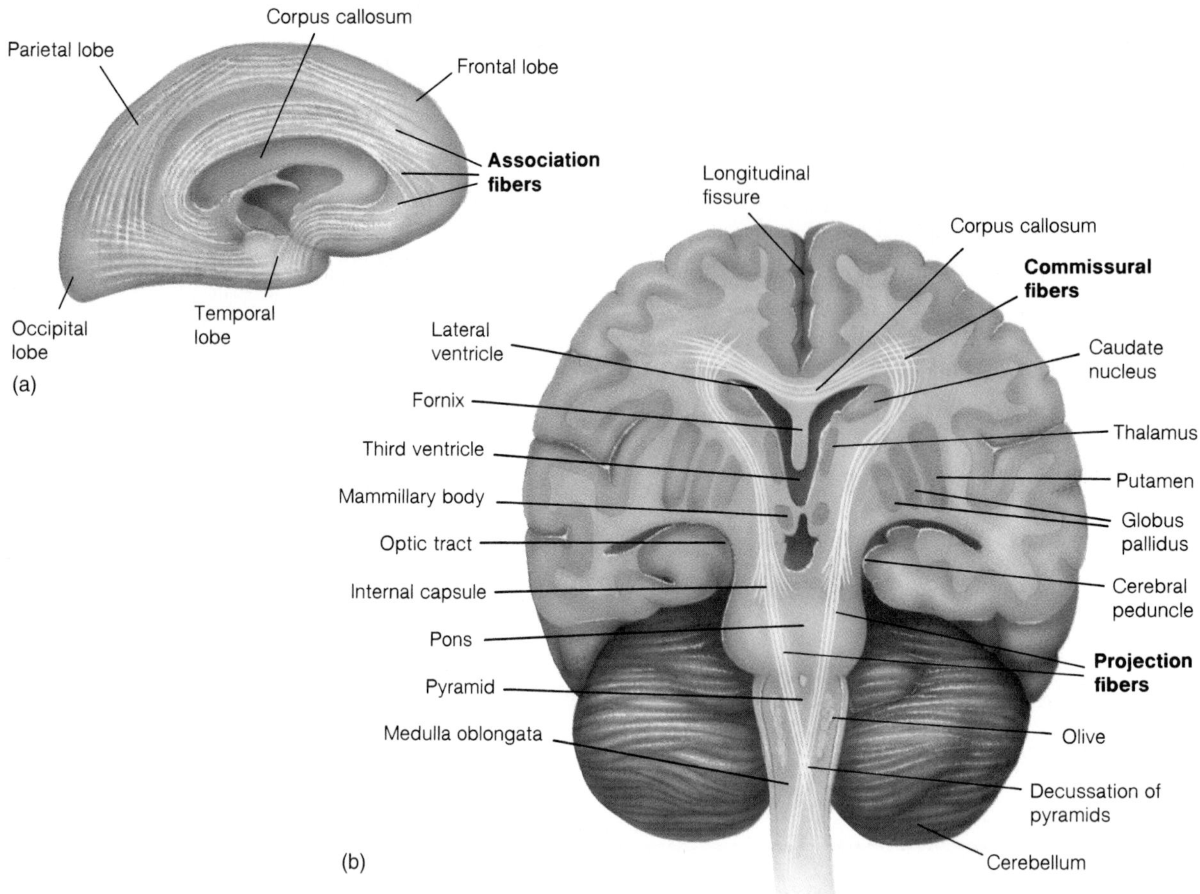

Basal Nuclei

The **basal nuclei** are specialized paired masses of gray matter located deep within the white matter of the cerebrum (fig. 15.10). The most prominent of the basal nuclei is the **corpus striatum,** so-named because of its striped appearance. The corpus striatum is composed of several masses of nuclei. The **caudate nucleus** is the upper mass. A thick band of white matter lies between the caudate nucleus and the next two masses underneath, collectively called the **lentiform nucleus.** The lentiform nucleus consists of a lateral portion, called the **putamen** (*pyoo-ta'men*), and a medial portion, called the **globus pallidus.** The **claustrum** (*klos'trum*) is another portion of the basal nuclei. It is a thin layer of gray matter just deep to the cerebral cortex of the insula.

The basal nuclei are associated with other structures of the brain, particularly within the mesencephalon. The caudate nucleus and putamen of the basal nuclei control unconscious contractions of certain skeletal muscles, such as those of the upper extremities involved in involuntary arm movements during walking. The globus pallidus regulates the muscle tone necessary for specific, intentional body movements. Neural diseases or physical trauma to the basal nuclei generally cause a variety of motor movement dysfunctions, including rigidity, tremor, and rapid and aimless movements.

Language

Knowledge of the brain regions involved in language has been gained primarily by the study of *aphasias*—speech and language disorders caused by damage to specific language areas of the brain. These areas (fig. 15.11) are generally located in the cerebral cortex of the left hemisphere in both right-handed and left-handed people.

The **motor speech area** (Broca's area) is located in the left inferior gyrus of the frontal lobe. Neural activity in the

lentiform: L. *lentis*, elongated
putamen: L. *putare*, to cut, prune
globus pallidus: L. *globus*, sphere; *pallidus*, pale

aphasia: L. *a*, without; Gk. *phasis*, speech
Broca's area: from Pierre P. Broca, French neurologist, 1824–80

Figure 15.9

A midsagittal section through the brain.

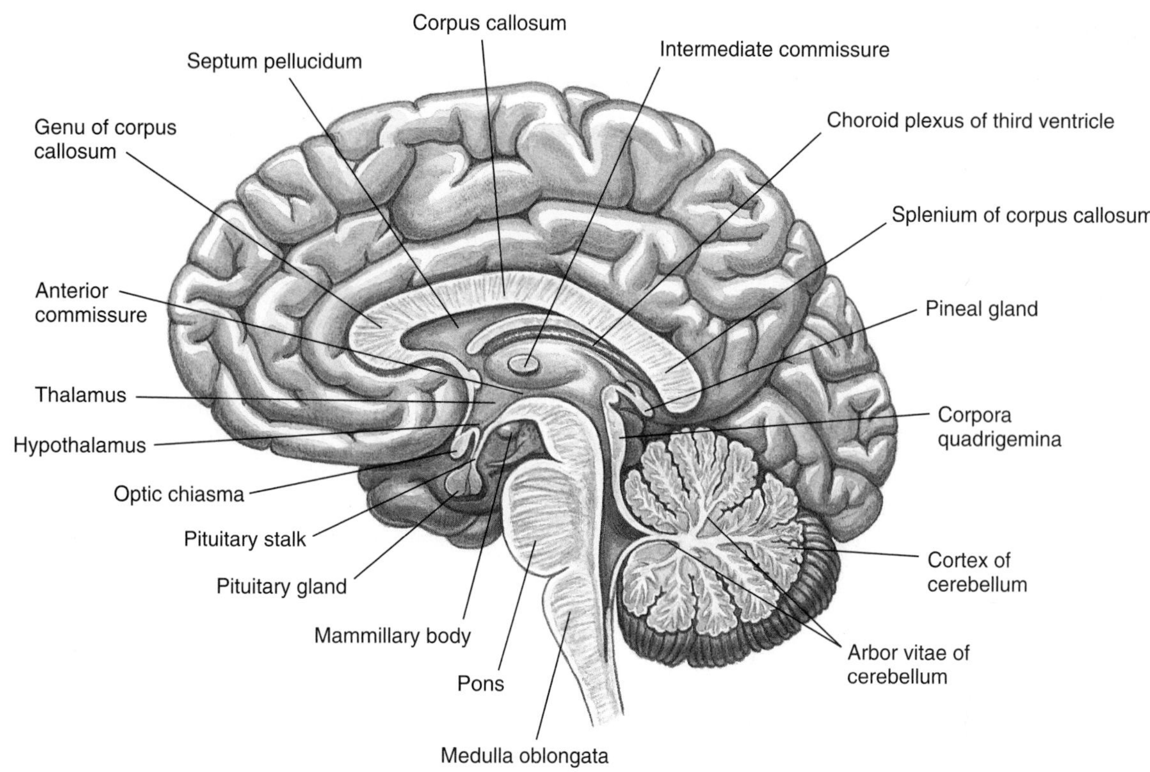

Figure 15.10

Structures of the cerebrum containing nuclei involved in the control of skeletal muscles.
The thalamus is a relay center between the motor cerebral cortex and the other brain areas.

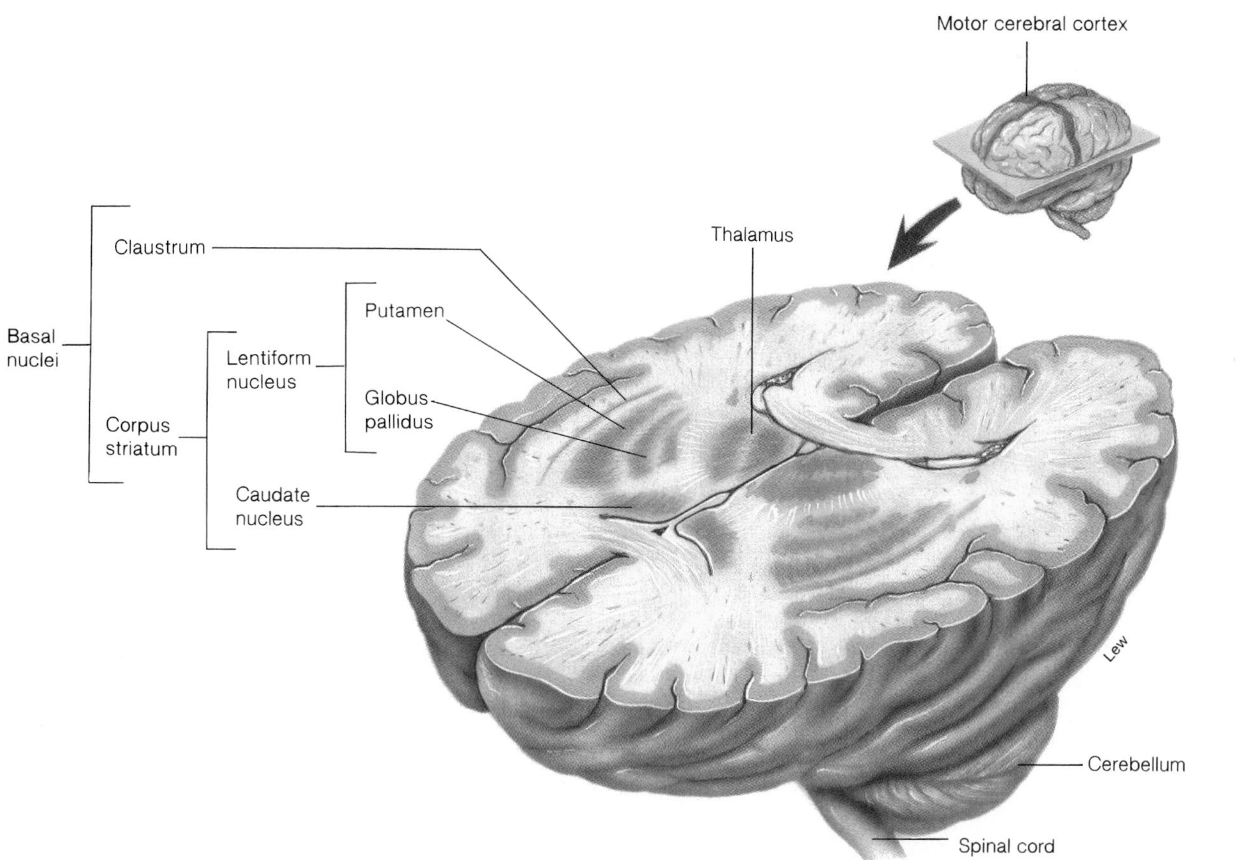

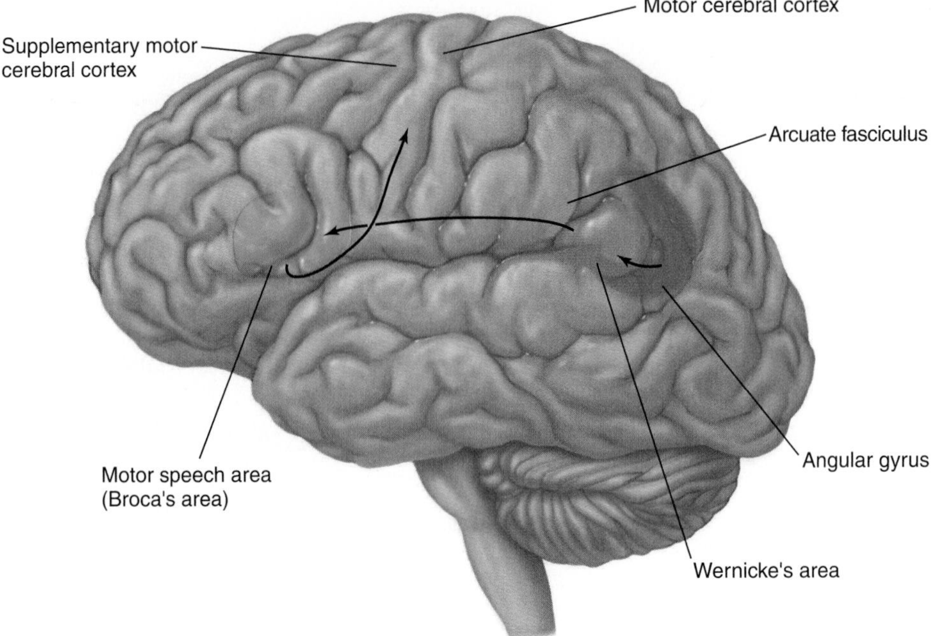

Figure 15.11

Brain areas involved in the control of speech.
Arrows indicate the direction of communication between these areas.

motor speech area causes selective stimulation of motor impulses in motor centers elsewhere in the frontal lobe, which in turn causes coordinated skeletal muscle movement in the pharynx and larynx. At the same time, motor impulses are sent to the respiratory muscles to regulate air movement across the vocal folds (cords). The combined muscular stimulation translates thought patterns into speech.

Wernicke's (*ver'ni-kez*) **area** is located in the superior gyrus of the temporal lobe and is directly connected to the motor speech area by a fiber tract called the **arcuate fasciculus.** People with *Wernicke's aphasia* produce speech that has been described as a "word salad." The words used may be real words that are randomly mixed together, or they may be made-up words. Language comprehension has been destroyed in people with Wernicke's aphasia; they cannot understand either spoken or written language.

It appears that the concept of words to be spoken originates in Wernicke's area and is then communicated to the motor speech area through the arcuate fasciculus. Damage to the arcuate fasciculus produces *conduction aphasia,* which is fluent but nonsensical speech as in Wernicke's aphasia, even though both the motor speech area and Wernicke's area are intact.

The **angular gyrus,** located at the junction of the parietal, temporal, and occipital lobes, is believed to be a center for the integration of auditory, visual, and somatesthetic information. Damage to the angular gyrus produces aphasias, which suggests that this area projects to Wernicke's area. Some patients with damage to the left angular gyrus can speak and understand spoken language but cannot read or write. Other patients can write a sentence but cannot read it, presumably due to damage to the projections from the occipital lobe (involved in vision) to the angular gyrus.

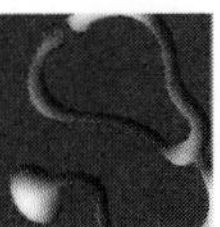

Recovery of language ability, by transfer to the right hemisphere after damage to the left hemisphere, is very good in children but decreases after adolescence. Recovery is reported to be faster in left-handed people, possibly because language ability is more evenly divided between the two hemispheres in left-handed people. Some recovery usually occurs after damage to the motor speech area, but damage to Wernicke's area produces more severe and permanent aphasias.

Memory

Clinical studies of amnesia suggest that several different brain regions are involved in memory storage and retrieval. Amnesia has been found to result from damage to the temporal lobe of the cerebral cortex, hippocampus, caudate nucleus (in Huntington's disease), or the dorsomedial thalamus (in alcoholics suffering from Korsakoff's syndrome with thiamine deficiency). Clinical studies also suggest that there are two major categories of memory: **short-term memory** and **long-term memory.** People with head trauma, for example, and patients with suicidal depression who are treated by *electroconvulsive shock* (*ECS*) *therapy* may lose their memory of recent events but retain their older memories.

The **hippocampus** (see fig. 15.12) appears to be required for short-term memory and for the consolidation of that memory into a long-term form. The surgical removal of the left hippocampus due to the presence of a tumorous growth impairs the consolidation of short-term verbal memories, and removal of the right hippocampus impairs the consolidation of nonverbal memories. The surgical removal of both the left and the right hippocampi leave a patient totally without short-term memory.

Wernicke's area: from Karl Wernicke, German neurologist, 1848–1905

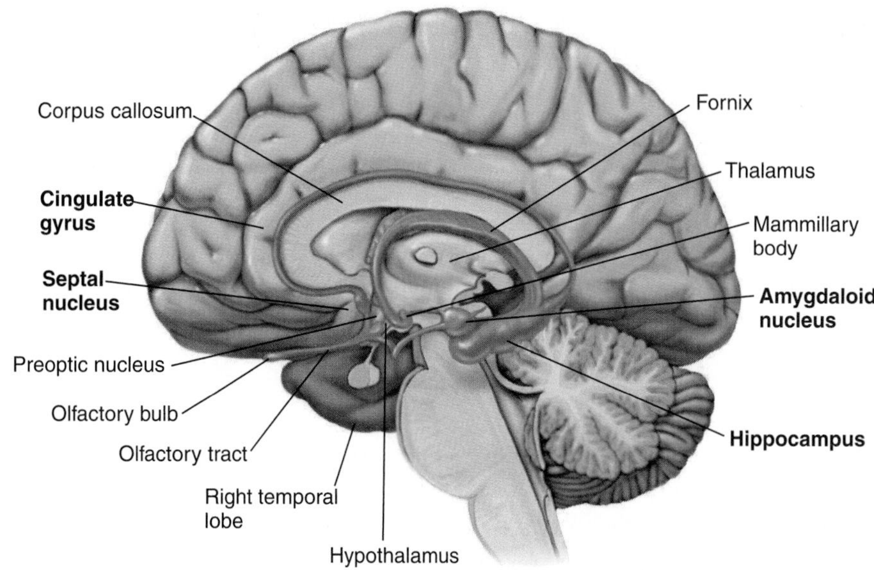

Figure 15.12

The limbic system and its interconnecting neural pathway. (*Note:* The cerebral cortex of the left temporal lobe has been removed. The structures of the limbic system are indicated in boldface type.)

Memories are stored in the cerebral cortex, with verbal information lateralized to the left hemisphere and visuospatial information to the right hemisphere. Electrical stimulation of various regions in the cerebrum of awake patients often evokes visual or auditory memories that are extremely vivid. Electrical stimulation of specific points in the temporal lobe evokes specific memories in such detail that patients feel as if the events were currently being experienced. Surgical removal of these regions does not, however, eradicate the memory. The amount of memory destroyed by ablation of brain tissue appears to depend more on the amount of brain tissue removed than on the location of the surgery. On the basis of these observations, it appears that memory may be diffusely located in the cerebrum, and that stimulation of the correct location of the cerebral cortex then retrieves the memory.

Since long-term memory is not eradicated by electroconvulsive shock, it seems reasonable to conclude that the consolidation of memory depends on relatively permanent changes in the chemical structure of neurons and their synapses. Experiments suggest that protein synthesis is required for the consolidation of the "memory trace." According to one theory, these proteins may be secreted into the intracellular environment, where they influence synaptic connections. Another theory holds that new receptor proteins in the membrane of the postsynaptic neuron are made available as a result of high-frequency stimulation of the presynaptic neuron. This would help to account for the increased sensitivity of postsynaptic neurons to neurotransmitters, as seen in long-term potentiation (discussed in chapter 14). Much more research is obviously needed in this exciting area of physiology before memory can be fully explained at a cellular and molecular level.

Emotion

The parts of the brain that appear to be of paramount importance in the neural basis of emotional states are the *hypothalamus* (in the diencephalon) and the **limbic system.** The limbic system consists of a group of forebrain nuclei and fiber tracts that form a ring around the brain stem. The structures of the limbic system include the **cingulate gyrus** (part of the cerebral cortex), **amygdaloid nucleus** (or *amygdala*), **hippocampus,** and the **septal nuclei** (fig. 15.12).

The limbic system was once called the *rhinencephalon*, or "smell brain," because it is involved in the central processing of olfactory information. This may be its primary function in lower vertebrates, whose limbic system may constitute the entire forebrain. It is now known however, that the limbic system in humans is a center for basic emotional drives. The limbic system was derived early in the course of vertebrate evolution, and its tissue is phylogenetically older than the cerebral cortex. There are thus few synaptic connections between the cerebral cortex and the structures of the limbic system, which perhaps helps to explain why we have so little conscious control over our emotions.

There is a closed circuit of information flow between the limbic system and the thalamus and hypothalamus. Through these interconnections, the limbic system and the hypothalamus appear to cooperate in the neural basis of emotional states.

Studies of the functions of these regions include electrical stimulation of specific locations, destruction of tissue (producing *lesions*) in particular sites, and surgical removal, or *ablation*, of specific structures. These studies suggest that the hypothalamus and limbic system are involved in the following feelings and behaviors:

1. **Aggression.** Stimulation of certain areas of the amygdala produces rage and aggression, and lesions of the amygdala can produce docility in experimental animals. Stimulation of particular areas of the hypothalamus can produce similar effects.

limbic: L. *limbus*, edge or border

2. **Fear.** Fear can be produced by electrical stimulation of the amygdala and hypothalamus, and surgical removal of the limbic system can result in an absence of fear. Monkeys are normally terrified of snakes, for example, but with their limbic system removed, they can handle snakes without fear.
3. **Feeding.** The hypothalamus contains both a *feeding center* and a *satiety center.* Electrical stimulation of the former causes overeating, and stimulation of the latter will stop feeding behavior in experimental animals.
4. **Sex.** The hypothalamus and limbic system are involved in the regulation of the sexual drive and sexual behavior, as shown by stimulation and ablation studies in experimental animals. The cerebral cortex, however, is also critically important for the sex drive in vertebrate animals, and especially so in humans.
5. **Goal-directed behavior (reward and punishment system).** Electrodes placed in particular sites between the frontal cortex and the hypothalamus can deliver shocks that function as a reward. In rats, this reward is more powerful than food or sex in motivating behavior. Similar studies have been done in humans, who report feelings of relaxation and relief from tension, but not of ecstasy. Electrodes placed in slightly different positions apparently stimulate a punishment system in experimental animals, who stop their behavior when stimulated in these regions.

Diencephalon

The diencephalon is a major autonomic region of the brain that consists of such vital structures as the thalamus, hypothalamus, epithalamus, and pituitary gland.

The **diencephalon** (*di″en-sef′a-lon*) is the second subdivision of the forebrain and is almost completely surrounded by the cerebral hemispheres of the telencephalon. The third ventricle (see fig. 15.20) is a narrow midline cavity within the diencephalon. The most important structures of the diencephalon are the thalamus (*thal′ă-mus*), hypothalamus (*hi″po-thal′ă-mus*), epithalamus, and pituitary (*pĭ-too′ĭ-ter-e*) gland.

Thalamus

The **thalamus** is a large ovoid mass of gray matter, constituting nearly four-fifths of the diencephalon. It is actually a paired organ, with each portion positioned immediately below the lateral ventricle of its respective cerebral hemisphere (see fig. 15.5). The principal function of the thalamus is to act as a relay center for all sensory impulses, except smell, to the cerebral cortex. Specialized masses of nuclei relay the incoming impulses to precise locations within the cerebral lobes for interpretation.

The thalamus also performs some sensory interpretation. The cerebral cortex discriminates pain and other tactile stimuli, but the thalamus responds to general sensory stimuli and provides crude awareness. The thalamus probably plays a role in the initial autonomic response of the body to intense pain, and is therefore, partially responsible for the physiological shock that frequently follows serious trauma.

Hypothalamus

The **hypothalamus,** named for its position below the thalamus, is the most inferior portion of the diencephalon. It forms the floor and part of the lateral walls of the third ventricle (figs. 15.9 and 15.13) and contains several masses of nuclei that are interconnected with other parts of the nervous system. Despite its small size, the hypothalamus performs numerous vital functions, most of which relate directly or indirectly to the regulation of visceral activities. It also participates with the limbic system in emotional states, as previously discussed.

The hypothalamus acts as an autonomic nervous center in accelerating or decelerating certain body functions. It secretes eight hormones, including two released from the posterior pituitary. These hormones and their functions are discussed in chapter 19. The principal autonomic and limbic (emotional) functions of the hypothalamus are as follows:

1. **Cardiovascular regulation.** Although the heart has an innate pattern of contraction, impulses from the hypothalamus indirectly cause autonomic acceleration or deceleration of the heart. Impulses from the posterior hypothalamus produce a rise in arterial blood pressure and an increase of the heart rate. Impulses from the anterior portion have the opposite effect. Rather than traveling directly to the heart, impulses from these regions pass first to the cardiovascular centers of the medulla oblongata.
2. **Body-temperature regulation.** Specialized nuclei within the anterior portion of the hypothalamus are sensitive to changes in body temperature. If the arterial blood flowing through this portion of the hypothalamus is above normal temperature, the hypothalamus initiates impulses that cause heat loss through sweating and vasodilation of cutaneous vessels of the skin. A below-normal blood temperature causes the hypothalamus to relay impulses that result in heat production and retention through shivering, contraction of cutaneous blood vessels, and cessation of sweating.
3. **Regulation of water and electrolyte balance.** Specialized *osmoreceptors* in the hypothalamus continuously monitor the osmotic concentration of the blood. An increased osmotic concentration resulting from lack of water causes the production of antidiuretic hormone (ADH) by the hypothalamus and its release from the posterior pituitary. At the same time, a *thirst center* within the hypothalamus produces feelings of thirst.

---
thalamus: L. *thalamus*, inner room

Figure 15.13

The pituitary gland.
This endocrine gland is positioned within the sella turcica of the sphenoid bone and is attached to the brain by the pituitary stalk.

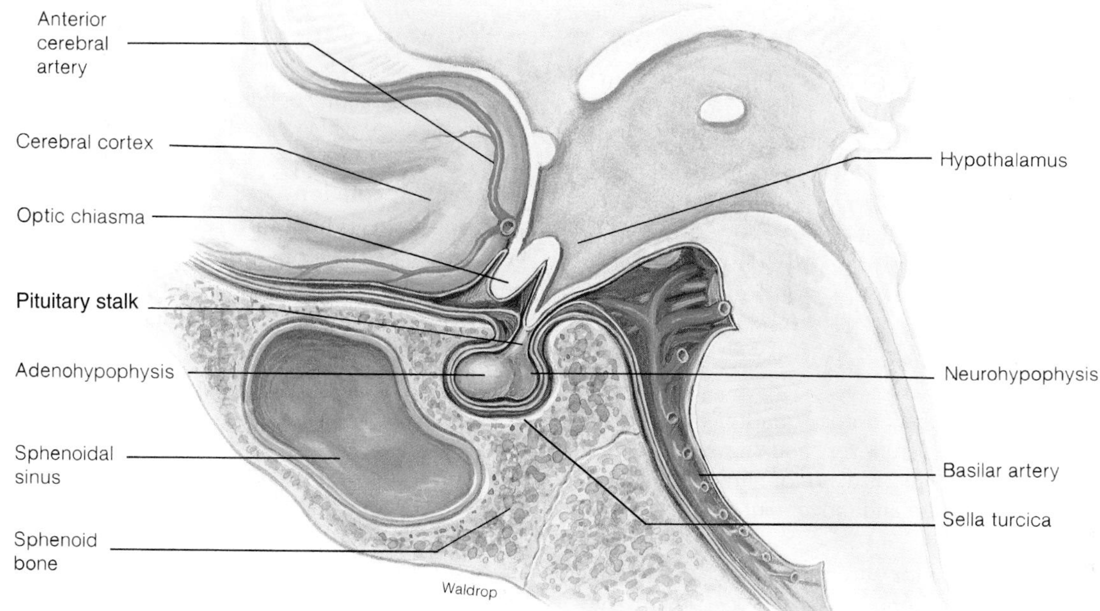

4. **Regulation of hunger and control of gastrointestinal activity.** The *feeding center* is a specialized portion of the lateral hypothalamus that monitors the blood glucose, fatty acid, and amino acid levels. Low levels of these substances in the blood are partially responsible for a sensation of hunger elicited from the hypothalamus. When enough food has been eaten, the *satiety* (*să-ti'ĭ-te*) *center* in the midportion of the hypothalamus inhibits the feeding center. The hypothalamus also receives sensory impulses from the abdominal viscera and regulates glandular secretions and the peristaltic movements of the GI tract.
5. **Regulation of sleeping and wakefulness.** The hypothalamus has both a *sleep center* and a *wakefulness center* that function with other parts of the brain to determine the level of conscious alertness.
6. **Sexual response.** Specialized *sexual center* nuclei within the superior portion of the hypothalamus respond to sexual stimulation of the tactile receptors within the genital organs. The experience of orgasm involves neural activity with the sexual center of the hypothalamus.
7. **Emotions.** A number of nuclei within the hypothalamus are associated with specific emotional responses, including anger, fear, pain, and pleasure.
8. **Control of endocrine functions.** The hypothalamus produces neurosecretory chemicals that stimulate the anterior and posterior pituitary to release various hormones.

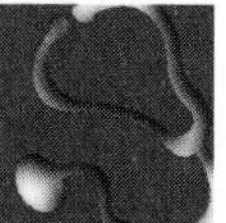

The hypothalamus is a vital structure in maintaining overall body homeostasis. Dysfunction of the hypothalamus may seriously affect autonomic, somatic, or psychic body functions. Not surprisingly, this organ is implicated as a principal factor in psychosomatic illness. Insomnia, peptic ulcers, palpitation of the heart, diarrhea, and constipation are a few symptoms of psychophysiologic disorders.

Epithalamus

The **epithalamus** is the posterior portion of the diencephalon that forms a thin roof over the third ventricle (see fig. 15.21). The inside lining of the roof consists of a vascular **choroid plexus** where cerebrospinal fluid is produced (see fig. 15.19). A small cone-shaped mass called the **pineal** (*pin'e-al*) **gland,** named for its resemblance to a small pine cone, extends outward from the posterior end of the epithalamus (see fig. 15.19). It is thought to have a neuroendocrine function. The **posterior commissure,** located inferior to the pineal gland, is a tract of commissural fibers that connects the right and left superior colliculi of the midbrain (see fig. 15.16).

pineal: L. *pinea*, pine cone

Pituitary Gland

The rounded, pea-shaped **pituitary gland,** or **cerebral hypophysis** (*hi-pof'ĭ-sis*), is positioned on the inferior aspect of the diencephalon (fig. 15.13). It is attached to the hypothalamus by the **pituitary stalk** and is supported by the sella turcica of the sphenoid bone (fig. 15.12). The cerebral arterial circle (circle of Willis) (see fig. 21.21), surrounds the highly vascular pituitary gland, providing it with a rich blood exchange. The pituitary, which has an endocrine function, is structurally and functionally divided into an anterior portion, called the **adenohypophysis** (*ad"ĕ-no-hi-pof'ĭ-sis*), and a posterior portion, called the **neurohypophysis** (see chapter 19).

Mesencephalon

The mesencephalon contains the corpora quadrigemina, concerned with visual and auditory reflexes, and the cerebral peduncles, composed of fiber tracts. It also contains specialized nuclei that help to control posture and movement.

The *brain stem* contains nuclei for autonomic functions of the body and their connecting tracts. It is that portion of the brain that attaches to the spinal cord and includes the midbrain, pons, and medulla oblongata. The **mesencephalon** (*mes"en-sef'ă-lon*), or **midbrain,** is the short section of the brain stem between the diencephalon and the pons (see fig. 15.16). Within the midbrain is the **mesencephalic aqueduct** (aqueduct of Sylvius) (see fig. 15.21), which connects the third and fourth ventricle. The midbrain also contains the corpora quadrigemina (see fig. 15.9), cerebral peduncles (see fig. 15.8), red nucleus, and substantia nigra.

The **corpora quadrigemina** (*kwad"rĭ-jem'ĭ-nă*) are the four rounded elevations on the posterior portion of the midbrain (see fig. 15.16). The two upper eminences, the **superior colliculi** (*ko-lik'yu-li*) are concerned with visual reflexes. The two posterior eminences, the **inferior colliculi,** are responsible for auditory reflexes. The **cerebral peduncles** (*pĕ-dung'k'lz*) are a pair of cylindrical structures composed of ascending and descending projection fiber tracts that support and connect the cerebrum to the other regions of the brain.

The **red nucleus** is deep within the midbrain between the cerebral peduncle and the cerebral aqueduct. It connects the cerebral hemispheres and the cerebellum and functions in reflexes concerned with motor coordination and maintenance of posture. Its reddish color is due to its rich blood supply and an iron-containing pigment in the cell bodies of its neurons.

pituitary: L. *pituita*, phlegm (this gland was originally thought to secrete mucus into the nasal cavity)
aqueduct of Sylvius: Sylvius, French anatomist, 1478–1555
corpora quadrigemina: L. *corpus*, body; *quadri*, four; *geminus*, twin
colliculus: L. *colliculus*, small mound
peduncle: L. *peduncle*, diminutive of *pes*, foot

Another nucleus, the **substantia nigra** (*ni'gra*), lies inferior to the red nucleus. The substantia nigra is thought to inhibit involuntary movements. Its dark color reflects its high content of melanin pigment.

Metencephalon

The metencephalon contains the pons, which relays impulses, and the cerebellum, which coordinates skeletal muscle contractions.

The **metencephalon** (*met"en-sef'ă-lon*) is the most superior portion of the hindbrain. Two vital structures of the metencephalon are the pons and cerebellum. The mesencephalic aqueduct of the mesencephalon enlarges to become the **fourth ventricle** (see fig. 15.21) within the metencephalon and myelencephalon.

Pons

The **pons** can be observed as a rounded bulge on the underside of the brain, between the midbrain and the medulla oblongata (fig. 15.14). It consists of white fiber tracts that course in two principal directions. The surface fibers extend transversely to connect with the cerebellum through the middle

pons: L. *pons*, bridge

Figure 15.14

The respiratory center.
Nuclei within the pons and medulla oblongata that constitute the respiratory center.

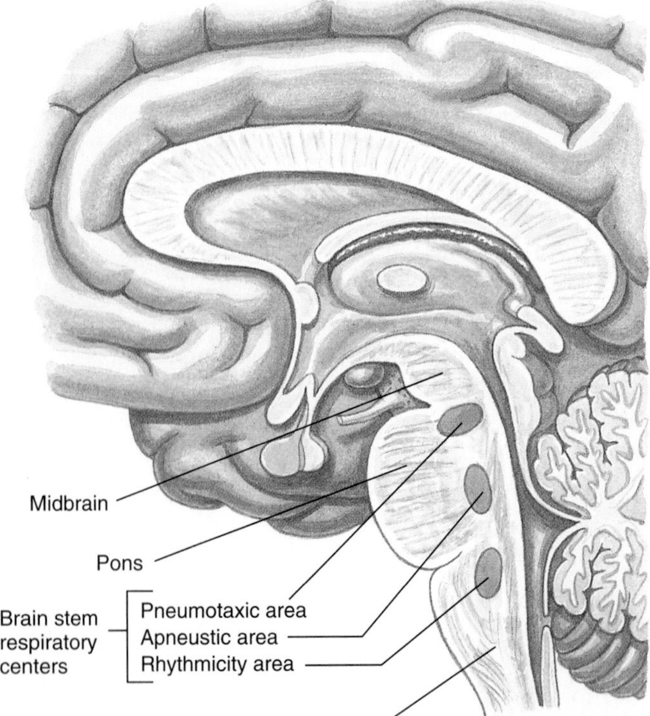

Figure 15.15

The structure of the cerebellum.
(*a*) a superior view, (*b*) an inferior view, and (*c*) a sagittal view.

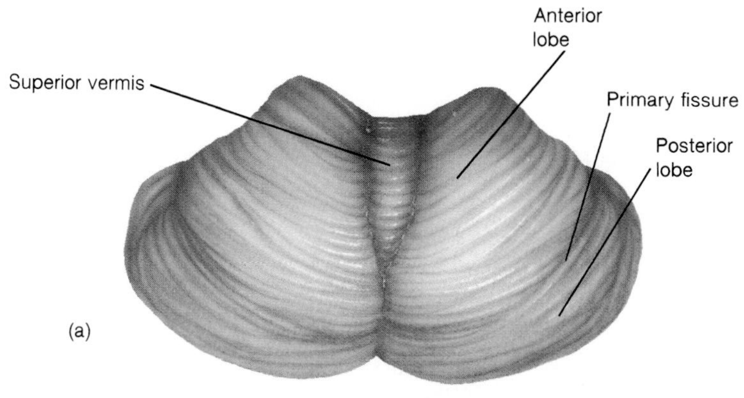

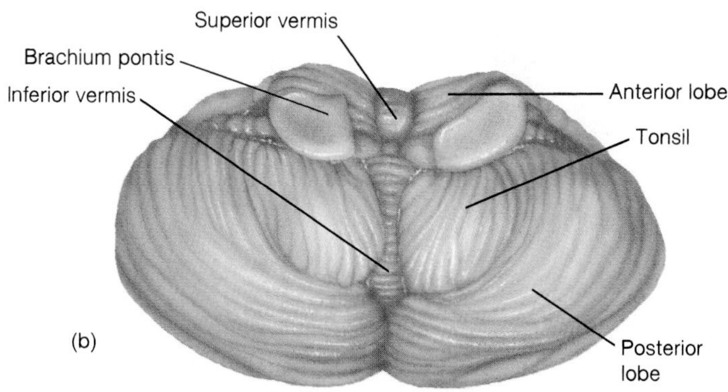

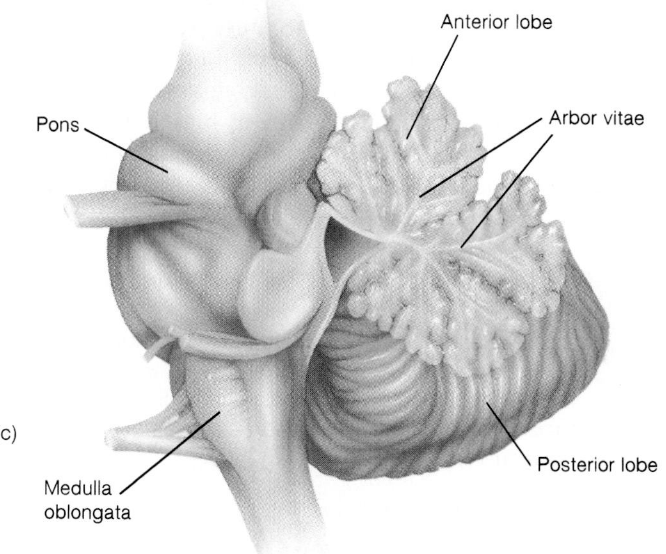

cerebellar peduncles. The deeper longitudinal fibers are part of the motor and sensory tracts that connect the medulla oblongata with the tracts of the midbrain.

Scattered throughout the pons are several nuclei associated with specific cranial nerves (see fig. 15.17). The cranial nerves that have nuclei within the pons include the trigeminal (V), which transmits impulses for chewing and sensory sensations from the head; the abducens (VI), which controls certain movements of the eyeball; the facial (VII), which transmits impulses for facial movements and sensory sensations from the taste buds; and the vestibular branches of the vestibulocochlear (VIII), which maintain equilibrium.

Other nuclei of the pons function with nuclei of the medulla oblongata to regulate the rate and depth of breathing. The two respiratory centers of the pons are called the **apneustic** and the **pneumotaxic areas** (fig. 15.14).

Cerebellum

The **cerebellum** (*ser″ĕ-bel′um*), the second largest structure of the brain, is located in the metencephalon and occupies the inferior and posterior aspect of the cranial cavity. The cerebellum is separated from the overlying cerebrum by a **transverse fissure** (see fig. 15.4). A portion of the meninges called the **tentorium cerebelli** extends into the transverse fissure. The cerebellum consists of two **hemispheres** and a central constricted area called the **vermis** (fig. 15.15). The **falx cerebelli** is the portion of the meninges that partially extends between the hemispheres (table 15.4).

cerebellum: L. *cerebellum*, diminutive of cerebrum, brain
vermis: L. *vermis*, worm

Table 15.4

Septa of the Cranial Dura Mater

Septa	Location
Falx cerebri	Extends downward into the longitudinal fissure to partition the right and left cerebral hemispheres; anchored anteriorly to the crista galli of the ethmoid bone and posteriorly to the tentorium
Tentorium cerebelli	Separates the occipital and temporal lobes of the cerebrum from the cerebellum; anchored to the tentorium, petrous parts of the temporal bones, and occipital bone
Falx cerebelli	Partitions the right and left cerebellar hemispheres; anchored to the occipital crest
Diaphragma sellae	Forms the roof of the sella turcica

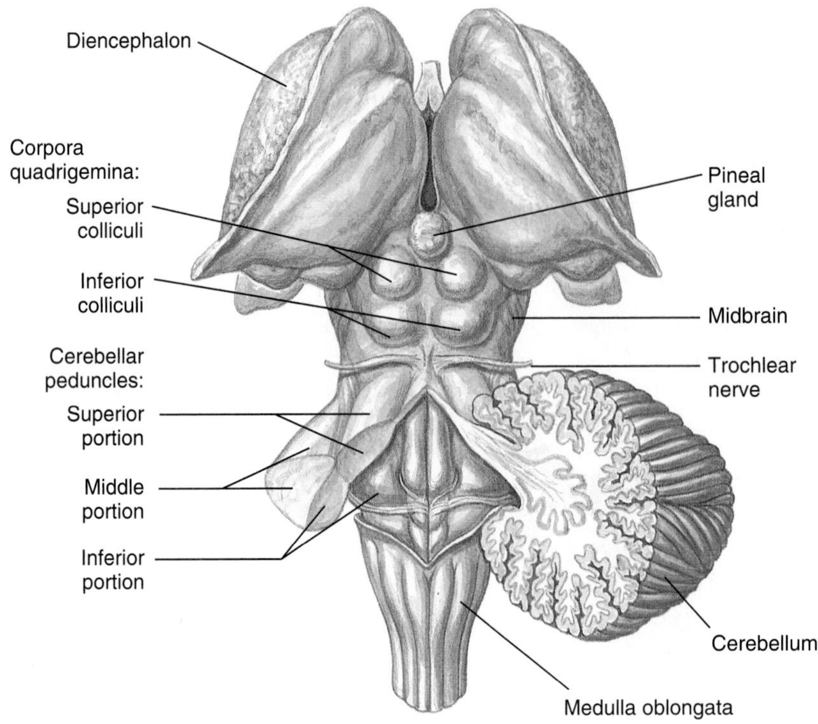

Figure 15.16

The cerebellar peduncles.
These fiber tracts can be seen when the cerebellum has been removed from its attachment to the brain stem.

Like the cerebrum, the cerebellum has a thin, outer layer of gray matter, the **cerebellar cortex,** and a thick, deeper layer of white matter. The cerebellum is convoluted into a series of slender, parallel *folia*. The tracts of white matter within the cerebellum have a distinctive branching pattern called the **arbor vitae** (*vi′te*) that can be seen in the sagittal view (fig. 15.15*c*).

Three paired bundles of nerve fibers called **cerebellar peduncles** support the cerebellum and provide it with tracts for communicating with the rest of the brain (fig. 15.16). Following is a description of the cerebellar peduncles.

1. **Superior cerebellar peduncles** connect the cerebellum with the midbrain. The fibers within these peduncles originate primarily from specialized **dentate nuclei** within the cerebellum and pass through the red nucleus to the thalamus, and then to the motor areas of the cerebral cortex. Impulses through the fibers of these peduncles provide feedback to the cerebrum.
2. **Middle cerebellar peduncles** convey impulses of voluntary movement from the cerebrum through the pons and to the cerebellum.
3. **Inferior cerebellar peduncles** connect the cerebellum with the medulla oblongata and the spinal cord. They contain both incoming vestibular and proprioceptive fibers and outgoing motor fibers.

The principal function of the cerebellum is coordinating skeletal muscle contractions by recruiting precise motor units within the muscles. Impulses for voluntary muscular movement originate in the cerebral cortex and are coordinated by the cerebellum. The cerebellum constantly initiates impulses to selective motor units for maintaining posture and muscle tone. The cerebellum also adjusts to incoming impulses from *proprioceptors* (*pro″pre-o-sep′torz*) within muscles, tendons, joints, and special sense organs to refine learned movement patterns. A proprioceptor is a sensory nerve ending that is sensitive to changes in the tension of a muscle or tendon.

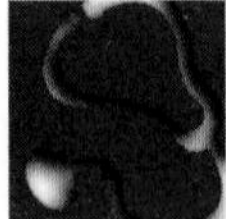

Trauma or diseases of the cerebellum, such as cerebral palsy or a stroke, frequently cause an impairment of skeletal muscle function. Movements become jerky and uncoordinated in a condition known as *ataxia.* There is also a loss of equilibrium, resulting in a disturbance of gait. *Alcohol intoxication* causes similar uncoordinated body movements.

Myelencephalon

The medulla oblongata, contained within the myelencephalon, connects to the spinal cord and contains nuclei for the cranial nerves and vital autonomic functions.

arbor vitae: L. *arbor*, tree; *vitae*, life

Medulla Oblongata

The **medulla oblongata** (*me-dul'a ob"long-ga'ta*), a bulbous structure about 3 cm (1 in.) long, is the most inferior structure of the brain stem. It is continuous with the pons anteriorly and the spinal cord posteriorly at the level of the foramen magnum (see figs. 15.8 and 15.9). Externally, the medulla oblongata resembles the spinal cord, except for two triangular elevations called **pyramids** on the inferior side and an oval enlargement called the **olive** on each lateral surface. The fourth ventricle, the space within the medulla oblongata, is continuous posteriorly with the central canal of the spinal cord and anteriorly with the mesencephalic aqueduct (see fig. 15.21).

The medulla oblongata is composed of vital nuclei and white matter that form all of the descending and ascending tracts communicating between the spinal cord and various parts of the brain. Most of the fibers within these tracts cross over to the opposite side through the pyramidal region of the medulla oblongata, permitting one side of the brain to receive information from and send information to the opposite side of the body (see fig. 15.8).

The gray matter of the medulla oblongata contains several important nuclei for cranial nerves and sensory relay (fig. 15.17). The **nucleus ambiguus** (*am-big'-yoo-us*) and the **hypoglossal nucleus** are the centers from which arise the vestibulocochlear (VIII), glossopharyngeal (IX), accessory (XI), and hypoglossal (XII) nerves. The vagus nerves (X) arise from **vagus nuclei,** one on each lateral side of the medulla oblongata, adjacent to the fourth ventricle. The **nucleus gracilis** (*gras'i-lis*) and the **nucleus cuneatus** (*kyoo-ne-a'tus*) relay sensory information to the thalamus, and the impulses are then relayed to the cerebral cortex via the thalamic nuclei (not illustrated). The **inferior olivary nuclei** and the **accessory olivary nuclei** of the olive mediate impulses passing from the forebrain and midbrain through the inferior cerebellar peduncles to the cerebellum.

Three other nuclei within the medulla oblongata function as autonomic centers for controlling vital visceral functions.

1. **Cardiac center.** Both *inhibitory* and *accelerator fibers* arise from nuclei of the cardiac center. Inhibitory impulses constantly travel through the vagus nerves to slow the heartbeat. Accelerator impulses travel through the spinal cord and eventually innervate the heart through fibers within spinal nerves T1–T5.
2. **Vasomotor center.** Nuclei of the vasomotor center send impulses via the spinal cord and spinal nerves to the smooth muscles of arteriole walls, causing them to constrict and elevate arterial blood pressure.
3. **Respiratory center.** The respiratory center of the medulla oblongata controls the rate and depth of breathing and functions in conjunction with the respiratory nuclei of the pons (see fig. 15.14) to produce rhythmic breathing.

medulla: L. *medulla*, marrow

Figure 15.17

Nuclei of the brain stem.
A sagittal section of the pons and medulla oblongata showing the nuclei of the cranial nerves that arise from this portion of the brain stem.

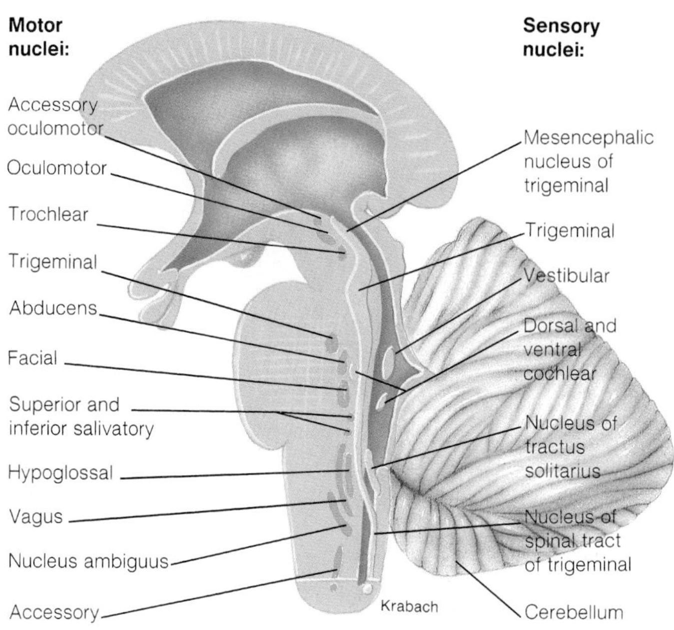

Other nuclei of the medulla oblongata function as centers for reflexes involved in sneezing, coughing, swallowing, and vomiting. Some of these activities (swallowing, for example) may be initiated voluntarily, but once they progress to a certain point they become involuntary and cannot be stopped.

Reticular Formation

The **reticular formation** is a complex network of nuclei and nerve fibers within the brain stem that functions as the *reticular activating system* (*RAS*) in arousing the cerebrum. Portions of the reticular formation are located in the spinal cord, pons, midbrain, and parts of the thalamus and hypothalamus (fig. 15.18) The reticular formation contains ascending and descending fibers from most of the structures within the brain.

Nuclei within the reticular formation generate a continuous flow of impulses unless they are inhibited by other parts of the brain. The principal functions of the RAS are to keep the cerebrum in a state of alert consciousness and to selectively monitor the sensory impulses perceived by the cerebrum. The RAS also helps the cerebellum activate selected motor units to maintain muscle tonus and produce smooth, coordinated contractions of skeletal muscles.

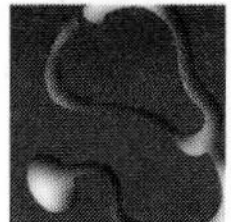

The RAS is sensitive to changes in and trauma to the brain. The sleep response is thought to occur because of a decrease in activity within the RAS, perhaps due to the secretion of specific neurotransmitters. A blow to the head or certain drugs and diseases may damage the RAS, causing unconsciousness. A *coma* is a state of unconsciousness and inactivity of the RAS that even the most powerful external stimuli cannot disturb.

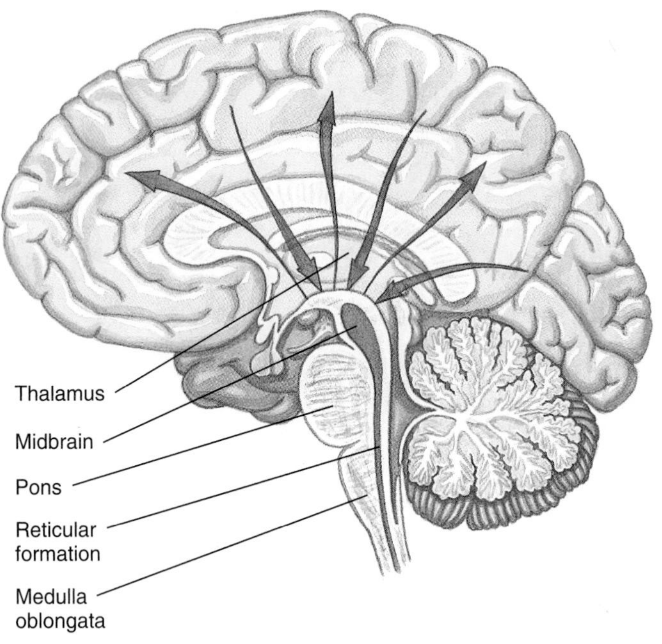

Figure 15.18

The reticular activating system.
The arrows indicate the direction of impulses along nerve pathways that connect with the RAS.

Meninges of the Central Nervous System

The CNS is covered by protective meninges, namely the dura mater, the arachnoid, and the pia mater.

As mentioned previously at the beginning of this chapter, the entire delicate CNS is protected by a bony encasement—the cranium surrounding the brain and the vertebral column surrounding the spinal cord. It is also protected by three membranous connective tissue coverings called the **meninges** (figs. 15.19 and 15.20). Individually, from the outside in, they are known as the dura mater, the arachnoid, and the pia mater.

Dura Mater

The **dura mater** (*door'a ma'ter*) is in contact with the bone and is composed primarily of dense connective tissue. The **cranial dura mater** is a double-layered structure. The outer **periosteal** (*per"e-os'te-al*) **layer** adheres lightly to the inner surface of the cranium, where it constitutes the periosteum (fig. 15.19*a*). The inner **meningeal layer,** which is thinner, follows the general contour of the brain. The **spinal dura mater** is not double layered. It is similar to the meningeal layer of the cranial dura mater.

The two layers of the cranial dura mater are generally fused and cover most of the brain. In certain regions, however, the layers are separated, enclosing **dural sinuses** (fig. 15.19*a*) that collect venous blood and drain it to the internal jugular veins of the neck.

In four locations, the meningeal layer of the cranial dura mater forms distinct septa to partition major structures on the surface of the brain and anchor the brain to the inside of the cranial case. These septa were identified earlier and are reviewed in table 15.4.

The spinal dura mater forms a tough, tubular **dural sheath** that continues into the vertebral canal and surrounds the spinal cord. There is no connection between the dural sheath and the vertebrae forming the vertebral canal, but instead there is a potential cavity called the **epidural space** (fig. 15.19*b*). The epidural space is highly vascular and contains loose and adipose connective tissues that form a protective pad around the spinal cord.

Arachnoid

The **arachnoid** (*a-rak'noid*) is the middle of the three meninges. This delicate, netlike membrane spreads over the CNS but generally does not extend into the sulci or fissures of the brain. The **subarachnoid space,** located between the arachnoid and the deepest meninx, the pia mater, contains cerebrospinal fluid. The subarachnoid space is maintained by weblike strands that connect the arachnoid and pia mater (fig. 15.19).

Pia Mater

The thin **pia mater** is attached to the surfaces of the CNS and follows the irregular contours of the brain and spinal cord. Composed of modified loose connective tissue, the pia mater supports the vascular network that nourishes the underlying cells of the brain and spinal cord. The pia mater is specialized over the roofs of the ventricles, where it contributes to the formation of the choroid plexuses along with the arachnoid. Lateral extensions of the pia mater along the spinal cord form the **ligamentum denticulatum,** which attaches the spinal cord to the dura mater (fig. 15.20).

Meningitis, an inflammation of the meninges, is usually caused by bacteria or viruses. The arachnoid and the pia mater are the two meninges most frequently affected. Meningitis is accompanied by high fever and severe headache. Complications may cause sensory impairment, paralysis, or mental retardation. Untreated meningitis generally results in coma and death.

meninges: L. plural form of *meninx,* membrane
dura mater: L. *dura,* hard; *mater,* mother
arachnoid: L. *arachnoides,* like a cobweb
pia mater: L. *pia,* soft or tender; *mater,* mother

Figure 15.19

Meninges and associated structures (*a*) surrounding the brain and (*b*) surrounding the spinal cord. The epidural space in the lower lumbar region is of clinical importance as a site for an epidural block that may be administered to facilitate parturition (childbirth). (The meninges are labeled in boldface type.)

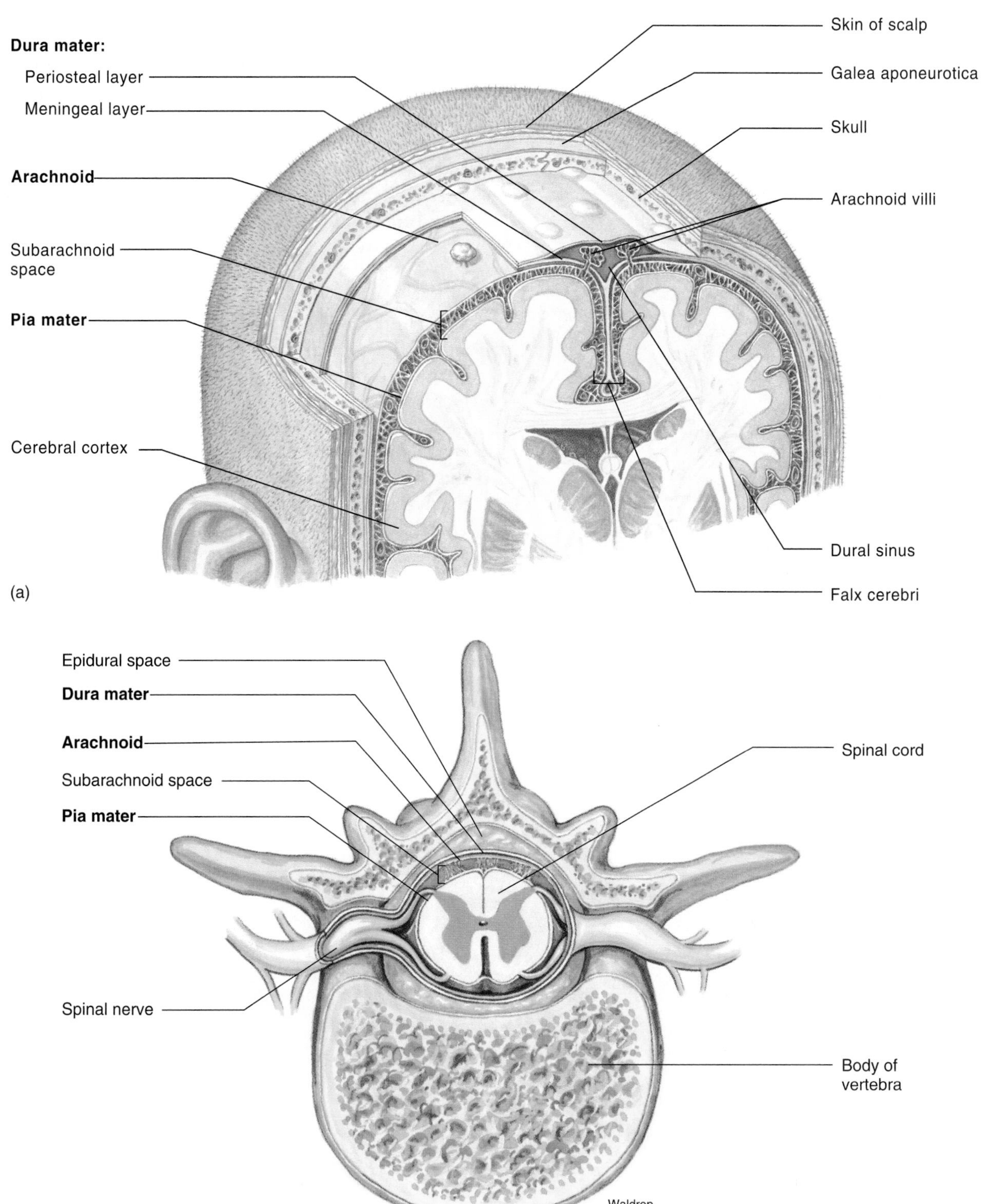

Figure 15.20

Meninges and associated structures surrounding the spinal cord.
(The meninges are labeled in boldface type.)

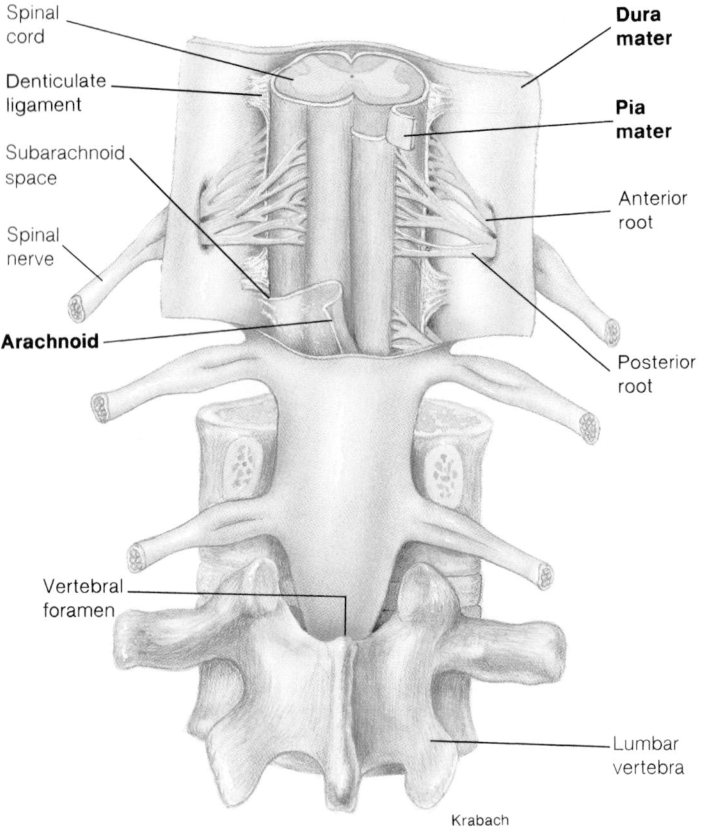

Ventricles and Cerebrospinal Fluid

The ventricles, central canal, and subarachnoid space contain cerebrospinal fluid, formed by the active transport of substances from blood plasma in the choroid plexuses.

Cerebrospinal fluid (CSF) is a clear, lymphlike fluid that forms a protective cushion around and within the CNS. The fluid also buoys the brain. CSF circulates through the various ventricles of the brain, the **central canal** of the spinal cord, and the subarachnoid space around the entire CNS. The CSF returns to the circulatory system by draining through the walls of the **arachnoid villi,** which are venous capillaries.

Ventricles of the Brain

The ventricles of the brain are connected to one another and to the central canal of the spinal cord (fig. 15.21). Each of the two **lateral ventricles** (first and second ventricles) is located in a cerebral hemisphere, inferior to the corpus callosum. The **third ventricle** is located in the diencephalon, between the thalami. Each lateral ventricle is connected to the third ventricle by a narrow, oval opening called the **interventricular foramen** (foramen of Monro). The **fourth ventricle** is located in the brain stem, within the pons and medulla oblongata. The **mesencephalic aqueduct** (cerebral aqueduct) passes through the midbrain to link the third and

foramen of Monro: from Alexander Monro Jr., Scottish anatomist, 1733–1817

Figure 15.21

The ventricles of the brain.
(*a*) An anterior view and (*b*) a lateral view.

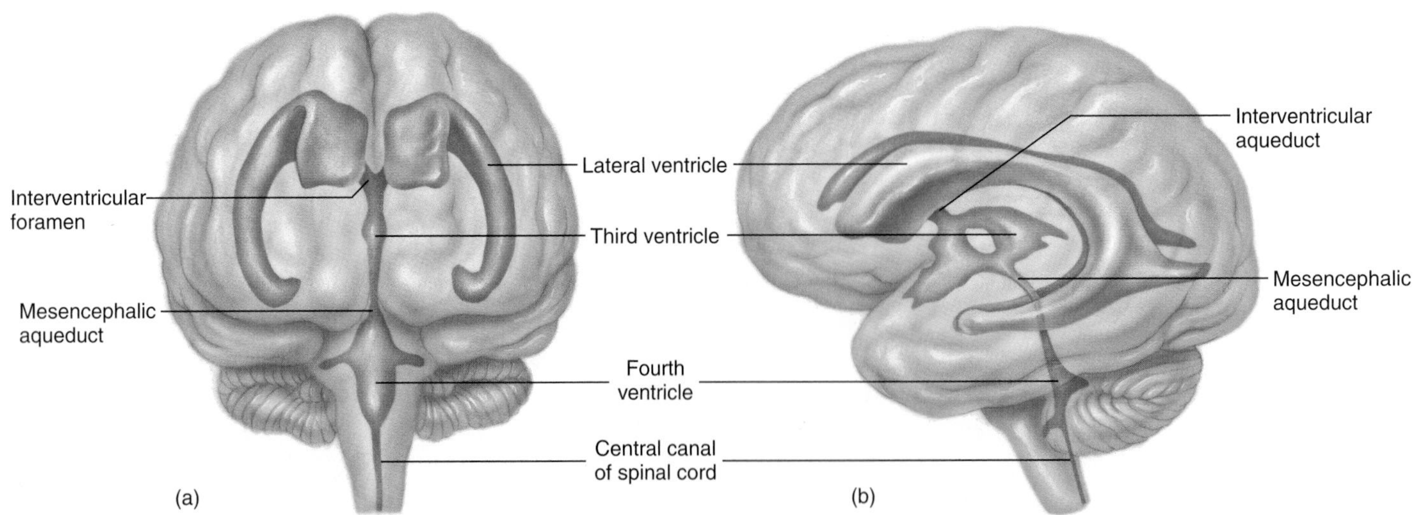

fourth ventricles. The fourth ventricle also communicates posteriorly with the **central canal** of the spinal cord. Cerebrospinal fluid exits from the fourth ventricle into the subarachnoid space (fig. 15.22) through three foramina: the **median aperture** (foramen of Magendie), a medial opening, and two **lateral apertures** (foramina of Luschka) (not illustrated). Cerebrospinal fluid returns to the venous blood through the **arachnoid villi** (see fig. 15.18).

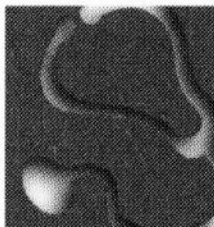

Internal hydrocephalus (*hi"dro-sef'a-lus*) is a condition in which cerebrospinal fluid builds up within the ventricles of the brain. It is more common in infants, whose cranial sutures have not yet strengthened or ossified, than in older individuals. If the pressure is excessive, the condition may have to be treated surgically.

External hydrocephalus, an accumulation of fluid within the subarachnoid space, usually results from an obstruction of drainage at the arachnoid villi. The external pressure compresses neural tissue and is likely to cause brain damage.

Cerebrospinal Fluid

Cerebrospinal fluid buoys the CNS and protects it from mechanical injury. The brain weighs about 1,500 grams, but suspended in CSF its buoyed weight is about 50 grams. This means that the brain has a near neutral buoyancy; at a true neutral buoyancy, an object does not float or sink but is suspended in its fluid environment.

In addition to buoying the CNS, CSF reduces the damaging effect of an impact to the head by spreading the force over a larger area. It also helps to remove metabolic wastes from nervous tissue. Since the CNS lacks lymphatic circulation, the CSF moves cellular wastes into the venous return at its places of drainage.

The clear, watery CSF is continuously produced by the filtration of blood plasma through masses of specialized capillaries called choroid plexuses and, to a lesser extent, by secretions of the ependymal cells. The ciliated ependymal cells cover the choroid plexuses, as well as line the central canal, and presumably aid the movement of the CSF. The tight junctions between the ependymal cells also help to form a *blood–cerebrospinal fluid barrier* that prohibits certain potentially harmful substances in the blood from entering the CSF.

CSF is formed mainly by the active transport and ultrafiltration of substances within the blood plasma. CSF has more sodium, chloride, magnesium, and hydrogen ions than blood plasma and less calcium, potassium, and glucose. In addition, CSF contains some proteins, urea, and white blood cells.

Up to 800 ml of CSF are produced each day, although only 140–200 ml are bathing the CNS at any given moment. A person lying in a horizontal position has a slow but continuous circulation of CSF, with a fluid pressure of about 10 mm of mercury.

foramen of Magendie: from François Magendie, French physiologist, 1783–1855

foramen of Luschka: from Hubert Luschka, German anatomist, 1820–75

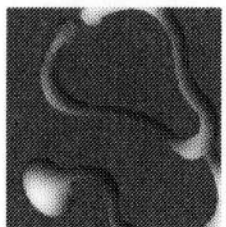

The homeostatic consistency of the CSF composition is critical, and a chemical imbalance may have marked effects on CNS functions. An increase in glycine (an amino acid) concentration, for example, produces hypothermia and hypotension as temperature and blood pressure regulatory mechanisms are disrupted. A slight change in pH may affect the respiratory rate and depth.

Spinal Cord

The spinal cord consists of centrally located gray matter involved in reflexes, and peripherally located ascending and descending tracts of white matter, which conduct impulses to and from the brain.

The **spinal cord** is the portion of the CNS that extends through the vertebral canal of the vertebral column (fig. 15.23). It is continuous with the brain through the foramen magnum of the skull. The spinal cord has two principal functions.

1. **Impulse conduction.** It provides a means of neural communication to and from the brain through tracts of white matter. **Ascending tracts** conduct impulses from the peripheral sensory receptors of the body to the brain. **Descending tracts** conduct motor impulses from the brain to the muscles and glands.
2. **Reflex integration.** It serves as a center for spinal reflexes. Specific nerve pathways enable some movements to be reflexive rather than initiated voluntarily by the brain. Movements of this type are not confined to skeletal muscles; reflexive movements of cardiac and smooth muscles control heart rate, breathing rate, blood pressure, and digestive activities. Spinal nerve pathways are also involved in swallowing, coughing, sneezing, and vomiting.

Structure of the Spinal Cord

The spinal cord extends inferiorly from the position of the foramen magnum of the occipital bone to the level of the first lumbar vertebra (L1). It is somewhat flattened anteroposteriorly, making it oval in cross section. Two prominent enlargements can be seen in a posterior view (fig. 15.23). The **cervical enlargement** is located between the third cervical (C3) and the second thoracic vertebrae (T2). Nerves emerging from this region serve the upper extremities. The **lumbar enlargement** lies between the ninth and twelfth thoracic vertebrae. Nerves from the lumbar enlargement supply the lower extremities.

The embryonic spinal cord develops more slowly than the associated vertebral column; thus, in the adult, the spinal cord does not extend beyond L1. The tapering, terminal portion of the spinal cord is called the **conus medullaris** (*med-yoo-lar'is*). The **filum terminale** (*fi'lum ter-mi-nal'e*), a fibrous

filum terminalis: L. *filum*, filament; *terminus*, end

Figure 15.22

The flow of cerebrospinal fluid.

Cerebrospinal fluid is secreted by choroid plexuses in the ventricular walls. The fluid circulates through the ventricles and central canal, enters the subarachnoid space, and is reabsorbed into the blood of the dural sinuses through the arachnoid villi.

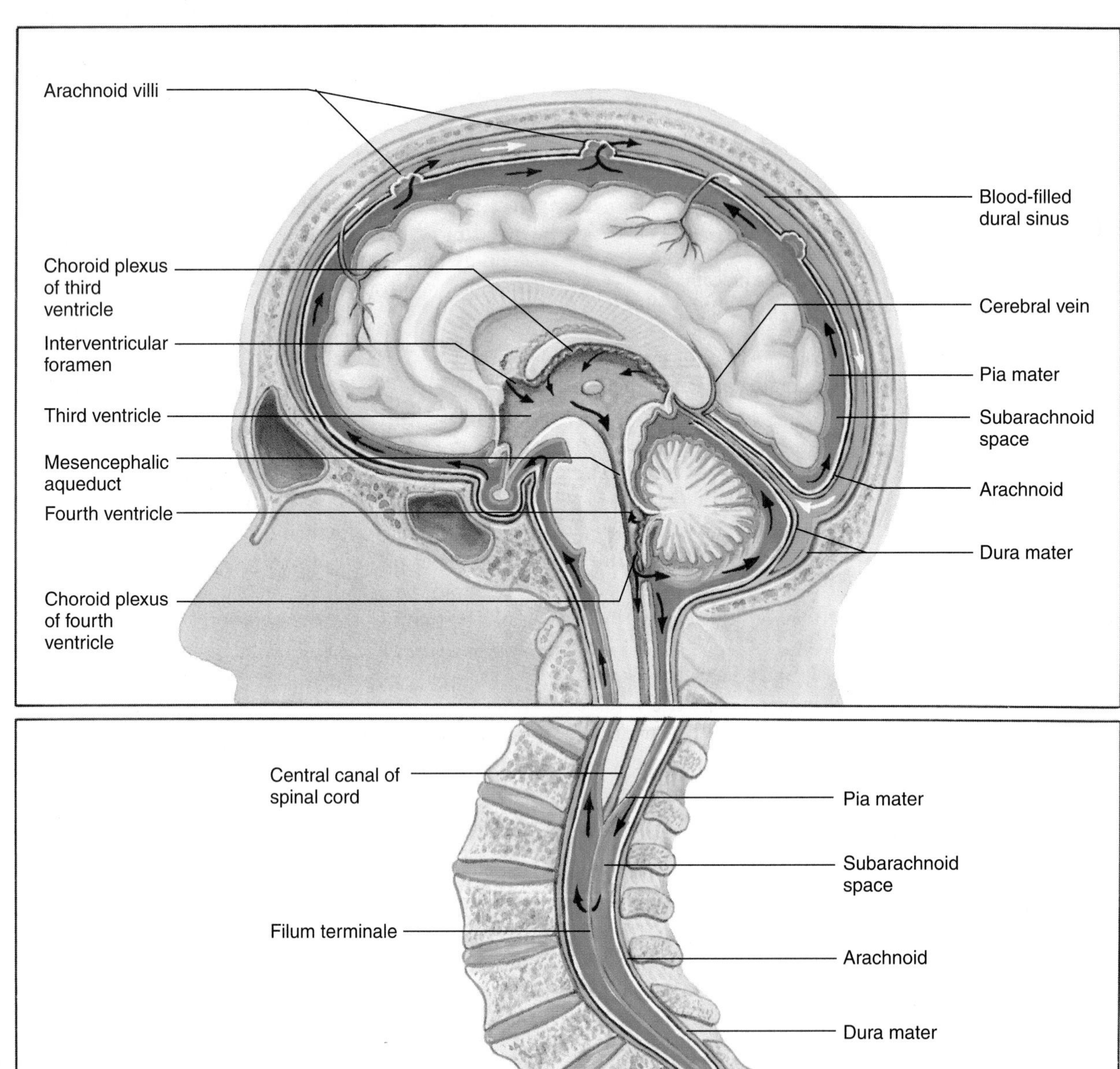

Figure 15.23

The spinal cord and plexuses.
(The plexuses are indicated in boldface type.)

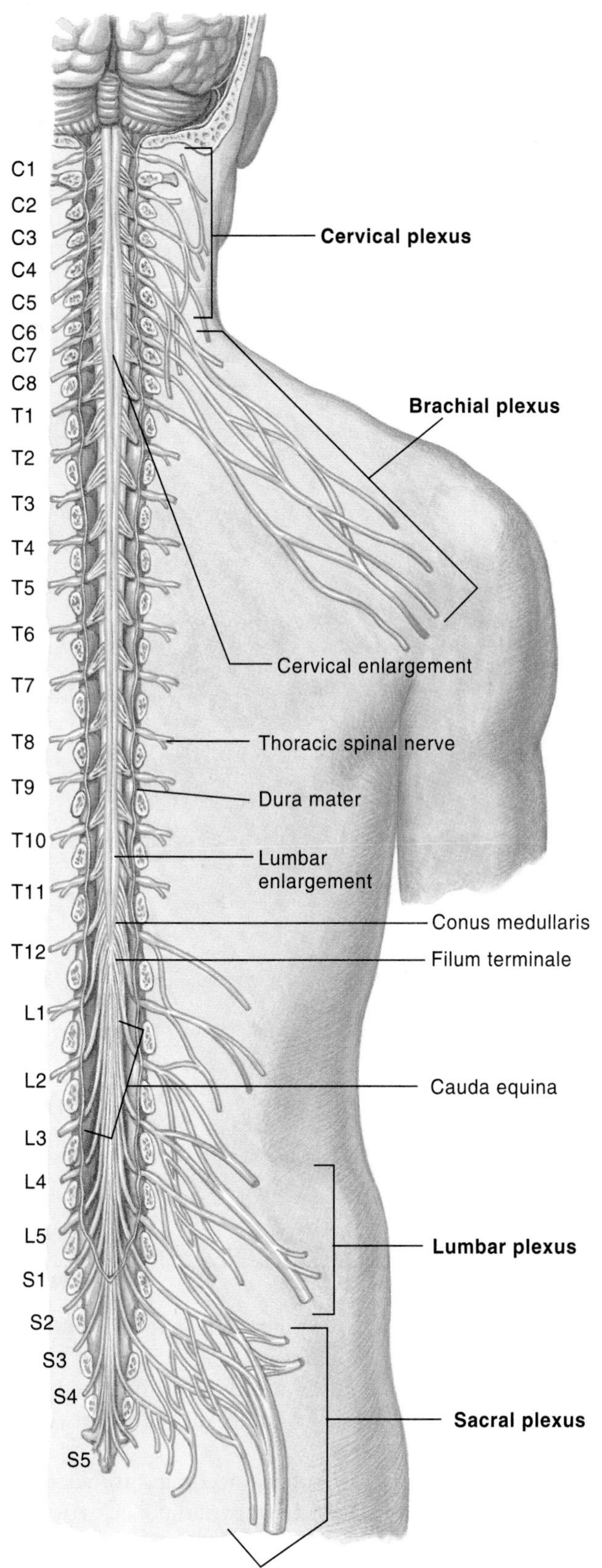

strand composed mostly of pia mater, extends inferiorly from the conus medullaris at the level of L1 to the coccyx (see figs. 15.22 and 15.23). Nerve roots also radiate inferiorly from the conus medullaris through the vertebral canal. These nerve roots are collectively referred to as the **cauda equina** (*kaw'da e-qui'na*) because they resemble a horse's tail.

The spinal cord develops as 31 segments, each of which gives rise to a pair of **spinal nerves** that emerge from the cord through the intervertebral foramina. Two grooves, an **anterior median fissure** and a **posterior median sulcus,** extend the length of the spinal cord and partially divide the cord into right and left portions. Like the brain, the spinal cord is protected by three distinct meninges and is cushioned by cerebrospinal fluid. The pia mater contains an extensive vascular network.

The **gray matter** of the spinal cord is centrally located and surrounded by white matter. It is composed of nerve cell bodies, neuroglia, and unmyelinated association neurons (interneurons). The **white matter** consists of bundles, or tracts, of myelinated fibers of sensory and motor neurons.

The relative size and shape of the gray and white matter varies throughout the spinal cord. The amount of white matter increases toward the brain as the nerve tracts become thicker. More gray matter is found in the cervical and lumbar enlargements where innervations from the upper and lower extremities, respectively, make connections.

The core of gray matter roughly resembles the letter H (fig. 15.24). Projections of the gray matter within the spinal cord are called horns and are named according to the direction in which they project. The paired **posterior horns** extend posteriorly and the paired **anterior horns** project anteriorly. Between the posterior and anterior horns, the short paired **lateral horns** extend to the sides. Lateral horns are prominent only in the thoracic and upper lumbar regions. The transverse bar of gray matter that connects the paired horns across the center of the spinal cord is called the **gray commissure.** Within the gray commissure is the **central canal.** It is continuous with the ventricles of the brain and is filled with cerebrospinal fluid.

Spinal Cord Tracts

Impulses are conducted through the ascending and descending tracts of the spinal cord within the columns of white matter. The spinal cord has six columns of white matter called **funiculi** (*fyoo-nik'-yŭ-li*), which are named according to their relative position within the spinal cord. The two **anterior funiculi** are located between the two anterior horns of gray matter, to either side of the anterior median fissure (fig. 15.24). The two **posterior funiculi** are located between the two posterior horns of gray matter to either side of the posterior median sulcus. Two **lateral funiculi** are located between the anterior and posterior horns of gray matter.

---

cauda equina: L. *cauda*, tail; *equus*, horse
commissure: L. *commissura*, a joining
funiculus: L. diminutive of *funis*, cord, rope

Under Development

Development of the Spinal Cord

The spinal cord, like the brain, develops as the neural tube undergoes differentiation and specialization. Throughout the developmental process, the hollow central canal persists while the specialized white and gray matter forms (fig. 1). Changes in the neural tube become apparent during the sixth week as the lateral walls thicken to form a groove, called the **sulcus limitans,** along each lateral wall of the central canal. A pair of **alar plates** forms dorsal to the sulcus limitans, and a pair of **basal plates** forms ventrally. By the ninth week, the alar plates have specialized to become the **posterior horns,** containing fibers of the sensory cell bodies, and the basal plates have specialized to form the **anterior** and **lateral horns,** containing motor cell bodies. Sensory neurons of spinal nerves conduct impulses toward the spinal cord, whereas motor neurons conduct impulses away from the spinal cord.

Figure 1 The development of the spinal cord. (*a*) A dorsal view of an embryo at 23 days with the position of a transverse cut indicated in (*b*). (*c*) The formation of the alar and basal plates is evident in a cross section through the spinal cord at 6 weeks. (*d*) The central canal has reduced in size, and functional posterior and anterior horns have formed at 9 weeks.

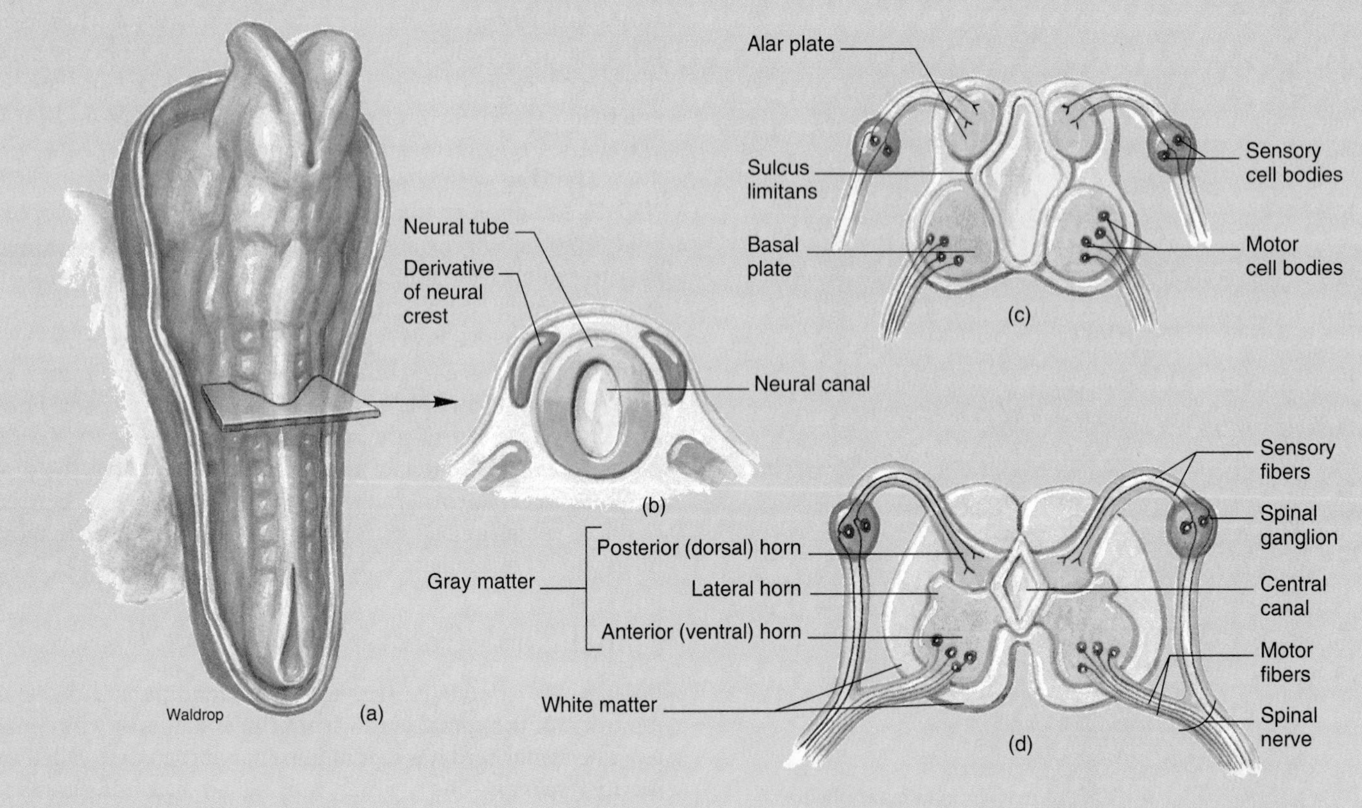

Each funiculus consists of both ascending and descending tracts. The nerve fibers within the tracts are generally myelinated and are named according to their origin and termination. The fibers of the tracts either remain on the same side of the brain and spinal cord or cross over within the medulla oblongata or the spinal cord. The crossing over of nerve tracts is referred to as *decussation* (*de″kŭ-sa′shun*). Illustrated in figures 15.25 and 15.26 are descending and ascending tracts, respectively, that decussate within the medulla oblongata.

The principal ascending and descending tracts within the funiculi are summarized in table 15.5 and illustrated in figure 15.27.

Descending tracts are grouped according to place of origin as either corticospinal or extrapyramidal. **Corticospinal** (pyramidal) **tracts** descend directly, without synaptic interruption, from the cerebral cortex to the lower motor neurons. The cell bodies of the neurons that contribute fibers to these tracts

decussation: L. *decussare*, to form an X intersection

Figure 15.24

The spinal cord in cross section.
(*a*) A diagram and (*b*) a photomicrograph.

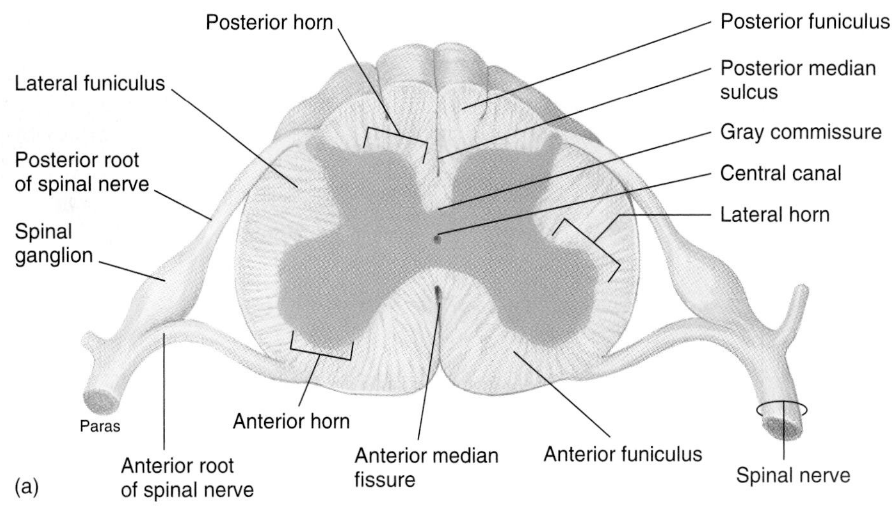

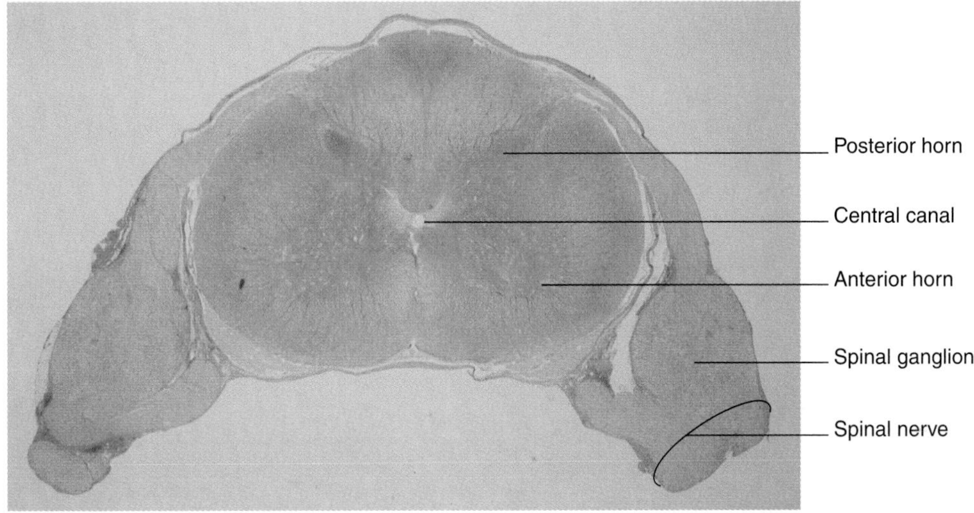

are located primarily in the precentral gyrus of the frontal lobe. Most (about 85%) of the corticospinal fibers decussate in the pyramids of the medulla oblongata (see fig. 15.8). The remaining 15% do not cross from one side to the other. The fibers that cross comprise the **lateral corticospinal tracts,** and the remaining uncrossed fibers comprise the **anterior corticospinal tracts.** Because of the crossing of fibers from higher motor neurons in the pyramids, the right hemisphere primarily controls the musculature on the left side of the body, whereas the left hemisphere controls the right musculature.

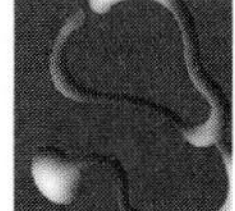
The corticospinal tracts appear to be particularly important in voluntary movements that require complex interactions between the motor cortex and sensory input. Speech, for example, is impaired when the corticospinal tracts are damaged in the thoracic region of the spinal cord, whereas involuntary breathing continues. Damage to the pyramidal motor system can be detected clinically by the presence of Babinski reflex, in which stimulation of the sole of the foot causes extension (upward movement) of the great toe and fanning out of the other toes. Babinski reflex is normally present in infants because neural control is not yet fully developed.

The remaining descending tracts are **extrapyramidal tracts** that originate in the brain stem region. Electrical stimulation of the cerebral cortex, the cerebellum, and the basal nuclei indirectly evokes movements because of synaptic connections within extrapyramidal tracts.

The **reticulospinal** (*rĕ-tik″yŭ-lo-spi′nal*) **tracts** are the major descending pathways of the extrapyramidal system. These tracts originate in the reticular formation of the brain stem. Neurostimulation of the reticular formation by the cerebrum either facilitates or inhibits the activity of lower motor neurons (depending on the area stimulated) (fig. 15.28).

There are no descending tracts from the cerebellum. The cerebellum can influence motor activity only indirectly, through the vestibular nuclei, red nucleus, and basal nuclei. These structures, in turn, affect lower motor neurons via the **vestibulospinal tracts, rubrospinal tracts,** and **reticulospinal tracts.** Damage to the cerebellum disrupts the coordination of movements with spatial judgment. Underreaching or overreaching for an object may occur, followed by *intention tremor*, in which the limb moves back and forth in a pendulumlike motion.

The basal nuclei, acting through synapses in the reticular formation in particular, appear normally to exert an inhibitory

Figure 15.25

A descending corticospinal tract.

This type of neural pathway is composed of motor fibers that decussate (cross-over) in the medulla oblongata of the brain stem.

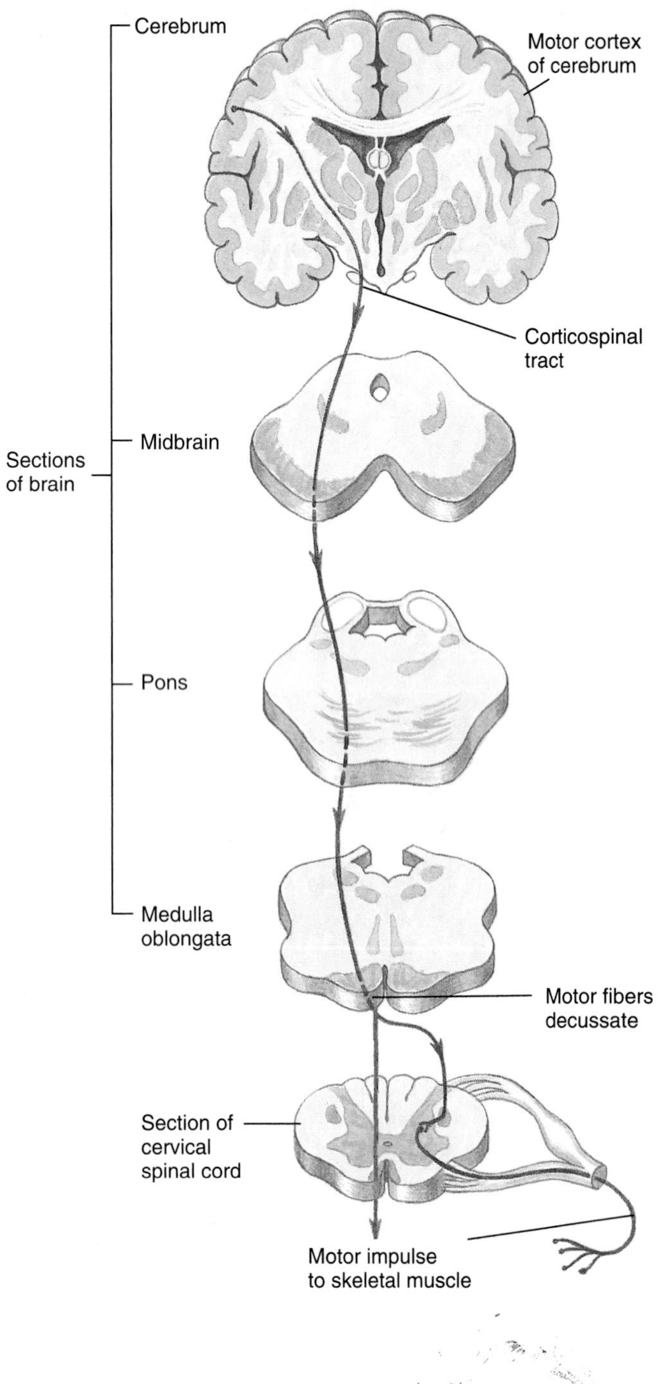

Figure 15.26

An ascending tract.

This type of neural pathway is composed of sensory fibers that decussate in the medulla oblongata of the brain stem.

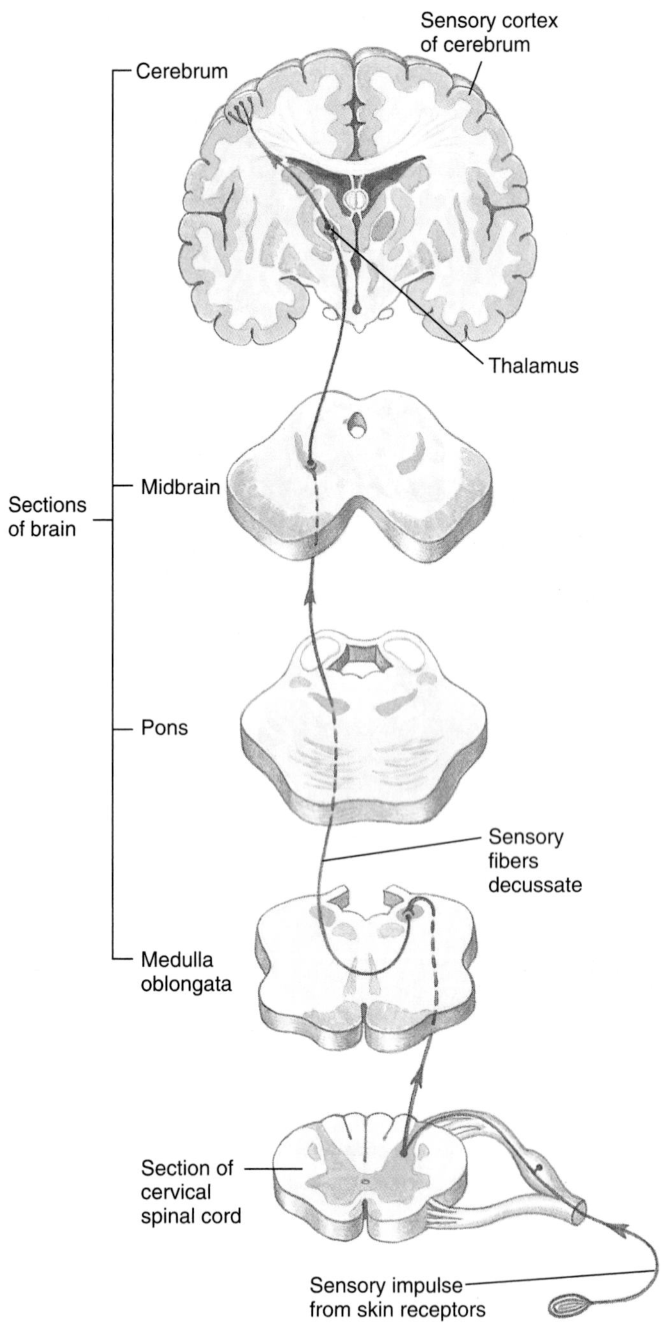

Table 15.5

Principal Ascending and Descending Tracts of the Spinal Cord

Tract	Funiculus	Origin	Termination	Function
Ascending Tracts				
Anterior spinothalamic	Anterior	Posterior horn on one side of spinal cord; crosses to opposite side	Thalamus, then cerebral cortex	Conducts sensory impulses for crude touch and pressure
Lateral spinothalamic	Lateral	Posterior horn on one side of spinal cord; crosses to opposite side	Thalamus, then cerebral cortex	Conducts pain and temperature impulses that are interpreted within cerebral cortex
Fasciculus gracilis and fasciculus cuneatus	Posterior	Peripheral sensory neurons; does not cross over	Nucleus gracilis and nucleus cuneatus of medulla oblongata; crosses to opposite side; eventually thalamus, then cerebral cortex	Conducts sensory impulses from skin, muscles, tendons, and joints, which are interpreted as sensations of fine touch, precise pressures, and body movements
Posterior spinocerebellar	Lateral	Posterior horn; does not cross over	Cerebellum	Conducts sensory impulses from one side of body to same side of cerebellum for subconscious proprioception required for coordinated muscular contractions
Anterior spinocerebellar	Lateral	Posterior horn; some fibers cross, others do not	Cerebellum	Conducts sensory impulses from both sides of body to cerebellum for subconscious proprioception required for coordinated muscular contractions
Descending Tracts				
Anterior corticospinal	Anterior	Cerebral cortex on one side of brain; crosses to opposite side of spinal cord	Anterior horn	Conducts motor impulses from cerebrum to spinal nerves, and outward to cells of anterior horns for coordinated, precise voluntary movements of skeletal muscle
Lateral corticospinal	Lateral	Cerebral cortex on one side of brain; crosses in base of medulla oblongata to opposite side of spinal cord	Anterior horn	Conducts motor impulses from cerebrum to spinal nerves, and outward to cells of anterior horns for coordinated, precise voluntary movements
Tectospinal	Anterior	Mesencephalon; crosses to opposite side of spinal cord	Anterior horn	Conducts motor impulses to cells of anterior horns, and eventually to muscles that move the head in response to visual, auditory, or cutaneous stimuli
Rubrospinal	Lateral	Mesencephalon (red nucleus); crosses to opposite side of spinal cord	Anterior horn	Conducts motor impulses concerned with muscle tone and posture
Vestibulospinal	Anterior	Medulla oblongata; does not cross over	Anterior horn	Conducts motor impulses that regulate body tone and posture (equilibrium) in response to movements of head
Anterior and medial reticulospinal	Anterior	Reticular formation of brain stem; does not cross over	Anterior horn	Conducts motor impulses that control muscle tone and sweat gland activity
Bulboreticulospinal	Lateral	Reticular formation of brain stem; does not cross over	Anterior horn	Conducts motor impulses that control muscle tone and sweat gland activity

Figure 15.27

A cross section showing the principal ascending and descending tracts within the spinal cord.

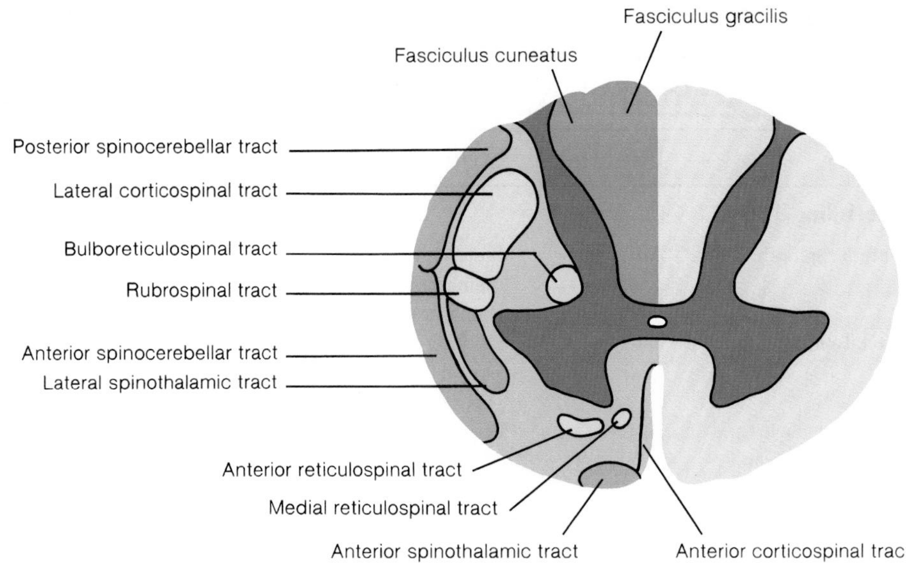

Figure 15.28

Pathways involved in the higher motor neuron control of skeletal muscles.
(The pyramidal [corticospinal] tracts are shown in red, and the extrapyramidal tracts are shown in black.)

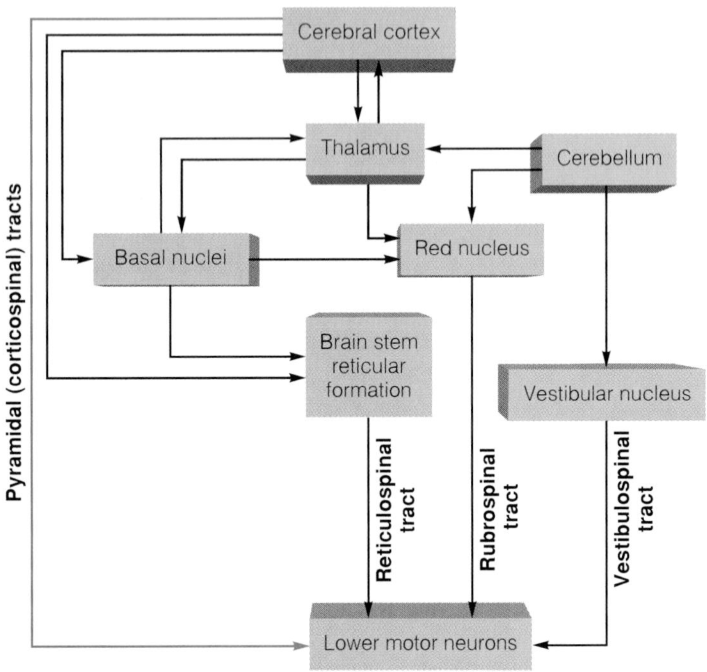

influence on the activity of lower motor neurons. Damage to the basal nuclei thus results in decreased muscle tone. People with such damage display *akinesia* (*a″kĭ-ne′zha*) (complete or partial loss of muscle movement) and *chorea* (*ko-re′a*) (sudden and uncontrolled random movements).

akinesia: Gk. *a*, without; *kinesis*, movement
chorea: Fr. *choros*, a dance

Clinical Considerations

The clinical aspects of the central nervous system are extensive and usually complex. Numerous diseases and developmental problems directly involve the nervous system, and the nervous system is indirectly involved with most of the diseases that afflict the body because of the location and activity of sensory pain receptors. Pain receptors are free nerve endings that are present throughout living tissue. The pain sensations elicited by disease or trauma are important in localizing and diagnosing specific diseases or dysfunctions.

Only a few of the many clinical considerations of the central nervous system will be discussed here. These include neurological assessment and drugs, developmental problems, injuries, infections and diseases, and degenerative disorders.

Neurological Assessment and Drugs

Neurological assessment has become exceedingly sophisticated and accurate in the past few years. In a basic physical examination, only the reflexes and sensory functions are assessed. But if the physician suspects abnormalities involving the nervous system, further neurological tests may be done, employing the following techniques.

A **lumbar puncture** is performed by inserting a fine needle between the third and fourth lumbar vertebrae and withdrawing a sample of CSF from the subarachnoid space (fig. 15.29). A **cisternal puncture** is similar to a lumbar puncture, except that the CSF is withdrawn from a cisterna at the base of the skull, near the foramen magnum. The pressure of the CSF, which is normally about 10 mm of mercury, is measured with a *manometer*. Samples of CSF may also be examined for abnormal constituents. In addition, excessive fluid, accumulated as a result of disease or trauma, may be drained.

Figure 15.29

(*a*) A lumbar puncture is performed by inserting a needle between the third and fourth lumbar vertebrae (L3–L4) and (*b*) withdrawing cerebrospinal fluid from the subarachnoid space.

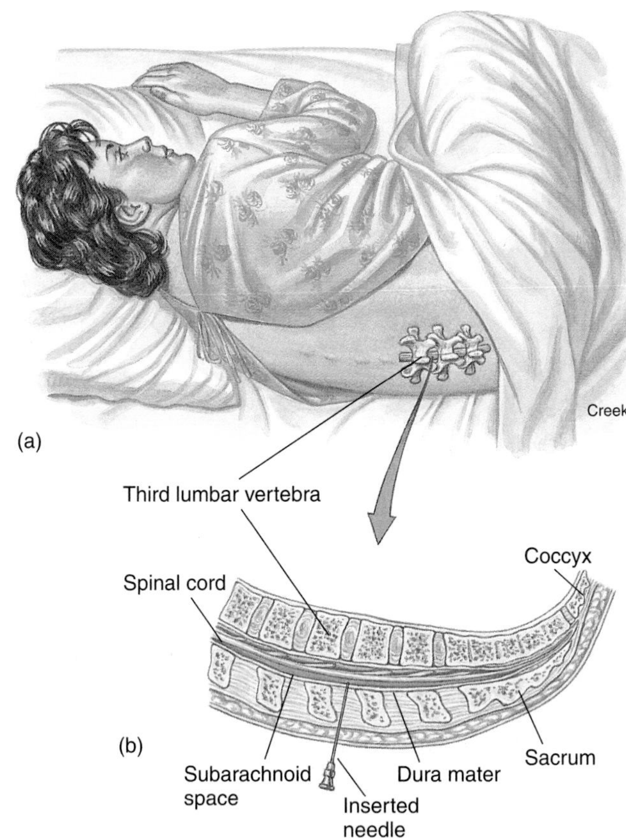

The condition of the arteries of the brain can be determined through a **cerebral angiogram** (*an'je-o-gram*). In this technique, a radiopaque substance is injected into the common carotid arteries and allowed to disperse through the cerebral vessels. Aneurysms and vascular constrictions or displacements by tumors may then be revealed on radiographs.

The development of the **CT scanner,** or **computerized axial tomographic scanner,** revolutionized the diagnosis of brain disorders. The CT scanner projects a sharply focused, detailed tomogram, or cross section, of a patient's brain onto a television screen. The versatile CT scanner allows quick and accurate diagnoses of tumors, aneurysms, blood clots, and hemorrhage. The CT scanner may also be used to detect certain types of birth defects, brain damage, scar tissue, and evidence of old or recent strokes.

A machine with even greater potential than the CT scanner is the **DSR,** or **dynamic spatial reconstructor.** Like the CT scanner, the DSR is computerized to transform radiographs into composite video images. However, with the DSR, a three-dimensional view is obtained, and the image is produced much faster than with the CT scanner. The DSR can produce 75,000 cross-sectional images in 5 seconds, whereas CT can produce only one. With that speed, body functions as well as structures may be studied. Blood flow through vessels of the brain can be observed. This type of data is important in detecting early symptoms of a stroke or other disorders.

Certain disorders of the brain may be diagnosed more simply by examining brain-wave patterns using an **electroencephalogram.** Sensitive electrodes placed on the scalp record particular EEG patterns being emitted from evoked cerebral activity. EEG recordings are used to monitor epileptic patients to predict seizures and determine proper drug therapy, and also to monitor comatose patients.

The fact that the nervous system is extremely sensitive to various drugs is fortunate; at the same time, this sensitivity has potential for disaster. *Drug abuse* is a major clinical concern because of the addictive and devastating effect that certain drugs have on the nervous system. Much has been written on drug abuse, and it is beyond the scope of this text to elaborate on the effects of drugs. A positive aspect of drugs is their administration in medicine to temporarily interrupt the passage or perception of sensory impulses. Injecting an anesthetic drug near a nerve, as in dentistry, desensitizes a specific area and causes a *nerve block*. Nerve blocks of a limited extent occur if an appendage is cooled or if a nerve is compressed for a period of time. Before the discovery of pharmacological drugs, physicians frequently cooled an affected appendage with ice or snow before performing surgery. **General anesthetics** affect the brain and render a person unconscious. A **local anesthetic** causes a nerve block by desensitizing a specific area.

Injuries

Although the brain and spinal cord seem to be well protected within a bony encasement, they are sensitive organs, highly susceptible to injury.

Certain symptomatic terms are used when determining possible trauma within the CNS. **Headaches** are the most common ailment of the CNS. Most headaches are due to dilated blood vessels within the meninges of the brain. Headaches are generally asymptomatic of brain disorders; rather, they tend to be associated with physiological stress, eyestrain, or fatigue. Persistent and intense headaches may indicate a more serious problem, such as a brain tumor. A **migraine** is a specific type of headache that is commonly preceded or accompanied by visual impairments and GI unrest. It is not known why only 5%–10% of the population periodically suffer from migraines or why they are more common in women. Fatigue, allergy, and emotional stress tend to trigger migraines.

Fainting is a brief loss of consciousness that may result from a rapid pooling of blood in the lower extremities. It may occur when a person rapidly arises from a reclined position, receives a blow to the head, or experiences an intense psychologic stimulus, such as viewing a cadaver for the first time. Fainting is of more concern when it is symptomatic of a particular disease.

A **concussion** is an injury resulting from a violent jarring of the brain, usually by a forceful blow to the head. Bones of the skull may or may not be fractured. A concussion usually results in a brief period of unconsciousness, followed by mild **delirium** in which the patient is in a state of confusion. **Amnesia** is a more intense disorientation in which the patient suffers varying degrees of memory loss.

A person who survives a severe head injury may be **comatose** for a short or an extended period of time. A coma is a state of unconsciousness from which the patient cannot be aroused, even by the most intense external stimuli. Severe injury to the reticular activating system is likely to result in irreversible coma. Although a head injury is the most common cause of coma, chemical imbalances associated with certain diseases (e.g., diabetes) or the ingestion of drugs or poisons may also be responsible.

The flexibility of the vertebral column is essential for body movements, but because of this flexibility the spinal cord and spinal nerves are somewhat vulnerable to trauma. Falls or severe blows to the back are a common cause of injury. A skeletal injury, such as a fracture, dislocation, or compression of the vertebrae, usually traumatizes nervous tissue as well. Other frequent causes of trauma to the spinal cord include gunshot wounds, stabbings, herniated discs, and birth injuries. The consequences of the trauma depend on the location and severity of the injury and the medical treatment the patient receives. If nerve fibers of the spinal cord are severed, motor or sensory functions will be permanently lost.

Paralysis is a permanent loss of motor control, usually resulting from disease or a lesion of the spinal cord or specific nerves. Paralysis of both lower extremities is called **paraplegia.** Paralysis of both the upper and lower extremity on the same side is called **hemiplegia,** and paralysis of all four extremities is **quadriplegia.** Paralysis may be flaccid or spastic. **Flaccid** (*flak'sid*) **paralysis** generally results from a lesion of the anterior horn cells and is characterized by noncontractible muscles that atrophy. **Spastic paralysis** results from lesions of the corticospinal tracts of the spinal cord and is characterized by hypertonicity of the skeletal muscles.

Whiplash is a sudden hyperextension and flexion of the cervical vertebrae (fig. 15.30) as may occur during a rear-end automobile collision. Recovery from a minor whiplash (muscle and ligament strains) is generally complete, albeit slow. Severe whiplash (spinal cord compression) may cause permanent paralysis to the structures below the level of injury.

Figure 15.30

Whiplash.

This type of neck trauma varies in severity from muscle and ligament strains to dislocation of the vertebrae and compression of the spinal cord. Injuries such as this may cause permanent loss of some or all of the spinal cord functions.

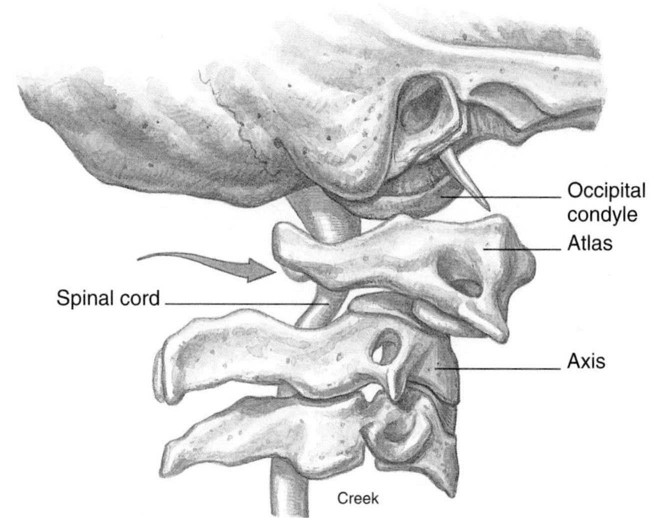

Disorders of the Nervous System

Mental Illness

Mental illness is a major clinical consideration of the nervous system and is perhaps the least understood. Traditionally, mental disorders have been grouped into two broad categories: neurosis and psychosis. In **neurosis,** a maladjustment to certain aspects of life interferes with normal functioning, but contact with reality is maintained. An irrational fear (*phobia*) is an example of neurosis. Neurosis frequently causes intense anxiety or abnormal distress that brings about increased sympathetic stimulation. **Psychosis,** a more serious mental condition, is typified by personality disintegration and a loss of contact with reality. The more common forms of psychosis include *schizophrenia,* in which a person withdraws into a world of fantasy; *paranoia,* in which a person has systematized delusions, often of a persecutory nature; and *manic-depressive psychosis,* in which a person's moods swing widely from intense elation to deepest despair.

Epilepsy

Epilepsy is a relatively common brain disorder with a strong hereditary basis, but it also can be caused by head injuries, tumors, and childhood infectious diseases. It is sometimes idiopathic (without demonstrable cause). A person with epilepsy may periodically suffer from an *epileptic seizure,* which has various symptoms depending on the type of epilepsy.

The most common kinds of epilepsy are petit mal, psychomotor epilepsy, and grand mal. **Petit mal** (*pet'e mal'*) occurs almost exclusively in children between the ages of 3 and 12. A child experiencing a petit mal seizure loses contact with reality for 5 to 30 seconds but does not lose consciousness or

amnesia: L. *amnesia,* forgetfulness
comatose: Gk. *koma,* deep sleep
paralysis: Gk. *paralysis,* loosening
paraplegia: Gk. *para,* beside; *plessein,* to strike

epilepsy: Gk. *epi,* upon; *lepsis,* seize
petit mal: L. *pitinnus,* small child; *malus,* bad

display convulsions. There may, however, be slight uncontrollable facial gestures or eye movements, and the child will stare, as if in a daydream. During a petit mal seizure, the thalamus and hypothalamus produce an extremely slow EEG pattern of 3 waves per second. Children with petit mal usually outgrow the condition by age 10 and generally require no medication.

Psychomotor epilepsy is often confused with mental illness because of the symptoms characteristic of the seizure. During such a seizure, EEG activity accelerates in the temporal lobes, causing a person to become disoriented and lose contact with reality. Occasionally during a seizure, specific cerebral motor areas will cause involuntary lip smacking or hand clapping. If motor areas in the brain are not stimulated, a person having a psychomotor epileptic seizure may wander aimlessly until the seizure subsides.

Grand mal is a more serious form of epilepsy characterized by periodic convulsive seizures that generally render a person unconscious. Grand mal epileptic seizures are accompanied by rapid EEG patterns of 25 to 30 waves per second. This sudden increase from the norm of about 10 waves per second may cause an extensive stimulation of motor units and, therefore, uncontrollable skeletal muscle activity. During a grand mal seizure, a person loses consciousness, convulses, and may lose urinary bladder and bowel control. After a few minutes, the muscles relax and the person awakes but remains disoriented for a short time.

Epilepsy almost never affects intelligence and can be effectively treated with drugs in about 85% of the patients.

Cerebral Palsy

Cerebral palsy is a motor nerve disorder characterized by paresis (partial paralysis) and lack of muscular coordination. It is caused by damage to the motor areas of the brain during prenatal development, birth, or infancy. For example, radiation or bacterial toxins (such as from German measles) transferred through the placenta of the mother to the fetus during neural development may cause cerebral palsy. Oxygen deprivation resulting from complications at birth and hydrocephalus in a newborn may also cause cerebral palsy. The three areas of the brain most severely affected by this disease are the cerebral cortex, the basal nuclei, and the cerebellum. The type of cerebral palsy is determined by the particular region of the brain that is affected.

Some degree of mental retardation occurs in 60% to 70% of cerebral palsy victims. Partial blindness, deafness, and speech problems frequently accompany this disease. Cerebral palsy is nonprogressive (that is, impairments do not worsen as a person ages). However, neither are there physical improvements.

Neoplasms of the CNS

Neoplasms of the CNS are either **intracranial tumors,** which affect brain cells or cells associated with the brain, or they are **intravertebral** (intraspinal) **tumors,** which affect cells within or near the spinal cord. **Primary neoplasms** develop within the CNS. Approximately half of these are benign, but they may become lethal because of the pressure they exert on vital centers as they grow. Patients with **secondary,** or **metastatic, neoplasms** within the brain have a poor prognosis because the cancer has already established itself in another body organ—frequently the liver, lung, or breast—and has only secondarily spread to the brain. The symptoms of a brain tumor include headache, convulsions, pain, paralysis, or a change in behavior.

Neoplasms of the CNS are classified according to the tissues in which the cancer occurs. Tumors arising in neuroglia are called **gliomas** (*gle-o'maz*) and account for about half of all primary neoplasms within the brain. Gliomas are frequently spread throughout cerebral tissue, develop rapidly, and usually cause death within a year after diagnosis. **Astrocytoma** (*as"tro-si-to'mă*), **oligodendroglioma** (*ol"ĭ-go-den"drog-li-o'mă*), and ependymoma (*ĕ-pen"dĭ-mo'mă*) are common types of gliomas.

Meningiomas arise from meningeal coverings of the brain and account for about 15% of primary intracranial tumors. Meningiomas are usually harmless if they are treated readily.

Intravertebral tumors are classified as **extramedullary** when they develop on the outside of the spinal cord and as **intramedullary** when they develop within the substance of the spinal cord. Extramedullary neoplasms may cause pain and numbness in body structures distant from the tumor as the growing tumor compresses the spinal cord. An intramedullary neoplasm causes a gradual loss of function below the spinal–segmental level of the affliction.

Methods of detecting and treating cancers within the CNS have greatly improved in the last few years. Early detection and competent treatment have lessened the likelihood of death from these cancers and have reduced the probability of physical impairment.

Dyslexia

Dyslexia is a defect in the language center within the brain, causing otherwise intelligent people to reverse the order of letters in syllables, of syllables in words, and of words in sentences. The sentence, "The man saw a red dog," for example, might be read by the dyslexic as "A red god was the man." Dyslexia is believed to result from the failure of one cerebral hemisphere to respond to written language, perhaps due to structural defects. Dyslexia can usually be overcome by intense remedial instruction in reading and writing.

Meningitis

The nervous system is vulnerable to a variety of organisms and viruses that may cause abscesses or infections. Meningitis is an infection of the meninges. It may be confined to the spinal cord, in which case it is referred to as **spinal meningitis,** or it may involve the brain and associated meninges, in which case it is known as **encephalitis.** When both the brain and spinal cord are involved, the correct

---

dyslexia: Gk. *dys*, bad; *lexis*, speech

term is **encephalomyelitis** (*en-sef″ă-lo-mi″ĕ-li′-tis*). The microorganisms that most commonly cause meningitis are meningococci, streptococci, pneumococci, and tubercle bacilli. Viral meningitis is more serious than bacterial meningitis; nearly 20% of viral encephalitides are fatal. The organisms that cause meningitis probably enter the body through respiratory passageways.

Poliomyelitis

Poliomyelitis, or infantile paralysis, is primarily a childhood disease caused by a virus that destroys nerve cell bodies within the anterior horn of the spinal cord, especially those within the cervical and lumbar enlargements. This degenerative disease is characterized by fever, severe headache, stiffness and pain in the head and back, and the loss of certain somatic reflexes. Muscle paralysis follows within several weeks, and eventually the muscles atrophy. Death results if the virus invades the vasomotor and respiratory nuclei within the medulla oblongata or anterior horn cells controlling respiratory muscles. Poliomyelitis has been effectively controlled with immunization.

Syphilis

Syphilis is a sexually transmitted disease that, if untreated, progressively destroys body organs. When syphilis causes organ degeneration, it is said to be in the *tertiary stage* (10 to 20 years after the primary infection). The organs of the nervous system are frequently infected, causing a condition called **neurosyphilis.** Neurosyphilis is classified according to the tissue involved, and the symptoms vary correspondingly. If the meninges are infected, the condition is termed **chronic meningitis. Tabes** (*ta′bez*) **dorsalis** is a form of neurosyphilis in which there is a progressive degeneration of the posterior funiculi of the spinal cord and posterior roots of spinal nerves. Motor control is gradually lost, and patients eventually become bedridden, unable even to feed themselves.

Degenerative Diseases of the Nervous System

Degenerative diseases of the CNS are characterized by a progressive, symmetrical deterioration of vital structures of the brain or spinal cord. The etiologies of these diseases are poorly understood, but it is thought that most of them are genetic.

Cerebrovascular Accident (CVA)

Cerebrovascular accident is the most common disease of the nervous system. It is the third most frequent cause of death in the United States, and perhaps the number one cause of disability. The term **stroke** is frequently used as a synonym for CVA, but actually a stroke refers to the sudden and dramatic appearance of a neurological defect. *Cerebral thrombosis*, in which a thrombus, or clot, forms in an artery of the brain, is the most common cause of CVA. Other causes of CVA include intracerebral hemorrhages, aneurysms, atherosclerosis, and arteriosclerosis of the cerebral arteries.

Patients who recover from CVA frequently suffer partial paralysis and mental disorders, such as loss of language skills. The dysfunction depends upon the severity of the CVA and the regions of the brain that were injured. Patients surviving a CVA can often be rehabilitated, but approximately two-thirds die within 3 years of the initial damage.

Syringomyelia

Syringomyelia (*sĭ-ring″go-mi-e′le-ă*) is a relatively uncommon condition characterized by the appearance of cystlike cavities, called *syringes*, within the gray matter of the spinal cord. These syringes progressively destroy the spinal cord from the inside out. As the spinal cord deteriorates, the patient experiences muscular weakness and atrophy and sensory loss, particularly of the senses of pain and temperature. The cause of syringomyelia is unknown.

Clinical Investigation Answer

The affected upper motor neurons have cell bodies that reside in the right cerebral hemisphere. They give rise to fibers that, as they course downward, cross in the medulla oblongata to the left side of the brain stem and spinal cord. They continue on the left side, eventually synapsing at the appropriate level with lower motor neurons in the anterior horn of the spinal cord. The lower motor neurons then give rise to fibers that travel to the periphery, where they innervate end organs; namely, muscle cells in the left side of the body. Since the patient's neurological defects are all motor as opposed to sensory, the tumor is most likely located in the right frontal lobe. The parietal lobe contains sensory neurons. The persistent headache is due to the pressure of the tumorous mass on the meninges, which are heavily sensory innervated.

Chapter Summary

Characteristics of the Central Nervous System (pp. 408–409)

1. The central nervous system (CNS) consists of the brain and spinal cord and contains gray and white matter. It is covered with meninges and is bathed in cerebrospinal fluid.
2. The tremendous metabolic rate of the 1.5-kg brain requires a continuous flow of blood, amounting to approximately 20% of the total cardiac output.

Cerebrum (pp. 412–424)

1. The cerebrum, consisting of two convoluted hemispheres, is concerned with higher brain functions, such as the perception of sensory impulses, the instigation of voluntary movement, the storage of memory, thought processes, and reasoning ability.
2. The cerebral cortex of the cerebral hemispheres is convoluted with gyri and sulci.
3. Each cerebral hemisphere contains frontal, parietal, temporal, occipital, and insula lobes.
4. Brain waves generated by the cerebral cortex are recorded as an electroencephalogram and may provide valuable diagnostic information.
5. The white matter of the cerebrum consists of association, commissural, and projection fibers.

6. Basal nuclei are specialized masses of gray matter located within the white matter of the cerebrum.
7. The motor speech area, Wernicke's area, the arcuate fasciculus, and the angular gyrus are the language areas of the brain and are generally located in the cerebral cortex of the left hemisphere.
8. Consisting of the cingulate gyrus, amygdaloid nucleus, hippocampus, and the septal nucleus, the limbic system is the center for basic emotional drives.
9. The consolidation of memory requires protein synthesis and probably involves changes in the chemical structure and function of synapses.

Diencephalon (pp. 424–426)

1. The diencephalon is a major autonomic region of the brain.
2. The thalamus is an ovoid mass of gray matter that functions as a relay center for sensory impulses and responds to pain.
3. The hypothalamus is an aggregation of specialized nuclei that regulate many visceral activities. It also performs emotional and instinctual functions.
4. The epithalamus contains the pineal gland and the vascular choroid plexus over the roof of the third ventricle.

Mesencephalon (p. 426)

1. The mesencephalon contains the corpora quadrigemina, the cerebral peduncles, and specialized nuclei that help to control posture and movement.
2. The superior colliculi of the corpora quadrigemina are concerned with visual reflexes; the inferior colliculi are concerned with auditory reflexes.
3. The red nucleus and the substantia nigra are concerned with motor activities.

Metencephalon (pp. 426–428)

1. The pons consists of fiber tracts connecting the cerebellum and medulla oblongata to other structures of the brain. The pons also contains nuclei for certain cranial nerves and the regulation of respiration.
2. The cerebellum consists of two hemispheres connected by the vermis and supported by three paired cerebellar peduncles.
   (a) The cerebellum is composed of a white matter tract called the arbor vitae, surrounded by a thin convoluted cerebellar cortex of gray matter.
   (b) The cerebellum is concerned with coordinated contractions of skeletal muscle.

Myelencephalon (pp. 428–429)

1. The medulla oblongata is composed of the ascending and descending tracts of the spinal cord and contains nuclei for several autonomic functions.
2. The reticular formation functions as the reticular activating system in arousing the cerebrum.

Meninges of the Central Nervous System (pp. 430–431)

1. The cranial dura mater consists of an outer periosteal layer and an inner meningeal layer. The spinal dura mater is a single layer surrounded by the vascular epidural space.
2. The arachnoid is a netlike meninx surrounding the subarachnoid space. The subarachnoid space contains cerebrospinal fluid.
3. The thin pia mater adheres to the contour of the CNS.

Ventricles and Cerebrospinal Fluid (pp. 432–433)

1. The lateral (first and second), third, and fourth ventricles are interconnected chambers within the brain that are continuous with the central canal of the spinal cord.
2. These chambers are filled with cerebrospinal fluid, which also flows throughout the subarachnoid space.
3. Cerebrospinal fluid is continuously secreted by the choroid plexuses and is absorbed into the blood at the arachnoid villi.

Spinal Cord (pp. 433–440)

1. The spinal cord is composed of 31 segments, each of which gives rise to a pair of spinal nerves.
   (a) It is characterized by a cervical enlargement, a lumbar enlargement, and two longitudinal grooves that partially divide it into right and left halves.
   (b) The conus medullaris is the terminal portion of the spinal cord, and the cauda equina are nerve roots that radiate inferiorly from that point.
2. Ascending and descending spinal cord tracts are referred to as funiculi.
   (a) Descending tracts are grouped as either corticospinal (pyramidal) or extrapyramidal.
   (b) Many of the fibers in the funiculi decussate (cross over) in the spinal cord or in the medulla oblongata.

Review Activities

Objective Questions

1. Which of the following is *not* a lobe of the cerebrum?
   (a) parietal (d) insula
   (b) sphenoid (e) occipital
   (c) temporal
2. The principal connection between the cerebral hemispheres is
   (a) the corpus callosum.
   (b) the pons.
   (c) the intermediate mass.
   (d) the vermis.
   (e) the precentral gyrus.
3. Which of the following structures of the brain is most directly involved in the autonomic response to pain?
   (a) pons
   (b) hypothalamus
   (c) medulla oblongata
   (d) thalamus
4. Which statement concerning the basal nuclei is *false?*
   (a) They are located within the cerebrum.
   (b) They regulate the basal metabolic rate.
   (c) They consist of the caudate nucleus, lentiform nucleus, putamen, and globus pallidus.
   (d) They indirectly exert an inhibitory influence on lower motor neurons.
5. In which region of the brain are the corpora quadrigemina, red nucleus, and substantia nigra located?
   (a) diencephalon
   (b) metencephalon
   (c) mesencephalon
   (d) myelencephalon
6. The fourth ventricle is contained within
   (a) the cerebrum.
   (b) the cerebellum.
   (c) the midbrain.
   (d) the metencephalon.
7. The right cerebral cortex controls voluntary movements on the left side of the body because
   (a) most people are right-handed.
   (b) the right hemisphere dominates.
   (c) many of the fibers in the funiculi decussate in the medulla oblongata.
   (d) there is distinct cerebral specialization of hemispheres.
8. A patient experiencing a fluctuating body temperature, lack of hunger and thirst, and psychosomatic disorders may have a malfunctioning
   (a) hypothalamus.
   (b) midbrain.
   (c) cerebellum.
   (d) medulla oblongata.

9. Spinal cord tracts that descend from the cerebral cortex to the lower motor neurons without synaptic interruption are called
   (a) reticulospinal tracts.
   (b) corticospinal tracts.
   (c) rubrospinal tracts.
   (d) vestibulospinal tracts.
10. The disease characterized by the destruction of the myelin sheaths of neurons in the CNS and the formation of plaques is
   (a) syringomyelia.
   (b) neurosyphilis.
   (c) poliomyelitis.
   (d) multiple sclerosis.

Essay Questions

1. List the types of brain waves recorded on an electroencephalogram, and explain the diagnostic value of each.
2. List the functions of the hypothalamus. Why is the hypothalamus considered a major part of the autonomic nervous system?
3. What structures are found within the midbrain? List the nuclei located in the midbrain, and give the function of each.
4. Describe the location and structure of the medulla oblongata. List the nuclei found within this structure. What are the functions of the medulla oblongata?
5. What is cerebrospinal fluid? Where is it produced, and what is its pathway of circulation?
6. What do the following abbreviations stand for? EEG, ANS, CSF, PNS, RAS, CT scan, MS, DSR, and CVA.
7. Describe the various techniques available for conducting a neurological assessment.
8. Define the various psychological terms used to describe mental illness.
9. What is epilepsy? What causes it, and how is it controlled?
10. What do meningitis, poliomyelitis, and neurosyphilis have in common? How do these conditions differ?

Critical Thinking Questions

1. In a patient's case study, any observed or suspected structural/functional abnormalities of body structures are reported. Prepare a brief case study of a patient who has suffered severe trauma to the medulla oblongata from a blow to the back of the skull.
2. A young man develops weakness in his legs over the course of several days, which worsens until he can no longer walk. He also loses urinary bladder control and complains of loss of sensation from the umbilicus down. Locate the probable site of neurological impairment.
3. Electrical stimulation of the cerebellum or the basal nuclei can produce skeletal movements. How would damage to these two regions of the brain affect skeletal muscle function differently?
4. If an entire cerebral hemisphere is destroyed, a person can still survive. Yet, damage to the medulla oblongata, a much smaller mass of tissue, can be fatal. Explain this difference.
5. A seizure occurs when abnormal electrical activity overwhelms the brain's normal function. A facial seizure originates from a specific irritable tissue, such as a tumorous mass, and causes isolated muscle jerking or sensory abnormalities. Immediately before a focal seizure, the sufferer frequently perceives a fleeting sensation, called an aura, that suggests the origin of the electrical burst. Using your knowledge of cerebral lobe function, predict the origin of a seizure that was preceded by the following: (1) perception of a foul odor, (2) a flash of light, (3) a painful hand, and (4) a familiar song.
6. Explain how meningitis contracted through the meninges over the roof of the nose may be detected from a spinal tap performed in the lumbar region.

Related Web Sites

For a listing of the most current web sites related to this chapter, please visit the *Concepts of Human Anatomy and Physiology* home page at http://www.mhhe.com/biosci/abio/.

SIXTEEN

Peripheral Nervous System

OBJECTIVES

- Define *peripheral nervous system* and distinguish between sensory, motor, and mixed nerves.
- List the 12 pairs of cranial nerves and describe the location and functions of each.
- Describe the clinical methods for determining cranial nerve dysfunction.
- Explain how the spinal nerves are grouped.
- Describe the general distribution of a spinal nerve.
- List the spinal nerve composition of each of the plexuses arising from the spinal cord.
- List the principal nerves that emerge from the plexuses and describe their general innervation.
- Define *reflex arc* and list its five components.
- Distinguish between the various kinds of reflexes.

Clinical Investigation

Following an auto accident, a 23-year-old male was brought to the emergency room for treatment of a fractured right humerus. Although the skin was not broken, there was an obvious deformity caused by an angulated fracture at the midshaft. While conducting an examination on the patient's injured arm, the attending orthopedist noticed that the patient was unable to extend the joints of his hand.

What structure could be injured in the brachial region of this patient that would account for his inability to extend his hand? List the muscles that would be affected and describe the movements that would be diminished. Do you think there might be other neurological defects? Explain.

Clues: Because the nervous system functions to coordinate body movement, nerve trauma may be expressed in structures far removed from the site of injury. Carefully read the section dealing with the brachial plexus.

Introduction to the Peripheral Nervous System

The peripheral nervous system consists of all of the nervous tissue outside the central nervous system, including sensory receptors, nerves and their associated ganglia, and nerve plexuses. It provides a communication pathway for impulses traveling between the CNS and the rest of the body.

The **peripheral nervous system (PNS)** is that portion of the nervous system outside the central nervous system. The PNS conveys impulses to and from the brain and spinal cord. Sensory receptors within the sensory organs, nerves, ganglia, and plexuses are all part of the PNS, which serves virtually every part of the body (fig. 16.1). The sensory receptors are discussed in chapter 18.

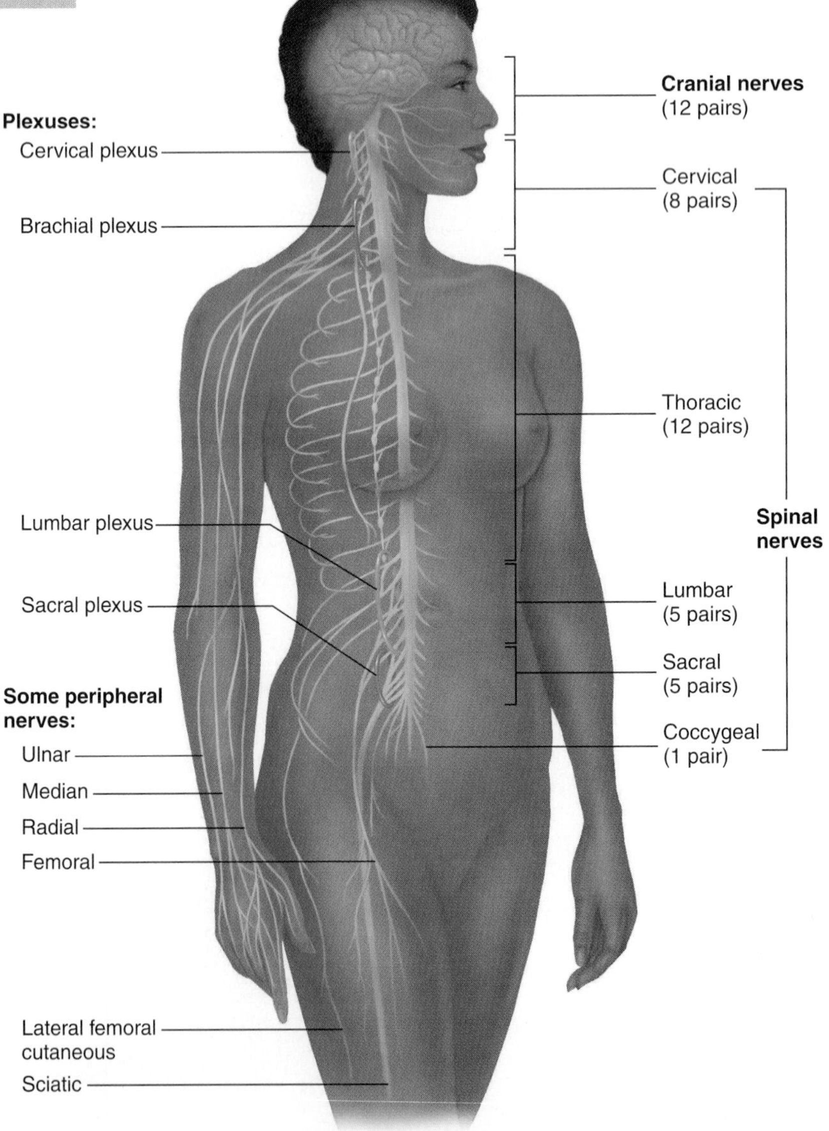

Figure 16.1

The peripheral nervous system.
This division of the nervous system includes cranial nerves and spinal nerves, and the nerves that arise from them. Plexuses and ganglia (not shown) are also part of the peripheral nervous system.

Figure 16.2

A scanning electron micrograph of a spinal nerve seen in cross section (about 1000×).

From Tissues and Organs: A Text-Atlas of Scanning Electron Microscopy, *by R. G. Kessel and R. Kardon. © 1979 W. H. Freeman and Company.*

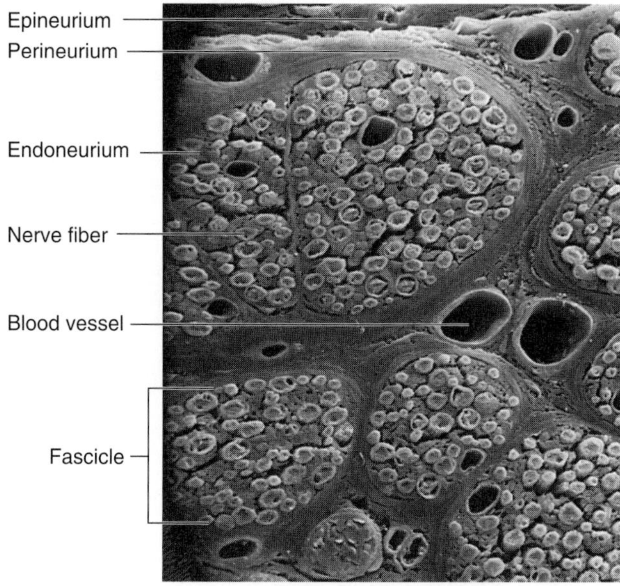

The nerves of the PNS are classified as cranial nerves or spinal nerves, depending on whether they arise from the brain or the spinal cord. A cross section of a spinal nerve is shown in figure 16.2. The terms *sensory nerve*, *motor nerve*, and *mixed nerve* relate to the direction in which the nerve impulses are being conducted. **Sensory nerves** consist of sensory (afferent) neurons that convey impulses toward the CNS. **Motor nerves** consist primarily of motor (efferent) neurons that convey impulses away from the CNS. (Technically speaking, there are no nerves that are motor only; all motor nerves contain some proprioceptor fibers that convey sensory information to the CNS.) **Mixed nerves** are composed of both sensory and motor neurons in about equal numbers, and they convey impulses both to and from the CNS.

Cranial Nerves

Twelve pairs of cranial nerves emerge from the inferior surface of the brain and pass through the foramina of the skull to innervate structures in the head, neck, and visceral organs of the trunk.

Structure and Function of the Cranial Nerves

Of the 12 pairs of cranial nerves, 2 pairs arise from the forebrain, and 10 pairs arise from the midbrain and brain stem (fig. 16.3). The cranial nerves are designated by Roman numerals and names. The Roman numerals refer to the order in which the nerves are positioned from the front of the brain to the back. The names indicate the structures innervated or the principal functions of the nerves. A summary of the cranial nerves is presented in table 16.1.

Figure 16.3

The cranial nerves.

With the exception of the olfactory nerves, each cranial nerve is composed of a bundle of nerve fibers. The olfactory nerves are minute and diffuse strands of nerve fibers that attach to the olfactory bulbs (see fig. 16.4).

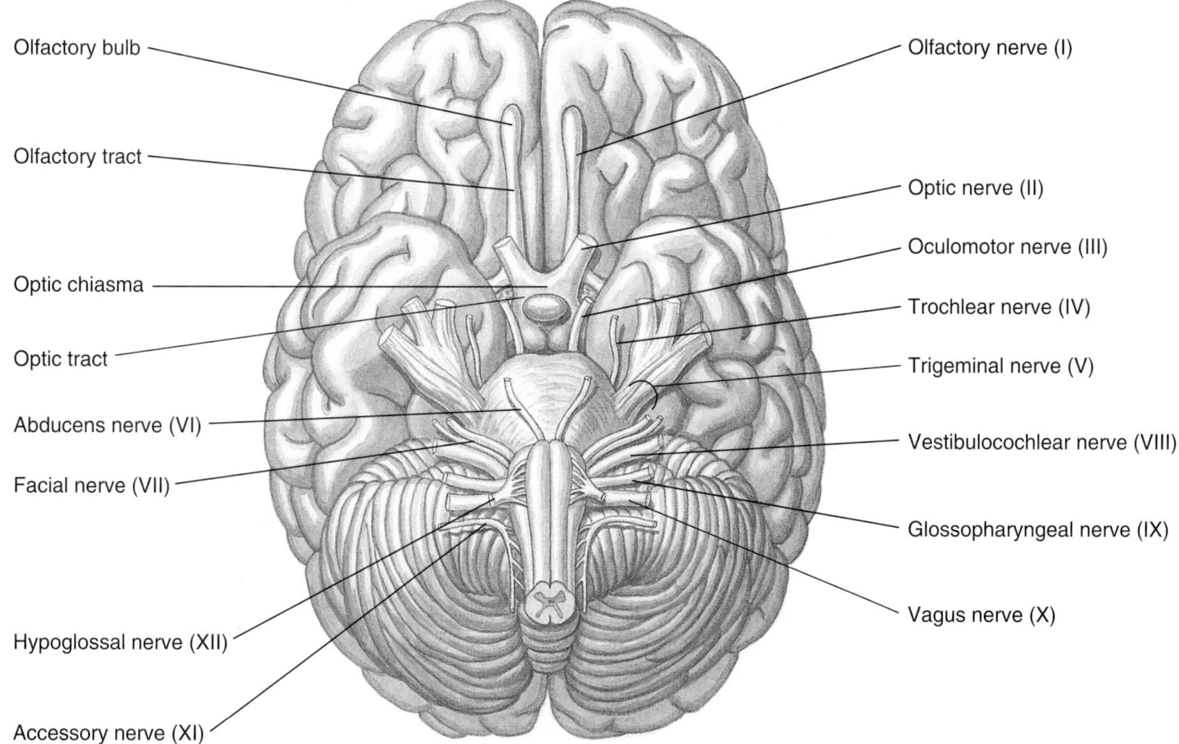

Table 16.1

Summary of Cranial Nerves

Number and Name	Foramen Transmitting	Composition	Location of Cell Bodies	Function
I Olfactory	Foramina in cribriform plate of ethmoid bone	Sensory	Bipolar cells in nasal mucosa	Olfaction
II Optic	Optic foramen	Sensory	Ganglion cells of retina	Vision
III Oculomotor	Superior orbital fissure	Motor	Oculomotor nucleus	Motor impulses to levator palpebrae superioris and extrinsic eye muscles except superior oblique and lateral rectus
		Motor: parasympathetic		Innervation to muscles that regulate amount of light entering eye and that focus the lens
		Sensory: proprioception		Proprioception from muscles innervated with motor fibers
IV Trochlear	Superior orbital fissure	Somatic motor	Trochlear nucleus	Motor impulses to superior oblique muscle of eyeball
		Sensory: proprioception		Proprioception from superior oblique muscle of eyeball
V Trigeminal				
Ophthalmic nerve	Superior orbital fissure	Sensory	Trigeminal ganglion	Sensory impulses from cornea, skin of nose, forehead, and scalp
Maxillary nerve	Foramen rotundum	Sensory	Trigeminal ganglion	Sensory impulses from nasal mucosa, upper teeth and gums, palate, upper lip, and skin of cheek
Mandibular nerve	Foramen ovale	Sensory	Trigeminal ganglion	Sensory impulses from temporal region, tongue, lower teeth and gum, and skin of chin and lower jaw
		Sensory: proprioception		Proprioception from muscles of mastication
		Somatic Motor	Motor trigeminal nucleus	Motor impulses to muscles of mastication and muscle that tenses tympanum
VI Abducens	Superior orbital fissure	Somatic motor	Abducens nucleus	Motor impulses to lateral rectus muscle of eyeball
		Sensory: proprioception		Proprioception from lateral rectus muscle of eyeball
VII Facial	Stylomastoid foramen	Somatic motor	Motor facial nucleus	Motor impulses to muscles of facial expression and muscle that tenses the stapes
		Motor: parasympathetic	Superior salivatory nucleus	Secretion of tears from lacrimal parasympathetic gland and salivation from sublingual and submandibular glands
		Sensory	Geniculate ganglion	Sensory impulses from taste buds on anterior two-thirds of tongue; nasal and palatal sensation
		Sensory: proprioception		Proprioception from muscles of facial expression

Although most cranial nerves are mixed, some are associated with special senses and consist of sensory neurons only. The cell bodies of sensory neurons are located in ganglia outside the brain.

Generations of anatomy students have used a mnemonic device to help them remember the order in which the cranial nerves emerge from the brain: "On old Olympus's towering top, a Finn and German viewed a hop." The initial letter of each word in this jingle corresponds to the initial letter of each pair of cranial nerves. A problem with this classic verse is that the eighth cranial nerve represented by *and* in the jingle, which used to be referred to as "auditory," is currently recognized as the vestibulocochlear nerve. Hence, the following topical mnemonic: "On old Olympus's towering top, a fat vicious goat vandalized a hat."

Table 16.1 Continued

Number and Name	Foramen Transmitting	Composition	Location of Cell Bodies	Function
VIII Vestibulocochlear	Internal acoustic meatus	Sensory	Vestibular ganglion Spiral ganglion	Sensory impulses associated with equilibrium Sensory impulses associated with hearing
IX Glossopharyngeal	Jugular foramen	Somatic motor	Nucleus ambiguous	Motor impulses to muscles of pharynx used in swallowing
		Sensory: proprioception	Petrosal ganglion	Proprioception from muscles of pharynx
		Sensory	Petrosal ganglion	Sensory impulses from taste buds on posterior one-third of tongue, pharynx, middle-ear cavity, and carotid sinus
		Parasympathetic	Inferior salivatory nucleus	Salivation from parotid gland
X Vagus	Jugular foramen	Somatic motor	Nucleus ambiguous	Contraction of muscles of pharynx (swallowing) and larynx (phonation)
		Sensory: proprioception		Proprioception from visceral muscles
		Sensory	Nodose ganglion	Sensory impulses from taste buds on rear of tongue; sensations from auricle of ear; general visceral sensations
		Motor: parasympathetic	Dorsal motor nucleus	Motor impulses to visceral muscles
XI Accessory	Jugular foramen	Somatic motor	Nucleus ambiguous Accessory nucleus	Laryngeal movement; soft palate Motor impulses to trapezius and sternocleidomastoid muscles for movement of head, neck, and shoulders
		Sensory: proprioception		Proprioception from muscles that move head, neck, and shoulders
XII Hypoglossal	Hypoglossal canal	Somatic motor	Hypoglossal nucleus	Motor impulses to intrinsic and extrinsic muscles of tongue and infrahyoid muscles
		Sensory: proprioception		Proprioception from muscles of tongue

I Olfactory Nerve

Actually, numerous olfactory nerves relay sensory impulses of smell from the mucous membranes of the nasal cavity (fig. 16.4). Olfactory nerves are composed of bipolar neurons that function as *chemoreceptors*, responding to volatile chemical particles breathed into the nasal cavity. The dendrites and cell bodies of olfactory neurons are positioned within the mucosa, primarily that which covers the superior nasal conchae and adjacent nasal septum. The axons of these neurons pass through the cribriform plate of the ethmoid bone to the **olfactory bulb** where synapses are made, and the sensory impulses are passed through the **olfactory tract** to the primary olfactory area in the cerebral cortex.

olfactory: L. *olfacere*, smell out

II Optic Nerve

The optic nerve, another sensory nerve, conducts impulses from the *photoreceptors* (rods and cones) in the retina of the eye. Each optic nerve is composed of an estimated 1.25 million nerve fibers that converge at the back of the eyeball and enter the cranial cavity through the optic foramen. The two optic nerves unite on the floor of the diencephalon to form the **optic chiasma** (*ki-as'mă*) (fig. 16.5). Nerve fibers that arise from the medial half of each retina cross at the optic chiasma to the opposite side of the brain, whereas fibers arising from the lateral half remain on the same side of the brain. The optic nerve fibers pass posteriorly from the optic chiasma to the thalamus via the **optic tracts.** In the thalamus, a majority of

optic: L. *optica*, see
chiasma: Gk. *chiasma*, an X-shaped arrangement

Figure 16.4
The olfactory nerve.

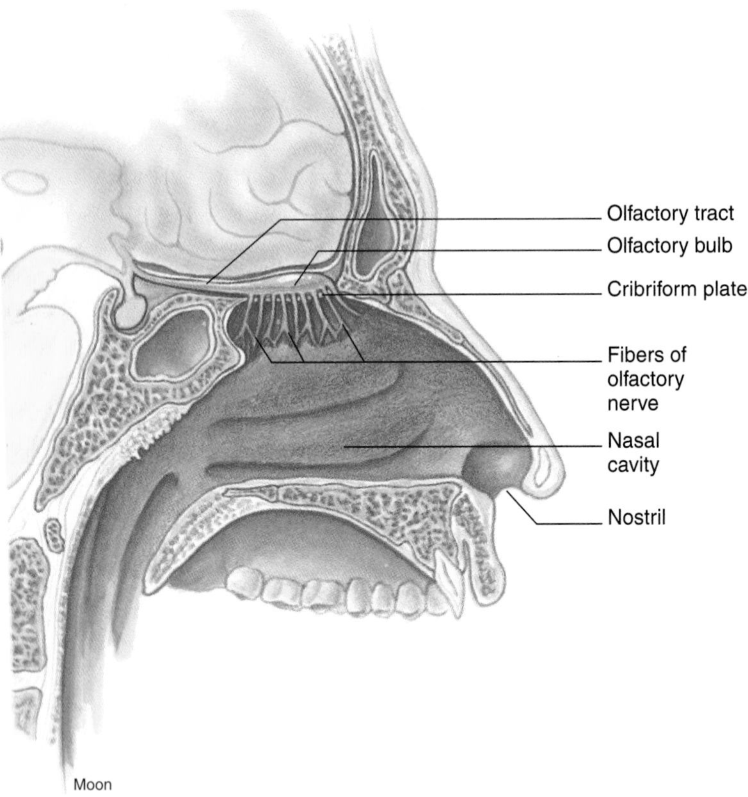

Figure 16.5
The optic nerve and visual pathways.

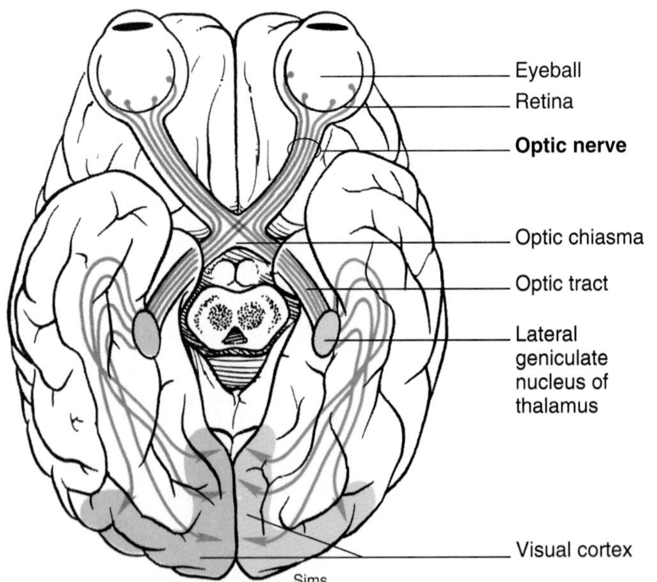

the fibers terminate within certain thalamic nuclei. A few of the ganglion-cell axons that reach the thalamic nuclei have collaterals that convey impulses to the superior colliculi. Synapses within the thalamic nuclei, however, permit impulses to pass through neurons to the **visual cortex** within the occipital lobes. Other synapses permit impulses to reach the nuclei for the oculomotor, trochlear, and abducens nerves, which regulate intrinsic (internal) and extrinsic (from orbit to eyeball) eye muscles. The visual pathway into the eyeball functions reflexively to produce motor responses to light stimuli. If an optic nerve is damaged, the eyeball served by that nerve is blinded.

III Oculomotor Nerve

Nerve impulses through the oculomotor nerve produce certain extrinsic and intrinsic movements of the eyeball. The oculomotor is primarily a motor nerve that arises from nuclei within the midbrain. It divides into superior and inferior branches as it passes through the superior orbital fissure in the orbit (fig. 16.6). The superior branch innervates the **superior rectus** eye muscle, which moves the eyeball superiorly, and the **levator palpebrae** (*le-va'tor pal'pĕ-bre*) **superioris** muscle, which raises the upper eyelid. The inferior branch innervates the **medial rectus, inferior rectus,** and **inferior oblique** eye muscles for medial, inferior, and superior and lateral movement of the eyeball, respectively. In addition, fibers from the inferior branch of the oculomotor nerve enter the eyeball to supply autonomic motor innervation to the intrinsic smooth muscles of the iris for pupil constriction and to the muscles within the ciliary body for lens accommodation.

A few sensory fibers of the oculomotor nerve originate from proprioceptors within the intrinsic muscles of the eyeball. These fibers convey impulses that affect the position and activity of the muscles they serve. A person whose oculomotor nerve is damaged may have a drooping upper eyelid or a dilated pupil, or be unable to move the eyeball in the directions permitted by the four extrinsic muscles innervated by this nerve.

IV Trochlear Nerve

The trochlear (*trok'le-ar*) nerve is a very small mixed nerve that emerges from a nucleus within the midbrain and passes from the cranium through the superior orbital fissure of the orbit. The trochlear nerve innervates the **superior oblique** muscle of the eyeball with both motor and sensory fibers (fig. 16.6). Motor impulses to the superior oblique muscle cause the eyeball to rotate downward and away from the midline. Sensory impulses originate in proprioceptors of the superior oblique muscle and provide information about its position and activity. Damage to the trochlear nerve impairs movement in the direction permitted by the superior oblique eye muscle.

---

trochlear: Gk. *trochos*, a wheel

Figure 16.6

The nerves associated with the eye.
The optic nerve provides sensory innervation to the eye. The oculomotor, trochlear, and abducens nerves provide motor innervation to the extrinsic eye muscles.

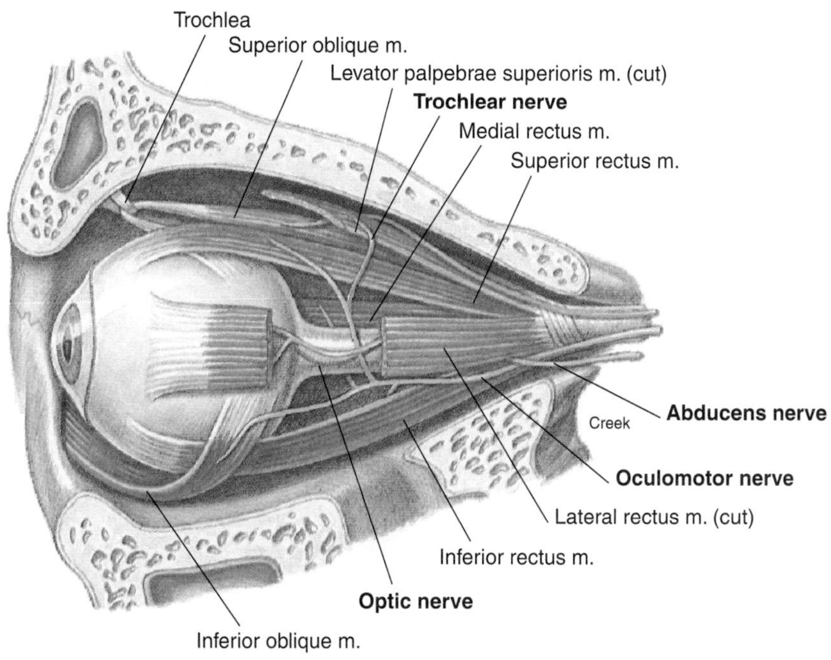

V *Trigeminal Nerve*

The large trigeminal (*tri-jem'in-al*) nerve is a mixed nerve with motor functions originating from the nuclei within the pons and sensory functions terminating in nuclei within the midbrain, pons, and medulla oblongata. Two roots of the trigeminal nerve are apparent as they emerge from the anterolateral side of the pons (see fig. 16.3). The larger **sensory root** immediately enlarges into a swelling called the **trigeminal** (semilunar) **ganglion,** located in a bony depression on the inner surface of the petrous part of the temporal bone. Three large nerves arise from the trigeminal ganglion (fig. 16.7): the **ophthalmic nerve** enters the orbit through the superior orbital fissure; the **maxillary nerve** extends through the foramen rotundum; and the **mandibular nerve** passes through the foramen ovale. The smaller **motor root** consists of motor fibers of the trigeminal nerve that accompany the mandibular nerve through the foramen ovale and innervate the muscles of mastication and certain

trigeminal: L. *trigeminus*, three born together
ophthalmic: L. *ophthalmia*, region of the eye

Figure 16.7

The trigeminal nerve and its distribution.

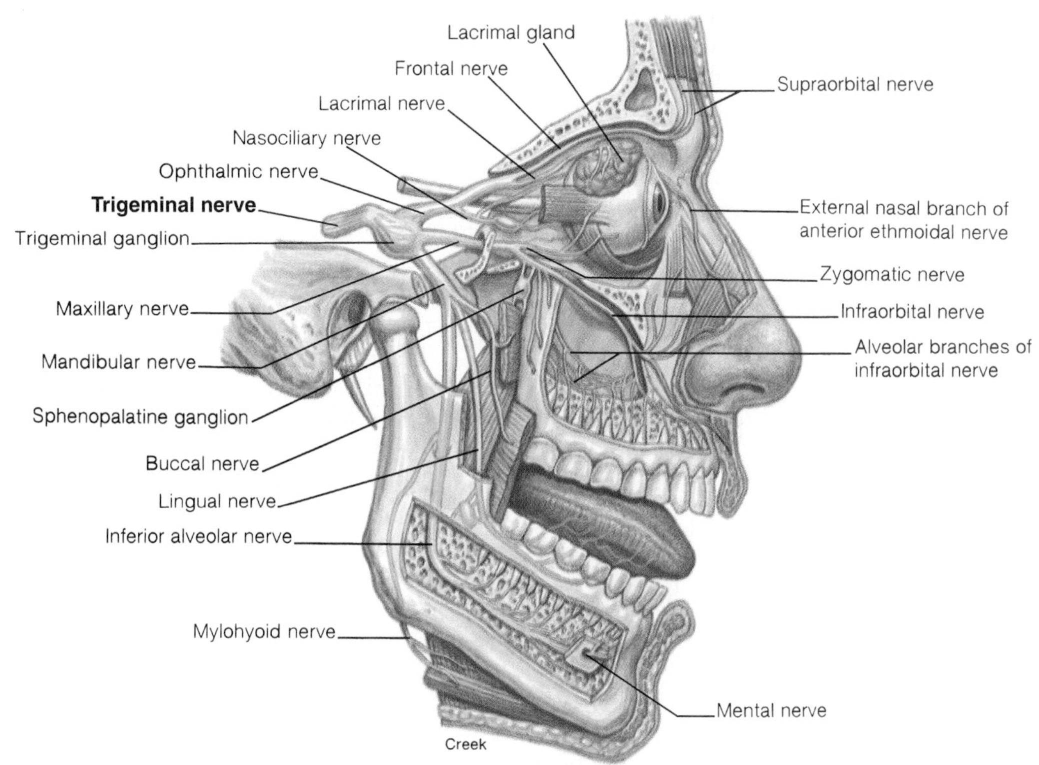

muscles in the floor of the mouth. Impulses through the motor portion of the mandibular nerve of the trigeminal ganglion stimulate contraction of the muscles involved in chewing, including the *medial* and *lateral pterygoid, masseter, temporalis,* and *mylohyoid muscles* and the anterior belly of the *digastric muscle*.

Although the trigeminal is a mixed nerve, its sensory functions are much more extensive than its motor functions. The three sensory nerves of the trigeminal ganglion respond to touch, temperature, and pain sensations from the face. More specifically, the ophthalmic nerve consists of sensory fibers from the anterior half of the scalp, skin of the forehead, upper eyelid, surface of the eyeball, lacrimal (tear) gland, side of the nose, and upper mucosa of the nasal cavity. The maxillary nerve is composed of sensory fibers from the lower eyelid, lateral and inferior mucosa of the nasal cavity, palate and portions of the pharynx, teeth and gums of the upper jaw, upper lip, and skin of the cheek. Sensory fibers of the mandibular nerve transmit impulses from the teeth and gums of the lower jaw, anterior two-thirds of the tongue (not taste), mucosa of the mouth, auricle of the ear, and lower part of the face. Trauma to the trigeminal nerve results in a lack of sensation from specific facial structures. Damage to the mandibular nerve impairs chewing.

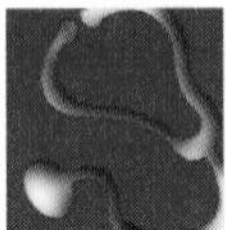

The trigeminal nerve is the principal nerve relating to the practice of dentistry. Before teeth are filled or extracted, anesthetic is injected near the appropriate nerve to block sensation. A *maxillary,* or *second-division, nerve block,* achieved by injecting near the sphenopalatine ganglion (see fig. 16.7), desensitizes the teeth in the upper jaw. A *mandibular,* or *third-division, nerve block* desensitizes the lower teeth. This is performed by injecting anesthetic near the inferior alveolar nerve, which branches off the mandibular nerve as it enters the mandible through the mandibular foramen.

VI Abducens Nerve

The small abducens (*ab-doo'senz*) nerve originates from a nucleus within the pons and emerges from the lower portion of the pons and the anterior border of the medulla oblongata. It is a mixed nerve that traverses the superior orbital fissure of the orbit to innervate the **lateral rectus muscle of the eye** (see fig. 16.6). Impulses through the motor fibers of the abducens nerve cause the lateral rectus muscle to contract and the eyeball to move away from the midline laterally. Sensory impulses through the abducens nerve originate in proprioceptors in the lateral rectus muscle and are conveyed to the pons, where muscle contraction is mediated. If the abducens nerve is damaged, not only will the patient be unable to move the eyeball laterally, but because of the lack of muscle tonus to the lateral rectus muscle, the eyeball will be pulled medially.

VII Facial Nerve

The facial nerve arises from nuclei within the lower portion of the pons, traverses the petrous part of the temporal bone (see fig. 16.9), and emerges on the side of the face near the parotid (salivary) gland. The facial nerve is mixed. Impulses through the motor fibers cause contraction of the posterior belly of the digastric muscle and the muscles of facial expression, including the scalp and platysma muscles (fig. 16.8). The submandibular and sublingual (salivary) glands also receive some autonomic motor innervation from the facial nerve, as does the lacrimal gland.

Figure 16.8

The facial nerve and its distribution to superficial structures.

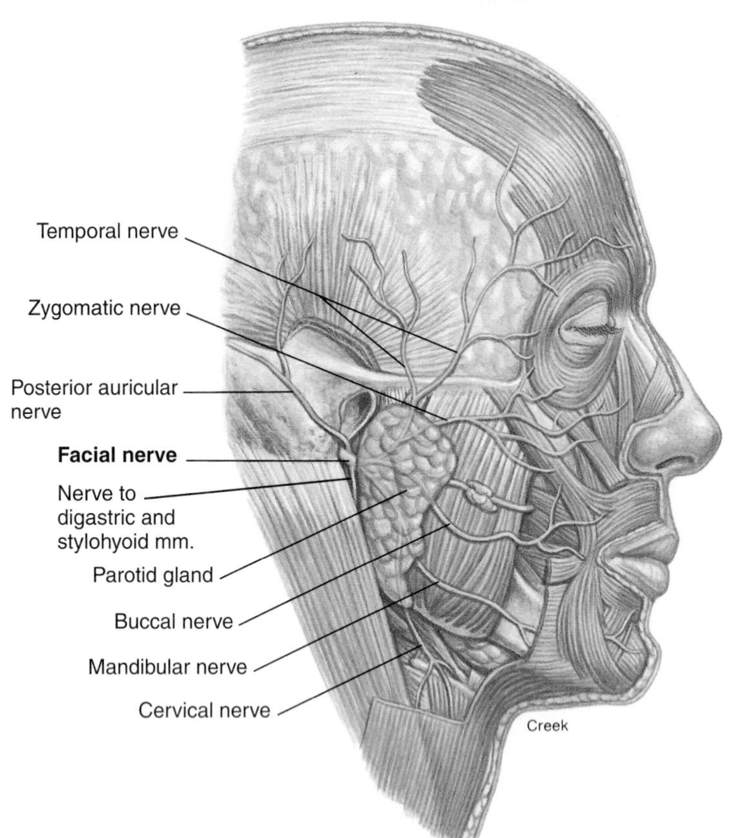

Sensory fibers of the facial nerve arise from taste buds on the anterior two-thirds of the tongue. Taste buds function as *chemoreceptors* because they respond to specific chemical stimuli.

The **geniculate** (*je-nik'yoo-lat*) **ganglion** is the enlargement of the facial nerve just before the entrance of the sensory portion into the pons. Sensations of taste are conveyed to nuclei within the medulla oblongata through the thalamus, and ultimately to the gustatory (taste) area in the parietal lobe of the cerebral cortex.

Trauma to the facial nerve results in inability to contract facial muscles on the affected side of the face and distorts taste perception, particularly of sweets. The affected side of the face tends to sag because muscle tonus is lost. *Bell's palsy* is a functional disorder (probably of viral origin) of the facial nerve.

Bell's palsy: from Sir Charles Bell, Scottish physician, 1774–1842

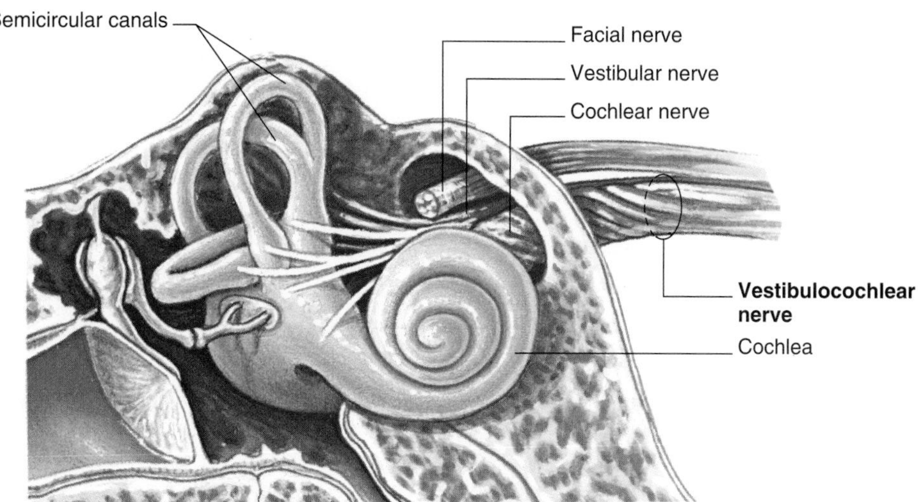

Figure 16.9

The vestibulocochlear nerve.
The structures of the inner ear are served by this sensory nerve. The semicircular ducts, concerned with balance and equilibrium, form the membranous labyrinth within the semicircular canals. The cochlea contains the structures concerned with hearing.

VIII Vestibulocochlear Nerve

The vestibulocochlear (*ves-tib"yu-lo-kok'le-ar*) nerve, also referred to as the *auditory, acoustic,* or *statoacoustic nerve,* serves structures contained within the skull. It is the only cranial nerve that does not exit the cranium through a foramen. A purely sensory nerve, the vestibulocochlear is composed of two nerves that arise within the inner ear (fig. 16.9). The **vestibular nerve** arises from the **vestibular organs** associated with equilibrium and balance. Bipolar neurons from the vestibular organs (saccule, utricle, and semicircular ducts) extend to the **vestibular ganglion,** where cell bodies are contained. From there, fibers convey impulses to the **vestibular nuclei** within the pons and medulla oblongata. Fibers from there extend to the thalamus and the cerebellum.

The **cochlear nerve** arises from the **spiral organ** (organ of Corti) within the cochlea and is associated with hearing. The cochlear nerve is composed of bipolar neurons that convey impulses through the **spiral ganglion** to the **cochlear nuclei** within the medulla oblongata. From there, fibers extend to the thalamus and synapse with neurons that convey the impulses to the auditory areas of the cerebral cortex.

Injury to the cochlear nerve results in perception deafness, whereas damage to the vestibular nerve causes dizziness and loss of balance.

IX Glossopharyngeal Nerve

The glossopharyngeal (*glos"o-fa-rin'je-al*) nerve is a mixed nerve that innervates part of the tongue and pharynx (fig. 16.10). The motor fibers of this nerve originate in a nucleus within the medulla oblongata and pass through the jugular foramen. The motor fibers innervate the muscles of the pharynx and the parotid gland to stimulate the swallowing reflex and the secretion of saliva.

Figure 16.10

The glossopharyngeal nerve.

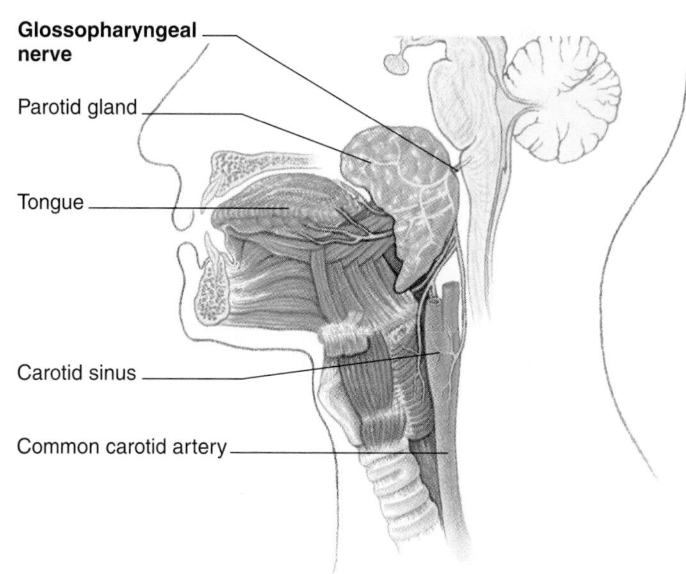

The sensory fibers of the glossopharyngeal nerve arise from the pharyngeal region, the parotid gland, the middle-ear cavity, and the taste buds on the posterior one-third of the tongue. These taste buds, like those innervated by the facial nerve, are *chemoreceptors.* Some sensory fibers also arise from sensory receptors within the carotid sinus of the neck and help regulate blood pressure. Impulses from the glossopharyngeal nerve travel through the medulla oblongata and into the thalamus, where they synapse with fibers that convey the impulses to the gustatory area of the cerebral cortex.

Damage to the glossopharyngeal nerve results in the loss of perception of bitter and sour taste from taste buds on the posterior portion of the tongue. If the motor portion of this nerve is damaged, swallowing becomes difficult.

---

vestibulocochlear: L. *vestibulum,* chamber; *cochlea,* snail shell
organ of Corti: from Alfonso Corti, Italian anatomist, 1822–88
glossopharyngeal: L. *glossa,* tongue; Gk. *pharynx,* throat

Figure 16.11

Distribution of the vagus nerves.

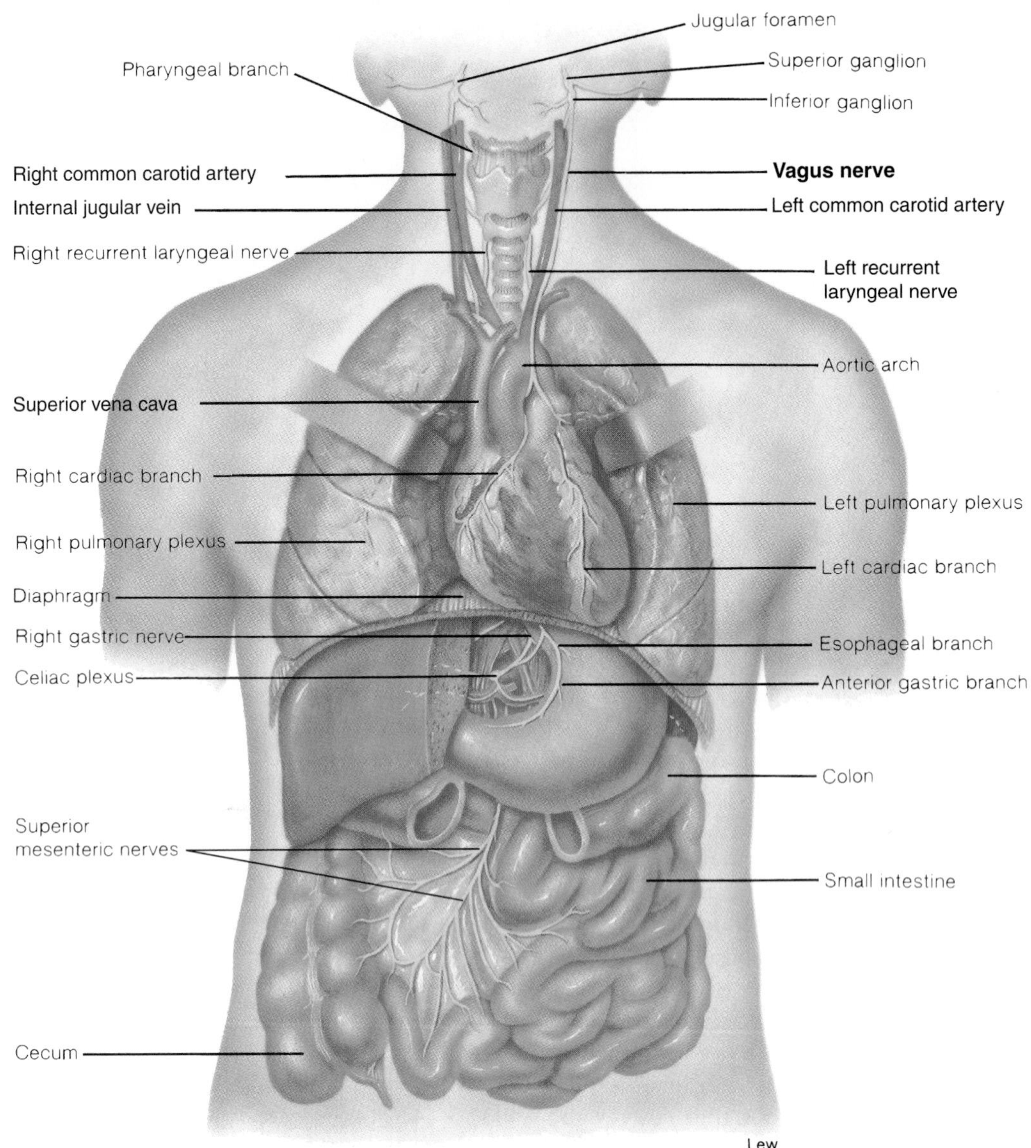

X Vagus Nerve

The vagus (*va'gus*) nerve has motor and sensory fibers that innervate visceral organs of the thoracic and abdominal cavities (fig. 16.11). The motor portion arises from the **nucleus ambiguus** and **dorsal motor nucleus** of the vagus within the medulla oblongata and passes through the jugular foramen. The vagus is the longest of the cranial nerves, and through various branches it innervates the muscles of the pharynx, larynx, respiratory tract, lungs, heart, esophagus, and those of the abdominal viscera, with the exception of the lower portion of the large intestine. One motor branch of the vagus nerve, the **recurrent laryngeal nerve,** innervates the larynx, enabling speech.

Sensory fibers of the vagus nerve convey impulses from essentially the same organs served by motor fibers. Impulses through the sensory fibers relay specific sensations, such as hunger pangs, distension, intestinal discomfort, or laryngeal movement. Sensory fibers also arise from proprioceptors in the muscles innervated by the motor fibers of this nerve.

If both vagus nerves are seriously damaged, death ensues rapidly because vital autonomic functions stop. The injury of one nerve causes vocal impairment, difficulty in swallowing, or other visceral disturbances.

XI Accessory Nerve

The accessory nerve is principally a motor nerve, but it does contain some sensory fibers from proprioceptors within the muscles it innervates. The accessory nerve is unique in that it

vagus: L. *vagus*, wandering

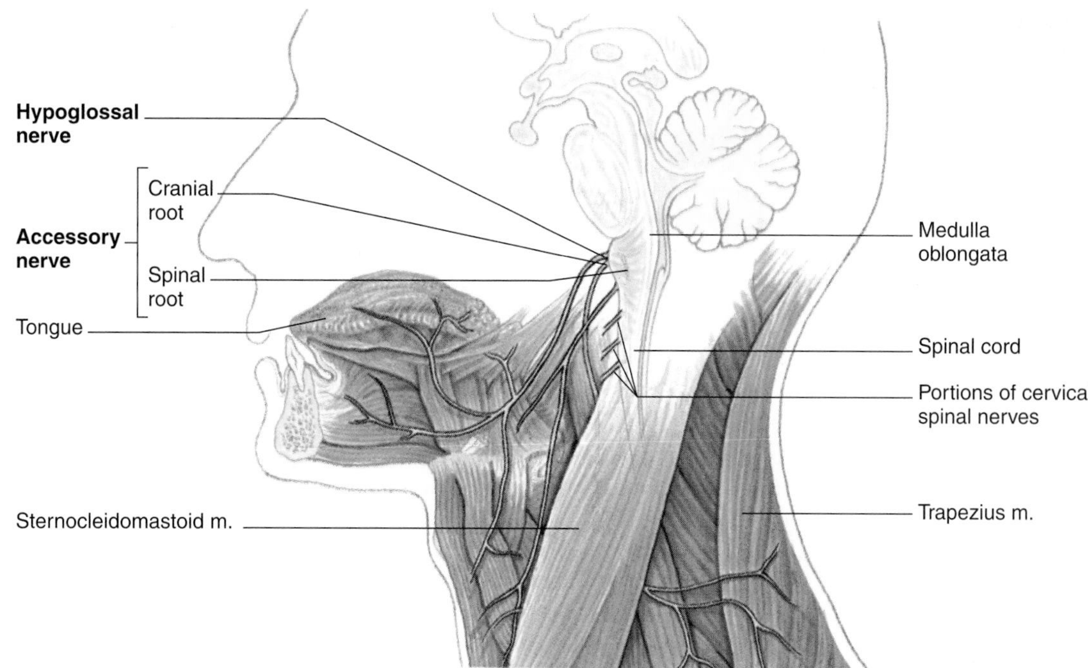

Figure 16.12
The accessory and hypoglossal nerves.

arises from both the brain and the spinal cord (fig. 16.12). The **cranial root** arises from nuclei within the medulla oblongata (ambiguus and accessory), passes through the jugular foramen with the vagus nerve, and innervates the skeletal muscles of the soft palate, pharynx, and larynx, which contract reflexively during swallowing. The **spinal root** arises from the first five segments of the cervical portion of the spinal cord, passes superiorly through the foramen magnum to join with the cranial root, and passes through the jugular foramen. The spinal root of the accessory nerve innervates the sternocleidomastoid and the trapezius muscles that move the head, neck, and shoulders. Damage to an accessory nerve makes it difficult to move the head or shrug the shoulders.

XII Hypoglossal Nerve

The hypoglossal nerve is a mixed nerve. The motor fibers arise from the hypoglossal nucleus within the medulla oblongata and pass through the hypoglossal canal of the skull to innervate both the extrinsic and intrinsic muscles of the tongue (fig. 16.12). Motor impulses along these fibers account for the coordinated contraction of the tongue muscles that is needed for such activities as food manipulation, swallowing, and speech.

The sensory portion of the hypoglossal nerve arises from proprioceptors within the same tongue muscles and conveys impulses to the medulla oblongata regarding the position and function of the muscles.

If a hypoglossal nerve is damaged, a person will have difficulty in speaking, swallowing, and protruding the tongue.

hypoglossal: Gk. *hypo*, under; L. *glossa*, tongue

Neurological Assessment of the Cranial Nerves

Head injuries and brain concussions are common occurrences in automobile accidents. The cranial nerves would seem to be well protected on the inferior side of the brain. But the brain, immersed in and filled with cerebrospinal fluid, is like a water-sodden log; a blow to the top of the head can cause a serious rebound of the brain from the floor of the cranium. Routine neurological examinations involve testing for cranial nerve dysfunction.

Commonly used clinical methods for determining cranial nerve dysfunction are presented in table 16.2.

Spinal Nerves

Each of the 31 pairs of spinal nerves is formed by the union of a posterior and an anterior spinal root that emerges from the spinal cord through an intervertebral foramen to innervate a body dermatome.

The 31 pairs of **spinal nerves** (see fig. 16.1) are grouped as follows: 8 cervical, 12 thoracic, 5 lumbar, 5 sacral, and 1 coccygeal. With the exception of the first cervical nerve, the spinal nerves leave the spinal cord and vertebral canal through intervertebral foramina. The first pair of cervical nerves emerges between the occipital bone of the skull and the atlas. The second through the seventh pairs of cervical nerves emerge above the vertebrae for which they are named, whereas the eighth pair of cervical nerves passes between the seventh cervical and first thoracic vertebrae. The remaining pairs of spinal nerves emerge below the vertebrae for which they are named.

Table 16.2

Methods of Determining Cranial Nerve Dysfunction

Nerve	Techniques of Examination	Comments
Olfactory	Patient asked to differentiate odors (tobacco, coffee, soap, etc.) with eyes closed.	Nasal passages must be patent and tested separately by occluding the opposite side.
Optic	Retina examined with ophthalmoscope; visual acuity tested with eye charts.	Visual acuity must be determined with lenses on, if patient wears them.
Oculomotor	Patient follows examiner's finger movement with eyes—especially movement that causes eyes to cross; pupillary change observed by shining light into each eye separately.	Examiner should note rate of pupillary change and coordinated constriction of pupils. Light in one eye should cause a similar pupillary change in other eye, but to a lesser degree.
Trochlear	Patient follows examiner's finger movement with eyes—especially lateral and downward movement.	
Trigeminal	Motor portion: Temporalis and masseter muscles palpated as patient clenches teeth; patient asked to open mouth against resistance applied by examiner.	Muscles of both sides of the jaw should show equal contractile strength.
	Sensory portion: Tactile and pain receptors tested by lightly touching patient's entire face with cotton and then with pin stimulus.	Patient's eyes should be closed, and innervation areas for all three nerves branching from the trigeminal nerve should be tested.
Abducens	Patient follows examiner's finger movement—especially lateral movement.	Motor functioning of cranial nerves III, IV, and VI may be tested simultaneously through selective movements of eyeball.
Facial	Motor portion: Patient asked to raise eyebrows, frown, tightly constrict eyelids, smile, puff out cheeks, and whistle.	Examiner should note lack of tonus expressed by sagging regions of face.
	Sensory portion: Sugar placed on each side of tip of patient's tongue.	Not reliable test for specific facial-nerve dysfunction because of tendency to stimulate taste buds on both sides of tip of tongue.
Vestibulocochlear	Vestibular portion: Patient asked to walk a straight line.	Not usually tested unless patient complains of dizziness or balance problems.
	Cochlear portion: Tested with tuning fork.	Examiner should note ability to discriminate sounds.
Glossopharyngeal and vagus	Motor: Examiner notes disturbances in swallowing, talking, and movement of soft palate; gag reflex tested.	Visceral innervation of vagus cannot be examined except for innervation to larynx, which is also served by glossopharyngeal.
Accessory	Patient asked to shrug shoulders against resistance of examiner's hand and to rotate head against resistance.	Sides should show uniformity of strength.
Hypoglossal	Patient requested to protrude tongue; tongue thrust may be resisted with tongue blade.	Tongue should protrude straight out; deviation to side indicates ipsilateral-nerve dysfunction; asymmetry, atrophy, or lack of strength should be noted.

A spinal nerve is a mixed nerve attached to the spinal cord by a **posterior** (dorsal) **root** composed of sensory fibers, and an **anterior** (ventral) **root,** composed of motor fibers (fig. 16.13). The posterior root contains an enlargement called the **spinal** (or dorsal root) **ganglion,** where the cell bodies of sensory neurons are located. The axons of sensory neurons convey sensory impulses through the posterior root and into the spinal cord, where synapses occur with dendrites of other neurons. The anterior root consists of axons of motor neurons, which convey motor impulses away from the CNS. A spinal nerve is formed as the fibers from the posterior and anterior roots converge and emerge through an intervertebral foramen.

The disease *herpes zoster,* also known as *shingles,* is a viral infection of the spinal ganglia. Herpes zoster causes painful, often unilateral, clusters of fluid-filled vesicles in the skin along the paths of the affected peripheral sensory neurons. The disease develops in adults who were first exposed to the virus as children, and is usually self-limited. Treatment may involve large doses of the antiviral drug acyclovir (Zorivax).

A spinal nerve divides into several branches immediately after it emerges through the intervertebral foramen. The small **meningeal branch** reenters the vertebral canal to innervate the meninges, vertebrae, and vertebral ligaments.

Figure 16.13

A section of the spinal cord and thoracic spinal nerves.

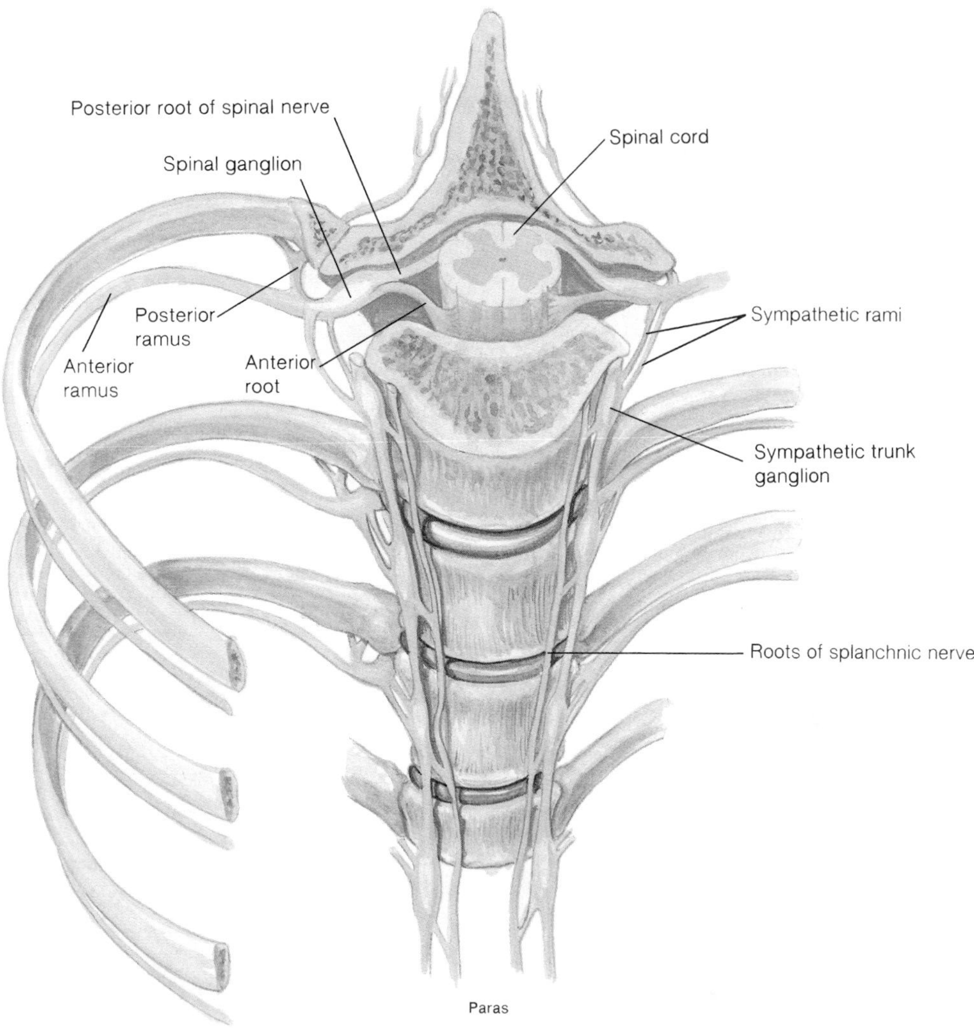

A larger branch, called the **posterior ramus,** innervates the muscles, joints, and skin of the back along the vertebral column (fig. 16.13). An **anterior ramus** of a spinal nerve innervates the muscles and skin on the lateral and anterior side of the trunk. Combinations of anterior rami innervate the limbs.

The **rami communicantes** are two branches from each spinal nerve that connect to a **sympathetic trunk ganglion,** which is part of the autonomic nervous system. The rami communicantes are composed of a **gray ramus,** containing unmyelinated fibers, and a **white ramus,** containing myelinated fibers. This arrangement is described in more detail in chapter 17.

Nerve Plexuses

Except in the thoracic nerves T2–T12, the anterior rami of the spinal nerves combine and then split again as networks of nerves referred to as nerve plexuses. There are four plexuses of spinal nerves: the cervical, the brachial, the lumbar, and the sacral. Nerves emerging from the plexuses are named according to the structures they innervate or the general course they take.

Cervical Plexus

The **cervical plexus** (*plek'sus*) is positioned deep on the side of the neck, lateral to the first four cervical vertebrae (fig. 16.14). It is formed by the anterior rami of the first four cervical nerves (C1–C4) and a portion of C5. Branches of the cervical plexus innervate the skin and muscles of the neck and portions of the head and shoulders. Some fibers of the cervical plexus also combine with the accessory and hypoglossal cranial nerves to supply dual innervation to some specific neck and pharyngeal muscles. Fibers from the third, fourth, and fifth cervical nerves unite to become the **phrenic** (*fren'ik*) **nerve,** which innervates the diaphragm. Motor impulses through the paired phrenic nerves cause the diaphragm to contract, moving air into the lungs.

The nerves of the cervical plexus are summarized in table 16.3.

Brachial Plexus

The **brachial plexus** is positioned to the side of the last four cervical vertebrae and first thoracic vertebra. It is formed by the anterior rami of C5 through T1, with occasional contributions

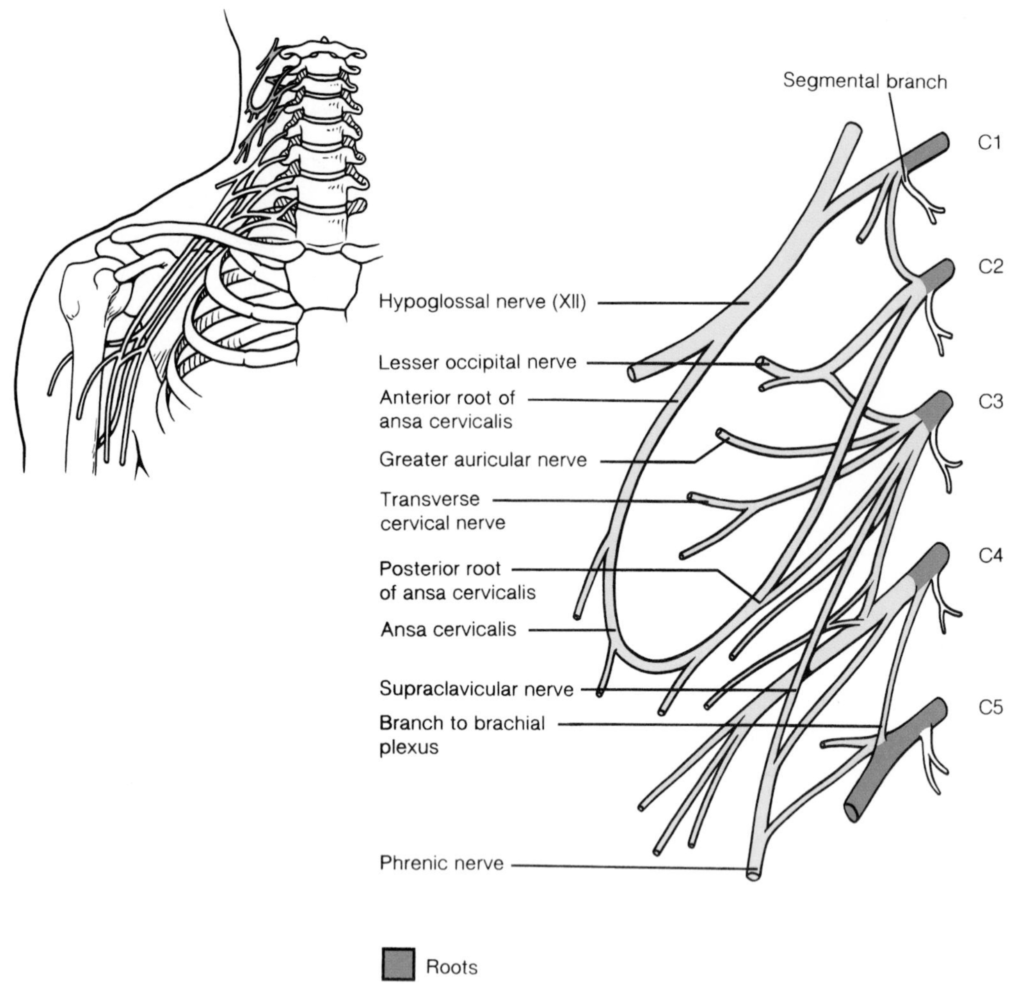

Figure 16.14
The cervical plexus.

Table 16.3
Branches of the Cervical Plexus

Nerve	Spinal Component	Innervation
Superficial Cutaneous Branches		
Lesser occipital	C2,C3	Skin of scalp above and behind ear
Greater auricular	C2,C3	Skin in front of, above, and below ear
Transverse cervical	C2,C3	Skin of anterior aspect of neck
Supraclavicular	C3,C4	Skin of upper portion of chest and shoulder
Deep Motor Branches		
Ansa cervicalis		
Anterior root	C1,C2	Geniohyoid, thyrohyoid, and infrahyoid muscles of neck
Posterior root	C3,C4	Omohyoid, sternohyoid, and sternothyroid muscles of neck
Phrenic	C3–C5	Diaphragm
Segmental branches	C1–C5	Deep muscles of neck (levator scapulae ventralis, trapezius, scalenus, and sternocleidomastoid)

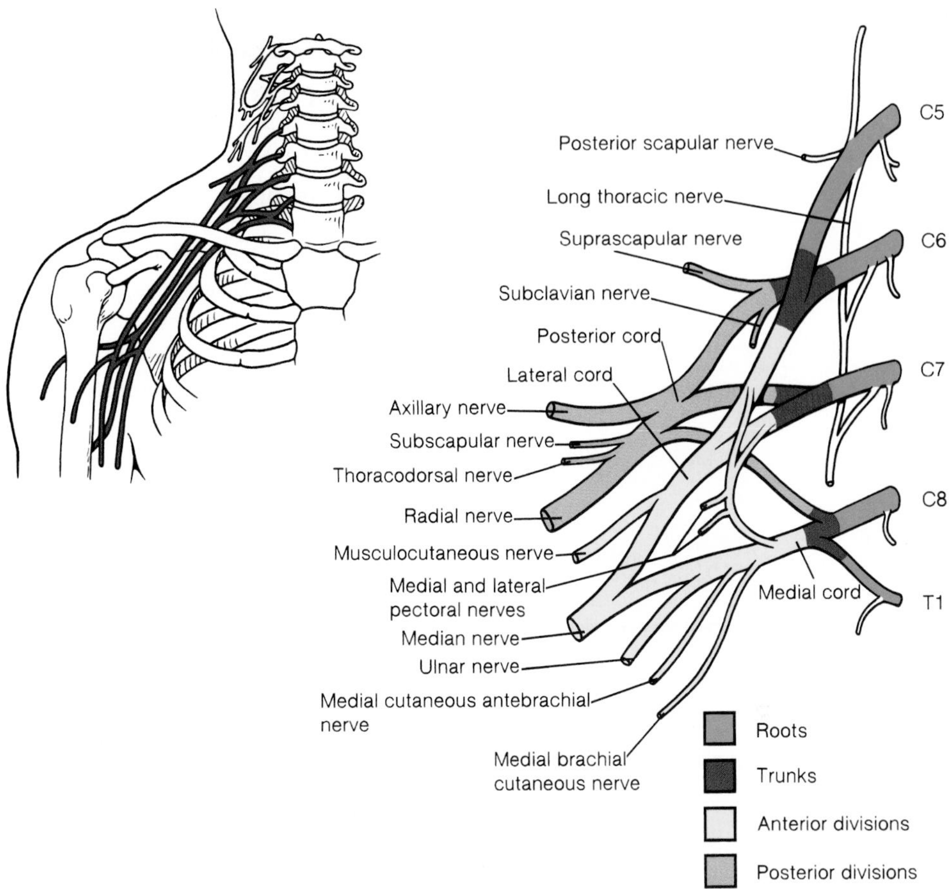

Figure 16.15
The brachial plexus.

from C4 and T2. From its emergence, the brachial plexus extends downward and laterally, passes over the first rib behind the clavicle, and enters the axilla. Each brachial plexus innervates the entire upper extremity of one side, as well as a number of shoulder and neck muscles.

Structurally, the brachial plexus is divided into *roots*, *trunks*, *divisions*, and *cords* (fig. 16.15). The roots of the brachial plexus are simply continuations of the anterior rami of the cervical nerves. The anterior rami of C5 and C6 converge to become the **superior trunk,** the C7 ramus becomes the **middle trunk,** and the anterior rami of C8 and T1 converge to become the **inferior trunk.** Each of the three trunks immediately divides into an **anterior division** and a **posterior division.** The divisions then converge to form three cords. The **posterior cord** is formed by the convergence of the posterior divisions of the upper, middle, and lower trunks, and hence contains fibers from C5 through C8. The **medial cord** is a continuation of the anterior division of the lower trunk and primarily contains fibers from C8 and T1. The **lateral cord** is formed by the convergence of the anterior division of the upper and middle trunk and consists of fibers from C5 through C7.

In summary, the brachial plexus is composed of nerve fibers from spinal nerves C5 through T1 and a few fibers from C4 and T2. Roots are continuations of the anterior rami. The roots converge to form trunks, and the trunks branch into divisions. The divisions in turn form cords, and the nerves of the upper extremity arise from the cords.

The brachial plexus may suffer trauma, especially if the clavicle, upper ribs, or lower cervical vertebrae are seriously fractured. Occasionally, the brachial plexus of a newborn is severely strained during a difficult delivery when the baby is pulled through the birth canal. In such cases, the arm of the injured side is paralyzed and eventually withers as the muscles atrophy in relation to the extent of the injury.

The entire upper extremity can be anesthetized in a procedure called a *brachial block* or *brachial anesthesia*. The site for injection of the anesthetic is midway between the base of the neck and the shoulder, posterior to the clavicle. At this point, the anesthetic can be injected in close proximity to the brachial plexus.

Five major nerves—the axillary, radial, musculocutaneous, ulnar, and median—arise from the three cords of the brachial plexus to supply cutaneous and muscular innervation to the upper extremity (table 16.4). The **axillary nerve** arises from the posterior cord and provides sensory innervation to the skin of the shoulder and shoulder joint, and motor innervation to the deltoid and teres minor muscles (fig. 16.16). The **radial nerve** arises from the posterior cord and extends along the posterior aspect of the brachial region to the radial side of the forearm. It provides sensory innervation to the skin of the posterior lateral surface of the upper extremity, including the posterior surface of the hand (fig. 16.17), and motor innervation to all of the extensor muscles of the upper extremity, the supinator muscle, and two muscles that flex the elbow joint.

Table 16.4

Selected Branches of the Brachial Plexus

Nerve	Cord and Spinal Components	Innervation
Axillary	Posterior cord (C5,C6)	Skin of shoulder; shoulder joint, deltoid and teres minor muscles
Radial	Posterior cord (C5–C8,T1)	Skin of posterior lateral surface of arm, forearm, and hand; posterior muscles of brachium and antebrachium (triceps brachii, supinator, anconeus, brachioradialis, extensor carpi radialis brevis, extensor carpi radialis longus, extensor carpi ulnaris)
Musculocutaneous	Lateral cord (C5–C7)	Skin of lateral surface of forearm; anterior muscles of brachium (coracobrachialis, biceps brachii, brachialis)
Ulnar	Medial cord (C8,T1)	Skin of medial third of hand; flexor muscles of anterior forearm (flexor carpi ulnaris, flexor digitorum), medial palm, and intrinsic flexor muscles of hand (profundus, third and fourth lumbricales)
Median	Medial cord (C6–C8,T1)	Skin of lateral two-thirds of hand; flexor muscles of anterior forearm, lateral palm, and first and second lumbricales

Figure 16.16

Muscular and cutaneous distribution of the axillary nerve.

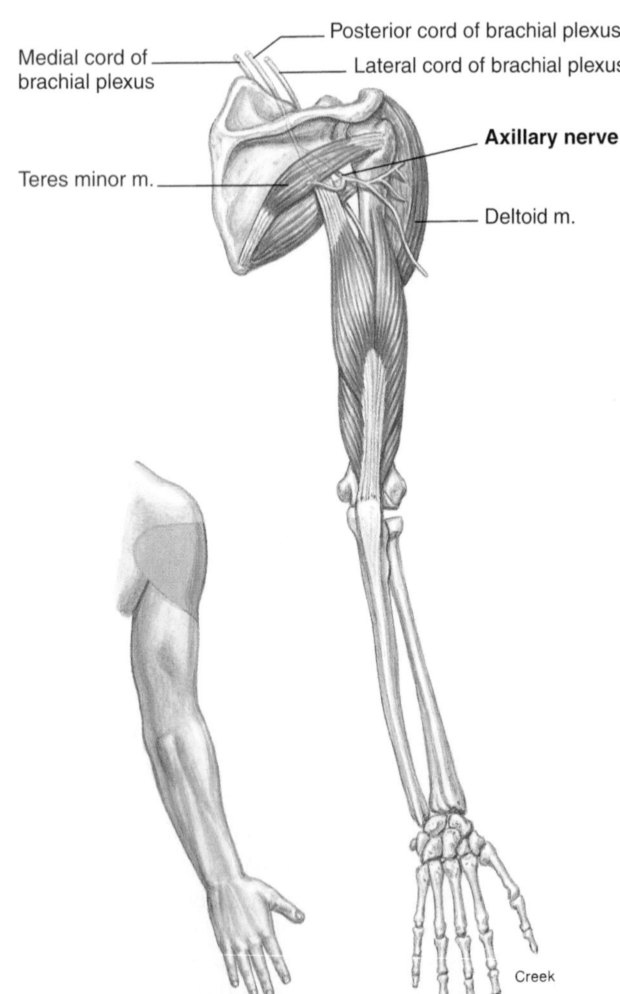

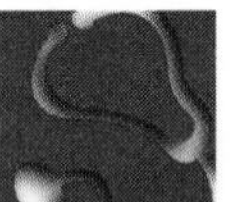

The radial nerve is vulnerable to several types of trauma. *Crutch paralysis* may result when a person improperly supports the weight of the body for an extended period of time with a crutch pushed tightly into the axilla. Compression of the radial nerve between the top of the crutch and the humerus may result in radial nerve damage. Likewise, dislocation of the shoulder frequently traumatizes the radial nerve. Children are particularly at risk as adults yank on their arms. A fracture to the body of the humerus may damage the radial nerve that parallels the bone at this point. The principal symptom of radial nerve damage is *wristdrop,* in which the extensor muscles of the fingers and wrist do not function. As a result, the joints of the fingers, wrist, and elbow are in a constant state of flexion.

The **musculocutaneous nerve** arises from the lateral cord. It provides sensory innervation to the skin of the posterior lateral surface of the arm and motor innervation to the anterior muscles of the brachium (fig. 16.18). The **ulnar nerve** arises from the medial cord and provides sensory innervation to the skin on the medial (ulnar side) third of the hand (fig. 16.19). The motor innervation of the ulnar nerve is to two muscles of the forearm and the intrinsic muscles of the hand (except some that serve the thumb).

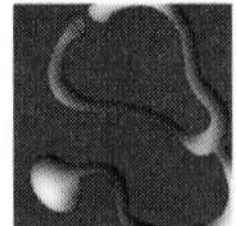

The ulnar nerve can be palpated in the ulnar sulcus (see fig. 10.4) behind the medial epicondyle of the humerus and the olecranon of the ulna. This area is commonly known as the "funny bone" or "crazy bone." Damage to the ulnar nerve may occur as the medial side of the elbow is banged against a hard object. The immediate perception of this trauma is a painful tingling that extends down the ulnar side of the forearm and into the hand and medial two digits. Although common, ulnar nerve damage is generally not serious.

Figure 16.17

Muscular and cutaneous distribution of the radial nerve.

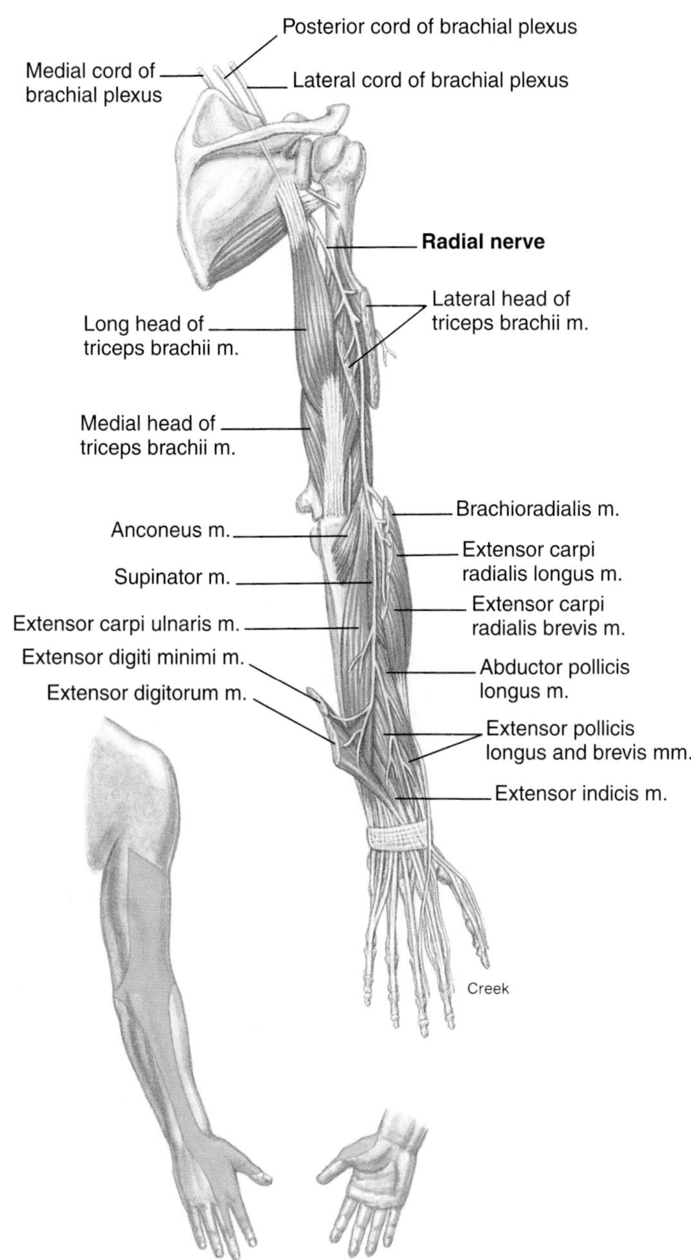

The **median nerve** arises from the medial cord. It provides sensory innervation to the skin on the radial portion of the palm of the hand (fig. 16.20) and motor innervation to all but one of the flexor muscles of the forearm and most of the hand muscles of the thumb (thenar muscles).

The median nerve, which serves the thumb, is the nerve in the forearm most commonly injured by stab wounds or the penetration of glass. If this nerve is severed, the muscles of the thumb are paralyzed and waste away, resulting in an inability to oppose the thumb in grasping.

Figure 16.18

Muscular and cutaneous distribution of the musculocutaneous nerve.

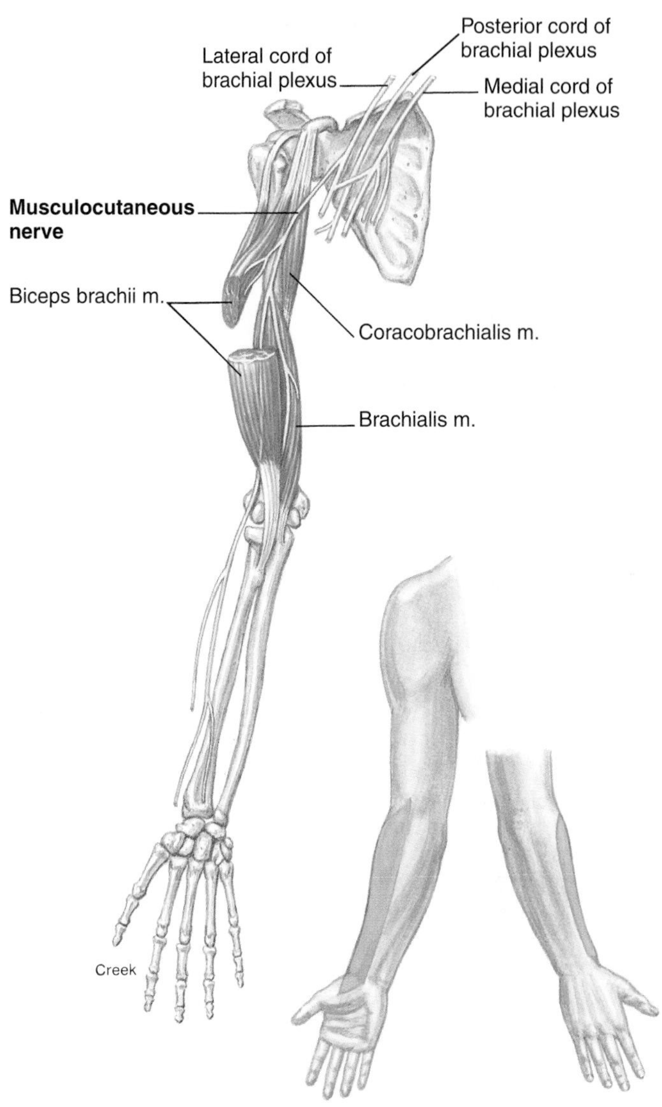

Lumbar Plexus

The **lumbar plexus** is positioned to the side of the first four lumbar vertebrae. It is formed by the anterior rami of spinal nerves L1 through L4 and some fibers from T12 (fig. 16.21). The nerves that arise from the lumbar plexus innervate structures of the lower abdomen and anterior and medial portions of the lower extremity. The lumbar plexus is not as complex as the brachial plexus, having only roots and divisions rather than roots, trunks, divisions, and cords of the brachial plexus.

Structurally, the **posterior division** of the lumbar plexus passes obliquely outward, deep to the psoas major muscle, whereas the **anterior division** is superficial to the quadratus lumborum muscle. The nerves that arise from the lumbar plexus are summarized in table 16.5. Because of their extensive innervation, the femoral nerve and the obturator nerve are illustrated in figures 16.22 and 16.23, respectively.

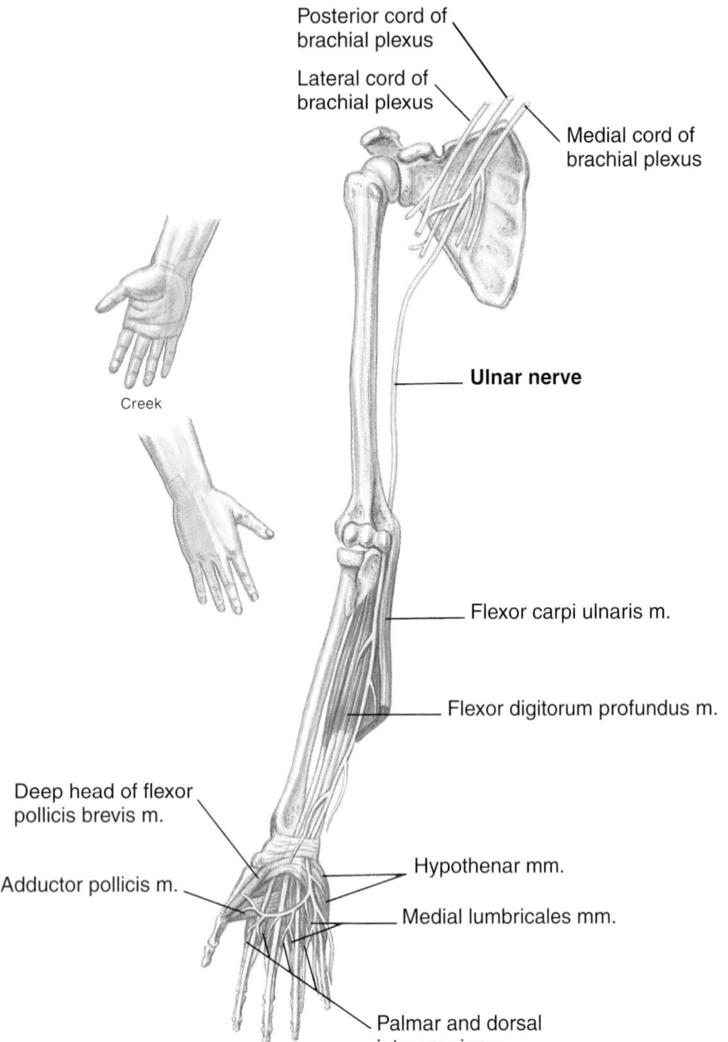

Figure 16.19

Muscular and cutaneous distribution of the ulnar nerve.

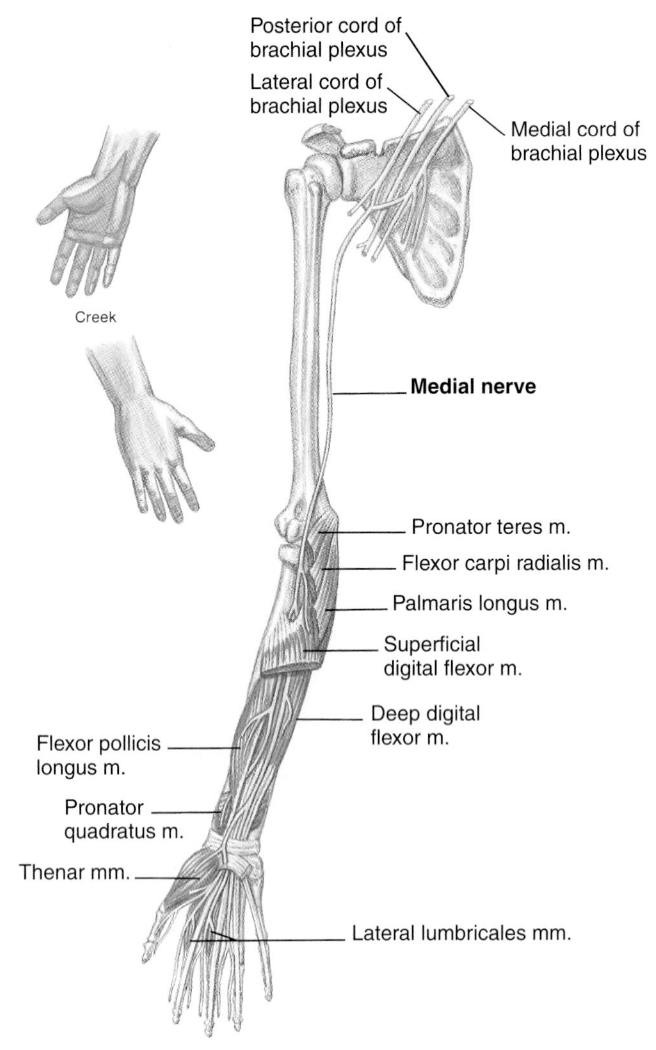

Figure 16.20

Muscular and cutaneous distribution of the median nerve.

The **femoral nerve** arises from the posterior division of the lumbar plexus and provides cutaneous innervation to the anterior and lateral thigh and the medial leg and foot (fig. 16.22). The motor innervation of the femoral nerve innervates the anterior muscles of the thigh, including the iliopsoas and sartorius muscles and the quadriceps femoris group.

The **obturator** (*ob'too-ra"tor*) **nerve** arises from the anterior division of the lumbar plexus. It provides cutaneous innervation to the medial thigh and motor innervation to the adductor muscles of the thigh (fig. 16.23).

Sacral Plexus

The **sacral plexus** lies immediately inferior to the lumbar plexus. It is formed by the anterior rami of spinal nerves L4, L5, and S1 through S4 (fig. 16.24). The nerves arising from the sacral plexus innervate the lower back, pelvis, perineum, posterior surface of the thigh and leg, and the superior and inferior (plantar) surfaces of the foot (table 16.6). Like the lumbar plexus, the sacral plexus consists of *roots* and *anterior* and *posterior divisions* from which nerves arise. Because some of the nerves of the sacral plexus also contain fibers from the nerves of the lumbar plexus through the **lumbosacral trunk,** these two plexuses are frequently described collectively as the **lumbosacral plexus.**

The **sciatic** (*si-at'ik*) **nerve** is the largest nerve arising from the sacral plexus and is the largest nerve in the body. The sciatic nerve passes from the pelvis through the greater sciatic notch of the os coxae and extends down the posterior aspect of the thigh. It is actually composed of two nerves—the *tibial nerve* and *common fibular nerve*—wrapped in a connective tissue sheath.

---

sacral: L. *sacris*, sacred

sciatic: L. *sciaticus*, hip joint

Figure 16.21
The lumbar plexus.

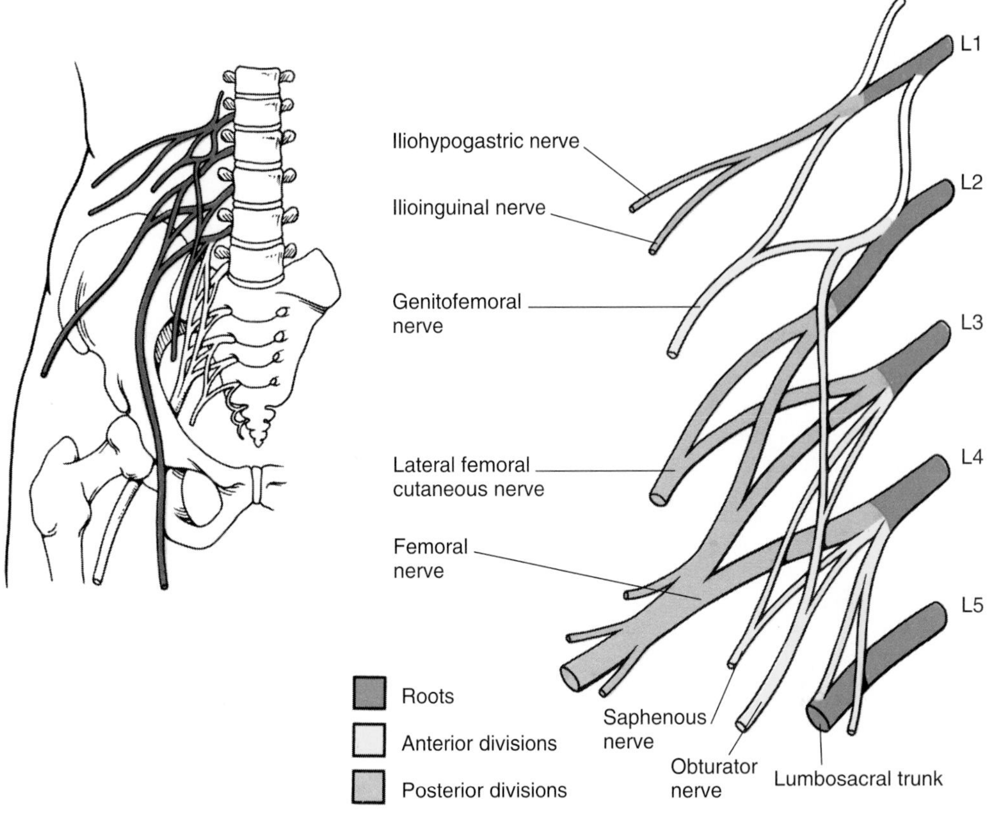

Table 16.5
Branches of the Lumbar Plexus

Nerve	Spinal Components	Innervation
Iliohypogastric	T12-L1	Skin of lower abdomen and buttock; muscles of anterolateral abdominal wall (external abdominal oblique, internal abdominal oblique, transversus abdominis)
Ilioinguinal	L1	Skin of upper median thigh, scrotum and root of penis in male and labia majora in female; muscles of anterolateral abdominal wall with iliohypogastric nerve
Genitofemoral	L1,L2	Skin of middle anterior surface of thigh, scrotum in male and labia majora in female; cremaster muscle in male
Lateral cutaneous femoral	L2,L3	Skin of anterior, lateral, and posterior aspects of thigh
Femoral	L2–L4	Skin of anterior and medial aspect of thigh and medial aspect of leg and foot; anterior muscles of thigh (iliacus, psoas major, pectineus, rectus femoris, sartorius) and extensor muscles of leg (rectus femoris, vastus lateralis, vastus medialis, vastus intermedius)
Obturator	L2–L4	Skin of medial aspect of thigh; adductor muscles of lower extremity (external obturator, pectineus, adductor longus, adductor brevis, adductor magnus, gracilis)
Saphenous	L2–L4	Skin of medial aspect of lower extremity

Figure 16.22

Muscular and cutaneous distribution of the femoral nerve.

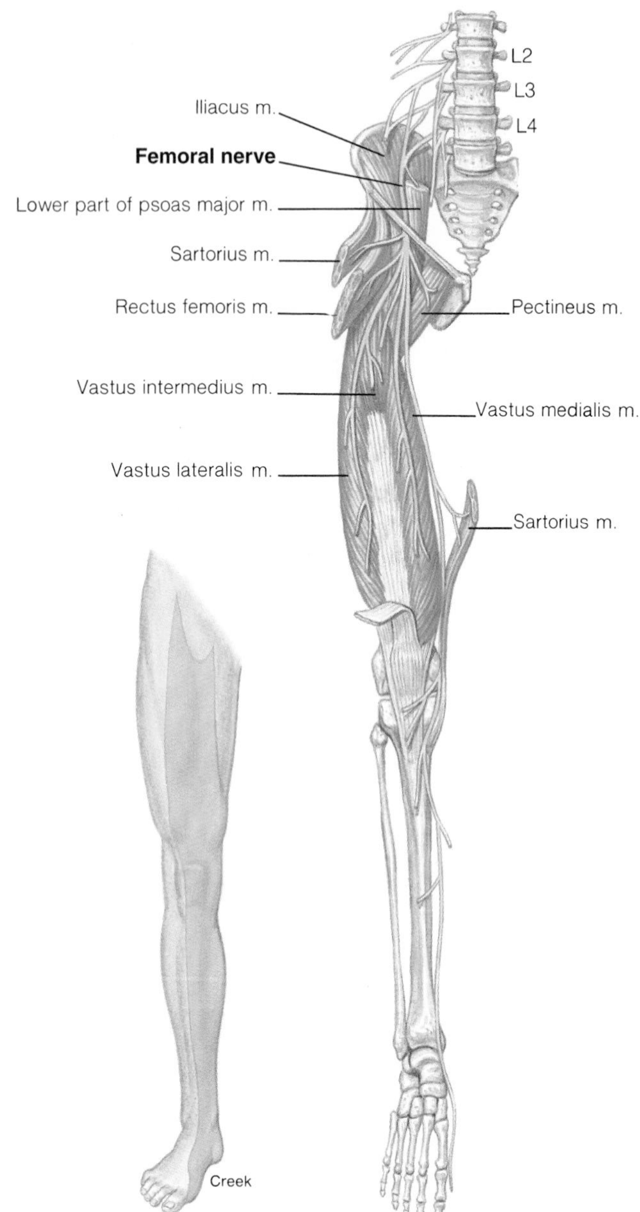

Figure 16.23

Muscular and cutaneous distribution of the obturator nerve.

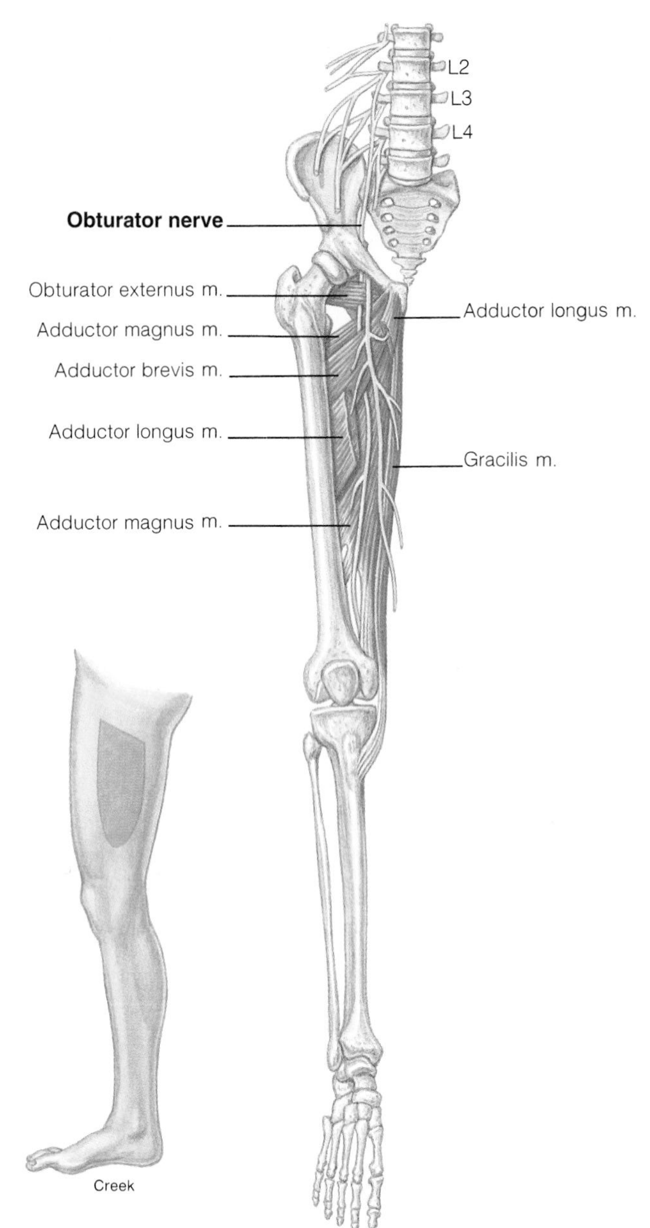

The **tibial nerve** arises from the anterior division of the sacral plexus, extends through the posterior regions of the thigh and leg, and branches in the foot to form the **medial** and **lateral plantar nerves** (fig. 16.25). The cutaneous innervation of the tibial nerve is to the calf of the leg and the plantar surface of the foot. The motor innervation of the tibial nerve is to most of the posterior thigh and leg muscles and to many of the intrinsic muscles of the foot.

The **common fibular nerve** (peroneal nerve) arises from the posterior division of the sacral plexus, extends through the posterior region of the thigh, and branches in the upper portion of the leg into the **deep** and **superficial fibular nerves** (fig. 16.26). The cutaneous innervation of the common fibular nerve and its branches is to the anterior and lateral leg and to the superior surface (dorsum) of the foot. The motor innervation is to the anterior and lateral muscles of the leg and foot.

Figure 16.24
The sacral plexus.

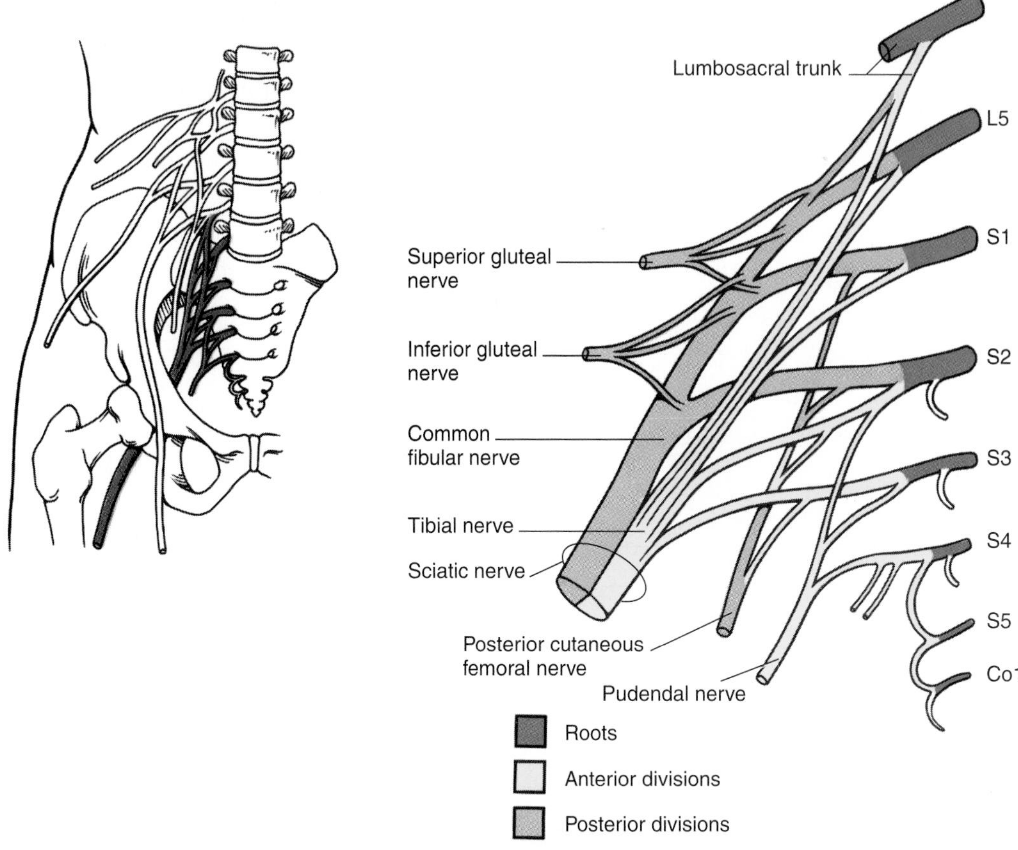

Table 16.6
Branches of the Sacral Plexus

Nerve	Spinal Components	Innervation
Superior gluteal	L4,L5,S1	Abductor muscles of thigh (gluteus minimus, gluteus medius, tensor fasciae latae)
Inferior gluteal	L5–S2	Extensor muscle of hip joint (gluteus maximus)
Nerve to piriformis	S1,S2	Abductor and rotator of thigh (piriformis)
Nerve to quadratus femoris	L4,L5,S1	Rotators of thigh (gemellus inferior, quadratus femoris)
Nerve to internal obturator	L5–S2	Rotators of thigh (gemellus superior, internal obturator)
Perforating cutaneous	S2,S3	Skin over lower medial surface of buttock
Posterior cutaneous femoral	S1–S3	Skin over lower lateral surface of buttock, anal region, upper posterior surface of thigh, upper aspect of calf, scrotum in male and labia majora in female
Sciatic	L4–S3	Composed of two nerves (tibial and common fibular); splits into two portions at popliteal fossa; branches from sciatic in thigh region to "hamstring muscles" (biceps femoris, semitendinosus, semimembranosus) and adductor magnus muscle
Tibial (sural, medial and lateral plantar)	L4–S3	Skin of posterior surface of leg and sole of foot; muscle innervation includes gastrocnemius, soleus, flexor digitorum longus, flexor hallucis longus, tibialis posterior, popliteus, and intrinsic foot muscles
Common fibular (superficial and deep fibular)	L4–S2	Skin of anterior surface of the leg and dorsum of foot; muscle innervation includes peroneus tertius, peroneus brevis, peroneus longus, tibialis anterior, extensor hallucis longus, extensor digitorum longus, extensor digitorum brevis
Pudendal	S2–S4	Skin of penis and scrotum in male and skin of clitoris, labia majora, labia minora, and lower vagina in female; muscles of perineum

Figure 16.25

Muscular and cutaneous distribution of the tibial nerve.

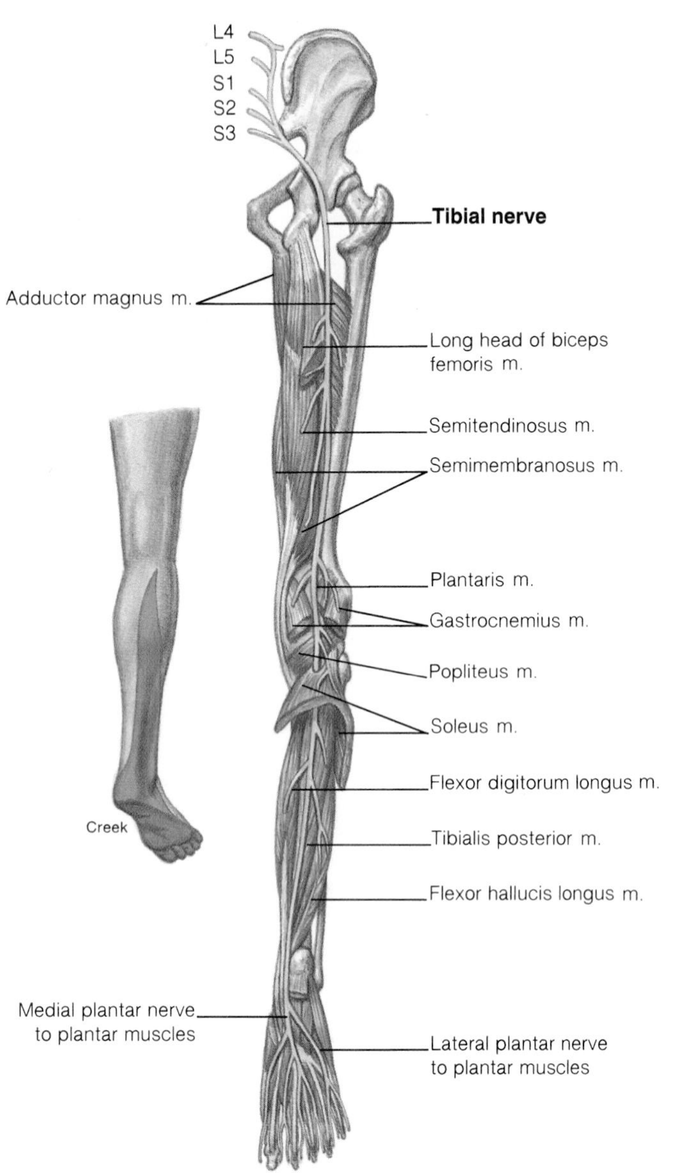

Figure 16.26

Muscular and cutaneous distribution of the common fibular nerve.

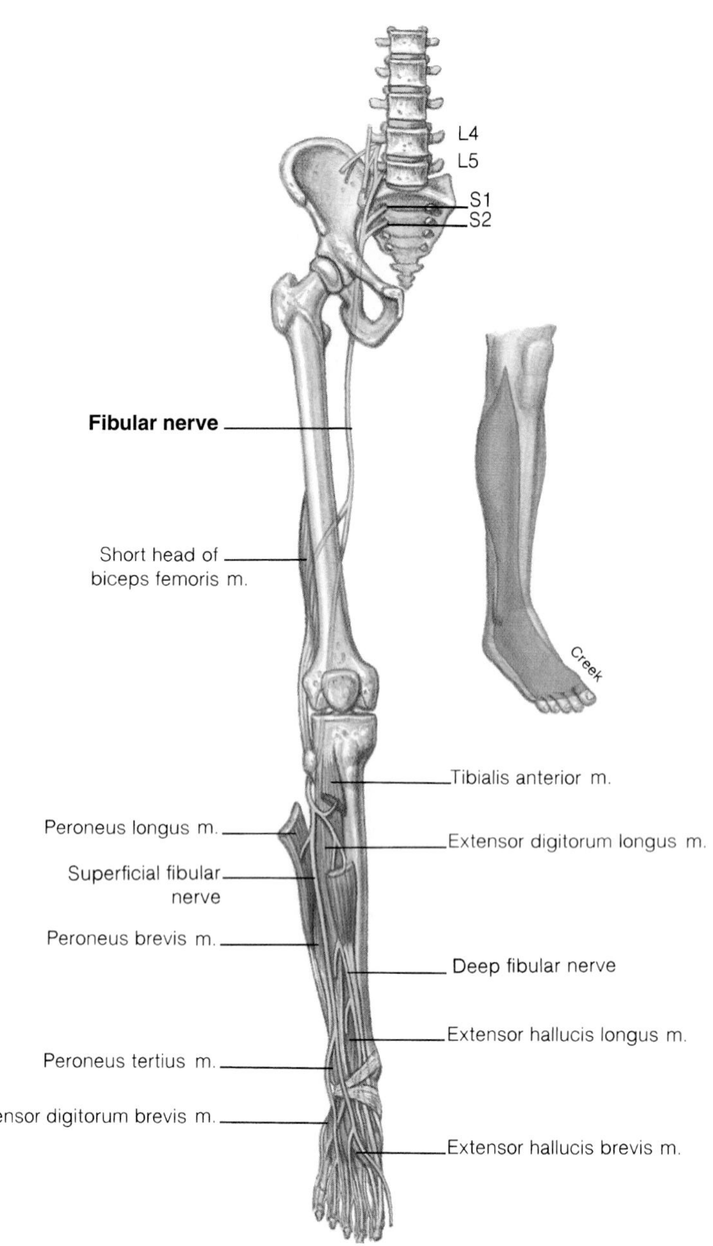

The sciatic nerve in the buttock lies deep to the gluteus maximus muscle, midway between the greater trochanter and the ischial tuberosity. Because of its position, the sciatic nerve is of tremendous clinical importance. A posterior dislocation of the hip joint will generally injure the sciatic nerve. A herniated disc (fig. 16.27) or pressure from the uterus during pregnancy may damage the nerve roots, resulting in a condition called *sciatica* (*si-at'ĭ-kă*). Sciatica is characterized by a sharp pain in the gluteal region that extends down the posterior side of the thigh. An improperly administered injection into the buttock may injure the sciatic nerve itself. Even a temporary compression of the sciatic nerve as a person sits on a hard surface for a period of time may result in the perception of tingling throughout the limb as the person stands up. The limb is said to have "gone to sleep."

Reflex Arcs and Reflexes

The conduction pathway of a reflex arc consists of a receptor, a sensory neuron, a motor neuron and its innervation in the PNS, and an association neuron in the CNS. The reflex arc provides the mechanism for a rapid, automatic response to a potentially threatening stimulus.

Specific **nerve pathways** provide routes by which impulses travel through the nervous system. Frequently, a nerve pathway begins with the conduction of impulses to the CNS

Figure 16.27

An MR scan of a herniated disc (arrow) in the lumbar region.

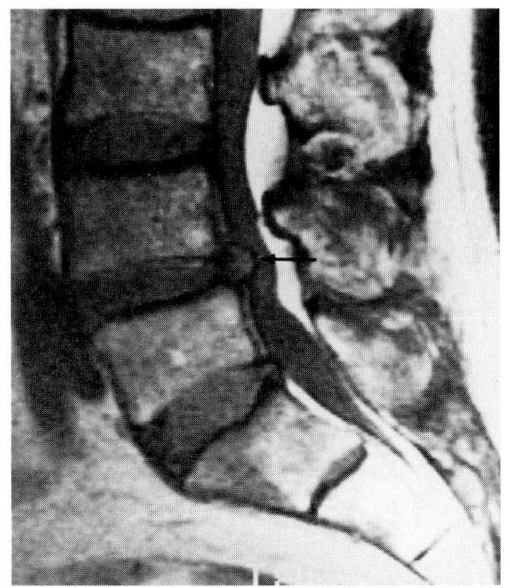

through sensory receptors and sensory neurons of the PNS. Once within the CNS, impulses may immediately travel back through motor portions of the PNS to activate specific skeletal muscles, glands, or smooth muscles. Impulses may also be sent simultaneously to other parts of the CNS through ascending tracts within the spinal cord.

Components of the Reflex Arc

The simplest type of nerve pathway is a **reflex arc** (fig. 16.28). A reflex arc implies an automatic, unconscious, protective response to a situation in an attempt to maintain body homeostasis. Impulses are conducted over a short route from sensory to motor neurons, and only two or three neurons are involved. The five components of a reflex arc are the receptor, sensory neuron, center, motor neuron, and effector. The **receptor** includes the dendrite of a sensory neuron and the place where the electrical impulse is initiated. The **sensory neuron** relays the impulse through the posterior root to the CNS. The **center** is located within the CNS and usually involves an association neuron (interneuron). It is here that the arc is made and other impulses are sent through synapses to other parts of the

Figure 16.28

The reflex arc.

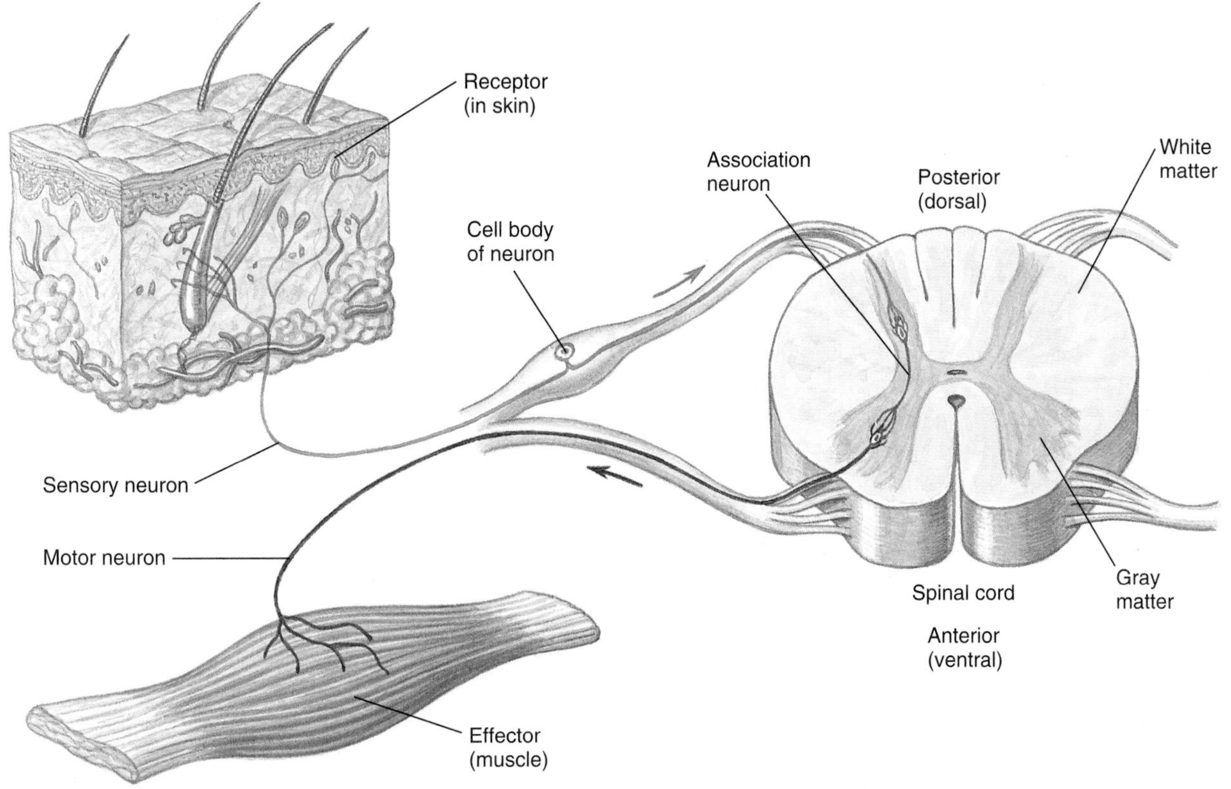

Figure 16.29

The knee-jerk reflex.
This is an ipsilateral reflex because the receptor and effector organs are on the same side of the spinal cord. The knee-jerk reflex is also a monosynaptic reflex because it involves only two neurons and one synapse.

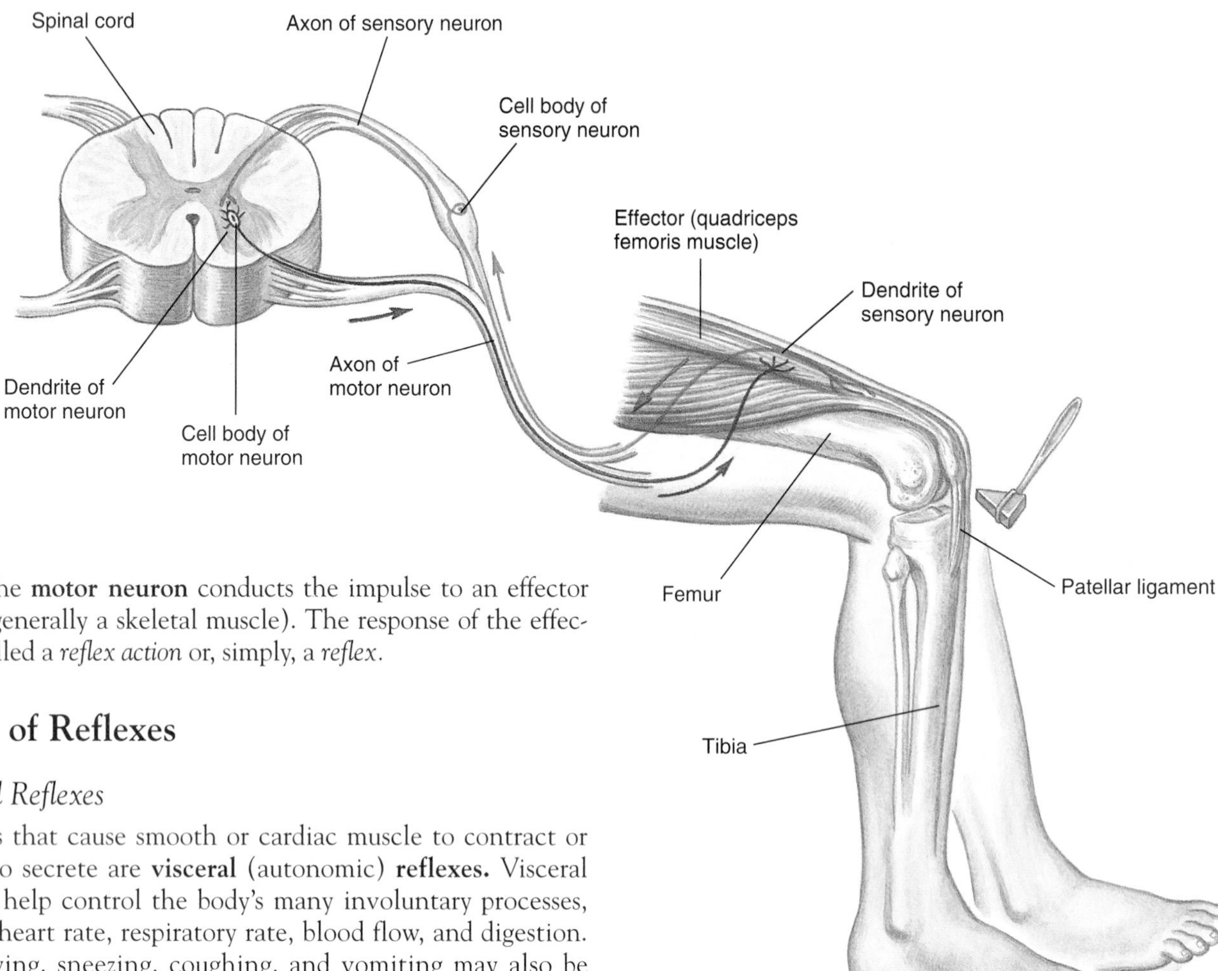

body. The **motor neuron** conducts the impulse to an effector organ (generally a skeletal muscle). The response of the effector is called a *reflex action* or, simply, a *reflex*.

Kinds of Reflexes

Visceral Reflexes

Reflexes that cause smooth or cardiac muscle to contract or glands to secrete are **visceral** (autonomic) **reflexes.** Visceral reflexes help control the body's many involuntary processes, such as heart rate, respiratory rate, blood flow, and digestion. Swallowing, sneezing, coughing, and vomiting may also be reflexive, although they involve the autonomic action of skeletal muscles.

Somatic Reflexes

Somatic reflexes are those that result in the contraction of skeletal muscles. The three principal kinds of somatic reflexes are named according to the response they produce.

The **stretch reflex** involves only two neurons and one synapse in the pathway; it is therefore called a *monosynaptic reflex arc*. Slight stretching of neuromuscular spindle receptors (described in chapter 18) within a muscle initiates an impulse along a sensory neuron to the spinal cord. A synapse with a motor neuron occurs in the anterior gray column, and a motor unit is activated, resulting in contraction of specific muscle fibers. Since the receptor and effector organs of the stretch reflex involve structures on the same side of the spinal cord, the reflex arc is an *ipsilateral reflex arc*. The knee-jerk reflex is an ipsilateral reflex (fig. 16.29), as are all monosynaptic reflex arcs.

A **flexor reflex,** or **withdrawal reflex,** consists of a *polysynaptic reflex arc* (fig. 16.30). Flexor reflexes involve association neurons in addition to the sensory and motor neurons. A flexor reflex is initiated as a person encounters a painful stimulus, such as a hot or sharp object. As a receptor organ is stimulated, sensory neurons transmit the impulse to the spinal cord where association neurons are activated. There, the impulses are directed through motor neurons to flexor muscles, which contract in response. Simultaneously, antagonistic muscles are inhibited (relaxed) so that the traumatized extremity can be quickly withdrawn from the harmful source of stimulation.

Several additional reflexes may be activated while a flexor reflex is in progress. In an *intersegmental reflex arc,* motor units from several segments of the spinal cord are activated by impulses coming in from the receptor organ. In an intersegmental reflex arc, more than one effector organ is stimulated. Frequently, sensory impulses from a receptor organ cross over through the spinal cord to activate effector organs in the opposite (*contralateral*) limb. This type of reflex is called a **crossed extensor reflex** (fig. 16.31) and is important for maintaining body balance while a flexor reflex is in progress. For example,

Figure 16.30

The flexor reflex.
Also called a withdrawal reflex, the flexor reflex is a disynaptic (two-synapse) reflex because it involves association neurons in addition to sensory and motor neurons.

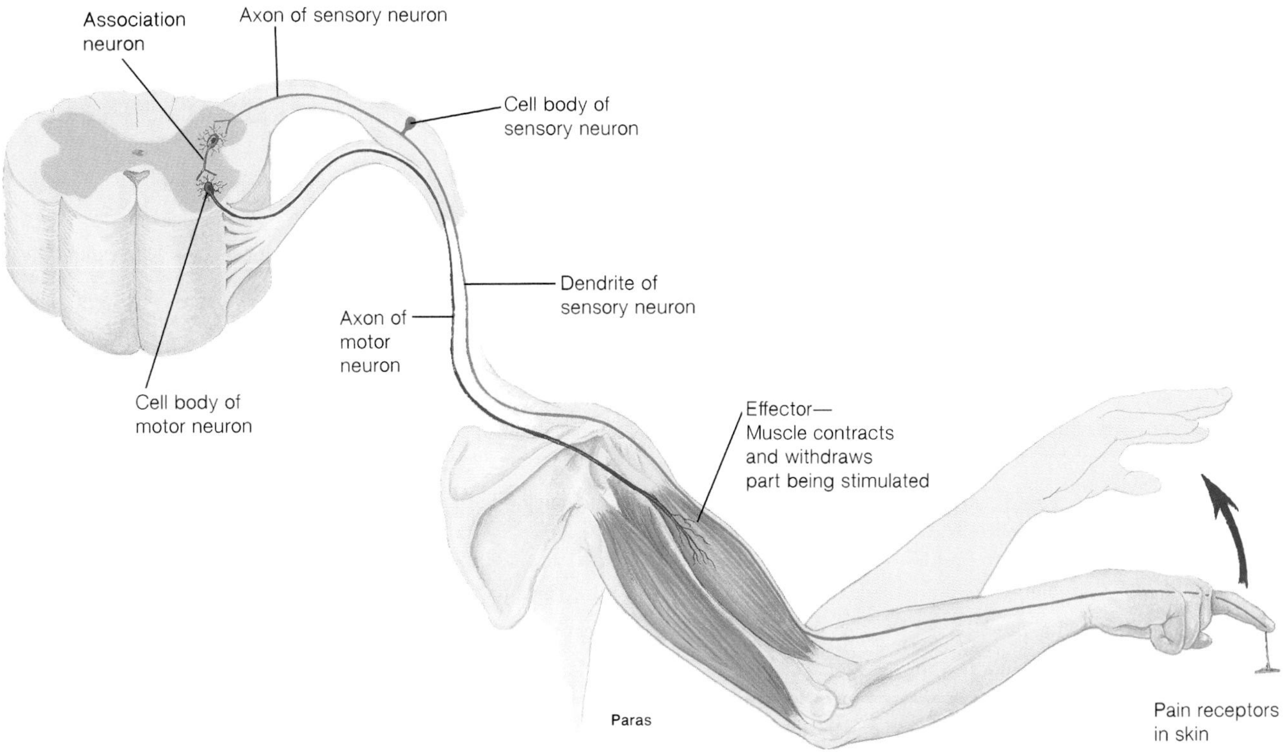

Figure 16.31

The crossed extensor reflex.
This type of reflex causes a reciprocal inhibition of muscles of the opposite appendage, which is important in maintaining balance.

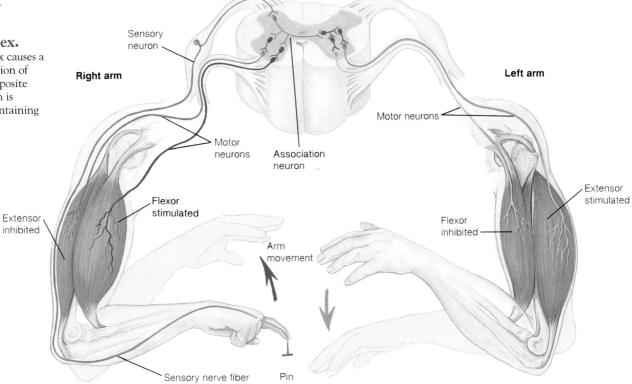

Table 16.7
Selected Reflexes of Clinical Importance

Reflex	Spinal Segment	Site of Receptor Stimulation	Effector Action
Biceps reflex	C5,C6	Tendon of biceps brachii muscle, near attachment on radial tuberosity	Contracts biceps brachii muscle to flex elbow
Triceps reflex	C7,C8	Tendon of triceps brachii muscle, near attachment on olecranon	Contracts triceps brachii muscle to extend elbow
Supinator or brachioradialis reflex	C5,C6	Radial attachment of supinator and brachioradialis muscles	Supinates forearm and hand
Knee-jerk	L2–L4	Patellar ligament, just below patella	Contracts quadriceps muscle to extend the knee
Ankle reflex	S1, S2	Tendo calcaneus, near attachment on calcaneus	Plantar flexes ankle
Plantar reflex	L4, L5, S1, S2	Lateral aspect of sole, from heel to ball of foot	Plantar flexes foot and flexes toes
Babinski reflex*	L4, L5, S1, S2	Lateral aspect of sole, from heel to ball of foot	Extends great toe and fans other toes
Abdominal reflexes	T8–T10 above umbilicus and T10–T12 below umbilicus	Sides of abdomen, above and below level of umbilicus	Contract abdominal muscles and deviate umbilicus toward stimulus
Cremasteric reflex	L1, L2	Upper inside of thigh in males	Contracts cremasteric muscle and elevates testis on same side of stimulation

*If the Babinski reflex rather than the plantar reflex occurs as the sole of the foot is stimulated, it may indicate damage to the corticospinal tract within the spinal cord. However, the Babinski reflex is present in infants up to 12 months of age because of the immaturity of their corticospinal tracts.

withdrawal of one leg, after stepping on broken glass, requires extension of the other in order to keep from falling. The reflexive inhibition of certain muscles to contract, called *reciprocal inhibition*, also helps maintain balance while either flexor or crossed extensor reflexes are in progress.

Certain reflexes are important for physiological functions, while others are important for avoiding injury. Some common reflexes are described in table 16.7 and illustrated in figure 16.32.

Babinski reflex: from Joseph F. Babinski, French neurologist, 1857–1932.

Part of a routine physical examination involves testing a person's reflexes. Several reflexes are used to assess certain neurological conditions, including functioning of the synapses. If some portion of the nervous system has been injured, the testing of certain reflexes may indicate the location and extent of the injury. Also, an anesthesiologist may try to initiate a reflex to ascertain the effect of an anesthetic.

anesthesia: Gk. *an*, without; *aisthesis*, sensation

Figure 16.32
Some reflexes of clinical importance.

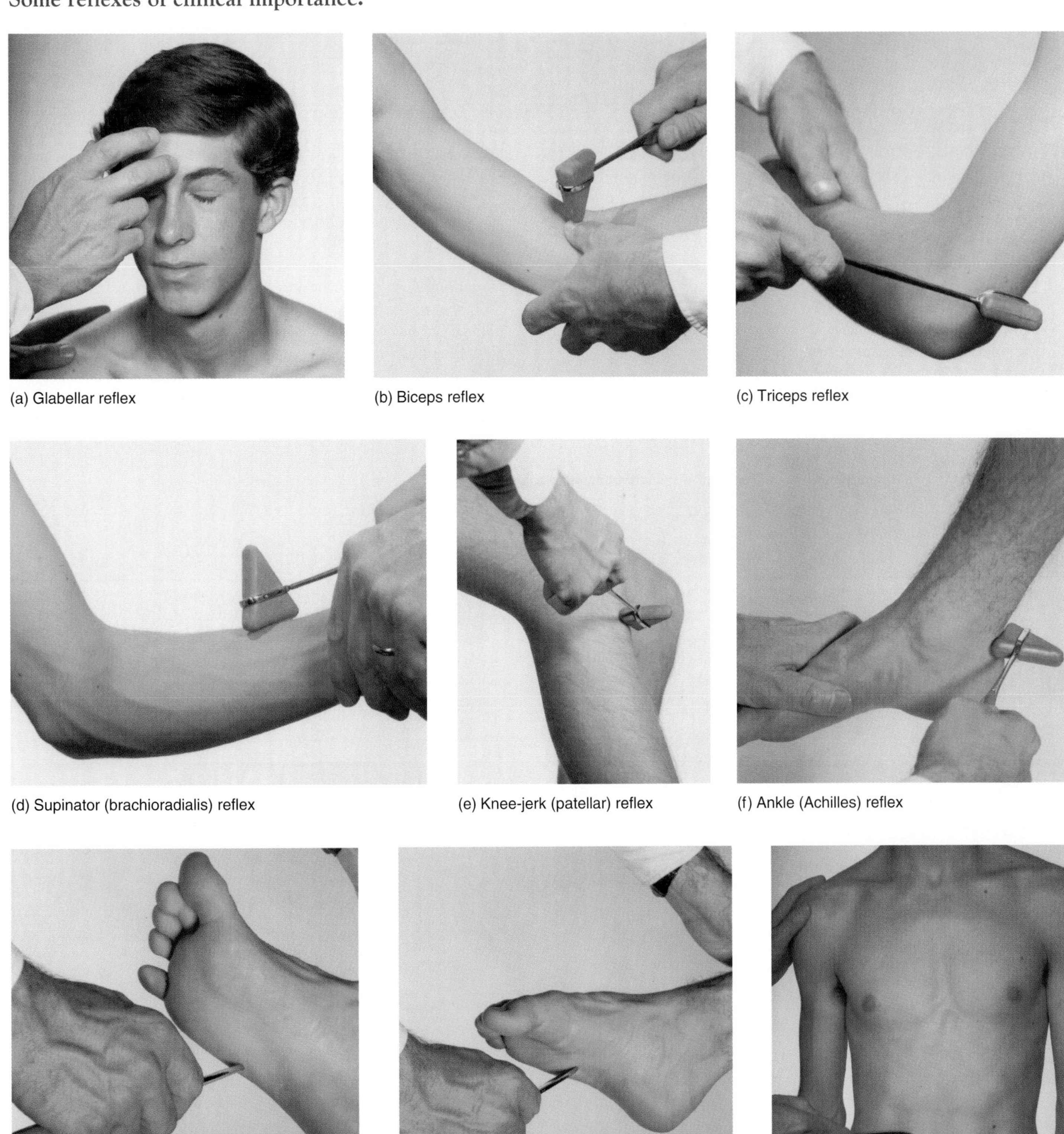

(a) Glabellar reflex
(b) Biceps reflex
(c) Triceps reflex
(d) Supinator (brachioradialis) reflex
(e) Knee-jerk (patellar) reflex
(f) Ankle (Achilles) reflex
(g) Babinski's reflex
(h) Plantar reflex
(i) Abdominal reflex

UNDER DEVELOPMENT

Development of the Peripheral Nervous System

Development of the peripheral nervous system produces the pattern of dermatomes within the body (fig. 1). A **dermatome** (*der′mă-tōm*) is an area of the skin innervated by all the cutaneous neurons of a certain spinal or cranial nerve. Most of the scalp and face is innervated by sensory neurons from the trigeminal nerve. With the exception of the first cervical nerve (C1), all of the spinal nerves are associated with specific dermatomes. Dermatomes are consecutive in the neck and trunk regions. In the appendages, however, adjacent dermatome innervations overlap. The apparently uneven dermatome arrangement in the appendages is due to the uneven rate of nerve growth into the limb buds. Actually, the limbs are segmented, and dermatomes overlap only slightly.

The pattern of dermatome innervation is of clinical importance when a physician wants to anesthetize a particular portion of the body. Because adjacent dermatomes overlap in the appendages, at least three spinal nerves must be blocked to produce complete anesthesia in these regions. Abnormally functioning dermatomes provide clues about injury to the spinal cord or specific spinal nerves. If a dermatome is stimulated but no sensation is perceived, the physician can infer that the injury involves the innervation to that dermatome.

dermatome: Gk. *derma*, skin; *tomia*, a cutting

Figure 1 The pattern of dermatomes and the peripheral distribution of spinal nerves. (*a*) An anterior view and (*b*) a posterior view.

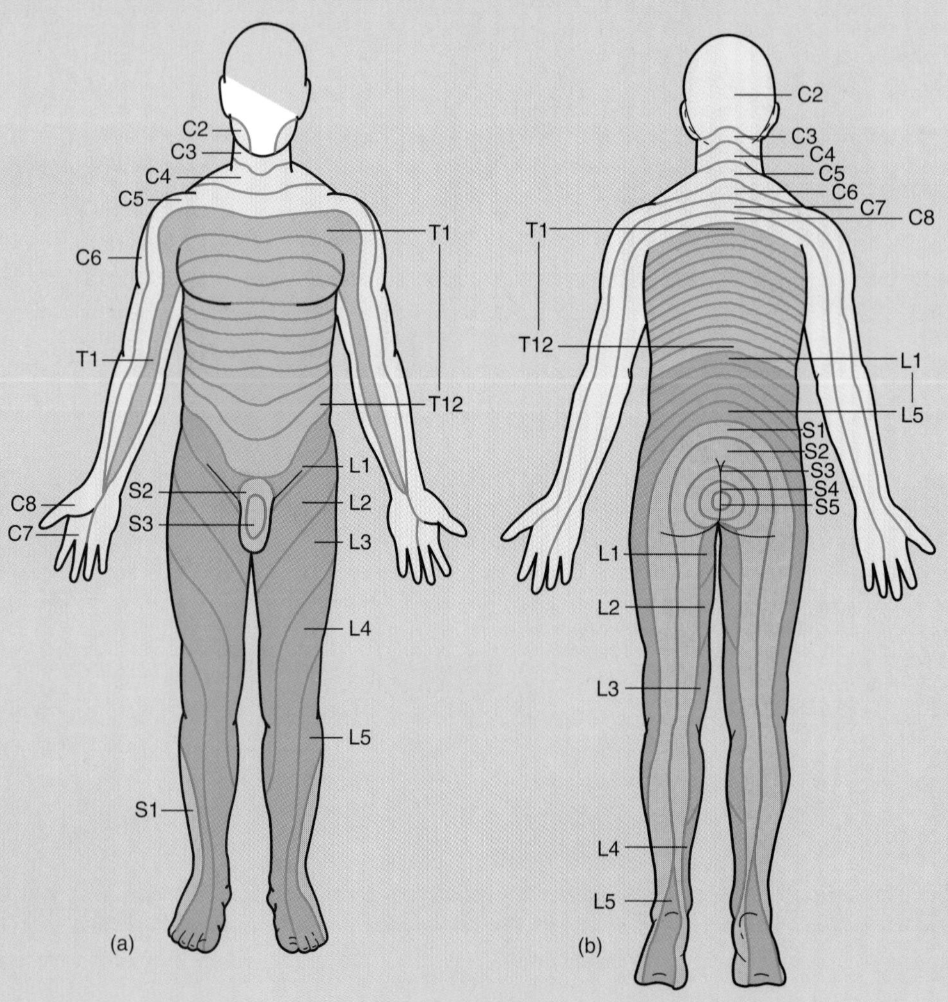

Clinical Investigation Answer

The radial nerve lies in the radial groove of the humerus as it extends through the brachial region toward the arm and hand. In this position, the radial nerve is susceptible to injury in the case of a fracture to the midshaft of the humerus. The muscles innervated by the radial nerve include those of wrist extension (extensor carpi radialis longus and brevis, extensor carpi ulnaris); wrist adduction (extensor carpi ulnaris); wrist abduction (extensor carpi radialis brevis and longus); supination (supinator); and finger extension (finger extensors). A detectable weakness in flexion of the elbow could also result from impairment of the brachioradialis muscle. The triceps brachii muscle would not be affected, however, because of the more proximal branching point of its motor nerve supply. The radial nerve is a mixed nerve, carrying both motor and sensory fibers; therefore, a sensory deficit would be present. Decreased sensation would be detectable on the posterolateral aspect of the hand (see fig. 16.17).

Chapter Summary

Introduction to the Peripheral Nervous System (pp. 448–449)

1. The peripheral nervous system consists of sensory receptors and the nerves that convey impulses to and from the central nervous system. Ganglia and nerve plexuses are also part of the PNS.
2. The cranial nerves arise from the brain and the spinal nerves arise from the spinal cord.
3. Sensory (afferent) nerves convey impulses toward the CNS, whereas motor (efferent) nerves convey impulses away from the CNS. Mixed nerves are composed of both sensory and motor fibers.

Cranial Nerves (pp. 449–457)

1. Twelve pairs of cranial nerves emerge from the inferior surface of the brain and, with the exception of the vestibulocochlear nerve, pass through foramina of the skull to innervate structures in the head, neck, and visceral organs of the trunk.
2. The names of the cranial nerves indicate their primary function or the general distribution of their fibers.
3. The olfactory, optic, and vestibulocochlear nerves are sensory only; the trigeminal, glossopharyngeal, and vagus are mixed; and the others are primarily motor, with only a few proprioceptive sensory fibers.
4. Some of the cranial nerve fibers are somatic; others are visceral.
5. Tests for cranial-nerve dysfunction are clinically important in a neurological examination.

Spinal Nerves (pp. 457–459)

1. Each of the 31 pairs of spinal nerves is formed by the union of an anterior (ventral) and posterior (dorsal) spinal root that emerges from the spinal cord through an intervertebral foramen to innervate a body dermatome.
2. The spinal nerves are grouped according to the levels of the spinal column from which they arise, and they are numbered in sequence.
3. Each spinal nerve is a mixed nerve consisting of a posterior root of sensory fibers and an anterior root of motor fibers.
4. Just beyond its intervertebral foramen, each spinal nerve divides into several branches.

Nerve Plexuses (pp. 459–468)

1. Except in the thoracic nerves T2 through T12, the anterior rami of the spinal nerves combine and then split again as networks of nerves called plexuses.
   (a) There are four plexuses of spinal nerves: the cervical, the brachial, the lumbar, and the sacral.
   (b) Nerves that emerge from the plexuses are named according to the structures they innervate or the general course they take.
2. The cervical plexus is formed by the anterior rami of C1 through C4 and by a portion of C5.
3. The brachial plexus is formed by the anterior rami of C5 through T1, and occasionally by some fibers from C4 and T2.
   (a) The brachial plexus is divided into roots, trunks, divisions, and cords.
   (b) The axillary, radial, musculocutaneous, ulnar, and median are the five largest nerves arising from the brachial plexus.
4. The lumbar plexus is formed by the anterior rami of L1 through L4 and by some fibers from T12.
   (a) The lumbar plexus is divided into roots and divisions.
   (b) The femoral and obturator are two important nerves arising from the lumbar plexus.
5. The sacral plexus is formed by the anterior rami of L4, L5, and S1 through S4.
   (a) The sacral plexus is divided into roots and divisions.
   (b) The sciatic nerve, composed of the common fibular nerve and tibial nerve, arises from the sacral plexus.
   (c) The lumbar plexus and the sacral plexus are collectively referred to as the lumbosacral plexus.

Reflex Arcs and Reflexes (pp. 468–473)

1. The conduction pathway of a reflex arc consists of a receptor, a sensory neuron, a motor neuron and its innervation in the PNS, and a center containing an association neuron in the CNS. The reflex arc enables a rapid, automatic response to a potentially threatening stimulus.
2. A reflex arc is the simplest type of nerve pathway.
3. Visceral reflexes cause smooth or cardiac muscle to contract or glands to secrete.
4. Somatic reflexes cause skeletal muscles to contract.
   (a) The stretch reflex is a monosynaptic reflex arc.
   (b) The flexor reflex is a polysynaptic reflex arc.

Review Activities

Objective Questions

1. Which of the following statements concerning the peripheral nervous system is *false?*
   (a) It consists of cranial and spinal nerves only.
   (b) It contains components of the autonomic nervous system.
   (c) Sensory receptors, nerves, ganglia, and plexuses are all part of the PNS.
2. An inability to cross the eyes would most likely indicate a problem with which cranial nerve?
   (a) the optic nerve
   (b) the oculomotor nerve
   (c) the abducens nerve
   (d) the facial nerve
3. Which cranial nerve innervates the muscle that raises the upper eyelid?
   (a) the trochlear nerve
   (b) the oculomotor nerve
   (c) the abducens nerve
   (d) the facial nerve
4. The inability to walk a straight line may indicate damage to which cranial nerve?
   (a) the trigeminal nerve
   (b) the facial nerve
   (c) the vestibulocochlear nerve
   (d) the vagus nerve

5. Which cranial nerve passes through the stylomastoid foramen?
   (a) the facial nerve
   (b) the glossopharyngeal nerve
   (c) the vagus nerve
   (d) the hypoglossal nerve
6. Which of the following cranial nerves does *not* contain parasympathetic fibers?
   (a) the oculomotor nerve
   (b) the accessory nerve
   (c) the vagus nerve
   (d) the facial nerve
7. Which of the following is *not* a spinal nerve plexus?
   (a) the cervical plexus
   (b) the brachial plexus
   (c) the thoracic plexus
   (d) the lumbar plexus
   (e) the sacral plexus
8. Roots, trunks, divisions, and cords are characteristic of
   (a) the sacral plexus.
   (b) the thoracic plexus.
   (c) the lumbar plexus.
   (d) the brachial plexus.
9. Which of the following nerve-plexus associations is *incorrect?*
   (a) median/sacral
   (b) phrenic/cervical
   (c) axillary/brachial
   (d) femoral/lumbar
10. Extending the leg when the patellar ligament is tapped is an example of
   (a) a visceral reflex.
   (b) a flexor reflex.
   (c) an ipsilateral reflex.
   (d) a crossed extensor reflex.

Essay Questions

1. Explain the structural and functional relationship between the central nervous system, the autonomic nervous system, and the peripheral nervous system.
2. List the cranial nerves, and describe the major function(s) of each. How is each cranial nerve tested for dysfunction?
3. Describe the structure of a spinal nerve.
4. List the roots of each of the spinal plexuses. Describe where each plexus is located and state the nerves that originate from it.
5. What is a reflex arc? Explain how reflexes are important in maintaining body homeostasis.
6. Distinguish between monosynaptic, polysynaptic, ipsilateral, stretch, and flexor reflexes.

Critical Thinking Questions

1. A person with quadriplegia from a spinal cord injury at the level of C5 can speak, digest food, breathe, and regulate his or her heartbeat, yet the person cannot move muscles from the shoulder down. Explain why.
2. The doctor taps your patellar ligament with a reflex hammer and can't elicit a kick. What's the matter with you? If it's good to have your leg jump a little, is it better to have your leg jump a lot?
3. A 63-year-old truck driver made an appointment with the company doctor because of pain and numbness in his left leg. Following a routine physical exam, the physician scheduled the man for a magnetic resonance imaging of his lumbar region. The MR scan indicated a herniated disc in the lumbar region. Explain how such a condition could develop and account for the man's symptoms. Discuss the possible treatment of this condition. (You may have to refer to medical textbooks in your library to find the answer.)
4. After a boxer was revived from a knockout punch during the world featherweight championship fight, the ringside physician determined that he had sustained oculomotor and facial nerve damage when his right zygomatic bone was shattered from a hard left hook. What symptoms might the boxer have displayed that would cause the physician to come to this conclusion?

Related Web Sites

For a listing of the most current web sites related to this chapter, please visit the *Concepts of Human Anatomy and Physiology* home page at http://www.mhhe.com/biosci/abio/.

Seventeen

Autonomic Nervous System

Objectives

- Describe the preganglionic and postganglionic neurons in the motor autonomic pathway.
- Describe the characteristics of the visceral effector organs that are innervated by the autonomic nervous system.
- Describe the origin of preganglionic sympathetic neurons and the location of sympathetic ganglia.
- Explain what is meant by the mass activation of the sympathetic division and describe the relationship between the sympathetic division and the adrenal medulla.
- Describe the origin of preganglionic parasympathetic neurons and the location of parasympathetic ganglia.
- Identify the neurotransmitters of each of the autonomic neurons.
- Distinguish between the different types of adrenergic receptors and describe the effects of sympathetic innervation.
- Distinguish between the different types of cholinergic receptors and describe the actions of the drug atropine.
- Describe the antagonistic, complementary, and cooperative actions of sympathetic and parasympathetic nerves on different body organs.
- Explain how the autonomic nervous system is controlled by the brain.

Clinical Investigation

A student laboratory assistant, studying for final examinations, found that her pulse rate was faster than normal and her blood pressure was elevated. Earlier that day, after having mixed some drugs for a laboratory exercise dealing with autonomic control, she had developed a severe headache and a very dry mouth, and her pupils were markedly dilated. What might be responsible for the student's symptoms?

Clues: Study the sections "Responses to Adrenergic Stimulation" and "Responses to Cholinergic Stimulation." Examine table 17.6 and the boxed information on the effects of atropine.

Neural Control of Involuntary Effectors

The autonomic nervous system helps to regulate the activities of cardiac muscle, smooth muscle, and glands. In this regulation, impulses are conducted from the CNS by an axon that synapses with a second autonomic neuron. It is the axon of this second neuron in the pathway that innervates the involuntary effectors.

Autonomic motor nerves innervate organs whose functions are not usually under voluntary control. The effectors that respond to autonomic regulation include **cardiac muscle** (the heart), **smooth** (visceral) **muscles,** and **glands.** These effectors are part of the *visceral organs* (organs within the body cavities) and of blood vessels. The involuntary effects of autonomic innervation contrast with the voluntary control of skeletal muscles by way of somatic motor neurons.

Autonomic Neurons

As discussed in chapter 16, neurons of the peripheral nervous system (PNS) that conduct impulses away from the central nervous system (CNS) are known as *motor,* or *efferent, neurons.* There are two major categories of motor neurons: somatic and autonomic. Somatic motor neurons have their cell bodies within the CNS and send axons to skeletal muscles, which are usually under voluntary control. This was described in chapter 16 and is reviewed in the left half of figure 17.1. The control of skeletal muscles by somatic motor neurons is discussed in depth in chapter 12.

Unlike somatic motor neurons, which conduct impulses along a single axon from the spinal cord to the neuromuscular junction, autonomic motor control involves two neurons in the motor pathway (table 17.1). The first of these neurons has its cell body in the gray matter of the brain or spinal cord. The axon of this neuron does not directly innervate the effector organ but instead synapses with a second neuron within an *autonomic ganglion* (a ganglion is a collection of cell bodies outside the CNS). The first neuron is thus called a **preganglionic neuron.** The second neuron in this pathway, called a **postganglionic neuron,** has an axon that extends from the autonomic ganglion to an effector organ, where it synapses with its target tissue (fig. 17.1, right side).

Preganglionic autonomic fibers originate in the midbrain and hindbrain and in the upper thoracic to the fourth sacral levels of the spinal cord. Autonomic ganglia are located in the head, neck, and abdomen; chains of autonomic ganglia also parallel the right and left sides of the spinal cord. The origin of the preganglionic fibers and the location of the autonomic ganglia help to distinguish the *sympathetic* and *parasympathetic divisions of the autonomic nervous system,* discussed in later sections of this chapter.

Visceral Effector Organs

Since the autonomic nervous system helps to regulate the activities of glands, smooth muscles, and cardiac muscle, autonomic control is an integral aspect of the physiology of most of the body systems. Autonomic regulation, then, partly explains endocrine regulation (chapter 19), smooth muscle function (chapter 12), functions of the heart and circulation (chapters 21 and 22), and, in fact, all the remaining systems to be discussed. Although the functions of the target organs of autonomic innervation are described in subsequent chapters, at this point we will consider some of the common features of autonomic regulation.

Unlike skeletal muscles, which enter a state of flaccid paralysis and atrophy when their motor nerves are severed, the involuntary effectors are somewhat independent of their innervation. Smooth muscles maintain a resting tone (tension) in the absence of nerve stimulation, for example. In fact, damage to an autonomic nerve makes its target cells more sensitive than normal to stimulating agents. This phenomenon is called denervation (*de″ner-va′shun*) **hypersensitivity.** Such compensatory changes can explain why, for example, the ability of the mucosa of the stomach to secrete acid may be restored after its neural supply from the vagus nerve has been severed. (This procedure is called vagotomy, and is sometimes performed as a treatment for ulcers.)

In addition to their intrinsic ("built-in") muscle tone, cardiac muscle and many smooth muscles take their autonomy a step further. These muscles can contract rhythmically, even in the absence of nerve stimulation, in response to electrical waves of depolarization initiated by the muscles themselves. Autonomic innervation simply increases or decreases this intrinsic activity. Autonomic nerves also maintain a resting tone in the sense that they maintain a baseline firing rate that can

---

autonomic: Gk. *auto,* self; *nomos,* law
viscera: L. *viscera,* internal organs

ganglion: Gk. *ganglion,* a swelling

Figure 17.1

Comparison of a somatic motor reflex with an autonomic motor reflex.
In a skeletal muscle reflex, a single somatic motor neuron passes from the CNS to the skeletal muscle. In an autonomic reflex, a preganglionic neuron passes from the CNS to an autonomic ganglion, where it makes a synapse with a second autonomic neuron. It is that second, or postganglionic neuron, which innervates the smooth muscle, cardiac muscle, or gland.

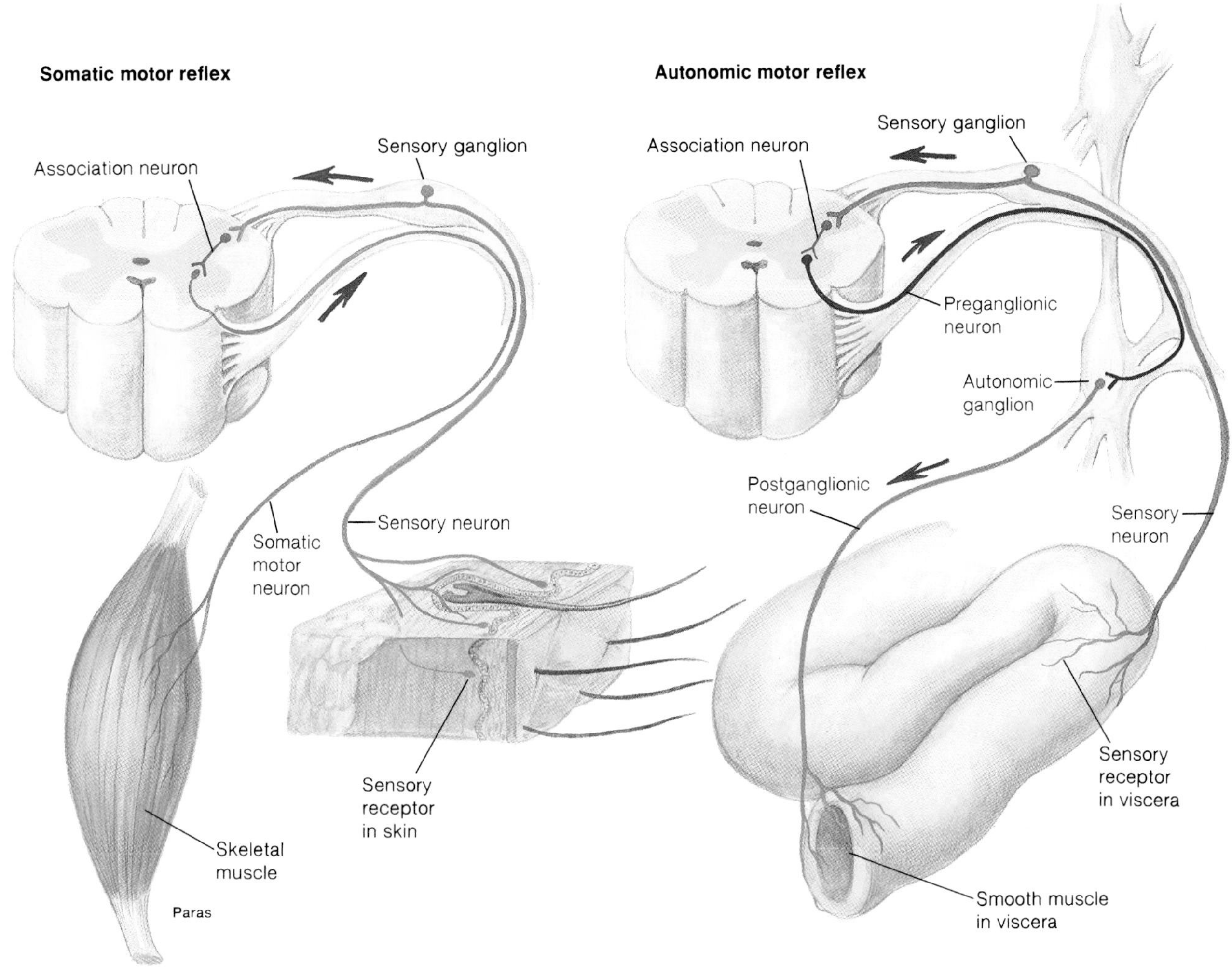

Table 17.1

Comparison of the Somatic Motor and Autonomic Motor Systems

Feature	Somatic Motor	Autonomic Motor
Effector organs	Skeletal muscles	Cardiac muscle, smooth muscle, and glands
Presence of ganglia	No ganglia	Cell bodies of postganglionic autonomic fibers located in paravertebral, prevertebral (collateral), and terminal ganglia
Number of neurons from CNS to effector	One	Two
Type of neuromuscular junction	Specialized motor end plate	No specialization of postsynaptic membrane; all areas of smooth muscle cells contain receptor proteins for neurotransmitters
Effect of nerve impulse on muscle	Excitatory only	Either excitatory or inhibitory
Type of nerve fibers	Fast-conducting, thick (9–13 μm), and myelinated	Slow-conducting; preganglionic fibers, lightly myelinated but thin (3 μm); postganglionic fibers, unmyelinated and very thin (about 1.0 μm)
Effect of denervation	Flaccid paralysis and atrophy	Muscle tone and function persist; target cells show denervation hypersensitivity

be either increased or decreased. A decrease in the excitatory input to the heart, for example, will slow its rate of beat.

The release of the neurotransmitter ACh from somatic motor neurons always stimulates the effector organ (skeletal muscles). By contrast, some autonomic nerves release transmitters that inhibit the activity of their effectors. An increase in the activity of the vagus, a nerve that supplies inhibitory fibers to the heart, for example, will slow the heart rate, whereas a decrease in this inhibitory input will increase the heart rate.

Divisions of the Autonomic Nervous System

Preganglionic neurons of the sympathetic division of the autonomic nervous system originate in the thoracic and lumbar levels of the spinal cord and send axons to sympathetic ganglia, which parallel the spinal cord. Preganglionic neurons of the parasympathetic division, by contrast, originate in the brain and in the sacral level of the spinal cord, and send axons to ganglia located in or near the effector organs.

The sympathetic and parasympathetic divisions of the autonomic nervous system have some structural features in common. Both consist of preganglionic neurons that originate in the CNS and postganglionic neurons that originate outside of the CNS in ganglia. However, the specific origin of the preganglionic fibers and the location of the ganglia differ in the two divisions of the autonomic system.

Sympathetic (Thoracolumbar) Division

The **sympathetic division** is also called the *thoracolumbar division* of the autonomic nervous system because its preganglionic fibers exit the spinal cord from the first thoracic (T1) to the second lumbar (L2) levels. Most sympathetic nerve fibers, however, separate from the somatic motor fibers and synapse with postganglionic neurons within a double row of sympathetic ganglia, called **paravertebral ganglia,** located on either side of the spinal cord (fig. 17.2). Ganglia within each row are interconnected, forming a **sympathetic chain of ganglia** that parallels the spinal cord on each lateral side.

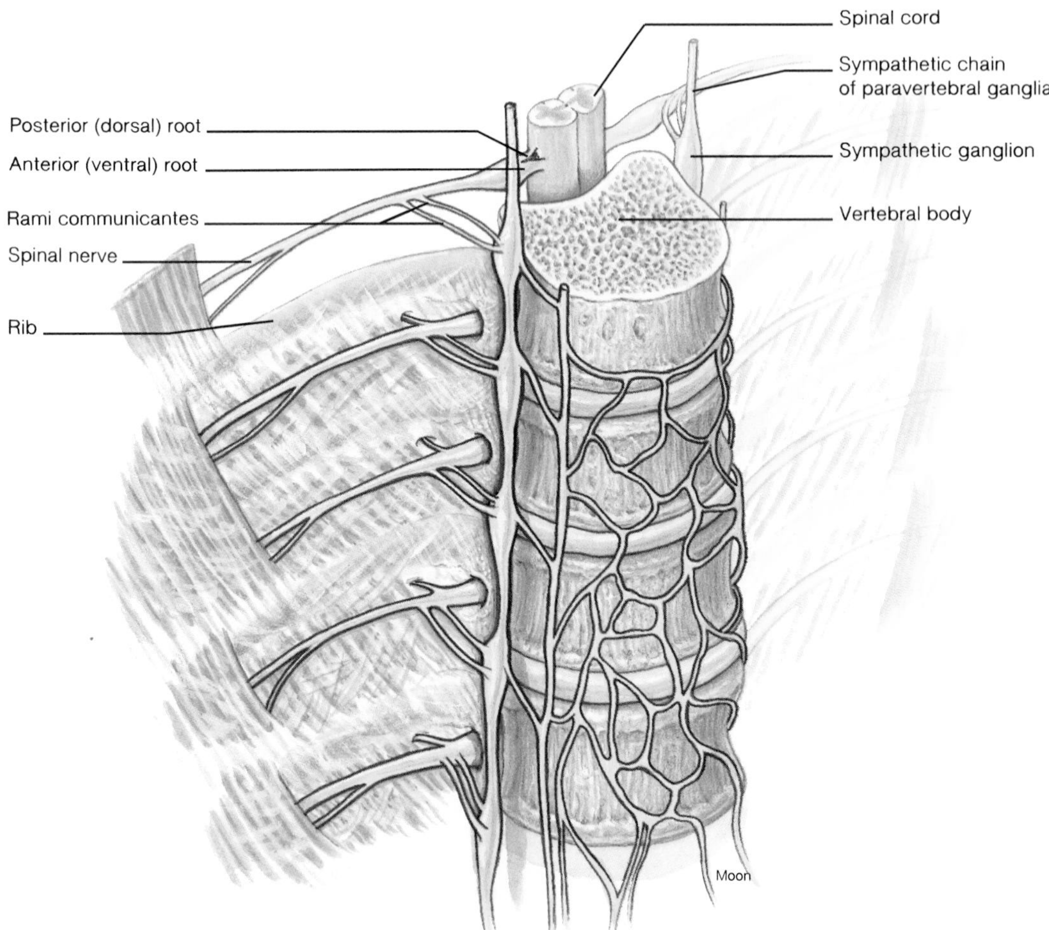

Figure 17.2

The sympathetic chain of paravertebral ganglia. This diagram shows the anatomical relationship between the sympathetic ganglia and the vertebral column and spinal cord.

The myelinated preganglionic sympathetic axons exit the spinal cord in the anterior roots of spinal nerves, but they soon diverge from the spinal nerves within *white rami communicantes* (*ra'mi kŏ"myoo-nĭ-kan'tēz*; singular, *ramus communicans*). The axons within each ramus enter the sympathetic chain of ganglia, where they can travel to ganglia at different levels and synapse with postganglionic sympathetic neurons. The axons of the postganglionic sympathetic neurons are unmyelinated and form the *gray rami communicantes* as they return to the spinal nerves and travel as part of the spinal nerves to their effector organs (fig. 17.3). Since sympathetic axons form a component of spinal nerves, they are widely distributed to the skeletal muscles and skin of the body where they innervate blood vessels and other involuntary effectors.

Divergence occurs within the sympathetic chain of ganglia as preganglionic fibers branch to synapse with numerous postganglionic neurons located in ganglia at different levels in the chain. *Convergence* also occurs here when a postganglionic neuron receives synaptic input from a large number of preganglionic fibers. The divergence of impulses from the spinal cord to the ganglia and the convergence of impulses within the ganglia usually results in the *mass activation* of almost all of the postganglionic sympathetic neurons. This explains why the sympathetic system is usually activated as a unit, so that it affects all of its effector organs at the same time.

Many preganglionic fibers that exit the spinal cord from the upper thoracic level travel into the neck, where they synapse in cervical sympathetic ganglia (fig. 17.4). Postganglionic fibers from here innervate the smooth muscles and glands of the head and neck.

Collateral Ganglia

Many preganglionic fibers that exit the spinal cord below the level of the diaphragm pass through the sympathetic chain of ganglia without synapsing. Beyond the sympathetic chain, these preganglionic fibers form **splanchnic** (*splangk'nik*) **nerves**. Preganglionic fibers in the splanchnic nerves synapse

ramus: L. *ramus*, a branch

splanchnic: Gk. *splankhna*, viscera

Figure 17.3

Pathway of sympathetic neurons.

The preganglionic neurons enter the sympathetic chain of ganglia on the white ramus (one of the two rami communicantes). Some synapse there, and the postganglionic axon leaves on the grey ramus to rejoin a spinal nerve. Others pass through the ganglia without synapsing. These ultimately synapse in a collateral ganglion, such as the celiac ganglion.

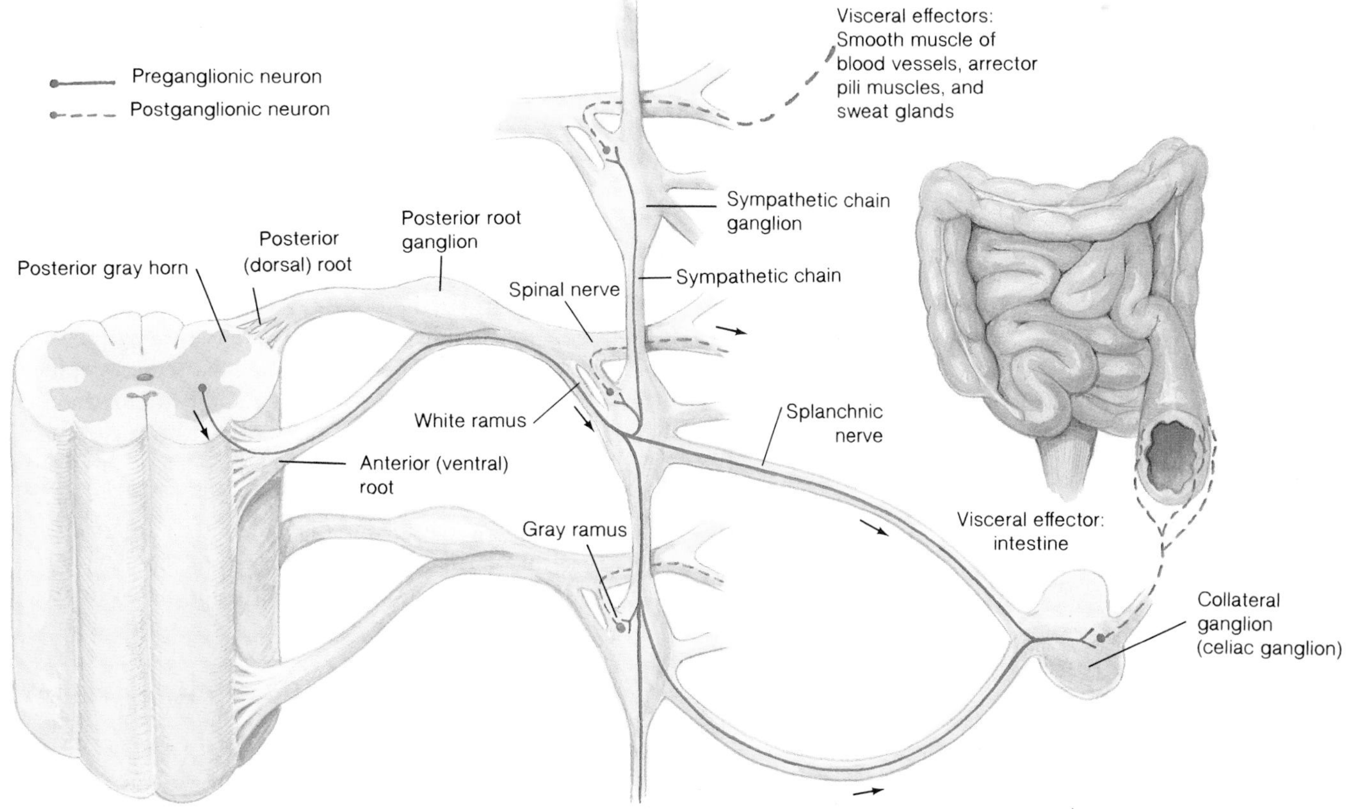

Figure 17.4

The cervical sympathetic ganglia. These ganglia provide important sympathetic innervation to the head and neck.

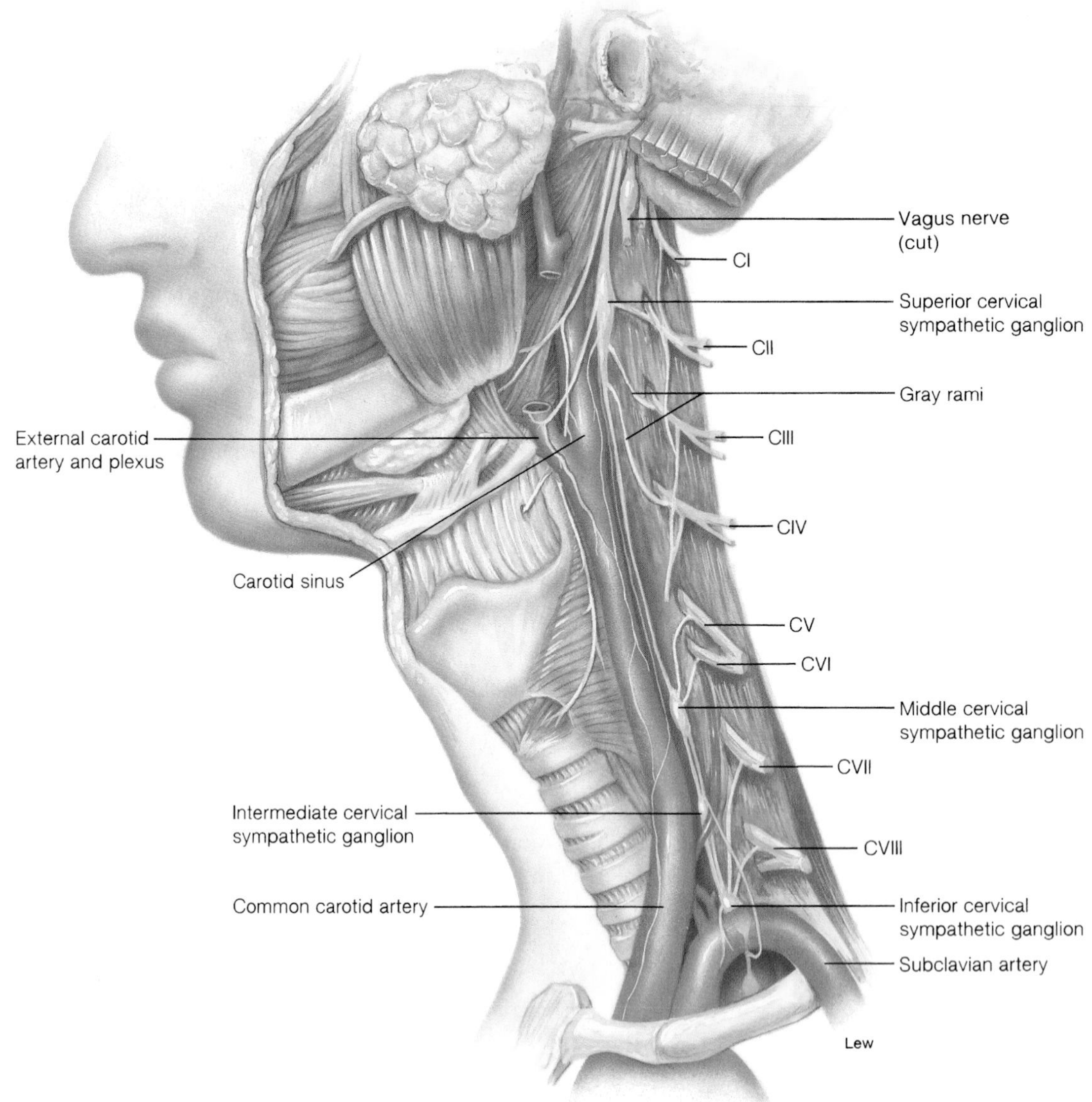

in **collateral,** or **prevertebral, ganglia.** These include the *celiac* (*se′le-ak*), *superior mesenteric*, and *inferior mesenteric ganglia* (figs. 17.5 and 17.6). Postganglionic fibers that arise from the collateral ganglia innervate organs of the digestive, urinary, and reproductive systems.

Adrenal Glands

The paired **adrenal glands** are located above each kidney. Each adrenal is composed of two parts: an outer **adrenal cortex** and an inner **adrenal medulla.** These two parts are really two functionally different glands, with different embryonic origins, different hormones, and different regulatory mechanisms. The adrenal cortex secretes steroid hormones; the adrenal medulla secretes the hormone **epinephrine** (adrenaline) and, to a lesser degree, **norepinephrine,** when it is stimulated by the sympathetic system.

The adrenal medulla can be likened to a modified sympathetic ganglion; its cells are derived from the same embryonic tissue (the neural crest, chapter 15) that forms postganglionic sympathetic neurons. Like a sympathetic ganglion, the cells of the adrenal medulla are innervated by preganglionic sympathetic fibers. The adrenal medulla secretes epinephrine into the blood in response to this neural stimulation. The effects of epinephrine are complementary to those of the neurotransmitter norepinephrine, which is released from postganglionic sympathetic nerve endings. For this reason, and because the adrenal medulla is stimulated as part of the mass activation of the sympathetic system, the two are often grouped together as a single **sympathoadrenal system.**

Parasympathetic (Craniosacral) Division

The **parasympathetic division** is also called the *craniosacral division* of the autonomic nervous system. This is because its preganglionic fibers originate in the brain (specifically, in the midbrain, medulla oblongata, and pons) and in the second

---

adrenal: L. *ad*, to; *renes*, kidney
cortex: L. *cortex*, bark
medulla: L. *medulla*, marrow

Figure 17.5

The collateral sympathetic ganglia.
These include the celiac and the superior and inferior mesenteric ganglia.

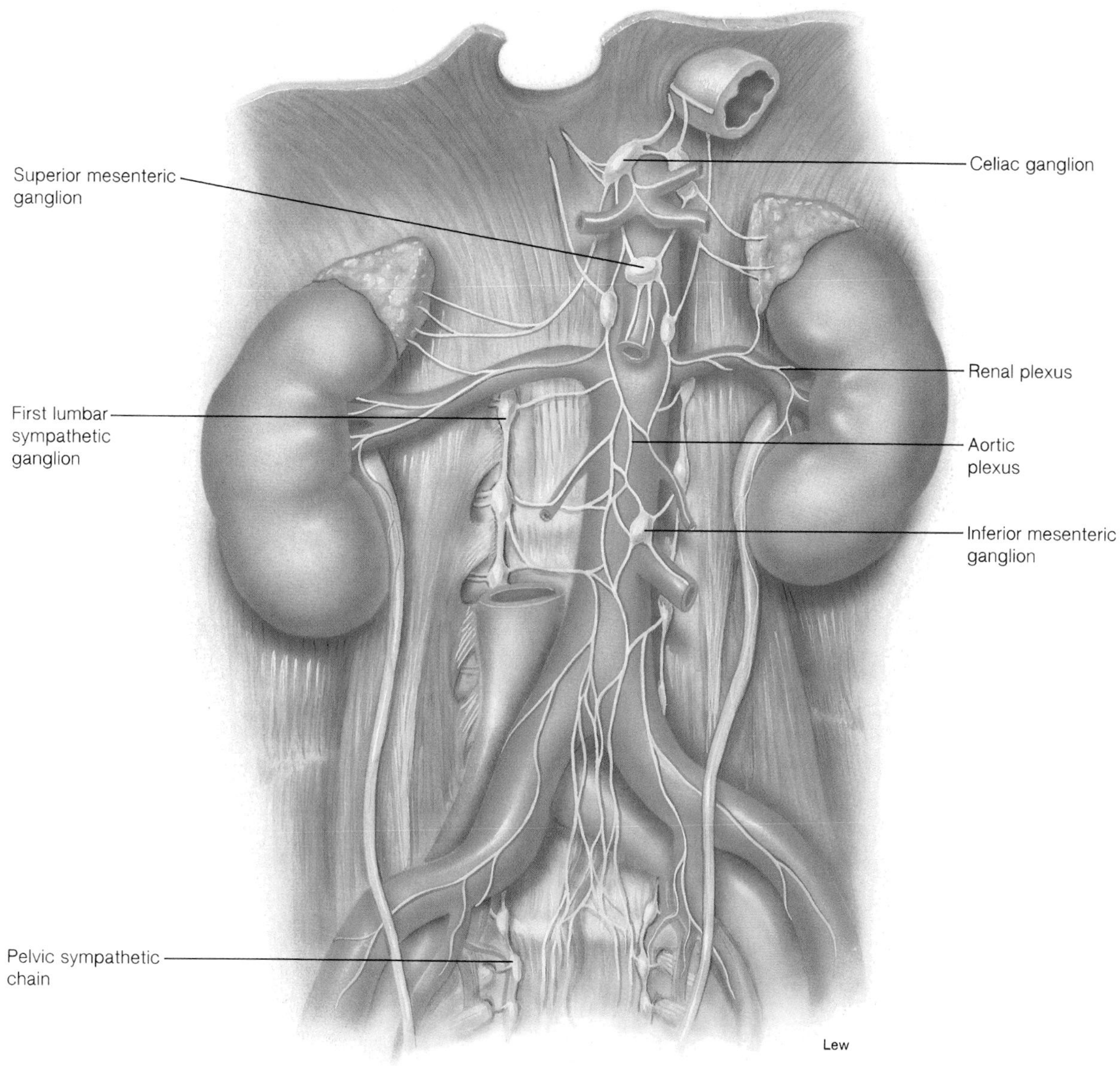

through fourth sacral levels of the spinal column. These preganglionic parasympathetic fibers synapse in ganglia that are located next to—or actually within—the organs innervated. These parasympathetic ganglia, called **terminal ganglia,** supply the postganglionic fibers that synapse with the effector cells.

The comparative structures of the sympathetic and parasympathetic divisions are listed in tables 17.2 and 17.3. It should be noted that most parasympathetic fibers, unlike sympathetic fibers, do not travel within spinal nerves. As a result, cutaneous effectors (blood vessels, sweat glands, and arrector pili muscles) and blood vessels in skeletal muscles receive sympathetic but not parasympathetic innervation.

Four of the 12 pairs of cranial nerves (described in chapter 16) contain preganglionic parasympathetic fibers. These are the oculomotor (III), facial (VII), glossopharyngeal (IX), and vagus (X) nerves. Parasympathetic fibers within the first three of these cranial nerves synapse in ganglia located in the head; fibers in the vagus nerve synapse in terminal ganglia located in widespread regions of the body.

The oculomotor nerve contains somatic motor and parasympathetic fibers that originate in the oculomotor nuclei of the midbrain. These parasympathetic fibers synapse in the *ciliary ganglion,* whose postganglionic fibers innervate the ciliary muscle and constrictor fibers in the iris of the eye. Preganglionic fibers that originate in the pons travel in the facial nerve to the *pterygopalatine* (*ter″ĭ-go-pal′ă-tēn*) *ganglion,* which sends postganglionic fibers to the nasal mucosa, pharynx,

---

vagus: L. *vagus*, wandering

Figure 17.6

The path of the vagus nerves.
The vagus nerves and their branches provide parasympathetic innervation to most organs within the thoracic and abdominal cavities.

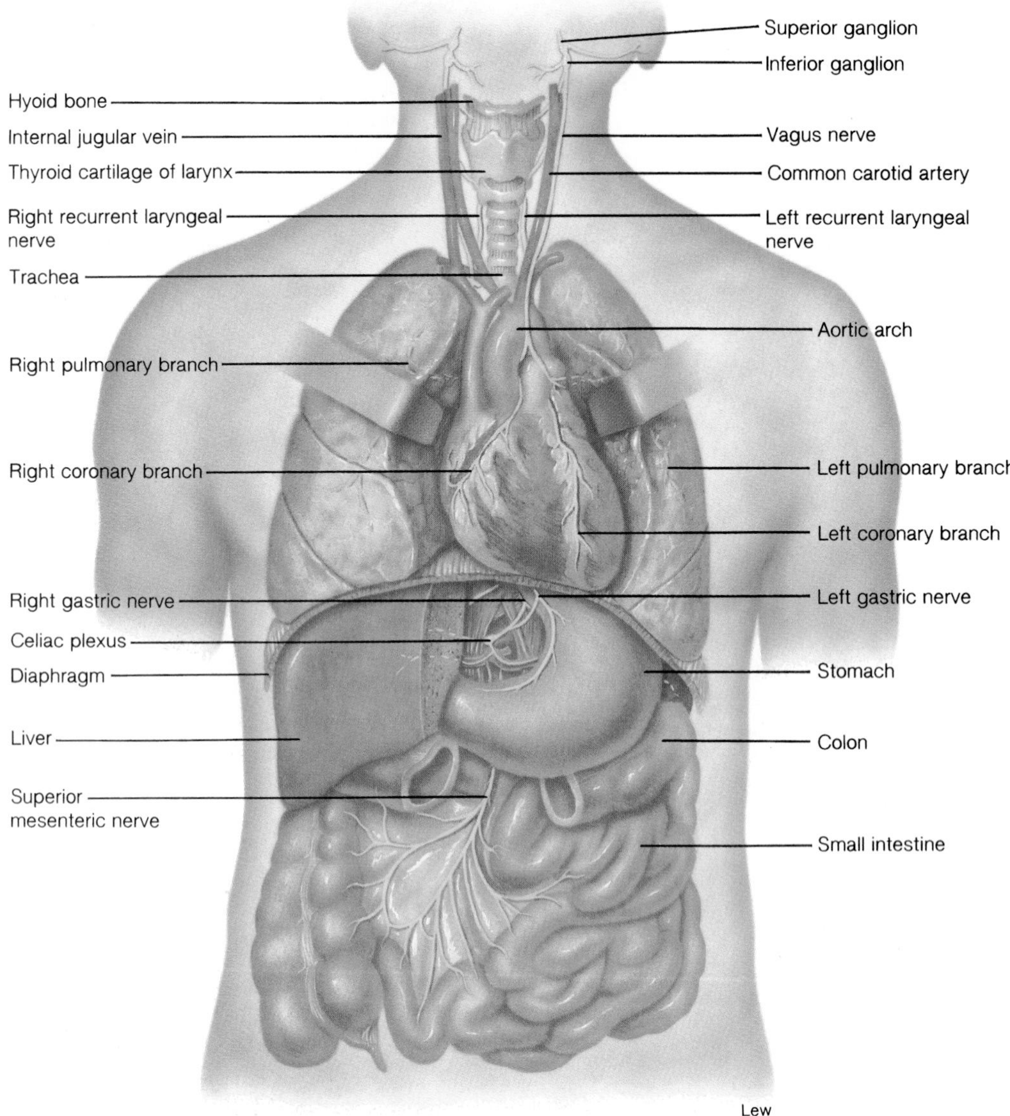

palate, and lacrimal glands. Another group of fibers in the facial nerve terminates in the *submandibular ganglion,* which sends postganglionic fibers to the submandibular and sublingual salivary glands. Preganglionic fibers of the glossopharyngeal nerve synapse in the *otic ganglion,* which sends postganglionic fibers to innervate the parotid (salivary) gland.

Nuclei in the medulla oblongata contribute preganglionic fibers to the very long tenth cranial, or vagus nerves (the "vagrant" or "wandering" nerves). These preganglionic fibers travel through the neck to the thoracic cavity and through the esophageal opening in the diaphragm to the abdominal cavity (fig. 17.6). In each region, some of these preganglionic fibers branch from the main trunks of the vagus nerves and synapse with postganglionic neurons located *within* the innervated organs. The preganglionic vagus fibers are thus quite long; they provide parasympathetic innervation to the heart, lungs, esophagus, stomach, pancreas, liver, small intestine, and the upper half of the large intestine. Postganglionic parasympathetic fibers arise from terminal ganglia within these organs and synapse with effector cells (smooth muscles and glands).

Preganglionic fibers from the sacral levels of the spinal cord provide parasympathetic innervation to the lower half of the large intestine, the rectum, and to the urinary and reproductive systems. These fibers, like those of the vagus, synapse with terminal ganglia located within the effector organs.

Parasympathetic nerves to the visceral organs thus consist of preganglionic fibers, whereas sympathetic nerves to these organs contain postganglionic fibers. A composite view of the sympathetic and parasympathetic systems is provided in figure 17.7, and the comparisons are summarized in table 17.4.

Table 17.2
The Sympathetic (Thoracolumbar) Division

Parts of Body Innervated	Spinal Origin of Preganglionic Fibers	Origin of Postganglionic Fibers
Eye	C8, T1	Cervical ganglia
Head and neck	T1–T4	Cervical ganglia
Heart and lungs	T1–T5	Upper thoracic (paravertebral) ganglia
Upper extremities	T2–T9	Lower cervical and upper thoracic (paravertebral) ganglia
Upper abdominal viscera	T4–T9	Celiac and superior mesenteric (collateral) ganglia
Adrenal medulla	T10, T11	Not applicable
Urinary and reproductive systems	T12–L2	Celiac and inferior mesenteric (collateral) ganglia
Lower extremities	T9–L2	Lumbar and upper sacral (paravertebral) ganglia

Functions of the Autonomic Nervous System

The sympathetic division of the autonomic nervous system activates the body to "fight or flight," largely through the release of norepinephrine from postganglionic fibers and the secretion of epinephrine from the adrenal medulla. The parasympathetic division often produces antagonistic effects through the release of acetylcholine from its postganglionic fibers. The actions of the two divisions must be balanced in order to maintain homeostasis.

The sympathetic and parasympathetic divisions of the autonomic system affect the visceral organs in different ways. Mass activation of the sympathetic system prepares the body for intense physical activity in emergencies; the heart rate increases, the blood glucose level rises, and blood is diverted to the skeletal muscles (away from the visceral organs and skin). These and other effects are listed in table 17.5. The theme of the sympathetic system has been aptly summarized in a phrase: **"fight or flight."**

The effects of parasympathetic nerve stimulation are in many ways opposite to the effects of sympathetic stimulation. The parasympathetic system, however, is not normally activated as a whole. Stimulation of separate parasympathetic nerves can result in slowing of the heart, dilation of visceral blood vessels, and increased activity of the digestive tract (table 17.5). Visceral organs respond differently to sympathetic and parasympathetic nerve activity because the postganglionic fibers of these two divisions release different neurotransmitters.

Table 17.3
The Parasympathetic (Craniosacral) Division

Nerve	Origin of Preganglionic Fibers	Location of Terminal Ganglia	Effector Organs
Oculomotor nerve (III)	Midbrain (cranial)	Ciliary ganglion	Eye (smooth muscle in iris and ciliary body)
Facial nerve (VII)	Pons (cranial)	Pterygopalatine and submandibular ganglia	Lacrimal, mucous, and salivary glands
Glossopharyngeal nerve (IX)	Medulla oblongata (cranial)	Otic ganglion	Parotid gland
Vagus nerve (X)	Medulla oblongata (cranial)	Terminal ganglia in or near organ	Heart, lungs, gastrointestinal tract, liver, pancreas
Pelvic spinal nerves	S2–S4 (sacral)	Terminal ganglia near organs	Lower half of large intestine, rectum, urinary bladder, and reproductive organs

Figure 17.7

The autonomic nervous system.
The sympathetic division is shown in red; the parasympathetic is in blue. The solid lines indicate preganglionic fibers, and the dashed lines indicate postganglionic fibers.

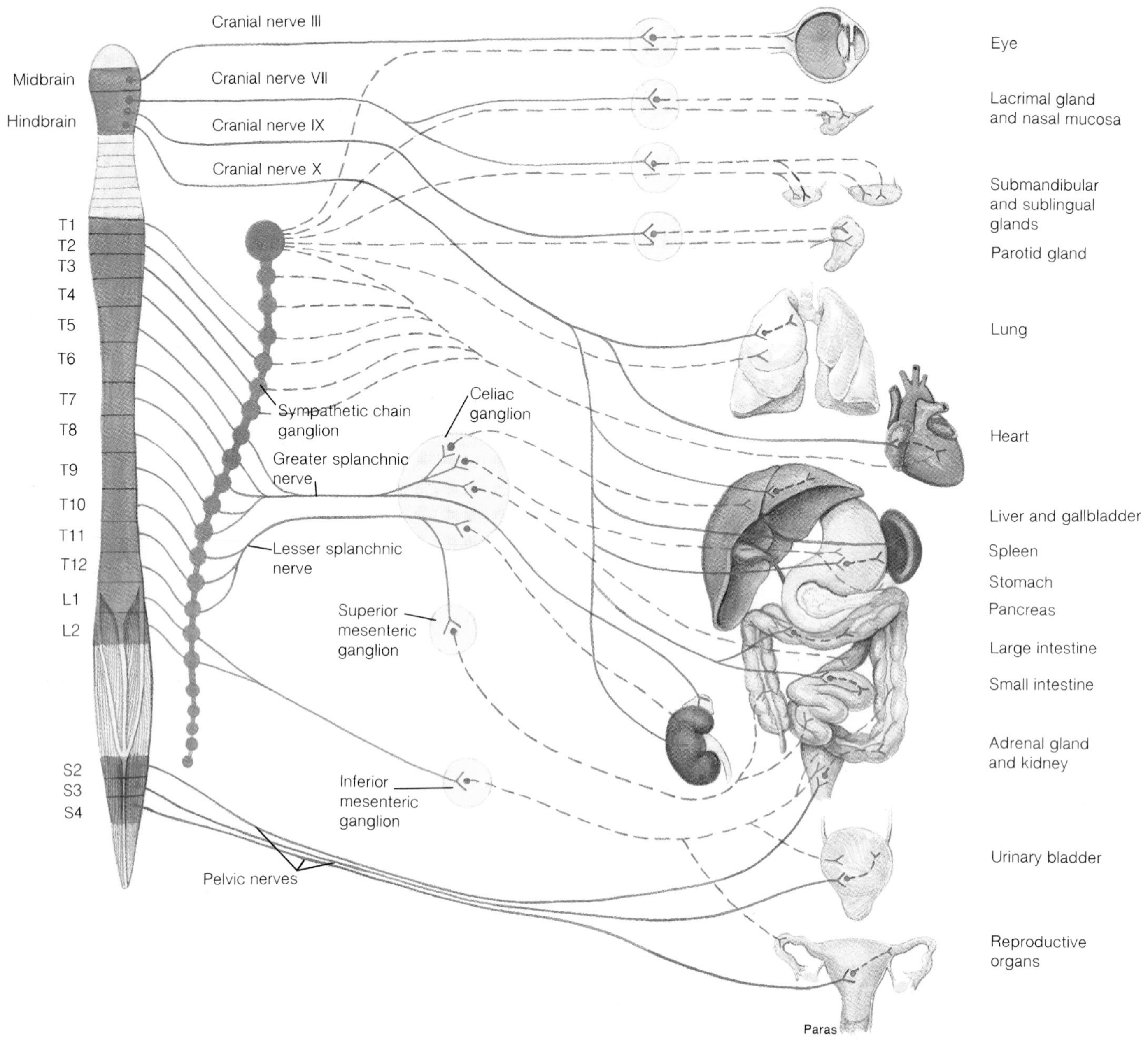

Table 17.4

Comparison of the Sympathetic and Parasympathetic Divisions of the ANS

Feature	Sympathetic	Parasympathetic
Origin of preganglionic outflow	Thoracolumbar portion of spinal cord	Midbrain, hindbrain, and sacral portion of spinal cord
Location of ganglia	Chain of paravertebral ganglia and prevertebral (collateral) ganglia	Terminal ganglia in or near effector organs
Distribution of postganglionic fibers	Throughout the body	Mainly limited to the head and viscera
Divergence of impulses from pre- to postganglionic fibers	Great divergence (1 preganglionic may activate 20 postganglionic fibers)	Little divergence (1 preganglionic activates only a few postganglionic fibers)
Mass discharge of system as a whole	Usually	Not normally

Table 17.5
Effects of Autonomic Nerve Stimulation on Various Effector Organs

Effector Organ	Sympathetic Effect	Parasympathetic Effect
Eye		
Iris (pupillary dilator muscle)	Dilation of pupil	—
Iris (pupillary sphincter muscle)	—	Constriction of pupil
Ciliary muscle	Relaxation (for far vision)	Contraction (for near vision)
Glands		
Lacrimal (tear)	—	Stimulation of secretion
Sweat	Stimulation of secretion	—
Salivary	Decreased secretion; saliva becomes thick	Increased secretion; saliva becomes thin
Stomach	—	Stimulation of secretion
Intestine	—	Stimulation of secretion
Adrenal medulla	Stimulation of hormone secretion	—
Heart		
Rate	Increased	Decreased
Conduction	Increased rate	Decreased rate
Strength	Increased	—
Blood vessels	Mostly constriction; affects all organs	Dilation in a few organs (e.g., penis)
Lungs		
Bronchioles (tubes)	Dilation	Constriction
Mucous glands	Inhibition of secretion	Stimulation of secretion
Gastrointestinal tract		
Motility	Inhibition of movement	Stimulation of movement
Sphincters	Closing stimulated	Closing inhibited
Liver	Stimulation of glycogen hydrolysis	—
Adipocytes (fat cells)	Stimulation of fat hydrolysis	—
Pancreas	Inhibition of exocrine secretions	Stimulation of exocrine secretions
Spleen	Stimulation of contraction	—
Urinary bladder	Muscle tone aided	Stimulation of contraction
Arrector pili muscles	Stimulation of hair erection, causing goosebumps	—
Uterus	If pregnant, contraction; if not pregnant, relaxation	—
Penis	Ejaculation	Erection (due to vasodilation)

Adrenergic and Cholinergic Synaptic Transmission

Acetylcholine (ACh) is the neurotransmitter of all preganglionic fibers (both sympathetic and parasympathetic). Acetylcholine is also the transmitter released by most parasympathetic postganglionic fibers at their synapses with effector cells (fig. 17.8). Transmission at these synapses is thus said to be **cholinergic** (*ko″lĭ-ner′jik*).

The neurotransmitter released by most postganglionic sympathetic nerve fibers is **norepinephrine** (*noradrenaline*). Transmission at these synapses is thus said to be **adrenergic** (*ad″rĕ-ner′jik*). There are a few exceptions, however. Some sympathetic fibers that innervate blood vessels in skeletal muscles, as well as sympathetic fibers to sweat glands, release ACh (are cholinergic).

In view of the fact that the cells of the adrenal medulla are embryologically related to postganglionic sympathetic

Figure 17.8

Neurotransmitters of the autonomic motor system. ACh = acetylcholine; NE = norepinephrine; E = epinephrine. Those nerves that release ACh are called cholinergic; those nerves that release NE are called adrenergic. The adrenal medulla secretes both epinephrine (85%) and norepinephrine (15%) as hormones into the blood.

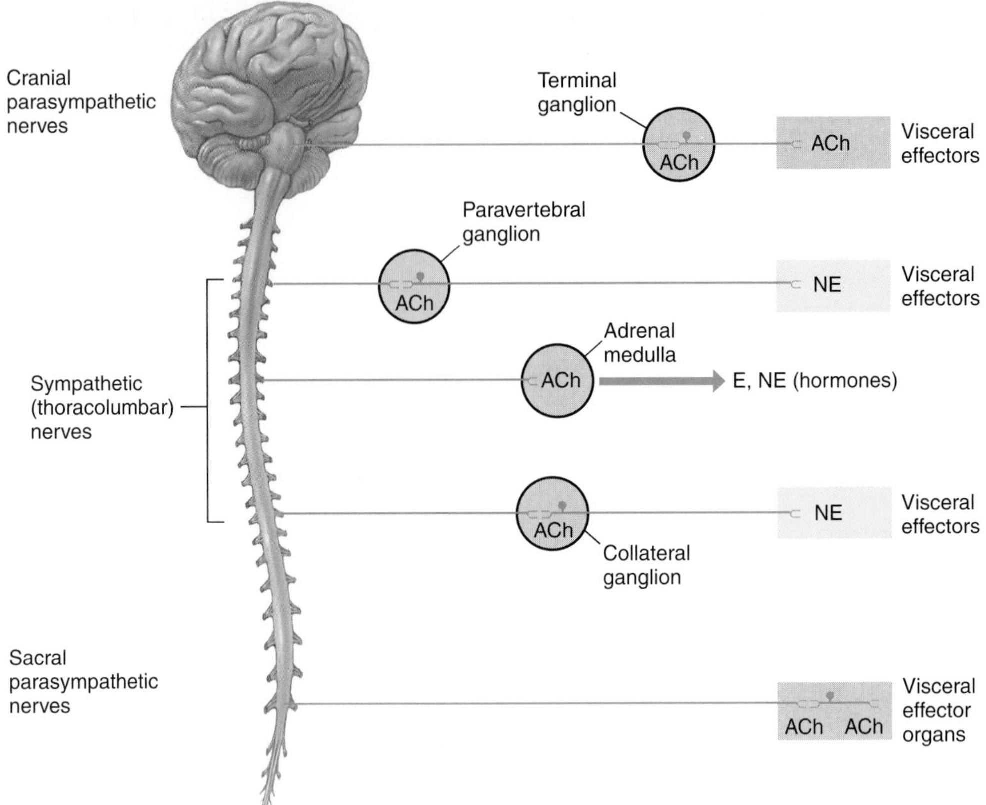

neurons, it is not surprising that the hormones they secrete should consist of epinephrine (about 85%) and norepinephrine (about 15%). Epinephrine differs from norepinephrine only in that the former has an additional methyl (CH_3) group, as shown in figure 17.9. As described in chapter 14, epinephrine, norepinephrine, and dopamine (a transmitter within the CNS) are all derived from the amino acid tyrosine and are collectively termed **catecholamines.**

Figure 17.9

The structure of norepinephrine and epinephrine. Both are in the chemical family known as catecholamines. Epinephrine has an additional methyl group (CH_3) compared to norepinephrine.

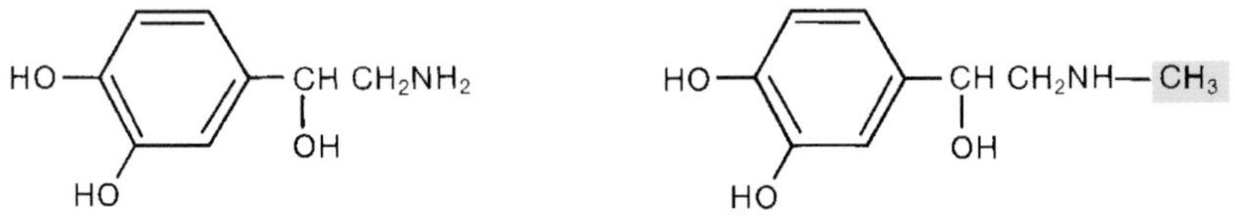

HO — CH CH_2NH_2 — OH — HO

Norepinephrine

HO — CH CH_2NH— CH_3 — OH — HO

Epinephrine

Responses to Adrenergic Stimulation

Adrenergic stimulation—by epinephrine in the blood and by norepinephrine released from sympathetic nerve endings—has both excitatory and inhibitory effects. The heart, dilatory muscles of the iris, and the smooth muscles of many blood vessels are stimulated to contract. The smooth muscles of the bronchioles and of some blood vessels, however, are inhibited from contracting; adrenergic chemicals, therefore, cause these structures to dilate.

Since excitatory and inhibitory effects can be produced in different tissues by the same neurotransmitter, the responses must depend on the characteristics of the tissue cells. To some degree, this is due to the presence of different membrane *receptor proteins* for the catecholamine neurotransmitters. (The interaction of neurotransmitters and receptor proteins in the postsynaptic membrane was described in chapter 14.) The two major classes of these receptor proteins are designated **alpha- (α)** and **beta- (β) adrenergic receptors.**

Experiments have revealed that each class of adrenergic receptor has two subtypes. These are designated by subscripts: α_1 and α_2; β_1 and β_2. Compounds have been developed that selectively bind to one or the other type of adrenergic receptor and, by this means, either promote or inhibit the normal action produced when epinephrine or norepinephrine binds to the receptor. As a result, the drug may either promote or inhibit the adrenergic effect. Also, by using these selective compounds, it has been possible to determine which subtype of adrenergic receptor is present in each organ (table 17.6).

Both subtypes of beta receptors produce their effects by stimulating the production of cyclic AMP (discussed in chapter 14) within the target cells. The activation of the α_2 receptors has the opposite effect—cyclic AMP production is blocked and the cAMP concentration within the target cell is lowered, inhibiting the effects of beta-adrenergic-receptor

Table 17.6

Adrenergic and Cholinergic Effects of Sympathetic and Parasympathetic Nerves

	Effect of			
	Sympathetic		Parasympathetic	
Organ	Action	Receptor*	Action	Receptor*
Eye				
Iris				
Pupillary dilator muscle	Contracts	α	—	—
Pupillary sphincter muscle	—	—	Contracts	M
Heart				
Sinoatrial node	Accelerates	β_1	Decelerates	M
Contractility	Increases	β_1	Decreases (atria)	M
Vascular smooth muscle				
Skin, splanchnic vessels	Contracts	α	—	—
Skeletal muscle vessels	Relaxes	β_2	—	—
	Relaxes	M**	—	—
Bronchiolar smooth muscle	Relaxes	β_2	Contracts	M
Gastrointestinal tract				
Smooth muscle				
Walls	Relaxes	β_2	Contracts	M
Sphincters	Constricts	α	Relaxes	M
Secretion	Decreases	α	Increases	M
Myenteric plexus	Inhibits	α	—	—
Genitourinary smooth muscle				
Wall of urinary bladder	Relaxes	β_2	Contracts	M
Urethral sphincter	Constricts	α	Relaxes	M
Uterus, pregnant	Relaxes	β_2	—	—
	Contracts	α	—	—
Penis, seminal vesicles	Ejaculation	α	Erection	M
Skin				
Arrector pili muscle	Contracts	α	—	—
Sweat gland activity				
Eccrine	Increases	M	—	—
Apocrine (stress)	Increases	α	—	—

Source: Reproduced and modified, with permission, from Katzung, BG: *Basic and Clinical Pharmacology*, 4th edition, copyright Appleton & Lange, 1989.

*Adrenergic receptors are indicated as alpha (α) or beta (β); cholinergic receptors are indicated as muscarinic (M).

**Vascular smooth muscle in skeletal muscle has sympathetic cholinergic dilator fibers.

stimulation. The response of a target cell when norepinephrine binds to the α_1 receptors is mediated by a different second-messenger system—a rise in the cytoplasmic concentration of Ca^{2+}. This Ca^{2+} second-messenger system is similar in many ways to the cAMP system, and is discussed together with endocrine regulation in chapter 19. It should be remembered that each of the intracellular changes following the binding of norepinephrine to its receptor ultimately results in the characteristic response of the tissue to the neurotransmitter.

A review of table 17.6 reveals certain generalities about the actions of adrenergic receptors. The stimulation of alpha-adrenergic receptors consistently causes contraction of smooth muscles. We can thus state that the vasoconstrictor effect of sympathetic nerves always results from the activation of alpha-adrenergic receptors. The effects of beta-adrenergic activation are more complex; stimulation of beta receptors promotes the relaxation of smooth muscles (in the digestive tract, bronchioles, and uterus, for example) but increases the force of contraction of cardiac muscle and promotes an increase in cardiac rate.

The diverse effects of epinephrine and norepinephrine can be understood in terms of the "fight-or-flight" theme. Adrenergic stimulation wrought by activation of the sympathetic division produces an increase in cardiac pumping (a β_1 effect), vasoconstriction and thus reduced blood flow to the visceral organs (an α_1 effect), dilation of pulmonary bronchioles (a β_2 effect), and so on, preparing the body for physical exertion.

Responses to Cholinergic Stimulation

Somatic motor neurons, all preganglionic autonomic neurons, and most postganglionic parasympathetic neurons are cholinergic—they release acetylcholine as a neurotransmitter. The cholinergic effects of somatic motor neurons and preganglionic autonomic neurons are always excitatory. The cholinergic effects of postganglionic parasympathetic fibers are usually excitatory, but there are notable exceptions. The parasympathetic fibers innervating the heart, for example, cause slowing of the heart rate. It is useful to remember that the effects of parasympathetic stimulation are, in general, opposite to the effects of sympathetic stimulation.

Just as adrenergic receptors are divided into alpha and beta subtypes, cholinergic receptors are divided into nicotinic and muscarinic subtypes (described in chapter 14). The drug *muscarine* (*mus'kă-ren*), derived from certain poisonous mushrooms, stimulates the cholinergic receptors in the heart, digestive system, and other target organs of postganglionic parasympathetic nerve fibers. These axons must thus exert their effects on the target organs by stimulating the muscarinic subtype of cholinergic receptors (table 17.6). Muscarine, however, does not stimulate ACh receptor proteins in autonomic ganglia or at the neuromuscular junction of skeletal muscle fibers. The drug nicotine, derived from the tobacco plant, specifically stimulates these cholinergic receptors, which must therefore be the nicotinic subtype of ACh receptors. The drug *curare*, used clinically to cause skeletal muscle relaxation, specifically blocks nicotinic receptors but has little effect on muscarinic receptors.

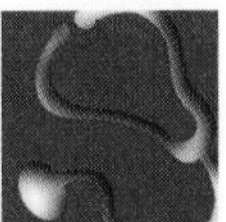

The muscarinic effects of ACh are specifically inhibited by the drug *atropine,* (*at'rŏ-pēn*), derived from the deadly nightshade plant (*Atropa belladonna*). Indeed, extracts of this plant were used by women during the Middle Ages to dilate their pupils (atropine inhibits parasympathetic stimulation of the iris). This was thought to enhance their beauty (in Italian, *bella* = beautiful, *donna* = woman). Atropine is used clinically today to dilate pupils during eye examinations, to reduce secretions of the respiratory tract prior to general anesthesia, to inhibit spasmodic contractions of the lower digestive tract, and to inhibit stomach acid secretion in a person with gastritis.

Other Autonomic Neurotransmitters

Certain postganglionic autonomic axons produce their effects through mechanisms that do not involve either norepinephrine or acetylcholine. This can be demonstrated experimentally by the inability of drugs that block adrenergic and cholinergic effects from inhibiting the actions of those autonomic axons. These axons, consequently, have been termed "nonadrenergic noncholinergic fibers." Proposed neurotransmitters for these axons include ATP, a polypeptide called vasoactive intestinal peptide (VIP), and nitric oxide (NO).

The nonadrenergic noncholinergic parasympathetic axons that innervate the blood vessels of the penis cause the smooth muscles of these vessels to relax, thereby producing vasodilation and a consequent erection of the penis (see chapter 28). These parasympathetic axons have been shown to use the gas nitric oxide (chapter 14) as their neurotransmitter. In a similar manner, nitric oxide appears to function as the autonomic neurotransmitter that causes vasodilation of cerebral arteries. Studies suggest that nitric oxide is not stored in synaptic vesicles, as are other neurotransmitters, but instead is produced immediately when Ca^{2+} enters the axon terminal in response to action potentials. This Ca^{2+} indirectly activates nitric oxide synthetase, the enzyme that forms nitric oxide from the amino acid L-arginine. Nitric oxide then diffuses across the synaptic cleft and promotes relaxation of the postsynaptic smooth muscle cells.

Nitric oxide can produce relaxation of smooth muscles in many organs, including the stomach, small intestine, large intestine, and urinary bladder. There is some controversy, however, about whether the nitric oxide functions as a neurotransmitter in each case. It has been suggested that, in some cases, nitric acid could be produced in the organ itself in response to autonomic stimulation. The fact that different tissues, such as the endothelium of blood vessels, can produce nitric oxide leads support to this suggestion. Indeed, nitric oxide is a member of a class of local tissue regulatory molecules called *paracrine regulators* (see chapter 19). Regulation can therefore be a complex process involving the interacting effects of different neurotransmitters, hormones, and paracrine regulators.

Organs with Dual Innervation

Most visceral organs receive dual innervation—they are innervated by both sympathetic and parasympathetic fibers. The effects of the two divisions of the autonomic nervous system may be antagonistic, complementary, or cooperative.

Antagonistic Effects

The effects of sympathetic and parasympathetic innervation of the pacemaker region of the heart is the best example of the antagonism of these two systems. In this case, sympathetic and parasympathetic fibers innervate the same cells. Adrenergic stimulation from sympathetic fibers increases the heart rate, whereas the release of acetylcholine from parasympathetic fibers decreases the heart rate. A reverse of this antagonism is seen in the GI tract, where sympathetic nerves inhibit and parasympathetic nerves stimulate intestinal movements and secretions.

The effects of sympathetic and parasympathetic stimulation on the diameter of the pupil of the eye are analogous to the reciprocal innervation of flexor and extensor skeletal muscles by somatic motor neurons (chapter 12). This is because the iris contains antagonistic muscle layers. Contraction of the radial muscles, which are innervated by sympathetic nerves, causes dilation; contraction of the circular muscles, which are innervated by parasympathetic nerve endings, causes constriction of the pupils (see chapter 18, fig. 18.35).

Complementary and Cooperative Effects

The effects of sympathetic and parasympathetic nerves are generally antagonistic, but in a few cases they can be complementary or cooperative. The effects are complementary when sympathetic and parasympathetic stimulation produce similar effects. The effects are cooperative, or *synergistic*, when sympathetic and parasympathetic stimulation produce different effects that work together to promote a single action.

The effects of sympathetic and parasympathetic stimulation on salivary gland secretion are complementary. The secretion of watery saliva is stimulated by parasympathetic nerves, which also stimulate the secretion of other exocrine glands in the GI tract. Sympathetic nerves stimulate the constriction of blood vessels throughout the digestive tract. The resultant decrease in blood flow to the salivary glands causes the production of a thicker, more viscous saliva.

The effects of sympathetic and parasympathetic stimulation on the reproductive and urinary systems are cooperative. Erection of the penis, for example, is due to vasodilation resulting from parasympathetic nerve stimulation; ejaculation is due to stimulation through sympathetic nerves. The two divisions of the autonomic system thus cooperate to enable sexual function in the male. There is also cooperation between the two divisions in the *micturition* (*mik″tŭ-rish′un*), or urination, reflex. Although the contraction of the urinary bladder is largely independent of nerve stimulation, it is promoted in part by the action of parasympathetic nerves. This reflex is also enhanced by sympathetic nerve activity, which increases the tone of the bladder muscles. Emotional states that are accompanied by high sympathetic nerve activity (such as extreme fear) may thus result in reflex urination at bladder volumes that are normally too low to trigger this reflex.

micturition: L. *micturire*, to urinate

Organs without Dual Innervation

Although most organs are innervated by both sympathetic and parasympathetic nerves, some—including the adrenal medulla, arrector pili muscles, sweat glands, and blood vessels in the skin and skeletal muscles—receive only sympathetic innervation. In these cases, regulation is achieved by increases or decreases in the tone (firing rate) of the sympathetic fibers. Constriction of cutaneous blood vessels, for example, is produced by increased sympathetic activity that stimulates alpha-adrenergic receptors, and vasodilation results from decreased sympathetic nerve stimulation.

The sympathoadrenal system is required for *nonshivering thermogenesis*: animals deprived of their sympathetic system and adrenals cannot tolerate cold stress. The sympathetic system itself is required for proper thermoregulatory responses to heat. In a hot room, for example, decreased sympathetic stimulation produces dilation of the blood vessels in the skin, which increases cutaneous blood flow and provides better heat radiation. During exercise, by contrast, sympathetic activity increases, causing constriction of the blood vessels in the skin of the limbs and stimulation of sweat glands in the trunk.

The sweat glands in the trunk secrete a watery fluid in response to cholinergic sympathetic stimulation. Evaporation of this dilute sweat helps to cool the body. The sweat glands also secrete a chemical called *bradykinin* (*brad″ĭ-ki′nin*) in response to sympathetic stimulation. Bradykinin stimulates dilation of the surface blood vessels near the sweat glands, helping to radiate some heat despite the fact that other cutaneous blood vessels are constricted. At the conclusion of exercise sympathetic stimulation is reduced, causing cutaneous blood vessels to dilate. This increases blood flow to the skin, helping to eliminate metabolic heat. Notice that all of these thermoregulatory responses are achieved without the direct involvement of the parasympathetic division.

Control of the Autonomic Nervous System by Higher Brain Centers

Visceral functions are largely regulated by autonomic reflexes. In most autonomic reflexes, sensory input is transmitted to brain centers that integrate this information and respond by modifying the activity of preganglionic autonomic neurons. The neural centers that directly control the activity of autonomic nerves are influenced by higher brain areas, as well as by sensory input.

Table 17.7

Effects Stimulated by Sensory Input from Afferent Fibers in the Vagus Nerves, Which Transmit This Input to Centers in the Medulla Oblongata

Organs	Type of Receptors	Reflex Effects
Lungs	Stretch receptors	Further inhalation inhibited; increase in cardiac rate and vasodilation stimulated
	Type J receptors	Stimulated by pulmonary congestion—produces feelings of breathlessness and causes a reflex fall in cardiac rate and blood pressure
Aortic arch	Chemoreceptors	Stimulated by rise in CO_2 and fall in O_2—produces increased rate of breathing, rise in heart rate, and vasoconstriction
	Baroreceptors	Stimulated by increased blood pressure—produces a reflex decrease in heart rate
Heart	Atrial stretch receptors	Antidiuretic hormone secretion inhibited, thus increasing the volume of urine excreted
	Stretch receptors in ventricles	Produces a reflex decrease in heart rate and vasodilation
Gastrointestinal tract	Stretch receptors	Feelings of satiety, discomfort, and pain

The **medulla oblongata** of the brain stem is the area that most directly controls the activity of the autonomic system. Almost all autonomic responses can be elicited by experimental stimulation of the medulla, where centers for the control of the cardiovascular, pulmonary, urinary, reproductive, and digestive systems are located. Much of the sensory input to these centers travels in the afferent fibers of the vagus nerve—a mixed nerve containing both sensory and motor fibers. The reflexes that result are listed in table 17.7.

Although the medulla oblongata directly regulates the activity of autonomic motor fibers, the medulla is itself responsive to regulation by higher brain areas. One of these is the **hypothalamus,** the brain region that contains centers for the control of body temperature, hunger, and thirst; for regulation of the pituitary gland; and—together with the limbic system and cerebral cortex—for various emotional states.

As described in chapter 15, the **limbic system** is a group of fiber tracts and nuclei that form a ring around the brain stem. It includes the cingulate gyrus of the cerebral cortex, the hypothalamus, the fornix (a fiber tract), the hippocampus, and the amygdaloid nucleus (fig. 17.10). The limbic system is involved in basic emotional drives, such as anger, fear, sex, and hunger. The involvement of the limbic system with the control of autonomic function is responsible for the visceral responses that characterize these emotional states. Blushing, pallor, fainting, breaking out in a cold sweat, a racing heartbeat, and "butterflies in the stomach" are only some of the many visceral reactions that accompany emotions as a result of autonomic activation.

The autonomic correlates of motion sickness—nausea, sweating, and cardiovascular changes—are eliminated by cutting the motor tracts from the **cerebellum.** This demonstrates that impulses from the cerebellum to the medulla oblongata influence activity of the autonomic nervous system. Experimental and clinical observations have also demonstrated that the frontal and temporal lobes of the **cerebral cortex** influence lower brain areas as part of their involvement in emotion and personality.

limbic: L. *limbus*, edge or border

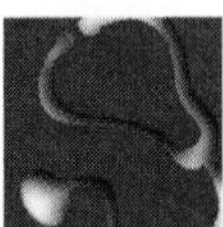

Traditionally, the distinction between the somatic system and the autonomic nervous system was drawn on the basis that the former is under conscious control, whereas the latter is not. Recently, however, we have learned that conscious processes in the cerebrum can influence autonomic activity. In *biofeedback* techniques, data obtained from devices that detect and amplify changes in blood pressure and heart rate, for example, are "fed back" to patients in the form of light signals or audible tones. The patients can often be trained to consciously reduce the frequency of the signals and, eventually, to control visceral activities without the aid of a machine. Biofeedback has been used successfully to treat hypertension, stress, and migraine headaches.

One of the most dramatic examples of the role of higher brain areas in personality and emotion is the famous crowbar accident of 1848. A 25-year-old railroad foreman, Phineas P. Gage, was tamping gunpowder into a hole in a rock with a metal rod when the gunpowder suddenly exploded. The rod—three feet, seven inches long and one and one-fourth inches thick—was driven through his left eye and through his brain, finally emerging through the back of his skull.

After a few minutes of convulsions, Gage got up, rode a horse three-quarters of a mile into town, and walked up a long flight of stairs to see a doctor. He recovered well, with no noticeable sensory or motor deficits. His associates, however, noted striking personality changes. Before the accident, Gage was a responsible, capable, financially prudent man. Afterward, he was much less inhibited socially, engaging, for example, in gross profanity (which he had never done previously).

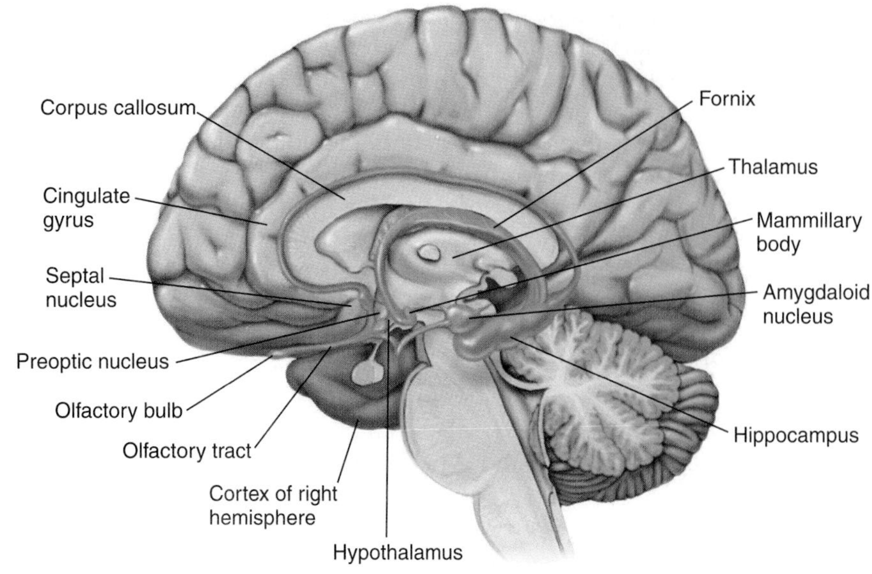

Figure 17.10

The limbic system.
The structures of the limbic system and the pathways that interconnect these structures are shown. In order to visualize the limbic system, the left temporal lobe has been removed in this figure.

He also seemed to be tossed about by chance whims. Eventually Gage was fired from his job, and his old friends remarked that he was "no longer Gage."

Clinical Considerations

Autonomic Dysreflexia

Autonomic dysreflexia, a serious condition producing rapid elevations in blood pressure that can lead to stroke (cerebrovascular accident), occurs in 85% of people with quadriplegia and others with spinal cord lesions above the sixth thoracic level. Lesions to the spinal cord first produce the symptoms of spinal shock, characterized by the loss of both skeletal muscle and autonomic reflexes. After a period of time, both types of reflexes return in an exaggerated state. The skeletal muscles may become spastic in the absence of higher inhibitory influences, and the visceral organs experience denervation hypersensitivity. Patients in this state have difficulty emptying their urinary bladders and often must be catheterized.

Noxious stimuli, such as overdistension of the urinary bladder, can result in reflex activation of the sympathetic nerves below the spinal cord lesion. This produces goose bumps, cold skin, and vasoconstriction in the regions served by the spinal cord below the level of the lesion. The rise in blood pressure resulting from this vasoconstriction activates pressure receptors that transmit impulses along sensory nerve fibers to the medulla oblongata. In response to this sensory input, the medulla oblongata directs a reflex slowing of the heart and vasodilation. Since descending impulses are blocked by the spinal lesion, however, the skin above the lesion is warm and moist (due to vasodilation and sweat gland secretion), but it is cold below the level of spinal cord damage.

Pharmacology of the Autonomic Nervous System

A drug that binds to receptors for a neurotransmitter and that enhances the effects produced by that neurotransmitter is said to be an **agonist** of the neurotransmitter. A drug that blocks the action of a neurotransmitter, by contrast, is said to be an **antagonist** of the neurotransmitter. The use of specific drugs that selectively stimulate or block α_1, α_2, β_1, and β_2 receptors has proven extremely useful in many medical applications.

Adrenergic Drugs

Epinephrine, norepinephrine, acetylcholine, and some chemicals that are not normally found in the body can either enhance or block the physiological effects of the autonomic nervous system. Those drugs that enhance sympathetic action are called **sympathomimetic** (*sim″pă-tho-mĭ-met′ik*) **drugs;** those that suppress sympathetic action are called **sympatholytic** (*sim″pă-tho-lit′ik*) **drugs.**

Many people with hypertension were once treated with a beta-blocking drug known as *propranolol.* This drug blocks β_1 receptors, which are located in the heart, and thus produces the desired effect of lowering the cardiac rate and blood pressure. Propranolol, however, also blocks β_2 receptors, which are located in the bronchioles of the lungs. This reduces the bronchodilation effect of epinephrine, producing bronchoconstriction and asthma in susceptible people. A more specific β_1 antagonist, *atenolol,* is now used instead to slow the cardiac rate and lower blood pressure. At one time, asthmatics inhaled an epinephrine spray, which stimulates β_1 receptors in the heart as well as β_2 receptors in the airways. Now, drugs such as *terbutaline* that selectively function as β_2 agonists are more commonly used.

Drugs that function as α_1 agonists, such as *phenylephrine,* are often part of nasal sprays because they promote vasoconstriction in the nasal mucosa. *Clonidine* is a drug that selectively stimulates α_2 receptors located on presynaptic axon terminals in the brain. (Activation of these receptors inhibits the release of neurotransmitter from the axon terminals.) As a consequence of its action, clonidine suppresses the activation of the sympathoadrenal system and thereby helps to lower the blood pressure. For reasons that are poorly understood, this drug is also helpful in treating patients with an addiction to opiates who are experiencing withdrawal symptoms.

Cholinergic Drugs

Acetylcholine is used as a neurotransmitter by somatic motor neurons, preganglionic autonomic fibers, all postganglionic parasympathetic nerve fibers, and some sympathetic nerve fibers. Drugs that are similar in structure and action to acetylcholine (*methacholine* and *bethanecholine*) can enhance the effects of these neurons. Drugs that inhibit the action of acetylcholinesterase (an enzyme that degrades ACh) enhance the action of ACh. *Physostigmine* and *neostigmine* are examples of such drugs. Because of their effects on the neuromuscular junction, these drugs are used in the treatment of myasthenia gravis and other muscular disorders.

Some of these drugs are also used as **parasympathomimetics**—drugs that duplicate the action of parasympathetic neurons—in the treatment of glaucoma. Atropine blocks the action of ACh released by postganglionic parasympathetic neurons; it is a **parasympatholytic drug.** Atropine is used clinically to block the actions of the parasympathetic division, as previously described.

Clinical Investigation Answer

The prolonged stress associated with final examinations had overstimulated the student's sympathoadrenal system. The increased sympathoadrenal activity could account for her rapid pulse (due to increased heart rate) and her hypertension (due to increased heart rate and vasoconstriction). Her headache could be attributed to dilated pupils, which admitted excessive amounts of light. Since she had been preparing drugs for a laboratory exercise on autonomic control, she may have been exposed to atropine, which would have caused dilation of her pupils. This possibility is strengthened by the fact that she felt her mouth to be excessively dry.

Chapter Summary

Neural Control of Involuntary Effectors (pp. 478–480)

1. There are two basic categories of neurons in the autonomic motor pathway: preganglionic and postganglionic.
   (a) Preganglionic autonomic neurons originate in the brain or spinal cord.
   (b) Postganglionic neurons originate in ganglia located outside the CNS.
2. Cardiac muscle, smooth muscle, and glands receive autonomic innervation.

Divisions of the Autonomic Nervous System (pp. 480–484)

1. Preganglionic neurons of the sympathetic division originate in the spinal cord, between the thoracic and lumbar levels.
   (a) Many of these fibers synapse with postganglionic neurons, whose cell bodies are located in a double chain of sympathetic (paravertebral) ganglia outside the spinal cord.
   (b) Some preganglionic fibers synapse in collateral (prevertebral) ganglia. These include the celiac, superior mesenteric, and inferior mesenteric ganglia.
   (c) Some preganglionic fibers innervate the adrenal medulla, which secretes epinephrine (and some norepinephrine) into the blood in response to stimulation.
2. Preganglionic parasympathetic fibers originate in the brain and in the spinal cord at the sacral levels.
   (a) Preganglionic parasympathetic fibers contribute to the oculomotor, facial, glossopharyngeal, and vagus nerves.
   (b) Preganglionic fibers of the vagus nerves are long; they synapse in terminal ganglia located next to or within the innervated organ. Short postganglionic fibers then innervate the effector organs.

Functions of the Autonomic Nervous System (pp. 485–493)

1. The sympathetic division of the autonomic nervous system activates the body to "fight or flight" through adrenergic effects; the parasympathetic division often exerts antagonistic actions through cholinergic effects.
   (a) All preganglionic autonomic nerve fibers are cholinergic (use ACh as a neurotransmitter).
   (b) All postganglionic parasympathetic fibers are cholinergic.
   (c) Most postganglionic sympathetic fibers are adrenergic (use norepinephrine as a neurotransmitter).
2. Adrenergic effects include stimulation of the heart, vasoconstriction in the viscera and skin, bronchodilation, and glycogenolysis in the liver.
   (a) The two main classes of adrenergic receptor proteins are alpha and beta.
   (b) There are two subtypes of alpha receptors (α_1 and α_2) and two subtypes of beta receptors (β_1 and β_2). These subtypes can be selectively stimulated or blocked by therapeutic drugs.
3. In organs with dual innervation, the actions of the sympathetic and parasympathetic divisions can be antagonistic, complementary, or cooperative.
4. The medulla oblongata of the brain stem is the structure that most directly controls the activity of the autonomic nervous system.
   (a) The medulla oblongata is in turn influenced by sensory input and by input from the hypothalamus.
   (b) The hypothalamus is influenced by input from the limbic system, cerebellum, and cerebrum. These interconnections provide an autonomic component to some of the visceral responses that accompany emotions.

Review Activities

Objective Questions

1. When a visceral organ is denervated,
   (a) it ceases to function.
   (b) it becomes less sensitive to subsequent stimulation by neurotransmitters.
   (c) it becomes hypersensitive to subsequent stimulation.

2. Parasympathetic ganglia are located
   (a) in a chain parallel to the spinal cord.
   (b) in the posterior roots of spinal nerves.
   (c) next to or within the organs innervated.
   (d) in the brain.
3. The neurotransmitter of preganglionic sympathetic fibers is
   (a) norepinephrine.
   (b) epinephrine.
   (c) acetylcholine.
   (d) dopamine.
4. Which of the following results from stimulation of alpha-adrenergic receptors?
   (a) constriction of blood vessels
   (b) dilation of bronchioles
   (c) decreased heart rate
   (d) sweat gland secretion
5. Which of the following fibers release norepinephrine?
   (a) preganglionic parasympathetic fibers
   (b) postganglionic parasympathetic fibers
   (c) postganglionic sympathetic fibers in the heart
   (d) postganglionic sympathetic fibers in sweat glands
   (e) all of the above
6. The actions of sympathetic and parasympathetic fibers are cooperative in
   (a) the heart.
   (b) the reproductive system.
   (c) the digestive system.
   (d) the eyes.
7. Propranolol is a beta blocker. It would therefore cause
   (a) vasodilation.
   (b) slowing of the heart rate.
   (c) increased blood pressure.
   (d) secretion of saliva.
8. Atropine blocks parasympathetic nerve effects. It would therefore cause
   (a) dilation of the pupils.
   (b) decreased mucus secretion.
   (c) decreased movements of the GI tract.
   (d) increased heart rate.
   (e) all of the above.
9. Which area of the brain is most directly involved in the reflex control of the autonomic system?
   (a) hypothalamus
   (b) cerebral cortex
   (c) medulla oblongata
   (d) cerebellum
10. The two subtypes of cholinergic receptors are
    (a) adrenergic and nicotinic.
    (b) dopaminergic and muscarinic.
    (c) nicotinic and muscarinic.
    (d) nicotinic and dopaminergic.
11. A fall in cyclic AMP within the target cell occurs when norepinephrine binds to which of the following adrenergic receptors?
    (a) α_1
    (b) α_2
    (c) β_1
    (d) β_2
12. A drug that serves as an agonist for β_2 receptors can be used to
    (a) increase the heart rate.
    (b) decrease the heart rate.
    (c) dilate the bronchioles.
    (d) constrict the bronchioles.
    (e) constrict the blood vessels.

Essay Questions

1. Compare the sympathetic and parasympathetic systems in terms of the location of their ganglia and the distribution of their nerves.
2. Explain the anatomical and physiological relationship between the sympathetic nervous system and the adrenal glands.
3. Compare the effects of adrenergic and cholinergic stimulation on the cardiovascular and digestive systems.
4. Explain how effectors that receive only sympathetic innervation are regulated by the autonomic nervous system.
5. Distinguish between the different types of adrenergic receptors and state where these receptors are located in the body.
6. Give examples of drugs that selectively stimulate or block different adrenergic receptors and explain how these drugs are used clinically.
7. Explain what is meant by nicotinic and muscarinic ACh receptors and describe the distribution of these receptors in the body.
8. Give examples of drugs that selectively stimulate and block the nicotinic and muscarinic receptors and explain how these drugs are used clinically.

Critical Thinking Questions

1. Shock is the medical condition that occurs when body tissues do not receive enough oxygen-carrying blood. It is characterized by low blood flow to the brain, leading to decreased levels of consciousness. Why would a patient with a cervical spinal cord injury be at risk of going into shock?
2. A person in shock may have pale, cold, and clammy skin and a rapid and weak pulse. What is the role of the autonomic nervous system in producing these symptoms? Discuss how drugs that influence autonomic activity might be used to treat this condition.
3. Imagine yourself at the starting block of the 100-meter dash of the Olympics. The gun is about to go off in the biggest race of your life. What is the autonomic nervous system doing at this point? How are your organs reacting?
4. Some patients with hypertension (high blood pressure) are given beta-blocking drugs to lower their blood pressure. How does this effect occur? Explain why these drugs are not administered to patients with a history of asthma. Why might drinking coffee help asthma?
5. Why do many cold medications contain an alpha-adrenergic agonist and atropine (belladonna)? Why is there a label warning for people with hypertension? Why would a person with gastritis be given a prescription for atropine? Explain how this drug might affect the ability to digest and absorb food.

Related Web Sites

For a listing of the most current web sites related to this chapter, please visit the *Concepts of Human Anatomy and Physiology* home page at http://www.mhhe.com/biosci/abio/.

NEXUS

Some Interactions of the Nervous System with the Other Body Systems

Integumentary System
- Sensory neurons conduct impulses from cutaneous receptors (p. 501).
- Sympathetic neurons to the skin help to regulate cutaneous blood flow (p. 676).
- Autonomic neurons innervate integumentary glands and arrector pili muscles (p. 485).

Skeletal System
- The skeleton supports and protects the brain and spinal cord (p. 183).
- Bones store calcium needed for neural function (p. 185).
- Sensory neurons from sensory receptors monitor movements of joints (p. 504).

Endocrine System
- Autonomic neurons innervate some endocrine glands, such as the adrenal medulla and pancreatic islets (p. 482).
- The brain controls the secretion of the anterior pituitary gland (p. 569).
- The brain controls the secretion of the posterior pituitary gland (p. 569).
- Many hormones, including sex steroids, act on the brain (p. 571).
- Hormones and neurotransmitters, such as epinephrine and norepinephrine, can have synergistic actions on target cells (p. 557).

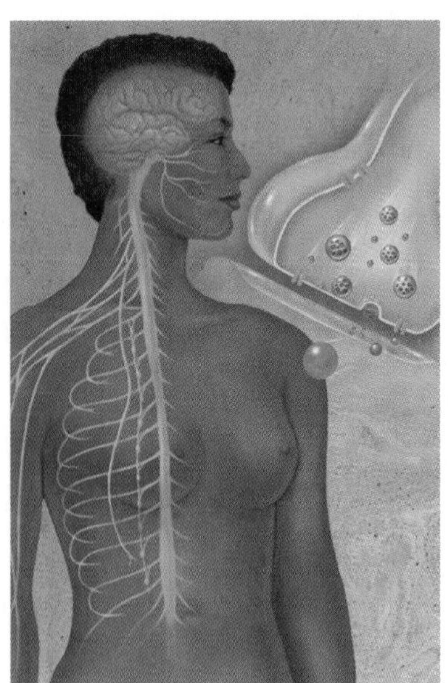

Muscular System
- Sensory neurons from muscle spindles transmit impulses to the CNS (p. 505).
- Somatic motor neurons innervate skeletal muscles (p. 287).
- Autonomic motor neurons innervate cardiac and smooth muscles (p. 478).

Circulatory System
- The circulatory system transports O_2 and CO_2, nutrients,and fluids to and from all organs, including the brain and spinal cord (p. 592).
- Autonomic nerves help to regulate the cardiac output (p. 656).
- Autonomic nerves promote constriction and dilation of blood vessels, helping to regulate the blood flow and blood pressure (p. 668).

Lymphatic and Immune Systems
- Chemical factors called cytokines, released by cells of the immune system, act on the brain to promote a fever (p. 698).
- Cytokines from the immune system act on the brain to modify its regulation of pituitary gland secretion (p. 718).

Respiratory System
- The lungs provide O_2 for all body systems, and eliminate CO_2 (p. 729).
- Neural centers within the brain control breathing (p. 752).

Urinary System
- The kidneys eliminate metabolic wastes and help maintain homeostasis of the blood plasma (p. 779).
- The kidneys regulate the plasma concentrations of Na^+, K^+, and other ions needed for the functioning of neurons (p. 801).
- The nervous system innervates organs of the urinary system to control urination (p. 807).
- Autonomic nerves help to regulate renal blood flow (p. 785).

Digestive System
- The GI tract provides nutrients for all body organs, including those of the nervous system (p. 817).
- Autonomic nerves innervate digestive organs (p. 822).
- The GI tract contains a complex neural system, called an enteric brain, that regulates the motility and secretions of the GI tract (p. 851).
- Secretions of gastric juice can be stimulated through activation of brain regions (p. 849).
- Hunger is controlled by centers in the hypothalamus in the brain (p. 872).

Reproductive System
- Gonads produce sex hormones that influence brain development (p. 897).
- The brain helps to regulate the secretions of gonadotrophic hormones from the anterior pituitary (p. 896).
- Autonomic nerves regulate blood flow into the external genitalia, contributing to the male and female sexual response (p. 930).
- The nervous and endocrine systems cooperate in the control of lactation (p. 942).

EIGHTEEN

Sensory Organs

OBJECTIVES

- Discuss the different categories of sensory receptors and the selectivity of receptors for specific stimuli.
- Explain the law of specific nerve energies and distinguish between tonic and phasic receptors.
- Describe the structure, functions, and locations of the tactile receptors and the neural pathway for somatic sensation.
- Describe the receptors and neural pathways for pain and explain what is meant by referred pain and phantom pain.
- Discuss the receptors and neural pathways that mediate proprioception.
- Describe the location and structure of taste buds and the distribution of the different kinds of taste buds.
- Describe the olfactory receptors and the neural pathways for olfaction.
- Distinguish between the membranous and bony labyrinths and describe the structure of the vestibular apparatus.
- Explain how mechanical movements are transduced into nerve impulses in the semicircular canals and in the otolith organs.
- Describe the neural pathways for the sense of equilibrium and explain how the vestibular apparatus can influence eye movements.
- Describe the structure of the outer ear, middle ear, and cochlea, and explain how they function in hearing.
- Explain how different pitches of sounds affect the cochlea and how different pitches are coded in the neural pathways for hearing.
- Describe the structures of the eyeball, trace the path of light through the eye, and explain how the focus of the eye adjusts for viewing different distances.
- Describe the structure of the retina and compare the structure and function of rods and cones.
- Explain how light affects the rods and ultimately results in the production of action potentials in ganglion cells.
- Explain how the cones provide color vision and greater visual acuity than the rods.
- Describe the neural pathways for visual perception.
- Describe the receptive fields of ganglion cells and explain how the different types of cortical neurons are best stimulated.

Clinical Investigation

During a routine eye exam, a 45-year-old woman mentioned to her optometrist that her glasses no longer allowed her to read small print, although they still helped her read street signs and other distant objects while driving. She explained further that she had been experiencing this problem for about a year, but that it had gotten worse in the past few months. The optometrist recommended bifocals and also tested her eyes with a device that blows a puff of air on the cornea to measure the intraocular pressure. The results of this test were normal. What can you conclude about this woman's vision?

Clues: Study the sections on refraction and accommodation under the main heading "The Eyes and Vision."

Characteristics of Sensory Receptors

Each type of sensory receptor responds to a particular modality of environmental stimulus by causing the production of action potentials in a sensory neuron. These impulses are conducted to parts of the brain that provide the proper interpretation of the sensory information when that particular neural pathway is activated.

Our perceptions of the world—its textures, colors, and sounds; its warmth, smells, and tastes—are created by the brain from electrochemical nerve impulses delivered to it from sensory receptors. These receptors **transduce** (change) different forms of energy in the "real world" into the energy of nerve impulses that are conducted into the central nervous system by sensory neurons. Different *modalities* (forms) of sensation—sound, light, pressure, and so forth—result from differences in neural pathways and synaptic connections. The brain thus interprets impulses arriving from the auditory nerve as sound and from the optic nerve as sight, even though the impulses themselves are identical in the two nerves.

We know, through the use of scientific instruments, that our senses act as energy filters that allow us to perceive only a narrow range of energy. Vision, for example, is limited to light in the visible spectrum; ultraviolet and infrared light, X rays and radio waves, which are the same type of energy as visible light, cannot normally excite the photoreceptors in the eyes. The perception of cold is entirely a product of the nervous system—there is no such thing as cold in the physical world, only varying degrees of heat. The perception of cold, however, has obvious survival value. Although filtered and distorted by the limitations of sensory function, our perceptions of the world allow us to interact effectively with the environment.

Categories of Sensory Receptors

Sensory receptors can be categorized on the basis of structure or various functional criteria. Structurally, the sensory receptors may be the dendritic endings of sensory neurons. Sensory nerve endings are either free—such as those in the skin that mediate pain and temperature—or encapsulated within nonneural structures, such as pressure receptors in the skin (fig. 18.1). The photoreceptors in the retina of the eyes (rods and cones) are highly specialized neurons that synapse with other neurons in the retina. In the case of taste buds and of hair cells in the inner ears, modified epithelial cells respond to an environmental stimulus and activate sensory neurons.

Functional Categories

Sensory receptors can be grouped according to the type of stimulus energy they transduce. These categories include (1) **chemoreceptors** (*ke″mo-re-sep′torz*), such as the taste buds, olfactory epithelium, and the aortic and carotid bodies, which sense chemical stimuli in the external environment or the blood; (2) **photoreceptors**—the rods and cones in the retina of the eye; (3) **thermoreceptors,** which respond to heat and cold; and (4) **mechanoreceptors** (*mek″ă-no-re-sep′torz*), which are stimulated by mechanical deformation of the receptor cell membrane—these include touch and pressure receptors in the skin and hair cells within the inner ear.

Nociceptors (*no″sĭ-sep′torz*)—or pain receptors—have a higher threshold for activation than the other cutaneous receptors; thus, a more intense stimulus is required for their activation. Their firing rate then increases with stimulus intensity. Receptors that subserve other sensations may also become involved in pain transmission when the stimulus is prolonged, particularly when tissue damage occurs.

Receptors also can be grouped according to the type of sensory information they deliver to the brain. **Proprioceptors** (*pro″pre-o-sep′torz*) include the muscle spindles, neurotendinous receptors, and joint receptors. These provide a sense of body position and allow fine control of skeletal movements. **Cutaneous** (skin) **receptors** include (1) touch and pressure receptors, (2) warm and cold receptors, and (3) pain receptors. The receptors that mediate sight, hearing, and equilibrium are grouped together as the **special senses.**

Tonic and Phasic Receptors: Sensory Adaptation

Some receptors respond with a burst of activity when a stimulus is first applied, but then quickly decrease their firing rate—adapt to the stimulus—if the stimulus is maintained. Receptors with this response pattern are called *phasic receptors*. Receptors that produce a relatively constant rate of firing as long as the stimulus is maintained are known as *tonic receptors* (fig. 18.2).

---

nociceptor: L. *nocco*, to injure; *ceptus*, taken
proprioceptor: L. *proprius*, one's own; *ceptus*, taken

Figure 18.1

Different types of sensory receptors.

Free nerve endings (*a*) mediate many cutaneous sensations, including heat. Some nerve endings are encapsulated within associated structures: the lamellated corpuscle (*b*) for deep pressure and the tactile (Meissner's) corpuscle (*c*) for light touch. Some receptors, such as the taste bud (*d*), are modified epithelial cells that are innervated by sensory neurons.

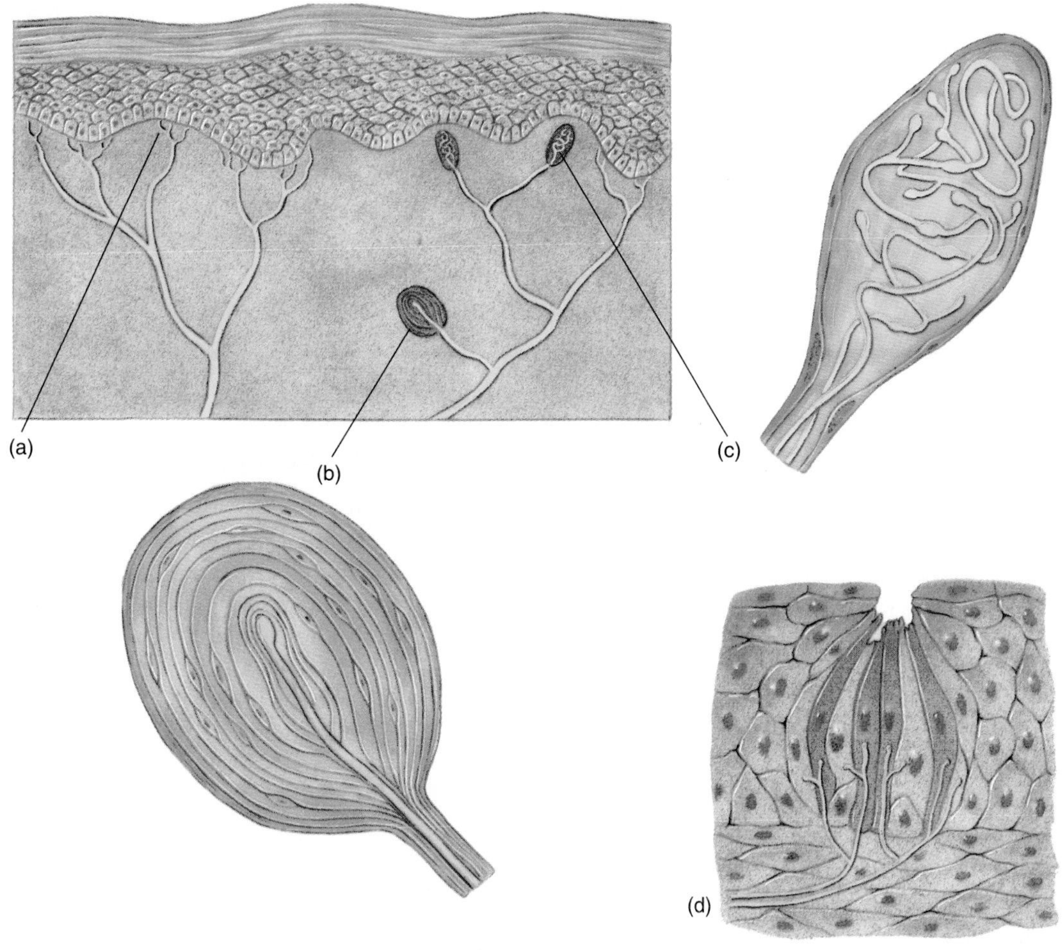

Figure 18.2

Comparison of tonic and phasic receptors.

Tonic receptors (*a*) continue to fire at a relatively constant rate as long as the stimulus is maintained. These produce slow-adapting sensations. Phasic receptors (*b*) respond with a burst of action potentials when the stimulus is first applied, but then quickly reduce their rate of firing if the stimulus is maintained. This produces fast-adapting sensations.

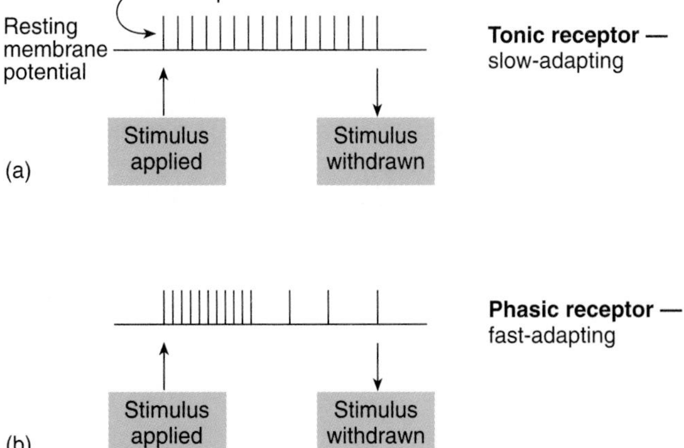

Table 18.1

Classification of Receptors Based on Their Normal (or "Adequate") Stimulus

Receptor	Normal Stimulus	Mechanisms	Examples
Mechanoreceptors	Mechanical force	Deforms cell membrane of sensory dendrites or deforms hair cells that activate sensory nerve endings	Cutaneous touch and pressure receptors; vestibular apparatus and spiral organ in cochlea of ear
Pain receptors	Tissue damage	Damaged tissues release chemicals that excite sensory endings	Cutaneous pain receptors
Chemoreceptors	Dissolved chemicals	Chemical interaction affects ionic permeability of sensory cells	Smell and taste (exteroreceptors); osmoreceptors and carotid body chemoreceptors (interoreceptors)
Photoreceptors	Light	Photochemical reaction affects ionic permeability of receptor cell	Rods and cones in retina of eye

Phasic receptors alert us to changes in sensory stimuli and are in part responsible for the fact that we can cease paying attention to constant stimuli. This ability is called **sensory adaptation.** Odor, touch, and temperature, for example, adapt rapidly; bathwater feels hotter when we first enter it. Sensations of pain, by contrast, adapt little if at all.

Law of Specific Nerve Energies

Stimulation of a sensory nerve fiber produces only one sensation—touch, cold, pain, and so on. According to the **law of specific nerve energies,** the sensation characteristic of each sensory neuron is that produced by its normal stimulus, or *adequate stimulus* (table 18.1). Also, although a variety of different stimuli may activate a receptor, the adequate stimulus requires the least amount of energy to do so. The adequate stimulus for the photoreceptors of the eye, for example, is light, where a single photon can have a measurable effect. If these receptors are stimulated by some other means—such as by the high pressure produced by a punch to the eye—a flash of light (the adequate stimulus) may be perceived.

The effect of *paradoxical cold* provides another example of the law of specific nerve energies. When the tip of a cold metal rod is touched to the skin, the perception of cold gradually disappears as the rod warms to body temperature. Then, when the tip of a rod heated to 45° C is applied to the same spot, the sensation of cold is perceived once again. This paradoxical cold is produced because the heat slightly damages receptor endings, and by this means produces an "injury current" that stimulates the receptor.

Regardless of how a sensory neuron is stimulated, therefore, only one sensory modality will be perceived. This specificity is due to the synaptic pathways within the brain that are activated by the sensory neuron. The ability of receptors to function as sensory filters so that they are stimulated by only one type of stimulus (the adequate stimulus) allows the brain to perceive the stimulus accurately under normal conditions.

Generator (Receptor) Potential

The electrical behavior of sensory nerve endings is similar to that of the dendrites of other neurons. In response to an environmental stimulus, the sensory endings produce local graded changes in the membrane potential. In most cases, these potential changes are depolarizations analogous to the excitatory postsynaptic potentials (EPSPs) described in chapter 14. In the sensory endings, however, these potential changes in response to environmental stimulation are called **receptor,** or **generator, potentials** because they serve to generate action potentials in response to the sensory stimulation. Since sensory neurons are pseudounipolar (chapter 14), the action potentials produced in response to the generator potential are conducted continuously from the periphery into the CNS.

The *lamellated,* or *pacinian, corpuscle,* a cutaneous receptor for pressure (see fig. 18.1), is an example of sensory transduction. When a light touch is applied to the receptor, a small depolarization (the generator potential) is produced. Increasing the pressure on the lamellated corpuscle increases the magnitude of the generator potential until it reaches the threshold depolarization required to produce an action potential (fig. 18.3). The lamellated corpuscle, however, is a phasic receptor; if the pressure is maintained, the size of the generator potential produced quickly diminishes. It is interesting to note that this phasic response is a result of the onionlike covering around the dendritic nerve ending; if the layers are peeled off and the nerve ending is stimulated directly, it will respond in a tonic fashion.

When a tonic receptor is stimulated, the generator potential it produces is proportional to the intensity of the stimulus. After a threshold depolarization is produced, increases in the amplitude of the generator potential result in increases in

pacinian corpuscle: from Filippo Pacini, Italian anatomist, 1812–93
corpuscle: L. *corpusculum,* diminutive of *corpus,* body

Figure 18.3

The receptor (generator) potential.
Sensory stimuli result in the production of local graded potential changes known as receptor, or generator, potentials (numbers 1–4). If the receptor potential reaches a threshold value of depolarization, it generates action potentials (number 5) in the sensory neuron.

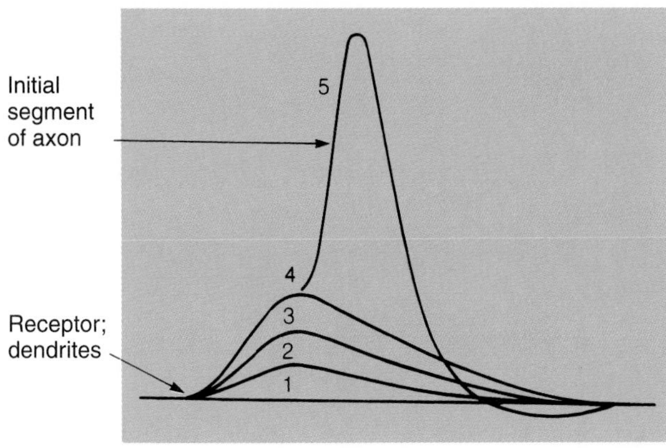

Figure 18.4

The response of tonic receptors to stimuli.
Three successive stimuli of increasing strengths are delivered to a receptor. The increasing amplitude of the generator potential results in increases in the frequency of action potentials, which last as long as the stimulus is maintained.

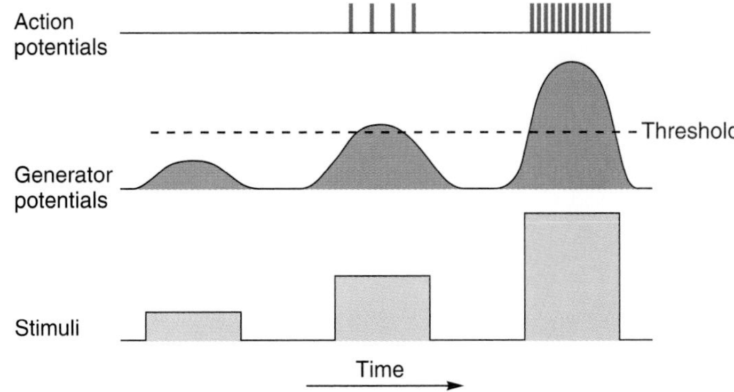

the *frequency* with which action potentials are produced (fig. 18.4). In this way, the frequency of action potentials that are conducted into the central nervous system serves as the code for the strength of the stimulus. As described in chapter 14, this frequency code is needed because the amplitude of action potentials is constant (all-or-none). Acting through changes in action potential frequency, tonic receptors thus provide information about the relative intensity of a stimulus.

Somatic Senses

Somatic senses include cutaneous receptors and proprioceptors. There are several types of sensory receptors in the skin, each of which is specialized to be maximally sensitive to one modality of sensation. Proprioceptors provide sensory information about muscles, joints, and tendons.

Cutaneous Receptors

The cutaneous sensations of touch, pressure, hot and cold, and pain are mediated by the dendritic nerve endings of different sensory neurons. The receptors for heat, cold, and pain are simply the naked endings of sensory neurons. Sensations of touch are mediated by naked dendritic endings surrounding hair follicles and by expanded dendritic endings, called Ruffini endings and Merkel's discs. The sensations of touch and pressure are also mediated by dendrites that are encapsulated within various structures (table 18.2); these include corpuscles of touch (Meissner's corpuscles) Krause's end bulbs, and lamellated corpuscles. In lamellated corpuscles, for example, the dendritic endings are encased within 30 to 50 onionlike layers of connective tissue (fig. 18.5). These layers absorb some of the pressure when a stimulus is maintained, and thus help to accentuate the phasic response of this receptor. The encapsulated touch receptors thus adapt rapidly, in contrast to the more slowly adapting Ruffini endings and Merkel's discs.

There are far more free dendritic endings that respond to cold than to heat. The *cold receptors* are located in the upper region of the dermis, just below the epidermis. These receptors are stimulated by cooling and inhibited by warming. The *warm receptors* are located somewhat deeper in the dermis and are excited by warming and inhibited by cooling. Nociceptors are

Table 18.2

Cutaneous Receptors

Anatomical Class (Structure)	Type	Sensation
Encapsulated (dendrites within associated structures)	Lamellated (pacinian) corpuscle; corpuscles of touch; Krause's end bulbs; Ruffini's endings	All serve touch and pressure
Free nerve endings	—	Touch, pressure, heat, cold, pain

somatic: Gk. *somatikos*, body
Ruffini endings: from Angelo Ruffini, Italian anatomist, 1864–1929
Merkel's discs: from Friedrich S. Merkel, German anatomist and physiologist, 1845–1911

Meissner's corpuscles: From George Meissner, German histologist, 1829–1905
Krause's end bulbs: From Wilhelm J. F. Krause, German anatomist, 1833–1910

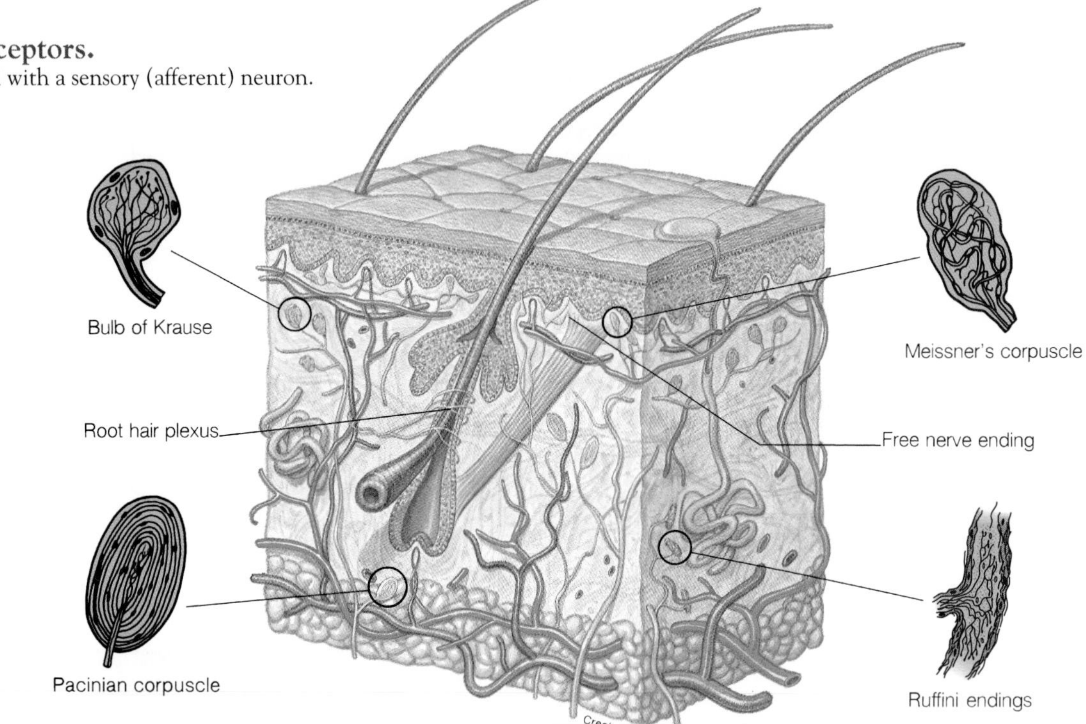

Figure 18.5

The cutaneous sensory receptors.
Each of these structures is associated with a sensory (afferent) neuron.

also free sensory nerve endings of either myelinated or unmyelinated fibers. The initial sharp sensation of pain, as from a pinprick, is transmitted by rapidly conducting myelinated axons, whereas a dull, persistent ache is transmitted by slower conducting unmyelinated axons. These afferent neurons synapse in the spinal cord using substance P (an 11–amino acid polypeptide) and glutamate as neurotransmitters.

Hot temperatures produce sensations of pain through the action of a particular membrane protein in sensory dendrites. This protein, called a *capsaicin receptor*, serves as both an ion channel and a receptor for capsaicin—the molecule in chili peppers that causes sensation of heat and pain. In response to a noxiously high temperature, or to capsaicin in chili peppers, these ion channels open. This allows Ca^{2+} and Na^{+} to diffuse into the neuron, producing depolarization and resulting action potentials that are transmitted to the CNS and perceived as heat and pain.

The sensation of pain can be clinically classified as *somatic pain* or *visceral pain*. Stimulation of the cutaneous pain receptors results in the perception of superficial somatic pain. Deep somatic pain comes from stimulation of receptors in skeletal muscles, joints, and tendons. Stimulation of the receptors within the viscera causes the perception of visceral pain. The sensation of pain from certain visceral organs, however, may not be perceived as arising from those organs but from other somatic locations. This phenomenon is known as **referred pain** (fig. 18.6). The pain of a heart attack, for example, may be perceived subcutaneously over the heart and down the medial side of the left arm. Ulcers of the stomach may cause pain that is perceived as coming from the upper central (epigastric) region of the trunk. Pain from problems of the liver or gallbladder may be perceived as localized visceral pain or as referred pain arising from the right neck and shoulder regions.

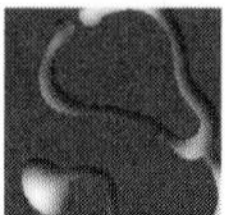

The phenomenon of the *phantom limb* was first described by a neurologist during the Civil War. In this account, a veteran with amputated legs asked for someone to massage his cramped leg muscle. It is now known that this phenomenon is common in amputees, who may experience complete sensations from the missing limbs. They perceive the phantom as being very real, especially with their eyes closed, and they sense it moving in accordance with the way the limb would naturally move if it were real. These sensations are sometimes useful (for example, in fitting prostheses into which the phantom has seemingly entered). However, pain in the phantom is experienced by 70% of amputees, and the pain can be severe and persistent.

One explanation for phantom limbs is that the nerves remaining in the stump can grow into nodules called neuromas, and these may generate nerve impulses that are transmitted to the brain and interpreted as arising from the missing limb. However, phantom limbs may occur in cases where the limb has not been amputated, but the nerves that normally enter from the limb have been severed. Or it may occur in individuals with spinal cord injuries above the level of the limb, so that sensations from the limb do not enter the brain. In these cases, the phantom limb phenomenon requires a different explanation. Current theories propose that the phantom may be produced by brain reorganization caused by the absence of the sensations that would normally arise from the missing limb.

Figure 18.6

Sites of referred pain.
Referred pain is perceived cutaneously but actually originates from specific visceral organs.

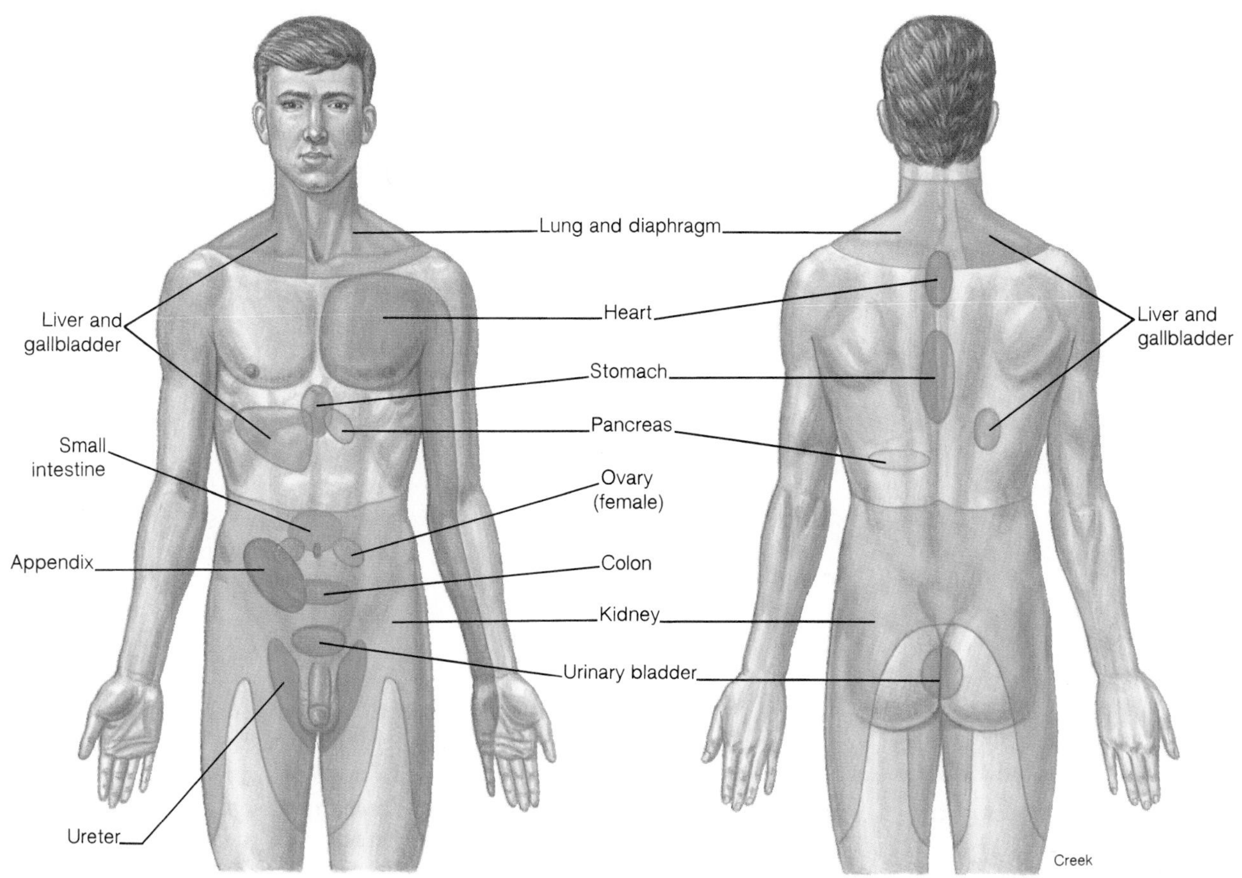

Receptive Fields and Sensory Acuity

The **receptive field** of a neuron serving cutaneous sensation is the area of skin whose stimulation results in changes in the firing rate of the neuron. Changes in the firing rate of primary sensory neurons affect the firing of second- and third-order neurons, which in turn affects the firing of those neurons in the postcentral gyrus of the cerebral cortex that receive input from the third-order neurons. Indirectly, therefore, neurons in the postcentral gyrus can be said to have receptive fields in the skin.

The area of each receptive field in the skin varies inversely with the density of receptors in the region. In the back and legs, where a large area of skin is served by relatively few sensory endings, the receptive field of each neuron is correspondingly large. In the fingertips, where a large number of cutaneous receptors serve a small area of skin, the receptive field of each sensory neuron is correspondingly small.

Two-Point Touch Threshold

The approximate size of the receptive fields serving light touch can be measured by the *two-point touch threshold test*. In this procedure, two points of a pair of calipers are lightly touched to the skin at the same time. If the distance between the points is sufficiently great, each point will stimulate a different receptive field and a different sensory neuron—two separate points of touch will thus be felt. If the distance is sufficiently small, both points will touch the receptive field of only one sensory neuron, and only one point of touch will be felt (fig. 18.7).

The **two-point touch threshold,** which is the minimum distance that can be distinguished between two points of touch, is a measure of the distance between receptive fields. If the distance between the two points of the calipers is less than this minimum distance, only one "blurred" point of touch can be felt. The two-point touch threshold is thus an indication of tactile *acuity*, or the sharpness of touch perception.

The tactile acuity of the fingertips is exploited in the reading of *braille*. Braille symbols are formed by raised dots on the page that are separated from each other by 2.5 mm, which is slightly greater than the two-point touch threshold in the fingertips (table 18.3). Experienced braille readers can scan words at about the same speed that a sighted person can read aloud—a rate of about 100 words per minute.

---

acuity: L. *acuo*, sharpen
braille: from Louis Braille, French teacher of the blind, 1809–52

Figure 18.7

The two-point touch threshold test.
If each point touches the receptive fields of different sensory neurons, two separate points of touch will be felt. If both caliper points touch the receptive field of one sensory neuron, only one point of touch will be felt.

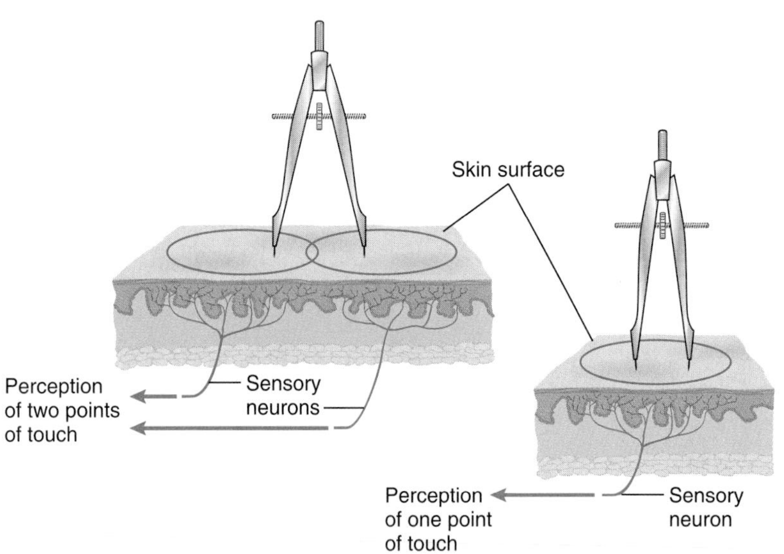

Lateral Inhibition

When a blunt object touches the skin, a number of receptive fields are stimulated—some more than others. The receptive fields in the central areas where the touch is strongest will be stimulated more than those in the neighboring fields where the touch is lighter. Stimulation will gradually diminish from the point of greatest contact, without a clear, sharp boundary. What we perceive, however, is not the fuzzy sensation that might be predicted. Instead, only a single touch with well-defined borders is felt. This sharpening of sensation is due to a process called **lateral inhibition.**

Lateral inhibition and the resultant sharpening of sensation occur within the central nervous system. Those sensory neurons whose receptive fields are stimulated most strongly inhibit—via interneurons that pass "laterally" within the CNS—sensory neurons that serve neighboring receptive fields. Lateral inhibition similarly plays a prominent role in the ability of the ears and brain to discriminate sounds of different pitch.

Table 18.3

The Two-Point Touch Threshold for Different Regions of the Body

Body Region	Two-Point Touch Threshold (mm)
Big toe	10
Sole of foot	22
Calf	48
Thigh	46
Back	42
Abdomen	36
Upper arm	47
Forehead	18
Palm of hand	13
Thumb	3
First finger	2

Source: From S. Weinstein and D. R. Kenshalo, the editors. *The Skin Senses*, © 1968. Courtesy of Charles C. Thomas, Publisher, Ltd., Springfield, Illinois.

Proprioceptors

Proprioceptors are located in and around joints, within skeletal muscle, and between tendons and muscles. Some of the sensory impulses from proprioceptors reach the level of consciousness as the **kinesthetic sense,** by which the position of the body parts is perceived. Other proprioceptor information, not consciously interpreted, is used to adjust the intensity and timing of muscle contractions that permit coordinated movements. With the kinesthetic sense, the position and movement of the limbs can be determined without visual sensations, such as when dressing or walking in the dark.

High-speed transmission is a vital characteristic of the kinesthetic sense, since rapid feedback to various body parts is essential for quick, smooth, coordinated body movements. There are three types of proprioceptors: joint kinesthetic receptors, neuromuscular spindles, and neurotendinous receptors.

Joint Kinesthetic Receptors

Joint kinesthetic receptors are located in the connective tissue capsule in synovial joints, where they are stimulated by changes in position caused by movement at the joints.

Neuromuscular Spindles

Neuromuscular spindles are located in skeletal muscle, particularly in the muscles of the extremities. Each neuromuscular spindle (*spindle apparatus*) contains several thin muscle cells, called **intrafusal fibers,** packaged within a connective

intrafusal: L. *intra*, within; *fusus*, spindle

Figure 18.8

The location and structure of a muscle spindle.
A muscle spindle is located within a skeletal muscle, and runs parallel to the muscle fibers. In the close-up view, the structure and innervation of a muscle spindle can be seen.

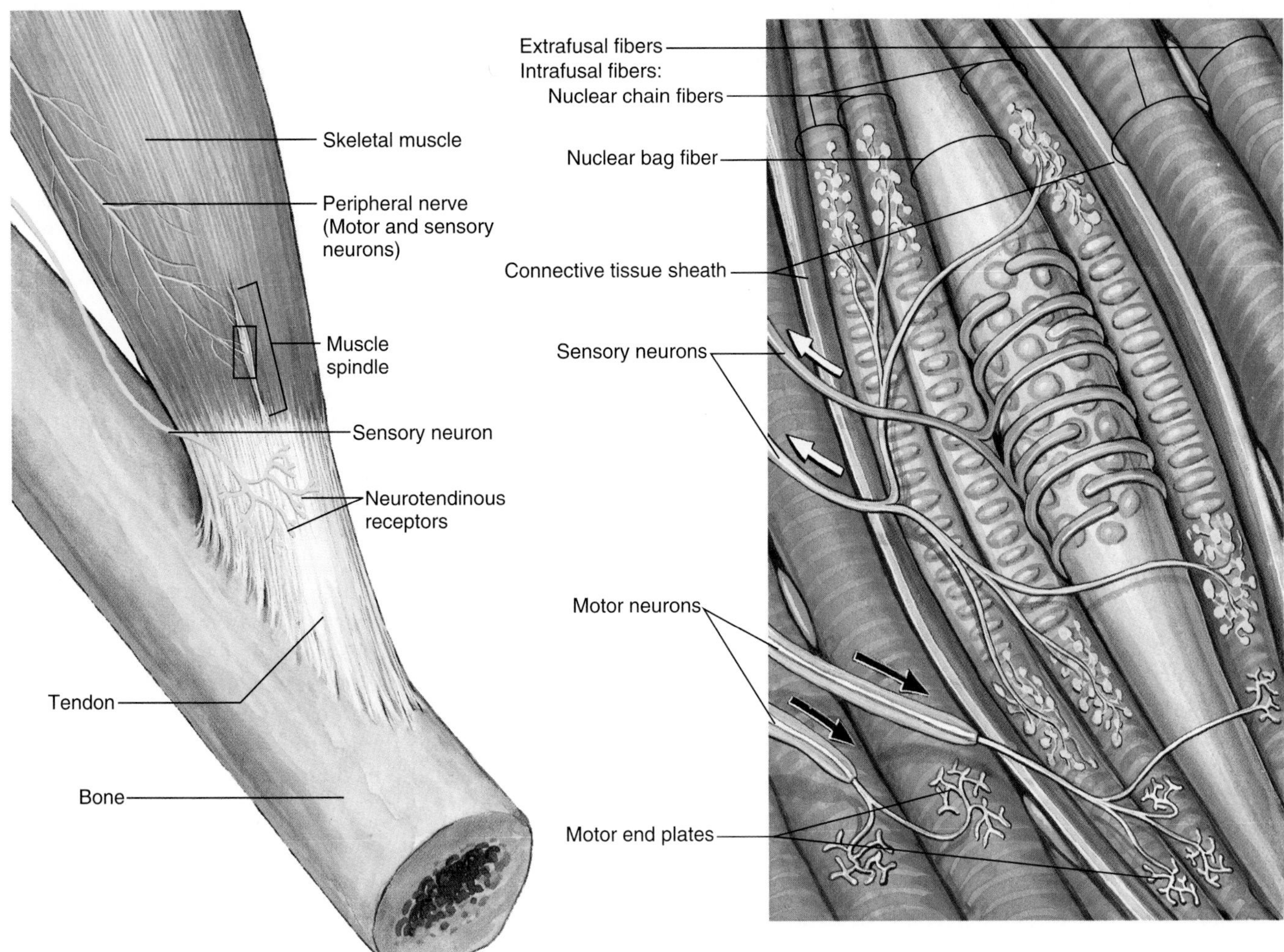

tissue sheath. Like the stronger and more numerous "ordinary" muscle fibers outside the spindles—the **extrafusal fibers**—the spindles insert into tendons on each end of the muscle. Spindles are therefore said to be parallel with the extrafusal fibers.

The extrafusal fibers contain myofibrils along their entire length, but the intrafusal fibers have no contractile apparatus in their central regions. The central, noncontracting part of an intrafusal fiber contains nuclei. There are two types of intrafusal fibers. One type, the *nuclear bag fibers*, have their nuclei arranged in a loose aggregate in the central regions of the fibers. The other type of intrafusal fibers, called *nuclear chain fibers* have their nuclei arranged in rows. Two types of sensory neurons serve these intrafusal fibers. **Primary,** or **annulospiral, sensory endings** wrap around the central regions of the nuclear bag and chain fibers (fig. 18.8) and **secondary,** or **flower-spray, endings** are located over the contracting poles of the nuclear chain fibers.

Since the spindles are arranged in parallel with the extrafusal muscle fibers, stretching a muscle causes its spindles to stretch. This stimulates both the primary and secondary sensory endings. The spindle thus serves as a length detector because the frequency of impulses produced in the primary and secondary endings is proportional to the length of the muscle. The primary endings, however, are most stimulated at the onset of stretch, whereas the secondary endings respond in a more tonic (sustained) fashion as the stretch is maintained. Sudden, rapid stretching of a muscle activates both types of sensory endings, and is thus a more powerful stimulus for the spindles than a slower, more gradual stretching that has less of an effect on the primary sensory endings. Since activation of the sensory endings in neuromuscular spindles produces a reflex contraction, the force of this reflex contraction is greater in response to rapid stretch than to gradual stretch.

In the spinal cord, two types of **lower motor neurons** innervate skeletal muscles. The motor neurons that innervate the extrafusal muscle fibers are called **alpha motoneurons;** those that innervate the intrafusal fibers are called **gamma motoneurons** (fig. 18.8). The alpha motoneurons are faster conducting (60–90 meters per second) than the thinner gamma motoneurons (10–40 meters per second). Since only the extrafusal muscle fibers are sufficiently strong and numerous to cause a muscle to shorten, only stimulation by the alpha motoneurons can cause muscle contraction that results in skeletal movements. These are the motor nerve fibers involved in the knee-jerk reflex and other stretch reflexes (fig. 18.9).

The intrafusal fibers of the muscle spindle are stimulated to contract by gamma motoneurons, which represent one-third of all motor fibers in spinal nerves. However, because the intrafusal fibers are too few in number and their contraction too weak to cause a muscle to shorten, stimulation by gamma motoneurons results only in isometric contraction of the spindles. Since myofibrils are present in the poles but absent in the central regions of intrafusal fibers, the more distensible central region of the intrafusal fiber is pulled toward the ends in response to stimulation by gamma motoneurons. As a result, the spindle is tightened. This effect of gamma motoneurons, which is sometimes termed *active stretch* of the spindles, serves to increase the sensitivity of the spindles when the entire muscle is passively stretched by external forces. The activation of gamma motoneurons thus enhances the stretch reflex and is an important factor in the voluntary control of skeletal movements.

Under normal conditions, the activity of gamma motoneurons is maintained at the level needed to keep the spindles under proper tension while the muscles are relaxed. Undue relaxation of the muscles is prevented by stretch and activation of the spindles, which in turn elicits a reflex contraction. This mechanism produces a normal resting muscle length and state of tension, or *muscle tone*.

Higher motor neurons—neurons in the brain that contribute fibers to descending motor tracts—usually stimulate both alpha and gamma motoneurons simultaneously. Such stimulation is known as *coactivation*. Stimulation of alpha motoneurons results in muscle contraction and shortening; stimulation of gamma motoneurons stimulates contraction of the intrafusal fibers and thus "takes out the slack" that would otherwise be present in the spindles as the muscles shorten. In this way, the spindles remain under tension and provide information about the length of the muscle even while the muscle is shortening.

Neurotendinous Receptors

Neurotendinous receptors (also called *Golgi tendon organs*) are located where muscles attach to tendons. They continuously monitor the tension in the tendons produced by muscle contraction or passive stretching of a muscle. Sensory neurons from these receptors synapse with association neurons in the spinal cord; these association neurons, in turn, have *inhibitory synapses* (via IPSPs and postsynaptic inhibition—chapter 14) with motor neurons that innervate the muscle (fig. 18.10). This helps to prevent excessive muscle contractions or excessive passive muscle stretching. Indeed, if a muscle is stretched extensively it will actually relax as a result of the inhibitory effects of the neurotendinous receptors.

Rapid stretching of skeletal muscles produces very forceful muscle contractions as a result of the activation of primary and secondary endings in the muscle spindles and the monosynaptic stretch reflex. This can result in painful muscle spasms, as may occur, for example, when muscles are forcefully pulled in the process of setting broken bones. Painful muscle spasms may be avoided in physical exercise by stretching slowly and thereby stimulating mainly the secondary endings in the muscle spindles. A slower rate of stretch also allows time for the inhibitory neurotendinous receptor reflex to occur and promote muscle relaxation.

Neural Pathways for Somatic Sensations

The conduction pathways for the somatic senses are shown in figure 18.11. Sensory information from proprioceptors and pressure receptors are carried by large, myelinated nerve fibers that ascend in the dorsal columns of the spinal cord on the same (ipsilateral) side. These fibers do not synapse until they reach the medulla oblongata of the brain stem; hence, fibers that carry these sensations from the feet are remarkably long. After the fibers synapse in the medulla oblongata with other second-order sensory neurons, information in the latter neurons crosses over to the contralateral side as it ascends via a fiber tract, called the **medial lemniscus** (*lem-nis'kus*), to the thalamus. Third-order sensory neurons in the thalamus that receive this input in turn project to the **postcentral gyrus** (the precentral gyrus, as described in chapter 15).

Sensations of heat, cold, and pain are carried mostly by thin, unmyelinated sensory neurons into the spinal cord. These synapse with second-order association neurons within the spinal cord that cross over to the contralateral side and ascend to the brain in the **lateral spinothalamic tract.** Fibers that mediate touch and pressure ascend in the **anterior spinothalamic tract.** Fibers of both spinothalamic tracts synapse with third-order neurons in the thalamus, which in turn project to the postcentral gyrus. Notice that somatesthetic information is always carried to the postcentral gyrus in third-order neurons. Also, because of crossing-over, somatesthetic information from each side of the body is projected to the postcentral gyrus of the contralateral cerebral hemisphere.

Since all somatic information from the same area of the body projects to the same area of the postcentral gyrus, a "map" of the body can be drawn on the postcentral gyrus to represent sensory projection points (see fig. 15.7). This map is distorted, however, because it shows larger areas of cerebral cortex devoted to sensation in the face and hands than in other areas in the body. This disproportionately larger area of the cerebral cortex devoted to the face and hands reflects the fact that the density of sensory receptors is higher in these regions.

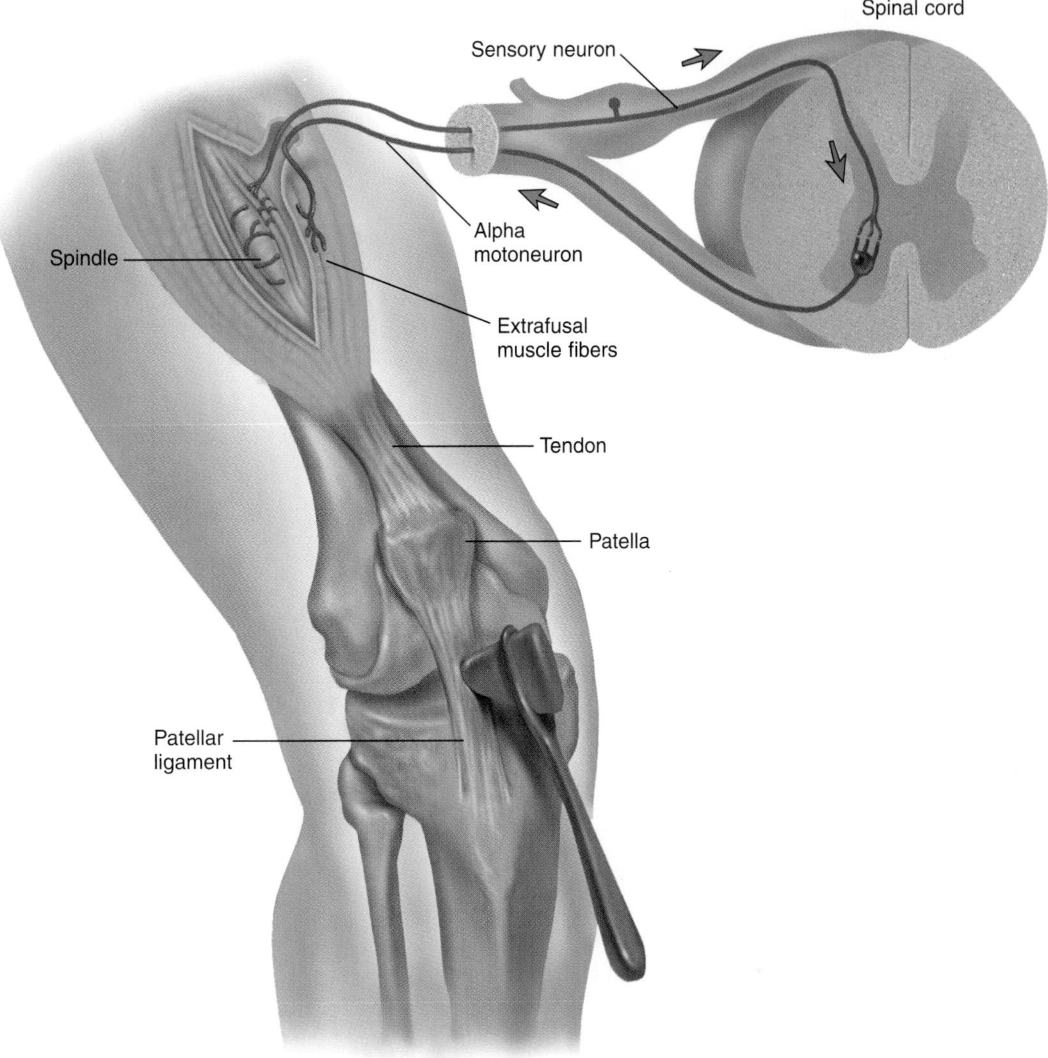

Figure 18.9

The knee-jerk reflex.
This is an example of a monosynaptic stretch reflex.

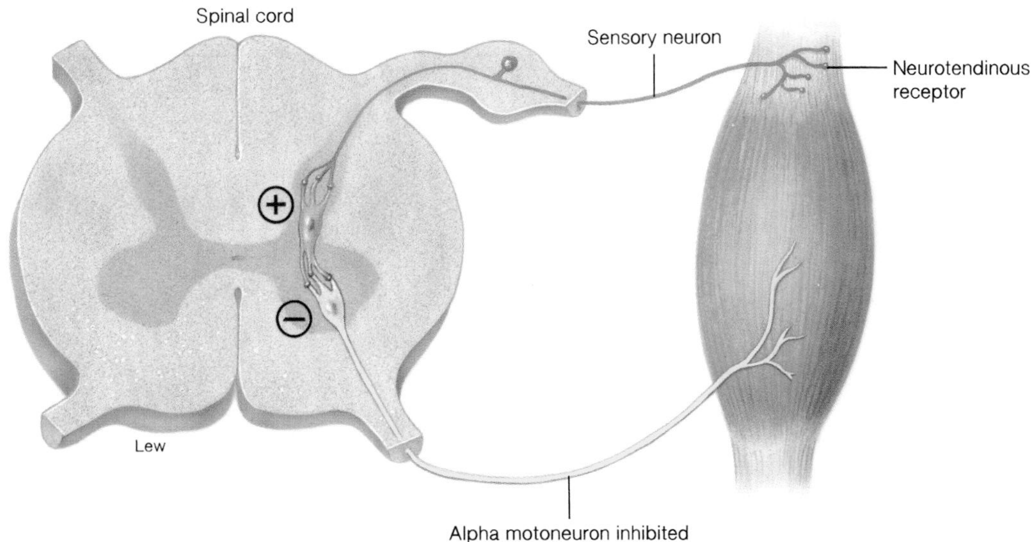

Figure 18.10

The action of the neurotendinous receptor.
An increase in muscle tension stimulates the activity of sensory nerve endings in the neurotendinous receptor (Golgi tendon organ). This sensory input stimulates an associ ynaptic reflex.

Figure 18.11

Somatic sensory pathways.
These neural pathways lead from the cutaneous receptors and proprioceptors into the postcentral gyrus in the cerebral cortex.

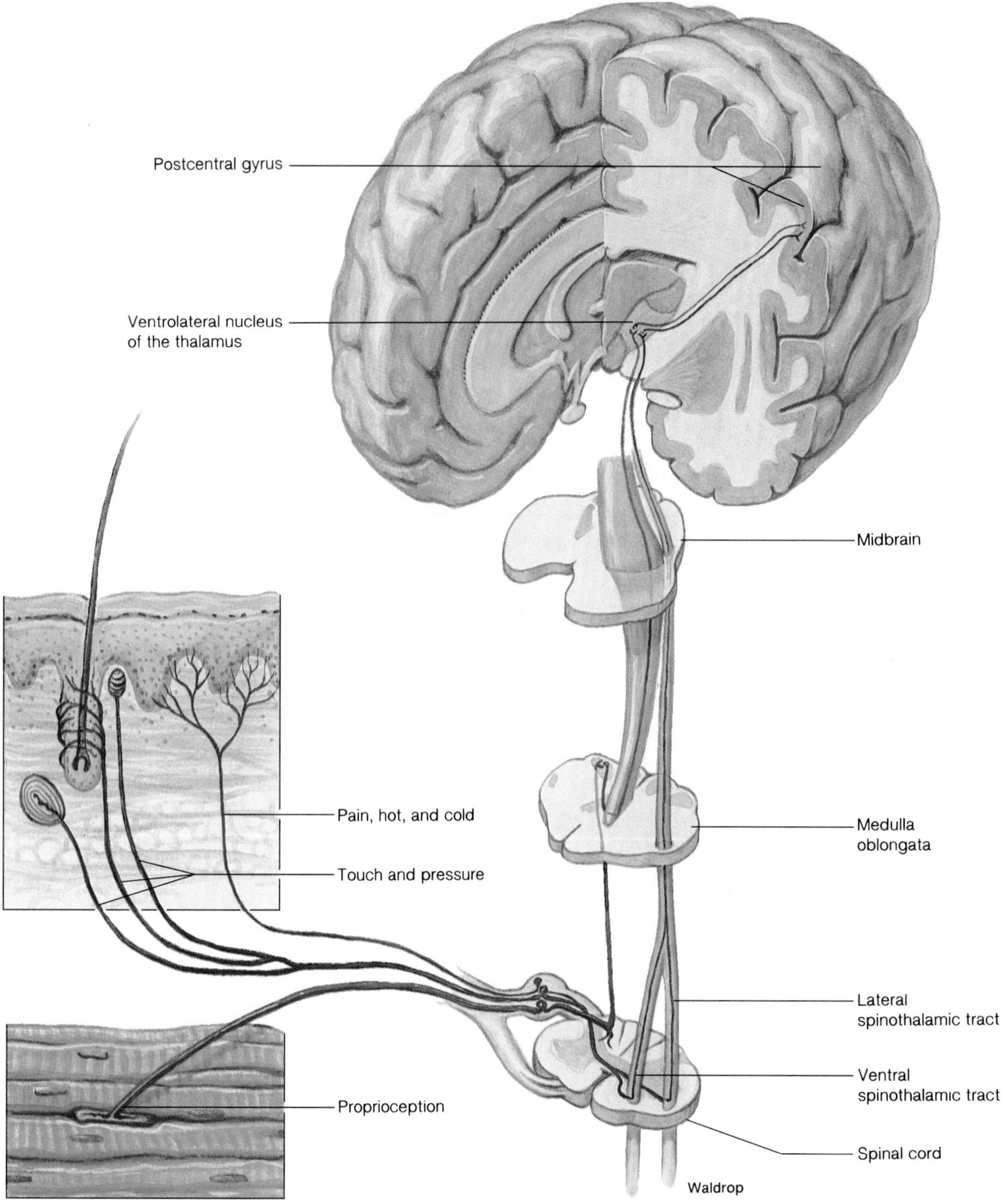

Taste and Smell

The receptors for taste and smell respond to molecules that are dissolved in fluid; hence, they are classified as chemoreceptors. Although there are only four basic modalities of taste, they combine in various ways and are influenced by the sense of smell, thus permitting a wide variety of different sensory experiences.

Chemoreceptors that respond to chemical changes in the internal environment are called **interoceptors,** or **visceroceptors** (*vis″er-o-sep′torz*); those that respond to chemical changes in the external environment are **exteroceptors** (*ek″ste-ro-sep′torz*). Included in the latter category are *taste* (*gustatory*) *receptors*, which respond to chemicals dissolved in food or drink, and *smell* (*olfactory*) *receptors*, which respond to gaseous molecules in the air. This distinction is somewhat arbitrary, however, because odorant molecules in air must first dissolve in fluid within the olfactory mucosa before the sense of smell can be stimulated. Also, the sense of olfaction strongly influences the sense of taste, as can easily be verified by eating an onion (or almost anything else) with the nostrils pinched together.

Taste

Gustation, the sense of taste, is evoked by receptors that consist of specialized epithelial cells clustered together in **taste buds.** Taste buds are most numerous on the surface of the tongue but are also present on the soft palate and on the walls of the oropharynx. The cylindrical taste bud is composed of numerous sensory *gustatory cells* that are encapsulated by supporting cells (fig. 18.12). Projecting from the tip of each gustatory cell is a long microvillus called a *gustatory hair* that projects to the surface through an opening in the

gustatory: L. *gustare*, to taste

Figure 18.12

Taste buds.

(*a*) Three types of papillae (labeled in boldface type) cover the surface of the tongue. (*b*) A magnified view of the papillae and the associated taste buds. (*c*) The structure of a taste bud.

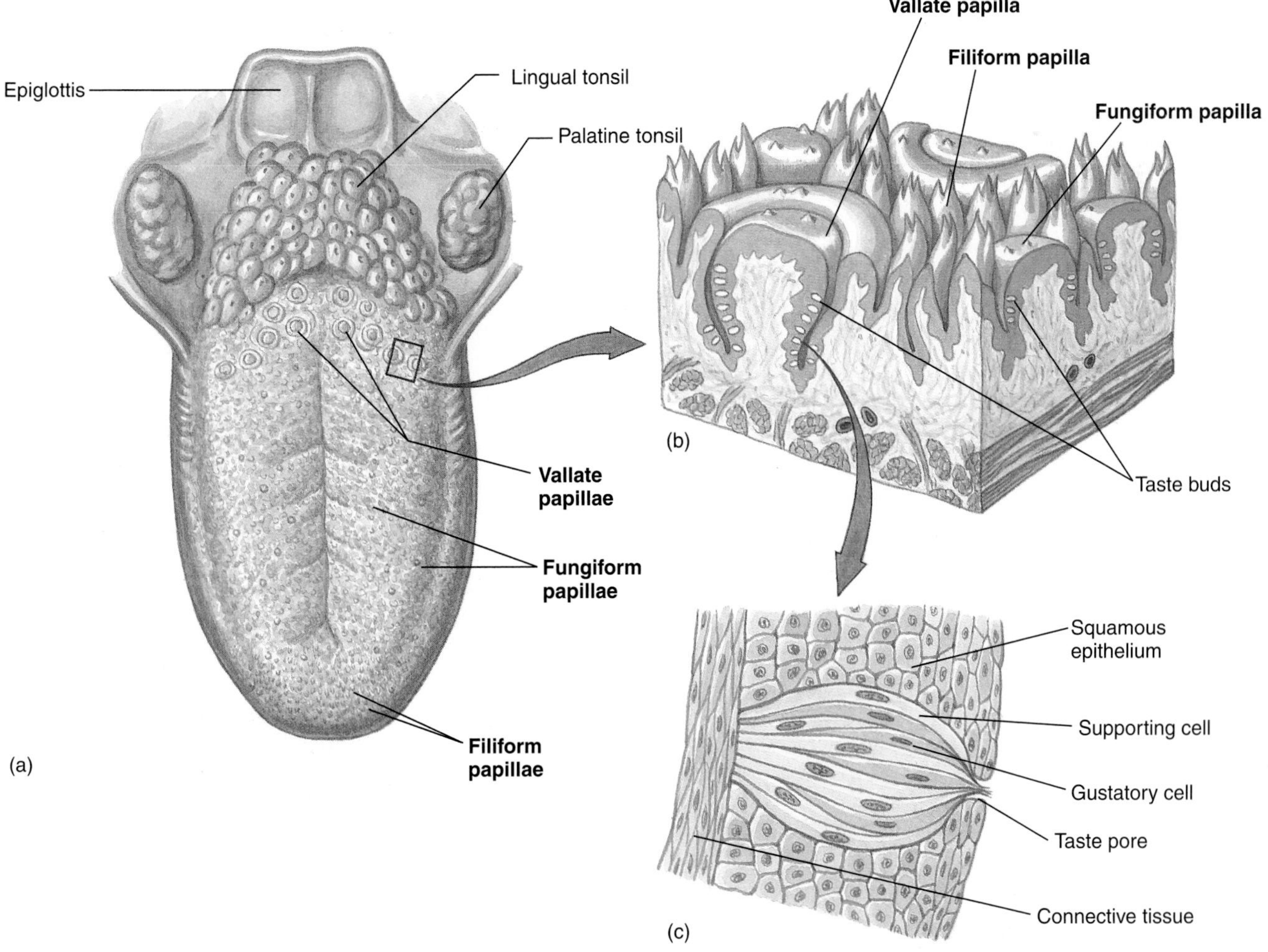

Figure 18.13

Patterns of taste receptor distribution on the surface of the tongue.
This diagram indicates the tongue regions that are maximally sensitive to different tastes.

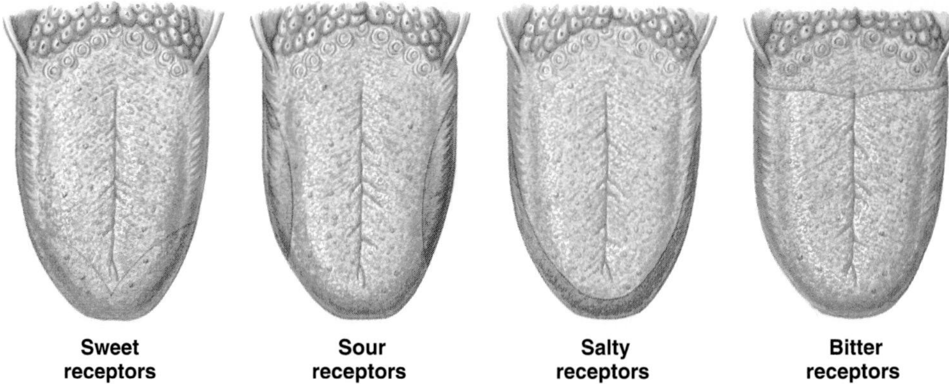

taste bud called the *taste pore*. The gustatory hairs make up the sensitive portion of the receptor cells. Saliva provides the moistened environment that is needed for a chemical stimulus to activate the gustatory cells.

Taste buds are elevated by surrounding connective tissue and epithelium to form **papillae** (*pă-pil′e*) (fig. 18.12). Three types of papillae can be identified: *vallate*, *fungiform*, and *filiform* (see chapter 26). Taste buds are found only in the vallate and fungiform papillae. The filiform papillae, although the most numerous, are not involved in the perception of taste. Taste buds in the posterior third of the tongue are innervated by the *glossopharyngeal nerve (IX)*; those in the anterior two-thirds of the tongue are innervated by the chorda tympani branch of the *facial nerve (VII)*.

There are four basic modalities of taste, each of which is sensed most acutely in a particular region of the tongue. These are *sweet* (tip of the tongue), *sour* (sides of the tongue), *bitter* (back of the tongue), and *salty* (over most of the tongue, but concentrated on the sides). This distribution is illustrated in figure 18.13. All the different tastes that we can perceive are combinations of these four, together with nuances provided by the sense of smell. There is also some evidence that a fifth taste modality—for water—also may be present in humans.

The salty taste of food is due to the presence of sodium ions (Na^+) along with some other cations, which activate specific receptor cells for the salty taste. Different substances taste salty to the degree that they activate these particular receptor cells. The Na^+ passes into the sensitive receptor cells through channels in the apical membranes. This depolarizes the cells, causing them to release their transmitter. The anion associated with the Na^+, however, modifies the perceived saltiness to a surprising degree: NaCl tastes much saltier than other sodium salts (such as sodium acetate). There is evidence to suggest that the anions can pass through the tight junctions between the receptor cells, and that the Cl^- anion passes through this barrier more readily than the other anions. This is presumably related to the ability of Cl^- to impart a saltier taste to the Na^+ than the other anions.

papilla: L. *papilla*, nipple

Sour taste, like salty taste, is produced by ion movement through membrane channels. Sour taste, however, is due to the presence of hydrogen ions (H^+); all acids therefore taste sour. In contrast to the salty and sour tastes, the sweet and bitter tastes are produced by interaction of taste molecules with specific membrane receptor proteins.

Most organic molecules, particularly sugars, taste sweet to varying degrees. Bitter taste is evoked by quinine and seemingly unrelated molecules. It is the most acute taste sensation and is generally associated with toxic molecules (although not all toxins taste bitter). Both sweet and bitter sensations are mediated by receptors that are coupled to G-proteins (chapter 14). The particular type of G-protein involved in taste has recently been identified and termed **gustducin.** This term is used to emphasize the similarity to a related group of G-proteins, of a type called *transducin*, associated with the photoreceptors in the eye (discussed in a later section). Dissociation of the gustducin G-protein subunit activates second-messenger systems, leading to depolarization of the receptor cell. The stimulated receptor cell, in turn, activates an associated sensory neuron that transmits impulses to the brain, where they are interpreted as the corresponding taste perception.

Smell

The receptors responsible for **olfaction,** the sense of smell, consist of the dendritic endings of several million bipolar sensory neurons of the *olfactory nerve (I)*, located within a pseudostratified epithelium in the superior portion of the nasal cavity. Unique among the neurons of an adult, these sensory neurons divide mitotically and replace themselves every 1 to 2 months. Each bipolar sensory neuron has one unmyelinated axon that projects up into the olfactory bulb of the cerebrum, where it synapses with second-order neurons, and one dendrite that projects into the nasal cavity, where it terminates in a knob containing cilia (figs. 18.14 and 18.15). Therefore, unlike other sensory modalities that are relayed to the cerebrum from the thalamus, the sense of smell is transmitted directly to the cerebral cortex. The olfactory bulb is part of the limbic system (see fig. 17.10), which has an important role in generating emotions and in memory. The human amygdala, in particular, has been implicated in the emotional responses to olfactory stimulation. Perhaps this explains why the smell of a

Figure 18.14

The olfactory neuron pathway.
(*a*) The olfactory epithelium contains receptor neurons that synapse with neurons in the olfactory bulb of the brain. (*b*) Olfaction is the only sense that goes directly to the cerebral cortex without first synapsing the thalamus.

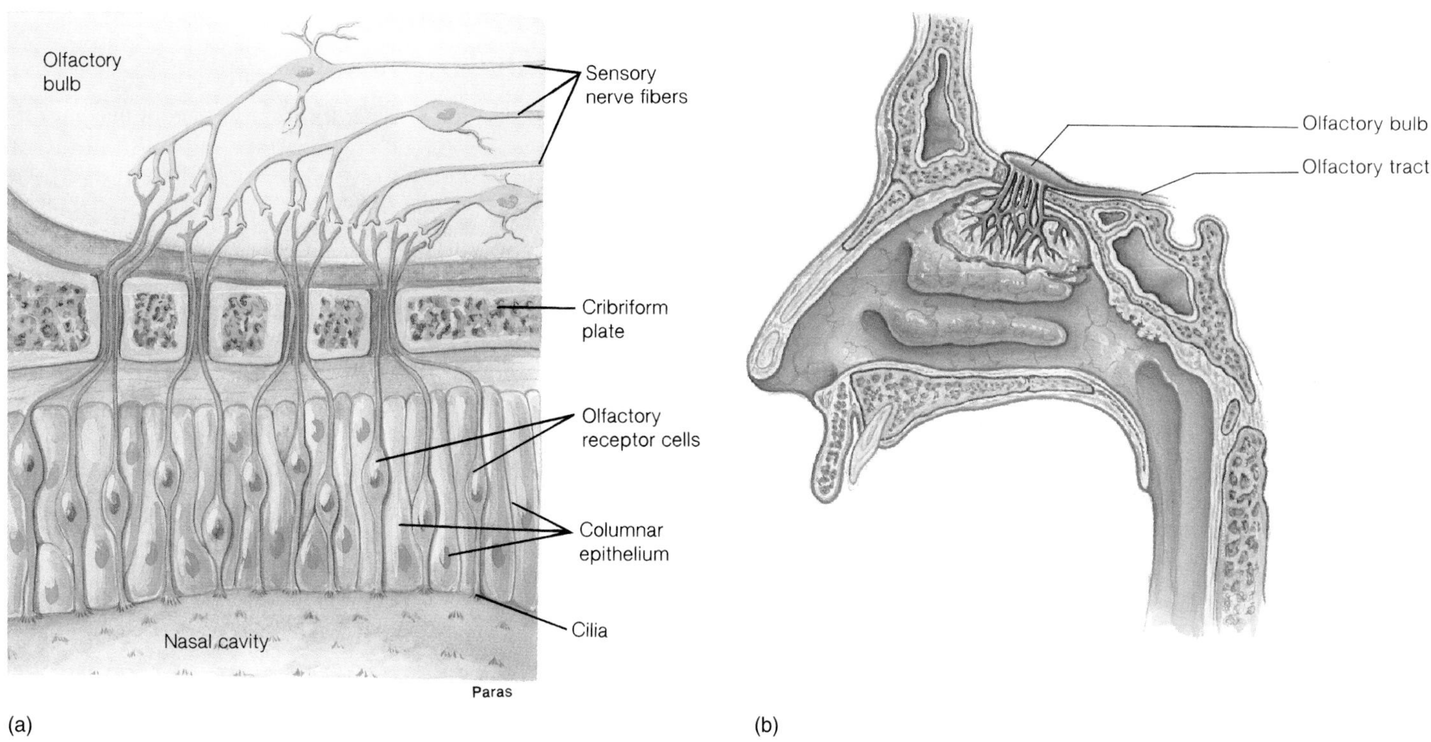

particular odor, more powerfully than other sensations, can evoke emotionally charged memories.

The molecular basis of olfaction is complex. At least in some cases, odorant molecules bind to receptors and act through G-proteins to increase the cyclic AMP within the cell. This, in turn, opens membrane channels and causes the depolarization of the generator potential, which then stimulates the production of action potentials. Up to 50 G-proteins may be associated with a single receptor protein. Dissociation of these G-proteins releases many G-protein subunits, thereby amplifying the effect many times. This amplification could account for the extreme sensitivity of the sense of smell: the human nose can detect a billionth of an ounce of perfume in air. Even at that, our sense of smell is not nearly as keen as that of many other mammals.

A family of genes that codes for the olfactory receptor proteins has been discovered. This is a large family that may include as many as a thousand genes. The large number may reflect the importance of the sense of smell to mammals in general. Even a thousand different genes coding for a thousand different receptor proteins, however, cannot account for the fact that humans can distinguish up to 10,000 different odors. Clearly, the brain must integrate the signals from several sensory neurons that have different olfactory receptor proteins and then interpret the pattern as a characteristic "fingerprint" for a particular odor.

Figure 18.15

A scanning electron micrograph of an olfactory neuron.
The tassel of cilia is clearly visible.

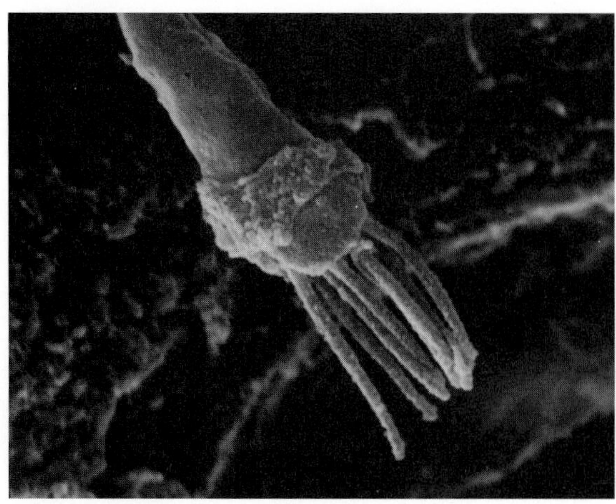

Vestibular Apparatus and Equilibrium

The sense of equilibrium is provided by structures in the inner ear, collectively known as the vestibular apparatus. Movements of the head cause fluid within these structures to bend extensions of sensory hair cells, and this mechanical bending results in the production of action potentials.

The sense of equilibrium, which provides orientation with respect to gravity, is due to the function of an organ called the **vestibular apparatus.** The vestibular apparatus and a snail-like structure called the *cochlea*, which is involved in hearing, form the **inner ear** within the temporal bones of the skull. The vestibular apparatus consists of two parts: (1) the *otolith organs*, which include the *utricle* and *saccule*, and (2) the *semicircular canals* (fig. 18.16).

The sensory structures of the vestibular apparatus and cochlea are located within a tubular structure called the **membranous labyrinth** (*lab'ĭ-rinth*), which is filled with a fluid that is similar in composition to intracellular fluid. This fluid is called *endolymph*. The membranous labyrinth is located within a bony cavity in the skull that forms a **bony labyrinth** (fig. 18.17). Within this cavity, between the membranous labyrinth and the bone, is a fluid called *perilymph*. Perilymph is similar in composition to cerebrospinal fluid.

Sensory Hair Cells of the Vestibular Apparatus

The utricle and saccule provide information about *linear acceleration*—changes in velocity when traveling horizontally or vertically. We therefore have a sense of acceleration and deceleration when riding in a car or when skipping rope. A sense of *rotational*, or *angular, acceleration* is provided by the semicircular canals, which are oriented in three planes like the faces of a cube. This helps us maintain balance when turning the head, spinning, or tumbling.

The receptors for equilibrium are modified epithelial cells. They are known as **hair cells** because they contain 20 to 50 hairlike extensions. All but one of these hairlike extensions are **stereocilia**—processes containing filaments of protein surrounded by part of the cell membrane. One larger extension has the structure of a true cilium (chapter 3), and is known as a **kinocilium** (*ki''no-sil'e-um*) (fig. 18.18). When the stereocilia bend in the direction of the kinocilium, the cell membrane is depressed and becomes depolarized. This causes the hair cell to

Figure 18.16

The cochlea and vestibular apparatus of the inner ear.
The vestibular apparatus consists of the utricle and saccule (together called the otolith organs) and the three semicircular canals. The base of each semicircular canal is expanded into an ampulla that contains sensory hair cells.

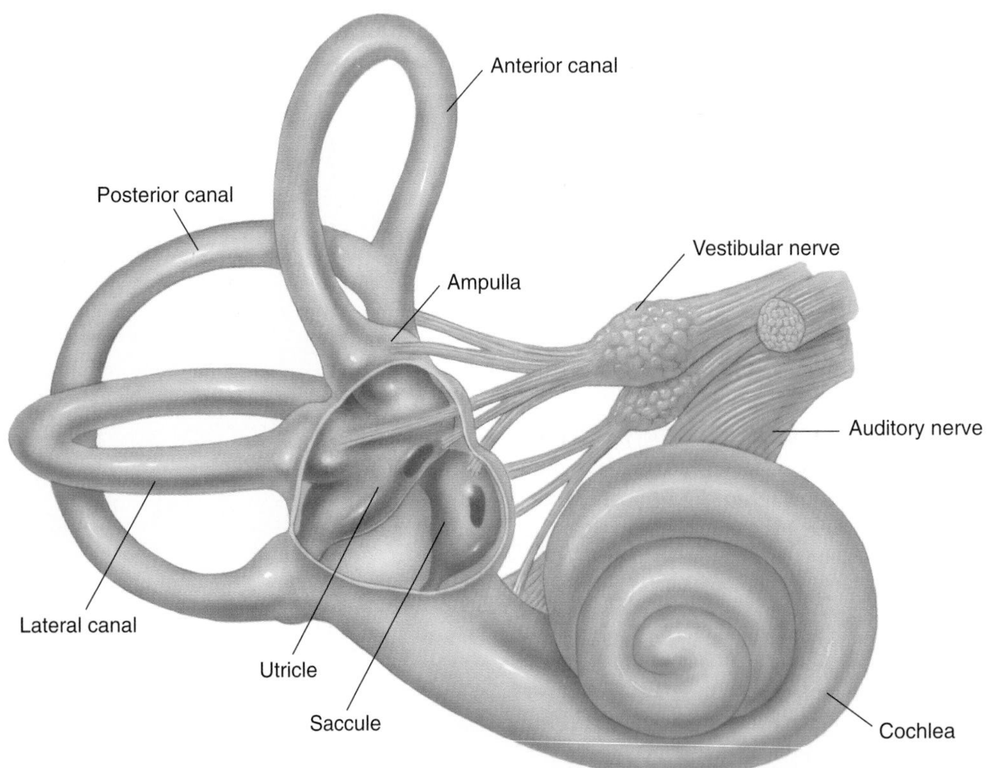

Figure 18.17

The labyrinths of the inner ear.

The membranous labyrinth (darker color) is contained within the bony labyrinth.

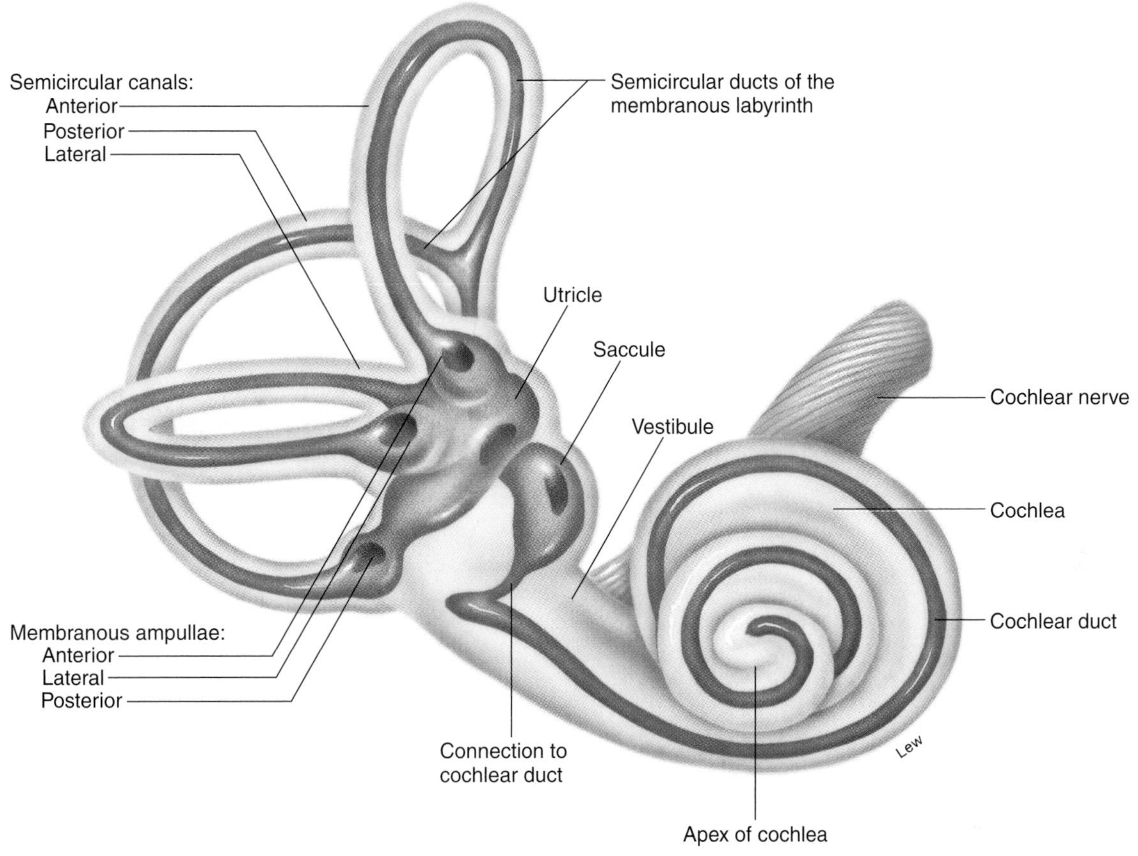

Figure 18.18

Sensory hair cells within the vestibular apparatus.

(*a*) A scanning electron photograph of a kinocilium and stereocilia. (*b*) Each sensory hair cell contains a single kinocilium and several stereocilia. (*c*) When stereocilia are displaced toward the kinocilium (arrows), the cell membrane is depressed, and the sensory neuron innervating the hair cell is stimulated. (*d*) When the stereocilia bend in the opposite direction, away from the kinocilium, the sensory neuron is inhibited.

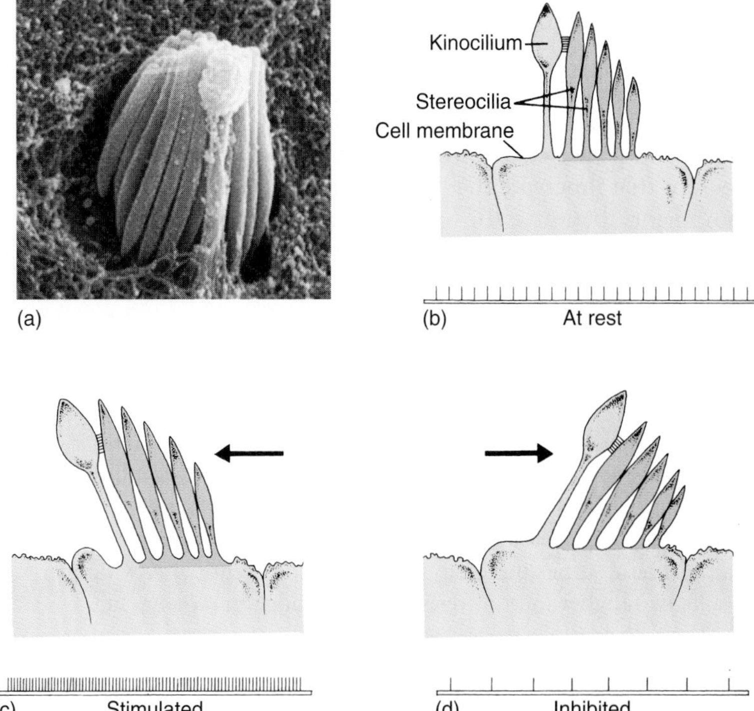

Figure 18.19

The otolith organ.
(*a*) When the head is in an upright position, the weight of the otoliths applies direct pressure to the sensitive cytoplasmic extensions of the hair cells.
(*b*) As the head is tilted forward, the extensions of the hair cells bend in response to gravitational force, stimulating the sensory nerve fibers.

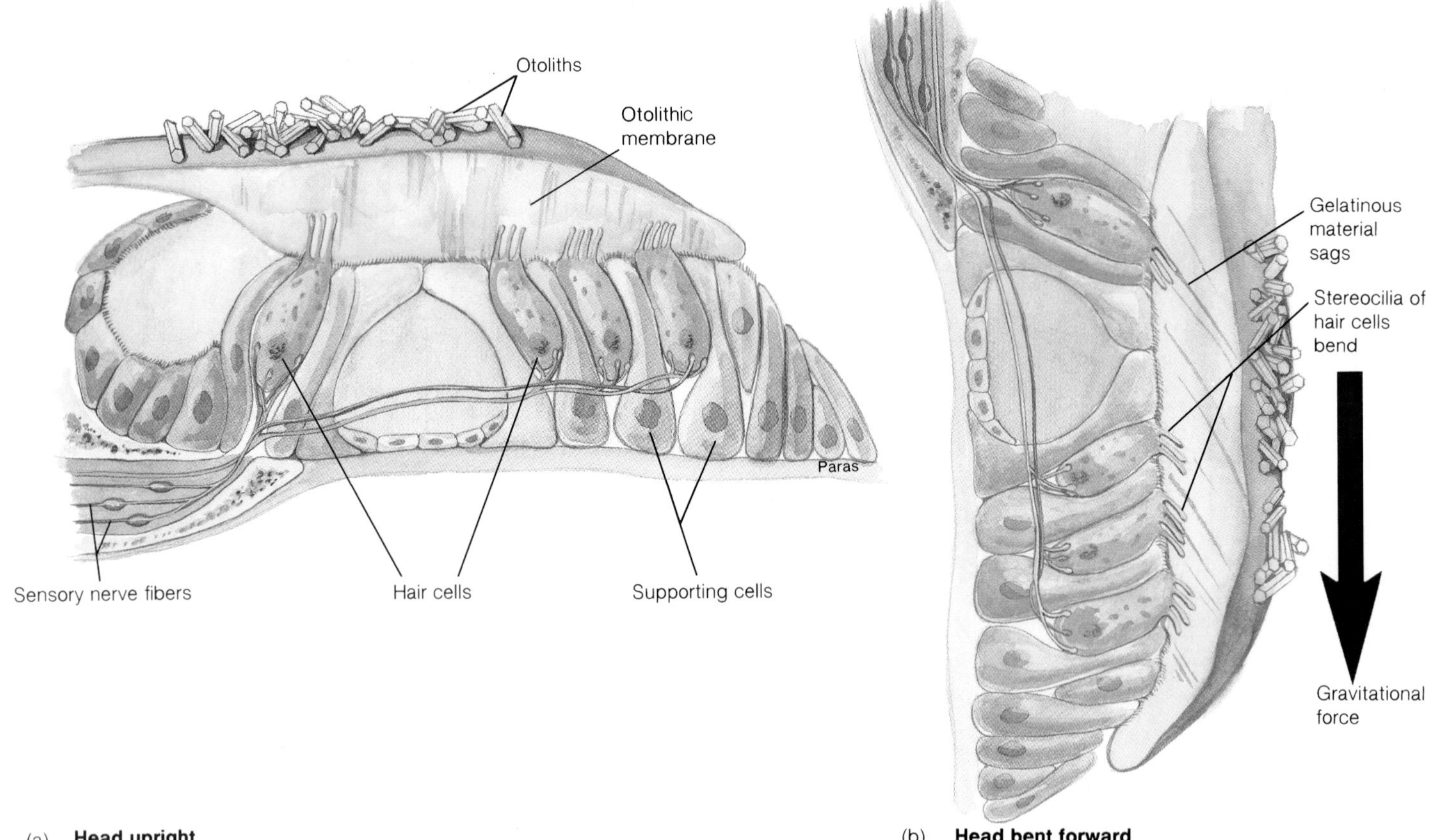

release a synaptic transmitter that stimulates the dendrites of sensory neurons that are part of the vestibulocochlear nerve (VIII). When the stereocilia bend in the opposite direction, the membrane of the hair cell becomes hyperpolarized (fig. 18.18) and, as a result, releases less synaptic transmitter. In this way, the frequency of action potentials in the sensory neurons that innervate the hair cells carries information about movements that cause the hair cell processes to bend.

Utricle and Saccule

The otolith organs, the **utricle** (*yoo'trĭ-ku'l*) and **saccule** (*sak'yool*), each have a patch of specialized epithelium called a **macula** (*mak'yŭ-lă*), consisting of hair cells and supporting cells. The hair cells project into the endolymph-filled membranous labyrinth, with their hairs embedded in a gelatinous **otolithic membrane.** The otolithic membrane contains microscopic crystals of calcium carbonate (*otoliths*), from which it derives its name. These stones increase the mass of the membrane, which results in a higher inertia (resistance to change in movement).

Because of the orientation of their hair cell processes into the otolithic membrane, the utricle is more sensitive to horizontal acceleration and the saccule is more sensitive to vertical acceleration. During forward acceleration, the otolithic membrane lags behind the hair cells, so the hairs of the utricle are pushed backward (fig. 18.19). This is similar to the backward thrust of the body when a car quickly accelerates forward. The inertia of the otolithic membrane similarly causes the hairs of the saccule to be pushed upward when a person descends rapidly in an elevator. These effects, and the opposite ones that occur when a person accelerates backward or upward, produce a changed pattern of action potentials in sensory nerve fibers that allows us to maintain our equilibrium with respect to gravity during linear acceleration.

Semicircular Canals

The three **semicircular canals** project in three different planes (frontal, sagittal, and transverse) at nearly right angles to each other. Each canal contains an inner extension of the membranous labyrinth called a **semicircular duct,** and at the base of each duct is an enlarged swelling called the **ampulla.** The sensory hair cells are located in an elevated area of the ampulla called the **crista ampullaris** (*kris'tă am"poo-lar'is*).

---

macula: L. *macula*, spot
otolith: Gk. *otos*, ear; *lithos*, stone

Figure 18.20

The cupula and hair cells within the semicircular canals.
(*a*) The structures at rest or at a constant velocity. (*b*) Movement of the endolymph during rotation causes the cupula to bend, stimulating the hair cells.

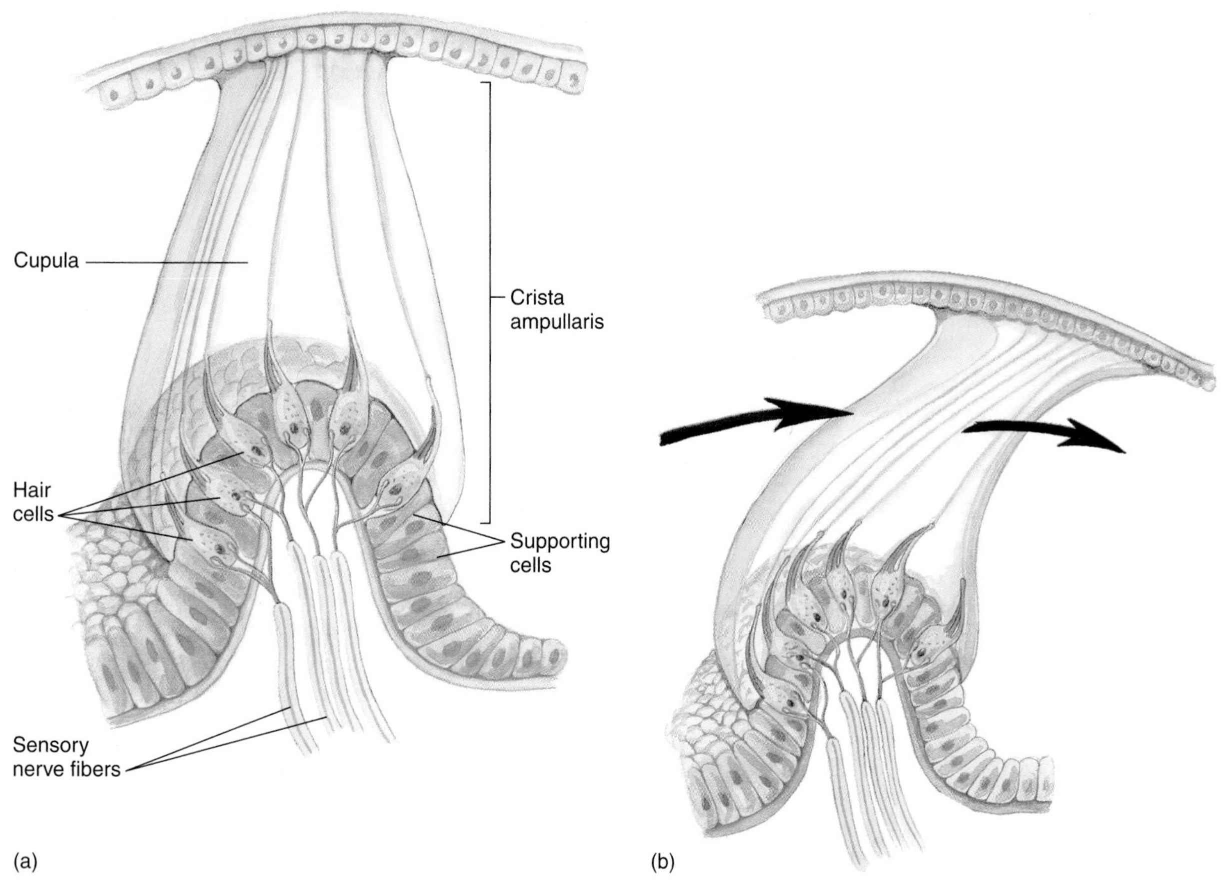

The processes of these sensory cells are embedded in a gelatinous membrane, the **cupula** (*kup'yoo-lă*) (fig. 18.20), which has a higher density than that of the surrounding endolymph. Like a sail in the wind, the cupula can be pushed in one direction or the other by movements of the endolymph.

The endolymph of the semicircular canals serves a function analogous to that of the otolithic membrane—it provides inertia so that the sensory processes will bend in a direction opposite to that of the angular acceleration. As the head rotates to the right, for example, the endolymph causes the cupula to bend toward the left, thereby stimulating the hair cells.

Neural Pathways

Stimulation of hair cells in the vestibular apparatus activates sensory neurons of the *vestibulocochlear nerve (VIII)*. These fibers transmit impulses to the cerebellum and to the vestibular nuclei of the medulla oblongata. The vestibular nuclei, in turn, send fibers to the oculomotor center of the brain stem and to the spinal cord (fig. 18.21). Neurons in the oculomotor center

Figure 18.21

Neural pathways involved in the maintenance of equilibrium and balance.
Sensory input enters the vestibular nuclei and the cerebellum, which coordinate motor responses.

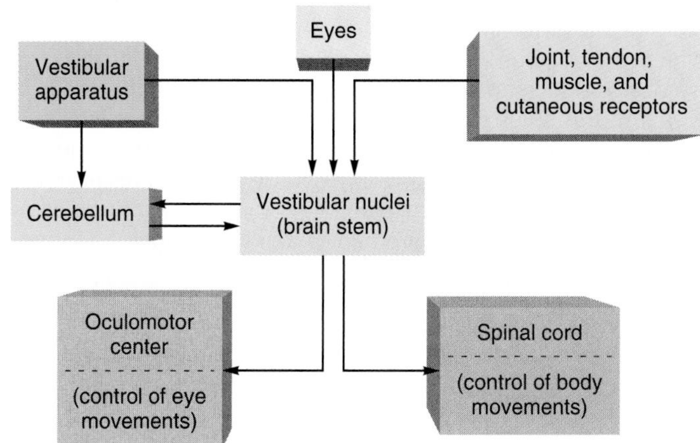

cupula: L. *cupula*, cup-shaped

control eye movements, and neurons in the spinal cord stimulate movements of the head, neck, and limbs. Movements of the eyes and body produced by these pathways serve to maintain balance and "track" the visual field during rotation.

Nystagmus and Vertigo

When a person first begins to spin, the inertia of endolymph within the semicircular ducts causes the cupula to bend in the opposite direction. As the spin continues, however, the inertia of the endolymph is overcome and the cupula straightens. At this time the endolymph and the cupula are moving in the same direction and at the same speed. If a person's movement suddenly stops, the greater inertia of the endolymph causes it to continue moving in the previous direction of spin and to bend the cupula in that direction.

Bending of the cupula after movement has stopped affects muscular control of the eyes and body through the neural pathways previously discussed. The eyes slowly drift in the direction of the previous spin and then are rapidly jerked back to the midline position, producing involuntary oscillations. These movements are called **vestibular nystagmus** (*nĭ'stag'mus*), and people experiencing this effect may feel that they, or the room, are spinning. The loss of equilibrium that results is called *vertigo*. If the vertigo is sufficiently severe or the person particularly susceptible, the autonomic system may become involved. This can produce dizziness, pallor, sweating, and nausea.

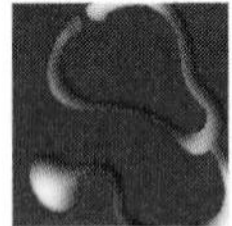

Vestibular nystagmus is one of the symptoms of an inner-ear disease called *Ménière's (mān-e-ārz) disease*. An early symptom of this disease is often "ringing in the ears," or *tinnitus* (*tĭ-nĭ'tus*). Since the endolymph of the cochlea and the endolymph of the vestibular apparatus are continuous through a tiny canal, the duct of Hensen, vestibular symptoms of vertigo and nystagmus often accompany hearing problems in this disease (see "Clinical Considerations" at the end of the chapter).

The Ears and Hearing

Sound causes vibrations of the tympanic membrane. These vibrations, in turn, produce movements of the auditory ossicles, which press against the oval window. Movements of the oval window produce pressure waves within the fluid of the cochlea, which in turn cause movements of the basilar membrane. Sensory hair cells are located on the basilar membrane, and the movements of this membrane in response to sound result in the bending of the hair cell processes. This stimulates action potentials in sensory fibers that are transmitted to the brain and interpreted as sound.

vertigo: L. *vertigo*, dizziness
Ménière's disease: from Prosper Ménière, French physician, 1799–1862
tinnitus: L. *tinnitus*, ring or tingle
duct of Hensen: From Viktor Hensen, German physiologist, 1835–1924

Figure 18.22

Surface anatomy of the auricle of the ear.
The external auditory canal leads to the eardrum, which cannot be seen without a specialized instrument (an otoscope).

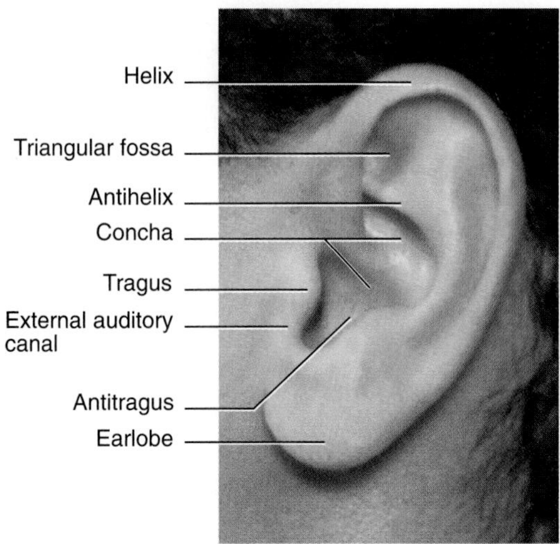

Sound waves travel in all directions from their source, like ripples in a pond where a stone has been dropped. These waves are characterized by their frequency and intensity. The **frequency** is measured in *hertz* (*Hz*), which is the modern designation for *cycles per second* (*cps*). The *pitch* of a sound is directly related to its frequency—the greater the frequency of a sound, the higher its pitch.

The **intensity,** or loudness, of a sound is directly related to the amplitude of the sound waves and is measured in units known as *decibels* (*dB*). A sound that is barely audible—at the threshold of hearing—has an intensity of 0 decibels. Every 10 decibels indicates a tenfold increase in sound intensity; a sound is 10 times louder than threshold at 10 dB, 100 times louder at 20 dB, a million times louder at 60 dB, and 10 billion times louder at 100 dB.

The ear of a trained, young individual can hear sound over a frequency range of 20,000 to 30,000 Hz, yet still can distinguish between two pitches that have only a 0.3% difference in frequency. The human ear can detect differences in sound intensities of only 0.1 to 0.5 dB, although the range of audible intensities covers 12 orders of magnitude (10^{12}), from the barely audible to the limits of painful loudness.

Outer Ear

Sound waves are funneled by the **auricle** (*or'ĭ-kul*), or **pinna** (fig. 18.22), into the external auditory canal, which is the fleshy tube within the bony external acoustic meatus (fig. 18.23). The auricle and external auditory canal constitute the **outer (external) ear.** The external auditory canal channels the sound waves to the **tympanic** (*tim-pan'ik*) **membrane,** or eardrum. The tympanic membrane, approximately 1 cm in diameter, is composed of an outer concave layer of stratified squamous epithelium and an inner convex layer of low columnar

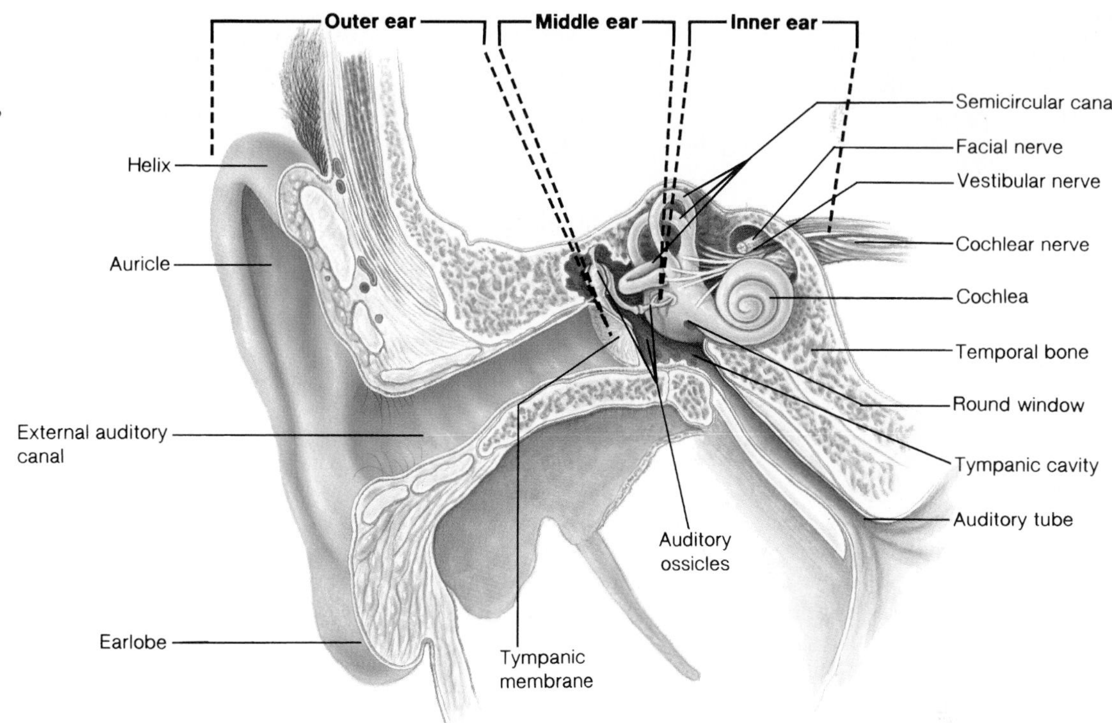

Figure 18.23

The ear.
Note the structures of the outer, middle, and inner ear.

epithelium. Sound waves in the external auditory canal produce extremely small vibrations of the tympanic membrane; movements of the tympanic membrane during speech (with an average sound intensity of 60 dB) are estimated to be about equal to the diameter of a molecule of hydrogen!

The **external auditory canal** is a slightly S-shaped canal about 2.5 cm (1 in.) long, extending slightly upward from the auricle to the tympanic membrane (see fig. 18.22). The skin that lines the canal contains fine hairs and sebaceous glands near the entrance. Specialized wax-secreting glands, called **ceruminous** (*sĕ-roo′mĭ-nus*) **glands,** are located in the skin, deep within the canal. *Cerumen* (earwax) secreted from ceruminous glands keeps the tympanic membrane soft and waterproof. Cerumen and the hairs also help to prevent small foreign objects from reaching the tympanic membrane. The bitter cerumen is probably an insect repellent as well.

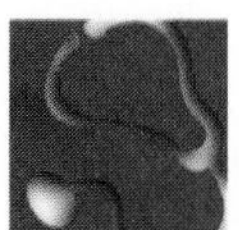
Inspecting the tympanic membrane with an *otoscope* during a physical examination yields significant information about the condition of the middle ear. The color, curvature, presence of lesions, and position of the malleus of the middle ear are particularly important. If ruptured, the tympanic membrane generally will regenerate and readily heal itself.

Middle Ear

The **middle ear** is the cavity between the tympanic membrane on the outer side and the cochlea on the inner side (fig. 18.24). It is located within the petrous part of the temporal bone and contains three *auditory ossicles*—the **malleus** (*mal′e-us*) (hammer), **incus** (*ing′kus*) (anvil), and **stapes** (*sta′pēz*) (stirrup). The malleus is attached to the tympanic membrane, so that vibrations of this membrane are transmitted via the malleus and incus to the stapes. The stapes, in turn, is attached to a membrane in the cochlea called the **oval (vestibular) window,** which thus vibrates in response to vibrations of the tympanic membrane. When the stapes presses the oval window into the cochlea, another flexible membrane—called the **round window** (fig. 18.24)—bulges outward to relieve the pressure.

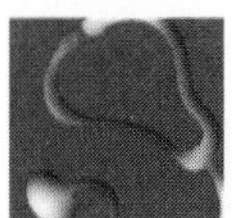
The *auditory* (*eustachian*) *tube* is a passageway leading from the middle ear to the nasopharynx (a cavity positioned behind the nasal cavity and extending down to the soft palate). The auditory tube is usually collapsed, so that debris and infectious agents cannot travel from the oral cavity to the middle ear. In order to open the auditory tube, the tensor tympani muscle, attaching to the auditory tube and the malleus (fig. 18.24), must contract. This occurs during swallowing, yawning, and sneezing. People sense a "popping" sensation in their ears as they swallow when driving up a mountain because the opening of the auditory canal permits air to move from the region of higher pressure in the middle ear to the region of lower pressure in the nasopharynx.

The fact that vibrations of the tympanic membrane are transferred through three bones instead of just one affords protection. If a sound is too intense, the auditory ossicles may buckle. This protection is enhanced by the action of the *stapedius* (*stă-pe′de-us*) *muscle*, the smallest of all the skeletal

otoscope: Gk. *otikos*, ear; *skopein*, to examine

eustachian tube: From Bartolommeo E. Eustachio, Italian anatomist, 1520–74

Under Development

Development of the Ear

The ear begins to develop at the same time as the eye, early during the fourth week. All three embryonic germ layers—ectoderm, mesoderm, and endoderm—are involved in the formation of the ear. Both types of ectoderm (neuroectoderm and surface ectoderm) play a role.

The ear of an adult is structurally and functionally divided into an outer ear, a middle ear, and an inner ear. Although each of these regions has a separate embryonic origin, by the end of the eighth week each of the ear's component parts is in place and formation of the ear is complete.

The first indication of ear formation is the appearance of a plate of surface ectoderm called the **otic** (*o'tik*) **placode** lateral to the developing embryonic hindbrain. The otic placode soon invaginates and forms an **otic fovea.** Toward the end of the fourth week, the outer edges of the invaginated otic fovea come together and fuse to form an **otocyst** (fig. 1). The otocyst soon pinches off and separates from the surface ectoderm. The otocyst further differentiates to form a posterior *utricular portion* and an anterior *saccular portion.* Structures of the inner ear form from these two portions of the otocyst. Three separate diverticula extend outward from the utricular portion and develop into the **semicircular canals** (not illustrated), which later function in balance and equilibrium. A tubular diverticulum, called the **cochlear duct** (not illustrated), extends in a coiled fashion from the saccular portion and forms the membranous portion of the cochlea (*kok'le-ă*), the organ of hearing. The sensory nerves that innervate the inner ear derive from neuroectoderm of the developing brain and grow toward the developing structures of the inner ear so that the nerve tracts will be in place when the rest of the ear has completed its development.

The **auditory ossicles** (bones) have an interesting developmental origin from the first and second pharyngeal arches. Specialized cells known as **mesenchymal condensations** migrate from their sites of origin to a location just below the developing otocyst. Going first through a cartilaginous stage, they soon ossify to bone and are positioned and structured to amplify sound waves that will pass through the middle ear.

The middle-ear chamber is referred to as the **tympanic cavity** and derives from the first pharyngeal pouch. As the tympanic cavity enlarges, it surrounds and encloses the developing auditory ossicles. The connection of the tympanic cavity to the pharynx gradually elongates to develop into the **auditory** (eustachian) **tube,** which remains patent throughout life and is important in maintaining an equilibrium of air pressure between the pharyngeal and tympanic cavities.

The **external auditory canal** forms from the surface ectoderm that covers the posterior end of the first branchial groove. This canal permits sound waves to pass from the outer ear to contact the **tympanic membrane.** The tympanic membrane actually derives from the tissues that contributed to the formation of the tympanic cavity and from tissues that contributed to the formation of the external auditory canal.

otic: Gk. *otikos*, ear

muscles, which attaches to the neck of the stapes (fig. 18.25). When sound becomes too loud, the stapedius muscle contracts and dampens the movements of the stapes against the oval window. This action helps to prevent nerve damage within the cochlea. If sounds reach high amplitudes extremely rapidly, however—as in gunshots—the stapedius muscle may not respond fast enough to prevent nerve damage.

Cochlea

Encased within the dense petrous part of the temporal bone is an organ called the **cochlea** (*kok'le-ă*), about the size of a pea and shaped like the shell of a snail. Together with the vestibular apparatus (previously described), it comprises the inner ear.

Vibrations of the stapes and oval window displace perilymph fluid within a part of the bony labyrinth known as the **scala vestibuli** (*ska'lă vĕ-stib'yŭ-li*), which is the upper of three chambers within the cochlea. The lower of the three chambers is also a part of the bony labyrinth and is known as the **scala tympani** (*tim'pă-ni*). The middle chamber of the cochlea is a part of the membranous labyrinth called the **cochlear duct,** or **scala media.** Like the cochlea as a whole, the cochlear duct coils to form three levels (fig. 18.25), similar to the basal, middle, and apical portions of a snail shell. Since the cochlear duct is a part of the membranous labyrinth, it contains endolymph rather than perilymph.

The perilymph of the scala vestibuli and scala tympani is continuous at the apex of the cochlea because the cochlear duct ends blindly, leaving a small space called the **helicotrema** (*hel''ĭ-kŏ-tre'ma*) between the end of the cochlear duct and the wall of the cochlea. Vibrations of the oval window produced by movements of the stapes cause pressure waves within the scala vestibuli, which pass to the scala tympani. Movements of perilymph within the scala tympani, in turn, travel to the

cochlea: L. *cochlea*, snail shell

scala: Gk. *scala*, staircase
helicotrema: Gk. *helix*, a spiral; *trema*, a hole

Figure 1 Development of the outer- and middle-ear regions and the auditory ossicles. (*a*) A lateral view of a 4-week-old embryo showing the position of the cut depicted in the sequential development (*b–e*). (*b*) The embryo at 4 weeks illustrating the invagination of the surface ectoderm and the evagination of the endoderm at the level of the first pharyngeal pouch. (*c*) During the fifth week, mesenchymal condensations are apparent, from which the auditory ossicles will be derived. (*d*) Further invagination and evagination at 6 weeks correctly positions the structures of the outer- and middle-ear regions. (*e*) By the end of the eighth week, the auditory ossicles, tympanic membrane, auditory tube, and external auditory canal have formed.

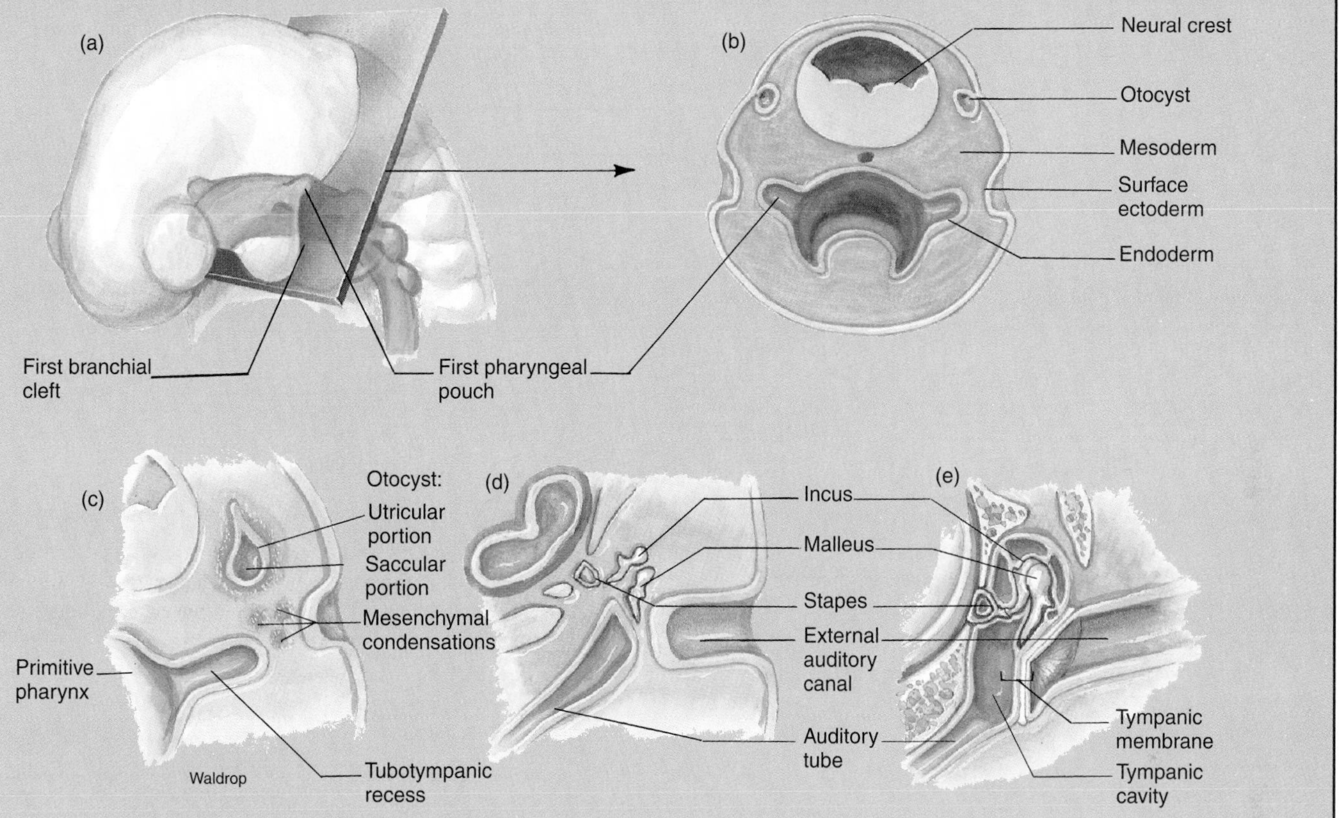

base of the cochlea where they cause displacement of the round window into the middle-ear cavity (see fig. 18.24). This occurs because fluid, such as perilymph, cannot be compressed; an inward movement of the oval window is thus compensated for by an outward movement of the round window.

When the sound frequency (pitch) is sufficiently low, there is adequate time for the pressure waves of perilymph within the upper scala vestibuli to travel through the helicotrema to the scala tympani. As the sound frequency increases, however, pressure waves of perilymph within the scala vestibuli do not have time to travel all the way to the apex of the cochlea. Instead, they are transmitted through the **vestibular membrane,** which separates the scala vestibuli from the cochlear duct, and through the **basilar membrane,** which separates the cochlear duct from the scala tympani, to the perilymph of the scala tympani (fig. 18.25). The distance that these pressure waves travel, therefore, decreases as the sound frequency increases.

Movements of perilymph from the scala vestibuli to the scala tympani thus produce displacement of the vestibular membrane and the basilar membrane. Movement of the vestibular membrane does not directly contribute to hearing, whereas displacement of the basilar membrane is central to pitch discrimination. Each sound frequency produces maximum vibrations at a different region of the basilar membrane. Sounds of higher frequency (pitch) cause maximum vibrations of the basilar membrane closer to the stapes, as illustrated in figure 18.26.

Spiral Organ (Organ of Corti)

The sensory *hair cells* are located on the basilar membrane. These are similar to the hair cells of the vestibular apparatus, with hair processes (actually stereocilia) projecting into the cochlear duct but without a kinocilium. These hair cells are arranged to form one row of inner cells, which extends the

organ of Corti: from Alfonso Corti, Italian anatomist, 1822–88

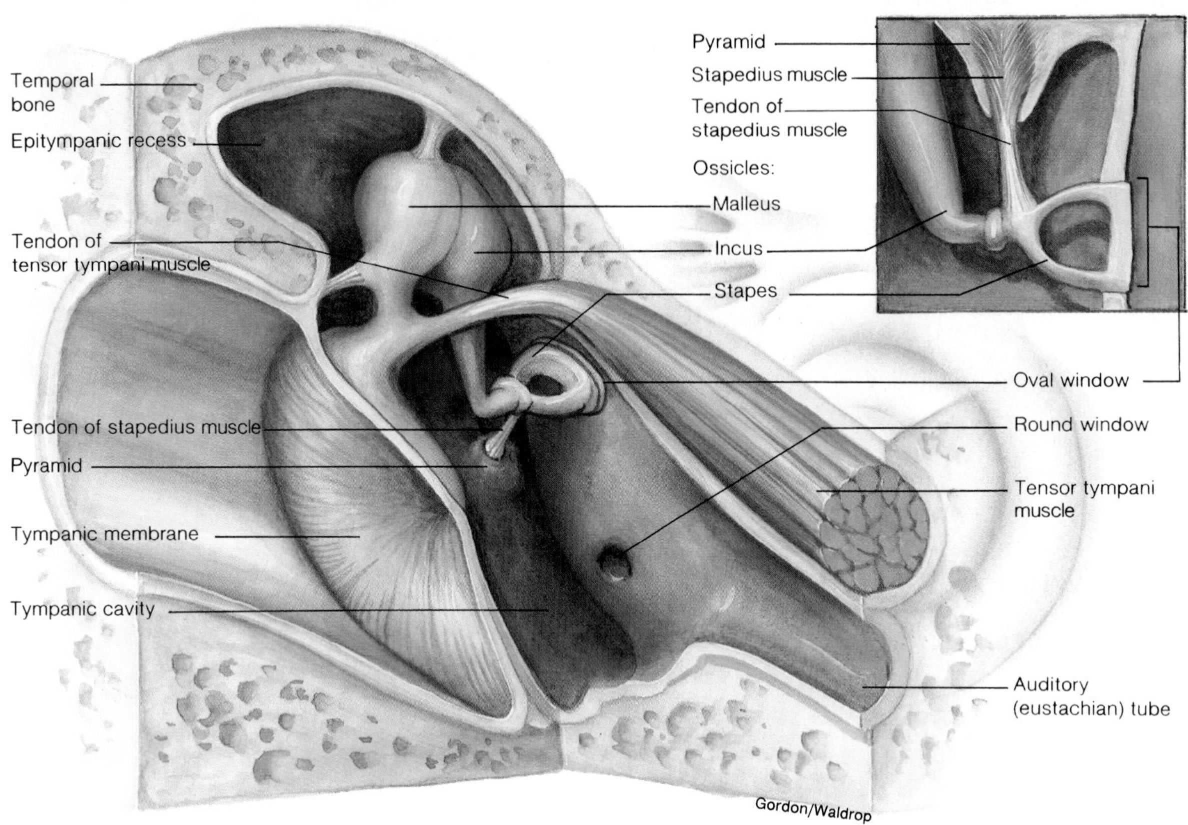

Figure 18.24

A medial view of the middle ear.
The position of auditory muscles, attached to the middle ear ossicles, is shown.

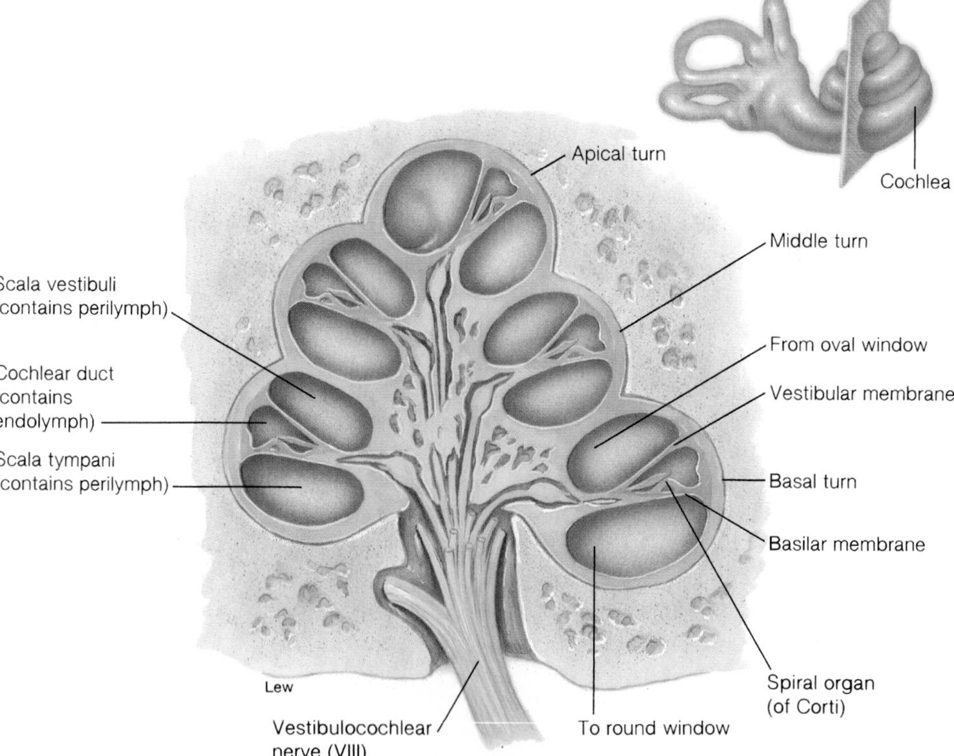

Figure 18.25

A cross section of the cochlea.
This view demonstrates its three turns and its three compartments—the scala vestibuli, cochlear duct (scala media), and scala tympani.

Figure 18.26

The effect of sounds of different frequency on the basilar membrane.

(The cochlea is shown "unwound" in this diagram.) Sounds of low frequency cause pressure waves of perilymph to pass through the helicotrema. Sounds of higher frequency cause pressure waves to "shortcut" through the cochlear duct. This causes displacement of the basilar membrane, which is central to the transduction of sound waves into nerve impulses. Maximum displacement of the basilar membrane occurs closer to its base as the sound frequency is increased. (The frequency of sound waves is measured in hertz [Hz], or cycles per second.)

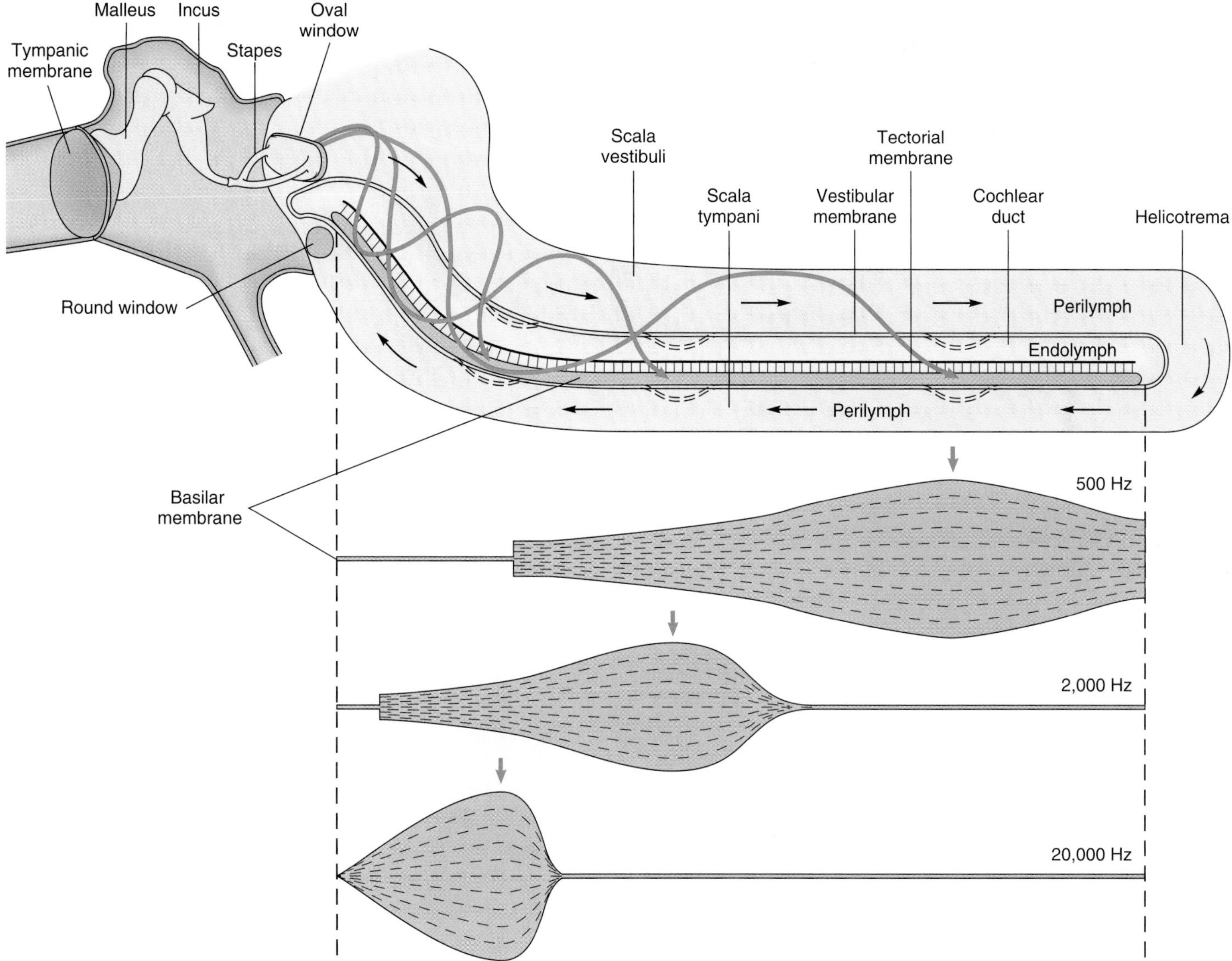

length of the basilar membrane, and multiple rows of outer hair cells: three rows in the basal turn, four in the middle turn, and five in the apical turn of the cochlea.

The stereocilia of the outer hair cells are embedded in a gelatinous **tectorial** (*tek-to're-al*) **membrane,** which overhangs the hair cells within the cochlear duct (fig. 18.27). The association of the basilar membrane, hair cells with sensory fibers, and tectorial membrane forms a functional unit called the **spiral organ,** or **organ of Corti** (fig. 18.27). When the cochlear duct is displaced by pressure waves of perilymph, a shearing force is created between the basilar membrane and the tectorial membrane. This causes the stereocilia to move and bend. Such movement causes ion channels in the membrane to open, which in turn depolarizes the hair cells. Each depolarized hair cell then releases a transmitter chemical (probably glutamate) that stimulates an associated sensory neuron.

The greater the displacement of the basilar membrane and the bending of the stereocilia, the greater the amount of transmitter released by the hair cell, and therefore the greater the generator potential produced in the sensory neuron. In other words, a greater bending of the stereocilia will increase the frequency of action potentials produced by the fibers of the cochlear nerve that are stimulated by the hair cells. Experiments suggest that the stereocilia need bend only

Figure 18.27

The spiral organ (organ of Corti).
This is depicted (*a*) within the cochlear duct and (*b*) isolated to show greater detail.

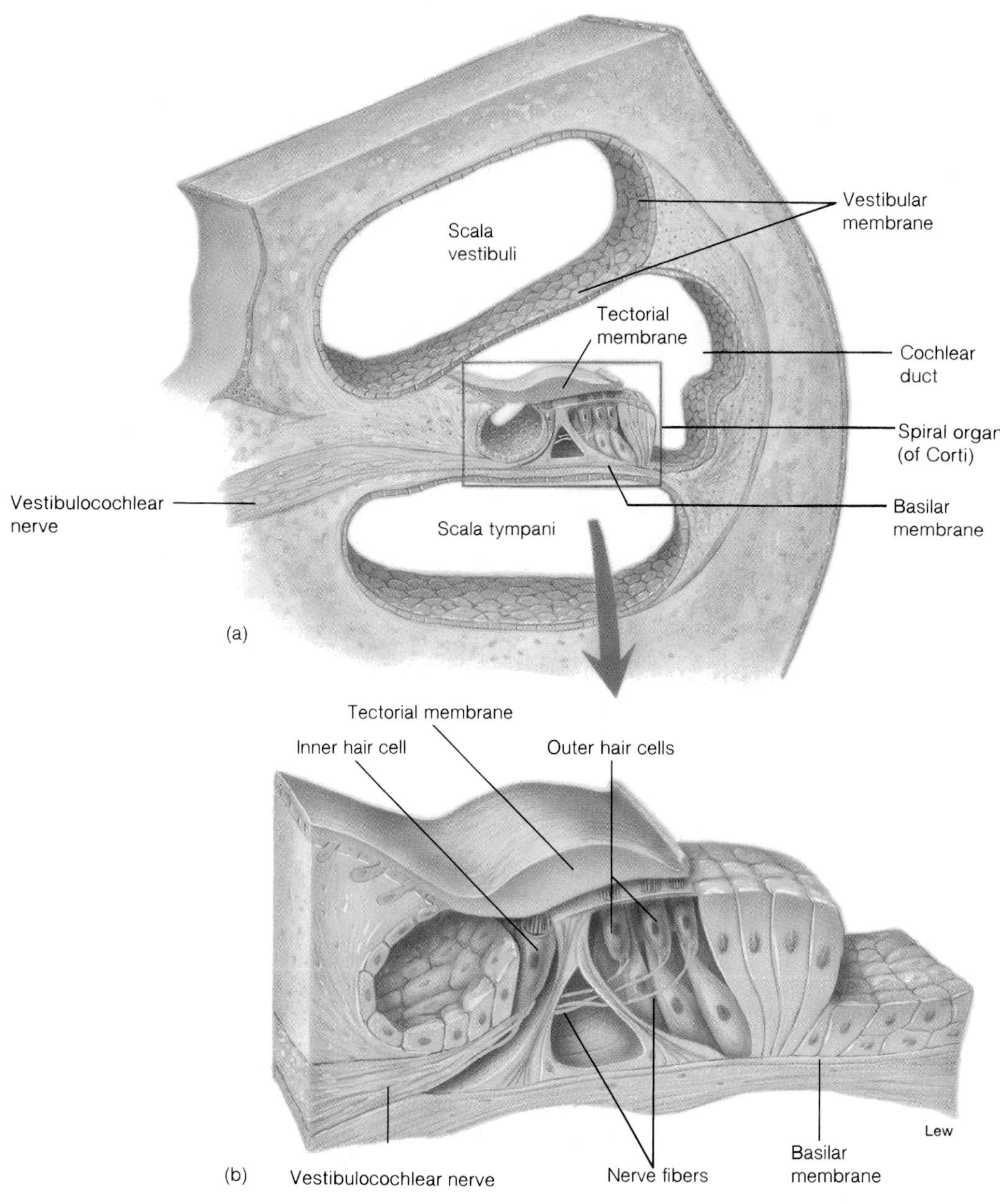

0.3 nanometers to be detected at the threshold of hearing! A greater bending will result in a higher frequency of action potentials, which will be perceived as a louder sound.

As mentioned earlier, traveling waves in the basilar membrane reach a peak in different regions, depending on the pitch of the sound. High-pitched sounds produce a peak displacement closer to the base, while sounds of lower pitch cause peak displacement further toward the apex (see fig. 18.26). Those neurons that originate in hair cells located where the displacement is greatest will be stimulated more than neurons that originate in other regions. This mechanism provides a neural code for **pitch discrimination.**

Neural Pathways for Hearing

Sensory neurons in the vestibulocochlear nerve (VIII) synapse with neurons in the medulla oblongata that project to the inferior colliculus of the midbrain (fig. 18.28). Neurons in this area, in turn, project to the medial geniculate nucleus of the thalamus, which sends axons to the auditory cortex of the temporal lobe. By means of this pathway, neurons in different regions of the basilar membrane stimulate neurons in corresponding auditory areas of the cerebral cortex. Each area of this cortex thus represents a different part of the basilar membrane and a different pitch (fig. 18.29).

Figure 18.28

Neural pathways of hearing.

These pathways extend from the spiral organ in the cochlea to the auditory area of the cerebral cortex.

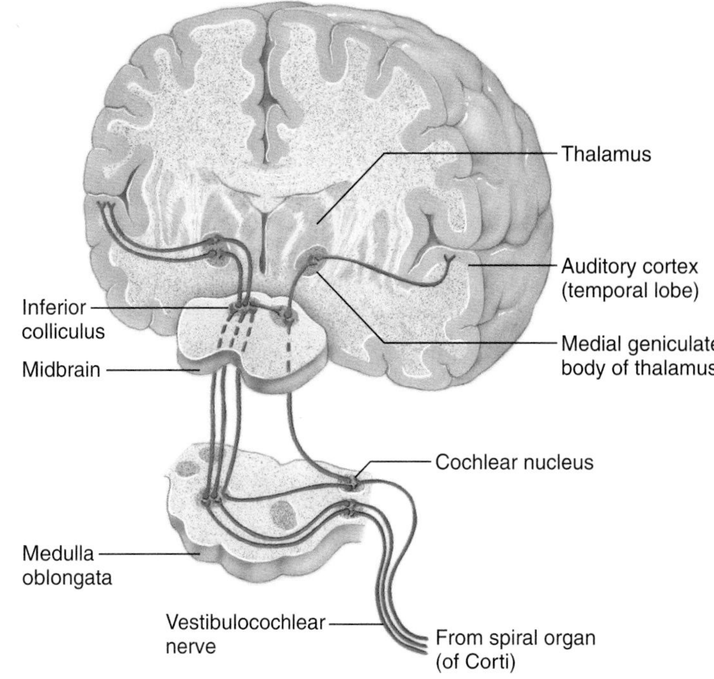

Figure 18.29

Correlation of pitch location in the cochlea and auditory area of the cerebral cortex.

Sounds of different frequencies (pitches) cause vibration of different parts of the basilar membrane, exciting different sensory neurons in the cochlea. These in turn send their input to different regions of the cerebral cortex.

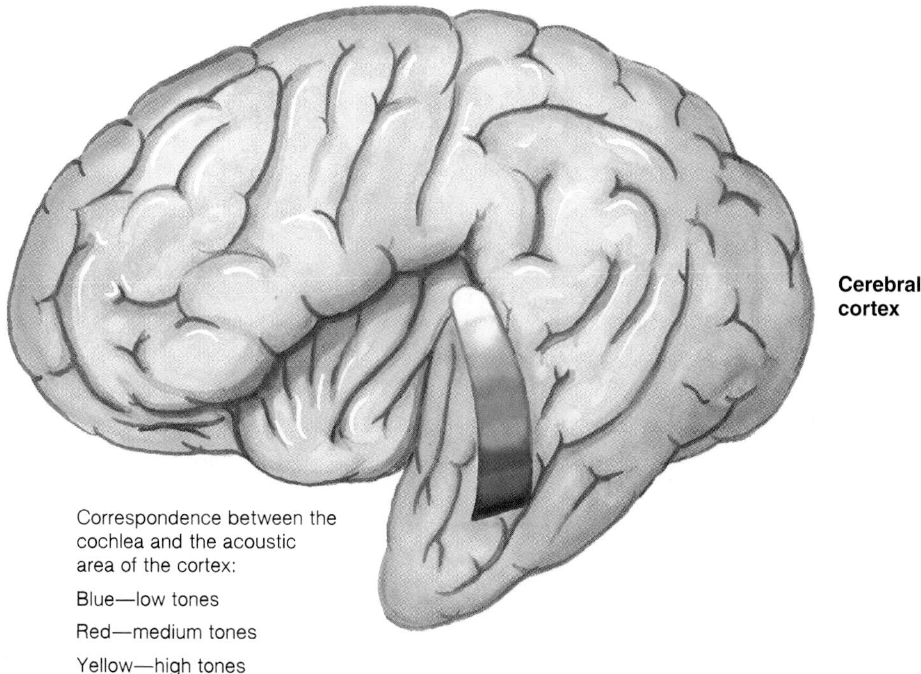

The Eyes and Vision

Light from an observed object is focused by the cornea and lens onto the photoreceptive retina at the back of the eye. The focus is maintained on the retina at different distances between the eyes and the object by muscular contractions that change the thickness and degree of curvature of the lens.

The eyes transduce energy in the electromagnetic spectrum (fig. 18.30) into nerve impulses. Only a limited part of this spectrum can excite the photoreceptors—electromagnetic energy with wavelengths between 400 and 700 nanometers (1 nm = 10^9 m, or one-billionth of a meter) constitute *visible light*. Light of longer wavelengths, which is in the infrared regions of the spectrum, does not have sufficient energy to excite the receptors but is felt as heat. Ultraviolet light, which has shorter wavelengths and more energy than visible light, is filtered out by the yellow color of the eye's lens. Honeybees—and people who have had their lenses surgically removed—can see light in the ultraviolet range.

Structures Associated with the Eye

Accessory structures of the eye either protect the eyeball or provide eye movement. Protective structures include the bony orbit, eyebrows, facial muscles, eyelids, eyelashes, the conjunctiva, and the lacrimal gland that produces tears. Eyeball movements are enabled by the actions of the extrinsic eye muscles that arise from the orbit and insert on the outer layer of the eyeball.

Orbit

Each eyeball is positioned in a bony depression in the skull called the **orbit** (see fig. 9.8 and table 9.2). Seven bones of the skull (frontal, lacrimal, ethmoid, zygomatic, maxilla, sphenoid, and palatine) form the walls of the orbit that support and protect the eye.

Eyebrows

Eyebrows consist of short, thick hairs positioned transversely above the eyes along the supraorbital margins of the skull. In this position, they effectively shade the eyes from the sun and prevent perspiration or falling particles from getting into the eyes.

Eyelids and Eyelashes

Eyelids, or **palpebrae** (*pal'pĕ-bre*), develop as reinforced folds of skin with attached skeletal muscle that make them movable. In addition to the orbicularis oculi muscle attached to the skin that surrounds the front of the eye, the *levator palpebrae superioris* muscle attaches along the upper eyelid, making this eyelid

palpebra: L. *palpebra*, eyelid (related to *palpare*, to pat gently)

Figure 18.30

The electromagnetic spectrum.
Different regions of the electromagnetic spectrum (top) are shown in Anstrom units (1 Å = 10^{-10} meter). The visible spectrum comprises only a small range of this spectrum (bottom), shown in nanometer units (1 nm = 10^{-9} meter).

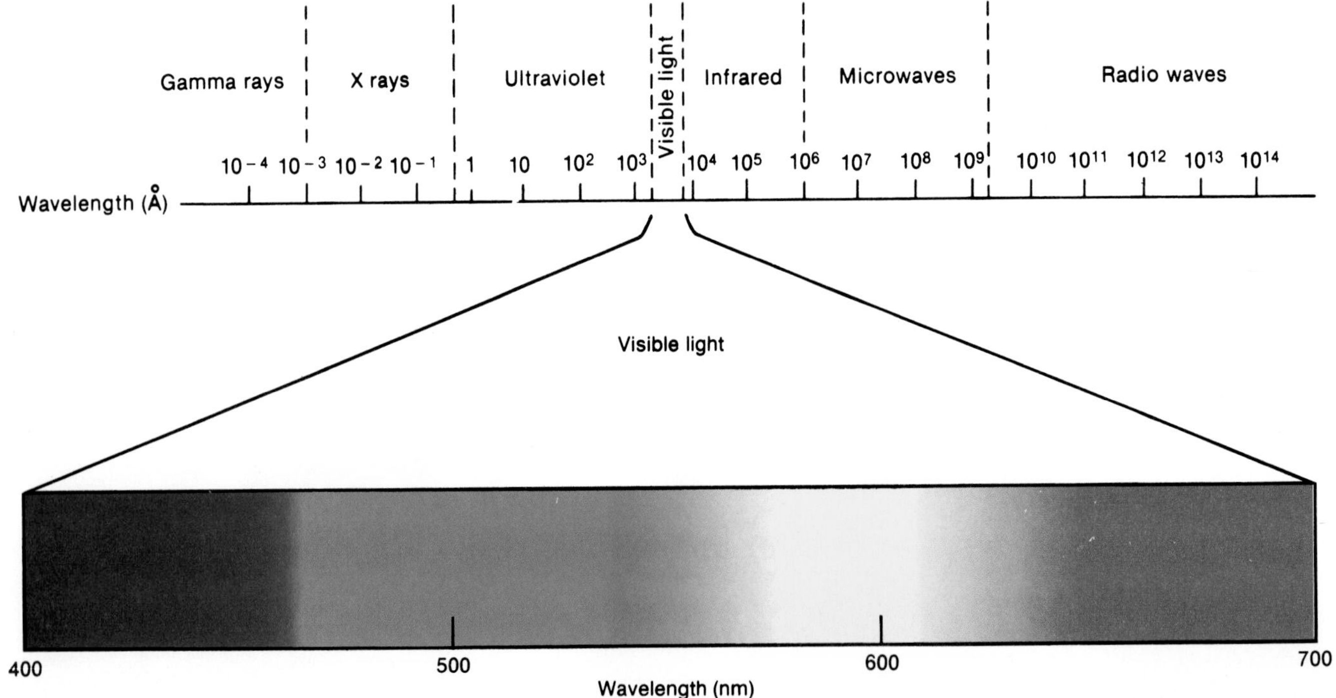

Figure 18.31

Surface anatomy of the eye.
The white of the eye is the sclera, the colored portion is the muscular iris, and the pupil is the opening of the iris that admits light into the eye.

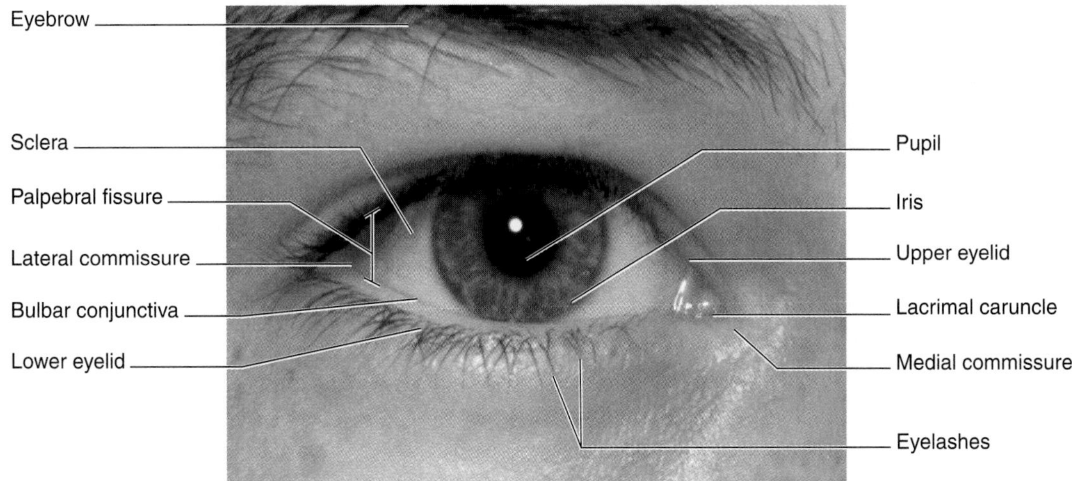

much more mobile than the lower eyelid. Contraction of the *orbicularis oculi* closes the eyelids over the eye, and contraction of the levator palpebrae superioris elevates the upper eyelid to expose the eye. The eyelids protect the eyeball from desiccation by reflexively blinking about every 7 seconds and moving lacrimal fluid (tears) across the anterior surface of the eyeball.

The **palpebral fissure** (fig. 18.31) is the interval between the upper and lower eyelids. The **commissures,** or **canthi** (*kan'thi*), of the eye are the medial and lateral angles where the eyelids come together. The **medial commissure** is broader than the **lateral commissure** and is characterized by a small, reddish, fleshy elevation called the **lacrimal caruncle** (*kar'ung-kul*) (fig. 18.32). The lacrimal caruncle contains sebaceous and sudoriferous glands; it produces the whitish secretion commonly called "sleep dust" that sometimes collects during sleep. Each eyelid supports a row of numerous **eyelashes,** which protect the eye from airborne particles.

In people of Asian descent, a fold of skin of the upper eyelid called the *epicanthic fold* may normally cover part of the medial commissure. An epicanthic fold may also be present in some infants with Down syndrome.

In addition to the layers of the skin and the underlying connective tissue and orbicularis oculi muscle fibers, each eyelid contains a tarsal plate, tarsal glands, and conjunctiva. The **tarsal plates,** composed of dense connective tissue, are important in maintaining the shape of the eyelids (fig. 18.33). Specialized sebaceous glands called **tarsal glands** are embedded in the tarsal plates along the exposed inner surfaces of the eyelids. The ducts of the tarsal glands open onto the edges of the eyelids, and their oily secretions help to keep the eyelids

Figure 18.32

The lacrimal apparatus.
The lacrimal apparatus consists of the lacrimal (tear) gland and the drainage pathway of lacrimal fluid into the nasal cavity.

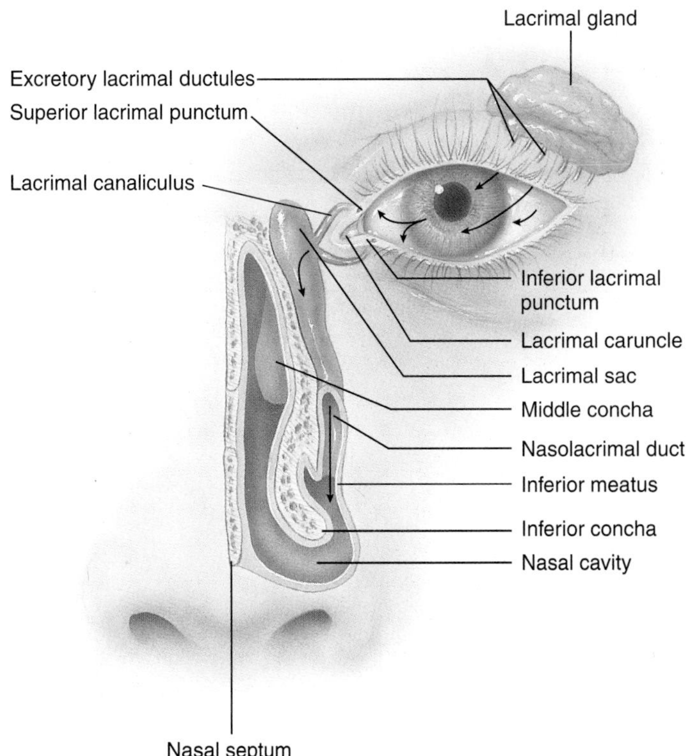

commissure: L. *commissura*, a joining
caruncle: L. *caruncula*, diminutive of *caro*, flesh
tarsal: Gk. *tarsos*, flat basket

from adhering to each other. Modified sweat glands called **ciliary** (*sil'e-er"e*) **glands** are also located within the eyelids, along with additional sebaceous glands at the bases of the hair follicles of the eyelashes.

Conjunctiva

The **conjunctiva** (*con"jungk-ti'vă*) is a delicate mucus-secreting epithelial membrane that lines the interior surface of each eyelid and exposed anterior surface of the eyeball (fig. 18.33). It consists of stratified squamous epithelium that varies in thickness in different regions. The *palpebral conjunctiva* is thick and adheres to the tarsal plates of the eyelids. As the conjunctiva reflects onto the anterior surface of the eyeball, it is known as the *bulbar conjunctiva.* This portion is transparent and especially thin where it covers the cornea. Because the conjunctiva is continuous and reflects from the eyelids to the anterior surface of the eyeball, a space called the **conjunctival sac** is present when the eyelids are closed. The conjunctival sac protects the eyeball by preventing foreign objects from passing beyond the confines of the sac. The conjunctiva can repair itself rapidly if scratched.

Lacrimal Apparatus

The **lacrimal** (*lak'rĭ-mal*) **apparatus** consists of the lacrimal gland, which secretes the lacrimal fluid (tears), and a series of ducts that drain the secretion into the nasal cavity (see fig. 18.32). The **lacrimal glan** shape of an almond, is located in the superolateral portion of the orbit. It is a compound tubuloacinar gland that secretes lacrimal fluid through several excretory lacrimal ductules into the conjunctival sac of the upper eyelid. With each blink of the eyelids, lacrimal fluid passes medially and downward and drains into two small openings, called **lacrimal puncta,** on both sides of the lacrimal caruncle. From here the lacrimal fluid drains through two small ducts, the **lacrimal canaliculi** (*kan"ă-lik'yŭ-li*), into the lacrimal sac, and continues through the nasolacrimal duct to the inferior meatus of the nasal cavity.

Lacrimal fluid is a watery secretion that moistens and lubricates the conjunctival sac. Because it contains a bactericidal enzyme called *lysozyme,* lacrimal fluid also reduces the likelihood of eye infections. Normally about 1 milliliter of lacrimal fluid is produced each day by the lacrimal gland of each eye. If irritating substances, such as particles of sand or chemicals from onions, come in contact with the conjunctiva, the lacrimal glands are stimulated to oversecrete. The extra lacrimal fluid protects the eye by diluting and washing away the irritating substance.

Extrinsic Eye Muscles

The movements of the eyeball are controlled by six extrinsic eye muscles called the **extrinsic ocular muscles** (see fig. 13.6 and table 13.4). Four **recti muscles** maneuver the eyeball in the

Figure 18.33

Accessory structures of the eyeball.
The eyeball and associated structures in sagittal section.

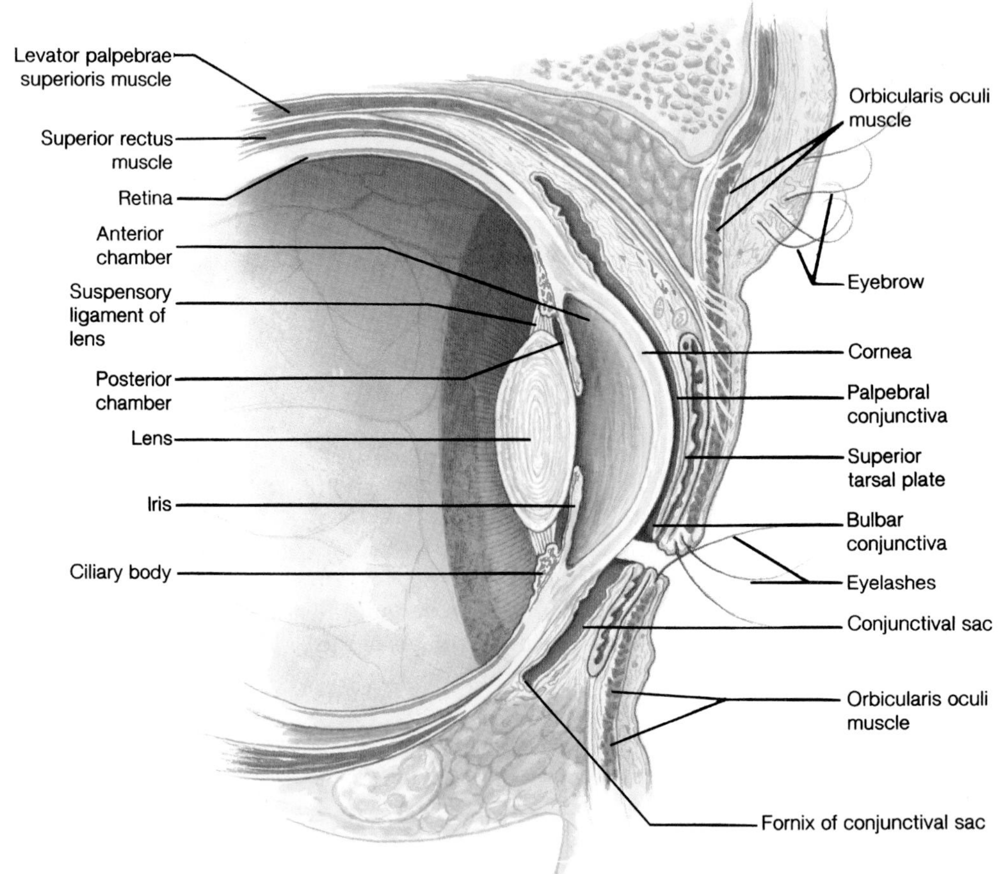

and **medial**), and two **oblique muscles** (**superior** and **inferior**) rotate the eyeball on its axis. One of the extrinsic ocular muscles, the superior oblique, passes through a pulleylike cartilaginous loop, the *trochlea* (*trok'le-ă*), before attaching to the eyeball. Although stimulation of each muscle causes a precise movement of the eyeball, most of the movements involve the combined contraction of two muscles or more.

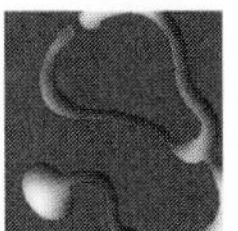

A physical examination may include an eye movement test. As the patient's eyes follow the movement of a physician's finger, the physician can assess weaknesses in specific muscles or dysfunctions of specific cranial nerves. The patient experiencing *double vision* or *diplopia* (*dĭ-plo'pe-ă*), when moving the eyes may be suffering from muscle weakness. Looking laterally tests the abducens cranial nerve; looking inferiorly and laterally tests the trochlear cranial nerve; and crossing the eyes tests the oculomotor and trochlear nerves of both eyes.

Amblyopia (*am"ble-o'pe-ă*), commonly called "lazy eye," is a condition of ocular muscle weakness causing a deviation of one eye. Because of this, two images are received by the optic cerebral cortex, and one is suppressed to avoid diplopia. A person who has amblyopia will experience dimness of vision and partial loss of sight. Young children are tested for amblyopia because little can be done to strengthen the afflicted muscle if it has not been treated before age 6.

trochlea: Gk. *trochos*, a wheel
amblyopia: Gk. *amblys*, dull; *ops*, vision

Structure of the Eyeball

The wall of the eyeball contains three layers, or tunics. The outer layer—the **fibrous tunic**—is divided into two regions: the posterior five-sixths is the opaque *sclera* and the anterior one-sixth is the transparent *cornea*. The middle layer is the **vascular tunic,** or **uvea** (*yoo've-ă*), and consists of the *choroid*, *ciliary body*, and the *iris*. The innermost layer of the eyeball—the **internal tunic**—consists of the *retina*. These structures are summarized in table 18.4.

The fibrous tunic of the eye consists of a tough coat of connective tissue called the **sclera** (*skler'ă*) (fig. 18.34), which can be seen externally as the white of the eye. The tissue of the sclera is continuous with the transparent **cornea.** The **choroid** (*kor'oid*) is a thin, highly vascular layer that lines most of the internal surface of the sclera. The choroid contains numerous pigment-producing melanocytes, which give it a brownish color that prevents light from being reflected out of the eyeball. Light passes through the cornea to enter the **anterior chamber** of the eye. Light then passes through an opening called the **pupil,** surrounded by a pigmented (colored) smooth muscle known as the **iris.** After

uvea: L. *uva*, grape
sclera: Gk. *skleros*, hard
cornea: L. *cornu*, horn
choroid: Gk. *chorion*, membrane
iris: Gk. *irid*, rainbow

Table 18.4
Structures of the Eyeball

Tunic and Structure	Location	Composition	Function
Fibrous tunic	Outer layer of eyeball	Avascular connective tissue	Gives shape to the eyeball
Sclera	Posterior outer layer; white of the eye	Tightly bound elastic and collagen fibers	Supports and protects the eyeball
Cornea	Anterior surface of eyeball	Tightly packed dense connective tissue—transparent and convex	Transmits and refracts light
Vascular tunic (uvea)	Middle layer of eyeball	Highly vascular pigmented tissue	Supplies blood; prevents reflection
Choroid	Middle layer in posterior portion of the eyeball	Vascular layer	Supplies blood to the eyeball
Ciliary body	Anterior portion of vascular tunic	Smooth muscle fibers and glandular epithelium	Supports the lens through suspensory ligament and determines its thickness; secretes aqueous humor
Iris	Anterior portion of vascular tunic, continuous with ciliary body	Pigment cells and smooth muscle fibers	Regulates the diameter of the pupil, and hence the amount of light entering the vitreous chamber
Internal tunic	Inner layer of the eyeball	Tightly packed photoreceptors, neurons, blood vessels, and connective tissue	Provides location and support for rods and cones
Retina	Principal portion of internal tunic	Photoreceptor neurons (rods and cones), bipolar neurons, and ganglion neurons	Photoreception; transmits impulses
Lens (not part of any tunic)	Between posterior and vitreous chambers; supported by suspensory ligament of ciliary body	Tightly arranged protein fibers; transparent	Refracts light and focuses onto fovea centralis

Figure 18.34

The internal anatomy of the eyeball.
Light enters the eye from the right side of this figure and is focused on the retina.

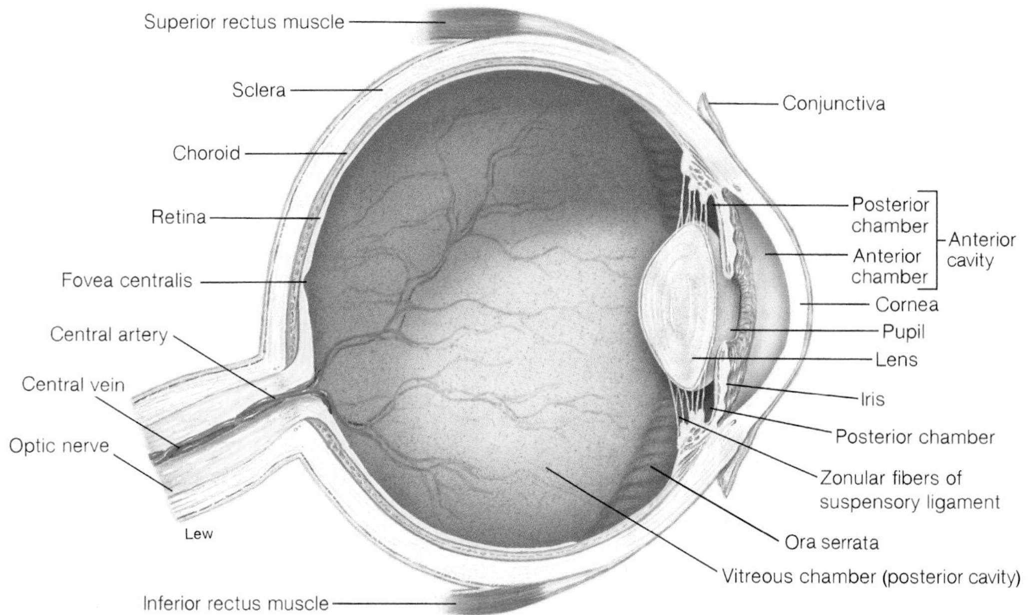

passing through the pupil, light enters the **posterior chamber** and then passes through the **lens** as it enters the **vitreous chamber** (*posterior cavity*) (fig. 18.34).

The iris is like the diaphragm of a camera; it can increase or decrease the diameter of its aperture (the pupil) to admit more or less light. Constriction of the pupils is produced by contraction of the circularly arranged fibers within the iris; dilation is produced by contraction of the radially arranged fibers. Constriction of the pupils results from parasympathetic stimulation, whereas dilation results from sympathetic stimulation (fig. 18.35). Variations in the diameter of the pupil are similar in effect to variations in the f-stop of a camera.

The posterior part of the iris contains a pigmented epithelium that gives the eye its color. The color of the eye is determined by the amount of pigment—blue eyes have the least pigment, brown eyes have more, and black eyes have the greatest amount of pigment. In the condition of *albinism* (a congenital absence of normal pigmentation due to a defect in the ability to produce melanin pigment), the eyes appear pink because the absence of pigment allows blood vessels to be seen.

Enclosed within a *lens capsule,* the lens is suspended from a muscular process called the **ciliary body,** which is connected to the sclera. Numerous extensions of the ciliary body attach to *zonular fibers* that in turn attach to the lens capsule. The zonular fibers constitute the **suspensory ligament.** The space between the cornea and iris is the *anterior chamber,* and the space between the iris and the ciliary body and lens is the *posterior chamber* (fig. 18.36).

The anterior and posterior chambers (together, called the *anterior cavity*) are filled with a watery fluid called the **aqueous humor.** This fluid is secreted by the ciliary body into the posterior chamber and passes through the pupil into the anterior chamber. From here, it drains into the **scleral venous sinus** (*canal of Schlemm*), which returns this fluid to the venous blood (fig. 18.36). Inadequate drainage of aqueous humor can lead to excessive accumulation of fluid, which in turn results in increased intraocular pressure. This condition, called *glaucoma* (see "Clinical Considerations"), may seriously damage the retina and cause loss of vision.

The portion of the eye located behind the lens (the posterior cavity) is filled with a clear gel known as the **vitreous humor.** Light passing through the lens and vitreous humor enters the retina. While passing through the retina, some of this light stimulates photoreceptors, which in turn activate other neurons. Neurons in the retina contribute fibers that are gathered together at a region called the **optic disc** (fig. 18.37), where they exit the retina as the optic nerve. This region lacks photoreceptors and is therefore a blind spot. The optic disc is also the site of entry and exit of blood vessels.

Refraction

Light that passes from a medium of one density into a medium of a different density is *refracted,* or bent. The degree of refraction depends on the comparative densities of the two media,

zonular: L. *zona*, girdle

canal of Schlemm: From Friedrich S. Schlemm, German anatomist, 1795–1858
vitreous: L. *vitreus*, glassy

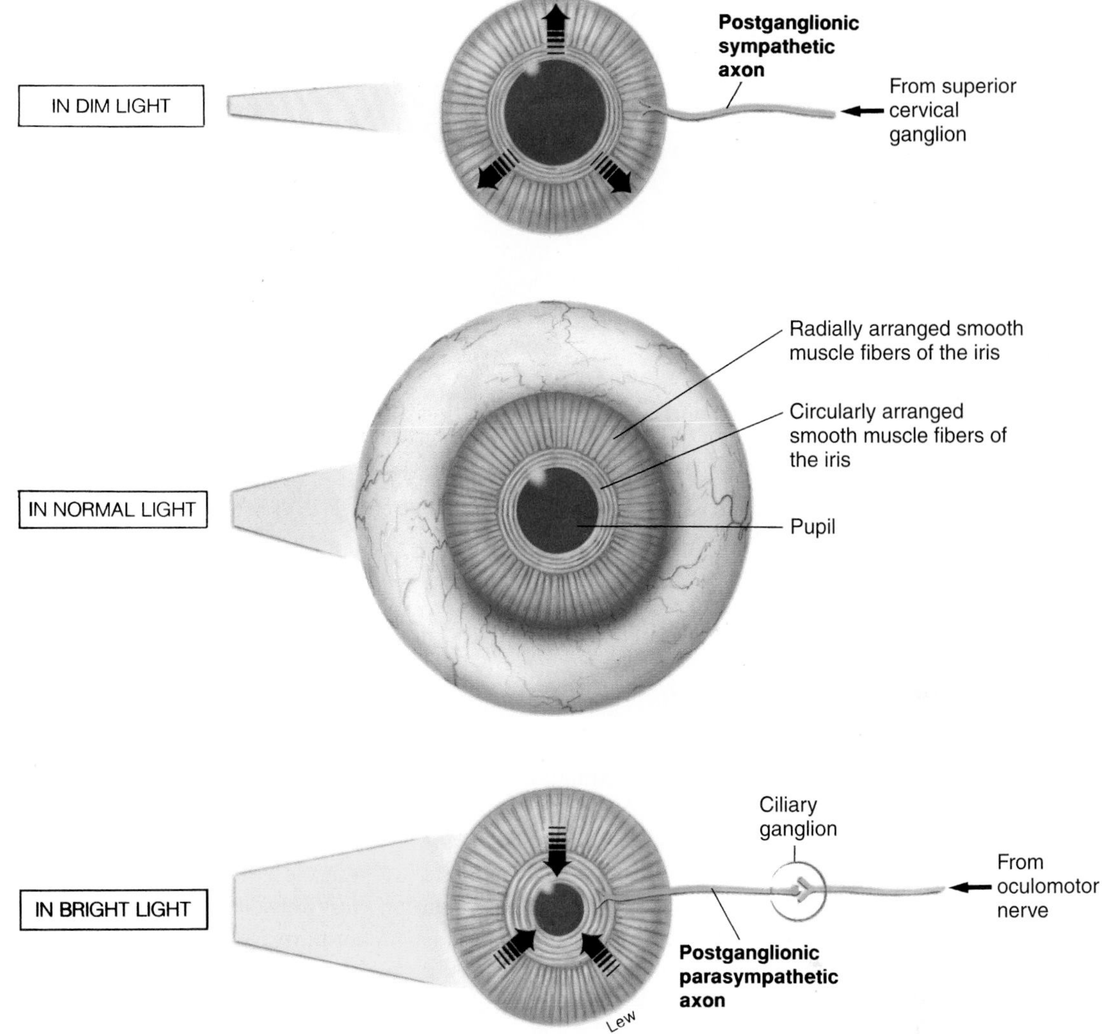

Figure 18.35

Dilation and constriction of the pupil.

In dim light, the radially arranged smooth muscle fibers are stimulated to contract by sympathetic stimulation, dilating the pupil. In bright light, the circularly arranged smooth muscle fibers are stimulated to contract by parasympathetic stimulation, constricting the pupil.

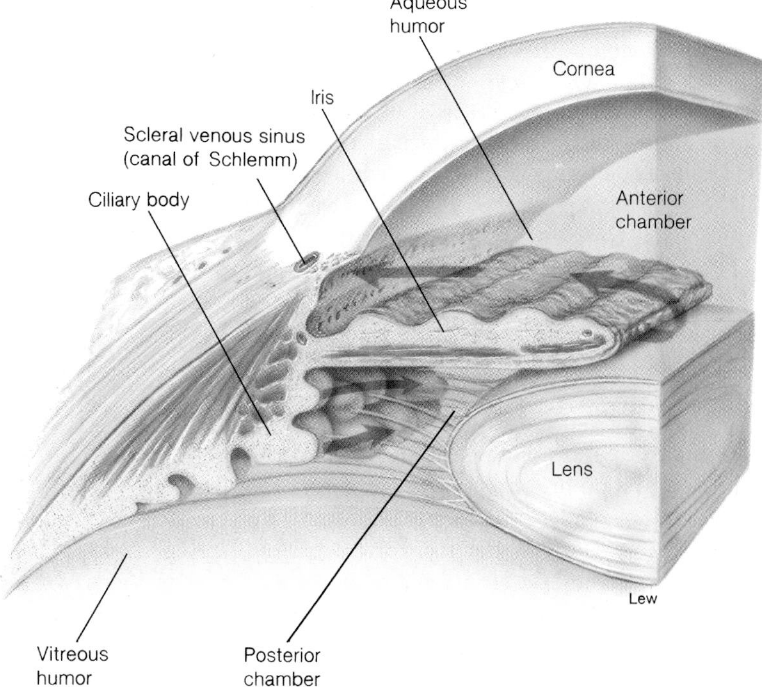

Figure 18.36

Production and drainage of aqueous humor.

Aqueous humor maintains the intraocular pressure within the anterior and posterior chambers. It is secreted into the posterior chamber, flows through the pupil into the anterior chamber, and drains from the eyeball through the scleral venous sinus.

Figure 18.37

A view of the retina as seen with an ophthalmoscope.
Optic nerve fibers leave the eyeball at the optic disc to form the optic nerve. (Note the blood vessels that can be seen entering the eyeball at the optic disc.)

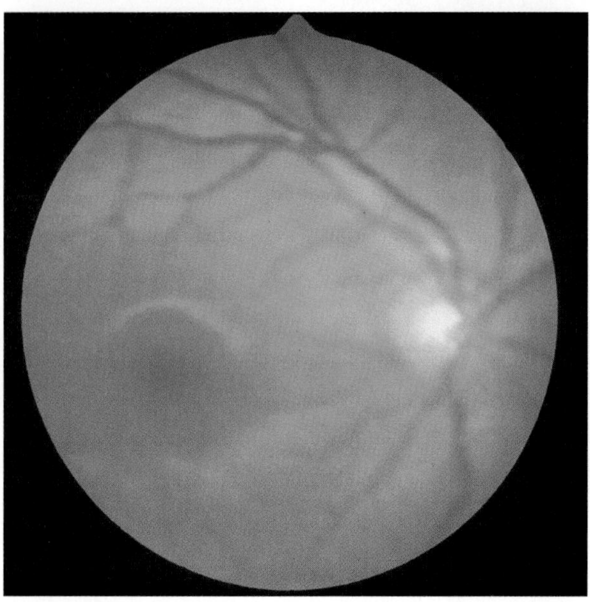

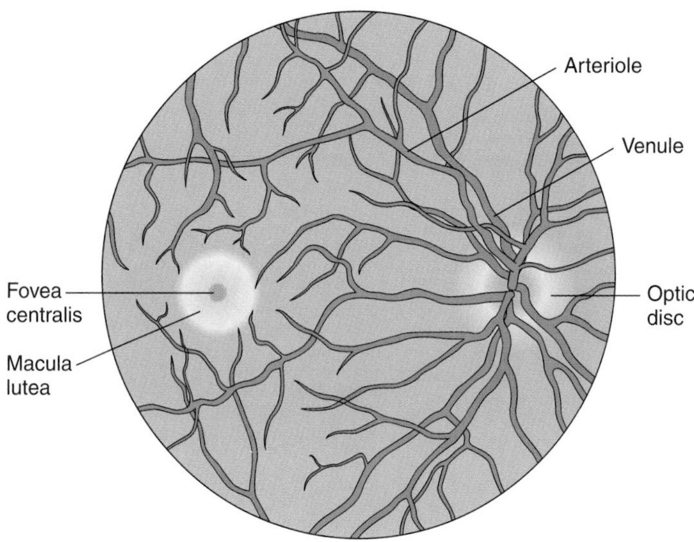

as indicated by their *refractive index*. The refractive index of air is set at 1.00; the refractive index of the cornea, by comparison, is 1.38; and the refractive indices of the aqueous humor and lens are 1.33 and 1.40, respectively. Since the greatest difference in refractive index occurs at the air-cornea interface, light is refracted most at the cornea.

The degree of refraction also depends on the curvature of the interface between two media. The curvature of the cornea is constant, but the curvature of the lens can be varied. The refractive properties of the lens can thus provide fine control for focusing light on the retina. As a result of light refraction, the image formed on the retina is upside down and right to left (fig. 18.38).

The *visual field*—which is the part of the external world projected onto the retina—is thus reversed in each eye. The cornea and lens focus the right part of the visual field on the left half of the retina of each eye, while the left half of the visual field is focused on the right half of each retina (fig. 18.39). The medial (or nasal) half-retina of the left eye therefore receives the same image as the lateral (or temporal) half-retina of the right eye. The nasal half-retina of the right eye receives the same image as the temporal half-retina of the left eye.

Accommodation

When a normal eye views an object, parallel rays of light are refracted to a point, or *focus*, on the retina (see fig. 18.55). If the degree of refraction remained constant, movement of the object closer to or farther from the eye would cause corresponding movement of the focal point, so that the focus would be either behind or in front of the retina.

The ability of the eyes to keep the image focused on the retina as the distance between the eyes and object changes is called **accommodation.** Accommodation results from contraction of the ciliary muscle, which is like a sphincter muscle that can vary its aperture (fig. 18.40). When the ciliary muscle is relaxed, its aperture is wide. Relaxation of the ciliary muscle thus places tension on the zonular fibers of the suspensory ligament and pulls the lens taut. These are the conditions when viewing an object that is 20 feet or more from a normal eye; the image is focused on the retina, and the lens is in its most flat, least convex form. As the object moves closer to the eyes, the muscles of the ciliary body contract. This muscular contraction narrows the aperture of the ciliary body and thus reduces the tension on the zonular fibers that suspend the lens. When the tension is reduced, the lens becomes more rounded and convex as a result of its inherent elasticity (fig. 18.41).

The ability of a person's eyes to accommodate can be measured by the near-point-of-vision test. The *near point of vision* is the minimum distance from the eyes that an object can be maintained in focus. This distance increases with age; indeed, accommodation in almost everyone over the age of 45 is significantly impaired. Loss of accommodating ability with age is known as *presbyopia* (*prez-be-o'pe-ă*). This loss appears to have a number of causes, including thickening of the lens and a forward movement of the attachments of the zonular fibers to the lens. As a result of these changes, the zonular fibers and lens are pulled taut even when the ciliary muscle contracts. The lens is thus not able to thicken and increase its refraction when, for example, a printed page is brought close to the eyes.

---

presbyopia: Gr. *presbys*, old; *ops*, eye

Figure 18.38

The image is inverted on the retina.

Refraction of light, which causes the image to be inverted, occurs to the greatest degree at the air/cornea interface. Changes in the curvature of the lens, however, provide the required fine focusing adjustments.

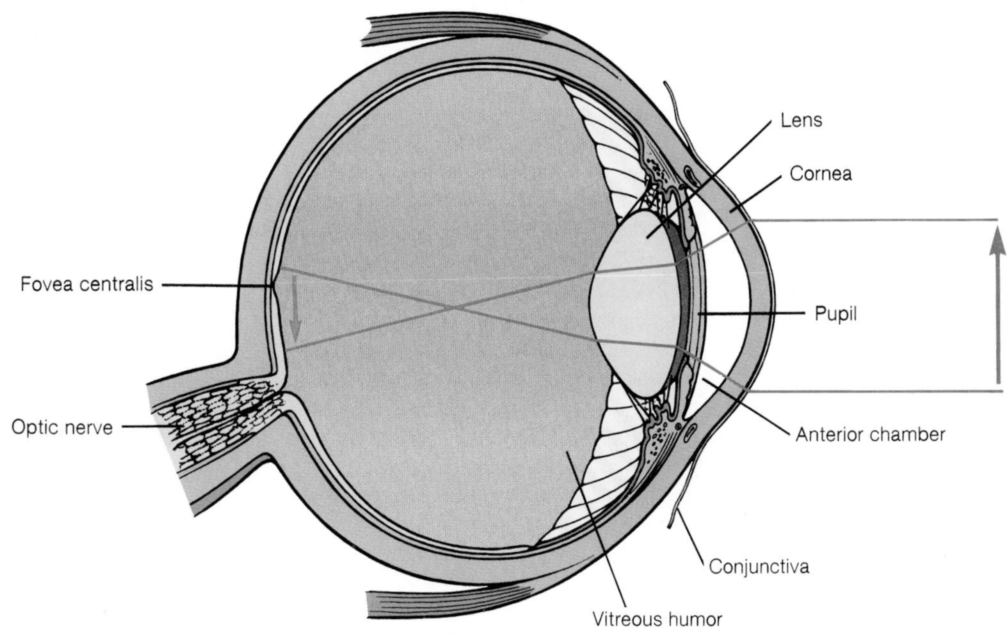

Figure 18.39

The image is switched right-to-left on the retina.

The left side of the visual field is projected to the right half of each retina, while the right side of each visual field is projected to the left half of each retina.

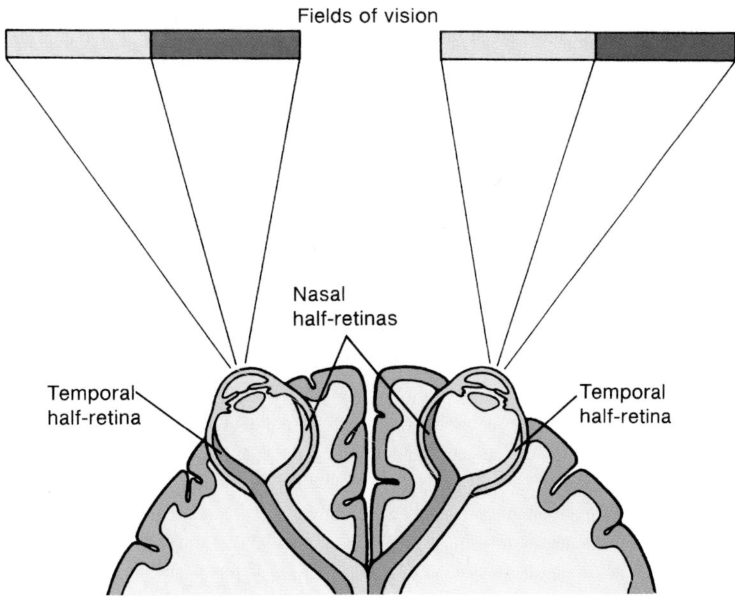

Figure 18.40

The relationship between the ciliary muscle and the lens.

(*a*) A diagram and (*b*) scanning electron micrograph (from the eye of a 17-year-old-boy) showing the relationship between the lens, zonular fibers, and ciliary muscle of the eye.

(Part [b] from "How the Eye Focuses" by James F. Koretz and George H. Handleman. Copyright © 1988 by Scientific American, Inc. All rights reserved.)

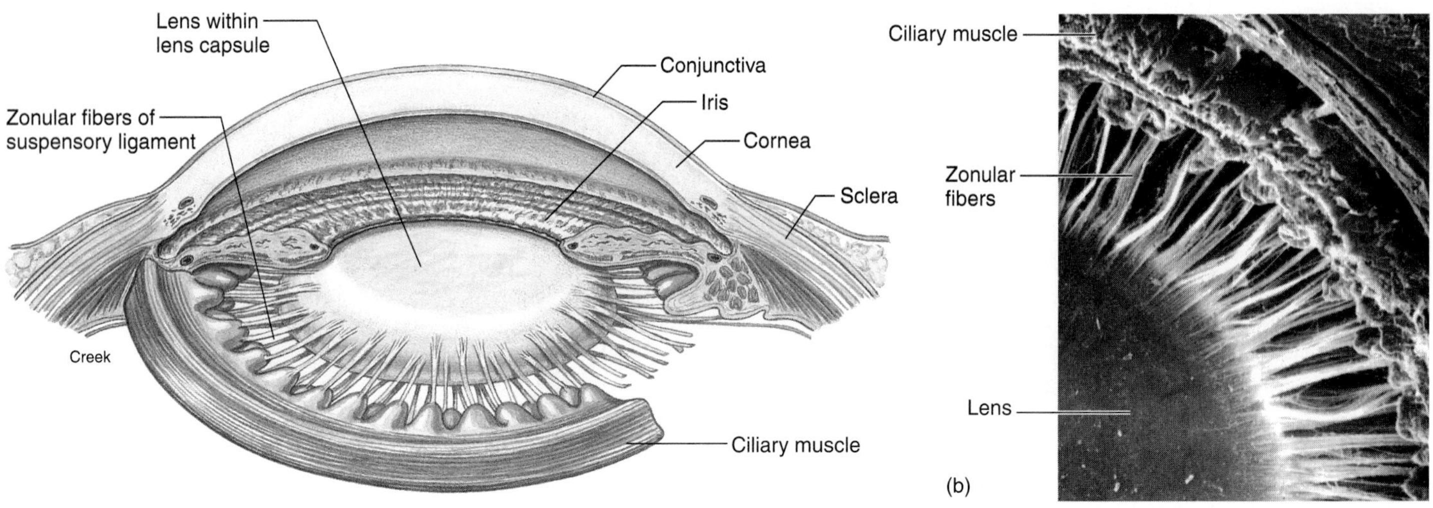

Figure 18.41

Changes in the shape of the lens produce accommodation.

(*a*) The lens is flattened for distant vision when the ciliary muscle fibers are relaxed and the suspensory ligament is taut. (*b*) The lens is more spherical for close-up vision when the ciliary muscle fibers are contracted and the suspensory ligament is relaxed.

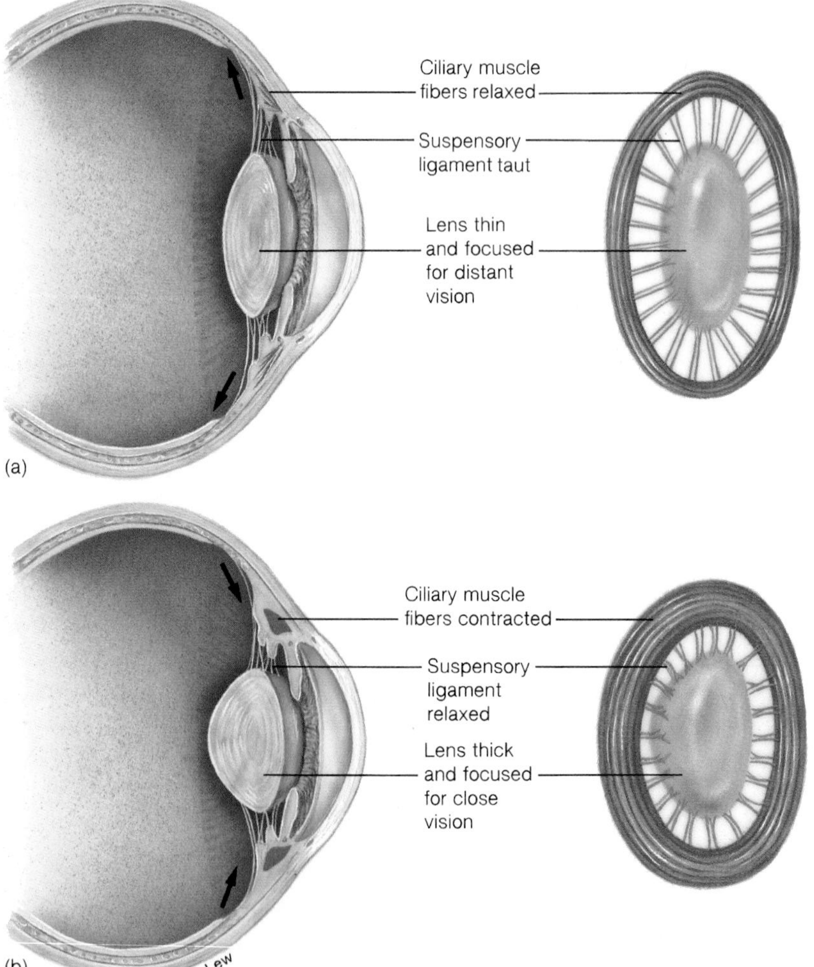

Retina

There are two types of photoreceptor neurons: rods and cones. Both receptor cell types contain pigment molecules that undergo dissociation in response to light, and it is this photochemical reaction that eventually results in the production of action potentials in the optic nerve. Rods provide black-and-white vision under conditions of low light intensities, whereas cones provide sharp color vision when light intensities are greater.

The **retina** is the principal structure of the internal tunic. The retina consists of an outer *pigmented layer* (in contact with the choroid of the vascular tunic) and an inner *nervous layer*, or *visual portion*. The retina is principally in the posterior part of the eyeball (see fig. 18.34). The thick nervous layer of the retina terminates in a jagged margin near the ciliary body called the **ora serrata** (*or'ă sĕ-ra'tă*). The thin pigmented layer extends anteriorly over the back of the ciliary body and iris.

The retina contains photoreceptor neurons called **rods** and **cones,** along with layers of other neurons (fig. 18.42). The neural layers of the retina are actually a forward extension of the brain. In this sense, the optic nerve can be considered a tract, and indeed the myelin sheaths of its fibers are derived from oligodendrocytes (like other CNS axons) rather than from Schwann cells.

Since the retina is an extension of the brain, the neural layers face outward, toward the incoming light. Light, therefore, must pass through several neural layers before striking the photoreceptors (fig. 18.43). The photoreceptors then synapse with other neurons, so that nerve impulses are conducted outward in the retina.

ora serrata: L. *ora*, margin; *serra*, saw

The neurons that contribute axons to the optic nerve are called **ganglion cells.** These neurons receive synaptic input from **bipolar cells** in the layer underneath, which in turn receive input from rods and cones. In addition to the flow of information from photoreceptors to bipolar cells to ganglion cells, neurons called *horizontal cells* synapse with several photoreceptors (and possibly also with bipolar cells), and neurons called *amacrine* (*am'ă-krin*) *cells* synapse with several ganglion cells.

Effect of Light on the Rods

The photoreceptors—rods and cones (fig. 18.44)—are activated when light produces a chemical change in molecules of pigment contained within the membranous lamellae of the outer segments of the receptor cells. Rods contain a purple pigment known as **rhodopsin** (*ro-dop'sin*). The pigment appears purple (a combination of red and blue), because it transmits light in the red and blue regions of the spectrum, while absorbing light energy in the green region. The wavelength of light that is absorbed best—the *absorption maximum*—is about 500 nm (a green-colored light).

Green cars (and other green objects) are seen more easily at night—when rods are used for vision—than red objects. This is because red light is not absorbed well by rhodopsin, and only absorbed light can produce the photochemical reaction that results in vision. In response to absorbed light, rhodopsin dissociates into its two components: the pigment **retinaldehyde** (also called **retinene** or **retinal**), which is derived from vitamin A, and a protein called **opsin.** This reaction is known as the **bleaching reaction**.

rhodopsin: Gk. *rhodon*, rose; *ops*, eye

Figure 18.42

Photomicrograph of the retina.
The retina is composed of successive layers. These layers are inverted, so that light must pass through various layers of nerve cells before reaching the photoreceptors (rods and cones).

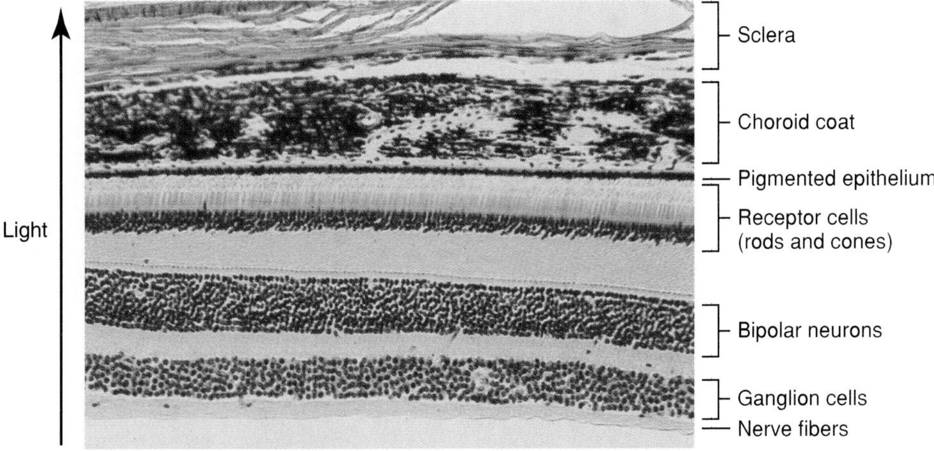

Figure 18.43

The layers of the retina.

A diagram that shows the layers of the retina and the direction that light travels through it. Nerve impulses are transmitted in the opposite direction.

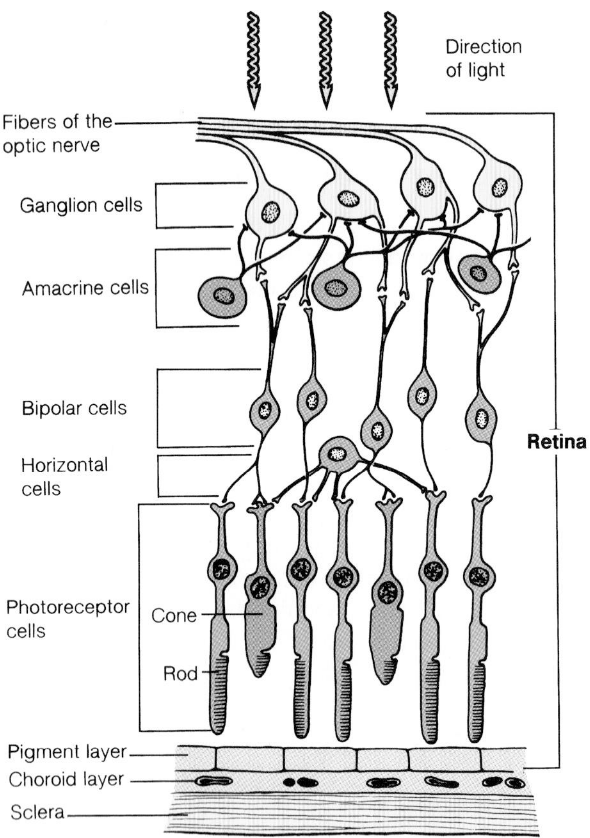

Figure 18.44

Rods and cones.

(*a*) A diagram showing the structure of a rod and a cone. (*b*) A scanning electron micrograph of rods and cones. Note that each photoreceptor contains an outer and inner segment.

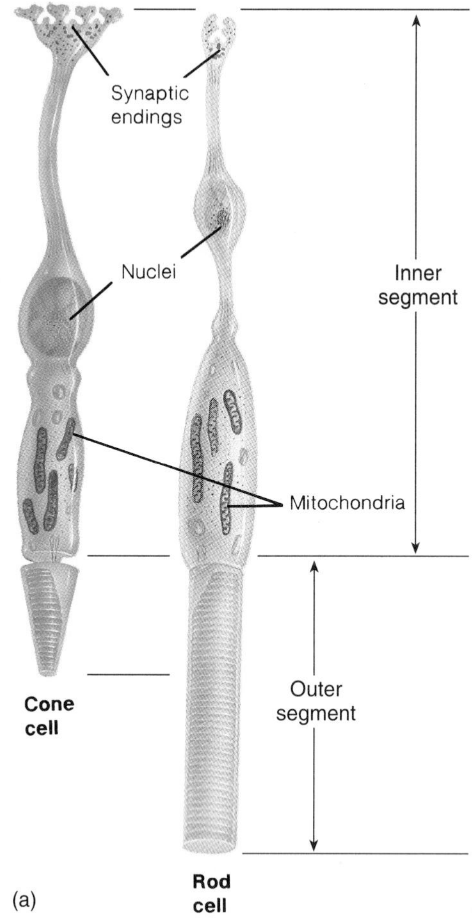

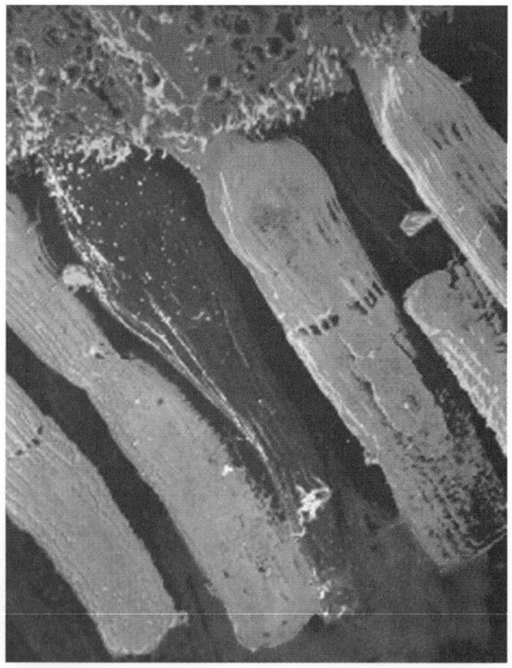

Retinene can exist in two possible configurations (shapes)—one known as the all-*trans* form and one called the 11-*cis* form (fig. 18.45). The all-*trans* form is more stable, but only the 11-*cis* form is found attached to opsin. In response to absorbed light energy, the 11-*cis* retinene is converted to the all-*trans* isomer, causing it to dissociate from the opsin. This dissociation reaction in response to light initiates changes in the ionic permeability of the rod cell membrane and ultimately results in the production of nerve impulses in the ganglion cells. As a result of these effects, rods provide black-and-white vision under conditions of low light intensity.

Dark Adaptation

The bleaching reaction that occurs in the light results in a lowered amount of rhodopsin in the rods and lowered amounts of visual pigments in the cones. When a light-adapted person first enters a darkened room, therefore, sensitivity to light is low and vision is poor. A gradual increase in photoreceptor sensitivity, known as **dark adaptation,** then occurs, reaching maximal sensitivity in about 20 minutes. The increased sensitivity to low light intensity is due partly to increased amounts of visual pigments produced in the dark.

Figure 18.45

The photodissociation of rhodopsin.
(*a*) The photopigment rhodopsin consists of the protein opsin combined with 11-*cis* retinene. (*b*) Upon exposure to light, the retinene is converted to a different form, called all-*trans*, and dissociates from the opsin. This photochemical reaction induces changes in ionic permeability that ultimately result in stimulation of ganglion cells in the retina.

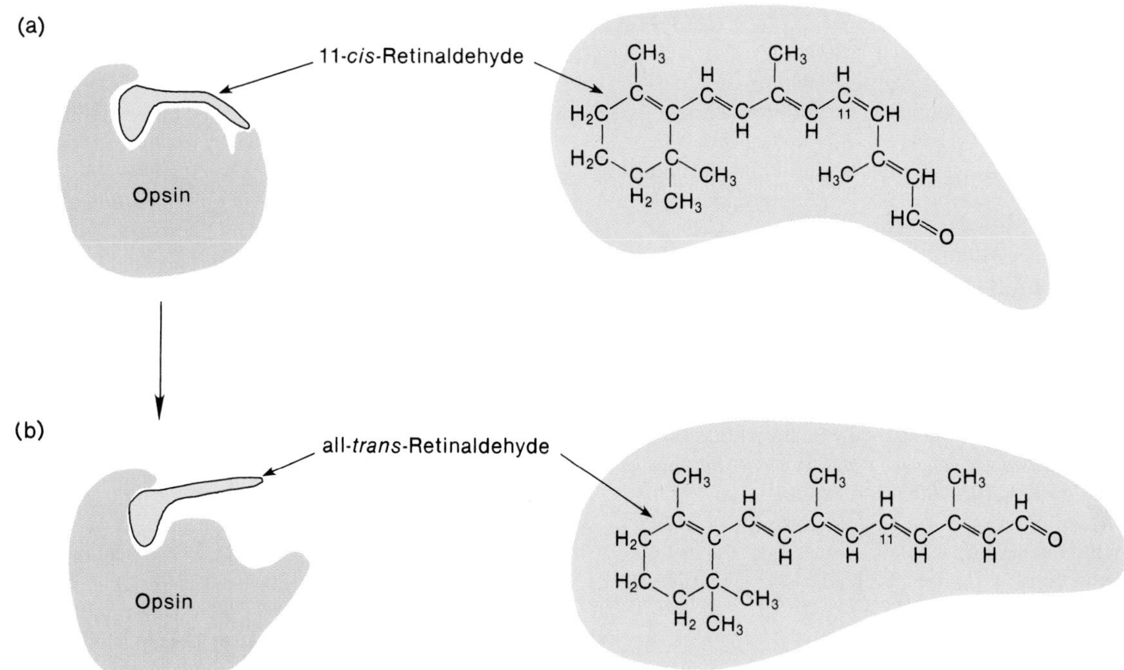

Increased pigments in the cones produce a slight dark adaptation in the first 5 minutes. Increased rhodopsin in the rods produces a much greater increase in sensitivity to low light levels and is partly responsible for the adaptation that occurs after about 5 minutes in the dark. In addition to the increased concentration of rhodopsin, other more subtle (and less well understood) changes occur in the rods that ultimately result in a 100,000-fold increase in light sensitivity in dark-adapted as compared to light-adapted eyes.

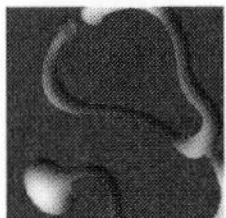

Scientists have recently discovered the genetic basis for blindness in the disease *dominant retinitis pigmentosa.* People with this disease inherit a gene for the opsin protein in which a single base change in the gene (substitution of adenine for cytosine) causes the amino acid histidine to be substituted for proline at a specific point in the polypeptide chain. This abnormal opsin leads to degeneration of the photoreceptors.

Electrical Activity of Retinal Cells

The only neurons in the retina that produce all-or-none action potentials are ganglion cells and amacrine cells. The photoreceptors, bipolar cells, and horizontal cells instead produce only graded depolarizations or hyperpolarizations, analogous to EPSPs and IPSPs.

The transduction of light energy into nerve impulses follows a cause-and-effect sequence that is the inverse of the usual way in which sensory stimuli are detected. This is because, in the dark, the photoreceptors release an inhibitory neurotransmitter that hyperpolarizes the bipolar neurons. Thus inhibited, the bipolar neurons do not release excitatory neurotransmitter to the ganglion cells. Light *inhibits* the photoreceptors from releasing their inhibitory neurotransmitter, and by this means, *stimulates* the bipolar cells and thus the ganglion cells, which transmit action potentials to the brain.

A rod or cone contains many Na^+ channels in the cell membrane of its outer segment (see fig. 18.44), and in the dark, many of these channels are open. As a consequence, Na^+ continuously diffuses into the outer segment and across the narrow stalk to the inner segment. This small flow of Na^+ that occurs in the absence of light stimulation is called the **dark current,** and it causes the membrane of a photoreceptor to be somewhat depolarized in the dark. The Na^+ channels in the outer segment rapidly close in response to light, reducing the dark current and causing the photoreceptor to hyperpolarize.

It has been discovered that cyclic GMP (cGMP) is required to keep the Na^+ channels open, and that the channels will close if the cGMP is converted into GMP. Light causes this conversion and consequent closing of the Na^+ channels. When a photopigment absorbs light, 11-*cis* retinene is converted into its isomer, all-*trans* retinene (fig. 18.45), and dissociates from the opsin, causing the opsin protein to change shape. Each opsin is associated with over a hundred regulatory G-proteins (chapter 14) known as **transducins,** and the change in the opsin induced by light causes the alpha subunits

UNDER DEVELOPMENT

Development of the Eye

The development of the eye is a complex, rapid process involving the precise interaction of neuroectoderm, surface ectoderm, and mesoderm. It begins early in the fourth week as the neuroectoderm forms a lateral diverticulum on each side of the prosencephalon (forebrain). As the diverticulum increases in size, the distal portion dilates to become the **optic vesicle,** and the proximal portion constricts to become the **optic stalk** (fig. 1). Once the optic vesicle has formed, the overlying surface ectoderm thickens and invaginates. The thickened portion is the **lens placode** (*plak'ōd*), and the invagination is the **lens fovea.**

During the fifth week, the lens placode is depressed and eventually cut off from the surface ectoderm, causing the formation of the **lens vesicle.** Simultaneously, the optic vesicle invaginates and differentiates into the two-layered **optic cup.** A groove called the **optic fissure** along the inferior surface of the optic cup allows for passage of the *hyaloid artery* and *hyaloid vein* that serve the developing eyeball. The walls of the optic fissure eventually close so that the hyaloid vessels are within the tissue of the optic stalk. They become the **central vessels** of the retina of the mature eye. The optic stalk eventually becomes the optic nerve, composed of sensory axons from the retina.

By the early part of the seventh week, the optic cup differentiates into two sheets of epithelial tissue that become the sensory and pigmented layers of the retina. Both of these layers also line the entire vascular coat, including the ciliary body, iris, and the choroid. A proliferation of cells in the lens vesicle leads to the formation of the lens. The **lens capsule** forms from the mesoderm surrounding the lens, as does the vitreous humor. Mesoderm surrounding the optic cup differentiates into two distinct layers of the developing eyeball. The inner layer of mesoderm becomes the vascular choroid; the outer layer becomes the toughened sclera posteriorly and the transparent cornea anteriorly. Once the cornea forms, additional surface ectoderm gives rise to the thin conjunctiva covering the anterior surface of the eyeball. Epithelium of the eyelids and the lacrimal glands and duct develop from surface ectoderm, whereas the extrinsic eye muscles and all connective tissues associated with the eye develop from mesoderm. These accessory structures of the eye gradually develop during the embryonic period and into the fetal period as late as the fifth month.

hyaloid: Gk. *hyalos*, glass; *oiodos*, form

of the G-proteins to dissociate. These G-protein subunits then bind to and activate hundreds of molecules of the enzyme *phosphodiesterase*. This enzyme converts cGMP to GMP, thus closing the Na^+ channels at a rate of about 1,000 per second and inhibiting the dark current. The absorption of a single photon of light can block the entry of more than a million Na^+ ions, thereby causing the photoreceptor to hyperpolarize and release less inhibitory neurotransmitter. Freed from inhibition, the bipolar cells activate ganglion cells, which transmit action potentials to the brain so that light can be perceived.

Cones and Color Vision

Cones are less sensitive than rods to light, but the cones provide color vision and greater visual acuity, as described in the next section. During the day, therefore, the high light

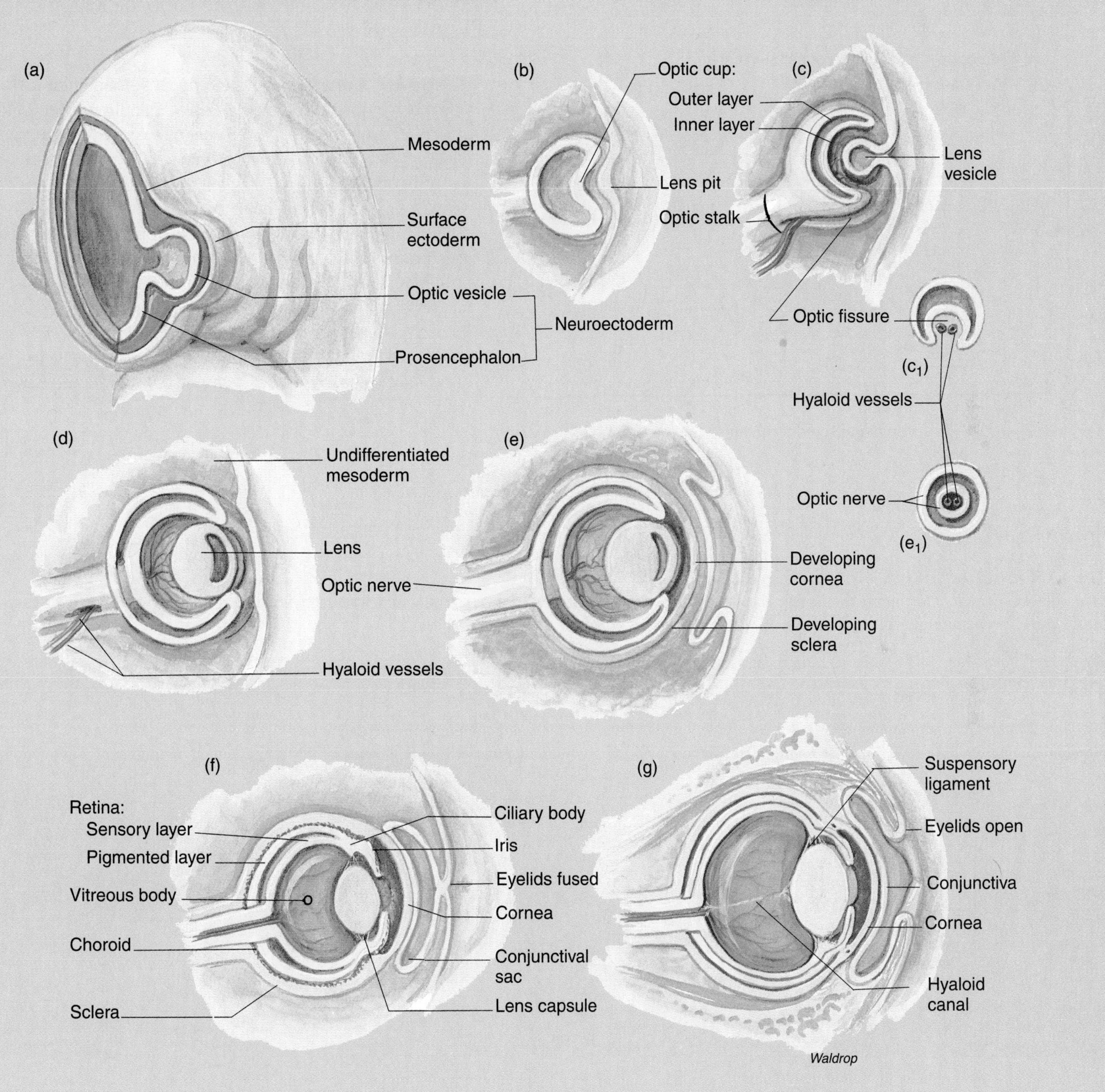

Figure 1 The development of the eye. (*a*) An anterior view of the developing head of a 22-day-old embryo and the formation of the optic vesicle from the neuroectoderm of the prosencephalon (forebrain). (*b*) The development of the optic cup. The lens vesicle is formed (*c*) as the ectodermal lens placode invaginates during the fourth week. The hyaloid vessels become enclosed within the optic nerve (c_1 and e_1) as there is fusion of the optic fissure. (*d*) The basic shape of the eyeball and the position of its internal structures are established during the fifth week. The successive development of the eye is shown at 6 weeks (*e*) and at 20 weeks (*f*), respectively. (*g*) The eye of the newborn.

intensity bleaches out the rods, and color vision with high acuity is provided by the cones. According to the **trichromatic theory of color vision,** our perception of a multitude of colors is due to stimulation of only three types of cones. Each type of cone contains retinene, as in rhodopsin, but the retinene in the cones is associated with proteins called **photopsins,** which are different from the opsin in rods. The photopsin protein is unique for each of the three cone pigments, and so each of the pigments absorbs different wavelengths of light (corresponding to colors) to a different degree. The three types of cones are designated *blue*, *green*, and *red*, according to the region of the visible spectrum in which each cone pigment absorbs light maximally (fig. 18.46). Our perception of any given color is produced by the relative degree to which each cone is stimulated by any given wavelength of visible light.

Figure 18.46

The three types of cones.

Each type contains retinene, but the protein with which the retinene is combined is different in each case. Thus, each different pigment absorbs light maximally at a different wavelength. Color vision is produced by the activity of these blue cones, green cones, and red cones.

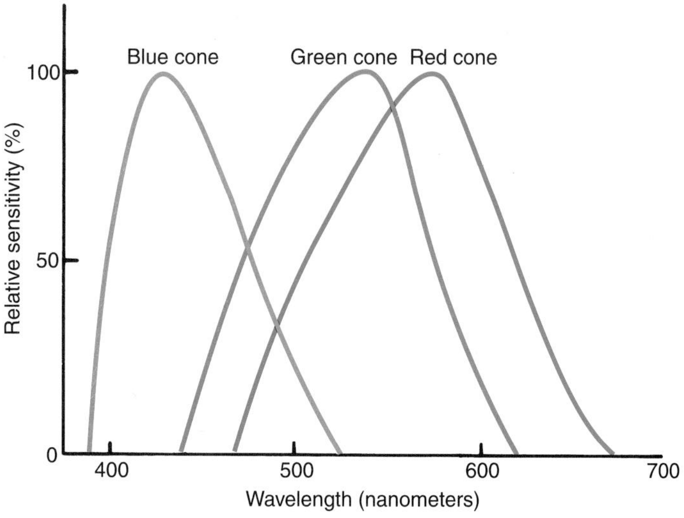

Suppose a person has become dark-adapted in a photographic darkroom over a period of 20 minutes or longer but needs light to examine some prints. Since rods do not absorb red light but red cones do, a red light in a photographic darkroom allows vision (because of the red cones) but does not cause bleaching of the rods. When the light is turned off, therefore, the rods will still be dark-adapted and the person will still be able to see.

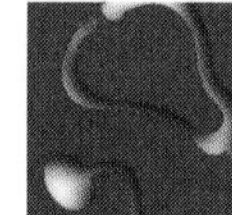

Color blindness is due to a congenital lack of one or more types of cones. People with normal color vision are *trichromats;* those with only two types of cones are *dichromats.* Dichromats may be missing red cones (have *protanopia*), or green cones (have *deuteranopia*), or blue cones (have *tritanopia*). They may have difficulty distinguishing red from green, for example. People who are *monochromats* have only one cone system and can see only black, white, and shades of gray. Color blindness is a trait carried on the X chromosome. Since men have only one X chromosome per cell, whereas women have two X chromosomes (see chapter 30), men are far more likely to be color blind than women (who can carry this trait in a recessive state).

Figure 18.47

The fovea centralis.

When the eyes "track" an object, the image is cast upon the fovea centralis of the retina. The fovea centralis is literally a "pit" formed by parting of the neural layers. In this region, light falls directly in the photoreceptors (cones).

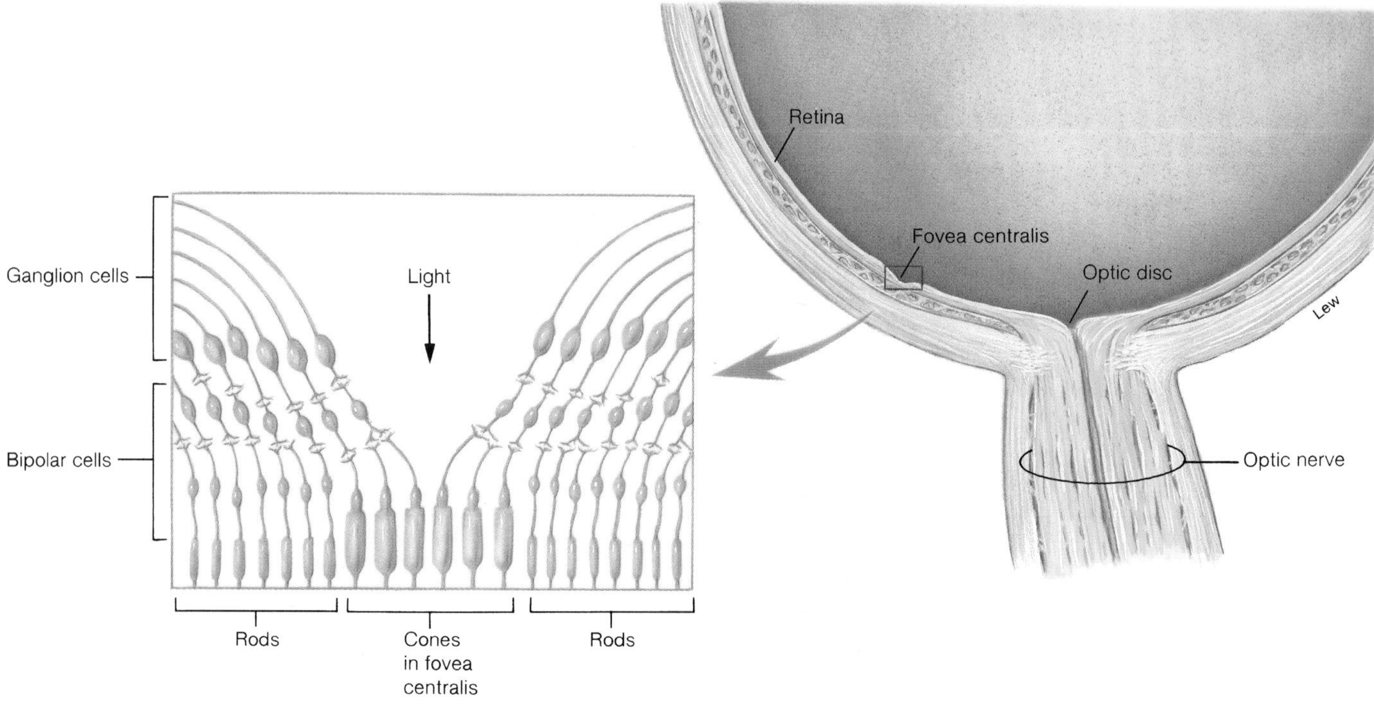

Visual Acuity and Sensitivity

While reading or similarly viewing objects in daylight, each eye is oriented so that the image falls within a tiny area of the retina called the **fovea centralis** (*fo've-ă sen-tra'lis*). The fovea centralis is a pinhead-sized pit within a yellow area of the retina called the *macula lutea* (*mak'yŭ-lă loo'te-ă*). The pit forms as a result of the displacement of neural layers around the periphery; therefore, light falls directly on photoreceptors in the center (fig. 18.47). Light falling on other areas, by contrast, must pass through several layers of neurons, as previously described.

There are approximately 120 million rods and 6 million cones in each retina, but only about 1.2 million nerve fibers enter the optic nerve of each eye. This gives an overall convergence ratio of photoreceptors on ganglion cells of about 105:1. This ratio is misleading, however, because the degree of convergence is much lower for cones than rods. In the fovea centralis, the ratio is 1:1.

---

fovea: L. *fovea*, small pit
macula lutea: L. *macula*, spot; *luteus*, yellow

The photoreceptors are distributed in such a way that the fovea centralis contains only cones, whereas more peripheral regions of the retina contain a mixture of rods and cones. Approximately 4,000 cones in the fovea centralis provide input to approximately 4,000 ganglion cells; each ganglion cell in this region, therefore, has a private line to the visual field. Each ganglion cell thus receives input from an area of retina corresponding to the diameter of one cone (about 2 μm). Peripheral to the fovea centralis, however, many rods synapse with a single bipolar cell, and many bipolar cells synapse with a single ganglion cell. A single ganglion cell outside the fovea centralis may thus receive input from large numbers of rods, corresponding to an area of about 1 mm^2 on the retina (fig. 18.48).

Since each cone in the fovea centralis has a private line to a ganglion cell, and since each ganglion cell receives input from only a tiny region of the retina, visual acuity is greatest and sensitivity to low light is poorest when light falls on the fovea. In dim light, only the rods are activated and vision is best out of the corners of the eye, where the image falls away from the fovea. Under these conditions, the convergence of large numbers of rods on a single bipolar cell

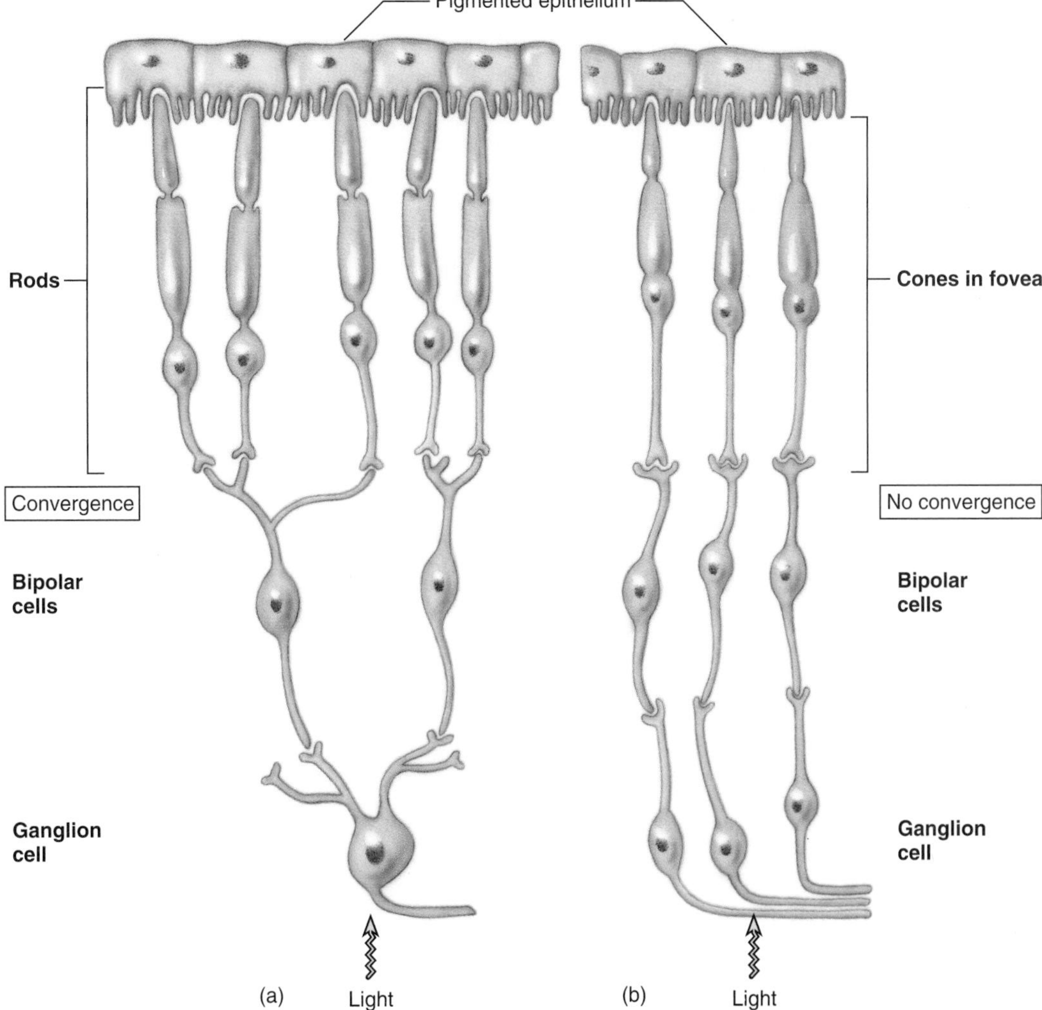

Figure 18.48

Convergence in the retina and light sensitivity. Since bipolar cells receive input from the convergence of many rods (*a*), and since a number of such bipolar cells converge on a single ganglion cell, rods maximize sensitivity to low levels of light at the expense of visual acuity. By contrast, the 1:1:1 ratio of cones to bipolar cells to ganglion cells in the fovea centralis (*b*) provides high visual acuity, but sensitivity to light is reduced.

and the convergence of large numbers of bipolar cells on a single ganglion cell increase sensitivity to dim light at the expense of visual acuity. Night vision is therefore less distinct than day vision.

The difference in visual sensitivity between cones in the fovea centralis and rods in the periphery of the retina can be easily demonstrated using a technique called *averted vision*. If you go out on a clear night and stare hard at a very dim star, it will disappear. This is because the light falls on the fovea centralis and is not sufficiently bright to activate the cones. If you then look slightly off to the side, the star will reappear because the light falls on rods away from the fovea.

Neural Pathways from the Retina

As a result of light refraction by the cornea and lens, the right half of the visual field projects to the left half of the retina of both eyes (the temporal half of the left retina and the nasal half of the right retina). The left half of the visual field projects to the right half of the retina of both eyes. The temporal half of the left retina and the nasal half of the right retina therefore see the same image. Axons from ganglion cells in the left (temporal) half of the left retina pass to the left **lateral geniculate nucleus** of the thalamus. Axons from ganglion cells in the nasal half of the right retina cross over (decussate) in the X-shaped *optic chiasma*, also to synapse in the left lateral geniculate nucleus. The left lateral geniculate, therefore, receives input from both eyes that relates to the right half of the visual field (fig. 18.49).

The right lateral geniculate nucleus, similarly, receives input from both eyes relating to the left half of the visual field. Neurons in both lateral geniculate nuclei of the thalamus in turn project to the **striate cortex** of the occipital lobe in the cerebral cortex (fig. 18.49). This area is also called area 17, in reference to a numbering system developed by K. Brodmann in 1906. Neurons in area 17 synapse with neurons in areas 18 and 19 of the occipital lobe (fig. 18.50).

Approximately 70% to 80% of the axons from the retina pass to the lateral geniculate nuclei and to the striate cortex. This **geniculostriate system** is involved in perception of the visual field. Put another way, the geniculostriate system is needed to answer the question, What is it? Approximately 20% to 30% of the fibers from the retina, however, follow a different path to the **superior colliculus** of the midbrain (also called the *optic tectum*). Axons from the superior colliculus activate motor pathways leading to eye and body movements. The **tectal system,** in other words, is needed to answer the question, Where is it?

Figure 18.49

The neural pathway of vision.
The neural pathway leading from the retina, to the lateral geniculate nucleus, and then to the visual cortex is needed for visual perception. As a result of the crossing of optic fibers, the visual cortex of each cerebral hemisphere receives input from the opposite (contralateral) visual field.

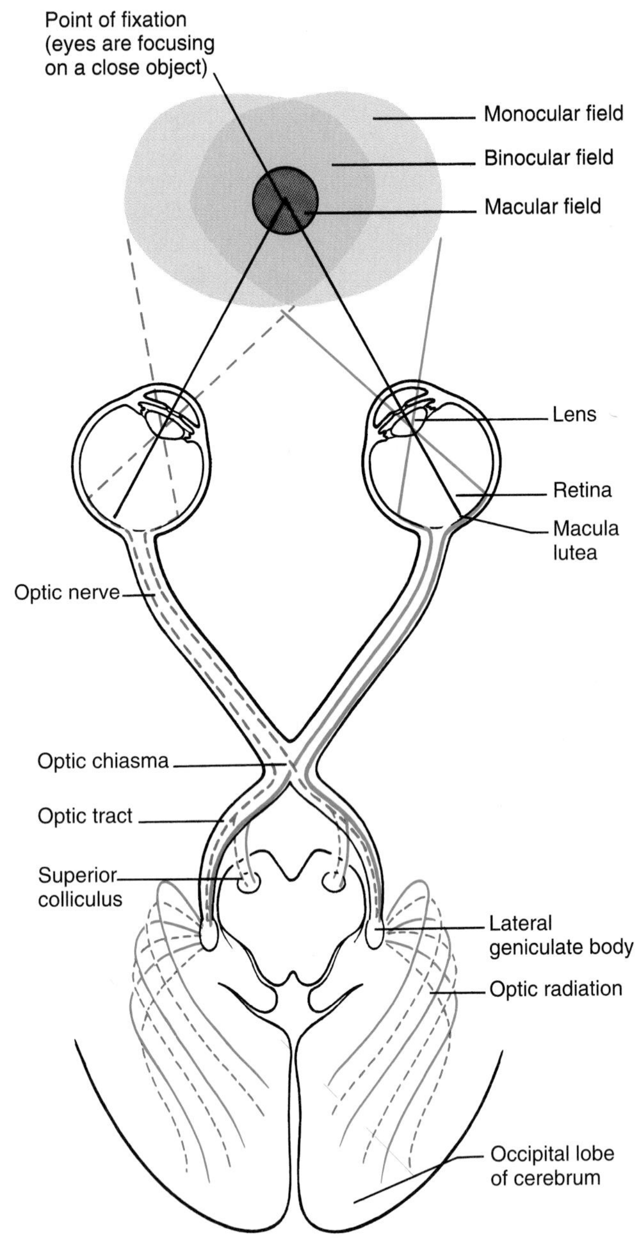

Superior Colliculus and Eye Movements

Neural pathways from the superior colliculus to motor neurons in the spinal cord help mediate the startle response to the sight of an unexpected intruder. Other nerve fibers from the superior colliculus stimulate the extrinsic eye muscles, which are the skeletal muscles that move the eyes.

Two types of eye movements are coordinated by the superior colliculus. **Smooth pursuit movements** track moving objects and keep the image focused on the fovea centralis. **Saccadic** (*să-kad'ik*) **eye movements** are quick (lasting 20 to 50 msec), jerky movements that occur while the eyes appear to be still. These saccadic movements continuously move the image to different photoreceptors; if they were to stop, the image would disappear as the photoreceptors became bleached.

saccadic: Fr. *saccade*, a sudden jerk

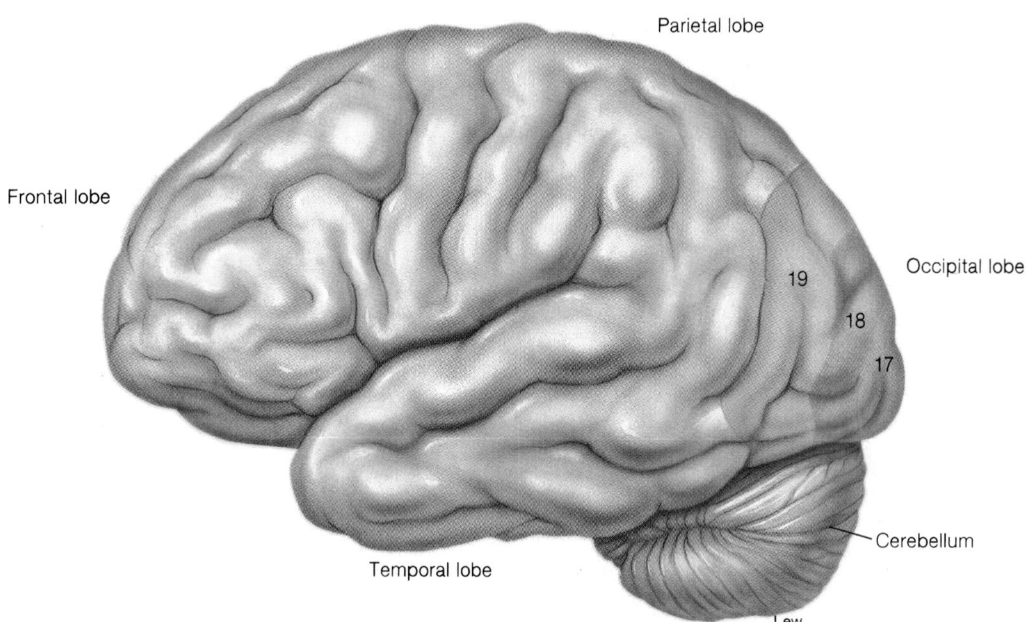

Figure 18.50

The striate cortex (area 17) and the visual association areas (18 and 19). Neural communication between the striate cortex, the visual association areas, and other brain regions is required for normal visual perception.

The tectal system is also involved in the control of the intrinsic eye muscles—the iris and the muscles of the ciliary body. Shining a light into one eye stimulates the **pupillary reflex** in which both pupils constrict. This is caused by activation of parasympathetic neurons from the superior colliculus. Postganglionic axons from the ciliary ganglia behind the eyes, in turn, stimulate constrictor fibers in the iris. Contraction of the ciliary body during accommodation for near vision also involves parasympathetic stimulation by the superior colliculus.

Neural Processing of Visual Information

Electrical activity in ganglion cells of the retina and in neurons of the lateral geniculate nucleus and cerebral cortex is evoked in response to light on the retina. The way in which each type of neuron responds to light at a particular point on the retina provides information about how the brain interprets visual information.

Light cast on the retina directly affects the activity of photoreceptors and indirectly affects the neural activity in bipolar and ganglion cells. The part of the visual field that affects the activity of a particular ganglion cell can be considered its **receptive field.** As previously mentioned, each cone in the fovea centralis has a private line to a ganglion cell, and thus the receptive fields of these ganglion cells are equal to the width of one cone (about 2 μm). By contrast, ganglion cells in more peripheral parts of the retina receive input from hundreds of photoreceptors and are therefore influenced by a larger area of the retina (about 1 mm in diameter).

Ganglion Cell Receptive Fields

Studies of the electrical activity of ganglion cells have yielded some surprising results. In the dark, each ganglion cell discharges spontaneously at a slow rate. When the room lights are turned on, the firing rate of many (but not all) ganglion cells increases slightly. A small spot of light that is directed at the center of some ganglion cells' receptive fields, however, stimulates a large increase in firing rate. Surprisingly, then, a small spot of light can be a more effective stimulus than a larger spot of light!

When the spot of light is moved only a short distance away from the center of the receptive field, the ganglion cell responds in the opposite manner. The ganglion cell that was stimulated with light at the center of its receptive field is inhibited by light in the periphery of its field. The responses produced by light in the center and by light in the "surround" of the visual field are *antagonistic*. Those ganglion cells that are stimulated by light at the center of their visual fields are said to have **on-center fields;** those that are inhibited by light in the center and stimulated by light in the surround have **off-center fields.**

The reason wide illumination of the retina has less effect than pinpoint illumination is now clear; diffuse illumination gives each ganglion cell conflicting orders—on and off. Because of the antagonism between the center and surround of ganglion cell receptive fields, the activity of each ganglion cell is a result of the *difference in light intensity* between the center and surround of its visual field. This is a form of *lateral inhibition* that helps to accentuate the contours of images, thereby sharpening edges and improving visual acuity.

Lateral Geniculate Nuclei

Each of the two lateral geniculate nuclei receives input from ganglion cells in both eyes. The right lateral geniculate receives input from the right half of each retina (corresponding to the left half of the visual field); the left lateral geniculate receives input from the left half of each retina (corresponding to the right half of the visual field). Each neuron in the lateral geniculate, however, is activated by input from only one eye. Neurons

Figure 18.51

Electron micrograph of the lateral geniculate nucleus.
Each lateral geniculate consists of six layers (numbered 1 through 6 in this figure). Each of these layers receives input from only one eye, with right and left eyes alternating. For example, the long arrow through these six layers of the left lateral geniculate encounters corresponding projections from a part of the right visual field in right and left eyes, alternately, as it passes from the outer to the inner layers.

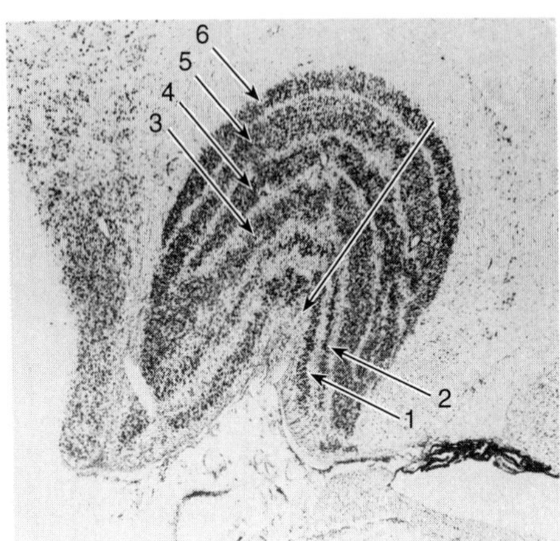

that are activated by ganglion cells from the left eye and those that are activated by ganglion cells from the right eye are in separate layers within the lateral geniculate (fig. 18.51).

The receptive field of each ganglion cell, as previously described, is the part of the retina it "sees" through its photoreceptor input. The receptive field of lateral geniculate neurons, similarly, is the part of the retina it "sees" through its ganglion cell input. Experiments in which the lateral geniculate receptive fields are mapped with a spot of light reveal that they are circular, with an antagonistic center and surround like the ganglion cell receptive fields.

Cerebral Cortex

Projections of nerve fibers from the lateral geniculate nuclei to area 17 of the occipital lobe form the *optic radiation* (see fig. 18.49). Because these fiber projections give area 17 a striped or striated appearance, this area is also known as the *striate cortex*. As mentioned earlier, neurons in area 17 project to areas 18 and 19 of the occipital lobe. Cortical neurons in areas 17, 18, and 19 are thus stimulated indirectly by light on the retina. On the basis of their stimulus requirements, these cortical neurons are classified as simple, complex, and hypercomplex.

Simple Cortical Neurons

The receptive fields of **simple cortical neurons** are rectangular rather than circular. This is because they receive input from lateral geniculate neurons whose receptive fields are aligned in a particular way (as illustrated in fig. 18.52). Simple cortical neurons are best stimulated by a slit or bar of light located in a precise part of the visual field (of either eye) at a precise orientation (fig. 18.53).

Figure 18.52

Stimulus requirements for simple cortical neurons.
Cortical neurons called simple cells have rectangular receptive fields that are best stimulated by slits of light of particular orientations. This may be due to the fact that these simple cells receive input from ganglion cells that have circular receptive fields along a particular line.

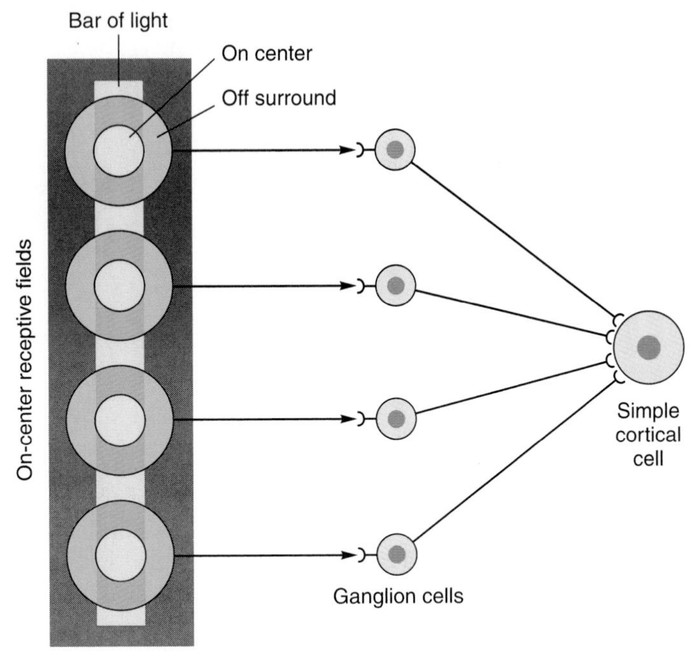

Figure 18.53

The effect of light orientation on simple cortical cells.
Simple cells are best stimulated by a slit of bar of light along a particular orientation within a particular region of the receptive field. The behavior of two different cortical cells is illustrated in (*a*) and (*b*).

Source: Data from Gunther Stent, "Cellular Communication" in Scientific American, *1972.*

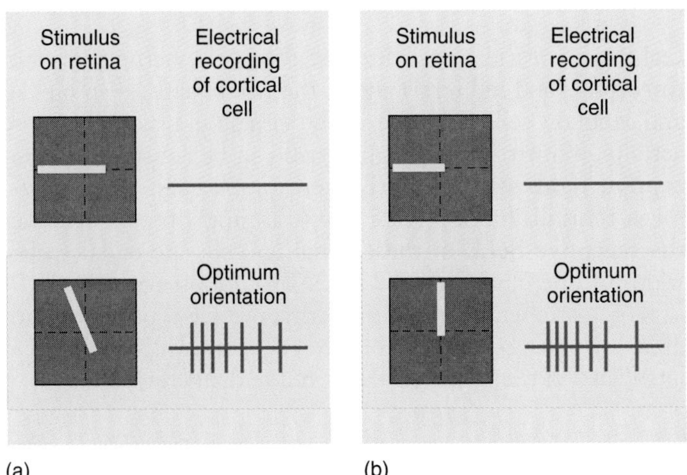

The striate cortex (area 17) contains simple, complex, and hypercomplex neurons. The other visual association areas, designated areas 18 and 19, contain only complex and hypercomplex cells. Complex neurons receive input from simple cells, and hypercomplex neurons receive input from complex cells.

Complex and Hypercomplex Cortical Neurons

Complex cortical neurons respond best to straight lines with a specific orientation that move in a particular direction through the receptive field. Unlike simple neurons, complex neurons do not require that the stimulus have a particular position within the receptive field. **Hypercomplex cortical neurons** require that the stimulus be of a particular length or have a particular bend or corner.

The dotlike information from ganglion and lateral geniculate cells is thus transformed in the occipital lobe into information about edges—their position, length, orientation, and movement. Although this information is highly abstract, the visual association areas of the occipital lobes probably represent only an early stage in the integration of visual information. Other areas of the brain receive input from the visual association areas and provide meaning to visual perception.

Clinical Considerations

Numerous disorders and diseases afflict the sensory organs. Some of these occur during the sensitive period of prenatal development and are the result of hereditary influences. Other sensory impairments, some of which are avoidable, can be acquired at any time of life. Still others result from changes associated with the natural aging process. The loss of a sense is traumatic and frequently involves a long adjustment period. Fortunately, however, when a sensory function is impaired or lost, the other senses seem to become keener, lessening the extent of the handicap. A blind person, for example, compensates somewhat for the loss of sight by developing a remarkable hearing ability.

Entire specialties within medicine are devoted to treating the disorders of specific sensory organs. It is beyond the scope of this text to attempt a comprehensive discussion of the numerous diseases and dysfunctions of these organs. We will comment generally, however, on the diagnosis of sensory disorders and on developmental problems that can affect the ears and eyes. In addition, we will discuss the more common diseases and dysfunctions of these organs.

Diagnosis of Sensory Organs

Ear

Otorhinolaryngology (*o″to-ri″no-lar″ing-gol′ ŏ-je*) is the specialty of medicine dealing with the diagnosis and treatment of diseases or conditions of the ear, nose, and throat. *Audiology* is the study of hearing, particularly assessment of the ear and its functioning.

Three common instruments or techniques are used in examining the ears to assess auditory function: (1) an *otoscope* is an instrument used to examine the tympanic membrane of the ear; abnormalities of this membrane are informative when diagnosing specific auditory problems, including middle-ear infections; (2) *tuning fork tests* are useful in determining hearing acuity and especially for discriminating the various kinds of hearing loss; and (3) *audiometry* is a functional examination for hearing sensitivity and speech discrimination.

Eye

There are two distinct professional specialties concerned with the structure and function of the eye. *Optometry* is the paramedical profession concerned with assessing vision and treating visual problems. An *optometrist* prescribes corrective lenses or visual training but is not a medical doctor and does not treat eye diseases. *Ophthalmology* (*of″thal-mol′ ŏ-je*) is the specialty of medicine concerned with diagnosing and treating eye diseases.

Although the eyeball is an extremely complex organ, it is quite accessible to examination. The following devices are frequently employed: (1) a *cycloplegic drug,* which is instilled into the eyes to dilate the pupils and temporarily inactivate the ciliary muscles; (2) a *Snellen's chart,* which determines the visual acuity of a person standing 20 feet from the chart (a reading of 20/20 is considered normal for the test); (3) an *ophthalmoscope,* which contains a light, mirrors, and lenses to illuminate and magnify the interior of the eyeball so that the structures within may be examined; and (4) a *tonometer,* which measures ocular tension, important in detecting glaucoma.

Developmental Problems of the Ears and Eyes

Although there are many congenital abnormalities of the ears and eyes, most of them are rare. The sensitive period of development for these organs is between 24 and 45 days after conception. Indeed, 85% of newborns suffer anomalies if infected during this interval. Most congenital disorders of the eyes and ears are caused by genetic factors or by intrauterine infections such as *rubella virus*.

If a pregnant woman contracts rubella (German measles), there is a 90% probability that the embryo or fetus will contract it also. An embryo afflicted with rubella is 30% more likely to be aborted, stillborn, or congenitally deformed than one that is not afflicted. Rubella interferes with the mitotic process and thus causes underdeveloped organs. An embryo with rubella may suffer from a number of physical disorders, including cataracts and *glaucoma,* which are common disorders of the eye.

Snellen's chart: From Herman Snellen, Dutch ophthalmologist, 1834–1908

glaucoma: Gk. *glaukos,* gray

Ear

Congenital deafness is generally caused by an autosomal recessive gene but may also be caused by a maternal rubella infection. The actual functional impairment is generally either a defective set of auditory ossicles or improper development of the neurosensory structures of the inner ear.

Although the shape of the auricle varies widely, **auricular abnormalities** are not uncommon, especially in infants with chromosomal syndromes causing mental deficiencies. In addition, the external auditory canal frequently does not develop in these children, resulting in a condition called **atresia** (*ă-tre′se-ă*) of the external auditory canal.

Eye

Most **congenital cataracts** are hereditary, but they may also be caused by maternal rubella infection during the critical fourth to sixth week of eye development. In this condition, the lens is opaque and frequently appears grayish white.

Cyclopia is a rare condition in which the eyes are partially fused into a median eye enclosed by a single orbit. Other severe malformations, which are incompatible with life, are generally expressed with this condition.

Infections, Diseases, and Functional Impairments of the Ear

Disorders of the ear are common and may affect both hearing and the vestibular functions. The ear is afflicted by numerous infections and diseases, some of which can be prevented.

Infections and Diseases

External otitis (*o-ti′tis*) is a general term for infections of the outer ear. The causes of external otitis range from dermatitis to fungal and bacterial infections.

Acute purulent otitis media is a middle-ear infection. Pathogens of this disease usually enter through the auditory tube, most often following a cold or tonsillitis. Children frequently have middle-ear infections because of their susceptibility to infections and their short and straight auditory tubes. As a middle-ear infection progresses to the inflammatory stage, the auditory tube closes, and drainage is prohibited. An intense earache is a common symptom of a middle-ear infection. The pressure from the inflammation may eventually rupture the tympanic membrane to permit drainage.

Repeated middle-ear infections, particularly in children, usually require an incision of the tympanic membrane known as a **myringotomy** (*mir″ing-got′ŏ-me*) and the implantation of a tiny tube within the tympanic membrane (fig. 18.54) to assist the patency of the auditory tube. The tube, which is eventually sloughed out of the ear, permits the infection to heal and helps prohibit further infections by keeping the auditory tube open.

Perforation of the tympanic membrane may occur as the result of infections or trauma. Sudden, intense noise can rupture the membrane. Spontaneous perforation of the membrane usually heals rapidly, but scar tissue may form and lessen the sensitivity to sound vibrations.

Figure 18.54

An implanted ventilation tube in the tympanic membrane.
This procedure followed a myringotomy.

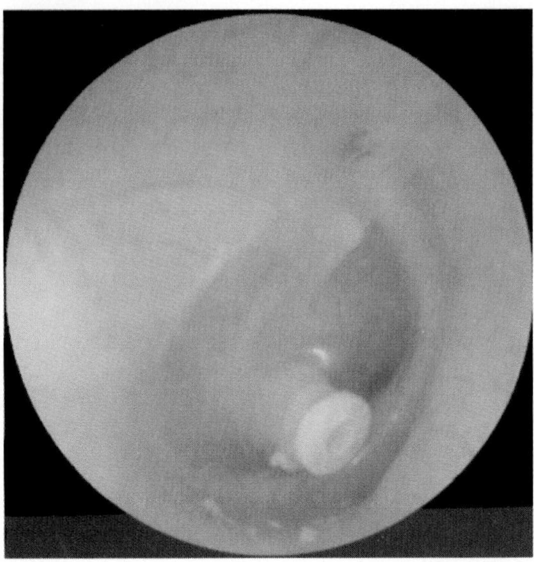

Otosclerosis (*o″to-sklĕ-ro′sis*) is a progressive deterioration of the normal bone of the bony labyrinth and its replacement with vascular spongy bone. This frequently causes hearing loss as the auditory ossicles are immobilized. Surgical scraping of the bone growth and replacement of the stapes with a prosthesis is the most frequent treatment for otosclerosis.

Ménière's disease afflicts the inner ear and may cause hearing loss as well as equilibrium disturbance. The causes of Ménière's disease vary and are not completely understood, but they are thought to be related to a dysfunction of the autonomic nervous system that causes a vasoconstriction within the inner ear. The disease is characterized by recurrent periods of **vertigo** (dizziness and a sensation of rotation), **tinnitus** (ringing in the ear), and progressive deafness in the affected ear. Ménière's disease is chronic and affects both sexes equally. It is more common in middle-aged and elderly people.

Auditory Impairment

Loss of hearing results from disease, trauma, or developmental problems involving any portion of the auditory apparatus, cochlear nerve and auditory pathway, or areas of auditory perception within the brain. Hearing impairment varies from slight disablement, which may or may not worsen, to total deafness. Some types of hearing impairment, including deafness, can be mitigated through hearing aids or surgery.

Based on the structures involved, there are two types of deafness. **Conduction deafness** is caused by interference with the sound waves through the outer or middle ear. Conduction problems include impacted cerumen (wax), a ruptured tympanic membrane, a severe middle-ear infection, or adhesions of one or more auditory ossicles (otosclerosis). Medical treatment usually improves the hearing loss from conduction deafness.

Perceptive (sensorineural) deafness results from disorders that affect the inner ear, the cochlear nerve or nerve pathway, or auditory centers within the brain. Perceptive impairment ranges in severity from the inability to hear certain frequencies to total deafness. Perceptive deafness may be caused by diseases, trauma, or genetic or developmental problems. Elderly people frequently experience some perceptive deafness. The ability to perceive high-frequency sounds is generally affected first. Hearing aids may help patients with perceptive deafness. This type of deafness is permanent, however, because it involves destruction of sensory structures that cannot regenerate.

Functional Impairments of the Eye

Few people have perfect vision. Slight variations in the shape of the eyeball or curvature of the cornea or lens result in an imperfect focal point of light waves onto the retina. Most variations are slight, however, and the error of refraction goes unnoticed. Severe deviations that are not corrected may cause blurred vision, fatigue, chronic headaches, and depression.

The primary clinical considerations associated with defects in the refractory structures or general shape of the eyeball are myopia, hyperopia, presbyopia, and astigmatism. **Myopia** (nearsightedness) is an elongation of the eyeball that causes light waves to focus at a point in the vitreous humor in front of the retina (fig. 18.55). Only light waves from close objects can be focused clearly on the retina; distant objects appear blurred. **Hyperopia** (farsightedness) is a condition in which the eyeball is too short; consequently, light waves are brought to a focal point behind the retina. Although visual accommodation aids a hyperopic person, it generally does not help enough for the person to clearly see very close or distant objects. **Presbyopia** is a condition in which the lens tends to lose its elasticity and ability to accommodate. It is relatively common in people over the age of 50. In order to read print on a page, a person with presbyopia must hold the page farther than the normal reading distance from the eyes. **Astigmatism** is a condition in which an irregular curvature of the cornea or lens of the eye distorts the refraction of light waves. If a person with astigmatism views a circle, the image will not appear clear in all 360°; the parts of the circle that appear blurred can be used to map the astigmatism.

Various glass or plastic lenses frequently benefit people with the visual impairments just described. Myopia may be corrected with a biconcave lens; hyperopia with a biconvex lens; and presbyopia with bifocals, or a combination of two lenses adjusted for near and distant vision. Correction for astigmatism requires a careful assessment of the irregularities and the prescription of a specially ground corrective lens.

Figure 18.55

Refraction problems in the eyes and their correction.
In a normal eye (*a*), parallel rays of light are brought to a focus on the retina by refraction in the cornea and lens. If the eye is too long, as in myopia (*b*), the focus is in front of the retina. This can be corrected by a concave lens. If the eye is too short, as in hyperopia (*c*), the focus is behind the retina. This is corrected by a convex lens. In astigmatism (*d*), light refraction is uneven due to an abnormal shape of the cornea or lens.

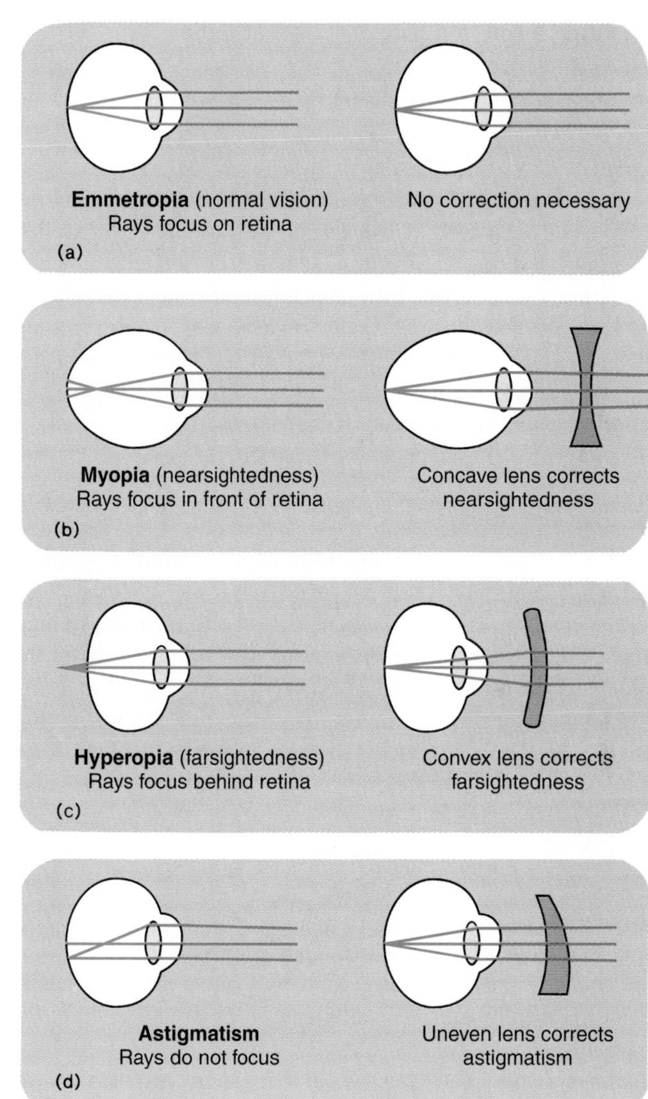

The condition of myopia may also be treated by a surgical procedure called **radial keratotomy** (*ker-ă-tot'ŏ-me*). In this technique, 8 to 16 microscopic slashes, like the spokes of a wheel, are made in the cornea from the center to the edge. The ocular pressure inside the eyeball bulges the weakened cornea and flattens its center, changing the focal length of the eyeball. Side effects may include vision fluctuations, sensitivity to glare, and incorrect corneal alteration. In a relatively new laser surgery called **photorefractive keratectomy** (*ker-ă-tek'tŏ-me*), the cornea is flattened by vaporizing microscopic slivers from its surface.

astigmatism: Gk. *a*, without; *stigma*, point

Cataracts

A **cataract** is a clouding of the lens that leads to a gradual blurring of vision and the eventual loss of sight. A cataract is not a growth within or upon the eye, but rather a chemical change in the protein of the lens. It is caused by injury, poisons, infections, or age degeneration. Recent evidence indicates that even excessive UV light may cause cataracts.

Cataracts are the leading cause of blindness. A cataract can be removed surgically, however, and vision is restored by implanting a tiny intraocular lens that either clips to the iris or is secured into the vacant lens capsule. Special contact lenses or thick lenses for glasses are other options.

Detachment of the Retina

Retinal detachment is a separation of the nervous or visual layer of the retina from the underlying pigment epithelium. It generally begins as a minute tear in the retina that gradually extends as vitreous humor accumulates between the layers. Retinal detachment may result from hemorrhage, a tumor, degeneration, or trauma from a violent blow to the eye. A detached retina may be repaired by using laser beams, cryoprobes, or intense heat to destroy the tissue beneath the tear and rejoin the layers.

Macular Degeneration

Another retinal disorder, **macular degeneration,** is common after age 70, but its cause is not fully known. It occurs when the blood supply to the macular area of the retina is reduced, often eventually resulting in hemorrhages, or when other fluid collects in this area, reducing central vision sharpness. People with macular degeneration retain peripheral vision but have difficulty focusing on objects directly in front of them. A simple test called the *Amsler grid* can be used to detect this disorder in its early stages.

Strabismus

Strabismus is a condition in which both eyes do not focus upon the same axis of vision. This prevents stereoscopic vision, and the afflicted individuals will have varied visual impairments. Strabismus is usually caused by a weakened extrinsic eye muscle.

Strabismus is assessed while the patient attempts to look straight ahead. If the afflicted eye turns toward the nose, the condition is called convergent strabismus **(esotropia).** If the eye turns outward, it is called divergent strabismus **(exotropia).** Disuse of the afflicted eye causes a visual impairment called **amblyopia.** Visual input from the normal eye and the eye with strabismus results in **diplopia,** or double vision. A normal, healthy person who has overindulged in alcoholic beverages may experience diplopia.

Amsler grid: From Marc Amsler, Swiss ophthalmologist, 1891–1968

Infections and Diseases of the Eye

Infections

Infections and inflammation can occur in any of the accessory structures of the eye and in structures within or on the eyeball itself. The causes of infections are usually microorganisms, mechanical irritation, or sensitivity to particular substances.

Conjunctivitis (inflammation of the conjunctiva) may result from sensitivity to light, allergens, or an infection caused by viruses or bacteria. Bacterial conjunctivitis is commonly called "pinkeye."

Keratitis (inflammation of the cornea) may develop secondarily from conjunctivitis or be caused by diseases such as tuberculosis, syphilis, mumps, or measles. Keratitis is painful and may cause blindness if untreated.

A **chalazion** (*kă-la'ze-on*) is a tumor or cyst on the eyelid that results from infection of the tarsal glands and a subsequent blockage of the ducts of these glands.

Styes (hordeola) are relatively common mild, infections of the follicle of an eyelash or the sebaceous gland of the follicle. Styes may spread readily from one eyelash to another if untreated. Poor hygiene and the excessive use of cosmetics may contribute to development of styes.

Diseases

Trachoma (*tră-ko'mă*) is a highly contagious bacterial disease of the conjunctiva and cornea. Although rare in the United States, trachoma afflicts an estimated 500 million people worldwide. Trachoma may be treated readily with sulfonamides and some antibiotics, but if untreated it will spread progressively until it covers the cornea. At this stage, vision is lost and the eye undergoes degenerative changes.

Glaucoma is the second leading cause of blindness and is particularly common in underdeveloped countries. Although it can afflict individuals of any age, 95% of the cases involve people over the age of 40. Glaucoma is an abnormal increase in the intraocular pressure of the eyeball where aqueous humor does not drain through the scleral venous sinus as quickly as it is produced. Accumulation of fluid causes compression of the blood vessels in the eyeball and compression of the optic nerve. Retinal cells die and the optic nerve may atrophy, producing blindness.

chalazion: Gk. *chalazion*, hail; a small tubercle

Clinical Investigation Answer

The woman apparently has myopia (with or without astigmatism), because her glasses allow her to see distant objects that would otherwise be blurry. In the past year or so, she seems also to have developed presbyopia, which would be expected at her age. Her inability to focus on small print held close to her eyes while wearing her glasses indicates a loss of ability to accommodate. Bifocals contain two different lenses—one to help focus on distant objects, and one for close vision—and would help compensate for this woman's visual impairments. The fact that her introocular pressure is normal indicates that she does not have glaucoma.

Chapter Summary

Characteristics of Sensory Receptors (pp. 498–501)

1. Receptors may be classified as chemoreceptors, mechanoreceptors, thermoreceptors, or photoreceptors; they may be further classified as exteroceptors or interoceptors (visceroceptors).
   (a) Receptors that fire only with a change in the stimulus are phasic; those that continue to fire as long as the stimulus is maintained are tonic.
   (b) According to the law of specific nerve energies, a receptor is sensitive to only one modality of stimulus, and the brain interprets action potentials originating from that receptor as a perception of that modality of stimulus.
2. Stimuli produce graded generator potentials in the sensory organs, analogous to EPSPs.

Somatic Senses (pp. 501–508)

1. Somatic senses are those in which the receptors are localized within the body wall. They include cutaneous receptors and proprioceptors.
   (a) Cutaneous receptors may be either free dendritic endings or dendrites encased in supporting structures.
   (b) Receptors for heat, cold, and pain are naked dendrites; lamellated (pacinian) corpuscles are receptors for deep pressure.
   (c) Proprioceptors include joint kinesthetic receptors, neuromuscular spindles, and neurotendinous receptors.
2. Joint kinesthetic receptors provide information about joint position and movement.
3. Neuromuscular spindles elicit the muscle stretch reflex and provide information about muscle length.
   (a) Motor neurons that stimulate extrafusal (outside the neuromuscular spindle) fibers to contract are called alpha motoneurons; those that cause the intrafusal muscle fibers of the spindles to contract are called gamma motoneurons.
   (b) Only stimulation by the alpha motoneurons can cause muscle contraction that results in skeletal movements.
   (c) Activation of gamma motoneurons enhances the stretch reflex.
4. Neurotendinous receptors (Golgi tendon organs) monitor tension in tendons. When activated, they inhibit contracting muscles, causing them to relax.

Taste and Smell (pp. 509–511)

1. Taste buds are located on the papillae of the tongue and respond to either sweet, sour, bitter, or salty tastes, depending upon their location on the tongue.
   (a) Sour is evoked by H^+ and salty tastes by the presence of Na^+.
   (b) Bitter and sweet tastes involve the interaction of organic molecules with receptor proteins that are linked to a family of G-proteins of the type called gustducin.
2. The olfactory receptors of the olfactory nerve transmit their impulses directly through the olfactory bulb to the cerebral cortex.
   (a) Olfaction involves the interaction of an odorant molecule with a specific receptor protein, and the effect is amplified many times by the release of numerous G-protein subunits.
   (b) As many as 1,000 genes may code for different olfactory receptor proteins, allowing humans to distinguish up to 10,000 different odors.

Vestibular Apparatus and Equilibrium (pp. 512–516)

1. The sense of equilibrium is provided by the vestibular apparatus of the inner ear. The vestibular apparatus includes the otolith organs (utricle and saccule) and the three semicircular canals.
   (a) The sensory structures for equilibrium and hearing are hair cells contained within a membranous labyrinth filled with endolymph.
   (b) The membranous labyrinth is contained within a bony cavity called the bony labyrinth, which is filled with perilymph.
2. Sensory hair cells of the otolith organs have hair processes, or stereocilia, and one kinocilium that project into an otolithic membrane. Bending of the stereocilia in the direction of the kinocilium evokes a depolarization of the hair cell and the release of a synaptic transmitter, which stimulates associated sensory neurons.
3. In the semicircular canals, the hair processes are embedded in a gelatinous cupula, which is bent by movements of the head along the axis of each canal. Bending of the hairs results in depolarization of the hair cell and the release of a synaptic transmitter, which stimulates the associated neurons of the eighth cranial nerve.

The Ears and Hearing (pp. 516–523)

1. The outer ear, consisting of the auricle and the external auditory canal, is bounded by the tympanic membrane.
2. The middle-ear cavity contains the malleus, incus, and stapes. The malleus is attached to the tympanic membrane, and the stapes is attached to the oval window of the cochlea.
3. Vibrations of the oval window in response to sound produce displacements of fluid within the cochlear duct, which is a part of the membranous labyrinth of the inner ear.
4. Displacements of fluid within the cochlea in response to sound cause the lower part of the cochlear duct, known as the basilar membrane, to vibrate.
   (a) These vibrations cause hair processes of hair cells on this membrane to bend.
   (b) The basilar membrane, hair cells, and the tectorial membrane in which the hair processes are embedded together comprise the spiral organ, or organ of Corti.
   (c) Bending of hair cells in the spiral organ produces impulses that are transmitted to the brain in the eighth cranial nerve and interpreted as sound.
   (d) The pitch of the sound is determined by the location of the stimulated hair cells on the basilar membrane.

The Eyes and Vision (pp. 524–532)

1. Accessory structures of the eyes include the eyebrows, eyelids, eyelashes, conjunctiva, and lacrimal apparatus.

2. The fibrous tunic of the eyeball includes the opaque sclera, or white of the eye, and the transparent cornea, which admits light into the eye.
   (a) The middle layer, or vascular tunic, consists of the choroid, ciliary body, and iris.
   (b) The iris is composed of two layers of muscle that can dilate or constrict the pupil to admit more or less light into the eye.
3. The ciliary body contains muscles that, through attachments of the suspensory ligament to the lens, can alter the degree to which the lens is stretched and thereby vary the refractive power of the lens.
   (a) Contraction of the ciliary muscles releases tension on the lens, allowing it to increase its convexity and focusing power as an object moves closer to the eyes.
   (b) Relaxation of the ciliary muscles places tension on the lens and flattens it, allowing an image to remain in focus as an object moves farther away from the eyes.
4. Light passes through the lens and vitreous humor to reach the inner layer of the eye—the retina.

Retina (pp. 533–541)

1. The retina consists of a pigmented layer and a nervous layer; the latter includes photoreceptor rods and cones.
   (a) Light passes through a ganglion cell layer and a bipolar cell layer before reaching the photoreceptors in the retina.
   (b) Impulses are conducted from the photoreceptors to bipolar neurons, and then to ganglion neurons that provide axons to the optic nerve, which exits the eye at the optic disc.
2. Cones provide daylight color vision and are responsible for visual acuity.
   (a) They are most concentrated in the fovea centralis, where they have 1:1 synapses with bipolar neurons.
   (b) There are three types of cones, described by their maximal sensitivity as red, green, and blue.
3. Rods respond to dim light for black-and-white vision.
   (a) They are stimulated by the photodissociation of rhodopsin, the retinene part of which is derived from vitamin A.
   (b) Rods converge onto bipolar cells, thus providing high visual sensitivity at low levels of illumination.
4. In the dark, Na^+ enters the photoreceptors to produce a "dark current" that is inhibited by the presence of light; this reduces the release of inhibitory neurotransmitter, thereby activating the bipolar and ganglion cells.

Neural Processing of Visual Information (pp. 541–543)

1. Visual information is transmitted from the ganglion cells in the retina to the lateral geniculate bodies, and then to the striate cortex of the occipital lobes of the cerebral cortex.
2. The receptive field of a ganglion cell is roughly circular, with an "on" or "off" center and an antagonistic surround.
3. Each lateral geniculate body receives input from both eyes relating to the same part of the visual field.
4. Cortical neurons involved in vision are either simple, complex, or hypercomplex.
   (a) Simple neurons receive input from neurons in the lateral geniculate; complex neurons receive input from simple cells; and hypercomplex neurons receive input from complex cells.
   (b) Simple neurons are best stimulated by a slit or bar of light located in a precise part of the visual field at a precise orientation; complex and hypercomplex neurons have different stimulus requirements.

Review Activities

Objective Questions

Match the vestibular organ on the left with its correct component on the right.

1. utricle and saccule — (a) cupula
2. semicircular canals — (b) ciliary body
3. cochlea — (c) basilar membrane
   (d) otolithic membrane

4. Which of the following is an avascular ocular tissue?
   (a) sclera
   (b) cornea
   (c) ciliary body
   (d) iris
5. The middle ear is separated from the inner ear by
   (a) the round window.
   (b) the tympanic membrane.
   (c) the oval window.
   (d) both *a* and *c*.
6. The dissociation of rhodopsin in the rods in response to light causes
   (a) the Na^+ channels to become blocked.
   (b) the rods to secrete less neurotransmitter.
   (c) the bipolar cells to become either stimulated or inhibited.
   (d) all of the above.
7. Tonic receptors
   (a) are fast-adapting.
   (b) do not fire continuously to a sustained stimulus.
   (c) produce action potentials at a greater frequency as the generator potential increases.
   (d) are described by all of the above.
8. Cutaneous receptive fields are smallest in
   (a) the fingertips.
   (b) the back.
   (c) the thighs.
   (d) the arms.
9. The process of lateral inhibition
   (a) increases the sensitivity of receptors.
   (b) promotes sensory adaptation.
   (c) increases sensory acuity.
   (d) prevents adjacent receptors from being stimulated.
10. The receptors for taste are
    (a) naked sensory nerve endings.
    (b) encapsulated sensory nerve endings.
    (c) modified epithelial cells.
11. Which of the following statements about the utricle and saccule are *true?*
    (a) They are otolith organs.
    (b) They are located in the middle ear.
    (c) They provide a sense of linear acceleration.
    (d) Both *a* and *c* are true.
    (e) Both *b* and *c* are true.
12. Since fibers of the optic nerve that originate in the nasal halves of each retina cross at the optic chiasma, each lateral geniculate receives input from
    (a) both the right and left sides of the visual field of both eyes.
    (b) the ipsilateral visual field of both eyes.
    (c) the contralateral visual field of both eyes.
    (d) the ipsilateral field of one eye and the contralateral field of the other eye.
13. When a person with normal vision views an object from a distance of at least 20 feet,
    (a) the ciliary muscles are relaxed.
    (b) the suspensory ligament is tight.
    (c) the lens is in its most flat, least convex shape.
    (d) all of the above apply.

14. Parasympathetic nerves that stimulate constriction of the iris (in the pupillary reflex) are activated by neurons in
    (a) the lateral geniculate.
    (b) the superior colliculus.
    (c) the inferior colliculus.
    (d) the striate cortex.
15. A bar of light in a specific part of the retina, with a particular length and orientation, is the most effective stimulus for
    (a) ganglion cells.
    (b) lateral geniculate cells.
    (c) simple cortical cells.
    (d) complex cortical cells.
16. The ability of the lens to increase its curvature and maintain a focus at close distances is called
    (a) convergence.
    (b) accommodation.
    (c) astigmatism.
    (d) ambylopia.
17. Which of the following sensory modalities is transmitted directly to the cerebral cortex without being relayed through the thalamus?
    (a) taste
    (b) sight
    (c) smell
    (d) hearing
    (e) touch
18. Stimulation of membrane protein receptors by binding to specific molecules is *not* responsible for
    (a) the sense of smell.
    (b) the sense of sweet taste.
    (c) the sense of sour taste.
    (d) the sense of bitter taste.
19. Epithelial cells release transmitter chemicals that excite sensory neurons in all of the following senses *except*
    (a) taste.
    (b) smell.
    (c) equilibrium.
    (d) hearing.

Essay Questions

1. Explain what is meant by lateral inhibition and give examples of its effects in three sensory systems.
2. Describe the nature of the generator potential and explain its relationship to stimulus intensity and to frequency of action potential production.
3. Describe the phantom limb phenomenon and give a possible explanation for its occurrence.
4. Draw a muscle spindle within a muscle and show the sensory and motor innervations. Explain the steps involved in a knee-jerk reflex and then compare the functions of alpha and gamma motoneurons.
5. Explain the relationship between smell and taste. How are these senses similar? How do they differ?
6. Diagram the structure of the eyeball and label the sclera, cornea, choroid, macula lutea, ciliary body, suspensory ligament, lens, iris, pupil, retina, optic disc, and fovea centralis.
7. Diagram the ear and label the structures of the outer, middle, and inner ear.
8. Explain how the vestibular apparatus provides information about changes in the position of our body in space.
9. Describe the sequence of changes that occur during accommodation. Why is it more of a strain on the eyes to look at a small nearby object than at large objects far away?
10. Describe the effects of light on the photoreceptors and explain how these effects influence the bipolar cells.
11. Explain why images that fall on the fovea centralis are seen more clearly than images that fall on the periphery of the retina. Why are the "corners of the eyes" more sensitive to light than the fovea centralis?
12. Explain why rods provide only black-and-white vision. Include a discussion of different types of color blindness in your answer.
13. Explain why green objects can be seen better at night than objects of other colors. What effect does red light in a darkroom have on a dark-adapted eye?
14. Describe the receptive fields of ganglion cells and explain how the nature of these fields helps to improve visual acuity.
15. How many genes code for the sense of color vision? How many for taste? How many for smell? What does this information say about the level of integration required by the brain for the perception of these senses?

Critical Thinking Questions

1. You know your contact lens is somewhere in your eye, but you can't seem to find it and you're worried that it might be displaced into your orbit. Considering the anatomy of the eye, do you have cause for concern? Why or why not?
2. People with conduction deafness often speak quietly. By contrast, people with sensorineural deafness tend to speak more loudly than normal. Explain these differences.
3. Opioid drugs reduce the sensation of dull, persistent pain but have little effect on the initial sharp pain of a noxious stimulus (e.g., a pinprick). What do these different effects imply? What conclusion can be drawn from the fact that aspirin (a drug that inhibits the formation of prostaglandins) functions as a pain reliever?
4. Compare the role of G-proteins in the senses of taste and sight. What is the advantage of having G-proteins mediate the effect of a stimulus on a receptor cell?
5. Discuss the role that inertia plays in the physiology of the vestibular apparatus. Why is there no sensation of movement in an airplane once it has achieved cruising speed?

Related Web Sites

For a listing of the most current web sites related to this chapter, please visit the *Concepts of Human Anatomy and Physiology* home page at http://www.mhhe.com/biosci/abio/.

NEXUS

Some Interactions of the Sensory System with the Other Body Systems

Integumentary System
- The skin helps to protect the body from pathogens (p. 696).
- The skin helps to regulate body temperature (p. 676).
- Cutaneous receptors provide sensations of touch, pressure, pain, and heat and cold (p. 501).

Skeletal System
- The skull provides protection and support for the eye and ear (p. 183).
- Proprioceptors provide sensory information about joint movement and the tension of tendons (p. 504).

Nervous System
- Sensory neurons transduce graded receptor potentials into action potentials (p. 500).
- Sensory neurons conduct action potentials from sensory receptors into the CNS for processing (p. 498).

Endocrine System
- Stimulation of stretch receptors in the heart causes secretion of atrial natriuretic hormone (p. 679).
- Stimulation of receptors in the GI tract causes secretion of particular hormones (p. 850).
- Sensory stimulation from a baby evokes the secretion of hormones involved in lactation (p. 942).

Muscular System
- Sensory information from the heart helps to regulate the heart beat (p. 679).
- Sensory information from certain arteries helps to regulate the blood pressure (p. 677).
- Muscle spindles within skeletal muscles monitor the length of the muscle (p. 504).

Circulatory System
- The blood delivers oxygen and nutrients to sensory organs and removes metabolic wastes (p. 592).
- Sensory stimuli from the heart provides information for neural regulation of the heart beat (p. 679).
- Sensory stimuli from certain blood vessels provides information for the neural regulation of blood flow and blood pressure (p. 677).

Lymphatic and Immune System
- The immune system protects against infections of sensory organs (p. 696).
- Pain sensations may arise from inflammation, alerting our bodies to infection (p. 707).
- The detection of particular chemicals in the brain evokes a fever, which may help to defeat infections (p. 698).

Respiratory System
- The lungs provide oxygen for the blood and provide for the elimination of carbon dioxide (p. 729).
- Chemoreceptors in the aorta, carotid, and medulla oblongata provide sensory information for the regulation of breathing (p. 752).

Urinary System
- The kidneys regulate the volume, pH, and electrolyte balance of the blood, and eliminate wastes (p. 779).
- Stretch receptors in the atria of the heart cause the secretion of natriuretic factor, which helps to regulate the kidneys (p. 679).
- Receptors in renal blood vessels contribute to the regulation of renal blood flow (p. 785).

Digestive System
- The GI tract provides nutrients for all body organs, including those of the sensory system (p. 817).
- Stretch receptors in the GI tract participate in reflex control of the digestive system (p. 851).
- Chemoreceptors in the GI tract contribute to the regulation of digestive activities (p. 850).

Reproductive System
- Gonads produce sex hormones that influence sensations involved in the male and female sexual response (p. 895).
- Sensory receptors provide information for erection and orgasm, as well as for other aspects of the sexual response (p. 912).

NINETEEN

Endocrine System

OBJECTIVES

- Describe the nature of endocrine glands and the chemical classification of hormones.
- Explain how hormones may act as neurotransmitters and describe the interactions that can occur between different hormones.
- Describe the structure of the pituitary gland and identify the hormones of the adenohypophysis and neurohypophysis.
- Describe the actions of the anterior pituitary hormones.
- Explain how the secretions of the posterior pituitary and anterior pituitary are regulated by the hypothalamus.
- Explain how the secretions of the hypothalamus and anterior pituitary are regulated by negative feedback.
- Relate the embryonic origin of the adrenal medulla to the hormones it secretes and describe the regulation of its secretions.
- Describe the functional categories of corticosteroids and explain how the secretion of these hormones is regulated.
- Explain how the secretions of the adrenal cortex are related to stress.
- Describe the structure of the thyroid, the formation of thyroxine and triiodothyronine, and the regulation of thyroid function.
- Describe the location of the parathyroid glands and discuss their significance.
- Describe the structure of the pancreatic islets and the actions of insulin and glucagon.
- Describe the location and endocrine function of the pineal gland and thymus.
- Identify the hormones of the gonads and placenta.
- Describe the mechanism of action of steroid hormones and thyroxine.
- Describe the mechanism of action of hormones that use cAMP as a second messenger and explain how Ca^{2+} can act as a second messenger.
- Discuss the concept of prehormones and the influence of hormone concentration on tissue responsiveness.
- Explain why regulation by prostaglandins is called autocrine rather than endocrine.
- Describe the effects of various prostaglandins and explain how nonsteroidal anti-inflammatory drugs work.

Clinical Investigation

A male patient was found to have hypertension and hyperglycemia. The results of an oral glucose tolerance test for insulin action were within the normal range. Blood tests revealed that the patient had normal catecholamine levels and normal levels of T_4 and T_3, but his blood cortisol levels were abnormally high. He had a generalized "puffiness," but not myxedema. He did not have a history of chronic inflammation and had not been taking immunosuppressive drugs. Assay of the blood ACTH concentration revealed levels that were only about one-fiftieth of normal. What might account for this patient's symptoms?

Clues: Review the sections "Feedback Control of the Anterior Pituitary," "Functions of the Adrenal Cortex," "Functions of the Adrenal Medulla," "Production and Action of Thyroid Hormones," and "Pancreatic Islets (Islets of Langerhans)."

Endocrine Glands and Hormones

Hormones are regulatory molecules secreted into the blood by endocrine glands. Chemical categories of hormones include steroids, amines, polypeptides, and glycoproteins. Interactions occur between the various hormones to produce effects that may be synergistic, permissive, or antagonistic.

Endocrine glands lack the ducts that are present in exocrine glands (chapter 6). The endocrine glands secrete their products, which are biologically active molecules called **hormones,** into the blood. The blood carries these hormones to target organs, which respond to them in a specific fashion. Many endocrine glands are discrete organs (fig. 19.1*a*) whose primary functions are the production and secretion of hormones. The pancreas functions as both an exocrine and an endocrine gland; the endocrine portion of the pancreas is composed of clusters of cells called the pancreatic islets (islets of Langerhans)

endocrine: Gk. *endon*, within; *krinein*, to separate
hormone: Gk. *hormone*, to set in motion

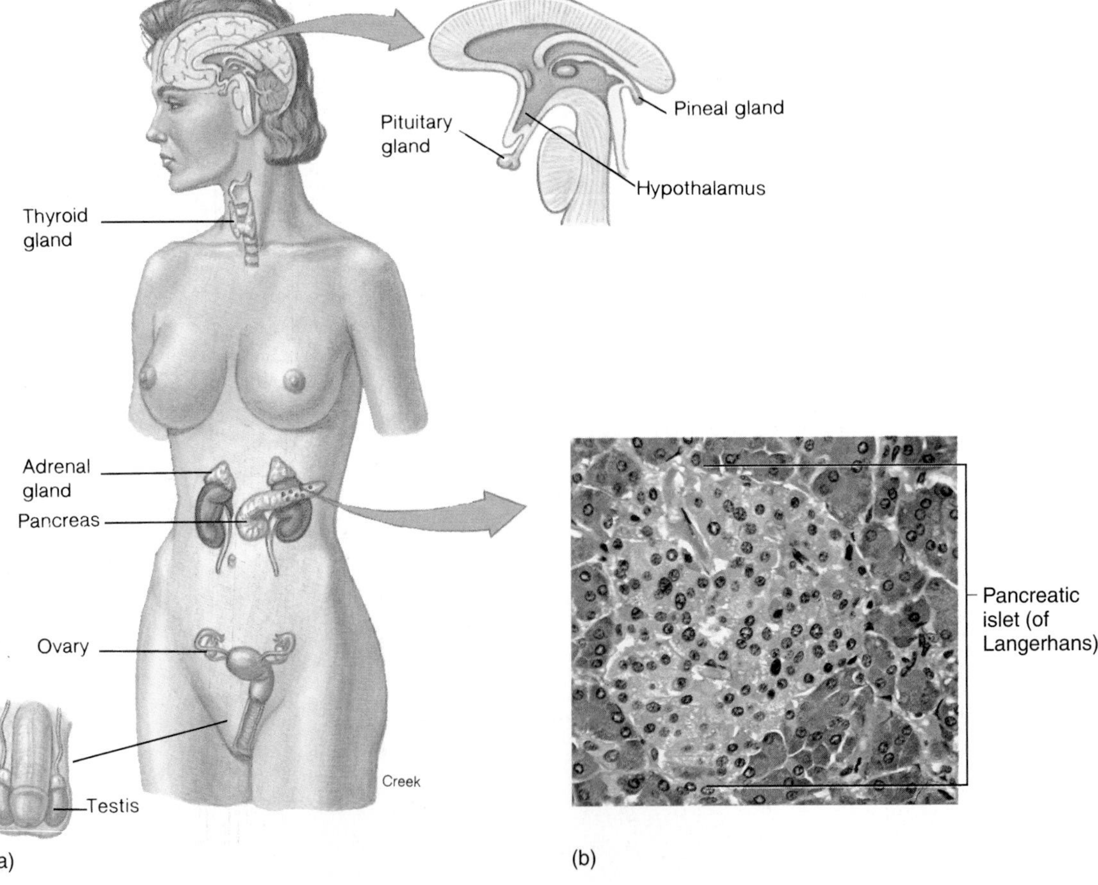

Figure 19.1 **The major endocrine glands.** *(a)* The anatomical location of some of the endocrine glands. *(b)* A photomicrograph of a pancreatic islet (islet of Langerhans) within the pancreas.

Table 19.1
A Partial Listing of the Endocrine Glands

Endocrine Gland	Major Hormones	Primary Target Organs	Primary Effects
Adipose tissue	Leptin	Hypothalamus	Suppresses appetite
Adrenal cortex	Glucocorticoids Aldosterone	Liver and muscles Kidneys	Glucocorticoids influence glucose metabolism; aldosterone promotes Na^+ retention, K^+ excretion
Adrenal medulla	Epinephrine	Heart, bronchioles, and blood vessels	Causes adrenergic stimulation
Heart	Atrial natriuretic hormone	Kidneys	Promotes excretion of Na^+ in the urine
Hypothalamus	Releasing and inhibiting hormones	Anterior pituitary	Regulates secretion of anterior pituitary hormones
Small intestine	Secretin and cholecystokinin	Stomach, liver, and pancreas	Inhibits gastric motility and stimulates bile and pancreatic juice secretion
Pancreatic islets	Insulin Glucagon	Many organs Liver and adipose tissue	Insulin promotes cellular uptake of glucose and formation of glycogen and fat; glucagon stimulates hydrolysis of glycogen and fat
Kidneys	Erythropoietin	Bone marrow	Stimulates red blood cell production
Liver	Somatomedins	Cartilage	Stimulates cell division and growth
Ovaries	Estradiol-17β and progesterone	Female reproductive tract and mammary glands	Maintains structure of reproductive tract and promotes secondary sex characteristics
Parathyroid glands	Parathyroid hormone	Bone, small intestine, and kidneys	Increases Ca^{2+} concentration in blood
Pineal gland	Melatonin	Hypothalamus and anterior pituitary	Affects secretion of gonadotrophic hormones
Pituitary, anterior	Trophic hormones	Endocrine glands and other organs	Stimulates growth and development of target organs; stimulates secretion of other hormones
Pituitary, posterior	Antidiuretic hormone Oxytocin	Kidneys and blood vessels Uterus and mammary glands	Antidiuretic hormone promotes water retention and vasoconstriction; oxytocin stimulates contraction of uterus and mammary secretory units
Skin	1,25-Dihydroxyvitamin D_3	Small intestine	Stimulates absorption of Ca^{2+}
Stomach	Gastrin	Stomach	Stimulates acid secretion
Testes	Testosterone	Prostate, seminal vesicles, and other organs	Stimulates secondary sexual development
Thymus	Thymosin	Lymph nodes	Stimulates white blood cell production
Thyroid gland	Thyroxine (T_4) and triiodothyronine (T_3); calcitonin	Most organs	Thyroxine and triiodothyronine promote growth and development and stimulate basal rate of cell respiration (basal metabolic rate or BMR); calcitonin may participate in the regulation of blood Ca^{2+} levels

(fig. 19.1*b*). The concept of the **endocrine system,** however, must be extended beyond these organs. In recent years, scientists have discovered that many other organs in the body secrete hormones. When these hormones can be demonstrated to have significant physiological functions, the organs that produce them may be categorized as endocrine glands, although they serve other functions as well. It is appropriate, then, that a partial list of the endocrine glands (table 19.1) should include the heart, liver, adipose tissue, and kidneys.

Some specialized neurons, particularly in the hypothalamus, secrete chemical messengers into the blood rather than into a narrow synaptic cleft. In these cases, the chemical that the neurons secrete is sometimes called a *neurohormone*. In addition, a number of chemicals—norepinephrine, for example—are secreted both as a neurotransmitter and a hormone. Thus, a sharp distinction between the nervous and endocrine systems cannot always be drawn on the basis of the chemicals they release.

Hormones affect the metabolism of their target organs and, by this means, help to regulate total body metabolism, growth, and reproduction. The effects of hormones on body metabolism and growth are discussed in chapter 27; the regulation of reproductive functions by hormones is included in chapters 28 and 29.

Table 19.2
Examples of Polypeptide and Glycoprotein Hormones

Hormone	Structure	Gland	Primary Effects
Antidiuretic hormone	8 amino acids	Posterior pituitary	Water retention and vasoconstriction
Oxytocin	8 amino acids	Posterior pituitary	Uterine and mammary contraction
Insulin	21 and 30 amino acids	Beta cells in pancreatic islets	Cellular glucose uptake, lipogenesis, and glycogenesis
Glucagon	29 amino acids	Alpha cells in pancreatic islets	Hydrolysis of stored glycogen and fat
ACTH	39 amino acids	Anterior pituitary	Stimulation of adrenal cortex
Parathyroid hormone	84 amino acids	Parathyroid glands	Increase in blood Ca^{2+} concentration
FSH, LH, TSH	Glycoproteins	Anterior pituitary	Stimulation of growth, development, and secretory activity of target glands

Chemical Classification of Hormones

Hormones secreted by different endocrine glands vary widely in chemical structure. All hormones, however, can be divided into a few chemical classes.

1. **Amines.** These are hormones derived from the amino acids tyrosine and tryptophan. They include the hormones secreted by the adrenal medulla, thyroid, and pineal gland.
2. **Polypeptides.** These are hormones composed of chains of amino acids that contain fewer than 100 amino acids. Examples are antidiuretic hormone and insulin (table 19.2).
3. **Glycoproteins.** These molecules consist of a long polypeptide (containing more than 100 amino acids) bound to one or more carbohydrate groups. Examples are follicle-stimulating hormone (FSH) and luteinizing hormone (LH) (table 19.2).
4. **Steroids.** These are lipids derived from cholesterol. They include the hormones testosterone, estradiol, progesterone, and cortisol (fig. 19.2).

In terms of their actions in target cells, hormone molecules can be divided into those that are polar, and therefore water-soluble, and those that are nonpolar, and thus insoluble in water. Since the nonpolar hormones are soluble in lipids, they are often referred to as **lipophilic** (*lip"ŏ-fil'ik*) **hormones.** Unlike the polar hormones, which cannot pass through cell membranes, lipophilic hormones can gain entry into their target cells. The lipophilic hormones include the steroid hormones and thyroid hormones.

Steroid hormones are secreted by only two endocrine glands: the adrenal cortex and the gonads. The gonads secrete *sex steroids;* the adrenal cortex secretes *corticosteroids* (including cortisol and aldosterone) and small amounts of sex steroids.

The major thyroid hormones are composed of two derivatives of the amino acid tyrosine bonded together (fig. 19.3). When the thyroid hormone molecule contains four iodine atoms, it is called *tetraiodothyronine* (T_4), or *thyroxine*. When it contains three atoms of iodine, it is called *triiodothyronine* (T_3). Although these hormones are not steroids, they are like steroids in that they are relatively small, nonpolar molecules. Steroid and thyroid hormones are the only hormones that are active when taken orally (as a pill). Sex steroids are the active agents in contraceptive pills, and thyroid hormone pills are taken by people whose thyroid is deficient (who are hypothyroid). By contrast, polypeptide and glycoprotein hormones cannot be taken orally because they would be digested into inactive fragments before being absorbed into the blood. Thus, insulin-dependent diabetics must inject themselves with this hormone.

The pineal gland secretes melatonin, a hormone derived from the amino acid tryptophan. Melatonin has properties that are similar in some ways to both the lipophilic hormones and the water-soluble hormones. The adrenal medulla secretes the *catecholamines* epinephrine and norepinephrine (chapter 17, see fig. 17.9), which are derived from the amino acid tyrosine. Like polypeptide and glycoprotein hormones, the catecholamine hormones are too large and polar to pass through cell membranes.

Prohormones and Prehormones

Hormone molecules that affect the metabolism of target cells are often derived from less active "parent," or *precursor,* molecules. In the case of polypeptide hormones, the precursor may be a longer chained **prohormone** that is cut and spliced together to make the hormone. Insulin, for example, is produced from *proinsulin* within the endocrine beta cells of the pancreatic islets (chapter 3, see fig. 3.26). In some cases, the prohormone itself is derived from an even larger precursor molecule; in the case of insulin, this molecule is called *preproinsulin*. The term *prehormone* is sometimes used to indicate such precursors of prohormones.

In some cases, the molecule secreted by the endocrine gland (and considered to be the hormone of that gland) is

Figure 19.2

Simplified biosynthetic pathways for steroid hormones.

Notice that progesterone (a hormone secreted by the ovaries) is a common precursor of all other steroid hormones and that testosterone (the major androgen secreted by the testes) is a precursor of estradiol-17β, the major estrogen secreted by the ovaries.

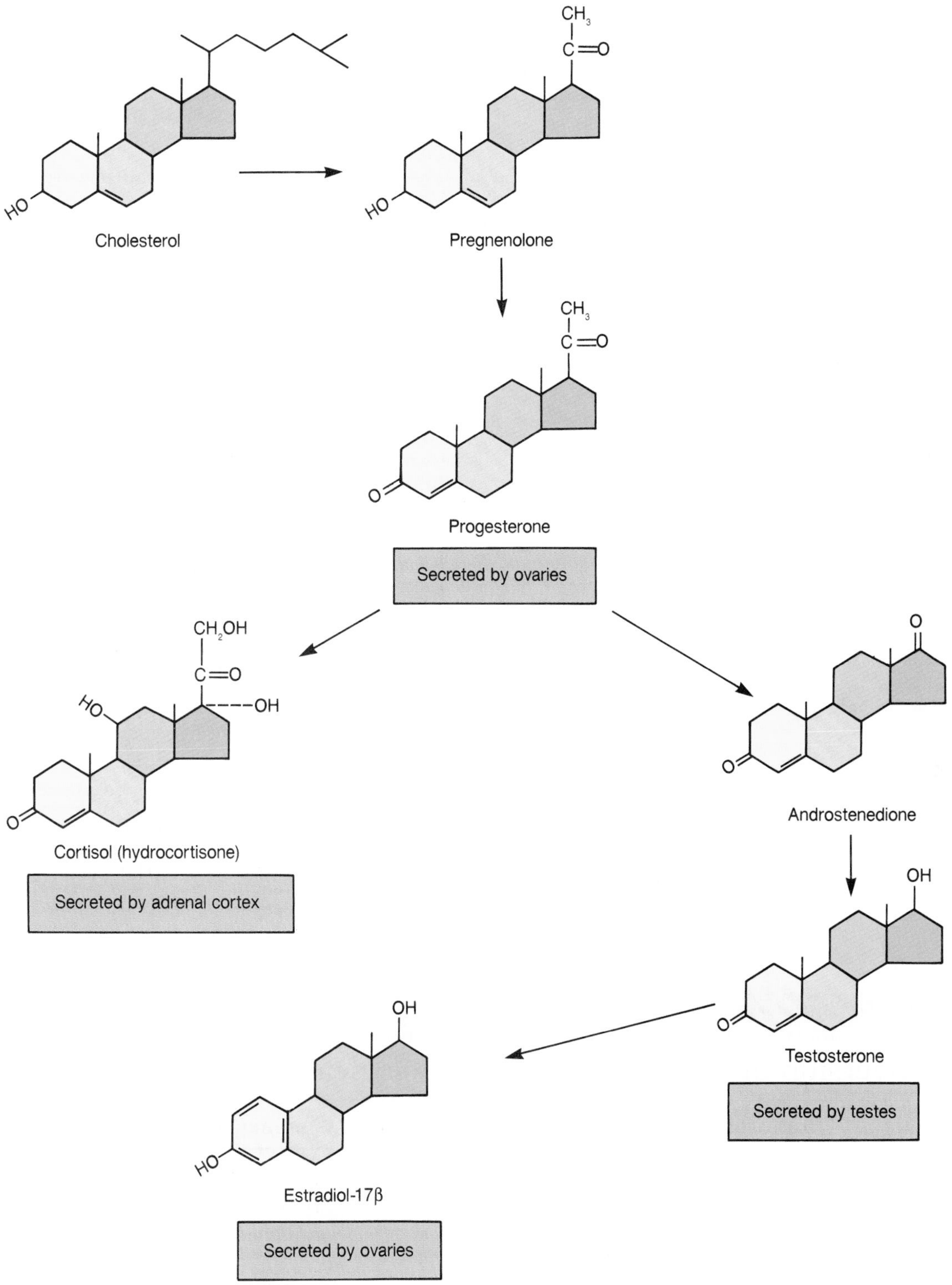

actually inactive in the target cells. In order to become active, the target cells must modify the chemical structure of the secreted hormone. Thyroxine (T_4), for example, must be changed into T_3 within the target cells to affect their metabolism. Similarly, testosterone (secreted by the testes) and vitamin D_3 (secreted by the skin) are converted into more active molecules within their target cells (table 19.3). In this text, the term **prehormone** will be used to designate those molecules secreted by endocrine glands that are inactive until changed by their target cells.

Common Aspects of Neural and Endocrine Regulation

The fact that endocrine regulation is chemical in nature might lead one to believe that it differs fundamentally from neural control systems, which depend on the electrical properties of cells. This assumption is incorrect. As explained in chapter 14, electrical nerve impulses are, in fact, chemical events produced by the diffusion of ions through the neuron cell membrane. Interestingly, the action of some hormones (such as insulin) is accompanied by ion diffusion and electrical changes in the target cells, so changes in membrane potential are not unique to the nervous system. Also, most nerve fibers stimulate the cells they innervate through the release of a chemical neurotransmitter. Neurotransmitters do not travel in the blood as do hormones; instead, they diffuse across a very narrow synaptic cleft to the membrane of the postsynaptic cell. In other respects, however, the actions of neurotransmitters are very similar to the actions of hormones.

Indeed, many polypeptide hormones, including those secreted by the pituitary gland and by the digestive tract, have been discovered in the brain. In certain locations in the brain, some of these compounds are produced and secreted as hormones. In other brain locations, some of these compounds apparently serve as neurotransmitters. The discovery of polypeptide hormones in unicellular organisms, which of course lack a nervous and endocrine system, suggests that these regulatory molecules appeared early in evolution and were incorporated into the function of nervous and endocrine tissue as these systems evolved. This fascinating theory would help to explain, for example, why insulin, a polypeptide hormone produced in the pancreas of vertebrates, is found in neurons of invertebrates (which lack a distinct endocrine system).

Regardless of whether a particular chemical is acting as a neurotransmitter or as a hormone, in order for it to function in physiological regulation: (1) target cells must have specific **receptor proteins** that combine with the regulatory molecule; (2) the combination of the regulatory molecule with its receptor proteins must cause a specific sequence of changes in the target cells; and (3) there must be a mechanism to quickly turn off the action of the regulator. This mechanism, which involves rapid removal and/or chemical inactivation of the regulator molecules, is essential because without an "off-switch" physiological control would be impossible.

Figure 19.3

The structure of the thyroid hormones.
Thyroxine, also called tetraiodothyronine (T_4), and triiodothyronine (T_3) are secreted in a ratio of 9 to 1.

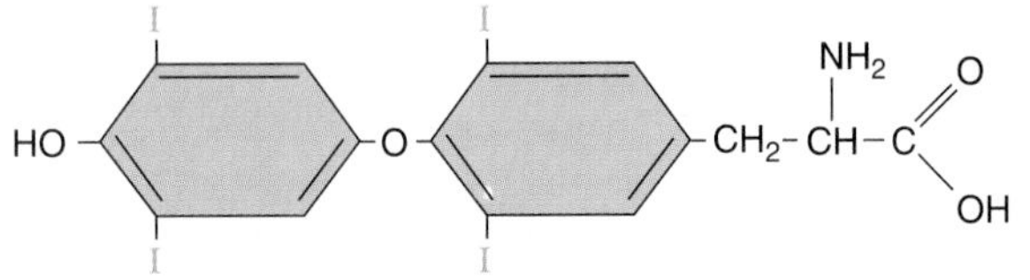

I I NH2 O
HO O CH2-CH-C OH
I I
Thyroxine, or tetraiodothyronine (T_4)

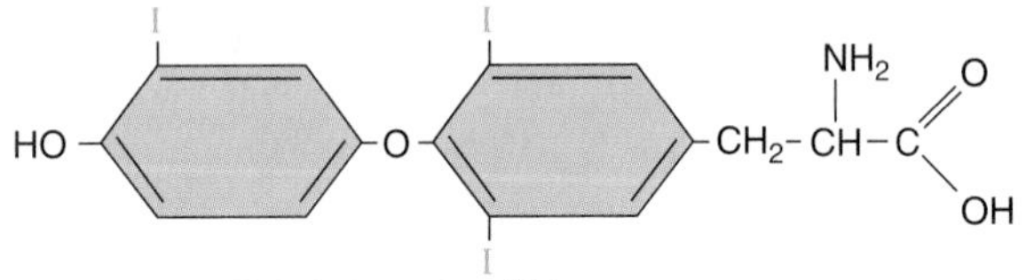

I I NH2 O
HO O CH2-CH-C OH
I
Triiodothyronine (T_3)

Table 19.3

Conversion of Prehormones into Biologically Active Derivatives

Endocrine Gland	Prehormone	Active Products	Comments
Skin	Vitamin D_3	1,25-dihydroxyvitamin D_3	Hydroxylation reactions occur in the liver and kidneys.
Testes	Testosterone	Dihydrotestosterone (DHT)	DHT and other 5α-reduced androgens form in most androgen-dependent tissue.
		Estradiol-17β (E_2)	E_2 forms in the brain from testosterone, where it is believed to affect both endocrine function and behavior; small amounts of E_2 are also produced in the testes.
Thyroid gland	Thyroxine (T_4)	Triiodothyronine (T_3)	Conversion of T_4 to T_3 occurs in almost all tissues.

Hormone Interactions

A given target tissue is usually responsive to a number of different hormones. These hormones may antagonize each other or work together to produce effects that are additive or complementary. The responsiveness of a target tissue to a particular hormone is thus affected not only by the concentration of that hormone, but also by the effects of other hormones on that tissue. Terms used to describe hormone interactions include synergistic, permissive, and antagonistic.

Synergistic and Permissive Effects

When two or more hormones work together to produce a particular result, their effects are said to be **synergistic** (*sin"er-jis'tik*). These effects may be *additive* or *complementary*. The action of epinephrine and norepinephrine on the heart is an example of an additive effect. Each of these hormones separately produces an increase in cardiac rate; acting together in the same concentrations, they stimulate an even greater increase in cardiac rate. The synergistic action of FSH and testosterone is an example of a complementary effect; each hormone separately stimulates a different stage of spermatogenesis during puberty, so that both hormones together are needed at that time to complete sperm development. Likewise, the ability of mammary glands to produce and secrete milk requires the synergistic action of many hormones—estrogen, cortisol, prolactin, oxytocin, and others.

A hormone is said to have a **permissive effect** on the action of a second hormone when it enhances the responsiveness of a target organ to the second hormone or when it increases the activity of the second hormone. Prior exposure of the uterus to estrogen, for example, induces the formation of receptor proteins for progesterone, which improves the response of the uterus when it is subsequently exposed to progesterone. Estrogen thus has a permissive effect on the responsiveness of the uterus to progesterone. Glucocorticoids (a class of corticosteroids including cortisol) exert permissive effects on the actions of catecholamines (epinephrine and norepinephrine). When these permissive effects are absent because of abnormally low glucocorticoids, the catecholamines will not be as effective as they are normally. One symptom of this condition may be an abnormally low blood pressure.

Vitamin D_3 is a prehormone that must be modified by enzymes in the kidneys and liver, where two hydroxyl (OH^-) groups are added to form the active hormone 1,25-dihydroxyvitamin D_3. This hormone helps to raise blood calcium levels. Parathyroid hormone (PTH) has a permissive effect on the actions of vitamin D_3 because it stimulates the production of the hydroxylating enzymes in the kidneys and liver. By this means, an increased secretion of PTH has a permissive effect on the ability of vitamin D_3 to stimulate the intestinal absorption of calcium.

Antagonistic Effects

In some situations, the actions of one hormone antagonize the effects of another. Lactation during pregnancy, for example, is inhibited because the high concentration of estrogen in the blood inhibits the secretion and action of prolactin. Another example of antagonism is the action of insulin and glucagon (two hormones from the pancreatic islets) on adipose tissue; the formation of fat is promoted by insulin, whereas glucagon promotes fat breakdown.

Effects of Hormone Concentrations on Tissue Response

The concentration of hormones in the blood primarily reflects the rate of secretion by the endocrine glands. Hormones do not generally accumulate in the blood because they are rapidly removed by target organs and by the liver. The **half-life** for most hormones—the time required for the plasma concentration of a given amount of the hormone to be reduced to half its reference level—ranges from minutes to hours. (Thyroid hormone, however, has a half-life of several days.) Hormones removed from the blood by the liver are converted by enzymatic reactions into less active products. Steroids, for example, are converted into more water-soluble polar derivatives that are released into the blood and are excreted in the urine and bile.

The effects of hormones are very dependent on concentration. Normal tissue responses are produced only when the hormones are present within their normal, or *physiological*, range of concentrations. When some hormones are taken in abnormally high, or *pharmacological*, concentrations (as when they are taken as drugs), their effects may be different from those produced by lower, more physiological concentrations. The fact that abnormally high concentrations of a hormone may cause the hormone to bind to tissue receptor proteins of different but related hormones may account, in part, for these different effects. Also, since some steroid hormones can be converted by their target cells into products that have different biological effects (as in the conversion of androgens into estrogens), the administration of large quantities of one steroid can result in the production of a significant quantity of other steroids with different effects.

Pharmacological doses of hormones, particularly steroids, can thus have widespread and often damaging side effects. People with inflammatory diseases who are treated with high doses of cortisone over long periods of time, for example, may develop osteoporosis and characteristic changes in soft tissue structure. Contraceptive pills, which contain sex steroids, have a number of potential side effects that could not have been predicted in 1960, when "the pill" was first introduced. At that time, the concentrations of sex steroids were much higher than they are in the pills presently being marketed.

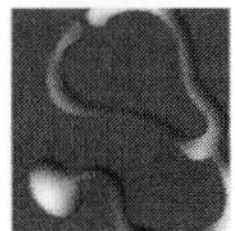

Anabolic (an″ă-bol′ik) steroids are synthetic androgens (male hormones) that promote protein synthesis in muscles and other organs. Use of these drugs by bodybuilders, weightlifters, and other athletes became widespread in the 1960s and, although prohibited by most athletic organizations, the practice is still common today. Although exogenous androgens do promote muscle growth, they can also cause a number of undesirable side effects. Since the liver and some other organs can change androgens into estrogens, male athletes who take exogenous androgens often develop *gynecomastia*—an abnormal growth of femalelike mammary tissue. For some, this tissue is so excessive as to warrant surgical removal. Damaging effects attributed to anabolic steroids include liver cancer, shrinkage of the testes and temporary sterility, stunted growth in teenage users, masculinization in female users, and antisocial behavior. Anabolic steroids may also raise the blood levels of cholesterol and LDL, thus predisposing users to atherosclerosis of the coronary arteries and heart disease.

Priming Effects

Variations in hormone concentration within the normal, physiological range can affect the responsiveness of target cells. This is due in part to the effects of polypeptide and glycoprotein hormones on the number of their receptor proteins in target cells. More receptors may be formed in the target cells in response to particular hormones. Small amounts of gonadotropin-releasing hormone (GnRH), secreted by the hypothalamus, for example, increase the sensitivity of anterior pituitary cells to further GnRH stimulation. This is a *priming effect*, sometimes also called **upregulation.** Subsequent stimulation by GnRH thus causes a greater response from the anterior pituitary.

Desensitization and Downregulation

Prolonged exposure to high concentrations of polypeptide hormones has been found to *desensitize* the target cells. Subsequent exposure to the same concentration of the same hormone thus produces less of a target tissue response. This desensitization may be partly due to the fact that high concentrations of these hormones cause a decrease in the number of receptor proteins in their target cells—a phenomenon called **downregulation.** Such desensitization and downregulation of receptors has been shown to occur, for example, in adipose cells exposed to high concentrations of insulin and in testicular cells exposed to high concentrations of luteinizing hormone (LH).

In order to prevent desensitization under normal conditions, many polypeptide and glycoprotein hormones are secreted in spurts rather than continuously. This *pulsatile secretion* is an important aspect, for example, in the hormonal control of the reproductive system. The pulsatile secretion of GnRH and LH is needed to prevent desensitization; when these hormones are artificially presented in a continuous fashion, they produce a decrease (rather than the normal increase) in gonadal function. This effect has important clinical implications, as will be described in chapter 28.

Mechanisms of Hormone Action

Each hormone exerts its characteristic effects on target organs by acting on the cells of these organs. Hormones of the same chemical class have similar mechanisms of action. Lipid-soluble hormones pass through the target cell membrane, bind to intracellular receptor proteins, and act directly within the target cell. Polar hormones do not enter the target cells, but instead bind to receptors on the cell membrane. This results in the activation of intracellular second-messenger systems that mediate the actions of the hormone.

Although each hormone exerts its own characteristic effects on specific target cells, hormones in the same chemical category have similar mechanisms of action. These similarities involve the location of cellular receptor proteins and the events that occur in the target cells after the hormone has combined with its receptor protein.

Hormones are delivered by the blood to every cell in the body, but only the **target cells** are able to respond to these hormones. In order to respond to any given hormone, a target cell must have specific receptor proteins for that hormone. Receptor protein–hormone interaction is highly specific. In addition to this property of *specificity*, hormones bind to receptors with a *high affinity* (high bond strength) and a *low capacity*. The latter characteristic refers to the possibility of saturating receptors with hormone molecules because of the limited number of receptors per target cell (usually a few thousand). Notice that the characteristics of specificity and saturation that apply to receptor proteins are similar to the characteristics of enzyme and carrier proteins discussed in previous chapters.

The location of a hormone's receptor proteins in its target cells depends on the chemical nature of the hormone. Since the lipophilic hormones (steroids and thyroxine) can pass through the cell membrane and enter their target cells, the receptor proteins for lipophilic hormones are located within the target cells. The receptor proteins for many steroid hormones are located in the cytoplasm; when they bind to the steroid hormone, the receptor protein–steroid hormone complex moves into the nucleus (as described below). The receptor proteins for thyroid hormones and other steroid hormones are located in the cell nucleus but are inactive until they bind to their hormone ligands. Since the water-soluble hormones (catecholamines, polypeptides, and glycoproteins) cannot pass through the cell membrane, their receptors are located on the outer surface of the membrane. In these cases, hormone action requires the activation of second messengers within the cell.

Hormones That Bind to Nuclear Receptor Proteins

Unlike the water-soluble hormones, the lipophilic steroid and thyroid hormones do not travel dissolved in the aqueous portion of the plasma; rather, they are transported to their target cells attached to plasma *carrier proteins*. These hormones must then dissociate from their carrier proteins in the blood in order to pass through the lipid component of the target cell membrane and enter the target cell, where their receptor proteins are located.

The receptors for the lipophilic hormones are known as **nuclear hormone receptors** because they function within the cell nucleus to activate genetic transcription (production of mRNA). The nuclear hormone receptors thus function as **transcription factors** that first must be activated by binding to their hormone ligands. The newly formed mRNA produced by the activated genes directs the synthesis of specific enzyme proteins that change the metabolism of the target cell in ways that are characteristic of the effects of that hormone on that target cell.

Each nuclear hormone receptor has two regions, or *domains:* a *ligand (hormone)-binding domain* and a *DNA-binding domain* (fig. 19.4). The receptor must be activated by binding to its hormone ligand before it can bind to a specific region of the DNA, which is called a **hormone-response element.** This is a short DNA span, composed of characteristic nucleotide bases, located adjacent to the gene that will be transcribed when the nuclear receptor binds to the hormone-response element.

Figure 19.4

Receptors for steroid hormones.
(a) Each nuclear hormone receptor protein has a ligand-binding domain that binds to a hormone molecule, and has a DNA-binding domain that binds to the hormone-response element of DNA. *(b)* Binding to the hormone causes the receptor to dimerize on the half-sites of the hormone-response element. This stimulates genetic transcription (synthesis of RNA).

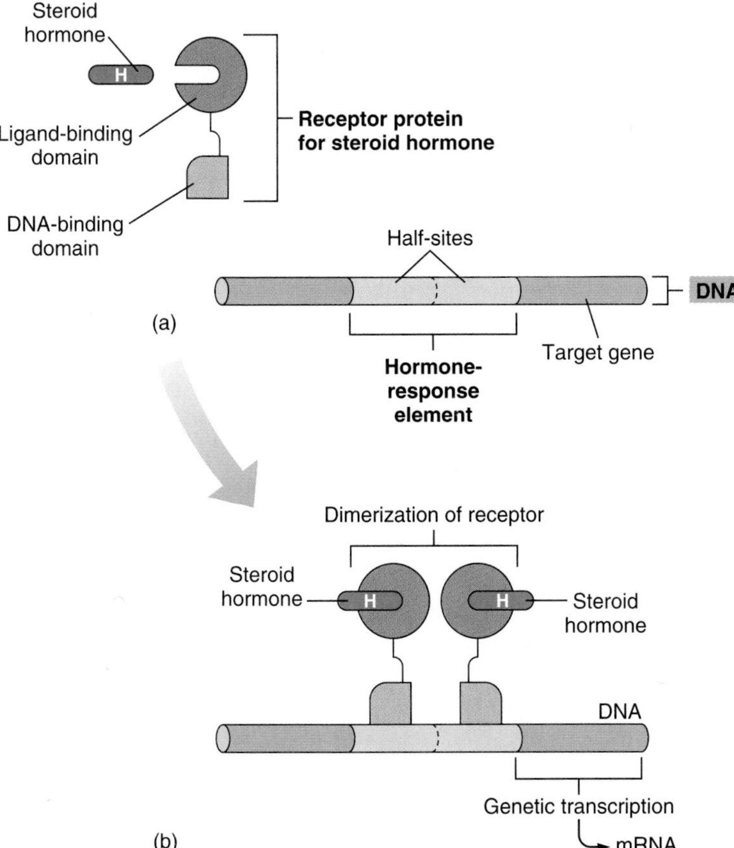

The nuclear hormone receptors are said to constitute a superfamily composed of two major families: the *steroid family* and the *thyroid hormone* (or *nonsteroid*) *family*. In addition to the receptor for thyroid hormone, the latter family also includes the receptors for the active form of vitamin D and for retinoic acid (derived from vitamin A, or retinol). Vitamin D and retinoic acid, like the steroid and thyroid hormones, are lipophilic molecules that play important roles in the regulation of cell function and organ physiology.

Mechanism of Steroid Hormone Action

Before they bind to the steroid hormones, many steroid receptors are located in the cytoplasm of the target cells. Once they bind to their steroid hormone ligand, the receptor-steroid complex moves (or *translocates*) to the nucleus, where its DNA-binding domain binds to the specific hormone-response element of the DNA (fig. 19.5). The hormone-response element of DNA consists of two *half-sites*, each six nucleotide bases long, separated by a three-nucleotide spacer segment. One steroid receptor, bound to one molecule of the steroid hormone, attaches as a single unit to one of the half-sites. Another steroid receptor, bound to another steroid hormone, attaches to the other half-site of the hormone-response element. The process of two receptor units coming together at the two half-sites is called **dimerization** (see fig. 19.4). Since both receptor units of the pair are the same, the steroid receptor is said to form a *homodimer*. (The situation is different for the nonsteroid family of receptors, as will be described.) Once dimerization has occurred, the activated nuclear hormone receptor stimulates transcription of particular genes and thus hormonal regulation of the target cell (fig. 19.5).

It should be noted that receptors for estradiol, as well as those for some other steroids, are like the receptor for thyroid hormone (discussed below) in that they are located in the nucleus rather than the cytoplasm. These receptors, then, do not have to translocate to the nucleus; instead, the steroid must move through the cytoplasm into the nucleus to find the receptor. Once the steroid binds to its receptor in the nucleus, the DNA-binding domain can attach to its hormone-response element of DNA.

Mechanism of Thyroid Hormone Action

As previously discussed, the major hormone secreted by the thyroid gland is thyroxine, or tetraiodothyronine (T_4). Like steroid hormones, thyroxine travels in the blood attached to carrier proteins (primarily attached to *thyroxine-binding globulin*, or *TBG*). The thyroid also secretes a small amount of triiodothyronine, or T_3. The carrier proteins have a higher affinity for T_4 than for T_3,

Figure 19.5

The mechanism of action of a steroid hormone on the target cells.

Some steroids bind to a cytoplasmic receptor, which then translocates to the nucleus. Other steroid hormones enter the nucleus and then bind to their receptor. In both cases, the steroid-receptor complex can then bind to a specific area of DNA to activate specific genes.

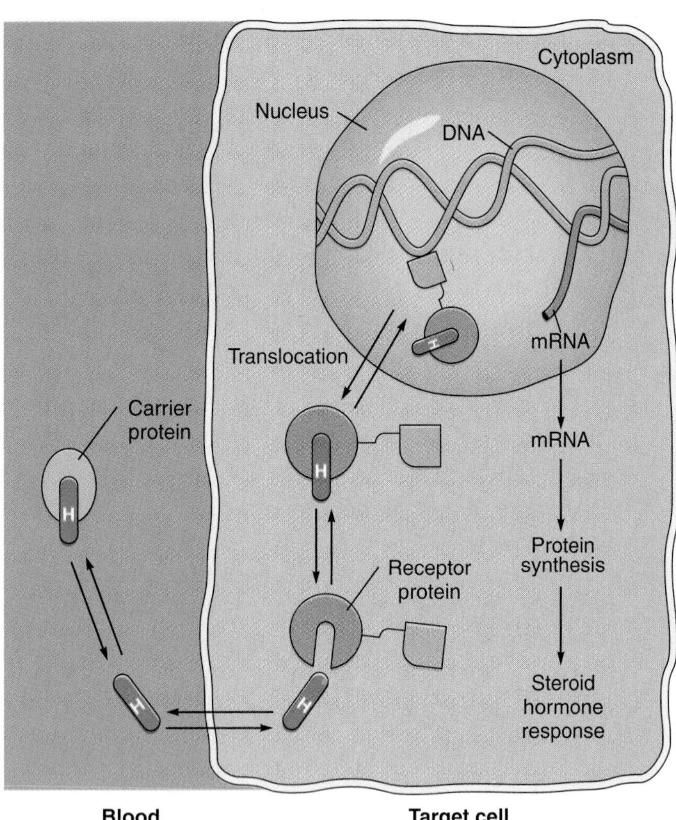

Figure 19.6

The mechanism of action of thyroid hormones on the target cells.

First, T_4 is converted into T_3 within the cytoplasm of the target cell. Then T_3 enters the nucleus and binds to its nuclear receptor. The hormone-receptor complex can then bind to a specific area of DNA to activate specific genes.

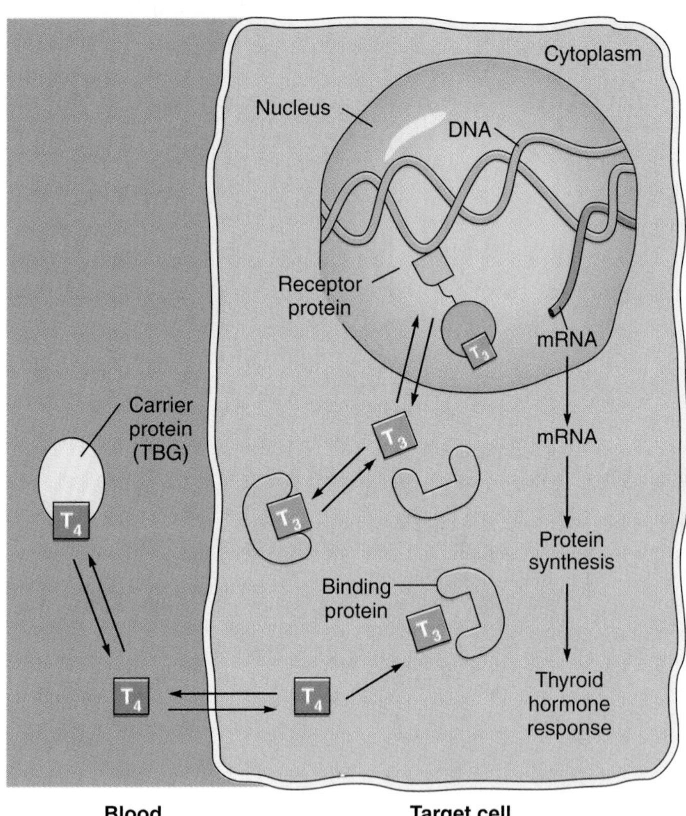

however, and, as a result, the amount of unbound (or "free") T_3 in the plasma is about 10 times greater than the amount of free T_4.

Approximately 99.96% of the thyroxine in the blood is attached to carrier proteins in the plasma; the rest is free. Only the free thyroxine and T_3 can enter target cells; the protein-bound thyroxine serves as a reservoir of this hormone in the blood (this is why it takes a couple of weeks after surgical removal of the thyroid for the symptoms of hypothyroidism to develop). Once the free thyroxine passes into the target cell cytoplasm it is enzymatically converted into T_3. As previously discussed, it is the T_3 rather than T_4 that is active within the target cells.

Unlike many of the steroid receptors, the inactive receptor proteins for T_3 (and some steroids) are located in the nucleus. Until they bind to T_3, however, the receptors are incapable of stimulating transcription. The T_3 may enter the cell from the plasma, or it may be produced in the cell by conversion from T_4. In either case, it uses some nonspecific binding proteins as "stepping stones" to enter the nucleus, where it binds to the ligand-binding domain of the receptor (fig. 19.6). Once the receptor binds to T_3, its DNA-binding domain can attach to the half-site of the DNA hormone-response element.

The other half-site, however, does *not* bind to another T_3 receptor protein. Unlike the steroid hormone receptors, the nuclear receptors in the nonsteroid family bind to DNA as *heterodimers*. The thyroid hormone receptor (abbreviated *TR*) is one partner in the heterodimer; the other partner is a receptor (abbreviated *RXR*) for the vitamin A derivative, 9-*cis*-retinoic acid. Once bound to their different ligands, the two partners in the heterodimer can bind to the DNA to activate the hormone-response element for thyroid hormone (fig. 19.7). In this way, thyroid hormones stimulate transcription of genes, production of specific mRNA, and therefore the production of specific enzymes (see fig. 19.6).

Interestingly, the receptor for 1,25-dihydroxyvitamin D_3, the active form of vitamin D, also forms heterodimers with the receptor for 9-*cis*-retinoic acid (the RXR receptor) when it binds to DNA and activates genes. The RXR receptor and its vitamin A derivative ligand thus form a link between the mechanisms of action of thyroid hormone, vitamin A, and vitamin D, along with those of some other molecules that are important regulators of genetic expression.

Figure 19.7

Receptor for triiodothyronine (T_3).
The nuclear receptor protein for T_3 forms a dimer with the receptor protein for 9-cis-retinoic acid, a derivative of vitamin A. This occurs when each bind to their ligand and to the hormone-response element of DNA. 9-cis-retinoic acid is thus required for the action of T_3. The heterodimer formed on the DNA stimulates genetic transcription.

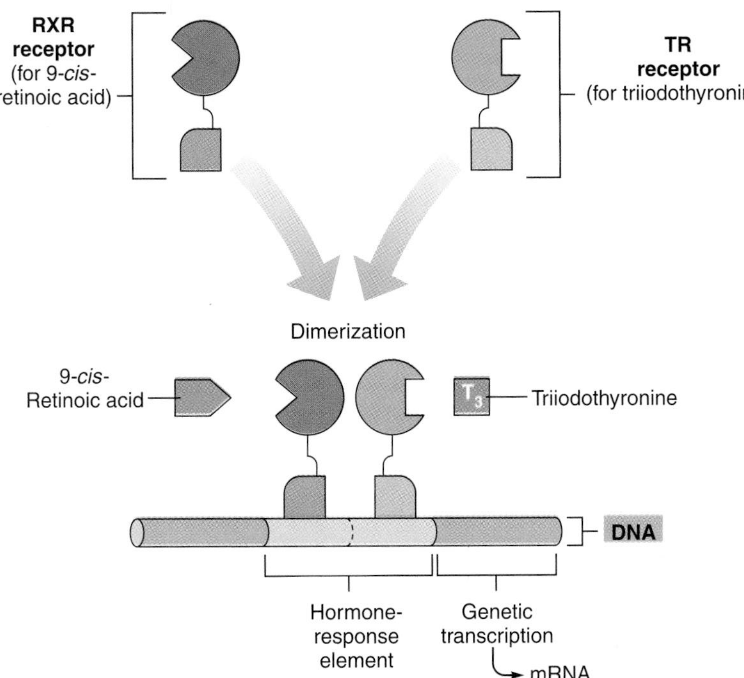

Hormones That Use Second Messengers

Hormones that are catecholamines (epinephrine and norepinephrine), polypeptides, and glycoproteins cannot pass through the lipid barrier of the target cell membrane. Although some of these hormones may enter the cell by pinocytosis, most of their effects result from their binding to receptor proteins on the outer surface of the target cell membrane. Since they exert their effects without entering the target cells, the actions of these hormones must be mediated by other molecules within the target cells. If you think of hormones as "messengers" from the endocrine glands, the intracellular mediators of the hormone's action can be called **second messengers.** (The concept of second messengers was introduced in connection with synaptic transmission in chapter 14.) Second messengers are thus part of *signal transduction* mechanisms, since extracellular signals (hormones) are transduced into intracellular signals (second messengers).

When these hormones bind to membrane receptor proteins, they must activate specific proteins in the cell membrane in order to produce the second messengers required to exert their effects. On the basis of the membrane enzyme activated, we can distinguish second-messenger systems that involve the activation of: (1) adenylate cyclase, (2) phospholipase C, and (3) tyrosine kinase.

Adenylate Cyclase–Cyclic AMP Second-Messenger System

Cyclic adenosine monophosphate (*ă-den'ŏ-sēn mon"o-fos'fāt*) (abbreviated **cAMP**) was the first "second messenger" to be discovered and is the best understood. The β-adrenergic effects (chapter 17) of epinephrine and norepinephrine are due to cAMP production within the target cells. Researchers later discovered that the effects of many (but not all) polypeptide and glycoprotein hormones are also mediated by cAMP.

When one of these hormones binds to its receptor protein, it causes the dissociation of a subunit from the complex of G-proteins (discussed in chapter 14). This G-protein subunit moves through the membrane until it reaches the enzyme **adenylate cyclase** (*ă-den'l-it si'klās*). The G-protein subunit then binds to and activates this enzyme, which catalyzes the following reaction within the cytoplasm of the cell:

$$\text{ATP} \rightarrow \text{cAMP} + \text{PP}_i$$

Adenosine triphosphate (ATP) is thus converted into cyclic AMP (cAMP) and two inorganic phosphates (*pyrophosphate*, abbreviated PP_i). As a result of the interaction of the hormone with its receptor and the activation of adenylate cyclase, therefore, the intracellular concentration of cAMP increases. Cyclic AMP activates a previously inactive enzyme in the cytoplasm called **protein kinase** (*ki'nās*). The inactive form of this enzyme consists of two subunits: a catalytic subunit and an inhibitory subunit. The enzyme is produced in an inactive form and becomes active only when cAMP attaches to the inhibitory subunit. Binding of cAMP to the inhibitory subunit causes it to dissociate from the catalytic subunit, which then becomes active (fig. 19.8). In summary, the hormone—acting through an increase in cAMP production—causes an increase in protein kinase enzyme activity within its target cells.

Active protein kinase catalyzes the phosphorylation of (attachment of phosphate groups to) different proteins in the target cells. This causes some enzymes to become activated and others to become inactivated. Cyclic AMP, acting through protein kinase, thus modulates the activity of enzymes that are already present in the target cell. This alters the metabolism of the target tissue in a manner characteristic of the actions of that specific hormone (table 19.4).

Like all biologically active molecules, cAMP must be rapidly inactivated for it to function effectively as a second messenger in hormone action. This inactivation is accomplished by **phosphodiesterase** (*fos"fo-di-es'tĕ-rās*), an enzyme within the target cells that hydrolyzes cAMP into inactive fragments. Through the action of phosphodiesterase, the stimulatory effect of a hormone that uses cAMP as a second messenger depends on the continuous generation of new cAMP molecules, and thus depends on the level of secretion of the hormone.

Figure 19.8

The adenylate cyclase–cyclic AMP second-messenger system.

The hormone causes the production of cAMP within the target cell cytoplasm, and cAMP activates protein kinase. The activated protein kinase then causes the activation or inactivation of a number of specific enzymes. These changes lead to the characteristic effects of the hormone on the target cell.

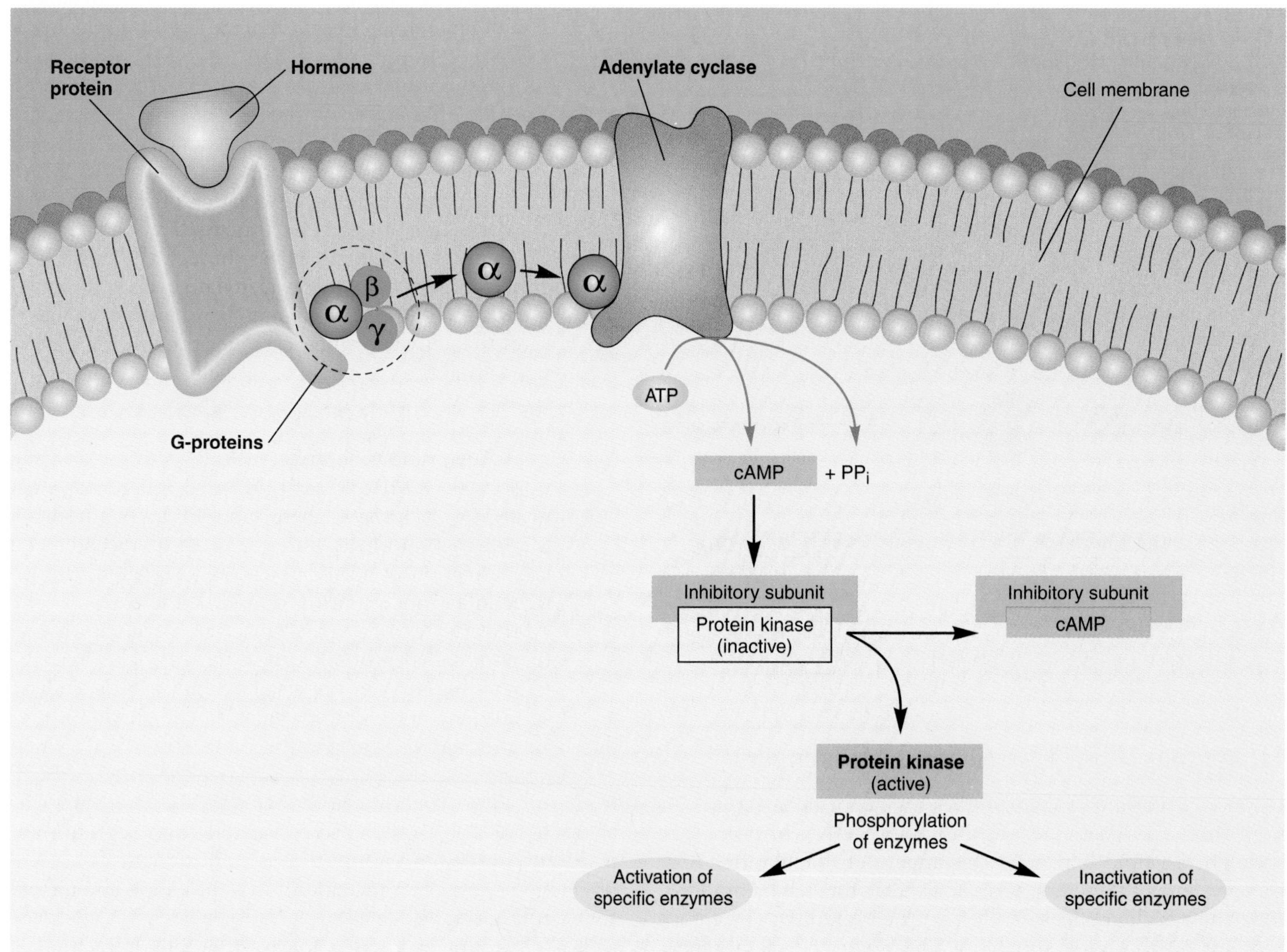

Table 19.4

Sequence of Events Involving Cyclic AMP as a Second Messenger

1. The hormones combine with their receptors on the outer surface of target cell membranes.
2. Hormone-receptor interaction stimulates (through G-protein subunits) the activation of adenylate cyclase on the cytoplasmic side of the membranes.
3. Activated adenylate cyclase catalyzes the conversion of ATP to cyclic AMP (cAMP) within the cytoplasm.
4. Cyclic AMP activates protein kinase enzymes that were already present in the cytoplasm in an inactive state.
5. Activated cAMP-dependent protein kinase transfers phosphate groups to (phosphorylates) other enzymes in the cytoplasm.
6. The activity of specific enzymes is either increased or inhibited by phosphorylation.
7. Altered enzyme activity mediates the target cell's response to the hormone.

In addition to cyclic AMP, **cyclic guanosine** (*gwan'ŏ-sēn*) **monophosphate (cGMP)** functions as a second messenger in certain cases. For example, the regulatory molecule nitric oxide (discussed in chapter 14 and later in this chapter) exerts its effects on smooth muscle by stimulating the production of cGMP in its target cells. In different regulatory systems cGMP and cAMP may interact, producing effects that are either antagonistic or complementary. For example, the control of cell division and the cell cycle (chapter 3) is related to the ratio of cAMP to cGMP in the cell.

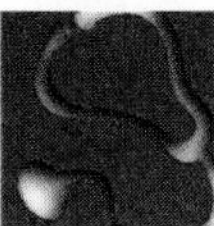

Drugs that inhibit the activity of phosphodiesterase prevent the breakdown of cAMP and thus result in increased concentrations of cAMP within the target cells. The drug *theophylline* and its derivatives, for example, are used clinically to raise cAMP levels within bronchiolar smooth muscle. This duplicates and enhances the effect of epinephrine on the bronchioles (producing dilation) in people who suffer from asthma. *Caffeine,* a compound related to theophylline, is also a phosphodiesterase inhibitor, and thus exerts its effects by raising the cAMP concentrations within cells.

Phospholipase C–Ca^{2+} Second-Messenger System

The concentration of Ca^{2+} in the cytoplasm is kept very low by the action of active transport carriers—calcium pumps—in the cell membrane. Through the action of these pumps, the concentration of calcium is about 10,000 times lower in the cytoplasm than in the extracellular fluid. In addition, the endoplasmic reticulum (chapter 3) of many cells contains calcium pumps that actively transport Ca^{2+} from the cytoplasm into the cisternae of the endoplasmic reticulum. The steep concentration gradient for Ca^{2+} that results allows various stimuli to evoke a rapid, though brief, diffusion of Ca^{2+} into the cytoplasm that can serve as a signal in different control systems.

At the terminal boutons of axons, for example, the entry of Ca^{2+} through voltage-regulated Ca^{2+} channels in the cell membrane serves as a signal for the release of neurotransmitters (chapter 14). Similarly, when muscles are stimulated to contract, Ca^{2+} couples electrical excitation of the muscle cell to the mechanical processes of contraction (chapter 12). Additionally, it is now known that Ca^{2+} serves as a part of a second-messenger system in the action of a number of hormones.

When epinephrine stimulates its target organs, it must first bind to adrenergic receptor proteins in the membrane of its target cells. As discussed in chapter 17, there are two types of adrenergic receptors—alpha and beta. Stimulation of the beta-adrenergic receptors by epinephrine results in activation of adenylate cyclase and the production of cAMP. Stimulation of alpha-adrenergic receptors by epinephrine, in contrast, activates the target cell via the Ca^{2+} second-messenger system.

The binding of epinephrine to its alpha-adrenergic receptor activates, via a G-protein intermediate, an enzyme in the cell membrane known as **phospholipase C.** The substrate of this enzyme, a particular membrane phospholipid, is split by the active enzyme into **inositol** (*in-o'sĭ-tol*) **triphosphate (IP_3)** and another derivative, **diacylglycerol** (*di″ă-sil-glis'er-ol*) **(DAG).** Both derivatives serve as second messengers, but the action of IP_3 is somewhat better understood and will be discussed in this section.

The IP_3 leaves the cell membrane and diffuses through the cytoplasm to the endoplasmic reticulum. The membrane of the endoplasmic reticulum contains receptor proteins for IP_3, so that the IP_3 is a second messenger in its own right, carrying the hormone's message from the cell membrane to the endoplasmic reticulum. Binding of IP_3 to its receptors causes specific Ca^{2+} channels to open, so that Ca^{2+} diffuses out of the endoplasmic reticulum and into the cytoplasm (fig. 19.9).

As a result of these events there is a rapid and transient rise in the cytoplasmic Ca^{2+} concentration. This signal is augmented, through mechanisms that are incompletely understood, by the opening of Ca^{2+} channels in the cell membrane. This may be due to the action of yet a different (and currently unknown) messenger sent from the endoplasmic reticulum to the cell membrane. The Ca^{2+} that enters the cytoplasm from the endoplasmic reticulum and extracellular fluid binds to a cytoplasmic protein called **calmodulin** (*kal″mod'yŭ-lin*). Once Ca^{2+} binds to calmodulin, the now-active calmodulin in turn activates specific protein kinase enzymes (those that add phosphate groups to proteins) that modify the actions of other enzymes in the cell (fig. 19.10). Activation of specific calmodulin-dependent enzymes is analogous to the activation of enzymes by cAMP-dependent protein kinase. The steps of the Ca^{2+} second-messenger system are summarized in table 19.5.

Tyrosine Kinase Second-Messenger System

Insulin promotes glucose and amino acid transport into cells and stimulates glycogen, fat, and protein synthesis in its target organs—primarily the liver, skeletal muscles, and adipose tissue. These effects are achieved by means of a mechanism of action that is quite complex and in some ways still incompletely understood. Nevertheless, it is known that insulin's mechanism of action bears similarities to the mechanism of action of regulatory molecules known as **growth factors.** These growth factors, examples of which are *epidermal growth factor (EGF), platelet-derived growth factor (PDGF),* and *insulin-like growth factors,* are autocrine regulators (described at the end of this chapter).

In the case of insulin and the growth factors, the receptor protein is located in the cell membrane and is itself a kind of enzyme known as a **tyrosine** (*ti'rŏ-sēn*) **kinase.** A *kinase* is an enzyme that adds phosphate groups to proteins, and a *tyrosine* kinase specifically adds these phosphate groups to the amino acid tyrosine within the proteins. The insulin receptor consists of two units that come together (dimerize) when they bind with insulin to form an active tyrosine kinase enzyme (fig. 19.11). Each unit of the receptor contains a site on the outside of the cell that binds to insulin (termed the *ligand-binding site*) and a part that spans the cell membrane with an *enzymatic site* in the cytoplasm. The enzymatic site is inactive until insulin binds to the ligand-binding site and causes dimerization of the receptor. When insulin binding and dimerization occur, the enzymatic site is activated in each

Figure 19.9

The phospholipase C–Ca^{2+} second-messenger system.

Some hormones, when they bind to their membrane receptors, activate phospholipase C (PLC). This enzyme catalyzes the formation of inositol triphosphate (IP_3), which causes Ca^{2+} channels to open in the endoplasmic reticulum. Ca^{2+} is released and acts as a second messenger in the action of the hormone.

Figure 19.10

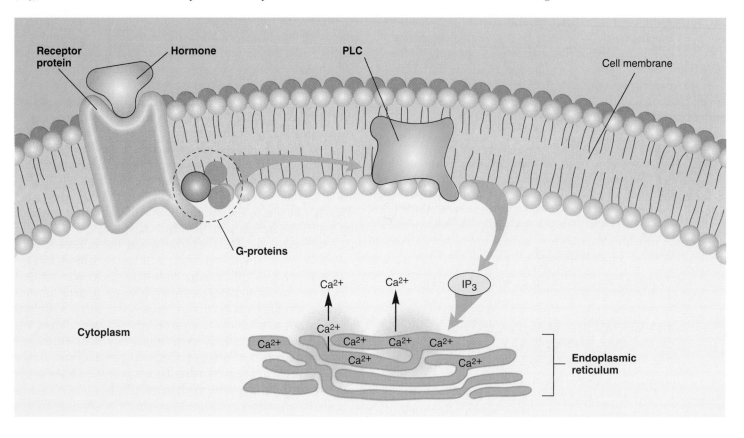

Epinephrine can act through two second-messenger systems.

The stimulation of β-adrenergic receptors invokes the cAMP second-messenger system, and the stimulation of the α-adrenergic receptors invokes the Ca^{2+} second-messenger system.

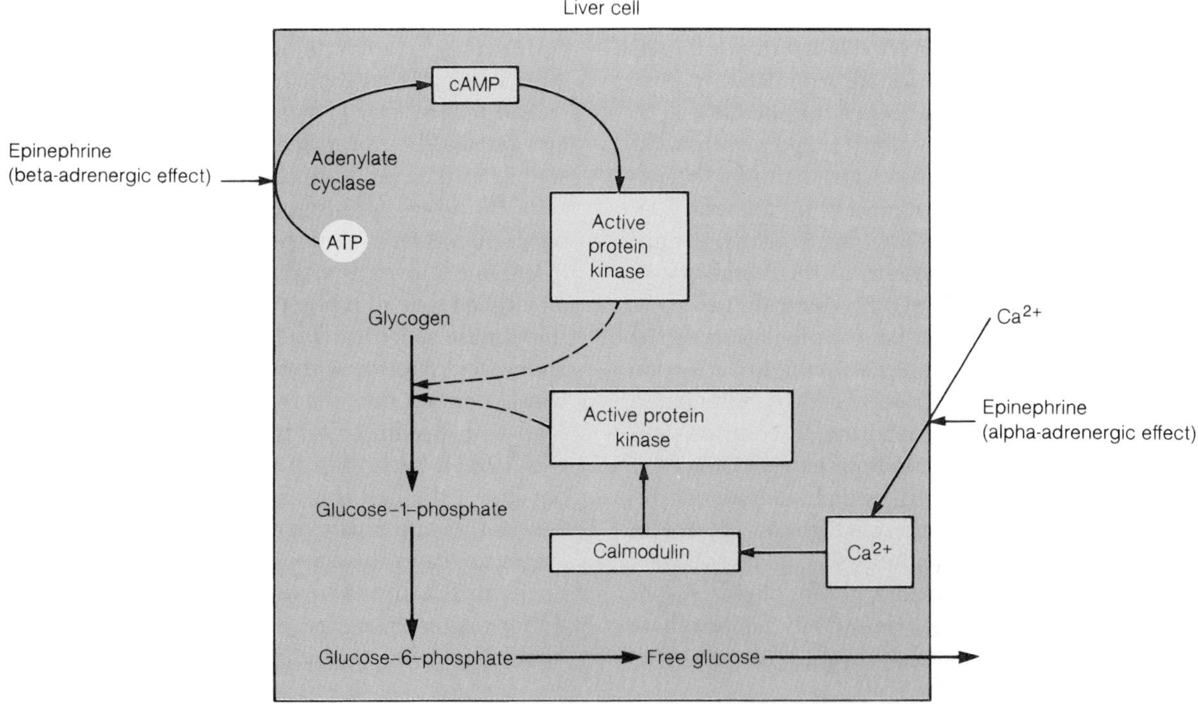

Table 19.5
Sequence of Events Involving Ca^{2+} as a Second Messenger

1. The hormone binds to its receptor on the outer surface of the target cell membrane.
2. Hormone-receptor interaction stimulates the activity of a membrane enzyme, phospholipase C.
3. Phospholipase C catalyzes the conversion of particular phospholipids in the membrane to inositol triphosphate (IP_3) and another derivative, diacylglycerol.
4. Inositol triphosphate enters the cytoplasm and diffuses to the endoplasmic reticulum, where it binds to its receptor proteins and causes the opening of Ca^{2+} channels.
5. Since the endoplasmic reticulum accumulates Ca^{2+} by active transport, a steep Ca^{2+} concentration gradient exists that favors the diffusion of Ca^{2+} into the cytoplasm.
6. Ca^{2+} that enters the cytoplasm binds to and activates a protein called calmodulin.
7. Activated calmodulin, in turn, activates protein kinase, which phosphorylates other enzyme proteins.
8. Altered enzyme activity mediates the target cell's response to the hormone.

Figure 19.11

Receptor for insulin.
Insulin binds to two units of its receptor protein, causing these units to dimerize (come together) on the cell membrane. This activates the tyrosine kinase enzyme portion of the receptor. As a result, the receptor phosphorylates itself and thereby makes the enzyme even more active. The receptor then phosphorylates a number of cytoplasmic "signal molecules," which exert a cascade of effects in the target cell.

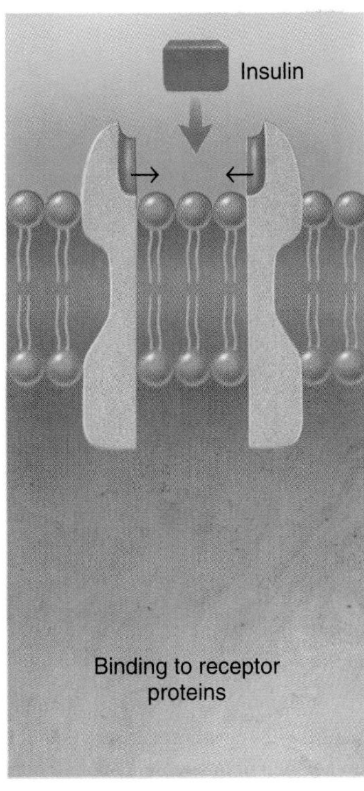

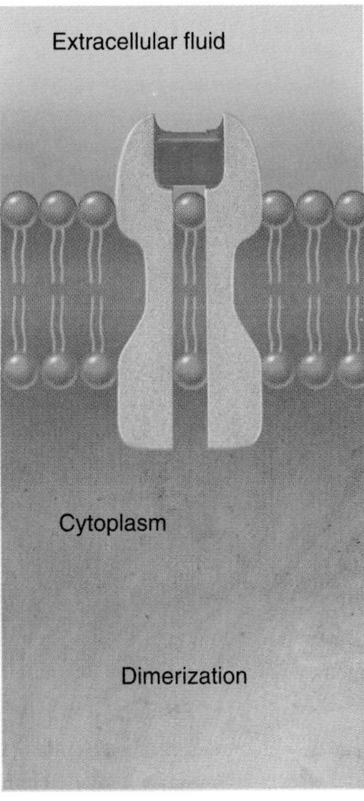

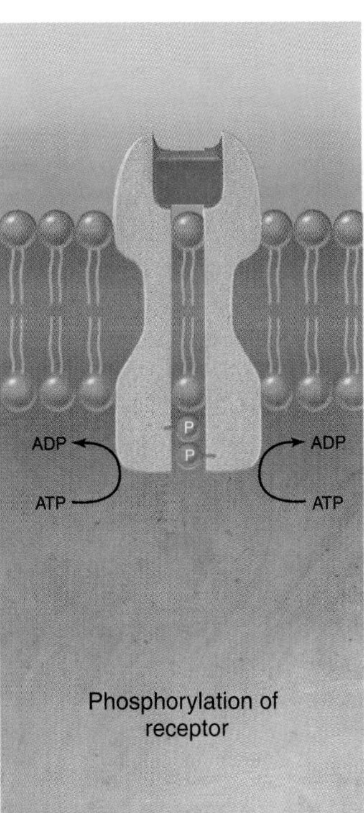

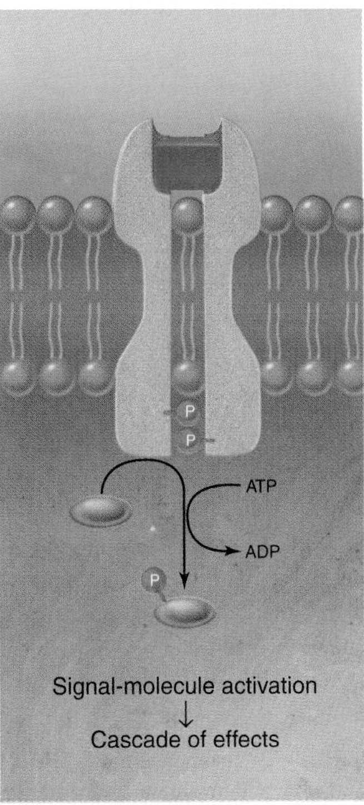

unit of the receptor, and one unit phosphorylates the other. This process, termed *autophosphorylation*, increases the tyrosine kinase activity of the dimerized receptor.

The activated tyrosine kinase receptor then phosphorylates other, cytoplasmic proteins that serve as **signaling molecules.** Some of these signaling molecules are themselves kinase enzymes that phosphorylate and activate other molecules. Some may activate other second-messenger systems; for example, they may activate phospholipase C and, via the action of IP_3, result in the release of Ca^{2+} and the activation of calmodulin. Still other signaling molecules can function as kinases that enter the nucleus and activate transcription factors, thereby turning on specific genes. As a result of this complex series of activations, insulin and the different growth factors regulate the metabolism of their target cells.

Such complexity is required for proper regulation. Different hormones must be able to act on the same target cell and produce different, even antagonistic, effects. For example, insulin stimulates the synthesis of fat and the hormone glucagon stimulates hydrolysis of fat in adipose cells. Clearly, these two hormones must use different second-messenger systems to achieve different results in the same target cells.

Pituitary Gland

The pituitary gland includes the anterior pituitary and the posterior pituitary. The posterior pituitary stores and releases hormones that are actually produced by the hypothalamus, whereas the anterior pituitary produces and secretes its own hormones. The anterior pituitary, however, is regulated by hormones secreted by the hypothalamus, as well as by feedback from the target gland hormones.

The **pituitary gland,** or **hypophysis** (*hi-pof′ĭ-sis*), is located on the inferior aspect of the brain in the region of the diencephalon (chapter 15). Roughly the size of a pea measuring about 1.3 cm (0.5 in.) in diameter, it is attached to the brain by the **pituitary stalk** (fig. 19.12).

The pituitary gland is structurally and functionally divided into an anterior lobe, or **adenohypophysis** (*ad″n-o-hi-pof′ĭ-sis*), and a posterior lobe called the **neurohypophysis** (*nour″o-hi-pof′ĭ-sis*). These two parts have different embryonic origins. The adenohypophysis is derived from epithelial tissue, whereas the neurohypophysis is formed as a downgrowth of the brain (see "Under Development" on page 568). The adenohypophysis consists of two parts in adults: (1) the *pars distalis*, also known as the **anterior pituitary,** is the rounded portion and the major endocrine part of the gland, and (2) the *pars tuberalis* is the thin extension in contact with the *infundibulum* (*in″fun-dib′yŭ-lum*), the portion of the pituitary stalk that connects to the hypothalamus. These parts are illustrated in figure 19.12. A *pars intermedia*, a strip of tissue between the anterior and posterior lobes, exists in the fetus. During fetal development, its cells mingle with those of the anterior lobe, and in adults they no longer constitute a separate structure.

The neurohypophysis is the neural part of the pituitary gland that consists of the *pars nervosa*, also called the **posterior pituitary.** The posterior pituitary connects to the hypothalamus through the infundibulum, the connecting stalk to the hypothalamus. Nerve fibers extend through the infundibulum along with small neuroglialike cells called *pituicytes* (*pĭ-too′ĭ-sīts*).

---

pituitary: L. *pituita*, phlegm (this gland was originally thought to secrete mucus into the nasal cavity)
adenohypophysis: Gk. *adeno*, gland; *hypo*, under; *physis*, a growing
infundibulum: L. *infundibulum*, a funnel

Figure 19.12

The structure of the pituitary gland in sagittal view. The anterior and posterior lobes are clearly seen and comprise most of the gland. The intermediate lobe in humans is poorly distinguished.

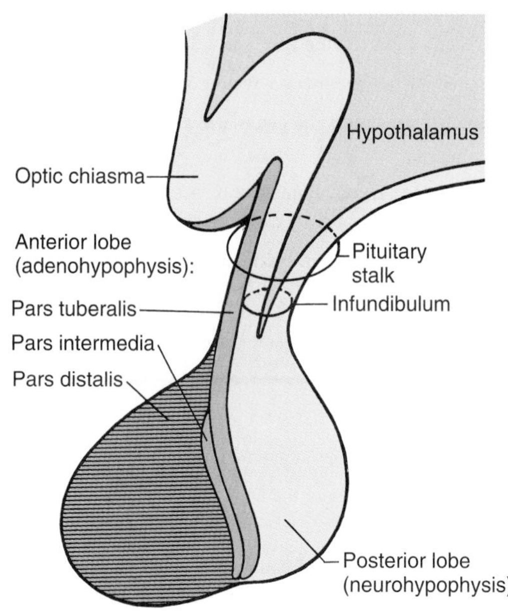

Pituitary Hormones

The hormones secreted by the anterior pituitary (the pars distalis of the adenohypophysis) are called **trophic hormones.** The term *trophic* means "food." Although the anterior pituitary hormones are not food for their target organs, this term is used because high concentrations of the anterior pituitary hormones cause their target organs to hypertrophy, whereas low levels cause their target organs to atrophy. When names are applied to the hormones of the anterior pituitary, "trophic" (conventionally shortened to *tropic*, meaning "attracted to") is incorporated into them. This is why the shortened forms of the names for the anterior pituitary hormones end in the suffix *-tropin*. The hormones of the pars distalis, listed below, are summarized in table 19.6.

1. **Adrenocorticotropic hormone (ACTH,** or **corticotropin).** ACTH stimulates the adrenal cortex to secrete the glucocorticoids, such as hydrocortisone (cortisol).
2. **Thyroid-stimulating hormone (TSH,** or **thyrotropin).** TSH stimulates the thyroid gland to produce and secrete thyroxine (tetraiodothyronine, or T_4).
3. **Follicle-stimulating hormone (FSH,** or **folliculotropin).** FSH stimulates the growth of ovarian follicles in females and the production of sperm in the testes of males.
4. **Luteinizing hormone (LH,** or **luteotropin).** LH and FSH are collectively called **gonadotropic** (*go″nad-ŏ-trop′ik*) **hormones.** In females, LH stimulates ovulation and the conversion of the ovulated ovarian follicle into

Table 19.6
Anterior Pituitary Hormones

Hormone	Target Tissue	Principal Actions	Regulation of Secretion
ACTH (adrenocorticotropic hormone)	Adrenal cortex	Stimulates secretion of glucocorticoids	Stimulated by CRH (corticotropin-releasing hormone); inhibited by glucocorticoids
TSH (thyroid-stimulating hormone)	Thyroid gland	Stimulates secretion of thyroid hormones	Stimulated by TRH (thyrotropin-releasing hormone); inhibited by thyroid hormones
FSH (follicle-stimulating hormone) and LH (luteinizing hormone)	Gonads	Promotes gamete production and sex steroid hormone secretion	Stimulated by GnRH (gonadotropin-releasing hormone); inhibited by sex steroids
LH (luteinizing hormone)	Gonads	Stimulates sex hormone secretion and ovulation and corpus luteum formation in females	Stimulated by GnRH
GH (growth hormone)	Most tissue	Promotes protein synthesis and growth, lipolysis, and increased blood glucose	Inhibited by somatostatin; stimulated by growth-hormone–releasing hormone
PRL (prolactin)	Mammary glands and other accessory sex organs	Promotes milk production in lactating females; controversial actions in other organs	Inhibited by PIH (prolactin-inhibiting hormone)

an endocrine structure called a corpus luteum. In males, LH is sometimes called *interstitial cell–stimulating hormone*, or *ICSH*; it stimulates the secretion of male sex hormones (mainly testosterone) from the interstitial (Leydig) cells in the testes.

5. **Growth hormone (GH,** or **somatotropin).** GH promotes the movement of amino acids into cells and the incorporation of these amino acids into proteins, thus promoting overall tissue and organ growth.
6. **Prolactin (PRL).** PRL is secreted in both males and females. Its best known function is the stimulation of milk production by the mammary glands of women after the birth of a baby. Prolactin plays a supporting role in the regulation of the male reproductive system by the gonadotropins (FSH and LH) and acts on the kidneys to help regulate water and electrolyte balance.

As mentioned earlier, the pars intermedia of the adenohypophysis ceases to exist as a separate lobe in the adult human pituitary, but it is present in the human fetus and in other animals. Until recently, it was thought to secrete **melanocyte-stimulating hormone (MSH),** as it does in fish, amphibians, and reptiles, where it causes darkening of the skin. In humans, however, plasma concentrations of MSH are insignificant. Some cells of the adenohypophysis, derived from the fetal pars intermedia, produce a large polypeptide called *pro-opiomelanocortin (POMC)*. POMC is a prohormone whose major products are beta-endorphin (chapter 14), MSH, and ACTH. Since part of the ACTH molecule contains the amino acid sequence of MSH, elevated secretions of ACTH (as in Addison's disease) can cause a marked darkening of the skin.

The posterior pituitary, or pars nervosa, stores and releases two hormones, both of which are produced in the hypothalamus.

1. **Antidiuretic hormone (ADH),** also known as **arginine** (*ar′jĭ-nēn*) **vasopressin (AVP)**. Antidiuretic (*an″te-di″yŭ-ret-ik*) hormone promotes the retention of water by the kidneys, so that less water is excreted in the urine and more water is retained in the blood. At high doses this hormone also has a "pressor" effect; that is, it causes vasoconstriction in experimental animals. The physiological significance of this pressor effect in humans is controversial, however.
2. **Oxytocin.** In females, oxytocin (*ok″sĭ-to′sin*) stimulates contractions of the uterus during labor and for this reason is needed for parturition (childbirth). Oxytocin also stimulates contractions of the mammary gland alveoli and ducts, which result in the milk-ejection reflex in a lactating woman. In men, a rise in oxytocin secretion at the time of ejaculation has been measured, but the physiological significance of this hormone in males remains to be demonstrated.

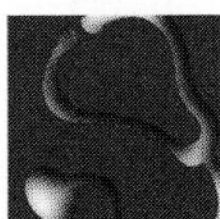

Injections of oxytocin may be given to a pregnant woman to induce labor if the pregnancy is prolonged, or if the fetal membranes have ruptured and there is a danger of infection. Labor may also be induced by injections of oxytocin in the case of severe pregnancy-induced hypertension, or *preeclampsia* (*pre″ĭ-klamp′se-ă*). Oxytocin administration after delivery promotes shrinkage of the uterus and squeezes the blood vessels, thus minimizing the danger of hemorrhage.

UNDER DEVELOPMENT

Development of the Pituitary Gland

The adenohypophysis begins to develop during the third week as a diverticulum (*di"ver-tik'yŭ-lum*), or pouchlike extension, called the **hypophyseal** (Rathke's) **pouch** (fig. 1). It arises from the roof of the primitive oral cavity and grows toward the brain. At the same time, another diverticulum called the **infundibulum** forms from the diencephalon on the inferior aspect of the brain. As the two diverticula come in contact, the hypophyseal pouch loses its connection with the oral cavity, and the primordial tissue of the adenohypophysis forms.

The neurohypophysis develops as the infundibulum extends inferiorly from the diencephalon to come in contact with the developing adenohypophysis. The fully formed neurohypophysis consists of the infundibulum and the pars nervosa. Specialized nerve fibers that connect the hypothalamus with the pars nervosa develop within the infundibulum.

Rathke's pouch: from Martin H. Rathke, German anatomist, 1793–1860

Figure 1 The development of the pituitary gland. (*a*) the head end of an embryo at 4 weeks showing the position of a midsagittal cut seen in the developmental sequence (*b–e*). The pituitary gland arises from a specific portion of the neuroectoderm, called the neurohypophyseal bud, which evaginates downward during the fourth and fifth weeks, respectively, in (*b*) and (*c*), and from a specific portion of the oral ectoderm, called the hypophyseal (Rathke's) pouch, which evaginates upward from a specific portion of the primitive oral cavity. At 8 weeks (*d*), the hypophyseal pouch is no longer connected to the pharyngeal roof of the oral cavity. During the fetal stage (*e*), the development of the pituitary gland is completed.

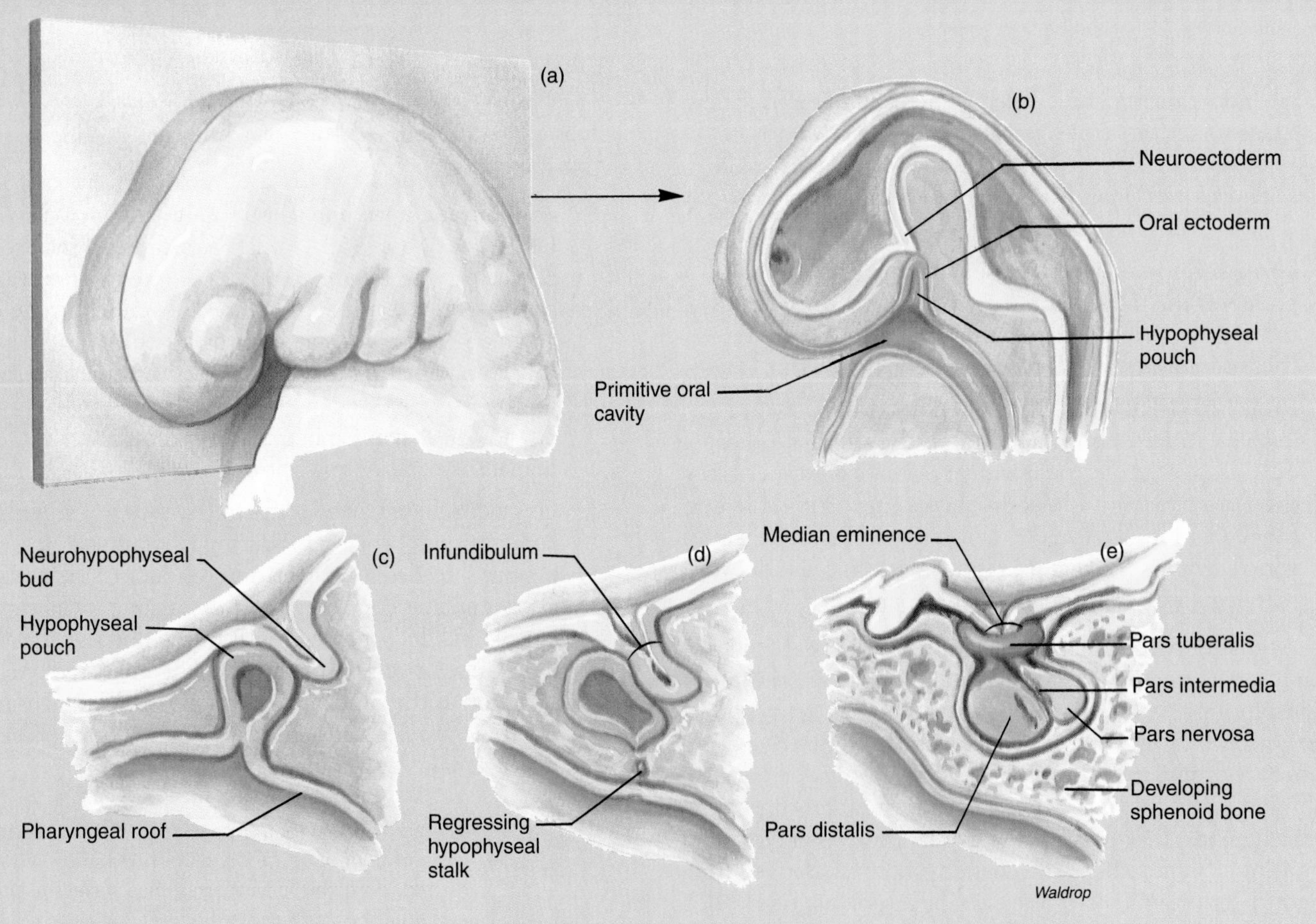

Hypothalamic Control of the Posterior Pituitary

Both of the posterior pituitary hormones—antidiuretic hormone and oxytocin—are actually produced in neuron cell bodies of the *supraoptic nuclei* and *paraventricular nuclei* of the hypothalamus. These nuclei within the hypothalamus are thus endocrine glands. The hormones they produce are transported along axons of the **hypothalamo-hypophyseal** (*hi″po-thă-lam′o-hi-pof″ĭ-se′al*) **tract** (fig. 19.13) to the posterior pituitary, where they are stored and later released. The posterior pituitary is thus more a storage organ than a true gland.

The release of ADH and oxytocin from the posterior pituitary is controlled by **neuroendocrine reflexes.** In nursing mothers, for example, the mechanical stimulus of suckling acts via sensory nerve impulses to the hypothalamus to stimulate the reflex secretion of oxytocin. The secretion of ADH is stimulated by osmoreceptor neurons in the hypothalamus in response to a rise in blood osmotic pressure (chapter 5); its secretion is inhibited by sensory impulses from stretch receptors in the left atrium of the heart in response to a rise in blood volume. These reflexes are discussed in more detail in later chapters.

Figure 19.13

Hypothalamic control of the posterior pituitary. The posterior pituitary, or neurohypophysis, stores and secretes hormones—vasopressin and oxytocin—that are actually produced in neurons within the supraoptic and paraventricular nuclei of the hypothalamus. These hormones are transported to the posterior pituitary by axons in the hypothalamo-hypophyseal tract.

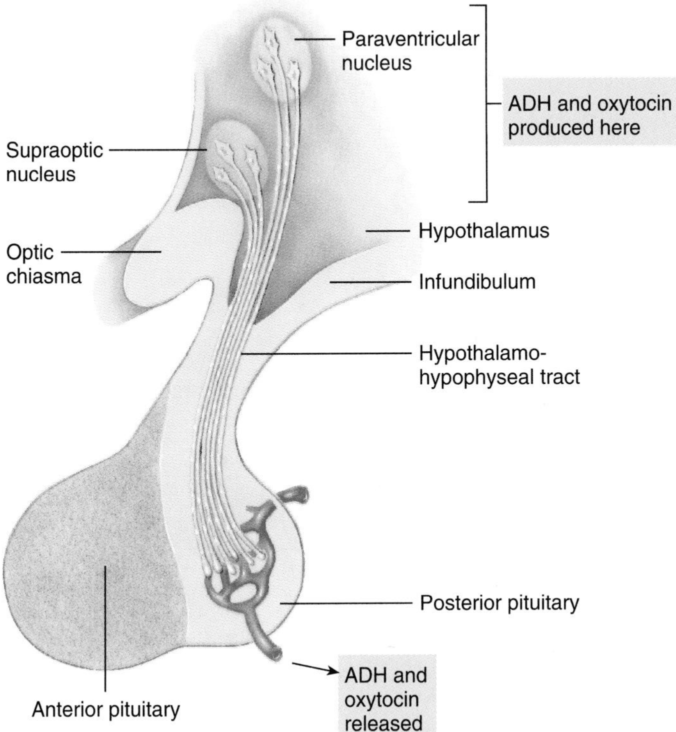

Hypothalamic Control of the Anterior Pituitary

At one time the anterior pituitary was called the "master gland" because it secretes hormones that regulate some other endocrine glands (fig. 19.14 and table 19.6). Adrenocorticotropic hormone (ACTH), thyroid-stimulating hormone (TSH), and the gonadotropic hormones (FSH and LH) stimulate the adrenal cortex, thyroid, and gonads, respectively, to secrete their hormones. The anterior pituitary hormones also have a "trophic" effect on their target glands in that the health of these glands depends on adequate stimulation by anterior pituitary hormones. The anterior pituitary, however, is not really the master gland, since secretion of its hormones is in turn controlled by hormones secreted by the hypothalamus.

Releasing and Inhibiting Hormones

Since axons do not enter the anterior pituitary, hypothalamic control of the anterior pituitary is achieved through hormonal rather than neural regulation. Neurons in the hypothalamus produce releasing and inhibiting hormones, which are transported to axon endings in the basal portion of the hypothalamus. This region, known as the *median eminence*, contains blood capillaries that are drained by venules in the stalk of the pituitary.

The venules that drain the median eminence deliver blood to a second capillary bed in the anterior pituitary. Since this second capillary bed is downstream from the capillary bed in the median eminence and receives venous blood from it, the vascular link between the median eminence and the anterior pituitary forms a *portal system*. (This is analogous to the hepatic portal system that delivers venous blood from the intestine to the liver, as described in chapter 26.) The vascular link between the hypothalamus and the anterior pituitary is thus called the **hypothalamo-hypophyseal portal system.**

Polypeptide hormones are secreted into the hypothalamo-hypophyseal portal system by neurons of the hypothalamus. These hormones regulate the secretions of the anterior pituitary (fig. 19.15 and table 19.7). **Thyrotropin-releasing hormone (TRH)** stimulates the secretion of TSH, and **corticotropin-releasing hormone (CRH)** stimulates the secretion of ACTH from the anterior pituitary. A single releasing hormone, **gonadotropin-releasing hormone,** or **GnRH,** stimulates the secretion of both gonadotropic hormones (FSH and LH) from the anterior pituitary. The secretion of prolactin and of growth hormone from the anterior pituitary is regulated by hypothalamic inhibitory hormones, known as **prolactin-inhibiting hormone (PIH)** and **somatostatin** (*so″mă-tŏ-stat′n*), respectively.

A specific **growth hormone–releasing hormone (GHRH)** that stimulates growth hormone secretion has been identified as a polypeptide consisting of 44 amino acids. Experiments suggest that a releasing hormone for prolactin may also exist, but no such specific releasing hormone has yet been discovered.

Figure 19.14

The hormones secreted by the anterior pituitary and their target organs.
Notice that the anterior pituitary controls some (but by no means all) of the other endocrine glands.

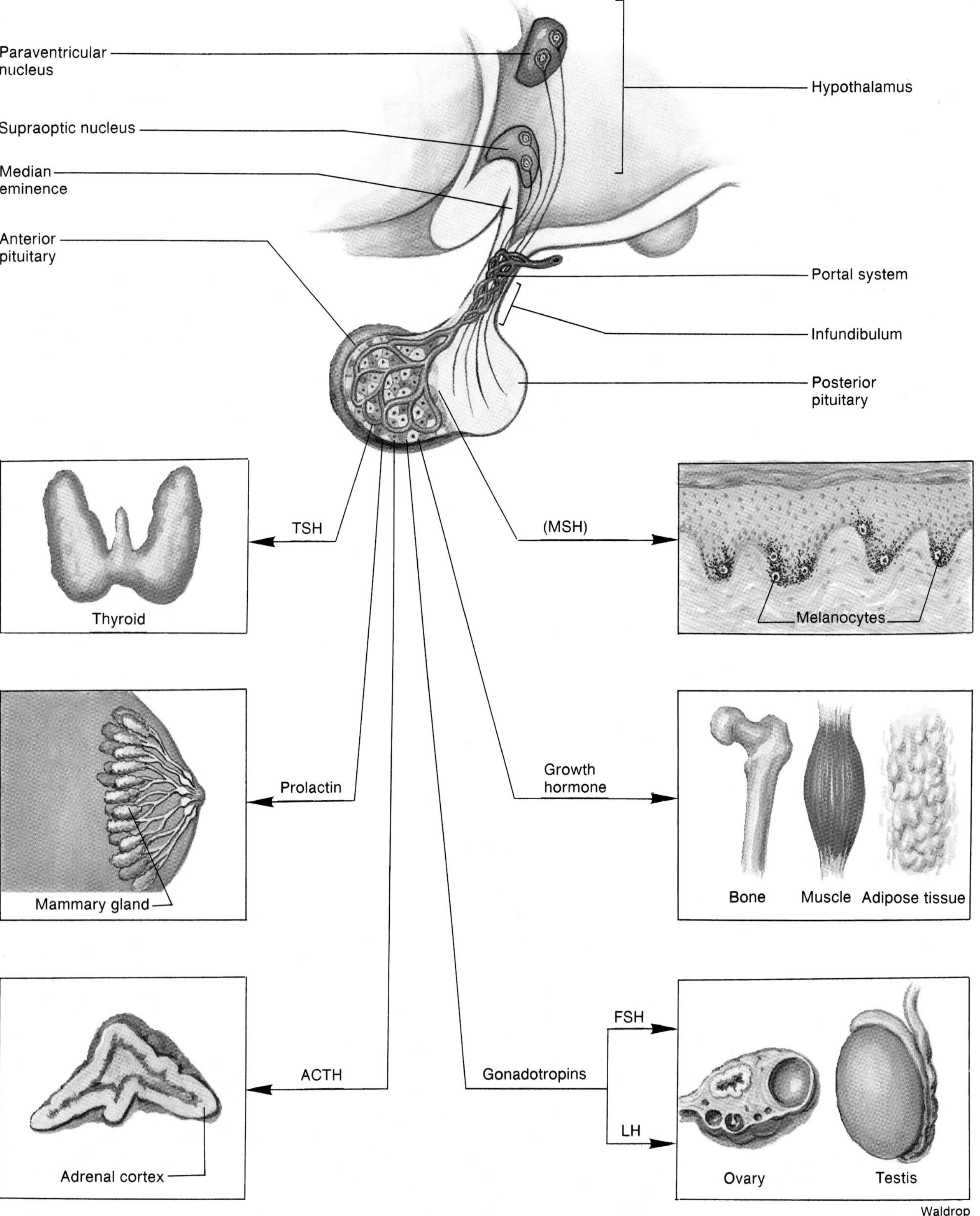

Figure 19.15

Hypothalamic control of the anterior pituitary. Neurons in the hypothalamus secrete releasing hormones (shown as dots) into the blood vessels of the hypothalamo-hypophyseal portal system. These releasing hormones stimulate the anterior pituitary to secrete its hormones (labeled "tropic hormones") into the general circulation.

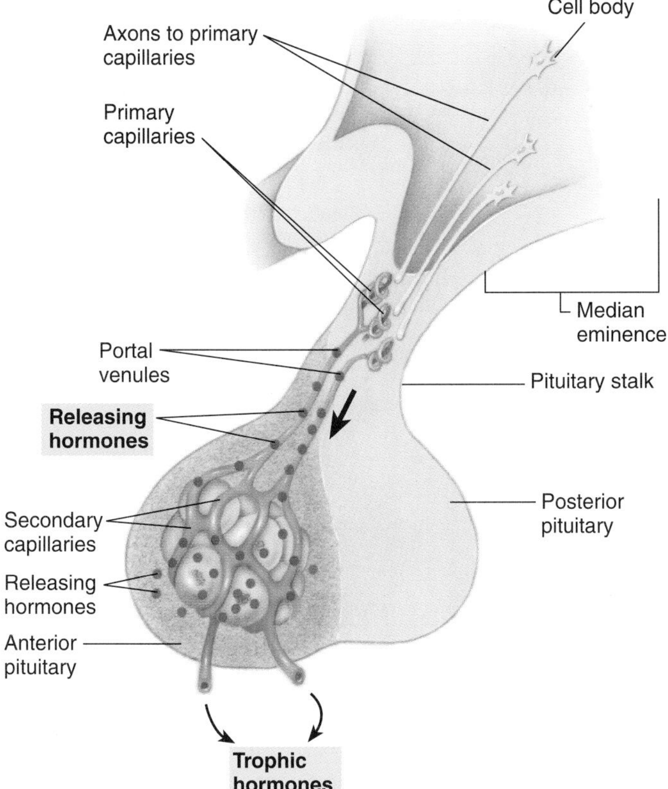

Feedback Control of the Anterior Pituitary

In view of its secretion of releasing and inhibiting hormones, the hypothalamus might be considered the "master gland." The chain of command, however, is not linear; the hypothalamus and anterior pituitary are controlled by the effects of their own actions. In the endocrine system, to use an analogy, the general takes orders from the private. The hypothalamus and anterior pituitary are not master glands because their secretions are controlled by the target glands they regulate.

Anterior pituitary secretion of ACTH, TSH, and the gonadotropins (FSH and LH) is controlled by **negative feedback inhibition** from the target gland hormones. Secretion of ACTH is inhibited by a rise in corticosteroid secretion, for example, and TSH is inhibited by a rise in the secretion of thyroxine from the thyroid. These negative feedback relationships are easily demonstrated by removal of the target glands. Castration (surgical removal of the gonads), for example, produces a rise in the secretion of FSH and LH. In a similar manner, removal of the adrenals or the thyroid results in an abnormal increase in ACTH or TSH secretion from the anterior pituitary.

The effects of removal of the target glands demonstrate that, under normal conditions, these glands exert an inhibitory effect on the anterior pituitary. This inhibitory effect can occur at two levels: (1) the target gland hormones can act on the hypothalamus and inhibit the secretion of releasing hormones, and (2) the target gland hormones can act on the anterior pituitary and inhibit its response to the releasing hormones. Thyroxine, for example, appears to inhibit the response of the anterior pituitary to TRH and thus acts to reduce TSH secretion (fig. 19.16). Sex steroids, by contrast,

Table 19.7

Hypothalamic Hormones Involved in the Control of the Anterior Pituitary

Hypothalamic Hormone	Structure	Effect on Anterior Pituitary
Corticotropin-releasing hormone (CRH)	41 amino acids	Stimulates secretion of adrenocorticotropic hormone (ACTH)
Gonadotropin-releasing hormone (GnRH)	10 amino acids	Stimulates secretion of follicle-stimulating hormone (FSH) and luteinizing hormone (LH)
Prolactin-inhibiting hormone (PIH)	Dopamine	Inhibits prolactin secretion
Somatostatin	14 amino acids	Inhibits secretion of growth hormone
Thyrotropin-releasing hormone (TRH)	3 amino acids	Stimulates secretion of thyroid-stimulating hormone (TSH)
Growth-hormone–releasing hormone (GHRH)	44 amino acids	Stimulates growth hormone secretion

Figure 19.16

The hypothalamus-pituitary-thyroid axis (control system).

The secretion of thyroxine from the thyroid is stimulated by thyroid-stimulating hormone (TSH) from the anterior pituitary. The secretion of TSH is stimulated by thyrotropin-releasing hormone (TRH) secreted from the hypothalamus. This stimulation is balanced by negative feedback inhibition (blue arrow) from thyroxine, which decreases the responsiveness of the anterior pituitary to stimulation by TRH.

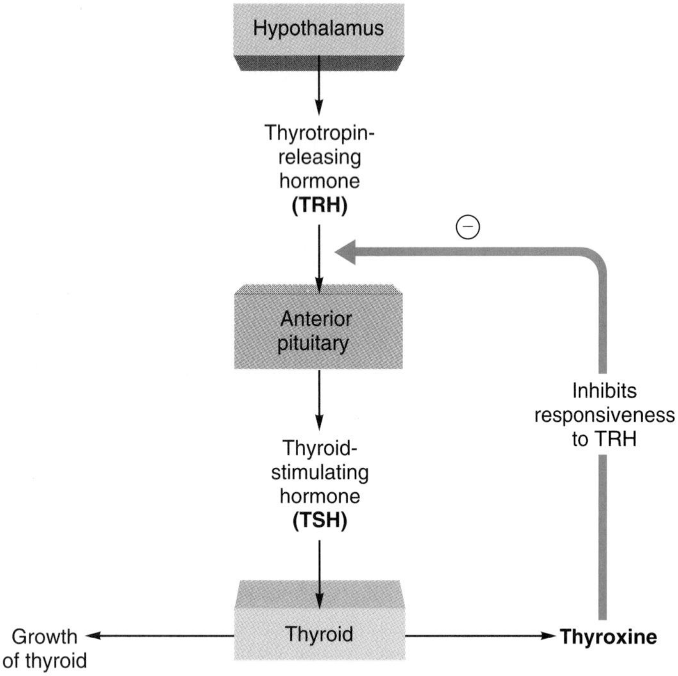

Figure 19.17

The hypothalamus-pituitary-gonad axis (control system).

The hypothalamus secretes GnRH, which stimulates the anterior pituitary to secrete the gonadotropins (FSH and LH). These, in turn, stimulate the gonads to secrete the sex steroids. The secretions of the hypothalamus and anterior pituitary are themselves regulated by negative feedback inhibition (blue arrows) from the sex steroids.

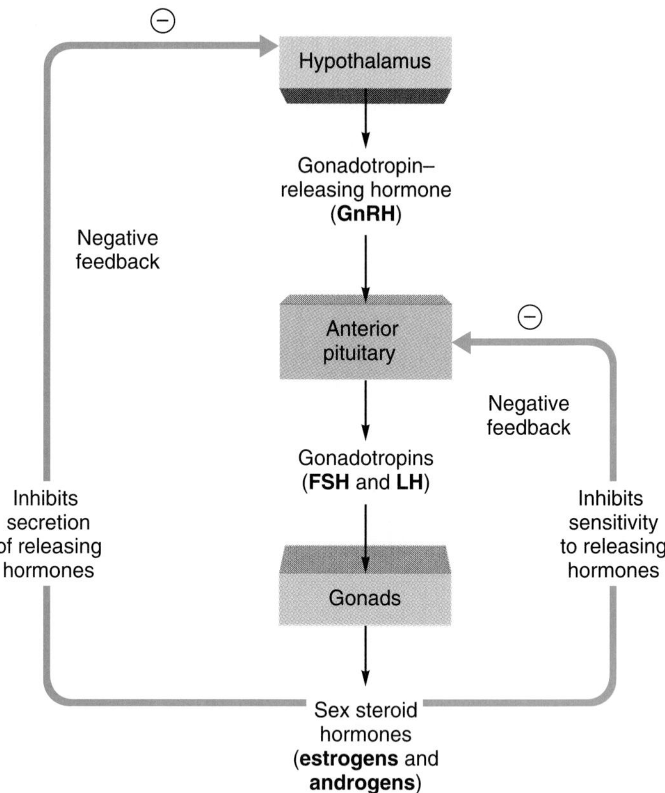

reduce the secretion of gonadotropins by inhibiting both GnRH secretion and the ability of the anterior pituitary to respond to stimulation by GnRH (fig. 19.17).

Evidence suggests that there may be retrograde transport of blood from the anterior pituitary to the hypothalamus. This may permit a *short feedback loop* in which a particular trophic hormone inhibits the secretion of its releasing hormone from the hypothalamus. A high secretion of TSH, for example, may inhibit further secretion of TRH by this means.

In addition to negative feedback control of the anterior pituitary, there is one instance of a hormone from a target gland that actually stimulates the secretion of an anterior pituitary hormone. Toward the middle of a menstrual cycle, the rising secretion of estradiol from the ovaries stimulates the anterior pituitary to secrete a "surge" of LH, which results in ovulation. This case is commonly described as a *positive feedback effect* to distinguish it from the more usual negative feedback inhibition of target gland hormones on anterior pituitary secretion. Interestingly, higher levels of estradiol at a later stage of the menstrual cycle exert the opposite effect—negative feedback inhibition—on LH secretion. The control of gonadotropin secretion is discussed in more detail in chapter 28.

Higher Brain Function and Pituitary Secretion

The relationship between the anterior pituitary and a particular target gland is described as an *axis;* the pituitary-gonad axis, for example, refers to the action of gonadotropic hormones on the testes and ovaries. This axis is stimulated by GnRH from the hypothalamus, as previously described. Since the hypothalamus receives neural input from "higher brain centers," however, it is not surprising that the pituitary-gonad axis can be affected by emotions. Indeed, the ability of intense emotions to alter the timing of ovulation or menstruation is well known. Psychological stress, as another example, also stimulates another axis—the pituitary-adrenal axis (described in the next section).

Stressors, as described later in this chapter, produce an increase in CRH secretion from the hypothalamus, which in turn results in elevated ACTH and corticosteroid secretion. In addition, the influence of higher brain centers produces *circadian* ("about a day") *rhythms* in the secretion of many anterior pituitary hormones. The secretion of growth hormone, for example, is highest during sleep and decreases during wakefulness, although its secretion is also stimulated by the absorption of particular amino acids following a meal.

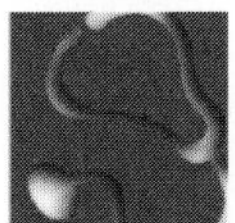

The influence of higher brain centers on the pituitary-gonad axis helps to explain the "dormitory effect"—that is, the tendency for the menstrual cycles of female roommates to synchronize. This synchronization will not occur in a new roommate if her nasal cavity is plugged with cotton, suggesting that the dormitory effect is due to the action of chemicals called *pheromones* (*fer'ŏ-mōnz*). These chemicals are secreted to the outside of the body and act through the olfactory sense to modify the physiology or behavior of another member of the same species. Pheromones are important regulatory molecules in the urine, vaginal fluid, and other secretions of most mammals, and help to regulate their reproductive cycles and behavior. The role of pheromones in human biology is difficult to assess. Recently, however, scientists discovered two pheromones produced in the axillae (underarms) of women that may cause the dormitory effect.

Adrenal Glands

The adrenal cortex and adrenal medulla are structurally and functionally different. The adrenal medulla secretes catecholamine hormones, which complement the sympathetic nervous system in the "fight-or-flight" reaction. The adrenal cortex secretes steroid hormones that participate in the regulation of mineral and energy balance.

The **adrenal glands** are paired organs that cap the superior borders of the kidneys (fig. 19.18). Each adrenal gland consists of an outer adrenal cortex and an inner adrenal medulla, which function as separate glands. The differences in function of the adrenal cortex and adrenal medulla are related to the differences in their embryonic derivation. The adrenal medulla is derived from embryonic neural crest ectoderm (the same tissue that produces the sympathetic ganglia), whereas the adrenal cortex is derived from a different embryonic tissue (mesoderm).

As a consequence of its embryonic derivation, the adrenal medulla, composed of **chromaffin** (*kro-maf'in*) **cells,** secretes catecholamine hormones (mainly epinephrine, with lesser amounts of norepinephrine) into the blood in response to stimulation by preganglionic sympathetic nerve fibers (chapter 17). The adrenal cortex does not receive neural innervation, and so must be stimulated hormonally (by ACTH secreted from the anterior pituitary). The adrenal cortex consists of three zones: an outer **zona glomerulosa** (*glo-mer"yoo-lo'să*), a middle **zona fasciculata** (*fa-sik"yoo-lă'tă*), and an inner

Figure 19.18

The structure of the adrenal gland, showing the three zones of the adrenal cortex.
The zona glomerulosa secretes the mineralocorticoids (including aldosterone), whereas the other two zones secrete the glucocorticoids (including cortisol).

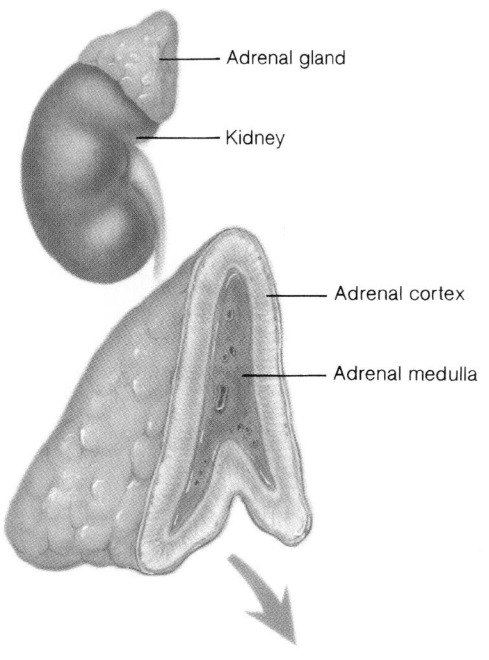

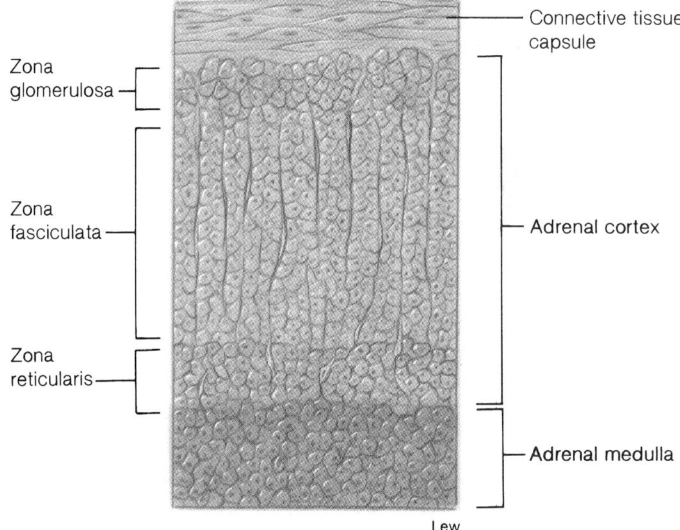

zona reticularis (fig. 19.18). These zones are believed to have different functions.

Functions of the Adrenal Cortex

The adrenal cortex secretes steroid hormones called **corticosteroids,** or **corticoids,** for short. There are three functional categories of corticosteroids: (1) **mineralocorticoids,** which regulate Na^+ and K^+ balance; (2) **glucocorticoids,** which regulate the metabolism of glucose and other organic molecules; and (3) **sex steroids,** which are weak androgens (such as

adrenal: L. *ad,* to; *renes,* kidney

Figure 19.19

Simplified pathways for the synthesis of steroid hormones in the adrenal cortex.
The adrenal cortex produces steroids that regulate Na^+ and K^+ balance (mineralocorticoids), steroids that regulate glucose balance (glucocorticoids), and small amounts of sex steroid hormones. (DHEA = dehydroepiandrosterone.)

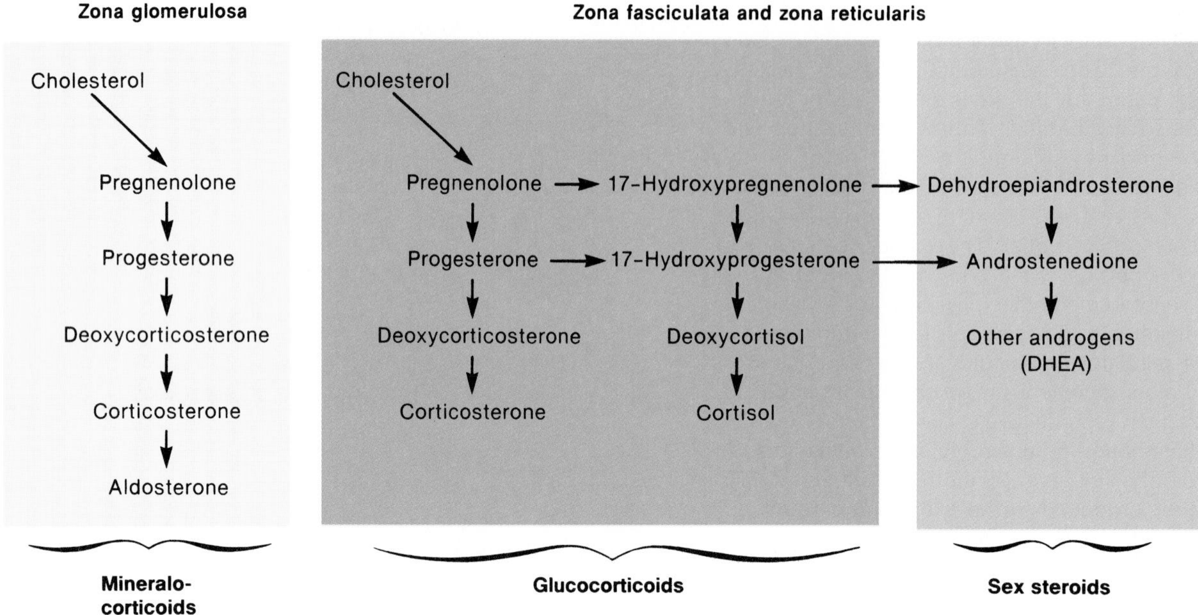

dehydroepiandrosterone, or *DHEA*) that supplement the sex steroids secreted by the gonads. These hormones are secreted by the different zones of the adrenal cortex.

Aldosterone is the most potent mineralocorticoid. The mineralocorticoids are produced in the zona glomerulosa (fig. 19.19). The predominant glucocorticoid in humans is *cortisol* (*hydrocortisone*), which is secreted by the zona fasciculata and perhaps also by the zona reticularis. The secretion of cortisol by the zona fasciculata is stimulated by ACTH (fig. 19.20). The secretion of aldosterone is controlled by other mechanisms related to blood volume and electrolyte balance, as described in chapter 22.

Functions of the Adrenal Medulla

The cells of the adrenal medulla secrete **epinephrine** (*ep″ĭ-nef′rin*) and **norepinephrine** in an approximate ratio of 4 to 1, respectively. The effects of these hormones are similar to those caused by stimulation of the sympathetic division of the autonomic nervous system, except that the hormonal effect lasts about 10 times longer. The hormones from the adrenal medulla increase the cardiac output and heart rate, dilate coronary blood vessels, increase mental alertness, increase the respiratory rate, and elevate the metabolic rate. The metabolic effects of these hormones are discussed in chapter 27.

Figure 19.20

The activation of the pituitary-adrenal axis by nonspecific stress.
Negative feedback control of the adrenal cortex (blue arrows) is also shown.

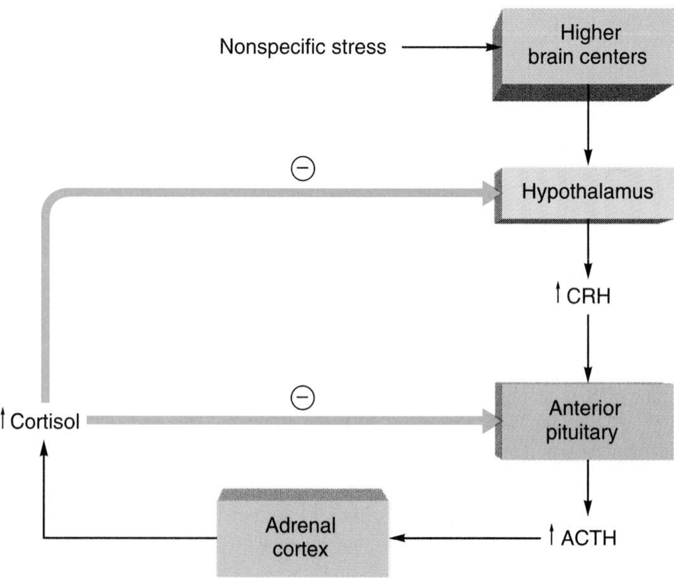

The adrenal medulla is innervated by sympathetic nerve fibers. Many stressors, therefore, activate the adrenal medulla as well as the adrenal cortex. Activation of the adrenal medulla as part of the mass activation of the sympathetic division of the autonomic nervous system prepares the body for greater physical performance—the fight-or-flight response (chapter 17).

Stress and the Adrenal Gland

In 1936, a Canadian physiologist, Hans Selye, discovered that injections of a cattle ovary extract into rats (1) stimulated growth of the adrenal cortex; (2) caused atrophy of the lymphoid tissue of the spleen, lymph nodes, and thymus; and (3) produced bleeding peptic ulcers. At first he attributed these effects to the action of a specific hormone in the extract. However, subsequent experiments revealed that a variety of substances—including foreign chemicals such as formaldehyde—could produce the same effects. Indeed, the same pattern occurred when Selye subjected rats to cold environments or when he dropped them into water and made them swim until they were exhausted.

The specific pattern of effects produced by these procedures suggested that the effects were due to something the procedures had in common. Selye reasoned that all of the procedures were stressful. Stress, according to Selye, is the reaction of an organism to stimuli called *stressors*, which may produce damaging effects. The pattern of changes he observed represented a specific response to any stressful agent. He later discovered that stressors produce these effects because they stimulate the pituitary-adrenal axis. Under stressful conditions there is increased secretion of ACTH from the anterior pituitary, and thus there is increased secretion of glucocorticoids from the adrenal cortex.

On this basis, Selye stated that there is "a nonspecific response of the body to readjust itself following any demand made upon it." A rise in the plasma glucocorticoid levels results from the demands of the stressors. Selye termed this nonspecific response the **general adaptation syndrome (GAS).** Stress, in other words, produces GAS. There are three stages in the response to stress: (1) the *alarm reaction*, when the adrenal glands are activated; (2) the *stage of resistance*, in which readjustment occurs; and (3) if the readjustment is not complete, the *stage of exhaustion*, which may lead to sickness and possibly death.

Selye's concept of stress has been refined by subsequent research. These investigations demonstrate that the sympathoadrenal system is activated, with increased secretion of epinephrine and norepinephrine, in response to stressors that challenge the organism to respond physically. This is the "fight-or-flight" reaction described in chapter 17. Different emotions, however, are accompanied by different endocrine responses. The pituitary-adrenal axis, with rising levels of glucocorticoids, becomes more active when the stress is of a chronic nature and when the person is more passive and feels less in control.

Glucocorticoids such as hydrocortisone can inhibit the immune system. For this reason, these steroids are often administered to treat various inflammatory diseases and to suppress the immune rejection of a transplanted organ. It seems reasonable, therefore, that the elevated glucocorticoid secretion that can accompany stress may inhibit the ability of the immune system to protect against disease. Indeed, studies suggest that prolonged stress results in an increased incidence of cancer and other diseases.

Thyroid and Parathyroid Glands

The thyroid secretes thyroxine (T_4) and triiodothyronine (T_3), which are needed for proper growth and development and that are primarily responsible for determining the basal metabolic rate (BMR). The parathyroid glands secrete parathyroid hormone, which helps to raise the blood Ca^{2+} concentration.

The **thyroid gland** is located just below the larynx (fig. 19.21). Its two lobes are positioned on either side of the trachea and are connected anteriorly by a medial mass of thyroid tissue called the *isthmus*. The thyroid is the largest of the endocrine glands, weighing between 20 and 25 grams.

On a microscopic level, the thyroid gland consists of many spherical hollow sacs called **thyroid follicles** (fig. 19.22). These follicles are lined with a simple cuboidal epithelium composed of *follicular cells* that synthesize the principal thyroid hormone, thyroxine. The interior of the follicles contains **colloid,** a protein-rich fluid. Between the follicles are epithelial cells called *parafollicular cells;* these cells produce a hormone called *calcitonin* (or *thyrocalcitonin*).

Production and Action of Thyroid Hormones

The thyroid follicles actively accumulate iodide (I^-) from the blood and secrete it into the colloid. Once the iodide has entered the colloid, it oxidizes to become iodine and attaches to specific amino acids (tyrosines) within the polypeptide chain of a protein called **thyroglobulin.** The attachment of one iodine to tyrosine produces *monoiodotyrosine (MIT)*; the attachment of two iodines produces *diiodotyrosine (DIT)*.

Within the colloid, enzymes modify the structure of MIT and DIT and couple them together, producing a molecule of **tetraiodothyronine** (*tet"ră-i-o"do-thi'rŏ-nēn*)—**(T_4),** or **thyroxine** (fig. 19.23). The combination of one MIT with one DIT forms **triiodothyronine** (*tri"i-o"do-thi'rŏ-nēn*), or **T_3.** Note that within the colloid T_4 and T_3 are still attached to thyroglobulin. Upon stimulation by TSH, the cells of the follicle take up a small volume of colloid by pinocytosis, hydrolyze the T_3 and T_4 from the thyroglobulin, and secrete the free hormones into the blood.

The transport of thyroid hormones through the blood and their mechanism of action at the cellular level was described earlier in this chapter. Through the activation of

Figure 19.21

The thyroid gland.
(*a*) Its relationship to the larynx and trachea. (*b*) A scan of the thyroid gland 24 hours after the intake of radioactive iodine.

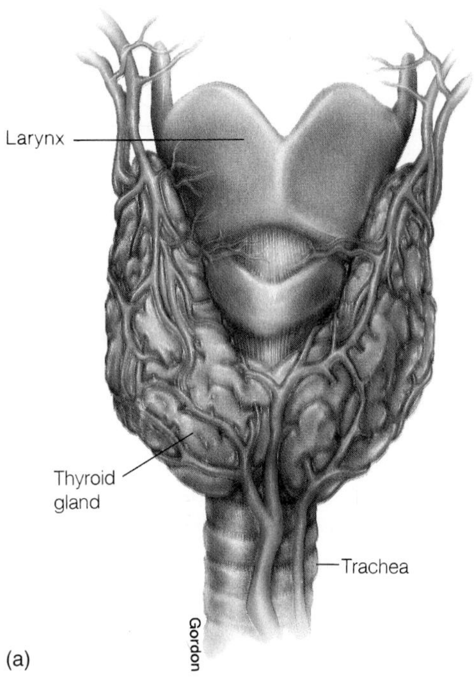

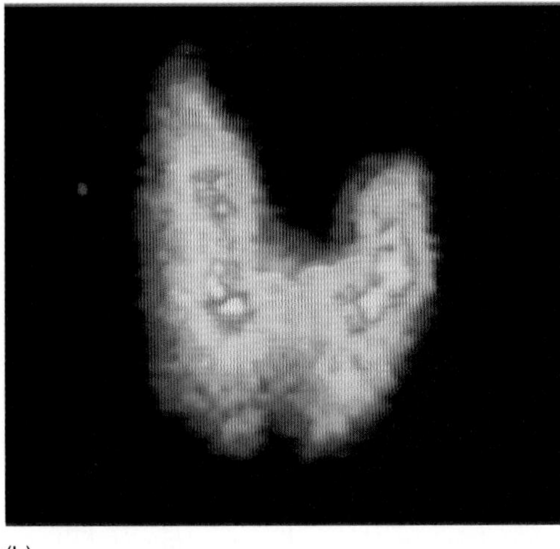

Figure 19.22

Photomicrograph (250×) of a thyroid gland.
Numerous thyroid follicles are visible. Each follicle consists of follicular cells surrounding the fluid known as colloid, which contains thyroglobulin.

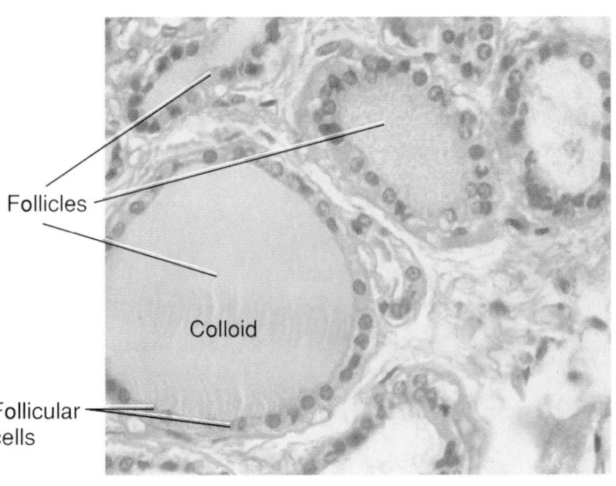

Figure 19.23

Production and storage of thyroid hormones.
Iodide is actively transported into the follicular cells. In the colloid, it is converted into iodine and attached to tyrosine amino acids within the thyroglobulin protein. MIT (monoiodotyrosine) and DIT (diiodotyrosine) are used to produce T_3 and T_4 within the colloid. Upon stimulation by TSH, the thyroid hormones, bound to thyroglobulin, are taken into the follicular cells by pinocytosis. Hydrolysis reactions within the follicular cells release the free T_4 and T_3, which are secreted.

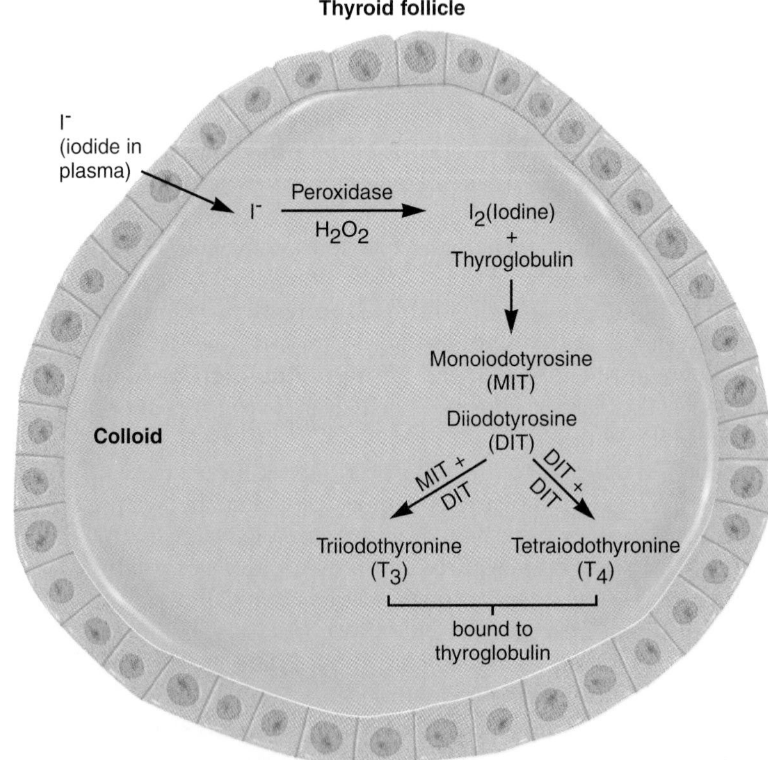

genes, thyroid hormones stimulate protein synthesis, promote maturation of the nervous system, and increase the rate of energy utilization by the body.

The development of the central nervous system is particularly dependent on thyroid hormones, and a deficiency of these hormones during development can cause serious mental retardation. The basal metabolic rate (BMR)—which is the minimum rate of caloric expenditure by the body—is determined to a large degree by the level of thyroid hormones in the blood.

Calcitonin, released by the parafollicular cells of the thyroid, works in concert with parathyroid hormone (discussed shortly) to regulate calcium levels in the blood. Calcitonin inhibits the breakdown of bone tissue and stimulates the excretion of calcium by the kidneys. Both actions result in a lowering of blood calcium levels.

The physiological functions of thyroid hormones are described in more detail in chapter 27.

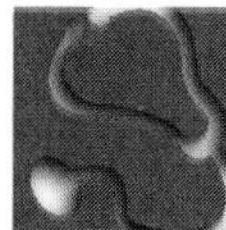

Thyroid-stimulating hormone (TSH) from the anterior pituitary stimulates the thyroid to secrete thyroxine and exerts a trophic effect on the thyroid gland. This trophic effect is immediately apparent in people who develop an *iodine-deficiency (endemic) goiter.* In the absence of sufficient dietary iodine, the thyroid cannot produce adequate amounts of T_4 and T_3. The resulting lack of negative feedback causes abnormally high levels of TSH secretion, which in turn stimulate the abnormal growth of the thyroid (a goiter). These events are summarized in figure 19.24.

Radiotherapy for Thyroid Cancer

The treatment of thyroid cancer illustrates how the principles previously introduced find practical application. Thyroid cancer, particularly when it occurs in younger adults, has a more optimistic prognosis than most other forms of cancer. It is extremely slow growing and usually spreads only to the lymph nodes of the neck (although other sites can be invaded). Surgical treatment involves removing most of the thyroid and the affected cervical lymph nodes. This surgical treatment is followed by having the patient swallow solutions containing radioactive iodine (^{131}I), which is selectively transported only into cells of the thyroid gland and into cancerous thyroid cells that have metastasized (traveled) to other regions of the body. These cells are then killed by the radioactive iodine.

Thyroid-stimulating hormone (TSH) stimulates the active accumulation of iodine in thyroid cells so that these cells can produce and secrete thyroxine. Therefore, in order for the ingested radioactive iodine to be maximally effective in the treatment of thyroid cancer, the blood levels of TSH must be raised to high levels.

When most of the patient's thyroid gland is surgically removed the blood levels of thyroxine gradually decline. Owing to the fact that most of the thyroxine is bound to protein in the plasma, it has an extremely long half-life. As the thyroxine levels in the blood decline, inhibition of TSH secretion lessens and TSH levels consequently rise. A period of

Figure 19.24

How iodine deficiency causes a goiter.
Lack of adequate iodine in the diet interferes with the negative feedback control of TSH secretion, resulting in the formation of an endemic goiter.

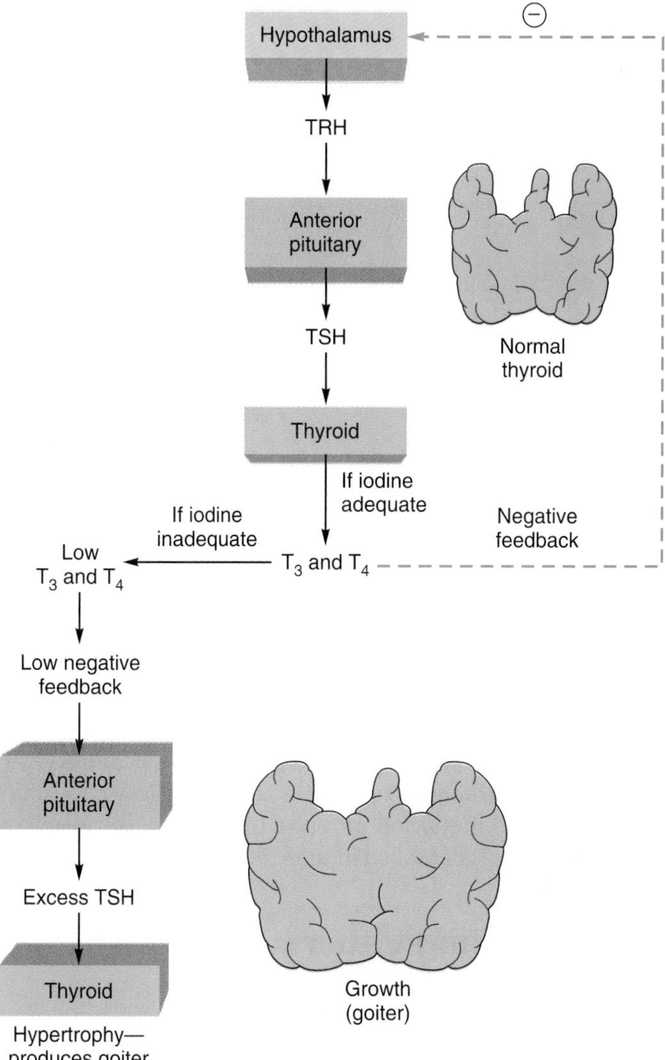

about 4 to 5 weeks is required after removal of the thyroid for the TSH levels to rise sufficiently. At this point, there is enough TSH to stimulate accumulation of radioactive iodine into thyroid cells and promote the destruction of both normal and metastasized cells.

Parathyroid Glands

The small, flattened **parathyroid glands** are embedded in the posterior surfaces of the lateral lobes of the thyroid gland, as shown in figure 19.25. There are usually four parathyroid glands (a *superior pair* and an *inferior pair*), although the precise number can vary. Each parathyroid gland is a small yellowish-brown body 3 to 8 mm (0.1 to 0.3 in.) long, 2 to 5 mm (0.07 to 0.2 in.) wide, and about 1.5 mm (0.05 in.) deep.

Figure 19.25

A posterior view of the parathyroid glands.
The parathyroids are embedded within the tissue of the thyroid gland.

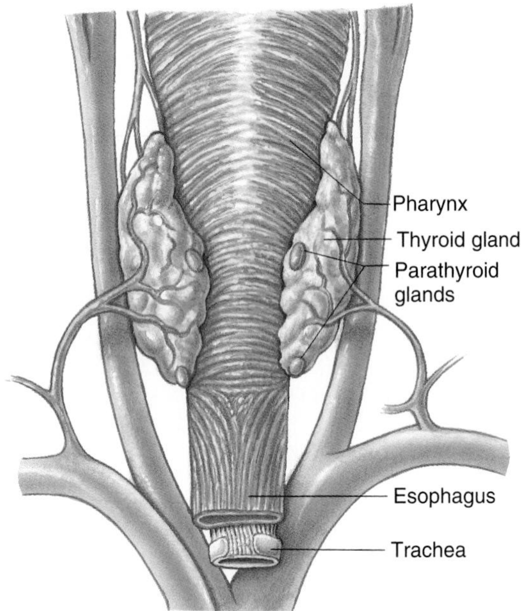

Figure 19.26

The actions of parathyroid hormone and the control of its secretion.
An increased level of parathyroid hormone causes the bones to release calcium and the kidneys to conserve calcium that would otherwise be lost through the urine. A rise in blood Ca^{2+} can then exert negative feedback inhibition on parathyroid hormone secretion.

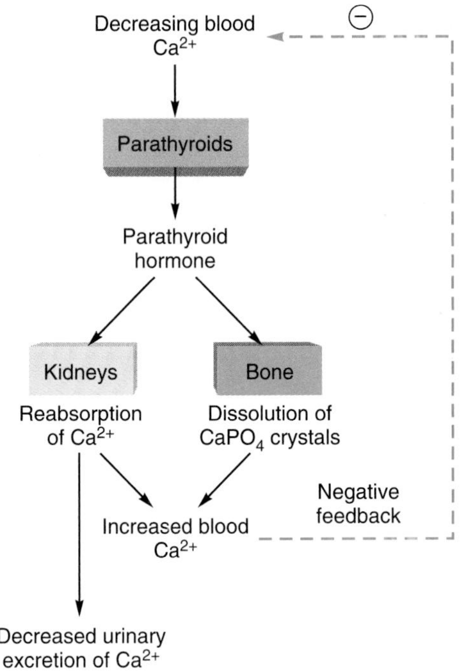

Parathyroid hormone (PTH) is the only hormone secreted by the parathyroid glands. PTH, however, is the single most important hormone in the control of the calcium levels of the blood. It promotes a rise in blood calcium levels by acting on the bones, kidneys, and intestine (fig. 19.26). Regulation of calcium balance is described in more detail in chapter 27.

Pancreas and Other Endocrine Glands

The pancreatic islets secrete two hormones, insulin and glucagon. Insulin promotes the lowering of blood glucose and the storage of energy in the form of glycogen and fat. Glucagon has antagonistic effects that act to raise the blood glucose concentration. Additionally, many other organs secrete hormones that help to regulate digestion, metabolism, growth, immune function, and reproduction.

The **pancreas** is both an endocrine and an exocrine gland. The gross structure of this gland and its exocrine functions in digestion are described in chapter 26. The endocrine portion of the pancreas consists of scattered clusters of cells called the **pancreatic islets,** or **islets of Langerhans** (*i'lets of lang'er-hanz*). These endocrine structures are most common in the body and tail of the pancreas (fig. 19.27).

islets of Langerhans: from Paul Langerhans, German anatomist, 1847–88

Pancreatic Islets (Islets of Langerhans)

On a microscopic level, the most conspicuous cells in the islets are the *alpha* and *beta cells* (fig. 19.27). The alpha cells secrete the hormone **glucagon** (*gloo'kă-gon*), and the beta cells secrete **insulin.**

Alpha cells secrete glucagon in response to a fall in the blood glucose concentrations. Glucagon stimulates the liver to hydrolyze glycogen to glucose (*glycogenolysis*), which causes the blood glucose level to rise. This effect thus represents the completion of a negative feedback loop. Glucagon also stimulates the hydrolysis of stored fat (*lipolysis*) and the consequent release of free fatty acids into the blood. This effect helps to provide energy substrates for the body during fasting, when blood glucose levels decrease. Glucagon, together with other hormones, also stimulates the conversion of fatty acids to ketone bodies, which can be secreted by the liver into the blood and used by other organs as an energy source. Glucagon is thus a hormone that helps to maintain homeostasis during times of fasting, when the body's energy reserves must be utilized (see chapter 27).

insulin: L. *insula*, island

Figure 19.27

The pancreas and the associated pancreatic islets (islets of Langerhans).
Alpha cells secrete glucagon, and beta cells secrete insulin. The pancreas is also exocrine, producing pancreatic juice for transport via the pancreatic duct to the duodenum of the small intestine.

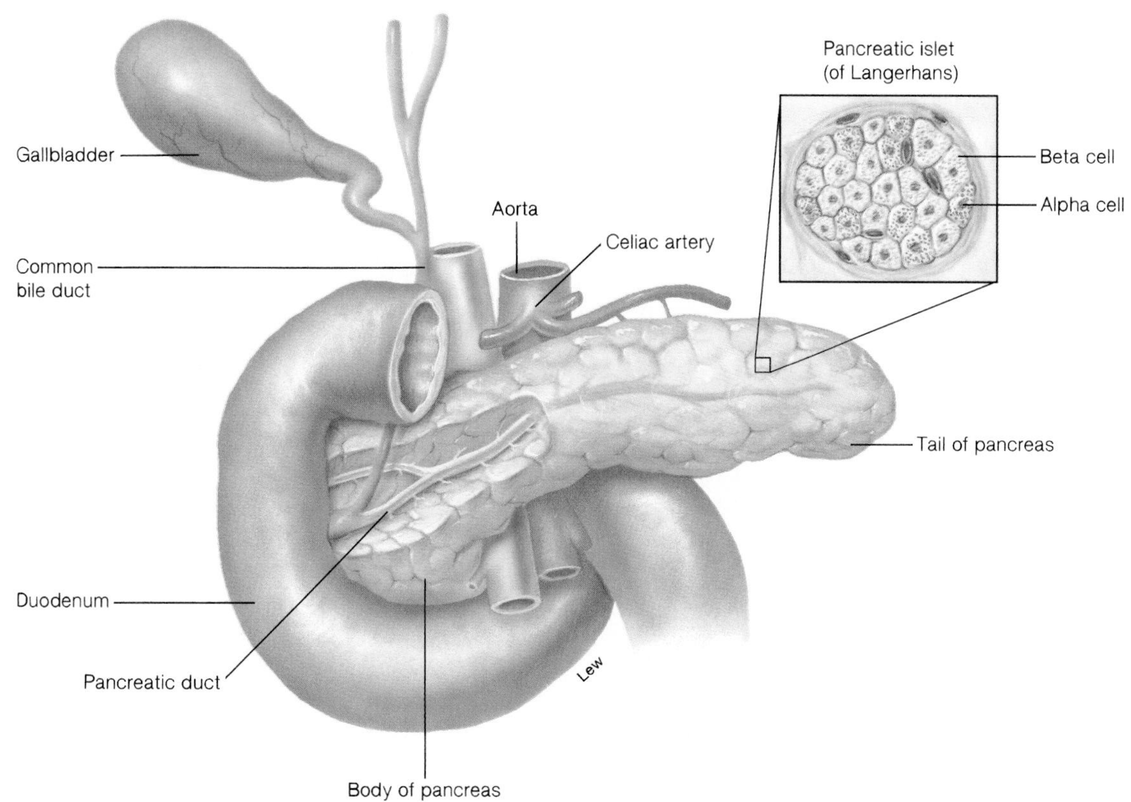

Beta cells secrete insulin in response to a rise in blood glucose concentrations. Insulin promotes the entry of glucose into tissue cells, and the conversion of this glucose into energy-storage molecules of glycogen and fat. Insulin also aids the entry of amino acids into cells and the production of cellular protein. The actions of insulin and glucagon are thus antagonistic. After a meal, insulin secretion increases and glucagon secretion decreases; fasting, by contrast, causes a rise in glucagon and a fall in insulin secretion.

Pineal Gland

The small, cone-shaped **pineal gland** is located in the roof of the third ventricle near the corpora quadrigemina, where it is encapsulated by the meninges covering the brain. The pineal gland of a child weighs about 0.2 g and is 5 to 8 mm (0.2 to 0.3 in.) long and 9 mm wide. The gland begins to regress in size at about age 7 and in the adult appears as a thickened strand of fibrous tissue. Although the pineal gland lacks direct nervous connections to the rest of the brain, it is highly innervated by the sympathetic nervous system from the superior cervical ganglion.

pineal: L. *pinea,* pine cone

The principal hormone of the pineal gland is **melatonin** (*mel″ă-to′nin*). Production and secretion of this hormone is stimulated by activity of the **suprachiasmatic nucleus (SCN)** in the hypothalamus of the brain via activation of sympathetic neurons to the pineal gland (fig. 19.28). The SCN is the primary center for **circadian** (*ser″kă-de′an* or *ser-ka′de-an*) **rhythms** in the body—rhythms of physiological activity that follow a 24-hour pattern. The circadian activity of the SCN is automatic, but environmental light/dark changes are required to entrain (synchronize) this activity to a day/night cycle. Activity of the SCN, and thus secretion of melatonin, increases with darkness and peaks by the middle of the night. During the day, neural pathways from the retina of the eyes to the hypothalamus (fig. 19.28) act to depress the activity of the SCN, reducing sympathetic stimulation of the pineal and thus decreasing melatonin secretion.

The pineal gland has been implicated in a variety of physiological processes. One of the most widely studied is the ability of melatonin to inhibit the pituitary-gonad axis (inhibiting GnRH secretion or the response of the anterior pituitary to GnRH, depending on the species of animal). Indeed, a decrease in melatonin secretion in many species is required for the maturation of the gonads during the reproductive season of seasonal breeders. Although there is evidence to support an antigonadotropic effect in humans, this possibility has

Figure 19.28

The secretion of melatonin.
The secretion of melatonin by the pineal gland is stimulated by sympathetic axons originating in the superior cervical ganglion. Activity of these neurons is regulated by the cyclic activity of the suprachiasmatic nucleus of the hypothalamus, which sets a circadian rhythm. This rhythm is entrained to light/dark cycles by neurons in the retina.

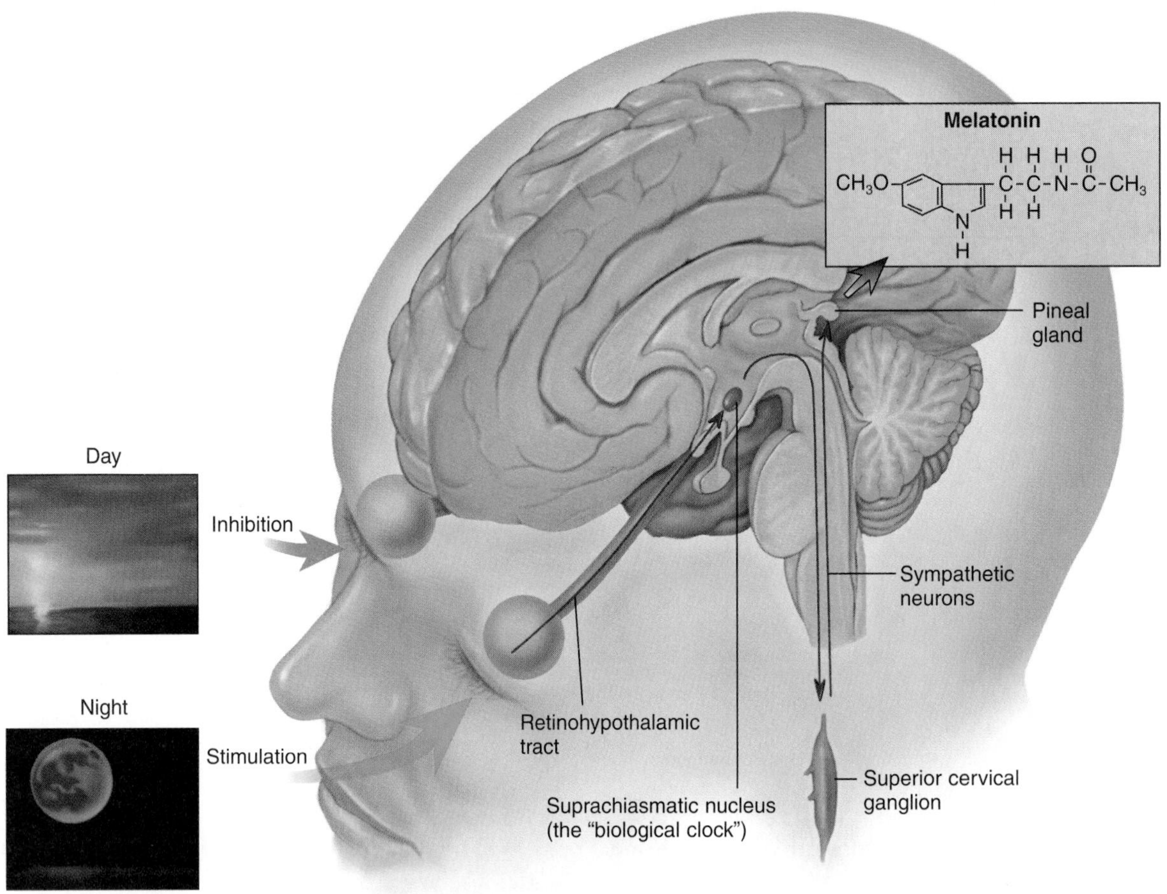

not yet been proven. For example, excessive melatonin secretion in humans is associated with a delay in the onset of puberty. Research findings indicate that melatonin secretion is highest in children between the ages of 1 and 5 and decreases thereafter, reaching its lowest levels at the end of puberty, when concentrations are 75% lower than during early childhood. This suggests a role for melatonin in the onset of human puberty. However, because of much conflicting data, the importance of melatonin in human reproduction is still highly controversial.

The pattern of melatonin secretion is altered when a person works night shifts or flies across different time zones. There is evidence that exogenous melatonin (taken as a pill) may be beneficial in the treatment of jet lag, but the optimum dosage is not currently known. Phototherapy using bright fluorescent lamps, which act like sunlight to inhibit melatonin secretion, has been used effectively in the treatment of *seasonal affective disorder* (*SAD*), or "winter depression."

Melatonin pills increase the speed of falling asleep and the duration of rapid eye movement (REM) sleep; for these reasons, they may be useful in the treatment of insomnia. This is particularly significant for elderly people with insomnia, who have the lowest nighttime levels of endogenous melatonin secretion. Melatonin can also act, much like vitamin E, as a scavenger of hydroxyl and other free radicals that can cause oxidative damage to cells. This antioxidant effect of melatonin, however, occurs only at pharmacological doses. The purported beneficial effects of exogenous melatonin (other than for insomnia and jet lag) are not yet proven, and the consensus of current medical opinion is against the uncontrolled use of melatonin pills.

Thymus

The **thymus** is a bilobed organ positioned in front of the aorta and behind the manubrium of the sternum (fig. 19.29). Although the size of the thymus varies considerably from person

Figure 19.29

The thymus is a bilobed organ within the mediastinum of the thorax.
The thymus secretes hormones that help to regulate the immune system.

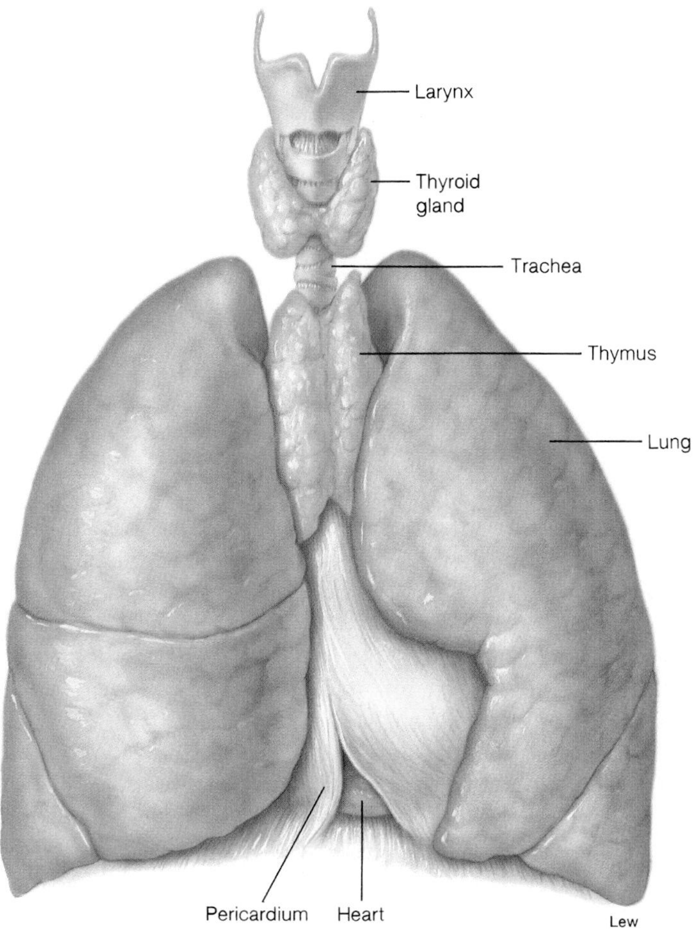

to person, it is relatively large in newborns and children, and sharply regresses in size after puberty. Besides decreasing in size, the thymus of adults becomes infiltrated with strands of fibrous and fatty connective tissue.

The thymus is the site of production of **T cells** (*thymus-dependent cells*), which are the lymphocytes involved in cell-mediated immunity (see chapter 23). In addition to providing T cells, the thymus secretes a number of hormones that are believed to stimulate T cells after they leave the thymus.

Gastrointestinal Tract

The stomach and small intestine secrete a number of hormones that act on the gastrointestinal tract itself and on the pancreas and gallbladder (see chapter 26). The effects of these hormones, in conjunction with regulation by the autonomic nervous system, coordinate the activities of different regions of the gastrointestinal tract and the secretions of pancreatic juice and bile.

Gonads and Placenta

The **gonads** (**testes** and **ovaries**) secrete sex steroids. These include male sex hormones, or *androgens*, and female sex hormones—*estrogens* and *progestogens*. The principal hormones in each of these categories are *testosterone*, *estradiol-17β*, and *progesterone*, respectively.

The testes consist of two compartments: *seminiferous tubules*, which produce sperm, and *interstitial tissue* between the convolutions of the tubules. The interstitial cells (**Leydig cells**) within the interstitial tissue secrete testosterone. Testosterone is needed for the development and maintenance of the male genitalia (penis and scrotum) and the male accessory sex organs (prostate, seminal vesicles, epididymides, and ductus [vas] deferens), as well as for the development of male secondary sex characteristics.

During the first half of the menstrual cycle estrogen is secreted by small structures within the ovary called *ovarian follicles*. These follicles contain the egg cell, or *ovum*, and **granulosa cells** that secrete estrogen. By about midcycle, one of these follicles grows very large and, in the process of ovulation, extrudes its ovum from the ovary. The empty follicle, under the influence of luteinizing hormone (LH) from the anterior pituitary, then becomes a new endocrine structure called a *corpus luteum*. The corpus luteum secretes progesterone as well as estradiol-17β.

The **placenta**—the organ responsible for nutrient and waste exchange between the fetus and mother—is also an endocrine gland in that it secretes large amounts of estrogens and progesterone. In addition, it secretes a number of polypeptide and protein hormones that are similar to some hormones secreted by the anterior pituitary. These hormones include *human chorionic gonadotropin (hCG)*, which is similar to LH, and *somatomammotropin*, which is similar in action to both growth hormone and prolactin. The physiology of the placenta and other aspects of reproductive endocrinology are considered in chapter 30.

Autocrine and Paracrine Regulation

Many regulatory molecules produced throughout the body act within the organs that produce them. These molecules may regulate different cells within one tissue, or they may be produced within one tissue and regulate a different tissue within the same organ.

Thus far in this text, two types of regulatory molecules have been considered—neurotransmitters (chapter 14) and hormones (this chapter). These two classes of regulatory molecules cannot be defined simply by differences in chemical structure, since on this basis the same molecule (such as norepinephrine) could be included in both categories. Rather,

Leydig cells: from Franz von Leydig, German anatomist, 1821–1908 they must be defined by function. Neurotransmitters are released by axons, travel across a narrow synaptic cleft, and affect a postsynaptic cell. Hormones are secreted into the blood by an endocrine gland and, through transport in the blood, influence the activities of one or more target organs.

There are yet other classes of regulatory molecules. These molecules are distinguished by the fact that they are produced in many different organs and are active within the organ in which they are produced. Molecules of this type are called **autocrine** (*aw'tŏ-krin*) **regulators** if they are produced and act within the same tissue of an organ. They are called **paracrine regulators** if they are produced within one tissue and regulate a different tissue of the same organ (table 19.8). In the following discussion, for the sake of simplicity and because the same chemical can function as an autocrine or a paracrine regulator in different situations, the term *autocrine* will be used in a generic sense to refer to both types of local regulation.

Examples of Autocrine Regulation

Many autocrine regulatory molecules are also known as **cytokines,** particularly if they regulate different cells of the immune system, and as **growth factors,** if they promote growth and cell division in any organ. This distinction is somewhat blurred, however, because some cytokines may also function as growth factors. Cytokines produced by lymphocytes (the type of white blood cell involved in specific immunity—see chapter 23) are also known as **lymphokines** (*lim'fŏ-kinz*), and the specific molecules involved are called *interleukins*. The terminology can be confusing because new regulatory molecules, and new functions for previously named regulatory molecules, are being discovered at a rapid pace. Cytokines secreted by macrophages (phagocytic cells found in connective tissues) and by lymphocytes stimulate proliferation of specific cells involved in the immune response.

Neurotrophins, such as *nerve growth factor,* guide regenerating peripheral neurons that have been injured (chapter 14). *Nitric oxide,* which can function as a neurotransmitter in memory processes (chapters 14 and 15) and in other functions, is also produced by the endothelium of blood vessels. In this context it is a paracrine regulator because it diffuses to the smooth muscle layer of the blood vessel and promotes smooth muscle relaxation, leading to dilation of the blood vessel. Neural and paracrine regulation interact in this case, since autonomic axons that release acetylcholine in blood vessels cause dilation by stimulating the synthesis of nitric oxide in those vessels.

The endothelium of blood vessels also produces other paracrine regulators. These include the *endothelins* (specifically *endothelin-1* in humans), which directly promote vasoconstriction, and *bradykinin,* which promotes vasodilation. These regulatory molecules are thus very important in the control of blood flow and blood pressure (see chapter 22). They are also involved in the development of atherosclerosis, the leading cause of heart disease and stroke (see chapter 21). In addition, endothelin-1 is produced by the epithelium of the airways and may be important in the embryological development and function of the respiratory system.

All autocrine regulators control gene expression in their target cells to some degree. This is very clearly the case with the various growth factors. These include *platelet-derived*

Table 19.8
Examples of Autocrine and Paracrine Regulators

Autocrine or Paracrine Regulator	Major Sites of Production	Major Actions
Insulinlike growth factors (somatomedins)	Many organs, particularly the liver and cartilages	Growth and cell division
Nitric oxide	Endothelium of blood vessels; neurons of CNS; macrophages	Dilation of blood vessels; neural messenger; antibacterial
Endothelins	Endothelium of blood vessels; other organs	Constriction of blood vessels; other effects
Platelet-derived growth factor	Platelets; macrophages; vascular smooth muscle cells	Cell division within blood vessels
Epidermal growth factors	Epidermal tissues	Cell division in wound healing
Neurotrophins	Neuroglial cells, Schwann cells, neurons	Regeneration of peripheral nerves
Bradykinin	Endothelium of blood vessels	Dilation of blood vessels
Interleukins	Macrophages; lymphocytes	Regulation of immune system
Prostaglandins	Many tissues	Wide variety (see text)
TNF_{α} (tumor necrosis factor)	Macrophages; adipocytes	Wide variety

growth factor, epidermal growth factor, and the *insulinlike growth factors* that stimulate cell division and proliferation of their target cells. Regulators in the last group interact with the endocrine system in a number of ways, as will be described in chapter 27.

Prostaglandins

The most diverse group of autocrine regulators are the **prostaglandins.** A prostaglandin is a 20-carbon–long fatty acid that contains a 5-membered carbon ring. Prostaglandins are members of a family called the **eicosanoids** (*i-ko'să-noidz*), which are molecules derived from the precursor *arachidonic acid.* Upon stimulation by hormones or other agents, arachidonic acid is released from phospholipids in the cell membrane and may then enter one of two possible metabolic pathways. In one case, the arachidonic acid is converted by the enzyme *cyclo-oxygenase* into a prostaglandin, which can then be changed by other enzymes into other prostaglandins. In the other case, arachidonic acid is converted by the enzyme *lipoxygenase* into **leukotrienes** (*loo"kŏ-tri'ēnz*), which are eicosanoids that are closely related to the prostaglandins (fig. 19.30).

Prostaglandins are produced in almost every organ and have been implicated in a wide variety of regulatory functions. The study of prostaglandin function can be confusing, in part because a given prostaglandin may have opposite effects in different organs. Prostaglandins of the E series (PGE), for example, cause smooth muscle to relax in the bladder, bronchioles, intestine, and uterus, but the same molecules cause vascular smooth muscle to contract. A different prostaglandin, designated $PGF_{2\alpha}$, has the opposite effects.

The antagonistic effects of prostaglandins on blood clotting make good physiological sense. Blood platelets, which are required for blood clotting, produce *thromboxane* A_2. This prostaglandin promotes clotting by stimulating platelet aggregation and vasoconstriction. The endothelial cells of blood vessels, by contrast, produce a different prostaglandin known as PGI_2, or *prostacyclin,* whose effects are the opposite—it inhibits platelet aggregation and causes vasodilation. These antagonistic effects ensure that, while clotting is promoted, the clots will not normally form on the walls of intact blood vessels.

Figure 19.30

The formation of leukotrienes and prostaglandins.
The actions of these autocrine regulators (PG = prostaglandin; TX = thromboxane) are also summarized.

Source: From L. M. Demers, "The Effects of Prostaglandins," Diagnostic Medicine, *September 1984, Medical Economics Company, Inc. Reprinted by permission.*

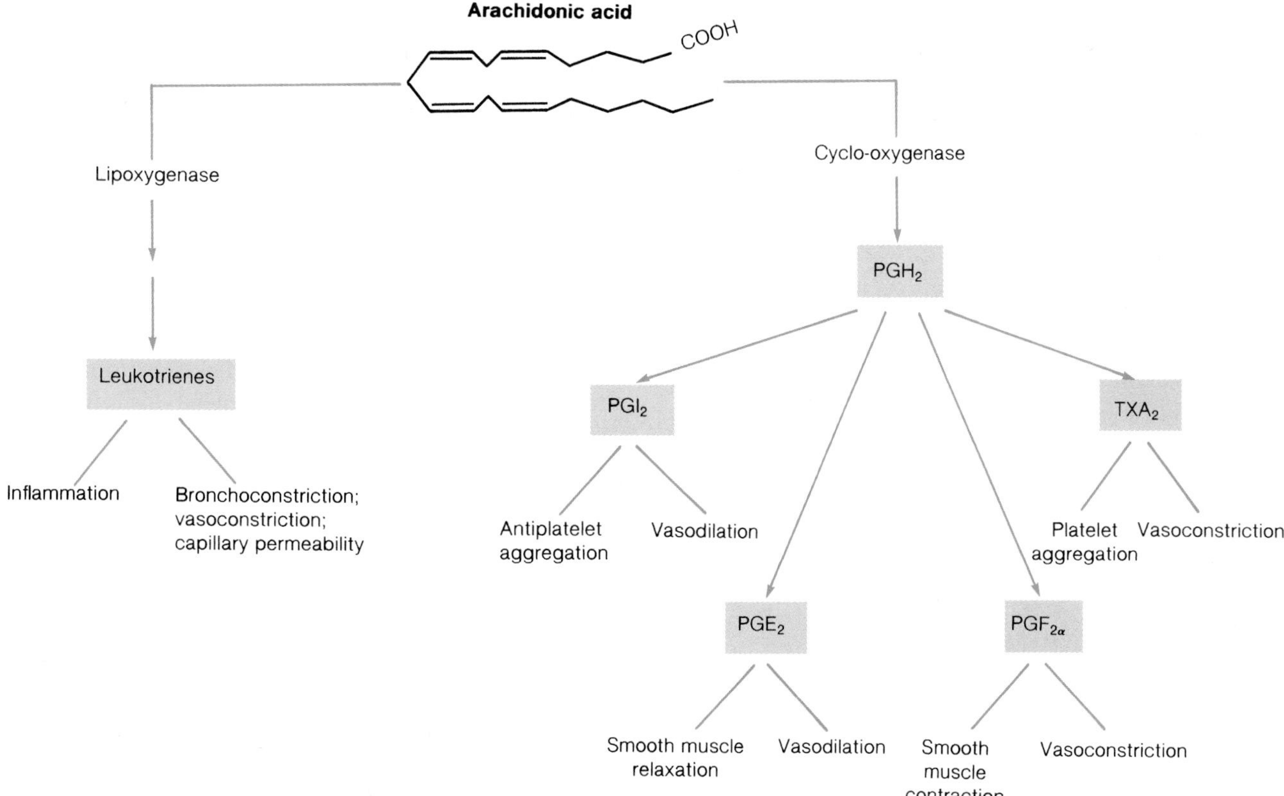

Examples of Prostaglandin Actions

Some of the regulatory functions proposed for prostaglandins in different systems of the body are as follows:

1. **Immune system.** Prostaglandins promote many aspects of the inflammatory process, including the development of pain and fever. Drugs that inhibit prostaglandin synthesis help to alleviate these symptoms.
2. **Reproductive system.** Prostaglandins may play a role in ovulation and corpus luteum function in the ovaries and in contraction of the uterus. Excessive prostaglandin production may be involved in premature labor, endometriosis, dysmenorrhea (painful menstrual cramps), and other gynecological disorders.
3. **Digestive system.** The stomach and intestines produce prostaglandins, which are believed to inhibit gastric secretions and influence intestinal motility and fluid absorption. Since prostaglandins inhibit gastric secretion, drugs that suppress prostaglandin production may make a patient more susceptible to peptic ulcers.
4. **Respiratory system.** Some prostaglandins cause constriction whereas others cause dilation of blood vessels in the lungs and of bronchiolar smooth muscle. The leukotrienes are potent bronchoconstrictors, and these compounds, together with some prostaglandins, may cause respiratory distress and contribute to bronchoconstriction in asthma.
5. **Circulatory system.** Some prostaglandins are vasoconstrictors, and others are vasodilators. Thromboxane A_2, a vasoconstrictor, and prostacyclin, a vasodilator, play a role in blood clotting, as previously described. In a fetus, PGE_2 is believed to promote dilation of the *ductus arteriosus*—a short vessel that connects the pulmonary artery with the aorta. After birth, the ductus arteriosus normally closes as a result of a rise in blood oxygen when the baby breathes. If the ductus remains patent (open), however, it can be closed by the administration of drugs that inhibit prostaglandin synthesis.
6. **Urinary system.** Prostaglandins are produced in the renal medulla and cause vasodilation, resulting in increased renal blood flow and increased excretion of water and electrolytes in the urine.

Clinical Considerations

Disorders of the Pituitary Gland

Panhypopituitarism

A reduction in the activity of the pituitary gland, called *hypopituitarism* (*hi″po-pĭ-too′ĭ-tă-riz″em*), can result from intracranial hemorrhage, a blood clot, prolonged steroid treatments, or a tumor. Total pituitary impairment, termed **panhypopituitarism,** brings about a progressive and general loss of hormonal activity. For example, the gonads stop functioning and the person suffers from amenorrhea (lack of menstruation) or aspermia (no sperm production) and loss of pubic and axillary hair. The thyroid and adrenal glands also eventually stop functioning. People with this condition, and those who have had their pituitary gland surgically removed—a procedure called *hypophysectomy* (*hi-pof″ĭ-sek′tŏ-me*)—receive thyroxine, cortisone, growth hormone, and gonadal hormones throughout life to maintain normal body function.

Abnormal Growth Hormone Secretion

Inadequate growth hormone secretion during childhood causes **pituitary dwarfism.** Hyposecretion of growth hormone in an adult produces a rare condition called **pituitary cachexia** (*kă-kek′se-ă*) (*Simmonds' disease*). One of the symptoms of this disease is premature aging caused by tissue atrophy. By contrast, oversecretion of growth hormone during childhood causes **gigantism.** Excessive growth hormone secretion in an adult does not cause further growth in height because the epiphyseal plates of the long bones have ossified. Hypersecretion of growth hormone in an adult causes **acromegaly** (*ak″ro-meg′ă-le*), in which the person's appearance gradually changes as a result of thickening of bones and growth of soft tissues, particularly in the face, hands, and feet.

Inadequate ADH Secretion

A dysfunction of the neurohypophysis results in a deficiency in ADH secretion, causing a condition called **diabetes insipidus.** Symptoms of this disease include polyuria (excessive urination), polydipsia (excessive thirst), and severe ionic imbalances. Diabetes insipidus is treated by injections of ADH.

Disorders of the Adrenal Glands

Tumors of the Adrenal Medulla

Tumors of the chromaffin cells of the adrenal medulla are referred to as **pheochromocytomas** (*fe-o-kro″mo-si-to′maz*). These tumors cause hypersecretion of epinephrine and norepinephrine, which produce an effect similar to continuous sympathetic nervous stimulation. The symptoms of this condition are hypertension, elevated metabolism, hyperglycemia and sugar in the urine, nervousness, digestive problems, and sweating. It does not take long for the body to become totally fatigued under these conditions, making the patient susceptible to other diseases.

---

cachexia: Gk. *kakos*, bad; *hexis*, habit
Simmonds' disease: from Morris Simmonds, German physician 1855–1925
acromegaly: Gk. *akron*, extremity; *megas*, large
diabetes: Gk. *diabetes*, to pass through a siphon
pheochromocytoma: Gk. *phaios*, dusky; *chroma*, color; *oma*, tumor

Addison's Disease

Addison's disease is caused by inadequate secretion of both glucocorticoids and mineralocorticoids, which results in hypoglycemia, sodium and potassium imbalance, dehydration, hypotension, rapid weight loss, and generalized weakness. A person with this condition who is not treated with corticosteroids will die within a few days because of the severe electrolyte imbalance and dehydration. Another symptom of this disease is darkening of the skin. This is caused by excessive secretion of ACTH, which results from lack of the inhibitory effects of corticosteroids on the pituitary. President John F. Kennedy had Addison's disease, but few knew of it because it was well controlled by corticosteroids.

Cushing's Syndrome

Hypersecretion of corticosteroids results in **Cushing's syndrome.** This is generally caused by a tumor of the adrenal cortex or by oversecretion of ACTH from the adenohypophysis. Cushing's syndrome is characterized by changes in carbohydrate and protein metabolism, hyperglycemia, hypertension, and muscular weakness. Metabolic problems give the body a puffy appearance and can cause structural changes characterized as "buffalo hump" and "moon face." Similar effects are also seen when people with chronic inflammatory diseases receive prolonged treatment with corticosteroids, which are given to reduce inflammation and inhibit the immune response.

Adrenogenital Syndrome

Usually associated with Cushing's syndrome, **androgenital syndrome** is caused by hypersecretion of adrenal sex hormones, particularly the androgens. Adrenogenital syndrome in young children causes premature puberty and enlarged genitals, especially the penis in males and the clitoris in females. An increase in body hair and a deepening of the voice are other characteristics. This condition in a mature woman can cause growth of a beard.

Disorders of the Thyroid and Parathyroid Glands

Hypothyroidism

The infantile form of hypothyroidism is known as **cretinism** (*krēt'n-iz"em*). The clinical symptoms of cretinism are stunted growth, thickened facial features, abnormal bone development, mental retardation, low body temperature, and general lethargy. If cretinism is diagnosed early, it may be successfully treated by administering thyroxine.

Addison's disease: from Thomas Addison, English physician, 1793–1860
Cushing's syndrome: from Harvey Cushing, American physician, 1869–1939

Figure 19.31

Endemic goiter is caused by insufficient iodine in the diet.
The lack of iodine causes hypothyroidism, and the resulting elevation in TSH secretion stimulates the excessive growth of the thyroid.

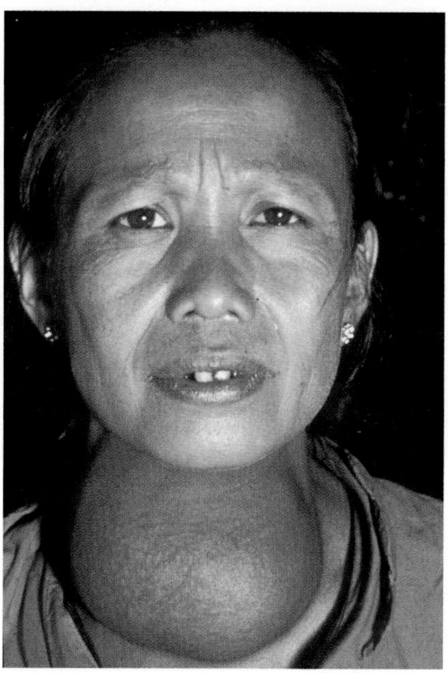

Myxedema

Hypothyroidism in an adult causes **myxedema** (*mik"sĕ-de'mă*). This disorder affects body fluids, causing edema and increasing blood volume, hence increasing blood pressure. Symptoms of myxedema include a low metabolic rate, lethargy, and a tendency to gain weight. This condition is treated with thyroxine or with triiodothyronine, which are taken orally (as pills).

Endemic Goiter

A *goiter* is an abnormal growth of the thyroid gland. When this condition results from inadequate dietary intake of iodine, it is termed **endemic goiter** (fig. 19.31). In this case, growth of the thyroid is due to excessive TSH secretion, which results from low levels of thyroxine secretion. Endemic goiter is thus associated with hypothyroidism.

Graves' Disease

Graves' disease, also called **toxic goiter,** involves growth of the thyroid associated with hypersecretion of thyroxine. This hyperthyroidism is produced by antibodies that act like TSH and stimulate the thyroid; it is an autoimmune disease. As a consequence of high levels of thyroxine secretion, the metabolic rate and heart rate increase, the person loses weight, and

myxedema: Gk. *myxa*, mucus; *oidema*, swelling
Graves' disease: from Robert James Graves, Irish physician, 1796–1853

Figure 19.32

A symptom of hyperthyroidism.
Hyperthyroidism is characterized by an increased metabolic rate, weight loss, muscular weakness, and nervousness. The eyes may also protrude (exophthalmos), as in this photograph.

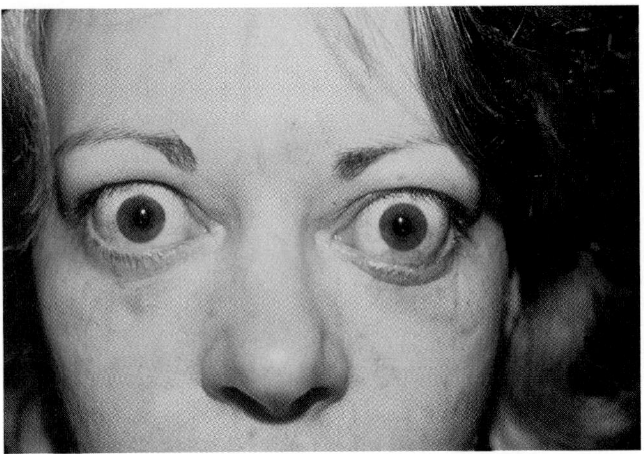

the autonomic nervous system induces excessive sweating. In about half of the cases, **exophthalmos** (*ek″sof-thal′mos*) (bulging of the eyes) also develops (fig. 19.32) because of edema in the tissues of the eye sockets and swelling of the extrinsic eye muscles. Graves' disease is more common in women than in men, and in smokers than in nonsmokers.

Disorders of the Parathyroid Glands

Surgical removal of the parathyroid glands sometimes unintentionally occurs when the thyroid is removed because of a tumor or the presence of Graves' disease. The resulting fall in parathyroid hormone (PTH) causes a decrease in plasma calcium levels, which can lead to severe muscle tetany. Hyperparathyroidism is usually caused by a tumor that secretes excessive amounts of PTH. This stimulates demineralization of bone, which makes the bones soft and raises the blood levels of calcium and phosphate. As a result of these changes, bones are subject to deformity and fracture, and stones (renal calculi) composed of calcium phosphate are likely to develop in the urinary tract.

Disorders of the Pancreatic Islets

Diabetes Mellitus

Diabetes mellitus is characterized by fasting hyperglycemia and the presence of glucose in the urine. There are two forms of this disease. **Type I,** or **insulin-dependent diabetes mellitus,** is caused by destruction of the beta cells and the resulting lack of insulin secretion. **Type II,** or **noninsulin-dependent diabetes mellitus** (which is the more common form) is caused by decreased tissue sensitivity to the effects of insulin, so that increasingly larger amounts of insulin are required to produce a normal effect. Both types of diabetes mellitus are also associated with abnormally high levels of glucagon secretion. Diabetes mellitus is discussed in more detail in chapter 27.

Reactive Hypoglycemia

People with a genetic predisposition for type II diabetes mellitus often first develop **reactive hypoglycemia.** In this condition, the rise in blood glucose that follows the ingestion of carbohydrates stimulates excessive secretion of insulin, which in turn causes the blood glucose levels to fall below the normal range. This can result in weakness, changes in personality, and mental disorientation.

Inhibitors of Prostaglandin Synthesis

Aspirin is the most widely used member of a class of drugs known as **nonsteroidal anti-inflammatory drugs (NSAIDs).** Other members of this class are indomethacin and ibuprofen. These drugs produce their effects because they specifically inhibit the cyclo-oxygenase enzyme that is needed for prostaglandin synthesis. Although they inhibit inflammation through this action, NSAIDs also produce some unwanted side effects, including gastric bleeding, possible kidney problems, and prolonged clotting time.

It is now known that there are two isoenzyme forms (chapter 4) of cyclo-oxygenase. The type I isoenzyme form, or *isoform* (abbreviated *COX1*), is produced constitutively (that is, in a constant fashion) by cells of the stomach and kidneys and by blood platelets, which are cell fragments involved in blood clotting (chapter 20). The type II isoform (*COX2*) is induced in a number of cells in response to cytokines involved in inflammation, and the prostaglandins produced by this isoenzyme promote the inflammatory condition.

When aspirin and the other drugs of its class inhibit the type I isoform of cyclo-oxygenase, the synthesis of prostacyclin is inhibited. Since prostacyclin protects the stomach lining, inhibition of its synthesis may be responsible for the stomach irritation caused by aspirin and indomethacin, the two most potent inhibitors of the type I isoform. Inhibition of the type I isoform is thus responsible for the negative side effects of these drugs. Inhibition of the type II isoform is responsible for the intended anti-inflammatory benefits of the drugs, and research is currently underway to develop new drugs that more selectively inhibit the type II isoenzyme of cyclo-oxygenase.

There is, however, one important benefit derived from the inhibition of the type I isoform by aspirin. The type I isoform is the form of cyclo-oxygenase present in blood platelets, where it is needed for the production of thromboxane A_2. Since this prostaglandin is needed for platelet aggregation, inhibition of its synthesis by aspirin reduces the ability of the blood to clot. While this can have negative consequences in some circumstances, low doses of aspirin have been shown to significantly reduce the risk of heart attacks and strokes by reducing platelet function. It should be noted that this beneficial effect is produced by lower doses of aspirin than are commonly taken to reduce inflammation.

exophthalmos: Gk. *ex,* cut; *opthalmos,* eyeball

Clinical Investigation Answer

The hyperglycemia cannot be attributed to diabetes mellitus because insulin activity is normal. The symptoms might be due to hyperthyroidism or to excessive catecholamine action (as in pheochromocytoma), but these possibilities are ruled out by the blood tests. The high blood levels of corticosteroids are not the result of ingestion of these compounds as drugs. However, the patient might have Cushing's syndrome, in which case an adrenal tumor could be responsible for the hypersecretion of corticosteroids and, as a result of negative feedback, a decrease in blood ACTH levels. Excessive corticosteroid levels cause the mobilization of glucose from the liver, thus increasing the blood glucose to hyperglycemic levels.

Chapter Summary

Endocrine Glands and Hormones (pp. 552–558)

1. Hormones are chemicals, including steroids, catecholamines, and polypeptides, that are secreted into the blood by endocrine glands.
   (a) The chemical classes of hormones include amines, polypeptides, glycoproteins, and steroids.
   (b) Nonpolar hormones, which can pass through the cell membrane of their target cells, are called lipophilic hormones.
2. Precursors of active hormones may be classified as either prohormones or prehormones.
   (a) Prohormones are relatively inactive precursor molecules made in the endocrine cells.
   (b) Prehormones are the normal secretions of an endocrine gland that must be converted to other derivatives by target cells in order to be active.
3. Hormones can interact in permissive, synergistic, or antagonistic ways.
4. The effects of a hormone in the body depend on its concentration.

Mechanisms of Hormone Action (pp. 558–566)

1. The lipophilic hormones (steroids and thyroid hormones) bind to nuclear receptor proteins.
   (a) Some steroid hormones bind to cytoplasmic receptors, which then move into the nucleus. Other steroids and thyroxine bind to receptors that are already in the nucleus.
   (b) Once it is bound to the hormone, the receptor can bind to a region of DNA called a hormone-response element and activate a gene, leading to the production of mRNA and protein within the target cell.
2. The polar hormones bind to receptors located on the outer surface of the target cell. This activates enzymes that enlist second-messenger molecules.
   (a) Many hormones activate adenylate cyclase, which leads to the production of cyclic AMP (cAMP) as a second messenger.
   (b) Other hormones may activate phospholipase C, leading to the production of inositol triphosphate (IP_3) as a second messenger, which stimulates the endoplasmic reticulum to release Ca^{2+}, in turn activating calmodulin.
   (c) The receptors for insulin and various growth factors are tyrosine kinase enzymes. Once activated by binding to the hormone, the receptor kinase phosphorylates various signaling molecules within the target cell.

Pituitary Gland (pp. 566–573)

1. The pituitary gland secretes eight hormones.
   (a) The anterior pituitary secretes growth hormone, thyroid-stimulating hormone, adrenocorticotropic hormone, follicle-stimulating hormone, luteinizing hormone, and prolactin.
   (b) The posterior pituitary releases antidiuretic hormone (also called vasopressin) and oxytocin.
2. The hormones of the posterior pituitary are produced in the hypothalamus and transported to the posterior pituitary via the hypothalamo-hypophyseal tract.
3. Secretions of the anterior pituitary are controlled by hypothalamic hormones that stimulate or inhibit these secretions.
   (a) Hypothalamic hormones include TRH, CRH, GnRH, PIH, somatostatin, and a growth-hormone–releasing hormone.
   (b) These hormones are carried to the anterior pituitary by means of the hypothalamo-hypophyseal portal system.
4. Secretions of the anterior pituitary are also regulated by the feedback (usually negative feedback) exerted by target gland hormones.
5. Higher brain centers, acting through the hypothalamus, can influence pituitary secretion.

Adrenal Glands (pp. 573–575)

1. The adrenal cortex secretes mineralocorticoids (mainly aldosterone), glucocorticoids (mainly cortisol), and sex steroids (primarily weak androgens).
   (a) The glucocorticoids help to regulate energy balance. They also can inhibit inflammation and suppress immune function.
   (b) The pituitary-adrenal axis is stimulated by stress as part of the general adaptation syndrome.
2. The adrenal medulla secretes epinephrine and lesser amounts of norepinephrine. These hormones complement the action of the sympathetic nervous system.

Thyroid and Parathyroid Glands (pp. 575–578)

1. The thyroid follicles secrete tetraiodothyronine (T_4, or thyroxine) and lesser amounts of triiodothyronine (T_3).
   (a) These hormones are formed within the colloid of the thyroid follicles.
   (b) The parafollicular cells of the thyroid secrete the hormone calcitonin, which may act to lower blood calcium levels.
2. The parathyroids are small structures embedded within the thyroid gland. They secrete parathyroid hormone (PTH), which promotes a rise in blood calcium levels.

Pancreas and Other Endocrine Glands (pp. 578–581)

1. Beta cells in the pancreatic islets (islets of Langerhans) secrete insulin and alpha cells secrete glucagon.
   (a) Insulin lowers blood glucose and stimulates the production of glycogen, fat, and protein.
   (b) Glucagon raises blood glucose by stimulating the breakdown of liver glycogen. It also promotes lipolysis and the formation of ketone bodies.
   (c) The secretion of insulin is stimulated by a rise in blood glucose following meals. The secretion of glucagon is stimulated by a fall in blood glucose during periods of fasting.
2. The pineal gland, located on the roof of the third ventricle of the brain, secretes melatonin.
   (a) Melatonin secretion is regulated by the suprachiasmatic nucleus of the hypothalamus, which is the major center for the control of circadian rhythms.

   (b) Melatonin secretion is highest at night, and this hormone has a sleep-promoting effect. It also may play a role in timing the onset of human puberty, but this is as yet unproven.
3. The thymus is the site of T cell lymphocyte production and secretes a number of hormones that may help to regulate the immune system.
4. The gastrointestinal tract secretes a number of hormones that help to regulate digestive functions.
5. The gonads secrete sex steroid hormones.
   (a) Interstitial (Leydig) cells of the testes secrete testosterone and other androgens.
   (b) Granulosa cells of the ovarian follicles secrete estrogen.
   (c) The corpus luteum of the ovaries secretes progesterone, as well as estrogen.
6. The placenta secretes estrogen, progesterone, and a variety of polypeptide hormones that have actions similar to some anterior pituitary hormones.

Autocrine and Paracrine Regulation (pp. 581–584)

1. Autocrine and paracrine regulators act within the same organ in which they are produced. They are thus local regulators, rather than blood-borne hormonal regulators.
2. Prostaglandins are special 20-carbon–long fatty acids produced by many different organs. They usually have regulatory functions within the organ in which they are produced.

Review Activities

Objective Questions

Match the gland to its embryonic origin.

1. adenohypophysis
2. neurohypophysis
3. adrenal medulla

(a) endoderm of pharynx
(b) diverticulum from brain
(c) endoderm of foregut
(d) neural crest ectoderm
(e) hypophyseal pouch

4. Hypothalamic-releasing hormones
   (a) are secreted into capillaries in the median eminence.
   (b) are transported by portal veins to the anterior pituitary.
   (c) stimulate the secretion of specific hormones from the anterior pituitary.
   (d) all of the above apply
5. The hormone primarily responsible for setting the basal metabolic rate and for promoting the maturation of the brain is
   (a) cortisol.
   (b) ACTH.
   (c) TSH.
   (d) thyroxine.
6. Which of the following statements about the adrenal cortex is *true?*
   (a) It is not innervated by nerve fibers.
   (b) It secretes some androgens.
   (c) The zona granulosa secretes aldosterone.
   (d) The zona fasciculata is stimulated by ACTH.
   (e) All of the above are true.
7. The hormone insulin
   (a) is secreted by alpha cells in the pancreatic islets.
   (b) is secreted in response to a rise in blood glucose.
   (c) stimulates the production of glycogen and fat.
   (d) both *a* and *b* apply
   (e) both *b* and *c* apply

Match the hormone with the primary agent that stimulates its secretion.

8. epinephrine
9. thyroxine
10. corticosteroids
11. ACTH

(a) TSH
(b) ACTH
(c) growth hormone
(d) sympathetic nerves
(e) CRH

12. Steroid hormones are secreted by
   (a) the adrenal cortex.
   (b) the gonads.
   (c) the thyroid.
   (d) both *a* and *b*.
   (e) both *b* and *c*.
13. The secretion of which hormone would be *increased* in a person with endemic goiter?
   (a) TSH
   (b) thyroxine
   (c) triiodothyronine
   (d) all of the above
14. Which hormone uses cAMP as a second messenger?
   (a) testosterone
   (b) cortisol
   (c) insulin
   (d) epinephrine
15. Which of the following terms best describes the interactions of insulin and glucagon?
   (a) synergistic
   (b) permissive
   (c) antagonistic
   (d) cooperative

Essay Questions

1. Explain how regulation of the neurohypophysis and adrenal medulla are related to the embryonic origins of these organs.
2. Explain the mechanism of action of steroid hormones and thyroxine.
3. Explain why polar hormones cannot regulate their target cells without using second messengers. Also explain how cyclic AMP is used as a second messenger in hormone action.
4. Describe the sequence of events by which a hormone can cause an increase in the Ca^{2+} concentration within a target cell. How can this increased Ca^{2+} affect the metabolism of the target cell?
5. Explain the significance of the term *trophic* with respect to the actions of anterior pituitary hormones.
6. Suppose a drug blocks the conversion of T_4 to T_3. Explain what the effects of this drug would be on (a) TSH secretion, (b) thyroxine secretion, and (c) the size of the thyroid gland.
7. Explain why the anterior pituitary is sometimes referred to as the "master gland" and why this reference is misleading.
8. Suppose a person's immune system made antibodies against insulin receptor proteins. What effect might this condition have on carbohydrate and fat metabolism?
9. Explain how light affects the function of the pineal gland. What is the relationship between pineal gland function and circadian rhythms?
10. Distinguish between endocrine and autocrine/paracrine regulation. List some of these autocrine/paracrine regulators and describe their functions.

Critical Thinking Questions

1. Brenda, your roommate, has been having an awful time lately. She can't even muster enough energy to go out on a date. She's been putting on weight, she's always cold, and every time she pops in the "Buns of Steel" workout video she complains of weakness. When she finally goes to the doctor, he finds her to have a slow pulse and a low blood pressure. Laboratory tests reveal that her T_4 is low and that her TSH is high. What is the matter with Brenda? Why are her symptoms typical of this disorder, and what type of treatment will the doctor most likely prescribe?

2. You decide it's time that your college basketball team make it to the Final Four. You figure your pal Steve has the competitive spirit needed to step in as the new star center—if only he weren't 5′8″. Confronting the problem head-on, you start injecting him with growth hormone as he sleeps each night. You think this is a clever strategy, but after a time you notice that he hasn't grown an inch. Instead, his jaw and forehead seem to have gotten disproportionately large, and his hands and feet are swollen. Explain why the growth hormone didn't make Steve grow taller and why it had the effect it did. What disease state do these changes mimic?
3. You see your friend Joe for the first time in over a year. When you last saw him he had been trying to bulk up by working out daily at the gym, but he was getting discouraged because his progress seemed so slow. Now, however, he's very muscular. In a frank discussion he tells you that he's been getting into trouble because he's become very aggressive. He also tells you, in strict confidence, that his testes have gotten smaller and that he's been developing breasts! What might Joe be doing to cause these changes? Explain how these changes came about.
4. Distinguish between the steroid and nonsteroid group of nuclear hormone receptors. Explain the central role of vitamin A in the actions of the nonsteroid group of receptors.
5. Suppose, in an experiment, you incubate isolated rat testes with hCG. What would be the effect, if any, of the hCG on the testes? Explain your answer. If there was an effect, discuss its potential significance in research and clinical settings.

Related Web Sites

For a listing of the most current web sites related to this chapter, please visit the *Concepts of Human Anatomy and Physiology* home page at http://www.mhhe.com/biosci/abio/.

NEXUS

Some Interactions of the Endocrine System with the Other Body Systems

Integumentary System
- The skin helps to protect the body from pathogens (p. 696).
- The skin produces vitamin D, which acts as a prehormone (p. 194).

Skeletal System
- Bones store calcium, which is needed for the action of many hormones (p. 185).
- Anabolic hormones, including growth hormone, stimulate bone development (p. 882).
- Parathyroid hormone and calcitonin regulate calcium deposition and resorption in bones (p. 194)

Nervous System
- The hypothalamus secretes hormones that control the anterior pituitary (p. 569).
- The hypothalamus produces the hormones secreted by the posterior pituitary (p. 569).
- Sympathetic nerves stimulate the secretion of the adrenal medulla (p. 575).
- Parasympathetic nerves stimulate the secretion of the pancreatic islets (p. 876).
- Sex hormones from the gonads regulate the hypothalamus (p. 902).
- Neurons stimulate the secretion of melatonin from the pineal gland, which in turn regulates parts of the brain (p. 579).

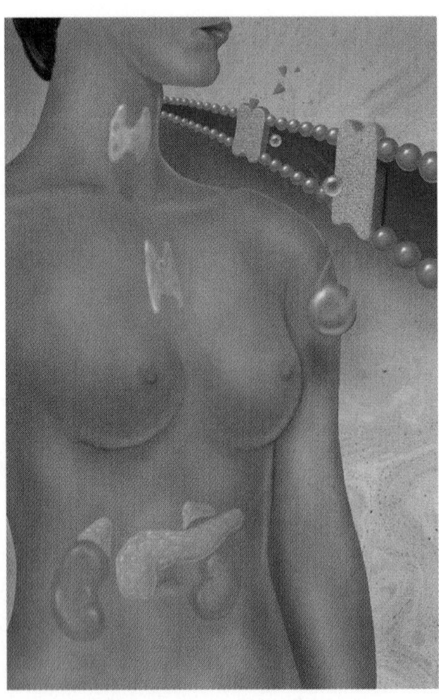

Muscular System
- The cardiac muscle of the heart and the smooth muscles of vessels help deliver blood to the endocrine system (p. 616).
- Anabolic hormones promote muscle growth (p. 558).
- Insulin stimulates the uptake of blood glucose into muscles (p. 877).

Circulatory System
- The blood transports oxygen, nutrients, and regulatory molecules to endocrine glands and removes wastes (p. 592).
- The blood transports hormones from endocrine glands to target cells (p. 552).
- Epinephrine and norepinephrine from the adrenal medulla stimulate the heart (p. 656).
- Thyroxine and other hormones have permissive effects on autonomic regulation of the cardiovascular system (p. 882).

Lymphatic and Immune Systems
- The immune system protects against infections that could damage endocrine glands (p. 696).
- Hormones from the thymus gland help regulate lymphocytes (p. 712).
- Adrenal corticosteroids have a suppressive effect on the immune system (p. 718).

Respiratory System
- The lungs provide oxygen for transport by the blood and eliminate carbon dioxide (p. 729).
- Thyroxine and epinephrine stimulate the rate of cell respiration in the body (p. 881).
- Epinephrine promotes bronchodilation, reducing airway resistance (p. 488).

Urinary System
- The kidneys eliminate metabolic wastes produced by body organs, including endocrine glands (p. 779).
- The kidneys release renin, which participates in the renin–angiotensin–aldosterone system (p. 799).
- The kidneys secrete erythropoietin, which serves as a hormone that regulates red blood cell production (p. 597).
- Antidiuretic hormone, aldosterone, and atrial natriuretic hormone regulate kidney functions (p. 791).

Digestive System
- The GI tract provides nutrients to the body organs, including those of the endocrine system (p. 817).
- Hormones of the stomach and small intestine help to coordinate activities of different regions of the GI tract (p. 849).
- Hormones from adipose tissue contribute to the regulation of hunger (p. 872).

Reproductive System
- Gonadal hormones help to regulate the secretions of the anterior pituitary (p. 896).
- Pituitary hormones regulate the ovarian cycle (p. 934).
- Testicular androgens regulate the accessory male sex organs (p. 894).
- Ovarian hormones regulate the uterus during the menstrual cycle (p. 938).
- Oxytocin is needed for labor and delivery (p. 973).
- The placenta secretes several hormones needed for the health of the pregnancy (p. 966).
- Several hormones are needed for lactation in a nursing mother (p. 942).

TWENTY

Circulatory System: Blood

OBJECTIVES

- Describe the major components of the circulatory system.
- List the functions of the circulatory system.
- Describe the composition of plasma and the functions of its constituents.
- Discuss the origin, structure, function, and life span of erythrocytes, leukocytes, and platelets.
- Discuss the origin and function of erythropoietin, thrombopoietin, and the cytokines involved in the regulation of leukocyte development.
- Discuss the ABO and Rh blood systems, comment on their clinical significance, and explain how the agglutination reaction occurs.
- Describe the platelet release reaction and the formation of a platelet plug.
- Compare the extrinsic and intrinsic clotting pathways.
- Describe the mechanisms of action of some commonly used anticoagulants.
- Describe the physiological mechanisms that promote the dissolution of a blood clot.

Clinical Investigation

A 40-year-old man went to his doctor complaining that he seemed to tire more quickly than usual. His job at an alpine ski resort was physically demanding, but he believed himself to be in good health. He stated that his muscles were very sore, and that he had been taking large amounts of aspirin and acetaminophen for the pain. Laboratory results revealed a somewhat lengthened clotting time, and a red blood cell count and hemoglobin concentration that were significantly below normal. Other laboratory results suggested that there was decreased kidney function, and kidney disease was confirmed by subsequent testing. Explain the symptoms of a prolonged clotting time and quick fatigue. What are the probable causes of these symptoms?

Clues: Read the sections on the regulation of erythropoiesis and the functions of platelets in blood clotting.

Functions and Components of the Circulatory System

Blood serves numerous functions, including the transport of respiratory gases, nutritive molecules, hormones and metabolic wastes. Blood is transported through the body in a system of vessels leading from and returning to the heart.

A unicellular organism can provide for its own maintenance and continuity by performing the wide variety of functions needed for life. By contrast, the complex human body is composed of trillions of specialized cells that demonstrate a division of labor. The specialized cells of a multicellular organism depend on one another for the very basis of their existence; since most are firmly implanted in tissues, they must have their oxygen and nutrients brought to them and their waste products removed. Therefore, a highly effective means of transporting materials within the body is needed.

The blood serves this transportation function. An estimated 60,000 miles of vessels throughout the body of an adult ensures that continued sustenance reaches each of the trillions of living cells. But then, too, the blood can serve to transport disease-causing viruses, bacteria, and their toxins. To guard against this, the circulatory system has protective mechanisms—the white blood cells and lymphatic system. In order to perform its various functions the circulatory system works together with the respiratory, urinary, digestive, endocrine, and integumentary systems in maintaining homeostasis.

Functions of the Circulatory System

The functions of the circulatory system can be divided into three broad areas: transportation, regulation, and protection.

1. **Transportation.** All of the substances essential for cellular metabolism are transported by the circulatory system. These substances can be categorized as follows:
   a. *Respiratory.* Red blood cells, or *erythrocytes*, transport oxygen to the cells. In the lungs, oxygen from the inhaled air attaches to hemoglobin molecules within the erythrocytes and is transported to the cells for aerobic respiration. Carbon dioxide produced by cell respiration is carried by the blood to the lungs for elimination in the exhaled air.
   b. *Nutritive.* The digestive system is responsible for the mechanical and chemical breakdown of food so that it can be absorbed through the intestinal wall into the blood vessels of the circulatory system. The blood then carries these absorbed products of digestion through the liver and to the cells of the body.
   c. *Excretory.* Metabolic wastes (such as urea), excess water and ions, and other molecules not needed by the body are carried by the blood to the kidneys and excreted in the urine.
2. **Regulation.** The circulatory system serves as a means for hormonal regulation and thermoregulation.
   a. *Hormonal regulation.* The blood carries hormones from their site of origin to distant target cells, where they perform a variety of regulatory functions.
   b. *Thermoregulation.* Maintenance of a constant body temperature is aided by the diversion of blood from deeper to more superficial cutaneous vessels or vice versa. When the ambient temperature is high, diversion of blood from deep to superficial vessels helps to cool the body, and when the ambient temperature is low, the diversion of blood from superficial to deeper vessels helps to keep the body warm.
3. **Protection.** The circulatory system protects against blood loss from injury and against foreign microbes or toxins introduced into the body.
   a. *Clotting.* The clotting mechanism protects against blood loss when vessels are damaged.
   b. *Immunity.* The immune function of the blood is performed by the *leukocytes* (white blood cells) that protect against many disease-causing agents (pathogens).

Major Components of the Circulatory System

The circulatory system consists of two divisions: the cardiovascular system and the lymphatic system. The **cardiovascular system** consists of the heart and blood vessels, and the **lymphatic system** consists of lymphatic vessels and lymphoid tissues within the spleen, thymus, tonsils, and lymph nodes.

The **heart** is a four-chambered double pump. Its pumping action creates the pressure head needed to push blood in the vessels to the lungs and body cells. At rest, the heart of an adult pumps about 5 liters of blood per minute. At this rate it takes about 1 minute for blood to be circulated to the most distal extremity and back to the heart.

Blood vessels form a tubular network that permits blood to flow from the heart to all the living cells of the body and then back to the heart. *Arteries* carry blood away from the heart, and *veins* return blood to the heart. Arteries and veins are continuous with each other through smaller blood vessels.

Arteries branch extensively to form a "tree" of progressively smaller vessels. The smallest of the arteries are called *arterioles* (*ar-tir'e-olz*). Blood passes from the arterial to the venous system in microscopic *capillaries* (*kap'ĭ-lar"ēz*), which are the thinnest and most numerous of the blood vessels. All exchanges of fluid, nutrients, and wastes between the blood and tissues occur across the walls of capillaries. Blood flows through capillaries into microscopic veins called *venules* (*ven'yoolz*), which deliver blood into progressively larger veins that eventually return the blood to the heart.

As blood *plasma* (the fluid portion of the blood) passes through capillaries, the hydrostatic pressure of the blood forces some of this fluid out of the capillary walls. Fluid derived from plasma that passes out of capillary walls into the surrounding tissues is called *interstitial* (*in"ter-stish'al*) *fluid* or *tissue fluid*. Some of this fluid returns directly to capillaries, and some enters into **lymphatic vessels** located in the connective tissues around the blood vessels. The fluid within the lymphatic vessels is called *lymph* (*limf*). This fluid is returned to the venous blood at specific sites. **Lymph nodes,** positioned along the way, cleanse the lymph prior to its return to the venous blood. The lymphatic system is discussed as part of the immune system in chapter 23.

Composition of the Blood

Blood consists of formed elements that are suspended and carried in a fluid called plasma. The formed elements—erythrocytes, leukocytes, and platelets—function, respectively, in oxygen transport, immune defense, and blood clotting. Plasma contains various types of proteins and many water-soluble molecules.

The total blood volume in an average-sized adult is about 5 liters, constituting about 8% of the total body weight. Blood leaving the heart is referred to as *arterial blood*. Arterial blood, with the exception of that going to the lungs, is bright red because of a high concentration of oxyhemoglobin (the combination of oxygen and hemoglobin) in the red blood cells. *Venous blood* is blood returning to the heart. Except for the venous blood from the lungs, it contains less oxygen and is therefore a darker red than the arterial blood because of the color of hemoglobin without oxygen (deoxyhemoglobin).

Blood is composed of a cellular portion, called **formed elements,** and a fluid portion, called **plasma.** When a blood sample is centrifuged, the heavier formed elements pack into the bottom of the tube, leaving plasma at the top (fig. 20.1). The formed elements constitute approximately 45% of the total blood volume—a measurement called the *hematocrit* (*hĭ-mat'ō-krit*)—and the plasma accounts for the remaining 55%.

Plasma

Plasma is a straw-colored liquid consisting of water and dissolved solutes. The major solute of the plasma in terms of its concentration is Na^+. In addition to Na^+, plasma contains

hematocrit: Gk. *haima*, blood; *krino*, to separate

Figure 20.1

The constituents of blood.
Blood cells become packed at the bottom of the test tube when whole blood is centrifuged, leaving the fluid plasma at the top of the tube. Red blood cells are the most abundant of the blood cells—white blood cells and platelets form only a thin, light-colored "buffy coat" at the interface between the packed red blood cells and the blood plasma.

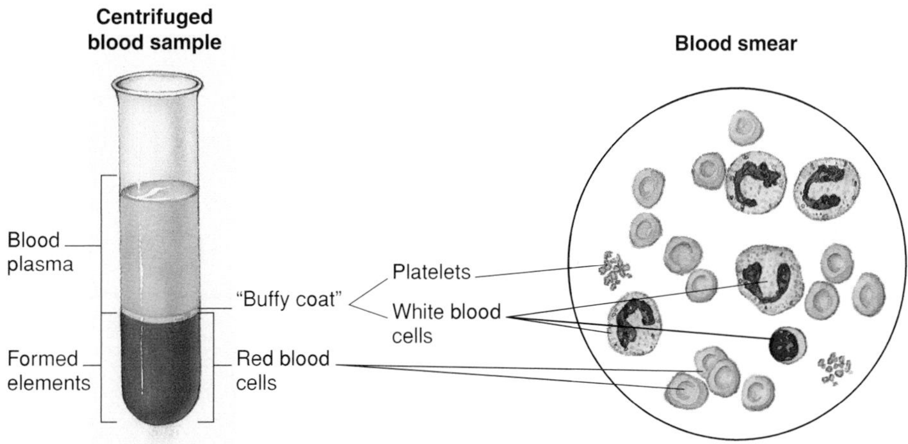

many other ions, as well as organic molecules such as metabolites, hormones, enzymes, antibodies, and other proteins. The concentrations of some of these constituents of plasma are shown in table 20.1.

Plasma Proteins

Plasma proteins constitute 7% to 9% of the plasma. The three types of proteins are albumins, globulins, and fibrinogen. **Albumins** (*al-byoo'minz*) account for most (60% to 80%) of the plasma proteins and are the smallest in size. They are produced by the liver and provide the osmotic pressure needed to draw water from the surrounding interstitial fluid into the capillaries. This action is needed to maintain blood volume and pressure. **Globulins** (*glob'yoo-linz*) are grouped into three subtypes: **alpha globulins, beta globulins,** and **gamma globulins.** The alpha and beta globulins are produced by the liver and function in transporting lipids and fat-soluble vitamins. Gamma globulins are antibodies produced by lymphocytes (one of the formed elements found in blood and lymphoid tissues) and function in immunity. **Fibrinogen** (*fi-brin'ŏ-jen*), which accounts for only about 4% of the total plasma proteins, is an important clotting factor produced by the liver. During the process of clot formation, fibrinogen is converted into insoluble threads of *fibrin*. Thus the fluid from clotted blood, called **serum,** does not contain fibrinogen but it is otherwise identical to plasma.

Plasma Volume

A number of regulatory mechanisms in the body maintain homeostasis of the plasma volume. If the body should lose water, the remaining plasma becomes excessively concentrated—its osmolality increases (chapter 5). This is detected by osmoreceptors in the hypothalamus, resulting in a sensation of thirst and the release of antidiuretic hormone (ADH) from the posterior pituitary (chapter 19). This hormone promotes water retention by the kidneys, which—together with increased intake of fluids—helps to compensate for the dehydration and lowered blood volume. This regulatory mechanism together with others that influence plasma volume are very important in maintaining blood pressure, as described in chapter 22.

Formed Elements of Blood

The formed elements of blood include two types of blood cells: erythrocytes, or *red blood cells,* and leukocytes, or *white blood cells.* Erythrocytes are by far the most numerous of the two. A cubic millimeter of blood contains 5.1 million to 5.8 million erythrocytes in males and 4.3 million to 5.2 million erythrocytes in females. The same volume of blood, by contrast, contains only 5,000 to 9,000 leukocytes.

albumin: L. *albumen*, white
globulin: L. *globulus*, small globe
serum: L. *serum*, liquid
erythrocytes: Gk. *erythros*, red; *kytos*, hollow (cell)
leukocytes: Gk. *leukos*, white; *kytos*, hollow (cell)

Table 20.1
Representative Normal Plasma Values

Measurement	Normal Range
Blood volume	80–85 ml/kg body weight
Blood osmolality	280–296 mOsm
Blood pH	7.35–7.45
Enzymes	
Creatine phosphokinase (CPK)	Female: 10–79 U/L Male: 17–148 U/L
Lactic dehydrogenase (LDH)	45–90 U/L
Phosphatase (acid)	Female: 0.01–0.56 Sigma U/ml Male: 0.13–0.63 Sigma U/ml
Hematology Values	
Hematocrit	Female: 37%–48% Male: 45%–52%
Hemoglobin	Female: 12–16 g/100 ml Male: 13–18 g/100 ml
Red blood cell count	4.2–5.9 million/mm^3
White blood cell count	4,300–10,880/mm^3
Hormones	
Testosterone	Male: 300–1,100 ng/100 ml Female: 25–90 ng/100 ml
Adrenocorticotropic hormone (ACTH)	15–70 pg/ml
Growth hormone	Children: over 10 ng/ml Adult male: below 5 ng/ml
Insulin	6–26 μU/ml (fasting)
Ions	
Bicarbonate	24–30 mmol/l
Calcium	2.1–2.6 mmol/l
Chloride	100–106 mmol/l
Potassium	3.5–5.0 mmol/l
Sodium	135–145 mmol/l
Organic Molecules (Other)	
Cholesterol	120–220 mg/100 ml
Glucose	70–110 mg/100 ml (fasting)
Lactic acid	0.6–1.8 mmol/l
Protein (total)	6.0–8.4 g/100 ml
Triglyceride	40–150 mg/100 ml
Urea nitrogen	8–25 mg/100 ml
Uric acid	3–7 mg/100 ml

Source: Excerpted from material originally appearing in *The New England Journal of Medicine*, "Case Records of the Massachusetts General Hospital," Vol. 302, No. 1, pp. 37–48, January 3, 1980 and "Case Records of the Massachusetts General Hospital," Vol. 314, pp. 39–49. Copyright © 1980, 1986 by *The New England Journal of Medicine*. Reprinted by permission.

Figure 20.2

Red blood cells: (*a*) diagram and (*b*) scanning electron micrograph.
They are seen here clinging to a hypodermic needle. Note the shape of the red blood cells, sometimes described as a "biconcave disc."

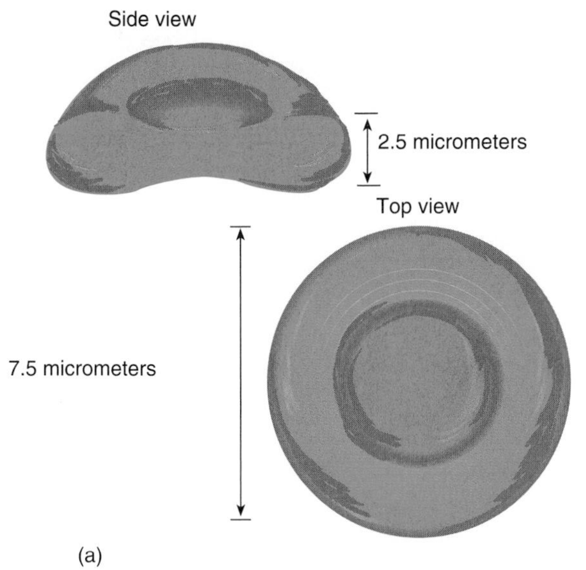

Erythrocytes

Erythrocytes (*ĕ-rith'rŏ-sīts*) are flattened, biconcave discs, about 7 µm in diameter and 2.2 µm thick. Their unique shape relates to their function of transporting oxygen; it provides an increased surface area through which gas can diffuse (fig. 20.2). Erythrocytes lack nuclei and mitochondria (they obtain energy through anaerobic respiration). Because of these deficiencies, erythrocytes have a relatively short circulating life span of only about 120 days. Older erythrocytes are removed from the circulation by phagocytic cells in the liver, spleen, and bone marrow.

Each erythrocyte contains approximately 280 million **hemoglobin** (*he'mŏ-glo"bin*) molecules, which give blood its red color. Each hemoglobin molecule consists of four polypeptide chains, each of which is bound to one *heme* (*hēm*), an iron-containing pigment. The iron group of heme is able to combine with oxygen in the lungs and release oxygen in the tissues.

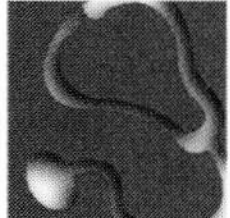

Anemia refers to any condition in which there is an abnormally low hemoglobin concentration and/or red blood cell count. The most common type is *iron-deficiency anemia,* caused by a deficiency of iron, which is an essential component of the hemoglobin molecule. In *pernicious anemia* there is inadequate availability of vitamin B_{12}, which is needed for red blood cell production. This is usually due to atrophy of the glandular mucosa of the stomach, which normally secretes a protein called *intrinsic factor.* In the absence of intrinsic factor, the vitamin B_{12} obtained in the diet cannot be absorbed by intestinal cells. *Aplastic anemia* is anemia due to destruction of the bone marrow, which may be caused by chemicals (including benzene and arsenic) or by radiation.

Leukocytes

Leukocytes (*loo'kŏ-sīts*) differ from erythrocytes in several respects. Leukocytes contain nuclei and mitochondria and can move in an amoeboid fashion (erythrocytes are not able to move independently). Because of their amoeboid ability, leukocytes can squeeze through pores in capillary walls and move to a site of infection, whereas erythrocytes usually remain confined within blood vessels. The movement of leukocytes through capillary walls is referred to as *diapedesis* (*di"ă-pĕ-de'sis*), or *extravasation.*

White blood cells are almost invisible under the microscope unless they are stained, and hence they are classified according to their staining properties. Those leukocytes that have granules in their cytoplasm are called **granular leukocytes;** those without clearly visible granules are called **agranular** (or **nongranular**) **leukocytes.**

The stain used to identify white blood cells is usually a mixture of a pink-to-red stain called *eosin* (*e'ŏ-sin*) and a blue-to-purple stain called a "basic stain." Granular leukocytes with pink-staining granules are therefore called **eosinophils** (*e"ŏ-sin'ŏ-filz*), and those with blue-staining granules are called **basophils.** Those with granules that have little affinity for either stain are **neutrophils** (fig. 20.3). Neutrophils are the most abundant type of leukocyte, accounting for 50% to 70% of the leukocytes in the blood. Immature neutrophils have sausage-shaped nuclei and are called *band cells.* As the band cells mature, their nuclei become lobulated, with three to five lobes connected by thin strands. At this stage, the neutrophils are also known as *polymorphonuclear* (*pol"e-mor"fo-noo'kle-ar*) *leukocytes* (*PMNs*).

---

diapedesis: Gk. *dia,* through; *pedester,* on foot

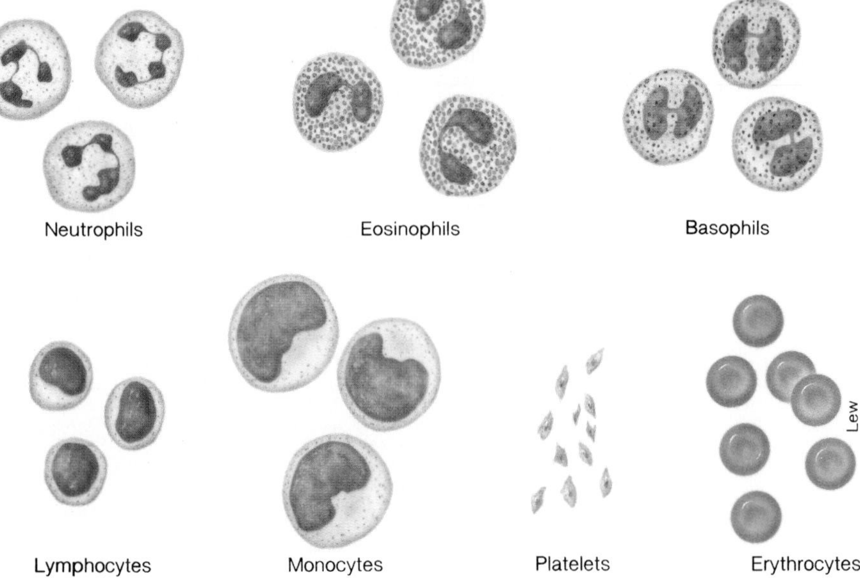

Figure 20.3

The blood cells and platelets.
The white blood cells depicted here are granular leukocytes, whereas lymphocytes and monocytes are nongranular leukocytes.

There are two types of agranular leukocytes: lymphocytes and monocytes. **Lymphocytes** are usually the second most numerous type of leukocyte; they are small cells with round nuclei and little cytoplasm. **Monocytes,** by contrast, are the largest of the leukocytes and generally have kidney- or horseshoe-shaped nuclei. In addition to these two cell types, there are smaller numbers of *plasma cells,* which are derived from lymphocytes. Plasma cells produce and secrete large amounts of antibodies. The immune functions of the different white blood cells are described in more detail in chapter 23.

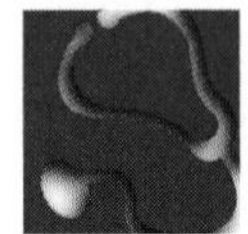

Blood cell counts are an important source of information in assessing a person's health. An abnormal increase in erythrocytes, for example, is termed *polycythemia* (*pol″e-si-the′me-ă*), and is indicative of several dysfunctions. As previously mentioned, an abnormally low red blood cell count is termed *anemia.* (Polycythemia and anemia are described in detail in chapter 24.) An elevated leukocyte count, called *leukocytosis,* is often associated with infection (see chapter 23). A large number of immature leukocytes in a blood sample is diagnostic of the disease *leukemia.* A low white blood cell count, called *leukopenia* (*loo′kŏ-pe′ne-ă*), may be due to a variety of factors; low numbers of lymphocytes, for example, may result from poor nutrition or from whole-body irradiation treatment for cancer.

Platelets

Platelets, or **thrombocytes,** are the smallest of the formed elements and are actually fragments of large cells called *megakaryocytes* (*meg″ă-kar′e-ŏ-sīts*), found in bone marrow. (This is why the term *formed elements* is used instead of *blood cells* to describe erythrocytes, leukocytes, and platelets.) The fragments that enter the circulation as platelets lack nuclei but, like leukocytes, are capable of amoeboid movement. The platelet count per cubic millimeter of blood ranges from 130,000 to 400,000, but this count can vary greatly under different physiological conditions. Platelets survive for about 5 to 9 days before being destroyed by the spleen and liver.

Platelets play an important role in blood clotting. They constitute most of the mass of the clot, and phospholipids in their cell membranes activate the clotting factors in plasma. This results in threads of fibrin, which reinforce the platelet plug. Platelets that attach together in a blood clot release *serotonin,* a chemical that stimulates constriction of the blood vessels and thus reduces the flow of blood to the injured area. Platelets also secrete growth factors (autocrine regulators—chapter 19), which are important in maintaining the integrity of blood vessels.

The formed elements of the blood are illustrated in figure 20.3, and their characteristics are summarized in table 20.2.

Hematopoiesis

Blood cells are constantly formed through a process called **hematopoiesis** (*he″mă-to-poi-e′sis*). The hematopoietic **stem cells**—those that give rise to blood cells—originate in the yolk sac of the human embryo and then migrate to the liver. Hematopoiesis thus occurs in the liver of the fetus. The stem cells then migrate to the bone marrow, and shortly after birth the liver ceases to be a source of blood cell production.

The term **erythropoiesis** (*ĕ-rith″ro-poi-e′sis*) refers to the formation of erythrocytes, and **leukopoiesis** to the formation of leukocytes. These processes occur in two classes of tissues after birth, myeloid and lymphoid. **Myeloid tissue** is the red bone marrow of the long bones, ribs, sternum, pelvis, bodies of the vertebrae, and portions of the skull. **Lymphoid tissue**

thrombocyte: Gk. *thrombos,* clot; *kytos,* hollow (cell)

hematopoiesis: Gk. *haima,* blood; *poiesis,* production

Table 20.2
Formed Elements of Blood

Component	Description	Approximate Number Present	Function
Erythrocyte (red blood cell)	Biconcave disc without nucleus; contains hemoglobin; survives 100 to 120 days	4,000,000 to 6,000,000/mm^3	Transports oxygen and carbon dioxide
Leukocytes (white blood cells)	5,000 to 10,000/mm^3		Aid in defense against infections by microorganisms
Granular leukocytes	About twice the size of red blood cells; cytoplasmic granules present; survive 12 hours to 3 days		
1. Neutrophil	Nucleus with 2 to 5 lobes; cytoplasmic granules stain slightly pink	54% to 62% of white cells present	Phagocytic
2. Eosinophil	Nucleus bilobed; cytoplasmic granules stain red in eosin stain	1% to 3% of white cells present	Helps to detoxify foreign substances; secretes enzymes that break down clots
3. Basophil	Nucleus lobed; cytoplasmic granules stain blue in hematoxylin stain	Less than 1% of white cells present	Releases anticoagulant heparin
Agranula leukocytes	Cytoplasmic granules not visible; survive 100 to 300 days (some much longer)		
1. Monocyte	2 to 3 times larger than red blood cell; nuclear shape varies from round to lobed	3% to 9% of white cells present	Phagocytic
2. Lymphocyte	Only slightly larger than red blood cell; nucleus nearly fills cell	25% to 33% of white cells present	Provides specific immune response (including antibodies)
Platelet (thrombocyte)	Cytoplasmic fragment; survives 5 to 9 days	130,000 to 400,000/mm^3	Enables clotting; releases serotonin, which causes vasoconstriction

includes the lymph nodes, tonsils, spleen, and thymus. The bone marrow produces all of the different types of blood cells, while the lymphoid tissue produces lymphocytes derived from cells that originated in the bone marrow.

Hematopoiesis begins the same way in both myeloid and lymphoid tissue (fig. 20.4). A population of undifferentiated (unspecialized) cells gradually differentiate (specialize) to become stem cells, which give rise to the blood cells. At each step along the way stem cells can duplicate themselves by mitosis, thus ensuring that the parent population will never become depleted. As the cells become differentiated, they develop membrane receptors for chemical signals that cause further development along particular lines. The earliest cells that can be distinguished under a microscope are the *erythroblasts* (which become erythrocytes), *myeloblasts* (which become granular leukocytes), *lymphoblasts* (which form lymphocytes), and *monoblasts* (which form monocytes).

Erythropoiesis is an extremely active process. It is estimated that about 2.5 million erythrocytes are produced every second in order to replace those that are continuously destroyed by the spleen and liver. The life span of an erythrocyte is approximately 120 days. Agranular leukocytes remain functional for 100 to 300 days under normal conditions. Granular leukocytes, by contrast, have an extremely short life span of 12 hours to 3 days.

The production of different subtypes of leukocytes is stimulated by chemicals called **cytokines** (*si'to-kīnz*). These are autocrine regulators (chapter 19) secreted by different cells of the immune system. The particular cytokines involved in leukopoiesis are discussed below. The production of red blood cells is stimulated by the hormone **erythropoietin** (*ĕ-rith"ro-poi'ĕ-tin*), secreted by the kidneys. The gene for erythropoietin has been commercially cloned, so that this hormone is now available for the treatment of the anemia that results from kidney disease in patients undergoing dialysis.

Scientists have identified a specific cytokine that stimulates proliferation of megakaryocytes and their maturation into platelets. By analogy with erythropoietin, they named this regulatory molecule **thrombopoietin.** The gene that codes for thrombopoietin also has been cloned, so that recombinant thrombopoietin is now available for medical research and applications. In clinical trials, thrombopoietin has been used to treat the *thrombocytopenia* (low platelet count) that occurs as a result of bone marrow depletion in patients undergoing chemotherapy for cancer.

Figure 20.4

The processes of hematopoiesis.
Formed elements begin as hemocytoblasts and differentiate into the various kinds of blood cells in accordance with the needs of the body.

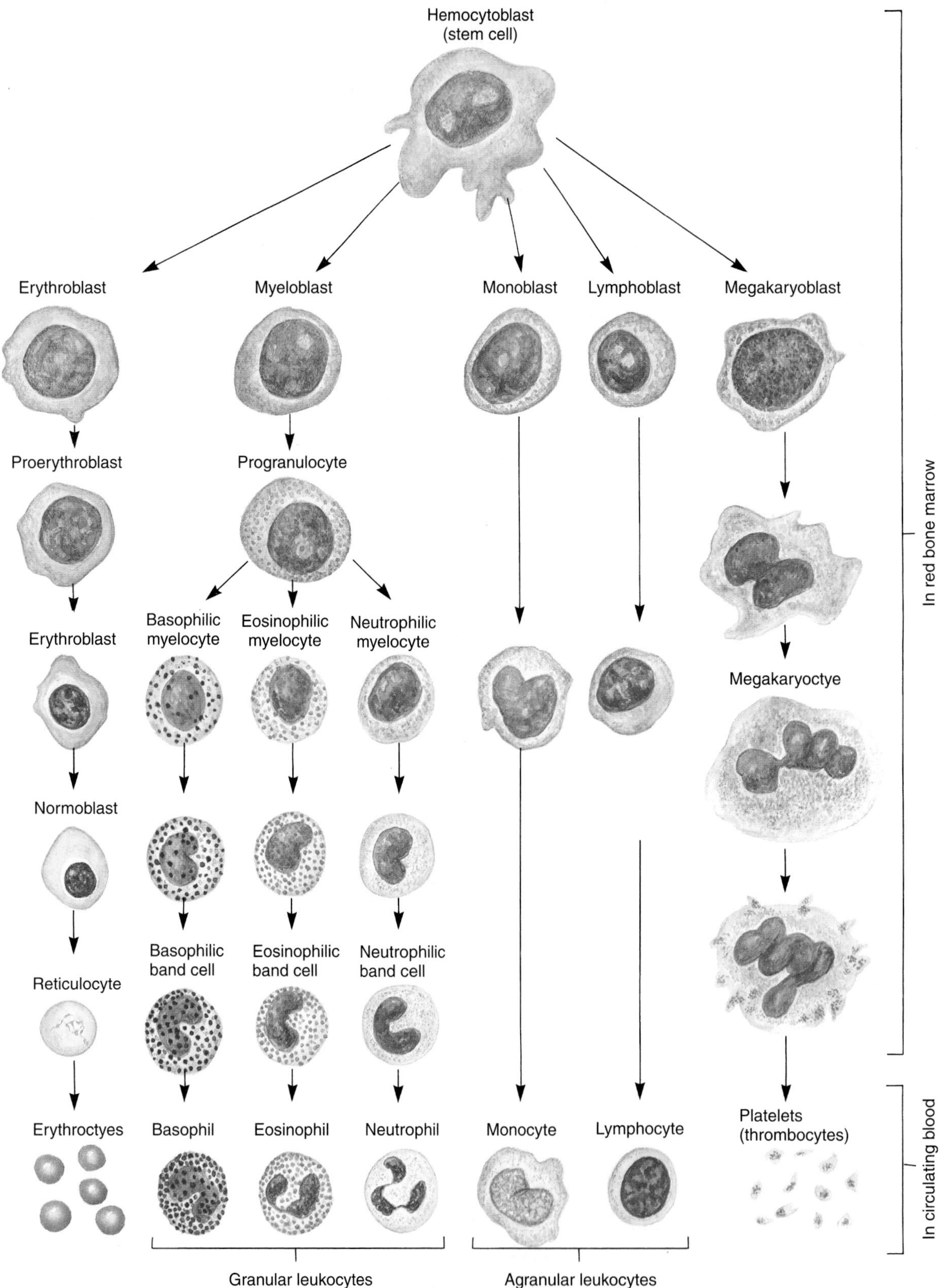

Regulation of Leukopoiesis

A variety of cytokines stimulate different stages of leukocyte development. The cytokines known as *multipotent growth factor-1*, *interleukin-1*, and *interleukin-3* have general effects, stimulating the development of different types of white blood cells. *Granulocyte colony-stimulating factor* (G-CSF) acts in a highly specific manner to stimulate the development of neutrophils, whereas *granulocyte-monocyte colony-stimulating factor* (GM-CSF) stimulates the development of monocytes and eosinophils. The genes for the cytokines G-CSF and GM-CSF have been cloned, making these cytokines available for medical applications.

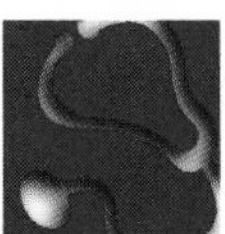

Approximately 10,000 *bone marrow transplants* are performed worldwide each year. This procedure generally involves the aspiration of marrow from the iliac crest and separation of the hematopoietic stem cells, which constitute only about 1% of the nucleated cells in the marrow. Stem cells also have been isolated from peripheral blood when the donor is first injected with G-CSF and GM-CSF, which stimulate the marrow to release more stem cells. Another recent technology involves the storage, or "banking," of hematopoietic stem cells obtained from the placenta or umbilical cord of a neonate. These cells may be used later in life if the person needs them for transplantation.

Regulation of Erythropoiesis

The primary regulator of erythropoiesis is erythropoietin, secreted by the kidneys whenever blood oxygen levels are decreased. One of the possible causes of decreased blood oxygen levels is a decreased red blood cell count. Because of erythropoietin stimulation, the daily production of new red blood cells compensates for the daily destruction of old red blood cells, preventing a decrease in the blood oxygen content. An increased secretion of erythropoietin and production of new red blood cells occurs when a person is at a high altitude or has lung disease, which are conditions that reduce the oxygen content of the blood.

Erythropoietin acts by binding to membrane receptors on cells that will become erythroblasts. The erythropoietin-stimulated cells undergo cell division and differentiation, leading to the production of erythroblasts and then *normoblasts*, which lose their nuclei to become *reticulocytes*. The reticulocytes then change into fully mature erythrocytes. This process takes 3 days; the reticulocyte normally stays in the bone marrow for the first 2 days and then circulates in the blood on the third day. At the end of the erythrocyte life span of 120 days, the old red blood cells are removed by phagocytic cells of the spleen, liver, and bone marrow. Most of the iron contained in hemoglobin molecules of the destroyed red blood cells is recycled back to the myeloid tissue to be used in the production of hemoglobin for new red blood cells. The production of red blood cells and synthesis of hemoglobin depends on the supply of iron along with that of vitamin B_{12} and folic acid.

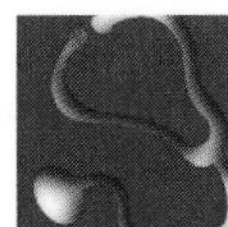

Because of the recycling of iron, dietary requirements for iron are usually quite small. Males (and women after menopause) require a dietary iron supplement of only about 1 mg/day. Women with average menstrual blood loss need 2 mg/day, and pregnant women need 4 mg/day. Because of these relatively small dietary requirements, iron-deficiency anemia in adults is usually not due to a dietary deficiency but rather to blood loss, which reduces the amount of iron that can be recycled.

Red Blood Cell Antigens and Blood Typing

Certain molecules on the surfaces of all cells in the body can be recognized as foreign by the immune system of another individual. These molecules are known as *antigens* (*an'tĭ-jenz*). As part of the immune response, particular lymphocytes secrete a class of proteins called *antibodies* that bond in a specific fashion with antigens. The specificity of antibodies for antigens is analogous to the specificity of enzymes for their substrates, and of receptor proteins for neurotransmitters and hormones. A complete description of antibodies and antigens is provided in chapter 23.

ABO System

The distinguishing antigens on other cells are far more varied than the antigens on red blood cells. Red blood cell antigens, however, are of extreme clinical importance because their types must be matched between donors and recipients for blood transfusions. There are several groups of red blood cell antigens, but the major group is known as the **ABO system.** In terms of the antigens present on the red blood cell surface, a person may be *type* A (with only A antigens), *type* B (with only B antigens), *type* AB (with both A and B antigens), or *type* O (with neither A nor B antigens). It should again be noted that the blood type denotes the class of antigens found on the red blood cell surface.

Each person inherits two genes (one from each parent) that control the production of the ABO antigens. The genes for A or B antigens are dominant to the gene for O, since O simply means the absence of A or B. The genes for A and B are usually shown as I^A and I^B, and the recessive gene for O is shown as the lowercase *i*. A person who is type A, therefore, may have inherited the A gene from each parent (may have the genotype I^AI^A), or the A gene from one parent and the O gene from the other parent (and thus have the genotype I^Ai). Likewise, a person who is type B may have the genotype I^BI^B or I^Bi. It follows that a type O person inherited the O gene from each parent (has the genotype *ii*), whereas a type AB person inherited the A gene from one parent and the B gene from the other (there is no dominant-recessive relationship between A and B).

The immune system exhibits tolerance to its own red blood cell antigens. People who are type A, for example, do not produce anti-A antibodies. Surprisingly, however, they do make antibodies against the B antigen and conversely, people with blood type B make antibodies against the A antigen.

This may be due to the fact that antibodies made in response to some common bacteria cross-react with the A or B antigens. People who are type A, therefore, acquire antibodies that can react with B antigens by exposure to these bacteria, but they do not develop antibodies that can react with A antigens because tolerance mechanisms prevent this.

People who are type AB develop tolerance to both of these antigens, and thus do not produce either anti-A or anti-B antibodies. Those who are type O, by contrast, do not develop tolerance to either antigen; therefore, they have both anti-A and anti-B antibodies in their plasma (table 20.3).

Table 20.3
The ABO System of Red Blood Cell Antigens

Genotype	Antigen on RBCs	Antibody in Plasma
I^AI^A; I^Ai	A	Anti-B
I^BI^B; I^Bi	B	Anti-A
ii	O	Anti-A and anti-B
I^AI^B	AB	Neither anti-A nor anti-B

Transfusion Reactions

Before transfusions are performed, a *major crossmatch* is made by mixing serum from the recipient with blood cells from the donor. If the types do not match—if the donor is type A, for example, and the recipient is type B—the recipient's antibodies attach to the donor's red blood cells and form bridges that cause the cells to clump together, or **agglutinate** (*ă-gloot'n-āt*) (fig. 20.5). Because of this agglutination reaction, the A and B antigens are sometimes called *agglutinogens*, and the antibodies against them are called *agglutinins*. Transfusion errors that result in such agglutination can lead to blockage of small blood vessels and cause *hemolysis* (rupture of red blood cells), which may damage the kidneys and other organs.

In emergencies, type O blood has been given to people who are type A, B, AB, or O. Since type O red blood cells lack A and B antigens, the recipient's antibodies cannot cause agglutination of the donor red blood cells. Type O is therefore a *universal donor*—but only as long as the volume of plasma donated is small, since plasma from a type O person would agglutinate type A, type B, and type AB red blood cells. Likewise, type AB people are *universal recipients* because they lack anti-A and anti-B antibodies, and thus cannot agglutinate donor red blood cells. (Donor plasma could agglutinate recipient red blood cells if the transfusion volume were too large.) Because of the dangers involved, use of the universal donor and recipient concept is strongly discouraged in practice.

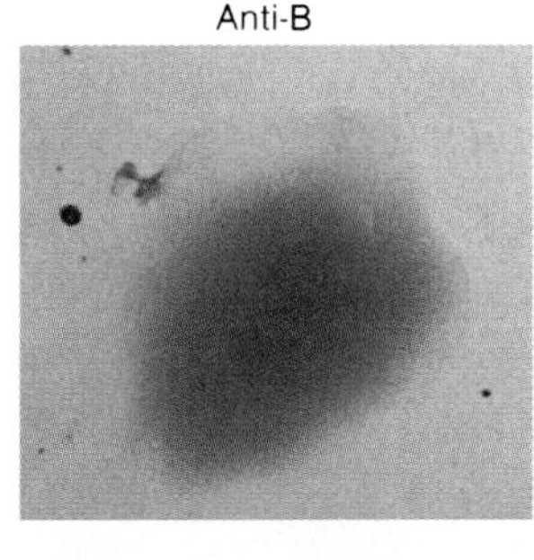
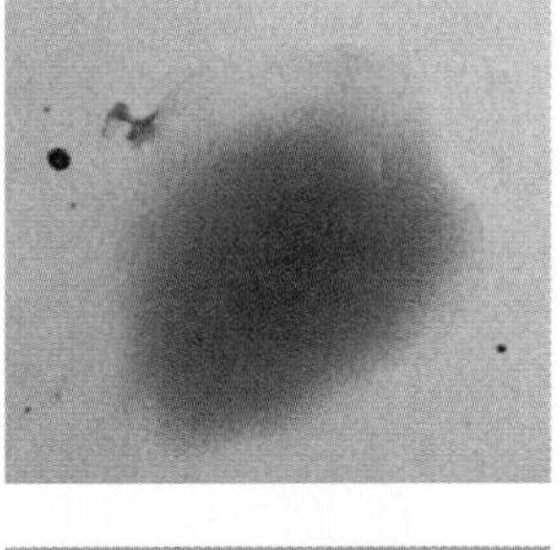
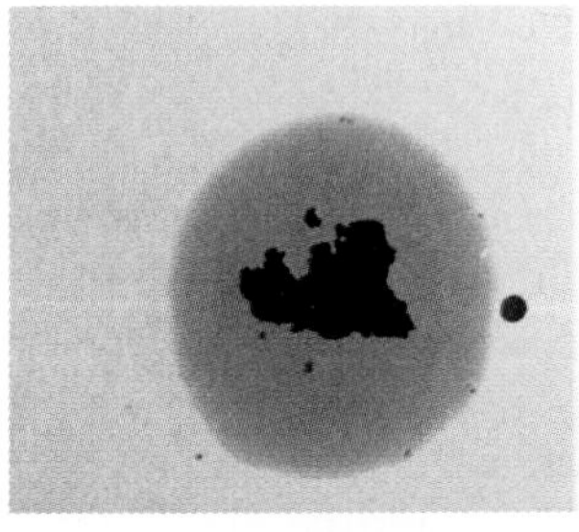
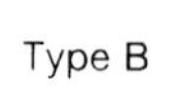

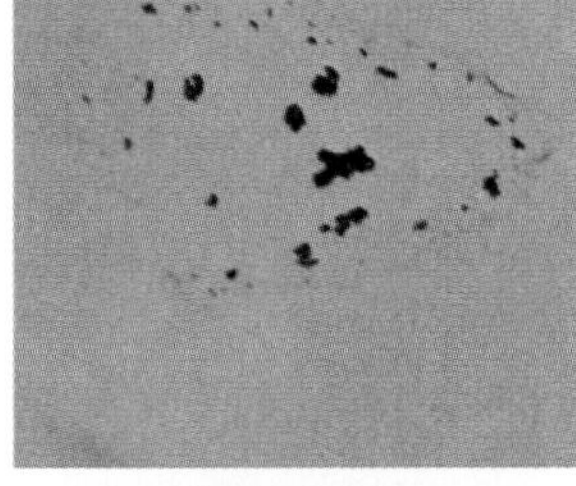
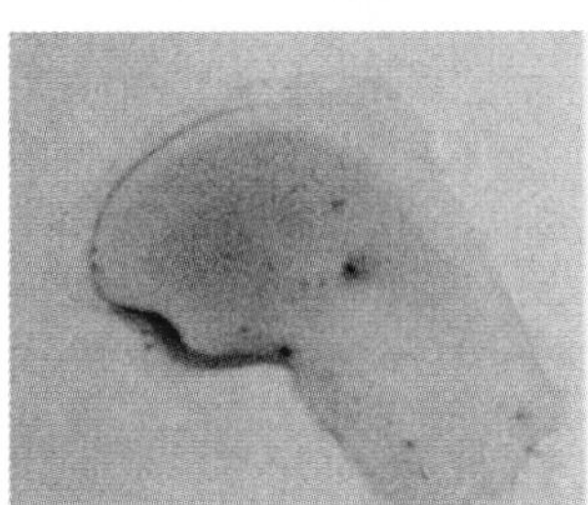

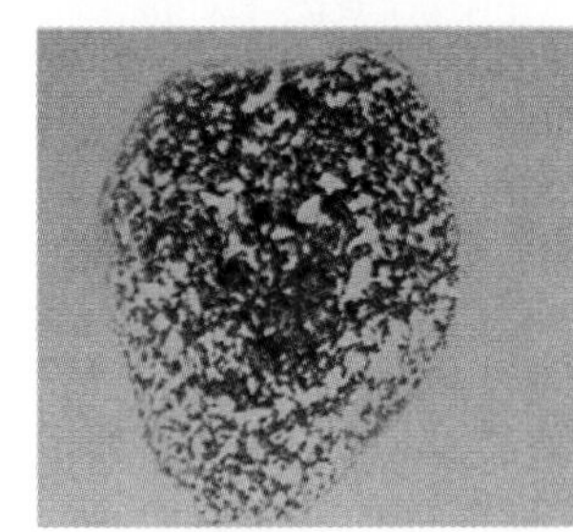
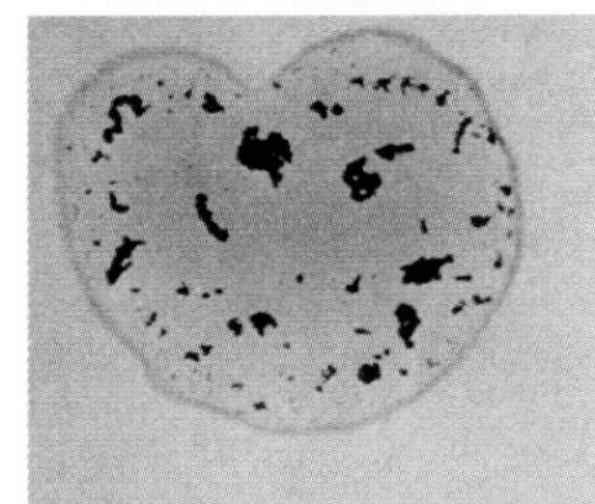

Figure 20.5

Blood typing.
Agglutination (clumping) of red blood cells occurs when cells with A-type antigens are mixed with anti-A antibodies and when cells with B-type antigens are mixed with anti-B antibodies. No agglutination would occur with type O blood (not shown).

Rh Factor

Another group of antigens found on the red blood cells of most people is the *Rh factor* (Rh is named for the rhesus monkey, in which these antigens were first discovered). People who have these antigens are said to be **Rh positive,** whereas those who do not are **Rh negative.** There are fewer Rh-negative people because this condition is recessive to Rh positive. The Rh factor is of particular significance when Rh-negative mothers give birth to Rh-positive babies.

Since the fetal and maternal blood are normally kept separate across the placenta (see chapter 29), the Rh-negative mother is not usually exposed to the Rh antigen of the fetus during the pregnancy. At the time of birth, however, a variable degree of exposure may occur, and the mother's immune system may become sensitized and produce antibodies against the Rh antigen. This does not always occur, however, because the exposure may be minimal and because Rh-negative women vary in their sensitivity to the Rh factor. If the woman does produce antibodies against the Rh factor, these antibodies may cross the placenta in subsequent pregnancies and cause hemolysis of the Rh-positive red blood cells of the fetus. Therefore the baby may be born anemic, with a condition called *erythroblastosis fetalis*, or *hemolytic disease of the newborn*.

Erythroblastosis fetalis can be prevented by injecting the Rh-negative mother with an antibody preparation against the Rh factor (a trade name for this preparation is RhoGAM—the "GAM" is short for gamma globulin, the class of plasma proteins in which antibodies are found) within 72 hours after the birth of each Rh-positive baby. This is a type of passive immunization in which the injected antibodies inactivate the Rh antigens and thus prevent the mother from becoming actively immunized to them. Some physicians now give RhoGAM throughout the Rh-positive pregnancy of any Rh-negative woman.

Blood Clotting

When a blood vessel is injured, a number of physiological mechanisms are activated that promote **hemostasis,** or the cessation of bleeding. Breakage of the endothelial lining of a vessel exposes collagen proteins from the subendothelial connective tissue to the blood. This initiates separate but overlapping hemostatic mechanisms: vasoconstriction, the formation of a platelet plug, and the production of a web of fibrin proteins around the platelet plug.

Functions of Platelets

In the absence of vessel damage platelets are repelled from each other and from the endothelial lining of vessels. The repulsion of platelets from an intact endothelium is believed to be due to *prostacyclin*, a derivative of prostaglandins produced within the endothelium. Mechanisms that prevent platelets from sticking to the blood vessels and to each other are obviously needed to prevent inappropriate blood clotting.

hemostasis: Gk. *haima*, blood; *stasis*, a standing

Damage to the endothelium of vessels exposes subendothelial tissue to the blood. Platelets are able to stick to exposed collagen proteins that have become coated with a protein (*von Willebrand factor*) secreted by endothelial cells. Platelets contain secretory granules; when platelets stick to collagen, they *degranulate* as the secretory granules release their products. These products include *adenosine diphosphate (ADP)*, *serotonin*, and a prostaglandin called *thromboxane* A_2. This event is known as the **platelet release reaction.**

Serotonin and thromboxane A_2 stimulate vasoconstriction, which helps to decrease blood flow to the injured vessel. Phospholipids that are exposed on the platelet membrane participate in the activation of clotting factors.

The release of ADP and thromboxane A_2 from platelets that are stuck to exposed collagen makes other platelets in the vicinity "sticky," so that they adhere to those stuck to the collagen. The second layer of platelets, in turn, undergoes a platelet release reaction, and the ADP and thromboxane A_2 that are secreted cause additional platelets to aggregate at the site of injury. This produces a **platelet plug** in the damaged vessel, which is then strengthened by the activation of plasma clotting factors (discussed below).

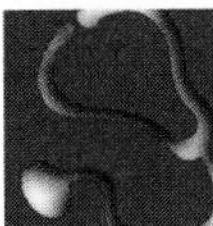

In order to undergo a release reaction, the production of prostaglandins by the platelets is required. *Aspirin* inhibits the cyclo-oxygenase enzyme that catalyzes the conversion of arachidonic acid (a cyclic fatty acid) into prostaglandins (chapter 19), thereby inhibiting the release reaction and consequent formation of a platelet plug. Since platelets lack nuclei and are not complete cells, they cannot regenerate new enzymes. Therefore, the inhibited enzymes persist for the life of the platelets. The ingestion of excessive amounts of aspirin can thus significantly prolong bleeding time for several days, which is why blood donors and women in the last trimester of pregnancy are advised to avoid aspirin. Slight inhibition of platelet aggregation by low doses of aspirin, however, can reduce the risk for atherosclerotic heart disease, and such a regimen is often recommended for patients diagnosed with this condition.

Clotting Factors: Formation of Fibrin

The platelet plug is strengthened by a meshwork of insoluble protein fibers known as **fibrin** (fig. 20.6). Blood clots therefore contain platelets and fibrin, and they usually contain trapped red blood cells that give the clot a red color. Clots formed in arteries, where the blood flow is more rapid, generally lack red blood cells and thus appear gray. Finally, contraction of the platelet mass in the process of *clot retraction* forms a more compact and effective plug. Fluid squeezed from the clot as it retracts is called *serum*, which is plasma without fibrinogen, the soluble precursor of fibrin. (Serum is obtained in laboratories by allowing blood to clot in a test tube and then centrifuging the tube so that the clot and blood cells become packed at the bottom of the tube.)

von Willebrand factor: from E. A. von Willebrand, Finnish physician, 1870–1949

Figure 20.6

A scanning electron micrograph showing threads of fibrin (2,740 ×).
Note that red blood cells are trapped within the clot.

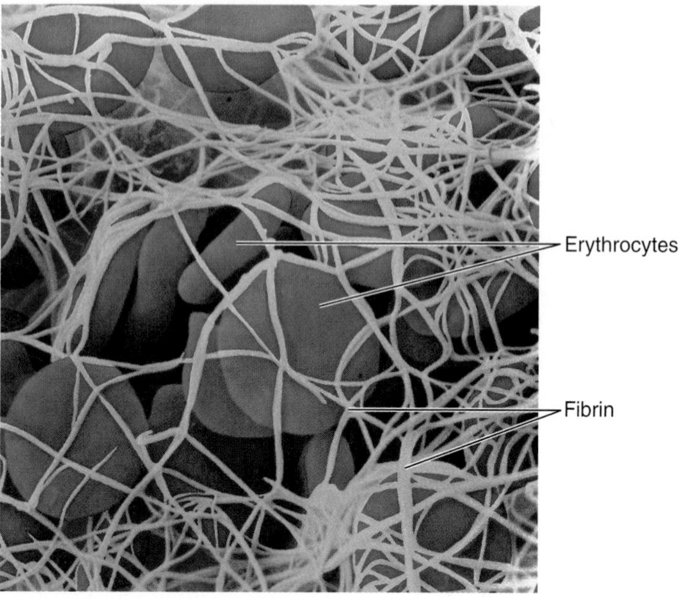

The conversion of fibrinogen into fibrin may occur via either of two pathways. Blood left in a test tube will clot without the addition of any external chemicals; the pathway that produces this clot is thus called the **intrinsic pathway.** The intrinsic pathway also produces clots in damaged blood vessels when collagen is exposed to plasma. Damaged tissues, however, release a chemical that initiates a "shortcut" to the formation of fibrin. Since this chemical is not part of blood, the shorter pathway is called the **extrinsic pathway.**

The intrinsic pathway is initiated by the exposure of plasma to a negatively charged surface, such as that provided by collagen at the site of a wound or by the glass of a test tube. This activates a plasma protein called factor XII (table 20.4), which is a protein-digesting enzyme, or *protease* (*pro′te-ās*). Active factor XII in turn activates another clotting factor, which activates yet another. The plasma clotting factors are numbered in order of their discovery, which does not reflect the actual sequence of reactions.

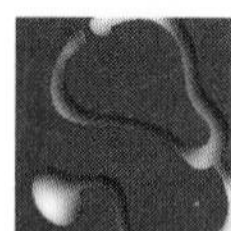

A number of hereditary diseases involve the clotting system. Examples of hereditary clotting disorders include two different genetic defects in factor VIII. A defect in one subunit of factor VIII prevents this factor from participating in the intrinsic clotting pathway. This genetic disease, called *hemophilia A,* is an X-linked recessive trait that is prevalent in the royal families of Europe. A defect in another subunit of factor VIII results in *von Willebrand's disease.* In this disease, rapidly circulating platelets are unable to stick to collagen and a platelet plug cannot be formed. Some acquired and inherited defects in the clotting system are summarized in table 20.5.

Table 20.4

The Plasma Clotting Factors

Factor	Name	Function	Pathway
I	Fibrinogen	Converted to fibrin	Common
II	Prothrombin	Enzyme	Common
III	Tissue thromboplastin	Cofactor	Extrinsic
IV	Calcium ions (Ca^{2+})	Cofactor	Intrinsic, extrinsic, and common
V	Proaccelerin	Cofactor	Common
VII*	Proconvertin	Enzyme	Extrinsic
VIII	Antihemophilic factor	Cofactor	Intrinsic
IX	Plasma thromboplastin component; Christmas factor	Enzyme	Intrinsic
X	Stuart–Prower factor	Enzyme	Common
XI	Plasma thromboplastin antecedent	Enzyme	Intrinsic
XII	Hageman factor	Enzyme	Intrinsic
XIII	Fibrin stabilizing factor	Enzyme	Common

*Factor VI is no longer referenced; it is now believed to be the same substance as activated factor V.

The next steps in the sequence require the presence of Ca^{2+} and of phospholipids, which are provided by platelets. These steps result in the conversion of an inactive plasma enzyme, called **prothrombin,** into the active enzyme **thrombin.** Thrombin converts the soluble protein **fibrinogen** into **fibrin** monomers. These monomers join together to produce the insoluble fibrin polymers that form a meshwork supporting the platelet plug. The intrinsic clotting sequence is shown on the right side of figure 20.7.

The formation of fibrin can occur more rapidly as a result of the release of **tissue thromboplastin** from damaged tissue cells. This extrinsic clotting pathway is shown on the left side of figure 20.7. Notice that the intrinsic and extrinsic clotting pathways eventually merge, forming a final common pathway that results in the formation of insoluble fibrin polymers.

Dissolution of Clots

As the damaged blood vessel wall is repaired, activated factor XII promotes the conversion of an inactive molecule in plasma into the active form called *kallikrein* (*ka″lĭ-kre′in*). Kallikrein, in turn, catalyzes the conversion of inactive *plasminogen* into the active molecule **plasmin.** Plasmin is an enzyme that digests fibrin into "split products," thus promoting dissolution of the clot.

Table 20.5

Acquired and Inherited Clotting Disorders and Some Anticoagulant Drugs

Category	Cause of Disorder	Comments
Acquired clotting disorders	Vitamin K deficiency	Inadequate formation of prothrombin and other clotting factors in the liver
Inherited clotting disorders	Hemophilia A (defective factor VIII_{AHF})	Recessive trait carried on X chromosome; results in delayed formation of fibrin
	von Willebrand's disease (defective factor VIII_{VWF})	Dominant trait carried on autosomal chromosome; impaired ability of platelets to adhere to collagen in subendothelial connective tissue
	Hemophilia B (defective factor IX); also called Christmas disease	Recessive trait carried on X chromosome; results in delayed formation of fibrin

Anticoagulants

Aspirin	Inhibits prostaglandin production, resulting in a defective platelet release reaction
Coumarin	Inhibits the activation of vitamin K
Heparin	Inhibits activity of thrombin
Citrate	Combines with Ca^{2+}, and thus inhibits the activity of many clotting factors

Figure 20.7

The extrinsic and intrinsic clotting pathways.
Both pathways lead to the formation of insoluble threads of fibrin polymers

Source: Adapted from A. Marchand, "Case of the Month, Circulating Anticoagulants: Chasing the Diagnosis" in Diagnostic Medicine, *June 1983, page 14. Copyright 1984. Reprinted with permission of Medical Economic Company, Inc. Oradell, NJ.*

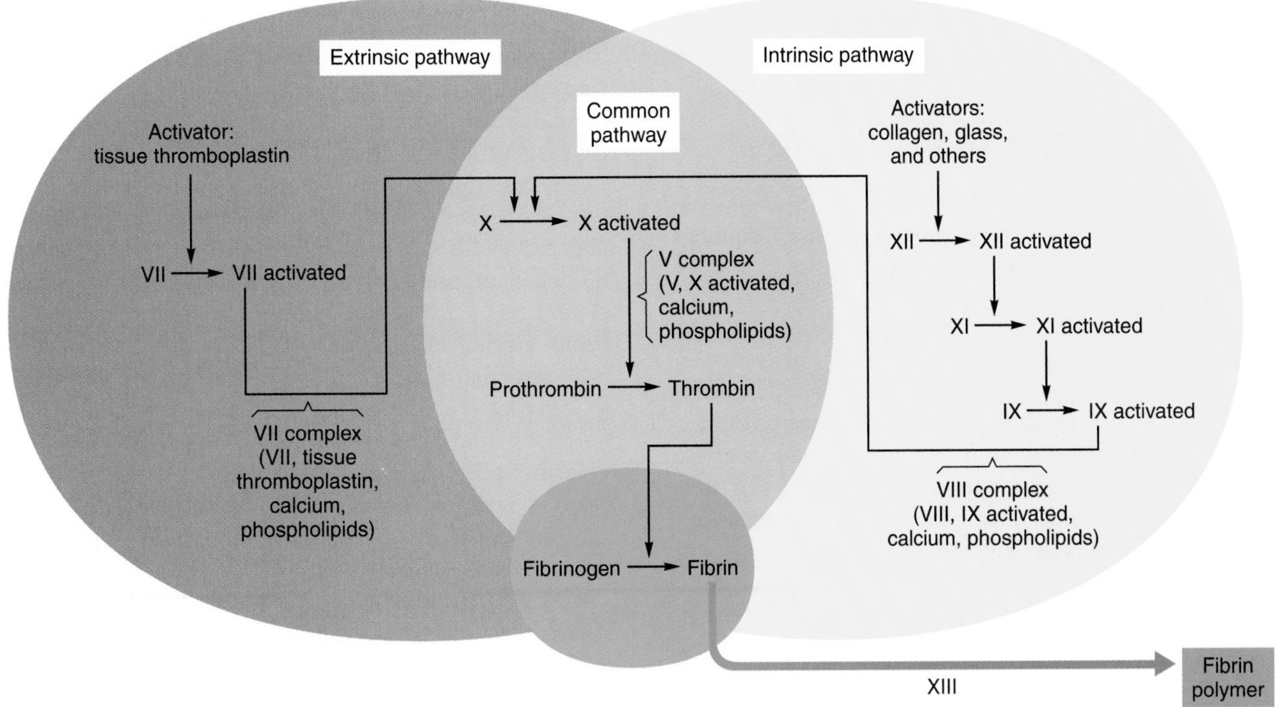

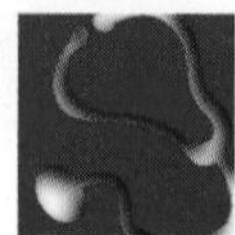

In addition to kallikrein, a number of other plasminogen activators are used clinically to promote dissolution of clots. An exciting development in genetic engineering technology is the commercial availability of an endogenous compound, *tissue plasminogen activator (tPA),* which is the product of human genes introduced into bacteria. *Streptokinase,* a natural bacterial product, is a potent and more widely used activator of plasminogen. Streptokinase and tPA may be injected into the general circulation or injected specifically into a coronary vessel that has become occluded by a thrombus (blood clot).

Anticoagulants

Clotting of blood in test tubes can be prevented by the addition of *sodium citrate* or *ethylenediaminetetraacetic acid (EDTA)*, both of which chelate (bind to) calcium. By this means, Ca^{2+} levels in the blood that can participate in the clotting sequence decrease, and clotting is inhibited. A mucoprotein called *heparin* can also be added to the tube to prevent clotting. Heparin activates *antithrombin III*, a plasma protein that combines with and inactivates thrombin. Heparin is also given intravenously during certain medical procedures to prevent clotting. The *coumarin* (*koo'mă-rin*) drugs, whose mechanism of action is different from that of heparin, are also used as anticoagulants. These drugs (dicumarol and warfarin) prevent blood clotting by inhibiting the cellular activation of vitamin K, thereby causing a vitamin K deficiency at the cellular level.

Vitamin K is needed for the conversion of glutamate, an amino acid found in many of the clotting factor proteins, into a derivative called *gamma-carboxyglutamate*. This derivative is more effective than glutamate at binding to Ca^{2+}, and such binding is needed for proper function of clotting factors II, VII, IX, and X. Because of the indirect action of vitamin K on blood clotting, coumarin must be given to a patient for several days before it becomes effective as an anticoagulant.

Acid-Base Balance of the Blood

The pH of blood plasma is maintained within a narrow range of values through the functions of the lungs and kidneys. The lungs regulate the carbon dioxide concentration of the blood, and the kidneys regulate the bicarbonate concentration.

The blood plasma within arteries normally has a pH between 7.35 and 7.45, with an average of 7.40. Using the definition of pH described in chapter 2, this means that arterial blood has a H^+ concentration of about $10^{-7.4}$ molar. Some of these hydrogen ions are derived from carbonic acid, which is formed in the blood plasma from carbon dioxide and which can ionize, as indicated in the following equations:

$$CO_2 + H_2O \leftrightharpoons H_2CO_3$$
$$H_2CO_3 \leftrightharpoons H^+ + HCO_3^-$$

Carbon dioxide is produced by cells through aerobic cell respiration and is transported by the blood to the lungs, where it can be exhaled. As will be described in more detail in chapter 24, carbonic acid can be reconverted to carbon dioxide, which is a gas. Because it can be converted to a gas, carbonic acid is referred to as a *volatile acid*, and its concentration in the blood is controlled by the lungs through proper ventilation (breathing). All other acids in the blood—including lactic acid, fatty acids, ketone bodies, and so on—are *nonvolatile acids*.

Under normal conditions, the H^+ released by nonvolatile acids do not affect the blood pH because these hydrogen ions are bound to molecules that function as *buffers*. The major buffer in the plasma is *bicarbonate* (HCO_3^-), and it buffers H^+ as described by the following equation:

$$HCO_3^- + H^+ \rightarrow H_2CO_3$$

This buffering reaction could not go on forever because the free HCO_3^- would eventually disappear. If this were to occur, the H^+ concentration would increase and the pH of the blood would decrease. Under normal conditions, however, excessive H^+ is eliminated in the urine by the kidneys. Through this action, and through their ability to produce bicarbonate, the kidneys function to maintain a normal concentration of free bicarbonate in the plasma. The role of the kidneys in acid-base balance is described in chapter 25.

A fall in blood pH below 7.35 is called **acidosis** (*as"ĭ-do'sis*) because the pH is to the acid side of normal. Acidosis does not mean acidic (pH less than 7); a blood pH of 7.2, for example, represents serious acidosis. Similarly, a rise in blood pH above 7.45 is called **alkalosis** (*al"kă-lo'sis*). Both of these conditions are categorized into respiratory and metabolic components of acid-base balance, as indicated in table 20.6.

Respiratory acidosis is caused by inadequate ventilation (hypoventilation), which results in a rise in the plasma concentration of carbon dioxide and thus carbonic acid. **Respiratory alkalosis,** in contrast, is caused by excessive ventilation (hyperventilation). **Metabolic acidosis** can result from excessive production of nonvolatile acids; for example, it can result from excessive production of ketone bodies in uncontrolled diabetes mellitus (see chapter 27). It can also result from the loss of bicarbonate, in which case there would not be sufficient free bicarbonate to buffer the nonvolatile acids. (This occurs in diarrhea because of the loss of bicarbonate derived from pancreatic juice—see chapter 26.) **Metabolic alkalosis,** in contrast, can be caused by either too much bicarbonate (perhaps from an intravenous infusion) or inadequate nonvolatile acids (perhaps as a result of excessive vomiting). Excessive vomiting may cause metabolic alkalosis through loss of the acid in gastric juice, which is normally absorbed from the intestine into the blood.

Since the *respiratory component* of acid-base balance is represented by the plasma carbon dioxide concentration and

Table 20.6
Terms Used to Describe Acid-Base Balance

Term	Definition
Acidosis, respiratory	Increased CO_2 retention (due to hypoventilation), which can result in the accumulation of carbonic acid and thus a fall in blood pH to below normal
Acidosis, metabolic	Increased production of "nonvolatile" acids such as lactic acid, fatty acids, and ketone bodies, or loss of blood bicarbonate (such as by diarrhea), resulting in a fall in blood pH to below normal
Alkalosis, respiratory	A rise in blood pH due to loss of CO_2 and carbonic acid (through hyperventilation)
Alkalosis, metabolic	A rise in blood pH produced by loss of nonvolatile acids (as in excessive vomiting) or by excessive accumulation of bicarbonate base
Compensated acidosis or alkalosis	Metabolic acidosis or alkalosis are partially compensated for by opposite changes in blood carbonic acid levels (through changes in ventilation). Respiratory acidosis of alkalosis are partially compensated for by increased retention or excretion of bicarbonate in the urine.

Table 20.7
Classification of Metabolic and Respiratory Components of Acidosis and Alkalosis

Plasma CO_2	Plasma HCO_3^-	Condition	Causes
Normal	Low	Metabolic acidosis	Increased production of "nonvolatile" acids (lactic acid, ketone bodies, and others), or loss of HCO_3^- in diarrhea
Normal	High	Metabolic alkalosis	Vomiting of gastric acid; hypokalemia; excessive steroid administration
Low	Low	Respiratory alkalosis	Hyperventilation
High	High	Respiratory acidosis	Hypoventilation

the *metabolic component* is represented by the free bicarbonate concentration, the study of acid-base balance can be simplified. A normal arterial blood pH is obtained when there is a proper ratio of bicarbonate to carbon dioxide. Indeed, the pH can be calculated given these values, and a normal pH is obtained when the ratio of HCO_3^- to CO_2 is 20 to 1. This is given by the **Henderson–Hasselbalch** equation:

$$pH = 6.1 + \log \frac{[HCO_3^-]}{[CO_2]}$$

Henderson–Hasselbalch equation: from Lawrence Joseph Henderson, American chemist, 1878–1942, and Karl A. Hasselbalch, Dutch scientist, 1874–1962

Respiratory acidosis or alkalosis occurs when the carbon dioxide concentrations are abnormal. Metabolic acidosis and alkalosis occur when the bicarbonate concentrations are abnormal (table 20.7). Often, however, a primary disturbance in one area (for example, metabolic acidosis), will be accompanied by secondary changes in another area (for example, respiratory alkalosis). It is important for hospital personnel to identify and treat the area of primary disturbance, but such analysis lies outside the scope of this discussion.

A more complete description of the respiratory and metabolic components of acid-base balance requires the study of pulmonary and renal function, and so will be presented with these topics in chapters 24 and 25.

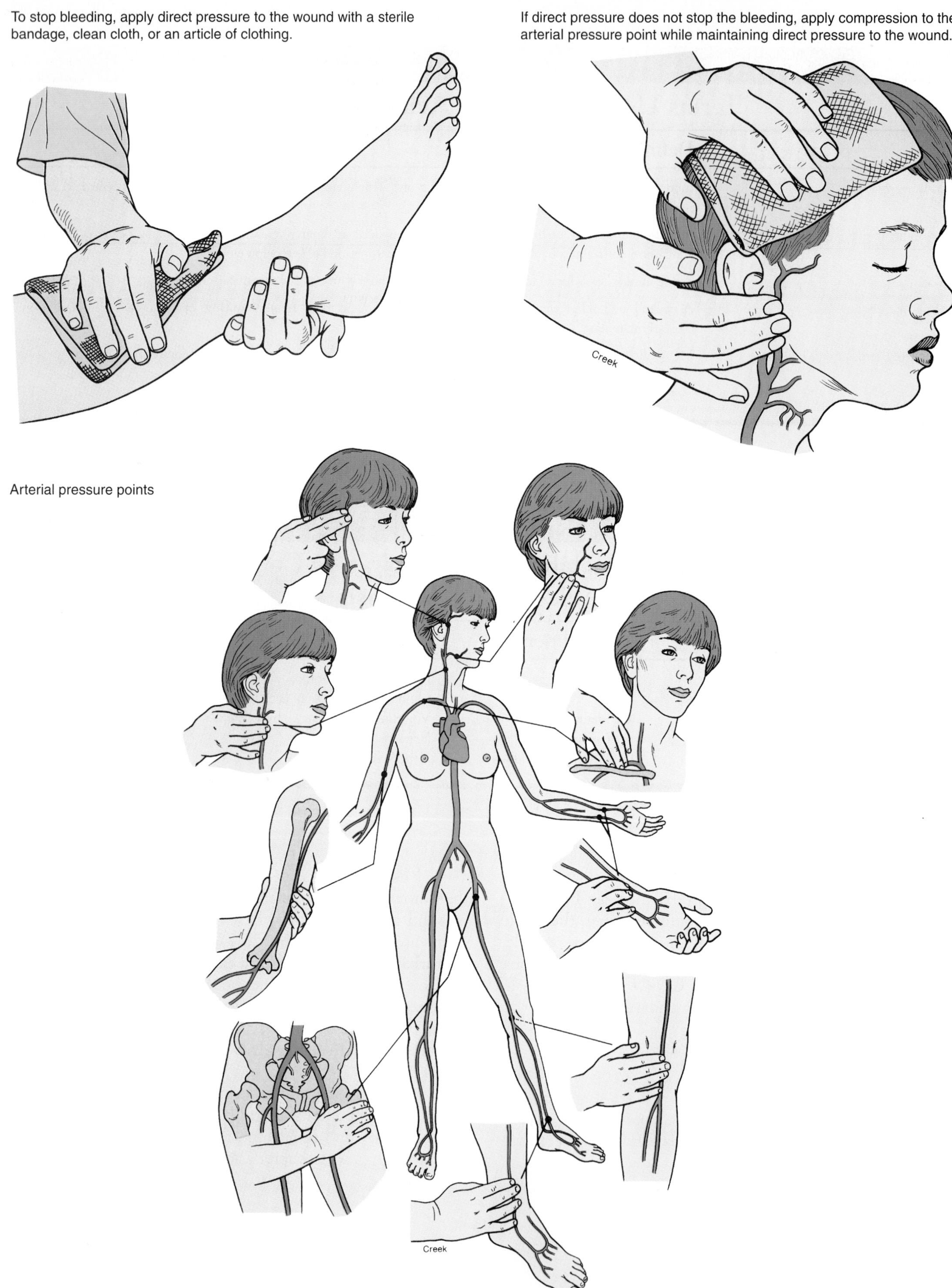
To stop bleeding, apply direct pressure to the wound with a sterile bandage, clean cloth, or an article of clothing.
If direct pressure does not stop the bleeding, apply compression to the arterial pressure point while maintaining direct pressure to the wound.
Creek
Arterial pressure points
Creek

Clinical Considerations

Arterial Pressure Points and Control of Bleeding

Because serious bleeding is life threatening, the principal first-aid concern is to stop loss of blood. The following are recommended steps in treating a victim who is hemorrhaging.

1. To reduce the chance that the victim will faint, lay the person down and slightly elevate his or her legs. If possible, elevate the site of bleeding above the level of the trunk. To minimize the chance of shock, cover the victim with a blanket.
2. Without causing further trauma, carefully remove any dirt or debris from the wound. *Do not remove any impaling objects*. This should be done at the hospital by trained personnel.
3. To stop bleeding, apply direct pressure to the wound with a sterile bandage, clean cloth, or an article of clothing (see upper left illustration on page 606).
4. Maintain direct pressure until the bleeding stops. Dress the wound with clean bandages or cloth lightly bound in place.
5. If the bleeding does not stop and continues to seep through the dressing, *do not remove the dressing*. Rather, place additional absorbent material on top of it and continue to apply direct pressure.
6. If direct pressure does not stop the bleeding, a pressure point proximal to the wound site may need to be compressed (see illustrations on page 606). In the case of a severe wound to the hand, for example, compress the brachial artery against the humerus. This should be done while pressure continues to be applied to the wound itself.
7. Once the bleeding has stopped, leave the bandage in place and immobilize the injured body part. Get the victim to the hospital or medical treatment center at once.

Recognizing and Treating Victims of Shock

***Shock** is the medical condition that occurs when body tissues do not receive enough oxygen-carrying blood. It is often linked with crushing injuries, heat stroke, heart attacks, poisoning, severe burns, and other life-threatening conditions. Symptoms of patients experiencing shock include the following.*

1. **Skin.** The skin is pale or gray, cool, and clammy.
2. **Pulse.** The heartbeat is weak and rapid. Blood pressure is reduced, frequently to below measurable values.
3. **Respiration.** The respiratory rate is hurried, shallow, and irregular.
4. **Eyes.** The eyes are staring and lusterless. The pupils may be dilated.
5. **State of consciousness.** The victim may be conscious or unconscious. If conscious, he or she is likely to feel faint, weak, and confused. Frequently, the victim is anxious and excited.

Most trauma victims will experience some degree of shock, especially if there has been considerable blood loss. Immediate first-aid treatment for shock is essential and includes the following steps.

1. Get the victim to lie down (see illustration below). Lay the person on his or her back with the feet elevated. This position maintains blood flow to the brain and may relieve faintness and mental confusion. Keep movement to a minimum. If the victim has sustained an injury in which raising the legs causes additional pain, leave the person flat on his or her back.
2. Keep the victim warm and comfortable. If the weather is cold, place a blanket under and over the person. If the weather is hot, position the person in the shade on top of a blanket. Loosen the victim's tight collars, belts, or other restrictive clothing. *Do not give the person anything to drink*, even if he or she complains of thirst.
3. Take precautions for internal bleeding or vomiting. If blood is coming from the victim's mouth, or if there is indication that the victim may vomit, position the person on his or her side to prevent choking or inhaling the blood or vomitus.
4. Treat injuries appropriately. If the victim is bleeding, treat accordingly (see "Arterial Pressure Points and Control of Bleeding"). Immobilize fractures and sprains. Always be alert to the possibility of spinal injuries and take the necessary precautions.
5. See that hospital care is provided as soon as possible.

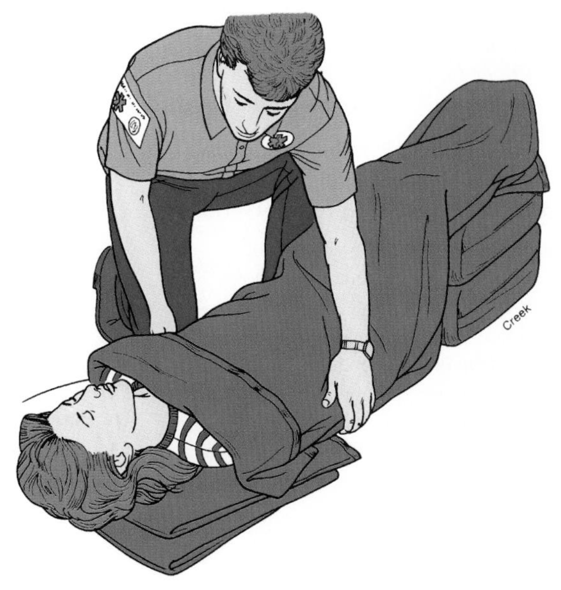

Clinical Investigation Answer

A person who lives and works at high altitude would be expected to have a somewhat higher-than-normal red blood cell count and hemoglobin concentration; this is an adaptation to the "thinner air" that allows the blood to carry more oxygen. Despite the fact that the man worked at an alpine ski resort, his red blood cell count and hemoglobin concentration were low. A low hemoglobin concentration, and a consequently low blood oxygen content, would explain his tendency to tire easily. The large amounts of aspirin and acetaminophen could have caused inflammation of the kidneys, resulting in the reduced kidney function observed.

Chapter Summary

Functions and Components of the Circulatory System (pp. 592–593)

1. Blood transports oxygen and nutrients to all the cells of the body and removes waste products from the tissues. It also serves regulatory and protective functions.
   - (a) Erythrocytes, or red blood cells, transport oxygen.
   - (b) Leukocytes, or white blood cells, protect the body from disease.
   - (c) Platelets, or thrombocytes, help to form blood clots.
2. The circulatory system consists of the cardiovascular system (heart and blood vessels) and the lymphatic system.
   - (a) Veins carry blood to the heart; arteries carry blood away from the heart.
   - (b) Blood flows from arterioles to capillaries, which transport the blood into venules.

Composition of the Blood (pp. 593–601)

1. Plasma is the fluid part of the blood, containing dissolved ions and various organic molecules.
   - (a) Hormones are found in the plasma portion of the blood.
   - (b) The three types of plasma proteins are albumins, globulins (alpha, beta, and gamma) and fibrinogen.
2. Hematopoiesis is the production of blood cells.
   - (a) Blood cells form from proerythroblasts, myeloblasts, lymphoblasts, and megakaryoblasts.
   - (b) Production of erythrocytes is stimulated by the hormone erythropoietin. Development of platelets is stimulated by thrombopoietin, and the different kinds of leukocytes develop under stimulation by a variety of cytokines.
3. The major blood-typing groups are the ABO system and the Rh system.
   - (a) Blood type refers to the kind of antigens found on the surface of erythrocytes.
   - (b) When different types of blood are mixed, antibodies against the erythrocytic antigens cause the erythrocytes to agglutinate.
   - (c) A person with type O blood is the universal donor because that person's plasma does not contain antibodies against the A or B antigen. This concept is of limited usefulness, however.
   - (d) Because of potential agglutination reactions, blood types must be matched for blood transfusions.

Blood Clotting (pp. 601–604)

1. When a blood vessel is damaged, platelets adhere to the exposed subendothelial collagen proteins.
   - (a) Platelets that stick to collagen undergo a release reaction whereby they secrete ADP, serotonin, and thromboxane A_2.
   - (b) Serotonin and thromboxane A_2 cause vasoconstriction; ADP and thromboxane A_2 attract other platelets and cause them to adhere to the growing mass of platelets that are stuck to the collagen in the broken vessel.
   - (c) In the formation of a blood clot, a soluble protein called fibrinogen is converted by the enzyme thrombin into insoluble threads of fibrin.
   - (d) Thrombin is derived from its inactive precursor prothrombin by either an intrinsic or an extrinsic pathway.
2. Dissolution of the clot eventually occurs by the digestive action of plasmin, which cleaves fibrin into split products.

Acid-Base Balance of the Blood (pp. 604–605)

1. The normal pH of arterial blood is 7.40, with a range of 7.35 to 7.45.
   - (a) Carbonic acid is formed from carbon dioxide and contributes to the blood pH. Its a volatile acid because it can be eliminated in the exhaled breath.
   - (b) Nonvolatile acids, such as lactic acid and the ketone bodies, are buffered by blood bicarbonate.
2. The blood pH is maintained by a proper ratio of carbon dioxide to bicarbonate.
   - (a) The lungs maintain the correct carbon dioxide concentration. An increase in carbon dioxide, as a result of inadequate ventilation, produces respiratory acidosis.
   - (b) The kidneys maintain the free bicarbonate concentration. An abnormally low plasma bicarbonate concentration produces metabolic acidosis.

Review Activities

Objective Questions

1. Which of the following is the correct sequence of blood flow?
   - (a) arteries-capillaries-venules-arterioles-veins
   - (b) veins-venules-capillaries-arterioles-arteries
   - (c) capillaries-arteries-arterioles-venules-veins
   - (d) arteries-arterioles-capillaries-venules-veins
2. Which of the following is the safest to use in a transfusion of a person with type B blood?
   - (a) type O blood
   - (b) type A blood
   - (c) type B blood
   - (d) type AB blood
3. Which of the following stem cells give rise to granular leukocytes?
   - (a) myeloblasts
   - (b) erythroblasts
   - (c) lymphoblasts
   - (d) megakaryoblasts
4. Which of the following is an agranular leukocyte?
   - (a) eosinophil
   - (b) neutrophil
   - (c) basophil
   - (d) monocyte
5. Hemoglobin is normally found in
   - (a) erythrocytes.
   - (b) plasma.
   - (c) granular leukocytes.
   - (d) agranular leukocytes.
   - (e) all of the above.
6. Factor X is activated
   - (a) in the intrinsic pathway only.
   - (b) in the extrinsic pathway only.
   - (c) in both the intrinsic and extrinsic pathways.
   - (d) in neither the intrinsic nor the extrinsic pathway.

7. Platelets
   (a) form a plug by sticking to each other.
   (b) release chemicals that stimulate vasoconstriction.
   (c) provide phospholipids needed for the intrinsic pathway.
   (d) all of the above apply
8. Antibodies against both type-A and type-B antigens are found in the plasma of a person who is
   (a) type A.
   (b) type B.
   (c) type AB.
   (d) type O.
   (e) all of the above apply.
9. Production of which of the following blood cells is stimulated by a hormone secreted by the kidneys?
   (a) lymphocytes
   (b) monocytes
   (c) erythrocytes
   (d) neutrophils
   (e) platelets
10. Which of the following statements about plasmin is *true?*
    (a) It is involved in the intrinsic clotting system.
    (b) It is involved in the extrinsic clotting system.
    (c) It functions in fibrinolysis.
    (d) It promotes the formation of emboli.
11. Anemia may be caused by
    (a) inadequate secretion of erythropoietin.
    (b) loss of iron from blood loss.
    (c) excessive destruction of red blood cells.
    (d) all of the above.
12. Vitamin K is needed
    (a) to produce hemoglobin.
    (b) to produce platelets.
    (c) to convert fibrinogen to fibrin.
    (d) to activate numerous clotting factors.
13. The blood concentration of carbon dioxide will increase and the blood pH will decrease when someone
    (a) hyperventilates (breathes excessively).
    (b) hypoventilates (breathes inadequately).
    (c) has prolonged diarrhea.
    (d) vomits excessively.

Essay Questions

1. Describe hematopoiesis in general terms and state where the different types of blood cells are formed.
2. Describe the stages of erythrocyte production and the characteristics of a mature erythrocyte. How is erythropoiesis regulated?
3. Describe the series of events that may culminate in hemolytic disease of the newborn.
4. Explain why agglutination occurs when different blood types are mixed.
5. Explain how a cut in the skin initiates both the intrinsic and extrinsic clotting pathways. Which pathway finishes first? Why?
6. Explain how aspirin, coumarin, EDTA, and heparin function as anticoagulants. Which of these are effective when added to a test tube? Which are not? Why?
7. Explain how bicarbonate helps to maintain acid-base balance and describe the conditions that may result in metabolic acidosis or alkalosis.

Critical Thinking Questions

1. Hematopoietic stem cells account for less than 1% of the cells in the bone marrow. These cells can be separated from the others prior to bone marrow transplantation, but it is better to first inject the donor with recombinant cytokines. Identify the cytokines that might be used and describe their effects.
2. A patient has a low red blood cell count, and microscopic examination of his blood reveals that there is an abnormally high proportion of circulating reticulocytes. Upon subsequent examination the patient is diagnosed with a bleeding ulcer. This is surgically corrected and in due course his blood measurements return to normal. What was the reason for the low red blood cell count and high proportion of reticulocytes?
3. A chemical called EDTA, like citrate, binds to (or "chelates") Ca^{2+}. Suppose a person had EDTA infused into the blood. What effect would this have on the intrinsic and extrinsic clotting pathways? How would these effects differ from the effects of aspirin on blood clotting?

Related Web Sites

For a listing of the most current web sites related to this chapter, please visit the *Concepts of Human Anatomy and Physiology* home page at http://www.mhhe.com/biosci/abio/.

Twenty-One

Circulatory System

Objectives

- Describe the location of the heart and its associated membranes. Also, describe the characteristics of the three layers of the heart wall.
- Discuss the nature and significance of the fibrous skeleton of the heart.
- List the chambers, valves, and associated vessels of the heart.
- Trace the flow of blood through the heart and distinguish between the systemic and pulmonary circulations.
- Describe the cycle of contraction and relaxation of the atria and ventricles.
- Explain how pressure changes in the heart affect the closing and opening of its valves and how these events produce the heart sounds.
- Describe the electrical activity of the SA node and explain why it functions as the normal pacemaker.
- Describe the conducting tissue of the heart, noting the pathway of electrical conduction.
- State the events that produce the ECG waves and correlate the ECG waves with the events of the cardiac cycle and the production of the heart sounds.
- Describe the structure of muscular and elastic arteries and the structure of veins. Explain how the structures of arteries and veins are functionally adaptive.
- Describe the structures of different types of capillaries and explain the functions of these capillaries.
- List the arterial branches of the ascending aorta and aortic arch and describe the arterial supply to the brain.
- Describe the arterial pathways that supply blood to the upper extremity, thorax, abdomen, and lower extremity.
- Describe the venous drainage of the head, neck, and upper extremity.
- Describe the venous drainage of the thorax, lower extremity, and abdomen.
- State the vessels involved in the hepatic portal system and explain the significance of this system.
- Describe fetal circulation and explain the changes that occur upon birth of the baby.

Clinical Investigation

A 65-year-old woman who had been discharged from the hospital following a myocardial infarction (heart attack) returned to the emergency room several weeks later because of the sudden onset of pain in her right lower extremity. The attending physician noticed that the patient's leg was also pale and cool from the knee down. Furthermore, he was unable to detect a pulse over the dorsal pedal and popliteal arteries. A good femoral pulse, however, was palpable in the inguinal region. Upon questioning, the patient stated that she was sent home with blood-thinning medication because tests had revealed that the heart attack had caused a blood clot to form within her heart. The doctors were worried, she added, that "a piece of the clot could break off and go to other parts of the body."

Explain how the patient's heart attack could have led to her current leg complaints. In which side of her heart (right or left) did the previously diagnosed blood clot most likely reside? Explain anatomically where the cause of her new problem is located and describe how it came to arrive there.

Structure of the Heart

The structure of the heart enables it to serve as a transport system pump that keeps blood continuously circulating through the blood vessels of the body.

The hollow, four-chambered muscular **heart** is roughly the size of a clenched fist. It averages 255 grams in adult females and 310 grams in adult males. The heart contracts an estimated 42 million times a year, pumping 700,000 gallons of blood.

The heart is located within the thoracic cavity between the lungs in the mediastinum (fig. 21.1). About two-thirds of the heart is located left of the midline, with its *apex*, or cone-shaped end, pointing downward in contact with the diaphragm. The *base* of the heart is the broad superior end, where the large vessels attach.

The **parietal pericardium** (*per"ĭ-kar'de-um*) is a loose-fitting, serous sac of dense irregular connective tissue that encloses and protects the heart (fig. 21.1). It separates the heart from the other thoracic organs and forms the wall of the *pericardial cavity* (table 21.1), which contains a watery,

Figure 21.1

The position of the heart and associated serous membranes within the thoracic cavity.

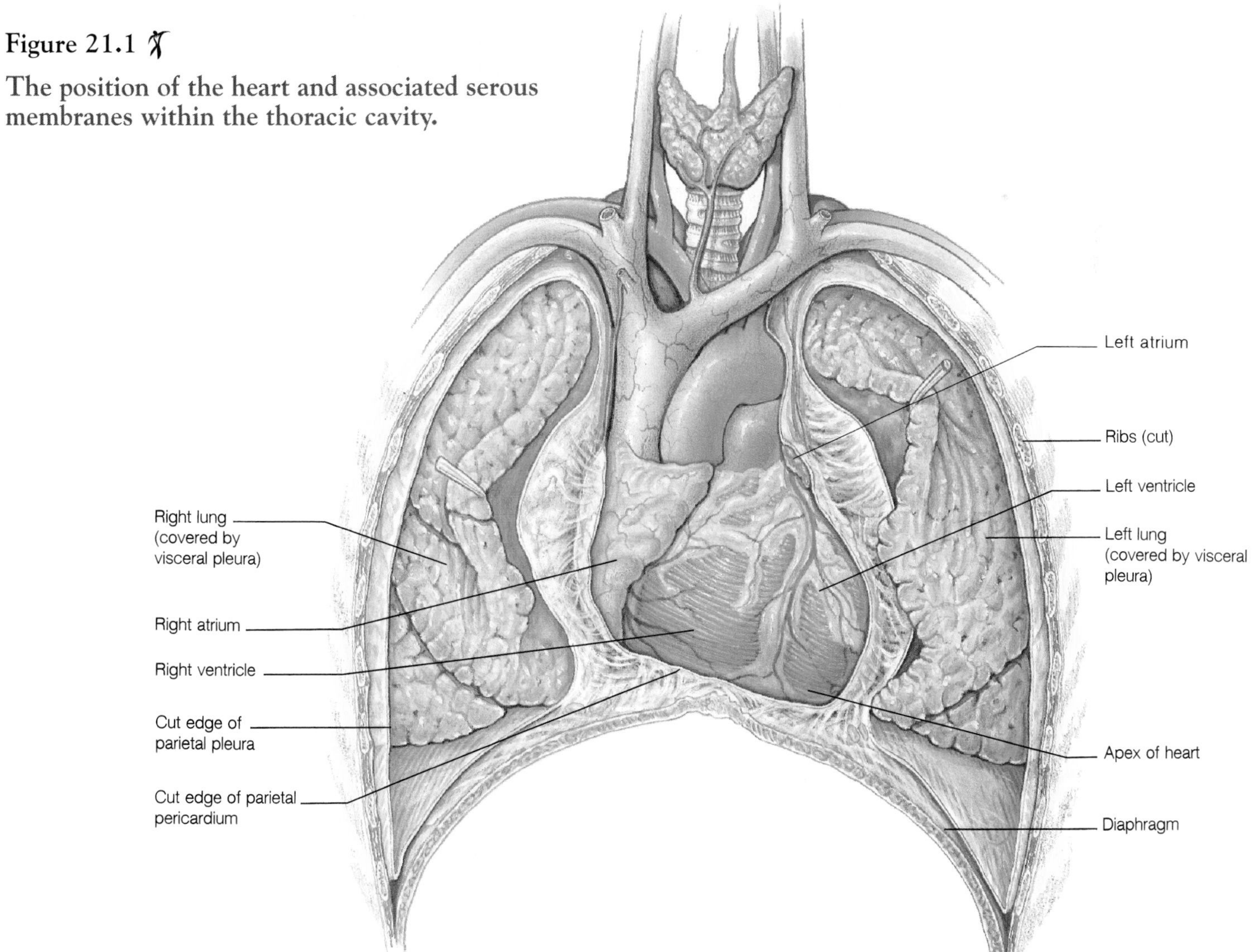

Table 21.1

Layers of the Heart Wall

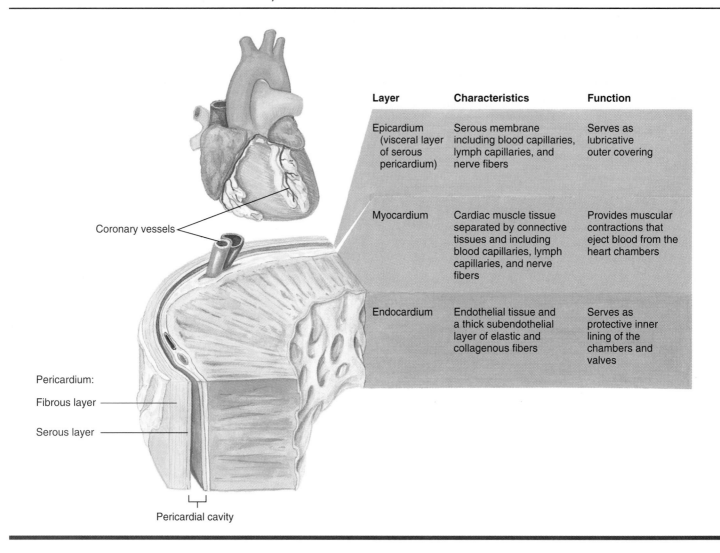

Layer	Characteristics	Function
Epicardium (visceral layer of serous pericardium)	Serous membrane including blood capillaries, lymph capillaries, and nerve fibers	Serves as lubricative outer covering
Myocardium	Cardiac muscle tissue separated by connective tissues and including blood capillaries, lymph capillaries, and nerve fibers	Provides muscular contractions that eject blood from the heart chambers
Endocardium	Endothelial tissue and a thick subendothelial layer of elastic and collagenous fibers	Serves as protective inner lining of the chambers and valves

lubricating *pericardial fluid*. The parietal pericardium is actually composed of an outer *fibrous pericardium* and an inner *serous pericardium*. The serous pericardium produces the lubricating pericardial fluid that allows the heart to beat in a kind of frictionless bath.

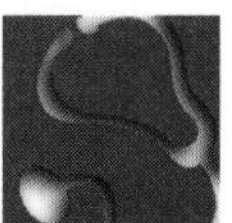

Pericarditis is an inflammation of the parietal pericardium that results in an increased secretion of fluid into the pericardial cavity. Because the tough, fibrous portion of the parietal pericardium is inelastic, an increase in fluid pressure impairs the movement of blood into and out of the chambers of the heart. Some of the pericardial fluid may be withdrawn for analysis by injecting a needle to the left of the xiphoid process to pierce the parietal pericardium.

Heart Wall

The wall of the heart is composed of three distinct layers (table 21.1). The outer layer is the **epicardium,** also called the *visceral pericardium*. The space between this layer and the parietal pericardium is the pericardial cavity, just described. The thick middle layer of the heart is the **myocardium.** It is composed of cardiac muscle tissue, and arranged in such a way that the contraction of the muscle bundles results in squeezing or wringing of the heart chambers. The thickness of the myocardium varies in accordance with the force needed to eject blood from the particular chamber. Thus, the thickest portion of the myocardium surrounds the left ventricle, and the atrial walls are relatively thin. The inner layer of the heart, called the **endocardium,** is continuous with the endothelium of blood vessels. The endocardium also covers the valves of the heart. Inflammation of the endocardium is called *endocarditis*.

Chambers and Valves

The interior of the heart is divided into four chambers: upper right and left **atria** (*a'tre-a*; singular, *atrium*) and lower right and left **ventricles.** The atria contract and empty simultaneously into the ventricles (fig. 21.2), which also contract in unison. Each atrium has an ear-shaped expandable appendage called the **auricle** (*or'i-kul*). The atria are separated by the thin, muscular **interatrial septum;** the ventricles are separated by the thick, muscular **interventricular septum. Atrioventricular valves** (*AV valves*) lie between the atria and ventricles, and **semilunar valves** are located at the bases of the two large vessels leaving the heart. Heart valves (table 21.2) maintain a one-way flow of blood.

Grooved depressions on the surface of the heart indicate the partitions between the chambers and also contain **cardiac vessels** that supply blood to the muscular wall of the heart. The most prominent groove is the *coronary sulcus* that encircles the heart and marks the division between the atria and ventricles. The partition between right and left ventricles is denoted by two (anterior and posterior) *interventricular sulci*.

The following discussion presents the sequence in which blood flows through the atria, ventricles, and valves. It is important to keep in mind that the right side of the heart (right atrium and right ventricle) receives deoxygenated blood (blood low in oxygen) and pumps it to the lungs. The left side of the heart (left atrium and left ventricle) receives oxygenated blood (blood rich in oxygen) from the lungs and pumps it throughout the body.

Right Atrium

The right atrium receives venous blood from the **superior vena cava,** which drains the upper portion of the body, and from the **inferior vena cava,** which drains the lower portion (fig. 21.2). The **coronary sinus** is an additional venous return into the right atrium that receives venous blood from the myocardium of the heart itself.

Right Ventricle

Blood from the right atrium passes through the **right atrioventricular (AV) valve** (also called the *tricuspid valve*) to fill the right ventricle. The right AV valve is characterized by three valve leaflets, or *cusps*. Each cusp is held in position by strong tendinous cords called **chordae tendineae** (*kor'de ten-din'e-e*). The chordae tendineae are secured to the ventricular wall by cone-shaped **papillary muscles.** These structures prevent the valves from everting, like an umbrella in a strong wind, when the ventricles contract and the ventricular pressure increases.

Ventricular contraction causes the right AV valve to close and the blood to leave the right ventricle through the **pulmonary trunk** and to enter the capillaries of the lungs via the **right** and **left pulmonary arteries.** The **pulmonary valve** (also called the *pulmonary semilunar valve*) lies at the base of the pulmonary trunk, where it prevents the backflow of ejected blood into the right ventricle.

Left Atrium

After gas exchange has occurred within the capillaries of the lungs, oxygenated blood is transported to the left atrium through two **right** and two **left pulmonary veins.**

Left Ventricle

The left ventricle receives blood from the left atrium. These two chambers are separated by the **left atrioventricular (AV) valves** (also called the *bicuspid*, or *mitral* [*mi'tral*], *valve*). When the left ventricle is relaxed, the valve is open, allowing blood to flow from the atrium into the ventricle; when the left ventricle contracts, the valve closes. Closing of the valve during ventricular contraction prevents the backflow of blood into the atrium.

The wall of the left ventricle is thicker than that of the right ventricle because the left ventricle bears a greater workload, pumping blood through the entire body. The endocardium of both ventricles is characterized by distinct ridges called **trabeculae carneae** (*tră-bek'yŭ-le kar'ne-e*) (see fig. 21.2*c*). Oxygenated blood leaves the left ventricle through the **ascending portion of the aorta.** The **aortic valve** (also called the *aortic semilunar valve*), located at the base of the ascending portion of the aorta, closes as a result of the pressure of the blood when the left ventricle relaxes, and thus prevents backflow of blood into the relaxed ventricle.

The valves are illustrated and their actions summarized in table 21.2.

Fibrous Skeleton of the Heart

The layer of dense connective tissue between the atria and ventricles is known as the **fibrous skeleton of the heart.** Bundles of myocardial cells (described in chapter 12) in the atria attach to the upper margin of this fibrous skeleton and form a single functioning unit, or **myocardium.** The myocardial cell bundles of the ventricles attach to the lower margin and form a different myocardium. As a result, the myocardia of the atria and ventricles are structurally and functionally separated from each other so that special conducting tissue is needed to carry action potentials from the atria to the ventricles, as will be described in a later section. The connective tissue of the fibrous skeleton also forms rings, called **annuli fibrosi,** around the four heart valves, providing a foundation for the support of the valve cusps.

auricle: L. *auricula*, a little ear
chordae tendineae: L. *chorda*, string; *tendere*, to stretch

semilunar: L. *semi*, half; *luna*, moon
mitral: L. *mitra*, like a bishop's mitre
trabeculae carneae: L. *trabecula*, small beams; *carneus*, flesh

Figure 21.2

The structure of the heart.
(*a*) An anterior view, (*b*) a posterior view, and (*c*) an internal view.

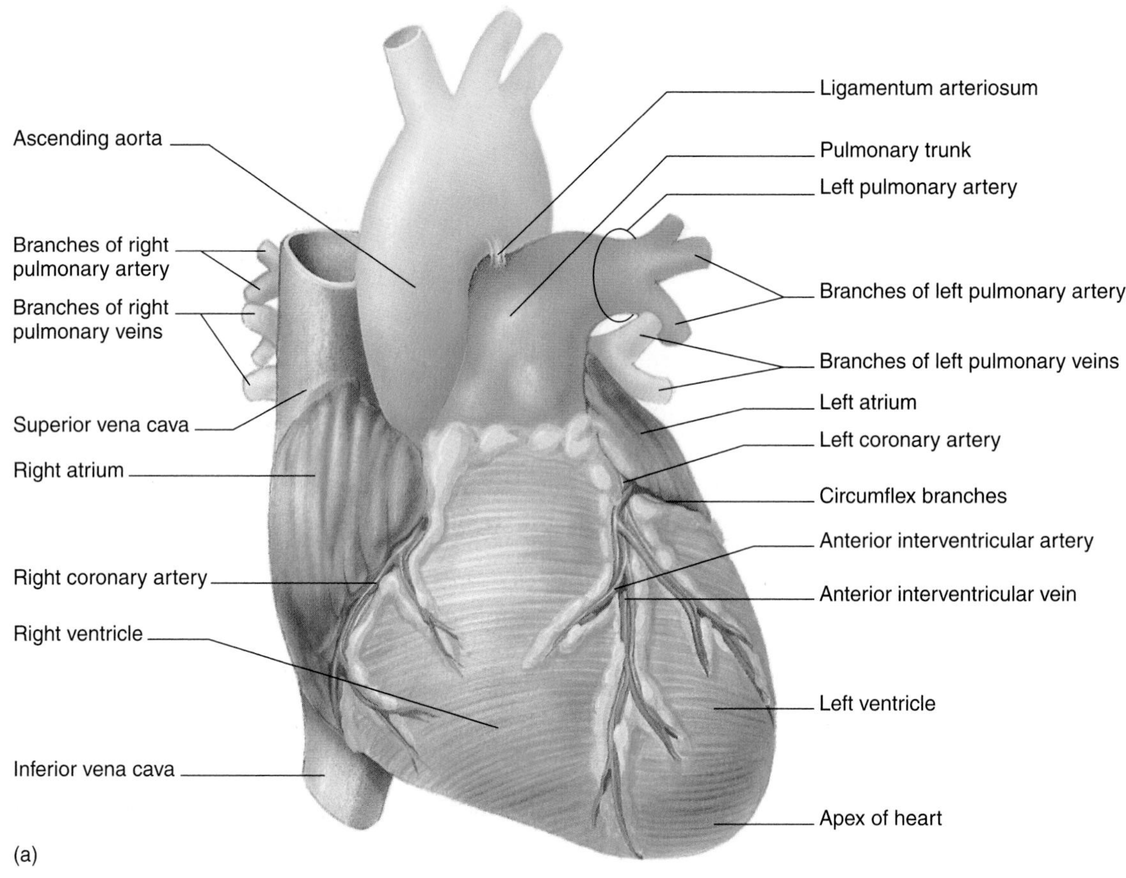
Ascending aorta
Branches of right pulmonary artery
Branches of right pulmonary veins
Superior vena cava
Right atrium
Right coronary artery
Right ventricle
Inferior vena cava
Ligamentum arteriosum
Pulmonary trunk
Left pulmonary artery
Branches of left pulmonary artery
Branches of left pulmonary veins
Left atrium
Left coronary artery
Circumflex branches
Anterior interventricular artery
Anterior interventricular vein
Left ventricle
Apex of heart
(a)

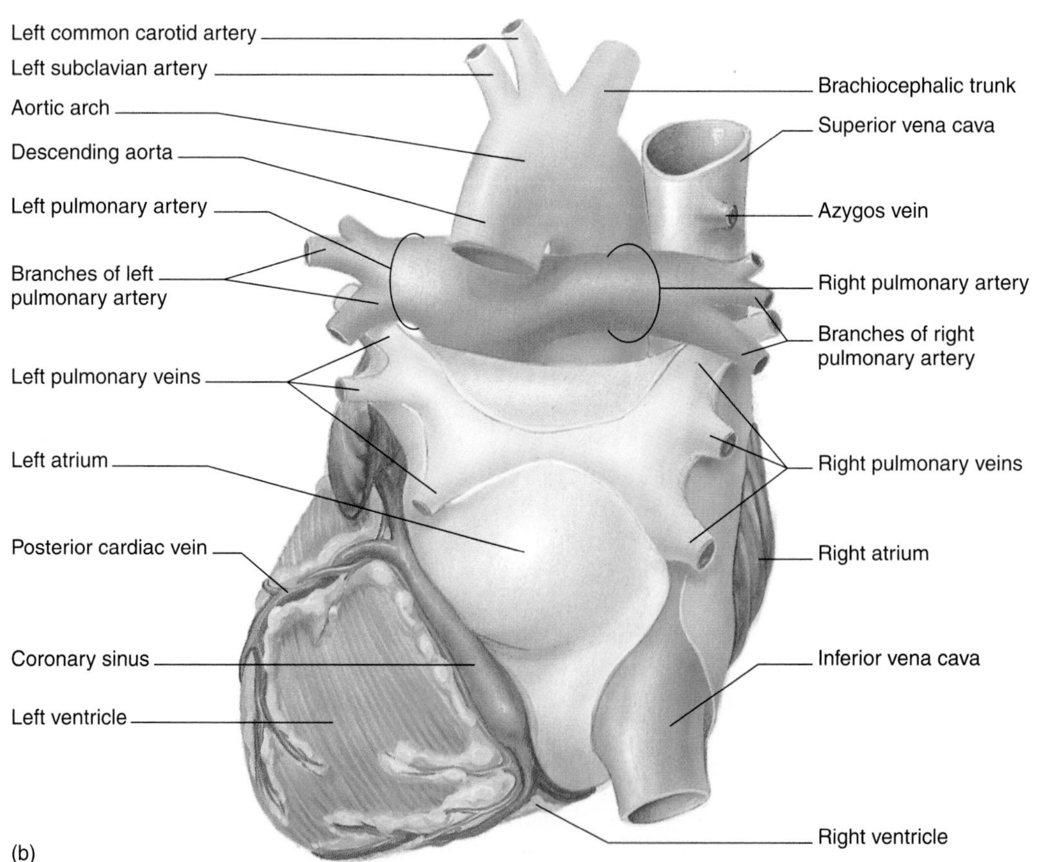
Left common carotid artery
Left subclavian artery
Aortic arch
Descending aorta
Left pulmonary artery
Branches of left pulmonary artery
Left pulmonary veins
Left atrium
Posterior cardiac vein
Coronary sinus
Left ventricle
Brachiocephalic trunk
Superior vena cava
Azygos vein
Right pulmonary artery
Branches of right pulmonary artery
Right pulmonary veins
Right atrium
Inferior vena cava
Right ventricle
(b)

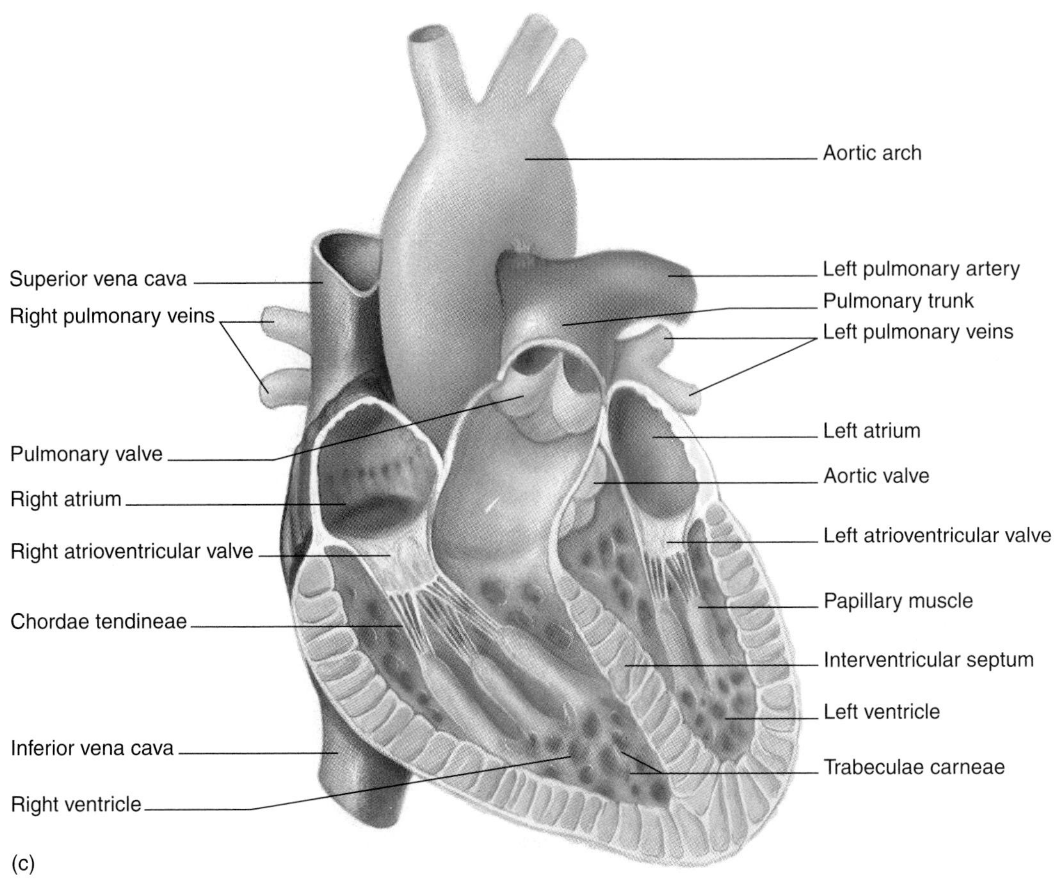

Table 21.2
Valves of the Heart

Valve	Location	Comments
Right atrioventricular (tricuspid) valve	Between right atrium and right ventricle	Composed of three cusps that prevent a backflow of blood from right ventricle into right atrium during ventricular contraction
Pulmonary valve	Entrance to pulmonary trunk	Composed of three half-moon–shaped flaps that prevent a backflow of blood from pulmonary trunk into right ventricle during ventricular relaxation
Left atrioventricular (bicuspid) valve; also called *mitral valve*	Between left atrium and left ventricle	Composed of two cusps that prevent a backflow of blood from left ventricle to left atrium during ventricular contraction
Aortic valve	Entrance to ascending aorta	Composed of three half-moon–shaped flaps that prevent a backflow of blood from aorta into left ventricle during ventricular relaxation

Figure 21.3
A schematic diagram of the circulatory system.

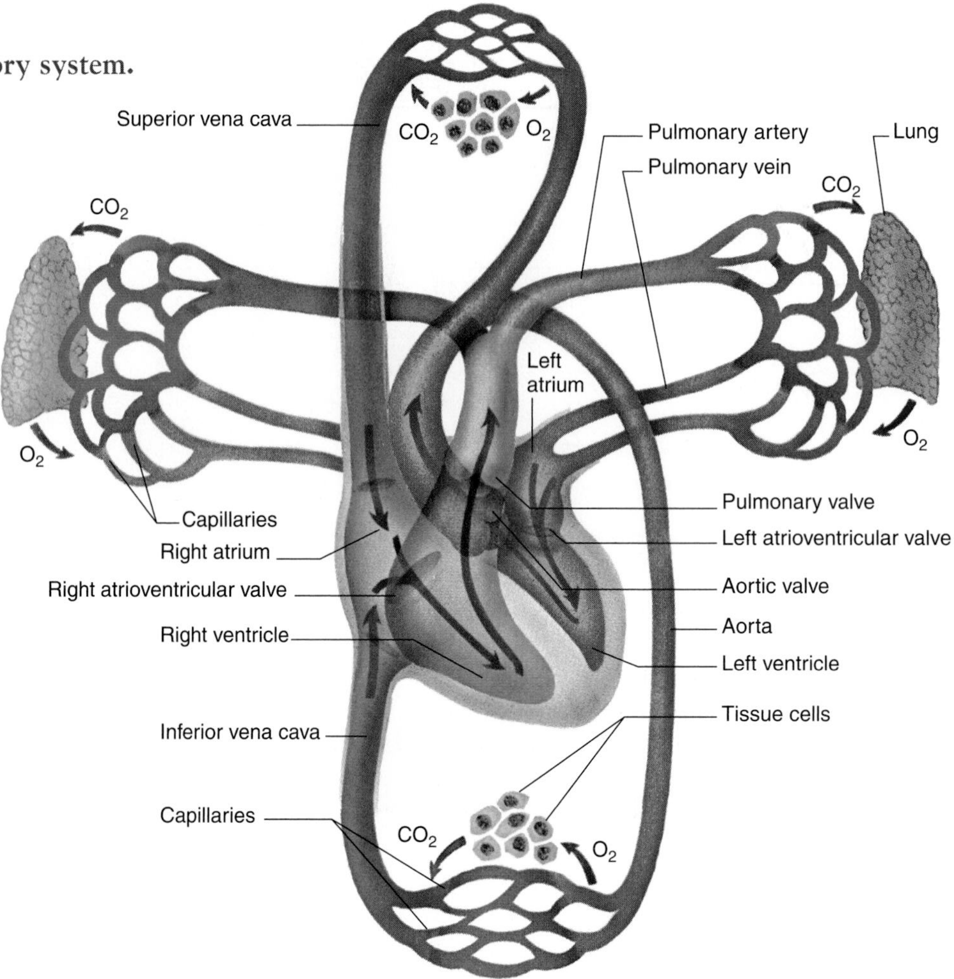

Circulatory Routes

The circulatory routes of the blood are illustrated in figure 21.3. The principal divisions of the circulatory blood flow are the pulmonary and systemic circulations.

The **pulmonary circulation** includes blood vessels that transport blood to the lungs for gas exchange and then back to the heart. It consists of the right ventricle that ejects the blood, the pulmonary trunk with its pulmonary valve, the pulmonary arteries that transport oxygen-depleted blood to the lungs, the pulmonary capillaries within each lung, the pulmonary veins that transport oxygenated blood back to the heart, and the left atrium that receives the blood from the pulmonary veins.

The **systemic circulation** involves all of the vessels of the body that are not part of the pulmonary circulation. It includes the right atrium, the left ventricle, the aorta with its aortic valve, all of the branches of the aorta, all capillaries other than those in the lungs, and all veins other than the pulmonary veins. The right atrium receives all the venous return of oxygen-depleted blood from the systemic veins.

Coronary Circulation

The wall of the heart has its own supply of blood vessels to meet its vital needs. The myocardium is supplied with blood by the **right** and **left coronary arteries** (fig. 21.4*a*). These two vessels arise from the ascending part of the aorta at the level of the aortic (semilunar) valve. The coronary arteries encircle the heart within the atrioventricular sulcus, the depression between the atria and ventricles. Two branches arise from both the right and left coronary arteries to serve the atrial and ventricular walls. The left coronary artery gives rise to the **anterior interventricular artery,** which courses within the anterior interventricular sulcus to serve both ventricles, and the **circumflex artery,** which supplies oxygenated blood to the walls of the left atrium and left ventricle. The right coronary artery gives rise to the **right marginal artery,** which serves the walls of the right atrium and right ventricle, and the **posterior interventricular artery,** which courses through the posterior interventricular sulcus to serve the two ventricles. The main trunk of the right and left coronaries *anastomose* (join together) on the posterior surface of the heart (figs. 21.4*a* and 21.5).

From the capillaries in the myocardium, the blood enters the **cardiac veins.** The course of these two vessels parallels that of the coronary arteries. The cardiac veins, however, have thinner walls and are more superficial than the arteries.

circumflex: L. *circum,* around; *flectere,* to bend
anastomosis: Gk. *anastomoo,* to furnish with a mouth (coming together)

Figure 21.4

Flowchart depicting the coronary circulation.
(*a*) The arterial supply and (*b*) the venous drainage from the myocardium.

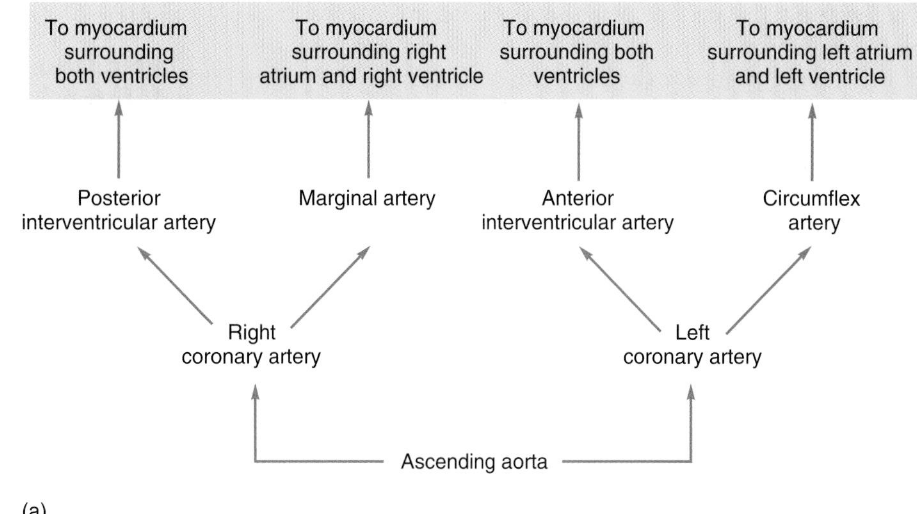

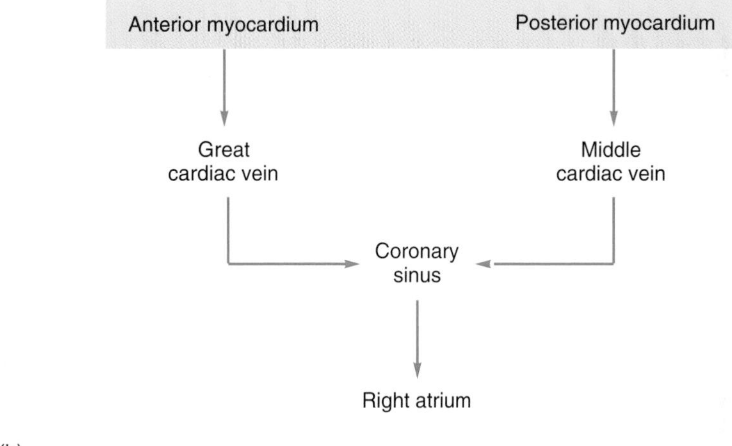

Figure 21.5

Coronary circulation.
(*a*) An anterior view of the arterial supply to the heart and (*b*) an anterior view of the venous drainage.

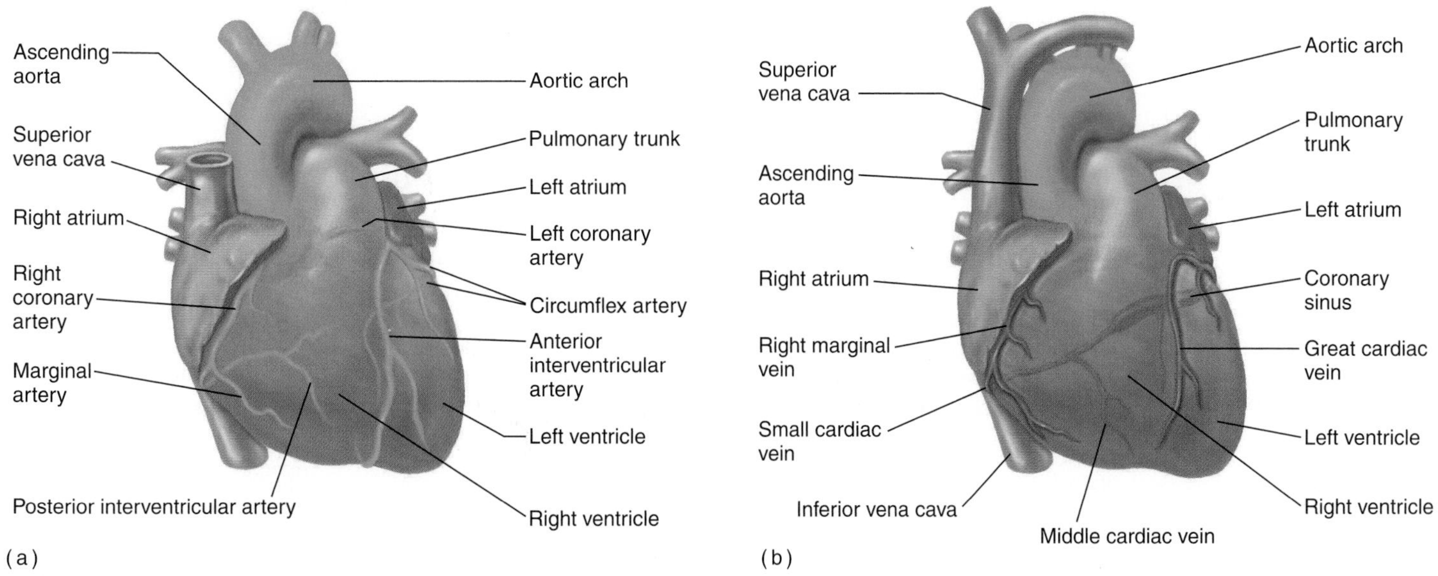

The two principal cardiac veins are the **anterior interventricular vein,** which returns blood from the anterior aspect of the heart, and the **posterior cardiac vein,** which drains the posterior aspect of the heart. These cardiac veins converge to form the **coronary sinus** channel on the posterior surface of the heart. The coronary venous blood then enters the heart through an opening into the right atrium.

Cardiac Cycle, Heart Sounds, and the Electrocardiogram

The two atria fill with blood and then contract simultaneously. This is followed by the simultaneous contraction of both ventricles, which sends blood through the pulmonary and systemic circulations. The AV valves close when the ventricles contract, and the semilunar valves close when the ventricles relax, producing the heart sounds. Contraction of the myocardium results from electrical excitation, which can be recorded as an electrocardiogram.

The *cardiac cycle* refers to the repeating pattern of contraction and relaxation of the heart. The contraction phase is called **systole** and the relaxation phase is called **diastole.** When these terms are used without reference to specific chambers, they refer to contraction and relaxation of the ventricles. It should be noted, however, that the atria also contract and relax. There is an atrial systole and diastole. Atrial contraction occurs toward the end of diastole, when the ventricles are relaxed; when the ventricles contract during systole, the atria are relaxed.

The heart thus has a two-step pumping action. The right and left atria contract almost simultaneously, followed by contraction of the right and left ventricles about 0.1 to 0.2 second later. During the time when both the atria and ventricles are relaxed, the venous return of blood fills the atria. The buildup of pressure that results causes the AV valves to open and blood to flow from atria to ventricles. It has been estimated that the ventricles are about 80% filled with blood even before the atria contract. Contraction of the atria adds the final 20% to the **end-diastolic volume (EDV),** which is the total volume of blood in the ventricles at the end of diastole.

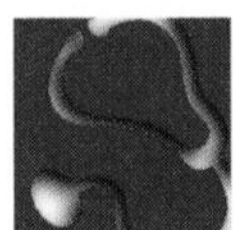

Interestingly, the blood contributed by contraction of the atria does not appear to be essential for life. Elderly people who have atrial fibrillation (a condition in which the atria fail to contract) do not appear to have a higher mortality than those who have normally functioning atria. People with atrial fibrillation, however, become fatigued more easily during exercise because the reduced filling of the ventricles compromises the ability of the heart to sufficiently increase its output and blood flow during exercise. Cardiac output and blood flow during rest and exercise are discussed in chapter 22.

systole: Gk. *systole*, contraction
diastole: Gk. *diastole*, prolonged or expansion

Figure 21.6

The relationship between the heart sounds and the left intraventricular pressure and volume.
Closing of the AV valves occurs during the early part of contraction, when the intraventricular pressure rises prior to ejection of blood. Closing of the semilunar valves occurs at the beginning of ventricular relaxation, just prior to filling. The first and second sounds thus appear during the stages of isovolumetric contraction and isovolumetric relaxation (*iso-* = same).

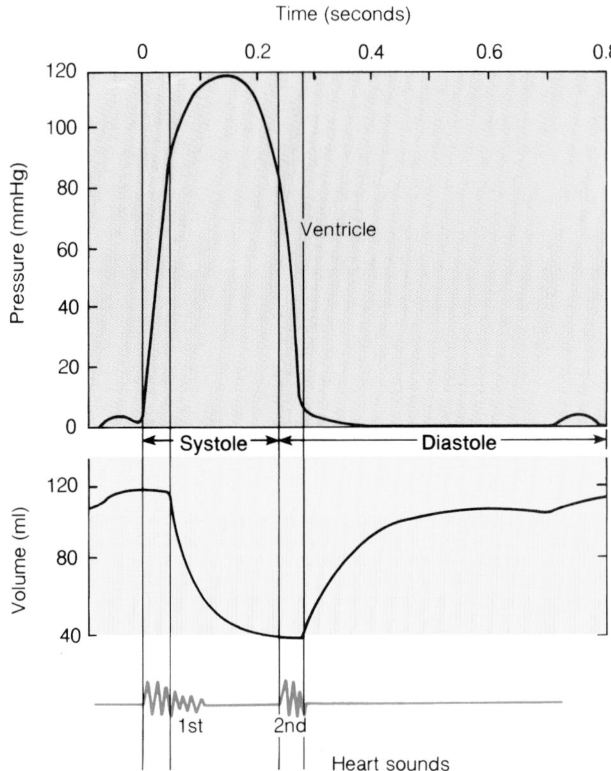

Contraction of the ventricles in systole ejects about two-thirds of the blood that they contain—an amount called the **stroke volume**—leaving one-third of the initial amount left in the ventricles as the **end-systolic volume.** The ventricles then fill with blood during the next cycle. At an average **cardiac rate** of 75 beats per minute, each cycle lasts 0.8 second; 0.5 second is spent in diastole, and systole takes 0.3 second (fig. 21.6).

Pressure Changes during the Cardiac Cycle

When the heart is in diastole, pressure in the systemic arteries averages about 80 mmHg (millimeters of mercury). The following events in the cardiac cycle then occur.

1. As the ventricles begin their contraction, the intraventricular pressure rises, causing the AV valves to snap shut. At this time, the ventricles are neither being filled with blood (because the AV valves are closed) nor ejecting blood (because the intraventricular pressure has not risen sufficiently to open the semilunar valves). This is the phase of *isovolumetric contraction*.

2. When the pressure in the left ventricle becomes greater than the pressure in the aorta, the phase of *ejection* begins as the semilunar valves open. The pressure in the left ventricle and aorta rises to about 120 mmHg (fig. 21.6) as the ventricular volume decreases.
3. As the pressure in the left ventricle falls below the pressure in the aorta, the back pressure causes the semilunar valves to snap shut. The pressure in the aorta falls to 80 mmHg, while pressure in the left ventricle falls to 0 mmHg.
4. During *isovolumetric relaxation*, the AV and semilunar valves are closed and the ventricles are expanding. This phase lasts until the pressure in the ventricles falls below the pressure in the atria.
5. When the pressure in the ventricles falls below the pressure in the atria, a phase of *rapid filling* of the ventricles occurs.
6. *Atrial contraction* (*atrial systole*) empties the final amount of blood into the ventricles immediately prior to the next phase of isovolumetric contraction of the ventricles.

Similar events occur in the right ventricle and pulmonary circulation, but the pressures are lower. The maximum pressure produced at systole in the right ventricle is 25 mmHg, which falls to a low of 8 mmHg at diastole.

Heart Sounds

Closing of the AV and semilunar valves produces sounds that can be heard at the surface of the chest with a stethoscope. These sounds are often described as "lub-dub." The "lub," or **first sound,** is produced by closing of the AV valves during isovolumetric contraction of the ventricles. The "dub," or **second sound,** is produced by closing of the semilunar valves when the pressure in the ventricles falls below the pressure in the arteries. The first sound is thus heard when the ventricles contract at systole; the second sound is heard when the ventricles relax at the beginning of diastole.

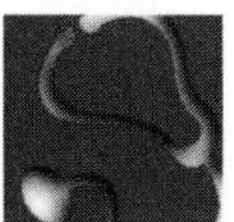

Heart sounds are of clinical importance because they provide information about the condition of the heart valves and other heart problems. Abnormal sounds are referred to as *heart murmurs.* Murmurs are caused by valvular leakage or other structural abnormalities that produce turbulence of the blood as it passes through the heart.

The valves of the heart are positioned directly deep to the sternum, which tends to obscure and dissipate valvular sounds. For this reason, a physician will listen with a stethoscope for the heart sounds at locations designated as **valvular auscultatory areas,** which are named according to the valve that can be detected (fig. 21.7). The *aortic area* is at the right second intercostal space near the sternum. The *pulmonic area* is directly across from the aortic area at the left second intercostal space near the sternum. The *tricuspid* and *bicuspid* (mitral) *areas* are located at the fifth and fourth intercostal spaces respectively, with the bicuspid area more laterally placed. Surface landmarks are extremely important in identifying auscultatory areas.

Figure 21.7

The valvular auscultatory areas.
These areas are the routine stethoscope positions for listening to the heart sounds.

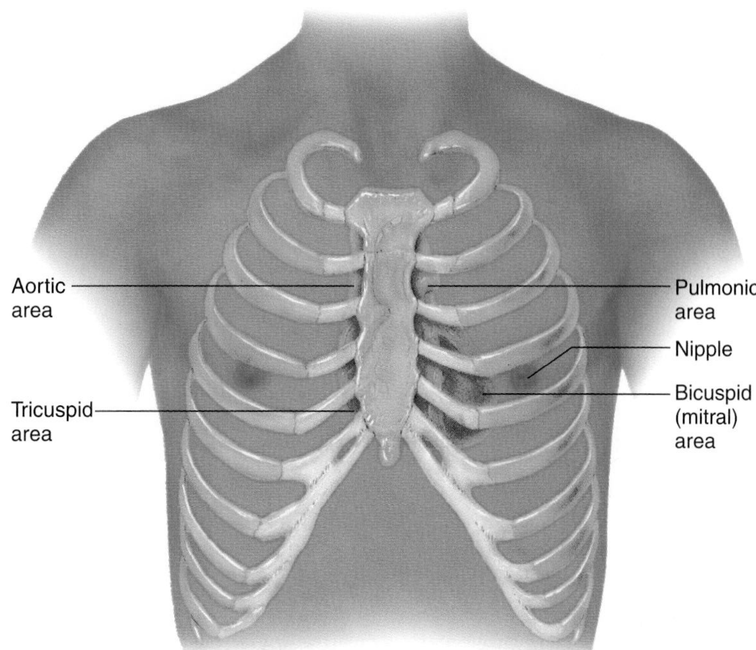

Electrical Activity of the Heart

If the heart of a frog is removed from the body and all neural connections are severed, the heart will still continue to beat as long as the myocardial cells remain alive. The automatic nature of the heartbeat is referred to as *automaticity*. As a result of experiments with isolated myocardial cells and clinical experience with patients who have specific heart disorders, many regions within the heart have been shown to be capable of originating action potentials and functioning as pacemakers.

In a normal heart, however, only one region demonstrates spontaneous electrical activity and by this means functions as a pacemaker. This pacemaker region is called the **sinoatrial (SA) node.** The SA node is located in the right atrium near the opening of the superior vena cava.

Figure 21.8

Pacemaker potentials and action potentials in the SA node.

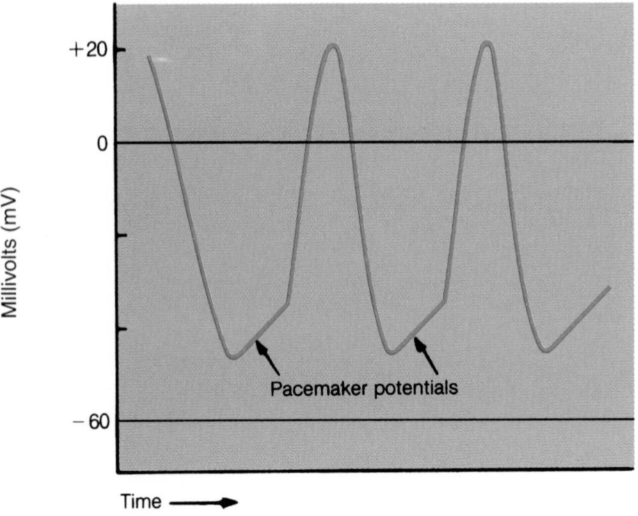

Figure 21.9

An action potential in a myocardial cell from the ventricles.

The plateau phase of the action potential is maintained by a slow inward diffusion of Ca^{2+}. As a result, the duration of the cardiac action potential is about 100 times longer than that of the "spike potential" of an axon.

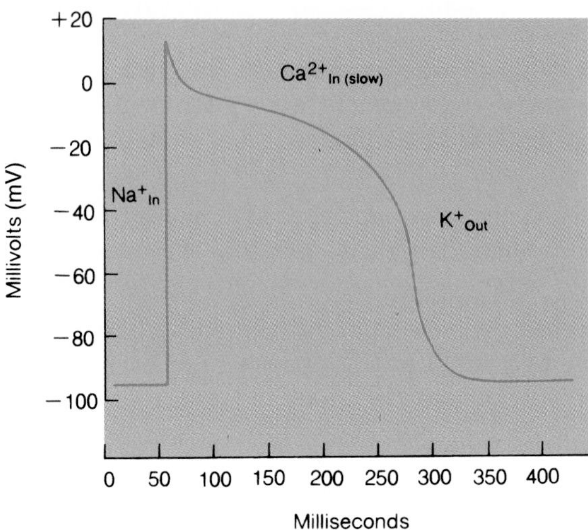

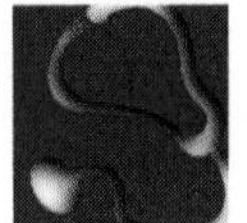

A number of abnormal conditions require the surgical implantation of an artificial pacemaker under the skin. This battery-powered device, about the size of a locket, stimulates or regulates contractions of the heart. The electrodes from the pacemaker are guided by means of a fluoroscope through a vein to the right atrium, through the right AV (tricuspid) valve, and into the right ventricle. The electrodes are fixed to the trabeculae carneae and are in contact with the myocardium of the right ventricle. When these electrodes deliver shocks, either at a continuous pace or on demand (when the heart's own impulse doesn't arrive on time), both ventricles depolarize and contract and then repolarize and relax—just as they do in response to endogenous stimulation.

The cells of the SA node do not keep a resting membrane potential in the manner of resting neurons or skeletal muscle cells. Instead, during the period of diastole, the SA node exhibits a spontaneous depolarization called the **pacemaker potential.** The membrane potential begins at about –60 mV and gradually depolarizes to –40 mV, which is the threshold for producing an action potential in these cells. This spontaneous depolarization is produced by the diffusion of Ca^{2+} through openings in the membrane called *slow calcium channels*. At the threshold level of depolarization, other channels—*fast calcium channels*—open, and Ca^{2+} rapidly diffuses into the cells. The opening of voltage-regulated Na^+ gates, and the inward diffusion of Na^+ that results, may also contribute to the upshoot phase of the action potential in pacemaker cells (fig. 21.8). Repolarization is produced by the opening of K^+ gates and outward diffusion of K^+. Once repolarization to –60 mV has been achieved, a new pacemaker potential begins, culminating again, at the end of diastole, with a new action potential.

Other regions of the heart, including the area around the SA node and the atrioventricular bundle (see fig. 21.10), can potentially produce pacemaker potentials. The rate of spontaneous depolarization of these regions, however, is slower than that of the SA node. Thus, potential pacemaker cells are stimulated by action potentials from the SA node before they can stimulate themselves through their own pacemaker potentials. If action potentials from the SA node are prevented from reaching these regions (through blockage of conduction), they will generate pacemaker potentials at their own rate and serve as sites for the origin of action potentials—they will function as pacemakers. A pacemaker other than the SA node is called an *ectopic pacemaker*, or an *ectopic focus*. From this discussion, it is clear that the rhythm set by an ectopic pacemaker is usually slower than that normally set by the SA node.

Once another myocardial cell has been stimulated by action potentials originating in the SA node, it produces its own action potential. The majority of myocardial cells have resting membrane potentials of about –90 mV. When stimulated by action potentials from a pacemaker region, these cells become depolarized to threshold, at which point their voltage-regulated Na^+ gates open. The upshoot phase of the action potential of nonpacemaker cells is due to the inward diffusion of Na^+. Following the rapid reversal of the membrane polarity, the membrane potential quickly declines to about –10 to –20 mV. Unlike the action potential of other cells, however, this level of depolarization is maintained for 200–300 msec before repolarization (fig. 21.9). This *plateau phase* results from a slow inward diffusion of Ca^{2+}, which balances a slow outward diffusion of cations. Rapid repolarization at the end of the plateau phase is achieved, as in other cells, by the opening of K^+ gates and the rapid outward diffusion of K^+ that results.

Figure 21.10

(a) The conduction system of the heart and (b) the electrocardiogram.
The electrocardiogram indicates the conduction of electrical impulses through the heart and measures and records both the intensity of this electrical activity (in millivolts) and the time intervals involved.

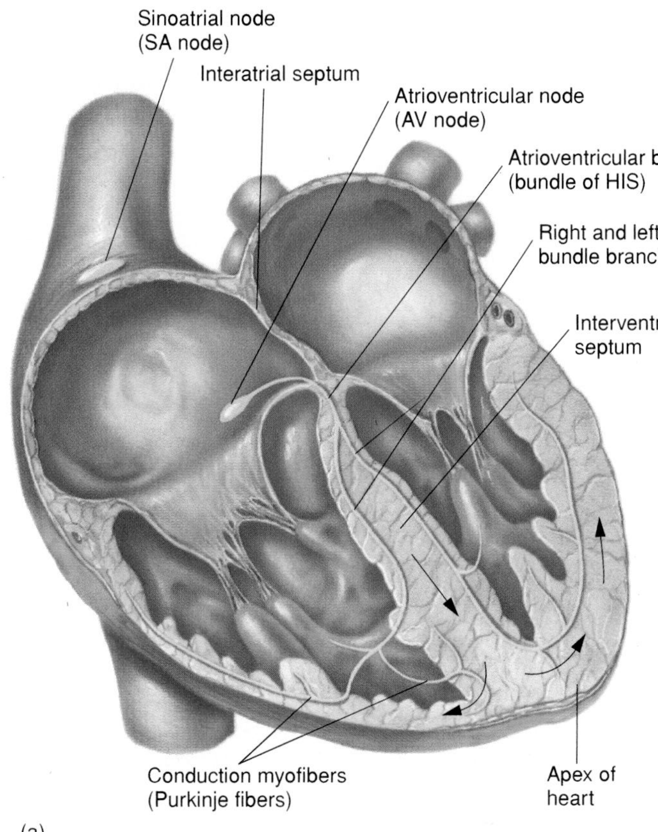

(a)

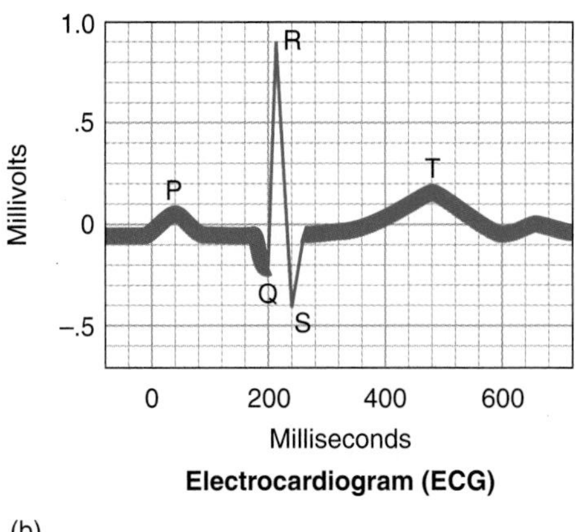

(b)

Conducting Tissue of the Heart

Action potentials that originate in the SA node spread to adjacent myocardial cells of the right and left atria through the gap junctions between these cells. Since the myocardium of the atria are separated from the myocardium of the ventricles by the fibrous skeleton of the heart, however, the impulse cannot be conducted directly from the atria to the ventricles. Specialized conducting tissue, composed of modified myocardial cells, is thus required. These specialized myocardial cells form the AV node, atrioventricular bundle, and conduction myofibers.

Once the impulse spreads through the atria, it passes to the **atrioventricular (AV) node,** which is located on the inferior portion of the interatrial septum (fig. 21.10). The AV node is smaller than the SA node, averaging 5 mm in length, 2 mm in width, and 0.5 mm in depth. From here, the impulse continues through the **atrioventricular bundle** (bundle of His), beginning at the top of the interventricular septum. This conducting tissue pierces the fibrous skeleton of the heart and continues to descend along the interventricular septum. The atrioventricular bundle divides into right and left bundle branches, which are continuous with the **conduction myofibers** (Purkinje fibers) within the ventricular walls. Stimulation of these fibers causes both ventricles to contract simultaneously and eject blood into the pulmonary and systemic circulation.

Conduction of the Impulse

Action potentials from the SA node spread very quickly—at a rate of 0.8 to 1.0 m/sec—across the myocardial cells of both atria. The conduction rate then slows considerably as the impulse passes into the AV node. Slow conduction of impulses (0.03–0.05 m/sec) through the AV node accounts for over half of the time delay between excitation of the atria and ventricles. After the impulses spread through the AV node, the conduction rate increases greatly in the atrioventricular bundle and reaches very high velocities (5 m/sec) in the conduction myofibers. As a result of this rapid conduction of impulses, ventricular contraction follows the contraction of the atria by only about 0.1 to 0.2 second, as mentioned earlier in the chapter.

Unlike skeletal muscles, the heart cannot sustain a contraction. This is because the atria and ventricles behave as if each were composed of only one muscle cell. The entire myocardium of each is electrically stimulated as a single unit and contracts as a unit. This contraction, which corresponds in time to the long action potential of myocardial cells and lasts almost 300 msec, is analogous to the twitch produced by a single skeletal muscle fiber (which lasts only 20 to 100 msec in comparison). The heart normally cannot be stimulated again until after it has relaxed from its previous contraction,

bundle of His: from Wilhelm His, Jr., Swiss physician, 1863–1934

Purkinje fibers: from Johannes E. von Purkinje, Bohemian anatomist, 1787–1869

Figure 21.11

The myocardial action potential and the duration of heart contraction.

The time course for the myocardial action potential (*A*) is compared with the duration of contraction (*B*). Notice that the long action potential results in a correspondingly long absolute refractory period (ARP) and relative refractory period (RRP). These refractory periods last almost as long as the contraction, so that the myocardial cells cannot be stimulated again until they have completed their contraction from the first stimulus.

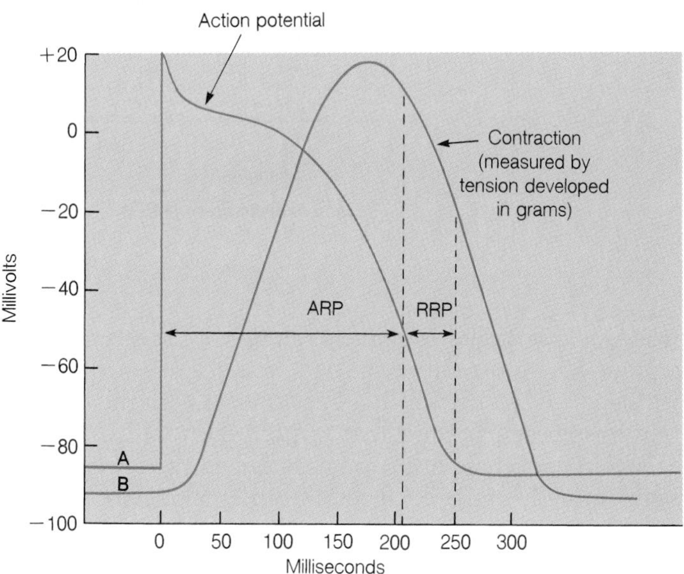

since myocardial cells have *long refractory periods* (fig. 21.11) that correspond to the long duration of their action potentials. Summation of contractions is thus prevented, and the myocardium must relax after each contraction. The rhythmic pumping action of the heart is thus ensured.

The Electrocardiogram

A pair of surface electrodes placed directly on the heart will record a repeating pattern of potential changes. As action potentials spread from the atria to the ventricles, the voltage measured between these two electrodes will vary in a way that provides a "picture" of the electrical activity of the heart. By changing the position of the recording electrodes, the observer can gain a more complete picture of the electrical events.

The body is a good conductor of electricity because tissue fluids contain a high concentration of ions that move (creating a current) in response to potential differences. Potential differences generated by the heart are thus conducted to the body surface, where they can be recorded by surface electrodes placed on the skin. The recording is called an **electrocardiogram (ECG** or **EKG)** (fig. 21.10); the recording device is called an **electrocardiograph.**

Each cardiac cycle produces three distinct ECG waves, designated P, QRS, and T. It should be noted that these waves are not action potentials; they represent changes in potential between two regions on the surface of the heart. These potential differences are produced by the composite effects of action potentials in many myocardial cells. For example, the spread of depolarization through the atria causes a potential difference that is indicated by an upward deflection of the ECG line. When about half the mass of the atria is depolarized, this upward deflection reaches a maximum value because the potential difference between the depolarized and unstimulated portions of the atria is at a maximum. When the entire mass of the atria is depolarized, the ECG returns to baseline because all regions of the atria have the same polarity. The spread of atrial depolarization thus creates the **P wave.**

Conduction of the impulse into the ventricles similarly creates a potential difference that results in a sharp upward deflection of the ECG line, which then returns to the baseline as the entire mass of the ventricles becomes depolarized. The spread of the depolarization into the ventricles is thus represented by the **QRS wave.** The atria repolarize during this period, but the event is hidden by the greater depolarization occurring in the ventricles. Finally, repolarization of the ventricles produces the **T wave** (fig. 21.12).

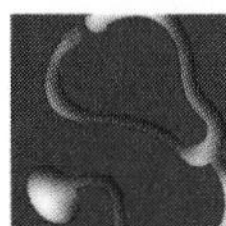

The time interval from the beginning of the P wave to the beginning of the Q wave, which is the beginning of the QRS complex, is known as the PR interval. This is a measure of the time required for the impulse to pass through the AV node. Damage to the AV node, therefore, can cause this interval to be prolonged. A PR interval greater than 0.20 second indicates *first-degree AV node block.* If the PR interval becomes too great, the QRS and T waves may not follow a given P wave, and may only appear with every second or third P wave. This condition is called *second-degree AV node block.* The condition of *third-degree AV node block* occurs when none of the atrial waves enter the ventricles. In this case, the ventricles beat according to a slower rhythm set by an ectopic pacemaker in the ventricles.

A depression in the ST segment of the ECG may indicate *myocardial ischemia,* or lack of sufficient blood flow to the heart muscle. This may not be evident at rest but may appear when the metabolism of the heart is increased during exercise. For this reason, an ECG may be performed while a person is walking on a treadmill. Some other abnormal conditions that may be detected by an ECG are described in "Clinical Considerations" at the end of this chapter.

Correlation of the ECG with Heart Sounds

Depolarization of the ventricles, as indicated by the QRS wave, stimulates contraction by promoting the uptake of Ca^{2+} into the regions of the sarcomeres. The QRS wave is thus seen to occur at the beginning of systole. The rise in intraventricular pressure that results causes the AV valves to close, so that the first heart sound (S_1, or "lub") is produced immediately after the QRS wave (fig. 21.13).

Repolarization of the ventricles, as indicated by the T wave, occurs at the same time that the ventricles relax at the beginning of diastole. The resulting fall in intraventricular pressure causes the aortic and pulmonary semilunar valves to close, so that the second heart sound (S_2, or "dub") is produced shortly after the T wave in an electrocardiogram begins.

Figure 21.12

The conduction of electrical impulses in the heart, as indicated by the electrocardiogram.
The direction of the arrows in (*e*) indicates that depolarization of the ventricles occurs from the inside (endocardium) out (to the epicardium), whereas the arrows in (*g*) indicate that repolarization of the ventricles occurs in the opposite direction.

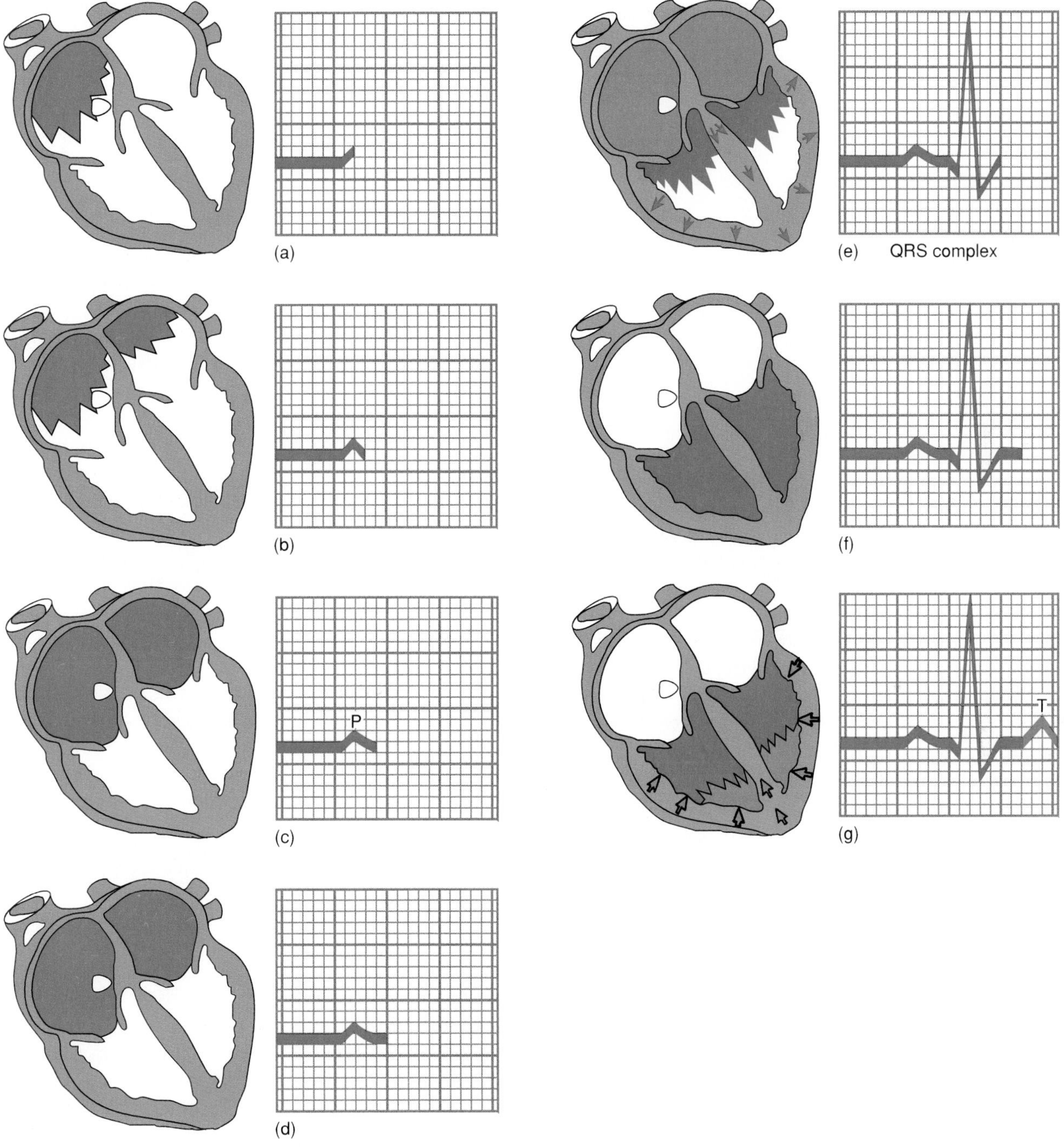

Figure 21.13

The relationship between the heart sounds and the electrocardiogram during the cardiac cycle.
The QRS wave (representing depolarization of the ventricles) occurs at the beginning of systole, whereas the T wave (representing repolarization of the ventricles) occurs at the beginning of diastole.

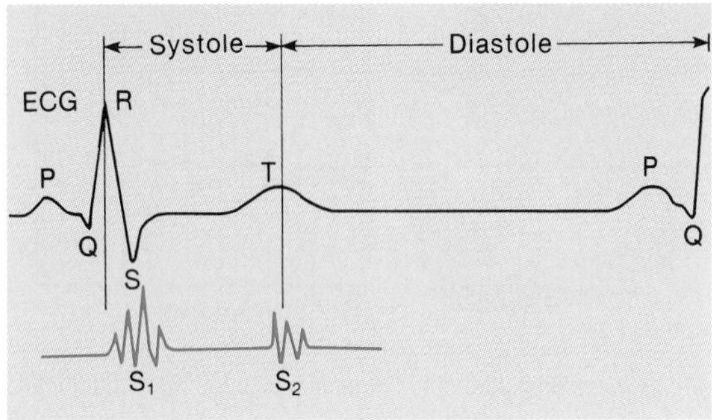

Blood Vessels

The structure of arteries and veins allows them to transport blood from the heart to the capillaries and from the capillaries back to the heart. The structure of capillaries permits the exchange of blood plasma and dissolved molecules between the blood and surrounding tissues.

Blood vessels form a closed tubular network that permits blood to flow from the heart to all the living cells of the body and then back to the heart. Blood leaving the heart passes through vessels of progressively smaller diameters referred to as arteries, arterioles, and capillaries. Capillaries are microscopic vessels that join the arterial flow to the venous flow. Blood returning to the heart from the capillaries passes through vessels of progressively larger diameters called venules and veins.

The walls of arteries and veins are composed of three layers, or tunics, as shown in figure 21.14.

1. The **tunica externa,** or *adventitia*, is the outermost layer composed of loose connective tissue.
2. The **tunica media,** the middle layer, is composed of smooth muscle. The tunica media of arteries has variable amounts of elastic fibers.
3. The **tunica interna,** the innermost layer, is composed of simple squamous epithelium and elastic fibers composed of *elastin*.

The simple squamous epithelium of the tunica interna is referred to as the **endothelium.** This thin coat of cells lines the inner wall of all blood vessels. Capillaries consist of only endothelium, supported by a basement membrane.

Although arteries and veins have the same basic structure, there are some important differences between the two types of vessels. Arteries have more muscle in proportion to their diameter than comparably sized veins. Also they appear rounder than veins in cross section. Veins are usually partially collapsed because they are not usually filled to capacity. They can stretch when they receive more blood, and thus function as reservoirs, or capacitance, vessels. In addition, many veins have valves, which are absent in arteries.

Arteries

In the tunica media of large arteries, there are numerous layers of elastin fibers between the smooth muscle cells. Thus, the large arteries expand when the pressure of the blood rises as a result of the heart's contraction; they recoil, like a stretched rubber band, when blood pressure falls during relaxation of the heart. This elastic recoil helps to produce a smoother, less pulsatile flow of blood through the smaller arteries and arterioles.

Small arteries and arterioles are less elastic than the larger arteries and have a thicker layer of smooth muscle in proportion to their diameter. Unlike the larger *elastic arteries*, therefore, the smaller *muscular arteries* retain a relatively constant diameter as the pressure of the blood rises and falls during the heart's pumping activity. Since small muscular arteries and arterioles have narrow lumina, they provide the greatest resistance to blood flow through the arterial system.

Small muscular arteries that are 100 µm or less in diameter branch to form smaller arterioles (20 to 30 mm in diameter). In some tissues, blood from the arterioles can enter the venules directly through **thoroughfare channels** (metarterioles) that form *vascular shunts* (fig. 21.15). In most cases, however, blood from arterioles passes into capillaries. Capillaries are the narrowest of blood vessels (7 to 10 µm in diameter), and serve as the functional units of the circulatory system. It is across their walls that exchanges of gases, (O_2 and CO_2), nutrients, and wastes between the blood and the tissues take place.

Capillaries

The arterial system branches extensively (table 21.3) to deliver blood to over 40 billion capillaries in the body. So extensive is this branching that scarcely any cell in the body is more than a fraction of a millimeter away from any capillary; moreover, the tiny capillaries provide a total surface area of 1,000 square miles for exchanges between blood and tissue fluid.

Despite their large number, at any given time capillaries contain only about 250 ml of blood out of a total blood volume of about 5,000 ml (most blood is contained in the venous system). The amount of blood flowing through a particular capillary bed is determined in part by the action of the **precapillary sphincter muscles** (fig. 21.15). These muscles allow only 5% to 10% of the capillary beds in skeletal muscles, for

tunic: L. *tunica*, covering or coat
endothelium: Gk. *endo*, within; *thelium*, to cover

Figure 21.14

Relative thickness and composition of the tunics in comparable arteries and veins.

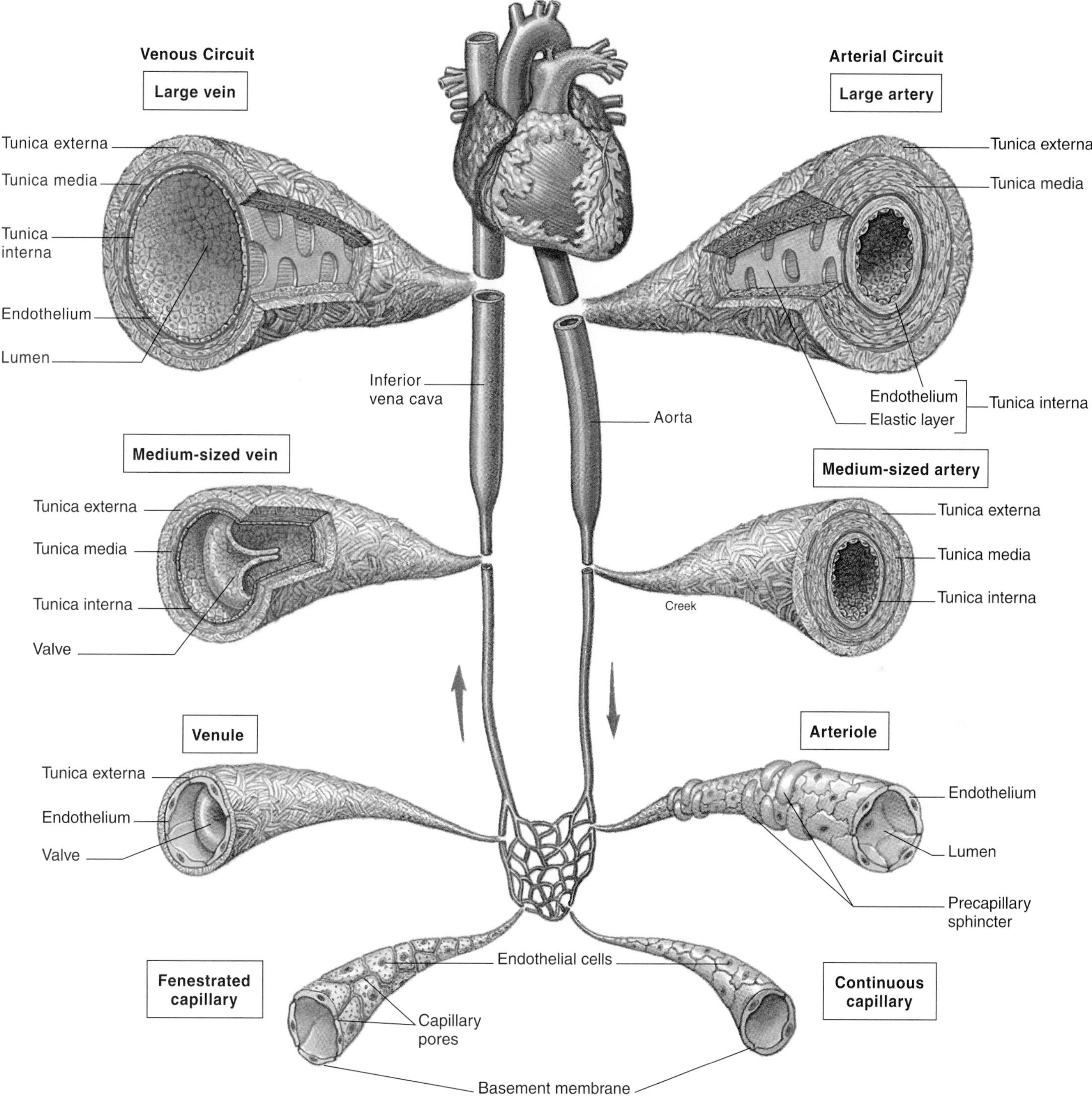

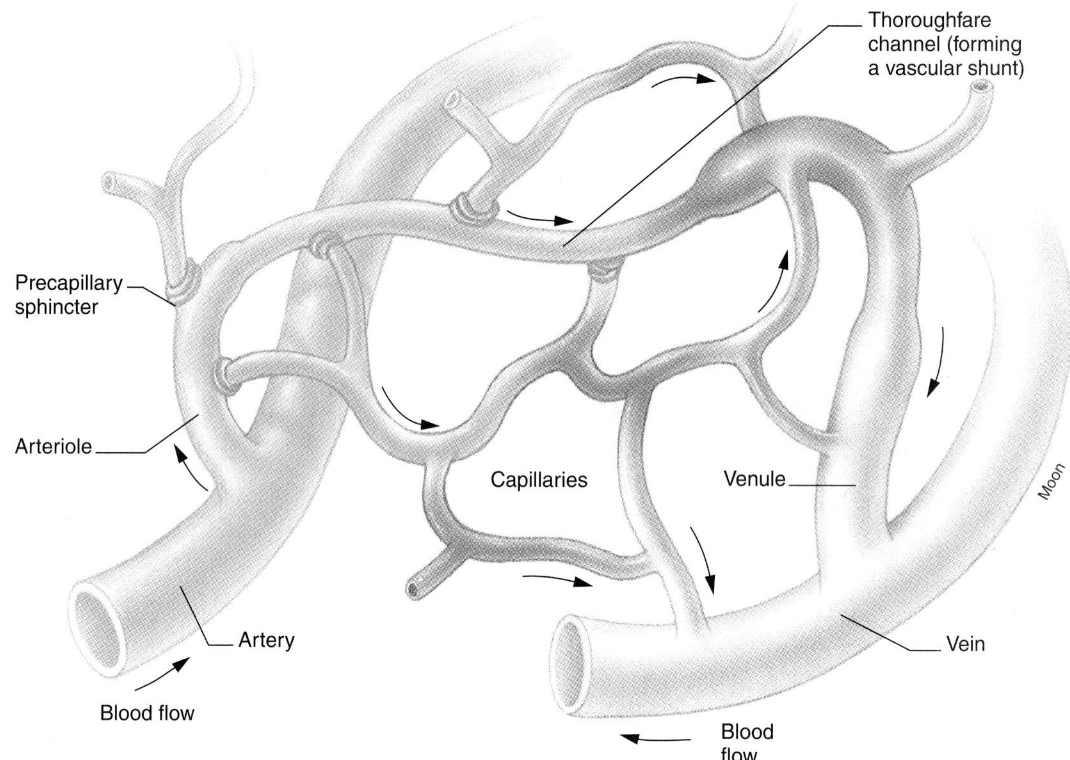

Figure 21.15

Microcirculation at the capillary level.
Thoroughfare channels form vascular shunts, providing paths of least resistance between arterioles and venules. Precapillary sphincter muscles regulate the flow of blood through the capillaries.

Table 21.3

Characteristics of the Vascular Supply to the Mesenteries in a Dog (Pattern Is Similar in a Human)

Kind of Vessel	Diameter (µm)	Number	Total cross-sectional area (cm^2)
Aorta	10.0	1	0.8
Large arteries	3.0	40	3.0
Main artery branches	1.0	600	5.0
Terminal branches	0.6	1,800	5.0
Arterioles	0.02	40,000,000	125.0
Capillaries	0.008	1,200,000,000	600.0
Venules	0.03	80,000,000	570.0
Terminal veins	1.5	1,800	30.0
Main venous branches	2.4	600	27.0
Large veins	6.0	40	11.0
Vena cavae	12.5	2	1.2

Source: Data from Malcolm S. Gordon, *Animal Physiology: Principles and Adaptations*, 3d ed., Macmillan Publishing Company. Copyright © 1977 by Malcolm S. Gordon.

example, to be open at rest. Blood flow to an organ is regulated by the action of these precapillary sphincters and by the degree of resistance to blood flow provided by the small arteries and arterioles in the organ.

Unlike the vessels of the arterial and venous systems, the walls of capillaries are composed of just one cell layer—a simple squamous epithelium, or endothelium (fig. 21.16). The absence of smooth muscle and connective tissue layers allows for a more rapid exchange of materials between the blood and the tissues.

Types of Capillaries

There are several different types of capillaries, distinguished by significant differences in structure. In terms of their endothelial lining, these capillary types include those that are continuous, those that are discontinuous, and those that are fenestrated (*fen′ĭ-stra″tid*).

Continuous capillaries are those in which adjacent endothelial cells are tightly joined together. These are found in muscles, lungs, adipose tissue, and in the central nervous system. The fact that continuous capillaries in the CNS lack intercellular channels contributes to the blood-brain barrier. Continuous capillaries in other organs have narrow intercellular channels (about 40 to 45 Å wide) that permit the passage of molecules other than protein between the capillary blood and tissue fluid.

The examination of endothelial cells with an electron microscope has revealed the presence of pinocytotic vesicles, which suggests that the intracellular transport of material may occur across the capillary walls. This type of transport appears

Figure 21.16

An electron micrograph of a capillary in the heart.
Notice the thin intercellular channel (middle left) and the capillary wall, composed of only one cell layer. Arrows show some of the many pinocytic vesicles.

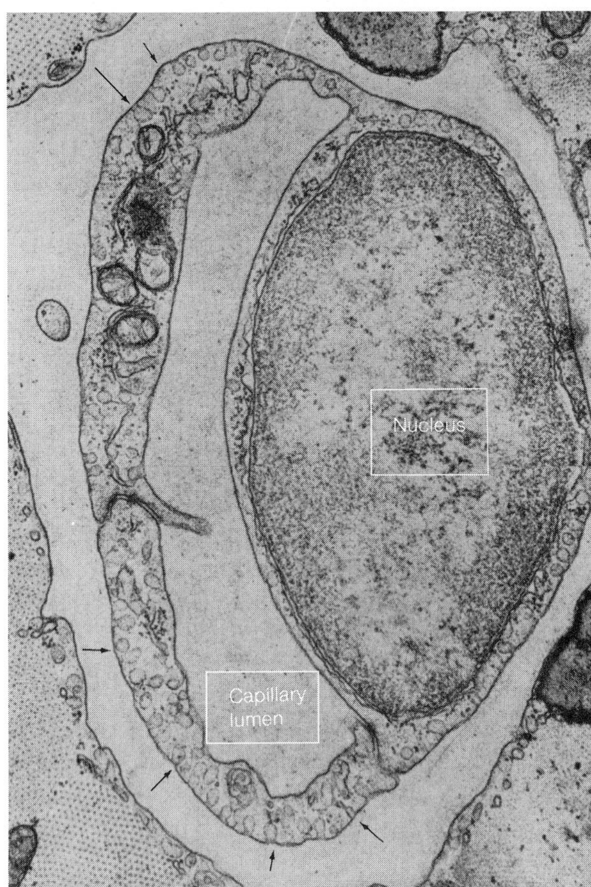

Figure 21.17

The action of the one-way venous valves.
The contraction of skeletal muscles helps to pump blood toward the heart, but the flow of blood away from the heart is prevented by closure of the venous valves.

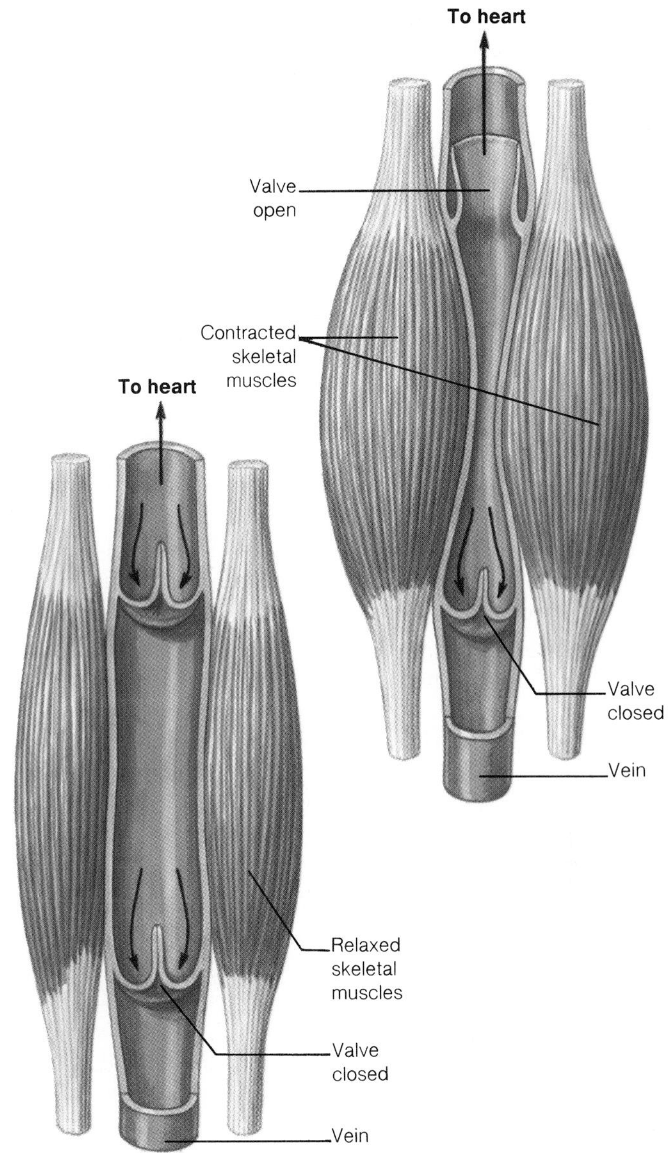

to be the only available mechanism of capillary exchange within the central nervous system and may account, in part, for the selective nature of the blood-brain barrier.

Discontinuous capillaries are found in the bone marrow, liver, and spleen. The space between endothelial cells is so great that these capillaries look like little cavities (*sinusoids*) in the organ. **Fenestrated capillaries** occur in the kidneys, endocrine glands, and small intestine. These capillaries are characterized by wide intercellular pores (800 to 1000 Å) that are covered by a layer of mucoprotein, which may serve as a diaphragm.

Veins

Veins are vessels that carry blood from capillaries back to the heart. The blood is delivered from microscopic vessels called venules into progressively larger vessels that empty into the large veins. The average pressure in the veins is only 2 mmHg, compared to a much higher average arterial pressure of about 100 mmHg. These pressures represent the hydrostatic pressure that the blood exerts on the walls of the vessels.

The low venous pressure is insufficient to return blood to the heart, particularly from the lower limbs. Veins, however, pass between skeletal muscle groups that produce a massaging action as they contract (fig. 21.17). As the veins are squeezed by contracting skeletal muscles, a one-way flow of blood to the heart is ensured by the presence of **venous valves.** The ability of these valves to prevent the flow of blood away from the heart was demonstrated in the seventeenth century by William Harvey (see fig. 1.4). After applying a tourniquet to a subject's arm, Harvey found that he could push the blood in a bulging vein toward the heart but not in the reverse direction.

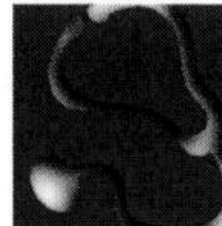

The effect of the massaging action of skeletal muscles on venous blood flow is often described as the *skeletal muscle pump.* The rate of venous return to the heart is dependent, in large part, on the action of skeletal muscle pumps. When these pumps are less active—for example, when a person stands still or is bedridden—blood accumulates in the veins and causes the veins to bulge. When a person is more active, blood returns to the heart at a faster rate and less is left in the venous system.

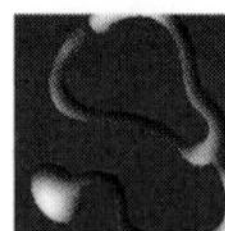

The accumulation of blood in the veins of the legs over a long period of time, as may occur in people with occupations that require standing still all day, can cause the veins to stretch to the point where the venous valves are no longer efficient. This can produce *varicose* (*var-ĭ-kōs*) *veins.* During walking, the movements of the foot activate the soleus muscle pump. This effect can be produced in bedridden people by extending and flexing the ankle joints.

Action of the skeletal muscle pumps aid the return of venous blood from the lower limbs to the large abdominal veins. Movement of venous blood from abdominal to thoracic veins, however, is aided by an additional mechanism—breathing. When a person inhales, the diaphragm—a dome-shaped muscular sheet separating the thoracic and abdominal cavities—contracts. As it contracts, it moves inferiorly and flattens out, so that it protrudes more into the abdomen. This has the dual effect of increasing the pressure in the abdomen, thus squeezing the abdominal veins, and decreasing the pressure in the thoracic cavity. The pressure difference in the veins created by this inspiratory movement of the diaphragm forces blood into the thoracic veins that return the venous blood to the heart.

Principal Arteries of the Body

The aorta ascends from the left ventricle to a position just above the heart, where it arches to the left and then descends through the thorax and abdomen. Branches of the aorta carry oxygenated blood to all the cells of the body.

Contraction of the left ventricle forces oxygenated blood into the arteries of the systemic circulation. The principal arteries of the body are shown in figure 21.18 and are described by region and identified in order from largest to smallest, or as the blood flows through the system. The major systemic artery is the **aorta** (*a-or'ta*), from which all the primary systemic arteries arise.

Aortic Arch

The systemic vessel that ascends from the left ventricle of the heart is called the **ascending portion of the aorta.** The **right** and **left coronary arteries,** which serve the myocardium of the heart with blood, are the only branches that arise from the ascending aorta. The aorta arches to the left and posteriorly over the pulmonary arteries as the **aortic arch** (fig. 21.19). Three vessels arise from the aortic arch: the **brachiocephalic** (*bra"ke-o-sĕ-fal'ik*) **trunk,** the **left common carotid** (*kă-rot'id*) **artery,** and the **left subclavian artery.**

The brachiocephalic trunk is the first vessel to branch from the aortic arch and, as its name suggests, supplies blood to the structures of the shoulder, upper extremity, and head on the right side of the body. It is a short vessel, rising superiorly through the mediastinum to a point near the junction of the sternum and the right clavicle. There it branches into the **right common carotid artery,** which extends to the right side of the neck and head, and the **right subclavian artery,** which carries blood to the right shoulder and upper extremity.

The remaining two branches from the aortic arch are the **left common carotid** and **left subclavian arteries.** The left common carotid artery transports blood to the left side of the neck and head, while the left subclavian artery supplies the left shoulder and upper extremity.

Arteries of the Neck and Head

The common carotid arteries course upward in the neck along lateral sides of the trachea (fig. 21.20). Each common carotid artery branches into the **internal** and **external carotid arteries** slightly below the angle of the mandible. By pressing gently in this area, a pulse can be detected. At the base of the internal carotid artery is a slight dilation called the **carotid sinus.** The carotid sinus contains *baroreceptors,* which monitor blood pressure. Surrounding the carotid sinus are the **carotid bodies,** small neurovascular organs that contain *chemoreceptors* (*ke"mo-re-sep'torz*), which respond to chemical changes in the blood.

Blood Supply to the Brain

The brain is supplied with arterial blood that arrives through four vessels. These vessels eventually unite on the inferior surface of the brain in the area surrounding the pituitary gland (fig. 21.21). The four vessels are the paired *internal carotid arteries* and the paired *vertebral arteries.* The value of having four separate vessels coming together at one location is that if one becomes occluded, the three alternate routes may still provide an adequate blood supply to the brain.

The **vertebral arteries** arise from the subclavian arteries at the base of the neck (see fig. 21.20). They pass superiorly through the transverse foramina of the cervical vertebrae and enter the skull through the foramen magnum. Within the cranium, the two vertebral arteries unite to form the **basilar artery** at the level of the pons. The basilar artery ascends along the inferior surface of the brain stem and terminates by forming two **posterior cerebral arteries** that supply the posterior portion of the cerebrum. The **posterior communicating arteries** are branches that arise from the posterior cerebral arteries

carotid: Gk. *karotikos,* stupefying (a state that can be induced by finger pressure in the region of the carotid sinus)

Figure 21.18

Principal arteries of the body.
(*a.* = artery; *aa.* = arteries.)

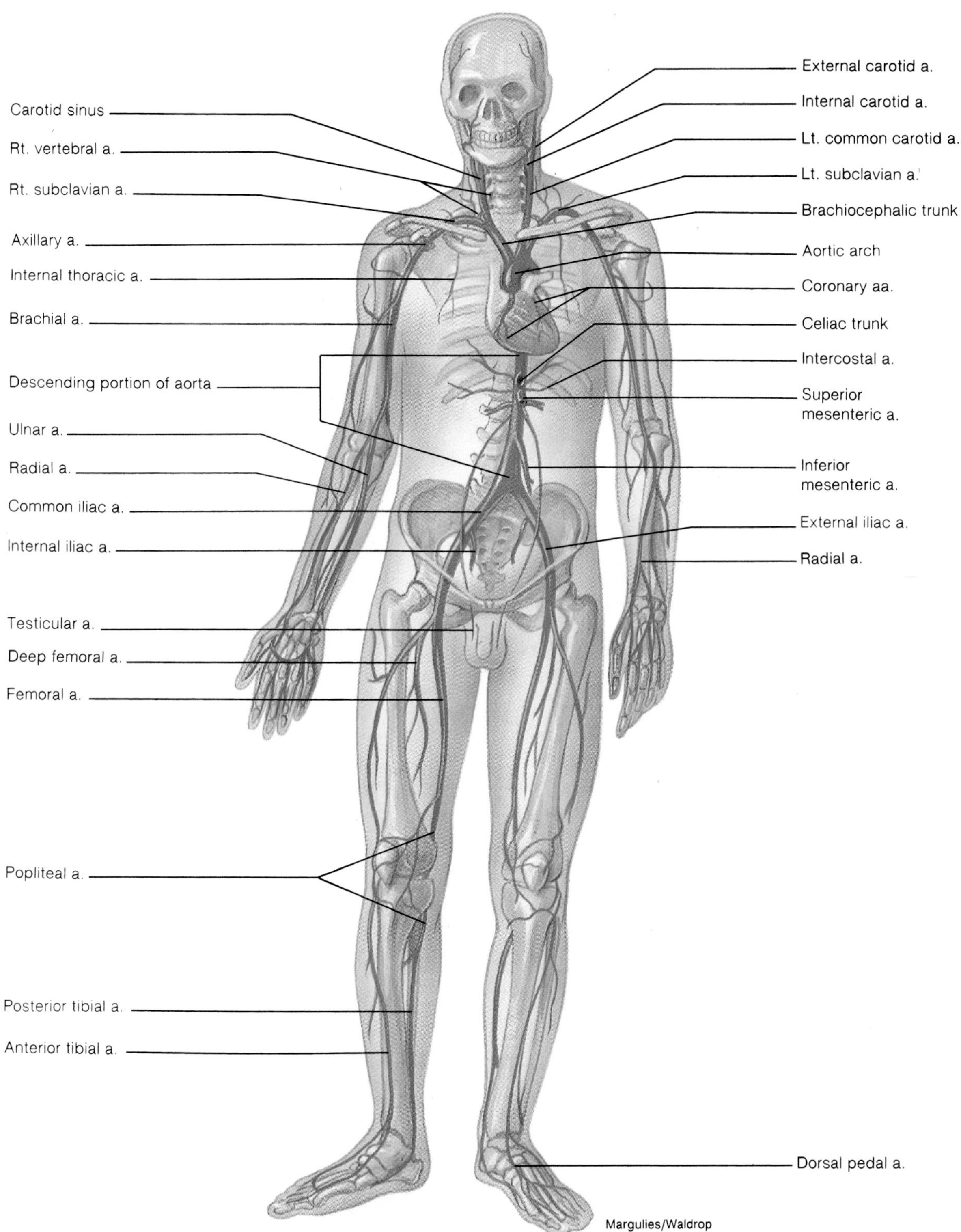

Figure 21.19

The structural relationship between the major arteries and veins to and from the heart.
(*v.* = vein; *vv.* = veins.)

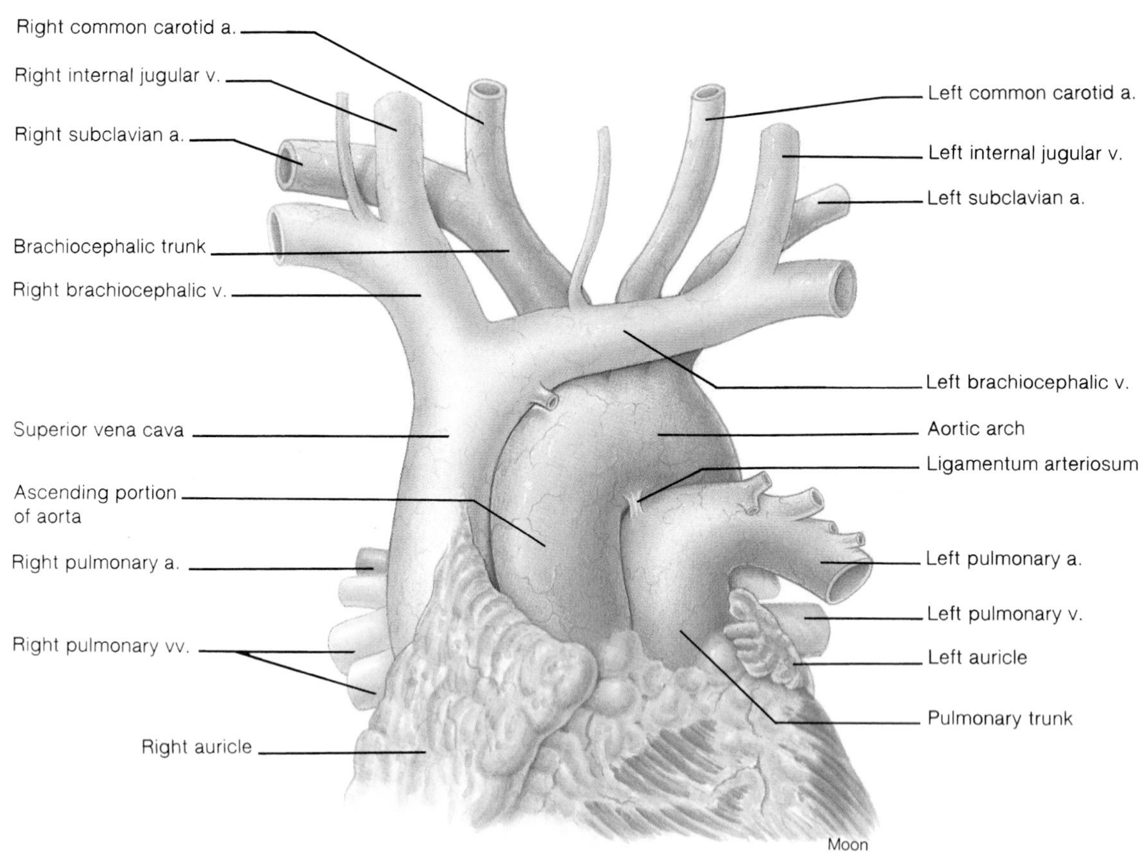

and participate in forming the **cerebral arterial circle** (*circle of Willis*) around the pituitary gland.

Each **internal carotid artery** arises from the common carotid artery and ascends in the neck until it reaches the base of the skull, where it enters the carotid canal of the temporal bone. Several branches arise from the internal carotid artery once it is on the inferior surface of the brain. Three of the more important branches are the **ophthalmic** (*of-thal'mik*) **artery** (see fig. 21.20), which supplies the eye and associated structures, and the **anterior** and **middle cerebral arteries,** which provide blood to the cerebrum. The internal carotid arteries are connected to the posterior cerebral arteries at the cerebral arterial circle (fig. 21.22).

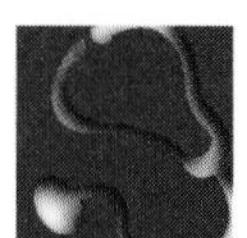

Capillaries within the pituitary gland receive both arterial and venous blood. The venous blood arrives from venules immediately superior to the pituitary gland, which drain capillaries in the hypothalamus of the brain. This arrangement of two capillary beds in series—whereby the second capillary bed receives venous blood from the first—is called a *portal system.* The venous blood that travels from the hypothalamus to the pituitary contains hormones from the hypothalamus that help regulate pituitary hormone secretion.

Blood Supply to the Head and Neck

The external carotid artery gives off several branches as it extends upward along the side of the neck and head (fig. 21.20). The names of these branches are determined by the areas or structures that they serve. The principal vessels that arise from the external carotid artery are the **superior thyroid, ascending pharyngeal, lingual, facial, occipital,** and **posterior auricular arteries.**

The external carotid artery terminates at a level near the condylar process of the mandible by dividing into **maxillary** and **superficial temporal arteries.** Pulsations through the temporal artery can be easily detected by placing the fingertips immediately in front of the ear at the level of the eye. This vessel is frequently used by anesthesiologists to check a patient's pulse rate during surgery.

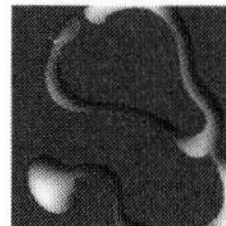

Headaches are usually caused by vascular pressure on the sensitive meninges covering the brain. The two principal vessels serving the meninges are the meningeal branches of the occipital and maxillary arteries. Vasodilation of these vessels creates excessive pressure on the sensory receptors within the meninges, resulting in a headache.

Figure 21.20

Arteries of the neck and head.
(*a*) Major branches of the right common carotid and right subclavian arteries. (*b*) A radiograph (X ray) of the head following a radiopaque injection of the arteries.

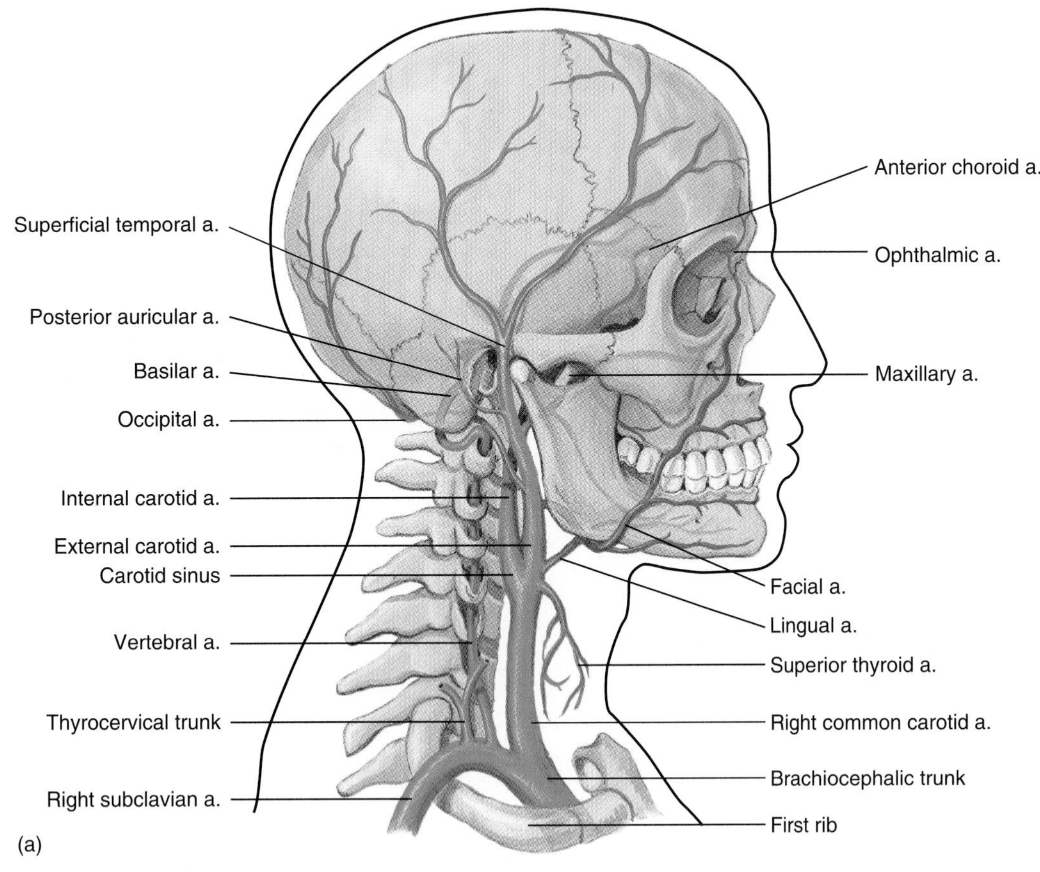

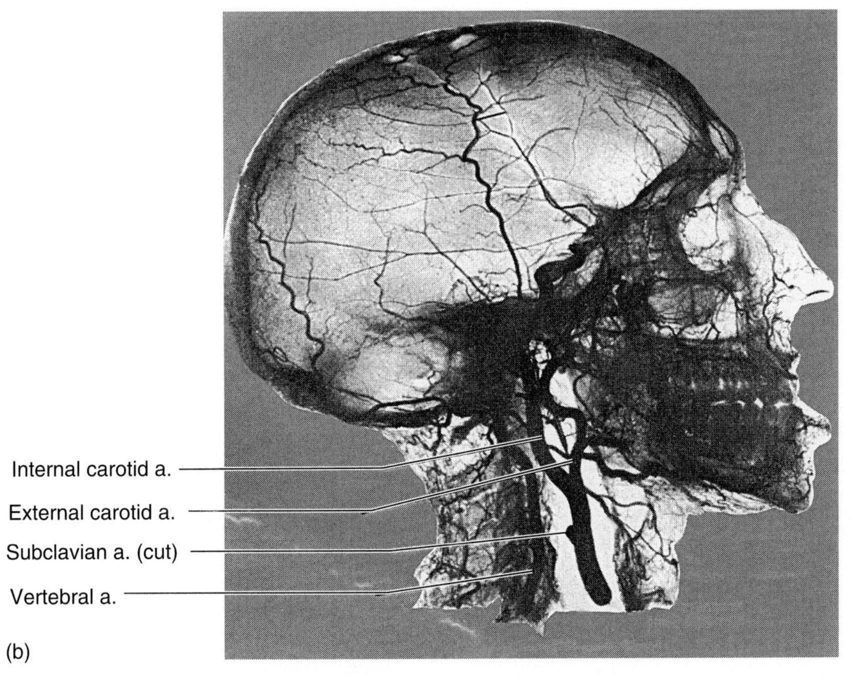

Figure 21.21

Arteries that supply blood to the brain.
(*a*) An inferior view of the brain and (*b*) a close-up view of the region of the pituitary gland. The cerebral arterial circle (circle of Willis) consists of the arteries that ring the pituitary gland.

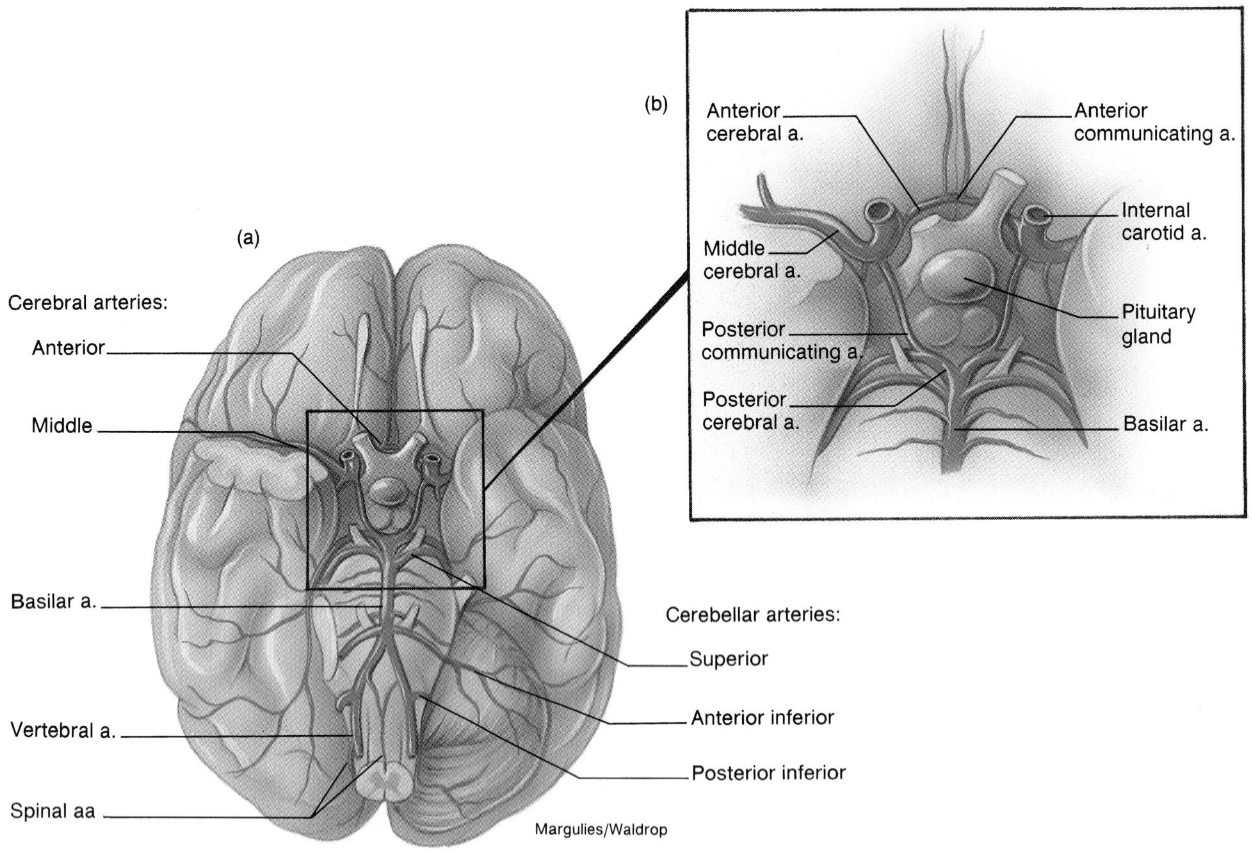

Arteries of the Shoulder and Upper Extremity

As mentioned earlier, the right subclavian artery arises from the brachiocephalic trunk, and the left subclavian artery arises directly from the aortic arch (see fig. 21.18). Each subclavian artery passes laterally deep to the clavicle, carrying blood toward the arm (figs. 21.23 and 21.24). From each subclavian artery arises a **vertebral artery** that carries blood to the brain (already described); a short **thyrocervical trunk** that serves the thyroid gland, trachea, and larynx; and an **internal thoracic artery** that descends into the thorax to serve the thoracic wall, thymus, and pericardium. The **costocervical trunk** branches to serve the upper muscles of the rib cage, posterior neck muscles, and the spinal cord and its meninges.

The **axillary** (*ak′sĭ-lar″e*) **artery** is the continuation of the subclavian artery as it passes into the axillary region. The axillary artery is that portion of the major artery of the upper extremity between the outer border of the first rib and the lower border of the teres major muscle. Several small branches arise from the axillary artery and supply blood to the tissues of the upper thorax and shoulder region.

The **brachial** (*bra′ke-al*) **artery** is the continuation of the axillary artery through the brachial region. The brachial artery courses on the medial side of the humerus, where it is a major pressure point and the most common site for determining blood pressure. A **deep brachial artery** branches from the brachial artery and curves posteriorly near the radial nerve to supply the triceps brachii muscle. Two additional branches from the brachial, the **anterior** and **posterior humeral circumflex arteries,** form a continuous ring of vessels around the proximal portion of the humerus.

Just proximal to the cubital fossa, the brachial artery branches into the **radial** and **ulnar arteries,** which supply blood to the forearm and a portion of the hand and digits. The radial artery courses down the lateral, or radial, side of the forearm, and the ulnar artery courses down the medial, or ulnar, side. Both vessels provide numerous small branches to the muscles of the forearm. The **radial recurrent artery** serves the region of the elbow and is the first and largest branch of the radial artery. The radial artery is important as a site for recording the pulse near the wrist.

At the wrist, the ulnar and radial arteries anastomose to form the **superficial** and the **deep palmar arch.** The metacarpal arteries of the hand (not shown) arise from the deep palmar arch, and the **digital arteries** of the fingers arise from the superficial palmar arch.

Figure 21.22

The path of arterial blood flow to the brain.
(Note the ring of vessels surrounding the pituitary gland to form the cerebral arterial circle.)

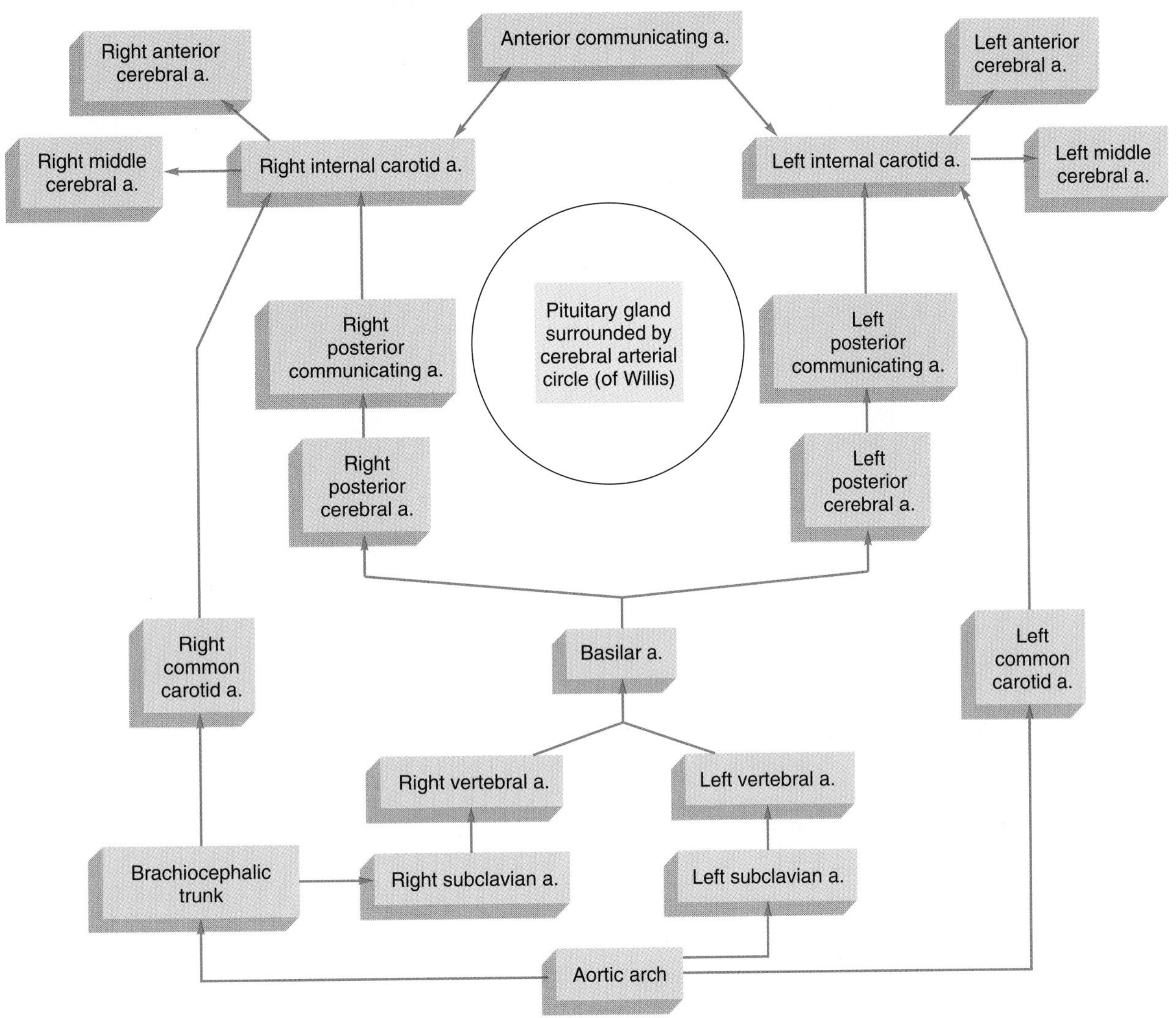

Branches of the Thoracic and Abdominal Portions of the Aorta

The **thoracic portion of the aorta** is a continuation of the aortic arch as it descends through the thoracic cavity to the diaphragm. This large vessel gives off many branches, including the **pericardial, bronchial, esophageal** (*ĕ-sof'ă-je'al*), **posterior intercostal,** and **superior phrenic** (*fren'ik*) **arteries.** These vessels are summarized according to their location and function in table 21.4.

The **abdominal portion of the aorta** is the segment of the aorta between the diaphragm and the level of the fourth lumbar vertebra, where it divides into the **right** and **left common iliac** (*il'e-ak*) **arteries.** The first branches of the abdominal aorta are the paired **inferior phrenic arteries.** Next, the large **celiac** (*se'le-ak*) **trunk** arises and divides immediately into three arteries: the **splenic, left gastric,** and the **common hepatic arteries** (fig. 21.25).

Other unpaired arteries are the **superior mesenteric** and the **inferior mesenteric arteries.** Other paired arteries include the **renal, suprarenal,** and **gonadal (testicular** or **ovarian) arteries.** These vessels are summarized according to their location and function in table 21.4.

Arteries of the Pelvis and Lower Extremity

The abdominal portion of the aorta terminates in the posterior pelvic area as it divides into the **right** and **left common iliac arteries.** These vessels pass downward approximately 5 cm on their respective sides and terminate by dividing into the internal and external iliac arteries.

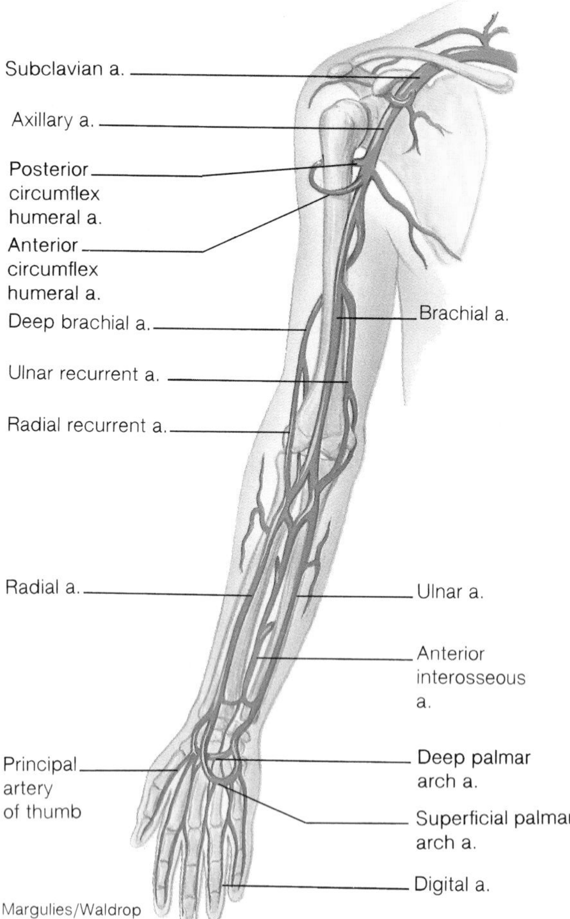

Figure 21.23

An anterior view of the arteries of the right shoulder and upper extremity.

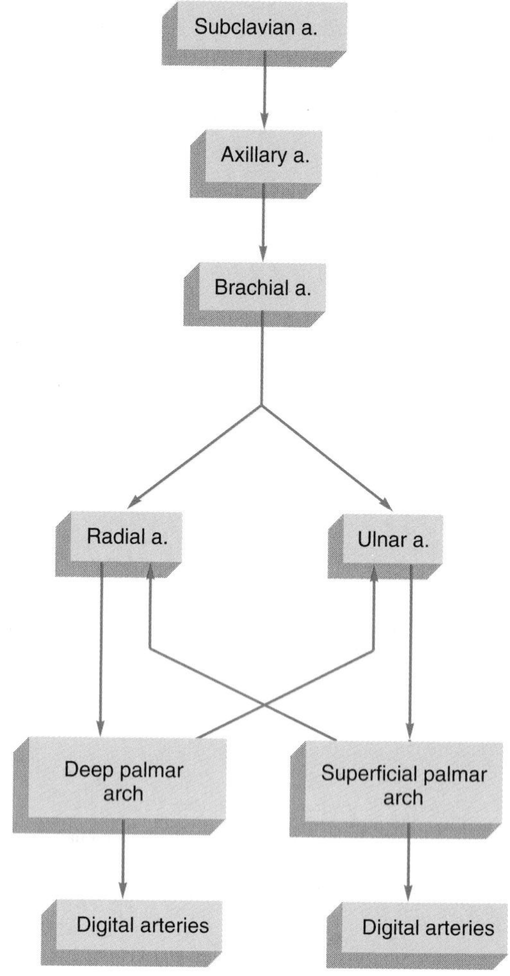

Figure 21.24

The path of arterial blood flow from the subclavian artery to the digital arteries of the fingers.

The **internal iliac artery** has extensive branches to supply arterial blood to the gluteal muscles and the organs of the pelvic region (fig. 21.26). Arteries that arise from the internal iliac artery include the **iliolumbar, lateral sacral, middle rectal, vesicular** (to the urinary bladder), **uterine** and **vaginal, gluteal, obturator,** and **internal pudendal** (to the external genitalia) **arteries.**

The **external iliac artery** gives off the **inferior epigastric artery** and **deep circumflex iliac artery** before exiting the pelvic cavity beneath the inguinal ligament (fig. 21.27). Once through the inguinal canal, the external iliac artery becomes the **femoral artery.**

The femoral artery passes through an area called the **femoral triangle** on the upper medial portion of the thigh (fig. 21.28). At this point, the femoral artery is close to the surface and is an important arterial pressure point (see page 606). Several vessels arise from the femoral artery, including the **deep femoral** and the lateral and **medial femoral circumflex arteries.** The femoral artery becomes the **popliteal** (*pop"lĭ-te'al*) **artery** as it passes across the posterior aspect of the knee.

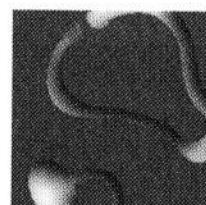

Hemorrhage can be a serious problem in many accidents. To prevent a victim from bleeding to death, it is important to know where to apply pressure to curtail the flow of blood. The arterial pressure points for the appendages are the brachial artery on the medial side of the arm and the femoral artery in the groin (see the section "Arterial Pressure Points and Control of Bleeding" in chapter 20). Firmly applied pressure to these regions greatly diminishes blood flow to traumatized areas below. A tourniquet may have to be applied if bleeding is life threatening and if other, safer methods have proved ineffective.

The popliteal artery supplies small branches to the knee joint and then divides into an **anterior tibial artery** and **posterior tibial artery** (fig. 21.27). At the ankle, the anterior tibial artery becomes the **dorsal pedal artery** that serves the ankle and superior surface of the foot and then contributes to the formation of the **plantar arch.** The posterior tibial artery gives off the large **peroneal** (*per"ŏ-ne'al*) **artery** and then, at the ankle, the posterior tibial bifurcates into the **lateral** and **medial**

Table 21.4
Segments and Branches of the Aorta

Segment of Aorta	Arterial Branch	General Region or Organ Served
Ascending portion of aorta	Right and left coronary aa.*	Heart
Aortic arch	Brachiocephalic trunk	
	Right common carotid a.	Right side of head and neck
	Right subclavian a.	Right shoulder and right upper extremity
	Left common carotid a.	Left side of head and neck
	Left subclavian a.	Left shoulder and left upper extremity
Thoracic portion of aorta	Pericardial aa.	Pericardium of heart
	Posterior intercostal aa.	Intercostal and thoracic muscles, and pleurae
	Bronchial aa.	Bronchi of lungs
	Superior phrenic aa.	Superior surface of diaphragm
	Esophageal aa.	Esophagus
Abdominal portion of aorta	Inferior phrenic aa.	Inferior surface of diaphragm
	Celiac trunk	
	Common hepatic a.	Liver, upper pancreas, and duodenum
	Left gastric a.	Stomach and esophagus
	Splenic a.	Spleen, pancreas, and stomach
	Superior mesenteric a.	Small intestine, pancreas, cecum, appendix, ascending colon, and transverse colon
	Suprarenal aa.	Adrenal (suprarenal) glands
	Lumbar aa.	Muscles and spinal cord of lumbar region
	Renal aa.	Kidneys
	Gonadal aa.	
	Testicular aa.	Testes
	Ovarian aa.	Ovaries
	Inferior mesenteric a.	Transverse colon, descending colon, sigmoid colon, and rectum
	Common iliac aa.	
	External iliac aa.	Lower extremities
	Internal iliac aa.	Genital organs and gluteal muscles

*"aa." = "arteries"; "a." = "artery."

plantar arteries. The lateral plantar artery anastomoses with the dorsal pedal artery to form the plantar arch in an arterial arrangement similar to that of the hand. **Digital arteries** arise from the plantar arch to supply the toes with blood.

Principal Veins of the Body

After systemic blood has passed through the tissues, this oxygen-depleted blood is returned through veins of progressively larger diameters to the right atrium of the heart.

In the venous portion of the systemic circulation, blood flows from smaller vessels into larger ones, so that a vein receives smaller tributaries instead of giving off branches as an artery does. The veins from all parts of the body (except the lungs and myocardium of the heart) converge into two major vessels that empty into the right atrium: the **superior vena cava** (*ve′na ka′va*) and **inferior vena cava** (fig. 21.29). Veins are more numerous than arteries and are both superficial and deep. Superficial veins generally can be seen just beneath the skin and are clinically important in drawing blood and giving injections. Deep veins are close to the principal arteries and are usually similarly named. As with arteries, veins are named according to the region in which they are found or the organ that they serve. (Note that when a vein serves an organ, it drains blood away from the organ.)

Figure 21.25

An anterior view of the abdominal aorta and its principal branches.
In (*a*) the abdominal viscera have been removed; in (*b*) they are intact.

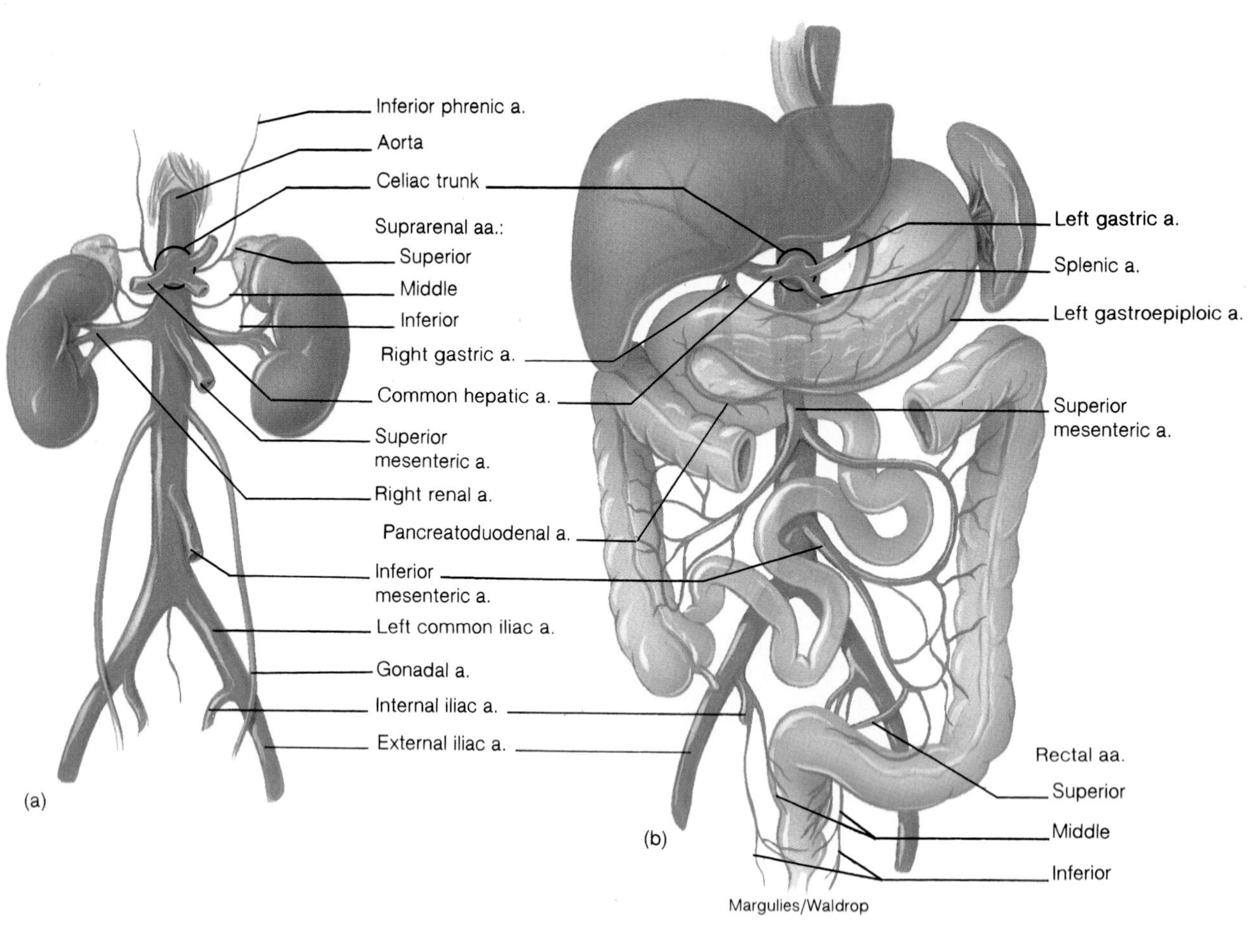

Figure 21.26

Arteries of the pelvic region.

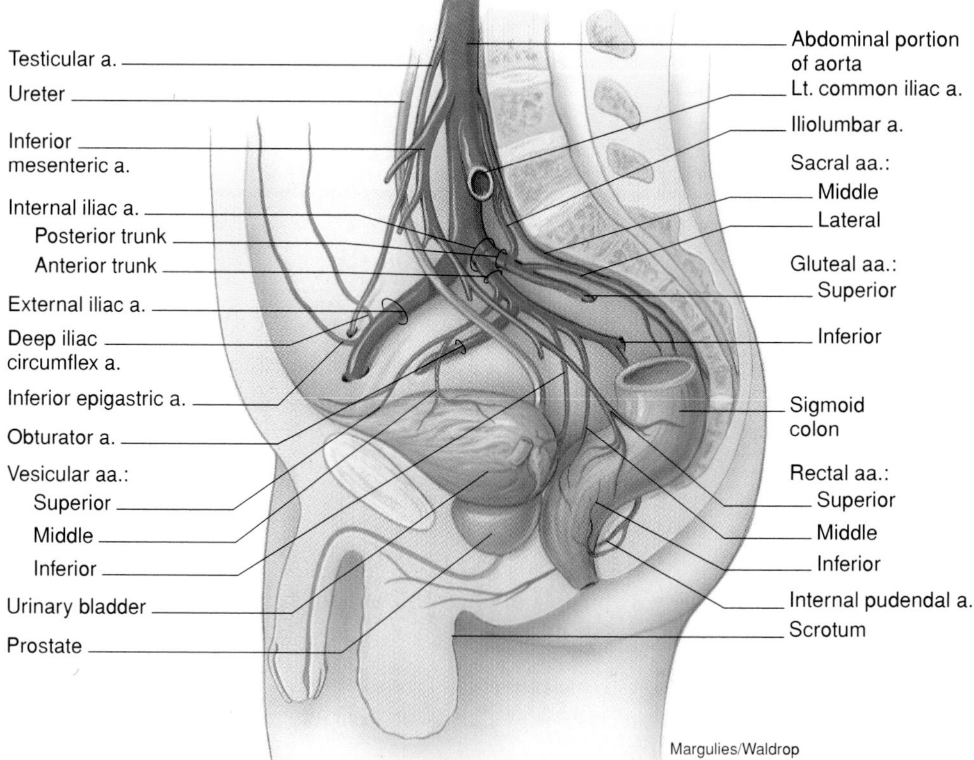

Figure 21.27

Arteries of the right hip and lower extremity.
(*a*) An anterior view and (*b*) a posterior view.

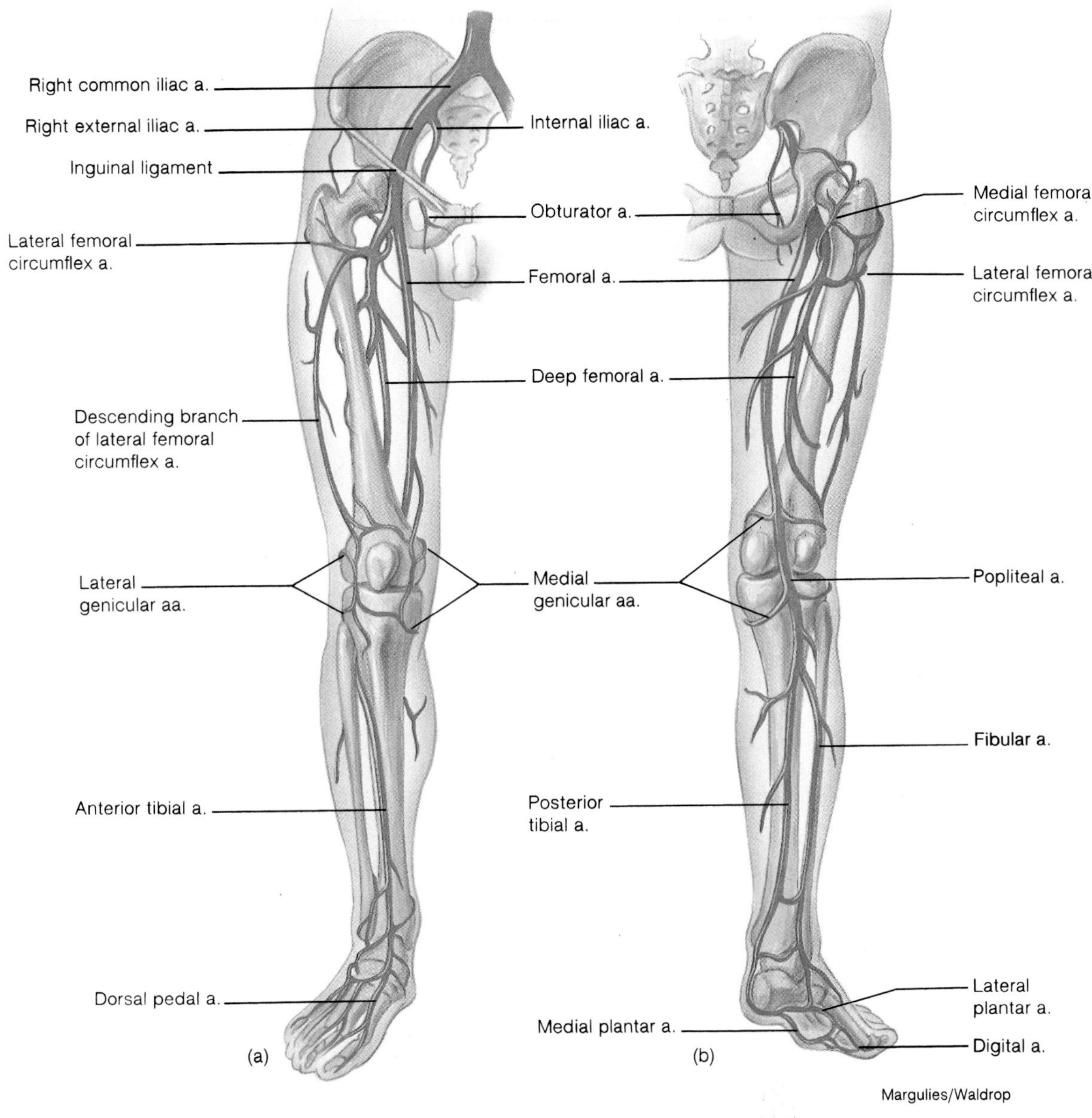

Veins Draining the Head and Neck

Blood from the scalp, portions of the face, and the superficial neck regions is drained by the **external jugular veins** (fig. 21.30). These vessels descend on either lateral side of the neck and drain into the right and left **subclavian veins,** which are located just behind the clavicles.

The paired **internal jugular veins** drain blood from the brain, meninges, and deep regions of the face and neck. The internal jugular veins are larger and deeper than the external jugular veins. They arise from numerous cranial **venous sinuses,** which constitute a series of both paired and unpaired channels within the dura mater. The venous sinuses, in turn, receive venous blood from the **cerebral,** the **cerebellar,** the **ophthalmic,** and the **meningeal veins.**

The internal jugular vein passes inferiorly down the neck, adjacent to the common carotid artery and the vagus nerve. The internal jugular on each side empties into the subclavian vein, and the union of these two vessels forms the large **brachiocephalic vein** on each side. The two brachiocephalic veins merge to form the superior vena cava, which drains into the right atrium of the heart (see fig. 21.29).

jugular: L. *jugulum*, throat or neck

Figure 21.28

The femoral triangle.
The structures within the femoral triangle are shown in (*a*); the boundaries of the triangle are shown in (*b*).

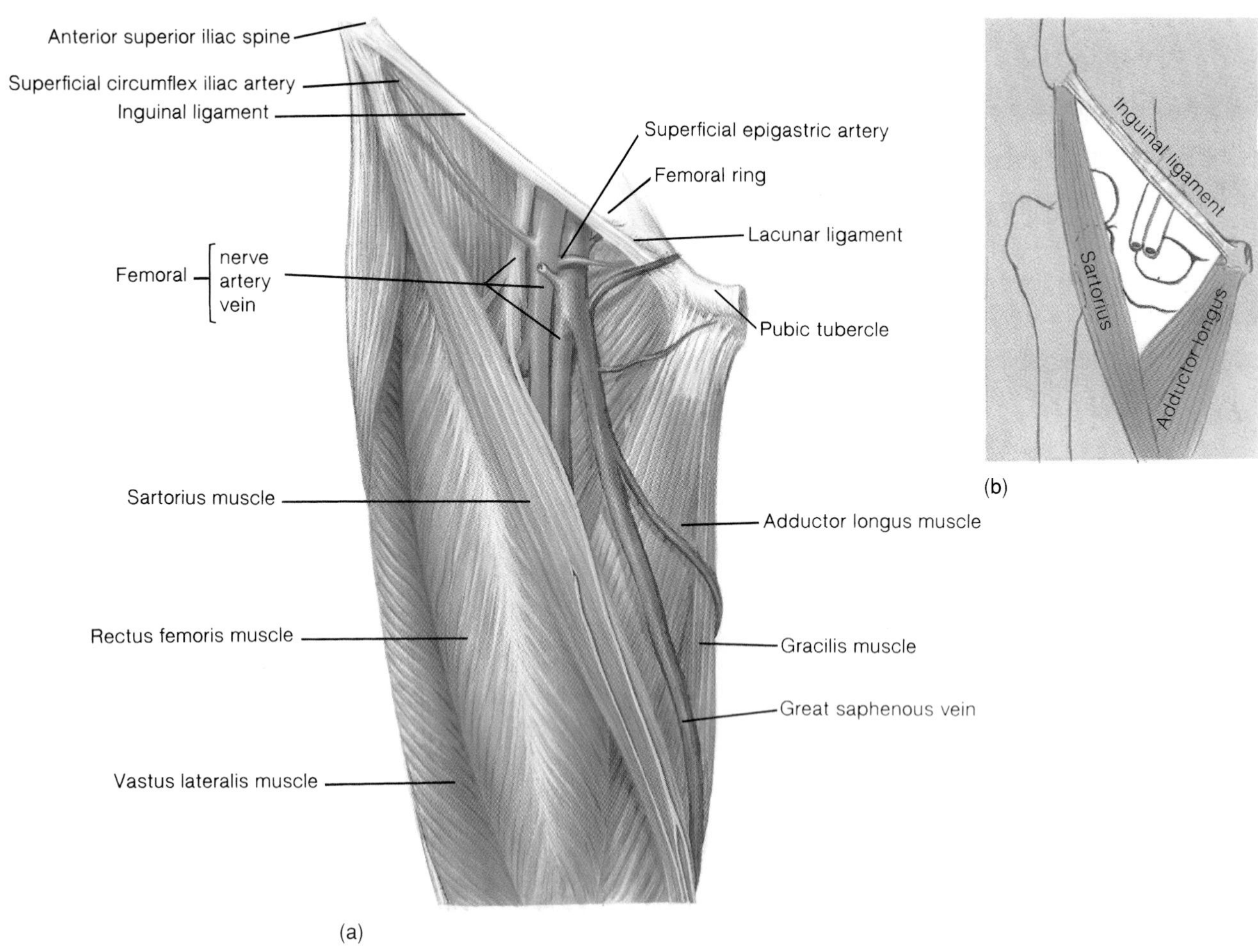

Veins of the Upper Extremity

The upper extremity has both superficial and deep venous drainage (figs. 21.31 and 21.32). The superficial veins are highly variable and form an extensive network just below the skin. The deep veins accompany the arteries of the same region and are given similar names: the **radial, ulnar, brachial,** and **axillary veins** are the major examples.

The main superficial vessels of the upper extremity are the **basilic vein** and the **cephalic vein.** In the cubital fossa of the elbow, the superficial **median cubital vein** connects the cephalic vein on the lateral side with the basilic vein on the medial side. The median cubital vein is commonly punctured to obtain blood samples for clinical tests. Both the basilic and the cephalic veins drain into the axillary vein in the shoulder region.

Veins of the Thorax

The superior vena cava, formed by the union of the two brachiocephalic veins, empties venous blood from the head, neck, and upper extremities directly into the right atrium of the heart. These large vessels lack the valves that are characteristic of most other veins in the body. In addition to receiving blood from the brachiocephalic veins, the superior vena cava collects blood from the *azygos* (*az'ĭ-gos*) *system of veins* arising from the posterior thoracic wall (fig. 21.33). This system includes the **azygos, ascending lumbar, intercostal, accessory hemiazygos,** and **hemiazygos veins.**

Veins of the Lower Extremity

The lower extremities, like the upper extremities, have both a deep and a superficial group of veins (fig. 21.34). The deep veins accompany corresponding arteries and have more valves than the superficial veins.

The deep veins include the **posterior** and **anterior tibial veins** that originate in the foot and course upward to the back of the knee, where they merge to form the **popliteal vein.** Just

azygos: Gk. *a*, without; *zygon*, yoke

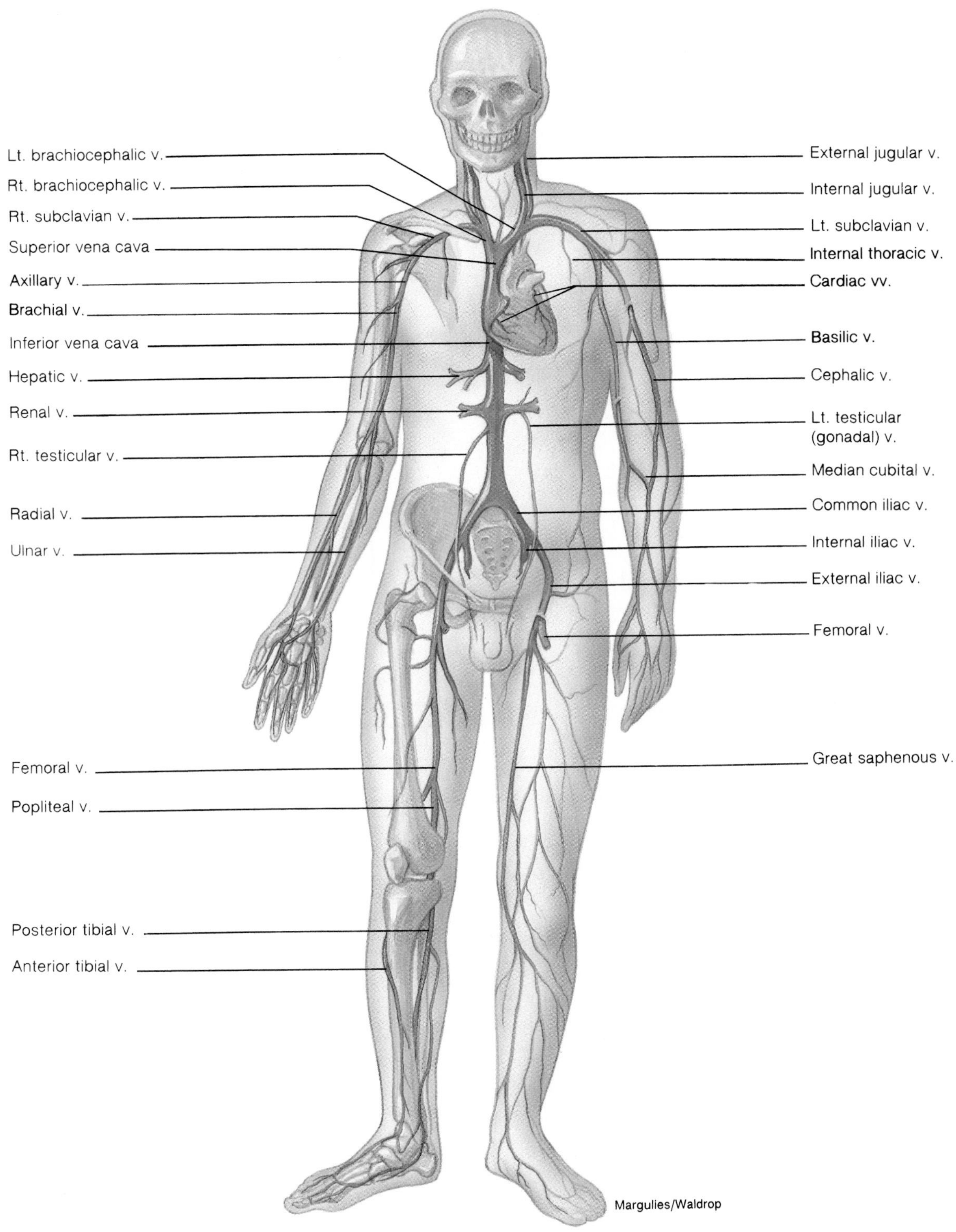

Figure 21.29

Principal veins of the body.
Superficial veins are depicted in the left extremities and deep veins in the right extremities.

Figure 21.30

Veins that drain the head and neck.

(Note the cranial venous sinuses that drain blood from the brain into the internal jugular vein.)

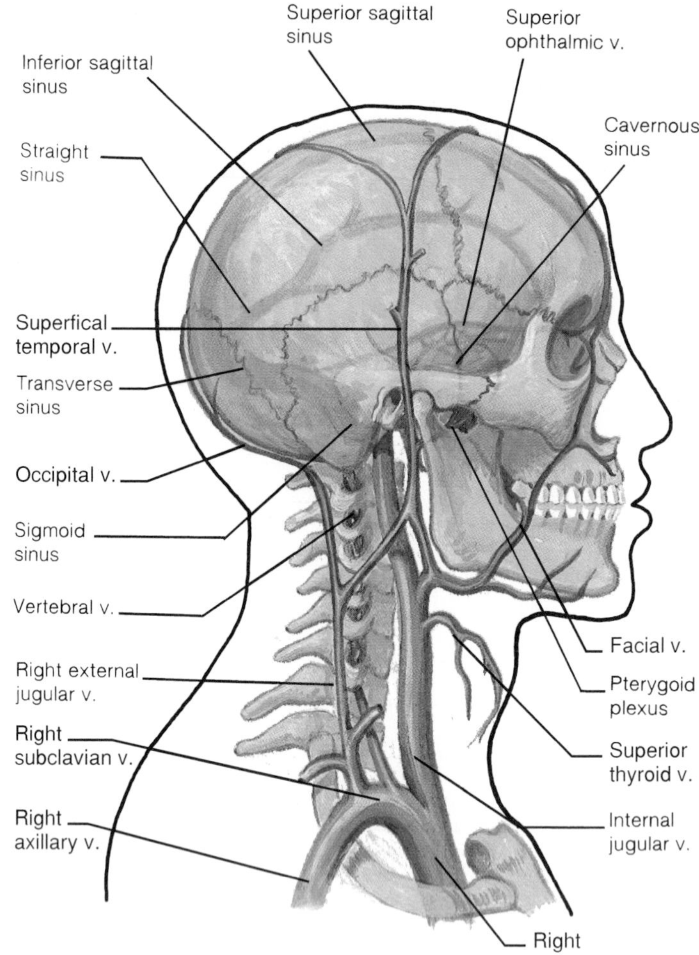

Figure 21.31

An anterior view of the veins that drain the upper right extremity.

(*a*) Superficial veins and (*b*) deep veins.

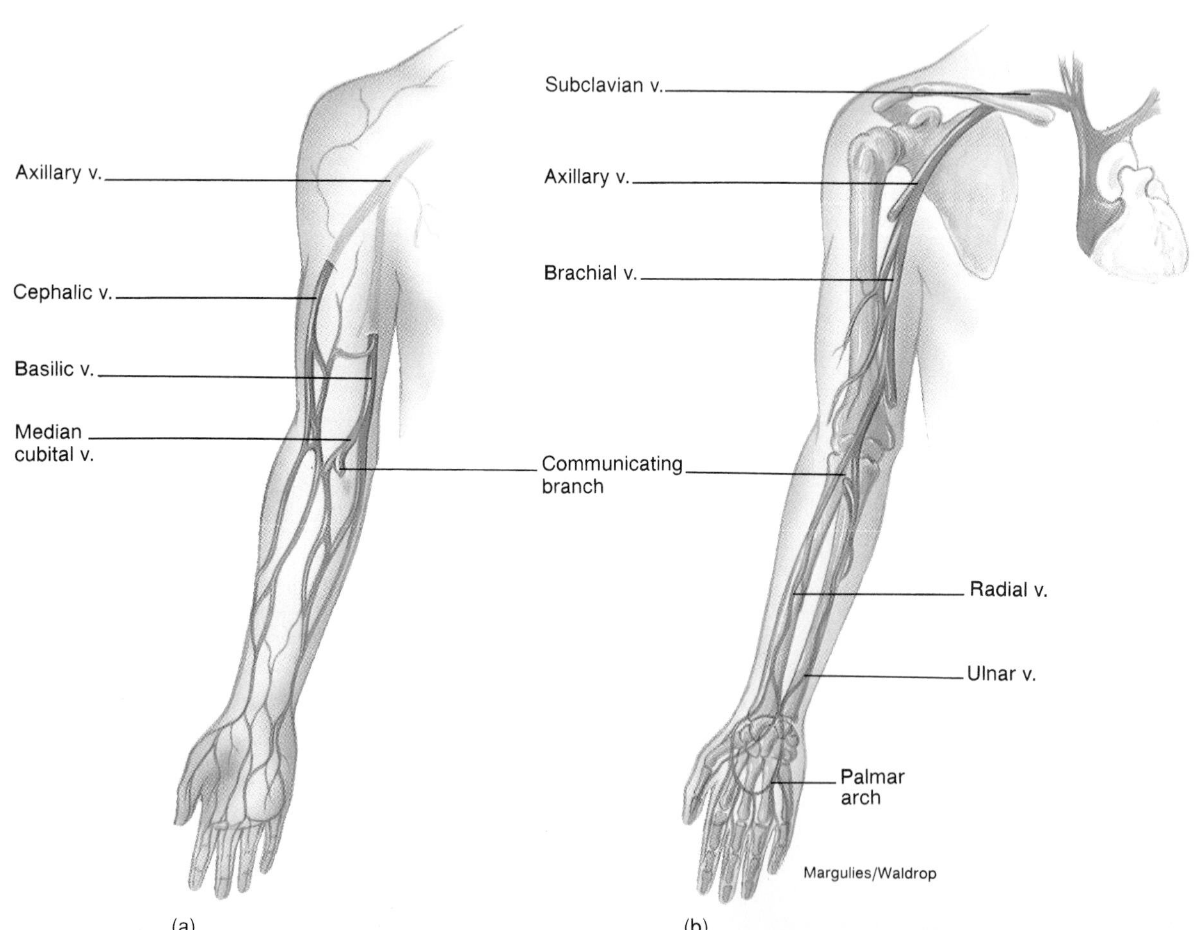

Figure 21.32

Venous return of blood from the head and the upper extremity to the heart.

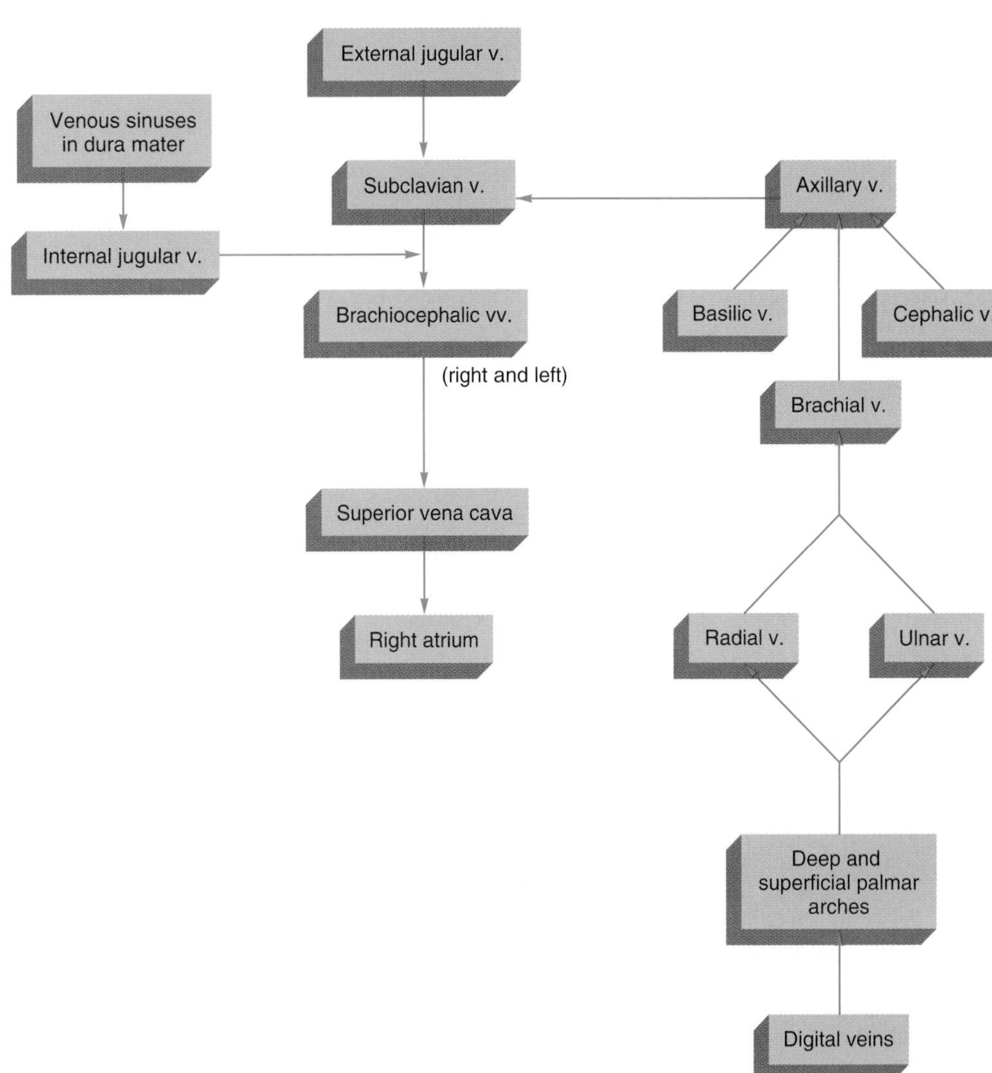

Figure 21.33

Veins of the thoracic region.
(The lungs and heart have been removed.)

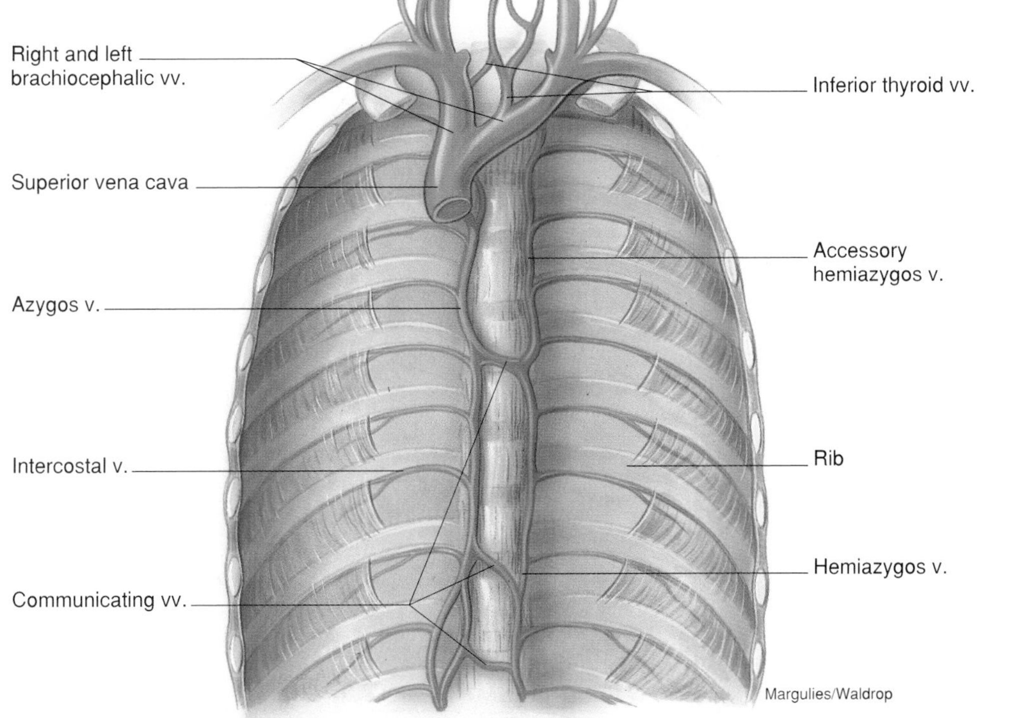

Figure 21.34

Veins of the lower extremity and hip.

(*a*) Superficial veins, medial and posterior aspects, and (*b*) deep veins, a medial view.

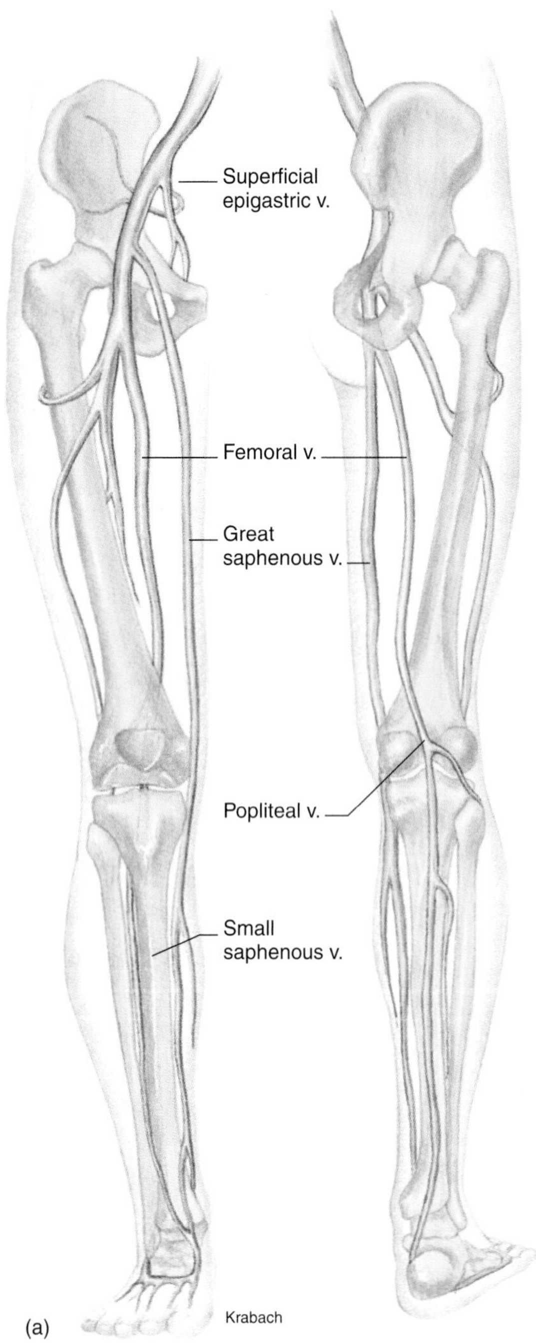

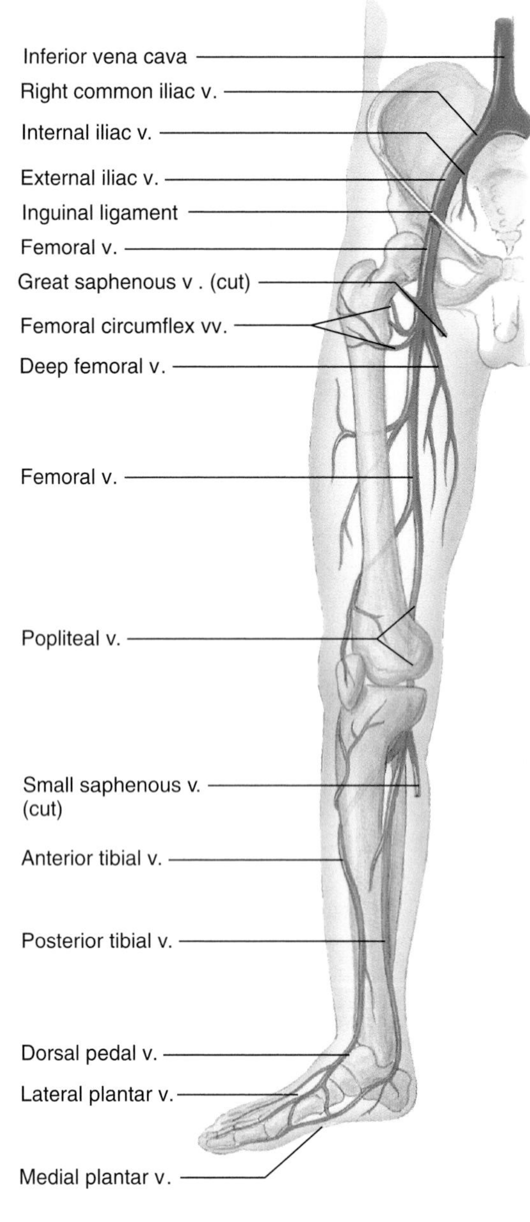

above the knee, this vessel becomes the **femoral vein.** The femoral vein receives blood from the **deep femoral vein** near the groin and then becomes the **external iliac vein** as it passes under the inguinal ligament. The external iliac merges with the **internal iliac vein** at the pelvic and genital regions to form the **common iliac vein.** At the level of the fifth lumbar vertebra, the right and left common iliacs unite to form the large inferior vena cava (see figs. 21.29 and 21.34).

The superficial veins of the lower extremity are the **small** and **great saphenous** (*să-fe'nus*) **veins.** The small saphenous vein arises from the lateral side of the foot and empties into the popliteal vein behind the knee. The great saphenous vein is the longest vessel in the body. It originates from the medial side of the foot and ascends superiorly along the medial aspect of the leg and thigh before draining into the femoral vein.

saphenous: L. *saphena,* the hidden one

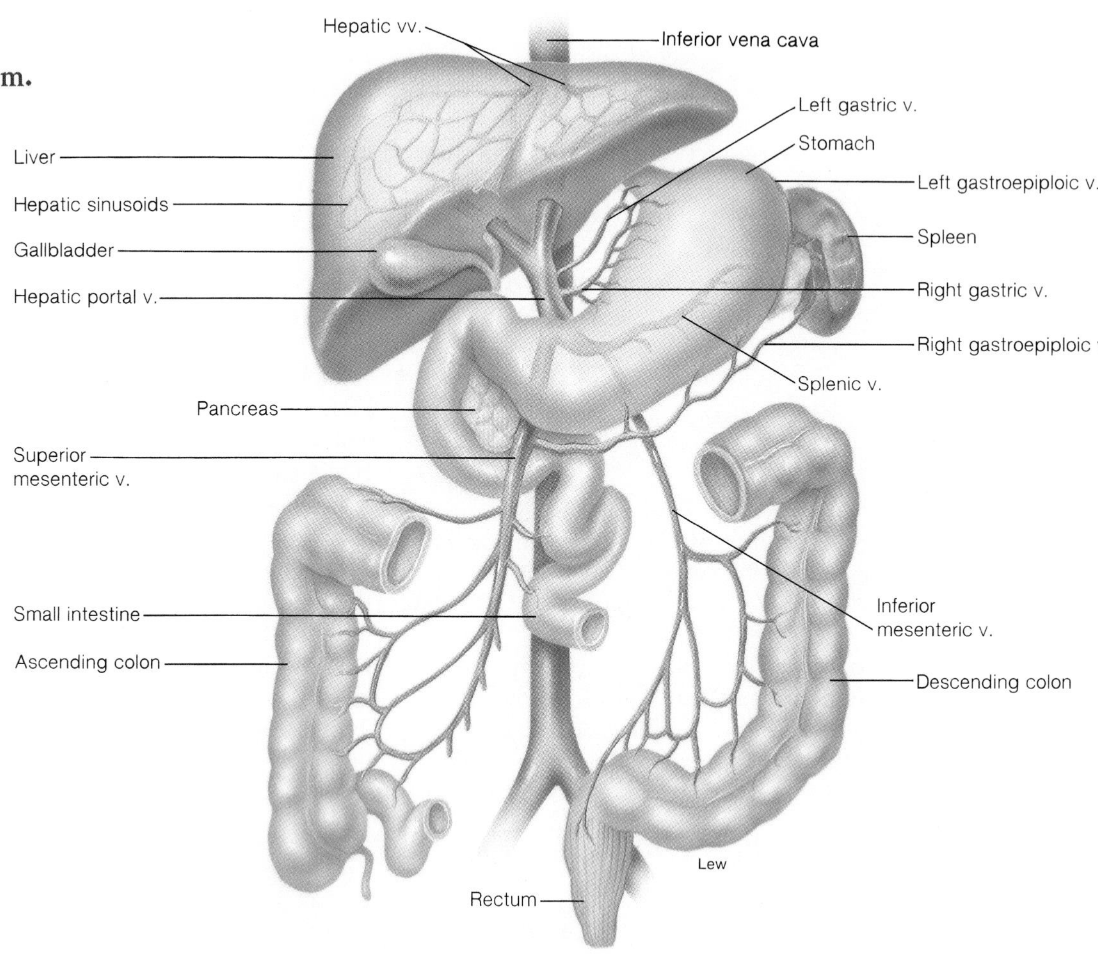

Figure 21.35 The hepatic portal system.

Veins of the Abdominal Region

The inferior vena cava parallels the abdominal aorta on the right side as it ascends through the abdominal cavity to penetrate the diaphragm and enter the right atrium (see fig. 21.29). It is the largest in diameter of the vessels in the body and is formed by the union of the two common iliac veins that drain the lower extremities. As the inferior vena cava ascends through the abdominal cavity, it receives tributaries from veins that correspond in name and position to arteries previously described.

Note that the inferior vena cava does not receive blood directly from the GI tract, pancreas, or spleen. Instead, the venous outflow from these organs first passes through capillaries in the liver.

Hepatic Portal System

As previously described, in a portal system, one capillary bed is located downstream from another. Drainage from the first capillary bed is delivered into the second, and drainage from the second is delivered finally into the general venous circulation. Therefore, there are two capillary beds in series. The **hepatic** (*he-pat′ik*) **portal system** is composed of veins that drain blood from capillaries in the intestines, pancreas, spleen, stomach, and gallbladder into capillaries in the liver (called *sinusoids*) and of the **right** and **left hepatic veins** that empty into the inferior vena cava (fig. 21.35). As a consequence of the hepatic portal system, the absorbed products of digestion must first pass through the liver before entering the general circulation.

The **hepatic portal vein** is the large vessel that receives blood from the digestive organs. It is formed by a union of the **superior mesenteric vein,** which drains nutrient-rich blood from the small intestine, and the **splenic vein,** which drains the spleen. The splenic vein is enlarged because of a convergence of the following three tributaries: (1) the **inferior mesenteric vein,** from the large intestine, (2) the **pancreatic vein,** from the pancreas, and (3) the **left gastroepiploic** (*gas″tro-ep″i-plo′ik*) **vein,** from the stomach. The **right gastroepiploic vein,** also from the stomach, drains directly into the superior mesenteric vein.

Three additional veins empty into the hepatic portal vein. The **right** and **left gastric veins** drain the lesser curvature of the stomach, and the **cystic vein** drains blood from the gallbladder.

gastroepiploic: Gk. *gastros,* stomach; *epiplein,* to float on (referring to greater omentum)

In summary, it is important to note that the sinusoids of the liver receive blood from two sources. The hepatic artery supplies oxygen-rich blood to the liver, and the hepatic portal vein transports nutrient-rich blood from the small intestine for processing. These two blood sources become mixed in the liver sinusoids. Liver cells exposed to this blood obtain nourishment from it and are uniquely qualified (because of their anatomical position and enzymatic ability) to modify the chemical nature of the venous blood that enters the general circulation from the GI tract. The nature of these modifications is discussed in detail in chapter 26.

Fetal Circulation

All of the respiratory, excretory, and nutritional needs of the fetus are provided for by diffusion across the placenta instead of by the fetal lungs, kidneys, and gastrointestinal tract. Fetal circulation is adaptive to these conditions.

The circulation of blood through a fetus is necessarily different from blood circulation in a newborn (fig. 21.36). Respiration, the procurement of nutrients, and the elimination of metabolic wastes occur through the maternal blood instead of through the organs of the fetus. The capillary exchange between the maternal and fetal circulation occurs within the *placenta* (*pla-sen'ta*) (see fig. 30.14). This remarkable structure, which includes maternal and fetal capillary beds, is discharged following delivery as the afterbirth.

The **umbilical cord** is the connection between the placenta and the fetal umbilicus. It includes one **umbilical vein** and two **umbilical arteries,** surrounded by a gelatinous substance. Oxygenated and nutrient-rich blood flows through the umbilical vein toward the inferior surface of the liver. At this point, the umbilical vein divides into two branches. One branch merges with the hepatic portal vein, while the other branch, called the **ductus venosus** (*ve-no'sus*) enters the inferior vena cava. Thus, oxygenated blood is mixed with venous blood returning from the lower extremities of the fetus before it enters the heart. The umbilical vein is the only vessel of the fetus that carries fully oxygenated blood.

The inferior vena cava empties into the right atrium of the fetal heart. Most of the blood passes from the right atrium into the left atrium through the **foramen ovale** (*o-val'e*), an opening between the two atria. Here, it mixes with a small quantity of blood returning from the pulmonary circulation. The blood then passes into the left ventricle, from which it is pumped into the aorta and through the body of the fetus.

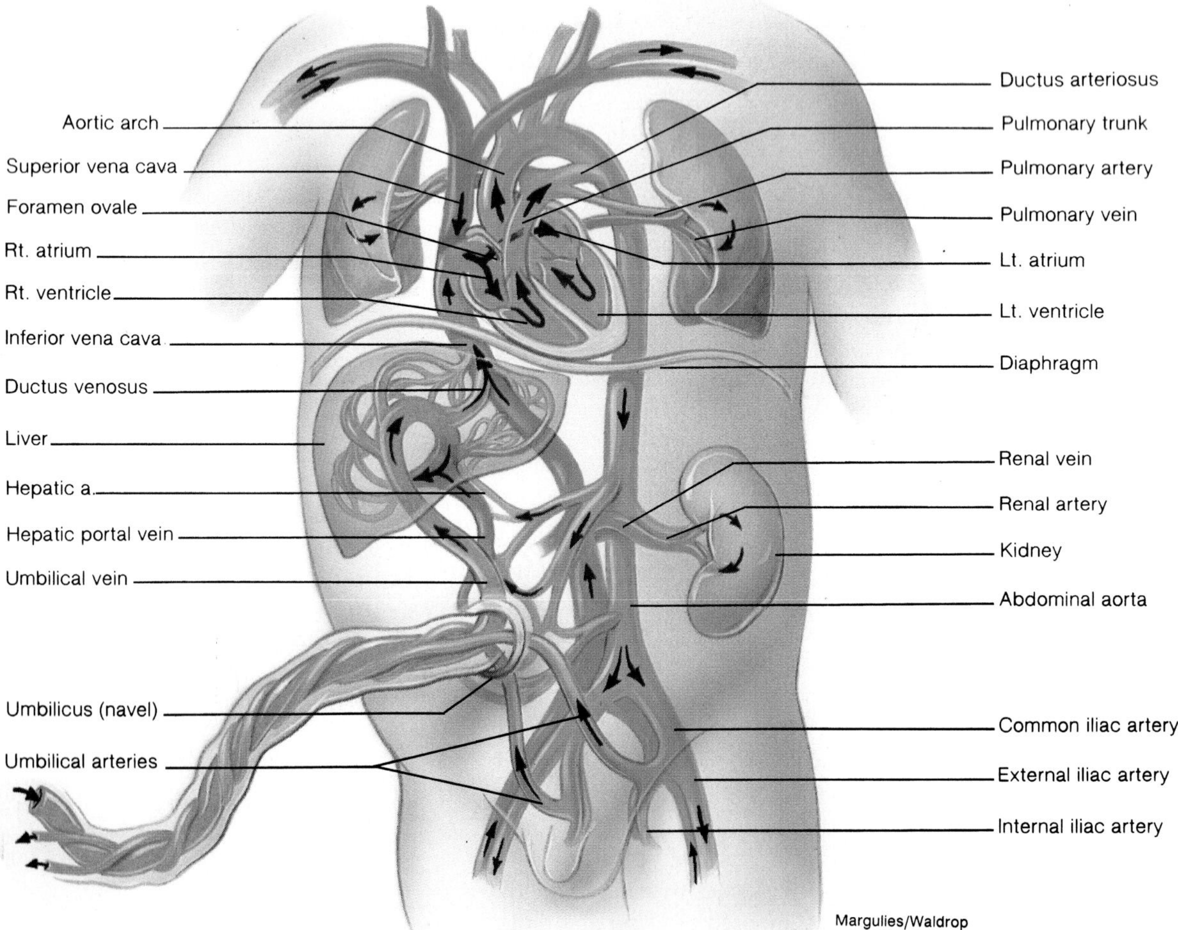

Figure 21.36

Fetal circulation.
(Arrows indicate the direction of blood flow.)

UNDER DEVELOPMENT

Development of the Heart

The development of the heart from two separate segments of mesoderm requires only 6 to 7 days. Heart development is first apparent at day 18 or 19 in the *cardiogenic* (*kar″de-o-jen′ik*) *area* of the mesoderm layer. A small paired mass of specialized cells called **heart cords** form here. Shortly after, a hollow center develops in each heart cord, and each structure is then referred to as a **heart tube** (fig. 1). The heart tubes begin to migrate toward each other on day 21 and soon fuse to form a single median endocardial heart tube. During this time, the endocardial heart tube undergoes dilations and constrictions, so that when fusion is completed during the fourth week, five distinct regions of the heart can be identified. These are the **truncus arteriosus, bulbus cordis, ventricle, atrium,** and **sinus venosus.**

After the fusion of the heart tubes and the formation of distinct dilations, the heart begins to pump blood. Partitioning of the heart chambers begins during the middle of the fourth week and is complete by the end of the fifth week. During this crucial time, many congenital heart problems develop.

Major changes occur in each of the five primitive dilations of the developing heart during the week-and-a-half embryonic period beginning in the middle of the fourth week. The truncus arteriosus differentiates to form a partition between the aorta and the pulmonary trunk. The bulbus cordis is incorporated in the formation of the walls of the ventricles. The sinus venosus forms the **coronary sinus** and a portion of the wall of the right atrium. The ventricle is divided into the right and left chambers by the growth of the **interventricular septum.** The atrium is partially partitioned into right and left chambers by the **septum secundum.** An opening between the two atria called the **foramen ovale** persists throughout fetal development. This opening is covered by a flexible valve that permits blood to pass from the right to the left side of the heart.

---

cardiogenic: Gk. *kardia*, heart; *genesis*, be born (origin)
ventricle: L. *ventriculus*, diminutive of *venter*, belly
atrium: L. *atrium*, chamber

Figure 1 Formation of the heart chambers. (*a*) The heart tubes fuse during days 21 and 22. (*b*) The developmental chambers are formed during day 23. (*c*) Differential growth causes folding between the chambers during day 24, and vessels are developed to transport blood to and from the heart. The embryonic heart generally has begun rhythmic contractions and pumping blood by day 25.

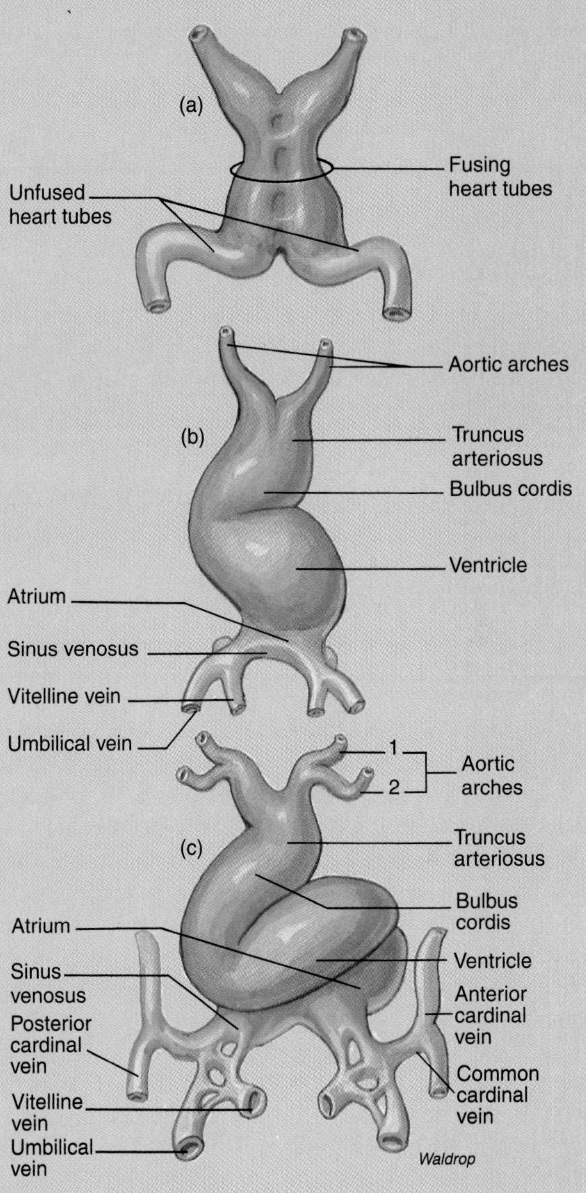

Table 21.5
Cardiovascular Structures of the Fetus and Changes in the Neonate

Structure	Location	Function	Neonate Transformation
Umbilical vein	Connects the placenta to the liver; forms a major portion of the umbilical cord	Transports nutrient-rich oxygenated blood from the placenta to the fetus	Forms the round ligament of the liver
Ductus venosus	Venous shunt within the liver that connects the umbilical vein and the inferior vena cava	Transports oxygenated blood directly into the inferior vena cava	Forms the ligamentum venosum, a fibrous cord in the liver
Foramen ovale	Opening between the right and left atria	Acts as a shunt to bypass the pulmonary circulation	Closes at birth and becomes the fossa ovalis, a depression in the interatrial septum
Ductus arteriosus	Connects the pulmonary trunk and the aortic arch	Acts as a shunt to bypass the pulmonary circulation	Closes shortly after birth, atrophies, and becomes the ligamentum arteriosum
Umbilical arteries	Arise from internal iliac arteries and form a portion of the umbilical cord	Transport blood from the fetus to the placenta	Atrophies to become the lateral umbilical ligaments

Some blood entering the right atrium passes into the right ventricle and out of the heart via the pulmonary trunk. Since the lungs of the fetus are not functional, only a small portion of blood continues through the pulmonary circulation (the resistance to blood flow is very high in the collapsed fetal lungs). Most of the blood in the pulmonary trunk passes through the **ductus arteriosus** (*ar-te″re-o′sus*) into the aortic arch, where it mixes with blood coming from the left ventricle. Blood is returned to the placenta by the two umbilical arteries that arise from the internal iliac arteries.

Notice that, in the fetus, oxygen-rich blood is transported by the inferior vena cava to the heart, and via the foramen ovale and ductus arteriosus to the systemic circulation.

Important changes occur in the cardiovascular system at birth. The foramen ovale, ductus arteriosus, ductus venosus, and the umbilical vessels are no longer necessary. The foramen ovale abruptly closes with the first breath of air because the reduced pressure in the right side of the heart causes a flap to cover the opening. The reduction in pressure occurs because the vascular resistance to blood flow in the pulmonary circulation falls far below that of the systemic circulation when the lungs fill with air. The pressure in the inferior vena cava and right atrium falls as a result of the loss of the placental circulation.

The constriction of the ductus arteriosus occurs gradually over a period of about 6 weeks after birth as the vascular smooth muscle fibers constrict in response to the higher oxygen concentration in the postnatal blood. The remaining structure of the ductus arteriosus gradually atrophies and becomes the *ligamentum arteriosum* (see fig. 21.19). Transformation of the unique fetal cardiovascular system is summarized in table 21.5.

Clinical Considerations

Electrocardiograph Leads

There are two types of electrocardiograph recording electrodes, or "leads." The **bipolar limb leads** record the voltage between electrodes placed on the wrists and legs. These bipolar leads include lead I (right arm to left arm), lead II (right arm to left leg), and lead III (left arm to left leg). In the **unipolar leads,** voltage is recorded between a single "exploratory electrode" placed on the body and an electrode that is built into the electrocardiograph and maintained at zero potential (ground).

The unipolar limb leads are placed on the right arm, left arm, and left leg, and are abbreviated AVR, AVL, and AVF, respectively. The unipolar chest leads are labeled 1 through 6, starting from the midline position (fig. 21.37). Thus, there are a total of 12 standard ECG leads that "view" the changing pattern of the heart's electrical activity from different perspectives (table 21.6). Using many leads at one time is important because certain abnormalities are best seen with some and may not be visible at all with others.

Arrhythmias Detected by the Electrocardiogram

Arrhythmias, or abnormal heart rhythms, can be detected and described by the abnormal ECG patterns they produce. Although the proper clinical interpretation of electrocardiograms requires knowledge of technical information not covered in this chapter, some knowledge of abnormal rhythms is interesting in itself and is useful in understanding normal physiology.

Figure 21.37

The placement of the bipolar limb leads and the exploratory electrode for the unipolar chest leads in an electrocardiogram.
(RA = right arm; LA = left arm; LL = left leg.)

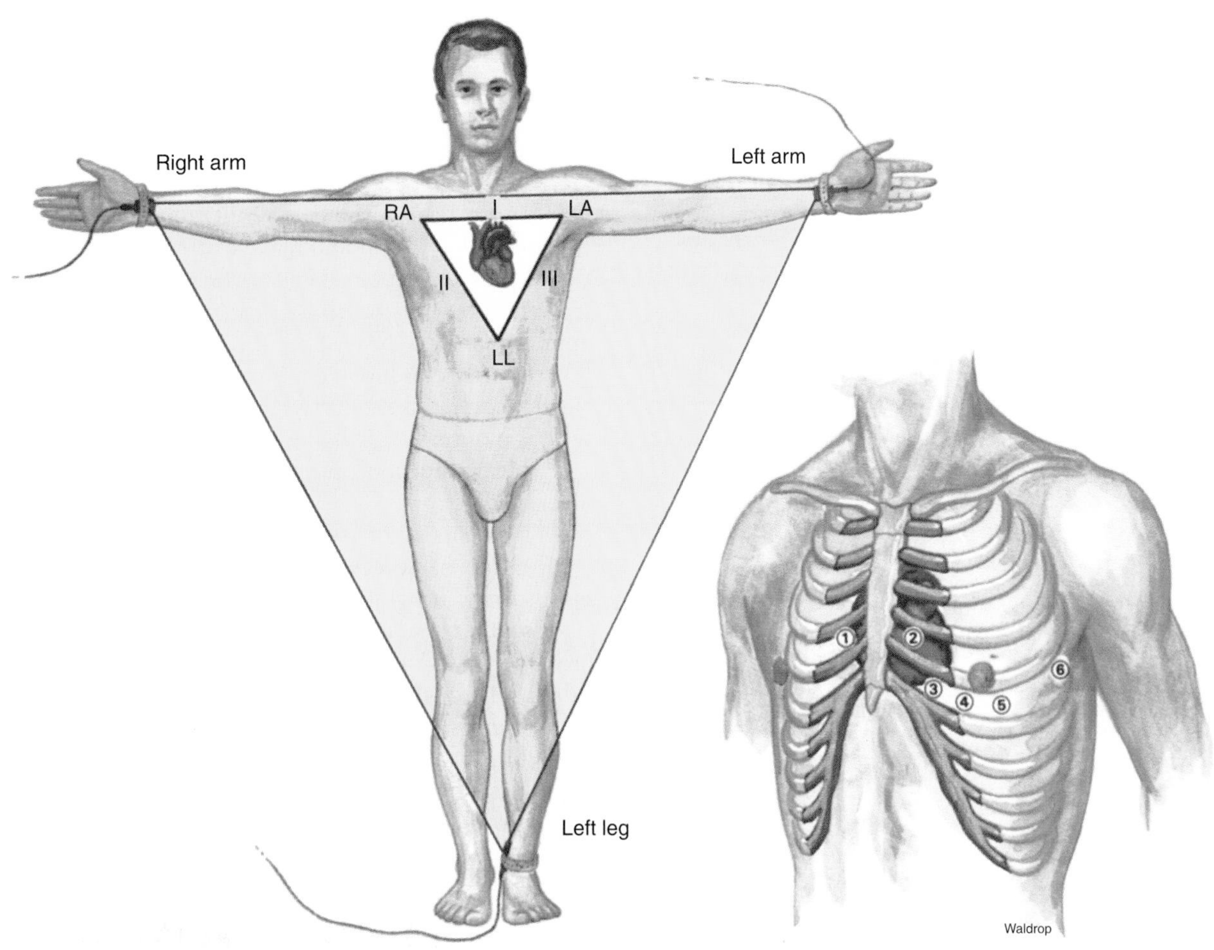

Since a heartbeat occurs whenever a normal QRS complex is seen, and since the ECG chart paper moves at a known speed so that its *x*-axis indicates time, the cardiac rate (beats per minute) can easily be obtained from the ECG recording. A cardiac rate slower than 60 beats per minute indicates **bradycardia** (*brad″ĭ-kar′de-ă*); a rate faster than 100 beats per minute is described as **tachycardia** (*tak″ĭ-kar′de-ă*).

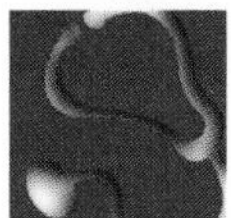

Both bradycardia and tachycardia can occur normally. Endurance-trained athletes, for example, commonly have slower heart rates than the general population. *Athlete's bradycardia* occurs as a result of higher levels of parasympathetic inhibition of the SA node and is a beneficial adaptation. Activation of the sympathetic division of the ANS during exercise or emergencies causes a normal tachycardia to occur.

bradycardia: Gk. *bradys*, slow; *kardia*, heart
tachycardia: Gk. *tachys*, rapid; *kardia*, heart

Abnormal tachycardia occurs when a person is at rest. This may result from abnormally fast pacing by the atria due to drugs, or it may result from the development of abnormally fast *ectopic pacemakers*—cells located outside the SA node that assume a pacemaker function. This abnormal atrial tachycardia thus differs from normal "sinus" (SA node) tachycardia. *Ventricular tachycardia* results when abnormally fast ectopic pacemakers in the ventricles cause them to beat rapidly and independently of the atria (fig. 21.38). This is very dangerous because it can quickly degenerate into a lethal condition known as *ventricular fibrillation*.

Ventricular Fibrillation

Fibrillation is caused by a continuous recycling of electrical waves through the myocardium. Normally, recycling is prevented because the myocardium simultaneously enters a refractory period at all regions. If some cells emerge from their refractory periods before others, however, electrical waves can be continuously regenerated and conducted. The recycling of electrical waves along continuously changing pathways produces uncoordinated contraction and an impotent pumping action. These effects can be produced by damage to the myocardium.

Fibrillation can sometimes be stopped by a strong electric shock delivered to the chest, a procedure called **electrical defibrillation.** The electric shock depolarizes all the myocardial cells at the same time, causing them to enter a refractory state. The conduction of random, recirculating impulses thus stops, and—within a short time—the SA node can begin to stimulate contraction in a normal fashion. Although this does not correct the initial problem that caused the abnormal electrical patterns, it can keep a person alive long enough so that other corrective measures can be taken.

Table 21.6
Electrocardiograph (ECG) Leads

Name of Lead	Placement of Electrodes
Bipolar Limb Leads	
I	Right arm and left arm
II	Right arm and left leg
III	Left arm and left leg
Unipolar Limb Leads	
AVR	Right arm
AVL	Left arm
AVF	Left leg
Unipolar Chest Leads	
V1	4th intercostal space right of sternum
V2	4th intercostal space left of sternum
V3	5th intercostal space left of sternum
V4	5th intercostal space in line with the middle of the clavicle
V5	5th intercostal space to the left of V4
V6	5th intercostal space in line with the middle of the axilla

Structural Heart Disorders

Congenital heart problems result from abnormalities in embryonic development and may be attributed to heredity, nutritional problems (poor diet) of the pregnant mother, or viral infections such as rubella. Congenital heart diseases occur in approximately 3 of every 100 births and account for about 50% of early childhood deaths. Many congenital heart defects can be corrected surgically, however, and others are not of a serious nature.

Heart murmurs can be congenital and acquired. Generally, they are of no clinical significance; nearly 10% of all people have heart murmurs, ranging from slight to severe. In general, three basic conditions cause murmurs: (1) *valvular insufficiency,* in which the cusps of the valves do not form a tight seal; (2) *stenosis,* in which the walls surrounding a valve are roughened or constricted; and (3) *turbulence of the blood* moving through the heart during heavy exercise. This last condition produces functional murmurs, which are common in children; they are not considered pathological.

A **septal defect** is the most common type of congenital heart problem. An **atrial septal defect,** or **patent foramen**

stenosis: Gk. *stenosis*, a narrowing

Figure 21.38

ECG patterns.
In (*a*) the heartbeat is paced by the normal pacemaker—the SA node (hence the name "sinus rhythm"). This can be abnormally slow (bradycardia—46 beats per minute in this example) or fast (tachycardia—136 beats per minute in this example). Compare the pattern of tachycardia in (*a*) with the tachycardia in (*b*). Ventricular tachycardia is produced by an ectopic pacemaker in the ventricles. This dangerous condition can quickly lead to ventricular fibrillation, also shown in (*b*).

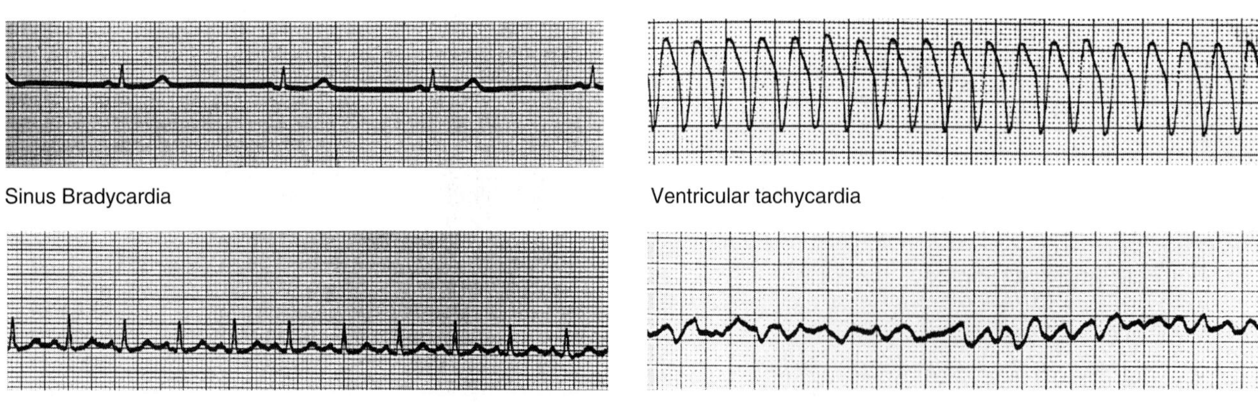

ovale, is a failure of the fetal foramen ovale to close after birth. A ventricular septal defect is caused by an abnormal development of the interventricular septum. **Pulmonary stenosis** is a narrowing of the opening into the pulmonary trunk from the right ventricle. It may lead to a pulmonary embolism and is usually recognized by extreme lung congestion. A **patent ductus arteriosus** is a failure of the ductus arteriosus to close after birth, allowing a backflow of blood into the pulmonary circulation from the aortic arch.

The **tetralogy of Fallot** (fig. 21.39) is a combination of four defects in a newborn and immediately causes a cyanotic condition, leading to the newborn being termed a "blue baby." The four characteristics of this anomaly are (1) a ventricular septal defect, (2) an ascending aorta that has shifted in position so that it overrides the interventricular septum and thus receives blood from the right as well as left ventricle, (3) pulmonary stenosis, and (4) right ventricular hypertrophy (fig. 21.39). Pulmonary stenosis obstructs blood flow to the lungs and causes hypertrophy of the right ventricle. Open-heart surgery is necessary to correct this condition, and the overall mortality rate is about 5%.

Acquired heart disease may develop suddenly or gradually. Heart attacks are included in this category and are the leading cause of death in the United States. The American Heart Association estimates that one in five individuals over the age of 60 will succumb to a heart attack. The immediate cause of a heart attack is generally one of the following: an inadequate coronary blood supply, an anatomical disorder, or a conduction disturbance.

Other types of acquired heart diseases affect the layers of the heart. **Bacterial endocarditis** is a disease of the lining of the heart, especially of the cusps of the valves. It is caused by infectious organisms that enter the bloodstream. **Myocardial disease** is an inflammation of the heart muscle followed by cardiac enlargement and congestive heart failure. **Pericarditis** causes an inflammation of the pericardium—the covering membrane of the heart. Its distinctive feature is pericardial friction rub, a transitory scratchy sound heard during auscultation.

A tissue is said to be **ischemic** (*ĭ-ske′mik*) when it receives an inadequate supply of oxygen because of an inadequate blood flow. The most common cause of myocardial ischemia is atherosclerosis of the coronary arteries. The adequacy of blood flow is relative—it depends on the metabolic requirements of the tissue for oxygen. An obstruction in a coronary artery, for example, may allow sufficient blood flow at rest but may produce ischemia when the heart is stressed by exercise or emotional factors. In patients with this condition, angioplasty or coronary artery bypass surgery may be performed.

Myocardial ischemia is associated with increased concentrations of blood lactic acid produced by anaerobic respiration of the ischemic tissue. This condition often causes substernal pain, which may also be referred to the left shoulder and arm, as well as to other areas. This referred pain is called **angina pectoris.** People with angina frequently take nitroglycerin or related drugs that help to relieve the ischemia and pain. These drugs are effective because they stimulate vasodilation, which improves circulation to the heart and decreases the work that the heart must perform to eject blood into the arteries. Myocardial cells are adapted to respire aerobically and cannot respire anaerobically for more than a few minutes. If ischemia and anaerobic respiration continue for more than a few minutes, *necrosis* (cellular death) may occur in the areas most deprived of oxygen. A sudden, irreversible injury of this kind is called a **myocardial infarction (MI).** The lay term "heart attack," though imprecise, usually refers to a myocardial infarction.

Myocardial ischemia may be detected by characteristic changes in the electrocardiogram. The diagnosis of myocardial infarction is aided by measurement of the concentration of enzymes in the blood that are released by the infarcted tissue. Plasma concentrations of *creatine phosphokinase* (*CPK*), for example, increase within 3 to 6 hours after the onset of symptoms and return to normal after 3 days. Plasma levels of *lactate dehydrogenase* (*LDH*) reach a peak within 48 to 72 hours after the onset of symptoms and remain elevated for about 11 days.

tetralogy of Fallot: from Étienne-Louis A. Fallot, French physician, 1850–1922

Figure 21.39

The tetralogy of Fallot.
The four defects of this anomaly are (1) ventricular septal defect, (2) an overriding aorta, (3) pulmonary stenosis, and (4) right ventricular hypertrophy.

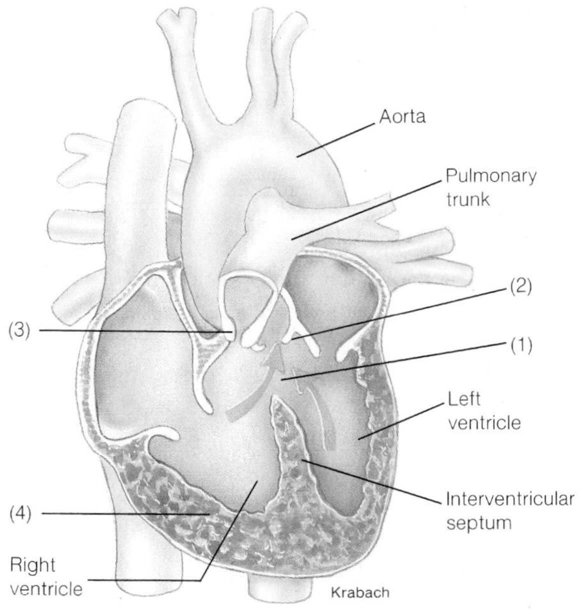

Atherosclerosis

Atherosclerosis is the most common form of arteriosclerosis (hardening of the arteries) and, through its contribution to heart disease and stroke, is responsible for about 50% of the

Figure 21.40

Atherosclerosis.

(*a*) The lumen of a human coronary artery is almost completely occluded by an atheroma. (*b*) A close-up view of the cleared left anterior descending coronary artery containing calcified atherosclerotic plaques. (*c*) The structure of an atheroma is diagrammed.

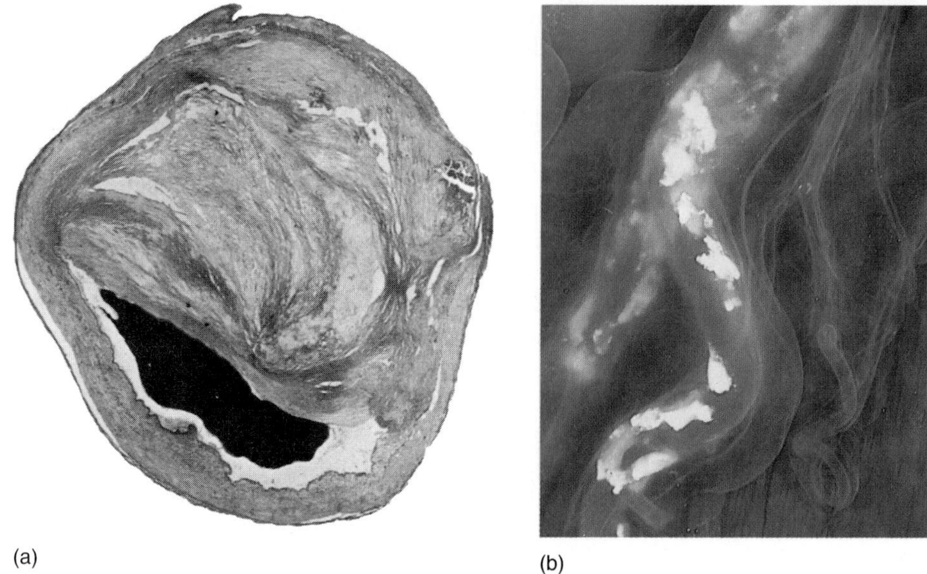

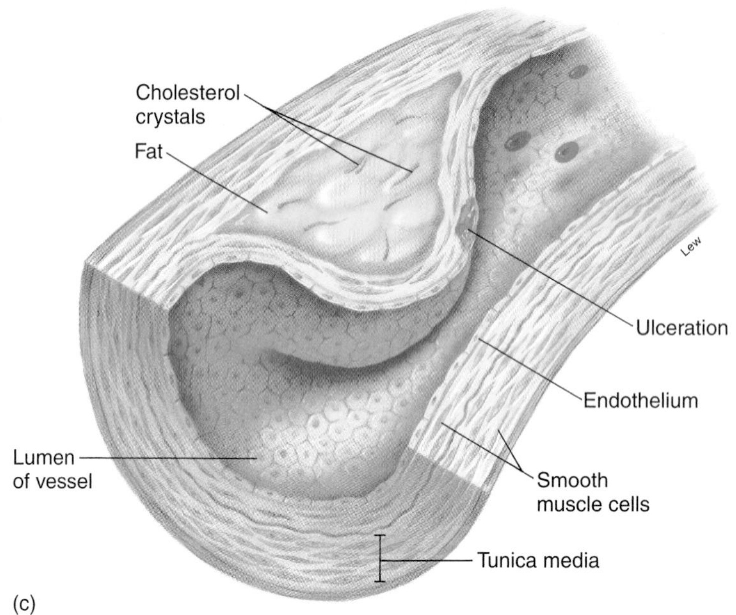

deaths in North America, Europe, and Japan. In atherosclerosis, localized plaques, or **atheromas,** protrude into the lumen of the artery and thus reduce blood flow. The atheromas additionally serve as sites for *thrombus* (blood clot) *formation*, which can further occlude the blood supply to an organ (fig. 21.40).

It is currently believed that the process of atherosclerosis begins as a result of damage, or "insult," to the endothelium. Such insults are produced by smoking, hypertension (high blood pressure), high blood cholesterol, and diabetes. The first anatomically recognized change is the appearance of "fatty streaks," which are gray-white areas that protrude into the lumina of arteries, particularly at arterial branch points. These are aggregations of lipid-filled macrophages and lymphocytes within the tunica intima. They are present to a small degree in the aorta and coronary arteries of children aged 10 to 14, but progress to more advanced stages at different rates in different people. In the intermediate stage, the area contains layers of macrophages and smooth muscle cells. The more advanced lesions are called *fibrous plaques* and consist of a cap of connective tissue with smooth muscle cells over accumulated lipid and debris, macrophages that have been derived from monocytes (chapter 23), and lymphocytes.

The process may be instigated by damage to the endothelium, but its progression appears to be a result of a wide variety of cytokines and other autocrine regulators (chapter 19) secreted by the endothelium and by the other participating cells, including platelets, macrophages, and lymphocytes. Some of these regulators attract monocytes and

lumen: L. *lumen*, opening

lymphocytes to the damaged endothelium and cause them to penetrate into the tunica intima. The monocytes then become macrophages, engulf lipids, and take on the appearance of "foamy cells." Smooth muscle cells change from a contractile state to a "synthetic" state where they produce and secrete connective tissue matrix proteins. (This is unique; in other tissues, connective tissue matrix is produced by cells called fibroblasts.) The changed smooth muscle cells respond to chemical attractants and migrate from the tunica media to the tunica intima, where they can proliferate.

Endothelial cells normally prevent this progression by presenting a physical barrier to the penetration of monocytes and lymphocytes, and by secreting autocrine regulators. Hypertension, smoking, and high blood cholesterol, among other risk factors, interfere with this protective function. The role of cholesterol in this process is described in the next section.

Cholesterol and Plasma Lipoproteins

There is good evidence that high blood cholesterol is associated with an increased risk of atherosclerosis. This high blood cholesterol can be produced by a diet rich in cholesterol and saturated fat, or it may be the result of an inherited condition known as **familial hypercholesteremia.** This condition is inherited as a single dominant gene; individuals who inherit two of these genes have extremely high cholesterol concentrations (regardless of diet) and usually suffer heart attacks during childhood.

Lipids, including cholesterol, are carried in the blood attached to protein carriers, as discussed in detail in chapter 26. Cholesterol is carried to the arteries by plasma proteins called **low-density lipoproteins (LDLs).** These particles, produced by the liver, consist of a core of cholesterol surrounded by a layer of phospholipids and a protein (to make the particle water-soluble). Cells in various organs contain receptors for the protein in LDL. When LDL attaches to its receptors, the cell engulfs the LDL by receptor-mediated endocytosis (described in chapter 3) and utilizes the cholesterol for different purposes. Most of the LDL in the blood is removed in this way by the liver.

When endothelial cells engulf LDL, they oxidize it to a product called oxidized LDL. Recent evidence suggests that oxidized LDL contributes to endothelial cell injury, migration of monocytes and lymphocytes into the tunica intima, conversion of monocytes into macrophages, and other events that occur in the progression of atherosclerosis. Since oxidized LDL appears to be so important in the progression of atherosclerosis, it seems logical that antioxidant compounds may aid in the prevention or treatment of this condition. The antioxidant drug *probucol,* as well as vitamin C, vitamin E, and beta-carotene, which are antioxidants (chapter 27), have been shown to be effective in this regard in experimental animals.

People who eat a diet high in cholesterol and saturated fat, and people with familial hypercholesteremia, have a high blood LDL concentration because their livers have a low number of LDL receptors. With fewer LDL receptors, the liver is less able to remove the LDL from the blood and thus more LDL is available to enter the endothelial cells of arteries.

Excessive cholesterol may be released from cells and travel in the blood as **high-density lipoproteins (HDLs),** which are removed by the liver. The cholesterol in HDLs is not taken into the artery wall because these cells lack the membrane receptor required for endocytosis of HDL, and therefore this cholesterol does not contribute to atherosclerosis. Indeed, in contrast to LDL, a high proportion of cholesterol in HDL is beneficial, since it indicates that cholesterol may be traveling away from the blood vessels to the liver. The concentration of HDL-cholesterol appears to be higher and the risk of atherosclerosis lower in people who exercise regularly. The HDL-cholesterol concentration, for example, is higher in marathon runners than in joggers and is higher in joggers than in sedentary individuals. Women in general have higher HDL-cholesterol concentrations and a lower risk of atherosclerosis than men.

Most people can significantly lower their blood cholesterol concentration through a regimen of exercise and diet. Since saturated fat in the diet raises blood cholesterol, such foods as fatty meat, egg yolks, and internal animal organs (liver, brain, etc.) should be eaten only sparingly. The American Heart Association recommends that fat contribute less than 30% to the total calories of a diet, and many experts argue for an even lower percentage. By way of comparison, the typical fast-food meal contains from 40% to 58% of its calories as fat. The single most effective action that smokers can take to lower their risk of atherosclerosis, however, is to stop smoking.

Other Vascular Disorders

An **aneurism** (*an-yŭ-riz-em*) is an expansion or bulging of the heart, aorta, or any other artery. A **coarctation** is a constriction in a segment of a vessel, usually the aorta, and is frequently caused by a remnant of the ductus arteriosus tightening around the vessel. **Varicose veins** are weakened veins that become stretched and swollen. They are most common in the legs because the force of gravity tends to weaken the valves and overload the veins. Varicose veins can also occur in the rectum, in which case they are called **hemorrhoids.** *Vein stripping* is the surgical removal of superficial weakened veins. **Phlebitis** (*flĕ-bi'tis*), an inflammation of a vein, may develop as a result of trauma or be an aftermath of surgery. Frequently, however, it appears for no apparent reason. Phlebitis interferes with normal venous circulation.

Clinical Investigation Answer

Mural thrombus (a blood clot adherent to the inner surface of one of the heart's chambers) is a fairly common complication of myocardial infarction. Once a thrombus forms within the heart, a piece may break off and travel throughout the body. This is the most likely cause of the symptoms in the right leg of our patient. Because the embolus traveled and lodged in the systemic circulation (as opposed to the pulmonary circulation), the mural thrombus was probably located in the left side of the heart. The embolus traveled to a point in the femoral artery and lodged there, thus occluding blood flow to the popliteal artery and its distal branches. The route of travel was as follows: left side of heart → ascending aorta → descending thoracic aorta → abdominal aorta → right common iliac artery → right external iliac artery → femoral artery.

The standard treatment for this problem is emergency surgery to extract the clot from the leg. Anticoagulation (blood-thinning) therapy is then continued or instituted.

Chapter Summary

Structure of the Heart (pp. 611–618)

1. The heart is enclosed within a pericardial sac. The wall of the heart consists of the epicardium, myocardium, and endocardium.
   (a) The right atrium receives blood from the superior and inferior venae cavae, and the right ventricle pumps blood through the pulmonary trunk into the pulmonary arteries.
   (b) The left atrium receives blood from the pulmonary veins, and the left ventricle pumps blood into the ascending aorta.
2. The heart contains right and left atrioventricular (AV) valves (the tricuspid and bicuspid valves, respectively), a pulmonary (semilunar) valve, and an aortic (semilunar) valve. Closing of the AV valves produces the "lub" sound at the beginning of systole; closing of the semilunar valves produces the "dub" sound at the beginning of diastole.
3. The two principal circulatory divisions are the pulmonary and the systemic; in addition, the coronary system serves the heart.
   (a) The pulmonary circulation includes the vessels that carry blood from the right ventricle through the lungs, and from there to the left atrium.
   (b) The systemic circulation includes all other arteries, capillaries, and veins in the body. These vessels carry blood from the left ventricle through the body and return blood to the right atrium.
   (c) The myocardium of the heart is served by right and left coronary arteries that branch from the ascending portion of the aorta. The coronary sinus collects and empties the blood into the right atrium.

Cardiac Cycle, Heart Sounds, and the Electrocardiogram (pp. 618–623)

1. The heart is a two-step pump; first the atria contract, and then the ventricles contract.
   (a) During diastole, first the atria and then the ventricles fill with blood.
   (b) The ventricles are about 80% filled before the atria contract and add the final 20% to the end-diastolic volume.
2. When the ventricles contract at systole, the pressure within them first rises sufficiently to close the AV valves and then rises sufficiently to open the semilunar valves.
   (a) Blood is ejected from the ventricles until the pressure within them falls below the pressure in the arteries; at this point, the semilunar valves close and the ventricles begin relaxation.
   (b) When the pressure in the ventricles falls below the pressure in the atria, a phase of rapid filling of the ventricles occurs, followed by the final filling caused by contraction of the atria.
3. The electrical impulse begins in the sinoatrial (SA) node and spreads through both atria by electrical conduction from one myocardial cell to another.
   (a) The impulse then excites the atrioventricular (AV) node, from which it is conducted by the atrioventricular bundle into the ventricles.
   (b) The conduction myofibers (Purkinje fibers) transmit the impulse into the ventricular muscle and cause it to contract.
4. The regular pattern of conduction in the heart produces a changing pattern of potential differences between two points on the body surface.
   (a) A recording of the voltage between two points on the surface of the body caused by the electrical activity of the heart is called an electrocardiogram (ECG).
   (b) The P wave is caused by depolarization of the atria; the QRS wave is caused by depolarization of the ventricles; the T wave is produced by repolarization of the ventricles.

Blood Vessels (pp. 624–628)

1. Arteries contain three layers, or tunics: the tunica intima, tunica media, and tunica externa.
   (a) The tunica intima consists of a layer of endothelium that is separated from the tunica media by a band of elastin fibers.
   (b) The tunica media consists of smooth muscle.
   (c) The tunica externa consists of loose connective tissue, which strengthens the arteries and provides elasticity.
2. Capillaries are the narrowest but the most numerous of the blood vessels; they provide for the exchange of molecules between the blood and the surrounding tissues.
   (a) Capillaries within most organs of the body have pores between adjacent endothelial cells, so that fluid derived from blood plasma can be filtered to produce interstitial, or tissue, fluid.
   (b) The capillaries in the brain are continuous and lack pores, thus contributing to the blood-brain barrier.
3. Veins have the same three tunics as arteries, but veins generally have a thinner muscular layer than comparably sized arteries.
   (a) Veins are more distensible than arteries and can expand to hold a larger quantity of blood.
   (b) Many veins have venous valves that permit a one-way flow of blood to the heart.
   (c) Contractions of skeletal muscles surrounding veins can squeeze the veins and aid the return of venous blood to the heart. This action is known as the skeletal muscle pump.

Principal Arteries of the Body (pp. 628–635)

1. Three arteries arise from the aortic arch: the brachiocephalic trunk, the left common carotid artery, and the left subclavian artery. The brachiocephalic trunk divides into the right common carotid artery and the right subclavian artery.
   (a) Each common carotid artery divides into internal and external carotid arteries.
   (b) Each subclavian artery gives rise to a vertebral artery and then continues through the shoulder region as the axillary artery.
2. The neck and head receive an arterial supply from branches of the internal and external carotid arteries and the vertebral arteries.
   (a) Each internal carotid artery enters the skull through the carotid foramen; each vertebral artery enters through the foramen magnum.
   (b) Branches of the internal carotid arteries and vertebral arteries form the cerebral arterial circle (circle of Willis), which supplies the brain.
3. The shoulder and upper extremity are served by the subclavian artery and its derivatives.
4. The abdominal portion of the aorta has the following branches: the inferior phrenic, celiac trunk, superior mesenteric, renal, suprarenal, testicular (or ovarian), and inferior mesenteric arteries.
5. The abdominal aorta terminates in the posterior pelvic area as it splits into the right and left common iliac arteries. These vessels terminate by dividing into the internal and external iliac arteries, which supply blood to the pelvis and lower extremities.

Principal Veins of the Body (pp. 635–644)

1. Blood from the head and neck is drained by the external and internal jugular veins; blood from the brain is drained by the internal jugular vein.
2. The upper extremity and shoulder region is drained by superficial and deep veins.
3. In the thorax, the superior vena cava is formed by the union of the two brachiocephalic veins and also collects blood from the azygos vein.
4. The lower extremity is drained by both superficial and deep veins. At the level of the fifth lumbar vertebra, the right and left iliac veins unite to form the inferior vena cava.
5. Blood from capillaries in the GI tract and accessory digestive organs is drained via the hepatic portal vein to the liver.
   (a) A portal system is one in which there is a second capillary bed downstream from the first; in this case, the capillary bed in the liver is downstream from the GI tract.
   (b) The liver can modify the chemical composition of the blood arriving from the GI tract before this blood goes back to the heart to enter the general circulation.

Fetal Circulation (pp. 644–646)

1. Fully oxygenated blood is carried only in the umbilical vein, which drains the placenta. This blood is carried via the ductus venosus to the inferior vena cava of the fetus.
2. Partially oxygenated blood is shunted from the right atrium to the left atrium via the foramen ovale and from the pulmonary trunk to the aortic arch via the ductus arteriosus.
   (a) In this way, blood is diverted away from the lungs, which are not active in oxygenating the blood, to the systemic circulation, where it can be delivered to the placenta via the umbilical arteries.
   (b) The foramen ovale normally closes immediately following a newborn's first breath; the ductus arteriosus closes by the sixth week following birth.

Review Activities

Objective Questions

1. All arteries in the body contain oxygen-rich blood with the exception of
   (a) the aorta.
   (b) the pulmonary arteries.
   (c) the renal arteries.
   (d) the coronary arteries.
2. Most blood from the coronary circulation directly enters
   (a) the inferior vena cava.
   (b) the superior vena cava.
   (c) the right atrium.
   (d) the left atrium.
3. The second heart sound immediately follows the occurrence of which event?
   (a) P wave
   (b) QRS wave
   (c) T wave
   (d) U wave
4. Which of the following arteries does *not* arise from the aortic arch?
   (a) brachiocephalic trunk
   (b) coronary artery
   (c) left common carotid artery
   (d) left subclavian artery
5. Which of the following arteries does *not* supply blood to the brain?
   (a) external carotid artery
   (b) internal carotid artery
   (c) vertebral artery
   (d) basilar artery
6. The maxillary and superficial temporal arteries are derived from
   (a) the external carotid artery.
   (b) the internal carotid artery.
   (c) the vertebral artery.
   (d) the facial artery.
7. Which of the following statements is *false?*
   (a) Most of the total blood volume is contained in veins.
   (b) Capillaries have a greater total surface area than any other type of vessel.
   (c) Exchanges between blood and tissue fluid occur across the walls of venules.
   (d) Small arteries and arterioles present great resistance to blood flow.
8. The "lub," or first heart sound, is produced by the closing of
   (a) the aortic semilunar valve.
   (b) the pulmonary semilunar valve.
   (c) the tricuspid valve.
   (d) the bicuspid valve.
   (e) both AV valves.
9. The first heart sound is produced at
   (a) the beginning of systole.
   (b) the end of systole.
   (c) the beginning of diastole.
   (d) the end of diastole.
10. Changes in the cardiac rate primarily reflect changes in the duration of
    (a) systole.
    (b) diastole.
11. The QRS wave of an ECG is produced by
    (a) depolarization of the atria.
    (b) repolarization of the atria.
    (c) depolarization of the ventricles.
    (d) repolarization of the ventricles.
12. The cells that normally have the fastest rate of spontaneous diastolic depolarization are located in
    (a) the SA node.
    (b) the AV node.
    (c) the atrioventricular bundle.
    (d) the conduction myofibers.
13. Which of the following statements is *true?*
    (a) The heart can produce a graded contraction.
    (b) The heart can produce a sustained contraction.
    (c) All of the myocardial cells in the ventricles are normally in a refractory period at the same time.

14. During the phase of isovolumetric relaxation of the ventricles, the pressure in the ventricles is
    (a) rising.
    (b) falling.
    (c) first rising, then falling.
    (d) constant.

Essay Questions

1. Explain why the beat of the heart is automatic and why the SA node functions as the normal pacemaker.
2. Compare the duration of the heart's contraction with those of the myocardial action potential and refractory period. Explain the significance of these relationships.
3. Describe the pressure changes that occur during the cardiac cycle and relate these changes to the occurrence of the heart sounds.
4. Describe the causes of the P, QRS, and T waves of an ECG and indicate when each of these waves occurs in the cardiac cycle. Explain why the first heart sound occurs immediately after the QRS wave and why the second sound occurs at the time of the T wave.
5. Can a defective valve be detected by an ECG? Can a partially damaged AV node be detected by auscultation with a stethoscope? Explain.
6. Describe the functions of the foramen ovale and ductus arteriosus in a fetus and explain why the newborn is at risk if they remain patent after birth.
7. Trace the flow of blood from the left ventricle to the upper teeth.
8. Trace the flow of blood from the small intestine, to the heart, and back to the small intestine.
9. What is significant about the hepatic portal system? What do we mean when we say that the liver has two blood supplies?
10. Name the vessels of the thoracic and shoulder regions that are not symmetrical (do not have a counterpart on the opposite side of the body).

Critical Thinking Questions

1. Examine figure 21.2*c* and predict the structures that might be harmed in the event of endocarditis involving the endocardium lining the interventricular septum or ventricular walls.
2. The walls of the ventricles are thicker than those of the atria, and the wall of the left ventricle is the thickest of all. How do these structural differences relate to differences in function?
3. An endurance-trained athlete will typically have a lower resting cardiac rate and a greater stroke volume than a person who is out of shape. Explain why these adaptations are beneficial.
4. Some passenger planes are now equipped with defibrillators for use as emergency life-saving devices. Why is it critical that ventricular fibrillation receive attention within minutes? Is atrial fibrillation less urgent? Explain.
5. A hospitalized 45-year-old man developed a thrombus (blood clot) in his lower thigh following severe trauma to his knee. The patient's physician explained that although the clot was near the great saphenous vein, the main concern was a pulmonary embolism. Explain the physician's reasoning and list, in sequence, the vessels the clot would have to pass through to cause a heart problem.

Related Web Sites

For a listing of the most current web sites related to this chapter, please visit the *Concepts of Human Anatomy and Physiology* home page at http://www.mhhe.com/biosci/abio/.

Twenty-Two

Circulatory System: Cardiac Output and Blood Flow

Objectives

- Define *cardiac output* and explain how the cardiac output is calculated.
- Explain how autonomic nerves regulate the cardiac rate and the strength of ventricular contraction.
- Explain the Frank–Starling law of the heart.
- Explain how the venous return of blood to the heart is regulated.
- Describe how tissue fluid forms and how it returns to blood capillaries.
- Describe the regulation of ADH secretion and the effects of ADH on blood volume.
- Describe the regulation of aldosterone secretion and the effects of aldosterone on blood volume and pressure.
- Explain how blood flow is affected by blood pressure and by vascular resistance.
- Discuss the regulation of vascular resistance by the autonomic nervous system and by intrinsic regulatory mechanisms.
- Describe the relationship between resistance and the radius of a vessel and explain how blood flow can be diverted from one organ to another.
- Explain how resistance and blood flow are regulated by sympathetic and parasympathetic innervation.
- Discuss autoregulation and explain how it is accomplished.
- Describe the mechanisms that control cerebral blood flow.
- Describe the cutaneous circulation and explain how it is regulated.
- Describe the changes in blood pressure as blood passes through the arterial system to capillaries, and then to the venous system.
- State the factors that directly influence blood pressure.
- Describe the baroreceptor reflex and comment on its significance.
- Describe the auscultatory method of blood pressure measurement.

Clinical Investigation

A young man who was participating in a class project in the Mojave Desert wandered off on his own and lost his way. Thirty-six hours later, he was found crawling along a seldom-used one-lane road. He was very weak, his skin was cold, and he was found to have low blood pressure and a rapid pulse. Intravenous albumin was administered in the hospital, where it was further observed that he had a low urine output. Analysis of his urine revealed a high total solute concentration (osmolality), but a virtual absence of sodium. What could account for this man's symptoms and laboratory findings?

Clues: Study the sections "Exchange of Fluid between Capillaries and Tissues," "Regulation of Blood Volume by the Kidneys," and "Baroreceptor Reflex." Also note the description of hypovolemic shock in the last section of the chapter.

Cardiac Output

The pumping ability of the heart is a function of the number of beats per minute (cardiac rate) and the volume of blood ejected per beat (stroke volume). The cardiac rate and stroke volume are regulated by autonomic nerves and by mechanisms intrinsic to the cardiovascular system.

The **cardiac output** is equal to the volume of blood pumped per minute by each ventricle. The average resting **cardiac rate** in an adult is 70 beats per minute; the average **stroke volume** (volume of blood pumped per beat by each ventricle) is 70 to 80 ml per beat. The product of these two variables gives an average cardiac output of 5,500 ml (5.5 L) per minute:

$$\frac{\text{Cardiac output}}{\text{(ml/min)}} = \frac{\text{stroke volume}}{\text{(ml/beat)}} \times \frac{\text{cardiac rate}}{\text{(beats/min)}}$$

The **total blood volume** also averages about 5.5 L. This means that each ventricle pumps the equivalent of the total blood volume each minute under resting conditions. Put another way, it takes about a minute for a drop of blood to complete the systemic and pulmonary circuits. An increase in cardiac output, as occurs during exercise, must thus be accompanied by an increased rate of blood flow through the circulation. This is accomplished by factors that regulate the cardiac rate and stroke volume.

Regulation of Cardiac Rate

In the complete absence of neural influences, the heart will continue to beat according to the rhythm set by the SA node. This automatic rhythm is produced by the spontaneous depolarization of the resting membrane potential to a threshold level, at which point voltage-regulated membrane gates are opened and action potentials are produced. As described in chapter 21, Ca^{2+} enters the myocardial cytoplasm during the action potential, attaches to troponin, and causes contraction.

Figure 22.1

Effect of autonomic nerves on the pacemaker potentials in the SA node.
The heart's rhythm is set by the rate of spontaneous depolarization in the SA node. This spontaneous depolarization is known as the pacemaker potential, and its rate is increased by sympathetic nerve stimulation and decreased by parasympathetic nerve inhibition.

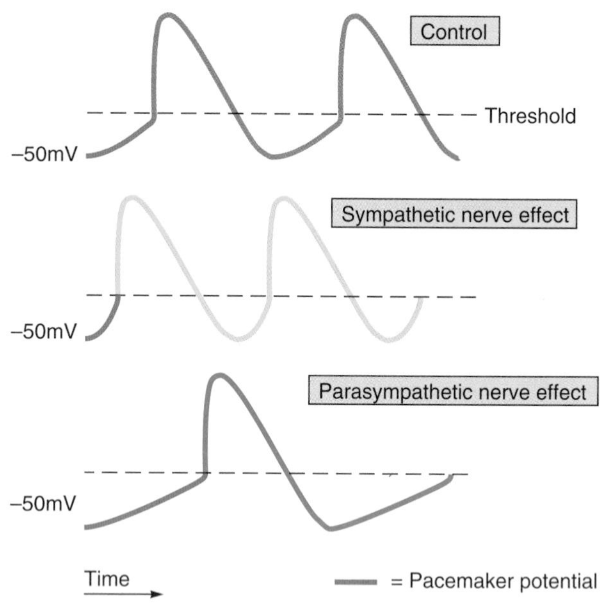

Normally, however, sympathetic and vagus (parasympathetic) nerve fibers to the heart are continuously active and modify the rate of spontaneous depolarization of the SA node. Norepinephrine, released primarily by sympathetic nerve endings, and epinephrine, secreted by the adrenal medulla, stimulate an increase in the spontaneous rate of firing of the SA node. Acetylcholine, released from parasympathetic endings, hyperpolarizes the SA node and thus decreases the rate of its spontaneous firing (fig. 22.1). The actual pace set by the SA node at any time depends on the net effect of these antagonistic influences. Mechanisms that affect the cardiac rate are said to have a **chronotropic** (*kron"ŏ-trop'ik*) **effect.** Those that increase cardiac rate have a positive chronotropic effect; those that decrease the rate have a negative chronotropic effect.

Autonomic innervation of the SA node represents the major means by which cardiac rate is regulated. However, other autonomic control mechanisms also affect cardiac rate to a lesser degree. Sympathetic endings in the musculature of the atria and ventricles increase the strength of contraction and cause a slight decrease in the time spent in systole when the cardiac rate is high (table 22.1).

During exercise, the cardiac rate first increases as a result of decreased inhibition of the SA node by the vagus nerves. Further increases in cardiac rate are achieved by increased

chronotropic: Gk. *chronos*, time; *trope*, turn, change

Table 22.1

Effects of Autonomic Nerve Activity on the Heart

Region Affected	Sympathetic Nerve Effects	Parasympathetic Nerve Effects
SA node	Increased rate of diastolic depolarization; increased cardiac rate	Decreased rate of diastolic depolarization; decreased cardiac rate
AV node	Increased conduction rate	Decreased conduction rate
Atrial muscle	Increased strength of contraction	Decreased strength of contraction
Ventricular muscle	Increased strength of contraction	No significant effect

sympathetic nerve stimulation. The resting bradycardia (slow heart rate) of endurance-trained athletes is due largely to high vagal activity.

The activity of the autonomic innervation of the heart is coordinated by **cardiac control centers** in the medulla oblongata of the brain stem. The question of whether there are separate cardioaccelerator and cardioinhibitory centers in the medulla is currently controversial. These cardiac control centers, in turn, are affected by higher brain areas and by sensory feedback from pressure receptors, or *baroreceptors*, in the aorta and carotid arteries. In this way, a rise in blood pressure can produce a reflex slowing of the heart. This *baroreceptor reflex* is discussed in more detail in relation to blood pressure regulation later in this chapter.

Regulation of Stroke Volume

The stroke volume is regulated by three variables: (1) the **end-diastolic volume (EDV),** which is the volume of blood in the ventricles at the end of diastole; (2) the **total peripheral resistance,** which is the frictional resistance, or impedance to blood flow in the arteries; and (3) the **contractility,** or strength, of ventricular contraction.

The end-diastolic volume is the amount of blood in the ventricles immediately before they begin to contract. This is a workload imposed on the ventricles prior to contraction, and thus is sometimes called a **preload.** The stroke volume is directly proportional to the preload; an increase in EDV results in an increase in stroke volume. The stroke volume is also directly proportional to contractility; when the ventricles contract more forcefully, they pump more blood.

In order to eject blood, the pressure generated in a ventricle when it contracts must be greater than the pressure in the arteries (since blood flows only from a location of higher pressure to one of lower pressure). The pressure in the arterial system before the ventricle contracts is, in turn, a function of the total peripheral resistance—the higher the peripheral resistance, the higher the pressure. As blood begins to be ejected from the ventricle, the added volume of blood in the arteries causes a rise in mean arterial pressure against the "bottleneck" presented by the peripheral resistance; ejection of blood stops shortly after the aortic pressure becomes equal to the intraventricular pressure. The total peripheral resistance thus presents an impedance to the ejection of blood from the ventricle, or an **afterload** imposed on the ventricle after contraction has begun.

In summary, the stroke volume is inversely proportional to the total peripheral resistance; the greater the peripheral resistance, the lower the stroke volume. It should be noted that this lowering of stroke volume in response to a raised peripheral resistance occurs for only a few beats. Thereafter, a healthy heart is able to compensate for the increased peripheral resistance by beating more strongly. This compensation occurs by means of a mechanism called the Frank–Starling law, to be described shortly.

The proportion of the end-diastolic volume that is ejected against a given afterload depends on the strength of ventricular contraction. Normally, contraction strength is sufficient to eject 70 to 80 ml of blood out of a total end-diastolic volume of 110 to 130 ml. The *ejection fraction* is thus about 60%. More blood is pumped per beat as the EDV increases, and thus the ejection fraction remains relatively constant over a range of end-diastolic volumes. In order for this to be true, the strength of ventricular contraction must increase as the end-diastolic volume increases.

Frank–Starling Law of the Heart

Two physiologists, Otto Frank and Ernest Starling, demonstrated in 1918 that the strength of ventricular contraction varies directly with the end-diastolic volume (fig. 22.2). Even in experiments where the heart is removed from the body (and is thus not subject to neural or hormonal regulation), and where the still-beating heart is filled with blood flowing from a reservoir, an increase in EDV within the physiological range results in increased contraction strength and, therefore, in increased stroke volume. This relationship between EDV, contraction strength, and stroke volume is thus a built-in, or *intrinsic*, property of heart muscle, and is known as the **Frank–Starling law of the heart.**

Frank–Starling law of the heart: from Otto Frank, German physiologist, 1865–1944, and Ernest Starling, English physiologist, 1866–1927

Figure 22.2

The Frank–Starling law and sympathetic nerve effects.

The graphs demonstrate the Frank–Starling law: as the end-diastolic volume is increased, the stroke volume is increased. The graphs also demonstrate, by comparing the three lines, that the stroke volume is higher at any particular end-diastolic volume when the ventricle is stimulated by sympathetic nerves. This is shown by the steeper curves to the left (see the red arrow).

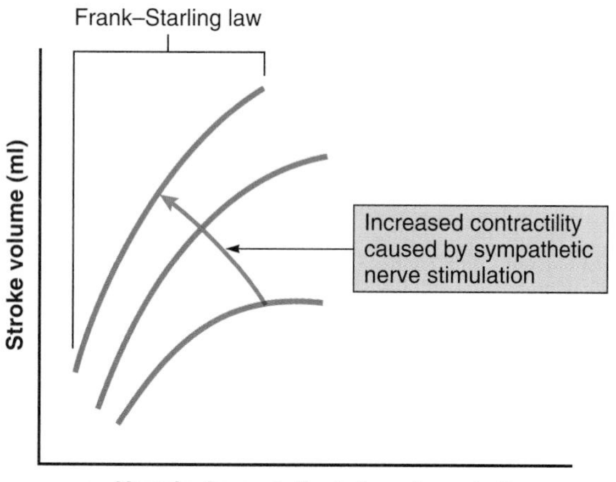

Intrinsic Control of Contraction Strength

The intrinsic control of contraction strength and stroke volume is due to variations in the degree to which the myocardium is stretched by the end-diastolic volume. As the EDV rises within the physiological range the myocardium is increasingly stretched, and as a result, contracts more forcefully.

As discussed in chapter 12, stretch can also increase the contraction strength of skeletal muscles (see fig. 12.16). The resting length of skeletal muscles is close to ideal, however, so that significant stretching decreases contraction strength. This is not true of the heart. Prior to filling with blood during diastole, the sarcomere lengths of myocardial cells are only about 1.5 µm. At this length, the actin filaments from each side overlap in the middle of the sarcomeres and the cells can contract only weakly (fig. 22.3).

As the ventricles fill with blood, the myocardium stretches so that the actin filaments overlap with myosin only at the edges of the A bands (fig. 22.3). This allows more force to be developed during contraction. Since this more advantageous overlapping of actin and myosin is produced by stretching of the ventricles, and since the degree of stretching is controlled by the degree of filling (the end-diastolic volume), the strength of contraction is intrinsically adjusted by the end-diastolic volume.

The Frank–Starling law explains how the heart can adjust to a rise in total peripheral resistance: (1) a rise in peripheral resistance causes a decrease in the stroke volume of the ventricle, so that (2) more blood remains in the ventricle and the end-diastolic volume is greater for the next cycle; as a result, (3) the ventricle is stretched to a greater degree in the next cycle and contracts more strongly to eject more blood. This allows a healthy ventricle to sustain a normal cardiac output.

A very important consequence of these events is that the cardiac output of the left ventricle, which pumps blood into the systemic circulation with its ever-changing resistances, can be adjusted to match the output of the right ventricle, which pumps blood into the pulmonary circulation. Clearly, the rate of blood flow through the pulmonary and systemic circulations must be equal in order to prevent fluid accumulation in the lungs and to deliver fully oxygenated blood to the body.

Extrinsic Control of Contractility

The *contractility* is the strength of contraction at any given fiber length. At any given degree of stretch, the strength of ventricular contraction depends on the activity of the sympathoadrenal system. Norepinephrine from sympathetic nerve endings and epinephrine from the adrenal medulla produce an increase in contraction strength (see fig. 22.2). This **positive inotropic effect** is believed to result from an increase in the amount of Ca^{2+} available to the sarcomeres.

The cardiac output is thus affected in two ways by the activity of the sympathoadrenal system: (1) through a positive inotropic effect on contractility and (2) through a positive chronotropic effect on cardiac rate (fig. 22.4). Stimulation through parasympathetic nerve endings to the SA node and conducting tissue has a negative chronotropic effect but does not directly affect the contraction strength of the ventricles. However, the increased EDV that results from a slower cardiac rate can increase contraction strength through the Frank–Starling mechanism.

Venous Return

The end-diastolic volume—and thus the stroke volume and cardiac output—is controlled by factors that affect the **venous return,** which is the return of blood to the heart via veins. The rate at which the atria and ventricles fill with venous blood depends on the total blood volume and the venous pressure (pressure in the veins). It is the venous pressure that serves as the driving force for the return of blood to the heart.

Veins have thinner, less muscular walls than arteries; thus, they have a higher **compliance.** This means that a given amount of pressure will cause more distension (expansion) in veins than in arteries, so that the veins can hold more blood. Approximately two-thirds of the total blood volume is located in the veins (fig. 22.5). Veins are therefore called *capacitance vessels*, after electronic devices called capacitors that store electrical charges. Muscular arteries and arterioles expand less under pressure (are less compliant), and thus are called *resistance vessels*.

---

inotropic: Gk. *inos*, fiber; *trope*, turn, change

Figure 22.3

The Frank–Starling mechanism (law of the heart).

When the heart muscle is subjected to an increasing degree of stretch, it contracts more forcefully. As a result of the increased contraction strength (shown as tension), the time required to reach maximum contraction remains constant, regardless of the degree of stretch.

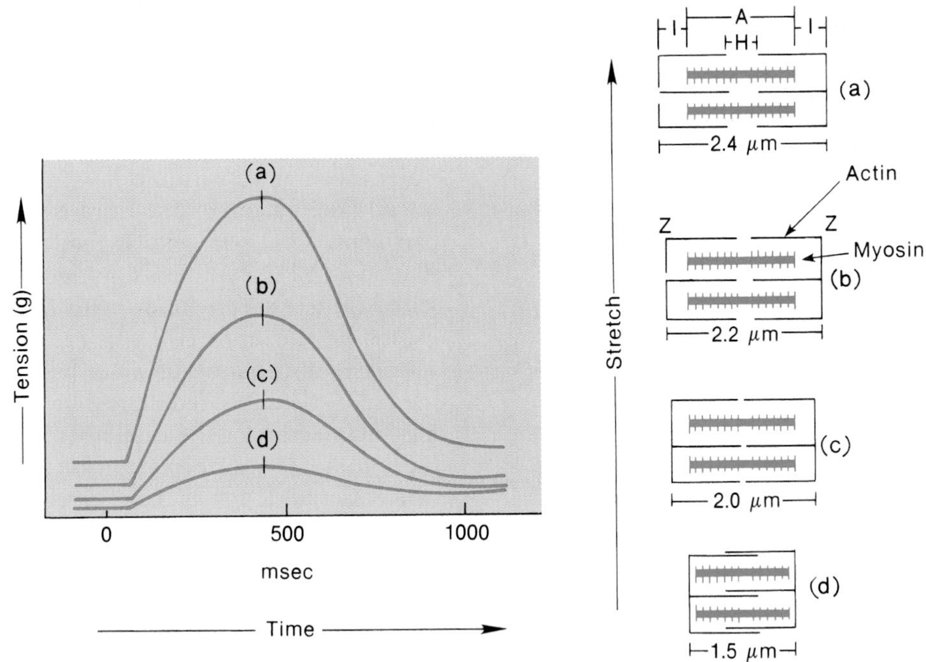

Figure 22.4

The regulation of cardiac output.

Factors that stimulate cardiac output are shown as solid arrows; factors that inhibit cardiac output are shown as dashed arrows.

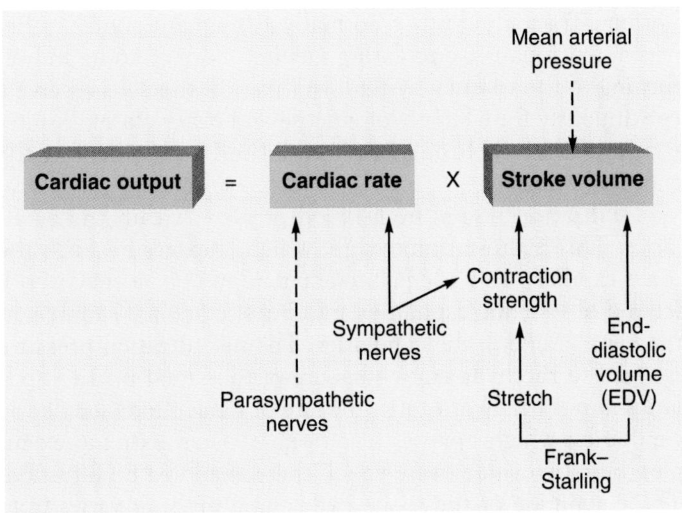

Figure 22.5

The distribution of blood within the circulatory system at rest.

Note that the venous system contains most of the blood; it functions as a reservoir that can add more blood to the circulation under appropriate conditions (such as exercise).

Source: Data from Bjorn Folkow and Eric Neil, *Circulation*. Copyright © 1971 Oxford University Press.

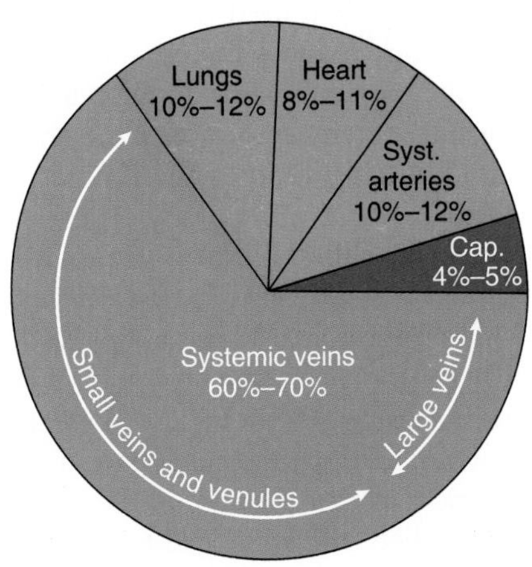

Figure 22.6

Variables that affect venous return and thus end-diastolic volume.
Direct relationships are indicated by solid arrows; inverse relationships are shown with dashed arrows.

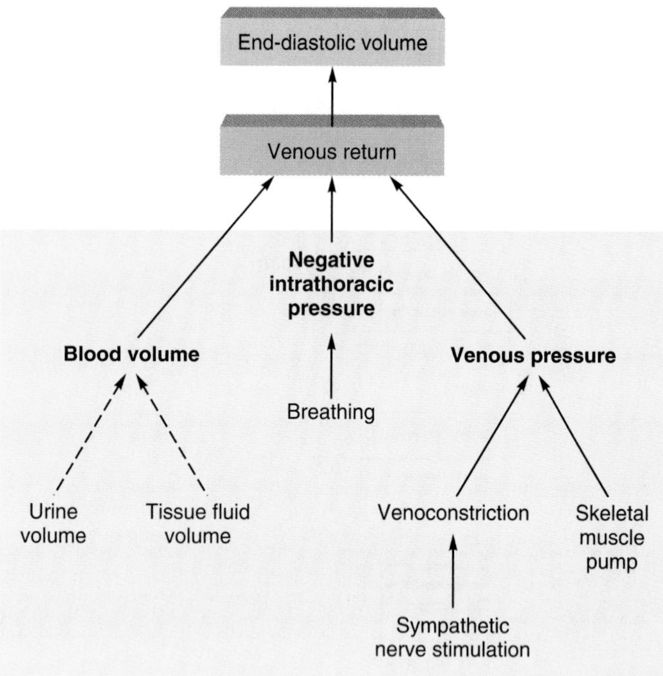

Although veins contain almost 70% of the total blood volume the mean venous pressure is only 2 mmHg, compared to a mean arterial pressure of 90 to 100 mmHg. The lower venous pressure is due in part to a pressure drop between arteries and capillaries and in part to the high venous compliance.

The venous pressure is highest in the venules (10 mmHg) and lowest at the junction of the venae cavae with the right atrium (0 mmHg). In addition to this pressure difference, the venous return to the heart is aided by (1) sympathetic nerve activity, which stimulates smooth muscle contraction in the venous walls and thus reduces compliance; (2) the skeletal muscle pump, which squeezes veins during muscle contraction; and (3) the pressure difference between the thoracic and abdominal cavities, which promotes the flow of venous blood back to the heart.

Contraction of the skeletal muscles functions as a "pump" by virtue of its squeezing action on veins (described in chapter 21). Contraction of the diaphragm during inhalation also improves venous return. The diaphragm lowers as it contracts, thus increasing the thoracic volume and decreasing the abdominal volume. This creates a partial vacuum in the thoracic cavity and a higher pressure in the abdominal cavity. The pressure difference thus produced favors blood flow from abdominal to thoracic veins (fig. 22.6).

Blood Volume

Fluid in the extracellular environment of the body is distributed between the blood and the tissue fluid compartments by filtration and osmotic forces acting across the walls of capillaries. The function of the kidneys influences blood volume because urine is derived from blood plasma. The hormones ADH and aldosterone act on the kidneys to help regulate the blood volume.

Blood volume represents one part, or compartment, of the total body water. Approximately two-thirds of the total body water is contained within cells—in the intracellular compartment. The remaining one-third is in the **extracellular compartment.** This extracellular fluid is normally distributed so that about 80% is contained in the tissues—as **tissue** or **interstitial** (*in″ter-stish′al*) **fluid**—with the blood plasma accounting for the remaining 20% (fig. 22.7).

The distribution of water between the tissue fluid and the blood plasma is determined by a balance between opposing forces acting at the capillaries. Blood pressure, for example, promotes the formation of tissue fluid from plasma, whereas osmotic forces draw water from the tissues into the vascular system. The total volume of intracellular and extracellular fluid is normally maintained constant by a balance between water loss and water gain. Mechanisms that affect water intake, urine volume, and the distribution of water between plasma and tissue fluid thus help to regulate blood volume and, by this means, help to regulate cardiac output and blood flow.

Exchange of Fluid between Capillaries and Tissues

The distribution of extracellular fluid between the plasma and interstitial compartments is in a state of dynamic equilibrium. Tissue fluid is not normally a "stagnant pond"; rather it is a continuously circulating medium, formed from and returning to the vascular system. In this way, the cells receive a continuously fresh supply of glucose and other plasma solutes that are filtered through tiny endothelial channels in the capillary walls.

Filtration results from blood pressure within the capillaries. This hydrostatic pressure, which is exerted against the inner capillary wall, is equal to about 37 mmHg at the arteriolar end of systemic capillaries and drops to about 17 mmHg at the venular end of the capillaries. The **net filtration pressure** is equal to the hydrostatic pressure of the blood in the capillaries minus the hydrostatic pressure of tissue fluid outside the capillaries, which opposes filtration. If, as an extreme example, these two values were equal, there would be no filtration. The magnitude of the tissue hydrostatic pressure varies from organ to organ. With a hydrostatic pressure in the tissue fluid of 1 mmHg, as it is outside the capillaries of skeletal muscles,

Figure 22.7

The distribution of body water between the intracellular and extracellular compartments.
The extracellular compartment includes the blood plasma and the interstitial (tissue) fluid.

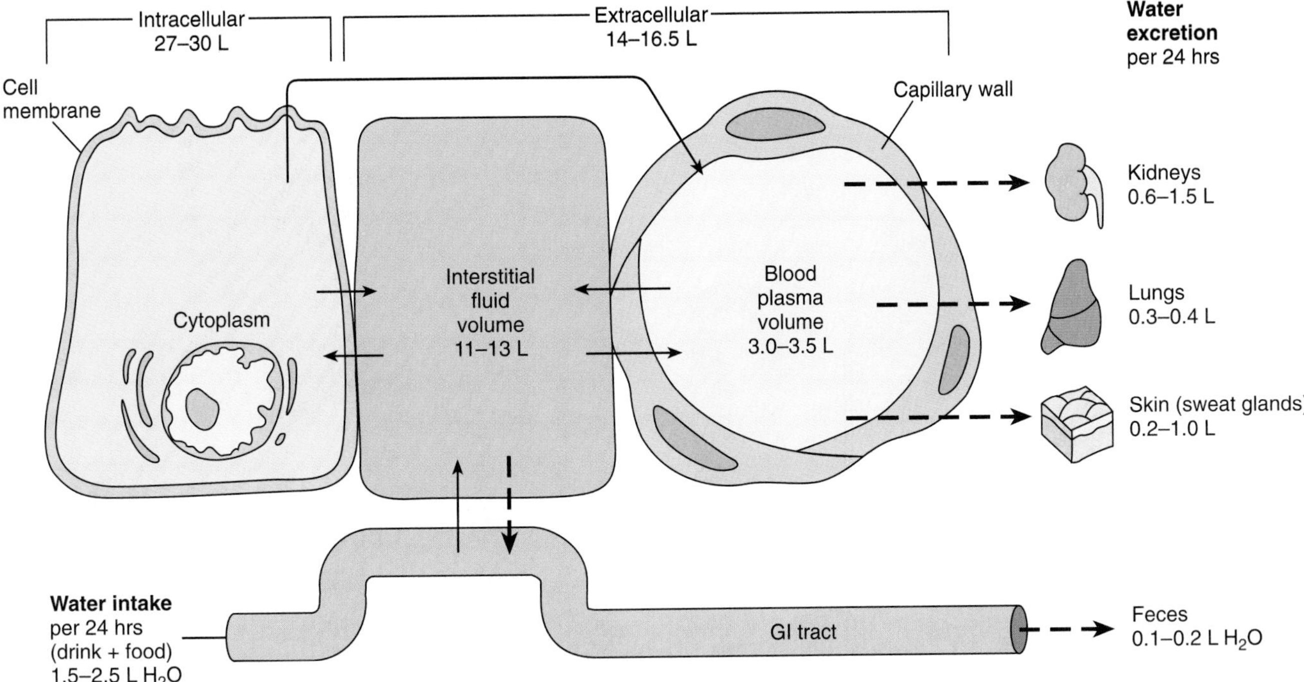

the net filtration pressure would be 37 – 1 = 36 mmHg at the arteriolar end of the capillary and 17 – 1 = 16 mmHg at the venular end.

Glucose, comparably sized organic molecules, inorganic salts, and ions are filtered along with water through the capillary channels. The concentrations of these substances in tissue fluid are thus the same as in plasma. The protein concentration of tissue fluid (2 g/100 ml), however, is less than the protein concentration of plasma (6 to 8 g/100 ml). This difference is due to the restricted filtration of proteins through the capillary pores. The osmotic pressure exerted by plasma proteins—called the **colloid osmotic pressure** of the plasma (because proteins are present as a colloidal suspension)—is therefore much greater than the colloid osmotic pressure of tissue fluid. The difference between these two pressures is called the **oncotic pressure.** Since the colloid osmotic pressure of the tissue fluid is sufficiently low to be neglected, the oncotic pressure is essentially equal to the colloid osmotic pressure of the plasma. This value has been estimated to be 25 mmHg. Since water will move by osmosis from the solution of lower to the solution of higher osmotic pressure (chapter 5), this oncotic pressure favors the movement of water into the capillaries.

Whether fluid will move out of or into the capillary depends on the magnitude of the net filtration pressure, which varies from the arteriolar to the venular end of the capillary, and on the oncotic pressure. These opposing forces that affect the distribution of fluid across the capillary are known as **Starling forces,** and their effects can be calculated according to the following equation:

Fluid movement is proportional to:

$$\underbrace{(P_c + \pi_i)}_{\text{(Fluid out)}} - \underbrace{(P_i + \pi_p)}_{\text{(Fluid in)}}$$

where

P_c = hydrostatic pressure in the capillary
π_i = colloid osmotic pressure of the interstitial (tissue) fluid
P_i = hydrostatic pressure of interstitial fluid
π_p = colloid osmotic pressure of the blood plasma

The expression to the left of the minus sign represents the sum of forces acting to move fluid out of the capillary. The expression to the right represents the sum of forces acting to move fluid into the capillary. In the capillaries of skeletal muscles, the values are as follows: At the arteriolar end of the capillary, (37 + 0) – (1 + 25) = 11 mmHg; at the venular end of the capillary, (17 + 0) – (1 + 25) = –9 mmHg. The positive value at the arteriolar end indicates that the force favoring the extrusion of fluid from the capillary predominates. The negative value at the venular end indicates that the net Starling forces favor the return of fluid to the capillary. Fluid thus leaves the capillaries at the arteriolar end and returns to the capillaries at the venular end (fig. 22.8).

This "classic" view of capillary dynamics has been modified in recent years by the realization that the balance of filtration and reabsorption varies in different tissues and under different conditions in a particular capillary. For example, a capillary may be open or closed off by precapillary muscles that function as sphincters. When the capillary is open, blood

Figure 22.8

The distribution of fluid across the walls of a capillary.

Tissue, or interstitial, fluid is formed by filtration (*orange arrows*) as a result of blood pressures at the arteriolar ends of capillaries and is returned to venular ends of capillaries by the colloid osmotic pressure of plasma proteins (*yellow arrows*).

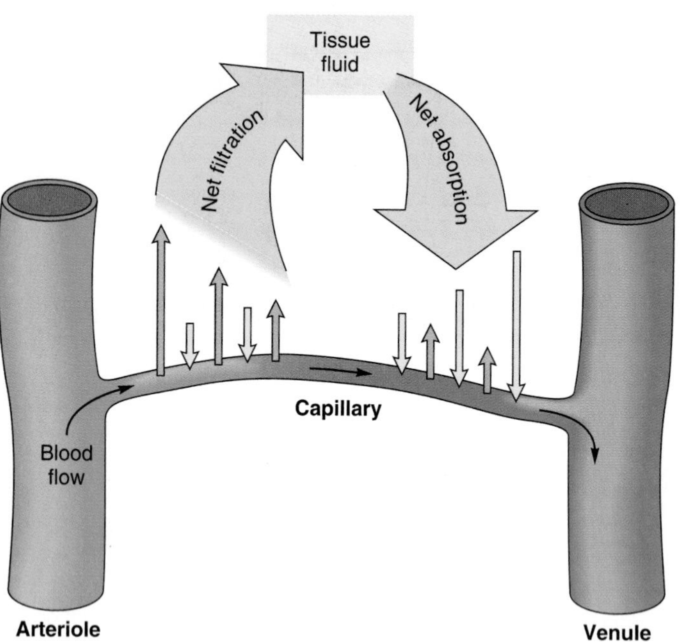

Figure 22.9

The relationship between blood capillaries and lymph capillaries.

Note that lymphatic capillaries are blind-ended. They are, however, highly permeable, so excess fluid and protein within the interstitial space can drain into the lymphatic system.

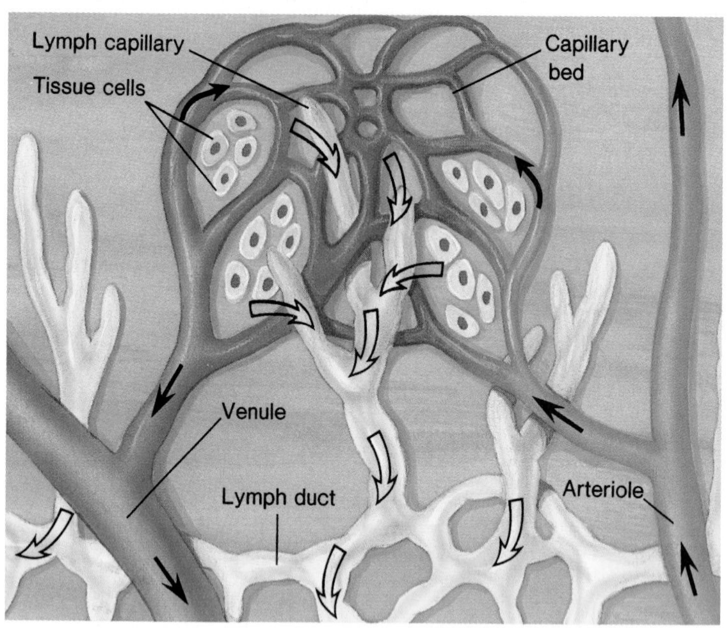

flow is high and the net filtration force exceeds the force for the osmotic return of water throughout the length of the capillary. The opposite is true if the precapillary sphincter closes and the blood flow through the capillary is reduced.

Lymphatic Drainage

Through the action of the Starling forces, plasma and tissue fluid are continuously interchanged. The return of fluid to the vascular system at the venular ends of the capillaries, however, does not exactly equal the amount filtered at the arteriolar ends. According to some estimates, approximately 85% of the capillary filtrate is returned directly to the capillaries; the remaining 15% (amounting to at least 2 L per day) is returned to the vascular system by way of the lymphatic system.

Excessive accumulation of interstitial fluid and filtered proteins is normally prevented by drainage of interstitial fluid into highly permeable, blind-ended **lymphatic capillaries** (fig. 22.9). Interstitial fluid that enters these lymphatic capillaries is known as **lymph.** Lymph is transported by lymphatic ductules into two large lymphatic vessels that drain the lymph into the right and left subclavian veins. In this way, interstitial fluid is ultimately returned to the circulatory system from which it was originally derived. The structure of the lymphatic system is described in detail in chapter 23.

Causes of Edema

Excessive accumulation of tissue fluid is known as **edema** (*ĕ-de′mă*). This condition is normally prevented by a proper balance between capillary filtration and osmotic uptake of water and by proper lymphatic drainage. Edema may thus result from (1) *high arterial blood pressure*, which increases capillary pressure and causes excessive filtration; (2) *venous obstruction*—as in phlebitis (where a thrombus forms in a vein) or mechanical compression of veins (during pregnancy, for example)—which produces a congestive increase in capillary pressure; (3) *leakage of plasma proteins into tissue fluid*, which causes reduced osmotic flow of water into the capillaries (this occurs during inflammation and allergic reactions as a result of increased capillary permeability); (4) *myxedema*—the excessive production of particular glycoproteins (mucin) in the interstitial spaces caused by hypothyroidism; (5) *decreased plasma protein concentration* as a result of liver disease (the liver makes most of the plasma proteins) or kidney disease, in which proteins are excreted in the urine; and (6) *obstruction of the lymphatic drainage* (table 22.2).

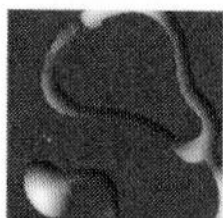

In the tropical disease *filariasis* (*fil″ă-ri′ă-sis*), mosquitoes transmit a nematode worm parasite to humans. The larvae of these worms invade lymphatic vessels and block lymphatic drainage. The edema that results can be so severe that the tissues swell to produce an elephantlike appearance, with thickening and cracking of the skin. This condition is thus aptly named *elephantiasis* (fig. 22.10). A new drug regimen is highly effective against the filariasis parasite, and a worldwide effort is now underway to treat this disease.

Table 22.2

Causes of Edema

Cause	Comments
Increased blood pressure or venous obstruction	Increases capillary filtration pressure so that more interstitial fluid is formed at the arteriolar ends of capillaries.
Increased tissue protein concentration	Decreases osmosis of water into the venular ends of capillaries. Usually a localized tissue edema due to leakage of blood plasma proteins through capillaries during inflammation and allergic reactions. Myxedema due to hypothyroidism is also in this category.
Decreased plasma protein concentration	Decreases osmosis of water into the venular ends of capillaries. May be caused by liver disease (which can be associated with insufficient blood plasma protein production), kidney disease (due to leakage of blood plasma protein into urine), or protein malnutrition.
Obstruction of lymphatic vessels	Infections by filaria roundworms (nematodes) transmitted by a certain species of mosquito block lymphatic drainage, causing edema and tremendous swelling of the affected areas.

Figure 22.10

The severe edema of elephantiasis.
Parasitic larvae that block lymphatic drainage produce tissue edema and the tremendous enlargement of the limbs and external genitalia in elephantiasis.

Regulation of Blood Volume by the Kidneys

The formation of urine begins in the same manner as the formation of tissue fluid—by filtration of plasma through capillary pores. These capillaries are known as *glomeruli*, and the filtrate they produce enters a system of tubules that transports and modifies the filtrate (by mechanisms discussed in chapter 25). The kidneys produce about 180 L per day of blood filtrate, but since there is only 5.5 L of blood in the body, it is clear that most of this filtrate must be returned to the vascular system and recycled. Only about 1.5 L of urine is excreted daily; 98% to 99% of the amount filtered is **reabsorbed** back into the vascular system.

The volume of urine excreted can be varied by changes in the reabsorption of filtrate. If 99% of the filtrate is reabsorbed, for example, 1% must be excreted. Decreasing the reabsorption by only 1%—from 99% to 98%—would double the volume of urine excreted (an increase to 2% of the amount filtered). Carrying the logic further, a doubling of urine volume from, for example, 1 to 2 liters, would result in the loss of an additional liter of blood volume. The percentage of the glomerular filtrate reabsorbed—and thus the urine volume and blood volume—is adjusted according to the needs of the body by the action of specific hormones on the kidneys. Through their effects on the kidneys and the resulting changes in blood volume, these hormones serve important functions in the regulation of the cardiovascular system.

Regulation by Antidiuretic Hormone (ADH)

One of the major hormones involved in the regulation of blood volume is **antidiuretic** (*an″te-di″yŭ-ret′ik*) **hormone (ADH),** also known as *vasopressin*. As described in chapter 19, this hormone is produced by neurons in the hypothalamus, transported by axons into the posterior pituitary, and released from this storage gland in response to hypothalamic stimulation. The release of ADH from the posterior pituitary occurs when neurons in the hypothalamus called **osmoreceptors** detect an increase in plasma osmolality (osmotic pressure).

An increase in plasma osmolality occurs when the plasma becomes more concentrated (chapter 5). This can occur either through *dehydration* or through *excessive salt intake*. Stimulation of osmoreceptors produces sensations of thirst, leading to increased water intake and an increase in the amount of ADH released from the posterior pituitary. Through mechanisms that will be discussed in conjunction with kidney physiology in chapter 25, ADH stimulates water reabsorption from the filtrate. A smaller volume of urine is thus excreted as a result of the action of ADH (fig. 22.11).

A person who is dehydrated or who consumes excessive amounts of salt thus drinks more and urinates less. This raises the blood volume and, in the process, dilutes the plasma to lower its previously elevated osmolality. The rise in blood volume that results from these mechanisms is extremely important in stabilizing the condition of a dehydrated person with low blood volume and pressure.

Drinking excessive amounts of water without excessive amounts of salt does not result in a prolonged increase in blood volume and pressure. The water does enter the blood from the intestine and momentarily raises the blood volume; at the same time, however, it dilutes the blood. Dilution of the blood decreases the plasma osmolality and thus inhibits the release of ADH. With less ADH there is less reabsorption of filtrate in the kidneys—a larger volume of urine is excreted. Water is therefore a *diuretic* (*di″yŭ-ret′ik*)—a substance that promotes urine formation—because it inhibits the release of antidiuretic hormone.

In addition to the activity of osmoreceptors, another mechanism operates to inhibit the release of ADH when fluid intake is excessive. An excessively high blood volume stimulates stretch receptors located in the left atrium of the heart. Stimulation of these stretch receptors, in turn, activates a reflex inhibition of ADH release, which promotes a lowering of blood volume through increased urine production.

During prolonged exercise, particularly on a warm day, a substantial amount of water (up to 900 ml per hour) may be lost from the body through sweating. The lowering of blood volume that results decreases the ability of the body to dissipate heat, and the consequent overheating of the body can cause ill effects and put an end to the exercise. The need for athletes to remain well hydrated is commonly recognized, but drinking pure water may not be the answer. This is because blood sodium is lost in sweat, so that a lesser amount of water is required to dilute the blood osmolality back to normal. When the blood osmolality is normal the urge to drink is extinguished. For these reasons, athletes performing prolonged endurance exercise should drink solutions containing sodium (as well as carbohydrates for energy), and they should drink at a predetermined rate rather than at a rate determined only by thirst.

Regulation by Aldosterone

From the preceding discussion, it is clear that a certain amount of dietary salt is required to maintain blood volume and pressure. Since Na^+ and Cl^- are easily filtered in the kidneys, a mechanism must exist to promote the reabsorption and retention of salt when the dietary salt intake is too low. **Aldosterone** (*al-dos′ter-ōn*), a steroid hormone secreted by the adrenal cortex, stimulates the reabsorption of salt by the kidneys. Aldosterone is thus a "salt-retaining hormone." Retention of salt indirectly promotes retention of water (in part, by the action of ADH as previously discussed). The action of aldosterone thus produces an increase in blood volume, but, unlike ADH, it does not produce a change in plasma osmolality. This is because aldosterone promotes the reabsorption of salt and water in proportionate amounts, whereas ADH promotes only the reabsorption of water. Thus, unlike ADH, aldosterone does not act to dilute the blood.

The secretion of aldosterone is stimulated during salt deprivation, when the blood volume and pressure are reduced. The adrenal cortex, however, is not directly stimulated to secrete aldosterone by these conditions. Instead, a decrease in blood volume and pressure activates an intermediate mechanism described in the next section.

diuretic: Gk. *dia*, through; *ouresis*, urination

Figure 22.11

The negative feedback control of blood volume and blood osmolality.

Thirst and ADH secretion are triggered by a rise in plasma osmolality. Homeostasis is maintained by countermeasures, including drinking and the increased retention of water by the kidneys.

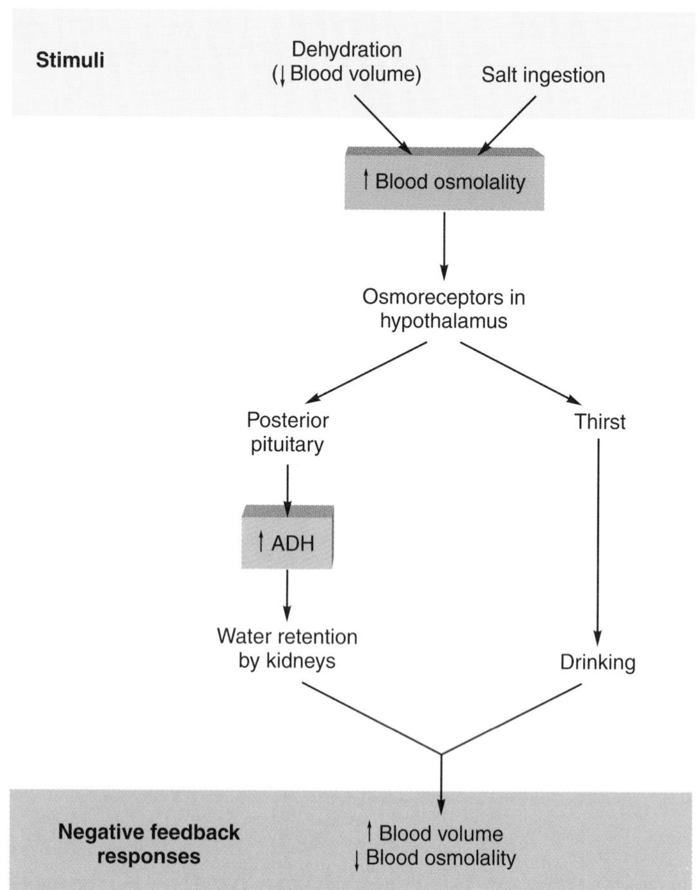

Figure 22.12

The renin-angiotensin-aldosterone system.
This system helps to maintain homeostasis through the negative feedback control of blood volume and pressure. (ACE = angiotensin converting enzyme.)

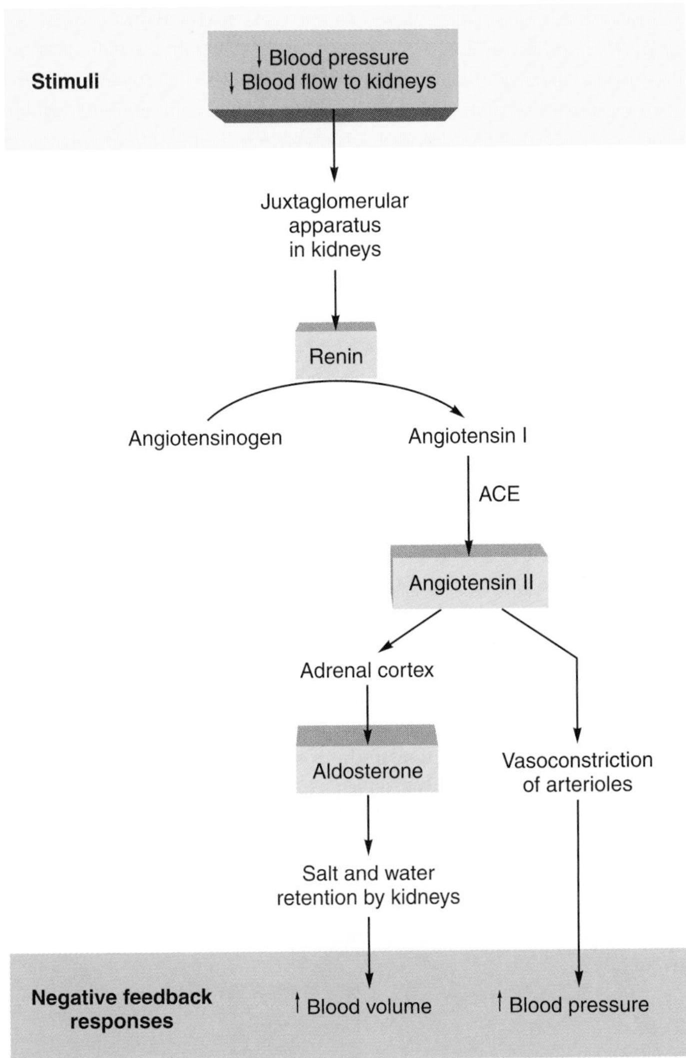

Throughout most of human history, salt was in short supply and was therefore highly valued. Moorish merchants in the sixth century traded an ounce of salt for an ounce of gold, and salt cakes were used as money in Abyssinia. Part of a Roman soldier's pay was given in salt—a practice from which the word *salary* (*sal* = salt) derives. Salt was also used to purchase slaves—hence the phrase "worth his salt."

Renin-Angiotensin System

When the blood flow and pressure are reduced in the renal artery (as they would be in the low-blood-volume state of salt deprivation), a group of cells in the kidneys called the **juxtaglomerular** (*juk″stă-glo-mer′yŭ-lar*) **apparatus** secretes the enzyme **renin** into the blood. This enzyme cleaves a 10-amino-acid polypeptide called *angiotensin I* from a plasma protein called *angiotensinogen*. As angiotensin I passes through the capillaries of the lungs and other organs, an *angiotensin-converting enzyme (ACE)* removes two amino acids. This leaves an 8-amino-acid polypeptide called **angiotensin II** (fig. 22.12). Conditions of salt deprivation, low blood volume, and low blood pressure, in summary, cause increased production of angiotensin II in the blood.

Angiotensin II exerts numerous effects that produce a rise in blood pressure. This rise in pressure is partly due to vasoconstriction and partly to increases in blood volume. Vasoconstriction of arterioles and small muscular arteries is produced directly by the effects of angiotensin II on the smooth muscle layers of these vessels. The increased blood volume is an indirect effect of angiotensin II.

Angiotensin II promotes a rise in blood volume by means of two mechanisms: (1) thirst centers in the hypothalamus are stimulated by angiotensin II, and thus more water is ingested, and (2) secretion of aldosterone from the adrenal cortex is stimulated by angiotensin II, and higher aldosterone secretion causes more salt and water to be retained by the kidneys. The relationship between angiotensin II and aldosterone is sometimes described as the *renin-angiotensin-aldosterone system*.

The renin-angiotensin-aldosterone system can also work in the opposite direction: high salt intake, leading to high blood volume and pressure, normally inhibits renin secretion. With less angiotensin II formation and less aldosterone secretion, less salt is retained by the kidneys and more is excreted in the urine. Unfortunately, many people with chronically high blood pressure may have normal or even elevated levels of renin secretion. In these cases, the intake of salt must be lowered to match the impaired ability to excrete salt in the urine.

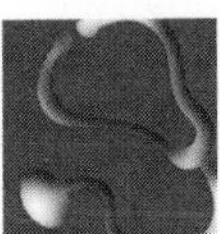

One of the newer classes of drugs that can be used to treat hypertension (high blood pressure) are the *angiotensin-converting enzyme,* or *ACE, inhibitors.* These drugs (such as *captopril*) block the formation of angiotensin II, thus reducing its vasoconstrictor effect. The ACE inhibitors also increase the activity of bradykinin, a polypeptide that promotes vasodilation. The reduced formation of angiotensin II and increased action of bradykinin result in vasodilation, which decreases the total peripheral resistance. Because this reduces the afterload of the heart, the ACE inhibitors are also used to treat left ventricular hypertrophy and congestive heart failure. Another new class of antihypertensive drugs allows angiotensin II to be formed but selectively blocks the angiotensin II receptors.

Atrial Natriuretic Factor

As described in the previous section, a fall in blood volume is compensated for by renal retention of fluid through activation of the renin-angiotensin-aldosterone system. An increase in blood volume, conversely, is compensated for by renal excretion of a larger volume of urine. Experiments suggest that the increase in water excretion under conditions of high blood volume is at least partly due to an increase in the excretion of Na^+ in the urine, or *natriuresis*.

natriuresis: L. *natrium*, sodium; Gk. *ouresis*, urination

Increased Na^+ excretion (natriuresis) may be produced by a decline in aldosterone secretion, but there is evidence that there is a separate hormone that stimulates natriuresis. This *natriuretic* (*na″trĭ-yoo-ret′ik*) *hormone* would thus be antagonistic to aldosterone and would promote Na^+ and water excretion in the urine in response to a rise in blood volume. A polypeptide hormone with these properties, identified as **atrial natriuretic factor (ANF),** is produced by the atria of the heart. By promoting salt and water excretion in the urine, ANF can act to lower the blood volume and pressure. This is analogous to the action of diuretic drugs taken by people with hypertension, as described in the section "Clinical Considerations," later in this chapter.

In addition to its stimulation of salt and water excretion by the kidneys, ANF also antagonizes various actions of angiotensin II. As a result of this action, ANF decreases the secretion of aldosterone and promotes vasodilation.

Vascular Resistance to Blood Flow

The rate of blood flow to an organ is related to the resistance to flow in the small arteries and arterioles that serve the organ. Vasodilation decreases resistance and increases flow, whereas vasoconstriction increases resistance and decreases flow. Vasodilation and vasoconstriction occur in response to intrinsic and extrinsic regulatory mechanisms.

The amount of blood that the heart pumps per minute is equal to the rate of venous return, and thus is equal to the rate of blood flow through the entire circulation. The cardiac output of 5 to 6 L per minute is distributed unequally to the different organs. At rest, blood flow is about 2,500 ml/min through the liver, kidneys, and gastrointestinal tract; 1,200 ml/min through the skeletal muscles; 750 ml/min through the brain; and 250 ml/min through the coronary arteries of the heart. The balance of the cardiac output (500 to 1,100 ml/min) is distributed to the other organs, as indicated in table 22.3.

Physical Laws Describing Blood Flow

The flow of blood through the vascular system, like the flow of any fluid through a tube, depends in part on the difference in pressure at the two ends of the tube. If the pressure at both ends of the tube is the same, there will be no flow. If the pressure at one end is greater than at the other, blood will flow from the region of higher to the region of lower pressure. The rate of blood flow is proportional to the pressure difference ($P_1 - P_2$) between the two ends of the tube. The term **pressure difference** is abbreviated ΔP, in which the Greek letter Δ (delta) means "change in."

If the systemic circulation is pictured as a single tube leading from and back to the heart (fig. 22.13), blood flow through this system would occur as a result of the pressure difference between the beginning of the tube (the aorta) and the end of the tube (the junction of the venae cavae with the right atrium). The average, or mean, arterial pressure is about 100 mmHg; the pressure at the right atrium is 0 mmHg. The "pressure head," or driving force (ΔP), is therefore about $100 - 0 = 100$ mmHg.

Blood flow is directly proportional to the pressure difference between the two ends of the tube (ΔP) but is *inversely proportional* to the frictional resistance to blood flow through the vessels. Inverse proportionality is expressed by showing

Table 22.3

Estimated Distribution of the Cardiac Output at Rest

Organs	Blood Flow	
	Milliliters per minute	Percent total
Gastrointestinal tract and liver	1,400	24
Kidneys	1,100	19
Brain	750	13
Heart	250	4
Skeletal muscles	1,200	21
Skin	500	9
Other organs	600	10
Total	5,800	100

Source: From O. L. Wade and J. M. Bishop, *Cardiac Output and Regional Blood Flow*. Copyright © 1962 Blackwell Science, Ltd., England. Used with permission.

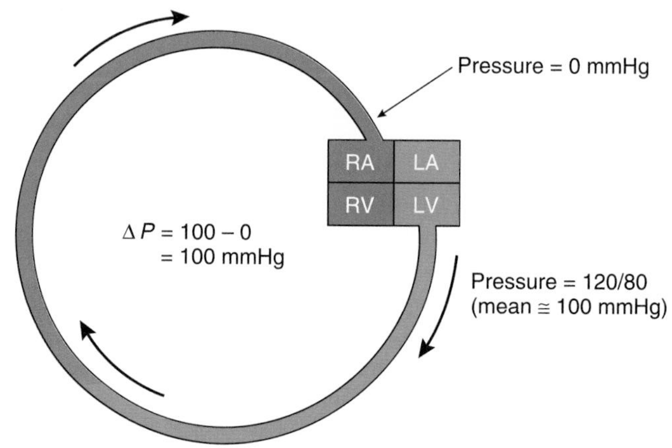

Figure 22.13

Blood flow is produced by a pressure difference. The flow of blood in the systemic circulation is ultimately dependent on the pressure difference (ΔP) between the mean pressure of about 100 mmHg at the origin of flow in the aorta and the pressure at the end of the circuit—0 mmHg in the vena cava, where it joins the right atrium (RA). (LA = left atrium; RV = right ventricle; LV = left ventricle.)

Figure 22.14

The relationships between blood flow, vessel radius, and resistance.
(*a*) The resistance and blood flow are equally divided between two branches of a vessel. (*b*) A doubling of the radius of one branch and halving of the radius of the other produces a sixteenfold increase in blood flow in the former and a sixteenfold decrease of blood flow in the latter.

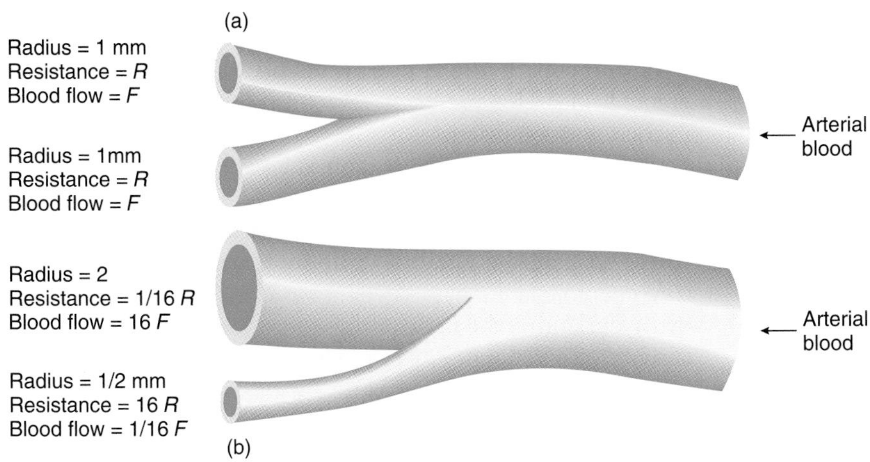

Vessel length and blood viscosity do not vary significantly in normal physiology, although blood viscosity is increased in severe dehydration and in the polycythemia (high red blood cell count) that occurs as an adaptation to life at high altitudes. The major physiological regulators of blood flow through an organ are the mean arterial pressure (driving the flow) and the vascular resistance to flow. At a given mean arterial pressure, blood can be diverted from one organ to another by variations in the degree of vasoconstriction and vasodilation. Vasoconstriction in one organ and vasodilation in another result in a diversion, or *shunting*, of blood to the second organ. Since arterioles are the smallest arteries and can become narrower by vasoconstriction, they provide the greatest resistance to blood flow (fig. 22.15). Blood flow to an organ is thus largely determined by the degree of vasoconstriction or vasodilation of its arterioles. The rate of blood flow to an organ can be increased by dilation of its arterioles and can be decreased by constriction of its arterioles.

one of the factors in the denominator of a fraction, since a fraction decreases when the denominator increases:

$$\textbf{Blood flow} \propto \frac{\Delta P}{\textbf{resistance}}$$

The **resistance** to blood flow through a vessel is directly proportional to the length of the vessel and to the viscosity of the blood (the "thickness," or ability of molecules to "slip over" each other). Of particular physiological importance, the vascular resistance is inversely proportional to the fourth power of the radius of the vessel:

$$\textbf{Resistance} \propto \frac{L\eta}{r^4}$$

where

L = length of vessel
η = viscosity of blood
r = radius of vessel

For example, if one vessel has half the radius of another and if all other factors are the same, the smaller vessel will have 16 times (2^4) the resistance of the larger vessel. Blood flow through the larger vessel, as a result, will be 16 times greater than in the smaller vessel (fig. 22.14).

When physical constants are added to this relationship, the rate of blood flow can be calculated according to **Poiseuille's** (*pwă-zuh'yez*) **law:**

$$\textbf{Blood flow} = \frac{\Delta P r^4 (\pi)}{\eta L(8)}$$

Poiseuille's law: from Jean Poiseuille, French physiologist, 1799–1869

Total Peripheral Resistance

The sum of all the vascular resistances within the systemic circulation is called the **total peripheral resistance.** The arteries that supply blood to the organs are generally in parallel rather than in series with each other. That is, arterial blood passes through only one set of resistance vessels (arterioles) before returning to the heart (fig. 22.16). Since one organ is not "downstream" from another in terms of its arterial supply, changes in resistance within one organ directly affect blood flow in that organ only.

Vasodilation in a large organ might, however, significantly decrease the total peripheral resistance and, by this means, might decrease the mean arterial pressure. In the absence of compensatory mechanisms, the driving force for blood flow through all organs might be reduced. This situation is normally prevented by an increase in the cardiac output and by vasoconstriction in other areas. During exercise of the large muscles, for example, the arterioles in the exercising muscles are dilated. This would cause a great fall in mean arterial pressure if there were no compensations. The blood pressure actually rises during exercise, however, because the cardiac output is increased and because there is constriction of arterioles in the viscera and skin.

Extrinsic Regulation of Blood Flow

The term *extrinsic regulation* refers to control by the autonomic nervous system and endocrine system. Angiotensin II, for example, directly stimulates vascular smooth muscle to produce generalized vasoconstriction. Antidiuretic hormone (ADH) also has a vasoconstrictor effect at high concentrations; this is

Figure 22.15

Blood pressure in different vessels of the systemic circulation.

Note that the pressure generated by the beating of the ventricles is largely dissipated by the time the blood gets into the venous system.

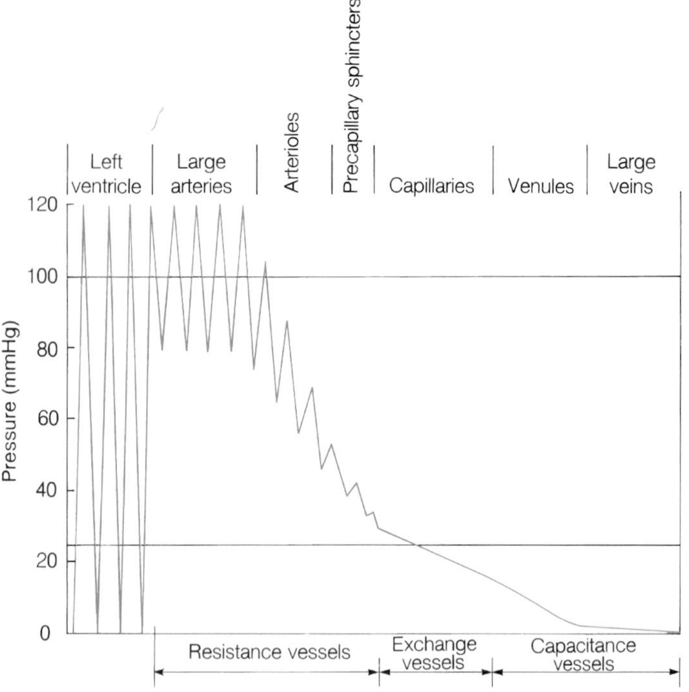

why it is also called *vasopressin*. This vasopressor effect of ADH is not believed to be significant under physiological conditions in humans.

Regulation by Sympathetic Nerves

Stimulation of the sympathoadrenal system produces an increase in the cardiac output (as previously discussed) and an increase in total peripheral resistance. The latter effect is due to alpha-adrenergic stimulation (chapter 17) of vascular smooth muscle by norepinephrine and, to a lesser degree, by epinephrine. This produces vasoconstriction of the arterioles in the viscera and skin.

Even when a person is calm, the sympathoadrenal system is active to a certain degree and helps set the "tone" of vascular smooth muscles. In this case, **adrenergic sympathetic fibers** (those that release norepinephrine) activate alpha-adrenergic receptors to cause a basal level of vasoconstriction throughout the body. During the fight-or-flight reaction, an increase in the activity of adrenergic fibers produces vasoconstriction in the digestive tract, kidneys, and skin.

Cocaine inhibits the reuptake of norepinephrine into the adrenergic axons, resulting in enhanced sympathetic-induced vasoconstriction. Chest pain, as a result of myocardial ischemia produced in this way, is a common cocaine-related problem. The nicotine from cigarette smoke acts synergistically with cocaine to induce vasoconstriction.

Figure 22.16

A diagram of the systemic and pulmonary circulations.

Notice that with few exceptions (such as blood flow in the renal circulation), the flow of arterial blood is in parallel rather than in series (arterial blood does not usually flow from one organ to another).

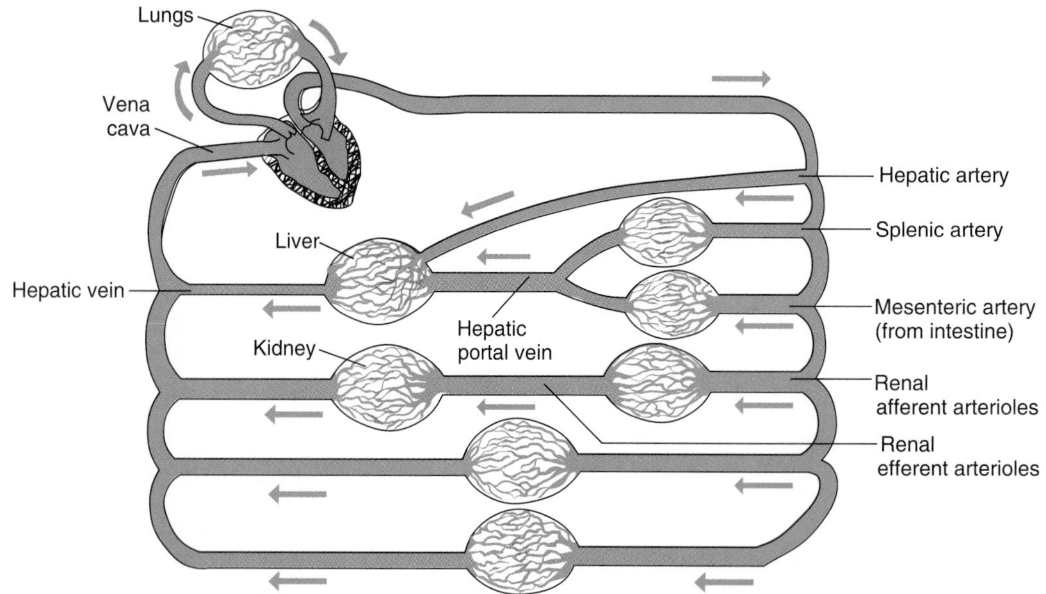

Table 22.4

Extrinsic Control of Vascular Resistance and Blood Flow

Extrinsic Agent	Effect	Comments
Sympathetic nerves		
Alpha-adrenergic	Vasoconstriction	The effect occurs throughout the body and is the dominant effect of sympathetic nerve stimulation on the circulatory system.
Beta-adrenergic	Vasodilation	There is some activity in arterioles in skeletal muscles and in coronary vessels, but effects are masked by dominant alpha-receptor-mediated constriction.
Cholinergic	Vasodilation	Effects are localized to arterioles in skeletal muscles and are produced only during defense (fight-or-flight) reactions.
Parasympathetic nerves	Vasodilation	Effects are primarily restricted to the gastrointestinal tract, external genitalia, and salivary glands and have little effect on total peripheral resistance.
Angiotensin II	Vasoconstriction	A powerful vasoconstrictor produced as a result of secretion of renin from the kidneys, it may function to help maintain adequate filtration pressure in kidneys when systemic blood flow and pressure are reduced.
ADH (vasopressin)	Vasoconstriction	Although the effects of this hormone on vascular resistance and blood pressure in anesthetized animals are well documented, the importance of these effects in conscious humans is controversial.
Histamine	Vasodilation	Histamine promotes localized vasodilation during inflammation and allergic reactions.
Bradykinins	Vasodilation	Bradykinins are polypeptides secreted by sweat glands and by the endothelium of blood vessels; they promote local vasodilation.
Prostaglandins	Vasodilation or vasoconstriction	Prostaglandins are cyclic fatty acids that can be produced by most tissues, including blood vessel walls. Prostaglandin I_2 is a vasodilator, whereas thromboxane A_2 is a vasoconstrictor. The physiological significance of these effects is presently controversial.

Arterioles in skeletal muscles receive **cholinergic sympathetic fibers,** which release acetylcholine as a neurotransmitter. During the fight-or-flight reaction the activity of these cholinergic fibers increases. This causes vasodilation. Vasodilation in skeletal muscles is also produced by epinephrine secreted by the adrenal medulla, which stimulates beta-adrenergic receptors. During the fight-or-flight reaction, therefore, blood flow decreases to the viscera and skin because of the alpha-adrenergic effects of vasoconstriction in these organs, whereas blood flow to the skeletal muscles increases. This diversion of blood flow to the skeletal muscles during emergency conditions may give these muscles an "extra edge" in responding to the emergency.

Parasympathetic Control of Blood Flow

Parasympathetic endings in arterioles are always cholinergic and always promote vasodilation. Parasympathetic innervation of blood vessels, however, is limited to the digestive tract, external genitalia, and salivary glands. Because of this limited distribution, the parasympathetic system is less important than the sympathetic system in the control of total peripheral resistance.

The extrinsic control of blood flow is summarized in table 22.4.

Paracrine Regulation of Blood Flow

Paracrine regulators, as described in chapter 19, are molecules produced by one tissue that help to regulate another tissue of the same organ. Blood vessels are particularly subject to paracrine regulation. Specifically, the endothelium of the tunica interna produces a number of paracrine regulators that cause the smooth muscle of the tunica media to either relax or contract.

The endothelium produces several molecules that promote smooth muscle relaxation, including nitric oxide, bradykinin, and prostacyclin (chapter 19). The endothelium-derived relaxation factor that earlier research had shown to be required for the vasodilation response to nerve stimulation appears to be nitric oxide. The steps in this process are as follows: (1) the parasympathetic axons release ACh, which stimulates the opening of Ca^{2+} channels in the endothelial cell membrane; (2) the Ca^{2+} then binds to and activates calmodulin (chapter 19); (3) calmodulin, in turn, activates nitric oxide synthetase, the enzyme that converts L-arginine into nitric oxide; and (4) the nitric oxide then diffuses into the smooth muscle cells of the vessel to produce the vasodilation response to the nerve stimulation. It is interesting in this regard that vasodilator drugs often given to treat angina pectoris—including nitroglycerin and sodium nitroprusside—promote vasodilation indirectly through their conversion into nitric oxide.

The endothelium also produces paracrine regulators that stimulate vasoconstriction. Notable among these is the polypeptide *endothelin-1*. Although the precise physiological role of this regulator is incompletely understood, it is currently believed that it works together with the vasodilator regulators to help maintain normal vessel diameter and blood pressure.

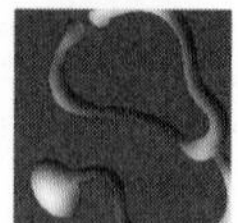

In people with hypertension (high blood pressure), there is evidence that the ability of the endothelium to produce nitric oxide may be compromised, and that the reduced production of nitric oxide may contribute to the hypertension. Drugs that increase nitric oxide, therefore, may help to lower blood pressure. Inhalation therapy using nitric oxide has been shown to relieve pulmonary hypertension in infants.

Intrinsic Regulation of Blood Flow

Intrinsic, or "built-in," mechanisms within individual organs provide a localized regulation of vascular resistance and blood flow. Intrinsic mechanisms are classified as *myogenic* or *metabolic*. Some organs, the brain and kidneys in particular, utilize these intrinsic mechanisms to maintain relatively constant flow rates despite wide fluctuations in blood pressure. This ability is termed **autoregulation.**

Myogenic Control Mechanisms

If the arterial blood pressure and flow through an organ are inadequate—if the organ is inadequately *perfused* with blood—the metabolism of the organ cannot be maintained beyond a limited time period. Excessively high blood pressure can also be dangerous, particularly in the brain, because this may result in the rupture of fine blood vessels (causing cerebrovascular accident—CVA, or stroke).

Changes in systemic arterial pressure are compensated for in the brain and some other organs by the appropriate responses of vascular smooth muscle. A decrease in arterial pressure causes cerebral vessels to dilate, so that adequate rates of blood flow can be maintained despite the decreased pressure. High blood pressure, by contrast, causes cerebral vessels to constrict, so that finer vessels downstream are protected from the elevated pressure. These responses are myogenic; they are direct responses by the vascular smooth muscle to changes in pressure.

Metabolic Control Mechanisms

Local vasodilation within an organ can occur as a result of the chemical environment created by the organ's metabolism. The localized chemical conditions that promote vasodilation include (1) *decreased oxygen concentrations* that result from increased metabolic rate; (2) *increased carbon dioxide concentrations;* (3) *decreased tissue pH* (due to CO_2, lactic acid, and other metabolic products); and (4) the *release of adenosine or K^+* from the cells. Through these chemical changes, the organ signals its blood vessels of its need for increased oxygen delivery.

The vasodilation that occurs in response to tissue metabolism can be demonstrated by constricting the blood supply to an area for a short time and then removing the constriction. The constriction allows metabolic products to accumulate by preventing venous drainage of the area. When the constriction is removed and blood flow resumes, the metabolic products that have accumulated cause vasodilation. The tissue thus appears red. This response is called **reactive hyperemia** (*hi″pĕ-re′me-ă*). A similar increase in blood flow occurs in skeletal muscles and other organs as a result of increased metabolism. This is called **active hyperemia.** A few minutes after exercise ends, blood flow can fall to pre-exercise levels because the increased blood flow can wash out the vasodilator metabolites.

Blood Flow to the Heart and Skeletal Muscles

Blood flow to the heart and skeletal muscles is regulated by both extrinsic and intrinsic mechanisms. These mechanisms provide increased blood flow when the metabolic requirements of these tissues are raised during exercise.

Survival requires that the heart and brain receive an adequate supply of blood at all times. The ability of skeletal muscles to respond quickly in emergencies and to maintain continued high levels of activity also may be critically important for survival. During such times, high rates of blood flow to the skeletal muscles must be maintained without compromising blood flow to the heart and brain. This is accomplished by mechanisms that increase the cardiac output and divert the blood away from the viscera and skin so that the heart, skeletal muscles, and brain receive a greater proportion of the cardiac output.

Aerobic Requirements of the Heart

The coronary arteries supply an enormous number of capillaries, which are packed within the myocardium at a density ranging from 2,500 to 4,000 per cubic millimeter of tissue. Fast-twitch skeletal muscles, by contrast, have a capillary density of 300 to 400 per cubic millimeter of tissue. Each myocardial cell, as a consequence, is within 10 μm of a capillary (compared to an average distance in other organs of 70 μm). The exchange of gases by diffusion between myocardial cells and capillary blood thus occurs very quickly.

Contraction of the myocardium squeezes the coronary arteries. Unlike blood flow in all other organs, flow in the coronary vessels thus decreases in systole and increases during diastole. The myocardium, however, contains large amounts of *myoglobin*, a pigment related to hemoglobin (the molecules in red blood cells that carry oxygen). Myoglobin in the myocardium stores oxygen during diastole and releases its oxygen during systole. In this way, the myocardial cells can receive a continuous supply of oxygen even though coronary blood flow is temporarily reduced during systole.

In addition to containing large amounts of myoglobin, heart muscle contains numerous mitochondria and aerobic respiratory enzymes. This indicates that—even more than slow-twitch skeletal muscles—the heart is extremely specialized for aerobic respiration. The normal heart always respires aerobically, even during heavy exercise when the metabolic demand for oxygen can rise to five times resting levels. This increased oxygen requirement is met by a corresponding increase in coronary blood flow, from about 80 ml at rest to about 400 ml per minute per 100 g tissue during heavy exercise.

Regulation of Coronary Blood Flow

Sympathetic nerve fibers, through stimulation of alpha-adrenergic receptors in the coronary arterioles, produce a relatively high vascular resistance in the coronary circulation at rest. Vasodilation of coronary vessels may be produced in part by sympathoadrenal activation of beta-adrenergic receptors. Most of the vasodilation that occurs during exercise, however, is due to intrinsic metabolic control mechanisms. The intrinsic mechanisms occur as follows: (1) as the metabolism of the myocardium increases, there are local accumulations of carbon dioxide, K^+, and adenosine in the tissue, together with depletion of oxygen; (2) these localized changes act directly on the vascular smooth muscle to cause relaxation and vasodilation.

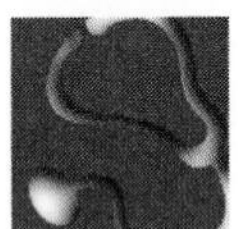

Under abnormal conditions the blood flow to the myocardium may be inadequate, resulting in myocardial ischemia (chapter 21). The inadequate flow may be due to blockage by atheromas and/or blood clots or to muscular spasm of a coronary artery (fig. 22.17). Occlusion of a coronary artery can be visualized by inserting a catheter (plastic tube) into a brachial or femoral artery all the way to the opening of the coronary arteries in the aorta, and then injecting a radiographic contrast material. The picture thus obtained is called an *angiogram.*

In a technique called *balloon angioplasty,* an inflatable balloon is used to open the coronary arteries. However, *restenosis* (recurrence of narrowing) often occurs. If the occlusion is sufficiently great, a *coronary bypass* may be performed. In this procedure a length of blood vessel, usually taken from the saphenous vein in the leg, is sutured to the aorta and to the coronary artery at a location beyond the site of the occlusion (fig. 22.18).

Regulation of Blood Flow through Skeletal Muscles

The arterioles in skeletal muscles, like those of the coronary circulation, have a high vascular resistance at rest as a result of alpha-adrenergic sympathetic stimulation. This produces a relatively low blood flow. Because muscles have such a large mass, however, they still receive from 20% to 25% of the total blood flow in the body at rest. Also, as in the heart, blood flow in a skeletal muscle decreases when the muscle contracts and squeezes its arterioles, and in fact blood flow stops entirely when the muscle contracts beyond about 70% of its

Figure 22.17

Angiograms of the left coronary artery of one patient.
These angiograms were taken (*a*) when the ECG was normal and (*b*) when the ECG showed evidence of myocardial ischemia. Notice that a coronary artery spasm (see arrow in [*b*]) appears to accompany the ischemia.

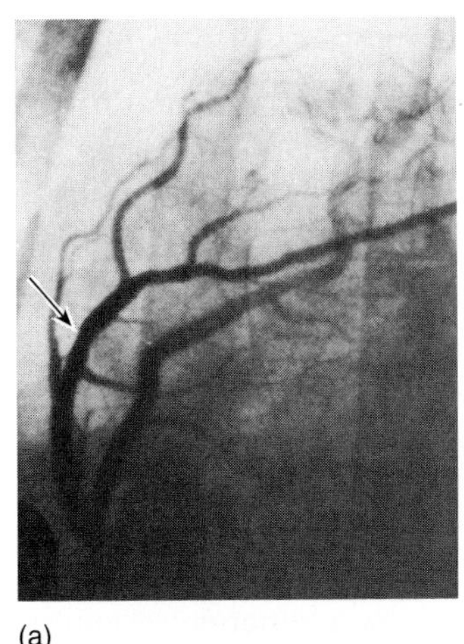

(a)

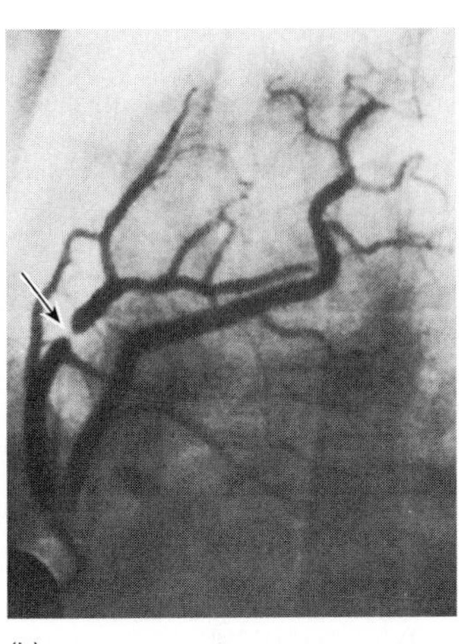

(b)

Figure 22.18

A diagram of coronary artery bypass surgery.
Segments of the saphenous vein of the patient are commonly used in the graft.

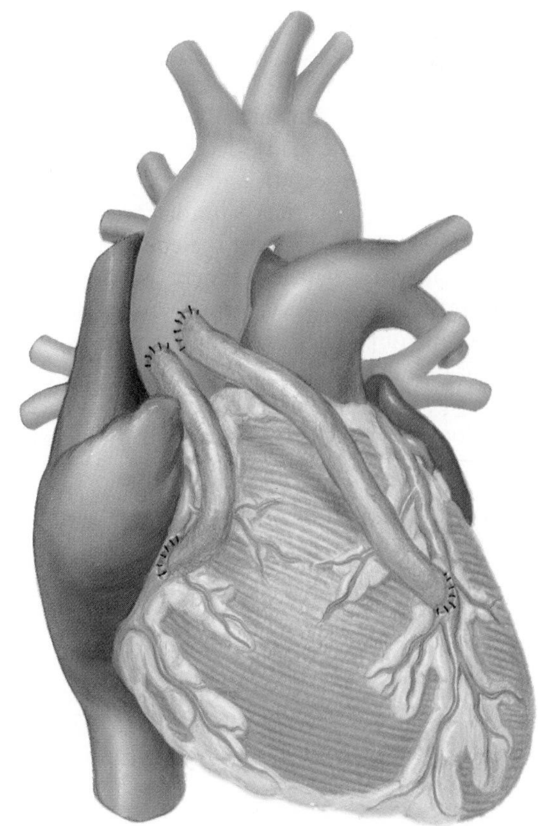

Table 22.5

Changes in Skeletal Muscle Blood Flow under Conditions of Rest and Exercise

Condition	Blood Flow (ml/min)	Mechanism
Rest	1,000	High adrenergic sympathetic stimulation of vascular alpha-receptors, causing vasoconstriction
Beginning exercise	Increased	Dilation of arterioles in skeletal muscles due to cholinergic sympathetic nerve activity and stimulation of beta-adrenergic receptors by the hormone epinephrine
Heavy exercise	20,000	Decreased alpha-adrenergic activity Increased sympathetic cholinergic activity Increased metabolic rate of exercising muscles, producing intrinsic vasodilation

maximum. Pain and fatigue thus occur much more quickly when an isometric contraction is sustained than when rhythmic isotonic contractions are performed.

In addition to adrenergic fibers, which promote vasoconstriction by stimulation of alpha-adrenergic receptors, there are also sympathetic cholinergic fibers in skeletal muscles. These cholinergic fibers, together with the stimulation of beta-adrenergic receptors by the hormone epinephrine, stimulate vasodilation as part of the fight-or-flight response to any stressful state, including that existing just prior to exercise (table 22.5). These extrinsic controls have been previously discussed and function to regulate blood flow through muscles at rest and upon anticipation of exercise.

As exercise progresses, the vasodilation and increased skeletal muscle blood flow that occur are almost entirely due to intrinsic metabolic control. The high metabolic rate of skeletal muscles during exercise causes local changes, such as increased carbon dioxide concentrations, decreased pH (due to carbonic acid and lactic acid), decreased oxygen, increased extracellular K^+, and the secretion of adenosine. As in the intrinsic control of the coronary circulation, these changes cause vasodilation of arterioles in skeletal muscles. This decreases the vascular resistance and increases the rate of blood flow. This effect is combined with the recruitment of capillaries by the opening of precapillary sphincters (only 5% to 10% of the skeletal muscle capillaries are open at rest). As a result of these changes, skeletal muscles can receive as much as 85% of the total blood flow in the body during maximal exercise.

Circulatory Changes during Exercise

While the vascular resistance in skeletal muscles decreases during exercise, the resistance to flow through visceral organs and skin increases. This increased resistance occurs because of vasoconstriction stimulated by adrenergic sympathetic fibers, and it results in decreased rates of blood flow through these organs. During exercise, therefore, the blood flow to skeletal muscles increases because of three simultaneous changes: (1) increased total blood flow (cardiac output), (2) metabolic vasodilation in the exercising muscles, and (3) the diversion of blood away from the viscera and skin. Blood flow to the heart also increases during exercise, whereas blood flow to the brain does not appear to change significantly (fig. 22.19).

During exercise the cardiac output can increase fivefold—from about 5 L per minute to about 25 L per minute. This is primarily due to an increase in cardiac rate. The cardiac rate, however, can increase only up to a maximum value (table 22.6), which is determined mainly by a person's age. In well-trained athletes the stroke volume can also increase significantly, allowing these individuals to achieve cardiac outputs during strenuous exercise up to six or seven times greater than their resting values. This high cardiac output results in increased oxygen delivery to the exercising muscles; this is the major reason for the much higher than average maximal oxygen uptake of elite athletes (chapter 12).

In most people the increase in stroke volume that occurs during exercise will not exceed 35%. The fact that the stroke volume can increase at all during exercise may at first be surprising, given that the heart has less time to fill with blood between beats when it is pumping faster. Despite the faster beat, however, the end-diastolic volume during exercise does not decrease. This is because the venous return is aided by the improved action of the skeletal muscle pumps and by increased respiratory movements during exercise (fig. 22.20). Since the end-diastolic volume does not significantly change during exercise, any increase in stroke volume that occurs must be due to an increase in the proportion of blood ejected per stroke.

The proportion of the end-diastolic volume ejected per stroke can increase from 60% at rest to as much as 90% during heavy exercise. This increased **ejection fraction** is produced by the increased contractility that results from sympathoadrenal stimulation. There also may be a decrease in total peripheral resistance as a result of vasodilation in the exercising skeletal muscles, which decreases the afterload and thus further augments the increase in stroke volume.

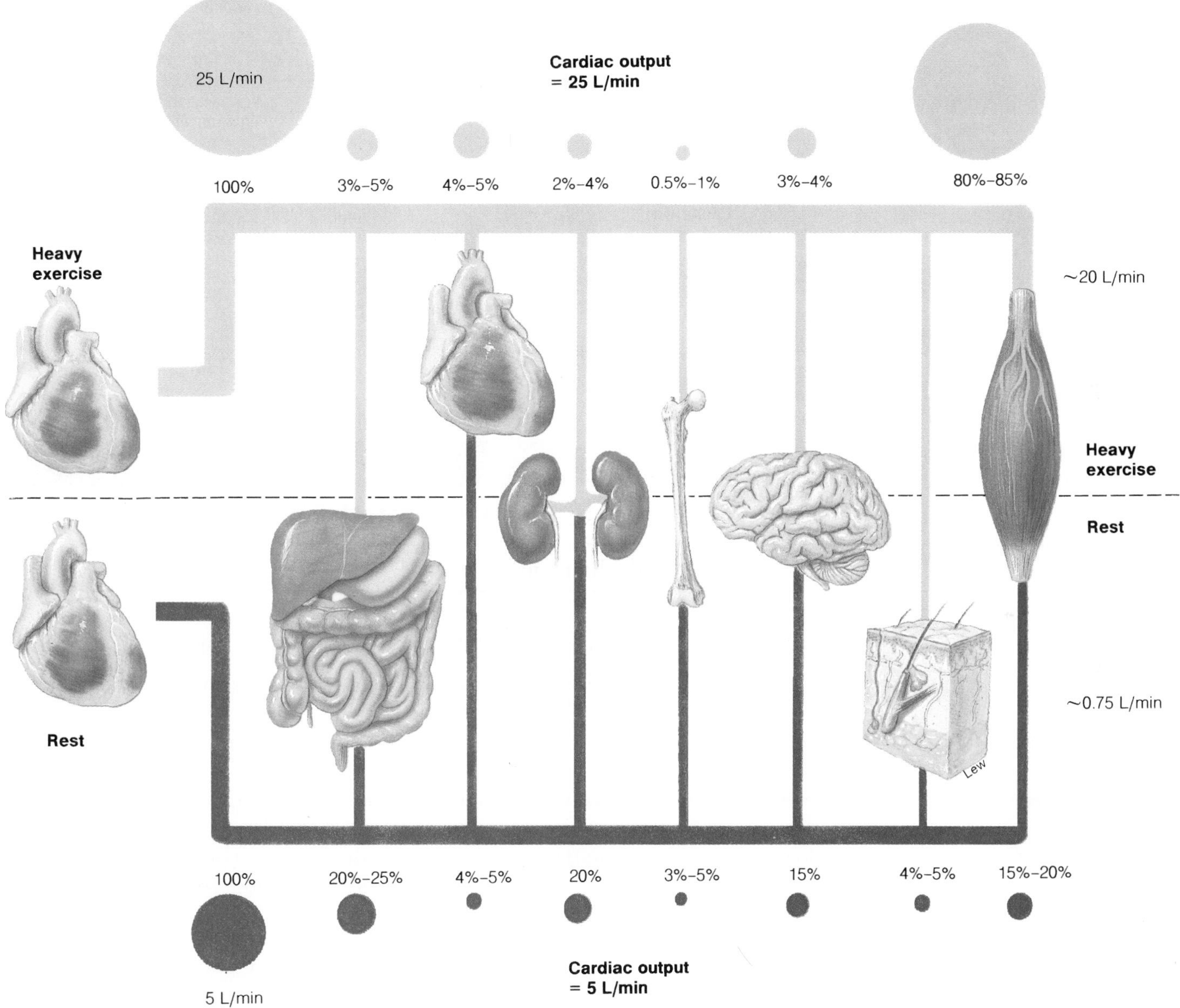

Figure 22.19

The distribution of blood flow (cardiac output) during rest and heavy exercise.
At rest, the cardiac output is 5 L per minute (*bottom of figure*); during heavy exercise the cardiac output increases to 25 L per minute (*top of figure*). At rest, for example, the brain receives 15% of 5 L per minute (= 750 ml/min), whereas during exercise it receives 3% to 4% of 25 L per minute ($0.03 \times 25 = 750$ ml/min). Flow to the skeletal muscles increases more than twentyfold because the total cardiac output increases (from 5 L/min to 25 L/min) and because the percentage of the total received by the muscles increases from 15% to 80%.

Source: Adapted from P. Astrand and K. Rodahl, *Textbook of Work Physiology*, 3rd edition, copyright 1986 McGraw-Hill, Inc., New York. Used by permission of the author.

The cardiovascular changes that occur during exercise are summarized in table 22.7.

Endurance training often results in a lowering of the resting cardiac rate and an increase in the resting stroke volume. The lowering of the resting cardiac rate results from a greater degree of inhibition of the SA node by the vagus nerve. The increased resting stroke volume is believed to be due to an increase in blood volume; indeed, studies have shown that the blood volume can increase by about 500 ml after only 8 days of training. These adaptations enable the trained athlete to produce a larger proportionate increase in cardiac output and achieve a higher absolute cardiac output during exercise. This large cardiac output is the major factor in the improved oxygen delivery to skeletal muscles that occurs as a result of endurance training.

Table 22.6

Relationship between Age and Average Maximum Cardiac Rate

Age	Maximum Cardiac Rate
20–29	190 beats/min
30–39	160 beats/min
40–49	150 beats/min
50–59	140 beats/min
60+	130 beats/min

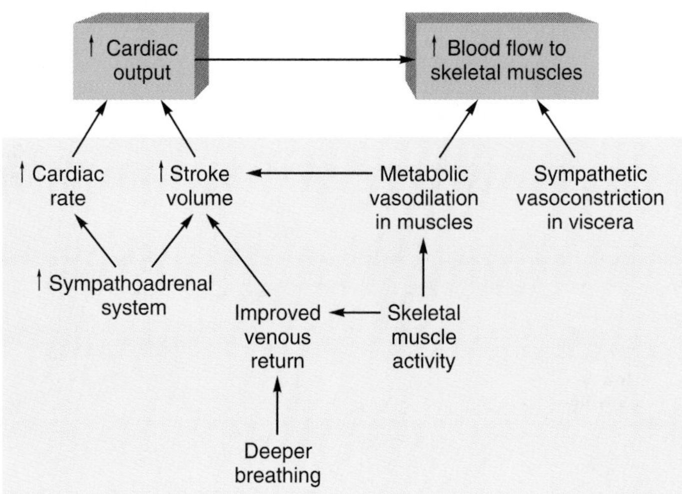

Figure 22.20

Cardiovascular adaptations to exercise. These adaptations (1) increase the cardiac output, and thus the total blood flow, and (2) cause vasodilation in the exercising muscles, thereby diverting a higher proportion of the blood flow to those muscles.

Table 22.7

Cardiovascular Changes during Moderate Exercise

Variable	Change	Mechanisms
Cardiac output	Increases	Increased cardiac rate and stroke volume
Cardiac rate	Increases	Increased sympathetic nerve activity; decreased activity of the vagus nerve
Stroke volume	Increases	Increased myocardial contractility due to stimulation by sympathoadrenal system; decreased total peripheral resistance
Total peripheral resistance	Decreases	Vasodilation of arterioles in skeletal muscles (and in skin when thermoregulatory adjustments are needed)
Arterial blood pressure	Increases	Increased systolic and pulse pressure due primarily to increased cardiac output; diastolic pressure rises less due to decreased total peripheral resistance
End-diastolic volume (EDV)	Unchanged	Decreased filling time at high cardiac rates is offset by increased venous pressure, increased activity of the skeletal muscle pump, and decreased intrathoracic pressure aiding the venous return
Blood flow to heart and muscles	Increases	Increased muscle metabolism produces intrinsic vasodilation; aided by increased cardiac output and increased vascular resistance in visceral organs
Blood flow to visceral organs	Decreases	Vasoconstriction in GI tract, liver, and kidneys due to sympathetic nerve stimulation
Blood flow to skin	Increases	Metabolic heat produced by exercising muscles produces reflex (involving hypothalamus) that reduces sympathetic constriction of arteriovenous shunts and arterioles
Blood flow to brain	Unchanged	Autoregulation of cerebral vessels, which maintains constant cerebral blood flow despite increased arterial blood pressure

Blood Flow to the Brain and Skin

Intrinsic control mechanisms help to maintain a relatively constant blood flow to the brain. Blood flow to the skin, by contrast, can vary tremendously in response to regulation by sympathetic nerve stimulation.

The examination of cerebral and cutaneous blood flow is a study in contrasts. Cerebral blood flow is regulated primarily by intrinsic mechanisms; cutaneous blood flow is regulated by extrinsic mechanisms. Cerebral blood flow is relatively constant; cutaneous blood flow exhibits more variation than blood flow in any other organ. The brain is the organ that can least tolerate low rates of blood flow; the skin is the organ that can tolerate low rates the most.

Figure 22.21

Changing patterns of blood flow in the brain.
A computerized picture of blood-flow distribution in the brain after injecting the carotid artery with a radioactive isotope. In (*a*), on the left, the subject followed a moving object with his eyes. High activity is seen over the occipital lobe of the brain. In (*a*), on the right, the subject listened to spoken words. Notice that the high activity is seen over the temporal lobe (the auditory cortex). In (*b*), on the left, the subject moved his fingers on the side of the body opposite to the cerebral hemisphere being studied. In (*b*), on the right, the subject counted to 20. High activity is shown over the mouth area of the motor cortex, the supplementary motor area, and the auditory cortex.

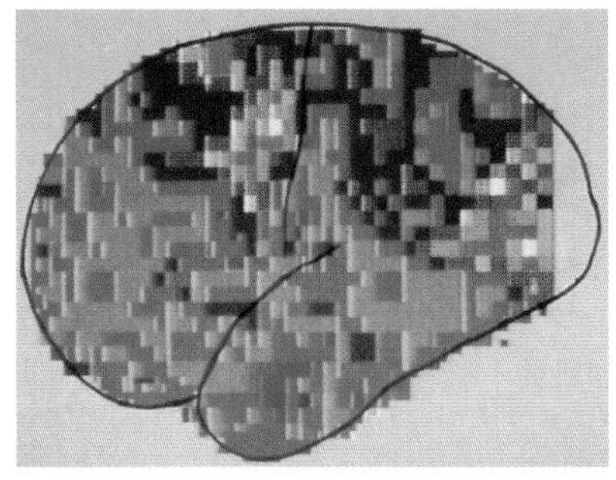

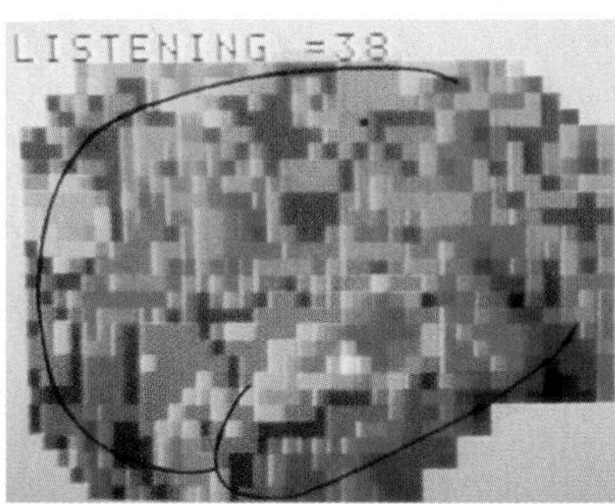

(a)

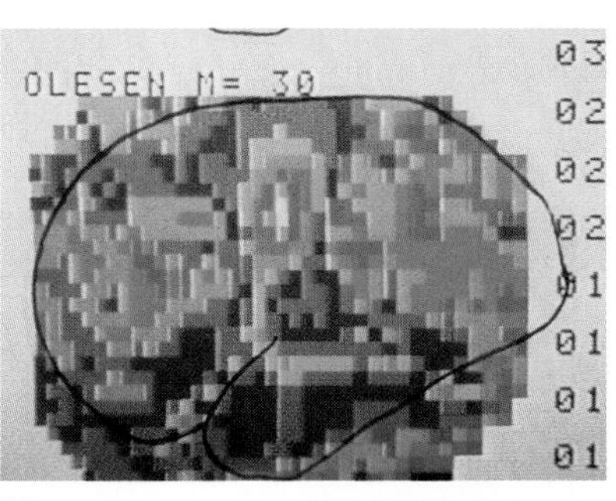

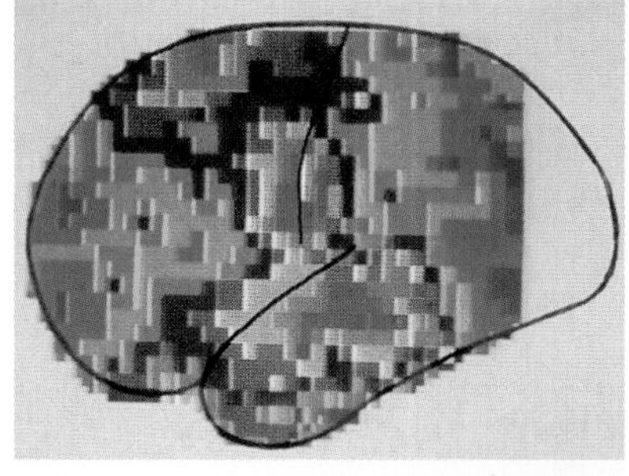

(b)

Cerebral Circulation

When the brain is deprived of oxygen for just a few seconds a person loses consciousness; irreversible brain injury may occur after a few minutes. For these reasons, the cerebral blood flow is held remarkably constant at about 750 ml per minute. This amounts to about 15% of the total cardiac output at rest.

Unlike the coronary and skeletal muscle blood flow, cerebral blood flow is not normally influenced by sympathetic nerve activity. Only when the mean arterial pressure rises to about 200 mmHg do sympathetic nerves cause a significant degree of vasoconstriction in the cerebral circulation. This vasoconstriction helps to protect small, thin-walled arterioles from bursting under the pressure, and thus helps to prevent cerebrovascular accident (stroke).

In the normal range of arterial pressures, cerebral blood flow is regulated almost exclusively by local, intrinsic mechanisms—a process called autoregulation, as previously mentioned. These mechanisms help to ensure a constant rate of blood flow despite changes in systemic arterial pressure. The autoregulation of cerebral blood flow is achieved by both myogenic and metabolic mechanisms.

Myogenic Regulation

Myogenic regulation occurs when there is variation in systemic arterial pressure. When the blood pressure falls, the cerebral arteries automatically dilate; when the blood pressure rises, they constrict. This helps to maintain a constant flow rate during the normal pressure variations that occur during rest, exercise, and emotional states.

The cerebral vessels are also sensitive to the carbon dioxide concentration of arterial blood. When the carbon dioxide concentration rises as a result of inadequate ventilation (hypoventilation), the cerebral arterioles dilate. This is believed to be due to decreases in the pH of cerebrospinal fluid rather than to a direct effect of CO_2 on the cerebral vessels. Conversely, when the arterial CO_2 falls below normal during hyperventilation, the cerebral vessels constrict. The resulting decrease in cerebral blood flow is responsible for the dizziness that occurs during hyperventilation.

Metabolic Regulation

The cerebral arterioles are exquisitely sensitive to local changes in metabolic activity, so that those brain regions with the highest metabolic activity receive the most blood. Indeed, areas of the brain that control specific processes have been mapped by the changing patterns of blood flow that result when these areas are activated. Visual and auditory stimuli, for example, increase blood flow to the appropriate sensory areas of the cerebral cortex, whereas motor activities, such as movements of the eyes, arms, and organs of speech, result in different patterns of blood flow (fig. 22.21).

The exact mechanisms by which increases in neural activity in a particular area of the brain elicit local vasodilation are not completely understood. There is evidence, however, that local cerebral vasodilation may be caused by K^+, which is released from active neurons during repolarization. It has been proposed that astrocytes may take up this extruded K^+ near the active neurons and then release the K^+ through their vascular processes (chapter 14) that surround arterioles, thereby causing the arterioles to dilate.

Cutaneous Blood Flow

The skin is the outer covering of the body and as such serves as the first line of defense against invasion by disease-causing organisms. The skin, as the interface between the internal and external environments, also helps to maintain a constant deep-body temperature despite changes in the ambient (external) temperature—a process called **thermoregulation.** The thinness and extensiveness of the skin (1.0–1.5 mm thick; 1.7–1.8 square meters in surface area) make it an effective radiator of heat when the body temperature rises above the ambient temperature. The transfer of heat from the body to the external environment is aided by the flow of warm blood through capillary loops near the surface of the skin.

Blood flow through the skin is adjusted to maintain deep-body temperature at about 37° C (98.6° F). These adjustments are made by variations in the degree of constriction or dilation of ordinary arterioles and of unique **arteriovenous anastomoses** (*ă-nas″tŏ-mo′sēz*) (fig. 22.22). These latter vessels, found predominantly in the fingertips, palms of the hands, toes, soles of the feet, ears, nose, and lips, shunt (divert) blood directly from arterioles to deep venules, thus bypassing superficial capillary loops. Both the ordinary arterioles and the arteriovenous anastomoses are innervated by sympathetic nerve fibers. When the ambient temperature is low, sympathetic nerves stimulate cutaneous vasoconstriction; cutaneous blood flow is thus decreased, so that less heat will be lost from the body. Since the arteriovenous anastomoses also constrict, the skin may appear rosy because the blood is diverted to the superficial capillary loops. In spite of this rosy appearance, however, the total cutaneous blood flow and rate of heat loss is lower than under usual conditions.

Skin can tolerate an extremely low blood flow in cold weather because its metabolic rate decreases when the ambient temperature decreases. In cold weather, therefore, the skin requires less blood. As a result of exposure to extreme cold, however, blood flow to the skin can be so severely restricted that the tissue dies—a condition known as *frostbite*. Blood flow to the skin can vary from less than 20 ml per minute at maximal vasoconstriction to as much as 3 to 4 L per minute at maximal vasodilation.

As the temperature rises, cutaneous arterioles in the hands and feet dilate as a result of decreased sympathetic nerve activity. Continued warming causes dilation of arterioles in other areas of the skin. If the resulting increase in cutaneous blood flow is not sufficient to cool the body, secretion of the sweat glands may be stimulated. Sweat helps to cool the body as it evaporates from the surface of the skin. The sweat glands also secrete **bradykinin** (*brad″ĭ-ki′nin*), a polypeptide that stimulates vasodilation. This increases blood flow to the skin and to the sweat glands, so that larger volumes of more dilute sweat are produced.

Under the usual conditions of ambient temperature, the cutaneous vascular resistance is high and the blood flow is low when a person is not exercising. In the pre-exercise state of fight or flight, sympathetic nerve activity further reduces cutaneous blood flow. During exercise, however, the need to maintain a deep-body temperature takes precedence over the need to maintain an adequate systemic blood pressure. As the body temperature rises during exercise, vasodilation in cutaneous vessels occurs together with vasodilation in the exercising muscles. This can produce an even greater lowering of total peripheral resistance. If exercise is performed in hot and humid weather and if restrictive clothing increases skin temperature and cutaneous vasodilation, a dangerously low blood pressure may be produced after exercise has ceased and the cardiac output has declined. People have lost consciousness and have even died as a result.

Changes in cutaneous blood flow occur as a result of changes in sympathetic nerve activity. Since the activity of the sympathetic nervous system is controlled by the brain,

Figure 22.22

Circulation in the skin showing arteriovenous anastomoses.
These vessels function as shunts, allowing blood to be diverted directly from the arteriole to the venule and thus bypass superficial capillary loops.

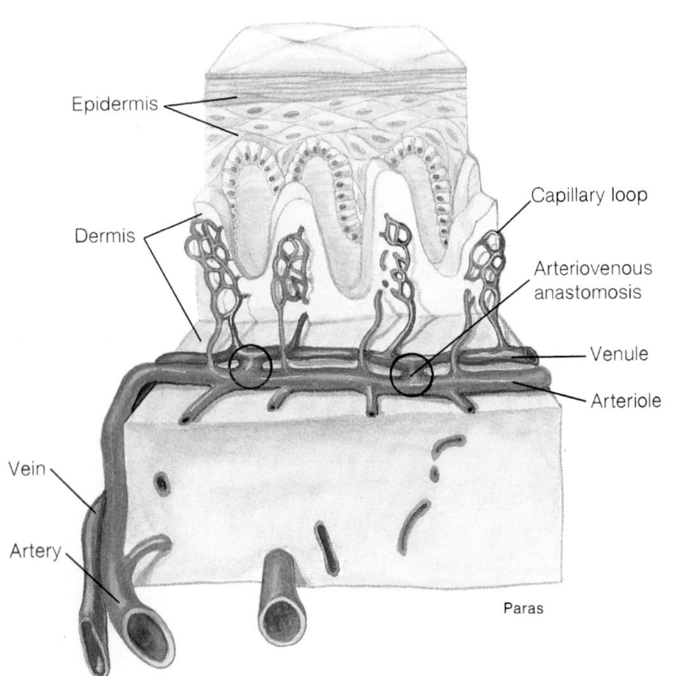

anastamosis: Gk. *anastamosis*, opening or outlet

emotional states, acting through control centers in the medulla oblongata, can affect sympathetic activity and cutaneous blood flow. During fear reactions, for example, vasoconstriction in the skin, along with activation of the sweat glands, can produce a pallor and a "cold sweat." Other emotions may cause vasodilation and blushing.

Blood Pressure

The pressure of the arterial blood is regulated by the blood volume, the total peripheral resistance, and the cardiac rate. Regulatory mechanisms adjust these factors in a negative feedback manner to compensate for deviations. Arterial pressure rises and falls as the heart goes through systole and diastole.

Resistance to flow in the arterial system is greatest in the arterioles because these vessels have the smallest diameters. Although the total blood flow through a system of arterioles must be equal to the flow in the larger vessel that gave rise to those arterioles, the narrow diameter of each arteriole reduces the flow rate in each according to Poiseuille's law. Blood flow rate and pressure are thus reduced in the capillaries, which are located downstream of the high resistance imposed by the arterioles. The blood pressure upstream of the arterioles—in the medium and large arteries—is correspondingly increased (fig. 22.23).

The blood pressure and flow rate within the capillaries are further reduced by the fact that their total cross-sectional area is much greater, due to their large number, than the cross-sectional areas of the arteries and arterioles (fig. 22.24). Thus, although each capillary is much narrower than each arteriole, the capillary beds served by arterioles do not provide as great a resistance to blood flow as do the arterioles.

Variations in the diameter of arterioles as a result of vasoconstriction and vasodilation thus simultaneously affect both blood flow through capillaries and the *arterial blood pressure* "upstream" from the capillaries. In this way, an increase in total peripheral resistance due to vasoconstriction of arterioles can raise arterial blood pressure. Blood pressure can also be raised by an increase in the cardiac output. This may be due to elevations in cardiac rate or stroke volume, which in turn are affected by other factors. The three most important variables affecting blood pressure are the **cardiac rate, stroke volume** (determined primarily by the **blood volume**), and **total peripheral resistance.** An increase in any of these, if not compensated for by a decrease in another variable, will result in an increased blood pressure.

Arterial blood pressure $\propto$ cardiac output × total peripheral resistance

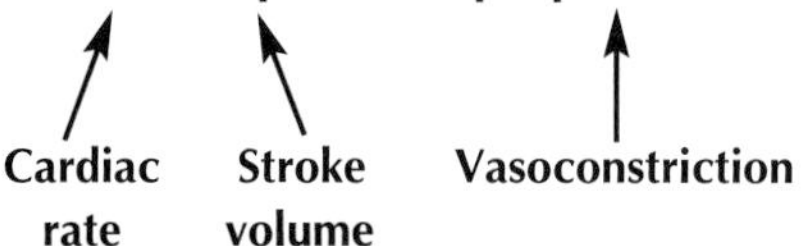

Blood pressure can thus be regulated by the kidneys, which control blood volume and thus stroke volume, and by the sympathoadrenal system. Increased activity of the sympathoadrenal system can raise blood pressure by stimulating vasoconstriction of arterioles (thus raising total peripheral resistance) and by promoting an increased cardiac output. Sympathetic stimulation can also affect blood volume indirectly, by stimulating constriction of renal blood vessels and thus reducing urine output.

Baroreceptor Reflex

In order for blood pressure to be maintained within limits, specialized receptors for pressure are needed. These **baroreceptors** are stretch receptors located in the *aortic arch* and in the *carotid sinuses*. An increase in pressure causes the walls of these arterial regions to stretch, increasing the frequency of action potentials along sensory nerve fibers (fig. 22.25). A fall in pressure below the normal range, by contrast, causes a decrease in the frequency of action potentials produced by these sensory nerve fibers.

Sensory nerve activity from the baroreceptors ascends, via the vagus and glossopharyngeal nerves, to the medulla oblongata, which directs the autonomic system to respond appropriately. The **vasomotor control center** in the medulla controls vasoconstriction/vasodilation, and hence helps to regulate total peripheral resistance. The **cardiac control center** in the medulla regulates the cardiac rate (fig. 22.26). Acting through the activity of motor fibers within the vagus and sympathetic nerves controlled by these brain centers, the baroreceptors function to counteract blood pressure changes so that fluctuations in pressure are minimized.

The baroreceptor reflex is activated whenever blood pressure increases or decreases. The reflex is somewhat more sensitive to decreases in pressure than to increases, and is more sensitive to sudden changes in pressure than to more gradual

baroreceptor: Gk. *baros*, pressure; L. *receiver*, to receive

Figure 22.23

The effect of vasoconstriction on blood pressure. A constriction increases blood pressure upstream (analogous to the arterial pressure) and decreases pressure downstream (analogous to capillary and venous pressure).

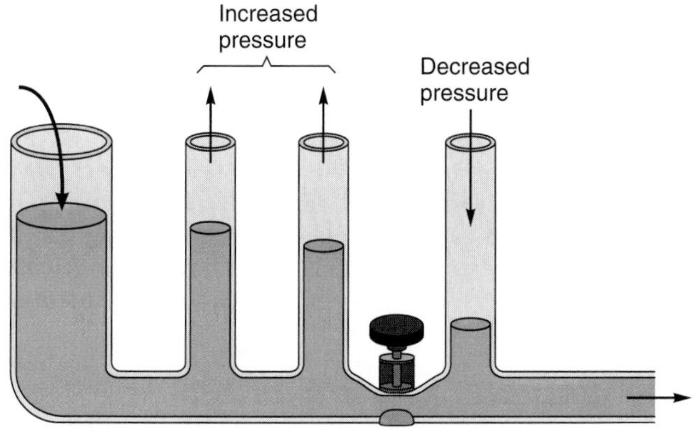

Figure 22.24

The relationship between vessels' cross-sectional area and blood pressure.
As blood passes from the aorta to the smaller arteries, arterioles, and capillaries, the cross-sectional area increases as the pressure decreases.

Source: Redrawn from E. O. Feigel, "*Physics in the Cardiovascular System*," Physiology and Biophysics, Vol. II, 20th ed., ed. by T. C. Ruch and H. D. Patton, © 1974 W. B. Saunders, Philadelphia, PA. Used by permission.

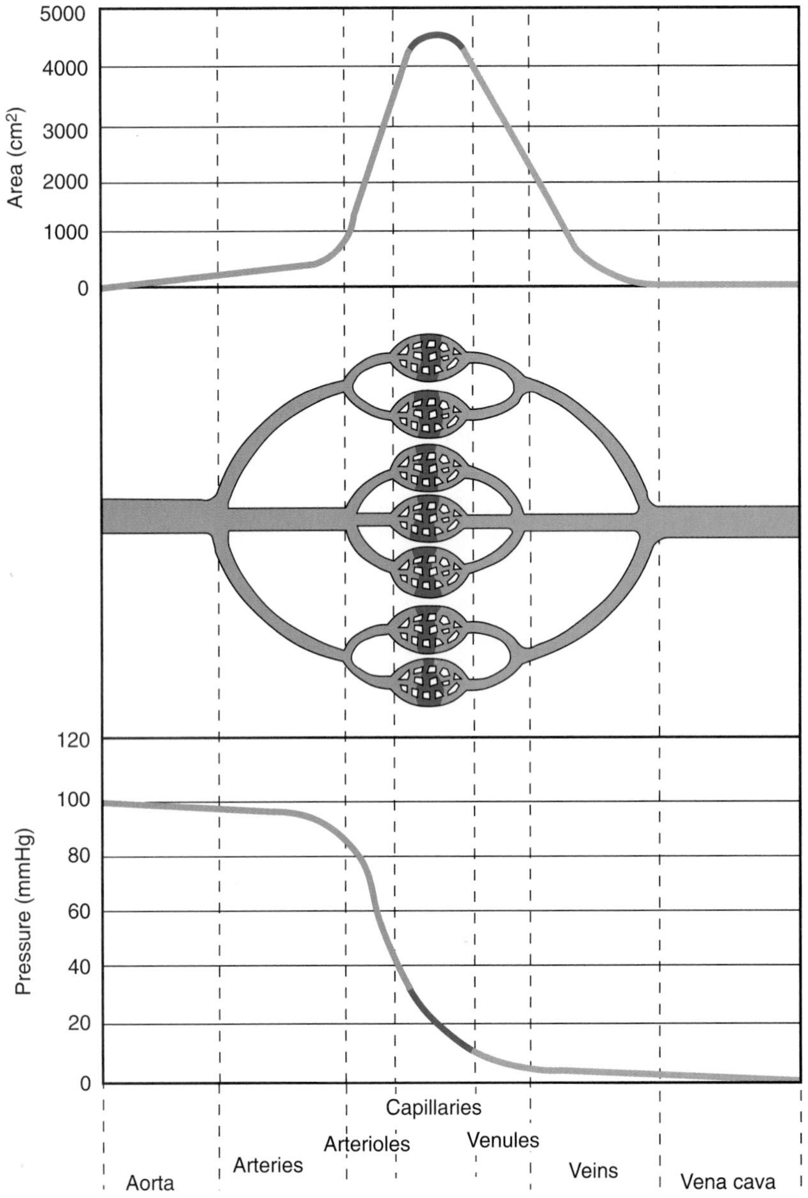

changes. A good example of the importance of the baroreceptor reflex in normal physiology is its activation whenever a person goes from a lying to a standing position; at this time, there is a shift of 500 to 700 ml of blood from the veins of the thoracic cavity to veins in the lower extremities, which expand to contain the extra volume of blood. This pooling of blood in the lower extremities reduces the venous return and cardiac output, but the resulting fall in blood pressure is almost immediately compensated for by the baroreceptor reflex. A decrease in baroreceptor sensory information, traveling in the glossopharyngeal nerve (IX) and the vagus nerve (X) to the medulla oblongata, inhibits parasympathetic activity and promotes sympathetic nerve activity. This produces an increase in cardiac rate and vasoconstriction, which help to maintain an adequate blood pressure upon standing (fig. 22.27).

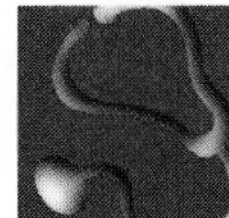

Since the baroreceptor reflex may require a few seconds before it is fully effective, many people feel dizzy and disoriented if they stand up too quickly. If the baroreceptor sensitivity is abnormally reduced, perhaps by atherosclerosis, an uncompensated fall in pressure may occur upon standing. This condition—called *postural,* or *orthostatic, hypotension* (hypotension = low blood pressure)—can make a person feel extremely dizzy or even faint because of inadequate perfusion of the brain.

Input from baroreceptors can also mediate the opposite response. When the blood pressure rises above an individual's normal range, the baroreceptor reflex causes a slowing of the cardiac rate and vasodilation. Manual massage of the carotid sinus, a procedure sometimes employed by physicians to reduce tachycardia and lower blood pressure, also evokes this reflex. Such carotid massage should be used cautiously, however, because the intense vagus nerve–induced slowing of the cardiac rate could cause loss of consciousness (as occurs in emotional fainting). Manually massaging both carotid sinuses simultaneously can even cause cardiac arrest in susceptible people.

Valsalva's (val-sal'vaz) maneuver is the term used to describe an expiratory effort against a closed glottis (which prevents the air from escaping). This maneuver, commonly performed during forceful defecation or when lifting heavy weights, increases the intrathoracic pressure. Compression of the thoracic veins causes a fall in venous return and cardiac output, thus lowering arterial blood pressure. The lowering of arterial pressure then stimulates the baroreceptor reflex, resulting in tachycardia and increased total peripheral resistance. When the glottis is finally opened and the air is exhaled, the cardiac output returns to normal. The total peripheral resistance is still elevated, however, causing a rise in blood pressure. The blood pressure is then brought back to normal by the baroreceptor reflex, which causes a slowing of the heart rate. These fluctuations in cardiac output and blood pressure can be dangerous in people with cardiovascular disease. Even healthy people are advised to exhale normally when lifting weights.

Valsalva's maneuver: from Antonio Valsalva, Italian anatomist, 1666–1723

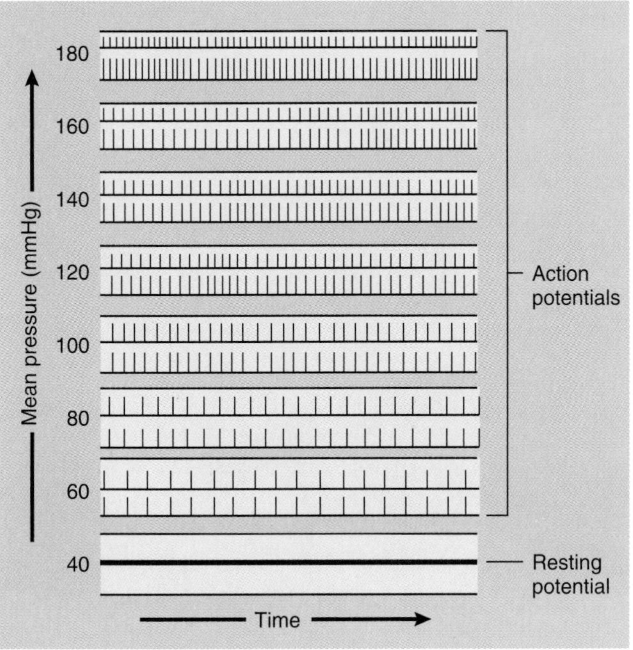

Figure 22.25

Effect of blood pressure on baroreceptor response. This is a recording of the action potential frequency in sensory nerve fibers from baroreceptors in the carotid sinus and aortic arch. As the blood pressure increases, the baroreceptors become increasingly stretched. This results in an increase in the frequency of action potentials that are transmitted to the cardiac and vasomotor control centers in the medulla oblongata.

Atrial Stretch Reflexes

In addition to the baroreceptor reflex, several other reflexes help to regulate blood pressure. The reflex control of ADH release by osmoreceptors in the hypothalamus and the control of angiotensin II production and aldosterone secretion by the juxtaglomerular apparatus of the kidneys have been previously discussed. Antidiuretic hormone and aldosterone increase blood pressure by increasing blood volume, and angiotensin II stimulates vasoconstriction to cause an increase in blood pressure.

Other reflexes important to blood pressure regulation are initiated by **atrial stretch receptors** located in the atria of the heart. These receptors are activated by increased venous return to the heart and, in response (1) stimulate reflex tachycardia, as a result of increased sympathetic nerve activity; (2) inhibit ADH release, resulting in the excretion of larger volumes of urine and a lowering of blood volume; and (3) promote increased secretion of atrial natriuretic factor (ANF). The ANF, as previously discussed, lowers blood volume by increasing urinary salt and water excretion and by antagonizing the actions of angiotensin II.

Measurement of Blood Pressure

The first documented measurement of blood pressure was accomplished by the Englishman Stephen Hales. Hales inserted a cannula into the artery of a horse and measured the heights

Stephen Hales: English physiologist and clergyman, 1677–1761

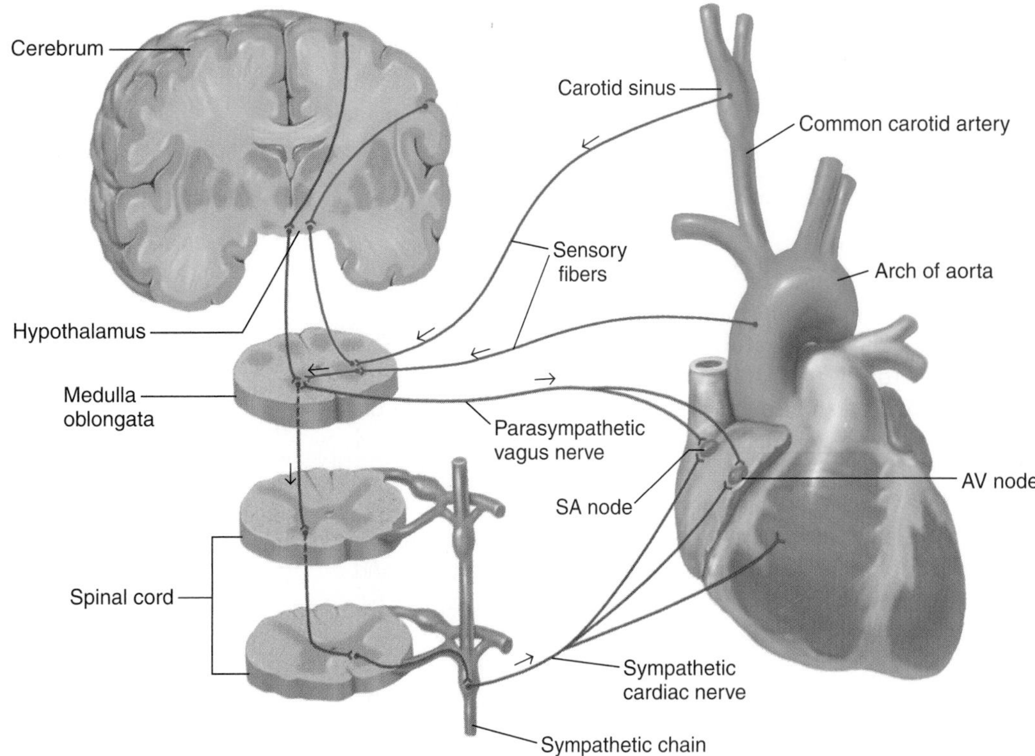

Figure 22.26

The structures involved in the baroreceptor reflex. Sensory stimuli from baroreceptors in the carotid sinus and the aortic arch, acting via control centers in the medulla oblongata, affect the activity of sympathetic and parasympathetic nerve fibers in the heart.

Figure 22.27

The negative feedback control of blood pressure by the baroreceptor reflex.
This reflex helps to maintain an adequate blood pressure upon standing.

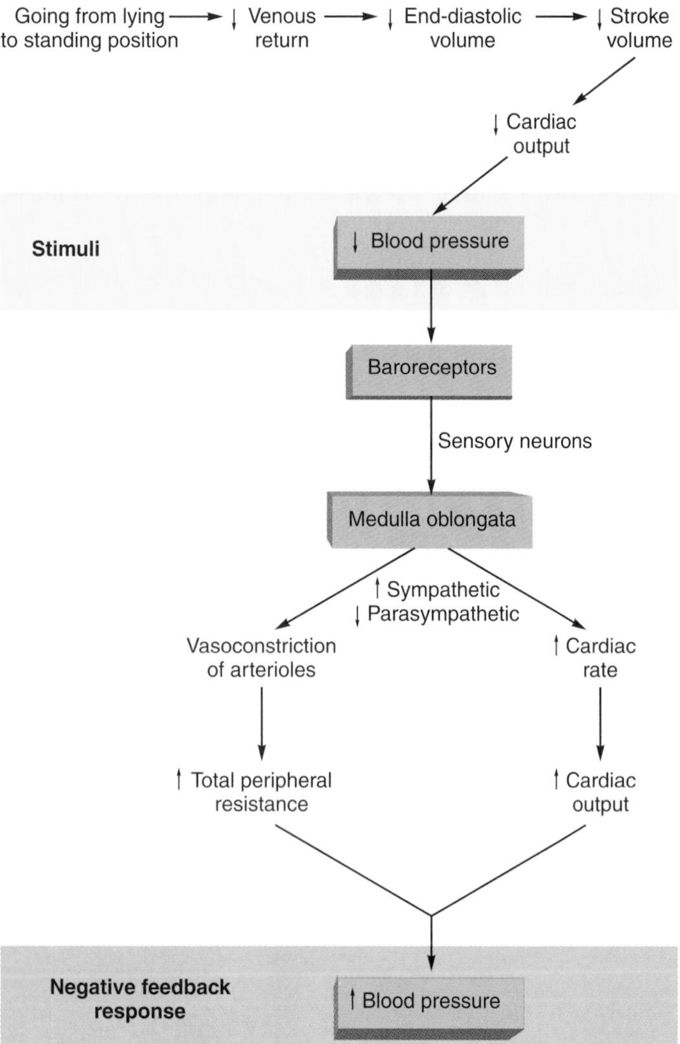

Figure 22.28

The use of a pressure cuff and sphygmomanometer to measure blood pressure.
The examiner is listening for the Korotkoff sounds.

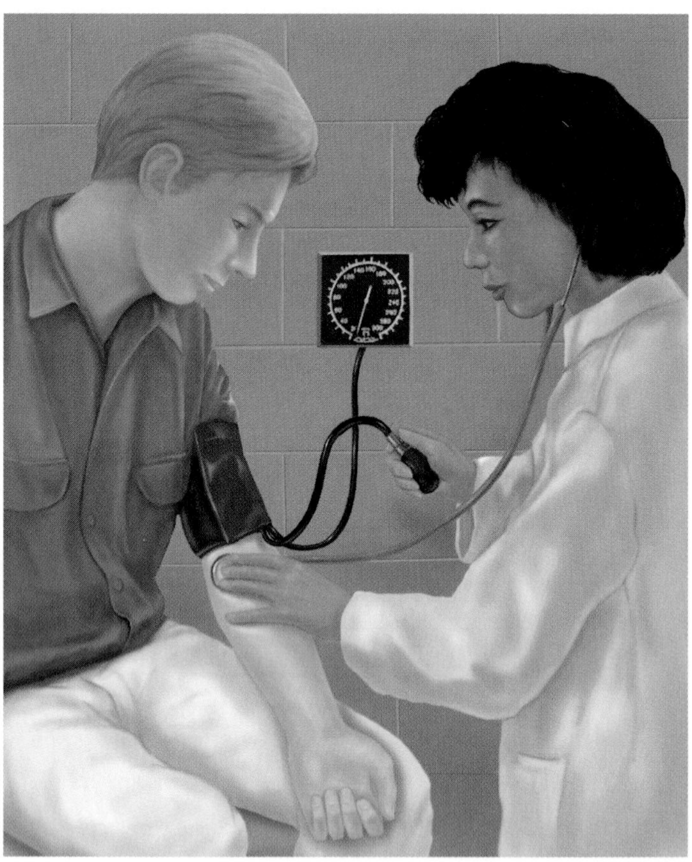

to which blood would rise in the vertical tube. The height of this blood column bounced between the **systolic pressure** at its highest and the **diastolic pressure** at its lowest, as the heart went through its cycle of systole and diastole. Modern clinical blood pressure measurements, fortunately, are less direct. The indirect, or **auscultatory** (*aw-skul'tă-tor"e*), method is based on the correlation of blood pressure and arterial sounds.

In the auscultatory method, a cloth cuff containing an inflatable rubber bladder is wrapped around the upper arm and a stethoscope is applied over the brachial artery (fig. 22.28). The artery is normally silent before inflation of the cuff because blood normally travels in a smooth *laminar flow* through the arteries. The term *laminar* means "layered"—blood in the central axial stream moves the fastest, and blood flowing closer to the artery wall moves more slowly. There is little transverse movement between these layers that would produce mixing.

The laminar flow that normally occurs in arteries produces little vibration and is thus silent. When the artery is pinched, however, blood flow through the constriction becomes turbulent. This causes the artery to vibrate and produce sounds, much like the sounds produced by water flowing through a kink in a garden hose. The tendency of the cuff pressure to constrict the artery is opposed by the blood pressure. Thus, in order to constrict the artery, the cuff pressure must be greater than the diastolic blood pressure. If the cuff pressure is also greater than the systolic blood pressure the artery will be pinched off and silent. Therefore, *turbulent flow* and the sounds produced by vibrations as a result of this flow occur only when the cuff pressure is greater than the diastolic pressure but lower than the systolic pressure.

Let's say that a person has a systolic pressure of 120 mmHg and a diastolic pressure of 80 mmHg (the average normal values). When the cuff pressure is between 80 and 120 mmHg the artery will be closed during diastole and open during systole. As the artery begins to open with every systole, turbulent flow of blood through the constriction will create

---

auscultatory: L. *auscultare*, to listen to

Figure 22.29

The blood flow and Korotkoff sounds during a blood pressure measurement.
When the cuff pressure is above the systolic pressure, the artery is constricted. When the cuff pressure is below the diastolic pressure, the artery is open and flow is laminar. When the cuff pressure is between the diastolic and systolic pressure, blood flow is turbulent and the Korotkoff sounds are heard with each systole.

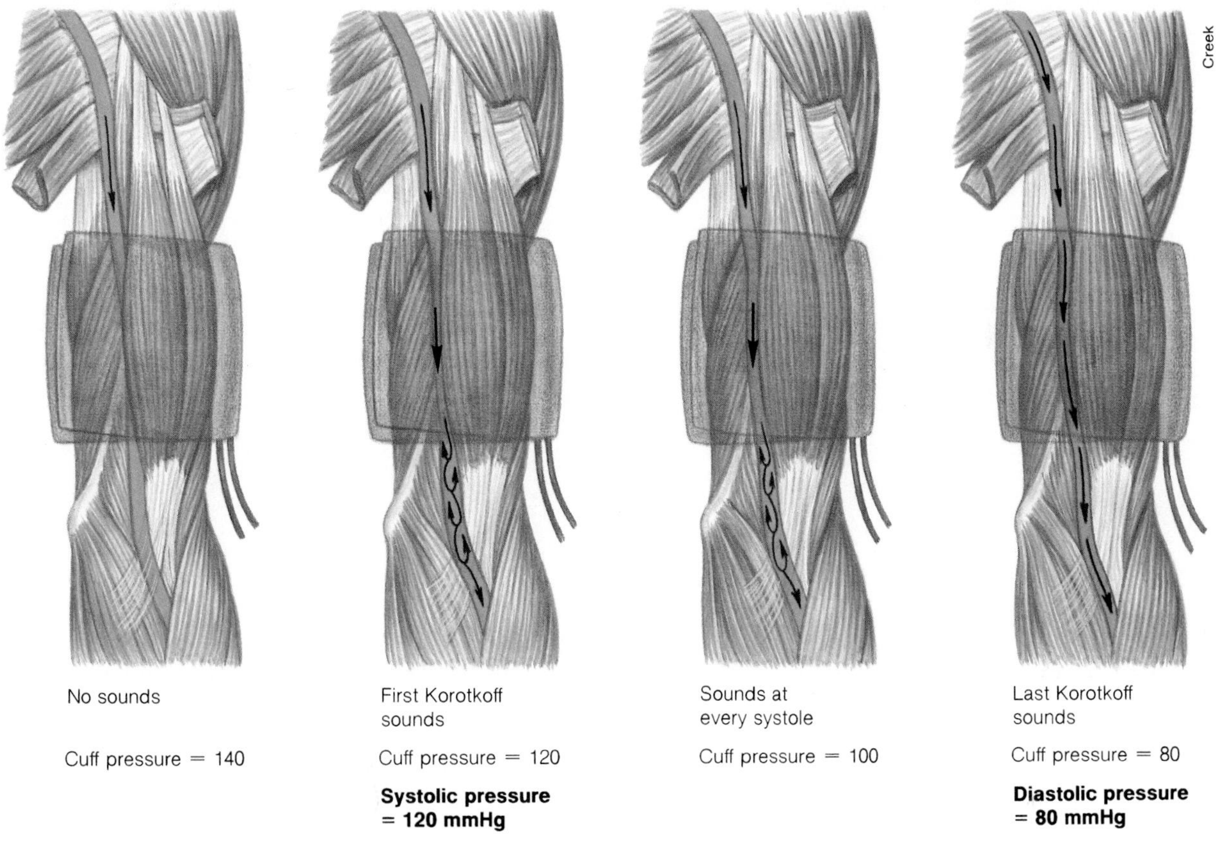

vibrations that are known as the **sounds of Korotkoff** (*kŏ-rot'kof*), as shown in figure 22.29. These are usually "tapping" sounds because the artery becomes constricted, blood flow stops, and silence is restored with every diastole. It should be understood that the sounds of Korotkoff are *not* "lub-dub" sounds produced by closing of the heart valves (those sounds can be heard only on the chest, not on the brachial artery).

Initially the cuff is inflated to produce a pressure greater than the systolic pressure, so that the artery is pinched off and silent. The pressure in the cuff is read from an attached meter called a *sphygmomanometer* (*sfig"mo-mă-nom'ĭ-ter*). A valve is then turned to allow the release of air from the cuff, causing a gradual decrease in cuff pressure. When the cuff pressure is equal to the systolic pressure, the **first Korotkoff sound** is heard as blood passes in a turbulent flow through the constricted artery.

sounds of Korotkoff: from Nicolai S. Korotkoff, Russian physician, 1874–1920

sphygmomanometer: Gk. *sphygmos*, pulse; *manos*, thin; *metro*, measure

Korotkoff sounds will continue to be heard at every systole as long as the cuff pressure remains greater than the diastolic pressure. When the cuff pressure becomes equal to or less than the diastolic pressure the sounds disappear; the artery remains open, laminar flow occurs, and the vibrations of the artery stop (fig. 22.30). The **last Korotkoff sound** thus occurs when the cuff pressure is equal to the diastolic pressure.

Different phases in the measurement of blood pressure are identified on the basis of the quality of the Korotkoff sounds (fig. 22.31). In some people, the Korotkoff sounds do not disappear even when the cuff pressure is reduced to zero (zero pressure means that it is equal to atmospheric pressure). In these cases—and often routinely—the onset of muffling of the sounds (phase 4 in fig. 22.31) is used as an indication of diastolic pressure rather than the onset of silence (phase 5). Normal blood pressure values are shown in table 22.8.

The average arterial blood pressure in the systemic circulation is 120/80 mmHg, whereas the average pulmonary arterial blood pressure is only 22/8 mmHg. Because of the Frank–Starling relationship, the cardiac output from the right ventricle into the pulmonary circulation is matched to that of

Figure 22.30

The indirect, or auscultatory, method of blood pressure measurement.
The first Korotkoff sound is heard when the cuff pressure is equal to the systolic blood pressure, and the last sound is heard when the cuff pressure is equal to the diastolic pressure. The dashed line indicates the falling cuff pressure.

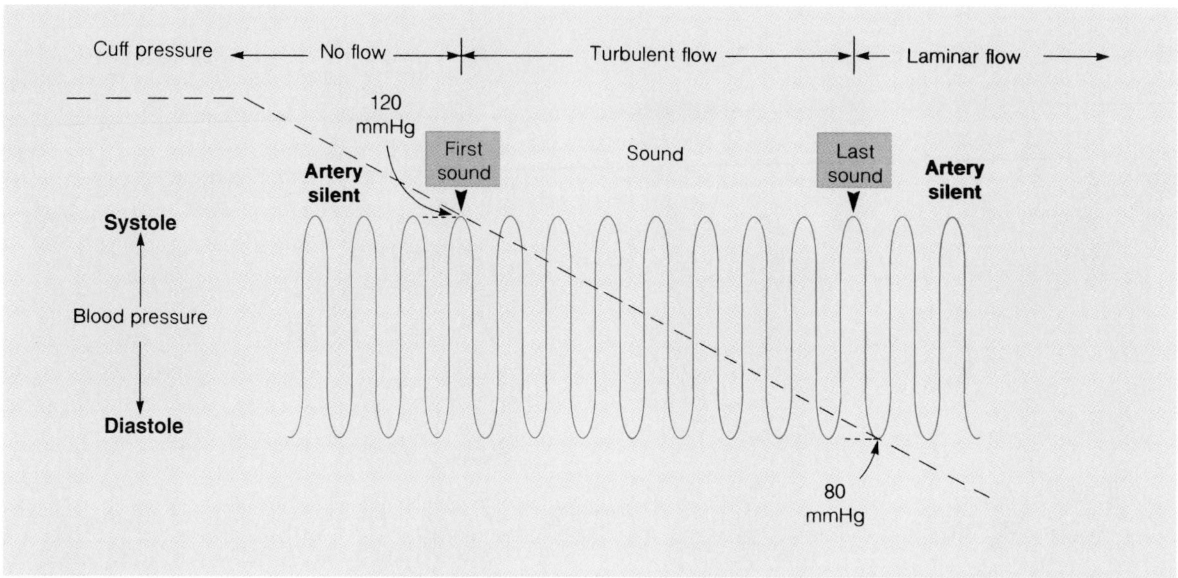

Figure 22.31

The five phases of blood pressure measurement.
Not all phases are heard in all people.

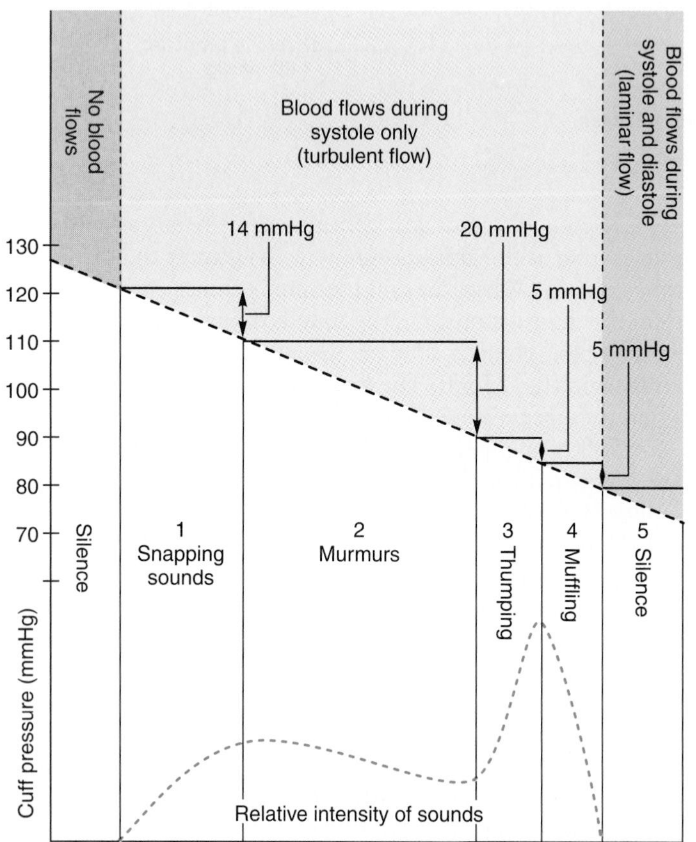

the left ventricle into the systemic circulation. Since the cardiac outputs are the same, the lower pulmonary pressure must be caused by a lower total peripheral resistance in the pulmonary circulation. Because the right ventricle pumps blood against a lower resistance, it has a lighter workload and its walls are thinner than those of the left ventricle.

Pulse Pressure and Mean Arterial Pressure

When someone "takes a pulse," he or she palpates an artery (for example, the radial artery) and feels the expansion of the artery occur in response to the beating of the heart; the pulse rate is thus a measure of the cardiac rate. The expansion of the artery with each pulse occurs as a result of the rise in blood pressure within the artery as the artery receives the volume of blood ejected by a stroke of the left ventricle.

The pulse is thus produced by the **pulse pressure,** which is equal to the difference between the systolic and diastolic pressures. If a person has a blood pressure of 120/80 (systolic/diastolic), therefore, the pulse pressure would be 40 mmHg.

Pulse pressure = systolic pressure – diastolic pressure

At diastole in this example the aortic pressure equals 80 mmHg. When the left ventricle contracts, the intraventricular pressure rises above 80 mmHg and ejection begins. As a result, the amount of blood in the aorta increases by the amount ejected from the left ventricle (the stroke volume). Due to the increase in volume, there is an increase in blood pressure. The pressure in the brachial artery, where blood

Table 22.8

Normal Arterial Blood Pressure at Different Ages

	Systolic		Diastolic			Systolic		Diastolic	
Age	Male	Female	Male	Female	Age	Male	Female	Male	Female
1 day	70				16 years	118	116	73	72
3 days	72				17 years	121	116	74	72
9 days	73				18 years	120	116	74	72
3 weeks	77				19 years	122	115	75	71
3 months	86				20–24 years	123	116	76	72
6–12 months	89	93	60	62	25–29 years	125	117	78	74
1 year	96	95	66	65	30–34 years	126	120	79	75
2 years	99	92	64	60	35–39 years	127	124	80	78
3 years	100	100	67	64	40–44 years	129	127	81	80
4 years	99	99	65	66	45–49 years	130	131	82	82
5 years	92	92	62	62	50–54 years	135	137	83	84
6 years	94	94	64	64	55–59 years	138	139	84	84
7 years	97	97	65	66	60–64 years	142	144	85	85
8 years	100	100	67	68	65–69 years	143	154	83	85
9 years	101	101	68	69	70–74 years	145	159	82	85
10 years	103	103	69	70	75–79 years	146	158	81	84
11 years	104	104	70	71	80–84 years	145	157	82	83
12 years	106	106	72	72	85–89 years	145	154	79	82
13 years	108	108	72	73	90–94 years	145	150	78	79
14 years	110	110	73	74	95–106 years	145	149	78	81
15 years	112	112	75	76					

Source: Documenta Geigy Scientific Tables, edited by K. Diem and C. Lentner, 7th ed. Copyright © 1970 J. R. Geigy S. A., Basle, Switzerland.

pressure measurements are commonly taken, therefore increases to 120 mmHg in this example. The rise in pressure from diastolic to systolic levels (pulse pressure) is thus a reflection of the stroke volume.

The **mean arterial pressure** represents the average arterial pressure during the cardiac cycle. This value is significant because it is the difference between this pressure and the venous pressure that drives blood through the capillary beds of organs. The mean arterial pressure is not a simple arithmetic average because the period of diastole is longer than the period of systole. Mean arterial pressure can be approximated by adding one-third of the pulse pressure to the diastolic pressure. For a person with a blood pressure of 120/80, for example, the mean arterial pressure would be approximately 80 + 1/3 (40) = 93 mmHg.

Mean arterial pressure = diastolic pressure + 1/3 pulse pressure

A rise in total peripheral resistance and cardiac rate increases the diastolic pressure more than it increases the systolic pressure. When the baroreceptor reflex is activated by going from a lying to a standing position, for example, the diastolic pressure usually increases by 5 to 10 mmHg, whereas the systolic pressure either remains unchanged or is slightly reduced (as a result of decreased venous return). People with hypertension (high blood pressure), who usually have elevated total peripheral resistance and cardiac rates, likewise have a greater increase in diastolic than in systolic pressure. Dehydration or blood loss results in decreased cardiac output, and thus also produces a decrease in pulse pressure.

An increase in cardiac output, by contrast, raises the systolic pressure more than it raises the diastolic pressure (although both pressures do rise). This occurs during exercise, for example, when the blood pressure may rise to values as high as 200/100 (yielding a pulse pressure of 100 mmHg).

Clinical Considerations

Hypertension

Approximately 20% of all adults in the United States have *hypertension*—blood pressure in excess of the normal range for a person's age and sex. Hypertension that is a result of (secondary to) known disease processes is logically called **secondary hypertension.** Of the hypertensive population, secondary hypertension accounts for only about 5%. Hypertension that is the result of complex and poorly understood processes is not so logically called **primary,** or **essential, hypertension.** Hypertension in adults is defined by a systolic pressure greater than 140 mmHg and/or a diastolic pressure greater than 90 mmHg (table 22.9).

Table 22.9

Classification of Blood Pressure for Adults Age 18 Years and Older

Category	Systolic mmHg	Diastolic mmHg
Normal	<130	<85
High normal	130–139	85–89
Hypertension*		
Stage 1 (Mild)	140–159	90–99
Stage 2 (Moderate)	160–179	100–109
Stage 3 (Severe)	180–209	110–119
Stage 4 (Very Severe)	≥210	≥120

Source: Fifth Report of the Joint National Committee on Detection, Evaluation, and Treatment of High Blood Pressure, National Institutes of Health, Washington, D.C., 1993.

*Based on the average of two or more readings taken at each of two or more visits following an initial screening.

Diseases of the kidneys and arteriosclerosis of the renal arteries can cause secondary hypertension because of high blood volume. More commonly, the reduction of renal blood flow can raise blood pressure by stimulating the secretion of vasoactive chemicals from the kidneys. Experiments in which the renal artery is pinched, for example, produce hypertension that is associated (at least initially) with elevated renin secretion. These and other causes of secondary hypertension are summarized in table 22.10.

Essential Hypertension

The vast majority of people with hypertension have essential hypertension. An increased total peripheral resistance is a universal characteristic of this condition. Cardiac rate and the cardiac output are elevated in many, but not all, of these cases.

The secretion of renin, which is correlated with angiotensin II production and aldosterone secretion, is likewise variable. Although some people with essential hypertension have low renin secretion, most have either normal or elevated levels of renin secretion. Renin secretion in the normal range is inappropriate for people with hypertension, since high blood pressure should inhibit renin secretion and, through a lowering of aldosterone, result in greater excretion of salt and water. Inappropriately high levels of renin secretion could thus contribute to hypertension by promoting (via stimulation of aldosterone secretion) salt and water retention and high blood volume.

Sustained high stress (acting via the sympathetic nervous system) and high salt intake appear to act synergistically in the development of hypertension. There is some evidence that Na^+ enhances the vascular response to sympathetic stimulation. Further, sympathetic nerve stimulation can cause constriction of the renal blood vessels and thus decrease the excretion of salt and water.

As an adaptive response to prolonged high blood pressure, the arterial wall thickens. This response can lead to arteriosclerosis and results in an even greater increase in total peripheral resistance, thus raising blood pressure still more in a positive feedback fashion.

Table 22.10

Possible Causes of Secondary Hypertension

System Involved	Examples	Mechanisms
Kidneys	Kidney disease Renal artery disease	Decreased urine formation Secretion of vasoactive chemicals
Endocrine	Excess catecholamines (tumor of adrenal medulla) Excess aldosterone (Conn's syndrome)	Increased cardiac output and total peripheral resistance Excess salt and water retention by the kidneys
Nervous	Increased intracranial pressure Damage to vasomotor center	Activation of sympathoadrenal system Activation of sympathoadrenal system
Cardiovascular	Complete heart block; patent ductus arteriosus Arteriosclerosis of aorta; coarctation of aorta	Increased stroke volume Decreased distensibility of aorta

The interactions between salt intake, sympathetic nerve activity, cardiovascular responses to sympathetic nerve activity, kidney function, and genetics make it difficult to sort out the cause-and-effect sequence that leads to essential hypertension. Current evidence suggests that the inability of the kidneys to properly eliminate salt and water is a shared characteristic in all cases of essential hypertension. Further, there is evidence that salt intake may be the single most important factor. Chimpanzees, with their natural low-salt diet, have low blood pressure. When given human levels of dietary salt, however, their blood pressure rises. "Pre-literate" people whose diet is natural and low in salt similarly exhibit low blood pressure that does not rise with age. Even though some people may be more salt-sensitive than others, these findings suggest that everyone with hypertension should restrict their intake of dietary salt.

Dangers of Hypertension

If other factors remain constant, blood flow increases as arterial blood pressure increases. The organs of people with hypertension are thus adequately perfused with blood until the excessively high pressure causes vascular damage. Because most patients are asymptomatic (without symptoms) until substantial vascular damage has occurred, hypertension is often referred to as a silent killer.

Hypertension is dangerous for a number of reasons. First, high arterial pressure increases the afterload, making it more difficult for the ventricles to eject blood. The heart, then, must work harder, which can result in pathological changes in heart structure and function, leading to congestive heart failure. Additionally, high pressure may damage cerebral blood vessels, leading to cerebrovascular accident, or "stroke." (Stroke is the third-leading cause of death in the United States.) Finally, hypertension contributes to the development of atherosclerosis, which can itself lead to heart disease and stroke as previously described.

Preeclampsia (*pre″ĭ-klamp′se-ă*) is a toxemia of late pregnancy characterized by high blood pressure, proteinuria (the presence of proteins in the urine), and edema of the lower extremities. For reasons discussed in chapter 25, only negligible amounts of proteins are normally found in urine, and the excretion of plasma proteins in the urine can cause edema. In preeclampsia, the sensitivity of blood vessels to pressor agents (which causes vasoconstriction) is increased, resulting in decreased organ perfusion and increased blood pressure. The danger of preeclampsia is that it can quickly degenerate into a state called *eclampsia,* in which seizures occur. This can be life threatening, and so the woman with preeclampsia is immediately treated for her symptoms and the fetus is delivered as quickly as possible.

Treatment of Hypertension

The first form of treatment that is usually attempted is modification of lifestyle. This modification includes cessation of smoking, moderation of alcohol intake, and weight reduction, if applicable. It can also include regular physical exercise and a reduction in sodium intake. People with essential hypertension may have a potassium deficiency, and there is evidence that eating food that is rich in potassium may help to lower blood pressure. There is also evidence that supplementing the diet with Ca^{2+} may be of benefit, but this is more controversial.

If lifestyle modifications alone are insufficient, various drugs may be prescribed. Most commonly, these are *diuretics* that increase urine volume, thus decreasing blood volume and pressure. Drugs that block β_1-adrenergic receptors (such as atenolol) lower blood pressure by decreasing the cardiac rate and are also frequently prescribed. ACE inhibitors, calcium antagonists, and various vasodilators (table 22.11) may also be used in particular situations. A new class of drugs, angiotensin II–receptor antagonists, is currently undergoing clinical trials.

Circulatory Shock

Circulatory shock occurs when there is inadequate blood flow and/or oxygen utilization by the tissues. Some of the signs of shock (table 22.12) are a result of inadequate tissue perfusion; other signs of shock are produced by cardiovascular responses that help to compensate for the poor tissue perfusion (table 22.13). When these compensations are effective, they (together with emergency medical care) are able to reestablish adequate tissue perfusion. In some cases, however, and for reasons that are not clearly understood, the shock may progress to an irreversible stage and death may result.

Hypovolemic Shock

The term **hypovolemic** (*hi″pŏ-vo-le′mik*) **shock** refers to circulatory shock due to low blood volume, as might be caused by hemorrhage (bleeding), dehydration, or burns. This is accompanied by decreased blood pressure and decreased cardiac output. In response to these changes, the sympathoadrenal system is activated by means of the baroreceptor reflex. As a result, tachycardia is produced and vasoconstriction occurs in the skin, digestive tract, kidneys, and muscles. Decreased blood flow through the kidneys stimulates renin secretion and activation of the renin-angiotensin-aldosterone system. A person in hypovolemic shock thus has low blood pressure; a rapid pulse; cold, clammy skin; and a reduced urine output.

Since the resistance in the coronary and cerebral circulations is not increased, blood is diverted to the heart and brain at the expense of other organs. Interestingly, a similar response occurs in diving mammals and, to a lesser degree, in Japanese pearl divers during prolonged submersion. These responses help to deliver blood to the two organs that have the highest requirements for aerobic metabolism.

Vasoconstriction in organs other than the brain and heart raises total peripheral resistance, which helps (along with the reflex increase in cardiac rate) to compensate for the drop in blood pressure due to low blood volume. Constriction of arterioles also decreases capillary blood flow and capillary filtration pressure. As a result, less filtrate forms. At the same time, the osmotic return of fluid to the capillaries is either

Table 22.11

Mechanisms of Action of Selected Antihypertensive Drugs

Category of Drugs	Examples	Mechanisms
Extracellular fluid volume depletors	Thiazide diuretics	Increase volume of urine excreted, thus lowering blood volume
Sympathoadrenal system inhibitors	Clonidine; alpha-methyldopa	Act on brain to decrease sympathoadrenal stimulation
	Guanethidine, reserpine	Deplete norepinephrine from sympathetic nerve endings
	Propranolol; Atenolol	Block beta-adrenergic receptors, decreasing cardiac output and/or renin secretion
	Phentolamine	Blocks alpha-adrenergic receptors, decreasing sympathetic vasoconstriction
Direct vasodilators	Hydralazine; sodium nitroprusside	Cause vasodilation by acting directly on vascular smooth muscle
Calcium channel blockers	Verapamil	Inhibits diffusion of Ca^{2+} into vascular smooth muscle cells, causing vasodilation and reduced peripheral resistance
Angiotensin-converting enzyme (ACE) inhibitors	Captopril; benazepril	Inhibit the conversion of angiotensin I into angiotensin II

Table 22.12

Signs of Shock

	Early Sign	Late Sign
Blood pressure	Decreased pulse pressure Increased diastolic pressure	Decreased systolic pressure
Urine	Decreased Na^{+} concentration Increased osmolality	Decreased volume
Blood pH	Increased pH (alkalosis) due to hyperventilation	Decreased pH (acidosis) due to "metabolic" acids
Effects of poor tissue perfusion	Slight restlessness; occasionally warm, dry skin	Cold, clammy skin; "cloudy" senses

Source: *Principles and Techniques of Critical Care,* edited by R. F. Wilson, Vol. 1. Copyright © 1977 Upjohn Company. Used by permission of F. A. Davis Company, Philadelphia, PA.

Table 22.13

Cardiovascular Reflexes That Help To Compensate for Circulatory Shock

Organ(s)	Compensatory Mechanisms
Heart	Sympathoadrenal stimulation increases cardiac rate and stroke volume due to "positive inotropic effect" on myocardial contractility
Gastrointestinal tract and skin	Decreased blood flow due to vasoconstriction as a result of sympathetic nerve stimulation (alpha-adrenergic effect)
Kidneys	Decreased urine production as a result of sympathetic nerve–induced constriction of renal arterioles; increased salt and water retention due to increased plasma levels of aldosterone and antidiuretic hormone (ADH)

unchanged or increased (during dehydration). The blood volume is thus raised at the expense of tissue fluid volume. Blood volume is also conserved by decreased urine production, which occurs as a result of vasoconstriction in the kidneys and the water-conserving effects of ADH and aldosterone, which are secreted in increased amounts during shock.

Septic Shock

Septic shock refers to a dangerously low blood pressure (hypotension) that may result from *sepsis*, or infection. This can occur through the action of a bacterial lipopolysaccharide called *endotoxin*. The mortality associated with septic shock is presently very high, estimated at 50% to 70%. According to recent information, endotoxin activates the enzyme nitric oxide synthetase within macrophages—cells that play an important role in the immune response (see chapter 23). As previously discussed, nitric oxide synthetase produces nitric oxide, which promotes vasodilation and, as a result, a fall in blood pressure. Septic shock has been treated effectively with drugs that inhibit the production of nitric oxide.

Other Causes of Circulatory Shock

A rapid fall in blood pressure occurs in **anaphylactic** (*ă″nă-fĭ-lak′tik*) **shock** as a result of a severe allergic reaction (usually to bee stings or penicillin). This results from the widespread release of histamine, which causes vasodilation and thus decreases total peripheral resistance. A rapid fall in blood pressure also occurs in **neurogenic shock,** in which sympathetic tone is decreased, usually because of upper spinal cord damage or spinal anesthesia. **Cardiogenic shock** results from cardiac failure, as defined by a cardiac output inadequate to maintain tissue perfusion. This commonly results from infarction that causes the loss of a significant proportion of the myocardium.

Congestive Heart Failure

Cardiac failure occurs when the cardiac output is insufficient to maintain the blood flow required by the body. This may be due to heart disease—resulting from myocardial infarction or congenital defects—or to hypertension, which increases the afterload of the heart. The most common causes of left ventricular heart failure are myocardial infarction, aortic valve stenosis, and incompetence of the aortic and bicuspid (mitral) valves. Failure of the right ventricle is usually caused by prior failure of the left ventricle.

Heart failure can also result from disturbance in the electrolyte concentrations of the blood. Excessive plasma K^+ concentration decreases the resting membrane potential of myocardial cells, and low blood Ca^{2+} reduces excitation-contraction coupling. High blood K^+ and low blood Ca^{2+} can thus cause the heart to stop in diastole. Conversely, low blood K^+ and high blood Ca^{2+} can arrest the heart in systole.

The term *congestive* is often used in describing heart failure because of the increased venous volume and pressure that results. Failure of the left ventricle, for example, raises the left atrial pressure and produces pulmonary congestion and edema. This causes shortness of breath and fatigue; if severe, pulmonary edema can be fatal. Failure of the right ventricle results in increased right atrial pressure, which produces congestion and edema in the systemic circulation.

The compensatory responses that occur during congestive heart failure are similar to those that occur during hypovolemic shock. Activation of the sympathoadrenal system stimulates cardiac rate, contractility of the ventricles, and constriction of arterioles. As in hypovolemic shock, renin secretion increases and urine output decreases. The increased secretion of renin and consequent activation of the renin-angiotensin-aldosterone system causes salt and water retention. This occurs despite an increased secretion of atrial natriuretic factor (which would have the compensatory effect of promoting salt and water excretion).

As a result of these compensations, chronically low cardiac output is associated with elevated blood volume and dilation and hypertrophy of the ventricles. These changes can themselves be dangerous. Elevated blood volume places a work overload on the heart, and the enlarged ventricles have a higher metabolic requirement for oxygen. These problems are often treated with drugs that increase myocardial contractility, drugs that are vasodilators, and diuretic drugs that lower blood volume by increasing the volume of urine excreted.

People with congestive heart failure are often treated with the drug **digitalis** (*dij″ĭ-tal′is*). Digitalis appears to bind to and inhibit the action of Na^+/K^+ pumps in the cell membranes, causing a rise in the intracellular concentrations of Na^+. The increased availability of Na^+, in turn, stimulates the activity of another membrane transport carrier that exchanges Na^+ for extracellular Ca^{2+}. As a result, the intracellular concentrations of Ca^{2+} increase, which strengthens the contractions of the heart.

Clinical Investigation Answer

The man was suffering from dehydration, which lowered his blood volume and thus lowered his blood pressure. This stimulated the baroreceptor reflex, resulting in intense activation of sympathetic nerves. Sympathetic nerve activation caused vasoconstriction in cutaneous vessels—hence the cold skin—and an increase in cardiac rate (hence the high pulse rate). The intravenous albumin solution was given in the hospital in order to increase his blood volume and pressure. His urine output was low as a result of (1) sympathetic nerve–induced vasoconstriction of arterioles in the kidneys, which decreased blood flow to the kidneys; (2) water reabsorption in response to high ADH secretion, which resulted from stimulation of osmoreceptors in the hypothalamus; and (3) water and salt retention in response to aldosterone secretion, which was stimulated by activation of the renin-angiotensin system. The absence of sodium in his urine resulted from the high aldosterone secretion.

Chapter Summary

Cardiac Output (pp. 656–660)

1. Cardiac rate is increased by sympathoadrenal stimulation and decreased by the effects of parasympathetic fibers that innervate the SA node.
2. Stroke volume is regulated both extrinsically and intrinsically.
   (a) The Frank–Starling law of the heart describes the way the end-diastolic volume, through various degrees of myocardial stretching, influences the contraction strength of the myocardium, and thus the stroke volume.
   (b) The end-diastolic volume is called the preload. The total peripheral resistance, through its effect on arterial blood pressure, provides an afterload that acts to reduce the stroke volume.
   (c) At a given end-diastolic volume, the amount of blood ejected depends on contractility. Strength of contraction is increased by sympathoadrenal stimulation.
3. The venous return of blood to the heart is largely dependent on the total blood volume and mechanisms that improve the flow of blood in veins.
   (a) The total blood volume is regulated by the kidneys.
   (b) The venous flow of blood to the heart is aided by the action of skeletal muscle pumps and the effects of breathing.

Blood Volume (pp. 660–666)

1. Tissue fluid is formed from and returns to the blood.
   (a) The hydrostatic pressure of the blood forces fluid from the arteriolar ends of capillaries into the interstitial spaces of the tissues.
   (b) Since the colloid osmotic pressure of plasma is greater than that of tissue fluid, water returns by osmosis to the venular ends of capillaries.
   (c) Excess tissue fluid is returned to the venous system by lymphatic vessels.
   (d) Edema occurs when excess fluid accumulates in the tissues.
2. The kidneys control the blood volume by regulating the amount of filtered fluid that will be reabsorbed.
   (a) Antidiuretic hormone stimulates reabsorption of water from the kidney filtrate, and thus acts to maintain the blood volume.
   (b) A decrease in blood flow through the kidneys activates the renin-angiotensin system.
   (c) Angiotensin II stimulates vasoconstriction and the secretion of aldosterone by the adrenal cortex.
   (d) Aldosterone acts on the kidneys to promote the retention of salt and water.

Vascular Resistance to Blood Flow (pp. 666–670)

1. According to Poiseuille's law, blood flow is directly related to the pressure difference between the two ends of a vessel and inversely related to the resistance to blood flow through the vessel.
2. Extrinsic regulation of vascular resistance is provided mainly by the sympathetic nervous system, which stimulates vasoconstriction of arterioles in the viscera and skin.
3. Intrinsic control of vascular resistance allows organs to autoregulate their blood flow rates.
   (a) Myogenic regulation occurs when vessels constrict or dilate as a direct response to a rise or fall in blood pressure.
   (b) Metabolic regulation occurs when vessels dilate in response to the local chemical environment within the organ.

Blood Flow to the Heart and Skeletal Muscles (pp. 670–674)

1. The heart normally respires aerobically because of its extensive capillary supply, myoglobin content, and enzyme content.
2. During exercise, when the heart's metabolism increases, intrinsic metabolic mechanisms stimulate vasodilation of the coronary vessels, and thus increase coronary blood flow.
3. Just prior to exercise and at the start of exercise, blood flow through skeletal muscles increases because of vasodilation resulting from stimulation of cholinergic sympathetic nerve fibers. During exercise, intrinsic metabolic vasodilation occurs.
4. Since cardiac output can increase fivefold during exercise, the heart and skeletal muscles receive an increased proportion of a higher total blood flow.
   (a) The cardiac rate increases because of decreased activity of the vagus nerve and increased activity of sympathetic nerves.
   (b) The venous return is greater because of greater activity of the skeletal muscle pumps and an increased respiratory movement.
   (c) Increased contractility of the heart, combined with a decrease in total peripheral resistance, can result in a higher stroke volume.

Blood Flow to the Brain and Skin (pp. 675–677)

1. Cerebral blood flow is regulated both myogenically and metabolically.
   (a) Cerebral vessels automatically constrict if the systemic blood pressure rises too high.
   (b) Metabolic products cause local vessels to dilate and supply more-active areas with more blood.

2. The skin contains unique arteriovenous anastomoses that can divert blood away from surface capillary loops.
   (a) Sympathetic nerve stimulation causes constriction of cutaneous arterioles.
   (b) As a thermoregulatory response, cutaneous blood flow and blood flow through surface capillary loops increase when the body temperature rises.

Blood Pressure (pp. 677–683)

1. Baroreceptors in the aortic arch and carotid sinuses affect the cardiac rate and the total peripheral resistance via the sympathetic nervous system.
   (a) The baroreceptor reflex causes pressure to be maintained when an upright posture is assumed. This reflex can cause a lowered pressure when the carotid sinuses are massaged.
   (b) Other mechanisms that affect blood volume help to regulate blood pressure.
2. Blood pressure is commonly measured indirectly by auscultation of the brachial artery when a pressure cuff is inflated and deflated.
   (a) The first sound of Korotkoff, caused by turbulent flow of blood through a constriction in the artery, occurs when the cuff pressure equals the systolic pressure.
   (b) The last sound of Korotkoff is heard when the cuff pressure equals the diastolic blood pressure.
3. The mean arterial pressure represents the driving force for blood flow through the arterial system.

Review Activities

Objective Questions

1. According to the Frank–Starling law, the strength of ventricular contraction is
   (a) directly proportional to the end-diastolic volume.
   (b) inversely proportional to the end-diastolic volume.
   (c) independent of the end-diastolic volume.
2. In the absence of compensations, the stroke volume will decrease when
   (a) blood volume increases.
   (b) venous return increases.
   (c) contractility increases.
   (d) arterial blood pressure increases.
3. Which of the following statements about tissue fluid is *false?*
   (a) It contains the same glucose and salt concentration as plasma.
   (b) It contains a lower protein concentration than plasma.
   (c) Its colloid osmotic pressure is greater than that of plasma.
   (d) Its hydrostatic pressure is lower than that of plasma.
4. Edema may be caused by
   (a) high blood pressure.
   (b) decreased plasma protein concentration.
   (c) leakage of plasma protein into tissue fluid.
   (d) blockage of lymphatic vessels.
   (e) all of the above.
5. Both ADH and aldosterone act to
   (a) increase urine volume.
   (b) increase blood volume.
   (c) increase total peripheral resistance.
   (d) produce all of the above effects.
6. The greatest resistance to blood flow occurs in
   (a) large arteries.
   (b) medium-sized arteries.
   (c) arterioles.
   (d) capillaries.
7. If a vessel were to dilate to twice its previous radius, and if pressure remained constant, blood flow through this vessel would
   (a) increase by a factor of 16.
   (b) increase by a factor of 4.
   (c) increase by a factor of 2.
   (d) decrease by a factor of 2.
8. The sounds of Korotkoff are produced by
   (a) closing of the semilunar valves.
   (b) closing of the AV valves.
   (c) the turbulent flow of blood through an artery.
   (d) elastic recoil of the aorta.
9. Vasodilation in the heart and skeletal muscles during exercise is primarily due to the effects of
   (a) alpha-adrenergic stimulation.
   (b) beta-adrenergic stimulation.
   (c) cholinergic stimulation.
   (d) products released by the exercising muscle cells.
10. Blood flow in the coronary circulation
    (a) increases during systole.
    (b) increases during diastole.
    (c) remains constant throughout the cardiac cycle.
11. Blood flow in the cerebral circulation
    (a) varies with systemic arterial pressure.
    (b) is regulated primarily by the sympathetic system.
    (c) is maintained constant within physiological limits.
    (d) increases during exercise.
12. Which of the following organs is able to tolerate the greatest reduction in blood flow?
    (a) brain
    (b) heart
    (c) skeletal muscles
    (d) skin
13. Which of the following statements about arteriovenous shunts in the skin is *true?*
    (a) They divert blood to superficial capillary loops.
    (b) They are closed when the ambient temperature is very cold.
    (c) They are closed when the deep-body temperature rises much above 37° C.
    (d) All of the above are true.
14. An increase in blood volume will cause
    (a) a decrease in ADH secretion.
    (b) an increase in Na^+ excretion in the urine.
    (c) a decrease in renin secretion.
    (d) all of the above.
15. The volume of blood pumped per minute by the left ventricle is
    (a) greater than the volume pumped by the right ventricle.
    (b) less than the volume pumped by the right ventricle.
    (c) the same as the volume pumped by the right ventricle.
    (d) either less or greater than the volume pumped by the right ventricle, depending on the strength of contraction.
16. Blood pressure is lowest in
    (a) arteries.
    (b) arterioles.
    (c) capillaries.
    (d) venules.
    (e) veins.
17. Stretch receptors in the aortic arch and carotid sinus
    (a) stimulate secretion of atrial natriuretic factor.
    (b) serve as baroreceptors that affect activity of the vagus and sympathetic nerves.
    (c) serve as osmoreceptors to stimulate secretion of ADH.
    (d) stimulate secretion of renin, thus increasing angiotensin II formation.
18. Angiotensin II
    (a) stimulates vasoconstriction.
    (b) stimulates the adrenal cortex to secrete aldosterone.
    (c) inhibits the action of bradykinin.
    (d) does all of the above.

19. Which of the following is a paracrine regulator that stimulates vasoconstriction?
    (a) nitric oxide
    (b) prostacyclin
    (c) bradykinin
    (d) endothelin-1
20. The pulse pressure is a measure of
    (a) the number of heartbeats per minute.
    (b) the sum of the diastolic and systolic pressures.
    (c) the difference between the systolic and diastolic pressures.
    (d) the difference between the arterial and venous pressures.

Essay Questions

1. Define the terms *contractility*, *preload*, and *afterload*, and explain how these factors affect the cardiac output.
2. Using the Frank–Starling law, explain how the stroke volume is affected by (*a*) bradycardia and (*b*) a "missed beat."
3. Which part of the cardiovascular system contains the most blood? Which part provides the greatest resistance to blood flow? Which part provides the greatest cross-sectional area? Explain.
4. Explain how the kidneys regulate blood volume.
5. A person who is dehydrated drinks more and urinates less. Explain the mechanisms involved.
6. Using Poiseuille's law, explain how arterial blood flow can be diverted from one organ system to another.
7. Describe the mechanisms that increase the cardiac output during exercise and that increase the rate of blood flow to the heart and skeletal muscles.
8. Explain why an anxious person may have a cold, clammy skin and why the skin becomes hot and flushed on a hot, humid day.
9. Explain the different ways in which a drug that acts as an inhibitor of angiotensin-converting enzyme (ACE) can lower the blood pressure. Also, explain how diuretics and β_1 adrenergic–blocking drugs work to lower the blood pressure.
10. Explain how hypotension may be produced in (*a*) hypovolemic shock and (*b*) septic shock. Also, explain the mechanisms whereby people in shock have a rapid but weak pulse, cold and clammy skin, and low urine output.

Critical Thinking Questions

1. One consequence of the Frank–Starling law is that the outputs of the right and left ventricles are matched. Explain why this is important, and how this matching is accomplished.
2. An elderly man who is taking digoxin for a weak heart complains that his feet hurt. Upon examination, his feet are swollen and discolored, with purple splotches and expanded veins. He is told to keep his feet raised and is given a prescription for Lasix, a powerful diuretic. Discuss this man's condition and the rationale for his treatment.
3. You are bicycling in a 100-mile benefit race because you want to help the cause, but you did not count on such a hot, humid day. You've gone through both water bottles, and in the last 10 miles, you are thirsty again. Should you accept the water that one bystander offers or the sports drink offered by another? Explain your choice.
4. As the leader of a revolution to take over a large country, you direct your followers to seize the salt mines. Why is this important? When the revolution succeeds and you become president, you ask your surgeon general to wage a health campaign urging the citizens to reduce their salt intake. Why?
5. Which type of exercise, isotonic contractions or isometric contractions, puts more of a "strain" on the heart? Explain.

Related Web Sites

For a listing of the most current web sites related to this chapter, please visit the *Concepts of Human Anatomy and Physiology* home page at http://www.mhhe.com/biosci/abio/.

NEXUS

Some Interactions of the Circulatory System with Other Body Systems

Integumentary System

- The skin helps to protect the body from pathogens (p. 696).
- The circulatory system delivers blood for exchange of gases, nutrients, and wastes with all of the body organs, including the skin (p. 592).
- Blood clotting occurs if the skin is broken (p. 601).

Skeletal System

- Hematopoiesis occurs in the bone marrow (p. 596).
- The rib cage protects heart and thoracic vessels (p. 183).
- The blood delivers calcium and phosphate for deposition of bone, and removes calcium and phosphate during bone resorption (p. 193).
- The blood delivers parathyroid hormone and other hormones that regulate bone (p. 194).

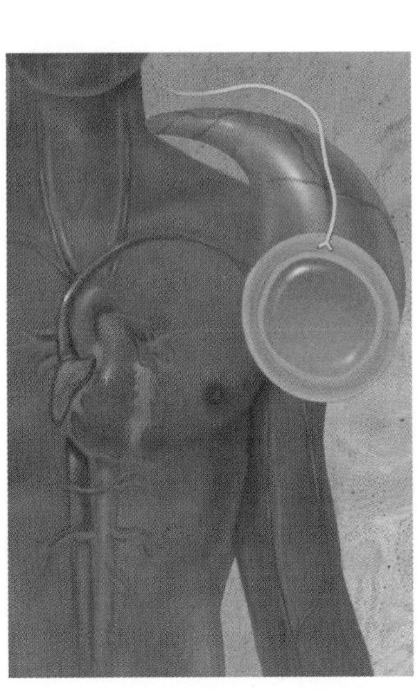

Nervous System

- Autonomic nerves help to regulate the cardiac output (p. 656).
- Autonomic nerves help to regulate the vascular resistance, blood flow, and blood pressure (p. 677).
- Cerebral capillaries participate in the blood-brain barrier (p. 378).

Endocrine System

- Epinephrine and norepinephrine from the adrenal medulla help to regulate cardiac function and vascular resistance (p. 668).
- Thyroxine and other hormones influence the blood pressure (p. 882).
- The blood transports hormones to their target organs (p. 552).

Muscular System

- Cardiac muscle function is central to the activity of the heart (p. 612).
- Smooth muscle function in blood vessels regulates the blood flow and blood pressure (p. 667).
- Skeletal muscle contractions squeeze veins and thus aid venous blood flow (p. 627).

Lymphatic and Immune Systems

- The immune system protects against infections (p. 696).
- Lymphatic vessels drain tissue fluid and returns it to the venous system (p. 693).
- Lymphocytes from the bone marrow and lymphoid organs circulate in the blood (p. 701).
- Neutrophils leave the vascular system to participate in aspects of the immune response (p. 698).
- The circulation carries chemical regulators of the immune response (p. 713).

Respiratory System

- The lungs provide O_2 for transport by blood and provides for elimination of CO_2 (p. 729).
- The blood transports gases between the lungs and tissue cells (p. 750).
- Ventilation helps to regulate the pH of the blood (p. 761).

Urinary System

- The kidneys regulate the volume, pH, and electrolyte balance of blood (p. 779).
- The kidneys excrete waste products, derived from blood plasma, in the urine (p. 792).

Digestive System

- Intestinal absorption of nutrients, including iron and particular B vitamins, is needed for red blood cell production (p. 595).
- The circulation transports nutrients from the GI tract to all the tissues in the body (p. 616).
- The hepatic portal vein allows some absorbed molecules to have an enterohepatic circulation (p. 842).

Reproductive System

- Gonadal hormones, particularly testosterone, stimulate red blood cell production (p. 903).
- The placenta permits exchanges of gases, nutrients, and waste products between the maternal and fetal blood (p. 965).
- Erection of the penis and clitoris results from vasodilation of blood vessels (p. 912).

Twenty-Three

Lymphatic System and Immunity

Objectives

- Describe the pattern of lymph flow from the lymphatic capillaries to the venous system.
- Describe the structure of the lymph nodes and lymphoid organs and state where they are located.
- Describe the mechanisms of nonspecific immunity.
- Discuss the nature of antigens and define *hapten* and *antigenic determinant site*.
- Discuss the origin and functions of B and T lymphocytes and distinguish between humoral and cell-mediated immunity.
- Describe the structure and origin of antibodies and explain how antibodies promote the destruction of invading pathogens.
- Discuss the complement system and its functions.
- Describe the events that occur in a local inflammation.
- Describe the nature of the primary and secondary immune responses.
- Explain the clonal selection theory and its relationship to the process of active immunization.
- Explain how passive immunizations are performed.
- Discuss the nature of monoclonal antibodies.
- Discuss the role of histocompatibility antigens in the function of T lymphocytes and antigen-presenting cells.
- Explain how interaction between T lymphocytes and macrophages leads to stimulation of both cell-mediated and humoral immunity.
- Discuss the nature of immunological tolerance and explain how it might be produced.
- Identify the cells and mechanisms involved in the immunological surveillance against cancer.
- Explain how stress and aging might result in increased susceptibility to cancer.
- Explain how allergic and autoimmune reactions may be produced.

Clinical Investigation

A 6-year-old girl was playing by crawling through the underbrush in the surrounding hills while her parents were picnicking. When she returned to their campsite, she tearfully showed them a bee sting—the first she had ever received. The next day she developed an itchy rash on her chest and abdomen, which was not relieved by antihistamines. The family physician prescribed oral corticosteroids, which alleviated her symptoms. Three weeks later she was again stung by a bee, and this time she developed severe swelling that responded to antihistamine treatment. What happened to the little girl?

Clues: Review the section "Active Immunity and the Clonal Selection Theory" and note the description of immediate and delayed hypersensitivity under the heading "Allergy." Also refer to table 23.11.

Lymphatic System

The lymphatic system, consisting of lymphatic vessels and various lymphoid tissues and organs, helps to maintain fluid balance in the tissues and to absorb fats from the gastrointestinal tract. It is also part of the body's defense system against disease.

The **lymphatic system** has three basic functions: (1) it transports interstitial (tissue) fluid, initially formed as a blood filtrate, back to the bloodstream; (2) it serves as the route by which absorbed fat from the small intestine is transported to the blood; and (3) its cells—called **lymphocytes**—help to provide immunological defenses against disease-causing agents.

Lymph and Lymphatic Capillaries

The smallest vessels of the lymphatic system are the **lymphatic capillaries** (chapter 22, see fig. 22.9). Lymphatic capillaries are microscopic closed-ended tubes that form vast networks in the intercellular spaces within most tissues. Within the villi of the small intestine, for example, lymphatic capillaries called *lacteals* (*lak'te-alz*) transport the products of fat absorption away from the GI tract. Because the walls of lymphatic capillaries are composed of endothelial cells with porous junctions, interstitial fluid, proteins, microorganisms, and absorbed fat (in the small intestine) can easily enter. Once the interstitial fluid enters the lymphatic capillaries, it is referred to as **lymph.**

lacteal: L. *lacteus*, milk
lymph: L. *lympha*, clear water

Lymph Ducts

From merging lymphatic capillaries, the lymph is carried into larger lymphatic vessels, or **lymph ducts.** The walls of lymph ducts are similar to those of veins. They have the same three layers and also contain valves to prevent the backflow of lymph. The pressure that pushes lymph through the lymph ducts comes from the massaging actions produced by skeletal muscle contractions, gravity, intestinal movements, and other body movements. The many valves keep lymph moving in one direction.

Interconnecting lymph ducts eventually empty into one of the two principal vessels: the **thoracic duct** and the **right lymphatic duct** (fig. 23.1). These ultimately drain the lymph into the left and right subclavian veins, respectively, so that it is returned to the circulatory system (fig. 23.2).

The larger thoracic duct drains lymph from the lower extremities, abdomen, left thoracic region, left upper extremity, and left side of the head and neck. The main trunk of this vessel ascends along the spinal column and drains into the left subclavian vein. In the abdominal area there is a saclike enlargement of the thoracic duct called the **cisterna chyli** (*sis-ter'nă ki'le*). The smaller right lymphatic duct drains lymph from the right upper extremity, right thoracic region, and right side of the head and neck. The right lymphatic duct empties into the right subclavian vein near the right internal jugular vein.

Lymph Nodes

Lymph filters through the reticular tissue of **lymph nodes** (fig. 23.3). The reticular tissue contains phagocytic cells that help to purify the fluid. A lymph node is a small oval body enclosed within a fibrous connective tissue *capsule*. Specialized connective tissue bands called *trabeculae* (*tră-bek'yŭ -le*) divide the node. Afferent lymphatic vessels carry lymph into the node, where it is circulated through sinuses in the *cortical tissue*. Lymph leaves the node through the efferent lymphatic vessel, which emerges from the *hilum*—a depression on the concave side. **Germinal centers** within the node are sites of lymphocyte production and are important in the development of an immune response.

Lymph nodes usually occur in clusters in specific regions of the body (fig. 23.4). Some of the principal groups of lymph nodes are the **popliteal** (not illustrated) and **inguinal nodes** of the lower extremity, the **lumbar nodes** of the pelvic region, the **cubital** and **axillary nodes** of the upper extremity, the **thoracic nodes** of the chest, and the **cervical nodes** of the neck. The wall of the small intestine contains numerous scattered lymphocytes and lymphatic nodules and larger aggregations of lymphoid tissue called **aggregated lymphatic follicles,** or **Peyer's** (*pi'erz*) **patches** (fig. 23.4).

cisterna chyli: L. *cisterna*, box; Gk. *chylos*, juice
Peyer's patches: from Johann K. Peyer, Swiss anatomist, 1653–1712

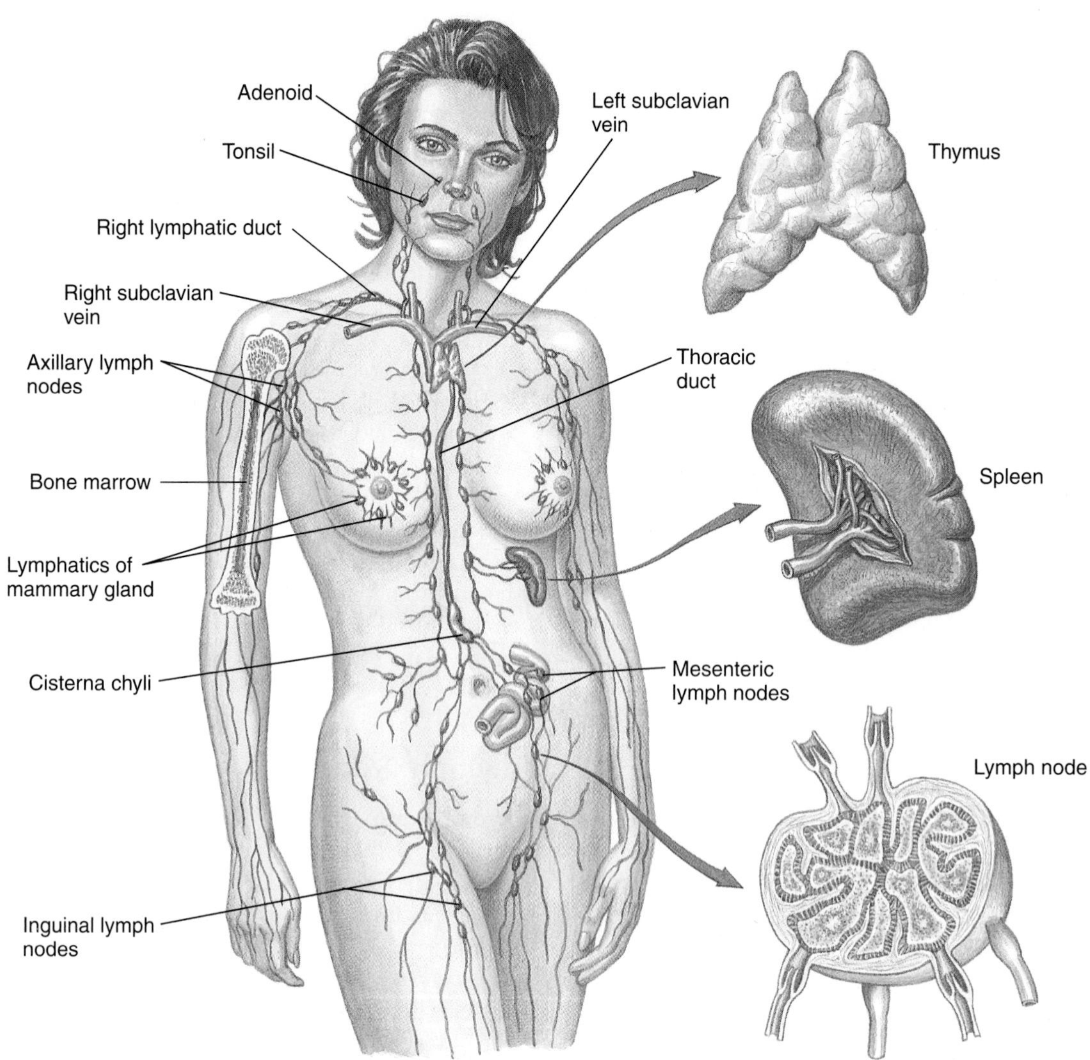

Figure 23.1

Principal lymphatic vessels and lymph nodes.
Lymph from the upper right extremity, the right side of the head and neck, and the right thoracic region drains through the right lymphatic duct into the right subclavian vein. Lymph from the remainder of the body drains through the thoracic duct into the left subclavian vein.

Migrating cancer cells (metastases) are especially dangerous if they enter the lymphatic system, which can disperse them widely. Once cancer invades the lymph nodes, the cancer cells can multiply and establish secondary tumors in organs far removed from the site of the primary tumor.

Other Lymphoid Organs

In addition to the lymph nodes just described, the *tonsils*, *spleen*, and *thymus* are lymphoid organs. The **tonsils** form a protective ring of lymphoid tissue around the openings between the nasal and oral cavities and the pharynx (chapter 24, see fig. 24.2).

The tonsils, of which there are three pairs, combat infection of the ear, nose, and throat regions. Because of the persistent infections that some children suffer, the tonsils may become so overrun with infections that they themselves become a source of infections that spread to other parts of the body. A *tonsillectomy* may then have to be performed. This operation is not as common as it was in the past because of the availability of powerful antibiotics and because the functional value of the tonsils is now appreciated to a greater extent.

The **spleen** is located on the left side of the abdominal cavity, posterior and lateral to the stomach to which it is

spleen: L. *splen*, low spirits (thought to cause melancholy)

Figure 23.2

Lymphatic drainage.

This is a schematic representation of the pattern of fluid movement from the vascular system, into the lymphatic system, and back to the vascular system.

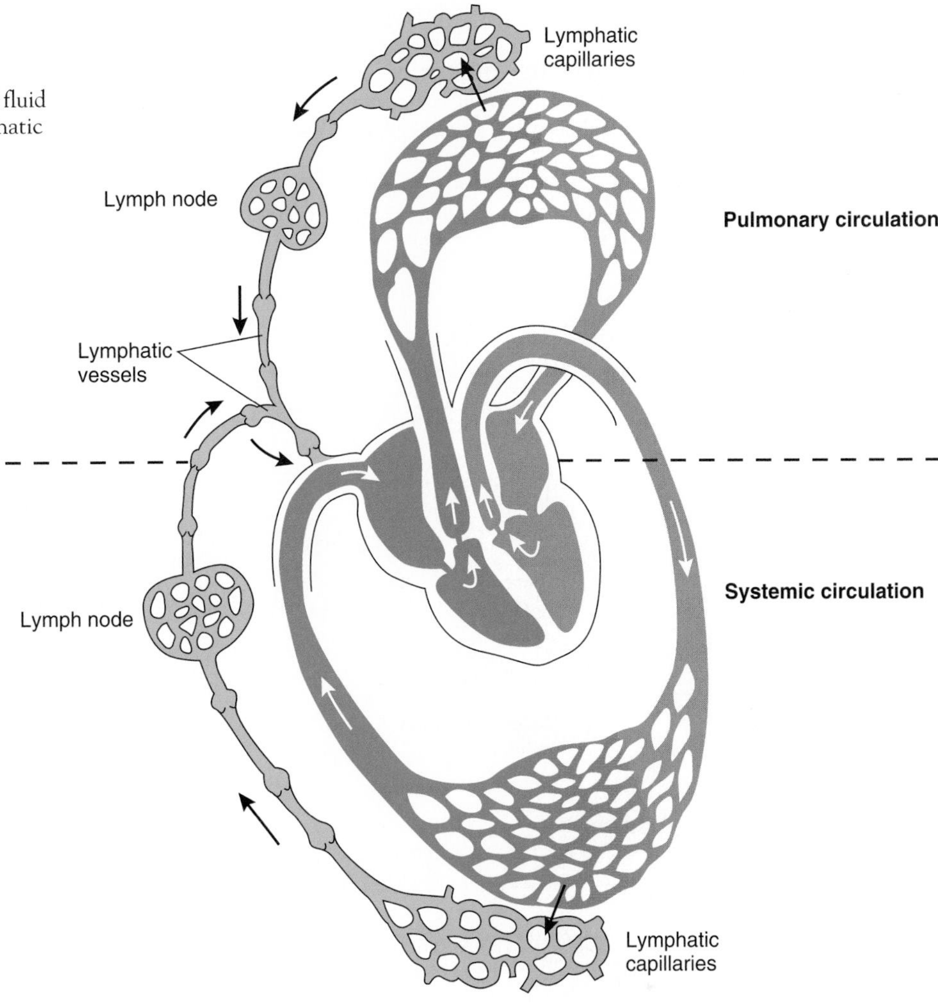

Figure 23.3

The structure of a lymph node.

(*a*) A schematic diagram of a sectioned lymph node and associated vessels and (*b*) a photograph of a lymph node positioned near a blood vessel.

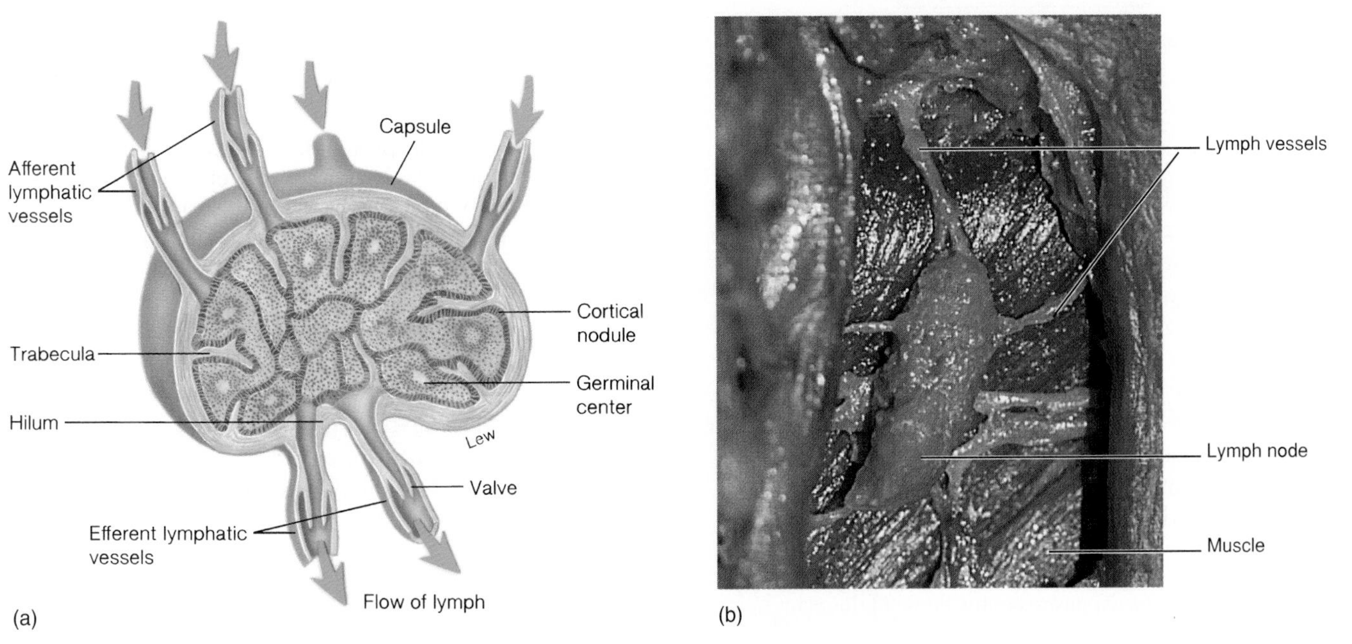

Figure 23.4

Lymph nodes associated with the intestine.
The aggregated lymphatic follicles, also called Peyer's patches, are located in the wall (specifically the submucosa) of the small intestine.

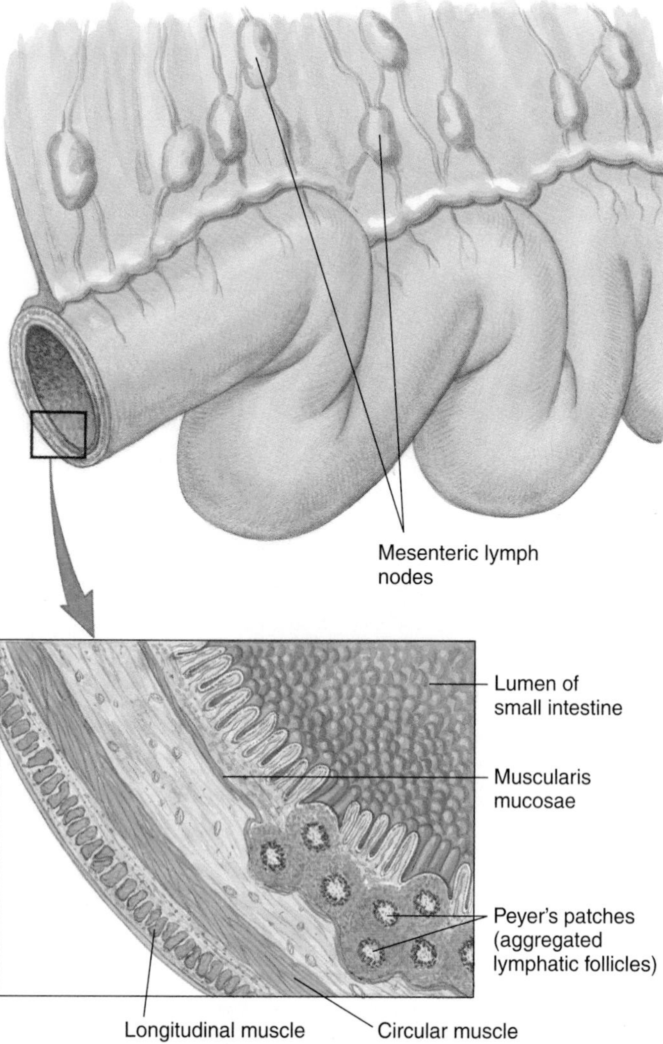

Figure 23.5

The structure of the spleen.
The spleen contains two major kinds of tissues that have different functions. These are the reticuloendothelial tissue, which helps to destroy old red blood cells and recirculate the iron, and the germinal centers, which produce lymphocytes and plasma cells.

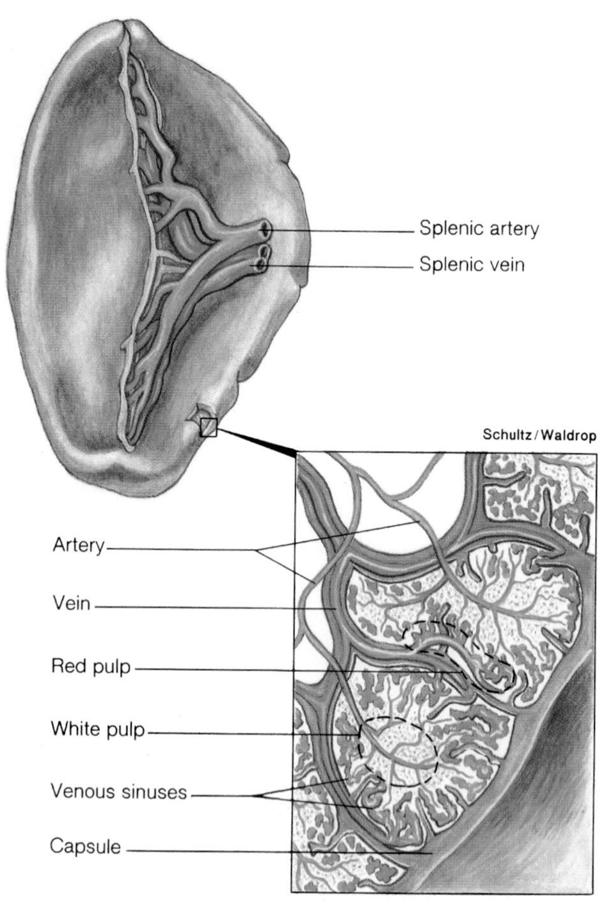

attached (fig. 23.5). The spleen is not a vital organ in an adult, but it does assist other body organs in producing lymphocytes, filtering the blood, and destroying old erythrocytes. In an infant it is an important site for the production of erythrocytes. In an adult the spleen contains *red pulp*, which serves to destroy old red blood cells, and *white pulp*, which contains germinal centers for the production of lymphocytes.

The **thymus** is located in the anterior thorax, deep to the manubrium of the sternum. Because it regresses in size during puberty, it is much larger in a fetus and child than in an adult. The thymus plays a key role in the immune system.

The lymphoid organs are summarized in table 23.1.

thymus: Gk. *thymos*, thyme (compared to the flower of this plant by Galen)

Defense Mechanisms

Nonspecific immune protection is provided by such mechanisms as phagocytosis, fever, and the release of interferons. Specific immunity, which involves the functions of lymphocytes, is directed at specific molecules, or parts of molecules, known as antigens.

The **immune system** includes all of the structures and processes that provide a defense against potential pathogens (disease-causing agents). These defenses can be grouped into *nonspecific* and *specific* categories. Nonspecific, or *innate*, defense mechanisms are inherited as part of the structure of each organism. Epithelial membranes that cover the body surfaces, for example, restrict infection by most pathogens. The

pathogen: Gk. *pathema*, suffering; *gen*, to produce

Table 23.1

Lymphoid Organs

Organ	Location	Function
Lymph nodes	In clusters or chains along the paths of larger lymphatic vessels	Sites of lymphocyte production; house T lymphocytes and B lymphocytes that are responsible for immunity; phagocytes filter foreign particles and cellular debris from lymph
Tonsils	In a ring at the junction of the oral cavity and pharynx	Protect against invasion of foreign substances that are ingested or inhaled
Spleen	In upper left portion of abdominal cavity, beneath the diaphragm and suspended from the stomach	Serves as blood reservoir; phagocytes filter foreign particles, cellular debris, and worn erythrocytes from the blood; houses lymphocytes
Thymus	Within the mediastinum, behind the manubrium	Important site of immunity in a child; houses lymphocytes; changes undifferentiated lymphocytes into T lymphocytes

strong acidity of gastric juice (pH 1–2) also helps to kill many microorganisms before they can invade the body. These external defenses are backed by internal defenses, such as phagocytosis, which function in both a specific and nonspecific manner (table 23.2).

Each individual can acquire the ability to defend against specific pathogens by prior exposure to those pathogens. This specific, or *acquired,* immune response is a function of lymphocytes. Internal specific and nonspecific defense mechanisms function together to combat infection, with lymphocytes interacting in a coordinated effort with phagocytic cells.

Nonspecific Immunity

Bacteria or other invading pathogens that have crossed epithelial barriers enter connective tissues. These invaders—or chemicals, called *toxins,* secreted from them—may enter blood or lymphatic capillaries and be carried to other areas of the body. Nonspecific immunological defenses are the first employed to counter the invasion and spread of infection. If these defenses are not sufficient to destroy the pathogens, lymphocytes may be recruited and their specific actions used to reinforce the nonspecific immune defenses.

Phagocytosis

The nonspecific defense mechanisms distinguish between the kinds of carbohydrates that are produced by mammalian cells and those produced by bacteria. The bacterial carbohydrates that "flag" the cell for phagocytic attack are part of the glycoproteins and lipopolysaccharides in the bacterial cell wall.

Table 23.2

Structures and Defense Mechanisms of Nonspecific Immunity

Structure	Mechanisms
External	
Skin	Physical barrier to penetration by pathogens; secretions contain lysozyme (enzyme that destroys bacteria)
GI tract	High acidity of stomach; protection by normal bacterial population of colon
Respiratory tract	Secretion of mucus; movement of mucus by cilia; alveolar macrophages
Urinary tract	Acidity of urine
Female reproductive tract	Vaginal lactic acid
Internal	
Phagocytic cells	Ingest and destroy bacteria, cellular debris, denatured proteins, and toxins
Interferons	Inhibit replication of viruses
Complement proteins	Promote destruction of bacteria and other effects of inflammation
Endogenous pyrogen	Secreted by leukocytes and other cells; produces fever

There are three major groups of phagocytic cells: (1) **neutrophils;** (2) the cells of the **mononuclear phagocyte system,** including *monocytes* in the blood and *macrophages* (derived from monocytes) in the connective tissues; and (3) **organ-specific phagocytes** in the liver, spleen, lymph nodes, lungs, and brain (table 23.3). Organ-specific phagocytes, such as the microglia of the brain, are embryologically and functionally related to macrophages and may be considered part of the mononuclear phagocyte system.

The *Kupffer* (*Koop'fer*) *cells* in the liver, as well as phagocytic cells in the spleen and lymph nodes, are **fixed phagocytes.** This term refers to the fact that these cells are immobile ("fixed") in the walls of the sinusoids (chapter 21) within these organs. As blood flows through these wide capillaries of the liver and spleen, foreign chemicals and debris are removed by phagocytosis and chemically inactivated within the phagocytic cells. Invading pathogens are very effectively removed in this manner, so that blood is usually sterile after a few passes through the liver and spleen. Fixed phagocytes in lymph nodes similarly help to remove foreign particles from the lymph.

Connective tissues have a resident population of all leukocyte types. Neutrophils and monocytes in particular can be highly mobile within connective tissues as they scavenge

Kupffer cell: from Karl Wilhelm von Kupffer, Bavarian anatomist, 1829–1902

Table 23.3

Phagocytic Cells and Their Locations

Phagocyte	Location
Neutrophils	Blood and all tissues
Monocytes	Blood
Tissue macrophages (histiocytes)	All tissue (including spleen, lymph nodes, bone marrow)
Kupffer cells	Liver
Alveolar macrophages	Lungs
Microglia	Central nervous system

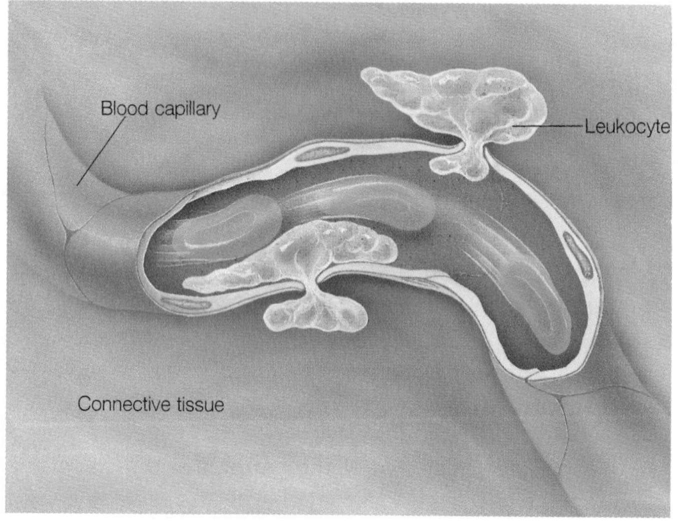

Figure 23.6

The process of diapedesis. White blood cells squeeze through openings between capillary endothelial cells to enter underlying connective tissues.

for invaders and cellular debris. These leukocytes are recruited to the site of an infection by a process known as **chemotaxis** (*ke″mo-tak′sis*)—movement toward chemical attractants. Neutrophils are the first to arrive at the site of an infection; monocytes arrive later and can be transformed into macrophages as the battle progresses.

If the infection has spread, new phagocytic cells from the blood may join those already in the connective tissue. These new neutrophils and monocytes are able to squeeze through the tiny gaps between adjacent endothelial cells in the capillary wall and enter the connective tissues. This process, called **diapedesis** (*di″ă-pĕ-de′sis*) or *extravasation*, is illustrated in figure 23.6.

Phagocytic cells engulf particles in a manner similar to the way an amoeba eats. The particle becomes surrounded by

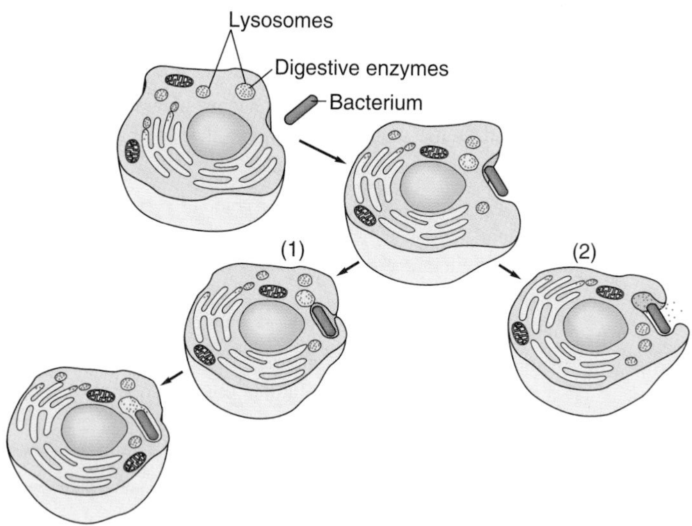

Figure 23.7

Phagocytosis by a neutrophil or macrophage. A phagocytic cell extends its pseudopods around the object to be engulfed (such as a bacterium). (Blue dots represent lysosomal enzymes.) (1) If the pseudopods fuse to form a complete food vacuole, lysosomal enzymes are restricted to the organelle formed by the lysosome and food vacuole. (2) If the lysosome fuses with the vacuole before fusion of the pseudopods is complete, lysosomal enzymes are released into the infected area of tissue.

cytoplasmic extensions called pseudopods, which ultimately fuse. The particle thus becomes surrounded by a membrane derived from the cell membrane (fig. 23.7) and contained within an organelle analogous to a food vacuole in an amoeba. This vacuole then fuses with lysosomes (organelles that contain digestive enzymes), so that the ingested particle and the digestive enzymes still are separated from the cytoplasm by a continuous membrane. Often, however, lysosomal enzymes are released before the food vacuole has completely formed. When this occurs, free lysosomal enzymes may be released into the infected area and contribute to inflammation.

Fever

The cell wall of gram-negative bacteria contains **endotoxin,** a lipopolysaccharide that stimulates monocytes and macrophages to release various cytokines. These cytokines, including *interleukin-1*, *interleukin-6*, and *tumor necrosis factor*, act to produce fever, drowsiness, and a fall in the plasma iron concentration.

Fever may be a component of the nonspecific defense system. Body temperature is regulated by the hypothalamus, which contains a thermoregulatory control center (a "thermostat") that coordinates skeletal muscle shivering and the activity of the sympathoadrenal system to maintain body temperature at about 37° C. This thermostat is reset upward in response to a chemical called **endogenous pyrogen.** In at least some infections, the endogenous pyrogen has been identified as interleukin-1β, which is first produced as a cytokine in response to endotoxin stimulation and is then produced by the brain itself.

diapedesis: Gk. *dia*, through; *pedesis*, a leaping

Figure 23.8

The life cycle of the human immunodeficiency virus (HIV).
This virus, like others of its family, contains RNA instead of DNA. Once inside the host cell, the viral RNA is transcribed by reverse transcriptase into complementary DNA (cDNA). The genes in the cDNA then direct the synthesis of new virus particles.

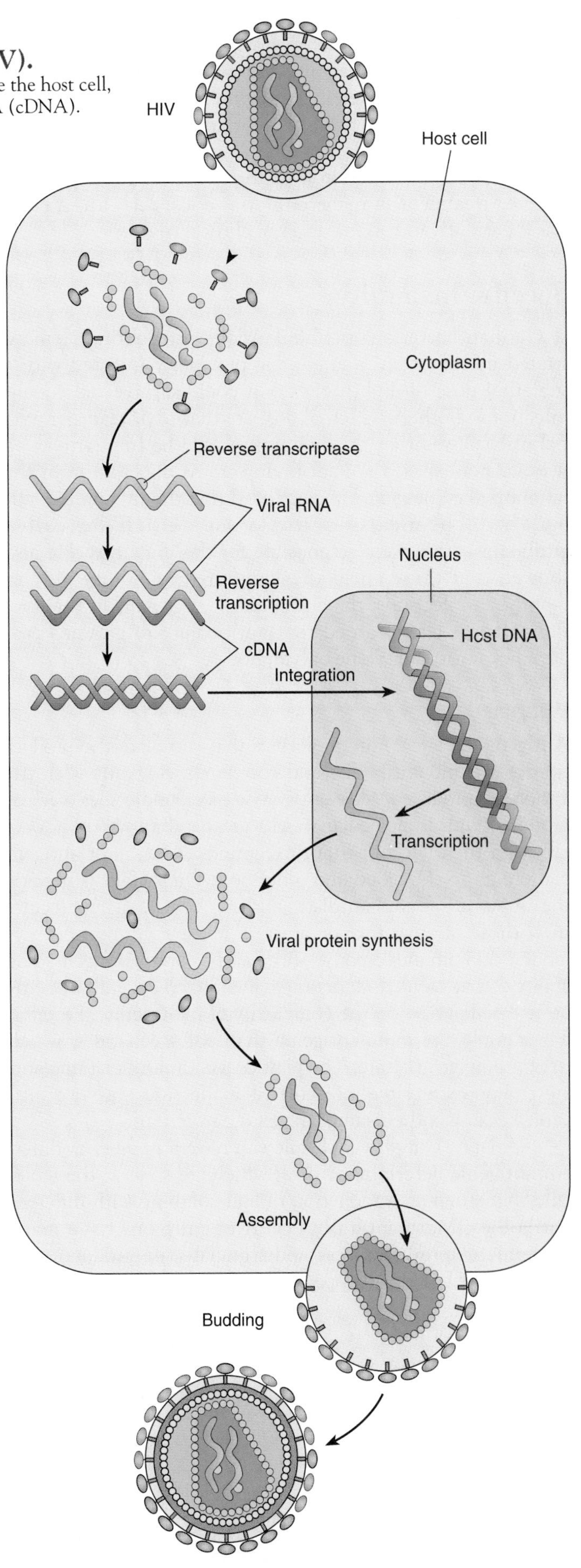

Although high fevers are definitely dangerous, a mild to moderate fever may be a beneficial response that aids recovery from bacterial infections. The fall in plasma iron concentrations that accompany a fever can inhibit bacterial activity and represents one possible benefit of a fever; others include increased activity of neutrophils and increased production of interferon.

Interferons

In 1957, researchers demonstrated that cells infected with a virus produced polypeptides that interfered with the ability of a second, unrelated strain of virus to infect other cells in the same culture. These **interferons** (*in"ter-fēr'onz*), as they were called, thus produced a nonspecific short-acting resistance to viral infection. Although this discovery generated a great deal of excitement, further research was hindered by the fact that human interferons could be obtained only in very small quantities; moreover, animal interferons were shown to have little effect in humans. In 1980, however, a technique called *genetic recombination* (chapter 3) made it possible to introduce human interferon genes into bacteria, enabling the bacteria to act as interferon factories.

There are three major categories of interferons: *alpha*, *beta*, and *gamma interferons*. Almost all cells in the body make alpha and beta interferons. These polypeptides act as messengers that protect other cells in the vicinity from viral infection. The viruses are still able to penetrate these other cells, but the ability of the viruses to replicate and assemble new virus particles is inhibited. Viral infection, replication, and dispersal are illustrated in figure 23.8, using the virus that causes AIDS as an example. Gamma interferon is produced only by particular lymphocytes and a related type of cell called a natural killer cell. The secretion of gamma interferon by these cells is part of the immunological defense against infection and cancer. Some of the effects of interferons are summarized in table 23.4.

Table 23.4
Effects of Interferons

Stimulation	Inhibition
Macrophage phagocytosis	Cell division
Activity of cytotoxic (killer) T lymphocytes	Tumor growth
Activity of natural killer cells	Maturation of adipose cells
Production of antibodies	Maturation of erythrocytes

The Food and Drug Administration (FDA) has currently approved the use of interferons to treat a number of diseases. Alpha interferon, for example, is now being used to treat hepatitis C, hairy-cell leukemia, virally induced genital warts, and Kaposi's sarcoma. The FDA has also approved the use of beta interferon to treat relapsing-remitting multiple sclerosis and the use of gamma interferon to treat chronic granulomatous disease. Treatment of numerous forms of cancer with interferons is currently in various stages of clinical trials.

Specific Immunity

A German bacteriologist, Emil Adolf von Behring, demonstrated in 1890 that a guinea pig previously injected with a sublethal dose of diphtheria toxin could survive subsequent injections of otherwise lethal doses of that toxin. Further, von Behring showed that this immunity could be transferred to a second, nonexposed animal by injections of serum from the immunized guinea pig. He concluded that the immunized animal had chemicals in its serum—which he called **antibodies**—that were responsible for the immunity. He also showed that these antibodies conferred immunity only to diphtheria infections; the antibodies were *specific* in their actions. It was later learned that antibodies are proteins produced by a particular type of lymphocyte.

Antigens

Antigens (*an'tĭ-jenz*) are molecules that stimulate the production of specific antibodies and combine specifically with the antibodies produced. Most antigens are large molecules (such as proteins) with a molecular weight greater than about 10,000, although there are important exceptions. Also, most antigens are foreign to the blood and other body fluids. This is because the immune system can distinguish its own "self" molecules from those of any other organism ("nonself"), and normally mounts an immune response only against nonself antigens. The ability of a molecule to function as an antigen depends not only on its size but also on the complexity of its structure. Proteins, for example, are more antigenic than polysaccharides, which have a simpler structure. The plastics used in artificial implants are composed of large molecules, but they are not very antigenic because of their simple, repeating structures.

A large, complex molecule can have a number of different **antigenic determinant sites,** which are areas of the molecule that stimulate production of and combine with different antibodies. Most naturally occurring antigens have many antigenic determinant sites and stimulate the production of different antibodies with specificities for these sites.

Haptens

Many small organic molecules are not antigenic in and of themselves but can become antigens if they bind to proteins (and thus become antigenic determinant sites on the proteins). This discovery was made by Karl Landsteiner, who is also credited with the discovery of the ABO blood groups (chapter 20). By bonding these small molecules—which Landsteiner called **haptens**—to proteins in the laboratory, new antigens could be created for research or diagnostic purposes. The bonding of foreign haptens to a person's own proteins can also occur in the body. By this means, derivatives of penicillin, for example, that would otherwise be harmless can produce fatal allergic reactions in susceptible people.

Immunoassays

When the antigen or antibody is attached to the surface of a cell or to particles of latex rubber (in commercial diagnostic tests), the antigen-antibody reaction becomes visible because the particles *agglutinate* (clump) as a result of antigen-antibody bonding (fig. 23.9). These agglutinated particles can be used to assay a variety of antigens, and tests that utilize this procedure are called **immunoassays** (*im"yŭ-no-as'az*). Blood typing and modern pregnancy tests are examples of such immunoassays. In order to increase their sensitivity, modern immunoassays generally use antibodies that exhibit specificity for just one antigenic determinant site. The technique for generating such uniformly specific antibodies is described in a later section on monoclonal antibodies.

Emil Adolph von Behring: German bacteriologist, 1854–1917

Karl Landsteiner: Austrian-born American pathologist and immunologist, 1868–1943

Figure 23.9

Immunoassay using the agglutination technique. Antibodies against a particular antigen are absorbed to latex particles. When these are mixed with a solution that contains the appropriate antigen, the formation of the antigen-antibody complexes produces clumping (agglutination) that can be seen with the unaided eye.

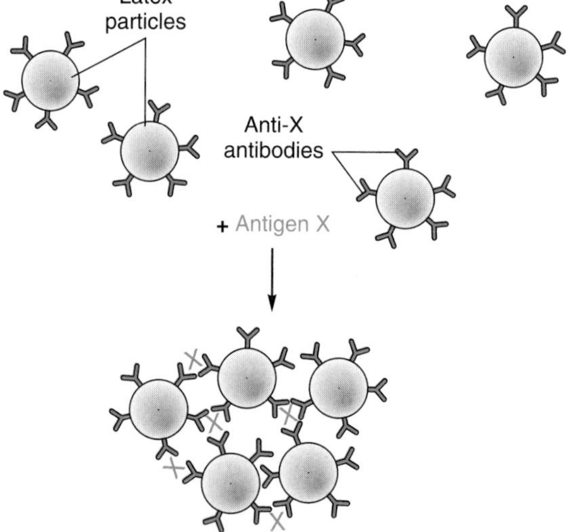

Lymphocytes

Leukocytes, erythrocytes, and blood platelets all ultimately derive from ("stem from") unspecialized cells in the bone marrow. These *stem cells* produce the specialized blood cells, and they replace themselves by cell division so that the stem cell population is not exhausted. Lymphocytes produced in this manner seed the thymus, spleen, and lymph nodes, producing self-replacing lymphocyte colonies in these organs.

The lymphocytes that seed the thymus become **T lymphocytes,** or **T cells** (the letter *T* stands for "thymus-dependent"). These cells have surface characteristics and an immunological function that differ from those of other lymphocytes. The thymus, in turn, seeds other organs; about 65% to 85% of the lymphocytes in blood and most of the lymphocytes in the germinal centers of the lymph nodes and spleen are T lymphocytes. T lymphocytes, therefore, either come from or had an ancestor that came from the thymus.

Most of the lymphocytes that are not T lymphocytes are called **B lymphocytes,** or **B cells.** The letter *B* is derived from immunological research performed in chickens. Chickens have an organ called the *bursa of Fabricius* that processes B lymphocytes. Since mammals do not have a bursa, the *B* is often translated as the "bursa equivalent" for humans and other mammals. It is currently believed that the B lymphocytes in mammals are processed in the bone marrow, which conveniently also begins with the letter *B*.

bursa of Fabricius: from Hieronymus Fabricius, Italian anatomist, 1533–1619

Both B and T lymphocytes function in specific immunity. The B lymphocytes combat bacterial infections, as well as some viral infections, by secreting antibodies into the blood and lymph. Because blood and lymph are body fluids (humors), the B lymphocytes are said to provide **humoral immunity,** although the term *antibody-mediated immunity* is also used. T lymphocytes attack host cells that have become infected with viruses or fungi, transplanted human cells, and cancerous cells. The T lymphocytes do not secrete antibodies; they must come in close proximity to the victim cell, or have actual physical contact with the cell, in order to destroy it. T lymphocytes are therefore said to provide **cell-mediated immunity** (table 23.5).

Functions of B Lymphocytes

B lymphocytes secrete antibodies that can bind to antigens in a specific fashion. This binding stimulates a cascade of reactions whereby a system of plasma proteins called complement is activated. Some of the activated complement proteins kill the cells containing the antigen; others promote phagocytosis and other activity that result in a more effective defense against pathogens.

Exposure of a B lymphocyte to the appropriate antigen results in cell growth followed by many cell divisions. Some of the progeny become **memory cells;** these are visually indistinguishable from the original cell and are important in active immunity. Others are transformed into **plasma cells** (fig. 23.10). Plasma cells are protein factories that produce about 2,000 antibody proteins per second.

Table 23.5
Comparison of B and T Lymphocytes

Characteristic	B Lymphocyte	T Lymphocyte
Site where processed	Bone marrow	Thymus
Type of immunity	Humoral (secretes antibodies)	Cell-mediated
Subpopulations	Memory cells and plasma cells	Cytotoxic (killer) T lymphocytes, helper cells, suppressor cells
Presence of surface antibodies	Yes—IgM or IgD	Not detectable
Receptors for antigens	Present—are surface antibodies	Present—are related to immunoglobulins
Life span	Short	Long
Tissue distribution	High in spleen, low in blood	High in blood and lymph
Percent of blood lymphocytes	10%–15%	75%–80%
Transformed by antigens to	Plasma cells	Activated lymphocytes
Secretory product	Antibodies	Lymphokines
Immunity to viral infections	Enteroviruses, poliomyelitis	Most others
Immunity to bacterial infections	*Streptococcus, staphylococcus,* many others	Tuberculosis, leprosy
Immunity to fungal infections	None known	Many
Immunity to parasitic infections	Trypanosomiasis, maybe to malaria	Most others

Figure 23.10

B lymphocytes are stimulated to become plasma cells and memory cells.
B lymphocytes have antibodies on their surface that function as receptors for specific antigens. The interaction of antigens and antibodies on the surface stimulates cell division and the maturation of the B cell progeny into memory cells and plasma cells. Plasma cells produce and secrete large amounts of the antibody. (Note the extensive rough endoplasmic reticulum in these cells.)

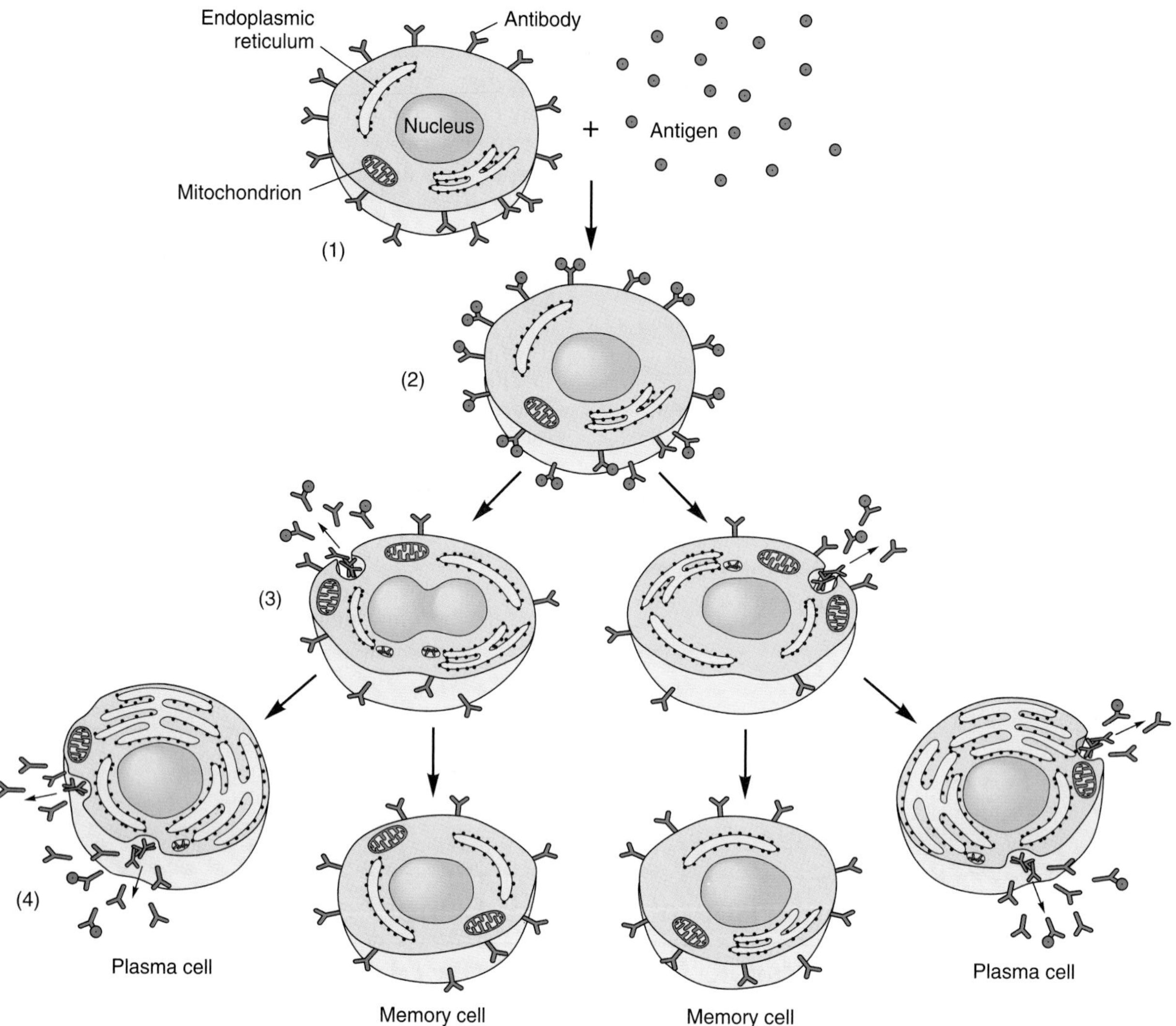

The antibodies that are produced by plasma cells when B lymphocytes are exposed to a particular antigen react specifically with that antigen. Such antigens may be isolated molecules, or they may be molecules at the surface of an invading foreign cell. The specific binding of antibodies to antigens serves to identify the enemy and to activate defense mechanisms that lead to the invader's destruction.

Antibodies

Antibody proteins are also known as **immunoglobulins** (*im″yŭ-no-glob′yŭ-linz*). These are found in the gamma globulin class of plasma proteins, as identified by a technique called *electrophoresis* (*ĕ-lek″tro-fŏ-re′sis*) in which different types of plasma proteins are separated by their movement in an electric field (fig. 23.11). The five distinct bands of proteins that appear are albumin, alpha-1 globulin, alpha-2 globulin, beta globulin, and gamma globulin.

The gamma globulin band is wide and diffuse because it represents a heterogeneous class of molecules. Since antibodies are specific in their actions, it follows that different types of antibodies should have different structures. An antibody against smallpox, for example, does not confer immunity to poliomyelitis and, therefore, must have a slightly different structure than an antibody against polio. Despite these differences, antibodies are structurally related and form only a few classes.

There are five immunoglobulin (Ig) subclasses: *IgG*, *IgA*, *IgM*, *IgD*, and *IgE*. Most of the antibodies in serum are in the IgG subclass, whereas most of the antibodies in external secretions (saliva and milk) are IgA (table 23.6). Antibodies in the IgE subclass are involved in certain allergic reactions, which are described in a later section.

Figure 23.11

The separation of serum protein by electrophoresis. This technique separates different groups of proteins on the basis of their electric charges and sizes. (A = albumin; α_1 = alpha-1 globulin; α_2 = alpha-2 globulin; β = beta globulin; γ = gamma globulin.)

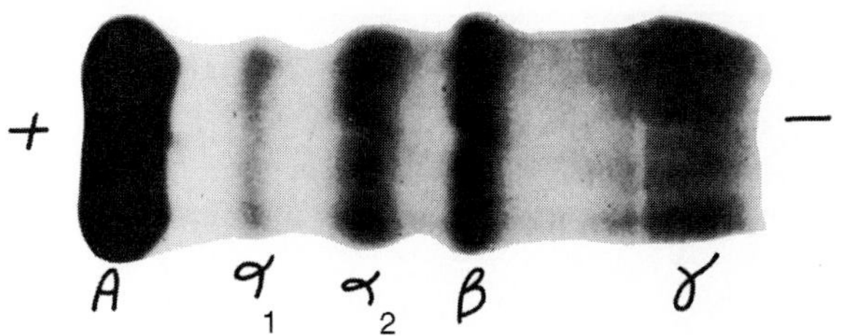

Antibody Structure

All antibody molecules consist of four interconnected polypeptide chains. Two long, heavy chains (the *H chains*) are joined to two shorter, lighter *L chains*. Research has shown that these four chains are arranged in the form of a Y. The stalk of the Y has been called the "crystallizable fragment" (abbreviated F_c), whereas the top of the Y is the "antigen-binding fragment" (F_{ab}). This structure is shown in figure 23.12.

The amino acid sequences of some antibodies have been determined through the analysis of antibodies sampled from people with multiple myelomas. These lymphocyte tumors arise from the division of a single B lymphocyte, forming a population of genetically identical cells (a clone) that secretes identical antibodies. Clones and the antibodies they secrete are different, however, from one patient to another. Analyses of these antibodies have shown that the F_c regions of different antibodies are the same (are constant), whereas the F_{ab} regions are variable. Variability of the antigen-binding regions is required for the specificity of antibodies for antigens. Thus, it is the F_{ab} region of an antibody that provides a specific site for bonding with a particular antigen (fig. 23.13).

B lymphocytes have antibodies on their cell membrane that serve as **receptors** for antigens. Combination of antigens with these antibody receptors stimulates the B cell to divide and produce more of these antibodies, which are secreted. Exposure to a given antigen thus results in increased amounts of the specific type of antibody that can attack that antigen. This provides active immunity, as described in the next major section.

Diversity of Antibodies

There are an estimated 100 million trillion (10^{20}) antibody molecules in each individual, representing a few million different specificities for different antigens. Considering that antibodies that bind to particular antigens can cross-react with closely related antigens to some extent, this tremendous antibody diversity usually ensures that there will be some antibodies that can combine with almost any antigen a person might encounter. These observations evoke a question that has long fascinated scientists: How can a few million different antibodies be produced? A person cannot possibly inherit a correspondingly large number of genes devoted to antibody production.

Two mechanisms have been proposed to explain antibody diversity. First, since different combinations of heavy and light chains can produce different antibody specificities, a person does not have to inherit a million different genes to code for a million different antibodies. If a few hundred genes code for different H chains and a few hundred code for different L chains, different combinations of these polypeptide chains could produce millions of different antibodies. The number of possible combinations is made even greater by the fact that different segments of DNA code for different segments of the heavy and light chains. Three segments in the antigen-combining region of a heavy chain and two in a light chain are coded by different segments of DNA and can be combined in different ways to make an antibody molecule.

Second, the diversity of antibodies could increase during development if, when some lymphocytes divided, the progeny received antibody genes that had been slightly altered by mutations. Such mutations are called *somatic mutations* because they occur in body cells rather than in sperm or ova. Antibody diversity would thus increase with age as the lymphocyte population increased.

Table 23.6

The Immunoglobulins

Immunoglobulin	Functions
IgG	Main form of antibodies in circulation; production increased after immunization; secreted during secondary response
IgA	Main antibody type in external secretions, such as saliva and mother's milk
IgE	Responsible for allergic symptoms in immediate hypersensitivity reactions
IgM	Function as antigen receptors on lymphocyte surface prior to immunization; secreted during primary response
IgD	Function as antigen receptors on lymphocyte surface prior to immunization; other functions unknown

The Complement System

The combination of antibodies with antigens does not itself cause destruction of the antigens or the pathogenic organisms that contain these antigens. Antibodies, rather, serve to identify the targets for immunological attack and to activate nonspecific immune processes that destroy the invader. Bacteria that are "buttered" with antibodies, for example, are better targets for phagocytosis by neutrophils and macrophages. The

Figure 23.12

The structure of antibodies.

Antibodies are composed of four polypeptide chains—two are heavy (H) and two are light (L). (*a*) A computer-generated model of antibody structure. (*b*) A simplified diagram showing the constant and variable regions. (The variable regions are abbreviated V, and the constant regions are abbreviated C.) Antigens combine with the variable regions. Each antibody molecule is divided into an F_{ab} (antigen-binding) fragment and an F_c (crystallizable) fragment.

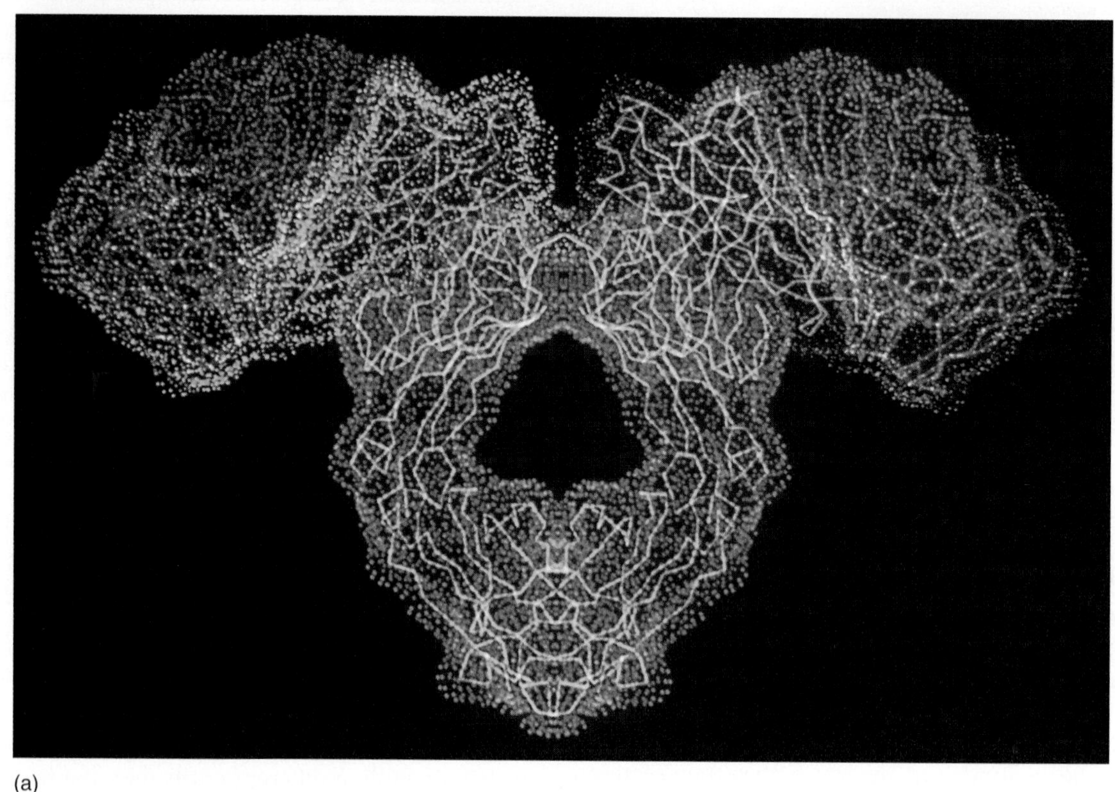

(a)

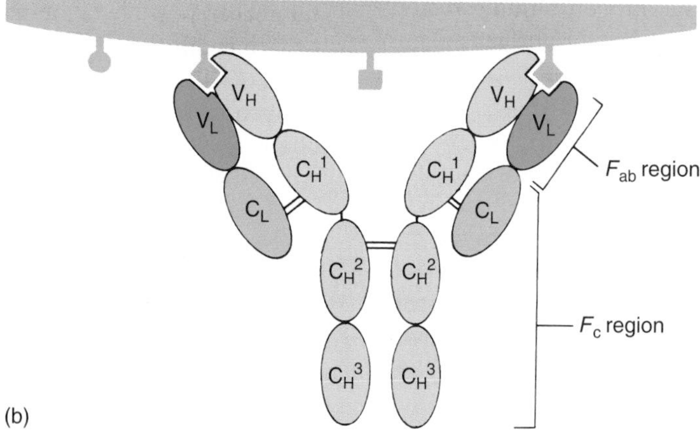

(b)

ability of antibodies to stimulate phagocytosis is termed **opsonization** (*op″sŏ-nĭ-za′shun*). Immune destruction of bacteria is also promoted by antibody-induced activation of a system of serum proteins known as *complement*.

In the early part of the twentieth century, it was learned that rabbit antibodies that bind to the red blood cell antigens of sheep could not lyse (destroy) these cells unless certain protein components of serum were present. These proteins, called **complement,** constitute a nonspecific defense system that is activated by the binding of antibodies to antigens, and by this means is directed against specific invaders that have been identified by antibodies.

There are 11 complement proteins, designated C1 (which has three protein components) through C9. These proteins are present in an inactive state within plasma and other body fluids and become activated by the attachment of

Figure 23.13

The antigen-binding site of an antibody.

The structure of the F_{ab} portion of an antibody molecule and the antigen with which it combines as determined by X-ray diffraction. (*a*) The heavy and light chains of the antibody are shown in blue and yellow, respectively, and the antigen is shown in green. Note the complementary shape at the region where the two join together in (*b*).

Photos from "Three-Dimensional Structure of an Antigen-Antibody Complex at 2.8A Resolution," *Science, vol. 233, p. 747.* © 1986 AAAS.

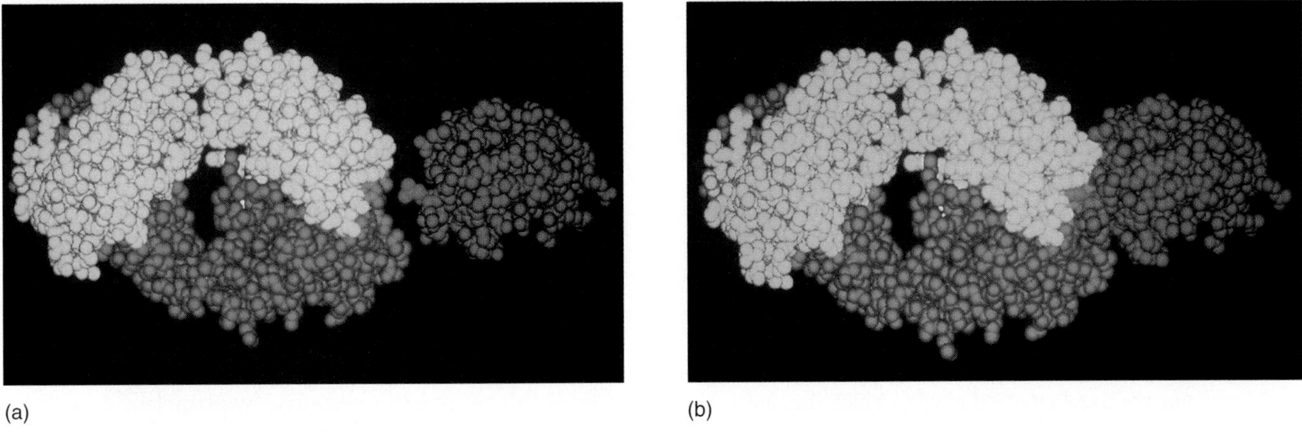

Figure 23.14

The fixation of complement proteins.

The formation of an antibody-antigen complex causes complement protein C4 to be split into two subunits—$C4_a$ and $C4_b$. The $C4_b$ subunit attaches (is fixed) to the membrane of the cell to be destroyed (such as a bacterium). This event triggers the activation of other complement proteins, some of which attach to the $C4_b$ on the membrane surface.

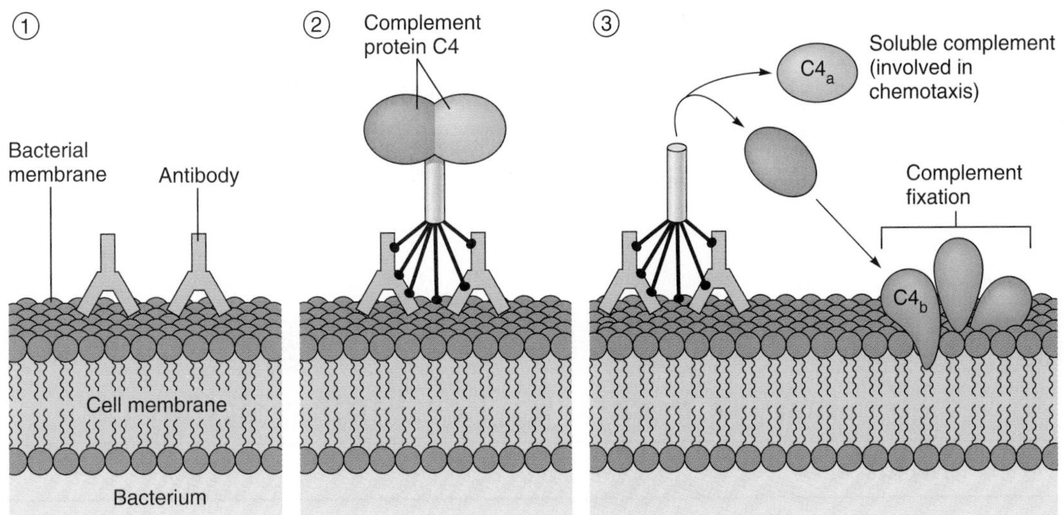

antibodies to antigens. In terms of their functions, the complement proteins can be subdivided into three components: (1) recognition (C1); (2) activation (C4, C2, and C3, in that order); and (3) attack (C5–C9). The attack phase consists of **complement fixation,** in which complement proteins attach to the cell membrane and destroy the victim cell.

Antibodies of the IgG and IgM subclasses attach to antigens on the invading cell's membrane, bind to C1, and by this means activate its enzyme activity. Activated C1 catalyzes the hydrolysis of C4 into two fragments (fig. 23.14), designated $C4_a$ and $C4_b$. The $C4_b$ fragment binds to the cell membrane (is "fixed") and becomes an active enzyme that splits C2 into two fragments, $C2_a$ and $C2_b$. The $C2_a$ becomes attached to $C4_b$ and cleaves C3 into $C3_a$ and $C3_b$. Fragment $C3_b$ becomes attached to the growing complex of complement proteins on the cell membrane. The $C3_b$ converts C5 to $C5_a$ and $C5_b$. The $C5_b$ and, eventually, C6 through C9 become fixed to the cell membrane.

Complement proteins C5 through C9 create large pores in the membrane (fig. 23.15). These pores permit the osmotic influx of water, so that the victim cell swells and bursts. Note that the complement proteins, not the antibodies directly, kill the cell; antibodies serve only as activators of this process. Other molecules can also activate the complement system in an alternate nonspecific pathway that bypasses the early phases of the specific pathway described here.

Complement fragments that are liberated into the surrounding fluid rather than becoming fixed have a number of effects. These effects include (1) *chemotaxis*—the liberated complement fragments attract phagocytic cells to the site of complement activation; (2) *opsonization*—phagocytic cells have receptors for $C3_b$, so that this fragment may form bridges between the phagocyte and the victim cell, thus facilitating phagocytosis; and (3) *stimulation of the release of histamine* from mast cells (a connective tissue cell type) and basophils by fragments $C3_a$ and $C5_a$. As a result of histamine release, blood flow to the infected area is increased because of vasodilation and increased capillary permeability. This helps to bring in more phagocytic cells to combat the infection, but the increased capillary permeability can also result in edema through leakage of plasma proteins into the surrounding tissue fluid.

Figure 23.15

Complement fixation.
Fixed complement proteins C5 through C9 (illustrated as a doughnut-shaped ring) puncture the membrane of the cell to which they are attached. This aids destruction of the cell.

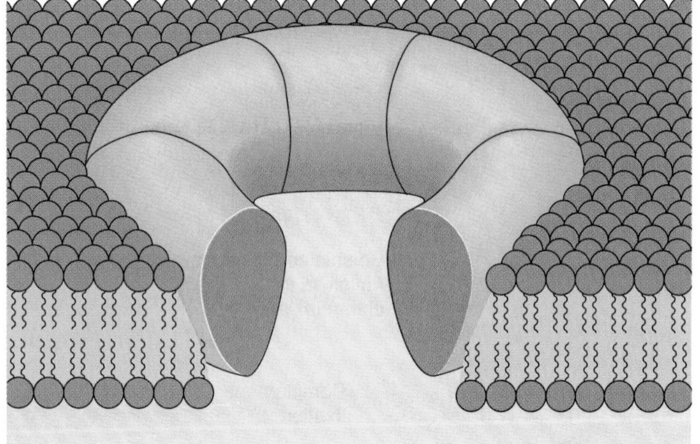

Local Inflammation

Aspects of the nonspecific and specific immune responses and their interactions are well illustrated by the events that occur when bacteria enter a break in the skin and produce a **local inflammation** (table 23.7). The inflammatory reaction is initiated by the nonspecific mechanisms of phagocytosis and complement activation. Activated complement further increases this nonspecific response by attracting new phagocytes to the area and by stimulating their activity.

After some time, B lymphocytes are stimulated to produce antibodies against specific antigens that are part of the invading bacteria. Attachment of these antibodies to antigens in the bacteria greatly amplifies the previously nonspecific response. This occurs because of greater activation of complement, which directly destroys the bacteria and which also—together with the antibodies themselves—promotes the phagocytic activity of neutrophils, macrophages, and monocytes (fig. 23.16).

Leukocytes within vessels in the inflamed area stick to the endothelial cells of the vessels through interactions between *adhesion molecules* on the two surfaces. The leukocytes can then roll along the wall of the vessel toward particular chemicals. As mentioned earlier, this movement toward chemoattractants is called *chemotaxis*. Complement proteins

Table 23.7

Summary of Events in a Local Inflammation

Category	Events
Nonspecific immunity	Bacteria enter through a break in the skin.
	Resident phagocytic cells—neutrophils and macrophages—engulf the bacteria.
	Nonspecific activation of complement protein occurs.
Specific immunity	B lymphocytes are stimulated to produce specific antibodies.
	Phagocytosis is enhanced by antibodies attached to bacterial surface antigens (opsonization).
	Specific activation of complement proteins occurs, which stimulate phagocytosis, chemotaxis of new phagocytes to the infected area, and secretion of histamine from tissue mast cells.
	Diapedesis allows new phagocytic leukocytes (neutrophils and monocytes) to invade the infected area.
	Vasodilation and increased capillary permeability (as a result of histamine secretion) produce redness and edema.

and bacterial products may serve as chemoattractants, drawing the leukocytes toward the site of infection.

The leukocytes squeeze between adjacent endothelial cells (the process of diapedesis, discussed earlier) and enter the subendothelial connective tissue. There, particular molecules on the leukocyte membrane interact with surrounding molecules that guide the leukocytes to the infection. The first to arrive are the neutrophils, followed by monocytes (which can change into macrophages) and T lymphocytes (fig. 23.17). Most of the phagocytic leukocytes (neutrophils and monocytes) die in the course of the infection, but lymphocytes can travel through the lymphatic system and re-enter the circulation.

As inflammation progresses, the release of lysosomal enzymes from macrophages causes the destruction of leukocytes and other cells. In addition, **mast cells** release *histamine* and other mediators of inflammation. Although mast cells, with their secretion of histamine, are associated with allergy and anaphylactic shock (see chapter 22), the mast cells have a beneficial effect in a local inflammation. The histamine released by mast cells causes vasodilation and increased capillary permeability, bringing more leukocytes to the infected area. Also, in response to a bacterial infection, the mast cells secrete *tumor necrosis factor* (TNF_{α}), which acts through chemotaxis to recruit neutrophils to the infected site.

These effects produce the characteristic symptoms of a local inflammation: *redness* and *warmth* (due to histamine-stimulated vasodilation), *swelling* (edema), *pus* (the accumulation of dead leukocytes), and *pain*. If the infection continues, the release of endogenous pyrogen from leukocytes and macrophages may also produce a fever, as previously discussed.

Figure 23.16

The events of a local inflammation.

In this inflammatory reaction, antigens on the surface of bacteria are coated with antibodies and ingested by phagocytic cells. Symptoms of inflammation are produced by the release of lysosomal enzymes and by the secretion of histamine and other chemicals from tissue mast cells.

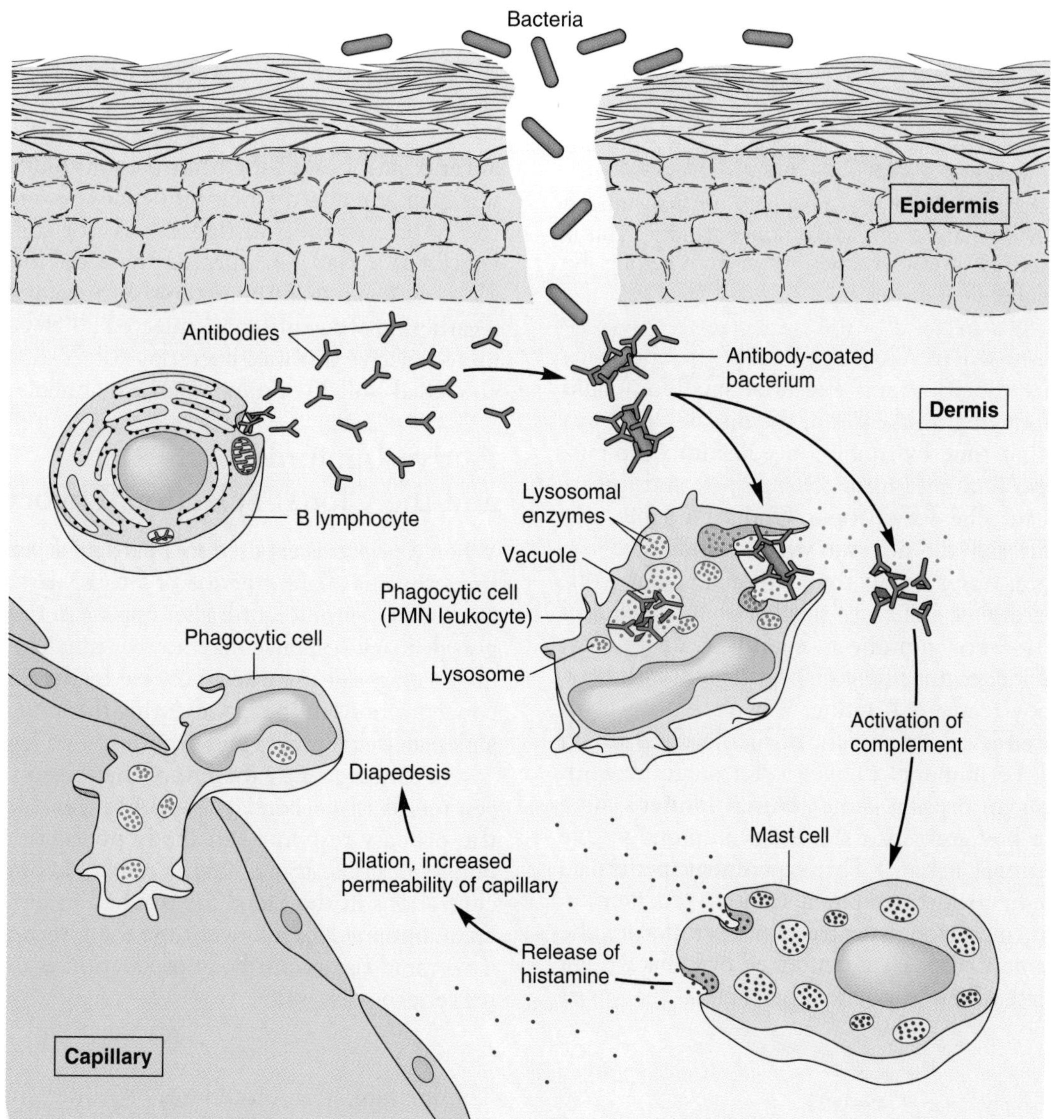

Figure 23.17

Leukocyte infiltration into an inflamed site.
Different types of leukocytes infiltrate into the site of a local inflammation. Neutrophils arrive first, followed by monocytes and then lymphocytes.

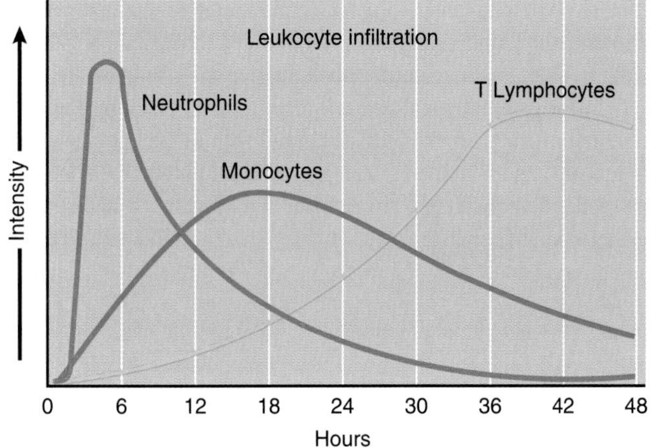

Figure 23.18

Virulence and antigenicity.
Active immunity to a pathogen can be gained by exposure to the fully virulent form or by inoculation with a pathogen whose virulence (ability to cause disease) has been attenuated (reduced) without altering its antigenicity (nature of its antigens).

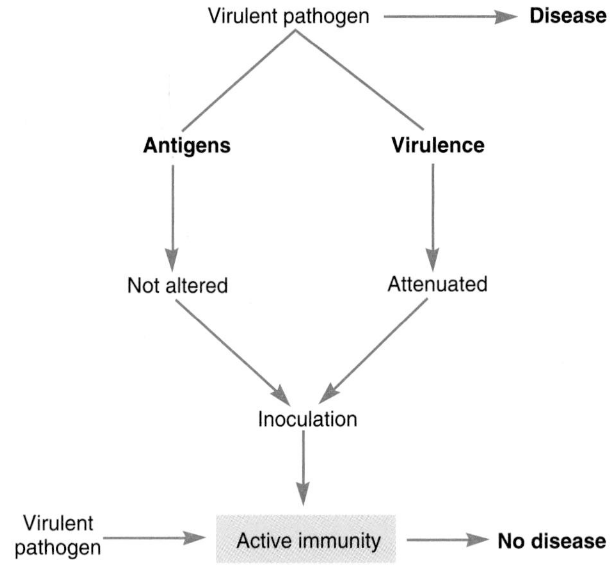

Active and Passive Immunity

When a person is first exposed to a pathogen, the immune response may be insufficient to combat the disease. In the process, however, the lymphocytes that have specificity for that antigen are stimulated to divide many times and produce a clone. This is active immunity, and it can protect the person from getting the disease upon subsequent exposures.

It first became known in Western Europe in the mid-eighteenth century that the fatal effects of smallpox could be prevented by inducing mild cases of the disease. This was accomplished at that time by rubbing needles into the pustules of people who had mild forms of smallpox and injecting these needles into healthy people. Understandably, this method of immunization did not gain wide acceptance.

Acting on the observation that milkmaids who contracted cowpox—a disease similar to smallpox but less *virulent* (less pathogenic)—were immune to smallpox, an English physician, Edward Jenner, inoculated a healthy boy with cowpox. When the boy recovered, Jenner inoculated him with what was considered a deadly amount of smallpox, to which the boy proved to be immune. (This was fortunate for both the boy—who was an orphan—and Jenner; Jenner's fame spread, and as the boy grew into manhood he proudly gave testimonials on Jenner's behalf.) This experiment, performed in 1796, began the first widespread immunization program.

A similar but more sophisticated demonstration of the effectiveness of immunizations was performed by Louis Pasteur almost a century later. Pasteur isolated the bacteria that cause anthrax and heated them until their ability to cause disease was greatly reduced (their virulence was *attenuated*), although the nature of their antigens was not significantly altered (fig. 23.18). He then injected these attenuated bacteria into 25 cows, leaving 25 unimmunized. Several weeks later, before a gathering of scientists, he injected all 50 cows with the completely active anthrax bacteria. All 25 of the unimmunized cows died—all 25 of the immunized animals survived.

Edward Jenner: English physician, 1749–1823
Louis Pasteur: French chemist and bacteriologist, 1822–95

Active Immunity and the Clonal Selection Theory

When a person is exposed to a particular pathogen for the first time, there is a latent period of 5 to 10 days before measurable amounts of specific antibodies appear in the blood. This sluggish **primary response** may not be sufficient to protect the individual against the disease caused by the pathogen. Antibody concentrations in the blood during this primary response reach a plateau in a few days and decline after a few weeks.

A subsequent exposure of that person to the same antigen results in a **secondary response** (fig. 23.19). Compared to the primary response, antibody production during the secondary response is much more rapid. Maximum antibody concentrations in the blood are reached in less than 2 hours and are maintained for a longer time than in the primary response. This rapid rise in antibody production is usually sufficient to prevent the disease.

Clonal Selection Theory

The immunization procedures of Jenner and Pasteur were effective because the people who were inoculated produced a

Figure 23.19

The primary and secondary immune responses.
A comparison of antibody production in the primary response (upon first exposure to an antigen) to antibody production in the secondary response (upon subsequent exposure to the antigen). The greater secondary response is believed to be due to the development of lymphocyte clones produced during the primary response.

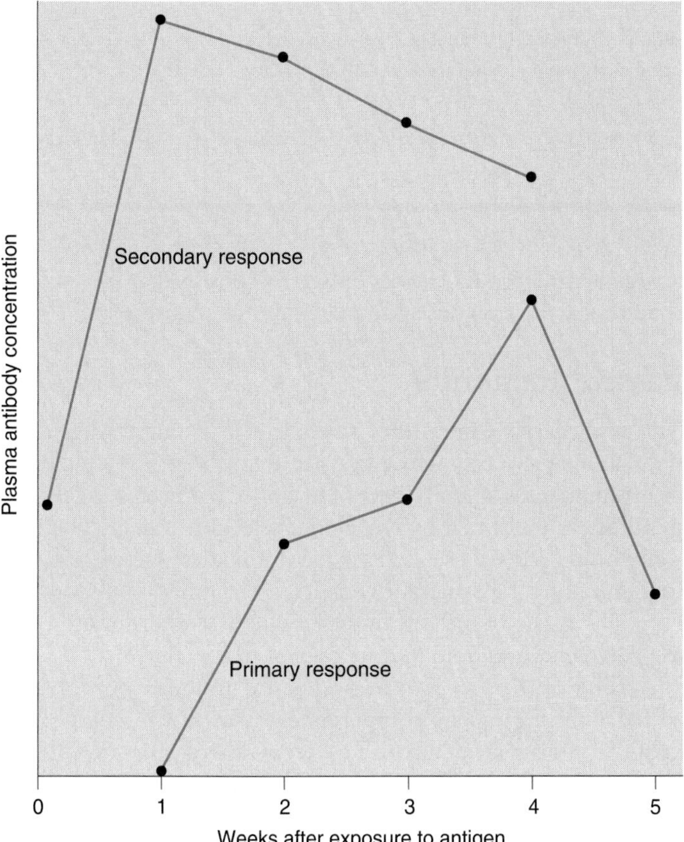

Figure 23.20

The clonal selection theory as applied to B lymphocytes.
Most members of the B lymphocytes clone become memory cells, but some become antibody-secreting plasma cells.

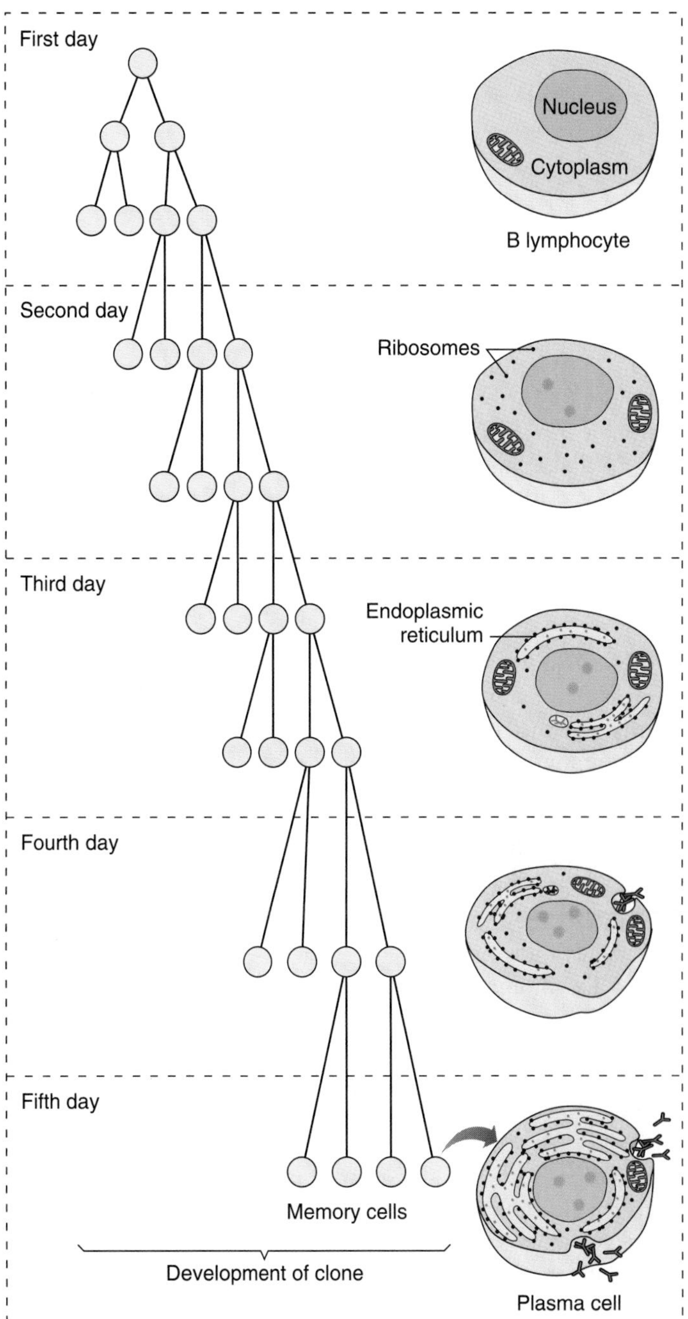

secondary rather than a primary response when exposed to the virulent pathogens. The type of protection they were afforded does not depend on accumulations of antibodies in the blood, since secondary responses occur even after antibodies produced by the primary response have disappeared. Immunizations, therefore, seem to produce a type of "learning" in which the ability of the immune system to combat a particular pathogen is improved by prior exposure.

The mechanisms by which secondary responses are produced are not completely understood; the **clonal selection theory,** however, appears to account for most of the evidence. According to this theory, B lymphocytes *inherit* the ability to produce particular antibodies (and T lymphocytes inherit the ability to respond to particular antigens). A given B lymphocyte can produce only one type of antibody, with specificity for one antigen. Since this ability is genetically inherited rather than acquired, some lymphocytes can respond to smallpox, for example, and produce antibodies against it even if the person has not been previously exposed to this disease.

The inherited specificity of each lymphocyte is reflected in the antigen receptor proteins on the surface of the lymphocyte's plasma membrane. Exposure to smallpox antigens thus stimulates these specific lymphocytes to divide many times until a large population of genetically identical cells—a *clone*—is produced. Some of these cells become plasma cells that secrete antibodies for the primary response; others become memory cells that can be stimulated to secrete antibodies during the secondary response (fig. 23.20).

Table 23.8

Summary of the Clonal Selection Theory (as Applied to B Lymphocytes)

Process	Results
Lymphocytes inherit the ability to produce specific antibodies.	Prior to antigen exposure, lymphocytes that can make the appropriate antibodies are already present in the body.
Antigens interact with antibody receptors on the lymphocyte surface.	Antigen-antibody interaction stimulates cell division and the development of lymphocyte clones that contain memory cells and plasma cells that secrete antibodies.
Subsequent exposure to the specific antigens produces a more efficient response.	Exposure of lymphocyte clones to specific antigens results in greater and more rapid production of specific antibodies.

Notice that, according to the clonal selection theory (table 23.8), antigens do not induce lymphocytes to make the appropriate antibodies. Rather, antigens select lymphocytes (through interaction with surface receptors) that are already able to make antibodies against that antigen. This is analogous to evolution by natural selection. An environmental agent (in this case, antigens) acts on the genetic diversity already present in a population of organisms (lymphocytes) to cause an increase in number of the individuals selected.

Active Immunity

The development of a secondary response provides **active immunity** against the specific pathogens. The development of active immunity requires prior exposure to the specific antigens, at which time the sluggishness of the primary response may cause the person to get the disease. Some parents, for example, deliberately expose their children to others who have measles, chickenpox, or mumps so that their children will be immune to these diseases in later life, when the diseases are potentially more serious.

Clinical immunization programs induce primary responses by inoculating people with pathogens whose virulence has been attenuated or destroyed (such as Pasteur's heat-inactivated anthrax bacteria) or by using closely related strains of microorganisms that are antigenically similar but less pathogenic (such as Jenner's cowpox inoculations). The name for these procedures—**vaccinations** (after the Latin word *vacca,* meaning "cow")—reflects the history of this technique. All of these procedures cause the development of lymphocyte clones that can combat the virulent pathogens by producing secondary responses.

The first successful polio vaccine (the Salk vaccine) was composed of viruses that had been inactivated by treatment with formaldehyde. These "killed" viruses were injected into the body, in contrast to the currently used oral (Sabin) vaccine. The oral vaccine contains "living" viruses that have attenuated virulence. These viruses invade the epithelial lining of the intestine and multiply but do not invade nerve tissue. The immune system can, therefore, become sensitized to polio antigens and produce a secondary response if polio viruses that attack the nervous system are later encountered.

Passive Immunity

The term **passive immunity** refers to the immune protection that can be produced by the transfer of antibodies to a recipient from a human or animal donor. The donor has been actively immunized, as explained by the clonal selection theory. The person who receives these ready-made antibodies is thus passively immunized to the same antigens. Passive immunity also occurs naturally in the transfer of immunity from mother to fetus during pregnancy and from mother to baby during nursing.

The ability to mount a specific immune response—called **immunological competence**—does not develop until about a month after birth. The fetus, therefore, cannot immunologically reject its mother. The immune system of the mother is fully competent but does not usually respond to fetal antigens for reasons that are not completely understood. Some IgG antibodies from the mother do cross the placenta and enter the fetal circulation, however, and these serve to confer passive immunity to the fetus.

The fetus and the newborn baby are thus immune to the same antigens as the mother. However, since the baby did not itself produce the lymphocyte clones needed to form these antibodies, such passive immunity disappears when the infant is about 1 month old. If the baby is breast-fed, it can receive additional antibodies of the IgA subclass in its mother's milk or *colostrum* (the secretion an infant feeds on for the first 2 or 3 days until the onset of true lactation).

Passive immunizations are used clinically to protect people who have been exposed to extremely virulent infections or toxins, such as tetanus, hepatitis, rabies, and snake venom. In these cases, the affected person is injected with *antiserum* (serum containing antibodies), also called *antitoxin,* from an animal that has been previously exposed to the pathogen. The animal develops the lymphocyte clones and active immunity, and thus has a high concentration of antibodies in its blood. Since the person who is injected with these antibodies does not develop active immunity, he or she must again be injected with antitoxin upon subsequent exposures.

Active and passive immunity are compared in table 23.9.

Salk vaccine: from Jonas Salk, American immunologist, 1914–95
Sabin vaccine: from Albert B. Sabin, American virologist, 1906–93

Table 23.9
Comparison of Active and Passive Immunity

Characteristic	Active Immunity	Passive Immunity
Injection of person with	Antigens	Antibodies
Source of antibodies	The person inoculated	Natural—the mother; artificial—injection with antibodies
Method	Injection with killed or attenuated pathogens or their toxins	Natural—transfer of antibodies across the placenta; artificial—injection with antibodies
Time to develop resistance	5 to 14 days	Immediately after injection
Duration of resistance	Long (perhaps years)	Short (days to weeks)
When used	Before exposure to pathogen	Before or after exposure to pathogen

Figure 23.21

The production of monoclonal antibodies.
These are antibodies produced by the progeny of a single B lymphocyte, so that all of the antibodies are directed against a specific antigen.

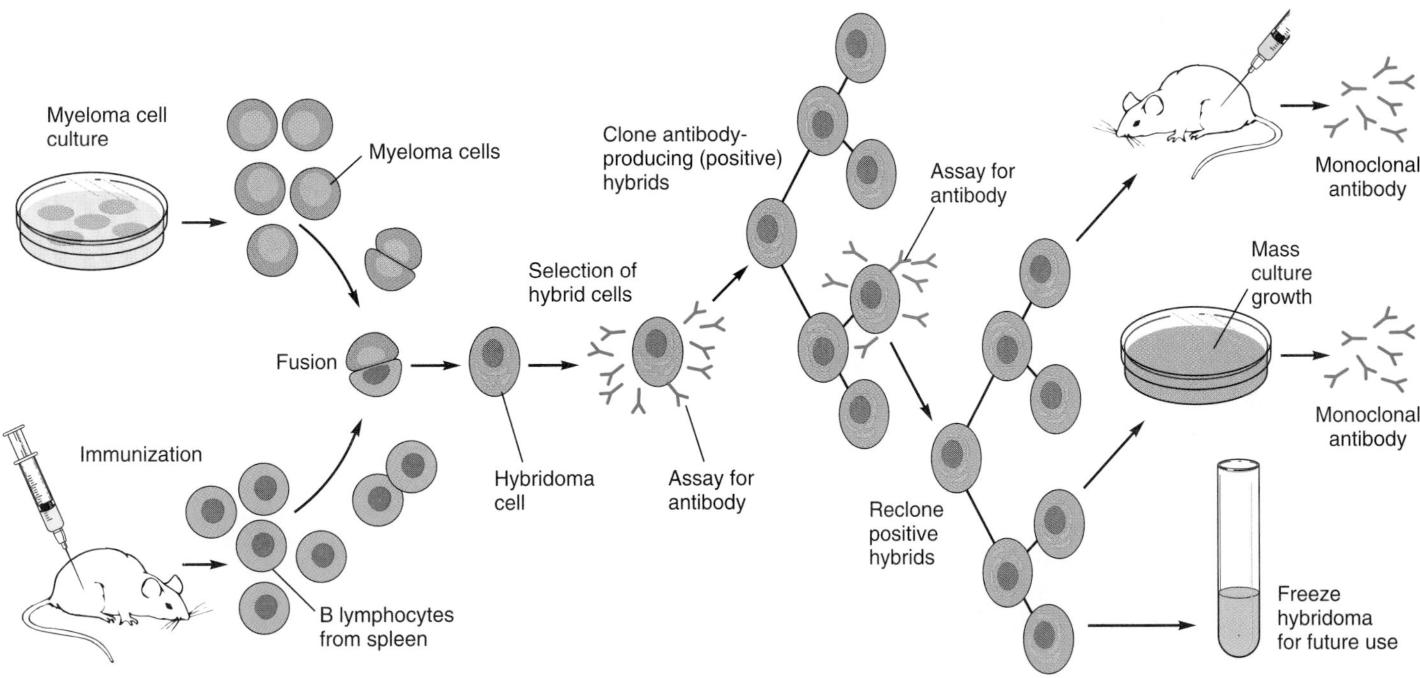

Monoclonal Antibodies

In addition to their use in passive immunity, antibodies are also commercially prepared for use in research and clinical laboratory tests. In the past, antibodies were obtained by chemically purifying a specific antigen and then injecting this antigen into animals. Since an antigen typically has many different antigenic determinant sites, however, the antibodies obtained by this method were polyclonal; they had different specificities. This decreased their sensitivity to a particular antigenic site and resulted in some degree of cross-reaction with closely related antigen molecules.

Monoclonal antibodies, by contrast, exhibit specificity for one antigenic determinant only. In the preparation of monoclonal antibodies, an animal (frequently, a mouse) is injected with an antigen and subsequently killed. B lymphocytes are then obtained from the animal's spleen and placed in thousands of different in vitro incubation vessels. These cells soon die, however, unless they are hybridized with cancerous multiple myeloma cells. The fusion of a B lymphocyte with a cancerous cell produces a potentially immortal hybrid that undergoes cell division and produces a clone, called a *hybridoma (hi''brĭ-do'mă)*. Each hybridoma secretes large amounts of identical monoclonal antibodies. From among the thousands of hybridomas produced in this way, the one that produces the desired antibody is cultured for large-scale production and the rest are discarded (fig. 23.21).

The availability of large quantities of pure monoclonal antibodies has resulted in the development of much more sensitive clinical laboratory tests (for pregnancy, for example). These pure antibodies have also been used to pick one molecule (the specific antigen interferon, for example) out of a solution of many molecules so as to isolate and concentrate it. In the future, monoclonal antibodies against specific tumor antigens may aid the diagnosis of cancer. Even more exciting, drugs that can kill normal as well as cancerous cells might be aimed directly at a tumor by combining these drugs with monoclonal antibodies against specific tumor antigens.

Functions of T Lymphocytes

Each subpopulation of T lymphocytes has specific immune functions. Killer T cells effect cell-mediated destruction of specific victim cells, and helper and supressor T cells play supporting roles. T cells are activated only by antigens presented to them on the surface of particular antigen-presenting cells. Activated helper T cells produce lymphokines that stimulate other cells of the immune system.

The thymus processes lymphocytes in such a way that their functions become quite distinct from those of B cells. Lymphocytes residing in the thymus or originating from the thymus, or those derived from cells that came from the thymus, are all T lymphocytes. These cells can be distinguished from B cells by specialized techniques. Unlike B cells, the T lymphocytes provide specific immune protection without secreting antibodies. This is accomplished in different ways by the three subpopulations of T lymphocytes.

Thymus

The **thymus** extends from below the thyroid in the neck into the thoracic cavity. As mentioned in chapter 19, this organ grows during childhood but gradually regresses after puberty. Lymphocytes from the fetal liver and spleen, and from the bone marrow postnatally, seed the thymus and become transformed into T cells. These lymphocytes, in turn, enter the blood and seed lymph nodes and other organs, where they divide to produce new T cells when stimulated by antigens.

Small T lymphocytes that have not yet been stimulated by antigens have very long life spans—months or perhaps years. Still, new T cells must be continuously produced to provide efficient cell-mediated immunity. Since the thymus atrophies after puberty, this organ may not be able to provide new T cells in later life. Colonies of T cells in the lymph nodes and other organs are apparently able to produce new T cells under the stimulation of various **thymus hormones.**

Two hormones that are believed to be secreted by the thymus—*thymopoietin I* and *thymopoietin II*—may promote the transformation of lymphocytes into T cells. Another thymus hormone, called *thymosin*, may promote the maturation of T lymphocytes.

Killer, Helper, and Suppressor T Lymphocytes

The **killer,** or **cytotoxic, T lymphocytes** destroy specific victim cells that are identified by specific antigens on their surface. In order to effect this *cell-mediated destruction*, the T lymphocytes must be in actual contact with their victim cells (B cells, by contrast, kill at a distance). Although the mechanisms by which the cytotoxic lymphocytes kill their victims are not completely understood, there is evidence that they accomplish this task by secreting certain molecules at the region of contact. Among these molecules, specific polypeptides called **perforins** have been identified. Perforins polymerize in the cell membrane of the victim cell and form cylindrical channels through the membrane. This process, which is similar to the formation of channels by complement proteins previously discussed, can result in osmotic destruction of the victim cell.

The killer T lymphocytes defend against viral and fungal infections and are also responsible for transplant rejection reactions and for immunological surveillance against cancer. Although most bacterial infections are fought by B lymphocytes, some are the targets of cell-mediated attack by killer T lymphocytes. This is the case with the tubercle bacilli that cause tuberculosis. Injections of some of these bacteria under the skin produce inflammation after a latent period of 48 to 72 hours. This *delayed hypersensitivity reaction* is cell-mediated rather than humoral, as shown by the fact that it can be induced in an unexposed guinea pig by an infusion of lymphocytes, but not of serum, from an exposed animal.

The **helper T lymphocytes** and **suppressor T lymphocytes** indirectly participate in the specific immune response by regulating the responses of the B cells (fig. 23.22) and the killer T cells. The activity of B cells and killer T cells is increased by helper T lymphocytes and decreased by suppressor T lymphocytes. The amount of antibodies secreted in response to antigens is thus affected by the relative numbers of helper to suppressor T cells that develop in response to a given antigen.

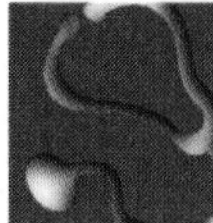

Acquired immune deficiency syndrome (AIDS) has caused the deaths of hundreds of thousands of people worldwide. Millions more are infected, and since AIDS has been shown to have a latency period of approximately 8 years, most will display symptoms of the disease in the near future. AIDS is caused by the *human immunodeficiency virus (HIV)* (see fig. 23.8), which specifically destroys the helper T lymphocytes. This results in decreased immunological function and greater susceptibility to opportunistic infections, including *Pneumocystis carinii pneumonia.* Many people with AIDS also develop a previously rare form of cancer known as *Kaposi's sarcoma.*

Kaposi's sarcoma: from Moritz Kaposi, Austrian physician, 1837–1902

Figure 23.22

Effect of an antigen on B and T lymphocytes. A given antigen can stimulate the production of both B and T lymphocyte clones. The ability to produce B lymphocyte clones, however, is also influenced by the relative effects of helper and suppressor T lymphocytes.

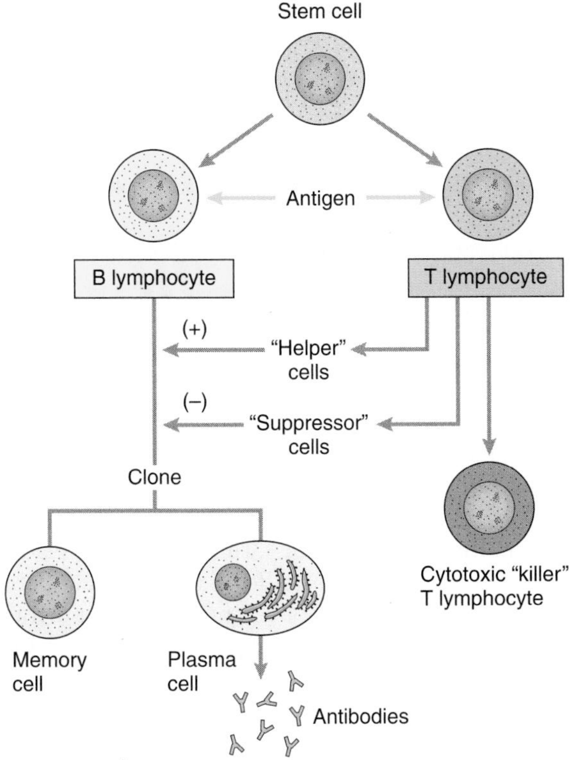

Current treatment for AIDS includes the use of drugs that inhibit reverse transcriptase, the enzyme used by the virus to replicate its RNA (see fig. 23.8). Recently two different reverse transcriptase inhibitors have been combined with a protease inhibitor (protease enzymes are needed to cut viral protein into segments for assembly of the viral coat) to produce a "cocktail" that has proved to be a highly effective treatment. The current therapies do not cure AIDS, however, and research on possible vaccines and other treatments is ongoing.

Lymphokines

The T lymphocytes, as well as some other cells such as macrophages, secrete a number of polypeptides that serve in an autocrine fashion (chapter 19) to regulate many aspects of the immune system. These products are generally called **cytokines;** the term **lymphokine** (*lim'fŏ-kīn*) is often used to refer to the cytokines of lymphocytes. When a cytokine is first discovered, it is named according to its biological activity (e.g., *B cell–stimulating factor*). Since each cytokine has many different actions (table 23.10), however, such names can be misleading. Scientists have thus agreed to use the name *interleukin*, followed by a number, to indicate a cytokine once its amino acid sequence has been determined.

Interleukin-1, for example, is secreted by macrophages and other cells and can activate the T cell system. B cell–stimulating factor, now called *interleukin-4*, is secreted by T lymphocytes and is required for the proliferation and clone development of B cells. *Interleukin-2* is released by helper T lymphocytes and is required for activation of killer T lymphocytes, among other functions. *Granulocyte colony-stimulating factor (G-CSF)* and *granulocyte-monocyte colony-stimulating factor (GM-CSF)* are lymphokines that promote leukocyte development; as mentioned in chapter 20, they are now available for use in medical treatments.

Current research has demonstrated that there are two subtypes of helper T lymphocytes, designated $\mathbf{T_H1}$ and $\mathbf{T_H2}$. Helper T lymphocytes of the T_H1 subtype produce interleukin-2 and gamma interferon. Because they secrete these lymphokines, T_H1 cells activate killer T cells and promote cell-mediated immunity. The lymphokines secreted by the T_H1 lymphocytes also stimulate nitric oxide production in macrophages, increasing their activity. The T_H2 lymphocytes secrete interleukin-4, interleukin-5, interleukin-10, and other lymphokines that stimulate B lymphocytes to promote humoral immunity. The lymphokines secreted by T_H2 cells can also activate mast cells and other agents that promote an allergic immune response.

Scientists have discovered that "uncommitted" helper T lymphocytes change into the T_H1 subtype in response to a cytokine called interleukin-12, which is secreted by macrophages under appropriate conditions. This process could thus provide a switch for determining how much of the immune response to an antigen will be cell-mediated and how much will be humoral.

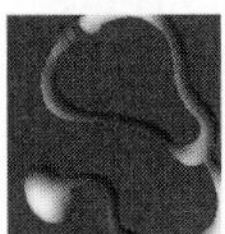

In response to endotoxin, a molecule released by bacteria, and to cytokines such as interleukin-1 and gamma interferon, production of the enzyme nitric oxide synthetase is induced within macrophages. As discussed in chapter 22, this enzyme catalyzes the formation of nitric oxide, which in excessive amounts may produce the hypotension of septic shock. A normal amount of nitric oxide, however, is required for macrophages to destroy bacteria and tumor cells.

T Cell Receptor Proteins

The antigens recognized by B lymphocytes may be either proteins or carbohydrates, but only protein antigens are recognized by most T lymphocytes. Unlike B cells, T cells do not make antibodies and thus do not have antibodies on their surfaces to serve as receptors for these antigens. The T cells do, however, have a different type of antigen receptor on their membrane surfaces, and these T cell receptors have recently been identified as molecules closely related to the immunoglobulins. The T cell receptors differ from the antibody receptors on B cells in a very important respect: the T cell receptors *cannot bind to free antigens*. In order for T lymphocytes to respond to foreign antigens, the antigens must be presented to the T cells on the membrane of **antigen-presenting cells.**

The chief antigen-presenting cells are macrophages and **dendritic** (*den-drit'ik*) **cells.** Dendritic cells are the most potent antigen-presenting cells for the activation of helper

Table 23.10

Some Cytokines That Regulate the Immune System

Lymphokine	Biological Functions	Secreted by
Interleukin-1	Activates resting T lymphocytes	Macrophages and others
Interleukin-2	Serves as growth factor for activated T lymphocytes; activates cytotoxic T lymphocytes	Helper T lymphocytes
Interleukin-3	Promotes growth of bone marrow stem cells; serves as growth factor for mast cells	Helper T lymphocytes
Interleukin-4 (B lymphocyte–stimulating factor)	Promotes growth of activated B lymphocytes; promotes growth of resting T lymphocytes; enhances activity of cytotoxic T lymphocytes	Helper T lymphocytes
B lymphocyte–differentiating factor	Induces the conversion of activated B lymphocytes into antibody-secreting plasma cells	T lymphocytes and others
Colony-stimulating factors	Different colony-stimulating factors stimulate the proliferation of granulocytic leukocytes and macrophages	T lymphocytes and others
Interferons	Activate macrophages; augment natural killer cell body activity; exhibit antiviral activity	T lymphocytes and others
Tumor necrosis factors	Exert direct cytotoxic effect on some tumor cells; stimulate production of other lymphokines	Macrophages and others

Source: Adapted from information appearing in C. A. Dinarello and J. W. Mier, *The New England Journal of Medicine*, Vol. 317, p. 940, 1987.

T lymphocytes. The dendritic cells originate in the bone marrow and migrate through the blood and lymph to temporarily reside in various lymphoid organs, as well as in other organs such as the lungs and skin. The antigen-presenting cells engulf protein antigens by pinocytosis, partially digest these proteins into shorter polypeptides, and then move these polypeptides to the cell surface. At the cell surface, the foreign polypeptides are associated with molecules called *histocompatibility antigens* and presented to the T lymphocytes. Some knowledge of the histocompatibility antigens is thus required before T cell functions can be understood.

Histocompatibility Antigens

Tissue that is transplanted from one person to another contains antigens that are foreign to the host. This is because all the cells of the body, with the exception of mature red blood cells, are genetically marked with a characteristic combination of **histocompatibility antigens** on the membrane surface. The greater the variance in these antigens between the donor and the recipient in a transplant, the greater the chance of transplant rejection. Prior to organ transplantation, therefore, the "tissue type" of the recipient is matched to that of potential donors. Since the person's white blood cells are used for this purpose, histocompatibility antigens in humans are also called **human leukocyte antigens (HLAs).** They are also called *MHC molecules*, after the name of the genes that code for them.

The histocompatibility antigens are proteins that are coded by a group of genes called the **major histocompatibility complex (MHC),** located on chromosome number 6. These four genes are labeled A, B, C, and D. Each of them can code for only one protein in a given individual, but because each gene has multiple alleles (forms), this protein can be different in different people. Two people, for example, could both have antigen A3, but one might have antigen B17 and the other antigen B21. The closer two people are related, the closer the match between their histocompatibility antigens.

Interactions between Antigen-Presenting Cells and T Lymphocytes

The major histocompatibility complex of genes produces two classes of MHC molecules, designated *class 1* and *class 2*, that are found on the cell surface. The class-1 molecules are made by all cells in the body except red blood cells. Class-2 MHC molecules are produced only by antigen-presenting cells (macrophages and dendritic cells) and B lymphocytes, where they promote the interactions with the T lymphocytes.

Killer T lymphocytes can interact only with antigens presented to them in conjunction with class-1 MHC molecules, whereas helper T lymphocytes can interact only with antigens presented with class-2 MHC molecules. These restrictions result from the presence of *coreceptors*, which are proteins associated with the T cell receptors. The coreceptor known as *CD8* is associated with the killer T lymphocyte receptor and interacts only with the class-1 MHC molecules; the coreceptor known as *CD4* is associated with the helper T lymphocyte receptor and interacts only with the class-2 MHC molecules. These structures are illustrated in figure 23.23.

T Lymphocyte Response to a Virus

When a foreign particle such as a virus infects the body, it is taken up by macrophages via phagocytosis and partially digested. Within the macrophage, the partially digested virus particles provide foreign antigens that move to the surface of

Figure 23.23

Coreceptors on helper and killer T cells.
A foreign antigen is presented to T lymphocytes in association with MHC molecules. The CD4, on helper T cells, and CD8 coreceptors, on killer T cells, permit each type of T cell to interact with only a specific class of MHC molecule.

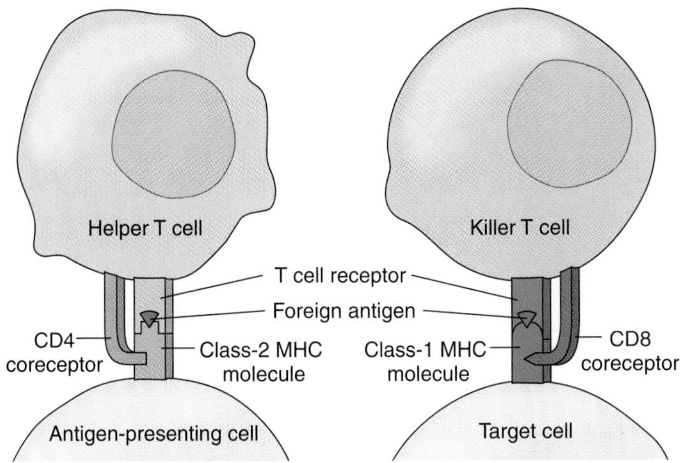

the cell membrane. At the membrane, these foreign antigens form a complex with the class-2 MHC molecules. This combination of MHC molecules and foreign antigens is required for interaction with the receptors on the surface of helper T cells. The macrophages thus "present" the antigens to the helper T cells and, in this way, stimulate activation of the T cells (fig. 23.24). It should be remembered that T cells are "blind" to free antigens; they can respond only to antigens presented to them by dentritic cells and macrophages in combination with class-2 MHC molecules.

The first phase of macrophage–T cell interaction then occurs: the macrophage is stimulated to secrete the cytokine known as interleukin-1. As previously discussed, interleukin-1 stimulates cell division and proliferation of T lymphocytes. The activated helper T cells, in turn, secrete macrophage colony-stimulating factor and gamma interferon, which promote the activity of macrophages. In addition, interleukin-2 is secreted by the T lymphocytes and stimulates the macrophages to secrete tumor necrosis factor, which is particularly effective in killing cancer cells.

Killer T cells can destroy infected cells only if those cells display the foreign antigen together with their class-1 MHC molecules (fig. 23.25). Such interaction of killer T cells with the foreign antigen–MHC class-1 complex also stimulates proliferation of those killer T cells. In addition, proliferation of the killer T lymphocytes is stimulated by interleukin-2 secreted by the helper T lymphocytes that were activated by macrophages, as previously described (fig. 23.26).

The network of interactions among the different cell types of the immune system now spreads outward. Helper T cells, activated to an antigen by macrophages or other antigen-presenting cells, can also promote the humoral immune response of B cells. In order to do this, the membrane receptor

Figure 23.24

Interactions between antigen-presenting cells, T cells, and B cells.
(*a*) An electron micrograph showing contact between a macrophage (left) and a lymphocyte (right). As illustrated in (*b*), such contact between a macrophage (or other antigen-presenting cell) and a T cell requires that the helper T cell interact with both the foreign antigen and the class-2 MHC molecule on the surface of the macrophage. In this figure, the helper T cell is now activated and able to interact with a B cell.

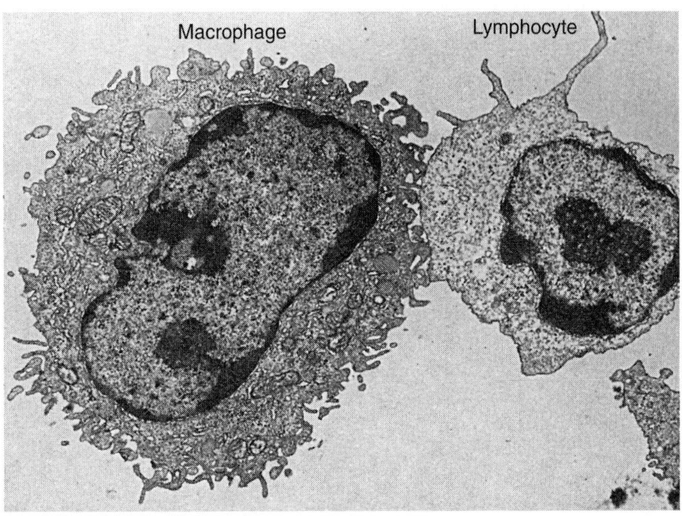

(a)

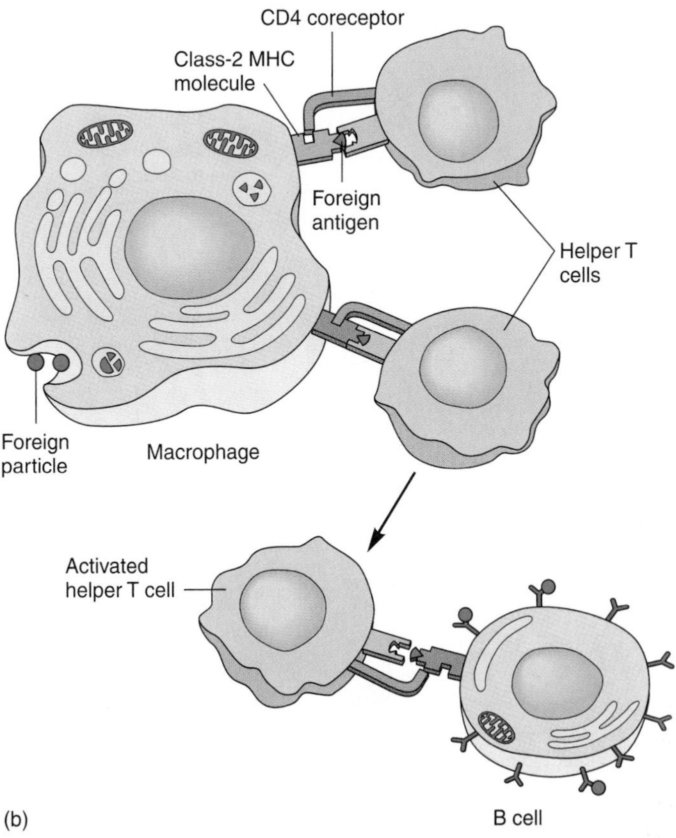

(b)

proteins on the surface of the helper T lymphocytes must interact with molecules on the surface of the B cells. This occurs when the foreign antigen attaches to the immunoglobulin receptors on the B cells, so that the B cells can present this antigen together with its class-2 MHC molecules to the receptors on the helper T cells (fig. 23.27). This interaction stimulates proliferation of the B cells, their conversion to plasma cells, and their secretion of antibodies against the foreign antigens.

Destruction of T Lymphocytes

The activated T lymphocytes must be destroyed after the infection has been cleared. This occurs because T cells produce a surface receptor called **FAS.** Production of FAS increases during the infection and, after a few days, the activated T lymphocytes begin to produce another surface molecule called **FAS ligand.** The binding of FAS to FAS ligand, on the same or on different cells, triggers the apoptosis (cell suicide) of the lymphocytes.

This mechanism also helps to maintain certain parts of the body—such as the inner region of the eye and the tubules of the testis—as *immunologically privileged sites.* These sites

Figure 23.25

A killer T cell destroys an infected cell.
In order for a killer T cell to destroy a tissue cell infected with viruses, the T cell must interact with both the foreign antigen and the class-1 MHC molecule on the surface of the infected cell.

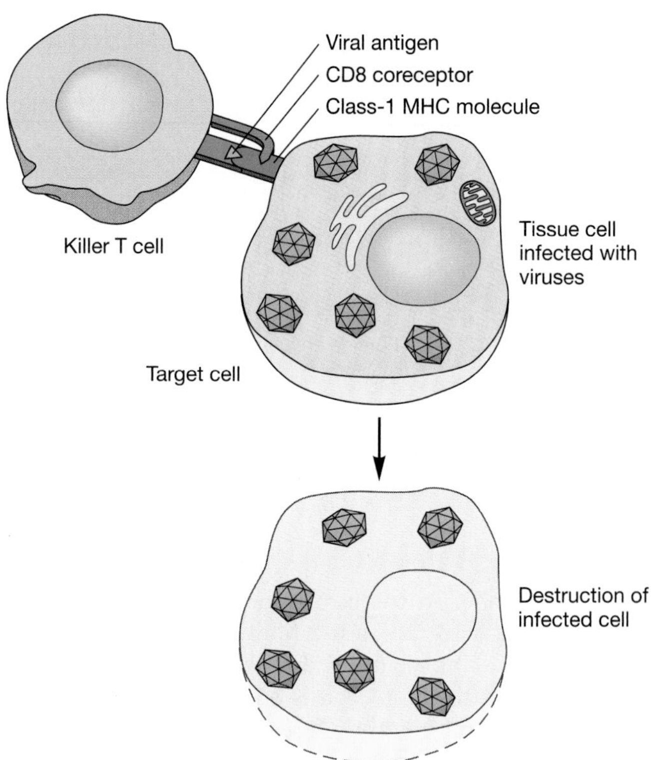

Figure 23.26

Interaction between macrophages, helper T lymphocytes, and killer T lymphocytes.
This sequence leads to the activation of killer T cells and thus the destruction of infected cells in the defense against viral infections.

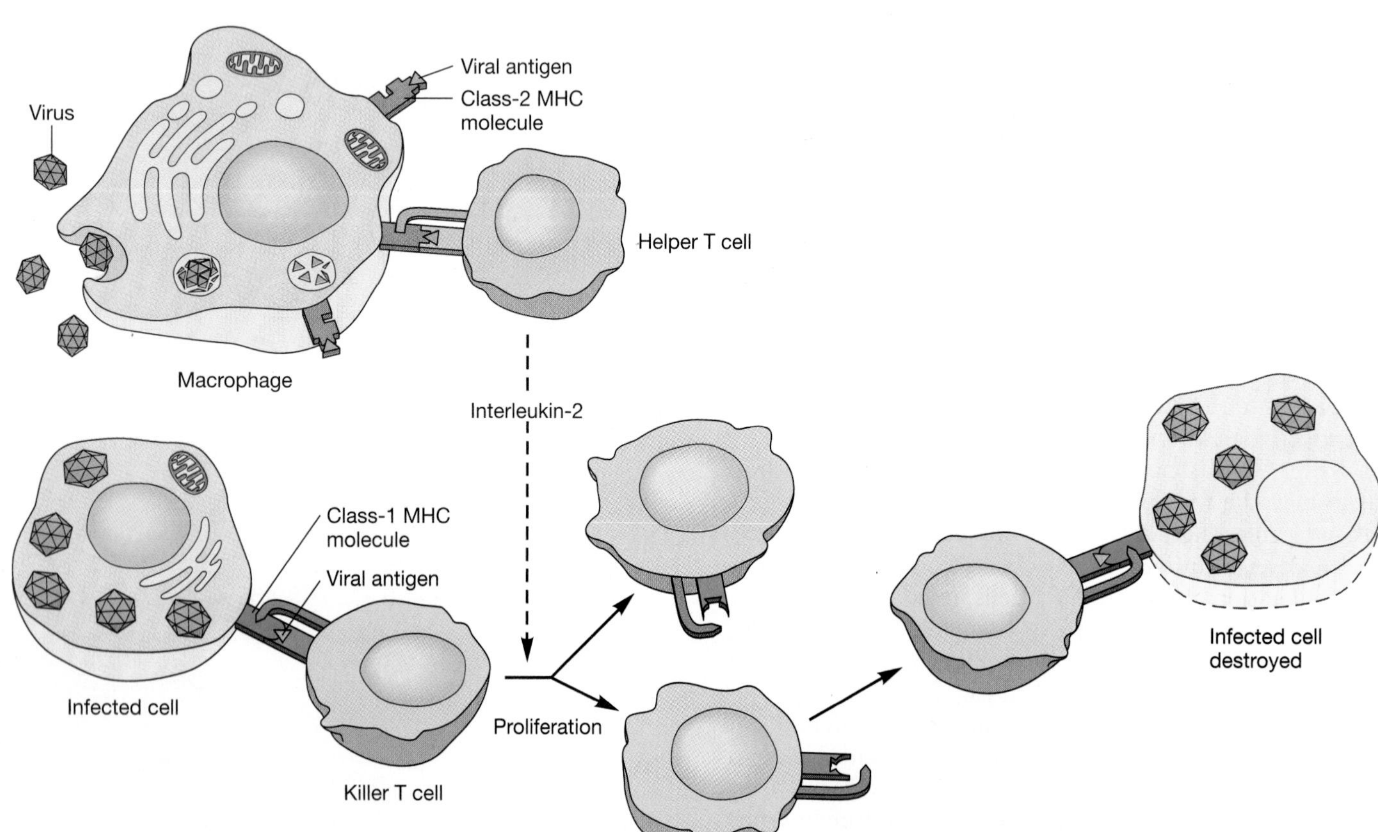

Figure 23.27

Interactions between macrophages, helper T lymphocytes, and B lymphocytes.
A schematic representation of the interactions that can be involved in the activation of B lymphocytes.

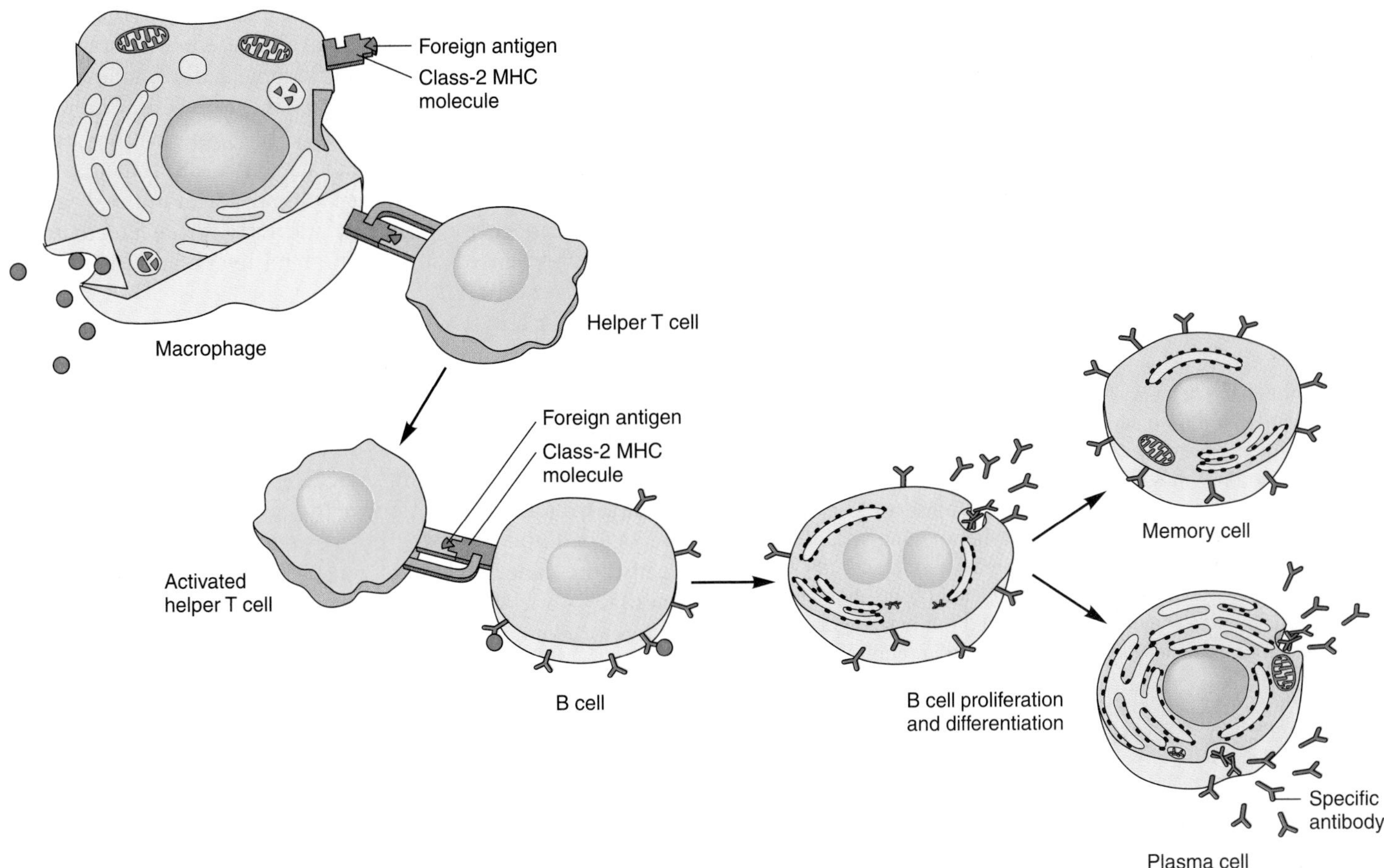

harbor molecules that the immune system would mistakenly treat as foreign antigens if the site were not somehow protected. The sustentacular (Sertoli) cells of the testicular tubules (see chapter 28), for example, protect the developing sperm from immune attack through two mechanisms. First, the tight junctions between adjacent sustentacular cells form a blood-testis barrier that normally prevents exposure of the immune system to the developing sperm. Second, these cells produce FAS ligand, which triggers apoptosis of any T lymphocytes that may enter the area.

Some tumor cells, unfortunately, have also been found to produce FAS ligand, which may defend the tumor from immune attack by triggering the apoptosis of lymphocytes. The role of the immune system in the defense against cancer is discussed in the next major section.

Immunological Tolerance

The ability to produce antibodies against **non-self** (foreign) **antigens,** while tolerating (not producing antibodies against) **self-antigens** occurs during the first month or so of postnatal life, when immunological competence is established. If a fetal mouse of one strain receives transplanted antigens from a different strain, therefore, it will not recognize tissue transplanted later in life from the other strain as foreign; consequently, it will not immunologically reject the transplant.

The ability of an individual's immune system to recognize and tolerate self-antigens requires continuous exposure of the immune system to those antigens. If this exposure begins when the immune system is weak—such as in fetal and early postnatal life—tolerance is more complete and long lasting than that produced by exposure beginning later in life. Some self-antigens, however, are normally hidden from the blood, such as thyroglobulin within the thyroid gland and lens protein in the eye. An exposure to these self-antigens results in antibody production just as if these proteins were foreign. Antibodies made against self-antigens are called **autoantibodies.** Killer T cells that attack self-antigens are called **autoreactive T cells.**

Although the mechanisms are not well understood, two general types of theories have been proposed to account for immunological tolerance: **clonal deletion** and **clonal anergy.** According to the clonal deletion theory, tolerance to self-antigens is achieved by destruction of the lymphocytes that recognize self-antigens. This occurs primarily during fetal life, when those lymphocytes that have receptors on their surface for self-antigens are recognized and destroyed. There is much evidence for clonal deletion in the thymus, and this mechanism is believed to be largely responsible for T cell tolerance. Anergy (which means "without working") occurs when lymphocytes directed against self-antigens are present throughout life but, for complex and poorly understood reasons, do not attack those antigens. Clonal anergy is believed to be largely responsible for tolerance in B cells, and there is some evidence that it may also contribute to tolerance in T cells.

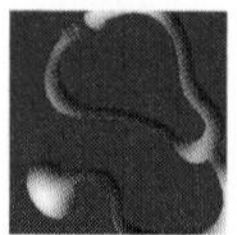

Glucocorticoids (such as hydrocortisone), secreted by the adrenal cortex, can act to suppress the activity of the immune system and inflammation. This is why *cortisone* and its analogues are used clinically to treat inflammatory disorders and to inhibit the immune rejection of transplanted organs. The immunosuppressive effect of these hormones may result from the fact that they inhibit the secretion of the cytokines. It is interesting in this regard that interleukin-1 (IL-1), which can be produced by microglia in the brain, has been shown to stimulate the pituitary-adrenal axis by promoting CRH, ACTH, and glucocorticoid secretion (chapter 19). In a negative feedback fashion, the glucocorticoids then inhibit the immune system and suppress the production of the inflammatory cytokines, including IL-1, IL-2, and TNF_{α}. These and related observations have opened up a new scientific field devoted to the study of interactions among the nervous, endocrine, and immune systems.

Tumor Immunology

Tumor cells can reveal antigens that stimulate the destruction of the tumor. When cancers develop, this immunological surveillance system—primarily the function of T cells and natural killer cells—has failed to prevent the growth and metastasis of the tumor.

Oncology (the study of tumors) has revealed that tumor biology is similar to and interrelated with the functions of the immune system. Most tumors appear to be clones of single cells that have become transformed in a process similar to the development of lymphocyte clones in response to specific antigens. Lymphocyte clones, however, are under complex inhibitory control systems—such as those exerted by suppressor T lymphocytes and negative feedback by antibodies. The division of tumor cells, by contrast, is not effectively controlled by normal inhibitory mechanisms. Tumor cells are also relatively unspecialized—they *dedifferentiate*, which means that they become similar to the less specialized cells of an embryo.

Tumors are described as *benign* when they are relatively slow growing and limited to a specific location (warts, for example). *Malignant* tumors grow more rapidly and undergo **metastasis** (*mĕ-tas′tă-sis*), a term that refers to the dispersion of tumor cells and the resultant seeding of new tumors in different locations. The term **cancer**, as it is generally applied, refers to malignant tumors.

As tumor cells dedifferentiate, they reveal surface antigens that can stimulate the immune destruction of the tumor. Consistent with the concept of dedifferentiation, some of these antigens are proteins produced in embryonic or fetal life and are not normally produced postnatally. Since they are absent at the time immunological competence is established, they are treated as foreign and fit subjects for immunological attack when they are produced by cancerous cells. The release of two such antigens into the blood has provided the basis for laboratory diagnosis of some cancers. *Carcinoembryonic* (*kar″sĭ-no-em″bre-on′ik*) *antigen tests* are useful in the diagnosis of colon cancer, for example, and tests for *alpha-fetoprotein* (normally produced only by the fetal liver) help in the diagnosis of liver cancer.

Tumor antigens activate the immune system, initiating an attack primarily by killer T lymphocytes (fig. 23.28) and natural killer cells (described in the next section). The concept of **immunological surveillance** against cancer was introduced in the early 1970s to describe the proposed role of the immune system in fighting cancer. According to this concept, tumor cells frequently appear in the body but are normally recognized and destroyed by the immune system before they can cause cancer. There is evidence that immunological surveillance does prevent some types of cancer; this explains why, for example, people with AIDS (who have a depressed immune system) have a high incidence of Kaposi's sarcoma. It is not clear, however, why all types of cancers do not appear with high frequency in AIDS patients and in others whose immune systems are suppressed. For these reasons, the generality of the immunological surveillance system concept is currently controversial.

Immunotherapy for Cancer

The production of human interferons by genetically engineered bacteria has made large amounts of these substances available for the experimental treatment of cancer. Thus far, interferons have proven to be a useful addition to the treatment of particular forms of cancer, including some types of lymphomas, renal carcinoma, melanoma, Kaposi's sarcoma, and breast cancer. They have not, however, proved to be the "magic bullet" against cancer (a term coined by Paul Ehrlich) as had previously been hoped.

A team of scientists at the National Cancer Institute has pioneered the use of another lymphokine that is now available through genetic engineering techniques. This is *interleukin-2 (IL-2)*, which activates both killer T lymphocytes and B lymphocytes. The investigators removed some of the blood from cancer patients who could not be successfully treated by conventional means and isolated a population of their lymphocytes. They treated these lymphocytes with IL-2 to produce *lymphokine-activated killer (LAK) cells* and then reinfused these cells, together with IL-2 and interferons, into the patients. Depending on the combinations and dosages,

cancer: L. *cancer*, a crab

Paul Ehrlich: German bacteriologist, 1854–1915

Figure 23.28

T cell destruction of a cancer cell.
A killer T cell (*a*) contacts a cancer cell (the larger cell) in a manner that requires specific interaction with antigens on the cancer cell. The killer T cell releases lymphokines, including toxins that cause the death of the cancer cell, as shown in (*b*).

Scanning electron micrographs © Andrejs Liepens.

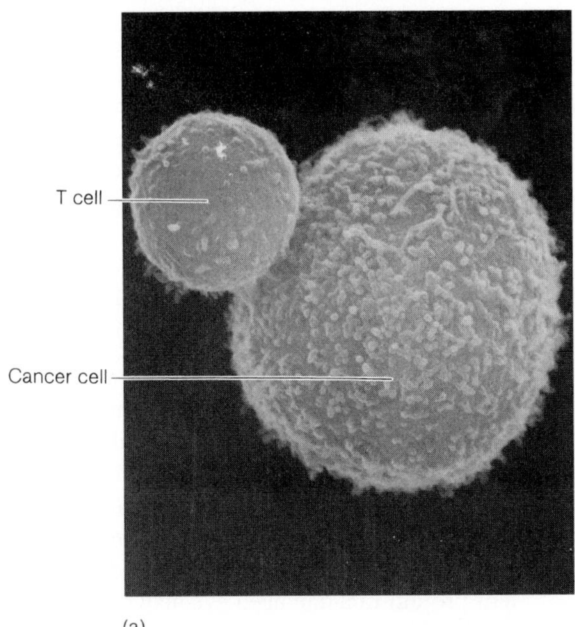

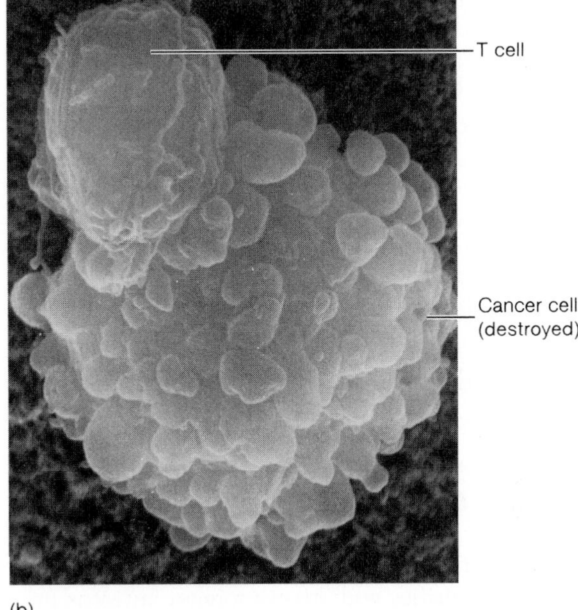

they achieved remarkable success (but not a complete cure for all cancers) in many of these patients.

The research group next identified a subpopulation of lymphocytes that had invaded solid tumors in mice. These *tumor-infiltrating lymphocyte (TIL) cells* were allowed to replicate in tissue culture, whereupon they were reintroduced into the mice with excellent results. Recently, the same techniques were used to treat an experimental group of people with metastatic melanoma, a cancer that claims the lives of 6,000 Americans annually. The patients were first given conventional chemotherapy and radiation therapy. They were then treated with their own TIL cells and interleukin-2. Some of the preliminary results of this treatment are encouraging, but IL-2, like gamma interferon, does not appear to be a magic bullet against cancer.

Besides interleukin-2 and gamma interferon, other cytokines may be useful in the treatment of cancer and are currently undergoing experimental investigations. Interleukin-12, for example, seems promising because it is needed for the changing of uncommitted helper T lymphocytes into the T_H1 subtype that bolsters cell-mediated immunity. Scientists are also attempting to identify specific antigens that may be uniquely expressed in cancer cells in an effort to help the immune system target cancer cells for destruction.

Natural Killer Cells

In a particular strain of hairless mice, a thymus and T lymphocytes are genetically lacking, yet these mice do not appear to have an especially high incidence of tumor production. This surprising observation led to the discovery of **natural killer (NK) cells,** which are lymphocytes that are related to, but different from, T lymphocytes. Unlike killer T cells, NK cells destroy tumors in a nonspecific fashion and do not require prior exposure for sensitization to the tumor antigens. The NK cells thus provide a first line of cell-mediated defense, which is subsequently backed up by a specific response mediated by killer T cells. These two cell types interact, however; the activity of NK cells is stimulated by interferon, released as one of the lymphokines from T lymphocytes.

Effects of Aging and Stress

Susceptibility to cancer varies greatly. The Epstein–Barr virus that causes Burkitt's lymphoma in some individuals (mainly in Africa), for example, can also be found in healthy people throughout the world. Most often the virus is harmless; in some cases, it causes mononucleosis (involving a limited proliferation of white blood cells). Only rarely does this virus cause the uncontrolled proliferation of leukocytes characteristic of Burkitt's lymphoma. The reasons for these differences in response to the Epstein–Barr virus, and indeed for the differing susceptibilities of people to other forms of cancer, are not well understood.

Epstein–Barr virus: from Michael A. Epstein, British physician, b. 1921, and Yvonne M. Barr, twentieth-century British virologist
Burkitt's lymphoma: from Denis P. Burkitt, Ugandan physician, b. 1911

Cancer risk increases with age. According to one theory, this is because aging lymphocytes gradually accumulate genetic errors that decrease their effectiveness. The secretion of thymus hormones also decreases with age in parallel with a decrease in cell-mediated immune competence. Both of these changes, and perhaps others not yet discovered, could increase susceptibility to cancer.

Numerous experiments have demonstrated that tumors grow faster in experimental animals subjected to stress than they do in unstressed control animals. This is generally attributed to the fact that stressed animals, including humans, exhibit increased secretion of corticosteroid hormones that act to suppress the immune system (which is why cortisone is given to people who receive organ transplants and to people with chronic inflammatory diseases). Some recent experiments, however, suggest that the stress-induced suppression of the immune system may also be due to other factors that do not involve the adrenal cortex. Future advances in cancer therapy may incorporate methods of strengthening the immune system into protocols aimed at directly destroying tumors.

Clinical Considerations

The ability of the normal immune system to tolerate self-antigens while it identifies and attacks foreign antigens provides a specific defense against invading pathogens. In every individual, however, this system of defense against invaders at times commits domestic offenses. This can result in diseases that range in severity from the sniffles to sudden death.

Diseases caused by the immune system can be grouped into three interrelated categories: (1) autoimmune diseases, (2) immune complex diseases, and (3) allergy, or hypersensitivity. It is important to remember that these diseases are not caused by foreign pathogens but by abnormal responses of the immune system.

Autoimmunity

Autoimmune diseases are those produced by failure of the immune system to recognize and tolerate self-antigens. This failure results in the activation of autoreactive T cells and the production of autoantibodies by B cells, causing inflammation and organ damage (table 23.11). There are over 40 known or suspected autoimmune diseases that affect 5% to 7% of the population. Two-thirds of those affected are women.

There are at least five reasons why self-tolerance may fail:

1. **An antigen that does not normally circulate in the blood may become exposed to the immune system.** Thyroglobulin protein that is normally trapped within the thyroid follicles, for example, can stimulate the production of autoantibodies that cause the destruction of the thyroid (fig. 23.29); this occurs in *Hashimoto's thyroiditis*. Similarly, autoantibodies developed against lens protein in a damaged eye may cause the destruction of a healthy eye (in *sympathetic ophthalmia*).

Hashimoto's thyroiditis: from Hakaru Hashimoto, Japanese surgeon, 1881–1934

Table 23.11

Examples of Autoimmune Diseases

Disease	Antigen	Ig and/or T Cell Response
Postvaccinal and postinfectious encephalomyelitis	Myelin, cross-reactive	T cell
Aspermatogenesis	Sperm	T cell
Sympathetic ophthalmia	Uvea	T cell
Hashimoto's thyroiditis	Thyroglobulin	IgG and T cell
Graves' disease	Receptor proteins for TSH	Thyroid-stimulating antibody (TSAb)
Autoimmune hemolytic disease	I, Rh, and others on surface of RBCs	IgM and IgG
Thrombocytopenic purpura	Hapten-platelet or hapten-absorbed antigen complex	IgG
Myasthenia gravis	Acetylcholine receptors	IgG
Rheumatic fever	Streptococcal, cross-reactive with heart	IgG and IgM
Glomerulonephritis	Streptococcal, cross-reactive with kidney	IgG and IgM
Rheumatoid arthritis	IgG	IgM to $F_c(\gamma)$
Systemic lupus erythematosus	DNA, nucleoprotein, RNA, etc.	IgG

Source: From James T. Barrett, Textbook of Immunology, 5th ed. Copyright © 1988 Mosby/Yearbook. Reprinted by permission.

2. **A self-antigen that is otherwise tolerated may be altered by combining with a foreign hapten.** The disease *thrombocytopenia* (low platelet count), for example, can be caused by the autoimmune destruction of platelets. This occurs when drugs such as aspirin, sulfonamides, antihistamines, digoxin, and others combine with platelet proteins to produce new antigens. The symptoms of this disease usually disappear when the person stops taking these drugs.
3. **Antibodies that are directed against other antibodies may be produced.** Such interactions may be necessary for the prevention of autoimmunity, but imbalances may actually cause autoimmune diseases. *Rheumatoid arthritis*, for example, is an autoimmune disease associated with the abnormal production of one group of antibodies (of the IgM type) that attack other antibodies (of the IgG type). This contributes to an inflammation reaction of the joints characteristic of the disease.
4. **Antibodies produced against foreign antigens may cross-react with self-antigens.** Autoimmune diseases of this sort can occur, for example, as a result of *Streptococcus* bacterial infections. Antibodies produced in response to antigens in this bacterium may cross-react with self-antigens in the heart and kidneys. The inflammation induced by such autoantibodies can produce heart damage (including the valve defects characteristic of *rheumatic fever*) and damage to the glomerular capillaries in the kidneys (*glomerulonephritis*).
5. **Self-antigens, such as receptor proteins, may be presented to the helper T lymphocytes together with class-2 MHC molecules.** Normally, only antigen-presenting cells (macrophages, dendritic cells, and antigen-activated B cells) produce class-2 MHC molecules, which are associated with foreign antigens and recognized by helper T cells. Perhaps as a result of viral infection, however, cells that do not normally produce class-2 MHC molecules may start to do so and, in this way, present a self-antigen to the helper T cells. In *Graves' disease* (chapter 19), for example, the thyroid cells produce class-2 MHC molecules, and the immune system produces autoantibodies against the TSH receptor proteins in the thyroid cells. These autoantibodies, called *TSAbs* for "thyroid-stimulating antibody," interact with the TSH receptors and overstimulate the thyroid gland. Similarly, in *type I diabetes mellitus*, the beta cells of the pancreatic islets abnormally produce class-2 MHC molecules, resulting in autoimmune destruction of the insulin-producing cells.

Immune Complex Diseases

The term *immune complexes* refers to antigen-antibody combinations that are free rather than attached to bacterial or other cells. The formation of such complexes activates complement proteins and promotes inflammation. This inflammation is normally self-limiting because the immune complexes are removed by phagocytic cells. When large numbers of immune complexes are continuously formed, however, the inflammation may be prolonged. Also, the dispersion of immune complexes to other sites can lead to widespread inflammation and organ damage. The damage produced by this inflammatory response is called **immune complex disease.**

Immune complex diseases can result from infections by bacteria, parasites, and viruses. In hepatitis B, for example, an immune complex that consists of viral antigens and antibodies can cause widespread inflammation of arteries (*periarteritis*). Arterial damage is not caused by the hepatitis virus itself but by the inflammatory process.

Figure 23.29

Autoimmune thyroiditis.
This was induced in a rabbit by injection with thyroglobulin. Compare the picture of a normal thyroid (*a*) with that of the diseased thyroid (*b*). The grainy appearance of the diseased thyroid is due to the infiltration of large numbers of lymphocytes and macrophages.

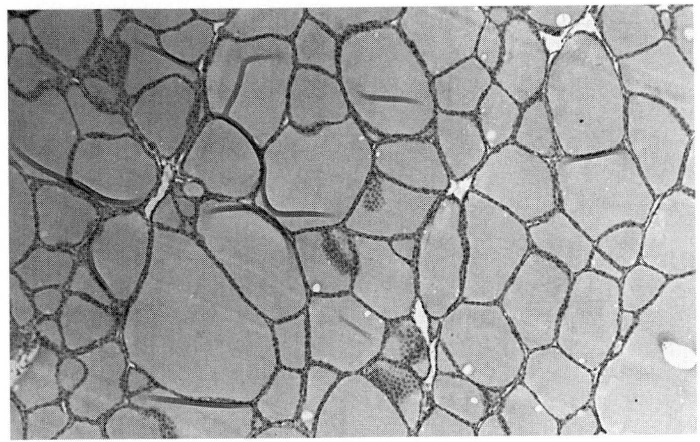

(a)

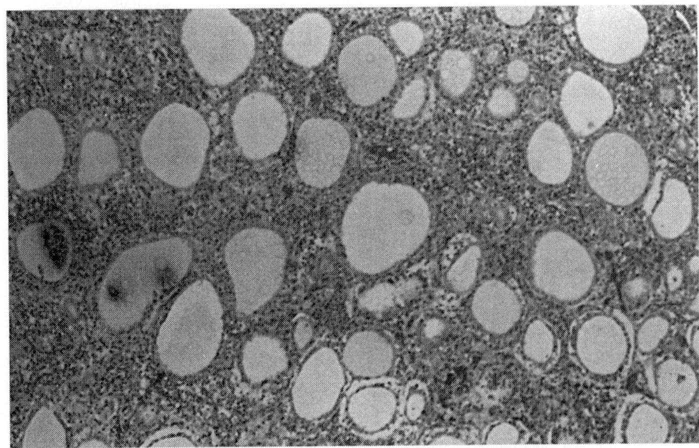

(b)

Immune complex diseases can also result from the formation of complexes between self-antigens and autoantibodies. This is the case in rheumatoid arthritis, where the inflammation is produced by complexes of altered IgG antibodies (the antigens in this case) and IgM antibodies. Another immune complex disease that has an autoimmune basis is *systemic lupus erythematosus (SLE)*. People with SLE produce antibodies against their own DNA and nuclear proteins. This can result in the formation of immune complexes throughout the body, including the glomerular capillaries, where glomerulonephritis may be produced.

Allergy

The term *allergy*, often used interchangeably with *hypersensitivity*, refers to particular types of abnormal immune responses to antigens, which in these cases are called *allergens*. There are two major forms of allergy: (1) immediate hypersensitivity, which is due to an abnormal B lymphocyte response to an allergen that produces symptoms within seconds or minutes, and (2) delayed hypersensitivity, which is an abnormal T cell response that produces symptoms between 24 and 72 hours after exposure to an allergen. These two types of hypersensitivity are compared in table 23.12.

Immediate Hypersensitivity

Immediate hypersensitivity can produce allergic rhinitis (chronic runny or stuffy nose), conjunctivitis (red eyes), allergic asthma, atopic dermatitis (urticaria, or hives), and other symptoms. These symptoms result from the production of antibodies of the IgE subclass instead of the normal IgG antibodies.

Table 23.12

Allergy: Comparison of Immediate and Delayed Hypersensitivity Reactions

Characteristic	Immediate Reaction	Delayed Reaction
Time for onset of symptoms	Within several minutes	Within a period of 1 to 3 days
Lymphocytes involved	B lymphocytes	T lymphocytes
Immune effector	IgE antibodies	Cell-mediated immunity
Allergies most commonly produced	Hay fever, asthma, and most other allergic conditions	Contact dermatitis (such as to poison ivy and poison oak)
Therapy	Antihistamines and adrenergic drugs	Corticosteroids (such as cortisone)

Unlike IgG antibodies, IgE antibodies do not circulate in the blood. Instead, they attach to tissue mast cells and basophils, which have membrane receptors for these antibodies. When the person is again exposed to the same allergen, the allergen binds to the antibodies attached to the mast cells and basophils. This stimulates these cells to secrete various chemicals, including **histamine** (fig. 23.30). During this process, leukocytes may also secrete **prostaglandin D** and related molecules called **leukotrienes** (*loo″kŏ-tri′ēnz*). These chemicals produce the symptoms of the allergic reactions. It should be noted that histamine stimulates smooth muscle contraction in the respiratory tract but causes smooth muscle relaxation in the walls of blood vessels. The different effects are due to differences in the histamine receptors of these target tissues.

The symptoms of hay fever (itching, sneezing, tearing, runny nose) are produced largely by histamine and can be treated effectively by antihistamine drugs. Food allergies, causing diarrhea and colic, are mediated primarily by prostaglandins and can be treated with aspirin, which inhibits prostaglandin synthesis (these are the only allergies that respond positively to aspirin). In a certain type of asthma, the difficulty in breathing is caused by inflammation and smooth muscle constriction in the bronchioles in the lungs as a result of leukotrienes and other molecules released in an allergic reaction. Asthma is treated with epinephrine and more specific β-adrenergic stimulating drugs (chapter 17), which cause bronchodilation, and with corticosteroids, which inhibit inflammation and leukotriene synthesis. Asthma and its treatment are discussed more fully in chapter 24.

Immediate hypersensitivity to a particular antigen is commonly tested for by injecting various antigens under the skin (fig. 23.31). Within a short time, a *flare-and-wheal reaction* is produced if the person is allergic to that antigen. This reaction is due to the release of histamine and other chemical mediators: the flare (spreading flush) is due to vasodilation, and the wheal (elevated area) results from local edema.

Allergens that provoke immediate hypersensitivity include various foods, bee stings, and pollen grains. The most common allergy of this type is seasonal hay fever, which may be provoked by ragweed (*Ambrosia*) pollen grains (fig. 23.32*a*). People who have chronic allergic rhinitis and asthma due to an allergy to dust or feathers are usually allergic to a tiny mite (fig. 23.32*b*) that lives in dust and eats the scales of skin that are constantly shed from the body. Actually, most of the antigens from the dust mite are not in its body but rather in its feces—tiny particles that can enter the nasal mucosa, much like pollen grains.

Delayed Hypersensitivity

In **delayed hypersensitivity,** as the name implies, symptoms take a longer time (hours to days) to develop than in immediate hypersensitivity. This may be because immediate hypersensitivity is mediated by antibodies, whereas delayed hypersensitivity is a cell-mediated T lymphocyte response. Since

Figure 23.30

The mechanism of immediate hypersensitivity.
Allergy (immediate hypersensitivity) is produced when antibodies of the IgE subclass attach to tissue mast cells. The combination of these antibodies with allergens (antigens that provoke an allergic reaction) causes the mast cell to secrete histamine and other chemicals that produce the symptoms of allergy.

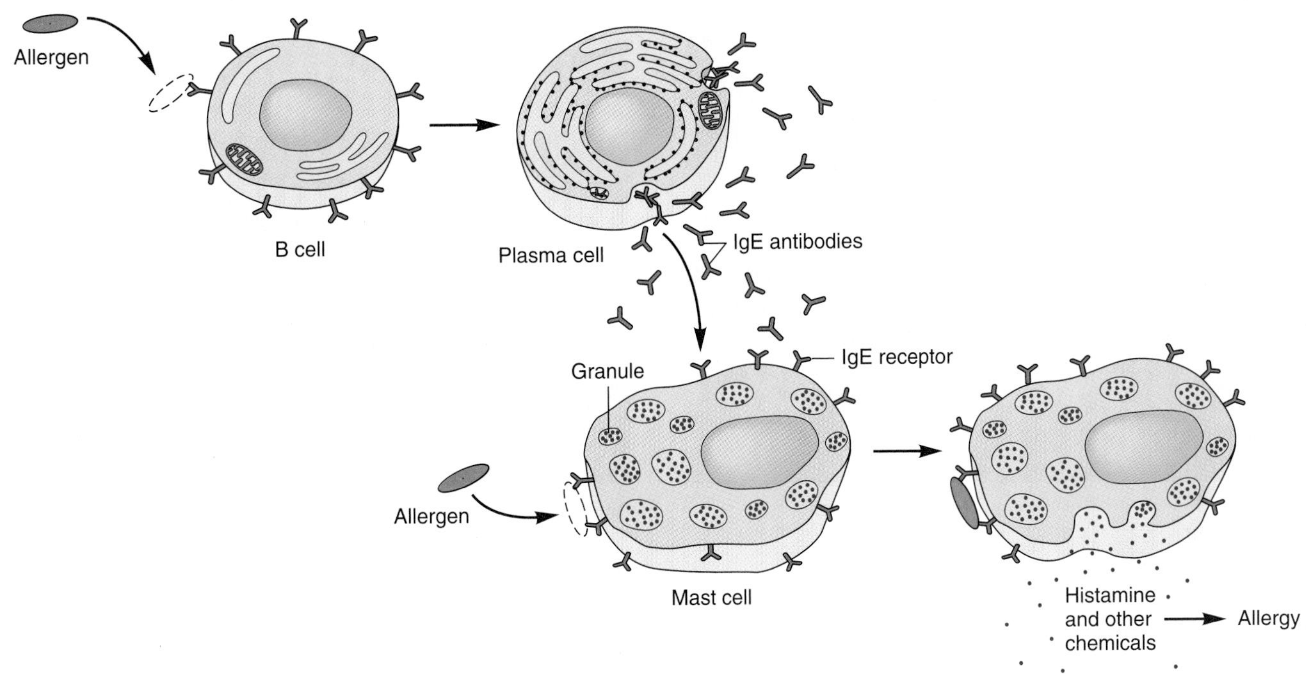

Figure 23.31

A skin test for allergy.
If an allergen is injected into the skin of a sensitive individual, a typical flare-and-wheal response occurs within several minutes.

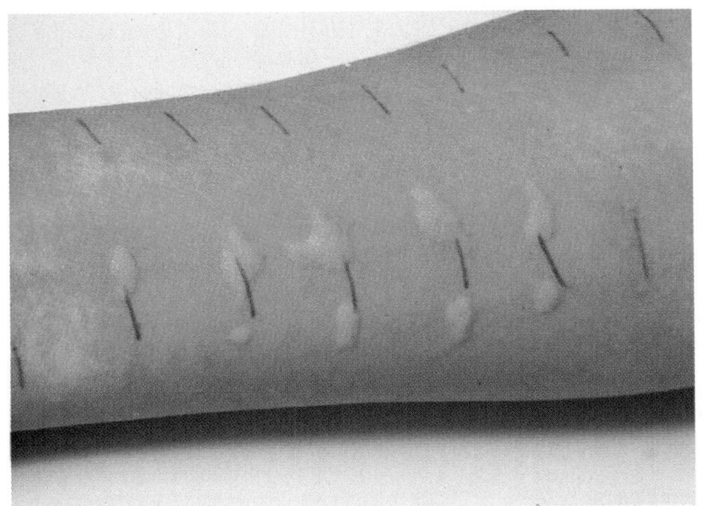

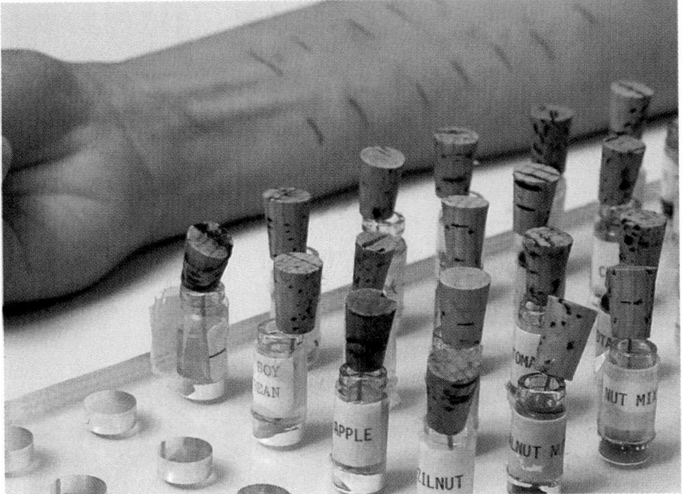

Figure 23.32

Common allergens.
(*a*) A scanning electron micrograph of ragweed (*Ambrosia* pollen), which is responsible for hay fever. (*b*) A scanning electron micrograph of the house dust mite (*Dermatophagoides farinae*). Waste-product particles produced by the dust mite are often responsible for chronic allergic rhinitis and asthma.
Part [a]: R. G. Kessel and C. Y. Shih. SCANNING ELECTRON MICROSCOPY IN BIOLOGY. A STUDENT'S ATLAS ON BIOLOGICAL ORGANIZATION. 1976 Springer-Verlag, Berlin.

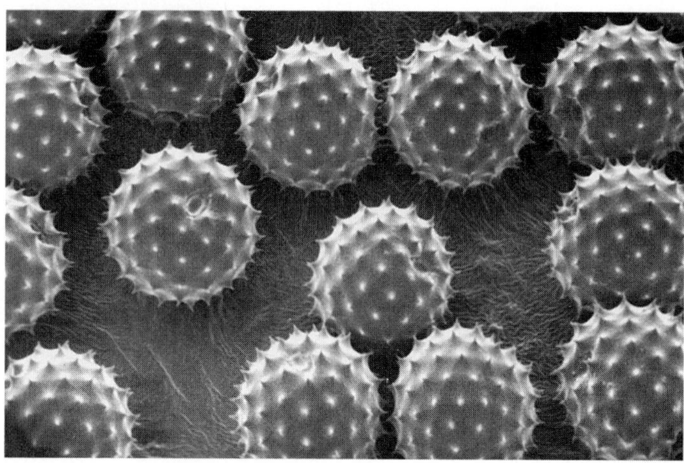
(a)

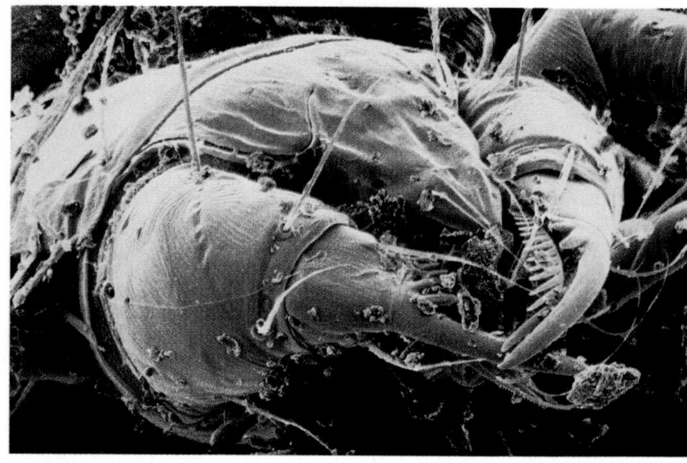
(b)

the symptoms are caused by the secretion of lymphokines rather than by the secretion of histamine, treatment with antihistamines provides little benefit. At present, corticosteroids are the only drugs that can effectively treat delayed hypersensitivity.

One of the best-known examples of delayed hypersensitivity is **contact dermatitis,** caused by poison ivy, poison oak, and poison sumac. The skin tests for tuberculosis—the tine test and the Mantoux (*man-too'*) test—also rely on delayed hypersensitivity reactions. If a person has been exposed to the tubercle bacillus and consequently has developed T cell clones, skin reactions appear within a few days after the tubercle antigens are rubbed into the skin with small needles (tine test) or are injected under the skin (Mantoux test).

Mantoux test: from Charles Mantoux, French physician, 1877–1947

Clinical Investigation Answer

While crawling through the underbrush the little girl may have been exposed to poison oak, causing a contact dermatitis. Since this is a delayed hypersensitivity response mediated by T cells, antihistamines would not have alleviated the symptoms. Cortisone helped, however, because of its immunosuppressive effect. The first bee sting did not have much of an effect, but it served to sensitize the girl (through the development of B cell clones) to the second bee sting. The second sting resulted in an immediate hypersensitivity response (mediated by IgE), which caused the release of histamine. This allergic reaction could thus be treated effectively with antihistamines.

Chapter Summary

Lymphatic System (pp. 693–696)

1. Lymphatic capillaries drain excess interstitial fluid and are highly permeable to proteins and particles such as microorganisms.
2. Lymph ducts receive lymph from the lymph capillaries and eventually empty into the thoracic duct and right lymphatic duct, which in turn empty into the left and right subclavian veins, respectively.
3. Lymph nodes filter the lymph and contain germinal centers for lymphocyte production.
4. Other lymphoid organs include the spleen, thymus, and tonsils.
   (a) The spleen also filters the blood and destroys old red blood cells.
   (b) The thymus plays a key role in the immune system.
   (c) The tonsils protect against foreign substances that are inhaled or ingested.

Defense Mechanisms (pp. 696–701)

1. Nonspecific defense mechanisms include barriers to penetration of the body, as well as internal defenses.
2. Specific immune responses are directed against antigens.
   (a) Antigens are molecules, or parts of molecules, that are usually large, complex, and foreign.
   (b) A given molecule can have a number of antigenic determinant sites that stimulate the production of different antibodies.

3. Specific immunity is a function of lymphocytes.
   (a) B lymphocytes secrete antibodies and provide humoral, or antibody-mediated, immunity.
   (b) T lymphocytes provide cell-mediated immunity.

Functions of B Lymphocytes (pp. 701–707)

1. There are five subclasses of antibodies, or immunoglobulins: IgG, IgA, IgM, IgD, and IgE.
2. Antigen-antibody complexes activate a system of proteins called the complement system.
   (a) Soluble complement proteins released into the area attract leukocytes by chemotaxis.
   (b) Fixed complement in the cell membrane forms pores that kill the victim cell.
3. Specific and nonspecific immune mechanisms cooperate in the development of a local inflammation.

Active and Passive Immunity (pp. 708–712)

1. A primary response is produced when a person is first exposed to a pathogen. A subsequent exposure results in a secondary response.
2. The secondary response is believed to be due to lymphocyte clones that develop after the first exposure as a result of the antigen-stimulated proliferation of appropriate lymphocytes.
3. Passive immunity is provided by transfer of antibodies from an immune to a nonimmune organism.
   (a) Passive immunity occurs naturally in the transfer of antibodies from mother to fetus.
   (b) Injections of antiserum provide passive immunity to some pathogenic organisms and toxins.
4. Monoclonal antibodies are made by hybridomas, which are formed artificially by the fusion of B lymphocytes with multiple myeloma cells.

Functions of T Lymphocytes (pp. 712–718)

1. The thymus processes T lymphocytes and secretes hormones that are believed to be required for an effective immune response by T lymphocytes throughout the body.
2. There are three subcategories of T lymphocytes.
   (a) Killer T lymphocytes are responsible for transplant rejection and for the immunological defense against fungal and viral infections, as well as for the defense against some bacterial infections.
   (b) Helper T lymphocytes stimulate, and suppressor T lymphocytes suppress, the function of B lymphocytes and killer T lymphocytes.
   (c) The T lymphocytes secrete a family of compounds called lymphokines, which promote the action of lymphocytes and macrophages.
   (d) Receptor proteins on the cell membrane of T lymphocytes must bind to a foreign antigen in combination with a MHC molecule in order for the T lymphocyte to become activated.
3. Macrophages and dendritic cells partially digest a foreign body, such as a virus, and present the antigens to the lymphocytes on the surface of the macrophage in combination with class-2 MHC molecules.
4. Interleukin-2 stimulates proliferation of killer T lymphocytes that are specific for the foreign antigen.
5. Tolerance to self-antigens may be due to the destruction of lymphocytes that can recognize the self-antigens, or it may be due to suppression of the immune response by the action of specific suppressor T lymphocytes.

Tumor Immunology (pp. 718–720)

1. Immunological surveillance against cancer is provided mainly by killer T lymphocytes and natural killer cells.
2. Natural killer cells are nonspecific, whereas T lymphocytes are directed against specific antigens on the cancer cell surface.

Review Activities

Objective Questions

1. Which of the following offers a nonspecific defense against viral infection?
   (a) antibodies
   (b) leukotrienes
   (c) interferon
   (d) histamine

Match the cell type with its secretion.

2. killer T cells	(a) antibodies
3. mast cells	(b) perforins
4. plasma cells	(c) lysosomal enzymes
5. macrophages	(d) histamine

6. Which of the following statements about the F_{ab} portion of antibodies is *true?*
   (a) It binds to antigens.
   (b) Its amino acid sequences are variable.
   (c) It consists of both H and L chains.
   (d) All of the above are true.
7. Which of the following statements about complement proteins $C3_a$ and $C5_a$ is *false?*
   (a) They are released during the complement fixation process.
   (b) They stimulate chemotaxis of phagocytic cells.
   (c) They promote the activity of phagocytic cells.
   (d) They produce pores in the victim cell membrane.
8. Mast cell secretion during an immediate hypersensitivity reaction is stimulated when antigens combine with
   (a) IgG antibodies.
   (b) IgE antibodies.
   (c) IgM antibodies.
   (d) IgA antibodies.
9. During a secondary immune response,
   (a) antibodies are made quickly and in great amounts.
   (b) antibody production lasts longer than in a primary response.
   (c) antibodies of the IgG class are produced.
   (d) lymphocyte clones are believed to develop.
   (e) all of the above apply.
10. Which of the following cell types aids the activation of T lymphocytes by antigens?
    (a) macrophages
    (b) neutrophils
    (c) mast cells
    (d) natural killer cells
11. Which of the following statements about T lymphocytes is *false?*
    (a) Some T cells promote the activity of B cells.
    (b) Some T cells suppress the activity of B cells.
    (c) Some T cells secrete interferon.
    (d) Some T cells produce antibodies.
12. Delayed hypersensitivity is mediated by
    (a) T cells.
    (b) B cells.
    (c) plasma cells.
    (d) natural killer cells.

13. Active immunity may be produced by
    (a) contracting a disease.
    (b) receiving a vaccine.
    (c) receiving gamma globulin injections.
    (d) both *a* and *b*.
    (e) both *b* and *c*.
14. Which of the following statements about class-2 MHC molecules is *false?*
    (a) They are found on the surface of B lymphocytes.
    (b) They are found on the surface of macrophages.
    (c) They are required for B cell activation by a foreign antigen.
    (d) They are needed for interaction of helper and killer T cells.
    (e) They are presented together with foreign antigens by macrophages.

Match the cytokine with its description.

15. interleukin-1
16. interleukin-2
17. interleukin-12

(a) stimulates formation of T_H1 helper T lymphocytes
(b) stimulates ACTH secretion
(c) stimulates proliferation of killer T lymphocytes
(d) stimulates proliferation of B lymphocytes

18. Which of the following statements about gamma interferon is *false?*
    (a) It is a polypeptide autocrine regulator.
    (b) It can be produced in response to viral infections.
    (c) It stimulates the immune system to attack infected cells and tumors.
    (d) It is produced by almost all cells in the body.

Essay Questions

1. Explain how antibodies help to destroy invading bacterial cells.
2. Identify the different types of interferons and describe their origin and actions.
3. Distinguish between the class-1 and class-2 MHC molecules in terms of their locations and functions.
4. Describe the role of macrophages in activating the specific immune response to antigens.
5. Distinguish between the two subtypes of helper T lymphocytes and explain how they may be produced.
6. Describe how plasma cells attack antigens and how they can destroy an invading foreign cell. Compare this mechanism with that by which killer T lymphocytes destroy a target cell.
7. Explain how tolerance to self-antigens may be produced. Also, give two examples of autoimmune diseases and explain their possible causes.
8. Use the clonal selection theory to explain how active immunity is produced by vaccinations.
9. Describe the nature of passive immunity and explain how antitoxins are produced and used.
10. Distinguish between immediate and delayed hypersensitivity. What drugs are used to treat immediate hypersensitivity, and how do these drugs work? Why don't these compounds work in treating delayed hypersensitivity?

Critical Thinking Questions

1. The specific T lymphocyte immune response is usually directed against proteins, whereas nonspecific immune mechanisms are generally directed against foreign carbohydrates in the form of glycoproteins and lipopolysaccharides. How might these differences in target molecules be explained?
2. Lizards are cold-blooded; their body temperature is largely determined by the ambient temperature. Devise an experiment using lizards to test whether an elevated body temperature, as in a fever, can be beneficial to an organism with an infection.
3. Why are antibodies composed of different chains, and why are there several genes that encode the parts of a particular antibody molecule? What would happen if each antibody were coded for by only one gene?
4. As a scientist trying to cure allergy, you are elated to discover a drug that destroys all mast cells. How might this drug help to prevent allergy? Would the drug have negative side effects? Explain.
5. The part of the placenta that invades the mother's uterine lining (the endometrium) has recently been found to produce FAS ligand. What might this accomplish, and why might this action be necessary?

Related Web Sites

For a listing of the most current web sites related to this chapter, please visit the *Concepts of Human Anatomy and Physiology* home page at http://www.mhhe.com/biosci/abio/.

NEXUS

Some Interactions of the Lymphatic and Immune Systems with the Other Body Systems

Integumentary System

- The skin serves as a first line of defense from invasion of pathogens (p. 696).
- Lymphatic vessels drain excess interstitial fluid from the dermis and other connective tissues (p. 693).

Skeletal System

- Hermatopoiesis, including formation of leukocytes involved in immunity, occurs in the bone marrow (p. 596).
- The immune system protects all systems, including the skeletal system, against infection (p. 696).

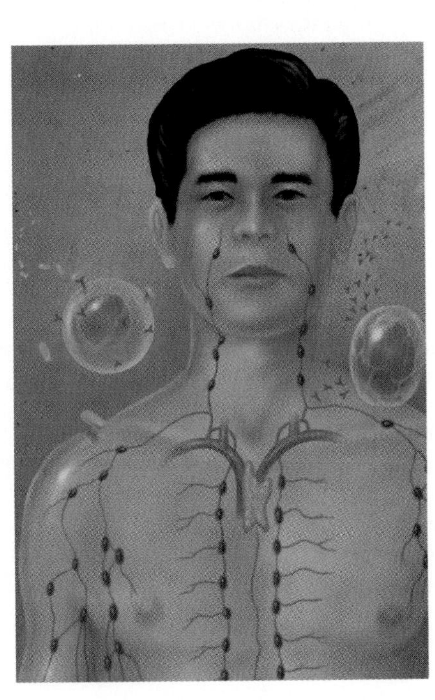

Nervous System

- Neural regulation of the pituitary and adrenal glands indirectly influences activity of the immune system (p. 718).
- Nerves regulate blood flow to most organs, including the lymphatic organs (p. 667).

Endocrine System

- The pituitary and adrenal glands influence immune function (p. 718).
- The thymus regulates production of T lymphocytes (p. 712).

Muscular System

- Cardiac muscle in the heart pumps blood to the body organs, including those of the immune system (p. 611).
- The smooth muscle of blood vessels helps to regulate the blood flow to areas of infection (p. 670).

Circulatory System

- The circulatory system transports neutrophils, monocytes, and lymphocytes to infected areas (p. 706).
- Hematopoiesis generates the cells required for the immune response (p. 596).

Respiratory System

- The lungs provide oxygen for transport by the blood and eliminate carbon dioxide from the blood (p. 729).
- Alveolar macrophages help to combat pulmonary infections (p. 735).

Urinary System

- The kidneys regulate the volume, pH, and electrolyte balance of the blood and eliminate wastes (p. 779).
- The immune system protects against infection of the urinary tract (p. 696).

Digestive System

- The GI tract provides nutrients to be transported by the circulatory system to all body cells, including those of the immune system (p. 817).
- Lymphatic vessels carry absorbed fat, which enters lymphatic capillaries called lacteals (p. 855).
- Areas of the GI tract contain many lymphocytes and lymphatic nodules (p. 696).

Reproductive System

- The blood-testes barrier prevents sperm cell antigens from provoking an autoimmune response (p. 906).
- The placenta is an immunologically privileged site that is normally protected against immune attack (p. 716).
- A mother's breast milk provides antibodies that passively immunize her baby (p. 710).

TWENTY-FOUR

Respiratory System

OBJECTIVES

- Discuss the functions of the respiratory system and describe and identify the organs associated with respiration.
- State the type of epithelium found in each region of the respiratory tract and describe the location of the paranasal sinuses.
- Describe the regions of the pharynx and the structure and functions of the larynx and conducting airways.
- Describe the structure and function of the pulmonary alveoli.
- Describe the gross structure of the lungs and discuss the significance of the thoracic serous membranes.
- Describe the changes in intrapulmonary and intrapleural pressures during breathing and discuss how Boyle's law relates to lung function.
- Define *compliance*, *elasticity*, and *surface tension* and explain how these lung properties affect ventilation.
- State the muscles involved in quiet and in forced inspiration and expiration, and describe their actions.
- Distinguish between restrictive and obstructive pulmonary disease and give examples of each type.
- Define the terms used to describe lung volumes and capacities, and describe how pulmonary function tests help in diagnosing lung disorders.
- Discuss the calculation of partial pressures of gases in the air and explain how they are measured in the blood.
- Explain the significance of the plasma P_{O_2} measurement.
- Discuss the functions of the pneumotaxic, apneustic, and rhythmicity centers in the brain.
- Explain the chemoreceptor reflex control of breathing and describe the locations of the chemoreceptors.
- Distinguish between the different forms of hemoglobin and describe the loading and unloading reactions.
- Explain how changes in pH, temperature, and 2,3-DPG influence the function of hemoglobin.
- Describe how carbon dioxide is transported in the blood and discuss the chloride shift.
- Explain how breathing helps to maintain acid-base balance and how changes in ventilation can produce respiratory acidosis and alkalosis.
- Explain the increased ventilation that occurs during exercise and describe the effect of exercise on blood P_{O_2}, P_{CO_2}, and pH.
- Discuss the effect of endurance training on the anaerobic threshold.
- Describe the respiratory system changes associated with acclimatization to a high altitude.

Clinical Investigation

A taxi driver was found parked at the curb of a downtown street, complaining of great pain in his right thoracic region where he had been stabbed by an irate passenger with the tip of an umbrella. The wound had punctured his chest, and radiographs revealed that his right lung was collapsed, although his left lung was still functional. His arterial blood had a high P_{CO_2} and a pH of 7.15. He was treated surgically, and upon recovery his blood gases were again analyzed. Although the arterial P_{CO_2} and pH had been restored to normal, he had a carboxyhemoglobin concentration of 20%. Pulmonary function tests revealed that his vital capacity was slightly low and that his $FEV_{1.0}$ was significantly lower than normal. What do the radiographs and laboratory tests disclose about the health of this man?

Clues: Examine the sections "Pulmonary Function Tests," "Partial Pressures of Gases in Blood," and "Ventilation and Acid-Base Balance." Read the boxed information pertaining to pneumothorax and to carboxyhemoglobin in the major sections "Physical Aspects of Ventilation" and "Hemoglobin and Oxygen Transport," respectively.

Structure of the Respiratory System

The respiratory system can be divided anatomically into upper and lower divisions and functionally into a conducting division and a respiratory division. Gas exchange between the air and blood occurs in the respiratory division, so that the blood leaving the lungs has a higher oxygen and a lower carbon dioxide concentration than the blood that enters the lungs.

The term *respiration* refers to three separate but related functions: (1) **ventilation** (breathing); (2) **gas exchange,** which occurs between the air and blood in the lungs and between the blood and other tissues of the body; and (3) **oxygen utilization** by the tissues in the energy-liberating reactions of cell respiration. Ventilation and the exchange of gases (oxygen and carbon dioxide) between the air and blood are collectively called **external respiration.** Gas exchange between the blood and other tissues and oxygen utilization by the tissues are collectively known as **internal respiration.**

Ventilation is the mechanical process that moves air into and out of the lungs. Since air in the lungs has a higher oxygen concentration than the blood, oxygen diffuses from air to blood. Carbon dioxide, conversely, moves from the blood to the air within the lungs by diffusing down its concentration gradient. As a result of this gas exchange, the inspired air contains more oxygen and less carbon dioxide than the expired air. More importantly, blood leaving the lungs (in the pulmonary veins) has a higher oxygen concentration and a lower carbon dioxide concentration than the blood delivered to the lungs in the pulmonary arteries. This is because the lungs function to bring the blood into gaseous equilibrium with the air.

The major passages and structures of the respiratory system (fig. 24.1) are the nasal cavity, pharynx, larynx, and trachea, and the bronchi, bronchioles, and pulmonary alveoli within the lungs. The respiratory system is frequently divided into the **conducting division** and the **respiratory division.** The conducting division includes all of the cavities and structures that transport gases to the respiratory division. The structures involved in the gas exchange between the air and blood constitute the respiratory division.

respiration: L. *re*, back; *spirare*, to breathe

Conducting Division

Nose and Pharynx

The **nose** includes an external portion that juts out from the face and an internal **nasal cavity** for the passage of air. The external portion of the nose is supported by paired **nasal bones,** forming the bridge (chapter 9), and pliable cartilage, forming the distal portions. The **septal cartilage** forms the anterior portion of the **nasal septum,** and the paired **lateral cartilages** and **alar cartilages** form the framework around the **nostrils.**

The **vomer** and the perpendicular plate of the **ethmoid bone,** together with the septal cartilage, constitute the supporting framework called the **nasal septum,** which divides the nasal cavity into two lateral halves. Each half opens anteriorly through the **nostril** (fig. 24.2) and communicates posteriorly with the **nasopharynx** through the **choana** (*ko-a'nă*). The roof of the nasal cavity is formed anteriorly by the frontal bone and paired nasal bones, medially by the cribriform plate of the ethmoid bone, and posteriorly by the sphenoid bone. The palatine and maxillary bones form the floor of the cavity. On the lateral walls are the **superior, middle,** and **inferior conchae** (*kong'ke*) or **turbinates.** Air passages between the conchae are referred to as **meatuses** (*me-a'tus-es*) (fig. 24.2). The anterior openings of the nasal cavity are lined with stratified squamous epithelium, whereas the conchae are lined with pseudostratified ciliated columnar epithelium (figs. 24.3 and 24.4). Mucus-secreting goblet cells are present in great abundance throughout both regions.

The three functions of the nasal cavity and its contents are as follows:

1. The nasal epithelium covering the conchae warms, moistens, and cleanses the air. The curved conchae enhance the turbulence of the air and extend the surface area of the nasal epithelium, which is highly vascular. This is important for warming the air but unfortunately also makes humans susceptible to

meatus: L. *meatus*, path

Figure 24.1

The gross anatomy of the respiratory system.
Air is conducted through airways into the lungs, which are located in the thoracic cavity.

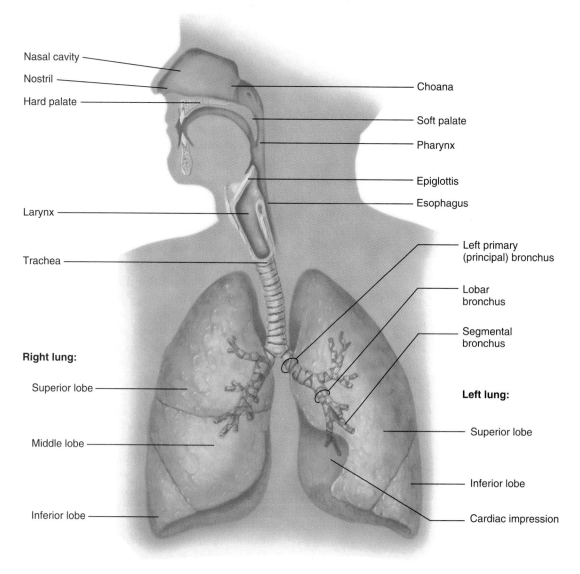

nosebleeds. Nasal hairs called **vibrissae** (*vi-bris'e*), which often extend from the nostrils, filter macroparticles that might otherwise be inhaled. Fine particles, such as dust, pollen, or smoke, are trapped along the moist mucous membrane lining the nasal cavity.

2. Olfactory epithelium in the upper medial portion of the nasal cavity responds to inhaled chemicals during olfaction.
3. The nasal cavity affects the quality of the voice by functioning as a resonating chamber.

There are several drainage openings into the nasal cavity (see fig. 24.2). The paranasal ducts drain mucus from the paranasal sinuses, and the nasolacrimal ducts drain tears from the eyes. An excessive secretion of tears causes the nose to run as the tears drain into the nasal cavity. The auditory tube from the middle-ear cavity enters the upper respiratory tract posterior to the nasal cavity in the nasopharynx. With all these accessory connections, it is no wonder that infections can spread so easily from one chamber to another throughout the facial area. To avoid causing damage or spreading infections to other areas, one must be careful not to blow the nose too forcefully.

Figure 24.2

The structures of the upper respiratory tract. These structures are revealed in a sagittal section of the head. There are several openings into the nasal cavity, including the openings of the paranasal sinuses, the nasolacrimal ducts that drain from the eyes, and the auditory tubes that drain from the middle ears.

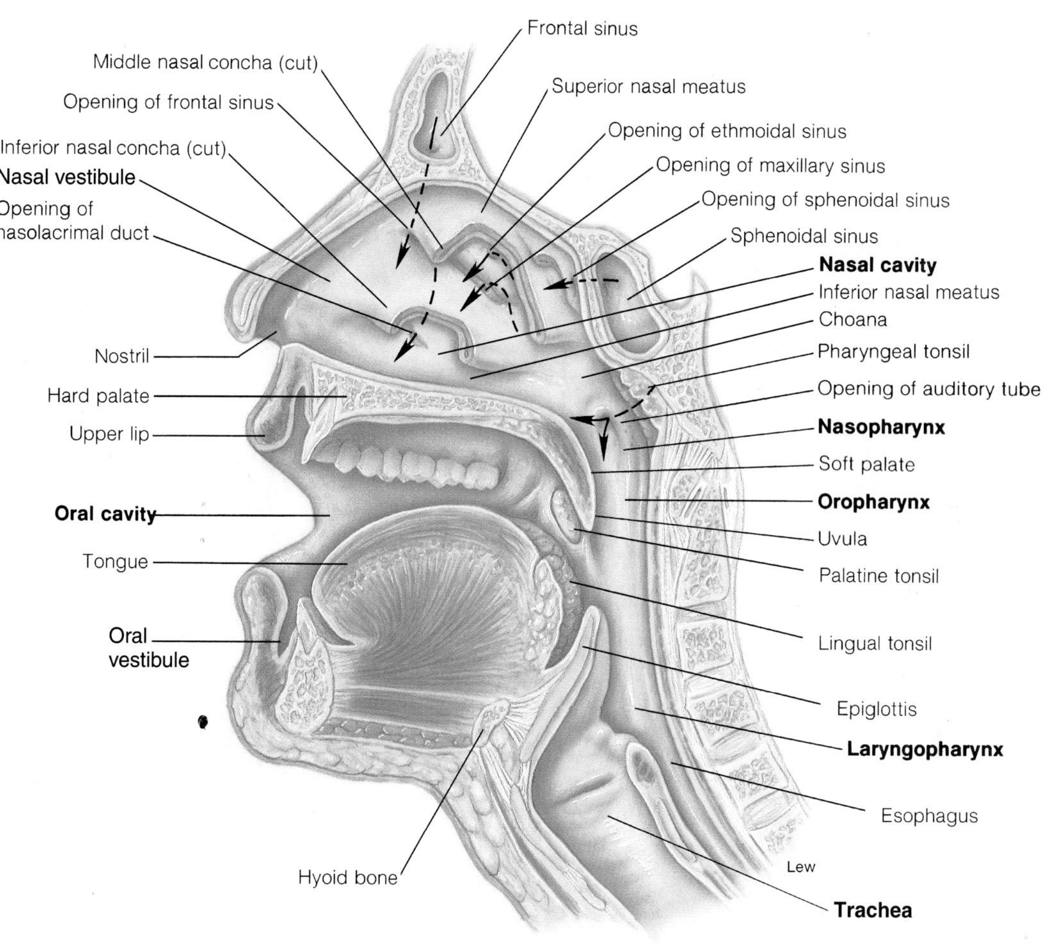

Figure 24.3

Types of epithelial tissues in the respiratory system. Note the thinness of the simple squamous epithelium of the lungs, which is specialized for a rapid rate of gas diffusion.

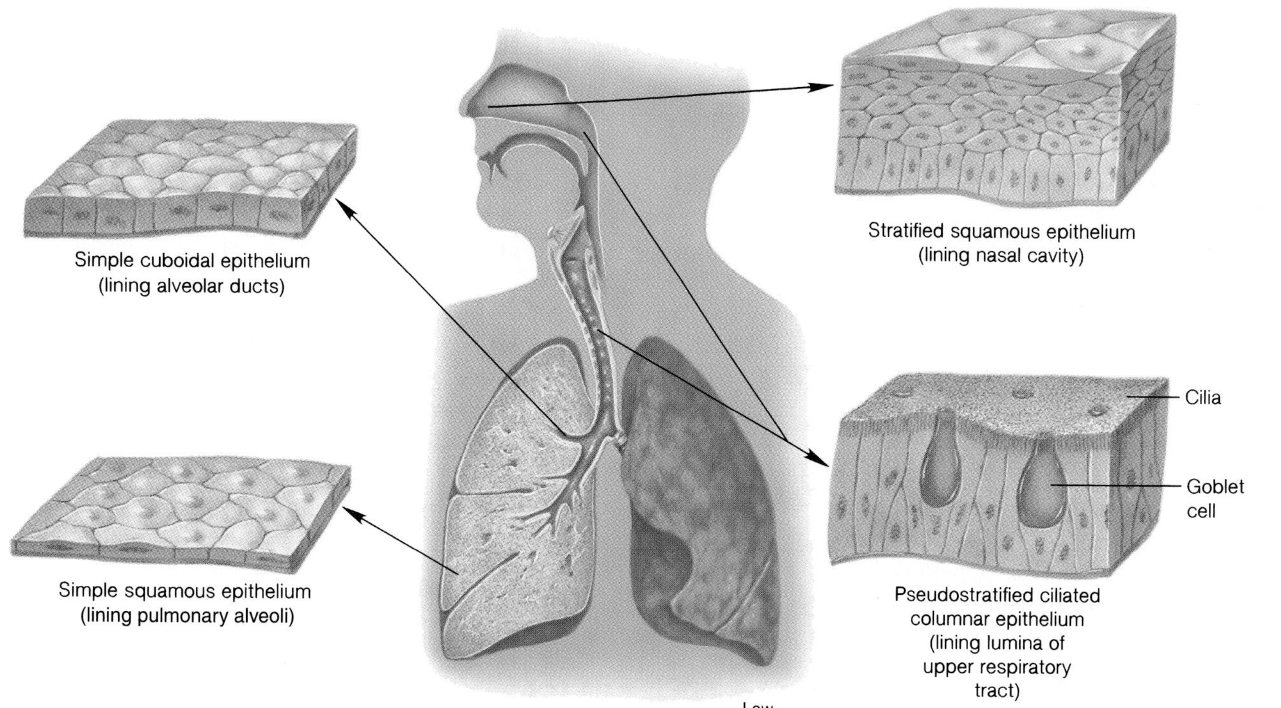

Figure 24.4

Color-enhanced scanning electron micrograph of a bronchial wall showing cilia.
In the trachea and bronchi, there are about 300 cilia per cell. The cilia move mucus-dust packages toward the pharynx, where they can either be swallowed or expectorated.

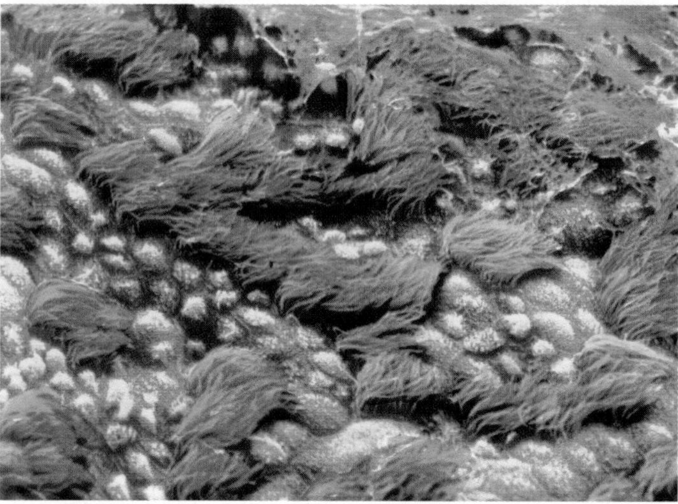

The **pharynx** (*far'ingks*) is a funnel-shaped passageway, approximately 13 cm (5 in.) long that connects the nasal and oral cavities to the larynx at the base of the skull. The supporting walls of the pharynx are composed of skeletal muscle, and the lumen is lined with a mucous membrane. Within the pharynx are several paired lymphoid organs, called **tonsils** (see fig. 24.2). The pharynx has both respiratory and digestive functions and is divided on the basis of location and function into three regions: nasal, oral, and laryngeal.

The **nasopharynx** (*na"zo-far'ingks*) has a respiratory function only. It is the uppermost portion of the pharynx, directly posterior to the nasal cavity and superior to the level of the soft palate. A pendulous **uvula** (*yoo'vyŭ-lă*) hangs from the middle lower border of the soft palate. The paired **auditory** (eustachian) **tubes** connect the nasopharynx with the middle-ear cavities. A collection of lymphoid tissue called the **pharyngeal tonsils,** or **adenoids,** (*ad'n-oidz*), are situated in the posterior wall of the nasopharynx. During the act of swallowing, the soft palate and uvula elevate to close off the nasopharynx and prevent food from entering the nasal cavity. Occasionally a person may suddenly exhale air (as with a laugh) while in the process of swallowing fluid. If this occurs before the uvula effectively blocks the nasopharynx, fluid will be discharged through the nasal cavity.

The **oropharynx** (*o"ro-far'ingks*) is the middle portion of the pharynx between the soft palate and the level of the hyoid bone. The base of the tongue forms the anterior wall of the oropharynx. Paired **palatine tonsils** are located along the posterior lateral wall of the oropharynx, and the **lingual tonsils** are found on the base of the tongue. This portion of the pharynx has both a respiratory and a digestive function.

The **laryngopharynx** (*lă-ring"go-far'ingks*) is the lowermost portion of the pharynx. It extends posteriorly from the level of the hyoid bone to the larynx and opens into the esophagus and larynx. It is at the lower laryngopharynx that the respiratory and digestive systems become distinct. Swallowed food and fluid are directed into the esophagus, whereas inhaled air is moved anteriorly into the larynx.

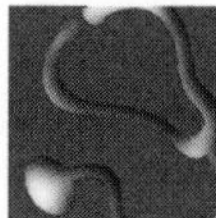

During a physical examination, a physician commonly depresses the patient's tongue and examines the palatine tonsils. Tonsils are pharyngeal lymphoid organs and tend to become swollen and inflamed after persistent infections. As mentioned in chapter 23, tonsils may have to be surgically removed when they become so overrun with pathogens that they themselves become the source of infection. The removal of the palatine tonsils is called a *tonsillectomy,* whereas the removal of the pharyngeal tonsils is called an adenoidectomy (*ad"n-oi-dek'tŏ-me*).

Larynx

The **larynx** (*lar'ingks*), or "voice box," forms the entrance into the lower respiratory system as it connects the laryngopharynx with the trachea. It is positioned in the anterior midline of the neck at the level of the fourth through sixth cervical vertebrae. The primary functions of the larynx are to prevent food or fluid from entering the trachea and lungs during swallowing and to permit passage of air while breathing. Additionally, it functions to produce sounds.

The larynx is shaped like a triangular box (fig. 24.5). It is composed of nine cartilages: three are large unpaired structures and six are smaller and paired. The largest of the unpaired cartilages is the anterior **thyroid cartilage.** The **laryngeal prominence** of the thyroid cartilage, commonly called the "Adam's apple," forms an anterior vertical ridge along the larynx that can be palpated on the midline of the neck. The thyroid cartilage is typically larger and more prominent in males than in females because of the effect of testosterone on the development of the larynx during puberty.

The spoon-shaped **epiglottis** (*ep"ĭ-glot'is*) has a cartilaginous framework. It is located behind the root of the tongue and aids in closing the **glottis,** or laryngeal opening, during swallowing. The lower end of the larynx is formed by the ring-shaped **cricoid** (*kri'koid*) **cartilage.** This third unpaired cartilage connects the thyroid cartilage above and the trachea below. The paired **arytenoid** (*ar"ĭ-te'noid*) **cartilages,** located above the cricoid cartilage and behind the thyroid cartilage, furnish the attachment of the *vocal folds*. The other paired **cuneiform cartilages** and **corniculate** (*kor-nik'yŭ-lāt*)

pharynx: L. *pharynx*, throat
uvula: L. *uvula*, small grape
adenoids: Gk. *adenoeides*, glandlike

thyroid: Gk. *thyreos*, shieldlike
arytenoid: Gk. *arytaina*, ladle or cup-shaped
cuneiform: L. *cuneus*, wedge-shaped
corniculate: L. *corniculum*, diminutive of *cornu*, horn

Figure 24.5

The structure of the larynx.
(*a*) An anterior view, (*b*) a lateral view, and (*c*) a sagittal view.

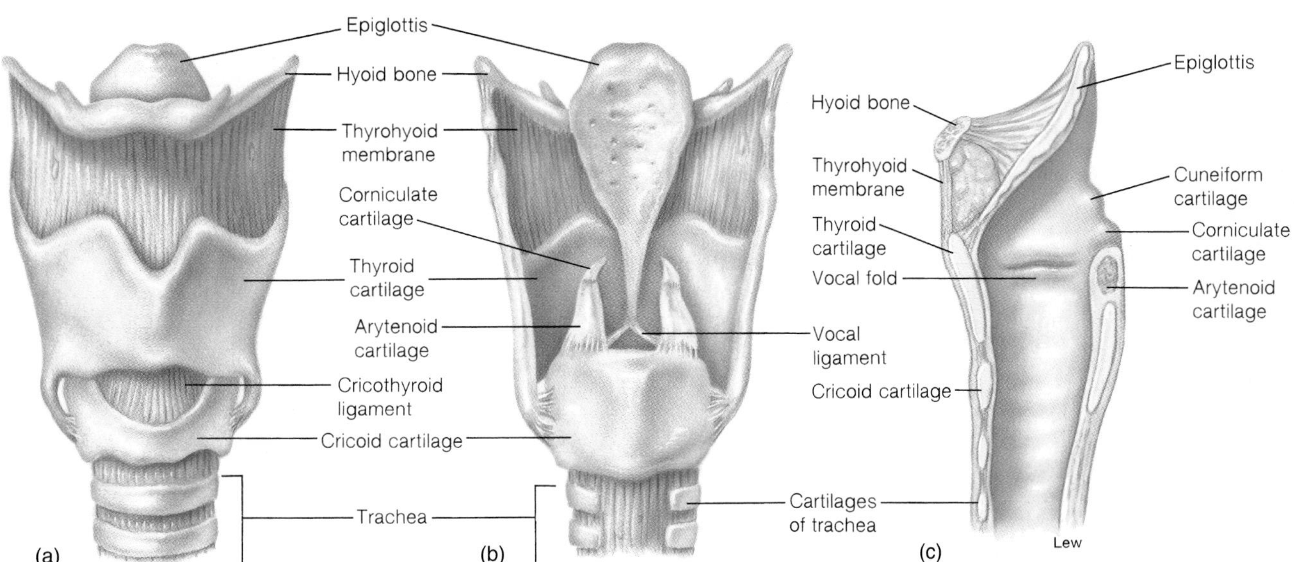

Figure 24.6

A superior view of the vocal cords.
In (*a*) the vocal cords are taut; in (*b*) they are relaxed and the glottis is opened. (*c*) A photograph through a laryngoscope showing the vocal folds (vocal cords), ventricular folds (false vocal cords), and the glottis.

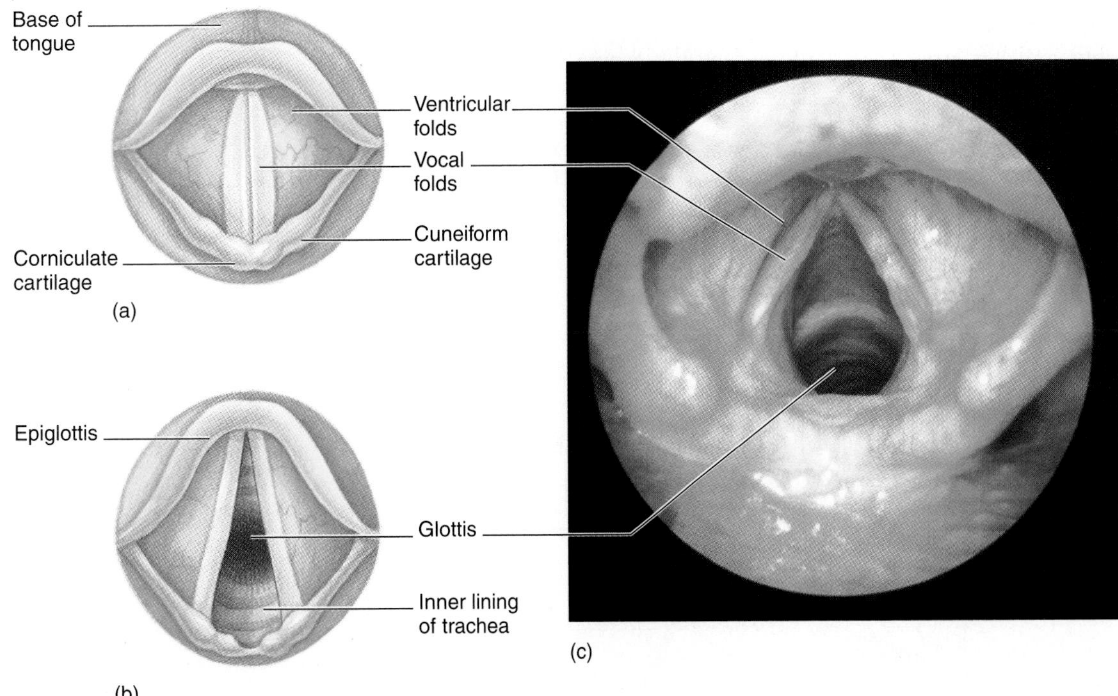

cartilages are small accessory cartilages that are closely associated with the arytenoid cartilages (fig. 24.6).

Two pairs of strong connective tissue bands are stretched across the upper opening of the larynx from the thyroid cartilage anteriorly to the paired arytenoid cartilages posteriorly. These are the **vocal folds,** or **true vocal cords,** and the **ventricular folds,** or **false vocal cords** (fig. 24.6). The ventricular folds support the vocal folds and are not used in sound production. Sound is produced by vibrations of the vocal folds.

The **laryngeal muscles** are extremely important in closing the glottis during swallowing and in speech. There are two groups of laryngeal muscles: **extrinsic muscles,** responsible for elevating the larynx during swallowing, and **intrinsic muscles** that, when contracted, change the length, position, and tension of the vocal folds. Sounds of various pitches are produced as air passes over the altered vocal folds. If the vocal folds are taut, vibration is more rapid and causes a higher pitch. Less tension on the folds produces a lower pitch. Mature males generally have thicker and longer vocal folds than females; therefore, the vocal folds of males vibrate more slowly and produce lower pitched sounds.

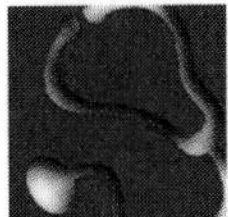

During the act of swallowing, the larynx is elevated to close the glottis against the epiglottis. This movement can be noted by cupping the fingers lightly over the larynx and then swallowing. Food may become lodged within the glottis if it is not closed as it should be. In this case, the *abdominal thrust* (Heimlich) *maneuver* (see page 771) can be used to prevent suffocation.

Trachea and Bronchial Tree

The **trachea** (*tra'ke-ă*), or "windpipe," is a rigid tube, approximately 12 cm (4 in.) long and 2.5 cm (1 in.) in diameter. It is positioned anterior to the esophagus and connects the larynx to the primary bronchi (fig. 24.7). A series of 16 to 20 C-shaped rings of hyaline cartilage form the walls of the trachea (fig. 24.8). The open part of the C is positioned posteriorly and is covered by fibrous connective tissue and smooth muscle. The cartilages provide a rigid but flexible tube that allows the airway to be permanently open. The division of the trachea into right and left bronchi is reinforced by the **carina** (*kă-ri'nă*), a keel-like cartilage plate (see fig. 24.7).

Heimlich maneuver: from Henry J. Heimlich, American surgeon, b. 1920
trachea: L. *trachia*, rough air vessel
carina: L. *carina*, keel

Figure 24.7

Anterior view of the larynx, trachea, and bronchi.
The hard tissue of these structures is composed of hyaline cartilage.

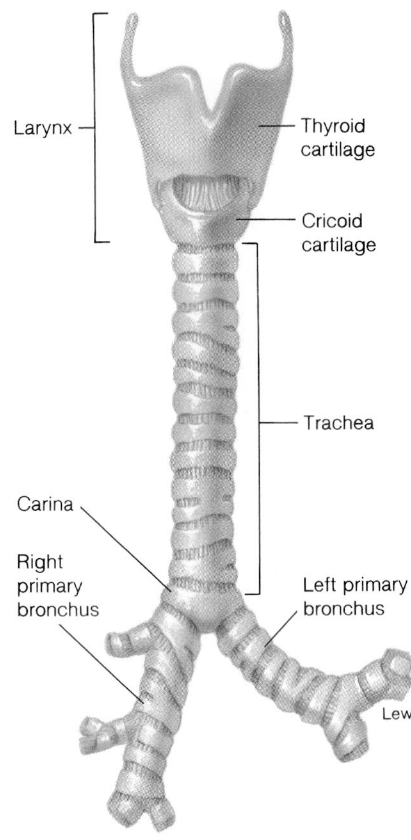

Figure 24.8

Histology of the trachea.
(*a*) A photomicrograph showing the relationship of the trachea to the esophagus (3×) and (*b*) a photomicrograph of tracheal cartilage (63×).

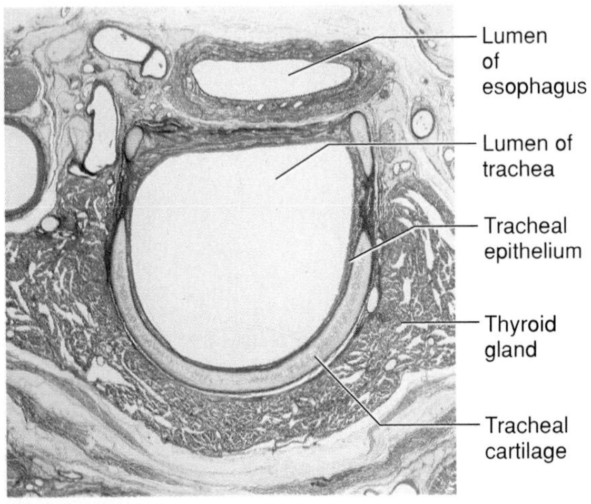

(a)

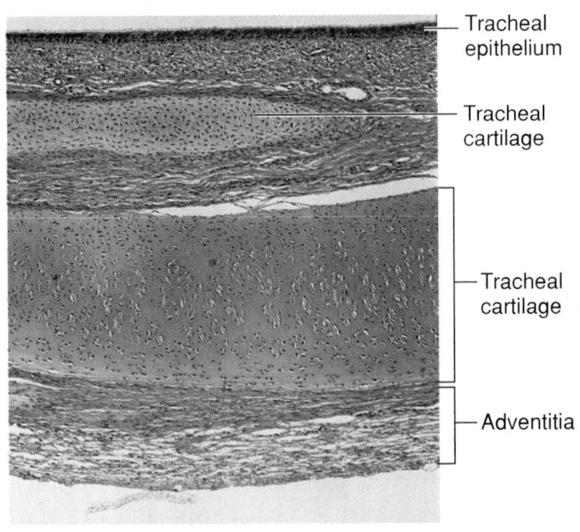

(b)

Figure 24.9

Conducting and respiratory zones of the respiratory system.
The conducting zone consists of airways that conduct the air to the respiratory zone, which is the region where gas exchange occurs.

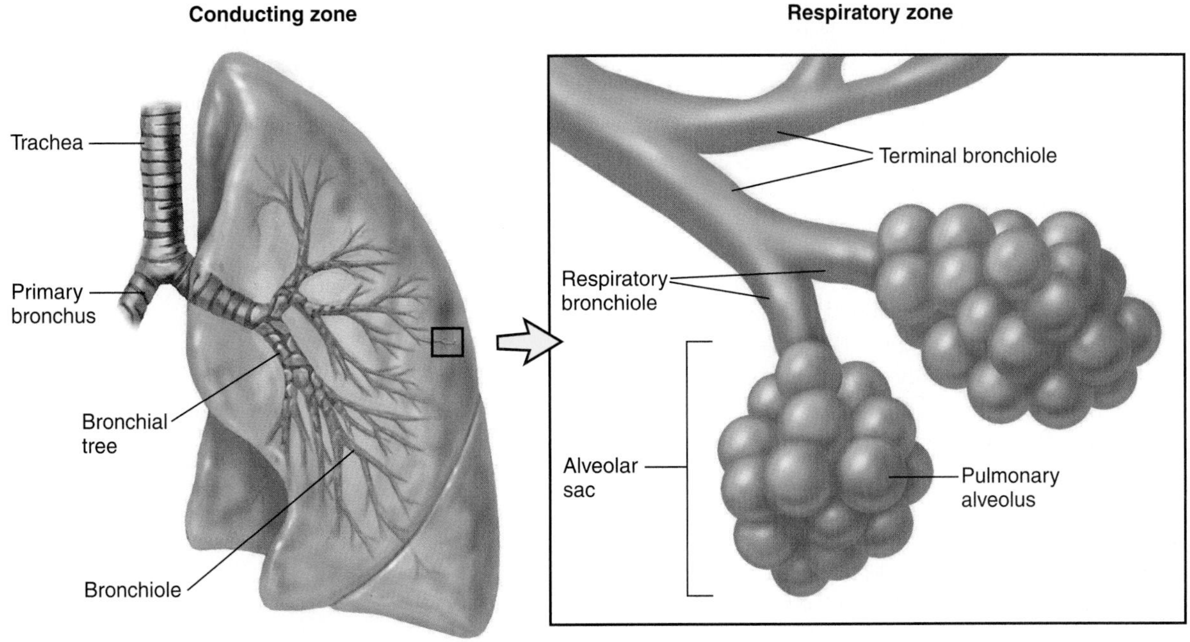

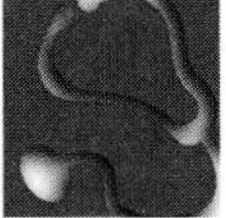

If the trachea becomes occluded, as by aspiration of a foreign object, it may be necessary to create an emergency opening so that ventilation can still occur. A *tracheotomy* is the procedure of surgically opening the trachea, and a *tracheostomy* involves the insertion of a tube into the trachea to permit breathing and to keep the passageway open. A tracheotomy should be performed only by a competent physician because of the great risk of cutting a recurrent laryngeal nerve or the common carotid artery.

The **bronchial tree** is so-named because it is composed of a series of respiratory tubes that branch into progressively narrower tubes as they extend into the lung. The trachea bifurcates into a **right** and a **left primary bronchus** (*brong'kus*). Each bronchus has hyaline cartilage rings surrounding its lumen to keep it open as it extends into the lung. Because of the more vertical position of the right bronchus, foreign particles are more likely to lodge here than in the left bronchus.

The primary bronchi divide deeper in the lungs to form **lobar (secondary) bronchi** and **segmental (tertiary) bronchi.** The bronchial tree continues to branch into yet smaller tubules called **bronchioles** (*brong'ke-ōlz*) (fig. 24.9). There is almost no cartilage in the bronchioles, and the smooth muscle in their walls can constrict or dilate these airways. Bronchioles provide the greatest resistance to air flow in the conducting passages, and thus their function is analogous to that of the arterioles in the vascular system. A simple cuboidal epithelium lines the bronchioles rather than the pseudostratified ciliated columnar epithelium that lines the bronchi (see fig. 24.3). Numerous **terminal bronchioles** mark the end of the air-conducting pathway to the pulmonary alveoli.

Regardless of the temperature and humidity of the ambient air, when the inspired air reaches the respiratory zone it is at a temperature of 37° C (body temperature), and it is saturated with water vapor as it flows over the warm, wet mucous membranes that line the respiratory airways. This ensures that a constant internal body temperature will be maintained and that delicate lung tissue will be protected from desiccation.

Mucus secreted by cells of the conducting zone structures serves to trap small particles in the inspired air and thereby performs a filtration function. This mucus is moved along at a rate of 1 to 2 centimeters per minute by cilia projecting from the tops of epithelial cells that line the conducting zone (fig. 15.7). About 300 cilia per cell beat in a coordinated fashion to move mucus toward the pharynx, where it can either be swallowed or expectorated.

As a result of this filtration function, particles larger than about 6 μm do not normally enter the respiratory zone of the lungs. The importance of this function is evidenced by *black lung*, a disease that occurs in miners who inhale large amounts of carbon dust, which causes them to develop pulmonary fibrosis. The pulmonary alveoli themselves are normally kept clean by the action of macrophages that reside within them. The cleansing action of cilia and macrophages in the lungs is diminished by cigarette smoke.

bronchus: L. *bronchus*, windpipe

Pulmonary Alveoli, Lungs, and Pleurae

Pulmonary Alveoli

Air from the terminal bronchioles enters the pulmonary **alveolar ducts.** The pulmonary alveolar ducts contain individual **pulmonary alveoli** (*al-ve'ŏ-li*) as outpouchings along their length and open into clusters of pulmonary alveoli called **alveolar sacs** at their ends (fig. 24.9). The last three structures make up the *respiratory division* of the lungs.

Gas exchange in the lungs occurs across an estimated 300 million tiny (0.25–0.50 mm in diameter) pulmonary alveoli. Their enormous number provides a large surface area (60–80 square meters, or about 760 square feet) for diffusion of gases. The diffusion rate is further increased because each alveolus is only one cell layer thick; hence, the total "air-blood barrier" is only two cells across (an alveolar cell and a capillary endothelial cell), or about 2 μm. This is an average distance because there are two types of cells in the alveolar wall, *type I* and *type II*, and the type II alveolar cells are thicker than the type I cells (fig. 24.10). Where the basement membranes of capillary endothelial cells fuse with those of type I alveolar cells, the diffusion distance is less than 1 μm.

Pulmonary alveoli are polyhedral and are usually clustered together, like the units of a honeycomb (see fig. 24.9). As mentioned previously, individual alveoli also occur as separate outpouchings along the length of the alveolar ducts. Although the distance between each alveolar duct and its terminal alveoli is only about 0.5 mm, these units together constitute most of the mass of the lungs.

The enormous surface area of alveoli and the short diffusion distance between alveolar air and the capillary blood quickly bring the blood into gaseous equilibrium with the alveolar air. This function is further aided by the capillary networks surrounding the pulmonary alveoli. Each alveolus is surrounded by so many capillaries that they form an almost continuous sheet of blood (see fig. 24.25).

Lungs

The large spongy **lungs** are paired organs contained within the thoracic cavity. They lie against the rib cage anteriorly and posteriorly and extend from the diaphragm to a point just above the clavicles. The lungs are separated from one another by the heart and other structures of the **mediastinum** (*me"de-ă-sti'num*), as shown in figure 24.11. The mediastinum is the area between the lungs. All structures of the respiratory system beyond the primary bronchi, including the bronchial tree and pulmonary alveoli, are contained within the lungs.

Each lung presents four surfaces that match the contour of the thoracic cavity. The mediastinal (medial) surface of each lung is slightly concave and contains a vertical slit, the **hilum** (*hi'lum*), through which pulmonary vessels, nerves, and bronchi pass. The inferior border of the lung, called the **base,** is concave as it fits over the convex dome of the diaphragm. Anteriorly, the portion of each lung that extends above the level of the clavicle is called the **apex.** Finally, the broad, rounded surface in contact with the rib cage is called the **costal surface.**

Although the right and left lungs are basically similar, they are not identical. The left lung is somewhat smaller than the right and has a **cardiac impression** on its medial surface to accommodate the heart. The left lung is subdivided into a **superior lobe** and an **inferior lobe** by a single fissure. The right lung is subdivided by two fissures into **superior, middle,** and **inferior lobes.** Each lobe of the lung is divided into many small lobules, which in turn contain the pulmonary alveoli. Lobular divisions of the lungs comprise specific *bronchial segments*. The right lung contains 10 bronchial segments, and the left lung contains 8 (fig. 24.12).

Pleurae

Pleurae (*ploor'e*) are serous membranes surrounding the lungs and lining the thoracic cavity (see fig. 24.11). The **visceral pleura** adheres to the outer surface of the lung and extends

Figure 24.10

Relationship between pulmonary alveoli and pulmonary capillaries.
Note that alveolar walls are quite narrow and lined with type I and type II alveolar cells. Pulmonary macrophages can phagocytose particles that enter the lungs.

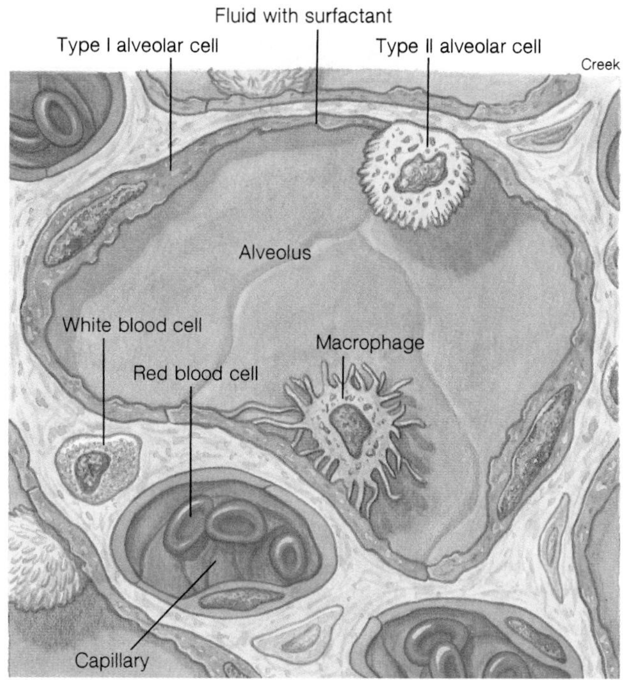

alveolus: L. diminutive of *alveus*, cavity
mediastinum: L. *mediastinus*, intermediate
hilum: L. *hilum*, a trifle (little significance)
pleura: Gk. *pleura*, side or rib

Figure 24.11

Cross section of the thoracic cavity.
In addition to the lungs, the mediastinum and pleural membranes are visible. The parietal pluera is shown in green, and the visceral pleura in blue.

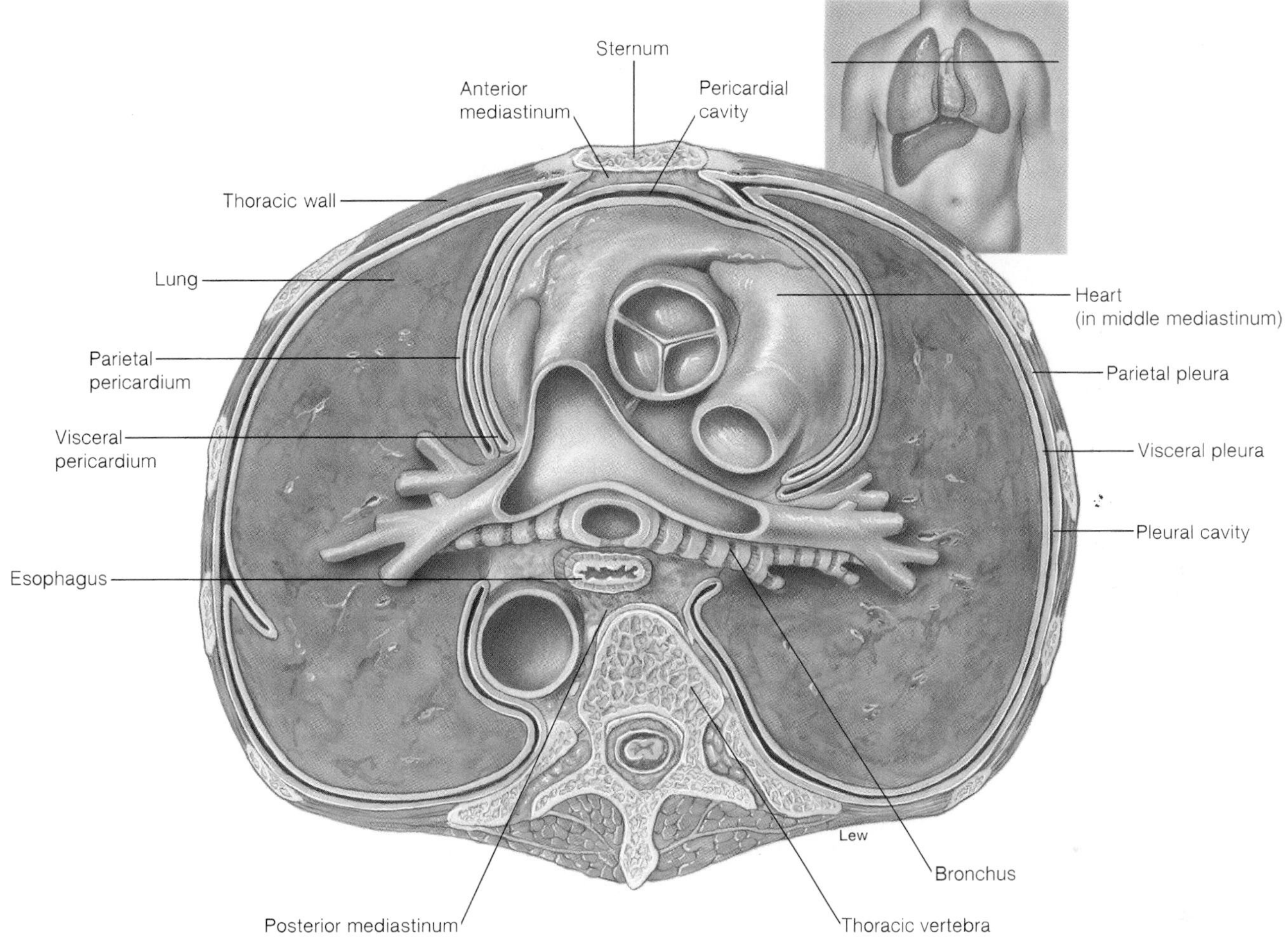

into each of the interlobar fissures. The **parietal pleura** lines the thoracic walls and the thoracic surface of the diaphragm. A continuation of the parietal pleura around the heart and between the lungs forms the boundary of the mediastinum. Between the visceral and parietal pleurae is a moistened space called the **pleural cavity.** An inferiorly extending reflection of the pleural layers around the roots of each lung is called the **pulmonary ligament.** The pulmonary ligaments help to support the lungs. The normal position of the lungs in the thoracic cavity is shown in the radiograph in figure 24.13.

The membranes of the thoracic cavity serve to compartmentalize the different organs. There are four distinct compartments: two pleural cavities (one surrounding each lung); a pericardial cavity surrounding the heart; and the mediastinum, within which the esophagus, thoracic duct, major vessels, various nerves, and portions of the respiratory tract are located. This compartmentalization has protective value because infections are usually confined to one compartment; moreover, damage to one organ will not usually involve another. Pleurisy, for example, which is an inflamed pleura, is generally confined to one side. A penetrating injury to one side, such as a knife wound, might cause one lung to collapse but not the other.

pulmonary: Gk. *pleumon*, lung

Physical Aspects of Ventilation

The movement of air into and out of the lungs occurs as a result of pressure differences induced by changes in lung volumes. Ventilation is thus influenced by the physical properties of the lungs, including their compliance, elasticity, and surface tension.

Movement of air, from higher to lower pressure, between the conducting zone and the terminal bronchioles occurs as a result of the pressure difference between the two ends of the airways. Air flow through bronchioles, like blood flow through blood vessels, is directly proportional to the pressure difference

Figure 24.12

Lobes and lobules of the lungs.

The bronchopulmonary segments are also indicated.

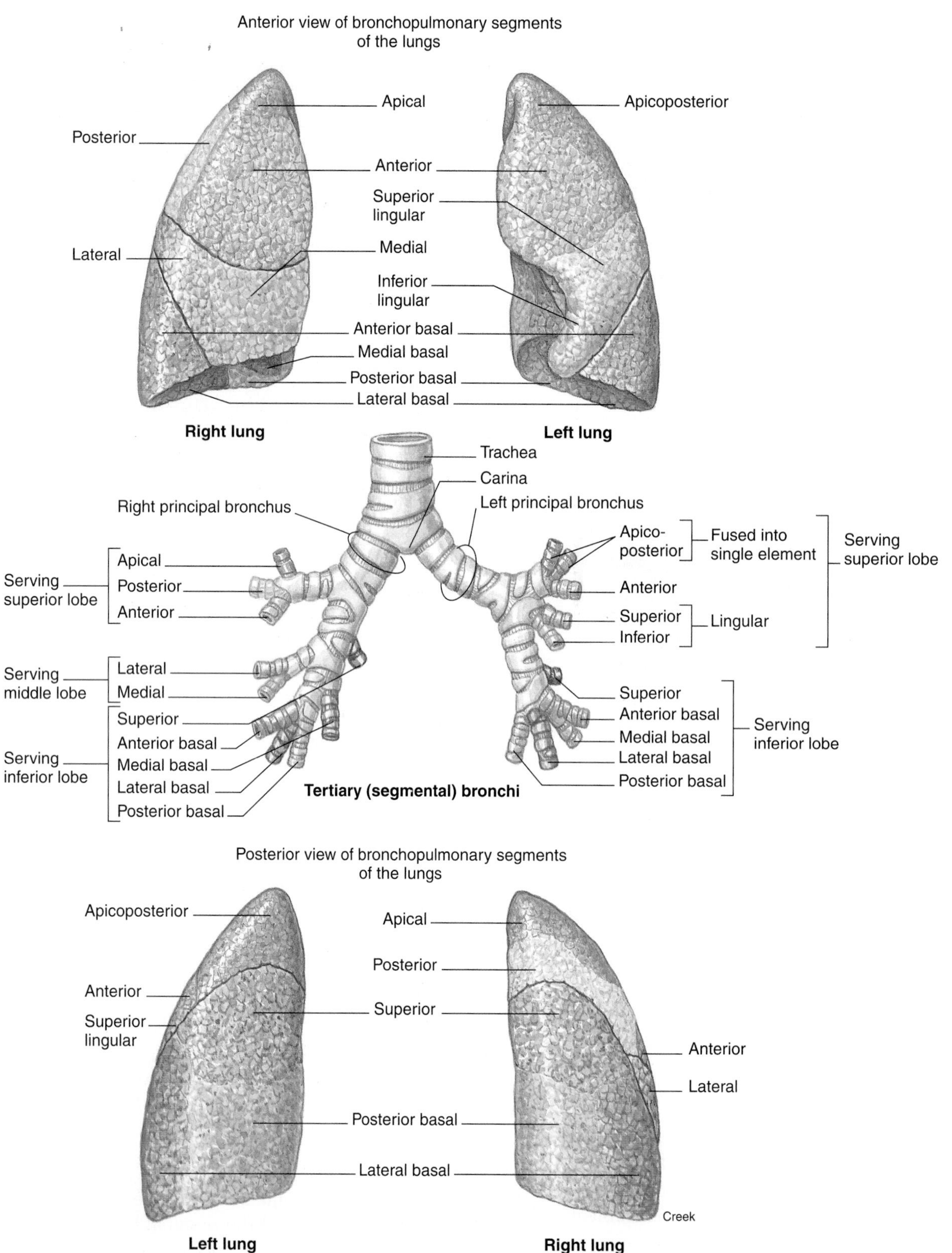

Figure 24.13

Radiographic (X-ray) views of the chest.
(*a*) A normal female and (*b*) a normal male.

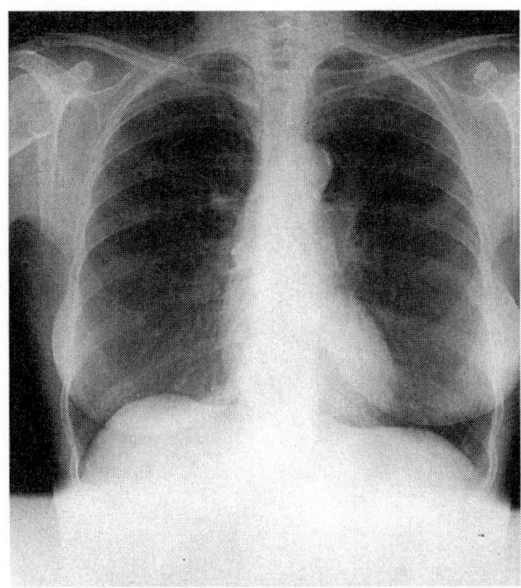
(a)

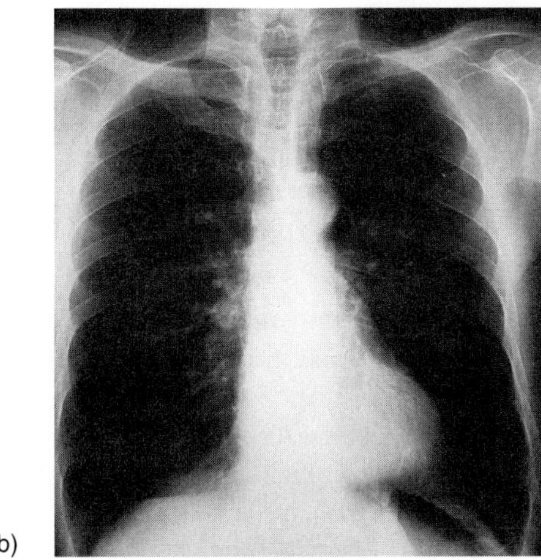
(b)

and inversely proportional to the frictional resistance to flow. The pressure differences in the pulmonary system are induced by changes in lung volumes. The compliance, elasticity, and surface tension of the lungs are physical properties that affect their functioning, as will be discussed shortly.

Intrapulmonary and Intrapleural Pressures

The moistened serous membranes of the visceral and parietal pleurae are normally flush against each other, so that the lungs are stuck to the chest wall in the same manner as two wet pieces of glass stick to each other. The *intrapleural space* between the two wet membranes contains only a thin layer of fluid secreted by the pleural membranes. The pleural cavity in a healthy person is thus potential rather than real; it can become real only in abnormal situations when air enters the intrapleural space. Since the lungs normally remain in contact with the chest wall, they get larger and smaller together with the thoracic cavity during respiratory movements.

Table 24.1

Pressure Changes during Normal, Quiet Breathing

	Inspiration	Expiration
Intrapulmonary pressure (mmHg)	–3	+3
Intrapleural pressure (mmHg)	–6	–3
Transpulmonary pressure (mmHg)	+3	+6

Note: Pressure indicates mmHg below or above atmospheric pressure. Intrapleural pressure is normally always negative (subatmospheric).

Air enters the lungs during inspiration because the atmospheric pressure is greater than the **intrapulmonary,** or **intra-alveolar, pressure.** Since the atmospheric pressure does not usually change, the intrapulmonary pressure must fall below atmospheric pressure to cause inspiration. A pressure below that of the atmosphere is called a *subatmospheric pressure*, or *negative pressure*. During quiet inspiration, for example, the intrapulmonary pressure may decrease to 3 mmHg below the pressure of the atmosphere. This subatmospheric pressure is shown as –3 mmHg. Conversely, expiration occurs when the intrapulmonary pressure is greater than the atmospheric pressure. During quiet expiration, for example, the intrapulmonary pressure may rise to at least +3 mmHg over the atmospheric pressure.

The lack of air in the intrapleural space produces a subatmospheric **intrapleural pressure** that is lower than the intrapulmonary pressure (table 24.1). There is thus a pressure difference across the wall of the lung—called the **transpulmonary pressure**—which is the difference between the intrapulmonary pressure and the intrapleural pressure. Since the pressure within the lungs (intrapulmonary pressure) is greater than that outside the lungs (intrapleural pressure), the difference in pressure (transpulmonary pressure) keeps the lungs against the chest wall. Thus, changes in lung volume parallel changes in thoracic volume during inspiration and expiration.

Boyle's Law

Changes in intrapulmonary pressure occur as a result of changes in lung volume. This follows from **Boyle's law,** which states that the pressure of a given quantity of gas is inversely proportional to its volume. An increase in lung volume during inspiration decreases intrapulmonary pressure to subatmospheric levels; air therefore goes in. A decrease in lung volume, conversely, raises the intrapulmonary pressure above that of the atmosphere, expelling air from the lungs. These changes in lung volume occur as a consequence of

Boyle's law: from Robert Boyle, Irish-born British physicist, 1627–91

changes in thoracic volume, as will be described in a later section on the mechanics of breathing.

Physical Properties of the Lungs

In order for inspiration to occur, the lungs must be able to expand when stretched; they must have high *compliance*. In order for expiration to occur, the lungs must get smaller when this tension is released; they must have *elasticity*. The tendency to get smaller is also aided by *surface tension* forces within the pulmonary alveoli.

Compliance

The lungs are very distensible (stretchable)—they are, in fact, about a hundred times more distensible than a toy balloon. Another term for distensibility is **compliance,** which here refers to the ease with which the lungs can expand under pressure. Lung compliance can be defined as the change in lung volume per change in transpulmonary pressure, expressed symbolically as $\Delta V/\Delta P$. A given transpulmonary pressure, in other words, will cause greater or lesser expansion depending on the compliance of the lungs.

The compliance of the lungs is reduced by factors that produce a resistance to distension. If the lungs were filled with concrete (as an extreme example), a given transpulmonary pressure would produce no increase in lung volume and no air would enter; the compliance would be zero. The infiltration of lung tissue with connective tissue proteins, a condition called *pulmonary fibrosis*, similarly decreases lung compliance.

Elasticity

The term **elasticity** refers to the tendency of a structure to return to its initial size after being distended. Because of their high content of elastin proteins, the lungs are very elastic and resist distension. Since the lungs are normally stuck to the chest wall, they are always in a state of elastic tension. This tension increases during inspiration when the lungs are stretched and is reduced by elastic recoil during expiration. The elasticity of the lungs and of other thoracic structures thus aids in pushing the air out during expiration.

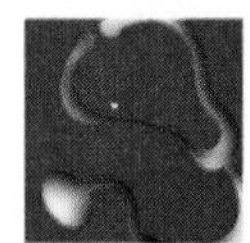

The elastic nature of lung tissue is revealed when air enters the intrapleural space (as a result of an open chest wound, for example). This condition, called a *pneumothorax,* is shown in figure 24.14. As air enters the intrapleural space, the intrapleural pressure rises until it is equal to the atmospheric pressure. When the intrapleural pressure is the same as the intrapulmonary pressure, the lung can no longer expand. Not only does the lung not expand during inspiration, it actually collapses away from the chest wall as a result of elastic recoil. Collapse of a lung is called *atelectasis* (*at′′l-ek′tă-sis*). Fortunately, a pneumothorax usually causes only one lung to collapse, since each lung is contained in a separate pleural compartment.

pneumothorax: Gk. *pneumon*, spirit (air); L. *thorax*, chest

Figure 24.14

Pneumothorax of the right lung.
The right side of the thorax appears uniformly dark because it is filled with air; the spaces between the ribs are also greater on the right due to release from the elastic tension of the lungs. The left lung appears denser (less dark) because of shunting of blood from the right to the left lung.

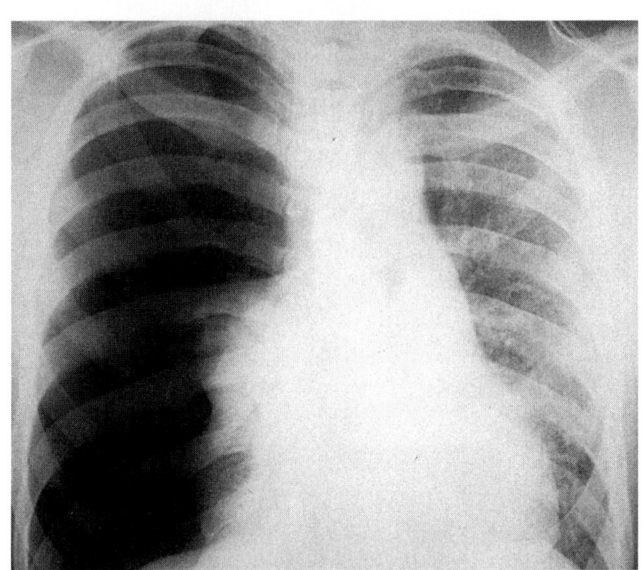

Surface Tension

The forces that act to resist distension include elastic resistance and the **surface tension** that is exerted by fluid in the pulmonary alveoli. The lungs both secrete and absorb fluid in two antagonistic processes that normally leave only a very thin film of fluid on the alveolar surface. Fluid absorption is driven (through osmosis) by the active transport of Na^+, whereas fluid secretion is driven by the active transport of Cl^- out of the alveolar epithelial cells. Research has demonstrated that people with cystic fibrosis have a genetic defect in one of the Cl^- carriers (called the *cystic fibrosis transmembrane regulator*, or *CFTR*, as described in chapter 6). This results in an imbalance of fluid absorption and secretion, so that the airway fluid becomes excessively viscous (with a lower water content) and difficult to clear.

The thin film of fluid normally present in the pulmonary alveolus has a surface tension because water molecules at the surface are attracted more to other water molecules than to air. As a result, the surface water molecules are pulled tightly together by attractive forces from underneath. This surface tension produces a force that is directed inward, raising the pressure within the alveolus. As described by the **law of Laplace,** the pressure thus created is directly proportional to the surface tension and inversely proportional to the radius of the alveolus (fig. 24.15). According to this law, the pressure in a smaller alveolus would be greater than in a

law of Laplace: from Pierre Simon, Marquis de Laplace, French astronomer and mathematician, 1749–1827

Figure 24.15

The law of Laplace.
According to the law of Laplace, the pressure created by surface tension should be greater in the smaller alveolus (left) than in the larger alveolus (right). This implies that (without surfactant) smaller alveoli would collapse and empty their air into larger alveoli.

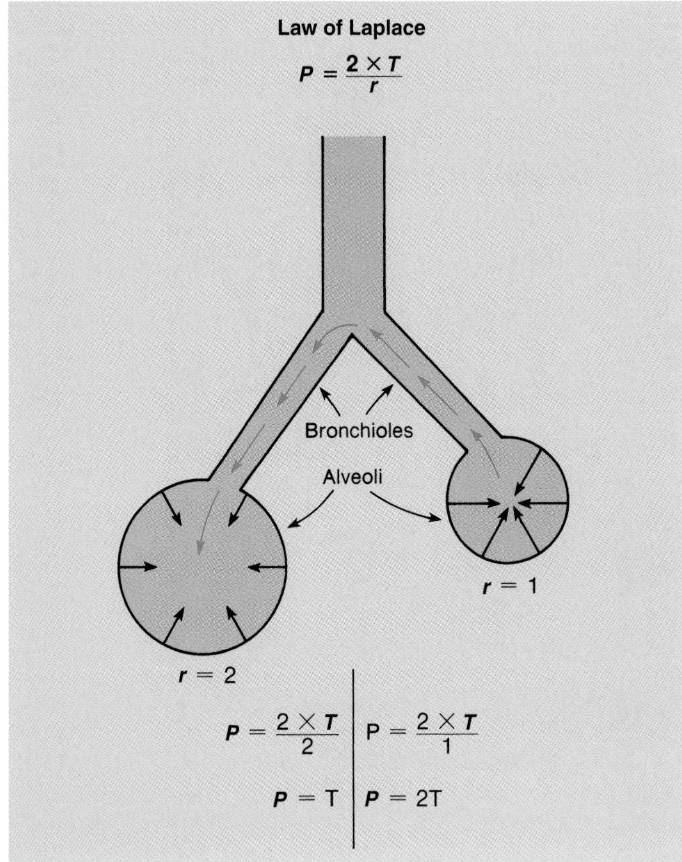

larger alveolus if the surface tension were the same in both. The greater pressure of the smaller alveolus would then cause it to empty its air into the larger one (fig. 24.15). This does not normally occur because, as an alveolus decreases in size, its surface tension (the numerator in the equation) is decreased at the same time that its radius (the denominator) is reduced. The reason for the decreased surface tension, which prevents the alveoli from collapsing, is described in the next section.

Surfactant and the Respiratory Distress Syndrome

Alveolar fluid contains a phospholipid known as *dipalmitoyl lecithin*, probably attached to a protein, which functions to lower surface tension. This compound is called **surfactant** (*sur-fak'tant*), a contraction of the term *surface active agent*. The surfactant molecules become interspersed between water molecules, thereby reducing the attractive forces (hydrogen bonds, described in chapter 2) between water molecules that produce the surface tension. Thus, because of surfactant, the surface tension in the pulmonary alveoli is reduced. Further, the ability of surfactant to lower surface tension improves as the alveoli get smaller during expiration. This might be because the surfactant molecules become more concentrated as the alveoli get smaller. Surfactant thus prevents the alveoli from collapsing during expiration, as would be predicted by the law of Laplace. Even after a forceful expiration, the alveoli remain open, and a *residual volume* of air remains in the lungs. Since the alveoli do not collapse, less surface tension has to be overcome to inflate them at the next inspiration.

Surfactant is produced by type II alveolar cells (fig. 24.16) in late fetal life. Because no surfactant is produced until about the eighth month, premature babies are sometimes born with lungs that lack sufficient surfactant, and their alveoli are collapsed as a result. This condition is called **respiratory distress syndrome (RDS).** It is also called **hyaline** (*hi'ă-līn*) **membrane disease** because the high surface tension causes plasma fluid to leak into the alveoli, producing a glistening "membrane" appearance (and pulmonary edema). This condition does not occur in all premature babies; the rate of lung development depends on hormonal conditions (thyroxine and cortisol primarily) and on genetic factors.

Even under normal conditions, the first breath of life is a difficult one because the newborn must overcome great surface tension forces in order to inflate its partially collapsed pulmonary alveoli. The transpulmonary pressure required for the first breath is 15 to 20 times that required for subsequent breaths, and an infant with respiratory distress syndrome must duplicate this effort with every breath. Fortunately, many babies with this condition can be saved by mechanical ventilators and by exogenous surfactant delivered to the baby's lungs by means of an endotracheal tube. The exogenous surfactant may be a synthetic mixture of phospholipids, or it may be surfactant obtained from bovine lungs. The mechanical ventilator and exogenous surfactant help to keep the baby alive long enough for its lungs to mature, so that it can manufacture sufficient surfactant on its own.

People with septic shock may develop a condition called **acute respiratory distress syndrome (ARDS).** In this condition, inflammation causes increased capillary and alveolar permeability that lead to accumulation of a protein-rich fluid in the lungs. This decreases lung compliance and is accompanied by a reduced surfactant, which further lowers compliance. The blood leaving the lungs, as a result, has an abnormally low oxygen concentration (a condition called *hypoxemia*).

Figure 24.16

The production of pulmonary surfactant.
Produced by type II alveolar cells, surfactant appears to be composed of a derivative of lecithin combined with protein.

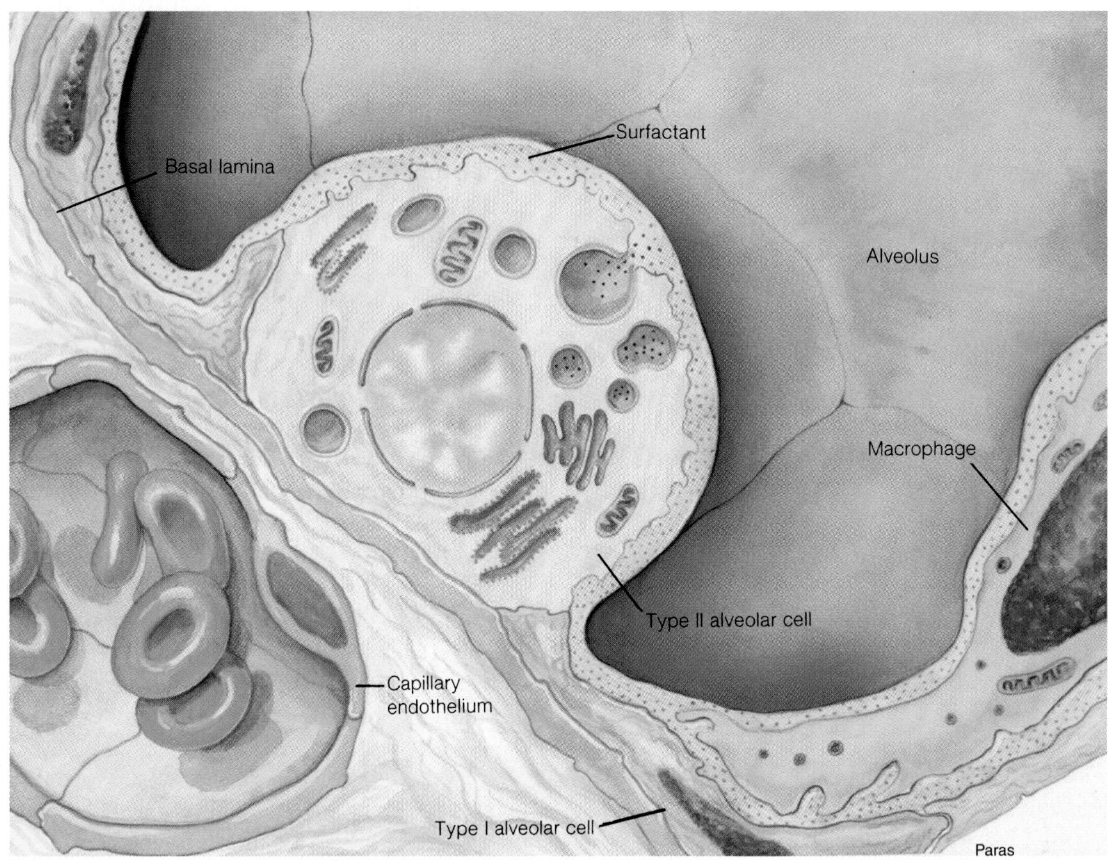

Mechanics of Breathing

Normal, quiet inspiration results from muscle contraction, and normal expiration results from muscle relaxation and elastic recoil. These actions can be forced by contractions of the accessory respiratory muscles. The amount of air inspired and expired can be measured in a number of ways to test pulmonary function.

The thorax must be sufficiently rigid to protect vital organs and provide attachments for a number of short, powerful muscles. However, breathing, or **pulmonary ventilation,** also requires a flexible thorax that can function as a bellows during the ventilation cycle. The structure of the rib cage and associated cartilages provides continuous elastic tension, so that when stretched by muscle contraction during inspiration, the rib cage can return passively to its resting dimensions when the muscles relax. This elastic recoil is greatly aided by the elasticity of the lungs.

Pulmonary ventilation consists of two phases: *inspiration* and *expiration*. Inspiration (inhalation) and expiration (exhalation) are accomplished by alternately increasing and decreasing the volumes of the thorax and lungs (fig. 24.17).

Inspiration and Expiration

Between the bony portions of the rib cage are two layers of intercostal muscles: the **external intercostal muscles** and the **internal intercostal muscles** (fig. 24.18). Between the costal cartilages, however, there is only one muscle layer, and its fibers are oriented in a manner similar to those of the internal intercostals. These muscles are therefore called the *interchondral part* of the internal intercostals. Another name for them is the **parasternal intercostals.**

An unforced, or quiet, inspiration results primarily from contraction of the dome-shaped **diaphragm,** which lowers and flattens when it contracts. This increases thoracic volume in a vertical direction. Inspiration is aided by contraction of the parasternal and external intercostals, which raises the ribs and increases thoracic volume laterally. Other thoracic muscles become involved in forced (deep) inspiration. The most important of these are the *scalenes*, followed by the *pectoralis minor*, and in extreme cases the *sternocleidomastoid muscles*. Contraction of these muscles elevates the ribs in an anteroposterior direction; at the same time, the upper rib cage is stabilized so

diaphragm: Gk. *dia*, across; *phragma*, fence

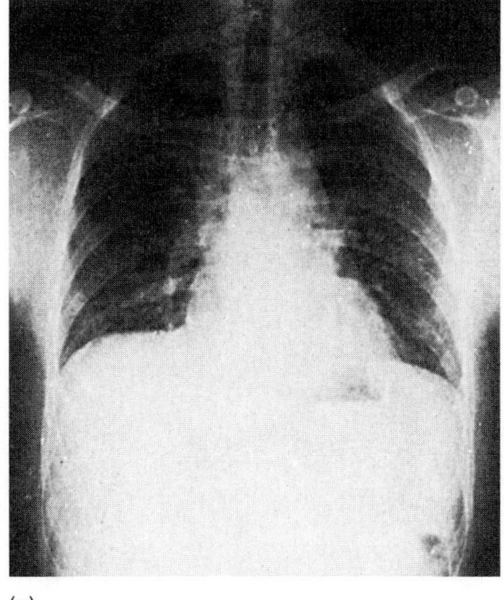
(a)

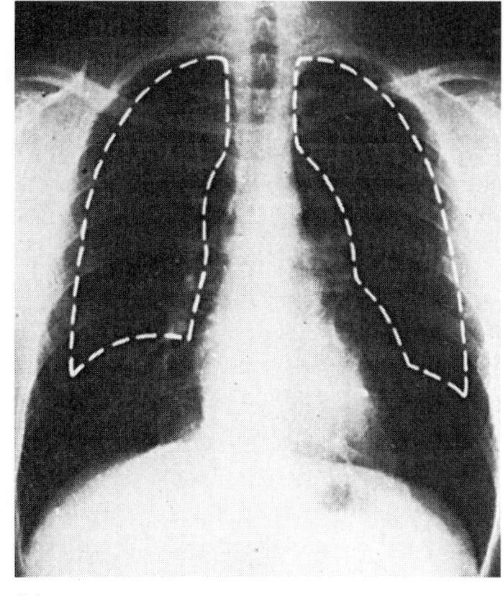
(b)

Figure 24.17

Changes in lung volume during breathing.
A change in lung volume, as shown by radiographs (*a*) during expiration and (*b*) during inspiration. The increase in lung volume during full inspiration is shown by comparison with the lung volume in full expiration (dashed lines).

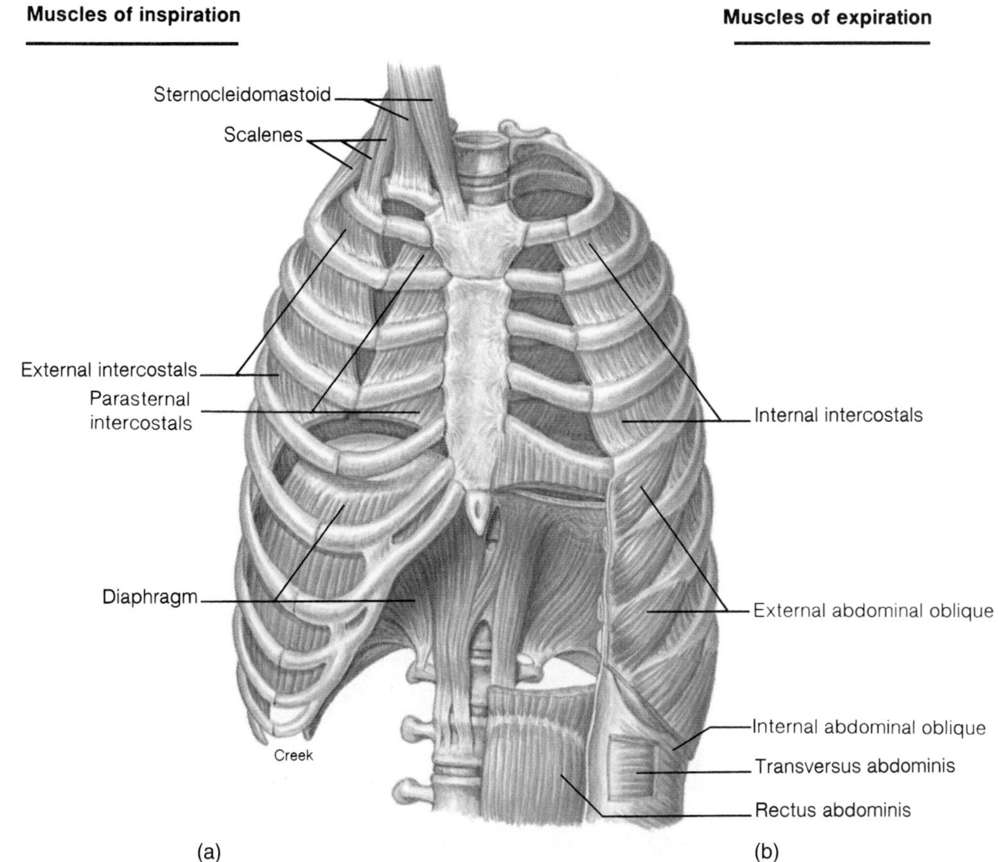

Figure 24.18

The muscles involved in breathing.
(*a*) The principal muscles of inspiration and (*b*) those of expiration.

that the intercostals become more effective. The increase in thoracic volume produced by these muscle contractions decreases intrapulmonary (intra-alveolar) pressure, thereby causing air to flow into the lungs.

Quiet expiration is a passive process. After becoming stretched by contractions of the diaphragm and thoracic muscles, the thorax and lungs recoil as a result of their elastic tension when the respiratory muscles relax. The decrease in lung volume raises the pressure within the alveoli above the atmospheric pressure and pushes the air out. During forced expiration, the internal intercostal muscles (excluding the interchondral part) contract and depress the rib cage. The abdominal muscles also aid expiration because, when they contract, they force abdominal organs up against the diaphragm and further decrease the volume of the thorax. By this means, the intrapulmonary pressure can rise 20 or 30 mmHg above the atmospheric pressure. The events that occur during inspiration and expiration are shown in figure 24.19.

Pulmonary Function Tests

Pulmonary function may be assessed clinically by means of a technique known as *spirometry* (*spi-rom'ĭ-tre*).In this procedure, a subject breathes in a closed system in which air is trapped within a light plastic bell floating in water. The bell moves up when the subject exhales and down when the subject inhales. The movements of the bell cause corresponding movements of a pen, which traces a record of

spirometry: L. *spiro*, breathe; Gk. *metron*, measure

Figure 24.19

The mechanics of pulmonary ventilation.
Pressures are shown (*a*) before inspiration, (*b*) during inspiration, and (*c*) during expiration. During inspiration, the intrapulmonary pressure is less than the atmospheric pressure, and during expiration it is greater than the atmospheric pressure.

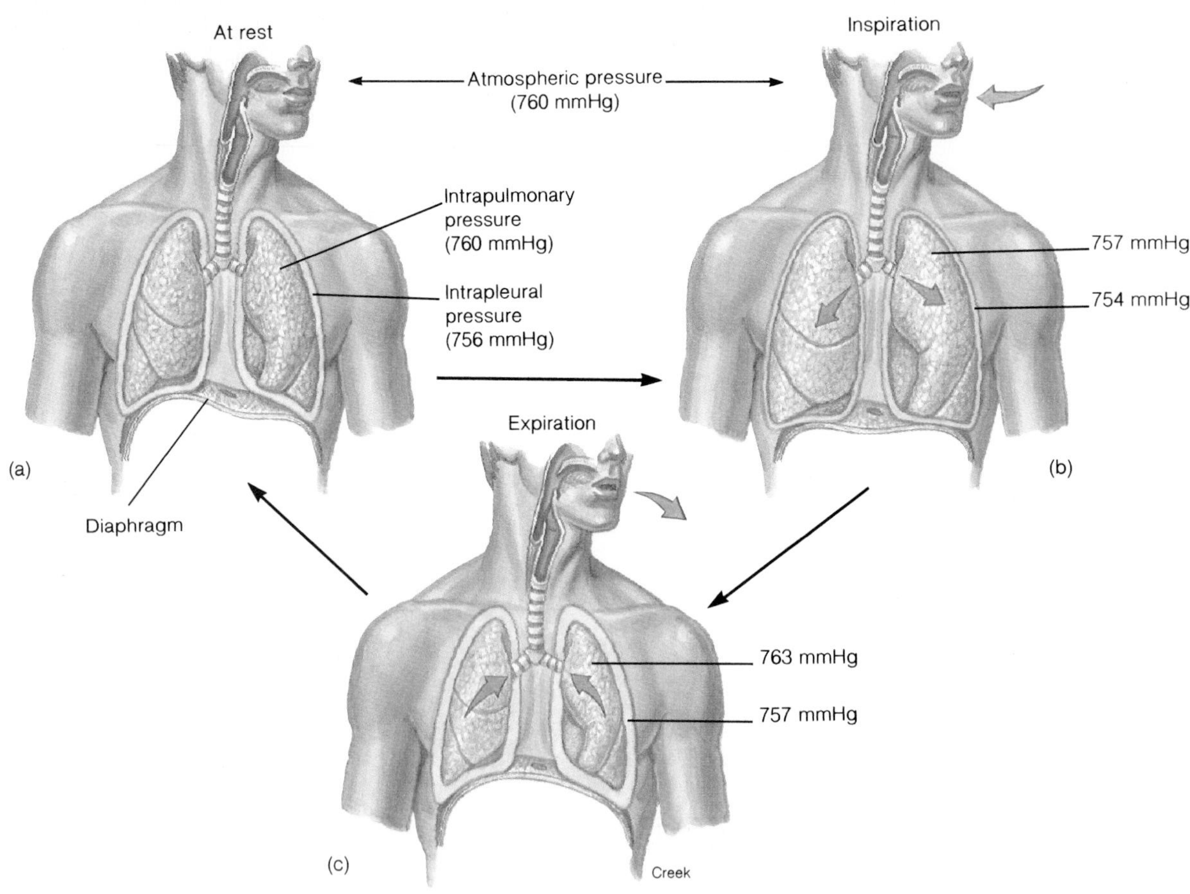

the breathing on a rotating drum recorder (fig. 24.20). More sophisticated computerized devices are also commonly employed to assess lung function.

Lung Volumes and Capacities

An example of a spirogram is shown in figure 24.21, and the various lung volumes and capacities are defined in table 24.2. A lung capacity is equal to the sum of two or more lung volumes. During quiet breathing, for example, the amount of air expired in each breath is the **tidal volume.** The maximum amount of air that can be forcefully exhaled after a maximum inhalation is called the **vital capacity,** which is equal to the sum of the **inspiratory reserve volume, tidal volume,** and **expiratory reserve volume** (fig. 24.21). Multiplying the tidal volume at rest by the number of breaths per minute yields a **total minute volume** of about 6 L per minute. During exercise, the tidal volume and the number of breaths per minute increase to produce a total minute volume as high as 100 to 200 L per minute.

Figure 24.20

A spirometer.
This instrument is used to measure lung volumes and capacities.

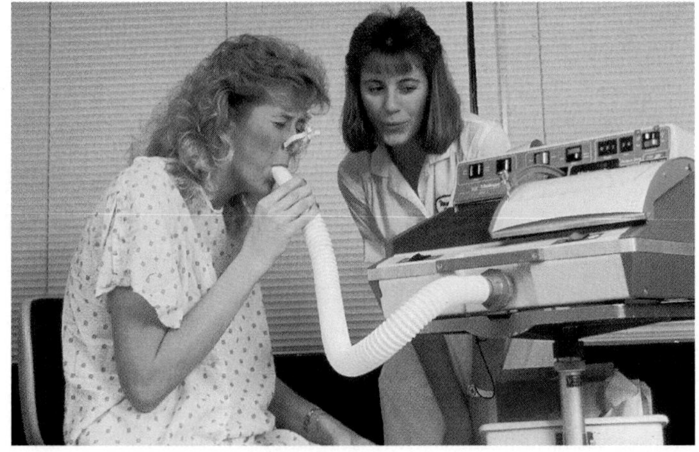

Figure 24.21

A spirogram showing lung volumes and capacities.
A lung capacity is the sum of two or more lung volumes. The vital capacity, for example, is the sum of the tidal volume, the inspiratory reserve volume, and the expiratory reserve volume. Note that residual volume cannot be measured with a spirometer, because it is air that cannot be exhaled. Therefore, the total lung capacity (the sum of the vital capacity and the residual volume) also cannot be measured with a spirometer.

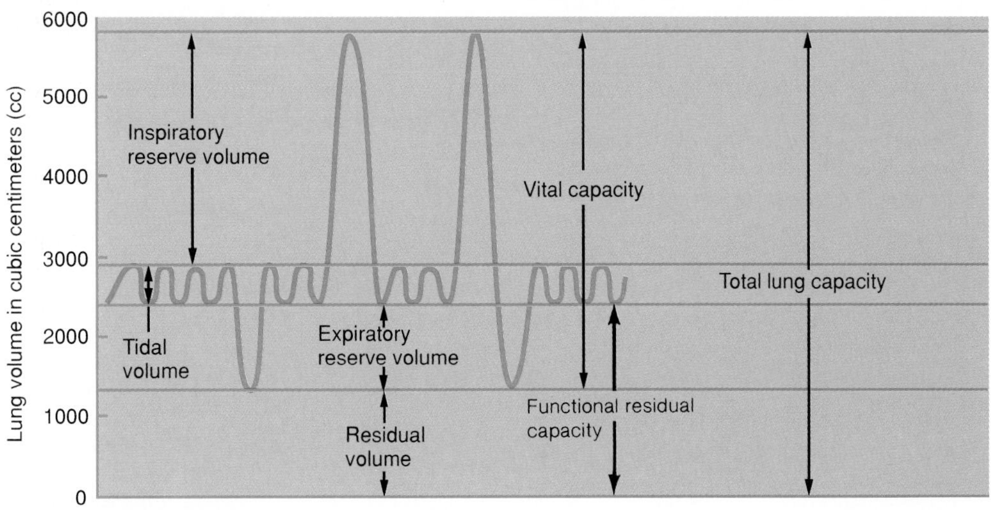

Table 24.2

Terms Used to Describe Lung Volumes and Capacities

Term	Definition	Term	Definition
Lung volumes	The four nonoverlapping components of the total lung capacity	**Lung capacities**	Measurements that are the sum of two or more lung volumes
Tidal volume	The volume of gas inspired or expired in an unforced respiratory cycle	Total lung capacity	The total amount of gas in the lungs at the end of a maximum inspiration
Inspiratory reserve volume	The maximum volume of gas that can be inspired during forced breathing in addition to tidal volume	Vital capacity	The maximum amount of gas that can be expired after a maximum inspiration
Expiratory reserve volume	The maximum volume of gas that can be expired during forced breathing in addition to tidal volume	Inspiratory capacity	The maximum amount of gas that can be inspired at the end of a tidal expiration
Residual volume	The volume of gas remaining in the lungs after a maximum expiration	Functional residual capacity	The amount of gas remaining in the lungs at the end of a tidal expiration

It should be noted that not all of the inspired volume reaches the alveoli with each breath. As fresh air is inhaled, it is mixed with air in the **anatomical dead space** (table 24.3). This dead space comprises the conducting zone of the respiratory system—nose, mouth, larynx, trachea, bronchi, and bronchioles—where no gas exchange occurs. Air within the anatomical dead space has a lower oxygen concentration and a higher carbon dioxide concentration than the external air. Since the air in the dead space enters the alveoli first, the amount of fresh air reaching the alveoli with each breath is less than the tidal volume. But, since the volume of air in the dead space is an anatomical constant, the percentage of fresh air entering the alveoli is increased with increasing tidal volumes. For example, if the anatomical dead space is 150 ml and the tidal volume is 500 ml, the percentage of fresh air reaching the alveoli is 350/500 × 100% = 70%. If the tidal volume is increased to 2,000 ml, and the anatomical dead space is still 150 ml, the percentage of fresh air reaching the alveoli is increased to 1,850/2,000 × 100% = 93%. An increase in tidal volume can thus be a factor in the respiratory adaptations to exercise and high altitude.

Table 24.3
Ventilation Terminology

Term	Definition
Air spaces	Alveolar ducts, alveolar sacs, and alveoli
Airways	Structures that conduct air from the mouth and nose to the respiratory bronchioles
Alveolar ventilation	Removal and replacement of gas in pulmonary alveoli; equal to the tidal volume minus the volume of dead space times the ventilation rate
Anatomical dead space	Volume of the conducting airways to the zone gas exchange occurs
Apnea	Cessation of breathing
Dyspnea	Unpleasant subjective feeling of difficult or labored breathing
Eupnea	Normal, comfortable breathing at rest
Hyperventilation	Alveolar ventilation that is excessive in relation to metabolic rate; results in abnormally low alveolar CO_2
Hypoventilation	An alveolar ventilation that is low in relation to metabolic rate; results in abnormally high alveolar CO_2
Physiological dead space	Combination of anatomical dead space and underventilated or underperfused alveoli that do not contribute normally to blood-gas exchange
Pneumothorax	Presence of gas in the intrapleural space (the space between the visceral and parietal pleurae), causing lung collapse
Torr	Unit of pressure very nearly equal to the millimeter of mercury (760 mmHg = 760 torr)

Spirometry is useful in the diagnosis of lung diseases. On the basis of pulmonary function tests, lung disorders can be classified as *restrictive* or *obstructive*. In restrictive disorders, such as pulmonary fibrosis, the vital capacity is reduced to below normal. The rate at which the vital capacity can be forcibly exhaled, however, is normal. In disorders that are exclusively obstructive, by contrast, the vital capacity is normal because lung tissue is not damaged. In asthma, for example, the vital capacity is normal, but expiration is more difficult and takes longer because bronchoconstriction increases the resistance to air flow. Obstructive disorders are therefore diagnosed by tests that measure the rate of expiration. One such test is the **forced expiratory volume (FEV),** in which the percentage of the vital capacity that can be exhaled in the first second ($FEV_{1.0}$) is measured (fig. 24.22). An $FEV_{1.0}$ that is significantly less than 80% suggests the presence of obstructive pulmonary disease.

Figure 24.22

The one-second forced expiratory volume ($FEV_{1.0}$) test.
The percentage in (*a*) is normal, whereas that in (*b*) may indicate an obstructive pulmonary disorder, including asthma and bronchitis.

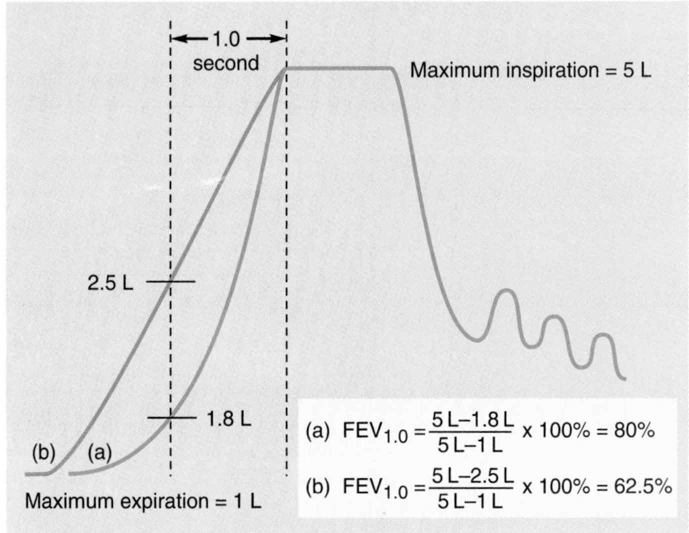

Bronchoconstriction often occurs in response to inhalation of noxious agents present in smoke or smog. The $FEV_{1.0}$ has therefore been used by researchers to determine the effects of inhaling various components of smog and passive cigarette smoke on pulmonary function. These studies have shown that it is unhealthy to exercise on very smoggy days and that inhaling smoke from other people's cigarettes in a closed environment can adversely affect pulmonary function.

There is normally a decline in the $FEV_{1.0}$ with age, but research suggests that this decline may be accelerated in cigarette smokers. Smokers under the age of 35 who quit have improved lung function; those who quit after the age of 35 slow their age-related decline in $FEV_{1.0}$ to normal rates.

People with pulmonary disorders frequently complain of **dyspnea** (*disp'ne-ă*), a subjective feeling of "shortness of breath." Dyspnea may occur even when ventilation is normal, however, and may not occur even when total minute volume is very high, as in exercise. Some of the terms associated with ventilation are defined in table 24.3, which also includes clinical terminology.

Gas Exchange in the Lungs

Gas exchange between the alveolar air and the blood in pulmonary capillaries results in an increased oxygen concentration and a decreased carbon dioxide concentration in the blood leaving the lungs. This blood enters the systemic arteries, where blood gas measurements are taken to assess the effectiveness of lung function.

dyspnea: Gk. *dys*, bad; *pnoe*, breathing

The atmosphere is an ocean of gas that exerts pressure on all objects within it. This pressure can be measured with a glass U-tube filled with fluid. One end of the U-tube is exposed to the atmosphere, while the other side is continuous with a sealed vacuum tube. Since the atmosphere presses on the open-ended side, but not on the side connected to the vacuum tube, atmospheric pressure pushes fluid in the U-tube up on the vacuum side to a height determined by the atmospheric pressure and the density of the fluid. Water, for example, will be pushed up to a height of 33.9 feet (10,332 mm) at sea level, whereas mercury (Hg)—which is more dense—will be raised to a height of 760 mm. As a matter of convenience, therefore, devices used to measure atmospheric pressure (barometers) use mercury rather than water. The atmospheric pressure at sea level is thus said to be equal to 760 mmHg (or 760 *torr*), which is also described as a pressure of *one atmosphere* (fig. 24.23).

Figure 24.23

Measurement of atmospheric pressure.
Atmospheric pressure at sea level can push a column of mercury to a height of 760 millimeters. This is also described as 760 torr, or one atmospheric pressure.

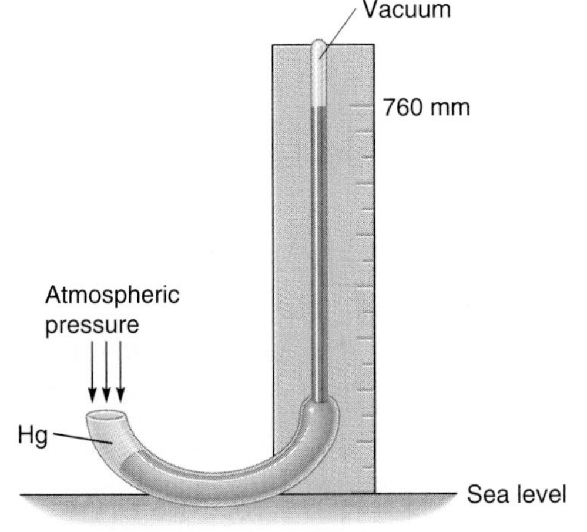

According to **Dalton's law,** the total pressure of a gas mixture (such as air) is equal to the sum of the pressures that each gas in the mixture would exert independently. The pressure that a particular gas in a mixture exerts independently is the **partial pressure** of that gas, which is equal to the product of the total pressure and the fraction of that gas in the mixture. The total pressure of the gas mixture is thus equal to the sum of the partial pressures of the constituent gases. Since oxygen constitutes about 21% of the atmosphere, for example, its partial pressure (abbreviated P_{O_2}) is 21% of 760, or about 159 mmHg. Since nitrogen constitutes about 78% of the atmosphere, its partial pressure is equal to $0.78 \times 760 =$ 593 mmHg. These two gases thus contribute about 99% of the total pressure of 760 mmHg:

$$P_{\text{dry atmosphere}} = P_{N_2} + P_{O_2} + P_{CO_2} = 760 \text{ mmHg}$$

Calculation of P_{O_2}

With increasing altitude, the total atmospheric pressure and the partial pressure of the constituent gases decrease (table 24.4). At Denver, for example (5,000 feet above sea level), the atmospheric pressure is decreased to 619 mmHg, and the P_{O_2} is therefore reduced to $619 \times 0.21 = 130$ mmHg. At the peak of Mount Everest (at 29,000 feet), the P_{O_2} is only 42 mmHg. As one descends below sea level, as in ocean diving, the total pressure increases by 1 atmosphere for every 33 feet. At 33 feet therefore, the pressure equals $2 \times 760 =$ 1,520 mmHg. At 66 feet, the pressure equals 3 atmospheres.

Inspired air contains variable amounts of moisture. By the time the air has passed into the respiratory zone of the

Dalton's law: from John Dalton, English chemist, 1766–1844

Table 24.4
The Effect of Altitude on P_{O_2}

Altitude (Feet above Sea Level)	Atmospheric Pressure (mmHg)	P_{O_2} in Air (mmHg)	P_{O_2} in Alveoli (mmHg)	P_{O_2} in Arterial Blood (mmHg)
0	760	159	105	100
2,000	707	148	97	92
4,000	656	137	90	85
6,000	609	127	84	79
8,000	564	118	79	74
10,000	523	109	74	69
20,000	349	73	40	35
30,000	226	47	21	19

lungs, however, it is normally saturated with water vapor (has a relative humidity of 100%). The capacity of air to contain water vapor depends on its temperature; since the temperature of the respiratory zone is constant at 37° C, its water vapor pressure is also constant (at 47 mmHg).

Water vapor, like the other constituent gases, contributes a partial pressure to the total atmospheric pressure. Since the total atmospheric pressure is constant (depending only on the height of the air mass), the water vapor "dilutes" the contribution of other gases to the total pressure:

$$P_{\text{wet atmosphere}} = P_{N_2} + P_{O_2} + P_{CO_2} + P_{H_2O}$$

When the effect of water vapor pressure is considered, the partial pressure of oxygen in the inspired air is decreased at sea level to

$$P_{O_2} \text{ (sea level)} = 0.21\ (760 - 47) = 150 \text{ mmHg}$$

As a result of gas exchange in the alveoli, the P_{O_2} of alveolar air is further diminished to about 105 mmHg. The partial pressures of the inspired air and the partial pressures of alveolar air are compared in figure 24.24.

Figure 24.24

Partial pressures of gases in inspired air and alveolar air.

Notice that, as air enters the alveoli, its oxygen content decreases and its carbon dioxide content increases. Also notice that air in the alveoli is saturated with water vapor (giving it a partial pressure of 47 mmHg), which dilutes the contribution of other gases to the total pressure.

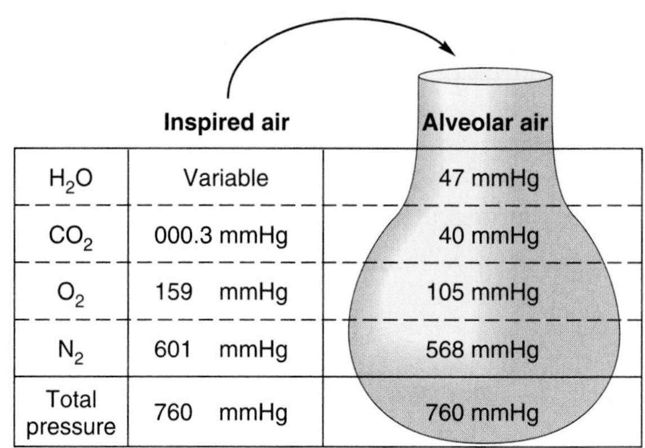

	Inspired air	Alveolar air
H_2O	Variable	47 mmHg
CO_2	000.3 mmHg	40 mmHg
O_2	159 mmHg	105 mmHg
N_2	601 mmHg	568 mmHg
Total pressure	760 mmHg	760 mmHg

Figure 24.25

Association between pulmonary alveoli and blood vessels.

The extensive surface area of contact between the pulmonary capillaries and the pulmonary alveoli allows for rapid exchange of gases between the air and blood.

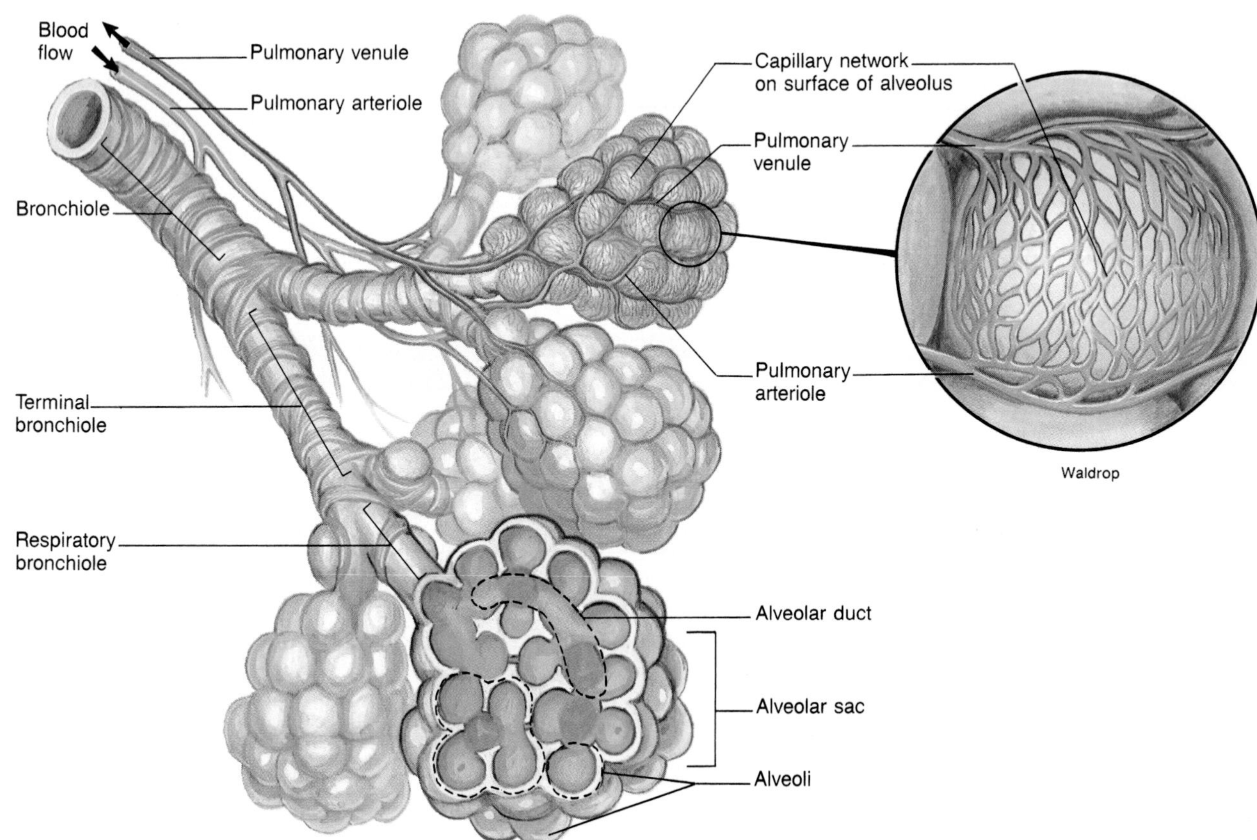

Figure 24.26

Blood gas measurements using the P_{O_2} electrode.
(*a*) The electrical current generated by the oxygen electrode is calibrated so that the needle of the blood gas machine points to the P_{O_2} of the gas with which the fluid is in equilibrium. (*b*) Once standardized in this way, the electrode can be inserted into a fluid, such as blood, and the P_{O_2} of this solution can be measured.

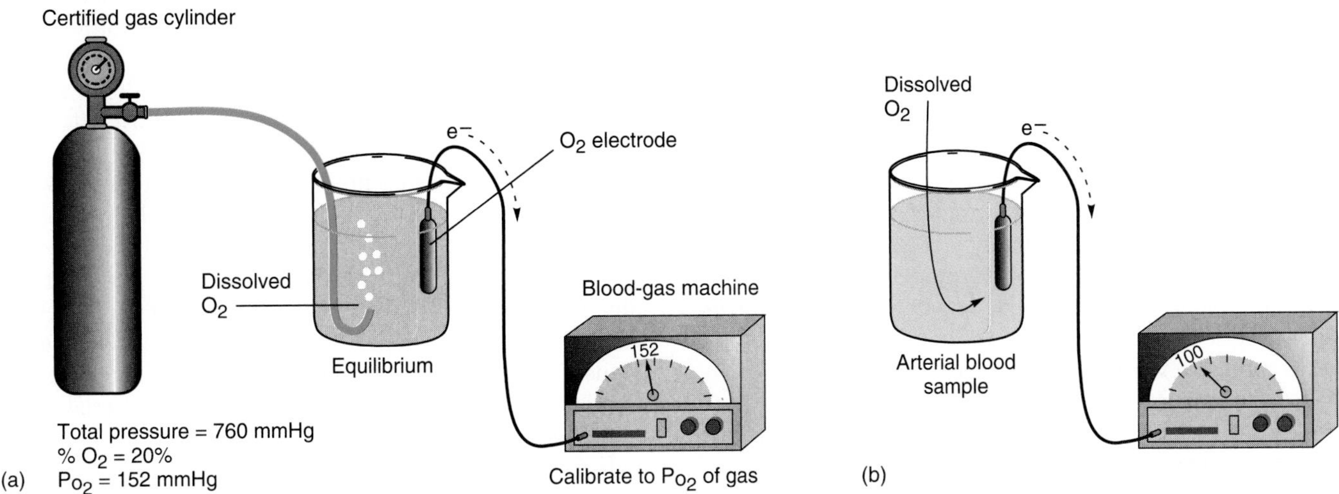

Partial Pressures of Gases in Blood

The enormous surface area of alveoli and the short diffusion distance between alveolar air and the capillary blood help to quickly bring the blood into gaseous equilibrium with the alveolar air. This function is further aided by the tremendous number of capillaries that form an almost continuous sheet of blood around each alveolus (fig. 24.25).

When a liquid and a gas, such as blood and alveolar air, are at equilibrium, the amount of gas dissolved in the fluid reaches a maximum value. According to **Henry's law,** this value depends on (1) the solubility of the gas in the fluid, which is a physical constant; (2) the temperature of the fluid—more gas can be dissolved in cold water than warm water; and (3) the partial pressure of the gas. Since the temperature of the blood does not vary significantly, *the concentration of a gas dissolved in a fluid (such as plasma) depends directly on its partial pressure in the gas mixture*. When water—or plasma—is brought into equilibrium with air at a P_{O_2} of 100 mmHg, for example, the fluid will contain 0.3 ml of O_2 per 100 ml fluid at 37° C. If the P_{O_2} of the gas were reduced by half, the amount of dissolved oxygen also would be reduced by half.

Blood Gas Measurements

Measurement of the oxygen content of blood (in ml O_2 per 100 ml blood) is a laborious procedure. Fortunately, an **oxygen electrode** that produces an electric current in proportion to the concentration of *dissolved oxygen* has been developed. If this electrode is placed in a fluid while oxygen is artificially bubbled into it, the current produced by the oxygen electrode will increase up to a maximum value. At this maximum value the fluid is saturated with oxygen—that is, all of the oxygen that can be dissolved at that temperature and P_{O_2} is dissolved. At a constant temperature, the amount dissolved, and thus the electric current, depend only on the P_{O_2} of the gas.

As a matter of convenience, it can now be said that *the fluid has the same P_{O_2} as the gas*. If it is known that the gas has a P_{O_2} of 152 mmHg, for example, the deflection of a needle by the oxygen electrode can be calibrated on a scale at 152 mmHg (fig. 24.26). The actual amount of dissolved oxygen under these circumstances is not particularly important (it can be looked up in solubility tables, if desired); it is simply a linear function of the P_{O_2}. A lower P_{O_2} indicates that less oxygen is dissolved; a higher P_{O_2} indicates that more oxygen is dissolved.

If the oxygen electrode is next inserted into an unknown sample of blood, the P_{O_2} of that sample can be read directly from the previously calibrated scale. Suppose, as illustrated in figure 24.26, the blood sample has a P_{O_2} of 100 mmHg. Since alveolar air has a P_{O_2} of about 105 mmHg, this reading indicates that the blood is almost in complete equilibrium with the alveolar air.

The oxygen electrode responds only to oxygen dissolved in water or plasma; it cannot respond to oxygen that is bound to hemoglobin in red blood cells. Most of the oxygen in blood, however, is located in the red blood cells attached to hemoglobin. The oxygen content of whole blood thus depends on both its P_{O_2} and its red blood cell and hemoglobin content. At a P_{O_2} of about 100 mmHg, whole blood normally contains almost 20 ml O_2 per 100 ml blood; of this amount, only 0.3 ml of O_2 is dissolved in the plasma and 19.7 ml of O_2 is found within the red blood cells. Since only the 0.3 ml of O_2 affects the P_{O_2} measurement, this measurement would be unchanged if the red blood cells were removed from the sample.

Henry's law: from William Henry, English chemist, 1775–1837

Significance of Blood P_{O_2} and P_{CO_2} Measurements

Since blood P_{O_2} measurements are not directly affected by the oxygen in red blood cells, the P_{O_2} does not provide a measurement of the total oxygen content of whole blood. It does, however, provide a good index of *lung function*. If the inspired air had a normal P_{O_2} but the arterial P_{O_2} was below normal, for example, you could conclude that gas exchange in the lungs was impaired. Measurements of arterial P_{O_2} thus provide valuable information in treating people with pulmonary diseases, in performing surgery (when breathing may be depressed by anesthesia), and in caring for premature babies with respiratory distress syndrome.

When the lungs are functioning properly, the P_{O_2} of systemic arterial blood is only 5 mmHg less than the P_{O_2} of alveolar air. At a normal P_{O_2} of about 100 mmHg, hemoglobin is almost completely loaded with oxygen. Thus, an increase in blood P_{O_2}—produced, for example, by breathing 100% oxygen from a gas tank—cannot significantly increase the amount of oxygen contained in the red blood cells. It can, however, significantly increase the amount of oxygen dissolved in the plasma (because the amount dissolved is directly determined by the P_{O_2}). If the P_{O_2} doubles, the amount of oxygen dissolved in the plasma also doubles, but the total oxygen content of whole blood increases only slightly. This is because the plasma contains relatively little oxygen compared to the red blood cells.

Since the oxygen carried by red blood cells must first dissolve in plasma before it can diffuse to the cells, however, a doubling of the blood P_{O_2} means that the *rate of oxygen diffusion* to the tissues would double under these conditions. For this reason, breathing from a tank of 100% oxygen (with a P_{O_2} of 760 mmHg) would significantly increase oxygen delivery to the tissues, although it would have little effect on the total oxygen content of blood.

An electrode that produces a current in response to dissolved carbon dioxide is also used, so that the P_{CO_2} of blood can be measured together with its P_{O_2}. Blood in the systemic veins, which is delivered to the lungs by the pulmonary arteries, usually has a P_{O_2} of 40 mmHg and a P_{CO_2} of 46 mmHg. After gas exchange in the alveoli of the lungs, blood in the pulmonary veins and systemic arteries has a P_{O_2} of about 100 mmHg and a P_{CO_2} of 40 mmHg (fig. 24.27). The values in arterial blood are relatively constant and are clinically significant because they reflect lung function. Blood gas measurements of venous blood are not as useful, since these values are far more variable. Venous P_{O_2} is much lower and P_{CO_2} much higher after exercise, for example, than at rest, whereas arterial values are not significantly affected by moderate physical activity.

Figure 24.27

Partial pressures of gases in blood.
The P_{O_2} and P_{CO_2} values of blood are a result of gas exchange in the pulmonary alveoli and gas exchange between systemic capillaries and body cells.

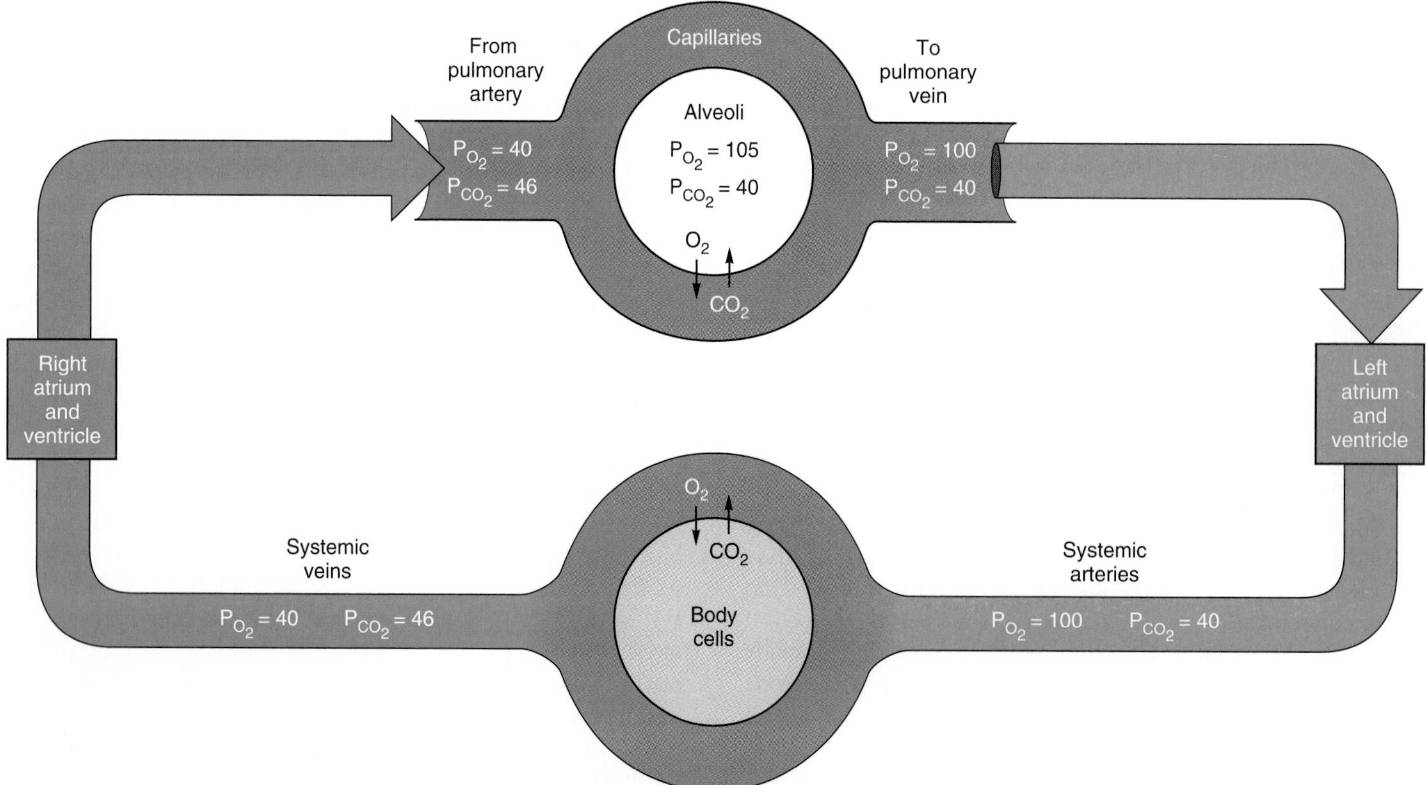

Pulmonary Circulation and Ventilation/Perfusion Ratios

In a fetus, the pulmonary circulation has a high vascular resistance because the lungs are partially collapsed. This high vascular resistance helps to shunt blood from the right to the left atrium through the foramen ovale, and from the pulmonary artery to the aorta through the ductus arteriosus (described in chapter 21). After birth, the foramen ovale and ductus arteriosus close, and the vascular resistance of the pulmonary circulation falls sharply. This fall in vascular resistance at birth is due to (1) opening of the vessels as a result of the subatmospheric intrapulmonary pressure and physical stretching of the lungs during inspiration and (2) dilation of the pulmonary arterioles in response to increased alveolar P_{O_2}.

In the adult, the right ventricle (like the left) has an output of about 5.5 L per minute. The rate of blood flow through the pulmonary circulation is thus equal to the flow rate through the systemic circulation. Blood flow, as described in chapter 22, is directly proportional to the pressure difference between the two ends of a vessel and inversely proportional to the vascular resistance. In the systemic circulation, the mean arterial pressure is 90 to 100 mmHg and the pressure of the right atrium is 0 mmHg; therefore, the pressure difference is about 100 mmHg. The mean pressure of the pulmonary artery, by contrast, is only 15 mmHg and the pressure of the left atrium is 5 mmHg. The driving pressure in the pulmonary circulation is thus 15 – 5, or 10 mmHg.

Since the driving pressure in the pulmonary circulation is only one-tenth that of the systemic circulation, and since the flow rates are equal, it follows that the pulmonary vascular resistance must be one-tenth that of the systemic vascular resistance. The pulmonary circulation, in other words, is a low-resistance, low-pressure pathway. The low pulmonary blood pressure produces less filtration pressure (chapter 22) than that produced in the systemic capillaries, and thus affords protection against *pulmonary edema.* This is a dangerous condition in which excessive fluid can enter the interstitial spaces of the lungs and then the alveoli, impeding ventilation and gas exchange. Pulmonary edema can occur when there is pulmonary hypertension, which may be produced by left ventricular heart failure.

Pulmonary arterioles constrict when the alveolar P_{O_2} is low and dilate as the alveolar P_{O_2} is raised. This response is opposite to that of systemic arterioles, which dilate in response to low tissue P_{O_2} (as described in chapter 22). Dilation of the systemic arterioles when the P_{O_2} is low helps to supply more blood and oxygen to the tissues; constriction of the pulmonary arterioles when the alveolar P_{O_2} is low helps to decrease blood flow to alveoli that are inadequately ventilated.

Constriction of the pulmonary arterioles where the alveolar P_{O_2} is low and their dilation where the alveolar P_{O_2} is high helps to *match ventilation to perfusion* (the term *perfusion* refers to blood flow). If this autoregulation of blood flow did not occur, blood from poorly ventilated alveoli would mix with blood from well-ventilated alveoli, and the blood leaving the lungs would have a lowered P_{O_2} as a result of this dilution effect.

Dilution of the P_{O_2} of pulmonary vein blood actually does occur to some degree despite these regulatory mechanisms. When a person stands upright, the force of gravity causes a greater blood flow to the base of the lungs than to the apex (top). Ventilation likewise increases from apex to base, but this increase is not proportionate to the increase in blood flow. The *ventilation/perfusion ratio* at the apex is thus high (0.24 L air divided by 0.07 L blood per minute gives a ratio of 3.4/1.0), while at the base of the lungs it is low (0.82 L air divided by 1.29 L blood per minute gives a ratio of 0.6/1.0). This is illustrated in figure 24.28.

Functionally, the alveoli at the apex of the lungs are thus overventilated (or underperfused) and are actually larger than alveoli at the base. This mismatch of ventilation/perfusion ratios is normal, but is largely responsible for the 5-mmHg difference in P_{O_2} between alveolar air and arterial blood. Abnormally large mismatches of ventilation/perfusion ratios can occur in cases of pneumonia, pulmonary emboli, edema, or other pulmonary disorders.

Figure 24.28

Lung ventilation/perfusion ratios.
The ventilation, blood flow, and ventilation/perfusion ratios are indicated for the apex and base of the lungs. The ratios indicate that the apex is relatively overventilated and the base underventilated in relation to their blood flows. As a result of such uneven matching of ventilation to perfusion, the blood leaving the lungs has a P_{O_2} that is slightly less (by about 4.5 mmHg) than the P_{O_2} of alveolar air.

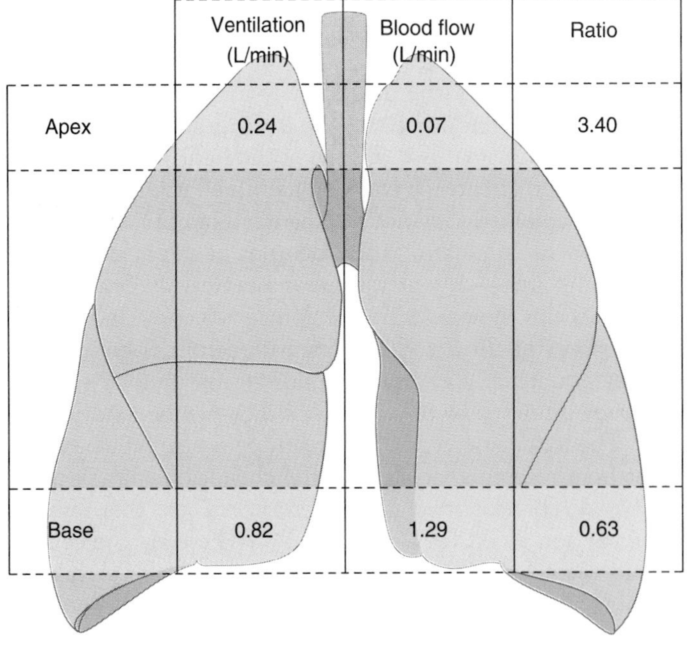

Ventilation (L/min)
Blood flow (L/min)
Ratio
Apex
0.24
0.07
3.40
Base
0.82
1.29
0.63

Regulation of Breathing

The motor neurons that stimulate the respiratory muscles are controlled by two major descending pathways: one that controls voluntary breathing, and another that controls involuntary breathing. The unconscious rhythmic control of breathing is influenced by sensory feedback from receptors sensitive to the P_{CO_2}, pH, and P_{O_2} of arterial blood.

Inspiration and expiration occur by means of the contraction and relaxation of skeletal muscles in response to activity in somatic motor neurons in the spinal cord. The activity of these motor neurons, in turn, is controlled by descending tracts from neurons in the respiratory control centers in the medulla oblongata and from neurons in the cerebral cortex.

Brain Stem Respiratory Centers

As described in chapter 15, a loose aggregation of neurons in the reticular formation of the *medulla oblongata* functions as a neural center to produce rhythmic breathing. This **rhythmicity** (*rith-mis'ĭ-te*) **center** consists of interacting pools of neurons that fire either during inspiration (*I neurons*) or expiration (*E neurons*). The I neurons project to and stimulate spinal motoneurons that innervate the respiratory muscles. Expiration is a passive process that occurs when the I neurons are inhibited, presumably by the activity of the E neurons. The activity of I and E neurons varies in a reciprocal way, so that a rhythmic pattern of breathing is produced.

The inspiratory neurons are located primarily in the *superior respiratory group*, and the expiratory neurons are in the *inferior respiratory group*. These form two parallel columns within the medulla oblongata. The superior group of neurons regulates the activity of the phrenic nerves to the diaphragm, and the inferior group controls the motor neurons to the internal intercostal muscles.

The activity of the medullary rhythmicity center is influenced by centers in the *pons*. As a result of research in which the brain stem is destroyed at different levels, two respiratory control centers have been identified in the pons. One area—the **apneustic** (*ap-noo'stik*) **center**—appears to promote inspiration by stimulating the I neurons in the medulla. The other pons area—called the **pneumotaxic** (*noo"mŏ-tak'sik*) **center**—seems to antagonize the apneustic center and inhibit inspiration (fig. 24.29). The apneustic center is believed to provide a tonic, or constant, stimulus for inspiration, which is cyclically inhibited by the activity of the pneumotaxic center.

The automatic control of breathing is also influenced by input from receptors sensitive to the chemical composition of the blood. There are two groups of *chemoreceptors* that respond to changes in blood P_{CO_2}, pH, and P_{O_2}. These are the **central chemoreceptors** in the medulla oblongata and the **peripheral chemoreceptors.** The peripheral chemoreceptors are contained within small nodules associated with the aorta and the carotid arteries, and they receive blood from these critical arteries via small arterial branches. The peripheral chemoreceptors include the **aortic bodies,** located near the aortic arch, and the **carotid bodies,** located in each common carotid artery at the point where it branches into the internal and external carotid arteries (fig. 24.30). The aortic and carotid bodies should not be confused with the aortic and carotid sinuses (chapter 22) that are located within these arteries. The aortic and carotid sinuses contain receptors that monitor the blood pressure.

The peripheral chemoreceptors control breathing indirectly via sensory nerve fibers to the medulla oblongata. The aortic bodies send sensory information to the medulla oblongata in the vagus (cranial nerve X); the carotid bodies stimulate sensory fibers in the glossopharyngeal (cranial nerve IX). The neural and sensory control of ventilation is summarized in figure 24.31.

Figure 24.29

Approximate locations of the brain stem respiratory centers.
The rhythmicity center in the medulla oblongata directly controls breathing, but it receives input from the control centers in the pons and from chemoreceptors.

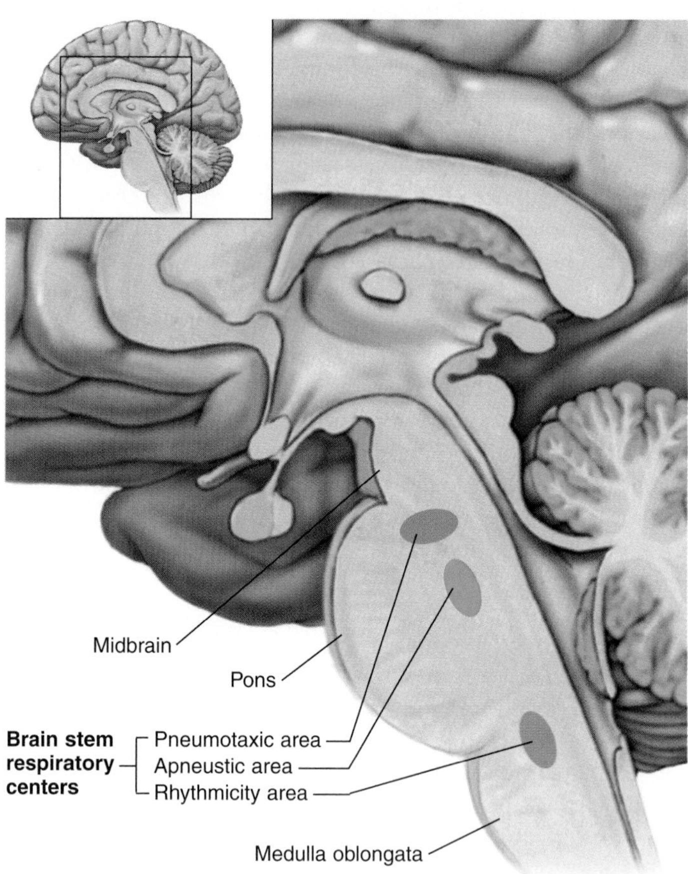

Figure 24.30

Sensory input from the aortic and carotid bodies.
The peripheral chemoreceptors (aortic and carotid bodies) regulate the brain stem respiratory centers by means of sensory nerve stimulation.

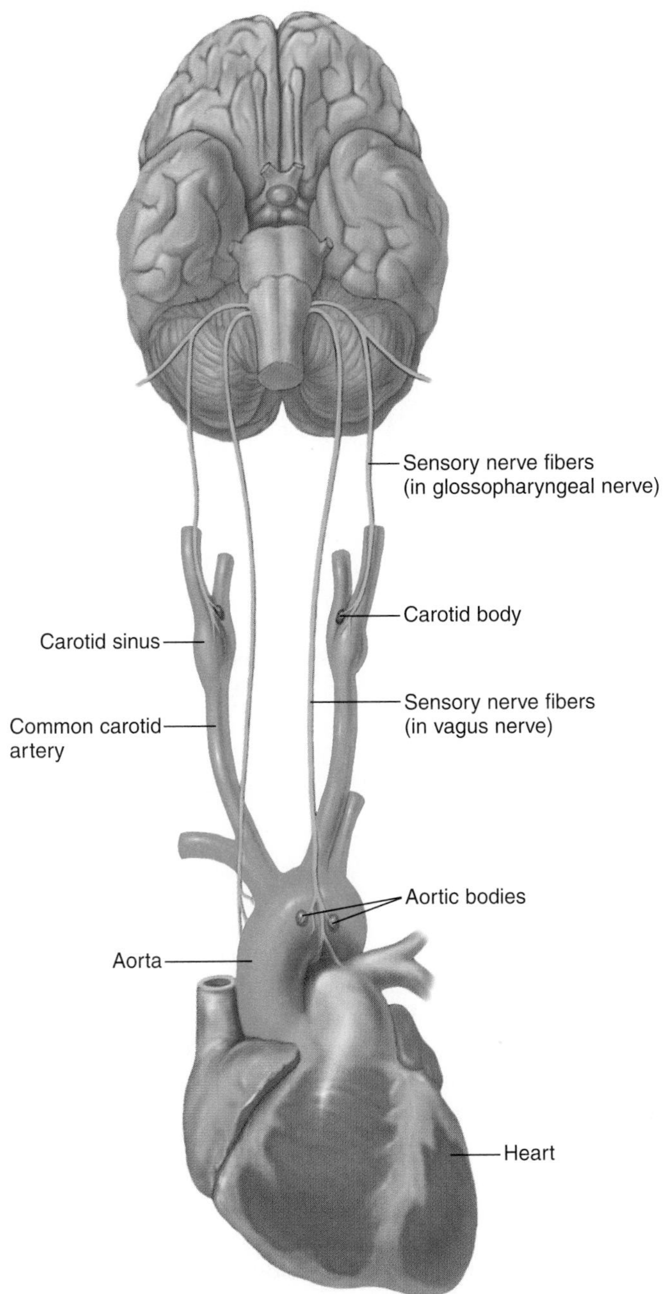

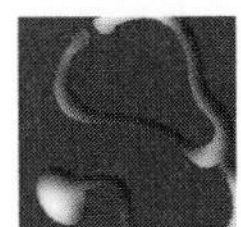

The automatic control of breathing is regulated by nerve fibers that descend from the medulla oblongata in the lateral and ventral white matter of the spinal cord. The voluntary control of breathing is a function of the cerebral cortex and involves nerve fibers that descend in the corticospinal tracts (chapter 15). The separation of the voluntary and involuntary pathways is dramatically illustrated in the condition named *Ondine's curse* (after a German fairy tale). In this condition, neurological damage abolishes the automatic control of breathing. People with Ondine's curse must remind themselves to breathe, and therefore they cannot sleep without the aid of a mechanical respirator.

Figure 24.31

Regulation of ventilation by the central nervous system.
Note that the feedback effects of pulmonary stretch receptors and "irritant" receptors on the control of breathing are not shown.

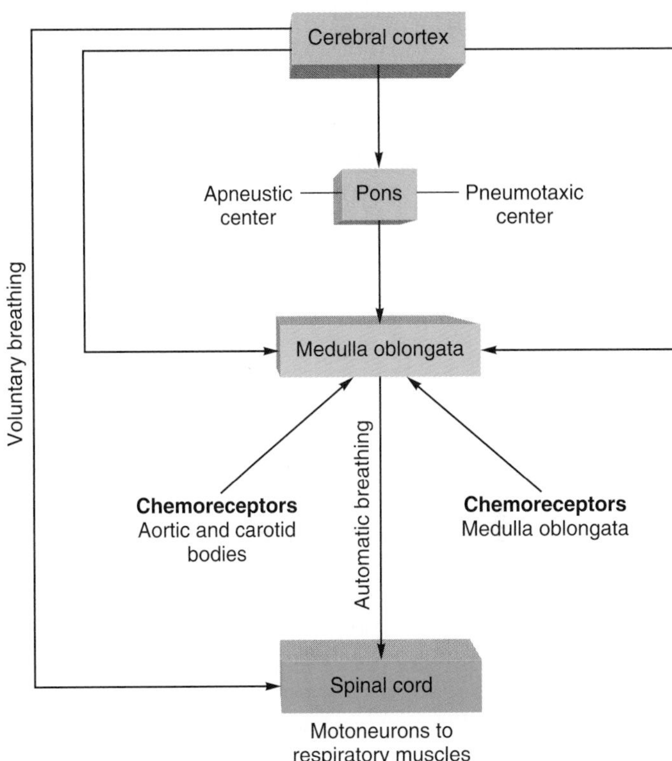

Effects of Blood P_{CO_2} and pH on Ventilation

Chemoreceptor input to the brain stem modifies the rate and depth of breathing so that, under normal conditions, arterial P_{CO_2}, pH, and P_{O_2} remain relatively constant. If hypoventilation (inadequate ventilation) occurs, P_{CO_2} quickly rises and pH falls. The fall in pH results from the fact that carbon dioxide can combine with water to form carbonic acid, which in turn can release H^+ into the solution. This is shown in the following equations:

$$CO_2 + H_2O \rightarrow H_2CO_3$$
$$H_2CO_3 \rightarrow HCO_3^- + H^+$$

The oxygen content of the blood decreases much more slowly because of the large "reservoir" of oxygen attached to hemoglobin. During hyperventilation, conversely, blood P_{CO_2} quickly falls and pH rises because of the excessive elimination of carbonic acid. The oxygen content of blood, on the other hand, is not significantly increased by hyperventilation (hemoglobin in arterial blood is 97% saturated with oxygen during normal ventilation).

For the reasons just stated, the blood P_{CO_2} and pH are more immediately affected by changes in ventilation than is the oxygen content. Indeed, changes in P_{CO_2} provide a sensitive index of ventilation, as shown in table 24.5. In view of these facts, it is not surprising that changes in P_{CO_2} provide

Table 24.5

The Effect of Ventilation, as Measured by Total Minute Volume (Breathing Rate × Tidal Volume), on the P_{CO_2} of Arterial Blood

Total Minute Volume	Arterial P_{CO_2}	Type of Ventilation
2 L/min	80 mmHg	Hypoventilation
4–5 L/min	40 mmHg	Normal ventilation
8 L/min	20–25 mmHg	Hyperventilation

the most potent stimulus for the reflex control of ventilation. Ventilation, in other words, is adjusted to maintain a constant P_{CO_2}; proper oxygenation of the blood occurs naturally as an extension of this reflex control.

The rate and depth of ventilation are normally adjusted to maintain an arterial P_{CO_2} of 40 mmHg. Hypoventilation causes a rise in P_{CO_2}—a condition called *hypercapnia* (*hi″per-kap′ne-ă*). Hyperventilation, conversely, results in *hypocapnia*. Chemoreceptor regulation of breathing in response to changes in P_{CO_2} is illustrated in figure 24.32.

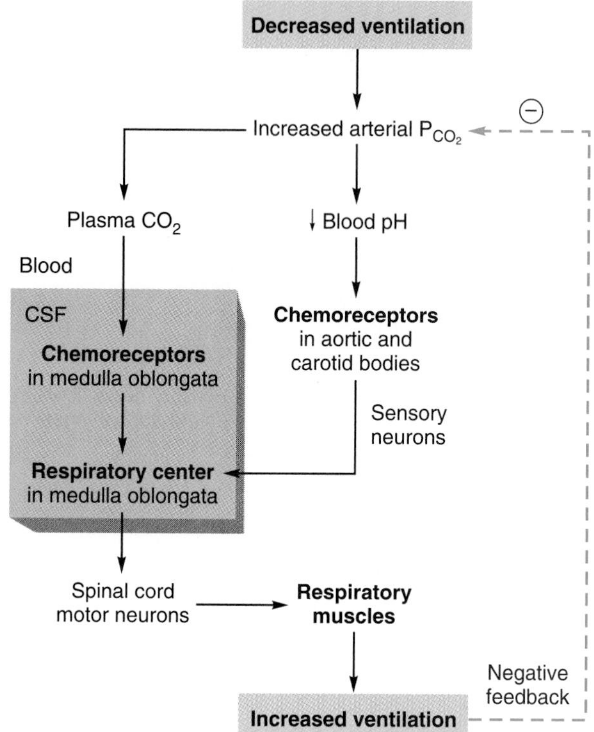

Figure 24.32

Chemoreceptor control of breathing. This figure depicts the negative feedback control of ventilation through changes in blood P_{CO_2} and pH. The orange box represents the blood-brain barrier, which allows CO_2 to pass into the cerebrospinal fluid but prevents the passage of H^+.

Chemoreceptors in the Medulla Oblongata

The chemoreceptors most sensitive to changes in the arterial P_{CO_2} are located in the inferior area of the medulla oblongata, near the exit of the ninth and tenth cranial nerves. These chemoreceptor neurons are anatomically separate from, but synaptically communicate with, the neurons of the rhythmicity control center in the medulla oblongata.

An increase in arterial P_{CO_2} causes a rise in the H^+ concentration of the blood as a result of increased carbonic acid concentrations. The H^+ in the blood, however, cannot cross the blood-brain barrier, and thus cannot influence the medullary chemoreceptors. Carbon dioxide in the arterial blood *can* cross the blood-brain barrier and, through the formation of carbonic acid, can lower the pH of cerebrospinal fluid (fig. 24.33). This fall in cerebrospinal fluid pH directly stimulates the chemoreceptors in the medulla oblongata when there is a rise in arterial P_{CO_2}.

The chemoreceptors in the medulla oblongata are ultimately responsible for 70% to 80% of the increased ventilation that occurs in response to a sustained rise in arterial P_{CO_2}. However, this response takes several minutes. The immediate increase in ventilation that occurs when P_{CO_2} rises is produced by stimulation of the peripheral chemoreceptors.

Peripheral Chemoreceptors

The aortic and carotid bodies are not stimulated directly by blood CO_2. Instead, they are stimulated by a rise in the H^+ concentration (fall in pH) of arterial blood, which occurs when the blood CO_2, and thus carbonic acid, rises. The retention of CO_2 during hypoventilation thus stimulates the medullary chemoreceptors through a lowering of cerebrospinal fluid pH and stimulates peripheral chemoreceptors through a lowering of blood pH.

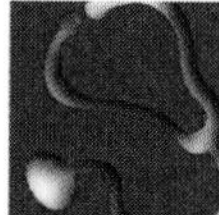

People who hyperventilate during psychological stress are sometimes told to breathe into a paper bag so that they rebreathe their expired air, enriched with CO_2. This procedure helps to raise their blood P_{CO_2} to the normal range. This is needed because hypocapnia causes cerebral vasoconstriction, reducing brain perfusion and producing ischemia. The cerebral ischemia causes dizziness and can lead to an acidotic condition in the brain. Through stimulation of the medullary chemoreceptors, this causes further hyperventilation. Breathing into a paper bag can thus relieve the hypocapnia and stop the hyperventilation.

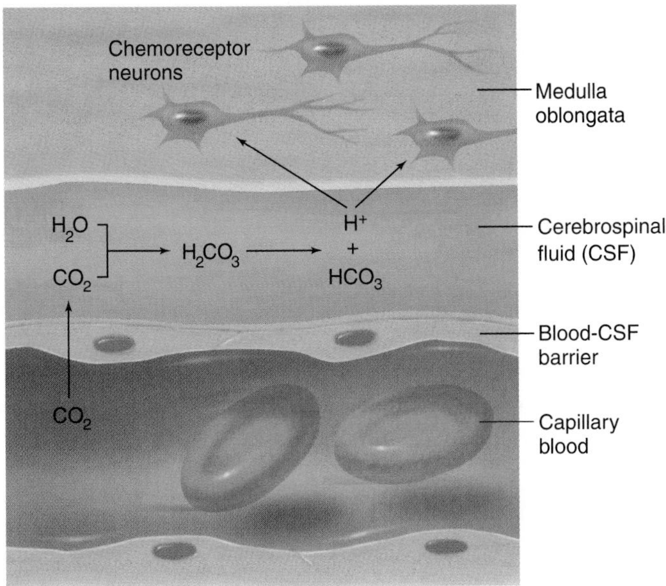

Figure 24.33

How blood CO_2 affects chemoreceptors in the medulla oblongata.
An increase in blood CO_2 stimulates breathing indirectly by lowering the pH of blood and cerebrospinal fluid (CSF). This figure illustrates how a rise in blood CO_2 increases the H^+ concentration (lowers the pH) of CSF and thereby stimulates chemoreceptor neurons in the medulla oblongata.

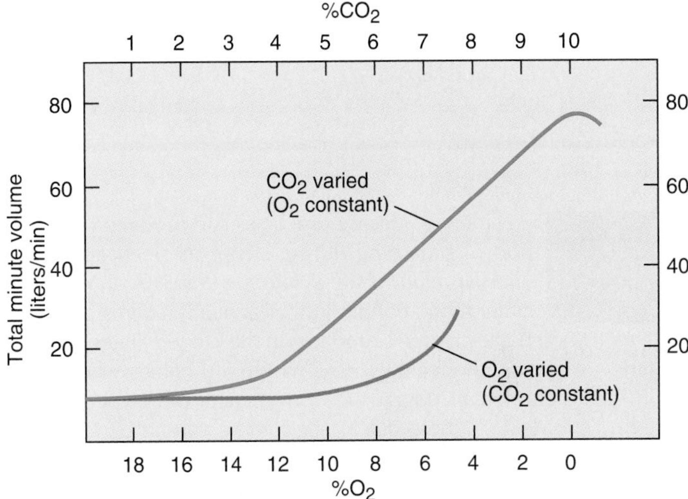

Figure 24.34

Comparing the effects of blood CO_2 and O_2 on breathing.
The graph depicts the effects of increasing blood concentrations of CO_2 (see the scale on the top of the graph) on breathing, as measured by the total minute volume. The effects of decreasing concentrations of blood O_2 (see the scale on the bottom of the graph) on breathing are also shown for comparison. Notice that respiration increases linearly with increasing CO_2 concentration, whereas O_2 concentrations must decrease to half the normal value before respiration is stimulated.

Effects of Blood P_{O_2} on Ventilation

Under normal conditions, blood P_{O_2} affects breathing only indirectly, by influencing the chemoreceptor sensitivity to changes in P_{CO_2}. Chemoreceptor sensitivity to P_{CO_2} is augmented by a low P_{O_2} (so ventilation is increased at a high altitude, for example) and is decreased by a high P_{O_2}. Therefore, if the blood P_{O_2} is raised by breathing 100% oxygen, the breath can be held longer because the response to increased P_{CO_2} is blunted.

When the blood P_{CO_2} is held constant by experimental techniques, the P_{O_2} of arterial blood must fall from 100 mmHg to below 50 mmHg before ventilation is significantly stimulated (fig. 24.34). This stimulation is apparently due to a direct effect of P_{O_2} on the carotid bodies. Since this degree of *hypoxemia* (*hi″pok-se′me-ă*), or low blood oxygen (table 24.6), does not normally occur at sea level, P_{O_2} does not normally exert this direct effect on breathing.

In emphysema, when there is a chronic retention of carbon dioxide, the chemoreceptor response to the carbon dioxide becomes blunted. This is because the choroid plexus in the brain (chapter 15) secretes more bicarbonate into the cerebrospinal fluid, buffering the fall in cerebrospinal fluid pH. The abnormally high P_{CO_2}, however, enhances the sensitivity of the carotid bodies to a fall in P_{O_2}. For people with emphysema, therefore, breathing may be stimulated by a *hypoxic drive* rather than by increases in blood P_{CO_2}. Over a long period, however, the chronic hypoxia reduces the sensitivity of the carotid bodies in people with emphysema or other forms of chronic obstructive pulmonary disease, exacerbating their breathing problems.

The effects of changes in the blood P_{CO_2}, pH, and P_{O_2} on chemoreceptors and the regulation of ventilation are summarized in table 24.7.

Table 24.6

Terms Used to Describe Blood Oxygen and Carbon Dioxide Levels

Term	Definition
Hypoxemia	A lower than normal oxygen content or P_{O_2} in arterial blood.
Hypoxia	A lower than normal oxygen content or P_{O_2} in the lungs, blood, or tissues. This is a more general term than hypoxemia. Tissues can be hypoxic, for example, even though there is no hypoxemia (as when the blood flow is occluded).
Hypercapnia, or hypercarbia	An increase in the P_{CO_2} of systemic arteries to above 40 mmHg. Usually this occurs when the ventilation is inadequate for a given metabolic rate (hypoventilation). Antonyms are *hypocapnia* and *hypocarbia* (usually produced by hyperventilation).

Table 24.7

Sensitivity of Chemoreceptors to Change in Blood Gases and pH

Stimulus	Chemoreceptor	Comments
↑P_{CO_2}	Medullary chemoreceptors; aortic and carotid bodies	Medullary chemoreceptors are sensitive to the pH of cerebrospinal fluid (CSF). Diffusion of CO_2 from the blood into the CSF lowers the pH of CSF by forming carbonic acid. Similarly, the aortic and carotid bodies are stimulated by a fall in blood pH induced by increases in blood CO_2.
↓pH	Aortic and carotid bodies	Peripheral chemoreceptors are stimulated by decreased blood pH independent of the effect of blood CO_2. Chemoreceptors in the medulla oblongata are not affected by changes in blood pH because H^+ cannot cross the blood-brain barrier.
↓P_{O_2}	Carotid bodies	Low blood P_{O_2} (hypoxemia) augments the chemoreceptor response to increases in blood P_{CO_2} and can stimulate ventilation directly when the P_{O_2} falls below 50 mmHg.

A variety of disease processes can produce cessation of breathing during sleep, or sleep apnea. *Sudden infant death syndrome* (*SIDS*) is an especially tragic condition that annually claims about 1 in 10,000 babies under 12 months in the United States. Victims are apparently healthy 2-to-5-month-old babies who die in their sleep for no obvious reason—hence, the layperson's term "crib death." These deaths seem to be caused by failure of the respiratory control mechanisms in the brain stem and/or by failure of the carotid bodies to be stimulated by reduced arterial oxygen. Since 1992, when the American Academy of Pediatrics began a campaign recommending that parents put infants to sleep on their backs rather than on their stomachs, the number of infants dying from SIDS has dropped by 38%.

Pulmonary Stretch and Irritant Reflexes

The lungs contain various types of receptors that influence the brain stem respiratory control centers via sensory fibers in the vagus nerves. These receptors are involved in both regulatory and defensive reflexes. Irritant receptors in the lungs, for example, stimulate reflex constriction of the bronchioles in response to smoke, ozone, and smog. A chemical called *capsaicin* (*kap-sa'ĭ-sin*)—the ingredient in hot peppers that creates the burning sensation—has been shown to be a potent stimulator of sensory fibers in the vagus nerves that promote reflex bronchosecretion and bronchoconstriction. These reflexes are presumed to be beneficial in situations where they improve the propulsive force, and thus the effectiveness, of coughing.

The **Hering–Breuer** (*her'ing broy'er*) **reflex** is stimulated by pulmonary stretch receptors. The activation of these receptors during inspiration inhibits the respiratory control centers, making further inspiration increasingly difficult. This helps to prevent undue distension of the lungs and may contribute to the smoothness of the ventilation cycles. A similar inhibitory reflex may occur during expiration. The Hering–Breuer reflex appears to be important in maintaining normal ventilation in the newborn. Pulmonary stretch receptors in adults, however, are probably not active at normal resting tidal volumes (500 ml per breath) but may contribute to respiratory control at high tidal volumes, as during exercise.

Hering–Breuer reflex: from Ewald Hering, German physiologist, 1834–1918, and Josef Breuer, Austrian physician, 1842–1925

Hemoglobin and Oxygen Transport

Hemoglobin without oxygen, or deoxyhemoglobin, can bond with oxygen to form oxyhemoglobin. This "loading" reaction occurs in the capillaries of the lungs. The dissociation of oxyhemoglobin, or "unloading" reaction, occurs in the tissue capillaries. The bond strength between hemoglobin and oxygen, and thus the extent of the unloading reaction, is adjusted by various factors to ensure an adequate delivery of oxygen to the tissues.

If the lungs are functioning properly, blood leaving in the pulmonary veins and traveling in the systemic arteries has a P_{O_2} of about 100 mmHg, indicating a plasma oxygen concentration of about 0.3 ml O_2 per 100 ml blood. The total oxygen content of the blood, however, cannot be determined if only the P_{O_2} of blood plasma is known. As mentioned earlier, the total oxygen content depends not only on the P_{O_2} but also on the hemoglobin concentration. If the P_{O_2} and hemoglobin concentration are normal, arterial blood contains about 20 ml of O_2 per 100 ml of blood (fig. 24.35).

Hemoglobin

Most of the oxygen in the blood is contained within the red blood cells, where it is chemically bonded to **hemoglobin.** As described in chapter 20, each hemoglobin molecule consists of a protein *globin* part, composed of four polypeptide chains, and four nitrogen-containing, disc-shaped organic pigment molecules called *hemes* (fig. 24.36).

The protein part of hemoglobin is composed of two identical *alpha chains*, each 141 amino acids long, and two identical *beta chains*, each 146 amino acids long. Each of the four polypeptide chains is combined with one heme group.

Figure 24.35

Oxygen content of blood.

Plasma and whole blood that are brought into equilibrium with the same gas mixture have the same P_{O_2} and thus the same number of dissolved oxygen molecules (shown as black dots). The oxygen content of whole blood, however, is much higher than that of plasma because of the binding of oxygen to hemoglobin.

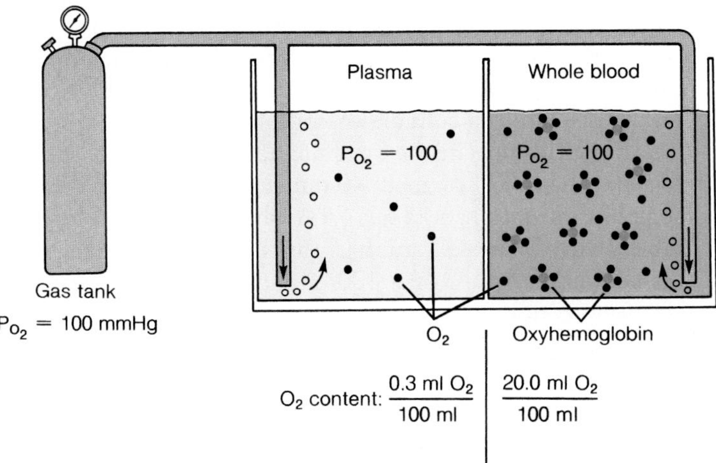

Figure 24.36

The structure of hemoglobin.

(*a*) An illustration of the three-dimensional structure of hemoglobin in which the two alpha and two beta polypeptide chains are shown. The four heme groups are represented as flat structures with iron (shown as spheres) in the centers. (*b*) The chemical structure of heme.

Hemoglobin

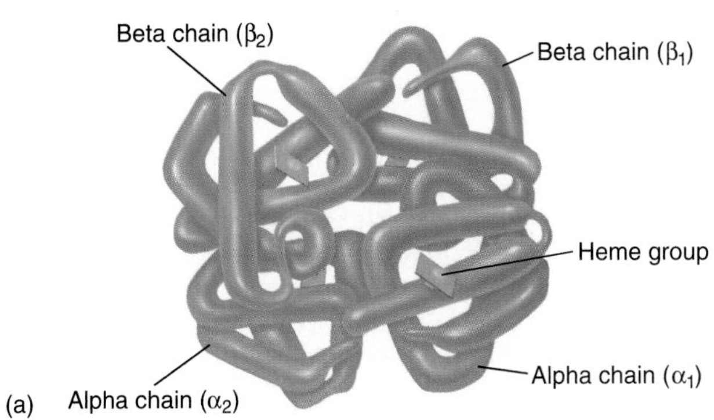

Heme

(b)

In the center of each heme group is one atom of iron, which can combine with one molecule of oxygen. One hemoglobin molecule can thus combine with four molecules of oxygen—and since there are about 280 million hemoglobin molecules per red blood cell, each red blood cell can carry over a billion molecules of oxygen.

Normal heme contains iron in the reduced form (Fe^{2+}, or ferrous iron). In this form, the iron can share electrons and bond with oxygen to form **oxyhemoglobin.** When oxyhemoglobin dissociates to release oxygen to the tissues, the heme iron is still in the reduced (Fe^{2+}) form, and the hemoglobin is called **deoxyhemoglobin,** or **reduced hemoglobin.** The term *oxyhemoglobin* is thus not equivalent to *oxidized* hemoglobin; hemoglobin does not lose an electron (and become oxidized) when it combines with oxygen. Oxidized hemoglobin, or **methemoglobin** (*met-he'mŏ-glo"bin*), has iron in the oxidized (Fe^{3+}, or ferric) state. Methemoglobin thus lacks the electron it needs to form a bond with oxygen and cannot participate in oxygen transport. Blood normally contains only a small amount of methemoglobin, but certain drugs can increase this amount.

In **carboxyhemoglobin,** another abnormal form of hemoglobin, the reduced heme is combined with *carbon monoxide* instead of oxygen. Since the bond with carbon monoxide is about 210 times stronger than the bond with oxygen, carbon monoxide tends to displace oxygen in hemoglobin and remains attached to hemoglobin as the blood passes through systemic capillaries. In carbon monoxide poisoning—which, in severe form results primarily from smoke inhalation and suicide attempts, and in milder forms from breathing smoggy air and smoking cigarettes—the transport of oxygen to the tissues is reduced.

According to federal standards, the percentage of carboxyhemoglobin in the blood of active nonsmokers should be no higher than 1.5%. However, concentrations of 3% in nonsmokers and 10% in smokers have been reported in some cities. Although these high levels may not cause immediate problems in healthy people, long-term adverse effects on health are possible. People with respiratory or cardiovascular diseases would be particularly vulnerable to the negative effects of carboxyhemoglobin on oxygen transport.

Hemoglobin Concentration

The *oxygen-carrying capacity* of whole blood is determined by its concentration of normal hemoglobin. If the hemoglobin concentration is below normal—a condition called **anemia**—the oxygen concentration of the blood falls below normal. Conversely, when the hemoglobin concentration rises above the normal range—as occurs in **polycythemia** (high red blood

anemia: Gk. *a*, negative; *haima*, blood

cell count)—the oxygen-carrying capacity of blood is increased accordingly. This can occur as an adaptation to life at a high altitude.

The production of hemoglobin and red blood cells in bone marrow is controlled by a hormone called **erythropoietin** (*ĕ-rith″ro-poi-e′tin*), produced by the kidneys. The secretion of erythropoietin—and thus the production of red blood cells—is stimulated when the amount of oxygen delivered to the kidneys and other organs is lower than normal. Red blood cell production is also promoted by androgens, which explains why the hemoglobin concentration in men is from 1 to 2 g per 100 ml higher than in women.

The Loading and Unloading Reactions

Deoxyhemoglobin and oxygen combine to form oxyhemoglobin; this is called the **loading reaction.** Oxyhemoglobin, in turn, dissociates to yield deoxyhemoglobin and free oxygen molecules; this is the **unloading reaction.** The loading reaction occurs in the lungs, and the unloading reaction occurs in the systemic capillaries.

Loading and unloading can thus be shown as a reversible reaction:

$$\text{Deoxyhemoglobin} + O_2 \underset{\text{(tissues)}}{\overset{\text{(lungs)}}{\rightleftarrows}} \text{Oxyhemoglobin}$$

The extent to which the reaction will go in either direction depends on two factors: (1) the P_{O_2} of the environment and (2) the *affinity*, or bond strength, between hemoglobin and oxygen. High P_{O_2} drives the equation to the right (favors the loading reaction); at the high P_{O_2} of the pulmonary capillaries, almost all the deoxyhemoglobin molecules combine with oxygen. Low P_{O_2} in the systemic capillaries drives the reaction in the opposite direction to promote unloading. The extent of this unloading depends on how low the P_{O_2} values are.

The affinity between hemoglobin and oxygen also influences the loading and unloading reactions. A very strong bond would favor loading but inhibit unloading; a weak bond would hinder loading but improve unloading. The bond strength between hemoglobin and oxygen is normally strong enough so that 97% of the hemoglobin leaving the lungs is in the form of oxyhemoglobin, yet the bond is sufficiently weak so that adequate amounts of oxygen are unloaded to sustain aerobic respiration in the tissues.

erythropoietin: Gk. *erythros*, red; *poiesis*, a making

The Oxyhemoglobin Dissociation Curve

Blood in the systemic arteries, at a P_{O_2} of 100 mmHg, has a *percent oxyhemoglobin saturation* of 97% (which means that 97% of the hemoglobin is in the form of oxyhemoglobin). This blood is delivered to the systemic capillaries, where oxygen diffuses into the cells and is consumed in aerobic respiration. Blood leaving in the systemic veins is thus reduced in oxygen; it has a P_{O_2} of about 40 mmHg and a percent oxyhemoglobin saturation of about 75% when a person is at rest (table 24.8). Expressed another way, blood entering the tissues contains 20 ml O_2 per 100 ml blood, and blood leaving the tissues contains 15.5 ml O_2 per 100 ml blood (fig. 24.37). Thus, 22%, or 4.5 ml of O_2 out of the 20 ml of O_2 per 100 ml blood, is unloaded to the tissues.

A graphic illustration of the percent oxyhemoglobin saturation at different values of P_{O_2} is called an **oxyhemoglobin dissociation curve** (fig. 24.37). The values in this graph are obtained by subjecting samples of blood in vitro to different partial oxygen pressures. The percent oxyhemoglobin saturations obtained, however, can be used to predict what the unloading percentages would be in vivo with a given difference in arterial and venous P_{O_2} values.

Figure 24.37 shows the difference between the arterial and venous P_{O_2} and the percent oxyhemoglobin saturation at rest. The relatively large amount of oxyhemoglobin remaining in the venous blood at rest functions as an oxygen reserve. If a person stops breathing, a sufficient reserve of oxygen in the blood will keep the brain and heart alive for about 4–5 minutes without requiring cardiopulmonary resuscitation (CPR) techniques. This reserve supply of oxygen can also be tapped when a tissue's requirements for oxygen are raised.

The oxyhemoglobin dissociation curve is S-shaped, or *sigmoidal*. The fact that it is relatively flat at high P_{O_2} values indicates that changes in P_{O_2} within this range have little effect on the loading reaction. One would have to ascend as high as 10,000 feet, for example, before the oxyhemoglobin saturation of arterial blood would decrease from 97% to 93%.

Table 24.8

The Relationship between Percent Oxyhemoglobin Saturation and P_{O_2} (at pH of 7.40 and Temperature of 37° C)

P_{O_2} (mmHg)	100	80	61	45	40	36	30	26	23	21	19
Percent oxyhemoglobin saturation	97	95	90	80	75	70	60	50	40	35	30
	Arterial blood				Venous blood						

At more common elevations, the percent oxyhemoglobin saturation would not be significantly different from the 97% value at sea level.

At the steep part of the sigmoidal curve, however, small changes in P_{O_2} values produce large differences in percent saturation. A decrease in *venous* P_{O_2} from 40 mmHg to 30 mmHg, as might occur during mild exercise, corresponds to a change in percent saturation from 75% to 58%. Since the arterial percent saturation is usually still 97% during exercise, the lowered venous percent saturation indicates that more oxygen has unloaded to the tissues. The difference between the arterial and venous percent saturations indicates the percent unloading. In the preceding example, 97% – 75% = 22% unloading at rest, and 97% – 58% = 39% unloading during mild exercise. During heavier exercise, the venous P_{O_2} can drop to 20 mmHg or lower, indicating a percent unloading of about 80%.

Effect of pH and Temperature on Oxygen Transport

In addition to changes in P_{O_2}, the loading and unloading reactions are influenced by changes in the affinity of hemoglobin for oxygen. Such changes ensure that active skeletal muscles will receive more oxygen from the blood than they do at rest. This occurs as a result of the lowered pH and increased temperature in exercising muscles.

The affinity is decreased when the pH lowers and is increased when the pH rises; this is called the **Bohr effect.** When the affinity of hemoglobin for oxygen is reduced, there is slightly less loading of the blood with oxygen in the lungs but greater unloading of oxygen in the tissues. The net effect is that the tissues receive more oxygen when the blood pH is lowered (table 24.9). Since the pH can be decreased by carbon dioxide (through the formation of carbonic acid), the Bohr effect helps to provide more oxygen to the tissues when their carbon dioxide output (and metabolism) is increased.

When oxyhemoglobin dissociation curves are graphed at different pH values, the dissociation curve is shown to be shifted to the right by a lowering of pH and shifted to the left by a rise in pH (fig. 24.38). If the percent unloading is calculated (by subtracting the percent oxyhemoglobin saturation for arterial and venous blood at a given P_{O_2}), it will be shown

Bohr effect: from Christian Bohr, Danish physiologist, 1855–1911

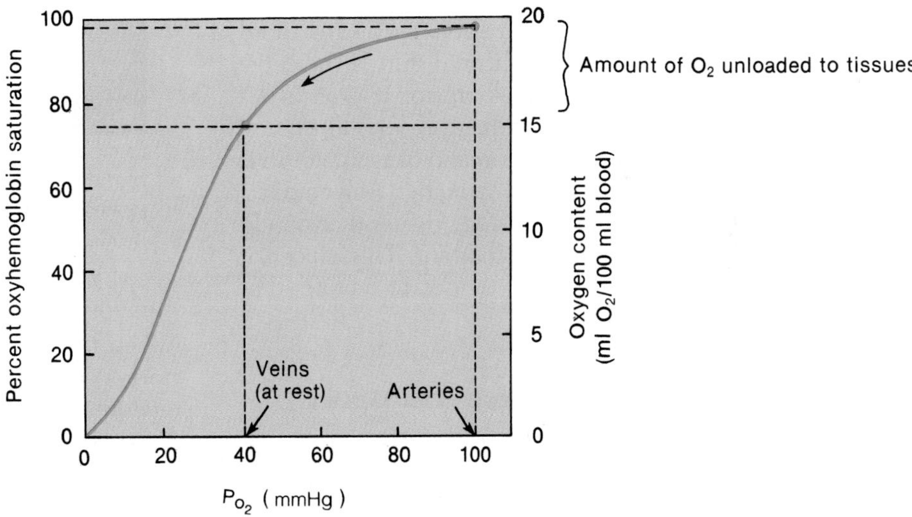

Figure 24.37

The oxyhemoglobin dissociation curve.
The percentage of oxyhemoglobin saturation and the blood oxygen content are shown at different values of P_{O_2}. Notice that oxyhemoglobin decreases by about 25% as the blood passes through the tissue from arteries to veins, resulting in the unloading of approximately 5 ml O_2 per 100 ml to the tissues.

Table 24.9

The Effect of pH on Hemoglobin Affinity for Oxygen and Unloading of Oxygen to the Tissues

pH	Affinity	Arterial O_2 Content per 100 ml	Venous O_2 Content per 100 ml	O_2 Unloaded to Tissues per 100 ml
7.40	Normal	19.8 ml O_2	14.8 ml O_2	5.0 ml O_2
7.60	Increased	20.0 ml O_2	17.0 ml O_2	3.0 ml O_2
7.20	Decreased	19.2 ml O_2	12.6 ml O_2	6.6 ml O_2

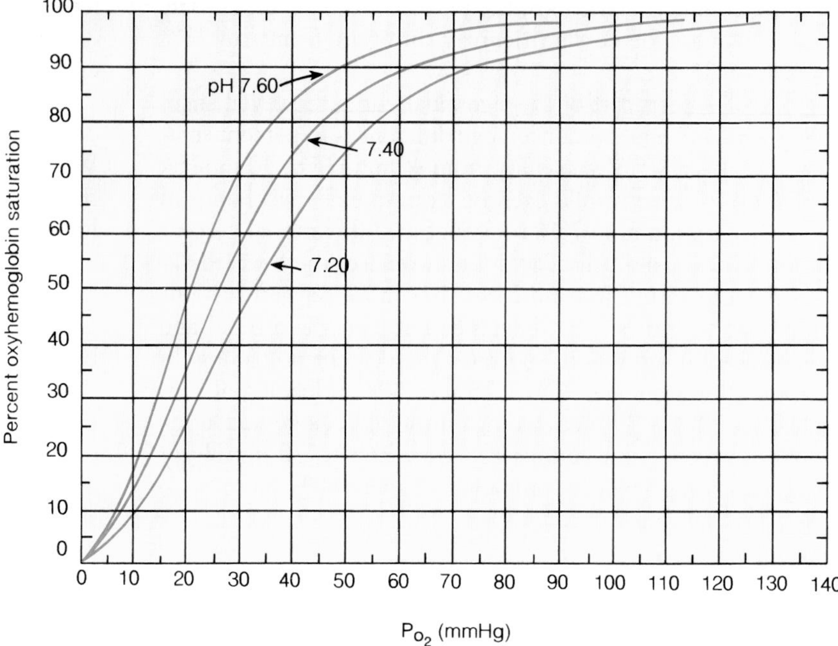

Figure 24.38

Effect of pH on the oxyhemoglobin dissociation curve.
A decrease in blood pH (an increase in H^+ concentration) decreases the affinity of hemoglobin for oxygen at each P_{O_2} value, resulting in a "shift to the right" of the oxyhemoglobin dissociation curve. A curve that is shifted to the right has a lower-percent oxyhemoglobin saturation at each P_{O_2}, but the effect is more marked at lower P_{O_2} values. This is called the Bohr effect.

that a *shift to the right* of the curve indicates a greater unloading of oxygen. A *shift to the left*, conversely, indicates less unloading but slightly more oxygen loading in the lungs.

When oxyhemoglobin dissociation curves are constructed at different temperatures, the curve moves rightward as the temperature increases. The rightward shift of the curve indicates that the affinity of hemoglobin for oxygen is decreased by a rise in temperature. An increase in temperature weakens the bond between hemoglobin and oxygen and thus has the same effect as a fall in pH. At higher temperatures, therefore, more oxygen is unloaded to the tissues than would be the case if the bond strength were constant. This effect can significantly enhance the delivery of oxygen to muscles that are warmed during exercise.

Effect of 2,3-DPG on Oxygen Transport

Mature red blood cells lack both nuclei and mitochondria. Without mitochondria, they cannot respire aerobically; the very cells that carry oxygen are the only cells in the body that cannot use it! Red blood cells, therefore, must obtain energy through the anaerobic respiration of glucose. At a certain point in the glycolytic pathway, a "side reaction" occurs in the red blood cells that results in a unique product—**2,3-diphosphoglyceric acid (2,3-DPG).**

The enzyme that produces 2,3-DPG is inhibited by oxyhemoglobin. When the oxyhemoglobin concentration is decreased, therefore, the production of 2,3-DPG is increased. This increase in 2,3-DPG production can occur when the total hemoglobin concentration is low (in anemia) or when the P_{O_2} is low (at a high altitude, for example). The bonding of 2,3-DPG with deoxyhemoglobin makes the deoxyhemoglobin more stable. Therefore, a higher proportion of the oxyhemoglobin will convert to deoxyhemoglobin by the unloading of its oxygen. An increased concentration of 2,3-DPG in red blood cells thus increases oxygen unloading and shifts the oxyhemoglobin dissociation curve to the right.

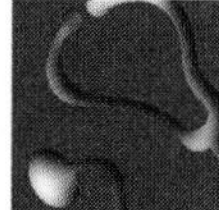

The importance of 2,3-DPG in red blood cells is now recognized in blood banking. Red blood cells that have been stored for some time can lose their ability to produce 2,3-DPG as they lose their ability to metabolize glucose. Modern techniques for blood storage, therefore, include the addition of energy substrates for respiration and phosphate sources needed for the production of 2,3-DPG.

Anemia

When the total blood hemoglobin concentration falls below normal in anemia, each red blood cell produces increased amounts of 2,3-DPG. A normal hemoglobin concentration of 15 g per 100 ml unloads about 4.5 ml O_2 per 100 ml at rest, as previously described. If the hemoglobin concentration were reduced by half, you might expect that the tissues would receive only half the normal amount of oxygen (2.25 ml O_2 per 100 ml). It has been shown, however, that an amount as great as 3.3 ml O_2 per 100 ml is unloaded to the tissues under these conditions. This occurs as a result of a rise in 2,3-DPG production that causes a decrease in the affinity hemoglobin for oxygen.

Hemoglobin F

The effects of 2,3-DPG are also important in the transfer of oxygen from maternal to fetal blood. In an adult, hemoglobin molecules are composed of two alpha and two beta chains as previously described. Fetal hemoglobin, by contrast, contains two alpha and two *gamma* chains in place of beta chains (gamma chains differ from beta chains in thirty-seven of their amino acids). Normal adult hemoglobin in the mother

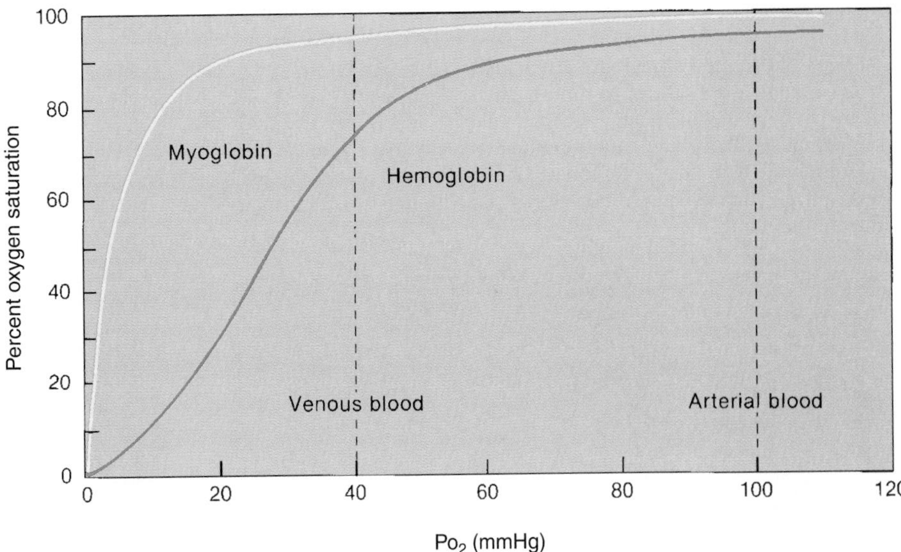

Figure 24.39

Comparison of the dissociation curves for hemoglobin and myoglobin.
Myoglobin is an oxygen-binding pigment in skeletal muscles. At the P_{O_2} of venous blood, the myoglobin retains almost all of its oxygen, indicating that it has a higher affinity for oxygen than hemoglobin. The myoglobin, however, does release its oxygen at the very low P_{O_2} values found inside the mitochondria.

(hemoglobin A) is able to bind to 2,3-DPG. Fetal hemoglobin, or **hemoglobin F,** cannot bind to 2,3-DPG, and thus has a higher affinity for oxygen than does hemoglobin A. Since hemoglobin F can have a higher percent oxyhemoglobin saturation than hemoglobin A, oxygen is transferred from the maternal to the fetal blood as these two come into close proximity in the placenta.

Muscle Myoglobin

As described in chapter 12, **myoglobin** is a red pigment found exclusively in striated muscle cells. In particular, slow-twitch, aerobically respiring skeletal fibers and cardiac muscle cells are rich in myoglobin. Myoglobin is similar to hemoglobin, but it has one heme rather than four; therefore, it can combine with only one molecule of oxygen.

Myoglobin has a higher affinity for oxygen than does hemoglobin, and its dissociation curve is therefore to the left of the oxyhemoglobin dissociation curve (fig. 24.39). The shape of the myoglobin curve is also different from the oxyhemoglobin dissociation curve. The myoglobin curve is rectangular, indicating that oxygen will be released only when the P_{O_2} becomes very low.

Since the P_{O_2} in mitochondria is very low (because oxygen is incorporated into water here), myoglobin may act as a "go-between" in the transfer of oxygen from blood to the mitochondria within muscle cells. Myoglobin may also have an oxygen-storage function, which is of particular importance in the heart. During diastole, when the coronary blood flow is greatest, myoglobin can load up with oxygen. This stored oxygen can then be released during systole, when the coronary arteries are squeezed closed by the contracting myocardium.

Carbon Dioxide Transport and Acid-Base Balance

Carbon dioxide is transported in the blood primarily in the form of bicarbonate (HCO_3^-), which is released when carbonic acid dissociates. Bicarbonate can buffer H^+ and thus helps to maintain a normal arterial pH. Hypoventilation raises, and hyperventilation lowers, the carbonic acid concentration of the blood.

Carbon dioxide is carried by the blood in three forms: (1) as *dissolved* CO_2—carbon dioxide is about 21 times more soluble than oxygen in water, and about one-tenth of the total blood CO_2 is dissolved in plasma; (2) as *carbaminohemoglobin*—about one-fifth of the total blood CO_2 is carried attached to an amino acid in hemoglobin (carbaminohemoglobin should not be confused with carboxyhemoglobin, which is a combination of hemoglobin and carbon monoxide); and (3) as *bicarbonate*, which accounts for most of the CO_2 carried by the blood.

Carbon dioxide is able to combine with water to form carbonic acid. This reaction occurs spontaneously in the blood plasma at a slow rate, but it occurs much more rapidly within the red blood cells because of the catalytic action of the enzyme **carbonic anhydrase.** Since this enzyme is confined to the red blood cells, most of the carbonic acid is produced there rather than in the plasma. The formation of carbonic acid from CO_2 and water is favored by the high P_{CO_2} found in tissue capillaries (this is an example of the *law of mass action*, described in chapter 4).

$$CO_2 + H_2O \xrightarrow[\text{high } P_{CO_2}]{\text{carbonic anhydrase}} H_2CO_3$$

Figure 24.40

Carbon dioxide transport and chloride shift.
Carbon dioxide is transported in three forms: as dissolved CO_2 gas, attached to hemoglobin as carbaminohemoglobin, and as carbonic acid and bicarbonate. Percentages indicate the proportion of CO_2 in each of the forms. Note that when bicarbonate (HCO_3^-) diffuses out of the red blood cells, Cl^- diffuses in to retain electrical neutrality. This exchange is the chloride shift.

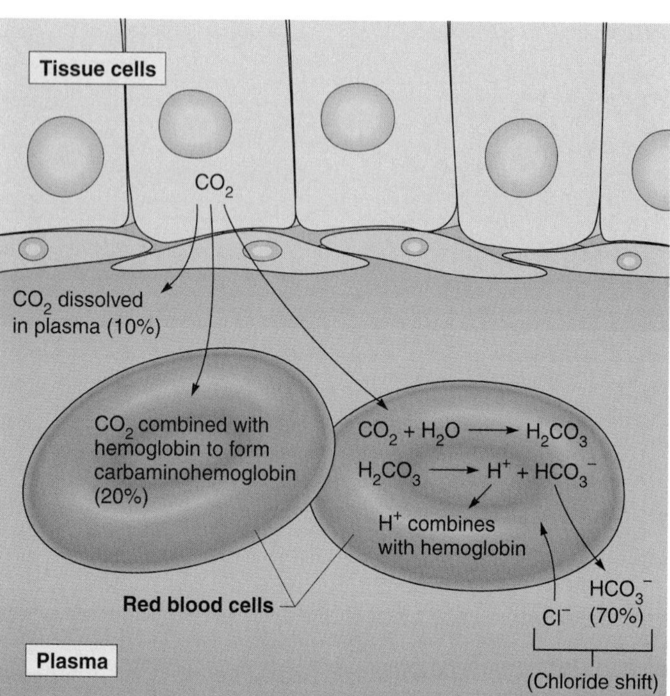

Figure 24.41

Reverse chloride shift in the lungs.
Carbon dioxide is released from the blood as it travels through the pulmonary capillaries. During this time a "reverse chloride shift" occurs, and carbonic acid is transformed into CO_2 and H_2O. The CO_2 is eliminated in the exhaled air.

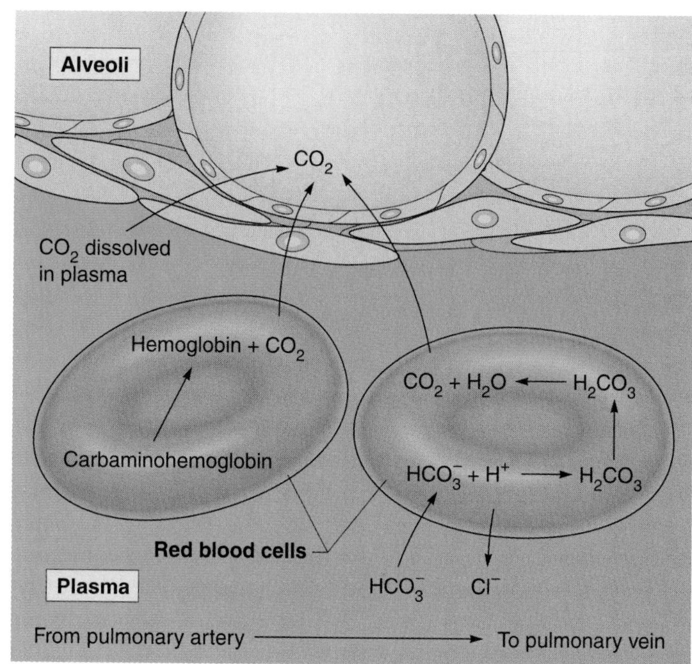

The Chloride Shift

As a result of catalysis by carbonic anhydrase within the red blood cells, large amounts of carbonic acid are produced as blood passes through the systemic capillaries. The buildup of carbonic acid concentrations within the red blood cells favors the dissociation of these molecules into hydrogen ions (protons, which contribute to the acidity of a solution) and HCO_3^- (bicarbonate), as shown by the following equation:

$$H_2CO_3 \rightarrow H^+ + HCO_3^-$$

The hydrogen ions (H^+) released by the dissociation of carbonic acid are largely buffered by their combination with deoxyhemoglobin within the red blood cells. Although the unbuffered hydrogen ions are free to diffuse out of the red blood cells, more bicarbonate diffuses outward into the plasma than does H^+. As a result of the "trapping" of hydrogen ions within the red blood cells by their attachment to hemoglobin and the outward diffusion of bicarbonate, the inside of the red blood cell gains a net positive charge. This attracts chloride ions (Cl^-), which move into the red blood cells as HCO_3^- moves out. This exchange of anions as blood travels through the tissue capillaries is called the **chloride shift** (fig. 24.40).

The unloading of oxygen is increased by the bonding of H^+, released from carbonic acid, to oxyhemoglobin. This is the Bohr effect, and it results in increased conversion of oxyhemoglobin to deoxyhemoglobin. Now, deoxyhemoglobin bonds H^+ more strongly than does oxyhemoglobin; therefore, the act of unloading its oxygen improves the ability of hemoglobin to buffer the H^+ released by carbonic acid. Removal of H^+ from solution by combining with hemoglobin (through the law of mass action), in turn, favors the continued production of carbonic acid and thereby improves the ability of the blood to transport carbon dioxide. Thus, carbon dioxide increases oxygen unloading, and oxygen unloading increases carbon dioxide transport.

When blood reaches the pulmonary capillaries, deoxyhemoglobin converts to oxyhemoglobin. Since oxyhemoglobin has a weaker affinity for H^+ than does deoxyhemoglobin, hydrogen ions are released within the red blood cells. This attracts HCO_3^- from the plasma, which combines with H^+ to form carbonic acid:

$$H^+ + HCO_3^- \Rightarrow H_2CO_3$$

In the condition of lower P_{CO_2}, as occurs in the pulmonary capillaries, carbonic anhydrase catalyzes the conversion of carbonic acid to carbon dioxide and water:

$$H_2CO_3 \xrightarrow[\text{low } P_{CO_2}]{\text{carbonic anhydrase}} CO_2 + H_2O$$

Figure 24.42

Effects of bicarbonate on blood pH.
Bicarbonate released into the plasma from red blood cells functions to buffer H^+ produced by the ionization of metabolic acids (lactic acid, fatty acids, ketone bodies, and others). Binding of H^+ to hemoglobin also promotes the unloading of O_2.

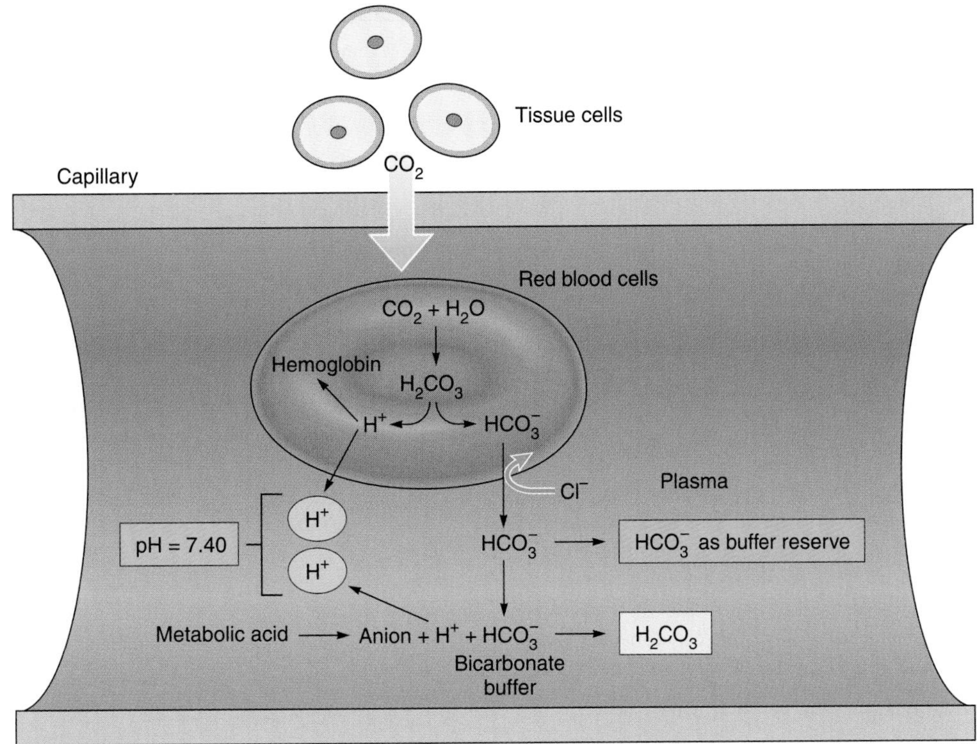

In summary, carbon dioxide produced by the cells is converted within the systemic capillaries, mostly through the action of carbonic anhydrase in the red blood cells, to carbonic acid. Once produced in the red blood cells, the carbonic acid dissociates into bicarbonate and H^+, which results in the chloride shift. A *reverse chloride shift* operates in the pulmonary capillaries to convert carbonic acid to H_2O and CO_2 gas, which is eliminated in the expired breath (fig. 24.41). The P_{CO_2}, carbonic acid, H^+, and bicarbonate concentrations in the systemic arteries are thus maintained relatively constant by normal ventilation. This is required to maintain the acid-base balance of the blood (fig. 24.42), as discussed in chapter 20 and in the next section.

Ventilation and Acid-Base Balance

The basic concepts and terminology relating to the acid-base balance of the blood were introduced in chapter 20. In brief review, *acidosis* refers to an arterial pH below 7.35, and *alkalosis* refers to an arterial pH above 7.45. There are two components of each: respiratory and metabolic. The *respiratory component* refers to the carbon dioxide concentration of the blood, as measured by the P_{CO_2}. As implied by its name, the respiratory component is regulated by the respiratory system. The *metabolic component* is controlled by the kidneys, and is discussed in chapter 25.

Ventilation is normally adjusted to keep pace with the metabolic rate, so that the arterial P_{CO_2} remains in the normal range. In **hypoventilation,** the ventilation is insufficient to "blow off" carbon dioxide and maintain a normal P_{CO_2}. Indeed, hypoventilation can be operationally defined as an abnormally high arterial P_{CO_2}. Under these conditions, carbonic acid production is excessively high and **respiratory acidosis** occurs.

In **hyperventilation,** conversely, the rate of ventilation is greater than the rate of CO_2 production. Arterial P_{CO_2} therefore decreases, so that less carbonic acid is formed than under normal conditions. The depletion of carbonic acid raises the pH, and **respiratory alkalosis** occurs.

A change in blood pH, produced by alterations in either the respiratory or metabolic component of acid-base balance, can be partially compensated for by a change in the other component. For example, a person with metabolic acidosis will hyperventilate. This is because the aortic and carotid bodies are stimulated by an increased blood H^+ concentration (fall in pH). As a result of the hyperventilation, a secondary respiratory alkalosis is produced. The person is still acidotic, but not as much so as would be the case without the compensation. People with partially compensated metabolic acidosis would thus have a low pH, which would be accompanied by a low blood P_{CO_2} as a result of the hyperventilation. Metabolic alkalosis, similarly, is partially compensated for by the retention of carbonic acid due to hypoventilation (table 24.10).

Table 24.10
The Effect of Lung Function on Blood Acid-Base Balance

Condition	pH	P_{CO_2}	Ventilation	Cause or Compensation
Normal	7.35–7.45	39–41 mmHg	Normal	Not applicable
Respiratory acidosis	Low	High	Hypoventilation	Cause of the acidosis
Respiratory alkalosis	High	Low	Hyperventilation	Cause of the alkalosis
Metabolic acidosis	Low	Low	Hyperventilation	Compensation for acidosis
Metabolic alkalosis	High	High	Hypoventilation	Compensation for alkalosis

Effect of Exercise and High Altitude on Respiratory Function

The arterial blood gases and pH do not significantly change during moderate exercise because ventilation increases to keep pace with increased metabolism. This increased ventilation requires neural feedback from the exercising muscles and chemoreceptor stimulation. Adjustments are made at high altitude in both the control of ventilation and the oxygen transport ability of the blood to permit adequate delivery of oxygen to the tissues.

Changes in ventilation and oxygen delivery occur during exercise and during acclimatization to a high altitude. These changes help to compensate for the increased metabolic rate during exercise and for the decreased arterial P_{O_2} at high altitudes.

Ventilation during Exercise

As soon as a person begins to exercise, breathing becomes deeper and more rapid to produce a total minute volume that is many times the resting value. This increased ventilation, particularly in well-trained athletes, is exquisitely matched to the simultaneous increase in oxygen consumption and carbon dioxide production by the exercising muscles. The arterial blood P_{O_2}, P_{CO_2}, and pH thus remain surprisingly constant during exercise (fig. 24.43).

It is tempting to suppose that ventilation increases during exercise as a result of the increased CO_2 production by the exercising muscles. Ventilation and CO_2 production increase simultaneously, however, so that blood measurements of P_{CO_2} during exercise are not significantly higher than at rest. The mechanisms responsible for the increased ventilation during exercise must therefore be more complex.

Two kinds of mechanisms—*neurogenic* and *humoral*—have been proposed to explain the increased ventilation that occurs during exercise. Possible neurogenic mechanisms include the following: (1) sensory nerve activity from the exercising

Figure 24.43

Effect of exercise on arterial blood gases and pH. Notice that there are no consistent and significant changes in these measurements during the first several minutes of moderate and heavy exercise and that only the P_{CO_2} changes (actually decreases) during more prolonged exercise.

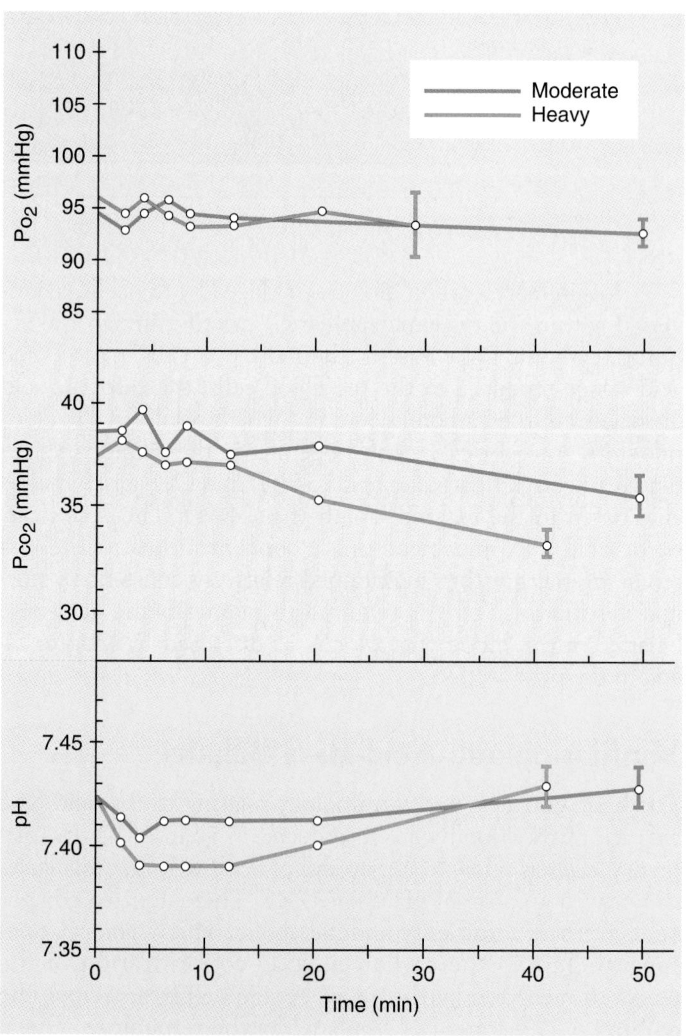

Table 24.11

Changes in Respiratory Function during Exercise

Variable	Change	Comments
Ventilation	Increased	In moderate exercise, ventilation is matched to increased metabolic rate. Mechanisms responsible for increased ventilation are not well understood.
Blood gases	No change	Blood gas measurements during light and moderate exercise show little change because ventilation is increased to match increased muscle O_2 consumption and CO_2 production.
O_2 delivery to muscles	Increased	Although the total O_2 content and P_{O_2} do not increase during exercise, there is an increased rate of blood flow to the exercising muscles.
O_2 extraction by muscles	Increased	Increased O_2 consumption lowers the tissue P_{O_2} and lowers the affinity of hemoglobin for O_2 (due to the effect of increased temperature). More O_2, as a result, is unloaded so that venous blood contains a lower oxyhemoglobin saturation than at rest. This effect is enhanced by endurance training.

limbs may stimulate the respiratory muscles, either through spinal reflexes or via the brain stem respiratory centers, and/or (2) input from the cerebral cortex may stimulate the brain stem centers to modify ventilation. These neurogenic theories help to explain the immediate increase in ventilation that occurs as exercise begins.

Rapid and deep ventilation continues after exercise has stopped, suggesting that humoral (chemical) factors in the blood may also stimulate ventilation during exercise. Since the P_{O_2}, P_{CO_2}, and pH of blood samples from exercising subjects are within the resting range, these humoral theories propose that (1) the P_{CO_2} and pH in the region of the chemoreceptors may be different from these values "downstream," where blood samples are taken, and/or (2) cyclic variations in these values that cannot be detected by blood samples may stimulate the chemoreceptors. The evidence suggests that both neurogenic and humoral mechanisms are involved in the **hyperpnea** (*hi″perp′ne-ă*), or increased total minute volume, of exercise. (Note that hyperpnea differs from hyperventilation in that the blood P_{CO_2} is decreased in hyperventilation.)

Lactate Threshold and Endurance Training

The ability of the cardiopulmonary system to deliver adequate amounts of oxygen to the exercising muscles at the beginning of exercise may be insufficient because of the time lag required to make proper cardiovascular adjustments. During this time, therefore, the muscles respire anaerobically and a "stitch in the side"—possibly due to hypoxia of the diaphragm—may develop. After numerous cardiovascular and pulmonary adjustments have been made, a person may experience a "second wind" when the muscles receive sufficient oxygen for their needs.

Continued heavy exercise can cause a person to reach the **lactate threshold,** which is the maximum rate of oxygen consumption that can be attained before blood lactic acid levels rise as a result of anaerobic respiration. This occurs when 50% to 70% of the person's maximal oxygen uptake has been reached. The rise in lactic acid levels is due to the aerobic limitations of the muscles; it is not due to a malfunction of the cardiopulmonary system. Indeed, the arterial oxygen hemoglobin saturation remains at 97%, and venous blood draining the muscles contains unused oxygen.

The lactate threshold, however, is higher in endurance-trained athletes than in other people. These athletes, because of their higher cardiac output, have a higher rate of oxygen delivery to their muscles. As mentioned in chapter 12, endurance training also increases the skeletal muscle content of mitochondria and Krebs cycle enzymes, enabling the muscles to utilize more of the oxygen delivered to them by the arterial blood. The effects of exercise and endurance training on respiratory function are summarized in table 24.11.

Acclimatization to High Altitude

When a person from a region near sea level moves to a significantly higher elevation, several adjustments in respiratory function are made to compensate for the decreased P_{O_2} at the higher altitude. These adjustments include changes in ventilation, in the hemoglobin affinity for oxygen, and in the total hemoglobin concentration.

Reference to table 24.12 indicates that at an altitude of 2,286 meters (7,500 feet), for example, the P_{O_2} of arterial blood is 69 to 74 mmHg (compared to 100 mmHg at sea level). This table also indicates that the percent oxyhemoglobin saturation at this altitude is between 92% and 93%, compared to about 96% at sea level. The amount of oxygen attached to hemoglobin, and thus the total oxygen content of blood, is therefore decreased. In addition, the rate at which oxygen can be delivered to the cells (by the plasma-derived tissue fluid) after it dissociates from oxyhemoglobin is reduced at the higher altitude. This is because the maximum concentration of oxygen that can be dissolved in the blood plasma decreases in a linear fashion with the fall in P_{O_2}. People may thus experience rapid fatigue even at more moderate elevations (for example, 5,000 to 6,000 feet), at

Table 24.12

Blood Gas Measurements at Different Altitudes

Altitude	Arterial P_{O_2} (mmHg)	Percent Oxyhemoglobin Saturation	Arterial P_{CO_2} (mmHg)
Sea level	90–95	96%	40
1,524 m (5,000 ft)	75–81	95%	32–33
2,286 m (7,500 ft)	69–74	92%–93%	31–33
4,572 m (15,000 ft)	48–53	86%	25
6,096 m (20,000 ft)	37–45	76%	20
7,620 m (25,000 ft)	32–39	68%	13
8,848 m (29,029 ft)	26–33	58%	9.5–13.8

Source: From P. H. Hackett et al., "High Altitude Medicine" in *Management of Wilderness and Environmental Emergencies*, 2d ed., edited by Paul S. Auerbach and Edward C. Geehr. Copyright © 1989 Mosby/Yearbook. Reprinted by permission.

which the oxyhemoglobin saturation is only slightly decreased. Compensations made by the respiratory system gradually reduce the amount of fatigue caused by a given amount of exertion at high altitudes.

Changes in Ventilation

Starting at altitudes as low as 1,500 meters (5,000 feet), the decreased arterial P_{O_2} stimulates an increase in ventilation. This *hypoxic ventilatory response* produces hyperventilation, which lowers the arterial P_{CO_2} (see table 24.5) and thus produces a respiratory alkalosis. The rise in arterial pH helps to blunt the hyperventilation, and within a few days the total minute volume becomes stabilized at a level 2.5 L/min higher than at sea level.

Hyperventilation at high altitude increases tidal volume, thus reducing the proportionate contribution of air from the anatomical dead space and increasing the proportion of fresh air brought to the alveoli. This improves the oxygenation of the blood over what it would be in the absence of the hyperventilation. Hyperventilation, however, cannot increase blood P_{O_2} above that of the inspired air. The P_{O_2} of arterial blood decreases with increasing altitude (table 24.12), regardless of the ventilation. In the Peruvian Andes, for example, the normal arterial P_{O_2} is reduced from 100 mmHg (at sea level) to 45 mmHg. The loading of hemoglobin with oxygen is therefore incomplete, producing an oxyhemoglobin saturation that is decreased from 97% (at sea level) to 81%.

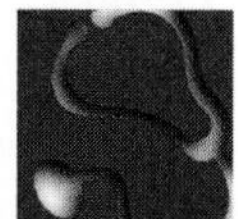

Acute Mountain Sickness (AMS) is common in people who arrive at altitudes in excess of 5,000 feet. Cardinal symptoms of AMS are headache, malaise, anorexia, nausea, and fragmented sleep. Headache, the most common symptom, may result from changes in blood flow to the brain. Low arterial P_{O_2} stimulates vasodilation of vessels in the pia mater (chapter 15), increasing blood flow and pressure within the skull. The hypocapnia produced by hyperventilation, however, causes cerebral vasoconstriction. Whether there is a net cerebral vasoconstriction or vasodilation depends on the balance between these two antagonistic effects. Pulmonary edema, common at altitudes above 9,000 feet, can produce shortness of breath, coughing, and a mild fever. Cerebral edema, which generally occurs above an altitude of 10,000 feet, can produce mental confusion and even hallucinations. Pulmonary and cerebral edema are potentially dangerous and should be alleviated by descending to a lower altitude.

The Affinity of Hemoglobin for Oxygen

Normal arterial blood at sea level unloads only about 22% of its oxygen to the tissues at rest; the percent saturation is reduced from 97% in arterial blood to 75% in venous blood. As a partial compensation for the decrease in oxygen content at high altitude, the affinity of hemoglobin for oxygen is reduced, so that a higher proportion of oxygen is unloaded. This occurs because the low oxyhemoglobin content of red blood cells stimulates the production of 2,3-DPG, which in turn decreases the affinity of hemoglobin for oxygen.

The capacity of 2,3-DPG to decrease the affinity of hemoglobin for oxygen thus predominates over the capacity of respiratory alkalosis (caused by the hyperventilation) to increase the affinity. At very high altitudes, however, the story becomes more complex. In a 1984 study, the very low arterial P_{O_2} (28 mmHg) of subjects at the summit of Mount Everest stimulated intense hyperventilation, so that the arterial P_{CO_2} was decreased to 7.5 mmHg. The resultant respiratory alkalosis (in this case, arterial pH greater than 7.7) caused the oxyhemoglobin dissociation curve to shift to the left (indicating greater affinity of hemoglobin for oxygen) despite the antagonistic effect of increased 2,3-DPG concentrations. It was suggested that the increased affinity of hemoglobin for oxygen

UNDER DEVELOPMENT

Embryological Development of the Respiratory System

The formation of the nasal cavity begins at 3½ to 4 weeks of embryonic life with the appearance of a region of thickened ectoderm called the **olfactory placode** (*plak'ōd*) on the front and inferior part of the head (fig. 1). The placode invaginates to form the **olfactory pit,** which extends posteriorly to connect with the **foregut.** The foregut, derived of endoderm, later develops into the pharynx.

The mouth, or oral cavity, develops at the same time as the nasal cavity, and for a short time a thin **oronasal** (*or''o-na'zal*) **membrane** separates the two cavities. This membrane ruptures during the seventh week, and a single large **oronasal cavity** forms. Shortly thereafter, tissue plates of mesoderm begin to grow horizontally across the cavity. At approximately the same time, a vertical plate develops inferiorly from the roof of the nasal cavity. These plates complete their formation by 3 months of development. The vertical plate forms the nasal septum, and the horizontal plates form the hard palate. A **cleft palate** is produced when the horizontal plates fail to meet in the midline.

The respiratory system begins to form during the fourth week of development as a diverticulum (*di''ver-tik'yŭ-lum*), or outpouching, called the **laryngotracheal** (*lă-ring''go-tra'ke-al*) **bud,** from the ventral surface of endoderm along the lower pharyngeal region. As the bud grows, the proximal portion forms the trachea, and the distal portion bifurcates (splits) into a right and left bronchus.

The buds continue to elongate and split until all the tubular network within the lower respiratory tract is formed. As the terminal portion forms air sacs, called **pulmonary alveoli,** at about 8 weeks of development, the supporting lung tissue begins to form. The complete structure of the lungs, however, is not fully developed until about 26 weeks of fetal development. Infants born prior to this time therefore require special mechanical respiratory equipment to live.

Figure 1 The development of the upper respiratory system. (*a*) An anterior view of the developing head of an embryo at 6 weeks showing the position of a sagittal cut depicted at different stages of development (b_1–b_4) through 14 weeks.

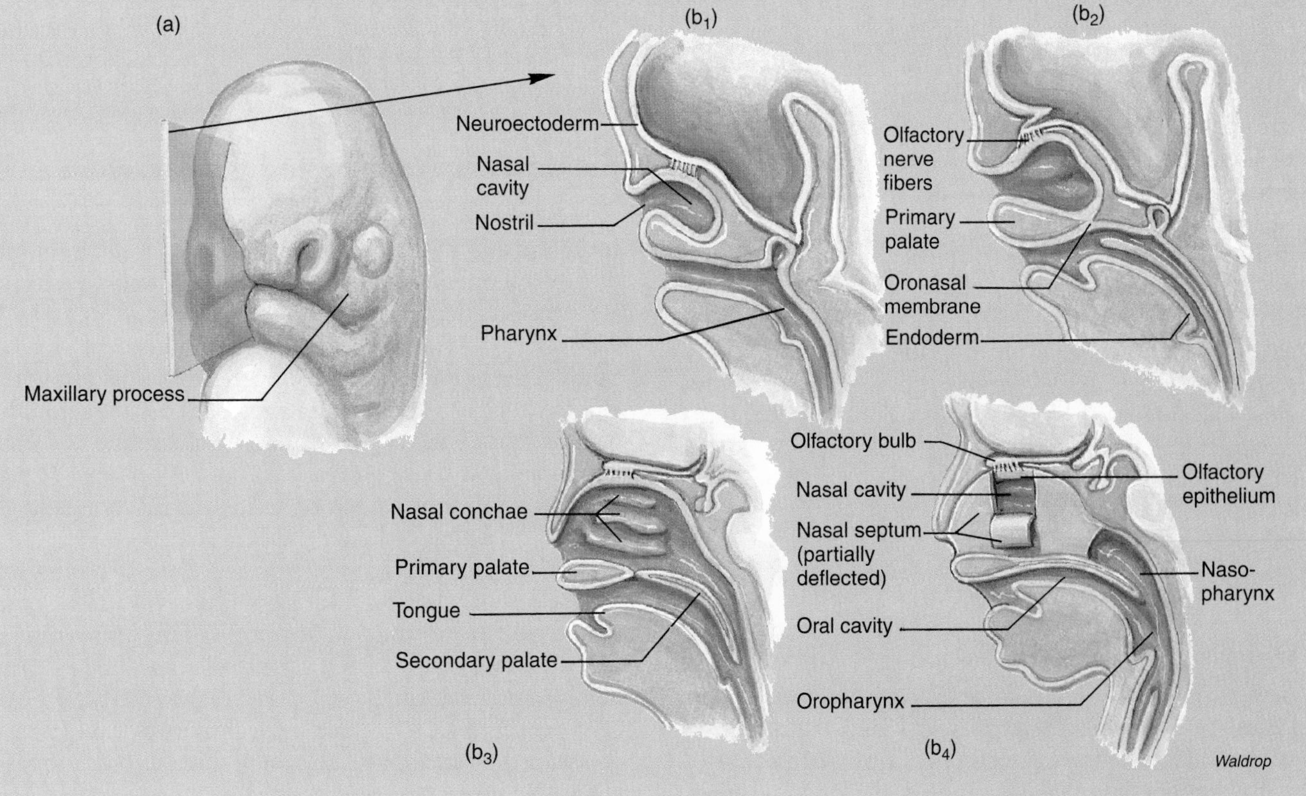

Table 24.13

Change in Respiratory Function during Acclimatization to High Altitude

Variable	Change	Comments
Partial pressure of O_2	Decreased	Due to decreased total atmospheric pressure
Partial pressure of CO_2	Decreased	Due to hyperventilation in response to low arterial P_{O_2}
Percent oxyhemoglobin saturation	Decreased	Due to lower P_{O_2} in pulmonary capillaries
Ventilation	Increased	Due to low arterial P_{O_2}; ventilation usually returns to normal after a few days
Total hemoglobin	Increased	Due to stimulation by erythropoietin; raises O_2 capacity of blood to partially or completely compensate for the reduced partial pressure
Oxyhemoglobin affinity	Decreased	Due to increased 2,3-DPG within the red blood cells; results in a higher unloading percentage of O_2 to the tissues, which may partially or completely compensate for the reduced arterial oxyhemoglobin saturation

caused by the respiratory alkalosis may have been beneficial at such a high altitude, since it increased the loading of hemoglobin with oxygen in the lungs.

Increased Hemoglobin and Red Blood Cell Production

In response to tissue hypoxia, the kidneys secrete the hormone erythropoietin (chapter 20). Erythropoietin stimulates the bone marrow to increase its production of hemoglobin and red blood cells. In the Peruvian Andes, for example, people have a total hemoglobin concentration that increases from 15 g per 100 ml (at sea level) to 19.8 g per 100 ml. Although the percent oxyhemoglobin saturation is still lower than at sea level, the total oxygen content of the blood is actually greater—22.4 ml O_2 per 100 ml compared to a sea-level value of about 20 ml O_2 per 100 ml. These adjustments of the respiratory system to high altitude are summarized in table 24.13.

It should be noted that these changes are not unalloyed benefits. Polycythemia (high red blood cell count) increases the viscosity of blood; hematocrits of 55% to 60% have been measured in people who live in the Himalayas, and higher values are reached if dehydration accompanies the polycythemia. The increased blood viscosity contributes to pulmonary hypertension, which can cause accompanying edema and ventricular hypertrophy that can lead to heart failure.

Clinical Considerations

Developmental Problems

Birth defects, inherited disorders, and premature births commonly cause problems in the respiratory system of infants. A **cleft palate** is a developmental deformity of the hard palate of the mouth. An opening persists between the oral and nasal cavities, making it difficult, if not impossible, for an infant to nurse. A cleft palate may be hereditary or a complication of some disease (e.g., German measles) contracted by the mother during pregnancy. A **cleft lip** is a genetically based developmental disorder in which the two sides of the upper lip fail to fuse. Cleft palates and cleft lips can be treated very effectively with cosmetic surgery.

Cystic fibrosis is the most common fatal inheritable disorder among Caucasians. Approximately 1 out of 20 Caucasians are carriers, and 1 in 2,000 children inherit the disease. The defective gene reduces the ability of epithelial cells to actively transport Cl^-, causing the production of a very thick, sticky mucus that blocks the airways of the lungs. It also affects the liver and other organs in potentially life-threatening ways.

Sickle-Cell Anemia and Thalassemia

A number of hemoglobin diseases are produced by inherited (congenital) defects in the protein part of hemoglobin. **Sickle-cell anemia**—a disease occurring almost exclusively in African Americans, and carried in a recessive state by 8% to 11% of the African American population of the United States—for example, is caused by the abnormal form of hemoglobin called *hemoglobin S*. Hemoglobin S differs from normal hemoglobin A in only one amino acid: valine is substituted for glutamic acid in position 6 on the beta chains. This amino acid substitution is caused by a single base change in the region of DNA that codes for the beta chains.

Under conditions of low blood P_{O_2}, hemoglobin S comes out of solution and cross-links to form a "paracrystalline gel" within the red blood cells. This causes the characteristic sickle shape of red blood cells and makes them less flexible and more fragile. Since red blood cells must be able to bend in the middle to pass through narrow capillaries, a decrease in their flexibility may cause them to block small blood channels and produce organ ischemia. The decreased solubility of hemoglobin S in solutions of low P_{O_2} is used in the diagnosis of sickle-cell anemia and sickle-cell trait (the carrier state, in which a person has the genes for both hemoglobin A and hemoglobin S). Sickle-cell anemia is treated with the drug *hydroxyurea*, which stimulates the production of gamma chains (characteristic of hemoglobin F) in place of the defective beta chains of hemoglobin S.

Thalassemia (*thal″ă-se′me-ă*) is any of a family of hemoglobin diseases found predominantly among people of Mediterranean ancestry. In *alpha thalassemia*, there is decreased synthesis of the alpha chains of hemoglobin, whereas in *beta thalassemia* the synthesis of the beta chains is impaired. One of the compensations for thalassemia is increased synthesis of gamma chains, resulting in the retention of large amounts of hemoglobin F (fetal hemoglobin) into adulthood.

Some types of abnormal hemoglobins have been shown to be advantageous in the environments in which they evolved. A person who is a carrier for sickle-cell anemia, for example (and who therefore has both hemoglobin A and hemoglobin S), has a high resistance to malaria. This is because the parasite that causes malaria cannot live in red blood cells that contain hemoglobin S.

Common Respiratory Disorders

A cough is the most common symptom of respiratory disorders. Acute problems may be accompanied by dyspnea or wheezing. Respiratory or circulatory problems may cause **cyanosis** (*si″ă-no′sis*), a blue discoloration of the skin resulting from blood with a low oxygen content.

Pulmonary disorders are classified as **obstructive** when there is increased resistance to air flow in the bronchioles; they are classified as **restrictive** when alveolar tissue is damaged. Asthma and acute bronchitis are usually exclusively obstructive, whereas chronic bronchitis and emphysema are both obstructive and restrictive. Pulmonary fibrosis, by contrast, is a purely restrictive disorder.

Asthma

The dyspnea, wheezing, and other symptoms of **asthma** are produced by an obstruction of air flow through the bronchioles that occurs in episodes or "attacks." This obstruction is caused by inflammation, secretion of mucus, and bronchoconstriction. Inflammation of the airways is characteristic of asthma and itself contributes to increased airway responsiveness to agents that promote bronchiolar constriction. Bronchoconstriction further increases airway resistance and makes breathing difficult. Constriction of bronchiolar smooth muscles is stimulated by leukotrienes and histamine released by mast cells and leukocytes (chapter 23), which can be provoked by an allergic reaction or by the release of acetylcholine from parasympathetic nerve endings.

Asthma is often treated with glucocorticoid drugs, which inhibit inflammation. A new antileukotriene drug also is now available to suppress the inflammatory response. Epinephrine and related compounds stimulate beta-adrenergic receptors in the bronchioles and by this means promote bronchodilation. Therefore, epinephrine was frequently used as an inhaled spray to relieve the symptoms of an asthma attack. It has since been learned that there are two subtypes of beta receptors for epinephrine, and that the subtype in the heart (called β_1) is different from the one in the bronchioles (β_2). Capitalizing on these differences, compounds such as *terbutaline* have been developed. These compounds can more selectively stimulate the β_2 adrenergic receptors and cause bronchodilation without affecting the heart to the extent that epinephrine does.

Emphysema

Alveolar tissue is destroyed in the chronic, progressive condition called **emphysema** (*em″fĭ-se′mă*), which results in fewer but larger pulmonary alveoli (fig. 24.44). This reduces the surface area for gas exchange and decreases the ability of the bronchioles to remain open during expiration. Collapse of the bronchioles as a result of the compression of the lungs during expiration produces *air trapping*, which further decreases the efficiency of gas exchange in the pulmonary alveoli.

Among the different types of emphysema, the most common occurs almost exclusively in people who have heavily smoked cigarettes over a period of years. A component of cigarette smoke apparently stimulates the macrophages and leukocytes to secrete proteolytic (protein-digesting) enzymes that destroy lung tissues. A less common type of emphysema results from the genetic inability to produce a plasma protein called α_1-antitrypsin. This protein normally inhibits proteolytic enzymes such as trypsin, and thus normally protects the lungs against the effects of enzymes that are released from alveolar macrophages.

Chronic bronchitis and emphysema, the two most common causes of respiratory failure, are together called **chronic obstructive pulmonary disease (COPD).** In addition to the more direct obstructive and restrictive aspects of these conditions, other pathological changes may occur. These include edema, inflammation, hyperplasia (an increase in the number of cells), zones of pulmonary fibrosis, pneumonia, and heart failure. Patients with severe chronic bronchitis or emphysema may develop *cor pulmonale* (*kor pool″mŏ-na′le*)—pulmonary hypertension with hypertrophy and the eventual failure of the right ventricle. COPD is the fifth leading cause of death in the United States.

Interstitial Lung Disease

Under certain conditions, for reasons that are poorly understood, lung damage leads to **interstitial lung disease** instead of emphysema. In this condition, the normal structure of the lungs is disrupted by the accumulation of fibrous connective

---

thalassemia: Gk. *thalassa*, sea
cyanosis: Gk. *kyanosis*, dark-blue color
asthma: Gk. *asthma*, panting

emphysema: Gk. *empxhysan*, blow up, inflate

Figure 24.44

Emphysema destroys lung tissue.
These are photomicrographs of tissue (*a*) from a normal lung and (*b*) from the lung of a person with emphysema. In emphysema, lung tissue is destroyed, resulting in fewer and larger pulmonary alveoli.

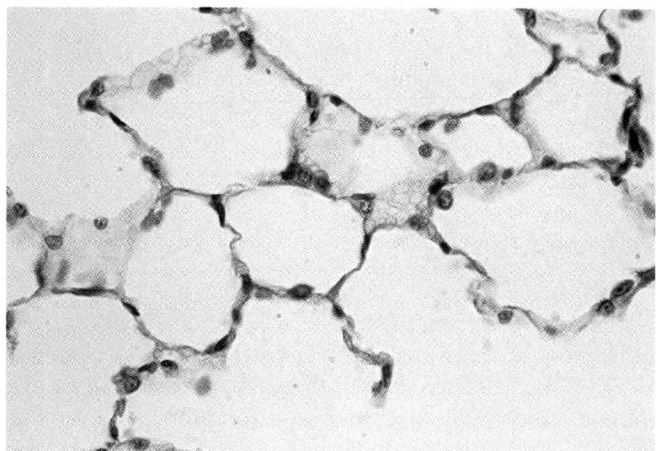

(a)

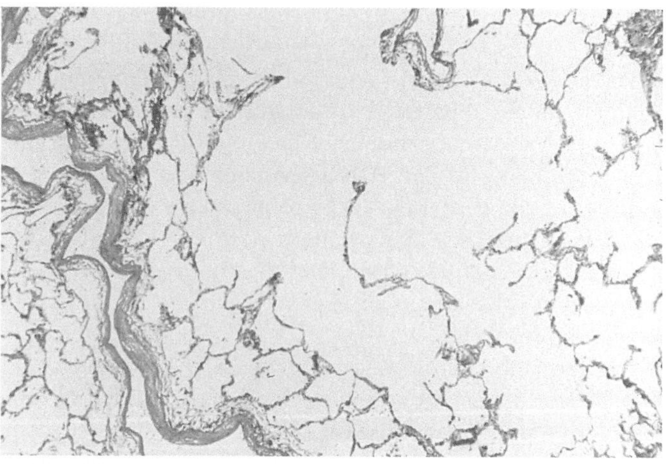

(b)

tissue proteins resulting in decreased lung elasticity. Fibrosis can result, for example, from the inhalation of particles less than 6 μm in size that can accumulate in the respiratory zone of the lungs. Included in this category is *anthracosis*, or black lung, which is produced by the inhalation of carbon particles from coal dust.

Other Respiratory Disorders

The **common cold** is the most widespread of all respiratory diseases. Colds occur repeatedly because acquired immunity to one virus does not protect against other viruses that cause colds. Cold viruses cause acute inflammation of the respiratory mucosa, resulting in a flow of mucus, a fever, and often a headache.

Nearly all the structures and regions of the respiratory tract can become infected and inflamed. **Influenza** is a viral disease that causes inflammation of the upper respiratory tract. **Sinusitis** (*si-nŭ-si'tis*) is an inflammation of the paranasal sinuses. **Tonsillitis** may involve one or all of the tonsils and frequently follows other lingering diseases of the oral or pharyngeal region. **Laryngitis** is inflammation of the larynx, which often produces a hoarse voice and limits the ability to talk. **Tracheobronchitis** and **bronchitis** are infections of the regions for which they are named. Severe inflammation of the bronchioles can cause significant airway resistance and dyspnea.

Diseases of the lungs are common and may be serious. **Pneumonia** is an acute infection and inflammation of lung tissue accompanied by exudation (accumulation of fluid). It is usually caused by bacteria, most commonly by the pneumococcus bacterium. Viral pneumonia is caused by a number of different viruses. **Tuberculosis** is an inflammatory disease of the lungs caused by the presence of tubercle bacilli. Tuberculosis softens lung tissue, which eventually becomes ulcerated. **Pleurisy** (*ploor'ĭ-se*) is an inflammation of the pleura and is usually secondary to some other respiratory disease. Inspiration may become painful, and fluid may collect within the pleural space resulting in pleural effusion.

Cancer in the respiratory system is often caused by repeated inhalation of irritating substances, such as cigarette smoke. Cancers of the lips, larynx, and lungs are especially common in smokers over the age of 50. The number of lung cancer cases has more than doubled for both men and women in the last 60 years, and lung cancer has now surpassed breast cancer as the leading form of cancer among women.

Disorders Caused by High Partial Pressures of Gases

The total atmospheric pressure increases by one atmosphere (760 mmHg) for every 10 meters (33 feet) below sea level. If a diver descends 10 meters below sea level, therefore, the partial pressures and amounts of dissolved gases in the plasma will be twice those at sea level. At 66 feet, they are three times, and at 100 feet, they are four times the values at sea level. The increased amounts of nitrogen and oxygen dissolved in the blood plasma under these conditions can have serious adverse effects on the body.

Oxygen Toxicity

Although breathing 100% oxygen at 1 or 2 atmospheres pressure can be safely tolerated for a few hours, higher partial oxygen pressures can be very dangerous. **Oxygen toxicity** may develop rapidly when the P_{O_2} rises above 2.5 atmospheres. This is apparently caused by the oxidation of enzymes and other

influenza: L. *influentia*, a flowing in

destructive changes that can damage the nervous system and lead to coma and death. For these reasons, deep-sea divers commonly use gas mixtures in which oxygen is diluted with inert gases such as nitrogen (as in ordinary air) or helium.

Hyperbaric (*hi″per-bar′ik*) **oxygen therapy,** in which a patient is given 100% oxygen gas at 2 to 3 atmospheres pressure to breathe for varying lengths of time, is used to treat such conditions as carbon monoxide poisoning, decompression sickness, severe traumatic injury (such as crush injury), infections that could lead to gas gangrene, and other conditions. While normal plasma oxygen concentration is 0.3 ml O_2 per 100 ml blood (as previously described), breathing 100% oxygen at a pressure of 3 atmospheres raises the plasma concentration to about 6 ml O_2 per 100 ml blood. This helps to kill anaerobic bacteria, such as those that cause gangrene; promote wound healing; reduce the size of gas bubbles (in the case of decompression sickness); and to quickly eliminate carbon monoxide from the body. Although hyperbaric oxygen was formerly used to treat premature infants for respiratory distress, the practice was discontinued because it caused a fibrotic deterioration of the retina that frequently resulted in blindness.

Nitrogen Narcosis

Although at sea level nitrogen is physiologically inert, larger amounts of dissolved nitrogen under hyperbaric (high-pressure) conditions have deleterious effects. Since it takes time for the nitrogen to dissolve, these effects usually do not appear until a diver has remained submerged for more than an hour. **Nitrogen narcosis** resembles alcohol intoxication; depending on the depth of the dive, the diver may experience what Jacques Cousteau termed "rapture of the deep." Dizziness and extreme drowsiness are other narcotizing effects.

Decompression Sickness

The amount of nitrogen dissolved in the plasma decreases as the diver ascends to sea level because of the progressive decrease in the P_{N_2}. If the diver surfaces slowly, a large amount of nitrogen can diffuse through the alveoli and be eliminated in the expired breath. If decompression occurs too rapidly, however, bubbles of nitrogen gas (N_2) can form in the tissue fluids and enter the blood. This process is analogous to the formation of carbon dioxide bubbles in a champagne bottle when the cork is removed. The bubbles of N_2 gas in the blood can block small blood channels, producing muscle and joint pain, as well as more serious damage. These effects are known as **decompression sickness,** commonly called "the bends." The primary treatment for decompression sickness is hyperbaric oxygen treatment.

Airplanes that fly long distances at high altitudes (30,000 to 40,000 ft) have pressurized cabins so that the passengers and crew do not experience the very low atmospheric pressures of these altitudes. If a cabin were to become rapidly depressurized at high altitude, much less nitrogen could remain dissolved at the greatly lowered pressure. People in this situation, like the divers who ascend too rapidly, would thus experience decompression sickness.

narcosis: Gk. *narkosis*, a numbing

Abdominal Thrust Maneuver

The abdominal thrust (Heimlich) maneuver can save the life of a person who is choking. This technique is performed as follows:

A. If the victim is standing or sitting:
   1. Stand behind the victim or the victim's chair and wrap your arms around his or her waist.
   2. Grasp your fist with your other hand and place the fist against the victim's abdomen, slightly above the navel and below the rib cage.
   3. Press your fist into the victim's abdomen with a quick upward thrust.
   4. Repeat several times if necessary.

B. If the victim is lying down:
   1. Position the victim on his or her back.
   2. Face the victim, and kneel astride his or her hips.
   3. With one of your hands on top of the other, place the heel of your bottom hand on the victim's abdomen, slightly above the navel and below the rib cage.
   4. Press into the victim's abdomen with a quick upward thrust.
   5. Repeat several times if necessary.

If you are alone and choking, use whatever is available to apply force just below your diaphragm. Press into a table or a sink, or use your own fist.

Reviving a Person Who Has Stopped Breathing

Mouth-to-Mouth Method

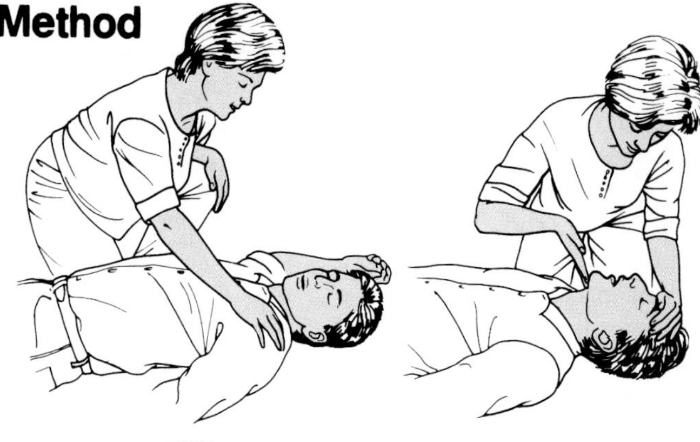

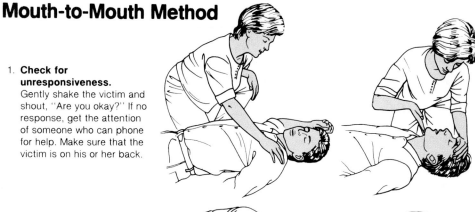

1. **Check for unresponsiveness.** Gently shake the victim and shout, "Are you okay?" If no response, get the attention of someone who can phone for help. Make sure that the victim is on his or her back.

2. **Open the airway.** Tilt the victim's head back by pushing on his or her forehead with your hand and lifting the chin with your fingers under his or her jaw. This will open the airway by moving the tongue away from the back of the victim's throat.

3. **Check for breathing.** Put your ear close to the victim's face to listen and feel for any return of air. At the same time, look to see if there is chest movement. Check for breathing for about 5 seconds.

4. **If no breathing, give two full breaths.** While maintaining the victim in the head-tilt position, pinch his or her nose to close off the nasal passageway. Take a deep breath, then seal your mouth around the victim's mouth and give two full breaths. (After the first breath, raise your head slightly to inhale quickly and then give the second breath.)

5. **Check for pulse.** While maintaining head tilt, feel for a carotid pulse for 5 to 10 seconds on the side of the victim's neck.

6. **Continue rescue breathing.** With the victim in the head-tilt position and his or her nostrils pinched, give one breath every 5 seconds. Observe for signs of breathing between breaths. For an infant, give one gentle puff every 3 seconds.

7. **Recheck for pulse.** Feel for a carotid pulse at 1-minute intervals. If the victim has a pulse but is not breathing, continue rescue breathing.

Mouth-to-Nose Method

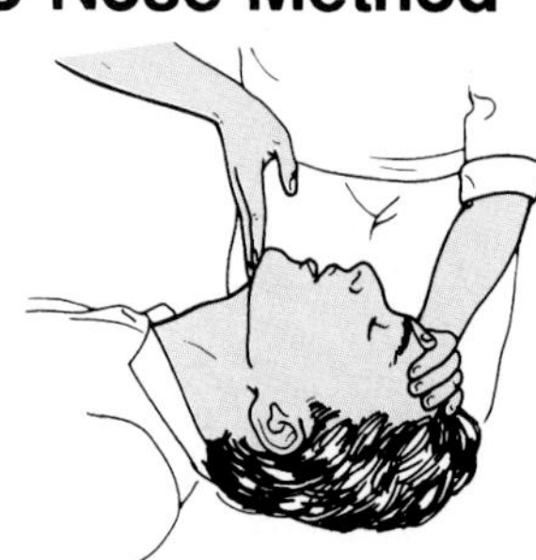

1. **Open the airway.** Place the victim in the head-tilt position as described above.

2. **Blow into the victim's nose.** Using the same sequence described above, blow into the victim's nose while holding his or her mouth closed.

3. **Feel and observe for breathing.** With the victim's mouth held open, detect for breathing between giving forced breaths.

Clinical Investigation Answer

The puncture wound must have produced a pneumothorax, thus raising the intrapleural pressure and causing collapse of the right lung. Since the left lung is surrounded by a separate pleural membrane, it was unaffected by the wound. As a result of the collapse of his right lung, the patient was hypoventilating. This caused retention of CO_2, thus raising his arterial P_{CO_2} and resulting in respiratory acidosis (as indicated by an arterial pH lower than 7.35). Upon recovery, analysis of his arterial blood revealed that he was breathing adequately but that he had a carboxyhemoglobin saturation of 20%. This very high level is probably due to a combination of smoking and driving in heavily congested areas, with much automobile exhaust. The high carboxyhemoglobin would reduce oxygen transport, thus aggravating any problems he might have with his cardiovascular or pulmonary system.

The significantly low $FEV_{1.0}$ indicates that the patient has an obstructive pulmonary problem, possibly caused by smoking and the inhalation of polluted air. A low $FEV_{1.0}$ could simply indicate bronchoconstriction, but the fact that this man's vital capacity was a little low suggests that he may have early-stage lung damage, possibly emphysema. He should be strongly advised to quit smoking, and further pulmonary tests should be administered at regular intervals.

Chapter Summary

Structure of the Respiratory System (pp. 729–737)

1. Respiration refers not only to breathing, but also to the exchange of gases between the atmosphere, the blood, and individual cells.
2. The respiratory system is divided into a respiratory division, which includes the alveoli of the lungs in which gas exchange occurs, and the conducting division, which includes all the structures that conduct air to the respiratory division.
3. The paranasal sinuses are located in the maxillary, frontal, sphenoid, and ethmoid bones.
4. The pharynx is a funnel-shaped passageway that connects the oral and nasal cavities with the larynx.
5. The larynx is composed of a number of cartilages that keep the passageway to the trachea open during breathing and that close the respiratory passageway during swallowing.
6. The trachea is a rigid tube, supported by incomplete rings of cartilage, that leads from the larynx to the bronchial tree.
7. The bronchial tree includes a right and a left primary bronchus, which divide to produce secondary bronchi, tertiary bronchi, and bronchioles; the conducting division ends with the terminal bronchioles, which connect to the pulmonary alveoli.
8. Pulmonary alveoli are the functional units of the lungs where the exchange of respiratory gases occur. They are small, thin-walled air sacs, and their abundance provides an immense surface area for gas exchange.
9. The lungs are covered by a visceral pleural membrane (visceral pleura), and the thoracic cavity is lined by a parietal pleural membrane (parietal pleura).
   (a) The potential space between the two pleural membranes is called the pleural cavity or intrapleural space.
   (b) The pleural membranes compartmentalize each lung and exclude the structures located in the mediastinum.

Physical Aspects of Ventilation (pp. 737–741)

1. The intrapleural and intrapulmonary pressures vary during ventilation.
   (a) The intrapleural pressure is always less than the intrapulmonary pressure.
   (b) The intrapulmonary pressure is subatmospheric during inspiration and greater than the atmospheric pressure during expiration.
2. Pressure changes in the lungs are produced by variations in lung volume in accordance with Boyle's law; that is, the pressure of a gas is inversely proportional to its volume.
3. The mechanics of ventilation are influenced by the physical properties of the lungs.
   (a) Lung compliance may be defined as the change in lung volume as a function of change in transpulmonary pressure.
   (b) The elasticity of the lungs refers to their tendency to recoil after distension.
   (c) The surface tension of the fluid in the pulmonary alveoli exerts a force directed inward, which acts to resist distension.
4. The action of lung surfactant to reduce surface tension in pulmonary alveoli prevents them from collapsing during expiration. Premature babies may lack sufficient surfactant and consequently suffer from respiratory distress syndrome.

Mechanics of Breathing (pp. 742–746)

1. Inspiration and expiration are accomplished by contraction and relaxation of skeletal muscles.
   (a) During quiet inspiration, the diaphragm and the external intercostals and internal intercostals (interchondral part) contract, increasing the volume of the thorax.
   (b) During quiet expiration these muscles relax, and the elastic recoil of the lungs and thorax decreases the thoracic volume.
   (c) The lungs are stuck to the wall of the thorax because the intrapulmonary pressure is greater than the intrapleural pressure.
2. Spirometry aids in the diagnosis of a number of pulmonary disorders.
   (a) An abnormally low vital capacity indicates restrictive lung disorders, including emphysema and pulmonary fibrosis.
   (b) An abnormally low forced expiratory volume indicates obstructive disorders, including asthma and bronchitis.

Gas Exchange in the Lungs (pp. 746–751)

1. According to Dalton's law, the total pressure of a gas mixture is equal to the sum of the pressures that each gas in the mixture would exert independently.

2. According to Henry's law, the amount of gas that can be dissolved in a fluid is directly proportional to the partial pressure of that gas in contact with the fluid.
3. The P_{O_2} and P_{CO_2} measurements of arterial blood provide information about lung function.
   (a) Inadequate ventilation (hypoventilation) causes a rise in arterial P_{CO_2} and a fall in arterial P_{O_2}.
   (b) Excessive ventilation (hyperventilation) causes a fall in arterial P_{CO_2}, but has little effect on arterial P_{O_2}.

Regulation of Breathing (pp. 752–756)

1. The rhythmicity center in the medulla oblongata directly controls the muscles of respiration.
   (a) Activity of the inspiratory and expiratory neurons varies in a reciprocal way to produce an automatic breathing cycle.
   (b) Activity in the medulla oblongata is influenced by the apneustic and pneumotaxic centers in the pons, as well as by sensory feedback information.
   (c) Conscious breathing involves direct control by the cerebral cortex via corticospinal tracts.
2. Breathing is affected by chemoreceptors sensitive to the P_{O_2}, pH, and P_{CO_2} of the blood.
   (a) The P_{CO_2} of the blood and consequent changes in pH are usually of greater importance than the blood P_{O_2} in the regulation of breathing.
   (b) Central chemoreceptors in the medulla oblongata are sensitive to changes in blood P_{CO_2} because of the resultant changes in the pH of cerebrospinal fluid.
   (c) The peripheral chemoreceptors in the aortic and carotid bodies are sensitive to changes in blood P_{CO_2} indirectly, because of consequent changes in blood pH.
3. Decreases in blood P_{O_2} directly stimulate breathing only when the blood P_{O_2} is less than 50 mmHg. A drop in P_{O_2} also stimulates breathing indirectly, by making the chemoreceptors more sensitive to changes in P_{CO_2} and pH.

Hemoglobin and Oxygen Transport (pp. 756–761)

1. Hemoglobin is composed of two alpha and two beta polypeptide chains and four heme groups that contain a central atom of iron.
2. When the iron is in the reduced form and not attached to O_2, the hemoglobin is called deoxyhemoglobin. When it is attached to O_2, it is called oxyhemoglobin.
3. Deoxyhemoglobin combines with O_2 in the lungs (the loading reaction) and breaks its bonds with O_2 in the tissue capillaries (the unloading reaction). The extent of each reaction is determined by the P_{O_2} and the affinity of hemoglobin for O_2.
4. An oxyhemoglobin dissociation curve is a graph of percent oxyhemoglobin saturation at different values of P_{O_2}.
5. The pH and temperature of the blood influence the affinity of hemoglobin for O_2 and the extent of loading and unloading.
   (a) In the Bohr effect, a fall in pH decreases the affinity of hemoglobin for O_2, and a rise in pH increases the affinity of hemoglobin for O_2.
   (b) A rise in temperature decreases the affinity of hemoglobin for O_2.
   (c) When the affinity is decreased, the oxyhemoglobin dissociation curve is shifted to the right. This indicates a greater percentage of O_2 unloaded to the tissues.
6. The affinity of hemoglobin for O_2 is also decreased by 2,3-diphosphoglyceric acid (2,3-DPG), an organic molecule in the red blood cells.

Carbon Dioxide Transport and Acid-Base Balance (pp. 761–763)

1. An enzyme in red blood cells called carbonic anhydrase catalyzes the reversible reaction whereby carbon dioxide and water are used to form carbonic acid.
   (a) This reaction is favored by the high P_{CO_2} in the tissue capillaries. As a result, CO_2 produced by the tissues is converted into carbonic acid in the red blood cells.
   (b) A reverse reaction occurs in the lungs. In this process, the low P_{CO_2} favors the conversion of carbonic acid to CO_2, which can be exhaled.
2. By adjusting the blood concentration of CO_2, and thus of carbonic acid, the process of ventilation helps to maintain proper acid-base balance of the blood.
   (a) Normal arterial blood pH is 7.40. A pH below 7.35 is termed acidosis; a pH above 7.45 is termed alkalosis.
   (b) Hyperventilation causes respiratory alkalosis and hypoventilation causes respiratory acidosis.

Effect of Exercise and High Altitude on Respiratory Function (pp. 764–768)

1. During exercise the increased ventilation, or hyperpnea, is matched to the increased metabolic rate so that the arterial blood P_{CO_2} remains normal.
2. Acclimatization to a high altitude involves changes that help to deliver O_2 more effectively to the tissues, despite reduced arterial P_{O_2}.

Review Activities

Objective Questions

1. Which is *not* a component of the nasal septum?
   (a) the palatine bone
   (b) the vomer
   (c) the ethmoid bone
   (d) septal cartilage
2. An adenoidectomy is the removal of
   (a) the uvula.
   (b) the pharyngeal tonsils.
   (c) the palatine tonsils.
   (d) the lingual tonsils.
3. Which is not a paranasal sinus?
   (a) the palatine sinus
   (b) the ethmoidal sinus
   (c) the sphenoidal sinus
   (d) the frontal sinus
   (e) the maxillary sinus
4. Which of the following is *not* characteristic of the left lung?
   (a) a cardiac impression
   (b) a superior lobe
   (c) a single fissure
   (d) an inferior lobe
   (e) a middle lobe
5. The epithelial lining of the wall of the thorax is called
   (a) the parietal pleura.
   (b) the pleural peritoneum.
   (c) the mediastinal pleura.
   (d) the visceral pleura.
   (e) the costal pleura.

6. Which of the following statements about intrapulmonary and intrapleural pressure is *true?*
   (a) The intrapulmonary pressure is always subatmospheric.
   (b) The intrapleural pressure is always greater than the intrapulmonary pressure.
   (c) The intrapulmonary pressure is greater than the intrapleural pressure.
   (d) The intrapleural pressure equals the atmospheric pressure.
7. If the transpulmonary pressure equals zero,
   (a) a pneumothorax has probably occurred.
   (b) the lungs cannot inflate.
   (c) elastic recoil causes the lungs to collapse.
   (d) all of the above apply.
8. The maximum amount of air that can be expired after a maximum inspiration is the
   (a) tidal volume.
   (b) forced expiratory volume.
   (c) vital capacity.
   (d) maximum expiratory flow rate.
9. If the blood lacked red blood cells but the lungs were functioning normally,
   (a) the arterial P_{O_2} would be normal.
   (b) the oxygen content of arterial blood would be normal.
   (c) both *a* and *b* would apply.
   (d) neither *a* nor *b* would apply.
10. If a person were to dive with scuba equipment to a depth of 66 feet, which of the following statements would be *false?*
    (a) The arterial P_{O_2} would be three times normal.
    (b) The oxygen content of plasma would be three times normal.
    (c) The oxygen content of whole blood would be three times normal.
11. Which of the following would be most affected by a decrease in the affinity of hemoglobin for oxygen?
    (a) arterial P_{O_2}
    (b) arterial percent oxyhemoglobin saturation
    (c) venous oxyhemoglobin saturation
    (d) arterial P_{CO_2}
12. If a person with normal lung function were to hyperventilate for several seconds, there would be a significant
    (a) increase in the arterial P_{O_2}.
    (b) decrease in the arterial P_{CO_2}.
    (c) increase in the arterial percent oxyhemoglobin saturation.
    (d) decrease in the arterial pH.
13. Erythropoietin is produced by
    (a) the kidneys.
    (b) the liver.
    (c) the lungs.
    (d) the bone marrow.
14. The affinity of hemoglobin for oxygen is decreased under conditions of
    (a) acidosis.
    (b) fever.
    (c) anemia.
    (d) acclimatization to a high altitude.
    (e) all of the above.
15. Most of the carbon dioxide in the blood is carried in the form of
    (a) dissolved CO_2.
    (b) carbaminohemoglobin.
    (c) bicarbonate.
    (d) carboxyhemoglobin.
16. The bicarbonate concentration of the blood would be decreased during
    (a) metabolic acidosis.
    (b) respiratory acidosis.
    (c) metabolic alkalosis.
    (d) respiratory alkalosis.
17. The chemoreceptors in the medulla are directly stimulated by
    (a) CO_2 from the blood.
    (b) H^+ from the blood.
    (c) H^+ in cerebrospinal fluid that is derived from blood CO_2.
    (d) decreased arterial P_{O_2}.
18. The rhythmic control of breathing is produced by the activity of inspiratory and expiratory neurons in
    (a) the medulla oblongata.
    (b) the apneustic center of the pons.
    (c) the pneumotaxic center of the pons.
    (d) the cerebral cortex.
19. Which of the following occur(s) during hypoxemia?
    (a) increased ventilation
    (b) increased production of 2,3-DPG
    (c) increased production of erythropoietin
    (d) all of the above
20. During exercise, which of the following statements is *true?*
    (a) The arterial percent oxyhemoglobin saturation is decreased.
    (b) The venous percent oxyhemoglobin saturation is decreased.
    (c) The arterial P_{CO_2} is measurably increased.
    (d) The arterial pH is measurably decreased.
21. All of the following can bond with hemoglobin *except*
    (a) HCO_3^-.
    (b) O_2.
    (c) H^+.
    (d) CO_2.
22. Which of the following statements about the partial pressure of carbon dioxide is *true?*
    (a) It is higher in the alveoli than in the pulmonary arteries.
    (b) It is higher in the systemic arteries than in the tissues.
    (c) It is higher in the systemic veins than in the systemic arteries.
    (d) It is higher in the pulmonary veins than in the pulmonary arteries.

Essay Questions

1. Using a flow diagram to show cause and effect, explain how contraction of the diaphragm produces inspiration.
2. Radiographic (X-ray) pictures show that the rib cage of a person with a pneumothorax is expanded, and the ribs are farther apart. Explain why this should be so.
3. Using a flowchart, explain how a rise in blood P_{CO_2} stimulates breathing. Include both the central and peripheral chemoreceptors in your answer.
4. Explain why a person with ketoacidosis may hyperventilate. What benefit might this provide? Also explain why hyperventilation can be stopped by an intravenous fluid containing bicarbonate.

5. What blood measurements can be performed to detect (*a*) anemia, (*b*) carbon monoxide poisoning, and (*c*) poor lung function?
6. Explain how measurements of blood P_{CO_2}, bicarbonate, and pH are affected by hypoventilation and hyperventilation.
7. Describe the changes in ventilation that occur during exercise. How are these changes produced and how do they affect arterial blood gases and pH?
8. How would an increase in the red blood cell content of 2,3-DPG affect the P_{O_2} of venous blood? Explain your answer.
9. Explain the mechanisms that produce changes in ventilation at a high altitude. Why are these changes beneficial? Under what conditions could they be detrimental? What other factors operate at a high altitude to improve oxygen delivery to the tissues?
10. Compare asthma and emphysema in terms of their characteristics and the effects they have on pulmonary function tests.
11. Explain the mechanisms involved in quiet inspiration and in forced inspiration, and in quiet expiration and forced expiration. What muscles are involved in each case?
12. Describe the formation, composition, and function of pulmonary surfactant. What happens when surfactant is absent? How is this condition treated?

Critical Thinking Questions

1. The nature of the sounds produced by percussion (tapping) a patient's chest can tell a physician a great deal about the condition of the organs within the thoracic cavity. Healthy, air-filled lungs resonate, or sound hollow. How do you think the lungs of a person with emphysema would sound in comparison to healthy lungs? What kind of sounds would be produced by a collapsed lung, or one that was partially filled with fluid?
2. Explain why the first breath of a healthy neonate is more difficult than subsequent breaths and why premature infants often require respiratory assistance (a mechanical ventilator) to keep their lungs inflated. How else is this condition treated?
3. Nicotine from cigarette smoke causes the buildup of mucus and paralyzes the cilia that line the respiratory tract. How might these conditions affect pulmonary function tests? If smoking has led to emphysema, how would the pulmonary function tests change?
4. Carbon monoxide poisoning from smoke inhalation and suicide attempts is the most common cause of death from poisoning in the United States. How would carbon monoxide poisoning affect a person's coloring, particularly of the mucous membranes? How would it affect the hemoglobin concentration, hematocrit, and percent oxyhemoglobin saturation? How would chronic carbon monoxide poisoning affect the person's red blood cell content of 2,3-DPG?
5. After driving from sea level to a trail head in the high Sierras, you get out of your car and feel dizzy. What do you suppose is causing your dizziness? How is this beneficial and how is it detrimental? What may eventually happen that would counteract the cause of the dizziness?

Related Web Sites

For a listing of the most current web sites related to this chapter, please visit the *Concepts of Human Anatomy and Physiology* home page at http://www.mhhe.com/biosci/abio/.

NEXUS

Some Interactions of the Respiratory System with the Other Body Systems

Skeletal System

- The respiratory system provides all organs, including the bones, with oxygen and eliminates carbon dioxide (p. 729).
- The lungs are protected by the rib cage, and bones of the rib cage serve as levers for the action of respiratory muscles (p. 742).
- Red blood cells, needed for oxygen transport, are produced in the bone marrow (p. 596).

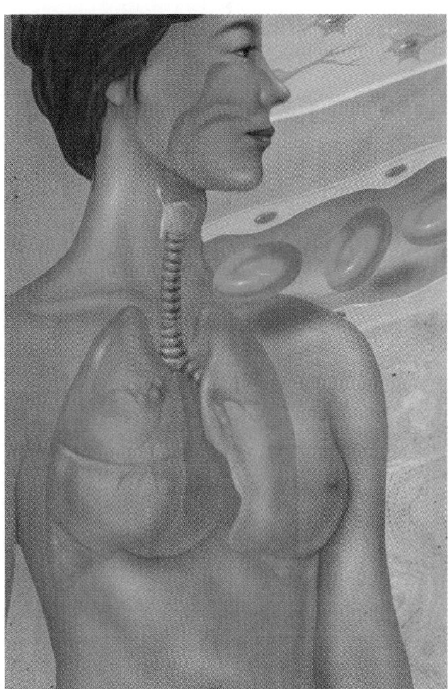

Nervous System

- The nervous system regulates the rate and depth of breathing (p. 752).
- Autonomic nerves regulate the blood flow, and hence the delivery of blood to tissues for gas exchange (p. 668).

Endocrine System

- Epinephrine dilates bronchioles, reducing airway resistance (p. 769).
- Thyroxine and epinephrine stimulate the rate of cell respiration (p. 880).

Muscular System

- Contractions of skeletal muscles are needed for ventilation (p. 742).
- Muscles consume large amounts of oxygen and produce large amounts of carbon dioxide during exercise (p. 670).

Circulatory System

- The heart and arteries deliver oxygen from the lungs to the body tissues, and veins transport carbon dioxide from the body tissues to the lungs (p. 592).
- Blood capillaries allow gas exchange for cellular respiration in the tissues and lungs (p. 626).

Lymphatic and Immune Systems

- The immune system protects against infections that could damage the respiratory system (p. 696).
- Alveolar macrophages and the action of cilia in the airways help protect the lungs from infection (p. 735).
- The tonsils contain lymphatic nodules (p. 732).

Urinary System

- The kidneys regulate the volume and electrolyte balance of the blood (p. 779).
- The kidneys participate with the lungs in the regulation of blood pH (p. 803).

Digestive System

- The GI tract provides nutrients to be used by cells of the lungs and other organs (p. 817).
- The respiratory system provides oxygen for cellular respiration of glucose and other nutrients brought into the blood by the digestive system (p. 729).

Reproductive System

- The lungs provide oxygen for cellular respiration of reproductive organs and eliminate carbon dioxide produced by these organs (p. 729).
- Changes in breathing and cellular respiration occur during sexual arousal (p. 930).

Twenty-Five

Urinary System: Fluid, Electrolyte, and Acid-Base Balance

Objectives

- Describe the position and gross structure of the kidney and the structure of a nephron.
- Explain how the nephrons are positioned in the kidney and trace the path of urine flow from the glomerulus to the renal pelvis.
- Describe the structural and functional relationships between the nephron tubules and their associated blood vessels.
- Describe the composition of glomerular ultrafiltrate and explain how it is produced.
- Define the term *glomerular filtration rate (GFR)* and explain how the GFR is regulated.
- Describe salt and water reabsorption in the proximal convoluted tubule.
- Describe the transport process in the nephron loop and explain how the countercurrent multiplier effect is produced.
- Discuss the structure and function of the vasa recta.
- Discuss the significance of a hypertonic renal medulla and describe the mechanism of ADH action.
- Define the term *renal plasma clearance* and explain how the processes of reabsorption and secretion affect the clearance of various substances.
- Describe how the GFR is measured by the clearance of inulin and explain why total renal blood flow can be measured by the clearance of PAH.
- Describe how the kidneys reabsorb glucose and amino acids and explain how glycosuria is produced.
- Describe the regulation of Na^+/K^+ balance by aldosterone and explain how aldosterone secretion is regulated.
- Explain how the interaction between plasma K^+ and H^+ concentrations affects the tubular secretions of these ions.
- Distinguish between the respiratory and metabolic components of acid-base balance and discuss the reabsorption of bicarbonate.
- Explain how the kidneys contribute to regulation of acid-base balance and how acid is excreted in the urine.
- Describe the structure of the urinary bladder and the ureters and explain the sequence of events involved in micturition.
- Compare the male urethra with that of the female.

Clinical Investigation

A teenage boy visited his family physician complaining of pain in his back between the twelfth rib and the lumbar vertebrae. The boy's urine was noticeably discolored, and he was referred to a urologist. Urinalysis revealed that he had hematuria (blood in urine); his urine sediment contained casts with associated red blood cells. Only trace amounts of protein were detected in the urine. Further analysis showed mild oliguria (reduced urine production) and an elevated plasma creatinine concentration. Edema was present. The boy mentioned that he was still competing on his cross-country running team, even though his throat had been sore for the past month. A throat culture demonstrated the presence of a streptococcus infection. The boy was placed on antibiotics and given hydrochlorothiazide. Within a few weeks, his symptoms were gone. What was responsible for the boy's symptoms, and why did they disappear with this treatment?

Clues: Examine the section "Renal Clearance of Inulin: Measurement of GFR" and the boxed information on creatinine. Look up hydrochlorothiazide and glomerulonephritis in the "Clinical Considerations" section.

Urinary System and Kidney Structure

Each kidney contains many tiny tubules that empty into a cavity drained by the ureter. Each of the tubules receives a blood filtrate from a capillary bed called the glomerulus. The filtrate is similar to tissue fluid but is modified as it passes through the tubules and is thereby changed into urine. The tubules and associated blood vessels thus form the functional units of the kidneys, which are known as nephrons.

The urinary system consists of two kidneys, two ureters, the urinary bladder, and the urethra (fig. 25.1). The primary function of the kidneys is regulation of the extracellular fluid (blood plasma and interstitial fluid) environment in the body. This is accomplished through the formation of urine, which is a modified filtrate of blood plasma. In the process of urine formation, the kidneys regulate (1) the volume of blood plasma (and thus contribute significantly to the regulation of blood pressure); (2) the concentration of waste products in the blood; (3) the concentration of electrolytes (Na^+, K^+, HCO_3^-, and other ions) in the plasma; and (4) the pH of plasma. In order to understand how these functions are performed by the kidneys, a knowledge of kidney structure is required.

Position and Gross Structure of the Kidney

The reddish-brown **kidneys** are positioned against the posterior abdominal wall, one on each side of the vertebral column,

Figure 25.1

Organs of the urinary system.
The urinary system of a female is shown. A male's urinary system is the same, except that the urethra extends through the penis.

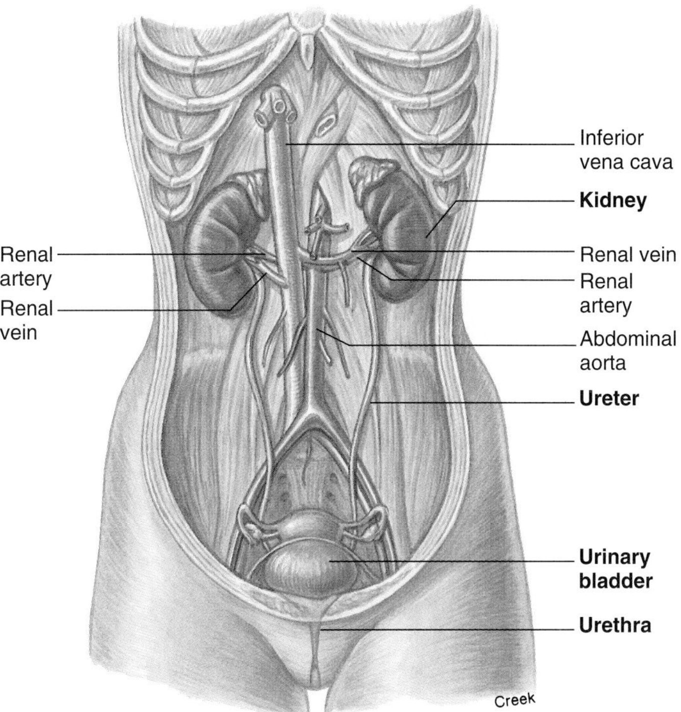

between the levels of the twelfth thoracic and the third lumbar vertebrae (fig. 25.2). The right kidney is usually 1.5 to 2.0 cm lower than the left because of the large area occupied by the liver above it.

The kidneys are *retroperitoneal,* which means that they are positioned behind the parietal peritoneum. Thus, strictly speaking, they are not within the peritoneal cavity. Each adult kidney is a lima bean–shaped organ about 11.25 cm (4 in.) long, 5.5 to 7.7 cm (2 to 3 in.) wide, and 3.0 cm (1.2 in.) thick. The **hilum** (*hi'lum*) of the kidney is the depression along the medial border through which the **renal artery** and nerves enter, and the **renal vein** and **ureter** (*yoo-re'ter*) exit.

Each kidney is embedded in a fatty fibrous pouch consisting of three layers. The **renal capsule**, the innermost layer, forms a strong, transparent fibrous attachment to the surface of the kidney. The renal capsule protects the kidney from trauma and the spread of infections. Surrounding the renal capsule is a firm protective mass of adipose tissue called the **adipose capsule.** The outermost layer, the **renal fascia,** is composed of dense irregular connective tissue. It anchors the kidney to the peritoneum and the abdominal wall.

A coronal section of the kidney shows two distinct regions and a major cavity (fig. 25.3*a*). The outer **renal cortex,** in contact with the real capsule, is reddish brown and granular in appearance because of its many capillaries. The deeper region, or **renal medulla,** is a darker color, and the presence of microscopic tubules and blood vessels gives it a

Figure 25.2

Pseudocolor radiograph of the urinary system.
In this photograph, shades of gray are assigned colors. The calyces of the kidneys, the renal pelvises, the ureters, and the urinary bladder are visible.

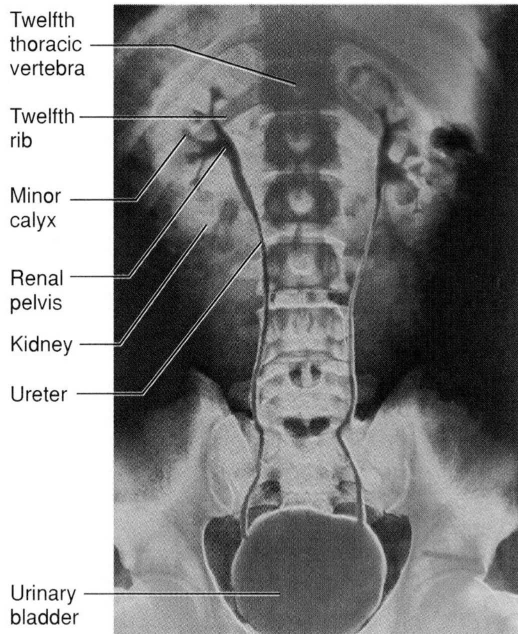

striped appearance. The renal medulla is composed of 8 to 15 conical **renal pyramids** separated by **renal columns.** The **renal papillae** (*pă-pil'e*) are the apexes of the renal pyramids. These nipplelike projections are directed toward the large cavity of the kidney called the **renal pelvis.**

The renal pelvis is divided into several portions. Each papilla of a renal pyramid projects into a small depression called the **minor calyx** (*ka'liks*)—in the plural, *calyces*. Several minor calyces unite to form a **major calyx.** In turn, the major calyces join to form the funnel-shaped renal pelvis. The renal pelvis serves to collect urine from the calyces and transport it to the ureter.

Microscopic Structure of the Kidney

The **nephron** (*nef'ron*) is responsible for the formation of urine and is thus the basic functional unit of the kidney. Each kidney contains more than a million nephrons. A nephron consists of small tubes, or **tubules,** and associated small blood vessels. Fluid formed by capillary filtration enters the tubules and is subsequently modified by transport processes; the resulting fluid that leaves the tubules is urine.

Renal Blood Vessels

Arterial blood enters the kidney at the hilum through the **renal artery,** which divides into **interlobar** (*in"ter-lo'bar*) **arteries** (fig. 25.4) that pass between the pyramids through the renal columns. **Arcuate** (*ar'kyoo-at*) **arteries** branch from the interlobar arteries at the boundary of the renal cortex and renal medulla. A number of **interlobular arteries** radiate from the arcuate arteries into the renal cortex and subdivide into numerous microscopic **afferent glomerular arterioles** (fig. 25.5). The afferent glomeruler arterioles transport the blood into the **glomeruli** (*glo-mer'yŭ-li*), the capillary networks that produce a blood filtrate that enters the urinary tubules. The blood remaining in the glomeruli leaves through **efferent glomerular arterioles,** which deliver the blood into another capillary network, the **peritubular capillaries** surrounding the tubules. The *vasa recta* (*va'ză rek'tă*) are specialized parts of the peritubular capillaries that extend deep into the renal medulla along with the nephron loop (discussed below).

This arrangement of blood vessels is unique. It is the only one in the body in which a capillary bed (the glomerulus) is drained by an arteriole rather than by a venule and delivered to a second capillary bed located downstream (the peritubular capillaries). Blood from the peritubular capillaries and the vasa recta is drained into veins that parallel the course of the arteries in the kidney. These veins are called the **interlobular veins, arcuate veins,** and **interlobar veins.** The interlobar veins descend between the pyramids, converge, and leave the kidney as a single **renal vein** that empties into the inferior vena cava.

Nephron Tubules

The tubular portion of a nephron consists of a glomerular capsule, a proximal convoluted tubule, a descending limb of the nephron loop, an ascending limb of the nephron loop, and a distal convoluted tubule (fig. 25.5).

The **glomerular (Bowman's) capsule** surrounds the glomerulus. The glomerular capsule and its associated glomerulus are located in the renal cortex of the kidney and together constitute the **renal corpuscle.** The glomerular capsule contains an inner visceral layer of epithelium, in contact with the glomerular capillaries, and an outer parietal layer. The space between these two layers is continuous with the lumen of the tubule and receives the glomerular filtrate, as will be described in the next section.

Filtrate that enters the glomerular capsule passes into the lumen of the **proximal convoluted tubule.** The wall of the proximal convoluted tubule consists of a single layer of cuboidal cells containing millions of microvilli; these microvilli increase the surface area for reabsorption. In the process of reabsorption, salt, water, and other molecules needed by the body are transported from the lumen, through the tubular cells, and into the surrounding peritubular capillaries.

The glomerulus, glomerular capsule, and proximal convoluted tubule are located in the renal cortex. Fluid passes from the proximal convoluted tubule to the **nephron loop**

arcuate: L. *arcuare*, to bend
glomerulus: L. diminutive of *glomus*, ball
Bowman's capsule: from Sir William Bowman, English anatomist, 1816–92

Figure 25.3

Structure of the kidney.
The figure depicts (*a*) a coronal section of the kidney and (*b*) a magnified view of the contents of a renal pyramid. (*c*) A single nephron tubule, microscopic in actual size, is shown isolated.

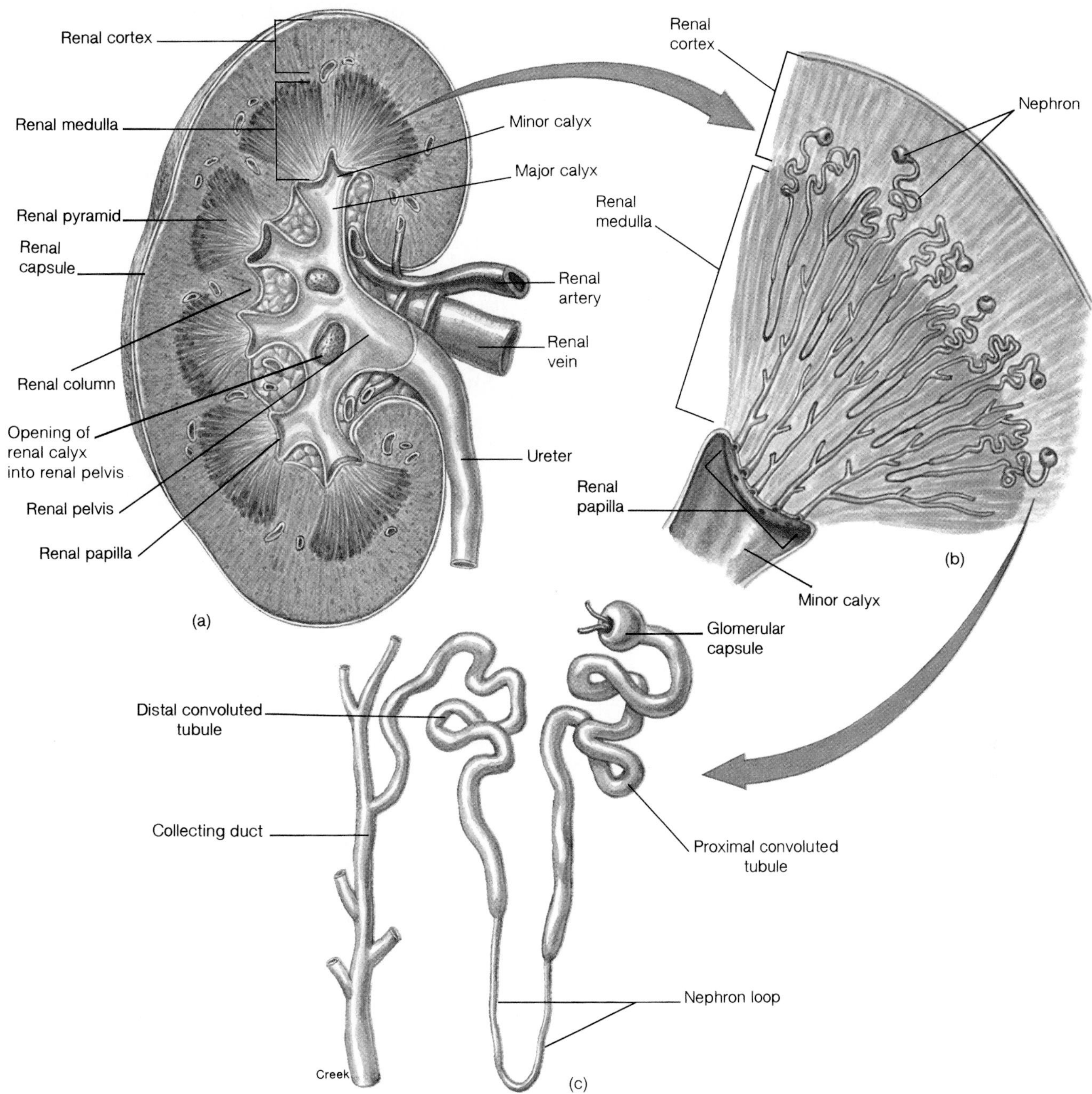

(loop of Henle). This fluid is carried into the renal medulla in the **descending limb** of the loop and returns to the renal cortex in the **ascending limb** of the loop. Back in the renal cortex, the tubule again becomes coiled and is called the **distal convoluted tubule.** The distal convoluted tubule is shorter than the proximal convoluted tubule and has relatively few microvilli. It is the last segment of the nephron and terminates as it empties into a collecting duct.

The two principal types of nephrons are classified according to their position in the kidney and the lengths of their nephron loops. Nephrons that originate in the inner one-third of the renal cortex—called **juxtamedullary** (*juk″stă-med′yŭ-ler-e*) **nephrons** because they are next to the renal medulla—have longer nephron loops than the more numerous **cortical nephrons,** which originate in the outer

loop of Henle: from Friedrich G. J. Henle, German anatomist, 1809–85

Figure 25.4

Vascular structure of the kidneys.

(*a*) An illustration of the major arterial supply and (*b*) a scanning electron micrograph of the glomeruli (300×).

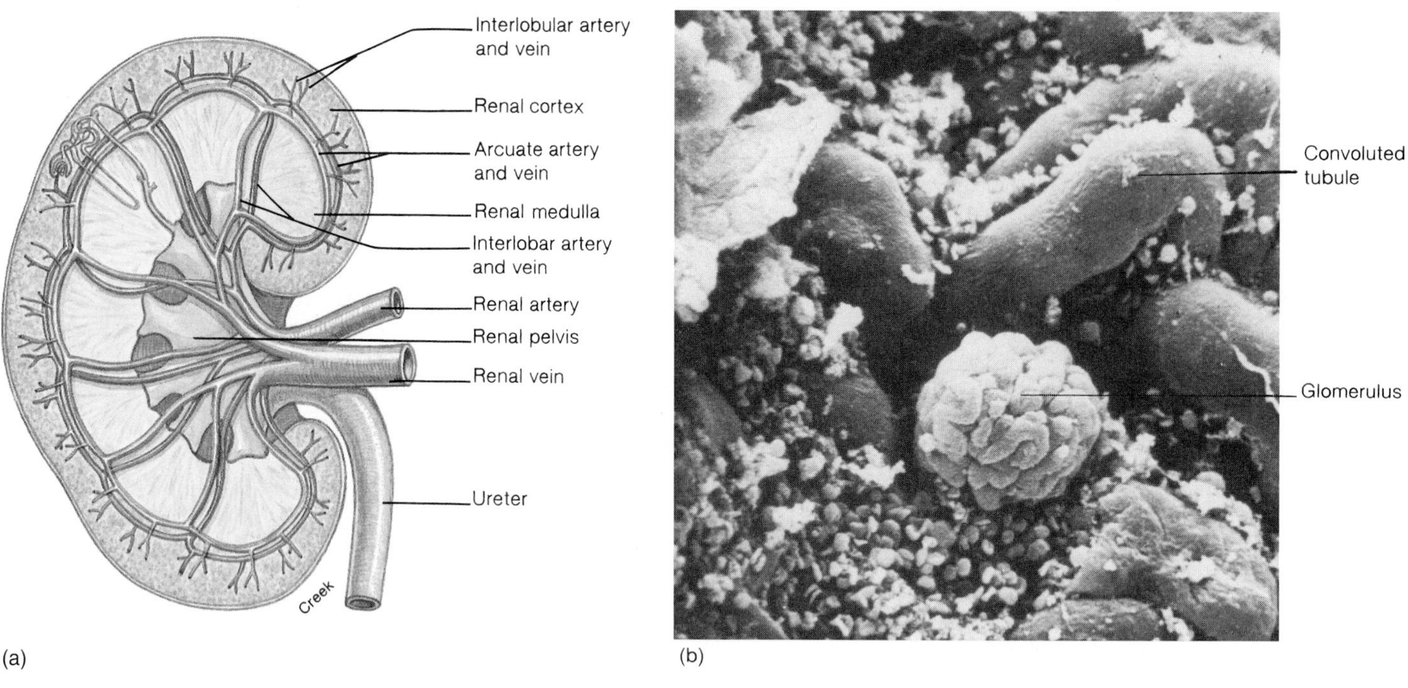

Figure 25.5

Renal tubules and associated blood vessels.

In this simplified illustration, the blood flow from a glomerulus to an efferent glomerular arteriole, to the peritubular capillaries, and to the venous drainage of the kidneys is indicated with arrows.

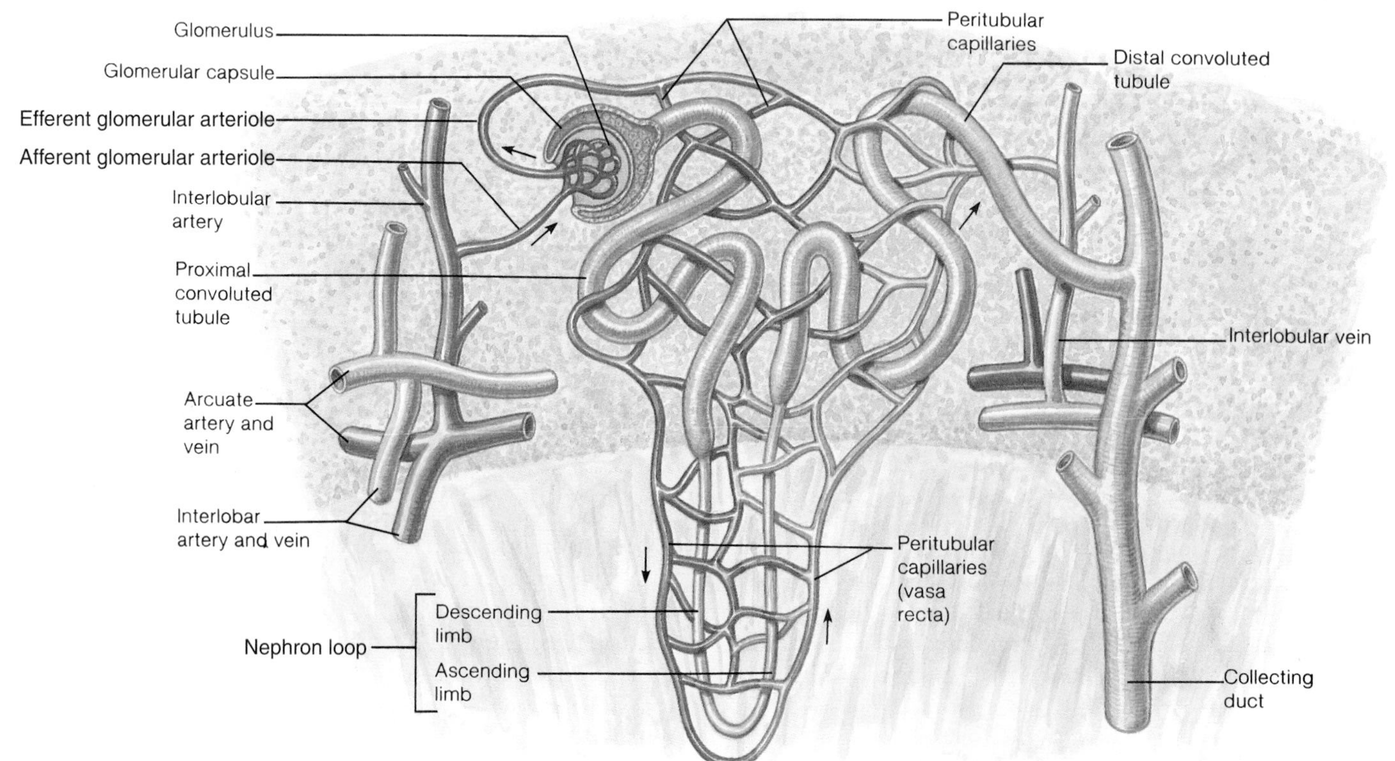

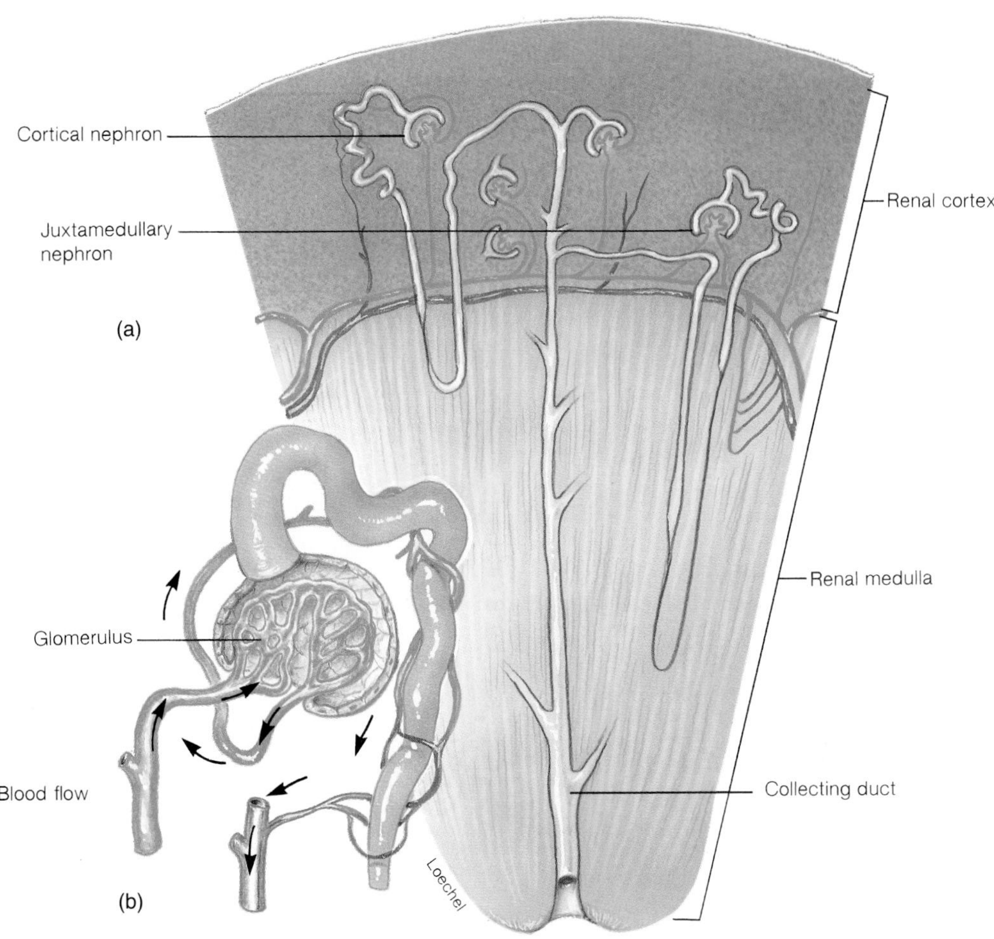

Figure 25.6

Contents of a renal pyramid. (*a*) The position of cortical and juxtamedullary nephrons is shown within the renal pyramid of the kidney. (*b*) The direction of blood flow in the vessels of the nephron is indicated with arrows.

two-thirds of the renal cortex (fig. 25.6). The juxtamedullary nephrons play an important role in the ability of the kidney to produce a concentrated urine.

A **collecting duct** receives fluid from the distal convoluted tubules of several nephrons. Fluid is then drained by the collecting duct from the renal cortex to the renal medulla as the collecting duct passes through a renal pyramid. This fluid, now called urine, passes into a minor calyx. Urine is then funneled through the renal pelvis and out of the kidney in the ureter.

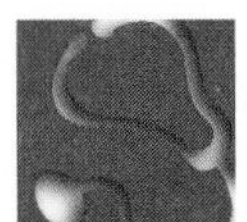

Polycystic kidney disease is a condition inherited as an autosomal dominant trait (see chapter 30) that affects 1 in every 600 to 1,000 people. This disease is thus more common than sickle-cell anemia, cystic fibrosis, and muscular dystrophy, which are also genetic diseases. In 50% of the people who inherit the defective gene (located on the short arm of chromosome 16), progressive renal failure develops during middle age to the point that dialysis or a kidney transplant is required. The cysts that develop are expanded portions of the renal tubule. Cysts that originate in the proximal convoluted tubule contain fluid that resembles glomerular filtrate and plasma. Cysts that originate in the distal tubule contain fluid with a lower NaCl concentration and a higher potassium and urea concentration than plasma as a result of the transport processes that occur during the passage of fluid through the tubules.

Glomerular Filtration

The glomerular capillaries have large pores in their walls, and the visceral layer of the glomerular capsule, in contact with the glomerulus, has filtration slits. Water, together with dissolved solutes (but not proteins), can thus pass from the blood plasma to the inside of the capsule and the nephron tubules. The volume of this filtrate produced by both kidneys each minute is called the glomerular filtration rate (GFR).

Endothelial cells of the glomerular capillaries have large pores (200–500 Å in diameter) called *fenestrae*, (*fĕ-nes'tre*), and are thus said to be *fenestrated*. As a result of these large pores, glomerular capillaries are 100 to 400 times more permeable to water in blood plasma and dissolved solutes than are the capillaries of skeletal muscles. Although the pores of glomerular capillaries are large, they are still small enough to prevent the passage of red blood cells, white blood cells, and platelets into the filtrate.

Before the filtrate can enter the lumen of the glomerular capsule, it must pass through the capillary pores, the basement membrane (a thin layer of glycoproteins lying immediately outside the endothelial cells), and the inner (visceral) layer of the glomerular capsule. The inner layer of the glomerular capsule is composed of unique cells called

Figure 25.7

Scanning electron micrograph of glomerular capillaries and capsule.
The inner (visceral) layer of the glomerular (Bowman's) capsule is composed of podocytes, as shown in this scanning electron micrograph. Very fine extensions of these podocytes form foot processes, or pedicels, that interdigitate around the glomerular capillaries. Spaces between adjacent pedicels form the "filtration slits." (See figure 25.8 also.)

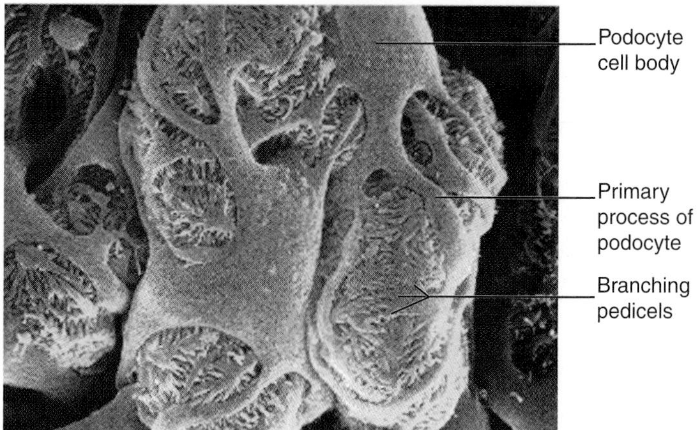

Figure 25.8

Structure of the glomerulus and capsule.
An illustration of the relationship between glomerular capillaries and the inner (visceral) layer of the glomerular (Bowman's) capsule. Note that filtered molecules pass out of the fenestrae of the capillaries and through the filtration slits to enter into the cavity of the capsule.

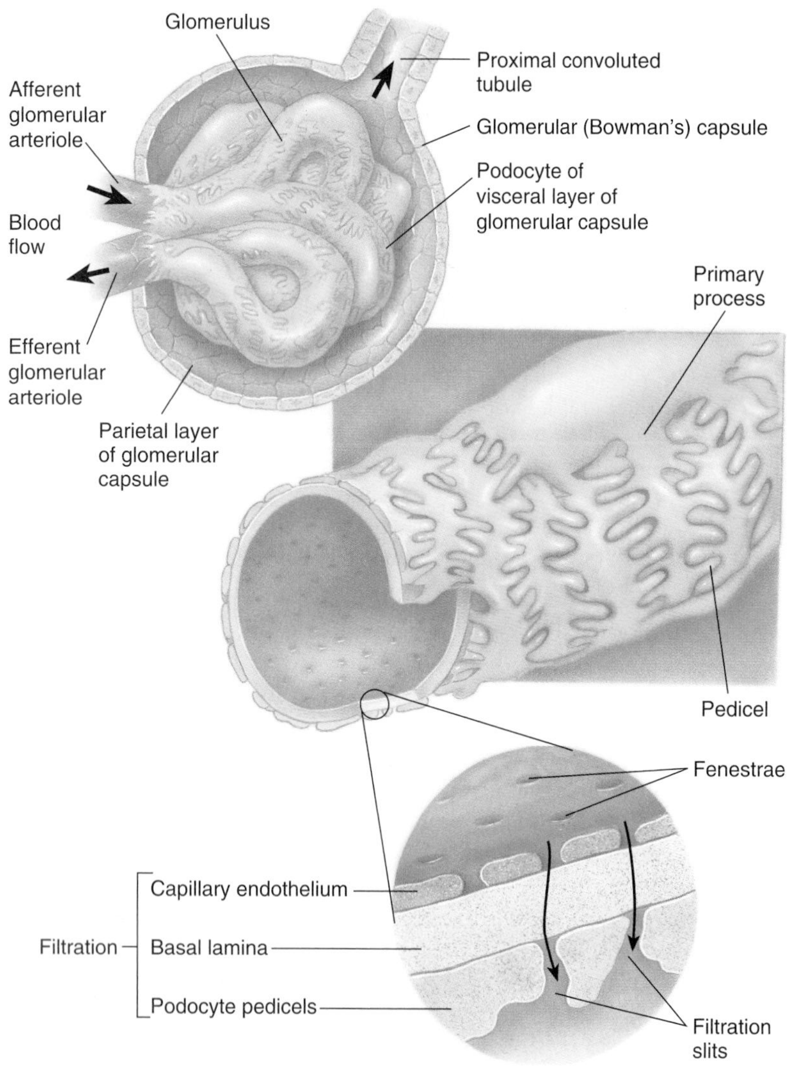

podocytes (*pod'ŏ-sīts*). The podocytes are shaped somewhat like octopi, with bulbous cell bodies and several thick arms. Each arm has thousands of cytoplasmic extensions known as **pedicels** (*ped'ĭ-selz*), or "foot processes" (fig. 25.7). These pedicels interdigitate, like the fingers of clasped hands, as they wrap around the glomerular capillaries. The narrow slits between adjacent pedicels provide the passageways through which filtered molecules must pass to enter the interior of the glomerular capsule (fig. 25.8).

Although the glomerular capillary pores are apparently large enough to permit the passage of proteins, the fluid that enters the capsular space contains only a small amount of plasma proteins. This relative exclusion of plasma proteins from the filtrate is partially a result of their negative charges, which hinder their passage through the negatively charged glycoproteins in the basement membrane of the capillaries (fig. 25.9). The large size and negative charges of plasma proteins may also restrict their movement through the filtration slits between pedicels.

Glomerular Ultrafiltrate

The fluid that enters the glomerular capsule is called **ultrafiltrate** (fig. 25.10) because it is formed under pressure—the hydrostatic pressure of the blood. This process is similar to the formation of tissue fluid by other capillary beds in the body in response to Starling forces (chapter 22). The force favoring filtration is opposed by a counterforce developed by the hydrostatic pressure of fluid in the glomerular capsule. Also, since the protein concentration of the tubular fluid is low (less than 2 to 5 mg per 100 ml) compared to that of plasma (6 to 8 g per 100 ml), the greater colloid osmotic pressure of blood plasma promotes the osmotic return of filtered water. When these opposing forces

podocyte: Gk. *pous*, foot; *kytos*, cell
pedicel: L. *peducekkus*, footplate

Figure 25.9

Electron micrograph of the filtration barrier. This electron micrograph shows the barrier separating the capillary lumen from the cavity of the glomerular (Bowman's) capsule.

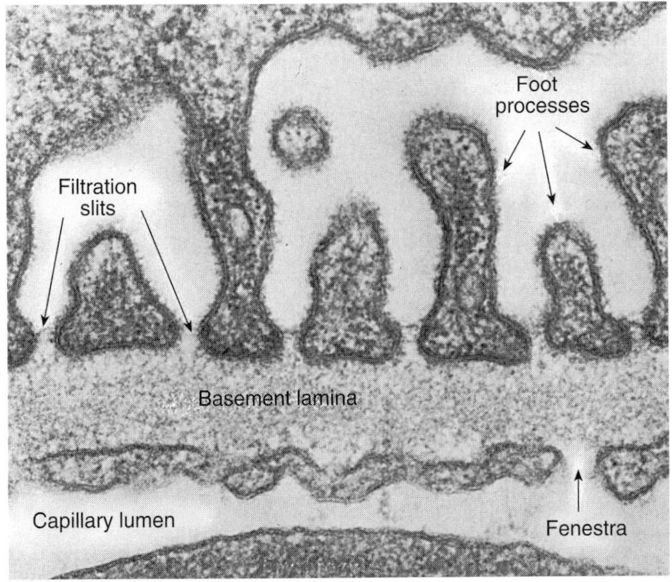

Figure 25.10

Formation of glomerular ultrafiltrate. Only a very small proportion of plasma proteins (green circles) are filtered, but smaller plasma solutes (purple dots) easily enter the glomerular ultrafiltrate. Arrows indicate the direction of filtration.

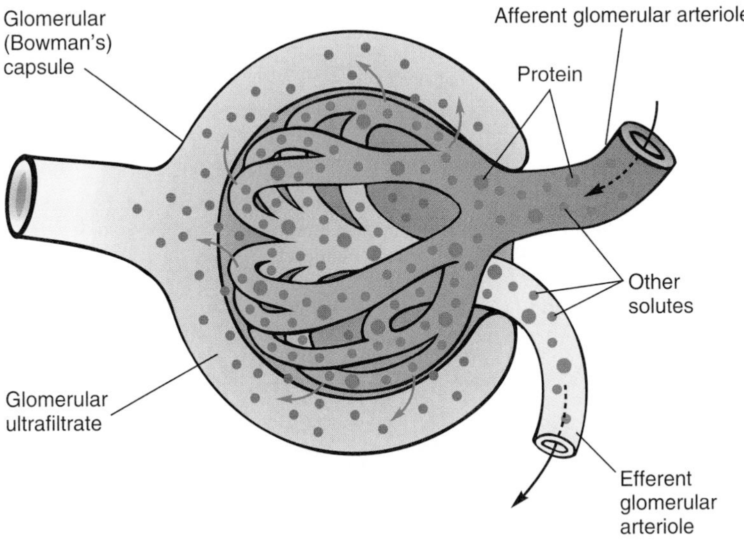

are subtracted from the hydrostatic pressure of the glomerular capillaries, a *net filtration pressure* of only about 10 mmHg is obtained.

Because glomerular capillaries are extremely permeable and have an extensive surface area, this modest net filtration pressure produces an extraordinarily large volume of filtrate. The **glomerular filtration rate (GFR)** is the volume of filtrate produced by both kidneys each minute. The GFR averages 115 ml per minute in women and 125 ml per minute in men. This is equivalent to 7.5 L per hour or 180 L per day (about 45 gallons)! Since the total blood volume averages about 5.5 L, this means that the total blood volume is filtered into the urinary tubules every 40 minutes. Most of the filtered water must obviously be returned immediately to the vascular system, or a person would literally urinate to death within minutes.

Regulation of Glomerular Filtration Rate

Vasoconstriction or dilation of afferent arterioles affects the rate of blood flow to the glomerulus and thus affects the glomerular filtration rate. Changes in the diameter of the afferent arterioles result from both extrinsic (sympathetic innervation) and intrinsic regulatory mechanisms. These mechanisms are needed to ensure that the GFR will be high enough to allow the kidneys to eliminate wastes and regulate blood pressure, but not so high as to cause excessive water loss.

Sympathetic Nerve Effects

An increase in sympathetic nerve activity, as occurs during the fight-or-flight reaction and exercise, stimulates constriction of the afferent arterioles. This helps to preserve blood volume and to divert blood to the muscles and heart. A similar effect occurs during cardiovascular shock when sympathetic nerve activity stimulates vasoconstriction. The decreased GFR and the resulting decreased rate of urine formation help to compensate for the rapid drop of blood pressure under these circumstances (fig. 25.11).

Renal Autoregulation

When the direct effect of sympathetic stimulation is experimentally removed, the effect of systemic blood pressure on GFR can be observed. Under these conditions, surprisingly, the GFR remains relatively constant despite changes in mean arterial pressure within a range of 70 to 180 mmHg (normal mean arterial pressure is 100 mmHg). The ability of the kidneys to maintain a relatively constant GFR in the face of fluctuating blood pressures is called **renal autoregulation.**

Renal autoregulation is achieved through the effects of locally produced chemicals on the afferent arterioles (effects on the efferent arterioles are believed to be of secondary importance). When systemic arterial pressure falls toward a mean of 70 mmHg, the afferent arterioles dilate, and when the pressure rises, the afferent arterioles constrict. Blood flow to the glomeruli and GFR can thus remain relatively constant within the autoregulatory range of blood pressure values. The effects of different regulatory mechanisms on the GFR are summarized in table 25.1.

Autoregulation is also achieved through a negative feedback relationship between the afferent arterioles and the volume of fluid in the filtrate. An increased flow of filtrate is sensed by a special group of cells called the *macula densa* in the thick portion of the ascending limb (see fig. 25.24). When the macula densa senses an increased flow of filtrate, it signals the afferent arterioles to constrict. This lowers the GFR, thereby decreasing the formation of filtrate in a process called **tubuloglomerular feedback.**

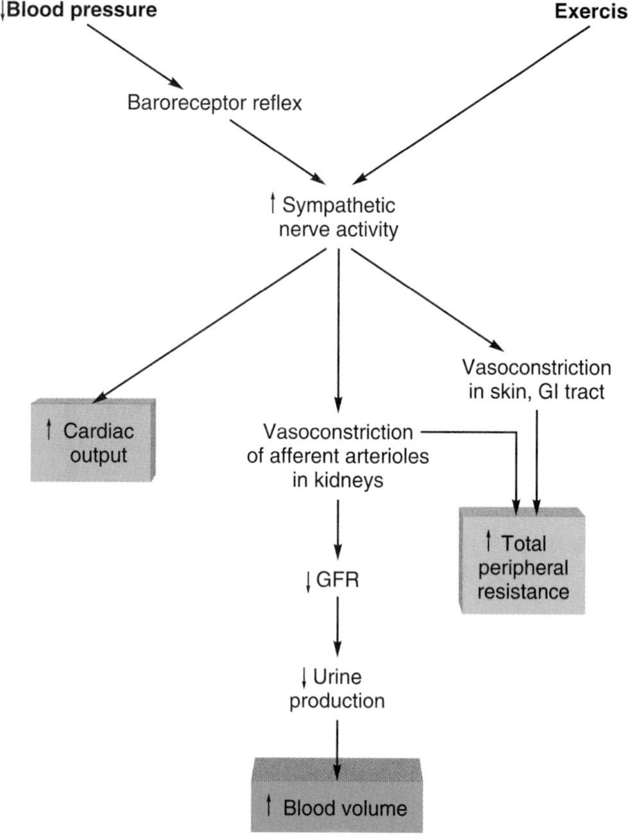

Figure 25.11

Sympathetic nerve effects.
The effect of increased sympathetic nerve activity on kidney function and other physiological processes is illustrated.

Table 25.1

Regulation of the Glomerular Filtration Rate (GFR)

Regulation	Stimulus	Afferent Arteriole	GFR
Sympathetic nerves	Activation by aortic and carotid baroreceptor reflex or by higher brain centers	Constricts	Decreases
Autoregulation	Decreased blood pressure	Dilates	No change
Autoregulation	Increased blood pressure	Constricts	No change

Reabsorption of Salt and Water

Most of the salt and water filtered from the blood is returned to the blood across the wall of the proximal convoluted tubule. The reabsorption of water occurs by osmosis, in which water follows the transport of NaCl from the tubule into the surrounding peritubular capillaries. Most of the water remaining in the filtrate is reabsorbed across the wall of the collecting duct in the renal medulla. This occurs as a result of the high osmotic pressure of the surrounding tissue fluid, which is produced by transport processes in the nephron loop.

Although about 180 L of glomerular ultrafiltrate are produced each day, the kidneys normally excrete only 1 to 2 L of urine in a 24-hour period. Approximately 99% of the filtrate must thus be returned to the blood, while 1% is excreted in the urine. The urine volume, however, varies according to the needs of the body. When a well-hydrated person drinks a liter or more of water, urine volume increases to 16 ml per minute (the equivalent of 23 L per day if this were to continue for 24 hours). In severe dehydration, when the body needs to conserve water, only 0.3 ml of urine per minute, or 400 ml per day, are produced. A volume of 400 ml of urine per day is the minimum needed to excrete the metabolic wastes produced by the body; this is called the *obligatory water loss*. When water in excess of this amount is excreted, the urine becomes increasingly diluted as its volume is increased.

Regardless of the body's state of hydration, it is clear that most of the filtered water must be returned to the vascular system to maintain blood volume and pressure. The return of filtered molecules from the tubules to the blood is called **reabsorption** (fig. 25.12). It is important to realize that the

Figure 25.12

Filtration and reabsorption.
Plasma water and its dissolved solutes (except proteins) enter the glomerular ultrafiltrate by filtration, but most of these filtered molecules are reabsorbed. The term *reabsorption* refers to the transport of molecules out of the tubular filtrate back into the blood.

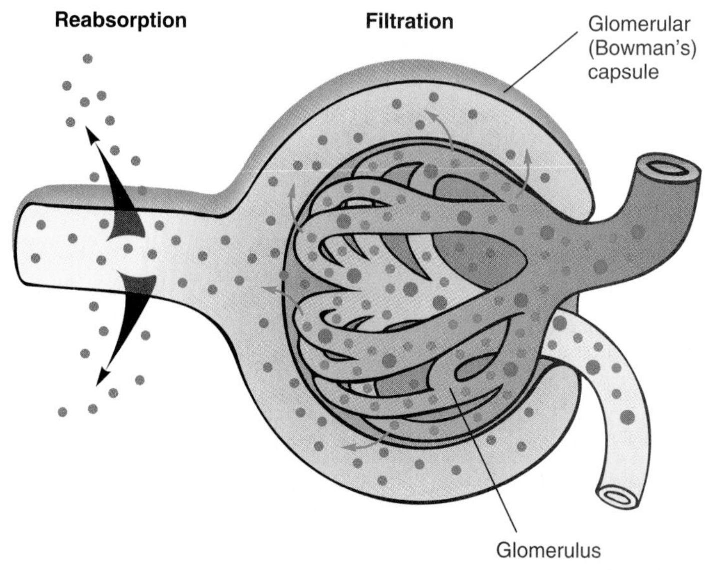

transport of water always occurs passively by *osmosis;* there is no such thing as active transport of water. A concentration gradient must thus be created between tubular fluid and blood that favors the osmotic return of water to the blood.

Reabsorption in the Proximal Convoluted Tubule

Since all plasma solutes, with the exception of proteins, are able to enter the glomerular ultrafiltrate freely, the total solute concentration (osmolality) of the filtrate is essentially the same as that of blood plasma. This total solute concentration is equal to 300 milliosmoles per liter (300 mOsm). The filtrate is thus said to be *isosmotic* with the plasma (chapter 5). Reabsorption by osmosis cannot occur unless the solute concentrations of plasma in the peritubular capillaries and the filtrate are altered by active transport processes. This is achieved by the active transport of Na^+ from the filtrate to the peritubular blood.

Active and Passive Transport

The epithelial cells that compose the wall of the proximal convoluted tubule are joined together by tight junctions only on their apical sides—that is, the sides of each cell that are closest to the lumen of the tubule (fig. 25.13). Each cell therefore has four exposed surfaces: the apical side facing the lumen, which contains microvilli; the basal side facing the peritubular capillaries; and the lateral sides facing the narrow clefts between adjacent epithelial cells.

The concentration of Na^+ in the glomerular ultrafiltrate—and thus in the fluid entering the proximal convoluted tubule—is the same as that in blood plasma. The epithelial cells of the tubule, however, have a much lower Na^+ concentration. This lower Na^+ concentration is partially due to the low permeability of the cell membrane to Na^+ and partially due to the active transport of Na^+ out of the cell by Na^+/K^+ pumps, as described in chapter 5. In the cells of the proximal convoluted tubule, the Na^+/K^+ pumps are located in the basal and lateral sides of the cell membrane but not in the apical membrane. As a result of the action of these active transport pumps, a concentration gradient is created that favors the diffusion of Na^+ from the tubular fluid across the apical cell membranes and into the epithelial cells of the proximal tubule. The Na^+ is then extruded into the surrounding interstitial (tissue) fluid by the Na^+/K^+ pumps.

The transport of Na^+ from the tubular fluid to the interstitial fluid surrounding the proximal convoluted tubule creates a potential difference across the wall of the tubule, with the lumen as the negative pole. This electrical gradient favors the passive transport of Cl^- toward the higher Na^+ concentration in the interstitial fluid. Chloride ions, therefore, passively follow sodium ions out of the filtrate into the interstitial fluid. As a result of the accumulation of NaCl, the osmolality and osmotic pressure of the interstitial fluid surrounding the epithelial cells are increased above those of the tubular fluid. This is particularly true of the interstitial fluid between the lateral membranes of adjacent epithelial cells, where the narrow spaces permit the accumulated NaCl to achieve a higher concentration.

An osmotic gradient is thus created between the tubular fluid and the tissue fluid surrounding the proximal convoluted tubule. Since the cells of the proximal convoluted tubule are permeable to water, water moves by osmosis from the tubular fluid into the epithelial cells and then across the basal and lateral sides of the epithelial cells into the interstitial fluid. The salt and water that were reabsorbed from the tubular fluid can then move passively into the surrounding peritubular capillaries, and in this way be returned to the blood (fig. 25.14).

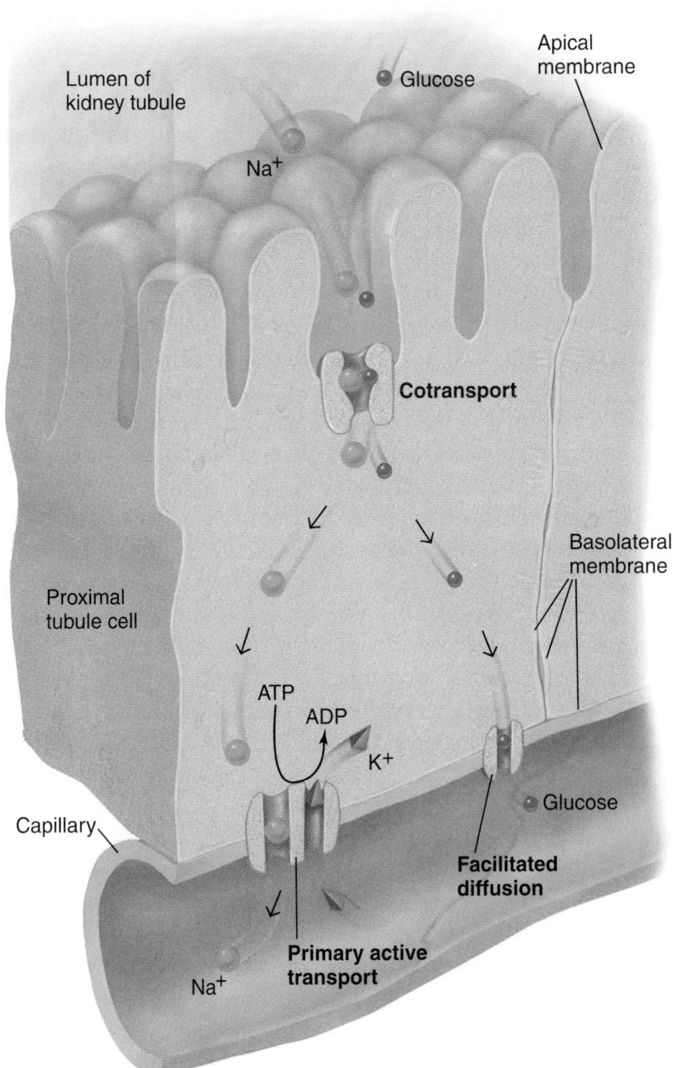

Figure 25.13

Mechanism of reabsorption in the proximal tubule. The appearance of proximal tubule cells in the electron microscope is illustrated. Molecules that are reabsorbed pass through the tubule cell from the apical membrane (facing the filtrate) to the basolateral membrane (facing the blood). There is coupled transport (a type of active transport) of glucose and Na^+ into the cytoplasm, and primary active transport of Na^+ across the basolateral membrane by the Na^+/K^+ pump.

Figure 25.14

Salt and water reabsorption in the proximal tubule. Sodium is actively transported out of the filtrate (see figure 25.13), and chloride follows passively by electrical attraction. Water follows the salt out of the tubular filtrate into the peritubular capillaries by osmosis.

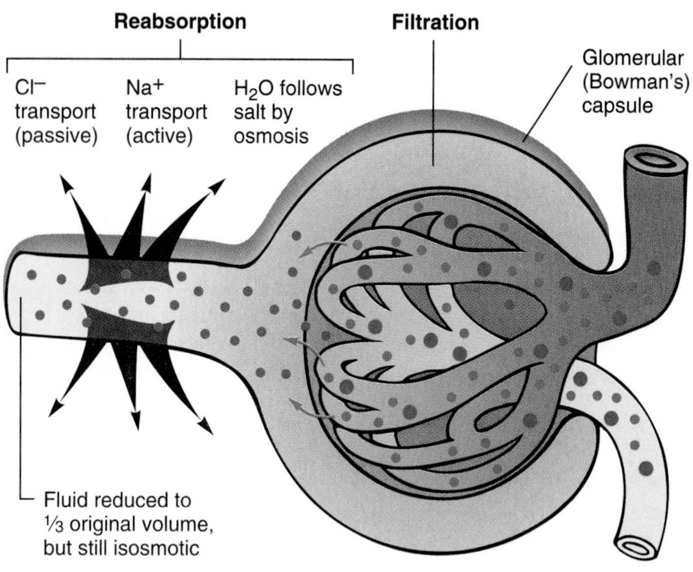

Significance of Proximal Tubule Reabsorption

Approximately 65% of the salt and water in the original glomerular ultrafiltrate is reabsorbed across the proximal convoluted tubule and returned to the blood. The volume of tubular fluid remaining is reduced accordingly, but this fluid is still isosmotic with the blood (has a concentration of 300 mOsm). This is because the cell membranes in the proximal convoluted tubule are freely permeable to water, so that water and salt are removed in proportionate amounts.

An additional smaller amount of salt and water (about 20%) is returned to the blood by reabsorption through the descending limb of the nephron loop. This reabsorption, like that in the proximal convoluted tubule, occurs constantly, regardless of the person's state of hydration. Unlike reabsorption in later regions of the nephron (distal convoluted tubule and collecting duct), it is not subject to hormonal regulation. Therefore, approximately 85% of the filtered salt and water is reabsorbed in a constant fashion in the early regions of the nephron (proximal convoluted tubule and nephron loop). This reabsorption is very costly in terms of energy expenditures, accounting for as much as 6% of the calories consumed by the body at rest.

Since 85% of the original glomerular ultrafiltrate is reabsorbed in the early region of the nephron, only 15% of the initial filtrate remains to enter the distal convoluted tubule and collecting duct. This is still a large volume of fluid—15% × GFR (180 L per day) = 27 L per day—that must be reabsorbed to varying degrees in accordance with the body's state of hydration. This "fine tuning" of the percentage of reabsorption and urine volume is accomplished by the action of hormones on the later regions of the nephron.

The Countercurrent Multiplier System

Water cannot be actively transported across the tubule wall, and osmosis of water cannot occur if the tubular fluid and surrounding interstitial fluid are isotonic to each other. In order for water to be reabsorbed by osmosis, the surrounding interstitial fluid must be hypertonic. The osmotic pressure of the interstitial fluid in the renal medulla is, in fact, raised to over four times that of blood plasma by the juxtamedullary nephrons. This results partly from the fact that the tubule bends; the geometry of the nephron loop permits interaction between the descending and ascending limbs. Since the ascending limb is the active partner in this interaction, its properties will be described before those of the descending limb.

Ascending Limb of the Nephron Loop

Salt (NaCl) is actively extruded from the ascending limb into the surrounding tissue fluid. This is not accomplished, however, by the same process that occurs in the proximal convoluted tubule. Instead, Na^+, K^+, and Cl^- passively diffuse from the filtrate into the cells of the thick portion of the ascending limb. This occurs in a ratio of 1 Na^+ to 1 K^+ to 2 Cl^-. The Na^+ is then actively transported across the basolateral membrane to the interstitial fluid by the Na^+/K^+ pumps. Cl^- follows the Na^+ passively because of electrical attraction, and K^+ diffuses back into the filtrate (fig. 25.15).

The ascending limb is structurally divisible into two regions: a *thin segment*, nearest to the tip of the loop, and a *thick segment* of varying lengths, which carries the filtrate outward into the renal cortex and into the distal convoluted tubule. It is currently believed that only the thick segments of the ascending limb are capable of actively transporting NaCl from the filtrate into the surrounding tissue fluid.

Although the mechanism of NaCl transport is different in the ascending limb than in the proximal convoluted tubule, the net effect is the same: salt (NaCl) is extruded into the surrounding interstitial fluid. Unlike the epithelial walls of the proximal convoluted tubule, however, the walls of the ascending limb of the nephron loop are *not permeable to water*. The tubular fluid thus becomes increasingly dilute as it ascends toward the renal cortex, whereas the tissue fluid around the nephron loops in the renal medulla becomes increasingly more concentrated. By means of these processes, the tubular fluid that enters the distal convoluted tubule in the cortex is made hypotonic (with a concentration of about 100 mOsm), whereas the tissue fluid in the renal medulla is made hypertonic.

Descending Limb of the Nephron Loop

The deeper regions of the renal medulla, around the tips of the loops of juxtamedullary nephrons, reach a concentration of 1,200 to 1,400 mOsm. In order to reach this high a concentration, the salt pumped out of the ascending limb must accumulate in the interstitial fluid of the medulla. This occurs because of the properties of the descending limb, to be discussed next, and because blood vessels around the nephron loop do not carry back all of the extruded salt to the general circula-

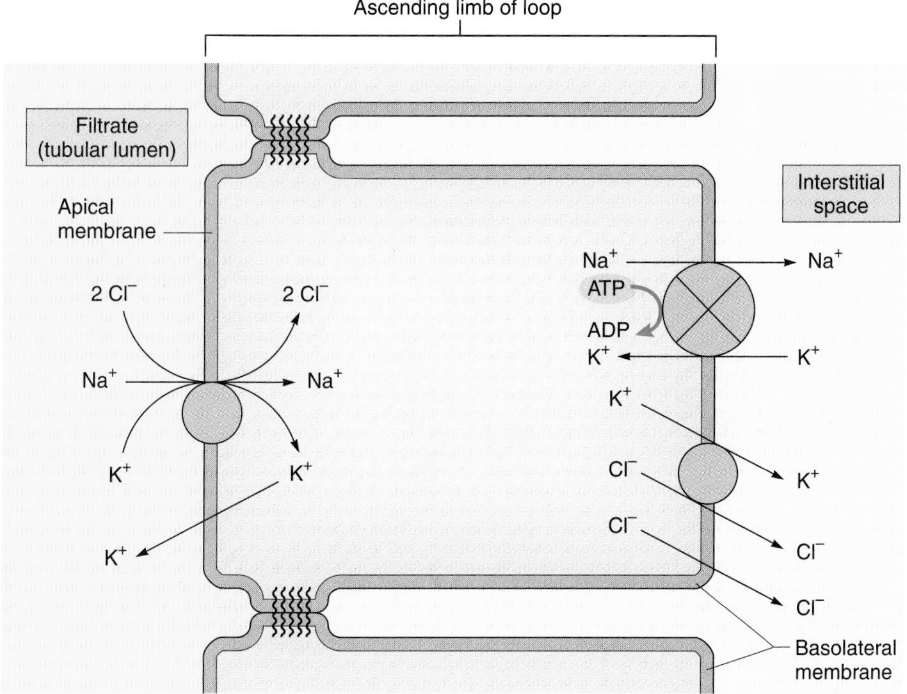

Figure 25.15

Transport of ions in the ascending limb.
In the thick segment of the ascending limb of the nephron loop, Na^+ and K^+, together with two Cl^-, enter the tubule cells. Na^+ is then actively transported out into the interstitial space, and Cl^- follows passively. The K^+ diffuses back into the filtrate, and some also enters the interstitial space.

tion. The capillaries in the medulla are uniquely arranged to trap NaCl in the tissue fluid, as will be discussed shortly.

The descending limb does not actively transport salt, and indeed is believed to be impermeable to the passive diffusion of salt. It is, however, permeable to water. Since the surrounding interstitial fluid is hypertonic to the filtrate in the descending limb, water is drawn out of the descending limb by osmosis and enters blood capillaries. The concentration of tubular fluid is thus increased, and its volume is decreased, as it descends toward the tips of the nephron loops.

As a result of these passive transport processes in the descending limb, the fluid that "rounds the bend" at the tip of the nephron loop has the same osmolality as that of the surrounding interstitial fluid (1,200 to 1,400 mOsm). There is, therefore, a higher salt concentration arriving in the ascending limb than there would be if the descending limb simply delivered isotonic fluid. Salt transport by the ascending limb is increased accordingly, so that the "saltiness" (NaCl concentration) of the interstitial fluid is multiplied (fig. 25.16).

Countercurrent Multiplication

Countercurrent flow (flow in opposite directions) in the ascending and descending limbs and the close proximity of the two limbs allow for interaction between them. Since the concentration of the tubular fluid in the descending limb reflects the concentration of surrounding tissue fluid, and since the concentration of this interstitial fluid is raised by the active extrusion of salt from the ascending limb, a *positive feedback mechanism* is created. The more salt the ascending limb extrudes, the more concentrated will be the fluid that is delivered to it from the descending limb. This positive feedback mechanism multiplies the concentration of tissue fluid and descending limb fluid, and is thus called the **countercurrent multiplier system.**

The countercurrent multiplier system recirculates salt and thus traps some of the salt that enters the nephron loop in the interstitial fluid of the renal medulla. This system results in a gradually increasing concentration of renal interstitial fluid from the renal cortex to the inner renal medulla; the osmolality of interstitial fluid increases from 300 mOsm (isotonic) in the renal cortex to between 1,200 and 1,400 mOsm in the deepest part of the renal medulla.

Vasa Recta

In order for the countercurrent multiplier system to be effective, most of the salt that is extruded from the ascending limbs must remain in the interstitial fluid of the renal medulla, while most of the water that leaves the descending limbs must be removed by the blood. This is accomplished by the **vasa recta**—long, thin-walled vessels that parallel the nephron loops of the juxtamedullary nephrons (see figs. 25.5 and 25.19). The descending vasa recta have characteristics of both capillaries and arterioles because their continuous endothelium is surrounded by smooth muscle remnants. These vessels have urea transporters and *aquaporin proteins*, which function as water channels through the membrane (chapter 5). The ascending vasa recta are capillaries with a fenestrated endothelium. As described in chapter 21, the wide gaps between endothelial cells in such capillaries permit rapid rates of diffusion.

The vasa recta maintain the hypertonicity of the renal medulla by means of a mechanism known as **countercurrent exchange.** Salt and other dissolved solutes (primarily urea, described in the next section) that are present at high concentrations in the medullary tissue fluid diffuse into the descending vasa recta. However, these same solutes then passively diffuse out of the ascending vasa recta and back into the medullary interstitial fluid to complete the countercurrent exchange. They do this because, at each level of the renal medulla, the concentration of solutes is higher in the ascending vessels than in the interstitial fluid, and higher in the interstitial fluid than in the descending vessels. Solutes are thus recirculated and trapped within the renal medulla.

The walls of the vasa recta are freely permeable to water and to dissolved NaCl and urea. Plasma proteins, however, do not easily pass through the capillary walls of the vasa recta.

Figure 25.16

The countercurrent multiplier system.
The extrusion of sodium chloride from the ascending limb makes the surrounding tissue fluid more concentrated. This concentration is multiplied by the fact that the descending limb is passively permeable so that its fluid increases in concentration as the surrounding tissue fluid becomes more concentrated. The transport properties of the nephron loop and their effect on tubular fluid concentration are shown in (*a*). The values of these changes in osmolality, together with the effect on surrounding tissue fluid concentration, are shown in (*b*).

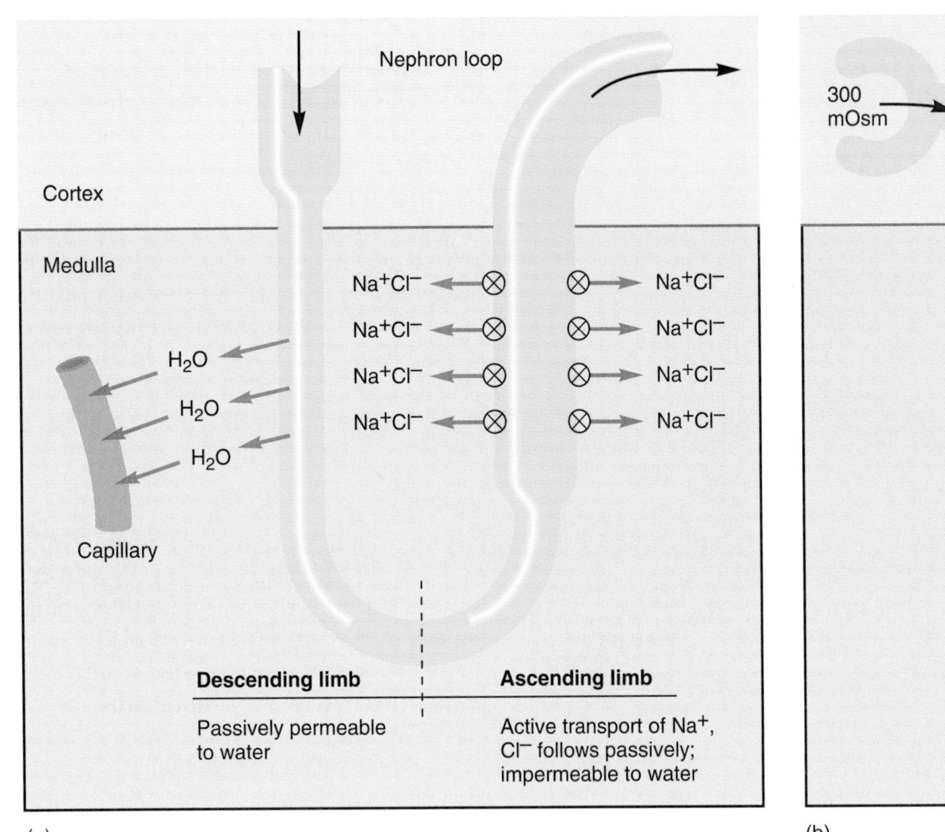

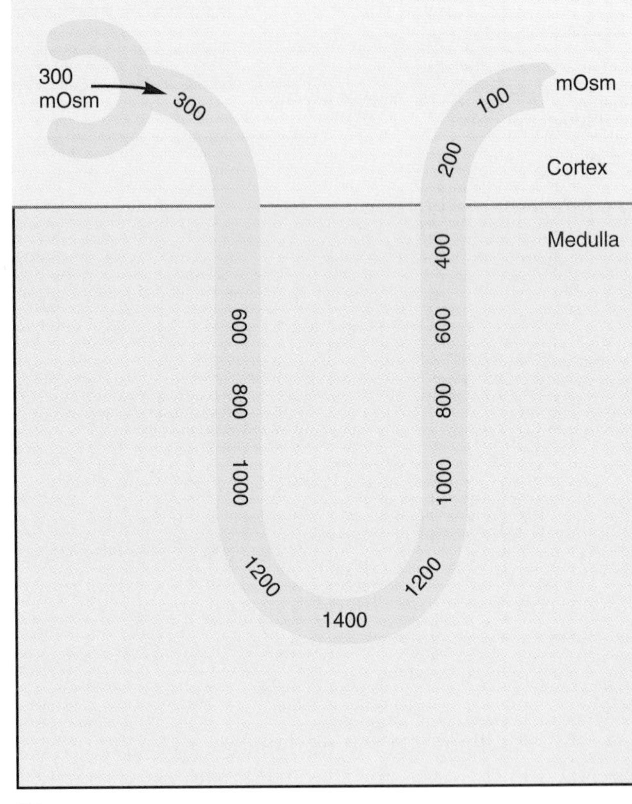

The colloid osmotic pressure (oncotic pressure) within the vasa recta, therefore, is higher than in the interstitial fluid. This is similar to the situation in other capillary beds (chapter 22) and results in the osmotic movement of water into both the descending and ascending limbs of the vasa recta. The vasa recta thus trap salt and urea within the interstitial fluid but transport water out of the renal medulla (fig. 25.17).

Effects of Urea

Countercurrent multiplication of the NaCl concentration is the mechanism that contributes most to the hypertonicity of the interstitial fluid in the renal medulla. However **urea,** a waste product of amino acid metabolism (chapter 4), also contributes significantly to the total osmolality of the interstitial fluid.

The role of urea was inferred from experimental evidence showing that active transport of Na^+ occurs only in the thick segments of the ascending limbs. The thin segments of the ascending limbs, which are located in the deeper regions of the medulla, are not able to extrude salt actively. But since salt does indeed leave the thin segments, a diffusion gradient for salt must exist, despite the fact that the surrounding interstitial fluid has the same osmolality as the tubular fluid. Investigators therefore concluded that molecules other than salt—specifically urea—contribute to the hypertonicity of the interstitial fluid.

It was later shown that the ascending limb of the nephron loop and the terminal portion of the collecting duct in the inner renal medulla are permeable to urea. Indeed, the region of the collecting duct in the inner renal medulla has specific urea transporters that permit a very high rate of diffusion into the surrounding tissue fluid. Urea can thus diffuse out of this portion of the collecting duct and into the ascending limb (fig. 25.18). In this way, a certain amount of urea is recycled through these two segments of the nephron; the urea is thereby trapped in the interstitial fluid where it can contribute significantly to the high osmolality of the medulla. This relates to the ability to produce a concentrated urine, as will be described in the next section.

The transport properties of different tubule segments are summarized in table 25.2.

Figure 25.17

Countercurrent exchange in the vasa recta.
The diffusion of salt and water first into and then out of these blood vessels helps to maintain the "saltiness" (hypertonicity) of the interstitial fluid in the renal medulla. (Numbers indicate osmolality.)

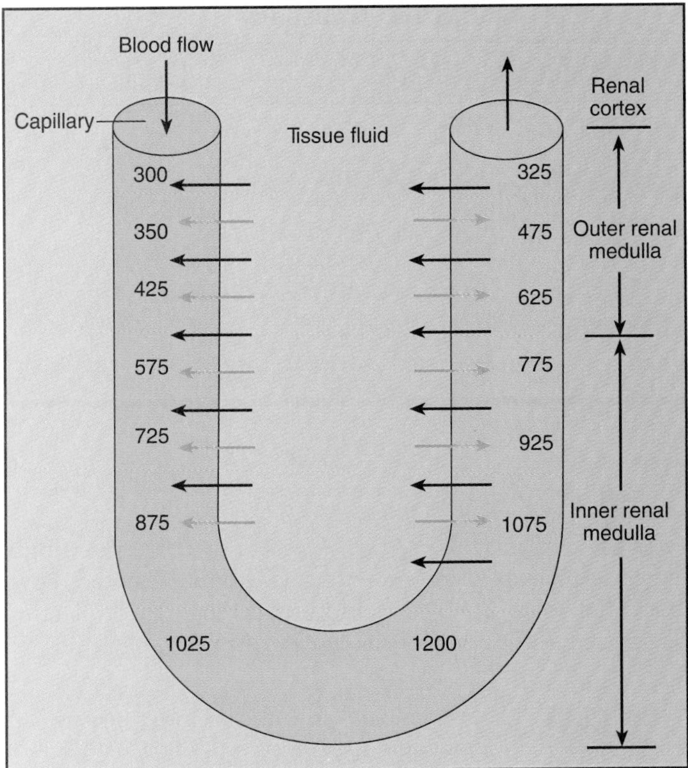

Figure 25.18

Role of urea in urine concentration.
Urea diffuses out of the inner collecting duct and contributes significantly to the concentration of the interstitial fluid in the renal medulla. The active transport Na^+ out of the thick segments of the ascending limbs also contributes to the hypertonicity of the renal medulla, so that water is reabsorbed by osmosis from the collecting ducts.

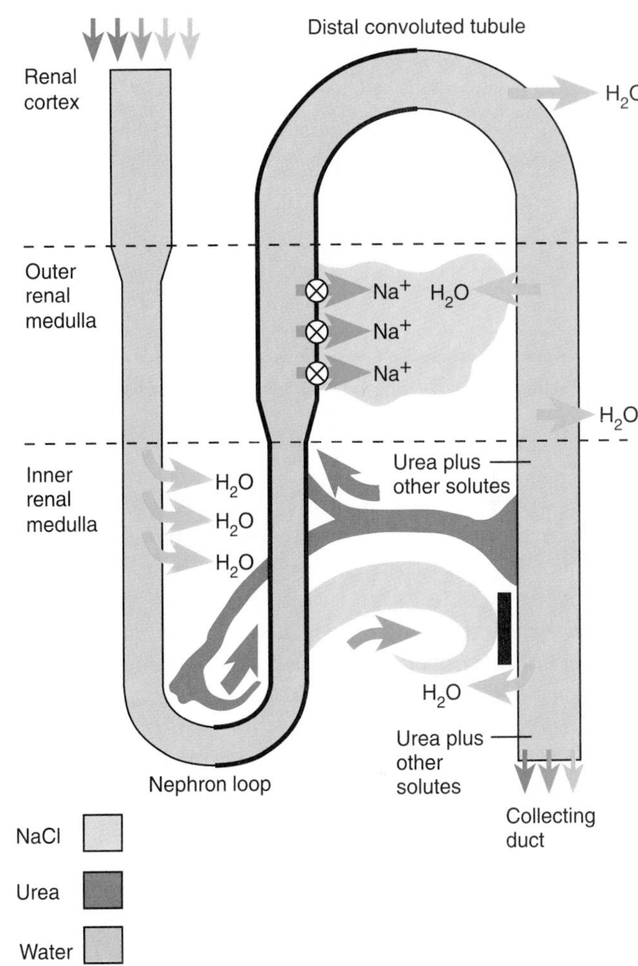

Collecting Duct: Effect of Antidiuretic Hormone (ADH)

As a result of the recycling of salt between the ascending and descending limbs and the recycling of urea between the collecting duct and the nephron loop, the medullary tissue fluid is made very hypertonic. The collecting ducts must channel their fluid through this hypertonic environment in order to empty their contents of urine into the calyces. Whereas the fluid surrounding the collecting ducts in the medulla is hypertonic, the fluid that passes into the collecting ducts in the cortex is hypotonic as a result of the active extrusion of salt by the ascending limbs of the nephron loops.

The medullary region of the collecting duct is impermeable to the high concentration of NaCl that surrounds it. The wall of the collecting duct, however, is permeable to water. Since the surrounding interstitial fluid in the renal medulla is very hypertonic because of the countercurrent multiplier system, water is drawn out of the collecting ducts by osmosis. This water does not dilute the surrounding interstitial fluid because it is transported by capillaries to the general circulation. In this way, most of the water remaining in the filtrate is returned to the vascular system (fig. 25.19).

Note that the osmotic gradient created by the countercurrent multiplier system provides the force for water reabsorption through the collecting ducts. The rate at which this osmotic movement occurs, however, is determined by the permeability of the collecting duct to water. This depends on the number of **aquaporins** (water channels) in the cell membranes of the collecting duct epithelial cells.

Aquaporins are protein channels within the membranes of vesicles that bud from the Golgi complex (chapter 3). In the absence of stimulation, these vesicles are present in the cytoplasm of the collecting duct cells. When **antidiuretic hormone (ADH)** binds to its membrane receptors on the collecting duct, it acts (via cAMP as a second messenger) to stimulate the fusion of these vesicles with the cell membrane. This is identical to exocytosis, except that here there is no secretion of product (see fig. 5.14, p. 119). The importance of this process in the collecting duct is that the water channels

Table 25.2

Transport Properties of Different Segments of the Renal Tubules and the Collecting Ducts

Nephron Segment	Active Transport	Passive Transport		
		Salt	Water	Urea
Proximal convoluted tubule	Na^+	Cl^-	Yes	Yes
Descending limb of nephron loop	None	Maybe	Yes	No
Thin segment of ascending limb	None	NaCl	No	Yes
Thick segment of ascending limb	Na^+	Cl^-	No	No
Distal convoluted tubule	Na^+	Cl^-	No	No
Collecting duct*	Slight Na^+	No	Yes (ADH) or slight (no ADH)	Yes

*The permeability of the collecting duct to water depends on the presence of ADH.

are incorporated into the cell membrane when the vesicles and membrane fuse. In response to ADH, therefore, the collecting duct becomes more permeable to water. When ADH is no longer available to bind to its membrane receptors, the water channels are removed from the cell membrane by a process of endocytosis. Endocytosis is the opposite of exocytosis; the cell membrane invaginates to re-form vesicles that again contain the water channels. Alternating exocytosis and endocytosis in response to the presence and absence of ADH, respectively, is believed to result in the recycling of water channels within the cell.

When the concentration of ADH is increased, the collecting ducts become more permeable to water, and more water is reabsorbed. A decrease in ADH, conversely, results in less reabsorption of water, and thus in the excretion of a larger volume of more dilute urine. ADH is produced by neurons in the hypothalamus and is released from the posterior pituitary (chapter 19). The secretion of ADH is stimulated when osmoreceptors in the hypothalamus respond to an increase in plasma osmotic pressure. During dehydration, therefore, when the plasma becomes more concentrated, increased secretion of ADH promotes increased permeability of the collecting ducts to water. In severe dehydration, only the minimal amount of water needed to eliminate the body's wastes is excreted. This minimum, about 400 ml per day, is limited by the fact that urine cannot become more concentrated than the medullary tissue fluid surrounding the collecting ducts. Under these conditions about 99.8% of the initial glomerular ultrafiltrate is reabsorbed.

A person in a state of normal hydration excretes about 1.5 L of urine per day, indicating that 99.2% of the glomerular ultrafiltrate volume is reabsorbed. Notice that small changes in percent reabsorption translate into large changes in urine volume. Drinking more water—and thus decreasing ADH secretion (table 25.3)—results in correspondingly larger volumes of urine excretion. It should be noted, however, that some water is reabsorbed through the collecting ducts even in the complete absence of ADH.

Diabetes insipidus is a disease associated with the inadequate secretion or action of ADH. When the secretion of ADH is adequate but a genetic defect in the ADH receptors or the aquaporin channels renders the kidneys incapable of responding to ADH, the condition is called *nephrogenetic diabetes insipidus.* Without proper ADH secretion or action, the collecting ducts are not very permeable to water, and so a large volume (5 to 10 L per day) of dilute urine is produced. The dehydration that results causes intense thirst, but a person with this condition has difficulty drinking enough to compensate for the large volumes of water lost in the urine.

Renal Plasma Clearance

As blood passes through the kidneys, some of the constituents of the plasma are removed and excreted in the urine. The blood is thus "cleared" of particular solutes in the process of urine formation. These solutes may be removed from the blood by filtration through the glomeruli or by secretion by the tubular cells into the filtrate. At the same time, certain molecules in the tubular fluid can be reabsorbed back into the blood.

One of the major functions of the kidneys is to eliminate excess ions and waste products from the blood by excreting them in the urine. These molecules are filtered through the glomerulus into the glomerular capsule along with water, salt, and other plasma solutes. In addition, some excess ions and waste products can gain access to the urine by a process called

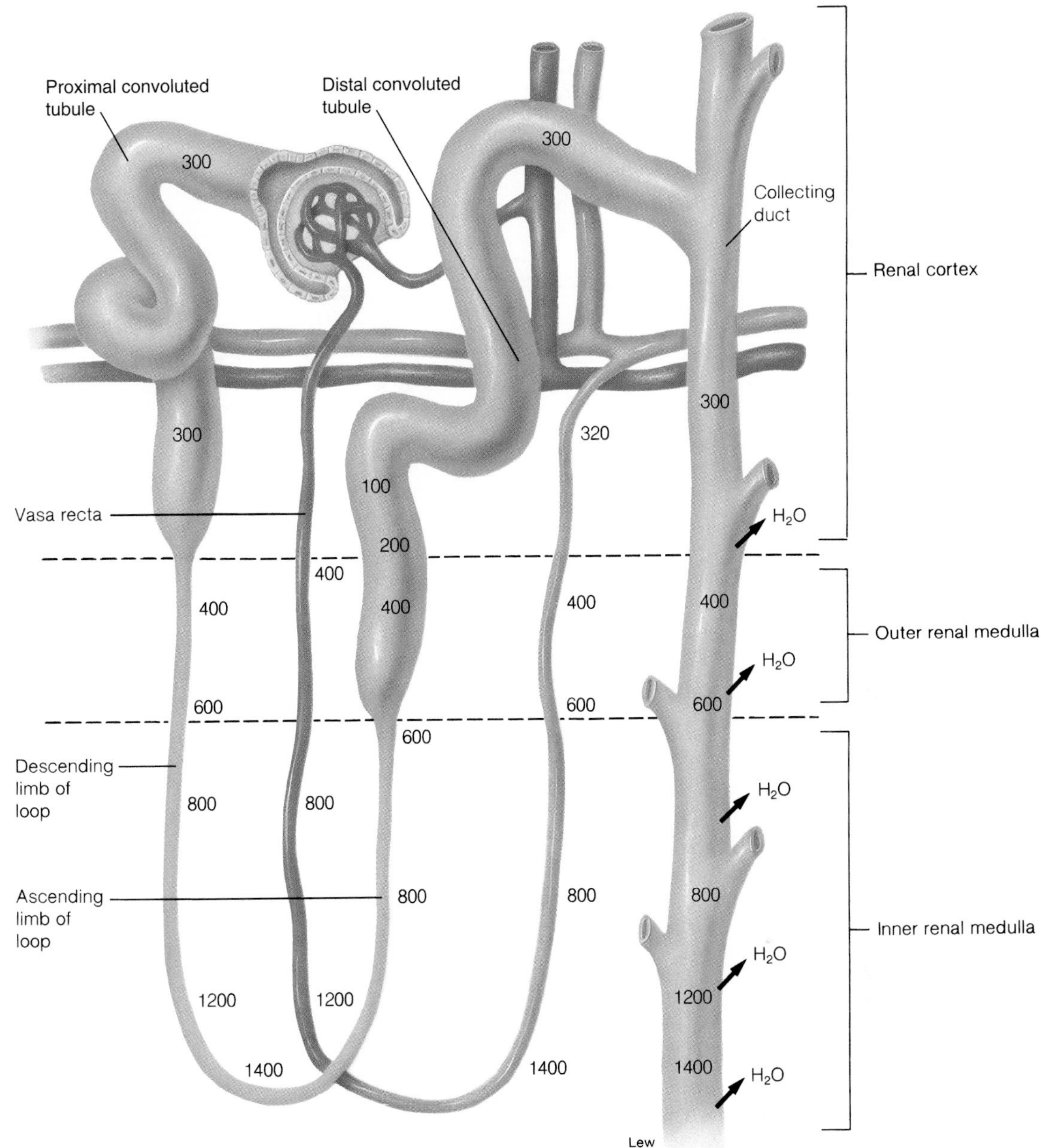

Figure 25.19

Osmolality of different regions of the kidney. The countercurrent multiplier system in the nephron loop (of Henle) and countercurrent exchange in the vasa recta help to create a hypertonic renal medulla. Under the influence of antidiuretic hormone (ADH) the collecting duct becomes more permeable to water, and thus more water is drawn by osmosis out into the hypertonic renal medulla and into the peritubular capillaries.

Table 25.3
Antidiuretic Hormone Secretion and Action

Stimulus	Receptors	Secretion of ADH	Effects on Urine Volume	Effects on Blood
↑Osmolality (dehydration)	Osmoreceptors in hypothalamus	Increased	Decreased	Increased water retention; decreased blood osmolality
↓Osmolality	Osmoreceptors in hypothalamus	Decreased	Increased	Water loss increases blood osmolality
↑Blood volume	Stretch receptors in left atrium	Decreased	Increased	Decreased blood volume
↓Blood volume	Stretch receptors in left atrium	Increased	Decreased	Increased blood volume

Figure 25.20

Secretion is the reverse of reabsorption.
Secretion refers to the active transport of substances from the peritubular capillaries into the tubular fluid. The direction of this transport is opposite to that of reabsorption.

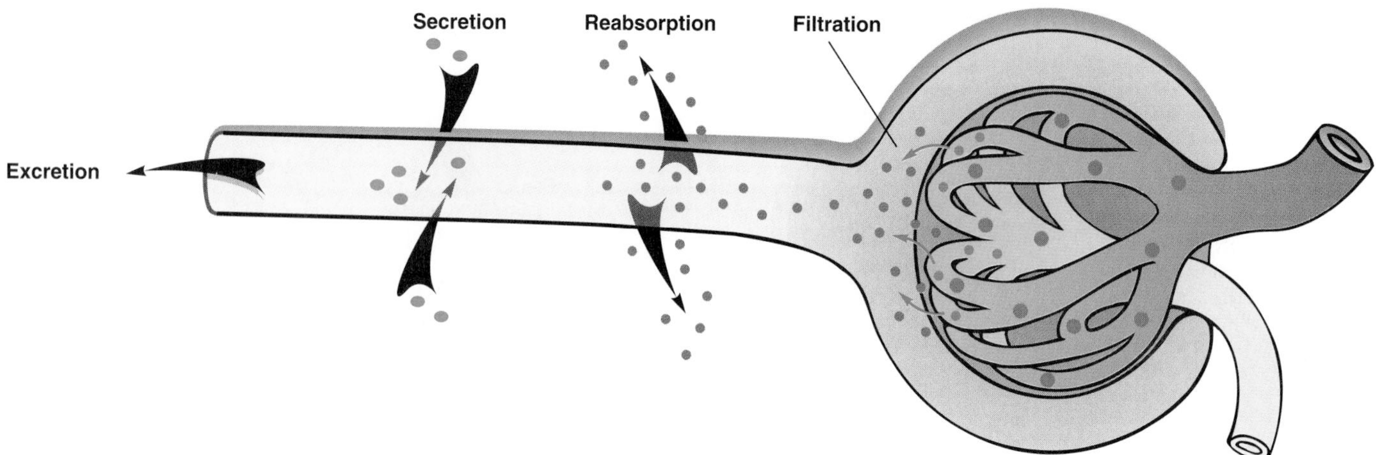

secretion (fig. 25.20). Secretion is the opposite of reabsorption. Molecules that are secreted move out of the peritubular capillaries and into the tubular cells, from which they are actively transported into the tubular lumen. In this way, molecules that were not filtered out of the blood in the glomerulus, but which instead passed through the efferent arterioles to the peritubular capillaries, can still be excreted in the urine. This is useful because it allows the kidneys to rapidly eliminate certain potential toxins. Also, K^+ and H^+ ions, which are first filtered and reabsorbed, are then secreted into the filtrate to varying degrees. This allows the kidneys to more finely control the amount of K^+ and H^+ excreted in the urine.

Although most (about 99%) of the filtered water is returned to the vascular system by reabsorption, most of the unneeded molecules that are filtered or secreted are eliminated in the urine. The concentration of these substances in the renal vein leaving the kidney is therefore lower than their concentrations in the blood entering the kidney in the renal artery. Some of the blood that passes through the kidneys, in other words, is "cleared" of these waste products.

Renal Clearance of Inulin: Measurement of GFR

If a substance is neither reabsorbed nor secreted by the tubules, the amount excreted in the urine per minute will be equal to the amount that is filtered out of the glomeruli per minute. There does not seem to be a single substance produced by the body, however, that is not reabsorbed or secreted to some degree. Plants such as artichokes, dahlias, onions, and garlic, fortunately, do produce such a compound. This compound, a polymer of the monosaccharide fructose, is **inulin** (*in'yŭ-lin*). Once injected into the blood, inulin is filtered by the glomeruli, and the amount of inulin excreted per minute is exactly equal to the amount that was filtered per minute (fig. 25.21).

If the concentration of inulin in urine is measured and the rate of urine formation is determined, the rate of inulin excretion can easily be calculated:

$$\underset{(\mathbf{mg/min})}{\textit{Quantity excreted per minute}} = \overset{V}{\left(\frac{\mathbf{ml}}{\mathbf{min}}\right)} \times \overset{U}{\left(\frac{\mathbf{mg}}{\mathbf{ml}}\right)}$$

where

V = rate of urine formation
U = inulin concentration in urine

The rate at which a substance is filtered by the glomeruli (in milligrams per minute) can be calculated by multiplying the milliliters of blood plasma filtered per minute—the **glomerular filtration rate,** or **GFR**—by the concentration of that substance in the plasma. This is shown in the following equation:

$$\underset{(\mathbf{mg/min})}{\textit{Quantity filtered per minute}} = \overset{GFR}{\left(\frac{\mathbf{ml}}{\mathbf{min}}\right)} \times \overset{P}{\left(\frac{\mathbf{mg}}{\mathbf{ml}}\right)}$$

where

P = inulin concentration in blood plasma

Since inulin is neither reabsorbed nor secreted, the amount filtered equals the amount excreted:

$$\underset{\text{(amount filtered)}}{GFR \times P} = \underset{\text{(amount excreted)}}{V \times U}$$

If the preceding equation is now solved for the glomerular filtration rate,

$$GFR_{(\text{ml/min})} = \frac{V_{(\text{ml/min})} \times U_{(\text{mg/ml})}}{P_{(\text{mg/ml})}}$$

Figure 25.21

Renal clearance of inulin.
(*a*) Inulin is present in the blood entering the glomeruli, and (*b*) some of this blood, together with its dissolved inulin, is filtered. All of this filtered inulin enters the urine, whereas most of the filtered water is returned to the vascular system (is reabsorbed). (*c*) The blood leaving the kidneys in the renal vein, therefore, contains less inulin than the blood that entered the kidneys in the renal artery. Since inulin is filtered but neither reabsorbed nor secreted, the inulin clearance rate equals the glomerular filtration rate (GFR).

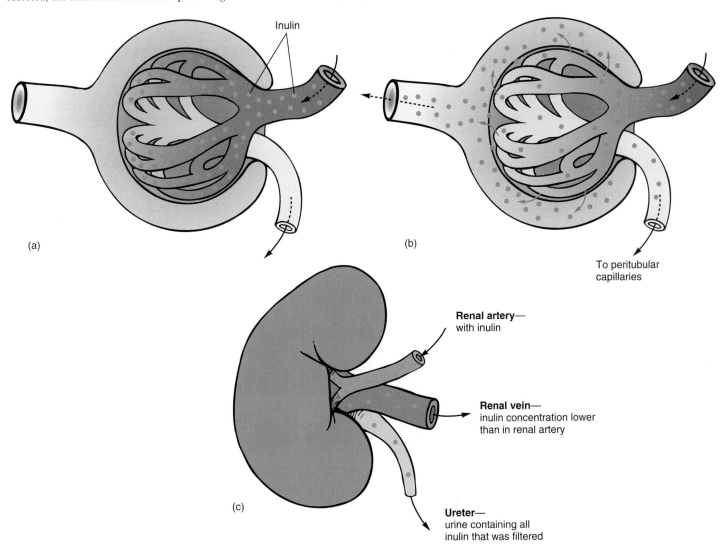

Suppose, for example, that inulin is infused into a vein and its concentrations in the urine and plasma are found to be 30 mg per ml and 0.5 mg per ml, respectively. If the rate of urine formation is 2 ml per minute, the GFR can be calculated as follows:

$$GFR = \frac{2 \text{ ml/min} \times 30 \text{ mg/ml}}{0.5 \text{ mg/ml}} = 120 \text{ ml/min}$$

This equation states that 120 ml of plasma must have been filtered each minute in order to excrete the measured amount of inulin that appeared in the urine. The glomerular filtration rate is thus 120 ml per minute in this example.

Measurements of the plasma concentration of *creatinine* (*kre'ă-tin*) are often used clinically as an index of kidney function. Creatinine, produced as a waste product of muscle creatine, is secreted to a slight degree by the renal tubules so that its excretion rate is a little above that of inulin. Since it is released into the blood at a constant rate, and since its excretion is closely matched to the GFR, an abnormal decrease in GFR causes the plasma creatinine concentration to rise. Thus, a simple measurement of blood creatinine concentration can indicate whether the GFR is normal and provide information about the health of the kidneys.

Table 25.4

Effects of Filtration, Reabsorption, and Secretion on Renal Plasma Clearance

Term	Definition	Effect on Renal Clearance
Filtration	A substance enters the glomerular ultrafiltrate	Some or all of a filtered substance may enter the urine and be cleared from the blood.
Reabsorption	A substance is transported from the filtrate, through tubular cells, and into the blood	Reabsorption decreases the rate at which a substance is cleared; clearance rate is less than the glomerular filtration rate (GFR).
Secretion	A substance is transported from peritubular blood through tubular cells and into the filtrate	When a substance is secreted by the nephrons, its clearance rate is greater than the GFR.

Table 25.5

Renal "Handling" of Different Plasma Molecules

If Substance Is	Example	Concentration in Renal Vein	Renal Clearance Rate
Not filtered	Proteins	Same as in renal artery	Zero
Filtered, not reabsorbed nor secreted	Inulin	Less than in renal artery	Equal to GFR (115–125 ml/min)
Filtered, partially reabsorbed	Urea	Less than in renal artery	Less than GFR
Filtered, completely reabsorbed	Glucose	Same as in renal artery	Zero
Filtered and secreted	PAH	Less than in renal artery; approaches zero	Greater than GFR; up to total plasma flow rate (~625 ml/min)
Filtered, reabsorbed, and secreted	K^+	Variable	Variable

Clearance Calculations

The **renal plasma clearance** is the volume of blood plasma from which a substance is completely removed in 1 minute by excretion in the urine. Notice that the units for renal plasma clearance are milliliters per minute. In the case of inulin, which is filtered but neither reabsorbed nor secreted, the amount of inulin that enters the urine is that which is contained in the volume of plasma filtered. The clearance of inulin is thus equal to the GFR (120 ml/min in the previous example). This volume of filtered blood plasma, however, also contains other solutes that may be reabsorbed to varying degrees. If a portion of a filtered solute is reabsorbed, the amount excreted in the urine is less than that which was contained in the 120 ml of plasma filtered. Thus, the renal plasma clearance of a substance that is reabsorbed must be less than the GFR (table 25.4).

If a substance is not reabsorbed, all of the filtered amount will be cleared. If this substance is, in addition, secreted by active transport into the renal tubules from the peritubular blood, an additional amount of plasma can be cleared of that substance. The renal plasma clearance of a substance that is filtered and secreted is therefore greater than the GFR (table 25.5). Thus, in order to compare the renal "handling" of various substances in terms of their reabsorption or secretion, the renal plasma clearance is calculated using the same formula used for the determination of the GFR:

$$\textit{Renal plasma clearance} = \frac{V \times U}{P}$$

where

V = urine volume per minute
U = concentration of substance in urine
P = concentration of substance in plasma

Clearance of Urea

Urea may be used as an example of how the clearance calculations can reveal the way the kidneys handle a molecule. Urea is a waste product of amino acid metabolism that is secreted by the liver into the blood and filtered into the glomerular capsules. Using the formula for renal clearance previously described and the following sample values, the following clearance may be calculated:

V = 2 ml/min
U = 7.5 mg/ml of urea
P = 0.2 mg/ml of urea

$$\textit{Urea clearance} = \frac{2 \text{ ml/min} \times 7.5 \text{ mg/ml}}{0.2 \text{ mg/ml}} = 75 \text{ ml/min}$$

Figure 25.22

Renal clearance of PAH.
Some of the para-aminohippuric acid (PAH) in glomerular blood (*a*) is filtered into the glomerular (Bowman's) capsules (*b*). The PAH present in the unfiltered blood is secreted from the peritubular capillaries into the nephron (*c*), so that all of the blood leaving the kidneys is free of PAH (*d*). The clearance rate of PAH therefore equals the total plasma flow to the glomeruli.

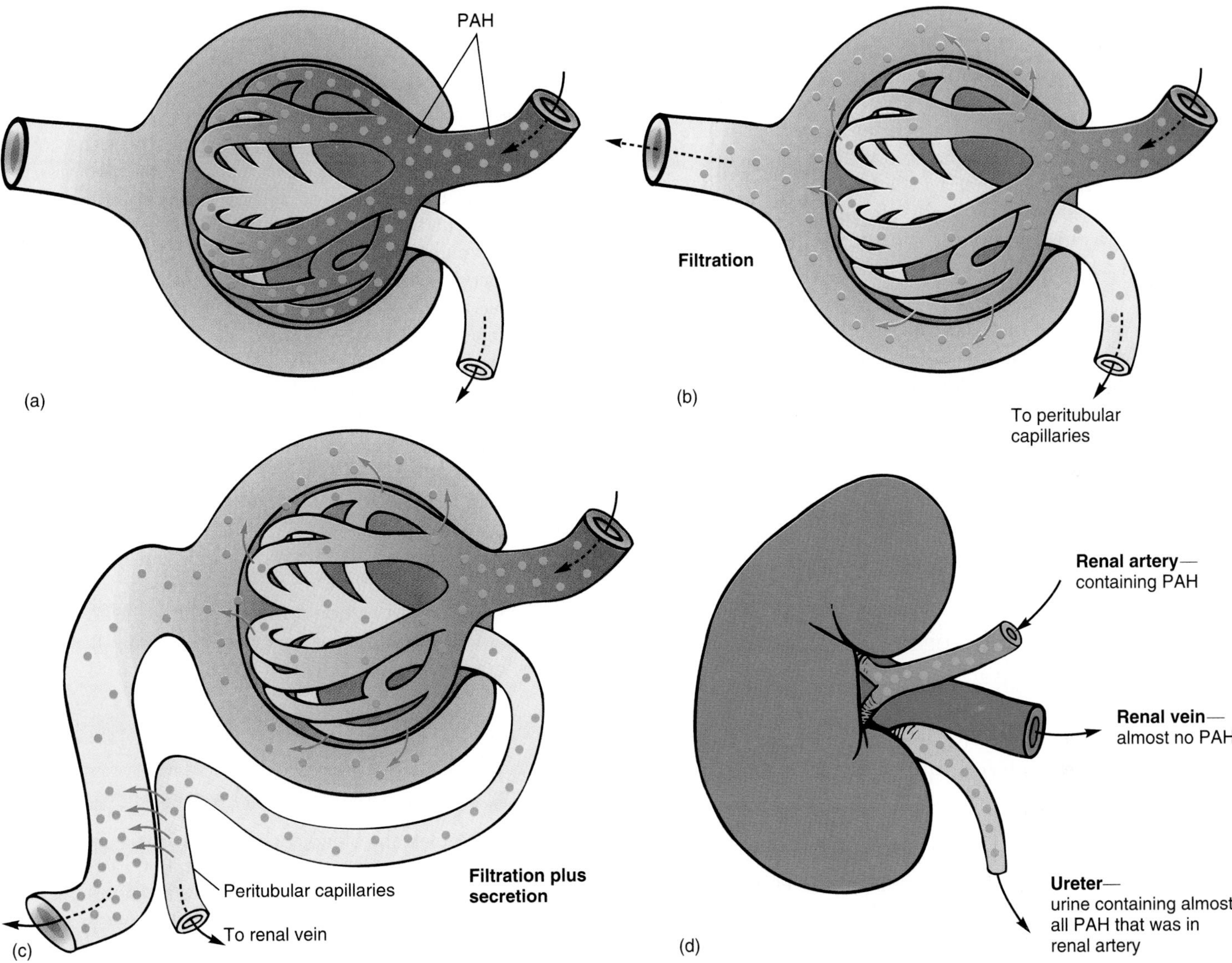

The clearance of urea in this example (75 ml/min) is less than the clearance of inulin (120 ml/min). Thus, even though 120 ml of plasma filtrate entered the nephrons per minute, only the amount of urea contained in 75 ml of filtrate is excreted. The kidneys therefore must reabsorb some of the urea that is filtered. Despite the fact that it is a waste product, a significant portion of the filtered urea (ranging from 40% to 60%) is always reabsorbed. This is a passive transport process that cannot be avoided because of the high permeability of cell membranes to urea.

Clearance of PAH: Measurement of Renal Blood Flow

Not all of the blood delivered to the glomeruli is filtered into the glomerular capsules; most of the glomerular blood passes through to the efferent arterioles and peritubular capillaries. The inulin and urea in this unfiltered blood are not excreted but instead return to the general circulation. Blood must therefore make many passes through the kidneys before it can be completely cleared of a given amount of inulin or urea.

In order for compounds in the unfiltered renal blood to be cleared, they must be secreted into the tubules by active transport from the peritubular capillaries. In this way, all of the blood going to the kidneys can potentially be cleared of a secreted compound in a single pass. This is the case for an exogenous molecule called **para-aminohippuric acid,** or **PAH** (fig. 25.22) that can be infused into the blood. The clearance (in milliliters per minute) of PAH can be used to measure the *total renal blood flow.* The normal PAH clearance has been found to average 625 ml/min. Since the glomerular filtration rate averages about 120 ml/min, this indicates that only about 120/625, or roughly 20%, of the renal plasma flow is filtered. The remaining 80% passes on to the efferent glomerular arterioles.

Since filtration and secretion clear only the molecules dissolved in blood plasma, the PAH clearance actually measures the renal plasma flow. In order to convert this to the total renal blood flow, the volume of blood occupied by erythrocytes must be taken into account. If the hematocrit (chapter 20) is 45, for example, erythrocytes occupy 45% of the blood volume and blood plasma accounts for the remaining 55%. The **total renal blood flow** is calculated by dividing the PAH clearance by the fractional blood volume occupied by plasma (0.55, in this example). The total renal blood flow in this example is thus 625 ml/min divided by 0.55, or 1.1 L/min.

Many antibiotics are secreted by the renal tubules and thus are cleared rapidly from the blood. *Penicillin,* for example, is rapidly removed from the blood by secretion into the tubular filtrate. Because of this, large amounts of penicillin must be administered to be effective. Many drugs and some hormones are inactivated in the liver by chemical transformations and, in these inactive and more water-soluble forms, are rapidly cleared from the blood by active secretion into the renal tubules.

Reabsorption of Glucose

Glucose and amino acids in the blood are easily filtered by the glomeruli into the renal tubules. These molecules, however, are usually not present in the urine. It can therefore be concluded that filtered glucose and amino acids are normally completely reabsorbed by the nephrons. This occurs by secondary active transport, which is mediated by membrane carriers that cotransport glucose and Na^+ (see fig. 25.13).

Carrier-mediated transport displays the property of *saturation*. This means that when the transported molecule (such as glucose) is present in sufficiently high concentrations, all of the carriers become occupied and the transport rate reaches a maximal value. The concentration of transported molecules needed to just saturate the carriers and achieve the maximal transport rate is called the **transport maximum** (abbreviated $\mathbf{T_m}$).

The carriers for glucose and amino acids in the renal tubules are not normally saturated and so are able to remove the filtered molecules completely. The T_m for glucose, for example, averages 375 mg per minute, which is well above the normal rate at which glucose is delivered to the renal tubules. The rate of glucose delivery can be calculated by multiplying the plasma glucose concentration (about 1 mg per ml in the fasting state) by the GFR (about 125 ml per minute). Approximately 125 mg per minute are thus delivered to the renal tubules, whereas a rate of 375 mg per minute is required to reach saturation.

Glycosuria

Glucose appears in the urine—a condition called **glycosuria** (*gli″kŏ-soor′e-a*)—when more glucose passes through the renal tubules than can be reabsorbed. This occurs when the plasma glucose concentration reaches 180 to 200 mg per 100 ml. Since the rate of glucose delivery under these conditions is still below the average T_m for glucose, we must conclude that some nephrons have considerably lower T_m values than the average.

The **renal plasma threshold** is the minimum plasma concentration of a substance that results in the excretion of that substance in the urine. The renal plasma threshold for glucose, for example, is 180 to 200 mg per 100 ml. Glucose is normally absent from urine because plasma glucose concentrations normally remain below this threshold value. Fasting plasma glucose is about 100 mg per 100 ml, for example, and the plasma glucose concentration following meals does not usually exceed 150 mg per 100 ml. The appearance of glucose in the urine (glycosuria) occurs only when the plasma glucose concentration is abnormally high (hyperglycemia) and exceeds the renal plasma threshold.

Fasting hyperglycemia is caused by the inadequate secretion or action of insulin. When this hyperglycemia results in glycosuria, the disease is called **diabetes mellitus.** A person with uncontrolled diabetes mellitus also excretes a large volume of urine because the excreted glucose carries water with it as a result of the osmotic pressure it generates in the renal tubules. This condition should not be confused with diabetes insipidus, in which a large volume of dilute urine is excreted as a result of inadequate ADH secretion.

Renal Control of Electrolyte and Acid-Base Balance

The kidneys regulate the blood concentration of Na^+, K^+, HCO_3^- and H^+. Aldosterone stimulates the reabsorption of Na^+ in exchange for K^+ in the tubule. Aldosterone thus promotes the renal retention of Na^+ and the excretion of K^+. Secretion of aldosterone from the adrenal cortex is stimulated directly by a high blood K^+ concentration and indirectly by a low Na^+ concentration via the renin-angiotensin system.

The kidneys help to regulate the concentrations of plasma electrolytes—sodium, potassium, chloride, bicarbonate, and phosphate—by matching the urinary excretion of these compounds to the amounts ingested. The control of plasma Na^+ is important in the regulation of blood volume and pressure; the control of plasma K^+ is required to maintain proper function of cardiac and skeletal muscles.

Role of Aldosterone in Na^+/K^+ Balance

Approximately 90% of the filtered Na^+ and K^+ is reabsorbed in the early part of the nephron before the filtrate reaches the distal convoluted tubule. This reabsorption occurs at a constant rate and is not subject to hormonal regulation. The final concentration of Na^+ and K^+ in the urine varies according to the needs of the body by processes that occur in the late distal convoluted tubule and in the cortical region of the collecting duct (the portion of the collecting duct within the renal medulla does not participate in this regulation). Renal reabsorption of Na^+ and secretion of K^+ are regulated by **aldosterone,** the principal mineralocorticoid secreted by the adrenal cortex (chapter 19).

Sodium Reabsorption

Although 90% of the filtered sodium is reabsorbed in the early region of the nephron, the amount left in the filtrate delivered to the distal convoluted tubule is still quite large. In the absence of aldosterone, 80% of this remaining amount is reabsorbed through the wall of the tubule into the peritubular blood; this is 8% of the amount filtered. The amount of sodium excreted without aldosterone is thus 2% of the amount filtered. Although this percentage seems small, the actual amount it represents is an impressive 30 grams of sodium excreted in the urine each day. When aldosterone is secreted in maximal amounts, by contrast, all of the sodium delivered to the distal convoluted tubule is reabsorbed. In this case urine contains no Na^+ at all.

The aldosterone-stimulated Na^+ reabsorption occurs to some degree in the late distal convoluted tubule, but the primary site of aldosterone action is in the cortical collecting duct. This is the initial portion of the collecting duct, located in the renal cortex, which has different permeability properties than the terminal portion of the collecting duct, located in the renal medulla.

Potassium Secretion

About 90% of the filtered potassium is reabsorbed in the early regions of the nephron (mainly from the proximal convoluted tubule). When aldosterone is absent, all of the remaining K^+ is also reabsorbed (principally from the cortical collecting duct). In the absence of aldosterone, therefore, no K^+ is excreted in the urine. The presence of aldosterone stimulates the secretion of K^+ from the peritubular blood into the cortical collecting duct (fig. 25.23). This aldosterone-induced secretion is thus the only means by which K^+ can be eliminated in the urine. When aldosterone secretion is maximal, as much as 50 times more K^+ is excreted in the urine than was originally filtered through the glomeruli.

In summary, aldosterone promotes sodium retention and potassium loss from the blood by stimulating the reabsorption of Na^+ and the secretion of K^+. Since aldosterone promotes the retention of Na^+, it contributes to an increased blood volume and pressure.

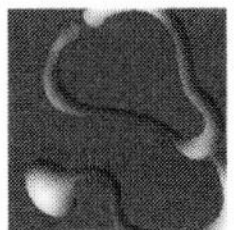

The body cannot get rid of excess K^+ in the absence of aldosterone-stimulated secretion of K^+ into the cortical collecting ducts. Indeed, when both adrenal glands are removed from an experimental animal, the *hyperkalemia* (high blood K^+) that results can produce fatal cardiac arrhythmias. Abnormally low plasma K^+ concentrations, as might result from excessive aldosterone secretion, can also produce arrhythmias, as well as muscle weakness.

Figure 25.23

Potassium is reabsorbed and secreted.
Potassium (K^+) is almost completely reabsorbed in the proximal convoluted tubule, but under aldosterone stimulation it is secreted into the cortical portion of the collecting duct. All of the K^+ in urine is derived from secretion rather than from filtration.

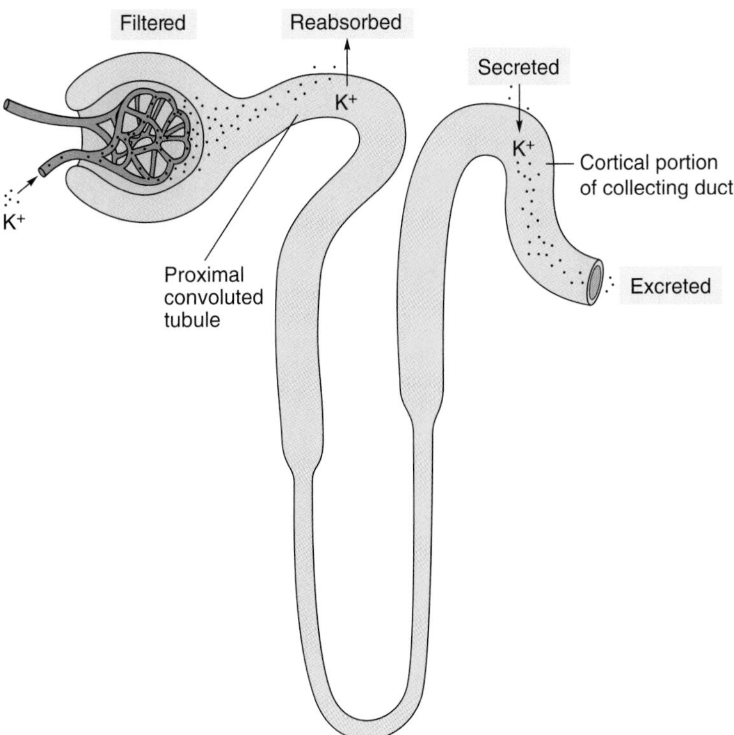

Control of Aldosterone Secretion

Since aldosterone promotes Na^+ retention and K^+ loss, one might predict (on the basis of negative feedback) that aldosterone secretion would be increased when there was a low Na^+ or a high K^+ concentration in the blood. This indeed is the case. A rise in blood K^+ *directly* stimulates the secretion of aldosterone from the adrenal cortex. A decrease in plasma Na^+ concentration, if it causes a fall in blood volume, also promotes aldosterone secretion. However, the stimulatory effect of a fall in blood volume on aldosterone secretion is indirect, as described in the next section.

Juxtaglomerular Apparatus

The **juxtaglomerular** (*juk"stă-glo-mer'yŭ-lar*) **apparatus** is the region in each nephron where the afferent glomerular arteriole comes into contact with the last portion of the thick ascending limb of the nephron loop (fig. 25.24). Under the microscope, the afferent arteriole and tubule in this small region have a different appearance than in other regions. *Granular cells* within the afferent arteriole secrete the enzyme **renin** into the blood; this enzyme catalyzes the conversion of *angiotensinogen* (a protein) into *angiotensin I* (a 10-amino-acid polypeptide).

Secretion of renin into the blood thus results in the formation of angiotensin I, which is then converted to **angiotensin II** (an 8-amino-acid polypeptide) by *angiotensin-converting*

Figure 25.24

The juxtaglomerular apparatus.
(*a*) The location of the juxtaglomerular apparatus. This structure includes the region of contact of the afferent glomerular arteriole with the distal tubule. The afferent glomerular arterioles in this region contain granular cells that secrete renin, and the distal tubule cells in contact with the granular cells form an area called the macula densa, seen in (*b*).

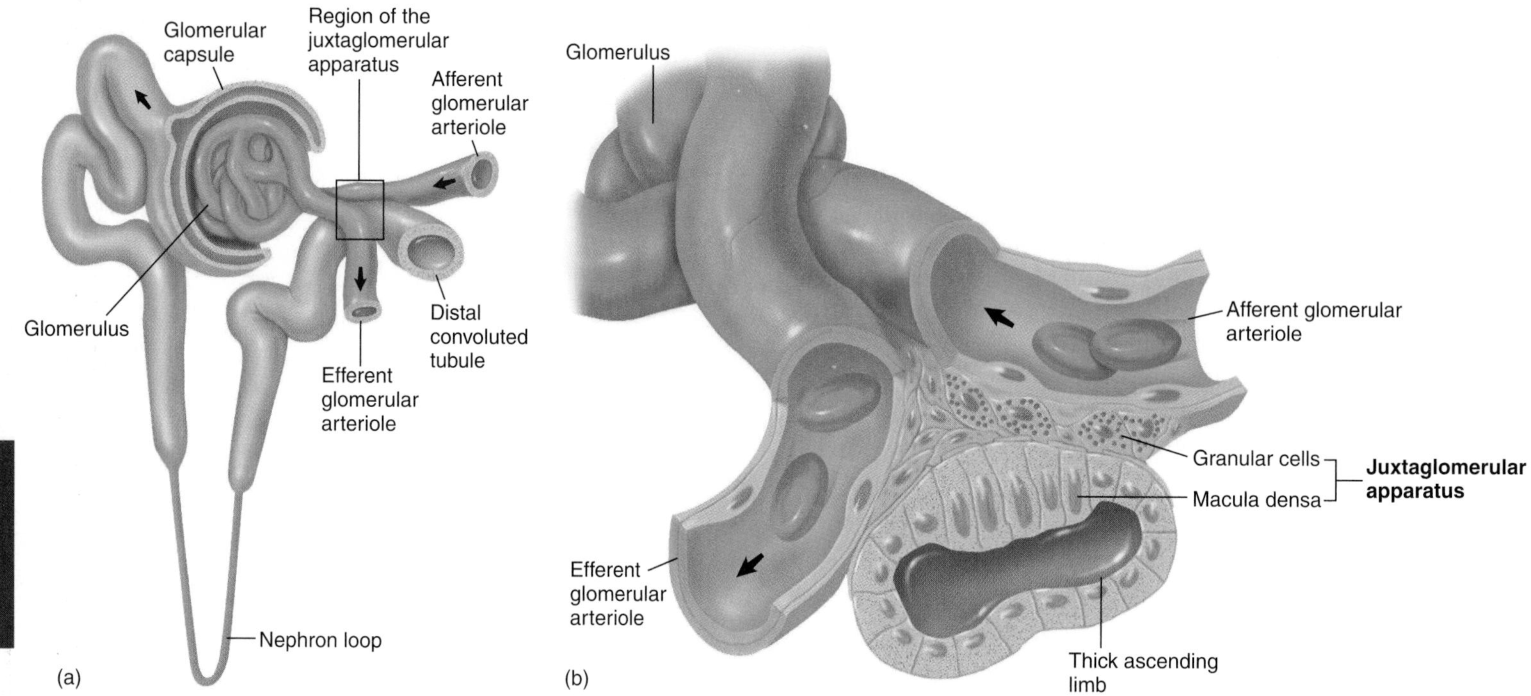

enzyme (ACE). This conversion occurs primarily as blood passes through the capillaries of the lungs, where most of the converting enzyme is present. Angiotensin II, in addition to its other effects (described in chapter 22), stimulates the adrenal cortex to secrete aldosterone. Thus, secretion of renin from the granular cells of the juxtaglomerular apparatus initiates the **renin-angiotensin-aldosterone system.** Conditions that result in increased renin secretion cause increased aldosterone secretion and, by this means, promote the reabsorption of Na^+ from the cortical collecting duct into the blood.

Regulation of Renin Secretion

An inadequate dietary intake of salt (NaCl) is always accompanied by a fall in blood volume. This is because the decreased plasma concentration (osmolality) inhibits ADH secretion. With less ADH, less water is reabsorbed through the collecting ducts and more is excreted in the urine. The fall in blood volume and the fall in renal blood flow that result cause increased renin secretion. Increased renin secretion is believed to be due in part to the direct effect of blood pressure on the granular cells, which may function as baroreceptors in the afferent arterioles. Renin secretion is also stimulated by sympathetic nerve activity, which is increased by the baroreceptor reflex (chapter 22) when the blood volume and pressure fall.

An increased secretion of renin acts, via the increased production of angiotensin II, to stimulate aldosterone secretion. Consequently, less sodium is excreted in the urine and more is retained in the blood. This negative feedback system is illustrated in figure 25.25.

Role of the Macula Densa

The region of the ascending limb in contact with the granular cells of the afferent glomerular arteriole is called the **macula densa** (*mak'yŭ-lă den'să*) (see fig. 25.24). There is evidence that this region helps to inhibit renin secretion when the blood Na^+ concentration is raised.

According to the proposed mechanism, the cells of the macula densa respond to Na^+ in the filtrate delivered to the distal convoluted tubule. When the plasma Na^+ concentration rises, or when the GFR increases, the rate of Na^+ delivered to the distal convoluted tubule also increases. Through an effect on the macula densa, this increase in filtered Na^+ inhibits the granular cells from secreting renin. Aldosterone secretion thus decreases, and since less Na^+ is reabsorbed in the cortical collecting duct, more Na^+ is excreted in the urine. The regulation of renin and aldosterone secretion is summarized in table 25.6.

---

macula densa: L. *macula,* a spot; *densitas,* thick

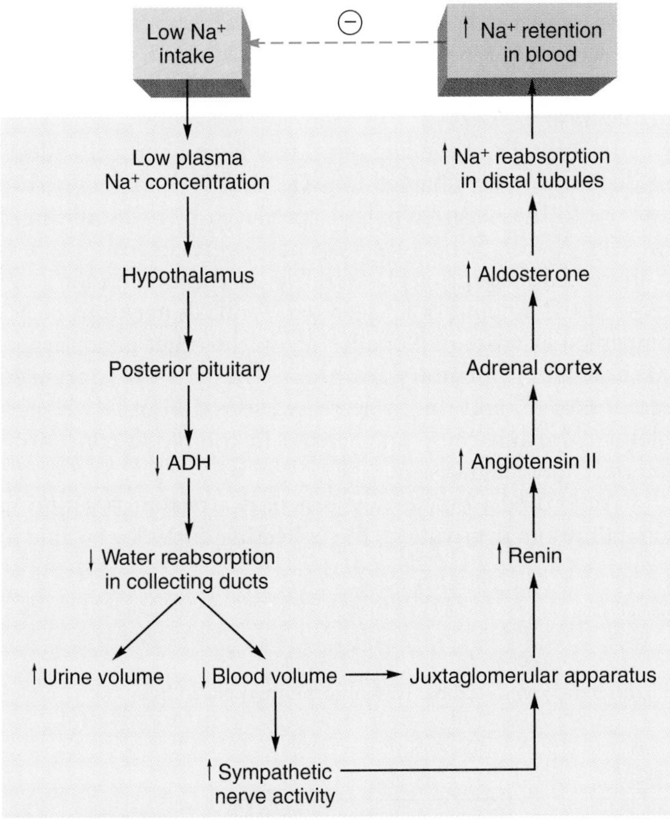

Figure 25.25

Homeostasis of plasma Na^+.
This is the sequence of events by which a low sodium (salt) intake leads to increased sodium reabsorption by the kidneys. The dashed arrow and negative sign show the completion of the negative feedback loop.

Atrial Natriuretic Peptide

Expansion of the blood volume causes increased salt and water excretion in the urine. This is due in part to an inhibition of aldosterone secretion, as previously described. In part, however, it is also caused by increased secretion of a *natriuretic* (*na″trĭ-yoo-ret′ik*) *hormone*, a hormone that stimulates salt excretion (an action opposite to that of aldosterone). The natriuretic hormone has been identified as a 28-amino-acid polypeptide called **atrial natriuretic peptide (ANP),** also called *atrial natriuretic factor.* Atrial natriuretic peptide is produced by the atria of the heart and secreted in response to the stretching of the atrial walls by increased blood volume. In response to ANP action, the kidneys lower the blood volume by excreting more of the salt and water filtered out of the blood by the glomeruli. Atrial natriuretic peptide thus functions as an endogenous diuretic.

Relationship between Na^+, K^+, and H^+

The reabsorption of Na^+ in the cortical collecting duct is accompanied by K^+ secretion. This occurs because the aldosterone-stimulated reabsorption of Na^+ creates a large potential difference between the two sides of the tubular wall, with the lumen side being very negative (–50 mV) compared to the basolateral side. The secretion of K^+ into the tubular fluid is driven by this electrical gradient. For example, if there is increased Na^+ reabsorption in the cortical collecting duct, the potential difference across the wall of the

natriuretic: L. *natrium*, sodium; *urina*, urine

Table 25.6

Regulation of Renin and Aldosterone Secretion

Stimulus	Effect on Renin Secretion	Angiotensin II Production	Aldosterone Secretion	Mechanisms
↓Blood volume	Increased	Increased	Increased	Low blood volume stimulates renal stretch receptors; granular cells release renin.
↑Blood volume	Decreased	Decreased	Decreased	Increased blood volume inhibits stretch receptors; increased Na^+ in distal convoluted tubule acts via macula densa to inhibit release of renin from granular cells.
↑K^+	None	Not changed	Increased	Direct stimulation of adrenal cortex.
↑Sympathetic nerve activity	Increased	Increased	Increased	α-adrenergic effect stimulates constriction of afferent glomerular arterioles; β-adrenergic effect stimulates renin secretion directly.

tubule will increase (with the lumen becoming more negatively charged). This will cause increased K^+ secretion.

Some diuretic drugs inhibit Na^+ reabsorption in the nephron loop and therefore increase the delivery of Na^+ to the distal convoluted tubule. This results in an increased reabsorption of Na^+ and secretion of K^+ in the distal tubule and cortical collecting duct. People who take these diuretics, therefore, tend to have excessive K^+ loss in the urine. The actions of different types of diuretics are discussed in the section "Clinical Considerations."

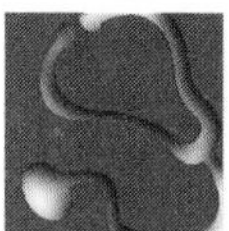

Complications may arise from the use of diuretics that cause an excessive loss of K^+ in the urine. If K^+ secretion into the tubules is significantly increased, *hypokalemia* (abnormally low blood K^+ levels) may result. This condition may lead to neuromuscular disorders and to electrocardiographic abnormalities. People who take diuretics for the treatment of high blood pressure are usually on a low-sodium diet, and they often must supplement their meals with potassium chloride (KCl) to offset the loss of K^+.

hypokalemia: Gk. *hypo,* under; L. *kalium,* potassium

The plasma K^+ concentration indirectly affects the plasma H^+ concentration (pH). Changes in plasma pH likewise affect the K^+ concentration of the blood. When the extracellular H^+ concentration increases, for example, some of the H^+ moves into the cells and causes cellular K^+ to diffuse outward into the extracellular fluid. The plasma concentration of H^+ is thus decreased while the K^+ increases, helping to reestablish the proper ratio of these ions in the extracellular fluid. A similar effect occurs in the cells of the distal region of the nephron.

In the cells of the late distal tubule and cortical collecting duct, positively charged ions (K^+ and H^+) are secreted in response to the negative polarity produced by reabsorption of Na^+ (fig. 25.26). When a person has severe acidosis, there is an increased amount of H^+ secretion at the expense of a decrease in the amount of K^+ secreted. Acidosis may thus be accompanied by a rise in blood K^+. If, on the other hand, hyperkalemia is the primary problem, there is an increased secretion of K^+ and thus a decreased secretion of H^+. Hyperkalemia can thus cause an increase in the blood concentration of H^+ and acidosis.

According to recent evidence, if a person is suffering from potassium deprivation, the collecting duct may be able to

Figure 25.26

Reabsorption of Na^+ and secretion of K^+.
In the distal convoluted tubule, K^+ and H^+ are secreted in response to the potential difference produced by the reabsorption of Na^+. High concentrations of H^+ may therefore decrease K^+ secretion, and vice versa.

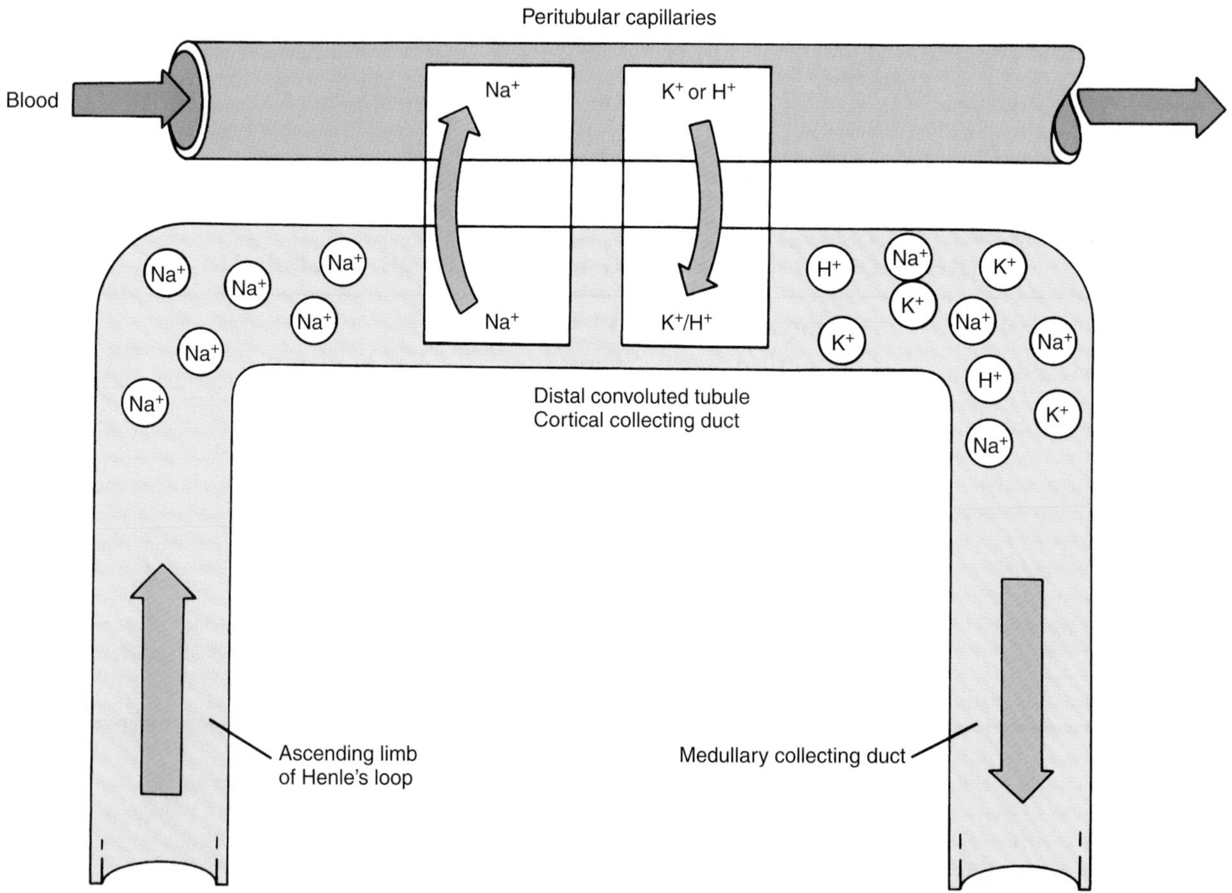

partially compensate by reabsorbing some K^+. This occurs in the outer renal medulla and results in the reabsorption of some of the K^+ that was secreted into the cortical collecting duct.

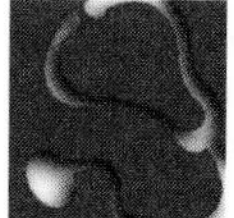

Aldosterone indirectly stimulates the secretion of H^+ as well as K^+ into the distal convoluted tubules. Abnormally high aldosterone secretion, as occurs in *primary aldosteronism* (*al-dos'tĕ-ro-niz'em*), or *Conn's syndrome,* results in both hypokalemia and metabolic alkalosis. Conversely, abnormally low aldosterone secretion, as occurs in *Addison's disease,* can produce hyperkalemia accompanied by metabolic acidosis.

Renal Acid-Base Regulation

The kidneys help to regulate the blood pH by excreting H^+ in the urine and by reabsorbing bicarbonate. The hydrogen ions enter the filtrate in two ways: by filtration through the glomeruli and by secretion into the tubules. Most of the H^+ secretion occurs across the wall of the proximal convoluted tubule in exchange for the reabsorption of Na^+. Since the kidneys normally reabsorb almost all of the filtered bicarbonate and excrete H^+, normal urine contains little bicarbonate and is slightly acidic (with a pH range of between 5 and 7). The mechanisms involved in acidification of the urine and reabsorption of bicarbonate are summarized in figure 25.27.

Conn's syndrome: from J. W. Conn, American physician, b. 1907
Addison's disease: from Christopher Addison, English anatomist, 1869–1951

Reabsorption of Bicarbonate in the Proximal Tubule

The apical membranes of the tubule cells (facing the lumen) are impermeable to bicarbonate. The reabsorption of bicarbonate must therefore occur indirectly. When the urine is acidic, HCO_3^- combines with H^+ to form carbonic acid. Carbonic acid in the filtrate is then converted to CO_2 and H_2O in a reaction catalyzed by **carbonic anhydrase.** This enzyme is located in the apical cell membrane of the proximal tubule in contact with the filtrate. Notice that the reaction that occurs in the filtrate is the same one that occurs within the red blood cells in pulmonary capillaries (as discussed in chapter 24).

The tubule cell cytoplasm also contains carbonic anhydrase. As CO_2 concentrations increase in the filtrate, the CO_2 diffuses into the tubule cells. Within the tubule cell cytoplasm, carbonic anhydrase catalyzes the reaction in which CO_2 and H_2O form carbonic acid. The carbonic acid then dissociates to HCO_3^- and H^+ within the tubule cells. (These are the same events that occur in the red blood cells of tissue capillaries.) The bicarbonate within the tubule cell can then

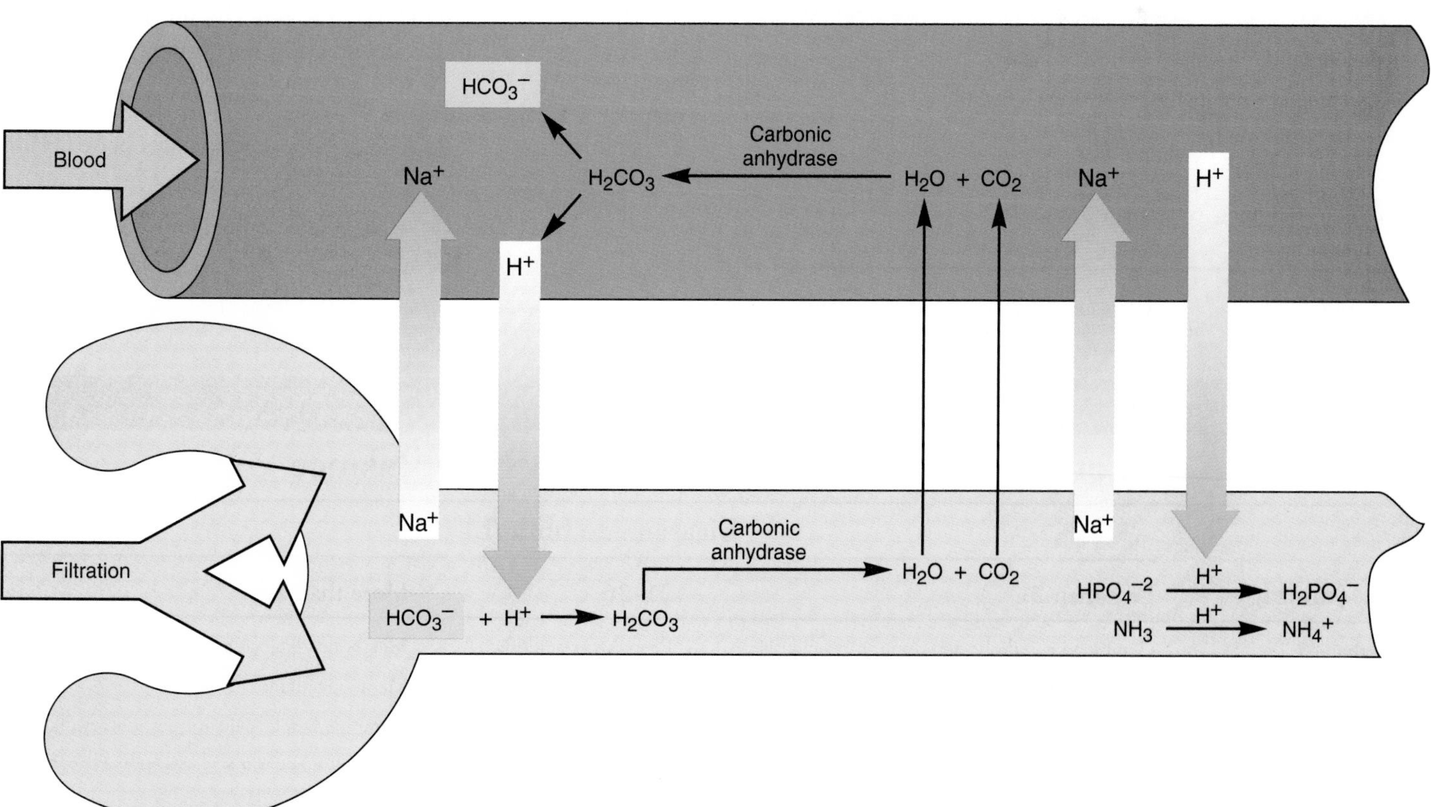

Figure 25.27

Acidification of the urine.
This diagram summarizes how the urine becomes acidified and how bicarbonate is reabsorbed from the filtrate. It also depicts the buffering of the urine by phosphate and ammonium buffers.

diffuse through the basolateral membrane and enter the blood (fig. 25.28). When conditions are normal, the same amount of HCO_3^- passes into the blood as was removed from the filtrate. The H^+, which was produced at the same time as HCO_3^- in the cytoplasm of the tubule cell, can either pass back into the filtrate or pass into the blood. Under acidotic conditions almost all of the H^+ goes back into the filtrate and is used to help reabsorb all of the filtered bicarbonate.

During alkalosis, less H^+ is secreted into the filtrate. Since the reabsorption of filtered bicarbonate requires that HCO_3^- combine with H^+ to form carbonic acid, less bicarbonate is reabsorbed. This results in urinary excretion of bicarbonate, which helps to partially compensate for the alkalosis.

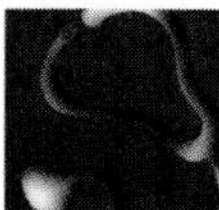

When people go to the high elevations of the mountains they hyperventilate (as discussed in chapter 24). This lowers the arterial P_{CO_2} and produces a respiratory alkalosis. The kidneys participate in this acclimatization by excreting a larger amount of bicarbonate. This helps to compensate for the alkalosis and bring the blood pH back down toward normal. It is interesting in this regard that the drug *acetazolamide,* which inhibits renal carbonic anhydrase, is often used to treat *acute mountain sickness* (chapter 24). The inhibition of renal carbonic anhydrase causes the loss of bicarbonate and water in the urine, producing a metabolic acidosis and diuresis that help to alleviate the symptoms.

By these mechanisms, disturbances in acid-base balance caused by respiratory problems can be partially compensated for by changes in plasma bicarbonate concentrations. Metabolic acidosis or alkalosis—in which changes in bicarbonate concentrations occur as the primary disturbance—similarly can be partially compensated for by changes in ventilation. These interactions of the respiratory and metabolic components of acid-base balance are summarized in table 25.7.

Figure 25.28

Mechanism of bicarbonate reabsorption.
Through this mechanism, the cells of the proximal convoluted tubule can reabsorb bicarbonate while secreting H^+. (CA = carbonic anhydrase.)

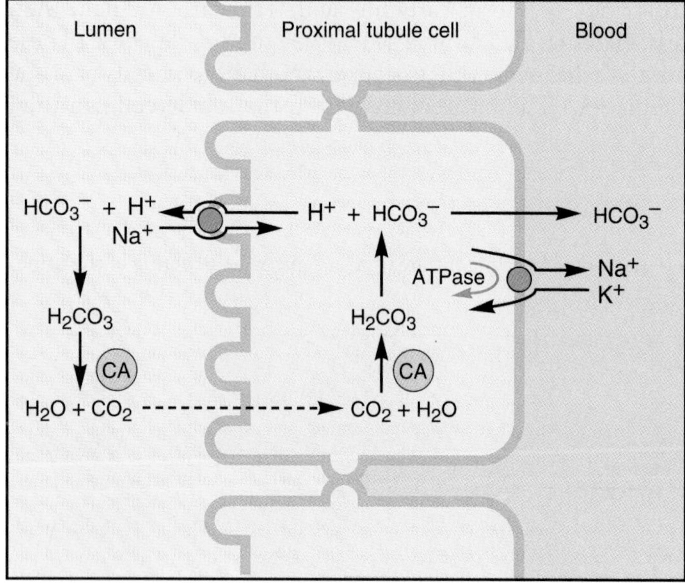

Urinary Buffers

When a person has a blood pH of less than 7.35 (acidosis), the urine pH almost always falls below 5.5. The nephron, however, cannot produce a urine pH that is significantly less than 4.5. In order for more H^+ to be excreted, the acid must be buffered. (Actually, even in normal urine, most of the H^+ excreted is in a buffered form.) Bicarbonate cannot serve this buffering function because it is normally completely reabsorbed. Instead, the buffering action of phosphates (mainly HPO_4^{2-}) and ammonia (NH_3) provide the means for excreting most of the H^+ in the urine. Phosphate enters the urine by filtration. Ammonia (whose presence is strongly evident in a diaper pail or kitty litter box) is produced in the tubule cells by deamination of amino acids. These molecules buffer H^+ as described by the following equations:

$$NH_3 + H^+ \rightarrow NH_4^+ \text{ (ammonium ion)}$$
$$HPO_4^{2-} + H^+ \rightarrow H_2PO_4^-$$

Table 25.7
Categories of Disturbances in Acid-Base Balance

	Bicarbonate (mEq/L)*		
P_{CO_2} (mmHg)	**Less than 21**	**21–26**	**More than 26**
More than 45	Combined metabolic and respiratory acidosis	Respiratory acidosis	Metabolic alkalosis and respiratory acidosis
35–45	Metabolic acidosis	Normal	Metabolic alkalosis
Less than 35	Metabolic acidosis and respiratory alkalosis	Respiratory alkalosis	Combined metabolic and respiratory alkalosis

*mEq/L = milliequivalents per liter. This is the millimolar concentration of HCO_3^- multiplied by its valence (× 1).

Ureters, Urinary Bladder, and Urethra

Urine is channeled from the kidneys to the urinary bladder by the ureters and expelled from the body through the urethra. The mucosa of the urinary bladder permits distension, and the muscles of the urinary bladder and urethra are used in the control of micturition.

Ureters

The **ureters** (*yoo're'terz*), like the kidneys, are retroperitoneal. These tubular organs, each about 25 cm (10 in.) long, begin at the renal pelvis and course inferiorly to enter the urinary bladder at the superior lateral angles of its base. A ureter is thickest—approximately 1.7 cm (0.5 in.) in diameter—near where it enters the urinary bladder.

The wall of the ureter consists of three layers, or tunics. The inner **mucosa** is continuous with the linings of the renal tubules and the urinary bladder. The mucosa consists of transitional epithelium (fig. 25.29). The cells of this layer secrete a mucus that coats the walls of the ureter with a protective film. The middle layer of the ureter is called the **muscularis.** It consists of an inner longitudinal and an outer circular layer of smooth muscle. In addition, the lower third of the ureter contains another longitudinal layer to the outside of the circular layer. Muscular peristaltic waves move the urine through the ureter. The peristaltic waves are initiated by the presence of urine in the renal pelvis, and their frequency is determined by the volume of urine. The waves force urine through the ureter and cause it to spurt into the urinary bladder. The outer layer of the ureter is called the **adventitia** (*ad"ven-tish'ă*). The adventitia is composed of loose connective tissue that covers and protects the underlying layers. In addition, extensions of the connective tissue anchor the ureter in place.

Figure 25.29

Photomicrograph of the ureter.
This is a transverse section.

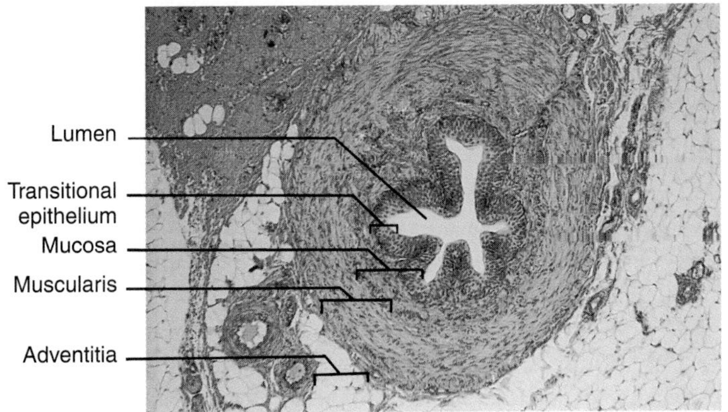

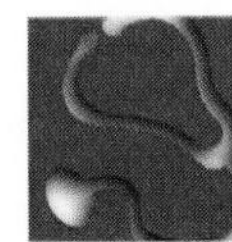

A *calculus,* or *renal stone,* may obstruct the ureter and greatly increase the frequency of peristaltic waves in an attempt to pass through. The pain from a lodged calculus is extreme and extends throughout the pelvic area. A lodged calculus also causes a sympathetic ureterorenal reflex that results in constriction of renal arterioles, thus reducing the production of urine in the kidney on the affected side.

Urinary Bladder

The **urinary bladder** is a storage sac for urine. It is located on the pelvic floor, posterior to the symphysis pubis and anterior to the rectum. In females, the urinary bladder is in contact with the uterus and vagina. In males, the prostate is positioned below the urinary bladder (fig. 25.30).

The shape of the urinary bladder is determined by the volume of urine it contains. An empty urinary bladder is pyramidal; as it fills, it becomes ovoid as the superior surface enlarges and bulges upward into the abdominal cavity. The apex of the urinary bladder is superior to the symphysis pubis and is secured to the **median umbilical ligament** by a fibrous cord called the **urachus** (*yoo'ră-kus*). The base of the urinary bladder receives the ureters along the superolateral angles, and the urethra exits at the neck. The urethra is a tubular continuation of the neck of the urinary bladder.

The wall of the urinary bladder consists of four layers: the mucosa, submucosa, muscularis, and serosa (adventitia). The **mucosa** is composed of transitional epithelium that becomes thinner as the urinary bladder distends and the cells stretch. Further distension is permitted by folds of the mucosa, called **rugae** (*roo'je*), which can be seen when the urinary bladder is empty. Fleshy flaps of mucosa located where the ureters pierce the urinary bladder act as valves to prevent a reverse flow of urine toward the kidneys as the urinary bladder fills. A triangular area known as the **trigone** (*tri'gōn*) is formed on the mucosa between the two ureteral openings and the single urethral opening (fig. 25.30). The internal trigone lacks rugae; it is therefore smooth in appearance and remains relatively fixed in position as the urinary bladder changes shape during distension and contraction.

The second layer of the urinary bladder, the **submucosa,** functions to support the mucosa. The **muscularis** consists of three interlaced smooth muscle layers and is referred to as the **detrusor** (*de-troo'sor*) **muscle.** At the neck of the urinary bladder, the detrusor muscle is modified to form the upper (the internal) sphincter of the two muscular sphincters surrounding the urethra. The outer covering of the urinary bladder is the **adventitia.** It appears only on the superior surface of the urinary bladder and is actually a continuation of the peritoneum.

---

calculus: L. *calculus*, small stone
trigone: L. *trigonum*, triangle

UNDER DEVELOPMENT

Development of the Urinary System

The urinary and reproductive systems originate from a specialized elevation of mesodermal tissue called the **urogenital ridge.** The two systems share common structures for part of the developmental period, but by the time of birth two separate systems have formed. The separation in the male is not totally complete, however, since the urethra serves to transport both urine and semen. The development of both systems is initiated during the embryonic stage, but the development of the urinary system starts and ends sooner than that of the reproductive system.

Three successive types of kidneys develop in the human embryo: the pronephros, mesonephros, and metanephros (fig. 1). The metanephric kidney persists as the permanent kidney.

The **pronephros** (*pro-nef'ros*) develops during the fourth week after conception and persists only through the sixth week. Of the three kidney types, it is the most superior in position on the urogenital ridge and is connected to the embryonic **cloaca** by the *pronephric duct.* Although the pronephros is nonfunctional and degenerates in humans, most of its duct is used by the mesonephric kidney (fig. 1), and part of it contributes significantly to the formation of the metanephros.

The **mesonephros** (*mez''ŏ-nef'ros*) develops toward the end of the fourth week as the pronephros degenerates. The mesonephros forms from an intermediate portion of the urogenital ridge and functions throughout the embryonic period of development.

Although the **metanephros** (*met''ă-nef'ros*) begins its formation during the fifth week, it does not become functional until immediately before the start of the fetal stage of development at the end of the eighth week. The paired metanephric kidneys produce urine throughout fetal development. The urine is expelled through the urinary system into the amniotic fluid.

The tubular drainage portion of the kidneys begins to form with the emergence of a diverticulum from the wall of the mesonephric duct near the cloaca. This outpouching expands into the metanephrogenic mass to form the drainage pathway for urine (fig. 1*d*). The stalk of the diverticulum develops into the ureter, whereas the expanded terminal portion forms the renal pelvis, calyces, and collecting tubules. Once the metanephric kidneys are formed, they begin to migrate from the pelvis to the upper posterior portion of the abdomen. The renal blood supply develops as the kidneys become positioned in the posterior body wall.

The urinary bladder develops from the **urogenital sinus,** which is connected to the embryonic umbilical cord by the fetal membrane called the **allantois** (*ă-lan'to-is*) (fig. 1*b*). By the twelfth week, the two ureters are emptying into the urinary bladder, the urethra is draining, and the connection of the urinary bladder to the allantois has been reduced to a supporting structure.

pronephros: Gk. *pro*, before; *nephros*, kidney
cloaca: L. *cloaca*, sewer

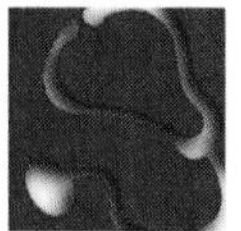

The urinary bladder becomes infected easily, and because a woman's urethra is so much shorter than a man's, women are particularly susceptible to these infections. A urinary bladder infection, called *cystitis,* may easily ascend from the urinary bladder to the ureters, since the mucous linings are continuous. An infection that involves the renal pelvis is called *pyelitis;* if it continues into the nephrons, it is known as *nephritis.* Urinary tract infections are discussed further in the section "Clinical Considerations."

Urethra

The tubular **urethra** (*yoo-re'thră*) conveys urine from the urinary bladder to the outside of the body. The urethral wall has an inside lining of mucous membrane surrounded by a relatively thick layer of smooth muscle, the fibers of which are directed longitudinally. Specialized **urethral glands** embedded in the urethral wall secrete mucus into the urethral canal.

Two muscular sphincters surround the urethra (fig. 25.30). The involuntary smooth muscle sphincter, the upper of the two, is the **internal urethral sphincter,** which is formed from the detrusor muscle of the urinary bladder. The lower sphincter, composed of voluntary skeletal muscle fibers, is called the **external urethral sphincter.**

The urethra of the female is a simple tube about 4 cm (1.5 in.) long that empties urine through the **urethral orifice** into the vestibule between the labia minora. The urethral orifice is positioned between the clitoris and the vaginal orifice.

Figure 1 Development of the urinary system. (*a, b*)Transverse section of embryo at 5 weeks. At 6 weeks (*c*) the metanephric kidney is forming. (*d*) A sagittal section of the kidney with magnified view shows the relationship between the urine collection system (gray) and the tissue derived from the metanephrogenic mass.

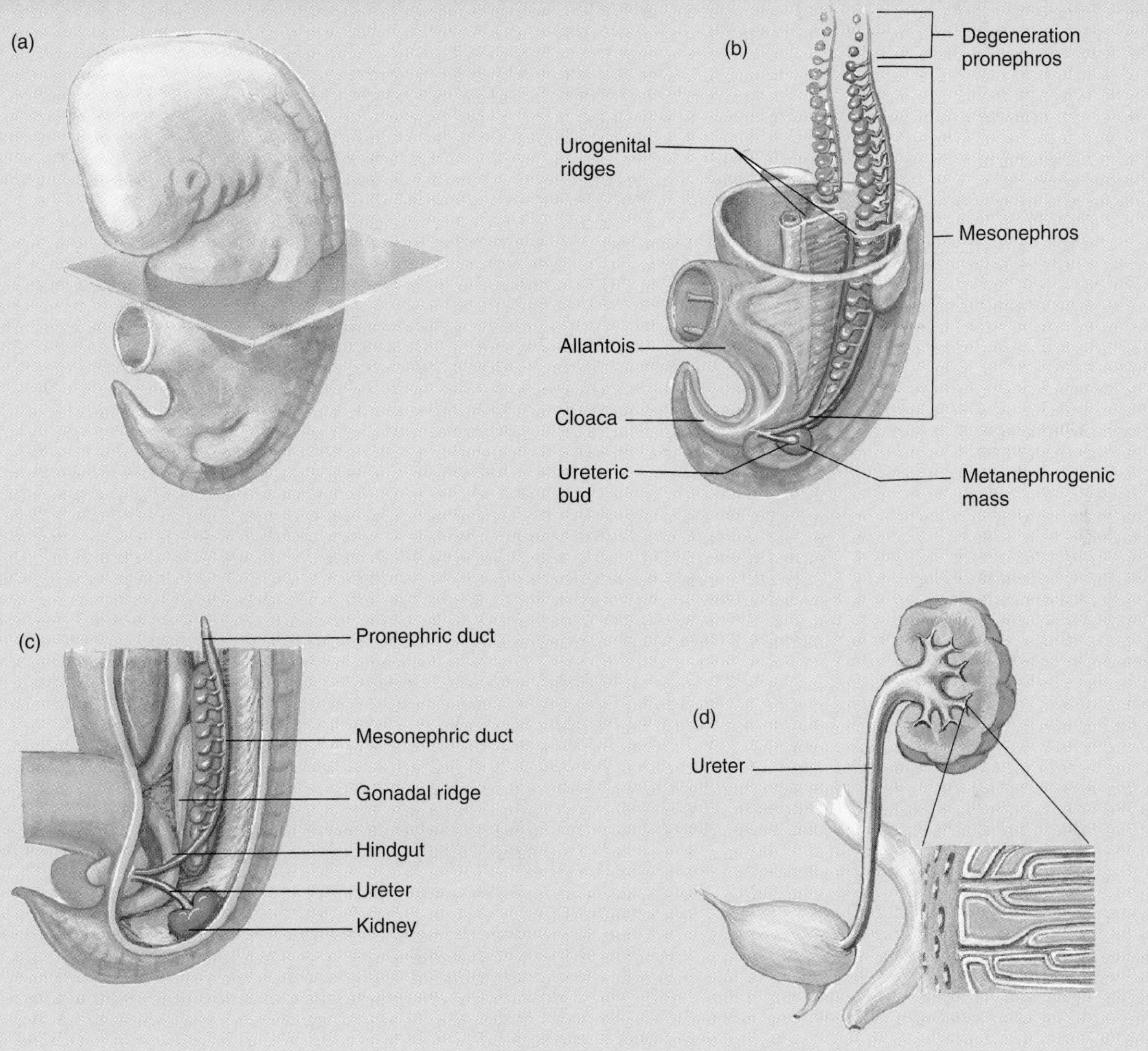

(a)

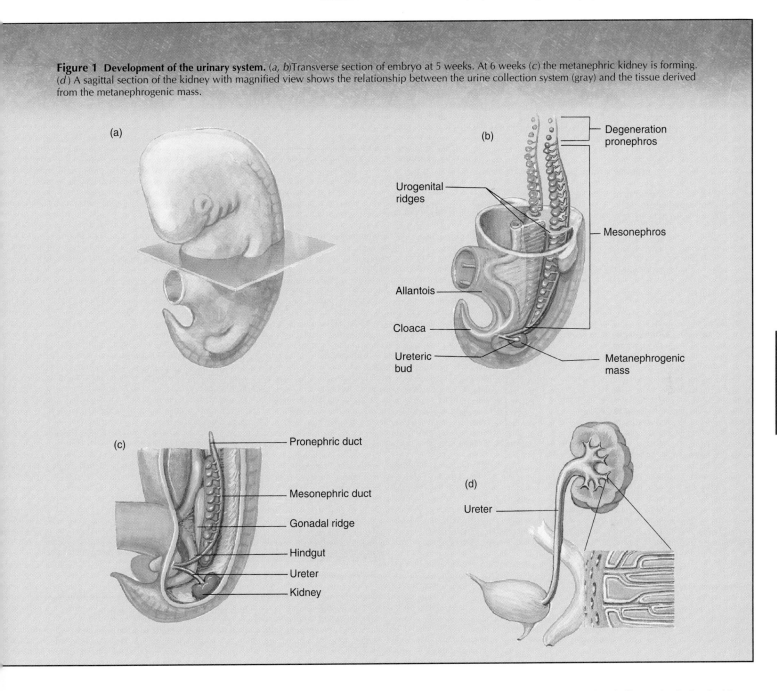

(b) Degeneration pronephros; Urogenital ridges; Mesonephros; Allantois; Cloaca; Ureteric bud; Metanephrogenic mass

(c) Pronephric duct; Mesonephric duct; Gonadal ridge; Hindgut; Ureter; Kidney

(d) Ureter

The female urethra has a single function: to transport urine to the exterior.

The urethra of the male serves both the urinary and reproductive systems. It is about 20 cm (8 in.) long and S-shaped because of the shape of the penis. As indicated in figure 25.30, three regions can be identified. The **prostatic urethra** is the proximal portion, about 2.5 cm long, that passes through the **prostate** near the neck of the urinary bladder. The **membranous urethra** is the short portion (0.5 cm) that passes through the urogenital diaphragm. The **spongy urethra** is the longest portion (15 cm), extending from the outer edge of the urogenital diaphragm to the external urethral orifice on the glans penis. This portion is surrounded by erectile tissue as it passes through the corpus spongiosum of the penis. As described in chapter 28, the paired ducts of the *bulbourethral glands* (Cowper's glands) of the reproductive system attach to the spongy urethra near the urogenital diaphragm.

Micturition

Micturition (*mik″tŭ-rish′un*), commonly called urination or voiding, is a reflex action that expels urine from the urinary bladder (table 25.8). It is a complex function that requires a stimulus from the urinary bladder and a combination of

Cowper's glands: from William Cowper, English surgeon, 1666–1709
micturition: L. *micturire*, to urinate

Figure 25.30

The urethra.

A longitudinal section of (*a*) the male urethra and (*b*) the female urethra.

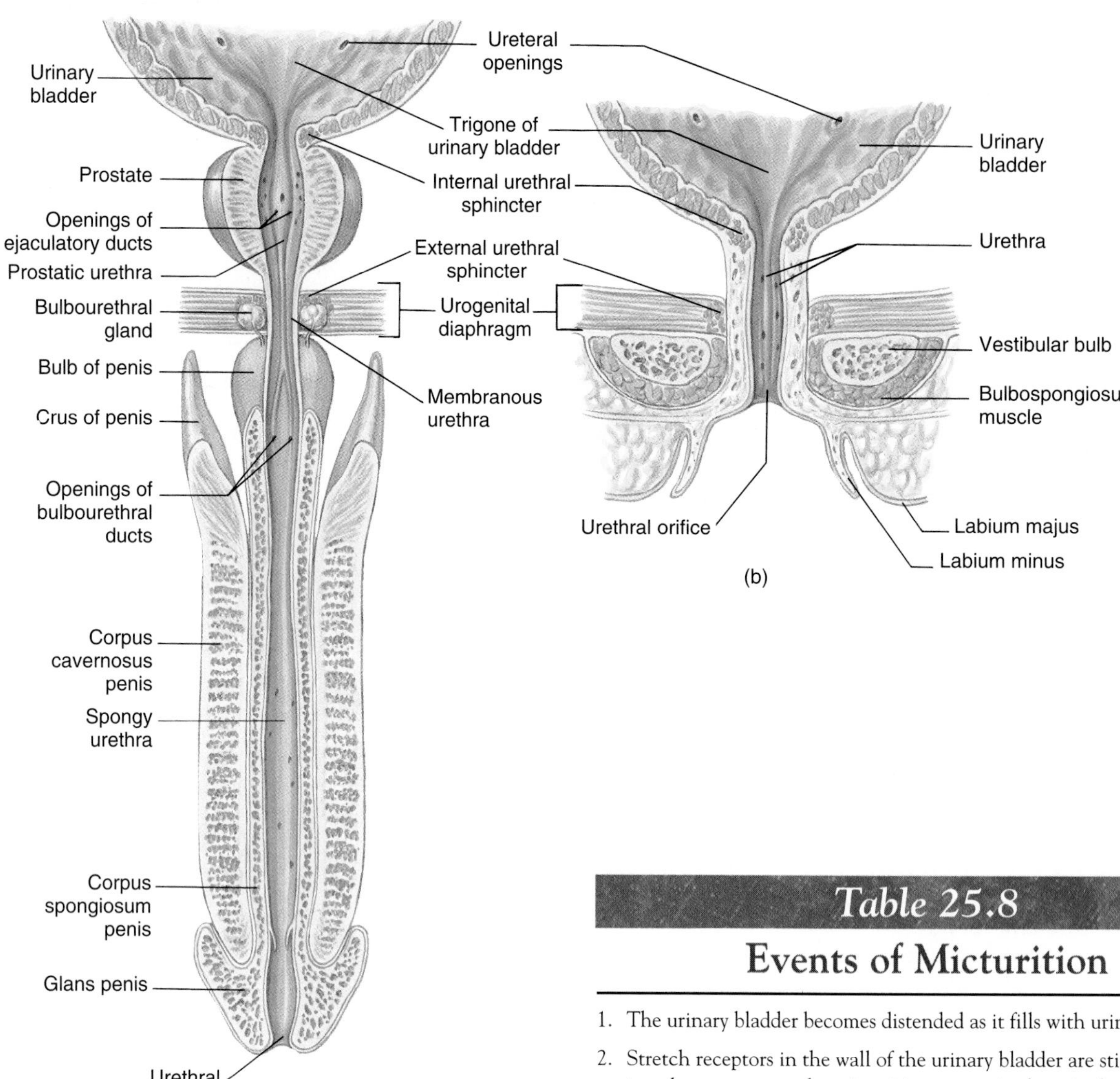

involuntary and voluntary nerve impulses to the appropriate muscular structures of the urinary bladder and urethra.

In young children, micturition is a simple reflex action that occurs when the urinary bladder becomes sufficiently distended. Voluntary control of micturition is normally established when a child is 2 or 3 years old. Voluntary control requires the development of inhibitory functioning by the cerebral cortex and a maturing of various portions of the spinal cord. The volume of urine produced by an adult averages about 1,200 ml per day, but it can range from 600 to 2,500 ml. The average capacity of the urinary bladder is 700 to 800 ml. A volume of 200 to 300 ml will distend the urinary bladder enough to stimulate stretch receptors and trigger the micturition reflex, creating a desire to urinate.

Table 25.8

Events of Micturition

1. The urinary bladder becomes distended as it fills with urine.
2. Stretch receptors in the wall of the urinary bladder are stimulated, and impulses are sent to the micturition center in the spinal cord.
3. Parasympathetic nerve impulses travel to the detrusor muscle and the internal urethral sphincter.
4. The detrusor muscle contracts rhythmically, and the internal urethral sphincter relaxes.
5. The need to urinate is sensed as urgent.
6. Urination is prevented by voluntary contraction of the external urethral sphincter and by inhibition of the micturition reflex by impulses from the midbrain and cerebral cortex.
7. Following the decision to urinate, the external urethral sphincter relaxes, and the micturition reflex is facilitated by impulses from the pons and the hypothalamus.
8. The detrusor muscle contracts, and urine is expelled through the urethra.
9. Neurons of the micturition reflex center are inactivated, the detrusor muscle relaxes, and the urinary bladder begins to fill with urine.

The micturition reflex center is located in the second, third, and fourth sacral segments of the spinal cord. Following stimulation of this center by impulses arising from stretch receptors in the urinary bladder, parasympathetic nerves that stimulate the detrusor muscle and the internal urethral sphincter are activated. Stimulation of these muscles causes a rhythmic contraction of the urinary bladder wall and a relaxation of the internal urethral sphincter. At this point, a sensation of urgency is perceived in the brain, but there is still voluntary control over the external urethral sphincter. At the appropriate time, the conscious activity of the brain activates the motor nerve fibers (S4) to the external urethral sphincter via the pudendal nerve (S2, S3, and S4), causing the sphincter to relax and urination to occur.

Clinical Considerations

The importance of kidney function in maintaining homeostasis and the ease with which urine can be collected and used as a mirror of the plasma's chemical composition make the clinical study of renal function and urine composition particularly useful. **Urology** (*yoo-rol'ă-je*) is the medical speciality concerned with dysfunctions of the urinary system. Urinary dysfunctions can be congenital or acquired; they may result from physical trauma or from conditions that secondarily involve the urinary organs.

Use of Diuretics

People who need to lower their blood volume because of hypertension, congestive heart failure, or edema take medications called **diuretics** (*di''yŭ-ret'iks*) that increase the volume of urine excreted. Diuretics directly lower blood volume (and hence, blood pressure) by increasing the proportion of the glomerular filtrate that is excreted as urine. These drugs also decrease the interstitial fluid volume (and hence, relieve edema) by a more indirect route. By lowering plasma volume, diuretic drugs increase the concentration, and thus the oncotic pressure, of the plasma within blood capillaries (chapter 22). This promotes the osmosis of interstitial fluid into the capillary blood, helping to reduce the edema.

The various diuretic drugs act on nephrons in different ways (table 25.9). On the basis of their chemical structure or aspects of their actions, commonly used diuretics are categorized as loop diuretics, thiazides, carbonic anhydrase inhibitors, osmotic diuretics, or potassium-sparing diuretics.

The most powerful diuretics, which inhibit salt and water reabsorption by as much as 25%, are the drugs that act to inhibit active salt transport out of the ascending limb of the nephron loop. Examples of these **loop diuretics** are *furosemide* and *ethacrynic acid*. The **thiazide diuretics,** (e.g., *hydrochlorothiazide*) inhibit salt and water reabsorption by as much as 8% through inhibition of salt transport by the first segment of the distal convoluted tubule. The **carbonic anhydrase inhibitors** (e.g., *acetazolamide*) are much weaker diuretics; they act primarily in the proximal convoluted tubule to prevent the water reabsorption that occurs when bicarbonate is reabsorbed.

When extra solutes are present in the filtrate, they increase the osmotic pressure of the filtrate and in this way decrease the reabsorption of water by osmosis. The extra solutes thus act as **osmotic diuretics.** *Mannitol* is sometimes used clinically for this purpose. Osmotic diuresis can occur in diabetes mellitus because glucose is present in the filtrate and urine; this extra solute causes the excretion of excessive amounts of water in the urine and can result in severe dehydration of a person with uncontrolled diabetes.

As mentioned in an earlier section, the use of diuretics can result in the excessive secretion of K^+ into the filtrate and its excessive elimination in the urine. For this reason, **potassium-sparing diuretics** are sometimes used. *Spironolactones* are aldosterone antagonists that compete with aldosterone for cytoplasmic receptor proteins in the cells of the

Table 25.9

Actions of Different Classes of Diuretics

Category of Diuretic	Example	Mechansim of Action	Major Site of Action
Loop diuretics	Furosemide	Inhibits sodium transport	Thick segments of ascending convoluted limbs
Thiazides	Hydrochlorothiazide	Inhibits sodium transport	Last part of ascending limb and first part of distal convoluted tubule
Carbonic anhydrase inhibitors	Acetazolamide	Inhibits reabsorption of bicarbonate	Proximal convoluted tubule
Osmotic diuretics	Mannitol	Reduces osmotic reabsorption of water by reducing osmotic gradient	Last part of distal convoluted tubule and cortical collecting duct
Potassium-sparing diuretics	Spironolactone	Inhibits action of aldosterone	Last part of distal convoluted tubule and cortical collecting duct
	Triamterene	Inhibits Na^+ reabsorption and K^+ secretion	Last part of distal convoluted tubule and cortical collecting duct

cortical collecting duct. These drugs thus block the aldosterone stimulation of Na^+ reabsorption and K^+ secretion. *Triamterene* is a different type of potassium-sparing diuretic that appears to act on the tubule more directly to block Na^+ reabsorption and K^+ secretion.

Symptoms and Diagnosis of Urinary Disorders

Normal micturition is painless. **Dysuria** (*dis-yur'e-ă*), or painful urination, is a sign of a urinary tract infection or obstruction of the urethra—as in an enlarged prostate in a male. **Hematuria** means blood in the urine and is usually associated with trauma. **Bacteriuria** means bacteria in the urine, and **pyuria** is the term for pus in the urine, which may result from a prolonged infection. **Oliguria** is an insufficient output of urine, whereas **polyuria** is an excessive output. Low blood pressure and kidney failure are two causes of oliguria. **Uremia** (*yoo-re'me-ă*) is a condition in which substances ordinarily excreted in the urine accumulate in the blood. **Enuresis** (*en"yŭ-re'sis*), or **incontinence,** is the inability to control micturition. It may be caused by psychological factors or by structural impairment.

The palpation and inspection of urinary organs is an important aspect of physical assessment. The right kidney is palpable in the supine position; the left kidney usually is not. The distended urinary bladder is palpable along the superior pelvic rim.

The urinary system may be examined using radiographic techniques. An **intravenous pyelogram** (*pi'ĕ-lŏ-gram*) **(IVP)** permits radiographic examination of the kidneys following the injection of radiopaque dye. In this procedure, the dye that has been injected intravenously is excreted by the kidneys so that the renal pelvises and the outlines of the ureters and urinary bladder can be observed in a radiograph.

Cystoscopy (*si-stos'kŏ-pe*) is the inspection of the inside of the urinary bladder by means of an instrument called a cystoscope. Tissue samples can be obtained with this technique, as well as urine samples from each kidney prior to mixing in the urinary bladder. Once the cystoscope is in the urinary bladder, the ureters and pelvis can be viewed through urethral catheterization and inspected for obstructions. A **renal biopsy** is a diagnostic test for evaluating certain types and stages of kidney diseases. The biopsy is performed either through a skin puncture (closed biopsy) or through a surgical incision (open biopsy).

In a **urinalysis,** the voided urine specimen is tested for color, specific gravity, chemical composition, and the presence of microscopic bacteria, crystals, and casts. *Casts* are accumulations of proteins that leaked through the glomeruli and that were pushed through the tubules like toothpaste through a tube. Casts may contain inclusions of red blood cells, white blood cells, bacteria, and other substances. Casts containing red blood cells are diagnostic of glomerulonephritis.

Infections of Urinary Organs

Urinary tract infections (UTIs) are a significant cause of illness and are also a major factor in the development of chronic renal failure. Females are more predisposed to urinary tract infections than males, and the incidence of infection increases directly with sexual activity and aging. The higher infection rate in females has been attributed to a shorter urethra, which is in close proximity to the rectum, and to the lack of protection provided by prostatic secretions in males. To reduce the risk of urinary infections, a female should wipe her anal region in a posterior direction, away from the urethral orifice, after a bowel movement.

Infections of the urinary tract are named according to the infected organ. An infection of the urethra is called **urethritis** (*yoo"re-thri'tis*) and involvement of the urinary bladder is **cystitis.** Cystitis is frequently a secondary infection from some other part of the urinary tract.

Nephritis is inflammation of the kidney tissue. **Glomerulonephritis** (*glo-mer"yŭ-lo-nĕ-fri'tis*) is inflammation of the glomeruli. Glomerulonephritis frequently occurs following an upper respiratory tract infection because antibodies produced against streptococci bacteria can produce an autoimmune inflammation in the glomeruli. This inflammation may permanently change the glomeruli and figure significantly in the development of chronic renal disease and renal failure.

Any interference with the normal flow of urine, such as from a renal stone or an enlarged prostate in a male, causes stagnation of urine in the renal pelvis and the development of **pyelitis**—an inflammation of the renal pelvis and its calyces. **Pyelonephritis** is inflammation involving the renal pelvis, the calyces, and the tubules of the nephron within one or both kidneys. Bacterial invasion from the blood or from the lower urinary tract is another cause of both pyelitis and pyelonephritis.

Reduced Renal Function

Renal function can be tested by techniques that include the renal plasma clearance of PAH, which measures total blood flow to the kidneys, and measurement of the GFR by the inulin clearance. The plasma creatinine concentration, as previously described, also provides an index of renal function. These tests aid the diagnosis of kidney diseases such as glomerulonephritis and renal insufficiency. The *urinary albumin excretion rate* is a commonly performed test that can detect an excretion rate of blood albumin that is slightly above normal. This condition, called **microalbuminuria,** is often the first manifestation of renal damage caused by diabetes or hypertension.

Acute Renal Failure

In **acute renal failure,** the ability of the kidneys to excrete wastes and regulate the homeostasis of blood volume, pH, and electrolytes deteriorates over a relatively short period of time (hours to days). There is a rise in blood creatinine concentration and a decrease in the renal plasma clearance of creatinine. This may be due to a reduced blood flow through the kidneys, perhaps as a result of atherosclerosis or inflammation of the renal rubules. The compromised kidney function may be the result of ischemia caused by the reduced blood flow, but it may also result from excessive use of nonsteroidal anti-inflammatory drugs (NSAIDs) such as phenacetin or from the use of other drugs.

The kidneys may sustain a 90% loss of their nephrons through tissue death and still continue to function. If a patient suffering acute renal failure is stabilized, the nephrons have an excellent capacity to regenerate.

Renal Insufficiency

When nephrons are destroyed—as in chronic glomerulonephritis, infection of the renal pelvis and nephrons (pyelonephritis), or loss of a kidney—or when kidney function is reduced by damage caused by diabetes mellitus, arteriosclerosis, or blockage by kidney stones, a condition of **renal insufficiency** may develop. This can cause hypertension, which is due primarily to the retention of salt and water, and **uremia** (high plasma urea concentrations). The inability to excrete urea is accompanied by an elevated plasma H^+ concentration (acidosis) and an elevated K^+ concentration, which are more immediately dangerous than the high levels of urea. Uremic coma appears to result from these associated changes. As the nephrons continue to deteriorate, the options for sustaining life are hemodialysis or kidney transplantation.

Hemodialysis

The equipment used in performing **hemodialysis** (*he″mo-di-al′ĭ-sis*) (fig. 25.31) is designed to filter the wastes from the blood of a patient who has chronic renal failure. During hemodialysis, the blood of a patient is pumped through a tube from the radial artery. In the machine, a selectively permeable cellophane membrane separates the blood from an isotonic solution containing molecules needed by the body (such as glucose). In a process called **dialysis,** waste products diffuse out of the blood through the membrane while glucose and other molecules needed by the body remain in the blood. After it has been cleansed, the blood is returned to the body through a vein. Hemodialysis is commonly performed three times a week for several hours each session.

More recent hemodialysis techniques include the use of the patient's own peritoneal membranes (which line the abdominal cavity) for dialysis. Dialysis fluid is introduced into the peritoneal cavity, and then, after a period of time, discarded after wastes have accumulated. This procedure, called *continuous ambulatory peritoneal dialysis* (CAPD), can be performed several times a day by the patients themselves on an outpatient basis. Although CAPD is more convenient and less expensive for patients than hemodialysis, it is less efficient in removing wastes and is more often complicated by infection.

Patients with kidney failure who are on hemodialysis frequently suffer from anemia. This is due to the lack of the hormone *erythropoietin*, secreted by the normal kidneys, which stimulates red blood cell production in the bone marrow (chapter 20). Such patients are now given recombinant erythropoietin produced by genetic engineering techniques.

Figure 25.31

Hemodialysis.
This schematic diagram (*a*) represents a device that uses a semipermeable membrane to filter the blood of a patient with (*b*) kidney failure.

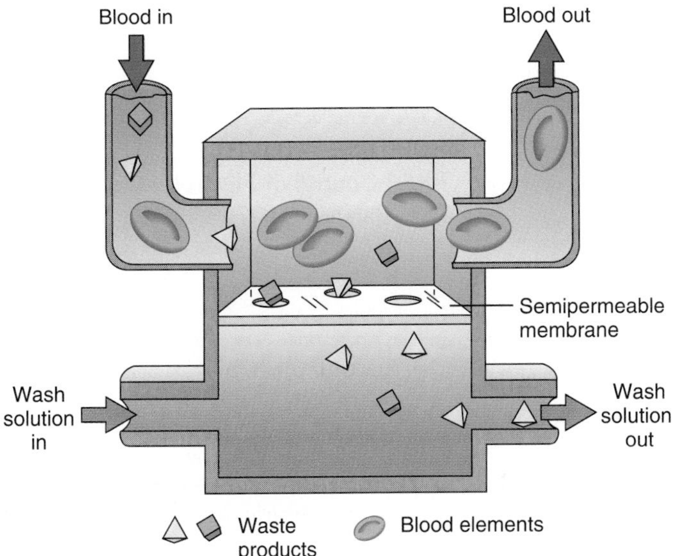

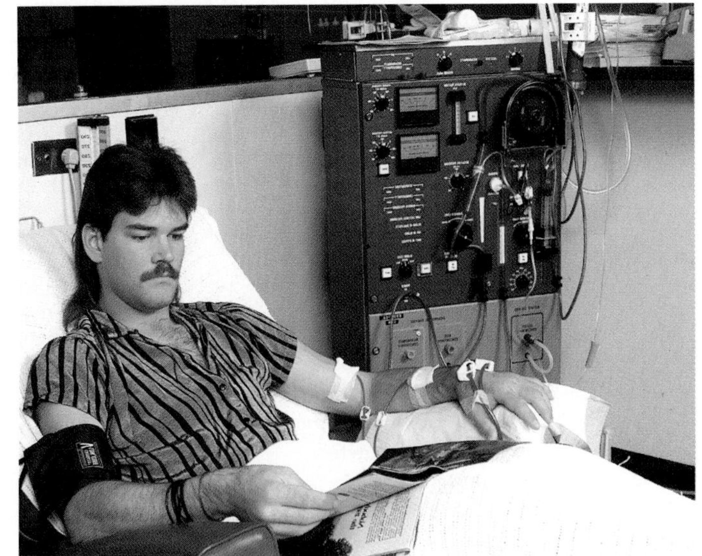

Clinical Investigation Answer

The location of the pain and the discoloration of the urine are indicative of a renal disorder. The hematuria (blood in the urine) was responsible for the discolored urine, and the presence of casts with associated red blood cells indicated glomerulonephritis. The elevated blood creatinine concentration indicated a reduction in the glomerular filtration rate (GFR) as a result of the glomerulonephritis, and this reduced GFR could have been responsible for the fluid retention and observed edema. The presence of only trace amounts of protein in the urine, however, was encouraging, and could be explained by the boy's running activity (proteinuria in this case would have been an ominous sign). The streptococcus infection, acting via an autoimmune reaction, was probably responsible for the glomerulonephritis. This was confirmed by the fact that the symptoms of glomerulonephritis disappeared after treatment with an antibiotic. Hydrochlorothiazide is a diuretic that helped to alleviate the edema by (1) promoting the excretion of larger amounts of urine and (2) shifting of edematous fluid from the interstitial to the vascular compartment.

Chapter Summary

Urinary System and Kidney Structure (pp. 779–783)

1. The urinary system consists of two kidneys, two ureters, the urinary bladder, and the urethra.
2. The urinary system maintains the composition and properties of the extracellular fluid. The end product of the urinary system is urine, which is voided from the body through the urethra.
3. The gross structure of the kidney includes the renal pelvis, calyces, renal medulla, and renal cortex.
   (a) The renal medulla is composed of the renal pyramids, which are separated by renal columns.
   (b) The renal pyramids empty urine into the calyces, which drain into the renal pelvis and out the ureter.
4. Each kidney contains more than a million microscopic functional units called nephrons. Nephrons have vascular and tubular components.
   (a) A capillary bed, the glomerulus, produces a filtrate that enters the first part of the nephron tubule, known as the glomerular (Bowman's) capsule.
   (b) Filtrate from the glomerular capsule enters, in turn, the proximal convoluted tubule, nephron loop (loop of Henle), distal convoluted tubule, and collecting duct.
   (c) The glomerulus, proximal convoluted tubule, and distal convoluted tubule are located in the renal cortex; the nephron loop can descend into the renal medulla.
   (d) The collecting ducts descend from the renal cortex through the renal medulla to empty their contents of urine into the calyces.

Glomerular Filtration (pp. 783–786)

1. A filtrate derived from blood plasma in the glomerulus must pass through a basement membrane of the glomerulae capillaries and through the inner layer of the glomerular capsule.
   (a) The glomerular ultrafiltrate, formed under the force of blood pressure, has a low protein concentration.
   (b) The glomerular filtration rate (GFR) is the volume of filtrate produced by both kidneys each minute. It ranges from 115 to 125 ml/min.
2. The GFR can be regulated by constriction or dilation of the afferent glomerular arterioles.
   (a) Sympathetic innervation causes constriction of the afferent glomerular arterioles.
   (b) Intrinsic mechanisms help to autoregulate the rate of renal blood flow and the GFR.

Reabsorption of Salt and Water (pp. 786–792)

1. Approximately 65% of the filtered salt and water is reabsorbed across the proximal convoluted tubules.
   (a) Na^+ is actively transported and Cl^- follows passively, going from the filtrate into the blood in the peritubular capillaries.
   (b) Water follows the NaCl by osmosis, so that the volume is reduced but the concentration of the filtrate that remains is unchanged.
2. The reabsorption of most of the remaining water occurs as a result of the action of the countercurrent multiplier system.
   (a) Sodium is actively extruded from the ascending limb of the nephron loop followed passively by chloride.
   (b) Since the ascending limb is impermeable to water, the remaining filtrate becomes hypotonic.
   (c) Because of this salt transport and because of countercurrent exchange in the vasa recta, the interstitial fluid of the renal medulla becomes hypertonic.
   (d) The hypertonicity of the renal medulla is multiplied by a positive feedback mechanism involving the descending limb, which is passively permeable to water.
3. The collecting duct is permeable to water but not to salt.
   (a) As the collecting ducts pass through the hypertonic renal medulla, water leaves by osmosis and is carried away in surrounding capillaries.
   (b) The permeability of the collecting ducts to water is stimulated by antidiuretic hormone (ADH).

Renal Plasma Clearance (pp. 792–798)

1. Inulin is filtered but neither reabsorbed nor secreted. Its clearance is thus equal to the glomerular filtration rate.
2. Some of the filtered urea is reabsorbed. Its clearance is therefore less than the glomerular filtration rate.
3. The PAH clearance is a measure of the total renal blood flow.
4. Normally, all of the filtered glucose and amino acids are reabsorbed. Glycosuria occurs when the transport carriers for glucose become saturated as a result of hyperglycemia.

Renal Control of Electrolyte and Acid-Base Balance (pp. 798–804)

1. Aldosterone stimulates Na^+ reabsorption and K^+ secretion in the distal convoluted tubule.
2. Aldosterone secretion is stimulated directly by a rise in blood K^+ and indirectly by a fall in blood volume.
   (a) Decreased blood flow and pressure through the kidneys stimulates the secretion of the enzyme renin from the juxtaglomerular apparatus.

(b) Renin catalyzes the formation of angiotensin I, which is then converted to angiotensin II.
(c) Angiotensin II stimulates the adrenal cortex to secrete aldosterone.

3. Aldosterone stimulates the secretion of H^+, as well as K^+, into the filtrate in exchange for Na^+.
4. The nephrons filter bicarbonate and reabsorb the amount required to maintain acid-base balance. Reabsorption of bicarbonate, however, in indirect.
   (a) Filtered bicarbonate combines with H^+ to form carbonic acid in the filtrate.
   (b) Carbonic anhydrase in the membranes of microvilli in the tubules catalyzes the conversion of carbonic acid to carbon dioxide and water.
   (c) Carbon dioxide is reabsorbed and converted in either the tubule cells or the red blood cells to carbonic acid, which dissociates to bicarbonate and H^+.
   (d) In addition to reabsorbing bicarbonate, the nephrons filter and secrete H^+, which is excreted in the urine buffered by ammonium and phosphate buffers.

Ureters, Urinary Bladder, and Urethra (pp. 805–809)

1. Each ureter contains three layers: the mucosa, muscularis, and adventitia.
2. The urinary bladder is lined by a transitional epithelium that is folded into rugae. The rugae enhance the ability of the urinary bladder to distend.
3. The urethra has an internal urethral sphincter of smooth muscle and an external urethral sphincter of skeletal muscle.
   (a) The male urethra conducts urine and seminal fluid. The much shorter female urethra conducts only urine.
   (b) The male urethra is composed of postatic, membranous, and spongy regions.
4. Micturition is controlled by reflex centers in the second through fourth segments of the spinal cord.

Review Activities

Objective Questions

1. Which of the following statements about metanephric kidneys is *true?*
   (a) They become functional at the end of the eighth week.
   (b) They are active throughout fetal development.
   (c) They are the third pair of kidneys to develop.
   (d) All of the above are true.
2. Which of the following statements about the renal pyramids is *false?*
   (a) They are located in the renal medulla.
   (b) They contain glomeruli.
   (c) They contain collecting ducts.
   (d) They empty urine into the calyces.

Match the following:

3. Active transport of sodium; water follows passively
4. Active transport of sodium; impermeable to water
5. Passively permeable to water only
6. Passively permeable to water and urea

(a) proximal convoluted tubule
(b) descending limb of nephron loop
(c) ascending limb of nephron loop
(d) distal convoluted tubule
(e) medullary collecting duct

7. Antidiuretic hormone promotes the retention of water by stimulating
   (a) the active transport of water.
   (b) the active transport of chloride.
   (c) the active transport of sodium.
   (d) the permeability of the collecting duct to water.
8. Aldosterone stimulates sodium reabsorption and potassium secretion in
   (a) the proximal convoluted tubule.
   (b) the descending limb of the nephron loop.
   (c) the ascending limb of the nephron loop.
   (d) the cortical collecting duct.
9. Substance X has a clearance greater than zero but less than that of inulin. What can you conclude about substance X?
   (a) It is not filtered.
   (b) It is filtered, but neither reabsorbed nor secreted.
   (c) It is filtered and partially reabsorbed.
   (d) It is filtered and secreted.
10. Substance Y has a clearance greater than that of inulin. What can you conclude about substance Y?
    (a) It is not filtered.
    (b) It is filtered, but neither reabsorbed nor secreted.
    (c) It is filtered and partially reabsorbed.
    (d) It is filtered and secreted.
11. About 65% of the glomerular ultrafiltrate is reabsorbed in
    (a) the proximal convoluted tubule.
    (b) the distal convoluted tubule.
    (c) the nephron loop.
    (d) the collecting duct.
12. Diuretic drugs that act in the nephron loop
    (a) inhibit active sodium transport.
    (b) cause an increased flow of filtrate to the distal convoluted tubule.
    (c) cause an increased secretion of potassium into the renal tubule.
    (d) promote the excretion of salt and water.
    (e) do all of the above.
13. The appearance of glucose in the urine
    (a) occurs normally.
    (b) indicates the presence of kidney disease.
    (c) occurs only when the transport carriers for glucose become saturated.
    (d) is a result of hypoglycemia.
14. Reabsorption of water through the tubules occurs by
    (a) osmosis.
    (b) active transport.
    (c) facilitated diffusion.
    (d) all of the above.
15. Which of the following factors oppose(s) filtration from the glomerulus?
    (a) plasma oncotic pressure
    (b) hydrostatic pressure in the glomerular (Bowman's) capsule
    (c) plasma hydrostatic pressure
    (d) both *a* and *b*
    (e) both *b* and *c*
16. The countercurrent exchange in the vasa recta
    (a) removes Na^+ from the extracellular fluid.
    (b) maintains high concentrations of NaCl in the extracellular fluid.
    (c) raises the concentration of Na^+ in the blood leaving the kidneys.
    (d) causes large quantities of Na^+ to enter the filtrate.
    (e) does all of the above.
17. The kidneys help maintain acid-base balance by
    (a) the secretion of H^+ in the distal regions of the nephron.
    (b) the action of carbonic anhydrase within the apical cell membranes.
    (c) the action of carbonic anhydrase within the cytoplasm of the tubule cells.
    (d) the buffering action of phosphates and ammonia in the urine.
    (e) all of the above means.
18. The detrusor muscle is located in
    (a) the kidneys.
    (b) the ureters.
    (c) the urinary bladder.
    (d) the urethra.

19. The internal urethral sphincter is innervated by
    (a) sympathetic nerve fibers.
    (b) parasympathetic nerve fibers.
    (c) somatic motor nerve fibers.
    (d) all of the above.

Essay Questions

1. Explain how glomerular ultrafiltrate is produced and why it has a low protein concentration.
2. Explain how the countercurrent multiplier system works and discuss its functional significance.
3. Explain how countercurrent exchange occurs in the vasa recta and discuss the functional significance of this mechanism.
4. Explain how an increase in ADH secretion promotes increased water reabsorption and how water reabsorption decreases when ADH secretion is decreased.
5. Explain how the structure of the epithelial wall of the proximal tubule and the distribution of Na^+/K^+ pumps in the epithelial cell membranes contribute to the ability of the proximal convoluted tubule to reabsorb salt and water.
6. Explain the action of different diuretic drugs. Which may cause an excessive loss of potassium? Also, explain how the potassium-sparing drugs work.
7. What happens to urinary bicarbonate excretion when a person hyperventilates? How might this response be helpful?
8. Describe the location of the macula densa and explain its role in the regulation of renin secretion and in tubuloglomerular feedback.
9. Describe how the nephron handles K^+, how the urinary excretion of K^+ changes under different conditions, and how this process is regulated by aldosterone.
10. Discuss the physiological and functional events of a voluntary micturition response.

Critical Thinking Questions

1. Recent studies suggest that the very high rates of urea transported by the collecting duct in the inner medulla are due to the presence of specific urea transporters that are stimulated by ADH. Suppose you collect urine from two patients who have been deprived of water overnight. One has normally functioning kidneys, and the other has a genetic defect in the urea transporters. How would the two urine samples differ? Explain.
2. Two men are diagnosed with diabetes insipidus. One didn't have the disorder until he suffered a stroke. The other had withstood the condition all his life, and it had never responded to exogenous ADH despite the presence of normal ADH receptors. What might be the cause of the diabetes insipidus in the two men?
3. Suppose a woman with a family history of polycystic kidney disease develops proteinuria. She has elevated blood creatinine levels and a reduced inulin clearance. What might these lab results indicate? Explain.
4. You love to spend hours fishing in a float tube in a lake, where the lower half of your body is submerged and the upper half is supported by an inner tube. However, you always have to leave the lake sooner than you'd like because you produce urine at a faster than usual rate. Using your knowledge about the regulation of urine volume, propose an explanation as to why people might produce more urine under these conditions.
5. You have an infection, and you see that the physician is about to inject you with millions of units of penicillin. What do you think will happen to your urine production as a result? Explain. In the hope of speeding your recovery, you gobble extra amounts of vitamin C. How will this affect your urine output?

Related Web Sites

For a listing of the most current web sites related to this chapter, please visit the *Concepts of Human Anatomy and Physiology* home page at http://www.mhhe.com/biosci/abio/.

NEXUS

Some Interactions of the Urinary System with the Other Body Systems

Integumentary System

- The kidneys maintain homeostasis of blood volume, pressure, and composition, which is needed for the health of the integumentary and other systems (p. 779).
- Evaporative water loss from the skin helps control body temperature, but effects on blood volume must be compensated by the kidneys (p. 664).

Skeletal System

- The pelvic girdle supports and protects some organs of the urinary system (p. 183).
- Bones store calcium and phosphate, and thus cooperate with the kidneys to regulate the blood levels of these ions. (p. 193).

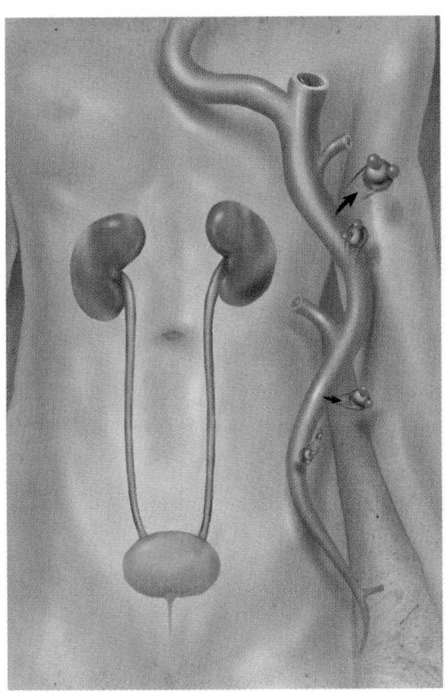

Nervous System

- Autonomic nerves help regulate renal blood flow, and hence glomerular filtration (p. 785).
- The nervous system provides motor control of micturition (p. 807).

Endocrine System

- Antidiuretic hormone stimulates reabsorption of water from the renal tubules (p. 791).
- Aldosterone stimulates sodium reabsorption and potassium secretion by the kidneys (p. 798).
- Natriuretic hormones stimulate sodium excretion by the kidneys (p. 801).
- The kidneys produce the hormone erythropoietin (p. 599).
- The kidneys secrete renin, which activates the renin-angiotensin-aldosterone system (p. 799).

Muscular System

- Muscles in the urinary tract assist storage and voiding of urine (p. 805).
- Smooth muscles in the renal blood vessels regulate renal blood flow and thus glomerular filtration rate (p. 785).

Circulatory System

- The blood transports oxygen and nutrients to all systems, including the urinary system, and removes wastes (p. 592).
- The kidneys filter the blood to produce urine, while regulating blood volume and composition (p. 779).
- The heart secretes atrial natriuretic hormone, which helps to regulate the kidneys (p. 801).

Lymphatic and Immune Systems

- The immune system protects all systems, including the urinary system against infections (p. 696).
- Lymphatic vessels help to maintain a balance between blood and interstitial fluid (p. 662).

Respiratory System

- The lungs provide oxygen and eliminates carbon dioxide for all systems, including the urinary system (p. 729).
- The lungs and kidneys cooperate in the regulation of blood pH (p. 763).

Digestive System

- The GI tract provides nutrients for all tissues, including those of the urinary system (p. 817).
- The GI tract, like the urinary system, helps to eliminate waste products (p. 842).

Reproductive System

- The urethra of a male passes through the penis and can eject either urine or semen (p. 806).
- Kidneys participate in the regulation of blood volume and pressure, which is required for functioning of the reproductive system (p. 779).

Twenty-Six

Digestive System

Objectives

- Briefly describe the activities of the digestive system and list its structures and regions.
- Locate and describe the serous membranes of the abdominal cavity.
- Describe the four tunics that compose the wall of the GI tract.
- Describe the anatomy of the oral cavity.
- Contrast the deciduous and permanent dentitions and describe the structure of a tooth.
- Describe the histological structure of the salivary glands and discuss the functions of saliva.
- Describe the location, gross structure, and functions of the stomach.
- Describe the histological structure of the esophagus and stomach, and list the cell types of the gastric mucosa and their secretions.
- Discuss the functions of hydrochloric acid and pepsin in digestion.
- State the regions of the small intestine and describe how bile and pancreatic juice are delivered to the small intestine.
- Describe the structure and functions of intestinal villi, microvilli, and intestinal crypts and discuss the nature and significance of the brush border enzymes.
- Describe the electrical activity that occurs in the intestine and discuss the nature of peristalsis and segmentation.
- State the regions of the large intestine and describe its structure.
- Describe the functions of the large intestine and explain how defecation is accomplished.
- Describe the structure of a liver lobule and trace the flow of blood and bile through the lobule.
- Discuss the functions of the liver and describe the enterohepatic circulation.
- Describe the anatomical relationship between the liver and gallbladder.
- Identify the endocrine and exocrine structures of the pancreas and describe the composition and functions of pancreatic juice.
- Explain how gastric secretion is regulated during the cephalic, gastric, and intestinal phases.
- Describe the structure and function of the enteric nervous system.
- Explain how pancreatic juice and bile secretion is regulated by nerves and hormones.
- Discuss the nature and actions of the various gastrointestinal hormones.
- Describe the enzymes involved in the digestion of carbohydrates, lipids, and proteins, and explain how monosaccharides and amino acids are absorbed.
- Describe the roles of bile and pancreatic lipase in fat digestion and trace the pathways and structures involved in the absorption of lipids.

Clinical Investigation

A male college student appeared at the student health center complaining of severe but transient pain. He said that the pain often occurred as soon as he had finished a meal. Upon questioning, it was learned that the pain was located below his right scapula, and that it occurred whenever he ate particular foods, such as peanut butter and bacon. The pain did not occur when he ate fish or drank milk, cola, or alcoholic beverages. The sclera of this student's eyes were markedly yellow. Laboratory tests revealed that he had fatty stools, a prolonged bleeding time, and elevated blood levels of conjugated bilirubin. His blood tests, however, showed normal levels of ammonia, urea, free bilirubin, and pancreatic amylase. His white blood cell count was normal, and he did not have a fever. What do these observations reveal about the possible cause of his symptoms?

Clues: Review the information on bile production and secretion in the section "Functions of the Liver" and study the boxed information on gallstones in the section entitled "Gallbladder." Also review the description of jaundice under "Clinical Considerations."

Introduction to the Digestive System

Within the lumen of the gastrointestinal (GI) tract, large food molecules are hydrolyzed into their monomers (subunits). These monomers pass through the inner layer, or mucosa, of the small intestine to enter the blood or lymph in a process called absorption. Digestion and absorption are aided by specializations of the mucosa and by characteristic movements caused by contractions of the muscle layers of the GI tract.

Unlike plants, which can form organic molecules using inorganic compounds such as carbon dioxide, water, and ammonia, humans and other animals must obtain their basic organic molecules from food. Some of the ingested food molecules are needed for their energy (caloric) value—obtained by the reactions of cellular respiration and used in the production of ATP—and the balance is used to make additional tissue.

Most of the organic molecules that are ingested are similar to the molecules that form human cells. These are generally large molecules (*polymers*), which are composed of subunits (*monomers*). Within the gastrointestinal tract, the **digestion** of these large molecules into their monomers occurs by means of *hydrolysis reactions* (reviewed in fig. 26.1). The monomers thus formed are transported across the wall of the small intestine into the blood and lymph in the process of **absorption.** Digestion and absorption are the primary functions of the digestive system.

Since the composition of food is similar to the composition of body tissues, enzymes that digest food are also capable of digesting a person's own cells. This does not normally occur, however, because a variety of protective devices inactivate digestive enzymes in the body and keep them away from the cytoplasm of tissue cells. The fully active digestive enzymes are normally confined to the lumen (cavity) of the GI tract.

The lumen of the GI tract is open at both ends (mouth and anus), and thus is continuous with the environment. In this sense, the harsh conditions required for digestion occur *outside* the body. Indigestible materials, such as cellulose from plant walls, pass from one end to the other without crossing the epithelial lining of the GI tract; since they are not absorbed, they do not enter the body.

In *planaria* (a type of flatworm), the GI tract has only one opening—the mouth is also the anus. Each cell that lines the GI tract is thus exposed to food, absorbable digestion products, and waste products. The two open ends of the GI tract of higher organisms, by contrast, permit one-way transport, which is ensured by wavelike muscle contractions and by the action of sphincter muscles. This one-way transport allows different regions of the GI tract to be specialized for different functions, as a "dis-assembly line." These functions of the digestive system include the following:

1. **Motility.** This refers to the movement of food through the GI tract through the processes of
   a. *Ingestion:* Taking food into the mouth.
   b. *Mastication:* Chewing the food and mixing it with saliva.
   c. *Deglutition:* Swallowing food.
   d. *Peristalsis:* Rhythmic, wavelike contractions that move food through the GI tract.
2. **Secretion.** This includes both exocrine and endocrine secretions.
   a. *Exocrine secretions:* Water, hydrochloric acid, bicarbonate, and many digestive enzymes are secreted into the lumen of the GI tract. The stomach alone, for example, secretes 2 to 3 liters of gastric juice a day.
   b. *Endocrine secretions:* The stomach and small intestine secrete a number of hormones that help to regulate the digestive system.
3. **Digestion.** This refers to the breakdown of food molecules into their smaller subunits, which can be absorbed.
4. **Absorption.** This refers to the passage of digested end products into the blood or lymph.

Anatomically and functionally, the digestive system can be divided into the tubular **gastrointestinal (GI) tract,** or *alimentary canal,* and **accessory digestive organs.** The GI tract is approximately 9 m (30 ft) long and extends from the mouth

mastication: Gk. *mastichan*, gnash the teeth
deglutition: L. *deglutire*, swallow down
peristalsis: Gk. *peri*, around; *stellain*, compress

Figure 26.1

Digestion of food molecules through hydrolysis reactions. These reactions ultimately release the subunit molecules of each food category.

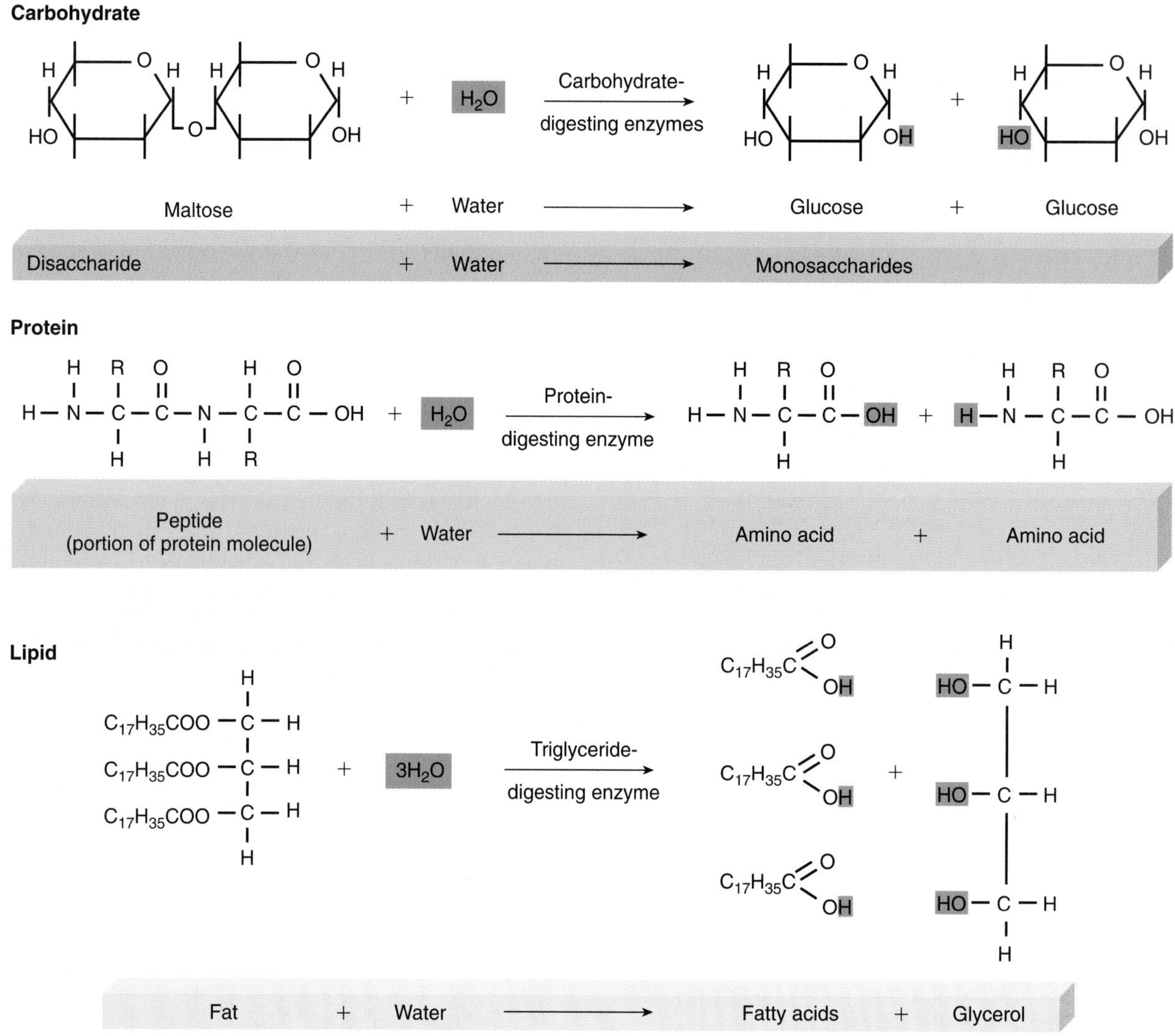

to the anus. It traverses the thoracic cavity and enters the abdominal cavity at the level of the diaphragm. The anus is located at the inferior portion of the pelvic cavity. The organs of the GI tract include the *oral cavity, pharynx, esophagus, stomach, small intestine,* and *large intestine* (fig. 26.2). The accessory digestive organs include the *teeth, tongue, salivary glands, liver, gallbladder,* and *pancreas*. The term *viscera* (*vis'er-ă*) is frequently used to refer to the abdominal organs of digestion, but it also can be used in reference to any of the organs in the thoracic and abdominal cavities.

Serous Membranes

Most of the GI tract and abdominal accessory digestive organs are located within the **abdominal cavity.** These organs are not firmly embedded in solid tissue but are supported and covered by **serous membranes.** Serous membranes line the body cavities and cover the organs that lie within these cavities. A serous membrane has a parietal portion lining the body wall and a visceral portion covering the internal organs. As described in chapter 24, the serous membranes associated with the lungs are called *pleurae*. The serous membranes of the abdominal cavity are called **peritoneal membranes,** or **peritoneum** (*per''ĭ-tŏ-ne'um*). The peritoneum is the largest serous membrane of the body. It is composed of simple squamous epithelium, portions of which are reinforced by connective tissue.

The **parietal peritoneum** lines the wall of the abdominal cavity (fig. 26.3). Along the posterior, or dorsal, aspect of the abdominal cavity, the parietal peritoneum comes together to form a double-layered peritoneal fold called the **mesentery** (*mes'en-ter''e*). The mesentery supports the GI tract, at the

mesentery: Gk. *mesos*, middle; *enteron*, intestine

Figure 26.2

The organs of the digestive system.

The digestive system includes the gastrointestinal tract and the accessory digestive organs.

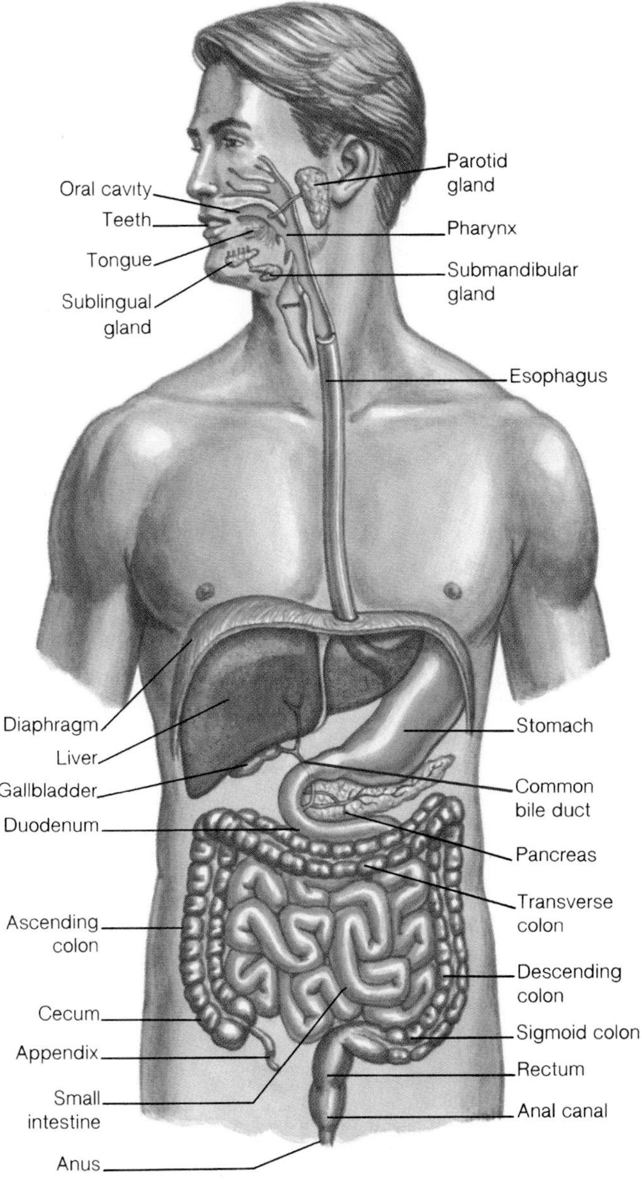

same time allowing for peristaltic movements of the small intestine and large intestine. It also provides a structure for the passage of intestinal nerves and vessels. The **mesocolon** is a specific portion of the mesentery that supports the large intestine (fig. 26.4*c*,*d*). The peritoneal covering continues around the intestinal viscera as the **visceral peritoneum.** The **peritoneal cavity** is the space between the parietal and visceral portions of the peritoneum.

Extensions of the parietal peritoneum, located in the peritoneal cavity, serve specific functions (fig. 26.4). The **falciform** (*fal'sĭ-form*) **ligament,** a serous membrane reinforced with connective tissue, attaches the liver to the diaphragm and anterior abdominal wall. The **lesser omentum** (*o-men'tum*) passes from the lesser curvature of the stomach and the upper duodenum to the inferior surface of the liver. The **greater omentum** extends from the greater curvature of the stomach to the transverse colon, forming an apronlike structure over most of the small intestine. Functions of the greater omentum include storing fat, cushioning visceral organs, supporting lymph nodes, and protecting against the spread of infections. In cases of localized inflammation, such as appendicitis, the greater omentum may compartmentalize the inflamed area, sealing it off from the rest of the peritoneal cavity.

Figure 26.3

Diagram of the abdominal serous membranes.

The parietal peritoneum lines the body wall, and the visceral peritoneum covers most of the internal organs of the abdominal cavity.

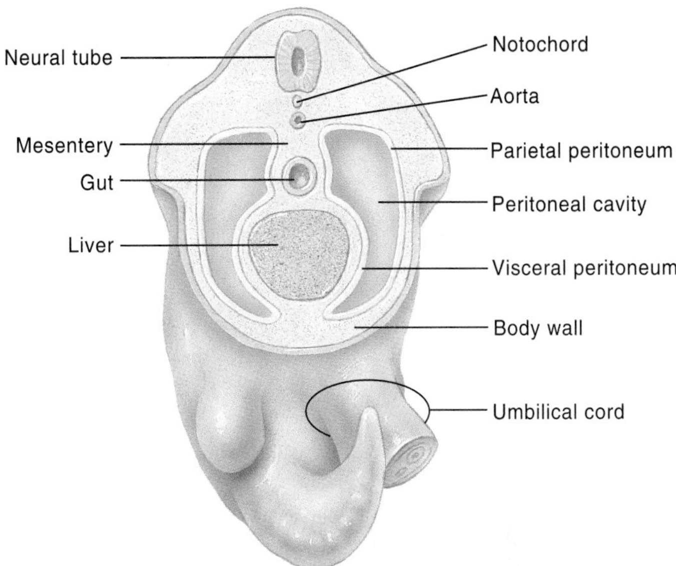

Certain of the abdominal organs closely associated with the posterior abdominal wall are not supported by mesentery and are covered only by parietal peritoneum. These organs are said to be *retroperitoneal* (behind the peritoneum). They include most of the pancreas, the kidneys, a portion of the duodenum, and the abdominal aorta.

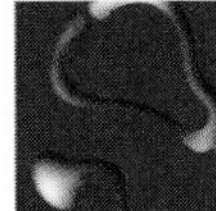

Peritonitis is a bacterial inflammation of the peritoneum. It may be caused by trauma, rupture of a visceral organ, or postoperative complications. Peritonitis is usually extremely painful and serious. Treatment involves the injection of massive doses of antibiotics and perhaps peritoneal intubation (insertion of a tube) to permit drainage.

---

omentum: L. *omentum*, apron

Figure 26.4

Arrangement of the abdominal organs and peritoneal membranes.
(*a*) The greater omentum, (*b*) the lesser omentum with the liver lifted, (*c*) the mesentery with the greater omentum lifted, and (*d*) the relationship of the peritoneal membranes to the visceral organs as shown in a sagittal view.

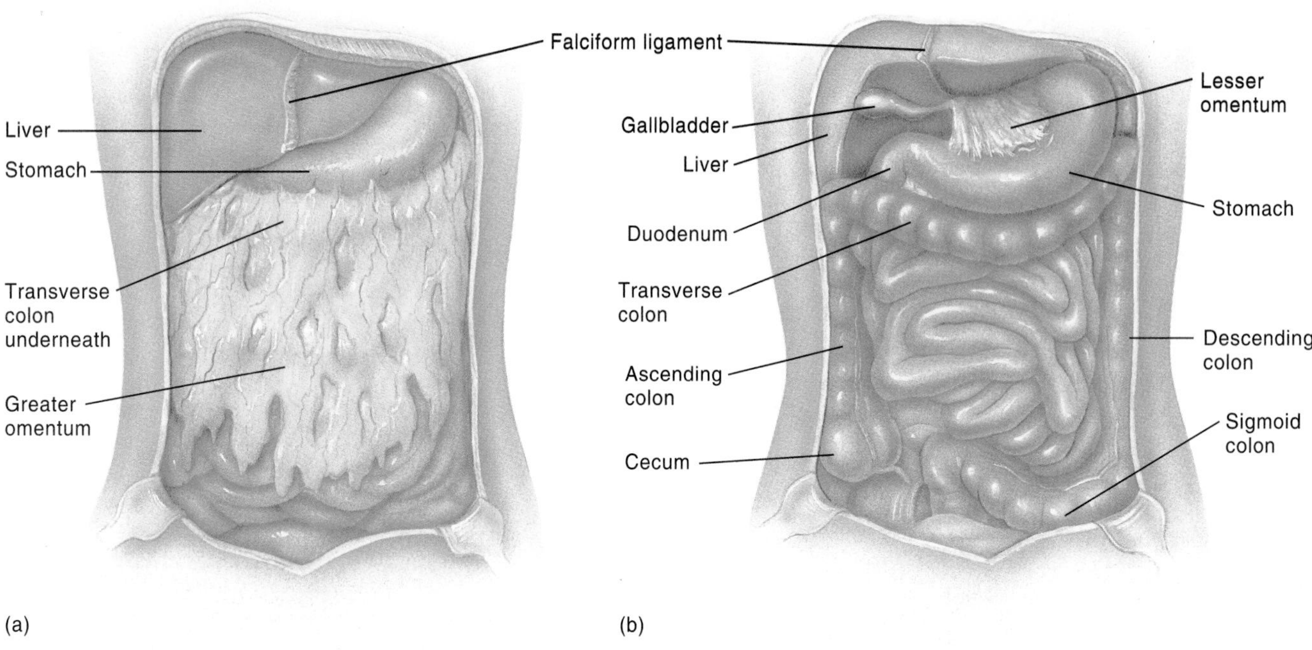

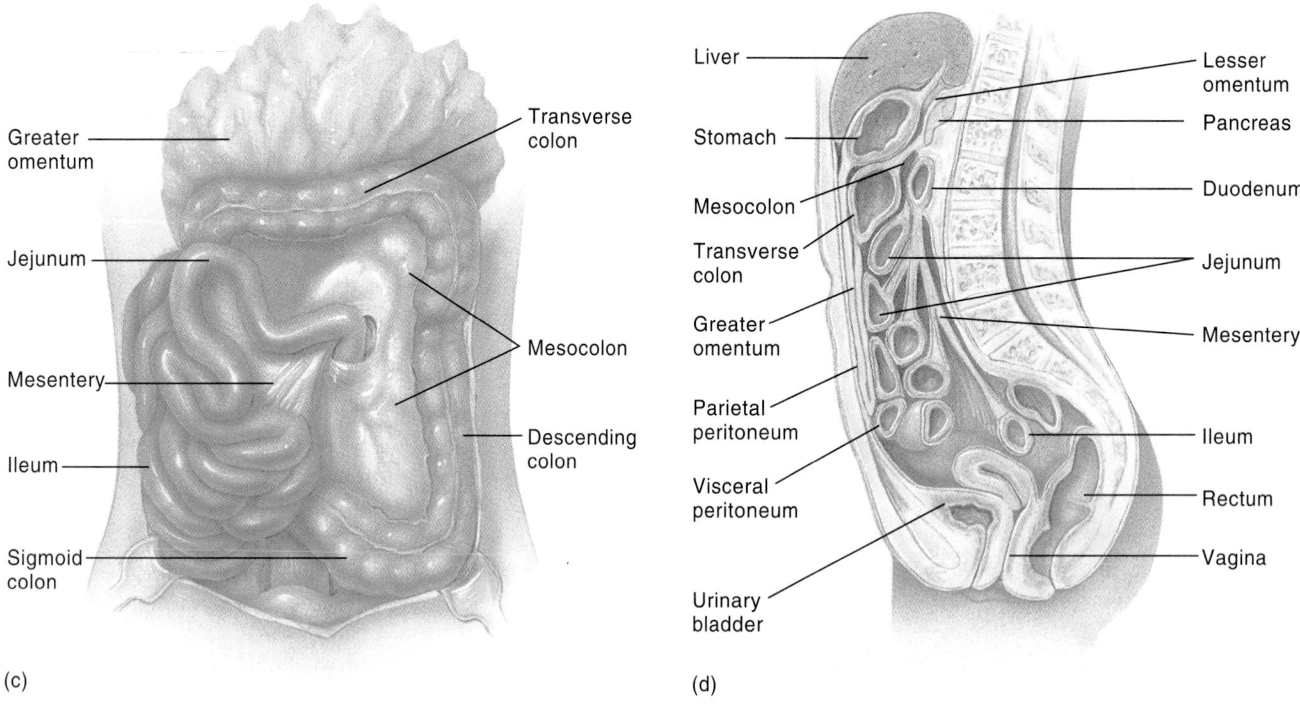

Figure 26.5

Layers of the digestive tract.
(*a*) An illustration of the major tunics, or layers, of the small intestine. The inset shows how folds of mucosa form projections called villi in the small intestine. (*b*) An illustration of a cross section of the small intestine showing layers and glands.

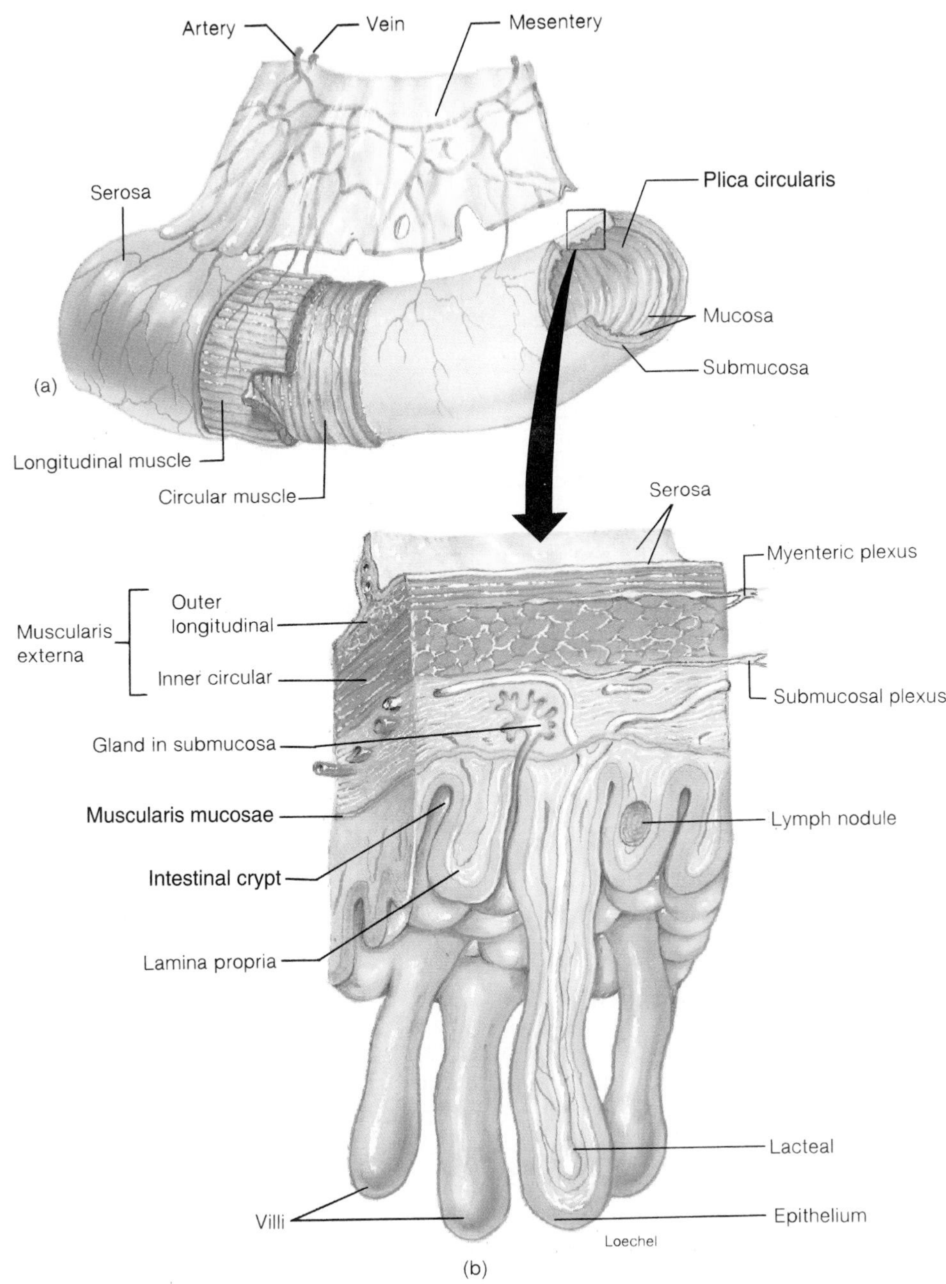

Mucosa

The **mucosa,** which lines the lumen of the GI tract, is the absorptive and major secretory layer. It consists of a simple columnar epithelium that also contains specialized *goblet cells* that secrete mucus. The epithelium is supported by a thin layer of loose connective tissue called the **lamina propria** (*lam'ĭ-nă pro'pre-ă*) (fig. 26.5*b*). The lamina propria contains numerous lymph nodules, which are important in protecting against disease. Deep to the lamina propria is a thin layer of smooth muscle called the **muscularis mucosae.**

Submucosa

The relatively thick **submucosa** is a highly vascular layer of connective tissue that serves the mucosa. Absorbed molecules that pass through the columnar epithelial cells of the mucosa enter into blood vessels or lymph ductules of the submucosa. In addition to blood vessels, the submucosa contains glands and nerve plexuses. The **submucosal** (*Meissner's*) **plexus** (fig. 26.5*b*), provides autonomic innervation to the muscularis mucosae.

Muscularis

The **muscularis** (also called the *muscularis externa*) is responsible for segmental contractions and peristaltic movement through the GI tract. The muscularis has an inner circular layer and an outer longitudinal layer of smooth muscle. Contractions of these layers move the food peristaltically through the GI tract and physically pulverize and churn the food with digestive enzymes. The **myenteric** (Auerbach's) **plexus** located between the two muscle layers provides the major nerve supply to the GI tract and includes fibers and ganglia from both the sympathetic and parasympathetic divisions of the autonomic nervous system.

Layers of the Gastrointestinal Tract

The GI tract from the esophagus to the anal canal is composed of four layers, or **tunics.** Each tunic contains a dominant tissue type that performs specific functions in the digestive process. The four tunics of the GI tract, from the inside out, are the mucosa, submucosa, muscularis, and serosa (fig. 26.5).

Serosa

The outer **serosa** completes the wall of the GI tract. It is a binding and protective layer consisting of loose connective

Meissner's plexus: from Georg Meissner, German histologist, 1829–1905
Auerbach's plexus: from Leopold Auerbach, German anatomist, 1828–97

tissue covered with a layer of simple squamous epithelium and subjacent connective tissue. The serosa is actually the visceral peritoneum of the abdominal cavity.

Innervation of the Gastrointestinal Tract

The GI tract is innervated by the sympathetic and parasympathetic divisions of the autonomic nervous system. The vagus nerves are the source of parasympathetic activity in the esophagus, stomach, pancreas, gallbladder, small intestine, and upper portion of the large intestine. The lower portion of the large intestine receives parasympathetic innervation from spinal nerves in the sacral region. The submucosal and myenteric plexuses are the sites where preganglionic fibers synapse with the postganglionic fibers that innervate the smooth muscle of the GI tract. Stimulation of the parasympathetic fibers promotes peristalsis and secretory activity of the GI tract.

Postganglionic sympathetic fibers pass through the submucosal and myenteric plexuses and innervate the GI tract. The effects of sympathetic nerve stimulation inhibit peristalsis and secretory activity and constrict the sphincter muscles along the GI tract; therefore, they are antagonistic to the effects of parasympathetic nerve stimulation.

Mouth, Pharynx, and Associated Structures

Ingested food is mechanically broken down by the action of teeth and chemically broken down by the activity of saliva. The resulting bolus is swallowed in the process of deglutition.

The **mouth** and associated structures initiate mechanical digestion of food through the process of mastication. The mouth is referred to as the **oral cavity** (fig. 26.6). It is formed by the cheeks, lips, hard palate, soft palate, and tongue. The *vestibule* is the depression between the cheeks and lips externally and the gums and teeth internally (fig. 26.7). The opening between the oral cavity and the pharynx is called the *fauces* (*faw'sēz*). The *pharynx* serves as a common passageway for both the respiratory and digestive systems. Both the mouth and pharynx are lined with nonkeratinized stratified squamous epithelium, which is constantly moistened by the secretion of saliva.

Cheeks and Lips

The **cheeks** form the lateral walls of the oral cavity. They consist of outer layers of skin, subcutaneous fat, facial muscles that assist in manipulating food in the oral cavity, and inner linings of moistened, stratified squamous epithelium. The anterior portion of the cheeks terminates in the superior and inferior lips.

The **lips** are fleshy, highly mobile organs whose principal function in humans is associated with speech. Each lip is attached from its inner surface to the gum by a midline fold of

Figure 26.6

Superficial structures of the oral cavity.
Note the position of the different teeth (also see figure 26.10).

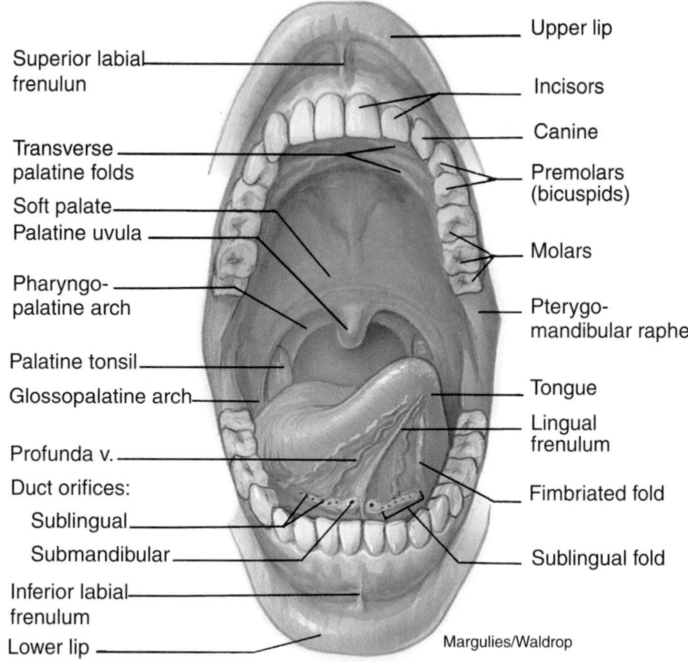

mucous membrane called the **labial frenulum** (*fren'yŭ-lum*) (see fig. 26.6). The lips are formed from the orbicularis oris muscle and associated connective tissue, and they are covered with soft, pliable skin. Between the outer skin and the mucous membrane of the oral cavity is a transition zone called the *vermilion*. This portion of the lips, where lipstick might be applied, is poorly keratinized and blood vessels close to the surface give it a reddish brown color.

Tongue

As a digestive organ, the **tongue** functions to move food around in the mouth during mastication and to assist in swallowing food. It contains **taste buds** (chapter 18) through which various food tastes are sensed, and it is also essential in producing speech. The tongue is a mass of skeletal muscle covered with a mucous membrane. Extrinsic tongue muscles (those that originate from bone and insert upon the tongue) move the tongue from side to side and in and out. Only the anterior two-thirds of the tongue lies in the oral cavity; the remaining one-third lies in the pharynx (fig. 26.7) and is attached to the hyoid bone. Rounded masses of **lingual tonsils** are located on the posterior surface of the tongue (fig. 26.8). The undersurface of the tongue is connected along the midline anteriorly to the floor of the mouth by the vertically positioned **lingual frenulum** (see fig. 26.6).

---

pharynx: L. *pharynx*, throat
tonsil: L. *toles*, swelling

Figure 26.7

Sagittal section of the facial region.
This section clearly reveals the relationship between the oral cavity, nasal cavity, and pharynx.

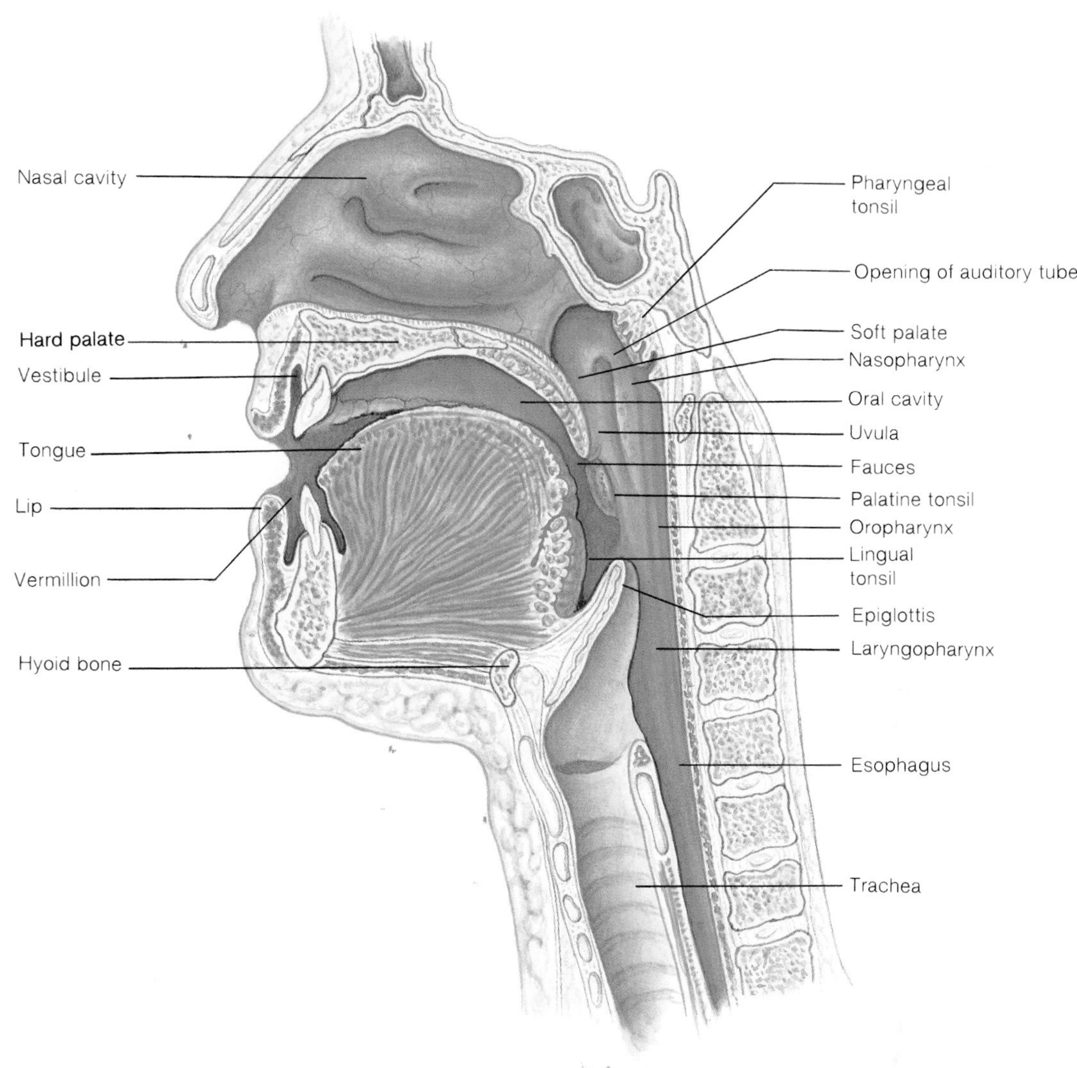

Figure 26.8

The surface of the tongue.
Note the position of the different papillae and tonsils.

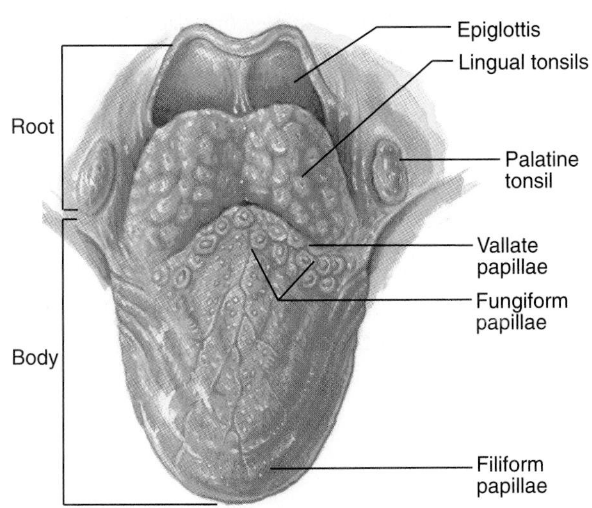

When a short lingual frenulum restricts tongue movements, the person is said to be "tongue-tied." If this developmental problem is severe, an infant may have difficulty suckling. Older children with this problem may have faulty speech. These functional problems can be easily corrected through surgery.

On the surface of the tongue are numerous small elevations called **papillae** (*pă pil'e*). The papillae give the tongue a distinct roughened surface that aids the handling of food. As described in chapter 18, some of them also contain taste buds that can distinguish sweet, salty, sour, and bitter sensations. The three types of papillae on the surface of the tongue are *filiform*, *fungiform*, and *vallate* (fig. 26.8). Filiform papillae, by far the most numerous, have tapered tips and are sensitive to touch. These papillae lack taste buds. The larger and rounded

---

papilla: L. *papula*, swelling or pimple
fungiform: L. *fungus*, fungus; *forma*, form

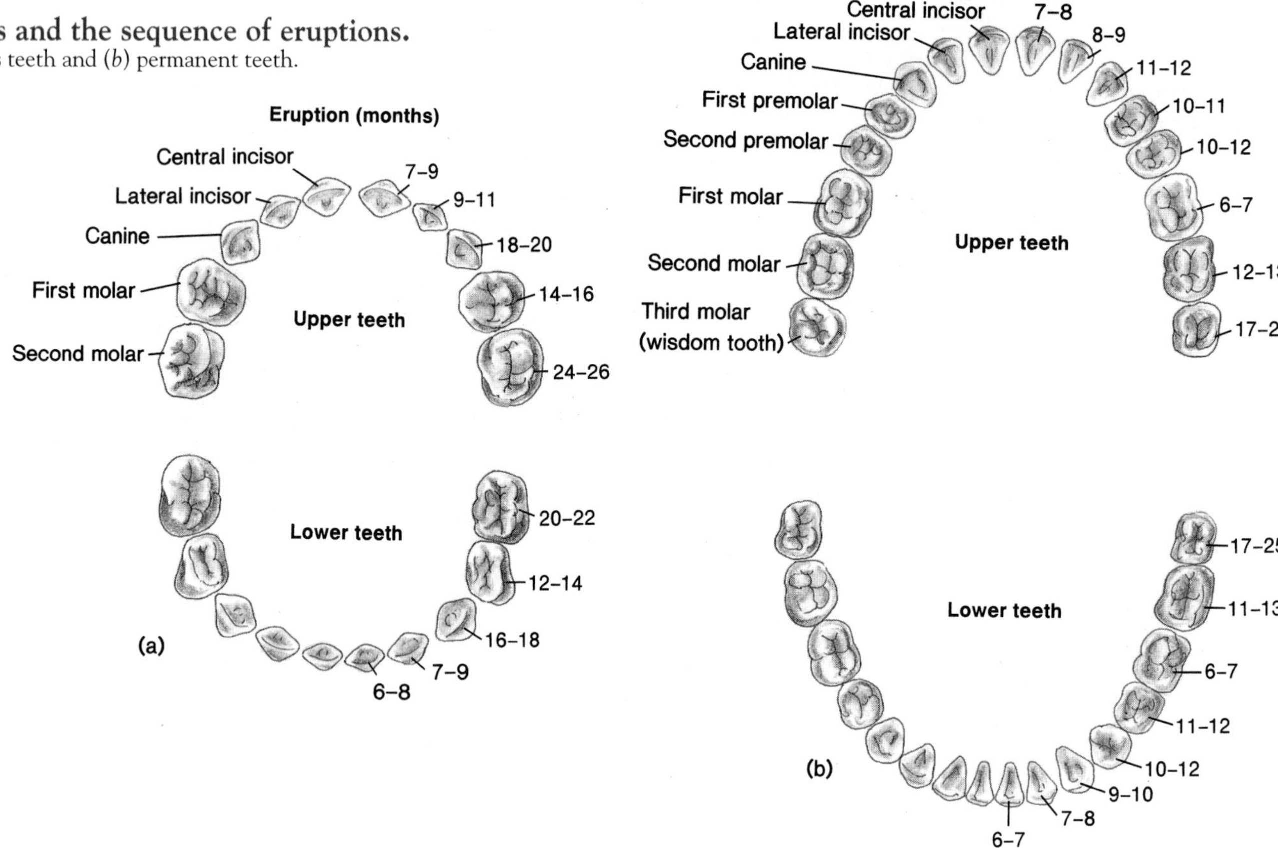

Figure 26.9

Dentitions and the sequence of eruptions. (*a*) Deciduous teeth and (*b*) permanent teeth.

fungiform papillae are scattered among the filiform type. The few vallate papillae are arranged in a V-shape on the posterior surface of the tongue.

Palate

The **palate,** which forms the roof of the oral cavity, consists of the bony hard palate anteriorly and the soft palate posteriorly (see figs. 26.6 and 26.7). The **hard palate,** formed by the palatine processes of the maxillae and the horizontal plates of the palatine bones, is covered with a mucous membrane. Transverse ridges called **palatal rugae** (*roo'je*) are located along the mucous membrane of the hard palate. These structures serve as friction ridges against which the tongue is placed during swallowing. The soft palate is a muscular arch covered with mucous membrane and is continuous with the hard palate anteriorly. Suspended from the middle lower border of the soft palate is a cone-shaped projection called the **uvula** (*yoo'vyŭ-lă*). During swallowing, the soft palate and uvula are drawn upward, closing the nasopharynx and preventing food and fluid from entering the nasal cavity.

Two muscular folds extend downward from both sides of the base of the uvula (see fig. 26.6). The anterior fold is called the **glossopalatine arch,** and the posterior fold is the **pharyngopalatine** (*fă-ring"go-pal'ă-tīn*) **arch.** The **palatine tonsils** are located between these two arches.

Teeth

The **teeth** of humans and other mammals vary in structure and are adapted to handle food in different ways (fig. 26.9). The four pairs (upper and lower jaws) of anteriormost teeth are the **incisors** (*in-si'sorz*). The chisel-shaped incisors are adapted for cutting and shearing food. The two pairs of cone-shaped **canines,** or **cuspids,** are located at the anterior corners of the mouth and are adapted for holding and tearing. Incisors and canines are further characterized by a single root on each tooth. Located behind the canines are the **premolars,** or *bicuspids*, and **molars.** These teeth have two or three roots, and their somewhat rounded, irregular surfaces, or **cusps,** are adapted for crushing and grinding food.

Two sets of teeth develop in a person's lifetime. Twenty **deciduous,** or **milk, teeth** begin to erupt at about 6 months of age (fig. 26.10), beginning with the incisors. All of the deciduous teeth have erupted by the age of 2. Thirty-two **permanent teeth** replace the deciduous teeth in a predictable sequence. This process begins at about age 6 and continues until about age 17. The **third molars,** or *wisdom teeth*, are the

incisor: L. *incidere*, to cut
canine: L. *canis*, dog
molar: L. *mola*, millstone
deciduous: L. *deciduus*, to fall away

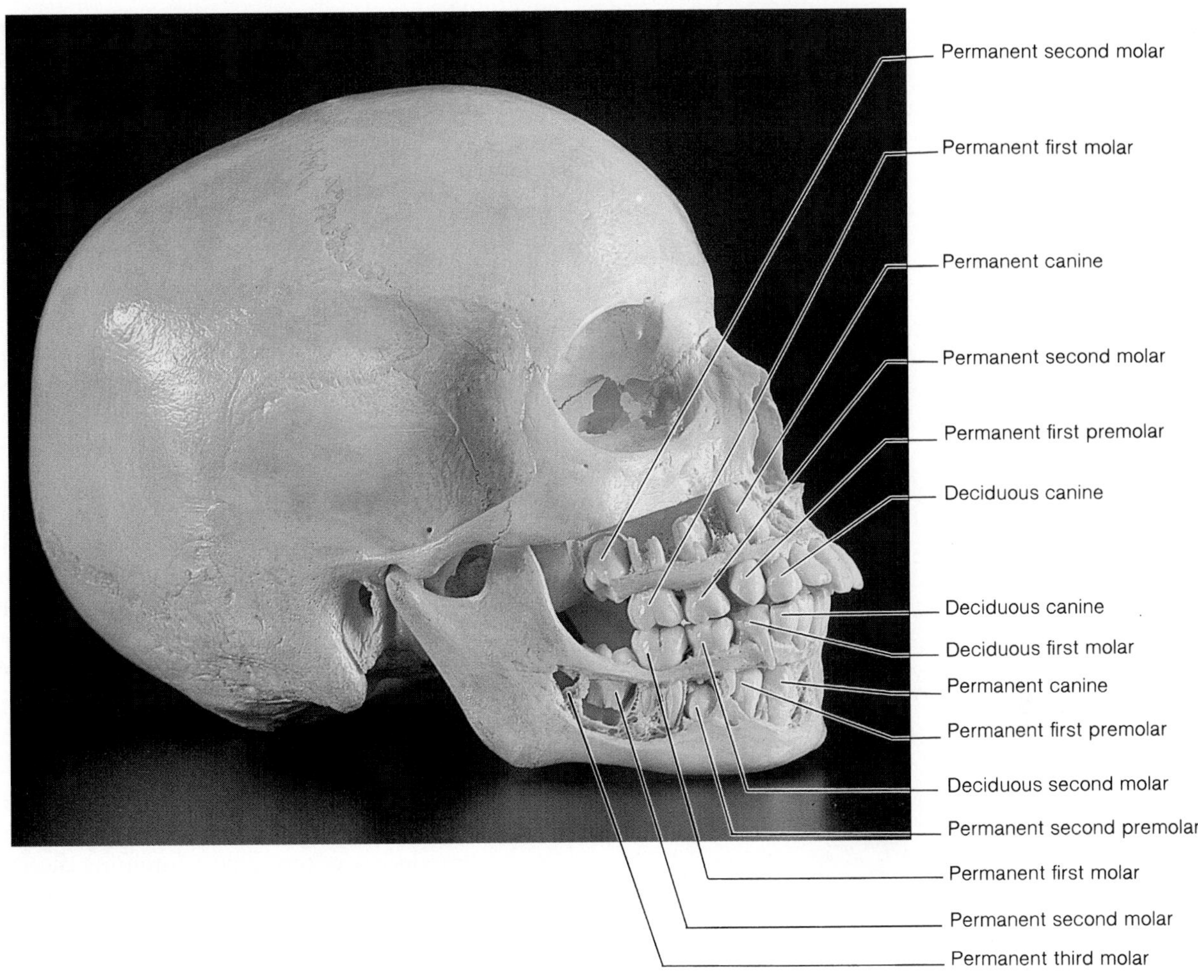

Figure 26.10

Eruption of teeth. This skull of a youth (9 to 12 years old) shows the eruptions of teeth.

last to erupt. Eruption of the wisdom teeth is less predictable. If they do erupt, it is between the ages of 17 and 25. Because the jaws are formed by this time and the other teeth are in place, the eruption of wisdom teeth may cause serious problems of crowding or impaction.

A **dental formula** is a graphic representation of the types, number, and position of teeth in the oral cavity. This formula gives the number of each tooth type in the left and right halves of each jaw; for example, **I** 2/2 means that there are 2 incisors on the left and 2 on the right halves of each jaw. Following are the deciduous and permanent dental formulae for humans:

Formula for deciduous dentition:

I 2/2, C 1/1, DM 2/2 = 10 × 2 = 20 teeth

Formula for permanent dentition:

I 2/2, C 1/1, P 2/2, M 3/3 = 16 × 2 = 32 teeth

where

I = incisor; **C** = canine; **P** = premolar;
DM = deciduous molar; **M** = molar.

The cusps of the upper and lower premolar and molar teeth occlude for chewing food, whereas the upper incisors normally form an overbite with the incisors of the lower jaw. An overbite of the upper incisors creates a shearing action as these teeth slide past one another. Masticated food is mixed with saliva, which initiates chemical digestion and facilitates swallowing. The soft mass of chewed food that is swallowed is called a *bolus* (*bo'lus*). A tooth consists of an exposed **crown,** which is supported by a **neck** that is anchored firmly into the jaw by one or more **roots** (fig. 26.11). The roots of teeth fit into sockets, called **dental alveoli,** in the alveolar processes of the mandible and maxillae. Each socket is lined with a connective tissue periosteum, specifically called the **periodontal ligament.** The root of a tooth is covered with a bonelike material called the **cementum;** fibers in the periodontal ligament insert into the cementum and fasten the tooth in its socket. The **gingiva** (*jin-ji'vă* or *jin'jĭ-vă*), also called the **gum,** is the mucous membrane surrounding the alveolar processes in the oral cavity.

The bulk of a tooth consists of **dentin** (*den'tin*)—a substance similar to bone, but harder. Covering the dentin on the outside and forming the crown is a tough, durable layer of

bolus: Gk. *bolos*, lump or mass
dentin: L. *dens*, tooth

Figure 26.11

Structure of a tooth.

This diagram illustrates a vertical section through a canine tooth.

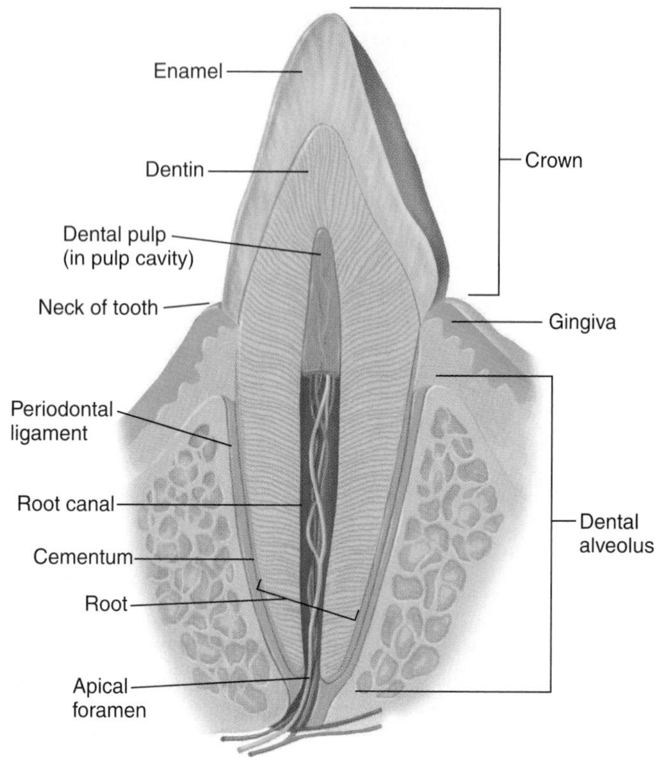

enamel. Enamel is composed primarily of calcium phosphate and is the hardest substance in the body. The central region of the tooth contains the **pulp cavity.** The pulp cavity contains the **pulp,** which is composed of connective tissue containing blood vessels, lymph vessels, and nerves. A **root canal** is continuous with the pulp cavity and opens to the connective tissue surrounding the root through an **apical foramen** at the tip of the root. The apical foramen permits blood vessels and nerves to enter the pulp.

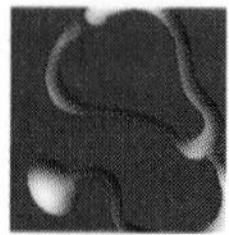

Although enamel is the hardest substance in the body, acidic conditions produced by bacterial activity can weaken it, possibly resulting in *dental caries* (*kar'ēz*), or *tooth decay.* The cavities produced by dental caries must be artificially filled because new enamel is not produced after a tooth erupts. The rate of tooth decay decreases after age 35, but then periodontal diseases may develop. *Periodontal diseases* result from plaque or tartar buildup at the gum line. This buildup pulls the gum away from the teeth, allowing bacterial infections to develop.

Salivary Glands

Salivary glands are accessory digestive glands that produce a fluid secretion called *saliva.* Saliva functions as a solvent in cleansing the teeth and dissolving food molecules so that they can be tasted. Saliva also contains a starch-digesting enzyme and mucus that lubricates the pharynx to facilitate swallowing. Numerous minor salivary glands are located throughout the mucous membranes of the oral cavity. Most of the saliva, however, is produced by three pairs of salivary glands that lie outside the oral cavity and is transported to the mouth via salivary ducts. The three major pairs of salivary glands are the parotid, submandibular, and sublingual glands (fig. 26.12).

The **parotid** (*pă-rot'id*) **gland** is the largest of the salivary glands. It is located below and in front of the ear, between the skin and the masseter muscle. The **parotid** (Stensen's) **duct** parallels the zygomatic arch across the masseter muscle, pierces the buccinator muscle, and drains into the oral cavity opposite the second upper molar. It is the parotid gland that becomes infected and swollen with the mumps.

The **submandibular gland** lies inferior to the body of the mandible, about midway along the inner side of the jaw. This gland is covered by the more superficial mylohyoid muscle. The **submandibular** (Wharton's) **duct** empties into the floor of the mouth on either side of the lingual frenulum.

The **sublingual gland** lies under the mucosa in the floor of the mouth on the side of the tongue. Associated with each sublingual gland are several small **sublingual** (Rivinus') **ducts** that empty into the floor of the mouth in an area posterior to the papilla of the submandibular duct.

Two types of secretory cells, *serous* and *mucous cells,* are found in all salivary glands in various proportions. Serous cells produce a watery fluid containing digestive enzymes; mucous cells secrete a thick, stringy mucus. Cuboidal epithelial cells line the lumina of the salivary ducts.

The salivary glands are innervated by both divisions of the autonomic nervous system. Sympathetic impulses stimulate the secretion of small amounts of viscous saliva. Parasympathetic stimulation causes the secretion of large volumes of watery saliva. Physiological responses of this type occur whenever a person sees, smells, tastes, or even thinks about desirable food. The amount of saliva secreted daily ranges from 1,000 to 1,500 ml. Information about the salivary glands is summarized in table 26.1.

Pharynx

The funnel-shaped **pharynx** (*far'ingks*) is a passageway approximately 13 cm (5 in.) long connecting the oral and nasal cavities to the esophagus and trachea. The external, circular layer of pharyngeal muscles, called *constrictor muscles* (fig. 26.13), compresses the lumen of the pharynx involuntarily during swallowing. The **superior constrictor muscle** attaches to bony processes of the skull and mandible and encircles the upper portion of the pharynx. The **middle constrictor muscle** arises from the hyoid bone and stylohyoid ligament and encircles the middle portion of the pharynx. The **inferior constrictor muscle**

parotid: Gk. *para,* beside; *otos,* ear

Stensen's duct: from Nicholaus Stensen, Danish anatomist, 1638–86

Wharton's duct: from Thomas Wharton, English anatomist, 1614–73

Rivinus' duct: from Augustus Rivinus, German anatomist, 1652–1723

Figure 26.12

The salivary glands.

These are exocrine glands with ducts that empty into the oral cavity.

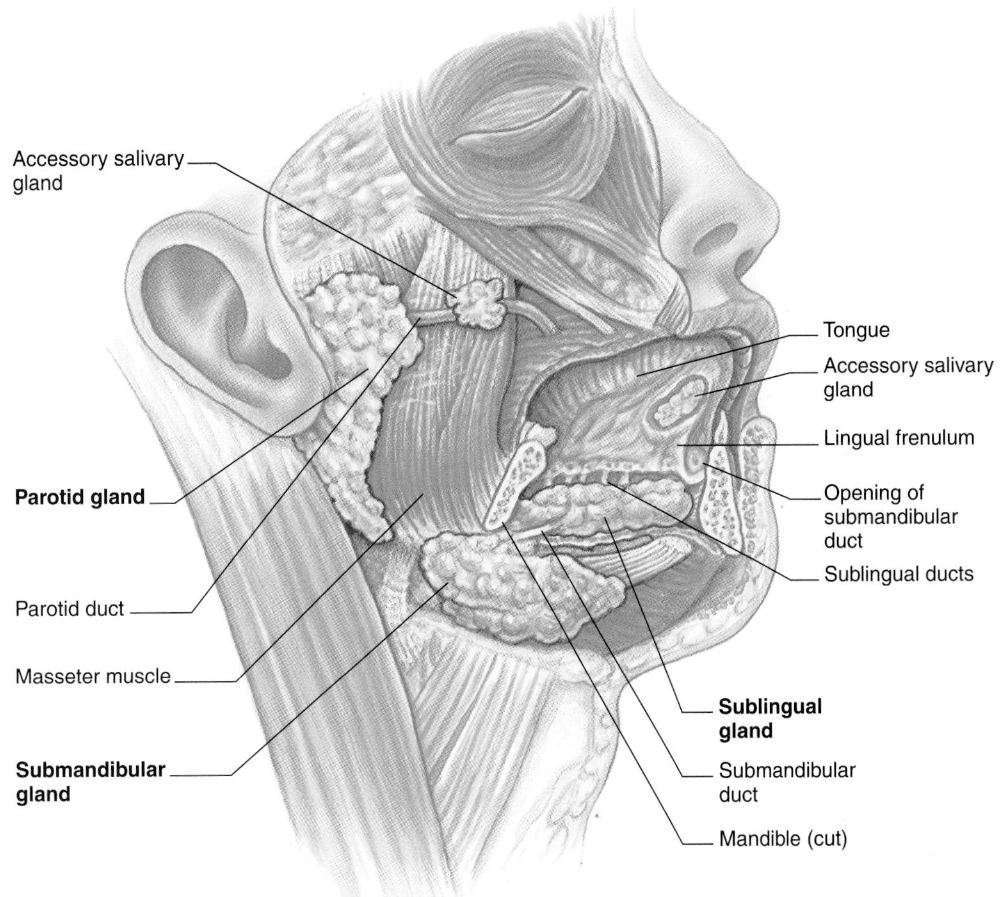

Table 26.1

Major Salivary Glands

Gland	Location	Entry into Oral Cavity	Type of Secretion
Parotid	Anterior and inferior to auricle; subcutaneous over masseter muscle	Lateral to upper second molar	Watery serous fluid, salts, and enzyme
Submandibular	Inferior to the base of the tongue	Papilla lateral to lingual frenulum	Watery serous fluid with some mucus
Sublingual	Anterior to submandibular gland; under the tongue	Ducts along the base of the tongue	Mostly thick, stringy mucus, salts, and enzyme

Figure 26.13

Posterior view of the constrictor muscles of the pharynx.
The right side has been cut away to illustrate the interior structures in the pharynx.

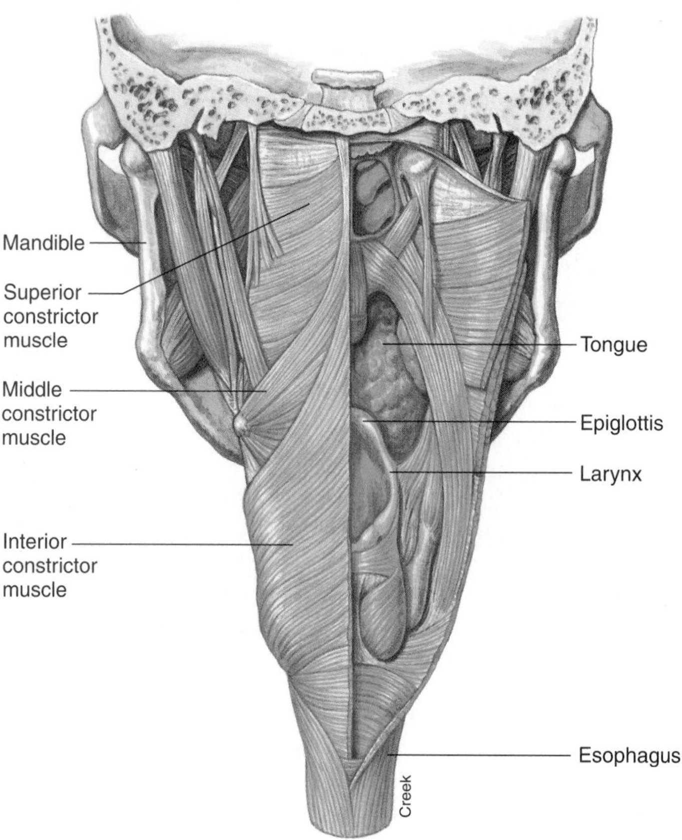

arises from the cartilages of the larynx and encircles the lower portion of the pharynx. During breathing, the lower portion of the inferior constrictor muscle is contracted, preventing air from entering the esophagus.

The three regions of the pharynx are described in detail in chapter 24 on the respiratory system.

Esophagus and Stomach

Swallowed food is passed through the esophagus to the stomach by wavelike contractions known as peristalsis. The mucosa of the stomach secretes hydrochloric acid and pepsinogen. Upon entering the lumen of the stomach, pepsinogen is converted into the active protein-digesting enzyme known as pepsin. The stomach partially digests proteins and functions to store its contents, called chyme, for later processing by the small intestine.

Esophagus

The **esophagus** (*ĕ-sof′ă-gus*) is that portion of the GI tract that connects the pharynx to the stomach. It is a collapsible muscular tube, approximately 25 cm (10 in.) long, located posterior to the trachea within the mediastinum of the thorax. The esophagus passes through the diaphragm by means of an opening called the **esophageal hiatus** (*ĕ-sof″ă-je′al hi-a′tus*) before terminating at the stomach. The esophagus is lined with a nonkeratinized stratified squamous epithelium; its walls contain either skeletal or smooth muscle, depending on the location. The upper third of the esophagus contains skeletal muscle, the middle third contains both skeletal and smooth muscle, and the terminal portion contains smooth muscle only.

Swallowing, or **deglutition** (*de″gloo-tish′un*), is the complex act of moving food or fluid from the oral cavity through the esophagus to the stomach. Swallowed food is pushed from one end of the esophagus to the other by a wavelike muscular contraction called **peristalsis** (*per″ĭ-stal′sis*) (fig. 26.14). Peristalsis is produced by a series of localized reflexes made in response to the distention of the GI tract walls by a bolus of food. Movement of the bolus along the GI tract occurs because the circular smooth muscle contracts behind, and relaxes in front of, the bolus. This is followed by shortening of the tube by longitudinal muscle contraction. These contractions progress from the superior end of the esophagus to the *gastroesophageal junction* at a rate of 2 to 4 cm per second as they empty the contents of the esophagus into the cardia of the stomach.

The lumen of the terminal portion of the esophagus is slightly narrowed because of a thickening of the circular muscle fibers in its wall. This portion is referred to as the **lower esophageal (gastroesophageal) sphincter.** The muscle fibers of this region constrict after food passes into the stomach to help prevent the stomach contents from regurgitating into the esophagus. Regurgitation would occur because the pressure in the abdominal cavity is greater than the pressure in the thoracic cavity as a result of respiratory movements. The lower esophageal sphincter must therefore remain closed until food is pushed through it by peristalsis into the stomach.

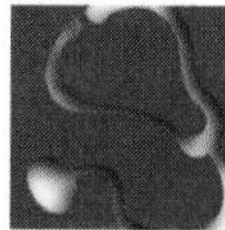

The lower esophageal sphincter is not a well-defined sphincter muscle compared to others located elsewhere along the GI tract, and it does at times permit the acidic contents of the stomach to enter the esophagus. This can create a burning sensation commonly called "heartburn," although the heart is not involved. In infants under a year of age, the lower esophageal sphincter may function erratically, causing them to "spit up" following meals. Certain mammals, such as rodents, have a true gastroesophageal sphincter and thus cannot regurgitate. This is why poison grains are effective in killing mice and rats.

esophagus: Gk. *oisein*, to carry; *phagema*, food

Figure 26.14

Peristalsis in the esophagus.
(*a*) A diagram and (*b*) a radiograph of peristalsis in the esophagus.

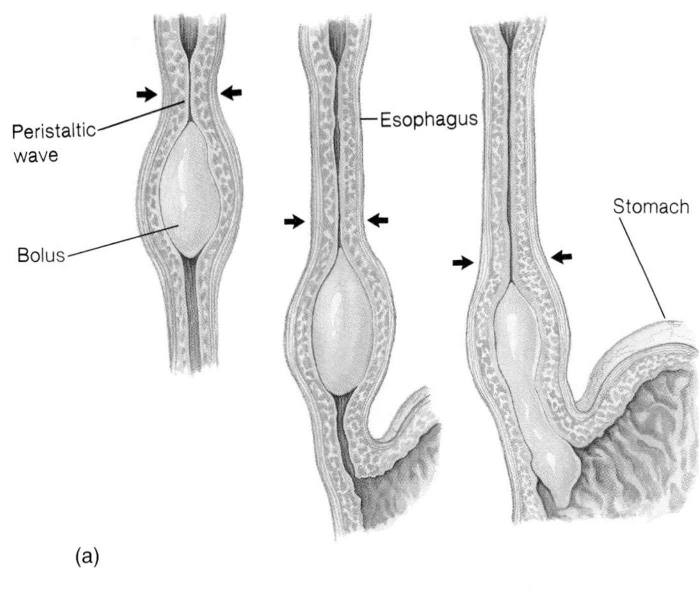

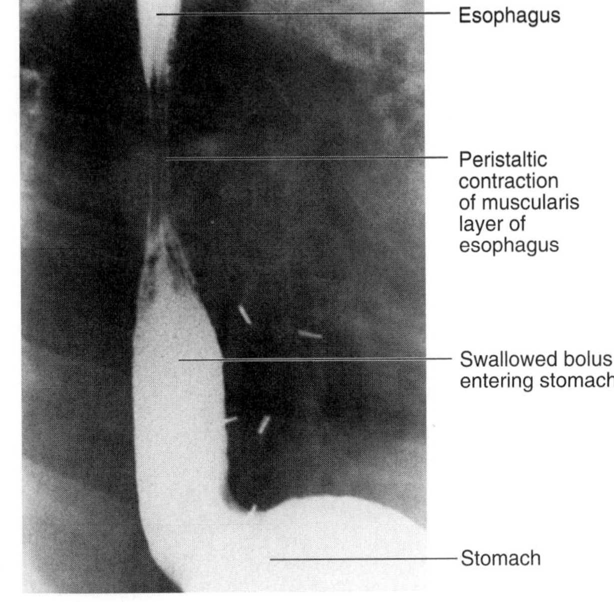

Stomach

The **stomach,** a J-shaped pouch in the left superior portion of the abdomen, is the most distensible part of the GI tract. It is continuous with the esophagus superiorly and empties into the duodenum of the small intestine inferiorly. The functions of the stomach are to store food, to initiate the digestion of proteins, and to move the food into the small intestine as a pasty material called **chyme** (*kīm*).

The stomach can be divided into four regions: the cardia, fundus, body, and pylorus (fig 26.15). The **cardia** is the narrow upper region immediately below the lower esophageal sphincter. The **fundus** is the dome-shaped portion to the left of the cardia and is in direct contact with the diaphragm. The **body** is the large central portion, and the **pylorus** is the funnel-shaped terminal portion. The pylorus begins in a somewhat widened area, the **antrum,** and ends at the **pyloric sphincter.** *Pylorus* is a Greek word meaning "gatekeeper," and this junction is just that, regulating the movement of chyme into the small intestine and prohibiting backflow.

The stomach has two surfaces and two borders. The broadly rounded surfaces are referred to as the *anterior* and *posterior surfaces*. The medial concave border is the **lesser curvature** (fig. 26.15), and the lateral convex border is the **greater curvature.** The lesser omentum extends between the lesser curvature and the liver; the greater omentum is attached to the greater curvature.

The wall of the stomach consists of the same four layers found in other regions of the GI tract, with certain modifications. The muscularis is composed of three layers of smooth muscle named according to the direction of fiber arrangement: an outer *longitudinal layer*, a middle *circular layer*, and an inner *oblique layer*. The circular muscle layer is further thickened at the gastroduodenal junction to form the pyloric sphincter.

The inner surface of the stomach is thrown into long folds called **gastric rugae,** which can be seen with the unaided eye. Microscopic examination of the gastric mucosa shows that it is likewise folded. The openings of these folds into the stomach lumen are called **gastric pits.** The cells that line the folds deeper in the mucosa secrete various products into the stomach; these form the exocrine **gastric glands** (figs. 26.16 and 26.17).

Gastric glands contain several types of cells that secrete different products: (1) **goblet cells,** which secrete *mucus*; (2) **parietal cells,** which secrete *hydrochloric acid (HCl)*; (3) **chief** (or **zymogenic**) **cells,** which secrete *pepsinogen*, an inactive form of the protein-digesting enzyme *pepsin*; (4) **enterochromaffin-like (ECL) cells,** which secrete *histamine*; (5) **G cells,** which secrete the hormone *gastrin*; and (6) **D cells,** which secrete the hormone *somatostatin*. In addition to these products (table 26.2), the gastric mucosa (probably the parietal cells) secretes a polypeptide called *intrinsic factor*, which is required for absorption of vitamin B_{12} in the small intestine.

The exocrine secretions of the gastric cells, together with a large amount of water (2 to 4 L/day), form a highly acidic solution known as **gastric juice.**

---

chyme: L. *chymus*, juice
fundus: L. *fundus*, bottom
pylorus: Gk. *pyloros*, gatekeeper

Figure 26.15

Primary regions and structures of the stomach.
Note the pylorus region includes the antrum (the wider portion of the pylorus) as well as the pyloric sphincter.

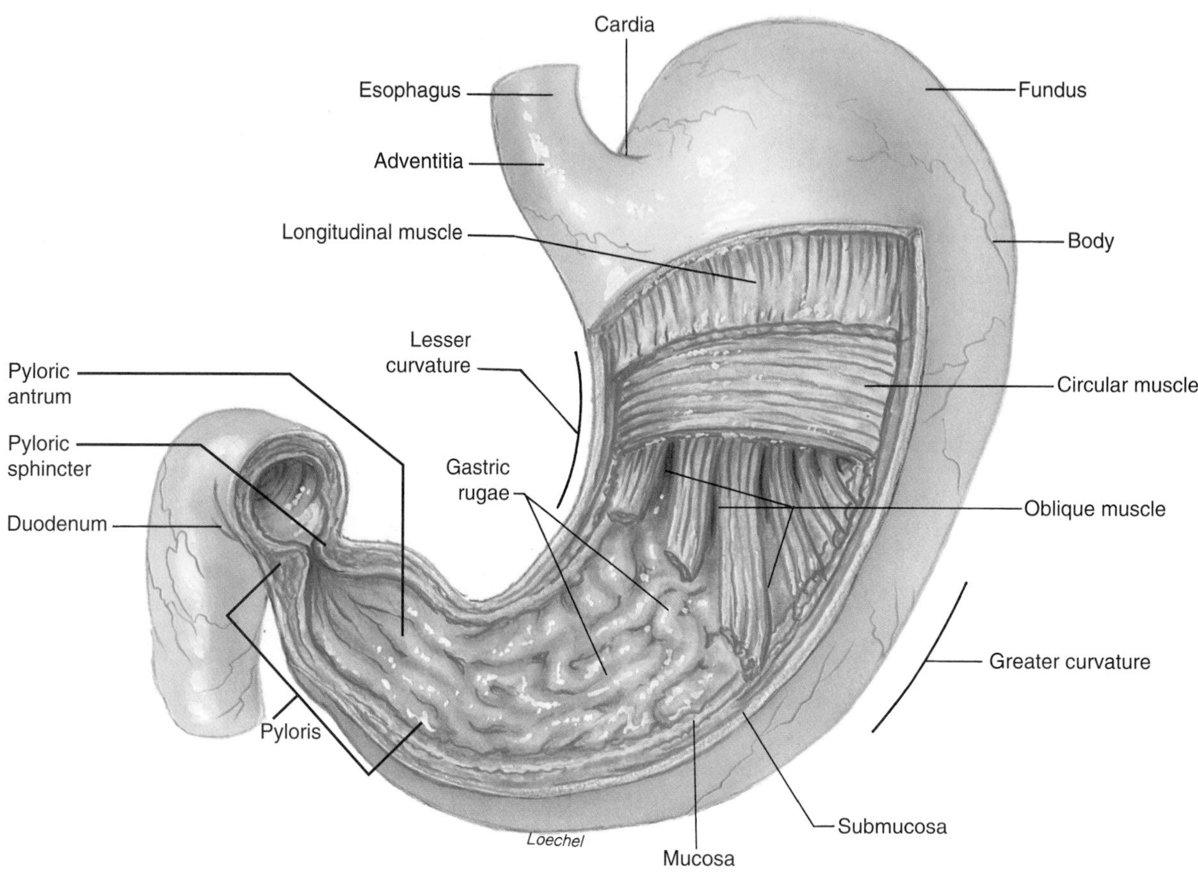

Figure 26.16

Microscopic view of the gastric mucosa.
The gastric glands secrete HCl and pepsinogen into the gastric pits.

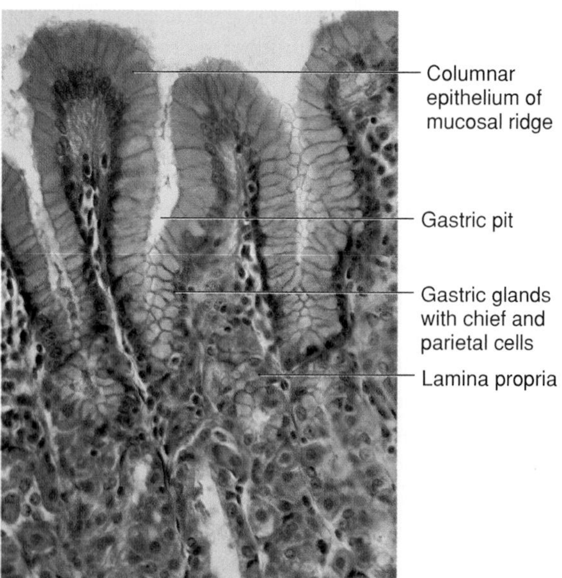

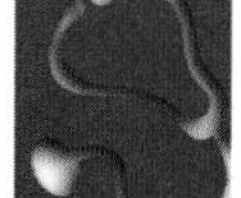

The only stomach function that appears to be essential for life is the secretion of *intrinsic factor.* This polypeptide is needed for the absorption of vitamin B_{12} in the terminal portion of the ileum in the small intestine, and vitamin B_{12} is required for maturation of red blood cells in the bone marrow. Following surgical removal of the stomach (gastrectomy), a patient has to receive B_{12} orally (together with intrinsic factor) or through injections to prevent the development of *pernicious anemia.*

Pepsin and Hydrochloric Acid Secretion

The parietal cells secrete H^+, at a pH as low as 0.8, into the gastric lumen by active transport against a million-to-one concentration gradient. This is accomplished by carriers in the parietal cell membrane that function as *H^+/K^+ ATPase pumps*, moving K^+ from the gastric lumen into the cell cytoplasm while an equivalent amount of H^+ is secreted into the gastric juice.

The secretion of HCl by the parietal cells is stimulated by a variety of factors, including the hormone gastrin, secreted by the G cells, and acetylcholine (ACh), released by axons of the vagus nerves. Most of the effects of gastrin and

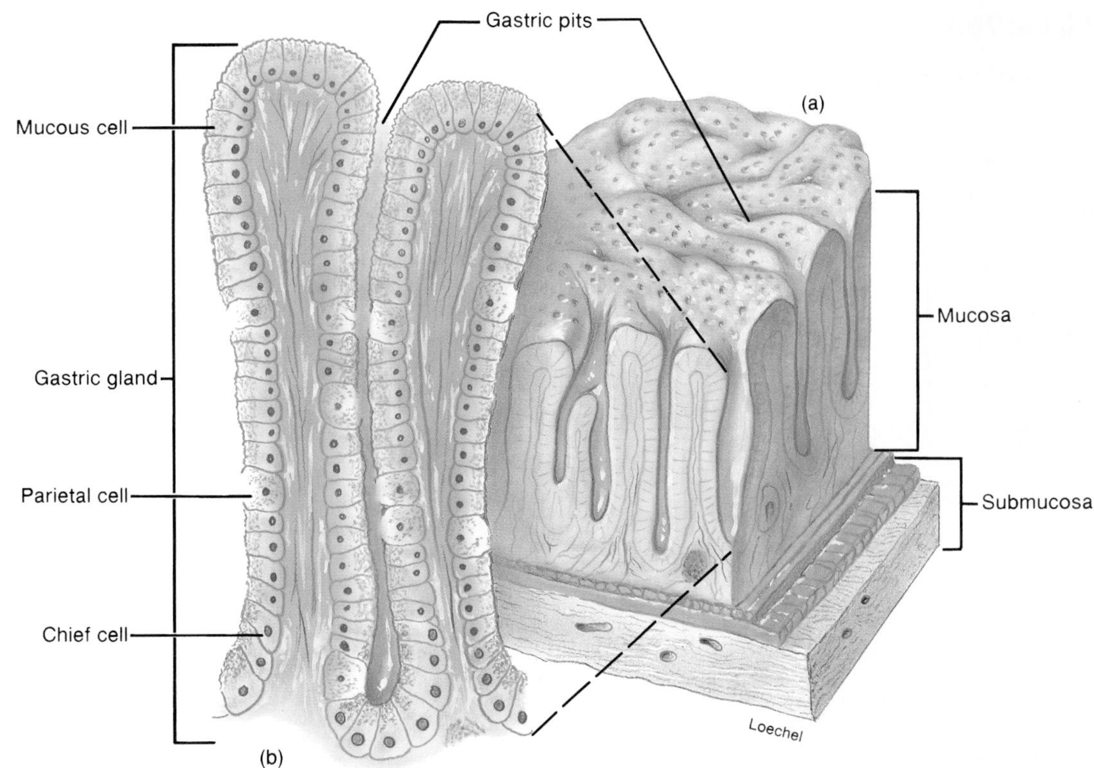

Figure 26.17

Gastric pits and gastric glands of the mucosa. (*a*) Gastric pits are the openings of the gastric glands. (*b*) Gastric glands consist of three types of cells—mucous cells, chief cells, and parietal cells—each of which produces a specific secretion.

Table 26.2

Secretions of the Fundus and Pylorus of the Stomach

Stomach Region	Cell Type	Secretions
Fundus	Parietal cells	Hydrochloric acid; intrinsic factor
	Chief cells	Pepsinogen
	Goblet cells	Mucus
	Enterochromaffin-like (ECL) cells	Histamine, serotonin
	D cells	Somatostatin
Pyloric antrum	G cells	Gastrin
	Chief cells	Pepsinogen
	Goblet cells	Mucus
	D cells	Somatostatin

ACh on acid secretion, however, are currently believed to be indirect. Gastrin and ACh from vagal axons stimulates the release of histamine from the ECL cells of the gastric mucosa, and histamine, in turn, acts as a paracrine regulator to stimulate the parietal cells to secrete HCl (see fig. 26.39). The endocrine regulation of the digestive system is discussed in detail later in this chapter.

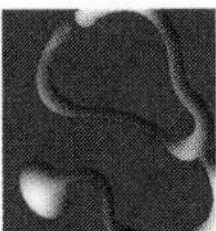

People with *gastroesophageal reflux disease,* a common disorder involving the reflux of acidic gastric juice into the esophagus, are often treated with specific drugs (e.g. *omeprazole*) that inhibit the K^+/H^+ pumps in the gastric mucosa. Since gastric acid secretion is stimulated by histamine released from the ECL cells, people with *peptic ulcers* can be treated with drugs that block histamine action. Drugs in this category, such as *Tagamet, Zantac,* and *Pepsid AC,* specifically block the H_2 histamine receptors in the gastric mucosa. This is a different receptor subtype than the one blocked by antihistamines commonly used to treat allergy symptoms. Peptic ulcers and their treatment are described in more detail in the section "Clinical Considerations" near the end of this chapter.

The high concentration of HCl from the parietal cells makes gastric juice very acidic, with a pH of less than 2. This strong acidity serves three functions: (1) ingested proteins are denatured at low pH—that is, their tertiary structure (chapter 2) is altered so that they become more digestible; (2) under acidic conditions, weak pepsinogen enzymes partially digest each other—this frees the active pepsin enzyme as small peptide fragments are removed (fig. 26.18); and (3) pepsin is more active under acidic conditions—it has a pH optimum (chapter 4) of about 2.0. The peptide bonds of ingested protein are broken (through hydrolysis reactions) by pepsin under acidic conditions; the HCl itself does not directly digest proteins.

Digestion and Absorption in the Stomach

Proteins are only partially digested in the stomach by the action of pepsin, while carbohydrates and fats are not digested at

Figure 26.18

Activation of pepsin.
The gastric mucosa secretes the inactive enzyme pepsinogen and hydrochloric acid (HCl). In the presence of HCl, the active enzyme pepsin is produced. Pepsin digests proteins into shorter polypeptides.

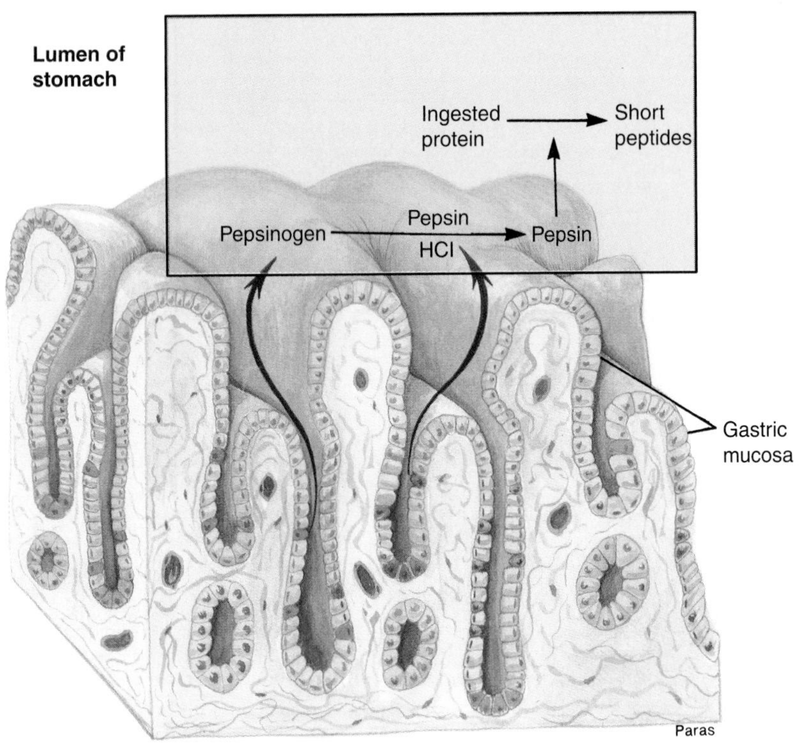

The **small intestine** is that portion of the GI tract between the pyloric sphincter of the stomach and the ileocecal valve, which opens into the large intestine. It is positioned in the central and lower portions of the abdominal cavity and is supported, except for the first portion, by mesentery (fig. 26.19). The fan-shaped attachment of mesentery to the small intestine permits movement but leaves little chance for the intestine to become twisted or kinked. Enclosed within the mesentery are blood vessels, nerves, and lymphatic vessels that supply the intestinal wall.

The small intestine is approximately 3 m (12 ft) long and 2.5 cm (1 in.) wide in a living person, but it will measure nearly twice this length in a cadaver when the muscle wall is relaxed. It is called the "small" intestine because of its relatively small diameter compared to that of the large intestine. The small intestine serves as the major site of digestion and absorption in the GI tract.

The small intestine is innervated by the superior mesenteric plexus. The branches of this plexus contain sensory fibers, postganglionic sympathetic fibers, and preganglionic parasympathetic fibers. The small intestine's arterial blood supply comes through the superior mesenteric artery and small branches from the celiac trunk and the inferior mesenteric artery. Venous drainage is through the superior mesenteric vein. This vein unites with the splenic vein to form the hepatic portal vein, which carries nutrient-rich blood to the liver.

all by pepsin. (Digestion of starch begins in the mouth with the action of salivary amylase and continues for a time when the food enters the stomach, but amylase soon becomes inactivated by the strong acidity of gastric juice.) The complete digestion of food molecules occurs later, when chyme enters the small intestine. Therefore, people who have had partial gastric resections—and even those who have had complete gastrectomies—can still adequately digest and absorb their food.

Almost all of the products of digestion are absorbed through the intestinal wall; the only commonly ingested substances that can be absorbed across the stomach wall are alcohol and aspirin. Absorption occurs as a result of the lipid solubility of these molecules. The passage of aspirin through the gastric mucosa has been shown to cause bleeding, which may be significant if aspirin is taken in large doses.

Small Intestine

The mucosa of the small intestine is folded into villi that project into the lumen. In addition, the cells that line these villi have foldings of their cell membrane called microvilli. This arrangement greatly increases the surface area for absorption. It also improves digestion, since the digestive enzymes of the small intestine are embedded within the cell membrane of the microvilli.

Regions of the Small Intestine

The small intestine is divided into three regions on the basis of function and histological structure. These three regions are the duodenum, jejunum, and ileum.

The **duodenum** (*doo"ŏ-de'num* or *doo-od'ĕ-num*) is a relatively fixed C-shaped tube measuring approximately 25 cm (10 in.) from the pyloric sphincter of the stomach to the **duodenojejunal** (*doo-od"ĕ-no"jĕ-joo'nal*) **flexure.** The concave surface of the duodenum faces to the left, where it receives bile secretions from the liver and gallbladder through the **common bile duct** and pancreatic secretions through the **pancreatic duct** of the pancreas (fig. 26.20). These two ducts unite to form a common entry into the duodenum called the **hepatopancreatic ampulla** (*hep"ă-to-pan"kre-at'ik am-pool'ă*) (*ampulla of Vater*), which pierces the duodenal wall and drains into the duodenum from an elevation called the **duodenal papilla.** It is here that bile and pancreatic juice enter the small intestine. The papilla can be opened or closed by the action of the **sphincter of ampulla** (*sphincter of Oddi*).

---

duodenum: L. *duodeni*, 12 each (distance of 12 fingers' breadth)
ampulla of Vater: from Abraham Vater, German anatomist, 1684–1751
sphincter of Oddi: from Ruggero Oddi, Italian physician, 1864–1913

Figure 26.19

The small intestine.

(*a*) The regions of the small intestine. (*b*) A section of the intestinal wall showing the tissue layers, plicae circulares, and villi.

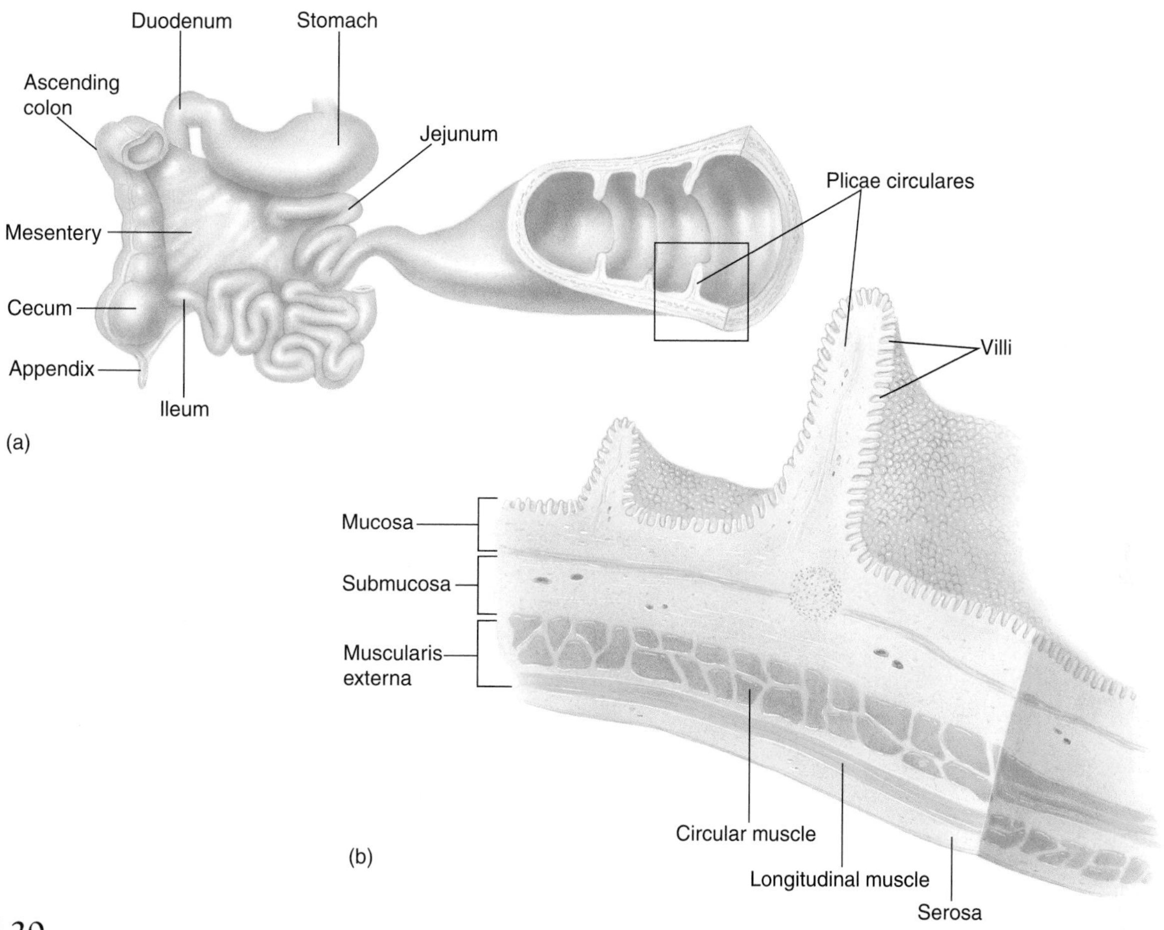

Figure 26.20

The duodenum and associated structures.

The pancreatic duct and the common bile duct merge to empty into the duodenum.

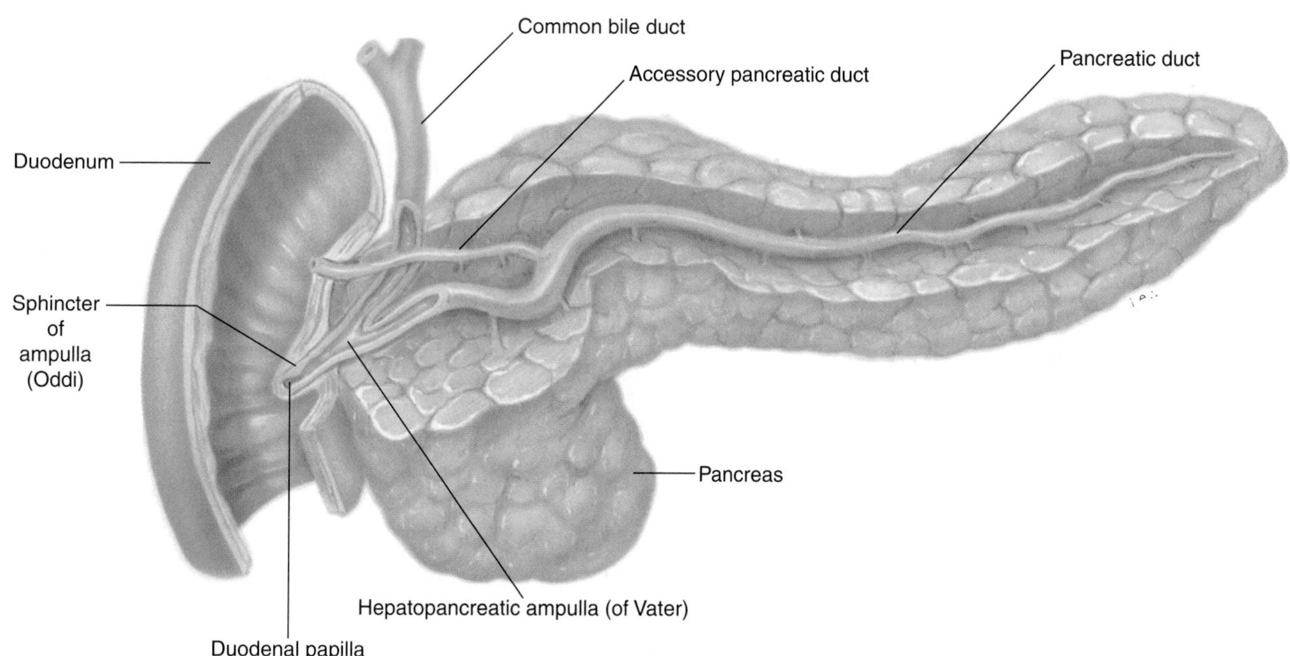

Figure 26.21

The histology of the duodenum.
Note the duodenal (Brunner's) glands, which extend into the submucosa and are unique to the duodenum. These glands produce a bicarbonate-rich, alkaline secretion.

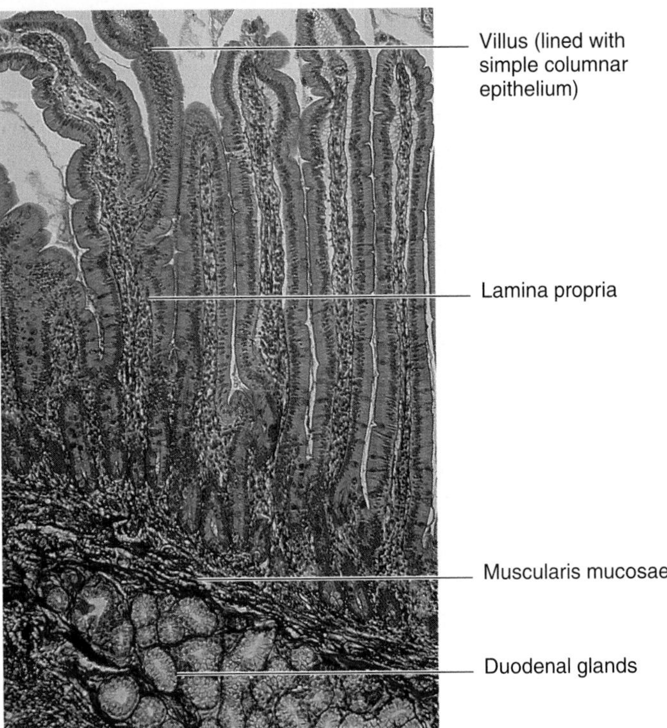

Figure 26.22

The structure of an intestinal villus.
The figure also depicts an intestinal crypt (of Lieberkühn), in which new epithelial cells are produced by mitosis.

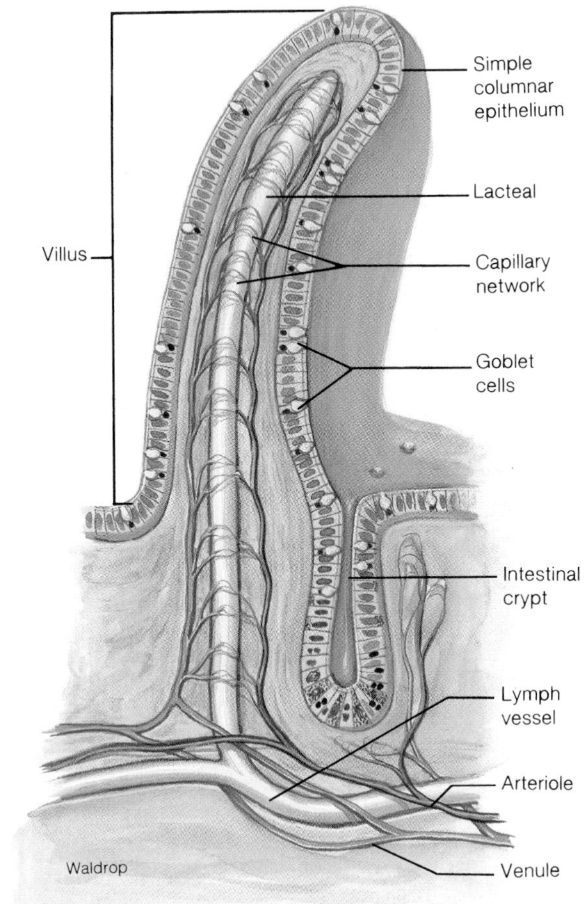

The duodenum is retroperitoneal, except for a short portion near the stomach. It differs histologically from the rest of the small intestine by the presence of **duodenal** (*Brunner's*) **glands** in the submucosa (fig. 26.21). These glands secrete alkaline mucus and are most numerous near the superior end of the duodenum, where their alkaline secretions help to protect the mucosa from damage by the acidic chyme.

The **jejunum** (*jĕ-joo'num*) is approximately 1 m (3 ft) long and extends from the duodenum to the ileum. The lumen of the jejunum is slightly larger than that of the ileum, and the jejunum has more internal folds. The two regions are similar in their histological structure.

The **ileum** (*il'e-um*) makes up the remaining 2 m (6–7 ft) of the small intestine. The terminal portion of the ileum empties into the medial side of the cecum through the **ileocecal** (*il''e-o-se'kal*) **valve.** Aggregations of lymphoid tissue called **Peyer's patches** (chapter 23) are abundant in the walls of the ileum.

Brunner's glands: from Johann C. Brunner, Swiss anatomist, 1653–1727
jejunum: L. *jejunus*, fasting (in dissection, it was always found empty)
Peyer's patches: from Johann K. Peyer, Swiss anatomist, 1653–1712

Structural Modifications for Absorption

The products of digestion are absorbed across the epithelial lining of the intestinal mucosa. Absorption of carbohydrates, lipids, protein, calcium, and iron occurs primarily in the duodenum and jejunum. Bile salts, vitamin B_{12}, water, and electrolytes are absorbed primarily in the ileum. The many folds in the small intestine and modifications in the structure of its walls provide a large mucosal surface area; therefore, absorption occurs at a rapid rate. The mucosa and submucosa form large folds called **plicae** (*pli'se*) **circulares,** which can be observed with the unaided eye. The surface area is further increased by microscopic folds of mucosa called *villi* (*vil'i*) and by foldings of the apical cell membrane of epithelial cells called *microvilli*.

plica: L. *plicatus*, folded
villus: L. *villosus*, shaggy

Figure 26.23

Electron micrographs of microvilli.
Microvilli are evident at the apical surface of the columnar epithelial cells in the small intestine, seen here at (*a*) lower magnification and (*b*) higher magnification. Microvilli increase the surface area for absorption and also have the brush border digestive enzymes embedded in their membranes.

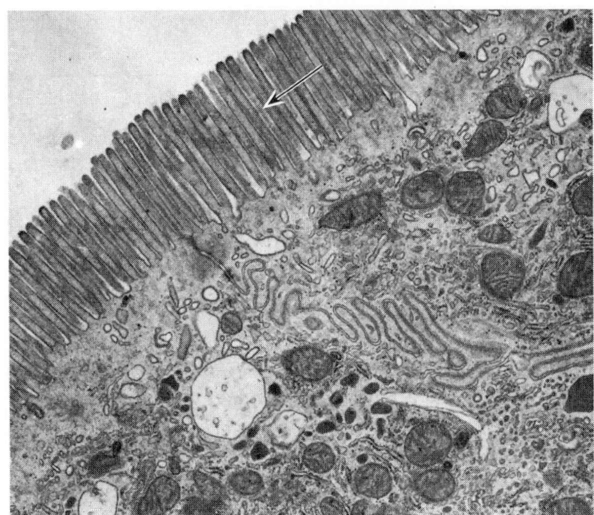
(a)

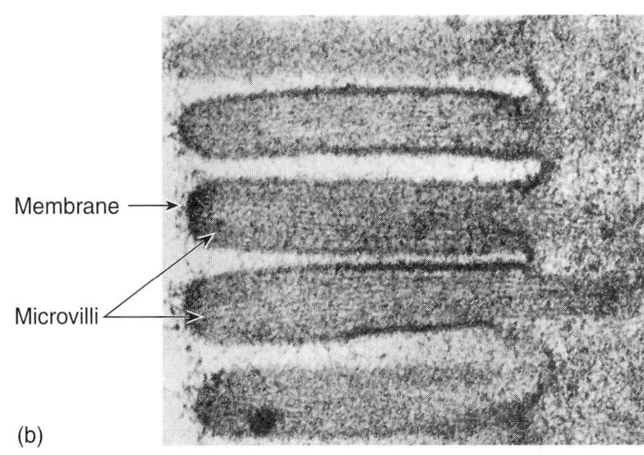

(b)

Each **villus** is a fingerlike fold of mucosa that projects into the lumen of the small intestine (fig. 26.21). The villi are covered with columnar epithelial cells, and interspersed among these are the mucus-secreting goblet cells. The lamina propria forms the connective core of each villus and contains numerous lymphocytes, blood capillaries, and a lymphatic vessel called the **lacteal** (*lak'te-al*) (fig. 26.22). Absorbed monosaccharides and amino acids enter the blood capillaries; absorbed fat enters the lacteals.

Epithelial cells at the tips of the villi are continuously exfoliated and replaced by cells that are pushed up from the bases of the villi. The epithelium at the base of each villus invaginates downward at various points to form narrow pouches that open through pores to the intestinal lumen. These structures are called the **intestinal crypts** (*crypts of Lieberkühn*) (fig. 26.22).

Microvilli, which can be seen clearly only in an electron microscope, are fingerlike projections formed by foldings of the cell membrane. In a light microscope, the microvilli display a somewhat vague **brush border** on the edges of the columnar epithelium. The terms *brush border* and *microvilli* are often used interchangeably in describing the small intestine (fig. 26.23).

Intestinal Enzymes

In addition to providing a large surface area for absorption, the cell membranes of the microvilli contain digestive enzymes that hydrolyze disaccharides, polypeptides, and other substrates (table 26.3). These **brush border enzymes** are not secreted into the lumen, but instead remain attached to the cell membrane with their active sites exposed to the chyme. One brush border enzyme, *enterokinase* (*en''tĕ-ro-ki'nās*), is required for activation of the protein-digesting enzyme *trypsin*, which enters the small intestine in pancreatic juice.

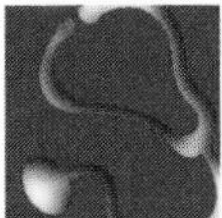
The ability to digest milk sugar, or lactose, depends on the presence of a brush border enzyme called *lactase.* This enzyme is present in all children under the age of 4 but becomes inactive to some degree in most adults (people of Asiatic or African heritage are more often lactase deficient than Caucasians). A deficiency of lactase can result in *lactose intolerance,* a condition in which too much undigested lactose in the small intestine causes diarrhea, gas, cramps, and other unpleasant symptoms. Yogurt is better tolerated than milk because it contains lactase (produced by the yogurt bacteria), which becomes activated in the duodenum and digests lactose.

Intestinal Contractions and Motility

Two major types of contractions occur in the small intestine: *peristalsis* and *segmentation*. Peristalsis is much weaker in the small intestine than in the esophagus and stomach. Intestinal motility—the movement of chyme through the small intestine—is relatively slow and occurs mainly because the pressure at the duodenal portion of the small intestine is greater than at the terminal end.

The major contractile activity of the small intestine is **segmentation.** This term refers to muscular constrictions of the lumen, which occur simultaneously at different intestinal segments (fig. 26.24). This action serves to mix the chyme more thoroughly.

crypts of Lieberkühn: from Johann N. Lieberkühn, German anatomist, 1711–56

Table 26.3

Brush Border Enzymes Attached to the Cell Membrane of Microvilli in the Small Intestine

Category	Enzyme	Comments
Disaccharidase	Sucrase	Digests sucrose to glucose and fructose; deficiency produces GI disturbances
	Maltase	Digests maltose to glucose
	Lactase	Digests lactose to glucose and galactose; deficiency produces GI disturbances (lactose intolerance)
Peptidase	Aminopeptidase	Produces free amino acids, dipeptides, and tripeptides
	Enterokinase	Activates trypsin (and indirectly other pancreatic juice enzymes); deficiency results in protein malnutrition
Phosphatase	Ca^{2+}, Mg^{2+}—ATPase	Needed for absorption of dietary calcium; enzyme activity regulated by vitamin D
	Alkaline phosphatase	Removes phosphate groups from organic molecules; enzyme activity may be regulated by vitamin D

Figure 26.24

Segmentation of the small intestine.
Simultaneous contraction of numerous segments of the small intestine help to mix the chyme with digestive enzymes and mucus.

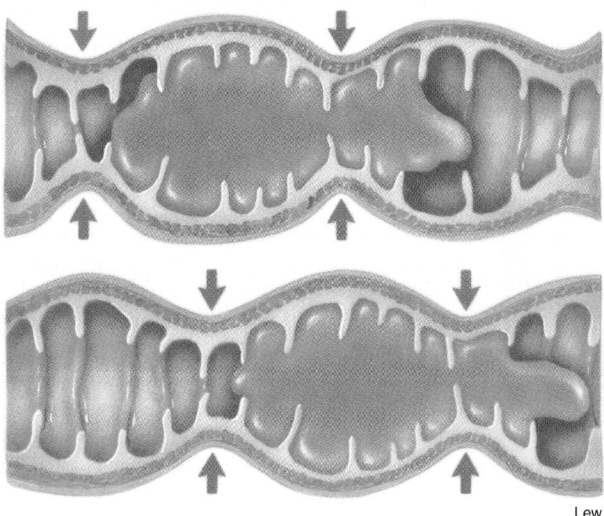

Figure 26.25

Electrical activity in intestinal smooth muscle.
The smooth muscle of the gastrointestinal tract produces and conducts spontaneous pacesetter (slow) potentials. As these potential changes reach a threshold level of depolarization, they stimulate the production of action potentials, which in turn stimulate smooth muscle contraction.

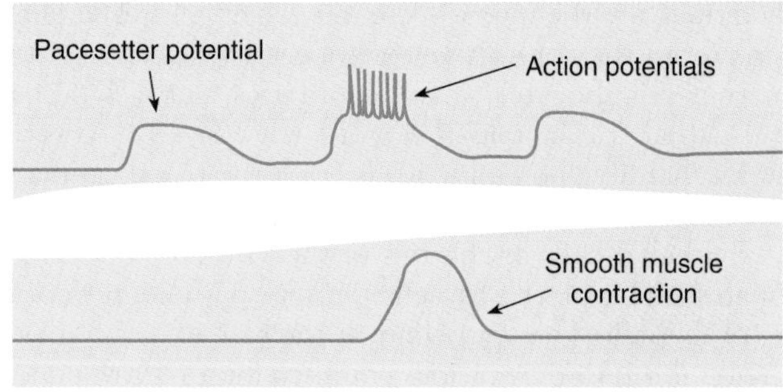

Like cardiac muscle, intestinal smooth muscle is capable of spontaneous electrical activity and automatic rhythmic contractions. Spontaneous depolarizations begin in pacemaker smooth muscle cells at the boundary of the circular muscle layer and spread through both the circular and longitudinal muscle layers across *nexuses*. The term *nexus* is used here to indicate an electrical synapse between smooth muscle cells. The spontaneous depolarizations, called **pacesetter potentials** or **slow waves,** decrease in amplitude as they are conducted from one muscle cell to another, much like excitatory postsynaptic potentials (EPSPs). The pacesetter potentials can stimulate the production of action potentials when they reach a plateau level of depolarization in the smooth muscle cells through which they are conducted (fig. 26.25).

The nexuses conduct the pacesetter potentials, not the action potentials. Action potentials are therefore limited to those smooth muscle cells that are depolarized to threshold by the spreading pacesetter potentials. These action potentials stimulate smooth muscle contraction in only limited regions of the small intestine, producing the localized contractions of segmentation. Although this activity is automatic in nature, the excitability of the intestinal smooth muscle cells is increased by parasympathetic stimulation and decreased by sympathetic nerves.

Intestinal peristalsis results from contraction of smooth muscles above and relaxation of smooth muscles below the chyme. These contractions and relaxations are local reflexes coordinated by neurons within the intestinal wall itself. Indeed, the neurons within the small intestine form a complex *enteric* (intestinal) *nervous system* (see fig. 26.40).

nexus: L. *nexus*, interconnection

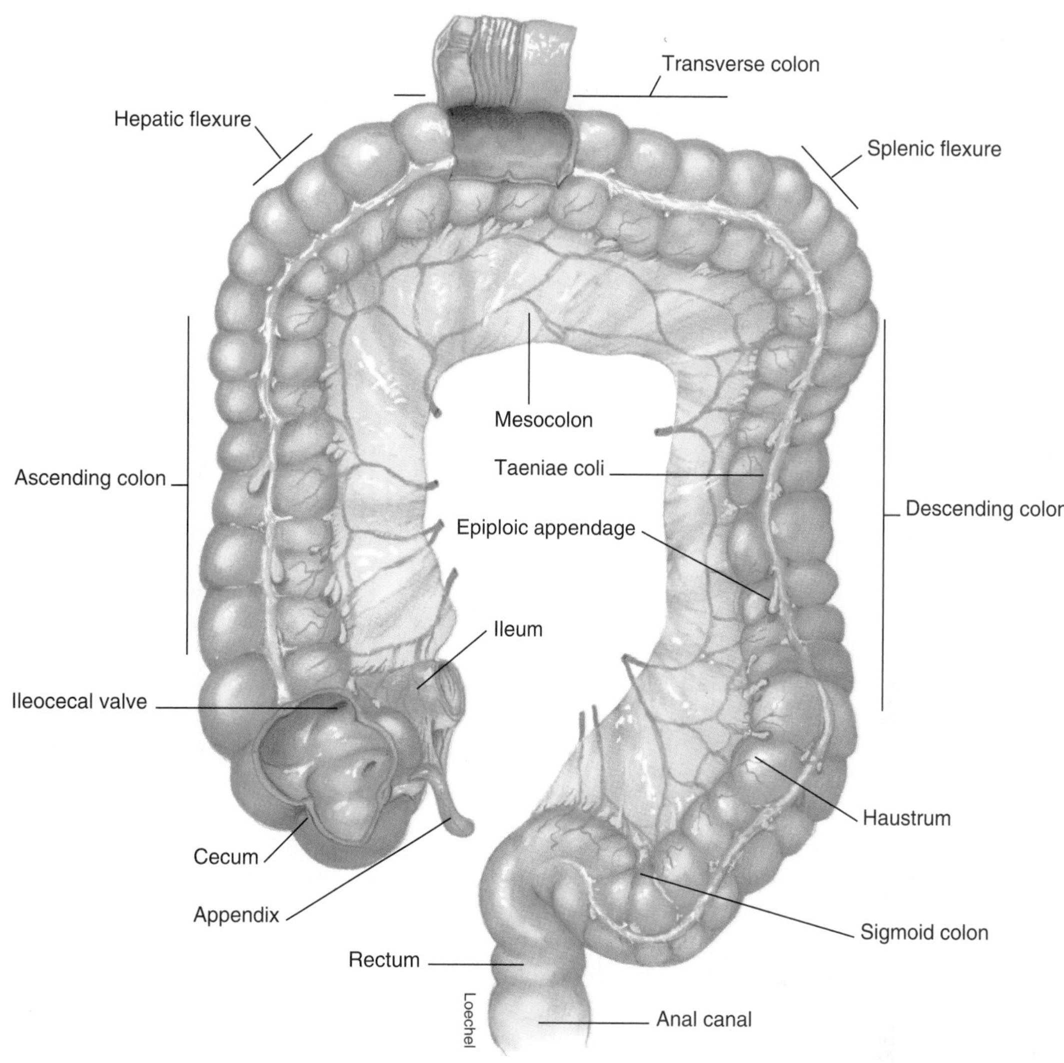

Figure 26.26

The large intestine. The regions of the large intestine are illustrated.

Large Intestine

The large intestine absorbs water , electrolytes, and certain vitamins from the chyme it receives from the small intestine. In a process regulated by the action of sphincter muscles, the large intestine then passes undigested waste products out of the GI tract through the rectum and anal canal.

The **large intestine** is about 1.5 m (5 ft) long and 6.5 cm (2.5 in.) in diameter. It is named the "large" intestine because its diameter is considerably larger than that of the small intestine. The large intestine begins at the terminal end of the ileum in the lower right quadrant of the abdominal cavity. From there, it leads superiorly on the right side to just below the liver; it then crosses to the left, descends into the pelvic region, and terminates at the anus. The **mesocolon,** a specialized portion of the mesentery, supports the large intestine along the posterior abdominal wall.

The large intestine has little or no digestive function, but it does absorb water and electrolytes from the remaining chyme, as well as several B complex vitamins and vitamin K. Resident bacteria in the large intestine also produce significant amounts of vitamin K and folic acid (see chapter 27), which are absorbed in the large intestine. In addition, the large intestine forms and stores *feces* and expels the feces from the GI tract.

Regions and Structures of the Large Intestine

The large intestine is structurally divided into the cecum, colon, rectum, and anal canal (fig. 26.26). The **cecum** (*se'kum*) is a dilated pouch hanging just below the *ileocecal valve*—a fold of mucous membrane at the junction of the small and large intestine. The ileocecal valve prohibits the backflow of undigested material into the ileum. The **appendix,** a fingerlike projection about 8 cm (3 in.) long, is attached to the inferior medial margin of the cecum. The appendix contains masses of lymphoid tissue that may serve to resist infection.

cecum: L. *caecum*, blind pouch
appendix: L. *appendix*, attachment

Figure 26.27

The anal canal.
The internal anal sphincter is composed of smooth muscle, whereas the external anal sphincter is composed of skeletal muscle.

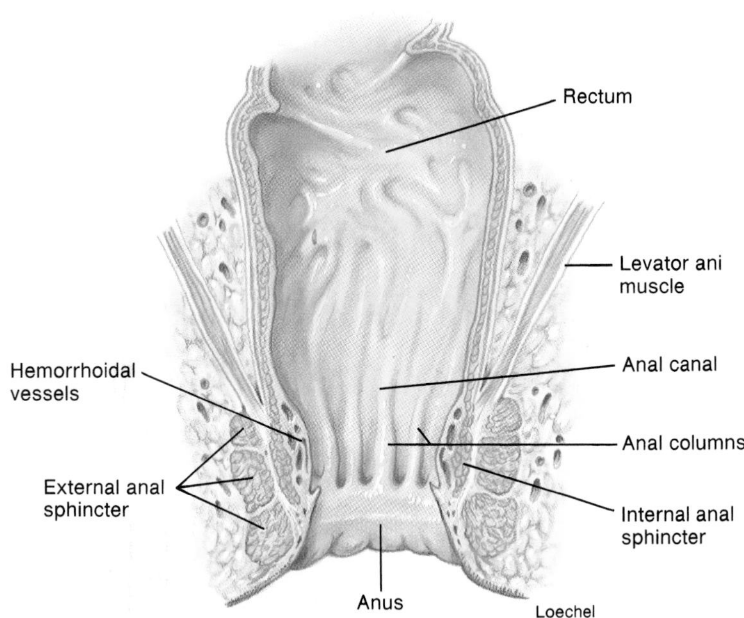

Figure 26.28

Radiograph of the large intestine.
The large intestine is seen after taking a barium enema; the haustra are clearly visible.

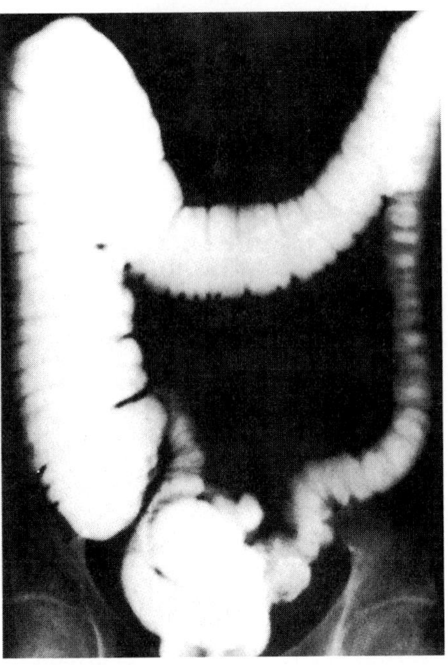

The superior portion of the cecum is continuous with the **colon,** which consists of ascending, transverse, descending, and sigmoid portions (fig. 26.26). The **ascending colon** extends superiorly from the cecum along the right abdominal wall to the inferior surface of the liver. Here the colon bends sharply to the left at the *hepatic flexure* and transversely crosses the upper abdominal cavity as the **transverse colon.** At the left abdominal wall, another right-angle bend called the *splenic (splen'ik) flexure* marks the beginning of the **descending colon.** The descending colon traverses inferiorly along the left abdominal wall to the pelvic region. The colon then angles medially from the brim of the pelvis to form an S-shaped bend known as the **sigmoid colon.**

The terminal 20 cm (7.5 in.) of the GI tract is the **rectum,** and the last 2 to 3 cm of the rectum is referred to as the **anal canal** (fig. 26.27). The **anus** is the external opening of the anal canal. Two sphincter muscles guard the anal opening: the **internal anal sphincter,** composed of smooth muscle fibers, and the **external anal sphincter,** composed of skeletal muscle. The mucous membrane of the anal canal is arranged into highly vascular longitudinal folds called **anal columns.** Masses of varicose veins in the anal area are known as *hemorrhoids (hem'ŏ-roidz)*; in reference to the condition in which they occur, they are also called *piles*.

As in the small intestine, the mucosa of the large intestine contains many scattered lymphocytes and lymphatic nodules and is covered by columnar epithelial cells and mucus-secreting goblet cells. Although this epithelium does form intestinal crypts, there are no villi in the large intestine—the intestinal mucosa therefore appears flat. The outer surface of the colon bulges outward to form pouches, or **haustra** (*haws'tră*) (figs. 26.26 and 26.28). The longitudinal muscle layer of the muscularis externa forms three distinct muscle bands, called **taeniae coli** (*te'ne-e ko'li*), which run the length of the large intestine. Numerous fat-filled pouches called **epiploic** (*ep-ĭ-plo'ik*) **appendages** (see fig. 26.26) are attached superficially to the taeniae coli in the serous layer.

The sympathetic innervation of the large intestine arises from the superior and inferior mesenteric plexuses, as well as from the celiac plexus. The parasympathetic innervation arises from the splanchnic and vagus nerves. Blood is supplied to the large intestine by branches of the superior and inferior mesenteric arteries. Venous blood is returned through the superior and inferior mesenteric veins, which in turn drain into the hepatic portal vein that enters the liver.

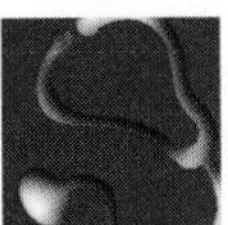

A common disorder of the large intestine is inflammation of the appendix, or *appendicitis.* Wastes that accumulate in the appendix cannot be moved easily by peristalsis, since the appendix has only one opening. Typical symptoms of appendicitis include muscular rigidity, localized pain in the lower right quadrant, and low-grade fever. Vomiting may or may not occur. The chief danger of appendicitis is that the appendix might rupture, resulting in peritonitis.

anus: L. *anus*, ring

haustrum: L. *haustrum*, bucket or scoop
epiploic: Gk. *epiplein*, to float on

Fluid and Electrolyte Absorption in the Large Intestine

Most of the fluid and electrolytes in the lumen of the GI tract are absorbed by the small intestine. Although a person may drink only about 1.5 L of water per day, the small intestine receives 7 to 9 L per day as a result of the fluid secreted into the GI tract by the salivary glands, stomach, pancreas, liver, and gallbladder. The small intestine absorbs most of this fluid and passes 1.5 to 2.0 L of fluid per day to the large intestine. The large intestine absorbs about 90% of this remaining volume, leaving less than 200 ml of fluid to be excreted in the feces.

Absorption of water in the intestine occurs passively as a result of the osmotic gradient created by the active transport of ions. The epithelial cells of the intestinal mucosa are joined together much like those of the renal tubules and, like the renal tubules, contain Na^+/K^+ pumps in the basolateral membrane. The analogy with renal tubules is emphasized by the observation that aldosterone, which stimulates salt and water reabsorption in the renal tubules, also appears to stimulate salt and water absorption in the ileum.

The handling of salt and water transport in the large intestine is made more complex by the fact that the large intestine can secrete, as well as absorb, water. The secretion of water by the mucosa of the large intestine occurs by osmosis as a result of the active transport of Na^+ or Cl^- out of the epithelial cells into the intestinal lumen. Secretion in this way is normally minor compared to the far greater amount of salt and water absorption, but this balance may be altered in some disease states.

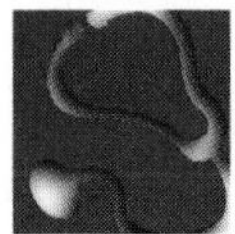

Diarrhea is characterized by excessive fluid excretion in the feces. Three different mechanisms, illustrated by three different diseases, can cause diarrhea. In *cholera,* severe diarrhea and dehydration result from *enterotoxin,* a chemical produced by the infecting bacteria. Release of enterotoxin stimulates active NaCl transport into the intestinal lumen, followed by the osmotic movement of water. In *celiac sprue* (*sproo*), diarrhea is caused by damage to the intestinal mucosa produced in susceptible people by eating foods that contain gluten (proteins from grains such as wheat). In *lactose intolerance,* diarrhea is produced by the increased osmolarity of the contents of the intestinal lumen as a result of the presence of undigested lactose.

Mechanical Activities of the Large Intestine

About 15 ml of pasty material enters the cecum with each rhythmic opening of the ileocecal valve. The ingestion of food intensifies peristalsis of the ileum and increases the frequency with which the ileocecal valve opens; this is called the **gastroileal reflex.** Undigested material entering the large intestine accumulates in the cecum and ascending colon.

Three types of movements occur throughout the large intestine: peristalsis, haustral churning, and mass movement. *Peristaltic movements* of the colon are similar to those of the small intestine, although they are usually more sluggish. In **haustral churning,** a relaxed haustrum is filled until it reaches a certain point of distension, and then the muscularis layer is stimulated to contract. Not only does this contraction move the material to the next haustrum, it also churns the material and exposes it to the mucosa, where water and electrolytes are absorbed. **Mass movement** is a very strong peristaltic wave, involving the action of the taeniae coli, that moves the colonic contents toward the rectum. Mass movements generally occur only two or three times a day, usually during or shortly after a meal. This response to eating, called the **gastrocolic reflex,** is best observed in infants who have a bowel movement during or shortly after feeding.

After electrolytes and water have been absorbed, the waste material that is left passes to the rectum, leading to an increase in rectal pressure and the urge to defecate. If the urge to defecate is denied, the feces are prevented from entering the anal canal by the internal anal sphincter. In this case, the feces remain in the rectum and may even back up into the sigmoid colon. The **defecation reflex** normally occurs when the rectal pressure rises to a particular level, determined largely by habit. At this point, the internal anal sphincter relaxes to admit the feces into the anal canal.

During the act of defecation, the longitudinal rectal muscles contract to increase rectal pressure, and the internal and external anal sphincter muscles relax. Excretion is aided by contractions of abdominal and pelvic skeletal muscles, which raise the intra-abdominal pressure (this is part of Valsalva's maneuver, described in chapter 22). The raised pressure helps to push the feces through the anal canal and out of the anus.

Liver, Gallbladder, and Pancreas

In addition to regulating the chemical composition of the blood in numerous ways, the liver produces and secretes bile, which is stored and concentrated in the gallbladder prior to its discharge into the duodenum. The pancreas produces pancreatic juice, an exocrine secretion containing bicarbonate and important digestive enzymes, which is passed into the duodenum via the pancreatic duct.

Structure of the Liver

The **liver** is the largest internal organ of the body, weighing about 1.7 kg (3.5 to 4.0 lbs) in an adult. It is positioned immediately beneath the diaphragm in the right hypochondrium of the abdominal cavity. Its reddish-brown color is due to its great vascularity.

The liver has two major lobes and two minor lobes. Anteriorly, the **right lobe** is separated from the smaller **left lobe** by the **falciform ligament** (fig. 26.29). Inferiorly, the **caudate** (*kaw'dāt*) **lobe** is positioned near the inferior vena cava, and the **quadrate lobe** is adjacent to the gallbladder. The falciform ligament attaches the liver to the anterior abdominal wall and the diaphragm. Continuous along the free

Figure 26.29

The liver.
(*a*) An anterior view of the gross structure. (*b*) An inferior view showing the lobes of the liver, gallbladder, hepatic vessels, and ducts. (*c*) A transverse section of the abdomen showing the relative position of the liver to other abdominal organs.

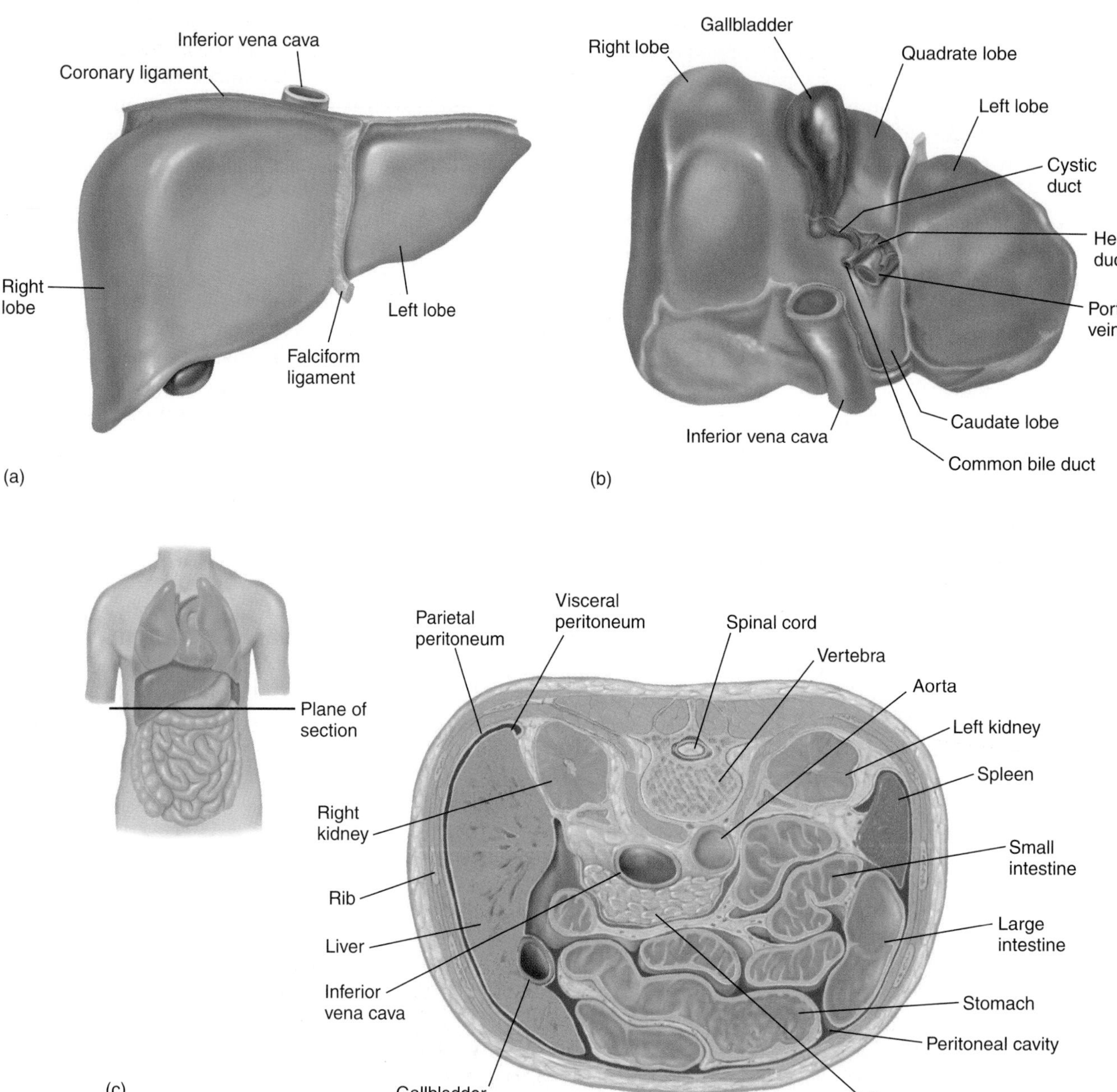

border of the falciform ligament to the umbilicus is a **ligamentum teres** (round ligament), which is the remnant of the umbilical vein of the fetus. The **porta** of the liver is where the hepatic artery, hepatic portal vein, lymphatics, and nerves enter the liver and where the **hepatic ducts** exit.

Although the liver is the largest internal organ, it is, in a sense, only one to two cells thick. This is because the liver cells, or **hepatocytes,** form **hepatic plates** that are one to two cells thick, and the hepatic plates are separated from each other by large capillary spaces called **sinusoids** (figs. 26.30 and 26.31). The sinusoids are lined with phagocytic **Kupffer** (*koop′fer*) **cells,** but the large intercellular gaps between adjacent Kupffer cells make these sinusoids more permeable than other capillaries. Because of the hepatic plate structure of the liver and the very permeable sinusoids, each hepatocyte is in close contact with the blood.

hepatic: Gk. *hepatos,* liver

Kupffer cells: from Karl W. Kupffer, Bavarian anatomist, 1829–1902

Figure 26.30

A liver lobule and the histology of the liver.
(*a*) A liver lobule seen in cross section and (*b*) longitudinal section. Blood enters a liver lobule through the vessels in a hepatic triad, passes through hepatic sinusoids, and leaves the lobule through a central vein. The central veins converge to form hepatic veins that transport venous blood from the liver. (*c*) A photomicrograph of a liver lobule in transverse section.

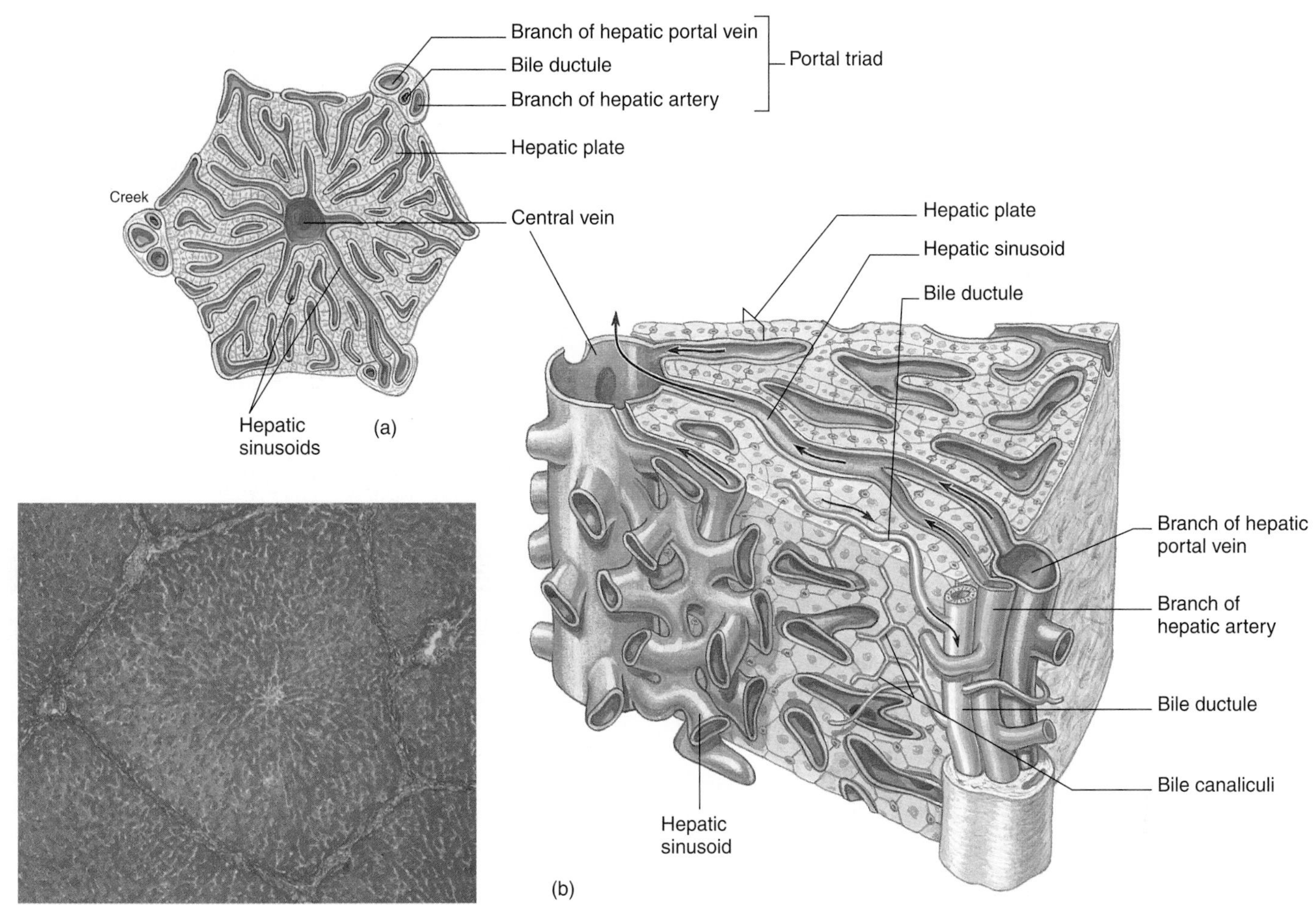

Hepatic Portal System

The products of digestion that are absorbed into blood capillaries in the GI tract do not enter the general circulation directly. Instead, this blood is delivered first to the liver. Capillaries in the stomach, small intestine, and large intestine drain venous blood into veins that converge to form the *hepatic portal vein*, which carries blood to capillaries in the liver (see fig. 21.35). It is not until the blood has passed through this second capillary bed that it enters the general circulation through the two *hepatic veins* that drain the liver. The term **hepatic portal system** is used to describe this unique pattern of circulation: capillaries ⇒ vein ⇒ capillaries ⇒ vein. In addition to receiving venous blood from the GI tract, the liver also receives arterial blood from the *hepatic artery*.

Liver Lobules

The hepatic plates are arranged to form functional units called **liver lobules** (figs. 26.30 and 26.31). In the middle of each lobule is a **central vein,** and at the periphery of each lobule are branches of the hepatic portal vein and of the hepatic artery. These branches open into the spaces *between* hepatic plates. The portal venous blood, containing molecules absorbed in the GI tract, mixes with the arterial blood within the sinusoids as it flows from the periphery of the lobule to the central vein. The central veins of different liver lobules converge to form the two hepatic veins that carry the blood from the liver to the inferior vena cava.

Bile is produced by the hepatocytes and secreted into thin channels called **bile canaliculi** (*kan″ă-lik′yŭ-li*) located *within* each hepatic plate. These bile canaliculi are drained at the periphery of each lobule by **bile ductules,** which in turn drain into **hepatic ducts** that carry bile away from the liver. Since blood travels in the sinusoids and bile travels in the opposite direction within the hepatic plates, blood and bile do not mix in the liver lobules.

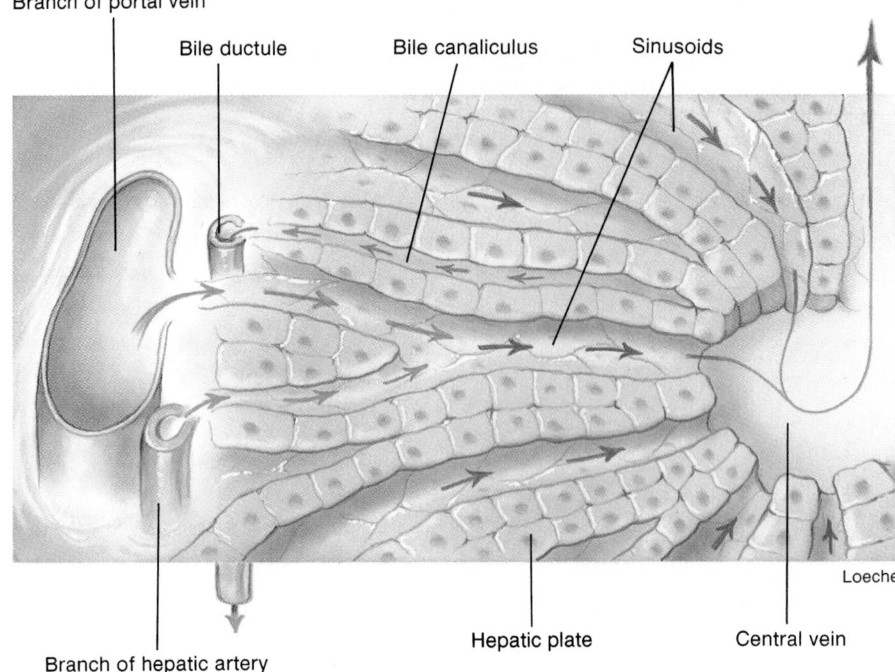

Figure 26.31

The flow of blood and bile in a liver lobule.
Blood flows within sinusoids from a portal vein to the central vein (from the periphery to the center of a lobule). Bile flows within hepatic plates from the center to bile ducts at the periphery of a lobule.

Enterohepatic Circulation

In addition to the normal constituents of bile, a wide variety of exogenous compounds (drugs) are secreted by the liver into the bile ducts (table 26.4). The liver can thus "clear" the blood of particular compounds by removing them from the blood and excreting them into the small intestine with the bile. Molecules that are cleared from the blood by secretion into the bile are eliminated in the feces; this is analogous to renal clearance of blood through excretion in the urine, as described in chapter 25.

Many compounds that are released with the bile into the small intestine are not eliminated with the feces, however. Some of these can be absorbed and enter the hepatic portal blood. These absorbed molecules are thus carried back to the liver, where they can be again secreted by hepatocytes into the bile ducts. Compounds that recirculate between the liver and small intestine in this way are said to have an **enterohepatic circulation** (fig. 26.32).

Functions of the Liver

As a result of its very large and diverse enzymatic content and its unique structure, and because it receives venous blood from the intestine, the liver has a wider variety of functions than any other organ in the body. The major categories of liver function are summarized in table 26.5.

Bile Production and Secretion

The liver produces and secretes 250 to 1,500 ml of bile per day. The major constituents of bile are bile pigment (bilirubin), bile salts, phospholipids (mainly lecithin), cholesterol, and inorganic ions.

Table 26.4

Compounds Excreted by the Liver into the Bile Ducts

	Compound	Comments
Endogenous (naturally occurring)	Bile salts, urobilinogen, cholesterol	High percentage is absorbed and has an enterohepatic circulation*
	Lecithin	Small percentage is absorbed and has an enterohepatic circulation
	Bilirubin	No enterohepatic circulation
Exogenous (drugs)	Ampicillin, streptomycin, tetracycline	High percentage is absorbed and has an enterohepatic circulation
	Sulfonamides, penicillin	Small percentage is absorbed and has an enterohepatic circulation

*Compounds with an enterohepatic circulation are absorbed to some degree by the intestine and are returned to the liver in the hepatic portal vein.

Bile pigment, or **bilirubin** (*bil′′ĭ-roo′bin*), is produced in the spleen, liver, and bone marrow as a derivative of the heme groups (minus the iron) from hemoglobin. The **free bilirubin** is not very water-soluble, and thus most is carried in the blood attached to albumin proteins. This protein-bound bilirubin can neither be filtered by the kidneys into the urine nor directly excreted by the liver into the bile.

Figure 26.32

The enterohepatic circulation.
Substances secreted in the bile may be absorbed by the intestinal epithelium and recycled to the liver via the hepatic portal vein.

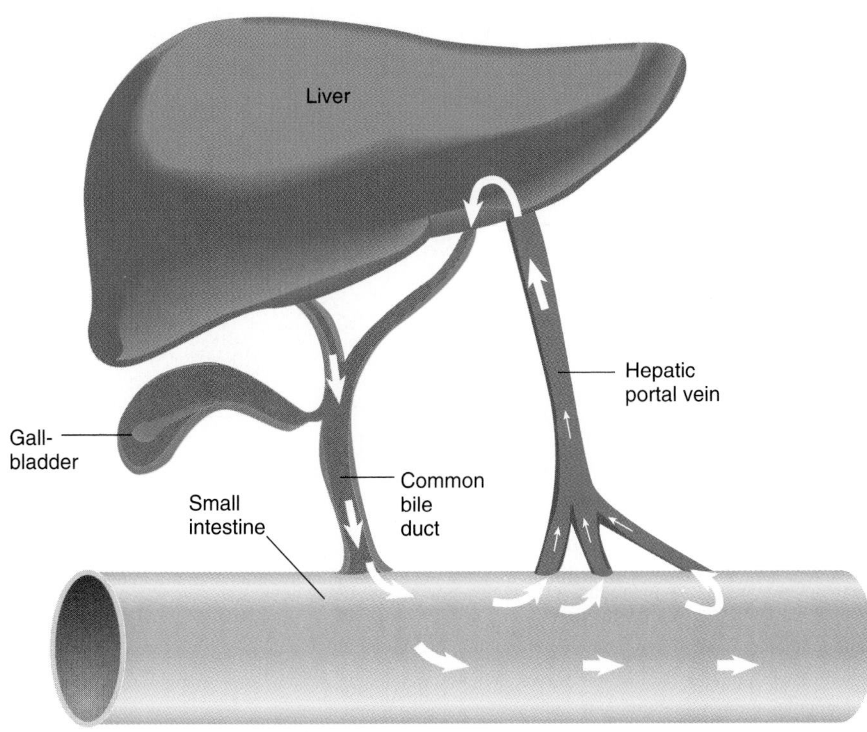

The liver can take some of the free bilirubin out of the blood and conjugate (combine) it with glucuronic acid. This **conjugated bilirubin** is water-soluble and can be secreted into the bile. Once in the bile, the conjugated bilirubin can enter the small intestine where it is converted by bacteria into another pigment—**urobilinogen** (*yoo″rŏ-bi-lin′o-jen*). Derivatives of urobilinogen impart a brown color to the feces. About 30% to 50% of the urobilinogen, however, is absorbed by the small intestine and enters the hepatic portal vein. Of the urobilinogen that enters the liver sinusoids, some is secreted into the bile and is thus returned to the small intestine in an enterohepatic circulation; the rest enters the general circulation (fig. 26.33). The urobilinogen in blood plasma, unlike free bilirubin, is not attached to albumin. Urobilinogen is therefore easily filtered by the kidneys into the urine, where its derivatives produce an amber color.

Bile salts are derivatives of cholesterol that have two to four polar groups on each molecule. The principal bile salts in humans are *cholic acid* and *chenodeoxycholic* (*ke″nŏ-de-ok-sĭ-ko′lik*) *acid* (fig. 26.34). In aqueous solutions these molecules "huddle" together to form aggregates known as **micelles.** As described in

Table 26.5
Major Categories of Liver Function

Functional Category	Actions
Detoxication of blood	Phagocytosis by Kupffer cells Chemical alteration of biologically active molecules (hormones and drugs) Production of urea, uric acid, and other molecules that are less toxic than parent compounds Excretion of molecules in bile
Carbohydrate metabolism	Conversion of blood glucose to glycogen and fat Production of glucose from liver glycogen and from other molecules (amino acids, lactic acid) by gluconeogenesis Secretion of glucose into the blood
Lipid metabolism	Synthesis of triglyceride and cholesterol Excretion of cholesterol in bile Production of ketone bodies from fatty acids
Protein synthesis	Production of albumin Production of plasma transport proteins Production of clotting factors (fibrinogen, prothrombin, and others)
Secretion of bile	Synthesis of bile salts Conjugation and excretion of bile pigment (bilirubin)

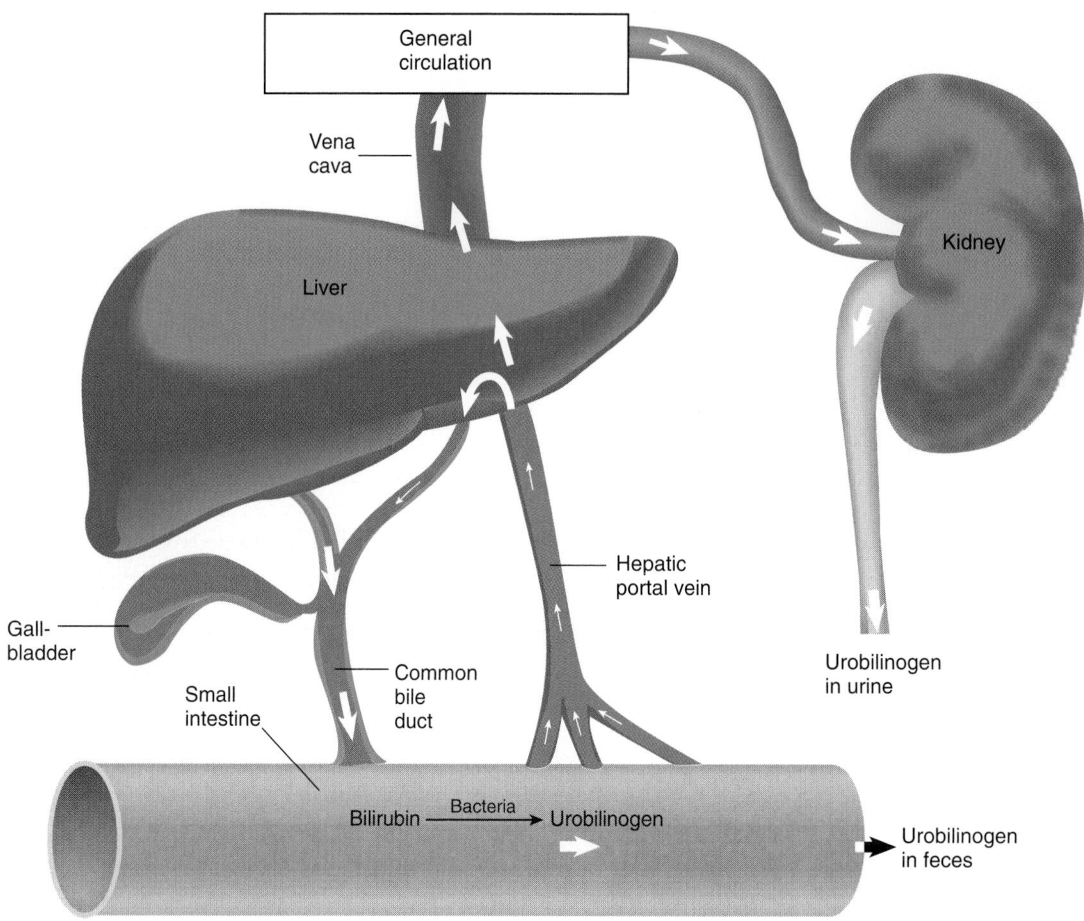

Figure 26.33

The enterohepatic circulation of urobilinogen. Bacteria in the intestine convert bile pigment (bilirubin) into urobilinogen. Some of this pigment leaves the body in the feces; some is absorbed by the intestine and is recycled through the liver. A portion of the urobilinogen that is absorbed enters the general circulation and is filtered by the kidneys into the urine.

chapter 2, the nonpolar parts are located in the central region of the micelle (away from water), whereas the polar groups face water around the periphery of the micelle (see fig. 2.20). Lecithin, cholesterol, and other lipids in the small intestine enter these micelles, and the dual nature of the bile salts (part polar, part nonpolar) allows them to emulsify fat in the chyme.

Detoxication of the Blood

The liver can remove hormones, drugs, and other biologically active molecules from the blood by (1) excretion of these compounds in the bile, as previously described; (2) phagocytosis by the Kupffer cells that line the sinusoids, and (3) chemical alteration of these molecules within the hepatocytes.

Ammonia, for example, is a very toxic molecule produced by deamination of amino acids in the liver and by the action of bacteria in the intestine. Since the ammonia concentration of blood in the hepatic portal vein is 4 to 50 times greater than that of blood in the hepatic veins, it is clear that the ammonia is removed by the liver. The liver has the enzymes needed to convert ammonia into less toxic **urea** molecules, which are secreted by the liver into the blood and excreted by the kidneys in the urine. Similarly, the liver converts toxic porphyrins into **bilirubin** and toxic purines into **uric acid.**

Steroid hormones and many drugs are inactivated in their passage through the liver by modifications of their chemical structures. The liver has enzymes that convert these nonpolar molecules into more polar (more water-soluble) forms by *hydroxylation* (the addition of OH^- groups) and by *conjugation* with highly polar groups such as sulfate and glucuronic acid. Polar derivatives of steroid hormones and drugs are less biologically active and, because of their increased water solubility, are more easily excreted by the kidneys into the urine.

Figure 26.34

The two major bile acids in humans.
These more polar derivatives of cholesterol form the bile salts.

OH CH$_3$
CH—CH$_2$—CH$_2$—COOH
HO OH
Cholic acid

CH$_3$
CH—CH$_2$—CH$_2$—COOH
HO OH
Chenodeoxycholic acid

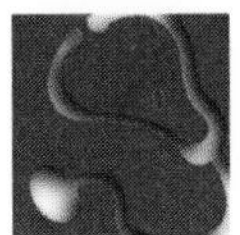

The liver cells contain enzymes for the metabolism of steroid hormones and other endogenous molecules, as well as for the detoxication of such exogenous toxic compounds as benzopyrene (a carcinogen from tobacco smoke and charbroiled meat), polychlorinated biphenyls (PCBs), and dioxin. The enzymes are members of a class called the *cytochrome P450 enzymes* (not related to the cytochromes of cell respiration) that is composed of a few dozen enzymes with varying specificities. Together, these enzymes can metabolize thousands of toxic compounds. Since there are individual differences in the hepatic content of the various cytochrome P450 enzymes, one person's sensitivity to a drug may be greater than another's because of a relative deficiency in the appropriate cytochrome P450 enzyme needed to metabolize that drug.

Secretion of Glucose, Triglycerides, and Ketone Bodies

As you may recall from chapter 4, the liver helps to regulate the blood glucose concentration by either removing glucose from the blood or adding glucose to it, according to the needs of the body. After a carbohydrate-rich meal, the liver can remove some glucose from the blood in the hepatic portal vein and convert it into glycogen and triglycerides through the processes of **glycogenesis** (*gli″kŏ-jen′ĭ-sis) and* **lipogenesis,** respectively. During fasting the liver secretes glucose into the blood. This glucose can be derived from the breakdown of stored glycogen in a process called **glycogenolysis** (*gli″kŏ-jĕ-nol′ĭ-sis*), or it can be produced by the conversion of noncarbohydrate molecules (such as amino acids) into glucose in a process called **gluconeogenesis** (*gloo″ko-ne″ŏ-jen′ĭ-sis*). The liver also contains the enzymes required to convert free fatty acids into ketone bodies **(ketogenesis),** which are secreted into the blood in large amounts during fasting. These processes are controlled by hormones and are explained further in chapter 27.

Production of Plasma Proteins

Plasma albumin and most of the plasma globulins (with the exception of immunoglobulins, or antibodies) are produced by the liver. Albumin constitutes about 70% of the total plasma protein and contributes most to the colloid osmotic pressure of the blood (chapter 22). The globulins produced by the liver have a wide variety of functions, including transport of cholesterol and triglycerides, transport of steroid and thyroid hormones, inhibition of trypsin activity, and blood clotting. Clotting factors I (fibrinogen), II (prothrombin), III, V, VII, IX, and XI, as well as angiotensinogen, are all produced by the liver.

Gallbladder

The **gallbladder** is a saclike organ attached to the inferior surface of the liver. This organ stores and concentrates bile, which drains to it from the liver by way of the hepatic ducts, bile ducts, and **cystic duct,** respectively. A sphincter valve at the neck of the gallbladder allows a storage capacity of about 35 to 100 ml. The inner mucosal layer of the gallbladder is arranged in rugae similar to those of the stomach. When the gallbladder fills with bile, it expands to the size and shape of a small pear. Bile is a yellowish green fluid containing bile salts, bilirubin, cholesterol, and other compounds, as previously discussed. Contraction of the muscularis ejects bile from the cystic duct into the **common bile duct,** through which it is conveyed to the duodenum (fig. 26.35).

Bile is continuously produced by the liver and drains through the hepatic and common bile ducts to the duodenum. When the small intestine is empty of food, the **sphincter of ampulla** (sphincter of Oddi) closes, and bile is forced up the cystic duct to the gallbladder for storage.

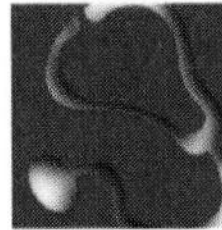

Approximately 20 million Americans have *gallstones*—small, hard mineral deposits (calculi) that can produce painful symptoms by obstructing the cystic or common bile ducts. Gallstones usually contain cholesterol as their major component. Cholesterol normally has an extremely low water solubility (20 μg/L), but it can be present in bile at 2 million times its water solubility (40 g/L) because cholesterol molecules cluster together with bile salts and lecithin in the hydrophobic centers of micelles. In order for gallstones to form, the liver must secrete enough cholesterol to create a supersaturated solution, and some substance within the gallbladder must serve as a nucleus for the formation of cholesterol crystals. The gallstone is formed from cholesterol crystals that become hardened by the precipitation of inorganic salts (fig. 26.36). Gallstones may be removed surgically or they may be dissolved by oral ingestion of bile acids. A newer treatment involves fragmentation of the gallstones by high-energy shock waves delivered to a patient immersed in a water bath. This procedure is called *extracorporeal shock-wave lithotripsy.*

cystic: Gk. *kystis*, pouch

Pancreas

The soft, lobulated **pancreas** (see fig. 26.20) has both exocrine and endocrine functions. The endocrine function, described in chapter 19, is performed by the **pancreatic islets** (*islets of Langerhans*), clusters of cells that secrete the hormones insulin and glucagon into the blood. As an exocrine gland, the pancreas secretes pancreatic juice through the pancreatic duct into the duodenum. The exocrine secretory units of the pancreas are called **acini** (*as'ĭ-ni*). Each acinus (fig. 26.37) consists of a single layer of epithelial cells surrounding a lumen into which pancreatic juice is secreted.

pancreas: Gk. *pan*, all; *kreas*, flesh

acinus: L. *acinus*, grape

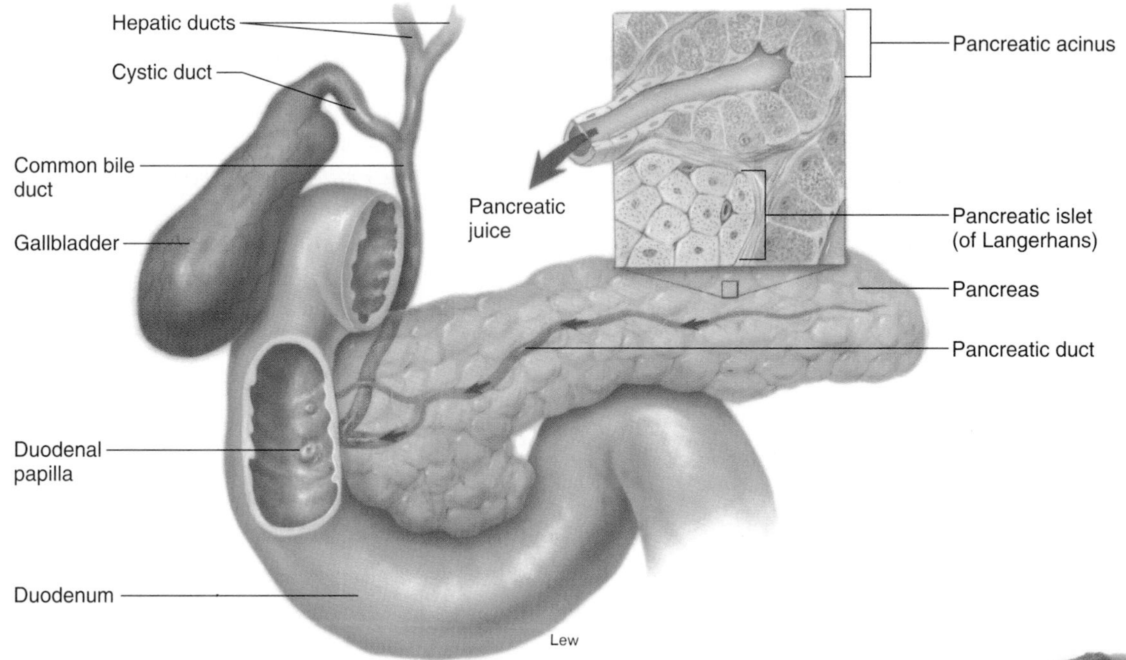

Figure 26.35

Pancreatic juice and bile are secreted into the duodenum.
The pancreatic duct joins the common bile duct to empty its secretions through the duodenal papilla into the duodenum.

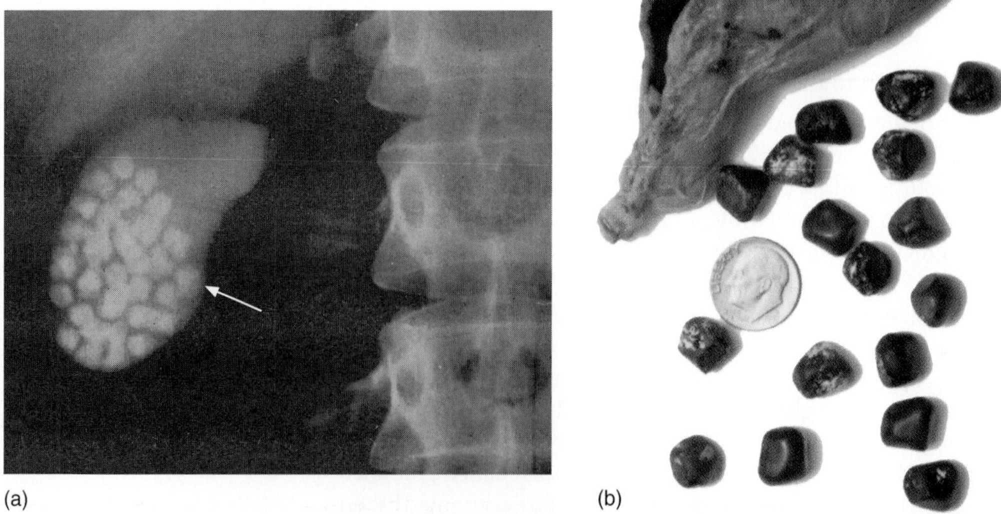

Figure 26.36

Gallstones.
(*a*) A radiograph of a gallbladder that contains gallstones. (*b*) A posterior view of a gallbladder that has been removed (cholecystectomy) and cut open to reveal its gallstones (bilary calculi). (A dime has been placed in the photo to show relative size.)

The pancreas is positioned horizontally along the posterior abdominal wall, adjacent to the greater curvature of the stomach. It is about 12.5 cm (6 in.) long and 2.5 cm (1 in.) thick and consists of an expanded **head,** associated with the C-shaped curve of the duodenum; a centrally located **body;** and a tapering **tail.** All but a portion of the head is positioned retroperitoneally.

Innervation of the pancreas is provided by branches of the celiac plexus. The glandular portion receives parasympathetic innervation, whereas the pancreatic blood vessels receive sympathetic innervation. The pancreas is supplied with blood by the pancreatic branch of the splenic artery arising from the celiac artery and by the pancreatoduodenal branches arising from the superior mesenteric artery. Venous blood is returned through the splenic and superior mesenteric veins into the hepatic portal vein.

Figure 26.37

Photomicrograph of the pancreas.
A microscopic view of the structure of the pancreas, showing exocrine acini and the pancreatic islet (of Langerhans).

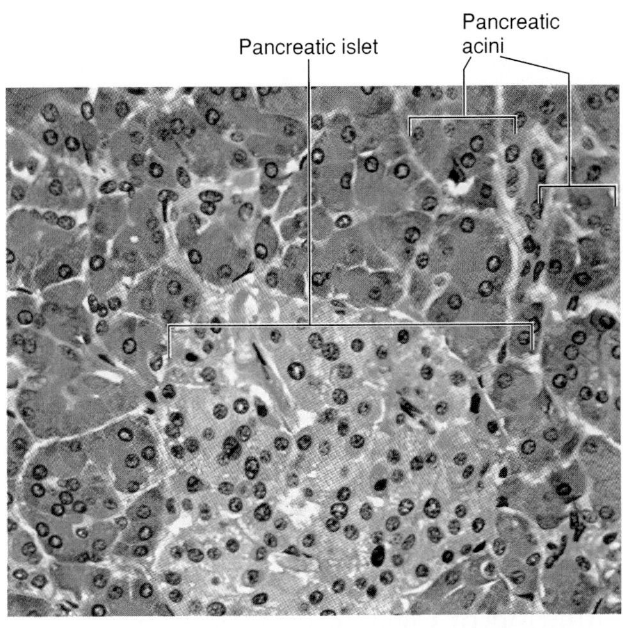

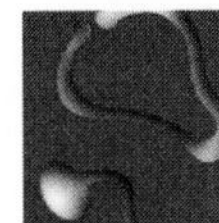

Pancreatic cancer has the worst prognosis of all types of cancer. This is probably because of the spongy, vascular nature of this organ and its vital exocrine and endocrine functions. Pancreatic surgery is a problem because the soft, spongy tissue is difficult to suture.

Pancreatic Juice

Pancreatic juice contains water, bicarbonate, and a wide variety of digestive enzymes. These enzymes include (1) **amylase,** which digests starch; (2) **trypsin,** which digests protein; and (3) **lipase,** which digests triglycerides. Other pancreatic enzymes are listed in table 26.6. It should be noted that the complete digestion of food molecules in the small intestine requires the action of both brush border enzymes and pancreatic enzymes.

Most pancreatic enzymes are produced as inactive molecules, or **zymogens** (*zi'mŏ-jenz*), so that the risk of self-digestion within the pancreas is minimized. The inactive form of trypsin, called trypsinogen, is activated within the small intestine by the catalytic action of the brush border enzyme **enterokinase.** Enterokinase converts trypsinogen to active trypsin. Trypsin, in turn, activates the other zymogens of pancreatic juice (fig. 26.38) by cleaving off polypeptide sequences that inhibit the activity of these enzymes.

The activation of trypsin serves as the trigger for the activation of other pancreatic enzymes. Actually, the pancreas

Table 26.6
Enzymes Contained in Pancreatic Juice

Enzyme	Zymogen	Activator	Action
Trypsin	Trypsinogen	Enterokinase	Cleaves internal peptide bonds
Chymotrypsin	Chymotrypsinogen	Trypsin	Cleaves internal peptide bonds
Elastase	Proelastase	Trypsin	Cleaves internal peptide bonds
Carboxypeptidase	Procarboxypeptidase	Trypsin	Cleaves last amino acid from carboxyl-terminal end of polypeptide
Phospholipase	Prophospholipase	Trypsin	Cleaves fatty acids from phospholipids such as lecithin
Lipase	None	None	Cleaves fatty acids from glycerol
Amylase	None	None	Digests starch to maltose and short chains of glucose molecules
Cholesterolesterase	None	None	Releases cholesterol from its bonds with other molecules
Ribonuclease	None	None	Cleaves RNA to form short chains
Deoxyribonuclease	None	None	Cleaves DNA to form short chains

Figure 26.38

Activation of pancreatic juice enzymes.
The pancreatic protein-digesting enzyme trypsin is secreted in an inactive form known as trypsinogen. This inactive enzyme (zymogen) is activated by a brush border enzyme, enterokinase (EN), located in the cell membrane of microvilli. Active trypsin in turn activates other zymogens in pancreatic juice.

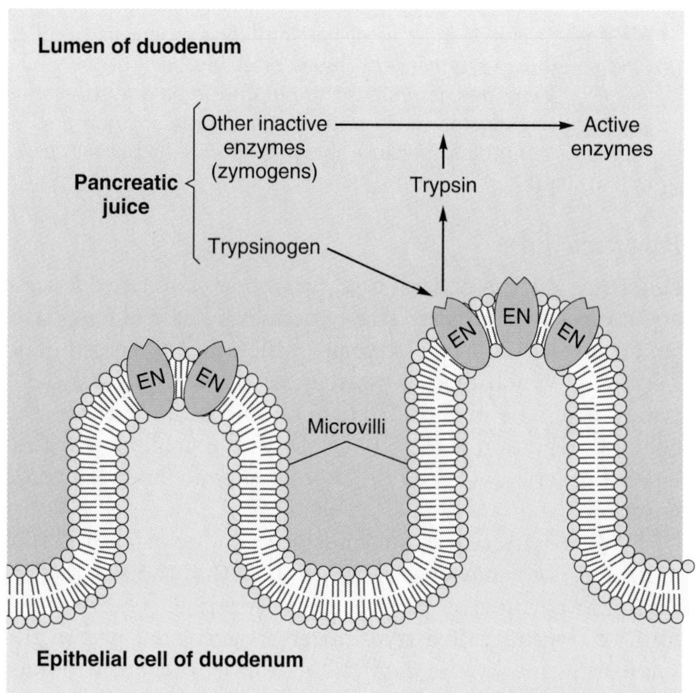

does produce small amounts of active trypsin, but the other enzymes are not activated until the pancreatic juice has entered the duodenum. This is because pancreatic juice also contains a small protein called *pancreatic trypsin inhibitor* that attaches to trypsin and inhibits its activity in the pancreas.

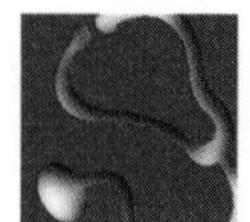

Pancreatitis (inflammation of the pancreas) may result when conditions such as alcoholism, gallstones, traumatic injury, infections, or toxicosis from various drugs provoke activation of digestive enzymes within the pancreas. Leakage of trypsin into the blood also occurs, but trypsin is inactive in the blood because of the inhibitory action of two plasma proteins, α_1*-antitrypsin* and α_2*-macroglobulin*. Pancreatic amylase may also leak into the blood, but it is not active because its substrate (starch) is not present in blood. Pancreatic amylase activity can be measured in vitro, however, and these measurements are commonly performed to assess the health of the pancreas.

Neural and Endocrine Regulation of the Digestive System

The activities of different regions of the GI tract are coordinated by the actions of the vagus nerves and by various hormones. The stomach begins to increase its secretion in anticipation of a meal, and further increases its activities in response to the arrival of food. The entry of chyme into the duodenum stimulates the secretion of hormones that promote contractions of the gallbladder, secretion of pancreatic juice, and inhibition of gastric activity.

Neural and endocrine control mechanisms modify the activity of the digestive system. The sight, smell, or taste of food, for example, can stimulate salivary and gastric secretions via activation of the vagus nerves, which helps to "prime" the digestive system in preparation for a meal. Stimulation of the vagus nerves, in this case, originates in the brain and is a conditioned reflex (as Pavlov demonstrated by training dogs to salivate in response to a bell). The vagus nerves are also involved in the reflex control of one part of the digestive system by another—these are "short reflexes," which do not involve the brain.

The GI tract is both an endocrine gland and a target for the action of various hormones. Indeed, the first hormones to be discovered were gastrointestinal hormones. In 1902 two English physiologists, Sir William Bayliss and Ernest Starling, discovered that the duodenum produced a chemical regulator. They named this substance **secretin** (*sĕ-kre′tin*) and proposed, in 1905, that it was but one of many yet undiscovered chemical regulators produced by the body. Bayliss and Starling coined the term *hormones* for this new class of regulators. In that same year, other investigators discovered that an extract from the cardia of the stomach stimulated gastric secretion. The hormone **gastrin** was thus the second hormone to be discovered.

The chemical structures of gastrin, secretin, and the duodenal hormone **cholecystokinin** (*ko″lĕ-sis″tŏ-ki′nin*), or **CCK,** were determined in the 1960s. More recently, a fourth hormone produced by the small intestine, **gastric inhibitory peptide (GIP),** has been added to the list of proven GI tract hormones. The effects of these and other gastrointestinal hormones are summarized in table 26.7.

Regulation of Gastric Function

Gastric motility and secretion are, to some extent, automatic. Waves of contraction that serve to push chyme through the pyloric sphincter, for example, are initiated spontaneously by pacesetter cells in the greater curvature of the stomach. Likewise, the secretion of HCl from parietal cells and pepsinogen from chief cells can be stimulated in the absence of neural and hormonal influences by the presence of cooked or partially digested protein in the stomach. This action involves other cells in the gastric mucosa—the G cells, which secrete the hormone gastrin; the enterochromaffin-like (ECL) cells, which secrete histamine; and the D cells, which secrete somatostatin.

The effects of autonomic nerves and hormones are superimposed on this automatic activity. The control of stomach function is conveniently divided into three phases: (1) the

Pavlov, Ivan Petrovich: Russian physiologist, 1849–1936
Bayliss, Sir William Maddock: English physiologist, 1860–1924
Starling, Ernest Henry: English physiologist, 1866–1927

Table 26.7
Physiological Effects of Gastrointestinal Hormones

Secreted by	Hormones	Effects
Stomach	Gastrin	Stimulates parietal cells to secrete HCl Stimulates chief cells to secrete pepsinogen Maintains structure of gastric mucosa
Small intestine	Secretin	Stimulates water and bicarbonate secretion in pancreatic juice Potentiates actions of cholecystokinin on pancreas
Small intestine	Cholecystokinin (CCK)	Stimulates contraction of gallbladder Stimulates secretion of pancreatic juice enzymes Inhibits gastric motility and secretion Maintains structure of exocrine pancreas (acini)
Small intestine	Gastric inhibitory peptide (GIP)	Inhibits gastric motility and secretion (?) Stimulates secretion of insulin from pancreatic islets
Ileum and colon	Glucagon-like peptide-1 (GLP-1)	Inhibits gastric motility and secretion Stimulates secretion of insulin from pancreatic islets
	Guanylin	Stimulates intestinal secretion of Cl^-, causing elimination of NaCl and water in the feces

cephalic phase, (2) the gastric phase, and (3) the intestinal phase. These are summarized in table 26.8.

Cephalic Phase

The **cephalic** (*sĕ-fal′ik*) **phase** of stomach regulation refers to control by the brain via the vagus nerves. As previously discussed, various conditioned stimuli can evoke gastric secretion. In humans, of course, this conditioning is more subtle than that exhibited by Pavlov's dogs in response to a bell. In fact, just talking about appetizing food is sometimes a more potent stimulus for gastric acid secretion than the actual sight and smell of food!

Activation of the vagus nerves (1) stimulates the chief cells to secrete pepsinogen and (2) indirectly stimulates the parietal cells to secrete HCl. The endings of the vagus nerves directly stimulate G cells to secrete gastrin and the ECL cells to secrete histamine. The gastrin secreted by G cells enters the systemic circulation and is carried back to the stomach, where it also stimulates the ECL cells to release histamine. Histamine, in turn, activates H_2 histamine receptors on the parietal cells to stimulate acid secretion (fig. 26.39).

This cephalic phase continues into the first 30 minutes of a meal, but then gradually declines in importance as the next phase becomes predominant.

Gastric Phase

The arrival of food into the stomach stimulates the **gastric phase** of regulation. Gastric secretion is stimulated in response to two factors: (1) distension of the stomach, which is determined by the amount of chyme, and (2) the chemical nature of the chyme.

Table 26.8
The Three Phases of Gastric Secretion

Phase of Regulation	Description
Cephalic phase	1. Sight, smell, and taste of food cause stimulation of vagal nuclei in brain 2. Vagus nerves stimulate acid secretion a. Direct stimulation of parietal cells (major effect) b. Stimulation of gastrin secretion; gastrin stimulates acid secretion (lesser effect)
Gastric phase	1. Distension of stomach stimulates vagus nerves; vagus nerves stimulate acid secretion 2. Amino acids and peptides in stomach lumen stimulate acid secretion a. Direct stimulation of parietal cells (lesser effect) b. Stimulation of gastrin secretion; gastrin stimulates acid secretion (major effect) 3. Gastrin secretion inhibited when pH of gastric juice falls below 2.5
Intestinal phase	1. Neural inhibition of gastric emptying and acid secretion a. Arrival of chyme in duodenum causes distension, increase in osmotic pressure b. These stimuli activate a neural reflex that inhibits gastric activity 2. In response to fat in chyme, duodenum secretes a hormone that inhibits gastric acid secretion

Figure 26.39

Regulation of gastric acid secretion.
Amino acids from partially digested proteins stimulate gastrin secretion. Gastrin secretion from G cells is also stimulated by vagus nerve activity. The secreted gastrin then acts as a hormone to stimulate histamine release from the ECL cells. The histamine, in turn, acts as a paracrine regulator to stimulate the parietal cells to secrete HCl (⊕ = stimulation; ⊖ = inhibition).

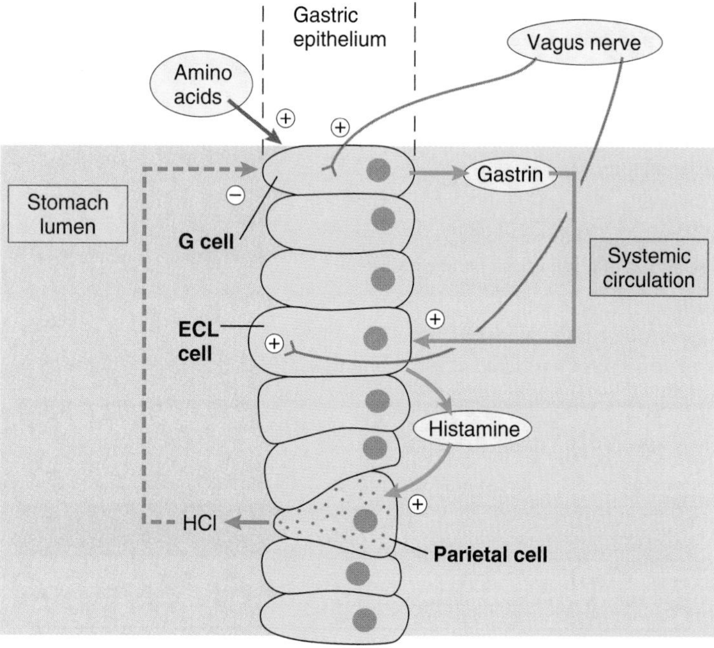

Although intact proteins in the chyme have little stimulatory effect, the partial digestion of proteins into shorter polypeptides and amino acids, particularly phenylalanine and tryptophan, stimulates the chief cells to secrete pepsinogen and the G cells to secrete gastrin. Gastrin, in turn, stimulates the secretion of pepsinogen from chief cells, but its effect on the parietal cells is primarily indirect. Gastrin stimulates the secretion of histamine from ECL cells, and the histamine then stimulates secretion of HCl from parietal cells as previously described (fig. 26.39). A *positive feedback mechanism* thus develops. As more HCl and pepsinogen are secreted, more short polypeptides and amino acids are released from the ingested protein. This stimulates additional secretion of gastrin and, therefore, additional secretion of HCl and pepsinogen. It should be noted that glucose in the chyme has no effect on gastric secretion, and the presence of fat actually inhibits acid secretion.

Secretion of HCl during the gastric phase is also regulated by a *negative feedback mechanism*. As the pH of gastric juice drops, so does the secretion of gastrin—at a pH of 2.5, gastrin secretion is reduced, and it shuts off entirely at a pH of 1.0. The secretion of HCl declines accordingly. This effect may be mediated by the hormone somatostatin, secreted by the D cells of the gastric mucosa. As the pH of gastric juice falls the D cells are stimulated to secrete somatostatin, which then acts as a paracrine regulator to inhibit the secretion of gastrin from the G cells.

The presence of proteins and polypeptides in the stomach helps to buffer the acid, thereby preventing a rapid fall in gastric pH. More acid can thus be secreted when proteins are present than when they are absent. The arrival of protein into the stomach thus stimulates acid secretion in two ways—by the positive feedback mechanism previously discussed and by inhibition of the negative feedback control of acid secretion. Through these mechanisms, the amount of acid secreted is closely matched to the amount of protein ingested. As the stomach empties the protein buffers exit, the pH falls, and the secretion of gastrin and HCl is accordingly inhibited.

Intestinal Phase

The **intestinal phase** of gastric regulation refers to the inhibition of gastric activity when chyme enters the small intestine. Investigators in 1886 demonstrated that the addition of olive oil to a meal inhibits gastric emptying, and in 1929 it was shown that the presence of fat inhibits gastric juice secretion. This inhibitory intestinal phase of gastric regulation is due to both a neural reflex originating from the duodenum and to a chemical hormone secreted by the duodenum.

The arrival of chyme into the duodenum increases its osmolality. This stimulus, together with the stretch of the duodenum and possibly other stimuli, produces a neural reflex that results in the inhibition of gastric motility and secretion. The presence of fat in the chyme also stimulates the duodenum to secrete a hormone that inhibits gastric function. The general term for such an inhibitory hormone is an **enterogastrone.**

In the past, **gastric inhibitory peptide (GIP)** was thought to function as an enterogastrone—hence the name for this hormone. Many researchers, however, now believe that other intestinal hormones may serve this function. Other polypeptide hormones secreted by the small intestine that can inhibit gastric activity include **somatostatin** (produced by the intestine, as well as by the brain and stomach); **cholecystokinin (CCK),** secreted by the duodenum in response to the presence of chyme; and **glucagon-like peptide-1 (GLP-1),** secreted by the ileum and colon. GLP-1 is one of a family of peptides produced by the intestine that structurally resemble the hormone glucagon (secreted by the alpha cells of the pancreatic islets).

It could be that the only physiological role of GIP is stimulation of insulin secretion from the pancreatic islets in response to the presence of glucose in the small intestine. Some scientists therefore propose that the name GIP be retained, but that it serve as an acronym for glucose-dependent *i*nsulinotropic *p*eptide. It should be noted, however, that GLP-1 is also a very potent stimulator of insulin secretion. These intestinal hormones therefore stimulate the pancreas to "anticipate" a rise in blood glucose by secreting insulin (which acts to lower the blood glucose concentration) even before the glucose has been absorbed into the blood.

Figure 26.40 The enteric nervous system.

Peristalsis is produced by local reflexes involving the enteric nervous system . Note that the enteric nervous system consists of motor neurons, association neurons, and sensory neurons. The neurotransmitters that stimulate smooth muscle contraction are indicated with a ⊕, those that produce smooth muscle relaxation are indicated with a ⊖. (NO = nitric oxide; VIP = vasoactive intestinal peptide.)

Regulation of Intestinal Function

The submucosal (Meissner's) and myenteric (Auerbach's) plexuses within the wall of the small intestine contain 100 million neurons, about as many as are in the spinal cord! These include preganglionic parasympathetic axons, the ganglion cell bodies of postganglionic parasympathetic neurons, postganglionic sympathetic axons, and sensory neurons. These plexuses also contain association neurons, as does the CNS. Also like the CNS, the **enteric nervous system** (or *enteric brain*) contains more neuroglial cells than neurons, and these neuroglial cells resemble the astrocytes of the brain.

Some sensory neurons within the intestinal plexuses send impulses all the way to the CNS, but many sensory neurons synapse with the association neurons in the wall of the small intestine. This allows for local reflexes that are controlled within the GI tract. Peristalsis can serve as an example. A mass of chyme stimulates sensory neurons that then activate intestinal association neurons, which in turn stimulate motor neurons within the intestinal wall. Smooth muscle contraction is stimulated by the neurotransmitters ACh and substance P above the chyme, and smooth muscle relaxation is stimulated by nitric oxide (NO), vasoactive intestinal peptide (VIP), and ATP below the chyme (fig. 26.40).

Several intestinal reflexes are controlled locally, by means of the enteric nervous system and paracrine regulators, and extrinsically through the actions of the nerves and hormones previously discussed. These reflexes include: (1) the **gastroileal reflex,** in which increased gastric activity causes increased motility of the ileum and increased movement of chyme through the ileocecal sphincter; (2) the **ileogastric reflex,** in which distension of the ileum causes a decrease in gastric motility; and (3) the **intestino-intestinal reflexes,** in which overdistension of one intestinal segment causes relaxation throughout the rest of the small intestine.

Guanylin (*gwa'nĭ-lin*) is a recently discovered paracrine regulator produced by the ileum and colon. It derives its name from its ability to activate the enzyme guanylate cyclase, and thus to cause the production of cyclic GMP (cGMP) within the cytoplasm of intestinal epithelial cells. Acting through cGMP as a second messenger, guanylin stimulates the intestinal epithelial cells to secrete Cl^- and water and inhibits their absorption of Na^+. These

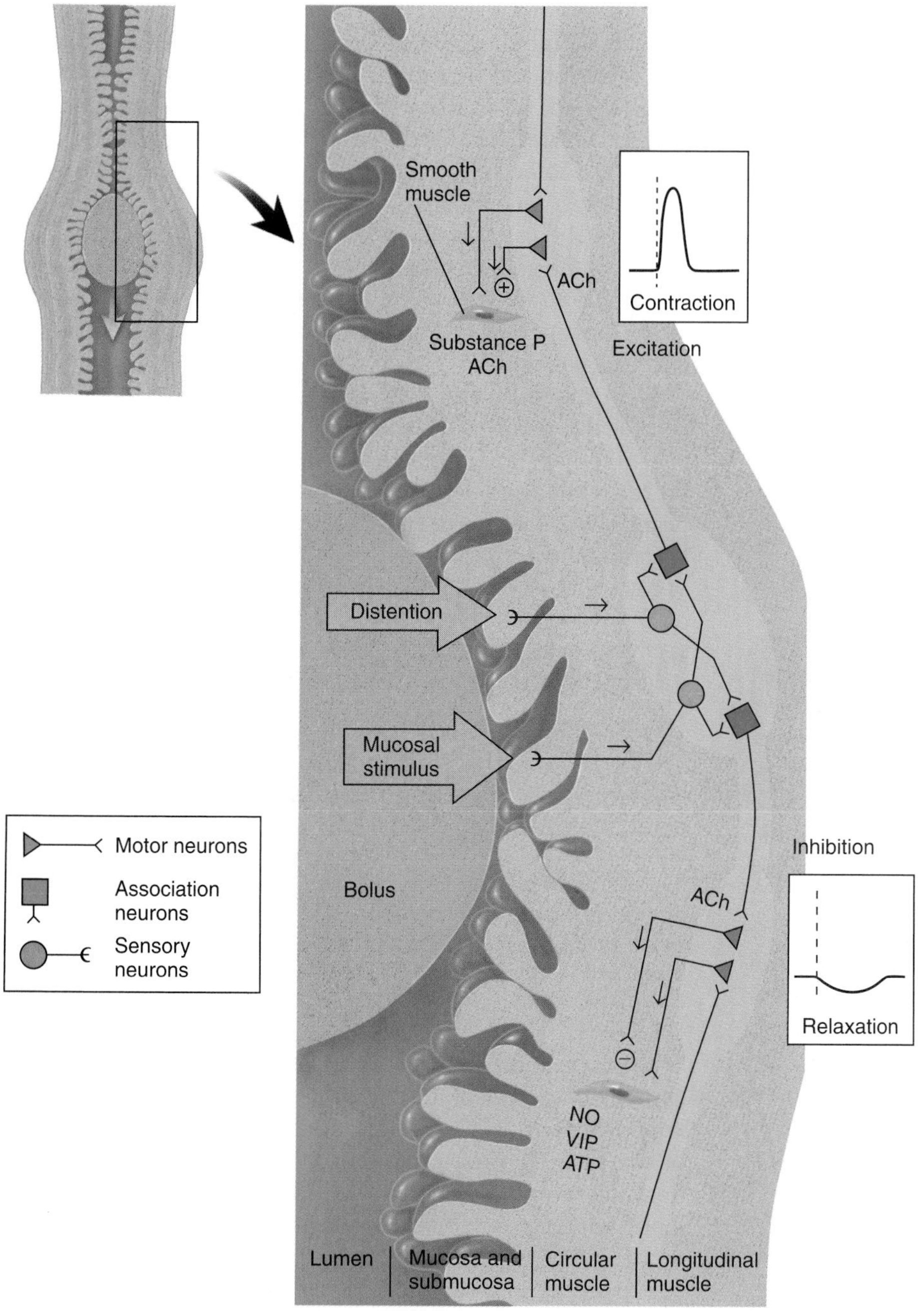

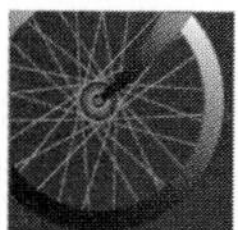

The inhibitory neural and endocrine mechanisms during the intestinal phase prevent the further passage of chyme from the stomach to the duodenum. This gives the duodenum time to process the load of chyme received previously. Since secretion of the enterogastrone is stimulated by fat in the chyme, a breakfast of bacon and eggs takes longer to pass through the stomach—and makes one feel "fuller" for a longer time—than does a breakfast of pancakes and syrup.

actions increase the amount of salt and water lost from the body in the feces. A related polypeptide, called **uroguanylin,** has been found in the urine. This polypeptide appears to be produced by the small intestine and may therefore function as a hormone that stimulates the kidneys to excrete salt in the urine.

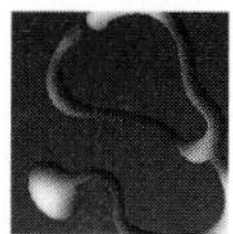

Particular *Escherichia coli* bacteria produce *heat-stable enterotoxins* that are responsible for "traveler's diarrhea." The enterotoxins act by stimulating the same receptors on the apical membranes of the intestinal epithelial cells that are activated by guanylin. By mimicking the actions of guanylin, the enterotoxins stimulate intestinal Cl^- and water secretion to produce the diarrhea.

Regulation of Pancreatic Juice and Bile Secretion

The arrival of chyme into the duodenum stimulates the intestinal phase of gastric regulation and, at the same time, stimulates reflex secretion of pancreatic juice and bile. The entry of new chyme is thus retarded as the previous load is digested. The secretion of pancreatic juice and bile is stimulated both by neural reflexes initiated in the duodenum and by secretion of the duodenal hormones cholecystokinin and secretin.

Pancreatic Juice

The secretion of pancreatic juice is stimulated by both secretin and CCK. These two hormones, however, are secreted in response to different stimuli, and they have different effects on the composition of pancreatic juice. The release of secretin occurs in response to a fall in duodenal pH to below 4.5; this pH fall occurs for only a short time, however, because the acidic chyme is rapidly neutralized by alkaline pancreatic juice. The secretion of CCK, by contrast, occurs in response to the protein and fat content of chyme in the duodenum.

Secretin stimulates the production of bicarbonate by the pancreas. Since bicarbonate neutralizes the acidic chyme, and since secretin is released in response to the low pH of chyme, a negative feedback loop is completed. CCK, by contrast, stimulates the production of pancreatic enzymes such as trypsin, lipase, and amylase. Partially digested protein and fat are the most potent stimulators of CCK secretion, and CCK secretion continues until the chyme has passed through the duodenum and early region of the jejunum.

Secretin and CCK can have different effects on the same cells (the pancreatic acinar cells) because their actions are mediated by different second messengers. The second messenger of secretin action is cyclic AMP, whereas the second messenger for CCK is Ca^{2+}.

Secretion of Bile

The liver secretes bile continuously, but this secretion is greatly augmented following a meal. The increased secretion is due to the release of secretin and CCK from the duodenum. Secretin stimulates the liver to secrete bicarbonate into the bile, and CCK enhances this effect. The arrival of chyme in the duodenum also causes the gallbladder to contract and eject bile. Contraction of the gallbladder occurs in response to neural reflexes from the duodenum and to hormonal stimulation by CCK.

Trophic Effects of Gastrointestinal Hormones

Patients with tumors of the stomach antrum exhibit excessive acid secretion and hyperplasia (growth) of the gastric mucosa. Surgical removal of the antrum reduces gastric secretion and prevents growth of the gastric mucosa. Patients with peptic ulcers are sometimes treated by vagotomy—cutting of the vagus nerve. Vagotomy also reduces acid secretion but has no effect on the gastric mucosa. These observations suggest that the hormone gastrin, secreted by the antrum, may exert stimulatory, or *trophic*, effects on the gastric mucosa. The structure of the gastric mucosa, in other words, is dependent upon the effects of gastrin.

In the same way, the structure of the acinar (exocrine) cells of the pancreas is dependent upon the trophic effects of CCK. Perhaps this explains why the pancreas, as well as the GI tract, atrophies during starvation. Since neural reflexes appear to be capable of regulating digestion, perhaps the primary function of the GI hormones is trophic—that is, maintenance of the structure of their target organs.

Digestion and Absorption of Carbohydrates, Lipids, and Proteins

Polysaccharides and polypeptides are hydrolyzed into their subunits. These subunits enter the epithelial cells of the intestinal villi and are secreted into blood capillaries. Fat is emulsified by the action of bile salts, hydrolyzed into fatty acids and monoglycerides, and absorbed into the intestinal epithelial cells. Once inside the cells, triglycerides are resynthesized and combined with proteins to form particles that are secreted into the lymphatic fluid.

The caloric (energy) value of food is derived mainly from its content of carbohydrates, lipids, and proteins. In the average American diet, carbohydrates account for approximately 50% of the total calories, protein accounts for 11% to 14%, and lipids make up the balance. These food molecules consist primarily of long combinations of subunits (monomers) that must be digested by hydrolysis reactions into free monomers before absorption can occur. The characteristics of the major digestive enzymes are summarized in table 26.9.

Digestion and Absorption of Carbohydrates

Most carbohydrates are ingested as starch, which is a long polysaccharide of glucose in the form of straight chains with occasional branchings. The most commonly ingested sugars are sucrose (table sugar, a disaccharide consisting of glucose and fructose) and lactose (milk sugar, a disaccharide consisting

Table 26.9
Characteristics of the Major Digestive Enzymes

Enzyme	Site of Action	Source	Substrate	Optimum pH	Product(s)
Salivary amylase	Mouth	Saliva	Starch	6.7	Maltose
Pepsin	Stomach	Gastric glands	Protein	1.6–2.4	Shorter polypeptides
Pancreatic amylase	Duodenum	Pancreatic juice	Starch	6.7–7.0	Maltose, maltriose, and oligosaccharides
Trypsin, chymotrypsin, carboxypeptidase			Polypeptides	8.0	Amino acids, dipeptides, and tripeptides
Pancreatic lipase			Triglycerides	8.0	Fatty acids and monoglycerides
Maltase	Small intestine	Epithelial membranes	Maltose	5.0–7.0	Glucose
Sucrase			Sucrose	5.0–7.0	Glucose + fructose
Lactase			Lactose	5.8–6.2	Glucose + galactose
Aminopeptidase			Polypeptides	8.0	Amino acids, dipeptides, tripeptides

of glucose and galactose). The digestion of starch begins in the mouth with the action of **salivary amylase,** or **ptyalin** (*ti'ă-lin*). This enzyme cleaves some of the bonds between adjacent glucose molecules, but most people don't chew their food long enough for sufficient digestion to occur in the mouth. The digestive action of salivary amylase stops some time after the swallowed bolus enters the stomach because this enzyme is inactivated at the low pH of gastric juice.

The digestion of starch, therefore, occurs mainly in the duodenum as a result of the action of **pancreatic amylase.** This enzyme cleaves the straight chains of starch to produce the disaccharide *maltose* and the trisaccharide *maltriose*. Pancreatic amylase, however, cannot hydrolyze the bond between glucose molecules at the branch points in the starch. As a result, short branched chains of glucose molecules, called *oligosaccharides*, are released together with maltose and maltriose by the activity of this enzyme (fig. 26.41).

Maltose, maltriose, and oligosaccharides are hydrolyzed to their monosaccharides by brush border enzymes located on the microvilli of the epithelial cells in the small intestine. The brush border enzymes also hydrolyze the disaccharides sucrose and lactose into their component monosaccharides. These monosaccharides are then moved across the epithelial cell membrane by secondary active transport, in which the glucose shares a common membrane carrier with Na^+ (chapter 5). Finally, glucose is secreted from the epithelial cells into blood capillaries within the intestinal villi.

Figure 26.41

Action of pancreatic amylase.
Pancreatic amylase digests starch into maltose, maltriose, and short oligosaccharides containing branch points in the chain of glucose molecules.

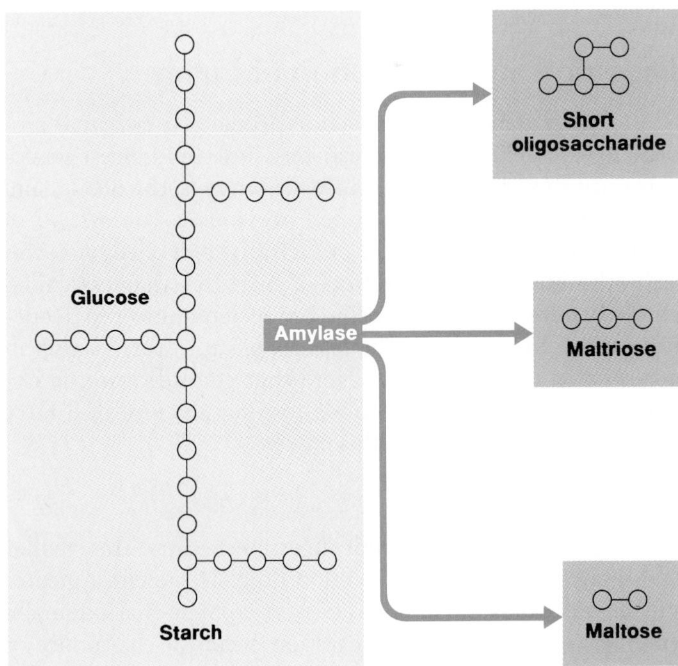

Digestion and Absorption of Proteins

Protein digestion begins in the stomach with the action of pepsin. Some amino acids are liberated in the stomach, but the major products of pepsin digestion are short-chain polypeptides. Pepsin digestion helps to produce a more homogenous chyme, but it is not essential for the complete digestion of protein that occurs in the small intestine.

Most protein digestion occurs in the duodenum and jejunum. The pancreatic juice enzymes **trypsin, chymotrypsin** (*ki"mŏ-trip'sin*), and **elastase** cleave peptide bonds in the interior of the polypeptide chains. These enzymes are thus grouped together as *endopeptidases* (*en″do-pep′tĭ-dās″ēz*). Enzymes that remove amino acids from the ends of polypeptide chains, by contrast, are *exopeptidases*. These include the pancreatic juice enzyme **carboxypeptidase,** which removes amino acids from the carboxyl-terminal end of polypeptide chains, and the brush border enzyme **aminopeptidase.** Aminopeptidase cleaves amino acids from the amino-terminal end of polypeptide chains.

As a result of the action of these enzymes, polypeptide chains are digested into free amino acids, dipeptides, and tripeptides. The free amino acids are absorbed by cotransport with Na^+ into the epithelial cells and secreted into blood capillaries. The dipeptides and tripeptides enter epithelial cells by the action of a single membrane carrier that has recently been characterized. This carrier functions in secondary active transport (chapter 5), using a H^+ gradient to transport dipeptides and tripeptides into the cell cytoplasm. Within the cytoplasm, the dipeptides and tripeptides are hydrolyzed into free amino acids, which are then secreted into the blood (fig. 26.42).

Newborn babies appear to be capable of absorbing a substantial amount of undigested proteins (hence they can absorb antibodies from their mother's milk); in adults, however, only the free amino acids enter the hepatic portal vein. Foreign food protein, which would be very antigenic, does not normally enter the blood. An interesting exception is the protein toxin that causes botulism, produced by the bacterium *Clostridium botulinum*. This protein is resistant to digestion, and is thus intact when it is absorbed into the blood.

Figure 26.42

Digestion and absorption of protein.
Polypeptide chains of proteins are digested into free amino acids, dipeptides, and tripeptides by the action of pancreatic juice enzymes and brush border enzymes. The amino acids, dipeptides, and tripeptides enter duodenal epithelial cells. Dipeptides and tripeptides are hydrolyzed into free amino acids within the epithelial cells, and these products are secreted into capillaries that carry them to the hepatic portal vein.

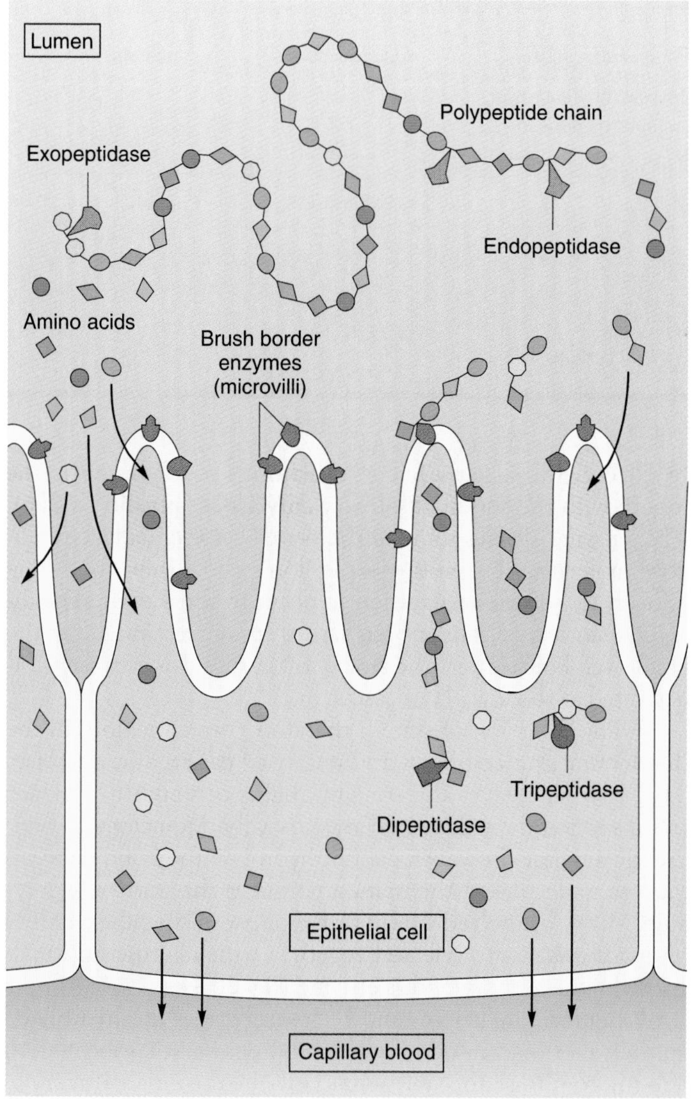

Digestion and Absorption of Lipids

The salivary glands and stomach of neonates (newborns) produce lipases. In adults, however, very little lipid digestion occurs until the lipid globules in chyme arrive in the duodenum. Through mechanisms described previously, the arrival of lipids (primarily triglyceride, or fat) in the duodenum serves as a stimulus for the secretion of bile. In a process called **emulsification,** bile salt micelles are secreted into the duodenum and act to break up the fat droplets into tiny *emulsification droplets* of triglycerides. Note that emulsification is not chemical digestion—the bonds joining glycerol and fatty acids are not hydrolyzed by this process.

Digestion of Lipids

The emulsification of fat aids digestion because the smaller and more numerous emulsification droplets present a greater surface area than the unemulsified fat droplets that originally entered the duodenum. Fat digestion occurs at the surface of the droplets through the enzymatic action of **pancreatic lipase,** which is aided in its action by a protein called *colipase* (also secreted by the pancreas) that coats the emulsification droplets and "anchors" the lipase enzyme to them. Through hydrolysis, lipase removes two of the three fatty acids from each triglyceride molecule and thus liberates *free fatty acids* and *monoglycerides* (fig. 26.43). **Phospholipase A** likewise digests phospholipids such as lecithin into fatty acids and *lysolecithin* (the remainder of the lecithin molecule after the removal of two fatty acids).

Free fatty acids, monoglycerides, and lysolecithin, which are more polar than the undigested lipids, quickly become associated with micelles of bile salts, lecithin, and cholesterol to form "mixed micelles" (fig. 26.44). These micelles then move to the brush border of the intestinal epithelium where absorption occurs.

Figure 26.43

Digestion of triglycerides.

Pancreatic lipase digest fat (triglycerides) by cleaving off the first and third fatty acids. This produces free fatty acids and monoglycerides. Sawtooth lines indicate hydrocarbon chains in the fatty acids.

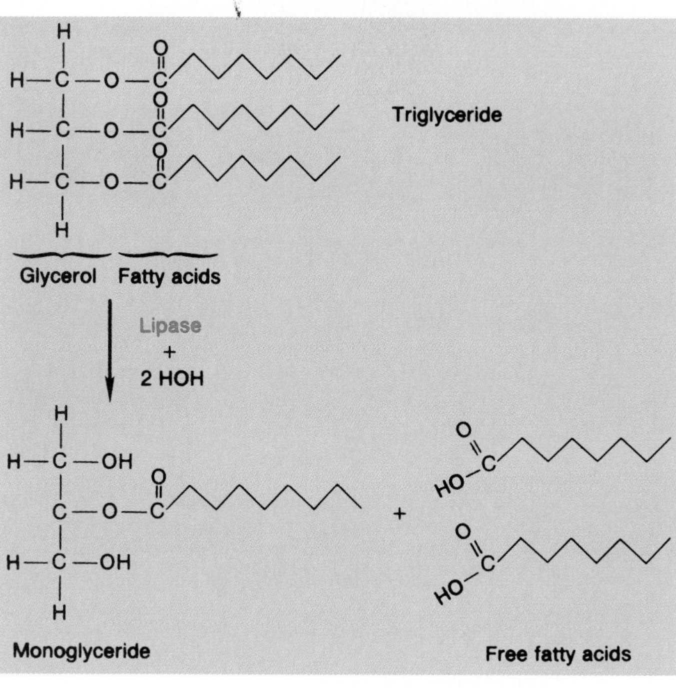

Absorption of Lipids

Free fatty acids, monoglycerides, and lysolecithin can leave the micelles and pass through the membrane of the microvilli to enter the intestinal epithelial cells. There is also some evidence that the micelles may be transported intact into the epithelial cells and that the lipid digestion products may be removed intracellularly from the micelles. In either event, these products are used to *resynthesize* triglycerides and phospholipids within the epithelial cells. This process is different from the absorption of amino acids and monosaccharides, which pass through the epithelial cells without being altered.

Triglycerides, phospholipids, and cholesterol then combine with protein inside the epithelial cells to form small particles called **chylomicrons** (*kī″lŏ-mi′kronz*). These tiny lipid and protein combinations are secreted into the lacteals (lymphatic capillaries) of the intestinal villi (fig. 26.45). Absorbed lipids thus pass through the lymphatic system, eventually entering the venous blood by way of the thoracic duct (chapter 23). By contrast, amino acids and monosaccharides enter the hepatic portal vein.

Transport of Lipids in the Blood

Once the chylomicrons are in the blood, their triglyceride content is removed by the enzyme *lipoprotein lipase*, which is attached to the endothelium of blood vessels. This enzyme hydrolyzes triglycerides and thus provides free fatty acids and

Figure 26.44

Steps in fat digestion and emulsification.

The three steps indicate the fate of fat in the small intestine. The digestion of fat (triglycerides) releases fatty acids and monoglycerides, which enter into micelles of bile salts secreted by the liver.

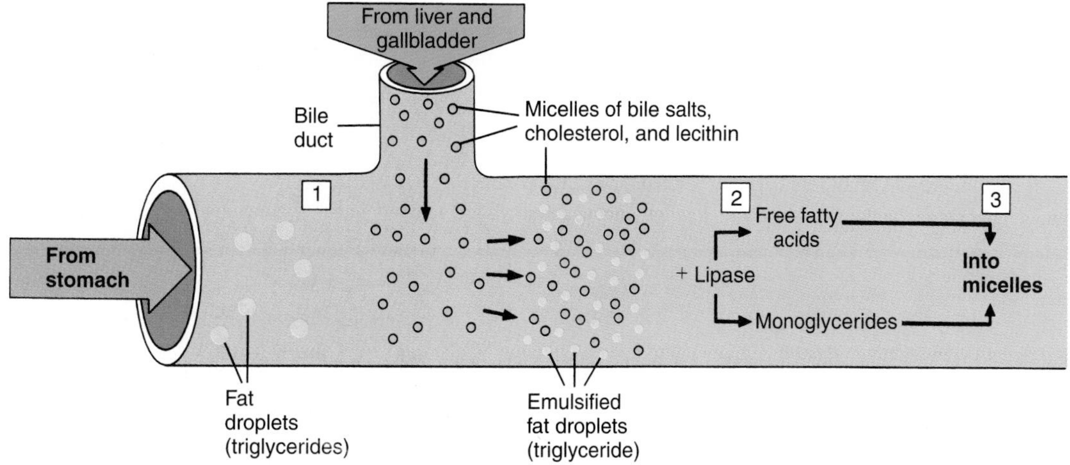

Figure 26.45

Absorption of fat.
Fatty acids and monoglycerides from the micelles within the small intestine are absorbed by epithelial cells and converted intracellularly into triglycerides. These are then combined with protein to form chylomicrons, which enter the lymphatic vessels (lacteals) of the villi. These lymphatic vessels transport the chylomicrons to the thoracic duct, which empties them into the venous blood (of the left subclavian vein).

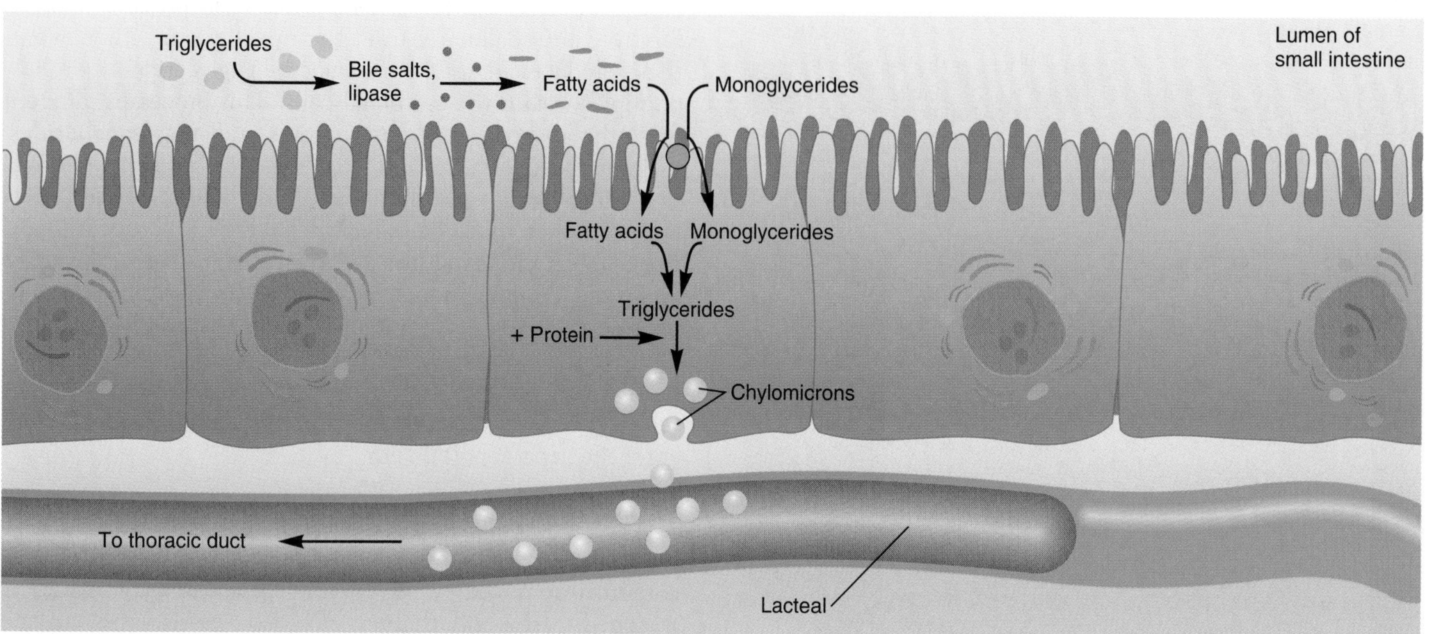

Table 26.10

Characteristics of the Lipid Carrier Proteins (Lipoproteins) Found in Plasma

Lipoprotein Class	Origin	Destination	Major Lipid(s)	Function
Chylomicrons	Small intestine	Many organs	Triglycerides, other lipids	Deliver lipids of dietary origin to body cells
Very-low-density lipoproteins (VLDLs)	Liver	Many organs	Triglycerides, cholesterol	Deliver endogenously produced triglycerides to body cells
Low-density lipoproteins (LDLs)	Intravascular removal of triglycerides from VLDL	Blood vessels, liver	Cholesterol	Deliver endogenously produced cholesterol to various organs
High-density lipoproteins (HDLs)	Liver and small intestine	Liver and steroid hormone producing glands	Cholesterol	Remove and degrade cholesterol

glycerol for use by the cells. The remaining *remnant particles*, containing cholesterol, are taken up by the liver. This is a type of endocytosis (chapter 3) that requires membrane receptors for the protein part (or *apoprotein*) of the remnant particle (see fig. 3.4).

Cholesterol and triglycerides produced by the liver are combined with other apoproteins and secreted into the blood as **very-low-density lipoproteins (VLDLs),** which deliver triglycerides to different organs. Once the triglycerides are removed, the VLDL particles are converted to **low-density lipoproteins (LDLs),** which transport cholesterol to various organs, including blood vessels. This can contribute to the development of atherosclerosis (chapter 21). Excess cholesterol is returned from these organs to the liver attached to **high-density lipoproteins (HDLs).** A high ratio of HDL-cholesterol to total cholesterol is believed to afford protection against atherosclerosis. The characteristics of these lipoproteins are summarized in table 26.10.

Clinical Considerations

Pathogens and Poisons

The GI tract presents a suitable environment for an array of microorganisms. Many of these are beneficial, but some bacteria and protozoa can cause diseases. The following discussion includes only a few examples of the pathogenic microorganisms.

Dysentery (*dis'en-ter"e*) is an inflammation of the intestinal mucosa characterized by the discharge of loose stools containing mucus, pus, and blood. The most common dysentery is **amoebic dysentery,** which is caused by the protozoan *Entamoeba histolytica*. Cysts from this organism are ingested in contaminated food, and after the protective coat has been removed by HCl in the stomach, the vegetative form invades the mucosal walls of the ileum and colon.

Food poisoning results from ingesting food contaminated with poisons or with bacteria containing toxins. *Salmonella* is a bacterium that commonly contaminates food. **Botulism,** the most serious type of food poisoning, is caused by ingesting food contaminated with the toxin produced by the bacterium *Clostridium botulinum*. This organism is widely distributed in nature, and the spores it produces are frequently found on food being processed by canning. For this reason, food must be heated to 120° C (248° F) before it is canned. It is the toxins produced by the bacteria growing in the food that are pathogenic, rather than the organisms themselves. The poison is a neurotoxin that is readily absorbed into the blood, at which point it can affect the nervous system.

Gastritis and Peptic Ulcers

Peptic ulcers are erosions of the mucous membranes of the stomach or duodenum produced by the action of HCl. In *Zollinger–Ellison syndrome*, ulcers of the duodenum are produced by excessive gastric acid secretion in response to very high levels of the hormone gastrin. Gastrin is normally produced by the stomach but, in this case, it may be secreted by a pancreatic tumor. This is a rare condition, but it does demonstrate that excessive gastric acid can cause ulcers of the duodenum. Ulcers of the stomach, however, are not believed to be due to excessive acid secretion, but rather to mechanisms that reduce the barriers of the gastric mucosa to self-digestion.

Experiments demonstrate that the cell membranes of the parietal and chief cells of the gastric mucosa are highly impermeable to the acid in the gastric lumen. Other protective mechanisms include a layer of alkaline mucus (containing bicarbonate) covering the gastric mucosa; tight junctions between adjacent epithelial cells, preventing acid from leaking into the submucosa; a rapid rate of cell division, allowing damaged cells to be replaced (the entire epithelium is replaced every 3 days); and several protective effects provided by prostaglandins produced by the gastric mucosa. Indeed, a common cause of gastric ulcers is believed to be the use of nonsteroidal anti-inflammatory drugs (NSAIDs). This class of drugs, including aspirin and ibuprofen, acts to inhibit the production of prostaglandins, as discussed in chapter 19.

When the gastric barriers to self-digestion are broken down, acid can leak through the mucosa to the submucosa, causing direct damage and stimulating inflammation. The histamine released from mast cells (chapter 23) during inflammation may stimulate further acid secretion and result in further damage to the mucosa. The inflammation that occurs during these events is called **acute gastritis.**

The duodenum is normally protected from gastric acid by the buffering action of bicarbonate in alkaline pancreatic juice, as well as by secretion of bicarbonate by duodenal (Brunner's) glands in the submucosa. However, people who develop duodenal ulcers produce excessive amounts of gastric acid that are not neutralized by the bicarbonate. People with gastritis and peptic ulcers must avoid substances that stimulate acid secretion, including coffee and wine, and often must take antacids.

It has been known for some time that most people who have peptic ulcers are infected with a bacterium known as *Helicobacter pylori*, which resides in the gastrointestinal tract. Also, clinical trials have demonstrated that antibiotics that eliminate this infection help in the treatment of the peptic ulcers. Indeed, modern antibiotic therapy can even cure peptic ulcers in many people and reduce the likelihood of recurrence.

Disorders of the Liver

Hepatitis

The liver is a remarkable organ that has the ability to regenerate even if up to 80% has been removed. The most serious diseases of the liver (hepatitis, cirrhosis, and hepatomas) affect the entire liver, so that it cannot repair itself. **Hepatitis** is inflammation of the liver. Certain chemicals may cause hepatitis, but generally it is caused by infectious viral agents. Scientists have identified seven kinds of hepatitis (A through G), but only the first three are common in the United States.

Hepatitis A (formerly termed infectious hepatitis) is a viral disease transmitted through contaminated foods and liquids. The fever, nausea, vomiting, and jaundice it causes can be severe but usually disappear within a week or two. The illness sometimes recurs, although generally in a milder form.

Hepatitis B (formerly termed serum hepatitis) is also caused by a virus and is transmitted primarily through sexual contact or infected blood. It is a more severe illness than hepatitis A. Some people with hepatitis B go on to develop chronic liver disease that can progress to cirrhosis and cancer. Treatment is not very effective, but vaccines produced through recombinant DNA technology are available to prevent hepatitis B infection.

dysentery: Gk. *dys*, bad; *entera*, intestine
Zollinger–Ellison syndrome: from Robert M. Zollinger, American surgeon, 1903–92, and Edwin H. Ellison, American physician, 1918–70

UNDER DEVELOPMENT

Development of the Digestive System

The entire digestive system develops from modifications of an elongated tubular structure called the **primitive gut.** These modifications are initiated during the fourth week of embryonic development. The primitive gut is composed solely of endoderm, and for descriptive purposes, can be divided into three regions: the *foregut, midgut,* and *hindgut.*

The **stomodeum** (*sto″mŏ-de′um*), or **oral pit,** is not part of the foregut but an invagination of ectoderm that breaks through a thin **oral membrane** to become continuous with the foregut and form the oral cavity, or mouth. Structures in the mouth, therefore, are ectodermal in origin. The esophagus, pharynx, stomach, pancreas, liver, gallbladder, and a portion of the duodenum are the organs that develop from the foregut (fig. 1). Along the GI tract, only the inside epithelial lining of the lumen is derived from endoderm. The vascular portion and smooth muscle layers are formed from mesoderm that develops from the surrounding mesenchyme.

The stomach first appears as an elongated dilation of the foregut. The caudal portion of the foregut and the cranial portion of the midgut form the duodenum. The liver and pancreas arise as small **hepatic** and **pancreatic buds,** respectively, from the wall of the duodenum. The hepatic bud experiences incredible growth to form the gallbladder, associated ducts, and the various lobes of the liver. By the sixth week, the liver is carrying out hematopoiesis (the formation of blood cells), and by the ninth week it has developed to the point where it represents 10% of the total weight of the fetus.

By the fifth week of embryonic development, the midgut has formed a ventral U-shaped loop, which projects into the umbilical cord. As development continues, the anterior limb of the midgut loop coils to form most of the small intestine. The posterior limb of the midgut loop expands to form the large intestine and a portion of the small intestine.

The hindgut extends from the midgut to the **cloacal** (*klo-a′kal*) **membrane.** This membrane is formed in part by the **proctodeum** (*prok″tŏ-de′um*), or **anal pit,** a depression in the anal region produced by an invagination of surface ectoderm. A partition develops to divide the cloacal membrane into an anterior **urogenital membrane** and a posterior **anal membrane.** Toward the end of the seventh week, the anal membrane perforates and forms the anal opening, which is lined with ectodermal cells.

stomodeum: Gk. *stoma,* mouth; *hodaios,* on the way to

cloaca: L. *cloaca,* sewer

Hepatitis C (formerly termed non-A, non-B hepatitis) is transmitted primarily through infected blood, rarely through sexual contact, and not through contaminated food. Some people develop a flulike illness that lasts several weeks. The hepatitis C virus remains in the liver and may eventually lead to liver failure or cancer. Patients are sometimes treated with alpha interferon, but this treatment is estimated to eliminate the virus in only about 20% of the cases.

Cirrhosis

In **cirrhosis** (*sĭ-ro′sis*), large numbers of liver lobules are destroyed and replaced with permanent connective tissue and "regenerative nodules" of hepatocytes. These regenerative nodules do not have the platelike structure of normal liver tissue, and are therefore less functional. One indication of their depressed function is the entry of ammonia from the hepatic blood into the general circulation. Other symptoms of cirrhosis are edema in the ankles and legs, uncontrolled bleeding, and increased sensitivity to drugs. Cirrhosis may be caused by chronic alcohol abuse, biliary obstruction, viral hepatitis, or by various chemicals that attack liver cells.

Jaundice

Jaundice is a yellow staining of the tissue produced by high blood concentrations of either free or conjugated bilirubin. Since free bilirubin is derived from hemoglobin, abnormally high concentrations of heme pigment may result from an unusually high rate of red blood cell destruction. This can occur, for example, as a result of *Rh disease* (*erythroblastosis fetalis*) in an Rh-positive baby born to a sensitized Rh-negative mother. Jaundice may also occur in otherwise healthy infants because red blood cells are normally destroyed at about the time of birth (hemoglobin concentrations decrease from 19 g per 100 ml to 14 g per 100 ml near the time of birth). This condition is called *physiological jaundice of the newborn* and is not indicative of disease. Premature infants may also develop jaundice

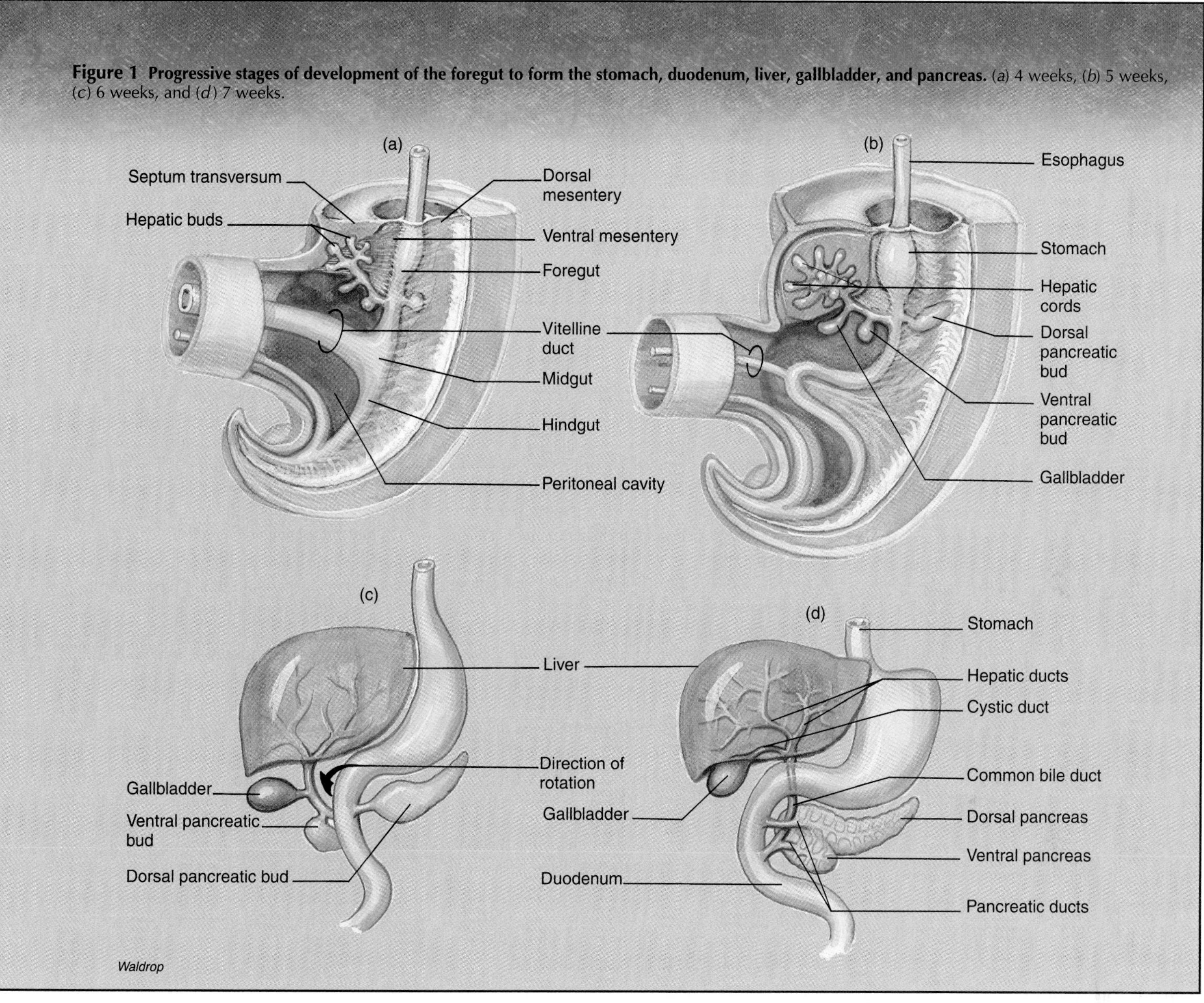

Figure 1 Progressive stages of development of the foregut to form the stomach, duodenum, liver, gallbladder, and pancreas. (*a*) 4 weeks, (*b*) 5 weeks, (*c*) 6 weeks, and (*d*) 7 weeks.

because the hepatic enzymes that conjugate bilirubin (a reaction needed for the excretion of bilirubin in the bile) mature late in gestation. Jaundice associated with high levels of conjugated bilirubin in the blood is commonly produced in adults when the excretion of bile is blocked by gallstones.

Newborn infants with jaundice are usually treated by exposing them to blue light in the wavelength range of 400 to 500 nm. This light is absorbed by bilirubin in cutaneous vessels and results in the conversion of the bilirubin into a form that can be dissolved in plasma without having to be conjugated with glucuronic acid. This more water-soluble photoisomer of bilirubin can then be excreted in the bile.

Intestinal Disorders

Enteritis is inflammation of the mucosa of the small intestine and is frequently referred to as intestinal flu. Causes of enteritis include bacterial or viral infections, irritating foods or fluids (including alcohol), and emotional stress. The symptoms are abdominal pain, nausea, and diarrhea. **Diarrhea** is the passage of watery, unformed stools. This condition is symptomatic of inflammation, stress, and other body dysfunctions.

A **hernia** is a protrusion of a portion of a visceral organ, usually the small intestine, through a weakened portion of the abdominal wall. Inguinal, femoral, umbilical, and hiatal hernias are the most common. With a **hiatal hernia,** a portion of the stomach pushes superiorly through the esophageal hiatus in the diaphragm and into the thorax. The potential dangers of a hernia are strangulation of the blood supply followed by gangrene, blockage of chyme, or rupture—each of which can threaten life.

Diverticulosis (*di″ver-tik″yŭ-lo′sis*) is a condition in which the intestinal wall weakens and an outpouching occurs. **Diverticulitis,** or inflammation of a diverticulum, can develop if fecal material becomes impacted in these pockets.

Clinical Investigation Answer

The student does not appear to have a peptic ulcer because the food and drinks that stimulate gastric acid secretion do not cause pain. The lack of fever and the normal white blood cell count suggest that the inflammation associated with appendicitis is absent. The yellowing of the sclera indicates jaundice, and this symptom—together with the prolonged clotting time—could be caused by liver disease. However, liver disease would elevate the blood levels of free bilirubin, which were found to be normal. The normal levels of urea and ammonia in the blood likewise suggest normal liver function. Similarly, the normal pancreatic amylase levels suggest that the pancreas is not affected.

This student's symptoms are most likely due to the presence of gallstones. Gallstones could obstruct the normal flow of bile, and thus prevent normal fat digestion. This would explain the fatty stools. The resulting loss of dietary fat could cause a deficiency in vitamin K, which is a fat-soluble vitamin required for the production of a number of clotting factors (chapter 20)—hence, the prolonged clotting time. The pain would be provoked by oily or fatty foods (peanut butter and bacon), which trigger a reflex contraction of the gallbladder once the fat arrives in the duodenum. Contraction of the gallbladder against an obstructed cystic duct or common bile duct often produces a severe referred pain below the right scapula.

Important Clinical Terminology

cheilitis (*ki-li'tis*) Inflammation of the lips.

colitis (*kŏ-li'tis*) Inflammation of the colon and rectum.

colostomy (*kŏ-los'tŏ-me*) The formation of an abdominal exit from the GI tract by bringing a loop of the colon to the surface of the abdomen. If the rectum is removed because of cancer, the colostomy provides a permanent outlet for the feces.

cystic fibrosis An inherited disease of the exocrine glands, particularly the pancreas. Pancreatic secretions that are too thick to drain easily cause inflammation of the ducts and promote connective tissue formation that occludes drainage from the ducts still further.

gingivitis (*jin-jĭ-vi'tis*) Inflammation of the gums. It may result from improper hygiene, poorly fitted dentures, improper diet, or certain infections.

halitosis (*hal-ĭ-to'sis*) Offensive breath odor. It may result from dental caries, certain diseases, or the ingestion of particular foods.

heartburn A burning sensation of the esophagus and stomach. It may result from the regurgitation of gastric juice into the lower portion of the esophagus.

hemorrhoids (*hem'ŏ-roidz*) Varicose veins of the rectum and anus.

jejunoileal (*jĕ-joo"no-il'e-al*) **bypass** A surgical procedure for creating a bypass of a considerable portion of the small intestine. It reduces the absorptive capacity of the small intestine and is thus used to control extreme obesity.

nausea Gastric discomfort and sensations of illness with a tendency to vomit. This feeling is symptomatic of motion sickness and many diseases and may occur during pregnancy.

pyorrhea (*pi"ŏ-re'ă*) The discharge of pus at the base of the teeth at the gum line because of periodontal disease.

regurgitation The forceful expulsion of gastric contents into the mouth. Nausea and vomiting are common symptoms of almost any dysfunction of the digestive system.

trench mouth A contagious bacterial infection that causes inflammation, ulceration, and painful swelling of the floor of the mouth. Generally it is contracted through direct contact by kissing an infected person. Trench mouth can be treated with penicillin and other medications.

vagotomy (*va-got'ŏ-me*) The surgical removal of sections of both vagus nerves where they enter the stomach in order to eliminate nerve impulses that stimulate gastric secretion. This procedure may be used to treat ulcers.

Chapter Summary

Introduction to the Digestive System (pp. 817–822)

1. The digestive system functions to break down food into its component monomers (digestion) and to move these monomers into the blood or lymph (absorption).
2. The digestive system consists of a GI tract and accessory digestive organs.
3. Peritoneal membranes line the abdominal wall and cover the visceral organs. The GI tract is supported by a double layer of peritoneum called the mesentery.
4. The layers (tunics) of the abdominal GI tract are, from the inside outward, the mucosa, submucosa, muscularis, and serosa.

Mouth, Pharynx, and Associated Structures (pp. 822–828)

1. The oral cavity is formed by the cheeks, lips, hard palate, soft palate, and tongue.
2. The four most anterior pairs of teeth are the incisors and canines, which have one root each; the premolars (bicuspids) and molars have two or three roots.
   (a) The tooth is anchored to a dental alveolus socket by a periodontal ligament.
   (b) The hard tooth tissues include enamel, dentin, and cementum. The pulp of the tooth receives blood vessels through the apical foramen.
3. The major pairs of salivary glands are the parotid, submandibular, and the sublingual glands.
4. The muscular pharynx is a passageway connecting the oral and nasal cavities to the esophagus and trachea.

Esophagus and Stomach (pp. 828–832)

1. Peristaltic waves of contraction push food through the lower esophageal sphincter into the stomach.
2. The stomach consists of a cardia, fundus, body, and pylorus. The pylorus terminates at the pyloric sphincter.
   (a) The lining of the stomach is thrown into folds, or gastric rugae, and the mucosal surface forms gastric pits that lead into gastric glands.
   (b) The parietal cells of the gastric glands secrete HCl; the chief cells secrete pepsinogen.
   (c) In the acidic environment of gastric juice, pepsinogen is converted into the active protein-digesting enzyme called pepsin.
3. The most important function of the stomach is the secretion of intrinsic factor, which is needed for the intestinal absorption of vitamin B_{12}.

Small Intestine (pp. 832–836)

1. The small intestine is divided into the duodenum, jejunum, and ileum. The common bile duct and pancreatic duct empty into the duodenum.
2. Fingerlike extensions of mucosa, called villi, project into the intestinal lumen.
   (a) Each villus is covered by a columnar epithelium and contains a connective tissue core of lamina propria.

(b) The apical surface (facing the intestinal lumen) of each columnar cell is folded to form microvilli. This brush border of the mucosa increases the absorptive surface area.

3. Digestive enzymes, called brush border enzymes, are located in the membranes of the microvilli.
   (a) Brush border enzymes include dipeptidases and disaccharidases, among others.
   (b) A deficiency of the disaccharidase known as lactase is responsible for lactose intolerance in many people.
4. The small intestine exhibits two major kinds of movements—peristalsis and segmentation.

Large Intestine (pp. 837–839)

1. The large intestine is divided into the cecum, colon, rectum, and anal canal.
   (a) The appendix is attached to the inferior medial margin of the cecum.
   (b) The colon consists of ascending, transverse, descending, and sigmoid portions.
2. The large intestine absorbs water and electrolytes.
3. Defecation occurs when the anal sphincters relax and contraction of other muscles raises the rectal pressure.

Liver, Gallbladder, and Pancreas (pp. 839–848)

1. The liver, the largest internal organ, is composed of functional units called lobules.
   (a) Liver lobules consist of plates of hepatic cells separated by capillary sinusoids.
   (b) Blood flows from the periphery of each lobule, where the hepatic artery and hepatic portal vein empty through the sinusoids and out the central vein.
   (c) Bile flows within the hepatocyte plates, in canaliculi, to the bile ducts.
   (d) Substances excreted in the bile can be returned to the liver in the hepatic portal blood. This is called an enterohepatic circulation.
   (e) The liver detoxifies the blood and modifies the blood plasma concentration of glucose, triglycerides, ketone bodies, and proteins.
2. The gallbladder stores and concentrates the bile. It releases bile through the cystic duct and common bile duct to the duodenum.
3. The pancreas performs both exocrine and endocrine functions.
   (a) The pancreatic islets, which secrete the hormones insulin and glucagon, perform the endocrine function.
   (b) The exocrine acini of the pancreas produce pancreatic juice, which contains various digestive enzymes and bicarbonate.

Neural and Endocrine Regulation of the Digestive System (pp. 848–852)

1. The regulation of gastric function occurs in three phases.
   (a) In the cephalic phase, the activity of higher brain centers, acting via the vagus nerves, stimulates gastric secretion.
   (b) In the gastric phase, the secretion of HCl and pepsin is controlled by the gastric contents and by the hormone gastrin, secreted by the gastric mucosa. Gastrin and vagus nerve stimulation occur via the action of histamine secreted from the gastric ECL cells.
   (c) In the intestinal phase, the activity of the stomach is inhibited by neural and endocrine reflexes from the duodenum. The intestinal hormone that inhibits gastric activity is known generically as an enterogastrone.
2. Intestinal activity is largely regulated intrinsically by the enteric nervous system, by a number of hormones secreted by the small intestine, and by paracrine regulators.
3. The secretion of the intestinal hormones secretin and cholecystokinin (CCK) regulate pancreatic juice and bile secretion.
4. Gastrointestinal hormones may be needed for the maintenance of the GI tract and accessory digestive organs.

Digestion and Absorption of Carbohydrates, Lipids, and Proteins (pp. 852–856)

1. The digestion of starch begins in the mouth with the action of salivary amylase.
   (a) Pancreatic amylase digests starch into disaccharides and short-chain oligosaccharides.
   (b) Complete digestion into monosaccharides is accomplished by brush border enzymes.
2. Protein digestion begins in the stomach with the action of pepsin.
   (a) Pancreatic juice contains protein-digesting enzymes, including trypsin, and chymotrypsin, among others.
   (b) The brush border contains digestive enzymes that help to complete the digestion of proteins into amino acids.
   (c) Amino acids, like monosaccharides, are absorbed and secreted into capillary blood entering the hepatic portal vein.
3. Lipids are digested in the small intestine after being emulsified by bile salts.
   (a) After digestion of triglycerides, free fatty acids and monoglycerides are absorbed into the epithelial cells of the small intestine.
   (b) Once inside the mucosal epithelial cells, these subunits are used to resynthesize triglycerides.
   (c) Triglycerides in the epithelial cells of the small intestine, together with proteins, form chylomicrons. These tiny particles are secreted into the lacteals of the villi.
   (d) Chylomicrons are transported by lymph to the thoracic duct, and from there enter the blood.

Review Activities

Objective Questions

1. Which of the following types of teeth are found in the permanent dentition but not in the deciduous dentition?
   (a) incisors
   (b) canines
   (c) premolars
   (d) molars
2. The double layer of peritoneum that supports the GI tract is called
   (a) the visceral peritoneum.
   (b) the mesentery.
   (c) the greater omentum.
   (d) the lesser omentum.
3. Which of the following tissue layers in the small intestine contains lacteals?
   (a) submucosa
   (b) muscularis mucosae
   (c) lamina propria
   (d) muscularis externa

4. Which of the following statements about intrinsic factor is *true?*
   (a) It is secreted by the stomach.
   (b) It is a polypeptide.
   (c) It promotes absorption of vitamin B_{12} in the intestine.
   (d) It helps prevent pernicious anemia.
   (e) All of the above are true.
5. Intestinal enzymes such as lactase are
   (a) secreted by the intestine into the chyme.
   (b) produced by the intestinal crypts (of Lieberkühn).
   (c) produced by the pancreas.
   (d) attached to the cell membrane of microvilli in the epithelial cells of the mucosa.
6. Which of the following statements about gastric secretion of HCl is *false?*
   (a) HCl is secreted by parietal cells.
   (b) HCl hydrolyzes peptide bonds.
   (c) HCl is needed for the conversion of pepsinogen to pepsin.
   (d) HCl is needed for maximum activity of pepsin.
7. Most digestion occurs in
   (a) the mouth.
   (b) the stomach.
   (c) the small intestine.
   (d) the large intestine.
8. Which of the following statements about trypsin is *true?*
   (a) Trypsin is derived from trypsinogen by the digestive action of pepsin.
   (b) Active trypsin is secreted into the pancreatic acini.
   (c) Trypsin is produced in the intestinal crypts.
   (d) Trypsinogen is converted to trypsin by the brush border enzyme enterokinase.
9. During the gastric phase, the secretion of HCl and pepsinogen is stimulated by
   (a) vagus nerve stimulation that originates in the brain.
   (b) polypeptides in the gastric lumen and by gastrin secretion.
   (c) secretin and cholecystokinin from the duodenum.
   (d) all of the above.
10. The secretion of HCl by the stomach mucosa is inhibited by
   (a) neural reflexes from the duodenum.
   (b) the secretion of an enterogastrone from the duodenum.
   (c) the lowering of gastric pH.
   (d) all of the above.
11. The first organ to receive the blood-borne products of digestion is
   (a) the liver.
   (b) the pancreas.
   (c) the heart.
   (d) the brain.
12. Which of the following statements about hepatic portal blood is *true?*
   (a) It contains absorbed fat.
   (b) It contains ingested proteins.
   (c) It is mixed with bile in the liver.
   (d) It is mixed with blood from the hepatic artery in the liver.
13. Absorption of salt and water is the principal function of which region of the GI tract?
   (a) esophagus
   (b) stomach
   (c) duodenum
   (d) jejunum
   (e) large intestine
14. Cholecystokinin (CCK) is a hormone that stimulates
   (a) bile production.
   (b) release of pancreatic enzymes.
   (c) contraction of the gallbladder.
   (d) both *a* and *b*.
   (e) both *b* and *c*.
15. Which of the following statements about vitamin B_{12} is *false?*
   (a) Lack of this vitamin can produce pernicious anemia.
   (b) Intrinsic factor is needed for absorption of vitamin B_{12}.
   (c) Damage to the gastric mucosa may lead to a deficiency in vitamin B_{12}.
   (d) Vitamin B_{12} is absorbed primarily in the jejunum.
16. Which of the following statements about starch digestion is *false?*
   (a) It begins in the mouth.
   (b) It occurs in the stomach.
   (c) It requires the action of pancreatic amylase.
   (d) It requires brush border enzymes for completion.
17. Which of the following statements about fat digestion and absorption is *false?*
   (a) Emulsification by bile salts increases the rate of fat digestion.
   (b) Triglycerides are hydrolyzed by the action of pancreatic lipase.
   (c) Triglycerides are resynthesized from monoglycerides and fatty acids in the intestinal epithelial cells.
   (d) Triglycerides, as particles called chylomicrons, are absorbed into blood capillaries within the villi.
18. Which of the following statements about contraction of intestinal smooth muscle is *true?*
   (a) It occurs automatically.
   (b) It is increased by parasympathetic nerve stimulation.
   (c) It produces segmentation.
   (d) All of the above are true.

Essay Questions

1. Explain how the gastric secretion of HCl and pepsin is regulated during the cephalic, gastric, and intestinal phases.
2. Describe how pancreatic enzymes become activated in the lumen of the small intestine. Why are these mechanisms needed?
3. Explain the function of bicarbonate in pancreatic juice. How may peptic ulcers in the duodenum be produced?
4. Describe the mechanisms that are believed to protect the gastric mucosa from self-digestion. What factors might be responsible for the development of a peptic ulcer in the stomach?
5. Explain why the pancreas is considered both an exocrine and an endocrine gland. Given this information, predict what effects tying of the pancreatic duct would have on pancreatic structure and function.
6. Explain how jaundice is produced when (*a*) the person has gallstones, (*b*) the person has a high rate of red blood cell destruction, and (*c*) the person has liver disease. In which case(s) would phototherapy for the jaundice be effective? Explain.
7. Describe the steps involved in the digestion and absorption of fat.
8. Distinguish between chylomicrons, very-low-density lipoproteins, low-density lipoproteins, and high-density lipoproteins.
9. Identify the different neurons present in the wall of the small intestine and explain how these neurons are involved in "short reflexes." Why is the enteric nervous system sometimes described as an "enteric brain"?
10. Trace the course of blood flow through the liver and discuss the significance of this pattern in terms of the detoxication of the blood. Describe the enzymes and the reactions involved in this detoxication.

Critical Thinking Questions

1. Which surgery do you think would have the most profound effect on digestion: (*a*) removal of the stomach (gastrectomy); (*b*) removal of the pancreas (pancreatectomy); or (*c*) removal of the gallbladder (cholecystectomy)? Explain your reasoning.
2. Describe the adaptations of the GI tract that make it more efficient by either increasing the surface area for absorption or increasing the contact between food particles and digestive enzymes.
3. Discuss how the ECL cells of the gastric mucosa function as a final common pathway for the neural, endocrine, and paracrine regulation of gastric acid secretion. What does this imply about the effectiveness of drug intervention to block excessive acid secretion?
4. Bacterial heat-stable enterotoxins can cause a type of diarrhea by stimulating the enzyme guanylate cyclase, which raises cyclic GMP levels within intestinal cells. Why might this be considered an example of mimicry? How does it cause diarrhea?
5. The hormone insulin is secreted by the pancreatic islets in response to a rise in blood glucose concentration. Surprisingly, however, the insulin secretion is greater in response to oral glucose than to intravenous glucose. Explain why this is so.

Related Web Sites

For a listing of the most current web sites related to this chapter, please visit the *Concepts of Human Anatomy and Physiology* home page at http://www.mhhe.com/biosci/abio/.

NEXUS

Some Interactions of the Digestive System with the Other Body Systems

Integumentary System

- The digestive system provides nutrients for all systems, including the integumentary system (p. 817).
- The skin produces vitamin D, which indirectly helps to regulate the intestinal absorption of Ca^{2+} (p. 194).
- Adipose tissue in the hypodermis of the skin stores triglycerides (p. 165).

Skeletal System

- The extracellular matrix of bones stores calcium phosphate (p. 193).
- The small intestine absorbs Ca^{2+} and PO_4^{-3}, which is needed for deposition of bone (p. 194).

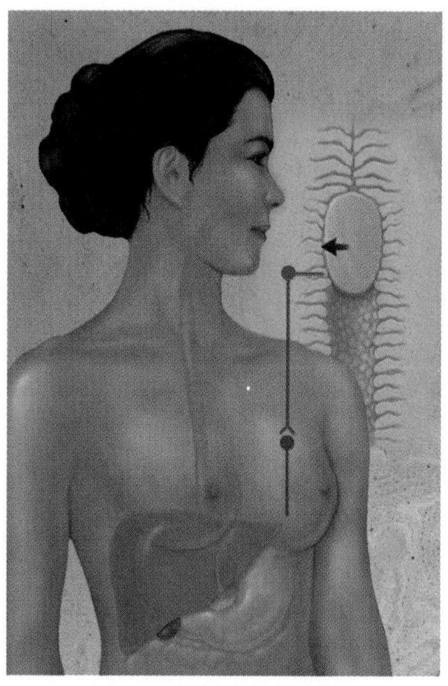

Nervous System

- Autonomic nerves help to regulate the digestive system (p. 822).
- The enteric nervous system functions like the CNS to regulate the intestine (p. 851).

Endocrine System

- Gastrin, produced by the stomach, helps to regulate the secretion of gastric juice (p. 830).
- Several hormones secreted by the small intestine regulate different aspects of the digestive system (p. 849).
- A hormone produced by the small intestine stimulates the pancreatic islets to secrete insulin (p. 850).
- Adipose tissue secretes leptin, which helps to regulate hunger (p. 873).
- The liver removes some hormones from the blood, changes them chemically, and excretes them in the bile (p. 842).

Muscular System

- Muscle contractions are needed for swallowing, peristalsis and segmentation (p. 835).
- Sphincter muscles help to regulate the passage of material along the GI tract (p. 829).

Lymphatic and Immune Systems

- The immune system protects against infections, including those of the digestive system (p. 696).
- Lymphatic vessels carry absorbed fat from the small intestine to the venous system (p. 855).
- The liver aids the immune system by metabolizing certain toxins and excreting them in the bile (p. 842).

Respiratory System

- The lungs provide oxygen for the metabolism of all organs, including those of the digestive system (p. 729).
- The oxygen provided by the respiratory system is used to metabolize food molecules brought into the body by the digestive system (p. 93).

Circulatory System

- The blood transports absorbed amino acids, monosaccharides, and other molecules from intestine to liver, and then to other organs (p. 643).
- The hepatic portal vein allows some absorbed molecules to have an enterohepatic circulation (p. 842).
- The intestinal absorption of vitamin B_{12} (needed for red blood cell production) requires intrinsic factor, secreted by the stomach (p. 830).
- Iron must be absorbed through the intestine to allow a normal rate of hemoglobin production (p. 599).

Urinary System

- The kidneys eliminate metabolic wastes from all organs, including those of the digestive system (p. 779).
- The kidneys help to convert vitamin D into the active form required for calcium absorption in the small intestine (p. 194).

Reproductive System

- Sex steroids, particularly androgens, stimulate the rate of fuel consumption by the body (p. 903).
- During pregnancy, the GI tract of the mother helps to provide nutrients that pass through the placenta to the embryo and fetus (p. 965).

Twenty-Seven

Regulation of Metabolism

Objectives

- Identify factors that influence the metabolic rate and explain the significance of the basal metabolic rate.
- Distinguish between the caloric and anabolic requirements for food and define the terms *essential amino acids* and *essential fatty acids*.
- Distinguish between fat-soluble and water-soluble vitamins and describe some of the functions of different vitamins.
- Define the terms *energy reserves* and *circulating energy substrates* and explain how these sources of energy interact during anabolism and catabolism.
- Describe the regulation of eating and discuss the endocrine control of metabolism in general terms.
- Describe the regulation of adipocyte development and the roles of adipocytes in the regulation of hunger and tissue responsiveness to insulin.
- Describe the actions of insulin and glucagon, and explain how the secretion of these hormones is regulated.
- Explain how insulin and glucagon regulate metabolism during feeding and fasting.
- Describe the symptoms of insulin-dependent and non-insulin-dependent diabetes mellitus and explain how these conditions are produced.
- Describe the metabolic effects of epinephrine and the glucocorticoids.
- Describe the effects of thyroxine on cell respiration and explain the relationship between thyroxine levels and the basal metabolic rate.
- Describe the symptoms of hypothyroidism and hyperthyroidism, and explain how these conditions are produced.
- Describe the metabolic effects of growth hormone and explain why growth hormone and thyroxine are needed for proper body growth.

Clinical Investigation

A middle-aged woman went to her physician complaining of nausea, headaches, frequent urination, and continuous thirst. In describing her medical history, she mentioned that both her mother and uncle were diabetics. The woman provided a sample of urine, which did not give evidence of glycosuria or ketonuria. She was told to return the next day to provide a fasting blood sample. When this was analyzed, a blood glucose concentration of 150 mg/dl was measured. An oral glucose tolerance test was subsequently performed, and a blood glucose concentration of 220 mg/dl was measured 2 hours following the ingestion of the glucose solution. The physician placed this patient on a weight-reduction program and advised her to begin a mild but regular exercise regimen. He told her that he would prescribe pills for her to take if the diet and exercise were not effective in relieving her symptoms.

What diagnosis did this physician make? Why did he make this diagnosis and the subsequent recommendations? What pills might he prescribe?

Clues: Read about the oral glucose tolerance test in the section "Regulation of Insulin and Glucagon Secretion." Study the information on non-insulin-dependent diabetes mellitus in "Clinical Considerations."

Nutritional Requirements

The body's energy requirements must be met by the caloric value of food to prevent catabolism of the body's own fat, carbohydrates, and protein. Additionally, food molecules—particularly the essential amino acids and fatty acids—are needed for replacement of molecules in the body that are continuously degraded. Vitamins and minerals do not directly provide energy but instead are required for diverse enzymatic reactions.

Living tissue is maintained by the constant expenditure of energy. This energy is obtained directly from ATP and indirectly from the cell respiration of glucose, fatty acids, ketone bodies, amino acids, and other organic molecules. These molecules are ultimately obtained from food, but they can also be obtained from the glycogen, fat, and protein stored in the body.

The energy value of food is commonly measured in **kilocalories** (*kil'ŏ-kal"ŏ-rēz*), which are also called "big calories" and spelled with a capital letter (Calories). One kilocalorie is equal to 1,000 calories; one calorie is defined as the amount of heat required to raise the temperature of 1 cubic centimeter of water from 14.5° to 15.5° C. As described in chapter 4, the amount of energy released as heat when a quantity of food is combusted in vitro is equal to the amount of energy released within cells through the process of aerobic respiration. This is 4 calories per gram for carbohydrates or proteins and 9 calories per gram for fat. When this energy is released by cell respiration, some is transferred to the high-energy bonds of ATP and some is lost as heat.

Metabolic Rate and Caloric Requirements

The total rate of body metabolism, or the **metabolic rate,** can be measured by either the amount of heat generated by the body or by the amount of oxygen consumed by the body per minute. This rate is influenced by a variety of factors. For example, the metabolic rate is increased by physical activity and by eating. The increased rate of metabolism that accompanies the assimilation of food can last more than 6 hours after a meal.

Temperature is also an important factor in determining metabolic rate. The reasons for this are twofold: (1) temperature itself influences the rate of chemical reactions and (2) the hypothalamus contains *temperature control centers*, as well as temperature-sensitive cells that act as sensors for changes in body temperature. In response to deviations from a "set point" for body temperature (chapter 1), the control areas of the hypothalamus can direct physiological responses that help to correct the deviations and maintain a constant body temperature. Changes in body temperature are thus accompanied by physiological responses that influence the total metabolic rate.

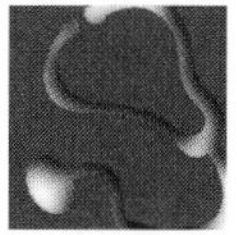

Hypothermia (low body temperature)—where the core body temperature is lowered to between 26° and 32° C (78° and 90° F)—is often induced during open heart or brain surgery. Compensatory responses to the lowered temperature are dampened by the general anesthetic, and the lower body temperature drastically reduces the needs of the tissues for oxygen. Under these conditions, the heart can be stopped and bleeding is significantly reduced.

The metabolic rate (measured by the rate of oxygen consumption) of an awake, relaxed person 12 to 14 hours after eating and at a comfortable temperature is known as the **basal metabolic rate (BMR).** The BMR is determined primarily by a person's age, sex, and body surface area, but it is also strongly influenced by the level of thyroid secretion. A person with hyperthyroidism has an abnormally high BMR, and a person with hypothyroidism has a low BMR. Interestingly, the BMR may be influenced by genetic inheritance; it appears that at least some families prone to obesity may have a genetically determined low BMR.

In general, however, individual differences in energy requirements are due primarily to differences in physical activity. Daily energy expenditures may range from 1,300 to 5,000 kilocalories per day. The average values for people not engaged in heavy manual labor but who are active during their leisure time are about 2,900 kilocalories per day for men and 2,100 kilocalories per day for women. People engaged in office work, the professions, sales, and comparable occupations consume up to 5 kilocalories per minute during work. More physically demanding occupations may require energy expenditures of 7.5 to 10 kilocalories per minute.

Table 27.1
Energy Consumed (in Kilocalories per Minute) in Various Activities

Activity	Weight in Pounds 105–115	127–137	160–170	182–192
Bicycling				
10 mph	5.41	6.16	7.33	7.91
Stationary, 10 mph	5.50	6.25	7.41	8.16
Calisthenics	3.91	4.50	7.33	7.91
Dancing				
Aerobic	5.83	6.58	7.83	8.58
Square	5.50	6.25	7.41	8.00
Gardening, weeding, and digging	5.08	5.75	6.83	7.50
Jogging				
5.5 mph	8.58	9.75	11.50	12.66
6.5 mph	8.90	10.20	12.00	13.20
8.0 mph	10.40	11.90	14.10	15.50
9.0 mph	12.00	13.80	16.20	17.80
Rowing, machine				
Easily	3.91	4.50	5.25	5.83
Vigorously	8.58	9.75	11.50	12.66
Skiing				
Downhill	7.75	8.83	10.41	11.50
Cross-country, 5 mph	9.16	10.41	12.25	13.33
Cross-country, 9 mph	13.08	14.83	17.58	19.33
Swimming, crawl				
20 yards per minute	3.91	4.50	5.25	5.83
40 yards per minute	7.83	8.91	10.50	11.58
55 yards per minute	11.00	12.50	14.75	16.25
Walking				
2 mph	2.40	2.80	3.30	3.60
3 mph	3.90	4.50	5.30	5.80
4 mph	4.50	5.20	6.10	6.80

When the caloric intake is greater than the energy expenditures, excess calories are stored primarily as fat. This is true regardless of the source of the calories—carbohydrates, protein, or fat—because these molecules can be converted to fat by the metabolic pathways described in chapter 4.

Weight is lost when the caloric value of the food ingested is less than the amount required in cell respiration over a period of time. Weight loss, therefore, can be achieved by dieting alone or in combination with an exercise program to raise the metabolic rate. A summary of the caloric expenditure associated with different forms of exercise is provided in table 27.1. Recent experiments, however, demonstrate why it is often so difficult to lose (or gain) weight. When subjects were maintained at 10% less than their usual weight, their metabolic rate decreased, and when they were maintained at 10% greater than their usual body weight, their metabolic rate increased. The body, it seems, tends to defend its usual weight by altering the energy expenditure as well as by regulating the food intake. This regulation is achieved, in part, by hormones secreted from adipose tissue (described in a later section).

Anabolic Requirements

In addition to providing the body with energy, food also supplies the raw materials for synthesis reactions—collectively termed **anabolism**—that occur constantly within the cells of the body. Anabolic reactions include those that synthesize DNA and

Table 27.2

Recommended Daily Allowances

Category	Age (Years) or Condition	Weight (kg)	Weight (lb)	Height (cm)	Height (in)	Protein (g)	Fat-Soluble Vitamins: Vitamin A (μg RE)[1]	Vitamin D (μg)[2]	Vitamin E (mg α-TE)[3]	Vitamin K (μg)
Infants	0.0–0.5	6	13	60	24	13	375	7.5	3	5
	0.5–1.0	9	20	71	28	14	375	10	4	10
Children	1–3	13	29	90	35	16	400	10	6	15
	4–6	20	44	112	44	24	500	10	7	20
	7–10	28	62	132	52	28	700	10	7	30
Males	11–14	45	99	157	62	45	1,000	10	10	45
	15–18	66	145	176	69	59	1,000	10	10	65
	19–24	72	160	177	70	58	1,000	10	10	70
	25–50	78	174	176	70	63	1,000	5	10	80
	51+	77	170	173	68	63	1,000	5	10	80
Females	11–14	46	101	157	62	46	800	10	8	45
	15–18	55	120	163	64	44	800	10	8	55
	19–24	58	128	164	65	46	800	10	8	60
	25–50	63	138	163	64	50	800	5	8	65
	51+	65	143	160	63	50	800	5	8	65
Pregnant						60	800	10	10	65
Lactating	1st 6 months					65	1,300	10	12	65
	2nd 6 months					62	1,200	10	11	65

Source: Reprinted with permission from *Recommended Dietary Allowances*, 10th Edition, © 1989 by the National Academy of Sciences. Courtesy National Academy Press, Washington, D.C.

[1]Retinol equivalent (1 RE = 1 μg retinol or 6 μg β-carotene)

[2]As cholecalciferol (10 μg cholecalciferol = 400 W of Vitamin D)

[3]α-tocopherol equivalents (1 mg α-tocopherol = 1 α-TE)

[4]Niacin equivalent (1 NE = 1 mg of niacin or 60 mg of dietary tryptophan)

RNA, protein, glycogen, triglycerides, and other polymers. These anabolic reactions must occur constantly to replace those molecules that are hydrolyzed into their component monomers. These hydrolysis reactions, together with the reactions of cell respiration that ultimately break down the monomers to carbon dioxide and water, are collectively termed **catabolism.**

Acting through changes in hormonal secretion, exercise and fasting increase the catabolism of stored glycogen, fat, and body protein. These molecules are also broken down at a certain rate in a person who is neither exercising nor fasting. Some of the monomers thus formed (amino acids, glucose, and fatty acids) are used immediately to resynthesize body protein, glycogen, and fat. However, some of the glucose derived from stored glycogen, for example, or fatty acids derived from stored triglycerides, are used to provide energy in the process of cell respiration. For this reason, new monomers must be obtained from food to prevent a continual decline in the amount of protein, glycogen, and fat in the body.

The *turnover rate* of a particular molecule is the rate at which it is broken down and resynthesized. For example, the average daily turnover for carbohydrates is 250 g/day. Since some of the glucose in the body is reused to form glycogen, the average daily dietary requirement for carbohydrate is somewhat less than this amount—about 150 g/day. The average daily turnover for protein is 150 g/day, but since many of the amino acids derived from the catabolism of body proteins can be reused in protein synthesis, a person needs only about 35 g/day of protein in the diet. It should be noted that these are average figures and will vary in accordance with individual differences in size, sex, age, genetics, and physical activity. The average daily turnover for fat is about 100 g/day, but very little is required in the diet (other than that which supplies fat-soluble vitamins and essential fatty acids), since fat can be produced from excess carbohydrates.

The minimal amounts of dietary protein and fat required to meet the turnover rate are adequate only if they supply sufficient amounts of the essential amino acids and fatty acids. As described in chapter 4, these molecules are termed *essential* because they are required for protein synthesis but cannot be synthesized by the body and must be obtained in the diet. The nine **essential amino acids** (see table 4.4) are lysine, tryptophan, phenylalanine, threonine, valine, methionine, leucine, isoleucine, and histidine. The **essential fatty acids** are linoleic acid and linolenic acid.

Water-Soluble Vitamins							Minerals						
Vitamin C (mg)	Thiamin (mg)	Riboflavin (mg)	Niacin (mg NE)[4]	Vitamin B_6 (mg)	Folate (μg)	Vitamin B_{12} (μg)	Calcium (mg)	Phosphorus (mg)	Magnesium (mg)	Iron (mg)	Zinc (mg)	Iodine (μg)	Selenium (μg)
30	0.3	0.4	5	0.3	25	0.3	400	300	40	6	5	40	10
35	0.4	0.5	6	0.6	35	0.5	600	500	60	10	5	50	15
40	0.7	0.8	9	1.0	50	0.7	800	800	80	10	10	70	20
45	0.9	1.1	12	1.1	75	1.0	800	800	120	10	10	90	20
45	1.0	1.2	13	1.4	100	1.4	800	800	170	10	10	120	30
50	1.3	1.5	17	1.7	150	2.0	1,200	1,200	270	12	15	150	40
60	1.5	1.8	20	2.0	200	2.0	1,200	1,200	400	12	15	150	50
60	1.5	1.7	19	2.0	200	2.0	1,200	1,200	350	10	15	150	70
60	1.5	1.7	19	2.0	200	2.0	800	800	350	10	15	150	70
60	1.2	1.4	15	2.0	200	2.0	800	800	350	10	15	150	70
50	1.1	1.3	15	1.4	150	2.0	1,200	1,200	280	15	12	150	45
60	1.1	1.3	15	1.5	180	2.0	1,200	1,200	300	15	12	150	50
60	1.1	1.3	15	1.6	180	2.0	1,200	1,200	280	15	12	150	55
60	1.1	1.3	15	1.6	180	2.0	800	800	280	15	12	150	55
60	1.0	1.2	13	1.6	180	2.0	800	800	280	10	12	150	55
70	1.5	1.6	17	2.2	400	2.2	1,200	1,200	300	30	15	175	65
95	1.6	1.8	20	2.1	280	2.6	1,200	1,200	355	15	19	200	75
90	1.6	1.7	20	2.1	260	2.6	1,200	1,200	340	15	16	200	75

Unsaturated fatty acids—those with double bonds between the carbons—are characterized by the location of the first double bond. Linoleic acid, found in corn oil, contains 18 carbons and two double bonds. It has its first double bond on the sixth carbon from the methyl (CH_3) end, and is therefore designated as an n-6 fatty acid. Linolenic acid, found in canola oil, also has 18 carbons, but it has three double bonds. More significantly for health, its first double bond is on the third carbon from the methyl end; linolenic acid is an n-3 (also called omega-3) fatty acid. Several studies suggest that n-3 fatty acids may offer protection from cardiovascular disease.

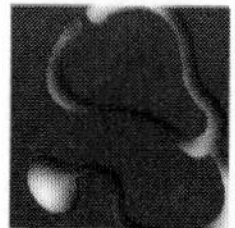

Eskimos who eat a traditional diet of meat and fish have a surprisingly low blood concentration of triglycerides and cholesterol and a low incidence of ischemic heart disease, despite the high fat and cholesterol content of their food. Several studies suggest that the n-3 fatty acids of the cold-water fish are the source of the apparent protective effect. The n-3 fatty acids of fish include *eicosapentaenoic acid,* or *EPA,* (with 20 carbons) and *docosahexaenoic acid,* or *DHA* (with 22 carbons). The n-3 fatty acids may help to inhibit platelet function in thrombus formation, the progression of atherosclerosis, and/or ventricular arrhythmias. Several studies have confirmed the protective effect of fish and fish oil in the diet, and on the basis of this evidence it seems prudent to eat fish at least once or twice a week on a continuing basis.

Vitamins and Minerals

Vitamins are small organic molecules that serve as coenzymes in metabolic reactions or that have other specific functions. They must be obtained in the diet because the body either doesn't produce them, or it produces them in insufficient quantities. (Vitamin D is produced in small quantities by the skin, and the B vitamins and vitamin K are produced by intestinal bacteria.) There are two classes of vitamins: fat-soluble and water-soluble. The **fat-soluble vitamins** include vitamins A, D, E, and K. The **water-soluble vitamins** include thiamine (B_1), riboflavin (B_2), niacin (B_3), pyridoxine (B_6), pantothenic acid, biotin, folic acid, vitamin B_{12}, and vitamin C (ascorbic acid). Recommended daily allowances for these vitamins are listed in table 27.2.

Water-Soluble Vitamins

Derivatives of water-soluble vitamins serve as coenzymes in the metabolism of carbohydrates, lipids, and proteins. **Thiamine** (*thi'ă-min* or *thi'ă-mēn*), for example, is needed for the activity of the enzyme that converts pyruvic acid to acetyl coenzyme A. **Riboflavin** and niacin are needed for the production of FAD and NAD, respectively. FAD and NAD serve as coenzymes that transfer hydrogens during cell respiration (chapter 4). **Pyridoxine** is a cofactor for the enzymes involved in amino acid metabolism. Deficiencies of the

Table 27.3
The Major Vitamins

Vitamin	Description/Comments	Deficiency Symptoms	Source
A	Constituent of visual pigment; strengthens epithelial membranes	Night blindness; dry skin	Yellow vegetables and fruit
B_1 (Thiamine)	Cofactor for enzymes that catalyze decarboxylation	Beriberi; neuritis	Liver, unrefined cereal grains
B_2 Riboflavin	Part of flavoproteins (such as FAD)	Glossitis; cheilosis	Liver, milk
B_6 (Pyridoxine)	Coenzyme for decarboxylase and transaminase enzymes	Convulsions	Liver, corn, wheat, and yeast
B_{12} (Cyanocobalamin)	Coenzyme for amino acid metabolism; needed for erythropoiesis	Pernicious anemia	Liver, meat, eggs, milk
Biotin	Needed for fatty acid synthesis	Dermatitis; enteritis	Egg yolk, liver, tomatoes
C	Needed for collagen synthesis in connective tissue	Scurvy	Citrus fruits, green leafy vegetables
D	Needed for intestinal absorption of calcium and phosphate	Rickets; osteomalacia	Fish, liver
E	Antioxidant	Muscular dystrophy	Milk, eggs, meat, leafy vegetables
Folates	Needed for reactions that transfer one carbon	Sprue; anemia	Green leafy vegetables
K	Promotes reactions needed for function of clotting factors	Hemorrhage; inability to form clot	Green leafy vegetables
Niacin	Part of NAD and NADP	Pellagra	Liver, meat, yeast
Pantothenic acid	Part of coenzyme A	Dermatitis; enteritis; adrenal insufficiency	Liver, eggs, yeast

water-soluble vitamins can thus have widespread effects in the body (table 27.3).

Free radicals are highly reactive molecules that carry an unpaired electron. An important example is the superoxide free radical, composed of two oxygen atoms with an unpaired electron. Such free radicals can damage tissues by removing an electron from (and thus oxidizing,) other molecules. **Vitamin C,** together with β-carotene (a precursor of vitamin A), and vitamin E (a fat-soluble vitamin) function as *antioxidants* through their ability to inactivate free radicals. These vitamins may afford protection against some of the causes of atherosclerosis and cancer.

Fat-Soluble Vitamins

Some fat-soluble vitamins have highly specialized functions. **Vitamin K,** for example, is required for the production of prothrombin and for clotting factors VII, IX, and X. Vitamins A and D also have functions unique to each, but these two vitamins overlap in their mechanisms of action.

Vitamin A is a collective term for a number of molecules that include *retinol* (the transport form of vitamin A), *retinal* (also known as *retinaldehyde*, used as the photopigment in the retina), and *retinoic acid*. Most of these molecules are ultimately derived from dietary β-carotene, present in such foods as carrots, leafy vegetables, and egg yolk. The β-carotene is converted by an enzyme in the intestine into two molecules of retinal. Most of the retinal is reduced to retinol, while some is oxidized to retinoic acid. It is the retinoic acid that binds to nuclear receptor proteins (chapter 19) and directly produces the effects of vitamin A. Retinoic acid is involved, for example, in regulating embryonic development; vitamin A deficiency interferes with embryonic development, while excessive vitamin A during pregnancy can cause birth defects. Retinoic acid is also needed for the maintenance of epithelial membrane structure and function. Indeed, retinoids are now widely used to treat acne and other skin conditions.

Vitamin D is produced by the skin under the influence of ultraviolet light, but usually it is not produced in sufficient amounts for all of the body's needs. That is why we must eat food containing additional amounts of vitamin D, and why it is classified as a vitamin even though it can be produced by the body. The vitamin D secreted by the skin or consumed in the diet is inactive in its original form; it must first be converted into a derivative by enzymes in the liver and kidneys

Table 27.4

Estimated Safe and Adequate Daily Dietary Intakes of Selected Vitamins and Minerals[1]

		Vitamins		Trace Elements[2]				
Category	Age (Years)	Biotin (μg)	Pantothenic Acid (mg)	Copper (mg)	Manganese (mg)	Fluorine (mg)	Chromium (μg)	Molybdenum (μg)
Infants	0–0.5	10	2	0.4–0.6	0.3–0.6	0.1–0.5	10–40	15–30
	0.5–1	15	3	0.6–0.7	0.6–1.0	0.2–1.0	20–60	20–40
Children and adolescents	1–3	20	3	0.7–1.0	1.0–1.5	0.5–1.5	20–80	25–50
	4–6	25	3–4	1.0–1.5	1.5–2.0	1.0–2.5	30–120	30–75
	7–10	30	4–5	1.0–2.0	2.0–3.0	1.5–2.5	50–200	50–150
	11+	30–100	4–7	1.5–2.5	2.0–5.0	1.5–2.5	50–200	75–250
Adults		30–100	4–7	1.5–3.0	2.0–5.0	1.5–4.0	50–200	75–250

Source: Reprinted with permission from *Recommended Dietary Allowances*, 10th Edition. Copyright 1989 by the National Academy of Sciences. Courtesy of the National Academy Press, Washington, D.C.

[1]Because there is less information on which to base allowances, these figures are not given in the main table of RDA and are provided here in the form of ranges of recommended intakes.

[2]Since the toxic levels for many trace elements may be only several times usual intakes, the upper levels for the trace elements given in this table should not be habitually exceeded.

before it can be active in the body. Once the active derivative is produced, vitamin D helps to regulate calcium balance.

As may be recalled from chapter 19, a nuclear receptor for the active form of thyroid hormone or of vitamin D consists of two different polypeptides. One polypeptide binds to thyroid hormone (thyroid receptor, or *TR*) or vitamin D (vitamin D receptor, or *DR*), and one polypeptide binds to retinoic acid (retinoic acid X receptor, or *RXR*). This overlapping of receptors may permit "cross-talk" between the actions of thyroid hormone, vitamin D, and vitamin A. In view of this, it is not surprising that thyroxine, vitamin A, and vitamin D have overlapping functions—all three are involved in regulating gene expression and promoting differentiation (specialization) of tissues.

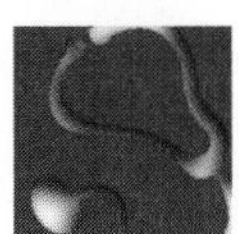

The best known function of vitamin D is regulation of calcium balance. As described in chapter 8, vitamin D promotes the intestinal absorption of Ca^{2+} and PO_4^{3-}, and thus is needed for proper calcification of the bones. However, a vitamin D derivative called *calcipotriene* is now widely used for the treatment of *psoriasis* (*sŏ-rī′ă-sis*), a skin condition characterized by inflammation and excessive proliferation of keratinocytes (the cells of the epidermis that produce keratin). In this case, the vitamin D analogue inhibits proliferation and promotes differentiation of the keratinocytes. It has been suggested that vitamin D produced in the skin may function as an autocrine regulator of the epidermis.

Minerals (Elements)

Minerals, or **elements,** are needed as cofactors for specific enzymes and for a wide variety of other critical functions. Those that are required daily in relatively large amounts include sodium, potassium, magnesium, calcium, phosphorus, and chlorine (see table 27.2). In addition, the following **trace elements** are recognized as essential: iron, zinc, manganese, fluorine, copper, molybdenum (*mŏ-lib′dĕ-num*), chromium, and selenium. These must be ingested in amounts ranging from 50 mg to 18 mg per day (table 27.4).

Regulation of Energy Metabolism

The blood plasma contains circulating glucose, fatty acids, amino acids, and other molecules that can be used by the body tissues for cell respiration. These circulating molecules may be derived from food or from the breakdown of the body's own glycogen, fat, and protein. The building of the body's energy reserves following a meal and the utilization of these reserves between meals are regulated by the action of a number of hormones that act to promote either anabolism or catabolism.

The molecules that can be oxidized for energy by the processes of cell respiration may be derived from **energy reserves** of glycogen, fat, or protein. Glycogen and fat function primarily as energy reserves; for proteins, by contrast, this represents a

Figure 27.1

A flowchart of energy pathways in the body.
The molecules indicated in the top and bottom rectangles are those found within cells, whereas the molecules in the middle rectangle are those that circulate in the blood.

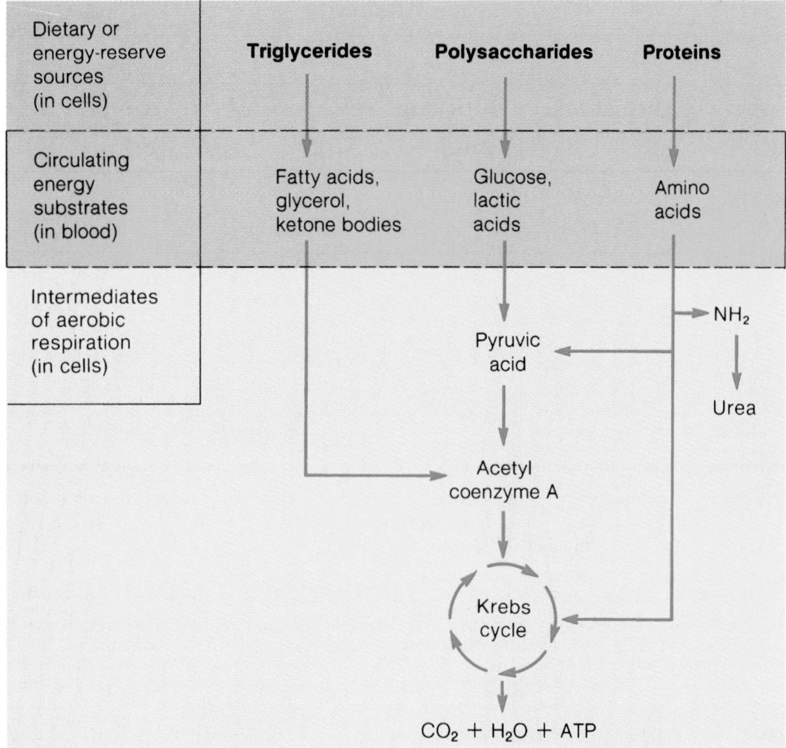

secondary emergency function. Although body protein can provide amino acids for energy, it can do so only through the breakdown of proteins needed for muscle contraction, structural strength, enzymatic activity, and other functions. Alternatively, the molecules used for cell respiration can be derived from the products of digestion that are absorbed through the small intestine. Since these molecules—glucose, fatty acids, amino acids, and others—are carried by the blood to the cells for use in cell respiration, they can be called **circulating energy substrates** (fig. 27.1).

Because of differences in cellular enzyme content, different organs have different *preferred energy sources*. The brain has an almost absolute requirement for blood glucose as its energy source, for example. A fall in the blood glucose concentration to below about 50 mg per 100 ml can thus "starve" the brain and have disastrous consequences. Resting skeletal muscles, by contrast, use fatty acids as their preferred energy source. Similarly, ketone bodies (derived from fatty acids), lactic acid, and amino acids can be used to different degrees as energy sources by various organs. The blood plasma normally contains adequate concentrations of all of these circulating energy substrates to meet the energy needs of the body.

Eating

Ideally, one should eat the kinds and amounts of foods that provide adequate vitamins, minerals, essential amino acids and fatty acids, and calories. Proper caloric intake maintains energy reserves (primarily fat and glycogen) and results in a body weight within an optimum range for health.

Eating behavior appears to be at least partially controlled by areas of the hypothalamus. Lesions (destruction) in the ventromedial area of the hypothalamus produce *hyperphagia* (*hi″per-fa′je-ă*), or overeating, and obesity in experimental animals. Lesions of the lateral hypothalamus, by contrast, produce *hypophagia* and weight loss. More recent experiments demonstrate that other brain regions are also involved in the control of eating behavior.

The chemical neurotransmitters that may be involved in neural pathways mediating eating behavior are being investigated. There is evidence, for example, that endorphins may be involved because injections of naloxone (a morphine-blocking drug) suppress overeating in rats. There is also evidence that the neurotransmitters norepinephrine and serotonin may be involved; injections of norepinephrine into the brain cause overeating in rats, whereas injections of serotonin have the opposite effect. Indeed, the diet pills *Redux* (D-fenfluramine) and *fen-phen* (a combination of L-fenfluramine and phentermine) were often prescribed to reduce hunger by elevating brain levels of serotonin. (Both pills were recently taken off the market because of their association with heart valve problems.)

Interestingly, the intestinal hormone cholecystokinin (CCK), which is also produced in the brain, has been found to promote satiety. Injections of CCK cause experimental animals and humans to stop eating. Also, glucagon-like peptide-1 (GLP-1), secreted by the ileum and colon, promotes satiety (at least in rodents). These two peptides may be a means by which the intestine, when it receives a load of chyme, can signal the brain to suppress appetite.

Regulatory Functions of Adipose Tissue

It is difficult for a person to lose (or gain) weight, many scientists believe, because the body has negative feedback loops that act to "defend" a particular body weight, or more accurately, the amount of adipose tissue. This regulatory system has been called an *adipostat* (*ad′ĭ-po-stat*). When a person eats more than is needed to maintain the set point of adipose tissue, the person's metabolic rate increases and hunger decreases, as previously described. Homeostasis of body weight implies negative feedback loops. Hunger and metabolism (acting through food and hormones) affect adipose cells, so in terms of negative feedback, it seems logical that adipose cells should influence hunger and metabolism.

Adipose cells, or **adipocytes** (*ad'ĭ-po-sīts*), store fat within large vacuoles during times of plenty and serve as sites for the release of circulating energy substrates, primarily free fatty acids, during times of fasting. Since the synthesis and breakdown of fat is controlled by hormones that act on these cells, the adipocytes traditionally have been viewed simply as passive storage depots of fat. Recent evidence suggests quite the opposite, however; adipocytes may themselves secrete hormones that play a pivotal role in the regulation of metabolism.

Development of Adipose Tissue

Some adipocytes appear during embryonic development, but their numbers increase greatly during a baby's first year. This increased number is due to both mitotic division of the adipocytes and the conversion of preadipocytes (derived from fibroblasts) into new adipocytes. This differentiation (specialization) is promoted by a high circulating level of fatty acids, particularly of saturated fatty acids. This represents a nice example of a negative feedback loop, where a rise in circulating fatty acids promotes processes that ultimately help to convert the fatty acids into stored fat.

The differentiation of adipocytes requires the action of a nuclear receptor protein—in the same family as the receptors for thyroid hormone, vitamin A, and vitamin D—known as **PPARγ.** (PPAR is an acronym for *p*eroxisome *p*roliferator *a*ctivated *r*eceptor, and the γ is the Greek letter gamma, indicating the subtype of PPAR.) Just as the thyroid receptor is activated when it is bound to its ligand, PPARγ is activated when it is bound to its own specific ligand, a type of prostaglandin abbreviated *15d-PGJ*$_2$. (The letter *d* stands for "deoxy-," and the letters *PG* for "prostaglandin.") This is a recently discovered autocrine regulator produced by adipocytes and some other tissues. When 15d-PGJ$_2$ binds to the PPARγ receptor, it stimulates adipogenesis by promoting the development of preadipocytes into mature adipocytes. This occurs primarily in children, since the development of new adipocytes is more limited in adults.

Regulation of Insulin Sensitivity

Adipocytes secrete a number of molecules that are also produced by other types of cells. One of the most interesting of these molecules is **tumor necrosis factor-alpha (TNF$_\alpha$)**. TNF$_\alpha$ is a cytokine that is also produced by macrophages and other cells of the immune system. Adipocyte secretion of TNF$_\alpha$ is increased in obesity, and there is evidence that increased secretion of TNF$_\alpha$ may contribute to the *insulin resistance* that is observed in obese people. "Insulin resistance" refers to a decreased sensitivity (primarily of skeletal muscles) to insulin, so that more insulin is required to maintain homeostasis. The possible role of adipocytes in insulin resistance has medical significance because of the association of insulin resistance with type II diabetes mellitus (see the section "Clinical Considerations" near the end of this chapter).

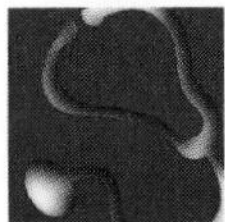

The insulin resistance of many diabetics may result, at least in part, from defective regulation by adipocytes. This is suggested by the action of *troglitazone,* one of a new class of thiazolidinedione drugs for the treatment of diabetes mellitus. These drugs have one known action: they specifically bind to PPARγ receptors in adipocytes and some other tissues. Through this action, the thiazolidinedione drugs lower the secretion of leptin and TNF$_\alpha$ from adipocytes. This may reduce the insulin resistance of skeletal muscles and thus help control the hyperglycemia of type II diabetes mellitus.

Regulation of Hunger

The possibility that adipose tissue secretes a hormonal *satiety factor* (a circulating chemical that decreases appetite) has been suspected for years on the basis of physiological evidence. According to this view, secretion of the satiety factor would increase following meals and decrease during fasting. Such a satiety factor could act through its regulation of the hunger centers in the hypothalamus.

The satiety factor secreted by adipose tissue has recently been identified. It is the product of a gene first observed in a strain of mice known as *ob/ob* (*ob* stands for "obese"; the double symbol indicates that the mice are homozygous for this gene—they inherit it from both parents). Mice of this strain display hyperphagia (they eat too much) and decreased energy consumption. The *ob* gene has been cloned in mice (and more recently in humans), and has been found to be expressed (produce mRNA) only in adipocytes. As expected, the expression of this gene is decreased during fasting and increased after feeding. The protein product of this gene, the presumed satiety factor, is a 167-amino-acid polypeptide now called **leptin.** The *ob* mice produce a mutated and ineffective form of leptin, and it is this defect that causes their obesity. When they are injected with normal leptin they stop eating and lose weight.

Scientists have also identified a few obese humans with defective leptin genes. However, studies show that in most obese people the activity of the *ob* gene and the blood concentrations of leptin are raised and that weight loss results in a lowering of plasma leptin concentrations. Thus, unlike the case of the *ob/ob* mice, the amount of leptin secretion in humans is correlated with body fat. It has therefore been suggested that most cases of obesity in humans may be associated with a reduced sensitivity of the brain to the actions of leptin.

Scientists have shown that three other genetically obese strains of rodents produce leptin but overeat anyway because they lack leptin receptors. In human obesity, however, the cause of the decreased sensitivity to leptin has not yet been determined.

In the *ob/ob* mice, it was observed that injections of leptin caused a decrease in the amount of neuropeptide Y in the hypothalamus. This observation provides a clue about how leptin might work. As discussed in chapter 14, neuropeptide Y is a potent stimulator of appetite. It functions as a neurotransmitter for axons that extend within the hypothalamus from the arcuate nucleus to the paraventricular nucleus, two regions implicated in the control of eating behavior. When weight is lost, a reduced secretion of leptin from the adipocytes may result in increased production of neuropeptide Y, which then stimulates hunger and food intake and decreases expenditure of energy.

When weight is gained, conversely, an increased secretion of leptin may reduce hunger by inhibiting neuropeptide Y release in the hypothalamus. The control of hunger, however, appears to be more complex than this. Scientists have discovered that appetite can be suppressed by melanocyte-stimulating hormone (MSH) or by a related neuropeptide of the *melanocortin* family that binds to a specific melanocortin receptor in the hypothalamus. It has thus been proposed that when weight is gained, the rising levels of leptin may increase the activity of these melanocortin pathways, suppressing appetite and increasing energy expenditure.

The course of this research is of interest to anyone concerned with weight control and obesity. Pharmaceutical companies, some with patents on the *ob* gene and the leptin receptor, have a particularly strong financial interest in the resolution of this fascinating area of human physiology.

Obesity

Obesity is a risk factor for cardiovascular diseases, diabetes mellitus, gallbladder disease, and some malignancies (particularly endometrial and breast cancer). The distribution of fat in the body is also important; there is a greater risk of cardiovascular disease when the distribution of fat produces a high waist-to-hip ratio, or an "apple shape," as compared to a "pear shape." This is because the amount of intra-abdominal fat in the mesenteries and greater omentum is a better predictor of a health hazard than is the amount of subcutaneous fat. In terms of the risk of diabetes mellitus, the larger adipocytes of the "apple shape" are less sensitive to insulin than the smaller adipocytes of the "pear shape."

Obesity in childhood is due to an increase in both the size and number of adipocytes; weight gain in adulthood is due mainly to an increase in adipocyte size, although the number of adipocytes may also increase to some degree. When weight is lost, the adipocytes get smaller but their number remains constant. It is thus important to prevent further increases in weight in all overweight people, but particularly so in children. This can best be achieved by a carefully chosen diet, low in saturated fat (because of the effect of fatty acids on adipocyte growth and differentiation, as previously described), and exercise. Prolonged exercise of low to moderate intensity promotes weight loss because, under these conditions, skeletal muscles use fatty acids as their primary source of energy.

Obesity is often diagnosed using a measurement called the **body mass index (BMI)**. This measurement is calculated using the formula shown below:

$$\mathbf{BMI} = \frac{w}{h^2}$$

where

w = weight in kilograms (pounds divided by 2.2)
h = height in meters (inches divided by 39.4)

Figure 27.2

Regulation of metabolic balance.
The balance of metabolism can be tilted toward anabolism (synthesis of energy reserves) or catabolism (utilization of energy reserves) by the combined actions of various hormones. Growth hormone and thyroxine have both anabolic and catabolic effects.

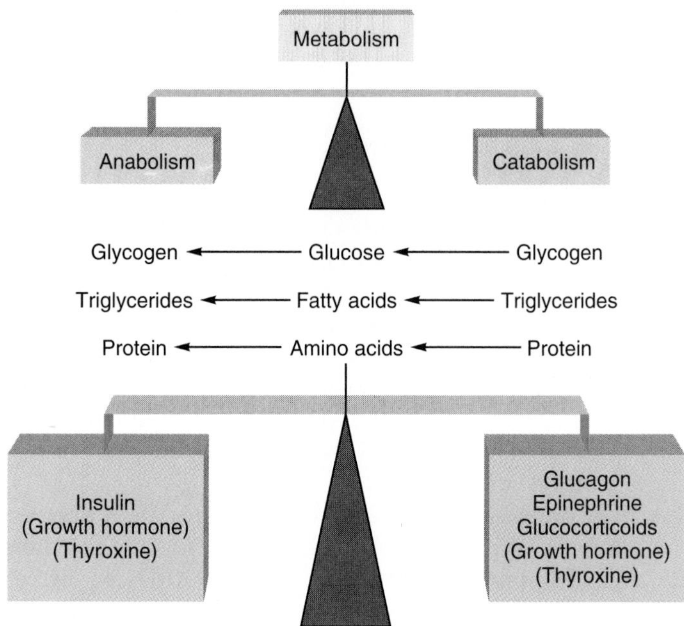

Obesity has been defined by different health agencies in different ways. The World Health Organization classifies people with a BMI of 30 or over as being at high risk for the diseases of obesity. According to the standards set by the National Institutes of Health, a healthy weight is indicated by a BMI of less than 25. Recent surveys, however, indicate that most Americans have a BMI above that level.

Hormonal Regulation of Metabolism

The absorption of energy carriers from the small intestine is not continuous; it rises to high levels during a 4-hour period following each meal (the **absorptive state**) and tapers toward zero between meals, after an absorptive state has ended (the **postabsorptive,** or **fasting, state**). Despite this fluctuation, the blood glucose concentration and other energy substrates does not remain high during periods of absorption, and it does not normally fall below a certain level during periods of fasting. During the absorption of digestion products from the small intestine, energy substrates are removed from the blood and deposited as energy reserves from which withdrawals can be made during times of fasting (fig. 27.2). This ensures an adequate plasma concentration of energy substrates to sustain tissue metabolism at all times.

The rate of deposit and withdrawal of energy substrates into and from the energy reserves and the conversion of one type of energy substrate into another are regulated by the actions of hormones. The balance between anabolism and catabolism is determined by the antagonistic effects of insulin,

Table 27.5

Endocrine Regulation of Metabolism

Hormone	Blood Glucose	Carbohydrate Metabolism	Protein Metabolism	Liquid Metabolism
Insulin	Decreased	↑Glycogen formation ↓Glycogenolysis ↓Gluconeogenesis	↑Amino acid transport	↑Lipogenesis ↓Lipolysis ↓Ketogenesis
Glucagon	Increased	↓Glycogen formation ↑Glycogenolysis ↑Gluconeogenesis	No direct effect	↑Lipolysis ↑Ketogenesis
Growth hormone	Increased	↑Glycogenolysis ↑Gluconeogenesis ↓Glucose utilization	↑Protein synthesis	↓Lipogenesis ↑Lipolysis ↑Ketogenesis
Glucocorticoids	Increased	↑Glycogen formation ↑Gluconeogenesis	↓Protein synthesis	↓Lipogenesis ↑Lipolysis ↑Ketogenesis
Epinephrine	Increased	↓Glycogen formation ↑Glycogenolysis ↑Gluconeogenesis	No direct effect	↑Lipolysis ↑Ketogenesis
Thyroxine	No effect	↑Glucose utilization	↑Protein synthesis	No direct effect

glucagon, growth hormone, thyroxine, and other hormones (fig. 27.2). The specific metabolic effects of these hormones are summarized in table 27.5, and some of their actions are illustrated in figure 27.3.

Energy Regulation by the Pancreatic Islets

Insulin secretion is stimulated by a rise in the blood glucose concentration, and insulin promotes the entry of blood glucose into tissue cells. Insulin thus increases the storage of glycogen and fat while causing the blood glucose concentration to fall. Glucagon secretion is stimulated by a fall in blood glucose, and glucagon acts to raise the blood glucose concentration by promoting glycogenolysis in the liver.

Scattered within a "sea" of pancreatic exocrine tissue (the acini) are islands of hormone-secreting cells. These **pancreatic islets,** or **islets of Langerhans** (fig. 27.4), contain three distinct cell types that secrete different hormones. The most numerous are the *beta cells,* which secrete the hormone **insulin.** About 60% of each islet consists of beta cells. The *alpha cells* form about 25% of each islet and secrete the hormone **glucagon** (*gloo'kă-gon*). The least numerous cell type, the *delta cells,* produce **somatostatin** (*so"mă-tŏ-stăt'n*), the composition of which is identical to the somatostatin produced by the hypothalamus and the small intestine.

All three pancreatic hormones are polypeptides. Insulin consists of two polypeptide chains—one that is 21 amino acids long and another that is 30 amino acids long—joined together by disulfide bonds. Glucagon contains 21 amino acids, and somatostatin contains 14. Insulin was the first of these hormones to be discovered (in 1921). The importance of insulin in diabetes mellitus was immediately recognized, and clinical use of insulin in the treatment of this disease began almost immediately after its discovery. The physiological role of glucagon was discovered later, but the physiological significance of islet-secreted somatostatin is still not well understood.

Regulation of Insulin and Glucagon Secretion

Insulin and glucagon secretion is largely regulated by the plasma concentrations of glucose and, to a lesser degree, of amino acids. The alpha and beta cells, therefore, act as both the sensors and effectors in this control system. Since the blood glucose concentration and amino acids rise during the absorption of a meal and fall during fasting, the secretion of insulin and glucagon likewise fluctuates between the absorptive and postabsorptive states. These changes in insulin and glucagon secretion, in turn, cause changes in blood glucose and amino acid concentrations and thus help to maintain homeostasis via negative feedback loops (fig. 27.5).

The mechanisms that regulate insulin and glucagon secretion and the actions of these hormones normally prevent the blood glucose concentration from rising above 170 mg per 100 ml after a meal or from falling below about 50 mg per 100 ml between meals. This regulation is important because abnormally high blood glucose can damage certain tissues (as may occur in diabetes mellitus), and abnormally low blood

Figure 27.3

Hormonal interactions in metabolic regulation.

Different hormones can work together synergistically or can have an antagonistic effect on metabolism. (⊕ = stimulatory effects; ⊖ = inhibitory effects.)

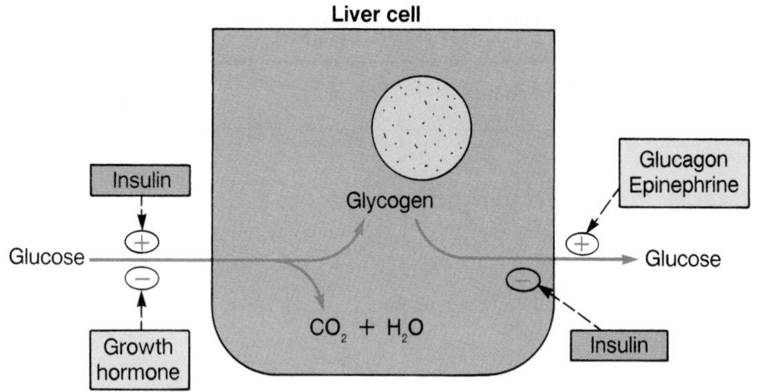

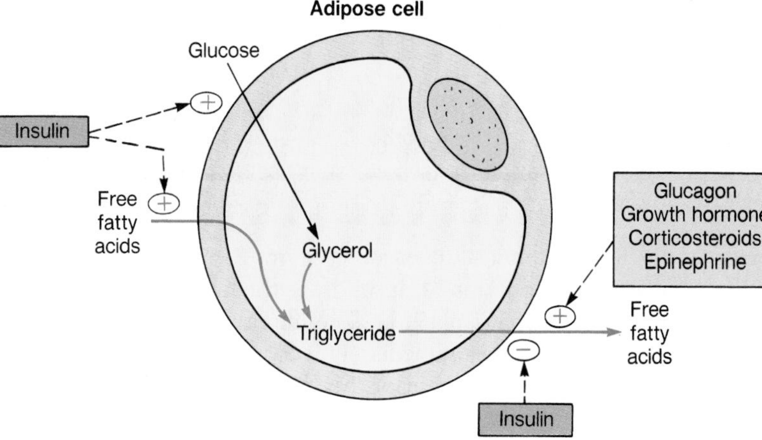

Figure 27.4

Photomicrograph of a pancreatic islet.

A normal pancreatic islet (of Langerhans) is visualized with the aid of fluorescently labeled antibodies that stain the cytoplasm green. The dark dots are nuclei.

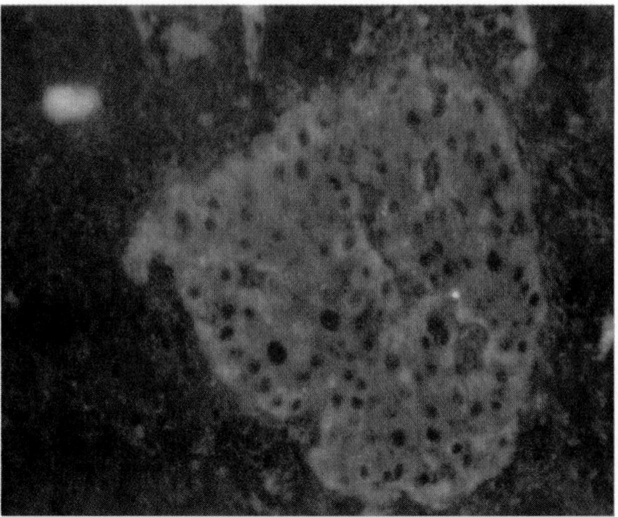

glucose can damage the brain. The later effect results from the fact that glucose enters the brain by facilitated diffusion; when the rate of this diffusion is too low, as a result of low blood glucose concentrations, the supply of metabolic energy for the brain may become inadequate. This can result in weakness, dizziness, personality changes, and ultimately in coma and death.

Effects of Glucose and Amino Acids

The fasting blood glucose concentration is 65 to 105 mg/dl. During the absorption of a meal, the blood glucose concentration usually rises to 140 to 150 mg/dl. This rise in blood glucose (1) stimulates the beta cells to secrete insulin and (2) inhibits the secretion of glucagon from the alpha cells. Insulin acts to stimulate the cellular uptake of blood glucose. A rise in insulin secretion therefore lowers the blood glucose concentration. Since glucagon has the antagonistic effect of raising the blood glucose concentration by stimulating glycogenolysis in the liver, the inhibition of glucagon secretion complements the effect of increased insulin during the absorption of a carbohydrate meal. A rise in insulin and a fall in glucagon secretion thus help to lower the high blood glucose concentration that occurs during periods of absorption.

During fasting, the blood glucose concentration falls. At this time, therefore, (1) insulin secretion decreases and (2) glucagon secretion increases. These changes in hormone secretion prevent the cellular uptake of blood glucose into organs such as the muscles, liver, and adipose tissue and promote the release of glucose from the liver (through the actions of glucagon). A negative feedback loop is therefore completed (fig. 27.5), helping to retard the fall in blood glucose concentration that occurs during fasting.

The **oral glucose tolerance test** (fig. 27.6) is a measure of the ability of the beta cells to secrete insulin and of the ability of insulin to lower blood glucose. In this procedure, a person drinks a glucose solution and then blood samples are taken periodically for blood glucose measurements. In a normal person, the rise in blood glucose produced by drinking this solution is reversed to normal levels within 2 hours following glucose ingestion.

Insulin secretion is also stimulated by particular amino acids derived from dietary proteins. Meals that are high in protein, therefore, stimulate the secretion of insulin; if the meal is high in protein and low in carbohydrates, glucagon secretion will be stimulated as well. The increased glucagon secretion acts to raise the blood glucose, and the increased insulin promotes the entry of amino acids into the cells.

Effects of Autonomic Nerves

The pancreatic islets have both parasympathetic and sympathetic innervation. The activation of the parasympathetic

Figure 27.5

Regulation of insulin and glucagon secretion.

The secretion from the β (beta) cells and α (alpha) cells of the pancreatic islet (of Langerhans) is regulated largely by the blood glucose concentration. (*a*) A high blood glucose concentration stimulates insulin and inhibits glucagon secretion. (*b*) A low blood glucose concentration stimulates glucagon and inhibits insulin secretion.

Source: From Mary L. Parker et al., "Juvenile Diabetes Mellitus, A deficiency in Insulin," in Diabetes *17:27–32. Copyright © 1968 American Diabetes Association, Inc. Reproduced with permission from the American Diabetes Association, Inc.*

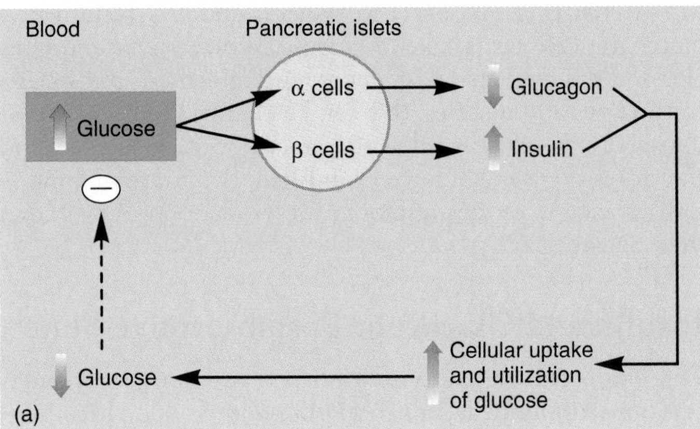

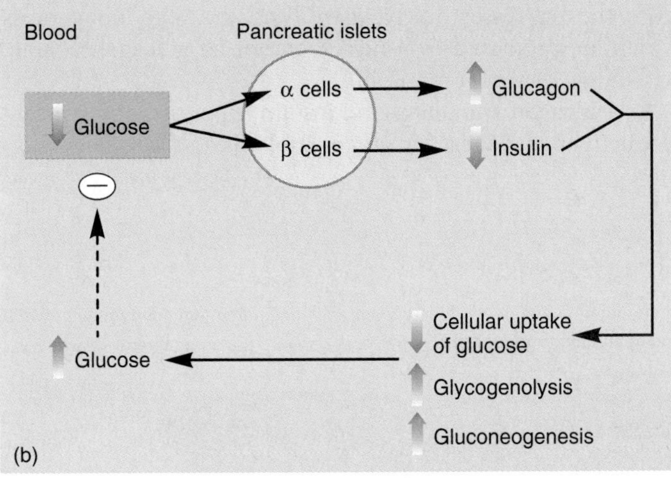

system during meals stimulates insulin secretion at the same time that gastrointestinal function is stimulated. The activation of the sympathetic system, by contrast, stimulates glucagon secretion and inhibits insulin secretion. The effects of glucagon, together with those of epinephrine, produce a "stress hyperglycemia" when the sympathoadrenal system is activated.

Effects of Intestinal Hormones

Surprisingly, insulin secretion increases more rapidly following glucose ingestion than it does following an intravenous injection of glucose. This is because the small intestine, in response to glucose ingestion, secretes hormones that stimulate insulin secretion before the glucose has been absorbed. Insulin secretion thus begins to rise "in anticipation" of a rise in blood glucose. One of the intestinal hormones that mediates this effect

Figure 27.6

Oral glucose tolerance test.

Changes in blood glucose and plasma insulin concentrations after the ingestion of 100 grams of glucose in an oral glucose tolerance test. The insulin is measured in activity units (U).

Source: From Mary L. Parker et al., "Juvenile Diabetes Mellitus, A deficiency in Insulin," in Diabetes *17:27–32. Copyright © 1968 American Diabetes Association, Inc. Reproduced with permission from the American Diabetes Association, Inc.*

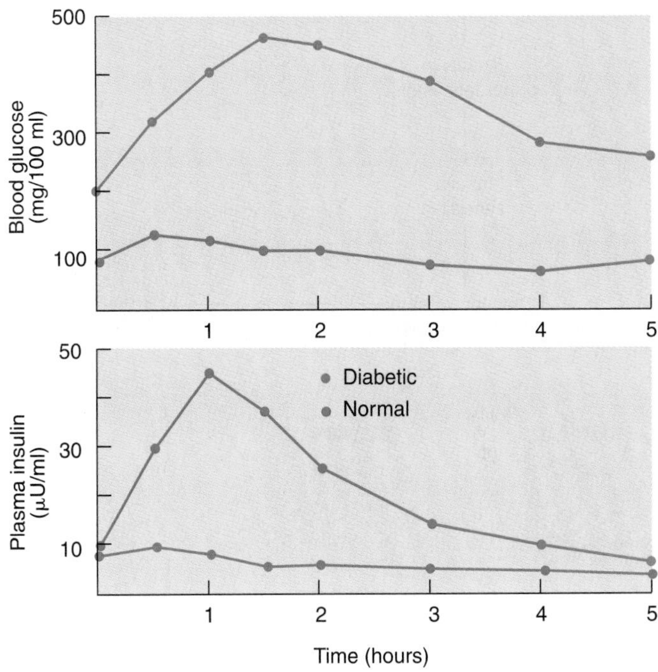

is GIP—gastric inhibitory peptide, or, more appropriately in this context, *glucose-dependent insulinotropic peptide* (chapter 26). Other polypeptide hormones secreted by the intestine that have similar effects are cholecystokinin (CCK) and glucagon-like peptide-1 (GLP-1), as described in chapter 26.

Insulin and Glucagon: Absorptive State

Insulin acts on skeletal muscle cells and adipocytes to stimulate the incorporation of a glucose facilitative diffusion carrier called **GLUT4** into their cell membranes. As a result of the facilitative diffusion of glucose into these cells, the blood glucose concentration will be decreased, thereby completing the negative feedback loop that stimulated insulin secretion (fig. 27.7).

The lowering of blood glucose by insulin is, in a sense, a side effect of the primary action of this hormone. Insulin is the major hormone that promotes anabolism in the body. During absorption of the products of digestion and the subsequent rise in the blood concentrations of circulating energy substrates, insulin promotes the cellular uptake of blood glucose and its incorporation into energy-reserve molecules of glycogen in the liver and muscles, and of triglycerides in adipose cells (fig. 27.7). Quantitatively, skeletal muscles are responsible for most of the insulin-stimulated glucose uptake.

Figure 27.7

Homeostasis of blood glucose.
A rise in blood glucose concentration stimulates insulin secretion. Insulin promotes a fall in blood glucose by stimulating the cellular uptake of glucose and the conversion of glucose to glycogen and fat.

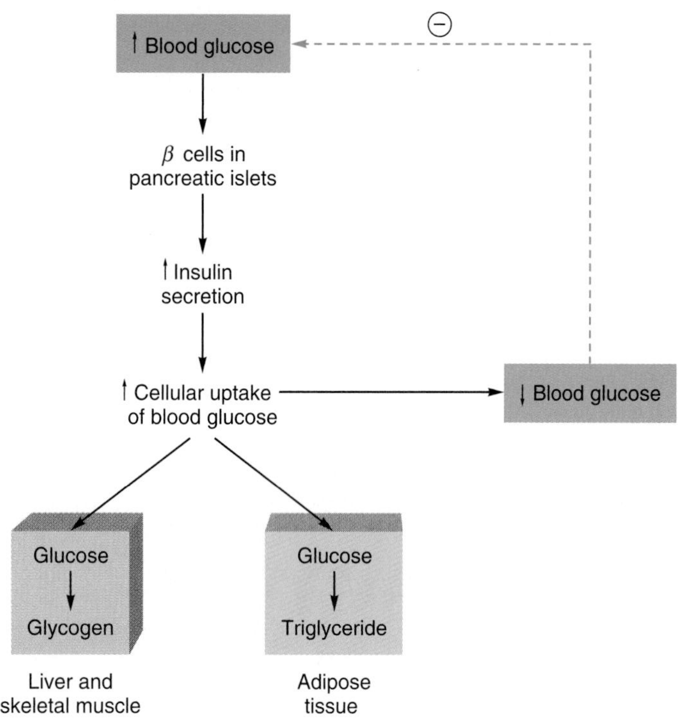

Insulin also promotes the cellular uptake of amino acids and their incorporation into proteins. The stores of large energy-reserve molecules are thus increased while the blood concentrations of glucose and amino acids are decreased.

A nonobese 70-kg (155-lb) man has approximately 10 kg (about 82,500 kcal) of stored fat. Since 250 g of fat can supply the energy requirements for 1 day, this reserve fuel is sufficient for about 40 days. Glycogen is less efficient as an energy reserve, and less is stored in the body; there are about 100 g (400 kcal) of glycogen stored in the liver and 375 to 400 g (1,500 kcal) in skeletal muscles. Insulin promotes the cellular uptake of glucose into the liver and muscles and the conversion of glucose into glucose-6-phosphate. In the liver and muscles, this can be changed into glucose-1-phosphate, which is used as the precursor of glycogen. Once the stores of glycogen have been filled, the continued ingestion of excess calories results in the production of fat rather than of glycogen.

Insulin and Glucagon: Postabsorptive State

The blood glucose concentration is maintained surprisingly constant during the fasting, or postabsorptive, state because of the secretion of glucose from the liver. This glucose is derived from the processes of glycogenolysis and gluconeogenesis, which are promoted by a high secretion of glucagon coupled with a low secretion of insulin.

Glucagon stimulates and insulin suppresses the hydrolysis of liver glycogen, or **glycogenolysis** (*gli″ko-jĕ-nol′ĭ-sis*).

Figure 27.8

Catabolism during fasting.
Increased glucagon secretion and decreased insulin secretion during fasting favor catabolism. These hormonal changes promote the release of glucose, fatty acids, ketone bodies, and amino acids into the blood. Notice that the liver secretes glucose that is derived both from the breakdown of liver glycogen and from the conversion of amino acids in gluconeogenesis.

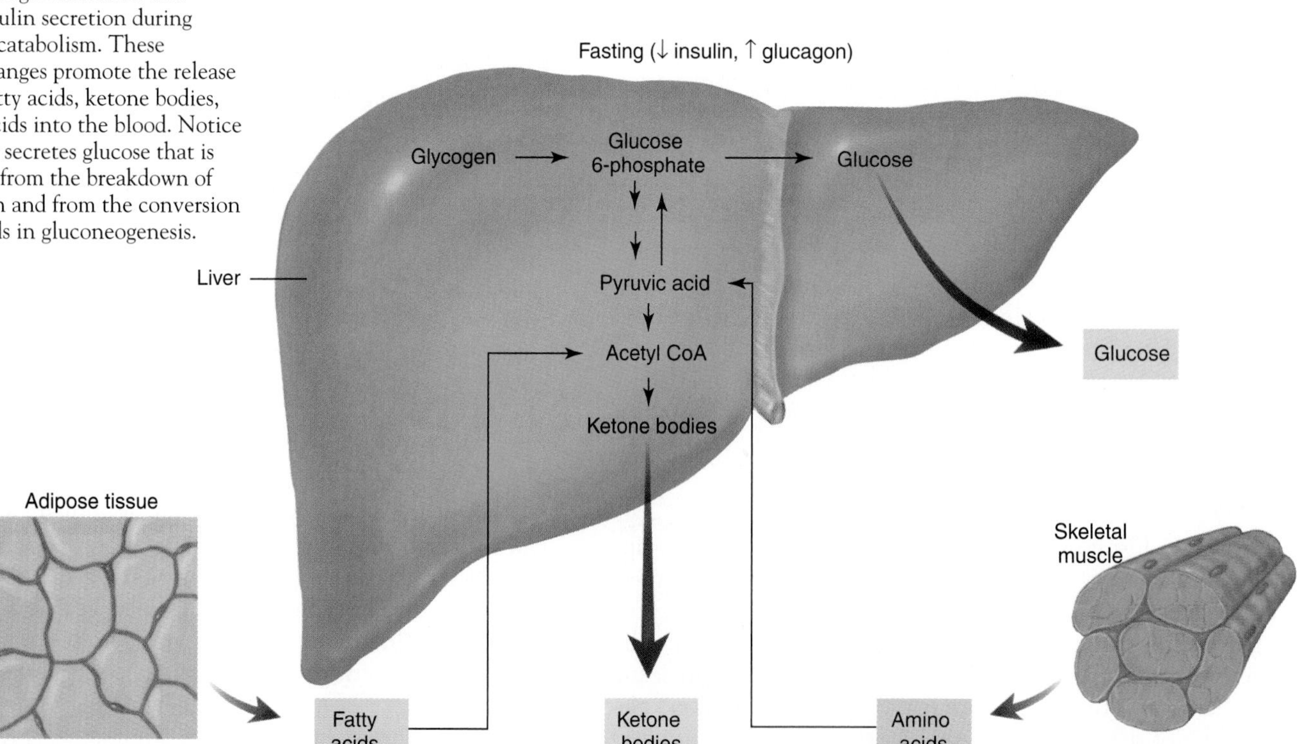

Figure 27.9

Effect of feeding and fasting on metabolism.
Metabolic balance is tilted toward anabolism by feeding (absorption of a meal) and towards catabolism by fasting. This occurs because of an inverse relationship between insulin and glucagon secretion. Insulin secretion rises and glucagon secretion falls during food absorption, whereas the opposite occurs during fasting.

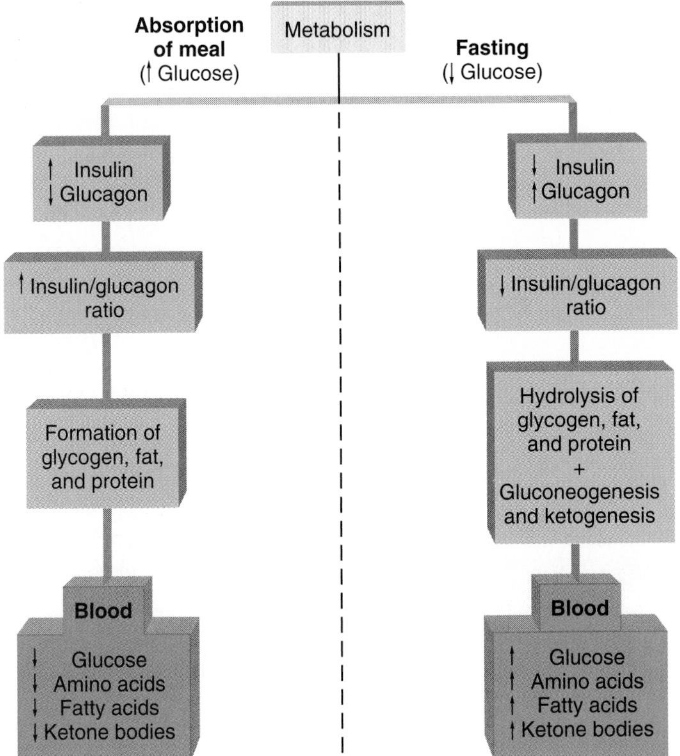

Thus, during times of fasting when glucagon secretion is high and insulin secretion is low, liver glycogen is used as a source of additional blood glucose. This results in the liberation of free glucose from glucose-6-phosphate by the action of an enzyme called *glucose-6-phosphatase* (chapter 4). Only the liver has this enzyme, and therefore only the liver can use its stored glycogen as a source of additional blood glucose. Since muscles lack glucose-6-phosphatase, the glucose-6-phosphate produced from muscle glycogen can be used for glycolysis only by the muscle cells themselves.

Since there are only about 100 grams of stored glycogen in the liver, adequate blood glucose levels could not be maintained for very long during fasting using this source alone. The low levels of insulin secretion during fasting, together with elevated glucagon secretion, however, promote **gluconeogenesis** (*gloo″ko-ne″ŏ-jen′ĭ-sis*), the formation of glucose from noncarbohydrate molecules. Low insulin allows the release of amino acids from skeletal muscles, while glucagon and cortisol (an adrenal hormone) stimulate the production of enzymes in the liver that convert amino acids to pyruvic acid, and pyruvic acid into glucose. During prolonged fasting and exercise, gluconeogenesis in the liver using amino acids from muscles may be the only source of blood glucose.

The secretion of glucose from the liver during fasting compensates for the low blood glucose concentrations and helps to provide the brain with the glucose that it needs. But because insulin secretion is low during fasting, skeletal muscles cannot utilize blood glucose as an energy source. Instead, skeletal muscles—as well as the heart, liver, and kidneys—use free fatty acids as their major source of fuel. This helps to "spare" glucose for the brain.

The free fatty acids are made available by the action of glucagon. In the presence of low insulin levels, glucagon activates an enzyme in adipose cells called *hormone-sensitive lipase*. This enzyme catalyzes the hydrolysis of stored triglycerides and the release of free fatty acids and glycerol into the blood. Glucagon also activates enzymes in the liver that convert some of these fatty acids into ketone bodies, which are secreted into the blood (fig. 27.8). Several organs in the body can use ketone bodies, as well as fatty acids, as a source of acetyl CoA in aerobic respiration.

Through the stimulation of **lipolysis** (*lĭ-pol′ĭ-sis*) and **ketogenesis** (the breakdown of fat and the formation of ketone bodies), the high glucagon and low insulin levels that occur during fasting provide circulating energy substrates for use by the muscles, liver, and other organs. Through liver glycogenolysis and gluconeogenesis, these hormonal changes help to provide adequate levels of blood glucose to sustain the metabolism of the brain. The antagonistic action of insulin and glucagon (fig. 27.9) thus promotes appropriate metabolic responses during periods of fasting and periods of absorption.

Metabolic Regulation by Adrenal Hormones, Thyroxine, and Growth Hormone

Epinephrine, the glucocorticoids, thyroxine, and growth hormone stimulate the catabolism of carbohydrates and lipids. These hormones are thus antagonistic to insulin in their regulation of carbohydrate and lipid metabolism. Thyroxine and growth hormone promote protein synthesis, however, and are needed for body growth and proper development of the central nervous system. The stimulatory effect of these hormones on protein synthesis is complementary to that of insulin.

The anabolic effects of insulin are antagonized by glucagon, as previously described, and by the actions of a variety of other hormones. The hormones of the adrenal glands, thyroid gland, and anterior pituitary (specifically growth hormone) antagonize the action of insulin on carbohydrate and lipid metabolism. The actions of insulin, thyroxine, and growth hormone, however, can act synergistically in the stimulation of protein synthesis.

Adrenal Hormones

As described in chapter 19, the adrenal gland consists of two different parts (of different embryonic origins) that function

Figure 27.10

Mechanism of epinephrine and glucagon effects on metabolism.
Cyclic AMP (cAMP) serves as a second messenger in the actions of epinephrine and glucagon on liver and adipose tissue metabolsim. (The mechansims of hormone action are discussed in more detail in chapter 19.)

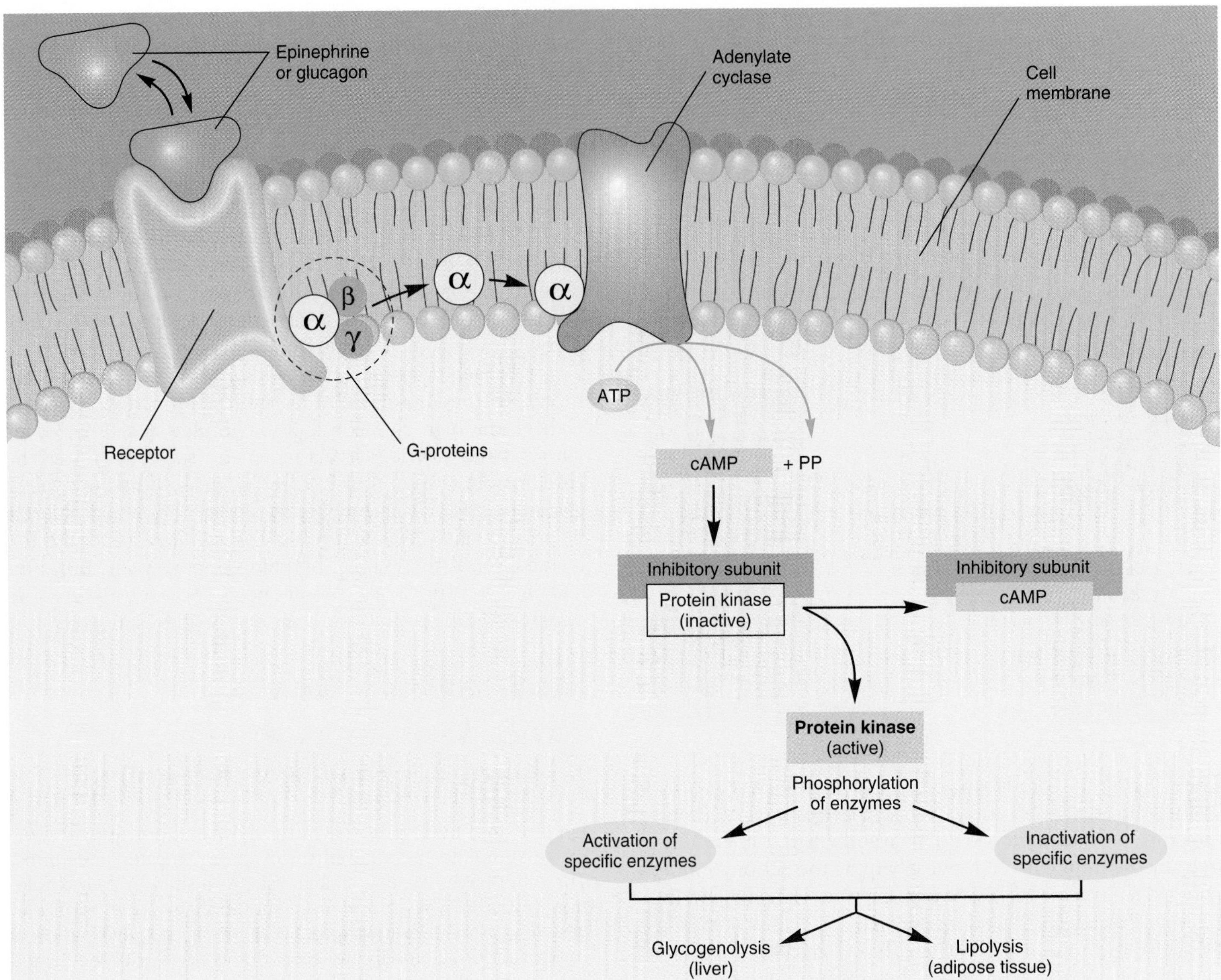

as separate glands. The two parts secrete different hormones and are regulated by different control systems. The **adrenal medulla** secretes **catecholamine** hormones—epinephrine and lesser amounts of norepinephrine—in response to sympathetic nerve stimulation. The **adrenal cortex** secretes corticosteroid hormones. These are grouped into two functional categories: **mineralocorticoids,** such as aldosterone, which act on the kidneys to regulate Na^+ and K^+ balance (chapter 25), and **glucocorticoids** (*gloo"ko-kor'tĭ-koidz*), such as hydrocortisone (cortisol), which participate in metabolic regulation.

Metabolic Effects of Catecholamines

The metabolic effects of catecholamines (epinephrine and norepinephrine) are similar to those of glucagon. They stimulate glycogenolysis and the release of glucose from the liver, as well as lipolysis and the release of fatty acids from adipocytes. These actions occur in response to glucagon during fasting, when low blood glucose stimulates glucagon secretion, and in response to catecholamines during the fight-or-flight reaction to stress. The latter effect provides circulating energy substrates in anticipation of the need for intense physical activity. Glucagon and epinephrine have similar mechanisms of action; the actions of both are mediated by cyclic AMP (fig. 27.10).

Sympathetic nerves, acting through the release of norepinephrine, can stimulate **β_3-adrenergic receptors** in brown adipose tissue (there also appears to be β_3 receptors in the ordinary, white fat of humans, but none in other tissues). As may be recalled from chapter 4, brown fat is a specialized tissue that contains a unique *uncoupling protein* that dissociates electron transport from the production of ATP. As a result, brown fat can have a very high rate of energy expenditure (unchecked by negative feedback from ATP) that is stimulated by catecholamines and thyroid hormones. Since human adults have less brown fat than is present in neonates, however, the significance of this effect is not fully understood. Even so, it has been shown that genetic defects in the β_3-adrenergic receptors of some people are associated with obesity and type II diabetes mellitus.

Metabolic Effects of Glucocorticoids

Hydrocortisone (cortisol) and other glucocorticoids are secreted by the adrenal cortex in response to ACTH stimulation. The secretion of ACTH from the anterior pituitary occurs as part of the *general adaptation syndrome* in response to stress (chapter 19). Since prolonged fasting or prolonged exercise certainly qualify as stressors, ACTH—and thus glucocorticoid secretion—is stimulated under these conditions. The increased secretion of glucocorticoids during prolonged fasting or exercise supports the effects of increased glucagon and decreased insulin secretion from the pancreatic islets.

Like glucagon, hydrocortisone promotes lipolysis and ketogenesis; it also stimulates the synthesis of hepatic enzymes that promote gluconeogenesis. Although hydrocortisone stimulates enzyme (protein) synthesis in the liver, it promotes protein breakdown in the muscles. This latter effect increases the blood levels of amino acids, and thus provides the substrates needed by the liver for gluconeogenesis. The release of circulating energy substrates—amino acids, glucose, fatty acids, and ketone bodies—into the blood in response to hydrocortisone (fig. 27.11) helps to compensate for a state of prolonged fasting or exercise. Whether these metabolic responses are beneficial in other stressful states is open to question.

Figure 27.11

Metabolic effects of glucocorticoids.
The catabolic actions of glucocorticoids help raise the blood concentration of glucose and other energy-carrier molecules.

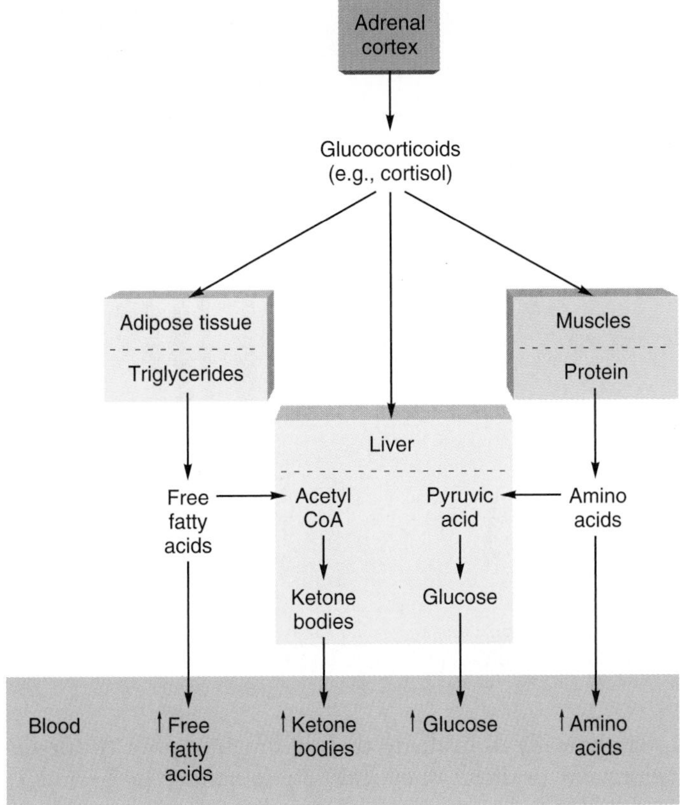

Thyroxine

The thyroid follicles secrete **thyroxine,** also called **tetraiodothyronine (T_4)**, in response to stimulation by thyroid-stimulating hormone (TSH) from the anterior pituitary. Almost all organs in the body are targets of thyroxine action. Thyroxine itself, however, is not the active form of the hormone within the target cells; thyroxine is a prehormone that must first be converted to triiodothyronine (T_3) within the target cells to be active (chapter 19). Acting via its conversion to T_3, thyroxine (1) regulates the rate of cell respiration and (2) contributes to proper growth and development, particularly during early childhood.

Thyroxine and Cell Respiration

Thyroxine (via its conversion to T_3) stimulates the rate of cell respiration in almost all cells in the body—an effect believed to be due to a lowering of cellular ATP concentrations. This effect is produced by (1) the production of uncoupling proteins (as in brown fat, discussed previously); and (2) stimulation of active transport Na^+/K^+pumps, which serve as an energy sink to further lower ATP concentrations. As discussed in chapter 4, ATP exerts an end-product inhibition of cell respiration, so that when ATP concentrations decrease the rate of cell respiration increases.

Much of the energy liberated during cell respiration escapes as heat, and uncoupling proteins increase the proportion of food energy that escapes as heat. Since thyroxine stimulates the production of uncoupling proteins and the rate of cell respiration, the actions of thyroxine increase the production of metabolic heat. The heat-producing, or *calorigenic* (*calor* = heat) *effects* of thyroxine are required for cold adaptation. This does not mean that people who are cold-adapted have high levels of thyroxine secretion. Rather, thyroxine levels in the normal range coupled with the increased activity of the sympathoadrenal system and other responses previously discussed are responsible for cold adaptation.

The metabolic rate under carefully defined resting conditions is known as the basal metabolic rate (BMR), as previously described. This rate of basal metabolism has two components—one that is independent of thyroxine action and one that is regulated by thyroxine. In this way, thyroxine acts to "set" the BMR. The BMR can thus be used as an index of thyroid function. Indeed, such measurements were

Figure 27.12

Cretinism.
Cretinism is a disease of infancy caused by an underactive thyroid gland.

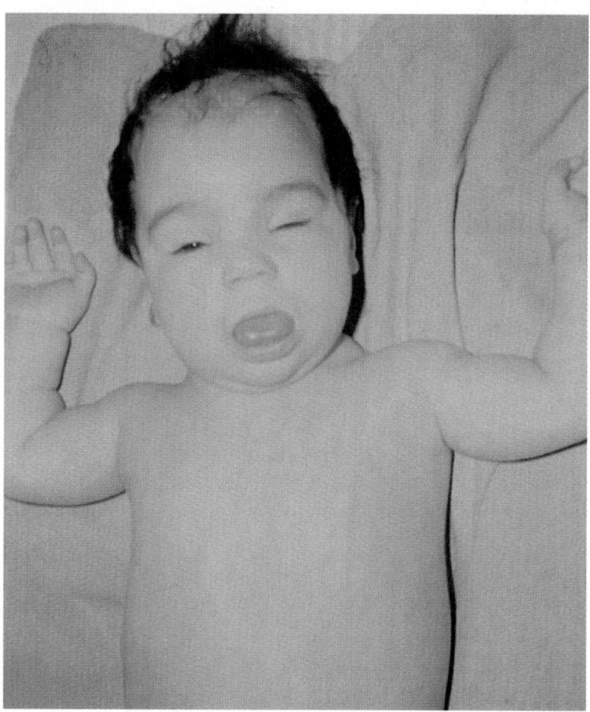

used clinically to evaluate thyroid function prior to the development of direct chemical determinations of T_4 and T_3 in the blood. A person who is hypothyroid may have a basal O_2 consumption about 30% lower than normal, while a person who is hyperthyroid may have a basal O_2 consumption up to 50% higher than normal.

Thyroxine in Growth and Development

Through its stimulation of cell respiration, thyroxine promotes the consumption of circulating energy substrates such as glucose, fatty acids, and other molecules. These effects, however, are mediated at least in part by the activation of genes; thyroxine thus stimulates both RNA and protein synthesis. As a result of its stimulation of protein synthesis throughout the body, thyroxine is considered to be an anabolic hormone like insulin and growth hormone.

Because of its stimulation of protein synthesis, thyroxine is needed for growth of the skeleton and, most importantly, for the proper development of the central nervous system. Recent evidence has demonstrated the presence of receptor proteins for T_3 in the neurons and astrocytes of the brain. This need for thyroxine is particularly great when the brain is undergoing its greatest rate of development—from the end of the first trimester of prenatal life to 6 months after birth. Hypothyroidism during this time may result in **cretinism** (fig. 27.12). Unlike people with dwarfism, who have normal thyroxine secretion but a low secretion of growth hormone, people with cretinism exhibit severe mental retardation. Treatment with thyroxine soon after birth, particularly before 1 month of age, has been found to completely or almost completely restore development of intelligence as measured by IQ tests administered 5 years later.

Hypothyroidism and Hyperthyroidism

As might be predicted from the effects of thyroxine, people who are hypothyroid have an abnormally low basal metabolic rate and experience weight gain and lethargy. A thyroxine deficiency also decreases the ability to adapt to cold stress. Hypothyroidism in adults causes **myxedema** (*mik″sĭ-de′mă*)—accumulation of mucoproteins and fluid in subcutaneous connective tissues. Symptoms of this disease include swelling of the hands, face, feet, and tissues around the eyes.

Hypothyroidism can result from a thyroid gland defect or secondarily from insufficient thyrotropin-releasing hormone (TRH) secretion from the hypothalamus, insufficient TSH secretion from the anterior pituitary, or insufficient iodine in the diet. In the case of lack of iodine, hypothyroidism is accompanied by excessive TSH secretion, which stimulates abnormal growth of the thyroid. This condition, an *endemic goiter*, can be reversed by iodine supplements.

A goiter can also be produced by another mechanism. In **Graves' disease,** apparently an autoimmune disease, autoantibodies (chapter 23) exert TSH-like effects on the thyroid. Since the production of these autoantibodies is not controlled by negative feedback, the thyroid is stimulated excessively to produce a *toxic goiter* associated with a hyperthyroid state. Hyperthyroidism produces a high BMR accompanied by weight loss, nervousness, irritability, and an intolerance to heat. There is also a significant increase in cardiac output and blood pressure. The symptoms of hypothyroidism and hyperthyroidism are compared in table 27.6.

Growth Hormone

The anterior pituitary secretes **growth hormone,** also called **somatotropin,** in larger amounts than any other of its hormones. As its name implies, growth hormone stimulates growth in children and adolescents. The continued high secretion of growth hormone in adults, particularly under the conditions of fasting and other forms of stress, implies that this hormone can have important metabolic effects even after the growing years have ended.

Regulation of Growth Hormone Secretion

The secretion of growth hormone is inhibited by somatostatin, which is produced by the hypothalamus and secreted into the hypothalamo-hypophyseal portal system (chapter 19). In addition, there is also a hypothalamic releasing hormone, GHRH, which stimulates growth hormone secretion. Growth hormone thus appears to be unique among the anterior pituitary hormones in that its secretion is controlled by both a releasing and an inhibiting hormone from the hypothalamus. The secretion of growth hormone follows a circadian ("about a day") pattern, increasing during sleep and decreasing during periods of wakefulness.

Graves' disease: from Robert James Graves, Irish physician, 1796–1853

Table 27.6
Comparison of Hypothyroidism and Hyperthyroidism

Symptom	Hypothyroid	Hyperthyroid
Growth and development	Impaired growth	Accelerated growth
Activity and sleep	Decreased activity; increased sleep	Increased activity; decreased sleep
Temperature tolerance	Intolerance to cold	Intolerance to heat
Skin characteristics	Coarse, dry skin	Normal skin
Perspiration	Absent	Excessive
Pulse	Slow	Rapid
Gastrointestinal symptoms	Constipation; decreased appetite; increased weight	Frequent bowel movements; increased appetite; decreased weight
Reflexes	Slow	Rapid
Psychological aspects	Depression and apathy	Nervousness and emotionality
Plasma T_4 levels	Decreased	Increased

Growth hormone secretion is stimulated by an increase in the plasma concentration of amino acids and by a decrease in the blood glucose concentration. These events occur during absorption of a high-protein meal when amino acids are absorbed. The secretion of growth hormone is also increased during prolonged fasting, when blood glucose is low and plasma amino acid concentration is raised by the breakdown of muscle protein.

Insulin-like Growth Factors

Insulin-like growth factors (IGFs), produced by many tissues, are polypeptides that are similar in structure to proinsulin (chapter 2). They have insulin-like effects and serve as mediators for some of growth hormone's actions. The term **somatomedins** (*so″mă-tŏ-med′nz*) is often used to refer to two of these factors, designated *IGF-1* and *IGF-2*, because they mediate the actions of somatotropin (growth hormone). The liver produces and secretes IGF-1 in response to growth hormone stimulation, and this secreted IGF-1 then functions as a hormone in its own right, traveling in the blood to the target tissue. A major target is cartilage, where IGF-1 stimulates cell division and growth. IGF-1 also functions as an autocrine regulator (chapter 19), since the chondrocytes (cartilage cells) themselves produce IGF-1 in response to growth hormone stimulation. The growth-promoting actions of IGF-1, acting as both a hormone and an autocrine regulator, thus directly mediate the effects of growth hormone on cartilage. These actions are supported by IGF-2, which has more insulin-like actions. The action of growth hormone in stimulating lipolysis in adipocytes and in decreasing glucose utilization is apparently not mediated by the somatomedins (fig. 27.13).

Effects of Growth Hormone on Metabolism

The fact that growth hormone secretion is increased during fasting and also during absorption of a protein meal reflects the complex nature of this hormone's action. Growth hormone has both anabolic and catabolic effects; it promotes protein synthesis (anabolism), and in this respect is similar to insulin. It also stimulates the catabolism of fat and the release of fatty acids from adipose tissue during periods of fasting (the postabsorptive state), as growth hormone secretion increases at night. A rise in the plasma fatty acid concentration induced by growth hormone results in decreased rates of glycolysis in many organs. This inhibition of glycolysis by fatty acids, perhaps together with a more direct action of growth hormone,

Figure 27.13

Metabolic effects of growth hormone. The growth-promoting, or anabolic, effects of growth hormone are mediated indirectly via stimulation of somatomedin production by the liver.

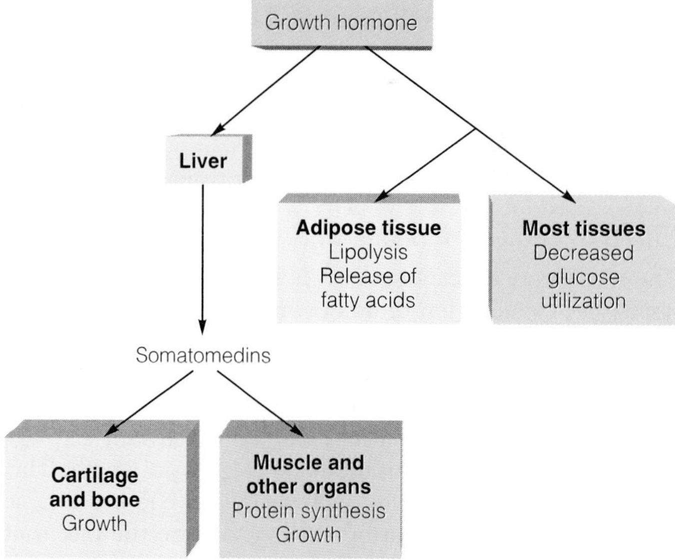

Figure 27.14

The progression of acromegaly in one individual.
The coarsening of features and disfigurement are evident by age 33 and severe at age 52.

Age 9

Age 16

Age 33

Age 52

results in decreased glucose utilization by the tissues. Growth hormone thus acts to raise the blood glucose concentration.

Growth hormone stimulates the cellular uptake of amino acids and protein synthesis in many organs of the body. These actions are useful when eating a protein-rich meal; amino acids are removed from the blood and used to form proteins, and the blood glucose concentration and fatty acids increase to provide alternate energy sources (fig. 27.13). The anabolic effect of growth hormone is particularly important during the growing years, when it contributes to increases in bone length and in the mass of many soft tissues.

Effects of Growth Hormone on Body Growth

The stimulatory effects of growth hormone on skeletal growth results from stimulation of mitosis in the epiphyseal plates of cartilage present in the long bones of growing children and adolescents. This action is mediated by the somatomedins, IGF-1 and IGF-2, which stimulate the chondrocytes to divide and secrete more cartilage matrix. Part of this growing cartilage converts to bone, enabling the bone to grow in length. This bone growth stops when the epiphyseal plates convert to bone after the growth spurt during puberty, despite the fact that growth hormone secretion continues throughout adulthood.

An excessive secretion of growth hormone in children can produce **gigantism.** These children may grow up to 8 feet tall, at the same time maintaining normal body proportions. An excessive growth hormone secretion that occurs after the epiphyseal discs have sealed, however, cannot produce increases in height. In adults, the oversecretion of growth hormone results in an elongation of the jaw and deformities in the bones of the face, hands, and feet. This condition, called **acromegaly,** is accompanied by the growth of soft tissues and coarsening of the skin (fig. 27.14). It is interesting that athletes who take growth hormone supplements to increase their muscle mass may also experience body changes that resemble those of acromegaly.

An inadequate secretion of growth hormone during the growing years results in **dwarfism.** An interesting variant of this condition is *Laron dwarfism*, in which there is a genetic insensitivity to the effects of growth hormone. This insensitivity is associated with a reduction in the number of growth hormone receptors in the target cells. Genetic engineering has made available recombinant IGF-1, which has recently been approved by the FDA for the medical treatment of Laron dwarfism.

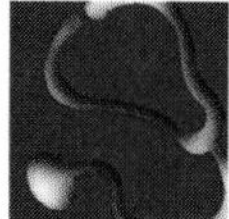

An adequate diet, particularly with respect to proteins, is required for the production of IGF-1. This helps to explain the common observation that many children are significantly taller than their parents, who may not have had an adequate diet in their youth. Children with severe protein deficiency, a condition called *kwashiorkor* (*kwă″she-or′kor*), have low growth rates and low levels of IGF-1 in the blood, despite the fact that their growth hormone secretion may be abnormally elevated. When these children are provided with an adequate diet, IGF-1 levels and growth rates increase.

acromegaly: Gk. *akron*, extremity; *mega*, large
Laron dwarfism: from Zui Laron, Israeli endocrinologist, b. 1927

Table 27.7

Comparison of the Two Major Forms of Diabetes Mellitus

Feature	Insulin-Dependent (Type I)	Non-Insulin-Dependent (Type II)
Usual age of onset	Under 20 years	Over 40 years
Development of symptoms	Rapid	Slow
Percentage of diabetic population	About 10%	About 90%
Development of ketoacidosis	Common	Rare
Association with obesity	Rare	Common
Beta cells of pancreatic islets (at onset of disease)	Destroyed	Usually not destroyed
Insulin secretion	Decreased	Normal or increased
Autoantibodies to islet cells	Present	Absent
Associated with particular MHC antigens	Yes	No
Usual treatment	Insulin injections	Diet; oral stimulators of insulin secretion

Clinical Considerations

Chronic high blood glucose, or *hyperglycemia,* is the hallmark of the disease **diabetes mellitus** (*di″ă-be′tēz mĕ-li′tus*). The name of this disease is derived from the fact that glucose "spills over" into the urine when the blood glucose concentration is too high—*mellitus* is derived from a Latin word meaning "honeyed" or "sweet." (The general term *diabetes* comes from a Greek word meaning "siphon"; it refers to the frequent urination associated with this condition.) The hyperglycemia of diabetes mellitus results from either the insufficient secretion of insulin by the beta cells of the pancreatic islets or the inability of secreted insulin to stimulate the cellular uptake of glucose from the blood. Diabetes mellitus, in short, results from the inadequate secretion or action of insulin.

There are two forms of diabetes mellitus. In **insulin-dependent diabetes mellitus (IDDM),** also called **type I diabetes,** the beta cells are progressively destroyed and secrete little or no insulin; injections of exogenous insulin are thus required to sustain the person's life. This form of the disease accounts for only about 10% of the known cases of diabetes. About 90% of the people who have diabetes have **non-insulin-dependent diabetes mellitus (NIDDM),** also called **type II diabetes.** Type I diabetes was once known as *juvenile-onset diabetes* because this condition is usually diagnosed in people under the age of 20. Type II diabetes has also been called *maturity-onset diabetes,* since it is usually diagnosed in people over the age of 40. The two forms of diabetes mellitus are compared in table 27.7. (It should be noted that only the early stages of type I and type II diabetes mellitus are compared; some people with severe type II diabetes may also require insulin injections to control the hyperglycemia.)

Insulin-Dependent Diabetes Mellitus

Insulin-dependent diabetes mellitus results when the beta cells of the pancreatic islets are progressively destroyed by autoimmune attack. Recent evidence in mice suggests that killer T lymphocytes (chapter 23) may target an enzyme known as glutamate decarboxylase in the beta cells. This autoimmune destruction of the beta cells may be provoked by an environmental agent, such as infection by viruses. In other cases, however, the cause is currently unknown. Removal of the insulin-secreting beta cells causes hyperglycemia and the appearance of glucose in the urine. Without insulin, glucose cannot enter the adipose cells; the rate of fat synthesis thus lags behind the rate of fat breakdown and large amounts of free fatty acids are released from the adipose cells.

In a person with uncontrolled IDDM, many of the fatty acids released from adipose cells are converted into ketone bodies in the liver. This may result in an elevated ketone body concentration in the blood (ketosis), and if the buffer reserve of bicarbonate is neutralized, it may also result in *ketoacidosis*. During this time, the glucose and excess ketone bodies that are excreted in the urine act as osmotic diuretics (chapter 25) and cause the excessive excretion of water in the urine. This can produce severe dehydration, which, together with ketoacidosis and associated disturbances in electrolyte balance, may lead to coma and death (fig. 27.15).

In addition to the lack of insulin, people with IDDM have an abnormally high secretion of glucagon from the alpha cells of the pancreatic islets. Glucagon stimulates glycogenolysis in the liver and thus helps to raise the plasma glucose concentration. Glucagon also stimulates the production of enzymes in the liver that convert fatty acids into ketone bodies. The full range of symptoms of diabetes may result from high glucagon

Figure 27.15

Consequences of insulin deficiency.
This is the possible sequence of events by which an insulin deficiency may lead to coma and death.

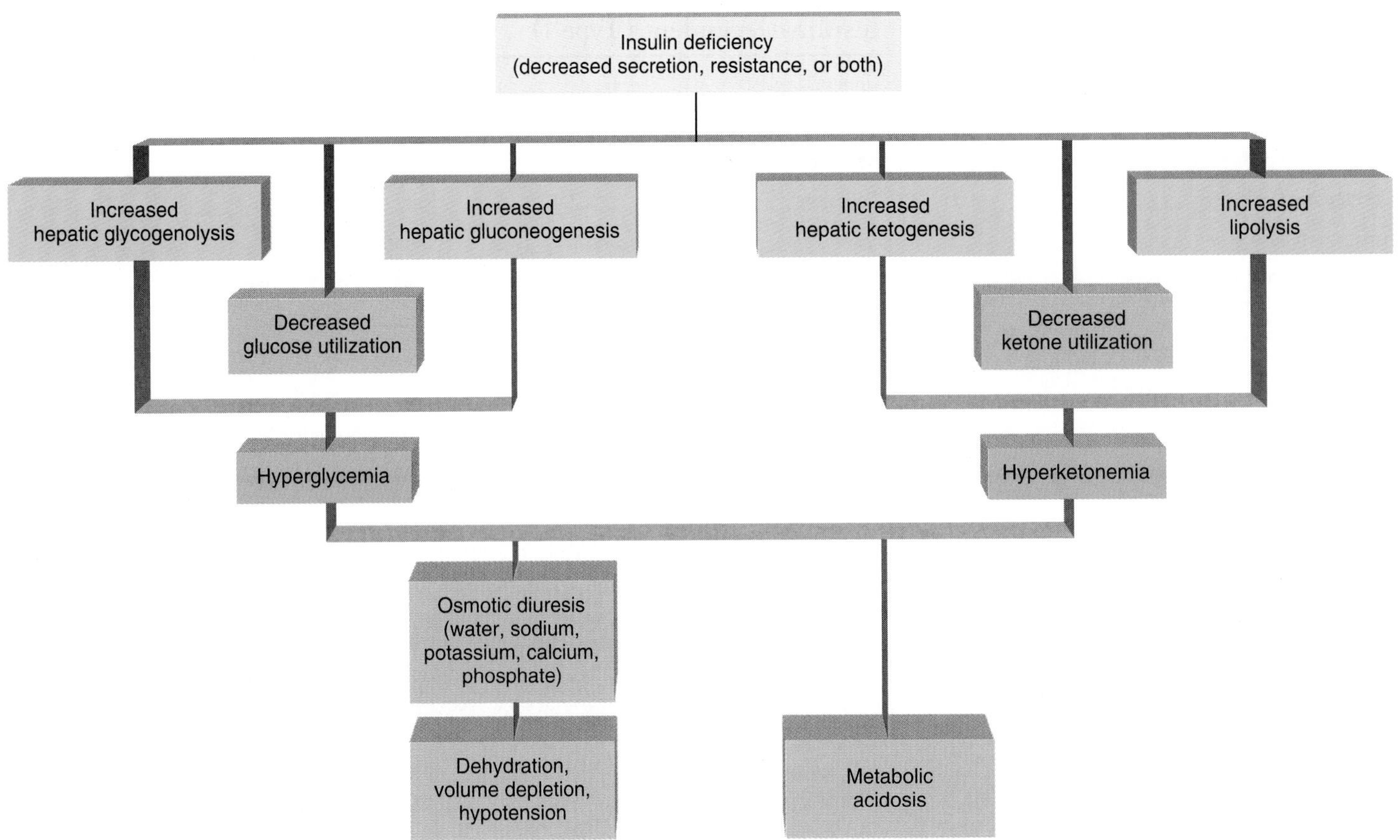

secretion as well as from the absence of insulin. The lack of insulin may be largely responsible for hyperglycemia and for the release of large amounts of fatty acids into the blood. The high glucagon secretion may contribute to the hyperglycemia and in large part cause the development of ketoacidosis.

Non-Insulin-Dependent Diabetes Mellitus

The effects produced by insulin, or any hormone, depend on the concentration of that hormone in the blood and on the sensitivity of the target cells to given amounts of the hormone. Cellular responsiveness to insulin, for example, varies under normal conditions. Exercise increases insulin sensitivity and obesity decreases insulin sensitivity of the target cells. The pancreatic islets of a nondiabetic obese person must therefore secrete high amounts of insulin to maintain the blood glucose concentration in the normal range. Conversely, nondiabetic people who are thin and who exercise regularly require lower amounts of insulin to maintain the proper blood glucose concentration.

Non-insulin-dependent diabetes is usually slow to develop, is hereditary, and occurs most often in people who are overweight. Genetic factors are very significant; people at highest risk are those whose parents both have NIDDM and those who are members of certain ethnic groups, particularly Native Americans of the Southwestern United States and Mexican-Americans. Unlike people with IDDM, those who have NIDDM can have normal or even elevated levels of insulin in their blood. Despite this, people with NIDDM have hyperglycemia if untreated. This must mean that the amount of insulin secreted is inadequate, even though insulin levels may be in the normal range.

Much evidence has been obtained to show that people with NIDDM have an abnormally low cellular sensitivity to insulin, or an *insulin resistance*. This is true even if the person is not obese, but the problem is compounded by the decreased cellular sensitivity that accompanies obesity, particularly of the "apple-shape" variety in which the adipocytes are enlarged. There is also evidence that the beta cells are not

Figure 27.16

Oral glucose tolerance in prediabetes and type II diabetes.

The oral glucose tolerance test showing (*a*) blood glucose concentrations and (*b*) insulin values following the ingestion of a glucose solution. Values are shown for people who are normal, prediabetic, and type II (non-insulin-dependent) diabetic. Prediabetics (those who demonstrate "insulin resistance") often show impaired glucose tolerance without fasting hyperglycemia. (NIDDM = non-insulin-dependent diabetes mellitus.)

Source: Data from Simeon I. Taylor, et al., "Insulin Resistance of Insulin Deficiency: Which is the Primary Cause of NIDDM?" in Diabetes, *vol. 43, June 1994, p. 735.*

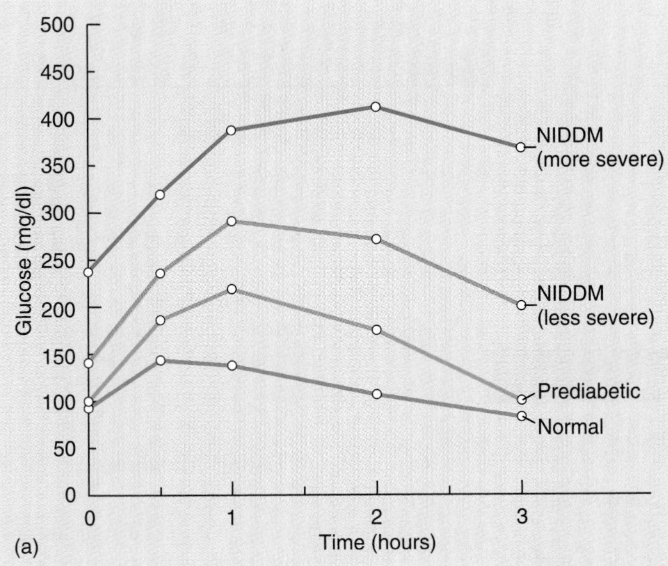

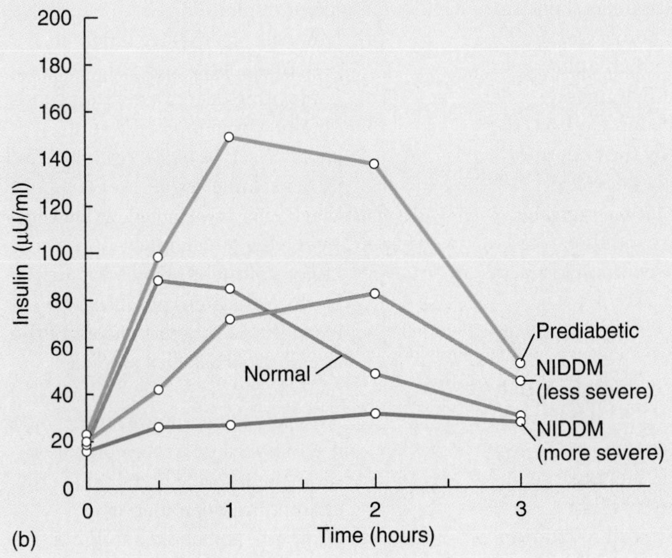

functioning correctly—whatever amount of insulin they secrete is inadequate to the task. People who are prediabetic (who have impaired glucose tolerance—see fig. 27.6) often have elevated levels of insulin without hypoglycemia, suggesting insulin resistance. People with chronic NIDDM have both an insulin resistance and insulin deficiency (fig. 27.16).

Since obesity decreases insulin sensitivity, people who are genetically predisposed to insulin resistance may develop symptoms of diabetes when they gain weight. Conversely, this type of diabetes mellitus can usually be controlled by increasing cellular sensitivity to insulin through diet and exercise. This is beneficial because exercise, like insulin, increases the amount of membrane GLUT4 carriers (for the facilitative diffusion of glucose) in skeletal muscle cells. If diet and exercise are insufficient, oral drugs are available that increase insulin secretion from the beta cells (e.g., *sulfonylureas*) and that decrease the insulin resistance of the target tissues (e.g., *troglitazone*).

People with type II diabetes do not usually develop ketoacidosis. The hyperglycemia itself, however, can be dangerous on a long-term basis. In the United States, diabetes is the leading cause of blindness, kidney failure, and amputation of the lower extremities. People with diabetes frequently have circulatory problems that increase the tendency to develop gangrene and increase the risk for atherosclerosis. The causes of damage to the retina and lens of the eyes and to blood vessels are not well understood. It is believed, however, that these problems result from a long-term exposure to high blood glucose, which may damage tissues through a variety of mechanisms. Scientists have recently discovered that injections of C-peptide, the polypeptide fragment left over when proinsulin is converted into insulin (chapter 2), prevents or reduces vascular and neural damage in diabetic rats.

Hypoglycemia

A person with type I diabetes mellitus depends on insulin injections to prevent hyperglycemia and ketoacidosis. If inadequate insulin is injected, the person may enter a coma as a result of the ketoacidosis, electrolyte imbalance, and dehydration that develop. An overdose of insulin, however, can also produce a coma as a result of the hypoglycemia (abnormally low blood glucose levels) produced. The physical signs and symptoms of diabetic and hypoglycemic coma are sufficiently different to allow hospital personnel to distinguish between these two types.

Less severe symptoms of hypoglycemia are usually produced by an oversecretion of insulin from the pancreatic islets after a carbohydrate meal. This **reactive hypoglycemia,** caused by an exaggerated response of the beta cells to a rise in blood glucose, is most commonly seen in adults who are genetically predisposed to type II diabetes. For this reason, people with reactive hypoglycemia must limit their intake of carbohydrates and eat small meals at frequent intervals rather than two or three meals per day.

The symptoms of reactive hypoglycemia include tremor, hunger, weakness, blurred vision, and mental confusion. The appearance of some of these symptoms, however, does not necessarily indicate reactive hypoglycemia, and a given level of blood glucose does not always produce these symptoms. To confirm a diagnosis of reactive hypoglycemia a number of tests must be performed. In the oral glucose tolerance test, for example, reactive hypoglycemia is shown when the initial rise in blood glucose produced by the ingestion of a glucose solution triggers excessive insulin secretion, so that the blood glucose levels fall below normal within 5 hours (fig. 27.17).

Figure 27.17

Reactive hypoglycemia.
An idealized oral glucose tolerance test in a person with reactive hypoglycemia. The blood glucose concentration falls below the normal range within 5 hours of glucose ingestion as a result of excessive insulin secretion.

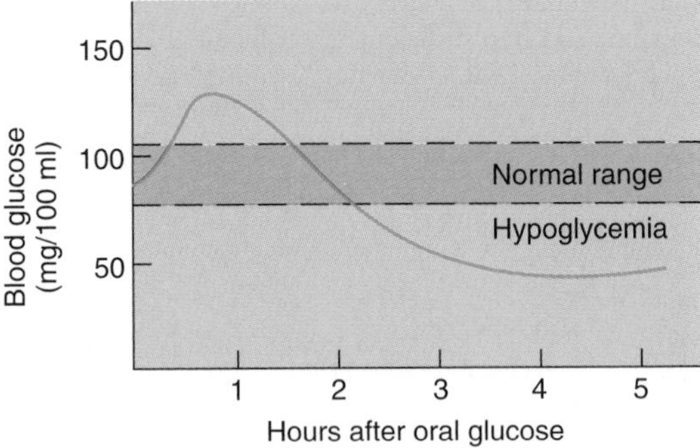

Clinical Investigation Answer

The woman's frequent urinations (polyuria) probably are causing her thirst and other symptoms. These symptoms and the fact that her mother and uncle were diabetics suggested that this woman might have diabetes mellitus. Indeed, polyuria and polydipsia (frequent drinking) are cardinal symptoms of diabetes mellitus. The fasting hyperglycemia (blood glucose concentration of 150 mg/dl) confirmed the diagnosis of diabetes mellitus. This abnormally high fasting blood glucose is too low to result in glycosuria. She could have glycosuria after meals, however, which would be responsible for her polyuria. The oral glucose tolerance test further confirmed the diagnosis of diabetes mellitus, and the observations that this condition appeared to have begun in middle age and that it was not accompanied by ketosis and ketonuria suggested that it was non-insulin-dependent diabetes mellitus. This being the case, she could increase her tissue sensitivity to insulin by diet and exercise. If this failed, she could probably control her symptoms with sulfonylurea drugs, which increase insulin secretion, or with troglitazone, which increases tissue sensitivity to the effects of insulin.

Chapter Summary

Nutritional Requirements (pp. 866–871)

1. Food provides molecules used in cell respiration for energy.
   - (a) The metabolic rate is influenced by physical activity, temperature, and food intake. The basal metabolic rate is measured as the rate of oxygen consumption when such influences are standardized and minimal.
   - (b) The energy provided in food and the energy consumed by the body are measured in units of kilocalories.
   - (c) When the caloric intake is greater than the energy expenditure over a period of time, the excess calories are stored primarily as fat.
2. Vitamins and minerals serve primarily as cofactors and coenzymes.
   - (a) Vitamins are divided into those that are fat-soluble (A, D, E, and K) and those that are water-soluble.
   - (b) Many water-soluble vitamins are needed for the activity of the enzymes involved in cell respiration.
   - (c) The fat-soluble vitamins A and D have specific functions but overlap in their mechanisms of action, activating nuclear receptors and regulating genetic expression.

Regulation of Energy Metabolism (pp. 871–875)

1. The body cells can use circulating energy substrates, including glucose, fatty acids, ketone bodies, lactic acid, and amino acid for cellular respiration.
   - (a) Different organs have different preferred energy sources.
   - (b) Circulating energy substrates can be obtained from food or from the energy reserves of glycogen, fat, and protein in the body.
2. Eating behavior is regulated, at least in part, by the hypothalamus.
   - (a) Lesions of the ventromedial area of the hypothalamus produce hyperphagia, whereas lesions of the lateral hypothalamus produce hypophagia.
   - (b) A variety of neurotransmitters have been implicated in the control of eating behavior. These include the endorphins, norepinephrine, serotonin, and cholecystokinin.

3. Adipose cells, or adipocytes, are the targets of hormonal regulation and are themselves endocrine in nature.
   (a) In children, circulating saturated fatty acids promote cell division and differentiation of new adipocytes. This activity involves the binding of a prostaglandin, 15d-PGJ_2, with a nuclear receptor known as PPARγ.
   (b) Adipocytes secrete leptin, which regulates food intake and metabolism, and TNF_α, which may help to regulate the sensitivity of skeletal muscles to insulin.
4. The control of energy balance in the body is regulated by the anabolic and catabolic effects of a variety of hormones.

Energy Regulation by the Pancreatic Islets (pp. 875–879)

1. A rise in blood glucose concentration stimulates insulin and inhibits glucagon secretion.
   (a) Amino acids stimulate the secretion of both insulin and glucagon.
   (b) Insulin secretion is also stimulated by parasympathetic innervation of the pancreatic islets and by the action of intestinal hormones such as gastric inhibitory peptide (GIP).
2. During the intestinal absorption of a meal, insulin promotes the uptake of blood glucose into skeletal muscle and other tissues.
   (a) This lowers the blood glucose concentration and increases the energy reserves of glycogen, fat, and protein.
   (b) Skeletal muscles are the major organs that remove blood glucose in response to insulin stimulation.
3. During periods of fasting, insulin secretion decreases, and glucagon secretion increases.
   (a) Glucagon stimulates glycogenolysis in the liver, gluconeogenesis, lipolysis, and ketogenesis.
   (b) These effects help to maintain adequate levels of blood glucose for the brain and provide alternate energy sources for other organs.

Metabolic Regulation by Adrenal Hormones, Thyroxine, and Growth Hormone (pp. 879–884)

1. The adrenal hormones involved in energy regulation include epinephrine from the adrenal medulla and glucocorticoids (mainly hydrocortisone) from the adrenal cortex.
   (a) The effects of epinephrine are similar to those of glucagon. Epinephrine stimulates glycogenolysis and lipolysis, and activates increased metabolism of brown fat.
   (b) Glucocorticoids promote the breakdown of muscle protein and the conversion of amino acids to glucose in the liver.
2. Thyroxine stimulates the rate of cellular respiration in almost all cells in the body.
   (a) Thyroxine sets the basal metabolic rate (BMR), which is the rate at which energy (and oxygen) is consumed by the body under resting conditions.
   (b) Thyroxine also promotes protein synthesis and is needed for proper body growth and development, particularly of the central nervous system.
3. The secretion of growth hormone is regulated by releasing and inhibiting hormones from the hypothalamus.
   (a) The secretion of growth hormone is stimulated by a protein meal and by a fall in glucose, as occurs during fasting.
   (b) Growth hormone stimulates catabolism of lipids and inhibits glucose utilization.
   (c) Growth hormone also stimulates protein synthesis, and thus promotes body growth.
   (d) The anabolic effects of growth hormone, including the stimulation of bone growth in childhood, are produced indirectly via polypeptides called insulin-like growth factors, or somatomedins.

Review Activities

Objective Questions

Match the following:

1. absorption of a carbohydrate meal	(a) rise in insulin; rise in glucagon
2. fasting	(b) fall in insulin; rise in glucagon
	(c) rise in insulin; fall in glucagon
	(d) fall in insulin; fall in glucagon

Match the following:

3. growth hormone	(a) increased protein synthesis; increased cell respiration
4. thyroxine	(b) protein catabolism in muscles; gluconeogenesis in liver
5. hydrocortisone	(c) protein synthesis in muscles; decreased glucose utilization
	(d) fall in blood glucose; increased fat synthesis

6. A lowering of blood glucose concentration promotes
   (a) decreased lipogenesis.
   (b) increased lipolysis.
   (c) increased glycogenolysis.
   (d) all of the above.
7. Glucose can be secreted into the blood by
   (a) the liver.
   (b) the muscles.
   (c) the liver and muscles.
   (d) the liver, muscles, and brain.
8. The basal metabolic rate is determined primarily by
   (a) hydrocortisone.
   (b) insulin.
   (c) growth hormone.
   (d) thyroxine.
9. Somatomedins are required for the anabolic effects of
   (a) hydrocortisone.
   (b) insulin.
   (c) growth hormone.
   (d) thyroxine.
10. At rest, about 12% of the total calories consumed are used for
   (a) protein synthesis.
   (b) cell transport.
   (c) the Na^+/K^+ pumps.
   (d) DNA replication.
11. Which of the following hormones stimulates anabolism of proteins and catabolism of fat?
   (a) growth hormone (d) glucagon
   (b) thyroxine (e) epinephrine
   (c) insulin
12. If a person eats 600 kilocalories of protein in a meal, which of the following statements will be *false?*
   (a) Insulin secretion will be increased.
   (b) The metabolic rate will be increased over basal conditions.

(c) The cells will use some of the amino acids for resynthesis of body proteins.
(d) The cells will obtain 600 kilocalories worth of energy.
(e) Body-heat production and oxygen consumption will be increased over basal conditions.

13. Ketoacidosis in untreated diabetes mellitus is due to
    (a) excessive fluid loss.
    (b) hypoventilation.
    (c) excessive eating and obesity.
    (d) excessive fat catabolism.
14. Which of the following statements about leptin is *false?*
    (a) It is secreted by adipocytes.
    (b) It increases the energy expenditure of the body.
    (c) It stimulates the release of neuropeptide Y in the hypothalamus.
    (d) It promotes feelings of satiety, decreasing food intake.

Essay Questions

1. Compare the metabolic effects of fasting to the state of uncontrolled insulin-dependent diabetes mellitus. Explain the hormonal similarities of these conditions.
2. Glucocorticoids stimulate the breakdown of protein in muscles but the synthesis of protein in the liver. Explain the significance of these different effects.
3. Describe how thyroxine affects cellular respiration. Why does a person who is hypothyroid have a tendency to gain weight and less tolerance for cold?
4. Compare and contrast the metabolic effects of thyroxine and growth hormone.
5. Why is vitamin D considered both a vitamin and a prehormone? In what ways are vitamin D similar to thyroxine?
6. Define the term *insulin resistance*. Explain the relationship between insulin resistance, obesity, exercise, and non-insulin-dependent diabetes mellitus.
7. Describe the chemical nature and origin of the somatomedins and explain the physiological significance of these growth factors.
8. Explain how the secretion of insulin and glucagon are influenced by (*a*) fasting, (*b*) a meal that is high in carbohydrate and low in protein, and (*c*) a meal that is high in protein and high in carbohydrate. Also, explain how the changes in insulin and glucagon secretion under these conditions function to maintain homeostasis.
9. How does adipose tissue regulate hunger and satiety?
10. Describe the conditions of gigantism, acromegaly, Laron dwarfism, and kwashiorkor, and explain how these conditions relate to blood levels of growth hormone and IGF-1.

Critical Thinking Questions

1. Your friend is trying to lose weight and at first is very successful. After a time, however, she complains that it seems to take more exercise and a far more stringent diet to lose even one more pound. What might explain her difficulties?
2. How can a high-fat diet in childhood lead to increased numbers of adipocytes? Explain how this process may be related to the ability of adipocytes to regulate the insulin sensitivity of skeletal muscles in adults.
3. Discuss the role of GLUT4 in glucose metabolism and use this concept to explain why exercise helps to control type II diabetes mellitus.
4. You are running in a 10-K race and, to keep your mind occupied, you try to remember which physiological processes regulate blood glucose levels during exercise. Step by step, what are these processes?
5. Discuss the location and physiological significance of the β_3-adrenergic receptors, and explain how a hypothetical β_3-adrenergic agonist drug might help in the treatment of obesity.

Related Web Sites

For a listing of the most current web sites related to this chapter, please visit the *Concepts of Human Anatomy and Physiology* home page at http://www.mhhe.com/biosci/abio/.

Twenty-Eight

Reproduction: Development and the Male Reproductive System

Objectives

- Explain how genetic sex determines whether testes or ovaries form in the embryo and describe the composition and function of the embryonic testes.
- Describe the descent of the testes into the scrotum.
- Describe the hormonal interactions between the hypothalamus, anterior pituitary, and gonads.
- Discuss the mechanisms that regulate the onset of puberty and describe the events that occur during puberty.
- Locate the structures of the male reproductive system and describe the structure and function of the scrotum.
- Describe the structure and function of the two compartments of the testes and discuss their hormonal regulation.
- Discuss the process of spermatogenesis and explain how this process is regulated.
- List the various spermatic ducts and describe the structure of the spermatic cord.
- Describe the location, structure, and function of the seminal vesicles, prostate, and bulbourethral glands.
- Describe the structure and function of the penis.
- Describe the composition of semen and explain the physiology of erection and ejaculation.
- Discuss the factors that affect male fertility.

Clinical Investigation

During a routine physical exam, a 27-year-old man mentioned to his family doctor that he and his wife had been unable to conceive a child after nearly 2 years of trying. He added that his wife had taken the initiative of having a thorough gynecological evaluation in an attempt to find out what was causing the problem. Her test findings revealed no physical conditions that could be linked to infertility. Upon palpating the patient's testes, the doctor found nothing unusual. When he examined the scrotal sac above the testes, however, the doctor appeared perplexed. He informed his patient that two tubular structures, one for each testis, appeared to be absent, and that they probably had been missing since birth. During a follow-up visit, the doctor told the patient that examination of his ejaculate revealed azoospermia (no viable sperm).

Explain how the result of the semenalysis relates to the patient's physical exam findings. What are the missing structures? Does it seem peculiar that the patient is capable of producing an ejaculate? Why or why not?

Clues: Review the section "Spermatic Ducts, Accessory Reproductive Glands, and the Penis." Also, carefully examine figure 28.10

Introduction to the Reproductive System

A gene on the Y chromosome causes the embryonic gonads to differentiate into testes. Females lack a Y chromosome, and the absence of this gene results in the development of ovaries. The embryonic testes secrete testosterone, which triggers the development of male accessory sex organs and external genitalia. The absence of testes (rather than the presence of ovaries) in a female embryo causes the development of female accessory sex organs.

The incredible complexity of structure and function in living organisms could not be produced in successive generations by chance; mechanisms must exist to transmit the genetic code from one generation to the next. This could be accomplished by either asexual or sexual reproduction. But sexual reproduction, in which genes from two individuals combine in random and novel ways with each new generation, offers the advantage of introducing great variability into a population. This variability of genetic constitution helps to ensure that some members of a population will survive changes in the environment over evolutionary time.

In sexual reproduction, **germ cells,** or **gametes** (*gam'ēts*) (sperm and ova) form within the **gonads** (testes and ovaries) by a process of reduction division, or *meiosis* (chapter 3). During this type of cell division, the number of chromosomes in most human cells—46—is halved, so that each gamete receives 23 chromosomes. Fusion of a sperm cell (spermatozoon) and an egg cell (ovum) in the act of **fertilization** results in restoration of the original chromosome number of 46 in the fertilized egg (the *zygote*). Growth of the zygote into an adult member of the next generation occurs by means of mitotic cell divisions, as described in chapter 3. When this individual reaches puberty, mature sperm or ova will form by meiosis within the gonads so that the life cycle can continue (fig. 28.1).

gamete: Gk. *gameta,* husband or wife

Figure 28.1

The human life cycle.
Numbers in parentheses indicate the haploid state (23 chromosomes) or diploid state (46 chromosomes),

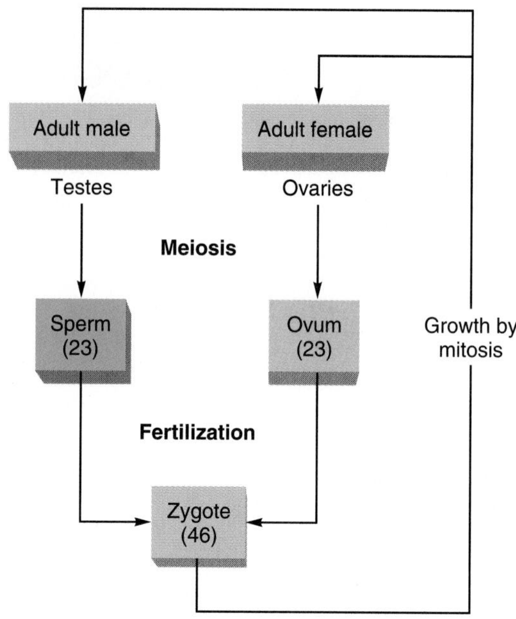

Sex Determination

Each zygote inherits 23 chromosomes from its mother and 23 chromosomes from its father. This does not produce 46 different chromosomes, but rather 23 pairs of *homologous chromosomes*. The members of a homologous pair, with the important exception of the sex chromosomes, look like each other and contain similar genes (such as those coding for eye color, height, and so on). These homologous pairs of chromosomes can be photographed and numbered (as shown in fig. 28.2). Each cell that contains 46 chromosomes (that is *diploid*) has two number 1 chromosomes, two number 2 chromosomes, and so on through pair number 22. The first 22 pairs of chromosomes are called **autosomal** (*aw"tŏ-so'mal*) **chromosomes.**

The twenty-third pair of chromosomes are the **sex chromosomes.** In a female these consist of two X chromosomes, whereas in a male there is one X chromosome and one Y chromosome. The X and Y chromosomes look different and

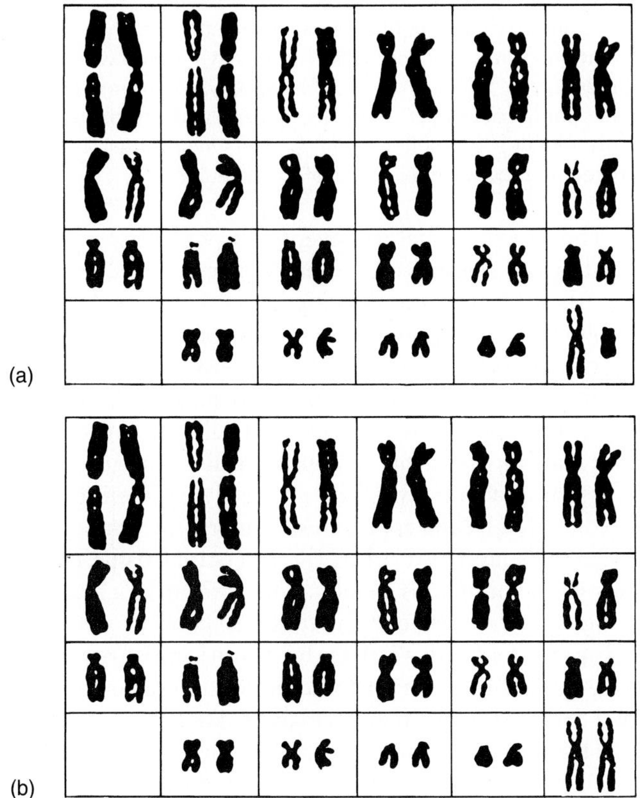

Figure 28.2

Homologous pairs of chromosomes.
These were obtained from diploid human cells. The first 22 pairs of chromosomes are called the autosomal chromosomes. The sex chromosomes are (*a*) XY for a male and (*b*) XX for a female.

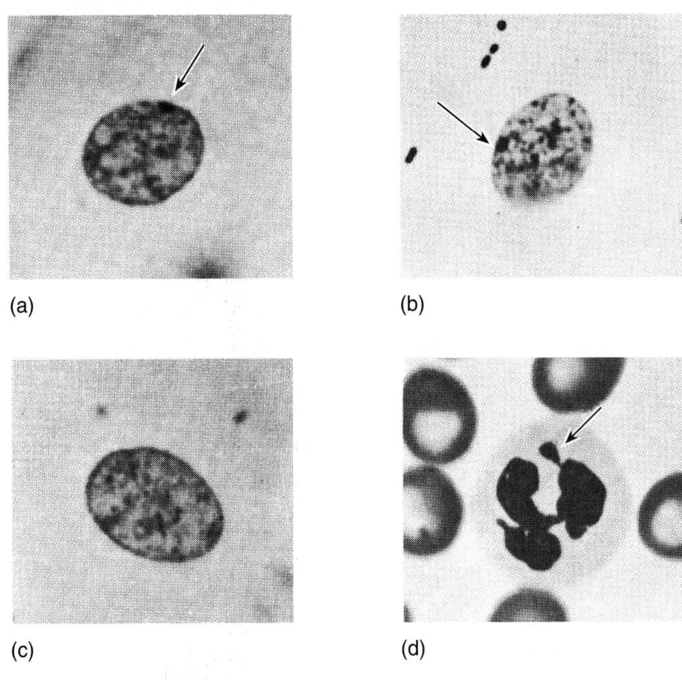

Figure 28.3

Barr bodies.
The nuclei of cheek cells from females (*a, b*) have Barr bodies (arrows). These are formed from one of the X chromosomes, which is inactive. No Barr body is present in the cell obtained from a male (*c*) because males have only one X chromosome, which remains active. Some neutrophils obtained from females (*d*) have a "drumsticklike" appendage (arrow) that is not found in the neutrophils of males.

contain different genes. This is the exceptional pair of homologous chromosomes mentioned earlier.

When a diploid cell (with 46 chromosomes) undergoes meiotic division, its daughter cells receive only one chromosome from each homologous pair of chromosomes. The gametes are therefore said to be *haploid* (they contain only half the number of chromosomes in the diploid parent cell). Each sperm cell, for example, will receive only one chromosome of homologous pair number 5—either the one originally contributed by the organism's mother, or the one originally contributed by the father (modified by the effects of crossing-over, as discussed in chapter 3). Which of the two chromosomes—maternal or paternal—ends up in a given sperm cell is completely random. This is also true for the sex chromosomes, so that approximately half of the sperm produced will contain an X and approximately half will contain a Y chromosome.

The egg cells (ova) in a woman's ovary will receive a similar random assortment of maternal and paternal chromosomes. Since the body cells of females have two X chromosomes, however, all of the ova will normally contain one X chromosome. Because all ova contain one X chromosome, whereas some sperm are X-bearing and others are Y-bearing, *the chromosomal sex of the zygote is determined by the fertilizing sperm cell.* If a Y-bearing sperm cell fertilizes the ovum, the zygote will be XY and male; if an X-bearing sperm cell fertilizes the ovum, the zygote will be XX and female.

Although each diploid cell in a woman's body inherits two X chromosomes, it appears that only one of each pair of X chromosomes remains active. The other X chromosome forms a clump of inactive "heterochromatin," which can often be seen as a dark spot, called a *Barr body,* at the edge of the nucleus of cheek cells (fig. 28.3). This provides a convenient test for chromosomal sex in cases where it is suspected that the chromosomal sex may differ from the apparent ("phenotypic") sex of the individual. Also, some of the nuclei in the neutrophils of females have a "drumstick" appendage not seen in neutrophils from males.

Formation of Testes and Ovaries

The gonads of males and females are similar in appearance for the first 40 or so days of development following conception. During this time, cells that will give rise to sperm (called *spermatogonia*) and cells that will give rise to ova (called *oogonia*) migrate from the yolk sac to the embryonic gonads. At this stage the embryonic structures have the

Barr body: from Murray L. Barr, Canadian anatomist, b. 1908
spermatogonia: Gk. *sperma*, seed; *gonos*, procreation
oogonia: Gk. *oion*, egg; *gonos*, procreation

Figure 28.4

The chromosomal sex and the development of embryonic gonads.
The very early embryo has "indifferent gonads" that can develop into either testes or ovaries. The testis-determining factor (TDF) is a gene located on the Y chromosome. In the absence of TDF, ovaries will develop.

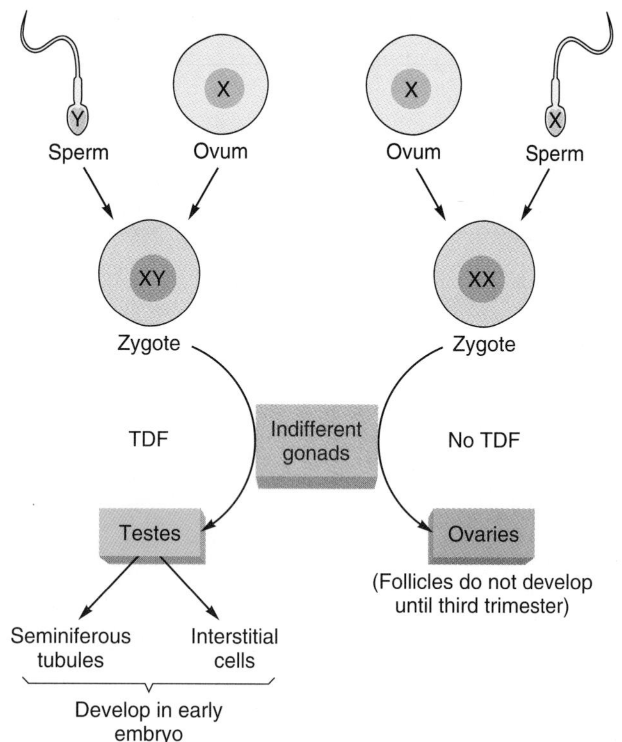

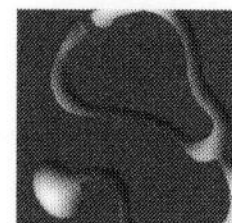

Notice that it is normally the presence or absence of the Y chromosome that determines whether the embryo will have testes or ovaries. This point is well illustrated by two genetic abnormalities. In *Klinefelter's syndrome,* the affected person has 47 instead of 46 chromosomes because of the presence of an extra X chromosome. This person, with an XXY genotype, will develop testes and have a male phenotype despite the presence of two X chromosomes. Patients with *Turner's syndrome,* who have the genotype XO (and therefore have only 45 chromosomes), have poorly developed ("streak") gonads and are phenotypically female. Klinefelter's and Turner's syndromes are discussed more fully under "Clinical Considerations."

potential to become either testes or ovaries. The hypothetical substance that promotes their conversion to testes has been called the **testis-determining factor (TDF).**

Although it has long been recognized that male sex is determined by the presence of a Y chromosome and female sex by the absence of the Y chromosome, the genes involved have only recently been localized. In rare male babies with XX genotypes, scientists have discovered that one of the X chromosomes contains a segment of the Y chromosome—the result of an error that occurred during the meiotic cell division that formed the sperm cell. Similarly, rare female babies with XY genotypes were found to be missing the same portion of the Y chromosome erroneously inserted into the X chromosome of XX males.

Through these and other observations, it has been shown that the gene for the testis-determining factor is located on the short arm of the Y chromosome (fig. 28.4). Evidence suggests that it may be a particular gene known as *SRY* (for sex-determining region of the Y). This gene is found in the Y chromosome of all mammals and is highly conserved, meaning that it shows little variation in structure over evolutionary time.

The structures that will eventually produce sperm within the testes, the **seminiferous** (*sem″ĭ-nif′er-us*) **tubules,** appear very early in embryonic development—between 43 and 50 days following conception. The tubules contain two major cell types: germinal and nongerminal. The **germinal cells** are those that will eventually become sperm, through meiosis and subsequent specialization. The nongerminal cells are called **sustentacular** (*sus″ten-tak′yŭ-lar*), or **Sertoli, cells.** The sustentacular cells appear at about day 42. At about day 65, the **interstitial,** or **Leydig** (*lī′dig*), **cells** appear in the embryonic testes. The interstitial cells are clustered in the interstitial tissue that surrounds the seminiferous tubules. The interstitial cells constitute the endocrine tissue of the testes. In contrast to the rapid development of the testes, the functional units of the ovaries—called the **ovarian follicles**—do not appear until the second trimester of pregnancy (at about day 105).

The early-appearing interstitial cells in the embryonic testes secrete large amounts of male sex hormones, or *androgens.* The major androgen secreted by these cells is **testosterone** (*tes-tos′tĕ-rōn*). Testosterone secretion begins as early as 8 weeks after conception, reaches a peak at 12 to 14 weeks, and thereafter declines to very low levels by the end of the second trimester (at about 21 weeks). Testosterone secretion during embryonic development in the male serves a very important function; similarly high levels of testosterone will not appear again in the life of the individual until the time of puberty.

Masculinization of the embryonic structures occurs as a result of the testosterone secreted by the embryonic testes. Testosterone itself, however, is not the active agent within all of the target organs. Once inside particular target cells, testosterone is converted by the enzyme *5α-reductase* into the active hormone known as **dihydrotestosterone (DHT)** (fig. 28.5). DHT is needed for the development and maintenance of the penis, spongy urethra, scrotum, and prostate. Evidence suggests that testosterone itself directly stimulates the development of the spermatic ducts and seminal vesicles.

Sertoli cells: from Enrico Sertoli, Italian histologist, 1842–1910
Leydig cells: from Franz von Leydig, German anatomist, 1821–1908
androgen: Gk. *andros,* male producing

Figure 28.5

Formation of DHT.
Testosterone, secreted by the interstitial (Leydig) cells of the testes, is converted into dihydrotestosterone (DHT) within the target cells. This reaction involves the addition of a hydrogen (and the removal of the double carbon bond) in the first (A) ring of the steroid.

Descent of the Testes

Developing on the outside surface of each testis is a fibromuscular cord called the **gubernaculum** (*goo''ber-nak'yŭ-lum*) (fig. 1, p. 915) that attaches to the inferior portion of the testis and extends to the labioscrotal swelling of the same side. At the same time, a portion of the embryonic mesonephric duct (see the discussion of embryonic development later in this chapter) attaches itself to the testis, becomes convoluted, and forms the epididymis. Another portion of the mesonephric duct becomes the ductus deferens.

The external genitalia of a male are completely formed by the end of the twelfth week. The descent of the testes from the site of development begins between the sixth and tenth week. Descent into the scrotal sac, however, does not occur until about week 28, when paired inguinal canals form in the abdominal wall to provide openings from the pelvic cavity to the scrotal sac. The process by which a testis descends is not well understood, but it seems to be associated with the shortening and differential growth of the gubernaculum, which is attached to the testis and extends through the inguinal canal to the wall of the scrotum. As the testis descends, it passes to the side of the urinary bladder and anterior to the symphysis pubis. It carries with it the ductus deferens, the testicular vessels and nerve, a portion of the internal abdominal oblique muscle, and lymphatic vessels. All of these structures remain attached to the testis and form what is known as the **spermatic cord.** By the time the testis has taken its position in the scrotal sac, the gubernaculum is no more than a remnant of scarlike tissue.

The temperature of the testes in the scrotum is maintained at about 35° C (95° F), or about 3.6° F below normal body temperature. This lower temperature is needed for proper sperm production and storage, as illustrated by the fact that spermatogenesis does not occur in males with undescended testes—a condition call *cryptorchidism* (see "Clinical Considerations").

During the physical examination of a neonatal male, a physician will palpate the scrotum to determine if the testes are in position. If one or both are not in the scrotal sac, it may be possible to induce descent by administering certain hormones. If this procedure is not effective, surgery is generally performed before the age of 5. Failure to correct the situation may result in sterility and possibly the development of a testicular tumor.

gubernaculum: L. *gubernaculum*, helm

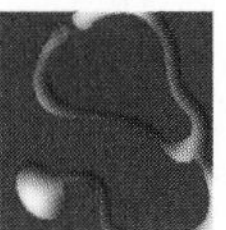

Associated with each spermatic cord is a strand of skeletal muscle called the cremaster muscle. In cold weather, the cremaster muscles contract and elevate the testes, bringing them closer to the warmth of the trunk. The *cremasteric* (*kre''mă-ster'ik*) *reflex* produces the same effect when the inside of a man's thigh is stroked. In a baby, however, this stimulation can cause the testes to be drawn up through the inguinal canal into the body cavity. Sumo wrestlers can also voluntarily draw up their testes into their body cavity.

Endocrine Regulation of Reproduction

The functions of the testes and ovaries are regulated by gonadotropic hormones secreted by the anterior pituitary. The gonadotropic hormones stimulate the gonads to secrete their sex steroid hormones, and these steroid hormones, in turn, have an inhibitory effect on the secretion of the gonadotropic hormones. This interaction between the anterior pituitary and the gonads forms a negative feedback loop.

During the first trimester of pregnancy the embryonic testes are active endocrine glands, secreting the high amounts of testosterone needed to masculinize the male embryo's external genitalia and accessory sex organs. Ovaries, by contrast, do not mature until the third trimester of pregnancy. Testosterone secretion in the male fetus declines during the second trimester of pregnancy, however, so that the gonads of both sexes are relatively inactive at the time of birth.

Before puberty, there are equally low blood concentrations of *sex steroids*—androgens and estrogens—in both males and females. Apparently, this is not due to deficiencies in the ability of the gonads to produce these hormones, but rather to lack of sufficient stimulation. During *puberty,* the gonads secrete increased amounts of sex steroid hormones as a result of increased stimulation by **gonadotropic hormones** from the anterior pituitary.

Interactions between the Hypothalamus, Pituitary Gland, and Gonads

The anterior pituitary produces and secretes two gonadotropic hormones—**follicle-stimulating hormone (FSH)** and **luteinizing hormone (LH).** Although these two hormones are named according to their actions in the female, the same hormones are secreted by the male's pituitary gland. The gonadotropic hormones of both sexes have three primary effects on the gonads: (1) stimulation of spermatogenesis or oogenesis (formation of sperm or ova); (2) stimulation of gonadal hormone secretion; and (3) maintenance of the structure of the gonads (the gonads atrophy if the pituitary gland is removed).

The secretion of both LH and FSH from the anterior pituitary is stimulated by a hormone produced by the hypothalamus and secreted into the hypothalamo-hypophyseal portal vessels (chapter 19). This releasing hormone is sometimes called *luteinizing hormone–releasing hormone* (*LHRH*). Since attempts to find a discrete FSH-releasing hormone have thus far failed, and since LHRH stimulates FSH as well as LH secretion, LHRH is often referred to as **gonadotropin-releasing hormone (GnRH).**

If a male or female animal is castrated (has its gonads surgically removed), the secretion of FSH and LH increases to much higher levels than those of the intact animal. This demonstrates that the gonads secrete products that exert a *negative feedback inhibition* on gonadotropin secretion. This negative feedback is exerted in large part by sex steroids: estrogen and progesterone in the female, and testosterone in the male. A biosynthetic pathway for these steroids is shown in figure 28.6.

The negative feedback effects of steroid hormones occurs by means of two mechanisms: (1) inhibition of GnRH secretion from the hypothalamus and (2) inhibition of the pituitary's response to a given amount of GnRH. In addition to steroid hormones, the testes and ovaries secrete a polypeptide hormone called **inhibin.** Inhibin is secreted by the sustentacular cells of the seminiferous tubules in males and by the granulosa cells of the ovarian follicles in females. This hormone specifically inhibits the anterior pituitary's secretion of FSH without affecting the secretion of LH.

Figure 28.7 illustrates the process of gonadal regulation. Although hypothalamus-pituitary-gonad interactions are similar in males and females, there are important differences. Secretion of gonadotropins and sex steroids is more or less constant in adult males. Secretion of gonadotropins and sex steroids in adult females, by contrast, shows cyclic variations (during the menstrual cycle). Also, as discussed in chapter 29, during one

Figure 28.6

A simplified biosynthetic pathway for the steroid hormones.

The sources of the sex hormones secreted in the blood are also indicated.

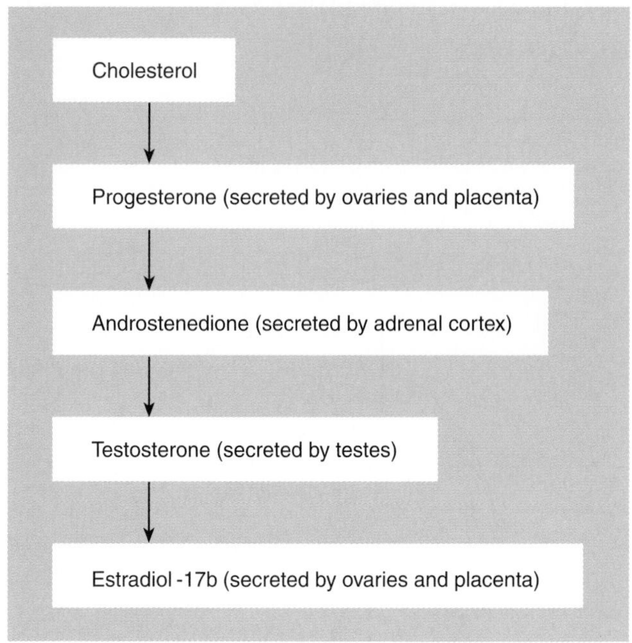

Figure 28.7

Interactions between the hypothalamus, anterior pituitary, and gonads.

Sex steroids secreted by the gonads have a negative feedback effect on the secretion of GnRH (gonadotropin-releasing hormone) and on the secretion of gonadotropins. The gonads may also secrete a polypeptide hormone called inhibin that functions in the negative feedback control of FSH secretion.

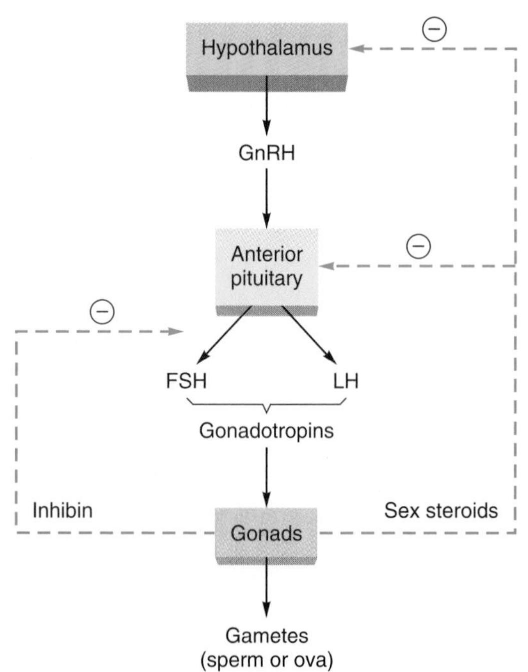

phase of the female cycle—shortly before ovulation—estrogen exerts a positive feedback effect on LH secretion.

Studies have shown that secretion of GnRH from the hypothalamus is *pulsatile* rather than continuous, and thus the secretion of FSH and LH follows this pulsatile pattern. Pulsatile patterns of secretion are needed to prevent desensitization and downregulation of the target glands (discussed in chapter 19). It appears that the frequency of the pulses of secretion, as well as their amplitude (how much hormone is secreted per pulse), affects the target gland's response to the hormone. For example, it has been proposed that a slow frequency of GnRH pulses in women preferentially stimulates FSH secretion, while faster pulses of GnRH favor LH secretion.

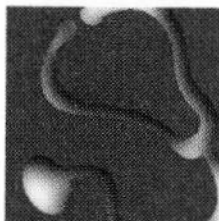

If a powerful synthetic analogue of GnRH (such as *nafarelin*) is administered, the anterior pituitary first increases and then decreases its secretion of FSH and LH. This decrease, which is contrary to the normal stimulatory action of GnRH, is due to a desensitization of the anterior pituitary evoked by continuous exposure to GnRH. The decrease in LH causes a fall in testosterone secretion from the testes, or of estradiol secretion from the ovaries. The decreased testosterone secretion is useful in the treatment of men who have *benign prostatic hyperplasia*. In this condition, common in men over the age of 60, testosterone supports abnormal growth of the prostate. The fall in estradiol secretion in women given synthetic GnRH analogues can be useful in the treatment of *endometriosis* (*en"do-me"tre-o'sis*). In this condition, ectopic endometrial tissue from the uterus (dependent on estradiol for growth) grows outside the uterus—for example, on the ovaries or on the peritoneum. These treatments illustrate the reasons why GnRH and the gonadotropins are normally secreted in a pulsatile fashion and are particularly beneficial clinically because they are reversible.

The Onset of Puberty

Secretion of FSH and LH is high in the newborn but falls to very low levels a few weeks after birth. Gonadotropin secretion remains low until the beginning of puberty, which is marked by rising levels of FSH followed by LH secretion. Experimental evidence suggests that this rise in gonadotropin secretion is a result of two processes: (1) maturational changes in the brain that result in increased GnRH secretion by the hypothalamus and (2) decreased sensitivity of gonadotropin secretion to the negative feedback effects of sex steroid hormones.

The maturation of the brain that leads to increased GnRH secretion at the time of puberty appears to be programmed—children without gonads show increased FSH secretion at the normal time. Also during this period of time, a given amount of sex steroids has less of a suppressive effect on gonadotropin secretion than the same dose would have if administered prior to puberty. This suggests that the sensitivity of the hypothalamus and the pituitary gland to negative feedback effects decreases at puberty, which would also help to account for rising gonadotropin secretion at this time.

During late puberty there is a pulsatile secretion of gonadotropins; FSH and LH secretion increase during periods of sleep and decrease during periods of wakefulness. These pulses of increased gonadotropin secretion during puberty stimulate a rise in sex steroid secretion from the gonads. Increased secretion of testosterone from the testes and of **estradiol-17β** (estradiol is the major *estrogen*, or female sex steroid) from the ovaries during puberty, in turn, produces changes in body appearance characteristic of the two sexes. Such **secondary sex characteristics** (tables 28.1 and 28.2) are the physical manifestations of the hormonal changes occurring during puberty. These changes are accompanied by a growth spurt, which begins at an earlier age in girls than in boys (fig. 28.8).

Table 28.1

Development of Secondary Sex Characteristics and Other Changes That Occur during Puberty in Girls

Characteristic	Age of First Appearance	Hormonal Stimulation
Appearance of breast buds	8–13	Estrogen, progesterone, growth hormone, thyroxine, insulin, cortisol
Pubic hair	8–14	Adrenal androgens
Menarche (first menstrual flow)	10–16	Estrogen and progesterone
Axillary (underarm) hair	About 2 years after the appearance of pubic hair	Adrenal androgens
Eccrine sweat glands and sebaceous glands; acne (from blocked sebaceous glands)	About the same time as axillary hair growth	Adrenal androgens

Table 28.2

Development of Secondary Sex Characteristics and Other Changes That Occur during Puberty in Boys

Characteristic	Age of First Appearance	Hormonal Stimulation
Growth of testes	10–14	Testosterone, FSH, growth hormone
Pubic hair	10–15	Testosterone
Body growth	11–16	Testosterone, growth hormone
Growth of penis	11–15	Testosterone
Growth of larynx (voice lowers)	Same time as growth of penis	Testosterone
Facial and axillary (underarm) hair	About 2 years after the appearance of pubic hair	Testosterone
Eccrine sweat glands and sebaceous glands; acne (from blocked sebaceous glands)	About the same time as facial and axillary hair growth	Testosterone

Figure 28.8

Pubescent growth rates in males and females.

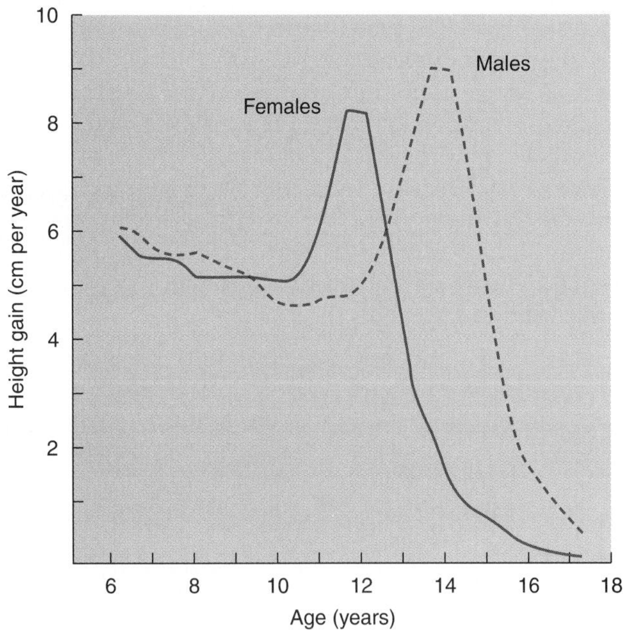

Figure 28.9

A simplified biosynthetic pathway for melatonin.
Melatonin is a hormone secreted by the pineal gland.

Tryptophane
(an amino acid)

Serotonin
(a biogenic amine)

Melatonin
(a pineal gland hormone)

The age at which puberty begins is related to the amount of body fat and level of physical activity of the child. The average age of *menarche* (*mĕ-nar'ke*)—the first menstrual flow—is later (age 15) in girls who are very active physically than in the general population (age 12.6). This appears to be due to a requirement for a minimum percentage of body fat for menstruation to begin; this may represent a mechanism favored by natural selection to ensure the ability to successfully complete a pregnancy and nurse the baby. Recent evidence suggests that the secretion of leptin from adipocytes (chapter 27) is required for puberty. Later in life, women who are very lean and physically active may have irregular cycles and *amenorrhea* (cessation of menstruation). This may also be related to the percentage of body fat. In addition, there is evidence that physical exercise may act to inhibit GnRH and gonadotropin secretion through activation of neural pathways involving endorphin neurotransmitters (chapter 14).

Pineal Gland

The role of the **pineal gland** in human physiology is poorly understood. However, it is known that this gland, located deep within the brain, secretes the hormone **melatonin** as a derivative of the amino acid tryptophane (fig. 28.9) and that production of this hormone is influenced by light-dark cycles.

The pineal glands of some vertebrates have photoreceptors that are directly sensitive to environmental light. Although no such photoreceptors are present in the pineal glands of mammals, the secretion of melatonin has been shown to increase at night and decrease during daylight hours. The inhibitory effect of light on melatonin secretion in mammals is indirect. Pineal secretion is stimulated by postganglionic sympathetic neurons that originate in the superior cervical ganglion; activity of these neurons, in turn, is inhibited by nerve tracts that are activated by light striking the retina. The physiology of the pineal gland was discussed in chapter 19 (see fig. 19.28).

There is abundant experimental evidence that melatonin can inhibit gonadotropin secretion and thus have an "antigonad" effect in many vertebrates. However, the role of melatonin in the regulation of human reproduction has not yet been clearly established.

Male Reproductive System

The interstitial (Leydig) cells of the testes are stimulated by LH to secrete testosterone—a potent androgen that acts to maintain the structure and function of the male secondary sex organs and to promote the development of male secondary sex characteristics. The sustentacular (Sertoli) cells in the seminiferous tubules of the testes are stimulated by FSH. The cooperative actions of FSH and testosterone are required to initiate spermatogenesis.

The structures of the male reproductive system are illustrated in figure 28.10. These include the *primary sex organs* of the male—the testes—and a number of *secondary sex organs*. In this section, we will describe the structure, function, and regulation of the testes. The secondary sex organs are covered in the next section.

Scrotum

The saclike **scrotum** is suspended immediately behind the base of the penis, anterior to the anal opening, in a region known as the **perineum** (*per″ĭ-ne′um*). The functions of the scrotum are to support and protect the testes and to regulate their position relative to the pelvic region of the body. The **dartos** (*dar′tos*) **muscle** consists of a layer of smooth muscle fibers in the subcutaneous tissue of the scrotum, and the **cremaster** (*krĕ-mas′ter*) **muscle** is a thin strand of skeletal muscle associated with the spermatic cord. Both muscles involuntarily contract in response to low temperatures to move the testes closer to the heat of the body in the pelvic region. The cremaster muscle is a continuation of the internal abdominal oblique muscle and is derived as the testes descend into the scrotum. High temperatures cause the dartos and cremaster muscles to relax and lower the testes away from the body cavity. The contraction and relaxation of the scrotal muscles help to maintain the temperature required by the testes for spermatogenesis, as previously described.

Although uncommon, male fertility may result from an excessively high temperature of the testes over an extended period of time. Normally the temperature of the testes is kept sufficiently low by their location outside of the body cavity, in the scrotum, and by a special arrangement of blood vessels in the scrotum. Each testicular artery, bringing warmer blood to the testis, is surrounded in the scrotum by a network of venules carrying cooler blood from the testis. This network of venules, the **pampiniform** (*pam-pin′i-form*) **plexus,** cools the arterial blood as it warms the venous blood. Since arterial and venous blood travel in opposite directions, this process is known as **countercurrent heat exchange.** Tight clothing that keeps the testes close to the body or frequent hot baths or saunas, however, may overcome these protective mechanisms and raise the testicular temperature sufficiently to impair sperm production and fertility.

The scrotum is subdivided into two longitudinal compartments by a fibrous **scrotal septum.** The site of the scrotal septum is apparent on the surface of the scrotum along a median longitudinal ridge called the *scrotal raphe* (*ra′fe*). The purpose of the scrotal septum is to compartmentalize each testis so that trauma or infection in one will not affect the other. Another protective feature is that the left testis is generally suspended lower than the right so that the two are less likely to be compressed forcefully together.

Structure of the Testes

The **testes** (*tes′tēz*) are paired organs each about 4 cm (1.5 in.) long and 2.5 cm (1 in.) in diameter. Each testis weighs between 10 and 14 g. Two tissue layers, or tunicas, cover the testis. The outer **tunica vaginalis** is a thin serous sac derived from the peritoneum during the descent of the testis. The **tunica albuginea** (*al″byoo-jin′e-ă*) is a tough fibrous membrane that directly encapsules each testis (fig. 28.11). Fibrous inward extensions of the tunica albuginea partition the testis into 250 to 300 wedge-shaped **testicular lobules.**

Each testicular lobule contains one to three tightly convoluted **seminiferous tubules** that may exceed 70 cm (28 in.) in length if uncoiled. It is in the seminiferous tubules that spermatogenesis occurs. Sperm are produced at the rate of thousands per second—more than 100 million per day—throughout the life of a healthy, sexually mature male.

Various stages of spermatogenesis can be observed in a histological section of seminiferous tubules (fig. 28.12). The germinal cells, called **spermatogonia** (*sper″mă-to-go′ne-ă*), are in contact with the basement membrane. Spermatogonia undergo

---

dartos: Gk. *dartos*, skinned or flayed
cremaster: Gk. *cremaster*, a suspender, to hang

pampiniform: L. *pampinus*, tendril; *forma*, shaped
septum: L. *septum*, a partition
raphe: Gk. *raphe*, a seam
tunica: L. *tunica*, a coat
vaginalis: L. *vagina*, a sheath
albuginea: L. *albus*, white

Figure 28.10

Organs of the male reproductive system.
(*a*) A sagittal view and (*b*) a posterior view.

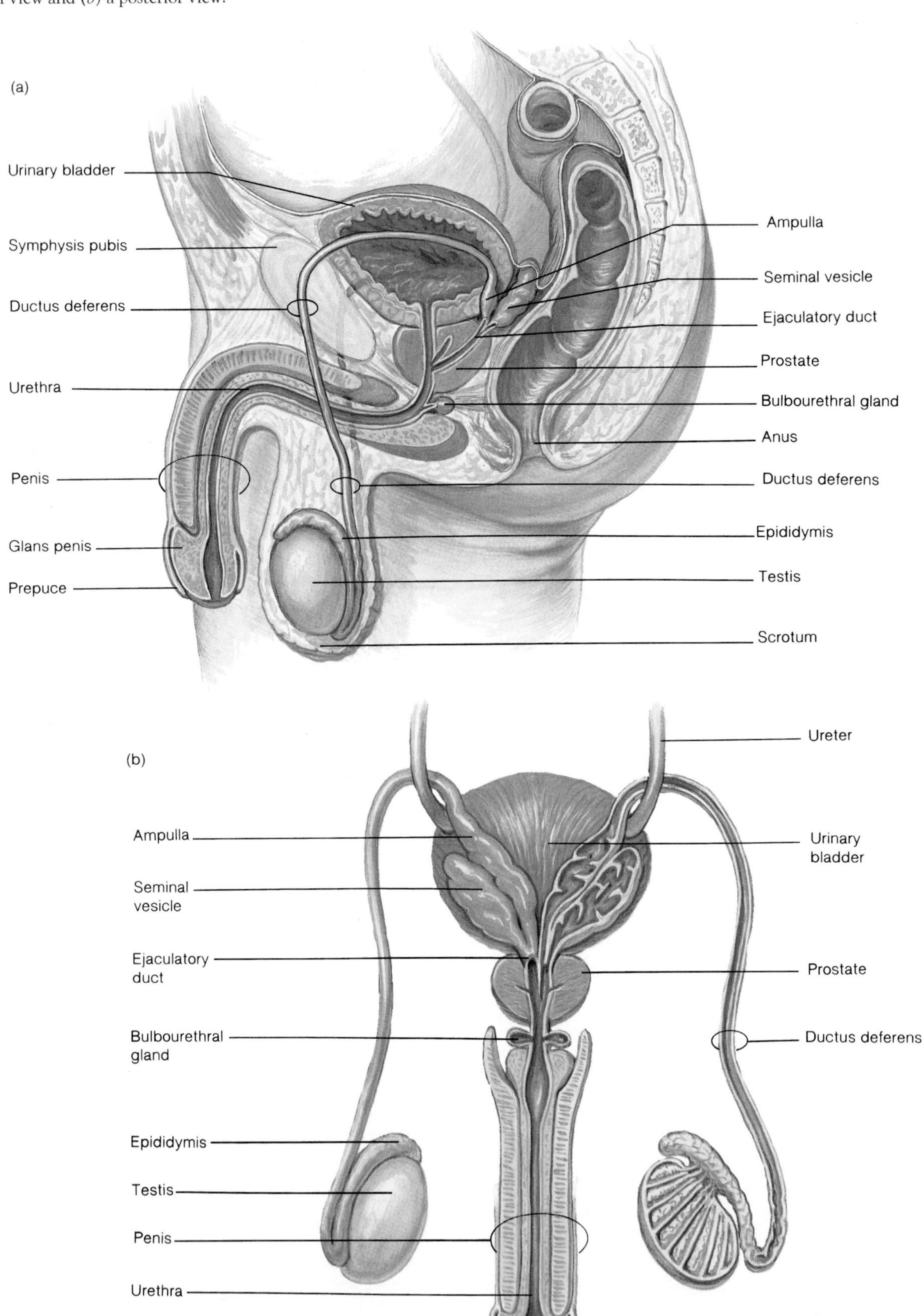

Figure 28.11

Male external genitalia.

The penis, scrotum, testis, and epididymis are visible in (*a*) a longitudinal view and (*b*) a transverse view of the scrotum and testis.

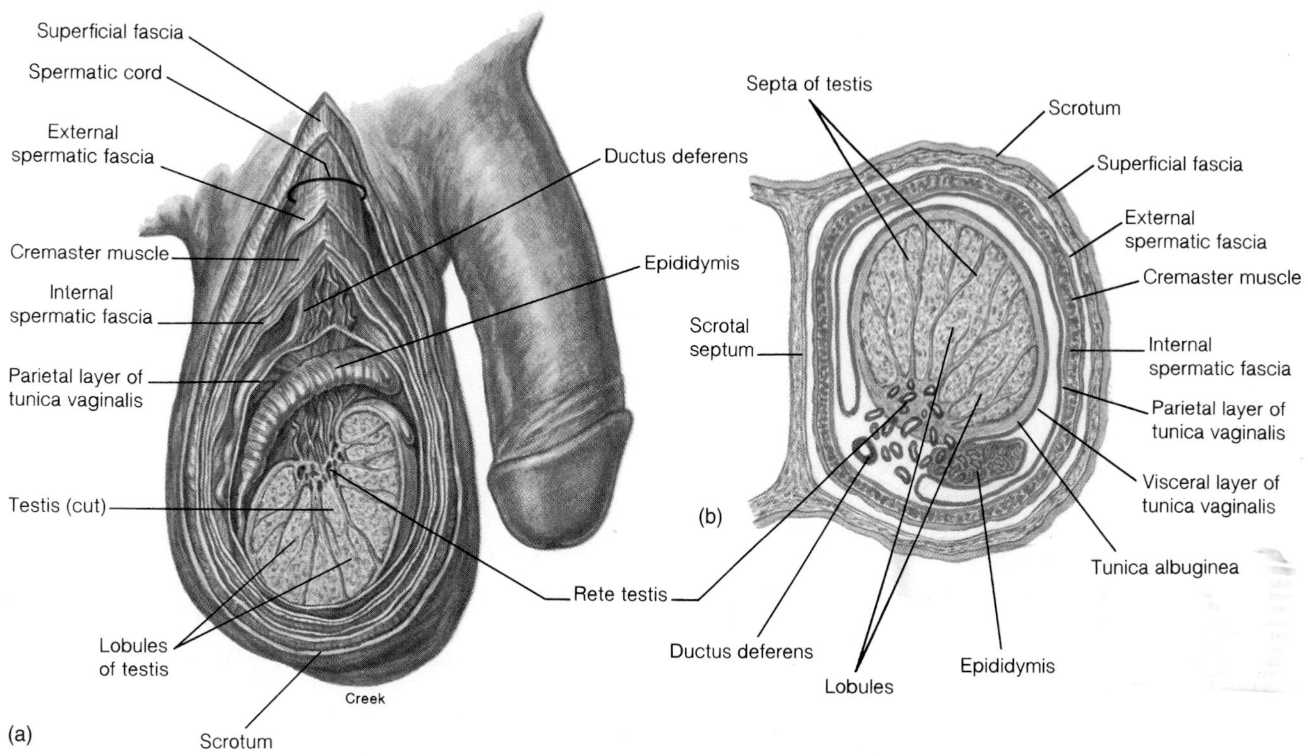

Figure 28.12

The seminiferous tubules.

(*a*) A sagittal section of a testis and (*b*) a transverse section of a seminiferous tubule.

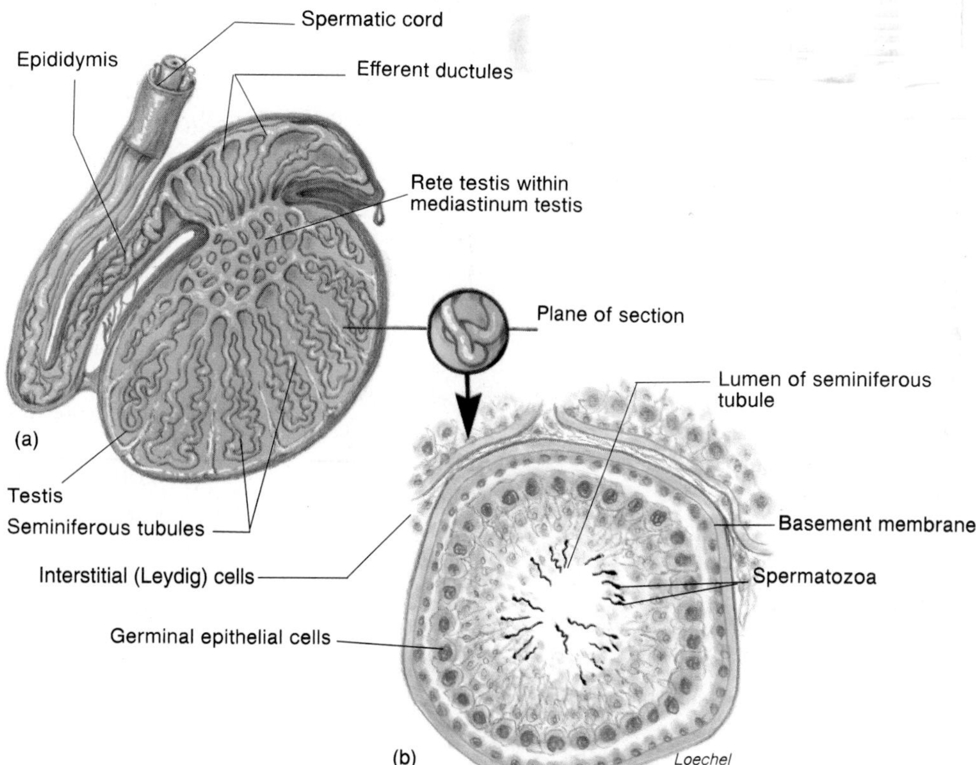

meiosis to produce, in order of advancing maturity, the primary spermatocytes, secondary spermatocytes, and spermatids (see fig. 28.16). Forming the walls of the seminiferous tubules are sustentacular (Sertoli) cells, which are sometimes referred to as *nurse cells*. The sustentacular cells are critically important to the developing spermatozoa that are embedded between them. The spermatozoa are formed, but not fully matured, by the time they reach the lumina of the seminiferous tubules.

Between the seminiferous tubules are specialized endocrine cells called interstitial (Leydig) cells. The function of these cells is to produce and secrete the male sex hormones. The testes are considered mixed exocrine and endocrine glands because they produce both sperm and androgens.

Once the sperm have been produced, they move through the seminiferous tubules and enter a tubular network called the **rete** (*re'te*) **testis** for further maturation. Cilia are located on some of the cells of the rete testis, presumably for moving sperm. The sperm are transported out of the testis and into the epididymis through a series of **efferent ductules.**

With regard to gonadotropin action, the testes are strictly compartmentalized. Cellular receptor proteins for FSH are located exclusively in the seminiferous tubules, where they are confined to the sustentacular cells. LH receptor proteins are located exclusively in the interstitial cells. Secretion of testosterone by the interstitial cells is stimulated by LH but not by FSH. Spermatogenesis in the tubules is stimulated by FSH.

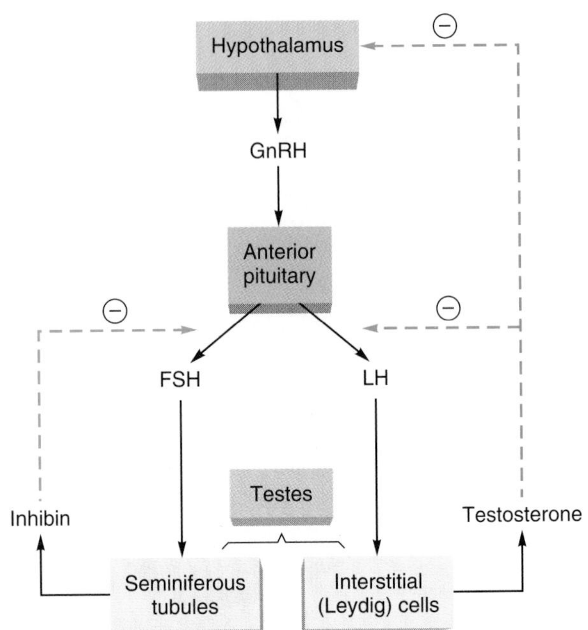

Figure 28.13

The anterior pituitary and testes.
The seminiferous tubules are the targets of FSH action; the interstitial (Leydig) cells are targets of LH action. Testosterone secreted by the interstitial cells inhibits LH secretion; inhibin secreted by the tubules may inhibit FSH secretion.

Control of Gonadotropin Secretion

Castration of a male animal results in an immediate rise in FSH and LH secretion. This demonstrates that hormones secreted by the testes exert negative feedback control of gonadotropin secretion. If testosterone is injected into the castrated animal, the secretion of LH can be returned to the previous (precastration) levels. This provides a classical example of negative feedback—LH stimulates testosterone secretion by the interstitial cells, and testosterone inhibits pituitary secretion of LH (fig. 28.13).

The amount of testosterone that is sufficient to suppress LH, however, is not sufficient to suppress the postcastration rise in FSH secretion in most experimental animals. In rams and bulls, a water-soluble (and, therefore, a peptide rather than a steroid) product of the seminiferous tubules specifically suppresses FSH secretion. This hormone, produced by the sustentacular cells, is called **inhibin.** The seminiferous tubules of the human testes also have been shown to produce inhibin, which inhibits FSH secretion in men. (There is also evidence that inhibin is produced by the ovaries, where it may function as a hormone and as a paracrine regulator of the ovaries.)

Testosterone Derivatives in the Brain

The brain contains testosterone receptors and is a target organ for this hormone. The effects of testosterone on the brain, such as the suppression of LH secretion, are not mediated directly by testosterone, however, but rather by its derivatives that are produced within the brain cells. Testosterone may be converted by the enzyme 5α-reductase to dihydrotestosterone (DHT), as previously described. The DHT, in turn, can be changed by other enzymes into other 5α-reduced androgens—abbreviated *3α-diol* and *3β-diol* (fig. 28.14). Alternatively, testosterone is also converted within the brain to estradiol-17β. Although usually regarded as a female sex steroid, estradiol is therefore an active compound in normal male physiology! Estradiol is formed from testosterone by the action of an enzyme called *aromatase*. This reaction is known as *aromatization*, a term that refers to the presence of an aromatic carbon ring (chapter 2). The estradiol formed from testosterone in the brain is required for the negative feedback effects of testosterone on LH secretion.

Testosterone Secretion and Age

The negative feedback effects of testosterone and inhibin help to maintain a relatively constant (that is, noncyclic) secretion of gonadotropins in males, resulting in relatively constant levels of androgen secretion from the testes. This contrasts with the cyclic secretion of gonadotropins and ovarian steroids in females. Women experience an abrupt cessation in sex steroid secretion during menopause. By contrast, the secretion of androgens declines only gradually and to varying degrees in men over 50 years of age. The causes of this age-related change in testicular function are not currently known. The decline in

rete: L. *rete*, a net
efferent ductules: L. *efferre*, to bring out; *ducere*, to lead

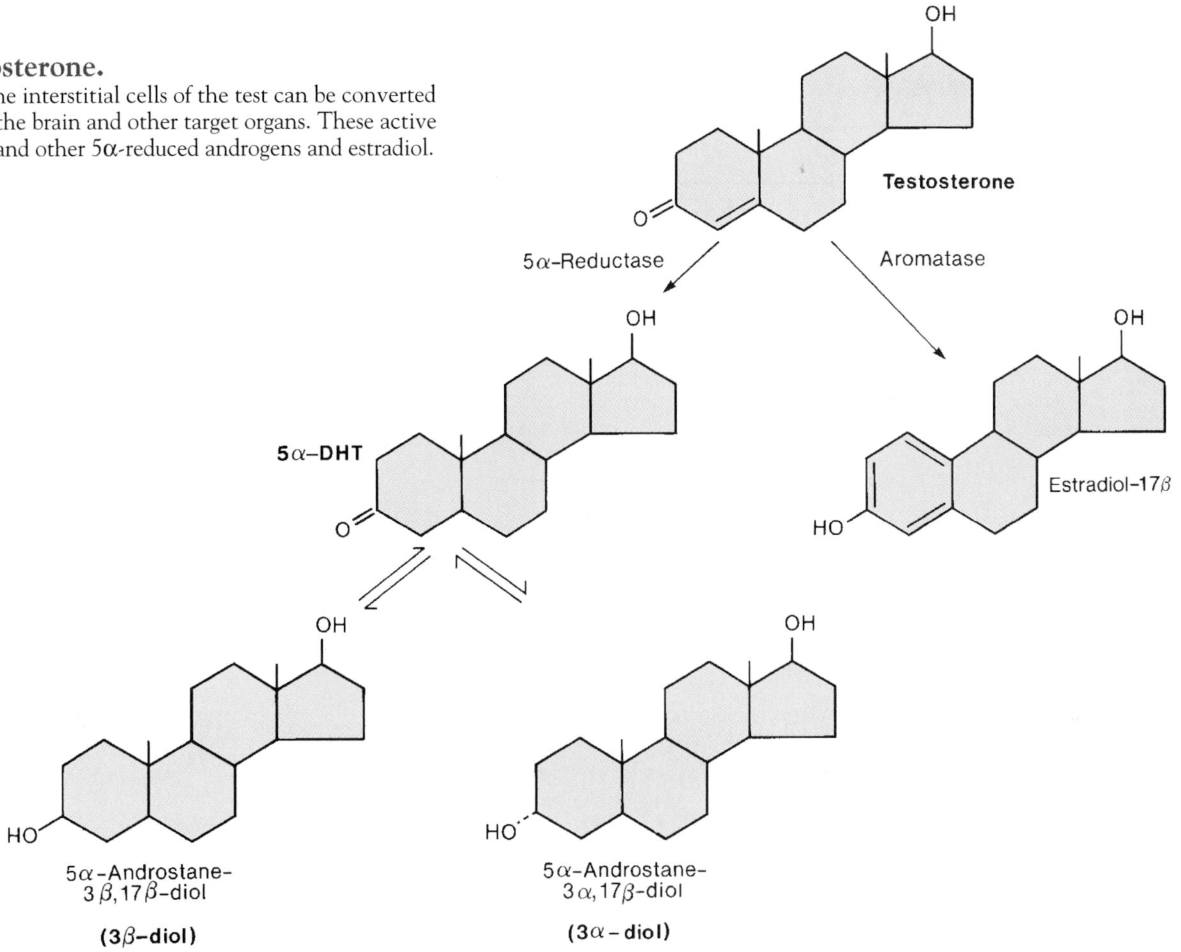

Figure 28.14

Derivatives of testosterone.
Testosterone secreted by the interstitial cells of the test can be converted into active metabolites in the brain and other target organs. These active metabolites include DHT and other 5α-reduced androgens and estradiol.

testosterone secretion cannot be due to decreasing gonadotropin secretion, since gonadotropin levels in the blood are, in fact, elevated (because of less negative feedback) at the time that testosterone levels are declining.

Endocrine Functions of the Testes

Testosterone is by far the major androgen secreted by the adult testes. This hormone and its derivatives (the 5α-reduced androgens) are responsible for initiation and maintenance of the body changes associated with puberty in males. Androgens are sometimes called *anabolic steroids* because they stimulate the growth of muscles and other structures (table 28.3). Increased testosterone secretion during puberty is also required for growth of the secondary sex organs—primarily the seminal vesicles and prostate. Removal of androgens by castration results in atrophy of these organs.

Androgens stimulate growth of the larynx (causing a lowering of the voice) and promote hemoglobin synthesis (males have higher hemoglobin levels than females) and bone growth. The effect of androgens on bone growth is self-limiting, however, because they ultimately cause replacement of cartilage by bone in the epiphyseal plates, thus "sealing" the plates and preventing further lengthening of the bones (as described in chapter 8).

Although androgens are by far the major endocrine products of the testes, there is evidence that the sustentacular cells and interstitial cells both secrete small amounts of estradiol. Further, receptors for estradiol are found in both of these cell types, as well as in the cells lining the male reproductive tract (efferent ductules and epididymides) and accessory sex organs (prostate and seminal vesicles). Indeed, knockout mice (chapter 3) missing an estrogen receptor gene are infertile! This suggests that estrogens have important regulatory functions in male reproduction.

Estradiol, either secreted by the testes or produced locally as a paracrine regulator, may be responsible for a number of effects in men that have previously been attributed to androgens. For example, the importance of the conversion of testosterone into estradiol in the brain for negative feedback control was described earlier. Estrogen also may be responsible for sealing of the epiphyseal plates of cartilage; this is suggested by observations that men who lack the ability to produce estrogen or who lack estrogen receptors (due to rare genetic defects only recently discovered) maintain their epiphyseal plates and continue to grow.

The two compartments of the testes interact with each other in a paracrine fashion (fig. 28.15). Paracrine regulation, as described in chapter 19, refers to chemical regulation that occurs among tissues within an organ. Testosterone from the

Table 28.3
Actions of Androgens in the Male

Category	Action
Sex determination	Growth and development of mesenephric ducts into epididymides, ductus deferens, seminal vesicles, and ejaculatory ducts Development of urogenital sinus into prostate Development of male external genitalia (penis and scrotum)
Spermatogenesis	At puberty: completion of meiotic division and early maturation of spermatids After puberty: maintenance of spermatogenesis
Secondary sex characteristics	Growth and maintenance of accessory sex organs Growth of penis Growth of facial and axillary hair Body growth
Anabolic effects	Protein synthesis and muscle growth Growth of bones Growth of other organs (including larynx) Erythropoiesis (red blood cell formation)

Figure 28.15

Interactions between the two compartments of the testes.
Testosterone secreted by the interstitial (Leydig) cells stimulates spermatogenesis in the seminiferous tubules. Interstitial cells may also secrete ACTH, MSH, and β-endorphin. Secretion of inhibin by the tubules may affect the sensitivity of the interstitial cells to LH stimulation.

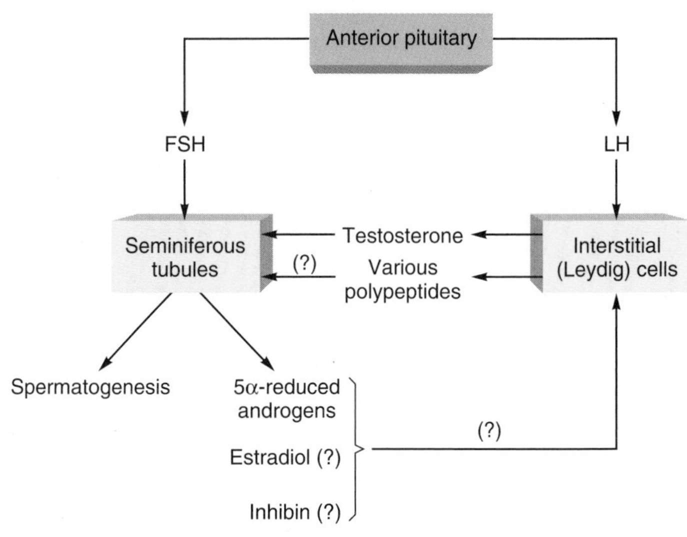

interstitial cells is metabolized by the tubules into other active androgens and is required for spermatogenesis (as described in the next section), for example. The tubules also secrete products that might influence interstitial cell function. Such interactions are suggested by evidence that exposure of pubertal male rats to FSH augments the responsiveness of the interstitial cells to LH. Since FSH can directly stimulate only the sustentacular cells of the seminiferous tubules, the FSH-induced enhancement of LH responsiveness must be mediated by products secreted from the sustentacular cells.

Inhibin, secreted by the sustentacular cells in response to FSH, can facilitate the interstitial cells' response to LH, as measured by the amount of testosterone secreted. Further, it has been shown that the interstitial cells are capable of producing a family of polypeptides previously associated only with the pituitary gland—ACTH, MSH, and β-endorphin. Experiments suggest that ACTH and MSH can stimulate sustentacular cell function, whereas β-endorphin can inhibit interstitial cell function. The physiological significance of these fascinating paracrine interactions between the two compartments of the testes remains to be demonstrated.

Spermatogenesis

The germ cells that migrate from the yolk sac to the testes during early embryonic development become "stem cells" or **spermatogonia,** within the outer region of the seminiferous tubules. Spermatogonia are diploid cells (with 46 chromosomes) that ultimately give rise to mature haploid gametes by a process of reductive cell division called *meiosis*, described in chapter 3.

Meiosis involves two nuclear divisions (see fig. 3.34). In the first part of this process, the DNA duplicates and homologous chromosomes separate into two daughter cells. Since each daughter cell contains only one of each homologous pair of chromosomes, the cells formed at the end of this first meiotic division contain 23 chromosomes each and are haploid. Each of the 23 chromosomes at this stage, however, consists of two strands (called chromatids) of identical DNA. During the second meiotic division, these duplicate chromatids separate into daughter cells. Meiosis of one diploid spermatogonium therefore produces four haploid cells.

Actually, only about 1,000 to 2,000 stem cells migrate from the yolk sac into the embryonic testes. In order to produce many millions of sperm throughout adult life, these spermatogonia duplicate themselves by mitotic division and only one of the two cells—now called a **primary spermatocyte**—undergoes meiotic division (fig. 28.16). In this way, spermatogenesis can occur continuously without exhausting the number of spermatogonia.

When a diploid primary spermatocyte completes the first meiotic division (at telophase I), the two haploid cells thus produced are called **secondary spermatocytes.** At the end of the second meiotic division, each of the two secondary spermatocytes produces two haploid **spermatids** (*sper′mă-tidz*). One primary spermatocyte therefore produces four spermatids.

Figure 28.16

Spermatogenesis.

Spermatogonia undergo mitotic division to replace themselves and produce a daughter cell that will undergo meiotic division. This cell is called a primary spermatocyte. Upon completion of the first meiotic division, the daughter cells are called secondary spermatocytes. Each of these completes a second meiotic division to form spermatids. Notice that the four spermatids produced by the meiosis of a primary spermatocyte are interconnected. Each spermatid forms a mature spermatozoon.

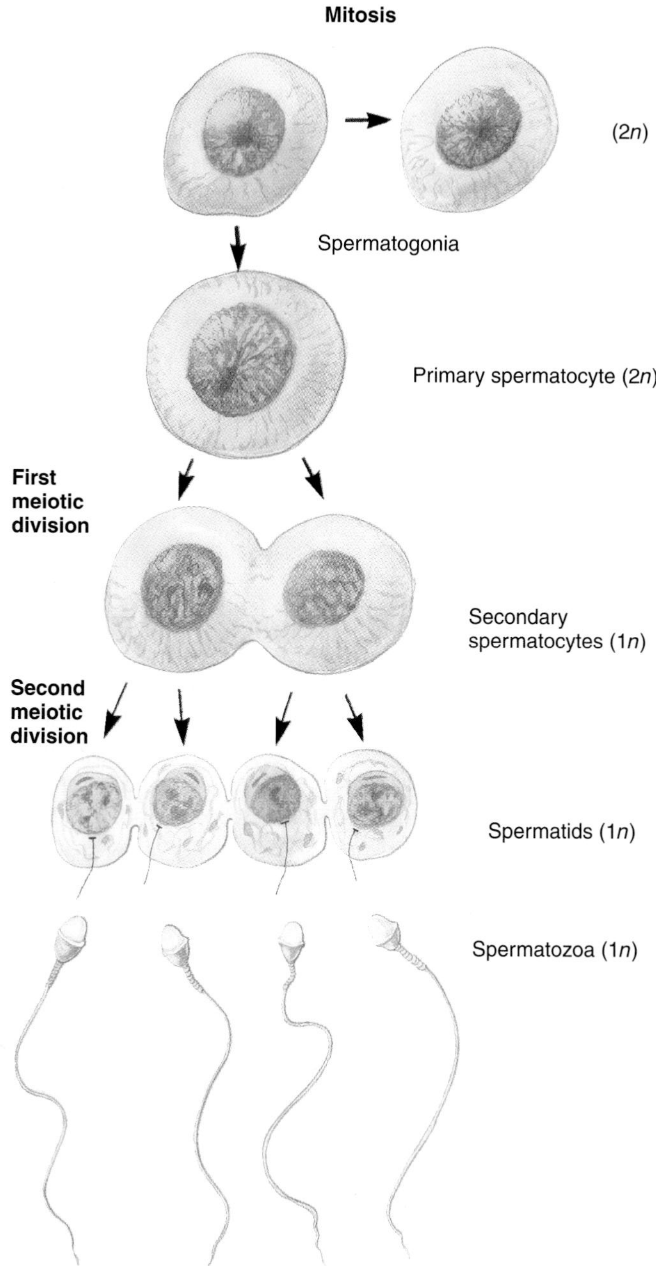

Figure 28.17

Photomicrograph and diagram of the seminiferous tubules.

(*a*) A cross section of the seminiferous tubules also show surrounding interstitial tissue. (*b*) The stages of spermatogenesis are indicated within the germinal epithelium of a seminiferous tubule. The relationship between sustentacular (Sertoli) cells and developing spermatozoa can also be seen.

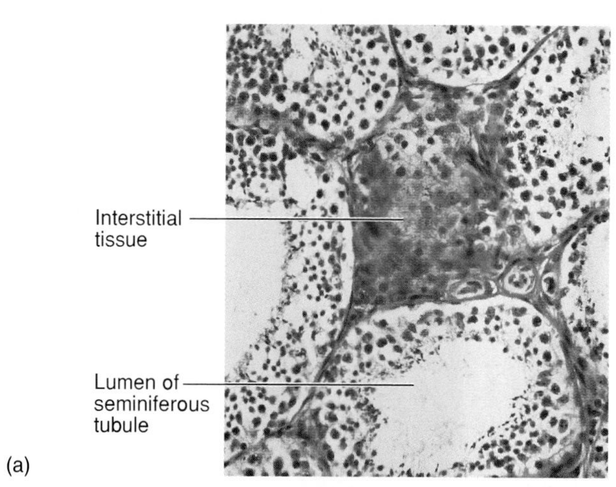

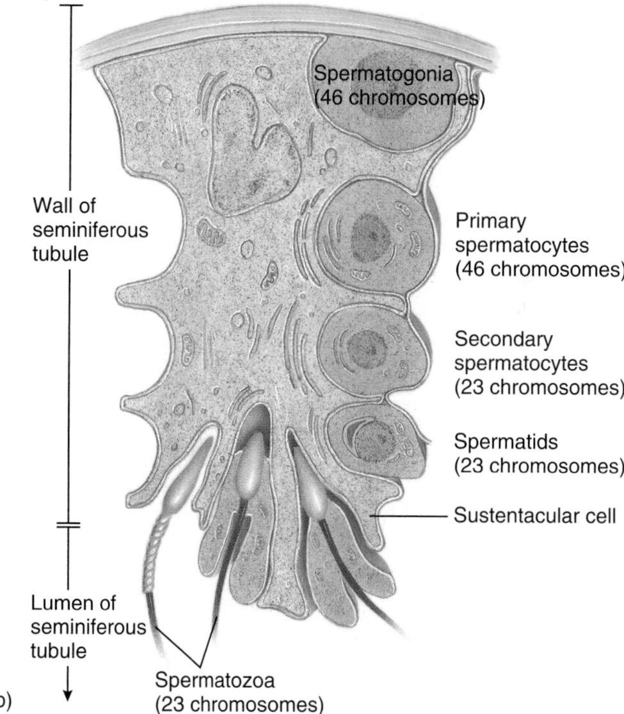

The sequence of events in spermatogenesis is reflected in the cellular arrangement of the wall of the seminiferous tubule. The spermatogonia and primary spermatocytes are located toward the outer side of the seminiferous tubule, whereas spermatids and mature spermatozoa are located on the side of the tubule facing the lumen.

At the end of the second meiotic division, the four spermatids produced by meiosis of one primary spermatocyte are interconnected—their cytoplasm does not completely pinch off at the end of each division. Development of these interconnected spermatids into separate, mature **spermatozoa** (singular, *spermatozoon*)—a process called **spermiogenesis**—requires the participation of the sustentacular cells (fig. 28.17).

spermatozoon: Gk. *sperma*, seed; *zoon*, animal

Figure 28.18

The processing of spermatids into spermatozoa (spermiogenesis).
As the spermatids develop into spermatozoa, most of their cytoplasm is pinched off as residual bodies and ingested by the surrounding sustentacular cell cytoplasm.

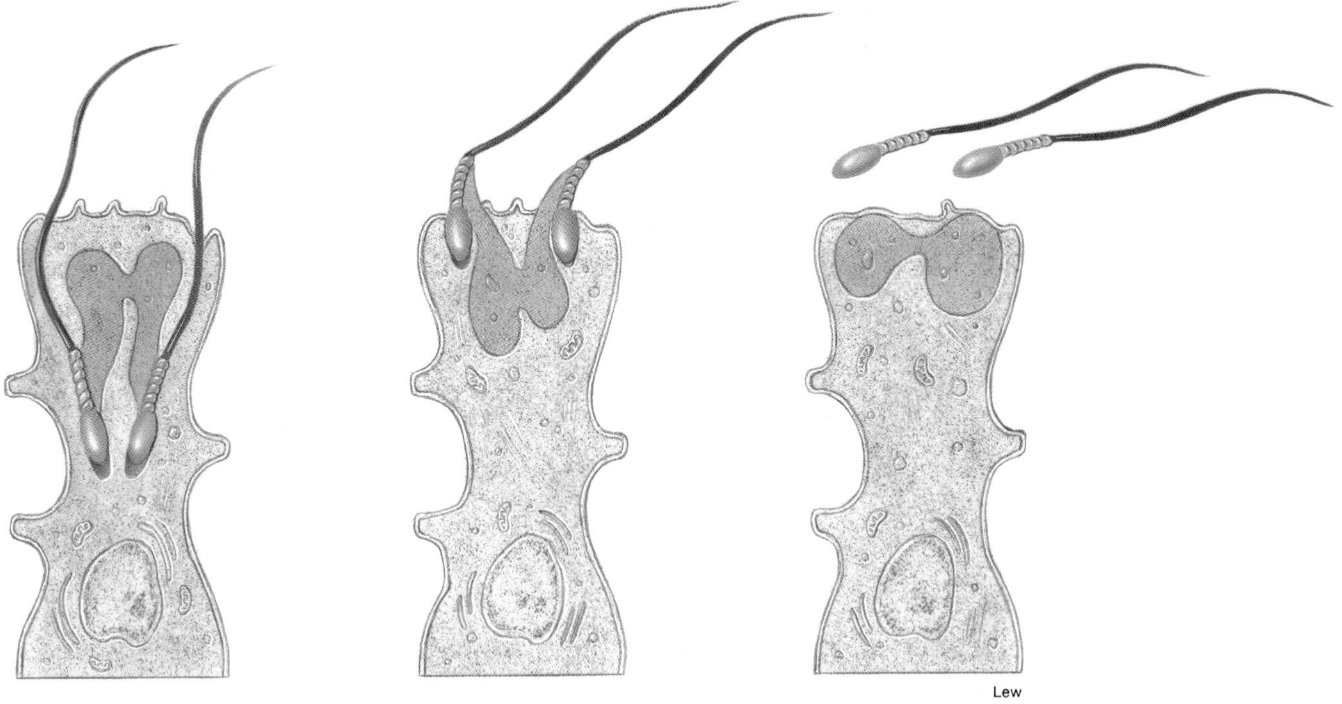

Sustentacular Cells

The nongerminal **sustentacular** (*Sertoli*) **cells** form a continuous layer connected by tight junctions around the circumference of each tubule. In this way the sustentacular cells constitute a **blood-testis barrier;** molecules from the blood must pass through the cytoplasm of the sustentacular cells before entering germinal cells. Similarly, this barrier normally prevents the immune system from becoming sensitized to antigens in the developing sperm, and thus prevents autoimmune destruction of the sperm. The cytoplasm of the sustentacular cells extends from the periphery to the lumen of the seminiferous tubule and envelops the developing germ cells, so that it is often difficult to tell where the cytoplasm of the sustentacular cells ends and that of germ cells begins.

In the process of spermiogenesis (conversion of spermatids to spermatozoa), most of the spermatid cytoplasm is eliminated. This occurs through phagocytosis by sustentacular cells of the "residual bodies" of cytoplasm from the spermatids (fig. 28.18). Phagocytosis of residual bodies may transmit informational molecules from germ cells to sustentacular cells. The sustentacular cells, in turn, may provide molecules needed by the germ cells. It is known, for example, that the X chromosome of germ cells is inactive during meiosis. Since this chromosome contains genes needed to produce many essential molecules, it is believed that these molecules are provided by the sustentacular cells during this time.

Sustentacular cells produce a protein called **androgen-binding protein (ABP)** into the lumen of the seminiferous tubules. This protein, as its name implies, binds to testosterone and thereby concentrates it within the tubules. The importance of sustentacular cells in tubular function is further evidenced by the fact that FSH receptors are confined to the sustentacular cells. Any effect of FSH on the tubules, therefore, must be mediated through the action of sustentacular cells. These include the FSH-induced stimulation of spermiogenesis and the autocrine interactions between sustentacular cells and interstitial cells previously described.

Hormonal Control of Spermatogenesis

The formation of primary spermatocytes and entry into early prophase I begins during embryonic development, but spermatogenesis is arrested at this point until puberty, when testosterone secretion rises. Testosterone is required for completion of meiotic division and for the early stages of spermatid maturation. This effect is probably not produced by testosterone directly, but rather by some of the molecules derived from testosterone in the tubules. The testes also produce a wide variety of paracrine regulators—transforming growth factor, insulin-like growth factor-1, inhibin, and others—that may help to regulate spermatogenesis.

The later stages of spermatid maturation during puberty appear to require stimulation by FSH (fig. 28.19). This FSH

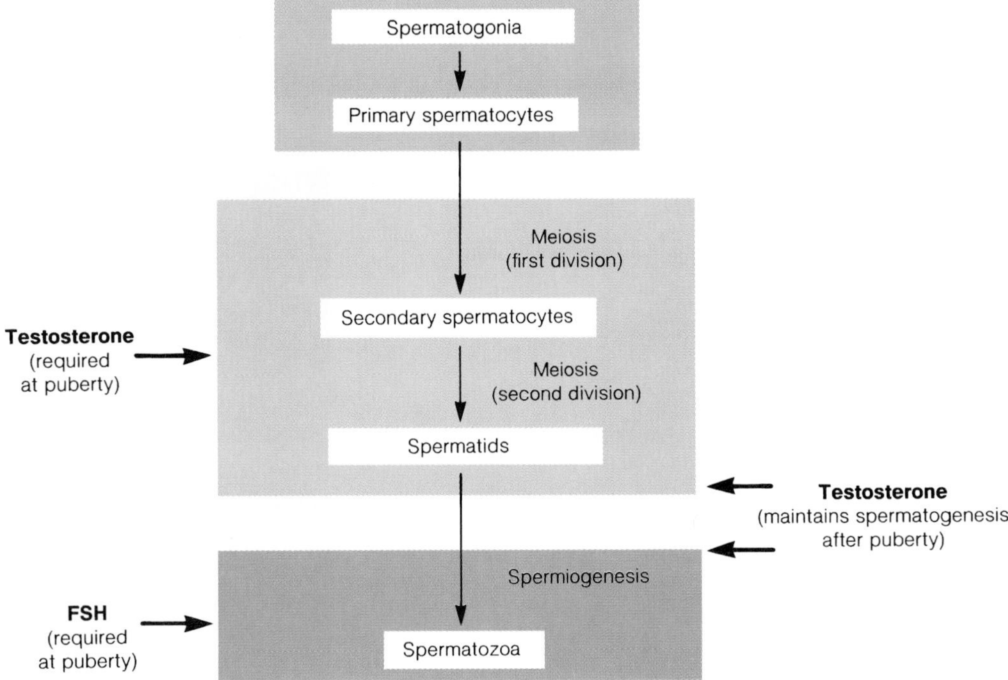

Figure 28.19

The endocrine control of spermatogenesis.
During puberty, both testosterone and FSH are required to initiate spermatogenesis. In the adult, however, testosterone alone can maintain spermatogenesis.

effect is mediated by the sustentacular cells, as previously described. During puberty, therefore, both FSH and androgens are needed for the initiation of spermatogenesis.

Experiments in rats and evidence in humans reveal that spermatogenesis within the adult testes can be maintained by androgens alone, in the absence of FSH. It appears, in other words, that FSH is needed to initiate spermatogenesis at puberty, but that it may not be needed for this function once spermatogenesis has begun.

At the conclusion of spermiogenesis, spermatozoa are released into the lumina of the seminiferous tubules. At this stage the sperm are nonmotile; they become motile and undergo other maturational changes outside the testis in the epididymis.

Structure of Spermatozoa

A mature sperm cell, or **spermatozoon** (*sper″mă-to-zo′on*), is a microscopic tadpole-shaped structure about 0.06 mm long (fig. 28.20). It consists of an oval-shaped **head** and an elongated **flagellum.** The head of a sperm cell contains a nucleus with 23 chromosomes. The tip of the head, called the **acrosome** (*ak′rŏ-sōm*), contains enzymes that help the sperm cell penetrate the ovum. The flagellum contains numerous mitochondria spiraled around a filamentous core. The mitochondria provide the energy necessary for locomotion. The flagellum propels the sperm cell with a lashing movement. The maximum unassisted rate of sperm movement is about 3 mm per hour.

acrosome: Gk. *akros*, extremity; *soma*, body

The life expectancy of ejaculated sperm is between 48 and 72 hours at body temperature. Many of the ejaculated sperm are defective, however, and are of no value. It is not uncommon for sperm to have enlarged heads, dwarfed and misshapen heads, two flagella, or a flagellum that is bent. Sperm such as these are unable to propel themselves adequately.

Spermatic Ducts, Accessory Reproductive Glands, and the Penis

The spermatic ducts store spermatozoa and transport them from the testes to the urethra. The accessory reproductive glands add secretions to the spermatozoa in the formation of semen. Semen is expelled from the erect penis during ejaculation.

Spermatic Ducts

The **spermatic ducts** store spermatozoa and transport them from the testes to the urethra. These ducts include the epididymides, the ductus deferentia, and the ejaculatory ducts.

Epididymis

The **epididymis** (*ep″ĭ-did′ĭ-mis*; plural, *epididymides* [*ep″ĭ-dĭ-dim′ĭ-dēz*])—is a long, flattened organ attached to the posterior surface of the testis (see figs. 28.10 and 28.11). The tubular portion of the epididymis is highly coiled and contains millions of sperm in their final stages of maturation (fig. 28.21). If the epididymis were uncoiled, it would measure an estimated 5 to 6 m (about 17 ft). The expanded upper portion of the epididymis is the **head,** the tapering middle section

Figure 28.20

A human spermatozoon.
(*a*) A diagrammatic representation and (*b*) a scanning electron micrograph in which a spermatozoon is in contact with an ovum.

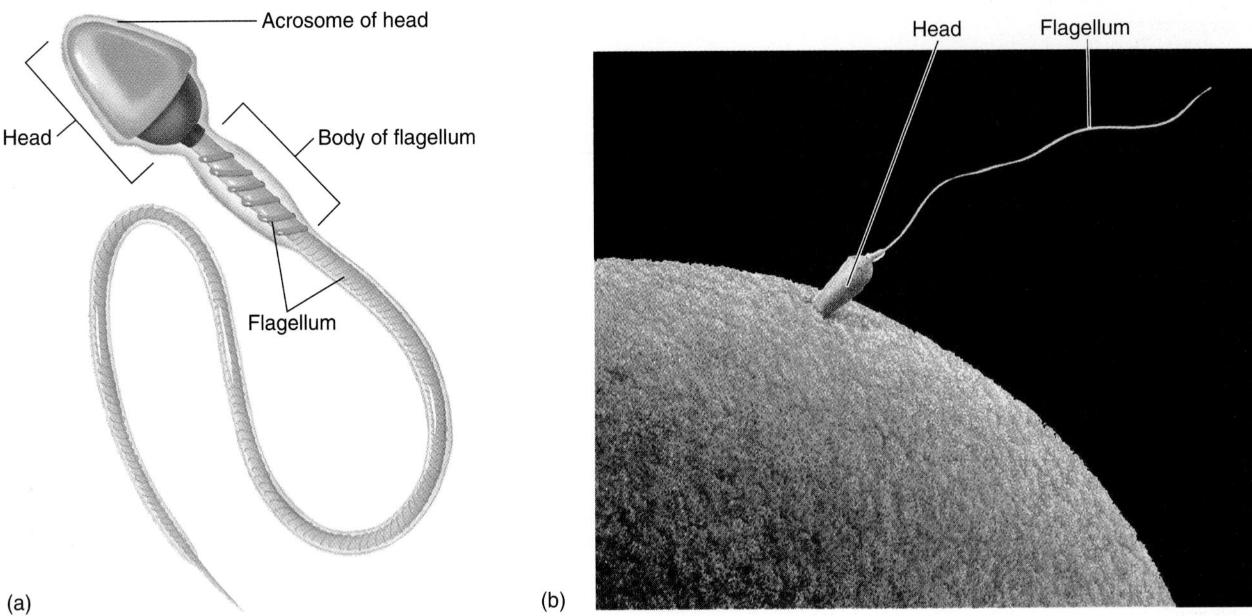

Figure 28.21

The histology of the epididymis (50×).
Sperm can be seen in the lumina. It is in the epididymis that sperm first become motile.

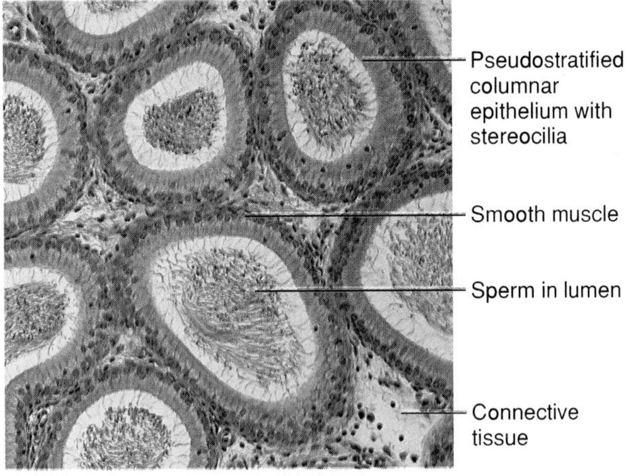

is the **body,** and the lower portion is the **tail.** The tail of the epididymis is continuous with the ductus deferens. The time required to produce mature sperm—from meiosis in the seminiferous tubules to storage of motile sperm in the ductus deferens—is approximately 2 months.

Ductus Deferens

The **ductus deferens** (*duk'tus def'er-enz;* plural, *ductus deferentia*) is a fibromuscular tube about 45 cm (18 in.) long and 2.5 mm thick (see fig. 28.10) that conveys sperm from the epididymis to the ejaculatory duct. Also called the **vas deferens** (plural, *vasa deferentia*), it originates at the tail of the epididymis and exits the scrotum as it ascends along the posterior border of the testis. From here, it penetrates the inguinal canal, enters the pelvic cavity, and passes to the side of the urinary bladder medial to the ureter. The **ampulla** (*am-pool'ă*) of the ductus deferens is the terminal portion that joins the ejaculatory duct.

The histological structure of the ductus deferens includes a layer of pseudostratified ciliated columnar epithelium lining the tubular lumen and three layers of tightly packed smooth muscle (fig. 28.22). Sympathetic nerves from the pelvic plexus serve the ductus deferens. Stimulation through these nerves causes peristaltic contractions of the muscular layer, which forcefully ejects the stored sperm toward the ejaculatory duct.

Much of the ductus deferens is located within a structure known as the **spermatic cord** (see fig. 28.26). The spermatic cord extends from the testis to the inguinal canal and consists of the ductus deferens, the testicular artery and venous plexus nerves, the cremaster muscle, lymphatic vessels, and connective tissue. As it passes anterior to the pubic bone, the spermatic cord can be easily palpated.

Ejaculatory Duct

The **ejaculatory** (*ĕ-jak'yoo-lă-tor-e*) **duct** is 2 cm (0.8 in.) long and is formed by the union of the ampulla of the ductus deferens and the duct of the seminal vesicle. The ejaculatory duct then pierces the capsule of the prostate on its posterior surface and continues through this gland (see fig. 28.10). Both ejacu-

deferens: L. *deferens*, conducting away

ampulla: L. *ampulla*, a two-handed bottle

Figure 28.22

The histology of the ductus deferens (250×).

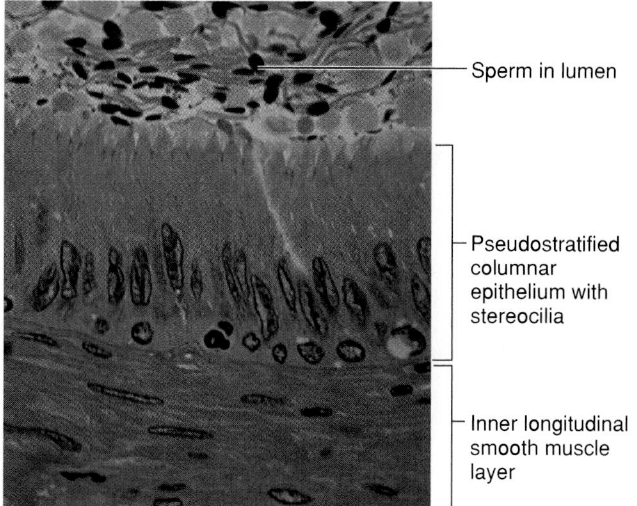

Figure 28.23

The histology of the seminal vesicle (10×).

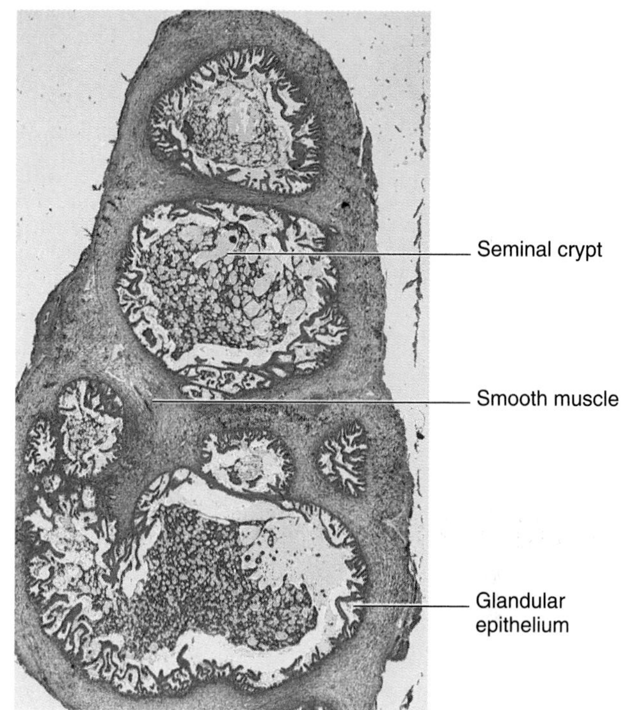

latory ducts receive secretions from the seminal vesicles and then eject the sperm and additives into the prostatic urethra to be mixed with secretions from the prostate. The urethra, discussed shortly, serves as a passageway for both semen and urine. It is the terminal duct of the male duct system.

Accessory Reprodictive Glands

Accessory reproductive glands include the seminal vesicles, the prostate, and the bulbourethral glands (see fig. 28.10). The contents of the seminal vesicles and the prostate mix with the sperm to form *semen* (*se′men*), or *seminal fluid*. The fluid from the bulbourethral glands is released in response to sexual stimulation prior to ejaculation.

Seminal Vesicles

The paired **seminal vesicles** (*sem′ĭ-nal ves′ĭ-k′lz*), each about 5 cm (2 in.) long, are convoluted, club-shaped glands positioned immediately posterior to the base of the urinary bladder, anterior to the rectum. They secrete a sticky, slightly alkaline, yellowish substance that contributes to the motility and viability of the sperm. The secretion from the seminal vesicles contains a variety of nutrients, including fructose, a monosaccharide that provides sperm with an energy source. It also contains citric acid, coagulation proteins, and prostaglandins. The discharge from the seminal vesicles makes up about 60% of the volume of semen.

Histologically, the seminal vesicle has an extensively coiled mucosal layer (fig. 28.23). This layer partitions the lumen into numerous intercommunicating spaces that are lined by pseudostratified columnar and cuboidal secretory epithelia (referred to as glandular epithelium). Sympathetic stimulation causes the contents of the seminal vesicles to empty into the ejaculatory ducts of their respective sides.

vesicle: L. *vesicula*, a blister; diminutive of *vesica*, bladder

Prostate

The firm **prostate** (*pros′tāt*) is the size and shape of a chestnut. It is about 4 cm (1.6 in.) across and 3 cm (1.2 in.) thick and lies immediately below the urinary bladder, where it surrounds the beginning portion of the urethra (see fig. 28.10). The prostate is enclosed by a fibrous capsule and divided into lobules formed by the urethra and the ejaculatory ducts that extend through the gland. Twenty to 30 small prostatic ducts from the lobules open into the urethra. Extensive bands of smooth muscular tissue course throughout the prostate to form a meshwork that supports the glandular tissue (fig. 28.24). Contraction of the smooth muscle expels the contents from the gland and provides part of the propulsive force needed to ejaculate the semen. The thin, milky-colored prostatic secretion assists sperm motility as a liquefying agent, and its alkalinity protects sperm in their passage through the acidic environment of the female vagina. The discharge from the prostate makes up about 40% of the volume of semen. (Spermatozoa account for less than 1% of the volume.) Clotting proteins in the prostatic fluid causes the semen to coagulate after ejaculation, but the hydrolytic action of *fibrinolysin* later causes the coagulated semen to again assume a more liquid form, thereby freeing the sperm. The prostate also secretes the enzyme *acid phosphatase*, which is often measured clinically to assess prostate function.

prostate: Gk. *prostate*, one standing before

Figure 28.24
The histology of the prostate (50×).

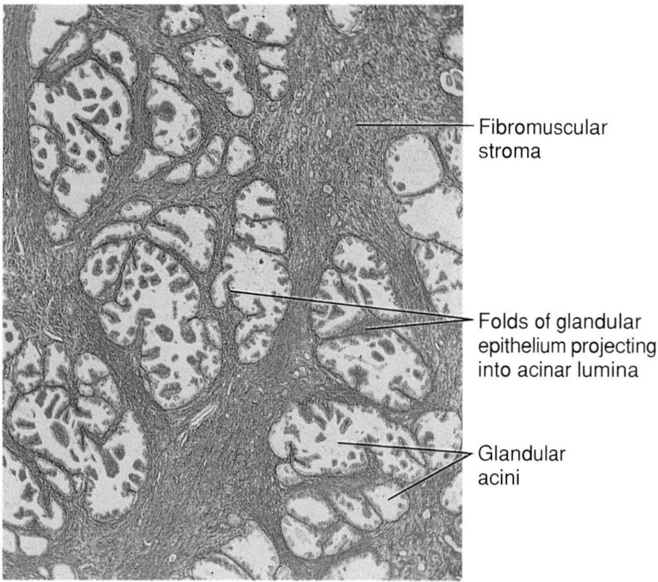

Figure 28.25
The histology of the urethra (10×).

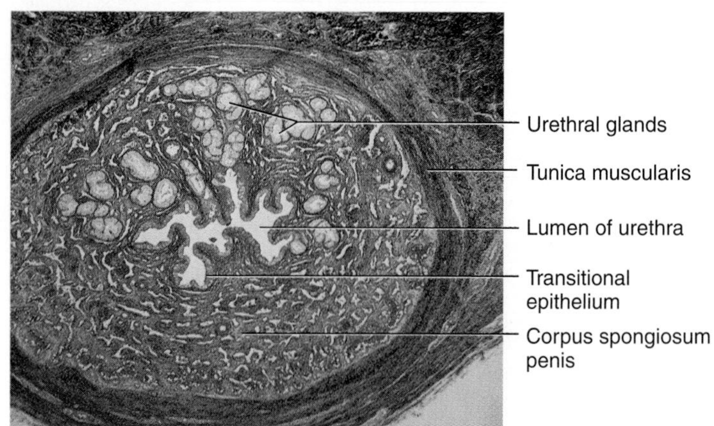

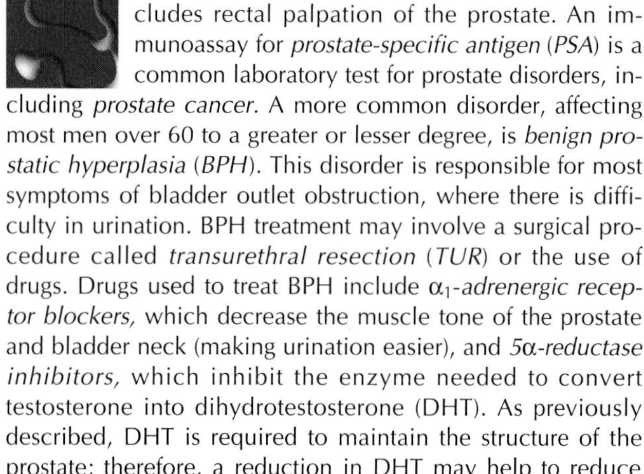

A routine physical examination of the male includes rectal palpation of the prostate. An immunoassay for *prostate-specific antigen* (*PSA*) is a common laboratory test for prostate disorders, including *prostate cancer.* A more common disorder, affecting most men over 60 to a greater or lesser degree, is *benign prostatic hyperplasia* (*BPH*). This disorder is responsible for most symptoms of bladder outlet obstruction, where there is difficulty in urination. BPH treatment may involve a surgical procedure called *transurethral resection* (*TUR*) or the use of drugs. Drugs used to treat BPH include *α_1-adrenergic receptor blockers,* which decrease the muscle tone of the prostate and bladder neck (making urination easier), and *5α-reductase inhibitors,* which inhibit the enzyme needed to convert testosterone into dihydrotestosterone (DHT). As previously described, DHT is required to maintain the structure of the prostate; therefore, a reduction in DHT may help to reduce the size of this gland.

Bulbourethral Glands

The paired, pea-sized, brownish **bulbourethral** (*bul″bo-yoo-re′thral*) (Cowper's) **glands** are located inferior to the prostate. Each bulbourethral gland is about 1 cm in diameter and drains by a 2.5-cm (1-in.) duct into the urethra (see fig. 28.10). Upon sexual excitement and prior to ejaculation, the bulbourethral glands are stimulated to secrete a mucoid substance. This secretion coats the lining of the urethra to neutralize the pH of the urine residue and lubricates the tip of the penis in preparation for coitus.

Cowper's glands: from William Cowper, English anatomist, 1666–1709

Urethra

The **urethra** of the male serves as a common tube for both the urinary and reproductive systems. However, urine and semen cannot simultaneously pass through the urethra because a neural reflex during ejaculation automatically inhibits micturition. The urethra of the male is about 20 cm (8 in.) long and S-shaped due to the shape of the penis. As described in chapter 25, three regions can be identified—the prostatic urethra, the membranous urethra, and the spongy urethra (see fig. 25.30).

The **prostatic urethra** is the 2.5-cm proximal portion that passes through the prostate. The prostatic urethra receives drainage from the small ducts of the prostate and the two ejaculatory ducts.

The **membranous urethra** is the 0.5-cm portion that passes through the urogenital diaphragm. The external urethral sphincter muscle is located in this region.

The **spongy,** or **penile, urethra** is the longest portion, extending from the outer edge of the urogenital diaphragm to the external urethral orifice on the glans penis. About 15 cm long, this portion is surrounded by erectile tissue as it passes through the corpus spongiosum of the penis. The paired ducts from the bulbourethral glands attach to the spongy urethra near the urogenital diaphragm.

The wall of the urethra has an inside lining of mucous membrane, composed of transitional epithelium (fig. 28.25) and surrounded by a relatively thick layer of smooth muscle tissue called *tunica muscularis*. Specialized **urethral glands,** embedded in the urethral wall, secrete mucus into the lumen of the urethra.

Penis

The **penis,** when distended, serves as the copulatory organ of the male reproductive system. It is composed mainly of **erectile tissue**—a spongy network of connective tissue and smooth muscle characterized by vascular spaces that become

Figure 28.26

Structure of the penis.

The illustration shows the attachment, blood and nerve supply, and the arrangement of the erectile tissue.

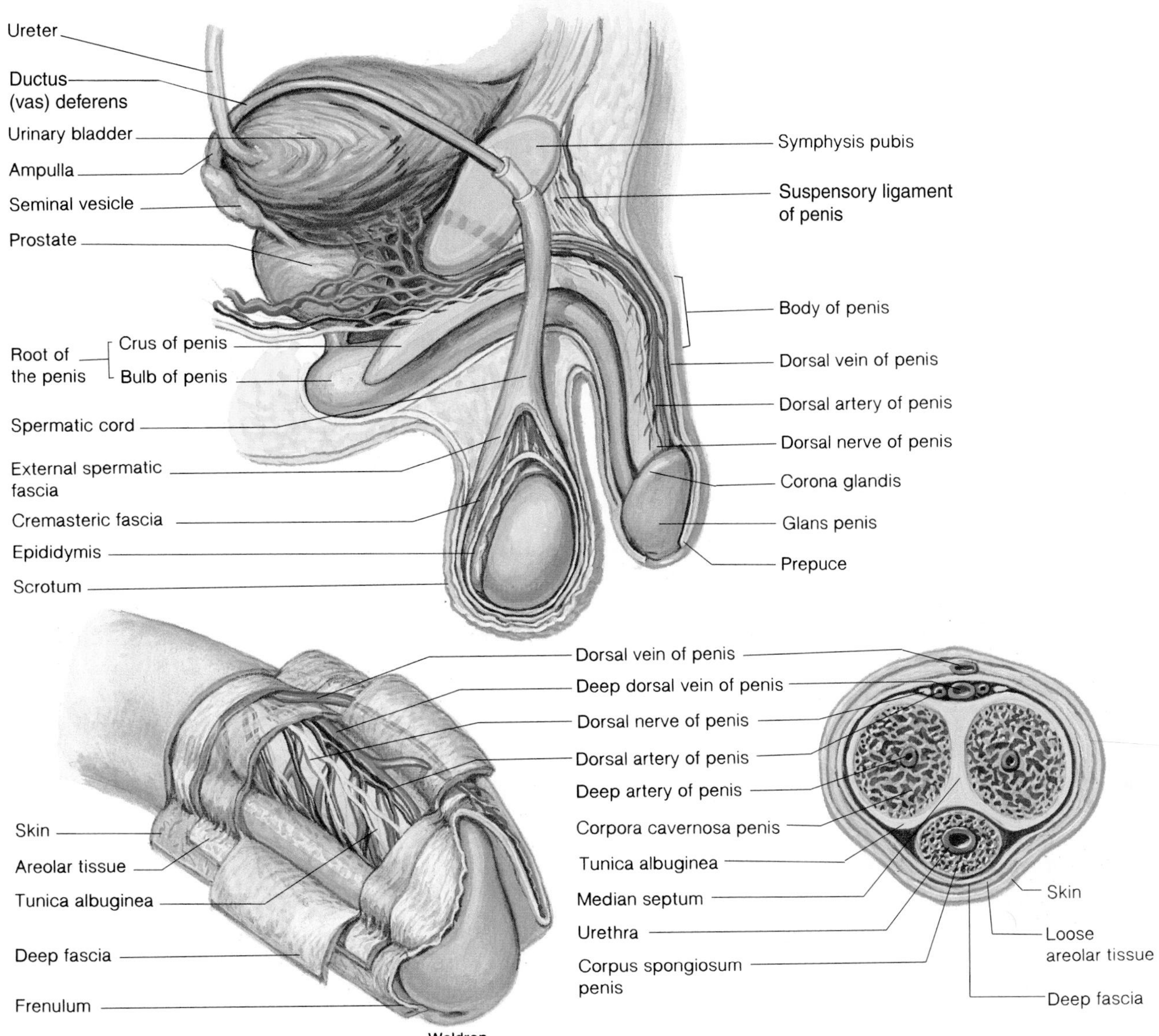

engorged with blood under the influence of sexual stimulation. The penis is a pendent structure, positioned anterior to the scrotum and attached to the pubic arch. It is divided into a proximally attached root; an elongated tubular body, or shaft; and a distal glans penis (fig. 28.26).

The **root of the penis** expands posteriorly to form the **bulb of the penis** and the **crus** (*krus*) **of the penis.** The bulb is positioned in the urogenital triangle of the perineum, where it is attached to the undersurface of the perineal membrane and enveloped by the bulbocavernosus muscle (see fig. 13.12). The crus attaches the root of the penis to the pubic arch (ischiopubic ramus) and to the perineal membrane. The crus, which is superior to the bulb, is enveloped by the ischiocavernosus muscle.

The **body,** or **shaft, of the penis** is composed of three cylindrical columns of erectile tissue, which are bound together by fibrous tissue and covered with skin (fig. 28.26). The paired columns that form the dorsum and sides of the penis are named the **corpora cavernosa** (*kor'por-ă kav"er-no'să*) **penis.** The fibrous tissue between the two corpora forms a **median septum.** The **corpus spongiosum** (*spon"je-o'sum*) **penis** is ventral to the other two and surrounds the spongy urethra. The penis is flaccid and relaxed when the spongelike tissue is not engorged with blood. In the sexually aroused male, it becomes firm and erect as the vascular spaces are filled.

crus: L. *crus,* leg, resembling a leg

The **glans penis** is the cone-shaped terminal portion of the penis, which is formed from the expanded corpus spongiosum. The opening of the urethra at the tip of the glans penis is called the **urethral orifice** (see fig. 25.30). The **corona glandis** is the prominent ridge of the glans penis. On the undersurface of the glans, a vertical fold of tissue called the **frenulum** (*fren'yŭ-lum*) attaches the skin covering the penis to the glans.

The skin covering the penis is hairless, lacks fat cells, and generally is more darkly pigmented than the other body skin. The skin of the body of the penis is loosely attached and is continuous over the glans penis as a retractable sheath called the **prepuce** (*pre'pyoos*), or **foreskin.** The prepuce is commonly removed in male infants on the third or fourth day after birth, or on the eighth day as part of a Jewish religious rite. This procedure is called a *circumcision*.

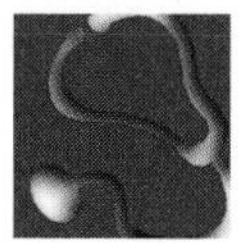

A *circumcision* is generally performed for hygienic purposes because the glans penis is easier to clean if exposed. A sebaceous secretion from the glans penis, called *smegma* (*smeg'ma*) will accumulate along the border of the corona glandis if good hygiene is not practiced. Smegma can foster bacteria that may cause infections and therefore should be removed through washing. Cleaning of the glans penis of an uncircumcised male requires retraction of the prepuce. Occasionally, a child is born with a prepuce that is too tight to permit retraction. This condition, called *phimosis* (*fĭ-mo'sis*) necessitates circumcision.

Erection of the Penis

Erection of the penis depends on the volume of blood that enters the arteries of the penis as compared to the volume that exits through venous drainage. Normally, constant sympathetic stimuli to the arterioles of the penis maintain a partial constriction of smooth muscles within the arteriole walls so that there is an even flow of blood throughout the penis. During sexual excitement, however, parasympathetic nerves cause marked vasodilation within the arterioles of the penis, resulting in more blood entering than venous blood draining. This causes the spongy tissue of the corpora cavernosa and the corpus spongiosum to become distended with blood and the penis to become turgid. Recent evidence has shown that the responses to parasympathetic stimulation that produce erection of the penis are mediated by nitric oxide (NO) as a neurotransmitter.

Erection is controlled by two portions of the central nervous system—the hypothalamus in the brain and the sacral portion of the spinal cord. The hypothalamus controls conscious sexual thoughts that originate in the cerebral cortex. Nerve impulses from the hypothalamus elicit parasympathetic responses from the sacral region that cause vasodilation of the arterioles within the penis. Conscious thought is not required for an erection, however, and stimulation of the penis can cause an erection because of a reflex response in the spinal cord. This reflexive action makes possible an erection in a sleeping male or in an infant—perhaps from the stimulus of a diaper.

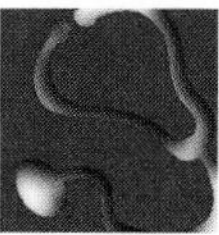

Nitric oxide, released from postganglionic parasympathetic neurons in the penis, diffuses into smooth muscle cells of blood vessels and stimulates the production of cyclic guanosine monophosphate (cGMP). The cGMP, in turn, causes the smooth muscle to relax so that blood can flow into the corpora cavernosa. This physiology is exploited by *Sildenafil,* a recently available drug that can be taken orally to treat *impotence* (also called *erectile dysfunction*). The drug blocks the enzyme which degrades cGMP (cGMP phosphodiesterase) and thereby enhances the action of cGMP in promoting erection.

Emission and Ejaculation of Semen

Continued sexual stimulation following erection of the penis causes the movement of sperm from the epididymides into the ejaculatory ducts and the discharge of accessory gland secretions into the ejaculatory ducts and urethra in the formation of semen. This process called **emission,** creates an urgent sensation that ejaculation must occur imminently. The first sympathetic response, which occurs prior to ejaculation, is the discharge of fluids from the bulbourethral glands into the urethra. These fluids are usually discharged before penetration of the penis into the vagina and serve to lubricate the urethra and the glans penis. Emission occurs as sympathetic stimulation from the pelvic plexus cause a rhythmic contraction of the smooth muscle layers of the testes, epididymides, ductus deferentia, ejaculatory ducts, seminal vesicles, and prostate.

Ejaculation immediately follows emission and is generally accompanied by *orgasm,* which is considered the climax of the human sexual response. Ejaculation occurs in a series of spurts of semen from the urethra. This takes place as parasympathetic impulses traveling through the pudendal nerves stimulate the bulbocavernosus muscles at the base of the penis and cause them to contract rhythmically. There is also sympathetic stimulation of the smooth muscles in the urethral wall that peristaltically contract to help eject the semen. Sexual function in the male thus requires the synergistic (rather than antagonistic) action of the parasympathetic and sympathetic nervous systems. The mechanism of emission and ejaculation is summarized in figure 28.27.

Immediately following ejaculation or cessation of sexual stimuli, sympathetic impulses cause vasoconstriction of the arterioles within the penis to reduce the inflow of blood. At the same time, cardiac output returns to normal, as does venous return of blood from the penis, and the penis again becomes flaccid. Another erection and ejaculation cannot be triggered for a period ranging from 10 minutes to a few hours. This is called a *refractory period,* and is not present in women (chapter 29).

Male Fertility

The approximate volume of semen for each ejaculation is 1.5 to 5.0 ml. The bulk of this fluid (about 60%) is produced by the seminal vesicles, and the rest (about 40%) is contributed

glans: L. *glans*, acorn
corona: L. *corona*, garland, crown
frenulum: L. diminutive of *frenum*, a bridle
prepuce: L. *prae*, before; *putium*, penis
phimosis: Gk. *phimosis*, a muzzling

Figure 28.27

The mechanism of emission and ejaculation.

Notice that sympathetic nerves, acting on different structures of the male reproductive system, stimulate both emission and ejaculation.

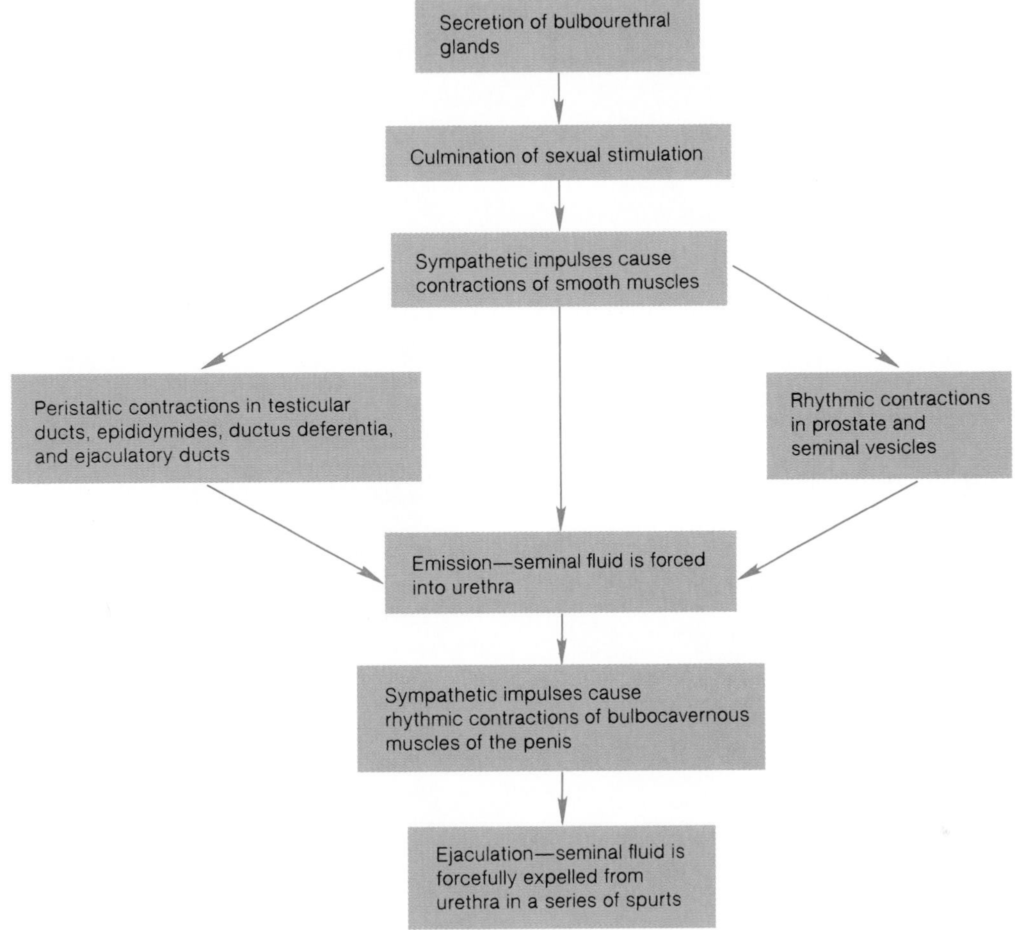

by the prostate. The sperm content in human males ranges between 60 and 150 million per milliliter in the ejaculated semen. Normal human semen values are summarized in table 28.4.

A sperm concentration below about 10 million per milliliter is termed *oligospermia,* and is associated with decreased fertility. A total sperm count below about 50 million per ejaculation is clinically significant in male infertility. Oligospermia may be caused by a variety of factors, including heat from a sauna or hot tub, various pharmaceutical drugs, lead and arsenic poisoning, and such illicit drugs as marijuana, cocaine, and anabolic steroids. In addition to low sperm counts as a cause of infertility, some men and women have antibodies against sperm antigens (this is very common in men with vasectomies). Such antibodies do not appear to affect health, but they do reduce fertility.

oligospermia: Gk. *oligos*, few; *sperma*, seed
fertility: L. *fere*, to bear

Table 28.4

Semen Analysis

Characteristic	Reference Value
Volume of ejaculate	1.5–5.0 ml
Sperm count	40–250 million/ml
Sperm motility	
Percentage of motile forms:	
1 hour after ejaculation	70% or more
3 hours after ejaculation	60% or more
Leukocyte count	0–2,000/ml
pH	7.2–7.8
Fructose concentration	150–600 mg/100 ml

Source: Modified from L. Glasser, "Seminal Fluid and Subfertility," *Diagnostic Medicine* July/August 1981, p. 28. Used by permission.

Under Development

Embryonic Development of the Reproductive System

The male and female reproductive systems follow a similar pattern of development, with sexual distinction resulting from the influence of hormones. A significant fact of embryonic development is that the sexual organs for both male and female derive from the same developmental tissues and are considered *homologous structures.*

The first sign of development of either the male or the female reproductive organs occurs during the fifth week as the medial aspect of each mesonephros (chapter 25) enlarges to form the **gonadal ridge** (fig. 1). The gonadal ridge continues to grow behind the developing peritoneal membrane. By the sixth week, stringlike masses called **primary sex cords** form within the enlarging gonadal ridge. The primary sex cords in the male will eventually mature to become the seminiferous tubules. In the female, the primary sex cords will contribute to nurturing tissue of developing ova. Each gonad develops near a **mesonephric** (wolffian) **duct** and a **paramesonephric** (müllerian) **duct.**

In the male embryo, each testis connects through a series of tubules to the mesonephric duct. During further development, the connecting tubules become the seminiferous tubules, and the mesonephric duct becomes the *efferent ductules, epididymis, ductus deferens, ejaculatory duct,* and *seminal vesicle.* The paramesonephric duct in the male degenerates without contributing any functional structures to the reproductive system.

In the female embryo, the mesonephric duct degenerates, and the paramesonephric duct contributes in large measure to structures of the female reproductive system. The distal ends of the paired paramesonephric ducts fuse to form the *vagina* and *uterus.* The proximal unfused portions become the *uterine tubes.*

By the sixth week, an external swelling called the **genital tubercle** appears anterior to the small embryonic tail (future coccyx). The mesonephric and paramesonephric ducts open to the outside through the genital tubercle. The genital tubercle consists of a *glans,* a *urethral groove,* paired *urethral folds,* and paired *labioscrotal swellings* (fig. 2). As the glans portion of the genital tubercle enlarges, it becomes known as the *phallus.* Early in fetal development (tenth through twelfth week), sexual distinction of the external genitalia becomes apparent. In the male, the phallus enlarges and develops into the *glans* of the penis, and the urethral folds fuse around the urethra to form the *body* of the penis. The urethra opens at the end of the glans as the *urethral orifice.* The labioscrotal swellings fuse to form the *scrotum* into which the testes will descend. In the female, the phallus gives rise to the *clitoris,* the urethral folds remain unfused as the *labia minora,* and the labioscrotal swellings become the *labia majora.* The urethral groove is retained as a longitudinal cleft known as the *vestibule.*

Development of Secondary Sex Organs

In addition to testes and ovaries, which are the primary sex organs, or gonads, the external genitalia and various internal accessory sex organs are needed for reproductive function. These are known as the secondary sex organs. Some of the male accessory sex organs derive from the mesonephric ducts, and female accessory organs derive from the paramesonephric ducts (fig. 2). Interestingly, from about day 25 to about day 50, male and female embryos alike have both duct systems, and therefore have the potential to form the accessory organs characteristic of either gender.

Experimental removal of the testes (castration) from male embryonic animals results in regression of the mesonephric ducts and development of the paramesonephric ducts into female accessory sex organs: the **uterus** and **uterine** (fallopian) **tubes.** Female accessory sex organs, therefore, develop as a result of the absence of testes (and their secretion of testosterone) rather than the presence of ovaries.

The developing seminiferous tubules within the testes secrete a polypeptide called *müllerian inhibition factor,* which causes regression of the paramesonephric ducts beginning about day 60. The secretion of testosterone by the interstitial cells of the testes subsequently causes growth and development of the mesonephric ducts into male accessory sex organs: the **epididymis, ductus deferens, seminal vesicles,** and **ejaculatory duct.**

Other structures that male and female embryos have in common are the *urogenital sinus, genital tubercle, urethral folds,* and *labiosacral swellings.* The secretions of the testes masculinize these structures to form the penis, prostate, and scrotum. The genital tubercle that forms the penis in a male will, in the absence of testes, become the clitoris in a female. The penis and clitoris are thus said to be homologous structures. Similarly, the labiosacral swellings form the scrotum in a male or the labia majora in a female; these structures are therefore also homologous (fig. 3).

In summary, the genetic sex is determined by whether a Y-bearing or an X-bearing spermatozoon fertilizes the ovum; the presence or absence of a Y chromosome in turn determines whether the gonads of the embryo will be testes or ovaries; and, finally, the presence or absence of testes determines whether the accessory sex organs and external genitalia will be male or female (table 1).

This regulatory pattern of sex determination makes sense in light of the fact that both male and female embryos develop within an environment high in estrogen, which is secreted by the mother's ovaries and the placenta. If the secretions of the ovaries determined the sex, all embryos would be female.

---

homologous: Gk. *homeo,* the same

wolffian duct: from Kaspar Friedrich Wolff, German embryologist, 1733–94

müllerian duct: from Johannes Peter Müller, German physiologist, 1801–58

Figure 1 Differentiation of the male and female gonads and genital ducts. (*a*) An embryo at 6 weeks showing the positions of a transverse cut depicted in (a_1), (*b*), and (*c*). (a_1) At 6 weeks, the developing gonads (primary sex cords) are still indifferent. By 4 months, the gonads have differentiated into male (*b*) or female (*c*). The oogonia have formed within the ovaries (c_1) by 6 months.

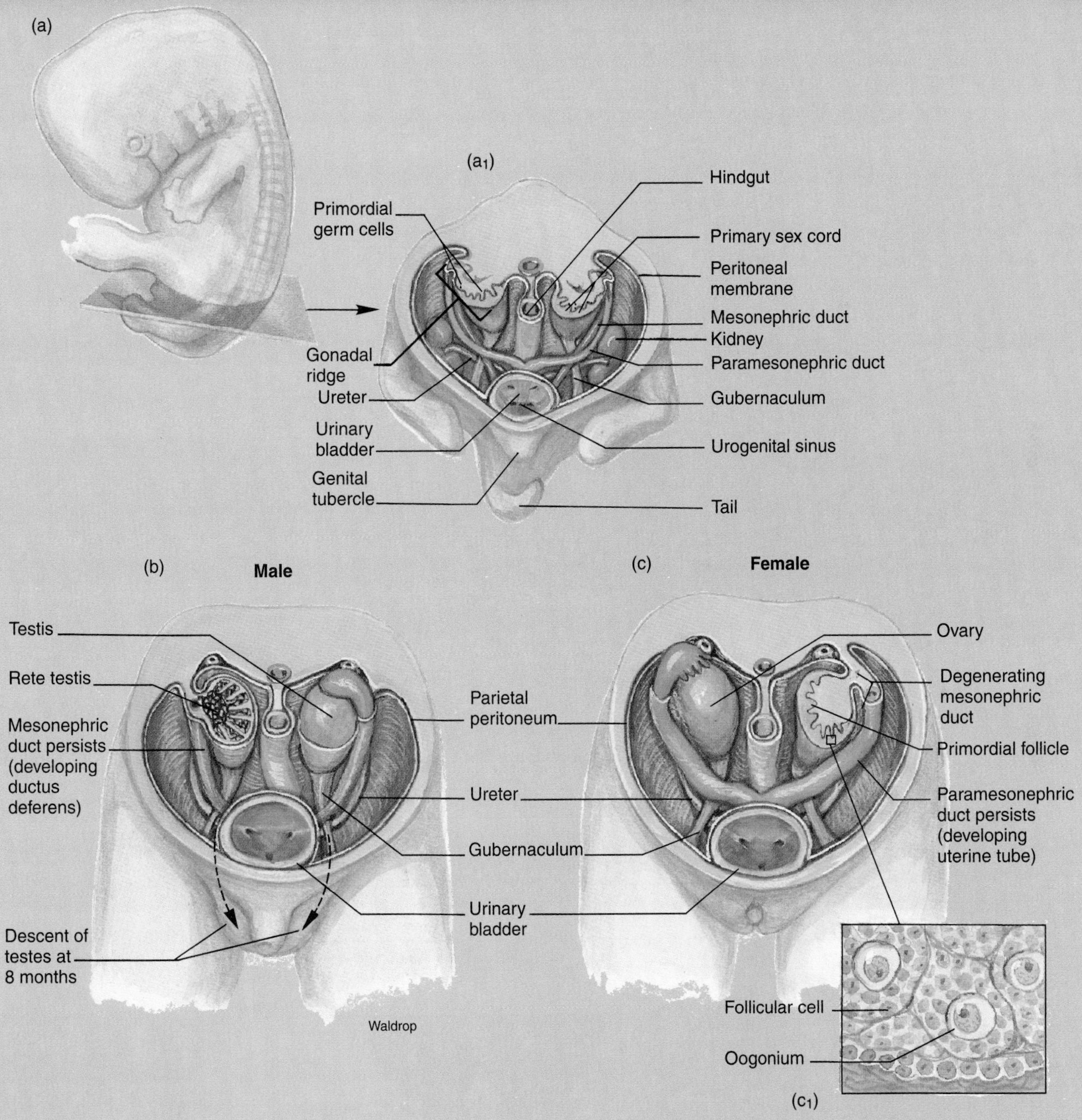

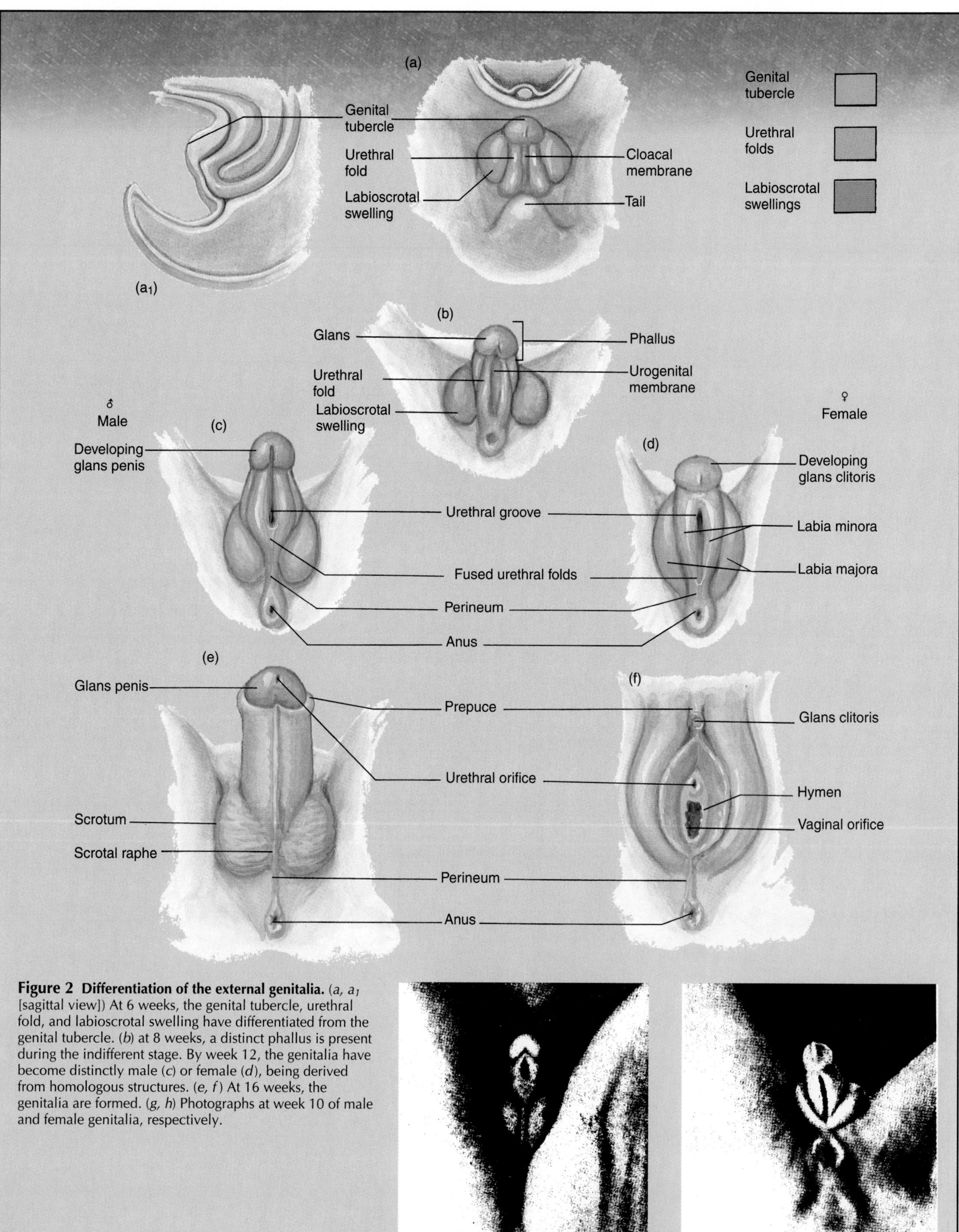

Figure 2 Differentiation of the external genitalia. (*a*, a_1 [sagittal view]) At 6 weeks, the genital tubercle, urethral fold, and labioscrotal swelling have differentiated from the genital tubercle. (*b*) at 8 weeks, a distinct phallus is present during the indifferent stage. By week 12, the genitalia have become distinctly male (*c*) or female (*d*), being derived from homologous structures. (*e, f*) At 16 weeks, the genitalia are formed. (*g, h*) Photographs at week 10 of male and female genitalia, respectively.

Figure 3 Regulation of embryonic sexual development. In the presence of testosterone and müllerian inhibition factor (MIF) secreted by the testes, male external genitalia and sex accessory organs develop. In the absence of these secretions, female structures develop.

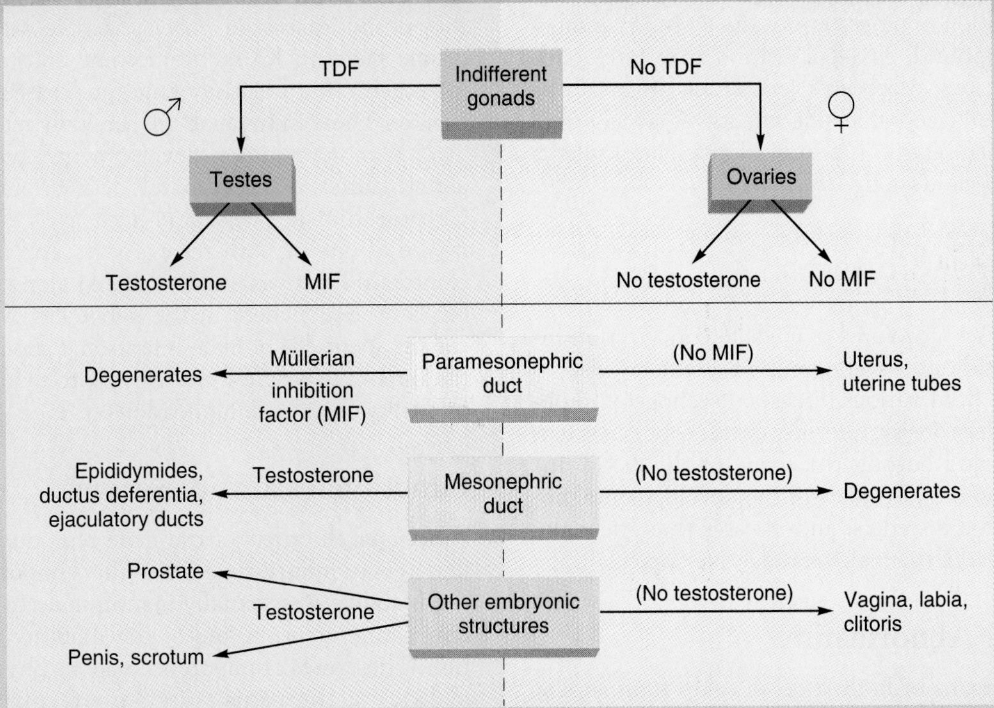

Table 1

A Developmental Timetable for the Reproductive System

Approximate Time after Fertilization		Developmental Changes		
Days	**Trimester**	**Indifferent**	**Male**	**Female**
19	First	Germ cells migrate from yolk sac.		
25–30		Mesonephric ducts begin development.		
44–48		Paramesonephric ducts begin development.		
50–52		Urogenital sinus and tubercle develop.		
43–60			Tubules and sustentacular cells appear. Paramesonephric ducts begin to regress.	
60–75			Interstitial cells appear and begin testosterone production.	Formation of vagina begins.
			Mesonephric ducts grow.	Regression of mesonephric ducts begins.
105	Second			Development of ovarian follicles begins.
120				Uterus forms.
160–260	Third		Testes descend into scrotum. Growth of external genitalia occurs.	Formation of vagina complete.

Source: Annual Review of Physiology, vol. 40, p. 279, 1978.

Attempts have been made to develop new methods of male contraception. These have generally involved compounds that suppress gonadotropin secretion, such as testosterone or a combination of progesterone and a GnRH antagonist. Another compound, *gossypol,* which interferes with sperm development, has also been tried. These drugs can be effective but have unacceptable side effects. A widely used method of male contraception is a surgical procedure called a **vasectomy** (*vă-sek'to-me*) (see fig. 28.29).

Clinical Considerations

Sexual dysfunction is a broad area of medical concern that includes developmental and psychogenic problems as well as conditions resulting from various diseases. Psychogenic problems of the reproductive system are extremely complex, poorly understood, and beyond the scope of this book. Included in this section are just a few of the developmental abnormalities, functional disorders, and diseases that affect the structure and function of the male reproductive system.

Developmental Abnormalities

The reproductive organs of both sexes develop from similar embryonic tissue that follows a consistent pattern of formation well into the fetal period. Because an embryo has the potential to differentiate into a male or a female, developmental errors can result in various degrees of intermediate sex, or **hermaphroditism** (*her-maf'rŏ-dĭ-tiz"em*). A person with undifferentiated or ambiguous external genitalia is called a **hermaphrodite.**

True hermaphroditism—in which both male and female gonadal tissues are present in the body—is a rare anomaly. True hermaphrodites usually have a 46, XX chromosome constitution. **Male pseudohermaphroditism** occurs more commonly and generally results from hormonal influences during early fetal development. This condition is caused either by inadequate secretion of androgenic hormones or by the delayed development of the reproductive organs after the period of tissue sensitivity has passed. These individuals have a 46, XY chromosome constitution and male gonads but intersexual and variable genitalia. The treatment of hermaphroditism varies, depending upon the extent of ambiguity of the reproductive organs. Although people with this condition are sterile, they may engage in normal sexual relations following hormonal therapy and plastic surgery.

Chromosomal anomalies result from the improper separation of the chromosomes during meiosis and are usually expressed in deviations of the reproductive organs. The two most frequent chromosomal anomalies cause Turner's syndrome and Klinefelter's syndrome. **Turner's syndrome** occurs when only one X chromosome is present. About 97% of embryos lacking an X chromosome die; the remaining 3% survive and appear to be females, but their gonads are rudimentary or absent, and they do not mature at puberty. People with **Klinefelter's syndrome** have an XXY chromosome constitution and develop male genitalia, but they have underdeveloped seminiferous tubules. These individuals are generally retarded.

A more common developmental problem than genetic abnormalities, and fortunately less serious, is cryptorchidism. **Cryptorchidism** (*krip-tor'kĭ-diz"em*) is characterized by the failure of one or both testes to descend into the scrotum. A cryptorchid testis is usually located along the path of descent but can be anywhere in the pelvic cavity (fig. 28.28). It occurs in about 3% of male infants and should be treated before the infant has reached the age of 5 to reduce the likelihood of infertility or other complications.

Functional Considerations

Functional disorders of the male reproductive system include impotence, infertility, and sterility. **Impotence** (*im'pŏ-tens*) is the inability of a sexually mature male to achieve and maintain penile erection and/or the inability to achieve ejaculation. The causes of impotence may be physical, such as abnormalities of the penis, vascular irregularities, neurological disorders, or certain diseases. Generally, however, the cause of impotence is psychological, and the patient requires skilled counseling by a sex therapist.

Infertility is the inability of the sperm to fertilize the ovum. Infertility problems may be due to factors originating in the male or female, or both. The term *impotence* should not be confused with infertility. In males, the most common cause of infertility is the inadequate production of viable sperm. This may result from alcoholism, dietary deficiencies, local injury, varicocele, excessive heat, or exposure to X rays. A hormonal imbalance may also contribute to infertility. Many of the causes of infertility can be treated through proper nutrition, gonadotropic hormone treatment, or microsurgery. If corrective treatment is not possible, however, it may be possible to concentrate the sperm obtained through *masturbation* (in males, self-stimulation to the point of ejaculation) and use this concentrate to artificially inseminate the female.

Sterility is similar to infertility, except that sterility is a permanent condition. Sterility may be genetically caused, or it may be the result of degenerative changes in the seminiferous tubules (for example, mumps in a mature male may secondarily infect the testes and cause irreversible tissue damage).

Voluntary sterilization of the male in a procedure called a **vasectomy** is a common technique of birth control and can be performed on an outpatient basis. In this procedure, a

---

hermaphrodite: Gk. (mythology) Hermaphrodites, son of Hermes (Mercury)
Turner's syndrome: from Henry H. Turner, American endocrinologist, 1892–1970

Klinefelter's syndrome: from Harry F. Klinefelter, Jr., American physician, b. 1912
cryptorchidism: Gk. *crypto,* hidden; *orchis,* testis
sterility: L. *sterlis,* barren
vasectomy: L. *vas,* vessel; Gk. *ektome,* excision

Figure 28.28

Cryptorchidism.
(*a*) Incomplete descent of a testis may involve a region (1) in the pelvic cavity, (2) in the inguinal canal, (3) at the superficial inguinal ring, or (4) in the upper scrotum. (*b and c*) An ectopic testis may be (1) in the superficial fascia of the anterior pelvic wall, (2) at the root of the penis, or (3) in the perineum, in the thigh alongside the femoral vessels.

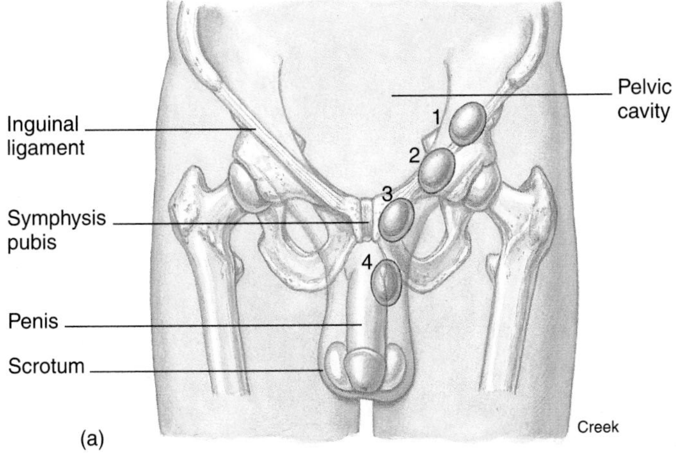

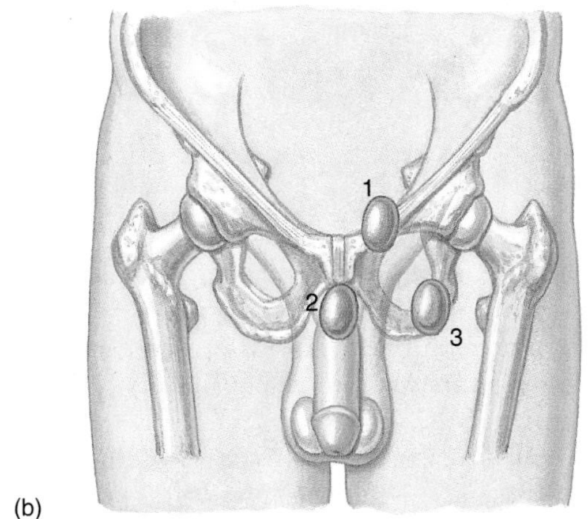

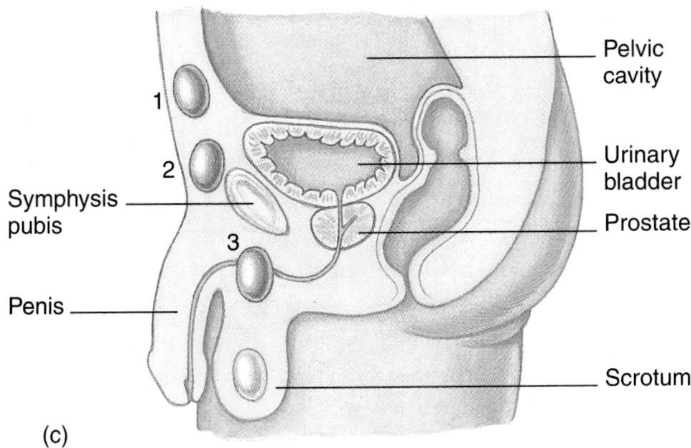

small section of each ductus deferens near the epididymis is surgically removed, and the cut ends of the ducts are tied (fig. 28.29). A vasectomy interferes with sperm transport but does not directly affect the secretion of androgens from interstitial cells in the interstitial tissue. Since spermatogenesis continues, the sperm cannot be drained from the testes and instead accumulate in the "crypts" that form in the seminiferous tubules and ductus deferens. These crypts present sites of inflammatory reactions in which spermatozoa are phagocytosed and destroyed by the immune system.

Diseases of the Male Reproductive System

Sexually Transmitted Diseases

Sexually transmitted diseases (STDs)—sometimes collectively called "VD" (venereal disease)—are contagious diseases that are transmitted during intimate sexual contact between individuals. Their frequency of occurrence in the United States is regarded by health authorities as epidemic. These diseases have not been eradicated mainly because humans cannot develop immunity to them and because increased sexual promiscuity increases the likelihood of infection and reinfection.

Gonorrhea (*gon″ŏ-re′ă*), commonly called "clap," is caused by the bacterium gonococcus, or *Neisseria gonorrhoeae*. Males with this disease suffer inflammation of the urethra, accompanied by painful urination and frequently the discharge of pus. In females, the condition is usually asymptomatic, and therefore many women may be unsuspecting carriers of the disease. Advanced stages of gonorrhea in females may infect the uterus and the uterine tubes. A pregnant woman with untreated gonorrhea may transmit the disease to the eyes of her newborn during its passage through the birth canal, possibly causing blindness.

Syphilis (*sif′ĭ-lis*) is caused by the bacterium *Treponema pallidum*. Syphilis is less common than gonorrhea but is more serious. During the *primary stage* of syphilis, a lesion called a *chancre* develops at the point where contact was made with the infectious syphilitic lesion of a carrier. The chancre—an ulcerated sore that has hard edges—endures for 10 days to 3 months. It is only during the primary stage that syphilis can be spread to another sexual partner. The chancre will heal with time, but if not treated it will be followed by secondary and tertiary stages of syphilis. During the initial contact, the bacteria enter the bloodstream and spread throughout the body. The *secondary stage* of syphilis is expressed by lesions or a rash of the skin and mucous membranes, accompanied by fever. This stage lasts from 2 weeks to 6 months, and the

venereal: L. (mythology) *Venus*, the goddess of love
gonorrhea: L. *gonos*, seed; *rhoia*, a flow
chancre: Fr. *chancre*, indirectly from L. *cancer*, a crab

Figure 28.29

A vasectomy.
This simplified diagram illustrates the way in which segments of the ductus deferens are removed through incisions in the scrotum. The procedure is then repeated on the opposite side.

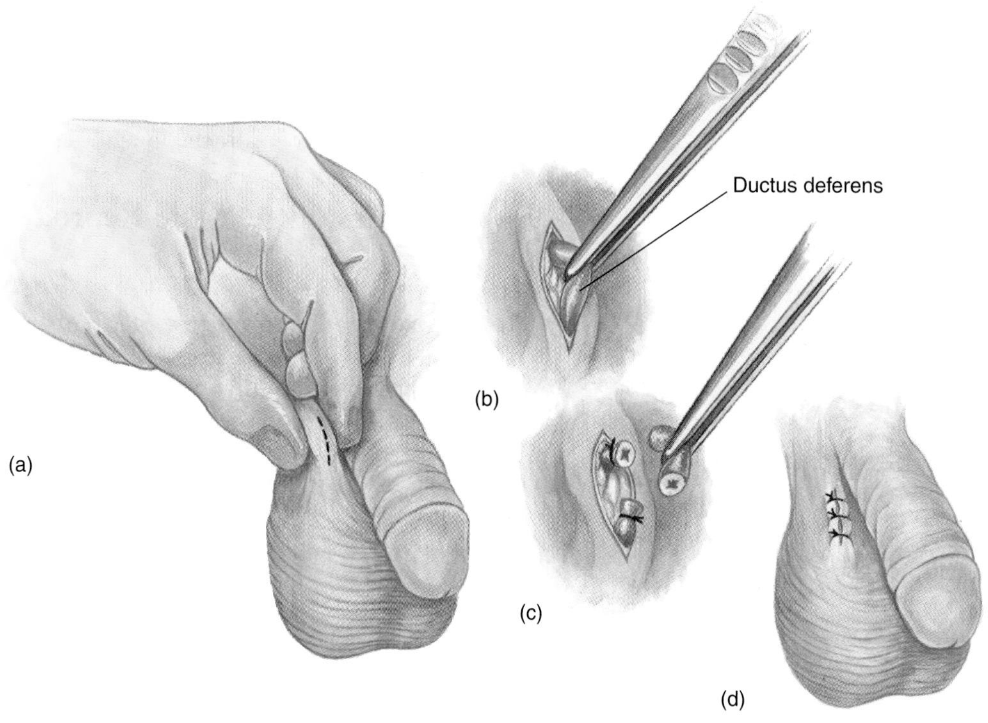

symptoms disappear of their own accord. The *tertiary stage* occurs 10 to 20 years following the primary infection. The circulatory, integumentary, skeletal, and nervous systems are particularly vulnerable to the degenerative changes caused by this disease. The end result of untreated syphilis is blindness, insanity, and eventual death.

AIDS, or **acquired immune deficiency syndrome,** is a viral disease that is transmitted primarily through intimate sexual contact and drug abuse (by sharing contaminated syringe needles). Additional information about this disease, for which there is currently no cure, is presented in chapter 23 and table 28.5.

Disorders of the Prostate

The prostate is subject to several disorders, most of which are common in older men. The four most frequent prostatic problems are acute prostatitis, chronic prostatitis, benign prostatic hyperplasia, and carcinoma of the prostate.

Acute prostatitis is common in sexually active young men through infections acquired from a gonococcus bacterium. The symptoms of acute prostatitis are a swollen and tender prostate, painful urination, and in extreme conditions, pus dripping from the penis. It is treated with penicillin, bed rest, and increased fluid intake.

Chronic prostatitis is one of the most common afflictions of middle-aged and elderly men. The symptoms of this condition range from irritation and slight difficulty in urinating to extreme pain and urine blockage, which commonly causes secondary renal infections. In this disease, several kinds of infectious microorganisms are believed to be harbored in the prostate and are responsible for inflammations elsewhere in the body, such as in the nerves (neuritis), the joints (arthritis), the muscles (myositis), and the iris (iritis).

Benign prostatic hyperplasia (BPH), or enlargement of the prostate, occurs in approximately one-third of all males over the age of 60. In this condition, an overgrowth of granular material compresses the prostatic urethra. The cause of BPH is not known. As the prostate enlarges, urination becomes painful and difficult. If the urinary bladder is not emptied completely, cystitis eventually occurs. People with cystitis may become incontinent and dribble urine continuously. BPH is usually treated by surgical removal of portions of the gland through transurethral curetting (cutting and removal of a small section) or the removal of the entire prostate, called **prostatectomy.**

Prostatic carcinoma, or cancer of the prostate, is the second leading cause of death from cancer in males in the United States. When prostatic cancer is confined to the prostate, it is generally small and asymptomatic. But as the cancer grows and invades surrounding nerve plexuses, it becomes extremely painful and is easily detected. The metastases of this cancer to the spinal column and brain are generally what kills the patient.

Table 28.5
Sexually Transmitted Diseases

Name	Organism	Resulting Condition	Treatment
Gonorrhea	*Gonococcus* (bacterium)	Adult: sterility due to scarring of epididymides and testicular ductules (in rare cases, blood poisoning); newborn: blindness	Penicillin injections; tetracycline tablets; eye drops (silver nitrate or penicillin) in newborns as preventative.
Syphilis	*Treponema pallidum* (bacterium)	Adult: gummy tumors, cardiovascular neurosyphilis; newborn: congenital syphilis (abnormalities, blindness)	Penicillin injections; tetracycline tablets
Chancroid (soft chancre)	*Hemophilus ducreyi* (bacterium)	Chancres, buboes	Tetracycline; sulfa drugs
Urethritis in males	Varius microorganisms	Clear discharge	Tetracycline
Vaginitis in females	*Trichomonas* (protozoan) *Candida albicans* (yeast)	Frothy white or yellow discharge Thick, white, curdy discharge (moniliasis)	Metronidazole Nystatin
Acquired immune deficiency syndrome (AIDS)	Human immunodeficiency virus (HIV)	Early symptoms include extreme fatigue, weight loss, fever, diarrhea; extreme susceptibility to pneumonia, rare infections, and cancer	Azidothymidine (AZT), dideoxyinosine (ddll), and dideoxycytidine (ddC); new drugs being developed, including protease inhibitors
Chlamydia	*Chlamydia trachomatis* (bacterium)	Whitish discharge from penis or vagina; pain during urination	Tetracycline and sulfonamides
Venereal warts	Virus	Warts	Podophyllin; cautery, cryosurgery, or laser treatment
Genital herpes	Herpes simplex virus	Sores	Palliative treatment
Crabs	Arthropod	Itching	Gamma benzene hexachloride
Hepatitis B	Virus	Liver injury and associated conditions	No specific treatment; each case treated symptomatically

As prostatic carcinoma develops, it has symptoms nearly identical to those of benign prostatic hyperplasia—painful urination and cystitis. When examined by rectal palpation with a gloved finger, however, a hard cancerous mass can be detected in contrast to the enlarged, soft, and tender prostate diagnostic of BPH. Prostatic carcinoma is treated by prostatectomy and frequently by the removal of the testes—called **orchiectomy** (*or″ke-ek′tŏ-me*)—as well. An orchiectomy inhibits metastases by eliminating testosterone secretion.

Disorders of the Testes and Scrotum

The most common cause of male infertility is a condition called **varicocele** (*var′ĭ-ko-sēl*). Varicocele occurs when one or both of the testicular veins draining from the testes becomes swollen, resulting in poor circulation in the testes. A varicocele generally occurs on the left side because the left testicular vein drains into the renal vein. The blood pressure is higher here than it is in the inferior vena cava, into which the right testicular vein empties.

A **hydrocele** (*hi′drŏ-sēl*) is a benign fluid mass within the tunica vaginalis that causes swelling of the scrotum. It is a frequent minor disorder in male infants, as well as in adults. The cause is unknown.

An infection in the testes is called **orchitis** (*or-ki′tis*). Orchitis may develop from a primary infection from a tubercle bacterium or as a secondary complication of mumps contracted after puberty. If orchitis from mumps involves both testes, it usually causes sterility.

Trauma to the testes and scrotum is common because of their pendent position. The testes are extremely sensitive to pain, and a male responds reflexively to protect the groin area.

varicocele: L. *varico*, a dilated vein; Gk. *kele*, tumor or hernia

Clinical Investigation Answer

The tubular structures that are apparently absent in the patient are the ductus (vasa) deferentia. This condition, known as *congenital bilateral absence of the ductus deferentia,* prevents the transport of spermatozoa from the testes to the ejaculatory ducts. This explains the absence of spermatozoa in the patient's ejaculate. His ejaculate consists only of the secretions from the seminal vesicles (which in many cases are also absent or nonfunctional in this deformity) and the prostate. Until recently, this patient's condition would have categorically prevented him from becoming a father. Microsurgical extraction of spermatozoa from the epididymides is now possible, however, and has allowed many afflicted men to father children.

Chapter Summary

Introduction to the Reproductive System (pp. 892–895)

1. Ova and sperm, collectively called gametes, each contain 23 chromosomes and are haploid.
   (a) The gametes are formed by meiosis.
   (b) The zygote that is formed as a result of fertilization is diploid, with 46 chromosomes.
   (c) The X and Y chromosomes are called the sex chromosomes; a female has the XX and a male has the XY genotype.
2. The indifferent embryonic gonads are converted into testes by the action of a testis-determining factor on the Y chromosome; in the absence of this factor, the gonads become ovaries.
3. Testosterone acts in its target cells through its conversion into derivatives, including dehydrotestosterone (DHT). Testosterone and DHT are responsible for the masculinization of the embryonic tissues to form male secondary sex organs.
4. The testes descend from the body cavity into the scrotum through the shortening and differential growth of the gubernaculum.

Endocrine Regulation of Reproduction (pp. 895–899)

1. The gonads of both sexes are stimulated by FSH and LH, which are secreted by the anterior pituitary in response to stimulation by GnRH from the hypothalamus.
2. At the time of puberty, a rise in the secretion of gonadal steroid hormones causes the development of secondary sex characteristics.
3. The pineal gland, with its secretion of melatonin, is believed by many to play a role in the initiation of puberty.

Male Reproductive System (pp. 899–907)

1. The testes are contained outside of the body cavity in the scrotum. The temperature of the testes is maintained at about 35° C (95° F) by the contraction or relaxation of the cremaster and dartos muscles and other mechanisms.
2. The testes are partitioned into wedge-shaped lobules. The lobules are composed of seminiferous tubules, which produce sperm, and of interstitial tissue, which produces androgens.
3. LH stimulates the interstitial cells to secrete testosterone, and FSH stimulates the tubules to secrete a polypeptide hormone called inhibin; testosterone and inhibin, in turn, exert feedback control of LH and FSH secretion, respectively.
4. Spermatogenesis begins with stem cells called spermatogonia. The diploid spermatogonia undergo meiosis to form haploid secondary spermatocytes at the end of the first division and spermatids at the end of the second division.
   (a) The four spermatids formed from the meiotic division of one parent cell are nurtured by the sustentacular cells, which aid the development of spermatids into spermatozoa.
   (b) Sustentacular cells are the targets of FSH action, and both FSH and testosterone are required for spermatogenesis at puberty.

Spermatic Ducts, Accessory Reproductive Glands, and the Penis (pp. 907–918)

1. Nonmotile sperm pass from the testes to the head of the epididymis, through its body, and out the tail of the epididymis to enter the ductus deferens as mature sperm.
   (a) The ductus deferens exits the scrotum, penetrates through the inguinal canal, and delivers sperm to the ejaculatory duct, which is formed by the union of the ductus deferens and the duct of the seminal vesicle.
   (b) The secretions of the seminal vesicles constitute about 60% of the semen; these secretions include fructose, citric acid, and coagulation proteins.
   (c) The ejaculatory ducts pass through the prostate, which contributes fluid and chemical agents to the semen. The secretions of the prostate constitute about 40% of the semen. (Spermatozoa account for less than 1% of the semen content.)
   (d) The bulbourethral glands secrete a mucoid substance that coats the urethra and lubricates the tip of the penis.
2. The male urethra, which serves as a common passageway for both the urinary and reproductive systems, is divided into prostatic, membranous, and spongy portions.
3. The penis contains three long columns of erectile tissue: two dorsal corpora cavernosa and one ventral corpus spongiosum surrounding the urethra.
   (a) Penile erection is produced by parasympathetic-induced vasodilation, which causes the spongy erectile tissue to become engorged with blood.
   (b) Emission is the movement of sperm into the ejaculatory ducts, together with fluid from the accessory reproductive glands.
   (c) Ejaculation is the forceful propulsion of semen from the male duct system as a result of muscular contractions of the bulbocavernosus muscles and sympathetic reflexes in the smooth muscles of the reproductive organs.

Review Activities

Objective Questions

1. An embryo with the genotype XY develops male accessory sex organs because of
   (a) androgens.
   (b) estrogens.
   (c) the absence of androgens.
   (d) the absence of estrogens.
2. Which of the following does not arise from the embryonic mesonephric duct?
   (a) epididymis
   (b) ductus deferens
   (c) seminal vesicle
   (d) prostate
3. The external genitalia of a male are completely formed by the end of
   (a) the embryonic period.
   (b) the ninth week.
   (c) the tenth week.
   (d) the twelfth week.

4. In the male,
   (a) FSH is not secreted by the pituitary.
   (b) FSH receptors are located in the interstitial cells.
   (c) FSH receptors are located in the spermatogonia.
   (d) FSH receptors are located in the sustentacular cells.
5. The secretion of FSH in a male is inhibited by negative feedback effects of
   (a) inhibin secreted from the seminiferous tubules.
   (b) inhibin secreted from the interstitial cells.
   (c) testosterone secreted from the seminiferous tubules.
   (d) testosterone secreted from the interstitial cells.
6. Which of the following is not a spermatic duct?
   (a) epididymis
   (b) spermatic cord
   (c) ejaculatory duct
   (d) ductus deferens
7. Spermatozoa are stored prior to emission and ejaculation in
   (a) the epididymides.
   (b) the seminal vesicles.
   (c) the penile urethra.
   (d) the prostate.
8. Urethral glands
   (a) secrete mucus.
   (b) produce nutrients.
   (c) secrete hormones.
   (d) regulate sperm production.
9. Which statement regarding erection of the penis is *false?*
   (a) It is a parasympathetic response.
   (b) It may be both a voluntary and an involuntary response.
   (c) It has to be followed by emission and ejaculation.
   (d) It is controlled by the hypothalamus of the brain and sacral portion of the spinal cord.
10. The condition in which one or both testes fails to descend into the scrotum is
    (a) cryptorchidism.
    (b) Turner's syndrome.
    (c) hermaphroditism.
    (d) Klinefelter's syndrome.
11. The androgen that is active within the cells of the prostate is
    (a) testosterone.
    (b) dehydrotestosterone.
    (c) estradiol.
    (d) none of these.
12. Which of the following statements regarding spermatozoa is *true?*
    (a) They are not motile at the conclusion of spermiogenesis.
    (b) They account for less than 1% of the volume of semen.
    (c) Prior to emission, they are stored in the epididymides.
    (d) All of the above are true.
13. If GnRH were secreted in large amounts and at a constant rate rather than in a pulsatile fashion, which of the following statements would be *true?*
    (a) LH secretion will increase at first and then decrease.
    (b) LH secretion will increase indefinitely.
    (c) Testosterone secretion in a male will be continuously high.
    (d) Estradiol secretion in a woman will be continuously high.

Essay Questions

1. Describe how development of the gonads and the secondary sex organs is determined by chromosomes and by the secretion of hormones.
2. Explain the sequence of events by which the male accessory reproductive organs and external genitalia develop. What occurs when a male embryo lacks receptor proteins for testosterone? What occurs when a male embryo lacks the enzyme 5α-reductase?
3. Explain why a testis is said to be composed of two separate compartments. Describe the interactions that may occur between these compartments.
4. Describe the roles of the sustentacular cells in the testes.
5. Describe the steps of spermatogenesis and explain its hormonal control.
6. Describe the interactions between the hypothalamus, anterior pituitary, and testes during puberty and discuss the possible role of the pineal gland in puberty.
7. Compare the seminal vesicles and the prostate in terms of location, structure, and function.
8. Describe the structure of the penis and explain the mechanisms that result in erection, emission, and ejaculation.
9. Define *semen*. How much semen is ejected during ejaculation? What does it consist of, and what are its properties?
10. Distinguish between impotence, infertility, and sterility.
11. Identify the conversion products of testosterone and describe their functions in the brain, prostate, and seminiferous tubules

Critical Thinking Questions

1. Suppose a woman is found to have testes rather than ovaries in her body cavity. Further study reveals that she has epididymides, vasa deferentia and seminal vesicles, but her external genitalia are more female than male in appearance, and she has female breast development. Laboratory tests reveal that she has an XX genotype. How might this condition be explained?
2. According to your friend, there is a female birth control pill and not a male birth control pill only because the medical establishment is run by men. Do you agree with her conspiracy theory? Provide physiological support for your answer.
3. Elderly men with benign prostatic hyperplasia are sometimes given estrogen treatments. How would this help the condition? What other types of drugs may be given, and what would you predict their possible side effects to be?

Related Web Sites

For a listing of the most current web sites related to this chapter, please visit the *Concepts of Human Anatomy and Physiology* home page at http://www.mhhe.com/biosci/abio/.

TWENTY-NINE

Female Reproductive System

OBJECTIVES

- List and briefly describe the primary and secondary female sex organs and define *secondary sex characteristics*.
- List the functions of the female reproductive system.
- Describe the structures of the uterine tubes, uterus, vulva, and vagina.
- Describe the changes that occur in the female reproductive structures during sexual excitement and coitus.
- Describe the position of the ovaries and of the ligaments associated with the ovaries, uterus, and uterine tubes.
- Discuss the changes that occur in the ovaries leading up to and following ovulation.
- Describe oogenesis and explain why meiosis of one primary oocyte results in the formation of only one mature ovum.
- Discuss the hormonal secretions of the ovaries during an ovarian cycle.
- Describe the hormonal changes that occur during the follicular phase and explain the hormonal control of ovulation.
- Discuss the formation, function, and fate of the corpus luteum.
- Discuss the structural changes that occur in the endometrium during the menstrual cycle and explain how these changes are hormonally controlled.
- Describe the structure of the breasts and mammary glands.
- Discuss the hormonal requirements for mammary gland development.
- Describe the action of prolactin and oxytocin on lactation and explain how secretion of these hormones is regulated.

Clinical Investigation

A first-year college student went to the student health center complaining of amenorrhea of several months' duration. Before that, her periods had always been normal. The young woman's body weight was significantly below average for her height, but she otherwise exhibited normal development of secondary sex characteristics. She had been diagnosed a few years earlier as hypothyroid, and thyroxine pills were prescribed for this condition. She stated that she exercises at least an hour a day and that she teaches aerobics at a local gym. A pregnancy test was performed and was negative. Blood tests were ordered for thyroxine, prolactin, androgens, estradiol, and gonadotropins. These all proved to be normal. What is the most likely diagnosis of this student's condition? What should she do about it?

Clue: Review the information about abnormal menstruations in Clinical Considerations, under the heading "Problems Involving the Uterus."

Structures and Functions of the Female Reproductive System

The structures of the female reproductive system include the ovaries, the secondary sex organs (vagina, uterine tubes, uterus, and mammary glands), and the external genitalia. The female reproductive system produces ova, secretes sex hormones, receives sperm from the male, and provides sites for fertilization and development of the embryo and fetus. Following parturition (birth), the secretions from the mammary glands provide nourishment for the baby.

The organs of the female and male reproductive systems are considered *homologous* because they develop from the same embryonic structures (see "Under Development" in this chapter and in chapter 28). The *primary sex organs*, called *gonads*, produce the *gametes*. Specifically, the **ovaries** are the gonads in females and the *ova* are the gametes that they produce. The ova of a female are completely formed, but not totally matured, during fetal development of the ovaries. The ova are generally discharged, or *ovulated*, one at a time in a cyclic pattern throughout the reproductive period of the female, which extends from puberty to menopause. *Menstruation* is the discharge of *menses* (*men'sēz*)—blood and solid tissue—from the uterus at the end of each menstrual cycle. Puberty in girls is heralded by the onset of menstruation, or *menarche* (*me-nar'ke*). *Menopause* is the termination of ovulation and menstruation. The reproductive period in females generally extends from about age 12 to age 50. The cyclic reproductive pattern of ovulation and the span of years over which a woman is fertile are determined by hormonal action.

The functions of the female reproductive system are (1) to produce ova; (2) to secrete sex hormones; (3) to receive the sperm from the male during sexual intercourse, or *coitus* (*ko'ĭ-tus*); (4) to provide sites for fertilization, implantation of the blastocyst (see chapter 30), and embryonic and fetal development; (5) to facilitate *parturition*, or delivery of the baby; and (6) to provide nourishment for the baby through the secretion of milk from the mammary glands in the breasts.

Secondary Sex Organs and Secondary Sex Characteristics

Secondary sex organs are those structures that are essential for successful fertilization, implantation of the embryo, development of the embryo and fetus, and parturition. The secondary sex organs include the *vagina*, which receives the penis and ejaculated semen during coitus and through which the baby passes during parturition; the uterine (fallopian) tubes, also called the *oviducts*, through which an ovum is transported toward the uterus after ovulation and in which fertilization normally occurs; and the uterus (womb), in which implantation and development occur and whose muscular walls play an active role in parturition (fig. 29.1). *Mammary glands* are also considered secondary sex organs because the milk they secrete after parturition provides nourishment for the baby. The structure and function of mammary glands will be discussed in a separate section.

Secondary sex characteristics are not essential for the reproductive process but are features considered to be sexual attractants. The female distribution of subcutaneous fat, broad pelvis, body hair pattern, and breast development are examples. Although the breasts contain the mammary glands, large breasts are not essential for nursing the young. Indeed, although all female mammals have mammary glands and are capable of nursing, only human females have breasts that are considerably larger than those of the male.

Uterine Tubes

The paired **uterine** (fallopian) **tubes** transport ova from the ovaries to the uterus. Fertilization normally occurs within the uterine tube. Each uterine tube is approximately 10 cm (4 in.) long and 0.7 cm (0.3 in.) in diameter and is positioned between the folds of the broad ligament of the uterus.

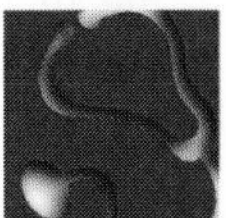

The term *salpinx* is occasionally used to refer to the uterine tubes. It is a Greek word meaning "trumpet" or "tube" and is the root of such clinical terms as *salpingitis* (*sal''pin-ji'tis*), or inflammation of the uterine tubes; *salpingography* (radiography of the uterine tubes); and *salpingolysis* (the breaking up of adhesions of the uterine tube to correct female infertility).

vagina: L. *vagina*, sheath or scabbard
fallopian tubes: from Gabriele Fallopius, Italian anatomist, 1523–62

Figure 29.1

Organs of the female reproductive system.

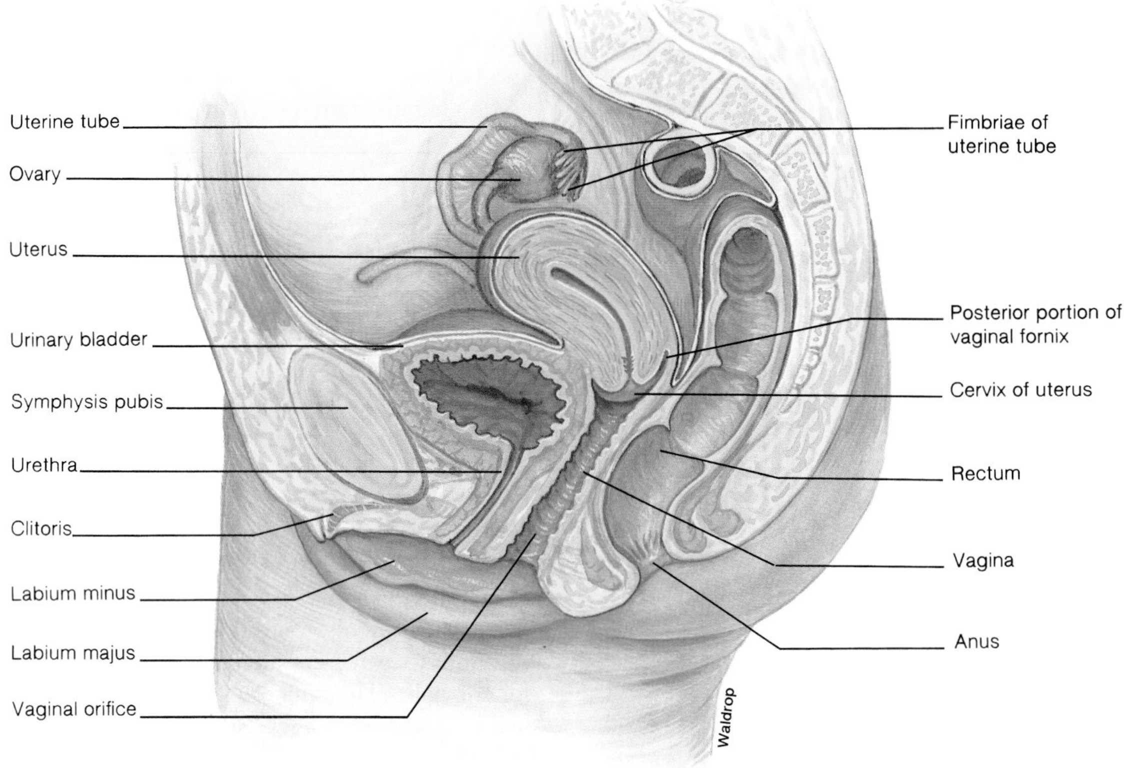

The funnel-shaped, open-ended portion of the uterine tube is called the **infundibulum.** Although the infundibulum is close to the ovary, it is not attached. A number of fringed, fingerlike processes called **fimbriae** (*fim'bre-e*) project from the margins of the infundibulum over the lateral surface of the ovary. The fimbriae are covered by ciliated columnar epithelium, which draws the ovum into the lumen of the uterine tube. From the infundibulum, the uterine tube extends medially and inferiorly to open into the cavity of the uterus at the uterine opening. The **ampulla** (*am-pool'a*) of the uterine tube is the longest and widest portion.

The wall of the uterine tube consists of three histological layers. The internal **mucosa** lines the lumen and is composed of a ciliated columnar epithelium (fig. 29.2). The mucosa is extensively folded, which delays the passage of the ovum, thus increasing the likelihood that fertilization will occur in the upper third of the tube. The **muscularis** is the middle layer, composed of a thick circular layer of smooth muscle and a thin outer layer of smooth muscle. Peristaltic contractions of the muscularis and ciliary action of the mucosa move the ovum through the lumen of the uterine tube. The outer lubricative **serous layer** of the uterine tube is part of the visceral peritoneum.

fimbriae: L. *fimbria*, fringe

Figure 29.2

Histology of the uterine tube.

The cilia of the uterine tube help to move an ovulated egg cell (oocyte) towards the uterus.

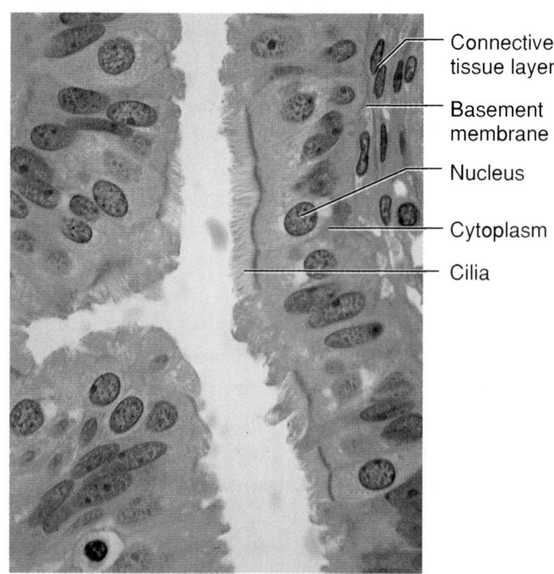

The ovum takes 4 to 5 days to move through the uterine tube. If enough viable sperm are ejaculated into the vagina during coitus, and if there is an oocyte in the uterine tube, fertilization will occur within hours after discharge of the semen. The zygote will move toward the uterus, where implantation occurs. If the developing embryo (called a *blastocyst*) implants into the uterine tube rather than the uterus, the pregnancy is termed an *ectopic* (*ek-top'ik*) *pregnancy*, meaning an implantation of the blastocyst in a site other than the uterus.

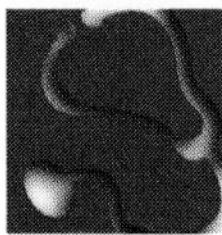

Since the infundibulum of the uterine tube is unattached, it provides a potential pathway for pathogens to enter the abdominopelvic cavity. The mucosa of the uterine tube is continuous with that of the uterus and vagina, and it is possible for infectious agents to enter the vagina and cause infections that may ultimately spread to the peritoneal linings, resulting in *pelvic inflammatory disease* (*PID*). There is no opening into the abdominopelvic cavity other than through the uterine tubes. The abdominopelvic cavity of a male is totally sealed from external contamination.

Uterus

The **uterus** (*yoo'ter-us*) is the normal site of implantation for the blastocyst that develops from a fertilized ovum. Prenatal development continues within the uterus until *gestation* (development) is completed, at which time the uterus plays an active role in the delivery of the baby.

Structure of the Uterus

The uterus is a hollow, thick-walled, muscular organ with the shape of an inverted pear. Although the shape and position of the uterus undergo enormous change during pregnancy (fig. 29.3), in its nonpregnant state it is about 7 cm (2.8 in.) long, 5 cm (2 in.) wide (through its broadest region), and 2.5 cm (1 in.) in diameter. The anatomical regions of the uterus include the uppermost dome-shaped portion above the entrance of the uterine tubes, called the **fundus;** the enlarged main portion, called the **body;** and the inferior constricted portion opening into the vagina, called the **cervix** (fig. 29.4). The uterus is located between the urinary bladder anteriorly and the rectum and sigmoid colon posteriorly. The fundus projects anteriorly and slightly superiorly over the urinary bladder. The cervix projects posteriorly and inferiorly, joining the vagina at nearly a right angle (see fig. 29.1).

The **uterine cavity** is the space within the regions of the fundus and body. The lumina of the uterine tubes open into the uterine cavity on the superior-lateral portions. The uterine cavity is continuous inferiorly with the **cervical canal,** which extends through the cervix and opens into the lumen of the vagina. The junction of the uterine cavity and the cervical canal is called the **isthmus of uterus,** and the opening of the cervical canal into the vagina is called the **uterine ostium.**

Figure 29.3

Uterus in a full-term pregnant woman.
The sagittal section demonstrates the size and position of the uterus during pregnancy.

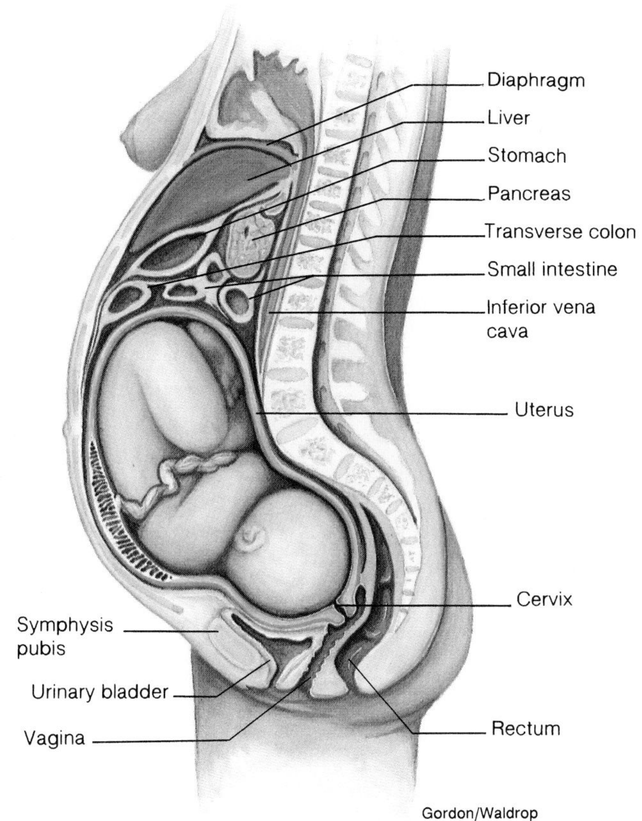

Support of the Uterus

The uterus is maintained in position by muscular support and by ligaments that extend from the pelvic girdle or body wall to the uterus. Muscles of the pelvic outlet (chapter 13), especially the levator ani muscle (see fig. 13.12), provide the principal muscular support. The ligaments that support the uterus undergo marked hypertrophy during pregnancy but regress in size after parturition. They atrophy after menopause, which may contribute to a condition called *uterine prolapse*, or downward displacement of the uterus.

Four paired ligaments support the uterus in position within the pelvic cavity. The paired **broad ligaments** (fig. 29.4) are folds of the peritoneum that extend from the pelvic walls and floor to the lateral walls of the uterus. The ovaries and uterine tubes are also supported by the broad ligaments. The **cardinal,** or **lateral cervical, ligaments** (not illustrated) are fibrous bands within the broad ligament that extend laterally from the cervix and vagina across the pelvic floor, where they attach to the wall of the pelvis. The cardinal ligaments contain some smooth muscle as well as vessels and

ectopic: Gk. *ex*, out; *topos*, place
fundus: L. *fundus*, bottom
cervix: L. *cervix*, neck

Figure 29.4

The female reproductive system and supporting ligaments.
The right portion of this figure depicts the release of an egg cell from the ovary, a process called ovulation.

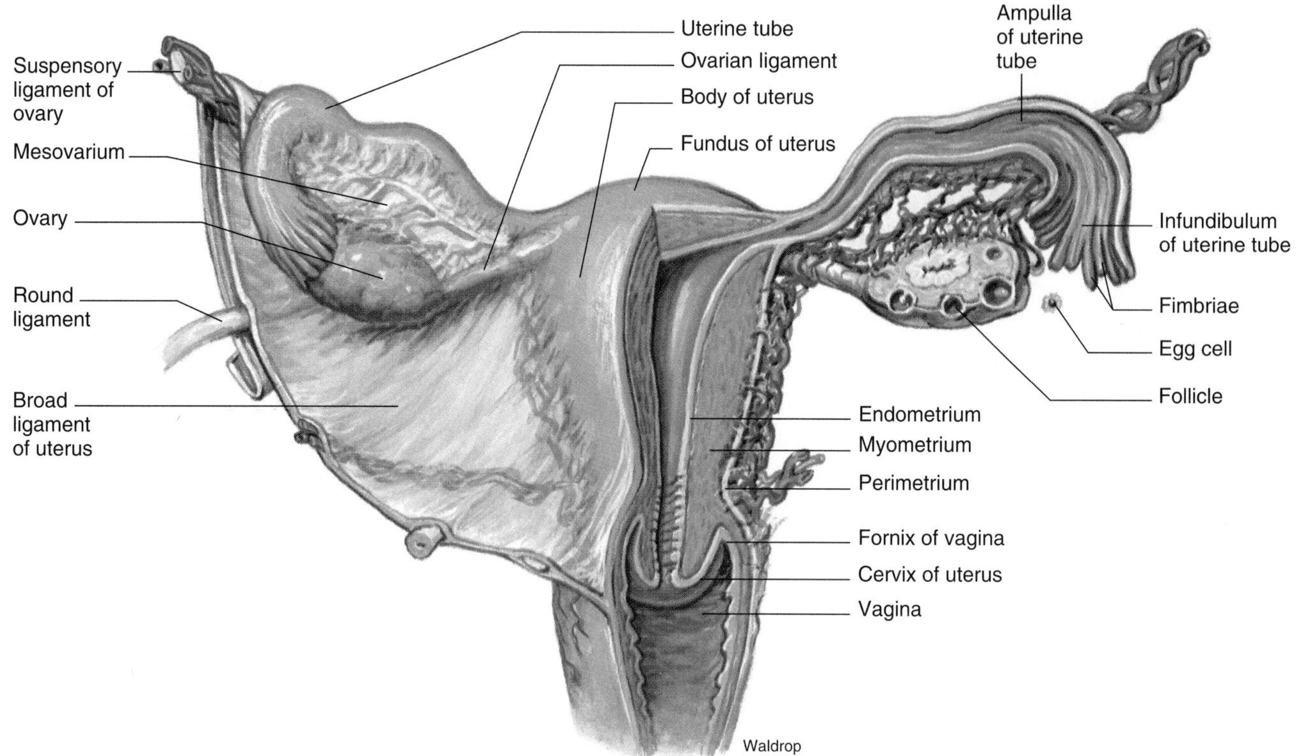

nerves that supply the cervix and vagina. The fourth paired ligaments are the **round ligaments.** Each round ligament extends from the lateral border of the uterus just below the point where the uterine tube attaches to the lateral pelvic wall. Similar to the course taken by the ductus deferentia in the male, the round ligaments continue through the inguinal canals of the abdominal wall, where they attach to the deep tissues of the labia majora.

Uterine Wall

The wall of the uterus is composed of three layers: the perimetrium, myometrium, and endometrium (fig. 29.4). The **perimetrium,** the outermost serosal layer, consists of the thin visceral peritoneum. The lateral portion of the perimetrium is continuous with the broad ligament. A shallow pouch called the **vesicouterine** (*ves″ĭ-ko-yoo′ter-in*) **pouch** is formed as the peritoneum is reflected over the urinary bladder. The **rectouterine pouch** (pouch of Douglas) is formed as the peritoneum is reflected onto the rectum. The rectouterine pouch is the lowest point of the pelvic cavity and provides a site for surgical entry into the peritoneal cavity.

The thick **myometrium** is composed of three poorly defined layers of smooth muscle arranged in longitudinal, circular, and spiral patterns. The myometrium is thickest in the fundus and thinnest in the cervix. During parturition, the muscles of this layer contract forcefully to expel the fetus.

The **endometrium,** the inner mucosal lining of the uterus, is composed of two distinct layers. The superficial **stratum functionale,** composed of columnar epithelium and containing secretory glands, is shed as *menses* during menstruation and is built up again under the stimulation of ovarian steroid hormones. The deeper **stratum basale** is highly vascular and serves to regenerate the stratum functionale after each menstruation.

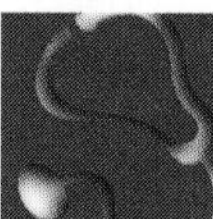

The uterus undergoes tremendous change during pregnancy. Its weight increases more than 16 times (from about 60 g to about 1,000 g), and its capacity increases from about 2.5 ml to over 5,000 ml. The principal change in the myometrium is a marked hypertrophy, or elongation, of the individual muscle cells to as much as 10 times their original length. There is some atrophy of the muscle cells after parturition, but the uterus may remain somewhat enlarged until menopause, at which time there is marked atrophy.

Uterine Blood Supply and Innervation

The uterus is supplied with blood through the **uterine arteries,** which arise from the internal iliac arteries, and from the

pouch of Douglas: from James Douglas, English anatomist and physician, 1675–1742

menses: L. *menses*, plural of *mensis*, monthly

Figure 29.5

Arteries serving the internal female reproductive organs.

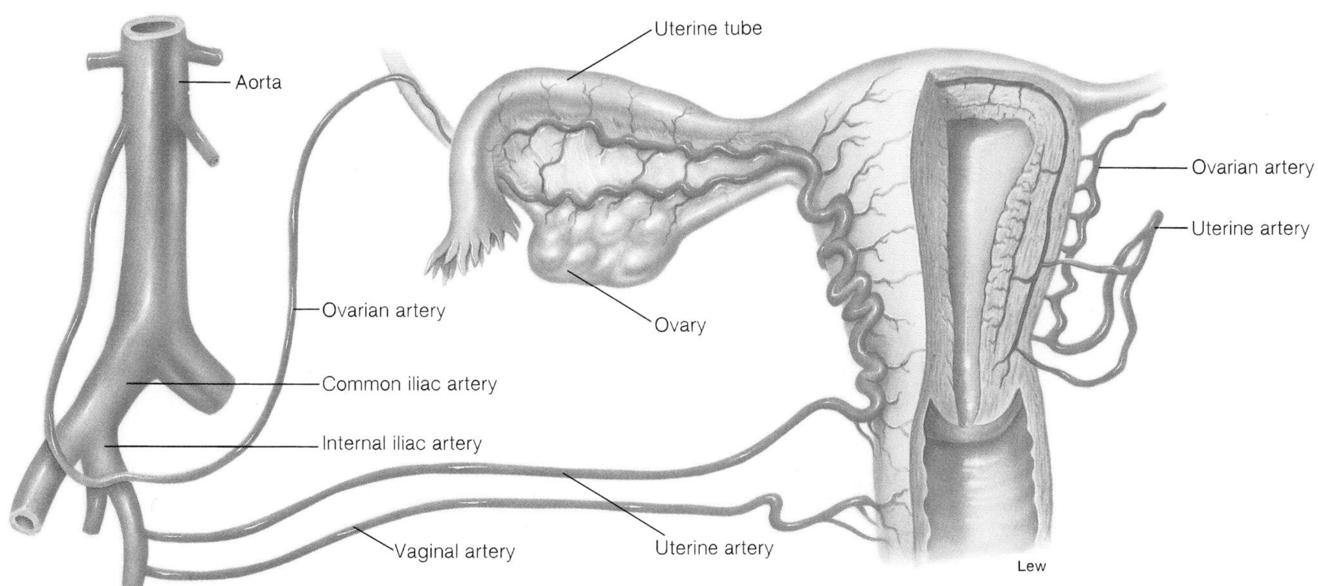

uterine branches of the **ovarian arteries** (fig. 29.5). Extensive interconnections between the arteries to the uterus help to ensure that it is adequately supplied with blood despite the changes it undergoes during pregnancy. The blood from the uterus returns through uterine veins that parallel the course of the arteries.

The uterus receives both sympathetic and parasympathetic innervation from the pelvic and hypogastric plexuses. Both autonomic innervations serve the arteries of the uterus, whereas the smooth muscle of the myometrium receives only sympathetic innervation.

Vagina

The **vagina** (*vă-ji'nă*) is the organ that receives the penis (and semen) during sexual intercourse. It also serves as the birth canal during parturition and provides for the passage of menses to the outside of the body. The vagina is a tubular organ about 9 cm (3.5 in.) long, passing from the cervix of the uterus to the vestibule. In its position between the urinary bladder and urethra anteriorly and the rectum posteriorly, it is continuous with the cervical canal of the uterus. The cervix attaches to the vagina at a nearly 90° angle. The deep recess surrounding the protrusion of the cervix into the vagina is called the **fornix** (see fig. 29.4). The exterior opening of the vagina, at its lower end, is the **vaginal orifice.** A thin fold of mucous membrane called the **hymen** (*hi'men*) may partially cover the vaginal orifice.

The vaginal wall is composed of three layers: an inner mucosa, a middle muscularis, and an outer fibrous layer. The **mucosal layer** consists of a nonkeratinized stratified squamous epithelium, which forms a series of transverse folds called **vaginal rugae** (*roo'je*) (fig. 29.6). The vaginal rugae provide friction ridges for stimulation of the penis during sexual intercourse. They also permit considerable distension of the vagina for penetration of the erect penis. The mucosal layer contains few glands; the acidic mucus that is present in the vagina comes primarily from glands within the uterus. This acidic environment of the vagina retards microbial growth. The semen, however, temporarily neutralizes the acidity of the vagina to ensure the survival of the ejaculated sperm.

Figure 29.6

Histology of a vaginal ruga.

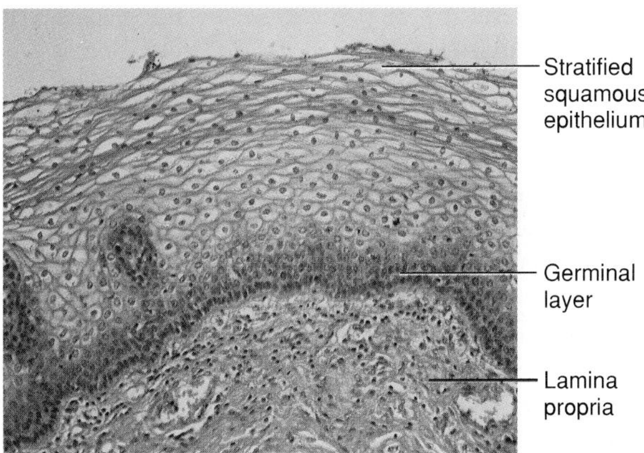

The **muscularis layer** consists of longitudinal and circular bands of smooth muscle interlaced with distensible connective tissue. The distension of this layer is especially important during parturition. Skeletal muscle strands near the vaginal orifice, including the levator ani muscle, partially constrict this opening. The **fibrous layer** covers the vagina and attaches it to surrounding pelvic organs. This layer consists of dense regular connective tissue interlaced with strands of elastic fibers.

Figure 29.7

The external female genitalia.
The labia majora and clitoris in a female are homologous to the scrotum and penis, respectively, in a male.

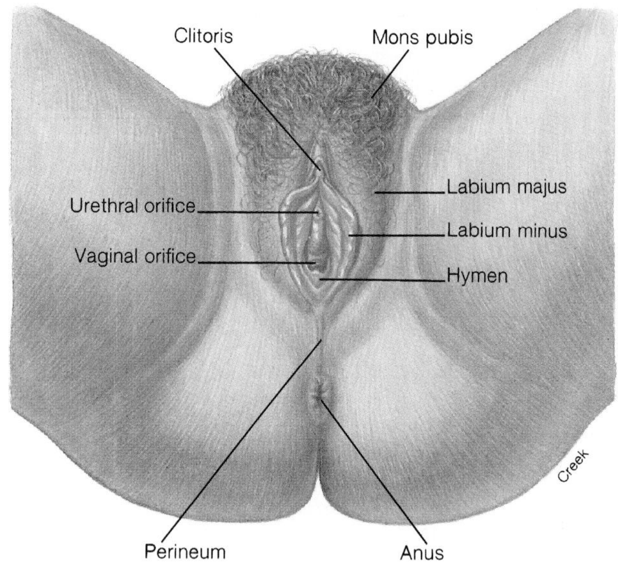

Vulva

The external genitalia of the female are referred to collectively as the **vulva** (*vul'vă*) (fig. 29.7). The structures of the vulva surround the vaginal orifice and include the mons pubis, labia majora, labia minora, clitoris, vaginal vestibule, vestibular bulbs, and vestibular glands.

The **mons pubis** is the subcutaneous pad of adipose connective tissue covering the symphysis pubis. At puberty, the mons pubis becomes covered with a pattern of coarse pubic hair that is somewhat triangular, usually with a horizontal upper border. The elevated and padded mons pubis cushions the symphysis pubis and vulva during sexual intercourse.

The **labia majora** (*la'be-ă mă-jor'ă*; singular, *labium majus*) are two thickened longitudinal folds of skin composed of adipose tissue and loose connective tissues, as well as some smooth muscle. After puberty their lateral surfaces are covered by pubic hair. The labia majora are continuous anteriorly with the mons pubis. They are separated longitudinally by the **pudendal** (*pyoo-den'dal*) **cleft** and converge again posteriorly on the **perineum** (*per"ĭ-ne'um*). The labia majora contain numerous sebaceous and sweat glands. They are homologous to the scrotum of the male and function to enclose and protect the other organs of the vulva.

Medial to the labia majora are two smaller longitudinal folds called the **labia minora** (singular, *labium minus*). The labia minora are hairless but do contain sebaceous glands. Anteriorly, the labia minora join to form the **prepuce** (*pre'pyoos*), a hoodlike fold that partly covers the clitoris. The labia minora further protect the vaginal and urethral openings.

The **clitoris** (*klit'or-is* or *kli-tor'is*) is a small rounded projection at the upper portion of the pudendal cleft, at the anterior junction of the labia minora. The clitoris corresponds in structure and origin to the penis in the male; it is smaller, however, and has no urethra. Although most of the clitoris is embedded, it does have an exposed **glans clitoris** of erectile tissue, richly innervated with sensory endings. The clitoris is about 2 cm (0.8 in.) long and 0.5 cm (0.2 in.) in diameter. The unexposed portion of the clitoris is composed of two columns of erectile tissue, the **corpora cavernosa,** which diverge posteriorly to form the **crura** and attach to the sides of the pubic arch.

The **vaginal vestibule** is the longitudinal cleft enclosed by the labia minora. The openings for the urethra and vagina are located in the vaginal vestibule. The external opening of the urethra is about 2.5 cm (1 in.) behind the glans clitoris and immediately in front of the vaginal orifice. The vaginal orifice is lubricated during sexual excitement by secretions from paired major and minor **vestibular** (Bartholin's) **glands** located within the wall of the region immediately inside the vaginal orifice. The ducts from these glands open into the vestibule near the lateral margins of the vaginal orifice. Bodies of vascular erectile tissue, called **vestibular bulbs,** are located immediately below the skin forming the lateral walls of the vestibule. The vestibular bulbs are separated from each other by the vagina and urethra, and extend from the level of the vaginal orifice to the clitoris.

The vulva has both sympathetic and parasympathetic innervation, as well as extensive somatic fibers that respond to sensory stimulation. Parasympathetic stimulation causes a response similar to that in the male: dilation of the arterioles of the genital erectile tissue and constriction of the venous return.

Mechanism of Erection and Orgasm

The homologous structures of the male and female reproductive systems respond to sexual stimulation in a similar fashion. The erectile tissues of a female, like those of a male, become engorged with blood and swollen during sexual arousal. During sexual excitement, the hypothalamus of the brain sends parasympathetic nerve impulses through the sacral segments of the spinal cord, which cause dilation of arteries serving the clitoris and vestibular bulbs. This increased blood flow causes the erectile tissues to swell. In addition, the erectile tissues in the areola of the breasts become engorged.

Simultaneous with the erection of the clitoris and vestibular bulbs, the vagina expands and elongates to accommodate the erect penis of the male, and parasympathetic impulses cause the vestibular glands to secrete mucus near the

vulva: L. *volvere*, to roll; wrapper
mons pubis: L. *mons*, mountain; *pubis*, genital area
vestibule: L. *vestibule*, an entrance, court
Bartholin's glands: from Caspar Bartholin, Jr., Danish anatomist, 1655–1738

vaginal orifice. The vestibular secretion moistens and lubricates the tissues of the vestibule, thus facilitating penetration of the erect penis into the vagina. Mucus continues to be secreted during intercourse so that the male and female genitalia do not become irritated as they would if the vagina became dry.

The position of the sensitive clitoris usually allows it to be stimulated during intercourse. If stimulation of the clitoris is of sufficient intensity and duration, a woman will experience a culmination of pleasurable psychological and physiological release called *orgasm*.

Associated with orgasm is a rhythmic contraction of the muscles of the perineum and the muscular walls of the uterus and uterine tubes. These reflexive muscular actions are thought to aid the movement of sperm through the female reproductive tract toward the upper end of a uterine tube, where an ovum might be located.

Ovaries and the Ovarian Cycle

The ovaries contain a large number of follicles, each of which encloses an ovum. Some of these follicles mature during the ovarian cycle, and the ova they contain progress to the secondary oocyte stage of meiosis. At ovulation, the largest follicle ruptures and expels its secondary oocyte from the ovary. The empty follicle then becomes a corpus luteum, which ultimately degenerates at the end of a nonfertile cycle.

The **ovaries** (*o'vă-rēz*) of sexually mature females are solid, ovoid structures about 3.5 cm (1.4 in.) long, 2 cm (0.8 in.) wide, and 1 cm (0.4 in.) thick. On the medial portion of each ovary is a **hilum** (*hi'lum*), which is the point of entrance for ovarian blood vessels and nerves. The lateral portion of the ovary is positioned near the open ends of the uterine tube (see fig. 29.4).

Position and Structure of the Ovaries

The ovaries are positioned in the upper pelvic cavity on the lateral sides of the uterus. Each ovary is situated in a shallow depression of the posterior body wall and is secured by several membranous attachments. The principal supporting membrane of the female reproductive tract is the **broad ligament.** The broad ligament is the parietal peritoneum that supports the uterine tubes and uterus. The **mesovarium** (*mes"ŏ-va're-um*) is a specialized posterior extension of the broad ligament that attaches to an ovary. Each ovary is additionally supported by an **ovarian ligament,** which is anchored to the uterus, and a **suspensory ligament,** which is attached to the pelvic wall (see fig. 29.4).

Each ovary consists of four layers. The **superficial epithelium** (see fig. 29.10) is the thin outermost layer composed of a simple cuboidal epithelium. A collagenous connective tissue layer called the **tunica albuginea** (*al"byoo-jin'e-ă*) is located immediately below the germinal epithelium. The principal substance of the ovary is divided into an outer **ovarian cortex** surrounding a vascular **ovarian medulla,** although the boundary between these layers is not distinct.

Blood is supplied by ovarian arteries that arise from the lateral sides of the abdominal aorta, just below the origin of the renal arteries. An additional supply comes from the ovarian branches of the uterine arteries. Venous return is through the ovarian veins. The right ovarian vein empties into the inferior vena cava, whereas the left ovarian vein drains into the left renal vein.

Ovarian Cycle

The germ cells that migrate into the ovaries during early embryonic development multiply, so that by about 5 months of gestation (prenatal life) the ovaries contain approximately 6 million to 7 million primordial oocytes called **oogonia** (*o"ŏ-go'ne-ă*). Most of these oogonia die prenatally through a process of apoptosis (cell suicide—see chapter 3). The production of new oogonia stops at this point and never resumes again. The oogonia begin meiosis toward the end of gestation, at which time they are called **primary oocytes** (*o'ŏ-sītz*). Like spermatogenesis in the prenatal male, oogenesis is arrested at prophase I of the first meiotic division. The primary oocytes are thus still diploid.

Primary oocytes decrease in number throughout a woman's life. The ovaries of a newborn girl contain about 2 million oocytes—all she will ever have. Each oocyte is contained within its own hollow ball of cells, the *ovarian follicles*. By the time a girl reaches puberty, the number of oocytes and follicles has been reduced to 400,000. Only about 400 of these oocytes will ovulate during the women's reproductive years, and the rest will die by apoptosis. Oogenesis ceases entirely at menopause (the time menstruation stops).

Primary oocytes that are not stimulated to complete the first meiotic division are contained within tiny **primary follicles** (fig. 29.8*a*). Immature primary follicles consist of only a single layer of follicle cells. In response to FSH stimulation, some of these oocytes and follicles get larger, and the follicular cells divide to produce numerous layers of **granulosa cells** that surround the oocyte and fill the follicle. Some primary follicles will be stimulated to grow still more, and they develop a number of fluid-filled cavities called *vesicles;* at this point, they are called **secondary follicles** (fig. 29.8*a*). Continued growth of one of these follicles will be accompanied by the fusion of its vesicles to form a single fluid-filled cavity called an *antrum*. At this stage, the follicle is known as a **vesicular ovarian,** or **graafian, follicle** (fig. 29.8*b*).

As the follicle develops, the primary oocyte completes its first meiotic division. This does not form two complete cells, however, because only one cell—the **secondary oocyte**—gets all the cytoplasm. The other cell formed at this

follicle: L. diminutive of *follis*, bag

graafian follicle: from Regnier de Graaf, Dutch anatomist and physician, 1641–73

Figure 29.8

Photomicrographs of the ovary.
(*a*) Primary follicles and one secondary follicle and (*b*) a vesicular ovarian (graafian) follicle are visible in these sections.

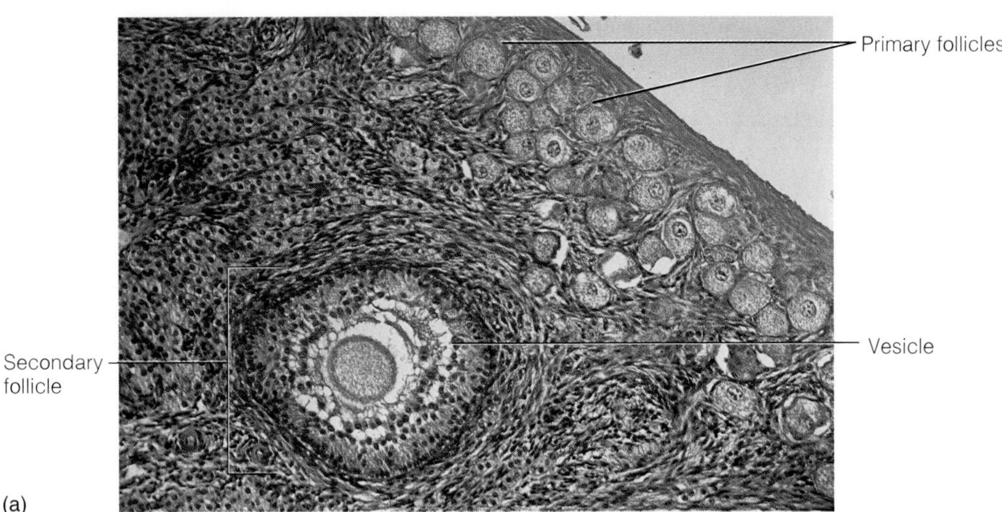

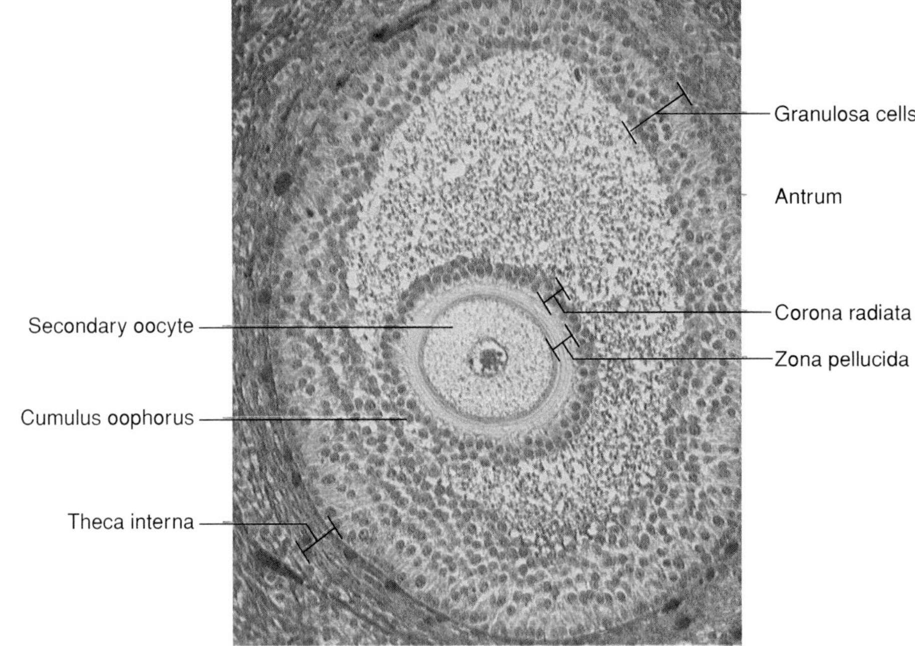

time becomes a small **polar body** (fig. 29.9), which eventually fragments and disappears. This unequal division of cytoplasm ensures that the ovum will be large enough to become a viable embryo should fertilization occur. The secondary oocyte then begins the second meiotic division, but meiosis is arrested at metaphase II. The second meiotic division is completed only by an ovum that has been fertilized.

The secondary oocyte, arrested at metaphase II, is contained within a graafian follicle. The granulosa cells of this follicle form a ring around the oocyte and form a mound that supports the oocyte. This mound is called the **cumulus oophorus** (*o-of'ŏ-rus*). The ring of granulosa cells surrounding the oocyte is the **corona radiata.** Between the oocyte and the corona radiata is a thin gel-like layer of proteins and polysaccharides called the **zona pellucida** (*pĕ-loo'sĭ-dă*) (see fig. 29.8*b*).

Under the stimulation of FSH from the anterior pituitary, the granulosa cells of the ovarian follicles secrete increasing amounts of estrogen as the follicles grow. Interestingly, the granulosa cells produce estrogen from its precursor testosterone, which is supplied by cells of the **theca interna,** the layer immediately outside the granulosa cells (fig. 29.8*b*).

Ovulation

Usually by about 10 to 14 days after the first day of menstruation only one follicle has continued its growth to become a fully mature graafian follicle (fig. 29.10). Other secondary follicles during that cycle regress and become *atretic* (*ă-tret'ik*). Follicle atresia, or degeneration, is a type of apoptosis that results from a complex interplay of hormones and paracrine regulators. The gonadotropins (FSH and LH), as well as various paracrine regulators and estrogen, act to protect follicles from atresia. By contrast, paracrine regulators that include androgens and FAS ligand (chapter 23) promote atresia of the follicles.

The follicle that is protected from atresia and that develops into a graafian follicle becomes so large that it forms a bulge on the surface of the ovary. Under proper hormonal stimulation, this follicle will rupture—much like the popping of a blister—and extrude its oocyte into the pelvic cavity in the process of **ovulation** (*ov-yŭ-la'shun*) (fig. 29.11).

cumulus oophorus: L. *cumulus*, a mound; Gk. *oophoros*, egg-bearing
zona pellucida: Gk. *zone*, girdle; L. *pellis*, skin; *lucere*, to shine
theca: Gk. *theke*, a box
atretic: Gk. *atretos*, not perforated

Figure 29.9

Photomicrographs of oocytes.

(*a*) A primary oocyte at metaphase I of meiosis. Notice the alignment of chromosomes (arrow). (*b*) A human secondary oocyte formed at the end of the first meiotic division and the first polar body (arrow).

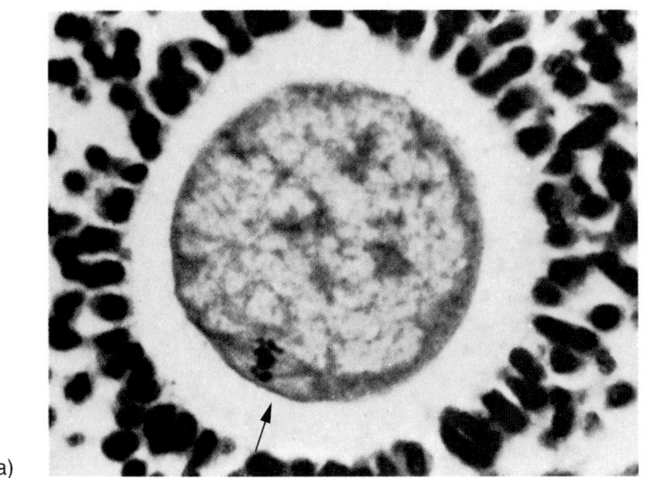

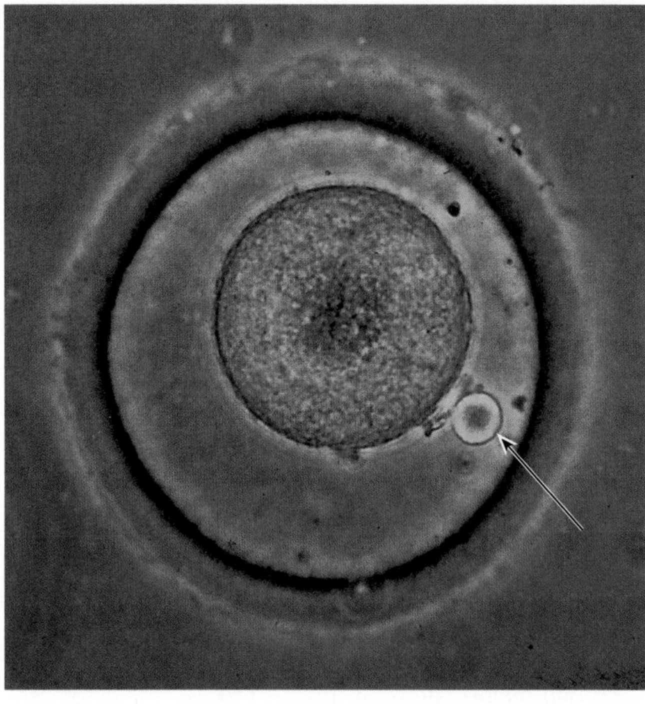

Figure 29.10

Ovary containing follicles at different stages of development.

An atretic follicle is one that is dying by apoptosis. It will eventually become a corpus albicans.

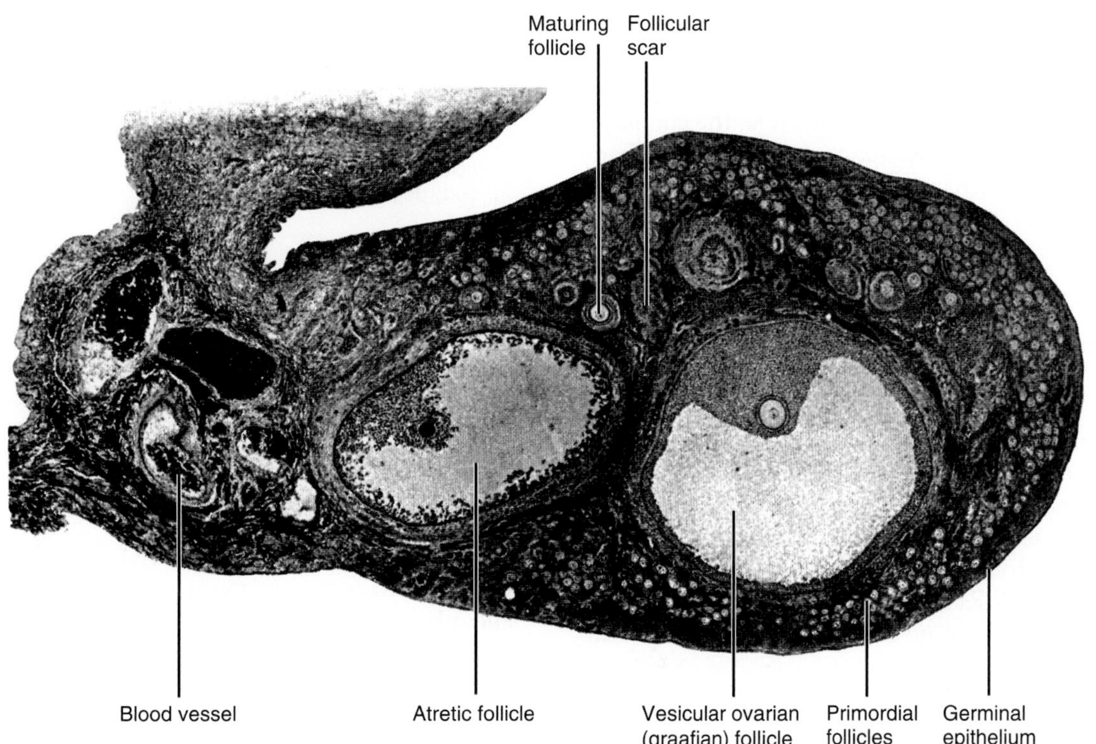

Figure 29.11

Ovulation from a human ovary.
Notice the cloud of fluid and granulosa cells surrounding the ovulated oocyte.

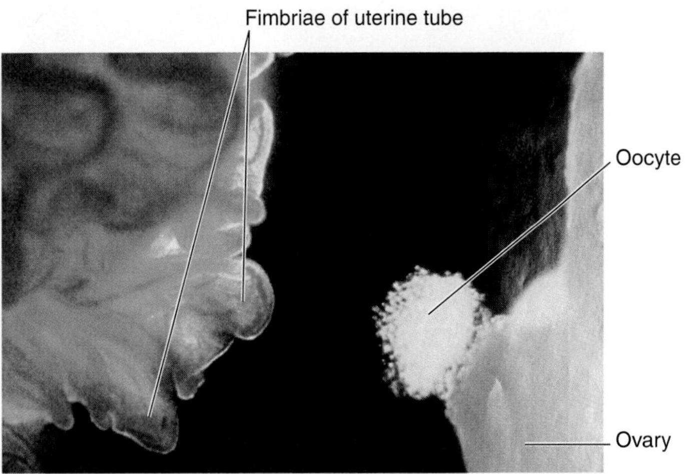

The released cell is a secondary oocyte, surrounded by the zona pellucida and corona radiata. If it is not fertilized it will degenerate in a couple of days. If a sperm passes through the corona radiata and zona pellucida and enters the cytoplasm of the secondary oocyte, the oocyte will then complete the second meiotic division. In this process, the cytoplasm is again not divided equally; most remains in the zygote (fertilized egg), leaving another polar body that, like the first, degenerates (fig. 29.12).

Changes continue in the ovary following ovulation. The empty follicle, under the influence of luteinizing hormone from the anterior pituitary, undergoes structural and biochemical changes to become a **corpus luteum** (*loo'te-um*). Unlike the ovarian follicles, which secrete only estrogen, the corpus luteum secretes two sex steroid hormones: estrogen and progesterone. Toward the end of a nonfertile cycle, the corpus luteum regresses to become a nonfunctional **corpus albicans** (*al'bĭ-kans*). These cyclic changes in the ovary are summarized in figure 29.13.

Figure 29.12

Oogenesis.
During meiosis, each primary oocyte produces a single haploid gamete. If the secondary oocyte is fertilized, it will form a secondary polar body and becomes a zygote.

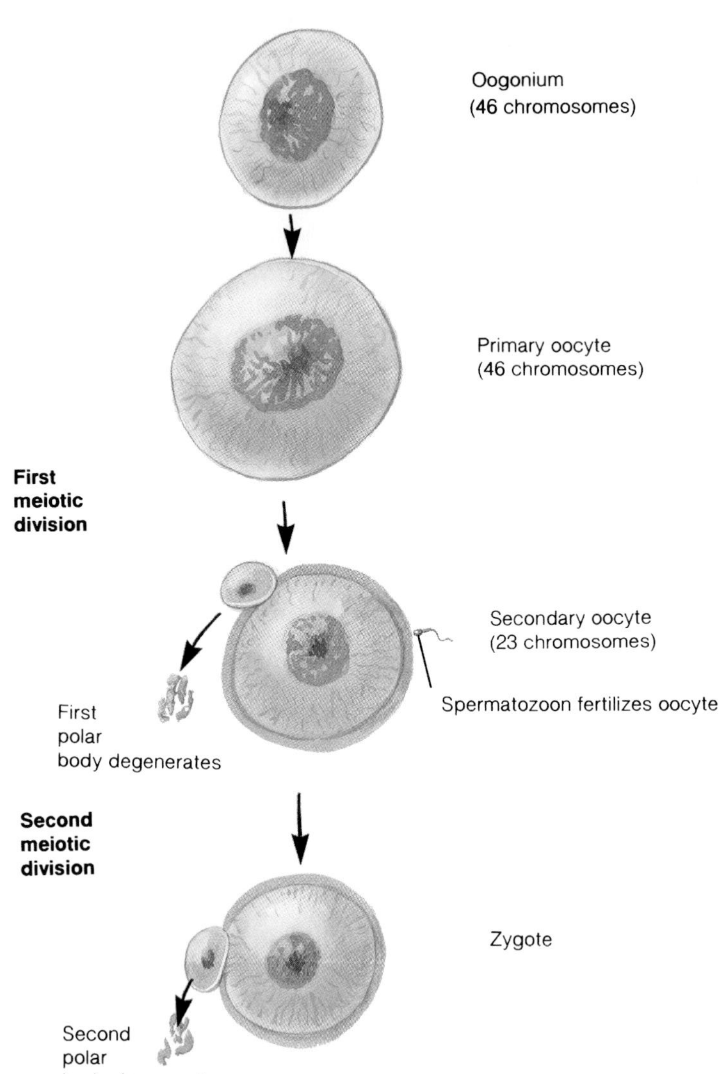

Pituitary-Ovarian Axis

The term **pituitary-ovarian axis** refers to the hormonal interactions between the anterior pituitary and the ovaries. The anterior pituitary secretes two gonadotropic hormones, follicle-stimulating hormone (FSH) and luteinizing hormone (LH), both of which promote cyclic changes in the structure and function of the ovaries. The secretion of both gonadotropic hormones, as discussed in chapter 28, is controlled by a single releasing hormone from the hypothalamus, gonadotropin-releasing hormone (GnRH), and by feedback effects from hormones secreted by the target organs—in the case of the female, the ovaries. The nature of these interactions will be described in detail in the next section.

Since one releasing hormone can stimulate the secretion of both FSH and LH, one might expect always to see parallel changes in the secretion of these gonadotropins. This, however, is not the case. FSH secretion is slightly greater than LH secretion during an early phase of the menstrual cycle, whereas LH secretion greatly exceeds FSH secretion just prior to ovulation. These differences are believed to result from the feedback effects of ovarian sex steroids, which can change the amount of GnRH secreted, the pulse frequency of GnRH secretion, and the ability of the anterior pituitary cells to secrete FSH and LH. These complex interactions result in a pattern of hormone secretion that regulates the phases of the menstrual cycle.

corpus luteum: L. *corpus*, body; *luteum*, yellow
albicans: L. *albicare*, to whiten

Figure 29.13

Stages of ovum and follicle development.
This diagram illustrates the stages that occur in an ovary during the course of a monthly cycle. The arrows indicate changes with time.

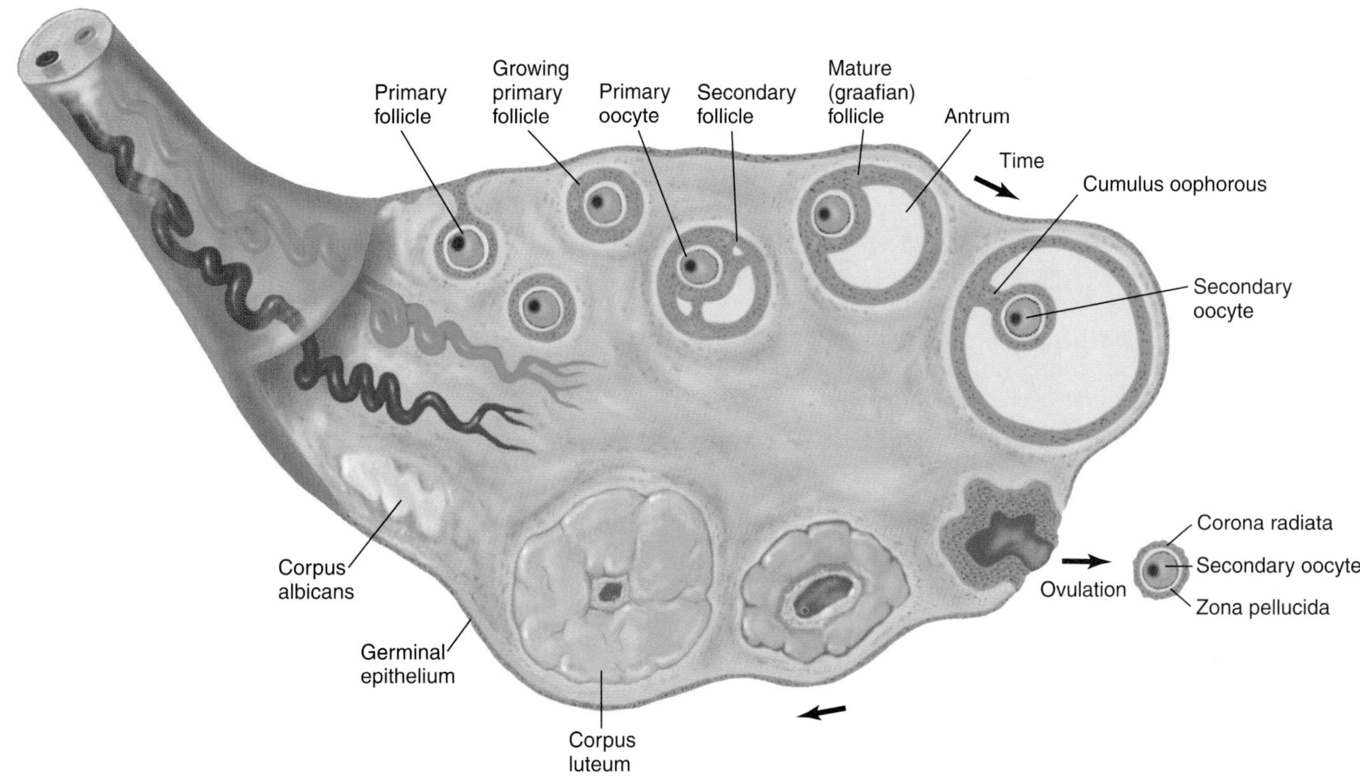

Menstrual Cycle

Cyclic changes in the secretion of gonadotropic hormones from the anterior pituitary cause the ovarian changes during a monthly cycle. The ovarian cycle is accompanied by cyclic changes in the secretion of estradiol and progesterone, which interact with the hypothalamus and pituitary to regulate gonadotropin secretion. The cyclic changes in ovarian hormone secretion also cause changes in the endometrium of the uterus during a menstrual cycle.

Humans, apes, and Old-World monkeys have cycles of ovarian activity that repeat at approximately one-month intervals; hence the name **menstrual cycle** (*menstru* = monthly). The term *menstruation* is used to indicate the periodic shedding of the stratum functionale of the endometrium, which becomes thickened prior to menstruation under stimulation by ovarian steroid hormones. In primates (other than New-World monkeys), this shedding of the endometrium is accompanied by bleeding. There is no bleeding when most other mammals shed their endometrium, and therefore their cycles are not called menstrual cycles.

In human females and other primates that have menstrual cycles, sexual intercourse (coitus) may be permitted at any time of the cycle. Nonprimate female mammals, by contrast, are sexually receptive (in "heat" or "estrus") only at a particular time in their cycles, shortly before or after ovulation. These animals are therefore said to have *estrous cycles*. Bleeding occurs in some animals (such as dogs and cats) that have estrous cycles shortly before they permit coitus. This bleeding is a result of high estrogen secretion and is not associated with shedding of the endometrium. The bleeding that accompanies menstruation, by contrast, is caused by a fall in estrogen and progesterone secretion.

Phases of the Menstrual Cycle: Cyclic Changes in the Ovaries

The duration of the menstrual cycle is typically about 28 days. Since it is a cycle, there is no beginning or end, and the changes that occur are generally gradual. It is convenient, however, to call the first day of menstruation "day one" of the cycle, since the flow of menstrual blood is the most apparent of the changes that occur. It is also convenient to divide the cycle into phases based on changes that occur in the ovary and in the endometrium. The ovaries are in the *follicular phase* from the first day of menstruation until the day of ovulation. After ovulation, the ovaries are in the *luteal phase* until the first day of menstruation. The cyclic changes that occur in the endometrium are called the menstrual, proliferative, and secretory phases. These will be discussed separately. It should be noted that the time frames used for the following discussion are only averages. Individual cycles may exhibit considerable variance.

Follicular Phase

Menstruation lasts from day 1 to day 4 or 5 of the average cycle. During this time the secretions of ovarian steroid hormones are at their lowest, and the ovaries contain only primary follicles. During the **follicular phase** of the ovaries, which lasts from day 1 to about day 13 of the cycle (this duration is highly variable), some of the primary follicles grow, develop vesicles, and become secondary follicles. Toward the end of the follicular phase, one follicle in one ovary reaches maturity and becomes a graafian follicle. As follicles grow, the granulosa cells secrete an increasing amount of **estradiol** (*es″tră-di′ol*)—the principal estrogen—which reaches its highest concentration in the blood at about day 12 of the cycle, 2 days before ovulation.

The growth of the follicles and the secretion of estradiol are stimulated by, and dependent upon, FSH secreted from the anterior pituitary. The amount of FSH secreted during the early follicular phase is believed to be slightly greater than the amount secreted in the late follicular phase (fig. 29.14), although this can vary from cycle to cycle. FSH stimulates the production of FSH receptors in the granulosa cells, so that the follicles become increasingly sensitive to a given amount of FSH. This increased sensitivity is augmented by estradiol, which also stimulates the production of new FSH receptors in the follicles. As a result, the stimulatory effect of FSH on the follicles increases, despite the fact that FSH levels in the blood do not increase throughout the follicular phase. Toward the end of the follicular phase, FSH and estradiol also stimulate the production of LH receptors in the graafian follicle. This prepares the graafian follicle for the next major event in the cycle.

The rapid rise in estradiol secretion from the granulosa cells during the follicular phase acts on the hypothalamus to increase the frequency of GnRH pulses. In addition, estradiol augments the ability of the pituitary to respond to GnRH with an increase in LH secretion. This stimulatory, or **positive feedback,** effect of estradiol on the pituitary causes an increase in LH secretion in the late follicular phase that culminates in an **LH surge** (fig. 29.14).

The LH surge begins about 24 hours before ovulation and reaches its peak about 16 hours before ovulation. This surge triggers ovulation. Since GnRH stimulates the anterior pituitary to secrete both FSH and LH, there is a simultaneous, smaller surge in FSH secretion. Some investigators believe that this midcycle peak in FSH acts as a stimulus for the development of new follicles for the next month's cycle.

Ovulation

Under the influence of FSH stimulation, the graafian follicle grows so large that it becomes a thin-walled "blister" on the surface of the ovary. The growth of the follicle is accompanied by a rapid increase in the rate of estradiol secretion. This rapid increase in estradiol, in turn, triggers the LH surge at about day 13. Finally, the surge in LH secretion causes the wall of the graafian follicle to rupture at about day 14 (fig. 29.15, *top*). In the course of ovulation a secondary oocyte, arrested at metaphase II of meiosis, is released into the pelvic cavity and is swept by cilia into a uterine tube. The ovulated oocyte is still surrounded by a zona pellucida and corona radiata as it begins its journey to the uterus.

Ovulation occurs, therefore, as a result of the sequential effects of FSH and LH on the ovarian follicles. By means of

Figure 29.14

Hormonal changes during the menstrual cycle. Sample values are indicated for LH, FSH, progesterone, and estradiol during the menstrual cycle. The midcycle peak of LH is used as a reference day.

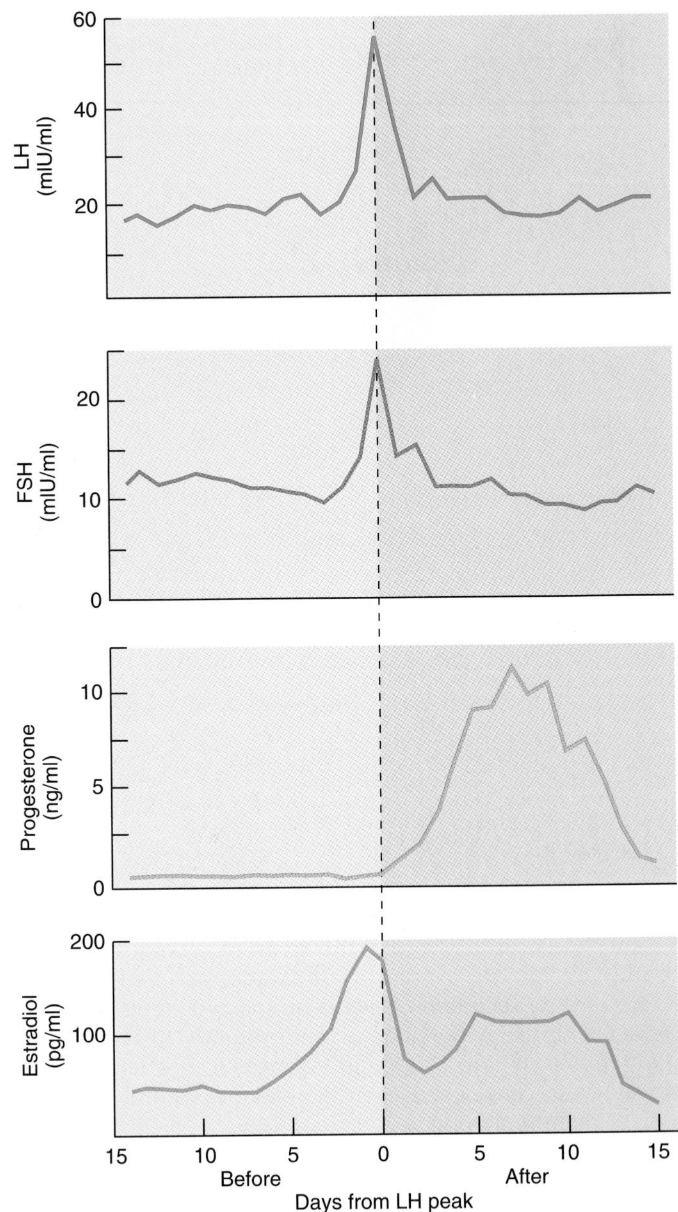

Figure 29.15

The cycle of ovulation and menstruation.
The downward arrows indicate the effects of the hormones.

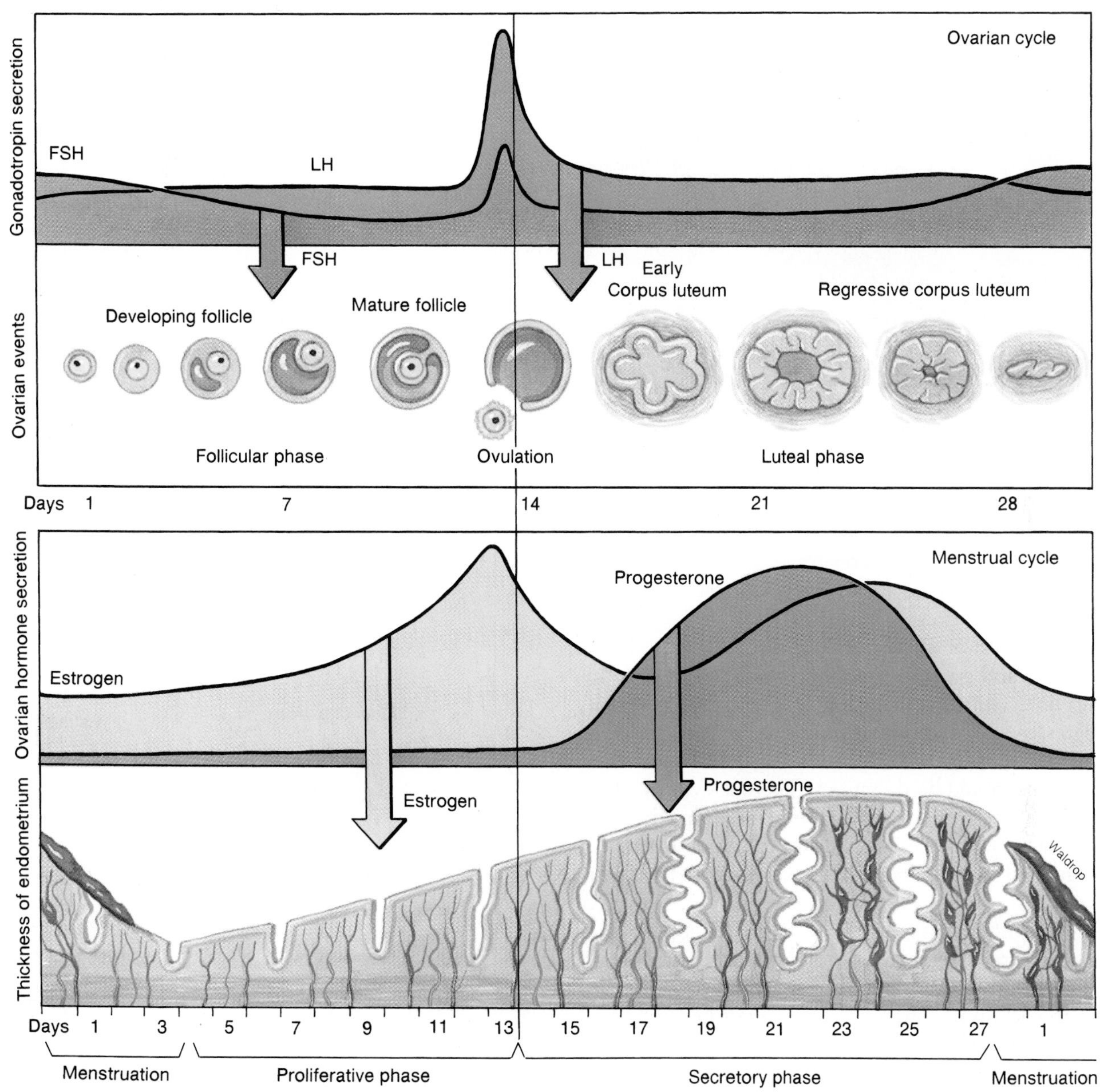

the positive feedback effect of estradiol on LH secretion, the follicle, in a sense, sets the time for its own ovulation. This is because ovulation is triggered by an LH surge, and the LH surge is triggered by increased estradiol secretion that occurs while the follicle grows. In this way, the graafian follicle is not normally ovulated until it has reached the proper size and degree of maturation.

Luteal Phase

After ovulation, the empty follicle is stimulated by LH to become a new structure—the **corpus luteum** (fig. 29.16). This change in structure is accompanied by a change in function. Whereas the developing follicles secrete only estradiol, the corpus luteum secretes both estradiol and *progesterone*. Progesterone levels in the blood are negligible before ovulation but rise rapidly to a peak level during the **luteal phase,** approximately 1 week after ovulation (see figs. 29.14 and 29.15).

The high levels of progesterone combined with estradiol during the luteal phase exert an inhibitory, or **negative feedback,** effect on FSH and LH secretion. There is also evidence that the corpus luteum produces inhibin during the luteal phase, which may help to suppress FSH secretion or action. This serves to retard development of new follicles, so that further ovulation does not normally occur during that cycle. In

Figure 29.16

A corpus luteum in a human ovary.

This structure is formed from the empty vesicular ovarian (graafian) follicle following ovulation.

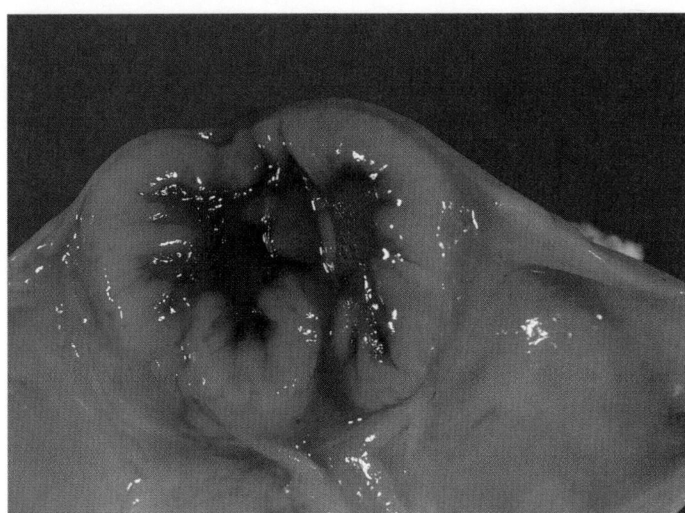

this way, multiple ovulations (and possible pregnancies) on succeeding days of the cycle are prevented.

However, new follicles start to develop toward the end of one cycle in preparation for the next. This may be due to a decreased production of inhibin toward the end of the luteal phase. Estrogen and progesterone levels also fall during the late luteal phase (starting about day 22) because the corpus luteum regresses and stops functioning. In lower mammals, the decline in corpus luteum function is caused by a hormone called *luteolysin* secreted by the uterus. There is evidence that the luteolysin in humans may be prostaglandin $F_{2\alpha}$ (chapters 2 and 19), but the mechanisms of corpus luteum regression in humans is still incompletely understood. Luteolysis (breakdown of the corpus luteum) can be prevented by high levels of LH, but LH levels remain low during the luteal phase as a result of negative feedback exerted by ovarian steroids. In a sense, therefore, the corpus luteum causes its own demise.

With the declining function of the corpus luteum, estrogen and progesterone fall to very low levels by day 28 of the cycle. The withdrawal of ovarian steroids causes menstruation and permits a new cycle of follicle development to progress.

Figure 29.17

The endocrine control of the ovarian cycle.

These changes are shown together with the associated phases of the endometrium during the menstrual cycle.

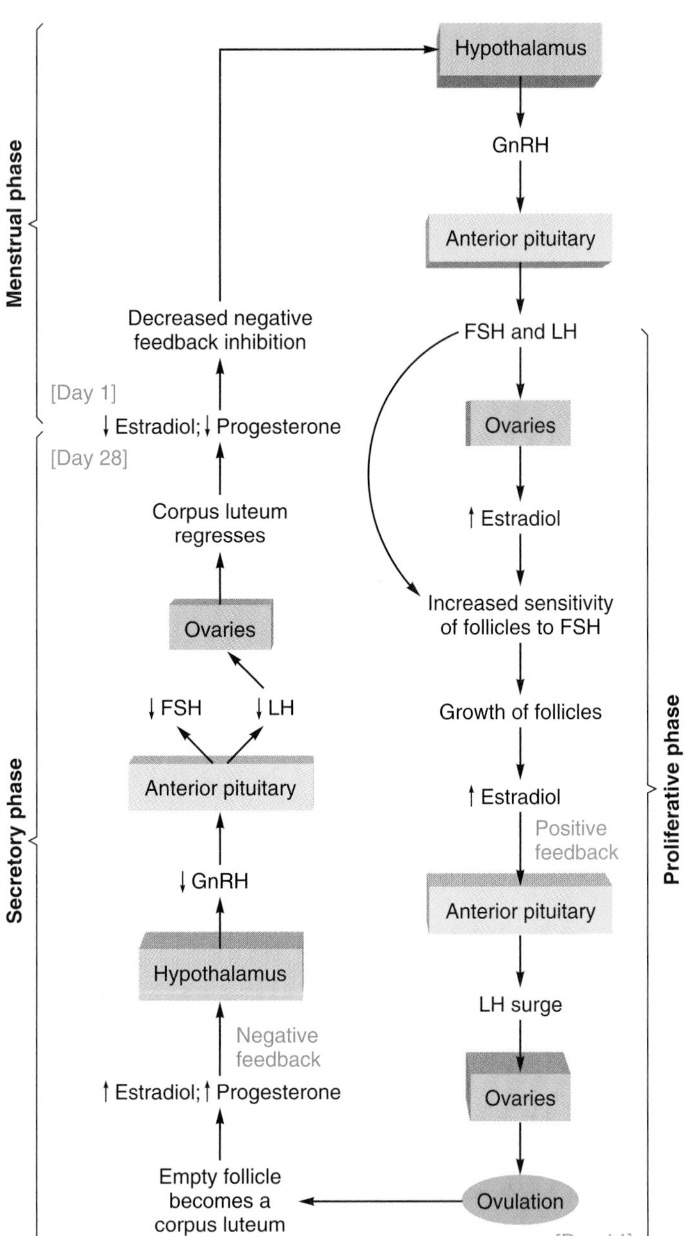

Cyclic Changes in the Endometrium

In addition to a description of the female cycle in terms of ovarian function, the cycle can also be described in terms of the changes in the endometrium of the uterus. These changes occur because the development of the endometrium is timed by the cyclic changes in the secretion of estradiol and progesterone from the ovarian follicles. Three phases can be identified on the basis of changes in the endometrium: (1) the proliferative phase, (2) the secretory phase, and (3) the menstrual phase (fig. 29.15, *bottom*).

The **proliferative phase** of the endometrium occurs while the ovary is in its follicular phase. The increasing amounts of estradiol secreted by the developing follicles stimulate growth (proliferation) of the stratum functionale of the endometrium. In humans and other primates, coiled blood vessels called *spiral arteries* develop in the endometrium during this phase. Estradiol may also stimulate the production of receptor proteins for progesterone at this time, in preparation for the next phase of the cycle.

Table 29.1

Phases of the Menstrual Cycle

Phases of Cycle		Hormonal Changes		Tissue Changes	
Ovarian	Endometrial	Pituitary	Ovarian	Ovarian	Endometrial
Follicular (days 1–4)	Menstrual	FSH and LH secretion low	Estradiol and progesterone remain low	Primary follicles grow	Outer two-thirds of endometrium is shed with accompanying bleeding
Follicular (days 5–13)	Proliferative	FSH slightly higher than LH secretion in early follicular phase	Estradiol secretion rises (due to FSH stimulation of follicles)	Follicles grow; graafian follicles develop (due to FSH stimulation)	Mitotic division increases thickness of endometrium; spiral arteries develop (due to estradiol stimulation)
Ovulatory (day14)	Proliferative	LH surge (and increased FSH) stimulated by positive feedback from estradiol	Estradiol secretion falls	Graafian follicle ruptures, and secondary oocyte is extruded into peritoneal cavity	No change
Luteal (days 15–28)	Secretory	LH and FSH decrease (due to negative feedback from steroids)	Progesterone and estrogen secretion increase, then fall	Development of corpus luteum (due to LH stimulation); regression of corpus luteum	Glandular development in endometrium (due to progesterone stimulation)

The **secretory phase** of the endometrium occurs when the ovary is in its luteal phase. In this phase, increased progesterone secretion stimulates the development of uterine glands. As a result of the combined actions of estradiol and progesterone, the endometrium becomes thick, vascular, and "spongy" in appearance, and the uterine glands become engorged with glycogen during the phase following ovulation. The endometrium is therefore well prepared to accept and nourish an embryo should fertilization occur.

The **menstrual phase** occurs as a result of the fall in ovarian hormone secretion during the late luteal phase. Necrosis (cellular death) and sloughing of the stratum functionale of the endometrium may be produced by constriction of the spiral arteries. It would seem that the spiral arteries are responsible for menstrual bleeding, since lower animals that lack these arteries do not bleed when they shed their endometrium. The phases of the menstrual cycle are summarized in figure 29.17 and in table 29.1.

The cyclic changes in ovarian secretion cause other cyclic changes in the female reproductive tract. High levels of estradiol secretion, for example, cause cornification of the vaginal epithelium (the upper cells die and become filled with keratin). High levels of estradiol also cause the production of a thin, watery cervical mucus that can easily be penetrated by spermatozoa. During the luteal phase of the cycle, the high levels of progesterone cause the cervical mucus to thicken and become sticky after ovulation has occurred.

Contraceptive Methods

Contraceptive Pill

About 10 million women in the United States and 60 million women worldwide currently use **oral contraceptives.** These contraceptives usually consist of a synthetic estrogen combined with a synthetic progesterone in the form of pills that are taken once each day for 3 weeks after the last day of a menstrual period. This procedure causes an immediate increase in blood levels of ovarian steroids (from the pill), which is maintained for the normal duration of a monthly cycle. As a result of *negative feedback inhibition* of gonadotropin secretion, *ovulation never occurs*. The entire cycle is like a false luteal phase, with high levels of progesterone and estrogen and low levels of gonadotropins.

Since the contraceptive pills contain ovarian steroid hormones, the endometrium proliferates and becomes secretory just as it does during a normal cycle. In order to prevent an abnormal growth of the endometrium, women stop taking the steroid pills after 3 weeks (placebo pills are taken during the fourth week). This causes estrogen and progesterone levels to fall, permitting menstruation to occur.

The side effects of earlier versions of the birth-control pill have been reduced through a decrease in the content of estrogen and through the use of newer generations of progestogens (analogues of progesterone). The newer contraceptive pills are very effective and have a number of beneficial side effects, including a reduced risk for endometrial and ovarian cancer, a reduced risk for cardiovascular disease, and a reduction in osteoporosis. However, there may be an increased risk for breast cancer, and possibly cervical cancer, with oral contraceptives. The current consensus is that the health benefits of oral contraceptives generally outweigh the risks, although each case should be evaluated individually.

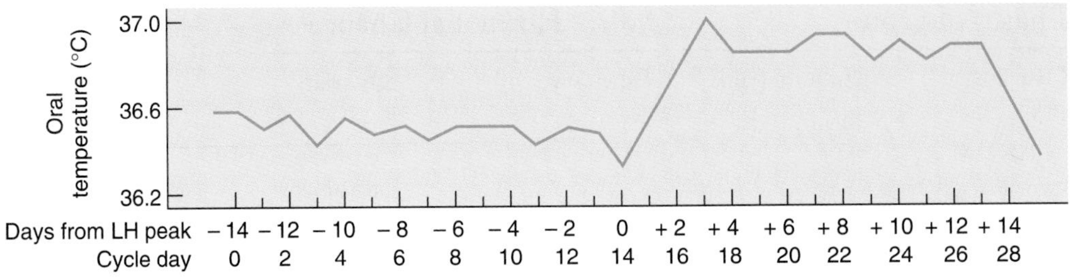

Figure 29.18

Changes in basal temperature during the menstrual cycle. Such changes can be used in the rhythm method of birth control.

Newer systems for delivery of contraceptive steroids are designed so that the steroids are not taken orally, and as a result do not have to pass through the liver before entering the general circulation. (All drugs taken orally pass from the hepatic portal vein to the liver before they are delivered to any other organ, as described chapter 26.) This permits lower doses of hormones to be effective. Such systems include a subcutaneous implant (Norplant), which need only be replaced after 5 years, and vaginal rings, which can be worn for 3 weeks. The long-term safety of these newer methods has not yet been established.

Rhythm Method

Studies have demonstrated that the likelihood of a pregnancy is close to zero if sexual intercourse occurs more than 3 days prior to ovulation, and that the likelihood is very low if it occurs more than a day following ovulation. Conception is most likely to result when intercourse takes place 1 to 2 days prior to ovulation. There is no evidence for differences in the sex ratio of the babies that were conceived at these different times.

Cyclic changes in ovarian hormone secretion also cause cyclic changes in basal body temperature. In the **rhythm method** of birth control, a woman measures her oral basal body temperature upon waking to determine when ovulation has occurred. On the day of the LH peak, when estradiol secretion begins to decline, there is a slight drop in basal body temperature. Starting about 1 day after the LH peak, the basal body temperature rises sharply as a result of progesterone secretion, and it remains elevated throughout the luteal phase of the cycle (fig. 29.18). The day of ovulation for that month's cycle can be accurately determined by this method, making the method useful if conception is desired. Since the day of the cycle on which ovulation occurs is quite variable in many women, however, the rhythm method is not very reliable for contraception by predicting when the next ovulation will occur. The contraceptive pill is a statistically more effective means of birth control.

Other contraceptive methods, including use of IUDs and various barrier methods, are discussed later in this chapter in the section "Clinical Considerations."

Menopause

The term **menopause** literally means "pause in the menses" and refers to the cessation of ovarian activity and menstruation that occurs at about the age of 50. During the postmenopausal years, which account for about a third of a woman's life span, the ovaries are depleted of follicles and stop secreting estradiol and inhibin. The fall in estradiol is due to changes in the ovaries, not in the pituitary; indeed, FSH and LH secretion by the pituitary is elevated because of a lack of negative feedback from estradiol and inhibin. Like prepubertal boys and girls, postmenopausal women only have estrogen that is formed by aromatization of the weak androgen *androstenedione*, which is secreted by the adrenal cortex and converted in the adipose tissue into a weak estrogen called *estrone*.

The withdrawal of estradiol secretion from the ovaries is most responsible for the symptoms of menopause. These include vasomotor disturbances and urogenital atrophy. Vasomotor disturbances produce the "hot flashes" of menopause, where a fall in core body temperature is followed by feelings of heat and profuse perspiration. Atrophy of the urethra, vaginal wall, and vaginal glands occurs, with loss of lubrication. There is also increased risk of atherosclerotic cardiovascular disease and increased progression of osteoporosis (chapter 8). These changes can be reversed to a significant degree by estrogen treatments.

Mammary Glands and Lactation

The structure and function of the mammary glands are dependent upon the action of a number of hormones. The secretion of prolactin and oxytocin is directly required for the production and delivery of milk to a suckling infant.

The **mammary glands,** located in the **breasts,** are modified sweat glands and thus part of the integumentary system (chapter 7). In function, however, these glands are associated with the reproductive system because they secrete milk for nourishment of the young. The size and shape of the breasts vary widely from female to female because of differences in

Figure 29.19

The structure of the breast and mammary glands.
(*a*) A sagittal section and (*b*) an anterior view partially sectioned.

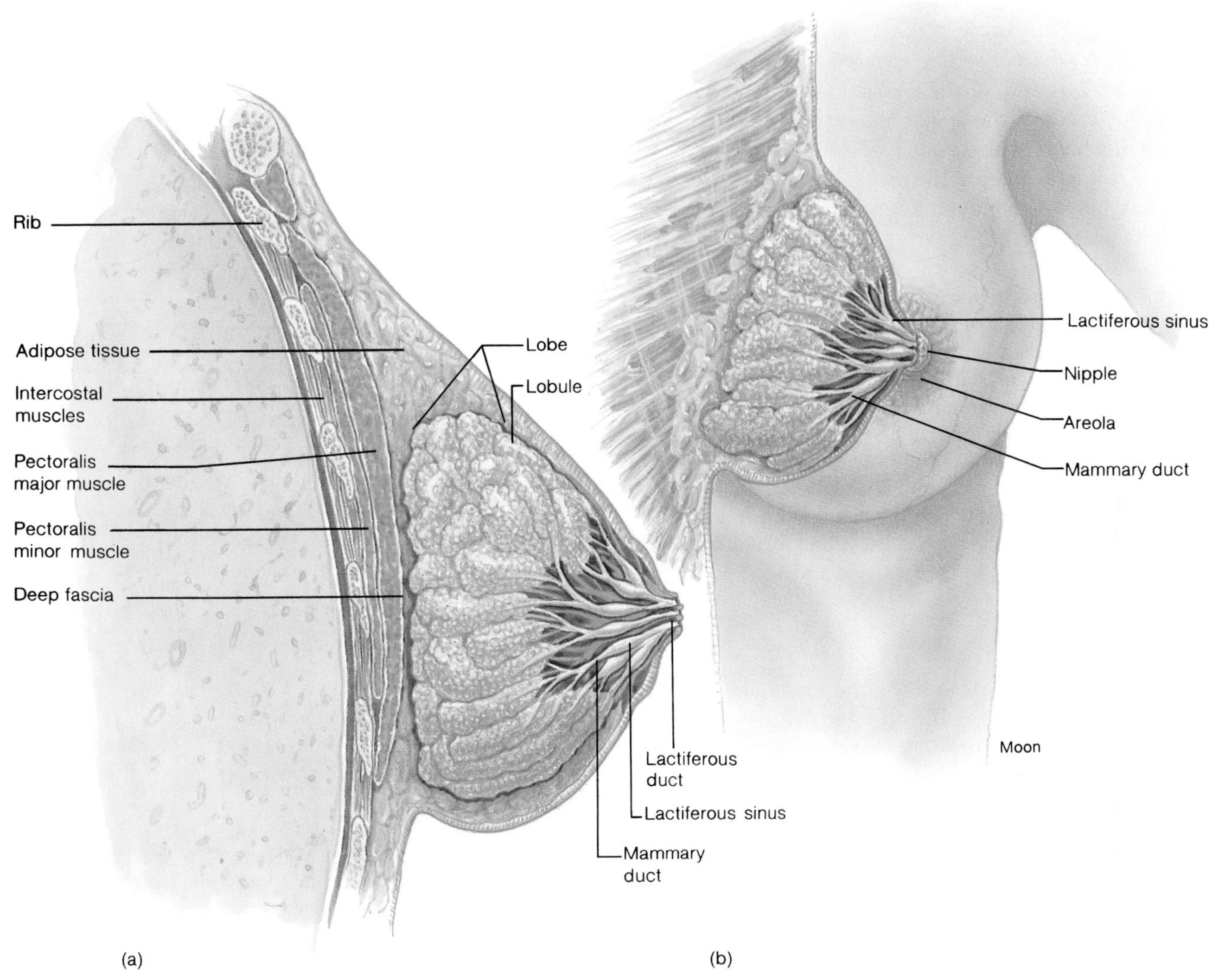

genetic makeup, age, and percentage of body fat. At puberty, estrogen from the ovaries stimulates growth of the mammary glands and the deposition of adipose tissue within the breasts. Mammary glands hypertrophy in pregnant and lactating women and usually undergo some degree of atrophy after menopause.

Structure of the Breasts and Mammary Glands

Each breast is positioned over ribs 2 through 6 and overlies the pectoralis major muscle and portions of the serratus anterior and external oblique muscles (fig. 29.19). The medial boundary of the breast overlies the lateral margin of the sternum, and the lateral margin of the breast follows the anterior border of the axilla. The **axillary process** of the breast extends upward and laterally toward the axilla, where it comes into close relationship with the axillary vessels. This region of the breast is clinically significant because of the high incidence of breast cancer within the lymphatic drainage of the axillary process.

Each mammary gland is composed of 15 to 20 **lobes,** each with its own drainage pathway to the outside. The lobes are separated by varying amounts of adipose tissue. The amount of adipose tissue determines the size and shape of the breast but has nothing to do with the ability of a woman to nurse. Each lobe is subdivided into **lobules.** The lobules contain the **mammary alveoli** that secrete the milk of a lactating female. **Suspensory ligaments** (ligaments of Cooper) between the lobules extend from the skin to the deep fascia overlying the pectoralis muscle and support the breasts. The clustered mammary alveoli secrete milk into a series of **mammary ducts,** which in turn converge to form **lactiferous** (*lak-tif'er-us*) **ducts.** The lumen of each lactiferous duct expands just deep to

ligaments of Cooper: from Sir Astley P. Cooper, English anatomist and surgeon, 1768–1841 the surface of the nipple to form a **lactiferous sinus.** Milk is stored in the lactiferous sinuses before drainning at the top of the nipple.

The **nipple** is a cylindrical projection containing some erectile tissue. A circular pigmented **areola** (*ă-re'o-lă*) surrounds the nipple. The surface of the areola may appear rough because of the presence of sebaceous **areolar glands** close to the surface. The secretions of the areolar glands keep the nipple pliable. The color of the areola and nipple varies according to the complexion of the woman and whether or not she is pregnant. During pregnancy, the areola becomes darker and enlarges somewhat, presumably to become more conspicuous to a nursing infant.

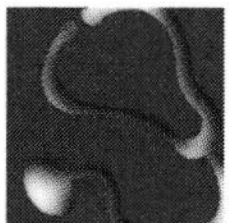

Lymphatic drainage and the location of lymph nodes within the breast are of considerable clinical importance because of the frequency of breast cancer and the high incidence of metastases. About 75% of the lymph drains through the axillary process of the breast into the axillary lymph nodes. Some 20% of the lymph passes toward the sternum to the internal thoracic lymph nodes. The remaining 5% of the lymph is subcutaneous and follows the lymph drainage pathway in the skin toward the back, where it reaches the intercostal nodes near the neck of the ribs.

Lactation

Lactation, the production of milk by the mammary glands, normally occurs in females following parturition and may continue for 2 to 3 years if sucking occurs at regular, frequent intervals. The changes that occur in the mammary glands during pregnancy and the regulation of lactation provide excellent examples of hormonal interactions and neuroendocrine regulation. Growth and development of the mammary glands during pregnancy require the permissive actions of insulin, cortisol, and thyroid hormones. In the presence of these hormones, high levels of progesterone stimulate the development of the mammary alveoli, and estrogen stimulates proliferation of the tubules and ducts (fig. 29.20).

The production of milk proteins, including casein and lactalbumin, is stimulated after parturition by **prolactin,** a hormone secreted by the anterior pituitary. The secretion of prolactin is controlled primarily by *prolactin-inhibiting hormone (PIH)*, which is believed to be dopamine produced by the hypothalamus and secreted into the portal blood vessels. The secretion of PIH is stimulated by high levels of estrogen. In addition, high levels of estrogen act directly on the mammary glands to block their stimulation by prolactin. During pregnancy, consequently, the high levels of estrogen prepare the breasts for lactation but prevent prolactin secretion and action.

After parturition, when the placenta is expelled as the *afterbirth*, (see chapter 30), declining levels of estrogen are accompanied by an increase in the secretion of prolactin. Milk production is therefore stimulated. If a woman does not wish to breast-feed her baby, she may take oral estrogens to inhibit prolactin secretion. A different drug commonly given in these circumstances, and in other conditions in which it is desirable to inhibit prolactin secretion, is *bromocriptine*. This drug binds to dopamine receptors and thus promotes the action of dopamine. The fact that this action inhibits prolactin secretion offers additional evidence that dopamine functions as the prolactin-inhibiting hormone (PIH).

The act of nursing helps to maintain high levels of prolactin secretion via a *neuroendocrine reflex* (fig. 29.21). Sensory endings in the breast, activated by the stimulus of suckling, relay impulses to the hypothalamus and inhibit the secretion of PIH. There is also indirect evidence that the stimulus of suckling may cause the secretion of a *prolactin-releasing hormone*, but this is controversial. Suckling thus results in the reflex secretion of high levels of prolactin that promotes the secretion of milk from the alveoli into the ducts. In order for the baby to get the milk, however, the action of another hormone is needed.

The stimulus of suckling also results in the reflex secretion of **oxytocin** (*ok"sĭ-to'sin*) from the posterior pituitary. This hormone is produced in the hypothalamus and stored in the posterior pituitary; its release results in the **milk-ejection reflex,** or milk letdown. This is because oxytocin stimulates contraction of the lactiferous ducts as well as that of the uterus. The effects of hormones on lactation are summarized in table 29.2.

Figure 29.20

Hormonal control of mammary gland development and lactation.

Notice that milk production is prevented during pregnancy by estrogen inhibition of prolactin secretion. This inhibition is accomplished by the stimulation of PIH (prolactin-inhibiting hormone) secretion from the hypothalamus.

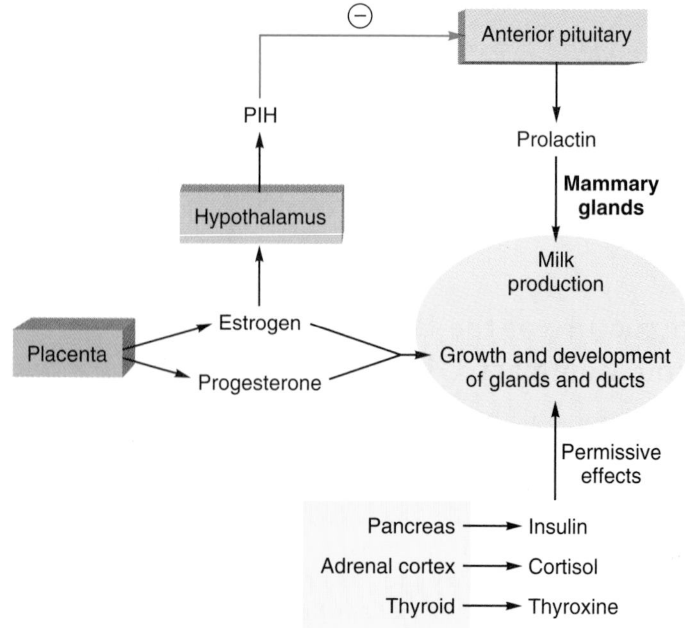

Infants who are breast-fed develop fewer infections than those who are bottle-fed on formula. This is because breast milk contains immunoglobulin A (IgA), in a form protected from digestive enzymes, gamma interferon, lysozyme, neutrophils and macrophages. The United Nations health organizations UNICEF and WHO recommend that babies be breast-fed until they are at least 2 years old.

Breast-feeding, acting through reflex inhibition of GnRH secretion, can also inhibit the secretion of gonadotropins from the mother's anterior pituitary and thus inhibit ovulation. Breast-feeding is thus a natural contraceptive mechanism that helps to space births. This mechanism appears to be most effective in women with limited caloric intake and in those who breast-feed their babies at frequent intervals throughout the day and night. In the traditional societies of the less industrialized countries, therefore, breast-feeding is an effective contraceptive. Breast-feeding has much less of a contraceptive effect in women who are well nourished and who breast-feed their babies at more widely spaced intervals.

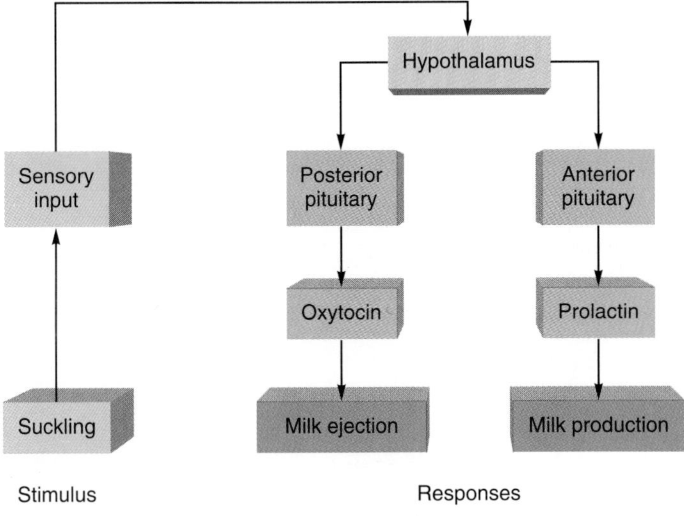

Figure 29.21

Milk production and the milk ejection reflex.
Lactation occurs in two stages: milk production (stimulated by prolactin) and milk ejection (stimulated by oxytocin). The stimulus of suckling triggers a neuroendocrine reflex that results in increased secretion of oxytocin and prolactin.

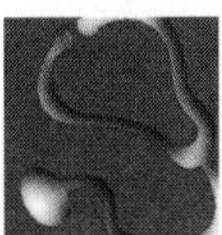

Milk letdown can become a conditioned reflex made in response to visual or auditory cues; the crying of a baby can elicit oxytocin secretion and the milk-ejection reflex. On the other hand, this reflex can be suppressed by the adrenergic effects produced in the fight-or-flight reaction. Thus, if a woman becomes nervous and anxious while breast-feeding, she will produce milk, but it will not flow (there will be no milk letdown). This can cause increased pressure, intensifying her anxiety and frustration and further inhibiting the milk-ejection reflex. It is therefore important for mothers to be in a quiet and calm environment while they nurse their babies. If needed, synthetic oxytocin can be given as a nasal spray to promote milk letdown.

Clinical Considerations

Females are more prone to dysfunctions and diseases of the reproductive organs than males because of cyclic changes in reproductive events, problems associated with pregnancy, and the susceptibility of the female breasts to infections and neoplasms. The termination of reproductive capabilities at menopause can also cause complications as a result of hormonal alterations. *Gynecology* (*gi″nĕ-kol′ŏ-je*) is the specialty of medicine concerned with dysfunction and diseases of the female reproductive system, whereas *obstetrics* is the specialty dealing with pregnancy and childbirth. Frequently a physician will specialize in both obstetrics and gynecology (OB-GYN).

A comprehensive discussion of the numerous clinical aspects of the female reproductive system is beyond the scope of this text. Only the most important conditions are discussed in the following sections, along with descriptions of the more popular methods of birth control.

Table 29.2

Hormonal Factors Affecting Lactation

Hormones	Major Source	Effects
Insulin, cortisol, thyroid hormones	Pancreas, adrenal cortex, and thyroid gland	Permissive effects—adequate amounts of these must be present for other hormones to exert their effects on mammary glands
Estrogen and progesterone	Placenta	Growth and development of secretory units (mammary alveoli) and ducts in mammary glands
Prolactin	Anterior pituitary	Production of milk proteins, including casein and lactalbumin
Oxytocin	Posterior pituitary	Stimulation of milk-ejection reflex

UNDER DEVELOPMENT

Development of the Female Reproductive System

Although genetic sex is determined at fertilization (XX for females and XY for males), both sexes develop similarly through the indifferent stage of the eighth week. The gonads of both sexes develop from gonadal ridges, and the genital tubercle develops during the sixth week as an external swelling.

The ovaries develop more slowly than the testes. Ovarian development begins at about the tenth week when **primordial follicles** begin to form within the medulla of the gonads. Each of the primordial follicles consists of an **oogonium** (*o″ŏ-go′ne-um*) surrounded by a layer of follicular cells. Mitosis of the oogonia occurs during fetal development, so that thousands of germ cells form. Unlike the male reproductive system, in which spermatogonia form by mitosis throughout life, all oogonia form prenatally, and their number continuously decreases after birth.

The uterus and uterine tubes develop from a pair of embryonic tubes called the **paramesonephric** (müllerian) **ducts,** which are so-called because they are located to the sides of the mesonephric ducts (which form temporary embryonic kidneys). As the mesonephric kidneys degenerate (chapter 25), the lower portions of the paramesonephric ducts fuse to form the uterus, and the upper portions give rise to the uterine tubes. As described in chapter 28, if the embryo is male, secretion of müllerian inhibition factor from the testes causes the paramesonephric ducts to degenerate.

The epithelial lining of the vagina develops from the endoderm of the urogenital sinus. The formation of a thin membrane called the **hymen** separates the lumen of the vagina from the urethral sinus. The hymen usually is perforated during later fetal development.

oogonium: Gk. *oion*, egg; *gonos*, procreation
hymen: Gk. (mythology) *Hymen*, god of marriage

The external genitalia of both sexes appear the same during the indifferent stage of the eighth week. A prominent **phallus** (*fal′us*) forms from the genital tubercle, and a **urethral groove** forms on the ventral side of the phallus. As described in chapter 28, the phallus becomes the penis in a male and the smaller clitoris in a female. Paired **urethral folds** surround the urethral groove on the lateral sides. In a male, these fuse to form the urethra of the penis; in a female, the urethral folds remain unfused and form the inner labia minora. Similarly, the **labioscrotal swellings** in a male fuse to form the scrotum; in a female, these remain unfused and form the prominent labia majora. The male and female structures that share a common embryological origin are said to be homologous structures (table 1).

Table 1
Homologous Reproductive Organs and the Undifferentiated Structures from Which They Develop

Indifferent Stage	Male	Female
Gonads	Testes	Ovaries
Urogenital groove	Membranous urethra	Vestibule
Genital tubercle	Glans penis	Clitoris
Urethral folds	Spongy urethra	Labia minora
Labioscrotal swelling	Scrotum Bulbourethral glands	Labia majora Vestibular glands

Diagnostic Procedures

A gynecological, or pelvic, examination is generally given in a thorough physical examination, especially prior to marriage, during pregnancy, or if problems involving the reproductive organs are suspected. In a gynecological examination the physician inspects the vulva for irritations, lesions, or abnormal vaginal discharge and palpates the vulva and internal organs. Most of the internal organs can be palpated through the vagina, especially if they are enlarged or tender. Inserting a lubricated *speculum* into the vagina permits visual examination of the cervix and vaginal walls. A speculum is an instrument for opening or distending a body opening to permit visual inspection.

In special cases, it may be necessary to examine the cavities of the uterus and uterine tubes by **hysterosalpingography** (*his″ter-o-sal″ping-gog′ră-fe*) (fig. 29.22). This technique involves injecting a radiopaque dye into the reproductive tract. The patency of the uterine tubes, irregular pregnancies, and various types of tumors may be detected using this technique. A **laparoscopy** (*lap″ă-ros′kŏ-pe*) permits in vivo visualization of the internal reproductive organs. The laparoscope may be inserted via the umbilicus, a small incision in the lower abdominal wall, or through the posterior fornix of the vagina

hysterosalpingography: Gk. *hystra*, uterus; *salpinx*, trumpet (uterine tube); *graphein*, to record

Figure 29.22

A hysterosalpingogram.
The photograph reveals the cavity of the uterus and lumina of the uterine tubes.

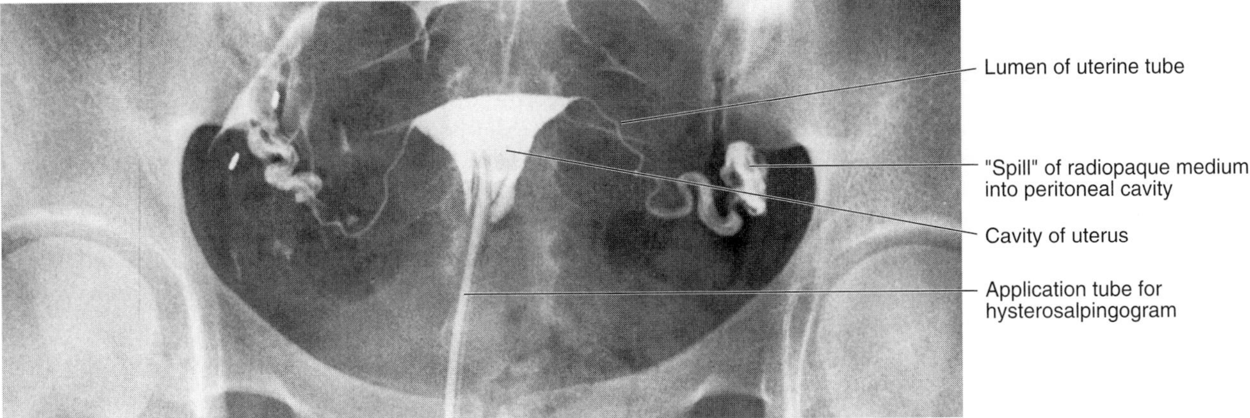

Figure 29.23

Tubal ligation.
This surgical procedure involves removal of a portion of each uterine tube.

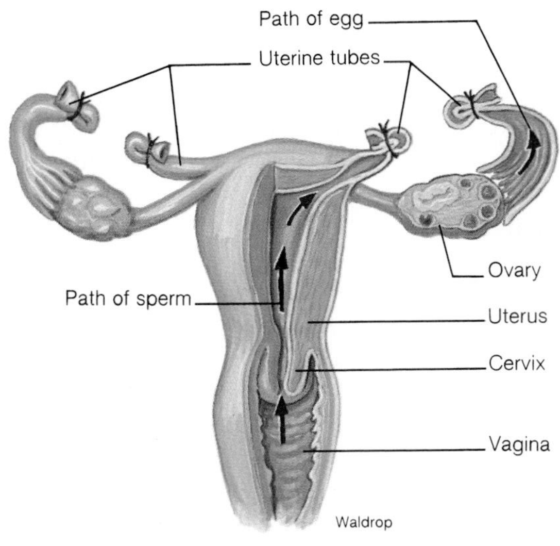

into the rectouterine pouch. Although a laparoscope is used primarily in diagnosis, it also can be used in performing a **tubal ligation** (fig. 29.23), a method of female sterilization in which the uterine tubes are surgically tied.

One diagnostic procedure performed routinely by many women is **breast self-examination** (BSE) (see page 950). The importance of a BSE is not to prevent diseases of the breast but to detect any problems before they become serious. A BSE should be performed monthly, 1 week after the cessation of menstruation so that the breast will not be swollen or especially tender.

Another important diagnostic procedure is a **Papanicolaou (Pap) smear.** The Pap smear permits microscopic examination of cells covering the tip of the cervix. Samples of cells are obtained by gently scraping the surface of the cervix with a specially designed wooden spatula. Women should have routine Pap smears for the early detection of cervical cancer.

Problems Involving the Ovaries and Uterine Tubes

Most **ovarian neoplasms** are nonmalignant ovarian cysts lined by cuboidal epithelium and filled with a serous albuminous fluid. These tumors often may be palpated during a gynecological examination and may require surgical removal if they exceed about 4 cm (1.5 in.) in diameter. They are generally removed as a precaution because it is impossible to determine by palpation whether the mass is malignant or benign.

Ovarian tumors, which occur most often in women over the age of 60, can grow to be massive. Ovarian tumors as heavy as 5 kg (14 lb) are not uncommon, and some weighing as much as 110 kg (300 lb) have been reported. Some ovarian tumors produce estrogen and thus cause feminization in elderly women, including the resumption of menstrual periods. The prognosis for women with ovarian tumors varies depending on the type of tumor, whether or not it is malignant, and if it is malignant, the stage of the cancer.

Two frequent problems involving the uterine tubes are salpingitis and ectopic pregnancies. **Salpingitis** is an inflammation of one or both uterine tubes. Infection of the uterine

Pap smear: from George N. Papanicolaou, American anatomist and physician, 1883–1962
neoplasm: Gk. *neos*, new; *plasma*, something formed

tubes is generally caused by a sexually transmitted disease, although secondary bacterial infections from the vagina may also cause salpingitis. Salpingitis may cause sterility if the uterine tubes become occluded.

Ectopic pregnancy results from implantation of the blastocyst in a location other than the body or fundus of the uterus. The most frequent ectopic site is in the uterine tube, where an implanted blastocyst causes what is commonly called a **tubular pregnancy** (see fig. 30.29). One danger of a tubular pregnancy is the enlargement, rupture, and subsequent hemorrhage of the uterine tube where implantation has occurred. A tubular pregnancy is frequently treated by removing the affected tube.

Infertility, or the inability to conceive, is a clinical problem that may involve the male or female reproductive system. Based on the number of people who seek help for this problem, it is estimated that 10% to 15% of couples have impaired fertility. Generally, when a male is infertile, it is because of inadequate sperm counts. Female infertility is frequently caused by an obstruction of the uterine tubes or abnormal ovulation.

Polycystic ovarian syndrome is characterized by chronic anovulation (lack of ovulation) combined with *hirsutism* (hairiness). The ovulation failure is due to hormone disturbances that originate elsewhere. The hirsutism is produced by an abnormally high secretion of androgens from the ovary. Progesterone therapy or birth control pills are generally used for treatment. Fertility drugs (e.g., Clomid) may also be used to stimulate ovulation in women with polycystic disease who want to become pregnant.

Problems Involving the Uterus

Abnormal menstruations are among the most common disorders of the female reproductive system. Abnormal menstruations may be directly related to problems of the reproductive organs and pituitary gland or associated with emotional and psychological stress.

Amenorrhea (*a-men-ŏ-re'ă*) is the absence of menstruation and can be categorized as normal, primary, or secondary. **Normal amenorrhea** follows menopause, occurs during pregnancy, and in some women may occur during lactation. **Primary amenorrhea** is the failure to have menstruated by the age at which menstruation normally begins. Primary amenorrhea is generally accompanied by lack of development of the secondary sex characteristics. Endocrine disorders may cause primary amenorrhea and abnormal development of the ovaries or uterus.

Secondary amenorrhea is the cessation of menstruation in women who previously have had normal menstrual periods and who are not pregnant and have not gone through menopause. Various endocrine disturbances, as well as psychological factors, may cause secondary amenorrhea. It is not uncommon, for example, for young women who are in the process of making major changes or adjustments in their lives to miss menstrual periods. Secondary amenorrhea is also frequent in female athletes during periods of intense training. A low percentage of body fat may be a contributing factor. Sickness, fatigue, poor nutrition, or emotional stress may also cause secondary amenorrhea.

Dysmenorrhea is painful or difficult menstruation that may be accompanied by severe menstrual cramps. The causes of dysmenorrhea are not totally understood but may include endocrine disturbances (inadequate progesterone levels), an abnormally positioned uterus, emotional stress, or some type of obstruction that prohibits menstrual discharge.

Abnormal uterine bleeding includes **menorrhagia** (*men"ŏ-ra'je-ă*), excessive bleeding during the menstrual period, and **metrorrhagia,** bleeding not associated with menstruation that occurs at irregular intervals. Other types of abnormal uterine bleeding are menstruations of excessive duration, menstruations that are too frequent, and postmenopausal bleeding. These abnormalities may be caused by hormonal irregularities, emotional factors, or various diseases and physical conditions.

Uterine neoplasms are an extremely common problem of the female reproductive tract. They include cysts, polyps, and smooth muscle tumors (leiomyomas), and most of them are benign. Any of these conditions may provoke irregular menstruations and may cause infertility if the neoplasms are massive.

Cancer of the uterus is the most common malignancy of the female reproductive tract. The most common site of uterine cancer is the cervix (fig. 29.24). Cervical cancer, second only to cancer of the breast in frequency of occurrence, is a disease of young women (ages 30 through 50), especially those who have had frequent intercourse with multiple partners during their teens and onward. If detected early through regular Pap smears, the disease can be cured before it metastasizes. The treatment of cervical cancer depends on the stage of the malignancy and the age and health of the woman. In the case of women for whom future fertility is not an issue, a **hysterectomy** (*his"tĕ-rek'tŏ-me*) (surgical removal of the uterus) is usually performed.

Endometriosis (*en"do-me"tre-o'sis*) is a condition characterized by the presence of endometrial tissue at sites other than the inner lining of the uterus. Frequent sites of ectopic endometrial cells are on the ovaries, on the outer layer of the uterus, on the abdominal wall, and on the urinary bladder. Although it is not certain how endometrial cells become established outside the uterus, it is speculated that some discharged endometrial tissue might be flushed backward from the uterus and through the uterine tubes during menstruation. Women with endometriosis will bleed internally with each menstrual period because the ectopic endometrial cells are stimulated along with the normal endometrium by ovarian hormones. The most common symptoms of endometriosis are extreme dysmenorrhea and a feeling of fullness during each menstrual period. Endometriosis can cause infertility. It is most often treated by suppressing the endometrial tissues with oral contraceptive pills or by surgery. An **ovariectomy** (*o-var"e-ek'tŏ-me*), or removal of the ovaries, may be necessary in extreme cases.

Figure 29.24

Sites of conditions and diseases of the female reproductive tract.
Each condition or disease could cause an abnormal discharge of blood.

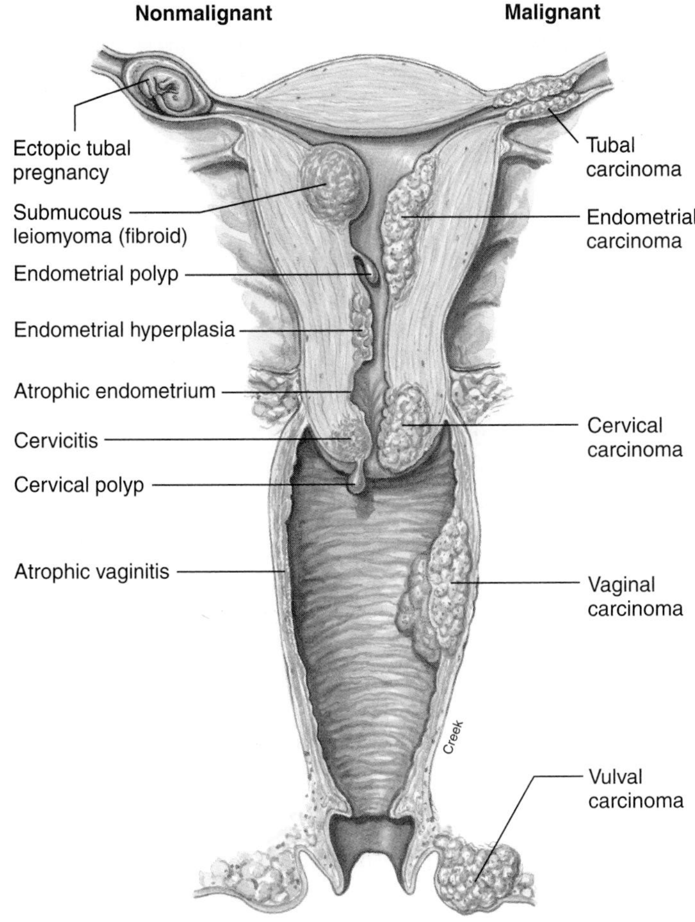

Diseases of the Vagina and Vulva

Pelvic inflammatory disease (PID) is a general term for inflammation of the female reproductive organs within the pelvis. The infection may be confined to a single organ, or it may involve all the internal reproductive organs. The pathogens generally enter through the vagina during coitus, induced abortion, or childbirth.

The vagina and vulva are generally resistant to infection because of the acidity of the vaginal secretions. Occasionally, however, localized infections and inflammations do occur; these are termed **vaginitis,** if confined to the vagina, or **vulvovaginitis,** if both the vagina and external genitalia are affected. The symptoms of vaginitis are a discharge of pus (*leukorrhea*) and itching (*pruritus*). The two most common organisms that cause vaginitis are the protozoan *Trichomonas vaginalis* and the fungus *Candida albicans*.

Diseases of the Breasts and Mammary Glands

The breasts and mammary glands of females are highly susceptible to infections, cysts, and tumors. Infections involving the mammary glands usually follow the development of a dry and cracked nipple during lactation. Bacteria enter the wound and establish an infection within the lobules of the gland. During an infection of the mammary gland, a blocked duct frequently causes a lobe to become engorged with milk. This localized swelling is usually accompanied by redness, pain, and fever. Administering specific antibiotics and applying heat are the usual treatments.

Nonmalignant cysts are the most frequent diseases of the breast. These masses are generally of two types, neither of which is life threatening. **Dysplasia,** or **fibrocystic disease,** is an inclusive designation referring to several nonmalignant diseases of the breast. All dysplasias are benign neoplasms of various sizes that may become painful during or prior to menstruation. Most of the masses are small and remain undetected. Dysplasia affects nearly 50% of women over the age of 30 prior to menopause.

A **fibroadenoma** (*fi″bro-ad-ĕ-no′-mă*) is a benign tumor of the breast that frequently occurs in women under the age of 35. Fibroadenomas are nontender rubbery masses that are easily moved about in the mammary tissue. A fibroadenoma can be excised in a physician's office under local anesthetics.

Carcinoma of the breast is the most common malignancy in women. One in nine women will develop breast cancer, and one-third of these will die from the disease. Breast cancer is the leading cause of death in women between 40 and 50 years of age. Men are also susceptible to breast cancer, but it is 100 times more frequent in women. Breast cancer in men is usually fatal.

The causes of breast cancer are not known, but women who are most susceptible are those who are over age 35, who have a family history of breast cancer, and who are nulliparous (never having given birth). The early detection of breast cancer is important because the prognosis worsens as the disease progresses.

One of the techniques geared specifically toward the early detection of breast cancer is *mammography* (fig. 29.25). A *mammogram* is a soft tissue X ray of the breast that shows the location and extent of any suspicious lesions. If the mammogram indicates possible breast cancer, the physician can obtain a biopsy (tissue sample) to be sent to a pathologist for microscopic examination. At present, biopsy analysis is considered the only definite way to determine whether a growth is cancerous.

The surgical treatment for breast cancer is usually some degree of *mastectomy* (*mas-tek′tŏ-me*), depending on the size of the tumor and whether or not metastasis has occurred. A simple *mastectomy* is removal of the entire breast but not the underlying lymph nodes. A *modified radical mastectomy* is the complete removal of the breast, the lymphatic drainage, and

Figure 29.25

Mammography.
(*a*) A mammogram of a patient with carcinoma of the upper breast. (Note the presence of a neoplasm indicated with an arrow.) (*b*) In mammography, the breasts are placed alternately on a metal plate and radiographed from the side and from above.

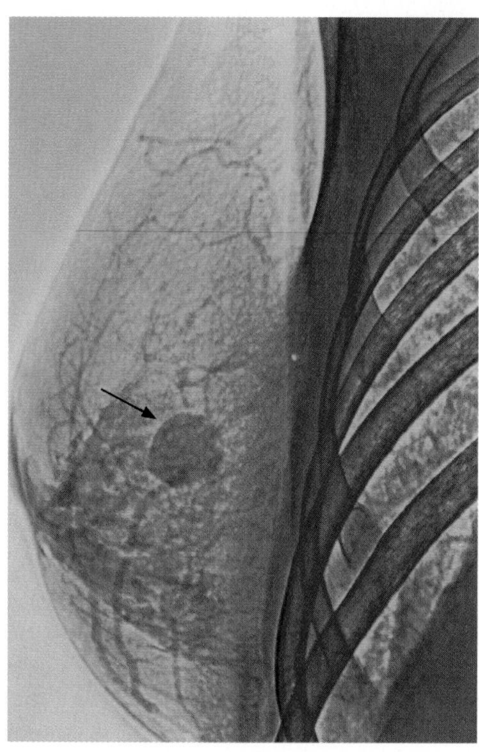

(a)

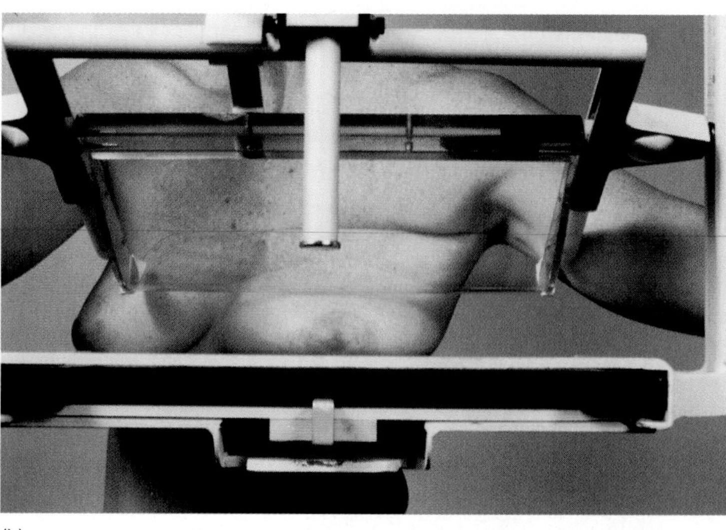

(b)

perhaps the pectoralis major muscle. A *radical mastectomy* is similar to a modified except that the pectoralis major muscle is always removed, as well as the axillary lymph nodes and adjacent connective tissue. A *lumpectomy*—removal of just the lump and a small amount of surrounding breast tissue—is currently used as an option for some small malignancies that are in the beginning stages.

Other Methods of Contraception

In addition to the rhythm method and the contraceptive (birth-control) pill, which have already been discussed, contraception may be accomplished by sterilization, intrauterine devices (IUDs), and barrier methods—including condoms, diaphragms, and spermicidal gels and foams (fig. 29.26).

Sterilization techniques include *vasectomy* for the male and *tubal ligation* for the female. In the latter technique (which currently accounts for over 60% of sterilization procedures performed in the United States), the uterine tubes are cut and tied. This is analogous to the procedure performed on the ductus deferens in a vasectomy and prevents fertilization of the ovulated ovum. Studies on the long-term effects of these procedures have failed to show deleterious side effects. With current procedures, tubal ligations (as well as vasectomies) should be considered essentially irreversible.

Intrauterine devices (IUDs) don't prevent ovulation, but instead prevent implantation of the blastocyst in the uterine wall should fertilization occur. The mechanisms by which their contraceptive effects are produced are not well understood, but the efficiency of different IUDs appears to be related to their ability to cause inflammatory reactions in the uterus. Uterine perforations are the foremost complication associated with the use of IUDs.

Barrier methods of birth control—physical or chemical barriers to keep sperm and ova apart—generally are quite effective for careful users and do not pose serious health risks. The failure rate for condoms—one of the oldest methods of contraception—is 12 to 20 pregnancies per 100 woman years of use, whereas the failure rate for the diaphragm is 12 to 18 pregnancies per 100 woman years. Latex condoms offer an additional benefit: they provide some protection against sexually transmitted diseases, including AIDS.

As an alternative to contraceptive pills, hormonal contraceptives may be delivered to a woman's body by means of **subdermal implants.** Implants are matchstick-sized rods filled with a synthetic progesterone (progestin); they are implanted just under the skin, usually on the upper arm, through a tiny incision. The hormone gradually leaches out through the walls of the rod and enters the bloodstream, preventing pregnancy for at least 5 years. Although their failure rate is less than 1%, up to 7% of women discontinue use because of side effects ranging from bleeding problems to depression.

Figure 29.26

Various types of birth control devices.
(*a*) IUD, (*b*) diaphragm, (*c*) birth control pills, (*d*) vaginal spermicide, (*e*) condom, (*f*) female condom, and (*g*) subdermal implants.

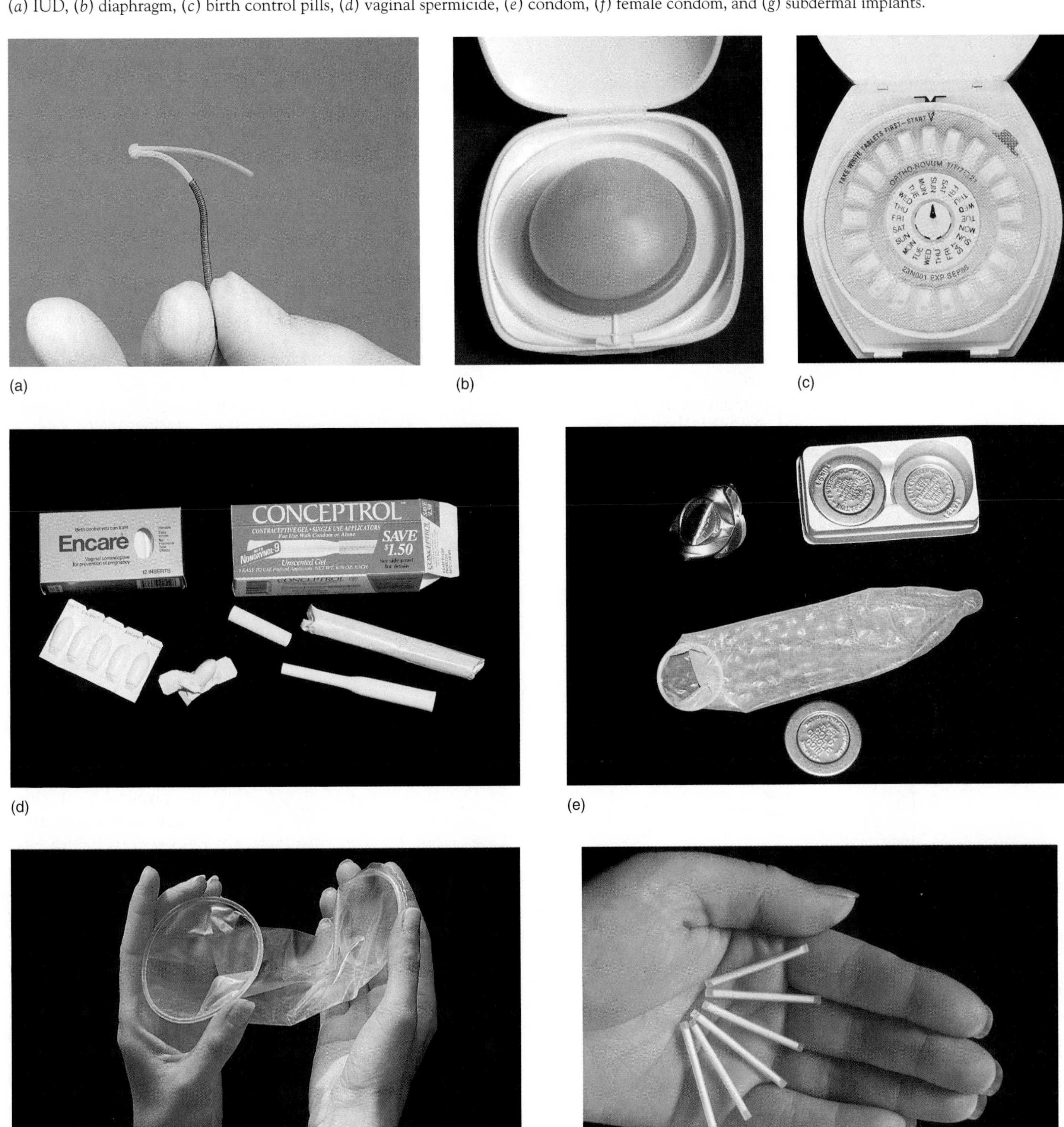

Breast Self-Examination (BSE)

One in nine women will develop breast cancer during her lifetime. Early detection of breast cancer and follow-up medical treatment minimizes the necessary surgical treatment and improves the patient's prognosis. Breast cancer is curable if it is caught early.

Many physicians recommend that women examine their breasts monthly. If a woman has not yet reached menopause, the ideal time for a BSE is 1 week after her period ends because the breasts are less likely to be swollen and tender at that time. A woman no longer menstruating should just pick any day of the month and do a BSE on that same day on a monthly basis. Visual inspection and palpation are equally important in doing a BSE. The steps involved in this procedure are the following:

1. **Observation before a mirror.** Inspect the breasts with the arms at the sides. Next, raise the arms high overhead. Look for any changes in the contour of each breast—a swelling, dimpling of skin, or changes in the nipple. The left and right breast will not exactly match—few women's breasts do. Finally, squeeze the nipple of each breast gently between the thumb and index finger. Any discharge from the nipple should be reported to a physician.
2. **Palpation during bathing.** Examine the breasts during a bath or shower when the hands will glide easily over wet skin. With the fingers flat, move gently over every part of each breast. Use the right hand to examine the left breast, and the left hand for the right breast. Palpate for any lump, hard knot, or thickening. If the breasts are normally fibrous or lumpy (fibrocystic tissue), the locations of these lumps should be noted and checked each month for changes in size and locations.
3. **Palpation while lying down.** To examine the right breast, put a pillow or folded towel under the right shoulder. Place the right hand behind the head—this distributes the breast tissue more evenly on the rib cage. With the fingers of the left hand held flat, press in small circular motions around an imaginary clock face. Begin at the outermost top of the right breast for 12 o'clock, then move to 1 o'clock, and so on around the circle back to 12 o'clock. A ridge of firm tissue in the lower curve of each breast is normal. Then move in an inch, toward the nipple, and keep circling to examine every part of the breast, including the nipple. Also examine the armpit carefully for enlarged lymph nodes. Repeat the procedure on the left breast.

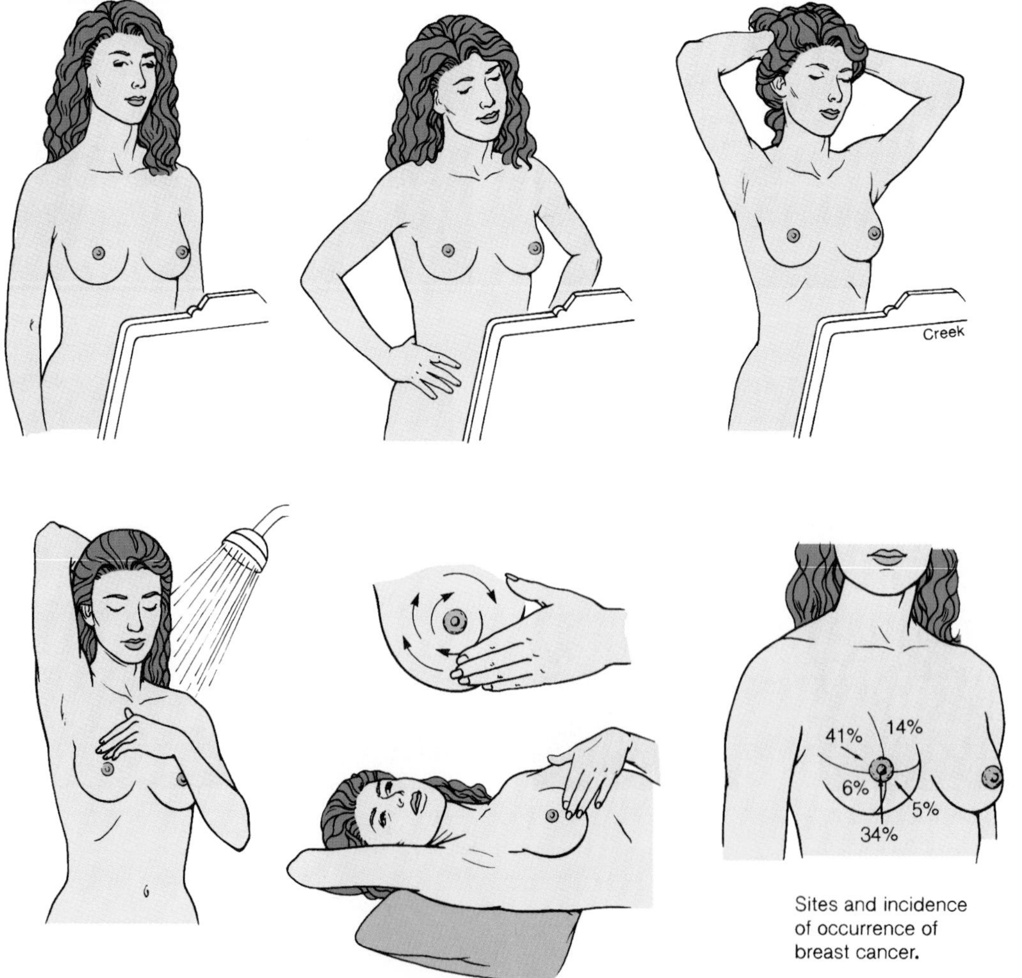

Sites and incidence of occurrence of breast cancer.

Clinical Investigation Answer

The young woman's lack of menstruation was not accompanied by pain, and she did not have a history of spotting or excessive menstrual bleeding. The fact that she had menstruated prior to her amenorrhea ruled out the possibility of primary amenorrhea. Her secondary amenorrhea could have been the result of pregnancy, but this was ruled out by the negative pregnancy test. The amenorrhea could have been caused by her hypothyroidism, but she stated that she took her thyroid pills regularly, and her blood test demonstrated normal thyroxine levels. Blood levels of estradiol and the gonadotropins were normal, suggesting that the amenorrhea was not secondary to a pituitary or ovarian neoplasm or other such problem.

This student most likely has a secondary amenorrhea that is due to emotional stress, low body weight, and/or her strenuous exercise program. She should take steps to alleviate these conditions if she wants to resume her normal menstrual periods. If she refuses to gain weight and reduce her level of physical activity, her physician might recommend the use of oral contraceptives to help regulate her cycles.

Chapter Summary

Structures and Functions of the Female Reproductive System (pp. 925–931)

1. The reproductive period of a female is the period between puberty (about age 12) and menopause (about age 50). In the course of this age span, cyclic ovulation and menstruation patterns occur in nonpregnant females.
2. The functions of the female reproductive system are to produce ova; secrete sex hormones; receive sperm from the male; provide sites for fertilization, implantation, and development of the embryo and fetus; facilitate parturition; and secrete milk from the mammary glands.
3. The female reproductive system consists of (*a*) primary sex organ—the ovaries; (*b*) secondary sex organs—those that are essential for sexual reproduction; and (*c*) secondary sex characteristics—features that are sexual attractants, expressed after puberty.
4. The female secondary sex organs—those that are essential for sexual reproduction—include the vagina, uterine tubes, uterus, and mammary glands.
5. The uterine (fallopian) tubes end in fimbriae, which project over the ovary and help to direct a secondary oocyte into the infundibulum of the tube.
6. The uterus consists of a fundus, body, and cervix, and is supported by four pairs of ligaments.
   (a) The wall of the uterus consists of a perimetrium, a muscular myometrium, and an epithelial lining called the endometrium.
   (b) The endometrium is stratified, with the layers divided into a stratum basale and a stratum functionale; the latter is shed during menstruation and rebuilt during the next cycle.
7. The vagina opens to the cervix of the uterus. The structures of the vulva (external genitalia) include the mons pubis, labia majora, labia minora, clitoris, vaginal vestibule, vestibular bulbs, and vestibular glands.
8. During sexual excitement, the clitoris and erectile tissue of the areola of the breasts swell with blood. During orgasm, the muscles of the perineum, uterus, and uterine tubes contract rhythmically.

Ovaries and the Ovarian Cycle (pp. 931–934)

1. The ovaries are supported by the mesovarium, which extends from the broad ligament, and by the ovarian and suspensory ligaments.
2. Primary oocytes, arrested at prophase I of the first meiotic division, are contained within primary follicles in the ovary.
   (a) Upon stimulation by gonadotropic hormones, granulosa cells divide and fill the follicle, forming a primary follicle.
   (b) Upon further stimulation, a fluid-filled cavity called the antrum begins to form to produce a secondary follicle.
   (c) As a secondary follicle develops, the primary oocyte completes its first meiotic division to form a secondary oocyte, arrested at metaphase II, and a polar body.
   (d) A single, fully mature follicle called the vesicular ovarian, or graafian, follicle releases its secondary oocyte at ovulation; the empty follicle then becomes a corpus luteum.
3. The pituitary gland secretes FSH and LH in response to the secretion of GnRH from the hypothalamus; secretion of FSH and LH, as well as GnRH, is modified by feedback from sex steroids secreted by the ovaries.

Menstrual Cycle (pp. 935–940)

1. The ovarian cycle can be divided into follicular and luteal phases, which are delimited by the event of ovulation.
   (a) During the follicular phase, FSH stimulates the growth and development of follicles. This is accompanied by increased secretion of estradiol from the granulosa cells of the follicles.
   (b) Estradiol exerts a positive feedback effect on the hypothalamus and pituitary gland, resulting in an LH surge that triggers ovulation.
   (c) At ovulation, the mature vesicular ovarian follicle ruptures, and its secondary oocyte is released from the ovary.
   (d) The empty vesicular ovarian follicle becomes a corpus luteum, which secretes estradiol and progesterone.
   (e) Estradiol and progesterone inhibit FSH and LH secretion during the luteal phase.
   (f) The corpus luteum regresses at the end of the cycle, and the resulting decline in estradiol and progesterone secretion causes menstruation.
2. In terms of the changes that occur in the endometrium, the cycle can be divided into the proliferative phase (following menstruation), secretory phase (corresponding to the luteal phase of the ovaries), and menstrual phase.
3. The contraceptive pill acts by duplicating the negative feedback control of FSH and LH secretion by estradiol and progesterone, which normally occurs during the luteal phase.

Mammary Glands and Lactation (pp. 940–943)

1. Mammary glands contain mammary alveoli, which drain into lactiferous ducts. The ducts in turn, open on the surface of the nipple of the breast.
2. Prolactin stimulates the production of milk proteins. Oxytocin stimulates contraction of the lactiferous ducts and ejection of milk from the nipple.
   (a) Secretion of oxytocin and prolactin occur in response to the stimulus of the baby's suckling, through the activation of a neuroendocrine reflex.

(b) Secretion of prolactin is stimulated by inhibiting the secretion of prolactin-inhibiting hormone from the hypothalamus.

Review Activities

Objective Questions

1. Which of the following statements about oogenesis is *true?*
   (a) Oogonia form continuously in postnatal life.
   (b) Primary oocytes are haploid.
   (c) Meiosis is completed prior to ovulation.
   (d) A secondary oocyte is released.
2. The paramesonephric (müllerian) ducts give rise to
   (a) the uterine tubes.
   (b) the uterus.
   (c) the pudendum.
   (d) both *a* and *b*.
   (e) both *b* and *c*.
3. In a female, the homologue of the male scrotum is/are
   (a) the labia majora.
   (b) the labia minora.
   (c) the clitoris.
   (d) the vestibule.
4. The cervix is a portion of
   (a) the vulva.
   (b) the vagina.
   (c) the uterus.
   (d) the uterine tubes.
5. Which of the following is shed as menses?
   (a) the perimetrial layer
   (b) the fibrous layer
   (c) the functionalis layer
   (d) the menstrual layer
6. The transverse folds in the mucosal layer of the vagina are called
   (a) perineal folds.
   (b) vaginal rugae.
   (c) fornices.
   (d) labia gyri.
7. Fertilization normally occurs in
   (a) the ovary.
   (b) the uterine tube.
   (c) the uterus.
   (d) the vagina.
8. Contractions of the mammary ducts are stimulated by
   (a) prolactin.
   (b) oxytocin.
   (c) estrogen.
   (d) progesterone.
9. The suspensory ligaments support
   (a) the ovary.
   (b) the uterus.
   (c) the uterine tube.
   (d) the breast.
10. Uterine contractions are stimulated by
   (a) oxytocin.
   (b) prostaglandins.
   (c) prolactin.
   (d) both *a* and *b*.
   (e) both *b* and *c*.

Match the following:

11. menstrual phase	(a) high estrogen and progesterone; low FSH and LH
12. follicular phase	(b) low estrogen and progesterone
13. luteal phase	(c) LH surge
14. ovulation	(d) increasing estrogen; low LH and low progesterone

Essay Questions

1. Describe the development of the genital ducts and explain why the external genitalia of males and females are considered homologous.
2. Describe the gross and histologic structure of the uterus and explain the significance of the strata functionale and basale of the endometrium.
3. Describe the hormonal interactions that control ovulation and cause it to occur at the proper time.
4. Compare menstrual bleeding and bleeding that occurs during the estrus cycle of a dog in terms of hormonal control mechanisms and the ovarian cycle.
5. "The contraceptive pill tricks the brain into thinking you're pregnant." Interpret this popularized explanation of how birth-control pills work in terms of physiological mechanisms.
6. Describe the mechanisms that trigger lactation in a nursing mother. By what mechanism might fertility be reduced in a women who is breast-feeding her baby?
7. Describe the endocrine changes that cause the endometrium to enter its proliferative and secretory phases during a normal menstrual cycle. How does this compare with the endometrial changes that occur in a woman taking the birth-control pill?
8. Why does menstruation normally occur? Under what conditions does menstruation not occur? Explain.
9. Describe the steps of oogenesis when fertilization occurs and when it does not occur. Why are polar bodies produced?
10. Describe the endocrine changes that occur at menopause and discuss the consequences of these changes. What are some of the benefits and risks of hormone replacement therapy?

Critical Thinking Questions

1. In a male, the gamete stem cells (spermatogonia) duplicate themselves throughout life, but this does not happen in a female; a girl is born with all the oogonia she will have. Why does this difference between male and female gamete development exist?
2. Discuss the role of apoptosis and follicle atresia in ovarian physiology and explain how this process may be regulated.
3. Is it true that estrogen is an exclusively female hormone and that testosterone is an exclusively male hormone? Explain your answer.
4. In a sense, it's the sperm's fault that the oocytes produce polar bodies. Explain this statement.
5. Surgical removal of a woman's ovaries (ovariectomy) can precipitate menstruation. Ovariectomy in a dog or cat, however, does not cause the discharge of uterine blood. How can you explain these different responses?

Related Web Sites

For a listing of the most current web sites related to this chapter, please visit the *Concepts of Human Anatomy and Physiology* home page at http://www.mhhe.com/biosci/abio/.

NEXUS

Some Interactions of the Reproductive System with Other Body Systems

Integumentary System
- The skin serves as a sexual stimulant (p. 160).
- Sex hormones affect the distribution of body hair, deposition of subcutaneous fat, and other secondary sex characteristics (p. 925).

Skeletal System
- The pelvic girdle supports and protects some reproductive organs (p. 183).
- Sex hormones stimulate bone growth and maintenance (p. 903).

Nervous System
- Autonomic nerves innervate the organs of male reproduction to stimulate erection and ejaculation (p. 912).
- Autonomic nerves promote parts of the female sexual response (p. 930).
- The CNS, acting through the pituitary, coordinates different aspects of reproduction (p. 896).
- The limbic system of the brain is involved in sexual drive (p. 424).
- Gonadal sex hormones influence brain activity (p. 902).

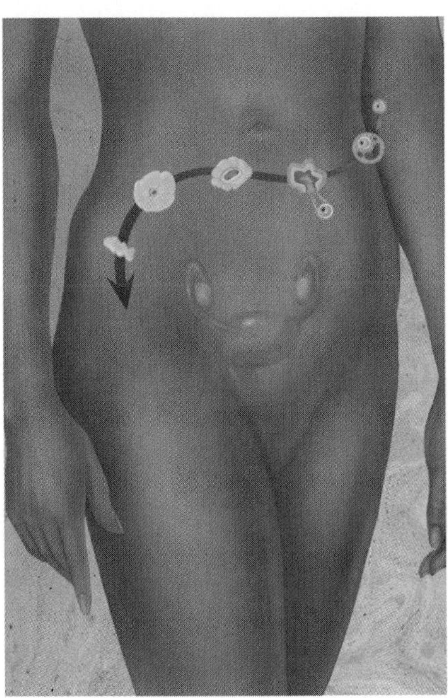

Endocrine System
- The anterior pituitary controls the activity of the gonads (p. 896).
- Testosterone secreted by the testes maintains the structure and function of the male reproductive system (p. 894).
- Estradiol and progesterone secreted by the ovaries regulates the female sex accessory organs, including the endometrium of the uterus (p. 938).
- Hormones secreted by the placenta are needed for maintenance of the pregnancy (p. 966).
- Prolactin and oxytocin are required for production for breast milk and the milk-ejection reflex (p. 942).

Muscular System
- Contractions of smooth muscles aids the movement of gametes (p. 912).
- Contractions of the myometrium aid labor and delivery (p. 973).
- Testosterone promotes an increase in muscle mass (p. 558).

Circulatory System
- The circulatory system transports oxygen and nutrients to the reproductive organs (p. 592).
- The fetal circulation permits the fetus to obtain oxygen and nutrients from the placenta (p. 644).
- Estrogen secreted by the ovaries helps to raise the level of HDL-cholesterol carries in the blood, lowering the risk of atherosclerosis (p. 651).

Lymphatic and Immune Systems
- The immune system protects the body, including the reproductive system, against infections (p. 696).
- The blood-testis barrier prevents the immune system from attacking sperm in the testes (p. 906).
- The placenta is an immunologically privileged site, which the mother's immune system is prevented from rejecting (p. 716).

Respiratory System
- The lungs provide oxygen for all body systems, including the reproductive system, and provide for the elimination of carbon dioxide (p. 729).
- The red blood cells of a fetus contain hemoglobin F, which has a high affinity for oxygen (p. 760).

Urinary System
- The kidneys regulate the volume, pH, and electrolytes balance of the blood and eliminate wastes (p. 779).

Digestive System
- The GI tract provides nutrients for all of the organs of the body, including those of the reproductive system (p. 817).
- Nutrients obtained from the GI tract of the mother can pass across the placenta to the embryo and fetus (p. 965).

THIRTY

Developmental Anatomy and Inheritance

OBJECTIVES

- Define *morphogenesis*, *capacitation*, and *fertilization*.
- Describe the changes that occur in the spermatozoon and ovum prior to, during, and immediately following fertilization.
- Describe the events of preembryonic development that result in the formation of the blastocyst.
- Discuss the role of the trophoblast in the implantation and development of the placenta.
- Explain how the primary germ layers develop and list the structures produced by each layer.
- Define *gestation* and explain how the parturition date is determined.
- Define *embryo* and describe the major events of the embryonic period of development.
- List the embryonic needs that must be met to avoid a spontaneous abortion.
- Describe the structure and function of each of the extraembryonic membranes.
- Describe the development and function of the placenta and umbilical cord.
- Define *fetus* and discuss the major events of the fetal period of development.
- Describe the various techniques available for examining the fetus or monitoring fetal activity.
- Describe the hormonal action that controls labor and parturition.
- Describe the three stages of labor.
- Define *genetics*.
- Discuss the variables that account for a person's phenotype.
- Explain how probability is involved in predicting inheritance and use a Punnett square to illustrate selected probabilities.

Clinical Investigation

A 27-year-old woman gave birth to twin boys, followed by an apparently single placenta. After examining the two infants, the pediatrician informed the mother that one of them had a cleft palate but the other was normal. She added that cleft palate could be hereditary and asked if any family members had the problem. The mother said that she knew of none. Further examination of the placenta revealed two amnions and only one chorion.

Does the presence of two amnions and one chorion indicate monozygotic or dizygotic twins? How would you account for the fact that one baby has a cleft palate, and the other does not?

Clues: Read the section on multiple pregnancy at the end of the chapter and carefully study figure 30.31. Can any conclusions be drawn regarding genetic similarities between the two infants? Note that identical twins are always the same gender.

Fertilization

Upon fertilization of an ovum by a spermatozoon in the uterine tube, meiotic development is completed, and a diploid zygote is formed.

Fertilization refers to the penetration of an ovum (egg) by a spermatozoon (fig. 30.1), with the subsequent union of their genetic material. It is this event that determines a persons' biological inheritance. Fertilization cannot occur, however, unless certain conditions are met. First of all, an ovum must be present in the uterine tube—it can be there for at most 24 hours before it becomes incapable of undergoing fertilization. Second, large numbers of spermatozoa must be ejaculated to ensure fertilization. Although a recent study has shown that sperm cells can remain viable 5 days after ejaculation, this still leaves a "window of fertility" of only 6 days each month—the day of ovulation and the 5 days leading up to it.

As described in chapter 29, a woman usually ovulates one ovum a month, totaling about 400 during her reproductive years. Each ovulation releases an ovum that is actually a secondary oocyte arrested at metaphase of the second meiotic division. An ovum is surrounded by a thin layer of protein and polysaccharides, called the **zona pellucida,** and a layer of granulosa cells, called the **corona radiata** (fig. 30.1). These layers provide a protective shield around the ovum as it enters the uterine tube.

During coitus, a male ejaculates between 100 million and 500 million spermatozoa into the female's vagina. This tremendous number is needed because of the high fatality rate of the sperm cells—only about 100 survive to encounter the ovum in the uterine tube. In addition to deformed sperm cells—up to 20% in the average fertile male—another 25% will perish as soon as they contact the vagina's acidic environment. Still others are destroyed by the woman's immune cells, which recognize them as foreign cells. Even if they manage to reach the cervix, many sperm cells will get stuck there and never make it through the uterus.

If it finally encounters an ovum, the spermatozoon must penetrate the protective corona radiata and zona pellucida for fertilization to occur. To do this, the head of each sperm cell is capped by an organelle called an **acrosome** (*ak'ro-som*) (figs. 30.1 and 30.2). The acrosome contains a trypsin-like protein-digesting enzyme and *hyaluronidase* (*hi''ă-loo-ron'ĭ-dās*), which digests hyaluronic acid, an important constituent of connective tissue. When a spermatozoon meets an ovum in the uterine tube, an *acrosomal reaction* occurs in which pores are created in the acrosomal membrane; the exposed digestive enzymes allow the spermatozoon to penetrate the corona radiata and the zona pellucida. A spermatozoon that comes along relatively late—after many others have undergone acrosomal reactions to expose the ovum membrane—is more likely to be the one that finally achieves penetration of the egg.

Experiments confirm that freshly ejaculated spermatozoa are infertile and must be in the female reproductive tract for at least 7 hours before they can fertilize an oocyte. Their membranes must become fragile enough to permit the release of the acrosomal enzymes—a process called *capacitation.* During in vitro fertilization, capacitation is induced artificially by treating the ejaculate with a solution of gamma globulin, free serum, follicular fluid, dextran, serum dialysate, and adrenal gland extract to chemically mimic the conditions of the female reproductive tract.

As soon as a spermatozoon penetrates the zona pellucida, a rapid chemical change in this layer prevents other spermatozoa from attaching to it. Therefore, only one spermatozoon is permitted to fertilize an oocyte. With the entry of a single spermatozoon through its cell membrane, the oocyte is stimulated to complete its second meiotic division (fig. 30.3). Like the first meiotic division, the second produces one cell that contains most of the cytoplasm and one polar body. The cell containing the cytoplasm is the mature ovum, and the second polar body, like the first, ultimately fragments and disintegrates.

At fertilization, the entire sperm cell enters the cytoplasm of the much larger ovum. Within 12 hours, the nuclear membrane in the ovum disappears, and the haploid number of chromosomes (23) in the ovum is joined by the *haploid number* of chromosomes from the spermatozoon. A fertilized egg,

zona pellucida: L. *zone*, a girdle; *pellis*, skin
corona radiata: Gk. *korone*, crown; *radiata*, radiate

acrosome: Gk. *akron*, extremity; *soma*, body
capacitation: L. *capacitas*, capable of
haploid: Gk. *haplous*, single; L. *ploideus*, multiple in form

Figure 30.1

The process of fertilization.

(*a, b*) As the head of the sperm cell (1) encounters the corona radiata of the oocyte (2), digestive enzymes are released from the acrosome (3, 4), clearing a path to the oocyte membrane. When the membrane of the sperm contacts the oocyte membrane (5), the membranes become continuous, and the nucleus and other content of the sperm move into the cytoplasm of the oocyte. (*c*) A scanning electron micrograph of a sperm cell bound to the surface of an oocyte.

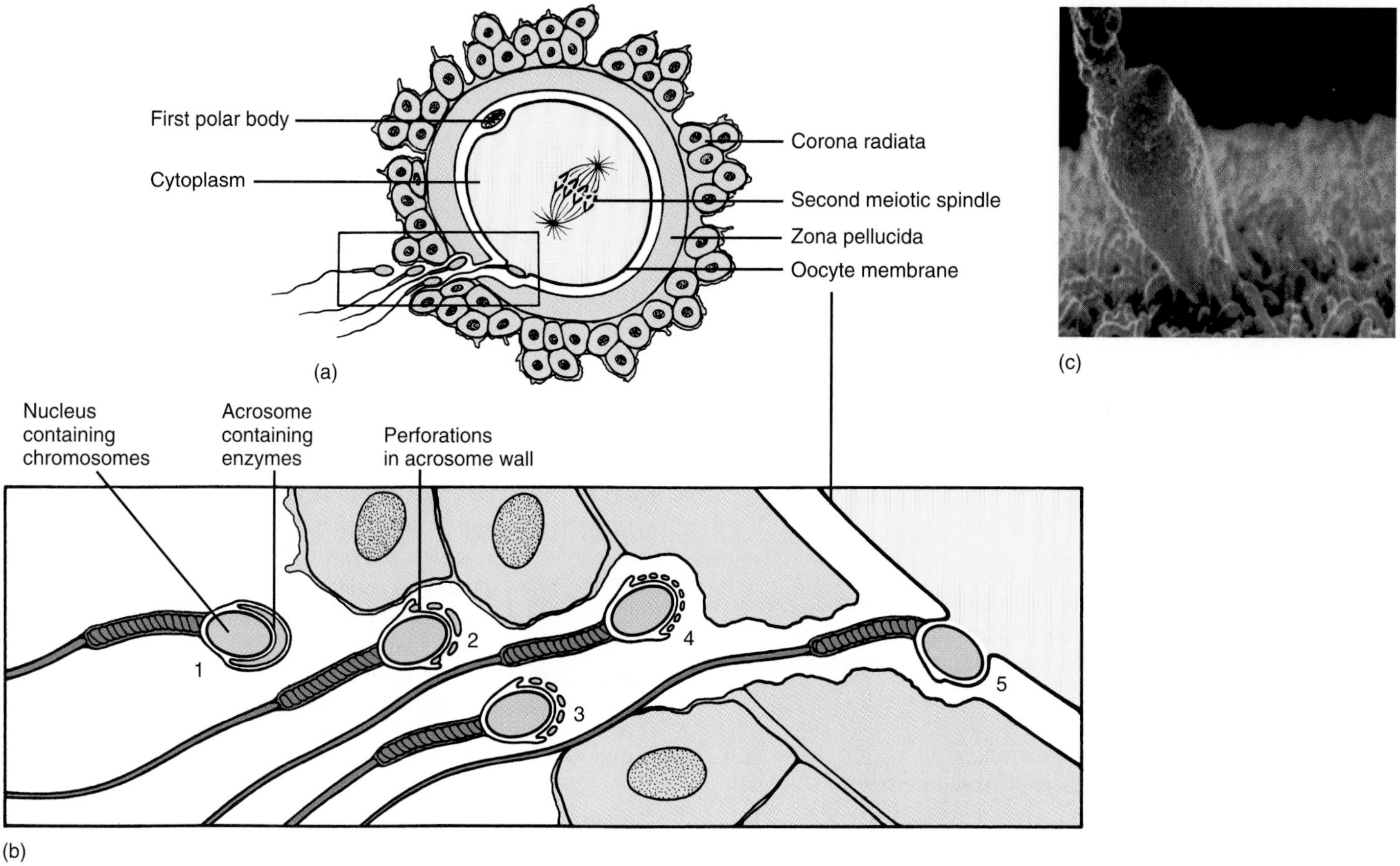

Figure 30.2

The head of a spermatozoon.

A transmission electron micrograph showing the head of a human spermatozoon with its nucleus and acrosome.

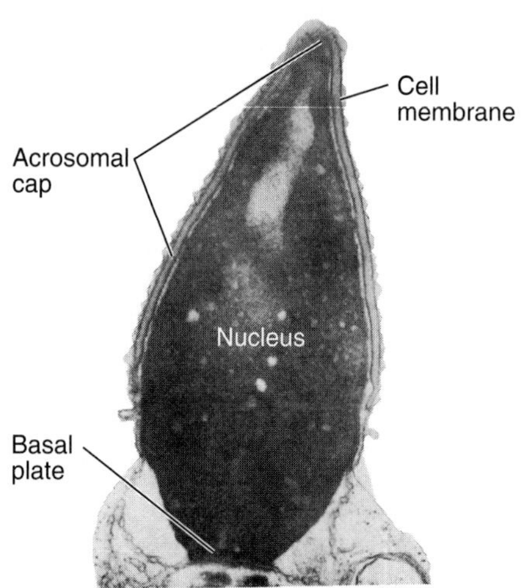

or **zygote** (*zi'got*), containing the *diploid number* of chromosomes (46), is thus formed.

Within hours after conception, the structure of the body begins to form from this single fertilized egg, culminating some 38 weeks later with the birth of a baby. The transformation involved in the growth and differentiation of cells and tissues is known as **morphogenesis** (*mor"fo-jen'ĕ-sis*), and it is through this awesome process that the organs and systems of the body are established in a functional relationship. Moreover, there are sensitive periods of morphogenesis for each organ and system, during which genetic or environmental factors may affect normal development.

Prenatal development can be divided into a preembryonic period, which is initiated by the fertilization of an ovum; an embryonic period, during which the body's organ systems are formed; and a fetal period, which culminates in parturition (childbirth).

---

zygote: Gk. *zygotos*, yolked, joined
diploid: Gk. *diplous*, double; L. *ploideus*, multiple in form
morphogenesis: Gk. *morphe*, form; *genesis*, beginning

Figure 30.3

Ovulation of a secondary oocyte.
A secondary oocyte, arrested at metaphase II of meiosis, is released at ovulation. If this cell is fertilized, it will become a mature ovum, complete its second meiotic division, and produces a second polar body. The fertilized ovum is known as a zygote.

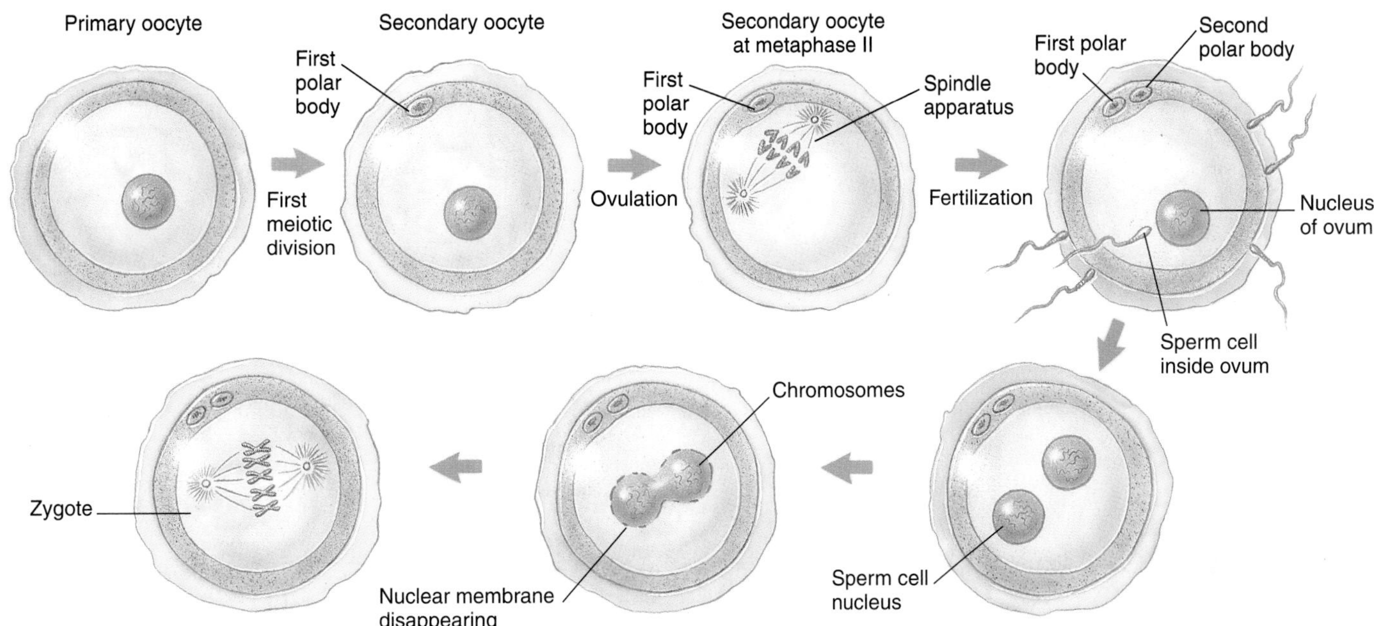

Preembryonic Period

The events of the 2-week preembryonic period include fertilization, transportation of the zygote through the uterine tube, mitotic divisions, implantation, and the formation of primordial embryonic tissue.

Cleavage and Formation of the Blastocyst

Within 30 hours following fertilization, the zygote undergoes a mitotic division called **cleavage.** This first division results in the formation of two identical daughter cells called *blastomeres* (fig. 30.4). Additional cleavages occur as the structure passes down the uterine tube and enters the uterus on about the third day. It is now composed of a ball of 16 or more cells called a **morula** (*mor'yu-la*). Although the morula has undergone several mitotic divisions, it is not much larger than the zygote because no additional nutrients necessary for growth have entered the cells. The morula floats freely in the uterine cavity for about 3 days. During this time, the center of the morula fills with fluid passing in from the uterine cavity. As the fluid-filled space develops within the morula, two distinct groups of cells form, and the structure becomes known as a **blastocyst** (*blas'to-sist*) (fig. 30.4). The hollow, fluid-filled center of the blastocyst is called the **blastocyst cavity.** The blastocyst is composed of an outer layer of cells, known as the **trophoblast,** and an inner aggregation of cells, called the **embryoblast** (*internal cell mass*) (see fig. 30.6). With further development, the trophoblast differentiates into a structure called the **chorion** (*kor'e-on*), which will later become a portion of the placenta. The embryoblast will become the embryo. A diagrammatic summary of the ovarian cycle, fertilization, and the morphogenic events of the first week is presented in figure 30.5.

Implantation

The process of **implantation** begins between the fifth and seventh day following fertilization. This is the process by which the blastocyst embeds itself into the endometrium of the uterine wall (fig. 30.6*a*). Implantation is made possible by the secretion of *proteolytic enzymes* by the trophoblast, which digest a portion of the endometrium. The blastula sinks into the depression, and endometrial cells move back to cover the defect in the wall. At the same time, the part of the uterine wall below the implanting blastocyst thickens, and specialized cells of the trophoblast produce fingerlike projections, called **syncytiotrophoblasts** (*sin-sit"e-ŏ-trof'ŏ-blasts*), into the thickened area. The syncytiotrophoblasts arise from a specific portion of the trophoblast called the **cytotrophoblast** (fig. 30.6*b*), located next to the embryoblast.

The blastocyst saves itself from being aborted by secreting a hormone that indirectly prevents menstruation. Even

morula: L. *morus*, mulberry

implantation: L. *im*, in; *planto*, to plant

Figure 30.4

Cleavage and formation of the blastocyst.
Sequential illustrations from the first cleavage of the zygote to the formation of the blastocyst. (Note the deterioration of the zona pellucida in the early blastocyst.)

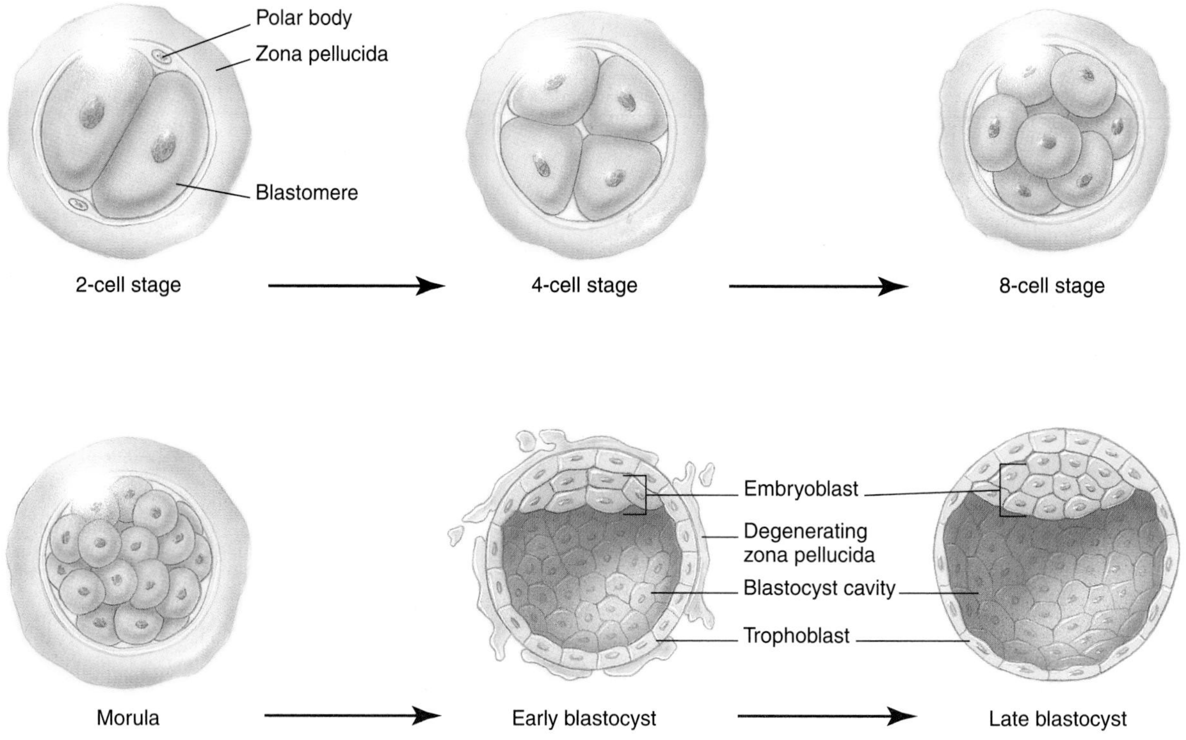

Polar body
Zona pellucida
Blastomere
2-cell stage
4-cell stage
8-cell stage
Morula
Embryoblast
Degenerating zona pellucida
Blastocyst cavity
Trophoblast
Early blastocyst
Late blastocyst

Figure 30.5

The ovarian cycle, fertilization, and the events of the first week.
Implantation of the blastocyst begins between the fifth and seventh days and is generally completed by the tenth day.

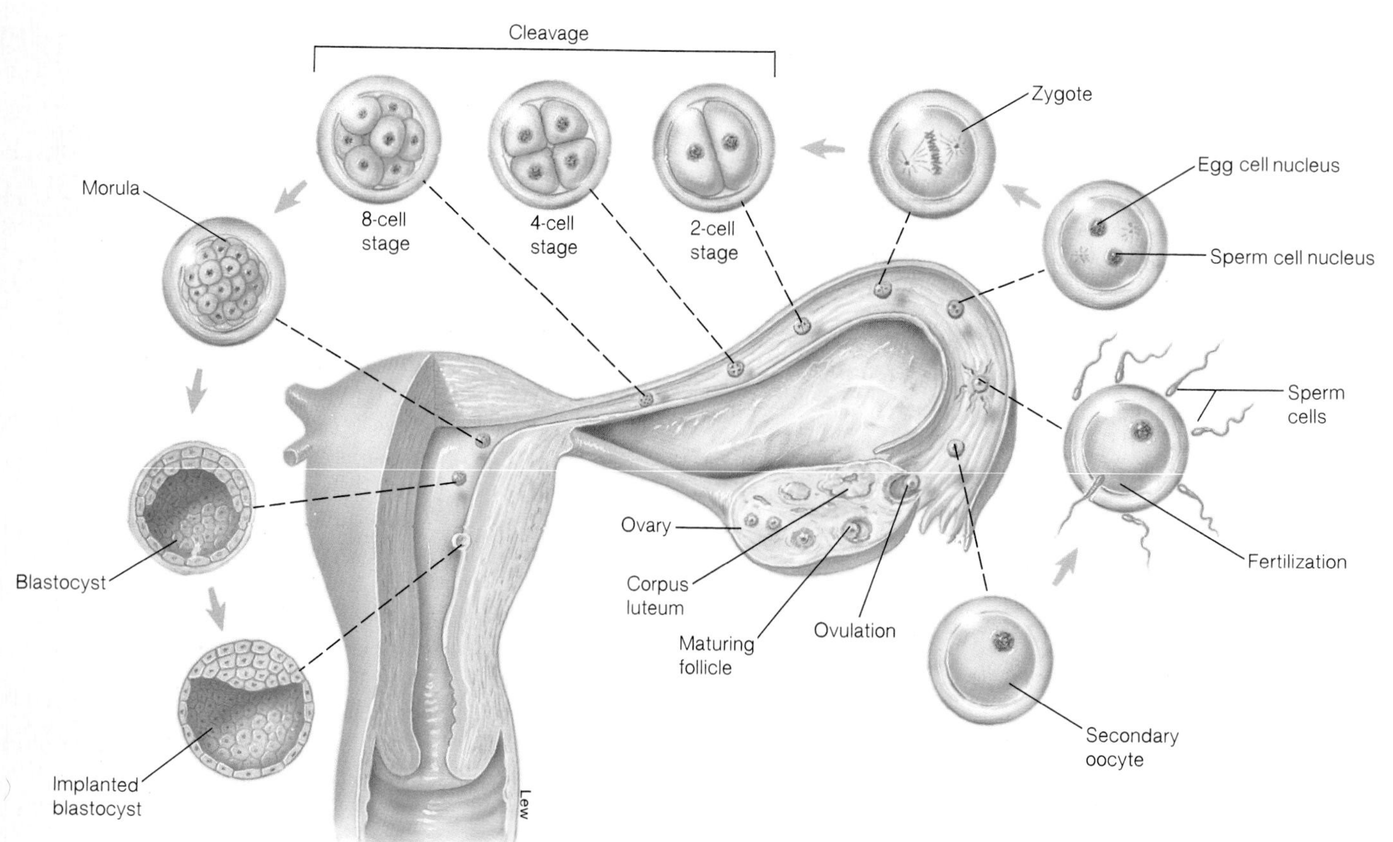

Figure 30.6

The blastocyst.

This preembryonic structure adheres to the endometrium on about the sixth day as seen in (*a*), a photomicrograph. By the seventh day (*b*), specialized syncytiotrophoblasts from the trophoblast have begun to invade the endometrium. The syncytiotrophoblasts secrete human chorionic gonadotropin (hCG) to sustain pregnancy and will eventually participate in the formation of the placenta from embryonic and fetal sustenance.

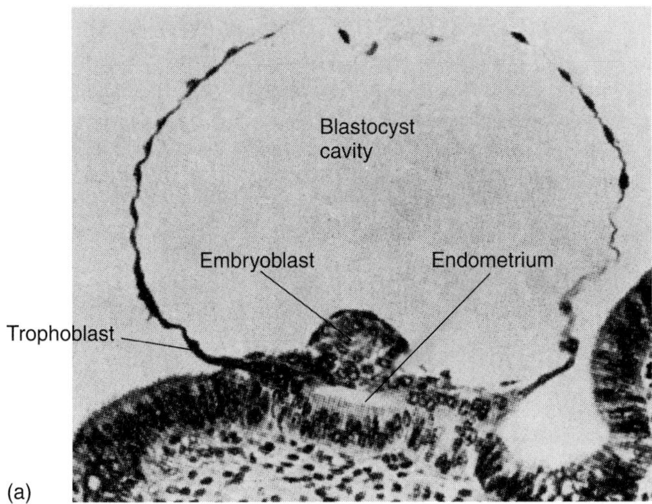

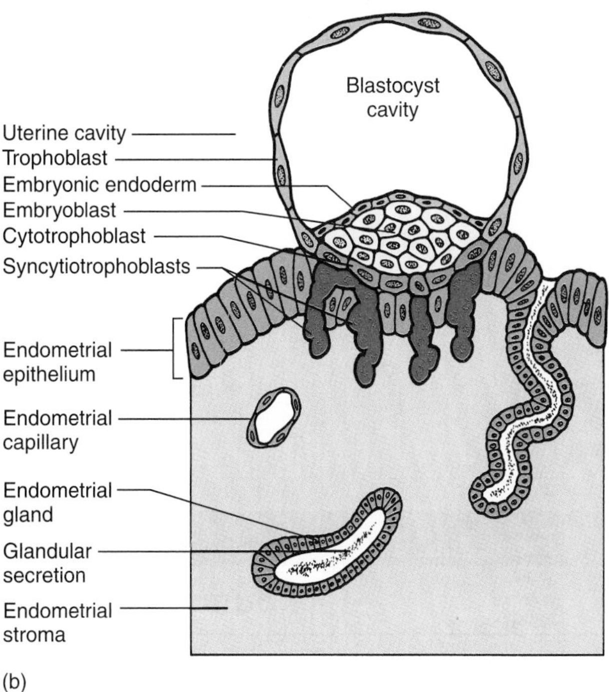

before the sixth day when implantation begins, the syncytiotrophoblasts secrete **human chorionic gonadotropin** (*kor″e-on′ik go-nad-ŏ-tro′pin*) **(hCG).** This hormone is identical to LH in its effects, and therefore is able to maintain the corpus luteum past the time when it would otherwise regress. The secretion of estrogen and progesterone is maintained, and menstruation is normally prevented (fig. 30.7).

The secretion of hCG declines by the tenth week of pregnancy. Actually, this hormone is required only for the first 5 to 6 weeks of pregnancy because the placenta itself becomes an active steroid-secreting gland by this time. At the fifth to sixth week, the mother's corpus luteum begins to regress (even in the presence of hCG), but the placenta secretes more than sufficient amounts of steroids to maintain the endometrium and prevent menstruation.

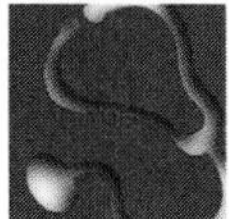

All pregnancy tests assay for the presence of hCG in the blood or urine because this hormone is secreted only by the blastocyst. Since there is no other source of hCG, the presence of this hormone confirms a pregnancy. Modern pregnancy tests detect the presence of hCG by use of antibodies against hCG or by the use of cellular receptor proteins for hCG.

Formation of Germ Layers

As the blastocyst completes implantation during the second week of development, the embryoblast undergoes marked differentiation. A slitlike space called the **amniotic** (*am″ne-ot′ik*) **cavity** forms between the embryoblast and the invading trophoblast (fig. 30.8). The embryoblast flattens into the **embryonic disc** (see fig. 30.10), which consists of two layers: an

Figure 30.7

Human chorionic gonadotropin (hCG).

This pituitary-like hormone is secreted by syncytiotrophoblasts during the first trimester of pregnancy. hCG maintains the mother's corpus luteum for the first 5½ weeks of pregnancy. The placenta then assumes the role of estrogen and progesterone production, and the corpus luteum degenerates.

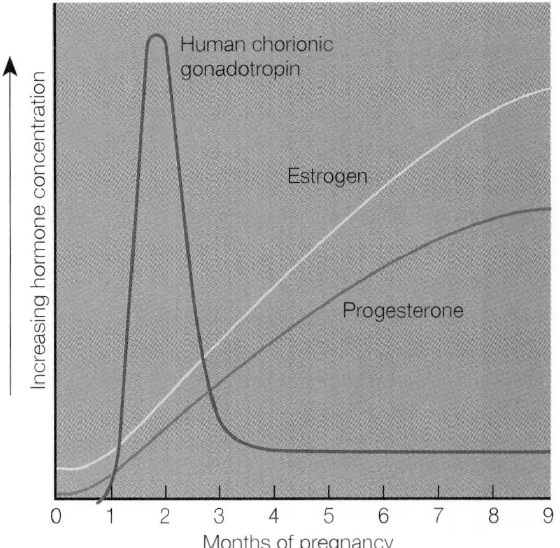

Figure 30.8

The primary germ layers. The completion of implantation occurs as the primary germ layers develop at the end of the second week.

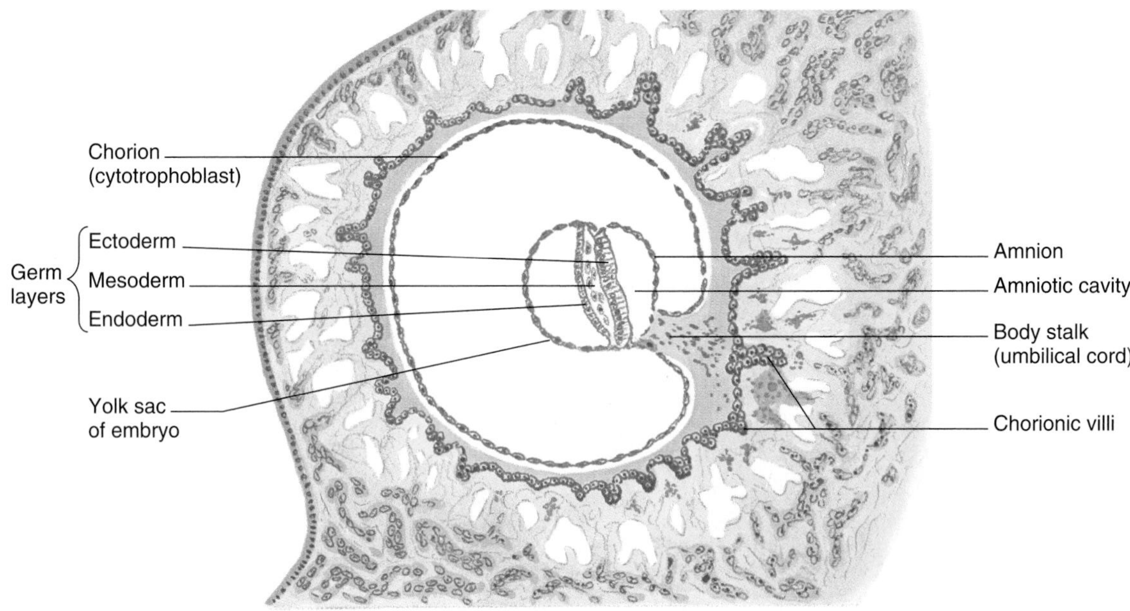

Table 30.1

Summary of Preembryonic Development

Morphogenic Stage	Time Period	Principal Events
Zygote	24 to 30 hours following ovulation	Egg is fertilized; zygote has 23 pairs of chromosomes (diploid) from haploid sperm and haploid egg and is genetically unique
Cleavage	30 hours to third day	Mitotic divisions produce increased number of cells
Morula	Third to fourth day	Solid ball-like structure forms, composed of 16 or more cells
Blastocyst	Fifth day to end of second week	Hollow ball-like structure forms, a single layer thick; embryoblast and trophoblast form; implantation occurs; embryonic disc forms, followed by primary germ layers

upper **ectoderm,** which is closer to the amniotic cavity, and a lower **endoderm,** which borders the blastocyst cavity. A short time later, a third layer called the **mesoderm,** forms between the endoderm and ectoderm. These three layers constitute the **primary germ layers** (fig. 30.8). Once they are formed, at the end of the second week, the preembryonic period is completed and the embryonic period begins.

The primary germ layers are especially significant because all of the cells and tissues of the body are derived from them. Ectodermal cells form the nervous system; the outer layer of skin (epidermis), including hair, nails, and skin glands; and portions of the sensory organs. Mesodermal cells form the skeleton, muscles, blood, reproductive organs, dermis of the skin, and connective tissue. Endodermal cells produce the lining of the GI tract, the digestive organs, the respiratory tract and lungs, and the urinary bladder and urethra.

The events of the preembryonic period are summarized in table 30.1. Refer to figure 30.9 for an illustration and a listing of the organs and body systems that derive from each of the primary germ layers.

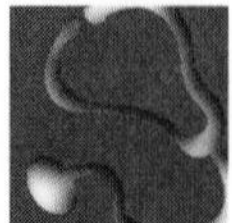

The period of prenatal development is referred to as *gestation.* Normal gestation for humans is 9 months. Knowing this and the pattern of menstruation make it possible to determine the baby's delivery date. In a typical reproductive cycle, a woman ovulates 14 days prior to the onset of the next menstruation and is fertile for approximately 20 to 24 hours following ovulation. Adding 9 months, or 38 weeks, to the time of ovulation gives one the estimated delivery date.

ectoderm: Gk. *ecto*, outside; *derm*, skin
endoderm: Gk. *endo*, within; *derm*, skin
mesoderm: Gk. *meso*, middle; *derm*, skin

gestation: L. *gestatus*, to bear

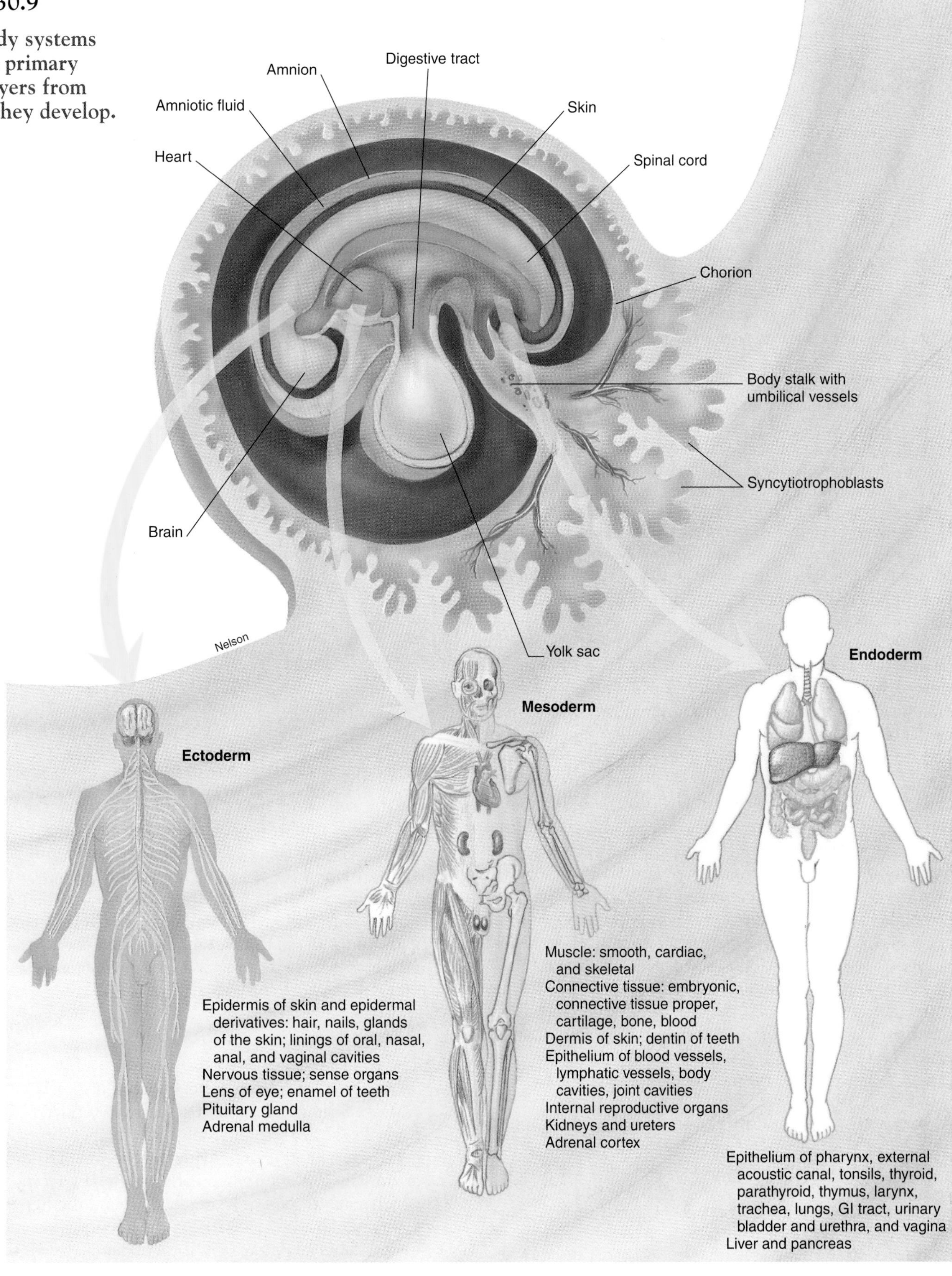
Digestive tract
Amnion
Amniotic fluid
Skin
Heart
Spinal cord
Chorion
Body stalk with umbilical vessels
Syncytiotrophoblasts
Brain
Nelson
Yolk sac
Endoderm
Mesoderm
Ectoderm
Muscle: smooth, cardiac, and skeletal
Connective tissue: embryonic, connective tissue proper, cartilage, bone, blood
Dermis of skin; dentin of teeth
Epithelium of blood vessels, lymphatic vessels, body cavities, joint cavities
Internal reproductive organs
Kidneys and ureters
Adrenal cortex
Epidermis of skin and epidermal derivatives: hair, nails, glands of the skin; linings of oral, nasal, anal, and vaginal cavities
Nervous tissue; sense organs
Lens of eye; enamel of teeth
Pituitary gland
Adrenal medulla
Epithelium of pharynx, external acoustic canal, tonsils, thyroid, parathyroid, thymus, larynx, trachea, lungs, GI tract, urinary bladder and urethra, and vagina
Liver and pancreas

Figure 30.9

The body systems and the primary germ layers from which they develop.

Embryonic Period

The events of the 6-week embryonic period include the differentiation of the germ layers into specific body organs and the formation of the placenta, the umbilical cord, and the extraembryonic membranes. Through these morphogenic events, the needs of the embryo are met.

During the embryonic period—from the beginning of the third week to the end of the eighth week—the developing organism is correctly called an **embryo.** During this period, all of the body tissues and organs form, as well as the placenta, umbilical cord, and extraembryonic membranes. The term *conceptus* refers to the embryo (or fetus later on) and all of the extraembryonic structures—the products of conception.

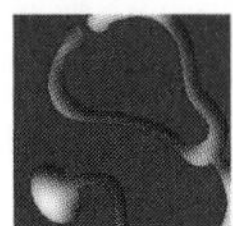

Embryology is the study of the sequential changes in an organism as the various tissues, organs, and systems develop. Chick embryos are frequently studied because of the easy access through the shell and their rapid development. Mice and pig embryos are also extensively studied as mammalian models. Genetic manipulation, induction of drugs, exposure to disease, radioactive tagging or dyeing of developing tissues, and X-ray treatments are some of the commonly conducted experiments that provide information that can be applied to human development and birth defects.

During the preembryonic period of cell division and differentiation, the developing structure is self-sustaining. The embryo, however, must derive sustenance from the mother. For morphogenesis to continue, certain immediate needs must be met. These needs include: (1) formation of a vascular connection between the uterus of the mother and the embryo so that nutrients and oxygen can be provided, and metabolic wastes and carbon dioxide can be removed; (2) establishment of a constant, protective environment around the embryo that is conducive to development; (3) establishment of a structural foundation for embryonic morphogenesis along a longitudinal axis; (4) provision for structural support for the embryo, both internally and externally; and (5) coordination of the morphogenic events through genetic expression. If these needs are not met, a spontaneous abortion will generally occur.

The first and second of these needs are provided for by extraembryonic structures; the last three are provided for intraembryonically. The extraembryonic membranes, the placenta, and the umbilical cord will be considered separately, prior to a discussion of the development of the embryo.

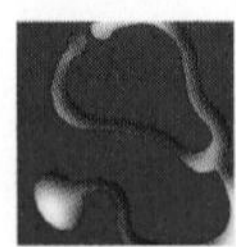

Serious developmental defects usually cause the embryo to be naturally aborted. About 25% of early aborted embryos have chromosomal abnormalities. Other abortions may be caused by environmental factors, such as infectious agents or teratogenic drugs (drugs that cause birth defects). In addition, an implanting, developing embryo is regarded as foreign tissue by the immune system of the mother, and is rejected and aborted unless maternal immune responses are suppressed.

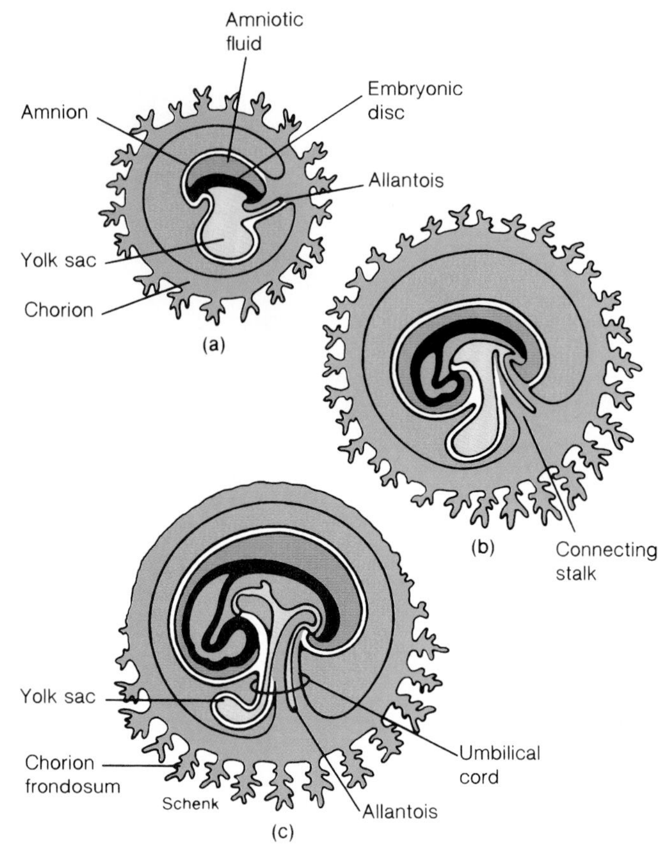

Figure 30.10

The formation of the extraembryonic membranes during a single week of rapid embryonic development. (*a*) At 3 weeks, (*b*) at 3½ weeks, and (*c*) at 4 weeks.

Extraembryonic Membranes

At the same time that the internal organs of the embryo are forming, a complex system of extraembryonic membranes is also developing (fig. 30.10). The **extraembryonic membranes** are the amnion, yolk sac, allantois, and chorion. These membranes are responsible for the protection, respiration, excretion, and nutrition of the embryo and subsequent fetus. At parturition, the placenta, umbilical cord, and extraembryonic membranes separate from the fetus and are expelled from the uterus as the *afterbirth*.

Amnion

The *amnion* (*am'ne-on*) is a thin extraembryonic membrane derived from ectoderm and mesoderm. It loosely envelops the embryo, forming an **amniotic sac** that is filled with *amniotic fluid* (fig. 30.11*b*). In later fetal development, the amnion expands to come in contact with the chorion. The development of the amnion is initiated early in the embryonic period, at which time its margin is attached around the free edge of the embryonic disc (fig. 30.10). As the amniotic sac enlarges during the late embryonic period (at about 8 weeks), the amnion gradually sheaths the developing umbilical cord with an epithelial covering (fig. 30.12).

Figure 30.11

An implanted embryo at approximately 4½ weeks.

(*a*) The interior of a uterus showing the implantation site. (*b*) The developing embryo, extraembryonic membranes, and placenta.

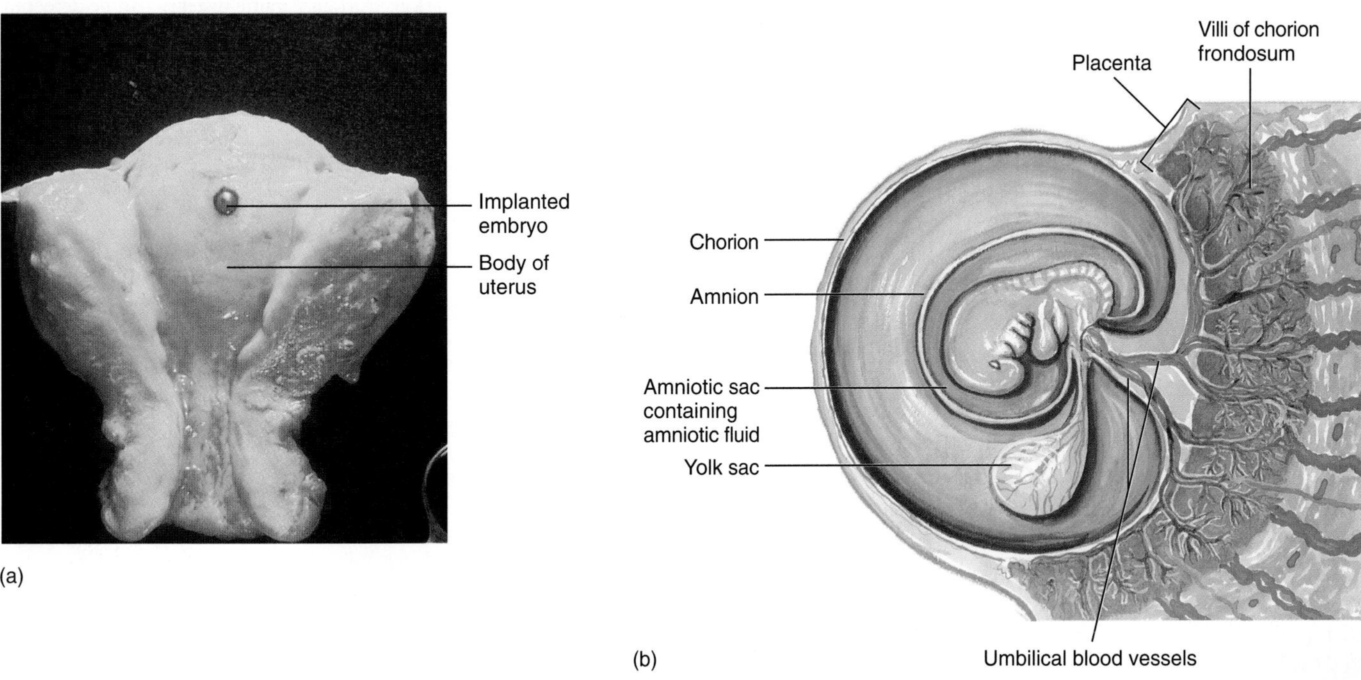

Figure 30.12

The embryo, extraembryonic membranes, and placenta at approximately 7 weeks of development.

Blood from the embryo is carried to and from the chorion frondosum by the umbilical arteries and vein. The maternal tissue between the chorionic villi is known as the decidua basalis; this tissue, together with the villi, forms the functioning placenta.

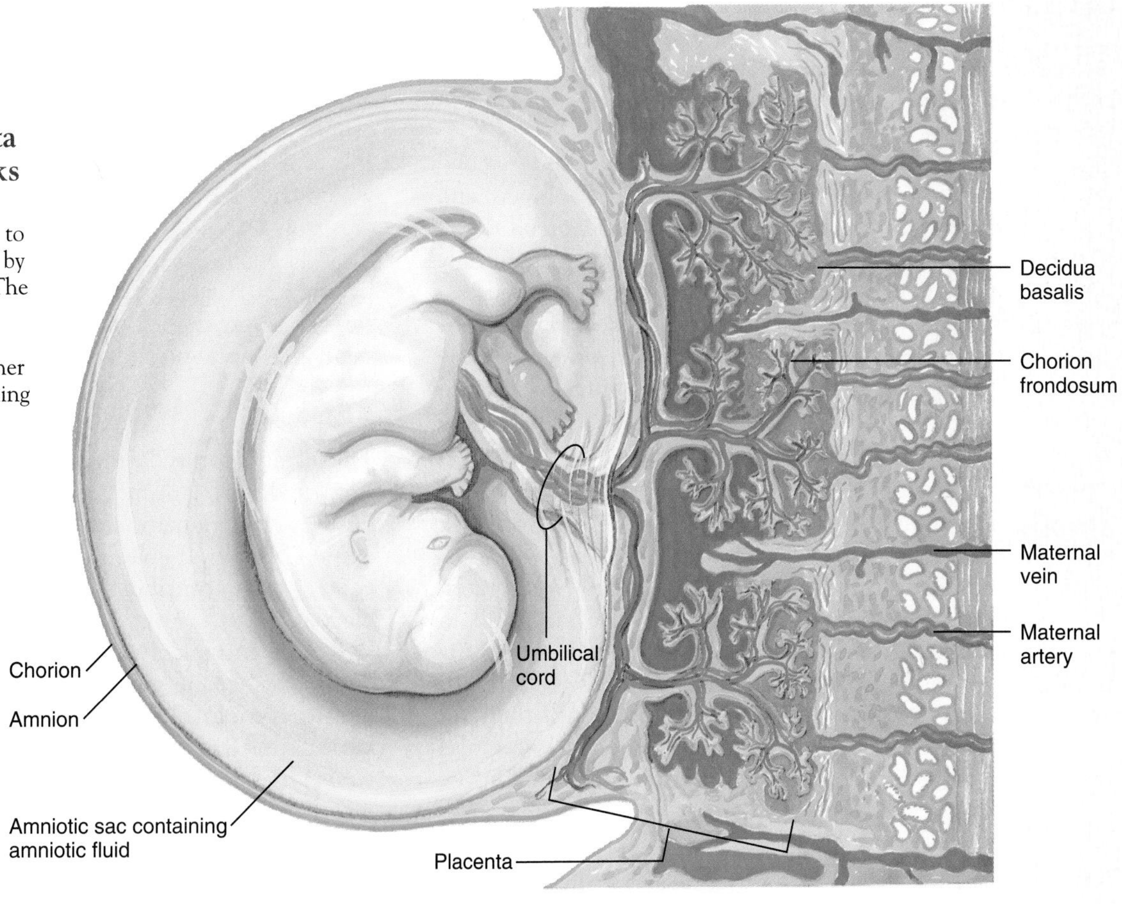

Decidua basalis
Chorion frondosum
Maternal vein
Maternal artery
Umbilical cord
Chorion
Amnion
Amniotic sac containing amniotic fluid
Placenta

As a buoyant medium, amniotic fluid performs four functions for the embryo and subsequent fetus.

1. It permits symmetrical structural development and growth.
2. It cushions and protects by absorbing jolts that the mother may receive.
3. It helps to maintain consistent pressure and temperature.
4. It allows the fetus to develop freely, which is important for musculoskeletal development and blood flow.

Amniotic fluid is formed initially as an isotonic fluid absorbed from the maternal blood in the endometrium surrounding the developing embryo. Later, the volume is increased and the concentration changed by urine excreted from the fetus into the amniotic sac. Amniotic fluid also contains cells that are sloughed off from the fetus, placenta, and amniotic sac. Since all of these cells are derived from the same fertilized egg, all have the same genetic composition. Many genetic abnormalities can be detected by aspirating this fluid and examining the cells obtained in a procedure called amniocentesis (*am″ne-o-sen-te′sis*).

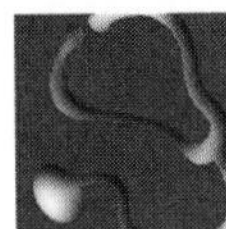

Amniocentesis (fig. 30.13) is usually performed at the fourteenth or fifteenth week of pregnancy, when the amniotic sac contains 175—225 ml of fluid. Genetic diseases, such as *Down syndrome,* or *trisomy 27* (in which there are three instead of two number 21 chromosomes), can be detected by examining chromosomes. Diseases such as *Tay–Sachs disease,* in which there is a defective enzyme involved in formation of myelin sheaths, can be detected by biochemical techniques from these fetal cells.

Amniotic fluid is normally swallowed by the fetus and absorbed in its GI tract. The fluid enters the fetal blood, and the waste products it contains enter the maternal blood in the placenta. Prior to delivery, the amnion is naturally or surgically ruptured, and the amniotic fluid (bag of waters) is released.

As the fetus grows, the amount of amniotic fluid increases. It is also continually absorbed and renewed. For the near-term baby, almost 8 liters of fluid are completely replaced each day.

Yolk Sac

The *yolk sac* is established during the end of the second week as cells from the trophoblast form a thin *exocoelomic* (*ek″so-sĕ-lo′mik*) *membrane*. Unlike the yolk sac of many vertebrates, the human yolk sac contains no nutritive yolk, but is still an essential structure during early embryonic development. Attached to the underside of the embryonic disc (see figs. 30.10 and 30.11), it produces blood for the embryo until the liver forms during the sixth week. A portion of the yolk sac is also involved in the formation of the primitive gut. In addition,

Figure 30.13

Amniocentesis.
In this procedure, amniotic fluid containing suspended cells is withdrawn for examination. Various genetic diseases can be detected prenatally by this means.

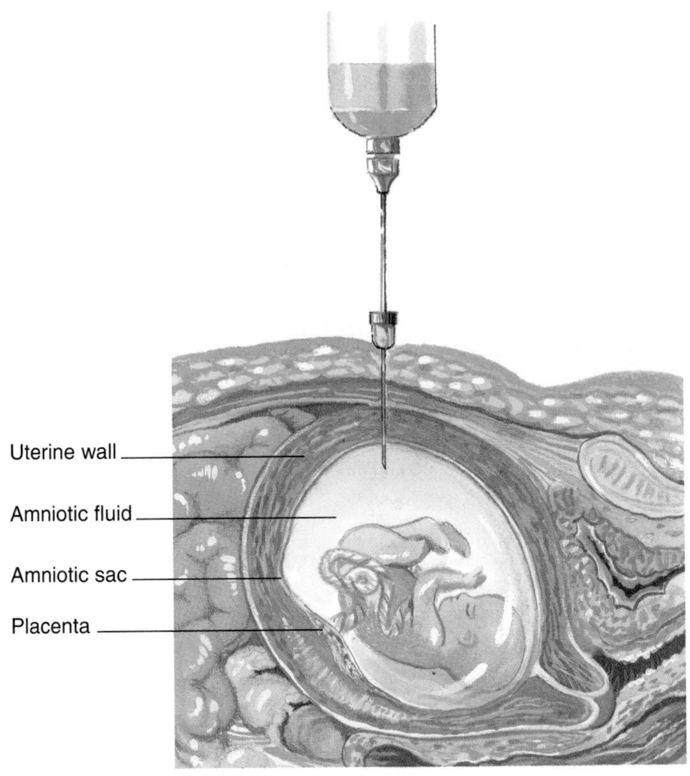

primordial germ cells form in the wall of the yolk sac. During the fourth week, they migrate to the developing gonads, where they become the primitive germ cells (spermatogonia or oogonia).

The stalk of the yolk sac usually detaches from the gut by the sixth week. Following this, the yolk sac gradually shrinks as pregnancy advances. Eventually, it becomes very small and serves no additional function.

Allantois

The *allantois* forms during the third week as a small outpouching, or diverticulum, near the base of the yolk sac (see fig. 30.10). It remains small but is involved in the formation of blood cells and gives rise to the fetal umbilical arteries and vein. It also contributes to the development of the urinary bladder.

The extraembryonic portion of the allantois degenerates during the second month. The intraembryonic portion involutes to form a thick urinary tube called the **urachus** (*yoo′ra-kus*). After birth, the urachus becomes a fibrous cord called the *median umbilical ligament* that attaches to the urinary bladder.

Down syndrome: from John L. H. Down, English physician, 1828–96

allantois: Gk. *allanto*, sausage; *iodos*, resemblance

Chorion

The *chorion* (*kor'e-on*) is the outermost extraembryonic membrane. It contributes to the formation of the placenta as small fingerlike extensions, called *choronic villi*, penetrate deeply into the uterine tissue (see fig. 30.10). Initially, the entire surface of the chorion is covered with choronic villi. But those villi on the surface toward the uterine cavity gradually degenerate, producing a smooth, bare area known as the **smooth chorion.** As this occurs, the choronic villi associated with the uterine wall rapidly increase in number and branch out. This portion of the chorion is known as the **villous chorion,** or **chorion frondosum** (*frondo'sum*). The villous chorion becomes highly vascular with embryonic blood vessels, and as the embryonic heart begins to function, embryonic blood is pumped in close proximity to the uterine wall.

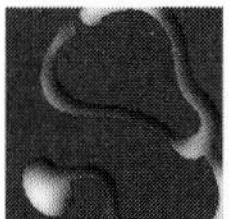

Chorionic villus biopsy is a technique used to detect genetic disorders much earlier than amniocentesis permits. In chorionic villus biopsy, a catheter is inserted through the cervix to the chorion, and a sample of chorionic villus is obtained by suction or cutting. Genetic tests can be performed directly on the villus sample, since this sample contains much larger numbers of fetal cells than does a sample of amniotic fluid. Chorionic villus biopsy can provide genetic information at 10 to 12 weeks' gestation.

Figure 30.14

The circulation of blood within the placenta.
Maternal blood is delivered to and drained from the spaces between the chorionic villi. Fetal blood is brought to blood vessels within the villi by branches of the umbilical arteries and is drained by branches of the umbilical vein.

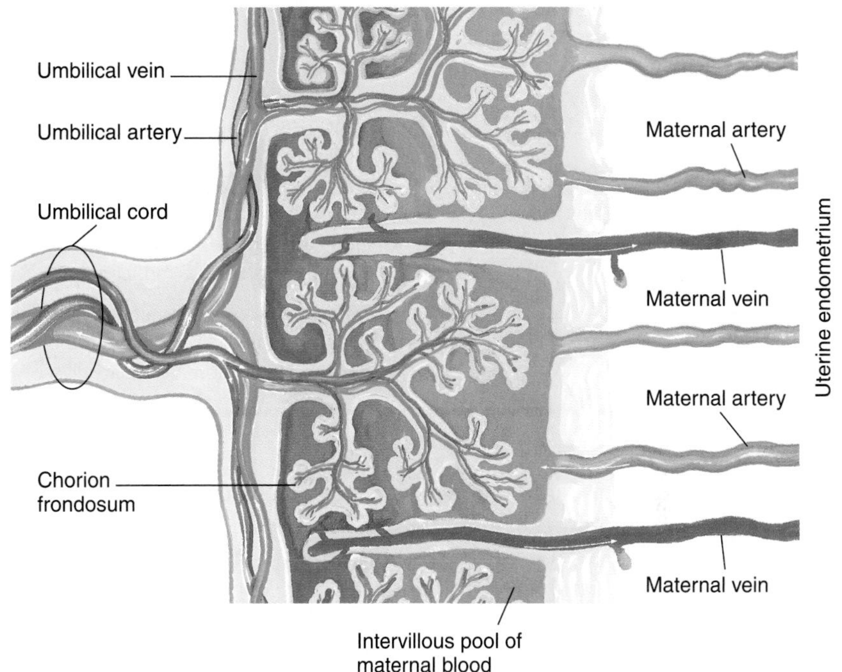

Placenta

The **placenta** (*plă-cen'tă*) is a vascular structure by which an unborn child is attached to its mother's uterine wall and through which metabolic exchange occurs (fig. 30.14). The placenta is formed in part from maternal tissue and in part from embryonic tissue. The embryonic portion of the placenta consists of the villous chorion, whereas the maternal portion is composed of the area of the uterine wall called the **decidua basalis** (see fig. 30.12), into which the choronic villi penetrate. Blood does not flow directly between these two portions, but because their membranes are in close proximity, certain substances diffuse readily.

When fully formed, the placenta is a reddish brown oval disc with a diameter of 15 to 20 cm (8 in.) and a thickness of 2.5 cm (1 in.). It weighs 500 to 600 gm, about one-sixth the weight of the fetus.

Exchange of Molecules across the Placenta

The two **umbilical arteries** deliver fetal blood to vessels within the villous chorion of the placenta. This blood circulates within the choronic villi and returns to the fetus via the **umbilical vein** (see fig. 21.36). Maternal blood is delivered to and drained from the cavities within the decidua basalis, which are located between the chorionic villi. In this way, maternal and fetal blood are brought close together but do not normally mix within the placenta.

The placenta serves as a site for the exchange of gases and other molecules between the maternal and fetal blood. Oxygen diffuses from mother to fetus, and carbon dioxide diffuses in the opposite direction. Nutrient molecules and waste products likewise pass between maternal and fetal blood.

The placenta has a high metabolic rate. It utilizes about one-third of all the oxygen and glucose supplied by the maternal blood. In fact, the rate of protein synthesis is actually higher in the placenta than in the fetal liver. Like the liver, the placenta produces a great variety of enzymes that are capable of converting biologically active molecules (such as hormones and drugs) into less active, more water-soluble forms. In this way, potentially dangerous molecules in the maternal blood are often prevented from harming the fetus.

Some substances ingested by a pregnant woman are able to pass through the placenta readily, to the detriment of the fetus. These include nicotine, heroin, and certain antidepressant drugs. Excessive nicotine will stunt the growth of the fetus; heroin can lead to fetal drug addiction, and certain antidepressants can cause respiratory problems.

chorion: Gk. *chorion*, external fetal membrane
villous: L. *villus*, tuft of hair
placenta: L. *placenta*, a flat cake

Table 30.2
Hormones Secreted by the Placenta

Hormones	Effects
Pituitary-like Hormones	
Chorionic gonadotropin (hCG)	Similar to LH; maintains mother's corpus luteum for first 5½ weeks of pregnancy; may be involved in suppressing immunological rejection of embryo; also exhibits TSH-like activity
Chorionic somatomammotropin (hCS)	Similar to prolactin and growth hormone; in the mother, hCS promotes increased fat breakdown and fatty acid release from adipose tissue and decreased glucose use by maternal tissues (diabetic-like effects)
Sex Steroids	
Progesterone	Helps maintain endometrium during pregnancy; helps suppress gonadotropin secretion; stimulates development of loose connective tissue in mammary glands
Estrogens	Help maintain endometrium during pregnancy; help suppress gonadotropin secretion; help stimulate mammary gland development; inhibit prolactin secretion: promote uterine sensitivity to oxytocin; stimulate duct development in mammary glands

Although the placenta is an effective barrier against diseases of bacterial origin, rubella and other viruses, as well as certain blood-borne diseases such as syphilis, they can diffuse through the placenta and affect the fetus. During parturition, small amounts of fetal blood may pass across the placenta to the mother. If the fetus is Rh positive and the mother Rh negative, the antigens of the fetal red blood cells elicit an antibody response in the mother. In a subsequent pregnancy, the maternal antibodies then cross the placenta and cause a breakdown of fetal red blood cells—a condition called *erythroblastosis fetalis*.

Endocrine Functions of the Placenta

The placenta functions as an endocrine gland in producing both glycoprotein and steroid hormones. The glycoprotein hormones have actions similar to those of some anterior pituitary hormones. This latter category of hormones includes **chorionic gonadotropin (hCG)** and **chorionic somatomammotropin (hCS)** (table 30.2). Chorionic gonadotropin has LH-like effects, as previously described; it also has thyroid-stimulating ability, like pituitary TSH. Chorionic somatomammotropin likewise has actions that are similar to two pituitary hormones: growth hormone and prolactin. The placental hormones hCG and hCS thus duplicate the actions of four anterior pituitary hormones.

Pituitary-Like Hormones from the Placenta

The importance of chorionic gonadotropin in maintaining the mother's corpus luteum for the first 5½ weeks of pregnancy has been previously discussed. There is also some evidence that hCG may in some way help to prevent immunological rejection of the implanting embryo. Chorionic somatomammotropin synergizes (acts together) with growth hormone from the mother's pituitary to produce a diabetic-like effect in the pregnant woman. The effects of these two hormones include (1) accelerated lipolysis, and therefore increased plasma fatty acid concentrations; (2) decreased maternal utilization of glucose and therefore increased blood glucose concentrations; and (3) polyuria (excretion of large volumes of urine), thereby producing a degree of dehydration and thirst. This diabetic-like effect in the mother helps to spare glucose for the placenta and fetus that, like the brain, use glucose as their primary energy source.

Steroid Hormones from the Placenta

After the first 5½ weeks of pregnancy, when the corpus luteum regresses, the placenta becomes the major sex-steroid–producing gland. The blood concentration of estrogens, as a result of placental secretion, rises to levels more than 100 times greater than those existing at the beginning of pregnancy. The placenta also secretes large amounts of progesterone, changing the estrogen/progesterone ratio in the blood from 100:1 at the beginning of pregnancy to a ratio of close to 1:1 toward full-term.

The placenta, however, is an incomplete endocrine gland because it cannot produce estrogen and progesterone without the aid of precursors supplied to it by both the mother and the fetus. The placenta, for example, cannot produce cholesterol from acetate and so must be supplied with cholesterol from the mother's circulation. Cholesterol, which is a steroid containing 27 carbons, can then be converted by enzymes in the placenta into steroids that contain 21 carbons—such as progesterone. The placenta, however, lacks the enzymes needed to convert progesterone into androgens (which have 19 carbons) and estrogens (which have 18 carbons).

In order for the placenta to produce estrogens, it needs to cooperate with steroid-producing tissues in the fetus. Fetus and placenta, therefore, form a single functioning system in terms of steroid hormone production. This system has been called the **fetal-placental unit** (fig. 30.15).

Figure 30.15

The hormonal relationship between the mother and the fetus.
The secretion of progesterone and estrogen from the placenta requires a supply of cholesterol from the mother's blood and the cooperation of fetal enzymes that convert progesterone to androgens.

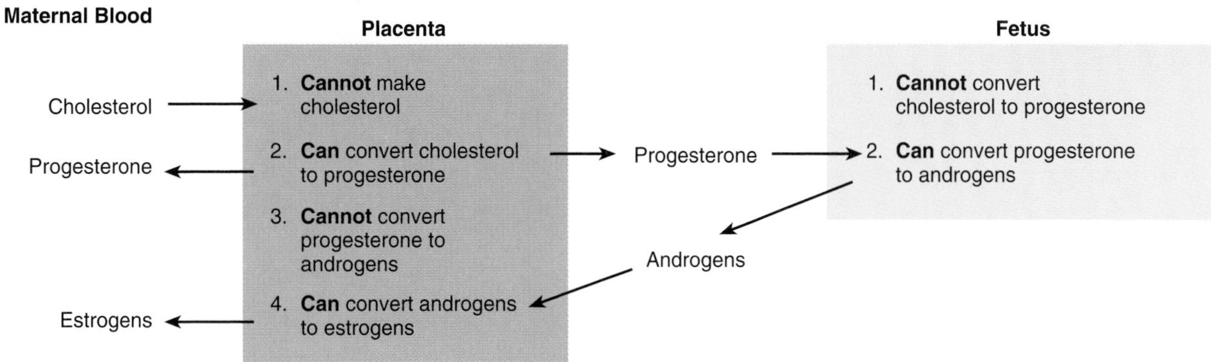

The ability of the placenta to convert androgens into estrogen helps to protect the female embryo from becoming masculinized by the androgens secreted from the mother's adrenal glands. In addition to producing estradiol, the placenta secretes large amounts of a weak estrogen called **estriol.** The production of estriol increases tenfold during pregnancy, so that by the third trimester estriol accounts for about 90% of the estrogens excreted in the mother's urine. Since almost all of this estriol comes from the placenta (rather than from maternal tissues), measurements of urinary estriol can be used clinically to assess the health of the placenta.

Umbilical Cord

The **umbilical cord** forms as the yolk sac shrinks and the amnion expands to envelop the tissues on the underside of the embryo (fig. 30.16). The umbilical cord usually attaches near the center of the placenta. When fully formed, it is between 1 and 2 cm (0.5 to 1 in.) in diameter and approximately 55 cm (2 ft) long. On average, the umbilical cords of male fetuses are approximately 5 cm (2 in.) longer than those of female fetuses. The umbilical cord contains two **umbilical arteries,** which carry deoxygenated blood from the embryo toward the placenta, and one **umbilical vein,** which carries oxygenated blood from the placenta to the embryo (see fig. 21.36). These vessels are surrounded by embryonic connective tissue called **mucoid connective tissue** (Wharton's jelly).

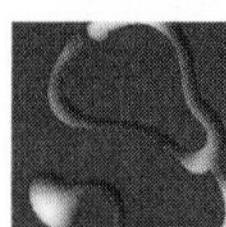

The umbilical cord has a helical, or screwlike, form that keeps it from kinking. The spiraling occurs because the umbilical vein grows faster and longer than the umbilical arteries. In about one-fifth of all deliveries, the cord is looped once around the baby's neck. If drawn tightly, the cord may cause death or serious perinatal problems.

Structural Changes of the Embryo by Weeks

Third Week

Early in the third week, a thick linear band called the **primitive line** (primitive streak) appears along the dorsal midline of the embryonic disc (figs. 30.10 and 30.17). Derived from mesodermal cells, the primitive line establishes a structural foundation for embryonic morphogenesis along a longitudinal axis. As the primitive line elongates, a prominent thickening called the **primitive node** appears at its cranial end (fig. 30.17). The primitive node later gives rise to the mesodermal structures of the head and to a rod of mesodermal cells called the **notochord.** The notochord forms a midline axis that is the basis of the embryonic skeleton. The primitive line also gives rise to loose embryonic connective tissue called **intraembryonic mesoderm** (mesenchyme). Mesenchyme differentiates into all the various kinds of connective tissue found in the adult. One of the earliest formed organs is the skin, which serves to support and maintain homeostasis within the embryo.

A tremendous amount of change and specialization occurs during the embryonic stage (fig. 30.18). The factors that cause precise, sequential change from one cell or tissue type to another are not fully understood. It is known, however, that the potential for change is programmed into the genetics of each cell and that under conducive environmental conditions the change takes place. The process of developmental change is referred to as **induction.** Induction occurs when one tissue, called the *inductor tissue*, has a marked effect on an adjacent tissue, causing it to become *induced tissue*, and stimulating it to differentiate.

Fourth Week

During the fourth week of development, the embryo increases about 4 mm (0.16 in.) in length. A **connecting stalk,** which is later involved in the formation of the umbilical

---

Wharton's jelly: from Thomas Wharton, English anatomist, 1614–73

induction: L. *inductus*, to lead in

Figure 30.16

The formation of the umbilical cord and other extraembryonic structures.
These structures are depicted in sagittal sections of a gravid uterus from week 4 to week 22. (*a*) A connecting stalk forms as the developing amnion expands around the embryo, finally meeting ventrally. (*b*) The umbilical cord begins to take form as the amnion ensheathes the yolk sac. (*c*) A cross section of the umbilical cord showing the embryonic vessels, mucoid connective tissue, and the tubular connection to the yolk sac. (*d*) By week 22, the amnion and chorion have fused, and the umbilical cord and placenta have become well-developed structures.

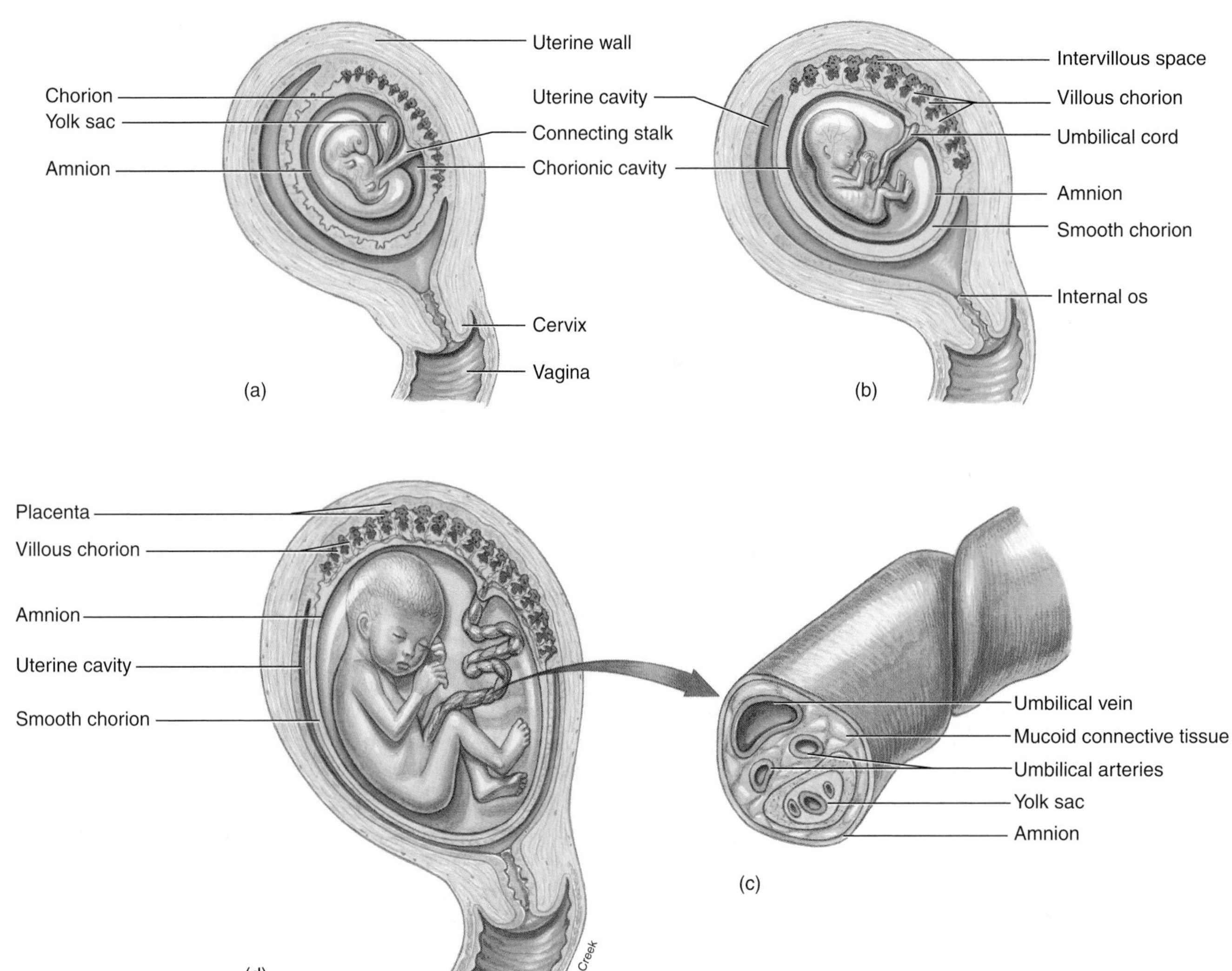

cord, is established from the body of the embryo to the developing placenta. By this time, the heart is already pumping blood to all parts of the embryo. The head and jaws are apparent, and the primordial tissue that will form the eyes, brain, spinal cord, lungs, and digestive organs has developed. The **superior** and **inferior limb buds** are recognizable as small swellings on the lateral body walls.

Fifth Week

Embryonic changes during the fifth week are not as extensive as those during the fourth week. The head enlarges, and the developing eyes, ears, and nasal pit become obvious. The appendages have formed from the limb buds, and paddle-shaped hand and foot plates develop.

Sixth Week

During the sixth week, the embryo is 16–24 mm (0.64–0.96 in.) long. The head is larger than the trunk, and the brain has undergone marked differentiation. This is the most vulnerable period of development for many organs. An interruption at this critical time can easily cause congenital damage. The limbs lengthen and are slightly flexed, and notches appear between the digital rays in the hand and foot plates. The gonads begin to produce hormones that will influence the development of the external genitalia.

Seventh and Eighth Weeks

During the last 2 weeks of the embryonic stage, the embryo, which is now 28–40 mm (1.12–1.6 in.) long, has distinct

Figure 30.17

The appearance of the primitive line and primitive node along the embryonic disc. These progressive changes occur through the process of induction.

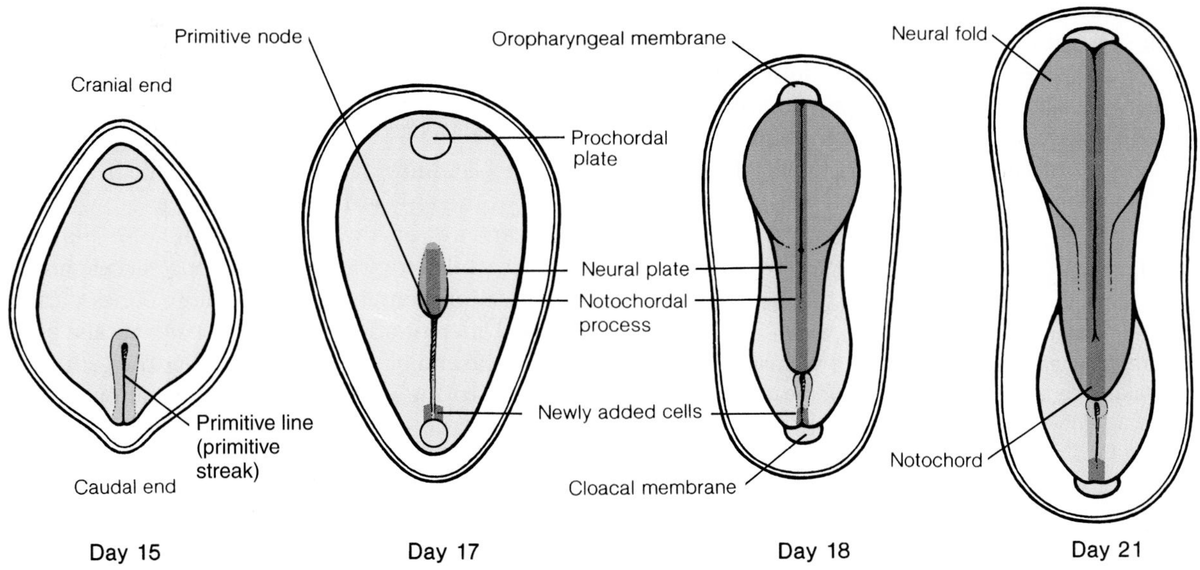

Figure 30.18

Structural changes in the embryo by weeks.

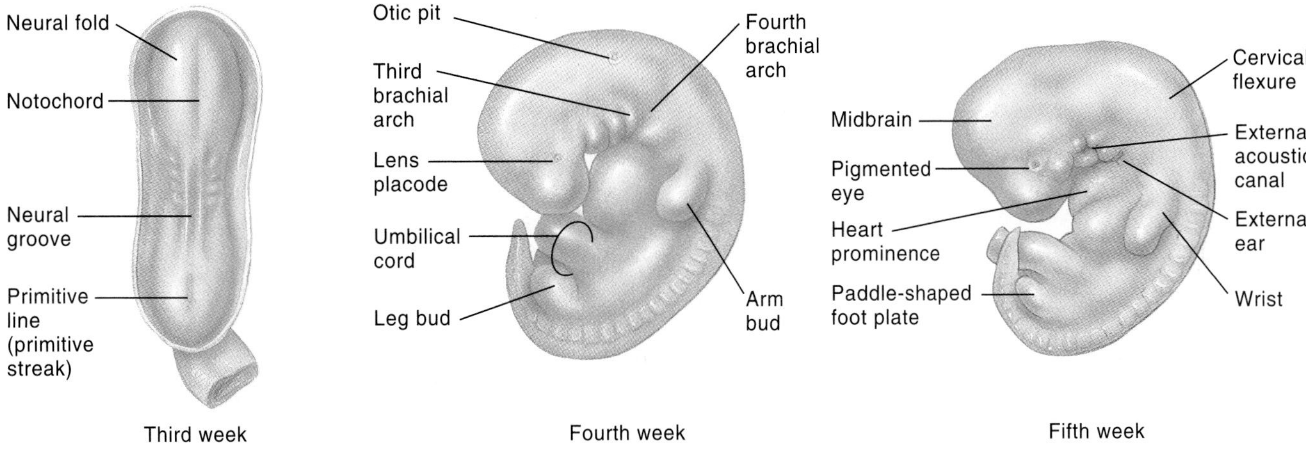

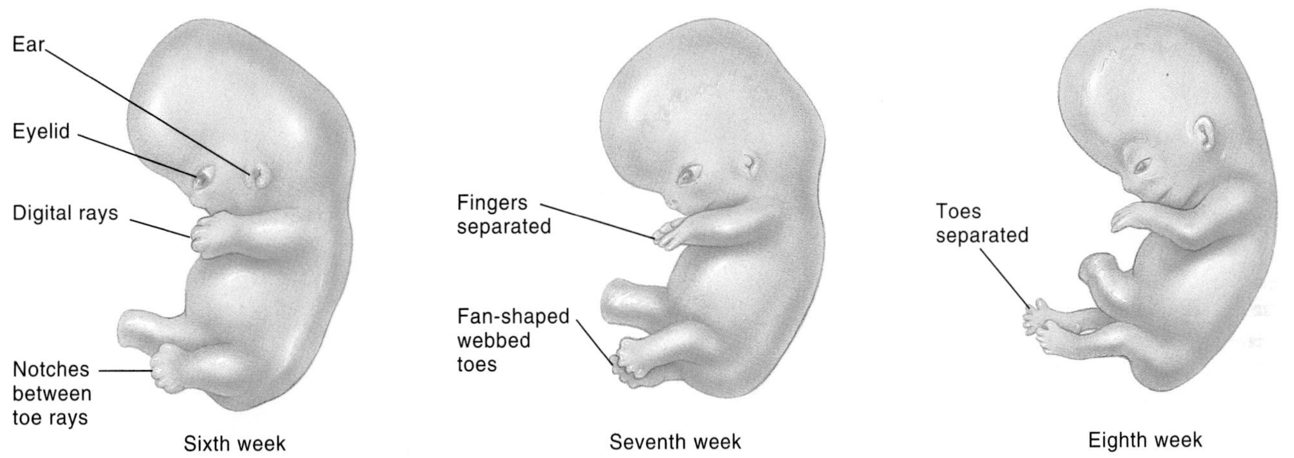

human characteristics (see fig. 30.18). The body organs are formed, and the nervous system starts to coordinate body activity. The neck region is apparent, and the abdomen is less prominent. The eyes are well developed, but the lids are stuck together to protect against probing fingers during muscular movement. The nostrils are developed but plugged with mucus. The external genitalia are forming but are still undifferentiated. The body systems are developed by the end of the eighth week, and from this time on the embryo is called a **fetus.**

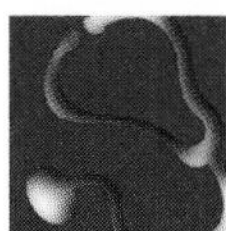

The most precarious time of prenatal development is the embryonic period—yet, well into this period many women are still unaware that they are pregnant. For this reason, a woman should abstain from taking certain drugs (including some antibiotics) if there is even a remote chance that she is pregnant or might become pregnant in the near future.

Fetal Period

The fetal period, beginning at week 9 and culminating at birth, is characterized by tremendous growth and the specialization of body structures.

Since most of the tissues and organs of the body form during the embryonic period, the fetus is recognizable as a human being at 9 weeks. The fetus is far less vulnerable than the embryo to the deforming effects of viruses, drugs, and radiation. Tissue differentiation and organ development continue during the fetal period, but to a lesser degree than before. For the most part, fetal development is primarily limited to body growth.

Structural Changes of the Fetus by Weeks

Changes in the external appearance of the fetus from the ninth through the thirty-eighth week are depicted in figure 30.19. A discussion of these structural changes follows.

Nine to Twelve Weeks

At the beginning of the ninth week, the head of the fetus is as large as the rest of the body. The eyes are widely spaced, and the ears are set low. Head growth slows during the next 3 weeks, while lengthening of the body accelerates (fig. 30.20). Ossification centers appear in most bones during the ninth week. Differentiation of the external genitalia becomes apparent at the end of the ninth week, but the genitalia are not developed to the point of sex determination until the twelfth week. By the end of the twelfth week, the fetus is 87 mm (3.5 in.) long and weighs about 45 g (1.6 oz). It can swallow, digest the fluid that passes through its system, and defecate and urinate into the amniotic fluid. Its nervous system and muscle coordination are developed to the point that it will withdraw its leg if tickled. The fetus begins inhaling through its nose but can only take in amniotic fluid.

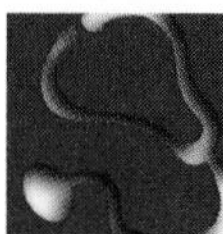

Major structural abnormalities, which may not be predictable from genetic analysis, can often be detected by *ultrasonography* (fig. 30.21). In this procedure, organs are bombarded with sound waves that reflect back in a certain pattern determined by tissue densities. For example, sound waves bouncing off amniotic fluid will produce an image much different from that produced by sound waves bouncing off the placenta or the mother's uterus. Ultrasonography is so sensitive that it can detect a fetal heartbeat several weeks before it can be heard by a stethoscope.

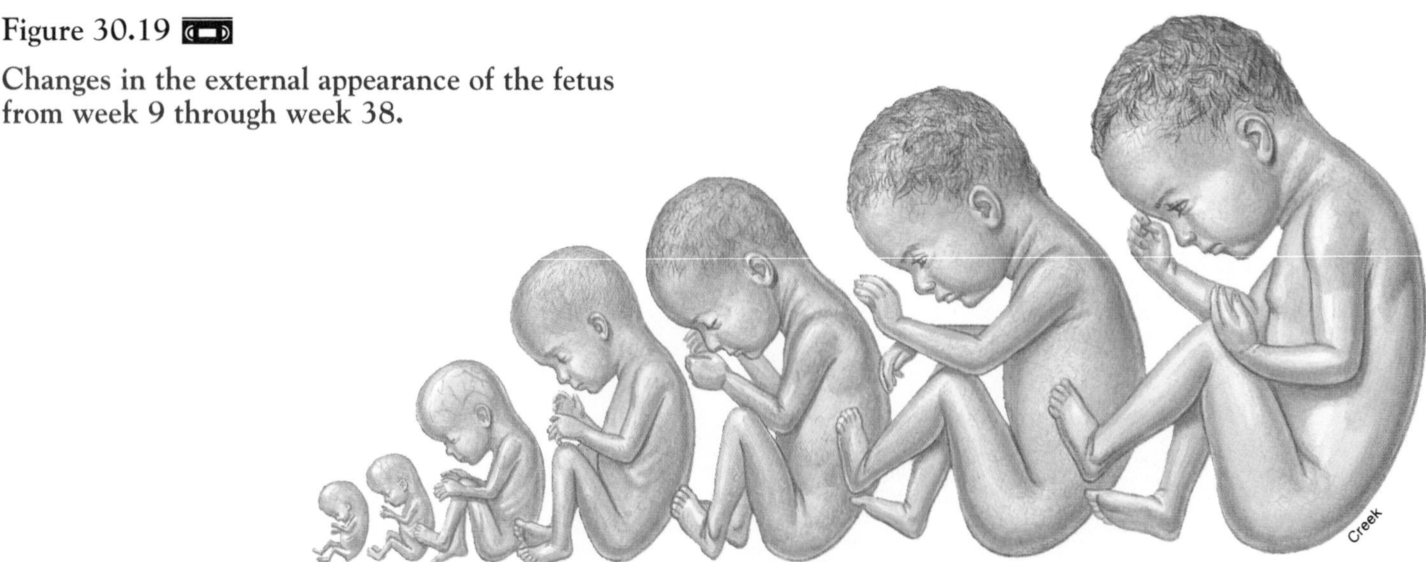

Figure 30.19
Changes in the external appearance of the fetus from week 9 through week 38.

Thirteen to Sixteen Weeks

By the thirteenth week, the facial features of the fetus are well formed, and epidermal structures such as eyelashes, eyebrows, hair on the head, fingernails, and nipples begin to develop. The appendages lengthen, and by the sixteenth week the skeleton is sufficiently developed to show up clearly on radiographs. During the sixteenth week, the fetal heartbeat can be heard by applying a stethoscope to the mother's abdomen. By the end of the sixteenth week, the fetus is 140 mm long (5.5 in.) and weighs about 200 g (7 oz).

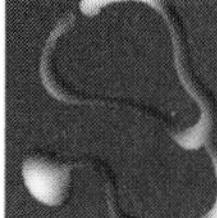

After the sixteenth week, fetal length can be determined from radiographs. The reported length of a fetus is generally derived from a straight-line measurement from the crown of the head to the developing ischium (crown-rump length). Measurements made on an embryo prior to the fetal stage, however, are not reported as crown-rump measurements but as total length.

Seventeen to Twenty Weeks

Between the seventeenth and twentieth weeks, the legs achieve their final relative proportions, and fetal movements, known as **quickening,** are commonly felt by the mother. The skin of the fetus is covered with a thin, white, cheeselike material known as **vernix caseosa** (*ka″se-o′sa*). It consists of fatty secretions from the sebaceous glands and dead epidermal cells. The function of vernix caseosa is to protect the fetus while it is bathed in amniotic fluid. Twenty-week-old fetuses usually have fine, silklike fetal hair called **lanugo** (*lă-noo′go*) covering the skin. Lanugo is thought to hold the vernix caseosa on the skin and produce a ciliarylike motion that moves amniotic fluid. A 20-week-old fetus is about 190 mm (7.5 in.) long, and it weighs about 460 g (16 oz). Because of cramped space, it develops a marked spinal

vernix caseosa: L. *vernix*, varnish; *caseus*, cheese
lanugo: L. *lana*, wool

Figure 30.20
A photographic summary of embryonic and early fetal development.

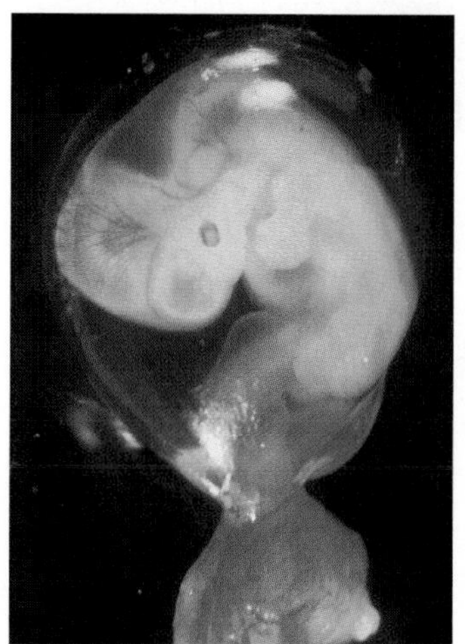
Five weeks

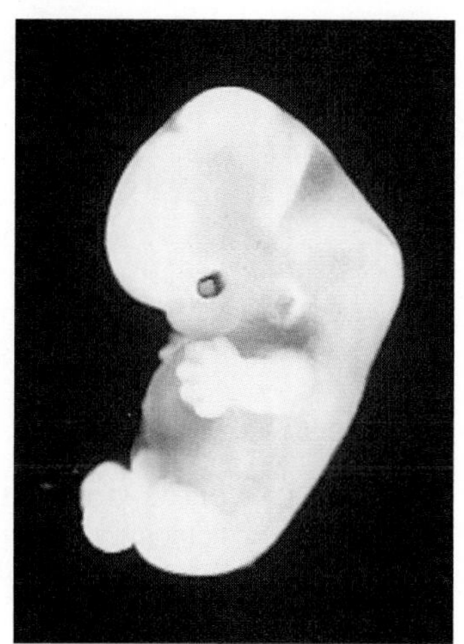
Six weeks

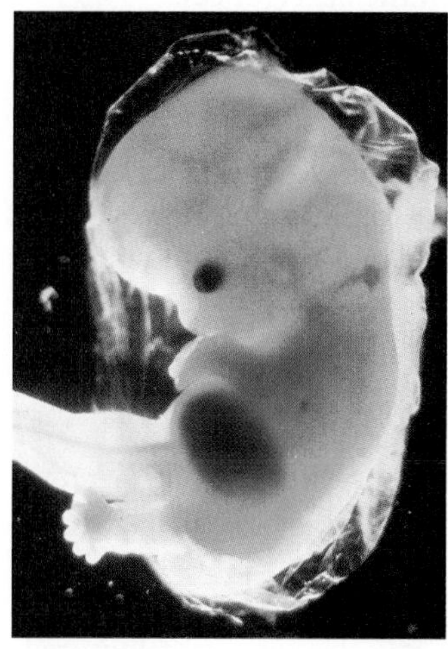
Seven weeks

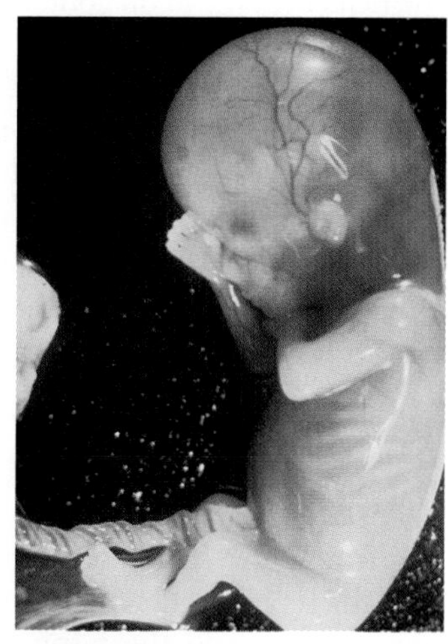
Eight weeks

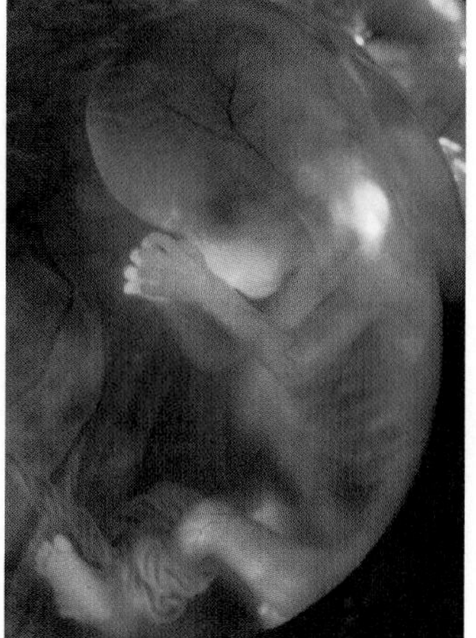
Ten weeks

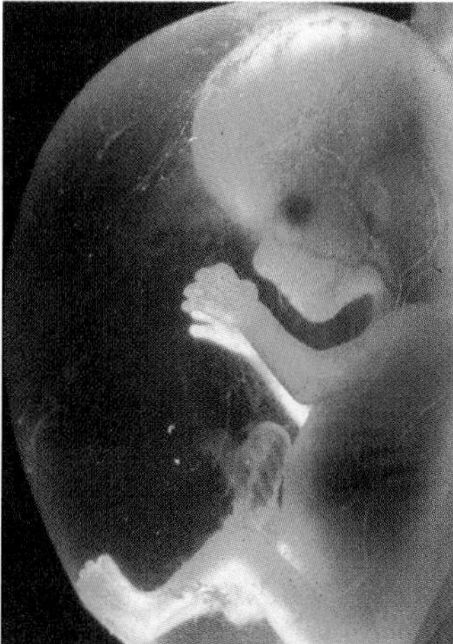
Twelve weeks

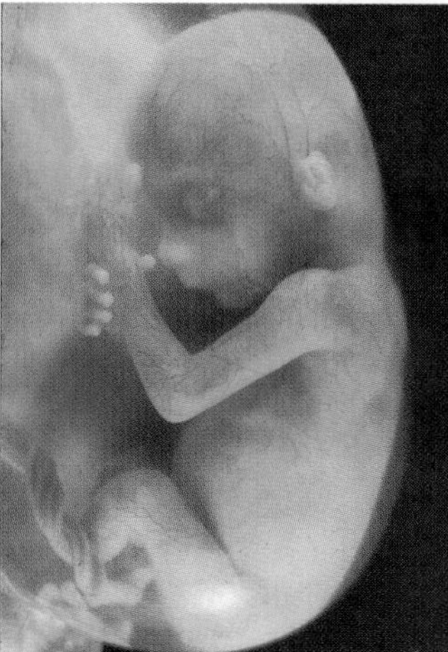
Fourteen weeks

Figure 30.21

Ultrasonography.

(*a*) Sound-wave vibrations are reflected from the internal tissues of a person's body. (*b*) Structures of the human fetus observed through an ultrasound scan.

(a)

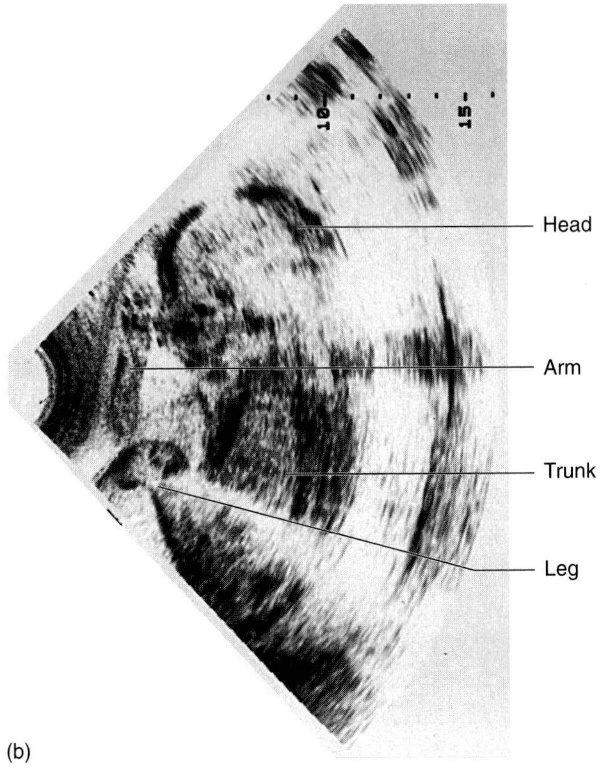

(b)

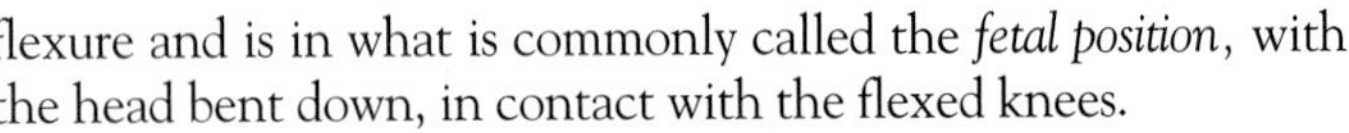
flexure and is in what is commonly called the *fetal position*, with the head bent down, in contact with the flexed knees.

Twenty-One to Twenty-Five Weeks

Between the twenty-first and twenty-fifth weeks, the fetus increases its weight substantially to about 900 g (32 oz). Body length increases only moderately (to 240 mm) however, so the weight is evenly proportioned. The skin is quite wrinkled and is translucent. Because the blood flowing in the capillaries is now visible, the skin appears pinkish.

Twenty-Six to Twenty-Nine Weeks

Toward the end of this period, the fetus will be about 275 mm (11 in.) long and will weigh about 1,300 g (46 oz). A fetus might now survive if born prematurely, but the mortality rate is high. Its body metabolism cannot yet maintain a constant temperature, and the respiratory muscles have not matured enough to provide a regular respiratory rate. If, however, the premature infant is put in an incubator, and a respirator is used to maintain its breathing, it may survive. The eyes open during this period, and the body is well covered with lanugo. If the fetus is a male, the testes should have begun descent into the scrotum. As the time of birth approaches, the fetus rotates to a **vertex position** in which the head is directed toward the cervix (fig. 30.22). The head repositions toward the cervix because of the shape of the uterus and because the head is the heaviest part of the body.

Thirty to Thirty-Eight Weeks

At the end of 38 weeks, the fetus is considered full-term. It has reached a crown-rump length of 360 mm (14 in.) and weighs about 3,400 g (7.5 lb). The average total length from

vertex: L. *vertex*, summit

Figure 30.22

A fetus in vertex position.

Toward the end of most pregnancies, the weight of the fetus' head causes the body to rotate, positioning the head against the cervix of the uterus.

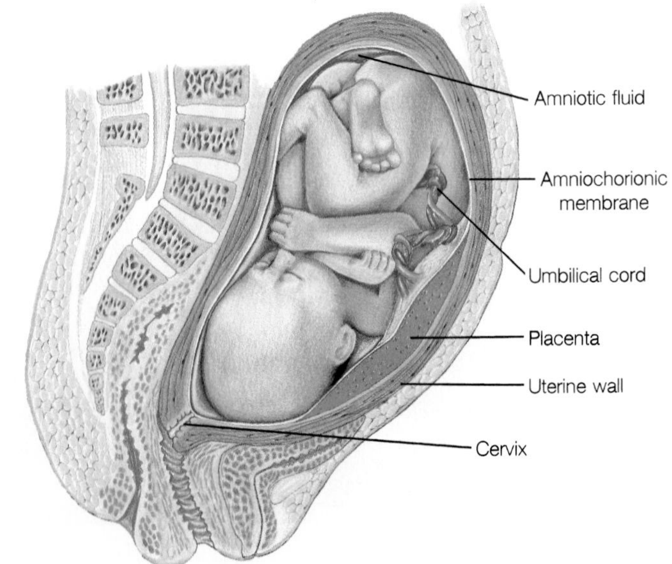

crown to heel is 50 cm (20 in.). Most fetuses are plump with smooth skin because of the accumulation of subcutaneous fat. The skin is pinkish-blue, even in fetuses of dark-skinned parents, because melanocytes do not produce melanin until the skin is exposed to sunlight. Lanugo is sparse and is generally found on the head and back. The chest is prominent, and the mammary area protrudes in both sexes. The external genitalia are somewhat swollen.

Labor and Parturition

Parturition, or childbirth, involves a sequence of events called labor. The uterine contractions of labor require the action of oxytocin, secreted by the posterior pituitary, and prostaglandins, produced in the uterus.

The time of prenatal development, or the time of pregnancy, is called **gestation.** In humans the average gestation time is usually 266 days, or about 280 days from the beginning of the last menstrual period to **parturition,** or birth. Most fetuses are born within 10 to 15 days before or after the calculated delivery date. Parturition is accompanied by a sequence of physiological and physical events called **labor.**

The *onset of labor* is denoted by rhythmic and forceful contractions of the myometrial layer of the uterus. In *true labor,* the pains from uterine contractions occur at regular intervals and intensify as the interval between contractions shortens. A reliable indication of true labor is dilation of the cervix and a "show," or discharge, of blood-containing mucus that has accumulated in the cervical canal. In *false labor,* abdominal pain is experienced at irregular intervals, and cervical dilation and "show" are absent.

The uterine contractions of labor are stimulated by two agents: (1) **oxytocin** (*ok″sĭ-to′sin*), a polypeptide hormone produced in the hypothalamus and released from the posterior pituitary, and (2) **prostaglandins** (*pros″tă-glan′dinz*), a class of fatty acids produced within the uterus itself. Labor can indeed be induced artificially by injections of oxytocin or by the insertion of prostaglandins into the vagina as a suppository.

Although the mechanisms operating during labor are well established, the factors responsible for initiating labor in humans are not completely understood. In sheep, the fetal adrenal cortex is known to play a critical role in timing the onset of labor and delivery. The fetal hypothalamus and anterior pituitary secrete corticotropin-releasing hormone (CRH) and ACTH, respectively, which stimulate the fetal adrenal cortex to secrete corticosteroids. The rising level of corticosteroids, in turn, stimulates the placenta to produce the enzymes that convert progesterone into estrogen. As a result, plasma progesterone levels fall and estrogen levels rise. Progesterone inhibits activity of the myometrium, while estrogen stimulates the ability of the myometrium to contract. As a result of these events, the birth process in sheep is known to be initiated by the fetal adrenal cortex.

gestation: L. *gestatus*, to bear

The initiation of labor in humans and other primates is more complex. The human placenta lacks the enzymes needed to convert progesterone into estrogen; it can only make estrogen when it is supplied with androgens from the fetus (see fig. 30.15). Therefore, despite a rise in fetal cortisol secretion, there is no accompanying fall in progesterone secretion in humans to trigger labor. There is, however, a rise in estradiol secretion from the human placenta in late pregnancy, probably as a result of a rise in the secretion of androgens (primarily *dehydroepiandrosterone*, or *DHEA*) from the fetal adrenal cortex. Indeed, injections of DHEA into monkeys can provoke labor, suggesting that this is a key event in the timing of birth in primates.

However it is timed, the DHEA secreted by the fetal adrenal cortex is converted into estrogens by the placenta, and the rising level of estrogen secretion in late pregnancy promotes the contractility of the myometrium. Estrogen may do this by (1) increasing the amount of oxytocin produced by the mother's hypothalamus and stored in the posterior pituitary, (2) increasing the production of oxytocin receptors in the myometrium, and (3) stimulating the production of prostaglandins in the uterus. As illustrated in figure 30.23, labor is divided into three stages:

1. **Dilation stage.** In this period, the cervix dilates to a diameter of approximately 10 cm. Contractions are regular during this stage, and the amniotic sac (bag of waters) generally ruptures. If the amniotic sac does not rupture spontaneously, it is broken surgically. The dilation stage generally lasts 8 to 24 hours.
2. **Expulsion stage.** This is the period of parturition, or actual childbirth. It consists of forceful uterine contractions and abdominal compressions to expel the fetus from the uterus and through the vagina. This stage may require 30 minutes in a first pregnancy but only a few minutes in subsequent pregnancies.
3. **Placental stage.** Generally within 10 to 15 minutes after parturition, the placenta separates from the uterine wall and is expelled as the *afterbirth*. Forceful uterine contractions characterize this stage, constricting uterine blood vessels to prevent hemorrhage. In a normal delivery, blood loss does not exceed 350 ml.

A *pudendal nerve block* may be administered during the early part of the expulsion stage to ease the trauma of delivery for the mother and to allow for an episiotomy. *Epidural anesthesia*, in which a local anesthetic is injected into the epidural space of the lumbar region of the spine, also may be used for these purposes.

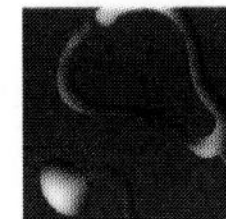

Five percent of newborns are born *breech.* In a breech birth, the fetus has not rotated and the buttocks are the presenting part. The principal concern of a breech birth is the increased time and difficulty of the expulsion stage of parturition. Attempts to rotate the fetus through the use of forceps may injure the infant. If an infant cannot be delivered breech, a *cesarean* (*sĭ-zar′e-an*) *section* must be performed. A cesarean section is delivery of the fetus through an incision made into the abdominal wall and the uterus.

Figure 30.23

The stages of labor and parturition.
(*a*) The position of the fetus prior to labor. (*b*) The ruptured amniotic sac during the *early dilation stage* of the cervix. (*c*) The *expulsion stage*, or period of parturition. (*d*) The *placental stage*, as the afterbirth is being expelled.

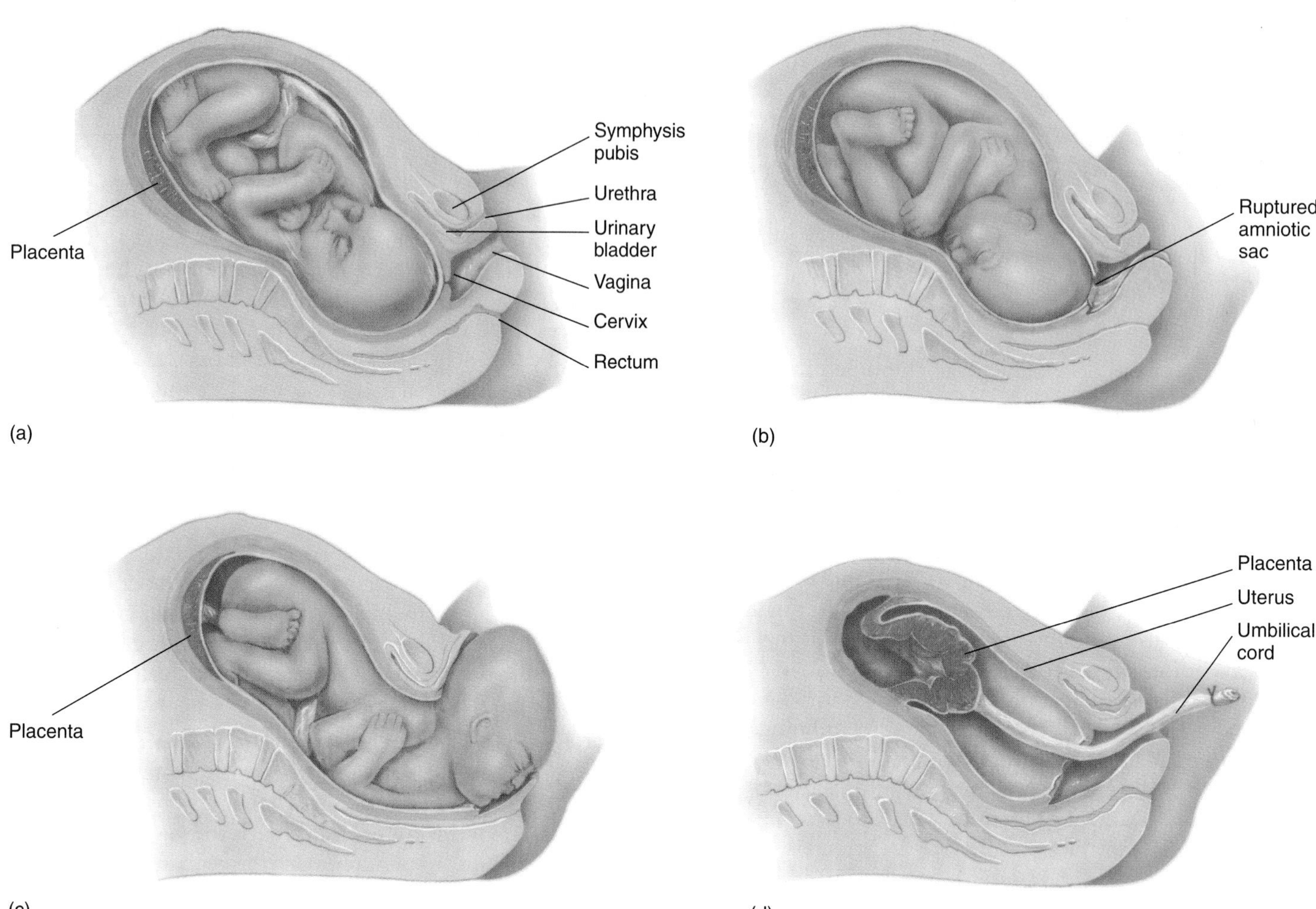

Inheritance

Inheritance is the acquisition of characteristics or qualities by transmission from parent to offspring. Hereditary information is transmitted by genes.

Genetics is the branch of biology that deals with inheritance. Genetics and inheritance are important in anatomy and physiology because of the numerous developmental and functional disorders that have a genetic basis. Knowledge of which disorders and diseases are inherited finds practical application in genetic counseling. The genetic inheritance of an individual begins with conception.

Each zygote inherits 23 chromosomes from the mother and 23 chromosomes from the father. This does not produce 46 different chromosomes; rather, it produces 23 pairs of *homologous chromosomes*. Each member of a homologous pair, with the important exception of the sex chromosomes, looks like the other and contains similar genes (such as those coding for eye color, height, and so on). These homologous pairs of chromosomes can be **karyotyped** (photographed or illustrated) and identified (as shown in fig. 30.24). Each cell that contains 46 chromosomes (that is, *diploid*) has two number 1 chromosomes, two number 2 chromosomes, and so on through chromosomes number 22. The first 22 pairs of chromosomes are called **autosomal** (*aw"to-so'mal*) **chromosomes.** The twenty-third pair of chromosomes are the **sex chromosomes,** which may look different and may carry different genes. In a female, these consist of two X chromosomes, whereas in a male there is one X chromosome and one Y chromosome.

Genes and Alleles

A **gene** is the portion of the DNA of a chromosome that contains the information needed to synthesize a particular protein molecule. Although each diploid cell has a pair of genes

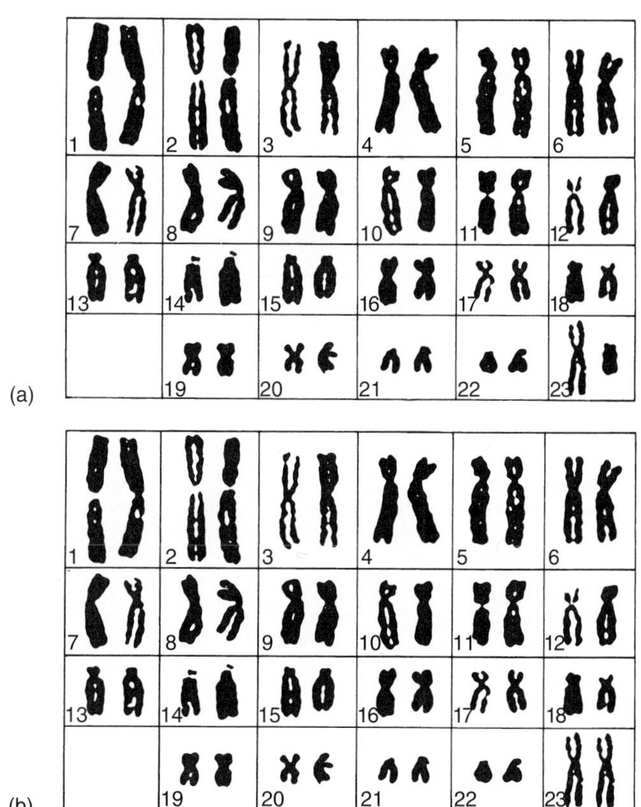

Figure 30.24

A karyotype of homologous pairs of chromosomes obtained from a human diploid cell. The first 22 pairs of chromosomes are called the autosomal chromosomes. The last pairs (number 23) are the sex chromosomes. The sex chromosomes are (*a*) XY for a male and (*b*) XX for a female.

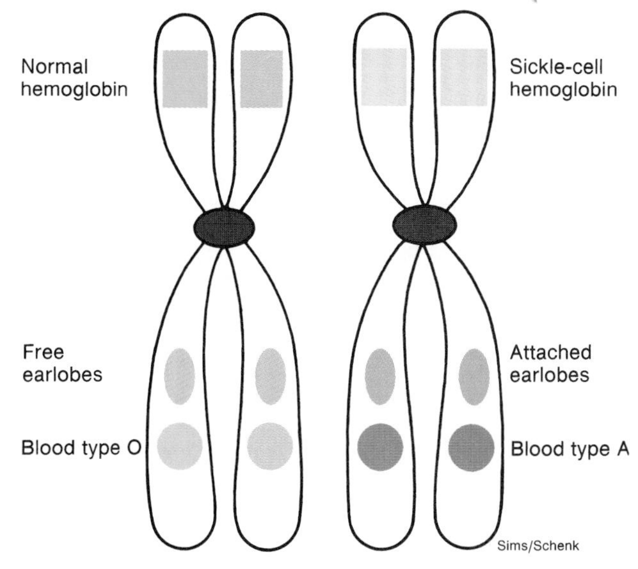

Figure 30.25

A pair of homologous chromosomes. Homologous chromosomes contain genes for the same characteristic at the same locus.

for each characteristic, these genes may be present in variant forms. Those alternate forms of a gene that affect the same characteristic but that produce different expressions of that characteristic are called **alleles** (*ă-lēlz*). One allele of each pair originates from the female parent and the other from the male. The shape of a person's ears, for example, is determined by the kind of allele received from each parent and how the alleles interact with one another. Alleles are always located on the same spot (called a **locus**) on homologous chromosomes (fig. 30.25).

For any particular pair of alleles in a person, the two alleles are either identical or not identical. If the alleles are identical, the person is said to be **homozygous** (*ho″mo-zi′gus*) for that particular characteristic. But if the two alleles are different, the person is **heterozygous** (*het″er-o-zi′gus*) for that particular trait.

Genotype and Phenotype

A person's DNA contains a catalog of genes known as the **genotype** (*jen′o-tip*) of that person. The expression of those genes results in certain observable characteristics referred to as the **phenotype** (*fe′no-tip*).

Table 30.3

Hereditary Traits in Humans Determined by Single Pairs of Dominant and Recessive Alleles

Dominant	Recessive	Dominant	Recessive
Free earlobes	Attached earlobes	Color vision	Color blindness
Dark brown hair	All other colors	Broad lips	Thin lips
Curly hair	Straight hair	Ability to roll tongue	Lack of this ability
Pattern baldness (♂♂)	Baldness (♀♀)	Arched feet	Flat feet
Pigmented skin	Albinism	A or B blood factor	O blood factor
Brown eyes	Blue or green eyes	Rh blood factor	No Rh blood factor

If the alleles for a particular trait are homozygous, the characteristic expresses itself in a specific manner (two alleles for attached earlobes, for example, results in a person with attached earlobes). If the alleles for a particular trait are heterozygous, however, the allele that expresses itself and the way in which the genes for that trait interact will determine the phenotype. The allele that expresses itself is called the **dominant allele;** the one that does not is the **recessive allele.** The various combinations of dominant and recessive alleles are responsible for a person's hereditary traits (table 30.3).

Figure 30.26

Inheritance of earlobe characteristics.
Two parents with unattached (free) earlobes can have a child with attached earlobes.

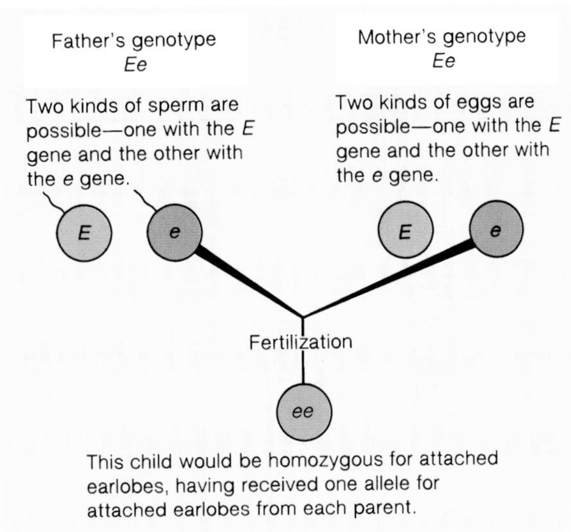

Figure 30.27

A Punnett square.
A Punnett square is used to determine genotypes and phenotypes that could result from the mating of two heterozygous parents.

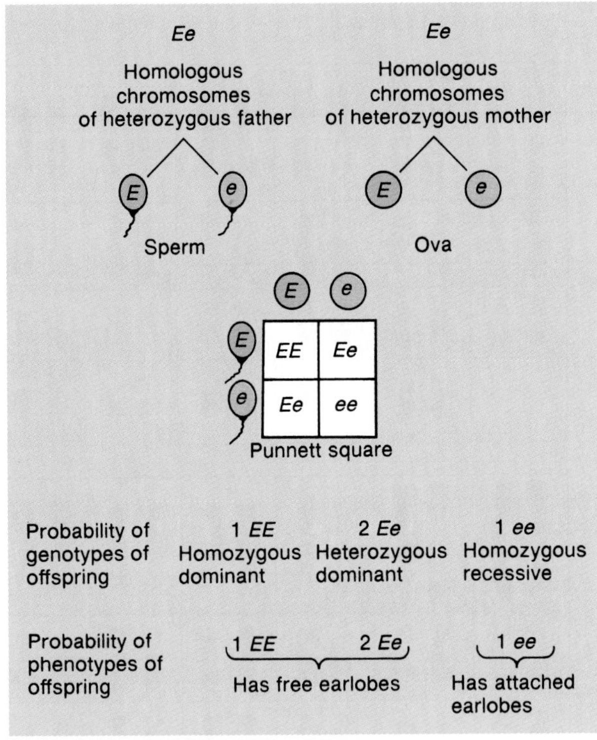

In describing genotypes, it is traditional to use letter symbols to refer to the alleles of an organism. The dominant alleles are symbolized by uppercase letters, and the recessive alleles are symbolized by lowercase letters. Thus, the *genotype* of a person who is homozygous for free earlobes as a result of a dominant allele is symbolized *EE*; a heterozygous pair is symbolized *Ee*. In both of these instances, the *phenotypes* of the individuals would be free earlobes resulting from the presence of a dominant allele in each genotype. A person who inherited two recessive alleles for earlobes would have the genotype *ee* and would have attached earlobes.

Thus, three genotypes are possible when gene pairing involves dominant and recessive alleles. They are *homozygous dominant* (*EE*), *heterozygous* (*Ee*), and *homozygous recessive* (*ee*). Only two phenotypes are possible, however, since the dominant allele is expressed in both the homozygous dominant (*EE*) and the heterozygous (*Ee*) individuals. The recessive allele is expressed only in the homozygous recessive (*ee*) condition. Refer to figure 30.26 for an illustration of how a homozygous recessive trait may be expressed in a child of parents who are heterozygous.

Probability

A **Punnett** (*pun'et*) **square** is a convenient way to express the probabilities of allele combinations for a particular inheritable trait. In constructing a Punnett square, the male gametes (spermatozoa) carrying a particular trait are placed at the side of the chart, and the female gametes (ova) are at the top, as in figure 30.27. The four spaces on the chart represent the possible combinations of male and female gametes that could form zygotes. The probability of an offspring having a particular genotype is 1 in 4 (0.25) for homozygous dominant and homozygous recessive and 1 in 2 (0.50) for heterozygous.

A genetic study in which a single characteristic (e.g., ear shape) is followed from parents to offspring is referred to as a **monohybrid cross.** A genetic study in which two characteristics are followed from parents to offspring is referred to as a **dihybrid cross** (fig. 30.28). The term *hybrid* refers to an offspring descended from parents who have different genotypes.

Sex-Linked Inheritance

Certain inherited traits are located on a sex-determining chromosome and are called **sex-linked** characteristics. The allele for *red-green color blindness*, for example, is determined by a recessive allele (designated c) found on the X chromosome but not on the Y chromosome. Normal color vision (designated C) dominates. The ability to discriminate red from green therefore, depends entirely on the X chromosomes. The genotype possibilities are as follows:

X^CY	Normal male
X^cY	Color-blind male
X^CX^C	Normal female
X^CX^c	Normal female carrying the recessive allele
X^cX^c	Color-blind female

In order for a female to be red-green color-blind, she must have the recessive allele on both of her X chromosomes. Her father would have to be red-green color-blind, and her mother would have to be a carrier for this condition. A male with

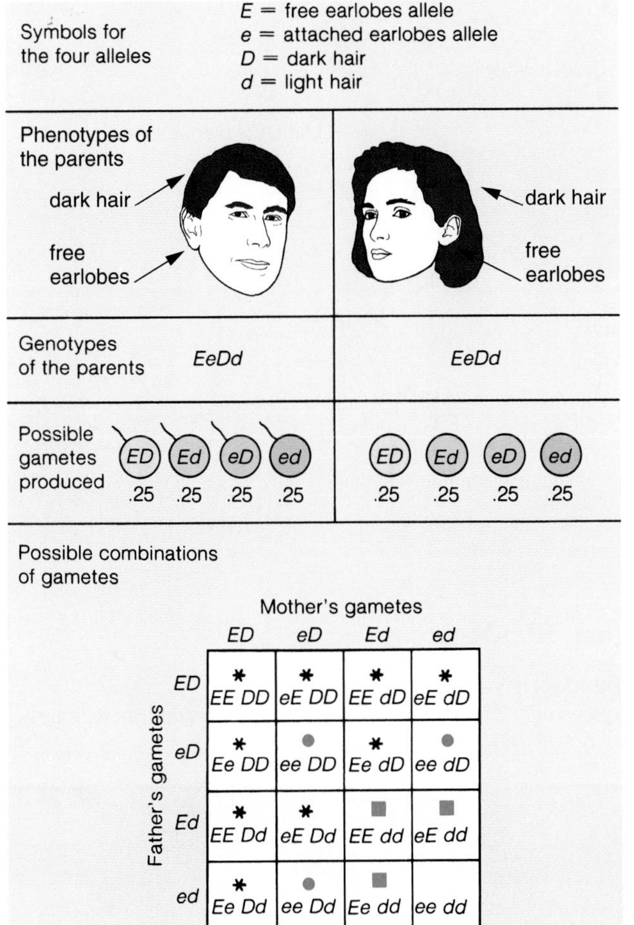

Figure 30.28

A dihybrid cross.
In a dihybrid cross, two pairs of traits are followed simultaneously. Any of the combinations of genes that have a *D* and an *E* (nine possibilities) will have free earlobes and dark hair. These are indicated with an asterisk (*). Three of the possible combinations have two alleles for attached earlobes (*ee*) and at least one allele for dark hair. They are indicated with a dot (•). Three of the combinations have free earlobes and light hair. These are indicated with a square (■). The remaining possibility has the genotype *eedd* for attached earlobes and light hair.

only one such allele on his X chromosome, however, will show the characteristic. Since a male receives his X chromosome from his mother, the inheritance of sex-linked characteristics usually passes from mother to son.

Hemophilia is a sex-linked condition caused by a recessive allele. The blood in a person with hemophilia fails to clot or clots very slowly after an injury. If *H* represents normal clotting and *h* represents abnormal clotting, then males with X^HY will be normal, and males with X^hY will be hemophiliac. Females with X^hX^h will have the disorder.

Clinical Considerations

Pregnancy and childbirth are natural events in human biology and generally progress smoothly without complications. Prenatal development is amazingly precise, and although traumatic, childbirth for most women in the world takes place without the aid of a physician. Occasionally, however, serious complications arise, and the knowledge of an obstetrician is required. The physician's knowledge of what constitutes normal development and what factors are responsible for congenital malformations ensures the embryo and fetus every possible chance to develop normally. Many of the clinical aspects of prenatal development involve what might be referred to as applied developmental biology.

In clinical terms, gestation is frequently divided into three phases, or **trimesters,** each lasting three calendar months. By the end of the **first trimester,** all of the major body systems are formed, the fetal heart can be detected, the external genitalia are developed, and the fetus is about the width of the palm of an adult's hand. During the **second trimester,** fetal quickening can be detected, epidermal features are formed, and the vital body systems are functioning. The fetus, however, would be still unlikely to survive if birth were to occur. At the end of the second trimester, fetal length is about equal to the length of an adult's hand. The fetus experiences a tremendous amount of growth and refinement in system functioning during the **third trimester.** A fetus of this age may survive if born prematurely, and of course, the chances of survival improve as the length of pregnancy approaches the natural delivery date.

Many clinical considerations are associated with prenatal development, some of which relate directly to the female reproductive system. Other developmental problems are genetically related and will be mentioned only briefly. Of clinical concern for developmental anatomy are such topics as ectopic pregnancies, so-called test-tube babies, multiple pregnancy, fetal monitoring, and congenital defects.

Abnormal Implantation Sites

In an **ectopic** (*ek-top'ik*) **pregnancy,** the blastocyst implants outside the uterus or in an abnormal site within the uterus (fig. 30.29). About 95% of the time, the ectopic location is within the uterine tube and is referred to as a **tubal pregnancy.** Occasionally, implantation occurs near the cervix, where development of the placenta blocks the cervical opening. This condition, called **placenta previa** (*pre've-a*), causes serious bleeding. Ectopic pregnancies will not develop normally in unfavorable locations, and the fetus seldom survives beyond the first trimester. Tubal pregnancies are terminated through medical intervention. If a tubal pregnancy is permitted to progress,

previa: L. *previa*, appearing before or in front of

Figure 30.29

Sites of ectopic pregnancies.
The normal implantation site is indicated by an X; the abnormal sites are indicated by letters in order of frequency of occurrence.

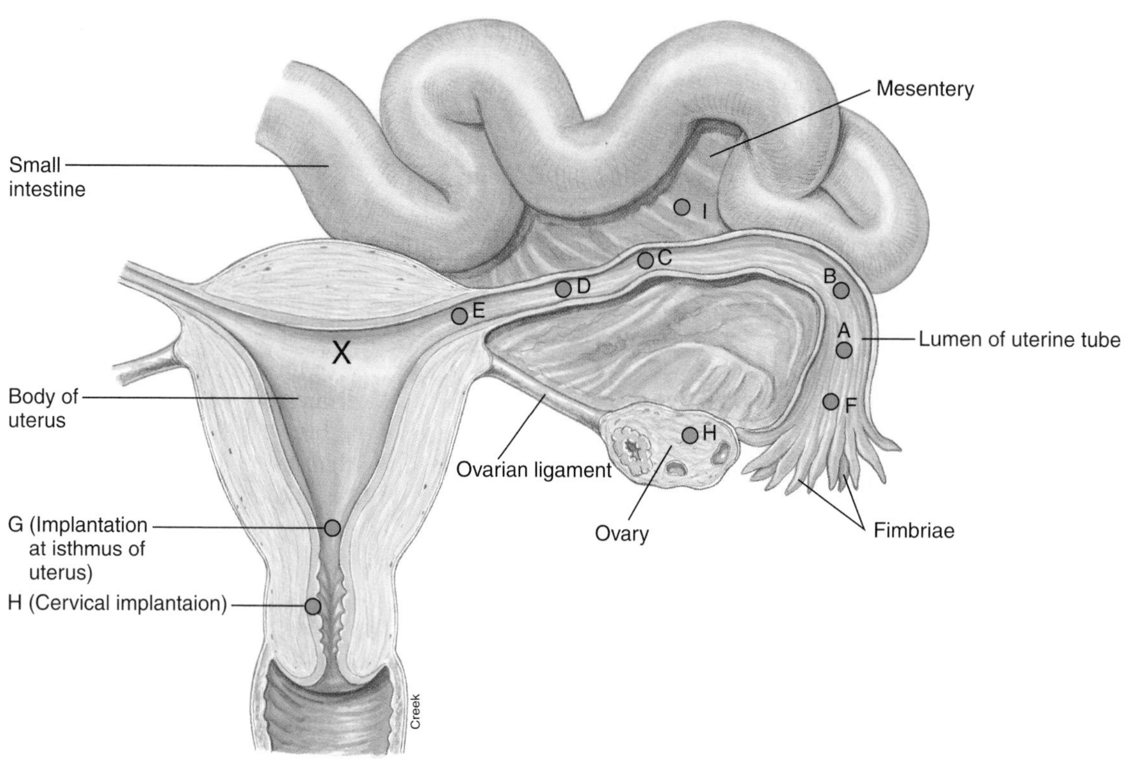

the uterine tube generally ruptures, followed by hemorrhaging. Depending on the location and the stage of development (hence vascularity) of a tubal pregnancy, it may or may not be life-threatening to the woman.

In Vitro Fertilization and Artificial Implantation

Reproductive biologists have been able to fertilize a human oocyte in vitro (outside the body), culture it to the blastocyst stage, and then perform artificial implantation, leading to a full-term development and delivery. This is the so-called test-tube baby. To obtain the oocyte, a specialized laparoscope (fig. 30.30) is used to aspirate the preovulatory egg from a graafian follicle. The oocyte is then placed in a suitable culture medium, where it is fertilized with spermatozoa. After the zygote forms, the sequential preembryonic development continues until the blastocyst stage, at which time implantation is performed. In vitro fertilization with artificial implantation is a means of overcoming infertility problems due to damaged, blocked, or missing uterine tubes in females or low sperm counts in males.

Multiple Pregnancy

Twins occur about once in 85 pregnancies. They can develop in two ways. **Dizygotic** (*di″zi-got′ik*) (fraternal) **twins** develop from two zygotes produced by the fertilization of two oocytes by two spermatozoa in the same ovulatory cycle (fig. 30.31). **Monozygotic** (identical) **twins** form from a single zygote (fig. 30.32). Approximately one-third of twins are monozygotic.

Figure 30.30

Laparoscopy.
A laparoscope, used for various abdominal operations, including the extraction of a preovulatory ovum.

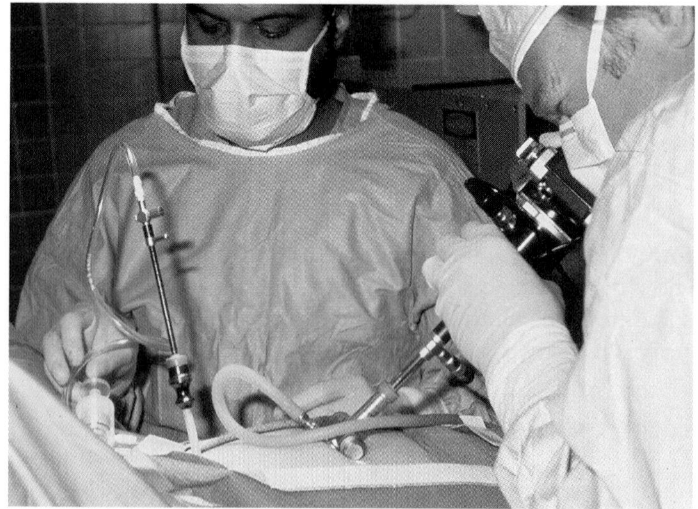

Figure 30.31

The formation of dizygotic twins.

Twins of this type are fraternal rather than identical and may have (*a*) separate or (*b*) fused placentas. (*c*) A photograph of fraternal twins at 11 weeks.

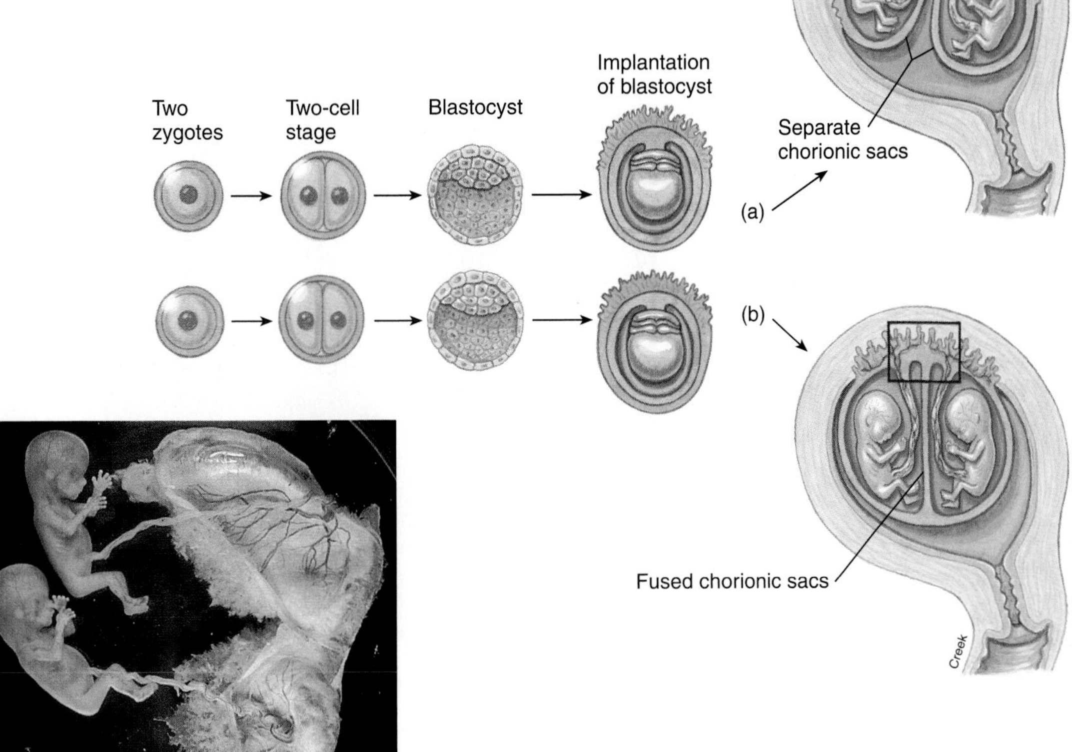

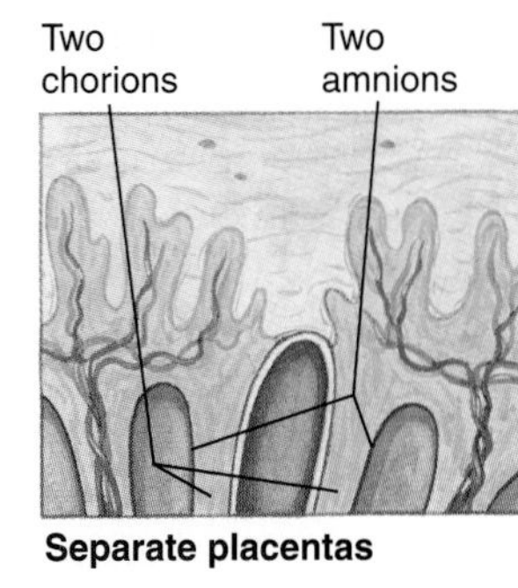

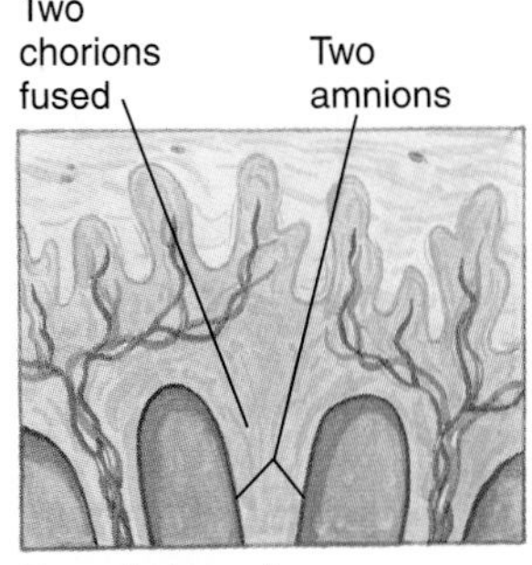

Figure 30.32

The formation of monozygotic twins.

Twins of this type develop from a single zygote and are identical. Such twins have two amnions but only one chorion and they share a common placenta.

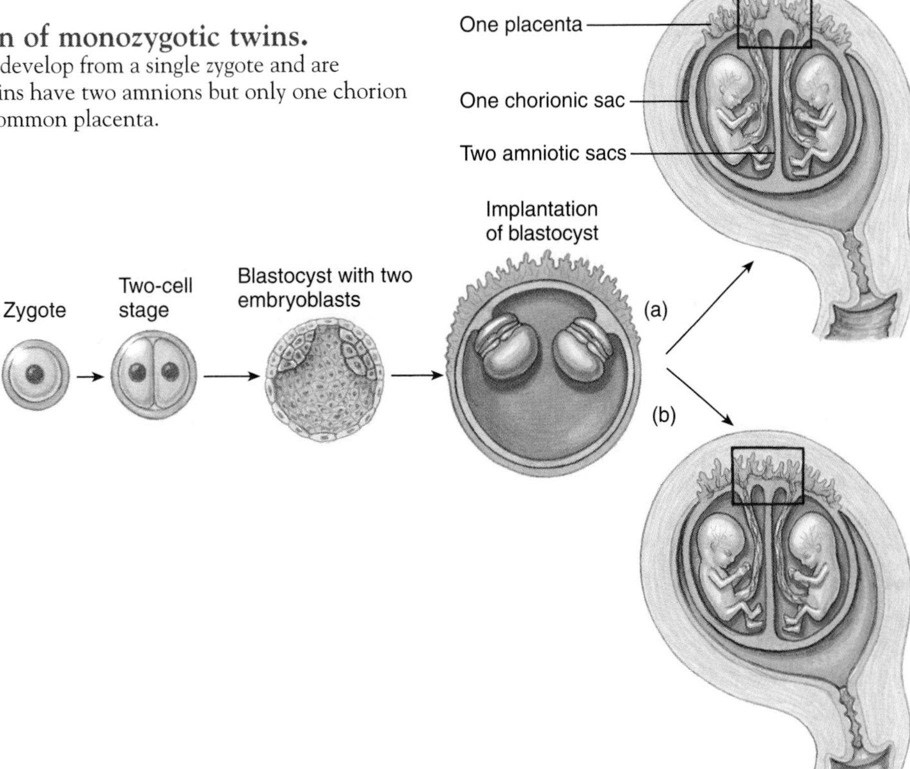

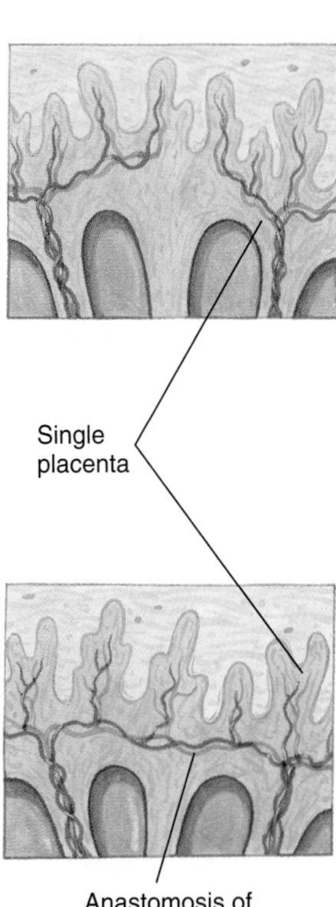

Dizygotic twins may be of the same sex or different sexes and are not any more alike than brothers or sisters born at different times. Dizygotic twins always have two chorions and two amnions, but the chorions and the placentas may be fused.

Monozygotic twins are of the same sex and are genetically identical. Any physical differences in monozygotic twins are caused by environmental factors during morphogenic development (e.g., there might be a differential vascular supply that causes slight differences to be expressed). Monozygotic twinning is usually initiated toward the end of the first week when the embryoblast divides to form two embryonic primordia. Monozygotic twins have two amnions but only one chorion and a common placenta. If the embryoblast fails to completely divide, **conjoined twins** (Siamese twins) may form.

Triplets occur about once in 7,600 pregnancies and may be (1) all from the same ovum and identical, (2) two identical and the third from another ovum, or (3) three zygotes from three different ova. Similar combinations occur in quadruplets, quintuplets, and so on.

Fetal Monitoring

Obstetrics has benefited greatly from the advancements made in fetal monitoring in the last two decades. Before modern techniques became available, physicians could only determine the welfare of the unborn child by auscultation of the fetal heart and palpation of the fetus. Currently, several tests may be used to gain information about the fetus during any stage of development. Fetal conditions that can now be diagnosed and evaluated include genetic disorders, hypoxia, blood disorders, growth retardation, placental functioning, prematurity, postmaturity, and intrauterine infections. These tests also help determine the advisability of an abortion.

Radiographs of the fetus were once commonly performed but were found harmful and have been replaced by other methods of evaluation that are safer and more informative. **Ultrasonography** employs a mechanical vibration of high frequency to produce a safe, high-resolution (sharp) image of fetal structure (fig. 30.33). Ultrasonic imaging is a reliable way to determine pregnancy as early as 6 weeks after ovulation. It can also be used to determine fetal weight, length, and position, as well as to diagnose multiple fetuses.

Amniocentesis is a technique used to obtain a small sample of amniotic fluid so that it can be assessed genetically and biochemically (see fig. 30.13). A wide-bore needle is inserted into the amniotic sac with guidance by ultrasound, and 5–10 ml of amniotic fluid is withdrawn with a syringe. Amniocentesis is most often performed to determine fetal maturity, but it can also help to predict such serious disorders as *Down syndrome* and *Gaucher's disease* (a metabolic disorder).

Fetoscopy (fig. 30.34) goes beyond amniocentesis by allowing direct examination of the fetus. Using fetoscopy, physicians scan the uterus with pulsed sound waves to locate fetal structures, the umbilical cord, and the placenta. Skin

amniocentesis: Gk. *amnion*, lamb (fetal membrane); *kentesis*, puncture
fetoscopy: L. *fetus*, offspring; *skopein*, to view

Figure 30.33

An ultrasonogram.
A color-enhanced ultrasonogram of a fetus during the third trimester. The left hand is raised, as if waving to the viewer.

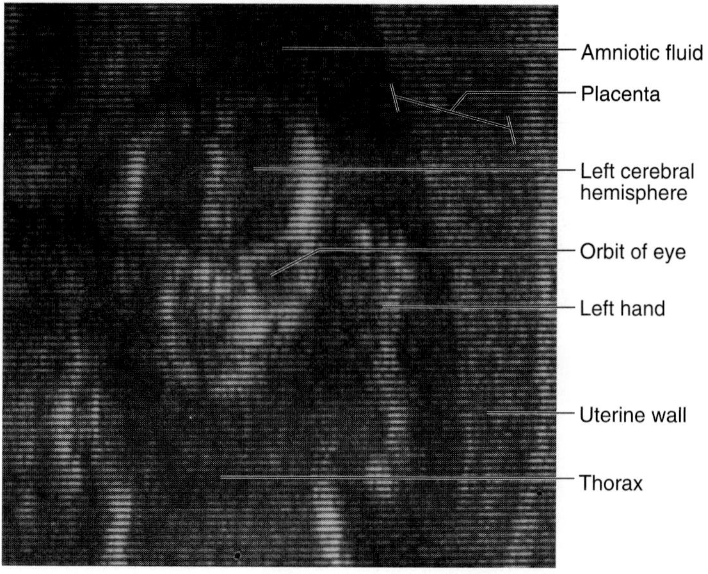

Figure 30.34

Fetoscopy.

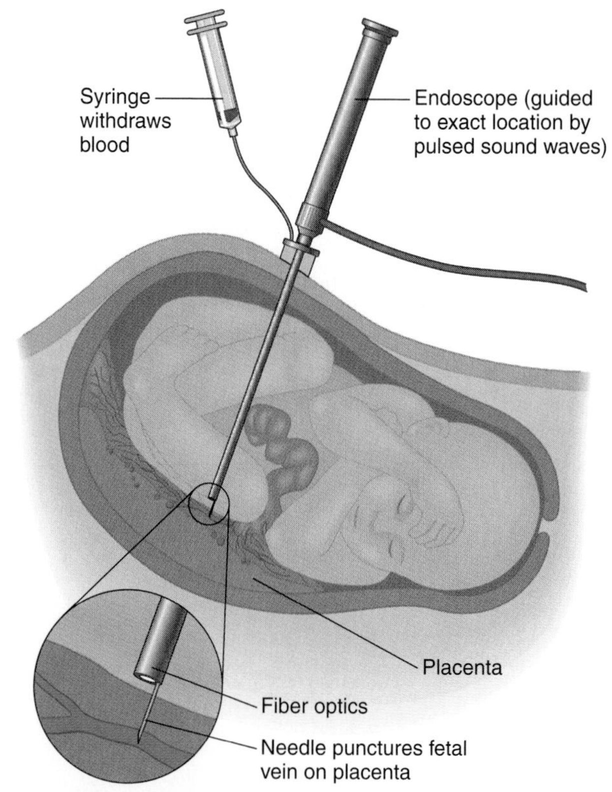

Figure 30.35

Monitoring the fetal heart rate and uterine contractions using an FHR-UC device.

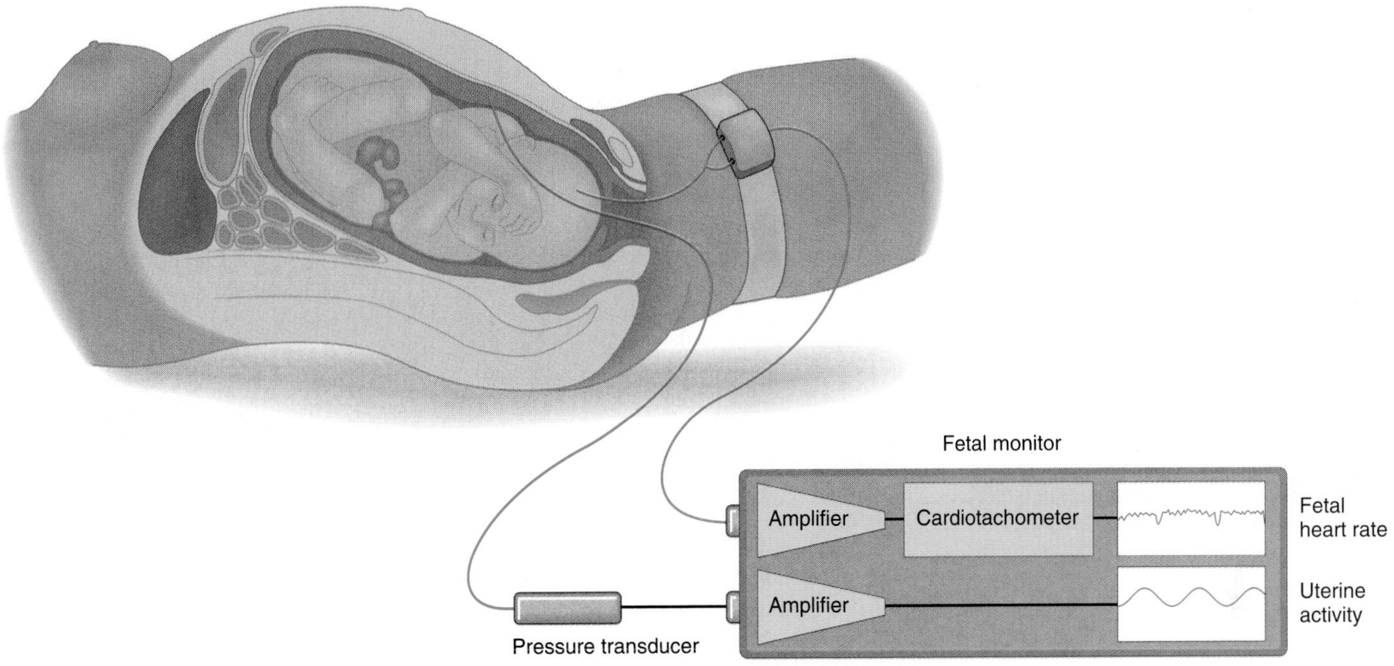

samples are taken from the head of the fetus, and blood samples are extracted from the placenta. The principal advantage of fetoscopy is that external features of the fetus (such as fingers, eyes, ears, mouth, and genitals) can be carefully observed. Fetoscopy is also used to determine several diseases, including hemophilia, thalassemia, and sickle-cell anemia, 40% of which are missed by amniocentesis.

Most hospitals are now equipped with instruments that monitor fetal heart rate and uterine contractions during labor. These instruments can detect any complication that may arise during the delivery. This procedure is called *Electronic Monitoring of Fetal Heart Rate and Uterine Contractions* (*FHR-UC Monitoring*). The extent of stress to the fetus from uterine contractions can be determined through FHR-UC monitoring (fig. 30.35). Long, arduous deliveries are taxing to both the mother and fetus. If the baby's health and vitality are presumed to be in danger because of a difficult delivery, the physician may decide to perform a cesarean section.

Congenital Defects

Major developmental problems called **congenital malformations** occur in approximately 2% of all newborn infants. They may be caused by genetic inheritance, mutation (genetic change), or environmental factors. About 15% of neonatal deaths are attributed to congenital malformations. The branch of developmental biology concerned with abnormal development and congenital malformations is called *teratology* (*ter″ah-tol′o-je*). Many congenital problems have been discussed in previous chapters, in connection with the body system in which they occur.

congenital: L. *congenitus*, born with
teratology: Gk. *teras*, monster; *logos*, study of

Clinical Investigation Answer

Two amions and one chorion in all but very unusual cases prove the twins to be monozygotic. The two infants, therefore, are genetically identical. Thus, when considering genetic disorders such as cleft palate, one would expect a high degree of concordance (both twins of a monozygotic pair exhibit a particular anomaly). Many such defects however can be present in only one twin—a consequence of nongenetic factors, such as intrauterine environment. An example would be inadequate blood supply to only one twin, resulting in a defect in that twin but not in the other.

Important Clinical Terminology

cystic fibrosis An autosomal recessive disorder characterized by the formation of thick mucus in the lungs and pancreas that interferes with normal breathing and digestion.

familial cretinism An autosomal recessive disorder characterized by a lack of thyroid secretion because of a defect in the iodine transport mechanism. Untreated children are dwarfed, sterile, and may be mentally retarded.

galactosemia (*ga-lak″te-se′me-a*) An autosomal recessive disorder characterized by an inability to metabolize galactose, a component of milk sugar. Patients with this disorder have cataracts, damaged livers, and mental retardation.

gout An autosomal dominant disorder characterized by an accumulation of uric acid in the blood and tissue resulting from an abnormal metabolism of purines.

hepatic porphyria (*por-fer′e-a*) An autosomal dominant disorder characterized by painful GI disorders and neurologic disturbances resulting from an abnormal metabolism of porphyrins.

hereditary hemochromatosis (*he″mo-kro″ma-to′sis*) A sex-influenced, autosomal dominant disorder characterized by an accumulation of iron in the pancreas, liver, and heart, resulting in diabetes, cirrhosis, and heart failure.

hereditary leukomelanopathy (*loo″ko-mel″a-nop′a-the*) An autosomal recessive disorder characterized by decreased pigmentation in the skin, hair, and eyes and abnormal white blood cells. Patients with this condition are generally susceptible to infections and early deaths.

Huntington's chorea An autosomal dominant disorder characterized by uncontrolled twitching of skeletal muscles and the deterioration of mental capacities. A latent expression of this disorder allows the mutant gene to be passed to children before the symptoms develop.

Marfan's syndrome An autosomal dominant disorder characterized by tremendous growth of the extremities, extreme looseness of the joints, dislocation of the lenses of the eyes, and congenital cardiovascular defects.

phenylketonuria (*fen″il-kēt″on-oor′e-ă*) (PKU) An autosomal recessive disorder characterized by an inability to metabolize the amino acid phenylalanine. It is accompanied by brain and nerve damage and mental retardation.

pseudohypertrophic muscular dystrophy A sex-linked recessive disorder characterized by progressive muscle atrophy. It usually begins during childhood and causes death in adolescence.

retinitis pigmentosa A sex-linked recessive disorder characterized by progressive atrophy of the retina and eventual blindness.

Tay–Sachs disease An autosomal recessive disorder characterized by a deterioration of physical and mental abilities, early blindness, and early death. It has a disproportionately high incidence in Jews of Eastern European origin.

Chapter Summary

Fertilization (pp. 955–956)

1. Upon fertilization of a secondary oocyte by a spermatozoon in the uterine tube, meiotic development is completed, and a diploid zygote forms.
2. Morphogenesis is the sequential formation of body structures during the prenatal period of human life. The prenatal period lasts 38 weeks and is divided into a preembryonic, an embryonic, and a fetal period.
3. A capacitated spermatozoon digests its way through the zona pellucida and corona radiata of the secondary oocyte to complete the fertilization process and formation of a zygote.

Preembryonic Period (pp. 957–961)

1. Cleavage of the zygote is initiated within 30 hours and continues until a morula forms; the morula enters the uterine cavity on about the third day.
2. A hollow, fluid-filled space forms within the morula, at which point it is called a blastocyst.
3. Implantation begins between the fifth and seventh day and is enabled by the secretion of enzymes that digest a portion of the endometrium.
   (a) During implantation, the trophoblast cells secrete human chorionic gonadotrophin (hCG), which prevents the breakdown of the endometrium and menstruation.
   (b) The secretion of hCG declines by the tenth week as the developed placenta secretes steroids that maintain the endometrium.
4. The embryoblast of the implanted blastocyst flattens into the embryonic disc, from which the primary germ layers of the embryo develop.
   (a) Ectoderm gives rise to the nervous system, the epidermis of the skin and epidermal derivatives, and portions of sensory organs.
   (b) Mesoderm gives rise to bones, muscles, blood, reproductive organs, the dermis of the skin, and connective tissue.
   (c) Endoderm gives rise to linings of the GI tract, digestive organs, the respiratory tract and lungs, and the urinary bladder and urethra.

Embryonic Period (pp. 962–970)

1. The events of the 6-week embryonic period include the differentiation of the germ layers into specific body organs and the formation of the placenta, the umbilical cord, and the extraembryonic membranes. These events make it possible for morphogenesis to continue.
2. The extraembryonic membranes include the amnion, yolk sac, allantois, and chorion.
   (a) The amnion is a thin membrane surrounding the embryo. It contains amniotic fluid that cushions and protects the embryo.
   (b) The yolk sac produces blood for the embryo.
   (c) The allantois also produces blood for the embryo and gives rise to the umbilical arteries and vein.
   (d) The chorion participates in the formation of the placenta.

Huntington's chorea: from George Huntington, American physician, 1850–1916
Marfan's syndrome: from Antoine Bernard-Jean Marfan, French physician, 1858–1942
Tay–Sachs disease: from Warren Tay, English physician, 1843–1927, and Bernard Sachs, American neurologist, 1858–1944

3. The placenta, formed from both maternal and embryonic tissue, has a transport role in providing for the metabolic needs of the fetus and in removing its wastes.
   (a) The placenta produces steroid and polypeptide hormones.
   (b) Nicotine, drugs, alcohol, and viruses can cross the placenta to the fetus.
4. The umbilical cord, containing two umbilical arteries and one umbilical vein, forms as the amnion envelops the tissues on the underside of the embryo.
5. From the third to the eighth week, the structure of all the body organs, except that of the genitalia, becomes apparent.
   (a) During the third week, the primitive node forms from the primitive line, which later gives rise to the notochord and intraembryonic mesoderm.
   (b) By the end of the fourth week, the heart is beating; the primordial tissues of the eyes, brain, spinal cord, lungs, and digestive organs are properly positioned; and the superior and inferior limb buds are recognizable.
   (c) At the end of the fifth week, the sense organs are formed in the enlarged head and the appendages have developed, with digital primordia evident.
   (d) During the seventh and eighth weeks, the body organs, except for the genitalia, are formed and the embryo appears distinctly human.

Fetal Period (pp. 970–973)

1. A small amount of tissue differentiation and organ development occurs during the fetal period, but for the most part fetal development is primarily limited to body growth.
2. Between weeks 9 and 12, ossification centers appear, the genitalia form, and the digestive, urinary, respiratory, and muscle systems show functional activity.
3. Between weeks 13 and 16, facial features form, and the fetal heartbeat can be detected with a stethoscope.
4. During the 17-to-20-week period, quickening can be felt by the mother, and vernix caseosa and lanugo cover the skin of the fetus.
5. During the 21-to-25-week period, substantial weight gain occurs, and the fetal skin becomes wrinkled and pinkish.
6. Toward the end of the 26-to-29-week period, the eyes have opened, the gonads have descended in a male, and the fetus is developed to the extent that it might survive if born prematurely.
7. At 38 weeks, the fetus is full-term; the normal gestation is 266 days.

Labor and Parturition (p. 973)

1. Labor and parturition are the culmination of gestation and require the action of oxytocin, secreted by the posterior pituitary, and prostaglandins, produced in the uterus.
2. Labor is divided into dilation, expulsion, and placental stages.

Inheritance (pp. 974–977)

1. Inheritance is the passage of hereditary traits carried on the genes of chromosomes from one generation to another.
2. Each zygote contains 22 pairs of autosomal chromosomes and 1 pair of sex chromosomes—XX in a female and XY in a male.
3. A gene is the portion of a DNA molecule that contains information for the production of one kind of protein molecule. Alleles are different forms of genes that occupy corresponding positions on homologous chromosomes.
4. The combination of genes in an individual's cells constitutes his or her genotype; the appearance of a person is his or her phenotype.
   (a) Dominant alleles are symbolized by uppercase letters, and recessive alleles are symbolized by lowercase letters.
   (b) The three possible genotypes are homozygous dominant, heterozygous, and homozygous recessive.
5. A Punnett square is a convenient means for expressing probability.
   (a) The probability of a particular genotype is 1 in 4 (0.25) for homozygous dominant and homozygous recessive, and 1 in 2 (0.50) for heterozygous dominant.
   (b) A single trait is studied in a monohybrid cross; two traits are studied in a dihybrid cross.
6. Sex-linked traits like color blindness and hemophilia are carried on the sex-determining chromosome.

Review Activities

Objective Questions

1. The preembryonic period is completed when
   (a) the blastocyst implants.
   (b) the placenta forms.
   (c) the blastocyst reaches the uterus.
   (d) the primary germ layers form.
2. The yolk sac produces blood for the embryo until
   (a) the heart is functional.
   (b) the kidneys are functional.
   (c) the liver is functional.
   (d) the baby is delivered.
3. Which of the following is a function of the placenta?
   (a) production of steroids and hormones
   (b) diffusion of nutrients and oxygen
   (c) production of enzymes
   (d) all of the above apply.
4. The decidua basalis is
   (a) a component of the umbilical cord.
   (b) the embryonic portion of the villous chorion.
   (c) the maternal portion of the placenta.
   (d) a vascular membrane derived from the trophoblast.
5. Which of the following could diffuse across the placenta?
   (a) nicotine
   (b) alcohol
   (c) heroin
   (d) all of the above apply.
6. During which week following conception does the embryonic heart begin pumping blood?
   (a) fourth week (c) sixth week
   (b) fifth week (d) eighth week
7. Twins that develop from two zygotes resulting from the fertilization of two ova by two sperm in the same ovulatory cycle are referred to as
   (a) monozygotic twins.
   (b) conjoined twins.
   (c) dizygotic twins.
   (d) identical twins.
8. Match the genotype descriptions in the left-hand column with the correct symbols in the right-hand column.
   homozygous recessive *Bb*
   heterozygous *bb*
   homozygous dominant *BB*
9. An allele that is not expressed in a heterozygous genotype is called
   (a) recessive. (c) genotypic.
   (b) dominant. (d) phenotypic.

10. If the genotypes of both parents are Aa and Aa, the offspring probably will be
    (a) ½ AA and ½ *aa*.
    (b) all *Aa*.
    (c) ¼ AA, ½ *Aa*, ¼ *aa*.
    (d) ¾ AA and ¼ *aa*.

Essay Questions

1. Describe the implantation of the trophoblast into the uterine wall and its involvement in the formation of the placenta.
2. Explain how the primary germ layers form. What major structures does each germ layer give rise to?
3. Explain why development during the embryonic period is so critical and list the embryonic needs that must be met during the embryonic period for morphogenesis to continue.
4. State the approximate time period (in weeks) for the following occurrences.
    (a) appearance of the arm and leg buds
    (b) differentiation of the external genitalia
    (c) perception of quickening by the mother
    (d) functioning of the embryonic heart
    (e) initiation of bone ossification
    (f) appearance of lanugo and vernix caseosa
    (g) survival of the fetus, if born prematurely
    (h) formation of all major body organs completed
5. State the features of a genetic disorder that would lead one to believe that it was a form of sex-linked inheritance.

Critical Thinking Questions

1. Write a short paragraph about pregnancy that includes the terms *ovum*, *blastocyst*, *implantation*, *embryo*, *fetus*, *gestation*, and *parturition*.
2. If a pregnant woman had a pack-a-day cigarette habit, what predictions could you make about the health of her baby? Justify your responses.
3. The sedative thalidomide was used by thousands of pregnant women in the 1960s to alleviate their morning sickness. This drug inhibited normal limb development and resulted in tragically deformed infants with flipperlike arms and legs. At what period of prenatal development did such abnormalities originate? What lessons were learned from the thalidomide tragedy?
4. Your friend just learned that she's pregnant, and she wants to name the baby either Louis or Louise. How soon can she know for sure whether she is carrying a boy or girl? Describe the technique by which the sex can be determined in vitro.
5. Cesarean section must be performed when a baby cannot be delivered breech. Can you think of some other reasons for a C-section?
6. Hemophilia used to be called a royal disease because it plagued many male members of the royal families of Europe and Russia. Why are female hemophilics rare, as their absence from royal pedigrees shows?
7. In many states, laws prohibit consanguineous marriages—those between blood relatives such as siblings or first cousins. Why is it likely that a geneticist would endorse such restrictions?

Related Web Sites

For a listing of the most current web sites related to this chapter, please visit the *Concepts of Human Anatomy and Physiology* home page at http://www.mhhe.com/biosci/abio/.

Appendix A

Answers to Objective Questions with Explanations

Chapter 1

1. (c) The concept of body humors was widely accepted by physicians in Greek and Roman times as an explanation for a person's disposition and general health. It was not until the Renaissance that this concept was gradually discarded in the light of scientific research.
2. (a) In William Harvey's time, there were many misconceptions regarding the flow of blood. Using the scientific method, Harvey demonstrated the flow of blood and described the function of valves within veins.
3. (c) The taxonomic scheme from general to specific is kingdom phylum, class, order, family, genus, and species.
4. (b) Although chordates have a distinct head, thorax, and abdomen, so do many invertebrate organisms.
5. (b) The cubital fossa is located on the anterior surface of the upper extremity, at the junction of the brachium and antebrachium. It is the anterior surface of the elbow.
6. (c) By definition, both the coronal and sagittal planes extend vertically through the body, but there is no vertical plane designation.
7. (c) The anatomical position provides a consistent frame of reference when describing structures of the body.
8. (b) Homeostasis is best thought of as a state of dynamic constancy. Adjustments in physiological processes are constantly being made to maintain the internal environment.
9. (b) The changes produced by the effector organ act to counter the original deviation from the set point.
10. (a) A fall in blood calcium will stimulate the parathyroid gland to secrete parathyroid hormone (PTH), which acts to increase blood calcium levels.

Chapter 2

1. (c) An atom contains an equal number of protons and electrons.
2. (b) In this bond, the oxygen side of the molecule is more negatively charged than the hydrogen side of the molecule.
3. (a) Covalent bonds between identical atoms are the strongest type of bond.
4. (d) The pH of a solution indicates the H^+ concentration of a solution.
5. (c) Glucose is a six-carbon monosaccharide.
6. (b) As the covalent bond between monosaccharides is broken, a molecule of water is split. The hydrogen atom and hydroxyl group from water are added to the ends of the subunits to form new molecules.
7. (c) Glycogen, or animal starch, consists of repeating glucose molecules and is highly branched.
8. (d) Two carbons of a three-carbon glycerol are attached to two fatty acids. The third carbon of the glycerol is attached to a phosphate group, and the phosphate group is bonded to a nitrogen-containing choline group.
9. (d) Steroids and prostaglandins are both involved in the regulation of ovulation.
10. (b) The folding and bending of the polypeptide chain is produced by chemical interactions between particular amino acids located in different regions of the chain.
11. (d) The hydrogen bonding between molecules of water is responsible for many of the physical properties of water, including its surface tension and capillary action.
12. (b) When amino acids are joined, the compound formed is called a peptide. The bond between two amino acids is called a peptide bond.

Chapter 3

1. (d) The transport properties of the membrane are believed to be due to its protein content.
2. (b) Each chromatid will become a separate chromosome once cell division is completed.
3. (a) Cells in the G1 phase perform the physiological functions characteristic of the tissue in which they are found, such as producing digestive enzymes or stomach acid.
4. (c) The aligning of chromosomes at the equator of the cell is believed to result from the action of the spindle fibers that are attached to the centromere of each chromosome.
5. (d) Each future cell gets one copy of each of the 46 chromosomes of the cell.
6. (b) In RNA, uracil bonds with adenine and cytosine bonds with guanine.
7. (a) There are three types of RNA made within the nucleus using DNA as a guide: mRNA, tRNA, and rRNA.
8. (c) An anticodon is a set of three tRNA nucleotides that are complementary to a specific codon in mRNA.
9. (a) Glycoproteins, formed by combining protein with carbohydrates, are components of cell membranes.
10. (b) The enzymes contained within a lysosome are called lysozymes.
11. (d) In DNA, adenine bonds with uracil or thymine.
12. (e) A tRNA containing a given anticodon will always bind to one specific amino acid.
13. (b) Genetic translation is the production of specific proteins according to the code contained in the mRNA base sequence.
14. (a) The anticodon is found in loop 2 of the tRNA molecule.

Chapter 4

1. (b) Enzymes act as biological catalysts to increase the rate of a reaction.
2. (d) When the reaction rate is maximal, the enzyme is said to be saturated.
3. (e) Endergonic reactions require an input of energy to occur; exergonic reactions produce energy.
4. (e) The release of energy from ATP drives the energy-requiring processes in the cell.
5. (d) This is an example of an oxidation-reduction reaction.
6. (d) Practically speaking, a reaction will occur if the enzyme is present, and will not occur if the enzyme is absent.
7. (a) The addition of two hydrogen molecules from NADH and H^+ to pyruvic acid produces lactic acid.
8. (c) Anaerobic respiration is a normal occurrence in skeletal muscle.
9. (d) Cyanide blocks the transfer of electrons from cytochrome a_3 to oxygen. This halts the production of ATP by oxidative phosphorylation.
10. (c) Aerobic respiration is the preferred means of ATP production. It is much more efficient than anaerobic respiration.
11. (a) An elevated level of ketone bodies, such as in uncontrolled diabetes mellitus, is called ketosis and is a very serious condition that can lead to coma and death.
12. (d) Glucose produced in the liver by gluconeogenesis can be used to replenish muscle glycogen stores after exercise.
13. (b) In starvation and prolonged exercise, gluconeogenesis is the only way to maintain adequate blood glucose levels for the brain.
14. (d) Amino acids can join metabolic pathways as pyruvic acid, acetyl CoA, or enter during the Krebs cycle.

Chapter 5

1. (c) Osmosis is the net diffusion of water (the solvent) across a membrane.
2. (b) Facilitated diffusion is powered by the thermal energy of the diffusing molecules, not ATP.

3. (a) ATP is required for the Na^+/K^+ pump, which transports two K^+ in and three Na^+ out of the cell.
4. (c) When red blood cells are placed in a hypotonic solution, they gain water and may burst (hemolysis).
5. (b) Isotonic saline has the same osmolality as plasma.
6. (d) 0.5 moles of NaCl will ionize to 0.5 mole of Na and 0.5 mole of Cl. Thus, a 0.5 m NaCl solution has a total concentration of 1.0 Osm.
7. (a) The membrane is most permeable to K^+. Na^+, Ca^{2+}, and Cl^- are less permeable, and therefore not as important in estimating membrane potential.
8. (a) An increase in blood osmolality causes an increase in blood osmotic pressure, which leads to an increase in ADH secretion.
9. (b) The diffusion gradient that causes K^+ to leave the cells is reduced, so more K^+ stays in the cell, neutralizing the negative charges, and bringing the membrane potential closer to zero.
10. (d) The Na^+/K^+ pump transports three Na^+ out for every two K^+ moved in.
11. (b) In facilitated diffusion, glucose moves by carrier proteins across the cell membrane when the concentration of glucose outside the cell in the blood is greater than the glucose concentration inside the cell.
12. (b) The Na^+/K^+ pump requires ATP for energy, an example of primary active transport. Cotransport is an example of secondary active transport.

Chapter 6

1. (b) The four principal types of tissues are epithelial tissue, connective tissue, muscular tissue, and nervous tissue. The integument, or skin, is an organ.
2. (c) Blood is a fluid tissue that flows through blood vessels.
3. (a) Many body structures derive from mesoderm, including cartilage, bone, and other connective tissue; smooth, cardiac, and skeletal muscle; and the dermis of the skin.
4. (b) Elastic and collagenous fibers are characteristic of certain kinds of connective tissues.
5. (d) The entire abdominal portion of the gastrointestinal (GI) tract contains simple columnar epithelium lining the lumen.
6. (a) Based on structural classification, the mammary glands are classified as compound acinar. Based on secretory classification, the mammary glands are classified as apocrine.
7. (c) Dense regular connective tissue is the principal component of tendons, which accounts for their great strength.
8. (a) Reticular tissue, which contains white blood cells, is phagocytic in cleansing body fluids.
9. (b) Cartilage tissue is slow to heal because it is avascular (without blood vessels). Cartilage generally derives its nutrients from surrounding fluids rather than from permeating blood vessels and capillaries.
10. (d) Like skeletal muscle, cardiac muscle is striated. As compared to skeletal muscle or smooth muscle tissues, however, cardiac muscle tissue is unique in containing intercalated discs and experiencing rhythmical involuntary contractions.

Chapter 7

1. (a) Ectoderm is the outermost of the three embryonic germ layers, and it is from this germ layer that hair, nail, integumentary glands, and the epidermis of the skin form.
2. (c) The appearance and general health of nails are indicators of general body health and of certain dietary deficiencies, such as iron deficiency.
3. (b) The stratum lucidum is an epidermal layer that occurs only in the skin of the palms of the hands and the soles of the feet.
4. (a) The dermal papillae may contribute to surface features of the skin, such as print patterns, but does not contribute to skin coloration.
5. (c) Mitosis occurs in the stratum basale, and to a limited extent, in the stratum spinosum.
6. (a) The sebum from sebaceous glands is emptied into a hair follicle, where it is dissipated along the shaft of the hair to the surface of the skin.
7. (c) Lanugo, or fetal hair, is thought to be important in the development of the hair follicles.
8. (b) Melanoma is an aggressive malignant skin cancer that is life threatening if untreated.
9. (b) A second-degree burn is serious, but it generally does not require skin grafts.
10. (d) A comedo is a blackhead or whitehead resulting from a small, localized skin infection.

Chapter 8

1. (b) A bone, such as the femur, is considered to be an organ because it is composed of more than two tissues integrated to perform a particular function.
2. (a) Skin, rather than bone, is involved in the synthesis of vitamin D.
3. (c) Bone tissue, derived from specialized mesodermal cells called mesenchyme, begins to ossify during the fourth week of development.
4. (d) An example of a fovea is the fovea centralis in the head of the femur.
5. (b) Perforating fibers are composed of bundles of collagenous fibers.
6. (a) The matrix of an osteon is laid down in columnlike rings, called lamellae, which surround a central canal.
7. (d) Osteoclasts are important in releasing stored minerals within bone tissue and in the continuous remodeling of bone.
8. (b) Only 1,25-dihydroxyvitamin D_3 stimulates intestinal absorption of calcium.
9. (b) Calcitonin acts to decrease blood calcium levels by inhibiting osteoclasts and inhibiting the reabsorption of Ca^{2+} in the kidneys.
10. (c) A person may gain some protection against osteoporosis by maintaining a healthy diet and a regular exercise program.

Chapter 9

1. (c) The maxillary nerve passes through foramen rotundum.
2. (d) The mental nerve and vessels pass through the mental foramen.
3. (e) The internal carotid artery passes through the carotid canal.
4. (a) The olfactory nerves pass through the ethmoid bone.
5. (b) The spinal cord passes through the foramen magnum to attach to the brain stem.
6. (c) The mandible articulates with the mandibular fossae of the temporal bones.
7. (a) The sella turcica of the sphenoid bone is located immediately superior to the sphenoidal sinus, where it supports the pituitary gland.
8. (a) The fontanels permit the skull to accommodate the rapid growth of the brain during infancy.
9. (d) A suture is a joint between two skull bones.
10. (c) The vomer forms the lower portion of the nasal septum and is considered to be a facial bone.
11. (b) The four parts of the temporal bone are the squamous, tympanic, mastoid, and petrous.
12. (a) Located in the squamous part of the temporal bone, the mandibular fossa is the depression for articulation of the condylar process of the mandible.
13. (c) The stylomastoid foramen is located in the mastoid part of the temporal bone.
14. (b) The crista galli is the portion of the ethmoid bone that contains numerous olfactory foramina for the passage of olfactory nerves from the nasal epithelium.
15. (c) Fovea are facets for articulation with the ribs found on the vertebrae.
16. (b) The first seven pair of ribs attach to the sternum by individual costal cartilages.

Chapter 10

1. (a) When in anatomical position, a person is erect and facing anteriorly. The subscapular fossa is found on the anterior side of the scapula. The supraspinatus and infraspinatus fossae are located on the posterior side.
2. (c) The sternal extremity of the clavicle articulates with the manubrium of the sternum. The acromial extremity of the clavicle articulates with the acromion of the scapula.
3. (d) The clavicle has a conoid tubercle near its acromial extremity.
4. (b) The bony prominence of the elbow is the olecranon of the ulna.
5. (b) Sesamoid bones are formed in tendons, and none of the carpal bones are sesamoid bones. Sesamoid bones are fairly common, however, at specific joints of the digits.
6. (d) Pelvimetry measures the dimension of the lesser pelvis in a pregnant woman to determine whether a cesarean delivery might be necessary.
7. (e) The linea aspera is a vertical ridge on the posterior surface of the body of the femur, where the posterior gluteal muscles of the hip attach.
8. (b) The intertrochanteric line is located on the anterior side of the femur between the greater trochanter and the lesser trochanter.

9. (d) Sex-related structural differences in the pelvis reflect modification for childbirth. For example, the female pelvis as compared to that of the male has a wider pelvic outlet, a shallower and shorter symphysis pubis, and a wider pubic arch.
10. (a) Talipes is a congenital malformation in which the sole of the foot is twisted medially.

Chapter 11

1. (b) The fused sutures in the skull and the ossified synchondroses of the long bones in the adult skeleton are immovable joints.
2. (c) Synchondroses are cartilaginous joints that have hyaline cartilage between the bone segments.
3. (d) Syndesmoses are fibrous joints in the antebrachium (forearm) and leg, where adjacent bones are held together by interosseous ligaments.
4. (d) Syndesmoses do not occur within the skull; rather, they are located only in the upper and lower extremities.
5. (c) The only saddle joints in the body are at the base of the thumb, where the trapezium articulates with the first metacarpal bone, and at the articulation of the malleus with the incus in the middle ear.
6. (d) Only the knee joints contain menisci.
7. (c) A pivotal joint is a synovial joint that permits rotational movement.
8. (b) The coxal (hip) joint has a wide range of movement, including hyperextention as the lower extremity is moved posteriorly beyond the vertical position of the body in anatomical position.
9. (b) The shoulder joint, with its relatively shallow socket and weak ligamentous support, is vulnerable to dislocation. In addition, we often place our arms in vulnerable positions as we engage in various activities.
10. (d) Rheumatoid arthritis is a chronic disease that frequently leaves the patient crippled. It occurs most commonly between the ages of 30 and 35.

Chapter 12

1. (b) In the body, graded contractions of whole muscles are produced by variations in the number of motor units that are activated.
2. (d) The series-elastic component absorbs some of the tension generated as a muscle contracts, and must be pulled tight before muscle contractions can result in muscle shortening.
3. (c) The muscles that innervate the fingers have fine neural control because the motor units are composed of a motor nerve and very few muscle fibers. The muscles of the leg, arm, and trunk have gross motor control.
4. (a) The A band, composed of thick myofilaments, does not shorten during contraction.
5. (c) Calcium must be released before tropomyosin moves and cross bridges attach to actin.
6. (b) ATP is the immediate source of energy for muscle contraction.
7. (b) ATP is required for cross-bridge detachment.
8. (d) Ca^{2+} binds to troponin and causes a conformational change that exposes tropomyosin for cross-bridge attachment.
9. (a) Muscle fatigue may decrease the amount of ATP available, but ATP will not totally disappear.
10. (e) Neither multiunit smooth muscle nor skeletal muscle are capable of spontaneous depolarization.
11. (c) Gap junctions join adjacent cells and allow electrical impulses to spread to all cells in the muscle mass.
12. (b) Muscle shortening occurs in an isotonic contraction in which the force for contraction remains constant.

Chapter 13

1. (c) The triceps brachii is antagonistic to the brachialis and biceps brachii at the elbow joint.
2. (c) An example of a pennate-type muscle is the flexor pollicis longus.
3. (e) Muscles are named on the basis of structural features, location, attachment, relative position, or function (action).
4. (a) The mandibular nerve innervates all of the muscles of mastication.
5. (a) A description of a muscle's contraction is always made in reference to a person in anatomical position. In anatomical position, the shoulder joint is positioned vertically at 180°. When the pectoralis muscle is contracted while in this position, the angle of the shoulder joint is decreased; therefore, the joint is flexed.
6. (d) The rectus abdominis flexes the vertebral column, whereas the longissimus thoracis extends it.
7. (b) Although it is positioned on the anterior surface of the humerus, the biceps brachii muscle arises from the coracoid process of the scapula and from the tuberosity above the glenoid fossa of the scapula. Both heads insert on the radial tuberosity.
8. (b) Contraction of the bulbospongiosus muscle is frequently associated with coitus (sexual intercourse) and micturition (urination).
9. (e) The vastus medialis muscle acts to extend the leg at the knee joint.
10. (c) As muscular dystrophy progresses, the muscle fibers atrophy and are replaced by adipose tissue.

Chapter 14

1. (c) Oligodendrocytes form myelin sheaths around axons in the CNS.
2. (d) A collection of cell bodies within the CNS are called nuclei.
3. (a) In a sensory neuron, one end of the longer process receives sensory stimuli and produces nerve impulses; the other end delivers these impulses to synapses within the brain or spinal cord.
4. (a) The Na^+ channels are voltage-regulated. When the axon voltage changes, the Na^+ channels open and allow Na^+ to diffuse into the cell.
5. (c) After an action potential is completed, the Na^+/K^+ pump extrudes the Na^+ that has entered the axon and recovers the K^+ that has diffused out of the cell.
6. (d) An increase in stimulus strength causes an increase in the frequency of action potentials, not an increase in action potential amplitude.
7. (d) Action potentials travel fastest in large, myelinated axons.
8. (a) The amplitude of a synaptic potential is dependent on the quanta of neurotransmitter released at the membrane.
9. (c) The opening of the K^+ channels is slower than the opening of the Na^+ channels.
10. (c) Acetylcholinesterase inactivates acetylcholine. Nerve gas inactivates acetylcholinesterase, which leads to spastic paralysis and then death.
11. (b) Hyperpolarizations produced by neurotransmitters are called inhibitory postsynaptic potentials (IPSPs).
12. (d) The binding of glycine to its receptor proteins causes the opening of Cl^- channels in the postsynaptic membrane.
13. (d) During the absolute refractory period, the voltage-regulated channels are inactivated and cannot be opened by depolarization. Thus, a membrane cannot respond to a stimulus during the absolute refractory period.
14. (b) Catechol-O-methyltransferase inactivates catecholamines in the postsynaptic neurons.
15. (a) In temporal summation, successive EPSPs from a single presynaptic axon come together to generate sufficient depolarization to cause an action potential.
16. (c) The ion channel of nicotinic receptors opens when the ligand, ACh, binds to the receptor. No G-protein is involved in nicotinic receptors.
17. (a) Glutamic acid is the major excitatory neurotransmitter in the brain.
18. (e) Nitric oxide is the first gas identified as a neurotransmitter. Once in its target cells, NO acts by stimulating cGMP production, which acts as a second messenger.

Chapter 15

1. (b) Each cerebral hemisphere contains frontal, parietal, temporal, occipital, and insula lobes.
2. (a) The corpus callosum is composed of commissural fibers that connect the two cerebral hemispheres.
3. (d) The thalamus autonomically responds to pain by activating the sympathetic nervous system. It also relays pain sensations to the parietal lobes of the cerebrum for perception.
4. (b) The basal nuclei consists of cell bodies of motor neurons that regulate contraction of skeletal muscles. Basal metabolic rate is regulated, for the most part, in the hypothalamus and medulla oblongata.
5. (c) Located within the mesencephalon, the corpora quadrigemina is concerned with visual and hearing reflexes, the red nucleus is concerned with motor coordination and posture maintenance, and the substantia

nigra is thought to inhibit forced involuntary movements.
6. (d) The fourth ventricle is located inferior to the cerebellum and contains cerebrospinal fluid.
7. (c) The crossing-over of nerve tracts is referred to as decussation.
8. (a) The hypothalamus has been implicated as a factor in psychosomatic illness.
9. (b) The cell bodies of the neurons that contribute fibers to the corticospinal tracts are primarily located in the precentral gyrus of the frontal lobe.
10. (d) The destruction of myelin prohibits normal conduction of impulses, resulting in a progressive loss of function.

Chapter 16

1. (a) Anatomically speaking, the PNS consists of all of the structures of the nervous system outside of the CNS. That means, then, that all nerves, sensory receptors, neurons, ganglia, and plexuses are part of the PNS.
2. (b) The oculomotor nerve innervates the medial rectus eye muscles that, when simultaneously contracted, cause the eyes to be directed medially.
3. (b) The oculomotor nerve innervates the levator palpebrae superioris muscle with motor fibers. This muscle elevates the upper eyelid when contracted.
4. (c) The vestibulocochlear nerve serves the vestibular organs of the inner ear with sensory fibers. These organs are associated with equilibrium and balance.
5. (a) Passing through the stylomastoid foramen, the facial nerve innervates the muscles of facial expression with motor fibers and the taste buds on the anterior two-thirds of the tongue with sensory fibers.
6. (b) The accessory nerve innervates several muscles that move the head and neck with motor fibers.
7. (c) The four spinal nerve plexuses are the cervical, brachial, lumbar, and sacral.
8. (d) Only the brachial plexus consists of roots, trunks, divisions, and cords. The nerves of the upper extremity arise from the cords.
9. (a) The median nerve arises from the brachial plexus.
10. (c) The knee-jerk reflex is an ipsilateral reflex because the receptor and effector organs are on the same side of the spinal cord.

Chapter 17

1. (c) Denervation hypersensitivity explains why stomach mucosa may resume secreting acid after a vagotomy.
2. (c) The sympathetic ganglia run parallel to the spinal cord.
3. (c) ACh is the neurotransmitter of all sympathetic and parasympathetic preganglionic fibers.
4. (a) Skin and splanchnic blood vessels constrict under adrenergic stimulation to divert blood flow to the brain, heart, lungs, and skeletal muscle.
5. (c) Norepinephrine will increase the heart rate.
6. (b) Erection of the penis is due to parasympathetic stimulation, and ejaculation is due to sympathetic stimulation.
7. (b) Propranalol blocks beta-1 receptors in the heart, which causes the heart rate to decrease.
8. (e) Atropine is a parasympatholytic drug, and is used clinically.
9. (c) Centers for control of the cardiovascular, pulmonary, urinary, reproductive, and digestive systems are located in the medulla oblongata.
10. (c) Nicotinic receptors contain ion channels, whereas muscarinic receptors operate through G-proteins.
11. (b) Binding of norepinephrine to alpha-2 receptors block cAMP production, so cAMP concentration decreases.
12. (c) Terbutaline is a specific agonist for beta-2 receptors. Drugs like terbutaline are used to treat bronchiolar constriction in asthmatics.

Chapter 18

1. (d) The hairs of the hair cells in the utricle and saccule are embedded in the gelatinous otolith membrane.
2. (a) The processes of the hair cells found in the semicircular canals are embedded in the cupula.
3. (c) The basilar membrane supports the hair cells inside the cochlea.
4. (b) The fibrous tunic, composed of the sclera and cornea, is avascular connective tissue.
5. (d) The stapes is attached to the oval window. The tympanic membrane separates the outer from the middle ear.
6. (d) Rods provide black and white vision.
7. (c) Tonic receptors produce slow-adapting sensations. Phasic receptors are fast-adapting and do not maintain a constant rate of firing.
8. (a) Cutaneous receptors are concentrated in the eyelids, lips, tongue, fingertips, palms, soles, nipples, and genitalia.
9. (c) Lateral inhibition helps to accentuate the contours of an image.
10. (c) Taste receptor cells release neurotransmitter onto sensory nerve fibers when they are stimulated.
11. (d) The utricle and saccule are located at the base of the semicircular canals.
12. (d) The left lateral geniculate nucleus receives input from the right visual field.
13. (d) As an object moves closer to the eyes, the ciliary muscles contract, reducing tension on the suspensory ligament, and allowing the lens to become more rounded and convex.
14. (b) Shining a light into one eye stimulates the pupillary reflex in which both pupils constrict.
15. (c) The receptive fields of simple cortical neurons are rectangular because they receive input from lateral geniculate neurons whose receptive fields are aligned in a particular way.
16. (b) Accommodation is caused by contraction of the ciliary muscles, which will then affect the shape of the lens.
17. (c) The sense of smell is transmitted directly to the olfactory bulb and then to the cerebral cortex.
18. (c) Ion movement through membrane channels produces sour taste.
19. (b) Receptors for olfaction consist of dendritic endings of bipolar sensory neurons of the olfactory nerve.

Chapter 19

1. (e) The adenohypophysis derives from the hypophyseal pouch in the roof of the oral cavity.
2. (b) The neurohypophysis derives from the neurohypophyseal bud of neuroectoderm within the developing brain.
3. (d) Because it is derived from neural crest ectoderm, the adrenal medulla secretes catecholamine hormones in response to stimulation.
4. (d) The hypothalamus maintains control of the anterior pituitary through hormonal regulation by hypothalamic-releasing hormones.
5. (d) TSH from the anterior pituitary causes the thyroxine to release from the thyroid gland.
6. (e) The adrenal cortex secretes steroid hormones that participate in the regulation of mineral balance, energy balance, and reproductive function.
7. (e) Insulin, secreted by beta cells, promotes the entry of glucose into tissue cells and the conversion of excess glucose into glycogen and fat.
8. (d) Epinephrine is the primary postganglionic neurotransmitter of the sympathetic nervous system.
9. (a) TSH released from the anterior pituitary stimulates thyroxine production and secretion from the thyroid.
10. (b) The adrenal cortex releases corticosteroids upon receiving ACTH stimulation from the adenohypophysis of the pituitary gland in response to stress.
11. (e) The adenohypophysis of the pituitary gland releases ACTH in response to receiving the CRF stimulation from the hypothalamus.
12. (d) Both the adrenal cortex and the gonads secrete steroid hormones.
13. (a) In the case of an endemic goiter, growth of the thyroid is due to excessive TSH secretion, which results from low levels of thyroxine secretion.
14. (d) Catecholamines cannot pass through the cell membrane, so they function through the second-messenger system.
15. (c) Insulin decreases blood glucose while glucagon increases blood glucose.

Chapter 20

1. (d) Capillaries are one cell thick and allow for exchange of nutrients and waste products.
2. (c) Transfusing blood of the same type will not cause an agglutination reaction.
3. (a) Hemocytoblasts differentiate into many cells, including myeloblasts.
4. (d) Monocytes and lymphocytes are both agranular leukocytes.
5. (a) Hemoglobin molecules in red blood cells bind oxygen for transport through the body.

6. (c) Once factor X is activated by either pathway, it converts prothrombin to thrombin.
7. (d) Platelets are essential for clotting reactions.
8. (d) Type-O blood does not contain type-A or type-B antigens.
9. (c) Erythropoietin is secreted by the kidneys in response to decreased oxygen delivery.
10. (c) Plasmin is an enzyme that digests fibrin to promote clot dissolution.
11. (d) Anemia refers to a condition in which there is an abnormally low hemoglobin concentration and/or erythrocyte count.
12. (d) Factors II, VII, IX, and X all require vitamin K for proper function.
13. (b) Prolonged diarrhea causes metabolic acidosis.

Chapter 21

1. (b) The pulmonary arteries carry oxygen-poor blood to the lungs to be oxygenated.
2. (c) The right atrium receives venous blood from the superior and inferior vena cavae and the coronary sinus. The coronary sinus collects venous blood from coronary circulation prior to delivery into an opening of the right atrium.
3. (c) The second heart sound is caused by the closure of the semilunar valves.
4. (b) The coronary arteries arise from the ascending aorta.
5. (a) The external carotid arteries supply blood to the entire head, excluding the brain. The paired internal carotid arteries and vertebral arteries (that unite to form the basilar artery) supply blood to the brain.
6. (a) Branching from the external carotid artery at the level of the mandibular condyle, the maxillary artery supplies blood to the teeth and gums of the upper jaw, and the superficial temporal artery supplies blood to the parotid gland and to the superficial temporal region.
7. (c) Exchanges between blood and tissue fluid occur across capillary walls.
8. (e) Closure of the atrioventricular valves causes the "lub" sound of the heart.
9. (a) The first heart sound is caused by closure of the AV valves.
10. (b) As heart rate increases, diastole (filling time) decreases.
11. (c) The QRS complex represents the depolarization of the ventricles. During this interval, the ventricles of the heart are in systole and blood is being ejected from the heart.
12. (a) The SA node is the pacemaker of the heart.
13. (c) Contraction of the heart is coordinated by the conduction system.
14. (b) Isovolumetric relaxation and filling time constitute diastole.

Chapter 22

1. (a) If diastolic filling is increased, then the strength of ventricular contraction will also increase.
2. (d) When arterial blood pressure increases, the afterload increases. If afterload is increased, the stroke volume will decrease.
3. (c) The colloid osmotic pressure of plasma is greater than the colloid osmotic pressure of tissue fluid.
4. (e) Edema is the excessive accumulation of interstitial fluid and can be caused by many factors.
5. (b) ADH causes water retention in the kidneys. Aldosterone promotes salt retention, which leads to water retention. Both ADH and aldosterone act to increase blood volume.
6. (c) The rate of blood flow to an organ can be increased by dilation of its arteries or decreased by constriction of its arteries.
7. (a) Poiseuille's law can be used to calculate the rate of blood flow.
8. (c) Korotkoff sounds are used to measure blood pressure by the ausculatory method.
9. (d) Nitric oxide, bradykinin, and prostacyclin are all released by the endothelium to promote smooth muscle relaxation to increase blood flow.
10. (b) Coronary blood flow during systole is almost completely occluded.
11. (c) Blood flow to the brain is autoregulated by myogenic and metabolic control mechanisms.
12. (d) Blood flow to the brain must remain relatively constant.
13. (b) When the arteriovenous shunts are closed, cutaneous blood flow decreases, so less body heat is lost.
14. (d) All of these mechanisms function to decrease blood volume by increasing the amount of water excreted through the kidneys.
15. (c) All of the chambers of the heart hold the same volume of blood.
16. (e) Because the pressure is the lowest, the veins contain most of the blood in the circulatory system at any given time.
17. (b) As the blood pressure rises, the stretch receptors are stimulated. They cause increased parasympathetic activity and decreased sympathetic activity to decrease the blood pressure.
18. (d) Angiotensin II promotes an increase in blood pressure, whereas bradykinin causes a decrease in blood pressure.
19. (d) Endothelin-1 helps to maintain normal vessel diameter and blood pressure.
20. (c) A blood pressure of 130/70 has a pulse pressure of 60 mmHg.

Chapter 23

1. (c) Interferons are proteins released by virally infected cells, which activate defense mechanisms in neighboring cells. These defense mechanisms prevent the spread of viral infections.
2. (b) Killer T cells release perforins, which polymerize on the cell membrane of the victim cell and form perforating channels.
3. (d) Mast cells release histamine when activated by IgE to cause vasodilation and chemoattraction.
4. (a) Plasma cells are mature B cells, which release antibodies in response to infection.
5. (c) Macrophages secrete lysosomal enzymes into their phagosomes to complete the killing process.
6. (d) The F_{ab} protein of antibodies is the variable portion that binds to antigens.
7. (d) The $C5_b$ through C9 proteins of the complement system produces pores in the victim cell membrane.
8. (b) IgE antibodies activate immediate hypersensitivity
9. (d) The presence of memory cells allows a secondary immune response to occur more quickly, last longer, and produce higher concentrations of antibodies.
10. (a) Macrophages are antigen presenting cells that help to activate T cells.
11. (d) Antibodies are produced only by B cells.
12. (a) Delayed hypersensitivity is mediated by T cells and takes longer to develop because the symptoms are caused by lymphokines rather than fast-acting histamine.
13. (d) Active immunity develops through the exposure of the body to the infectious agent.
14. (c) B cells may be activated by the presence of antigen-antibody complexes on the surface of the B cell.
15. (b) Interleukin-1 activates the T cell system.
16. (c) Interleukin-2 is released by T-helper cells and activates killer T cells.
17. (a) Interleukin-12 stimulates the differentiation of helper T cells to the T_H1 subtype.
18. (d) Only particular lymphocytes and a related type of cell called natural killer cells produce gamma interferon.

Chapter 24

1. (a) The paired palatine bones support the nasal septum but are not part of its structure.
2. (b) Adenoid is the common name of the pharyngeal tonsil. An adenoidectomy is the removal of both pharyngeal tonsils.
3. (a) There are four paranasal sinuses, each named according to the bone in which it is located. Hence, we have ethmoidal, sphenoidal, frontal, and maxillary sinuses.
4. (e) Unlike the right lung that has three lobes, the left lung has only a superior lobe and an inferior lobe.
5. (a) The parietal pleura lines the wall of the thoracic cavity, and the visceral pleura covers the lung. The space between the parietal pleura and the visceral pleura is referred to as the pleural cavity.
6. (c) The difference between the intrapulmonary pressure and intrapleural pressure is known as the transpulmonary pressure.
7. (d) When the transpulmonary pressure is zero, there is no pressure gradient to cause lung inflation.
8. (c) The vital capacity is equal to the sum of the inspiratory reserve volume, tidal volume, and expiratory reserve volume.
9. (a) Oxygen content of arterial blood will decrease because there are no red blood cells to carry oxygen. Arterial P_{O_2} would be normal because the lungs are properly functioning.
10. (c) The amount of oxygen carried in red blood cells would not change while scuba diving, so the whole blood oxygen cannot increase by three times.
11. (c) The venous portion of the oxyhemoglobin saturation curve is very steep, so a small change in hemoglobin affinity will cause a large difference in saturation.

12. (b) In hyperventilation, more carbon dioxide is blown off than is produced, so the P_{CO_2} will decrease.
13. (a) The kidneys produce and secrete erythropoietin to stimulate production of hemoglobin and red blood cells in response to tissue hypoxia.
14. (e) Oxygen delivery problems can result from all of these conditions.
15. (c) Carbon dioxide is carried in the blood in three forms: dissolved carbon dioxide, carbaminohemoglobin, and bicarbonate.
16. (a) A metabolic acidosis can occur when bicarbonate is lost as a result of diarrhea.
17. (c) Blood H^+ cannot cross the blood-brain barrier, so carbon dioxide crosses over and lowers the pH of cerebrospinal fluid.
18. (a) The medullary rhythmicity center is influenced by the apneustic and pneumotaxic centers of the pons.
19. (d) The body will respond to hypoxemia in ways to increase blood volume and flow.
20. (b) During heavier exercise, the venous P_{O_2} can drop to 20 mmHg or lower.
21. (a) Bicarbonate does not bind to hemoglobin, but H^+ does. The positive charge of the red blood cell attracts $C1^-$ into the red blood cell. This is known as the chloride shift.
22. (c) Carbon dioxide produced in the tissues by metabolism moves into the blood at the capillary level. Thus carbon dioxide levels in the veins are higher than in the arteries.

Chapter 25

1. (d) Metanephric kidneys are the third pair of kidneys to develop. They become functional at the end of the third week and are active throughout fetal development.
2. (b) Glomeruli are located in the renal cortex.
3. (a) In the proximal convoluted tubule, sodium is actively reabsorbed and water follows passively.
4. (c) Sodium is actively reabsorbed in the ascending loop of the nephron while water remains in the tubule to be reabsorbed later if the urine must be further concentrated.
5. (b) Because the descending limb of the nephron loop is permeable only to water, water can move out into the interstitium. This allows for concentration of the fluid in the descending limb.
6. (e) Urea moves out of the collecting duct to increase the concentration of the interstitium. Water will move out of the collecting duct and descending limb because the interstitium is more concentrated.
7. (d) Antidiuretic hormone acts on the collecting duct to increase its permeability to water and promote water reabsorption.
8. (d) Aldosterone acts on the cells of the cortical collecting duct to cause sodium reabsorption, potassium secretion, and ultimately, water retention.
9. (c) If substance X has a secretion greater than zero, it must be filtered. If it has a clearance less than inulin, a substance that is not reabsorbed, it must be somewhat reabsorbed.
10. (d) If substance Y has a clearance greater than that of inulin, a substance that is filtered but neither secreted nor absorbed, it must be filtered and secreted.
11. (a) The proximal convoluted tubule is the portion of the nephron where the majority of salt and water reabsorption occurs.
12. (e) Diuretic drugs that act in the nephron loop inhibit sodium reabsorption and increase the amount of potassium and water that reach the collecting systems.
13. (c) When the amount of glucose in the ultrafiltrate is increased because of increased blood glucose, the glucose transport system is saturated and glucose spills into the urine.
14. (a) Water is reabsorbed from renal tubules only by osmosis.
15. (d) Plasma oncotic pressure retains water in the plasma.
16. (b) The countercurrent exchange in the vasa recta maintains high concentrations of NaC1 in the extracellular fluid. That is necessary to remove water from the filtrate by osmosis.
17. (e) The kidneys can help maintain acid-base balance through the secretion of H^+, buffers in the urine, and the action of carbonic anhydrase.
18. (c) The detrusor muscle is found in the wall of the urinary bladder.
19. (b) The internal urethral sphincter is innervated by parasympathetic nerve fibers.

Chapter 26

1. (c) The deciduous dentition includes eight incisors, four canines, and eight molars. The permanent dentition includes eight incisors, four canines, eight premolars, and twelve molars. The third molars are called wisdom teeth.
2. (b) It is through the supporting mesentery that vessels and nerves supply the abdominal viscera.
3. (c) Lacteals are lymph ductules found within the lamina propria of villi.
4. (e) Secretion of intrinsic factor is the only stomach function essential for life.
5. (d) These types of enzymes are known as brush border enzymes.
6. (b) HC1 converts pepsinogen to pepsin. Pepsin is the enzyme that hydrolyzes peptide bonds.
7. (c) The mouth and stomach function in the mechanical breakdown of food in preparation for digestion. The large intestine functions in water and electrolyte absorption.
8. (d) Trypsin is secreted from the pancreas, but does not become active until it reaches the small intestine and is activated by enterokinase.
9. (b) Vagus nerve stimulation happens in the cephalic phase. Secretin and CCK stimulate the secretion of pancreatic juice.
10. (d) In the intestinal phase of gastric regulation, gastric activity is inhibited when chyme enters the small intestine.
11. (a) Venous blood from the GI tract passes into the liver through the hepatic portal system.
12. (d) Fat is absorbed into the lymphatic system through the lacteals, and proteins are broken down into amino acids in the small intestine. Blood and bile never mix in the liver.
13. (e) The large intestine also absorbs several B complex vitamins and vitamin K.
14. (e) CCK is secreted by the duodenum in response to the presence of chyme.
15. (d) Vitamin B_{12} is absorbed in the terminal portion of the ileum.
16. (b) Starch digestion occurs in the small intestine.
17. (d) Triglycerides are absorbed into central lacteals and enter the lymphatic system.
18. (d) Contraction of intestinal smooth muscle also causes peristalsis.

Chapter 27

1. (c) Absorption of a carbohydrate meal stimulates a rise in insulin and a fall in glucagon, because it causes a rise in the blood glucose levels.
2. (b) Fasting lowers blood glucose and causes the body to react by lowering insulin levels and increasing glucagon secretion.
3. (c) Growth hormone is secreted by the pituitary gland and causes an increase in protein synthesis in the muscle and decreased glucose utilization.
4. (a) Thyroxine, a hormone secreted by the thyroid gland, causes increased protein synthesis and increased cell respiration by stimulating metabolism.
5. (b) Hydrocortisone is released in response to low blood glucose and causes protein catabolism and gluconeogenesis in order to increase blood glucose.
6. (d) In response to low blood glucose, the body will respond by breaking down glycogen and creating substrates for gluconeogenesis.
7. (a) The liver is the only organ in the body that stores glucose for later secretion into the blood.
8. (d) The basal metabolic rate is determined mainly by thyroxine.
9. (c) The action of growth hormone is mediated through somatomedins.
10. (c) The Na^+/K^+ pump continually consumes energy to maintain membrane potentials.
11. (a) Growth hormones stimulate anabolism of protein to cause growth and catabolism of fat.
12. (d) There is not 100% absorption from the gastrointestinal tract.
13. (d) Increased levels of ketone bodies in the blood cause ketoacidosis. Ketone bodies are a product of fat catabolism, which is increased in diabetes mellitus.
14. (c) Leptin causes a decrease in the secretion of neuropeptide Y, which may cause a decrease in appetite.

Chapter 28

1. (a) The ovum from a female may be fertilized by a sperm cell containing either an X or a Y

chromosome. If the sperm cell contains an X chromosome, it will pair with the X chromosome of the ovum, and a female child will develop. A sperm cell carrying a Y chromosome results in an XY combination, and a male child will develop. The Y chromosome, therefore, is responsible for the subsequent production of androgens, which cause masculinization.

2. (d) The epididymides, ductus deferentia, and seminal vesicles derive from the mesonephric duct. The prostate arises from an endodermal outgrowth of the urogenital sinus.
3. (d) The testes do not descend into the scrotal sac until about week 28.
4. (d) FSH stimulates spermatogenesis. Sustentacular cells are critical to the developing spermatozoa.
5. (a) Inhibin feeds back to the anterior pituitary to decrease FSH secretion.
6. (b) The spermatic cord contains the ductus deferens, which is a spermatic duct. The spermatic cord, however, also contains spermatic vessels, a nerve, and the cremasteric muscle.
7. (a) The epididymides are long, flattened organs that store sperm in their last stage of maturation. Sperm are also stored in the ductus deferentia.
8. (a) Mucus secreted by the urethral glands lubricates the urethra and retards the entry of pathogens into the urinary bladder.
9. (c) Emission is movement of stored sperm from the epididymides and ductus deferentia to the ejaculatory ducts. Ejaculation is the forceful discharge of semen from the erect penis. Emission and ejaculation occur only if stimulation is sufficient, as in masturbation or coitus.
10. (a) Spermatogenesis does not occur in males with cryptorchidism because the temperature is too high for sperm production and storage.
11. (d) The discharge from the prostate makes up about 40% of the volume of semen.
12. (d) All of the statements are true.
13. (a) Pulsatile patterns of secretion are necessary to prevent desensitization and down-regulation of the target glands.

Chapter 29

1. (d) A secondary oocyte is discharged from an ovary during ovulation. If a spermatozoon passes through the corona radiata and zona pellucida and enters the cytoplasm of the secondary oocyte, the second meiotic division is completed and a mature ovum is formed.
2. (d) The paramesonephric ducts give rise to the genital tract of the female reproductive system, which includes the uterus and the uterine tubes.
3. (a) Both the labia majora of the female and the scrotum of the male derive from the embryonic labioscrotal swellings.
4. (c) The cervix of the uterus is the inferior constricted portion that opens into the vagina.
5. (c) The endometrium consists of a superficial stratum functionale layer that is shed as menses during menstruation and a deeper stratum basale layer that replenishes the stratum functionale layer after each menstruation.
6. (b) Vaginal rugae permit distension of the vagina during coitus and also during parturition.
7. (b) The usual site of fertilization is within a uterine tube. From there, the fertilized egg continues to develop and enters into the cavity of the uterus about three days after conception.
8. (b) Oxytocin also stimulates contraction of the uterus.
9. (d) The suspensory ligaments between the lobules of the breast extend from the skin to the deep fascia overlying the pectoralis muscles.
10. (d) Prolactin stimulates milk production.
11. (b) The stratum functionale is shed during the menstrual phase.
12. (d) During the follicular phase, some of the primary follicles grow, develop vesicles, and become secondary follicles.
13. (a) The corpus luteum secretes both estrogen and progesterone, which inhibits FSH and LH secretion and maintains the uterine lining.
14. (c) The LH surge is triggered by a rise in estrogen secreted by the mature follicle. This allows the follicle to cause ovulation when it is ready.

Chapter 30

1. (d) The preembryonic period of development lasts 14 days and is completed when the three primary germ layers have been formed and are in place to begin migration and differentiation.
2. (c) The yolk sac produces blood for only about a 2-week period (the third to the fifth week) prior to the formation of the liver. The liver then produces blood until the bone marrow is sufficiently developed to carry out the task.
3. (d) The placenta is an organ that serves many physiological functions, including the exchange of gases and other molecules between the maternal and fetal blood and the production of enzymes. In addition, it serves as a barrier against many harmful substances and as an endocrine gland in secreting steroid and glycoprotein hormones.
4. (c) The embryonic portion of the placenta is the chorion frondosum, and the maternal portion is the decidua basalis.
5. (d) Although the placenta is an effective barrier against diseases of bacterial origin, viruses and certain blood-borne diseases can diffuse through the vascular tissues. Furthermore, most drugs ingested by a pregnant woman can readily pass through the placenta, including nicotine, alcohol, and heroin.
6. (a) The embryonic heart begins pumping blood on about day 25, or during the fourth week of development.
7. (c) Dizygotic, or fraternal, twins may be of the same sex or of different sexes and are not any more alike than brothers or sisters born at different times.
8. homozygous recessive (*bb*)
heterozygous (*Bb*)
homozygous dominant (*BB*)
9. (c) A recessive allele is not expressed in a heterozygous genotype. This means that the particular recessive trait is not physically apparent.
10. (c) In a monohybrid cross, the probability of an offspring having a particular genotype is one in four for homozygous dominant and homozygous recessive and one in two for heterozygous dominant.

Appendix B

Some Laboratory Tests of Clinical Importance

Common Tests Performed on Blood

Test	Normal Values (Adult)	Clinical Significance
Acetone and acetoacetate (serum)	0.3–2.0 mg/100 ml	Values increase in diabetic acidosis, toxemia of pregnancy, fasting, and high-fat diet.
Albumin-globulin ratio or A/G ratio (serum)	1.5:1–2.5:1	Ratio of albumin to globulin is lowered in kidney diseases and malnutrition.
Albumin (serum)	3.2–5.5 gm/100 ml	Values increase in multiple myeloma and decrease with proteinuria and as a result of severe burns.
Ammonia (plasma)	50–170 μg/100 ml	Values increase in severe liver disease, pneumonia, shock, and congestive heart failure.
Amylase (serum)	80–160 Somogyi units/100 ml	Values increase in acute pancreatitis, intestinal obstructions, and mumps. They decrease in chronic pancreatitis, cirrhosis of the liver, and toxemia or pregnancy.
Bilirubin, total (serum)	0.3–1.1 mg/100 ml	Values increase in conditions causing red blood cell destruction or biliary obstruction.
Blood urea nitrogen or BUN (plasma or serum)	10–20 mg/100 ml	Values increase in various kidney disorders and decrease in liver failure and during pregnancy.
Calcium (serum)	9.0–11.0 mg/100 ml	Values increase in hyperparathyroidism, hypervitaminosis D, and respiratory conditions that cause a rise in CO_2 concentration. They decrease in hypoparathyroidism, malnutrition, and severe diarrhea.
Carbon dioxide (serum)	24–30 mEq/l	Values increase in respiratory diseases, intestinal obstruction, and vomiting. They decrease in acidosis, nephritis, and diarrhea.
Chloride (serum)	96–106 mEq/l	Values increase in nephritis, Cushing's syndrome, and hyperventilation. They decrease in diabetic acidosis, Addison's disease, diarrhea, and following severe burns.
Cholesterol, total (serum)	150–250 mg/100 ml	Values increase in diabetes mellitus and hypothyroidism. They decrease in pernicious anemia, hyperthyroidism, and acute infections.
Creatine phosphokinase or CPK (serum)	Men: 0–20 IU/l Women: 0–14 IU/l	Values increase in myocardial infarction and skeletal muscle diseases such as muscular dystrophy.
Creatine (serum)	0.2–0.8 mg/100 ml	Values increase in muscular dystrophy, nephritis, severe damage to muscle tissue, and during pregnancy.
Creatinine (serum)	0.7–1.5 mg/100 ml	Values increase in various kidney diseases.
Erythrocyte count or red blood cell count (whole blood)	Men: 4,600,000–6,200,000/cu mm Women: 4,200,000–5,400,000/cu mm Children: 4,500,000–5,100,000/cu mm (varies with age)	Values increase as a result of severe dehydration or diarrhea and decrease in anemia, leukemia, and following severe hemorrhage.
Fatty acids, total (serum)	190–420 mg/100 ml	Values increase in diabetes mellitus, anemia, kidney disease, and hypothyroidism. They decrease in hyperthyroidism.
Globulin (serum)	2.5–3.5 gm/100 ml	Values increase as a result of chronic infections.
Glucose (plasma)	70–115 mg/100 ml	Values increase in diabetes mellitus, liver diseases, nephritis, hyperthyroidism, and pregnancy. They decrease in hyperinsulinism, hypothyroidism, and Addison's disease.
Hematocrit (whole blood)	Men: 40–54 ml/100 ml	Values increase in polycythemia due to dehydration or shock. They

Test	Normal Values (Adult)	Clinical Significance
	Women: 37–47 ml/100 ml Children: 35–49 ml/100 ml (varies with age)	decrease in anemia and following severe hemorrhage.
Hemoglobin (whole blood)	Men: 14–18 gm/100 ml Women: 12–16 gm/100 ml Children: 11.2–16.5 gm/100 ml (varies with age)	Values increase in polycythemia, obstructive pulmonary diseases, congestive heart failure, and at high altitudes. They decrease in anemia, pregnancy, and as a result of severe hemorrhage or excessive fluid intake.
Iron (serum)	75–175 μg/100 ml	Values increase in various anemias and liver disease. They decrease in iron deficiency anemia.
Iron-binding capacity (serum)	250–410 μg/100 ml	Values increase in iron deficiency anemia and pregnancy. They decrease in pernicious anemia, liver disease, and chronic infections.
Lactic acid (whole blood)	6–16 mg/100 ml	Values increase with muscular activity and in congestive heart failure, severe hemorrhage, and shock.
Lactic dehydrogenase or LDH (serum)	90–200 milliunits/ml	Values increase in pernicious anemia, myocardial infarction, liver diseases, acute leukemia, and widespread carcinoma.
Lipids, total (serum)	450–850 mg/100 ml	Values increase in hypothyroidism, diabetes mellitus, and nephritis. They decrease in hyperthyroidism.
Oxygen saturation (whole blood)	Arterial: 94%–100% Venous: 60%–85%	Values increase in polycythemia and decrease in anemia and obstructive pulmonary diseases.
pH (whole blood)	7.35–7.45	Values increase due to vomiting, Cushing's syndrome, and hyperventilation. They decrease as a result of hypoventilation, severe diarrhea, Addison's disease, and diabetic acidosis.
Phosphatase, acid (serum)	1.0–5.0 King-Armstrong units/ml	Values increase in cancer of the prostate, hyperparathyroidism, certain liver diseases, myocardial infarction, and pulmonary embolism.
Phosphatase, alkaline (serum)	5–13 King-Armstrong units/ml	Values increase in hyperparathyroidism (and in other conditions that promote resorption of bone), liver diseases, and pregnancy.
Phospholipids (serum)	6–12 mg/100 ml as lipid phosphorus	Values increase in diabetes mellitus and nephritis.
Phosphorus (serum)	3.0–4.5 mg/100 ml	Values increase in kidney diseases, hypoparathyroidism, acromegaly, and hypervitaminosis D. They decrease in hyperparathyroidism.
Platelet count (whole blood)	150,000–350,000/cu mm	Values increase in polycythemia and certain anemias. They decrease in acute leukemia and aplastic anemia.
Potassium (serum)	3.5–5.0 mEq/l	Values increase in Addison's disease, hypoventilation, and conditions that cause severe cellular destruction. They decrease in diarrhea, vomiting, diabetic acidosis, and chronic kidney disease.
Protein, total (serum)	6.0–8.0 gm/100 ml	Values increase in severe dehydration and shock. They decrease in severe malnutrition and hemorrhage.
Protein-bound iodine or PBI (serum)	3.5–8.0 μg/100 ml	Values increase in hyperthyroidism and liver disease. They decrease in hypothyroidism.
Prothrombin time (serum)	12–14 sec (one stage)	Values increase in certain hemorrhagic diseases, liver disease, vitamin K deficiency, and following the use of various drugs.
Sedimentation rate, Westergren (whole blood)	Men: 0–15 mm/hr Women: 0–20 mm/hr	Values increase in infectious diseases, menstruation, pregnancy, and as a result of severe tissue damage.
Sodium (serum)	136–145 mEq/l	Values increase in nephritis and severe dehydration. They decrease in Addison's disease, myxedema, kidney disease, and diarrhea.
Thyroxine or T_4 (serum)	2.9–6.4 μg/100 ml	Values increase in hyperthyroidism and pregnancy. They decrease in hypothyroidism.
Thromboplastin time,	35–45 sec	Values increase in deficiencies of blood factors VIII, IX, and X.

continued

Common Tests Performed on Blood (*continued*)

Test	Normal Values (Adult)	Clinical Significance
partial (plasma)		
Transaminases or SGOT (serum)	5–40 units/ml	Values increase in myocardial infarction, liver disease, and diseases of skeletal muscles.
Uric acid (serum)	Men: 2.5–8.0 mg/100 ml Women: 1.5–6.0 mg/100 ml	Values increase in gout, leukemia, pneumonia, toxemia of pregnancy, and as a result of severe tissue damage.
While blood cell count, differential (whole blood)	Neutrophils 54%–62% Eosinophils 1%–3% Basophils 0%–1% Lymphocytes 25%–33% Monocytes 3%–7%	Neutrophils increase in bacterial diseases; lymphocytes and monocytes increase in viral diseases; eosinophils increase in collagen diseases, allergies, and in the presence of intestinal parasites.
White blood cell count, total (whole blood)	5000–10,000/cu mm	Values increase in acute infections, acute leukemia, and following menstruation. They decrease in aplastic anemia and as a result of drug toxicity.

Common Tests Performed on Urine

Test	Normal Values	Clinical Significance
Acetone and acetoacetate	0	Values increase in diabetic acidosis
Albumin, qualitative	0 to trace	Values increase in kidney disease, hypertension, and heart failure.
Ammonia	20–70 mEq/l	Values increase in diabetes mellitus and liver diseases.
Bacterial count	Under 10,000/ml	Values increase in urinary tract infection.
Bile and bilirubin	0	Values increase in melanoma and biliary tract obstruction.
Calcium	Under 250 mg/24 hr	Values increase in hyperparathyroidism and decrease in hypoparathyroidsim
Creatinine clearance	100–140 ml/min	Values increase in renal diseases.
Creatinine	1–2 gm/24 hr	Values increase in infections and decrease in muscular atrophy, anemia, leukemia, and kidney diseases.
Glucose	0	Values increase in diabetes mellitus and various pituitary gland disorders.
17-Hydroxycorticosteroids	2–10 mg/24 hr	Values increase in Cushing's syndrome and decrease in Addison's disease.
Phenylpyruvic acid	0	Values increase in phenylketonuria.
Urea clearance	Over 40 ml blood cleared of urea/mn	Values increase in renal diseases.
Urobilinogen	0–4 mg/24 hr	Values increase in liver disease and hemolytic anemia. They decrease in complete biliary obstruction and severe diarrhea.
Urea	25–35 gm/24 hr	Values increase as a result of excessive protein breakdown. They decrease as a result of impaired renal function.
Uric acid	0.6–1.0 gm/24 hr. as urate	Values increase in gout and decrease in various kidney diseases.

Glossary

Most of the words in this glossary are followed by a phonetic spelling that serves as a guide to pronunciation. The phonetic spellings reflect standard scientific usage and can be easily interpreted following a few basic rules.

Any unmarked vowel that ends a syllable or that stands alone as a syllable has the long sound. Any unmarked vowel that is followed by a consonant has the short sound.

If a long vowel appears in the middle of a syllable (followed by a consonant), it is marked with a macron (¯). Similarly, if a vowel stands alone or ends a syllable but should have a short sound, it is marked with a breve (˘).

Syllables that are emphasized are indicated by stress marks. A single stress mark (′) indicates the primary emphasis; a secondary emphasis is indicated by a double stress mark (″).

Page references are provided, except in the case of some adjectives and a few general terms.

A

abdomen (ab′dŏ-men, ab-do′men) The portion of the trunk between the diaphragm and pelvis. 15

abduction (ab-duk′shun) The movement of a body part away from the axis or midline of the body; movement of a digit away from the axis of the limb. **258**

ABO system The most common system of classification for red blood cell antigens. On the basis of antigens on the red blood cell surface, individuals can be type A, type B, type AB, or type O. 599

absorption (ab-sorp′shun) The transport of molecules across epithelial membranes into the body fluids. 765

accessory organs (ak-ses′ŏ-re) Organs that assist with the functioning of other organs within a system. 766

accommodation (ă-kom″ŏ-da′shun) A process whereby the focal length of the eye is changed by automatic adjustment of the curvature of the lens to bring images of objects from various distances into focus on the retina. **530**

acetabulum (as″ĕ-tab′yŭ-lum) A socket in the lateral surface of the hipbone (os coxa) with which the head of the femur articulates. 209

acetone (as′ĕ-tōn) A ketone body produced as a result of the oxidation of fats. 819

acetyl coenzyme A (acetyl CoA) (as′ĕ-tl, ă-set′l) A coenzyme derivative in the metabolism of glucose and fatty acids that contributes substrates to the Krebs cycle. 83

acetylcholine (ACh) (ă-set″l-ko′lēn) An acetic acid ester of choline—a substance that functions as a neurotransmitter in somatic motor nerve and parasympathetic nerve fibers. **287**

acetylcholinesterase (ă-set″l-ko″lĭ-nes′tĕ-rās) An enzyme in the membrane of postsynaptic cells that catalyzes the conversion of ACh into choline and acetic acid. This enzymatic reaction inactivates the neurotransmitter. 361

Achilles tendon (ă-kil′ēz) *See* tendo calcaneous. 316

acid (as′id) A substance that releases hydrogen ions when ionized in water. **29**

acidosis (as″ĭ-do′sis) An abnormal increase in the H^+ concentration of the blood that lowers the arterial pH to below 7.35. **604**

acromegaly (ak″ro-meg′ă-le) A condition caused by the hypersecretion of growth hormone from the pituitary gland after maturity and characterized by enlargement of the extremities, such as the nose, jaws, fingers, and toes. **196**

actin (ak′tin) A protein in muscle fibers that together with myosin is responsible for contraction. 263

action potential An all-or-none electrical event in an axon or muscle fiber in which the polarity of the membrane potential is rapidly reversed and reestablished. **381**

active immunity (ĭ-myoo′nĭ-te) Immunity involving sensitization, in which antibody production is stimulated by prior exposure to an antigen. 662

active transport The movement of molecules or ions across the cell membranes of epithelial cells by membrane carriers. An expenditure of cellular energy (ATP) is required. **111**

adduction (ă-duk′shun) The movement of a body part toward the axis or midline of the body; movement of a digit toward the axis of the limb. 233

adenohypophysis (ad″n-o-hi-pof′ĭ-sis) The anterior, glandular lobe of the pituitary gland that secretes FSH (follicle-stimulating hormone), LH (luteinizing hormone), ACTH (adrenocorticotropic hormone), TSH (thyroid-stimulating hormone), GH (growth hormone), and prolactin. Secretions of the adenohypophysis are controlled by hormones produced by the hypothalamus. **426**

adenoids (ad′ĕ-noidz) The tonsils located in the nasopharynx; pharyngeal tonsils. 686

adenylate cyclase (ă-den′l-it si′klās) An enzyme found in cell membranes that catalyzes the conversion of ATP to cyclic AMP and pyrophosphate (PP_I). This enzyme is activated by an interaction between a specific hormone and its membrane receptor protein. **561**

ADH Antidiuretic hormone; a hormone produced by the hypothalamus and released by the posterior pituitary that acts on the kidneys to promote water reabsorption; also known as *vasopressin.* **117**

ADP Adenosine diphosphate; a molecule that together with inorganic phosphate is used to make ATP (adenosine triphosphate). 76

adrenal cortex (ă-dre′nal kor′teks) The outer part of the adrenal gland. Derived from embryonic mesoderm, the adrenal cortex secretes corticosteroid hormones (such as aldosterone and hydrocortisone). **482**

adrenal medulla (mĕ-dul′ă) The inner part of the adrenal gland. Derived from embryonic postganglionic sympathetic neurons, the adrenal medulla secretes catecholamine hormones—epinephrine and (to a lesser degree) norepinephrine. **482**

adrenergic (ad″rĕ-ner′jik) A term used to describe the actions of epinephrine, norepinephrine, or other molecules with similar activity (as in *adrenergic receptor* and *adrenergic stimulation*). **487**

adventitia (ad″ven-tish′ă) The outermost epithelial layer of a visceral organ; also called *serosa.* 754

afferent (af′er-ent) Conveying or transmitting to. 346

afferent arteriole (ar-tir′e-ōl) A blood vessel within the kidney that supplies blood to the glomerulus. 730

afferent neuron (noor′on) *See* sensory neuron. 346

agglutinate (ă-gloot′n-āt) A clump of cells (usually erythrocytes) formed as a result of specific chemical interaction between surface antigens and antibodies. **600**

agranular leukocytes (ă-gran′yŭ-lar loo′kŏ-sītz) White blood cells (leukocytes) that do not contain cytoplasmic granules;

specifically, lymphocytes and monocytes. **595**

albumin (al-byoo′min) A water-soluble protein produced in the liver; the major component of the plasma proteins. **594**

aldosterone (al-dos′ter-ōn) The principal corticosteroid hormone involved in the regulation of electrolyte balance (mineralocorticoid). 527

alimentary canal The tubular portion of the digestive tract. *See also* gastrointestinal tract (GI tract). 766

allantois (ă-lan′to-is) An extraembryonic membranous sac involved in the formation of blood cells. It gives rise to the fetal umbilical arteries and veins and also contributes to the formation of the urinary bladder. 756

allergens (al′er-jenz) Antigens that evoke an allergic response rather than a normal immune response. 677

allergy (al′er-je) A state of hypersensitivity caused by exposure to allergens. It results in the liberation of histamine and other molecules with histaminelike effects. 677

all-or-none principle The statement of the fact that muscle fibers of a motor unit contract to their maximum extent when exposed to a stimulus of threshold strength. 355

allosteric (al″ŏ-ster′ik) A term used with reference to the alteration of an enzyme's activity as a result of its combination with a regulator molecule. Allosteric inhibition by an end product represents negative feedback control of an enzyme's activity. **84**

alveolar sacs (al-ve′ŏ-lar) A cluster of alveoli that share a common chamber or central atrium. 690

alveolus (al-ve′ŏ-lus) **1.** An individual air capsule within the lung. The alveoli are the basic functional units of respiration. 690 **2.** The socket that secures a tooth (tooth socket). 773

amniocentesis (am″ne-o-sen-te′sis) A procedure in which a sample of amniotic fluid is aspirated to examine suspended cells for various genetic diseases. 910

amnion (am′ne-on) A developmental membrane surrounding the fetus that contains amniotic fluid. 908

amphiarthrosis (am″fe-ar-thro′sis) A slightly movable articulation in a functional classification of joints. 224

amphoteric (am-fo-ter′ik) Having both acidic and basic characteristics; used to denote a molecule that can be positively or negatively charged, depending on the pH of its environment.

ampulla (am-pool′ă) A saclike enlargement of a duct or tube.

ampulla of Vater (Fă′ter) *See* hepatopancreatic ampulla. 781

anabolic steroids (an″ă-bol′ik ster′oidz) Steroids with androgenlike stimulatory effects on protein synthesis. 539

anabolism (ă-nab′ŏ-liz″em) A phase of metabolism involving chemical reactions within cells that result in the production of larger molecules from smaller ones; specifically, the synthesis of protein, glycogen, and fat. **89**

anaerobic respiration (an-ă-ro′bik res″pĭ-ra′shun) A form of cell respiration involving the conversion of glucose to lactic acid in which energy is obtained without the use of molecular oxygen. **90**

anal canal (a′nal) The terminal tubular portion of the large intestine that opens through the anus of the GI tract. 785

anaphylaxis (an″ă-fĭ-lak′sis) An unusually severe allergic reaction that can result in cardiovascular shock and death. 642

anastomosis (ă-nas′tŏ-mo′sis) An interconnecting aggregation of blood vessels or nerves that form a network plexus. 632

anatomical position (an″ă-tom′ĭ-kal) An erect body stance with the eyes directed interior, the arms at the sides, the palms of the hands facing interior, and the fingers pointing straight down. **11**

anatomy (ă-nat′ŏ-me) The branch of science concerned with the structure of the body and the relationship of its organs. 2

androgens (an′drŏ-jenz) Steroids containing 18 carbons that have masculinizing effects; primarily those hormones (such as testosterone) secreted by the testes, although weaker androgens are also secreted by the adrenal cortex. 840

anemia (ă-ne′me-ă) An abnormal reduction in the red blood cell count, hemoglobin concentration, or hematocrit, or any combination of these measurements. This condition is associated with a decreased ability of the blood to carry oxygen. 552

angina pectoris (an-ji′nă pek′tŏ-ris) A thoracic pain, often referred to the left pectoral and arm area, caused by myocardial ischemia. 608

angiotensin II (an″je-o-ten′sin) An 8-amino-acid polypeptide formed from angiotensin I (a 10-amino-acid precursor), which in turn is formed from cleavage of a protein (angiotensinogen) by the action of renin (an enzyme secreted by the kidneys). Angiotensin II is a powerful vasoconstrictor and a stimulator of aldosterone secretion from the adrenal cortex. 621

anions (an′i-onz) Ions that are negatively charged, such as chloride, bicarbonate, and phosphate. 24

antagonist (an-tag′ŏ-nist) A muscle that acts in opposition to another muscle. **493**

antebrachium (an″te-bra′ke-em) The forearm. **16**

anterior (ventral) Toward the front; the opposite of *posterior*, or *dorsal*. 12

anterior pituitary (pĭ-too′ĭ-ter-e) *See* adenohypophysis. **566**

anterior root The anterior projection of the spinal cord, composed of axons of motor neurons. **458**

antibodies (an′tĭ-bod″ēz) Immunoglobin proteins secreted by B lymphocytes that have transformed into plasma cells. Antibodies are responsible for humoral immunity. Their synthesis is induced by specific antigens, and they combine with these specific antigens but not with unrelated antigens. 657

anticodon (an″tĭ-ko′don) A base triplet provided by three nucleotides within a loop of transfer RNA that is complementary in its base-pairing properties to a triplet (the codon) in mRNA. The matching of codon to anticodon provides the mechanism for translating the genetic code into a specific sequence of amino acids. **61**

antigen (an′tĭ-jen) A molecule that can induce the production of antibodies and react in a specific manner with antibodies. 655

antigenic determinant site (an-tĭ-jen′ik) The region of an antigen molecule that specifically reacts with particular antibodies. A large antigen molecule may have a number of such sites. 655

antiserum (an′tĭ-sir″um) A serum that contains specific antibodies. 665

anus (a′nus) The terminal opening of the GI tract. 785

aorta (a-or′tă) The major systemic vessel of the arterial system of the body, emerging from the left ventricle. 583

aortic arch The superior left bend of the aorta between the ascending and descending portions. **628**

apex (a′peks) The tip or pointed end of a conical structure. 691

aphasia (ă-fa′zhă) Defects in speech, writing, or comprehension of spoken or written language caused by brain damage or disease. 388

apneustic center (ap-noo′stik) A collection of nuclei (nerve cell bodies) in the brain stem that participates in the rhythmic control of breathing. **427**

apocrine gland (ap′ŏ-krin) A type of sweat gland that functions in evaporative cooling. It may respond during periods of emotional stress. **141**

aponeurosis (ap″ŏ-noo-ro′sis) A fibrous or membranous sheetlike tendon. **281**

appendix A short pouch that attaches to the cecum. 785

aqueous humor (a′kwe-us) The watery fluid that fills the anterior and posterior chambers of the eye. **528**

arachnoid mater (ă-rak′noid) The weblike middle covering (meninx) of the central nervous system. **430**

arbor vitae (ar'bor vi'te) The branching arrangement of white matter within the cerebellum. 393

arm (brachium) The portion of the upper extremity from the shoulder to the elbow. 203

arrector pili muscle (ah-rek'tor pih'le) The smooth muscle attached to a hair follicle that, upon contraction, pulls the hair into a more vertical position, resulting in "goose bumps." **166**

arteriole (ar-tir'e-ōl) A minute arterial branch. 550

arteriosclerosis (ar-tir"e-o-sklĕ-ro'sis) Any one of a group of diseases characterized by thickening and hardening of the artery wall and in the narrowing of its lumen. 606

arteriovenous anastomoses (ar-tir"e-o-ve'nus ă-nas"tŏ-mo'sēz) Direct connections between arteries and veins that bypass capillary beds. 632

artery (ar'tĕ-re) A blood vessel that carries blood away from the heart. 579

arthrology (ar-throl'ŏ-je) The scientific study of the structure and function of joints. 224

articular cartilage (ar-tik'yŭ-lar kar'tĭ-lij) A hyaline cartilaginous covering over the articulating surface of the bones of synovial joints. **189**

articulation (ar-tik"yŭ-la'shun) A joint. **248**

arytenoid cartilages (ar"ĕ-te'noid) A pair of small cartilages located on the superior aspect of the larynx. 687

ascending colon (ko'lon) The portion of the large intestine between the cecum and the hepatic flexure. 785

association neuron (noor'on) A nerve cell located completely within the central nervous system. It conveys impulses in an arc from sensory to motor neurons; also called *interneuron* or *internuncial neuron*. **374**

astigmatism (ă-stig'mă-tiz"em) Unequal curvature of the refractive surfaces of the eye (cornea and/or lens), so that light entering the eye along certain meridians does not focus on the retina. **545**

atherosclerosis (ath"ĕ-ro-sklĕ-ro'sis) A common type of arteriosclerosis found in medium and larger arteries in which raised areas within the tunica intima are formed from smooth muscle cells, cholesterol, and other lipids. These plaques occlude arteries and serve as sites for the formation of thrombi. 606

atomic number The number of protons in the nucleus of an atom. **25**

atopic dermatitis (ă-top'ik der"mă-ti'tis) An allergic skin reaction to agents such as poison ivy and poison oak; a type of delayed hypersensitivity. 687

ATP Adenosine triphosphate; the universal energy donor of the cell. 76

atretic (ă-tret'ik) Without an opening. Atretic ovarian follicles are those that fail to ovulate. 878

atrioventricular bundle (a"tre-o-ven-trik'yŭ-lar) A group of specialized cardiac fibers that conduct impulses from the atrioventricular node to the ventricular muscles of the heart; also called the *bundle of His* or *AV bundle*. **621**

atrioventricular node A microscopic aggregation of specialized cardiac fibers located in the interatrial septum of the heart that are a part of the conduction system of the heart; *AV node*. **621**

atrioventricular valve A cardiac valve located between an atrium and a ventricle of the heart; *AV valve*. **613**

atrium (a'tre-um) Either of the two superior chambers of the heart that receive venous blood. 567

atrophy (at'rŏ-fe) A gradual wasting away or decrease in the size of a tissue or an organ. **156**

atropine (at'rŏ-pēn) An alkaloid drug obtained from a plant of the species *Belladonna* that acts as an anticholinergic agent. It is used medically to inhibit parasympathetic nerve effects, dilate the pupils of the eye, increase the heart rate, and inhibit intestinal movements. 457

auditory (aw'dĭ-tor-e) Pertaining to the structures of the ear associated with hearing. 482

auditory tube A narrow canal that connects the middle ear chamber to the pharynx; also called the *eustachian canal*. 483

auricle (or'ĭ-kul) **1.** The fleshy pinna of the ear. 482 **2.** An ear-shaped appendage of each atrium of the heart. **613**

autoantibodies (aw"to-an'tĭ-bod"ēz) Antibodies formed in response to, and that react with, molecules that are part of one's own body. 673

autonomic nervous system (aw"tŏ-nom'ik) The sympathetic and parasympathetic portions of the nervous system that function to control the actions of the visceral organs and skin; *ANS*. 446

autosomal chromosomes (aw"to-so'mal kro'mŏ-sōmz) The paired chromosomes; those other than the sex chromosomes. 839

axilla (ak-sil'ă) The depressed hollow commonly called the armpit. **14**

axon (ak'son) The elongated process of a nerve cell that transmits an impulse away from the cell body of a neuron. **155**

B

ball-and-socket joint The most freely movable type of synovial joint (e.g., the shoulder or hip joint). 231

baroreceptor (bar"o-re-sep'tor) A cluster of neuroreceptors stimulated by pressure changes. Baroreceptors monitor blood pressure. 634

basal metabolic rate (BMR) (ba'sal met"ă-bol'ik) The rate of metabolism (expressed as oxygen consumption or heat production) under resting or basal conditions (14 to 18 hours after eating). 812

basal nucleus (ba'sal noo'kle-us) A mass of nerve cell bodies located deep within a cerebral hemisphere of the brain; also called *basal ganglion*. 388

base A chemical substance that ionizes in water to release hydroxyl ions (OH^-) or other ions that combine with hydrogen ions. **29**

basement membrane A thin sheet of extracellular substance to which the basal surfaces of membranous epithelial cells are attached; also called the *basal lamina*. **133**

basophil (ba'sŏ-fil) A granular leukocyte that readily stains with basophilic dye. **595**

B cell lymphocytes Lymphocytes that can be transformed by antigens into plasma cells that secrete antibodies (and are thus responsible for humoral immunity). The B stands for *bursa equivalent*. 656

belly The thickest circumference of a skeletal muscle.

benign (bĭ-nīn') Not malignant. 674

bifurcate (bi'fur-kāt) Forked; divided into two branches.

bile A liver secretion that is stored and concentrated in the gallbladder and released through the common bile duct into the duodenum. It is essential for the absorption of fats. 791

bilirubin (bil"ĭ-roo'bin) Bile pigment derived from the breakdown of the heme portion of hemoglobin. 792

bipennate (bi-pen'āt) Denoting muscles that have a fiber architecture coursing obliquely on both sides of a tendon. 281

blastula (blas'tyoo-lă) An early stage of prenatal development between the morula and embryonic period. 903

blood The fluid connective tissue that circulates through the cardiovascular system to transport substances throughout the body. 551

blood-brain barrier A specialized mechanism that inhibits the passage of certain materials from the blood into brain tissue and cerebrospinal fluid. 352

bolus (bo'lus) A moistened mass of food that is swallowed from the oral cavity into the pharynx. 773

bone A solid, rigid, ossified connective tissue forming an organ of the skeletal system. 127

bony labyrinth (lab'ĭ-rinth) A series of chambers within the petrous part of the temporal bone associated with the vestibular organs and the cochlea. The bony labyrinth contains a fluid called perilymph. **512**

Bowman's capsule (bo'manz kap'sul) *See* glomerular capsule. 730

brachial plexus (bra′ke-al plek′sus) A network of nerve fibers that arise from spinal nerves C5–C8 and T1. Nerves arising from the brachial plexuses supply the upper extremities. **459**

bradycardia (brad″ĭ-kar′de-ă) A slow cardiac rate; fewer than 60 beats per minute. 603

bradykinins (brad″ĭ-ki′ninz) Short polypeptides that stimulate vasodilation and other cardiovascular changes. 633

brain The enlarged superior portion of the central nervous system located in the cranial cavity of the skull. 376

brain stem The portion of the brain consisting of the medulla oblongata, pons, and midbrain. 392

bronchial tree (brong′ke-al) The bronchi and their branching bronchioles. 689

bronchiole (brong′ke-ōl) A small division of a bronchus within the lung. 689

bronchus (brong′kus) A branch of the trachea that leads to a lung. 689

buccal cavity (buk′al) The mouth, or *oral cavity*. 770

buffer A molecule that serves to prevent large changes in pH by either combining with H^+ or by releasing H^+ into solution. **30**

bulbourethral glands (bul″bo-yoo-re′thral) A pair of glands that secrete a viscous fluid into the male urethra during sexual excitement; also called *Cowper's glands*. 855

bundle of His *See* atrioventricular bundle. 576

bursa (bur′sa) A saclike structure filled with synovial fluid. Bursae are located at friction points, as around joints, over which tendons can slide without contacting bone. **253**

buttocks (but′oks) The rump or fleshy masses on the posterior aspect of the lower trunk, formed primarily by the gluteal muscles. **16**

C

calcitonin (kal″sĭ-to′nin) Also called *thyrocalcitonin*. A polypeptide hormone produced by the parafollicular cells of the thyroid and secreted in response to hypercalcemia. It acts to lower blood calcium and phosphate concentrations and may serve as an antagonist of parathyroid hormones. **196**

calmodulin (kal″mod′yŭ-lin) A receptor protein for Ca^{2+} located within the cytoplasm of target cells. It appears to mediate the effects of this ion on cellular activities. **389**

calorie (kal′ŏ-re) A unit of heat equal to the amount needed to raise the temperature of one gram of water by 1°C. 76

calyx (ka′liks) A cup-shaped portion of the renal pelvis that encircles a renal papilla. 729

cAMP Cyclic adenosine monophosphate; a second messenger in the action of many hormones, including catecholamines, polypeptides, and glycoproteins. It serves to mediate the effects of these hormones on their target cells. 536

canal of Schlemm (shlem) *See* scleral venous sinus. 494

canaliculus (kan″ă-lik′yŭ-lus) A microscopic channel in bone tissue that connects lacunae. **150**

cancer A tumor characterized by abnormally rapid cell division and the loss of specialized tissue characteristics. This term usually refers to malignant tumors. 674

capacitation (kă-pas″ĭ-ta′shun) The process whereby spermatozoa gain the ability to fertilize ova. Sperm that have not been capacitated in the female reproductive tract cannot fertilize ova. 899

capillary (kap′ĭ-lar″e) A microscopic blood vessel that connects an arteriole and a venule; the functional unit of the circulatory system. 579

carbonic anhydrase (kar-bon′ik an-hi′drās) An enzyme that catalyzes the formation or breakdown of carbonic acid. When carbon dioxide concentrations are relatively high, this enzyme catalyzes the formation of carbonic acid from CO_2 and H_2O. When carbon dioxide concentrations are low, the breakdown of carbonic acid to CO_2 and H_2O is catalyzed. These reactions aid the transport of carbon dioxide from tissues to alveolar air. 711

cardiac muscle (kar′de-ak) Muscle of the heart, consisting of striated muscle cells. These cells are interconnected into a mass called the myocardium. **478**

cardiac output The volume of blood pumped per minute by either the right or left ventricle. 612

cardiogenic shock (kar″de-o-jen′ik) Shock that results from low cardiac output in heart disease. 642

carotid sinus (kă-rot′id) An expanded portion of the internal carotid artery located immediately above the point of branching from the external carotid artery. The carotid sinus contains baroreceptors that monitor blood pressure. **628**

carpus (kar′pus) The proximal portion of the hand that contains the eight carpal bones. 206

carrier-mediated transport The transport of molecules or ions across a cell membrane by means of specific protein carriers. It includes both facilitated diffusion and active transport. 97

cartilage (kar′tĭ-laj) A type of connective tissue with a solid elastic matrix. **149**

cartilaginous joint (kar″tĭ-laj′ĭ-nus) A joint that lacks a joint cavity, permitting little movement between the bones held together by cartilage. **248**

cast An accumulation of proteins molded from the kidney tubules that appears in urine sediment. 759

catabolism (kă-tab′o-liz-em) The metabolic breakdown of complex molecules into simpler ones, often resulting in a release of energy. **89**

catecholamines (kat″ĕ-kol′ă-mēnz) A group of molecules including epinephrine, norepinephrine, L-dopa, and related molecules with effects similar to those produced by activation of the sympathetic nervous system. **394**

cations (kat′i-onz) Positively charged ions, such as sodium, potassium, calcium, and magnesium. 24

cauda equina (kaw′dă e-kwi′nă) The lower end of the spinal cord where the roots of spinal nerves have a tail-like appearance. 402

cecum (se′kum) The pouchlike portion of the large intestine to which the ileum of the small intestine is attached. 785

cell The structural and functional unit of an organism; the smallest structure capable of performing all the functions necessary for life. 42

cell-mediated immunity (ĭ-myoo′nĭ-te) Immunological defense provided by T cell lymphocytes that come within close proximity of their victim cells (as opposed to humoral immunity provided by the secretion of antibodies by plasma cells). 656

cellular respiration (sel′yŭ-lar res″pĭ-ra′shun) The energy-releasing metabolic pathways in a cell that oxidize organic molecules such as glucose and fatty acids. 79

cementum (se-men′tum) Bonelike material that binds the root of a tooth to the periodontal membrane of the bony socket. 773

central canal An elongated longitudinal channel in the center of an osteon in bone tissue that contains branches of the nutrient vessels and a nerve; also called a *haversian canal*. **150**

central nervous system Part of the nervous system consisting of the brain and the spinal cord; *CNS*. **372**

centrioles (sen′trĭ-olz) Cell organelles that form the spindle apparatus during cell division. **68**

centromere (sen′trŏ-mēr) The central region of a chromosome to which the chromosomal arms are attached. 61

centrosome (sen′trŏ-sōm) A dense body near the nucleus of a cell that contains a pair of centrioles. **68**

cerebellar peduncle (ser″ĕ-bel′ar pĕ-dung′k′l) An aggregation of nerve fibers connecting the cerebellum with the brain stem. **428**

cerebellum (ser″ĕ-bel′um) The portion of the brain concerned with the coordination of skeletal muscle contraction. Part of the metencephalon, it consists of two hemispheres and a central vermis. **427**

cerebral arterial circle (ser′ĕ-bral) An arterial vessel that encircles the pituitary gland. It provides alternate routes for blood to reach the brain should a carotid or vertebral artery become occluded; also called the *circle of Willis*. **630**

cerebral peduncles A paired bundle of nerve fibers along the inferior surface of the midbrain that conducts impulses between the pons and the cerebral hemispheres. **426**

cerebrospinal fluid (ser″ĕ-bro-spi′nal) A fluid produced by the choroid plexus of the ventricles of the brain. It fills the ventricles and surrounds the central nervous system in association with the meninges. **432**

cerebrum (ser′ĕ-brum) The largest portion of the brain, composed of the right and left hemispheres. **412**

ceruminous gland (sĕ-roo′mĭ-nus) A specialized integumentary gland that secretes cerumen, or earwax, into the external auditory canal. **517**

cervical (ser′vĭ-kal) Pertaining to the neck or a necklike portion of an organ. **14**

cervical ganglion (gang′gle-on) A cluster of postganglionic sympathetic nerve cell bodies located in the neck, near the cervical vertebrae. 452

cervical plexus (plek′sus) A network of spinal nerves formed by the anterior branches of the first four cervical nerves. **459**

cervix (ser′viks) **1.** The narrow necklike portion of an organ. **2.** The inferior end of the uterus that adjoins the vagina (cervix of the uterus). 873

chemoreceptor (ke″mo-re-sep′tor) A neuroreceptor that is stimulated by the presence of chemical molecules. **498**

chemotaxis (ke″mo-tak′sis) The movement of an organism or a cell, such as a leukocyte, toward a chemical stimulus. 652

Cheyne–Stokes respiration (chān′stōkes′ res″pĭ-ra′shun) Breathing characterized by rhythmic waxing and waning of the depth of respiration, with regularly occurring periods of apnea (failure to breathe). 720

chiasma (ki-as′mă) A crossing of nerve tracts from one side of the CNS to the other; also called a *chiasm*. 420

choane (ko-a′ne) The two posterior openings from the nasal cavity into the nasal pharynx; also called the *internal nares*. 684

cholesterol (kŏ-les′ter-ol) A 27-carbon steroid that serves as the precursor of steroid hormones. 34

cholinergic (ko″lĭ-ner′jik) Denoting nerve endings that liberate acetylcholine as a neurotransmitter, such as those of the parasympathetic system. **394**

chondrocranium (kon″dro-kra′ne-um) The portion of the skull that supports the brain. It is derived from endochondral bone. 192

chondrocytes (kon′dro-sītz) Cartilage-forming cells. **149**

chordae tendineae (kor′de ten-din′e-e) Chordlike tendinous bands that connect papillary muscles to the leaflets of the atrioventricular valves within the ventricles of the heart. **613**

chorea (kŏ-re′ă) The occurrence of a wide variety of rapid, complex, jerky movements that appear to be well coordinated but that are performed involuntarily. 366

chorion An extraembryonic membrane that participates in the formation of the placenta. 910

choroid (kor′oid) The vascular, pigmented middle layer of the wall of the eye. **527**

choroid plexus A mass of vascular capillaries from which cerebrospinal fluid is secreted into the ventricles of the brain. 391

chromatids (kro′mă-tidz) Duplicated chromosomes, joined together at the centromere, that separate during cell division. **66**

chromatin (kro′mă-tin) Threadlike structures in the cell nucleus consisting primarily of DNA and protein. They represent the extended form of chromosomes during interphase. **58**

chromatophilic substances (kro″mă-to-fil′ik) Clumps of rough endoplasmic reticulum in the cell bodies of neurons; also called *Nissl bodies*. 345

chromosomes (kro′mŏ-sōmz) Structures in the nucleus that contain the genes for genetic expression. 64

chyme (kīm) The mass of partially digested food that passes from the pylorus of the stomach into the duodenum of the small intestine. 776

cilia (sil′e-ă) Microscopic hairlike processes that move in a wavelike manner on the exposed surfaces of certain epithelial cells. **50**

ciliary body (sil′e-er″e) A portion of the choroid layer of the eye that secretes aqueous humor. It contains the ciliary muscle. **528**

circadian rhythms (ser″kă-de′an) Physiological changes that repeat at about 24-hour intervals. These are often synchronized with changes in the external environment, such as the day-night cycles. **579**

circle of Willis *See* cerebral arterial circle. 585

circumduction (ser″kum-duk′shun) A movement of a body part that outlines a cone, such that the distal end moves in a circle while the proximal portion remains relatively stable. **259**

cirrhosis (sĭ-ro′sis) Liver disease characterized by loss of normal microscopic structure, which is replaced by fibrosis and nodular regeneration. 790

clitoris (klit′or-is, kli′tor-is) A small, erectile structure in the vulva of the female, homologous to the glans penis in the male. 876

clone (klōn) **1.** A group of cells derived from a single parent cell by mitotic cell division; since reproduction is asexual, the descendants of the parent cell are genetically identical. 62 **2.** A term used to refer to cells as separate individuals (as in white blood cells) rather than as part of a growing organ. 664

CNS *See* central nervous system. 345

coccygeal (kok-sij′e-al) Pertaining to the region of the coccyx; the caudal termination of the vertebral column. **216**

cochlea (kok′le-ă) The organ of hearing in the inner ear where nerve impulses are generated in response to sound waves. **518**

cochlear window *See* round window. 483

codon (ko′don) The sequence of three nucleotide bases in mRNA that specifies a given amino acid and determines the position of that amino acid in a polypeptide chain through complementary base pairing with an anticodon in tRNA. **61**

coelom (se′lom) The abdominal cavity. **16**

coenzyme (ko-en′zīm) An organic molecule, usually derived from a water-soluble vitamin, that combines with and activates specific enzyme proteins. **83**

cofactor (ko′fak-tor) A substance needed for the catalytic action of an enzyme; generally used in reference to inorganic ions such as Ca^{2+} and Mg^{2+}. **82**

collateral (kŏ-lat′er-al) A small branch of a blood vessel or nerve fiber.

colloid osmotic pressure (kol′oid oz-mot′ik) Osmotic pressure exerted by plasma proteins that are present as a colloidal suspension; also called *oncotic pressure*. 617

colon (ko′lon) The first portion of the large intestine. 785

common bile duct A tube formed by the union of the hepatic duct and cystic duct that transports bile to the duodenum. 793

compact bone Tightly packed bone that is superficial to spongy bone and covered by the periosteum; also called *dense bone*. **150**

compliance (kom-pli′ans) A measure of the ease with which a structure such as the lung expands under pressure; a measure of the change in volume as a function of pressure changes. 693

conduction myofibers Specialized large-diameter cardiac muscle fibers that conduct electrical impulses from the AV bundle into the ventricular walls; also called *Purkinje fibers*. **621**

condyle (kon′dīl) A rounded process at the end of a long bone that forms an articulation. 164

cone A color receptor cell in the retina of the eye. **533**

congenital (kon-jen′ĭ-tal) Present at the time of birth.

congestive heart failure (kon-jes′tiv) The inability of the heart to deliver an adequate blood flow as a result of heart disease or hypertension. This condition is associated with breathlessness, salt and water retention, and edema. 643

conjunctiva (kon″jungk-ti′vă) The thin membrane covering the anterior surface of the eyeball and lining the eyelids. **526**

conjunctivitis (kon-jungk″tĭ-vi′tis) Inflammation of the conjunctiva of the eye, which is sometimes called "pink eye." **546**

connective tissue One of the four basic tissue types within the body. It is a binding and supportive tissue with abundant matrix. **145**

Conn's syndrome (konz) Primary hyperaldosteronism; excessive secretion of aldosterone produces electrolyte imbalances. 750

contralateral (kon″tră-lat′er-al) Taking place or originating in a corresponding part on the opposite side of the body. 438

conus medullaris (kó nus med″yŭ-lăr′is) The inferior, tapering portion of the spinal cord. **433**

convolution (kon-vŏ-loo′shun) An elevation on the surface of a structure and an infolding of the tissue upon itself. 383

cornea (kor′ne-ă) The transparent, convex, anterior portion of the outer layer of the eyeball. **527**

coronal plane (kor′ŏ-nal, kŏ-ro′nal) A plane that divides the body into anterior and posterior portions; also called a *frontal plane*. **10**

coronary circulation (kor′ŏ-nar″e) The arterial and venous blood circulation to the wall of the heart. 571

coronary sinus A large venous channel on the posterior surface of the heart into which the cardiac veins drain. **613**

corpora quadrigemina (kor′por-ă kwad″rĭ-jem′ĭ-na) Four superior lobes of the midbrain concerned with visual and auditory functions. **426**

corpus callosum (kor′pus kă-lo′sum) A large tract of white matter within the brain that connects the right and left cerebral hemispheres. 380

corpuscle of touch (kor′pus′l) A touch sensory receptor found in the papillary layer of the dermis of the skin; also called *Meissner's corpuscle*. 467

cortex (kor′teks) **1.** The outer layer of an internal organ or body structure, as of the kidney or adrenal gland. 526 **2.** The convoluted layer of gray matter that covers the surface of each cerebral hemisphere. **168**

corticosteroids (kor″tĭ-ko-ster′oidz) Steroid hormones of the adrenal cortex, consisting of glucocorticoids (such as hydrocortisone) and mineralocortocoids (such as aldosterone). **38**

costal cartilage (kos′tal) The cartilage that connects the ribs to the sternum. **149**

cranial (kra′ne-al) Pertaining to the cranium.

cranial nerves One of 12 pairs of nerves that arise from the brain. 417

cranium (kra′ne-um) The bones of the skull that enclose or support the brain and the organs of sight, hearing, and balance. **14**

creatine phosphate (kre′ă-tin fos′fāt) An organic phosphate molecule in muscle cells that serves as a source of high-energy phosphate for the synthesis of ATP; also called *phosphocreatine*. **298**

crenation (krĭ-na′shun) A notched or scalloped appearance of the red blood cell membrane caused by the osmotic loss of water from these cells. 96

crest A thickened ridge of bone for the attachment of muscle.

cretinism (krēt′n-iz″em) A condition caused by insufficient thyroid secretion during prenatal development or early childhood. It results in stunted growth and inadequate mental development. **585**

cricoid cartilage (kri′koid) A ring-shaped cartilage that forms the inferior portion of the larynx. 687

crista (kris′tă) A crest, such as the crista galli that extends superiorly from the cribriform plate of the ethmoid bone. **209**

cryptorchidism (krip-tor′kĭ-diz″em) A developmental defect in which one or both testes fail to descend into the scrotum and, instead, remain in the body cavity. 863

cubital (kyoo′bĭ-tal) Pertaining to the antebrachium. The cubital fossa is the anterior aspect of the elbow joint. **16**

curare (koo-ră-re) A chemical derived from plant sources that causes flaccid paralysis by blocking ACh receptor proteins in muscle cell membranes. 362

Cushing's syndrome (koosh′ingz) Symptoms caused by the hypersecretion of adrenal steroid hormones as a result of tumors of the adrenal cortex or ACTH-secreting tumors of the anterior pituitary. **585**

cyanosis (si′ă-no′sis) A bluish discoloration of the skin or mucous membranes due to excessive concentration of deoxyhemoglobin; indicates inadequate oxygen concentration in the blood. 141

cystic duct (sis′tik dukt) The tube that transports bile from the gallbladder to the common bile duct. 793

cytochrome (si′tŏ-krōm) A pigment in mitochondria that transports electrons in the process of aerobic respiration. 84

cytokinesis (si″to-kĭ-ne′sis) The division of the cytoplasm that occurs in mitosis and meiosis, when a parent cell divides to produce two daughter cells. 62

cytology (si-tol′ŏ-je) The science dealing with the study of cells. 5

cytoplasm (si′tŏ-plaz″em) In a cell, the protoplasm located outside of the nucleus. **52**

cytoskeleton (si″to-skel′ĕ-ton) A latticework of structural proteins in the cytoplasm arranged in the form of microfilaments and microtubules. **52**

D

deciduous (dĭ-sij′oo-us) Pertaining to something shed or cast off in a particular sequence. Deciduous teeth are shed and replaced by permanent teeth during development. 771

decussation (dek″uh-sa′shun) A crossing of nerve fibers from one side of the CNS to the other. 403

defecation (def″ĕ-ka′shun) The elimination of feces from the rectum through the anal canal and out the anus. 787

deglutition (de″gloo-tish′un) The act of swallowing. 775

delayed hypersensitivity An allergic response in which the onset of symptoms may not occur until 2 or 3 days after exposure to an antigen. Produced by T cells, it is a type of cell-mediated immunity. 678

denaturation (de-na″chur-a′shun) Irreversible changes in the tertiary structure of proteins caused by heat or drastic pH changes. **41**

dendrite (den′drīt) A nerve cell process that transmits impulses toward a neuron cell body. **155**

dentin (den′tin) The main substance of a tooth, covered by enamel over the crown of the tooth and by cementum on the root. 773

dentition (den-tish′un) The number, arrangement, and shape of teeth. 773

depolarization (de-po″lar-ĭ-za′shun) The loss of membrane polarity in which the inside of the cell membrane becomes less negative in comparison to the outside of the membrane. The term is also used to indicate the reversal of membrane polarity that occurs during the production of action potentials in nerve and muscle cells. **379**

dermal papilla (pă-pil′ă) A projection of the dermis into the epidermis. **164**

dermis (der′mis) The second, or deep, layer of skin beneath the epidermis. **164**

descending colon The segment of the large intestine that descends on the left side from the level of the spleen to the level of the left iliac crest. 785

diabetes insipidus (di″ă-be′tēz in-sip′ĭ-dus) A condition in which inadequate amounts of antidiuretic hormone (ADH) are secreted by the posterior pituitary. It results in the inadequate reabsorption of water by the kidney tubules and, thus, in the excretion of a large volume of dilute urine. **584**

diabetes mellitus (mĕ-li′tus) The appearance of glucose in the urine due to the presence of high plasma glucose concentrations, even in the fasting state. This disease is caused by either lack of sufficient insulin secretion or inadequate responsiveness of the target tissues to the effects of insulin. **105**

diapedesis (di″ă-pĕ-de′sis) The migration of white blood cells through the endothelial walls of blood capillaries into the surrounding connective tissues. 652

diaphragm (di′ă-fram) A sheetlike dome of muscle and connective tissue that separates the thoracic and abdominal cavities. **316**

diaphysis (di-af′ĭ-sis) The shaft of a long bone. **188**

diarrhea (di″ă-re′ă) Abnormal frequency of defecation accompanied by abnormal liquidity of the feces. 786

diarthrosis (di″ar-thro′sis) A type of functionally classified joint in which the articulating bones are freely movable; also called a *synovial joint.* 224

diastole (di-as′tŏ-le) The sequence of the cardiac cycle during which a heart chamber wall is relaxed. **618**

diencephalon (di″en-sef′ă-lon) A major region of the brain that includes the third ventricle, thalamus, hypothalamus, and pituitary gland. **424**

diffusion (dĭ-fyoo′zhun) The net movement of molecules or ions from regions of higher to regions of lower concentration. 92

digestion The process by which larger molecules of food substance are broken down mechanically and chemically into smaller molecules that can be absorbed. 765

diploe (dip′lo-e) The spongy layer of bone positioned between the inner and outer layers of compact bone. 165

diploid (dip′loid) Denoting cells having two of each chromosome or twice the number of chromosomes that are present in sperm or ova. 839

disaccharide (di-sak′ă-rīd) Any of a class of double sugars; carbohydrates that yield two simple sugars, or monosaccharides, upon hydrolysis. **33**

distal (dis′tal) Away from the midline or origin; the opposite of *proximal.* 12

diuretic (di″yŭ-ret′ik) An agent that promotes the excretion of urine, thereby lowering blood volume and pressure. 620

DNA Deoxyribonucleic acid; composed of nucleotide bases and deoxyribose sugar. It is found in all living cells and contains the genetic code. **56**

dopamine (do′pă-mēn) A type of neurotransmitter in the central nervous system; also is the precursor of norepinephrine, another neurotransmitter molecule. **396**

dorsal (dor′sal) Pertaining to the back or posterior portion of a body part; the opposite of *ventral;* also called *posterior.* 12

dorsal root ganglion *See* posterior root ganglion. 427

dorsiflexion (dor″sĭ-flek′shun) Movement at the ankle as the dorsum of the foot is elevated. **257**

ductus arteriosus (duk′tus ar-tir″e-o′sus) The blood vessel that connects the pulmonary trunk and the aorta in a fetus. 601

ductus deferens (def′er-enz), pl. *ductus deferentia* A tube that carries spermatozoa from the epididymis to the ejaculatory duct; also called the *vas deferens* or *seminal duct.* 854

ductus venosus (ven-o′sus) A fetal blood vessel that connects the umbilical vein and the inferior vena cava. 601

duodenum (doo″ŏ-de′num, doo-od′ĕ-num) The first portion of the small intestine that leads from the pylorus of the stomach to the jejunum. 781

dura mater (door′ă ma′ter) The outermost meninx. **430**

dwarfism A condition in which a person is undersized due to inadequate secretion of growth hormone. **196**

dyspnea (disp-ne′ă) Subjective difficulty in breathing. 699

E

eccrine gland (ek′rin) A sweat gland that functions in thermoregulation. **171**

ECG *See* electrocardiogram. 577

ectoderm (ek′tŏ-derm) The outermost of the three primary germ layers of an embryo. 904

ectopic focus (ek-top′ik) An area of the heart other than the SA node that assumes pacemaker activity. 575

ectopic pregnancy Embryonic development that occurs anywhere other than in the uterus (as in the uterine tubes or body cavity). 890

edema (ĕ-de′mă) An excessive accumulation of fluid in the body tissues. 618

EEG *See* electroencephalogram. 386

effector (ĕ-fek′tor) An organ, such as a gland or muscle, that responds to a motor stimulation. 17

efferent (ef′er-ent) Conveying away from the center of an organ or structure. 346

efferent arteriole (ar-tir′e-ōl) An arteriole of the renal vascular system that conducts blood away from the glomerulus of a nephron. 730

efferent ductules (duk′toolz) A series of coiled tubules through which spermatozoa are transported from the rete testis to the epididymis. 848

efferent neuron (noor′on) *See* motor neuron. 346

ejaculation (ĕ-jak″yŭ-la′shun) The discharge of semen from the male urethra that accompanies orgasm. 858

ejaculatory duct (ĕ-jak′yŭ-lă-tor″-e) A tube that transports spermatozoa from the ductus deferens to the prostatic urethra. 854

elastic fibers (ĕ-las′tik) Protein strands that are found in certain connective tissue that have contractile properties. **145**

elbow The synovial joint between the brachium and the antebrachium. **266**

electrocardiogram (ĕ-lek″tro-kar′de-ŏ-gram″) A recording of the electrical activity that accompanies the cardiac cycle; ECG or EKG. **622**

electroencephalogram (ĕ-lek″tro-en-sef′ă-lŏ-gram) A recording of the brain-wave patterns or electrical impulses of the brain from electrodes placed on the scalp; EEG. **419**

electrolytes (ĕ-lek′tro-lītz) Ions and molecules that are able to ionize and thus carry an electric current. The most common electrolytes in the plasma are Na^+, HCO_3^-, and K^+. 728

electromyogram (ĕ-lek″tro-mi′ŏ-gram) A recording of the electrical impulses or activity of skeletal muscles using surface electrodes; EMG. 318

electrophoresis (ĕ-lek″tro-fŏ-re′sis) A biochemical technique in which different molecules can be separated and identified by their rate of movement in an electric field. 657

elephantiasis (el″ĕ-fan-ti′ă-sis) A disease caused by infection with a nematode worm in which the larvae block lymphatic drainage and produce edema; the lower areas of the body can become enormously swollen as a result. 618

embryology (em″bre-ol′ŏ-je) The study of prenatal development from conception through the eighth week in utero. 906

EMG *See* electromyogram. 318

emphysema (em″fĭ-se′mă, em″fĭ-ze′mă) A lung disease in which the alveoli are destroyed and the remaining alveoli become larger. It results in decreased vital capacity and increased airway resistance. 719

emulsification (ĕ-mul″sĭ-fĭ-ka′shun) The process of producing an emulsion or fine suspension; in the small intestine, fat globules are emulsified by the detergent action of bile. 798

enamel (ĕ-nam′el) The outer dense substance covering the crown of a tooth. 773

endergonic (en″der-gon′ik) Denoting a chemical reaction that requires the input of energy from an external source in order to proceed. **85**

endocardium (en″do-kar′de-um) The endothelial lining of the heart chambers and valves. **612**

endochondral bone (en″dŏ-kon′dral) Denoting bones that develop as hyaline cartilage models first and that are then ossified. 167

endocrine gland (en′dŏ-krin) A ductless, hormone-producing gland that is part of the endocrine system. **552**

endocytosis (en″do-si-to′sis) A general term for the cellular uptake of particles that are too large to cross the cell membrane. *See also* phagocytosis and pinocytosis. **49**

endoderm (en′dŏ-derm) The innermost of the three primary germ layers of an embryo. 904

endogenous (en-doj′ĕ-nus) Denoting a product or process arising from within the body (as opposed to exogenous products or influences from external sources).

endolymph (en′dŏ-limf) A fluid within the membranous labyrinth and cochlear duct of the inner ear that aids in the conduction of vibrations involved in hearing and the maintenance of equilibrium. 478

endometrium (en″do-me′tre-um) The inner lining of the uterus. 874

endomysium (en″do-mis′e-um) The connective tissue sheath that surrounds each skeletal muscle fiber, separating the muscle cells from one another. **281**

endoneurium (en″do-nyoo′re-um) The connective tissue sheath that surrounds each nerve fiber, separating the nerve fibers from another within a nerve. 347

endoplasmic reticulum (en-do-plaz′mik rĕ-tik′yŭ-lum) A cytoplasmic organelle composed of a network of canals running through the cytoplasm of a cell. **54**

endorphins (en-dor′finz) A group of endogenous opiate molecules that may act as a natural analgesic. 367

endothelium (en″do-the′le-um) The layer of epithelial tissue that forms the thin inner lining of blood vessels and heart chambers. **134**

endotoxin (en″do-tok′sin) A toxin found within certain types of bacteria that is able to stimulate the release of endogenous pyrogen and produce a fever. 653

enkephalins (en-kef′ă-linz) Short polypeptides, containing five amino acids, that have analgesic effects and that may function as neurotransmitters in the brain. The two known enkephalins (which differ in only one amino acid) are endorphins. 367

enteric (en-ter′ik) The term referring to the small intestine.

entropy (en′trŏ-pe) The energy of a system that is not available to perform work. A measure of the degree of disorder in a system, entropy increases whenever energy is transformed. 75

enzyme (en′zīm) A protein catalyst that increases the rate of specific chemical reactions. **79**

eosinophil (e″ŏ-sin′ŏ-fil) A type of white blood cell characterized by the presence of cytoplasmic granules that become stained by acidic eosin dye. Eosinophils normally constitute about 2% to 4% of the white blood cells. 553

epicardium (ep″ĭ-kar′de-um) A thin, outer layer of the heart; also called the *visceral pericardium*. **612**

epicondyle (ep″ĭ-kon′dīl) A projection of bone above a condyle. 164

epidermis (ep″ĭ-der′mis) The outermost layer of the skin, composed of several stratified squamous epithelial layers. **160**

epididymis (ep″ĭ-did′ĭ-mis) A highly coiled tube located along the posterior border of the testis. It stores spermatozoa and transports them from the seminiferous tubules of the testis to the ductus deferens. 853

epidural space (ep″ĭ-door′al) A space between the spinal dura mater and the bone of the vertebral canal. 396

epiglottis (ep″ĭ-glot′is) A leaflike structure positioned on top of the larynx. It covers the glottis during swallowing. 687

epimysium (ep″ĭ-mis′e-um) A fibrous outer sheath of connective tissue surrounding a skeletal muscle. **281**

epinephrine (ep″ĭ-nef′rin) A hormone secreted from the adrenal medulla resulting in actions similar to those resulting from sympathetic nervous system stimulation; also called *adrenaline*. **482**

epineurium (ep″ĭ-nyoo′re-um) A fibrous outer sheath of connective tissue surrounding a nerve. 347

epiphyseal plate (ep″ĭ-fiz′e-al) A hyaline cartilaginous layer located between the epiphysis and diaphysis of a long bone. It functions as a longitudinal growing region. **189**

epiphysis (ĕ-pif′ĭ-sis) The end segment of a long bone, separated from the diaphysis early in life by an epiphyseal plate but later becoming part of the larger bone. **188**

episiotomy (ĕ-pe″ze-ot′ŏ-me) An incision of the perineum at the end of the second stage of labor to facilitate delivery and to avoid tearing the perineum. 922

epithelial tissue (ep″ĭ-the′le-al) One of the four basic tissue types; the type of tissue that covers or lines all exposed body surfaces. 108

eponychium (ep″ŏ-nik′e-um) The thin layer of stratum corneum of the epidermis of the skin that overlaps and protects the lunula of the nail. **170**

EPSP Excitatory postsynaptic potential; a graded depolarization of a postsynaptic membrane in response to stimulation by a neurotransmitter chemical. EPSPs can be summated but can be transmitted only over short distances. They can stimulate the production of action potentials when a threshold level of depolarization has been attained. 361

erythroblastosis fetalis (ĕ-rith″ro-blas-to′sis fĭ-tal′is) Hemolytic anemia in an Rh-positive newborn caused by maternal antibodies against the Rh factor that have crossed the placenta. 558

erythrocyte (ĕ-rith′rŏ-sīt) A red blood cell. **153**

esophagus (ĕ-sof′ă-gus) A tubular portion of the GI tract that leads from the pharynx to the stomach as it passes through the thoracic cavity. 775

essential amino acids Those eight amino acids in adults or nine amino acids in children that cannot be made by the human body; therefore, they must be obtained in the diet. **101**

estrogens (es′tro-jenz) Any of several female sex hormones secreted from the ovarian (graafian) follicle. 533

estrus cycle (es′trus) Cyclic changes in the structure and function of the ovaries and female reproductive tract of mammals other than humans, accompanied by periods of "heat" (estrus) or sexual receptivity. Estrus is the equivalent of the human menstrual cycle but differs from the human menstrual cycle in that the endometrium is not shed with accompanying bleeding. 881

etiology (e″te-ol′ŏ-je) The study of cause, especially of disease, including the origin and what pathogens, if any, are involved.

eustachian canal (yoo-sta′ke-an) *See* auditory tube. 483

eversion (ĕ-ver′zhun) A movement of the foot in which the sole is turned outward. **259**

exergonic (ek″ser-gon′ik) Denoting chemical reactions that liberate energy. **85**

exocrine gland (ek′sŏ-krin) A gland that secretes its product to an epithelial surface, directly or through ducts. **141**

exocytosis (ek″so-si-to′sis) The process of cellular secretion in which the secretory products are contained within a membrane-enclosed vesicle. The vesicle fuses with the cell membrane so that the lumen of the vesicle is open to the extracellular environment. **50**

expiration (ek″spĭ-ra′shun) The process of expelling air from the lungs through breathing out; also called *exhalation*. 696

extension (ek-sten′shun) A movement that increases the angle between parts of a joint. **257**

extensor A muscle that, upon contraction, increases the angle of a joint.

external (superficial) Located on or toward the surface. 12

external acoustic meatus (ă-koo′stik me-a′tus) An opening through the temporal bone that connects with the tympanum and the middle-ear chamber and through which sound vibrations pass; also called the *external auditory meatus*. **207**

exteroceptors (ek″stĕ-ro-sep′torz) Sensory receptors that are sensitive to changes in the external environment (as opposed to interoceptors). **509**

extraocular muscles (ek″stră-ok′yŭ-lar) The muscles that insert into the sclera of the eye and that act to change the position of the eye in its orbit (as opposed to the intraocular muscles, such as those of the iris and ciliary body within the eye). 285

extrinsic (eks-trin′sik) Pertaining to an outside or external origin.

F

face **1.** The anterior aspect of the head not supporting or covering the brain. 12 **2.** The exposed surface of a structure.

facet (fas′et) A small, smooth surface of a bone where articulation occurs. 164

facilitated diffusion (fă-sil′ĭ-ta″tid) The carrier-mediated transport of molecules through the cell membrane along the direction of their concentration gradients. It does not require the expenditure of metabolic energy. **118**

FAD Flavin adenine dinucleotide; a coenzyme derived from riboflavin that participates in electron transport within the mitochondria. **87**

falciform ligament (fal′sĭ-form lig′ă-ment) An extension of parietal peritoneum that separates the right and left lobes of the liver. 767

fallopian tube (fă-lo′pe-an) *See* uterine tube. 872

false vocal cords The supporting folds of tissue for the true vocal cords within the larynx. 688

falx cerebelli (falks ser″ĕ-bel′e) A fold of the dura mater anchored to the occipital bone. It projects inward between the cerebellar hemispheres. 398

falx cerebri (ser′ĕ-bre) A fold of dura mater anchored to the crista galli of the ethmoid bone. It extends between the right and left cerebral hemispheres. 398

fascia (fash′e-ă) A tough sheet of fibrous tissue binding the skin to underlying muscles or supporting and separating muscles. **145**

fasciculus (fă-sik′yŭ-lus) A small bundle of muscle or nerve fibers. 255

fauces (faw′sēz) The passageway between the mouth and the pharynx. 770

feces (fe′sēz) Material expelled from the GI tract during defecation, composed of undigested food residue, bacteria, and secretions; also called *stool.* 787

fertilization (fer″tĭ-lĭ-za′shun) The fusion of an ovum and spermatozoon. 899

fetus (fe′tus) A prenatal human after 8 weeks of development. 917

fibrillation (fib″rĭ-la′shun) A condition of cardiac muscle characterized electrically by random and continuously changing patterns of electrical activity and resulting in the inability of the myocardium to contract as a unit and pump blood. It can be fatal if it occurs in the ventricles. 604

fibrin (fi′brin) The insoluble protein formed from fibrinogen by the enzymatic action of thrombin during the process of blood clot formation. **601**

fibrinogen (fi-brin′ŏ-jen) A soluble plasma protein that serves as the precursor of fibrin; also called *factor I.* **594**

fibroblast (fi′bro-blast) An elongated connective tissue cell with cytoplasmic extensions that is capable of forming collagenous fibers or elastic fibers. **145**

fibrous joint (fi′brus) A type of articulation bound by fibrous connective tissue that allows little or no movement (e.g., a syndesmosis). **248**

filiform papillae (fil′ĭ-form pă-pil′e) Numerous small projections over the entire surface of the tongue in which taste buds are absent. 771

filum terminale (fi′lum ter-mĭ-nal′e) A fibrous, threadlike continuation of the pia mater, extending inferiorly from the terminal end of the spinal cord to the coccyx. **433**

fimbriae (fim′bre-e) Fringelike extensions from the borders of the open end of the uterine tube. 782

fissure (fish′ur) A groove or narrow cleft that separates two parts, such as the cerebral hemispheres of the brain. 380

flagellum (flă-jel′um) A whiplike structure that provides motility for sperm. 45

flare-and-wheal reaction (hwēl, wēl) A cutaneous reaction to skin injury or the administration of antigens, produced by release of histamine and related molecules and characterized by local edema and a red flare. 679

flavoprotein (fla″vo-pro′te-in) A conjugated protein containing a flavin pigment that is involved in electron transport within the mitochondria. 84

flexion (flek′shun) A movement that decreases the angle between parts of a joint. **257**

flexor (flek′sor) A muscle that decreases the angle of a joint when it contracts.

fontanel (fon″tă-nel′) A membranous-covered region on the skull of a fetus or baby where ossification has not yet occurred; commonly called a *soft spot.* 192

foot The terminal portion of the lower extremity, consisting of the tarsal bones, metatarsal bones, and phalanges. **238**

foramen (fŏ-ra′men), pl. *foramina* An opening in an anatomical structure, usually in a bone, for the passage of a blood vessel or a nerve. 164

foramen ovale (o-val′e) An opening through the interatrial septum of the fetal heart. 601

forearm The portion of the upper extremity between the elbow and the wrist; also called the *antebrachium.* 205

fornix (for′niks) **1.** A recess around the cervix of the uterus where it protrudes into the vagina. 895 **2.** A tract within the brain connecting the hippocampus with the mammillary bodies. 460

fossa (fos′ă) A depressed area, usually on a bone. 164

fourth ventricle (ven′trĭ-k′l) A cavity within the brain, between the cerebellum and the medulla oblongata and the pons, containing cerebrospinal fluid. **432**

fovea centralis (fo′ve-ă sen-tra′ lis) A depression on the macula lutea of the eye, where only cones are located; the area of keenest vision. **539**

frenulum (fren′yŭ-lum) A membranous structure that serves to anchor and limit the movement of a body part. 771

frontal **1.** Pertaining to the region of the forehead. 12 **2.** A plane through the body, dividing the body into anterior and posterior portions; also called the *coronal plane.* **10**

FSH Follicle-stimulating hormone; one of the two gonadotropic hormones secreted from the anterior pituitary. In females, FSH stimulates the development of the ovarian follicles; in males, it stimulates the production of sperm in the seminiferous tubules. 520

fungiform papillae (fun′jĭ-form pă-pil′e) Flattened, mushroom-shaped projections interspersed over the surface of the tongue in which taste buds are present. 771

G

GABA Gamma-aminobutyric acid; believed to function as an inhibitory neurotransmitter in the central nervous system. **398**

gallbladder A pouchlike organ attached to the underside of the liver in which bile secreted by the liver is stored and concentrated. 793

gamete (gam′ēt) A haploid sex cell; either an egg cell or a sperm cell. 839

ganglion (gang′gle-on) An aggregation of nerve cell bodies occurring outside the central nervous system. **533**

gastric intrinsic factor (gas′trik) A glycoprotein secreted by the stomach that is needed for the absorption of vitamin B_{12}. 779

gastrin (gas′trin) A hormone secreted by the stomach that stimulates the gastric secretion of hydrochloric acid and pepsin. 778

gastrointestinal tract (GI tract) (gas″tro-in-tes′tĭ-nal) The portion of the digestive tract that includes the stomach and the small and large intestines. 766

gates Structures composed of one or more protein molecules that regulate the passage of ions through channels within the cell membrane. Gates may be chemically regulated (by neurotransmitters) or voltage regulated (in which case they open in response to a threshold level of depolarization). 380

genetic recombination (jĕ-net′ik re″kom-bĭ-na′shun) The formation of new combinations of genes, as by crossing-over between homologous chromosomes. **71**

genetic transcription (tran-skrip′shun) The process by which RNA is produced with a sequence of nucleotide bases that is complementary to a region of DNA. 54

genetic translation (trans-la′shun) The process by which proteins are produced with amino acid sequences specified by the sequence of codons in messenger RNA. 55

gigantism (ji-gan′tiz″em) Abnormal body growth as a result of the excessive secretion of growth hormone. **584**

gingiva (jin′jĭ-vă) The fleshy covering over the mandible and maxilla through which the teeth protrude within the mouth; also called the *gum*. 773

gland An organ that produces a specific substance or secretion.

glans penis (glanz pe′nis) The enlarged, sensitive, distal end of the penis. 856

gliding joint A type of synovial joint in which the articular surfaces are flat, permitting only side-to-side and back-and-forth movements. 228

glomerular capsule (glo-mer′yŭ-lar) The double-walled proximal portion of a renal tubule that encloses the glomerulus of a nephron; also called *Bowman's capsule*. 730

glomerular filtration rate (GFR) The volume of filtrate produced per minute by both kidneys. 733

glomerular ultrafiltrate (ul″tră-fil′trāt) Fluid filtered through the glomerular capillaries into the glomerular capsule of the kidney tubules. 733

glomerulonephritis (glo-mer″yŭ-lo-nĕ-fri′tis) Inflammation of the renal glomeruli, associated with fluid retention, edema, hypertension, and the appearance of protein in the urine. 759

glomerulus (glo-mer′yŭ-lus) A coiled tuft of capillaries surrounded by the glomerular capsule that filtrates urine from the blood. 730

glottis (glot′is) A slitlike opening into the larynx, positioned between the true vocal cords. 687

glucagon (gloo′kă-gon) A polypeptide hormone secreted by the alpha cells of the pancreatic islets. It acts primarily on the liver to promote glycogenolysis and raise blood glucose levels. **578**

glucocorticoids (gloo″ko-kor′tĭ-koidz) Steroid hormones secreted by the adrenal cortex (corticosteroids). They affect the metabolism of glucose, protein, and fat and also have anti-inflammatory and immunosuppressive effects. The major glucocorticoid in humans is hydrocortisone (cortisol). **573**

gluconeogenesis (gloo″ko-ne″ŏ-jen′ĭ-sis) The formation of glucose from noncarbohydrate molecules, such as amino acids and lactic acid. 82

glycerol (glis′ĕ-rol) A 3-carbon alcohol that serves as a building block of fats. 32

glycogen (gli′kŏ-jen) A polysaccharide of glucose—also called *animal starch*—produced primarily in the liver and skeletal muscles. Similar to plant starch in composition, glycogen contains more highly branched chains of glucose subunits than does plant starch. **33**

glycogenesis (gli″kŏ-jen′ĭ-sis) The formation of glycogen from glucose. 793

glycogenolysis (gli″kŏ-jĕ-nol′ĭ-sis) The hydrolysis of glycogen to glucose-1-phosphate, which can be converted to glucose-6-phosphate, which then may be oxidized via glycolysis or (in the liver) converted to free glucose. **92**

glycolysis (gli″kol′ĭ-sis) The metabolic pathway that converts glucose to pyruvic acid; the final products are two molecules of pyruvic acid and two molecules of reduced NAD, with a net gain of two ATP molecules. In anaerobic respiration, the reduced NAD is oxidized by the conversion of pyruvic acid to lactic acid. In aerobic respiration, pyruvic acid enters the Krebs cycle in mitochondria, and reduced NAD is ultimately oxidized to yield water. **89**

glycosuria (gli″kŏ-soor′e-ă) The excretion of an abnormal amount of glucose in the urine (urine normally contains only trace amounts of glucose). 746

goblet cell A unicellular mucus-secreting gland that is associated with columnar epithelia; also called a *mucous cell*. **134**

Golgi apparatus (gol′je) A network of stacked, flattened membranous sacs within the cytoplasm of cells. Its major function is to concentrate and package proteins for secretion from the cell. **63**

Golgi tendon organ A sensory receptor found near the junction of tendons and muscles. 473

gonad (go′nad) A reproductive organ, testis or ovary, that produces gametes and sex hormones. **581**

gonadotropic hormones (go-nad″ŏ-tro′pik) Hormones of the anterior pituitary that stimulate gonadal function—the formation of gametes and secretion of sex steroids. The two gonadotropins are FSH (follicle-stimulating hormone) and LH (luteinizing hormone), which are essentially the same in males and females. **566**

graafian follicle (graf′e-an) A mature ovarian follicle, containing a single fluid-filled cavity, with the ovum located toward one side of the follicle and perched on top of a hill of granulosa cells. 877

granular leukocytes (loo′kŏ-sītz) Leukocytes with granules in the cytoplasm; on the basis of the staining properties of the granules, these cells are classified as neutrophils, eosinophils, or basophils. **595**

Graves' disease A hyperthyroid condition believed to be caused by excessive stimulation of the thyroid gland by autoantibodies; it is associated with exophthalmos (bulging eyes), high pulse rate, high metabolic rate, and other symptoms of hyperthyroidism. **585**

gray matter The region of the central nervous system composed of nonmyelinated nerve tissue. **376**

greater omentum (o-men′tum) A double-layered peritoneal membrane that originates on the greater curvature of the stomach. It hangs inferiorly like an apron over the contents of the abdominal cavity. 767

gross anatomy The branch of anatomy concerned with structures of the body that can be studied without a microscope. 5

growth hormone A hormone secreted by the anterior pituitary that stimulates growth of the skeleton and soft tissues during the growing years and that influences the metabolism of protein, carbohydrate, and fat throughout life. 567

gustatory (gus′tă-tor″e) Pertaining to the sense of taste. 474

gut The GI tract or a portion thereof; generally used in reference to the embryonic digestive tube, consisting of the foregut, midgut, and hindgut. 766

gyrus (ji′rus) A convoluted elevation or ridge. 383

H

hair A threadlike appendage of the epidermis consisting of keratinized dead cells that have been pushed up from a dividing basal layer. **167**

hair cells Specialized receptor nerve endings for detecting sensations, such as in the spiral organ (organ of Corti). **512**

hair follicle (fol′lĭ-k′l) A tubular depression in the dermis of the skin in which a hair develops. **168**

hand The terminal portion of the upper extremity, containing the carpal bones, metacarpal bones, and phalanges. 206

haploid (hap′loid) A cell that has one of each chromosome type and therefore half the number of chromosomes present in most other body cells; only the gametes (sperm and ova) are haploid. 839

haptens (hap′tenz) Small molecules that are not antigenic by themselves, but which—when combined with proteins—become antigenic and thus capable of stimulating the production of specific antibodies. 655

hard palate (pal′it) The bony partition between the oral and nasal cavities, formed by the maxillae and palatine bones and lined by mucous membrane. 771

haustra (haws′tră) Sacculations or pouches of the colon. 786

haversian canal (hă-ver′shan) *See* central canal. 130

haversian system *See* osteon. 167

hay fever A seasonal type of allergic rhinitis caused by pollen; it is characterized by itching and tearing of the eyes, swelling of the nasal mucosa, attacks of sneezing, and often by asthma. 677

head The uppermost portion of a human that contains the brain and major sense organs. **14**

heart A four-chambered, muscular pumping organ positioned in the thoracic cavity, slightly to the left of midline. **593**

heart murmur An auscultatory sound of cardiac or vascular origin, usually caused by an abnormal flow of blood in the heart as a result of structural defects of the valves or septum. 605

helper T cells A subpopulation of T cells (lymphocytes) that helps to stimulate the antibody production of B lymphocytes by antigens. 668

hematocrit (hĭ-mat′ŏ-krit) The ratio of packed red blood cells to total blood volume in a centrifuged sample of blood, expressed as a percentage. 551

heme (hēm) The iron-containing red pigment that, together with the protein globin, forms hemoglobin. 552

hemoglobin (he′mŏ-glo″bin) The pigment of red blood cells constituting about 33% of the cell volume that transports oxygen and carbon dioxide. **164**

hemopoiesis (hem″ŏ-poi-e′sis) The production of red blood cells. **185**

heparin (hep′ar-in) A mucopolysaccharide found in many tissues, but most abundantly in the lungs and liver, that is used medically as an anticoagulant. 561

hepatic duct (hĕ-pat′ik) A duct formed from the fusion of several bile ducts that drain bile from the liver. The hepatic duct merges with the cystic duct from the gallbladder to form the common bile duct. 790

hepatic portal circulation The return of venous blood from the digestive organs and spleen through a capillary network within the liver before draining into the heart. 598

hepatitis (hep″ă-ti′tis) Inflammation of the liver. 803

hepatopancreatic ampulla (hep″ă-to-pan″kre-at′ik) A small, elevated area within the duodenum where the combined pancreatic and common bile duct empties; also called the *ampulla of Vater*. 781

Hering–Breuer reflex A reflex in which distension of the lungs stimulates stretch receptors, which in turn act to inhibit further distension of the lungs. 706

hermaphrodite (her-maf′rŏ-dīt) An organism having both testes and ovaries. 863

heterochromatin (het″ĕ-ro-kro′mă-tin) A condensed, inactive form of chromatin. 53

hiatal hernia (hi-a′tal her′ne-ă) A protrusion of an abdominal structure through the esophageal hiatus of the diaphragm into the thoracic cavity. 806

hiatus An opening or fissure; a foramen.

high-density lipoproteins (HDLs) (lip″o-pro′te-inz) Combinations of lipids and proteins that migrate rapidly to the bottom of a test tube during centrifugation. HDLs are carrier proteins for lipids, such as cholesterol, that appear to offer some protection from atherosclerosis. 608

hilum (hi′lum) A concave or depressed area where vessels or nerves enter or exit an organ; also called *hilus*. 691

hinge joint A type of synovial articulation characterized by a convex surface of one bone fitting into a concave surface of another such that movement is confined to one plane, as in the knee or interphalangeal joint. 228

histamine (his′tă-mēn) A compound secreted by tissue mast cells and other connective tissue cells that stimulates vasodilation and increases capillary permeability. It is responsible for many of the symptoms of inflammation and allergy. 677

histology (hĭ-stol′ŏ-je) Microscopic anatomy of the structure and function of tissues. **131**

homeostasis (ho″me-o-sta′sis) The dynamic constancy of the internal environment, the maintenance of which is the principal function of physiological regulatory mechanisms. The concept of homeostasis provides a framework for understanding most physiological processes. **18**

homologous chromosomes (hŏ-mol′ŏ-gus) The matching pairs of chromosomes in a diploid cell. **70**

horizontal (transverse) plane A directional plane that divides the body, organ, or appendage into superior and inferior or proximal and distal portions. 11

hormone (hor′mōn) A chemical substance produced in an endocrine gland and secreted into the bloodstream to cause an effect in a specific target organ. **552**

humoral immunity (hyoo′mor-al ĭ-myoo′nĭ-te) The form of acquired immunity in which antibody molecules are secreted in response to antigenic stimulation (as opposed to cell mediated immunity); also called *antibody-mediated immunity*. 656

hyaline cartilage (hi′ă-līn) A cartilage with a homogeneous matrix. It is the most common type, occurring at the articular ends of bones, in the trachea, and within the nose. Most of the bones in the body are formed from hyaline cartilage. 127

hyaline membrane disease A disease affecting premature infants who lack pulmonary surfactant, it is characterized by collapse of the alveoli (atelectasis) and pulmonary edema; also called *respiratory distress syndrome*. 695

hydrocortisone (hi″drŏ-kor′tĭ-sōn) The principal corticosteroid hormone secreted by the adrenal cortex, with glucocorticoid action; also called *cortisol*. 527

hydrophilic (hi″drŏ-fil′ik) Denoting a substance that readily absorbs water; literally, "water loving." **28**

hydrophobic (hi″drŏ-fo′bik) Denoting a substance that repels, and that is repelled by, water; "water fearing." **28**

hymen (hi′men) A developmental remnant (vestige) of membranous tissue that partially covers the vaginal opening. 875

hyperbaric oxygen (hi″per-bar′ik) Oxygen gas present at greater than atmospheric pressure. 721

hypercapnia (hi″per-kap′ne-ă) Excessive concentration of carbon dioxide in the blood. 705

hyperextension (hi″per-ek-sten′shun) Extension beyond the normal anatomical position or 180°. **257**

hyperglycemia (hi″per-gli-se′me-ă) An abnormally increased concentration of glucose in the blood. **105**

hyperkalemia (hi″per-kă-le′me-ă) An abnormally high concentration of potassium in the blood. **125**

hyperopia (hi″per-o′pe-ă) A refractive disorder in which rays of light are brought to a focus behind the retina as a result of the eyeball being too short; also called *farsightedness*. **545**

hyperplasia (hi″per-pla′zha) An increase in organ size due to an increase in cell numbers as a result of mitotic cell division (in contrast to hypertrophy). **68**

hyperpolarization (hi″per-po″lar-ĭ-za′shun) An increase in the negativity of the inside of a cell membrane with respect to the resting membrane potential. **379**

hypersensitivity (hi″per-sen″sĭ-tiv′ĭ-te) Another name for *allergy*; abnormal immune response that may be immediate (due to antibodies of the IgE class) or delayed (due to cell-mediated immunity). **478**

hypertension (hi″per-ten′shun) Elevated or excessive blood pressure. 640

hypertonic (hi″per-ton′ik) Denoting a solution with a greater solute concentration and thus a greater osmotic pressure than plasma. 96

hypertrophy (hi″per′trŏ-fe) Growth of an organ due to an increase in the size of its cells (in contrast to hyperplasia). **68**

hyperventilation (hi″per-ven″tĭ-la′shun) A high rate and depth of breathing that results in a decrease in the blood carbon dioxide concentration to below normal. 705

hypodermis (hi″pŏ-der′mis) A layer of fat beneath the dermis of the skin. **165**

hyponychium (hi″pŏ-nik′e-um) A thickened, supportive layer of stratum corneum at the distal end of a digit under the free edge of the nail. **170**

hypothalamic hormones (hi″po-thă-lam′ik) Hormones produced by the hypothalamus. These include antidiuretic hormone and oxytocin, which are secreted by the posterior pituitary, and both releasing and inhibiting hormones that regulate the secretions of the anterior pituitary. 523

hypothalamo-hypophyseal portal system (hi-pof″ĭs-e′al) A vascular system that transports releasing and inhibiting hormones from the hypothalamus to the anterior pituitary. **569**

hypothalamo-hypophyseal tract The tract of nerve fibers (axons) that transports antidiuretic hormone and oxytocin from the hypothalamus to the posterior pituitary. **569**

hypothalamus (hi″po-thal′ă-mus) A portion of the forebrain within the diencephalon that lies below the thalamus, where it functions as an autonomic nerve center and regulates the pituitary gland. **424**

hypovolemic shock (hi″po-vo-le′mik) A rapid fall in blood pressure as a result of diminished blood volume. 642

hypoxemia (hi″pok-se′me-ă) A low oxygen concentration of the arterial blood. 706

I

ileocecal valve (il″e-ŏ-se′kal) A modification of the mucosa at the junction of the small and large intestine that forms a one-way passage and prevents the backflow of food materials. 781

ileum (il′e-um) The terminal portion of the small intestine between the jejunum and cecum. **234**

immediate hypersensitivity (hi″per-sen″sĭ-tiv′ĭ-te) Hypersensitivity (allergy) mediated by antibodies of the IgE class that results in the release of histamine and related compounds from tissue cells. 677

immunization (im″yŭ-nĭ-za′shun) The process of increasing one's resistance to pathogens. In active immunity, a person is injected with antigens that stimulate the development of clones of specific B or T lymphocytes; in passive immunity, a person is injected with antibodies produced by another organism. 663

immunoassay (im″yŭ-no-as′a) Any of a number of laboratory or clinical techniques that employ the specific binding between an antigen and its homologous antibody in order to identify and quantify a substance in a sample. 655

immunoglobulins (im″yŭ-no-glob′yŭ-linz) Subclasses of the gamma globulin fraction of plasma proteins that have antibody functions, providing humoral immunity. 657

immunosurveillance (im″yŭ-no-ser-va′lens) The concept that the immune system recognizes and attacks malignant cells that produce antigens not recognized as "self." This function is believed to be cell mediated rather than humoral. 674

implantation (im″plan-ta′shun) The process by which a blastocyst attaches itself to and penetrates into the endometrium of the uterus. 903

incus (ing′kus) The middle of three auditory ossicles within the middle-ear chamber; commonly called the *anvil*. **214**

inferior vena cava (ve′nă ka′vă) A large systemic vein that collects blood from the body regions inferior to the level of the heart and returns it to the right atrium. **613**

infundibulum (in″fun-dib′yŭ-lum) The stalk that attaches the pituitary gland to the hypothalamus of the brain. 519

ingestion (in-jes′chun) The process of taking food or liquid into the body by way of the oral cavity. **568**

inguinal (ing′gwĭ-nal) Pertaining to the groin region.

inguinal canal The circular passageway in the abdominal wall through which a testis descends into the scrotum. 841

inhibin (in-hib′in) A polypeptide hormone secreted by the testes that is believed to specifically exert negative feedback inhibition of FSH secretion from the anterior pituitary. 842

inositol (ĭ-no′sĭ-tol) A sugarlike B-complex vitamin. Inositol triphosphate is believed to act as a second messenger in the action of some hormones. **563**

insertion The more movable attachment of a muscle, usually more distal. 255

inspiration (in″spĭ-ra′shun) The act of breathing air into the alveoli of the lungs; also called *inhalation*. 695

insula (in′sŭ-lă) A deep, paired cerebral lobe.

insulin (in′sŭ-lin) A polypeptide hormone secreted by the beta cells of the pancreatic islets that promotes the anabolism of carbohydrates, fat, and protein. Insulin acts to promote the cellular uptake of blood glucose and, therefore, to lower the blood glucose concentration; insulin deficiency results in hyperglycemia and diabetes mellitus. **578**

integument (in-teg′yoo-ment) The skin; the largest organ of the body. 138

intercalated disc (in-ter′kă-lāt-ed) A thickened portion of the sarcolemma that extends across a cardiac muscle fiber, indicating the boundary between cells. **153**

intercellular substance (in″ter-sel′yŭ-lar) The matrix or material between cells that largely determines tissue types. 107

interferons (in″ter-fēr′onz) A group of small proteins that inhibits the multiplication of viruses inside host cells and that also have antitumor properties. 654

internal (deep) Toward the center, away from the surface of the body. 12

internal ear The innermost portion or chamber of the ear, containing the cochlea and the vestibular organs. 483

interneurons (in″ter-noor′onz) Multipolar neurons interposed between sensory (afferent) and motor (efferent) neurons and confined entirely within the central nervous system; also called *association neurons*. 347

interoceptors (in″ter-o-sep′torz) Sensory receptors that respond to changes in the internal environment (as opposed to exteroceptors). **509**

interphase The interval between successive cell divisions, during which time the chromosomes are in an extended state and are active in directing RNA synthesis. **65**

interstitial cells (in″ter-stish′al) Cells located in the interstitial tissue between adjacent convolutions of the seminiferous tubules of the testes; they secrete androgens (mainly testosterone); also called *cells of Leydig*. 848

intervertebral disc (in″ter-ver′tĕ-bral) A pad of fibrocartilage located between the bodies of adjacent vertebrae. 194

intestinal crypt A simple tubular digestive gland opening onto the surface of the intestinal mucosa that secretes digestive enzymes; also called the *crypt of Lieberkühn*. 782

intrafusal fibers (in″tră-fyoo′sal) Modified muscle fibers that are encapsulated to form muscle spindle organs, which are muscle stretch receptors. **504**

intramembranous ossification *See* membranous bone. 163

intrapleural space (in″tră-ploor′al) An actual or potential space between the visceral pleural membrane covering the lungs and the somatic pleural membrane lining the thoracic wall. 693

intrinsic (in-trin′zik) Situated within or pertaining to internal origin.

inulin (in′yŭ-lin) A polysaccharide of fructose, produced by certain plants, that is filtered by the human kidneys but neither reabsorbed nor secreted. The clearance rate of injected insulin is thus used to measure the glomerular filtration rate. 742

inversion (in-ver′zhun) A movement of the foot in which the sole is turned inward. **259**

in vitro (in ve′tro) Occurring outside the body, in a test tube or other artificial environment. 257

in vivo (in ve′vo) Occurring within the body. 257

ion (i′on) An atom or group of atoms that has either lost or gained electrons and thus has a net positive or a net negative charge. 24

ionization (i-on-ĭ-za′shun) The dissociation of a solute to form ions. 26

ipsilateral (ip″sĭ-lat′er-al) On the same side (as opposed to contralateral). 438

IPSP Inhibitory postsynaptic potential; hyperpolarization of the postsynaptic membrane in response to a particular neurotransmitter chemical, which makes it more difficult for the postsynaptic cell to attain a threshold level of depolarization required to produce action potentials. It is responsible for postsynaptic inhibition. 366

iris (i'ris) The pigmented portion of the vascular tunic of the eye that surrounds the pupil and regulates its diameter. 492

ischemia (ĭ-ske'me-ă) A rate of blood flow to an organ that is inadequate to supply sufficient oxygen and maintain aerobic respiration in that organ. 82

islets of Langerhans (i'letz of lang'er-hanz) *See* pancreatic islets. **578**

isoenzymes (i"so-en'zīmz) Enzymes, usually produced by different organs, that catalyze the same reaction but that differ from each other in amino acid composition. **81**

isometric contraction (i"sŏ-met'rik) Muscle contraction in which there is no appreciable shortening of the muscle. **287**

isotonic contraction (i"sŏ-ton'ik) Muscle contraction in which the muscle shortens in length and maintains approximately the same amount of tension throughout the shortening process. **287**

isotonic solution A solution having the same total solute concentration, osmolality, and osmotic pressure as the solution with which it is compared; a solution with the same solute concentration and osmotic pressure as plasma. 96

isthmus (is'mus) A narrow neck or portion of tissue connecting two structures.

J

jaundice (jawn'dis) A condition characterized by high blood bilirubin levels and staining of the tissues with bilirubin, which imparts a yellow color to the skin and mucous membranes. 806

jejunum (jĕ-joo'num) The middle portion of the small intestine, located between the duodenum and the ileum. 781

joint capsule The fibrous tissue that encloses the joint cavity of a synovial joint. **251**

K

keratin (ker'ă-tin) An insoluble protein present in the epidermis and in epidermal derivatives, such as hair and nails. **162**

ketoacidosis (ke"to-ă-sĭ-do'sis) A type of metabolic acidosis resulting from the excessive production of ketone bodies, as in diabetes mellitus. **36**

ketogenesis (ke"to-jen'ĭ-sis) The production of ketone bodies. 793

ketone bodies (ke'tōn) The substances derived from fatty acids via acetyl coenzyme A in the liver; namely, acetone, acetoacetic acid, and β-hydroxybutyric acid. Ketone bodies are oxidized by skeletal muscles for energy. **36**

ketosis (ke-to'sis) An abnormal elevation in the blood concentration of ketone bodies that does not necessarily produce acidosis. **36**

kidney (kid'ne) One of a pair of organs of the urinary system that contains nephrons and that filters wastes from the blood in the formation of urine. 728

kilocalorie (kil'ŏ-kal"ŏ-re) A unit of measurement equal to 1,000 calories, which are units of heat (a kilocalorie is the amount of heat required to raise the temperature of 1 kilogram of water by 1°C). In nutrition, the kilocalorie is called a big calorie (Calorie). 812

kinesiology (kĭ-ne"se-ol'ŏ-je) The study of body movement. 224

Klinefelter's syndrome (klīn'fel-terz sin'drōm) An abnormal condition of male sex characteristics due to the presence of an extra X chromosome (genotype XXY). 863

knee A region in the lower extremity between the thigh and the leg that contains a synovial hinge joint. **16**

Krebs cycle (krebz) A cyclic metabolic pathway in the matrix of mitochondria by which the acetic acid part of acetyl CoA is oxidized and substrates provided for reactions that are coupled to the formation of ATP. **94**

Kupffer cells (koop'fer) Phagocytic cells lining the sinusoids of the liver that are part of the body immunity system. 789

L

labia majora (la'be-ă mă-jor'ă), sing. *labium majus* A portion of the external genitalia of a female consisting of two longitudinal folds of skin extending downward and backward from the mons pubis. 876

labia minora (mĭ-nor'ă), sing. *labium minus* Two small folds of skin, devoid of hair and sweat glands, lying between the labia major of the external genitalia of a female. 876

labial frenulum (la'be-al fren'yŭ-lum) A longitudinal fold of mucous membrane that attaches the lips to the gum along the midline of both the upper and lower lip. 770

labyrinth (lab'ĭ-rinth) An intricate structure consisting of interconnecting passages (e.g., the bony and membranous labyrinths of the inner ear). 478

lacrimal canaliculus (lak'rĭ-mal kan"ă-lik'yŭ-lus) A drainage duct for tears, located at the medial corner of an eyelid. It conveys the tears medially into the nasolacrimal sac. **526**

lacrimal gland A tear-secreting gland, located on the superior lateral portion of the eyeball underneath the upper eyelid. **526**

lactation (lak-ta'shun) The production and secretion of milk by the mammary glands. 887

lacteal (lak'te-al) A small lymphatic duct associated with a villus of the small intestine. 782

lactose (lak'tōs) Milk sugar; a disaccharide of glucose and galactose. 30

lactose intolerance A disorder resulting in the inability to digest lactose because of an enzyme (lactase) deficiency. Symptoms include bloating, intestinal gas, nausea, diarrhea, and cramps. 783

lacuna (lă-kyoo'nă) A small, hollow chamber that houses an osteocyte in mature bone tissue or a chondrocyte in cartilage tissue. **149**

lambdoidal suture (lam'doid-al soo'chur) The immovable joint in the skull between the parietal bones and the occipital bone. 185

lamella (lă-mel'ă) A concentric ring of matrix surrounding the central canal in an osteon of mature bone tissue. 167

lamellated corpuscle (lam'ĕ-la-ted) A sensory receptor for pressure, found in tendons, around joints, and in visceral organs; also called a *pacinian corpuscle*. 467

lamina (lam'ĭ-nă) A thin plate of bone that extends superiorly from the body of a vertebra to form either side of the arch of a vertebra. 194

lanugo (lă-noo'go) Short, silky fetal hair, which may be present for a short time on a premature infant. **169**

large intestine The last major portion of the GI tract, consisting of the cecum, colon, rectum, and anal canal. 784

laryngopharynx (lă-ring"go-far'ingks) The inferior or lower portion of the pharynx in contact with the larynx. 686

larynx (lar'ingks) The structure located between the pharynx and trachea that houses the vocal cords; commonly called the *voice box*. 687

lateral (lat'er-al) Pertaining to the side; farther from the midplane. **526**

lateral ventricle (ven'trĭ-k'l) A cavity within the cerebral hemisphere of the brain that is filled with cerebrospinal fluid. **432**

L-dopa Levodopa; a derivative of the amino acid tyrosine. It serves as the precursor for the neurotransmitter molecule dopamine and is given to patients with Parkinson's disease to stimulate dopamine production. 352

leg The portion of the lower extremity between the knee and ankle. **236**

lens (lenz) A transparent refractive organ of the eye positioned posterior to the pupil and iris. **528**

lesion (le'zhun) A wounded or damaged area. **171**

lesser omentum (o-men'tum) A peritoneal fold of tissue extending from the lesser curvature of the stomach to the liver. 767

leukocyte (loo'kŏ-sīt) A white blood cell; variant spelling, leucocyte. **153**

ligament (lig'ă-ment) A tough cord or fibrous band of connective tissue that binds bone to bone to strengthen and provide flexibility to a joint. It also may support viscera. **146**

limbic system (lim′bik) A portion of the brain concerned with emotions and autonomic activity. **423**

linea alba (lin′e-ă al′bă) A vertical fibrous band extending down the anterior medial portion of the abdominal wall. 291

lingual frenulum (ling′gwal fren′yŭ-lum) A longitudinal fold of mucous membrane that attaches the tongue to the floor of the oral cavity. 771

lipogenesis (lip″ŏ-jen′ĕ-sis) The formation of fat or triglycerides. **99**

lipolysis (lĭ-pol′ĭ-sis) The hydrolysis of triglycerides into free fatty acids and glycerol. **99**

liver A large visceral organ inferior to the diaphragm in the right hypochondriac region. The liver detoxifies the blood and modifies the blood plasma concentration of glucose, triglycerides, ketone bodies, and proteins. 788

low-density lipoproteins (LDLs) (lip″o-pro′te-inz) Plasma proteins that transport triglycerides and cholesterol. They are believed to contribute to arteriosclerosis. 606

lower extremity A lower appendage, including the hip, thigh, knee, leg, and foot. **14**

lumbar (lum′bar) Pertaining to the region of the loins. **16**

lumbar plexus (plek′sus) A network of nerves formed by the anterior branches of spinal nerves L1 through L4. **463**

lumen (loo′men) The space within a tubular structure through which a substance passes. 109

lung One of the two major organs of respiration positioned within the thoracic cavity on either side of the mediastinum. 690

lung surfactant (sur-fak′tant) A mixture of lipoproteins (containing phospholipids) secreted by type II alveolar cells into the alveoli of the lungs. It lowers surface tension and prevents collapse of the lungs as occurs in hyaline membrane disease, in which surfactant is absent. 694

lunula (loo′nyoo-lă) The half-moon–shaped whitish area at the proximal portion of a nail. **170**

luteinizing hormone (LH) (loo′te-ĭ-ni″zing) A hormone secreted by the adenohypophysis (anterior lobe) of the pituitary gland that stimulates ovulation and the secretion of progesterone by the corpus luteum. It also influences mammary gland milk secretion in females and stimulates testosterone secretion by the testes in males. **566**

lymph (limf) A clear, plasmalike fluid that flows through lymphatic vessels.

lymph node A small, ovoid mass of reticular tissue located along the course of lymph vessels. 593

lymphatic system (lim-fat′ik) The lymphatic vessels and lymph nodes. **592**

lymphocyte (lim′fŏ-sīt) A type of white blood cell characterized by agranular cytoplasm. Lymphocytes usually constitute about 20% to 25% of the white blood cell count. **596**

lymphokines (lim′fŏ-kīns) A group of chemicals released from T cells that contribute to cell-mediated immunity. **582**

lysosomes (li′sŏ-sōmz) Organelles containing digestive enzymes and responsible for intracellular digestion. **53**

M

macromolecules (mak″ro-mol′ĭ-kyoolz) Large molecules; a term that usually refers to protein, RNA, and DNA.

macrophage (mak′rŏ-fāj) A wandering phagocytic cell. 651

macula lutea (mak′yŭ-lă loo′te-ă) A yellowish depression in the retina of the eye that contains the fovea centralis, the area of keenest vision. 500

malignant Threatening to life; virulent. Of a tumor, cancerous, tending to metastasize. 674

malleus (mal′e-us) The first of three auditory ossicles that attaches to the tympanum; commonly called the *hammer*. **214**

mammary gland (mam′er-e) The gland of the female breast responsible for lactation and nourishment of the young. **171**

marrow (mar′o) The soft connective tissue found within the inner cavity of certain bones that produces red blood cells. 165

mast cell A type of connective tissue cell that produces and secretes histamine and heparin and promotes local inflammation. **145**

mastication (mas″tĭ-ka′shun) The chewing of food. 766

matrix (ma′triks) The intercellular substance of a tissue. 121

maximal oxygen uptake The maximum amount of oxygen that can be consumed by the body per unit time during heavy exercise. **297**

meatus (me-a′tus) A passageway or opening into a structure. 685

mechanoreceptor (mek″ă-no-re-sep′tor) A sensory receptor that responds to a mechanical stimulus. **498**

medial (me′de-al) Toward or closer to the midplane of the body. **527**

mediastinum (me″de-ă-sti′num) The partition in the center of the thorax between the two pleural cavities. **16**

medulla (mĕ-dul′ă) The center portion of an organ.

medulla oblongata (ob″long-gă′tă) A portion of the brain stem located between the spinal cord and the pons. **429**

medullary (marrow) cavity (med′u-l-er″e) The hollow core of the diaphysis of a long bone in which marrow is found. **188**

megakaryocyte (meg″ă-kar′e-o-sīt) A bone marrow cell that gives rise to blood platelets. 553

meiosis (mi-o′sis) A specialized type of cell division by which gametes or haploid sex cells are formed. **70**

Meissner's corpuscle (mīs′nerz) *See* corpuscle of touch. 467

melanin (mel′ă-nin) A dark pigment found within the epidermis or epidermal derivatives of the skin. **163**

melanocyte (mel′ă-no-sīt) A specialized melanin-producing cell found in the deepest layer of the epidermis. **162**

melanoma (mel″ă-no′mă) A dark, malignant tumor of the skin that frequently forms in moles. 143

melatonin (mel″ă-to′nin) A hormone secreted by the pineal gland that produces lightening of the skin in lower vertebrates and that may contribute to the regulation of gonadal function in mammals. Secretion follows a circadian rhythm and peaks at night. **579**

membrane potential The potential difference or voltage that exists between the inner and outer sides of a cell membrane. It exists in all cells but is capable of being changed by excitable cells (neurons and muscle cells). 102

membranous bone (mem′bră-nus) Bone that forms from membranous connective tissue rather than from cartilage. 163

membranous labyrinth (lab′ĭ-rinth) A system of communicating sacs and ducts within the bony labyrinth of the inner ear that includes the cochlea and vestibular apparatus. It is filled with endolymph and surrounded by perilymph and bone. **512**

menarche (mĕ-nar′ke) The first menstrual discharge. 843

Ménière's disease (mān-yarz′) Deafness, tinnitus, and vertigo resulting from a disorder of the labyrinth. **544**

meninges (mĕ-nin′jēz), sing. *meninx* A group of three fibrous membranes covering the central nervous system, composed of the dura mater, arachnoid mater, and pia mater. **430**

menisci (mĕ-nis′ke) Wedge-shaped fibrocartilages in certain synovial joints. **253**

menopause (men′ŏ-pawz) The period marked by the cessation of menstrual periods in the human female. 885

menstrual cycle (men′stroo-al) The rhythmic female reproductive cycle, characterized by changes in hormone levels and physical changes in the uterine lining. 881

menstruation (men″stroo-a′shun) The discharge of blood and tissue from the uterus at the end of the menstrual cycle. 884

mesencephalic aqueduct (mez″en-sĕ-fal′ik ak′wĕ-dukt) The channel that connects the third and fourth ventricles of the brain; also called the *aqueduct of Sylvius*. **426**

mesencephalon (mes″en-sef′ă-lon) The midbrain, which contains the corpora quadrigemina and the cerebral peduncles. **426**

mesenchyme (mez′en-kīm) An embryonic connective tissue that can migrate, and from which all connective tissues arise. **144**

mesenteric patches (mes″en-ter′ik) Clusters of lymph nodes on the walls of the small intestine; also called *Peyer's patches*. 650

mesentery (mes′en-ter″e) A fold of peritoneal membrane that attaches an abdominal organ to the abdominal wall. **17**

mesoderm (mes′ŏ-derm) The middle one of the three primary germ layers. 904

mesothelium (mes″ŏ-the′lium) A simple squamous epithelial tissue that lines body cavities and covers visceral organs; also called *serosa*. **134**

mesovarium (mes″ŏ-va′re-um) The peritoneal fold that attaches an ovary to the broad ligament of the uterus. 877

messenger RNA (mRNA) A type of RNA that contains a base sequence complementary to a part of the DNA that specifies the synthesis of a particular protein. 54

metabolism (mĕ-tab′ŏ-liz-em) The sum total of the chemical changes that occur within a cell. **89**

metacarpus (met″ă-kar′pus) The region of the hand between the wrist and the phalanges, including the five metacarpal bones that support the palm of the hand. 208

metarteriole (met″ar-tir′e-ōl) A small blood vessel that emerges from an arteriole, passes through a capillary network, and empties into a venule. 581

metastasis (mĕ-tas′tă-sis) The spread of a disease from one organ or body part to another. 674

metatarsus (met″ă-tar′sus) The region of the foot between the ankle and the phalanges that includes the five metatarsal bones. **238**

metencephalon (met″en-sef′ă-lon) The most superior portion of the hindbrain that contains the cerebellum and the pons. **426**

micelles (mi-selz′) Colloidal particles formed by the aggregation of many molecules. **36**

microglia (mi-krog′le-ă) Small phagocytic cells found in the central nervous system. **375**

microvilli (mi″kro-vil′i) Microscopic hairlike projections of cell membranes on certain epithelial cells. **51**

micturition (mik″tŭ-rish′un) The process of voiding urine; also called *urination*. 755

midbrain The portion of the brain between the pons and the forebrain. **426**

middle ear The middle of the three portions of the ear that contains the three auditory ossicles. **517**

midsagittal plane (mid-saj′ĭ-tal) A plane that divides the body into equal right and left halves; also called the *median plane* or *midplane*. 10

mineralocorticoids (min″er-al-o-kor′tĭ-koidz) Steroid hormones of the adrenal cortex (corticosteroids) that regulate electrolyte balance. **573**

mitochondria (mi″tŏ-kon′dre-ă), sing. mitochondrion Cytoplasmic organelles that serve as sites for the production of most of the cellular energy; the so-called powerhouses of the cell. **54**

mitosis (mi-to′sis) The process of cell division that results in two identical daughter cells, containing the same number of chromosomes. **66**

mitral valve (mi′tral) The left atrioventricular heart valve; also called the *bicuspid valve*. 569

mixed nerve A nerve that contains both motor and sensory nerve fibers. 347

molal (mo′lal) Pertaining to the number of moles of solute per kilogram of solvent. 95

molar (mo′lar) Pertaining to the number of moles of solute per liter of solution. 95

mole (mōl) The number of grams of a chemical that is equal to its formula weight (atomic weight for an element or molecular weight for a compound). 95

monoclonal antibodies (mon″ŏ-klōn′al an′tĭ-bod″ēz) Identical antibodies derived from a clone of genetically identical plasma cells. 666

monocyte (mon′o-sīt) A phagocytic type of white blood cell, normally constituting about 3% to 8% of the white blood cell count. **596**

monomer (mon′ŏ-mer) A single molecular unit of a longer, more complex molecule. Monomers are joined together to form dimers, trimers, and polymers; the hydrolysis of polymers eventually yields separate monomers. 765

monosaccharide (mon″ŏ-sak″ă-rīd) The monomer of the more complex carbohydrates, examples of which include glucose, fructose, and galactose; also called a *simple sugar*. **33**

mons pubis (monz pyoo′bis) A fatty tissue pad covering the symphysis pubis and covered by pubic hair in the female. 876

morula (mor′yŭ-lă) An early stage of embryonic development characterized by a solid ball of cells. 902

motile (mōt′l), mo′tīl) Capable of self-propelled movement.

motor area A region of the cerebral cortex from which motor impulses to muscles or glands originate. 388

motor nerve A nerve composed of motor nerve fibers. 347

motor neuron (noor′on) A nerve cell that conducts action potentials away from the central nervous system and innervates effector organs (muscle and glands). It forms the anterior roots of the spinal nerves; also called an *efferent neuron*. **374**

motor unit A single motor neuron and the muscle fibers it innervates. **287**

mucosa (myoo-ko′să) A mucous membrane that lines cavities and tracts opening to the exterior. 113

mucous cell (myoo′kus) *See* goblet cell. 111

mucous membrane A thin sheet consisting of layers of visceral organs that include the lining epithelium, submucosal connective tissue, and (in some cases) a thin layer of smooth muscle (the muscularis mucosa). **16**

multipolar neuron A nerve cell with many processes originating from the cell body. **374**

muscle (mus′el) A major type of tissue adapted to contract. The three kinds of muscle are cardiac, smooth, and skeletal. 130

muscle spindles Sensory organs within skeletal muscles composed of intrafusal fibers. They are sensitive to muscle stretch and provide a length detector within muscles. 471

muscularis (mus″kyŭ-la′ris) A muscular layer or tunic of an organ, composed of smooth muscle tissue. 753

myelencephalon (mi″ĕ-len-sef′ă-lon) The posterior portion of the hindbrain that contains the medulla oblongata. 394

myelin (mi′ĕ-lin) A lipoprotein material that forms a sheathlike covering around nerve fibers.

myelin sheath A sheath surrounding axons formed by successive wrappings of a neuroglial cell membrane. Myelin sheaths are formed by neurolemmocytes in the peripheral nervous system and by oligodendrocytes within the central nervous system. **375**

myenteric plexus (mi″en-ter′ik plek′sus) A network of sympathetic and parasympathetic nerve fibers located in the muscularis tunic of the small intestine; also called the *plexus of Auerbach*. 769

myocardial infarction (mi″ŏ-kar′de-al in-fark′shun) An area of necrotic tissue in the myocardium that is filled in by scar (connective) tissue. 608

myocardium (mi″ŏ-kar′de-um) The cardiac muscle layer of the heart. **612**

myofibril (mi″ŏ-fi′bril) A bundle of contractile fibers within muscle cells. **289**

myogenic (mi″ŏ-jen′ik) Originating within muscle cells; used to describe self-excitation by cardiac and smooth muscle cells. 275

myoglobin (mi″ŏ-glo′bin) A molecule composed of globin protein and heme pigment. It is related to hemoglobin but contains only one subunit (instead of the four in hemoglobin) and is found in skeletal and cardiac muscle cells where it serves to store oxygen. 272

myogram (mi′ŏ-gram) A recording of electrical activity within a muscle. 318

myology (mi-ol′ŏ-je) The science or study of muscle structure and function. 281

myometrium (mi″o-me′tre-um) The layer or tunic of smooth muscle within the uterine wall. 874

myoneural junction (mi″ŏ-noor′al) The site of contact between an axon of a motor neuron and a muscle fiber. **385**

myopia (mi-o′pe-ă) A visual defect in which objects may be seen distinctly only when very close to the eyes; also called *nearsightedness*. **545**

myosin (mi′ŏ-sin) A thick myofilament protein that together with actin causes muscle contraction. **290**

myxedema (mik″sĭ-de′mă) A type of edema associated with hypothyroidism. It is characterized by the accumulation of mucoproteins in tissue fluid. **585**

N

NAD Nicotinamide adenine dinucleotide; a coenzyme derived from niacin that helps to transport electrons from the Krebs cycle to the electron-transport chain within mitochondria. **87**

nail A hardened, keratinized plate that develops from the epidermis and forms a protective covering on the surface of the distal phalanges of fingers and toes. **169**

naloxone (nal′ok-sōn, nă-lok′sōn) A drug that antagonizes the effects of morphine and endorphins. 367

nasal cavity (na′zal) A mucosa-lined space above the oral cavity, divided by a nasal septum. It is the first chamber of the respiratory system. 684

nasal concha (kong′kă) A scroll-like bone extending medially from the lateral wall of the nasal cavity; also called a *turbinate bone*. 189

nasal septum (sep′tum) A bony and cartilaginous partition that separates the nasal cavity into two portions. 684

nasopharynx (na″zo-far′ingks) The first or uppermost chamber of the pharynx, positioned posterior to the nasal cavity and extending down to the soft palate. 684

natriuretic (na″trĭ-yoo-ret′ik) An agent that promotes the excretion of sodium in the urine. Atrial natriuretic hormone has this effect. 749

neck **1.** Any constricted portion, such as the neck of an organ. **2.** The cervical region of the body between the head and thorax.

necrosis (nĕ-kro′sis) Cellular death or tissue death due to disease or trauma. **14**

negative feedback A mechanism in the body for maintaining a state of internal constancy, or homeostasis; effectors are activated by changes in the internal environment, and the actions of the effectors serve to counteract these changes and maintain a state of balance. **19**

neonatal (ne″o-na′tal) The stage of life from birth to the end of 4 weeks.

neoplasm (ne′ŏ-plazm) A new, abnormal growth of tissue, as in a tumor. 151

nephron (nef′ron) The functional unit of the kidney, consisting of a glomerulus, convoluted tubules, and a nephron loop. 730

nerve A bundle of nerve fibers outside the central nervous system. **374**

neurilemma (noor″ĭ-lem′ă) A thin, membranous covering surrounding the myelin sheath of a nerve fiber. 349

neurofibril node A gap in the myelin sheath of a nerve fiber; also called a *node of Ranvier*. 350

neuroglia (noo-rog′le-ă) Specialized supportive cells of the central nervous system. **372**

neurohypophysis (noor″o-hi-pof′ĭ-sis) The posterior lobe of the pituitary gland derived from the brain. Its major secretions include antidiuretic hormone (ADH), also called vasopressin, and oxytocin, produced in the hypothalamus. **426**

neurolemmocyte (noor″ŏ-lem′ŏ-sīt) A specialized neuroglia cell that surrounds an axon fiber of a peripheral nerve and forms the neurilemmal sheath; also called a *Schwann cell*. **375**

neuron (noor′on) The structural and functional unit of the nervous system, composed of a cell body, dendrites, and an axon; also called a *nerve cell*. **372**

neurotransmitter (noor″o-trans′mit-er) A chemical contained in synaptic vesicles in nerve endings that is released into the synaptic cleft, where it stimulates the production of either excitatory or inhibitory postsynaptic potentials. **386**

neutrons (noo′tronz) Electrically neutral particles that exist together with positively charged protons in the nucleus of atoms. 22

neutrophil (noo′trŏ-fil) A type of phagocytic white blood cell, normally constituting about 60% to 70% of the white blood cell count. **595**

nexus (nek′sus) A bond between members of a group; the type of bonds present in single-unit smooth muscles. 784

nidation (ni-da′shun) Implantation of the blastocyst into the endometrium of the uterus. 903

nipple A dark pigmented, rounded projection at the tip of the breast. 887

Nissl bodies (nis′l) *See* chromatophilic substances. 345

node of Ranvier (ran′ve-a) *See* neurofibril node. **376**

norepinephrine (nor″ep-ĭ-nef″rin) A catecholamine released as a neurotransmitter from postganglionic sympathetic nerve endings and as a hormone (together with epinephrine) from the adrenal medulla. **397**

notochord (no′tŏ-kord) A flexible rod of tissue that extends the length of the back of an embryo. 7

nucleolus (noo-kle′ŏ-lus) A dark-staining area within a cell nucleus; the site where ribosomal RNA is produced. 53

nucleoplasm (noo′kle-ŏ-plaz″em) The protoplasmic contents of the nucleus of a cell. 54

nucleotide (noo′kle-ŏ-tīd) The subunit of DNA and RNA macromolecules. Each nucleotide is composed of a nitrogenous base (adenine, guanine, cytosine, and thymine or uracil); a sugar (deoxyribose or ribose); and a phosphate group. **56**

nucleus (noo′kle-us) A spheroid body within a cell that contains the genetic factors of the cell. **25**

nucleus pulposus (pul-po′sus) The soft, pulpy core of an intervertebral disc; a remnant of the notochord. 7

nystagmus (nĭ-stag′mus) Involuntary oscillary movements of the eye. 481

O

obese (o-bēs′) Excessively fat. 812

olfactory (ol-fak′tŏ-re) Pertaining to the sense of smell. 477

olfactory bulb An aggregation of sensory neurons of an olfactory nerve, lying inferior to the frontal lobe of the cerebrum on either lateral side of the crista galli of the ethmoid bone. **451**

olfactory tract The olfactory sensory tract of axons that conveys impulses from the olfactory bulb to the olfactory portion of the cerebral cortex. **451**

oligodendrocyte (ol″ĭ-go-den′drŏ-sīt) A type of neuroglial cell concerned with the formation of the myelin of nerve fibers within the central nervous system. **375**

oncology (on-kol′ŏ-je) The study of tumors. 674

oncotic pressure (on-kot′ik) The colloid osmotic pressure of solutions produced by proteins. In plasma, it serves to counterbalance the outward filtration of fluid from capillaries due to hydrostatic pressure. 617

oocyte (o′ŏ-sīt) A developing egg cell.

oogenesis (o″ŏ-jen′ĕ-sis) The process of female gamete formation. 877

opsonization (op″sŏ-nĭ-za′shun) The process by which antibodies enhance the ability of phagocytic cells to attack bacteria. 660

optic (op′tik) Pertaining to the eye.

optic chiasma (ki-az′mă) An X-shaped structure on the inferior aspect of the brain, anterior to the pituitary gland, where there is a partial crossing over of fibers in the optic nerves; also called the *optic chiasm*. **451**

optic disc A small region of the retina where the fibers of the ganglion neurons exit from the eyeball to form the optic nerve; also called the *blind spot*. **528**

optic tract A bundle of sensory axons located between the optic chiasma and the thalamus that functions to convey visual impulses from the photoreceptors within the eye. **451**

oral Pertaining to the mouth.

ora serrata The jagged peripheral margin of the retina. **533**

organ A structure consisting of two or more tissues that performs a specific function. 9

organ of Corti (kor′te) *See* spiral organ. **521**

organelle (or″gă-nel′) A minute living structure of a cell with a specific function. 9

organism An individual living creature. 10

orifice (or′ĭ-fis) An opening into a body cavity or tube. 281

origin The place of muscle attachment—usually the more stationary point or the proximal bone; opposite the insertion. 255

oropharynx (o″ro-far′ingks) The second portion of the pharynx, located posterior to the oral cavity and extending from the soft palate to the hyoid bone. 686

osmolality (oz″mŏ-lal′ĭ-te) A measure of the total concentration of a solution; the number of moles of solute per kilogram of solvent. **115**

osmoreceptors (oz″mŏ-re-cep′torz) Sensory neurons that respond to changes in the osmotic pressure of the surrounding fluid. 97

osmosis (oz-mo′sis) The passage of solvent (water) from a more dilute to a more concentrated solution through a membrane that is more permeable to water than to the solute. **113**

osmotic pressure (oz-mot′ik) A measure of the tendency of a solution to gain water by osmosis when separated by a membrane from pure water. Directly related to the osmolality of the solution, it is the pressure required to just prevent osmosis. 94

osseous tissue (os′e-us) Bone tissue. 127

ossicle (os′ĭ-kul) One of the three bones of the middle ear; also called the *auditory ossicle*. 483

ossification (os″ĭ-fĭ-ka′shun) The process of bone tissue formation. 168

osteoblast (os′te-ŏ-blast) A bone-forming cell. **189**

osteoclast (os′te-ŏ-klast) A cell that causes erosion and resorption of bone tissue. **189**

osteocyte (os′te-ŏ-sīt) A mature bone cell. **150**

osteology (os″te-ol′ŏ-je) The study of the structure and function of bone and the entire skeleton. 160

osteomalacia (os″te-o-mă-la′shă) Softening of bones due to a deficiency of vitamin D and calcium. **196**

osteon (os′te-on) A group of osteocytes and concentric lamellae surrounding a central canal, constituting the basic unit of structure in osseous tissue; also called a *haversian system*. **190**

osteoporosis (os″te-o-pŏ-ro′sis) Demineralization of bone, seen most commonly in postmenopausal women and patients who are inactive or paralyzed. It may be accompanied by pain, loss of stature, and other deformities and fractures. **197**

otoliths (o′tŏ-liths) Small, hardened particles of calcium carbonate in the saccule and utricle of the inner ear, associated with the receptors of equilibrium; also called *statoconia*. **514**

outer ear The outer portion of the ear, consisting of the auricle and the external auditory canal. **516**

oval window An oval opening in the bony wall between the middle and inner ear, into which the footplate of the stapes fits; also called the *vestibular window*. 483

ovarian follicle (o-var′e-an fol′ĭ-kul) A developing ovum and its surrounding epithelial cells. 878

ovarian ligament (lig′ă-ment) A cordlike connective tissue that attaches the ovary to the uterus. 877

ovary (o′vă-re) The female gonad in which ova and certain sexual hormones are produced. **581**

oviduct (o′vĭ-dukt) The tube that transports ova from the ovary to the uterus; also called the *uterine tube* or *fallopian tube*. 872

ovulation (ov-yŭ-la′shun) The rupture of an ovarian (graafian) follicle with the release of an ovum. 878

ovum (o′vum) A secondary oocyte capable of developing into a new individual when fertilized by a spermatozoon. 839

oxidative phosphorylation (ok″sĭ-da′tiv fos″for-ĭ-la′shun) The formation of ATP using energy derived from electron transport to oxygen. It occurs in the mitochondria. **98**

oxidizing agent (ok′sĭ-dīz-ing) An atom that accepts electrons in an oxidation-reduction reaction. **86**

oxyhemoglobin (ok″se-he″mŏ-glo′bin) A compound formed by the bonding of molecular oxygen to hemoglobin. 707

oxyhemoglobin saturation The ratio, expressed as a percentage, of the amount of oxyhemoglobin relative to the total amount of hemoglobin in blood. 709

oxytocin (ok″sĭ-to′sin) One of the two hormones produced in the hypothalamus and secreted by the posterior pituitary (the other hormone is vasopressin). Oxytocin stimulates the contraction of uterine smooth muscles and promotes milk ejection in females. **567**

P

pacemaker (pās′ma″ker) A group of cells that has the fastest spontaneous rate of depolarization and contraction in a mass of electrically coupled cells; in the heart, this is the sinoatrial, or SA, node. 275

pacinian corpuscle (pă-sin′e-an) *See* lamellated corpuscle. 467

PAH Para-aminohippuric acid; a substance used to measure total renal plasma flow because its clearance rate is equal to the total rate of plasma flow to the kidneys. PAH is filtered and secreted but not reabsorbed by the renal nephrons. 745

palate (pal′at) The roof of the oral cavity. 771

palatine (pal′ă-tīn) Pertaining to the palate.

palmar (pal′mar) Pertaining to the palm of the hand. **16**

palpebra (pal′pĕ-bră) An eyelid. **524**

pancreas (pan′kre-as) A mixed organ in the abdominal cavity that secretes pancreatic juices into the GI tract and insulin and glucagon into the blood. **578**

pancreatic duct (pan″kre-at′ik) A drainage tube that carries pancreatic juice from the pancreas into the duodenum of the hepatopancreatic ampulla. 794

pancreatic islets A cluster of cells within the pancreas that forms the endocrine portion and secretes insulin and glucagon; also called *islets of Langerhans*. **578**

papillae (pă-pil′e) Small, nipplelike projections. **510**

papillary muscle (pap′ĭ-ler″e) Muscular projections from the ventricular walls of the heart to which the chordea tendineae are attached. **613**

paranasal sinus (par″ă-na′zal si′nus) An air chamber lined with a mucous membrane that communicates with the nasal cavity. 685

parasympathetic (par″ă-sim″pă-thet′ik) Pertaining to the division of the autonomic nervous system concerned with activities that, in general, inhibit or oppose the physiological effects of the sympathetic nervous system. **482**

parathyroid hormone (PTH) A polypeptide hormone secreted by the parathyroid glands. PTH acts to raise the blood Ca^{2+} levels primarily by stimulating reabsorption of bone. **578**

parathyroids (par″ă-thi roidz) Small endocrine glands embedded on the posterior surface of the thyroid glands that are concerned with calcium metabolism. **577**

parietal (pă-ri′ĕ-tal) Pertaining to a wall of an organ or cavity. 12

parietal pleura (ploor′ă) The thin serous membrane attached to the thoracic walls of the pleural cavity. **17**

Parkinson's disease (par′kin-sunz) A tremor of the resting muscles and other symptoms caused by inadequate dopamine-producing neurons in the basal nuclei of the cerebrum; also called *paralysis agitans*. **397**

parotid gland (pă-rot′id) One of the paired salivary glands located on the side of the face over the masseter muscle just anterior to the ear and connected to the oral cavity through a salivary duct. 774

parturition (par″tyoo-rish′un) The process of giving birth; childbirth. 920

passive immunity (ĭ-myoo′nĭ-te) Specific immunity granted by the administration of antibodies made by another organism. 665

pathogen (path′ŏ-jen) Any disease-producing microorganism or substance. 803

pectoral (pek′tŏ-ral) Pertaining to the chest region. **14**

pectoral girdle The portion of the skeleton that supports the upper extremities. **183**

pedicle (ped′ĭ-k′l) The portion of a vertebra that connects and attaches the lamina to the body. 194

pelvic (pel′vik) Pertaining to the pelvis. **15**

pelvic girdle The portion of the skeleton to which the lower extremities are attached. **232**

pelvis (pel′vis) A basinlike bony structure formed by the sacrum and ossa coxae. **232**

penis (pe′nis) The male organ of copulation, used to introduce sperm into the female vagina and through which urine passes during urination. 856

pennate (pen′āt) Pertaining to a skeletal muscle fiber arrangement in which the fibers are attached to tendinous slips in a featherlike pattern. 281

pepsin (pep′sin) The protein-digesting enzyme secreted in gastric juice. 778

peptic ulcer (pep′tik ul′ser) An injury to the mucosa of the esophagus, stomach, or small intestine due to the action of acidic gastric juice. 779

perforating canal A minute duct through compact bone by which blood vessels and nerves penetrate to the central canal of an osteon; also called *Volkmann's canal*. **190**

pericardium (per″ĭ-kar′de-um) A protective serous membrane that surrounds the heart. 566

perichondrium (per″ĭ-kon′dre-um) A toughened connective sheet that covers some kinds of cartilage. **149**

perikaryon (per″ĭ-kar′e-on) The cell body of a neuron. **155**

perilymph (per′ĭ-limf) A fluid of the inner ear that provides a liquid-conducting medium for the vibrations involved in hearing and the maintenance of equilibrium. 478

perimysium (per″ĭ-mis′e-um) Fascia (connective tissue) surrounding a bundle of muscle fibers. **281**

perineum (per″ĭ-ne′um) The floor of the pelvis, which is the region between the anus and the symphysis pubis. It is the region that contains the external genitalia. **15**

perineurium (per″ĭ-noor′e-um) Connective tissue surrounding a bundle of nerve fibers. 347

periodontal membrane (per′e-ŏ-don′tal) A fibrous connective tissue lining the dental alveoli. 773

periosteum (per″e-os′te-um) A fibrous connective tissue covering the outer surface of bone. **189**

peripheral nervous system (pĕ-rif′er-al) The nerves and ganglia of the nervous system that lie outside of the brain and spinal cord; *PNS*. **372**

peristalsis (per″ĭ-stal′sis) Rhythmic contractions of smooth muscle in the walls of various tubular organs by which the contents are forced onward. 766

peritoneum (per″ĭ-tŏ-ne′um) The serous membrane that lines the abdominal cavity and covers the abdominal visceral organs. 767

Peyer's patches (pi′erz) *See* mesenteric patches. 650

pH A measure of the relative acidity or alkalinity of a solution, numerically equal to 7 for neutral solutions. The pH scale in common use ranges from 0 to 14. Solutions with a pH lower than 7 are acidic, and those with a higher pH are basic. 26

phagocytosis (fag″ŏ-si-to′sis) Cellular eating; the ability of some cells (such as white blood cells) to engulf large particles (such as bacteria) and digest these particles by merging the food vacuole in which they are contained with a lysosome containing digestive enzymes. **49**

phalanx (fa′langks), pl. *phalanges* A bone of a finger or toe. **232**

pharynx (far′ingks) The organ of the digestive system and respiratory system located at the back of the oral and nasal cavities that extends to the larynx anteriorly and to the esophagus posteriorly; also called the *throat*. 685

photoreceptor (fo″to-re-sep′tor) A sensory nerve ending that responds to the stimulation of light. **498**

physiology (fiz″e-ol′ŏ-je) The science that deals with the study of body functions. 2

pia mater (pi′ă ma′ter) The innermost meninx that is in direct contact with the brain and spinal cord. **430**

pineal gland (pin′e-al) A small cone-shaped gland located in the roof of the third ventricle. **579**

pinna (pin′ă) The outer, fleshy portion of the external ear; also called the *auricle*. **516**

pinocytosis (pin″ŏ-si-to′sis) Cell drinking; invagination of the cell membrane forming narrow channels that pinch off into vacuoles. This allows for cellular intake of extracellular fluid and dissolved molecules. **49**

pituitary gland (pĭ-too′ĭ-ter-e) A small, pea-shaped endocrine gland situated on the interior surface of the diencephalonic region of the brain, consisting of anterior and posterior lobes; also called the *hypophysis*. **426**

pivot joint (piv′ut) A synovial joint in which the rounded head of one bone articulates with the depressed cup of another to permit a rotational type of movement. 228

placenta (plă-sen′tă) The organ of metabolic exchange between the mother and the fetus. **581**

plantar (plan′tar) Pertaining to the sole of the foot. **16**

plasma (plaz′mă) The fluid, extracellular portion of circulating blood. **593**

plasma cells Cells derived from B lymphocytes that produce and secrete large amounts of antibodies. They are responsible for humoral immunity. 553

platelets (plāt-letz) Small fragments of specific bone marrow cells that function in blood coagulation; also called *thrombocytes*. **596**

pleural (ploor′al) Pertaining to the serous membranes associated with the lungs.

pleural cavity The potential space between the visceral pleura and parietal pleura. **16**

pleural membranes Serous membranes that surround the lungs and provide protection and compartmentalization. 116

plexus (plek′sus) A network of interlaced nerves or vessels.

plexus of Auerbach (ow′er-bak) *See* myenteric plexus. 769

plexus of Meissner (mīs′ner) *See* submucosal plexus. 769

plicae circulares (pli′ce sur-kyŭ-lar′ēz) Deep folds within the wall of the small intestine that increase the absorptive surface area. 782

pneumotaxic area (noo″mŏ-tak′sik) The region of the respiratory control center located in the pons of the brain. **427**

polar body A small daughter cell formed by meiosis that degenerates in the process of oocyte production. 877

polar molecule A molecule in which the shared electrons are not evenly distributed, so that one side of the molecule is negatively (or positively) charged in comparison with the other side. Polar molecules are soluble in polar solvents, such as water. **27**

polydipsia (pol″e-dip′se-ă) Excessive thirst.

polymer (pol′ĕ-mer) A large molecule formed by the combination of smaller subunits, or monomers. 765

polymorphonuclear leukocyte (pol″e-mor″fŏ-noo′kle-ar loo′kŏ-sīt) A granular leukocyte containing a nucleus with a number of lobes connected by thin, cytoplasmic strands. This type includes neutrophils, eosinophils, and basophils. 553

polypeptide (pol″e-pep′tīd) A chain of amino acids connected by covalent bonds called peptide bonds. A very large polypeptide is called a protein. **33**

polysaccharide (pol″e-sak′ă-rīd) A carbohydrate formed by covalent bonding of numerous monosaccharides. Examples include glycogen and starch. 30

polyuria (pol″e-yoor′e-ă) Excretion of an excessively large volume of urine in a given period. 758

pons (ponz) The portion of the brain stem just above the medulla oblongata and anterior to the cerebellum. **426**

popliteal (pop″lĭ-te′al, pop-lit′e-al) Pertaining to the concave region on the posterior aspect of the knee. **16**

posterior (pos-tēr′e-or) Toward the back; also called *dorsal*. 12

posterior pituitary (pĭ-too′ĭ-ter-e) *See* neurohypophysis. 566

posterior root An aggregation of sensory neuron fibers lying between a spinal nerve and the posterolateral aspect of the spinal cord; also called the *dorsal root* or *sensory root*. **458**

posterior root ganglion (gang′gle-on) A cluster of cell bodies of sensory neurons located along the posterior root of a spinal nerve. 427

postganglionic neuron (pōst″gang-gle-on′ik) The second neuron in an autonomic motor pathway. Its cell body is outside the central nervous system, and it terminates at an effector organ. **478**

postnatal (pōst-na′tal) After birth.

postsynaptic inhibition (pōst″sĭ-nap′tik) The inhibition of a postsynaptic neuron by axon endings that release a neurotransmitter that induces hyperpolarization (inhibitory postsynaptic potentials). **401**

preganglionic neuron (pre″gang-gle-on′ik) The first neuron in an autonomic motor pathway. Its cell body is inside the central nervous system, and it terminates on a postganglionic neuron. **478**

pregnancy A condition in which a female is carrying a developing offspring within the body. 925

prenatal (pre-na′tal) Pertaining to the period of offspring development during pregnancy; before birth.

prepuce (pre′pyoos) A fold of loose, retractable skin covering the glans of the penis or clitoris; also called the *foreskin*. 856

presynaptic inhibition (pre″sĭ-nap′tik) Neural inhibition in which axoaxonic synapses inhibit the release of neurotransmitter chemicals from the presynaptic axon terminal. **401**

prolactin (pro-lak′tin) A hormone secreted by the anterior pituitary that, in conjunction with other hormones, stimulates lactation in the postpartum female. It may also participate (along with the gonadotropins) in regulating gonadal function in some mammals. **567**

pronation (pro-na′shun) A rotational movement of the forearm in which the palm of the hand is turned posteriorly. **259**

proprioceptor (pro″pre-o-sep′tor) A sensory nerve ending that responds to changes in tension in a muscle or tendon. **498**

prostaglandin (pros″tă-glan′din) Any of a family of fatty acids that have numerous autocrine regulatory functions, including the stimulation of uterine contractions and of gastric acid secretion and the promotion of inflammation. **38**

prostate (pros′tāt) A walnut-shaped gland surrounding the male urethra just below the urinary bladder that secretes an additive to seminal fluid during ejaculation. 855

prosthesis (pros-the′sis) An artificial device to replace a diseased or worn body part. 250

proton (pro′ton) A unit of positive charge in the nucleus of atoms. **25**

protoplasm (pro′tŏ-plaz″em) A general term for the colloidal complex of protein that constitutes the living material of a cell. It includes cytoplasm and nucleoplasm. 9

protraction (pro-trak′shun) The movement of a body part, such as the mandible, forward on a plane parallel with the ground; the opposite of *retraction*. **259**

proximal (prok′-sĭ-mal) Closer to the midplane of the body or to the origin of an appendage; the opposite of *distal*. 12

pseudohermaphrodite (soo″dŏ-her-maf′rŏ-dīt) An individual with some of the physical characteristics of both sexes, but who lacks functioning gonads of both sexes; a true hermaphrodite has both testes and ovaries. 863

pseudopods (soo′dŏ-podz) Footlike extensions of the cytoplasm that enable some cells (with amoeboid motion) to move across a substrate. Pseudopods are also used to surround food particles in the process of phagocytosis. 44

ptyalin (ti′ă-lin) An enzyme in saliva that catalyzes the hydrolysis of starch into smaller molecules; also called *salivary amylase*. 796

puberty (pyoo′ber-te) The period of development in which the reproductive organs become functional. 843

pulmonary (pul′mŏ-ner″e) Pertaining to the lungs.

pulmonary circulation The system of blood vessels from the right ventricle of the heart to the lungs that transports deoxygenated blood and returns oxygenated blood from the lungs to the left atrium of the heart. **616**

pulp cavity A cavity within the center of a tooth that contains blood vessels, nerves, and lymphatics. 773

pupil The opening through the iris that permits light to enter the posterior cavity of the eyeball and be refracted by the lens through the vitreous chamber. **527**

Purkinje fibers (pur-kin′je) *See* conduction myofibers. 576

pyloric sphincter (pi-lor′ik sfingk′ter) A modification of the muscularis tunic between the stomach and the duodenum that functions to regulate the food material leaving the stomach. 777

pyramid (pir′ă-mid) Any of several structures that have a pyramidal shape (e.g., the renal pyramids in the kidney and the medullary pyramids on the anterior surface of the brain). 403

pyrogen (pi′rŏ-jen) A fever-producing substance. 653

Q

QRS complex The principal deflection of an electrocardiogram that is produced by depolarization of the ventricles. 577

R

ramus (ra′mus) A branch of a bone, artery, or nerve. **213**

raphe (ra′fe) A ridge or a seamlike structure between two similar parts of a body organ, as in the scrotum. 845

receptor (re-sep′tor) A sense organ or a specialized distal end of a sensory neuron that receives stimuli from the environment. **469**

rectum (rek′tum) The terminal portion of the GI tract, between the sigmoid colon and the anal canal. 785

red marrow (mar′o) A tissue that forms blood cells, located in the medullary cavity of certain bones. 165

red nucleus (noo′kle-us) An aggregation of gray matter of a reddish color located in the upper portion of the midbrain. It sends fibers to certain brain tracts. **426**

reduced hemoglobin (he′mŏ-glo″bin) Hemoglobin with iron in the reduced ferrous state. It is able to bond with oxygen but is not combined with oxygen. Also called *deoxyhemoglobin*. 707

reducing agent An electron donor in a coupled oxidation-reduction reaction. **86**

reflex (re′fleks) A rapid involuntary response to a stimulus. 438

reflex arc The basic conduction pathway through the nervous system, consisting of a sensory neuron, an association neuron, and a motor neuron. **469**

regional anatomy The division of anatomy concerned with structural arrangement in specific areas of the body, such as the head, neck, thorax, or abdomen. 323

renal (re′nal) Pertaining to the kidney.

renal corpuscle (kor′pus′l) The portion of the nephron consisting of the glomerulus and a glomerular capsule; also called the *malpighian corpuscle*. 730

renal cortex The outer portion of the kidney, primarily vascular. 728

renal medulla (mĕ-dul′ă) The inner portion of the kidney, including the renal pyramids and renal columns. 728

renal pelvis The inner cavity of the kidney formed by the expanded ureter and into which the calyces open. 729

renal plasma clearance rate The milliliters of plasma cleared of a particular solute per minute by the excretion of that solute in the urine. If there is no reabsorption or secretion of that solute by the nephron tubules, the plasma clearance rate is equal to the glomerular filtration rate. 744

renal pyramid A triangular structure within the renal medulla composed of nephron loops and the collecting ducts. 729

repolarization (re-po″lar-ĭ-za′shun) The reestablishment of the resting membrane potential after depolarization has occurred. **379**

respiration (res″pĭ-ra′shun) The exchange of gases between the external environment and the cells of an organism. **607**

respiratory acidosis (rĭ-spīr′ă-tor-e as″ĭ-do′sis) A lowering of the blood pH to below 7.35 due to accumulation of CO_2 as a result of hypoventilation. **604**

respiratory alkalosis (al″kă-lo′sis) A rise in blood pH to above 7.45 due to excessive elimination of blood CO_2 as a result of hyperventilation. **604**

respiratory center The structure or portion of the brain stem that regulates the depth and rate of breathing. 395

respiratory distress syndrome A lung disease of the newborn, most frequently occurring in premature infants, that is caused by abnormally high alveolar surface tension as a result of a deficiency in lung surfactant; also called *hyaline membrane disease*. 695

respiratory membrane A thin, moistened membrane within the lungs, composed of an alveolar portion and a capillary portion, through which gaseous exchange occurs. 690

rete testis (re′te tes′tis) A network of ducts in the center of the testis associated with the production of spermatozoa. 848

reticular formation (rĕ-tik′yŭ-lar) A network of nervous tissue fibers in the brain stem that arouses the higher brain centers. **429**

retina (ret′ĭ-nă) The principal portion of the internal tunic of the eyeball that contains the photoreceptors. **533**

retraction (re-trak′shun) The movement of a body part, such as the mandible, backward on a plane parallel with the ground; the opposite of *protraction*. **259**

retroperitoneal (ret″ro-per″ĭ-tŏ-ne′al) Positioned behind the parietal peritoneum. 117

rhodopsin (ro-dop′sin) A pigment in rod cells that undergoes a photochemical dissociation in response to light, and in so doing stimulates electrical activity in the photoreceptors. **533**

rhythmicity area (rith-mis′ĭ-te) A portion of the respiratory control center located in the medulla oblongata that controls inspiratory and expiratory phases. 703

ribosome (ri′bo-sōm) A cytoplasmic organelle composed of protein and RNA in which protein synthesis occurs. 55

rickets (rik′ets) A condition caused by a deficiency of vitamin D and associated with an interference of the normal ossification of bone. 173

right lymphatic duct (lim-fat′ik) A major vessel of the lymphatic system that drains lymph from the upper right portion of the body into the right subclavian vein. 648

rigor mortis (rig′or mor′tis) The stiffening of a dead body due to the depletion of ATP and the production of rigor complexes between actin and myosin in muscles. **156**

RNA Ribonucleic acid; a nucleic acid consisting of the nitrogenous bases adenine, guanine, cytosine, and uracil; the sugar ribose; and phosphate groups. There are three types of RNA found in cytoplasm: messenger RNA (mRNA), transfer RNA (tRNA), and ribosomal RNA (rRNA). **56**

rod A photoreceptor in the retina of the eye that is specialized for colorless, dim-light vision. **533**

root canal The hollow, tubular extension of the pulp cavity into the root of the tooth that contains vessels and nerves. 773

rotation (ro-ta′shun) The movement of a bone around its own longitudinal axis. **258**

round window A round, membrane-covered opening between the middle and inner ear, directly below the oval window; also called the *cochlear window*. **517**

rugae (roo′je) The folds or ridges of the mucosa of an organ. 754

S

saccadic eye movements (să-kad′ik) Very rapid eye movements that occur constantly and that change the focus on the retina from one point to another. **540**

saccule (sak′yool) A saclike cavity in the membranous labyrinth inside the vestibule of the inner ear that contains a vestibular organ for equilibrium. **514**

sacral (sa′kral) Pertaining to the sacrum. **16**

sacral plexus (plek′sus) A network of nerve fibers that arises from spinal nerves L4 through S3. Nerves arising from the sacral plexus merge with those from the lumbar plexus to form the lumbosacral plexus and supply the lower extremity. 434

saddle joint A synovial joint in which the articular surfaces of both bones are concave in one plane and convex or saddle shaped, in the other plane, such as in the distal carpometacarpal joint of the thumb. 231

sagittal plane (saj′ĭ-tal) A vertical plane, running parallel to the midsagittal plane, that divides the body into unequal right and left portions. 10

salivary gland (sal′ĭ-ver-e) An accessory digestive gland that secretes saliva into the oral cavity. 773

saltatory conduction (sal′tă-to″re) The rapid passage of action potentials from one node of Ranvier (neurofibril node) to another in myelinated axons. 357

sarcolemma (sar″kŏ-lem′ă) The cell membrane of a muscle fiber. 270

sarcomere (sar′kŏ-mēr) The portion of a striated muscle fiber between the two adjacent Z lines that is considered the functional unit of a myofibril. **290**

sacroplasm (sar′kŏ-plaz″em) The cytoplasm within a muscle fiber. 268

sarcoplasmic reticulum (sar″kŏ-plaz′mik rĕ-tik′yŭ-lum) The smooth or agranular endoplasmic reticulum of skeletal muscle cells. It surrounds each myofibril and stores Ca^{2+} when the muscle is at rest. **295**

scala tympani (ska′lă tim′pă-ne) The lower channel of the cochlea that is filled with perilymph. **518**

scala vestibuli (vĕ-stib′yŭ-le) The upper channel of the cochlea that is filled with perilymph. **518**

Schwann cell (schwahn) *See* neurolemmocyte. **375**

sclera (skler′ă) The outer white layer of fibrous connective tissue that forms the protective covering of the eyeball. **527**

scleral venous sinus (ve′nus) A circular venous drainage for the aqueous humor from the anterior chamber; located at the junction of the sclera and the cornea; also called the *canal of Schlemm*. **528**

scrotum (skro′tum) A pouch of skin that contains the testes and their accessory organs. 845

sebaceous gland (sĕ-ba′shus) An exocrine gland of the skin that secretes sebum. 147

sebum (se′bum) An oily, waterproofing secretion of the sebaceous glands. **170**

second messenger A molecule or ion whose concentration within a target cell is increased by the action of a regulatory compound (e.g., a hormone or neurotransmitter) and which stimulates the metabolism of that target cell in a way that mediates the intracellular effects of that regulatory compound. **394**

secretin (sĕ-kre′tin) A polypeptide hormone secreted by the small intestine in response to acidity of the intestinal lumen. Along with cholecystokinin, secretin stimulates the secretion of pancreatic juice into the small intestine. 800

semen (se′men) The thick, whitish secretion of the reproductive organs of the male, consisting of spermatozoa and additives from the prostate and seminal vesicles. 863

semicircular canals Tubule channels within the inner ear that contain receptors for equilibrium. *514*

semilunar valve (sem″e-loo′nar) Crescent- or half-moon–shaped heart valves positioned at the entrances to the aorta and the pulmonary trunk. **613**

seminal vesicles (sem′ĭ-nal ves′ĭ-k′lz) A pair of accessory male reproductive organs lying posterior and inferior to the urinary bladder that secrete additives to spermatozoa into the ejaculatory ducts. 854

seminiferous tubules (sem″ĭ-nif′er-us too′byoolz) Numerous small ducts in the testes, where spermatozoa are produced. 845

semipermeable membrane (sem″e-per′me-ă-b′l) A membrane with pores of a size that permits the passage of solvent and some solute molecules while restricting the passage of other solute molecules. 93

senescence (sĕ-nes′ens) The process of aging. 134

sensory area A region of the cerebral cortex that receives and interprets sensory nerve impulses. 385

sensory neuron (noor′on) A nerve cell that conducts an impulse from a receptor organ to the central nervous system; also called an *afferent neuron*. **374**

septum (sep′tum) A membranous or fleshy wall dividing two cavities.

serous membrane (ser′us) An epithelial and connective tissue membrane that lines body cavities and covers visceral organs within these cavities; also called *serosa*. **16**

Sertoli cells (ser-to′le) *See* sustentacular cells. 845

serum (ser′um) Blood plasma with the clotting elements removed. **594**

sesamoid bone (ses′ă-moid) A membranous bone formed in a tendon in response to joint stress (e.g., the patella). 161

sex chromosomes The X and Y chromosomes; the unequal pairs of chromosomes involved in sex determination (which is based on the presence or absence of a Y chromosome). Females lack a Y chromosome and normally have the genotype XX; males have a Y chromosome and normally have the genotype XY. 839

shock As it relates to the cardiovascular system, this term refers to a rapid, uncontrolled fall in blood pressure, which in some cases becomes irreversible and leads to death. **607**

shoulder The region of the body where the humerus articulates with the scapula. **16**

sickle-cell anemia A hereditary, autosomal recessive trait that occurs primarily in people of African ancestry, in which it evolved apparently as a protection (in the carrier state) against malaria. In the homozygous state, hemoglobin S is made instead of hemoglobin A; this leads to the characteristic sickling of red blood cells, hemolytic anemia, and organ damage. 718

sigmoid colon (sig′moid ko′lon) The S-shaped portion of the large intestine between the descending colon and the rectum. 785

sinoatrial node (sin″no-a′tre-al) A mass of specialized cardiac tissue in the wall of the right atrium that initiates the cardiac cycle; the SA node; also called the *pacemaker*. **619**

sinus (si′nus) A cavity or hollow space within a body organ, such as a bone.

sinusoid (si′nŭ-soid) A small, blood-filled space in certain organs, such as the spleen or liver. 789

skeletal muscle A specialized type of multinucleated muscle tissue that occurs in bundles, has crossbands of proteins, and contracts in either a voluntary or involuntary fashion. 131

sleep apnea (ap′ne-ă) A temporary cessation of breathing during sleep, usually lasting for several seconds. 720

sliding filament theory The theory that the thick and thin filaments of a myofibril slide past each other during muscle contraction, while maintaining their initial length. 263

small intestine The portion of the GI tract between the stomach and the cecum whose function is the absorption of food nutrients. 780

smooth muscle A specialized type of nonstriated muscle tissue composed of fusiform, single-nucleated fibers. It contracts in an involuntary, rhythmic fashion within the walls of visceral organs. **478**

sodium/potassium pump (so′de-um pŏ-tas′e-um) An active transport carrier with ATPase enzymatic activity that acts to accumulate K^+ within cells and extrude Na^+ from cells, thus maintaining gradients for these ions across the cell membrane. 99

soft palate (pal′at) The fleshy, posterior portion of the roof of the mouth, from the palatine bones to the uvula. 771

somatic (so-mat′ik) Pertaining to the nonvisceral parts of the body.

somatomedins (so″mat′ŏ-mēd-inz) A group of small polypeptides believed to be produced in the liver in response to growth hormone stimulation and to mediate the actions of growth hormone on the skeleton and other tissues. 831

somatostatin (so-mat″ŏ-stāt′in) A polypeptide produced in the hypothalamus that acts to inhibit the secretion of growth hormone from the anterior pituitary. Somatostatin is also produced in the pancreatic islets, but its function there has not been established. **569**

somatotropic hormone (so-mat″ŏ-trop′ik) Growth hormone; an anabolic hormone secreted by the anterior pituitary that stimulates skeletal growth and protein synthesis in many organs. 519

sounds of Korotkoff (kŏ-rot′kof) The sounds heard when pressure measurements are taken. These sounds are produced by the turbulent flow of blood through an artery that has been partially constricted by a pressure cuff. 637

spermatic cord (sper-mat′ik) The structure of the male reproductive system composed of the ductus deferens, spermatic vessels, nerves, cremaster muscle, and connective tissue. The spermatic cord extends from a testis to the inguinal ring. 841

spermatogenesis (sper-mat″ŏ-jen′ĭ-sis) The production of male sex gametes, or spermatozoa. 849

spermatozoon (sper-mat″ŏ-zo′on), pl. *spermatozoa* or, loosely, *sperm* A mature male sperm cell, or gamete. 853

spermiogenesis (sper″me-ŏ-jen′ĕ-sis) The maturational changes that transform spermatids into spermatozoa. 849

sphincter (sfingk′ter) A circular muscle that functions to constrict a body opening or the lumen of a tubular structure. 281

sphincter of ampulla The muscular constriction at the opening of the common bile and pancreatic ducts; also called the *sphincter of Oddi*. 794

sphincter of Oddi (o′de) *See* sphincter of ampulla. 794

sphygmomanometer (sfig″mo-mă-nom′ĭ-ter) A manometer (pressure transducer) used to measure the blood pressure. 637

spinal cord (spi′nal) The portion of the central nervous system that extends downward from the brain stem through the vertebral canal. 401

spinal ganglion A cluster of nerve cell bodies on the posterior root of a spinal nerve. **458**

spinal nerve One of the 31 pairs of nerves that arise from the spinal cord. 427

spindle fibers (spin′d′l) Filaments that extend from the poles of a cell to its equator and attach to the chromosomes during the metaphase stage of cell division. Contraction of the spindle fibers pulls the chromosomes to opposite poles of the cell. **67**

spinous process (spi′nus) A sharp projection of bone or a ridge of bone, such as on the scapula. 194

spiral organ The functional unit of hearing, consisting of a basilar membrane supporting receptor hair cells and a tectorial membrane within the endolymph of the cochlear duct; also known as the *organ of Corti*. **521**

spironolactones (spi″rŏ-no-lak′tōnz) Diuretic drugs that act as an aldosterone antagonist. 758

spleen (splēn) A large, blood-filled, glandular organ located in the upper left quadrant of the abdomen and attached by mesenteries to the stomach. 650

spongy bone Bone tissue with a latticelike structure; also called *cancellous bone*. **150**

squamous (skwa′mus) Flat or scalelike. **207**

stapes (sta′pēz) The innermost of the auditory ossicles that fits against the oval window of the inner ear; also called the *stirrup*. **214**

steroid (ster′oid) A lipid, derived from cholesterol, that has three 6-sided carbon rings and one 5-sided carbon ring. These form the steroid hormones of the adrenal cortex and gonads. **38**

stomach A pouchlike digestive organ located between the esophagus and the duodenum. 776

stratified (strat′ĭ-fīd) Arranged in layers, or strata. 110

stratum basale (stra′tum bă-să′le) The deepest epidermal layer, where mitotic activity occurs. **160**

stratum corneum (kor′ne-um) The outer, cornified layer of the epidermis of the skin. **162**

stroke volume The amount of blood ejected from each ventricle at each heartbeat. **618**

stroma (stro′mă) A connective tissue framework in an organ, gland, or other tissue. 122

subarachnoid space (sub″ă-rak′noid) The space within the meninges between the arachnoid mater and pia mater, where cerebrospinal fluid flows. 396

sublingual gland (sub-ling′gwal) One of the three pairs of salivary glands. It is located below the tongue, and its duct opens to the side of the lingual frenulum. 774

submandibular gland (sub″man-dib′yŭ-lar) One of the three pairs of salivary glands. It is located below the mandible, and its duct opens to the side of the lingual frenulum. 774

submucosa (sub″myoo-ko′sa) A layer of supportive connective tissue that underlies a mucous membrane. 767

submucosal plexus (sub″myoo-kōs′al plek′sus) A network of sympathetic and parasympathetic nerve fibers located in the submucosa tunic of the small intestine; also called the *plexus of Meissner*. 769

substrate (sub′strāt) In enzymatic reactions, the molecules that combine with the amino acids lining the active sites of an enzyme and are converted to products by catalysis of the enzyme. 70

sulcus (sul′kus) A shallow impression or groove. 164

superficial (soo″per-fish′al) Toward or near the surface. 12

superficial fascia (fash′e-ă) A binding layer of connective tissue between the dermis of the skin and the underlying muscle. 255

superior Toward the upper part of a structure or toward the head; also called *cephalic*. **526**

superior vena cava A large systemic vein that collects blood from regions of the body superior to the heart and returns it to the right atrium. **613**

supination (soo″pĭ-na′shun) Rotation of the arm so that the palm is directed forward or anteriorly; the opposite of *pronation*. **258**

suppressor T cell A subpopulation of T lymphocytes that acts to inhibit the production of antibodies against specific antigens by B lymphocytes. 668

surface anatomy The division of anatomy concerned with the structures that can be identified from the outside of the body. 322

surfactant (sur-fak′tant) A substance produced by the lungs that decreases the surface tension within the alveoli. **38**

suspensory ligament (sŭ-spen′sŏ-re) **1.** A portion of the peritoneum that extends laterally from the surface of the ovary to the wall of the pelvic cavity. 877 **2.** A ligament that supports an organ or body part, such as that supporting the lens of the eye. **528**

sustentacular cells (sus-ten-tak′yŭ-lar) Specialized cells within the testes that supply nutrients to developing spermatozoa; also called *Sertoli cells* or *nurse cells*. 845

sutural bone (soo′chur-al) A small bone positioned within a suture of certain cranial bones; also called a *wormian bone*. **208**

suture (soo′chur) A type of fibrous joint found between bones of the skull. **248**

sweat gland A skin gland that secretes a fluid substance for evaporative cooling. 148

sympathetic (sim″pă-thet′ik) Pertaining to the division of the autonomic nervous system concerned with activities that, in general, arouse the body for physical activity; also called the *thoracolumbar division*. **480**

symphysis (sim′fĭ-sis) A type of cartilaginous joint characterized by a fibrocartilaginous pad between the articulating bones, which provides slight movement. **250**

symphysis pubis (pyoo′bis) A slightly movable joint located anteriorly between the two pubic bones of the pelvic girdle. 209

synapse (sin′aps) A minute space between the axon terminal of a presynaptic neuron and a dendrite of a postsynaptic neuron. **385**

synarthrosis (sin″ar-thro′sis) A fibrous joint, such as a syndesmosis or a suture. 224

synchondrosis (sin″kon-dro′sis) A cartilaginous joint in which the articulating bones are separated by hyaline cartilage. **250**

syndesmosis (sin″des-mo′sis) A type of fibrous joint in which two bones are united by an interosseous ligament. **249**

synergist (sin′er-jist) A muscle that assists the action of the prime mover. 281

synergistic (sin″er-jis′tik) Pertaining to regulatory processes or molecules (such as hormones) that have complementary or additive effects. **57**

synovial cavity (sĭ-no′ve-al) A space between the two bones of a synovial joint, filled with synovial fluid. 227

synovial joint A freely movable joint in which there is a synovial cavity between the articulating bones; also called a *diarthrotic joint*. **248**

synovial membrane The inner membrane of a synovial capsule that secretes synovial fluid into the joint cavity. **253**

system A group of body organs that function together. 10

systemic (sis-tem′ik) Relating to the entire organism rather than to individual parts. 10

systemic anatomy The division of anatomy concerned with the structure and function of the various systems. 10

systemic circulation The portion of the circulatory system concerned with blood flow from the left ventricle of the heart to the entire body and back to the heart via the right atrium (in contrast to the pulmonary system, which involves the lungs). **616**

systole (sis′tŏ-le) The muscular contraction of a heart chamber during the cardiac cycle. **618**

systolic pressure (sis-tol′ik) Arterial blood pressure during the ventricular systolic phase of the cardiac cycle. 636

T

T cell A type of lymphocyte that provides cell-mediated immunity (in contrast to B lymphocytes, which provide humoral immunity through the secretion of antibodies). There are three subpopulations of T cells: cytotoxic, helper, and suppressor. **581**

tachycardia (tak″ĭ-kar′de-ă) An excessively rapid heart rate, usually in excess of 100 beats per minute (in contrast to bradycardia, in which the heart rate is very slow). 603

tactile (tak′til) Pertaining to the sense of touch.

taeniae coli (te′ne-e ko′li) The three longitudinal bands of muscle in the wall of the large intestine. 786

target organ The specific body organ that a particular hormone affects. 514

tarsal gland An oil-secreting gland that opens on the exposed edge of each eyelid; also called a *meibomian gland*. **525**

tarsus (tar′sus) The region of the foot containing the seven tarsal bones. **238**

taste bud An organ containing the chemoreceptors associated with the sense of taste. **509**

tectorial membrane (tek-to′re-al) A gelatinous membrane positioned over the hair cells of the spiral organ in the cochlea. **521**

telencephalon (tel″en-sef′ă-lon) The anterior portion of the forebrain, constituting the cerebral hemispheres and related parts. 379

tendo calcaneous (ten′do kal-ka′ne-us) The tendon that attaches the calf muscles to the calcaneous bone; also called the *Achilles tendon*. 316

tendon (ten′dun) A band of dense regular connective tissue that attaches muscle to bone. **146**

tendon sheath A covering of synovial membrane surrounding certain tendons. **253**

tentorium cerebelli (ten-to′re-um ser″ĕ-bel′e) An extension of dura mater that forms a partition between the cerebral hemispheres and the cerebellum and covers the cerebellum. **418**

teratogen (tĕ-rat′ŏ-jen) Any agent or factor that causes a physical defect in a developing embryo or fetus. 928

testis (tes′tis) The primary reproductive organ of a male that produces spermatozoa and male sex hormones. 845

testosterone (tes-tos′tĕ-rōn) The major androgenic steroid secreted by the interstitial cells of the testes after puberty. 840

tetanus (tet′n-us) A smooth contraction of a muscle (as opposed to muscle twitching). 259

thalamus (thal′ă-mus) An oval mass of gray matter within the diencephalon that serves as a sensory relay center. **424**

thalassemia (thal″ă-se′me-ă) Any of a group of hemolytic anemias caused by the hereditary inability to produce either the alpha or beta chain of hemoglobin. It is found primarily among Mediterranean people. 718

thigh The proximal portion of the lower extremity between the hip and the knee in which the femur is located. **16**

third ventricle (ven′trĭ-k′l) A narrow cavity between the right and left halves of the thalamus and between the lateral ventricles that contains cerebrospinal fluid. **432**

thoracic (tho-ras′ik) Pertaining to the chest region. **14**

thoracic duct The major lymphatic vessel of the body that drains lymph from the entire body, except for the upper right quadrant, and returns it to the left subclavian vein. 648

thorax (thor′aks) The chest. 14

threshold stimulus The weakest stimulus capable of producing an action potential in an excitable cell. 355

thrombocyte (throm′bŏ-sīt) A blood platelet formed from a fragmented megakaryocyte. **153**

thrombus (throm′bus) A blood clot produced by the formation of fibrin threads around a platelet plug. 606

thymus (thi′mus) A bilobed lymphoid organ positioned in the upper mediastinum, posterior to the sternum and between the lungs. **580**

thyroid cartilage (thi′roid kar′tĭ-lij) The largest cartilage in the larynx that supports and protects the vocal cords; commonly called the *Adam's apple*. 687

thyroxine (thi-rok′sin) Also called tetraiodothyronine, or T_4. The major hormone secreted by the thyroid gland, which regulates the basal metabolic rate and stimulates protein synthesis in many organs. A deficiency of this hormone in early childhood produces cretinism. **575**

tinnitus (tĭ-ni′tus) The spontaneous sensation of a ringing sound or other noise without sound stimuli. **544**

tissue An aggregation of similar cells and their binding intercellular substance, joined to perform a specific function. 9

tongue A protrusible muscular organ on the floor of the oral cavity. 771

tonsil (ton′sil) A node of lymphoid tissue located in the mucous membrane of the pharynx. 686

toxin (tok′sin) A poison. 665

trabeculae (tră-bek′yŭ-le) A supporting framework of fibers crossing the substance of a structure, as in the lamellae of spongy bone. **190**

trachea (tra′ke-ă) The airway leading from the larynx to the bronchi, composed of cartilaginous rings and a ciliated mucosal lining of the lumen; commonly called the *windpipe*. 688

tract A bundle of nerve fibers within the central nervous system. 403

transamination (trans″am-ĭ-na′shun) The transfer of an amino group from an amino acid to an alpha-keto acid, forming a new keto acid and a new amino acid without the appearance of free ammonia. **102**

transpulmonary pressure (trans″pul′mŏ-ner″e) The pressure difference across the wall of the lung, equal to the difference between intrapulmonary pressure and intrapleural pressure. 693

transverse colon (trans-vers′ ko′lon) A portion of the large intestine that extends from right to left across the abdomen between the hepatic and splenic flexures. 785

transverse fissure (fish′ur) The prominent cleft that horizontally separates the cerebrum from the cerebellum. 393

transverse plane A plane that divides the body into superior and inferior portions; also called a *horizontal*, or *cross-sectional, plane*. **10**

tricuspid valve (tri-kus′pid) The heart valve located between the right atrium and the right ventricle. 569

trigone (tri′gōn) A triangular area in the urinary bladder between the openings of the ureters and the urethra. 754

triiodothyronine (tri″i-o″do-thi′rŏ-nēn) Abbreviated T_3; a hormone secreted in small amounts by the thyroid; the active hormone in target cells, formed from thyroxine. 529

trochanter (tro-kan′ter) A broad, prominent process on the proximolateral portion of the femur. 212

trochlea (trok′le-ă) A pulleylike anatomical structure (e.g., the medial surface of the distal end of the humerus that articulates with the ulna). **228**

tropomyosin (tro″pŏ-mi′ŏ-sin) A filamentous protein that attaches to actin in the thin myofilaments and that acts, together with another protein called troponin, to inhibit and regulate the attachment of myosin cross bridges to actin. **294**

true vocal cords Folds of the mucous membrane in the larynx that produce sound as they are pulled taut and vibrated. 688

trunk The thorax and abdomen together. **14**

trypsin (trip′sin) A protein-digesting enzyme in pancreatic juice that is released into the small intestine. 795

tubercle (too′ber-k′l) A small, elevated process on a bone. 164

tuberosity (too′bĭ-ros′ĭ-te) An elevation or protuberance on a bone. 164

tunica albuginea (too′nĭ-kă al″byoo-jin′e-ă) A tough, fibrous tissue surrounding the testis. 845

tympanic membrane (tim-pan′ik) The membranous eardrum positioned between the external and middle ear. **516**

U

umbilical cord (um-bĭ′lĭ-kal) A cordlike structure containing the umbilical arteries and vein and connecting the fetus with the placenta. 601

umbilicus (um-bĭ-lĭ-kus) The site where the umbilical cord was attached to the fetus; commonly called the *navel*. **15**

unipolar neuron (yoo′nĭ-po-lar noor′on) A nerve cell that has a single nerve fiber extending from its cell body. 347

universal donor A person with blood type O who is able to donate blood to people with other blood types in emergency blood transfusions. 557

universal recipient A person with blood type AB who can receive blood of any type in emergency transfusions. 557

upper extremity The appendage attached to the pectoral girdle, consisting of the shoulder, brachium, elbow, antebrachium, and hand. **14**

urea (yoo-re′ă) The chief nitrogenous waste product of protein catabolism in the urine, formed in the liver from amino acids. 739

uremia (yoo-re′me-ă) The retention of urea and other products of protein catabolism as a result of inadequate kidney function. 759

ureter (yoo-re′ter) A tube that transports urine from the kidney to the urinary bladder. 753

urethra (yoo-re′thră) A tube that transports urine from the urinary bladder to the outside of the body. 754

urinary bladder (yoo′rĭ-ner″e) A distensible sac that stores urine, situated in the pelvic cavity posterior to the symphysis pubis. 754

urobilinogen (yoo″rŏ-bi-lin′ŏ-jen) A compound formed from bilirubin in the small intestine; some is excreted in the feces, and some is absorbed and enters the enterohepatic circulation, where it may be excreted either in the bile or in the urine. 792

uterine tube (yoo′ter-in) The tube through which the ovum is transported to the uterus and the site of fertilization; also called the *oviduct* or *fallopian tube*. 782

uterus (yoo'ter-us) A hollow, muscular organ in which a fetus develops. It is located within the female pelvis between the urinary bladder and the rectum; commonly called the *womb.* 873

utricle (yoo'trĭ-k'l) An enlarged portion of the membranous labyrinth, located within the vestibule of the inner ear. **514**

uvula (yoo'vyŭ-lă) A fleshy, pendulous portion of the soft palate that blocks the nasopharynx during swallowing. 771

V

vacuole (vak-yoo'ōl) A small space or cavity within the cytoplasm of a cell. 45

vagina (vă-ji'nă) A tubular organ leading from the uterus to the vestibule of the female reproductive tract that receives the male penis during coitus. 874

vallate papillae (val'āt pă-pil'e) The largest papillae on the surface of the tongue. They are arranged in an inverted V-shaped pattern at the posterior portion of the tongue. 771

vasectomy (vă-sek'tŏ-me, va-zek'tŏ-me) Surgical removal of portions of the ductus deferentia to induce infertility. 864

vasoconstriction (va"zo-kon-strik'shun) Narrowing of the lumen of blood vessels due to contraction of the smooth muscles in their walls. 624

vasodilation (va'zo-di-la'shun) Widening of the lumen of blood vessels due to relaxation of the smooth muscles in their walls. 624

vasomotor center (va"zo-mo'tor) A cluster of nerve cell bodies in the medulla oblongata that controls the diameter of blood vessels. It is therefore important in regulating blood pressure. 395

vein (vān) A blood vessel that conveys blood toward the heart. 582

vena cava (ve'nă ka'vă) One of two large vessels that return deoxygenated blood to the right atrium of the heart. 592

ventilation (ven"tĭ-la'shun) Breathing; the process of moving air into and out of the lungs. 692

ventral (ven'tral) Toward the front or facing surface; the opposite of *dorsal;* also called *inferior.* 12

ventricle (ven'trĭ-k'l) A cavity within an organ; especially those cavities in the brain that contain cerebrospinal fluid and those in the heart that contain blood to be pumped from the heart. **613**

venule (ven'yool) A small vessel that carries venous blood from capillaries to a vein. 615

vermis (ver'mis) The coiled middle lobular structure that separates the two cerebellar hemispheres. 393

vertebral canal (ver'tĕ-bral) The tubelike cavity extending through the vertebral column that contains the spinal cord; also called the *spinal canal.* 401

vertigo (ver'tĭ-go) A feeling of movement or loss of equilibrium. **544**

vestibular window *See* oval window.

vestibule (ves'tĭ-byool) A space or cavity at the entrance to a canal, especially that of the nose, inner ear, or vagina.

villus (vil'us) A minute projection that extends outward into the lumen from the mucosal layer of the small intestine. 782

virulent (vir'yŭ-lent) Pathogenic; able to cause disease.

viscera (vis'er-a) The organs within the abdominal or thoracic cavities. 663

visceral (vis'er-al) Pertaining to the membranous covering of the viscera. **16**

visceral peritoneum (per"ĭ-tŏ-ne'um) A serous membrane that covers the surfaces of abdominal viscera. 117

visceral pleura (ploor'ă) A serous membrane that covers the surfaces of the lungs. **17**

visceroceptor (vis"er-ŏ-sep'tor) A sensory receptor located within body organs that responds to information concerning the internal environment. **509**

vitreous humor (vit're-us hyoo' mer) The transparent gel that occupies the space between the lens and retina of the eyeball. **520**

Volkmann's canal (fōlk'manz) *See* perforating canal. 167

vulva (vul'vă) The external genitalia of the female that surround the opening of the vagina. 875

W

white matter Bundles of myelinated axons located in the central nervous system. **376**

wormian bone (wer'me-an) *See* sutural bone. 161

Y

yellow marrow (mar'o) Specialized lipid storage tissue within bone cavities. 165

Z

zygote (zi'gōt) A fertilized egg cell formed by the union of a sperm cell and an ovum. 899

zymogens (zi'mŏ-jenz) Inactive enzymes that become active when part of their structure is removed by the action of another enzyme or by some other means. 795

Credits

Photographs

Chapter 1

Fig 1.1: Bibliotheque Nationale; Fig 1.2: Fratelli Alinari; Fig 1.3: From the Works of Andrea Vesalins of Brussels by J. Bade, C. M. Saunders and Charley P. O'Malley, Pg. 1096, Dover Publications, Inc.; Fig 1.4: Stock Montage, Inc.; Fig 1.9a: Courtesy of Kodak; Fig 1.9b; © Carroll H. Weiss/Camera M. D. Studios; Fig 1.9c: Courtesy of Utah Valley Regional Medical Center, Dept. of Radiology; Fig 1.10a: L. V. Bergman & Assoc./ Project Masters, Inc.; Fig 1.10b: Hank Moorgan/Science Source/Photo Researchers, Inc.; Fig 1.10c: © Hank Morgan/Science Source/Photo Researchers, Inc.; Fig 1.12: Dr. Sheril D. Burton; Fig 1.13b: A. Glauberman/Photo Researchers, Inc.; Fig 1.13c: Martin M. Rotker/Photo Researchers, Inc.; Pg. 15: Dr. Sheril D. Burton.

Chapter 2

Fig 2.28: Ed Reschke.

Chapter 3

Fig 3.3a,b: Kwang W. Jeon; Fig 3.4 (all): M. M. Perry & A. B. Gilbert, *Journal of Cell Science* 39 257–272, 1979; Fig 3.5a: Ellen R. Dirksen/Visuals Unlimited; Fig 3.5b: Richard Chao; Fig 3.6a: Dr. Carolyn Chambers; Fig 3.6b: From R. G. Kessel and R. H. Kardon, *Tissues and Organs: A Text-Atlas of Scanning Electron Microscopy*, W. H. Freeman and Co.; Fig 3.7: K. G. Murti/Visuals Unlimited; Fig 3.9: Richard Chao; Fig 3.10a, 3.11a: Keith R. Porter; Fig 3.12: Richard Chao; Fig 3.13a: E. G. Pollack; Fig 3.21: Alexander Rich; Fig 3.27a: David M. Phillips/Visuals Unlimited; Fig 3.31a–e: Ed Reschke; Fig 3.32a: David M. Phillips/Visuals Unlimited; Fig 3.33: CNRI/SPL/Photo Researchers, Inc.

Chapter 4

Fig 4.27: Fernandes-Moran V. M. D. Ph.D.

Chapter 5

Fig 5.10: Richard Chao.

Chapter 6

Fig 6.1a: Ed Reschke; Fig 6.1b: Dr. Kerry L. Openshaw; Fig 6.2b: Ray Simons/Photo Researchers, Inc.; Fig 6.3b: © Ed Reschke/Peter Arnold, Inc.; Fig 6.4b–6.9b: Ed Reschke; Fig 6.11b: CNRI/SPL/Photo Researchers, Inc.; Fig 6.14b, 6.15b: Biophoto/Science Source/Photo Researchers, Inc.; Fig 6.16b: Ed Reschke; Fig 6.17b: Biology Media/Photo Researchers, Inc.; Fig 6.18b: © Ed Reschke/Peter Arnold, Inc; Fig 6.19b–6.22b: Ed Reschke; Fig 6.23b: © Ed Reschke/Peter Arnold, Inc; Fig 6.24b–6.26c: Ed Reschke; Fig 6.27b: © John D. Cunningham/Visuals Unlimited; Fig 6.27c: © 1984 Martin M. Rotker/Photo Researchers, Inc.

Chapter 7

Fig 7.2: © Ed Reschke/Peter Arnold, Inc.; Fig 7.3: J. Burgess/Photo Researchers, Inc.; Fig 7.4: Victor B. Eichler, Ph.D.; Fig 7.5: Dr. Sheril D. Burton; Fig 7.6: James M. Clayton; Fig 7.7: Lester V. Bergman & Assoc./ Project Masters, Inc.; Fig 7.8a: World Health Organization; Fig 7.8b: George P. Bogumill; Fig 7.9a: Michael Abbey/Photo Researchers, Inc.; Fig 7.9b: Dr. Kerry L. Openshaw; Fig 7.10b: © John D. Cunningham/Visuals Unlimited; Fig 7.13a: © Zeva Oelbaum/ Peter Arnold Inc.; Fig 7.13b,c: Dr. P. Marazz/SPL/Photo Researchers, Inc.; Fig 7.14a: Tierbild Okapia/Photo Researchers, Inc.; Fig 7.14b,c: SPL/Photo Researchers, Inc.; Fig 7.17: (Child): From *Science Year, The World Book Science Annual* © 1973 Field Enterprise Education Corporation. By permission of World Book, Inc. (Woman): Black Star Publishing Co.; Fig 7.18: Norman Lighfoot/Photo Researchers, Inc.

Chapter 8

Pg. 186, fig 1b: © Science VU/ Visuals Unlimited; Fig 8.5a,b: Biophoto Associates/Photo Researchers, Inc.; Fig 8.7a,b: From: *Tissues and Organs: A Text Atlas of Scanning Electron Microscopy* by R. G. Kessel and R. Kardon. © 1979, W. H. Freeman and Company; Fig 8.9: Ed Reschke; Fig 8.10: Courtesy of Utah Valley Regional Medical Center, Dept. of Radiology; Fig 8.13: CNRI/SPL/Photo Researchers, Inc.; Fig 8.14a,b: New England Journal of Medicine.

Chapter 9

Fig 9.9a–9.23a: Courtesy of Utah Valley Regional Medical Center, Dept. of Radiology.

Chapter 10

Fig 10.7b: Dr. Sheril D. Burton; Fig 10.9–10.16: Courtesy of Utah Valley Regional Medical Center, Dept. of Radiology; Fig 10.18a: Dr. Sheril D. Burton; Fig 10.18b: Courtesy of Utah Valley Regional Medical Center, Dept. of Radiology; Fig 10.20a,b: Blayne Hirshche; Fig 10.23e: Courtesy of Eastman Kodak Company.

Chapter 11

Fig 11.5: Courtesy of Utah Valley Regional Medical Center, Dept. of Radiology; Fig 11.17a–h: Dr. Sheril D. Burton; Fig 11.18b,c: Dr. Sheril D. Burton; Fig 11.18e,f: Dr. Sheril D. Burton; Fig 11.23–11.29: From *Clinical Anatomy Atlas*, R. A. Chase, et al., Mosby 1996 (Class Project, Stanford University); Fig 11.31b: Lester V. Bergman & Assoc./ Project Masters, Inc.; Fig 11.32a: James Stevenson/Science UV/Photo Researchers, Inc.; Fig 11.32b: Richard Anderson; Fig 11.33a–d: SIU, School of Medicine; Pg. 282 (all): Ed Reschke.

Chapter 12

Fig 12.2: Ed Reschke; Fig 12.3a: International Bio-Medical, Inc.; Fig 12.5b: © John D. Cunningham/Visuals Unlimited; Fig 12.8–12.9b: Dr. H. E. Huxley; Fig 12.9c: From R. G. Kessel and R. H. Kardon: *Tissues and Organs: A Text-Atlas of Scanning Electron Microscopy*, W. H. Freeman and Company © 1979; Fig 12.10a: Dr. H. E. Huxley; Fig 12.19: Hans Hoppler, *Respiratory Physiology* 44: 94 (1981); Fig 12.21: Avril V. Somlyo, University of Pennsylvania, Pennsylvania Muscle Institute.

Chapter 13

Reference Atlas

Fig 8: From B. Vidic and F. R. Suarez, "Photographic Atlas of the Human Body" (P1. 27, pg. 34), 1984, St. Louis, Mosby; Fig 9: From B. Vidic and F. R. Suarez, "Photographic Atlas of the Human Body" (P1. 34, pg. 48), 1984, St. Louis, Mosby; Fig 10: Wm. C. Brown Communcations/Karl Rubin, photographer; Fig 11: From B. Vidic and F. R.. Suarez, "Photographic Atlas of the Human Body" (P1. 113, pg. 182), 1984, St. Louis, Mosby; Fig 12: Wm. C. Brown Communcations/Karl Rubin, photographer; Fig 13: From B. Vidic and F. R. Suarez, "Photographic Atlas of the Human Body" (P1. 39, pg. 58), 1984, St. Louis, Mosby; Fig 14: From B. Vidic and F. R. Suarez, "Photographic Atlas of the Human Body" (P1. 148, pg. 240), 1984, St. Louis, Mosby; Fig 17: From B. Vidic and F. R. Suarez, "Photographic Atlas of the Human Body" (P1. 206, pg. 338), 1984, St. Louis, Mosby; Fig 18–20: Wm. C. Brown Communcations/Karl Rubin, photographer.

Chapter 14

Fig 14.6: H. Webster, from John Hubbard. *The Vertibrate Peripheral Nervous System* © 1974, Plenum Press; Fig 14.14: Bell et al., *Textbook of Physiology and Biochemistry* 10th ed. © Churchill Livingstone, Edinburgh; Fig 14.19a: From Gulula, Ranes and Steinbach "Metabolic Coupling and Cell Contracts", *Nature* 235: 262–265 © McMillan Journals Ltd.; Fig 14.20: John Heuser, Washington University School of Medicine, St. Louis, MO.

Chapter 15

Fig 15.2: Monte S. Buchsbaum; Fig 15.23b: Per H. Kjeldsen, University of Michigan, Ann Arbor.

Chapter 16

Fig 16.2: From R. G. Kessel and R. H. Kardon, *Tissues and Organs: A Text-Atlas of Scanning Electron Microscopy*, W. H. Freeman and Company © 1979; Fig 16.32a–i: Dr. Sheril D. Burton assisted by Dr. Douglas W. Hacking.

Chapter 18

Fig 18.15: 1993 *Discover Magazine*, reprinted with permission; Fig 18.18a: Dean E. Hillman; Fig 18.22, 18.31: Dr. Sheril D. Burton; Fig 18.37: Thomas Sims; Fig 18.40b: P. N. Farnsworth, University of Medicine and Dentistry, New Jersey, Medical School; Fig 18.42: Per H. Kjeldsen, University of Michigan Ann Arbor; Fig 18.44: Frank Werblin; Fig 18.51: D. Hubel, *American Scientist* 67 1979 534; Fig 18.54: Dr. Stephen Clark.

Chapter 19

Fig 19.1: Ed Reschke; Fig 19.21b: © SIU/Peter Arnold, Inc.; Fig 19.22: © Martin M. Rotker; Fig 19.31, 19.32: © Lester V. Bergman & Associates/ Project Masters, Inc.

Chapter 20

Fig 20.2b: Reginald J. Poole/Polaroid International Instant; Fig 20.6: CNRI/Science Photo Library/Photo Researchers, Inc.

Chapter 21

Fig 21.16: Don W. Fawcett; Fig 21.20b: *Practische Intleedkunde* from J. Dankemeyer, H. G. Lambers, and J. M. F. Landsmearr. Bohn, Scheltma and Holkema; Fig 21.38a–20.40b: Richard Menard.

Chapter 22

Fig 22.10: From E. K. Markell and M. Vogue, *Medical Parasitology*, 5th ed.,W. B. Saunders; Fig 22.17a,b: Donald S. Bain from Hurst et al. *The Heart*, 5/e © McGraw Hill Book; Fig 22.21a,b: Courtesy of Niels A. Lassen, Copenhagen, Denmark.

Chapter 23

Fig 23.12a: Courtesy Arthur J. Olson, Ph.D., The Scripps Research Institute, Molecular Graphics Laboratory; Fig 23.13a,b: Dr. D. A. G. Amit, "Three Dimensional Structure of an Antigen Antibody Complex"; Fig 23.24a: From Alan S. Rosenthal; Fig 23.28a,b: Dr. Andrejs Liepins; Fig 23.29a,b: Courtesy of Dr. Noel Rose; Fig 23.31 (both): SIU Biomed Comm/Custom Medical Stock; Fig 23.32b © Dr. Jeremy Burgess/SPL/ Photo Researchers, Inc.

Chapter 24

Fig 24.4: CNRI/SPL/Photo Researchers, Inc.; Fig 24.6c: © CNRI/Phototake, Inc.; Fig 24.8a: © John D. Cunningham/Visuals Unlimited; Fig 24.8b: Ed Reschke; Fig 24.13a–24.14: Edward C. Vasquez R. T. C. R. T./ Dept. of Radiologic Technology, Los Angeles; Fig 24.17a,b: From J. H. Comroe, Jr., *Physiology of Respiration*, © 1974, Yearbook Medical City College; Fig 24.20: SIU/Photo Researchers, Inc.; Fig 24.44a: © 1988 R. Calentine/Visuals Unlimited; Fig 24.44b: © M. Moore/Visuals Unlimited.

Chapter 25

Fig 25.2: SPL/Photo Researchers, Inc.; Fig 25.4b: Biophoto Assoc./Photo Researchers, Inc.; Fig 25.7: Professor P. M. Motta & M. Castellucci/Science Photo Library/ Photo Researchers, Inc.; Fig 25.9: © Daniel Friend from William Bloom and Don Fawcett, *Textbook of Histology*, 10th ed., W. B. Saunders, Co.; Fig 25.29: Per H. Kjeldsen, University of Michigan, Ann Arbor.

Chapter 26

Fig 26.14b: Courtesy of Utah Valley Regional Medical Center, Dept. of Radiology; Fig 26.16: Ed Reschke; Fig 26.21: © Manfred Kage/Peter Arnold, Inc.; Fig 26.23a,b: Keith R. Porter; Fig: 26.28: from W. A. Sodeman and T. M. Watson, *Pathologic Physiology*, 6th ed., W. B. Saunders Co., 1969; Fig 26.30c: Victor B. Eichler, Ph.D.; Fig 26.36a: © Carroll Weiss/Camera M. D. Studios; Fig 26.36b: Dr. Sheril D. Burton; Fig 26.37: Ed Reschke.

Chapter 27

Fig 27.4: Dr. Mark Atkinson; Fig 27.12: Lester V. Bergman & Assoc./ Project Masters, Inc.; Fig 28.2a,b: March of Dimes.

Chapter 28

Fig 28.3a–d: Williams, R. H.; *A Textbook of Endocrinology*, 6/e © W. B. Saunders, 1981; Fig 28.17a: Biophoto Assoc./Photo Researchers, Inc.; Fig 28.20b: Francis Leroy, Biocosmos/SPL/Photo Researchers, Inc.; Fig 28.21, 28.22: Ed Reschke; Fig 28.23, 28.24: © Manfred Kage/Peter Arnold, Inc.; Fig 28.25: Ed Reschke; Pg. 916 (both): Dr. Landrum Shettles.

Chapter 29

Fig 29.2–29.8a: Ed Reschke; Fig 29.8b: © Ed Reschke/Peter Arnold, Inc.; Fig 29.9a: Dr. Landrum Shettles; Fig 29.9b: R. J. Blandau, *A Textbook of Histology*, 10th ed. W. B. Saunders, Co.; Fig 29.10: From Bloom/Fawcett: *Textbook of Histology*, 10th ed. p. 859, W. B. Saunders; Fig 29.11: Dr. Landrum Shettles; Fig 29.16: Martin M. Rotker; Fig 29.22: Courtesy of Utah Valley Regional Medical Center, Dept. of Radiology; Fig 29.25a: SIU/Visuals Unlimited; Fig 29.25b: © John Kaprielian/Photo Researchers, Inc.; Fig 29.26a: © SIU/Visuals Unlimited; Fig 29.26b: © John Kaprielian/Photo Researchers, Inc.; Fig 29.26c: Ray Ellis/Photo Researchers, Inc.; Fig 29.26d: © Science VU-Ortho/Visuals Unlimited; Fig 29.26e,f: © M. Long/Visuals Unlimited; Fig 29.26g: John Kaprielian/Photo Researchers, Inc.; Fig 29.26h: © Hank Morgan/Science Source/Photo Researchers, Inc.

Chapter 30

Fig 30.1c: © David Phillips/Visuals Unlimited; Fig 30.2: Luciano Zamboni, from Greep, Roy, and Weiss, Leon, *Histology*, 3/e © McGraw-Hill Book Company, 1973; Fig 30.6a: Courtesy of Ronan O'Rahilly M. D., Carnegie Institute of Washington, Dept. of Embryology, Davis Division; Fig 30.11a, 30.20 (1): © Donald Yeager/Camera M. D. Studios; Fig 30.20 (2): Dr. Landrum Shettles; Fig 30.20 (3): Petit Format/ Nestle/Science Source/Photo Researchers, Inc.; Fig 30.20 (4–6): © Donald Yeager/Camera M. D. Studios; Fig 30.20 (7): Petit Format/Nestle/Science Source/Photo Researchers, Inc.; Fig 30.21a: A. Tsiaras/Photo Researchers, Inc.; Fig: 30.24a,b: March of Dimes; Fig 30.30: Lester V. Bergman & Assoc./ Project Masters, Inc.; Fig 30.31c: Dr. Landrum Shettles; Fig 30.33: Gregory Dellore, M. D. and Steven L. D. Clark, M. D.

Line Art

Chapter 7

Nexus: Rictor Lew

Chapter 10

Nexus: Rictor Lew

Chapter 11

Figure 11.3: Rictor Lew

Chapter 12

Figure 12.7: Rictor Lew

Chapter 15

Figure 15.1a,b (icon): Chris Creek; Figure 15.11: Rictor Lew

Chapter 17

Nexus: Rictor Lew

Chapter 21

Figure 21.5a,b: Rictor Lew

Chapter 23

Figure 23.4: Chris Creek

Chapter 26

Figures 26.11, 26.29: Rictor Lew; Figure 26.13: Chris Creek

Chapter 28

Nexus: Rictor Lew

Chapter 29

Figure 29.24: Chris Creek

Chapter 30

Figures 30.16, 30.29, 30.31, 30.32: Chris Creek; Figures 30.34, 30.35: Precision Graphics

Index

Page numbers followed by *f* and *t* refer to figures and tables, respectively.

A

B

C

D

E

F

I

J

K

M

Q

R

S

U

V

W

Z